23rd International Compressor Engineering Conference at Purdue 2016

AA006981

West Lafayette, Indiana, USA
11 - 14 July 2016

Volume 1 of 2

ISBN: 978-1-5108-2872-8

Printed from e-media with permission by:

Curran Associates, Inc.
57 Morehouse Lane
Red Hook, NY 12571

Some format issues inherent in the e-media version may also appear in this print version.

Copyright© (2016) by Perdue University
All rights reserved.

Printed by Curran Associates, Inc. (2016)

For permission requests, please contact Ray W. Herrick Laboratories
at the address below.

Ray W. Herrick Laboratories
School of Mechanical Engineering
Purdue University
177 S. Russell Street
West Lafayette, IN 47907-2099, USA.

Phone: +1 (765) 494-2132
Fax: +1 (765) 494-0787

rhlab@ecn.purdue.edu

Additional copies of this publication are available from:

Curran Associates, Inc.
57 Morehouse Lane
Red Hook, NY 12571 USA
Phone: 845-758-0400
Fax: 845-758-2633
Email: curran@proceedings.com
Web: www.proceedings.com

Ray W. Herrick Laboratories
Center for High Performance Buildings

Table of Content

C-01: Screw Compressors I

Time: Monday July 11, 2016: 1:00 PM - 3:00 PM — *Location*: 202
Session Chair: Matt Cambio

C-02: Valves I

Time: Monday July 11, 2016: 1:00 PM - 3:00 PM — *Location*: 206
Session Chair: Cesar J. Deschamps

C-03: Novel Compressors

Time: Monday July 11, 2016: 3:30 PM - 5:30 PM — *Location*: 202

Session Chair: Margaret Mathison

C-04: Heat Transfer in Compressors I

Time: Monday July 11, 2016: 3:30 PM - 5:30 PM — *Location*: 206
Session Chair: Vincent Lemort

C-05: Tribology and Lubrication I

Time: Tuesday July 12, 2016: 9:45 AM - 11:45 AM — *Location*: 202

Session Chair: Joseph Karnaz

C-06: Scroll Compressors I

Time: Tuesday July 12, 2016: 9:45 AM - 11:45 AM — *Location*: 206

Session Chair: Gene Fields

C-07: Expanders I

Time: Tuesday July 12, 2016: 1:30 PM - 3:30 PM — *Location*: 206

Session Chair: Martijn van den Broek

C-SPC: Student Paper Competition

Time: Tuesday July 12, 2016: 1:30 PM - 3:30 PM — *Location*: 278

Session Chair: Gene Fields

C-08: Heat Transfer in Compressors II

Time: Tuesday July 12, 2016: 4:00 PM - 6:00 PM — *Location*: 202
Session Chair: Joe Orosz

C-09: Linear and Reciprocating Compressors

Time: Tuesday July 12, 2016: 4:00 PM - 6:00 PM — *Location*: 206

Session Chair: Craig R Bradshaw

C-10: Noise, Vibration, & Harshness I

Time: Wednesday July 13, 2016: 9:45 AM - 12:00 PM — *Location*: 202

Session Chair: Stuart Bolton

C-11: Compressor Modeling

Time: Wednesday July 13, 2016: 9:45 AM - 12:00 PM — *Location*: 206

Session Chair: Craig R Bradshaw

C-12: Expanders II

Time: Wednesday July 13, 2016: 1:00 PM - 3:00 PM — *Location*: 202

Session Chair: Jason Hugenroth

C-13: Scroll Compressors II

Time: Wednesday July 13, 2016: 1:00 PM - 3:00 PM — *Location*: 206

Session Chair: Bryce Shaffer

C-14: Tribology and Lubrication II

Time: Wednesday July 13, 2016: 3:30 PM - 5:30 PM — *Location*: 202

Session Chair: Michael Perevozchikov

C-15: Rotary Compressors I

Time: Wednesday July 13, 2016: 3:30 PM - 5:30 PM — *Location*: 206

Session Chair: Doug Collings

C-16: Testing and Measurement

Time: Thursday July 14, 2016: 9:45 AM - 12:00 PM — *Location*: 202

Session Chair: Stefan Bertsch

C-17: Centrifugal Compressors

Time: Thursday July 14, 2016: 9:45 AM - 12:00 PM — *Location*: 206

Session Chair: Christian Bach

C-18: Noise, Vibration, & Harshness II

Time: Thursday July 14, 2016: 1:00 PM - 3:00 PM — *Location*: 202

Session Chair: Patricia Davies

C-19: Rotary Compressors II

Time: Thursday July 14, 2016: 1:00 PM - 3:00 PM — *Location*: 206

Session Chair: Margaret Mathison

C-20: Valves II

Time: Thursday July 14, 2016: 3:30 PM - 5:30 PM — *Location*: 202

Session Chair: Akash Bhatia

C-21: Scroll Compressors III

Time: Thursday July 14, 2016: 3:30 PM - 5:30 PM — *Location*: 206

Session Chair: Kirill M. Ignatiev

ID: 1079

Experimental Investigations on the Performance Improvement of Oil-Gas Separator in Electric Driven Scroll Compressor for Eco-Friendly Vehicles.....1066

Donglim Nam[1,4], Poyoung Lee[1], Seungbin Jung[1], Geonho Lee[2], Yunki Kwon[3], Jinho Lee[4]

[1]R&D Center of Doowon Heavy Industrial Co., Ltd., Korea, Republic of (South Korea); [2]Department of Building Engineering Services, Doowon Technical University College; [3]Department of Computer Aided Design, Doowon Technical University College; [4]School of Mechanical Engineering, Yonsei University

Keywords: Electric Driven Compressor, Oil Separator, Vortex Finder, Pressure Drop

ID: 1344

Influencing Factors Study of the Variable Speed Scroll Compressor With EVI Technology.....1076

Xiaoli Kang, Yusheng Hu, Caixia Shan, Yun Liu, Xiaojun Gao, Gang Lv

GREE ELECTRIC APPLIANCES, INC. OF ZHUHAI, China, People's Republic of

Keywords: Enhanced Vapor Injection; Scroll Compressor;the Area of Supplementary Channel;the Position of Supplementary Inlet;gas Force

ID: 1497

Lubrication Tests to Support Optimal Performance Design Guidelines for Thrust Slide-Bearings in Scroll Compressors.....1084

Noriaki Ishii[1], Eiji Nonoguchi[1], Keiko Anami[5], Atsushi Sakuda[2], Yuya Terada[1], Yusuke Imai[2], Kiyoshi Sawai[3], Charles William Knisely[4]

[1]Osaka Electro-Communication University, Japan; [2]Panasonic Corporation Appliances Company; [3]Hiroshima Institute of Technology; [4]Bucknell University, USA; [5]Ashikaga Institute of Technology, Japan

Keywords: Scroll Compressor, Thrust Slide-Bearing, Optimization, Dsign Guideline, Friction Power Loss

ID: 1613

Visualization and Simulation of Oil Flow in a Scroll Compressor Plenum.....1093

Jiu Xu[1], Pega Hrnjak[1,2]

[1]Air Conditioning and Refrigeration Center, University of Illinois at Urbana-Champaign; [2]Creative Thermal Solutions Inc., Urbana, Illinois

Keywords: Compressor, Oil, CFD, Flow Visulization, Droplet

ID: 1634

Numerical Simulation and Experimental Examination of an Oldham Coupling.....1101

Hang Ye[1], Zhigang Huang[1], Jinduo Ye[2]

[1]Danfoss (Tianjin) R&D, Tianjin, PR. China; [2]School of Mechanical, Tianjin University of Technology, Tianjin, PR.China

Keywords: Scroll Compressor, Oldham Coupling, Numercial, Experimental, Prediction

POLICY FOR PUBLISHING

POLICY FOR PUBLISHING CONFERENCE PAPERS IN ARCHIVAL JOURNALS

The organizing committee recommends that all authors who have papers that satisfy the criteria for archival journal articles submit them to such journals for publication. Because of the relatively small number of copies, publication in the proceedings of the International Compressor Engineering Conference at Purdue does not constitute prior publication as far as many journal editors are concerned. If in doubt, an inquiry should be made with the respective editor. In most cases, the limit of eight pages will require the creation of an expanded journal article version, which would qualify as a new publication anyway.

The criteria for archival papers are that they should contain original, quantitative, detailed scientific materials, which are placed in proper perspective relative to prior work and are supported by references to the appropriate literature. Thus, not all conference papers are suitable. Some describe present practice, or report on procedures that were applied first by someone else, or report design features which, although important to conference participants, have relatively short temporal interest.

The organizing committee requests that authors who choose to submit their papers to archival journals acknowledge that those papers were, in part, presented at the 2016 Conferences and send a copy of each submitted manuscript to the conference chairperson.

Purdue University sponsors this conference to provide a forum for the free exchange of ideas on the engineering state-of-the-art in the compressor field. The ideas expressed are those of the author(s) and do not necessarily represent the opinions or policies of the University or the cooperating organizations. The papers are not subject to the formal review procedures of the separate cooperating organizations as they would be if they were to be published by the cooperating societies. Instead, they are subject to the review procedures of the Organizing Committee.

CONTACT INFORMATION

Contact about the conferences and technical information:
Mrs. Kim Stockment
Conference Coordinator
Ray W. Herrick Laboratories
Purdue University, 177 S Russell St.
West Lafayette, IN 47907-2031 USA
Telephone: (765) 494-6078; Fax: (765) 494-0787
E-mail: hlconf16@purdue.edu
Website: www.engineering.edu/HerrickConf

Contact about the conference venue, hotel information and registration:
Mr. Ethan Kingery,
Conference Coordinator
Purdue University Conferences
128 Memorial Mall, Stewart Center, Room 116
West Lafayette, IN 47907-7220 USA
Telephone: (765) 496-3667; Fax: (765) 494-0567
E-mail: ekingery@purdue.edu
Website: www.conf.purdue.edu

FOREWORD

On behalf of the Organizing Committee, the Ray W. Herrick Laboratories, Purdue University, the Co-Sponsoring Organizations and the Endorsing Organizations, it is a great pleasure to present the Final Program of the 23rd International Compressor Engineering Conference, the 16th International Refrigeration and Air Conditioning Conference, and the 4th International High Performance Buildings Conference at Purdue 2016.

The Organizing Committee received more than 650 abstracts and accepted 476 papers for publication. Every effort was made to include papers of current engineering and scientific interest. In addition, an internal review was conducted of each paper. Nevertheless, the Organizing Committee takes no responsibility for the correctness or completeness of any papers published.

I would like to thank the authors for having chosen these conferences to present their work and for all their efforts in preparing and submitting papers. Thanks to them we present a conference program that is exciting and informative. I would also like to thank the session chairs and session co-chairs for their help in conducting the technical program. In addition, I would like to thank the members of the advisory committee, the Co-Sponsoring and Endorsing Organizations and their representatives for their support and helpful suggestions. In particular, I would like to thank the members of the Organizing Committee and the many graduate student helpers as well as Ms. Kim Stockment, our conference secretary, for their tremendous efforts in making our conferences a success.

We hope that you enjoy these conferences and return home with new ideas and professional contacts. The next Purdue Conferences are planned for July 16-19, 2018. We hope that we will see you again at that time.

Yours sincerely,

Eckhard A. Groll
General Chair

ORGANIZING COMMITTEE

General Conference Chair
Eckhard Groll

International Refrigeration and Air Conditioning Conference
Chair: James Braun
Co-Chair: Andrew Hjortland

International Compressor Engineering Conference
Chair: Travis Horton
Co-Chair: Harshad Inamdar

International High Performance Buildings Conference
Chair: Thanos Tzempelikos
Co-Chair: Jason Konstantzos

Short Course Chairs
Orkan Kurtulus – Compressor Short Course
Bill Murphy – Refrigeration Short Course
James Braun – 2016 IBO Workshops
Gregor Henze – 2016 IBO Workshops

Director, Herrick Laboratories
Patricia Davies

Student Organization Committee
Stephen Caskey
Domenique Lumpkin
Vatsal Shah

Student Paper Competition
Neera Jain

Conference Coordinator
Kim Stockment

ADVISORY BOARD

Representatives of Cooperating Organizations

Dr. Karim Amrane
Vice President, Regulatory & Research
Air-Conditioning, Heating and Refrigeration Institute, USA (AHRI)

Dr. Andreas Athienitis
Professor and Concordia Research Chair Tier I
Scientific Director of NSERC Solar Buildings Research Network, Concordia University, Canada

Chip Barnaby
IBPSA President
International Building Performance Simulation Association, USA

Mr. Van D. Baxter
Distinguished Research Staff Member, Building Equipment Research, Energy and Transportation Science Division
Oak Ridge National Laboratory, USA

Mr. Didier Coulomb
Director
International Institute of Refrigeration, France (IIR)

Ron Domitrovic
Program Manager, Energy Efficiency
Electric Power Research Insitute, USA

Dr. Ullrich Hesse
Professor, Technical University of Dresden, Germany
President, Deutscher Kaelte- und Klimatechnischer Verein e. V. (DKV)

Mr. Glenn Hourahan
Senior Vice President, Research & Technology
Air-Conditioning Contractors of America, USA

Dr. Ji-Hwan Jeong
Professor, Pusan National University, South Korea
Society of Air-Conditioning and Refrigeration Engineers of Korea (SAREK)

Akio Miyara, D. Eng.
Professor, Saga University
Japan Socity of Refrigerating & Air Conditioning Engineers

Mr. Brendan Owens
Vice President, Lead Technical Development
U.S. Green Building Research Program, USA

Mr. Michael Vaughn
Manager of Research and Technical Services
ASHRAE, USA

Adhoc Members

Dr. Stefan Bertsch
Professor and Head, Institute for Energy Systems
Interstate University of Applied Sciences Buchs, Switzerland

Mr. Sergio Bobbo
CNR
Instituo per le Tecnologie della Construzione, Sezione di Padova, Italy

Mr. Doug Collings
Director, N.A. Test Laboratories
Tecumseh Products Company, USA

Dr. Rob Comparin
Vice President, Research
Emerson Climate Technologies, USA

Prof. Cesar J. Deschamps
Professor
Federal University of Santa Catarina, Brazil

Dr. Ir. Carlos Infante Ferreira
Faculty of Mechanical, Maritime & Materials Engineering
Delft University of Technology, The Netherlands

Mr. Gene Fields
VP of Global Compressor Technologies
Johnson Controls, Inc., USA

Dr. Mitsuhiro Fukuta
Professor, Department of Mechanical Engineering
Shizuoka University, Japan

Dr. Andrea Gasparella
Faculty of Science & Technology
Libera Universita di Bolzano, Italy

Dr. Andrea Gernemann
Director of Advanced Engineering
Danfoss Commercial Compressors Division, Germany

Mr. Evandro Francisco Gon
R&D Director
Embraco S.A., Brazil

Dr. Zhaolin Gu
Professor and Executive Dean, School of Human Settlements and Civil Engineering
Xi'an Jiaotong Universoty, P.R. of China

ADVISORY BOARD

Mr. Bill Hooper
Research & Development Manager
Kawneer Company, Inc. (Alcoa Building & Construction Systems), USA

Mr. Joseph Karnaz
Global Technology Leader
CPI Fluid Engineering, USA

Dr. Vincent Lemort
Associate Professor
B49 Thermodynamics Laboratory, University of Liege, Belgium

Dr. Stephen B. Memory
Director, Thermal & Mechanical Technology
A.O. Smith Corporate Technology Center, USA

Mr. Kevin Mercer
Staff Engineer, Thermal Systems Technology Group
Carrier Corp., United Technologies, USA

Dr. William E. Murphy
Professor of Mechanical Engineering and Director of Engineering Extended Campus Programs
University of Kentucky - Paducah, USA

Mr. Andrew Pearson
Managing Director
Star Refrigeration Ltd., Scotland

Mr. Gordon Powell
Compressor Center of Excellence
Ingersoll Rand - Trane, USA

Dr. Brent Protzman
Manager - Energy Information and Analytics
Lutron Electronics, USA

Dr. Joaquim Rigola
Associate Professor, Heat and Mass Transfer Technological Center
Technical University of Catalonia, Spain

Dr. Frank Rinne
Director Application Engineering
Emerson Climate Technologies, GmbH, Germany

Dr.-Ing. Jurgen Suess
Leiter Operations (Leader of Operations)
Efficient Energy GmbH, Germany

Dr. Bob Turney
Engineering Fellow and Team Lead, Technology and Advanced Development Building Efficiency
Johnson Controls Inc., USA

Dr. Ruzhu Wang
Professor and Director, Engineering Research Center for Solar Energy
Shanghai Jiaotong University, P.R. of China

Mr. John Withouse
RAC Refrigerants - Sr. Principal Engineer
Sporlan Division of Parker Hannifin, USA

Mr. Jun Yang
Department Manager of Product Development, R&D Center
Shanghai Hitachi Electircal Appliances Co., Ltd., P.R. of China

SPEAKERS

MONDAY, JULY11 | 9:30 AM | LOEB THEATRE

Viraj Vithoontien

Program Leader, Montreal Protocol Program, World Bank

"Climate Action driving our future forward, key perspectives on Climate Change and Refrigeration and Air Conditioning linkages from the World Bank Montreal Protocol Program"
Climate Change considerations are transforming both our current and future development trajectory, largely driven by concerted actions taken at the international level. The 2016 International Refrigeration and Air Conditioning Conference is happening during a seminal year of key international decisions that are helping to define future development pathways. This session will present an overview of climate change issues at the international level and discuss the many inter-connections between this and the Refrigeration and Air Conditioning sectors today. The World Bank Montreal Protocol Program operates directly at this nexus, working with developing countries and their industries to help integrate climate change concerns into their industry and national and local development programs through a combination of technical advisory and financial support.

TUESDAY, JULY 12 | 8:30 AM | LOEB THEATRE

Ian Beausoleil-Morrison

Professor, Faculty of Engineering & Design, Carleton University

"Maximizing the use of solar energy to radically reduce the energy needs of housing?"
Heating, cooling, and ventilating the places we live in, and providing the hot water, lighting, and appliance services we need, consumes tremendous amounts of energy; this contributes significantly to environmental and energy security issues. For example, housing accounts for 30% of all electricity and 26% of all natural gas consumed in Canada, and produces 13.5% of the country's greenhouse gas (GHG) emissions. Although these figures vary from region to region, a similar situation exists in most OECD countries.

If we are to meet ambitious GHG emission reduction targets and reduce dependency on fossil fuels, innovative concepts are required for providing the energy services required by housing; incremental efficiency improvements will not be sufficient.

This talk will demonstrate new techniques aimed at maximizing the use of solar energy to radically reduce housing energy demands. This will include a hybrid active-passive system to increase passive solar gains whilst protecting against overheating. Concepts for the longterm storage of solar thermal energy at the single-house scale will also be explored.

TUESDAY, JULY 12 | 12:00 PM | PURDUE MEMORIAL UNION BALLROOMS

Manpreet Singh

Purdue University Food Science Department

"Refrigeration: A Key Strategy to Global Food Security"

Improved storage and transportation of food is an important aspect that can help in reducing food waste and is a key focus of several food security initiatives worldwide. Not only do transportation mechanisms need to be efficient, but also sustainable to be viable options for use by the industry. Refrigeration and advancing the technology to maintain a cold-chain can play a vital role in tackling the issue of 1.3bn tons of food wasted annually. Extreme weather calls for energy-intensive cooling technologies compared to milder climates, and unfortunately the compounding factor of developing and under-developed countries with weather extremes pose a bigger challenge. Additionally, the demand for chilled and frozen foods is limited due to access to effective refrigeration in developing countries. Therefore, prevention of food loss is a challenge that the food industry is faced with in addition to impact on quality (nutritional and microbial spoilage) and public health risks (due to pathogens).

WEDNESDAY, JULY13 | 8:30 AM | LOEB THEATRE

Drusilla Hufford

Environmental Protection Agency

"EPA's SNAP Program: Listing Safer Alternatives and the Climate Action Plan"

In the 1990 Clean Air Act Amendments, Congress tasked EPA with phasing out ozone depleting substances, and in addition, with the task of identifying safer substitutes. In response, EPA created the SNAP program, which in its over twenty years of existence has identified and listed more than 400 alternatives for use in major consumer and industrial uses. The President's June 2013 announcement of the Climate Action Plan specifically tasked SNAP with an important role in prioritizing listing of lower-global warming potential options, and in reviewing – and possibly prohibiting - options already listed as acceptable in light of the introduction of newer, safer alternatives. We will focus in this talk on EPA's activities under SNAP to respond to the CAP. In addition, we will focus and on how those activities support the broader engagement of the U.S. domestically and internationally in the effort to address rapid growth in the use of hydrofluorocarbons (HFCs). Also a focus of the CAP was the effort to amend the international treaty, the Montreal Protocol to phase down HFCs, and this talk will also update listeners on the status of this effort.

THURSDAY, JULY14 | 8:30 AM | LOEB THEATRE

Gene Fields

Vice President of Global Compressor Technologies, Johnson Controls

"The Past, Present, and Future of Scroll Compressors"

A colorful tour through the history of this great technology, including the key advancements since the introduction of the mechanism more than a century ago; as well as the many companies and people behind these things. The progress in manufacturing as well as design techniques made this story fascinating; and the Purdue Conference played a major role in the global evolution. From the complex mathematics of involute equations, displacements, volume ratios, and gas forces; to the various applications, including AC, refrigeration, homes/buildings, and automotive; the progress continues to inspire. The basics of simulation models will also be discussed, and how they led the way for very successful scroll products. Finally, future advancements with inverter technology, increased capacities, system integration, and expanders are discussed.

SPEAKER GUIDE

BREAKFAST FOR PRESENTERS, CHAIRPERSONS AND VICE CHAIRPERSONS

A complimentary lunch (on Monday) and breakfast (Tuesday – Thursday) is scheduled in the Purdue Memorial Union, West Faculty Lounge for presenting authors, session chairs & vice-chairs scheduled for each day. You can find your presentation time in the Conference Overview. **It is important to attend the speaker lunch or breakfasts if you are presenting a paper to meet your session chair and address any presentation questions you might have. Updated presentations can also be loaded at this time.** Instructions concerning audio/visual/projection systems and technical session organization will be addressed. **Please only attend the lunch or breakfast on the day in which you present.** If you have presentations scheduled for different days, you should attend the lunch or breakfasts **each day** so that you can meet your Session Chairperson and the other Presenters. If you are the Presenter in more than one session on the same day, please meet both Session Chairpersons. If you are a Presenter but not an author of a technical paper, you should be the person who attends this breakfast so that you may alert the Chairpersons that you are the Presenter.

PROGRAM FORMAT

*NEW Presentations are on a strict time schedule this year. **Each presenter gets ONLY 20 minutes total, for both presentation and question/answers.** Each presenter is allowed 15 minutes to present and 3-5 minutes for questions. Should there be a speaker who does not show up or there is a gap for whatever reason, **we will NOT move the speaker up, as has been done in the past.** This means some sessions will indeed end early, while others will run long.

PICTURES AND VIDEOTAPING DURING THE CONFERENCES & PICTURE PHONES

Conference attendees may **NOT** take pictures, videotapes, nor use picture phones at any presentation without the consent of the author or presenter. An official photographer will be taking pictures during the conferences. Attendees may **NOT** take pictures during the tours because the research is often highly sensitive.

PRESENTATIONS

If you have last minute changes to your presentation (different from what was uploaded into Conftool), please see your session co-chair during the daily breakfast Tuesday to Thursday or during the luncheon (Monday) to make arrangements to upload your new file.

FUTURE CONFERENCE DATES

July 16-19, 2018

SHORT COURSES

Short Courses are organized and presented by the faculty of Herrick Laboratories, in cooperation with our sponsoring organizations. Often, speakers from outside organizations and university serve as presenters for these courses. One CEU credit can be earned through attendance to any of the offered short courses.

The courses will meet from 8:00 a.m. until 5:00 p.m. on Sunday, July 10, 2016, and will meet concurrently.

OIL MANAGEMENT COMPRESSORS AND THEIR SYSTEMS

The short course focuses on oil management in hermetic, positive displacement compressors, such as scroll, rotary and reciprocating compressors, and their systems. It consists of eight 50-minute lectures and will provide ample time for discussion. In particular, the following topics will be part of the short course:

- Impact of lubricants on compressor and system performance and reliability.
- Thermodynamic and transport properties of lubricant and HFC/HFO/CO2/NH3/HC refrigerant mixtures, including miscibility, solubility, and viscosity.
- Impact of oil on heat transfer and pressure drop, and oil retention in heat exchangers.
- Review of components used for oil management in vapor compression systems, such as oil separators, accumulators, and filters.
- Oil management in simple and complex vapor compression systems, including supermarket applications using multiple compressors on a rack, air-source heat pump systems, VRF systems, systems with vapor injection, industrial CO2 and NH3 refrigeration systems.
- Performance of liquid-flooded compression with regeneration systems.

FINAL FRONTIERS IN VAPOR COMPRESSION CYCLE EFFICIENCY

The efficiency of vapor compression cycles used for refrigeration and comfort cooling/heating has steadily increased since the US Department of Energy initiated minimum efficiency requirements in the early 1980s. Some current offerings have SEER values that are more than double the initial DOE minimum SEER of 10.0. These efficiency improvements have been the result of better compressor designs, larger and more efficient heat exchangers, better expansion devices, and more efficient motor technology. While efficiencies have risen substantially, they are still just a fraction of the theoretical Carnot efficiency.

The short course speakers will evaluate the various components of the vapor compression cycle to identify all areas that still offer economic opportunities for even higher system efficiencies. Case studies using optimization methods will show how to "invent" new heat exchanger shapes that go beyond tube-fin and microchannel designs, as well as overall system optimization while considering system level performance metrics. In addition, an overview will be provided for other thermodynamic cycles that may hold promise to compete with, or even surpass, the traditional vapor compression cycle that is used in the majority of refrigeration or comfort cooling/heating applications.

2016 INTELLIGENT BUILDING OPERATIONS WORKSHOP

This will be the third (previously held in 2011 and 2013) in a series of workshops that brings together researchers and developers of intelligent building features and systems. The main goals of this one-day workshop are to

1. Review fundamentals of optimal control and automated diagnostics in intelligent buildings,
2. Understand the state-of-the-art in commercialized intelligent building technologies,
3. Identify existing gaps that should be addressed in future research and development.

The format of the workshop will include a review of relevant fundamentals, followed by brief case studies that highlight recent developments and demonstrations of model-predictive controls and automated diagnostics. Ample time will be allotted for open discussion to address what is necessary to move from research to widespread adoption of intelligent building technologies.

A single registration fee will cover participation in both the IBO Workshop and the 2016 Purdue Conferences. Special IBO paper sessions are incorporated within the High Performance Building and Refrigeration and Air Conditioning Conferences where current research critical for achieving scalable and cost effective intelligent building operations will be presented. Participants in the IBO Workshop are strongly encouraged to participate in these conferences.

SOCIAL EVENTS

MONDAY, JULY 11

Reception at the Lafayette Brewing Company

The reception is hosted by **UTC Carrier** and will be held from 6:00 to 8:00 p.m. Shuttle buses will depart from the Grant Street Parking Garage between 5:30 and 6:00 p.m. transporting attendees and guests to the reception. Attendees may also use CityBus and the Trolleys or walk to the Reception. There will be transportation available after the Reception. You may need to arrange shuttle service with your individual hotels. Some local restaurants and bars are open on Monday evenings in downtown Lafayette and the Purdue campus.

Driving Directions to the Lafayette Brewing Company: From Grant Street Parking Garage: Head south on Andrew Place toward W State Street. Turn right onto W State Street. Turn Right onto N Grant Street. Sharp right onto Northwestern Avenue. Use the two left lanes to turn left onto Wiggins St. Keep left to stay on Wiggins Street. Continue half mile onto Old U.S. 231. Turn right onto N 6th Street. Turn left onto Main street. Lafayette Brewing Company is on the left. Parking is available on the street, or in the lot directly south of the bar that will cost $1 as you exit.

TUESDAY, JULY 12

Luncheon – Purdue Memorial Union, North & South Ballrooms

The buffet luncheon will be held from 12:00 to 1:30 p.m. It is hosted by **Danfoss, Inc.** There will be a Keynote Address by Purdue University's Manpreet Singh of the Food Science Department.

WEDNESDAY, JULY 13

Steak BBQ – The Trails Banquet Facility, West Lafayette

The informal steak barbeque will be held from 6 p.m. to dusk and is hosted by **Emerson Climate Technologies.** Buses will begin departing from the Grant Street Parking Garage entrance across from the Union Club Hotel at 5:45 p.m. The last bus will leave at approximately 6:15 p.m. for the barbeque. There will be games, a beer wagon and a chance to visit outside. The buses will return attendees and guests to the Purdue Memorial Union, University Inn, Holiday Inn and the Hilton Garden Inn after the barbeque. You may also drive your own vehicle to this venue.

Driving Directions to The Trails: From Grant Street Parking Garage, exit on North Street, turn left. Turn right on Grant Street, turn right on Northwestern Avenue (stop light) immediately curve left on Wiggins Street. Follow US 231 (on right) down the hill to State Road 43 North. Turn left on State Road 43 North. Proceed 5-1/2 miles to Burnetts Road, turn right. The Trails will be on your right.

STUDENT EVENTS

Student Paper Competition: Winners of the Student Paper Competition from each conference will be announced and awarded at the Steak BBQ at the Trails. Plan to join us and congratulate these students on a job well done.

ASHRAE Student Meeting – Monday, July 11 from 3:30-5:30 pm, Stewart 279

Purdue Student Chapter of ASHRAE meeting with ASHRAE President, Tim Wentz. (Open to ALL student ASHRAE members, not just Purdue student members).

Student Mixer – Tuesday, July 12 from 8-10 pm, The Stacked Pickle

Gather with other students from around the world at The Stacked Pickle located at 516 Northwestern Avenue #1700, West Lafayette, IN 47906.

CONFERENCE OFFICE / HOSPITALITY ROOM

Conference Office is located in Stewart Center, Room 306

Monday	8:00 a.m. to 4:00 p.m.
Tuesday	8:00 a.m. to 4:00 p.m.
Wednesday	8:00 a.m. to 4:00 p.m.
Thursday	8:00 a.m. to 4:00 p.m.

E-Paper Center

The E-Paper Center is located in Room 302, Stewart Center. Attendees may print up to 10 papers via the computer stations set up in room 302. You may also use the e-stations to print airline boarding passes, check emails, etc. Please be considerate of other guests, and limit time on the computer to 15 minute intervals.

Internet Access

Wireless internet is available to all attendees via the Conference Division/ATT WIFI hotspot. Please choose ATT Wifi as your internet connection. When you open a browser, you will be automatically connected. Your will need to provide your email address and accept terms & conditions. If you have difficulties, please visit the Conference Office, Stewart Center, 306.

Display Center

There will be several tables available in the Conference Hospitality Room, Stewart Center, Room 306. Please check with Conference staff before displaying any material. No commercialism is allowed. There is also a Bulletin Board where messages can be posted for other attendees and the Organizing Committee will post any last minute schedule changes.

PRINTING, COPYING AND FAXING

Attendees may print from a CD/disk/memory stick, make copies, send faxes, and more at The BoilerCopy Maker in the Purdue Memorial Union, Second Floor. Charges vary for different services.

SMOKING POLICIES ON CAMPUS

Smoking is prohibited on the West Lafayette Campus, except in designated smoking areas. All smoking material shall be extinguished and disposed of in an appropriate receptacle at the perimeter of the campus. Designated smoking areas are published in maps around the Purdue campus. The closest smoking area to Stewart Center is located near State Road 26.

TEMPERATURES INSIDE & OUTSIDE DURING THE CONFERENCES

Outside summer temperatures in Indiana are usually between 75-95˚F and very humid. Room temperatures in technical session rooms are often cool for our European visitors; come prepared!

TIME DURING CONFERENCES

The time zone in Indiana is Eastern Daylight Time. This is the same time as New York City in the summer.

PRACTICAL GUIDE

BANKS AND CREDIT UNION LOCATED NEAR THE PURDUE CAMPUS (WITH ATM'S)

REGIONS BANK

Address: Stadium Square Center, 728 Northwestern Avenue, West Lafayette, IN
Telephone: (765) 476-8205
Hours: Monday-Thursday 9:00 a.m. – 5:00 p.m.; Friday 9:00 a.m. – 6:00 p.m.

CHASE BANK

Address: Chauncey Village, 210 West State Street, West Lafayette, IN
Telephone: (765) 423-0412
Hours: Monday – Friday 9:00 a.m. – 6:00 p.m.; Saturday 9:00 a.m. – 2:00 p.m.

PURDUE EMPLOYEES FEDERAL CREDIT UNION

ATMs available at Purdue Memorial Union in the Stewart Center Lobby, Northwestern Avenue and Garfield Streets
Hours: Monday – Friday 9:00 a.m. – 6:00 p.m.; Saturday 9:00 a.m. – 2:00 p.m.

TRANSPORTATION DURING THE CONFERENCE

There will be organized transportation available for the Conference Social Events or you may choose to use your own vehicles. The Conference Hotels will provide shuttle service between the hotels and the Purdue Memorial Union. They do not operate on a regular schedule so you may call for pick-up service. Please make sure that you obtain a contact number for your hotel shuttle. The Greater Lafayette Transit System's buses and trolleys will also be available.

TROLLEY ROUTE

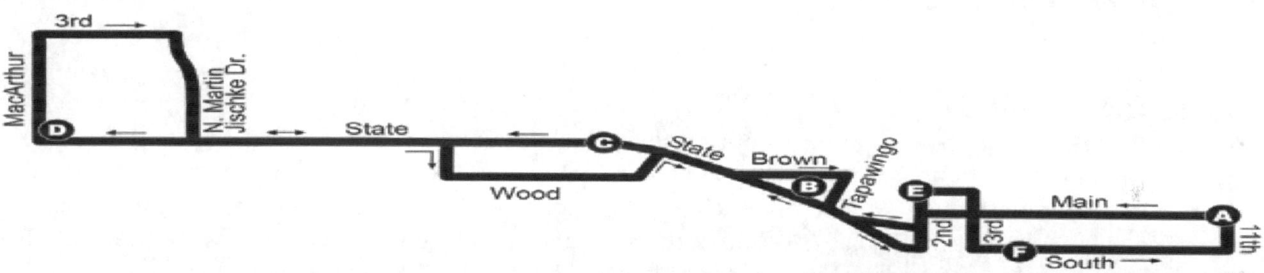

HOTEL LOCATIONS ON THE MAP & TELEPHONE CONTACT NUMBERS

- B **Hilton Garden Inn** .... (765) 743-2100
- C **The Union Club Hotel at Purdue University** .... (765) 494-8900
- D **First Street Towers** .... (765) 494-0926
- G **Holiday Inn -Lafayette City Centre** .... (765) 423-1000
- **Four Points by Sheraton** .... (765) 463-5511
  (Located on U.S. 52, West of campus)

LIMO SERVICE TO AND FROM INDIANAPOLIS AND CHICAGO AIRPORTS

LAFAYETTE LIMO SERVICE
Provides service to and from the Indianapolis International Airport
(765) 497-3828 or lafayettelimo.com

STAR OF AMERICA
Shuttle service to and from the Indianapolis International Airport
1-800-933-0097 or StarofAmerica.com

EXPRESS AIR COACH
Provides shuttle service to and from the Chicago O'Hare Airport
(765) 743-3120 or expressaircoach.com

Check their websites. Attendees must contact the companies themselves.

TRANSPORTATION SERVICES

AMTRAK – TRAIN
Riehle Plaza
Big Four Depot, 200 North 2nd Street
(800) 872-7245

GREYHOUND BUS
Riehle Plaza
Big Four Depot, 200 North 2nd Street
(765) 742-8836 or (800) 231-2222

PARKING ON CAMPUS
Parking is available in garages on the Purdue campus. The Grant Street Garage is the closest garage to the Purdue Memorial Union and Stewart Center. Attendees may park in that garage for a maximum price of $10. Guests at the Union Club Hotel can park free. Visitors can purchase "A" garage permits or multiple day passes at the Parking facilities office (494-9494) Monday-Friday 7:30-4:30. Price is $5.00/day and allows parking in any garage EXCEPT Grant Street. **Do not park in marked parking spaces or you will be ticketed. We are not able to get your ticket dismissed.**

CONFERENCE APP
Add the conference agenda on your mobile! Conference4me is a mobile app that facilitates participation in conferences and exhibitions. This app is offered free to conference attendees. Type Conference4me in Google Play / iTunes / WindowsPhone Store or scan the QR code below.

Google Play | **iTunes** | **WindowsPhone Store**

MONDAY, JULY 11

7:00am - 4:00 pm	Conference Registration – *East Foyer, Stewart Center*							
8:00-10:00 am	Welcome Continental Breakfast – *West Foyer, Stewart Center* – Hosted by Tecumseh							
9:30-11:30 am	**Opening Session, Welcome and Keynote Address** – *Loeb Playhouse, Stewart Center* VIRAJ VITHOONTIEN, Program Leader, Montreal Protocol Program, World Bank							
11:30-1:00 pm	Lunch Break (on your own)							
11:30-1:00 pm	Complimentary Lunch for Chairpersons & Presenting Authors for Monday's Sessions – *West Faculty Lounge, Second Floor, Purdue Memorial Union*							
	STEW 214AB	*STEW 214CD*	*STEW 218AB*	*STEW 218CD*	*STEW 310*	*STEW 278*	*STEW 202*	*STEW 206*
1:00-3:00 pm	B1: Supervisory Control Optimization (IBO)	B2: Building Envelope Systems	R1: Commercial & Industrial Refrigeration I	R2: Organic Rankine Cycle	R3: Heat Exchanger Design Improvements	R4: Heat Pump Design & Assessment	C1: Screw Compressors I	C2: Valves I
3:00-3:30 pm	Coffee Break – *Hospitality Room (STEW 302/306)* – Hosted by Tecumseh							
3:30-5:30 pm	B3: Model Predictive Controls (IBO)	B4: Net Zero Energy Buildings	R5: Commercial & Industrial Refrigeration II	R6: Lubricant Developments & Assessments	R7: Heat Exchanger Design Optimization	R8: Vapor Compression Cycle Enhancements	C3: Novel Compressors	C4: Heat Transfer in Compressors I
3:30-5:30 pm	ASHRAE Student Branch Meeting – *STEW 279* – ASHRAE Students meet with the ASHRAE President							
5:30-6:00 pm	Bus transportation provided from Grant Street Garage to Lafayette Brewing Company (LBC)							
6:00-8:00 pm	**Opening Night Reception** – *Lafayette Brewing Company (LBC) 622 Main Street, Lafayette, IN 47901* – Hosted by UTC Carrier							
8:00-9:00 pm	Bus transportation provided from Lafayette Brewing Company (LBC) to Grant Street Garage							

TUESDAY, JULY 12

8:00am - 4:00 pm	Conference Registration – *East Foyer, Stewart Center*							
7:15-8:15 am	Complimentary Breakfast for Chairpersons & Presenting Authors for Tuesday's Sessions – *West Faculty Lounge, 2nd Floor, Purdue Memorial Union*							
8:00-10:00 am	Continental Breakfast Available – *West Foyer, Stewart Center* – Hosted by Highly							
8:30-9:30 am	**Plenary Session** – *Loeb Playhouse, Stewart Center* IAN BEAUSOLEIL-MORRISON, Carleton University							
9:30-9:45 am	Coffee Break – *Hospitality Room (STEW 302/306)* – Hosted by Highly							
	STEW 214AB	*STEW 214CD*	*STEW 218AB*	*STEW 218CD*	*STEW 310*	*STEW 278*	*STEW 202*	*STEW 206*
9:45-11:45 am	B5: Building Data Analytics & Diagnostics (IBO)	B6: Facades, Lighting & Solar Radiation	R9: Advanced Equipment Controls I (IBO)	R10: Alternative Cooling & Heating Technologies	R11: Air-Side Heat Transfer Characterization & Enhancement	Student Paper Competition Session R	C5: Tribology & Lubrication I	C6: Scroll Compressors I
12:00-1:15 pm	**Conference Luncheon** – *North and South Ballrooms, Purdue Memorial Union* – Hosted by Danfoss (included in registration) MANPREET SINGH, Purdue University Food Science Department							
1:30-3:30 pm	B7: Demand Response & Load Mgmt (IBO)	B8: Building Analysis & Retrofits	R12: Advanced Equipment Controls II (IBO)	R13: Expansion Devices + Two Phase Separators	R14: Heat Exchanger Frost Formation	Student Paper Competition Session C		C7: Expanders I
3:30-4:00 pm	Coffee Break – *Hospitality Room (STEW 302/306)* – Hosted by Highly							
4:00-6:00 pm	B9: Control Oriented Modeling (IBO)		R15: Commercial & Industrial Refrigeration III	R16: Ejector/ Injector Analysis & Performance	R17: Heat Exchanger Re-frigerant Flow Distribution	Student Paper Competition Session B	C8: Heat Transfer in Compressors II	C9: Linear and Reciprocating Compressors
6:00-8:00	Tours of Herrick Laboratories							
7:00-8:30 pm	**Conference Advisory Committee Meeting (by invitation only)** – *Burton Morgan Café, near Herrick Laboratories*							
8:00-10:00 pm	**Student Mixer** – *The Stacked Pickle (516 Northwestern Ave. #1700, West Lafayette, IN 47906)*							

**IBO = Intelligent Building Operations Session*

WEDNESDAY, JULY 13

7:15-8:15 am	Complimentary Breakfast for Chairpersons & Presenting Authors for Wednesday's Sessions – *West Faculty Lounge, 2nd Floor, Purdue Memorial Union*							
8:00-10:00 am	Continental Breakfast Available – *West Foyer, Stewart Center* – Hosted by Kawneer							
8:30-9:30 am	**Plenary Session** – *Loeb Playhouse, Stewart Center* – DRUSILLA HUFFORD, Environmental Protection Agency							
9:30-9:45 am	Coffee Break – *Hospitality Room (STEW 302/306)* – Hosted by Kawneer							
	STEW 214AB	*STEW 214CD*	*STEW 218AB*	*STEW 218CD*	*STEW 310*	*STEW 278*	*STEW 202*	*STEW 206*
9:45-12:00pm	B10: Occupant Behavior and Predictive Models (IBO)	B11: Thermal Storage & Heat Pumps	R18: Rooftop Unit Diagnostics (IBO)	R19: Transient System Modeling	R20: Evaluation of R410A Alternatives	R21: Transportation Air-Conditioning	C10: Noise, Vibration & Harshness I	C11: Compressor Modeling
12:00-1:00 pm	Lunch Break							
1:00-3:00 pm	B12: Building Simulation and Optimization I	R22: Absorption / Adsorption Technology I	R23: Analysis of Systems with Ejectors	R24: Boiling Heat Transfer Enhancements	R25: Domestic Refrigeration I	R26: Heat Exchanger Design, Manufacturing & Operational Impacts	C12: Expanders II	C13: Scroll Compressors II
3:00-3:30 pm	Coffee Break – *Hospitality Room (STEW 302/306)* – Hosted by Kawneer							
3:30-5:30 pm	B13: Ventilation, Passive Cooling & IAQ	R27: Absorption Technology II	R28: Heat Exchanger Modeling and Characterization	R29: Desiccant and Other Heat/Mass Transfer Studies	R30: Domestic Refrigeration II	R31: Property Measurements, Modeling & Assessments I	C14: Tribology & Lubrication II	C15: Rotary Compressors I
5:45-6:15 pm	Shuttle buses will transport attendees from the Grant Street Parking Garage to the Steak Barbeque							
6:00-10:00 pm	**Steak Barbeque** – *The Trails (325 Burnetts Rd, West Lafayette, IN 47906)* – Hosted by Emerson Climate Technologies							
9:30-10:00 pm	Shuttle buses will transport attendees from the Steak Barbeque to the Grant Street Parking Garage and area hotels							

THURSDAY, JULY 14

7:15-8:15 am	Complimentary Breakfast for Chairpersons & Presenting Authors for Thursday's Sessions – *West Faculty Lounge, 2nd Floor, Purdue Memorial Union*							
8:00-10:00 am	Continental Breakfast Available – *West Foyer, Stewart Center* – Hosted by Parker Hannifin							
8:30-9:20 am	**Plenary Session** – *Loeb Playhouse, Stewart Center* – GENE FIELDS, Johnson Controls							
9:30-9:45 am	Coffee Break – *Hospitality Room (STEW 302/306)* – Hosted by Parker Hannifin							
	STEW 214AB	*STEW 214CD*	*STEW 218AB*	*STEW 218CD*	*STEW 310*	*STEW 278*	*STEW 202*	*STEW 206*
9:45-12:00pm	R32: Equipment Performance Measurements and Modeling	R33: Evaluation of Natural Refrigerants	R34: Flow Boiling	R35: Residential Heat Pumps	R36: Property Measurements, Modeling & Assessments II	R37: Plate Heat Exchangers	C16: Testing & Measurement	C17: Centrifugal Compressors
12:00-1:00 pm	Lunch Break – Boxed lunches available in the Conference Hospitality Room – Hosted by Rheem							
1:00-3:00 pm	R38: HVAC Equipment Performance Enhancements	R39: Alternative Refrigerant Evaluation Methods & Results	R40: Refigerant Heat Transfer & Pressure Drop I	R41: Heat Pump Water Heaters	R42: Thermal Storage		C18: Noise, Vibration & Harshness II	C19: Rotary Compressors II
3:00-3:30 pm	Coffee Break – *Hospitality Room (STEW 302/306)* – Hosted by Parker Hannifin							
3:30-5:30 pm	R43: Air Conditioning Equipment Assessments	R44: Electronics/ Thermoelectric Cooling	R45: Refrigerant Heat Transfer & Pressure Drop II	R46: Geothermal/ Ground Source Heat Pumps			C20: Valves II	C21: Scroll Compressors III
3:30-5:30pm	Conference Advisory Committee Meeting (by invitation only) – *STEW 307*							
5:30pm	End of Conference							

**IBO = Intelligent Building Operations Session*

STEWART CENTER

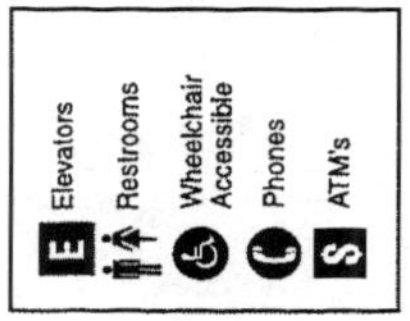

FIRST FLOOR

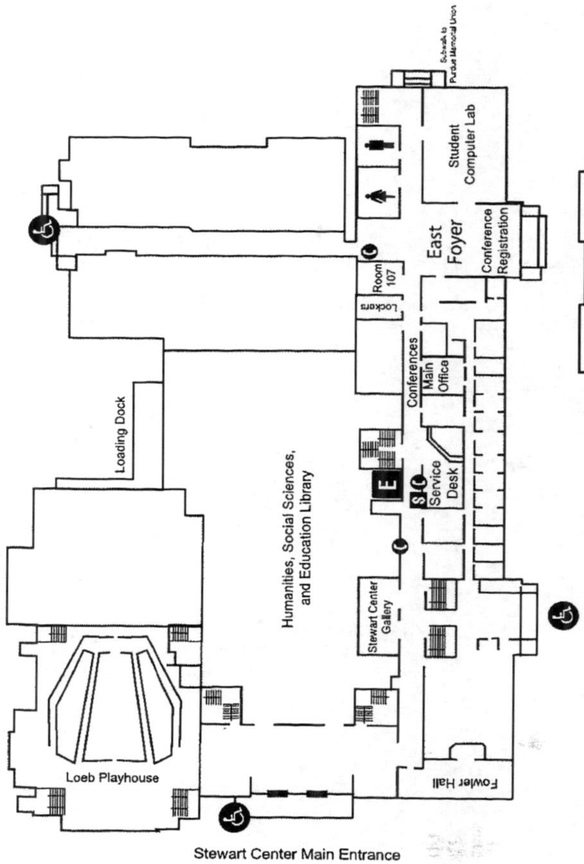

SECOND FLOOR

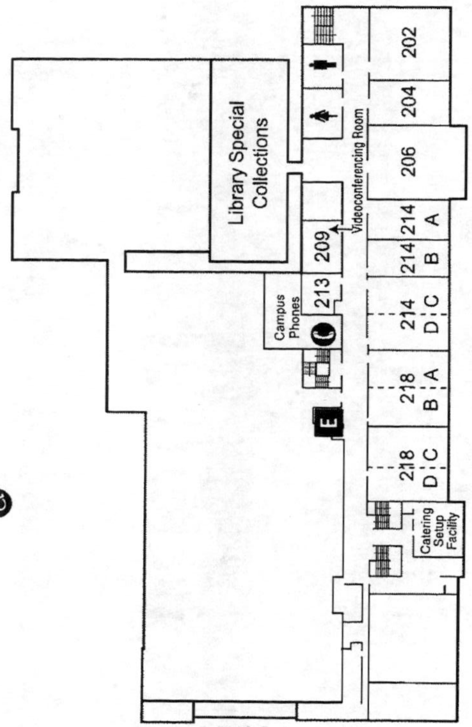

THIRD FLOOR

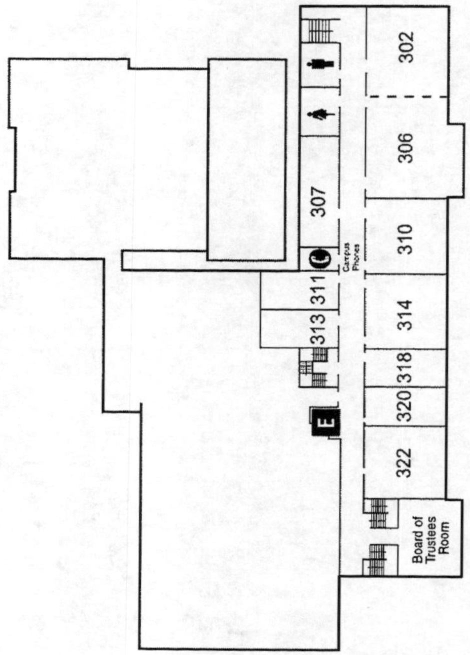

Carrier
turn to the experts
Danfoss
EMERSON
Climate Technologies
HIGHLY
KAWNEER
Parker
Rheem
The new degree of comfort.
Tecumseh

SPONSORS

CFD Analysis of Oil Flooded Twin Screw Compressors

Sham RANE*, Ahmed KOVACEVIC, Nikola STOSIC

City University London, Centre for Compressor Technology,
London, EC1V 0HB, UK
sham.rane@city.ac.uk

* Corresponding Author

ABSTRACT

Modelling of screw compressors using Computational Fluid Dynamics (CFD) offers better insight into the working chamber of twin screw machines when compared with chamber models. As shown by authors in earlier publications, CFD models predict performance of dry gas and refrigeration compressors fairly accurately. However numerical flow models used for modelling of oil flooded twin screw compressors are still at the development stage. This is mainly due to the lack of understanding of the flow complexity and the techniques used for solving coupled equations that represent interactions between the gas and the oil in such machines.

This paper presents the modelling approach used for calculation of the performance of an oil flooded screw compressor. It requires a structured numerical mesh which can represent all moving parts of the compressor in a single numerical domain. Such mesh is generated by SCORG^{TM} using novel boundary distribution technique called casing-to-rotor conformal boundary mapping. A test oil injected twin screw compressor with rotor configuration 4/5 and 127 mm main rotor diameter was measured in the compressor rig of the Centre for Compressor Technology at City University London. Measurements of the chamber pressure history and integral parameters of the compressor such as mass flow rate of gas and oil, indicated power and temperatures are used for the comparison with CFD results. The analysis showed a close match in the prediction of the mass flow rates of gas. The indicated power obtained by CFD predictions matched well with the measured shaft power. The model provided an exceptional visualization of the interaction of gas and oil inside the compression chamber. The mixing of the phases, distribution of oil, heat transfer between gas and oil and also effects on sealing due to high oil concentration in leakage gaps were well captured.

1. INTRODUCTION

Oil is injected into the working chamber of twin screw compressors mainly for three reasons. The main reason is to control the gas temperature during the compression. For high pressure ratios between the suction and the discharge the temperature rise in the absence of internal cooling may be significant and can result in thermal deformation of the rotors and the casing, eventually leading to seizure. Since early times, the adopted practice for twin screw compressors and many other positive displacement compressors is to inject oil in the compression chamber for cooling. Secondly, the injected oil reduces clearance gaps in the working chamber, thereby improving volumetric efficiency. Thirdly, the injected oil lubricates rotors and bearings. The rule of thumb in screw compressor industry is that oil contribution by mass to cooling, sealing and lubrication is in the proportion of 100:10:1 respectively. Although oil injection has several benefits and is essential for functioning of the compressor, the presence of oil has negative effects. Being of high viscosity, the shear of oil in clearances contributes to the additional mechanical losses. The bulk of the oil with density several orders of magnitude higher than the compressed gas has a high momentum and consumes some of the input power during injection and transport through the chambers. Deipenwisch and Kauder (1999) used a mathematical model for predicting oil induced losses in twin screw compressors shown in Figure 1 as a representative of contribution of the losses associated with oil injection in a typical screw compressor at a male rotor tip speed of 35 m/s. As such, oil injection needs to be controlled so that optimum quantity of oil is used and maximum effectiveness is achieved. 90% of the oil is injected through the oil ports that are timed accurately with the compression cycle such that for the given injection pressure and compressor internal pressure the desired quantity of oil mass moves into the compression process.

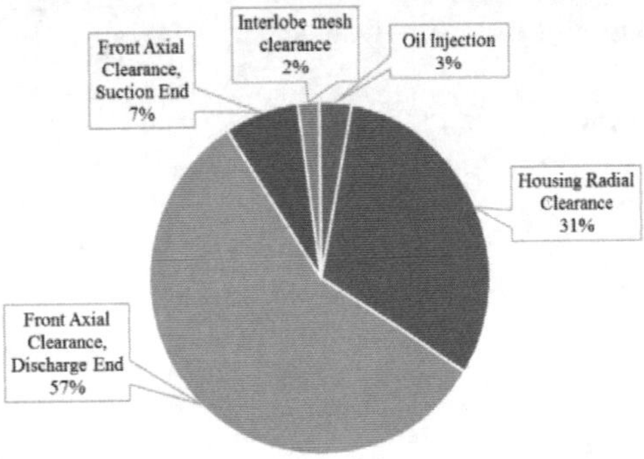

Figure 1: Oil power loss contributions in leakages (Reproduced from Deipenwisch and Kauder, 1999)

Figure 2 shows an example of the oil injection ports. It is possible to have oil injection systems where the compressor discharge pressure is utilized to inject the oil. In order to improve reliability or to account for high variations in loading many systems use an independent oil pumping system which often provides better control of oil injection but comes at the cost of additional power loss. It is essential to have a guideline for a compression system to vary the oil injection pressure for varying rotor speed or discharge pressure and a computational model can effectively provide this type of data. Additionally the oil temperature at the point of injection is important in achieving effective heat transfer.

If the oil temperature is very close to the gas temperature at injection point then there will be low convection; if the oil temperature is very low compared to gas temperature then due to a short residence time in the compression chamber, the heat transfer will continue in the discharge port thus increasing the load on the oil cooling heat exchanger downstream, without any benefit inside the compression chamber. It is required to develop a computational model that can provide a complete description of the flow process inside the oil flooded compression chamber for its design and optimisation.

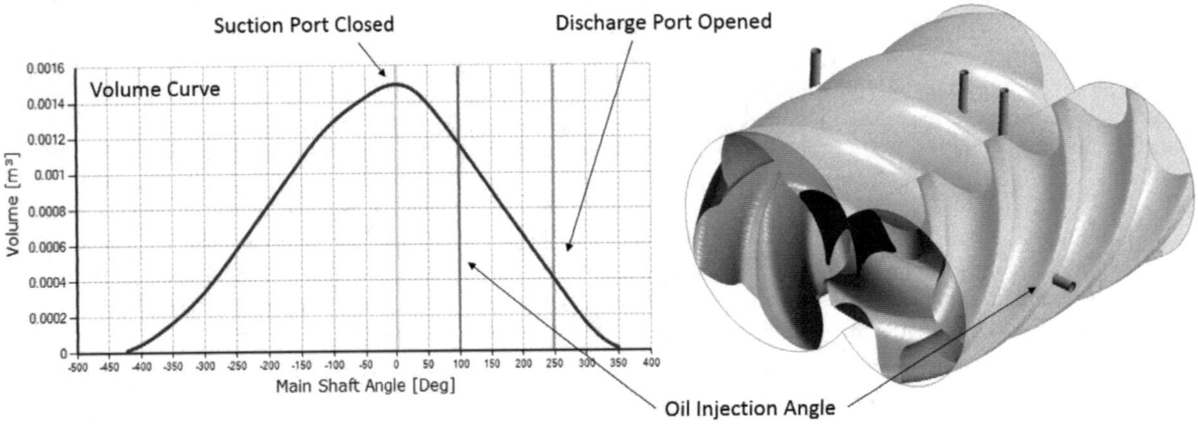

Figure 2: Oil injection ports and injection angle on a compression volume curve

CFD can provide a useful means of modelling the oil injection and study the physical phenomenon in detail in order to understand the complete behaviour of oil flooded twin screw compressors. Kovačević (2002, 2005) successfully used an algebraic grid generation method with boundary adaptation and transfinite interpolation which has been implemented in the program SCORG. Kovačević *et al.* (2007) have reported CFD simulations of twin screw machines to predict flow, heat transfer, fluid-structure interaction, etc. Kovačević (2002) also reported a test case study of an oil injected compressor using source terms in transport equations with a segregated pressure based solver. Vande Voorde *et al.* (2005) used an algorithm for generating block structured mesh from the solution of the Laplace equation for twin screw compressors and pumps using differential methods. Reports on analysis of dry air screw compressors and twin screw expanders with real gas models are available in literature using these techniques (Papes *et al.*, 2013). Recently, Arjeneh *et al.* (2014) have presented the analysis of flow through the suction port of a screw compressor with water injection. It was reported that it was difficult to stabilize the solver in a full 3D analysis with both deforming rotor domains and multiphase models. Although several attempts have been made in the recent past to extend the CFD technology to oil injected compressors, it has proven to be difficult to achieve the desired grid structure and the modelling conditions that can provide stability to the numerical solvers. Rane (2015) in his thesis proposed a new analytical approach for grid generation that can independently refine the interlobe region of the screw rotors. It was demonstrated that such grid refinement improves the prediction of mass flow rates. The same algorithm has been extended to produce a rotor grid that eliminates the interface between two rotors and thereby providing a desirable grid for oil injected or multiphase modelling. An attempt has been made in the presented work to utilize the latest developments in CFD technology for modelling oil injected twin screw compressors.

2. APPROACHES FOR MODELLING OIL INJECTION IN SCREW COMPRESSORS

2.1 Thermodynamic chamber model approach

Stošić *et al.* (1992, 2005) have presented an oil injected twin screw compressor model by considering the working chamber, suction and discharge process as an open thermodynamic system; mass flow varying with time and gas defined via equation of state. In such type of lumped parameter models, oil droplets are assumed to have a mean Sauter diameter and the heat exchange between the spherical droplets and the gas via convection can be balanced with the droplet temperature rise. The differential form of this energy balance is given as Equation 1.

$$\frac{dT_{oil}}{d\theta} = \frac{h_{oil}A_{oil}(T_{gas} - T_{oil})}{\omega m_{oil}c_{oil}} \quad (1)$$

Using, $Nu = 2 + 0.6\,Re^{0.6}Pr^{0.33}$, heat transfer coefficient for Stokes flow.

And integration of the equation gives the oil droplet temperature at each time step.

$$T_{oil} = \frac{T_{gas} - k\,T_{oil,p}}{1+k}$$

$T_{oil,p}$ is the oil droplet temperature at the previous time step.

$k = \frac{\tau}{\Delta t} = \frac{\omega\tau}{\Delta\theta}$, with $= \frac{m_{oil}c_{oil}}{h_{oil}A_{oil}}$, being the non-dimensional time constant of the droplet.

For a given mean Sauter diameter d_s, the non-dimensional time constant k is

$$k = \frac{\omega m_{oil}c_{oil}}{h_{oil}A_{oil}\Delta\theta} = \frac{\omega d_s c_{oil}}{6h_{oil}\Delta\theta}$$

If k tends to zero, the oil and gas temperatures will be equal, while for finite values of k, the gas and oil temperatures will differ. The above described approach is based on the assumption that the oil-droplet time constant τ is smaller than the droplet travelling time through the gas before it hits the rotor or casing wall, or reaches the compressor discharge port.

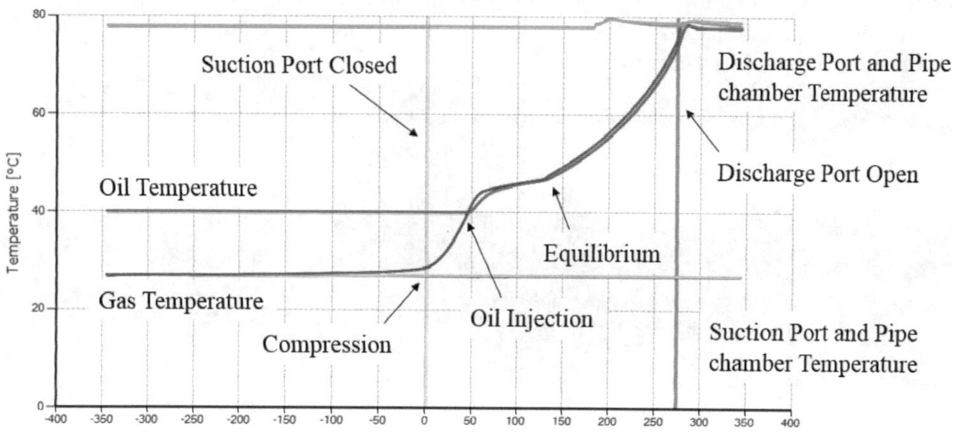

Figure 3: Oil injection modelling using chamber models

This means that the heat exchange is completed within the droplet residence time through the gas during the compression process. This prerequisite can be fulfilled by appropriate atomization of the injected oil which produces sufficiently small droplet sizes producing sufficiently small droplet time constant, as well as by choosing adequate nozzle angle and to some extent the initial oil spray velocity. Figure 3 shows a representative result of the gas and oil temperature variation calculated using a multi-chamber thermodynamic model.

2.2 CFD approach using source term formulations

Computational fluid dynamic solvers calculate the solution of the conservation of mass, momentum, energy and other transported quantities by numerically integrating the governing transport equations over the entire computational domain. Equation 2 is the general transport equation in which the last term is a generic source term and can be formulated in order to account for additional physicals effects that are not directly modelled in the solver.

$$\frac{\partial}{\partial t}\int_{\Omega}\rho\phi d\Omega + \int_{S}\rho\phi \boldsymbol{v}.\boldsymbol{n}dS = \int_{S}\Gamma\ grad\ \phi.\boldsymbol{n}dS + \boxed{\int_{\Omega} q_{\phi}d\Omega} \tag{2}$$

where, Ω is the volume of CV, S is the surface enclosing this CV, $\boldsymbol{n}$ is the unit vector orthogonal to S and directed outwards, $\boldsymbol{v}$ is the fluid velocity inside CV in a fixed coordinate system, ρ is fluid density and t is time. Here, the transient term represents rate of change of property ϕ in CV, the convection term accounts for the net advective transport of ϕ across surface S and diffusive transport is described by the gradient. Γ is the diffusivity for the quantity ϕ while q_{ϕ} represents source or sink of ϕ in the CV. In the approach of modelling oil injection using source term formulation, it is possible to model the thermal and sealing effects of the injected oil while still calculating the main flow field as a single gas phase. Even in conditions where the mass of injected oil is 10 times the mass of gas, the volume occupied by oil to that of gas is in the order of 1:100. Hence a passive scalar transport equation can be solved to model the distribution and mixing of oil in the gas. The numerical solver will solve for continuity of the gas, its x, y and z momentum components and total energy equation while the physical effects of oil injection will come from q_{ϕ} applied at the oil concentration regions calculated by passive scalar transport or at selected regions of the domain.

Thermal effect: The source term formulation for energy equation is such that if the gas temperature exceeds the injection oil temperature, an energy sink will become active. The assumption in this approach is that the heat transfer between gas and oil is instantaneous and the mass of oil is sufficient enough to completely get into thermal equilibrium with the gas. This energy source can be limited to the rotor domains and defined as in Equation 3.

$$q_e = \frac{\rho C_p}{\partial t}\left(T_{gas} - T_{oil}\right),\ q_{e\_c} = \frac{\partial}{\partial T}\left\{\frac{\rho C_p}{\partial t}\left(T_{gas} - T_{oil}\right)\right\} \tag{3}$$

q_{e_c} is a source coefficient and is required in order to linearize the source in discretized equations. T_{oil} is set at the oil injection temperature. The source term is such that if the gas temperature exceeds oil injection temperature, heat is removed from the computational cell. In this case a lower temperature limit source is required to stabilize the solution process. Figure 4a presents the temperature field in the rotor domain produced by such an energy source formulation. As seen the gas temperature is almost clipped to the oil injection temperature everywhere in the domain after the compression process begins. A careful calibration of such formulation is required to get basic results such as pressure rise and indicated power from the analysis. The accurate prediction of gas temperature rise and oil losses is challenging and dependant on the calibration of the source term.

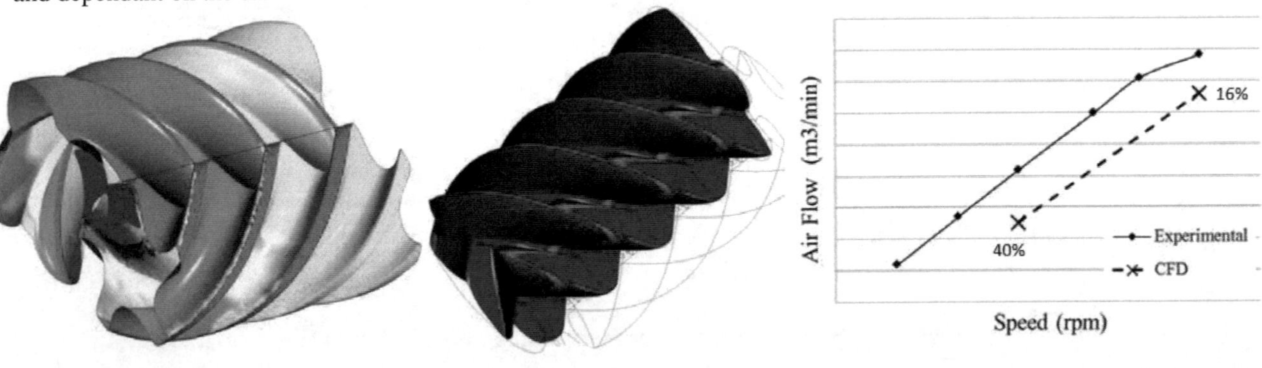

a. Temperature Result under an energy source formulation

b. Local Viscosity Function

c. Under prediction of flow under a local viscosity formulation

Figure 4: CFD approach for oil injection with source formulations

Sealing effect: A momentum source term in x, y, and z direction can be added to the momentum transport equations. The formulation is presented in Equation 4 and it can be applied in the leakage gaps to reduce the velocity of the gas.

$$\begin{aligned} q_u &= \frac{1}{2}\frac{\rho}{\partial t}(|u|\,u),\quad q_{u\_c} = \frac{\partial}{\partial u}\left\{\frac{1}{2}\frac{\rho}{\partial t}(|u|\,u)\right\} \\ q_v &= \frac{1}{2}\frac{\rho}{\partial t}(|u|\,v),\quad q_{v\_c} = \frac{\partial}{\partial v}\left\{\frac{1}{2}\frac{\rho}{\partial t}(|u|\,v)\right\} \\ q_w &= \frac{1}{2}\frac{\rho}{\partial t}(|u|\,w),\quad q_{w\_c} = \frac{\partial}{\partial w}\left\{\frac{1}{2}\frac{\rho}{\partial t}(|u|\,w)\right\} \end{aligned} \tag{4}$$

Cell volume parameter is utilized in order to mask the region that are in the leakage gaps using a dynamic masking function since the rotor domain is deforming and cell volumes move with time between the main domain and the leakage gaps. This method may make the solver unstable with the increase in the discharge pressure which results in the higher pressure gradients across the gaps. An alternative approach is to identify cells in the clearance gaps and modify the molecular viscosity of the gas such that it is close to that of the oil. The increase in viscosity results into the added flow resistance thus imitating the sealing effect of injected oil. This method has a limit to which the viscosity could be increased due to solver stability issues. Figure 4b shows the use of a local viscosity function. The resulting influence on the flow rate is shown in Figure 4c. The improvement in the air flow rate in comparison to a case without viscosity function was observed but it was not sufficient to match the measurements. Similar to the momentum source terms, the viscosity function affects solver stability when the increase in artificial viscosity exceeds 3-4 orders of the gas viscosity.

2.3 CFD approach using Multi-phase models

The complexity of the flow inside the compressor flow domain makes modelling challenging since variety of flow regimes can occur within the process. The oil injection itself starts as a spray formation. Before or after breakup the rotor surface is impinged with oil and oil film is formed. The film grows and spreads until the injection nozzle is cut-off from the compression chamber by the rotors trailing lobes. Further rotation causes flooding on the chamber whose volume is continuously reducing and squeezing the oil contained. In order to capture all these effects a top level model is required that can account for large changes in oil volume fraction. The applicable multiphase modelling treatment depends on the phase coupling effects that are necessary to be accounted for, as shown in Figure 5.

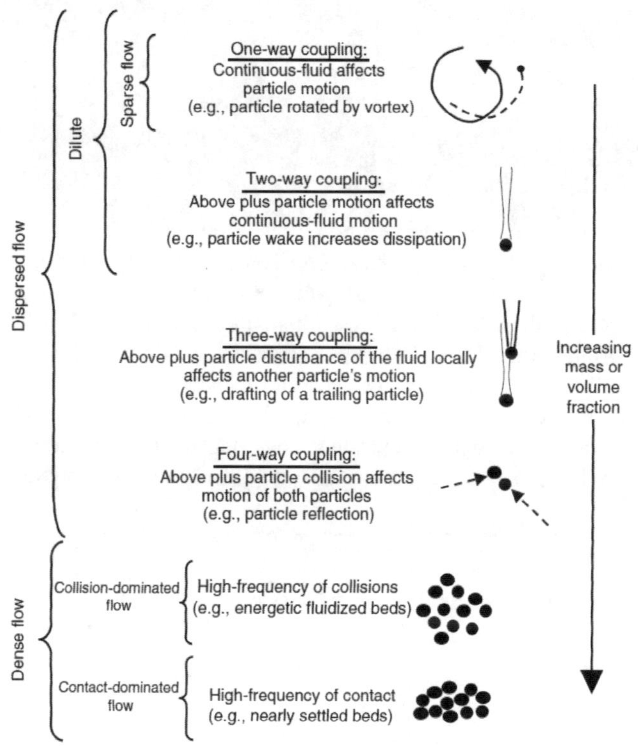

Figure 5: Flow conditions based on interphase couplings

Crowe *et al.* (2006) has presented a detailed characterization of multiphase flow regimes based on the various coupling effects between the continuous and the dispersed phases. If the dispersed phase quantity is very low and the phase particles are very fine with negligible momentum then the compressed gas is treated as the continuous phase and oil droplets as particles in the Lagrangian frame with just one way coupling, i.e. oil will move with the flow and heat transfer between phases will occur. Such a computational model has five transport equations for the gas phase and one momentum (motion kinematics) equation for the liquid phase that needs to be conserved. On the other end of the spectrum is the condition of heavily oil flooded operation. In addition to the oil droplets, oil film on the rotor and housing will occur resulting in oil being dragged along with the compressed gas. Under such conditions, Eulerian – Eulerian treatment for both the gas and oil phases is required; both can be continuous phases with energy, momentum interphase transfer effects. Such a computational model is the most expensive and has five transport equations for each of the gas and oil phases. The pressure field solution is shared by the two phases and the independent momentum equations can calculate relative slip and shear between the gas and oil.

In describing multiphase flow the concept of phase volume fractions is utilized. The phase volume fraction represents the space occupied by each phase in a computational cell in which case it is required that conservation of mass, momentum and energy is satisfied by each phase individually. As such Eulerian - Eulerian is the most accurate treatment for any multiphase flow regime and is the necessary model to be used for oil injected screw compressors. A simplification of the Eulerian – Eulerian approach is available for flows which have negligible slip between the gas and oil phases and this homogeneous multiphase treatment shares a common pressure as well as velocity field for both the phases. In the presented work a non-homogeneous Eulerian – Eulerian treatment has been used.

3. GRID GENERATION FOR OIL INJECTED SCREW ROTORS

By applying the principles of analytical grid generation through transfinite interpolation with adaptive meshing, the authors have derived a general, fast and reliable algorithm for automatic numerical mapping of arbitrary twin screw machine geometries. The procedure of analytical grid generation of screw machine working domain is explained in Kovačević (2002). In order to achieve a conformal single domain mesh, a new approach of background blocking has been presented in Rane (2015). In this procedure, outer boundary in each background block is defined as a combination of the rack segment and the casing circle segment. The rack segment stretches between the lower and higher cusp points and is closed by the casing as shown in Figure 6. The distribution obtained on the outer boundaries of the two blocks is used as the reference for the rotor profile distribution.

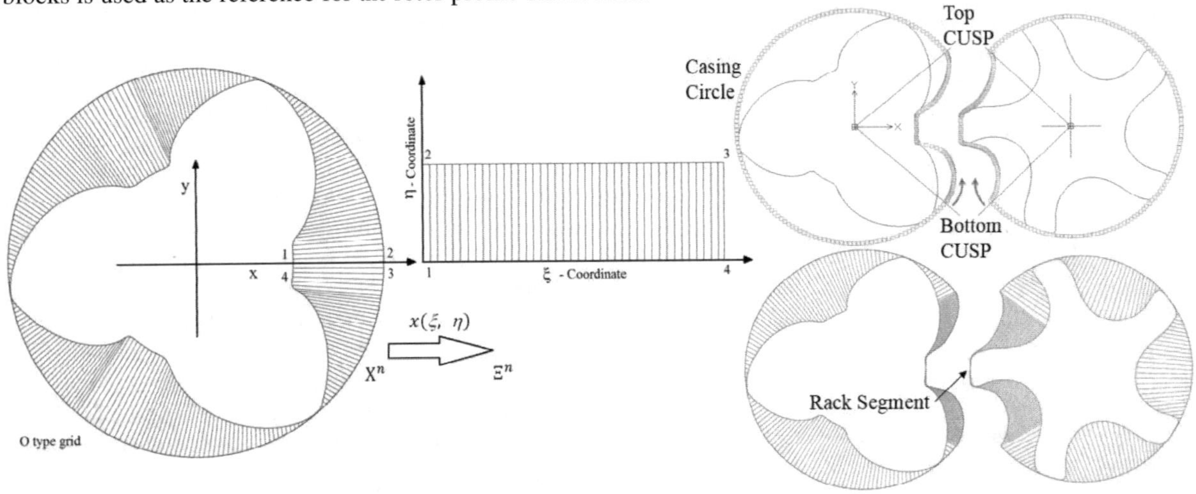

Figure 6: Background blocking and distribution on the main and gate rotors

Since the blocks of the main and gate rotors are obtained separately, the intersection points obtained on the common rack curve from the two blocks can be different. If different the resulting mesh has a non-conformal map between the two rotor blocks. Using the blocking approach it is possible to transform this boundary map into a conformal boundary map. The surface mesh on the interlobe surface mostly follows the axial grid lines with small transverse movements in the vicinity of the cusp points. The movements of points are along the surface of the interface and do not cause any irregularity of cells. This implementation allows for a fully conformal interface with the equal index around cusp points to ensure accurate capturing is the blow hole leakage area. The blocking approach is in more detail explained in Rane (2015). It allows both conformal and non-conformal boundary map to produce fully hexahedral 3D grid with both, the main and the gate rotor surfaces smoothly captured. The surface mesh on the casing is of the highest quality with the regular quadrilateral cells.

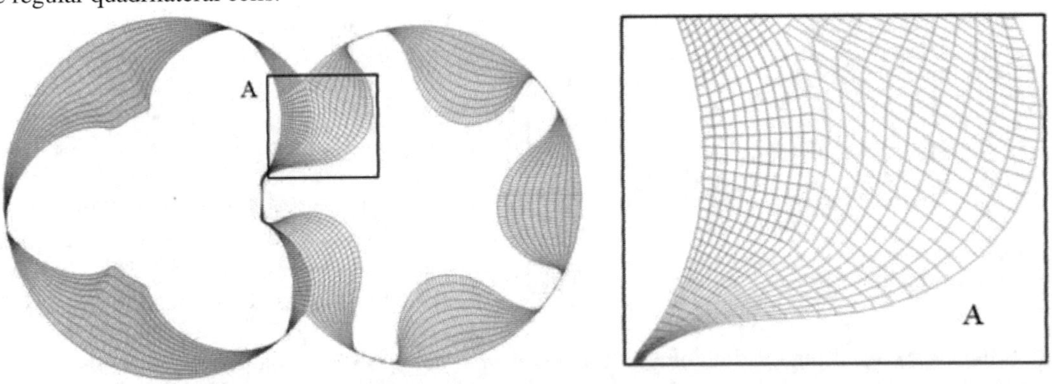

Figure 7: Grid generated with single domain for the main and gate rotors

Figure 7 shows the nodes on the rack segment between the main and the gate rotor grids with conformal boundary map. The 3D mesh generated from such 2D cross sections allows the rotor domains of the male and female rotors to be combined into a single rotor mesh. This avoids inaccuracies and instabilities that may arise due to the interface mismatch in non-conformal boundary map. The resultant grids are recommended for oil injected flow modelling.

4. CASE STUDY

The test compressor used for this study is an oil flooded twin screw compressor with a combined axial and radial suction and an axial discharge ports. The male rotor has four lobes and the female rotor has five lobes with 'N' profiles. The nominal interlobe, radial and axial leakage gaps are 60 μm. The built-in volume index V_i of the compressor is 3.6. The main rotor diameter is 127.0mm and the wrap angle is 306°.

The operating speed of the machine is between 3000 rpm and 6000 rpm, discharge pressure can vary between 4.0 bar and 12.0 bar. Measurements at few operating points were carried out at the air compressor test rig at City University London and the results have been used in this paper for comparison with CFD model predictions. The compressor is driven by a 75kW motor with a variable frequency drive for speed control. The discharge pressure is controlled by a valve and gas flow is measured using an orifice in the discharge line. In the current setup oil is injected using the discharge line pressure and oil temperature is controlled using a water cooled heat exchanger.

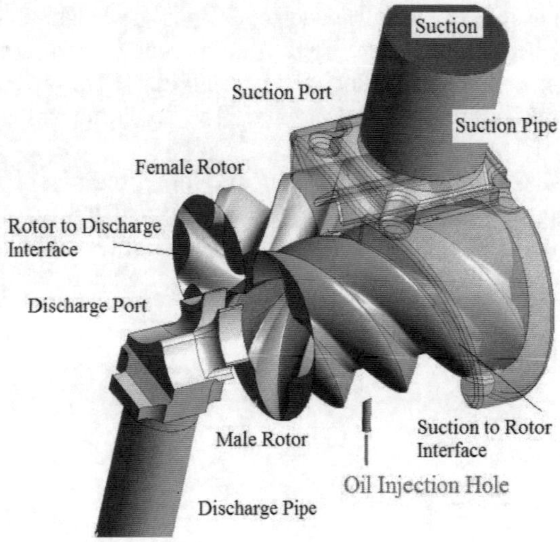

Figure 8: CFD Model of the compressor

4.1 CFD model

Figure 8 shows the CFD model domain that consists of the suction port and discharge port. Both the ports have an extended pipe to provide boundary pressure conditions. The oil injection is through four injection holes that are connected to the rotor domain by fluid-fluid interfaces. The rotor domain itself is a deforming grid, generated using the procedure described in Section 3. Three operating conditions have been calculated, listed in Table 1.

Table 1: CFD model cases and operating conditions

Case	Speed (rpm)	Suction Pressure (bar)	Suction Gas Temperature (K)	Discharge Pressure (bar)	Oil injection Pressure (bar)	Oil Injection Temperature (K)
1	3000	1.00	298.0	6.0	5.5	323.0
2	3000	1.00	298.0	8.0	7.5	323.0
3	6000	1.00	298.0	8.0	7.5	323.0

Table 2: Mesh statistics and cell quality report

Domain	Cell Structure	Node Count	Cell Count	Orthogonality Angle (Min)	Expansion Factor	Aspect Ratio
Rotor	Hexahedral	468677	406368	7.4	646	488
Suction Port	Tetra + Hex	119058	203255	30.2	279	9
Discharge Port	Tetra + Hex	98521	253095	19.6	53	28
Oil Injection Port	Hexahedral	28340	25144	55.7	4	4

The axial end clearances are not included in the CFD model for simplification. Eulerian – Eulerian multiphase model as described in Section 2.3 has been used. Air phase is defined as an ideal gas state with molar mass of 28.96 kg/kmol, specific heat capacity $1.0044e^{03}$ J/kg K, dynamic viscosity $1.831e^{-05}$ kg/m s and thermal conductivity $2.61e^{-02}$ W/m K. Oil phase is defined as constant density of 800 kg/m^3, specific heat capacity 1800.0 J/kg K, dynamic viscosity 0.0088 kg/m s and thermal conductivity 0.18 W/m K. Dynamic viscosity of oil is a function of temperature but it was defined constant in expected temperature range. Interphase heat transfer has been defined by assuming an interfacial length scale of 5.0 μm and Nu equal to 50. Same interfacial length scale is also used for interphase slip. The solver was set with first order upwind advection scheme, first order transient scheme and relaxation factors of the order of 15%. The model mesh statistics is presented in Table 2 along with the cell quality report. The rotor domain had the minimum orthogonality factor but it was below value of 10.0 in less than 1% cells. Both expansion factor and aspect ratio were in good control due to the fewer divisions in the radial direction of the rotor as well as the absence of a boundary layer mesh in the ports with tetrahedral grids. The turbulence model SST – k Omega was used.

The discharge and oil injection pressure had to be gradually increased over more than 4 full rotations of the main rotor. Calculations were continued until the flow and pressure pulsations showed cyclic repetitions. Mass imbalance for air phase was below 9% and for oil phase was about 11% in all three cases. This imbalance could be due to the grid quality, lower order discretization schemes used and the convergence level achieved which was in the range of $5e^{-4}$. On a 6 node parallel calculation with 3GHz processor and 32GB RAM, each computation took about 350 hours physical time for convergence. The flow time was equivalent of 15 rotations of the main rotor at full pressure.

4.2 Comparison of the flow and indicated power with measurements

Figure 9 shows a comparison of normalized mass flow rate of air and the normalized indicated power between measurements and CFD results for the three cases. Experimental flow and power at Case 1 conditions of 3000rpm, 6.0 bar discharge pressure have been used here for normalization.

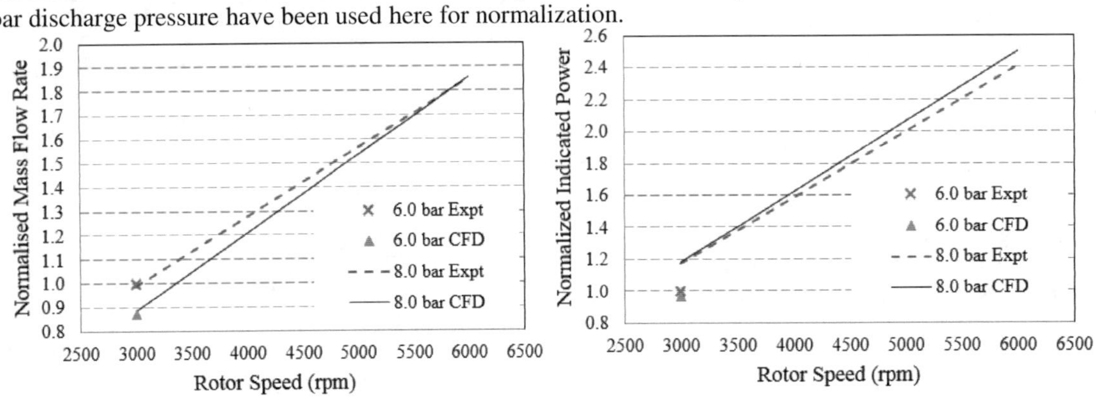

Figure 9: Comparison of Flow and Indicated Power between experimental and CFD results

It can be noted that at 3000 rpm at both discharge pressures the gas flow prediction is within around 12% lower than the measured values, while at 6000 rpm and 8.0 bar discharge, the predicted flow is within 2% of measurements. From 3000 to 6000 rpm, the flow has increased by 85%. Indicated power at 3000rpm is calculated within 4% of the measurements while at 6000 rpm it is about 10% higher than the measured value. It has to be noted that, in order to compare with CFD results, the measured shaft power was multiplied by the estimated constant mechanical efficiency of 70%. This may not be the most accurate estimate which could be overcome by the addition of a mechanical and bearing loss model to include additional mechanical losses into the CFD model instead. Power at 3000 rpm, 8.0 bar is 20% more than that at 6.0 bar. From 3000 to 6000rpm, indicated power increased by more than twice.

4.3 Pressure distribution

Figure 10 shows the variation of pressure in the compression chamber with the main rotor rotation angle. The distribution of pressure in rotors and the discharge with an iso-surface of oil volume fraction at 0.1 is also plotted. From the pressure angle plots it could be observed that the pressure during the compression part of the process aligns closely. At 3000 rpm when the discharge port opens, the flow pulsations in the port are low. These increase with the increase in pressure. Still for 3000 rpm pulsations and the maximum pressure are both lower that at the higher speed. At 8.0 bar discharge pressure and 6000 rpm the peak pressure reaches 10.0 bar visible in the contour plot in Figure 10. For the built-in volume ratio V_i=3.6, the 8.0 bar discharge pressure is in under-compression condition visible in the pressure curve. The peak over-compression at higher speed is due to the throttling and dynamic losses in the port.

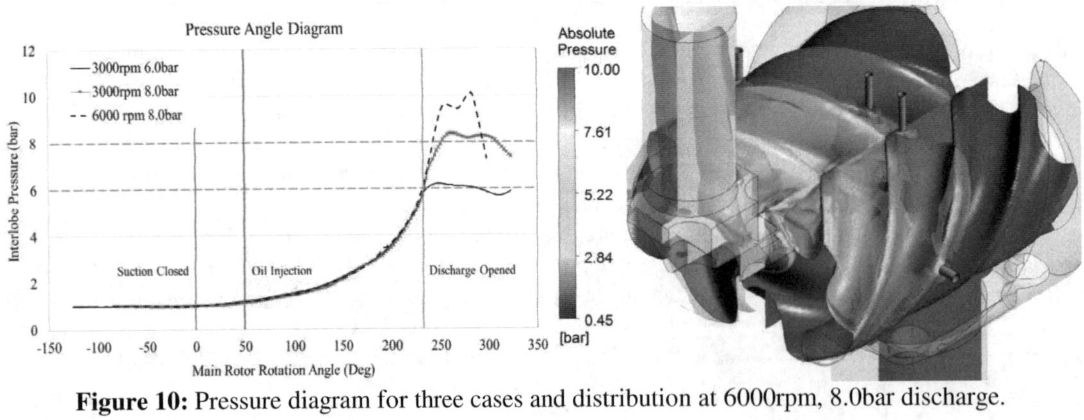

Figure 10: Pressure diagram for three cases and distribution at 6000rpm, 8.0bar discharge.

4.4 Oil distribution inside the compression chamber

Figure 11(a – h) shows a close interaction of the injected oil with the gate rotor lobes over one complete interlobe rotation. At position ‘a’ the injection hole is just exposed to the compression chamber when the formation of oil spray begins. Its direction is influenced by the passing leading gate rotor lobe. The velocity of oil depends on the injection pressure. At ‘c’ oil film forms in the root the rotor. At positions ‘d’, ‘e’ and ‘f’ the film spreads, spray continues and finally the intersection with the trailing lobe starts. At ‘g’ the hole is completely blocked by the trailing lobe width and oil velocity vectors show that an oil flow pulse is formed. At ‘h’ the position of injection coincides with that at ‘a’ and the cycle repeats itself. A large vortex remains in the trailing working chamber.

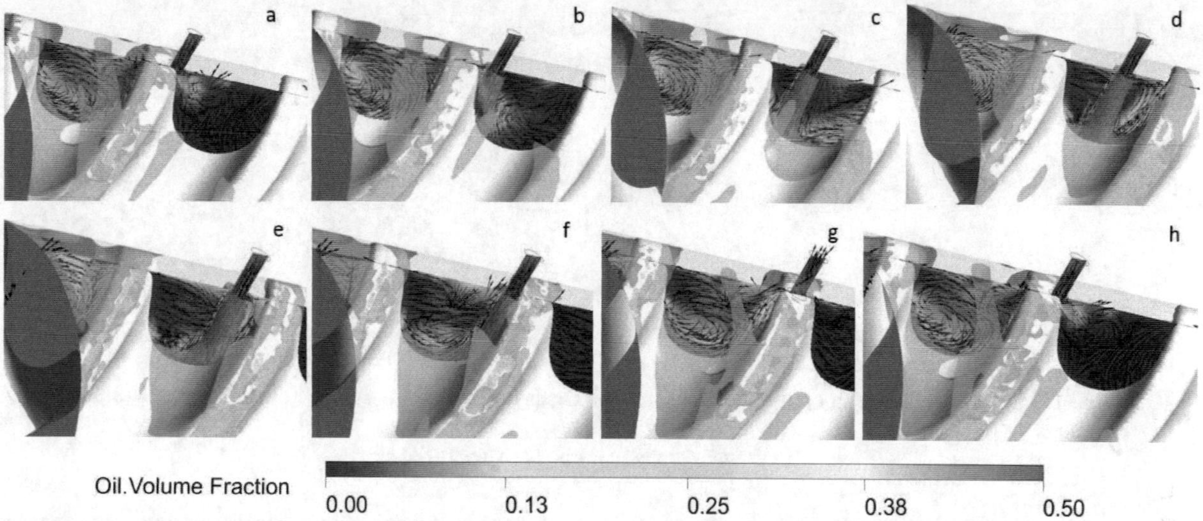

Figure 11: Oil injection interaction with the gate rotor lobes during one cycle

The pulsating nature of the oil injection can also be seen in Figure 12. Quantity of oil injected at 8.0bar discharge is similar at both speeds 3000 rpm and 6000 rpm. In the analysis, the system loss from separator vessel to the injection hole is assumed to be 0.5 bar in all the three cases. For a more accurate model, an injection pressure measurement would be required that can be specified as boundary condition. The oil flow rate at 6.0 bar is lower than at 8.0 bar discharge. Figure 12 also shows the distribution of oil and its interaction in the discharge port. Iso-surface of oil volume fraction 0.1 shows an accumulation of oil in the rotor tips and also high concentration on the discharge port walls. Similarly oil accumulates in the interlobe gaps and is transported by the rotors. The oil distribution in the discharge port is unsteady but cyclically repeating.

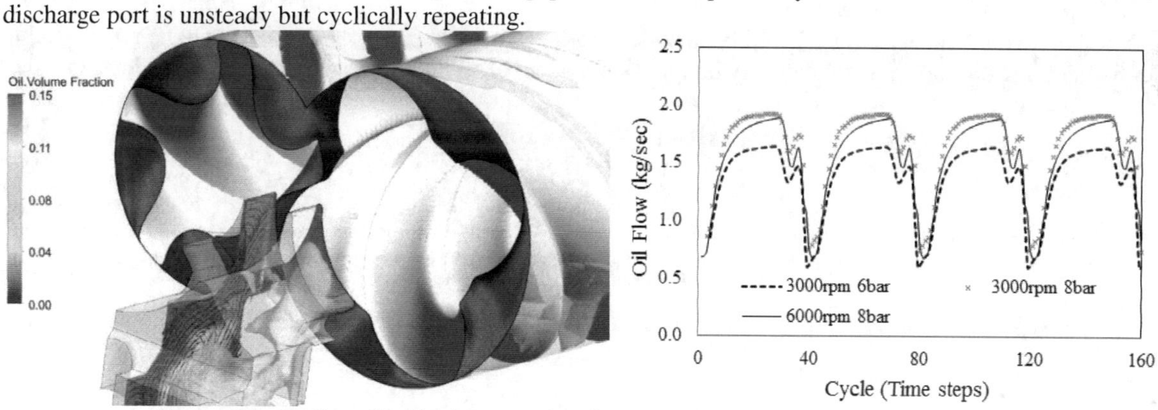

Figure 12: Oil flow distribution near the discharge port and flow cycles for three cases.

4.5 Temperature distribution

Figure 13 shows the cyclic variation of the discharge air temperature in the three cases. With built-in Vi of 3.6, the adiabatic discharge temperature without oil injection would be about 500 K but due to leakages and recompression effects, the temperature rise would be much higher. Due the injected oil at 323 K, the gas temperature is substantially lowered and does not exceed 340 K. At 3000 rpm and 8 bar discharge pressure, the peak temperature is about 10 degree lower than that at 6000rpm. This is because the amount of oil injected at the same discharge pressure and different rotational speeds is almost the same, as shown in Figure 12. This when combined with lower residence time of the oil at higher speed results in lower heat transfer from gas to oil and results in a higher discharge temperature. Similarly there is only a small difference in the gas temperature at the same speed and different discharge pressures because the quantity of oil injected at higher pressure increases, as shown in Figure 12. This higher amount of oil lowers the gas temperature and it is close to the injection temperature 323K. Figure 13 also shows the distribution of gas temperature at 8.0bar discharge pressure, 6000rpm with an iso-surface of oil volume fraction 0.1. It can be seen that the local maximum temperature reach 360 K. Also the gas temeperature is not uniformly distributed. This non-uniformity is due to the non-homogeneous distribution of oil in the compression chamber.

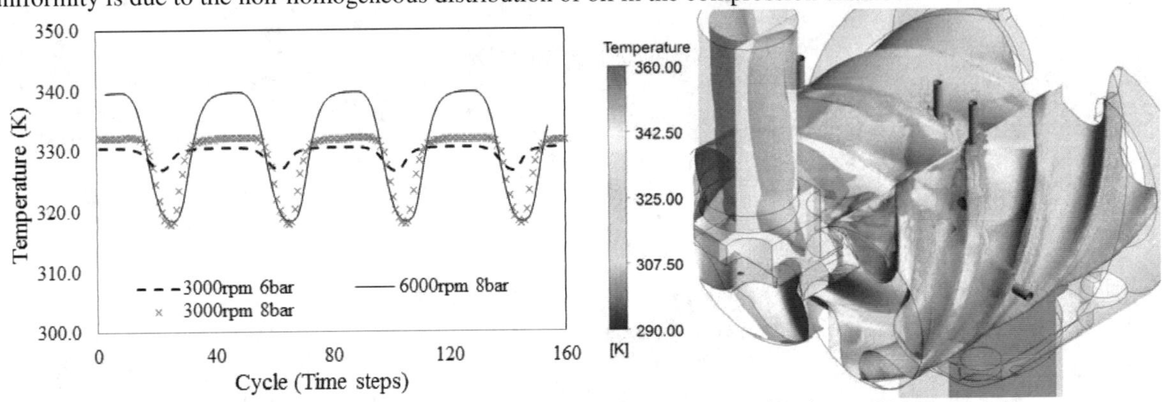

Figure 13: Discharge gas temperature cycles for three cases and distribution at 6000rpm, 8.0bar pressure.

5. CONCLUSIONS

This paper presents the CFD modelling approach used for the calculation of an oil flooded screw compressor. A single domain structured numerical mesh of the flow domain was generated by SCORG™ using recently developed boundary blocking and analytical grid generation procedure. A test oil injected compressor was measured and integral parameters such as flow rate of gas and indicated power were used for the comparison with CFD results.

- Test analysis showed a close match in the prediction of the mass flow rates of gas.
- The indicated power obtained by CFD predictions matched well with the measured shaft power.
- The mixing of the phases, distribution of oil, heat transfer between gas and oil and also effects on sealing due to high oil concentration in leakage gaps were well captured.
- Interaction of oil injection with the rotors, deposition on the rotor surface, discharge port walls and heavy accumulation in the interlobe region was observed.

In order to improve the model further it will be necessary to investigate impact of grid quality on the results and to test for higher accuracy solver settings along with the development of techniques for improving solver robustness.

REFERENCES

Arjeneh M., Kovačević A., Gavaises M., Rane S., (2014). Study of Multiphase Flow at the Suction of Screw Compressor, Proc. Int. Compressor Conf. at Purdue, Paper 1353.

Crowe, T. C., (2006). *Multiphase Flow Handbook*. Taylor and Francis, ISBN 0-8493-1280-9, CRC Press.

Deipenwisch R and Kauder K., (1999). Oil as a design parameter in screw type compressors: oil distribution and power losses caused by oil in the working chamber of a screw type compressor. *IMECHE Transactions*; **6**; 49-58; *Int. Conf. on Compressors and their systems*, London.

Kovačević A., (2002), Three-Dimensional Numerical Analysis for Flow Prediction in Positive Displacement Screw Machines, Thesis, City University London.

Kovačević A., (2005). Boundary Adaptation in Grid Generation for CFD Analysis of Screw Compressors, *Int. J. Numer. Methods Eng.*, Vol. **64**: 401-426.

Kovačević A., Stošić N. and Smith I. K., (2007). *Screw compressors - Three dimensional computational fluid dynamics and solid fluid interaction*, ISBN 3-540-36302-5, Springer-Verlag Berlin Heidelberg New York.

Papes, I., Degroote, J. and Vierendeels, J., (2013), 3D CFD analysis of an oil injected twin screw expander. Pro of the ASME 2013 International Mechanical Engineering Congress and Exposition, San Diego, USA: ASME IMECE 2013.

Rane S., Kovačević A., Stošić N. and Kethidi M., (2014). Deforming grid generation and CFD analysis of variable geometry screw compressors, Computers and Fluids, 99, p. 124–141.

Rane, S., (2015). Grid Generation and CFD analysis of Variable Geometry Screw Machines, Thesis, City University London.

Stošić N., Milutinovic Lj., Hanjalić K. and Kovačević A., (1992). Investigation of the influence of oil injection upon the screw compressor working process. Int. J. Refrig. vol. 15(4), pp. 206-220.

Stošić N., Smith I.K. and Kovačević A., (2005). Screw Compressors: Mathematical Modeling and Performance Calculation, Monograph, Springer Verlag, Berlin, June 2005, ISBN: 3-540-24275-9.

Voorde Vande J. and Vierendeels J., (2005). A grid manipulation algorithm for ALE calculations in screw compressors. *17th AIAA Computational Fluid Dynamics Conference*, Canada, AIAA 2005-4701.

Experimental investigation on the operating characteristics of a semi-hermetic twin screw refrigeration compressor by means of *p-V* diagram

Xiaokun Wu[1]*, Zhaorui Zhao[1], Wenqing Chen[2], Ziwen Xing[1]

[1]School of Energy and Power Engineering, Xi'an Jiaotong University,
Xi'an, Shaanxi, PR China
(Phone:+86-29-82675258, Fax:+86-29-82663783, E-mail:wxiaokun@foxmail.com)

[2]Suzhou Academy, Xi'an Jiaotong University,
Suzhou, Jiangsu, PR China
(Phone:+86-512-69562807, Fax:+86-512-69562807, E-mail:wqchen_xjtusz@126.com)

* Corresponding Author

ABSTRACT

In this paper, a comprehensive experimental investigation is carried out to evaluate the operating characteristics of a semi-hermetic twin screw refrigeration compressor at different oil flow rates and slide valve positions under various conditions. The working volume pressure of the compressor is recorded by a serial of sensors arranged in consecutive positions of the housing. These measured pressure data are then transformed into an indicator diagram. Based on the *p-V* diagrams, the effect mechanism of some factors such as evaporation temperature, condensation temperature, slide valve positions, oil flow rates for the suction and discharge end bearings lubricating and oil flow rate returned from the suction pipe on the compressor performance and working process is analyzed. These results can be useful for the optimum design of oil flow passage assembly and selection of optimal built-in volume ratio to improve the energy efficiency of refrigeration system with semi-hermetic twin screw compressor.

1. INTRODUCTION

Twin screw compressor is a kind of rotary positive displacement machine, the working cavity of which is enclosed by the housing bores, housing end plates and the helical surfaces of the male and female rotors. With advantages of efficient, reliable and compact, the twin screw compressor has been widely used in industrial refrigeration and central air-conditioning applications [1]. As a key component, the performance of which is very sensitive to a number of design parameters governing the thermodynamic and flow process. Thus the investigation on the operating characteristics of twin screw compressor is of great importance to improving the energy efficiency in refrigeration systems.

The analysis of thermodynamic processes within twin screw compressors by means of indicator diagram is well established and is frequently used as a tool in the design of high efficient compressors. In the case of semi-hermetic twin screw compressor, there are two approaches to access the indicator diagrams. The first is essentially theoretical. In this case a mathematical model is established, following basic thermodynamic laws, which is solved to obtain the performance parameters of the compressor as well as its indicator diagrams [2-5]. The second is essentially experimental. Here the compressor indicator diagrams are obtained from direct measurement and then used to analyze the compressor thermodynamic process. The helical structure of screw compressor is relatively complex and which makes it far more difficult to record the indicator diagrams. In general, in the case of oil-flooded screw compressors, there are two methods for recording indicator diagrams. The first involves the use of a serial of pressure sensors arranged in consecutive positions in the housing so that each is responsible for recording the pressure changes in a certain rotation degree. All the pressure signals are then combined to obtain continuous pressure data through the whole working cycle[6,7]. Alternatively, one pressure transducer can be embedded into a groove at the root of the female rotor on the discharge side to measure the whole working process [8,9].

In this paper, the working volume pressure of the compressor is recorded by a serial of sensors arranged in consecutive positions in the housing and the characteristics of the compressor is experimentally investigated when operating at different conditions, loads and especially at various lubricating oil flow rates by means of the indicator diagram. Based on the test data, performance parameters, such as the indicated power consumptions, the isentropic indicated efficiency and the mechanical efficiency, are calculated to evaluate the influence of structure and operating parameters on the working process and performance of the semi-hermetic twin screw refrigeration compressor. These results can be useful for optimum design of the compressor to improve the energy efficiency of refrigeration system. Also, the measurement data can supply some guidance for the development of mathematical model used to describe the compressor working process.

2. EXPERIMENTAL SET-UP

2.1 Modification of the Lubricating Oil Circuit

The lubricating oil circuit of a semi-hermetic twin screw refrigeration compressor is presented in the Figure 1(a). The oil outflowed from sump *Mo,out* can be divided into three parts: oil supplied for the suction end bearings *Mo,s*, oil supplied for the discharge end bearings *Mo,d* and the oil supplied for the capacity regulation *Mo,l*. The oil returned from discharge end bearings is then injected into the rotor cavity at the position with built-in volume ratio of 1.1. The oil returned from suction end bearings flows into the rotor cavity along with the intake refrigerant gas. The oil injected machine requires a separator to remove the oil from the high pressure discharged gas. The oil-gas stream from the rotor section exits the inner housing and then directly strikes the filter screens to remove the oil before the refrigerant gas leaves the compressor via the discharge valve. The separated oil is eventually collected in the oil storage sump of compressor assembly. Obviously, a little oil will still exit compressor along with the refrigerant gas because the oil separated efficiency η cannot reach 100%. As a result, additional oil needs to return back to the compressor for maintaining the oil level in the sump. This part of oil usually returns to the suction pipe and flows with refrigerant gas into the compressor. For the purpose of oil flow rate measurement, the screw compressor was modified as shown in Figure 1(b). External oil pipes are used to replace the inner oil loop arranged in compressor housing. Thus, oil flow meters and sight glasses can be installed to measure the oil supplied flow rates for the suction and discharge end bearings. The regulation of oil supplied flow rates are accomplished by the orifices plugs with various diameters and the variations of valves opening.

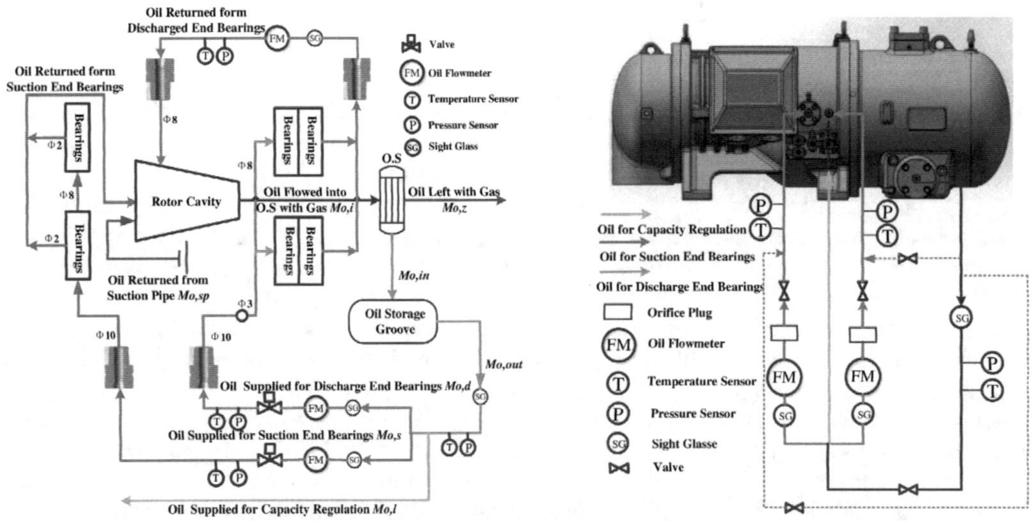

(a) (b)

Figure 1 Lubricating oil cyclic system of the screw compressor

2.2 Compressor Test Rig

Performance of the semi-hermetic twin screw refrigeration compressor with built-in volume ratio of 2.4 and R134a as the working fluid under various conditions is measured with the experimental apparatus shown in Figure 2. The gas-oil mixture is compressed in the compressor and discharged into an oil separator. The separated oil is cooled in the oil-cooler and then flows into the suction pipe to return back to the compressor. Part of the high pressure refrigerant gas then enters the condenser, liquid receiver and sub-cooler. The sub-cooled refrigerant liquid flow rate is measured by the Coriolos type mass flow meter. The rest high-pressure refrigerant gas directly flows into a regulating container. The liquid refrigerant from the sub-cooler is throttling. And pressures of the refrigerant gas and liquid are released by the controlling valves. Then heat and mass exchanges sufficiently between refrigerant gas and liquid in the regulating container to the specified low pressure state. The refrigerant flow rate is measured again by a vortex flow meter. Finally, the refrigerant gas flows into the compressor. The data acquisition, processing and archiving are accomplished by a computer and the steady state can be automatically estimated. Simultaneously experimental data is monitored and the working conditions such as the evaporation pressure and condensation pressure can be controlled precisely.

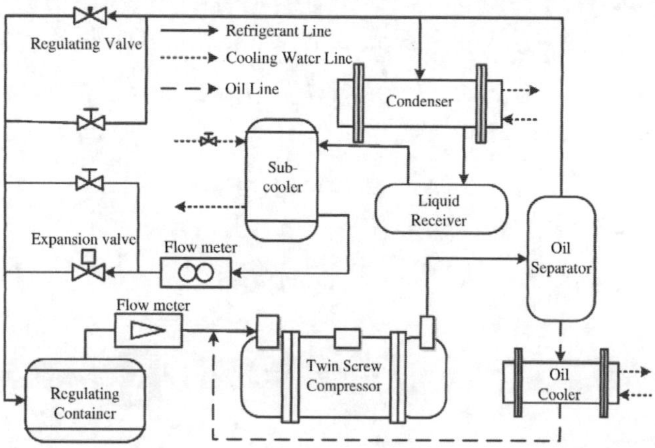

Figure 2 The screw refrigeration compressor test system

The method for recording indicator diagrams in this paper involves the use of five pressure sensors arranged in consecutive positions in the compressor housing as shown in Figure 3(a). The recording data of the five pressure sensors has a part of overlap and can be used to describe the pressure changes of the suction process, the compression process and part of discharge process. Because there is no position to arrange sensors in the compressor housing, pressure in the rest part of discharge process cannot be measured by this method. So this part of pressure is considered approximately equal to the condensation pressure. The pressure sensors are made by Kulite Group and the model is XTL-140M. The overall error of the sensor is less than ±0.3% and its response frequency is as high as 10 kHz. The vapor pressure changes along with time in the working chamber named *p-t* diagram can be achieved as shown in Figure 3(b) when the compressor operating under 5℃/36℃ with rotation speed of 2950 r/min and inlet temperature of 10℃. Then the *p-t* diagrams can be transformed to the *p-V* diagrams according to the rotation speed and variations of working volume with rotation angles.

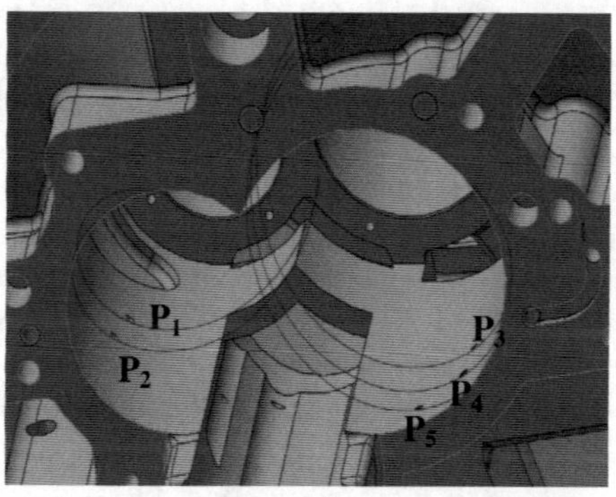

(a) Arrangement of pressure sensors

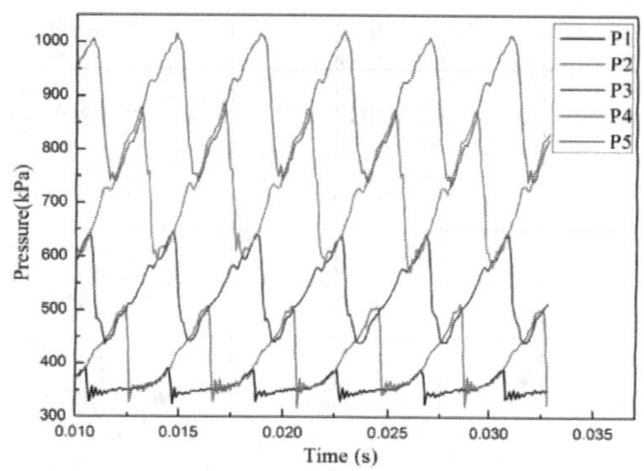

(b) *p-t* diagram under 5℃/36℃

Figure 3 Installation positions of pressure sensors and the tested *p-t* diagram

2.3 Performance Parameters

The performance parameters can be calculated from the measured pressures, temperatures, flow rates and the power consumptions. The theoretical volume flow rate q_{Vt} of semi-hermetic twin screw compressor investigated in this paper is 550 m^3/h. Then the volumetric efficiency η_V can be calculated as follows:

$$\eta_V = \frac{q_V}{q_{Vt}} \tag{1}$$

The indicated power P_{ind}, shaft power P_{out}, isentropic adiabatic power P_{ad} can also be calculated by the measured data:

$$P_{ind} = z\frac{w}{2\pi}\oint pdV \tag{2}$$

$$P_{out} = P_{in}\eta_m \tag{3}$$

$$\eta_m = \text{-}1.027\text{e-}07P_{in}^4 + 4.378\text{e-}05P_{in}^3 \text{-}0.007P_{in}^2 + 0.488P_{in} + 83.421 \tag{4}$$

$$P_{ad} = q_m(h_{ds} - h_s) \tag{5}$$

Furthermore, the indicated, mechanical, isentropic efficiencies are defined as follows:

$$\eta_{ind} = \frac{P_{ad}}{P_{ind}} \times 100\% \tag{5}$$

$$\eta_m = \frac{P_{ad}}{P_{out}} \times 100\% \tag{7}$$

$$\eta_{ad} = \frac{P_{ad}}{P_{ind}} \times 100\% \tag{8}$$

3. RESULTS AND DISCUSSION

3.1 Operating Characteristic at Various Conditions

Figure 4 shows the measured performance parameters and *p-V* diagrams of the screw compressor under different condensation temperature conditions. As the pressure difference between the condensation and evaporation side increases, external leakages (leaking to the suction pressure) cause the reduction of volumetric efficiency as presented in Figure 4(a). More refrigerant gas from the discharge process will leak into the following enclosed cavity due to the relative larger condensation pressure. The pressure in enclosed working volume will rise up as presented in Figure 4(b), especially for the working volume near to the discharge end. When operating under 5℃ /50℃ and 5℃/65℃ conditions, the pressure when the working volume connects with discharge port (with built-in volume ratio of 2.4) is smaller than the condensation pressure (under-compression condition), the refrigerant gas under higher pressure will flow reversely to the working chamber and the pressure in working volume will rapidly increase. Moreover, due to the existence of flow resistance during the discharge process, the pressure in discharge process needs to be larger than the condensation pressure so that the normal discharge process can begin. It is similar with the condition of 5℃/36℃. The pressure in the discharge process is also large than the condensation pressure even though the pressure when working volume is open to the discharge port is nearly to the condensation pressure in this condition. When operating under 5℃/29℃, the pressure when the working volume is open to the discharge port is larger than the condensation pressure (over-compression condition), the pressure drops as the discharge area increases. As the pressure line rises in the *p-V* diagrams, the indicated power and shaft power consumption of screw compressor increase. The isentropic power is also larger but the amplification is gradually to be slow down because the suction volume flow rate reduces as presented in Figure 4(c). Figure 4(d) gives the variation of efficiency of the screw compressor under different condensation temperature conditions. There exists an optimal operating condition for the isentropic and indicated isentropic efficiency. And under this condition the built-in compression end pressure is nearly equal to the condensation pressure. Moreover, the mechanical efficiency increases with the condensation temperature. It implies that the indicated power proportion of shaft power will continuously rise as the increase of condensation temperature.

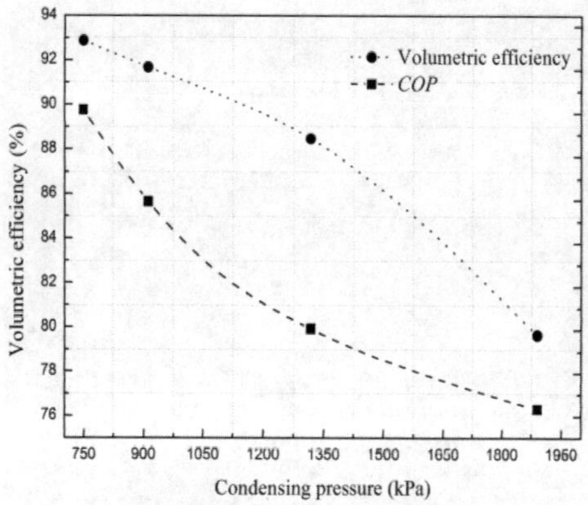

(a) Volumertic efficiency and *COP*

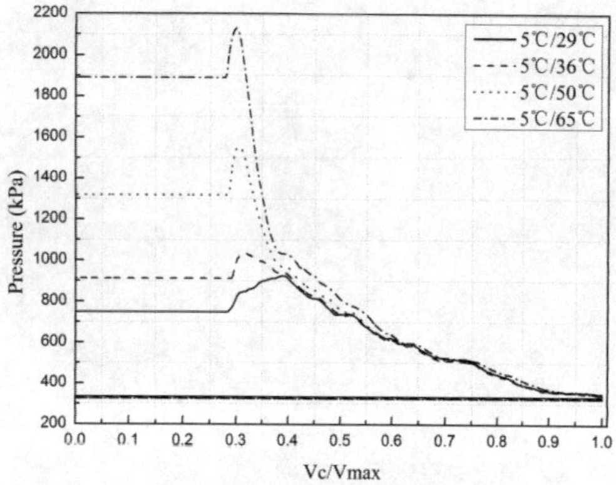

(b) *p-V* diagram

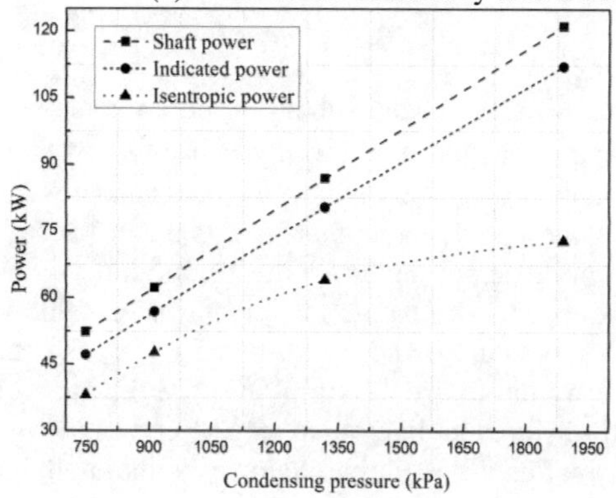

(c) Power

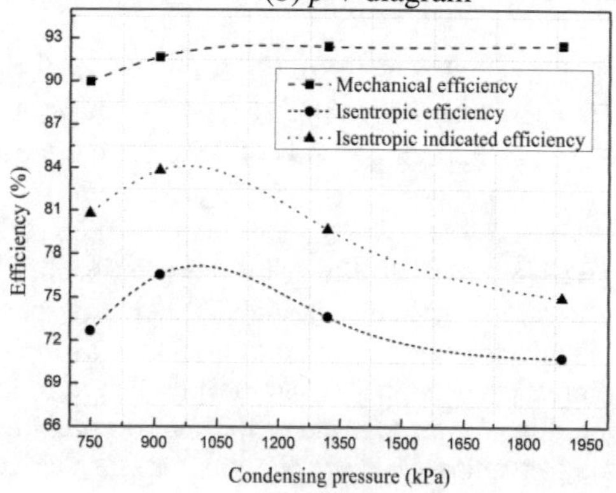

(d) Efficiency

Figure 4 Performance parameters and *p-V* diagrams under different condensing temperature

Figure 5 gives the performance parameters and *p-V* diagrams under different evaporation temperature. It can be observed that the volumetric efficiency increases as the evaporating pressure rises up due to the reduction of pressure difference through the external leakage paths. The pressure of compression process described in the *p-V* diagram ascends because the beginning pressure of the whole compression process increases with the evaporation pressure. The indicated power consumption calculated by the *p-V* diagram correspondingly increases. The density of R134a in the suction state increases so that the mass flow rate of R134a rises up which results in the increase of isentropic power even though the difference pressure between the evaporation and condensation reduces. The built-in compression end pressure is always smaller than the condensation pressure in the three testing conditions. However, the degree of under-compression is alleviated with the ascend of evaporation pressure which is beneficial to the isentropic and isentropic indicated efficiency as shown in Figure 5(d). Moreover, the mechanical efficiency is also increased with the evaporation pressure which is similar with the condition of condensation pressure increase.

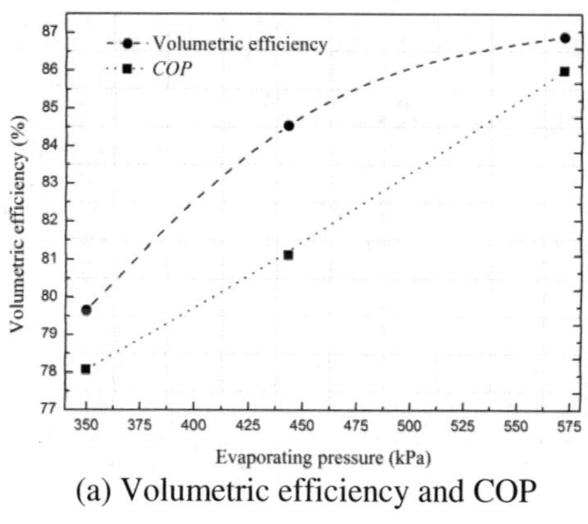

(a) Volumetric efficiency and COP

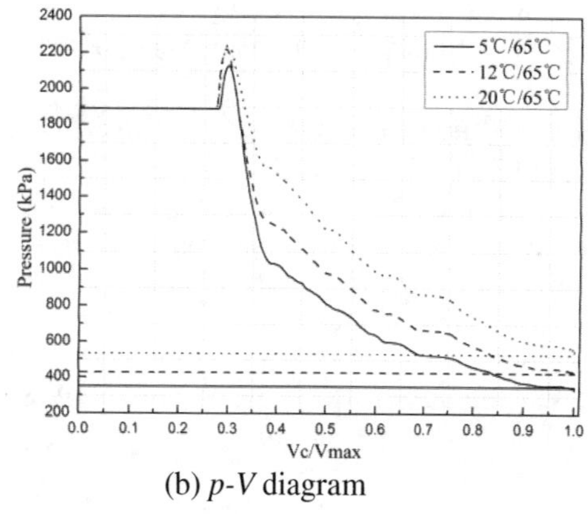

(b) *p-V* diagram

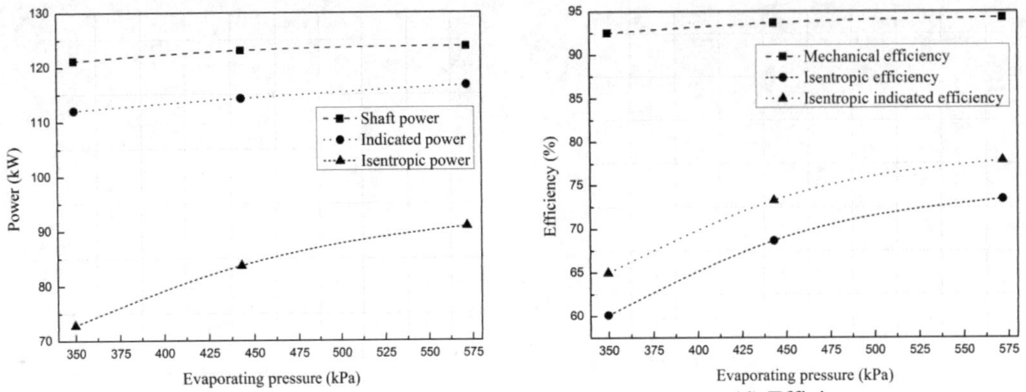

(c) Power (d) Efficiency

Figure 5 Performance parameters and *p-V* diagrams under different evaporating temperature

3.2 Operating Characteristic at Different Loads

Figure 6 shows the performance parameters and *p-V* diagrams under different loads with an evaporation temperature of 5℃ and a condensation temperature of 36℃. It can be observed that both the cooling capacity and the COP decrease as the load drops in Figure 6(a). That is due to the mechanism for slide valve to achieve capacity control. When the slide valve contacts with the slide stop to keep a seal between them, the compressor is under the full load condition. The R134a gas is drawn in by the rotation of the rotors, trapped between the rotors and the casing, compressed as the rotors meshed each other. When a by-pass port appears due to the movement of the slide valve, the compressor is under part-load condition. As soon as the rotors start to mesh and try to increase the gas pressure, the working volume exposes to the suction pressure through the by-pass port. Gas pressure rise in the working volume is slowed down, even held under the suction pressure until the contact line moves to the end of the by-pass port due to some gas venting back to the low pressure cavity, thereafter the working chamber is enclosed and the normal compression begins as shown in the *p-V* diagram of Figure 6(b).

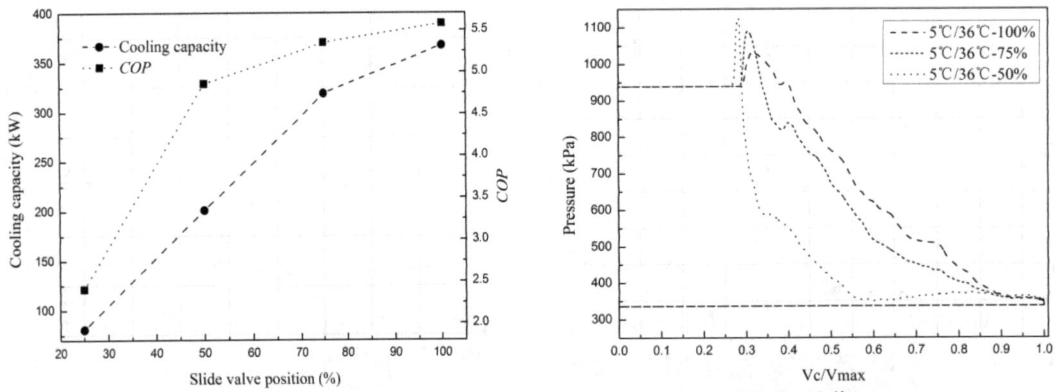

(a) Cooling capacity and *COP* (b) *p-V* diagram

Figure 6 Performance parameters and *p-V* diagram under different loads

3.3 Operating Characteristic under Various Oil Cyclic Flow Rates

For the purpose of maintaining the oil level in the sump, additional oil needs to compensate into the compressor and this part of oil usually returns to the suction pipe. Figure 7 gives the performance parameters and *p-V* diagrams with different oil return flow rates from suction pipe $M_{o,sp}$. The volumetric efficiency and *COP* reduce as the $M_{o,sp}$ increases. That is because that this part of oil flows along with gas from the suction valve and then is drawn in with gas by the rotation of rotors. As a result, part of the enclosed working volume is occupied by the oil. Moreover, the increase of oil flow rate causes extra flow resistance when gas-oil mixture flowed through the motor cavity which is detrimental to the volumetric efficiency. The pressure at the compression beginning position is also reduced with the increase of $M_{o,sp}$ as shown in the Figure 7(b). The higher volumetric efficiency means larger gas mass in the working volume which will cause higher working volume pressure as the reduction of oil flow rate. Another reason for increase of working volume pressure is the cooling effect of oil. This part of oil is firstly cooled in an oil-cooler installed at the test rig system to temperature of almost 20℃. In the compression process, this part of oil can absorb a certain amount of heat from refrigerant gas which causes the decline of pressure line in *p-V* diagram. Correspondingly, the indicated power calculated by the *p-V* diagram is larger as the oil flow rate increases as shown in Figure 7(c). The shaft power consumption also increases but the augment rate is small. That is because the increase of oil flow rate entering the rotor cavity which will alleviate the friction power loss to a certain degree. As a result, the mechanical efficiency is increased as shown in Figure 7(d). The isentropic power decreases with the increase of oil flow rate due to the reduction of gas volume flow rate. So the isentropic and isentropic indicated efficiency also reduce as the oil flow rate increases which implies that the increase of oil flow return flow rate from the suction pipe has adversely effect on the performance of screw compressor.

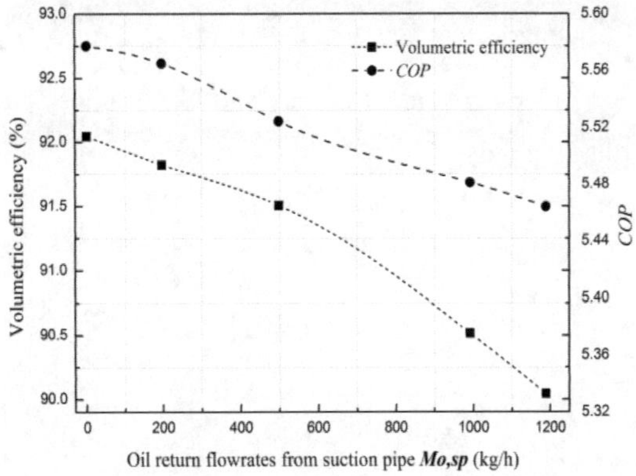

(a) Volumetric efficiency and *COP*

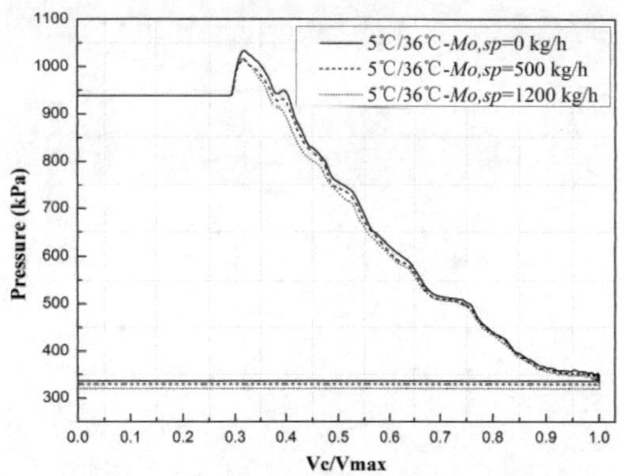

(b) *p-V* diagram

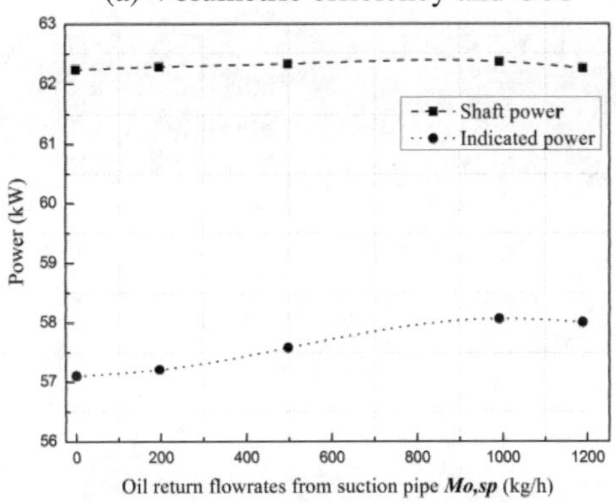

(c) Power

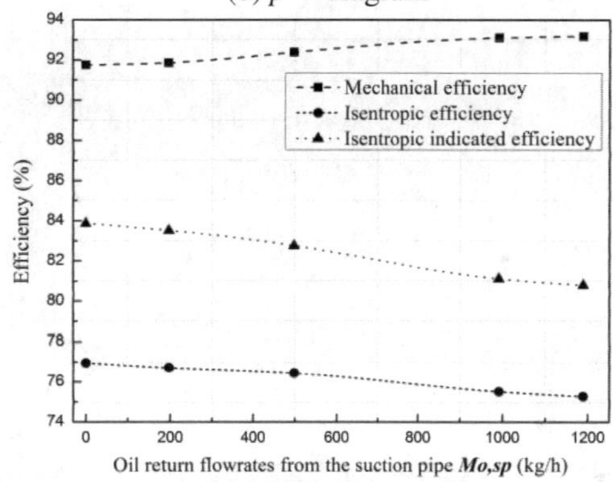

(d) Efficiency

Figure 7 Performance parameters and *p-V* diagrams under different oil return flow rates from suction pipe

Performance parameters and *p-V* diagrams at different oil flow rates for suction end bearings under conditions of 5℃/36℃ and 5℃/50℃ are shown in Figure 8. The oil flow rates are adjusted by replacement of orifice plugs installed in the oil supplied pipe. As the oil flow rates increase from zero to the maximum value, the volumetric efficiency reduces slightly. That is mainly because that the oil flows along with refrigerant gas into the working volume which is still at suction process, the oil will occupy part of working volume. Figure 8(c) and (d) give the *p-V* diagram at different oil flow rate. It can be observed that the pressure line with larger oil flow rate is higher than that with smaller oil flow rate. That is because that the oil temperature for suction end bearings lubricating is from the sump of discharge housing where the temperature is nearly to the discharge temperature. Once such high temperature oil enters the working volume, the gas should be heated and gas pressure will increase. The larger amount of oil with higher temperature enters into the working volume, the higher of gas pressure would be. That is why the increase rate of pressure when operating at 5℃/50℃ is larger compared with the condition of 5℃/36℃. The indicated power consumption calculated by the *p-V* diagram and the shaft power are also increased as shown in Figure 8(e) and (f). The isentropic power reduces due to the reduction of refrigerant volume flow rate. As a result, the isentropic and isentropic indicated efficiency also decline with the increase of oil flow rate as shown in Figure 8(g) and (h). The friction energy loss of suction end bearings is alleviated as the oil flow rate increasing. Thus, the mechanical efficiency increases.

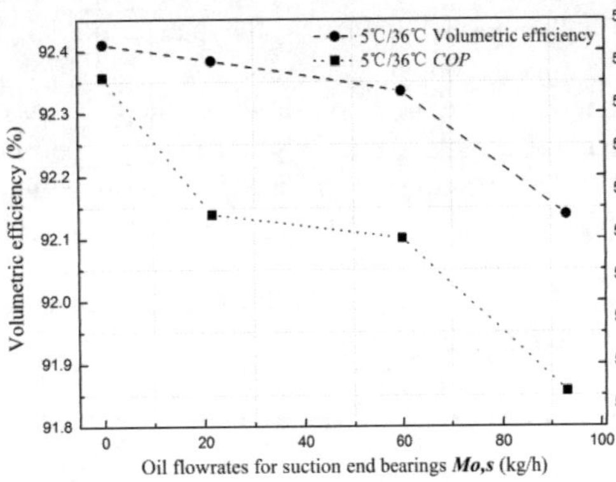

(a) Volumetric efficiency and *COP* under 5℃/36℃

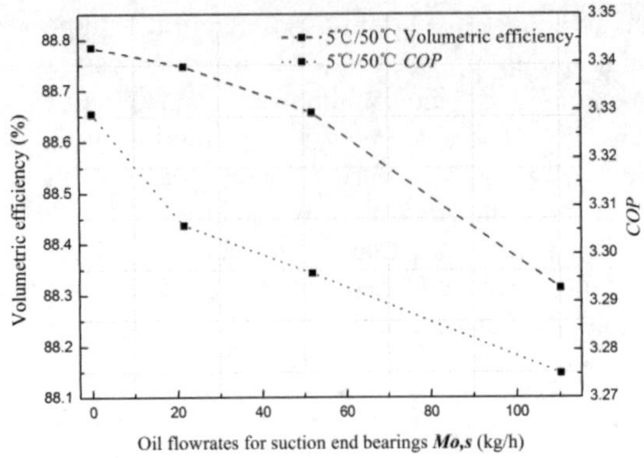

(b) Volumetric efficiency and *COP* under 5℃/50℃

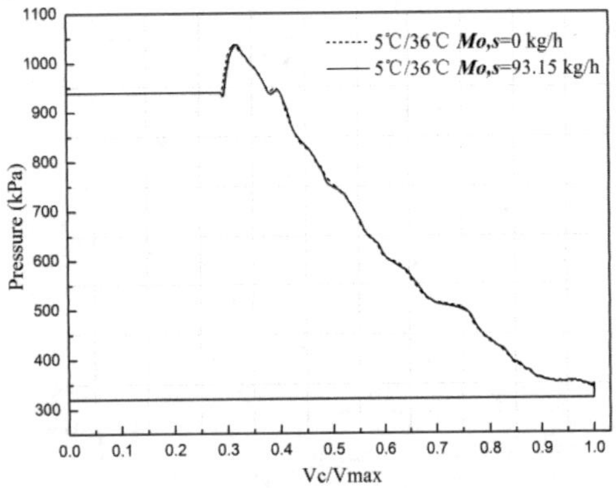

(c) *p-V* diagram under 5℃/36℃

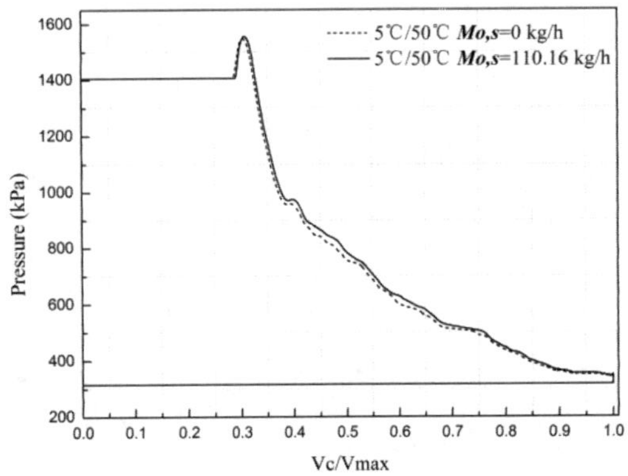

(d) *p-V* diagram under 5℃/50℃

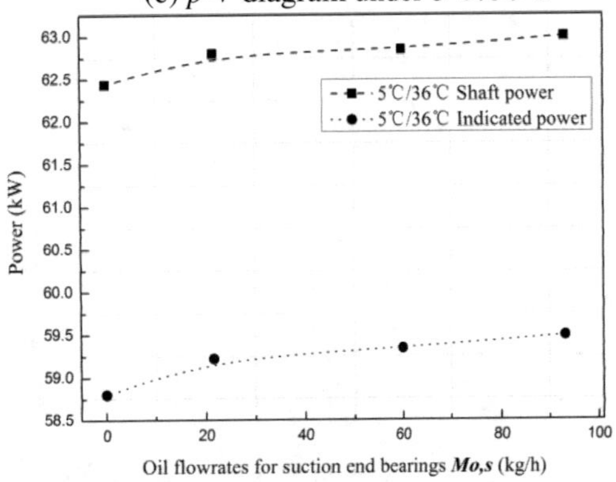

(e) Power under 5℃/36℃

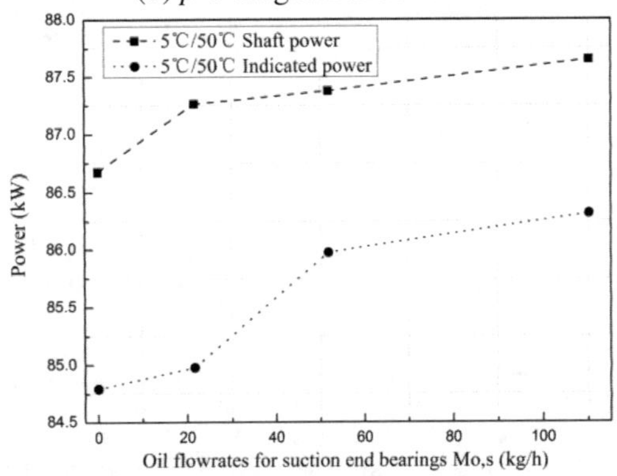

(f) Power under 5℃/50℃

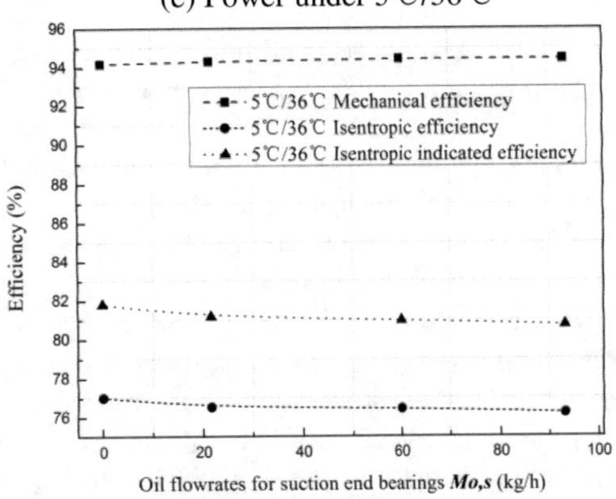

(g) Efficiency under 5℃/36℃

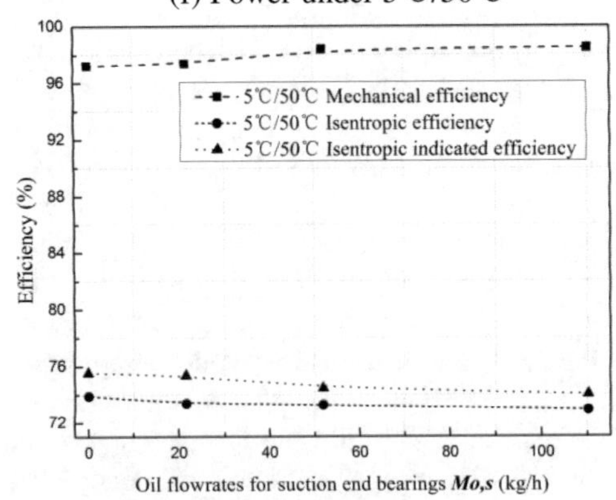

(h) Efficiency under 5℃/50℃

Figure 8 Performance parameters and *p-V* diagrams under different oil flow rates for suction end bearings

Performance parameters and *p-V* diagrams at different oil flow rates for discharge end bearings lubricating under conditions of 5℃/36℃ and 5℃/50℃ are shown in Figure 9. As the oil flow rate rises from zero to the maximum, the volumetric efficiency slightly increases. Different from the oil for suction end bearings lubricating, the oil for discharge end bearings returns to the working volume which has already enclosed. The oil will not occupy the working volume and this part of oil entering into the rotor cavity has a favorable effect on the sealing of leakage paths. However, the promote range is slight as the oil flow rate increases. It indicates that the oil flow rate coming from the suction pipe and suction end bearings is enough for the leakage path sealing effect. Extra oil flowing into the rotor cavity from the discharge end bearings has little promote effect on the volumetric efficiency. However, the pressure of working volume rises up with increase of the oil flow rate as shown in Figure 9(c) and (d). Similar with the oil for suction end bearings lubricating, the temperature of oil for discharge end bearings is also nearly equal to the discharge temperature. Once this part of oil flows into the rotor cavity, the gas of the working volume will be heated by the high temperature oil and correspondingly the pressure will increase. The indicated power calculated from the *p-V* diagram and the shaft power rise as the oil flow rate increase as shown in Figure 9(e) and (f). The mechanical efficiency has a slightly increase due to the reduction of friction loss of the discharge end bearings. And the isentropic indicated efficiency reduces which implies that additional oil entering from the discharge end bearings has an adverse effect on the screw performance once the oil flow rate entering into the rotor cavity is enough for the sealing of leakage paths.

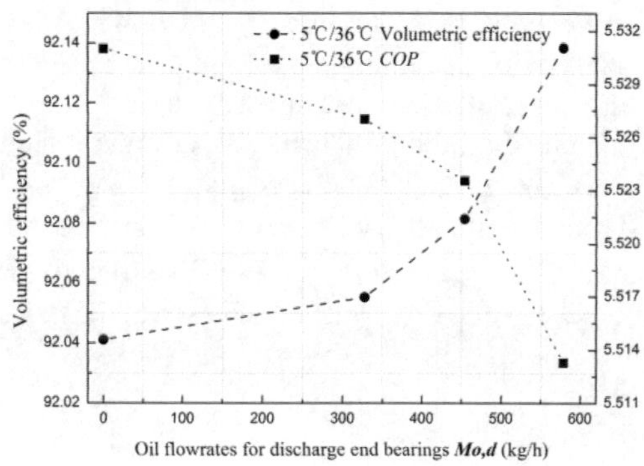

(a) Volumetric efficiency and *COP* under 5℃/36℃

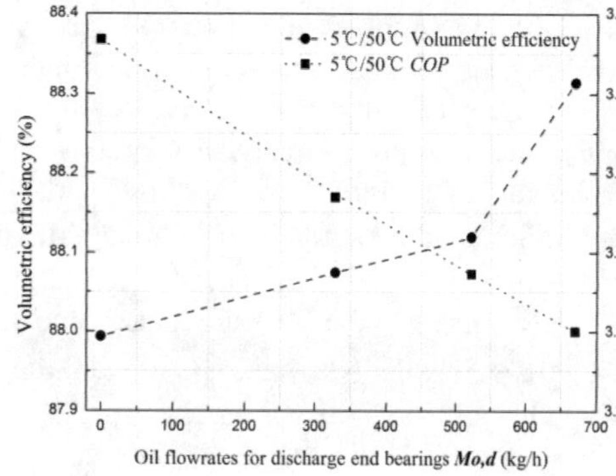

(b) Volumetric efficiency and COP under 5℃/50℃

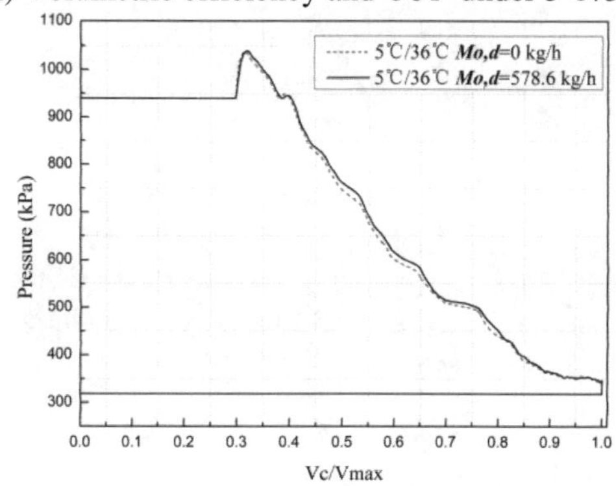

(c) *p-V* diagram under 5℃/36℃

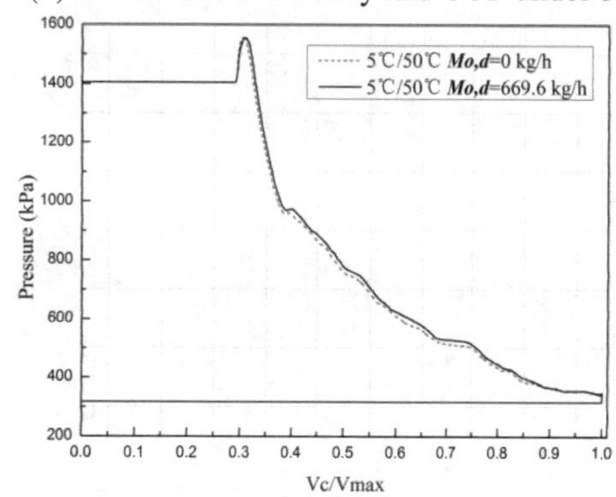

(d) *p-V* diagram under 5℃/50℃

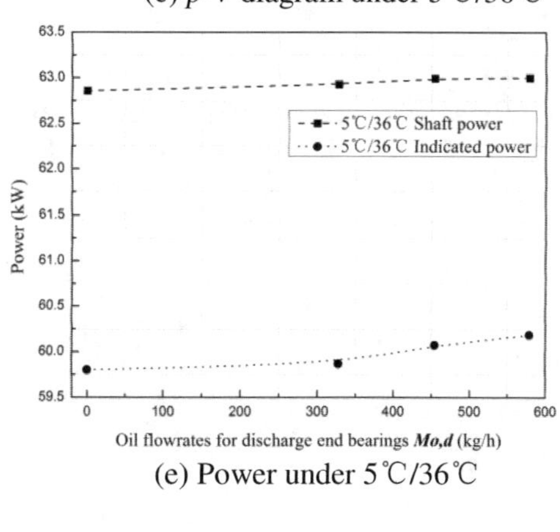

(e) Power under 5℃/36℃

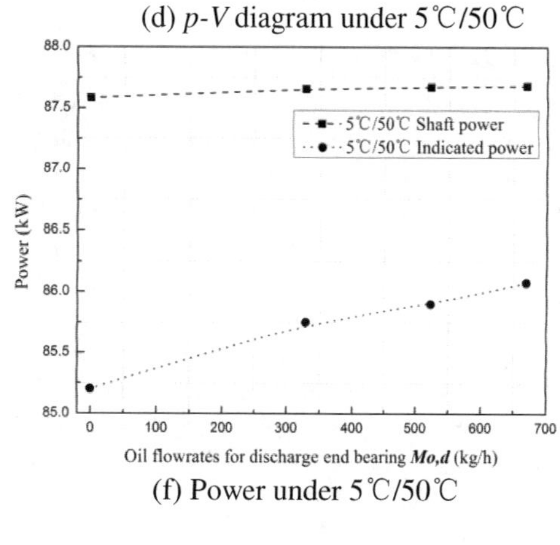

(f) Power under 5℃/50℃

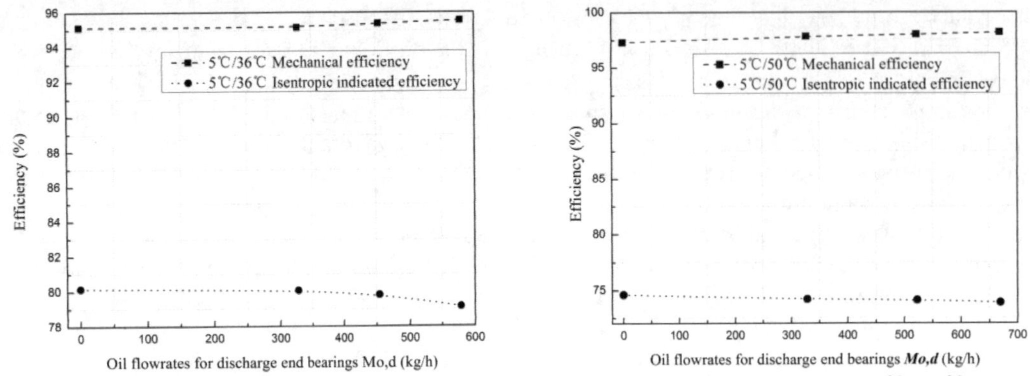

(g) Efficiency under 5℃/36℃ (h) Efficiency under 5℃/50℃

Figure 9 Performance parameters and *p-V* diagrams under different oil flow rates for discharge end bearings

Figure 10 gives the volumetric efficiency when different oil flow rates enter into the rotor cavity. This part of oil consists of the oil for suction and discharge end bearing lubricating and the returned oil from the suction pipe. The volumetric efficiency firstly rises up with the increase of this part oil due to the sealing effect of oil on the leakage paths. Then the increase rate becomes slight which indicates that the oil for sealing is enough. The reason why the volumetric efficiency eventually descends is because that in this stage the increase of oil entering into the rotor cavity is mainly caused by the increase of oil returned from the suction pipe which has an adverse effect on the volumetric efficiency. The minimum oil flow rate for sealing of leakage paths can be accessed. For the condition of 5℃/36℃, the minimum oil flow rate is 677.45 kg/h, the volume ratio of oil and gas is 0.15%. As for the condition of 5℃/50℃, the minimum oil flow rate needed for sealing is larger than the condition of 5℃/36℃ due to the larger pressure difference through the leakage paths and the minimum oil flow rate reaches 1334.51 kg/h, the corresponding oil-gas volume ratio is 0.3%.

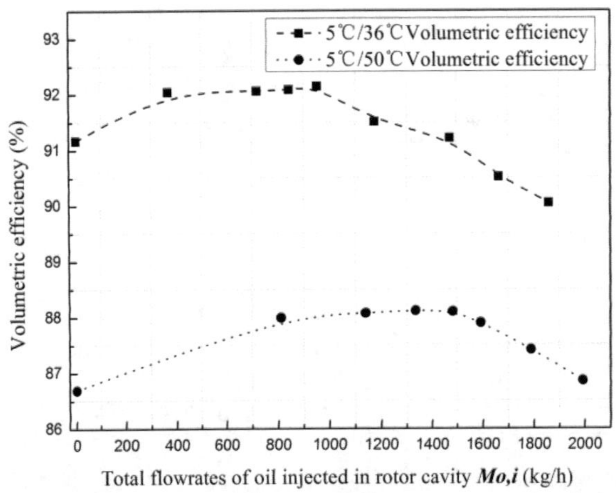

Figure 10 Volumetric efficiency at different oil flow rates entering into the rotor cavity

4. CONCLUSIONS

A comprehensive experimental investigation is carried out to evaluate the operating characteristic of a semi-hermetic twin screw refrigeration compressor under various conditions. The *p-V* diagram is also recorded to analyze the internal working process. The effects of operating conditions, slide valve positions and oil flow rates on the performance parameters are evaluated. The following conclusions are obtained from the results of the investigation:

- As the condensation temperature increases, the pressure line in the *p-V* diagram rises up. The shaft power, indicated power and isentropic power of the twin screw compressor increase while volumetric efficiency and the *COP* decrease. Isentropic efficiency and isentropic indicated efficiency reach the highest when the internal pressure ratio is matching with the external pressure ratio.
- When operating under part-load, an obvious pre-compression process and pressure reduction process exist in the *p-V* diagram recorded. As the slide valve position increases, the cooling capacity and *COP* increase.

- As the oil flow rate returned from suction pipe increases, the pressure line in the *p-V* diagram declines. The shaft power, indicated power and mechanical efficiency increase while the volumetric efficiency, isentropic efficiency, isentropic indicated efficiency and COP decrease.
- As the oil flow rate for suction end bearings lubricating increases, the pressure line in the *p-V* diagram slightly rises. The shaft power, indicated power and the mechanical efficiency increase while the volumetric efficiency, isentropic efficiency, isentropic indicated efficiency and COP decrease.
- As the oil flow rate for discharge end bearings lubricating increases, the pressure line in the *p-V* diagram rises. The shaft power, indicated power and the volumetric efficiency increase while the mechanical efficiency, isentropic indicated efficiency and the COP decrease.
- There exists a minimum oil flow rate for the sealing of leakage paths in the rotor cavity. The optimal volume ratio of oil and gas is 0.15% for condition of 5/36℃ and 0.3% for condition of 5/50℃.

NOMENCLATURE

η	efficiency	(–)
q_v	volume flow rate	(m^3/h)
q_m	mass flow rate	(kg/s)
P	power	(kW)
h	specific enthalpy	(kJ/kg)
Subscript		
ind	indicated	
ad	adiabatic	
m	motor	

REFERENCES

[1]. Z.W. Xing, Screw compressors: Theory, design and application. *China Machine Press*, 2000(In Chinese).

[2]. Stosic N, Milutinovic L, Hanjalic K and Kovacevic A, Investigation of the influence of oil injection upon the compressor working process, *Int.J.Refrig*, 1992,15(4),355-219.

[3]. Fleming J.S and Tang Y, The analysis of leakage in a twin screw compressor and its application to performance improvement. *Proc. Instn Mech. Engrs, Part C: J. Mechanical Engineering Science*, 1995, 212(5), 125-136.

[4]. H.G. Wu, Z.W. Xing, P.C. Shu, Theoretical and experimental study on indicator diagram of twin screw refrigeration compressor, *International Journal of Refrigeration* 27 (2004) 331-338.

[5]. W.Q. Chen, Z.W. Xing, H. Tang, H.G. Wu, Theoretical and experimental investigation on the performance of screw refrigeration compressor under part-load conditions, *International Journal of Refrigeration* 34(2011) 1141-1150.

[6]. J.B. Shen, H. Tang, Z. Zhang, Z.W. Xing, Experimental study of a water injected twin screw compressor for mechanical vapor compression system. *International Compressor Engineering Conference at Purdue*, USA, 2014.

[7]. Haugland K et al, Pressure indication of twin screw compressor. *Proc.10th International Compressor Engineering Conference at Purdue*, USA, pp. 450-456, 1990.

[8]. X.Y. Peng, Z.W. Xing, T.S. Cui, L.S. Li, Analysis of the working process in an oil-flooded screw compressor by means of an indicator diagram. *Proc. Instn Mech. Engrs, Part A: J. Power and Energy*, 2002, 465-470.

[9]. H.G. Wu, J.F. Li, Z.W. Xing, Theoretical and experimental research on the working process of screw refrigeration compressor under superfeed condition, International Journal of Refrigeration 30 (2007) 1329-1335.

ACKNOWLEDGEMENT

The authors would like to acknowledge the financial support from the National Nature Science Foundation of China (Grant No.51276134) and the National Nature Science Foundation of Jiangsu province of China (Grant No.BK20150380).

Research on Injected Effect and Heat-transfer Characteristics of Narrow-Slit Injection Orifice Used for the Screw Compressor

Jia Xie, Jian Li, Kang Lian, Quanke Feng*, Weifeng Wu

School of Energy and Power, Xi'an Jiaotong University,
Xi'an 710049, China

Corresponding author: Quanke Feng*
Tel (Fax): +86-029-82663783 Email: qkfeng@mail.xjtu.edu.cn

ABSTRACT

A narrow-slit injection orifice which is used for the screw compressor (mainly for single screw compressor) injection is described in this paper. Different from the traditional injection orifice, the orifice is slit shaped but not circular. Liquid which is injected by the narrow-slit injection orifice appears as a curtain shape so that the heat transfer area is increased. The cooling effect is enhanced and then the efficiency of the compressor is improved. The injected velocity of the orifice is simulated through the CFD software, and then the experiments are carried out. The study verified the accuracy of the simulation results. It will provide a theoretical foundation for the design of the narrow-slit injection orifice used for the screw compressor.

1. INTRODUCTION

Liquid injected screw compressors have been widely used in industry because of its high efficiency, stable performance and so on. While gas is compressed in the screw compressor, a certain amount of liquid with a certain pressure is injected into the compression cavity generally. The liquid contacts with compressed gas directly and it mainly to play a role of cooling and sealing. Usually it needs to spray 8-15L liquid into per cubic meter gas, the volume ratio is about 1:100(Deng and Shu, 1982). In general, the lubricant fluid flows in the loop depend on the pressure difference between the compressor's exhaust pressure and the spray pressure. For large screw compressor, the oil pump is added the circuit. The injected liquid droplets are mixed with the compressed gas. The extremely heat transfer surface absorbs the heat of gas rapidly, so the compression process is similar to the isothermal process. A reduction on the energy consumptions is up to 7% with a positive spray cooling (Li and Jin, 1992). The accuracy requirements for machining of screw rotors are reduced through a sprayed liquid injection technology, also the performance of the compressor is improved.

The main injection parameters have been investigated, such as injection pressure and temperature, liquid flow rate, orifice diameter, position of the injector, etc. Many scholars have studied the injection technology of screw compressor in the world. Stosic *et al.* (1988) presented some results of mathematical modeling of the oil injection into the screw compressor and its influence upon the thermodynamic process of the engine and analyzed the effect of various parameters upon the compressor physical cycle. Pawan J. and Bowman, J. L. (1986) described a mathematical model to calculate the oil-gas heat transfer, assuming that the oil is injected in the form of non-interacting spherical droplets. The model was used to calculate the effect of the heat transfer on compressor performance. N. Seshaiah *et al.*(2007) carried out the mathematical analysis of oil injected twin-screw compressor and effect of some of the compressor operating and design parameters on power and volumetric efficiencies had been analyzed and presented. Li *et al.* (1992) analyzed the effect of oil atomization and its effect on the exhaust gas temperature of single screw compressor. The influence of the main structural parameters of atomizer on lubricant oil atomization effect was analyzed, and the exhaust temperature of single screw compressor without atomization was compared. Finally, it was concluded that the performance of the compressor is improved effectively by the atomization of lubricant.

The section crosses of the injection orifice are all circle, and the liquid is sprayed into compression cavity in a column sharp. Because the injection path is so short, its effect on liquid injection optimization is limited.

2. INJECTION ORIFICE AND ITS POSITION

In the compressor without volume flow rate regulation, the position of injection orifice is on the side of compression gas and in the section of the compressor that volume has been out of the suction port and compression process starts. The lubricant liquid could be injected through a larger spray injection orifice or through a number of smaller orifices, usually it can obtain better performance of compressor by using a number of small injection orifice. In the large and medium sized injection screw refrigeration machine and process compressor, inner-placed volume flow refrigeration valve device is used usually. When the compressor working under different conditions, the working length of the rotor is changed and the inspiratory end position moves to the exhaust end with the slide valve. Generally, the injection orifice is opened in the slide valve in this type of compressor, the injection orifice will move with the slide valve together. So it is ensure that the lubricant liquid is always sprayed after the end of the suction. In order to improve the volumetric efficiency, sometimes the injection orifices are opened in the middle of the compression process for small and high pressure ratio compressor.

In twin-screw compressor, the injection orifice is opened in or near the intersection of two rotor's tooth addendum circle (Xing, 2000). Liquid holes can be arranged in one or two rows, as shown in Figure 1. It shows the top view and the left view of the casing.

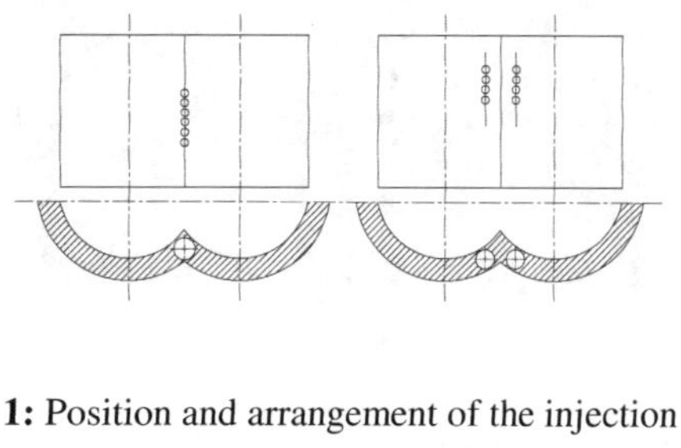

Figure 1: Position and arrangement of the injection orifice

Vertical liquid spray method is currently used in the single screw compressor, that is, the nozzle axis is perpendicular to the surface of the star wheel. The lubricant liquid is sprayed where the star-wheel teeth cutting into the depths of the screw, the liquid injection rate is equal to the circumferential speed of the screw. Another injection method is that lubricant liquid is sprayed into basic capability along the rotation direction of the screw tooth (Yu *et al.* 2012). The orifices are opened in front of projection of the outer edge of the spiral screw that is in front of the star wheel on cylinder when the basic capability is formed. The injection orifice are a pair or many pairs circular holes symmetrical arrangement in two regions of the cylinder， as shown in Figure. 2.

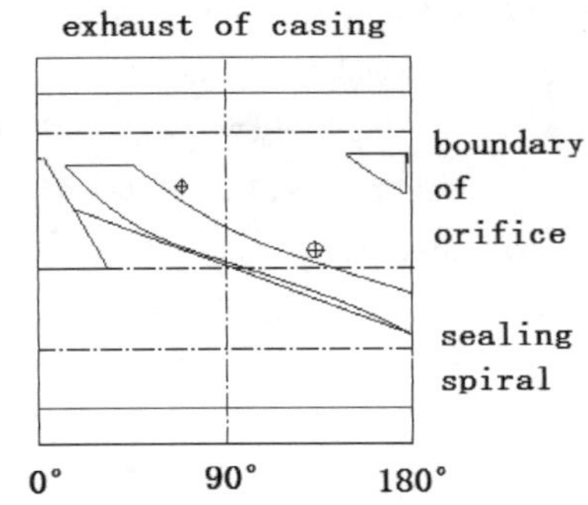

Figure 2: Diagram of the two pairs of injection orifices position

The injection orifices connect to the channel for injection liquid. Liquid channel is arranged on the casing by drilling hole along the axial direction. In part of the compressor, the orifice is opened along the radial of the casing directly.

3. NARROW-SLIT SHARPED INJECTION ORIFICE

Liquid is sprayed into the cogging space in column sharp by the above injection orifice, and the injection path in the cogging is so short. Column sharped liquid is too late to spread for atomization when it attaches to the metal wall of the casing. Therefore, the heat transfer area of the lubricant liquid and the compressed gas is very small. Due to the cooling effect is not ideal, the inner wall of the compression cavity is scaled. Scaling phenomenon is shown in Figure 3. It is a photo for the casing of a brand compressor when the compressor running some time, the cross section of the injection orifice is circular.

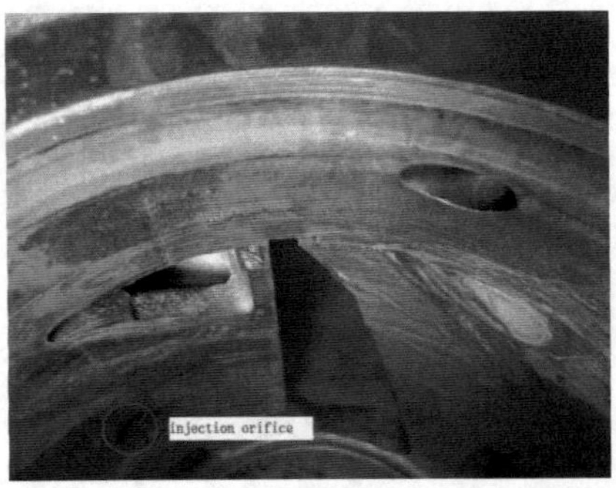

Figure 3: Scaling

The heat exchange between the compressed gas and the liquid is less when the size of the injection orifice is larger. That is because the large size of the injection orifice will lead to a lower injection rate and thus format large liquid droplets, heat transfer area decreases. In addition, the increase of liquid droplets will lead to a shorter contact time between compressed gas and lubricant liquid, and the heat transfer effect is weakened more. In order to enhance the heat transfer and promote that the compression process is more similar to isothermal compression of the most power saving theoretically, the lubricant liquid should have a larger surface area in the same amount of liquid. The narrow-slit injection orifice is proposed in this paper. It is opened by arranging appropriate width and length of the narrow slit on the air chamber, and injecting at an appropriate location. The lubricant liquid is sprayed into the cogging space in a curtain shaped, then the span of the lubricant liquid and gas mixed along the axial is larger. And the lubricant liquid in the casing is uniformly distributed in the compression cavity. As a result of the increase of the contact area between the gas and the liquid injection, cooling of the compression process is enhanced. Comparison of the spray effect between a narrow-slit orifice and circular orifice is shown in Figure 4.

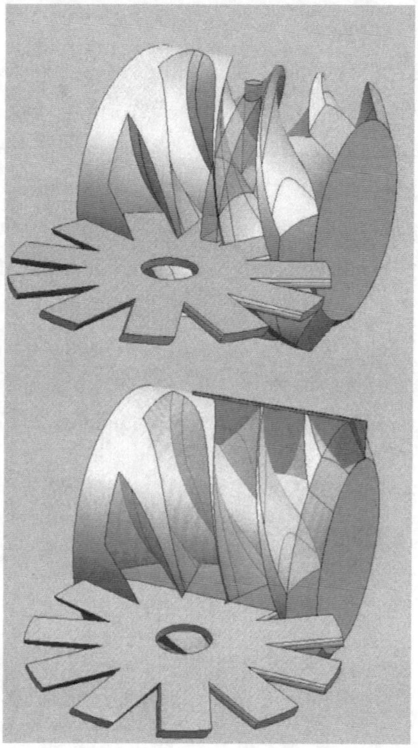

Figure 4: Comparison of liquid injection effects

It is clear that the axial span of lubricant liquid injected by the narrow-slit orifice is larger, and the span of screw groove is more.

In the process of injection, the viscous force of the lubricating liquid and the air resistance and the surface tension of the lubricating liquid play an important role in the crushing of lubricant liquid. Larger liquid droplets that are injected in compression cavity twist under air resistance, and generate friction under the relative velocity of liquid and air. Due to the inner wall of the liquid droplet, a part of the surface protrude out of the liquid droplet is devised and then a smaller liquid drop is formed under the action of the liquid surface tension. At the same time, the liquid droplets will distorted and deformed when the resultant force of surface tension and viscous force is overcome by the air resistance. It continues to that the air resistance and the internal force of the liquid drop regain the equilibrium state. Based on oil atomizing theory, the parameter of Weber number is put forward for evaluating the atomizing level of the oil in compression cavity. The flows of liquid atomization that resulted by different Weber number are divided into five forms (Bein and Hamilton, 1982): liquid drop (We<0.05), smooth liquid flow (0.5<We<5), ripple current (5<We<20), spray flow (We>20).

Weber number is estimated according to Equation (1).

$$W_e = \frac{\rho v^2 l}{\sigma} \tag{1}$$

Where ρ the density of the lubricant liquid, v is the liquid injection rate, l is the characteristic length of the injection orifice, σ is the surface tension coefficient of the lubricant liquid.

The better effect of atomization, the more compressor power consumption is saved. So it's the better that the atomization of the compressor working cavity is the form of spray flow, that is, liquid droplets are fine particles. The liquid droplets of the spray flow are smaller size droplets when Weber number is greater than 20. The compressor which speed is 3000r/min is the main object in reference (Lin *et al.* 2000), the experiment is carried out when the diameter is ϕ3mm、ϕ2mm and ϕ1.6mm. Weber number and the diameter of lubricant liquid drop is 43.4, 60.5, 118.3 and ϕ 3.8mm, ϕ 1.46mm, ϕ 0.75mm according the analysis of experimental data. The atomization of lubricant liquid injected by these three kinds of injection orifice is the spray form through Weber number. However, the diameter of liquid drop is too large when diameter of the injection orifice is ϕ 3.8mm through the drop diameter, quality of atomization is poor. Therefore the Sauter mean diameter (SMD) is introduced, also it is known as surface area mean diameter or Sauter mean diameter. SMD can be estimated by Equation (2).

,

$$SMD = 4.52\left(\frac{\sigma_0\mu_0^2}{\rho_a\Delta p}\right)^{0.25}\left[2.7\left(\frac{d_{or}\,m_0\mu_0}{\rho_a\Delta p}\right)\cos\gamma\right]^{0.25} + 0.39\left(\frac{\rho_0\mu_0}{\rho_a\Delta p}\right)^{0.25}\left[2.7\left(\frac{d_{or}\,m_0\mu_0}{\rho_a\Delta p}\right)\cos\gamma\right]^{0.75} \tag{2}$$

Where σ_0 is the surface tension coefficient of the lubricant liquid, u_0 is the velocity of the liquid jet, m_0 is the weight of the liquid per diameter range, ρ_0 is the density of the medium around, γ is half cone angle, d_{or} is the diameter of the injection orifice, Δp is pressure difference of nozzle.

The smaller the SMD value, the better the atomization quality is. Five different types of nozzles under the premise of the same mass flow rate were tested, including three diameter of holes with diameter of ϕ 3mm、ϕ 2.5mm and ϕ 2mm. The Sauter mean diameters of the drops are ϕ 3.53mm、ϕ 1.97mm、and ϕ 0.96mm.

It is concluded that the Sauter mean diameter of droplet will decrease with the decreasing of orifice diameter. Narrow-slit sharped injection orifice can be regarded as a lot of smaller circular injection orifice, and its role is equivalent to an atomizer.

4. INJECTION VELOCITY SIMULATION

The working process of screw compressor is complex, so the model must be assumed and simplified appropriately to receive results quickly in the premise of sufficient accuracy. Simplifications are as follows: 1. Lubricant liquid passage is simplified as a straight pipe; 2. Property of lubricating oil for working medium has not changed; 3. On-way resistance of the lubricant liquid flow process is negligible; 4. The screw cogging is a regular shape.

Steady flow field calculation is achieved by software Ansys 14.0. The grid division is achieved by CFD14.0 ICEM, the results shown in Figure 5:

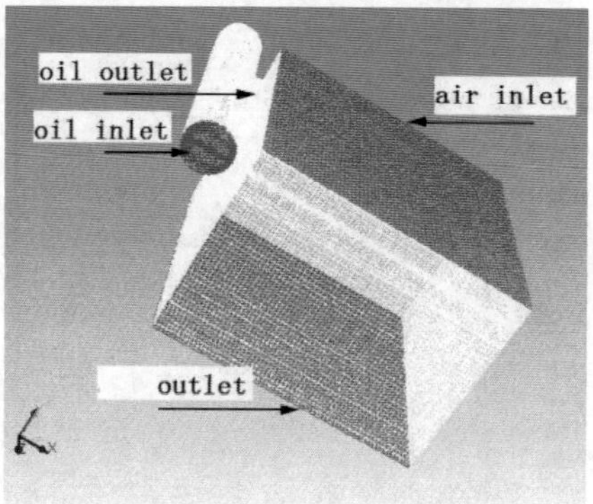

Figure 5: Grid division

The flow control equations are composed of incompressible flow Reynolds equation (3), turbulent kinetic energy equation (κ equation)(4) and the turbulent dissipation rate equation (ε equation)(5).

$$\frac{\partial}{\partial x}\left(\frac{ph^2}{\eta}\frac{\partial p}{\partial x}\right)+\frac{\partial}{\partial y}\left(\frac{ph^2}{\eta}\frac{\partial p}{\partial y}\right)=6\left(U\frac{\partial ph}{\partial x}+V\frac{\partial ph}{\partial y}+2\frac{\partial ph}{\partial t}\right) \tag{3}$$

$$\rho U_i\frac{\partial k}{\partial x_i}=\mu_t\left(\frac{\partial U_j}{\partial x_i}+\frac{\partial U_i}{\partial x_j}\right)\frac{\partial U_j}{\partial x_i}+\frac{\partial}{\partial x_i}\left\{\left(\mu_t/\sigma_k\right)\frac{\partial k}{\partial x_i}\right\}-\rho\varepsilon \tag{4}$$

$$\rho U_i\frac{\partial \varepsilon}{\partial x_i}=C_{1\varepsilon}\left(\frac{\varepsilon}{k}\right)\mu_t\left(\frac{\partial U_j}{\partial x_i}+\frac{\partial U_i}{\partial x_j}\right)\frac{\partial U_j}{\partial x_i}+\frac{\partial}{\partial x_i}\left\{\left(\mu_t/\sigma_k\right)\frac{\partial k}{\partial x_i}\right\}-C_{2\varepsilon}\rho\left(\frac{\varepsilon^2}{k}\right) \tag{5}$$

Where σ_κ, σ_ε, $C_{1\varepsilon}$, $C_{2\varepsilon}$ are experimental constant.

Firstly, the boundary conditions and the correlation of the model must be set up when the flow distribution of lubricating liquid is simulated by Fluent software. In the model, the setting of boundary at the inlet and outlet of the lubricant liquid is the key to influence whether the flow agree with actual condition of the lubricant liquid. Boundary conditions are set in Table 1.

Table 1: Boundary conditions set

Boundary definition	Oil inlet	Oil outlet	Air inlet	Outlet	Other wall
Boundary type	Pressure inlet	Interior	Velocity inlet	Pressure outlet	Wall

The simulation result is showed in Figure 6. It is a section of the simulation that liquid oil is injected into the screw cogging in Z axis direction.

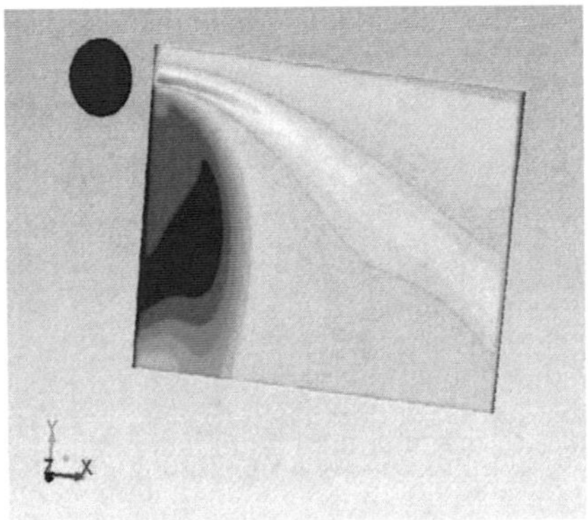

Figure 6 Simulation result

Simulation is carried out when the crack length is 40mm and width is 0.4mm. The injected velocity is shown in Table 2.

Table 2: Simulation results

Pressure(MPa)	0.05	0.1	0.15	0.2	0.25	0.3	0.35	0.4
Velocity(L/(mm^2 · min))	0.088	0.176	0.247	0.317	0.382	0.441	0.503	0.556

Maximum pressure condition is 0.4MPa because of the limit of the experimental equipment. It is indicated that the injected velocity increases with the increase of pressure in the range of pressure. The simulation results will be verified by experiment.

5. EXPERIMENT EQUIPMENT

Experiment equipment is shown in Figure 7. The experimental oil pump is a gear pump which type is KCB-33.3 with design flow 33.3L/min, pressure 1.45MPa and speed 1420r/min. The oil pump is mainly used for all kinds of machinery and equipment of lubricant system to convey lubricant oil. It is suitable for conveying the lubricating oil which viscosity for $5\times10^{-6}\sim1.5\times10^{-3}m^2/s$ (5-1500cSt) and the temperature under 300° C. The nozzle shape is obtained by machining and low tolerances need to be guaranteed. The accuracy is particularly important because of the small size of the slit. In the experiment, the narrow slit size of the nozzle is 40mm×0.4mm×3mm. The experimental medium is ANDEROL3032M, an ISO32 compressor lubricant formulated from synthetic basefluid utilizing synthetic hydrocarbon (PAO) with specially developed additivesy stems.

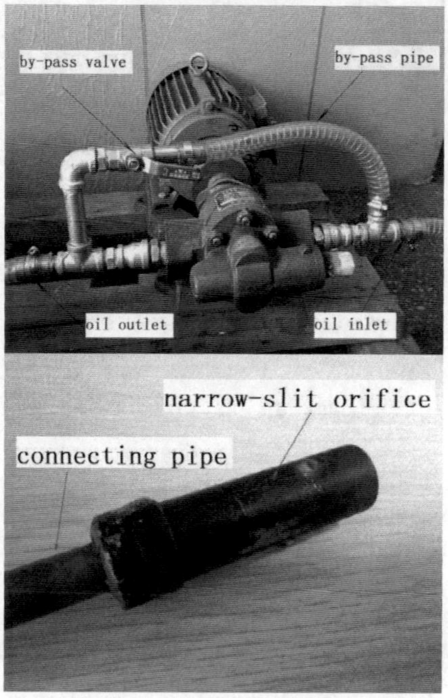

Figure 7: Experiment equipment

The injected velocity obtained by weighing method. Experiment results and errors are shown in Table 3.

Table 3: Experiment results and errors

Pressure(MPa)	0.05	0.1	0.15	0.2	0.25	0.3	0.35	0.4
Velocity(L/(mm^2 · min))	0.088	0.176	0.247	0.317	0.382	0.441	0.503	0.556
Error (%)	7.7	0.4	1.2	3.6	2.0	2.4	4.0	7.5

Error is the ratio of the difference between simulated results and experiment results to experiment results. It is shown that the simulation results is accordance with the real situation in the experiment because of all errors are less than 10%. The simulation lays a theoretical foundation for the design.

In order to observe the experimental results, the experimental media is water when the simulation of injection flow in the compressor is carried out. When the velocity of air is 35m/s (approximate value of average velocity in the compression chamber), the injection diagram is shown in Figure 8.

Figure 8: Injection diagram

It is shown that the injection is the spray flow. The atomization effect is better. The jet is blown into many small droplets; the heat exchange area increases a lot. The atomizing form is beneficial to the gas-liquid mixture uniformly, and the heat exchange effect is enhanced. It is verified the feasibility of narrow-slit injection orifice.

6 CONSLUSIONS

For screw compressor, the curtain jet that is formed by the narrow-slit orifice proposed in this paper promotes decrease of the heat transfer greatly. Heat transfer between compressed gas and liquid is strengthened so that the compressor temperature decreased. The thermal working process is similar to the isothermal process and the consumption of the compressor is further reduced, the efficiency is improved. And the manufacturing cost can be reduced; its application range is expanded. The comparison between circular injection orifice and narrow-slit injection orifice and the theoretical analysis lay a theoretical foundation for the design and optimization of the narrow-slit injection orifice.

Next the proper size of the narrow-slit injection orifice aiming at a volume model will be designed and the casing will be processed. The real machine experiments will be carried out to test and verify whether the narrow-slit orifice has its superiority.

REFERENCES

Bein, T. W., Hamilton, J. F. (1982). Computer modeling of an oil flooded single screw air compressor. In: *International Compressor Engineering Conference. Indiana,* Purdue University. Paper 383.

Dingguo Deng, Pengcheng Shu. (1982). Rotary compressor. Beijing, MA: China Machine Press, in China.

Hongqi Li, Junhong Feng, Guangxi Jin. (1992). Study on the effect of oil atomization and its effect on the exhaust gas temperature of single screw compressor. *Int. J. Fluid Engineering,* 20(8), 14-16.

N. Seshaiah, Subrata Kr. Ghosh, R.K. Sahoo, Sunil Kr. Sarangi. (2007). Mathematical modeling of the working cycle of oil injected rotary twin screw compressor. *Int. J. Applied Thermal Engineering* 27, 145-155.

Qiang Lin, Ling Fan, Guangxi Jin. (2000). Theoretical analysis and experimental study of oil atomizing in a single screw compressor. *Int. J. Journal of Xi'an Petroleum Institute(Natural Science Edition),* 15(2), 41-44.

Singh, P. J. and Bowman, J. L. (1986). Heat Transfer in Oil-Flooded Screw Compressors. In: *International Compressor Engineering Conference. Indiana,* Purdue University. Paper 521.

Stosic，N., Milutinovi, L., Hanjali, K., Kovaevi, A,A. (1992). Investigation of the influence of oil injection upon the screw compressor working process. *Int. J. Refrigeration 15*, 206-220.

Stosic, N., Kovaevi, Ac, A., Hanjali, K., Milutinovi, L. (1988). Mathematical modeling of the oil influence upon the working cycle of screw compressors. In: *Proceedings of the International Compressor Engineering Conference, Indiana,* Purdue University.

Yongzhang Yu, Peizheng Jiang, Siying Sun. (2012). Compressor engineering manual. Beijing, MA: China Petrochemical Press, in China.

Ziwen Xing. (2000). Theory, design and application of screw compressor. Beijing, MA: China Machine Press, in China.

Research on the performance of a high pressure 5.3MPa twin screw compressor

Zhao Zhaorui*, Wu Xiaokun, Hou Feng, Xing Ziwen

Xi'an Jiaotong University, School of Energy and Power Engineering,
Xi'an, Shaanxi, China
+86-18629596580 smile90613@163.com

ABSTRACT

High pressure twin screw compressors have been widely employed in fuel gas boosting and petrochemical industry. Recently, such compressors, whose maximum of discharging pressure is 5.3MPa, are also adopted in high temperature NH3 heat pumps and NH3/CO2 cascade refrigeration system. However, high pressure twin screw compressors are required to have large capacity, good performance and excellent stability at high operating pressure. In this paper, a semi-empirical model for open-type high pressure twin screw compressor is developed. Experimental research is conducted for identification of parameters, while validation is also made on the accuracy of the model. On the basis of theoretical and experimental research, the performance of the compressor, which includes volumetric efficiency, adiabatic efficiency, discharge temperature and lubricant oil flow rate/temperature, are illustrated. The change pattern of such features on the operating conditions, slide valve loadings and ambient features are then analyzed. Additionally, the stability test results of the high pressure twin screw compressor including the vibration and noise are also shown.

1. INTRODUCTION

With the substitution process of CFCs and HCFCs refrigerant, investigations on better choices for low temperature refrigerant, which is frequently employed in cold storage, instant freezer, pharmaceuticals and food industry, attracts global attention.

Natural refrigerants are proved to be promising in the low temperature refrigeration fields, among which CO2 stands out for its excellent mass and heat transfer properties, as well as large specific refrigeration capacity. CO2 has been widely employed in commercial or industrial refrigeration applications. However, due to the thermodynamic properties of CO2, specific requirement are made for the component of the system, especially the CO2 compressor. First, the CO2 refrigeration systems are usually working under specially high operating pressure. Therefore, the components of CO2 compressors should has excellent stability while working at extreme operating conditions. The moving parts as bearings or sealing must has enough loading capacity. Secondly, the compressing ration of CO2 refrigeration system is usually low, making the compressor to work under a condition of big pressure difference but low compressing ratio. The internal leakage of compressor appears to the key feature determining the performance of the system. Thus, the sealing capacity of moving components and the better control of leakage paths has direct influence on compressor performance. Additionally, CO2 has large specific refrigeration capacity, which is 5.12 times that of R22, so that the compressor must be made smaller. The manufacturing difficulty for either compressor body but also bearings or sealing are much higher. Moreover, CO2 has high ratio of specific heats, so that the discharge temperature might be high.

The CO2 refrigeration systems are divided into trans-critical and sub-critical systems. In trans-critical systems, the discharged CO2 is beyond critical point so that the refrigerant temperature change within the gas-cooler and the efficiency remains a big problem. On the other hand, the stability and efficiency of compressor is usually low at such extremely high pressure. In recently years, subcritical CO2 refrigeration systems are frequently employed in cascade refrigeration systems, including CO2/NH3, CO2/R22, CO2/R290 systems. The cascade refrigeration systems could have both high stability and efficiency for low temperature refrigeration. Moreover, the nature refrigerant employed in the systems have great effect on the substitution of CFCS/HCFCs refrigerant.

As the core component of refrigeration systems, high pressure CO_2 compressors attract much attention these years. The CO_2 compressors should has sufficient efficiency and reliability. Twin screw compressors have excellent efficiency and reliability in large capacity systems and high pressure situations, while refrigerant injection contributes to better performance and lower discharge temperature. In CO_2 subcritical refrigeration systems, the compressor may works at as high as operating pressure of 5.3 MPa, so that the performance and stability of components should be reevaluated. Conventional approaches for twin screw compressor performance prediction employs either CFD or working process modeling, but such approaches might be too complex and time consuming for applications in practical engineering. Recently, much investigation has been conducted on developing semi-empirical model for twin screws compressors. However, current model are not developed for high pressure twin screw compressors, so that the inter-stage leakage, heat transfer with oil and losses in compressing process are not profoundly considered.

In this paper, a novel high pressure 5.3 MPa CO_2 twin screw compressor set is designed and analyzed. Experimental test is conducted on the performance and stability of the compressor. Semi-empirical mathematical model is also built for the prediction of performance, while comparison between simulated and experimental results are compared on the accuracy of prediction for volumetric efficiency, adiabatic efficiency and COP. Their variation with change of operating conditions are also illustrated.

2. SEMI-EMPIRICAL COMPRESSOR MODEL

A series of semi-empirical model for the performance prediction of twin screw compressor and refrigeration system is developed. CO_2 is selected as the refrigerant. The condensing temperature ranges from -15 to 15 ℃, while the evaporating temperature ranges from -54 to -15 ℃.

The following assumptions are made to simplify the thermodynamic analysis.
(1) Steady-state assumption. All components are working at steady-state and dynamic characteristics are neglected.
(2) All throttling process are viewed as isenthalpic process.
(3) The heat and pressure loss in pipelines are neglected.

2.1 Model description

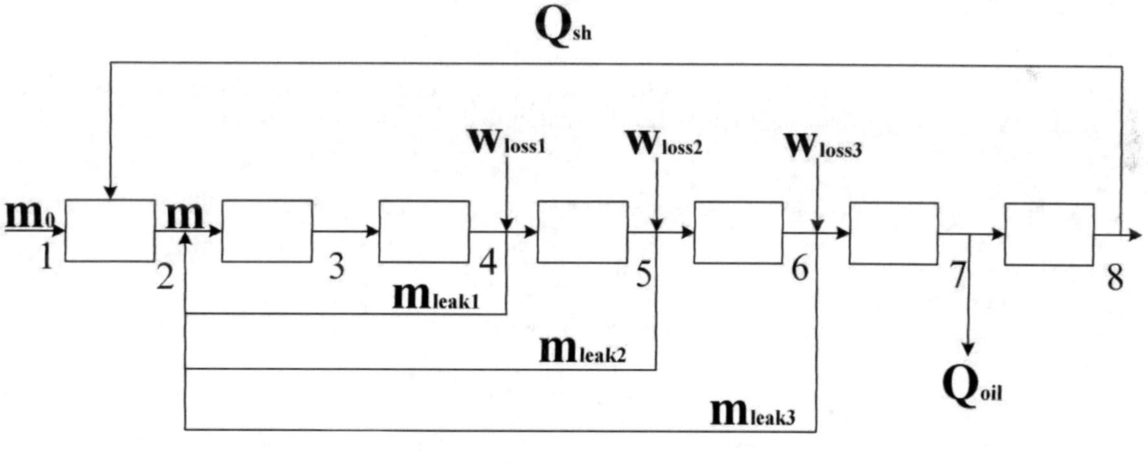

Fig.1 Schematic diagram of model

Conventional semi-empirical model for twin screw compressors are mainly concerning compressors whose operating pressure is ordinary .However, compressors have quite different working process and leakage characteristics at extremely high operating pressure. The inter-chamber leakage for high pressure twin screw compressor could not be neglected as it is in low pressure sets. Since as much as 30-40 percent of refrigerant may leak to suction side at extreme operating conditions, the calculation of power consumption for this part of compressing should be reconsidered. Therefore, in this paper, the compressing process is divided into three individual procedures, and the leakage and power losses is separately calculated after each of the three compressing process. Moreover, the heat exchange between refrigerant and oil for high pressure compressor will be more significant compared with the cooling effect of ambient condition. Therefore, in this paper, the cooling effect of oil, depending on the oil flow rate and oil suction temperature is calculated. Fig.1 illustrates the schematic processing of the semi-empirical model.

The refrigerant suction heating is considered in stage 1-2, while calculated the increased enthalpy of refrigerant contributing to the high temperature of compressor body or rotors. The amount of heat transfer from the rotor and compressor body may depend on the temperature difference between suction and discharge refrigerant as well as the time consumed.

$$Q_{1-2} = f_{1-2}\frac{am}{2\pi n}\left(t_{dis} - t_{suc}\right) \tag{1}$$

The heat transfer coefficient, as well as that between oil and refrigerant introduced later, is formulated as

$$f \propto \left(\frac{T_1 - T_2}{L}\right)^{\frac{1}{4}} \tag{2}$$

Stage 2-3 refers to the refrigerant mixing of refrigerant sucked and refrigerant leaked from higher pressure compressing chamber. The mass of refrigerant entering the compressor and the leakage mass mix altogether, which leads to a slight increase in refrigerant enthalpy after this isobarically mixing process.

$$m_0 h_2 + m_{leak1} h_4 + m_{leak2} h_5 + m_{leak3} h_6 = \left(m_0 + m_{leak1} + m_{leak2} + m_{leak3}\right) h_3 \tag{3}$$

Stage 3-4 refers to the first stage of imaginary compressing process. The refrigerant will firstly be isothermally compressed according to the built-in volumetric ratio in this stage. The ratio is calculated based on the relation between inter-rotor volume and rotating angle. Then, part of the refrigerant will be leaked to the suction chamber, after which the viscous friction torque losses are calculated. The compressing of refrigerant is viewed as isentropic.

$$s_3 = s_4 \tag{4}$$

$$v_4 = v_3 / BVR_{3-4} \tag{5}$$

The leakage of refrigerant in twin screw compressors are treated as leakage from leakage paths, which is one unique fictitious path connecting the refrigerant of high and low pressure. In this paper, the leakage mass flow rate is viewed as an isentropic flow through a nozzle.

$$\dot{m}_{leak1} = \frac{b}{v_4}\sqrt{p_4 - p_2} \tag{6}$$

Additionally, the power loss caused by viscous friction torque is calculated. The amount of shaft is consumed as proposed below, and equal amount of heat is viewed as transfer to the refrigerant at this point.

$$P_{loss3-4} = c\mu_{oil} V_g n^2 \tag{7}$$

Stages 4-5 and 5-6 are similarly calculated so that no redundant explanation is given.

Stage 6-7 is considering the discharge process of refrigerant after the compressing process. Either over-compression or under-compression may occur when the discharge port opens. The discharge process would be viewed as isentropic procedure where refrigerant is compressed or expanded by the backpressure refrigerant and then pushed out of the control volume by the rotor.

$$P_{6-7} = v_6 p_{dis} m \tag{8}$$

The heat transfer between oil and refrigerant is considered after the discharge process, because in practical process the time consumed for refrigerant to go through the compressing chamber is far less than enough for the profound heat transferring from high temperature refrigerant to low temperature oil. On the other hand, the heat transfer between compressor body and ambient temperature is neglected.

$$Q_{7-8} = \frac{d}{n} m q_{oil}\left(t_{dis} - t_{oil}\right)^{\frac{5}{4}} \tag{9}$$

Another shaft power loss necessary to be considered should be the transmission efficiency which is caused by the friction in bearings, sealing and gears.

$$P_{loss} = eP_{sp} \tag{10}$$

Therefore, the volumetric efficiency and adiabatic efficiency of the twin screw compressor could be calculated.

$$\eta_v = \frac{m}{m_0} \tag{11}$$

$$\eta_a = \frac{P_{th}}{P} \tag{12}$$

$$P_{th} = m\left(h_{dis}^{ic} - h_{suc}\right) \tag{13}$$

As for the evaporator, the amount of heat supply may depend on the refrigerant flow rate and the enthalpy difference between evaporator inlet and outlet.

$$\dot{Q}_e = \dot{m}_e\left(h_{out} - h_{in}\right) \tag{14}$$

$$COP = \frac{\dot{Q}_c}{P_t} \tag{15}$$

The thermodynamic property of ammonia refrigerant should be accurately calculated in the system modeling and performance prediction. In this paper, all properties as enthalpy, temperature, pressure, entropy are calculated via the NIST Standard Reference Database, which help determine the suction/discharge condition of compressor, and the status of refrigerant at each inlet or outlet of control volume.

2.2 Identification of parameters

As shown in the model description section, totally six parameter are necessarily to be calibrated, including a in the calculation of suction heating degree, b_4,b_5,b_6 in the calculation of refrigerant leakage mass flow rate of process 4,5,6, c in the calculation of viscous friction power loss and d in the calculation of transmission loss. Therefore, six points of experimental results are required in the calibration, where different suction pressure, discharge pressure, suction superheat degree, oil temperature and mass flow rate should be tested in the experiment.

In practical situations, the experiment is conducted on different operating conditions in order to better calibrate the parameters. The evaporating temperature ranges from 20 to 45 ℃ while the condensing temperature ranges from 60 to 85 ℃. The suction superheat degree ranges from 5.91 to 14.63 ℃ so that the suction temperature may be different in spite of constant suction pressure. The oil temperature are precisely controlled thus ranging from 52.7 to 65 ℃ in order to investigate the influence of oil temperature on performance. The oil flow rate ranges from 67 to 236 L/min. Therefore, totally fourteen points of experiment results are employed in this investigation, where six of them are used to calibrate the parameters while the eight points left are used to validate the accuracy.

The MATLAB optimization tools are employed to implement the model of compressor with the error target as follows.

$$err = \sqrt{\sum\left(\frac{\eta_a^{sl}}{\eta_a^{test}} - 1\right)^2 + \sum\left(\frac{\eta_v^{sl}}{\eta_v^{test}} - 1\right)^2 + \sum\left(\frac{t_{dis}^{sl}}{t_{dis}^{test}} - 1\right)^2} \tag{16}$$

3. EXPERIMENTAL FACILITY

In order to examine the performance and stability of high pressure twin screw compressor and CO2 refrigeration system, an experimental plant is built based on the RCH 16S high pressure compressor. The number of tips of the male rotor is 5 while that of the female rotor is 7. The length of rotors are 125mm. The diameter of male and female rotors are 189.5mm and 185mm respectively. Its theoretical discharge volume is 603.84 m^3/h. The schematic diagram is shown in Fig. 2. In application, the condensing temperature ranges from -15 to 15 ℃, and the evaporating temperature ranges from -54 ℃ to -15℃.

Fig.2 Experimental facility

The experiment content includes:

(1) Stability test. The stability of components is examined at high operating pressure, including the stability of bearings, slide valve reliability in loading control and the performance of oil pumps.

(2) Vibration and noise test. The intensity of vibration as well as noise is measured at different directions of the compressor in order to guarantee safe operation. The variations of vibration and noises under different operating conditions are also analyzed.

(3) Capacity test. The mass flow rate of refrigerant is measured at different operating condition, so that the volumetric efficiency could be calculated. At extreme high pressure difference condition, the capacity deterioration could then be calculated.

(4) Power consumption test. According to the measurement of power consumption on motors, the adiabatic efficiency could be calculated. Therefore, the variation of compressor performance, refrigeration system COP and cooling capacity could be analyzed with the change of operating conditions.

(5) Oil cooler test. The load and performance of oil cooler is examined in order to judge whether the requirement of oil cooling is satisfied, and to test the performance of oil cooler at high discharge temperature situation.

(6) Extreme operating condition test. All components are examined on the reliability at extreme operating condition of twin screw compressor.

4. RESULTS AND DICUSSION

The performance evaluation of compressor as well as refrigeration system are made by two steps. First, the volumetric efficiency, isentropic efficiency and COP of heat pump system is calculated based on the experimental data for the power consumption, refrigerant flow rate, heating capacity and operating conditions. Therefore, the parameters should be firstly calculated according to the variation of all parameters with the change of operating conditions, and then the compressor model is capable of predicting the performance of twin screw compressor and refrigeration system. The accuracy of simulated results are shown on volumetric efficiency, isentropic efficiency and COP. Second, with validated accuracy of the model, the refrigeration efficiency, as well as its variation with operating condition, is given.

4.1 Validation of modeling

4.1.1 Volumetric efficiency

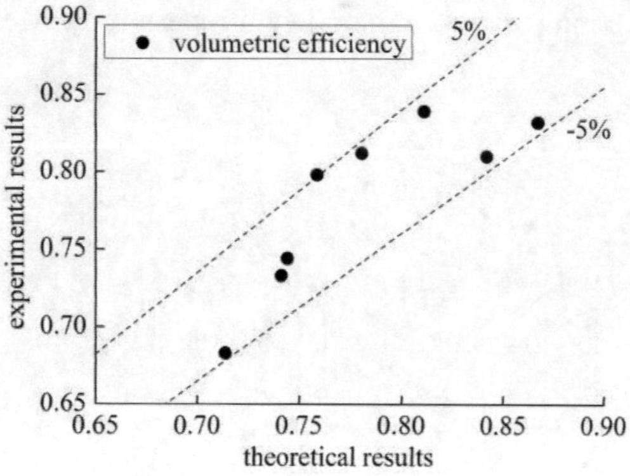

Fig.3 Accuracy of volumetric efficiency

The accuracy of volumetric efficiency prediction is shown in Fig.3. The X coordinate represents the simulated results while the Y coordinate represent the experimental results. At is shown in the figure, the simulated results of volumetric efficiency are within a maximum error of 5% compared with the experimental results in most situations. The volumetric efficiency is prominently higher when the pressure difference between suction and discharge is smaller. The difference of volumetric efficiency at different operating condition could be as high as 0.2. This is mainly the result of leakage which is much more significant when the pressure difference between the suction and discharge sides is larger.

4.1.2 Isentropic efficiency

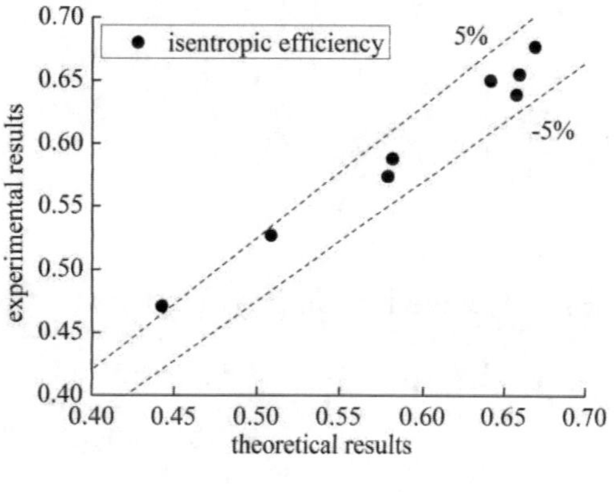

Fig.4 Accuracy of isentropic efficiency

The comparisons of simulated and experimental results of isentropic efficiency are shown in Fig.4. The error of simulated results is also controlled at the maximum of 5% for all operating situation. As it is similar with the trend of volumetric efficiency, the isentropic efficiency is higher when the compressing ratio is lower. The maximum difference at different operating condition is 0.25. The results may attribute to the inter-lobe leakage which is much more significant at bigger pressure difference.

4.1.3 COP

The comparisons of simulated and experimental results of system COP are shown in Fig. 5. The COP of the system ranges from 0.9 to 5 . Such trend is the combined result of theoretical COP increase and a higher isentropic efficiency of compressor at a smaller temperature difference between condensing and evaporating temperature.

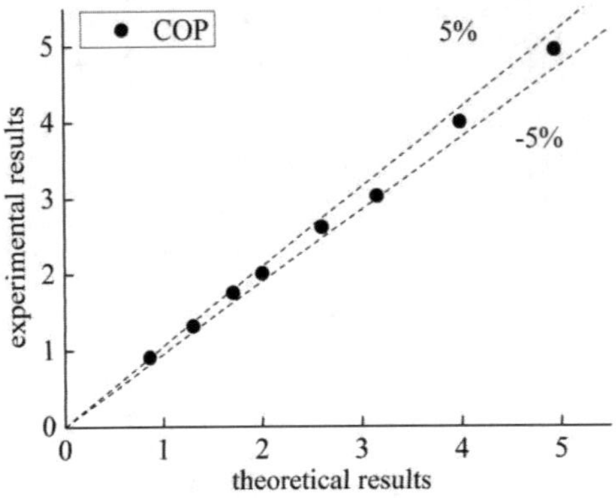

Fig.5 Accuracy of COP

4.2 Variation of performance

4.2.1 Volumetric efficiency

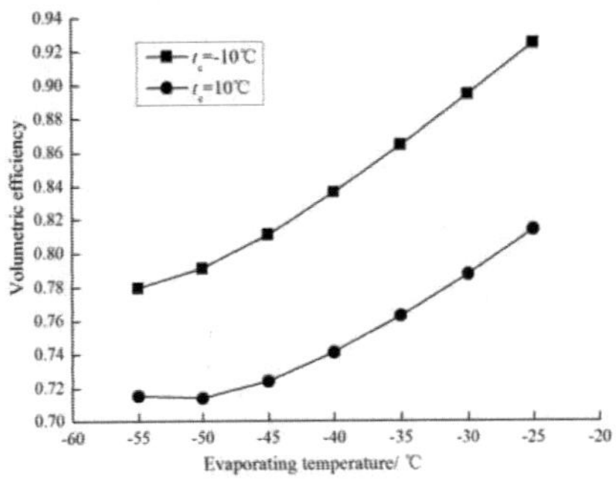

Fig. 6 Influence of evaporating temperature on volumetric efficiency

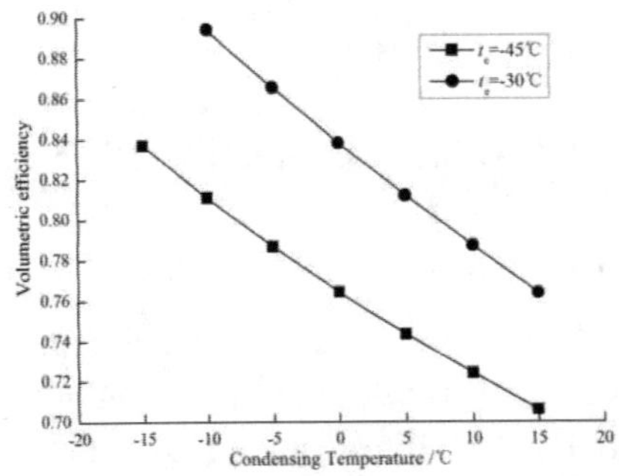

Fig.7 Influence of condensing temperature on volumetric efficiency

As shown in Fig.6 and Fig.7 , the volumetric efficiency decrease significantly with the increase of condensing temperature and the decrease of evaporating temperature. The compressor has 0.2 lower volumetric efficiency when the condensing temperature increase from -15 ℃ to 15℃ and the evaporating temperature decrease from -25℃ to -55℃, which reaches as high as 0.92. This would attribute to the bigger pressure difference between suction and discharge side, as well as more significant suction heating due to higher discharge temperature once the compressing ratio is higher.

4.2.2 Isentropic efficiency

The variations of isentropic efficiency with changing condensing/evaporating temperature are respectively illustrated in Fig.8 and Fig.9 . The adiabatic efficiency will first increase and then decrease with the increase of condensing temperature and the decrease of evaporating temperature. When the refrigerant evaporates at the temperature of -30℃, the adiabatic efficiency reaches its highest of 0.72 at 0℃ of condensing temperature. The efficiency ranges from 0.46 to 0.73 according to the operating condition. Such trend should attribute to the over-compression and owing-compression of screw compressors. If the compressing ratio is lower than the designed operating condition, the compressor will has lower adiabatic efficiency even if the pressure difference is lower.

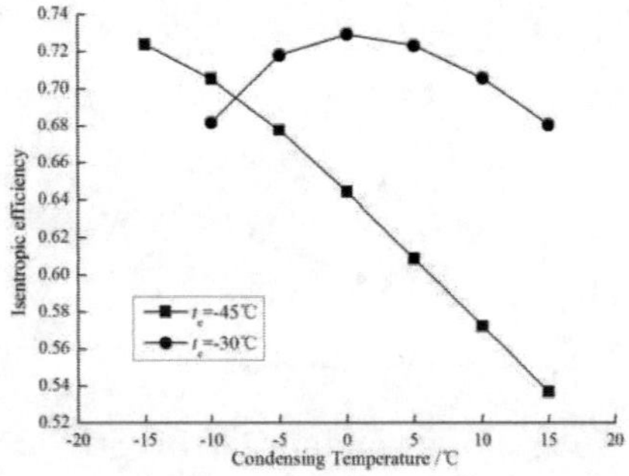

Fig. 8 Influence of condensing temperature on isentropic efficiency

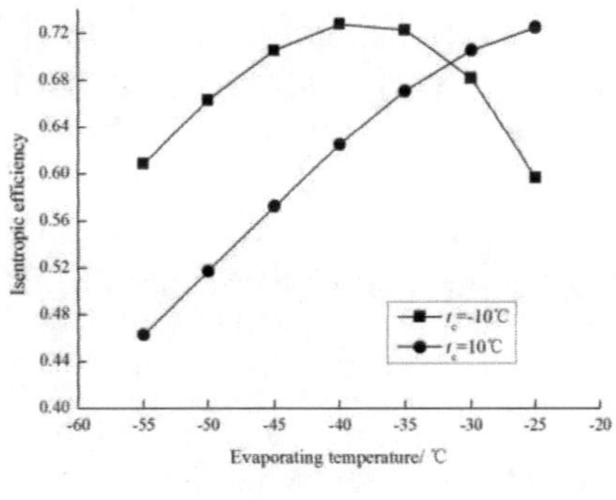

Fig. 9 Influence of evaporating temperature on isentropic efficiency

4.2.3 COP

Fig. 10 and Fig.11 show the variation of COP with changing operating condition. The COP decreases more rapidly when the evaporating temperature is higher or when the condensing temperature is lower. The COP ranges from 0.9 to 6.6 according to the operating condition. The COP difference is approximately 2.2 at 10℃ of condensing temperature, and is bigger with the decrease of heating temperature. Such may attribute to the increase of both the theoretical COP and isentropic efficiency at higher evaporating temperature or lower condensing temperature.

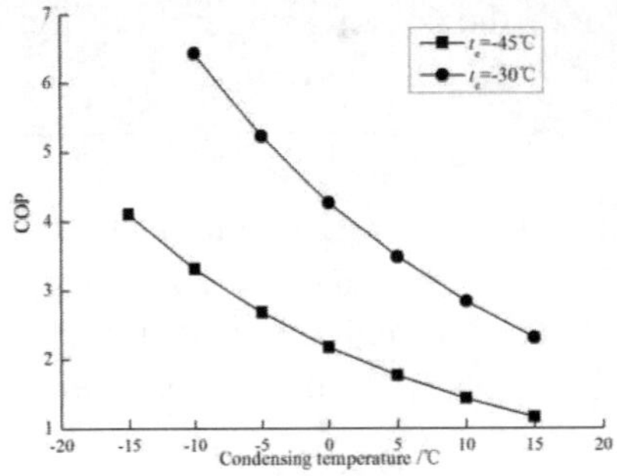

Fig. 10 Influence of condensing temperature on COP

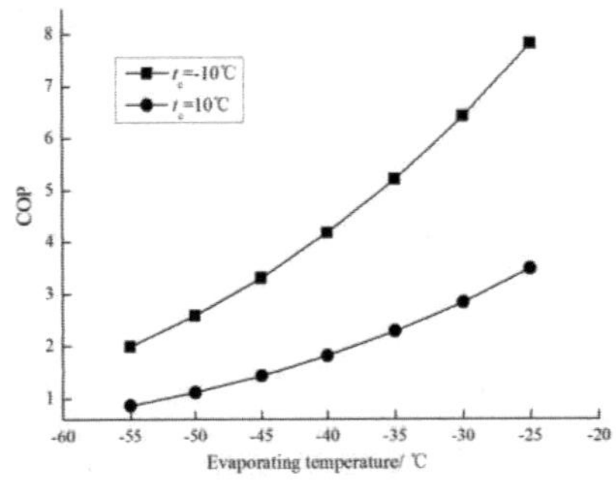

Fig. 11 Influence of evaporating temperature on COP

5. CONCLUSIONS

In this paper, the performance and stability of a high pressure 5.3MPa twin screw compressor is theoretically and experimentally analyzed. Novel semi-empirical model is developed to predict the performance of the CO2 subcritical refrigeration system, which has taken the inter-chamber leakage losses and cooling effect of oil into consideration. An experimental plant is set up to test both the performance of high pressure CO2 twin screw compressor and that of the subcritical refrigeration system. From the validation of compressor efficiency and COP of the refrigeration system, the accuracy of the model could be guaranteed, whose error is mainly controlled with 5%.

The high pressure CO2 twin screw compressor is guaranteed stability and reliability working at even extreme operating conditions. The COP decreases with the increase of condensing temperature, and decreases more rapidly when the evaporating temperature is higher. The COP ranges from 0.9 to 6.6 according to the operating condition. The adiabatic efficiency will first increase and then decrease while the volumetric efficiency may monotonously increase with the increase of condensing temperature and the decrease of evaporating temperature.

NOMENCLATURE

Q	quantity of heat	(kJ)
m	mass	(kg)
t	temperature	(K)
n	rotation speed	(r/min)
L	length	(m)

h	enthalpy	(kJ/kg)
s	entropy	(kJ/(kg·K))
v	specific volume	(m^3/kg)
p	pressure	(kPa)
P	power	(kW)
μ	viscosity	(Pa·s)
V	volume	(m^3)
q	mass flow rate	(kg/s)
η	efficiency	(-)
err	error	(-)
Subscript		
suc	suction	
dis	discharge	
sp	shaft power	
th	theoretical	
ic	isentropic compressing	
a	isentropic	
v	volumetric	
sl	simulate	
t	technical	
e	evaporator	
c	condenser	

REFERENCES

[1] Dopazo J A, Fernández-Seara J, Sieres J, et al. (2009) Theoretical analysis of a CO 2–NH 3 cascade refrigeration system for cooling applications at low temperatures[J]. *Applied Thermal Engineering*, 29(8):1577–1583.

[2] Lee T S, Liu C H, Chen T W. (2006) Thermodynamic analysis of optimal condensing temperature of cascade-condenser in CO2/NH3 cascade refrigeration systems[J]. *International Journal of Refrigeration*, 29(7):1100–1108.

[3] Wang B, Wu H, Li J, et al. (2009) Experimental investigation on the performance of NH3/CO2 cascade refrigeration system with twin-screw compressor[J]. *International Journal of Refrigeration*, 32:1358–1365.

[4] Sanz-Kock C, Llopis R, Sánchez D, et al. (2014) Experimental evaluation of a R134a/CO2 cascade refrigeration plant[J]. *Applied Thermal Engineering*, 73:41–50.

[5] Arias, J., (2005). Energy usage in supermarkets modelling and field measurements. Department of Energy Technology. Stockholm, Royal Institute of Technology. Doctoral thesis.

[6] Stosic N, Smith I K, Kovacevic A, et al. (2011) Review of Mathematical Models in Performance Calculation of Screw Compressors[J]. International Journal of Fluid Machinery & Systems, 4(2):271-288.Liu J, Li Q, Wang F, et al. A new model of screw compressor for refrigeration system simulation[J]. *International Journal of Refrigeration*, 2012, 35(4):861–870.

[7] Liu J, Li Q, Wang F, et al. (2012) A new model of screw compressor for refrigeration system simulation[J]. *International Journal of Refrigeration*, 35(4):861–870.

[8] Giuffrida A.(2015) A semi-empirical method for assessing the performance of an open-drive screw refrigeration compressor[J]. *Applied Thermal Engineering*, 93:813-823..

ACKNOWLEDGEMENT

The China authors would like to acknowledge the financial support from the National Natural Science Foundation of China (Grant No. 51276134).

Effect of Working Medium on Single Screw Refrigeration Compressor Performance

Zengli WANG[1]*, Yingfeng SHEN[1], Zhenbo WANG[1]

[1]College of Chemical Engineering, China University of Petroleum, Qingdao 266580, China

Tel: +86-0532-86983483 Fax: +86-0532-86983480 Email: wangzengli25@ upc.edu.cn

* Corresponding Author

ABSTRACT

The single screw refrigeration compressor (SSRC) is widely used in refrigeration and air conditioning systems due to the advantages of simple structure, balanced forces on the rotor, small vibration, low noise level, high volumetric efficiency and so on. In the SSRC, working medium is an important factor influencing the thermal dynamic performance of the compressor. Restricted by the application environment and the requirements of environmental protection, the types of the working medium used in the SSRC will change gradually in future, this to a certain extent, affect the compressor adaptability for medium and the operation stability and reliability of compressor. Thus the effects of the working medium on the geometric characteristics and dynamic performance of the SSRC must be studied. In this paper, the geometric model of the exhaust port and the fluid flow calculation model of the exhaust process were established for analyzing the influence of the working medium on the exhaust port opening position and the flow resistance loss during exhaust process in the SSRC. A dynamic characteristic analysis model was proposed to calculate the force act on the meshing pairs and the bearings under the working condition as well as the load conditions of the shaft system under the shutdown-starting condition. All analysis results obtained in this paper can be useful for optimum design of the SSRC to improve the operation stability and reliability and the energy efficiency of the compressor with different working medium. The first major section of the manuscript is an abstract. The abstract should describe the contents of the paper, discuss the contribution to the field as well as present the most important results. Authors are responsible for typing accuracy and proofreading the manuscript. If accepted, the manuscript must be submitted for reproduction without being edited or retyped by the staff before printing. The manuscript must look professional and be technically correct in order to be accepted.

1. INTRODUCTION

Single screw compressor was first invented by Zimmern (1963), and then it was introduced to the field of refrigeration soon in the mid of 1970s (Constant *et al.* 1981 and Zimmern 1984). The SSRC mainly consists of one screw rotor and two star-wheels which symmetrically located with the screw to double the swept volume and balance the thrusts. So it has advantages of simple structure, without valve, balanced forces on the rotor, small vibration, low noise level, high volumetric efficiency and so on. Currently, it has been widely used in refrigeration and air conditioning systems. In the SSRC, working medium is an important factor influencing the thermal dynamic performance of the compressor. But today, restricted by the application environment and the requirements of environmental protection, the new agreement about accelerating the phase-out of production and consumption of hydro-chlorofluorocarhons (HCFCs) have been established. So the types of the working medium used in the SSRC will change gradually, this to a certain extent, affect the compressor adaptability for medium and the operation stability and reliability of compressor.

Over the years, a series of studies have been carried out for the HCFCs replacement technologies, the properties of the alternative refrigerants and the current situation of refrigerants in different types of refrigeration compressors. Ma and Wang (2010) analyzed the refrigerant alternative technologies, summarized the advantages and disadvantages of various hydro-fluorocarbons (HFCs) and natural refrigerants for substitution. From their study, the HFCs alternative refrigerants such as R410A, R134a and R407 series were considered to be viable transition alternative refrigerants, and the natural refrigerants such as the R290, NH_3, CO_2 and H_2O were considered to be environmental friendly alternative refrigerants, but the R290 and NH_3 had the security problems, furthermore the requirement for the compressor and system of the CO_2 and H_2O became more strict. Li (2011) enumerated a kind of transition alternative refrigerant, such as R32, DR-2, HFO-1234yf in addition to natural refrigerants, and analyzed

the thermal physical properties of this transition alternative refrigerant in his paper. Huo (2015) analyzed the feasibility of using R1234ze(Z) and R152a mixture as alternative for R22. Chen *et al.* (2016) presented a review of the latest research and development of the refrigerants based on the new version of the Montreal Protocol 2014 Assessment report, the 11th HR Gustav Lorentzen Conference on Natural Refrigerants 2014 and the 25th international congress of refrigeration 2015. The current situation of refrigerants in different types of refrigeration units and the properties of the alternative refrigerants were discussed in their study. Through these studies, many kinds of alternative refrigerants had been proposed to replace HCFC, and been used in refrigeration systems.
But the properties of the alternative refrigerants are difference from that of the HCFC, which will affect the geometric characteristics and working characteristics of the compressor in the refrigeration system used these alternative refrigerants. So the scholars at home and abroad have carried out some analysis about the compressor operating characteristics with different alternative refrigerants. Jurgen, Mitsuhiro and Hubacher *et al.* (1998, 2000, and 2002) researched the performance of the reciprocating compressor, the twin rotary compressor and the vane compressor for carbon dioxide cycle respectively. Yang *et al.* (2008) developed the hermetic rotary CO_2 compressor and measured its performance. Li and Yang *et al.* (2008) analyzed the CO_2 expander used in refrigeration system. Weight *et al.* (2000) researched the efficiency of the water vapor compressor. Brandon *et al.* (2000) analyzed the commercial feasibility of the use of water vapor as a refrigerant. Sariibrahimoglu *et al.* (2009) studied the effect of R600a on tribological behavior of sintered steel in hermetic refrigeration compressor. Xu *et al.* (2016) compared the performance of the compressor used R22, R404A and R407F. Song *et al.* (2015) compared the performance of the open screw compressor using R717 and R22. These studies showed that the structure and performance of the compressor were greatly influenced by refrigerant. Although many scholars have done a lot of work, so far no work has been done about the effect of working medium on single screw refrigeration compressor performance. In the SSRC, working medium is an important factor influencing the thermal dynamic performance of the compressor. Thus the effects of the working medium on the geometric characteristics and dynamic performance of the SSRC must be studied.

In this paper, the geometric model of the exhaust port and the fluid flow calculation model of the exhaust process are established for analyzing the influence of the working medium on the exhaust port opening position and the flow resistance loss during exhaust process in the SSRC. A dynamic characteristic analysis model is proposed to calculate the force act on the meshing pairs and the bearings under the working condition as well as the load conditions of the shaft system under the shutdown-starting condition. All analysis results obtained in this paper can be useful for optimum design of the SSRC to improve the operation stability and reliability and the energy efficiency of the compressor with different working medium.

2. GEOMETRIC CHARATERISTICS ANANLYSIS

2.1 Geometric characteristics

In the SSRC, in order to increase the radial inlet area and reduce the friction area between the screw and the cylinder wall, the cylinder wall is usually designed to step as shown in Figure 1 (a), in which I is the closed spiral and II is the exhaust orifice. The spiral at the outside of the steps called close spiral. According to the geometric features of it, the closed helix refers to the spiral at the ventro-side of the tooth when the star wheel teeth form primitive volume. Because of that the properties of the refrigerants are difference from each other, the star wheel rotation Angle when the star wheel teeth form primitive volume is also different. So the location of the closed spiral will be affected by the refrigerants. In order to analysis the influence of refrigerants on the closed spiral, the closed spiral calculation model needs to be established.
Assuming that the star wheel Angle $\alpha = \alpha_{\text{in}}$ when the star-wheel tooth just forms a closed volume as shown in Figure 1 (b), then the equations of the closed spiral can be expressed as follows:

$$\begin{cases} Y_{\text{f}}(\alpha) = \left(A - R_{\text{1p}}\right)\tan\alpha - \dfrac{b}{2\cos\alpha} \\ S_{\text{f}}(\alpha) = PR_1(\alpha - \alpha_{\text{in}} - \dfrac{\alpha_\delta}{P}) \end{cases} \qquad \alpha_{\text{in}} \le \alpha \le \alpha_{\text{out}} \tag{1}$$

Where Y_{f} is the axial coordination of the closed spiral, A is the central distance between the star wheel and the screw rotor, R_{1p} is the equivalent radius in the meshing point of the star wheel, α is the rotation position angle of the star wheel, b is the tooth width of star-wheel, S_{f} is the circumferential coordination of the closed spiral, P is the tooth number ratio of the star-wheel to the screw rotor, R_1 is the radius of the screw rotor, α_{in} is the flow inlet angle, α_{out} is the angle when the star-wheel tooth meshing out of the spiral groove, and α_δ is the position angle of the star-wheel meshing line.

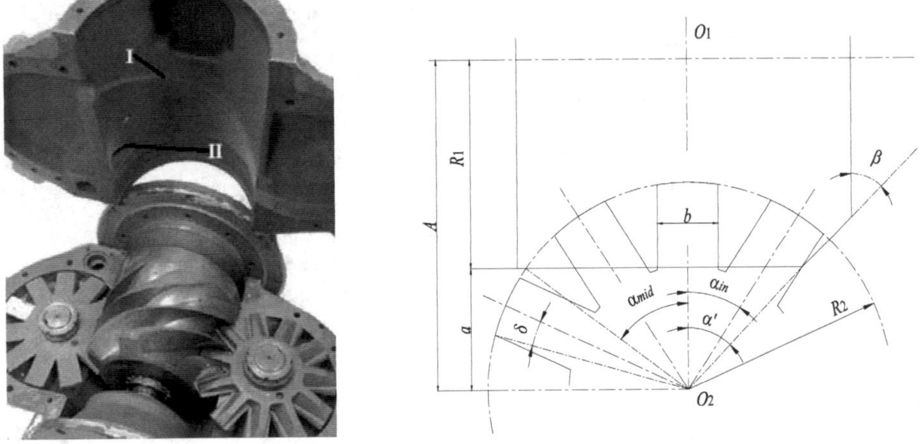

(a) The internal surface of the cylinder wall (b) Geometric parameters relationship

Figure 1: The schematic of the single-screw compressor

Different from reciprocating compressor, the exhaust process of the SSRC is the forced exhaust process. The location and shape of the exhaust orifice is the necessary condition to guarantee the gas realizing the designed internal compression ratio in the primitive volume, and is the important factor to influence the compressor performance. As a result of the forced exhaust process, the exhaust process will begin when the front screw groove flank of the closed working chamber connected to the exhaust orifice. At this time, the front screw groove flank spiral is set as the exhaust orifice spiral (As shown in Figure 2). Similarly, the refrigerants will also affect the location and shape of the exhaust orifice spiral, so the exhaust orifice spiral calculation model needs to be established to analysis the influence of refrigerants on the exhaust orifice spiral.

Set the star wheel angle as the α_d when the front screw groove flank of the closed working chamber connected to the exhaust orifice, and set the star wheel angle as the α_{mid} when the front star-wheel tooth meshing out of the spiral groove. Used the star wheel rotation Angle as the characterized angle, the exhaust orifice spiral equation is as follows:

$$\begin{cases} Y_p(\alpha) = (A - R_{1p})\tan\alpha + \dfrac{b}{2\cos\alpha} \\ S_p(\alpha) = PR_1(\alpha - \alpha_d - \dfrac{\alpha_\delta}{P}) \end{cases} \quad \alpha_d \le \alpha \le \alpha_{mid} \tag{2}$$

Where Y_p is the axial coordination of the exhaust orifice spiral, and S_p is the circumferential coordination of the exhaust orifice spiral.

2.2 Flow resistance loss

In the exhaust process of the SSRC, the high pressure gas flow through the exhaust orifice and be discharged to the exhaust pipe. There will be a big flow resistance loss when the gas flows through the exhaust orifice which will affect the performance of the SSRC. The exhaust orifice position and area is different in the SSRC using different refrigerants. If the exhaust orifice was designed improperly, the flow resistance loss will be increased and the performance of compressor will be reduced inevitably. In order to calculate the flow resistance loss, the area of the exhaust orifice needs to be gained. The exhaust orifice area can be calculated through the integration of the exhaust orifice spiral equation and exhaust end line, the calculation formula is:

$$S = \int_{\alpha}^{\alpha_{\text{mid}}} \left(y_{\text{p}} \left(\alpha_{\text{out}} \right) - y_{\text{p}} \left(\alpha \right) \right) PR_1 \text{d}\alpha \tag{3}$$

After calculating the exhaust orifice area, the flow velocity which used to measure the flow resistance loss can be obtained as following:

$$v = \dot{V} / S \tag{4}$$

Where $\dot{V}$ is the exhaust gas volume from the exhaust orifice per second.

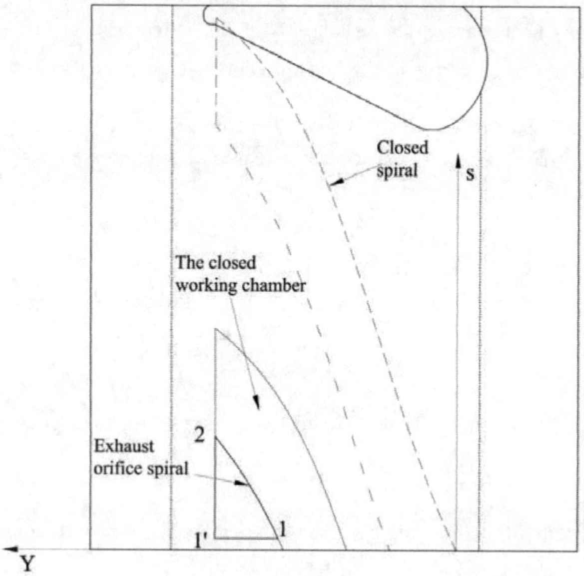

Figure 2: The expanded view of the internal surface of the cylinder wall

3. DYNAMIC CHARACTERISTICS ANALYSIS

3.1 Thermal process analysis

In the SSRC, the gas pressure in the compression chamber under different star wheel rotation angle can be expressed as:

$$p(\alpha) = \begin{cases} p_{\text{s}} & \alpha < \alpha_{\text{in}} \\ \dfrac{V_0^{\,n}}{V(\alpha)^n} p_{\text{s}} & \alpha_{\text{in}} \le \alpha \le \alpha_{\text{d}} \\ p_{\text{d}} & \alpha \ge \alpha_{\text{d}} \end{cases} \tag{5}$$

Where p_{s} is the suction pressure, V_0 is the groove volume at suction closure of the SSRC, $V(\alpha)$ the volume of the compression chamber under different star wheel rotation angle, which can be calculated by the equation listed in the Wang's study(2014), p_{d} is the exhaust pressure of the SSRC.

By using this equation, the gas temperature in the compression chamber under different star wheel rotation angle can be expressed as:

$$T(\alpha) = T_{\text{in}} \left(\frac{p(\alpha)}{p_{\text{s}}} \right)^{\frac{n-1}{n}} \qquad \alpha_{\text{in}} \le \alpha \le \alpha_{\text{d}} \tag{6}$$

3.2 Dynamic characteristics

In the SSRC, the symmetrical arranged star wheels make the radial gas force of the screw which caused by the compressed gas in the screw groove in balance with each other. On the other hand, the axial force of the screw also can be kept balance due to the compressed gas filled in the screw groove and the axial gas channel opened in the screw. Based on the above characteristics, the dynamic characteristics of the screw are very good. But the gas forces act on the star wheel can not be kept balance. The refrigerants will affect the dynamic characteristics of the star wheel, and then the operation reliability of the compressor will be affected. So in this study the dynamic characteristics of the star wheel must be analyzed.

The forces act on the star wheel and shaft system are shown in Figure 3, F_g is the resultant gas force acting on the multiple star wheel teeth at the same time. F_n is the normal force acting on the star wheel tooth flank. M_{fA} and M_{fB} are the torque of the star wheel bearing. The value of F_n is much smaller than that of the gas force under quasi-static equilibrium state, therefore its influence in the dynamic characteristics analysis of the star wheel shaft can be ignored.

For a standard gear ratio SSRC, there are three teeth engage in screw. Set the last tooth as the basis tooth to calculate resultant forces F_g. The gas force applied on the each star wheel tooth is expressed as follows:

$$F_g(\alpha_i)=\begin{cases}\int_{-\frac{b}{2}}^{\frac{b}{2}}\left(p(\alpha_i)-p_s\right)\cdot dS_{s,i} & \alpha_{in}\le\alpha_i\le\alpha_{mid}\\ \int_{-\frac{b}{2}}^{R_2\cdot\sin\gamma}\left(p(\alpha_i)-p_s\right)\cdot dS_{s,i} & \alpha_{mid}\le\alpha_i\le\alpha_{out}\end{cases}\tag{7}$$

$$\gamma_0=arc\cos\left(\frac{A-R_1}{R_2}\right)-\delta\tag{8}$$

So the resultant gas forces act on the multiple star wheel teeth at the same time can be calculated as following:

$$F_g(\alpha)=\sum F_{g,i}(\alpha_i)=\sum_{i=0}^{n}F_{g,i}(\alpha+i\lambda)\tag{9}$$

Where $p(\alpha_i)$ is the gas pressure of the working chamber at Arbitrary star wheel rotation Angle, p_s is the suction pressure, $S_{s,i}$ is the tooth area engage in screw, R_2 is the radius of the star wheel, δ is the half width tooth angle, λ is the intersection angle of the center line of the adjacent two star wheel.

As shown in Figure 3, the resultant gas force act on the tip position of the star wheel tooth, so an overturning moment will be produced by the gas force. In order to balance this overturning moment, bearings on both ends of the star wheel bracket need to be installed. The torque of the star wheel bearing can be obtained by the following equation:

$$M_i(\alpha_i)=\begin{cases}\int_{-\frac{b}{2}}^{\frac{b}{2}}\int_0^{\mu}\left(p(\alpha_i)-p_s\right)\cdot r(\alpha_i)\,d\sigma d\eta & \alpha_{in}\le\alpha\le\alpha_{mid}\\ \int_{-\frac{b}{2}}^{R_2\cdot\sin\gamma}\int_0^{\mu}\left(p(\alpha_i)-p_s\right)\cdot r(\alpha_i)\,d\sigma d\eta & \alpha_{mid}\le\alpha\le\alpha_{out}\end{cases}\tag{10}$$

$$W(\alpha)=\sum W_i(\alpha_i)=\sum_{i=0}^{n}W_i(\alpha+i\lambda)\tag{11}$$

Where $r(\alpha_i)$ is the distance between the gas force equivalent application point and the star wheel axis, which can be calculated by the equation listed in the Wang's study (2014).

Influence by the work environment, the SSRC alternate in downtime and normal work condition. When the compressor is in a state of shutdown, the refrigerant in the system will keep in an equilibrium state. In this case, the internal pressure of the system is much higher than suction pressure. So in the shutdown-starting condition, the suction gas is the refrigerant vapor in the equilibrium state. The pressure of the suction gas under the shutdown-starting condition is much higher than the normal suction pressure of the SSRC. So the forces act on the meshing pair will become higher than the normal working conditions. As the refrigerant used in the SSRC is different, the pressure of the refrigerant vapor under equilibrium state will be different too. So the refrigerants will affect the dynamic characteristics of the meshing pairs under shutdown-starting condition. In order to analyze the influence of the refrigerant on the dynamic characteristics under shutdown-starting condition, the forces acting on the star wheel under the shutdown-starting condition also have to be calculated in this study.

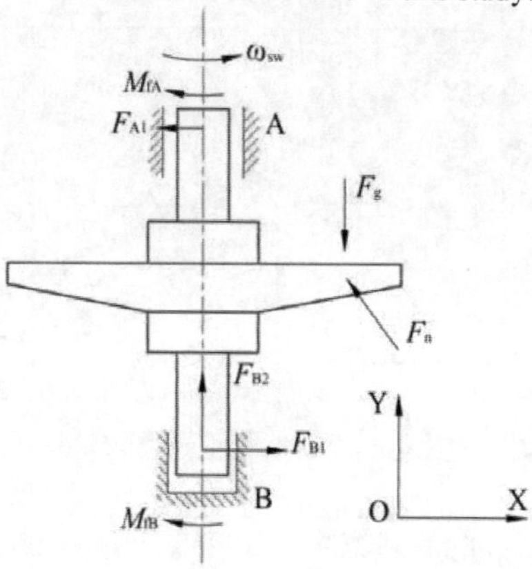

Figure 3: The forces act on the star wheel and shaft system

4. RESULTS AND DISCUSSION

In this paper, a typical SSRC used in refrigeration and air conditioning field with the structure and operation parameters as shown in Table 1 was chosen for the calculate example. Several refrigerants such as R22, R134a, R410A, R290, R717 and R32 were used in this study to contrastive analysis the influence of the refrigerant on the geometric characteristics and the dynamic performance of the SSRC.

Table 1: The main parameters

Parameters	Value	Parameters	Value
P	11/6	Length of the screw (mm)	155.7
R_1 (mm)	181	Thickness of the star wheel (mm)	7.25
R_2 (mm)	181	Lubricating oil	Suniso 4GS
A (mm)	144.8	Evaporation temperature (K)	278.15
b (mm)	26.5	Condensation temperature (K)	313.15
n (r.min^{-1})	2880	Superheat temperature (K)	5

4.1 Geometric characteristics analysis results

The location and shape of the exhaust orifice will be affected by the refrigerants. Figure 4 shows the opening location and the shape of the exhaust orifice in the SSRC with different refrigerant. The contrast of the exhaust orifice of the SSRC with different refrigerant shows that the opening location and the area of the exhaust orifice are greatly affected by the refrigerants because of that the pressure ratio and the adiabatic index of the refrigerants are very different from each other. The internal volume ratio increase by degress from curve 6 and curve 1.For the refrigerants used in this study, the exhaust orifice opening time of the SSRC with R410A and R32 which have a lower pressure ratio and higher adiabatic index is earlier than that of the SSRC with R22, R134a, R290 and R717 which have a higher pressure ratio and lower adiabatic index. And the exhaust orifice area of the SSRC with R410A and R32 is larger than that of the SSRC with R22, R134a, R290 and R717. So in the design phase, the exhaust orifice must be designed separately for each kind of the refrigerant. If the design is not changed, the over or under compression process will occur, then the additional energy losses will be caused.

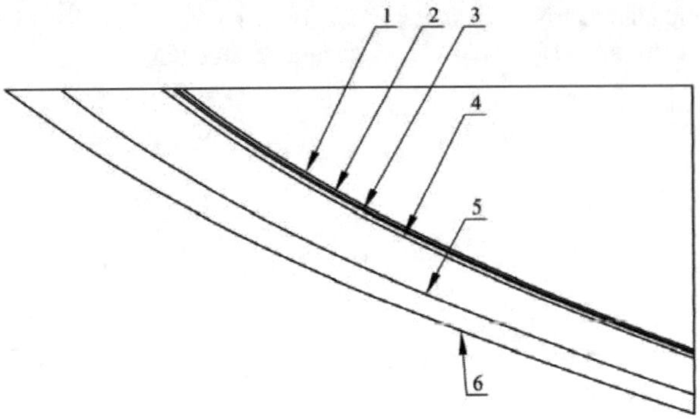

1-R134a, 2-R290, 3-R717, 4-R22, 5-R410A, 6-R32

Figure 4: The location and shape of the exhaust orifice

Figure 5 shows the closed spiral location of the SSRC with different refrigerant. As can be seen from the contrast of these closed spirals, the difference of closed spiral location of the SSRC with different refrigerant is very small. So the conclusion of that the influence of refrigerants on the closed spiral is very small and is smaller than that on the location and shape of the exhaust orifice. So in the design phase, only one closed spiral needs to be designed for all kinds of the refrigerant, but the closed spiral must be designed to meet the requirements of the refrigerant with the minimum inlet pressure.

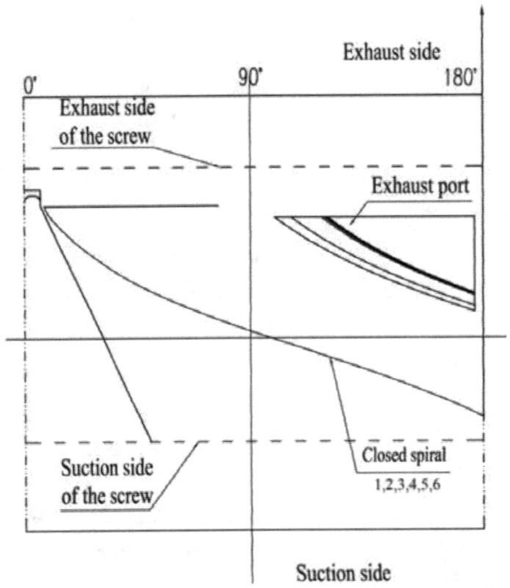

1-R134a, 2-R290, 3-R717, 4-R22, 5-R410A, 6-R32

Figure 5: The closed spiral location

Figure 6 shows the exhaust orifice flow speed ratio of the compressor using different refrigerants changing with the pressure ratio of the refrigerants. The variation tendency of the curves in Figure 6 shows that the exhaust orifice flow speed of the compressor will influence by the pressure ratio and the adiabatic index of the refrigerants rather than the evaporating pressure and condensing pressure of the refrigerants. For the refrigerants used in this study, the exhaust orifice flow speed of the compressor with R410A and R32 which have a lower pressure ratio and higher adiabatic index is smaller than that of the compressor with R22, R134a, R290 and R717 which have a higher pressure ratio and lower adiabatic index. That means the exhaust flow resistance loss of the compressor with R410A and R32 is lower than that of the compressor with R22, R134a, R290 and R717.

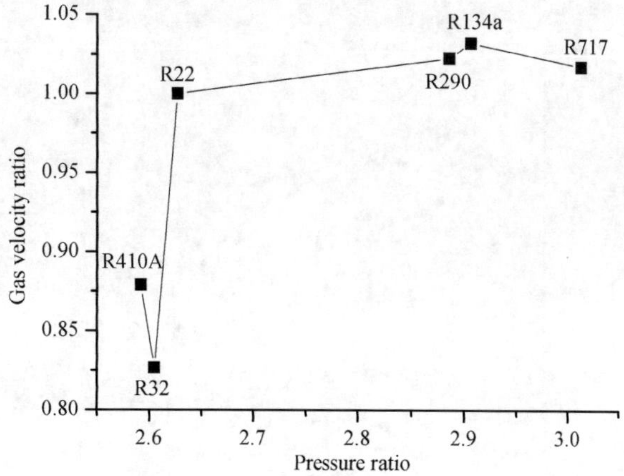

Figure 6: The exhaust orifice flow speed of the compressor with different refrigerants

4.2 Dynamic characteristics analysis results

The gas force acts on the star wheel and the torque acts on the shaft system are shown in Figure 7 and Figure 8 respectively. The contrasts of the gas force curves and the torque curves in Figure 7 and Figure 8 show that the gas force acts on the star wheel and the torque acts on the shaft system of the compressor will be affected by the evaporating pressure and condensing pressure of the refrigerant rather than the pressure ratio and the adiabatic index of the refrigerant. The gas force and the torque will increase with the increase of the evaporating pressure and condensing pressure of the refrigerant. So in the design phase, the dynamic characteristics of the meshing pair and the shaft system have to be checked if a refrigerant with high evaporation pressure and condensing pressure used in the SSRC. For each refrigerant, the gas force first increases and then decreases with the change of the star-wheel angle. Due to the change of work pressure ratio, the star wheel Angle is different to achieve the maximum gas force.

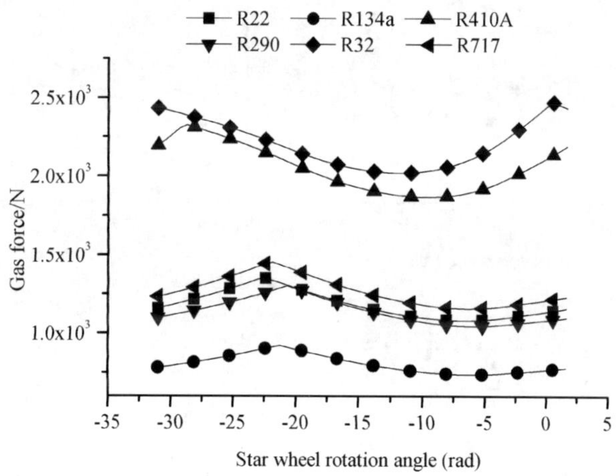

Figure 7: The gas force acts on the star wheel

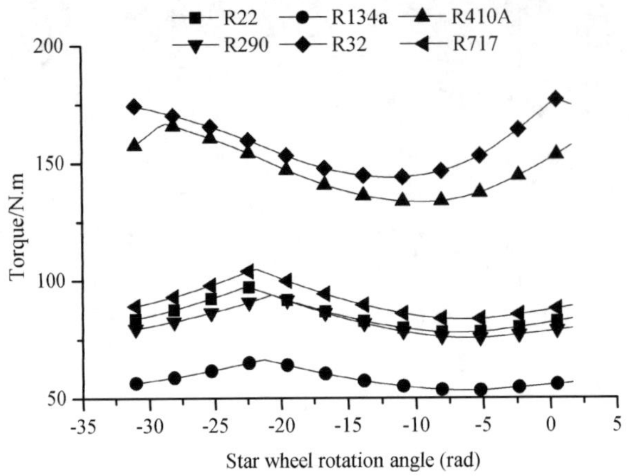

Figure 8: The torque acts on the shaft system

The gas force and the torque act on the shaft system under the shutdown-starting condition are shown in Figure 9. The contrasts of the gas force curves and the torque curves in Figure 7, Figure 8 and Figure 9 show that the influence law of the refrigerant on the gas force and the torque under the shutdown-starting condition are the same as that under the normal working condition. But the gas force and the torque act on the shaft system under the shutdown-starting condition are twice the gas force and the torque act on the shaft system under the normal working condition.

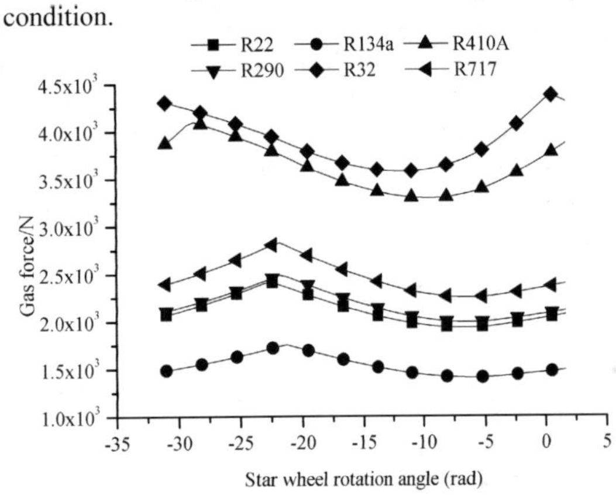

(a) Gas force

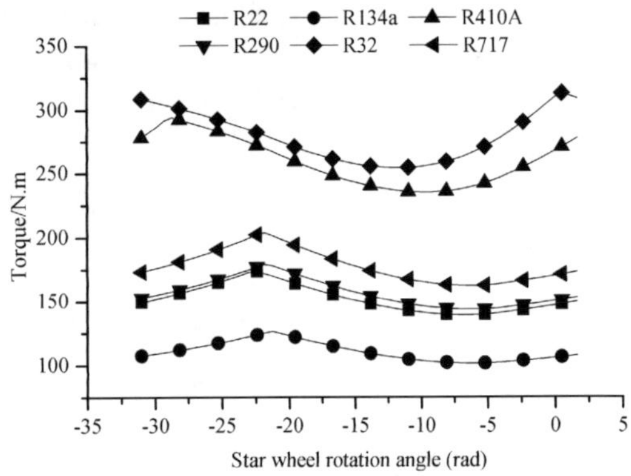

(b) Torque

Figure 9: The gas forces and torque under the shutdown-starting conditon

5. CONCLUSIONS

(1) The location and shape of the exhaust orifice will be affected by the refrigerants. The exhaust orifice opening time of the SSRC with refrigerant which have a lower pressure ratio and higher adiabatic index is earlier than that of the SSRC with refrigerant which have a higher pressure ratio and lower adiabatic index. And the exhaust orifice area is larger when the refrigerant has a lower pressure ratio and higher adiabatic index.
(2) The influence of refrigerants on the closed spiral is very small and is smaller than that on the location and shape of the exhaust orifice.
(3) The exhaust orifice flow speed of the compressor will be affected by the pressure ratio and the adiabatic index of the refrigerants rather than the evaporating pressure and condensing pressure of the refrigerants. The exhaust orifice flow speed of the compressor with refrigerant which have a lower pressure ratio and higher adiabatic index is smaller than that of the SSRC with refrigerant which have a higher pressure ratio and lower adiabatic index. That

means the exhaust flow resistance loss is lower when the refrigerant has a lower pressure ratio and higher adiabatic index.

(4) The gas force acts on the star wheel and the torque acts on the shaft system of the compressor will be affected by the evaporating pressure and condensing pressure of the refrigerant rather than the pressure ratio and the adiabatic index of the refrigerant. The gas force and the torque will increase with the increase of the evaporating pressure and condensing pressure of the refrigerant.

(5) The influence law of the refrigerant on the gas force and the torque under the shutdown-starting condition is the same as that under the normal working condition. But the gas force and the torque act on the shaft system under the shutdown-starting condition are twice the gas force and the torque under the normal working condition.

NOMENCLATURE

The nomenclature should be located at the end of the text using the following format:

A	The central distance between the star wheel and the screw rotor	(m)
b	The tooth width of star-wheel total cost	(m)
F	Force	(N)
n	Rotate speed	(r/min)
P	Tooth number ratio of the star-wheel to the screw rotor	
p	Pressure	(Pa)
R	Radius	(m)
r	The gas force equivalent application distance	(m)
R_{1p}	Equivalent radius in the meshing point of the star wheel	(m)
S	The circumferential coordination of the closed spiral	
$S_{s,i}$	The tooth area engages in screw	(m^2)
V_0	The primitive volume of the SSRC	(m^3)
$\dot{V}$	The exhaust gas volume from the exhaust orifice per second	(m^3/s)
W	Torque	(N.m)
Y	The axial coordination of the closed spiral	
α	Star-wheel rotation angle	(rad)
α_δ	The position angle of the star-wheel meshing line	(rad)
δ	The half width tooth angle	(rad)
λ	The intersection angle of the center line of the adjacent two star-wheel	(rad)

Subscript

1	Screw rotor
2	Star-wheel
d	Exhaust
f	Closed spiral
g	Gas
in	Inlet
mid	The front star-wheel tooth meshing out of the spiral groove
out	The star-wheel tooth meshing out of the spiral groove
p	Exhaust orifice spiral
s	Suction

REFERENCES

Zimmern B., Patel G. J. (1972). Design and operating characteristics of the zimmern single screw compressor, in: Proceedings of the 1972 Purdue Compressor technology Conference, West Lafayette, USA, pp. 96-99.

Constant L. C., Hundy G. F., Chan C.Y., et al. (1981). Single-screw compressor for refrigeration and other duties, London, UK: Mech Eng Publ Ltd, 1981: 117-126.

Zimmern B. (1984). From water to refrigerant: twenty years to develop the oil injection-free single screw compressor, in: Proceedings of the 1984 International Compressor Engineering Conference, West Lafayette, USA, pp. 513-518.

Ma Y. T., Wang W. (2010). Substitution and postponable technology of refrigerants, Journal of refrigeration, 2010, 31(05): 11-17.
Li L. S. (2011). Research progress on alternative refrigerants and their development trend, Journal of refrigeration, 2011, 32(06): 53-58.
Huo E. G., Dai Y. D., Gong P., et al. (2015). Feasibility of using R1234ze and R152a mixture as alternative for R22, CIESC Journal, 2015, 66(12): 4725-4729.
Chen G. M., Gao N., Piao C. C. (2016). State of the art of research and applications of low-carbon refrigerants, Journal of Refrigeration, 2016, 37(01):1-11.
Jurgen S., Horst K. (1998). Efficiency of the indicated process of CO_2-compressor. International Journal of Refrigeration, 1998, 21(3): 194-201.
Mitsuhiro F., Radermacher R. (2000). Performance of a vane compressor for CO_2 cycle// 4th IIR-Gustav Lorentzen Conference on Natural Working Fluids at Purdue University, USA, 2000: 339-346.
Hubacher B., Groll E. A. (2002). Measurement of performance of carbon dioxide compressors. http://www.arti-21cr.org/research/completed/.
Yang J., Lu P., Zhang L., et al. (2008). Development and performance measurements of a new hermetic rotary CO_2 compressor. Journal of Shanghai Jiaotong University, 2008, 42(3): 426-529.
Li M. X., Ma Y. T., An Q. S., et al. (2008). Hermetic swing expander in transcritical carbon dioxide cycle water source heat pump. Chinese Journal of Mechanical Engineering, 2008, 44(05):160-164.
Yang B. C., Guo B., Peng X. Y., et al. (2008). Experimental research on internal working process of rotary vane expander in CO_2 refrigeration system. Journal of Xi'an Jiaotong University, 2008, 42(3): 313-317.
Wight S. E., Yoshinaka T., LeDrew B. A., et al. (2000). The efficiency limits of water vapor compressors. Report for Air-Conditioning and Refrigeration Technology Institute, 2000.
Brandon F., Lacher Jr., Gregory F., et al. (2007), The commercial feasibility of the use of water vapor as a refrigerant. International Journal of Refrigeration, 2007, 30(4): 699-708.
Sariibrahimoglu K., Kizil H., Aksit M., et al. (2009). Effect of R600a on tribological behavior of sintered steel under starved lubrication. Tribology International, 2009, 43(5-6): 1054-1058.
Xu S., Gao F., Liu Z.S., et al. (2016). Performance comparison of R22, R404A and R407F used in compressors. Low Temperature and Specialty Gases, 2016, 34(01): 7-10.
Song Y. Q., Fan H. B., Xu J. D., et al. (2015). Performance experimental comparison of open screw compressor using R717 and R22 refrigerant. Refrigeration, 2015, 43(05): 62-64.
Wang Z. L. (2014). Research of thermal dynamic characteristics for variable load single screw refrigeration compressor with multicolumn compound enveloping meshing pairs. PhD thesis. Xi'an Jiaotong University.

ACKNOWLEDGEMENT

This work is supported by China Postdoctoral Science Foundation [NO.2015M572094], Qingdao Postdoctoral Sponsored Project [NO. T1504115] and Natural Science Foundation of Shandong Province, China [NO.ZR2015EL032].

Analysis on Lubricant Film Force for Two Types of Meshing Pair Profile in Single Screw Compressor

Feilong Liu [1], Jia Xie[2], Davide Ziviani[3], Quanke Feng[4]* , Martijn Van Den Broek[5], Michel De Paepe[6]

[1]School of Energy and Power Engineering, Xi'an Jiaotong University,
Xi'an, Shaanxi, China
[1]Department of Flow, Heat, and Combustion Mechanics, Ghent University,
Ghent, Belgium
liufeilong@stu.xjtu.edu.cn
[2]School of Energy and Power Engineering, Xi'an Jiaotong University,
Xi'an, Shaanxi, China
jiajiawa_1990@163.com
[3]Department of Flow, Heat, and Combustion Mechanics, Ghent University,
Ghent, Belgium
Davide.Ziviani@UGent.be
[4]School of Energy and Power Engineering, Xi'an Jiaotong University,
Xi'an, Shaanxi, China
qkfeng@mail.xjtu.edu.cn, Tel(Fax): +86-29-82663783
[5]Department of Flow, Heat, and Combustion Mechanics, Ghent University,
Ghent, Belgium
Martijn.vandenBroek@UGent.be
[6]Department of Flow, Heat, and Combustion Mechanics, Ghent University,
Ghent, Belgium
Michel.DePaepe@UGent.be
* Corresponding Author

ABSTRACT

Single screw compressors are widely used in various industrial fields, however the performance usually drops obviously after thousands of hours of operation due to the wear between the star-wheel and the gate rotor. A lot of meshing pair profiles and manufacturing methods are proposed to solve this problem. Thus, good oil lubrication conditions are conducive to improve the wear resistance of the meshing pair so as to increase the compressor's operation life. In this paper, the geometric models of the clearances for both profiles, a kind of multi-column profile and single-column profile for oil flooded compressors are carried out, and a mathematical lubrication model is used to analyses the oil lubrication. Results show that the torques caused by the oil film force on both sides of the star-wheel tooth are unbalanced. The total torques of the analyzed multi-column profile is smaller than that of the single column profile, which indicates a better lubrication condition. The analysis can also be used to optimize the profile design of the meshing pair of single screw compressors.

1. INTRODUCTION

Since Zimmern (1972) first developed the single screw compressor (SSC) with the profile of straight-line envelope type, it is now widely used in various fields, such as air compression, refrigeration, industrial progress, heat pump and etc.. However, due to the poor wear-resistance performance, the discharge capacity decreases sharply after thousands hours' operation. This is mainly because of the rapid wear of the star-wheel tooth flank surface while meshing with the screw groove, which can extends the clearance and then results in the leakage. A lot of works have been done by Zimmern (1976), Jing (1986), Boblitt (1987), Jensen (2000), Wu et al. (2009) in order to find a suitable structure profile for prolonging the operation life of the meshing pair, including both profile design and manufacturing methods. Feng et al. (2005) also proposed a multistraight-line envelope profile. Then, on the basis of column envelope meshing pair (CEMP) and multistraight-line envelope profile, a multicolumn envelope meshing pair (MEMP) was developed by Wu et al. (2009) aimed to reduce the friction.

According to Cameron's study (1983), a well formed hydrodynamic lubrication between two sliding surfaces can generate a lubricant film and avoid the wear. Similarly to the meshing pair in SSC, a good lubrication between the screw groove and the tooth flank surface can also reduce or even avoid the wear problem. In the study of Post and Zwaans (1986) about the hydrodynamic lubrication, the tooth will move to the smaller force side because of the unbalanced lubricant film forces generated on each side of the tooth flank. This will lead to the clearances' increasing on one side and decreasing on the other side. According to the research of Jisheng et al. (1997), Huang et al. (2007), Kim et al. (2014) and Larsson (2009), when the change reaches to the critical point, the flow regime will change form hydrodynamic lubrication to boundary lubrication, which will cause the metal contact between the meshing pair and lead to the wear out of the tooth flank. Thus, it is necessary to analyze the lubrication situation between the meshing pair. Huang et al. (2015) created a useful optimization method of the meshing pair profile based on the proposed lubrication simulation analysis about the CEMP.

In this paper, a simulation model is used to analyses the oil lubrication in two types of meshing pair, a kind of multi-column profile and single column profile for oil flooded compressors. Real design parameters of these two types meshing pair for the same single screw compressor are given as calculation examples. The oil lubrication analysis results are then compared and discussed.

2. GEOMETRY MODEL

2.1 Geometric And Kinematic Relations Between The Meshing Pair

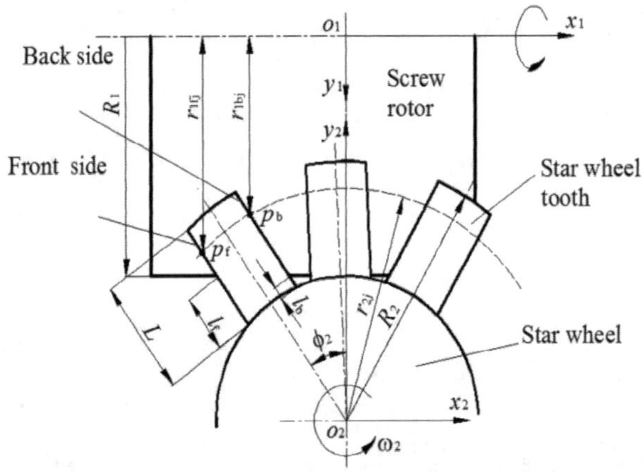

Figure 1: Geometric relations between the meshing pair

Fig.1 shows the relationships between screw rotor and star-wheel while operation. ϕ_2 is the rotation angle. R1 is the radius of the screw rotor. For any contact points P_f and P_b on each side along the tooth, the related swinging radius r_{2j} is

$$r_{2j} = \sqrt{\left(\sqrt{R_2^2-(b/2)^2}-(L-l)\right)^2-(b/2)^2} \tag{1}$$

Where R_2 is the radius of the star-wheel, b is the tooth width, l is the distance from the tooth root to the contact point, L is the tooth length.

The swinging radius from the contact point to the rotor center, r_{1fj} and r_{1bj} are

$$\begin{cases} r_{1fj} = a - r_{2j}\cos(\phi_2 + \delta) \\ r_{1bj} = a - r_{2j}\cos(\phi_2 - \delta) \end{cases} \tag{2}$$

Where a is the center distance between the screw rotor and the star-wheel, δ is the space width half angle

$$\delta = \arccos(\frac{b}{2R_2}) \tag{3}$$

The unengaged length on both sides of the tooth l_f and l_b are

$$\begin{cases} l_f = L - \dfrac{R_2 \cdot \cos(\delta + \phi_2) - a + R_1}{\cos\phi_2} \\ l_b = L - \dfrac{R_2 \cdot \cos(\delta - \phi_2) - a + R_1}{\cos\phi_2} \end{cases} \tag{4}$$

2.2 Double-column Meshing Pair

The double-column meshing pair profile is established mainly based on the MEMP, shown in Fig.2. The geometric and kinematic relations between the tooth and the groove can be described by three coordinates. Coordinate, $S\ (t,n)$ is used to describe the center coordinates of the designed columns, which could be given through the design process. The other two Cartesian coordinates $S_f\ (x_f,h_f)$ and $S_b\ (x_b,h_b)$ are used for expressing the geometry of the clearances in both sides (front and back) of the tooth. Axes x_f and x_b coincide with the front and back screw groove flank, h_f and h_b two orthogonal axes to x_f and x_b, respectively, which can express the clearances between the screw groove and the tooth flank. As shown in Fig.2, the tooth flanks of the double-column profile are consisted of two column segments and three tangent lines for each side. Points 1-8 are the tangent points of the arcs and lines. The groove inclined angles (α_{f1}, α_{b1}), equals to the inclined angles of the tangent lines at the meshing points (c,d). These angles can be obtained by the geometric and kinematic relations between the star wheel and the screw shown in Fig.1.

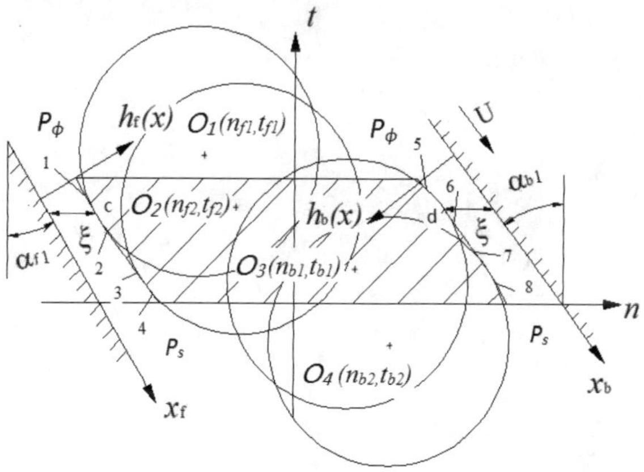

Figure 2: Double-column meshing pair

For this double-column meshing pair profile, columns O_2 and O_3 are the basic columns, which are parallel to the tooth center line. O_1 and O_4 are the auxiliary columns with inclined angles β_f and β_b. According to the MEMP theory proposed by Wu et al. (2009), the center coordinates of the auxiliary columns at any position along the tooth can be calculated once β_f, β_b and the center coordinates on the tooth top are given after design.

The groove inclined angles are expressed as

$$\begin{cases} \tan\alpha_{f1} = \dfrac{\omega_2 r_{2j}}{\omega_1 r_{1fj}} \\ \tan\alpha_{b1} = \dfrac{\omega_2 r_{2j}}{\omega_1 r_{1bj}} \end{cases} \tag{5}$$

The clearance functions for the front side of the tooth can be expressed as follows:

$$h_f(t) = \begin{cases} h_f(t_1) + \dfrac{(t-t_1)\sin(\alpha_{f1}-\alpha_1)}{\cos(\alpha_1)} & t_1 < t \le u \\ \xi\cos(\alpha_{f1}) + r - r\cos(\alpha_{f1} - \arcsin(\dfrac{t_{f1}-t}{r})) & t_2 \le t \le t_1 \\ h_f(t_2) + \dfrac{(t_2-t)\sin(\alpha_2-\alpha_{f1})}{\cos(\alpha_2)} & t_3 \le t < t_2 \\ h_f(t_3) + \dfrac{(t_3-t)\sin(\dfrac{\arcsin(\frac{t_{f2}-t}{r})+\alpha_2}{2} - \alpha_{f1})}{\cos(\dfrac{\arcsin(\frac{t_{f2}-t}{r})+\alpha_2}{2})} & t_4 \le t < t_3 \\ h_f(t_4) + \dfrac{(t_4-t)\sin(\alpha_4-\alpha_{f1})}{\cos(\alpha_4)} & 0 \le t < t_4 \end{cases} \tag{6}$$

$$x(t) = \begin{cases} \dfrac{(u-t)\cos(\alpha_{f1}-\alpha_1)}{\cos(\alpha_1)} & t_1 \le t \le u \\ x(t_1) + \dfrac{(t_1-t)\cos(\alpha_{f1} - \dfrac{\arcsin(\frac{t_{f1}-t}{r})+\alpha_1}{2})}{\cos(\dfrac{\arcsin(\frac{t_{f1}-t}{r})+\alpha_1}{2})} & t_c \le t < t_1 \\ x(t_c) + \dfrac{(t_c-t)\cos(\dfrac{\arcsin(\frac{t_{f1}-t}{r})-\alpha_{f1}}{2})}{\cos(\dfrac{\arcsin(\frac{t_{f1}-t}{r})+\alpha_{f1}}{2})} & t_2 \le t < t_c \\ x(t_2) + \dfrac{(t_2-t)\cos(\alpha_2-\alpha_{f1})}{\cos(\alpha_2)} & t_3 \le t < t_2 \\ x(t_3) + \dfrac{(t_3-t)\cos(\dfrac{\arcsin(\frac{t_{f2}-t}{r})+\alpha_2}{2} - \alpha_{f1})}{\cos(\dfrac{\arcsin(\frac{t_{f2}-t}{r})+\alpha_2}{2})} & t_4 \le t < t_3 \\ x(t_4) + \dfrac{(t_4-t)\cos(\alpha_4-\alpha_{f1})}{\cos(\alpha_4)} & 0 \le t < t_4 \end{cases} \tag{7}$$

Where r is the radius of the designed columns, ξ is the design clearance, u is the thickness of the tooth, α_1 and α_4 are angles of the tangent lines when points 1 and 4 act as the meshing points, which can be got as follows,

$$\begin{cases} \tan\alpha_1 = \dfrac{\omega_2 r_{2j}}{\omega_1 R_1} \\ \tan\alpha_4 = \dfrac{\omega_2 r_{2j}}{\omega_1 (a - r_{2j})} \end{cases} \quad (8)$$

Line 34 is the common tangent line of the two columns whose angle can be calculated according to the geometric relationships between the columns, therefore, α_2=α_3.t_1~t_4 are the related height of the points.

$$\begin{cases} t_i = t_{f1} - r\sin(\alpha_i) \quad i = 1,2 \\ t_i = t_{f2} - r\sin(\alpha_i) \quad i = 3,4 \end{cases} \quad (9)$$

Similarly to the front side, the clearance functions of the back side can be obtained as well.

2.3 Single-column Meshing Pair

The single-column meshing pair is introduced and studied by Wu and Feng (2009), shown in Fig.3.

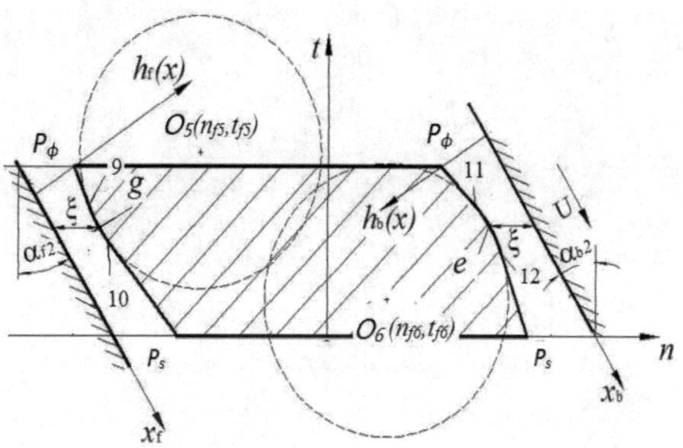

Figure 3: Single-column meshing pair

Three coordinates are also set similarly to the double-column meshing pair to describe the clearances between the crew groove and the tooth flank. The tooth flanks of the single-column profile are consisted of one column segments and two tangent lines for each side. Points 9-12 are the tangent points of the arcs and lines. The groove inclined angles (α_{f2}, α_{b2}), which are related to the meshing points (g,e), can be calculated by

$$\begin{cases} \tan\alpha_{f2} = \dfrac{\omega_2 r_{2j}}{\omega_1 r_{2fj}} \\ \tan\alpha_{b2} = \dfrac{\omega_2 r_{2j}}{\omega_1 r_{2bj}} \end{cases} \quad (10)$$

Then the clearance functions for the front side are

$$h_f(t) = \begin{cases} h_f(t_9) + \dfrac{(t - t_9)\sin(\alpha_{f2} - \alpha_9)}{\cos(\alpha_9)} & t_9 < t \le u \\ \xi\cos(\alpha_{f2}) + r - r\cos(\alpha_{f2} - \arcsin(\dfrac{t_{f5} - t}{r})) & t_{10} \le t \le t_9 \\ h_f(t_{10}) + \dfrac{(t_{10} - t)\sin(\alpha_{10} - \alpha_{f2})}{\cos(\alpha_{10})} & 0 \le t < t_{10} \end{cases} \quad (11)$$

$$
x_f(t)=\begin{cases}
\dfrac{(u-t)\cos(\alpha_{f2}-\alpha_9)}{\cos(\alpha_9)} & t_9\le t\le u \\[2ex]
x_f(t_9)+\dfrac{(t_{10}-t)\cos(\alpha_{f2}-\dfrac{\arcsin(\frac{t_{f5}-t}{r})+\alpha_9}{2})}{\cos(\dfrac{\arcsin(\frac{t_{f5}-t}{r})+\alpha_9}{2})} & t_g\le t<t_9 \\[2ex]
x_f(t_g)+\dfrac{(t_g-t)\cos(\alpha_{f2}-\dfrac{\arcsin(\frac{t_{f5}-t}{r})-\alpha_{f2}}{2})}{\cos(\dfrac{\arcsin(\frac{t_{f5}-t}{r})+\alpha_{f2}}{2})} & t_{10}\le t< t_g \\[2ex]
x_f(t_{10})+\dfrac{(t_{10}-t)\cos(\alpha_{10}-\alpha_{f2})}{\cos(\alpha_{10})} & 0\le t< t_{10}
\end{cases}
\tag{12}
$$

Where α_9, α_{10} are the angles of the tangent lines when points 9 and 10 act as the meshing points, which can be got by:

$$
\begin{cases}
\tan\alpha_9=\dfrac{\omega_2 r_{2j}}{\omega_1 R_1} \\[2ex]
\tan\alpha_{10}=\dfrac{\omega_2 r_{2j}}{\omega_1(a-r_{2j})}
\end{cases}
\tag{13}
$$

t_9 and t_{10} are the related height of the meshing points

$$
t_i=t_{f5}-r\sin(\alpha_i)\quad i=9,10 \tag{14}
$$

The clearance functions for the back side can also be obtained through the same way.

2.4 Lubrication Model

The whole path between the groove and the tooth flank can be divided into two parts by the meshing point, convergent region at the high pressure side, and divergent region at the low pressure side. For the divergent region, the pressure applied on the tooth flank can be seen as the suction pressure.

For the convergent region, the lubrication model introduced by Launder and Leschziner (1978) can be simplified as:

$$
\frac{\mathrm{d}P(x)}{\mathrm{d}x}=\frac{6\mu U}{h(x)^2}-\frac{12\mu M}{\rho h(x)^3}+\frac{1.2M^2}{\rho h(x)^3}\cdot\frac{\mathrm{d}h(x)}{\mathrm{d}x}-\frac{0.133\rho U^2}{h(x)}\cdot\frac{\mathrm{d}h(x)}{\mathrm{d}x} \tag{15}
$$

Where ρ is the density of the lubrication oil, $h(x)$ is the geometric model of the clearance.

The boundary condition is

$$
\begin{cases}
x=x_0, P(x)=P_\phi \\
x=x_n, P(x)=P_s
\end{cases}
\tag{16}
$$

Where P_ϕ is the gas pressure in the compressing chamber, P_s is the suction pressure. The pressure distribution along the tooth surface in the convergent region is

$$P(x) = P_\phi + \int_{x0}^{x} \frac{6\mu U}{h(x)^2}\mathrm{d}x - \frac{12\mu M}{\rho}\int_{x0}^{x}\frac{1}{h(x)^3}\mathrm{d}x$$
$$-\frac{1.2M^2}{2\rho}\left(\frac{1}{h(x)^2} - \frac{1}{h(x_0)^3}\right) - 0.133\rho U^2[\ln h(x) - \ln h(x_0)] \tag{17}$$

Where

$$M = \frac{\frac{12\mu}{\rho}\int_{x0}^{x_n}\frac{1}{h(x)^3}\mathrm{d}t - \sqrt{\left(\frac{12\mu}{\rho}\int_{x0}^{x_n}\frac{1}{h(x)^3}\mathrm{d}t\right)^2 + \frac{2\cdot 1.2}{\rho}\left(\frac{1}{h(x_n)^2} - \frac{1}{h(x_0)^2}\right)N}}{\frac{1.2}{\rho}\left(\frac{1}{h(x_0)^2} - \frac{1}{h(x_n)^2}\right)} \tag{18}$$

$$N = \int_{x0}^{x_n}\frac{6\mu U}{h(x)^2}\mathrm{d}x - 0.133\rho U^2[\ln h(x_n) - \ln h(x_0)] + P_s - P_\phi \tag{19}$$

U is the relative velocity between the meshing pair

$$U = \sqrt{(\omega_2 r_{2j})^2 + (\omega_1 r_{1(f,b)j})^2} \tag{20}$$

Then, the component forces caused by the oil film and the gas force, applied on the engaged element dl, which are parallel to the upper surface of the star-wheel are

$$\begin{cases} F_{oij}' = \cos\alpha \cdot \int_{x_0}^{x_n} P(x)dx \\ F_{gi} = \cos\alpha \cdot \int_{x_n}^{x_s} P_s dx \end{cases} \tag{21}$$

Where α is the groove inclined angle of the related meshing point.

The gas force, applied on the unengaged element dl, which is parallel to the upper surface of the star-wheel is

$$F_{gi}' = u \cdot P_s \tag{22}$$

Therefore, the torques applied on each side of the tooth are

$$\begin{cases} T_{fi} = \int_{l_f}^{L} F_{foij}' \cdot r_{2j}(l)dl + \int_{l_f}^{L} F_{fgi} \cdot r_{2j}(l)dl + \int_{0}^{l_f} F_{fgi}' \cdot r_{2j}(l)dl \\ T_{bi} = \int_{l_b}^{L} F_{oijb}' \cdot r_{2j}(l)dl + \int_{l_b}^{L} F_{bgi} \cdot r_{2j}(l)dl + \int_{0}^{l_b} F_{bgi}' \cdot r_{2j}(l)dl \end{cases} \tag{23}$$

Where l_f, l_b and $r_{2j}(l)$ can be got through equation (2~4).

The composite torque applied on one tooth is as follows:

$$T_i = T_{bi} - T_{fi} \tag{24}$$

During operation, there are usually three or four star-wheel teeth participating in the engaging at the same time. Thus, the total torque applied on the star-wheel is

$$T = \sum_{1}^{4} T_i \tag{25}$$

3. RESULTS

In this paper, two different designs of the two profiles for the same 2 m³/min oil-flooded single screw air compressor are taken as calculation examples. The Dimensions of the single screw compressor is shown in Tab. 1.

Table 1: Dimensions of the single screw compressor

Main Parameter	**Value**	**Main Parameter**	**Value**
Screw rotor radius R_1/mm	65	Tooth length L/mm	30
Star-wheel radius R_2/mm	61	Center distance a/mm	96
Tooth width b/mm	18	Envelope column radius r/mm	6
Rotating speed / $r{\cdot}min^{-1}$	2970	Discharge pressure $P_{out}/10^5 Pa$	10
Design clearance ξ/mm	0.04		

Table 2: Design parameters

Parameter	***n/mm***	***t/mm***
Tooth top of Column O_1	-4.754	7.292
Column O_2	-2.674	4.542
Column O_3	2.674	0.958
Tooth top of Column O_4	4.739	-1.792
Column O_5	-2.674	4.542
Column O_6	2.674	0.958

Tab. 2 is the design parameters related to the two different profiles. For the double-column, β_f=2.327°, β_b=-2.298°.

Fig.4 is the pressure distributions of the two types meshing pair under the discharge starting rotation angle (ϕ_2=35°) at the tooth top.

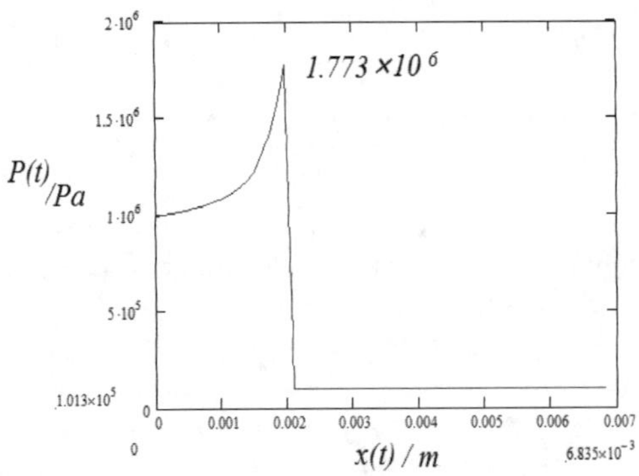

(a) Pressure distribution for double-column meshing pair

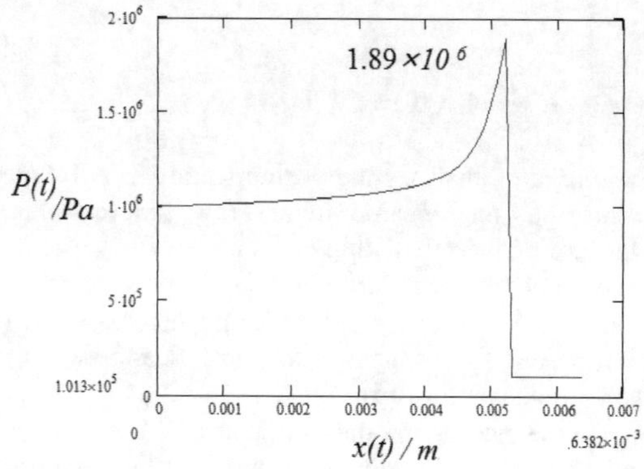

(b) Pressure distribution for single-column meshing pair

Figure 4: Pressure distributions at the top tooth

In both types of the meshing pair, the pressures increase in the convergent region, and will reach the peak value at the meshing point. For the double-column meshing pair, the peak pressure is 1.773×10^6 Pa. For single-column meshing pair, the peak pressure is 1.89×10^6 Pa.

The total torques of the two types meshing pair through the rotation angle ϕ_2 are shown in Fig.5. Obviously, the torques for both profiles have periodical changes during the operation. For the double-column, the total torque changes from 1.3 N·m to 3.7 N·m. For the single-column, the total torque changes from 2.92 N·m to 4.5 N·m. At any same rotation angle, the total torque applied on the double-column star-wheel is always smaller than that of the single-column. According to the study of Post and Zwaans (1986), there will be clearance changes between the meshing pair and the single-column meshing pair will have a larger clearance change because of the larger unbalanced torques. From the researches of Jisheng et al. (1997), Huang et al. (2007), Kim et al. (2014) and Larsson (2009), if the change reaches the critical point, the flow regime will change form hydrodynamic lubrication to boundary lubrication, which will cause a metal contact between the meshing pair and lead to the wear out of the tooth flank. That means the double-column have a better lubrication situation.

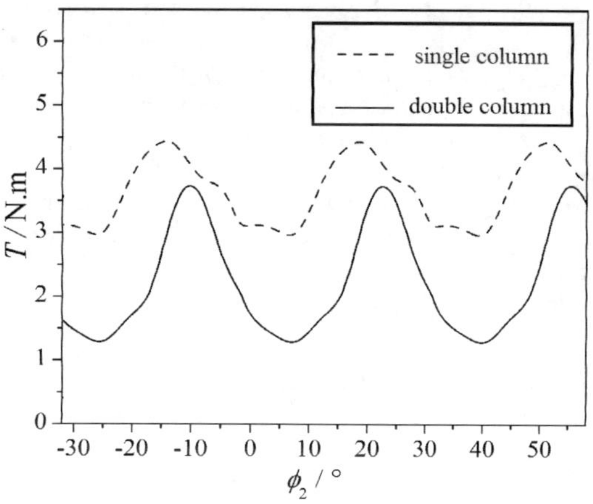

Figure 5: Total torques through the operation

4. CONCLUSIONS

In this paper, a simulation model is used to analyses the oil lubrication in two types of meshing pair. Real design parameters of these two types meshing pair for the same single screw compressor are given as calculation examples. Several conclusions can be made according the calculating results:

- The pressure increases to a peak value from the start of the convergent region to the meshing point.
- The total torques caused by the oil film forces are unbalanced, which would lead to a deflection of the tooth in order to regain balance. Furthering result in changes in the clearances between the meshing pair and may cause metal contact of the meshing pair reducing the operation life.
- The total torque generated by the double-column meshing pair profile is smaller than that of the single-column at any same rotation angle, which indicates the double-column meshing pair profile has a better hydrodynamic lubrication state and a longer operation life.
- The analysis can also be used to optimize the meshing pair profile design for single screw compressors.

REFERENCES

Zimmern, B., and Patel, G. C.(1972) "Design and Operating Characteristics of the Zimmern Single Screw Compressor,"(96–99). Compressor Technology Conference, Purdue

Zimmern B.(1976). Rotary inter-engaging worm and worm wheel with specific tooth shape. US Patent No. 3932077.

Jin GXTY(1986). A study of the single screw compressor profile(312–20). In: Proc. Purdue Compressor Technology Conference.

Boblitt, W.W.(1987). Method for Cutting Complex Tooth Profiles in a Cylindrical, Single-screw Gate-rotor. US Patent No. 4710076.

Jensen, D.(2000). Method for Manufacturing Fluid Compression/Compressor Rotor. US Patent No. 6122824.

Weifeng Wu, Quanke Feng, Column envelope meshing pair and its design method for single screw compressors. J Zhejiang Univ Sci A 2009 10(1):31-36.

Feng Q. K., Guo B., Zhao C., Xu J., Li Y.J., Cheng L., and Shu P.C.,2005, "A Design Method of Meshing Pair Profile of Single Screw Compressors Enveloped by Multi Straight Line," Compressor Technology, 191(3)pp.1–6(in Chinese).

Weifeng Wu, Quanke Feng, 2009, A Multicolumn Envelope Meshing Pair for Single Screw Compressors. ASME J. Mech. Jul., 131, p. 074505.

Cameron A. Basic lubrication theory. Brisbane, Queensland, Australia: Jacaranda-Wiley Ltd: Jacarand Press, 1983, p.17.

Post W and Zwaans M. (1986). Computer simulation of the hydrodynamic lubrication in a single screw compressor(334–348). In: Proceedings of Purdue compressor technology conference, West Lafayette, IN, USA.

Jisheng E and Gawne DT. Influence of lubrication regime on the sliding wear behaviour of an alloy steel. Wear 1997; 211: 1–8.

Huang P, Meng YG and Xu H. Tribology course. Beijing: China Higher Education Press, 2007, p.359.

Kim D-W and Kim K-W. Effects of sliding velocity and ambient temperature on the friction and wear of a boundary-lubricated, multi-layered DLC coating. Wear 2014; 315: 95–102.

Larsson R. Modelling the effect of surface roughness on lubrication in all regimes. Tribol Int 2009; 42: 512–516.

Huang R, Li T, et al. An optimization of the star-wheel profile in a single screw compressor [J]. Proceedings of the Institution of Mechanical Engineers, Part A: Journal of Power and Energy, 2015, 229(2): 139-150.

Lauder BE and Leschziner M. Flow in finite-width, thrust bearings including inertial effects 1-Laminar Flow. ASME J Lubric Technol 1978; 100: 330–338.

Linear Compressor Suction Valve Optimization

Rinaldo PUFF[1*], Dietmar E. B. LILIE[2], Marcelo KNIES[3], Marcos G. D. de BORTOLI[4]

Embraco – R&D
Joinville, Santa Catarina, Brazil

[1]++55 47 3441 2044, rinaldo.puff@embraco.com
[2]++55 47 3441 2348, dietmar.lilie@embraco.com
[3]++55 47 3441 2987, marcelo.knies@embraco.com
[4]++55 47 3441 2783, marcos.g.bortoli@embraco.com

* Corresponding Author

ABSTRACT

The Design approach of hermetic compressors for household refrigeration usually follows a sequence that considers an initial concept, normally based on existing families (goal here is incremental optimization or design itself). Those designs are then upgraded with new features aiming improvements in efficiency, sound quality or even cost reduction. Compressors usually have a cylinder head that fits discharge and suction valve. In a new and compact design for linear compressor, the suction valve is placed on the top of the piston, requiring new shapes in order to make it possible to comply with gas flow, fatigue and dynamic requirements. This article aims to present the use of an optimization procedure to develop the suction valve design for this type of compressor. The results obtained in terms of performance and reliability with the optimization process and final results in compressor are presented.

1. INTRODUCTION

In recent years, industry has been challenged to develop more efficient machines and equipments, in order to reduce consumption and save energy. This demand has been promoted by governments and other institutes with the final objective of reducing investments in energy generation and protect the planet. Governments have been placing restrictions to the commercialization of household electric machines with lower energetic efficiency, compelling the industries in this sector to invest more and more efforts to take care of these new requirements. In the household refrigeration industry for example, the efficiency targets for the next years are very compelling. Another positive effect of this increase in equipment's efficiency trend, is that it represents money saving for the end user, once the energy cost is becoming higher year over year.

Another significant factor in the design of machines and equipments is related to the size. Final user is increasingly concerned with the space optimization, and harmony with the environment in which the equipments are fit. The same trend can be observed in the household refrigeration industry, and particularly in the hermetic compressor industry. New technologies are constantly being developed with higher efficiency, lower size and weight, and that can offer more options for the installation in the systems.

As an attempt to comply with this rising demand, a technology of reciprocating compressor driven by a linear motor was developed. This type of compressor has its stroke controlled by an electronic central board called inverter, once it does not have a fixed stroke. This compressor is presented in Figure 1, which illustrates the machine in a cut view, as well as its main components.

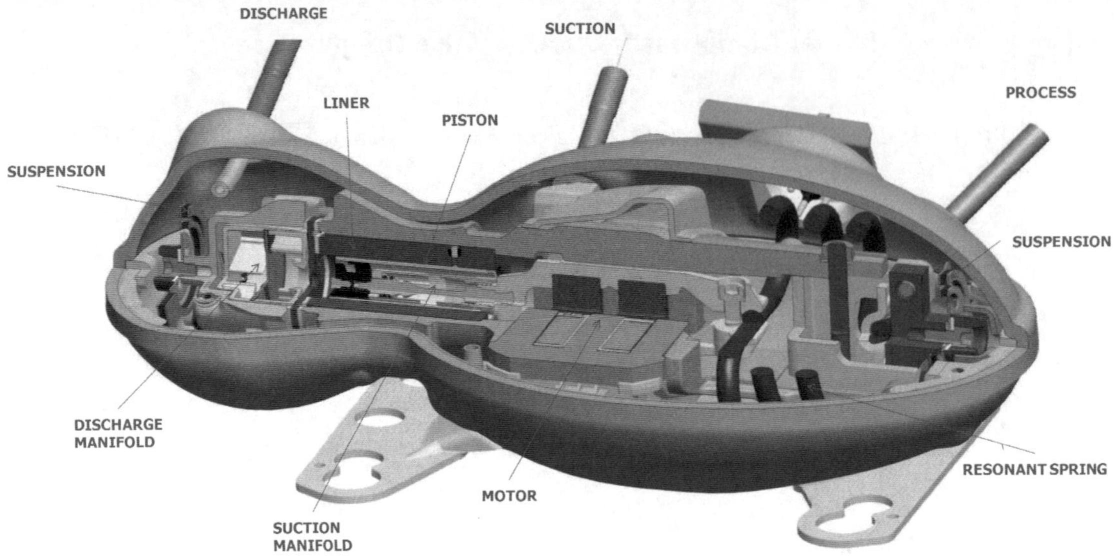

Figure 1: Linear compressor technology.

This technology also presents other innovative characteristics, which are listed below:

- A low height, that allows internal refrigerator space saving;
- It does not require lubricating oil, that allows a unique and total friendly characteristic related to the environment, as well as flexibility for installation;
- A larger capacity range, allowing lower temperature variation, and better food preservation;
- A better sound quality due to lower vibration and lower sound variation in-between starting, running and stopping,

The development of the new technology requires the use of new knowledge and methods, as well as new tests and approval criteria. One of the fundamental systems in this kind of compressor is the suction manifold. In this technology, this process is done through the piston, and the suction valve is placed at its top, as illustrated in Figure 2. This article aims to present the methodology used to develop the valve geometry, considering efficiency and reliability criteria necessary for this type of component. Final efficiency results are also presented in comparison with another valve geometry evaluated.

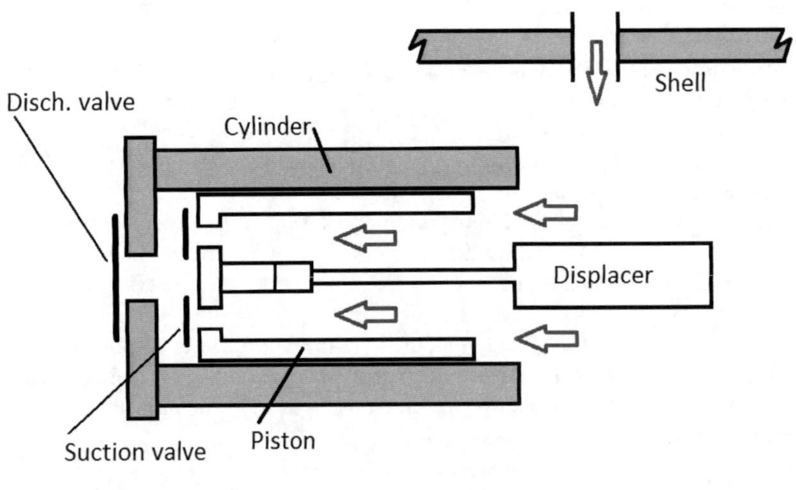

Figure 2: Schematic view of the linear compressor suction manifold

2. SUCTION VALVE OPTIMIZATION

According to Bortoli and Puff (1998), the use of design optimization together with FEM – Finite Element Method – is a powerful tool to improve the compressor components' design. It is not a magic box and depends a lot on the problem formulation quality. The optimum is just a function of the design variables, and the problem must be as simple as possible to reduce computer run time.

The main objective of this article is to perform the suction valve optimization, considering that it is assembled on the piston top. The main design variables were considered, although defining only one reduced thickness. The optimization software ModeFrontier was used together with the FEM software ANSYS to perform the optimization. The design variables are described in Table 1, and illustrated in Figure 3.

Table 1: Design Variables

Variable	Description
NP	Number of valve legs
E1	Width at the first leg region
E2	Width in the middle of the leg
E3	Width at the final portion of the leg
RR1	Fillet radius at the beginning of the leg
RR2	Fillet radius at the end of the leg

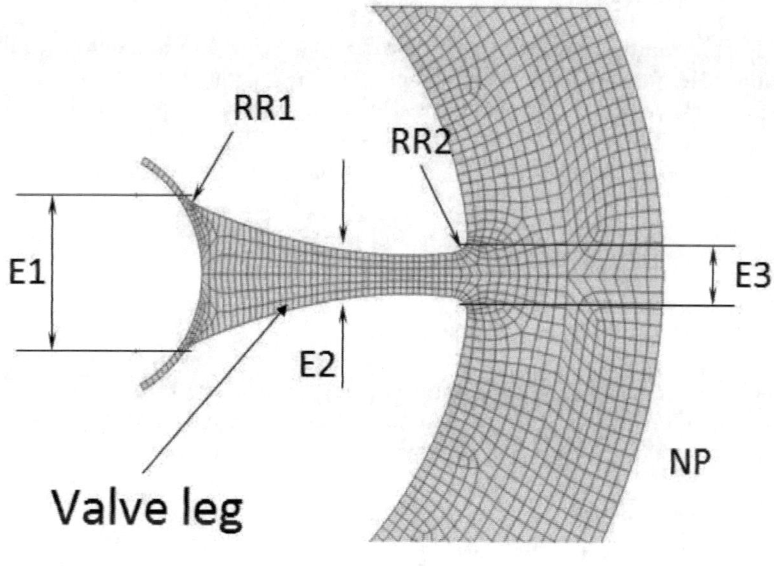

Figure 3: Illustration of the design variables

Once the design variables were defined, the optimization strategy was assembled, aiming to minimize the objective function, defined as the ratio between the maximum stress S_{max} and the valve stiffness K, as defined in Equation (1).

$$F_{obj} = \frac{S_{\max}}{K} \tag{1}$$

An APDL (2) – Ansys Parametric Design Language – was developed in order to have an automatic way to generate the designs according to the design variables. This application takes the design variables, builds the models and runs the FEM analysis. At the end, response variables are extracted. Figure 4 illustrates the optimization scheme assembled to perform the process.

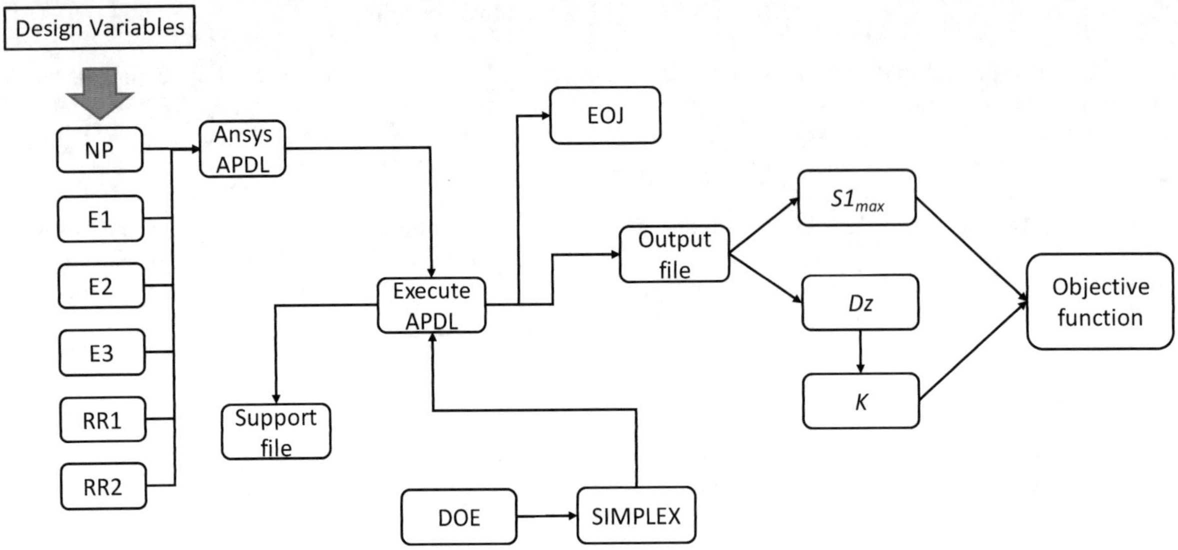

Figure 4: Optimization Scheme used in this analysis

3. STRESS AND FATIGUE ANALYSIS

After performing the optimization steps, one of the verifications needed to initiate the component certification is related to the reliability. Hermetic compressors for household refrigeration need by nature components that can support dynamic loads at the level of giga-cycles, as described by Berger and Kaiser (3) for mechanical springs. This is because compressor valves work in the VHCF – Very High Cycle Fatigue regime. To accomplish this requirement, a criterions evaluation must be performed, considering three king of loads to which valves are commonly submitted.

- Bending over the suction port due to the gas pressure;
- Bending due to valve opening;
- Impact velocity of the valve against the valve seat.

To evaluate bending stress due to valve opening at each cycle, a pressure is applied on the valve surface corresponding to the main pressure difference. Figure 5a represents the annular suction port on the piston top, and Figure 5b a typical stress result due to the valve opening.

The evaluation of the stresses resulting from the compression pressure over the valve orifice is done using FEM as well. The pressure load corresponding to the maximum compression pressure is applied on the annular area of the orifice. A typical result of this kind of load is presented in Figure 5c.

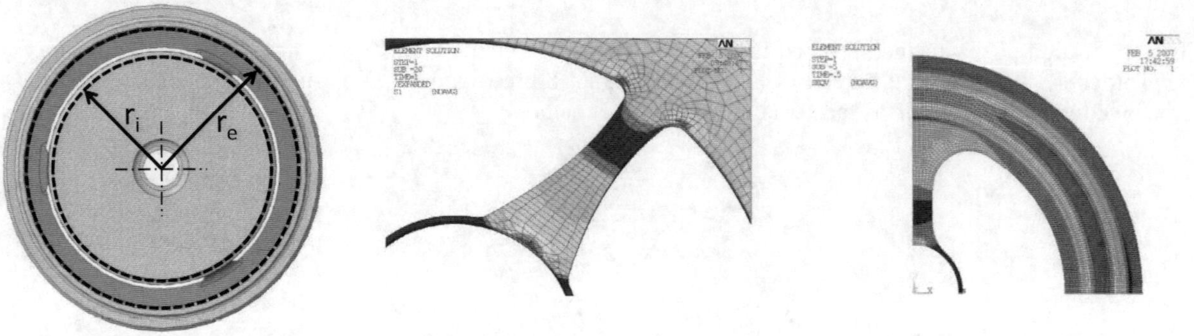

a) Piston top layout   b) Bending stress distribution   c) Orifice stress distribution

Figure 5: Models for stress evaluation

It is very important to emphasize that for both cases the load is cyclic and dynamic. Considering this, adequate fatigue criteria must be used to evaluate the probability of failure due to the real loads and real life.

The third type of load that is experienced by the valve is related to the collision velocity against the valve seat in each cycle. The determination of this velocity is extremely dependent on the application condition, and was not evaluated in this work using FEM. In substitution, an experimental approach was used by constructing a compressor with a glass cylinder and get images of the piston in movement. The valve movement was then evaluated with relation to the piston movement, using the treatment of the images. Figure 6 illustrates the experiment performed, which was only possible because the compressor does not use oil.

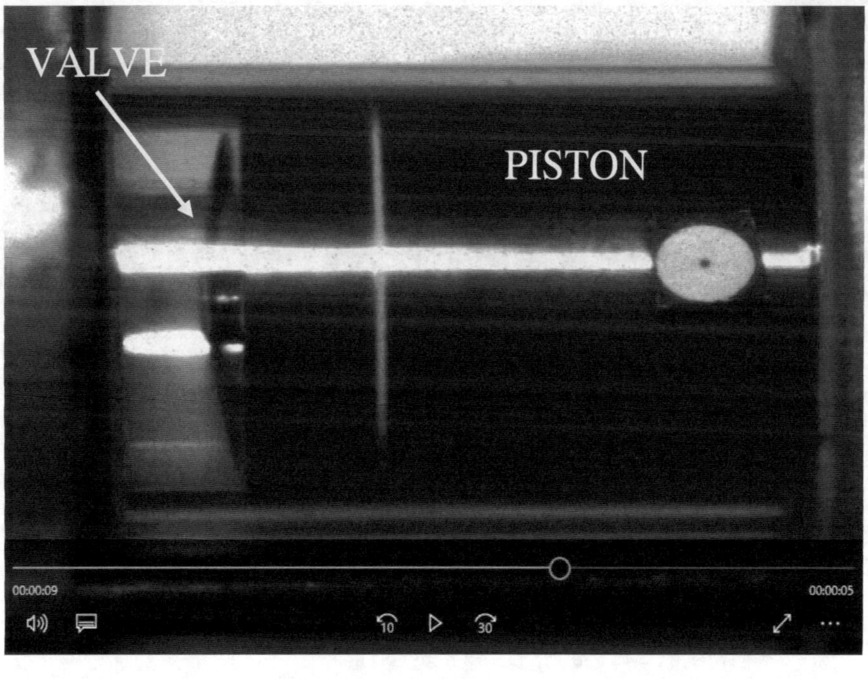

Figure 6: Illustration of the experimental approach used to evaluate the valve movement.

After all fatigue evaluations were performed, some compressors were assembled using different valve models, and based on them the performance was evaluated.

4. RESULTS

The application of this method resulted in the evaluation of different geometries for the suction valve. Two of the geometries evaluated are illustrated in Figure 7 and Figure 8 presents the optimization process result showing the objective function decrement as function of the use of the methodology.

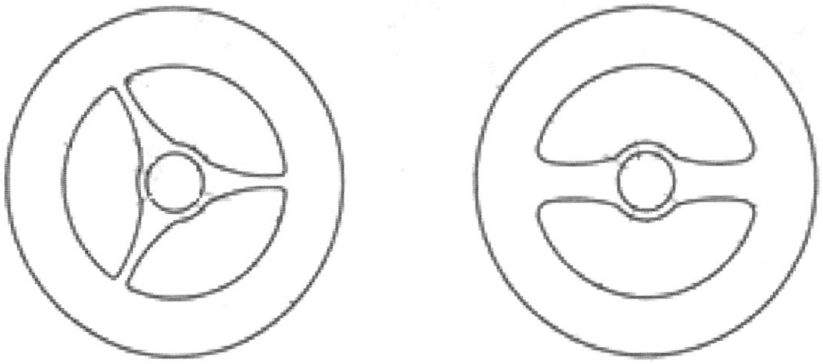

Figure 7: Two valve models obtained from the method usage.

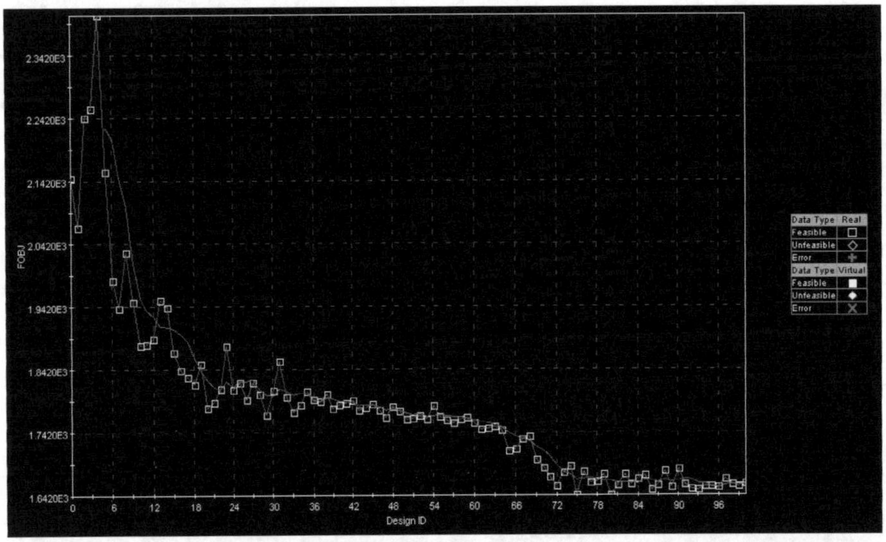

Figure 8: Path of the optimization process, showing the objective function minimization.

After performing the optimization process, a stress analysis was performed as per described in the previous section. The stress analysis was done considering two confidence levels according to Shigley and Mischke (4). These two levels consider the statistical confidence levels used as criteria for the valve orifice definition.

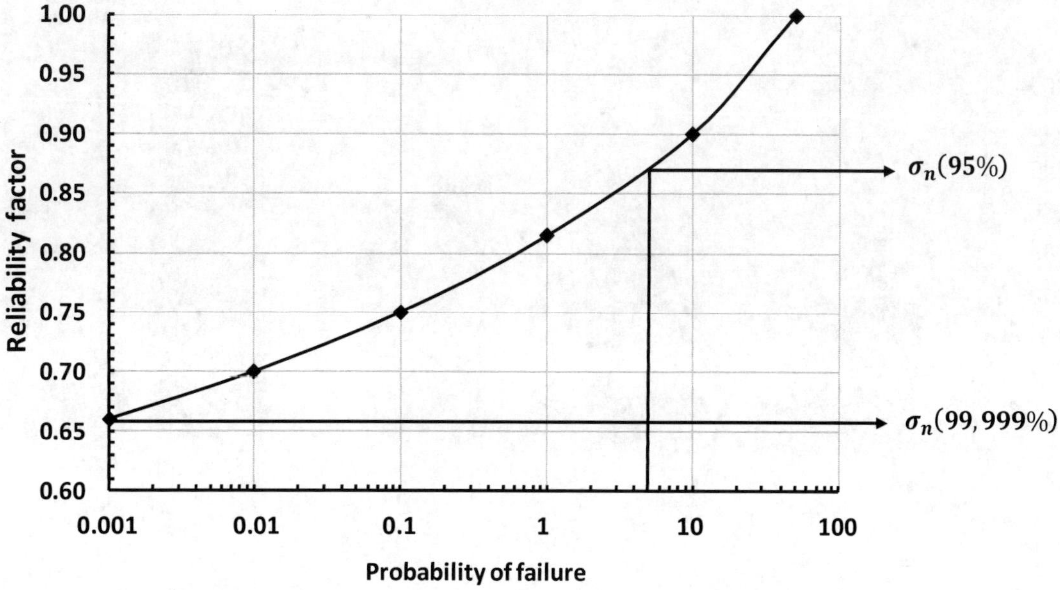

Figure 9: Reliability factor and confidence levels considered for the fatigue evaluation.

Figure 10 shows the results obtained for the valve movement.

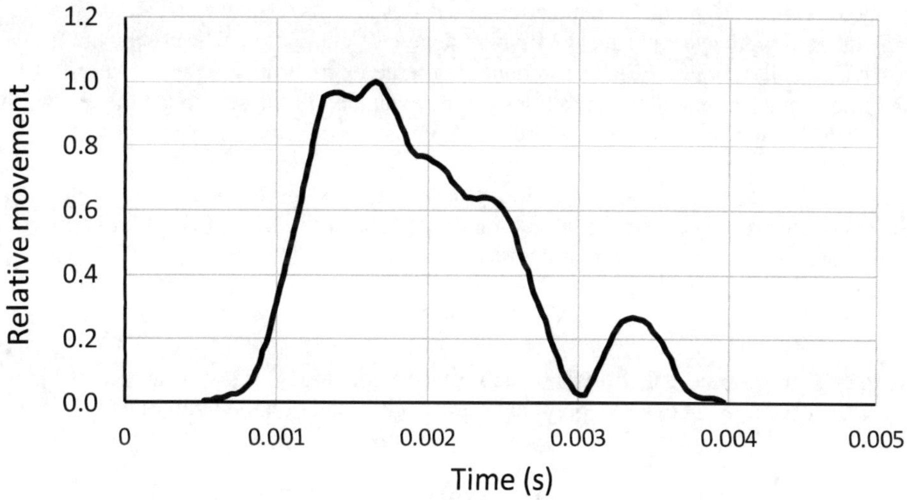

Figure 10: Suction valve movement measured using optical technique.

To close the analysis, resultant geometries were prototyped, some less optimized and other with the best theoretical responses. The prototypes were assembled in compressors, and those were tested in performance. The graphic of Figure 11 shows the comparison of efficiency (COP – Coefficient of Performance) obtained with the geometries tested. In this graphic one can observe the COP result for the options prototypes, taking as reference the one with lower performance. The result for the best one is 5.3% better than the worst one. In terms of compressor performance, it is a huge increase.

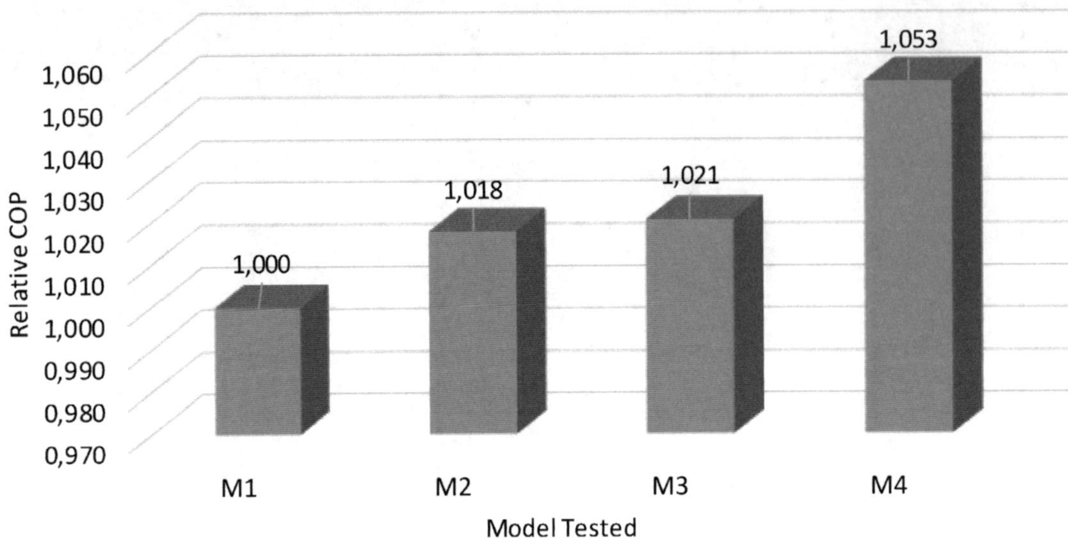

Figure 11: Relative COP measured for four models tested in the optimization process

5. CONCLUSION

The results obtained at the end of the optimization process show its efficacy. It is certainly a tool of major importance during the development of products with high degree of innovation. The combination of an optimization model with other numerical tools like FEM can reduce in a large degree the computational effort because through the optimization logic a larger number of possibilities can be evaluated in the search for the global optimum, within the boundaries stablished. The final result for COP, about 5.3% better than the reference.

For sure, an important check after the optimization process must be performed, taking into account reliability criteria, not considered during the objective function minimization. In this case, FEM tool is used together with an innovative experimental approach, on evaluating the valve movement using captured images. Combined, the methodologies were used to evaluate fatigue stresses and valve impact velocity, characteristics that are fundamental for the approval of the chosen model.

To finalize, it is important to observe that this new compressor technology, without the use of lubricating oil, adds new possibilities to the development of new approval criteria, and evaluation methodologies.

REFERENCES

Bortoli, M. G. D. de, Puff, R., 1998, Compressor Components Optimization With the Use of Finite Element Method, 14th *International Compressor Engineering Conference at Purdue*, Proceedings of 1998 International Compressor Engineering Conference at Purdue.

ANSYS, Release 14.0 – © SAS IP, Inc., 2011.

Berger, C., Kaiser, B. (2006). Results of very high cycle fatigue tests on helical compression springs. *International Journal of Fatigue,* 28, 1658–1663.

Shigley, J.E., Mischke, C.R. (1989), *Mechanical Engineering Design*, 5. ed., New York, McGraw-Hill.

Flap-X: Development of a new compressor reed valve material

Alexander LOF[1], Chris MILLWARD[1], Azhar NAWAZ[1,2]*

[1,2] voestalpine Precision Strip AB, Research & Development,
Munkfors, Sweden
[1](Phone: +46 732750268, E-mail: alexander.lof@voestalpine.com)
[1](Phone: +46 732717642, E-mail: chris.millward@voestalpine.com)
[1,2](Phone: +46 734618240, E-mail: azhar.nawaz@voestalpine.com)

* Corresponding Author

ABSTRACT

Flap-X is a new reed valve steel developed for compressor applications. The development work was undertaken after consistent feedback from customers requiring higher fatigue strength and improved damping properties. The renewed interest in improved valve strip is a direct result of the industry's latest trends towards energy efficient compressors. Different steel grades were investigated during process development and Flap-X material, which is an optimized version of the well-known compressor steel UHB SS716, showed the most promising results in terms of production feasibility and desired enhancements in alloy design, metallurgy and mechanical properties. Finished valve strip of each candidate material was used to quantify microstructure, tensile strength and fatigue life. Fracture investigations on the fatigue test samples were undertaken to detect any contributing non-metallic inclusions. Vibration damping was measured at temperatures and frequencies corresponding to operating conditions of reed valves in a compressor. Impact fatigue strength was measured in a recently developed test rig. Both fatigue life and damping properties are improved at increased tensile strength in all of the investigations.

1. INTRODUCTION

A majority of compressor reed or flapper valves for various refrigeration applications are made of UHB 20C (20C) or UHB SS716 (716) steel grades. Reed valves are also produced using other grades; however, most of the production volumes delivered by voestalpine Precision Strip AB is of these two steel grades. 20C is a high carbon steel similar to AISI 1095 whereas 716 is a modified version of AISI 420, which a martensitic stainless steel (Johansson and Nordberg, 1984)

The renewed interest in improved valve strip is a direct result of the industry's latest trends towards energy efficient compressors. This is reflected in recent market and our direct customer's feedback. Compressor OEMs are also required to develop better solutions as the result of National and international legislation for reducing energy consumption. For this reason, the manufacturers of refrigerators and air conditions are prioritizing the sourcing of high strength materials from multiple suppliers to continue to be competitive whilst meeting the designer's expectations. The designers of the compressors want a flapper valve strip with improved impact and bending fatigue properties in addition to material's noise reduction at operating frequencies. Flap-X is a new flapper valve material that fulfils designer's requirements of higher fatigue and damping properties. The material performance has also been verified in various compressor OEMs own tests.

2. FIRST PAHSE

The main focus of the work was to enhance the functional performance of the stainless valve alloy. Table 1 lists the type of materials investigated in the first phase of the work. A minimum of a pilot heat (approx. 60 tons) of each version was produced. In certain instances multiple heats of one specific version were produced to quantify the reproducibility of the process route. Standard 716 chemical compositions were modified by micro-alloying with certain alloying elements for improved hardening response in 716-A version. Secondary metallurgy treatment was fine tuned to change the chemistry of non-metallic inclusions from hard-brittle Al_2O_3 to soft-deformable SiO_2 in 716-B version. 716-C was ESR processed (electro slag re-melting) after ingot casting to achieve a homogenous microstructure with reduced non-metallic inclusions and segregation. Hardening trials were undertaken in production. Finished valve strip samples were used for all of the fatigue tests.

Table 1: List of investigated materials

Material	Process-Route
716-0	Standard
716-A	Modified chemistry
716-B	Modified Inclusions
716-C	Modified-ESR
716-D	Modified A+B

Lab hardening trials were undertaken for producing samples for the damping tests as per a standard (ASTM-E756), which required thick samples to guarantee the reproducibility of the thermal loss factor measurements. Tensile and bending fatigue tests3 for each specific material and tensile strength were aimed to be reproducible with ± 50 MPa tolerance to be able to spot the heat to heat variations.

Steel's functional properties for a specific application are dependent on large number of processing factors. Flapper valve steel in its finished form is essentially advanced high strength steel (AHSS), i-e: multiphase microstructure containing martensite, retained austenite and retained secondary carbides with tensile strength and elongation in the range of >1800 MPa & < 10 % respectively. Multiple strengthening mechanisms are used to achieve this range of strength and ductility resulting in optimal fatigue properties. Figure 1 in the following schematic illustrates the basic steps in valve strip processing and shows the level of complexity that is involved in achieving the desired set of flapper valve functional properties after strip production.

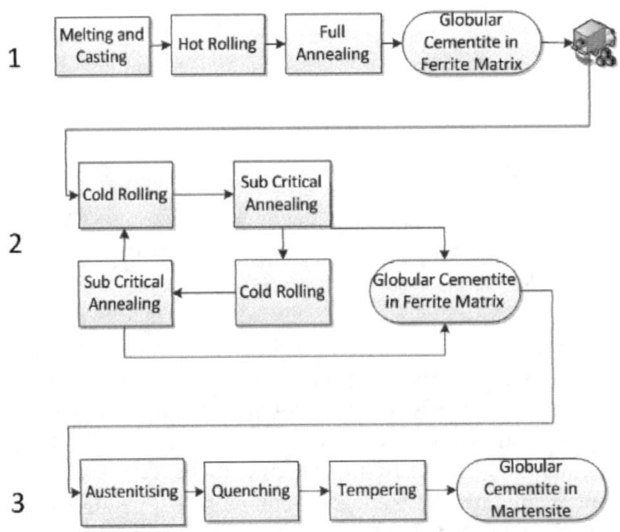

Figure 1: Basic metallurgical steps and desired microstructure during valve strip processing

In order to improve the functional properties of the new valve strip, Flap-X, different approaches were tried.

- Micro-alloying
- Modifying the chemistry of non-metallic inclusions
- Using advanced melting routes after secondary metallurgy processing
- Optimal hardening of the strip

The main consideration in trying different approaches to developing the new alloy was not to negatively affect the basic stainless flapper valve strip attributes. These attributes are described in the following list.

- Corrosion resistance
- Hardening response of each batch for volume production
- Precision rolling capability
- Flatness after hardening
- A fine surface finish

Following metallurgical parameters were investigated after each production trial of a specific thickness of the valve strip.

1. Matrix Microstructure
2. Secondary Carbides
3. Retained Austenite
4. Hardness and Tensile Strength
5. Flatness
6. Residual stresses

A martensitic matrix with secondary carbides and retained austenite was the typical microstructure in all of the trials as shown in the following figure 2.

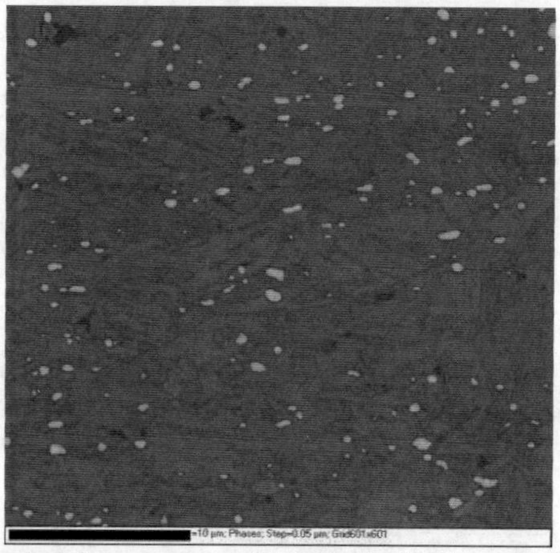

Figure 2: FEG-SEM EBSD image showing a martensitic matrix (red), secondary carbides (yellow) and retained austenite (blue).

Theoretically, after the final martempering or so called hardening treatment, the microstructure should be fully martensitic, however, due to the non-ideal cooling; it is always the case that some of austenite is not transformed into martensite. In flapper valve applications this retained austenite in the microstructure is beneficial for improving impact fatigue strength. The effort in the production trials has been to understand how to vary and control the retained austenite. As part of this work, strip with a range of tensile strength or hardness levels was produced. The retained austenite was measured in all of the strip samples by using X-Ray diffraction technique (XRD). A ratio, R, in the following equation 1 is defined to express the relationship of microstructure, hardening temperature and the resulting mechanical properties. It's important to note that actual martempering of the valve strip carried out in a continuous hardening line is far more complex process with many more variables than just austenitizing temperatures. Hardening speed, controlled cooling in the quenching zones (employing highly specialized oil bath), tempering temperatures and the furnace's online flatness monitoring constitute a few more of these variables. Furthermore, different sets of parameters are used for hardening different thicknesses in this continuous process.

$$R = UTS / T \quad (1)$$

Where
UTS is ultimate tensile strength
T is austenitizing temperature of the production furnace (°C)

Generally, retained austenite increased with an increase of hardening temperature and tensile strength as shown in the figure 3. The absolute value of retained austenite in a "fully martensitic" microstructure is difficult to establish due the low levels present and the resulting influence on measurement uncertainty (George F. Vander Voort, 2009). The average standard deviation of all the retained austenite measurements using XRD was ±2. To cross check the XRD measurements, the volume fraction of retained austenite was also validated by using FEG-SEM EBSD image analysis studies (Research & Development, 2013). Retained austenite in 716-0 material with a 1810MPa tensile strength was consistently lower than the higher tensile optimized versions. Furthermore, a lab simulation of the actual hardening process was undertaken using dilatometry (Research & Development, 2015) to verify the high tensile and retained austenite observation using the 716-A material. Here again, the samples from the study were measured for retained austenite using the SEM EBSD volume fraction method.

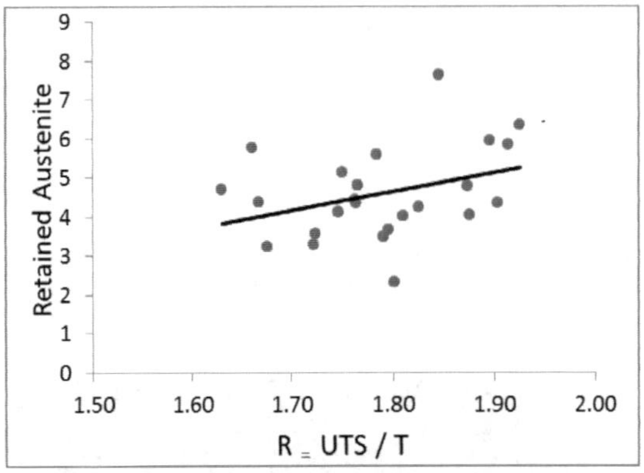

Figure 3: The effect of retained austenite on tensile strength

Figures 4 & 5, show a comparison of microstructure from two simulations where the retained austenite was varied from 6 % to 14% respectively. The increment of retained austenite in the martensitic matrix in this case was the function of higher hardening temperatures.

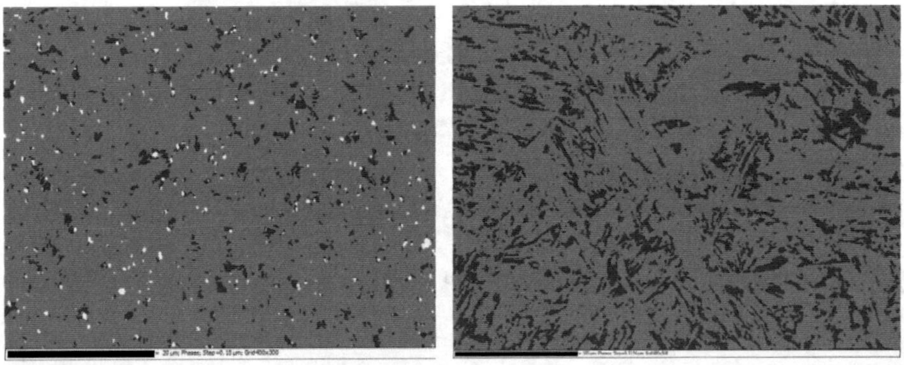

Figure 4 & 5: FEG-SEM EBSD image showing a martensitic matrix (red), secondary carbides (yellow) and retained austenite (blue).

A lab simulation of the optimized material also showed that increasing retained austenite in the matrix is more or less unaffected by the medium and lower hardening speeds, as shown in the figure 6. The maximum retained austenite in the lab and production trials differs due to the different maximum temperatures employed in lab simulation and production trials.

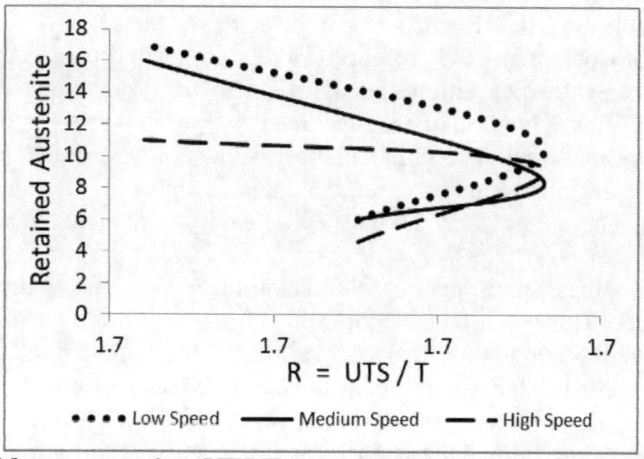

Figure 6: The effect of furnace speed on UTS/T ratio at different tempering and lead bath temperatures.

Secondary or retained carbides after martempering can be shown to be directly related the hardening response and mechanical properties of martensitic stainless steels for valves. It is of critical importance to control the density and the size of carbides in various processing stages after ingot casting. The density or the percentage area fraction, measured using LOM image analysis, was inversely proportional to the tensile strength for all materials after hardening. A minimum level of secondary carbides is aimed in the final microstructure for achieving a combination of higher tensile & yield strength in addition to increased retained austenite.

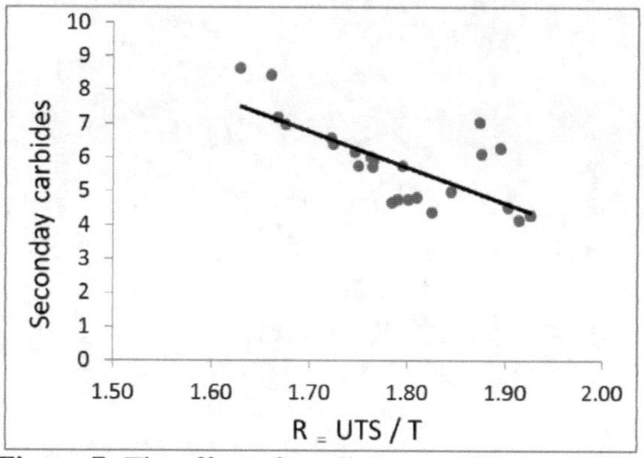

Figure 7: The effect of tensile on secondary carbides

The flatness of the valve strip after hardening is negatively influenced by an increase the tensile strength, mainly due to the higher thermal shock effect as shown in the following figure 8. The un-flatness of flapper valve is extremely undesirable as it may lead to refrigerant leakages in the end application. In theory (Pirouznia, 2014), the main factor influencing the flatness during martempering is quenching stage where it's critical to control and reduce uneven temperature gradients that lead to strip distortions. For the new material, Flap-X, both cold rolling and martempering were carefully optimized to produce strip within internal and customer specifications.

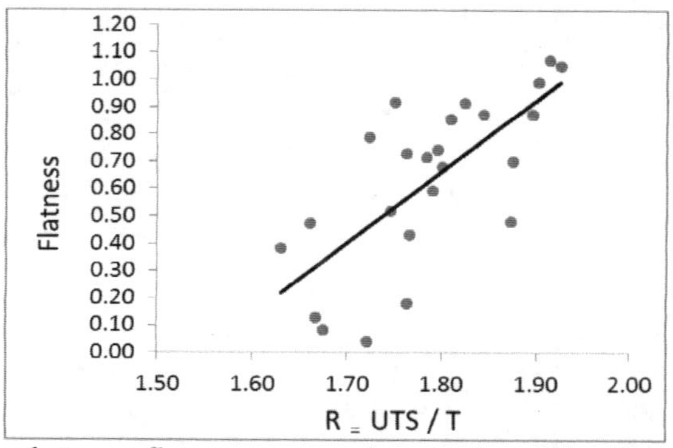

Figure 8: The correlation between flatness measured across the rolling direction and tensile strength

The fatigue strength of a steel alloy is influenced by a great many factors. Alloy type, microstructure, tensile strength, non-metallic inclusion content and the presence of other un-desired phases in the microstructure are a few material related factors that can influence fatigue strength. These factors are generally influenced positively by improvements in process and quality control of the steelmaking. For stainless flapper valve steels, the main the factors are steel cleanness, tensile strength, optimal microstructure, damping and compressive residual stresses. Three different types of fatigue regimes, namely: axial, bending and impact fatigue influence the material's failure. These different fatigue mechanisms are associated with specific modes of applied stresses and frequencies.

For the internal product development investigations, the quantification of fatigue life in thin martensitic strip is a detailed and painstaking process (Research & Development, 2014). There are so many factors that influence the results. The sample preparation, test conditions, analysis method and the sample thickness are as critical as material3 metallurgy and mechanical properties. Tensile or axial fatigue tests are a good starting point in determining the fatigue life improvement. The critical volume under stress, for "hour-glass" shape sample in axial fatigue tests is much larger than impact and bending fatigue tests. Therefore the probability of a non-metallic inclusion associated fracture becomes higher in this test. That leads to a quicker and definitive quantification of pilot or prototype materials fatigue strength which is helpful in early phase of the fatigue related alloy development. On average, tensile fatigue life increased with tensile strength as shown in the following figure 9 for all of the tested versions.

Intensive fracture studies were carried out in order to correctly identify the causes for fracture initiation. These investigations showed that most of the fractures started from some type of inclusion. For example, it was found that the batch with low fatigue results (section 1 in the figure 9) had more than twice the inclusion size compared to the batch with higher fatigue results (section 4 in the figure 9).

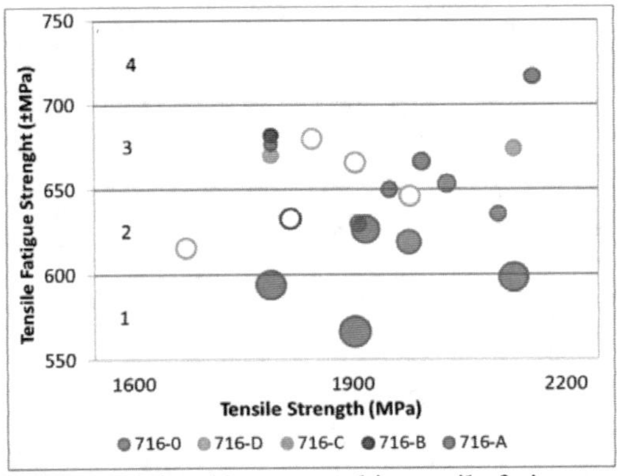

Figure 9: Tensile Vs Fatigue Limit for (each batch) measured in tensile fatigue tests at R- 0. The size of the bubble (Bubble diagram) shows the average inclusions size in the fracture samples according to Murakami √area. (Murakami, 2002)

3. SECOND PAHSE

The reverse bending fatigue tests were done using a Krouse type machine. It consists of a variable speed electric motor equipped with a variable pitch crank, connecting rod, specimen clamping device, motor controls, specimen break detector and a cycle counter (see figure 10)

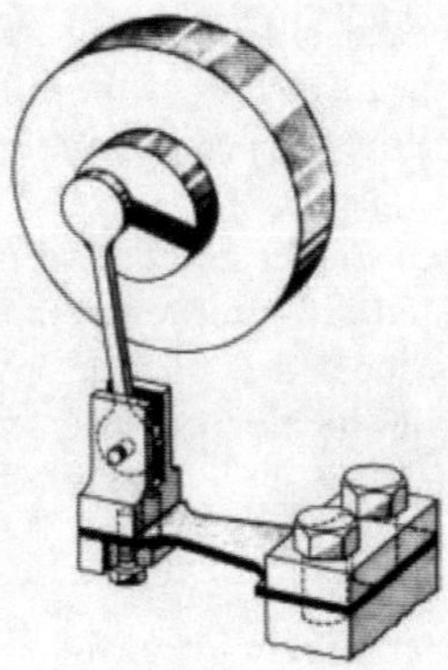

Figure 10: Reverse bending fatigue test setup (R=-1)

As the thickness of the tested samples were <1.5 mm, light weight nylon cranks were used to minimize any dynamic influence of a heavy connecting rod. Finished strip samples from each material were tested using a staircase approach. Evaluation of mean fatigue life was done according to the SS IS0-1207 2012(E) standard. The method attempts to identify the mean fatigue limit, upper and lower limit of the mean, at a prescribed confidence levels. The results are summarized in the following figure 11. Flap-X with the modified inclusions and chemistry showed the highest improvement compared to SS716 and the reference spring steel available in the market. Only the high alloy stainless grade AEB-L and S-Coat L+ tool steel had a higher mean fatigue than Flap-X.

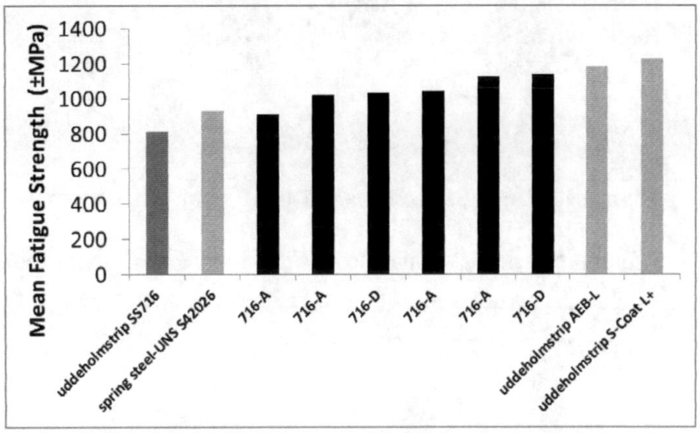

Figure 11: Mean reverse bending fatigue of different versions (heats) of Flap-X and the comparison with other materials.

It's also seen that bending fatigue increases with an increase of tensile strength as shown in the following figure 12.

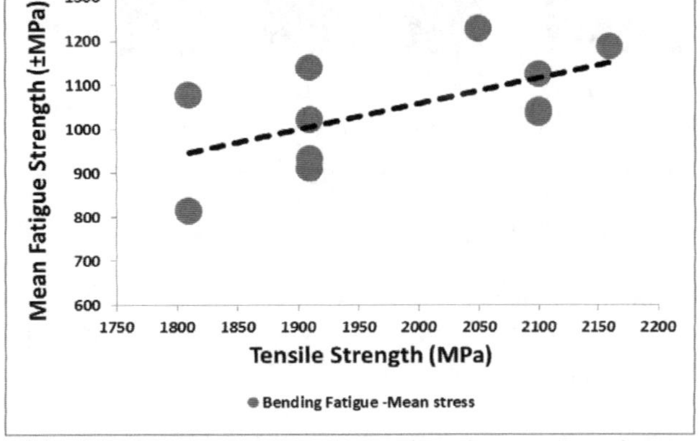

Figure 12: Mean Fatigue and Tensile Strength of flap-X and reference materials.

Material's thermal vibration damping and thermal stiffness properties are also a critical factor in choosing the optimal material for a specific application and in the design of the compressors. For compressor and reed designers, it's important to know the damping properties of the flapper valve material.

The standard ASTM E-756 test method was used to measure the vibration-damping properties of the hardened strip for the loss factor, η, and Young's modulus, E, at different temperatures. Each test strip of specific thickness was mounted in the cantilever beam test apparatus located inside an environmental chamber. It was then excited via random noise and the response was measured via the piezoelectric crystal. Frequency response was in the range of 70 Hz to 3000 Hz whereas the bending modes 2 to 6 were measured for each sample. Materials used for damping measurements were 716-A & 716-D.

Young's modulus, E, measured for Flap-X decreases with an increase of the strip temperature as shown in the following figure 13.

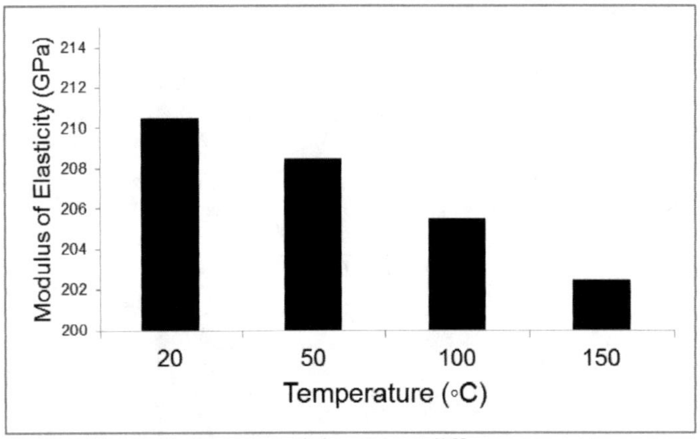

Figure 13: Young's Modulus (E) at different temperatures

The damping loss factor of different materials also increases with an increase of temperature shown in the following figures 14.

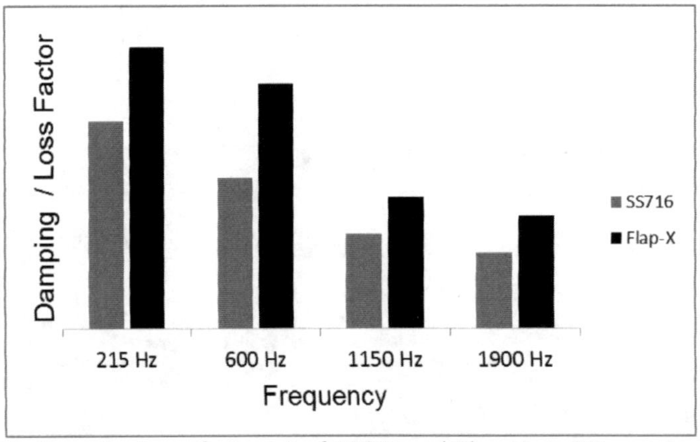

Figure 14: Damping Loss factor (η) of SS716 and Flap-X at different frequencies.

The thermal damping loss factor in the figure 15 is defined as the average loss factor (different temperatures) measured for each specific material at different frequencies. It was also compared with reference steel. For all the samples, average thermal damping value decreased with an increase of the applied frequency. Flap-X exhibited an increased thermal damping factor compared to the reference steel.

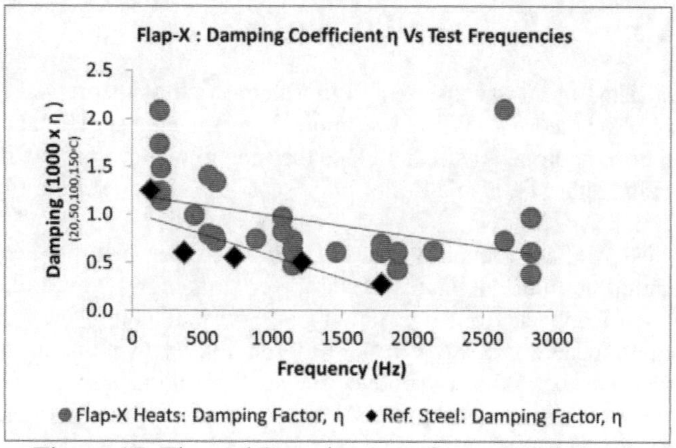

Figure 15: Thermal Damping Loss factor (η) of Flap-X

Impact fatigue tests were conducted in a newly developed test rig specifically designed for impact fatigue testing, in a similar manner to that outlined in experiments by previous work (Dusil & Johansson 1978). Reed movement was measured in terms of lift, impact velocity and force of each impact (Altunlu & Lazoglu 2012). The new rig also uses better control valves and regulators for precision control of the compressed air pulses in terms of duration, frequency and pressure. All this allows for the possibility of optimizing the impact testing for each material and reed design for repeatable tests and accurate comparisons.

Figure 16: Flap-X valve in new test rig.

The new material Flap-X exhibited a higher impact fatigue life compared to SS716 as shown in the following figure 17. This higher impact fatigue life performance is primarily associated with optimized chemical composition of the alloy and the non-metallic inclusion content in addition to higher tensile strength. Compressive residual stresses are also an important factor affecting the impact fatigue strength of flapper valve materials.

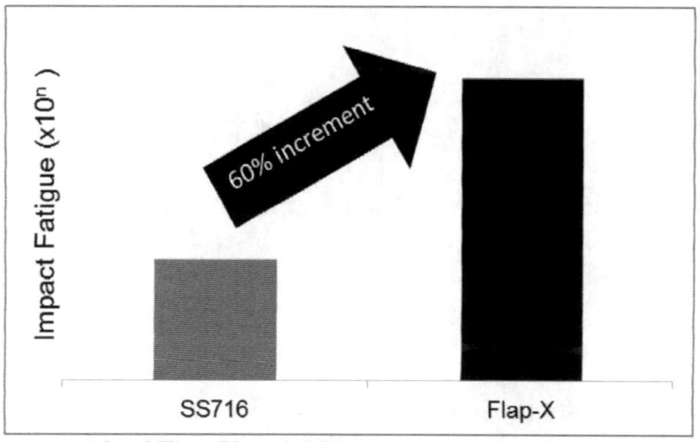

Figure 17: Impact fatigue strength of Flap-X and SS716 measured on finished valves of 0.203 mm thickness

4. DISCUSSION

The Flap-X development consisted of an exhaustive and thorough cycle of theoretical material design followed by full scale practical validation of the theories. While this approach is time consuming and expensive, in some ways it can be considered to be a superior approach that can reduce the time to market for new material. The reason for this is that the use of full scale trials, rather than small pilot plant scale trials, reduced the final industrialization stage of the development.

In the case of Flap-X, the design of the 'functional specification' for the new product was defined by talking to compressor manufacturers about the future demands of the material that they would like to be used for their future compressor designs. The main feedback from these compressor manufacturers was that fatigue life was the key improvement area. This requirement was related to both the possibility of reducing the probability of premature valve failure but also to give them the option to produce valves from thinner strip. Thinner reed valves are more energy efficient and also quieter. These are both properties of compressors that are both desirable and are required from a legislative perspective.

Another requirement was an improvement in the damping capacity of the material. This is also related to the suppression of noise (as well as being beneficial from a fatigue life perspective). These functional requirements were set against the conflicting desire to minimize the fundamental level of changes from the established and trusted materials they were currently using. In addition, there was a requirement to minimize material changes that might negatively influence the possibility to produce it in precision strip form.

The final constraint on the degree of changes that could be applied relates to the requirement to minimize changes that would have a detrimental effect on the reed valve manufacturer's process i.e. the final material should still be manufactured using the same basic tooling and tumbling process.

Against this backdrop, the development of Flap-X has managed to achieve significant improvements in the desired functional behavior of a material for reed valves without compromising either the processing constraints of the material during cold rolling to precision strip or during the production of the reed valves. This material should allow compressor designers and manufacturers to achieve their goals for more energy efficient and quieter compressors. This positive outcome should complement other approaches to achieving the recently agreed significant environmental improvements.

5. CONCLUSION

The fatigue life of reed valves is influenced by a number of metallurgical factors during strip processing and it is essential to have a good understanding and process control of the strip production. The new alloy, Flap-X, has an optimal chemical composition, microstructure and mechanical properties, making it significantly superior to SS716 for compressor application. The improved performance of Flap-X has been verified in both internal and external R&D tests, and also in compressor manufacturer's tests. Flap-X is being recommended by compressor manufacturers for demanding compressors reeds with long lifecycle requirements and low noise requirements.

ACKNOWLEDGEMENT

This paper is published by the permission of voestalpine Precision Strip AB, Munkfors, Sweden. The support of Stellan Eriksson in this work is greatly appreciated.

REFERENCESS

1. B. Johansson, B. and Nordberg, H. (1984) Properties of high strength steels, *International Compressor Engineering at Purdue, Paper 474.*
2. ASTM- E756, Standard Test Method for Measuring Vibration-Damping Properties of Materials (2010), http://www.astm.org/cgi-bin/resolver.cgi?E756
3. Fatigue Tests, Internal Reports (2014), Internal Reports, Research & Development, *voestalpine Precision Strip AB*
4. George F. Vander Voort (2009), Retained Austenite and Martensite, *Industrial Heating Magazine September 2009*
5. SEM-EBSD Investigations (2013), Internal Reports, Research & Development, *voestalpine Precision Strip AB*

6. SEM-EBSD Investigations (2015), Internal Reports, Research & Development, *voestalpine Precision Strip AB*
7. Pirouznia, P. and Jänsson P (2015) A Mathematical Model of Martempering of thin Martensitic Stainless Steel. *The 6th International Congress on the Science and Technology of Steelmaking (ICS2015).*
8. Y. Murakami, Y. Effects of small defects and non-metallic inclusions (2002), *Metal Fatigue, 1st Edition.*
9. Dusil, R. and Johansson, B. (1978). Material Aspects of Impact Fatigue of Valve Steels. *International Compressor Engineering Conference at Purdue Paper 254.*
10. A. C. Altunlu, I. Lazoglu ; 2012

An Investigation Into The Dynamics Of Self-Acting Compressor Valves

Carsten MÖHL[1]*, Christiane THOMAS[1], Ullrich HESSE[1]

[1]Institute of Power Engineering
Bitzer Chair of Refrigeration, Cryogenics and Compressor Technology
Technische Universität Dresden,
Dresden, Germany

* Corresponding Author
carsten.moehl@tu-dresden.de

ABSTRACT

In principle, a reciprocating compressor stage consists of a cylinder, the working volume which varies periodically due to a piston moving inside a liner. Latter is sealed by two sets of valves. On one hand the suction valve for admitting the gas to be compressed and on the other the discharge valve for allowing the high pressure gas to be delivered to the downstream process. The typical type of valves used within compressors for refrigeration plants are reed valves. State-of-the-art computation of these valves is of zero or one-dimensional order; however the number of publications regarding 3D numerical simulation is increasing. Among these two ways the user has to compromise between computation time and accuracy. This paper tries to address one possible alternative as its focusing on the application of two-way fluid-structure-interaction (FSI) in a two-dimensional representation of a reciprocating compressor. The results are validated against an established simulation software as well as measurement data. Furthermore, some modeling peculiarities will be elaborated enabling better insight to users which are new to this subject.

1. INTRODUCTION

Mathematical modeling as well as simulation is widely applied in the design of compressors. The in-depth analysis of underlying processes enables complicated problems to be solved particularly fast with the help of computers. A considerable amount of previous works has been published on the mathematical modeling and simulation, starting with Michael Costagliola (Costagliola, 1949). He developed the fundamental set of equations on valve dynamics affecting the performance of reciprocating compressors back in the days, when computers were difficult to access and programmed via perforated paper tape. Despite some simplifications, like omitting the rebound effect when the valve hits the limiter, his work was the starting point for many others to come.

20 years later, the first model with the need for solving two nonlinear differential equations simultaneously was introduced by MacLaren & Kerr (MacLaren & Kerr, 1969). The analytical model – assisted by several empirical coefficients – included plenum as well as pipeline pulsations and lead to some of the first studies regarding valve lifetime by using a mathematical model. These models, which were supported by empirical data, were consecutively improved until the end of the 20th century. Due to the large amount, only some of the major key influencers shall be named and referred to. Simon Touber developed and published his model on valve dynamics with the focus on solving with a computer (Touber, 1976). He even compared the computation times as well as costs of analog and digital computers. His extensive theory, decomposition of various valve influences plus comparison of different types of valves lead to a well-rounded set of equations for valve dynamics description. Concluding there is the work of Leopold Böswirth (Böswirth, 1994). His research of the effects of valve flutter resulted in an improvement of equations characterizing the flow and a similitude theory for transient processes within the valve as well as the cylinder. The second edition of his self-published work was amongst other things extended by one chapter focusing on the simulation of valve dynamics using the software KV-DYN, which will be used for comparison later in this paper.

Following these zero and one-dimensional modelling methods, making use of the parallel improvement of computer hardware and thus the reduction in the average calculating time, the application of 3D simulation started approximately with the turn of the millennium. At first the computations were often conducted in 2D, like seen by

Georg Flade as he studied the opening process of plate valves (Flade, 2006), later on even considering the structural reaction of the suction and discharge valve in a hermetic compressor (Rodrigues, 2014). To overcome several hurdles, some of them will be addressed in the later part of this paper, in modeling a complete working cycle, the extensive usage of interrupt control and 4 different sub-models had to be applied in ANSYS CFX. Since there were no measurements published to compare to, a qualitative analysis has not been conducted.

2. THE MODEL

As the main focus of this paper is on the application of two-way FSI simulation of reed valve dynamics and shall be the starting point for experiments to follow, the movement of the piston will be the driving force for changes of the working chamber volume and consequently the pressure. This enables the verification against well-established theories as well as current literature. Figure 1 depicts a sectional representation of a deformed reed valve within a reciprocating compressor. As shown in this drawing the limiter forces the reed, which is displaced due to the pressure difference between working and suction chamber, to bend and thus stress is induced.
Nowadays numerical simulation software makes it easy to setup a 3D representation of existing geometries. It is up to the user to decide, whether he tries to simplify the model utilizing symmetry, sectional representation or 2D slices. The resulting computation time biases the decision. As of now, there have been only a few published attempts to apply two-way fluid-structure-interaction (FSI) setups on reed valve dynamics. Considering that the one-dimensional models are still valid and often used, a 2D model could be the next best compromise between computation time and information gained. Thus – for this paper – the sectional representation of Figure 1 is extruded to a thickness of one mesh element in z-direction, leading to Figure 2 respectively Figure 4.

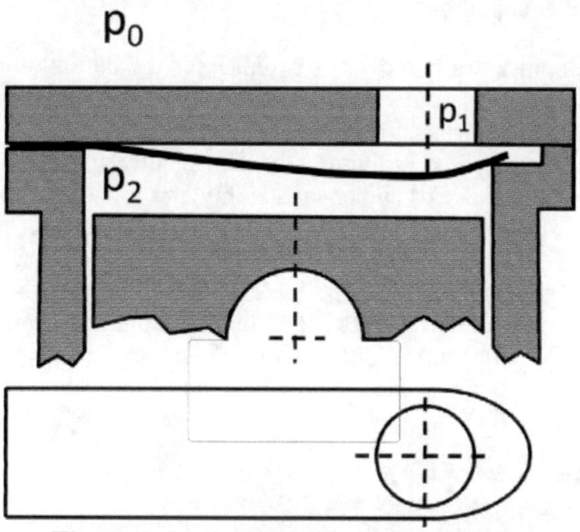

Figure 1: Sectional representation of reed valve deformation within a reciprocating compressor

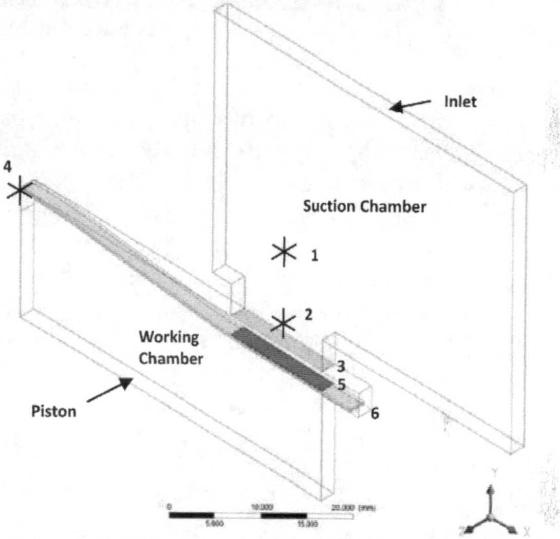

Figure 2: 2D representation of Figure 1 in CFX Post as well as definition of distinctive points, chambers and surfaces

As done in previous studies, the driving force for the necessary pressure difference will be the piston movement (Möhl, Langebach, & Hesse, 2014; Möhl, 2014). This enables to view the valve dynamics as a function of the crank angle und thus makes the data more comparable against literature. The losses due to streaming along the piston circumference will be neglected as there is no gap between the liner and the cylinder.
In consideration of the valve dynamics being a non-linear lift-off because of the fixed bearing as well as the fact that mesh elements cannot be created at run-time in ANSYS CFX, the initial position of the unbend lamella has to be shifted towards the piston. As depicted in Figure 3 a minimal gap (3, white border color) was introduced above the reed valve (2). This leads to streaming effects when the force is not yet able to move the lamella. Given that presently there is no oil fraction in the working fluid the flow can be interpreted as leakage.

Figure 3: Geometry conditioning for valve dynamics computation

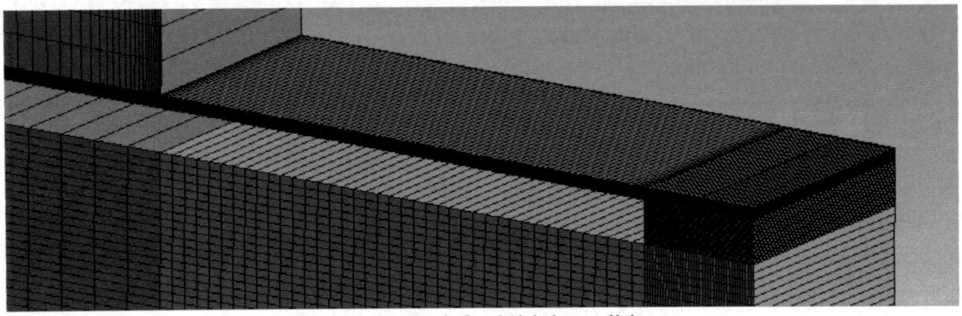

Figure 4: Mesh for initial conditions

Furthermore, the space at the tip of the reed valve should be split up in two distinct volumes for the definition of mesh movement, like shown in Figure 3 (1). The part connected to the outer wall shall only move parallel to its boundary, as the other one is defined as unspecified in motion. The drawback of this kind of modeling is preserving the mesh stiffness and quality. Figure 5 shows the coloring of the mesh according to the orthogonality angle at the tip. The reddening describes the angles in relation to the desirable 90 °. For numerical stability and reliable results one should aim for rectangular elements, even after deformation. Unfortunately, the non-linear movement complicates this task. Therefore, the user needs to influence the mesh stiffness. Advisable is the application of subdomains and the implicit addressing via Ansys CFX expression language (CEL). Table 1 shows example codes for the mesh stiffness interaction as well as the corresponding results. As they may vary according to the conditioned geometry the user should evaluate the results and evaluate different regimes. In this case the application of wall-based criteria should be avoided. This is due to the dimensions, in which this kind of boundary draws the mesh elements out of the deformation zone.

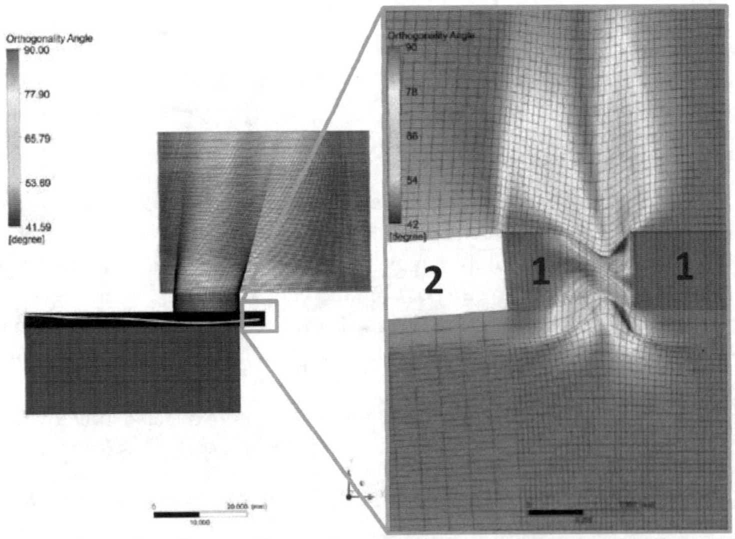

Figure 5: Mesh orthogonality at the reed tip (less reddening, smaller angle)

Table 1: 4 examples for mesh stiffness manipulation, equally scaled

#	Stiffness Parameter Regime	Result
(1)	$\text{Stiffness} = if(\text{inside()@SubdomainTip} == 1, 1e8\,[m\text{^}2/s], 1[m\text{^}2/s])$	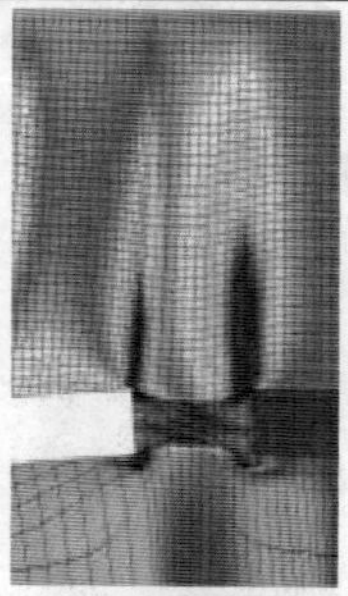
(2)	Increase near Boundaries	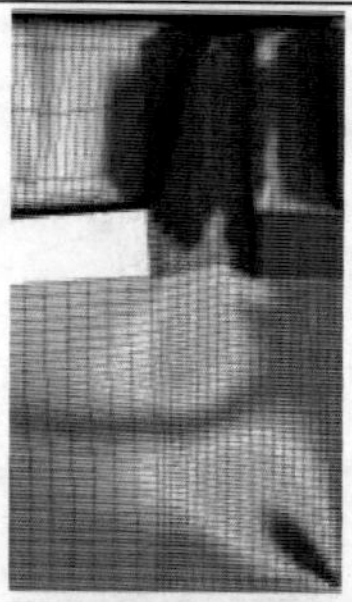
(3)	Increase near Small Volumes	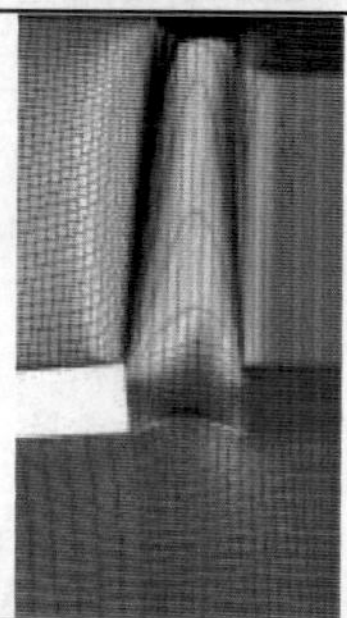
(4)	$\text{Stiffness} = if\left(\text{inside()@SubdomainTip} == 1, 1e8\left[\frac{m^2}{s}\right], \text{Stiffness2}\right)$ $\text{Stiffness2} = if\left(\text{inside()@SubdomainAroundReed} == WallFunction, 1\left[\frac{m^2}{s}\right]\right)$ $WallFunction = 1000\,[\text{m\^{}5 / s}] / (\max(0.1[\text{mm}], \text{Wall Distance})\text{\^{}}3)$	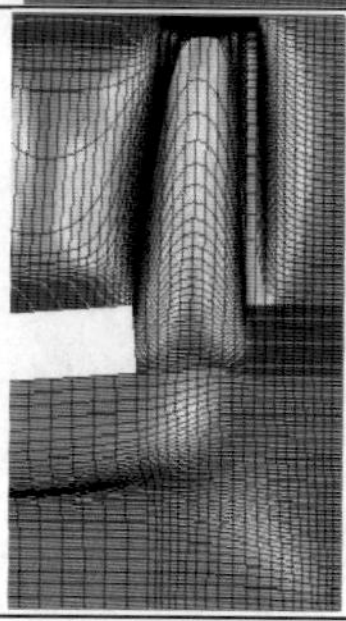

Subsequently to the fluid part of the simulation the structural interaction needs to be setup as it will be solved iteratively. The main focus is on limiting the lift so the reed valve does not hit the wall as the simulation will instantaneously stop. So the definition of a minimal gap between the contact pair is essential. In this study 0.1 mm were used between the limiter and the bottom side of the lamella. Furthermore, the correct material data for the solids should be applied as they tend to differ in tensile strength, fatigue strength as well as ductility (Chai, Zetterholm, & Walden, 2004). For the computations conducted here, the necessary parameters can be taken out of the corresponding datasheet (Sandvik, 2015).

3. RESULTS AND VALIDATION

Preceding the verification of the numerical data, the position of the bottom side (numbered as 6 and shown in turquois in Figure 2) is diagrammed in Figure 6. Due to the increasing pressure difference across the lamella, the reed valve moves and furthermore bends as it has a fixed bearing. The dashed perpendicular line shows the center of the suction port (point 2 in Figure 2). The time difference between the drawn curves is 50 time steps or 6.4° with respect to the crank angle (CA). While the upper four resemble convex bending functions, the lower two show a concave behavior due to the limiter that stops at 2.5 mm. As state-of-the-art valve dynamics measurements only capture points by design there is literally no data published to validate against. At our facilities at the Technische Universität Dresden currently a new measuring principle is evaluated, that enables to track the movement of a laser line on a moving object. Since the test rig dedicated to validate this computational data is still in production, only data acquired in preliminary tests is available at the moment. Figure 7 shows a provisional setup, where the lift of a discharge valve without a limiter was recorded. Nevertheless the measured data fits the computed ones well.

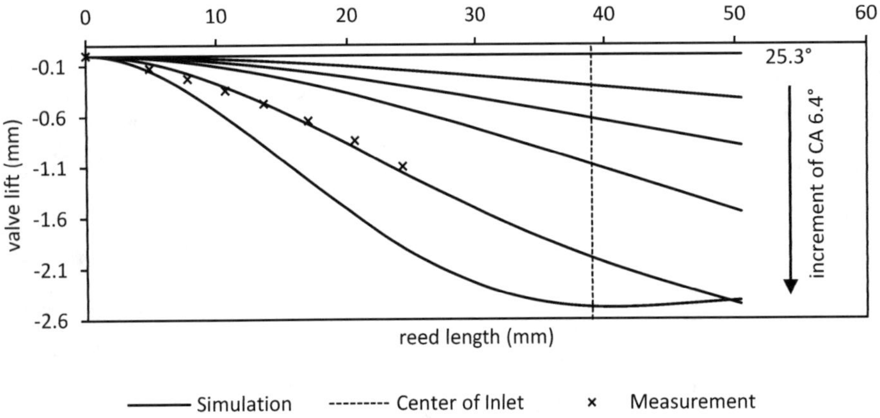

Figure 6: Valve lift vs. reed length

Figure 7: Application of a laser profile scanner for valve lift measurements

As the movement of the reed valve appears to be as indicated by our experiments, a more detailed analysis can be conducted. (Böswirth, 1994) presented the flow coefficient C_{D0} as a function of the ratio of valve lift to bore radius. Furthermore, he showed one function and a straight slope for approximation. Figure 8 shows the recreation of his chart extended to the data computed. As depicted in Figure 9 the deviation is sufficiently small, with a maximum of -0.37 %.

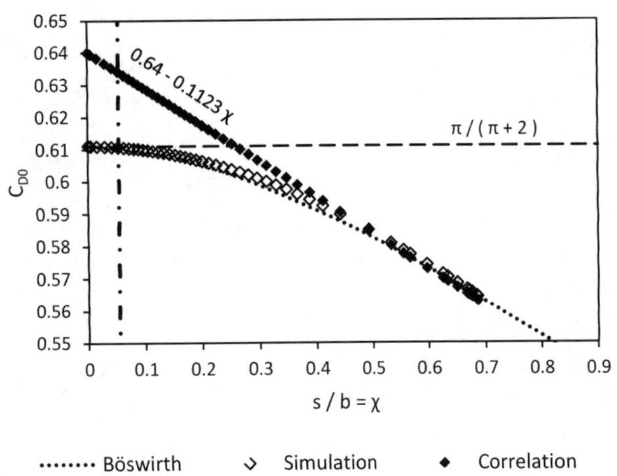

Figure 8: Flow coefficient vs. ratio of valve lift to bore radius (recreation of (Böswirth, 1994))

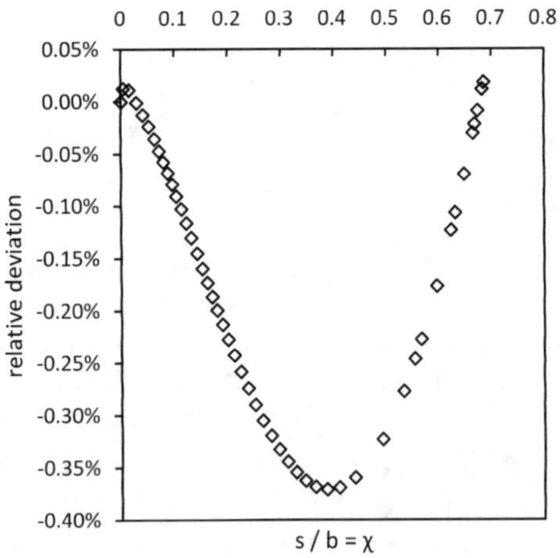

Figure 9: Relative deviation of C_{D0} computation vs. results presented in (Böswirth, 1994)

For further analysis, the steady state model implemented in KV-DYN will be used (Böswirth, 1994). This software enables the computation of valve dynamics according to the basic valve theory. The valve input parameters used are listed in Table 2. It appears that the coefficients are of a linear kind, even though Figure 8 showed that the flow coefficient C_{D0} for instance is a function of valve lift with respect to the time. Nonetheless the zero and one-dimensional computation is still state-of-the-art and widespread in the field of compressor design. The first nine parameters are pure hardware based and resemble the input data for ANSYS. The latter seven coefficients can be determined by rules of thumb, correlations or user experience. Applying these below-mentioned parameters yields to curves for valve lift, acceleration as well as the corresponding pressure difference as a function of the crank angle.

Table 2: Input parameters for KV-DYN

Symbol	Parameter	Value	Unit
b	bore diameter	7	mm
A_P	port area	38.48	mm²
L	circumference	21.99	mm
l	thickness of valve plate	4	mm
s_{max}	limiter position	2.50	mm
C	spring constant	1165.8	N/m
X_v	pretension	0	mm
m	mass	2.32	g
f_0	resonance frequency	112.90	Hz
Phi	angle for oil sticktion	0	°
PF	pocket factor	1	-
C_G	coefficient for gas spring effect	2	-
J	coefficient for gas inertia	3.94	-
C_D	flow coefficient	0.6	-
c_p	force coefficient the valve plate	1	-
d_Z	damping of the valve plate	0	-

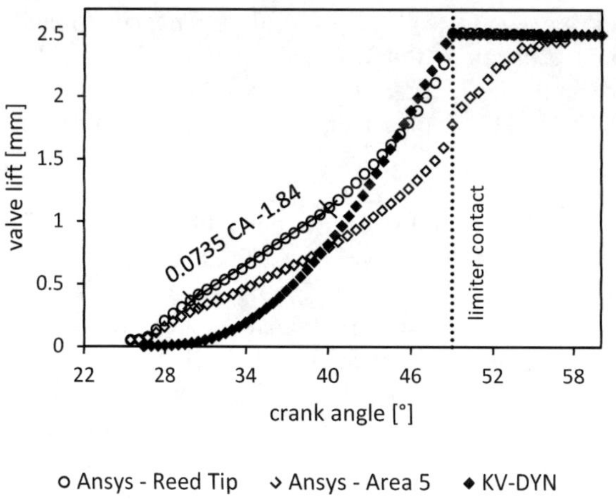

Figure 10: Comparison of simulated data – valve lift

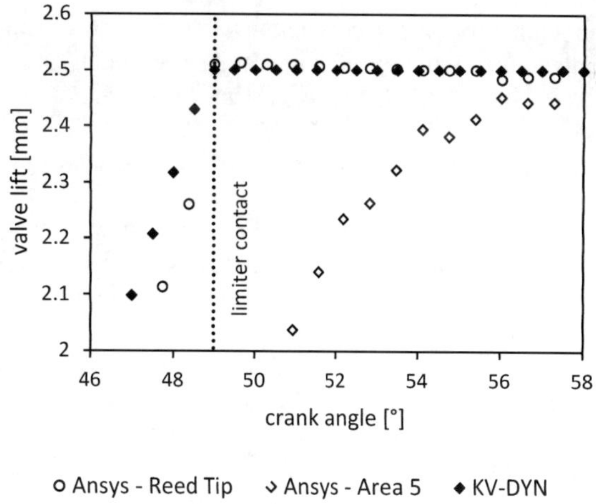

Figure 11: Detail of Figure 10 focusing on the limiter contact

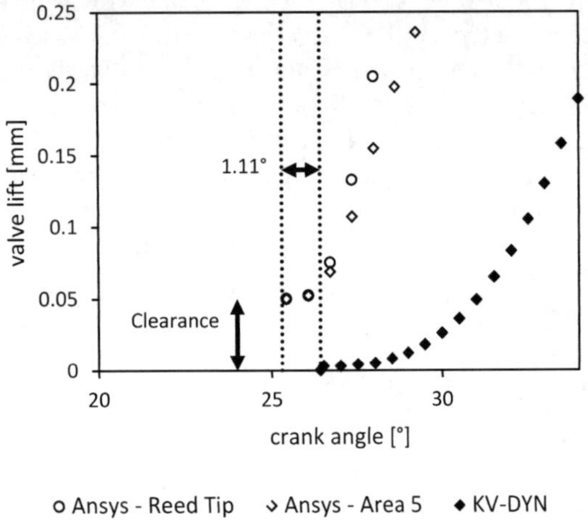

Figure 12: Detail of Figure 10 focusing on the lift-off, also depicted: clearance for mesh displacement and delay angle

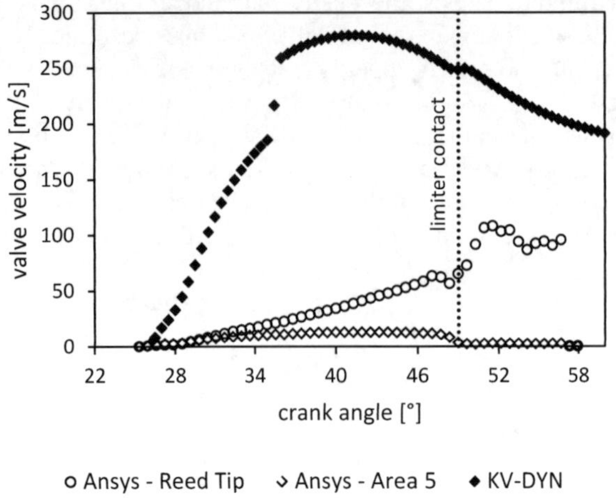

Figure 13: Comparison of simulated data – valve velocity

Figure 10 shows the simulation data for the valve lift created with both ANSYS CFX and KV-DYN. The first part shows the trend that is described by the reed tip movement. Secondly the movement of the valve can be observed analog to a simple valve plate. Therefore one should track a sample area approximately the size of the suction port. In this case the red surface labeled 5 in Figure 2 has been used. As it bends while moving, an average was used for the charts. While the movement of this duct-sized area was mainly introduced for information purposes, Figure 11 shows an oscillation close to the limiter between 54 ° and 57 °. Additionally as the time base was not the same, the limiter contact at 49 ° was used for synchronization purposes.
The curves of the reed tip and KV-DYN show a similar trend. However, they differ in acceleration during the opening process as shown in Figure 12. The valve lift-off in the ANSYS simulation starts approximately 1.1 ° in advance to KV-DYN, accelerates faster in the beginning, forms an almost linear zone from 30 ° to 40 ° and then speeds up again until it hits the limiter. At first glance this behavior conflicts with the often used and published curves (Aigner, Meyer, & Steinrück, 2013; Flade, 2006; Pereira, Deschamps, & Ribas, 2008). However (Flade, 2006) compared his data to other methods of calculation. The valve lift function of the 2D CFD computation showed the smallest gradient. Nonetheless, each model in his comparison has a similar slope. One possible explanation for this difference can be found in the quasi-stationary basis of KV-DYN. The emerging pressure gradients within each chamber caused by the prompt opening process require an accurate resolution in time and space. With the application of linear coefficients this accuracy cannot be achieved. Instead a valve lift dependency should be implemented which results in a higher computation time. By the utilization of higher mathematical method orders as realized in ANSYS, a proper time resolution is ensured due to the complexity of the simulation setup. This complexity results from the parallel solving of multiple mesh displacements, shifts and deformations, thermodynamic states as well as structural strains and stresses.

In contradiction to the state-of-the-art mathematical models a two-way fluid-structure-interaction was numerically solved in this paper. Therefore data describing stresses within the reed valve is available. Figure 14 depicts the case of a fully opened lamella that has hit the limiter as well as the corresponding stresses. Figure 15 presents the data as a function of time as well as the reed length. Thus it appears that there are two main relaxation areas. The first is at the limiter contact position while the second one shifts towards the mounting of the lamella as the forces increase. For a clear presentation of the loads the von-Mises stress criterion will be applied, given that the multiaxial stress state is reduced to an uniaxial equivalent stress. Thus the computed data has to be compared against the yield strength given in (Sandvik, 2015), namely 1450 MPa with a manufacturing tolerance of 60 MPa. As the peak value is approximately 495 MPa a safety factor of about 2.81 to 3.05, depending on the tolerance, can be calculated.

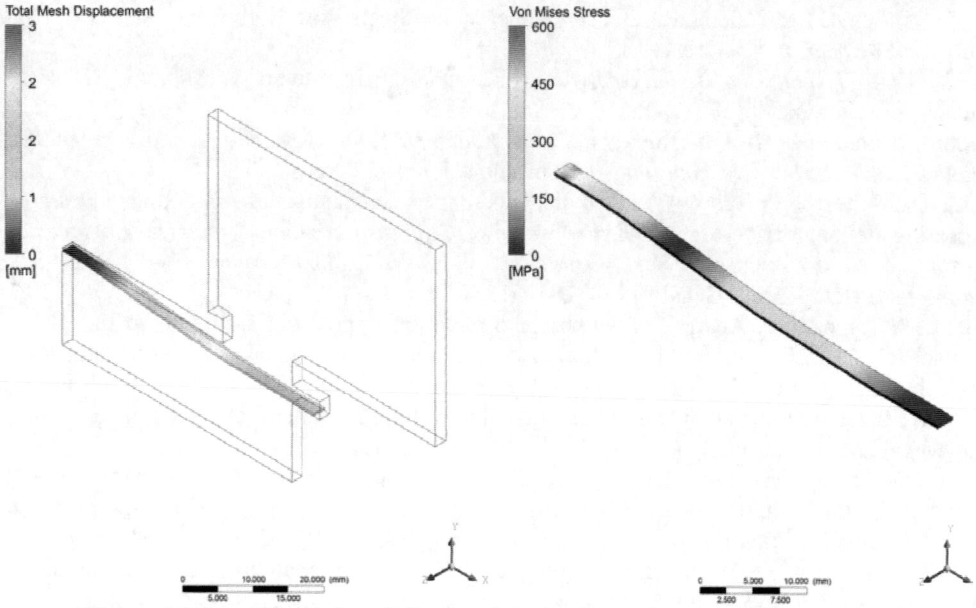

Figure 14: Parallel display of displaced reed valve and corresponding stresses

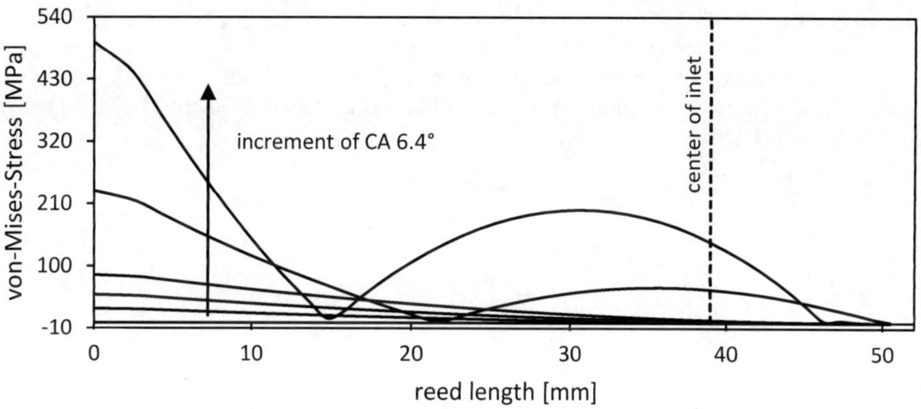

Figure 15: von-Mises-Stresses vs. reed length

4. CONCLUSIONS

Within this paper a two-dimensional two-way fluid-structure-interaction model has been presented. The computed data was then validated against literature as well as established methods. Further it could be shown that there was a fair agreement between the valve lift curves though the quasi-stationary evaluation lacks the transient resolution of an oscillation process of a reed valve. Although the bare simulation time of the data presented here was approximately 07:22 hours on an Intel® Core™ i5-2500 CPU @ 3.3 GHz with 16 GB of RAM, it enables the analysis of the entire lamella starting with the tip. Information like impact velocities in combination with different designs of seats for chamfer studies come to mind as well as the later inclusion of oil fractions in the refrigerant. Nonetheless, additional measurements should be conducted verifying the simulated data and possibly contribute in improvements regarding the lower dimensional computation methods.

REFERENCES

Aigner, R., Meyer, G., & Steinrück, H. (2013). Ventildynamik und interne Wellen in Kolbenkompressoren. *Workshop Kolbenverdichter*, *17*, 159–184.

Böswirth, L. (1994). *Strömung und Ventilplattenbewegung in Kolbenverdichterventilen*. Mödling: Eigenverlag.

Chai, G. C., Zetterholm, G., & Walden, B. (2004). Flapper Valve Steels With High Performance. *International Compressor Engineering Conference*, 8.

Costagliola, M. (1949). *Dynamics of a reed type valve*. Massachusetts Institute of Technology. Retrieved from http://dspace.mit.edu/handle/1721.1/12752

Flade, G. (2006). *Weiterentwicklung der Berechnungsmethoden für Kolbenverdichterventile auf der Basis zweidimensionaler Strömungssimulation*. Technische Universität Dresden.

MacLaren, J. F. T., & Kerr, S. V. (1969). An analytical and experimental study of self-acting valves in a reciprocating air compressor. *ARCHIVE: Proceedings of the Institution of Mechanical Engineers, Conference Proceedings 1964-1970 (vols 178-184), Various Titles Labelled Volumes A to S*, *184*(318), 24–33. http://doi.org/10.1243/PIME_CONF_1969_184_515_02

Möhl, C. (2014). A Comparative Analysis of Numerical Simulation Approaches for Reciprocating Compressors.

Möhl, C., Langebach, R., & Hesse, U. (2014). Numerische Simulation eines Hubkolbenverdichters unter Berücksichtigung der Ventilbewegung. *Kälte-, Luft- Und Klimatechnik*, 28–35. Retrieved from http://www.ki-portal.de/12305/numerische-simulation-eines-hubkolbenverdichters-unter-berucksichtigung-der-ventilbewegung/

Pereira, E. L. L., Deschamps, C. J., & Ribas, F. A. (2008). Performance analysis of reciprocating compressors through computational fluid dynamics. *Proceedings of the Institution of Mechanical Engineers Part E Journal of Process Mechanical Engineering*, *222*(4), 183–192. http://doi.org/10.1243/09544089JPME194

Rodrigues, T. T. (2014). ANSYS Mechanical Coupling Control in FSI Simulations to Reduce Computational Time. In *ESSS Conference & ANSYS Users Meeting*. Retrieved from http://www.ansys.com/staticassets/ANSYS/staticassets/resourcelibrary/presentation/brazil-2014ugm-mechanical-coupling-control-in-fsi-simulations.pdf

Sandvik. (2015). Sandvik 7C28Mo2 Compressor Valve Steel. Retrieved from http://smt.sandvik.com/en/materials-center/material-datasheets/strip-steel/sandvik-7c27mo2-flapper-valve-steel/

Touber, S. (1976). *A contribution to the improvement of compressor valve design*. TU Delft. Retrieved from http://repository.tudelft.nl/view/ir/uuid%3A4c12fa64-32c6-4fd5-9ebc-e733621c96ac/

Numerical Simulation of a Suction Valve: Comparison Between a 3D Complete Model and a 1D Model

José L. GASCHE[1]*, Allan D. S. L. DIAS[1], Douglas D. BUENO[1], Jônatas F. LACERDA[2]

[1]Unesp-State University of São Paulo, Department of Mechanical Engineering
Ilha Solteira, SP, Brazil
+55 18 3743 1035, gasche@dem.feis.unesp.br

[2]Tecumseh do Brasil LTDA, Research and Development Department,
São Carlos, SP, Brazil
+55 16 3633 7678, jonatas.lacerda@tecumseh.com

* Corresponding Author

ABSTRACT

Hermetic reciprocating compressors are widely used in small and medium size vapor compression refrigeration systems. One of the main parts of this type of compressor is the automatic valve system used to control the suction and discharge processes. As the opening and closing of the valves are caused by the forces produced by the refrigerant flow itself, the understanding of this fluid-structure interaction problem which characterizes the system dynamics is of fundamental importance in order to enhance the efficiency of the valve system. The numerical simulation of the flow has been used as an efficient method to perform this task. Most of the researchers simulated this problem by considering whether a geometric simplification for the valve (radial diffuser) or a simplification for the dynamics of the structure (one degree of freedom). In order to overcome these two drawbacks, in this work we simulate numerically the fluid structure interaction problem in a real geometry usually used in suction valves by considering a 3D complete model for both the flow and the dynamics of the structure. The Ansys CFX CFD-Computational Fluid Dynamics software was used for the simulation of the flow coupled with the structure module available in the Ansys Mechanical. In addition, a 1D model (one degree of freedom) was also implemented for simulating the dynamics of the structure for comparison reasons. The models were validated against experimental data for the aperture of the valve. We show that the 1D model saves small computational time and provides good results for general characteristics of the problem as the force acting on the valve and the aperture of the valve, while the 3D complete model allows us to predict the deformation of the structure. We suggest the use of the 3D complete model because the computational time is not significantly increased.

1. INTRODUCTION

Hermetic reciprocating compressors are widely used in small and medium size refrigeration cycles based on the vapor compression process. The thermodynamics losses (inlet superheating and flow through the suction and discharge systems) of high efficiency compressors can represent up to about 30% of the total losses - electrical, friction, thermodynamics, and cycle - (Possamai and Todescat, 2004). Among the thermodynamics losses, the suction and discharge flows through the suction and discharge valves can contribute with about 47% of the losses (Ribas *et al.*, 2010). Therefore, the flow through the valve system can represent about 15% of the total losses in this type of compressor, which means that small efficiency increase of the suction and discharge processes can provide significant increase of the global performance of the compressor.

In order to improve the efficiency of the suction and discharge processes, it is essential to investigate the flow through the suction and discharge valves. As the movement of the valves (reed type valves) depends on the forces of the flow acting on its surfaces and on the structural response of the reed, this is a very complex problem because it involves the solution of the Fluid-Structure Interaction (FSI) between the refrigerant flow and the reed.

The FSI problem in refrigeration compressor valves has been extensively investigated in the last decade. Several numerical modeling dealing with the full three-dimensional FSI problem have been developed (Kin *et al.*, 2008; Estruch *et al.*, 2014; Silva and Arceno, 2014; Tan *et al.*, 2014; Wu and Wang, 2014). Using this type of modeling we can obtain detailed characteristics of both fluid flow field and structural stress field, which can be very helpful for understanding the reasons of the thermodynamics losses. The main disadvantage of this modeling is the large computational cost. Using a hybrid modeling, in which whether the fluid flow or the structure dynamics is modeled in a simplified way (Kinjo *et al.*, 2010; Mistry *et al.*, 2012; Mayer *et al.*, 2014; Brancher and Deschamps, 2014; Yoshizumi *et al.*, 2014; Ding and Gao, 2014), one can diminish the computational cost. However, we do not obtain detailed results and we always need other information based on experimental or numerical results in order to complete the models. For optimization applications, in which several design have to be evaluated, it would be important to compare both type of modeling in order to evaluate their computational costs as well as the accuracy of the results.

We present here a comparison between two methodologies developed to simulate the flow through a real geometry usually used in suction valves. In the first methodology we consider a complete three-dimensional solution for both fluid flow and structural dynamics, in which a Finite Volume Method (FVM) is used to solve the fluid flow, and a FEM is used to solve the structure behavior of the reed. For the second methodology we consider the same model for the fluid flow but a simplified model to solve the structural dynamics, in which a one-degree of freedom mass-spring-damping (1D model) system is used to model the reed. The complete three-dimensional model (3D model) is validated by using experimental results for the lift of the reed obtained from an experimental setup design to study compressor valves. The main objective of the work is to evaluate which methodology is more appropriate for optimization purposes in industrial applications.

2. METHODOLOGY

2.1 Numerical Methodology

A schematic view of the geometry of the problem (solution domain) is shown in Fig. 1(a), which is the model for the same configuration experimentally studied by Gasche *et al.* (2014). In the solution domain, there is an entrance region representing the inlet tube of the experimental setup (aluminum tube, Fig. 2a) and an outlet region representing the test section (upper plate and reed, Fig. 2b). The diameter ratio and thickness of the reed are D/d=1.3 and 0.4 mm, respectively. Two parameters that play an important role on the dynamic of the reed are the stiffness, k, and the natural frequency, f_n. These two parameters were obtained numerically by using the commercial code Ansys-Mechanical®, resulting in k=214.7 N/m and f_n=31.4 Hz.

Two physical models were developed. In the first model, a three-dimensional formulation was considered for both fluid flow and structural dynamics. The governing equations for transient, compressible and isothermal flow of a Newtonian fluid (air) consisted of the mass conservation equation, momentum equation and an equation of state relating the pressure and the specific mass of the fluid (ideal gas was used here). The boundary conditions shown in Fig. 2 are: prescribed average velocity at the inlet of the flow (inlet), no-slip condition at all solid walls (wall), and ambient pressure at the outlet of the flow (outlet). The system of equations was solved by using the Finite Volume Method (FVM) implemented in the commercial code Ansys-CFX®. After performing a mesh independence test, we adopted a mesh containing 137,000 volumes (Fig. 1 b and c). A deforming mesh (expansion/contraction) was used near the reed in order to capture its movement. Based on preliminary tests we chose the k-Ω Shear-Stress-Transport (SST) in order to model the turbulence phenomenon, and a high order resolution scheme was used to discretize the convective terms of the momentum equation. The governing equation for the structure (reed) is given by:

$$[M]\{\ddot{x}\} + [C]\{\dot{x}\} + [k]\{x\} = \{F\} \tag{1}$$

where $[M]$, $[C]$, and $[k]$ are the mass, damping, and stiffness matrices, respectively; x, $\dot{x}$, and $\ddot{x}$ are the local position, velocity, and acceleration vectors, respectively; and $\{F\}$ is the force vector of the flow acting on the reed surfaces. Equation (1) was solved by using the FEM implemented in the commercial code Ansys-Mechanical® for a 5,600 elements mesh. The structural system was based on quadratic tetrahedral elements with ten nodes and three degree of freedom per node. The contact between the reed and the seat was modeled as zero velocity for the reed. The damping matrix was considered to be proportional to the mass and stiffness matrices:

$$[C] = \alpha_d[M] + \beta_d[k], \quad (2)$$

where the proportionality constants are α_d=3.63 and β_d=7.703x10^{-6}.

The fluid-structure interaction problem was solved by using an uncoupled method, in which the fluid flow and the structural dynamics are solved separately using the latest data provided from each other.

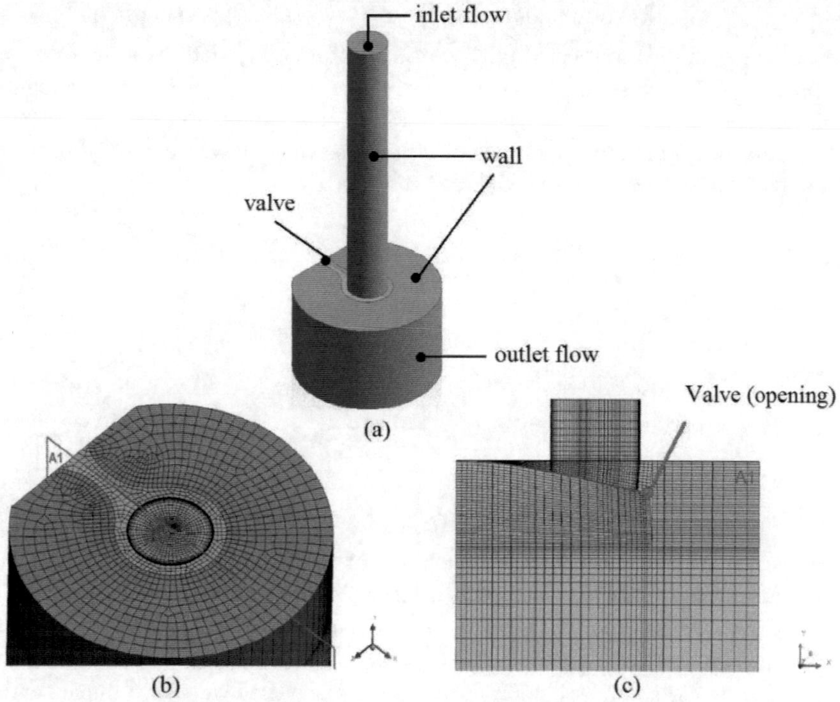

Figure 1: (a) schematic view of the solution domain; (b) top view of the mesh adopted for simulation; (c) longitudinal view of the mesh - plane A1

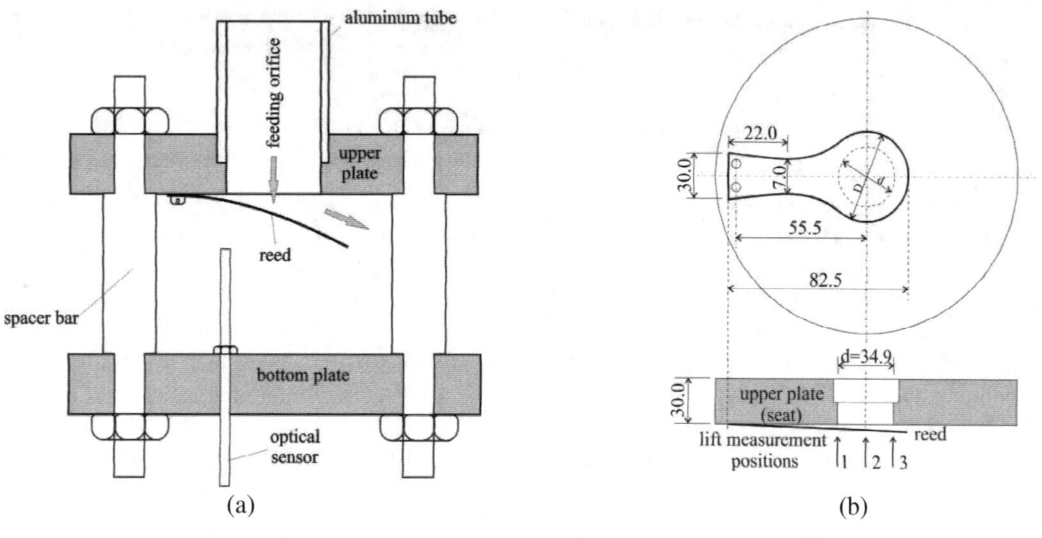

Figure 2: (a) test section; (b) upper plate and reed (Gasche *et al.*, 2014)

For the second methodology, the fluid flow was solved by using the same procedure adopted in the first methodology. However, the structural dynamics was solved by considering a simplified one-dimensional (1D) model, in which the reed was taken as a non-deforming rotating beam with a concentrated mass-spring-damping at the center of the circular part of the reed (Fig. 3). For this case, the structural dynamics is governed by a one-degree of freedom mass-spring model described by Eq. (3).

$$m_T\ddot{y} + c_{eq}\dot{y} + k_{eq}y = F_{eq}, \tag{3}$$

where m_T is the total mass of the valve, y is the position of the valve, and c_{eq} and k_{eq} are the equivalent damping and stiffness coefficients of the reed, respectively. The resultant force, F_{eq}, acting on the reed surfaces is calculated by integrating the pressure field just on the circular region of the reed in order to avoid computational expenses to treat the transfer of data between the fluid and structure meshes. In addition, considering that we have to use a very small time interval, Δt, for integrating the equations governing the fluid flow, we used a constant force (determined at the preceding time) in order to solve Eq. (3), yielding to:

$$y = e^{-\alpha^*\Delta t}\left[y_c\cos(\omega^*\Delta t) + \frac{\alpha^* y_c - \dot{y}^0}{\omega^*} sen(\omega^*\Delta t)\right] + \frac{F_{eq}}{k_{eq}} \tag{4}$$

where,

$$\alpha^* = \frac{c_{eq}}{2m_T}, \qquad \omega^* = \frac{\sqrt{4m_T k_{eq} - c_{eq}^2}}{2m_T} \qquad y_c = y^0 - \frac{F_{eq}}{k_{eq}}, \tag{4a}$$

where y^0 and $\dot{y}^0$are the valve displacement and valve velocity at the preceding time; m_T is the total mass of the reed; and the equivalent stiffness coefficient, k_{eq}, was determined using m_T and the first natural frequency obtained by a numerical modal analysis. The definition of the equivalent damping coefficient c_{eq} was based on the same procedure used to determined matrix $[C]$. Different values were not evaluated once that experimental tests showed undamped motions. Equation (3) and the position of the reed at each time step were implemented in the CFX code by using a user defined subroutine.

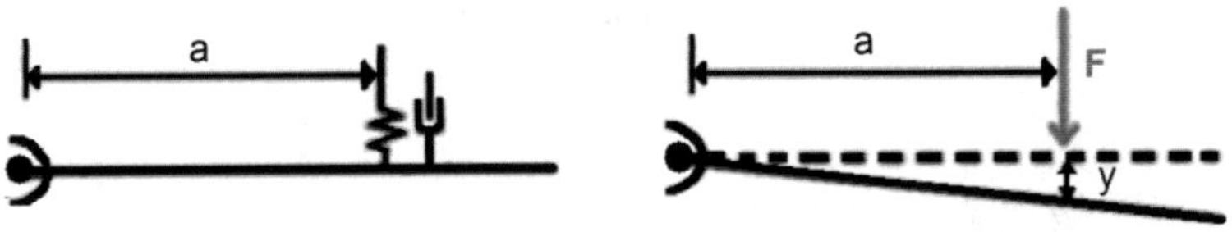

Figure 3: Schematic representation of the 1D structural dynamics

2.2 Validation of the Numerical Methodology

The 3D numerical methodology was validated against experimental data of the displacement of the reed for Reynolds number based on the inlet parameters of the flow (Re=Vd/ν) equal to 6,000. We measured the displacement of the reed at positions 1, 2, and 3 (Fig. 2b) using an optical sensor with measurement uncertainty equal to ±1μm. Details of the experimental apparatus and procedure can be seen in Gasche *et al.* (2014). The prediction of the reed displacement provided by the 3D model is qualitatively reasonable (Fig. 4). However, the model over-predicts the amplitude at the measurement positions 1 and 2 (Fig. 2b) by approximately 13% and 22%, respectively. In addition, the model over-predicts the frequency of the movement by a smaller scale, with a difference of approximately 5%. These differences may be due to the impact of the reed on the seat, which is not completely modeled here.

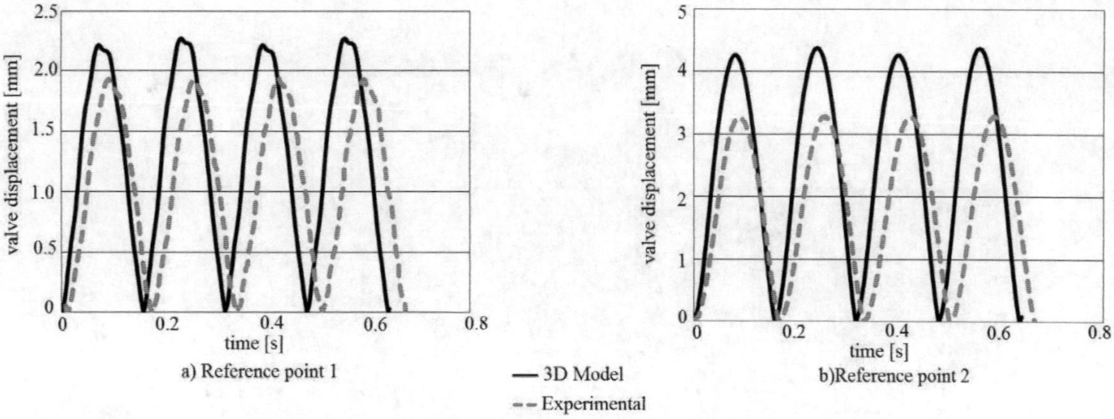

Figure 4: Displacement of the reed for Re=6,000

3. RESULTS AND DISCUSSIONS

In order to evaluate both methodologies (3D and 1D), we plotted the reed displacement and the resultant force (Fig. 5), and the pressure and velocity fields (Fig. 7 and Fig. 8) for Re=10,000. The best agreement for the valve displacement occurs at position 2. The disagreement at positions 1 and 3 is due to the deformation of the reed provided by the 3D model (Fig. 6). Figure 5d shows a very good agreement of the resultant force acting on the reed, except at the contact instant where a pressure pick is captured. Concerning the pressure and velocity fields (Fig. 7 and Fig.8), we do not see significant differences between the results provided by both models, which means that the simplified 1D model could be used without losing accuracy, mainly if global parameters as the resultant force and the mass flow rate (related to the valve aperture) are the concerns. Similar results were obtained for Re=4,000 and Re=6,000.

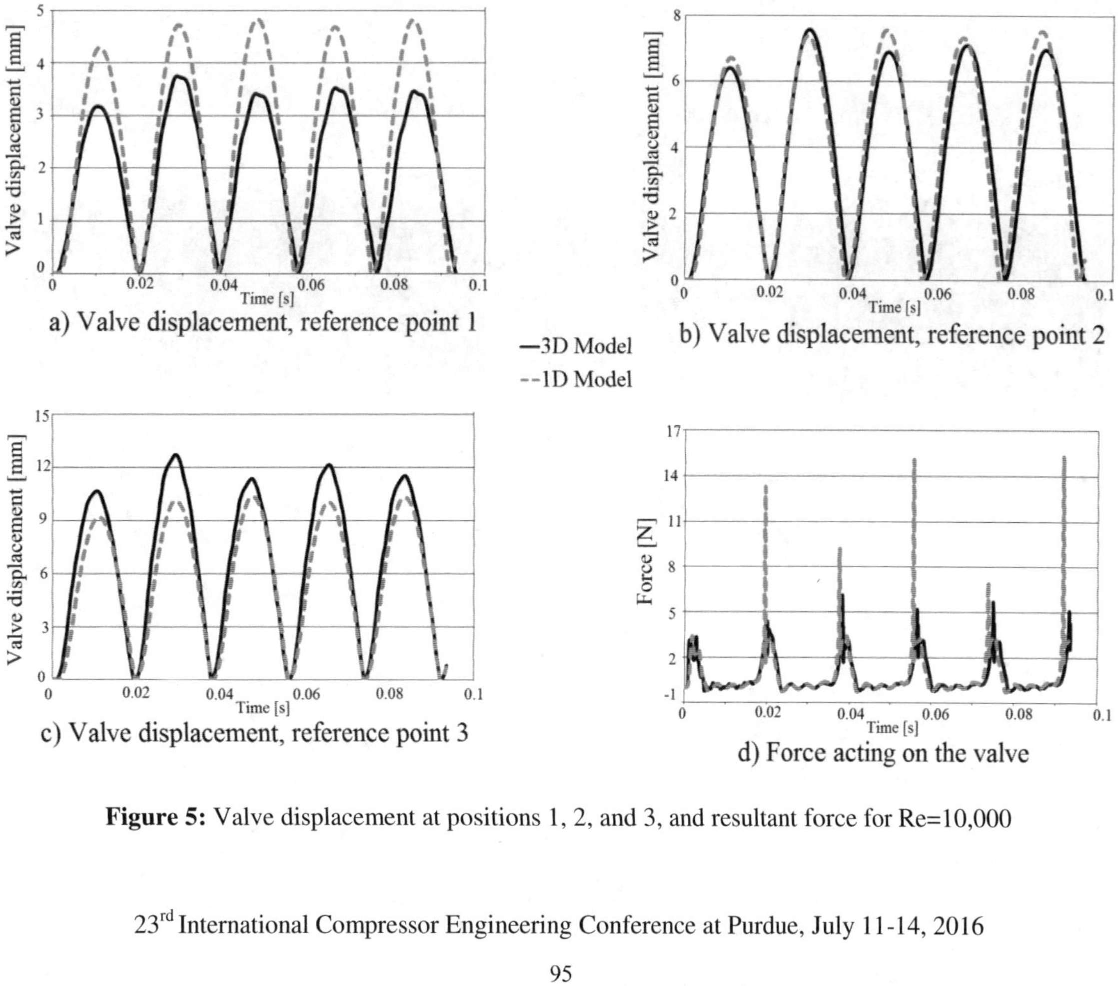

Figure 5: Valve displacement at positions 1, 2, and 3, and resultant force for Re=10,000

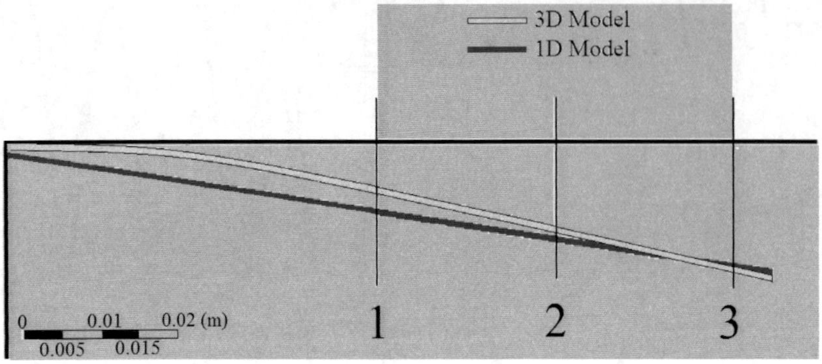

Figure 6: Positioning of the reed for the maximum aperture considering the 3D and 1D models

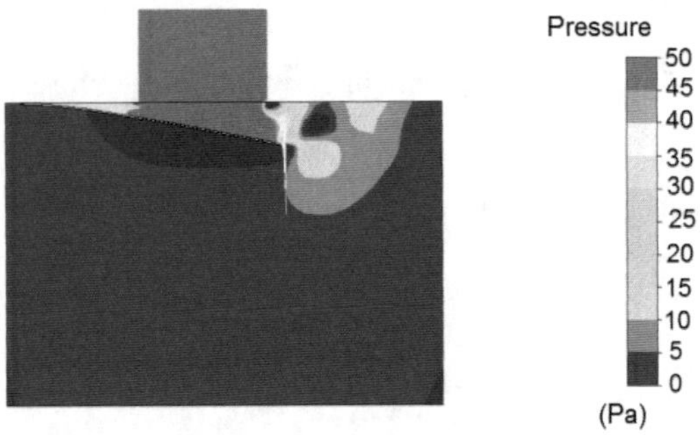

(a) Pressre field - 3D model

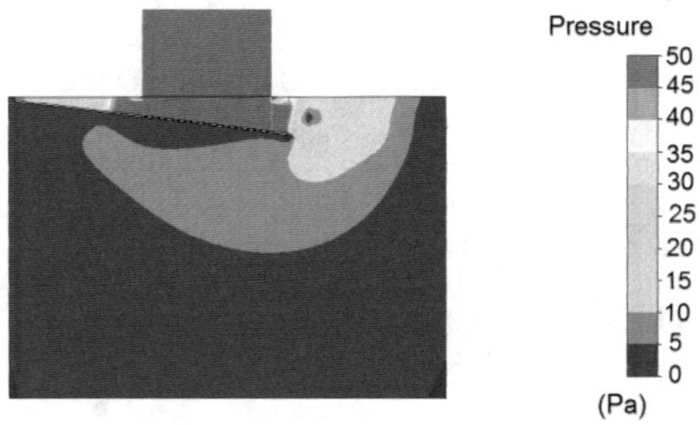

(b) Pressure field - 1D model

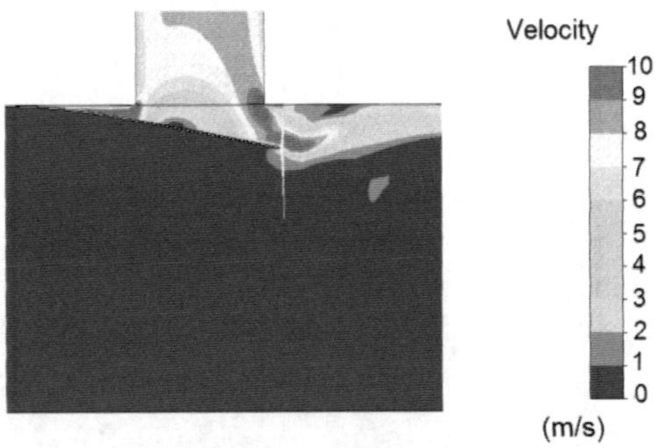

(c) Velocity field - 3D model

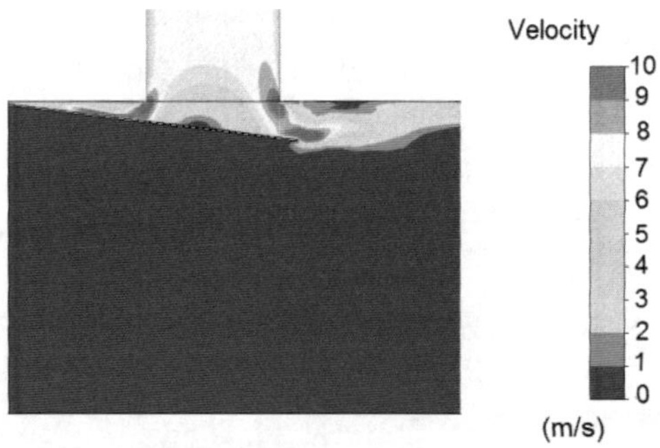

(d) Velocity field – 1D model

Figure 7: Pressure and velocity fields at the longitudinal view (plane A1) for Re=10,000

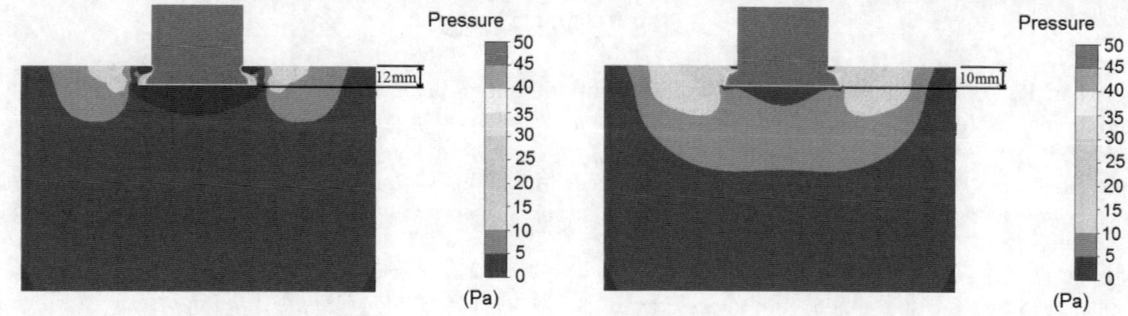

(a) Pressure field – 3D model (b) Pressure field – 1D model

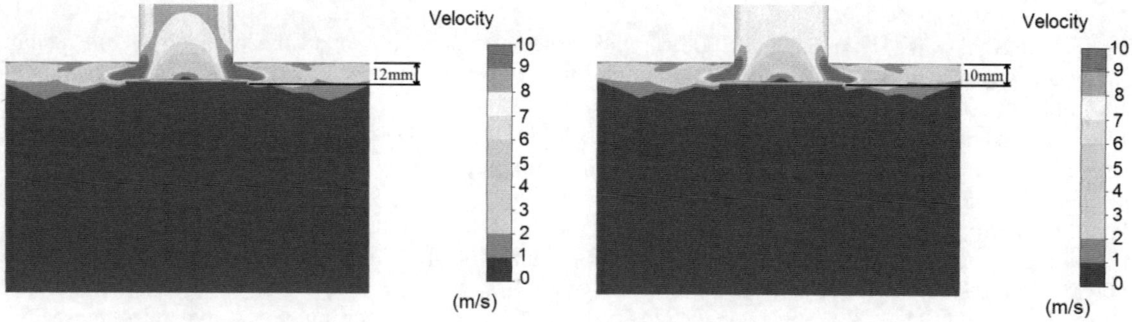

(c) Velocity field – 3D model (d) Velocity field – 1D model

Figure 8: Pressure and velocity fields at the cross-section view for Re=10,000

We could indicate the 1D model for compressor simulation purposes if the computational time for running each simulation was much lower than that necessary to run the 3D model. However, we save just approximately 15% (2 hours) of the time needed to simulate the same case by using the complete 3D model (Table 1). For all cases we used the single core Intel® processor i7-3770, with 3.40 GHz and 16.0 GB RAM.

Table 1: Real time simulation of the problem

Re	3D Model	1D Model	Difference (%)
4,000	13h 32min	11h 40min	13.8
6,000	13h 37min	11h 42min	14.1
10,000	13h 59min	11h 48min	15.6

4. CONCLUSIONS

In this paper we evaluate the possibility of using a simplified 1D model over a complete 3D model in order to represent the structural dynamics of reed type valves. The objective is to assess the global parameters (force and displacement) and the real computational time spent to run both models. Concerning the global parameters there are not significant differences between the results. However, only 15% of reduction for the computational time does not justify the use of the 1D model, mainly because the 3D model can provide detailed results that are essential for design purposes as for instance the stress field in the valve.

REFERENCES

Brancher, R. D., Deschamps, C. J., (2014). Modeling of rolling-piston compressors with special attention to the suction and discharge processes. In *Proceedings of the 22nd International Compressor Engineering Conference at Purdue*, Paper 1302. West Lafayette-IN, EUA.

Ding, H., Gao, H., (2014). 3-D Transiente CFD model for rolling piston compressor with dynamic reed valve. In *Proceedings of the International Compressor Engineering Conference at Purdue*, Paper 1548. West Lafayette-IN, EUA.

Estruch, O., Lehmkuhl, O., Rigola, J., Pérez Segarra, C. D., (2014). Fluid-structure interaction of a reed type valve subjected to piston displacement. In *Proceedings of the 22nd International Compressor Engineering Conference at Purdue*, Paper 1659. West Lafayette-IN, EUA.

Gasche, J. L., Arantes, D. M., Andreotti, T., (2014). Experimental analysis of the fluid structure interaction in a suction valve model. In *Proceedings of the 22nd International Compressor Engineering Conference at Purdue*, Paper 1217. West Lafayette-IN, EUA.

Kim, H.S., Ahn, J.W., Kim, D.H., (2008). Fluid structure interaction and impact analyses of reciprocating compressor discharge valves, In *Proceedings of the International Compressor Engineering Conference at Purdue*, Paper 1112. West Lafayette-IN, EUA.

Kinjo, K., Nakano, A., Hikichi, T., Morinishi, K., (2010). Study of CFD considering valve behavior in reciprocating compressor, In *Proceedings of the International Compressor Engineering Conference at Purdue*, Paper 1256. West Lafayette-IN, EUA.

Mayer, J., Bjerre, P., Brune, F., (2014). A comparative study of different numerical models for flapper valve motion. In *Proceedings of the 22nd International Compressor Engineering Conference at Purdue*, Paper 1315. West Lafayette-IN, EUA.

Mistry, H., Bhakta, A., Dhar, S., Bahadur, V., Dey, S., (2012). Capturing valve dynamics in reciprocating compressors through computational fluid dynamics, In *Proceedings of the International Compressor Engineering Conference at Purdue*, Paper 1210. West Lafayette-IN, EUA.

Possamai, F. C., Todescat, M. L., (2004). A review of household compressor energy, In *Proceedings of the International Compressor Engineering Conference at Purdue*, Paper 1711. West Lafayette-IN, EUA.

Silva, J. C., Arceno, E., (2014). Correlation between the fluid structure interaction method and experimental analysis of bending stress of a variable capacity compressor suction valve. *In Proceedings of the 22nd International Compressor Engineering Conference at Purdue*, Paper 1324. West Lafayette-IN, EUA

Tan, Q., Liu, Z., Cheng, J., Feng, Q., (2014). Effective flow and force areas of discharge valve in a rotary compressor. *In Proceedings of the 22nd International Compressor Engineering Conference at Purdue*, Paper 1159. West Lafayette-IN, EUA.

Wu, S., Wang, Z., (2014). A numerical simulation of fluid-structure interaction for refrigerator compressors suction and exhaust system performance analysis, *In Proceedings of the International Compressor Engineering Conference at Purdue*, Paper 1131. West Lafayette-IN, EUA.

Yoshizumi, F., Kondoh, Y., Moroi, T., Tamano, S., Morinishi, Y., (2014). Numerical simulation on the opening delay of a discharge valve in compressor. *In Proceedings of the 22nd International Compressor Engineering Conference at Purdue*, Paper 1428. West Lafayette-IN, EUA.

ACKNOWLEDGEMENT

The authors acknowledge the support provided by FAPESP (Grant n° 2010/16837-9), TECUMSEH do Brasil, and FUNDUNESP (Grant n° 01525/11).

Fluid-structure interaction of a reed type valve

Ignacio GONZALEZ[1*], Oriol LEHMKUHL[1], Alireza NASERI[1], Joaquim RIGOLA[1], Assensi OLIVA[1]

[1] Heat and Mass Transfer Technological Center (CTTC)
Universitat Politècnica de Catalunya – BarcelonaTech (UPC), ESEIAAT,
Colom 11, 08222, Terrassa (Barcelona), Spain
Tel. +34 937398192
Email: cttc@cttc.upc.edu, http://www.cttc.upc.edu

* Corresponding Author

ABSTRACT

This paper presents a complete numerical procedure to study the fluid-structure interaction problem of incompressible flow through reed valves, typically employed in hermetic reciprocating compressors. A partitioned semi-implicit coupling scheme is implemented, which only strongly couples the added-mass-effect (pressure term) of the fluid to the structure hence, assuring numerical stability and avoiding excessive computational cost. The fluid is solved by a three-dimensional CFD solver using large eddy simulation closures to model the turbulent flow, while the reed valve is described with the classical plate theory and the normal mode summation method. To showcase the potentiality of the proposed methodology, a sensitivity analysis regarding valve thickness is carried out for a given velocity in the feeding channel. Considerable differences, mainly in valve lift and pressure drop, are appreciated between the considered configurations.

1. INTRODUCTION

High efficiency of hermetic reciprocating compressors has become an essential requirement in the domestic refrigeration field as a result of their increasing worldwide electricity consumption and the environmental awareness. According to Ribas et al. (2008), almost half of the compressor thermodynamic losses are produced during suction and discharge processes. Therefore, special attention should be given to the flexible reed valves that regulate the gas flow in both processes. In addition, valves are determining not only in terms of efficiency but also regarding installation reliability, since they have been identified to be the most common cause of compressor failures (Leonard, 1996; Woo et al., 2010).

The study and optimization of these reed valves are challenging tasks due to the complexity of the fluid-structure interaction (FSI) problem. Significant work has been carried out in order to predict the flow field by means of computational fluid dynamics (CFD). Matos et al. (2000) and Barbi et al. (2012) performed a three-dimensional simulation of a laminar flow through a valve, in the former solved with a one-dimensional model and in the latter with a predefined motion. Besides, the flow was studied in two dimensions by Matos et al. (2002), Kim et al. (2006) and Gasche et al. (2010) resulting from an axisymmetric problem arrangement or from simplifications. Mayer et al. (2014) used a loosely-coupled strategy to combine a three-dimensional turbulent model for the refrigerant with a strain-stress structural solver for the valve. Encouraging agreement with experimental data was obtained.

This work arises from the effort made by Estruch et al. (2014), who focused on a three-dimensional turbulent flow through a simple rectangular reed valve. A newly developed FSI method allows strongly coupling of the fluid and structural solvers, assuring numerical stability with an acceptably moderate increase in computational cost. The CFD solver consists of a three-dimensional explicit finite volume model formulated in a second-order, conservative and collocated unstructured grid arrangement. The turbulent nature of the flow is modeled by the wall-adapting local eddy-viscosity (WALE) model. For the elastic reed valve, it is followed the classical flat plate theory based on the structure modal analysis in order to save considerable effort in time and cost. The geometric interaction between both dynamically moving media is managed with a fluid deformable mesh and the space conservation law, whereas the immerse boundary method is used to establish static bodies, i.e. the valve seat and the channel.

Two distinct objectives are pursued in this paper. The first one is to present a complete numerical procedure to analyze FSI applications, for instance the turbulent flow through reed valves. The second is to evaluate the sensitivity of these valves regarding their thickness. To do so, a generic reed valve is considered together with a feeding channel with a suction flow inlet condition.

2. MATHEMATICAL AND NUMERICAL MODEL

The developed FSI algorithm is built on a partitioned approach, which uses independent solvers for fluid and structural sub-problems and adopts a coupling scheme to account for the interaction of the domains. In order to precisely satisfy the coupling condition at the fluid-structure interface and to avoid numerical instabilities, a strongly coupled scheme is followed. It enforces the exact coupling condition by means of iterating between the fluid and structural solvers until the convergence criterion is met.

The change in the geometric domains –derived from the deformation of the structure– is taken into account by directly moving the solid mesh and updating the fluid grid according to the new solid interface position. The fluid mesh adaptation is governed by the equations of elasticity similarly to Stein and Tezduyar (2002) and Smith (2011). By introducing a stiffening in the smaller grid elements, large displacements can be accomplish preserving the mesh quality near the solid surfaces. Namely, not influencing the accurate calculation of the boundary layers.

Throughout the valve opening, new fluid elements will appear between the valve and the seat. Since there is no remeshing but the fluid mesh is just deforming, extra elements are needed under the valve, which first perform as rigid seat and later as fluid. The immerse boundary technique allows this change in the nature of an element thanks to a modification in its viscosity.

Firstly, the governing equations of both sub-domains, the fluid Ω_f and the structure Ω_s, are presented. Then, there is the description of the algorithm for solving the coupling problem on the interface $\Gamma = \Omega_f \cap \Omega_f$.

2.1 Governing equations

The governing equations for the fluid correspond with the incompressible Navier-Stokes and continuity equations within an arbitrary Lagrangian Eulerian formulation to consider dynamic mesh effects,

$$\nabla \cdot \vec{u} = 0 \quad (1)$$

$$\frac{\partial \vec{u}}{\partial t} + (\vec{u} - \vec{u}_w) \cdot \nabla \vec{u} = -\frac{1}{\rho_f} \nabla p + \frac{\mu}{\rho_f} \Delta \vec{u} \quad (2)$$

where the unknowns p and $\vec{u}$ are the fluid pressure and velocity, respectively. The velocity of the control volume's faces, $\vec{u}_w$, is nonzero as long as the mesh moves. Its value is calculated by means of the space conservation law (Estruch et al., 2013). The fractional step projection method is adopted to decouple velocity and pressure and solve equations 1 and 2. The expected turbulence of the flow will generate small scales of motion that need high grid refinement to be properly solved. The influence of these scales of turbulence can be modeled by Large Eddy Simulation (LES) techniques so as not to require that refinement. In this work, LES has been performed using the wall-adapting local-eddy viscosity model (WALE), available in TermoFluids code (Lehmkuhl et al., 2009).

On the fluid-structure interface Γ, there is a Dirichlet boundary condition determined by the displacement of the structure $\vec{d}$

$$\vec{u} = \frac{\partial \vec{d}}{\partial t} . \quad (3)$$

By contrast, the structure has a traction boundary condition on the common interface. This external force is represented by $\vec{\sigma}_\Gamma$ and is a function of fluid velocity and pressure distribution on Γ, $\vec{\sigma}_\Gamma = \vec{\sigma}_\Gamma (\vec{u}, p)_\Gamma$. Therefore, having adapted the mesh and updated the boundary conditions (Equation 3), the fluid solver is able to find p and $\vec{u}$ in Ω_f and, in this manner, evaluate a new traction distribution on the solid surface

$$\vec{\sigma}_\Gamma = \vec{F}(\vec{d}) . \quad (4)$$

In regards to the reed structure, a solver based on the Kirchhoff-Love theory for isotropic plates under transverse loading is employed, hence just the normal component of $\vec{\sigma}_\Gamma$ is taken into account. Likewise, only the displacement in the plate transverse direction is treated as unknown, since the rest can be derived easily from it,

$$d_x(\vec{x}) = -z\frac{\partial d_z}{\partial x} \tag{5}$$

$$d_y(\vec{x}) = -z\frac{\partial d_z}{\partial y} \tag{6}$$

$$d_z(\vec{x}) = d_z(x,y) \tag{7}$$

considering z as the plate normal direction and locating the mid-surface plane of the structure on $z = 0$.

Referring to Soedel (1992), the valve motion can be described from the superposition of its infinite normal modes of vibration. Each mode m has a natural radial frequency ω_m and a particular deformation pattern in z direction known as mode shape $\varphi_m(x,y)$. Thus, the transverse displacement is a linear combination of the normal modes of the system. For practical applications only the first few modes do actually participate and, consequently, only a finite number of modes should be computed:

$$d_z(x,y,t) = \sum_{m=1}^{\infty} q_m(t)\varphi_m(x,y) \tag{8}$$

where $\varphi_m(x,y)$ as well as ω_m of any given plate geometry can be found experimentally or numerically by a structural solver.

The combination of both assumptions, the plate theory and the assumed modes method, leads to one generalized momentum equation for each mode,

$$\ddot{q}_m(t) + 2\xi\omega_m\dot{q}_m(t) + \omega_m^2 q_m(t) = \frac{\int_\Gamma \varphi_m(x,y)\sigma_{\Gamma,z}(x,y,t)d\Gamma}{\rho_s h\int_\Gamma \varphi_m^2(x,y)d\Gamma} \tag{9}$$

where ξ is the damping ratio and h is the plate thickness. A Newmark time integration scheme is employed to find the generalized coordinates q_m, $\dot{q}_m$ and $\ddot{q}_m$.

Moreover, the impact between the valve and its seat is assumed to occur in such a short period of time that only the velocity is affected by the collision. Hence, a coefficient of restitution C_r has been advocated in order to model the impact. When the coefficient of restitution is set to one, all the kinetic energy is preserved after impact. Having solved Equation 9 at time step t^{n+1}, if contact is detected, the generalized velocity will be corrected with the approximation of Sheu and Hu (2000)

$$\dot{q}_m\left(t^{n+1}\right) = -C_r\dot{q}_m\left(t^n\right). \tag{10}$$

Similarly to Equation 4 for the fluid solver, the structural solver can be summarized by

$$\vec{d} = \vec{S}(\vec{\sigma}_\Gamma) \tag{11}$$

in other words, given the current fluid traction on the solid, the solver $\vec{S}$ will obtain the deformation or displacement of the structure.

2.2 Fluid-Structure Interaction coupling scheme

Defining the fluid and structural equations concisely as Equations 4 and 11 reduces the coupled fluid-structure system of equations into an interface problem of the form:

$$\vec{S}\circ\vec{F}\left(\vec{d}\right) - \vec{d} = 0 \tag{12}$$

with $\vec{d}$ and functions $\vec{F}$ and $\vec{S}$, all in the new time step t^{n+1}. Instead of solving the interface problem 12 as in implicit schemes, here a semi-implicit approach is followed. This approach couples only the pressure stress term of the fluid, implicitly to the structure, while the rest of the fluid terms are coupled explicitly. It has been shown that pressure stress term is responsible for the added-mass effect, and coupling it explicitly will cause numerical instability (Causin et al., 2005). By coupling this term implicitly, the stability issue is overcome, while explicitly coupling the other terms saves a significant computational cost.

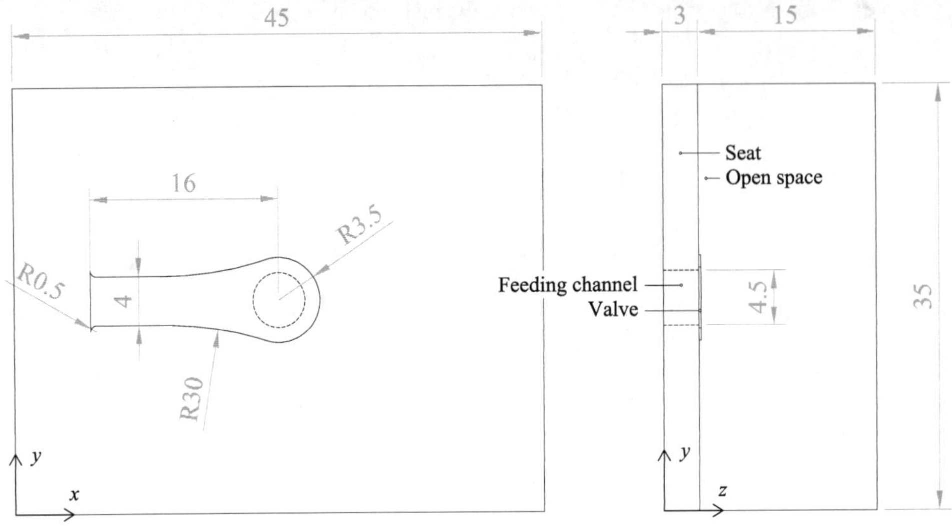

Figure 1: FSI problem domain. Dimensions in mm.

Thanks to the fractional step method, the pressure stress term can be split off from the fluid flow equations. At the beginning of a new time step, once the moving mesh function has updated the domain Ω_f and the velocities $\vec{u}_w$, a predictor velocity, which does not take into account the pressure gradient, is calculated. Then, a pressure field is derived from this intermediate velocity. The structure will deform facing this pressure field and, as a result, a new velocity boundary condition will arise in Γ (Equation 3). The implicit coupling is fixed for these two steps, the fluid pressure solver and the structural solver. So the new FSI interface problem is modified to

$$\vec{S} \circ \vec{f}(\vec{d}) - \vec{d} = 0 \tag{13}$$

where $\vec{f}$ is just the fluid pressure solver. To solve the nonlinear FSI interface problem, a fixed-point solver with a block Gauss-Seidel method and Aitken's relaxation has been chosen in this study. Aitken's dynamic relaxation mitigates possible convergence problems at the Gauss-Seidel method and reduces the number of iterations required. For further details of the FSI coupling scheme please refer to Naseri et al. (2016).

3. PROBLEM DEFINITION

Figure 1 shows the dimensions of the simulation domain and the reed valve chosen for the test. The domain contains a feeding channel, a rigid seat, a reed valve and an empty space downstream. Two different valve thicknesses have been considered: 0.18 mm and 0.20 mm.

The fluid properties are: ρ_f = 19 kg/m^3 and μ = 9.4 · 10^{-6} Pa·s. The valve material is steel and its properties are: ρ_s = 7870 kg/m^3, ν = 0.29 and E = 2.05 · 10^{11} Pa. The damping ratio of Equation 9 is taken from Prater and Hnat (2003) ξ = 0.04 and the coefficient of restitution from Habing and Peters (2006) C_r = 0.3.

There is an inlet velocity boundary condition at the beginning of the channel. It is a time dependent mass flow that belongs to the suction process of a one-dimensional simulation of a hermetic reciprocating compressor, Figure 2 (Lopez, 2016). The average fluid velocity along the channel results in a Reynolds number close to 3 · 10^4, hence a turbulent flow is developed indeed.

The natural modes of vibration of the reed valve ($\varphi_m(x,y)$ and ω_m) should be studied previously to the FSI simulation. A commercial structural solver has been used defining a clamped end, like in a cantilevered beam. For this problem only the first four modes has been considered. The normal mode shapes are illustrated for the thinnest valve in Figure 3 –the deformation is pretty similar for the thicker valve– whereas the natural frequencies are presented in Table 1.

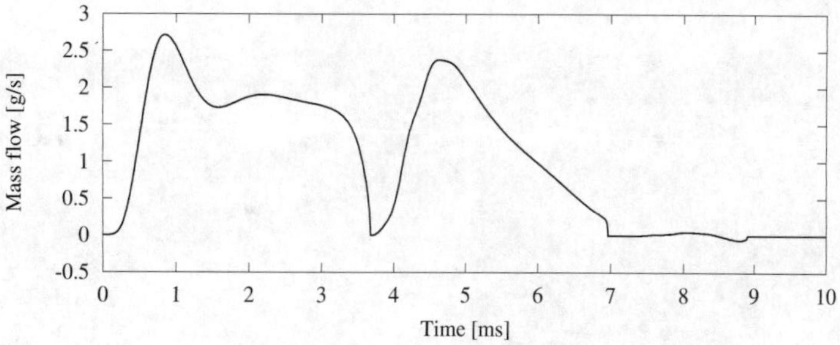

Figure 2: Suction mass flow profile extracted from Lopez (2016).

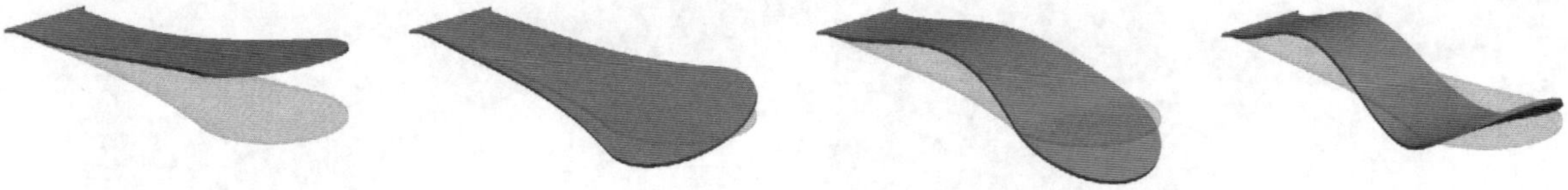

Figure 3: Normal mode shapes of the 0.18 mm-thick valve. From left to right: mode 1, 2, 3 and 4.

4. NUMERICAL RESULTS

The results presented in this section have been obtained with a coarse computational mesh of $4 \cdot 10^5$ control volumes. It would be appropriate to perform a mesh refinement analysis to ensure that the FSI solution has converged. Even so, the flow phenomena obtained is consistent and accomplishes a qualitative accurate transient simulation of the turbulent flow through the reed valve.

Figure 5 outlines the valve lift throughout the suction process, which is equivalent here to the displacement of the valve point coincident with the channel axis. The numerical results show that the dynamic action of the reed valve is definitely dependent on the thickness. Although the first peak force exerted by the fluid while impacting with the valve is almost equal in both cases, the maximum lift is 18 % lower in the thickest design. It is clearly due to the bending stiffness of the valve, which rises with the thickness. This rigidity is at the same time affecting the oscillation period, as was also derived from the modal analysis of the valves (Table 1). A higher thickness implies a major number of opening oscillations and impacts, what can eventually influence to structural and acoustic issues. Nevertheless, neither of them seems to be suitable for the suction process proposed, as their stiffness make them oscillate in excess for a regular refrigerant flow.

Since it is the mass flow that is imposed in the boundary condition, the parameter to compare suction process efficiency will be the pressure drop of the flow through the valve. Thus, the smaller is the pressure drop, the more fluid mass will be suctioned and placed in the compression chamber. The transient evolution of the pressure on the inlet orifice is presented in Figure 6. It seems to be generally higher with h = 0.20 mm. It has been observed that this pressure is qualitatively identical to the fluid pressure on the valve, specifically around the channel axis. As can be seen, there are five pressure peaks that emerge from those times when the valve is going to close and there is still a feeding flow along the channel. In fact, the last bounces of the valve, those beyond 7 ms, would not provide any benefit for the suction cycle because there is no feeding flow anymore.

Table 1: Natural frequencies of the valves.

	mode 1	mode 2	mode 3	mode 4
$h = 0.18$ mm	339.7 Hz	2349 Hz	2499 Hz	7234 Hz
$h = 0.20$ mm	377.4 Hz	2606 Hz	2777 Hz	8035 Hz

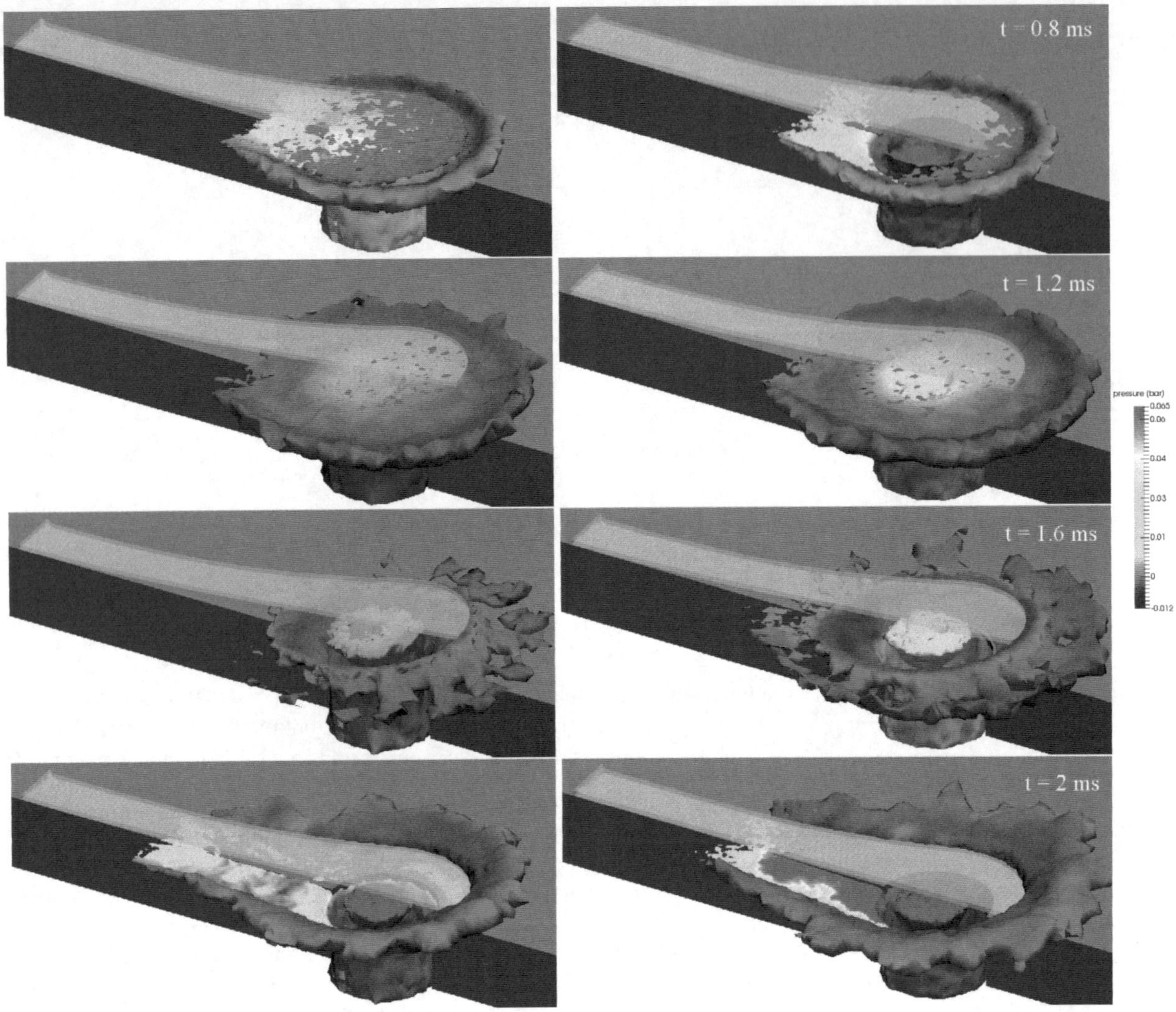

Figure 4: Fluid flow during the first valve oscillation for the 0.18 mm-thick valve (left) and for the 0.20 mm-thick valve (right). Contour surfaces of velocity at $|\vec{u}| = 5$ m/s.

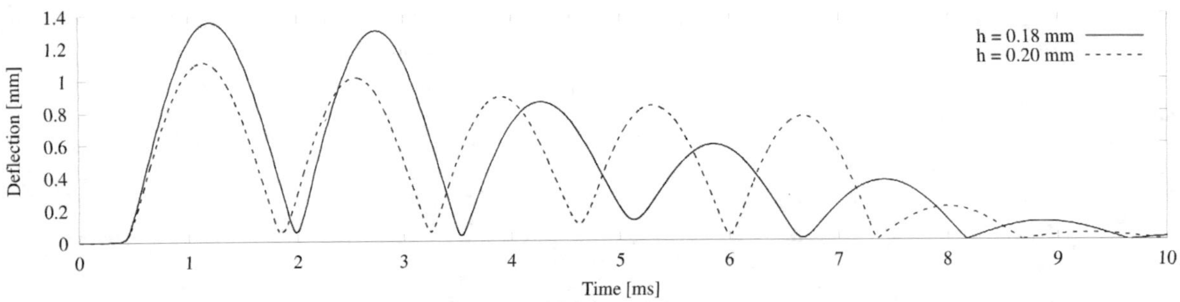

Figure 5: Valve lift.

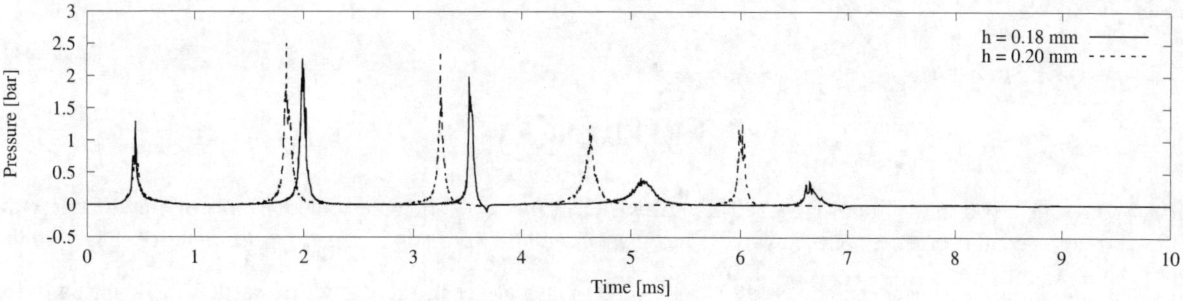

Figure 6: Pressure drop measured on the channel inlet orifice.

Finally, the flow behavior during the first opening oscillation is depicted in Figure 4. Just prior to having a wide outflow aperture (at t = 0.8 ms), a considerable increase of velocity is appreciated between the valve and the seat. Therefore, large velocity gradients and high vorticity appears in this area. Later, the velocity tends to slow down along with the feeding velocity (Figure 2) until the valve moves towards the seat again narrowing the passage.

5. CONCLUSIONS

A semi-implicit FSI coupling scheme, that only couples the pressure stress term of the fluid implicitly to the structure, is used to solve the fluid flow through a reed valve. The geometric arrangement of the different elements is treated with a moving mesh technique, for the flexible valve, and with the immerse boundary method, for the rigid seat. The flow is studied by means of a three-dimensional CFD solver with LES model whereas the valve response is evaluated according to the plate theory and the assumed modes method. The proposed methodology shows a good numerical performance and is proven to be appropriate for FSI applications as this one. All together is used to carry out a thickness sensitivity analysis considering a generic reed valve and a particular suction mass flow configuration. The results demonstrate that this parameter is a crucial aspect for the design of reed valves since it has a big influence in the valve and fluid dynamics during the suction process. In the particular case analyzed, the thinner plate of 0.18 mm seems to be better in terms of pressure drop through the valve, though it is clear that an even less-rigid valve would be more suitable for the suction conditions considered.

NOMENCLATURE

C_r	coefficient of restitution
$\vec{d}$	displacement vector of the structure
E	Young's modulus
h	valve thickness
p	pressure
$\vec{u}$	velocity vector
t	time
μ	dynamic viscosity
$\varphi_m(x,y)$	mth normal mode shape function
ν	Poisson's ratio
ξ	damping ratio
ρ	density
$\vec{\sigma}_\Gamma$	fluid load on the structure
ω_m	mth normal natural frequency

Subscript

f	fluid
s	structure

REFERENCES

Barbi, F., Gasche, J. L., and Villar, M. M. (2012). An efficient immersed boundary method for solving the unsteady flow through actual geometries of reed valves. In *International Compressor Engineering Conference*. Paper 2088.

Causin, P., Gerbeau, J. F., and Nobile, F. (2005). Added-mass effect in the design of partitioned algorithms for fluid–structure problems. *Computer Methods in Applied Mechanics and Engineering*, 194(42–44):4506 -- 4527.

Estruch, O., Lehmkuhl, O., Borrell, R., Pérez-Segarra, C. D., and Oliva, A. (2013). A parallel radial basis function interpolation method for unstructured dynamic meshes. *Computers & Fluids*, 80:44 -- 54.

Estruch, O., Lehmkuhl, O., Rigola, J., and Pérez-Segarra, C. D. (2014). Fluid-structure interaction of a reed type valve subjected to piston displacement. In *International Compressor Engineering Conference*. Paper 2380.

Gasche, J. L., Rodrigues, T. T., and Militzer, J. (2010). Flow simulation through moving hermetic compressor valves using the immersed boundary method. In *International Compressor Engineering Conference*. Paper 1999.

Habing, R. A. and Peters, M. C. A. M. (2006). An experimental method for validating compressor valve vibration theory. *Journal of Fluids and Structures*, 22(5):683 -- 697.

Kim, J., Wang, S., Park, S., Ryu, K., and La, J. (2006). Valve dynamic analysis of a hermetic reciprocating compressor. In *International Compressor Engineering Conference*. Paper 1805.

Lehmkuhl, O., Perez-Segarra, C. D., Borrell, R., Soria, M., and Oliva, A. (2009). Termofluids: A new parallel unstructured cfd code for the simulation of turbulent industrial problems on low cost pc cluster. In *Parallel Computational Fluid Dynamics 2007*, volume 67, pages 275 -- 282.

Leonard, S. M. (1996). Increase reliability of reciprocating hydrogen compressors. *Hydrocarbon Processing*, 75(1):67 -- 74.

Lopez, J. (2016). *Parallel object-oriented algorithms for simulation of multiphysics. Application to thermal systems.* PhD thesis, Universitat Politècnica de Catalunya.

Matos, F. F. S., Prata, A. T., and Deschamps, C. J. (2000). A numerical methodology for the analysis of valve dynamics. In *International Compressor Engineering Conference*. Paper 1411.

Matos, F. F. S., Prata, A. T., and Deschamps, C. J. (2002). Numerical simulation of the dynamics of reed type valve. In *International Compressor Engineering Conference*. Paper 1563.

Mayer, J., Bjerre, P., and Brune, F. (2014). A comparative study of different numerical models for flapper valve motion. In *International Compressor Engineering Conference*. Paper 2309.

Naseri, A., Lehmkuhl, O., Gonzalez, I., and Oliva, A. (2016). Partitioned semi-implicit methods for simulation of biomechanical fluid-structure interaction problems. In *7th European Thermal-Sciences Conference*.

Prater, G. and Hnat, W. P. (2003). Optical measurement of discharge valve modal parameters for a rolling piston refrigeration compressor. *Measurement*, 33(1):75 -- 84.

Ribas, F. A., Deschamps, C. J., Fagotti, F., Morriesen, A., and Dutra, T. (2008). Thermal analysis of reciprocating compressors - a critical review. In *International Compressor Engineering Conference*. Paper 1907.

Sheu, H. C. and Hu, Y. Z. R. (2000). Nonlinear vibration analysis of reed valves. In *International Compressor Engineering Conference*. Paper 1419.

Smith, R. (2011). A pde-based mesh update method for moving and deforming high reynolds number meshes. In *49th AIAA Aerospace Sciences Meeting*, Orlando. AIAA 2011-472.

Soedel, W. (1992). *Mechanics, Simulation and Design of Compressor Valves, Gas Passages and Pulsation Mufflers*. Ray W. Herrick Laboratories, School of Mechanical Engineering, Purdue University.

Stein, K. and Tezduyar, T. (2002). Advanced mesh update techniques for problems involving large displacements. In *Proceedings of the Fifth World Congress on Computational Mechanics*, Vienna, Austria.

Woo, S., O'Neal, D. L., and Pecht, M. (2010). Reliability design of a reciprocating compressor suction reed valve in a common refrigerator subjected to repetitive pressure loads. *Engineering Failure Analysis*, 17(4):979 -- 991.

ACKNOWLEDGEMENT

This work has been financially supported by the Ministerio de Educación, Cultura y Deporte, Spain (FPU13 fellowships).

Transient Experimental and 3D-FSI Investigation of Flapper Valve Dynamics for Refrigerant Compressors

Nicholas C. LEMKE[1]*, Michael KÖNIG[2], Jakob HENNIG[2], Sven FÖRSTERLING[1], Jürgen KÖHLER[1]

[1] Institute of Thermodynamics, Technical University of Braunschweig,
38106 Braunschweig, Niedersachsen, Germany
E-mail: n.lemke@tu-bs.de, j.koehler@tu-bs.de

[2] Volkswagen AG, Component Development,
38231 Salzgitter, Niedersachsen, Germany
E-mail: michael.koenig3@volkswagen.de, jakob.hennig@volkswagen.de

* Corresponding Author

ABSTRACT

The design of the valves for refrigerant compressors has a distinct influence on the efficiency of the refrigeration cycle. In order to predict the valve behavior for compressor optimization, flow characteristics and valve system dynamics are key factors. State of the art valve model approaches have to date ignored the retainer deflection and the influence of contact effects on fluid-structure interaction behavior. With this fact in mind, experimental valve lift investigations by means of laser vibrometry measurements were performed. In order to determine the dynamic flapper valve characteristics, oscillation frequency analyses for free oscillation of the flapper valve were carried out and compared to a FEA modal analysis.

Additionally this paper presents an extended three-dimensional numerical discharge valve model containing a fully-coupled fluid-structure interaction (FSI) approach. The two-way mechanical coupling is carried out by a commercial CFD code combined with a commercial FEM package. The numerical results of a 1D and the 3D-FSI model are validated using the presented experimental transient results. The numerical 3D-FSI results show good agreement with the experimental results with respect to the obtained oscillation frequencies. The 1D model allows error estimation of the mass flow considering retainer deflection. Neglecting the retainer deflection yields a maximum error of four percent for experimental boundary conditions.

1. INTRODUCTION

Pressure actuated reed valves (flapper valves) are widely used in the suction and discharge part of refrigerant compressors. Self-acting reed valves basically respond to pressure differences across the valve. Due to in-flow and outflow characteristics, geometrical and momentum effects reed valves cause 'valve flutter' and over-compression losses. Valve flutter describes the multiple opening and closing process during suction and discharge process through reed valves (Dhar *et al.*, 2014, Min *et al.*, 2014). The efficiency of reciprocating compressors is significantly influenced by the valve performance. General replacement of synthetic refrigerants by natural refrigerants considering high system absolute pressures – especially CO_2 (R744) – requires increased performance and reliability improvement of reed valves for both suction and discharge port. For that reason valve dynamics need to be deeply studied for various operating conditions. The present study employs experimental as well as 1D and 3D-FSI modeling for oil-free valve dynamics investigations. The study is particularly focusing on the extension of reed valve dynamics by retainer deflection and contact effects. The schematics of a flapper suction valve and flapper discharge valve with retainer can be seen in Figure 1. The flapper valve consists of a valve plate comprising the bore for refrigerant suction and discharge. The reed is excited by external pressure load. During the suction and discharge process the reed oscillates between the retainer and the valve plate. The discharge reed is clamped between the valve plate and the retainer which restricts to the maximum valve lift x_{max}. For positive pressure differences the refrigerant flows from the bore to the valve, as denoted by the direction of the mass flow $\dot{m}$ (cf. Figure 1). Pressure drop through the valve especially depends on geometrical parameters such as maximum reed lift and bore diameter. Both

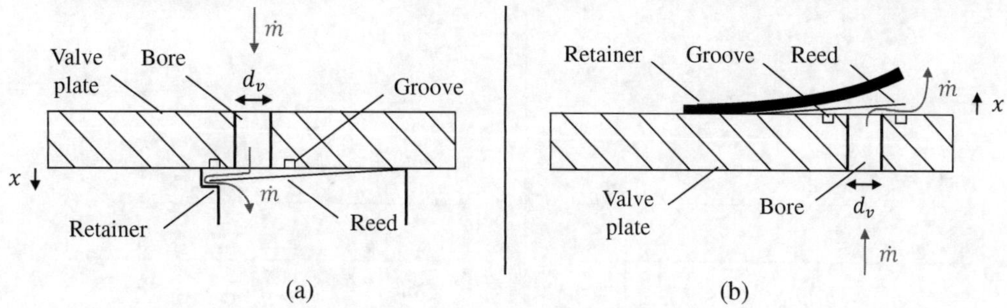

Figure 1: Schematic of flapper valve, (a) suction part, (b) discharge part

parameters significantly limit the flow area for the refrigerant in- and outflow between valve plate and reed. Maximum mechanical reed lifts of 1.05 mm for the suction valve and 0.388 mm and 0.642 mm for the discharge valve were allowed (measurement data at the axis of the bore). Both suction and discharge bore diameter have not been varied. Suction and discharge reeds with different thickness values (0.2 mm, 0.25 mm and 0.3 mm) have been investigated. A coated reed of 0.2 mm is additionally investigated. The coating is applied at both sides of the reed.

2. EXPERIMENTAL SETUP

For experimental investigation of a suction and discharge valve an experimental device with a reproduced reed valve from a refrigerant compressor was designed, operating in dry, oil-free conditions. Figure 2 shows the schematic of the test rig.

The test rig comprises a 300 bar storage nitrogen bottle. The decompression pressure is controlled by a manually actuated decompression valve. The valve pressure is set electronically by a LabVIEW-controlled solenoid valve. The outflow is detected by a Coriolis mass flow sensor. The decompression pressure (pressure transmitter) and the valve pressure (pressure transducer) as well as the valve temperature (thermocouple) are synchronously recorded. Constant outflow or single short pressure surges generated by the solenoid valve represent two different operating conditions which can be investigated at flapper valve test rig. The valve outflow is drained to the environment. For the reduction of valve unit vibrations a massive test housing was mounted around the valve. The valve lift is detected by a laser vibrometer orthogonally aligned on the reed for optimum signal quality. Figure 3 shows a schematic of the suction and discharge reed valve comprising a retainer bore for laser vibrometric lift measurements.

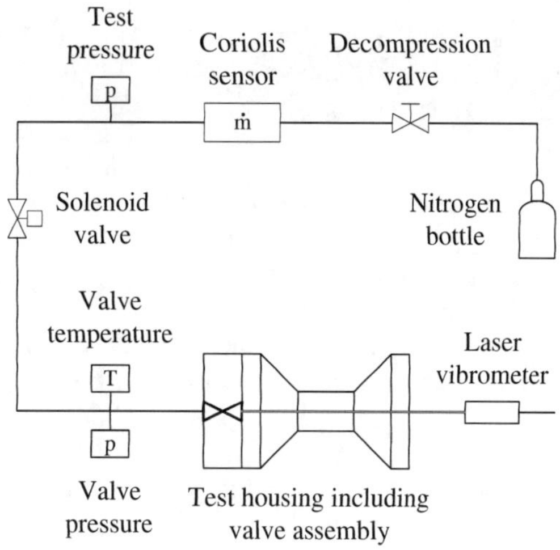

Figure 2: Schematic of test rig for dynamic investigation of flapper valves

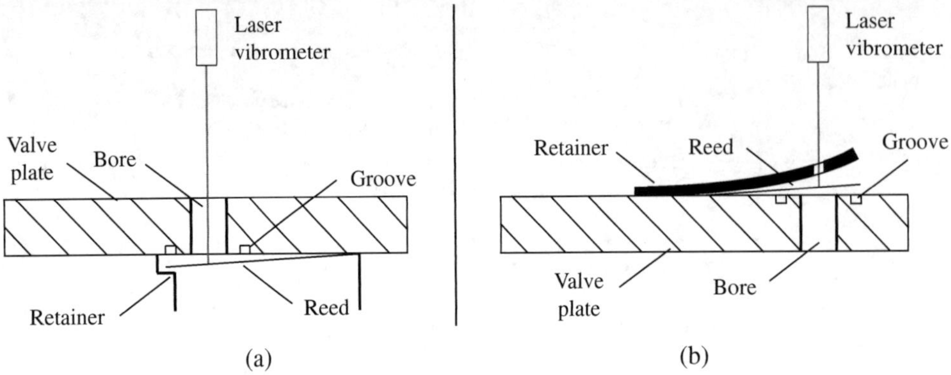

Figure 3: Schematic of reed valve with optical lift measurement, (a) suction valve, (b) discharge valve

3. NUMERICAL MODELING

3.1 1D valve dynamics and mass flow

The reduced model for a reed valve considers inertia, damping and stiffness effects on the one side and the pressure difference Δp as driving force on the other side of the force balance acting on the reed. The valve lift dynamics are described by a spring-mass system in the following general form as a one degree of freedom model:

$$F_{\mathrm{p}}(t) = m\ddot{x}_{\mathrm{v}} + b\dot{x}_{\mathrm{v}} + cx_{\mathrm{v}} = A_{\mathrm{eff}} \cdot \Delta p. \tag{1}$$

Where, m is the effective substitute valve mass, b is the damping factor and c is the substitute stiffness. The effective force area is considered as:

$$A_{\mathrm{eff}} = A_{\mathrm{geo}} \cdot c_{\mathrm{pv}}. \tag{2}$$

A_{geo} is the geometrical force area (equal to the bore area) corrected by an empirical force coefficient c_{pv}. The underlying equations and valve parameters are widely summarized by Touber (1976) and extended by Boeswirth (1994) and Baumgart (2010). Within the applied 1D model, the different valve parameters are calculated dependent on the valve geometry and for each discrete value of the valve displacement x_{v}. The resulting explicit non-linear second-order ordinary differential equation which is numerically solved using an implicit Runge-Kutta formula for stiff systems is derived as follows:

$$\frac{\mathrm{d}^2 x_{\mathrm{v}}}{\mathrm{d}t^2} = \frac{1}{m(x_{\mathrm{v}})}\left[c_{\mathrm{pv}}(x_{\mathrm{v}}) \cdot \frac{\pi}{4} \cdot d_{\mathrm{v}}^2 \cdot \Delta p - F_{\mathrm{s}}(x_{\mathrm{v}}) - b(x_{\mathrm{v}})\frac{\mathrm{d}x_{\mathrm{v}}}{\mathrm{d}t}\right]. \tag{3}$$

The normalized values ($\Phi^* = \Phi/\Phi_{\mathrm{max}}$) of the valve parameters m, F_{s}, b and c_{pv} of the 1D valve model depending on the normalized valve lift ($x^* = x/x_{\mathrm{max}}$) are given in Figure 4 for the investigated discharge valve. As soon as the valve interacts with the retainer and begins to align with the retainer shape ($x^* \approx 0.68$), the substitute valve mass m begins varying and the valve spring force F_{s} changes from linear to non-linear behavior. The damping coefficient b has local maximum values at $x^* \approx 0$ and $x^* \approx 1$ due to fluid damping effects at the valve seat and the retainer, whereby the damping effect at the valve seat is smaller. The pressure coefficient c_{pv} slightly increases due to decreasing pressure loss across the valve during the opening process. Since oil-free conditions are investigated, no additional adhesive effect is considered.

The resulting mass flow $\dot{m}_{\mathrm{v}}$ through the valve is calculated using a semi-empiric approach based on the Bernoulli equation for incompressible, frictionless, stationary and horizontal flows, corrected by the flow coefficient α and the expansion coefficient ε:

$$\dot{m}_{\mathrm{v}}(t) = A_{\mathrm{flow}}(x_{\mathrm{v}}) \cdot \alpha(x_{\mathrm{v}}) \cdot \epsilon(p_0, p_1) \cdot \sqrt{2\Delta p \cdot \rho_0}. \tag{4}$$

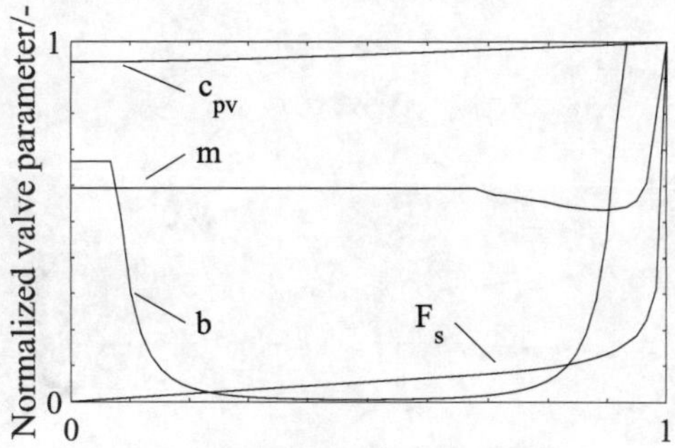

Figure 4: Valve lift dependent parameters of the 1D valve model (discharge valve, 0.25 mm valve thickness, 0.642 mm maximum theoretical valve lift)

Where, A_{flow} is the geometrical flow area, formed by the smallest shell surface between the valve seat and the valve. The retainer deflection is taken into account by a superposition approach. The valve displacement x_{v} calculated by solving Equation (3) is superposed by the retainer deflection x_{ret} leading to the total displacement x_{tot}:

$$x_{\text{tot}} = x_{\text{v}} + x_{\text{ret}}. \quad (5)$$

The retainer deflection x_{ret} is calculated assuming a linear-elastic material behavior, considering the pressure force F_{p} acting on the valve reed (cf. Equations (1) and (3)). The substitute stiffness of the retainer is determined experimentally in the present study.

$$x_{\text{ret}} = \frac{F_{\text{p}}}{k_{\text{ret}}} = \frac{c_{\text{pv}}(x_{\text{v}}) \cdot \frac{\pi}{4} \cdot d_{\text{v}}^2 \cdot \Delta p}{k_{\text{ret}}} \quad (6)$$

3.2 Fully-coupled 3D-FSI valve model

For a deeper understanding of the complex valve flow phenomena in consideration of the valve movement and deformation, a fully-coupled three-dimensional fluid-structure interaction (FSI) model for the investigated discharge valve is developed. The results are compared to the 1D results and the test rig measurements. The FSI model comprises the mechanical parts of the test housing (cf. Figure 2) and its connection to the test rig. The two-way mechanical coupling is realized by a commercial CFD code (STAR-CCM+, version 10.06) combined with a commercial FEA package (Abaqus, version 6.14). The pressure and stress data (CFD) and the geometrical displacement (FEA) are interchanged at any discrete time step. The 3D computational CFD grid consists of a fixed background mesh (3.5 M grid cells) including the boundary definitions and an overset mesh (1.9 M grid cells) which is morphed based on the displacement of the valve surface coupled with the FEA solver. The overset grid approach allows for modelling zero-gap configurations, especially if the valve is closed (cf. Figure 5 (a)) or in contact with the retainer. The CFD setup contains compressibility (ideal gas formulation) and turbulence modeling (SST k-omega model). The solution scheme is coupled for pressure and velocity and second-order implicit unsteady in time with a time step of $1.5 \cdot 10^{-5}$ s. In order to improve the numerical stability, the measured pressure data upstream of the valve have been smoothed and set up as inlet pressure boundary condition. Ambient pressure is set at the outlet boundary condition.

The FEA model consists of three parts: The reed, the retainer and the valve seat (cf. Figure 5 (b)), each of them consisting of about 3000 C3D20R-elements. All parts are modeled as solid bodies in order to be able to handle the contact events. Within the FEA model, the retainer is deformable, influencing the the valve lift. The CFD model, however, solely considers the movement of the reed (the retainer is assumed to have no influence on the flow field).

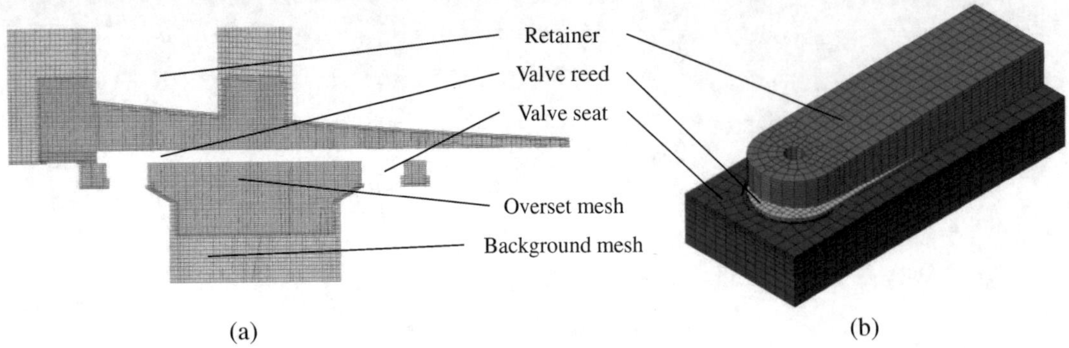

Figure 5: (a) Section of the computational CFD grid and (b) structural FEA model combinded for FSI coupling for the discharge valve

4. RESULTS AND DISCUSSION

4.1 Frequency analyses

In a first step external pressure surge measurements of oscillation frequency for suction and discharge valve were carried out. For determining oscillation frequencies the flapper valves are manually actuated while the valve lift is recorded. Figure 6 shows the absolute lift for the discharge valve over time, starting with maximum deflection amplitude of about 1 mm. Based on the measurements for the manual mechanical excitation, a frequency analysis was carried out.

Table 1 shows the identified oscillation frequency for the suction and discharge valves by varying reed thicknesses. The frequency values increase with higher thicknesses of the reed. For a non-damped mass-spring system increased reed stiffness leads to higher frequencies of the reed. Particularly the comparison of 0.20 mm and 0.20 (coated) for the suction valve and discharge valve, respectively, indicates a low effect of the coating on the natural frequency, compared to the influence of the effective metal valve thickness. This is confirmed by similar oscillation frequencies of the coated and non-coated 0.20 mm reeds although the coated valve has a ten percent higher absolute mass. A FEA modal analysis for the discharge valve with a thickness of 0.25 mm (cf. Figure 5 (b)) for free oscillation conditions yielded a natural frequency of 400 Hz for the first mode, which is about nine percent higher than the measured oscillation frequency (366 Hz). Although the FEA modal analysis does not consider physical fluid friction and damping due to ambient fluid the experimental and simulation results for the free valve oscillation fit well. The FEA modal analysis thus can be applied to predict natural frequency for the reed (cf. Section 4.2 (i)).

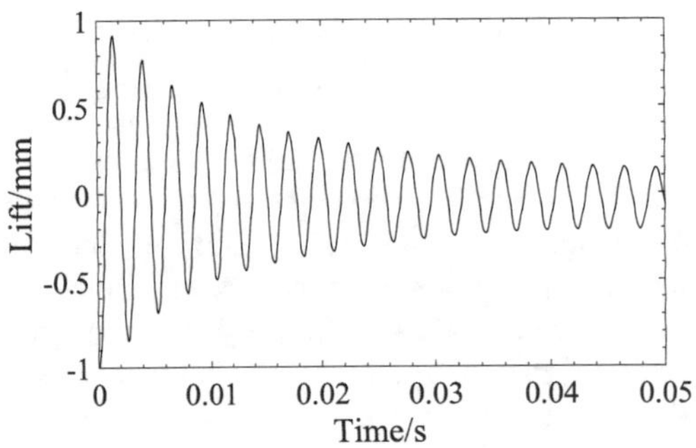

Figure 6: Free valve oscillation due to manual mechanical excitation (discharge reed valve thickness 0.25 mm)

Table 1: Mean oscillation frequency of reproduced suction and discharge valve

Reed thickness/mm	**Oscillation frequency/Hz**	
	Suction valve	Discharge valve
0.20	538	292
0.20 (coated)	557	274
0.25	695	366
0.30	765	449

4.2 Lift measurements

For further studies on valve dynamics during opening and closing events, lift measurements for short pressure surges for approximately 0.1 s were carried out. Figure 7 (a) shows the discharge reed lift and the retainer lift for a reed thickness of 0.25 mm with a maximum theoretical reed lift of 0.642 mm at the valve bore axis determined by the retainer. The reed lift and the retainer lift were recorded for consecutive measurements. The identified oscillation frequencies for the reed and the retainer are shown in Figure 7 (b) for sections (i), (ii) and (v) depicted in Figure 7 (a). A test pressure of 60 bar was applied. Due to throttle effects of the solenoid and the decompression valve (cf. Figure 2) the maximum absolute pressure upstream of the valve is significantly lower than the test pressure. The actual pressure difference between upstream pressure of the reed valve and ambient pressure is 7.6, 3.4, 1.4 and 0.47, respectively, for 120, 60, 30 and 10 bar test pressure. The evaluation of the following experimental results focusses on the discharge reed valve.

When the pressure surge upstream of the valve is applied at 0.006 s the opening process of the valve starts. The opening and closing process can be subdivided into six sections. Figure 7 illustrates the valve behavior for any section.

(i) The opening process of the valve starts as soon as the pressure signal is increasing. The reed valve oscillates between retainer and valve plate with decreasing oscillation amplitude. The reed is bent around the retainer resulting in smaller bending lengths for the reed. A retroactive effect between the reed and the retainer can be observed resulting in similar oscillation frequencies of the reed and the retainer. The oscillation frequencies of the reed and the retainer vary from about 800 Hz to 2000 Hz. The oscillation frequency for the first opening interval is close to the natural frequency from an FEA modal analysis of 829 Hz (first mode). The FEA analysis is carried out for the clamped reed. Since the retainer reduces the oscillation length of the reed (cf. Figure 5 (b)), the natural frequency for the mounted valve is higher than the free oscillation frequency of the total reed length (cf. Section 4.1).

(ii) The lift as well as the oscillation frequencies of the reed and the retainer are steadily increasing. The reed lift results in an interaction between reed, retainer and pressure with increasing amplitude of the lift, compatible with the behavior of a pressure forced oscillation of a coupled spring mass system. The obtained frequencies are determined by the oscillation length of the reed resulting from the valve lift.

(iii) The oscillation amplitude of the coupled system is approaching zero. A minimum bending length of the reed is obtained. At the end of section (iii) a maximum lift height is obtained when valve pressure gains a maximum value indicating the retainer deflection. No interaction between retainer and pressure is observed.

(iv) The valve pressure decreases, whereas the reed remains attached to the retainer. In consequence the reed valve and the retainer lift decrease while oscillation amplitude is low. At the end of section (iv) the reed begins again to interact with the retainer resulting in similar oscillation frequencies.

(v) The reed continuously detaches from the retainer. The reed oscillates with increasing amplitude while oscillation frequency is decreasing to values comparable to the end of section (i). Again, an interaction between reed, retainer and pressure is observed, however the retroactive effect is not as significant as in section (ii).

(vi) The reed completely loses from the retainer back to its initial position on the valve seat. The valve pressure drops to initial ambient pressure of around 1 bar.

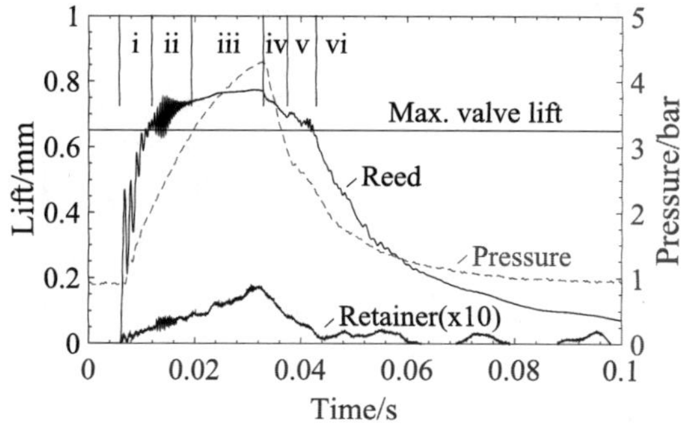

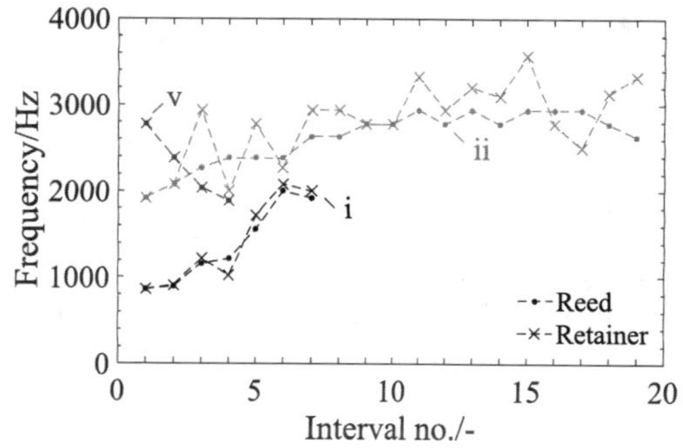

Figure 7: (a) Discharge valve lift and retainer lift (factor 10) with absolute pressure plot, (b) Frequency analysis for reed and retainer (valve thickness 0.25 mm, 0.642 mm retainer, 60 bar test pressure)

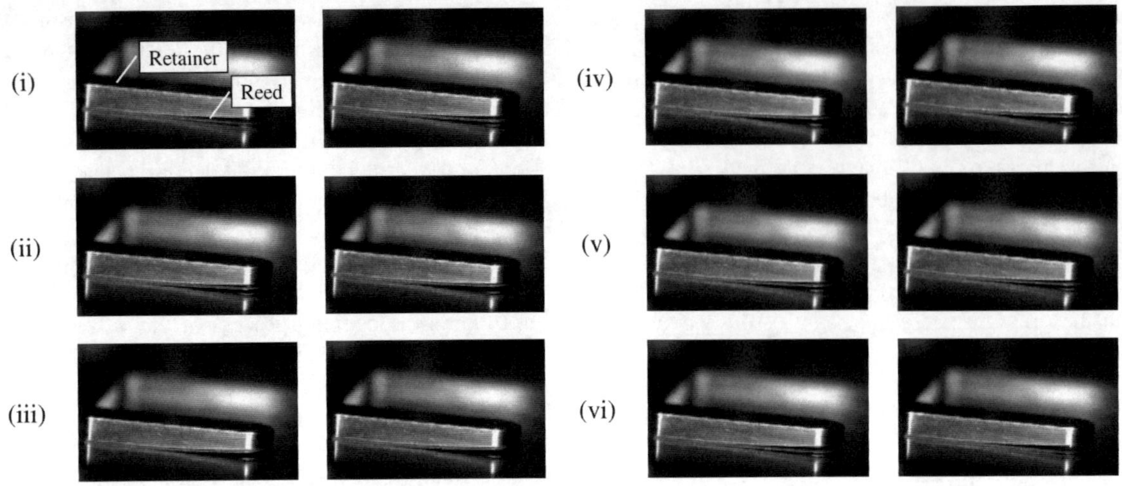

Figure 8: High speed video analysis for opening and closing process of discharge valve; left: beginning of section; right: end of section (valve thickness 0.25 mm, 0.642 mm retainer, 60 bar test pressure)

For the opening process in sections (i) and (ii) the oscillation frequencies of the reed and the retainer could clearly be determined. The oscillation frequencies of the reed and the retainer show similar values for the entire opening process due to interaction, cf. Figure 7 (b). Sections (iii) and (vi) do not show significant oscillations for a frequency analysis. For the closing process in section (v) the oscillation frequencies could again clearly be determined while oscillation frequencies are equal resulting from the reed and the retainer interaction. Figure 8 illustrates the reed lift for each section. The general opening and closing characteristics for varied discharge reed thicknesses, retainer geometries and varied suction reed thicknesses are similar. Due to possible geometrical laser vibrometer misalignment the measured valve lift values exceed the theoretical maximum valve lift, cf. Figure 7 (a). The maximum valve lift uncertainty is identified to be 20 % and can be adjusted by a scaling factor with the help of the numerical results, cf. Section 4.3. The general reproducibility error in lift measurement is identified to be one percent.

Figure 9 shows selected reed valve lift and pressure plots for different test pressure acting on discharge reed valve with 0.25 mm reed thickness. The valve pressure is increasing along with increasing test pressure. The valve lift measurements show longer open periods for increasing test pressure. The maximum valve lift is obtained for a maximum test pressure of 120 bar. The test pressures of 10 and 30 bar show continuous oscillation for the opening process. Sections (iii) and (iv) are not observed for 10 and 30 bar test pressure because the reed is not in complete contact with the retainer. For 10 bar even section (ii) is not present as a result of small lift amplitude. Resonance oscillation for the coupled reed and retainer spring mass system is distinct for 60 bar testing pressure. Oscillation amplitudes of more than 0.1 mm can be observed where no contact between reed and retainer is achieved. A higher valve lift is measured for 120 bar test pressure compared to 60 bar test pressure although both measurements show contact

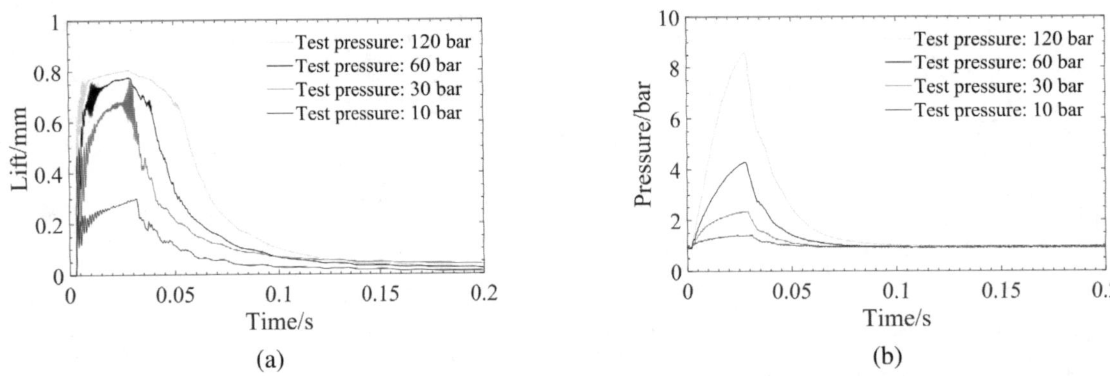

Figure 9: (a) Discharge valve lift and (b) pressure plot for different test pressures of 10, 30, 60 and 120 bar (valve thickness 0.25 mm, 0.642 mm retainer)

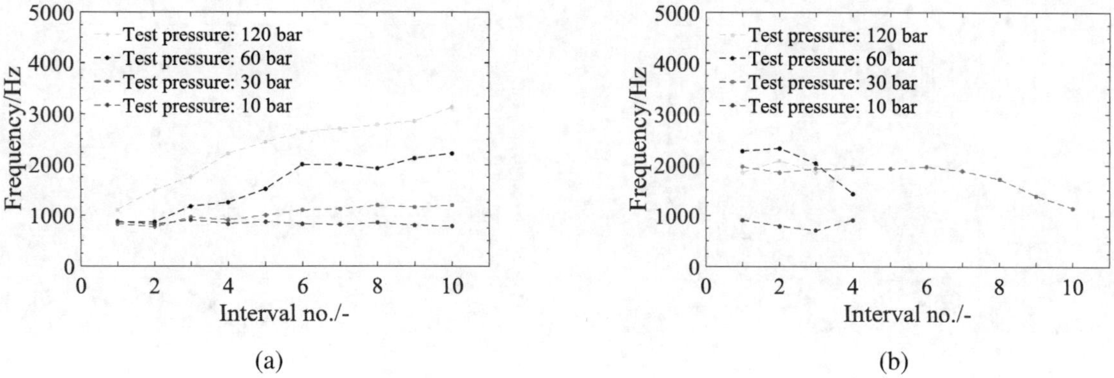

Figure 10: Frequency analysis of discharge valve lift: (a) opening process (cf. Figure 7, section (i)) and (b) closing process (cf. Figure 7, section (v)) (valve thickness 0.25 mm, 0.642 mm retainer)

behavior between valve and retainer. This demonstrates the effect of the retainer deflection on the total valve lift. Due to a higher density of nitrogen at higher test pressures, a higher mass of nitrogen needs to be expanded through the valve, resulting in a longer open period.

Figure 10 shows a frequency analyses for the discharge valve lift during opening and closing processes for different test pressures as far as oscillation lift maxima and minima can be determined. The frequency analyses of the valve lift measurements for any test pressure show similar frequencies for the first interval when opening (about 700 to 1100 Hz) and closing (about 1700 to 2300 Hz) for investigated test pressures from 10 to 120 bar. Further on the frequencies for the opening process show a similar progression for all test pressures as the frequency increases over the interval number up to 3000 Hz for 120 bar test pressure. For higher test pressures the frequency seems to be less damped. In contrast to the opening process the frequencies for the closing process decrease over the interval number from a frequency of 2000 Hz to a frequency of 1000 Hz for 30 bar test pressure; however the number of intervals is very low for 10, 60 and 120 bar test pressure. The measured frequencies due to pressure surges are significantly higher than measured free oscillation and natural frequencies for the discharge valve (cf. Table 1), caused by the reduced bending length of the clamped reed. In case of 10 bar test pressure, the reed shows lower frequencies, compared to higher test pressures, since it does not fully align with the retainer, cf. Figure 9 (a).

The frequency deviation between pressure surges and free oscillation as well as natural frequency can be explained by two physical facts: The first effect is the interaction between the reed and the retainer which results in a coupled spring-mass system with resonance phenomena and changed oscillation frequencies. The second effect is the interaction between the fluid and the reed which also leads to changed resonances and oscillation frequencies. Both effects can deeper be investigated by employing numerical models, as presented in the following.

4.3 Numerical results

The analysis of the numerical results covers the evaluation of the valve lift over time calculated by the above described 1D/3D-FSI models, see Figure 11. In order to compare them to the experimental results, the experimental lift data are adjusted by a constant scaling factor of 0.83, compensating the above mentioned measuring inaccuracy. The maximum lift is chosen as scaling reference value. The valve lift measurement for 60 bar (cf. Figure 7) is chosen as validation example comprising a representative valve behavior. For the 3D-FSI model, the valve lift in the bore axis has been analyzed. Due to high computational effort the 3D-FSI simulation is stopped after passing the maximum valve lift (at 0.03 s) and analyzed with focus on the valve opening process, as shown in Figure 11 (a). The overall shape of the valve lift fits well for both the 1D and the fully-coupled 3D-FSI model, compared to the scaled experimental data. The 1D model is unable to describe the identified oscillations in section (ii), cf. Figure 7. The 3D-FSI model, however, reproduces a characteristic oscillation behavior in the section of interest, even though the oscillation starts earlier and the calculated amplitude values are lower than the experimental values. The 3D-FSI model shows approximately ten characteristic oscillation intervals. The first oscillation shows a frequency of 1750 Hz, which is about twice the natural frequency of the clamped reed (first mode of FEA modal analysis: 829 Hz), whereas the tenth oscillation frequency is about 2780 Hz. This trend shows a good agreement to the experimentally determined frequencies at section (ii), cf. Figure 7 (b). Once the valve passed the section of distinct interaction of the reed and the retainer both models show good agreement with respect to the total displacement. The retainer deflection is also corresponding to the experimental results.

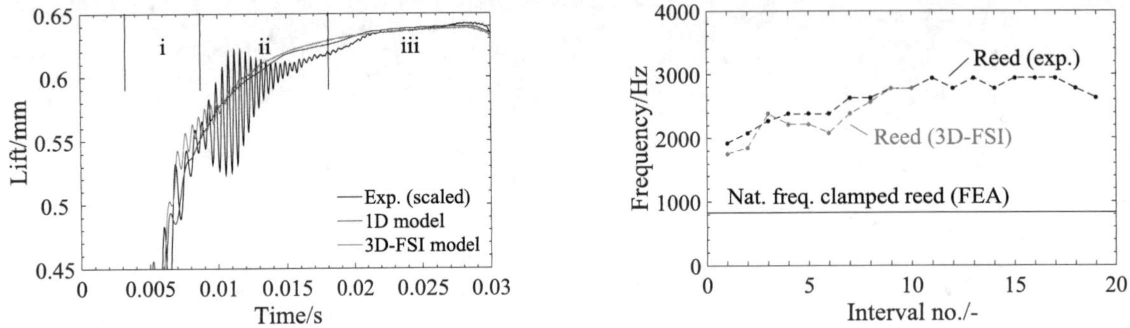

(a) (b)

Figure 11: (a) Calculated valve lift curves of 1D and 3D-FSI model at the opening sections (i) to (iii) and (b) frequency analysis for 3D-FSI lift data at section (ii) compared to experimental data, cf. Figure 7 (valve thickness 0.25 mm, 0.642 mm retainer, 60 bar test pressure)

The evaluation of the influence of the retainer deflection on the total valve lift and the resulting mass flow is carried out based on the 1D valve model for a test pressure of 120 bar (cf. Figure 9 (b)). As shown in Figure 12, the valve displacement x_v does not exceed the maximum theoretical valve lift of 0,642 mm, if the retainer is assumed to be infinite stiff. Taking into account the experimentally obtained retainer stiffness ($k_{ret} = 3.2 \cdot 10^5$ N/m), the retainer deflection x_{ret} achieves a maximum value of 0.024 mm, leading to a maximum total displacement of 0.663 mm. Thus, neglecting the retainer deflection results in a mass flow estimation error, see Equation (4). As illustrated in Figure 12, the maximum difference $\Delta\dot{m}_{v,max}$ between the normalized total mass flow in consideration of the retainer deflection $\dot{m}^*_{tot}$ and the normalized mass flow for a stiff retainer $\dot{m}^*_{stiffRet}$ achieves a value of almost four percent. The integration of the mass flow yields the cumulative mass $m_{v,0.1s}$ which passed the valve due to one pressure surge. For the maximum test pressure of 120 bar, the error $\Delta m_{v,0.1s}$ when neglecting the retainer deflection sums up to 2.2 %-, compared to the cumulative mass considering the retainer deflection. The calculated values for maximum retainer deflection $x_{ret,max}$, maximum mass flow difference $\Delta\dot{m}_{v,max}$ and cumulative mass difference $\Delta m_{v,0.1s}$ are summarized in Table 2. It is evident that the error when neglecting the retainer deflection increases by increasing pressure difference. For lower pressure differences the retainer difference can be neglected.

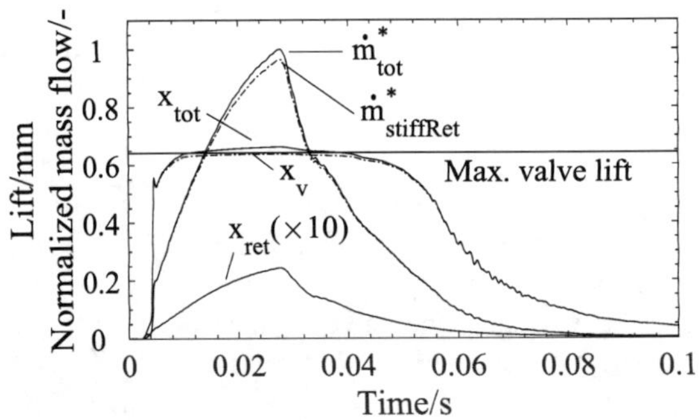

Figure 12: Valve lift and normalized mass flow calculated by the numerical 1D model with and without superposition of retainer deflection (valve thickness 0.25 mm, 0.642 mm retainer, test pressure 120 bar)

Table 2: Error estimation for mass/mass flow calculation at different pressure differences when neglecting the retainer deflection based on the 1D model

Test pressure/bar	**10**	**30**	**60**	**120**
$\Delta p_{v,max}$/bar	0.47	1.4	3.4	7.6
$x_{ret,max}$/mm	0.0014	0.0046	0.011	0.024
$\Delta\dot{m}_{v,max}$/%	0.65	0.67	1.4	3.6
$\Delta m_{v,0.1s}$/%	0.68	0.63	0.98	2.2

6. CONCLUSION

This study focuses on the dynamics of flapper valves for refrigerant compressors by means of laser vibrometry valve lift measurements for various operating conditions. A frequency analysis is carried out for suction and discharge valves with different thicknesses, which allows estimating the free oscillation behavior of a single valve. Further on, lift measurements for the discharge valve-retainer assembly excited by different pressure surges give a deeper understanding of the transient behavior and show distinct interaction and resonance effects between valve and retainer. Analyzing the oscillations of the coupled system shows varying frequencies which cover a range frequency in the order of 1000 Hz to 3000 Hz. Any experimental investigations have been carried out for dry conditions.
Additionally a 1D and a 3D-FSI valve model are presented and evaluated with focus on the valve lift. Compared to the experimental results, the 1D model reproduces well the generic valve lift behavior and mass flow characteristics. However it is incapable of describing the distinct oscillations of the system. The 3D-FSI model shows good agreement to the experimentally obtained oscillation frequencies of the coupled system, especially when valve and retainer begin to interact. To obtain deeper knowledge on the complexity of valve flows, further experimental and numerical investigations considering multiphysics phenomena such as adhesive effects should be carried out.

NOMENCLATURE

A	area	(m²)	ρ	density	(kg/m³)
b	damping coefficient	(Ns/m)	Φ^*	normalized physical quantity	(–)
c	stiffness	(N/m)			
CFD	computational fluid dynamics				
c_{pv}	pressure coefficient	(–)	**Subscript**		
d	(bore) diameter	(m)	0	upstream value	
F	force	(N)	1	downstream value	
FEA	finite element analysis		eff	effective	
FSI	fluid-structure interaction		geo	geometrical	
k	stiffness	(N/m)	max	maximum value	
m	mass	(kg)	p	pressure	
$\dot{m}$	mass flow	(kg/s)	ret	retainer	
p	pressure	(Pa)	s	spring force	
t	time	(s)	tot	total	
x	lift, displacement	(m)	v	valve	

REFERENCES

Baumgart, R. (2010). *Reduzierung des Kraftstoffverbrauches durch Optimierung von Pkw-Klimaanlagen. Doctoral thesis.* Chemnitz, Germany: Chemnitz University of Technology, Department of Mechanical Engineering.

Boeswirth, L. (1994). *Stroemung und Ventilplattenbewegung in Kolbenverdichterventilen*. Vienna: Self-published.

Dhar, S., Tamma, B., Bhakta, A., Krishna, M. (2014). An approach towards reed valve geometry design. International Compressor Engineering Conference, Purdue University, Indiana, USA, Paper 1144.

Min, B., Noh, K., Yang, J., Choi, G., Kim D. (2014). Prediction of Refrigerant Leakage for Discharge Valve System in a Rolling Piston Compressor. International Compressor Engineering Conference, Purdue University, Indiana, USA, Paper 1464.

Touber, S. (1976): *A Contribution to the Improvement of Compressor Valve Design. Doctoral Thesis.* Delft, Netherlands: Delft University of Technology, Department of Mechanical Engineering.

ACKNOWLEDGEMENT

The authors would like to thank our colleagues at the Volkswagen R&D department, B.Sc. Norman Welz, B.Sc. Maximilian Müller, Dr. Andreas Gitt-Gehrke and Dipl.-Wirtsch.-Ing. Mathias Möller, for their contribution and support to this paper.

An Integrated Model for an Oil Free Carbon Dioxide Compressor Using Sanderson-Rocker Arm Motion (S-RAM) Mechanism

Bin YANG[1*], Orkan KURTULUS[2], and Eckhard A. GROLL[3]

Purdue University, School of Mechanical Engineering,
Ray W. Herrick Laboratories, West Lafayette, IN, USA

[1]yang62@purdue.edu
[2]orkan@purdue.edu
[3]groll@purdue.edu

* Corresponding Author

ABSTRACT

The multi-piston axial reciprocating compressor using the Sanderson-Rocker Arm Motion (S-RAM) mechanism is expected to have high volumetric efficiencies due to the application of a new type of seal to prevent the in-cylinder refrigeration gas leaking through the clearance between piston and cylinder wall. The stroke of the compressor can be controlled by adjusting the inclination angle between the connecting shaft and machine driving shaft. This allows the control of the delivered refrigerant mass flow rate to match the capacity requirement in the field.

A comprehensive simulation model to predict the performance of a prototype reciprocating compressor using the S-RAM mechanism has been developed. The natural refrigerant carbon dioxide (CO_2) is used as the working fluid. The comprehensive model is comprised of a kinematics model, compression process model and an overall energy balance model. In the kinematics model, the movement of the piston is given including its displacement, velocity and acceleration. It is found that the moving path of the center of the ball in the ball-socket joint is moving around a corresponding cylinder centerline with a 'figure 8' motion instead of moving along the cylinder centerline. In the compression process model, the system of governing equations is solved, which incorporates a valve sub-model, leakage sub-model and gas pulsation sub-model. The classical 4th order Runge-Kutta method and Broyden's method are employed to solve the non-linear system of equations to find the in-cylinder refrigerant state (temperature, pressure) at each rotational angle of the machine driving shaft. The variations of suction and discharge valve movements with respect to drive shaft rotational angle are also given. The values of the cylinder wall temperature, the actual suction and discharge temperatures in the connecting pipes are required to initiate the solving of the compression process model. These temperatures are solved simultaneously by incorporating the overall energy balance model with the compression process model. A lumped temperature assumption is employed in the overall energy balance model to assume there is no temperature gradient in each compressor component at steady-state.

1. INTRODUCTION

Since Lorentzen and Pettersen presented experimental data on an automotive air conditioning system using CO_2 as the refrigerant in early the 90's last century (1992), tremendous research activities have been carried out on the application of CO_2 as an alternative refrigerant in refrigeration systems. Some of the activities were dedicated to the development of a CO_2 reciprocating compressor. Fagerli (1997) theoretically and experimentally investigated the possibility of the application of small hermetic reciprocating compressors in the CO_2 trans-critical cycle. The prototype single cylinder CO_2 compressor was retrofitted from an existing compressor using R22 as the working fluid. It was shown that the overall isentropic efficiency of the CO_2 compressor was lower than R22 compressor by 9% to 15%, the volumetric efficiency was lower by 5%. Süss and Kruse (1998) proposed that it was possible for a CO_2 compressor to achieve a higher performance by an appropriate design, since the pressure ratio was lower compared to that in a compressor using conventional refrigerant as the working fluid. Nekså et al. (1999) developed a series of semi-hermetic reciprocating compressors using CO_2 as the working fluid whose displacements expanded from 1.7 $m^3 \cdot h^{-1}$ to 10.7 $m^3 \cdot h^{-1}$. These compressors were single- or double-cylinders. It was shown in the experiments that the single-cylinder compressors could reach an overall isentropic efficiency of 0.69 and a volumetric efficiency of 0.77 if used in a water heating heat pump. Bauman and Conzett (2002) developed a non-lubricated four-cylinder prototype reciprocating compressor. Experiments showed an acceptable efficiency. By now, several commercial company mass produced single- and double-cylinder semi-hermetic reciprocating compressors used in trans-critical CO_2 systems (2015).

Meanwhile, simulation models were also developed to reveal the distinct characteristics of a reciprocating compressor using CO_2 as its working fluid. Yang et al. (2013) developed a comprehensive simulation model for a semi-hermetic two-cylinder reciprocating compressor. In this model, both the frictional power losses between piston rings and cylinder wall and that at the bearings were calculated. The simulation results were in a good agreement with the compressor performance measurements. Cavalcante et al. (2008) developed a transient model for a CO_2 automotive swash plate compressor in which the piston moved with a reciprocating motion. The in-cylinder thermodynamic phenomenon, valve dynamics, heat transfer, friction of swash plate and pistons, leakage through pistons were taken into account

The work presented in this paper focuses on a comprehensive simulation model for a novel prototype reciprocating compressor using the Sanderson-Rocker Arm Motion (S-RAM) driving mechanism, which converts the rotation of machine driving shaft into a reciprocating motion of a piston at high efficiency. The S-RAM compressor is expected to be promising in heat pump and cooling applications of 25 to 300 kW capacities. It has competitive volumetric efficiencies and overall isentropic efficiencies compared to other types of compressors used in this range of cooling capacities, such as scroll compressors and screw compressors. The schematic of the S-RAM mechanism can be seen in Figure 1. The machine driving shaft is coupled with an AC motor shaft. One end of the inclined shaft is fixed in a shoe. The other end of the connecting shaft forms conjunction with another shaft through a universal joint. The position of the universal joint is fixed relative to machine driving shaft when the compressor is running. The inclined angle between the connecting shaft and machine driving shaft keeps constant when the compressor is spinning. The wobble plate makes a nutation around the universal joint. There are five ball-socket joints connected the wobble plate and piston assembly. The centers of these balls form a wobble plane which goes through the universal joint. The ball-socket joint converts the rotation of machine driving shaft into a reciprocating motion of pistons.

There are several advantages using the S-RAM mechanism. First, the capacity of compressor can be changed by changing the stroke without varying the motor speed. This can be achieved by changing the inclined angle of the connecting shaft to the machine driving shaft. While the stroke is varying, the position of the top dead center (T.D.C.) relative to the machine itself is fixed, i.e., the clearance volume above the piston keeps constant which is one of the main advantages of this compressor. Second, the piston is strictly moving along the cylinder centerline, i.e., there is no side loads resulted from the contact between piston and cylinder wall, reducing the frictional power loss. Finally, a U-cup seal is employed instead of conventional piston rings to prevent the gas from leaking out of the in-cylinder control volume.

2. SIMULATION MODEL OF THE S-RAM COMPRESSOR

2.1 Geometric and kinematics model

Figure 1 shows a schematic of the driving system of the Sanderson-Rocker Arm Motion mechanism. A coordinate system is built to assist the establishment of the kinematics model. The $X_0Y_0Z_0$ coordinate system is located on the machine itself. The XYZ coordinate system is obtained by rotating $X_0Y_0Z_0$ counterclockwise (seen from up direction) around the Z_0-direction by a degree of θ. Rotating the XYZ coordinate system around the X-direction counterclockwise by a degree of α, forms the X'Y'Z' coordinate system, which is located on the wobble plane. The wobble plane is formed by the moving path of the ball centers. The three coordinate systems have the same origin, which is located at the center of the universal joint.

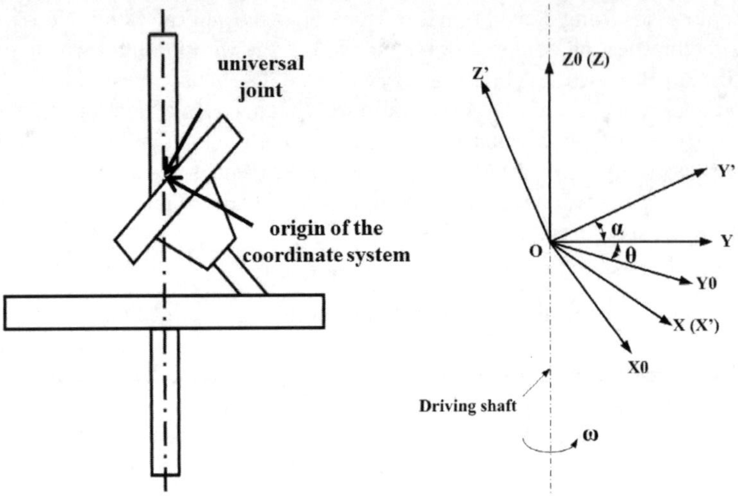

Figure 1: Coordinate system used in kinematics model

Through the transformation of the coordinate systems, the *i*-th ball center position in the XYZ system can be shown as

$$\begin{cases} X = R_w \cos(-\theta' + \theta_i) \\ Y = R_w \sin(-\theta' + \theta_i)\cos\alpha \\ Z = R_w \sin(-\theta' + \theta_i)\sin\alpha \end{cases} \quad (1)$$

Where, R_w is the radius of the wobble plate. θ' is the rotational angle of *i-th* ball center in X'Y'Z' coordinate system. θ_i is the *i*-th cylinder phase angle. α is the wobble plate tilting angle. The position of the *i*-th ball center in $X_0Y_0Z_0$ system is then determined using the following equations,

$$\begin{cases} X_0 = R_w \cos(-\theta' + \theta_i)\cos\theta - R_w \sin(-\theta' + \theta_i)\cos\alpha\sin\theta \\ Y_0 = R_w \cos(-\theta' + \theta_i)\sin\theta + R_w \sin(-\theta' + \theta_i)\cos\alpha\cos\theta \\ Z_0 = R_w \sin(-\theta' + \theta_i)\sin\alpha \end{cases} \quad (2)$$

The piston displacement is calculated using the third equation in Eq. (2). The mid-stroke point is designated as the starting point of the displacement. For convenience, the starting point can be specified to the bottom dead center (B.D.C.) or the top dead center (T.D.C.) by shifting calculated values correspondingly. The direction towards to the T.D.C. is assumed to be positive. The piston velocity and acceleration can be determined by further differentiation of Z_0 equation in Eq. (2).

2.2 Compression process model

2.2.1 Governing equations

The thermal properties of the refrigerant inside the cylinder are assumed to be uniform anywhere in the cylinder. According to the law of continuity, the mass change of refrigerant in the cylinder can be expressed by:

$$\frac{dm_c}{dt} = \frac{dm_{suc}}{dt} + \frac{dm_{li}}{dt} - \frac{dm_{dis}}{dt} - \frac{dm_{lo}}{dt}. \quad (3)$$

Where, m_c refers to the refrigerant mass in the cylinder.

The energy conservation equation can be derived by applying the first law of thermodynamics to the in-cylinder control volume:

$$\frac{dT_c}{dt} = \frac{1}{m_c C_v}[(h_{suc} - h_c)\frac{dm_{suc}}{dt} + (h_{li} - h_c)\frac{dm_{li}}{dt}] + \frac{1}{m_c C_v}\{[-pm_c - T(\frac{\partial p}{\partial T})_v + p]\frac{dv_c}{dt} + \frac{dQ}{dt}\}. \quad (4)$$

The in-cylinder refrigerant pressure in the above quation is determined by temperature T_c and specific volume v_c. The variation of specific volume with respect to time is evaluated by incorporating the kinematics model.

The instantaneous heat transfer between the refrigerant and the cylinder wall is determined using the heat transfer coefficient, proposed by by Todescat et al. (1992). The heat transfer coefficient is defined as follows:

$$h_{heat} = 3 \cdot 0.7 \cdot \frac{\tau_c}{D_p} \left(\frac{\rho_c U_{mean} D_p}{\mu_c}\right)^{0.7} . \tag{5}$$

Then, the heat transfer rate between refrigerant and cylinder wall is calculated:

$$\dot{Q}_{heat} = \frac{dQ}{dt} = h_{heat} A_{heat} (T_c - T_{wall}) . \tag{6}$$

Where, the cylinder wall temperature, T_{wall}, can be assumed to be constant (Adair et al., 1972).

2.2.2 Valve submodel

The mass flow rates through the valves are determined by assuming an isentropic, compressible flow:

$$\dot{m}_{valve} = \frac{dm_{valve}}{dt} = C_{valve} A_{valve} (2\rho_{high} p_{high})^{\frac{1}{2}} \left\{\frac{\kappa}{\kappa - 1}\left[\left(\frac{p_{low}}{p_{high}}\right)^{\frac{2}{\kappa}} - \left(\frac{p_{low}}{p_{high}}\right)^{\frac{\kappa+1}{\kappa}}\right]\right\}^{\frac{1}{2}} . \tag{7}$$

A discharge coefficient C_{valve} of 0.58 is used for the suction valve and a coefficient of 0.6 is used for the discharge valve (Lin and Sun, 2006). A_{valve} denotes the valve opening area for the suction and discharge valves. It can be determined by considering that there are two dominant regions, pressure dominant region and mass flux dominant region. Correspondingly, the motion of the valve plate which is modeled as a one degree-of-freedom problem can be expressed by the following equations (Kim and Groll, 2007),

In pressure dominant region:

$$m_{equivalent} \ddot{x} + k_{spring} x = (p_{high} - p_{low}) A_{vp} + \frac{1}{2} C_D \rho_{port} V_{port}^2 A_{vp} . \tag{8}$$

In mass flux dominant region:

$$m_{equivalent} \ddot{x} + k_{spring} x = \rho_{port} (V_{port} - x) A_{port} + \frac{1}{2} C_D \rho_{port} V_{port}^2 A_{vp} . \tag{9}$$

Where, $m_{equivalent}$ is the equivalent mass of the valve plate spring; k_{spring} is the spring stiffness; C_D is the drag coefficient, which is assumed to be 1.17; and A_{vp} is the area of valve plate.

2.2.3 Leakage submodel

Any gas leaking through the clearance between piston and cylinder wall is modeled as an isentropic, compressible fluid flowing through an orifice. The mass flow rate of the gas leakage is calculated by:

$$\begin{cases} \dot{m}_{gap} = \dfrac{dm_{gap}}{dt} = C_{gap} A_{gap} p_u \left(\dfrac{2}{ZRT_u}\right)^{\frac{1}{2}} \left\{\dfrac{\kappa}{\kappa-1}\left[\left(\dfrac{p_d}{p_u}\right)^{\frac{2}{\kappa}} - \left(\dfrac{p_d}{p_u}\right)^{\frac{\kappa+1}{\kappa}}\right]\right\}^{\frac{1}{2}}, \dfrac{p_d}{p_u} > 0.54 \\ \dot{m}_{gap} = \dfrac{dm_{gap}}{dt} = C_{gap} A_{gap} p_u \left(\dfrac{\kappa}{ZRT_u}\right)^{\frac{1}{2}} \left[\left(\dfrac{2}{\kappa+1}\right)^{\frac{\kappa+1}{\kappa-1}}\right]^{\frac{1}{2}}, \dfrac{p_d}{p_u} \le 0.54, choked \end{cases} . \tag{10}$$

Where the discharge coefficient C_{gap} is assumed to be 0.86 (Furuhama and Tada, 1961) and the compressible factor Z is determined by using REFPROP (Lemmon, 2012).

2.3 Gas pulsation model in the discharge pipes

When using a reciprocating compressor with multiple cylinders, gas pulsation is an inevitable issue. The gas pulsations in the discharge pipeline are closely related to the motion of the discharge valves. Moreover, they also depend on the distributed angle of the multiple cylinders. Elson and Soedel (1974) proposed a mathematical steady state model to investigate the gas pulsation in the discharge plenum and the response of the discharge valve by coupling a steady state solution for the acoustic impedance at the valve with the non-linear valve model. It is not necessary to simultaneously solve a number of equations as the classical method (Brablik, 1972) does if the proposed steady state method is used, making it more flexible to a complicated discharge system. In addition, the classical method makes an assumption that the oscillated back pressure in the piping does not affect the flow through the valve which is not a very good description of the real world. Therefore, the steady state method proposed by Elson and Soedel (1974) is employed.

The anechoic assumption is applied. It is assumed that the acoustic wave in the pipeline is a plane wave. Thus, the acoustic impedance at the valve is given as,

$$Z_{imp} = \frac{p_{pul}}{u_{pul}} = \rho_0 c \,. \qquad (11)$$

Where, p_{pul} is the oscillated the acoustic pressure in the discharge pipeline. U_{pul} is the gas velocity at the valve. P_0 is the refrigerant gas density in the pipeline and c is the speed of sound.

The gas velocity at the valve is determined first from the compression process model which incorporates the dynamic valve model. Using Eq. (11), the acoustic pressure is found. The actual pressure in the discharge pipeline should be the sum of the determined acoustic pressure and nominal discharge pressure. With the updated discharge pressure, the compression process model is solved again which results in a different gas velocity at the valve, then repeat this procedure until all the parameters are converged.

2.4 Overall energy balance model

In the abovementioned compression process model, the cylinder wall temperature is essential to calculate the heat transfer between refrigerant gas and cylinder wall. In addition, the suction gas temperature and discharge gas temperature are required in the valve model to accurately calculate the mass flow rate across the suction and discharge valves. Besides, the discharge temperature is also used to calculate the characteristic impedance of the gas after the discharge valve in the discharge gas pulsation model. However, all of these three temperatures (cylinder wall temperature, suction/discharge gas temperature) are undetermined. Hence, in this section, the overall energy balance of the compressor is incorporated into the compression process model to find these three temperatures and solve the compression process model simultaneously. Figure 2 shows the overall energy balance of the compressor.

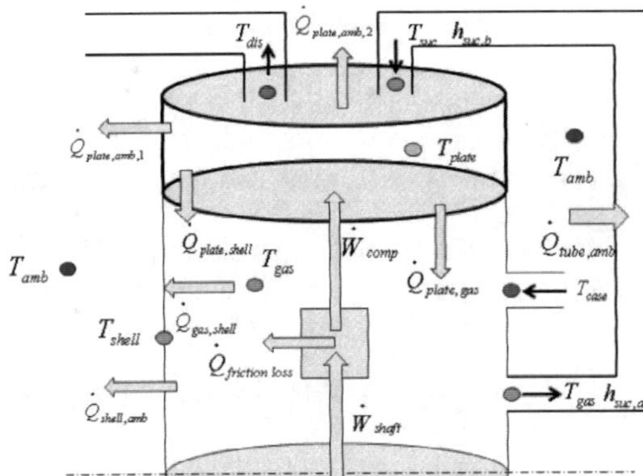

(a)

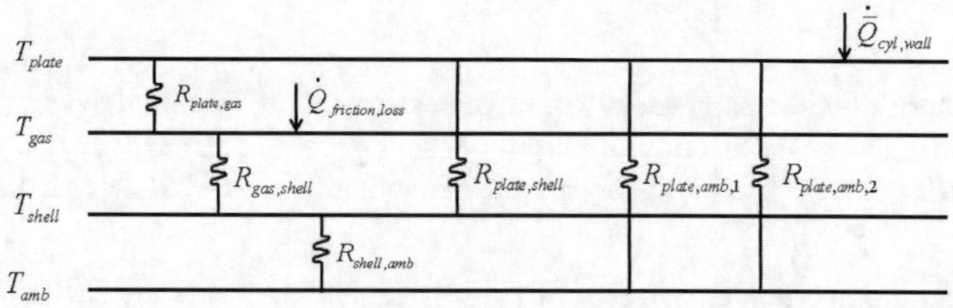

(b)

Figure 2: Energy balance of the compressor

A lumped temperature analysis method is used. There are 5 unknown temperatures: T_{amb}, ,T_{gas}, T_{plate}, T_{suc} and T_{dis}. It is noted here that the cylinder wall temperature is equal to the plate temperature since a lumped temperature analysis method is employed. Applying the energy conservation law to each component, the following equations can be derived:

$$\dot{m}_{comp}(h_{suc,a} - h_{case}) = \dot{Q}_{plate,gas} - \dot{Q}_{gas,shell} + \dot{Q}_{friction,loss}$$

$$\dot{Q}_{gas,shell} - \dot{Q}_{shell,amb} = 0$$

$$-\dot{Q}_{plate,amb,1} - \dot{Q}_{plate,amb,2} - \dot{Q}_{plate,gas} - \bar{\dot{Q}}_{cyl,wall} = 0 \quad . \tag{12}$$

$$\dot{Q}_{pipe,amb} + \dot{m}_{comp}(h_{suc,a} - h_{suc,b}) = 0$$

$$\dot{W}_{comp} + \bar{\dot{Q}}_{cyl,wall} = \dot{m}_{comp}(h_{dis} - h_{suc,b})$$

Where, the average heat transfer rate, $\bar{\dot{Q}}_{cyl,wall}$, is calculated by integrating the instantaneous heat transfer between the in-cylinder refrigerant gas and cylinder wall over one second.

The heat transfer rates, $\dot{Q}_{plate,gas}$, $\dot{Q}_{gas,shell}$, $\dot{Q}_{shell,amb}$, $\dot{Q}_{plate,amb,1}$, $\dot{Q}_{plate,amb,2}$, $\dot{Q}_{pipe,amb}$ can be expressed using the following general form,

$$\dot{Q} = \frac{T_a - T_b}{R_{ab}} \, . \tag{13}$$

Where, T_a-T_b is the temperature difference between two components. R_{ab} is the heat transfer thermal resistance. In order to find the thermal resistance, empirically correlated averaged Nusselt numbers are determined first considering different types of free convections shown in Figure 2. Since the ambient air and refrigerant gas inside the shell are both assumed to be quiescent, all heat transfer between the ambient air/refrigerant gas and compressor components are free convection.

3. NUMERICAL METHODOLOGY

The classical 4th-order Runge-Kutta method is used to solve the initial value problem formed by the combination of Eq. (3) and Eq. (4). As it is known, once the initial state of in-cylinder refrigerant together with other required information (such as initial suction/discharge valve plate positions) are given, the compression process model can be solved. The final in-cylinder refrigerant state and suction/discharge valve plate positions after one driving shaft rotation should match the initial values. However, since the initial values are arbitrarily assumed, it is not guaranteed that the final values would match the initial values. In this case, the problem becomes a root-finding problem for the following equation,

$$\vec{x} - g(\vec{x}) = 0. \tag{14}$$

Where $\vec{x}$ is a vector comprised of the parameters whose values are initially required to carry out the solution of the compression process model and g is a function of period of 2π. The root of the above equation is the initial values which would match the final values. Therefore, the Broyden's method is employed here (Bradie, 2006).

The cylinder wall temperature and the temperature of the refrigerant gas inside the shell are required to solve the compression process model. However, since the actual values of these two temperatures are unknown initially, arbitrary values are chosen. In order to get the actual cylinder wall temperature and refrigerant gas temperature inside the shell, the compressor overall energy balance model is incorporated with the compression process model. The bottom dead center (B.D.C.) is chosen as the starting point of the piston.

The converging criterion for the overall energy balance model is 10^{-4}. The converging criterion to find the initial values in the compression process model is also 10^{-4}. The flow chart of the algorithm is given in Figure 3.

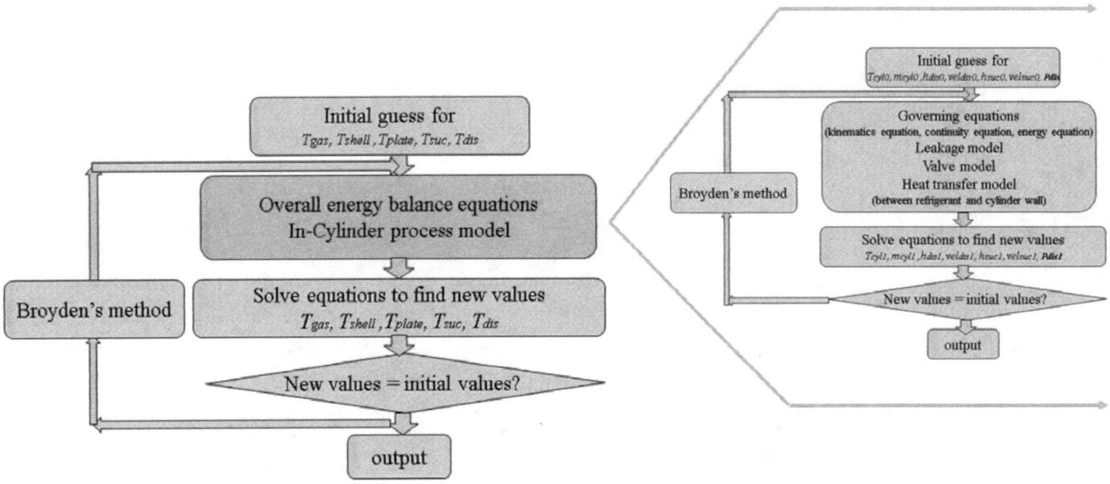

Figure 3: The algorithm flow chart of the whole compressor model

4. RESULTS AND DISCUSSION

4.1 Kinematics model

A prototype five-piston axial reciprocating compressor using the S-RAM mechanism was simulated using a suction pressure of 26 bars, discharge pressure of 90 bars, and rotational speed of 900 rpm. The variations of piston displacement, velocity and acceleration with respect to the machine driving shaft rotational angle are shown in Figure 4. It is clearly shown that all pistons in the five cylinders are moving similarly except for the phase difference. The piston starting point is located at the B.D.C.

Figure 5 provides the moving path of a ball center, which is clearly seen not to be located on the cylinder centerline leading to a side load to the restraints. The center of ball is moving with a 'figure 8' motion. The red vertical line is the centerline of one of the cylinders.

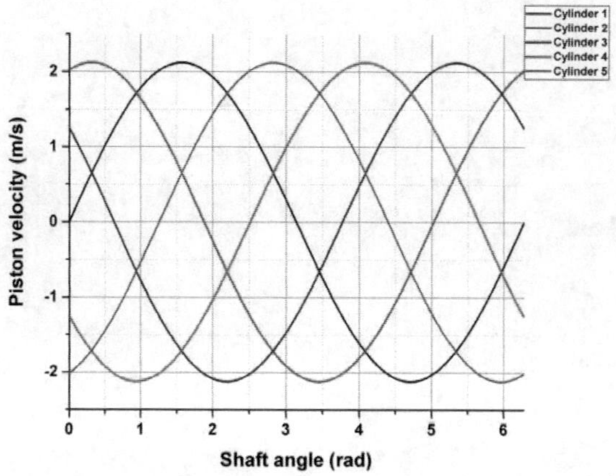

(a)

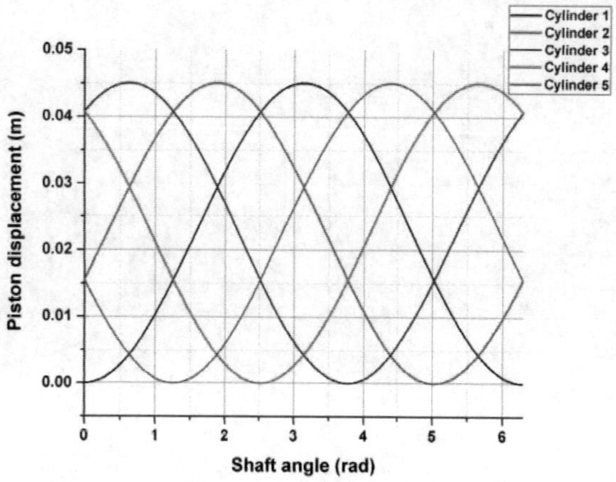

(b)

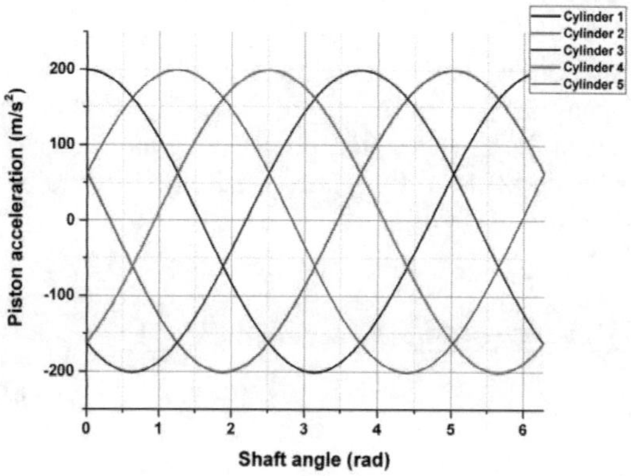

(c)

Figure 4: Piston displacement, velocity and acceleration (stroke = 0.04445 m) versus driving shaft angle (a) displacement; (b) velocity; (c) acceleration

4.2 Compression Process Model

The indicated P-V diagram for the prototype compressor is given in Figure 6. In the discharge process, the in-cylinder pressure reaches a value that is at first higher than the pressure in the discharge pipeline to overcome the valve spring force before the valve opens. As the discharge valve opens, some amount of refrigerant flows out of the cylinder decreasing the pressure inside the cylinder. The discharge valve starts to close when the in-cylinder pressure is not high enough to hold the resultant force of the valve spring and the discharge pressure behind valve. Hence, the discharge valve moves back and forth during the discharge process as shown in Figure 7.

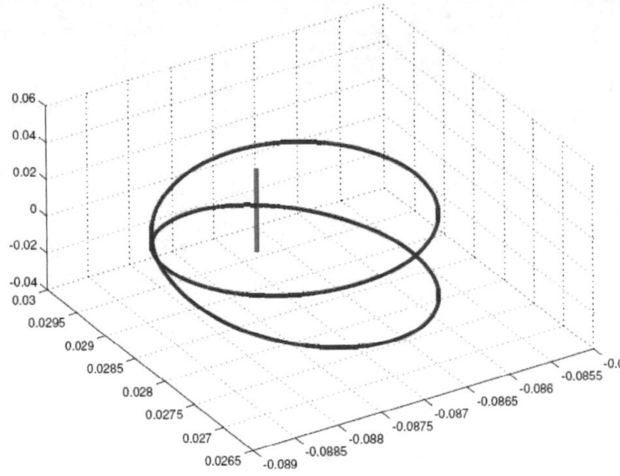

Figure 5: Moving path of a ball center ('figure 8' motion)

Figure 7 shows the variation of suction/discharge valve plate displacement with respect to the compressor driving shaft rotational angle. It can be seen that the suction valve is still open when the compression process starts. The discharge valve also stays open when the re-expansion process begins. Back flow may occur for both suction and discharge valves due to the late closing of the valve plates. The closing time of the valve plate can be adjusted by optimizing the valve spring stiffness and valve lift.

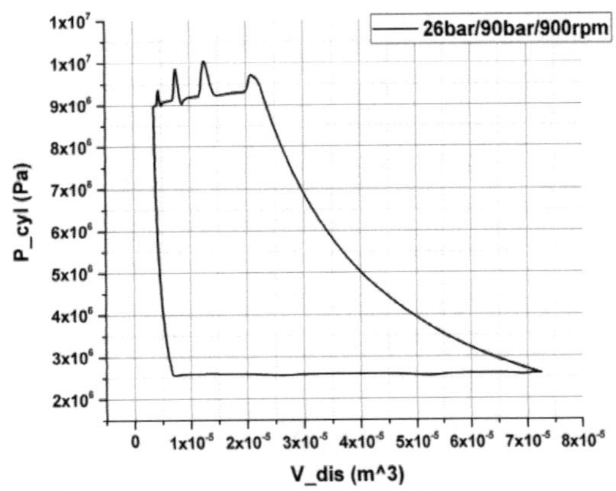

Figure 6: In-cylinder refrigerant pressure versus piston displacement volume

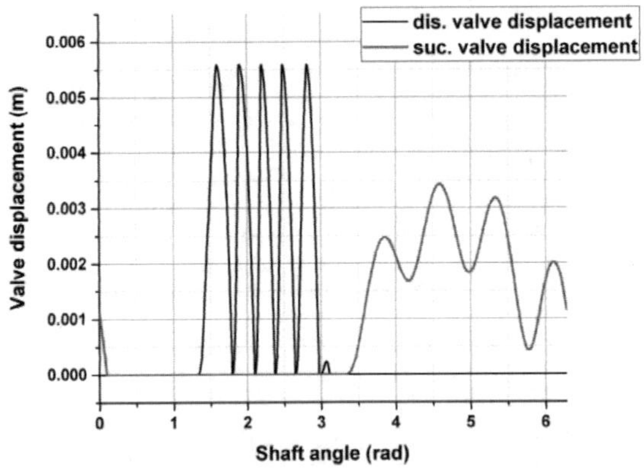

Figure 7: Valve displacement (26bar/42bar, 900 rpm)

4.3 Parametric Studies

4.3.1 Effect of Discharge Pressure

The effect of discharge pressure on the P-V diagram is shown in Figure 8. The suction pressure stays constant. As it is shown, the duration of the re-expansion process for a higher discharge pressure case is longer than that for a lower discharge pressure case. This is one of the reasons accounting for the decrease of the volumetric efficiency with an increment in pressure ratio, since less volume is available for the suction gas through the suction valve if a higher discharge pressure is applied. The discharge valve starts to open earlier for the low discharge pressure. Moreover, the in-cylinder pressure fluctuation is smoother for the higher discharge pressure case, which is consistent with the discharge valve movement shown in Figure 9.

Figure 10 shows the variation of the instantaneous heat transfer between in-cylinder refrigerant gas and cylinder wall with respect to the machine driving shaft angle. The shapes of the curves for both low and high pressure ratio cases are similar except that the magnitude of the heat transfer rate is larger for high pressure ratio case since the in-cylinder refrigerant temperature is higher, which can be seen in Figure 11. The trend of these two curves agrees well with that discovered by Recktenwald (1989).

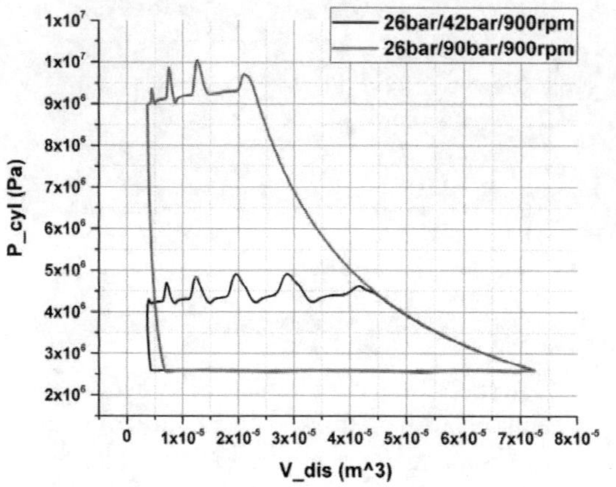

Figure 8: Effect of discharge pressure on P-V diagram

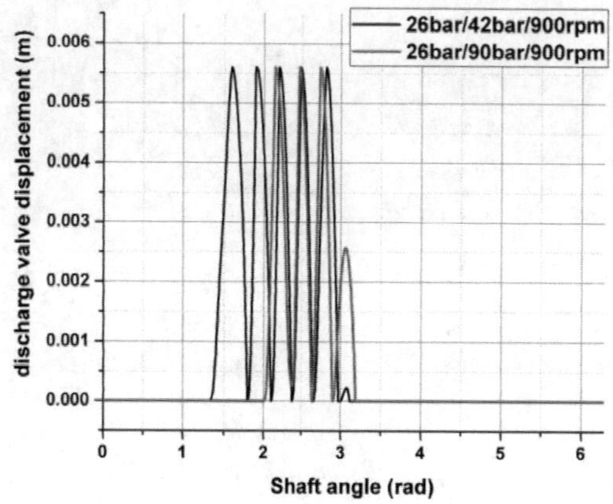

Figure 9: Effect of discharge pressure on discharge valve displacement

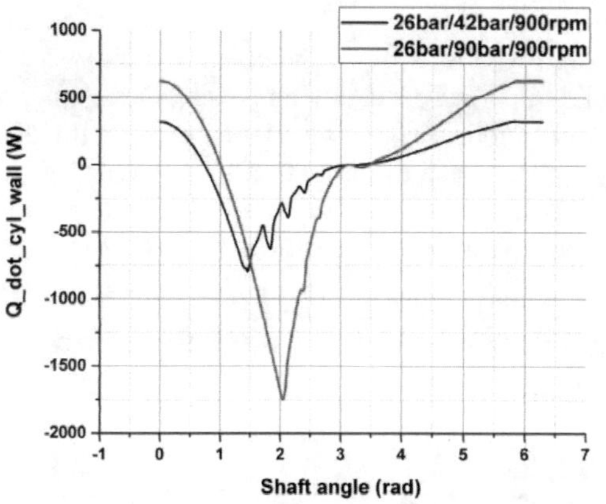

Figure 10: Effect of discharge pressure on the instantaneous heat transfer between in-cylinder refrigerant and cylinder wall

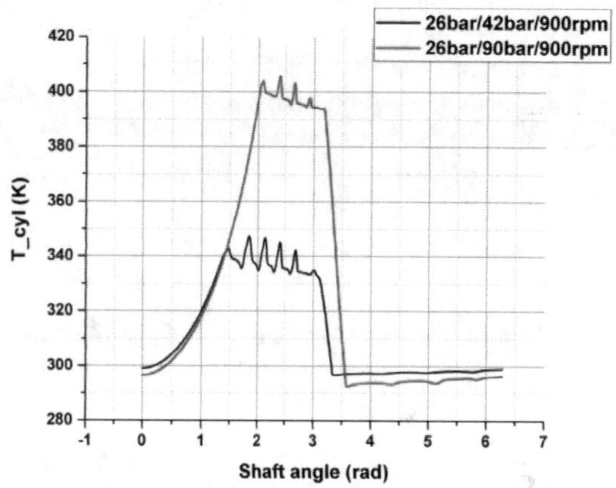

Figure 11: Effect of discharge pressure on the in-cylinder refrigerant temperature

4.3.2 Effect of Stroke-to-Bore Ratio

The stroke-to-bore ratio is an important parameter, which has a great influence on the reciprocating compressor performance. Figure 12 shows the variation of instantaneous leakage mass flow rate leaving from the in-cylinder control volume with respect to the stroke-to-bore ratio. It should be noted that the results presented in Figure 12 and Figure 13 are determined with the assumption that the compressor stroke is kept constant while the cylinder bore diameter is changing. It is clearly shown that the magnitude of the instantaneous leakage mass flow rate increases with the stroke-to-bore ratio decreases. A reduced stroke-to-bore ratio leads to a larger cylinder bore diameter when the stroke is constant. The leakage path is increased with a large cylinder bore diameter causing more leakage.

The effect of stroke-to-bore ratio on the instantaneous heat transfer rate between in-cylinder refrigerant and cylinder wall is given in Figure 13. It is found that the magnitude of the heat transfer rate has a higher value with a lower stroke-to-bore ratio. The heat transfer surface area is enlarged with a smaller stroke-to-bore ratio since the cylinder bore diameter is increased, causing the increment of the heat transfer rate.

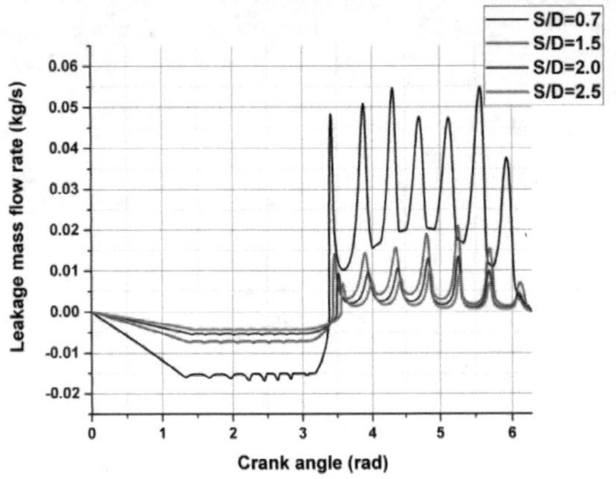

Figure 12: Effect of stroke-to-bore ratio on the instantaneous leakage mass flow rate (26bar/42bar/600rpm)

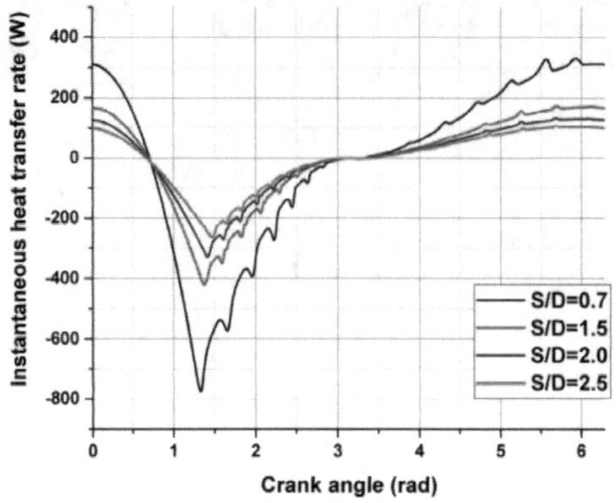

Figure 13: Effect of stroke-to-bore ratio on the instantaneous heat transfer rate between in-cylinder refrigerant and cylinder wall (26bar/42bar/600rpm)

5. CONCLUSIONS AND FUTURE WORK

A comprehensive simulation model to predict the performance of a prototype reciprocating compressor using the S-RAM mechanism is introduced. The comprehensive model is comprised of a kinematics model, compression process model and an overall energy balance model. In the kinematics model, the movement of the piston is revealed. The in-cylinder refrigerant pressure and temperature are given by solving the compression process model incorporated with kinematics model. The values of cylinder wall temperature, suction and discharge temperatures in their connection pipes are provided from the overall energy balance model. The gas pulsation issue in the discharge pipes is also considered. Parametric studies are conducted to investigate the effects of discharge pressure and stroke-to-bore ratio. It is shown that the stroke-to-bore ratio is critical for the compressor design to improve the volumetric efficiency by controlling the leakage through the leakage path between piston and cylinder wall.

It should be noted here that the dynamics model is not included in the current model, which is the focus of the future work. Therefore, currently only a certain portion of indicated power is assumed to be the frictional power loss.

NOMENCLATURE

A	area	h	unit enthalpy, distance between driving shaft center-line and cylinder centerline, heat transfer coefficient
C	discharge coefficient , specific heat capacity	m	mass
D	diameter	p	pressure
L	Length	v	specific volume
Q	quantity of heat transferred to control volume through boundary from its ambient	x	piston displacement
R	Length, or heat transfer thermal resistance	α	angle
S	piston stroke	θ	angle
T	temperature	κ	specific heat ratio
U	mean velocity of piston	μ	dynamic viscosity
W	work	ρ	density
Z	compact factor of carbon dioxide, or acoustic impedance at the discharge valve	τ	thermal conductivity of refrigerant in control volume
g	function	ω	angular speed of the main crankshaft

Subscript

0	clearance volume	p	piston
amb	ambient	plate	cylinder plate
c	refrigerant in control volume	port	valve port
comp	compressor	pul	gas pulsation
cyl	cylinder	s	surface
D	drag coefficient	shaft	crank shaft
d	Downstream	suc	suction
dis	Discharge	u	upstream
f	film	valve	valve
gap	leakage gap	vp	valve plate
heat	heat transfer	w	wobble plate
high	high pressure side	wall	cylinder wall
imp	impedence	∞	ambient
li, lo	leak in and out		
low	low pressure side		
mean	mean velocity of piston		

REFERENCES

Adair R P, Qvale E R, Pearson J T 1972. Instantaneous heat transfer to the cylinder wall in reciprocating compressors. Proc. of the Int. Comp. Eng. Conf. Purdue University, West Lafayette, IN USA, pp. 521-526.

Brablik J 1972. Gas pulsations as factor affecting operation of automatic valves in reciprocating compressors. Proc. of the Int. Comp. Eng. Conf. Purdue University, West Lafayette, IN USA.

Bradie B 2006. A friendly introduction to numerical analysis. Peason Prentice Hall, USA.

Baumannn H, Conzett M 2002. Small oil free piston type compressor for CO2. Proc. of the Int. Comp. Eng. Conf. Purdue University, West Lafayette, IN USA, pp. 861-868.

Cavalcante P, Försterling S, Tegethoff W, Stulgies N, Köhler J 2008. Transient Modeling and Sensitivity Analysis of a Controlled R744 Swash Plate Compressor. Proc. of the Int. Comp. Eng. Conf. Purdue University, West Lafayette, IN USA, paper 1880.

Elson J P, Soedel W 1974. Simulation of the interaction of compressor valves with acoustic back pressures in long discharge lines. J. Sound Vib 34 (2), pp. 211-220.

Fagerli B 1997. CO2 compressor development IEA/IIR Workshop on CO2 Technologies in Refrigeration, Heat Pump and Air Conditioning Systems: Trondheim, pp. 13-14.

Furuhama S, Tada T 1961. On the flow of the gas through the piston-rings (1st report, the discharge coefficient and temperature of leakage gas). Bulletin of JSME 4 (16), pp. 684-690.

Kim J H, Groll E A 2007. Feasibility study of a bowtie compressor with novel capacity modulation. Int. J. Refrigeration 30, pp. 1427-1438.

Lemmon, E. and Huber, M., Refprop 7.0, 2012, http://www.nist.gov/srd/nist23.htm

Lin M, Sun S Y 2006. Theory of the piston compressor. Xi'an Jiaotong University Press. Xi'an, China.

Lorentzen G, Pettersen J 1992. New possibilities for non-CFC refrigeration. IIR International Symposium on Refrigeration, Energy and Environment: Trondheim, pp. 147-63.

Nekså P, Rekstad H, Zakeri G, Schiefloe, P A 1999. Commercial heat pumps for water heating and heat recovery. IEA/IZWe.V./IIR Workshop on CO2 Technology in Refrigeration, Heat Pump and Air Conditioning System: Mainz.

Officine Mario Dorin. http://www.dorin.com/jsp/Template2/HomePage.jsp.

Recktenwald G W 1989. A study of heat transfer between the walls and gas inside the cylinder of a reciprocating compressor. PhD. Thesis. University of Minnesota. pp. 197-199.

Süss J, Kruse H 1998. Efficiency of the indicated process of CO2 compressors. International Journal of Refrigeration, 21(3), pp. 194-205.

Todescat M L, Fagotti F, Prata A T, Ferreira R T S 1992. Thermal energy analysis in reciprocating hermetic compressors. Proc. of the Int. Comp. Eng. Conf. Purdue University, West Lafayette, IN USA, pp. 1419-1428.

Yang B, Bradshaw C, Groll E A 2013. Modeling of a semi-hermetic CO2 reciprocating compressor including lubrication submodels for piston rings and bearings. Int J. Refrigeration 36 (7), pp.1925-1937.

ACKNOWLEDGEMENT

The authors would like to thank S-RAM Dynamics Inc. for the support to conduct this study.

Research of variable volume and gas injection DC inverter air conditioner compressor

Weimin Xiang[1], Bin Gao[2]*, Hualong Wu[3], Yangbo Yu[4]

Guandong Meizhi Compressor Co. Ltd.. R&D Section,
Foshan City, Guangdong Province, China, 528333

[1]Phone: +86 757 22329000, xiangwm@chinagmcc.com
[2] Phone: +86 757 22325565,gaob@chinagmcc.com
[3] Phone: +86 757 22325067, wuhl@chinagmcc.com
[4] Phone: +86 757 22329126, yuyb@chinagmcc.com

ABSTRACT

With the wide promotion of energy saving and environmental protection, the working efficiency of Room Air Conditioner (RAC for short) system become more and more important, in addition, to improve the function of RAC on the living comfort, the demand of heating capacity especially at low ambient temperature condition of RAC become a urgent issue needs to be resolved.

Firstly, the purpose of this paper was introduced. Then, the relative techniques like DC inverter, variable volume, gas injection were simply analyzed. Thirdly, a new coupling compressor with these three techniques to satisfy both high efficiency and heating capacity demands was indicated. Moreover, the detail design of the coupling compressor was described. At last, the prototype compressor was designed and assembled to validate the effect on RAC system.

According to the test result, this coupling compressor has significant effect on RAC system as we expectation. The maximum heating capacity increasing up to 185% by compare to the base system at -15℃ ambient temperature condition, which can reduce half time to achieve the same room temperature rising. Furthermore, the Annual Performance Factor (APF for short) efficiency of the RAC system increased by about 6%.

1. INTRODUCTION

Recently, with the advancement of RAC industry, people are focusing on the energy saving and the comfort function like high heating capacity of the RAC system. It is not too difficult to achieve high efficiency or high heating capacity for a RAC system. However, how to achieve both high efficiency and high heating capacity at one RAC system is still an important issue for the engineers.

There are different advanced techniques like DC inverter technique, variable volume technique and gas injection technique, etc. have been published for the RAC. Among these techniques, the DC inverter technique and gas injection technique have good effect on energy efficiency. For the heating capacity issue, when RAC system working at low ambient temperature condition, the evaporate temperature become very low which result in the low mass flow of refrigeration circulation, and then the heating capacity significantly decrease. In this case, the increase of running frequency and compressor displacement will make up the problem.

However, each single of these techniques has disadvantages like, the DC inverter technique has the limitation of maximum running frequency so that limit the heating capacity increase, the variable volume technique of the compressor has good heating capacity improvement but has the problem of low running efficiency, the gas injection technique has good efficiency but limited by the heating capacity increase and a uneconomic structure of the compressor and RAC system, etc.

2. THE PRINCIPLE AND SCHEMES OF COMPRESSOR

Refer to both the energy efficiency and heating capacity requirements of the RAC system, the target of the new solution has two parts, one is achieve good increase in system APF efficiency, the second one is the high heating capacity at low ambient temperature.

In order to satisfy both high efficiency and high heating capacity demands of the RAC system, a new compressor which we called coupling compressor will designed, it adopts all the DC invert technique, variable volume technique and gas injection technique in one compressor. This coupling compressor collecting the advantages while repairing the disadvantages of the three techniques which expected to has good effect on energy efficiency and significantly increases of heating capacity at low temperature at the same time.

2.1 The principle of DC inverter technique

There are two kinds of frequency conversion technique used in compressor, one is DC inverter technique, and another is AC inverter technique. The DC inverter technique is wildly used in compressor because of the better efficiency.

The principle of DC inverter technique is shown as figure 1. By the process of rectification, filtering and inverting, the power source 200V/50Hz will change to 3 phases with 120°difference rectangular wave current supply to the stator winding of BLDC compressor. With both the current and rotor location signals, the controller electrifies the 3 phase winding one by one to driving the motor in compressor.

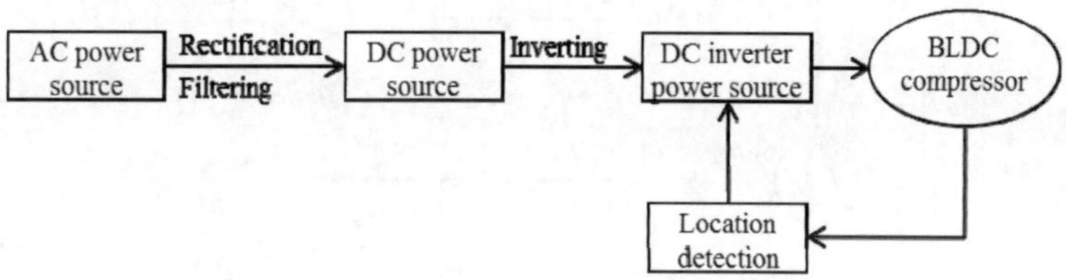

Figure 1: Principle of the DC inverter driver

The DC motor has higher efficiency than the fix speed motor. Furthermore, by the application of DC inverter technique, the compressor can adjust the power consumption by change the running frequency according to the RAC requirement to achieve high efficiency operation. This not only improves the efficiency of the air conditioner at different working conditions but also makes the room more comfortable with the stable room temperature.

2.2 The principle of gas injection technique

There are two popular gas injection designs for rolling piston compressor. The first one called Roller Rotation Control (RRC for short) as shown in figure 2: the gas injection valve set on the bearing surface, the open and close and anti-backflow of the gas injection is controlled by the position of the roller. The structure of RRC is very simply and without clearance volume makes it is easy application and high efficiency. However, the RRC structure is limited by the roller thickness.

The second one is the Check Valve Control (CVC for short). The gas injection valve is set on the cylinder nearby the discharge port. The valve open and close is controlled by the pressure difference of the valve. When the pressure in cylinder is lower than injection pressure, the valve open, the gas inject into cylinder. When the pressure in cylinder is higher than injection pressure, the valve close to prevent gas back-flow. The CVC has more injection quality than RRC. However, it has more complex structure, and the efficiency is affect by its clearance volume.

The principle of the gas injection is two-stage like refrigeration circulation as shown in Figure 3. The liquid refrigerant out from the condenser through the first expansion valve become the two-phase flow into the gas-liquid separator which is separated as gas and liquid. Hereinto, the liquid will go to the evaporator after the second expansion while the gas with middle pressure will directly inject into the compressive chamber of the cylinder.

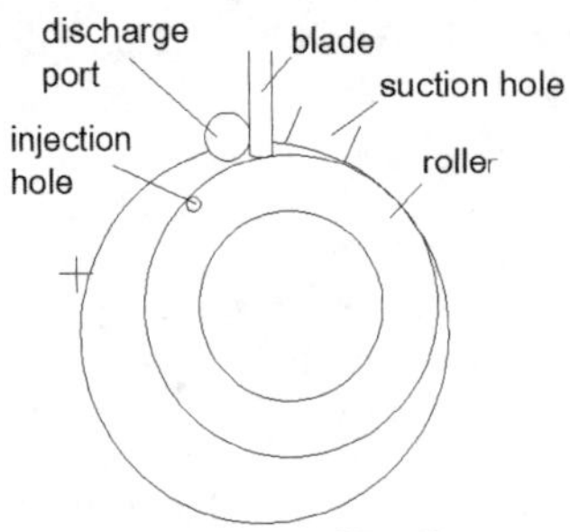

Figure 2: Diagram of RRC structure

In this gas injection process, the gas inject into the cylinder has middle pressure means the power consume to make this part of gas from the suction pressure to the middle pressure is avoid so that the circulation efficiency improving. In addition, because the gas from the gas-liquid separator is saturated gas which has lower temperature than the gas in compressive chamber, when these two kinds of gas mixed, the discharge temperature drops so that the compressor reliability improving.

According to the structure, by considering both the efficiency and reliability, we adopt the RRC and CVC design to the two cylinders in the compressor respectively.

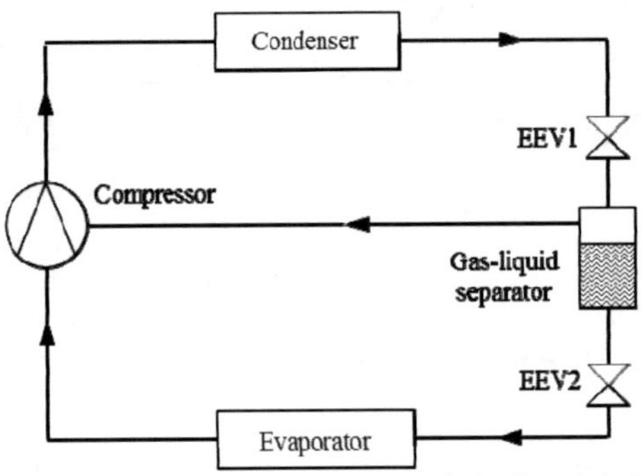

Figure 3: System schematic diagram and circling P-h drawing

By adjust the injection pressure and angle, the injection design is optimized to improve the gas quality inject into the cylinder. The injection quality can be calculated as the equation 1 below.

$$M_j = \alpha \times \frac{\pi d^2}{4} \times \sqrt{2 * \frac{p_j - p_0}{\rho}} \times \frac{\theta}{360 \times f} \tag{1}$$

Hereinto, M_j is the injection mass flow, p_j is the injection pressure, p_0 is the pressure in compressive chamber before injection, ρ is the gas density in compressive chamber before injection, α is the injection coefficient, θ is the injection angle, f is the running frequency of the compressor.

According to the equation 1, the injection quality has direct proportion to the injection port area.

The injection pressure calculates as below equation 2.

$$p_m = \sqrt{p_c \times p_e} \tag{2}$$

When $p_j > p_m$, the dryness in gas-liquid separator decrease, then the gas content reduce which may cause the risk of liquid injection. On the contrary, if $p_j < p_m$, the injection quality decrease which cause less effect on efficiency improve.

2.3 The principle of two cylinder variable volume technique

The variable volume technique means the compressor can adjust the working displacement at different conditions, it's usually using in a two cylinder rolling piston compressor. When system is working at low capacity condition, the compressor will working at one cylinder mode which we called partial volume model, here this working cylinder we called constant cylinder. When system is working at high capacity condition, the compressor will working at two cylinder model which we called full volume model, here the second cylinder we called alterable cylinder.

The alterable cylinder drawing is shown in Figure 4. There are two kinds of alterable cylinder control schemes, in both schemes, a pressure input which can switch to suction pressure or discharge pressure is necessary for the alterable cylinder.

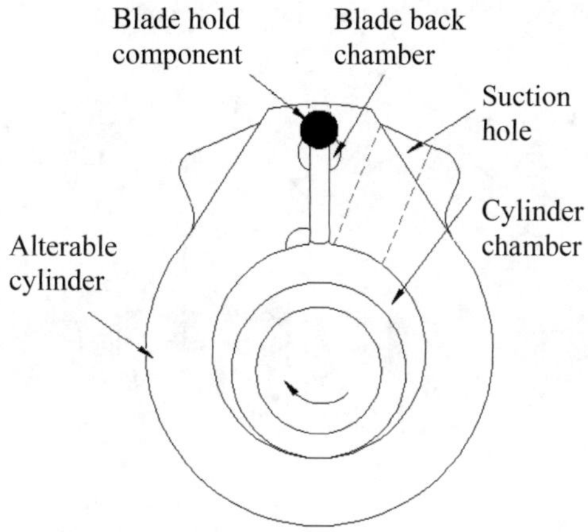

Figure 4: Drawing of alterable cylinder structure

For the first one, the alterable cylinder is controlled by the pressure change in the blade back chamber. Normally, it is necessary to add a blade hold component to fix the blade at the upper position. In this scheme, the blade back chamber must sealed from the pressure in the compressor case, when the pressure in the chamber is suction pressure, the alterable cylinder is idling while when the pressure in the chamber is discharge pressure, the alterable cylinder in operation condition.

In the second scheme, the alterable cylinder is controlled by the change the suction pressure, when suction normally, the alterable cylinder in operation condition while when suction hole connect to discharge pressure, the alterable cylinder unload. This scheme also needs a blade hold component.

2.4 Integrate design

According to the analysis above, the outlines of the integrate design include below points.

a. Adopted multi-increment design, combine the DC inverter technique, variable volume technique and gas injection technique together to improve energy efficiency and heating capacity at low ambient temperature conditions.

b. Different gas injection design for the two cylinders, the constant running cylinder use CVC injection design to enhance efficiency while the alterable cylinder applied the RRC structure to increase heating capacity. These two gas injections are using the same one injection supply path.

c. To achieve gas injection in alterable cylinder, the alterable cylinder adopted the variable volume scheme of change suction pressure as well as RRC injection with a non-return flap valve design. The non-return flap valve is to prevent back-flow in case of the alterable cylinder idling, when the pressure in alterable cylinder chamber is discharge pressure which is higher than injection pressure.

The two cylinders of the prototype compressor designed as equal displacement of 9.8 (cm^3/rev) to match a nominal 3500W cooling capacity room air conditioner using R410A refrigerant. This air conditioner used to applying a compressor with displacement of 10.8 (cm^3/rev).

The structure of the coupling compressor is shows in figure 5. The alterable cylinder is set as the lower cylinder. Compare to normal two cylinder compressor, there are two added small accumulator at the outside of compressor case, one is for the alterable cylinder and another one is for the gas injection. One head of the alterable cylinder accumulator connect to the cylinder suction hole and another head connect to the outlet of a three-way valve. The two input inlets of the three-way valve respectively connect to the main accumulator suction pipe and the compressor discharge pipe.

When the three-way valve input discharge pressure, the alterable cylinder is idling and its injection closed. However, the alterable cylinder and its gas injection are working if the three-way valve input suction pressure. The upper cylinder is a constant running cylinder so that the gas injection also at constant working status.

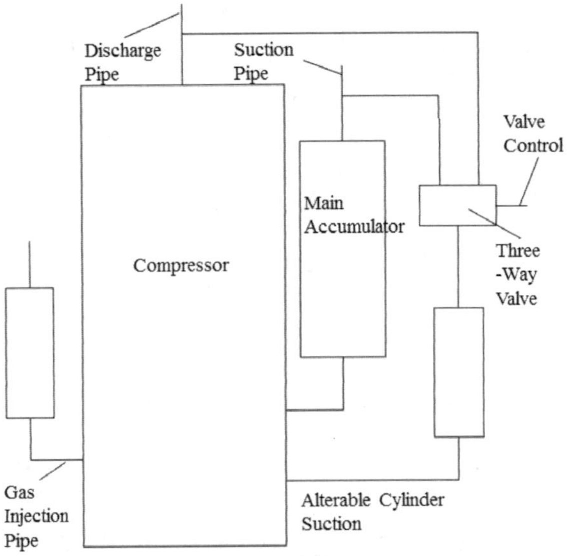

Figure 5: Diagram of coupling compressor structure

3. EXPERIMENTAL VERIFICATION

According to the design above, the prototype compressor is assembled. Firstly, the compressor performance was tested on the calorie meter. The result shows that this coupling compressor has equivalent performance (without gas injection) as the normal two cylinder compressor.

Then, we matched the coupling compressor and the normal two cylinder compressor on the same RAC system. The test conditions include the APF conditions and a super low temperature heating condition which are shown in table 1. At the RC, MC, RH, MH conditions, the compressor is working at partial model, at the LH and SLH conditions, the compressor is working at full volume model.

Two systems were tested, one is an original nominal 3500W air conditioner we called 35S which for the APF efficiency testing. When tested the SLH condition, because the prototype compressor has much bigger displacement than normal compressor used in 35S system, to release the heating capacity limit by the heat exchanger, we changed the condenser to a bigger one with 5100W capacity which was called 51S.

Table 1: APF testing conditions

Test Conditions		APF Conditions					Super Low temp. Heating (SLH)
		Rating Cooling (RC)	Middle Cooling (MC)	Rating Heating (RH)	Middle Heating (MH)	Low temp. Heating (LH)	
Indoor dry/wet temp.	℃	27/19	35/24	20/15	20/15	20/15	20/15
Outdoor dry/wet temp.	℃	35/24	27/19	7/6	7/6	2/1	-16/-15

The result of APF conditions as shown in figure 6. Compare to the result of normal two cylinder compressor one, the efficiency of all APF conditions were improved especially at the low temperature heating condition, the COP is about 22% higher than the base. After calculated, the APF efficiency improved 6% compare to the base one which means the coupling compressor has very good effect on the air conditioner system improvement.

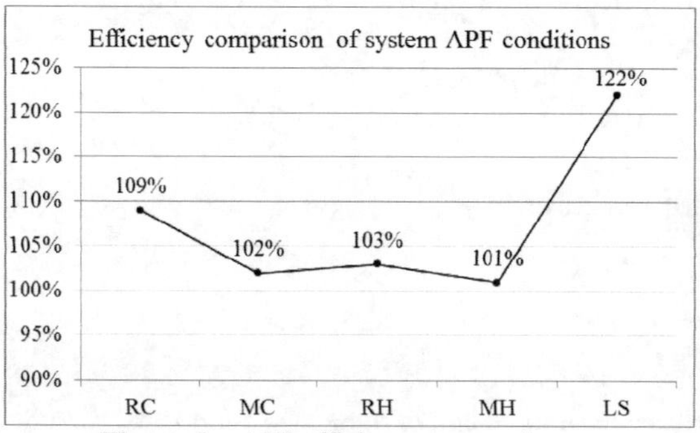

Figure 6: APF efficiency comparison

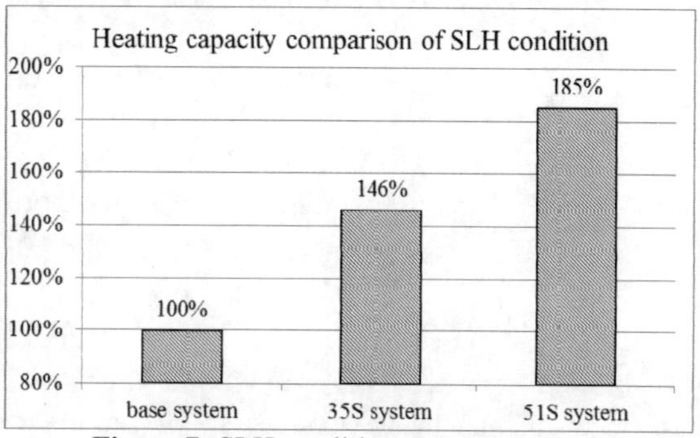

Figure 7: SLH condition comparison

The figure 7 shows the heating capacity result at SLH condition. Compare to the base system, when the system is 35S, the heating capacity improved about 46% while the 51S system was about 85%. This result means that for this coupling compressor, the condenser capacity of 35S is not big enough. In addition, when we test with the 51S system, the wind outlet temperature can up to 50℃ with enough air flow which will satisfy the heating requirement of the air conditioner customers. According to this result, we can apply this coupling compressor in a big system like 51S to obtain higher heating capacity, or match on the normal 35S system to achieve both high efficiency and high heating capacity.

4. CONCLUSIONS

According to the problem of the low heating capacity of room air conditioner especially at low temperature condition, as well as to satisfy the high efficiency requirement, this paper advanced a coupling compressor with DC inverter technique, variable volume technique and gas injection technique together, which can solve all the issues. The test result shows that, with this coupling compressor, the APF conditions efficiency of the room air conditioner system improves 6%, in addition, the maximum heating capacity increment of the system at -15℃ achieves about 85%, which has significant effect as the expectation.

REFERENCE

Ozu Masao, Bo Jiang(2009), *Patent of Gas injection compressor*, Guangdong Meizhi Compressor Ltd. China.

Yezheng Wu, Hongqi Li & Hua Zhang(2010),*Refrigeration Compressor*. Beijing: Mechanical Industry Publisher.

Liu Yang, Huang Zhimin, Liu Chunhui (2013). Impact of Injection Technology on Heating Performance of Rotary Comparssor (707-710). Beijing, China: *2013 China Household Electronic Conference*.

Updated Performance and Operating Characteristics of a Novel Rotating Spool Compressor

Joe OROSZ[1*], Craig R. BRADSHAW[2], Greg KEMP[3], Eckhard A. GROLL[4]

[1]Torad Engineering, 5192 Performance Drive
Cumming, GA 30040
Phone – (770-889-8400), Fax – (770-889-6900), E mail – joe.orosz@toradengineering.com

[2]Torad Engineering, 5192 Performance Drive
Cumming, GA 30040
Phone – (770-889-8400). Fax – (770-889-6900), E mail – craig.bradshaw@toradengineering.com

[3]Torad Engineering, 5192 Performance Drive
Cumming, GA 30040
Phone – (770-889-8400), Fax – (770-889-6900), E mail – greg.kemp@toradengineering.com

[4]Ray W. Herrick Laboratories, West Lafayette, IN 47907
Phone – (765-496-2201). Fax – (765-494-0787), E mail – groll@purdue.edu

* Corresponding Author

ABSTRACT

The basic mechanism of the novel rotary spool compressor has been described previously by Kemp et al. (2008, 2010). The device combines various aspects of rotary and reciprocating devices that are currently well understood. Due to increasing pressure in the global market for refrigerants with very low GWP levels extensive modeling was conducted to explore a spool compressor design for operation on medium pressure low GWP refrigerants in sizes applicable to the commercial air conditioning marker, specifically for application on air and water cooled chillers. The basis for a compressor design in this space is operation on R134a realizing that most low-GWP medium pressure gases are similar in nature to R134a from a compressor design point of view. A new compressor design with specific aspect ratio optimized for operation on R134a has been constructed and tested. The compressor was tested at a range of conditions suitable for evaluation in this application. The current compressor performance is comparable with today's screw compressors operating in this size range.

1. INTRODUCTION

The rotating spool compressor is a novel rotary compressor mechanism most similar to the sliding vane compressor. Primary differences are described by Kemp *et al.* (2008, 2010) and include three key differences from a sliding vane compressor, as shown in Figure 1.

- The vane is constrained by means of an eccentric cam allowing its distal end to be held in very close proximity to the housing bore while never contacting the bore.
- The rotor has affixed endplates that rotate with the central hub and vane forming a rotating spool.
- The use of dynamic sealing elements to minimize leakage between the suction and compression pockets as well as between the process pockets and the compressor containment.

The movement of the rotor is purely rotary with only the vane and tip seals performing any oscillating movement. The eccentric cam will force the movement of the vane to oscillate by twice the eccentricity during a single rotation. The tip seals will oscillate relative to the vane two times per rotation by an amount proportional to the ratio of diameters of the rotor to the housing bore (also known as the eccentricity ratio). The tip seal movement amount is roughly an order of magnitude smaller than the eccentricity and follows a sinusoidal path. Analytical details regarding the geometry, including the mathematical expressions describing the chamber volumes, is presented in Bradshaw and Groll (2013).

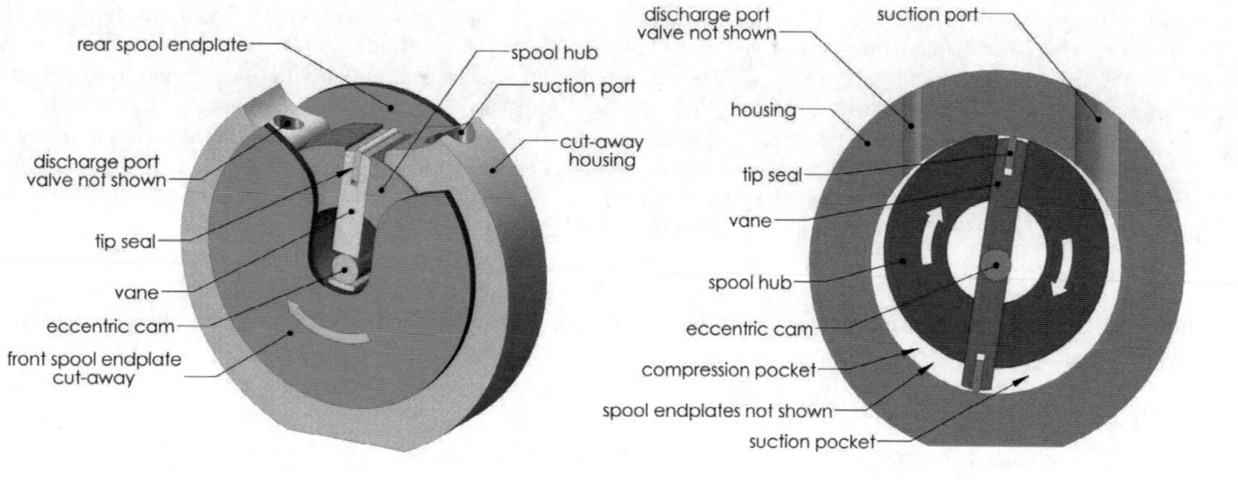

Figure 1: Cutaway view of rotating spool compressor mechanism with key components highlighted.

Since its introduction in 2008 the spool compressor has been improved and better understood. A comprehensive model was developed to deepen the understanding of the device holistically, as a system (Bradshaw and Groll, 2013). This holistic approach has allowed for better prediction of non-intuitive behavior of the spool compressor especially applied to applications and working fluids without experimental experience. The model includes 10 leak paths, 8 heat flow paths, port geometries, as well as oil solubility and frictional component sub-models.

Additionally, sub-components of the spool compressor have been studied in detail. Bradshaw (2013) presented a study on the analytical model developed to represent the tip seal in the spool compressor. The model was developed based on the tip seal dynamic model coupled with hydrodynamic lubrication theory. Kemp *et al.* (2012) presented iterations on the spool seal design and described the critical nature of the design of this component.

Bradshaw *et al.* (2016) recently presented the development of a loss analysis to describe the 5th generation spool compressor. This analysis utilized high-speed pressure measurements to determine the indicated, or flow, losses within the device. Coupled with analysis of the frictional losses a complete loss Pareto was generated. This analysis has resulted in a deep understanding of the location of potential improvements as well as an improved comprehensive modeling tool which accounts, more accurately, for the various frictional losses within the device. The resulting improvements using this analysis were also presented which showed a significant improvement in performance between the 5th and 6th generation compressor. These results echo the improved performance presented by Orosz *et al.* (2014).

This work presents an experimental characterization of a 141 kW (40 tonsR) prototype spool compressor for application to light-commercial air-conditioning. A custom-built hot-gas bypass load stand is described including an uncertainty analysis. The experimental performance data at two speeds is presented and discussed.

2. PERFORMANCE EVALUATION

A 141 kW (40 tonsR) R134a spool compressor was constructed to validate the modeling performed and presented by Bradshaw *et al.* (2016a). The compressor was mapped at two speeds with a range of suction pressures from 281.4 kPa (40.81 psia) to 497.5 kPa (72.2 psia) and a range of discharge pressures from 699.1 kPa (101.4 psia) to 1472.0 kPa (213.5 psia) utilizing R134a as the working fluid. This represents evaporating and condensing temperatures of -1.1 °C (30 °F) to 15.5 °C (60 °F) and 26.7 °C (80 °F) to 48.9 °C (120 °F), respectively. Intake gas superheat was fixed at 16.7 ºC (30 ºF) for all data points. These conditions represent a typical commercial air-conditioning application for both water and air cooled conditions.

2.2 Test Stand

Construction of a hot gas by-pass test stand was necessary to test the constructed prototype. This test stand needed accommodate the larger capacity compressor, up to 141 kW (40tonsR) of cooling. A schematic of this test stand is shown in Figure 2. The test stand includes the ability to separate and meter oil back into the compressor sump as well as into the intake for better simulation of actual system conditions. In addition, the test stand was designed to be ASHRAE 23.1 compliant and, where practical, includes redundant measurements on all critical functions including pressure, temperature and mass flow rate. Mass flow sensors are installed on the discharge as well as the intake of the compressor. The discharge mass flow meter is utilized to calculate the efficiencies because the gas is entering at a higher super heat with less risk for two-phase flow. The mass balance is considered acceptable at a level below 1.5%. Steady state is determined by calculating the total statistical variance of critical measurements over a 15 minute period. The test point is considered stable when the temperature variance is less than 0.2 K (0.4 R), pressure variances below 6.8 kPa (1 psi), speed variance is below 1 rpm, and mass flow rate variance is below 0.007 kg/sec (1 lb/min). The operating conditions of the compressor are controlled with software driven PID controllers which adjust the bypass valve, liquid line valve and condenser water flow controls. The power measurement is obtained with a rotary torque sensor and verified with the motor input power accounting for approximate motor efficiency.

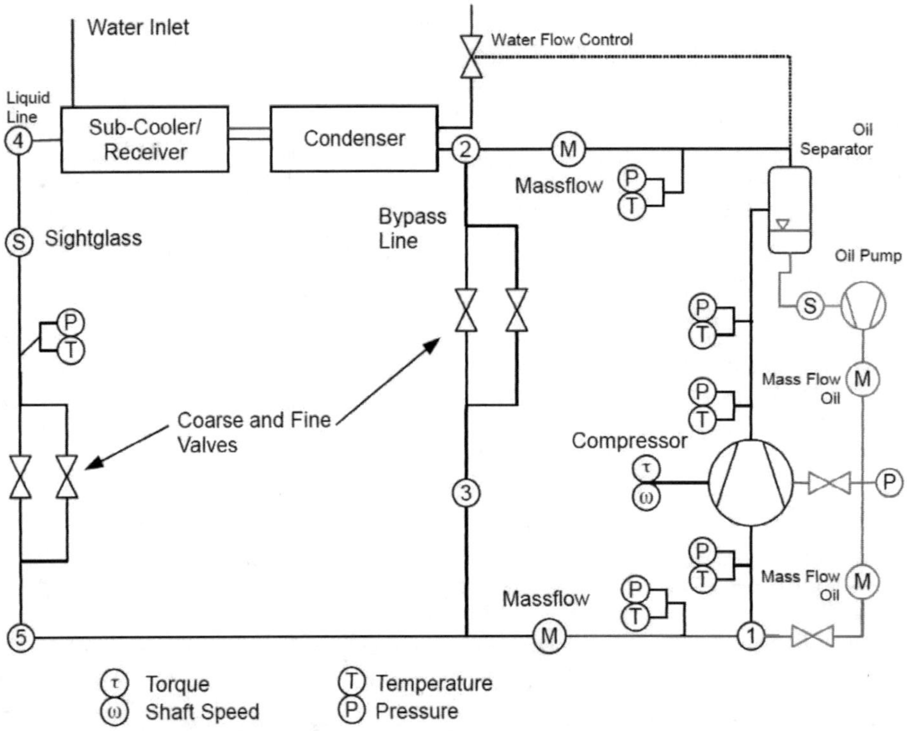

Figure 2: Schematic of 141 kW hot gas bypass test stand

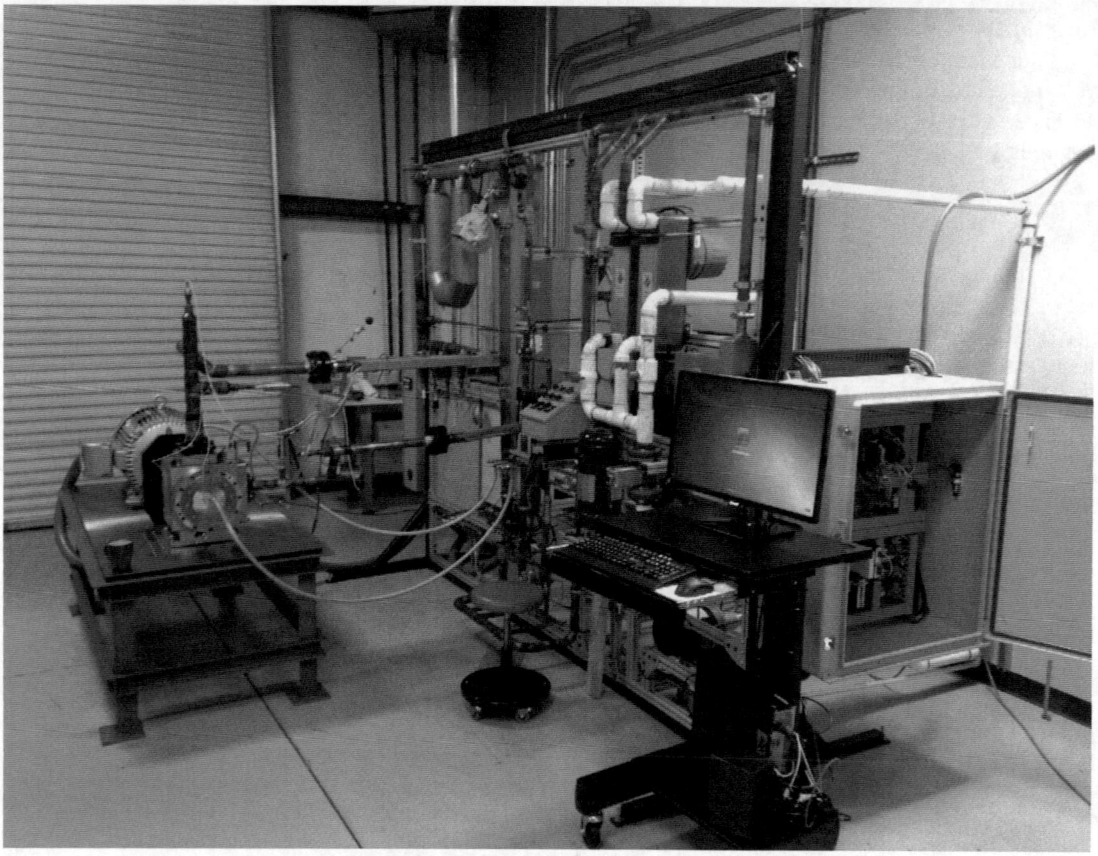

Figure 3: Photo of 141 kW hot gas bypass test stand

2.2 Efficiency Definitions

Test data is collected to calculate the Volumetric Efficiency and Overall Isentropic Efficiency as a function of pressure ratio and speed. The volumetric efficiency was determined using Equation (1), where the theoretical volume flow was obtained based on speed measurements and the displacement volume:

$$\eta_{vol} = \frac{\dot{m}_{act} \cdot v_1}{\dot{V}_{th}} \quad (1)$$

where $\dot{m}_{act}$ and v_1 are the measured mass flow rate and specific volume, respectively. The overall isentropic efficiency is a frequently used measure for the first law efficiency of compressors by using an overall control volume, i.e., an evaluation by using the thermodynamic states at the compressor inlet and outlet. The overall isentropic efficiency is obtained based on Equation (2):

$$\eta_{is,o} = \frac{\dot{m}_{act} \cdot (h_{2s} - h_1)}{\dot{W}_{shaft}} \quad (2)$$

where $\dot{W}_{shaft}$ is the shaft power input to the compressor mechanism only.

2.3 Shaft Power Adjustments

For an open drive prototype the true shaft power cannot be measured directly because of the power draw of the shaft seal. Similarly, for a hermetic compressor the motor inefficiencies need to be accounted for to obtain a true shaft power. The efficiency numbers presented here include the loses associated with the shaft seal, as a shaft seal map was unavailable. In the case of a hermetic application the shaft seal would be replaced with a hermetic motor and these loses would need to be accounted for.

2.4 Uncertainty Analysis

The uncertainties associated with the measurements collected from the load stand are presented in this section. These measurements include temperature, pressure, refrigerant mass flow, compressor speed and compressor torque. The absolute uncertainty is given in Table 1. As the uncertainty of the temperature measurements are defined by the following relationship,

$$\mu = (.15 + .002*T) \tag{3}$$

where T is the actual temperature measured, in °C. In Table 1 the accuracy stated for the suction and discharge temperature respectively are for the specific operating condition of 4.4 °C (40 °F) Saturated Suction, 37.8 °C (100 °F) Saturated Discharge and 16.7 °C (30 °F) Suction superheat. Using the absolute uncertainty, the propagation of uncertainty associated with the reported efficiency values is calculated used and uncertainty propagation analysis (Fox *et al.* 2004). The volumetric and overall isentropic efficiencies have average uncertainty values of 0.6% and 1%, respectively.

Table 1: Uncertainty of Experimentally Obtained Measurements

Measured Value	Manufacturer	Model Number	Accuracy
Suction Pressure	Setra	ASM1200P	± .05%
Discharge Pressure	Omega	PX01-C1	± .05%
Discharge Mass Flow	Micromotion	CMF100	± .35%
Compressor Speed	Encoder Products	702	± .01%
Shaft Torque	S Himmelstein	79061V	± .15%
Suction Temperature	Omega	PRTF-17-2-100	± .75%
Discharge Temperature	Omega	PRTF-17-2-100	± .54%

4. COMPRESSOR PERFORMANCE

The 141 kW (40 tonsR) rotary spool compressor was evaluated based on the above running conditions using refrigerant R134a. Figures 4 and Figure 5 represent the volumetric efficiency for speeds 900 rpm and 1300 rpm, respectively, plotted against the pressure ratio for various condensing temperatures. The data points shift to the right as the condensing temperatures increases due to the constant suction pressures. At 900 rpm and 26.7 °C (80°F), the volumetric efficiency degrades at a rate of approximately 2.5% per pressure ratio and increases with increasing condensing pressure. At a condensing pressure of 43.3 °C (110°F), the volumetric efficiency degrades at a rate of 3% per pressure ratio. At 48.9 °C (120°F), the volumetric efficiency line exhibits a slight increase in slope from the 43.3 °C (110°F) line to 3.5% volumetric loss per pressure ratio. From Figure 5, 1300 rpm, it can be seen that the slope of the volumetric efficiency degradation per pressure ratio is 3% for all condensing temperatures. The total loss of volumetric efficiency at 900 rpm for the range of 1.5 pressure ratios to 4.5 pressure ratios is 14.5%. At 1300 rpm this improves to 9% for the same range of pressure ratios.

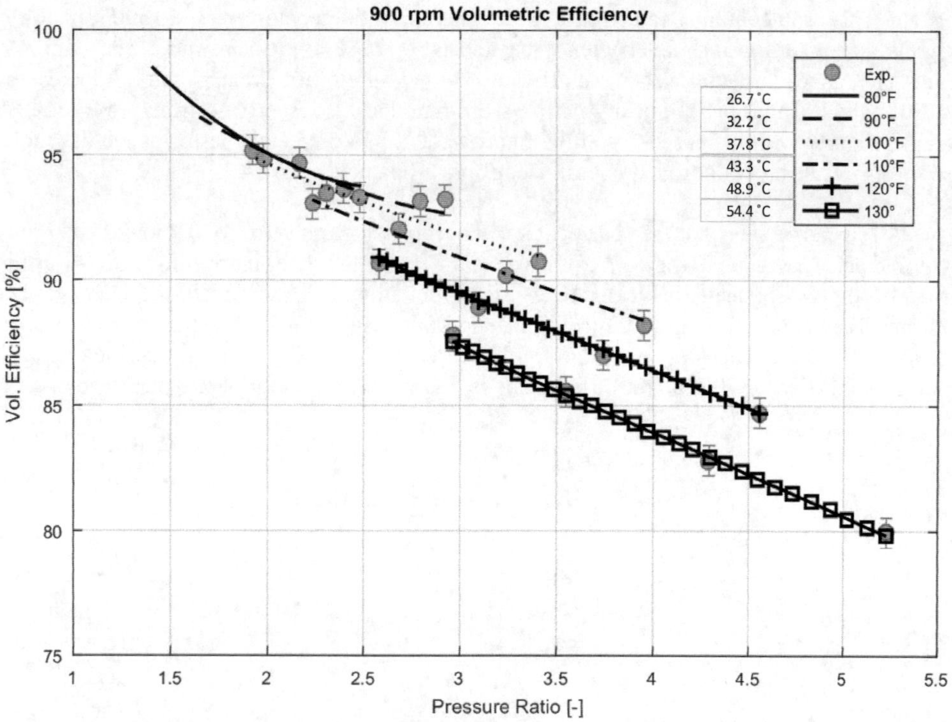

Figure 4: 141 kW spool compressor volumetric efficiency vs. pressure ratio at 900 rpm

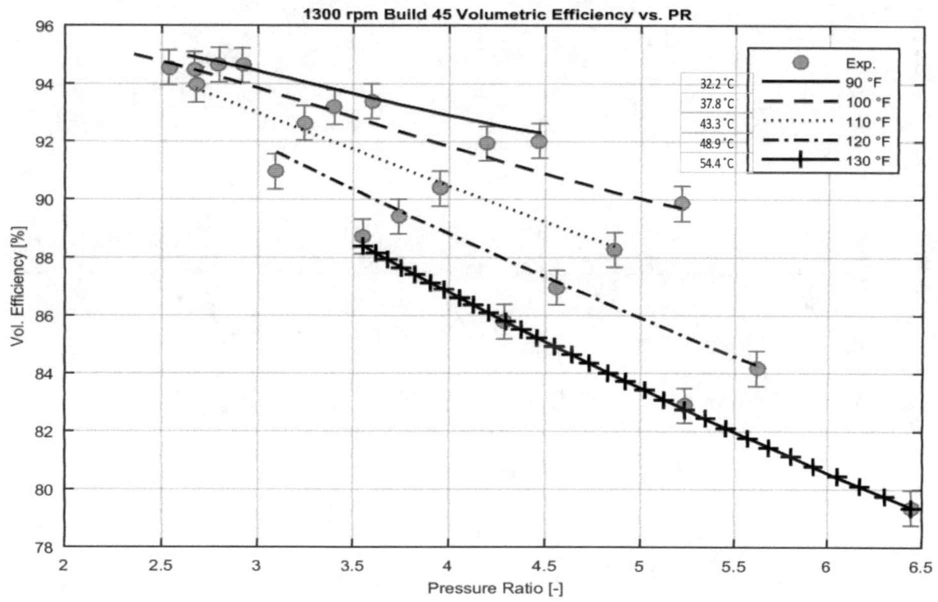

Figure 5: 141 kW (40 tonR) spool compressor volumetric efficiency vs. pressure ratio at 1300 rpm

Figures 6 and 7 present the overall isentropic efficiency at various pressure ratios at the two speeds, 900 and 1300 rpm, respectively. Looking at the 900 rpm curves in Figure 6, it can be seen that the spread between all condensing temperatures is in the range of 3.5% and the slope of the loss in overall isentropic efficiency per pressure ratio is 5.5%. Comparing that behavior to the 26.7ºC (80ºF) condensing line, it can be seen that this falls out of line with the balance of the data. Referring to Figure 8, it can be seen that the total shaft power required at 26.7ºC (80ºF) condensing is quite small and as such the frictional loses become disproportionately high causing the overall isentropic efficiency to drop at a faster rate. This drop continues to increase as the pressure ratio gets lower. The 1300 rpm curve in Figure 7 shows that the range of overall isentropic efficiency has closed to 1.5% at pressure ratios less than 3.5 down from 2.5% at 900 rpm. The range of overall isentropic efficiency at pressure ratios greater than 3.5 increases to 4%. The slope of the loss of overall isentropic efficiency per pressure ratio has also improved from 5.5% to 3.8%. While the torque has increase slightly with increased speed the improvement in Volumetric performance has more than offset the incremental torque resulting in improved overall isentropic efficiency.

From Figures 6 and 7, it can be seen that the peak efficiency occurs at a pressure ratio of 2 in the 900 rpm case at 78%. The efficiency at 2.5 pressure ratios in the 1300 rpm case is 77% with the trend increasing with lower pressure ratio. Lower pressure ratio points were unattainable due to test stand limitations. The volumetric efficiency is approximately the same at both speeds and thus, the overall isentropic efficiency is being degraded by the increase in torque. Looking at a pressure ratio of 4 and a condensing temperature of 48.9°C (120°F) it can be seen that the overall isentropic is 67% at 900 rpm but improved to 70% at 1300 rpm. This is due to the improving slope of the volumetric efficiency with speed.

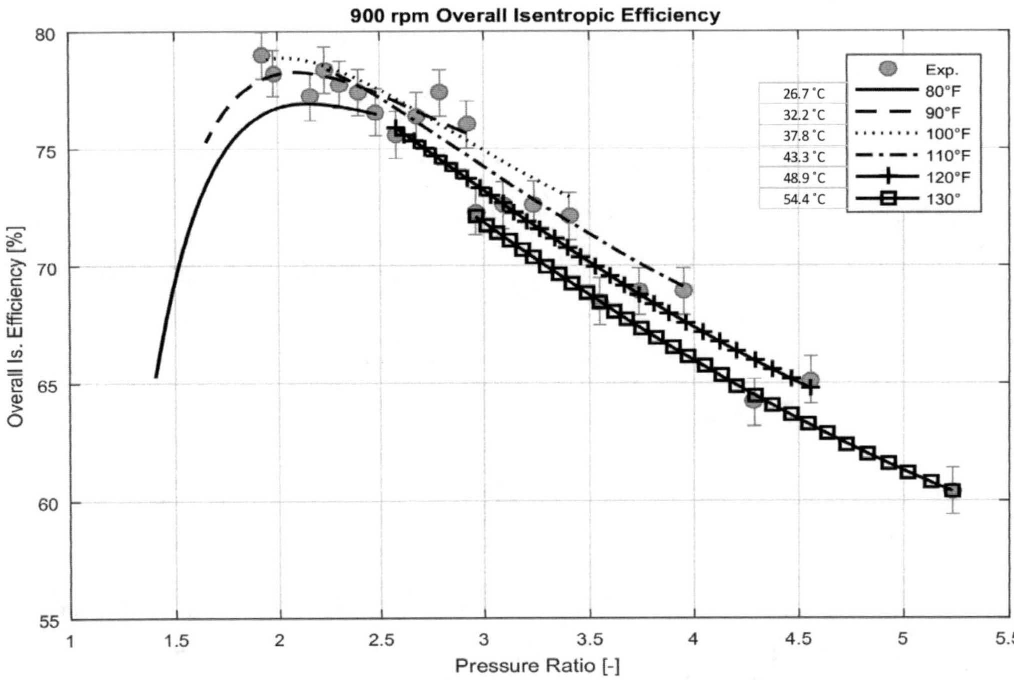

Figure 6: 141 kW spool compressor overall isentropic efficiency ($\eta_{is,0}$) vs. pressure ratio at 900 rpm

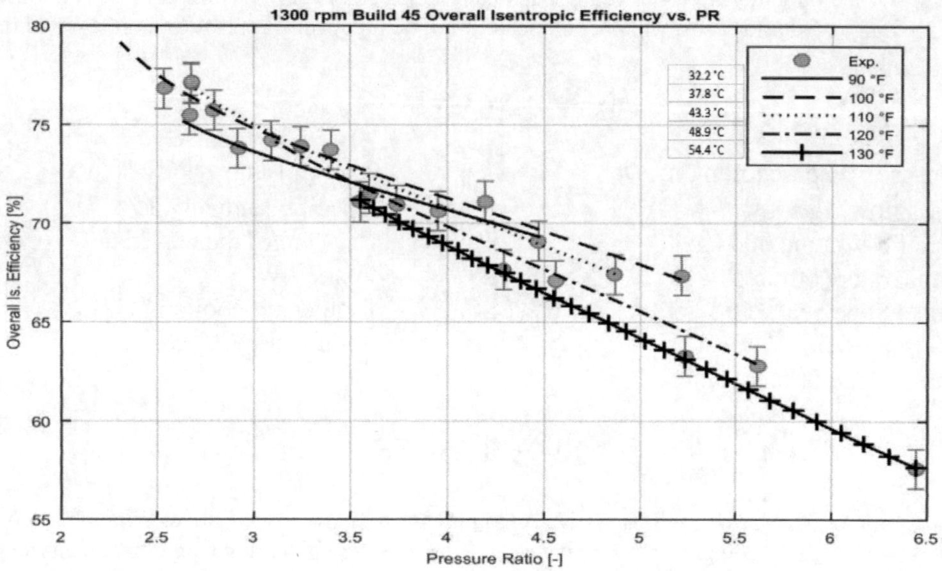

Figure 7: 141 kW spool compressor overall isentropic efficiency ($\eta_{is,0}$) vs. pressure ratio at1300 rpm

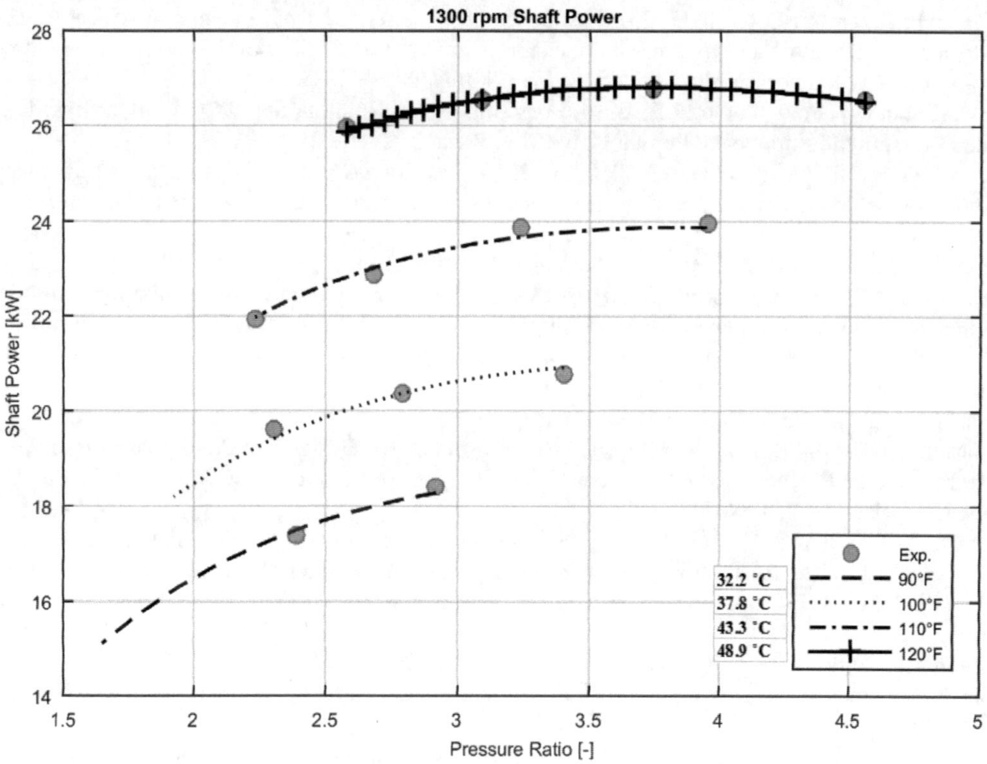

Figure 8: 141 kW spool compressor shaft power vs. pressure ratio

5. CONCLUSIONS

An experimental characterization of a prototype 141 kW (40 tonsR) R134a spool compressor is presented. This compressor design was based on the comprehensive modeling output presented in Bradshaw *et al.* (2016). A description of the custom-built hot-gas bypass test stand is included. The prototype compressor displayed good efficiency across a wide operating range. Additionally, the spool compressor displayed a very good capability to perform at a wide range of shaft speeds, which could lead to efficient capacity control when integrated into a unit.

NOMENCLATURE

V_{Th}	Compressor Displacement, m^3/h	$\eta_{is,o}$	Overall Isentropic Efficiency
$\dot{m}$	Mass flow, kg s-1	h	Specific Enthalpy, kJ kg-1
v	Specific Volume, m3 kg-1	P	Pressure of the gas, kPa
T	Temperature of the gas, K	$\dot{w}$	Work, kW
SSH	Intake Superheat, ºC	LSC	Liquid sub-cooling, ºC
Pr	Pressure Ratio	η_{vol}	Volumetric Efficiency
Rpm	Speed, Revolutions / min		

REFERENCES

Bradshaw, C.., Kemp, G., Orosz, J., Groll, E., 2012. A Comprehensive Model of a Novel Rotating Spool Compressor. In: Proceedings of the International Compressor Engineering Conference. Purdue University, West Lafayette, IN USA. No. 1142.

Bradshaw, C. R. and Groll, E. A. ,2013. A comprehensive model of a novel rotating spool compressor. International Journal of Refrigeration, 36(7):1974 – 1981. New Developments in Compressor Technology. Bradshaw, C.., Kemp, G., Orosz, J., Groll, E., 2014. Influence of Volumetric Displacement and Aspect Ratio on the Performance Metrics of the Rotating Spool Compressor. In: Proceedings of the International Compressor Engineering Conference. Purdue University, West Lafayette, IN USA. No. 1177.

Bradshaw, C.., Kemp, G., Orosz, J., Groll, E., 2014. Loss Analysis of Rotating Spool Compressor Based on High-Speed Pressure Measurements. Purdue University, West Lafayette, IN USA. No. 1178.

Bradshaw, C. R., Kemp, G., Orosz, J., and Groll, E. A., 2016. Development of a loss pareto for a rotating spool compressor using high-speed pressure measurements and friction analysis. Applied Thermal Engineering, 99:392–401.

Fox, R, McDonald, A., Pritchard, P., 2004. Introduction to Fluid Mechanics, 6th Edition, John Wiley and Sons.

Bradshaw, C.., Kemp, G., Orosz, J., Groll, E., 2016a. Design Methodology Improvements of a Rotating Spool Compressor using a Comprehensive Model. Purdue University, West Lafayette, IN USA. No. 1376.

Kemp, G., Elwood, L.., Groll, E., 2010. Evaluation of a Prototype Rotating Spool Compressor in Liquid Flooded Operation. In: Proceedings of the International Compressor Engineering Conference. Purdue University, West Lafayette, IN USA. No. 1389.

Kemp, G., Garrett, N., Groll, E., 2008. Novel Rotary Spool Compressor Design and Preliminary Prototype Performance. In: Proceedings of the International Compressor Engineering Conference. Purdue University, West Lafayette, IN USA. No. 1328.

Kemp, G., Orosz, J., Bradshaw, C., Groll, E., 2012. Spool Compressor Tip Seal Design Considerations and Testing. In: Proceedings of the International Compressor Engineering Conference. Purdue University, West Lafayette, IN USA. No. 1258.Kemp, G., Orosz, J., Bradshaw, C., Groll, E., 2012. Spool Seal Design and Testing for the Spool Compressor. In: Proceedings of the International Compressor Engineering Conference. Purdue University, West Lafayette, IN USA. No. 1259.

Orosz, J., Kemp, G., Bradshaw, C., Groll, E., 2012. Performance and Operating Characteristics of a Novel Rotating Spoll Compressor.

Orosz, J., Kemp, G., Bradshaw, C. R., and Groll, E. A. (2014). An update on the performance and operating characteristics of a novel rotating spool compressor. In International Compressor Engineering Conference at Purdue University, number 1378.

Theoretical Study of Seal Spring in a Wankel Compressor

De-lou Zhang[1,2],Yu-ting Wu[1,2]*,Jing-fu Wang[1,2],
Chun-xu Du[1,2],Xia Chen[1,2],Rui Ma[1,2],Chong-fang Ma[1,2]

[1]Key Laboratory of Enhanced Heat Transfer and Energy Conservation,
Ministry of Education and Key Laboratory of Heat Transfer and Energy Conversion,
Beijing municipality

[2]College of Environmental and Energy Engineering,
Beijing University of Technology, Beijing 100124, China

ABSTRACT

The successful fabrication of a miniature Wankel compressor relies on its seal and seal spring.The factors that influence the seal of a miniature Wankel compressor mainly include axial double mechanical seal and apex seals.Axial mechanical seal depends on machining precision on the end face.In the compression chamber, the springs for seal flake on the Wankel rotor put pressure on the internal face of the cylinder ,which generates the desired amount of pressing force.When springs fails,high-pressure chamber and low-pressure chamber will be connected.Spring performance will directly affect the efficiency of a Wankel compressor.In this paper,we aim to introduce kinematic analysis of apex seals and force analysis of springs in a miniature Wankel compressor.

Keywords: seal; seal spring; Wankel compressor; kinematic analysis; force analysis

1. INTRODUCTION

At present,with the increasing requirement for temperature control of personal air-conditioning system,the development of personal air-conditioning facility has become extremely urgent.Meso- and micro-compressors act as the core parts of a cooling system,and Wankel-type units have become prime choices for micro-cooling systems because of their unique advantages,including simplicity in structure,high efficiency,long life,low vibration,smaller and lighter.Several organizations have obtained some achievements in research on the micro-Wankel engine and Wankel compressor.Heppner et al. analyzed the leakage flow and the friction loss of rotary engine and compressor and estabished design parameters for micro engine sealing systems.Hsiao et al.presented the simulation of a rotary compressor and its performance comparison with measured results.Prater and William described the methodology and results of an experiment to measure the fundamental undamped natural frequency and damping ratio for the discharge reed valve in a rolling piston rotary compressor.

The Wankel compressor is a kind of rotary machine.It is similar to the Wankel engine and rotary compressor in terms of structure and operation;it consists of a rotor,a cylinder,a crankshaft,a pair of gears and apex seals,compression chambers. As shown in Figure1,a Wankel compressor consists of three compression chambers,and each chamber is bounded by the internal surface of the cylinder,rotor,two endplates,and apex seals.High-pressure chamber and low-pressure chamber are separated by the apex seals,to satisfy the demand for sealing in the working chamber.From the above discussion,it appears that the material and structure of seal is one of the key problems to research the miniature Wankel compressor at present.

In this paper,we aim at the research of all the seals and seal springs in a Wankel compressor,including the analysis of the motions and forces of seal springs,the design of seals.Finally, seals and springs are manufactured based on the theoretical analyses.

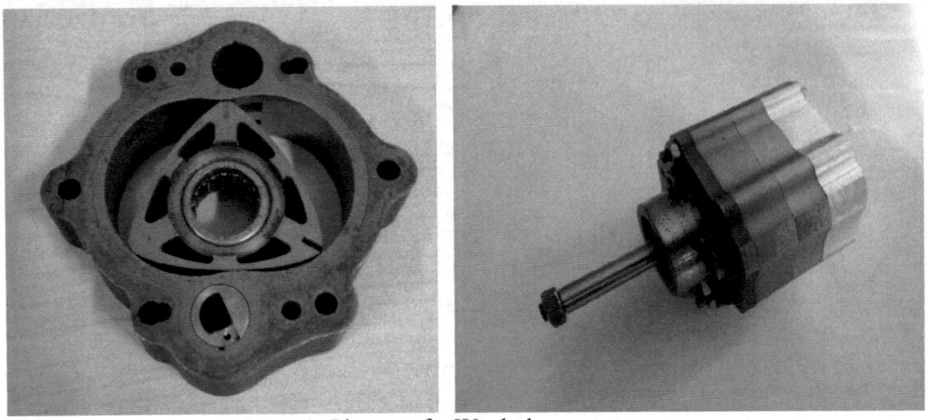

Figure1: Picture of a Wankel compressor

2. THE PRINCIPLE OF WANKEL ROTARY COMPRESSOR

The cylinder profile of a Wankel rotary compressor is epitrochoid, and the rotor is inner driving drum envelope of the epitrochoid.The eccentricity *e* is the distance bewteen cylinder center and rotor center.The operating principle of Wankel compressor is shown in Figure2. The rotor ,which slides around the internal face of the cylinder,spins round the axis of the cylinder,and on its own axis.A pair mesh gear is shown in Figure2, and the gear ratio is 3:2.
The diagram of the complete working process of a Wankel compressor is shown in the Figure2.Figure a shows the intake process,and the pressure remains unchanged until the volume of the intake chamber reaches its maximum. The compression process begins with the intake vavle closing,as illustrated in Figure b.The operation transforms into an exhaust process while the pressure reaches the piont of exhaust pressure,as demonstrated in Figure c.In Figure d,the gas pressure is kept stable until the minimum volume of the exhaust chamber is attained ,after which the exhaust value is shut.From the Figure 1,it is not difficult to find that there are three chambers working in parallel. That is to say,intake-compression-exhaust process are achieved in two chambers while the eccentric shaft completes a revolution.

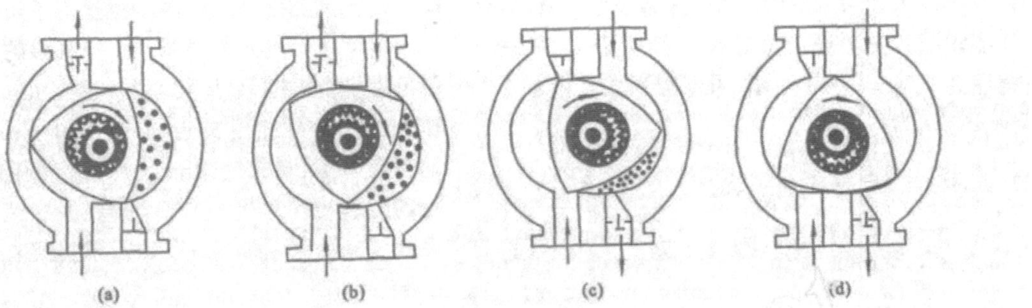

Figure 2: Schematic diagram of a Wankel compressor

3. KINEMATIC ANALYSIS OF THE SEAL IN A WANKEL COMPRESSOR

3.1Dynamics of a point on a body with translation and rotation

Figure3 illustrates the seal and the spring in a Wankel compressor,and the motion diagram of seals in the compression chamber is shown in Figure 4. With the effect of springs,the seals leans against the internal surface of the cylinder,which generates the desired amount of force.Once the spring fails,the performance of a Wankel compressor may be worse or lost.

As is shown in Figure 4,there are three apex seals equispaced on the rotor,120° apart.A Plane-Rectangular coordinate system will be established to describe a point at the apex of one of the rotor locations,including its acceleration, velocity,movement.

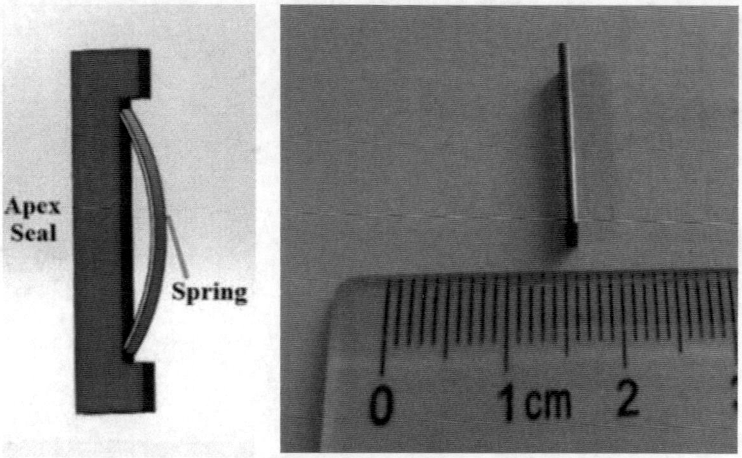

Figure 3: The three-dimensional diagram of the Apex seal

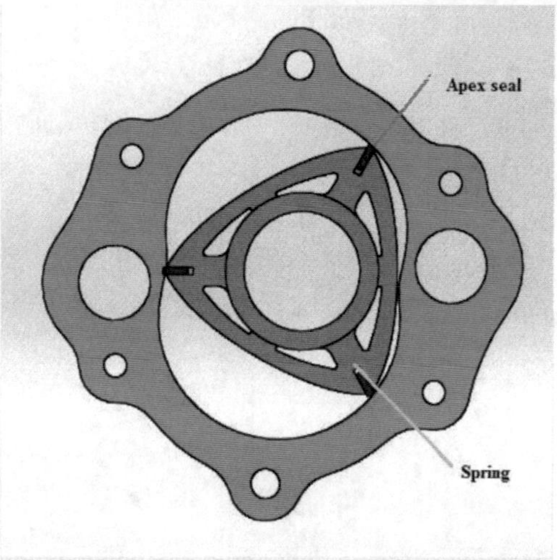

Figure 4: Position of Apex seal in the compressor

As one of the mainly reseach subjects,the piont P represents the location of the apex seals and coordinate system S_1 is the fixed frame of reference and coordinate system S_2 rotates with the rotor.O and O' are the centers for coordinate systems S_1 and S_2.In coordinate systems S_1 and S_2,θ_1 and θ_2 signifies the rotation angles of the eccentric shaft and rotor,respectively. $\vec{\omega}_1$ and $\vec{\omega}_2$ are the rotation velocities of the eccentric shaft and rotor,respectively.The position vector from coordinate system O to coordinate system O' is $\vec{R}$. $\vec{\rho}$ is the position vector from O' to the apex seal.

From the Figure 5,$\vec{r}$,the position vector,for point P is given as follows

$$\vec{r} = \vec{R} + \vec{\rho} \tag{1}$$

Differentialing Equation (1) with respect to time results in the velocity with respect to the fixed coordinate system O as

$$\vec{V} = \dot{\vec{r}} = \dot{\vec{R}} + \dot{\vec{\rho}} \tag{2}$$

where

$$\dot{\vec{\rho}} = \left(\dot{\vec{\rho}}\right)_r + \vec{\omega} \times \vec{\rho} \tag{3}$$

where ω is the absolute rotation rate of coordinate system O'.

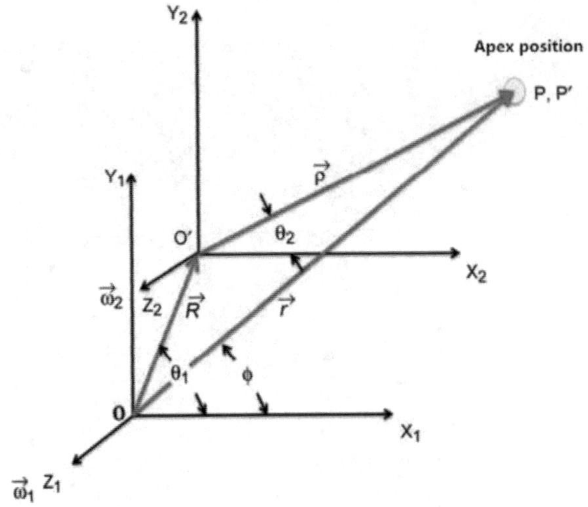

Figure 5: Coordinate system used for kinematic analysis

Therefore,the apex velocity $\vec{V}$ of the spring is given by

$$\vec{V} = \dot{\vec{R}} + \left(\dot{\vec{\rho}}\right)_r + \vec{\omega} \times \vec{\rho} \tag{4}$$

Piont P' is coincident with P in coordinate systems O' but fixed in this coordinate system.

$\dot{\vec{R}}$ illustrates the absolute velocity of O' and $\vec{\omega} \times \vec{\rho}$ is the velocity of P' relative to O' as viewed by a non-rotating observer.That is to say, $\dot{\vec{R}} + \vec{\omega} \times \vec{\rho}$ is the absolute velocity of P'.The term $\left(\dot{\vec{\rho}}\right)$ is the velocity of P relative to O' as seen by an observer rotating with the O' coordinate system.

Differentialing Equation (4)with respect to time can be expressed as

$$\frac{\partial}{\partial t}\left(\vec{V}\right) = \frac{\partial}{\partial t}\left(\dot{\vec{R}}\right) + \frac{\partial}{\partial t}\left(\left(\dot{\vec{\rho}}\right)_r\right) + \frac{\partial}{\partial t}\left(\vec{\omega} \times \vec{\rho}\right) \tag{5}$$

where the terms of equation become

$$\frac{\partial}{\partial t}\left(\dot{\vec{R}}\right) = \ddot{\vec{R}}$$

$$\frac{\partial}{\partial t}\left(\left(\dot{\vec{\rho}}\right)_r\right) = \left(\ddot{\vec{\rho}}\right)_r + \vec{\omega} \times \left(\left(\dot{\vec{\rho}}\right)_r\right)$$

$$\frac{\partial}{\partial t}\left(\vec{\omega} \times \vec{\rho}\right) = \dot{\vec{\omega}} \times \vec{\rho} + \vec{\omega} \times \dot{\vec{\rho}} = \dot{\vec{\omega}} \times \vec{\rho} + \vec{\omega} \times \left(\left(\dot{\vec{\rho}}\right)_r\right) + \vec{\omega} \times \left(\vec{\omega} \times \vec{\rho}\right)$$

taking these terms into Equation (5)results in the acceleration $\vec{a}$ being given as follows:

$$\vec{a} = \ddot{\vec{R}} + \dot{\vec{\omega}} \times \vec{\rho} + \vec{\omega} \times \left(\vec{\omega} \times \vec{\rho}\right) + \left(\left(\ddot{\vec{\rho}}\right)_r\right) + 2\vec{\omega} \times \left(\left(\dot{\vec{\rho}}\right)_r\right) \tag{6}$$

where

$\ddot{\vec{R}}$ represents the absolute acceleration of O', $\dot{\vec{\omega}} \times \vec{\rho}$ and $\vec{\omega} \times \left(\vec{\omega} \times \vec{\rho}\right)$ is the acceleration of P' relative to O' as viewed from a non-rotating observer.In other words with $\dot{\vec{\omega}} \times \vec{\rho}$ is the tangential acceleration and $\vec{\omega} \times \left(\vec{\omega} \times \vec{\rho}\right)$ is the centripetal acceleration of P towards the axis of rotation through O'.Vector $\left(\left(\ddot{\vec{\rho}}\right)_r\right)$ is the acceleration of point P relative to the O' coordinate system as viewed by an observer moving with O'.Vector $2\vec{\omega}\times\left(\left(\dot{\vec{\rho}}\right)_r\right)$ is the Coriolis acceleration.

3.2 Application to a rotary Wankel compressor

For the Wankel rotary compressor(Table 1),the following formulas are given as known due to the timing mechanism among the crankshaft,the motor and the internal-external gearing.In the micro-compressor, a 20-tooth external gear is fixed to the rear cover.A 30-tooth internal gear rotates about the fixed external gear with the crankshaft offset equal to the center distance difference of the meshing external- internal gears.This results in a 3:1 angular and rotational relationship between the crankshaft and the motor.

Table 1: Parameters of Wankel compressor

Parameter	Value	Parameter	Value
generation radius R	0.018 m	thickness of seal b	0.001 m
n_2	20	n_1	30
height of apex seal h	0.0035 m	eccentric distance e	0.003 m
mass of apex seal m	0.2873 g	length of apex seal l	0.015 m
displacement	8.4 cc	offset a	0.001 m

So the speed of the crankshaft and rotation relationships are given by

$$\theta_1 = 3\theta_2,\quad \omega_1 = 3\omega_2$$

Since $\left(\left(\dot{\vec{\rho}}\right)_r\right)$=0,point P does not move relative to O' for a rotating observer.The equation for the apex seal velocity can be described as

$$\vec{V} = \dot{\vec{R}} + \vec{\omega}\times\vec{\rho} \tag{7}$$

The acceleration a can be simplified as follows(since $\dot{\vec{\rho}}=0$,then $\ddot{\vec{\rho}}=0$):

$$\vec{a} = \ddot{\vec{R}} + \vec{\omega}\times\left(\vec{\omega}\times\vec{\rho}\right) \tag{8}$$

According to the Figure 5,the position vector for the apex seal is expressed as

$$\vec{r} = \begin{Bmatrix} |R|\cos\theta_1 + |\rho|\cos\theta_2 \\ |R|\sin\theta_1 + |\rho|\cos\theta_2 \\ 0 \end{Bmatrix} \tag{9}$$

So the velocity vector is given as

$$\vec{V} = \begin{Bmatrix} -\omega_1|R|\sin\theta_1 - \omega_2|\rho|\sin\theta_2 \\ \omega_1|R|\cos\theta_1 + \omega_2|\rho|\cos\theta_2 \\ 0 \end{Bmatrix} \tag{10}$$

The acceleration vector can be obtained by

$$\vec{a} = \begin{Bmatrix} -\omega_1^2|R|\cos\theta_1 - \omega_2^2|\rho|\cos\theta_2 \\ -\omega_1^2|R|\sin\theta_1 - \omega_2^2|\rho|\sin\theta_2 \\ 0 \end{Bmatrix} \tag{11}$$

In a Wankel compressor,the Equation (9) can be used to develop for the path of the apex seal for a given crankshaft offset and the distance from the center of the motor to the apex seal location.The path the seal takes for one revolution is shown in Figure 6.This is approximately the same geometry as the static cylinder that the seal is in contact with during operation and will be assumed to be the path of the apex seal in this paper.

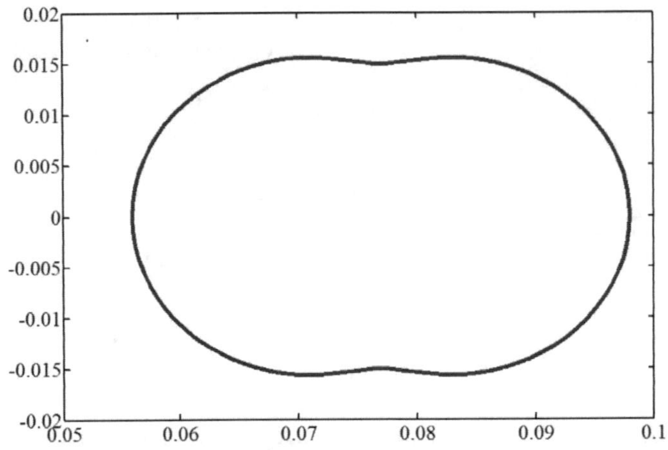

Figure6: Apex seal path

Utilizing the MATLAB software,the velocity of the seal as a function of crankshaft rotation angle found from Equation (10) is shown in Figure 5.And the seal velocity is presented as that in the fixed coordinate system.The velocity is given for the speed of crankshaft equal to 1500rpm.Meanwhile, the acceleration of the apex seal is shown in the fixed coordinate directions.As a function of crankshaft rotation angle at 1500rpm,the acceleration is presented by evaluating Equation (11) in Figure 8.

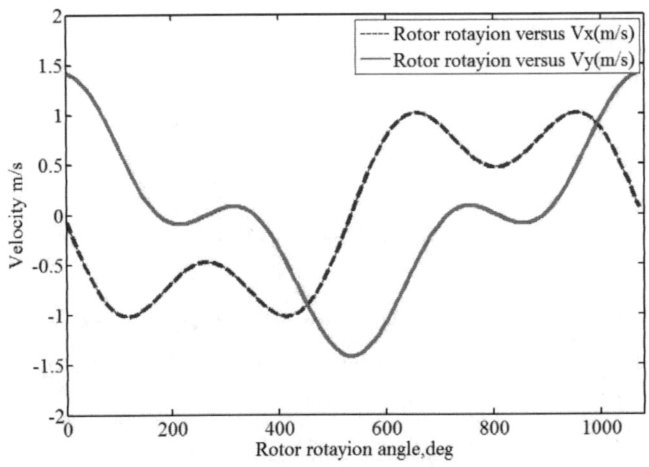

Figure7: Velocity of the apex seal in the coordinate axis

At this point, it would be useful to put the same calculation methodology into a coordinate system that is radial and transverse to that of the apex seal itself.This will be done via the coordinate system shown in Figure 9.The radial and transverse directions are along vector $\vec{\rho}$ and perpendicular to this vector, respectively(From Figure 9).In order to express the acceleration in terms of a radial and transverse direction,the acceleration vector should be transformed into these directions.The acceleration can be expressed as:

$$\begin{Bmatrix} a_r \\ a_t \end{Bmatrix} = \begin{vmatrix} \cos\theta_2 & \sin\theta_2 \\ -\sin\theta_2 & \cos\theta_2 \end{vmatrix} \begin{Bmatrix} a_x \\ a_y \end{Bmatrix} \tag{12}$$

Using Equation (12)the acceleration shown in Figure after transformation is shown in Figure 10.

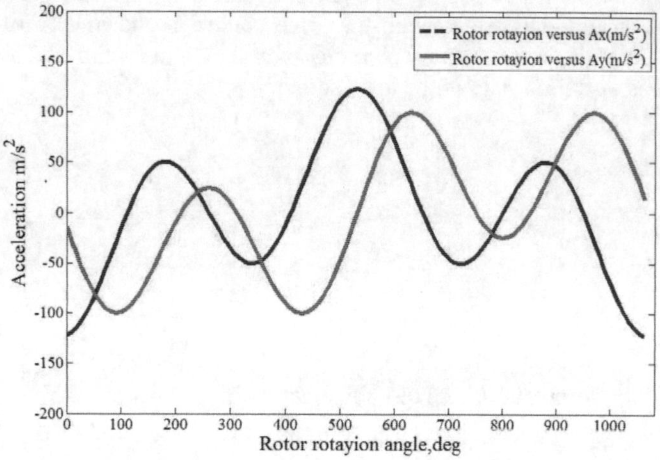

Figure 8: Acceleration of the apex seal in the coordinate axis

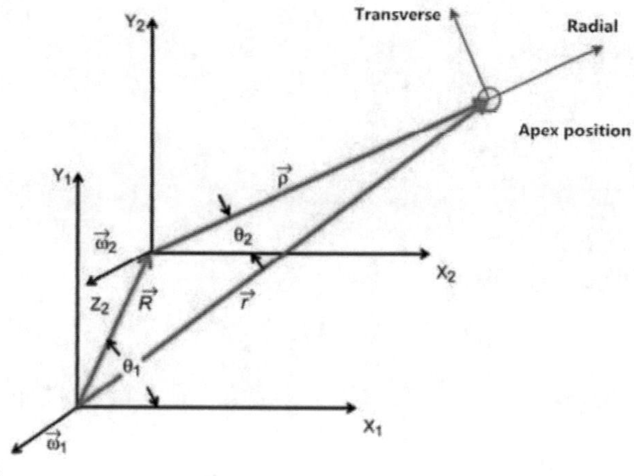

Figure 9: Coordinate system for radial and transverse components for the apex seal

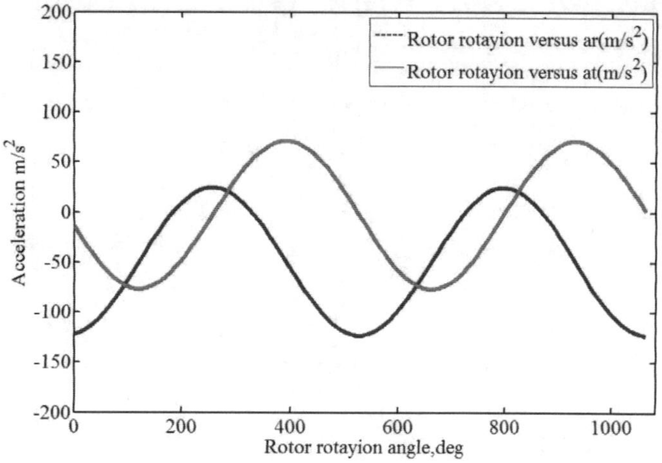

Figure 10: Acceleration components in the radial and transverse apex seal directions

4. FORCE ANALYSIS OF THE SEAL SPRING

The apex spring plays an vital role in compacting internal surface of the cylinder,which may lead to its plastic deformation and abrasion. It is important to research the force condition and motion of the springs.A schematic drawing of a plate-spring ,used in the Wankel rotary compressor,is shown in Figure 11.The black line represents the original state,and the red one is operating state of the apex spring.

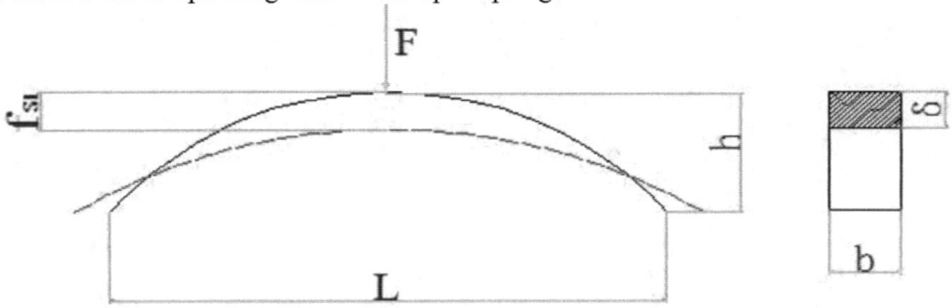

Figure11: The free-body diagram of apex spring

The stress condition of the spring is the worst,while the seal is along the short axis.As shown in Figure 12,the component force of spring is zero in the horizontal direction,and the resultant force F_R is worked out as follows:

$$F_R = F_r + F_{gr} + F_{sl} + F_m \tag{13}$$

where, F_r-- the radial inertial force of seal;F_{gr}--the back pressure ; F_{sl}--the force of spring; F_m--the friction bewteen the seal and the internal face of the cylinder.

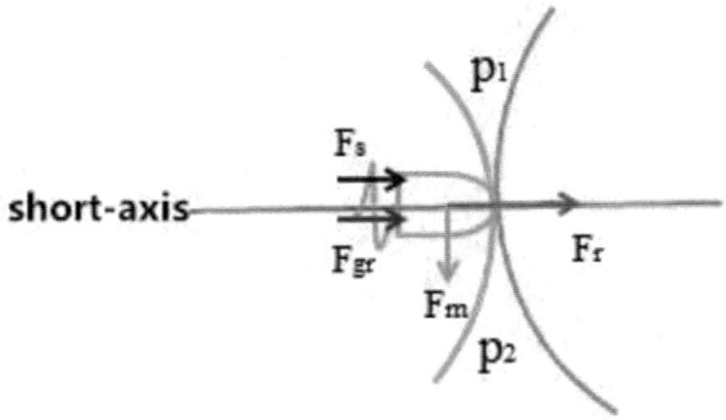

Figure 12: The free-body diagram of apex spring in short-axis direction

$$F_r = -ma_R = \frac{W}{g}\omega^2\left(\frac{R}{9} + e\cos\frac{2}{3}\alpha\right) \tag{14}$$

where,W-- the weight of the seal;g--acceleration of gravity;ω--the angular velocity of the eccentric shaft;R--the distance bewteen the center of gravity of seal and the rotor center;e--eccentric distance;α--the angle of the eccentric shaft.

$$F_{gr} = lBP_2 - lP_2\left(\frac{B}{2} - \alpha\sin\phi\right) - lP_1\left(\frac{B}{2} + \alpha\sin\phi\right) \tag{15}$$

where,l--the length of the seal;B--the width of the seal; ϕ--the swinging angle of the seal;P_2--the pressure of high-pressure chamber;P_1--the pressure of low-pressure chamber.

When the seal is along the short axis(*α=270° or 810°),* the radial inertial force of seal F_r has its minimum value:

$$F_{r\min} = \frac{W}{g}\omega^2\left(\frac{R}{9} - e\right) \tag{16}$$

At different speed of the motor,the F_{rmin} are shown in the following table2:

Table 2 : F_{rmin} at different motor speed

Speed (rpm)	*300*	*500*	*800*	*1000*	*1500*	*2000*	*2500*	*3000*
F_{rmin}	*0.003*	*0.008*	*0.02*	*0.031*	*0.072*	*0.125*	*0.196*	*0.288*

At the same time,F_{gr} is simplified as follows:

$$F_{gr} = l\frac{B}{2}(P_2 - P_1) = 0.075(P_2 - P_1) \quad (17)$$

The relationship between Frmin and Fgr will be discussed in this section:

(1)While the Wankel compressor is in the startup,the speed of the motor is about 300r/min:

$P_2=P_1$

Due to its low speed,the pressure in the high-pressure chamber is equal to that in the low-pressure chamber.The back pressure F_{gr} is zero.

(2)After the Wankel compressor start:

a.$P_2>>P_1$

The high-pressure chamber is in the compression process.The pressure in the high-pressure chamber is much greater than that in the low-pressure chamber.The absolute value of the back pressure is greater than the absolute value of the radial inertial force of seal.

b.$P_2>0,P_1<0$.

The high-pressure chamber is in the exhaust process,and the pressure is positive pressure($P_2>0$).At the same time,the gas pressure is negative pressure during the intake process in the low-pressure chamber($P_1<0$).The absolute value of the back pressure is greater than the absolute value of the radial inertial force of seal.

Moreover,the condition of contacting the seal with the internal surface of the cylinder is as follows:

$$F_R = F_r + F_{gr} + F_{sl} + F_m > 0 \quad (18)$$

In summary,only if the absolute value of the force of spring is greater than the radial inertial force of seal($F_{sl}>F_{rmin}$=0.0003N),the above relationship can be satisfied.And the back pressure F_{gr} plays a key role in achieving the effect of the seal,after which the Wankel compressor turns into the operation stage.The force of spring F_{sl} is so small that can be ignored.

In conclusion, in order to ensure the seal to press stably against the internal surface of the cylinder,the spring was designed when the Wankel compressor stops running.Meanwhile,the spring was used to counteract the centripetal force of the seal.Many factors should be concerned on the determination of the force of spring F_{sl},including the force of friction,the abrasion of the seal,we have increased the safety margin of the force.

the deflection of spring(initial amount of compression) is given by:

$$f_1 = \frac{F_{sl}l^3}{4Eb\delta^3} \quad (19)$$

So, the thickness of spring is worked out as follows:

$$\delta = \sqrt[3]{\frac{F_{sl}l^3}{4Ebf_1}} \quad (20)$$

In this study,65Mn is used to manufacture the spring,and the parameters of the spring is:

$$b = 0.8mm,\ h = 2.5mm,\ l = 9.2mm$$

the thickness of the spring is achieved:

$$\delta = 0.2mm$$

the maximum stress of the spring is expressed by:

$$\sigma_{\max} = \frac{3F_{sl}l}{2b\delta^2} \le [\sigma_{0.01}] \quad (21)$$

where: f_l--the deflection of spring(initial amount of compression);F_{sl}-- the initial force;l--the span of spring; b--the width of spring; E-- the elastic modulus(65Mn); $[\sigma_{0.01}]$--elastic limit(65Mn).
From the formula (21),the maximum stress of the spring is less than the limiting stress at given operational conditions. That is to say,the requirements of sealing system in our Wankelcompressor will be satisfied.As shown in Figure13,the seal and seal spring have been applied in our Wankel compressor.

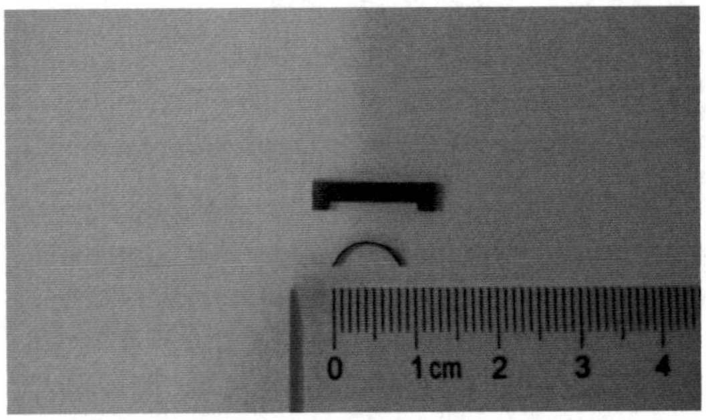

Figure13: Seal and Seal spring in our laboratory

5. CONCLUSIONS

This paper reviewed the research work of Wankel engine and Wankel compressor in recent years, provided a simple mathematical model of rotor, and discussed the influence of seal spring.Through those works, three conclusions are obtained:

(1)The dynamics of the rotor of a Wankel compressor have been described in this paper.The analytical treatment developed can be utilized for any rotary Wankel compressor given the crankshaft,rotor design parameters.The apex seal path(the cylinder profile), velocity and acceleration were determined.

(2)The seal spring plays a decisive role in seal system and in the performance of the Wankel compressor.If the seal spring failed,the low and high pressure chambers would be connected.And then the performance of the Wankel compressor will weaken or be lost.

(3)According to the analysis above,the factors causing seal spring in a Wankel compressor include its thickness (δ), span(*l*), material,and operational conditions.Among these parameters,the thickness and material are the most important factors that influence the performance of a Wankel compressor.

REFERENCES

[1]Fu,K.,Knobloch,F.Martinez,D.Walther,A.C.FernandezPello,A.Posano,and D.Liepmann.Design and fabrica-tion of a silicon-based MEMS rotary engine.*ASME International Mechanical Engineering Congress and Exposition*,New York,875-80

[2]Heppner,J.D.,D.C. Walther and A.P.Pisano.The design of ARCTIC:A rotary compressor thermally insulated μcooler. *Sensors and Actuators A*,134(1):47-56

[3]Lee,C.H.,K.C.Jiang,P.Jin,and P.D.Prewett.Design and fabrication of a micro Wankel engine using MEME technology.*Microelectronic Engineering*,73(4):529-34

[4]Zhang,Y.L.and W.Wang.Effects of leakage and friction on miniaturization of Wankel compressor.*Frontiers of Energy and Power Engineering in China*,5(1):83-92

[5]Zhang,Y.L.,and W.Wang.Factors affecting the cylinder displacement of a Wankel compressor in a micro-cooling system.*HVAC&R Research*,18(3):366-376

[6]Robert F. Handschuh,and A.Karl Owen.Analysis of apex seal friction power loss in rotary engines.2010

[7]Gordon R.Pennock,and John E.Beard.Force analysis of the apex seals in the Wankel rotary compressor including the influence of fluctuations in the crankshaft speed.*Mechanism and Machine Theory*,(32):349-361

[8]Ma G Y,Li H Q.*Rotary Compressor*.Beijing:Machinery Industry Press,2003

[9]Lu F,Yu N B.*Wankel Engine*.Beijing:National Defense Industry Prsee,1990

[10]Padhy S K.Dynamic analysis of a rotary compressor.*Journal of Mechanical Design*,116(2):639-646

[11]D.Greenwood.*Principles of Dynamics*.Prentice -Hall Inc,1988

[12]Hsiao W,Jiro Y,Takeshi A and Michio Y. Analysis of performance in a rotary compressor.*In:Proc of the 1982 ICECP, Purdue University, USA*,140-147

ACKNOWLEDGMENTS

The work was supported by the 973 Program' Project of China (2013CB228306).

A Novel Vapor Injection Structure on the Blade for Rotary Compressor

Baolong WANG[1]*, Xingru LIU[2], Wenxing SHI[3]

Department of Building and Science, School of Architecture,
Tsinghua University, Beijing, 100084, China
[1] Tel.: +86 10 62786571; Fax: +86 10 62773461; E-mail: wangbl@tsinghua.edu.cn
[2] Tel.: +86 18711346668; Fax: +86 10 62773461; E-mail: liuxr11@yeah.net
[3] Tel.: +86 10 62796114; Fax: +86 10 62773461; E-mail: wxshi@mail.tsinghua.edu.cn
* Corresponding Author

ABSTRACT

Heat pumps with rotary compressors operating in low temperature ambient will face heavy performance degradation. A novel vapor injection structure on the blade for the rotary compressor has been proposed in this paper to overcome the back-flowing drawback of the traditional cylinder injection structure and enlarge the injection area. Based on a verified numerical model, the performance of a rotary compressor with the blade injection structure has been investigated. The results indicate that: the proposed vapor injection structure can eliminate the back-flowing of the injected refrigerant; compared to the traditional injection structure, the new structure can enhance the volumetric efficiency and injection mass flow-rate by 1.8~2.7% and 26.6~57.2%, respectively; and compared to the traditional injection structure, the new structure can increase the heating capacity and COP by 23.1~48.9% and 3.2~8.0%, respectively.

1. INTRODUCTION

Rotary compressors have been extensively used in room air conditioners and household refrigerators for their advantages, including high efficiency, strong adaptability, and low cost. However, air source heat pumps with rotary compressors will face dramatically degradation as applied in cold regions [1]. Aimed at this problems, considerable research has been carried out and a series of solutions, including economizer technology [2], have been raised. At present, the economizer technology has proved an important technical solution to avoid the rapid performance degradation of heat pump in cold regions [3-4].

In the quasi two-stage compression system with an economizer, the refrigerant gas of the secondary loop is injected into the compression pocket through an injection port, which is also called vapor injection. Because only one compressor is used, the vapor injection technology is relatively cheap. So the vapor injection technology has been widely investigated and applied, especially in scroll compressors and screw compressors [5-9]. For the rotary compressor, most of the investigations of economizer technology focused on the twin-cylinder rotary compressor [10-11]. However, although the cost of a twin-cylinder rotary compressor is lower than two one-cylinder rotary compressors, there still exists an attempt to cut down the cost of the rotary compressor for low ambient temperature heating, which promotes the development of the gas injection technology in the one-cylinder rotary compressor.

Investigations on the gas injection into the one-cylinder rotary compressor is relatively seldom. Yan [12] and Jia [13] studied rotary compressors with vapor injection by experiments. Experimental results indicated that the system heating capacity was enhanced more than 12% by gas injection when outdoor temperatures was lower than -15°C. They also found that, compared to the twin-cylinder rotary compressor with economizer, the heating capacity and COP of the common one-cylinder rotary compressor with gas injection were both higher when ambient temperatures was higher than -15°C, which indicated the one-cylinder rotary compressor with gas injection will have a comparable even better performance than the twin-cylinder rotary compressor under mild working conditions.

In the one-cylinder rotary compressor with gas injection, the compressor injection port is often opened on the cylinder wall. In order to extend the injection time for injecting more refrigerant, the injection port is settled as close as possible to the discharge port. However, the limited area of the injection port and the unavoidable back-flowing of the injection refrigerant into the suction tube diminish the merits of gas injection on the one-cylinder rotary compressor. In this study, a new injection structure was put forward and the performance of the one-cylinder rotary compressor with the new injection structure was investigated.

2. THE NEW VAPOR INJECTION STRUCTURE

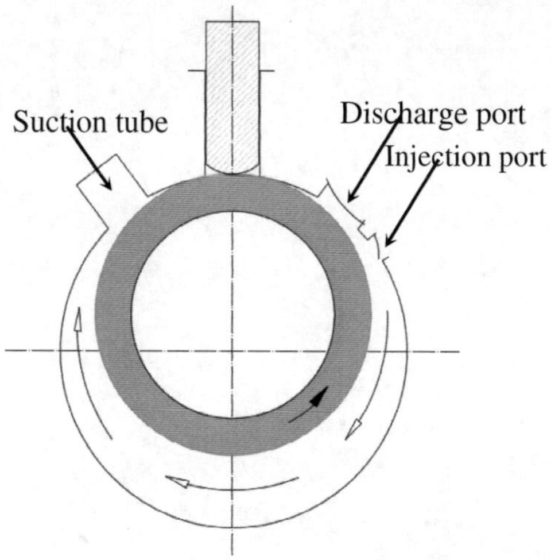

Fig. 1. Traditional injection structure

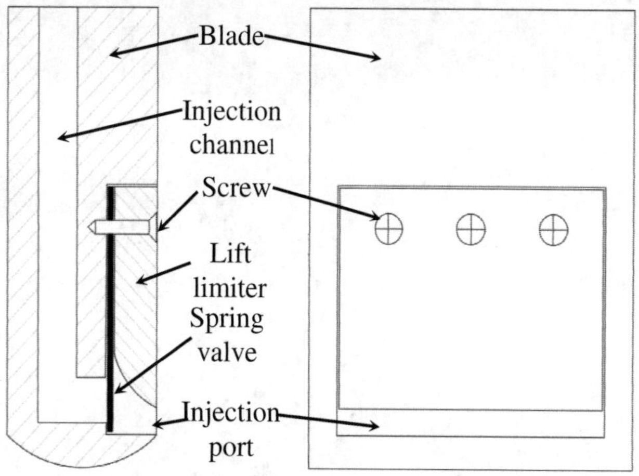

Fig. 2. Schematic of the blade with gas injection structure

Figure 1 shows a rotary compressor with traditional injection structure. The injection port is opened on the cylinder wall. In the beginning period of gas injection (just like the position in the Figure 1), the suction tube is still connected with the compression pocket. So the part of injected refrigerant will flow back to the suction tube, which will result in a decline of effective suction mass flow and compressor volumetric efficiency. At the same time, the area of the traditional injection port must be small in order to reduce the time of the back-flowing of the gas injection refrigerant.

In order to avoid the injected refrigeration gas flowing back into the suction tube and enlarge the area of injection port, a novel injection structure is proposed in Figure 2, in which the injection passage is screwed in the middle of the blade and the injection port is opened oriented to the discharge port. The new injection structure also need to mill a platform on the blade where the spring valve plate and the lift limiter are installed.

Figure 3 details the working process of the rotary compressor with the new injection structure. Before the suction pocket is isolated from the suction tube (Figure 3(a)), the blade don't protrude enough and the injection port is still sealed by the wall of the blade house. Under the press of the spring valve, the injection port is closed, and the injection process does not start. When the suction pocket is isolated from the suction tube, the suction process finishes (Figure 3(b)). At the same time, the blade moves down where the lower edge of the injection port exactly reveals, the injection port connects to the compression pocket and then the spring valve plate opens as the injection pressure is greater than the compression pocket pressure, starting injection. With increase of the rotation angle and increase of the compression pocket pressure, as the compression pocket pressure equals the injection pressure, the spring valve plate closes and the injection process ends, as shown Figure 3(c). Hereafter, the rolling piston rotates to the bottom center (Figure 3(d)), the initial discharge (Figure 3(e)) and the end of discharge (Figure 3(f)), respectively. The injection spring valve plate keeps close.

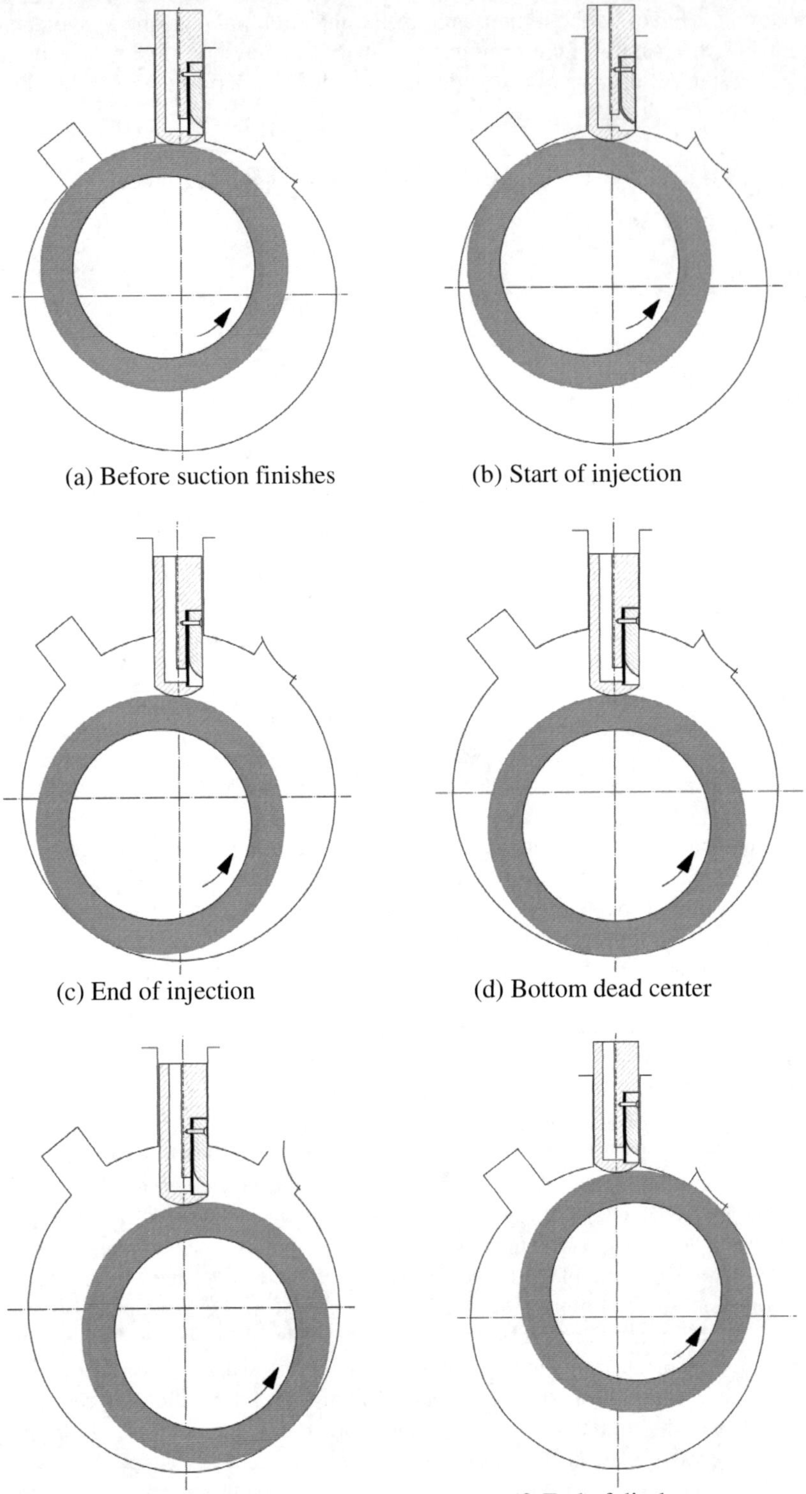

(a) Before suction finishes (b) Start of injection

(c) End of injection (d) Bottom dead center

(e) Initial discharge (f) End of discharge

Fig. 3. Working process of the rotary compressor with injection channel on blade

Therefore, the rotary compressor with the blade injection structure can totally avoid the back-flowing of the injection refrigerant into the suction tube. At the same time, it can achieve injection at the first time when the suction process finishes, and so it can achieve the maximum of injection mass flow. Moreover, the volume between the injection port and the spring valve plate at the injection structure has no influence upon the volumetric efficiency.

Because the volume does not connect to the suction tube, it has no influence on the compressor suction mass flow-rate from the suction tube.
The effects of setting gas injection structure in the blade on the pressure balance of the blade, lubrication and leakage also have been considered. Because the new injection port was designed to be narrow (1.0~2.0 mm in width) and long, and injection port is with a large area. After injection starts, the new injection structure can inject a lot of refrigerant rapidly at a short time and result in the injection process time being very short. The injection structure influence on the blade movement characteristic does not display a large difference compared to the traditional injection structure compressor, and the pressure balance can be achieved. Actually, the existence of the injection port and lift limiter can achieve better lubrication and leakage decrease. Besides, in order to ensure structural strength, the vertical injection channel could use multiple parallel round holes instead of a narrow and long injection channel.

3. METHODOLOGY

In order to investigate the performance of the proposed rotary compressor, a detailed distributed-parameter model of the rotary compressor with the new injection structure is built.

3.1 Modelling

The suction, compression and discharge process of the rotary compressor with gas injection is a typical thermodynamic process of the open system. The thermodynamic model of the rotary compressor could be deduced according to the energy-balance and mass-balance equations [14].
The geometric models include dynamic characteristics of the suction pocket, compression pocket and blade with the rotation angle change [14]. Before the rolling piston rotates from top dead center to the end of suction, the injection area of the blade injection structure equals 0. After the suction process finishes, the injection port opens and the injection area rapidly increases and reaches maximum at a short time. Afterward, the injection area remains unchanged until rotating to bottom dead center. The change rule of injection area with rolling piston rotating from bottom dead center to top dead center is symmetrical with the rolling piston rotating from top dead center to bottom dead center. The traditional injection structure (opening the injection port in cylinder wall close to the discharge port) needs to consider the throttling characteristic of the injection process. When the rolling piston is rotated to the injection port, the injection area is equal to the minimum value between the injection port area and three-dimensional surface area under the injection port and rolling piston.
A modified one-dimensional throttle nozzle model is used to calculate leakage process through radial clearance, blade clearance and clearance between the rolling piston and blade [15]. The researched compressor is with a high-pressure shell, which means the high-pressure gas discharged by compression pocket entries into the compressor shell, where the high-pressure gas exchanges heat with the motor, the outdoor environment and the suction tubes. The heat balance equation was applied to describe the heating exchange process in the compressor shell. The heat exchange between refrigerant gas and compression pocket in the compression process is also considered. The Dittus-Beolter equation is used to describe it under the assumption that the temperature change of the cylinder wall from the suction port to the discharge port is linear with rotation angle [16]. The Nusselt number in suction preheated process is assumed constant (Nu=3.66) as the suction tube is very short [17].

3.2 Solving procedure

The Visual Studio was used to solve equations because the solution procedure includes many submodules, and the method of variable step size was applied during calculating to reduce the calculation time. The method of 4th Order Runge-Kutta was used to improve calculation accuracy and accelerate the convergence process.

3.3 Model verification

In order to verify the accuracy of the model, the experimental data from two public papers [12] are collected for comparing. Figure 4(a) and (b) show the simulation results of a single-cylinder rotary compressor compared to experimental results [22], which indicates that the errors of the heating capacity and the power consumption are within 3%. Generally speaking, the developed models have relative high accuracy and can be used in the further research.

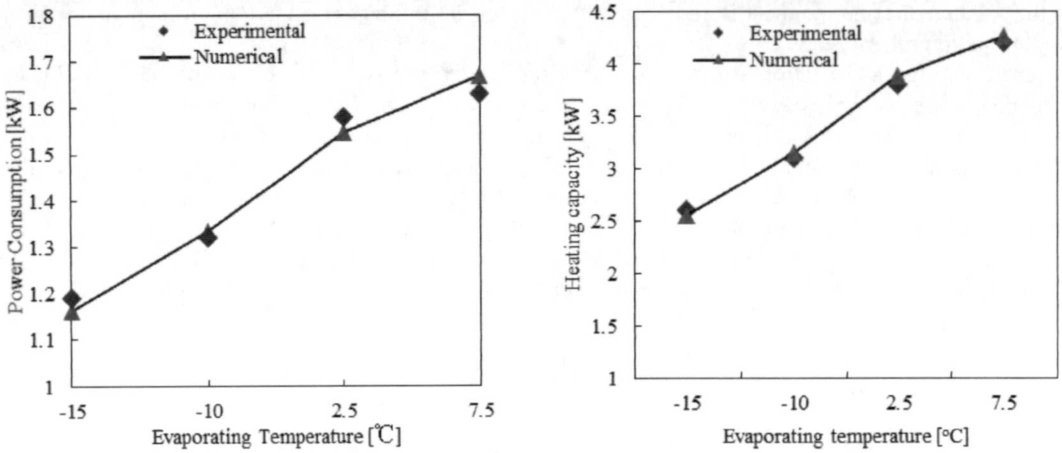

Fig. 4. Experimental verification for single rotary compressor

4. RESULTS AND DISGUSSION

4.1 Effect on inner compression process

The variations of the injection port areas of the traditional cylinder injection and the proposed blade injection with the change of the rotation angle are shown in Figure 5. According to Figure 5, the injection area of the traditional injection structure almost does not change when the rolling piston rotation angle is less than 300 degrees. With the rotation angle increase later, the rolling piston would be near to the injection port where the throttling effect appears, so the injection area decreased first, and then increased. When the rolling piston rotates from the position where the discharge process ends to the position where the suction process ends, the injection area of the traditional injection structure is not equal to 0, and the injection pressure is greater than the pressure of the suction pocket. Therefore, the process of the injection refrigerants back-flowing into the suction pocket is inevitable for rotary compressors with the traditional injection structure. However, the proposed injection structure opening injection port in the blade can totally avoid the back-flowing process, because its injection area equals 0 before the suction process finishes, and after the discharge process ends.

The non-back-flowing characteristic of the proposed compressor also can be observed by the instantaneous injection mass flow-rate through the injection port. Figure 6 shows the instantaneous injection mass flow-rate of different injection structures. Before the rolling piston rotates to the suction port, the injection mass flow-rate through the injection port of the traditional cylinder injection does not equal 0, then the injection refrigerant would flow back into the suction tube. However, when the injection port is opened in the blade, before the suction process is finished, the injection mass flow-rate keeps 0. After the rolling piston rotates over the suction port, the instantaneous injection mass flow-rate increases with the injection area increase rapidly, which means the blade injection structure can successfully avoid the injection refrigerant back-flowing into the suction tube.

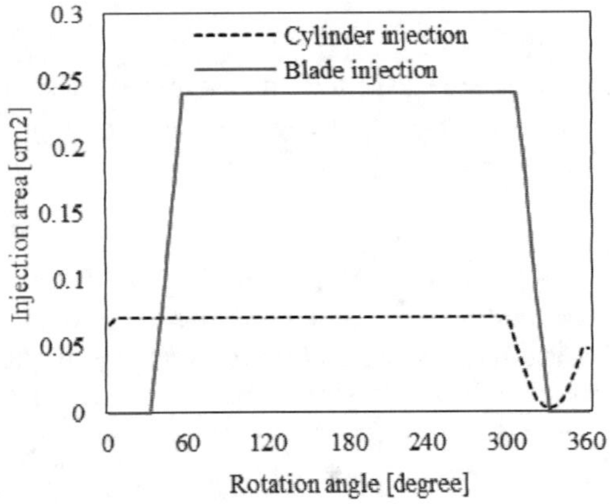

Fig. 5. Injection area

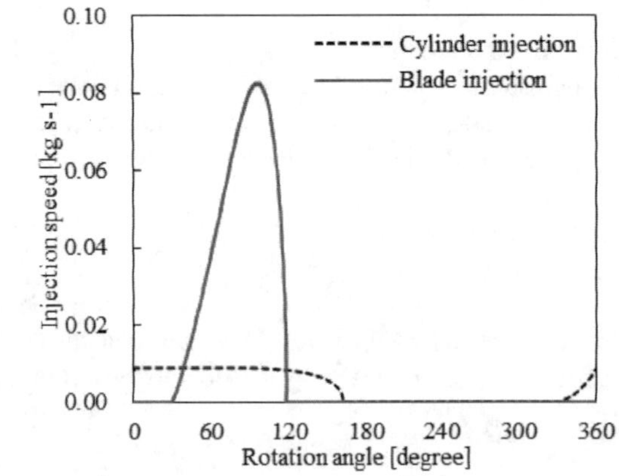

Fig. 6. Instantaneous injected mass flowrate

Moreover, it can found that the instantaneous injection mass flow-rate of the new injection structure is greater than the traditional injection structure, which can be attributed to the 2.5 times injection area of the new injection structure relative to the traditional injection structure. This much larger injection area also leads to the injection process finishes in a short time. After that, the pocket pressure is larger than the injection pressure and the injection spring valve plate closes.

To quantitatively describe the back-flowing, the back-flowing rate is defined, which is the ratio of the mass flow-rate of injection refrigerants directly back-flowing into suction tube divided by the total mass flow-rate of injection refrigerants and the leakage is not included. Figure 7 shows the back-flowing rates of the rotary compressor with different injection structures in condensing temperature 50°C. The back-flowing rates of the rotary compressor with the traditional injection structure increases from 23.3% to 29.3% with the increase of the evaporating temperature, but the back-flowing rate with the new injection structure remains 0.

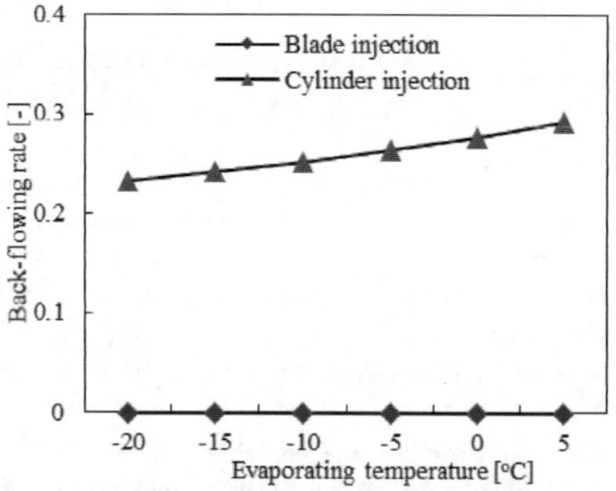

Fig. 7. Back-flowing rate

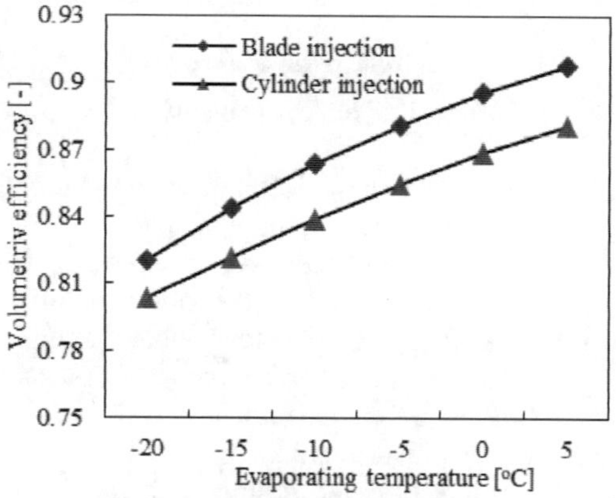

Fig. 8. Volumetric efficiency

Fig. 8 shows volumetric efficiency of the rotary compressor with different injection structures when the condensing temperature keeps 50°C. Compared to the rotary compressor with the traditional injection structure, the volumetric efficiency of the rotary compressor with the new injection structure can be enhanced by 1.8~2.7%.

4.2 Effect on compressor performance

The injection rate is the key factor affecting the performance of the injected compressor. Figure 9 shows injection rate of the rotary compressor with different injection structures under different working conditions. The injection rate with different injection structures both fall with the evaporating temperature increase. Compared to the traditional injection structure, the new blade structure can inject much larger refrigerant into the compressor.

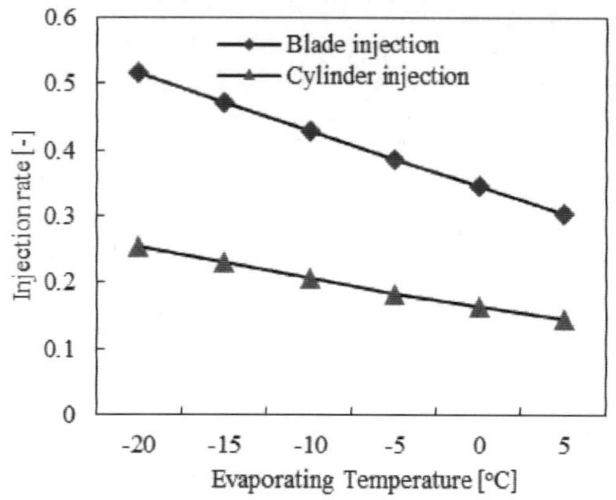

Fig. 9. Injection rate

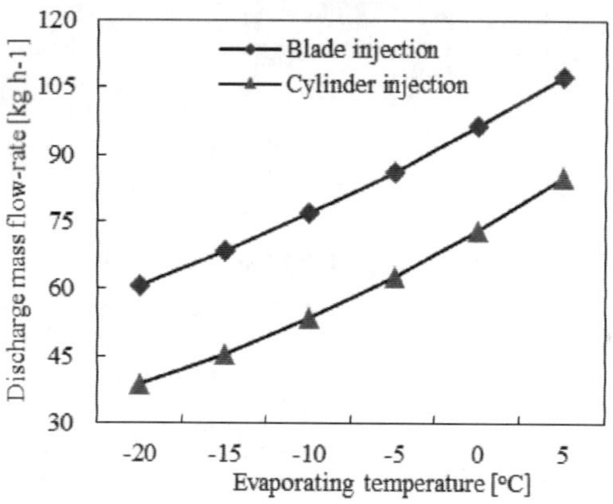

Fig. 10. Discharge mass flowrate

Figure 10 shows discharge mass flow-rate of a rotary compressor with different injection structures under different working conditions. Compared to the rotary compressor with the traditional injection structure, the discharge mass flow-rate of the rotary compressor with the new injection structure increased by 26.6~57.2%.

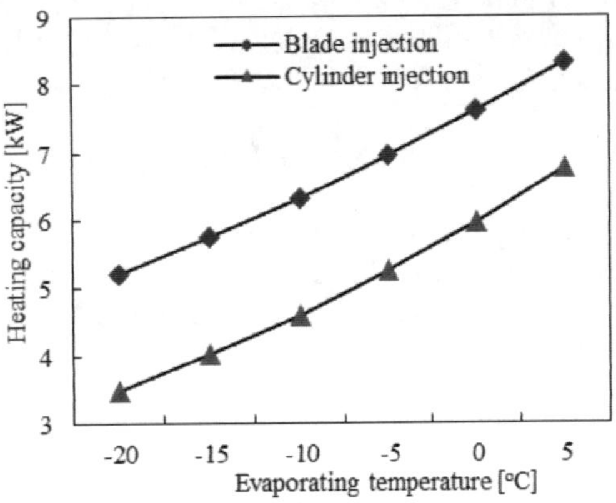

Fig. 11. Heating capacity

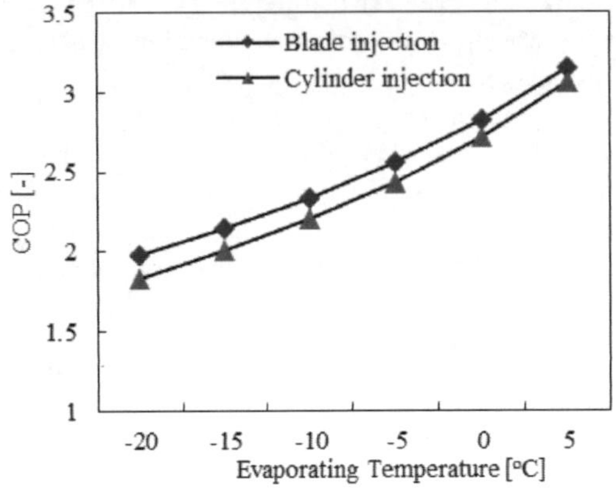

Fig. 12. COP

Figure 11 and Figure 12 show heating capacity and COP of a rotary compressor with different injection structures under different working conditions. Comparing to the rotary compressor with traditional injection structure, the heating capacity and COP of the rotary compressor with the blade injection structure are larger by 23.1~48.9% and 3.2~8.0%, respectively. That means the rotary compressor with the new injection structure can solve the degradation of the heating capacity and COP in low ambient temperature to a large extent.
However, it should be noticed that the performance here, including the injection rate, mass flowrate, heating capacity and COP, is just the performance of the compressor, not the performance of the injected heat pump system. Indeed, the real performance enhancement of injection on the heat pump system is much lower due to the limited gas refrigerant generated in the flash tank, higher condensing temperature, lower evaporating temperature, etc. The injection performance of the new injection structure will be investigated in the future.

5. CONCLUSIONS

Based on a verified numerical model, the performance of the proposed blade rotary compressor is investigated and the results indicate:
1) Compared to the traditional injection structure, the blade injection structure avoids about 23.3~29.3% of the injected refrigerant gas back-flowing into the suction tube;
2) The volumetric efficiency and the discharge mass flow-rate of the rotary compressor with the proposed injection structure is enhanced by 1.8~2.7% and 26.6~57.2%, respectively, relative to the traditional one.
3) Compared to the traditional injection structure, the heating capacity and COP of rotary compressor with the new injection structure is enhanced by 23.1~48.9% and 3.2~8.0%, respectively.

REFERENCES

[1] S. Bertsch Stefan, A. Groll Eckhard, Review of Air-source Heat Pumps for Low Temperature Climates. 8th International Energy Agency Heat Pump Conference, 2005.
[2] S. G.Kim, M. S.Kim, Experiment and simulation on the performance of an autocascade refrigeration system using carbon dioxide as a refrigeration. International Journal of Refrigeration 25 (2002) 1093–1101.
[3] B. L. Wang, W. X. Shi, X. T. Li, Numerical analysis on the effects of refrigerant injection on the scroll compressor. Applied Thermal Engineering 29 (2009) 37–46.
[4] E. L. Winandy, J. Lebrun, Scroll compressors using gas and liquid refrigerant: experimental analysis and modeling. International Journal of Refrigeration 25 (2002) 1143–1156.
[5] G. Y. Ma, Q.Chai, Characteristics of an improved heat-pump cycle for cold regions. International Journal of Refrigeration 26 (2003) 12–18.
[6] B. Wang, W. Shi, L. Han, Optimization of refrigerant system with gas-injection scroll compressor. International Journal of Refrigeration 32 (2009) 1544–1554
[7] J.Heo, M. W. Jeong,C. Baek, Comparison of the heating performance of air-source heat pumps using various types of refrigerant injection. International Journal of Refrigeration 34 (2011) 444–453.

[8]H. G. Wu, X. Y. Peng, Z.W. Xing, Experimental study on p–V indicator diagrams of twin-screw refrigeration compressor with economizer. Applied Thermal Engineering, 24 (2004) 1491–1500.
[9] S. X. Xu, G. Y. Ma, Experimental research on quasi-two stage compression heating pump system using HFC32. International Congress of Refrigeration, 2001.
[10] S. X. Xu, G. Y. Ma, Experiment study on two-stage compression refrigerant/heat pump system with dual-cylinder rolling piston compressor. Applied Thermal Engineering, 62 (2014) 803–808.
[11] Y. Ko, S. Park, S. Jin, The selection of volume ratio of two-stage rotary compressor and its effects on air-to-water heat pump with flash tank cycle. Applied Energy, 104 (2013) 187–196.
[12] Q. L. Jia, L.W. Feng, G. Yan, Experimental research on rotary compression system with vapor injection. Refrigeration and Air Conditioning 14 (2014) 128–132. (In Chinese)
[13] Q. L. Jia, L. W. Feng, G. Yan, Experimental Research on Heating Performance of Rotary Compression System with Vapor Injection. Journal of Refrigeration 36 (2015) 65–70. (In Chinese)
[14] Z. Liu, 1993. Simulation of a variable speed compressor with special attention to supercharging effects. Ph.D. dissertation, School of Mechanical Engineering, Purdue University.
[15] Senug-jun Lee, Jaesool Shim, Kyung Chun Kim, Development of capacity modulation compressor based on a two stage rotary compressor – part 1: Modeling and simulation of compressor performance. International Journal of Refrigeration. 54(2015) 22-37
[16] N. P. Halm. Mathematical model of scroll compressor. Master thesis, Purdue University, 1997.
[17] W. M. Kays, M. E. Crawford. Convective heat and mass transfer, 3rd ed. McGraw-Hill, New York, 1993.

ACKNOWLEDGEMENT

This research is funded by Innovative Research Groups of the National Natural Science Foundation of China (grant number 51521005).

Dynamic Modeling of a Poppet Valve for use in a Rotating Spool Compressor

Nathaniel Wood[1], Craig R. Bradshaw[1], Joe Orosz[1*], Greg Kemp[1], Eckhard A. Groll[2]

[1]Torad Engineering
Cumming, GA, USA
Phone – (770-889-8400), Fax – (770-889-6900), E mail – joe.orosz@toradengineering.com

[2]Ray W. Herrick Laboratories, West Lafayette, IN 47907
Phone – (765-496-2201). Fax – (765-494-0787), E mail – groll@purdue.edu

[*]Corresponding Author

ABSTRACT

The operational efficiency of the rotating spool compressor, first introduced by Kemp *et al.* (2008), has been improved in recent years through the modeling of various elements in the compression cycle. In order to continue the advancements of the design, a model for predicting the discharge valve dynamics is introduced. The model is created using high-pressure measurements both upstream and downstream of the valve. The model is able to predict the initial valve displacement with a mean absolute percent error of 6.91%. The model is also able to predict the amount of time the valve is open in one valve oscillation with a mean absolute error of 9.51%.

1. INTRODUCTION

The rotary spool compressor was first introduced by Kemp *et al.* (2008). Since that time, advancements have been made to the rotary spool compressor design. This includes improvements to tip seals, as seen in Kemp *et al.* (2012a), and Bradshaw (2013) and spool seals, as seen in Kemp *et al.* (2012b). Additional changes have come about through the comprehensive modeling of the rotary spool compressor (Bradshaw and Groll, 2013) and the losses through various elements in the compressor (Bradshaw *et al.*, 2016). The ability to model any losses in the compressor allows for a focused approach on which elements in the rotary spool compressor design need improvements.

The work in Bradshaw *et al.* (2016) identified the discharge process as a significant loss in the compression cycle of the rotary spool compressor. This includes any losses due to flow through the discharge port, any losses derived from the valve motion, and any losses caused by backflow through the discharge valve. The ability to understand how a valve will behave without testing it in the compressor would be useful as a design tool. This would allow for a fast and effective way to improve the discharge process and reduce the associated losses.

Reducing discharge losses in the compressor is assisted by modeling the initial opening displacement of the discharge valve. By modeling the initial displacement of a certain discharge valve system, the valve performance can be established for a certain set of discharge conditions. This is done by seeing how quickly the valve reaches the stop and how the valve reacts after hitting the stop. If the valve rapidly hits the stop and returns with a high velocity towards the seat, it would be expected that the valve would continue in this state of motion until the entire discharge process ended. On the other hand, if the model shows the valve hitting the stop and then coming to rest on the stop, this would indicate that the valve setup was performing in a nearly ideal manner.

Using a self-acting poppet valve in an HVAC&R compressor is uncommon. The poppet valve is used in the spool compressor to lower clearance volume and to decrease leakage around the tip seal. The most commonly used self-acting valve used is the reed valve. The functional difference between a reed valve and a poppet valve is the way in which the spring force is applied. A poppet valve consists of two main parts: the valve body and the external spring; whereas, the reed valve consists only of a valve body that includes the spring force by way of the deflection of the valve body material. This allows for less components for a reed valve configuration and ultimately a lower mass, leading to its wide use in the compressor industry. This becomes more evident when looking for previous documentation on modeling valve dynamics. Previous investigations into the discharge valve dynamics were conducted where pressures around the valve were found using an analytical approach. This approach can be seen in Benson and Ucert (1972), where the pressures acting on the valve are found using a combination of the energy and continuity equations. By using high-speed pressure measurements, a more accurate relationship between the pressures acting on the valve and its accompanying dynamics could be found. In order to fully understand the valve dynamics in a rotary spool compressor, this work developed an experimental environment to study the valve movement under various pressure loads.

2. EXPERIMENTAL SETUP

2.1 Discharge Valve Apparatus

The test apparatus was designed to simulate the discharge process of a rotary spool compressor in a more controlled fashion than the compressor itself can yield. The test apparatus simulates the two discharges of gas per shaft revolution, intrinsic to the spool compressor. The working fluid of the testing apparatus was compressed air instead of refrigerant. An assembly and section view of the final test apparatus is shown in Figure 1. The testing apparatus consists of four main parts: the rotor housing, the rotor, the valve seat plate, and the discharge plenum. The rotor housing includes an inlet port with an accompanying channel to supply compressed air to the valve. The inlet and discharge ports are located on opposite sides of the housing and offset from each other. The top of the rotor housing has a port to allow gas entering into the housing to travel into the valve seat plate. The rotor is manufactured of PEEK and includes two through holes placed perpendicular to each other and offset 29 mm in order for each hole to access either the inlet channel or the discharge channel in the housing. In the valve seat plate, there is a cavity upstream of the valve to allow for a larger volume of air. There is also a pressure tap located in the valve seat plate to allow for the upstream pressure measurement of the valve. The discharge plenum consists of the valve guide and stop, as well as a discharge port to allow flow through the valve and a pressure tap for measuring the downstream pressure.

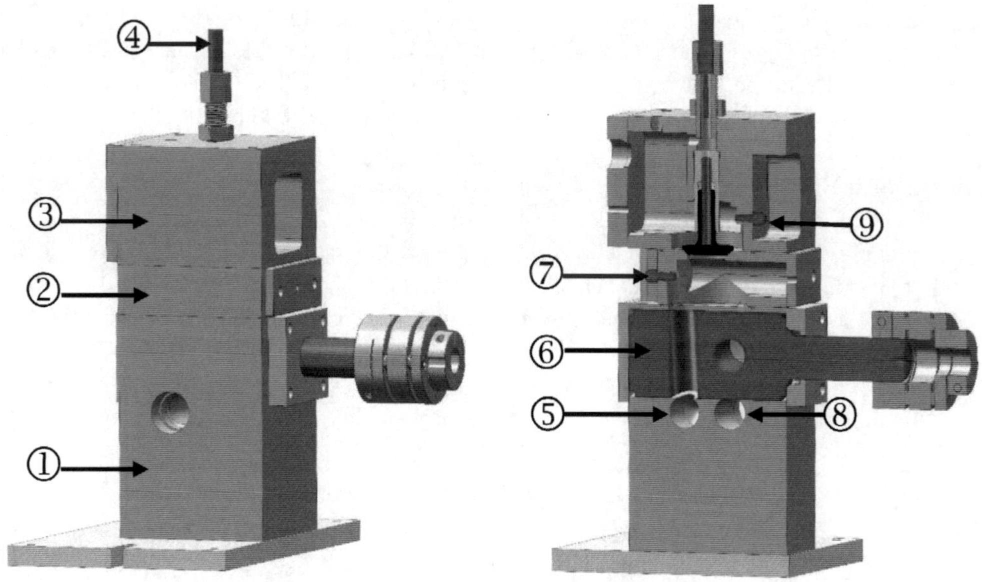

Figure 1: (left) Assembly view of the testing apparatus. (right) Section view of the testing apparatus. (1) rotor housing (2) valve seat plate (3) discharge plenum (4) LVDT (5) compressed air inlet (6) rotor (7) upstream pressure sensor (8) atmosphere vent (9) downstream pressure sensor

2.2 Experimental Measurements

The upstream and downstream pressures were measured using high-speed pressure sensors located in both the valve seat plate and the discharge plenum as shown in Figure 1, as described above. The two high-speed pressure sensors used in the testing apparatus were the Endevco Model 8530b-500 sensors. The upstream and downstream pressures were sampled at 50 kHz to capture any rapid changes in the pressures.

To capture the valve motion, a linear variable differential transformer (LVDT) was used. The armature was attached to the valve by the addition of a threaded hole to the valve face, where the armature for the LVDT could be threaded into the valve. To ensure the security of the armature to the valve, a jam nut was added to the threaded armature as shown in Figure 2. The LVDT used in this experiment was the Daytronic DS400SUHA. The LVDT was also sampled at a rate of 50 kHz.

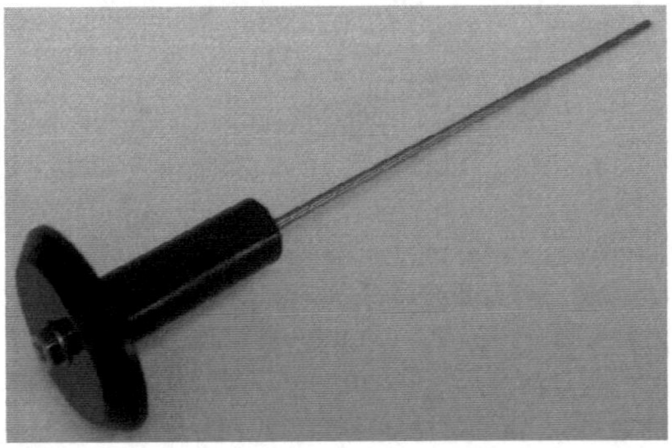

Figure 2: Valve body with attached LVDT armature

Along with the pressure measurements and displacement measurement described above, inlet pressure and temperature were collected. The pressure transducer and thermocouple were located in the inlet manifold adjacent to the apparatus. The pressure transducer used for the inlet was the Omega PX4202-600G5V and the thermocouple used was the Omega TMQSS-125U-6. The inlet temperature and pressure were sampled at 50 Hz.

2.2.1 Experimental Uncertainty

The absolute uncertainty for the Endevco Model 8530b-500 high-speed pressure sensor is ± 17 kPa. The absolute uncertainty associated with the LVDT, the DS400SUHA, is ± 0.05 mm. The absolute uncertainty for the PX4202-600G5V pressure transducer and the TMQSS-125U-6 thermocouple is ± 10 kPa and ± 0.50 °C, respectively.

2.3 Experimental Procedure

In order to fully capture the operating conditions that could be encountered during the discharge cycle of a spool compressor, inlet pressures and shaft speeds were varied. Five different inlet pressures were tested, ranging from 260 kPa to 430 kPa. In addition, five different shaft speeds were used ranging from 600 RPM to 1800 RPM in increments of 300 RPM. These testing points were repeated using five different spring rates. The full test matrix is shown in Table 1 below.

Table 1: Full test matrix; repeated for spring rates of 950, 1400, 1900, 2800, and 3300 N/m

	Shaft Speed (RPM)				
	600	**900**	**1200**	**1500**	**1800**
Inlet Pressure (kPa)	260	260	260	260	260
	300	300	300	300	300
	340	340	340	340	340
	390	390	390	390	390
	430	430	430	430	430

3. VALVE MODEL

3.1 Mechanical Model

The discharge valve system behaves as a single degree of freedom system due to the valve guide only allowing motion in a single direction. Damping forces were not directly accounted for even with the understanding that oil and friction between the valve post and valve guide were present. This allows for a simplification of the model and only a single empirically tuned coefficient. This is similar to the approach used when modeling reed valves in compressors as seen in Boswirth (1980) and Aigner *et al.* (2005). Bredesen (1974) came to the conclusion that any damping forces present in the valve system are negligible when compared to the gas and spring forces. Based on the aforementioned and the free-body diagram shown in Figure 3, the equation of motion for the valve is as shown in Equation (1).

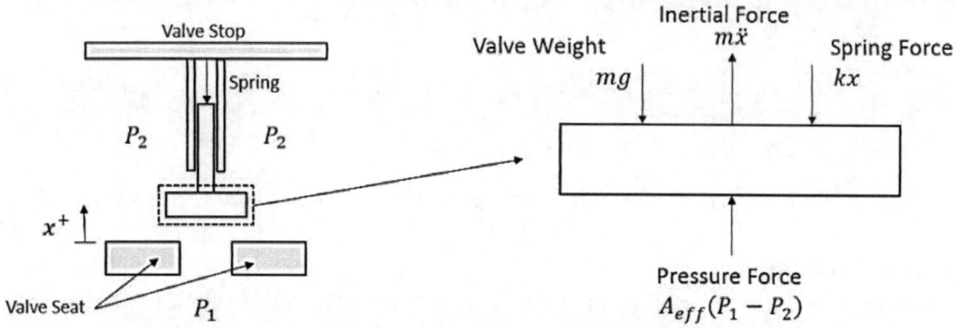

Figure 3: Valve Free Body Diagram

$$m\ddot{x} + k(x + x_0) = A_{eff}\Delta P - mg \quad (1)$$

The total mass, m, is a combination of the mass of the armature and valve as well as the effective spring mass, shown below in Equation (2). The effective spring mass is one-third of the total mass of the spring (Rao, 2004).

$$m = m_v + m_{arm} + \frac{1}{3}m_s \quad (2)$$

The ΔP term in Equation (1) is provided as an input which is found using the dynamic pressure measurements obtained from the test apparatus described in Section 2. The A_{eff} term in Equation (1) represents the effective force area. The effective area is commonly found in many reed valve models. Boswirth (1980) and Benson and Ucert (1972) represents the effective area as a drag coefficient multiplied by the valve disc area. The effective force area accounts for energy loss due to changes in the fluid flow as the valve opens.

3.2 Model Solver

The equation of motion was solved numerically using the fourth-order Runge-Kutta method. The collected pressure difference values were interpolated to use as continuous inputs into the equation of motion. The ODE describing the valve motion was solved in MATLAB (2015b) using the 'ode45' solver.

4. RESULTS

4.1 Mechanical Model Validation

To verify the valve model, a step response experiment was conducted on the valve system. The experiment consisted of supplying the valve with an initial displacement and then quickly releasing the valve to allow it to return to its equilibrium position. This test was conducted without any flow present to ensure correct functioning of the solver and to guarantee an accurate mechanical description of the valve motion. Without flow, there is confidence that any discrepancies in the model predictions is most likely related to the flow and/or pressure fields surrounding the valve, given the accurate prediction of valve displacement considering only the mechanical phenomena. The modeled displacement exhibited the same rise time and settling time as well as the same overshoot and number of oscillations as the experimentally collected displacement, shown in Figure 4.

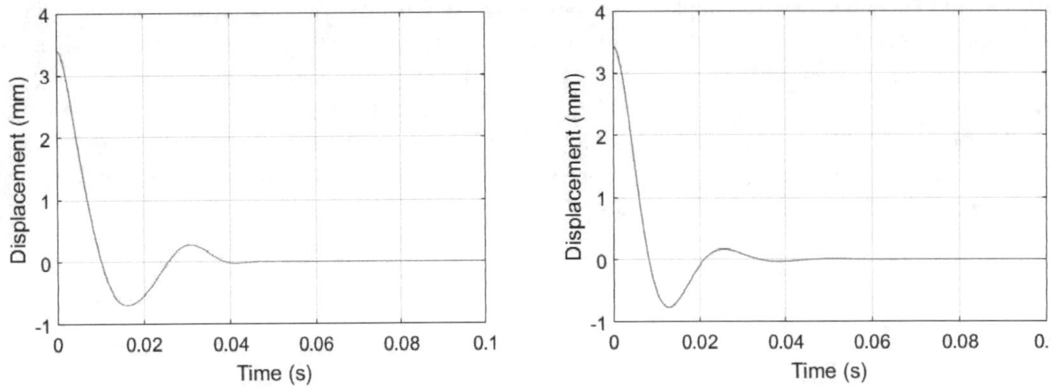

Figure 4: Mechanical step response of the actual system (left) and the model (right)

4.2 Effective Area Estimation

The effective force area for each collected point was found by tuning the model output to produce an initial valve displacement that matched the experimental valve displacement. Figure 5 shows the tuned effective area for each point plotted against the maximum pressure difference reached before the valve opened. This figure shows a wide variation in the effective area for each spring rate.

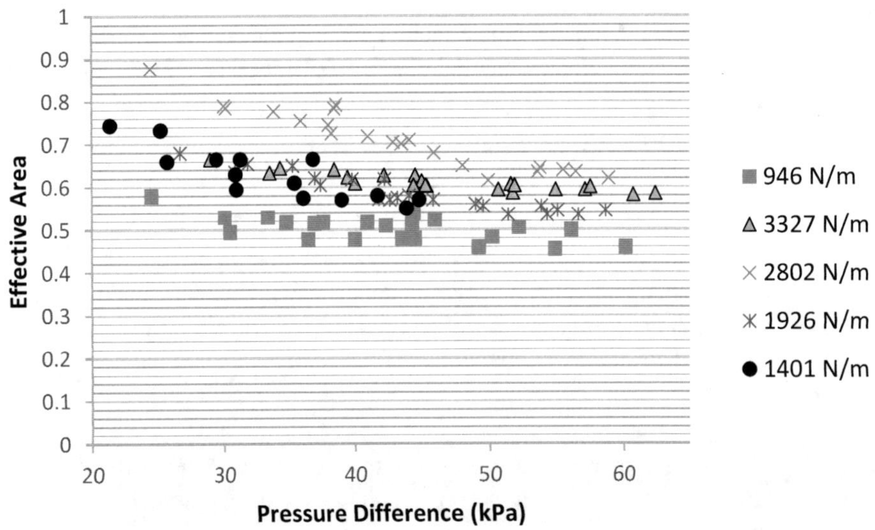

Figure 5: Effective area plotted against opening pressure difference

The values presented in Figure 5 were averaged based on the springs used. With the averaged effective areas, a linear equation was fit to the data and used to predict average effective areas based on the spring rate as shown in Figure 6. This trend allows for a reasonable approximation of the average effective area that would be expected for a given spring rate.

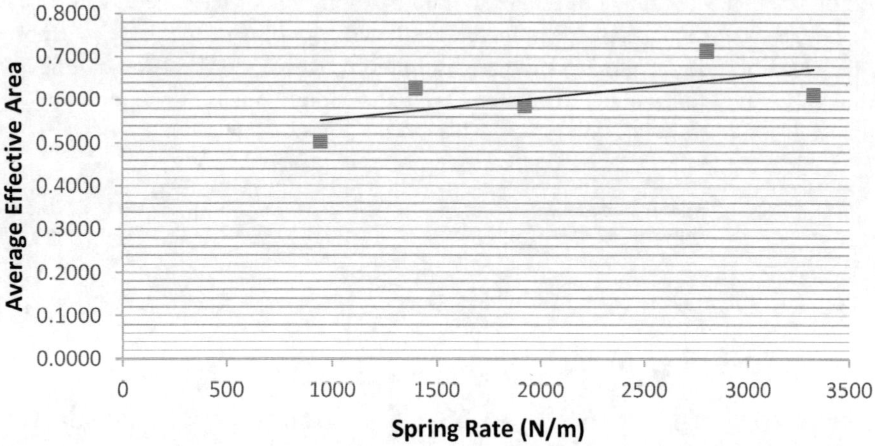

Figure 6: Average effective area plotted against spring stiffness

4.3 Model Validation

The mean absolute percent error of the model's predicted initial valve displacement came out to 6.91%. This value was calculated by taking the percent error between the predicted and the experimentally obtained displacements at each test point. These errors were then averaged to give the final error percentage.

Figure 7 shows the amount of time, for a single oscillation, the valve was open predicted by the model compared to the experimentally collected data. The parity plot includes all testing points introduced in Table 1 of Section 2. The opening time was defined as the difference in time between when the valve initially lifts off the seat and when the valve fully comes to rest on the seat. The model predicts the opening time within a 9.51% mean absolute error.

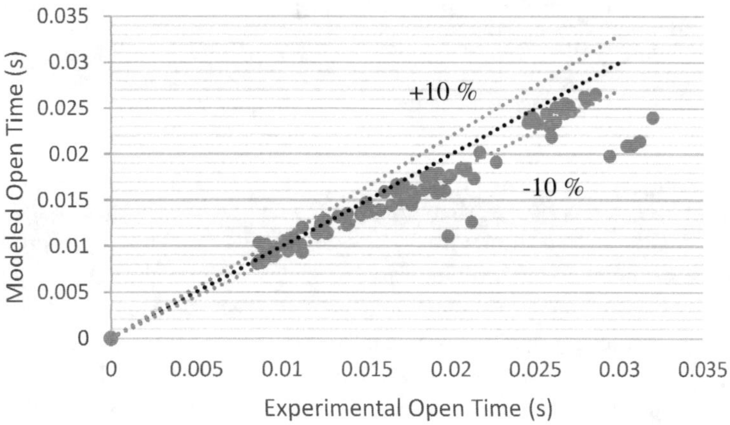

Figure 7: Parity plot of predicted and experimental open times for a single valve oscillation. The upper bound represents a +10% error and the lower bound represents a -10% error.

4.4 Model Strengths

The plots shown in Figure 8 represent points where the model can accurately predict the discharge valve displacement. Figure 8(a) shows the model predicted valve displacement with an inlet pressure of 300 kPa, a shaft speed of 1500 RPM, and a spring rate of 950 Newtons per meter. Figure 8(b) shows the model predicted valve displacement for an inlet pressure of 260 kPa, a shaft speed of 1200 RPM, and a spring rate of 1900 Newtons per meter. Figure 8(c) shows the model predicted valve displacement for an inlet pressure of 300 kPa, a shaft speed of 1200 RPM, and a spring rate of 1400 Newtons per meter. Figure 8(d) shows the model predicted valve displacement for an inlet pressure of 390 kPa, a shaft speed of 1800 RPM, and a spring rate of 2800 Newtons per meter. The model best predicts the valve motion at higher shaft speeds. As the shaft speed increases, the amount of time the valve is exposed to the pressure force decreases. This leads to a lower number of valve oscillations, where the flow does not have enough influence from the oscillating valve body to become as irregular. The valve simply opens due to the initial pressure force and then hits the stop, and then returns to the seat.

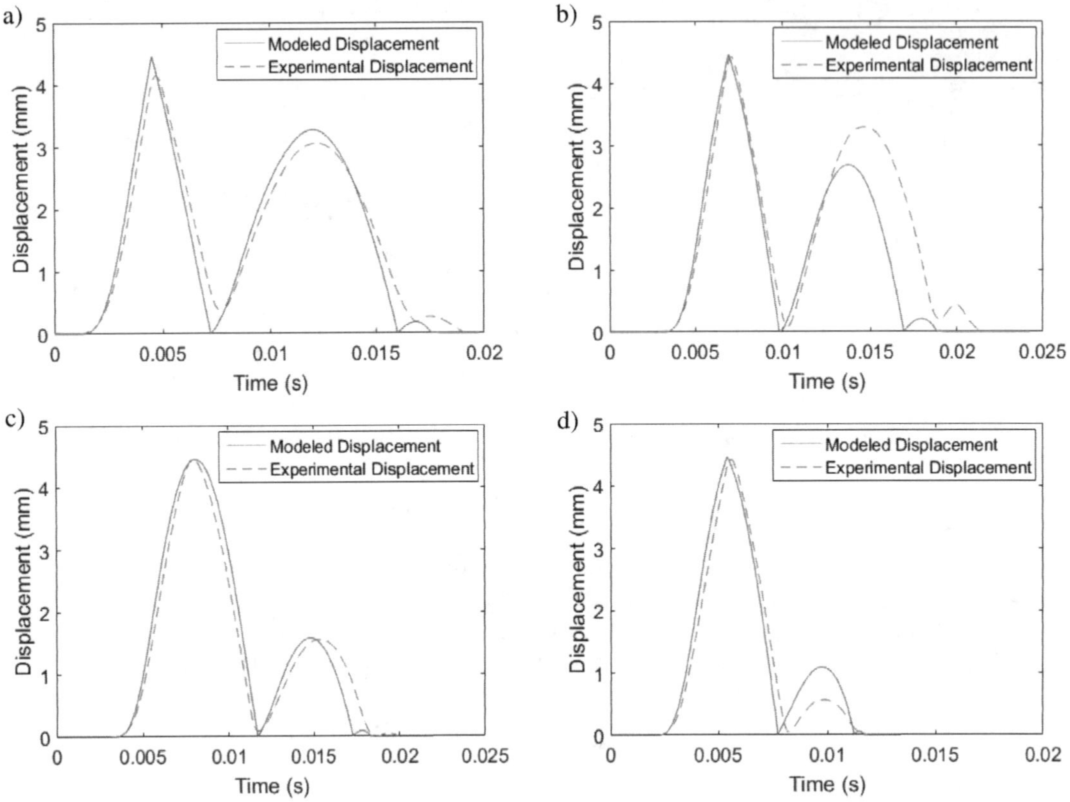

Figure 8: Model predicted valve displacement plotted against experimental valve displacement

4.5 Model Weaknesses

The plots shown in Figure 9 represent points where the model is not able to accurately predict the discharge valve displacement. Figure 9(a) shows the model predicted valve displacement with an inlet pressure of 260 kPa, a shaft speed of 600 RPM, and a spring rate of 2800 Newtons per meter. Figure 9(b) shows the model predicted valve displacement for an inlet pressure of 260 kPa, a shaft speed of 600 RPM, and a spring rate of 950 Newtons per meter. Figure 9(c) shows the model predicted valve displacement for an inlet pressure of 260 kPa, a shaft speed of 600 RPM, and a spring rate of 1400 Newtons per meter. Figure 9(d) shows the model predicted valve displacement for an inlet pressure of 300 kPa, a shaft speed of 600 RPM, and a spring rate of 1900 Newtons per meter.

These plots show the poor prediction capabilities of the model particularly when the shaft speed is below 900 RPM. At lower speeds, the valve is exposed to the pressure force for a relatively long duration. This forces the valve to oscillate, which is caused by the valve trying to reach its equilibrium position based on the pressure and flow acting on it. This oscillation, sometimes called fluttering, additionally causes rapid changes in the pressures surrounding the valve. This oscillation in the flow field becomes very irregular when compared to the flow experienced by the valve when it first opens. These irregular flow fields lead to difficulty in obtaining the measured pressures which can lead to an inaccurate prediction of the dynamics. This suggests, in order to fully describe the valve dynamics at such a shaft speed, more details about the flow around the valve would need to be explained.

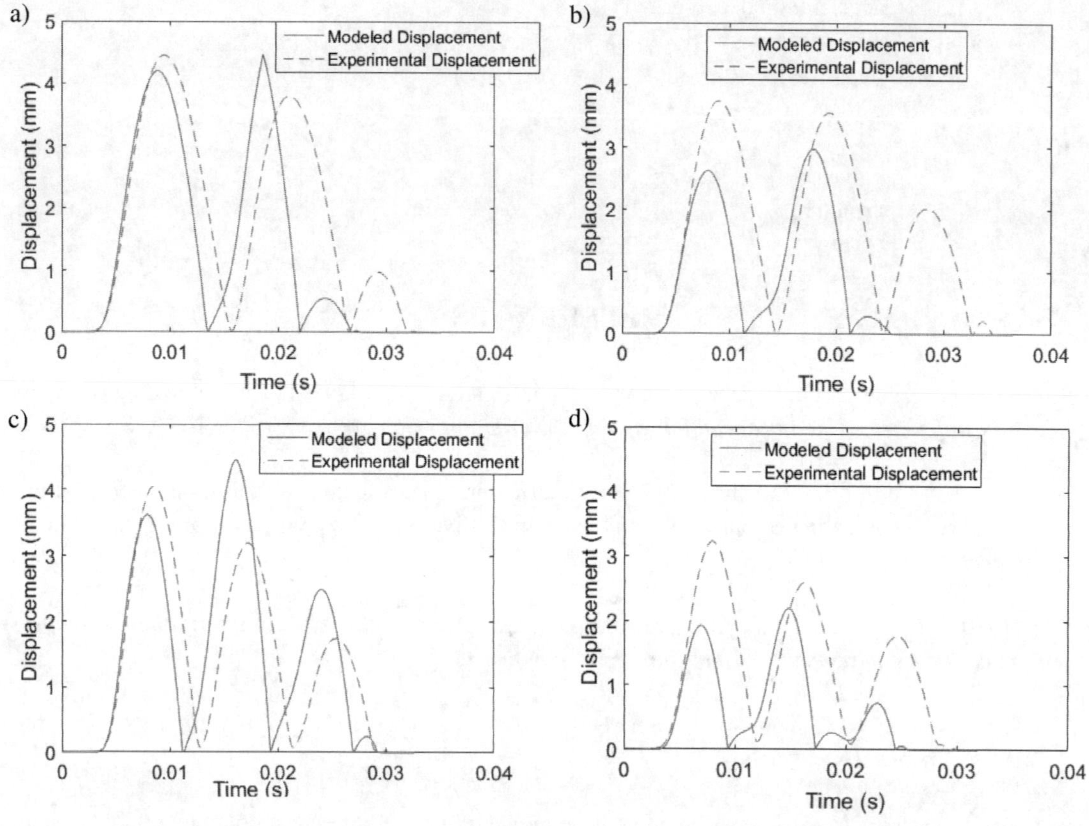

Figure 9: Model predicted valve displacement plotted against experimental valve displacement

5. CONCLUSIONS

5.1 Summary

A model predicting the discharge valve displacement has been presented. The model used high speed pressure measurements upstream and downstream of the valve and also the experimentally discovered relationship of the average effective area and spring rate to predict the discharge valve displacement. The model preforms reasonably well at higher shaft speeds. At lower speeds, the valve displacement is not accurately predicted due to flow forces that were not accounted for.

5.2 Future Work

In order for this model to be useful for future design considerations, the model must be verified by the experimentation of more valve masses. Flow characteristics must be accounted for to accurately predict valve displacements at lower shaft speeds. The addition of more spring rates would increase the confidence in the average effective area trend. Finally, this model will be integrated into the comprehensive model previously presented by Bradshaw and Groll (2013).

NOMENCLATURE

m	mass	(kg)
$\ddot{x}$	valve acceleration	(m/s^2)
x	valve displacement	(m)
k	spring rate	(N/m)

g	acceleration due to gravity	(m/s^2)
A_{eff}	effective area	(m^2)
ΔP	pressure difference	(kPa)

Subscript

v	valve
arm	armature
s	spring

REFERENCES

Aigner, R., Meyer, G., & Steinrück, H. (2005). Valve dynamics and internal waves in a reciprocating compressor. In *4th Conference of the European Forum for Reciprocating Compressors*. pages 169-178.

Benson, R. S., & Ucert, A. S. (1972). A theoretical and experimental investigation of a gas dynamic model for a single stage reciprocating compressor with intake and delivery pipe systems. *Journal Mechanical Engineering Science*, 14(4), 264–279.

Boswirth, L. (1980). Flow forces and the tilting of spring loaded valve plates - part I. In *International Compressor Engineering Conference at Purdue University*. number 330.

Bradshaw, C. R. (2013). Spool Compressor Tip Seal Design Considerations. In *2013 International Conference on Compressors and Their Systems.*

Bradshaw, C. R., & Groll, E. A. (2013). A comprehensive model of a novel rotating spool compressor. *International Journal of Refrigeration,* 36(7), 1974–1981.

Bradshaw, C. R., Kemp, G., Orosz, J., & Groll, E. A. (2016). Development of a loss pareto for a rotating spool compressor using high-speed pressure measurements and friction analysis. *Applied Thermal Engineering,* 99, 392–401.

Bradshaw, C. R., Kemp, G., Orosz, J., and Groll, E. A. (2014). Loss analysis of rotating spool compressor based on high-speed pressure measurements. In *International Compressor Engineering Conference at Purdue University*. number 2271.

Bredesen, A. M. (1974). Computer simulation of valve dynamics as an aid to design. In *International Compressor Engineering Conference at Purdue University*. number 117.

Kemp, G., Orosz, J., Bradshaw, C., and Groll, E. A. (2012a). Spool compressor tip seal design considerations and testing. In *International Compressor Engineering Conference at Purdue University*. number 2079.

Kemp, G., Orosz, J., Bradshaw, C., and Groll, E., A. (2012b). Spool seal design and testing for the spool compressor. In *International Compressor Engineering Conference at Purdue University*, number 2080.

Kemp, G.T., Garrett, N., Groll, E. A., (2008). Novel rotary spool compressor design and preliminary prototype performance, In *International Compressor Engineering Conference at Purdue University*, number 1328.

MATLAB, MATLAB version 8.6.0 (2015b), The MathWorks, Inc., Natick, Massachusetts, United States.

Rao, S. S., 2004. Mechanical Vibrations, 4th Edition. Prentice Hall.

Experimental and numerical investigation of the heat transfer inside a hollow piston rod

Konrad KLOTSCHE[1]*, Christiane THOMAS[2], Ullrich HESSE[3]

Technische Universität Dresden, Bitzer Chair of Refrigeration, Cryogenics and Compressor Technology
Dresden, Saxony, Germany

[1] phone: +49 351 463 32603, fax: +49 351 463 37247, konrad.klotsche@tu-dresden.de

[2] phone: +49 351 463 34048, fax: +49 351 463 37247, christiane.thomas@tu-dresden.de

[3] phone: +49 351 463 32548, fax: +49 351 463 37247, ullrich.hesse@tu-dresden.de

* Corresponding Author

ABSTRACT

The heat transfer in the system piston - piston rod - crosshead of reciprocating compressors can be influenced by a fluid enclosed in an internal cavity. Thus, heat is transferred by the fluid flow inside the hollow parts utilizing the reciprocating motion. Subsequently, the high working frequency of such compressors can lead to an improved cooling of the high temperature compressor parts in comparison with conventional solid material designs.

For the experimental examination of this cooling concept a test rig has been set up at the Technische Universität Dresden which comprises a vertically oscillating hollow rod that is heated at its upper end and cooled at its lower end. Among other factors the heat transfer of this cooling concept is affected by the internal diameter of the piston rod. In order to quantify the influence of the internal diameter on the axial heat transfer three different hollow rods with varying internal diameters where investigated by means of temperature measurements.

The experimental results are presented and the influence of the size of the cavity is discussed. Furthermore, the measured temperatures are used for numerical simulations of the test rig in order to derive the axial heat flux of the fluid inside the cavity. Based on the resulting values the heat transfer capability of the investigated cooling concept can be examined with respect to the internal diameter and the benefit for the cooling of different compressor components can be estimated.

1. INTRODUCTION

When there is a demand for oil-free compression typically dry-running reciprocating piston compressors are used. The applied sealing elements are usually made of polymer composite materials which do not require an additional lubricant, offering economic and environmental benefits. In this field not only a growing interest in research but also an increasing utilization can be observed over the last decades (Kriegel, 1977; Tomschi, 1995; Feistel *et al.*, 2003; Thomas, 2013).

Especially for higher capacities the advantages of crosshead-based compressor designs are typically preferred, requiring a sealing at the piston rod – the so-called packing – in order to minimize the leakage to the environment. Due to the lack of lubrication the mechanical and thermal stress of the sealing elements are significantly higher compared to oil-lubricated designs. Hence, on the one hand friction increases and on the other hand the omitted lubricant reduces the heat rejection. This leads to higher temperatures and to an increased wear of the sealing elements. As a result the service life of the sealing elements is reduced and the operating costs of the compressor increase.

According to Kriegel (1977) the temperature in the contact zone between the non-lubricated sealing and the counter surface is crucial to the wear rate of the polymer sealing elements. Their attrition increases with increasing temperature. Hence, their cooling is expedient.

Up to the present a coolant circuit provided in the housing of the packing is often applied for the rejection of heat (Feistel *et al.*, 2003). However, such an external packing cooling shows several drawbacks in terms of additional constructional and operating means giving rise to higher investment and operating costs. In addition, the risk of a blocking of the coolant ducts as well as leakage of coolant is well-known.

Based on these circumstances a different cooling approach was created for which a coolant is filled in an internal cavity in the piston rod. For the sake of simplicity this cavity can be designed as a cylindrical bore. The hereafter presented investigation and related results show the influence of the internal diameter on the heat transfer through a reciprocating cooling cavity within a reasonable range.

2. INTERNAL COOLING CONCEPT FOR RECIPROCATING COMPRESSORS

The idea of an internal cooling of the piston rod for reciprocating piston compressors was firstly delineated and patented by Quack and Nickl (2009). The following section gives an overview of the principle of operation and the latest state of the internal cooling concept as well as the consequential motivation for this publication.

2.1 Principle of Operation

The aim of the internal cooling concept is to extract heat from components that need to be cooled, *e.g.* the packing and the cylinder area. For this purpose, a coolant filled into a cavity inside the piston rod and/or the piston is used to absorb heat from the warmer regions, to transfer it through the piston rod and to reject it to the oil-lubricated crosshead (see figure 1).

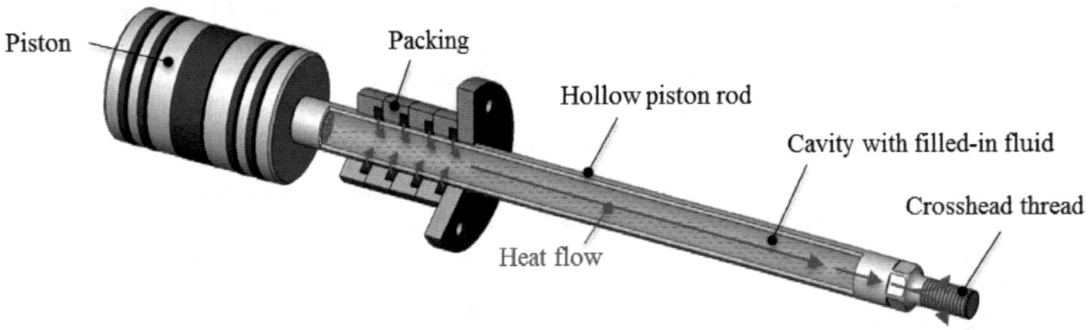

Figure 1: Design and principle of operation of the internal piston rod cooling concept

Although the thermodynamic drive is given by the temperature gradient, the main idea and benefit of this cooling technology is to exploit the reciprocating motion of the piston rod to cause an enhanced fluid flow inside the piston rod's cavity. In this way, the fluid absorbs heat by flowing through warmer areas and transfers it to colder areas where it is finally rejected. The typical high working frequency of such crosshead-type compressors (up to 1800 min^{-1}) enables an enhanced heat transfer through the piston rod and an increased cooling of high temperature compressor parts compared to conventional solid piston rods.

2.2 Previous Work and Results

Since the first reference of the described idea several experimental and theoretical studies were conducted focusing on its feasibility and general applicability. Hammer *et al.* (2012) measured a significant cooling effect at a test rig with a vertical oriented oscillating hollow rod induced by the internal cooling of the rod. In addition, results were presented indicating the possibility to model the heat transfer processes inside the rod's cavity by means of a Finite Element (FE) simulation. Thomas (2013) presented extensive results from both experimental and theoretical investigations which show the effectiveness of the internal cooling at three different test rigs, including a single-stage double-acting balanced-opposed compressor. Making use of the internal piston rod cooling the measured gas and solid body temperatures inside the packing of this compressor could be reduced by up to 70 K. In addition, the sealing quality could be improved by reducing the temperatures in the sealing gap.

Comparable approaches in terms of an enclosed internal cooling cavity supported by a reciprocating motion can be found in the field of combustion engines, for instance the cooling of engine valves (Bush and London, 1965) and the heat transfer from the piston to the crankcase oil via heat pipes molded into the piston (Cao and Wand, 1995). Unfortunately, different scales impede the transfer of these results to internally cooled compressor piston rods.

2.3 Open Tasks

Since the feasibility and effectiveness of the internal piston rod cooling has been demonstrated, the next target is to optimize the heat transfer. A variety of parameters, influencing the heat transfer of the discussed cooling technology, allow more detailed investigations. The results of these investigations are the basis to enhance the heat transfer capability. These parameters are:

- the fluid itself and the charge,
- the cross section (shape, size),
- the kinematics of the rod (speed, stroke),
- the orientation relative to gravity,
- the temperature difference between the heat source and sink, and
- the heat flow to be transferred

as well as other thermal resistances.

One of the most significant impacts of the enumerated parameters constitutes the cross section of the cavity in axial direction. It is represented by the internal diameter in case of a cylindrical bore. So the question arises how the internal diameter affects the heat transfer through a hollow piston rod.

3. TEST RIG AND EXPERIMENTAL INVESTIGATION

At present the theoretical examination of the fluid flow, *e.g.* by CFD simulations, seems not expedient since the transient und turbulent processes in the shaked-up cavity are not educible with reasonable effort. In contrast, experimental investigations offer promising results (Thomas, 2013). Hence, the general construction of the used test rig and the obtained results are shown in this section.

3.1 Test Rig Setup

For the experimental investigation of the heat transfer inside a reciprocating hollow rod a test rig is available at the Technische Universität Dresden (see figure 2). It comprises a vertically reciprocating hollow rod actuated by a crank drive that is driven by an electric motor. The stroke of the crank drive is 100 mm. The rod is mounted onto a temperature-controlled crosshead. It comprises a cooling duct connected to an external cooling circuit providing the entire heat to be completely rejected at the rod's lower end, *i.e.* the temperature at this location can be kept constant for all operation points. At the upper end of the rod a cylindrically shaped electrical heater is installed to provide a defined heat input and the rod is closed by an end cap. All relevant components – except the crosshead – are insulated against the ambient. By this, most of the heat input is transferred through the rod and the filled-in fluid. In order to connect all electrical signals to the measurement system and to transfer the coolant fluid from the cooling unit to the moved system an energy chain is attached to the rod's cap.

To accurately record the temperature field of the relevant components with reasonable extent 38 type K thermocouples (TCs) are installed whose signals are sampled by 1 Hz. Their positions are depicted in figure 2 (c). Five TCs positioned inside the cavity (FT1 … FT5) are distributed along the rod's axis measuring the fluid temperature. Twelve temperature sensors (WT1 … WT12) are located 2 mm below the rod's exterior radial surface capturing the wall temperature. They are distributed along the rod's length similar to the fluid TCs. For the striven thermal simulation of the experiments and therefore required boundary conditions two TCs are installed at the rod's cap and 3 TCs at the lateral surface of the cap bracket as well as 2 TCs inside the cooling duct of the crosshead.

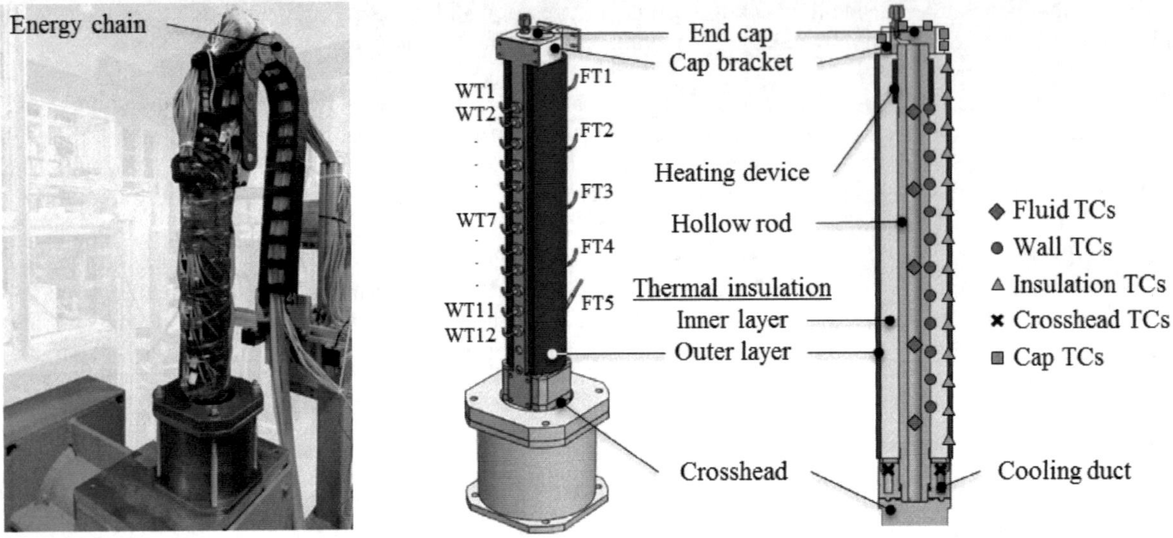

(a) (b) (c)

Figure 2: Setup of the test rig (a), CAD model (b), and sectional view with thermocouple positions (c)

The rod is equipped with two insulation layers. In between 14 TCs are placed measuring the temperature along the rod's axis. A former setup of the facility exhibits only one layer and the boundary condition at its outer surface was a convective heat transfer which is rather difficult to specify for a FE simulation (Hammer *et al.*, 2012). The TCs at the insulation are now used to provide the temperatures as boundary condition for the numerical simulation which is more accurate than the previous setup.

3.2 Measurement Procedure

At first the fluid to be investigated is filled into the cavity and the hollow rod is closed. This yields a coolant-air mixture at ambient conditions. Then the rotational speed (in this study only 600 min^{-1} are presented) is adjusted and the flow through the crosshead's cooling circuit as well as the power supply of the heater is switched on. After a transient period in which the temperatures are converging towards a steady state the following interval is used to record the final temperatures which are averaged over 5 minutes. In this way temporary fluctuations and phenomena are excluded. All measurements are carried out under laboratory conditions ensuring the ambient temperature to be sufficiently constant over time.

3.3 Measurement Series

For the experimental results shown here, 3 different hollow rods are investigated that differ only by their internal diameters (see test cases A, B, and C in table 1). For all measurements the same coolant-air-mixture was used as the filling. The power input of the heating device and the temperature of the crosshead coolant were constant for all measurements, thus the boundary conditions are consistent in terms of the heat sink and source. The cross sectional flow areas are referenced to the rod's value of test case A. Table 1 summarizes the relevant conditions.

Table 1: Relevant information on the test series

Parameter	Symbol	Unit	Test case		
			A	B	C
Rotational speed	N	[min^{-1}]	600	600	600
Heat input	P	[W]	70	70	70
Temperature of crosshead coolant	t_C	[°C]	10.3	10.3	10.3
Cross sectional flow area ratio	$A_{flow}/A_{flow,A}$	[-]	1.00	1.78	2.78

Since this paper aims for the determination of the heat transfer capability of the fluid, only the temperatures over time inside the cavity of the measurement's last six minutes are shown (see figure 3).

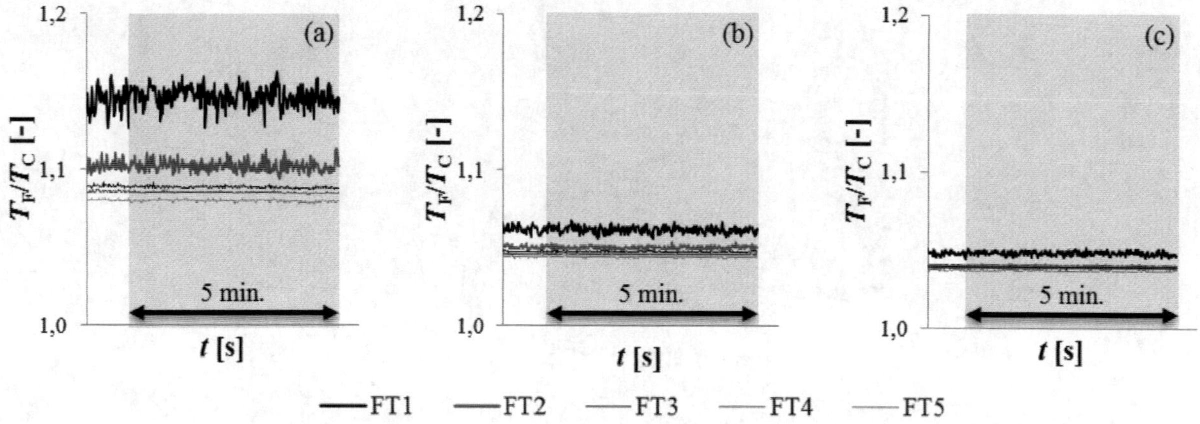

Figure 3: Fluid temperatures from the evaluation period (highlighted gray) of the test cases A (a), B (b), and C (c)

It can be seen that all fluid temperatures decrease with an increasing cross sectional flow area. The larger temperature reduction occurs between test case A and B. In comparison, the difference between test case B and C is less significant. Therefore it can be stated that increasing the internal diameter from test case A to B is considerably more effective than to increase it from test case B to C in terms of the axial heat transfer. Furthermore, the upper temperatures (FT1 and FT2) appear less steady for smaller internal diameters. This indicates a more fluctuating local heat transfer. Optical measurements of the fluid motion could help verifying how this is caused by or interrelated to the fluid's turbulent behavior. The temperatures exponentially increase with increasing height. Since the TCs are equally spaced along the rod the influence of the gravity cannot be neglected for the heat transfer in a vertical oriented internally cooled reciprocating piston rod even at 600 min^{-1}. The resulting values of the evaluation at the end of each test case measurement are summarized in table 2. The fluid temperatures are referenced to the temperature of the crosshead's coolant.

Table 2: Summary of measured steady-state temperatures of the test cases A, B, and C

Sensor	Quantity	Symbol	Unit	Test case		
				A	B	C
FT1	Mean referenced temperature	$\overline{T_{FT1}/T_C}$	[-]	1.147	1.062	1.049
	Relative deviation	$\sigma_{t_{FT1}}/t_{FT1}$	[%]	3.3	2.1	1.5
FT2	Mean referenced temperature	$\overline{T_{FT2}/T_C}$	[-]	1.102	1.051	1.040
	Relative deviation	$\sigma_{t_{FT2}}/t_{FT2}$	[%]	2.4	0.9	0.5
FT3	Mean referenced temperature	$\overline{T_{FT3}/T_C}$	[-]	1.089	1.048	1.040
	Relative deviation	$\sigma_{t_{FT3}}/t_{FT3}$	[%]	0.5	0.4	0.3
FT4	Mean referenced temperature	$\overline{T_{FT4}/T_C}$	[-]	1.085	1.047	1.039
	Relative deviation	$\sigma_{t_{FT4}}/t_{FT4}$	[%]	0.3	0.3	0.3
FT5	Mean referenced temperature	$\overline{T_{FT5}/T_C}$	[-]	1.080	1.045	1.038
	Relative deviation	$\sigma_{t_{FT5}}/t_{FT5}$	[%]	0.5	0.3	0.3

The measured values of both the fluid and the wall temperatures can be averaged over the steady-state period (highlighted gray in figure 3). Thereby, the thermal gradient along the rod represents the heat transfer capability according to the one-dimensional Fourier's law.

$$\hat{q}_z = -\lambda \frac{dt}{dz} \tag{1}$$

The mean fluid and wall temperatures for the three test cases are presented in figure 4.

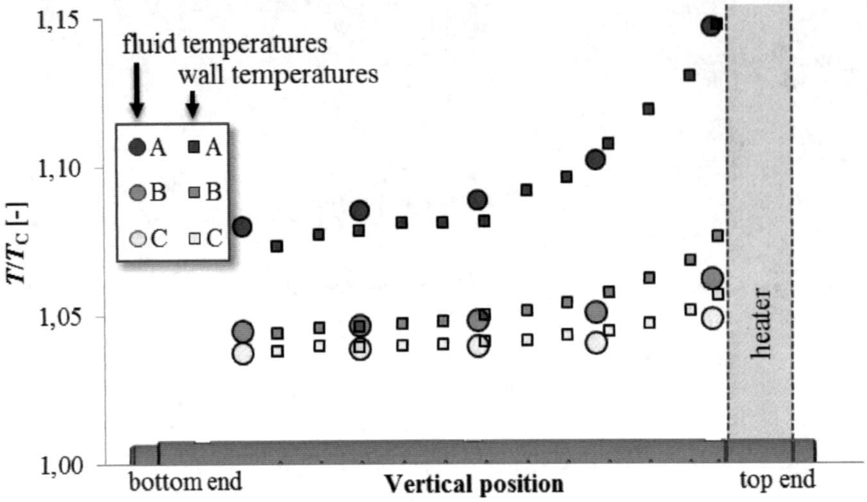

Figure 4: Fluid and wall temperatures from the evaluation period of the test cases A, B, and C

Besides some small differences it can be seen that the fluid and the wall temperatures show a similar distribution for each test case, respectively. As it is indicated in figure 3 larger internal diameters enhance the heat transfer and therefore lead to an improved cooling. Additionally, the larger differences between fluid and wall temperatures, *e.g.* at the top end of the rod for test case B and C, indicate a lowered radial heat transfer at this height. Those temperature differences can be seen at the bottom end of the rod for test case A as well. This is important to consider when it comes to optimize the heat transfer by minimizing the individual thermal resistance factors.

4. SIMULATION OF THE TEST RIG TEMPERATURE FIELD

The experimental results presented in the previous sections allow a qualitative evaluation of the axial heat transfer capability of reciprocating hollow rods depending on their internal diameter. To estimate the axial heat flux inside the cavity the recorded temperatures can be exploited to simulate the temperature distribution of the relevant parts. Therefore, the following section introduces the developed simulation model and the related assumptions as well as two derived quantities that are describing the axial heat transfer.

For calculating the thermal distribution of the test rig a steady-state thermal simulation model is used which is based on the finite element method. This simulation model consists only of solid elements, thus the fluid's thermal behavior (and not its flow characteristics) is represented by an equivalent thermal conductivity.

Since most of the test rig components are nearly rotationally symmetric the three-dimensional construction is transferred into a two-dimensional model (see figure 5). Moreover, the domain to be calculated is demarcated to an essential extension. This means that the outer layer of the rod's insulation and the insulation of the end cap are not adopted to the simulation model.

As shown in figure 5 (b) the area of the filled cavity (depicted white) is divided into several sections taking into account different dominant heat transfer mechanisms. According to different temperature gradients and corresponding to the design of the test rig this area is separated in the direction of the z axis. Additionally, these sub-areas are segmented into one part close to the rod's wall and into a second internal part which yields 12 areas representing the filling.

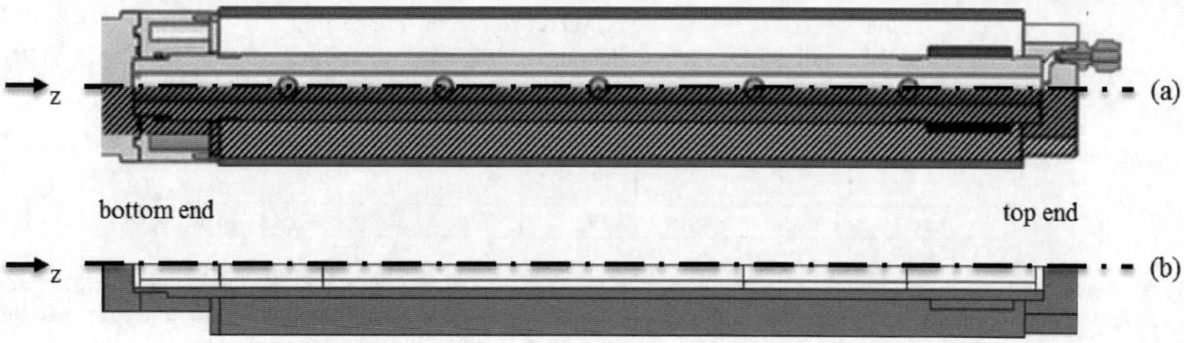

Figure 5: Cross sectional view of the CAD model (a) with areas adopted to the FE model (highlighted striped) and derived simplifying FE model (b) with white areas representing the cavity

After meshing the model with reasonable accuracy (Klotsche, 2012) all boundary conditions (temperature specifications and rotational symmetry) provided by the temperature measurements (see figure 2 (c)) and the thermal conductivity of all parts of the model except the fluid's one were predefined. The temperatures of the insulation are chosen to be piecewise linear functions between the thermocouple positions. Additionally, the heat input is specified as the heat generation rate for the cross sectional area of the heating device.

Then, the only unknown influence is the thermal behavior, *i.e.* the thermal conductivities of the filling areas. The idea here is to find a thermal behavior of the modelled filling that corresponds to the fluid's one in the experiments. This is done by an iterative procedure. In order to check this correspondence the calculated temperature field is compared to the measured fluid and wall temperatures. The result of such a comparison for test case A is presented in figure 6.

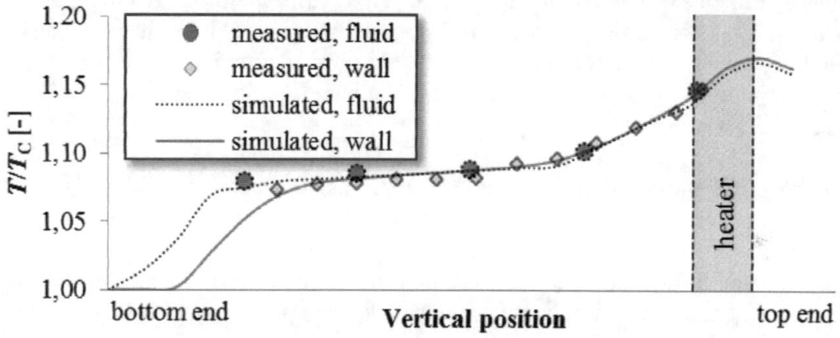

Figure 6: Measured and simulated steady-state temperatures of test case A

A sufficiently good correspondence between the measured and simulated fluid and wall temperatures can be seen in figure 6. Hence, the assumed equivalent thermal conductivities of the simulated filling areas accurately represent the thermal behavior of the fluid in the experiment. With this result the temperature field is exploited to deduce how much heat is actually transported by the fluid in axial direction. On the one hand, the power input of 70 W is not entirely transferred from the top end to the crosshead. On the other hand, some heat will be conducted by the rod's wall as well as by the insulation. To achieve an appropriate comparison between different measurements the axial heat flux at only one significant height is used. This position is chosen to be at the middle of the rod's length as it seems to be the most suitable representation of the axial heat transfer for the largest part of the rod.

To determine this parameter the axial heat flow through all finite elements at this height associated with the fluid are obtained at first. Then, each of these values is divided by the associated cross sectional area. This yields the axial heat flux for every circular ring of the fluid cross section. Finally, all values are added which gives the axial heat flux for the entire fluid cross section. The procedure of adjusting the simulation model to the experiment and ascertaining the axial heat flux was done for all measurements resulting in the values summarized in table 3.

Table 3: Axial heat flux and piston rod heat transfer capability at the middle of the rod's length of the test cases A, B, and C

Parameter	Symbol	Unit	Test case		
			A	B	C
Axial heat flux	$\lvert\hat{\dot{q}}_z\rvert$	W/mm²	0.285	0.191	0.109
Piston rod heat transfer capability	C	W/K	302.1	548.2	715.1

The axial heat flux is a useful indicator for the heat transfer inside of an internally cooled piston rod. However, for the designing process not only the transferred heat in relation to the cross sectional area is of importance but also the temperature level occurring at the heated end. Besides the heat flux also the temperature gradient over the relevant axial length have to be taken into account. Consequently, a novel parameter – the piston rod heat transfer capability C – is introduced covering most of the influences that affect the axial heat transfer:

$$C = \dot{Q}_z * (\Delta T)^{-1} * L * (d_i)^{-1} \tag{2}$$

where $\dot{Q}_z$ is the transferred heat in z direction, ΔT is the temperature difference between both ends of the piston rod, L is the length and d_i is the internal diameter of the piston rod. For the calculation of C (see table 3) the temperature difference and the distance between the thermocouples FT4 and FT3 is used. The cross section used for obtaining the axial heat flux shown in table 3 is located in-between those thermocouples.

When comparing both of the different parameters of table 3 two contrary behaviors can be identified. The test case A with the smallest internal diameter leads to the highest axial heat flux. In contrast, the piston rod heat transfer capability increases with an increasing internal diameter. This can be explained with the help of the simulation results. They show that nearly the same amount of heat is axially transferred by the fluid at the middle of the rod within all 3 test cases. If the heat flow is referenced to the cross sectional flow area the smallest internal diameter of test case A leads to the highest heat flux. The thermal gradient that causes the heat flow is not taken into account for the determination of the heat flux. Therefore, only the higher temperature difference between the ends of the rod for test case A enables a similar heat flow through the fluid compared to the other test cases B and C with larger cross sectional flow areas.

This illustrates the benefit of the self-defined piston rod heat transfer capability. This parameter allows the designing of an internally cooled piston rod based on the thermodynamic requirements. In particular, the necessary internal diameter can be calculated according to the heat flow that is to be transferred driven by a certain temperature gradient over a certain length. In contrast to the axial heat flux, the piston rod heat transfer capability of test case C is superior to the other test cases mainly because of the smallest temperature gradient.

5. CONCLUSIONS

Experimental and theoretical investigations have been carried out in order to verify the impact of different internal diameters in a vertically oriented internally cooled piston rod with regard to the axial heat transfer. The comparison of the resulting temperature distribution and a subsequent thermal FE simulation yield the axial heat flux inside the fluid for all test cases. Based on these heat fluxes, a new parameter is introduced – the piston rod heat transfer capability. This parameter characterizes the transferable heat flow as a function of the internal diameter and of the thermal gradient over the length of the piston rod as they are the most important influences.

The results constitute the basis for the selection of the internal diameter for future configurations and lead to an enhanced performance of the internal piston rod cooling. In a scientific perspective the attained results allow an improved insight into the thermal mechanisms of an internally cooled reciprocating hollow rod.

NOMENCLATURE

A	Area	(m²)
C	Piston rod heat transfer capability	(W/K)
d	Diameter	(m)
L	Piston rod length	(m)
N	Rotational speed	(min^{-1})
P	Power input of heater	(W)
$\dot{Q}$	Heat flow	(W)
$\hat{q}$	Heat flux	(W/mm²)
t	Temperature	(°C)
T	Temperature	(K)
z	z axis	(m)
λ	Thermal conductivity	(W/m/K)
σ	Standard deviation	

Subscript

A	test case A
C	Crosshead
F	Fluid
FT1 … FT5	Fluid temperature 1 … fluid temperature 5
flow	flow
i	internal
t	Temperature
W	Wall
z	z axis

REFERENCES

Bush, J. E., & London, A. L. (1965). *Design data for "cocktail shaker" cooled pistons and valves*. SAE Technical Paper No. 650727

Cao, Y., & Wang, Q. (1995). Thermal analysis of a piston cooling system with reciprocating heat pipes. *Heat Transfer Eng. 16*(2), 50–57.

Feistel, N., Samland, G. Fejzuli, B., & Spindler, R. (2003). Unterschiedliche Konzepte zur Wärmeabfuhr aus Kolbenstangen-Dichtsystemen unter Berücksichtigung der API618. *7. Workshop Kolbenverdichter, Rheine* (121–134)

Hammer, C., Will, G., & Hesse, U. (2012). Temperature Reduction of the Piston Rod in non-lubricated Piston Compressors by Internal Cooling. *Proceedings of the 8th EFRC Conference, Düsseldorf, Germany* (260–266)

Klotsche, K. (2012). *Simulation von stationärer und instationärer Wärmeleitung und Wärmeübergang in der Kolbenstange zur Bestimmung der Kolbenstangentemperatur mit ANSYS* (Großer Beleg). Technische Universität Dresden

Kriegel, G. (1978). *Beitrag zur Berechnung des Radialverschleißes von Trockenlaufkolbenringen* (Doctoral dissertation). Technische Universität Dresden

Quack, H., & Nickl, J. (2009). Patent DE 10 2007 00 652, Deutsches Patent- und Markenamt

Reay, D., McGlen, R., & Kew, P. (2013). *Heat pipes: theory, design and applications.* Butterworth-Heinemann.

Thomas (née Hammer), C. (2013). *Innenkühlung der Kolbenstange von trockenlaufenden Kolbenverdichtern* (Doctoral dissertation). Technische Universität Dresden

Tomschi, U. (1995). *Verschleißverhalten von Trockenlaufwerkstoffen für Abdichtelemente in Kolbenkompressoren* (Doctoral dissertation). Universität Erlangen-Nürnberg

ACKNOWLEDGEMENT

Sincere thanks I express to Prof. G. Will for his inspiring and motivating help.

Thermal Analysis of a Hermetic Reciprocating Compressor using Numerical Methods

Stefan POSCH[1]*, Johann HOPFGARTNER[1], Martin HEIMEL[1], Erwin BERGER[1], Raimund ALMBAUER[1], Stefan STANGL[2]

[1] Institute of Internal Combustion Engines and Thermodynamics, Graz University of Technology, Inffeldgasse 19, 8010 Graz, Austria

posch@ivt.tugraz.at, +43 316 873 30233
hopfgartner@ivt.tugraz.at, +43 316 873 30240
heimel@ivt.tugraz.at, +43 316 873 30235
berger@ivt.tugraz.at, +43 316 873 30234
almbauer@ivt.tugraz.at, +43 316 873 30230

[2] Secop Austria GmbH, Jahnstraße 30, 8280 Fürstenfeld, Austria
s.stangl@secop.com, +43 3382 5010 824

* Corresponding Author

ABSTRACT

Comprehensive knowledge about the heat transfer mechanisms and the temperature field inside hermetic compressors is very important for the thermal management and thus their performance. A numerical model to predict the temperature field in a hermetic reciprocating compressor for household refrigeration appliances is presented in this work. The model combines a high resolution three-dimensional heat conduction formulation of the compressor's solid parts, a three-dimensional computational fluid dynamics (CFD) approach for the gas line domain and lumped formulations of the shell gas and the lubrication oil. Heat transfer coefficients are determined by applying CFD to the gas line side and correlations from the literature on the shell gas and oil side, respectively. The valve in the gas line simulation is modelled as a parallel moving flat plate. By means of an iterative loop the temperature field of the solid parts acts as boundary condition for the CFD calculation of the gas line which returns a cycle averaged quantity of heat to the solid parts. Using an iteration method which is based on the temperature deviation between two iteration steps, the total number of iterations and consequently the computational time can be reduced. The loop is continued until a steady-state temperature field is obtained. Calculated temperatures of the solid parts are verified by temperature measurements of a calorimeter test bench. The numerical results show reasonable agreement with the measured data.

1. INTRODUCTION

Thermal management is one of the main topics in the development process of modern hermetic reciprocating compressors for household refrigeration application. Due to the hermetic design of the compressor, electrical and mechanical losses influence the thermodynamic efficiency of the compressor. Heat transfer mechanisms like convection and conduction determine the temperature field inside the compressor and consequently, the compressor performance. The knowledge of the temperature field inside the compressor is essential to quantify loss mechanisms like superheating.

One possibility to obtain the heat transfer and the temperature field inside the compressor is the experimental investigation using heat flux sensors (Dutra and Deschamps, 2010) or thermocouples (Kara and Oguz, 2010). The spatial resolution of these methods is low, they are not suitable for compressors in the design phase and the usage of sensors may affect the compressor behaviour. Another possibility to investigate the thermal performance of a hermetic reciprocating compressor is the use of simulation tools. Several strategies to model the temperature field and heat transfer mechanism inside the compressor have been developed and can be found in the open literature.

Simple approaches split the compressor into several lumped volumes using the first law of thermodynamics to calculate the temperature field (e.g. Meyer and Thompson, 1988; Todescat *et al.*, 1992). Such models use either experimental data or correlations for the convective heat transfer formulation. The lumped volumes are connected via thermal conductance adjusted to experimental data. The low spatial resolution of these models and the dependence on experimental data yield a very rough and inflexible estimation of the temperature field in a hermetic reciprocating compressor.

A more flexible modelling strategy is the Thermal Network approach (TNW). The usage of TNW for thermal modelling of a hermetic reciprocating compressor for refrigeration application can be found in e.g. Sim *et al.* (2000) and Ooi (2003). TNW uses mass points to model the considered compressor parts and the heat transfer between the mass points is represented with the Lumped Conductance Method. Convective heat transfer is modelled with correlations based on forced or natural convection Nusselt number. Due to the reduction of geometrical information, the heat transfer modelling is characterized by a high level of uncertainty especially for regions with transient 3d flows like in the suction or discharge line. Also the validation of TNW with experimental data can cause problems because it is not clear if the chosen measuring points represent the temperature of the lumped mass.

The use of Computational Fluid Dynamics (CFD) in the development process of hermetic reciprocating compressors leads to another approach for thermal modelling, the so called hybrid simulation models. Although the performance of CPUs increased significantly over the last years, an overall 3d simulation of a compressor is still not possible within a reasonable time. Hybrid models use different combinations of complex 3d formulations and simple correlations for convective and conductive heat transfer, respectively. Almbauer *et al.* (2006) applied 1d flow simulation of the gas line, 3d formulation of the cylinder solid domain and lumped formulation of the remaining compressor parts. Ribas (2007), Sanvezzo and Deschamps (2012) and Lohn *et al.* (2015) combined 3d heat conduction formulation for the solid parts of the compressor and lumped formulation of the gas path. The authors used either experimental data or correlations from the literature to model the heat transfer.

The model in the present study is based on the hybrid approach. Compared to the hybrid models found in the literature, this work contains 3d formulation of the full gas line and 3d formulation of solid components of the entire compressor in combination with lumped volume formulation of the gas inside the compressor shell and the lubrication oil. The simulation of the fluid flow in the gas line considers 3d phenomena in the suction and discharge mufflers, the flow in the cylinder as well as interactions with the compressor valves. Special focus is laid on the high geometric resolution of the solid compressor parts. A simulation algorithm is presented to combine the transient flow calculation and the steady-state heat conduction calculation to model the thermal behaviour of a hermetic reciprocating compressor running at calorimeter test conditions. The simulation results are verified by experimental data of a calorimeter test bench.

2. SIMULATION MODEL

The R600a (isobutane) hermetic reciprocating compressor used in the present study has a displacement of 5.5 ccm and the COP (ASHRAE test conditions -23.3 °C/55 °C) is approximately 1.8. A schematic view of the compressor is shown in Figure 1. The simulation model can basically be split up into three main parts: (i) simulation of the gas flow using commercial CFD software, (ii) heat conduction in the solid parts also using commercial CFD software and (iii) the energy balance for oil and refrigerant by assuming lumped control volumes. The three simulation parts exchange data in terms of heat flux and temperatures.

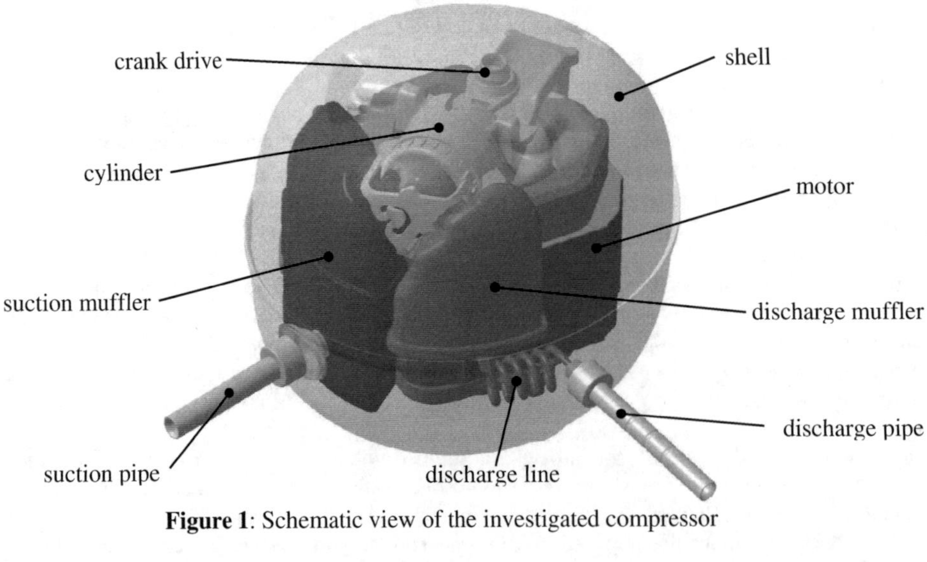

Figure 1: Schematic view of the investigated compressor

2.1 Gas flow

The simulation of the gas flow is done with the commercial CFD software package ANSYS Fluent. The simulation domain includes suction line, suction muffler, valves, cylinder, ports, discharge muffler and discharge line. To reduce the computational time the valves are assumed to be parallel moving flat plates. The valve motion is calculated via User Defined Functions solving the single-degree of freedom system considering oil stiction forces. The number of cells in the computational mesh varies between 3.1 and 5.5 million cells depending on the position of the piston. The k-ε turbulence model is used and 1st order spatial discretization of density, momentum, energy and turbulence is applied. Pressure-velocity coupling is used with coupled solver setting.

2.2 Conduction in the solid parts

Heat conduction in the solid parts of the hermetic compressor is also solved with the commercial software package ANSYS Fluent. The whole domain is meshed with about 2.4 million cells with refinements at the interfaces between the single parts. Heat flux boundary conditions are applied on the surfaces of the solid parts depending on the flow conditions. Surfaces that are in contact with the gas flow use area-weighted average heat flux values determined by the CFD calculation of the gas flow. Heat flux values for the convective heat transfer between the solid parts, the compressor oil and the gas inside the shell are calculated with correlations from the literature (e.g. flow over flat plates, flow inside pipes). The heat flux between the compressor shell and the ambient air is modelled by means of a natural convection correlation. Electrical power losses are considered as volumetric source terms in the energy equation of the corresponding part. Mechanical power losses are considered as heat input to the compressor oil (Zach, 2013).

2.3 Lumped control volumes

To complete the thermal modelling of the entire hermetic compressor, the refrigerant and the oil inside the compressor shell have to be taken into account. The gas caught inside the shell is regarded as one control volume. Leakage mass flow rates are neglected and the gas is assumed to be in thermal equilibrium. The sum of the heat fluxes between the shell gas and the adjacent parts $\dot{Q}_{SG}$ are calculated with the results of the solid part conduction simulation. The temperature of the gas can be calculated by applying the energy balance to the control volume:

$$T_{SG}^{n+1} = T_{SG}^{n} + \frac{\dot{Q}_{SG} \cdot \Delta t}{m_{SG} \cdot c_{v\,SG}} \tag{1}$$

The oil inside the compressor shell is divided into three control volumes, namely the oil in the sump, the oil transported by the oil pump and the expelled oil on the oil covered walls. The oil mass flow rate at steady-state operating conditions is determined by experiments (Posch *et al.*, 2015). Applying a first order upwind scheme for the calculation of the outlet enthalpy flow, the energy balance of the considered control volume yields the following equations for the oil temperatures:

$$T_{OS}^{n+1} = \frac{\dot{Q}_{OS} + c_{oil}\left(\dot{m}_{oil} \cdot T_{OW} + m_{OS}\frac{T_{OS}^{n}}{\Delta t}\right)}{c_{oil}\left(\dot{m}_{oil} + \frac{m_{OS}}{\Delta t}\right)} \tag{2}$$

$$T_{OC}^{n+1} = \frac{\dot{Q}_{OC} + c_{oil}\left(\dot{m}_{oil} \cdot T_{OS} + m_{OC}\frac{T_{OC}^{n}}{\Delta t}\right)}{c_{oil}\left(\dot{m}_{oil} + \frac{m_{OC}}{\Delta t}\right)} \tag{3}$$

$$T_{OW}^{n+1} = \frac{\dot{Q}_{OW} + c_{oil}\left(\dot{m}_{oil} \cdot T_{OC} + m_{OW}\frac{T_{OW}^{n}}{\Delta t}\right)}{c_{oil}\left(\dot{m}_{oil} + \frac{m_{OW}}{\Delta t}\right)} \tag{4}$$

2.4 Solution algorithm

At the beginning of the simulation process an initial temperature field in all solid parts, refrigerant and oil in the compressor is guessed. The solid part temperatures are used as boundary conditions for the following CFD simulation of the gas flow. The results of the CFD simulation are averaged over one rotation of the compressor and are considered as heat flux boundary conditions for the simulation of the solid part heat conduction. Since a steady-state temperature field of the compressor should be calculated, the solid part heat conduction model in ANSYS Fluent is set to steady. After calculating the temperature distribution in the solid parts, the energy balance for the refrigerant in the shell and the respective oil volumes are carried out using the CFD heat fluxes. The lumped control volumes are modelled as quasi-transient using a time step of 1s. The boundary conditions for the solid part conduction simulation are updated with the gas and oil temperatures. Lumped volume energy balance and solid part conduction forms an iteration loop which is continued until heat flux deviation between two consecutive iteration steps is below a certain limit. Using the new solid part temperature field the CFD simulation is carried out again. The iteration loop between CFD simulation of the gas line and the solid part heat conduction calculation is executed until heat flux deviations between two consecutive iteration steps are also below a certain convergence criteria.

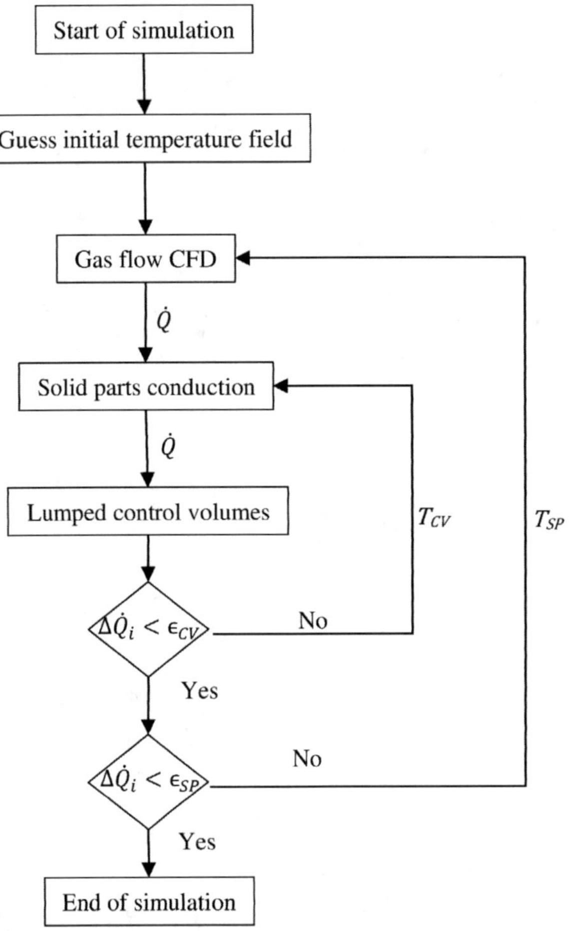

Figure 2: Flow chart of the solution algorithm

3. EXPERIMENTAL WORK

To validate the thermal modelling of the hermetic compressor, temperature measurements on several positions in the compressor are carried out. 22 thermocouples type T are distributed inside the compressor, on the shell and in the gas line. Furthermore, two pressure sensors are placed in the suction and discharge muffler. A special sealed feedthrough is used to pass the Teflon®-insulated sensor wires through the hermetic shell. Table 1 gives an overview of the thermocouples which are used for the validation of the present thermal model. A schematic overview of the sensor positions is illustrated in Figure 3.

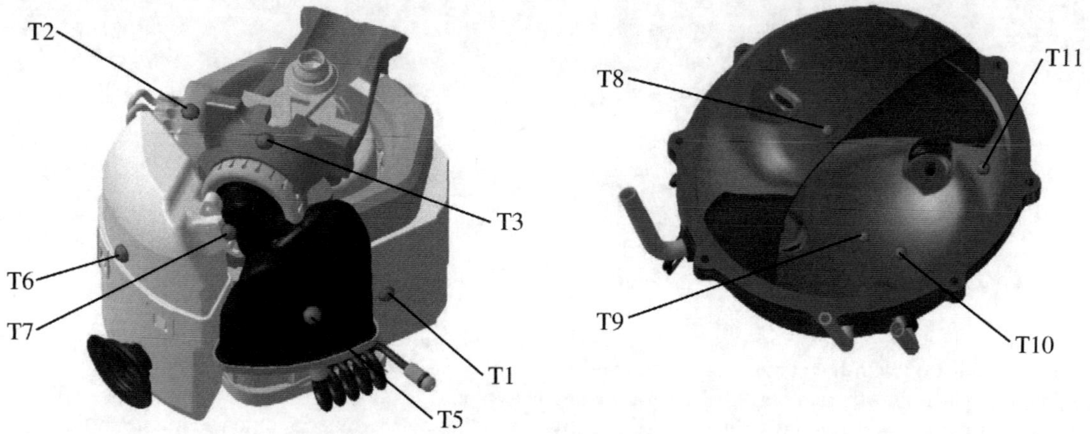

Figure 3: Schematic overview of the sensor positions

Table 1: Thermocouple measuring points

No.	Measuring point description
T1	Stator lamination (surface)
T2	Stator windings (surface)
T3	Cylinder (surface)
T4	Shell (gas)
T5	Discharge muffler (surface)
T6	Suction muffler (surface)
T7	Valve plate (surface)
T8	Shell top outside (surface)
T9	Shell bottom outside (surface)
T10	Oil sump
T11	Shell inside (surface)

4. RESULTS

The presented thermal model is used to simulate the temperature distribution in solid parts, gas and oil of a hermetic reciprocating compressor. The operating conditions of the compressor are set to -23 °C evaporating temperature, 45 °C condensing temperature and 32°C ambient temperature, respectively. The compressor works at constant speed of 2950 rpm.

Figure 4 shows exemplarily the velocity field in the cylinder of the gas flow obtained by CFD simulation. Regions with high local gradients of the heat transfer coefficient like suction and discharge port are treated as separated areas in the iteration loop between gas flow simulation and solid part conduction. Specific heat flux values for the individual regions can be seen in Table 2. The sign of the values is related to the heat flux direction from solid part to gas flow. The specific heat flux acts as boundary condition for the heat conduction simulation of the solid parts.

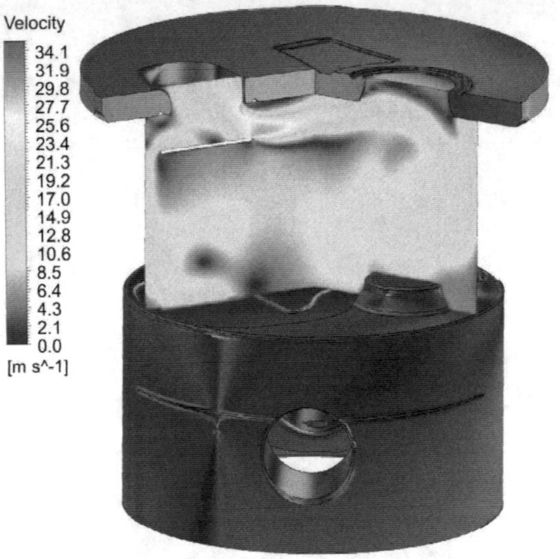

Figure 4: CFD simulation of the gas flow

Table 2: Specific heat flux from solid to gas [W/m²]

Solid	Value
piston	-7199
discharge muffler	-1289
suction muffler	351
valve plate cylinder	239
cylinder	2082
valve plate suction	414
valve plate discharge	-24421
discharge port	-10216
suction port	7485
serpentine	-3799

As described in chapter two the present model consists of two iteration loops. The outer iteration loop links the gas flow simulation with the simulation of the solid parts, shell gas and oil. The inner iteration loop links the solid part conduction with the energy balance of the lumped volumes of gas and oil in the compressor shell. The calculation time of the whole procedure is significantly depending on the outer iteration loop due to the time-consuming CFD simulation of the gas flow. To contain the number of iterations for the outer loop, the initial temperature field of the compressor has to be guessed well. In the present study, five iterations for the outer loop had to be carried out to fulfil the convergence criteria. Due to the fast calculation of the conduction in the solids and of the energy balance, the inner iteration loop has no significant impact on the overall computation time. The number of iterations for the inner loop in the present study is between 50 and 60. Figure 5 shows the temperature development over the number of iterations of the lumped volumes during the inner iteration loop.

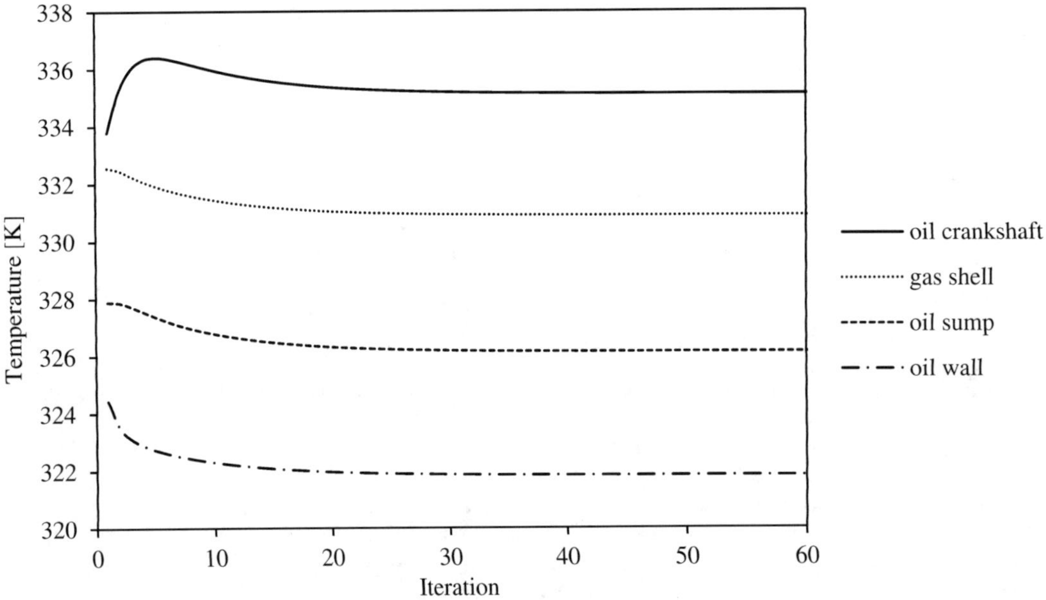

Figure 5: Temperature development during the inner iteration loop [K]

The comparison between the simulated and measured temperatures at the specific measuring points is illustrated in Table 3. The results of the thermal model are in agreement with calorimeter test data. Maximum absolute temperature difference between simulation and measurement is less than 4.2 K. Especially the temperatures of the solid parts inside the compressor and the lumped volumes are met very well. Although the thermal modeling of the hermetic compressor in the present study matches the experiment well, the uncertainties in the determination of the heat transfer coefficients should be discussed briefly. In general heat transfer can be modelled in two different ways: On the one hand by detailed CFD simulation of the complex gas flow inside the shell or by means of empirical correlations. With the CFD approach not only the model accuracy should increase but also the computation effort. If choosing empirical correlations a bad agreement between simulation and experimental data could turn out, that requires a slight calibration of the empirical correlations by experimental data. The temperature distribution in the solid parts of the compressor can be seen in Figure 6.

Table 3: Comparison between measured and simulated temperatures [K]

No.	Measuring point description	Measurement	Simulation	ΔT
T1	Stator lamination (surface)	332.9	333.4	0.5
T2	Stator windings (surface)	333.7	333.4	0.3
T3	Cylinder (surface)	337.7	338.3	0.6
T4	Shell (gas)	331.6	330.9	0.7
T5	Discharge muffler (surface)	350.9	350.9	0.0
T6	Suction muffler (surface)	328.2	326.8	1.4
T7	Valve plate (surface)	345.0	342.6	2.4
T8	Shell top outside (surface)	318.9	321.2	2.3
T9	Shell bottom outside (surface)	321.8	322.0	0.2
T10	Oil sump	327.8	326.1	1.7
T11	Shell inside (surface)	324.7	320.5	4.2

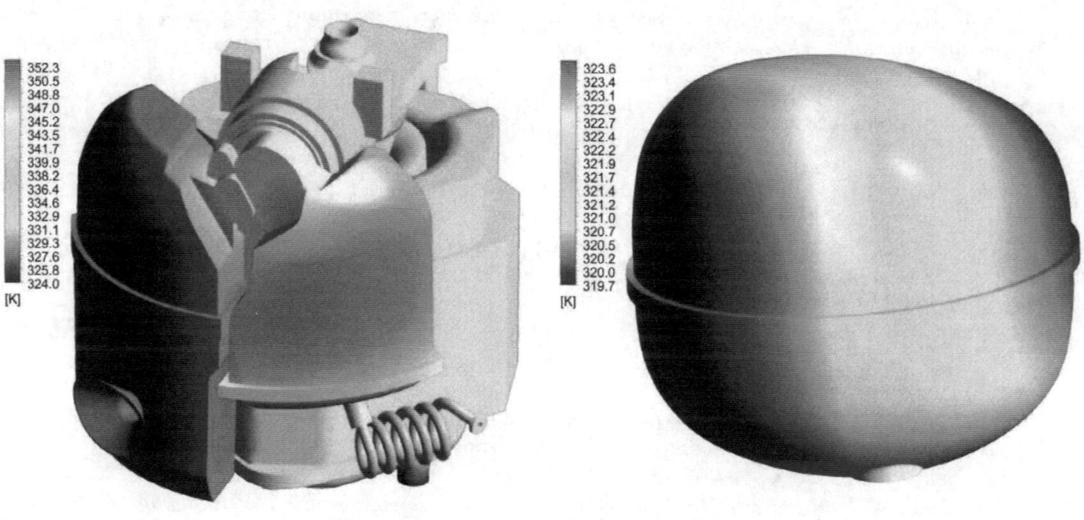

Figure 6: Temperature field of the solid parts

5. CONCLUSION

In this study a thermal model for the investigation of a hermetic reciprocating compressor is shown using mainly numerical methods. The method combines CFD simulation of the gas flow, numerical calculation of the heat conduction in the solid parts and simple lumped formulations of the refrigerant and oil in the compressor shell. Although some heat transfer coefficients are determined by simple correlations from literature, the results of the thermal modelling are in good agreement with experimental data gained by calorimeter tests. The present method is an advisable tool in the compressor development process along with detailed CFD simulation of the gas flow to validate thermodynamic compressor behaviour. Unavoidable uncertainties in the determination of the heat transfer coefficients between the solid parts and the fluids (refrigerant and oil) have to be considered in the evaluation of the results. To get a thermal model for the usage in thermodynamic parameter studies it is useful to calibrate the heat transfer coefficients with experimental data.

NOMENCLATURE

c	specific heat capacity	(J/kg K)	OS	oil in the sump	
$\dot{m}$	mass flow rate	(kg/s)	OW	oil on the walls	
m	mass	(kg)	oil	compressor oil	
n	iteration step		SG	shell gas	
$\dot{Q}$	heat flux	(W)	SP	solid parts	
T	temperature	(K)	v	isochoric	
t	time	(s)			
Subscripts			**Greek symbols**		
CV	control volumes		Δ	difference	(-)
OC	oil in the crankshaft		ε	convergence criteria	(-)

REFERENCES

Almbauer, R., Burgstaller, A., Abidin, Z., Nagy, D. (2006). 3-Dimensional Simulation for Obtaining the Heat Transfer Correlations of a Thermal Network Calculation for a Hermetic Reciprocating Compressor. *Proceedings of the International Compressor Engineering Conference at Purdue*, Purdue University, IN, USA, paper 1794.

Dutra, T., Deschamps, C. (2010). Experimental Investigation of Heat Transfer in Components of a Hermetic Reciprocating Compressor. *Proceedings of the International Compressor Engineering Conference at Purdue*, Purdue University, IN, USA, paper 1346.

Kara, S., Oguz, D.H. (2010). Thermal Analysis of a Small Hermetic Reciprocating Compressor, *Proceedings of the International Compressor Engineering Conference at Purdue*. Purdue University, IN, USA, paper 1307.

Lohn, S.K., Diniz, M.C., Deschamps, C.J. (2015). A Thermal Model for Analysis of Hermetic Reciprocating Compressors under the On-Off Cycling Operating Condition. *Proceedings of the International Conference on Compressors and their Systems*, City University, London, GB, paper 012068.

Meyer, W.A., Thompson, E. (2010). An Analytical Model of Heat Transfer to the Suction Gas in a Low-Side Hermetic Refrigeration Compressor. *Proceedings of the International Compressor Engineering Conference at Purdue*, Purdue University, IN, USA, paper 662.

Ooi, K.T. (2003). Heat Transfer Study of a Hermetic Refrigeration Compressor. *Applied Thermal Engineering*, Volume 23, pages 1931-1945.

Posch, S., Berger, E., Heimel, M., Hopfgartner, J., Almbauer, R., Schöllauf, P. (2015). Fluid Flow in the Oil Pumping System of a Hermetic Compressor. *Proceedings of the International Conference on Compressors and their Systems*, City University, London, GB, paper 012028.

Ribas Jr., F.A. (2007). Thermal Analysis of Reciprocating Compressors. *Proceedings of the International Conference on Compressors and their Systems*, City University, London, GB, pages 277-287.

Sim, Y.H., Youn, Y., Min, M.K. (2000). A Study on Heat Transfer and Performance Analysis of Hermetic Reciprocating Compressors for Refrigerators. *Proceedings of the International Compressor Engineering Conference at Purdue*, Purdue University, IN, USA, paper 1390.

Sanvezzo Jr, J., Deschamps, C.J. (2012). A Heat Transfer Model Combining Differential and Integral Formulations for Thermal Analysis of Reciprocating Compressors. *Proceedings of the International Compressor Engineering Conference at Purdue*, Purdue University, IN, USA, paper 2104.

Todescat, M.L., Fagotti, F., Prata, A.T., Ferreira, R.T.S. (1992). Thermal Energy Analysis in Reciprocating Hermetic Compressors. *Proceedings of the International Compressor Engineering Conference at Purdue*, Purdue University, IN, USA, paper 936.

Zach, D. (2013). Verlustteilung hermetischer Kältemittelkompressoren. *MSc Thesis*, Graz University of Technology, AUT.

ACKNOWLEDGEMENT

This work has been carried out within the framework of ECO-COOL, a research project initiated and funded by the FFG (Austrian Research Promotion Agency). Furthermore the authors particularly acknowledge the technical support by Secop Austria GmbH, formerly ACC Austria GmbH and Liebherr-Hausgeräte Lienz GmbH.

CFD Approach to Evaluate Heat Transfer in Reciprocating Compressors

Jônatas Ferreira LACERDA[1]*, Celso Kenzo TAKEMORI[2]

[1]Tecumseh Products Company, Research and Development,
São Carlos, São Paulo, Brazil
jonatas.lacerda@tecumseh.com

[2]Vibroacústica Development and Research
Joinville, Santa Catarina, Brazil
celso@vibroacustica.com.br

* Corresponding Author

ABSTRACT

How to deal with the heat exchange inside compressors is one of the main challenges on its design nowadays. The heating of the gas entering the compressors decreases its density reducing directly the mass flow and compressor efficiency, so should be avoid. To study this topic it was developed a full three-dimensional model of the gas and the solid parts inside a compressor. Prescribed steady state mass flow rate is imposed at suction and discharge tubes and heat sources due electrical and mechanical inefficiencies of the compressor are imposed. The portion of the suction and cavity gas are separated from the portion of the discharge gas and an isentropic compression from the suction muffler outlet temperature is assumed to define the temperature of the gas flowing through the discharge port. The main advantage of this approach is the directly calculation of the heat transfer coefficients from the solid parts to the fluid, however with a high computational cost. Comparison with experimental data shows good agreement is some parts, but some disagreement in other, probably due the non-consideration of the oil flowing inside the compressor, which are topics for future investigations.

1. INTRODUCTION

Understanding and predict the heat transfer phenomena and temperature distribution is crucial in a reciprocating compressors design. The heating of the gas entering the compressors decreases its density reducing directly the mass flow and compressor efficiency. Also, excessive increase of the temperature can degrade oil lubrication properties causing catastrophic fail of the compressor.

Several studies on modeling the heat transfer process inside the compressor has been conducted along the years. They can be characterized according to Tab. 1, with different levels of accuracy and complexity, making the choice of the method dependent on the accuracy and time required for the analysis. Some works falling in each category are presented, but there are more in current literature.

The steady state lumped models (0D) are based on energy balance equations. They are computationally inexpensive but are not able to describe the effect of drastic changes in the compressor layout. Besides, heat transfer coefficients and thermal resistances between solid parts must be obtained experimentally or using general heat transfer correlations that may lead to differences in the comparison with experimental results. However, even if this kind of model was used in the early stages of the studies (Meyer and Thompson, 1988; Todescat *et al.*, 1992 and Padhy, 1992), it is still widely used due to ease of integration of more physics into the model, as showed by Dutra and Deschamps (2015).

Advances were obtained with the combination of lumped models with differential models (0D & 3D), also known as hybrid models. Being less computationally demanding than full 3D models, they are more used in the combination of 3D discretization for solid parts and lumped formulation for flow parts. Thus, the temperature gradients present in solid parts are better captured, affecting positively the results. This kind of procedure was used for part of the compressor by Almbauer *et al.* (2006) and for the entire compressor as in the work of Lohn *et al.* (2015).

A third class of methods are those full 3D models for the fluid flow and heat transfer phenomena. They can consider both fluid and solid parts and are computationally expensive, so limited to perform few cases analysis or consider just part of the compressor as performed by Pereira *et al.* (2010) and Lacerda and Takemori (2014). On the other hand, there is no need to input heat transfer coefficients and strong modifications in compressor layout can be better evaluated with this procedure. Full compressor 3D models were performed by Chikurde *et al.* (2002) and Raja *et al.* (2003).

Table 1: Map of works on heat transfer according to a modeling classification

		0D (lumped model)	**0D & 3D**	**Full 3D**
Parts of the Compressor	**Steady State**		- Almbauer *et al.* (2006) - Kara and Oguz (2010)	- Birari *et al.* (2006) - Colmek (2014)
	Transient		- Abidin *et al.* (2006)	- Morriesen *et al.* (2009) - Pereira *et al.* (2010) - Lacerda and Takemori (2014)
Full Compressor	**Steady State**	- Meyer and Thompson (1988) - Todescat *et al.* (1992) - Padhy (1992) - Ooi (2003)	- Ribas Jr. (2007) - Schreiner *et al.* (2009) - Sanvezzo Jr. and Deschamps (2012)	- Chikurde *et al.* (2002) - Raja *et al.* (2003)
	Transient	- Dutra and Deschamps (2015)	- Lohn *et al.* (2015) - Rigola *et al.* (2014)	

In this work is presented a study of a compressor using a full compressor 3D model with a steady state mass flow rate. The compression process is not included, but the temperature of the discharged gas into the cylinder head is calculated considering an isentropic compression from the temperature of the gas leaving the suction muffler. More details on the model are further discussed. Comparisons with the experimental data is showed, as well as an energetic balance and heat transfer coefficient predictions.

2. METHODOLOGY

The thermal management of the compressor consists and understand and act on the heat transfer paths from thermal sources to a thermal "receiver". Usual thermal sources are heat generated due motor and friction losses, the compression process in the cylinder and the hot gas being discharged. All the solid parts, the oil and the gas inside the compressor can be considered as heat transfer paths and the receivers are the gas being suctioned at low temperature and the housing that exchange heat with the exterior ambient.

In the present work a full 3D discretization was adopted for the entire compressor: fluid and solid parts. The solid parts are the suction tube, the housing, the suction muffler, the valve plate, the cylinder head, the shockloop, the crankcase, the winding, the stator and the rotor. The fluid parts are the suction, the housing cavity and the discharge gas. The Tab. 2 shows a matrix of connections among all the parts in the model. The solid material properties of each part was considered, detaching that the only plastic part is the suction muffler.

A steady state mass flow rate is imposed at the suction tube considering ASHRAE-LBP conditions for a compressor that operates at 60Hz with R134a (suction pressure 97 kPa and suction temperature 32.2°C). Electric motor losses are estimated from their efficiency curves obtained in dynamometer tests and are imposed as volumetric heat sources in the winding, the stator and the rotor. Estimated losses of the bearings (Neto, 2006) are imposed locally as area heat sources. As mentioned before, the compression process is not included, but the temperature of the discharged gas into the cylinder head is calculated considering an isentropic compression from the temperature of the gas leaving the suction muffler up to the discharge pressure (discharge pressure 1,072 kPa). In the cavity gas volume is considered the buoyance effect causing some natural convection inside the housing. At the external part of the housing and the suction tube is imposed a natural convection heat transfer process considering the ambient temperature at 32.2 °C and heat transfer coefficient of 10.0 W/m².K. A graphical summary of the boundary conditions is shown in Fig. 1.

Table 2: Matrix of connections between parts

	Suction Tube	Housing	Shockloop	Cylinder Head	Stator	Muffler	Valve Plate	Crankcase	Rotor	Winding	Muffler Gas	Cavity Gas	Discharge Gas
Suction Tube													
Housing													
Shockloop													
Cylinder Head													
Stator													
Muffler													
Valve Plate													
Crankcase													
Rotor													
Winding													
Muffler Gas													
Cavity Gas													
Discharge Gas													

To perform this task the commercial code Ansys-CFX was used. Compressible transient mass conservation and Navier-Stokes equations were numerically solved. To close the problem were also solved the equation of energy balance without simplifications and a Peng Robinson real gas model for the R134a (Ansys CFX, 2012).

From the numeric point of view, the called 'High Resolution' method was employed as spatial discretization scheme. The SST turbulence model was used, which falls within the category of Unsteady Reynolds Averaged Navier-Stokes (URANS) equations and combines κ-ε and κ-ω models (Cezario, 2007), being the first used on turbulent free stream and the second on near-wall regions.

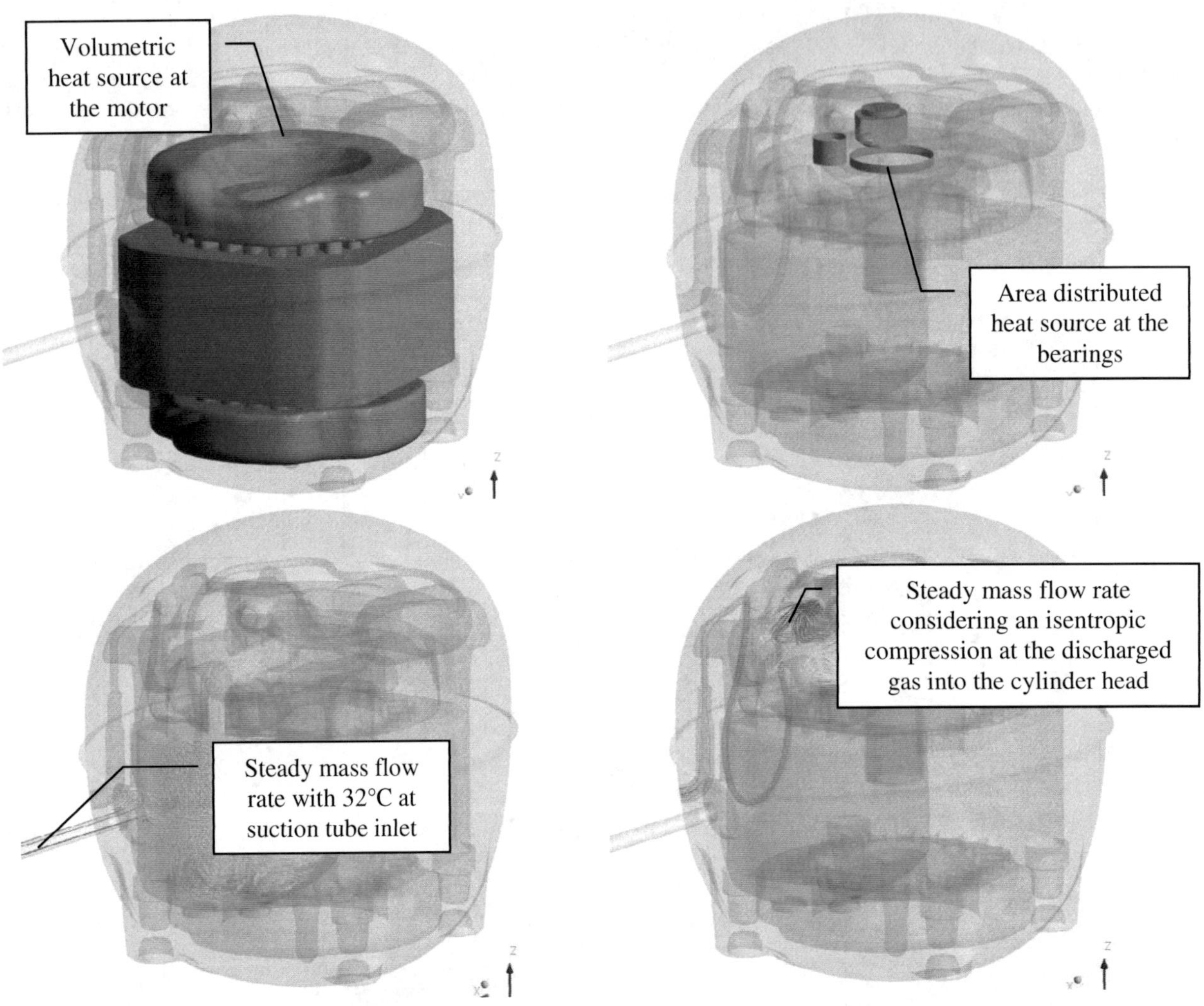

Figure 1: Boundary conditions used in the numerical simulation

The simulation was started on a coarser mesh and then finalized on a refined one with 840 thousand nodes. Around 2000 iterations was run taking around 24 hours using parallel simulation on a 12 Gb of RAM with four 3.33GHz computer processing unites.

3. RESULTS AND DISCUSSIONS

The same compressor model was instrumented with thermocouples in several parts to measure the temperature distribution inside the compressor and compare with numerical results. The compressor was mounted on a calorimeter facility and a test was carried out under ASHRAE-LBP conditions.

Table 3 and Fig. 2 shows a comparison between experimental and numerical data of the temperature distribution. The temperature are normalized by the maximum observed temperature for the solid and fluid parts, respectively. As can be noticed, in some parts it was observed a very good agreement as in the suction muffler wall close to the motor (S04, -2%) or the discharge muffler fluid (F06, 0%). However, in other parts the quantitative comparison was not so good, as for the crankcase close the cylinder (S08, 13%) or for the suction muffler outlet (F04, -12%). It was considered that most of these differences was obtained because of the absence of the lubricant oil in the model or the assumption of an isentropic compression process. The oil plays a major hole in the compressor temperature distribution because due its higher specific heat, compared to the gas, it is capable of absorb and reject a higher amount of heat wherever it goes. That can be the explanation of the temperature difference in solid parts as the motor, the crankcase, the shockloop and the housing, where a big amount of oil is splashed in during compressor run. The inclusion of the oil presence seems to be a great advance for future works using this technique.

Table 3: Temperature comparisons between experimental and numerical data

			Experimental	Numerical	% Difference
Solid Parts	S01	Lower Housing	0.52	0.49	-7
	S02	Upper Housing	0.61	0.53	-12
	S03	Suction Muffler (Housing Side)	0.49	0.45	-7
	S04	Suction Muffler (Motor Side)	0.56	0.55	-2
	S05	Suction Muffler Outlet	0.57	0.61	7
	S06	Cylinder Head	0.98	1.00	2
	S07	Discharge Muffler	0.97	1.00	3
	S08	Crankcase close to cylinder	0.73	0.82	13
	S09	Stator and Winding	0.72	0.79	10
Fluid Parts	F01	Suction Tube Outlet	0.28	0.28	-1
	F02	Suction Muffler Inlet	0.37	0.35	-4
	F03	Suction Muffler Volume	0.45	0.41	-11
	F04	Suction Muffler Outlet	0.48	0.42	-12
	F05	Cylinder Head	0.99	1.00	1
	F06	Discharge Muffler	0.97	0.97	0
	F07	Compressor Outlet	0.83	0.89	8
	F08	Oil	0.55	0.58*	7

* The numerical model did not considered the oil. It is used the gas temperature instead

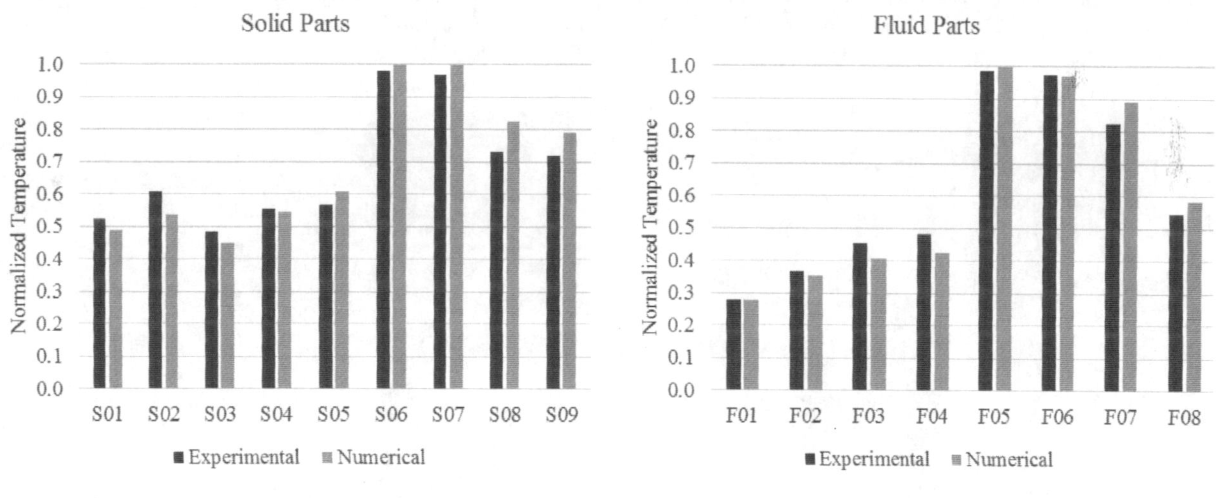

Figure 2: Normalized temperature distribution comparison between experimental and numerical results

Despite the observed differences in the temperature distribution, qualitatively it was observed a good agreement between numerical and experimental results. Therefore, an important information to understand how is the heat transfer process inside the compressor is the heat exchange from part to part that can be monitored numerically. For example, Fig. 3 shows the normalized heat exchange into and from the cavity gas (the maximum heat exchange value is used to normalize). Is noticed that a great amount of heat is coming from the motor (rotor, winding and stator) and from the crankcase. Other interesting aspect to point is how the heat coming from the discharge process is distributed, being a higher amount coming from the shockloop instead of the cylinder head. Thus, in this case, further work to decrease the contribution of the discharge process should be done in the shockloop.

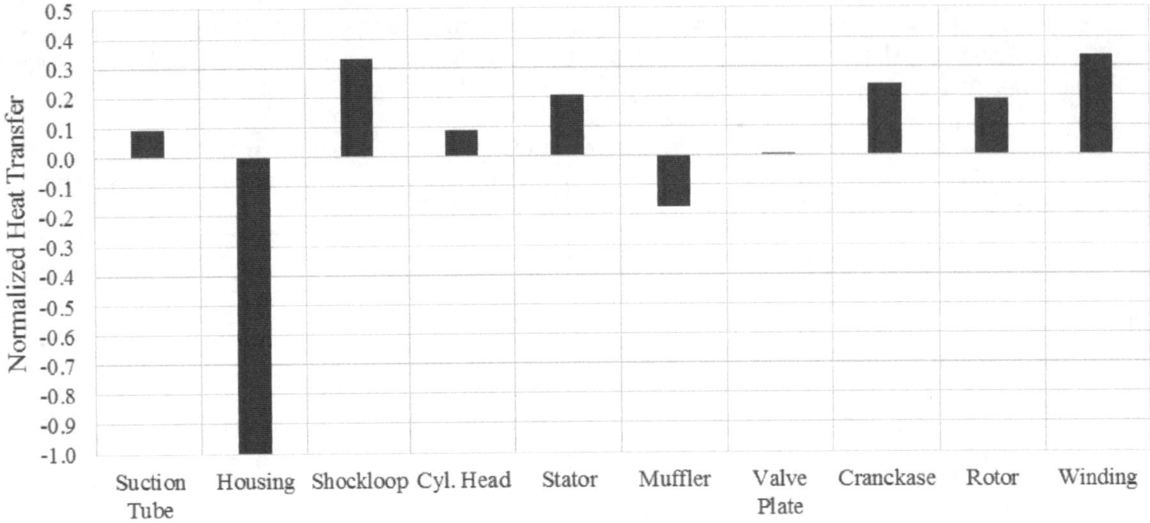

Figure 3: Heat exchange of the cavity gas

Since temperature distribution and heat flux between parts are easily accessed in this simulation, it is possible to calculate the heat transfer coefficient h, given by Eq. 1. This is an advantage to experimental analysis were heat flux is difficult to be measured and advantage to use of known heat transfer correlations that are suitable to more simple geometries whereas the geometries inside compressor are usually complex.

$$h = Q/A.\Delta T \tag{1}$$

In Figures 4 and 5 are presented the normalized temperature difference and the normalized heat transfer coefficients obtained for the cavity gas heat transfer. Again is detached the behavior for the shockloop, where is observed the highest value of the normalized heat transfer coefficient. This enhance in the heat transfer can be caused by an improved mixing of the gas inside the compressor due the consideration of buoyance effects.

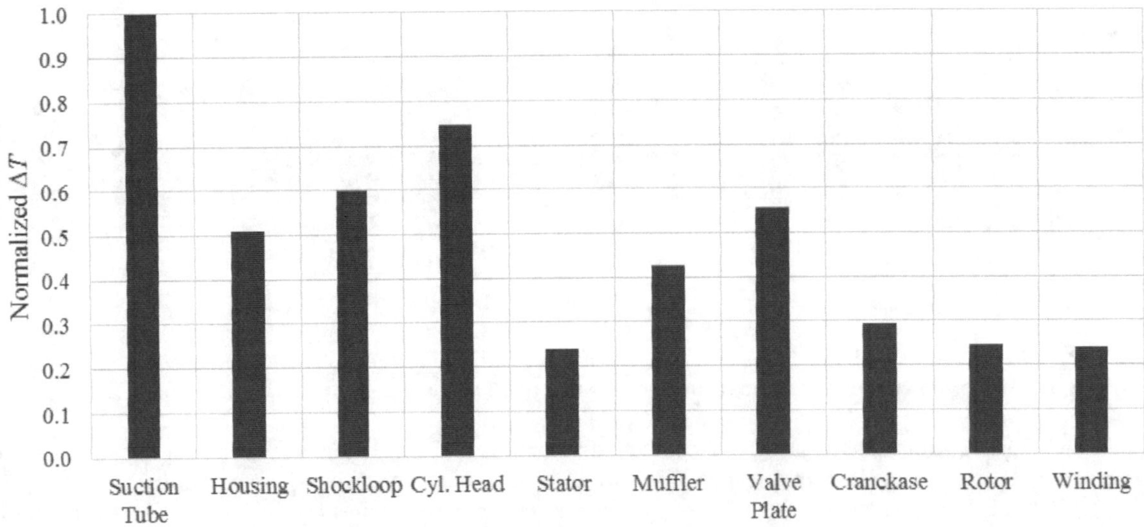

Figure 4: Normalized temperature difference between the cavity gas and solid parts

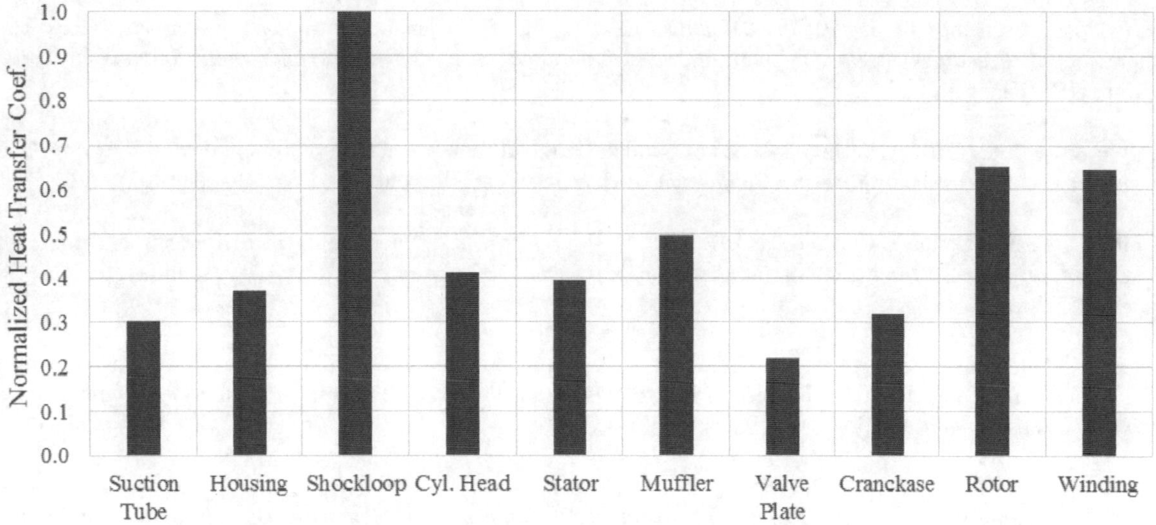

Figure 5: Normalized heat transfer coefficients between the cavity gas and solid parts

4. CONCLUSIONS

In this work the heat transfer phenomena happening inside a reciprocating compressor was investigated using a full 3D discretization model of the solid and fluid parts. This procedure is more computationally expensive than others described on literature, however it is not necessary to impose heat transfer coefficients and thermal resistances between solid parts. In fact, these information are results of the simulations. Comparison of the temperature distribution inside the compressor with experimental data showed good agreement is some parts (less than 1% difference), but some disagreement in other (more than 10%), probably due the non-consideration of the oil flowing inside the compressor or the non-consideration of transient aspects, which are topics for future investigations.

NOMENCLATURE

Q	heat transfer	(W)
ΔT	temperature difference	(°C)
A	surface area	(m^2)
h	heat transfer coefficient	($W/m^2.K$)

REFERENCES

Almbauer, R. A., Burgstaller, A., Abidin, Z., & Nagy, D. (2006). 3-dimensional simulation for obtaining the heat transfer correlations of a thermal network calculation for a hermetic compressor. *Proceedings of the 18th International Compressor Engineering Conference at Purdue*, West Lafayette, IN, USA. (C079).

Birari, Y. V., Gosavi, S. S. & Jorwekar, P. P. (2006). Use of CFD in design and development of R404a reciprocating compressor. *Proceedings of the 18th International Compressor Engineering Conference at Purdue*, West Lafayette, IN, USA. (C072).

Chikurde, R. C., Loganathan E., Dandekar D. P. & Manivasagam, S. (2002). Thermal mapping of hermetically sealed compressors using computational fluid dynamics technique. *Proceedings of the 16th International Compressor Engineering Conference at Purdue*, West Lafayette, IN, USA. (C6-4).

Colmek, S. (2014). The improvement of the compressor efficiency by reducing the heat transfer. *Proceedings of the 22nd International Compressor Engineering Conference at Purdue*, West Lafayette, IN, USA. (Paper 1168).

Dutra T. & Deschamps C. J. (2015). A thermal network model for induction motors of hermetic reciprocating compressors. *Proceedings of the 9th International Conference on Compressors and Their Systems at London*, London, UK. (Paper 012020).

Kara, S. & Oguz E. (2010) Thermal analysis of a small hermetic reciprocating compressor. *Proceedings of the 20th International Compressor Engineering Conference at Purdue*, West Lafayette, IN, USA. (Paper 1307).

Lacerda, J. F. & Takemori, C. K. (2014). Predicting the suction gas superheating in reciprocating compressors. *Proceedings of the 22nd International Compressor Engineering Conference at Purdue*, West Lafayette, IN, USA. (Paper 1412).

Lohn, S. K., Diniz, M. C. & Deschamps C. J. (2015). A thermal model for analysis of hermetic reciprocating compressors under the on-off cycling operation condition. *Proceedings of the 9th International Conference on Compressors and Their Systems at London*, London, UK. (Paper 012068).

Meyer, W. A. & Thompson, H. D. (1988). An analytical model of heat transfer to the suction gas in a low-side hermetic refrigeration compressor. *Proceedings of the 1998 International Compressor Engineering Conference at Purdue*, West Lafayette, IN, USA. (Paper 662).

Morriesen, A., Deschamps, C. J., Pereira, E. L. L. Dutra, T. & Ribas Jr., F. A. (2009). Numerical prediction of superheating in the suction muffler of a hermetic reciprocating compressor. *Proceedings of the 20th International Congress of Mechanical Engineering at Gramado/RS, Brazil.*

Neto, I. G., (2006). Modelagem de mancais hidrodinâmicos com movimento oscilatório, MSc dissertation (UNICAMP, Campinas-SP, Brazil)

Ooi, K.T. (2003). Heat transfer study of a hermetic refrigeration compressor. *Applied Thermal Engineering*, *23*, 1931–1945.

Padhy, S. K. (1992) Heat transfer model of a rotary compressor. *Proceedings of the 1992 International Compressor Engineering Conference at Purdue*, West Lafayette, IN, USA. (Paper 935).

Perereira, E. L. L., Deschamps C. J. & Ribas Jr., F. A. (2010). Numerical analysis of the heat transfer inside the cylinder of reciprocating compressors in the presence of the suction and discharge process. *Proceedings of the 20th International Compressor Engineering Conference at Purdue*, West Lafayette, IN, USA. (Paper 1310).

Raja, B., Sekhar, S. J., Lal, D. M. & Kalanidhi A. (2003). A numerical model for thermal mapping in a hermetically sealed reciprocating compressor. *Int. J. Refrig.*, *26*(6), 229-236.

Ribas Jr., F. A. (2007). Thermal analysis of reciprocating compressors. *Proceedings of the International Conference on Compressors and Their Systems at London*, London, UK. (277-287).

Schreiner, J. E., Ribas Jr., F. A., Deschamps C. J., Barbosa Jr. J. B. & da Rosa, V. H. P. (2009). Thermal management of a commercial reciprocating compressor through numerical simulation. *Proceeding of 2009 IIR/IIF International Congress on Compressors at Papiernicka*, Papiernicka, SK. (1-10).

Todescat, M. L., Fagotti F., Prata, A. T. & Ferreira R. T. (1992) Thermal energy analysis in reciprocating hermetic compressors. *Proceedings of the 1992 International Compressor Engineering Conference at Purdue*, West Lafayette, IN, USA. (Paper 936).

ACKNOWLEDGEMENT

To Tecumseh do Brasil Ltda. for the financial support.

Experimental Investigation of Heat Transfer in Components of a Hermetic Reciprocating Compressor under Thermal Transient

Gustavo L M SILVA[§], Thiago DUTRA, Cesar J DESCHAMPS*

POLO Research Laboratories for Emerging Technologies in Cooling and Thermophysics
Federal University of Santa Catarina
Florianopolis, SC, Brazil
gustavo.macedo@catolicasc.org.br, dutra@polo.ufsc.br, deschamps@polo.ufsc.br

*Corresponding Author

ABSTRACT

Heat transfer has an important role on the efficiency and reliability of hermetic reciprocating compressors adopted for household refrigeration. Most analyses of such compressors are carried out for steady-state operating conditions based on established procedures and standards. However, since household refrigeration systems work by alternating periods in which the compressor is on (ON) and idle (OFF), the performance assessment of compressors should take into account the ON/OFF cyclic operation. This paper presents an experimental study of the heat transfer in components of a hermetic reciprocating compressor during thermal transients due to ON/OFF cyclic operation. The compressor was instrumented with thin-film heat flux sensors and thermocouples to evaluate time-varying heat transfer coefficients. We found that heat flux in some regions is significantly affected by the cyclic operating condition. Moreover, the investigation revealed the presence of different time scales due to distinct thermal inertia and heat transfer associated with each component.

1. INTRODUCTION

Heat transfer can considerably affect the efficiency and reliability of hermetic reciprocating compressors. In one hand, the superheating of the suction gas reduces the compressor volumetric and isentropic efficiencies. On the other hand, reliability issues limit the electrical motor and lubricant oil temperatures. Hence, significant research has been developed to better understand the heat transfer between the compressor components.

Thermal simulation models have been developed based on lumped-parameter formulation (Todescat *et al.*, 1992; Ooi, 2003), in which the heat transfer between lumped elements is approached from measurements or correlations available in the literature. Other simulation models adopt differential formulation (Raja *et al.*, 2003; Birari *et al.*, 2006), solving the heat conduction in solid components and the fluid flow via the finite-volume method. Such models do not require convective heat transfer coefficients to estimate heat transfer between the gas and solid components. Although a full differential formulation provides a detailed thermal characterization of the compressor, the numerical solution of the governing equations is associated with high computation cost. Hybrid simulation models (Sanvezzo Jr. and Deschamps, 2012) formed by lumped and differential formulations for the gas and solid components, respectively, have been proposed to reduce the computational cost while retaining a detailed characterization of heat transfer in the solid components.

Experimental studies have considered different aspects of heat transfer in reciprocating compressors. Meyer and Thompson (1990) used thermocouples to measure the gas temperature along the flow path and, by applying energy balances, determined the heat transfer rate through components. Dutra and Deschamps (2013) carried out measurements of heat flux and temperature in a single-speed small reciprocating compressor by using thin-film heat flux sensors (HFSs) and thermocouples, and combined such measurements to estimate local heat transfer coefficients.

[§] Present address: Catholic University of Santa Catarina, Jaraguá do Sul, SC, Brazil.

All the aforementioned investigations were conducted for the compressor operating under steady-state condition, following well-established standards for evaluation and certification criteria of compressors (EN12900:2005). Nevertheless, household refrigeration systems operate under cyclic ON/OFF conditions and such transients can affect the compressor performance. Available studies in the literature concerning the ON/OFF operation of refrigeration systems are focused on heat exchangers and expansion devices (Hermes and Melo, 2008). Only recently, Lohn *et al.* (2015) presented a hybrid thermal model to predict the temperature distribution in a hermetic reciprocating compressor under a cyclic ON/OFF condition.

The present paper reports an experimental investigation aimed at characterizing the heat transfer in a hermetic reciprocating compressor during transient conditions. Evaporating and condensing temperatures were fixed at -21°C and 40°C, respectively. The compressor was tested under steady-state operating condition and two ON/OFF cyclic conditions: (i) 12 min. ON/15 min. OFF; and (ii) 25 min. ON/30min. OFF. Measurements of temperature and heat flux were carried out in different regions and components of the compressor, allowing estimates of local heat transfer coefficients as a function of time.

2. EXPERIMENTAL PROCEDURE

2.1 Compressor Instrumentation

Several HFSs and thermocouples were fitted onto the surface of the electrical motor, suction muffler and discharge system, so as to obtain a detailed thermal characterization of the compressor: (i) three on the electrical motor surfaces: em1, em2 and em3; (ii) two on the suction muffler: sm1 and sm2; and (iii) five on the discharge system: dp1, dm1, dm2, dm3 and dm4. The location of each HFS on these components is shown in Figures 1-3.

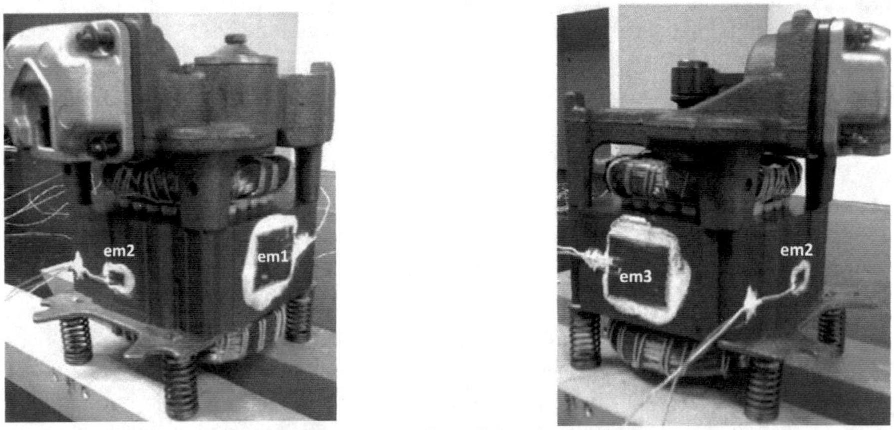

Figure 1: Instrumentation of the electrical motor.

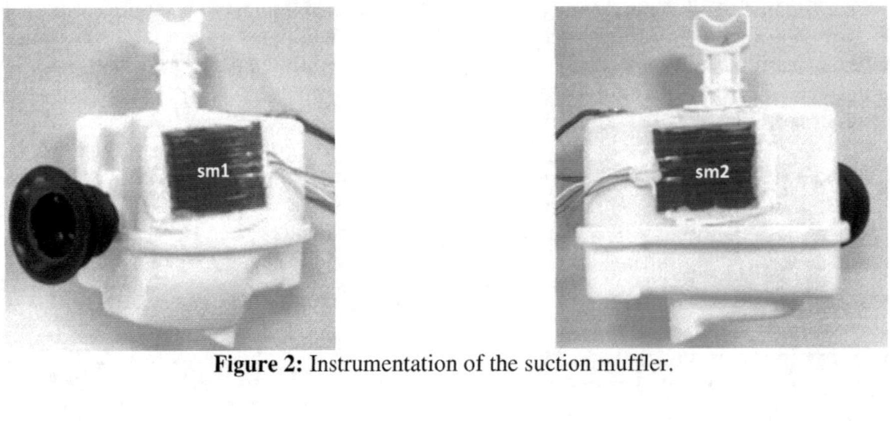

Figure 2: Instrumentation of the suction muffler.

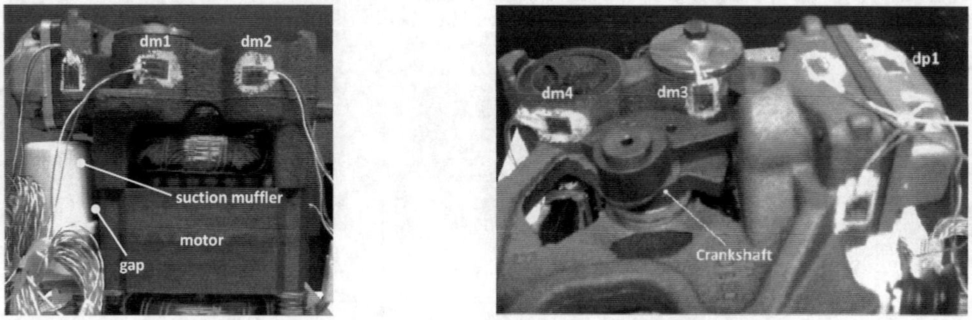

Figure 3: Instrumentation of the cylinder head and discharge muffler.

The operating principle of a thin-film HFS consists of a self-generated voltage output proportional to a heat flux through the sensor. The generated voltage (V) is the result of a temperature difference (ΔT) across the thickness of the HFS, which is measured by a thermocouple serial association (V = N α ΔT), being N and α the number of thermocouple junctions and the thermoelectric sensitivity of the materials, respectively. The heat flux through the HFS surfaces, q", is a function of its thermal conductivity, k, thickness, t, and temperature difference ΔT (q" = k ΔT / t). Therefore, a linear relationship between heat flux and voltage output is obtained (q" = V / S), where S (= N α t / k) is defined as the HFS sensitivity. Although the value of S can be theoretically determined, this is not a common practice. The HFS sensitivity is usually assessed from a standardized calibration procedure performed by the sensor manufacturer which supplies a datasheet containing all the technical specifications, including the HFS sensitivity.

The instrumentation of HFSs on the compressor components is not a simple task, due to the high temperature levels and the presence of lubricant oil. An epoxy-adhesive was necessary to attach the HFSs onto the surfaces and their wires were carefully positioned inside the compressor to minimize disturbances in the lubricant oil flow. This is a relevant aspect, because the lubricant oil significantly affects the heat transfer inside the compressor, increasing the heat loss to the external ambient (Dutra and Deschamps, 2013). After the installation of the HFSs, thermocouples were also positioned in the gas within the compressor shell, in order to establish suitable reference temperatures, T_∞, to estimate the local heat transfer coefficients:

$$h = \frac{q''}{T_s - T_\infty} \tag{1}$$

where T_s is the temperature of the sensor surface, measured by a thermocouple embedded in the HFS.

2.2 Experimental Setup

A calorimeter setup was employed to establish the operating conditions in which the compressor was tested. Figure 4 illustrates a schematic of the calorimeter composed of pipelines, control valves (CV), a mass flow meter (FM), heat exchangers (HX), a thermocouple (TC) and pressure transducers (PT). The compressor (C) raises the refrigerant pressure from point 1 up to point 2, which is then cooled by a heat exchanger (HX1) at point 3. After that, the refrigerant is throttled to an intermediate pressure level (point 4) by means of a control valve (CV1). The mass flow is measured by a Coriolis mass flow meter (FM) and then the refrigerant is cooled again with another heat exchanger (HX2) at point 5. The refrigerant is throttled again (CV2), reducing its pressure to the required evaporating pressure (point 6). Finally, an electrical heater (EH1) and a thermocouple (TC1) are used to adjust the level of superheating in the compressor suction line (point 1), completing the cycle. As can be seen in the pressure-enthalpy diagram of Figure 5, the refrigerant remains as superheated vapor along the entire thermodynamic cycle.

The first step in experimental test is to submit the compressor and the calorimeter pipeline to a suitable vacuum condition, in order to remove air and humidity within the system. After that, a refrigerant charge is supplied to the system and the operating condition is defined by setting the evaporating and condensing temperatures, as well as the ON/OFF period. The transient tests are carried out by switching on and off the compressor in an alternate mode for 8 ON/OFF cycles. The first 5 cycles are necessary to attain a periodically stationary condition and the last 3 are used to sample an average representative cycle. For the steady state condition, the compressor runs uninterruptedly for approximately 4 hours until no temperature variation higher than 1°C is observed in one hour. After this requirement is met, data are acquired during 45 minutes and the test is finished. For both cyclic and steady-state conditions the sampling rate was 0.25 Hz.

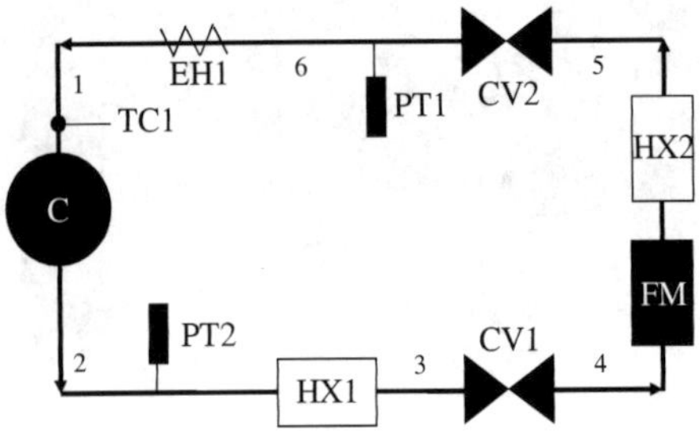

Figure 4: Schematic of the calorimeter facility.

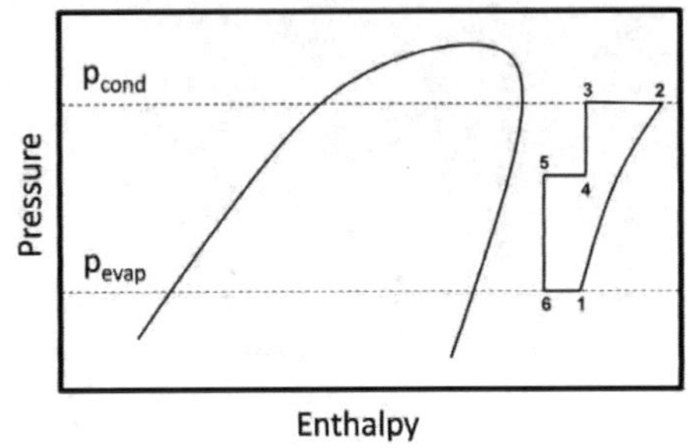

Figure 5: p-h diagram of the hot-cycle calorimeter.

3. RESULTS

A 60 Hz reciprocating compressor operating with R134a was adopted for the measurements. The suction and discharge pressures correspond to saturation pressures associated with the evaporating and condensing temperatures of -21°C and 40°C, respectively. These pressures were set and controlled within a range of ±1% whereas the temperatures of the gas in suction line and of the environment inside the calorimeter were controlled within ±0.5°C. Temperatures, heat fluxes and heat transfer coefficients are normalized in relation to steady-state measurements of the shell temperature, T_{shell}, (in °C), the heat transfer rate released through the shell, q''_{shell}, and the heat transfer coefficient between shell and environment, h_{shell}:

$$T^* = \frac{T}{T_{shell}} \qquad q''^* = \frac{q''}{q''_{shell}} \qquad h^* = \frac{h}{h_{shell}} \tag{2}$$

Results under steady-state condition are presented with uncertainty bars corresponding to a 95% confidence interval.

3.1 Gas inside the Compressor

Figures 6-8 show measurements for the gas temperature in four different positions inside the compressor shell under the cyclic ON/OFF conditions (12 min/15 min; 25 min/30 min) and the steady-state condition. The results show that the highest temperatures occur near the shell cover and discharge muffler, while the lowest temperatures are measured in the suction muffler and near the region em3 of the motor. Due to such temperature differences inside the compressor shell, the reference temperatures adopted to estimate the local heat transfer coefficients were those close to the surfaces of interest.

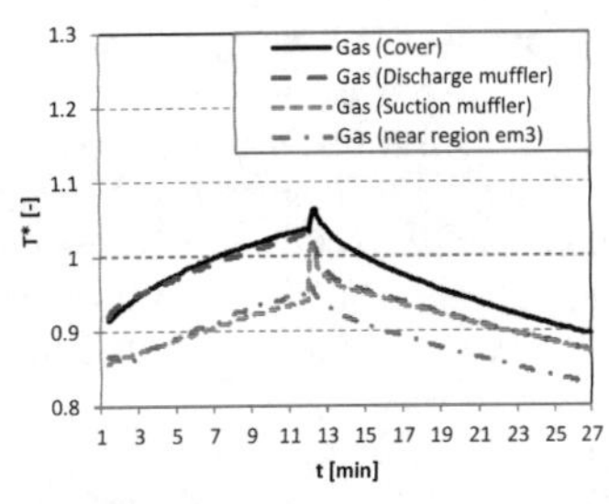

Figure 6: Gas temperatures - 12 min/15 min.

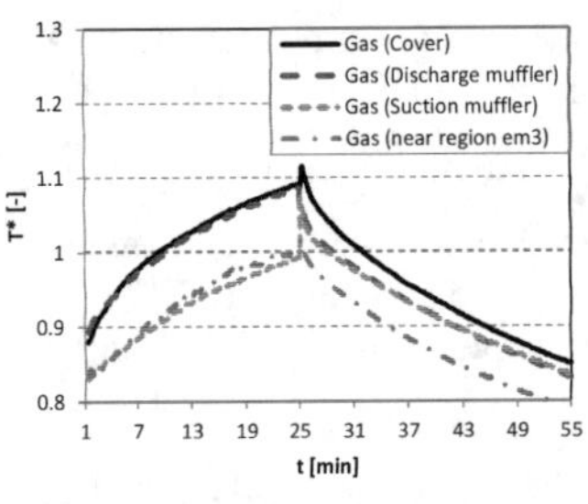

Figure 7: Gas temperatures - 25 min/30 min.

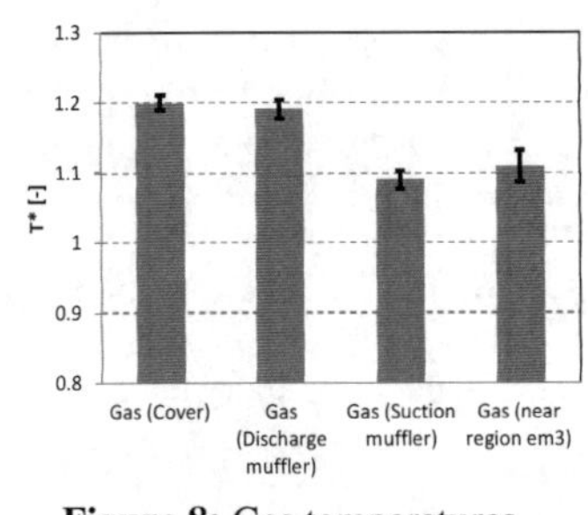

Figure 8: Gas temperatures - steady state.

3.2 Electrical Motor

Measurements of heat flux, temperature and local heat transfer coefficients for the electrical motor under the cyclic (12 min/15 min; 25 min/30 min) and the steady-state conditions are shown in columns (a), (b) and (c) of Figure 9.

During the ON period of the cyclic operation (12 min/15 min), the heat fluxes through the motor surfaces decrease in the first minutes and when this period is longer (25 min/30 min), the heat fluxes start to increase. This is a consequence of the temperature difference between the surfaces (em1, em2 and em3) and the gas inside the compressor shell, which is initially decreased and increased again in the case of a period of 25 min. Immediately after the compressor is switched off, the heat fluxes increase suddenly, since the gas temperature decreases at a much faster rate than the temperature of the solid components. The heat fluxes through the motor surfaces are on the same level when the compressor runs under the cyclic condition. On the other hand, the heat flux through surface em2 is much higher than the others when the compressor operates in steady-state condition. This is due to the proximity between the surface em2 and the suction muffler (Figure 3), which is the coldest component of the compressor under steady-state condition.

The temperature on the motor surfaces em1, em2 and em3 point out similar results, with a slightly higher value on em1, since this region is near the discharge muffler. The temperature of the gas close to the surface em2 was not measured, but estimated as the average of the measurements on the motor and suction muffler surfaces, em2 and sm2, respectively. This approach was adopted because the narrow gap between the electrical motor and the suction muffler (Figure 3) hinders the installation of a thermocouple in that location. Thus, the temperature in the gap is similar to that observed for the em2 region, making it difficult to correctly estimate the local heat transfer coefficient in the idle period. Since the difference between the surface and gas temperatures becomes very small when the compressor is switched off, the heat transfer coefficient suddenly increases and reaches excessive values. Despite this shortcoming, the estimates of heat transfer coefficients, h*, for the remaining regions are reasonable, in the range of 0.5 - 2, being on the same magnitude for ON/OFF and steady-state conditions.

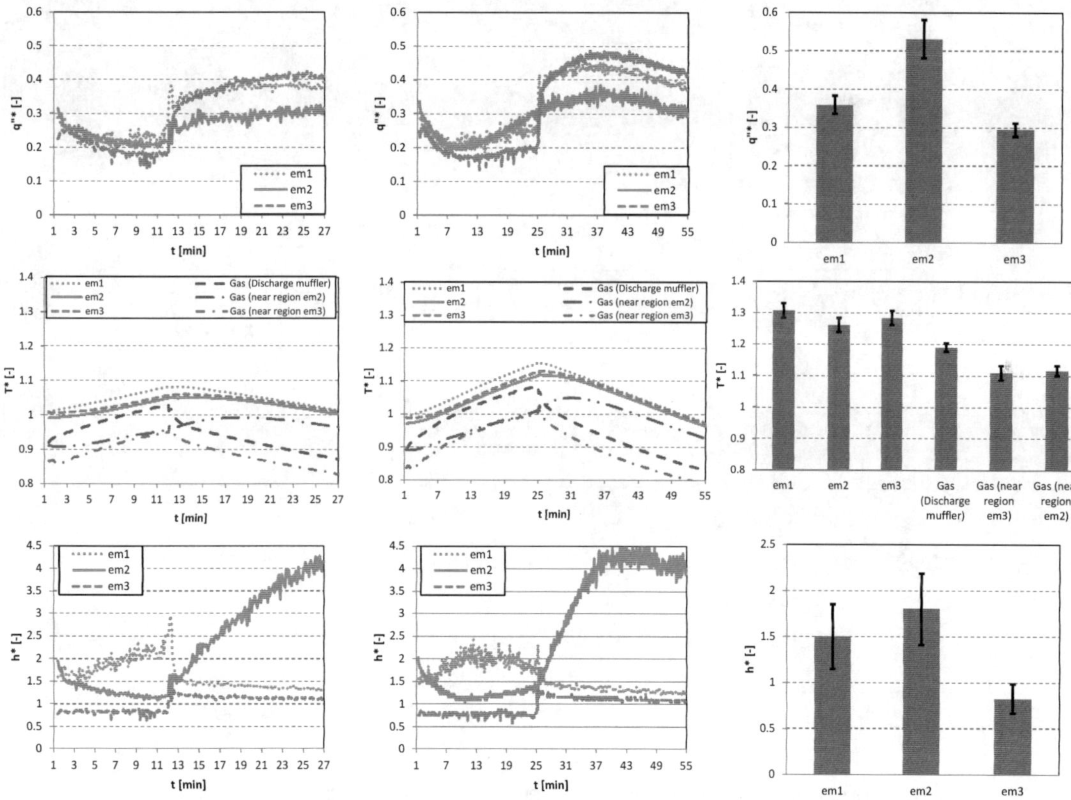

(a) (b) (c)

Figure 9: Electrical motor: Heat flux, temperatures and local heat transfer coefficients (a) 12min/15 min; (b) 25min/30 min; (c) steady state.

3.3 Suction Muffler

The results of heat flux, temperature and local heat transfer coefficients associated with the suction muffler surfaces are shown in Figure 10. The measurements of heat flux suggest that the electrical motor thermally interacts with the suction muffler, since the surface sm2 is subjected to higher heat flux than the surface sm1. Moreover, there is a sudden decrease of heat flux immediately after the compressor is switched off, reaching negative values. When the compressor is idle (OFF), heat transfer occurs first towards the suction muffler, since the gas temperature is still higher than the temperature of the muffler wall. However, eventually during the idle period the temperature of the solid surpasses the gas temperature, and heat flux has its direction changed. These temperature variations can be clearly observed in Figure 10, with the exception of the gas temperature near the region sm2 that is the reference temperature for the surface em2.

As the surface and gas temperatures approach each other, a sudden increase in the heat transfer coefficient is observed on the surface sm1. Naturally, this is not associated with any modification of the flow pattern in the suction muffler vicinity, but simply a result of the approach used to estimate the coefficient, represented by Equation (1). More representative results for the heat transfer coefficients in the idle period can be obtained just before the compressor is switched on, h* ~ 1. The heat transfer coefficients obtained for the ON period and the steady-state operation are on the same order of magnitude, 0.8 – 3.5, compared to those associated with the surface of the electrical motor.

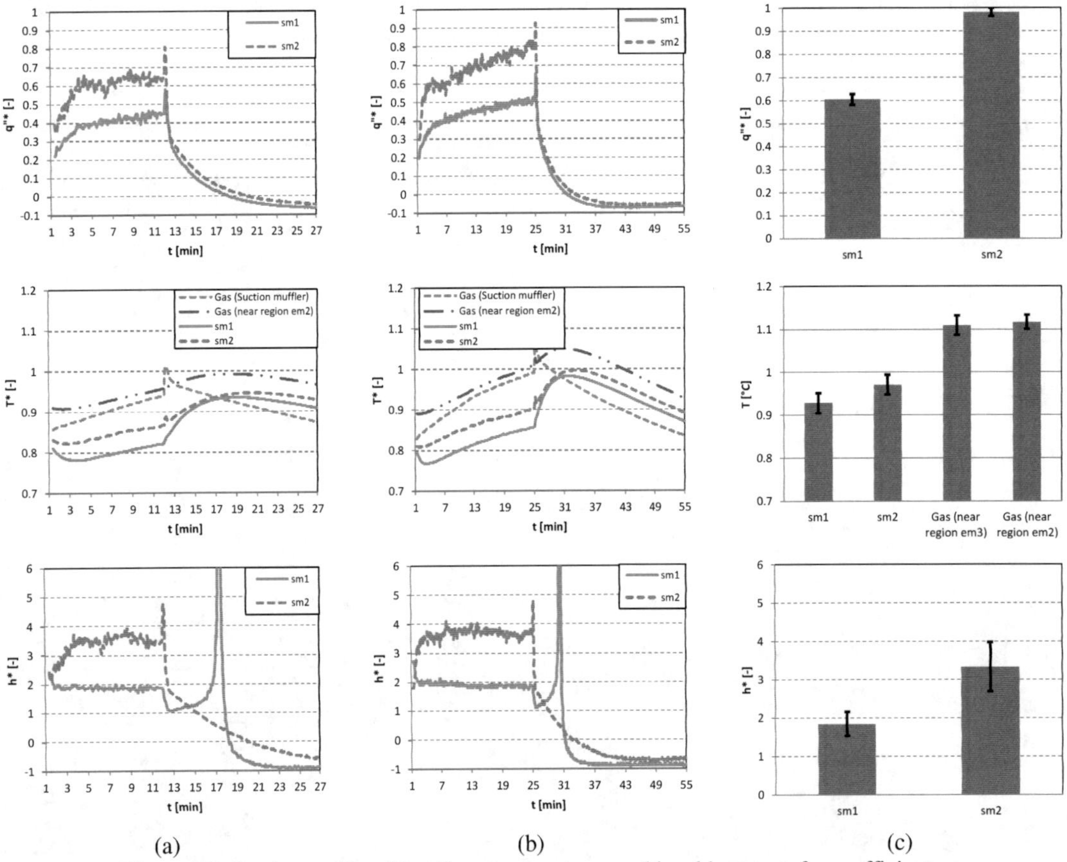

(a) (b) (c)

Figure 10: Suction muffler: Heat flux, temperatures and local heat transfer coefficients. (a) 12min/15 min; (b) 25min/30 min; (c) steady state.

Measurements of temperature were also carried out for the gas in the suction chamber (sc) depicted in Figure 11. This temperature is an important parameter since gas superheating along the suction path reduces the compressor volumetric and isentropic efficiencies. Figure 12 shows the temperature of the gas in suction chamber during the ON period for the two cyclic conditions and the steady-state condition. As can be seen, gas superheating varies according to how long the compressor runs before being switched off. For the 12min/15min ON/OFF condition, the non-dimensional temperature in the suction chamber reaches 0.72, whereas the temperature measured in the steady-state condition is 0.78.

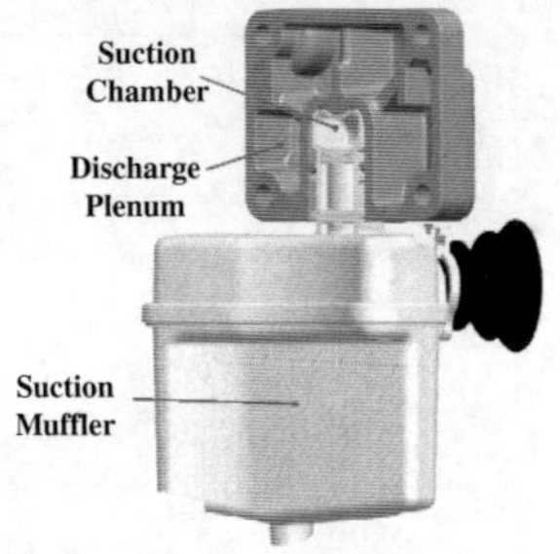

Figure 11: Suction chamber.

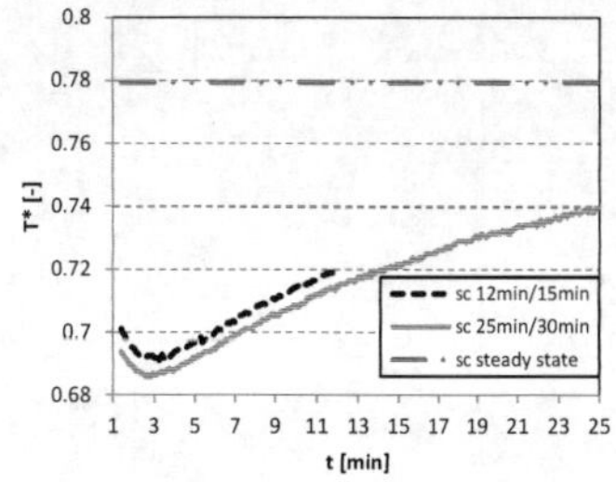

Figure 12: Suction chamber gas temperature.

3.4 Discharge System

Experimental data of temperature, heat flux and local heat transfer coefficients at the discharge system regions are shown in Figures 13-15 for the steady-state and ON/OFF operating conditions. As can be noted in Figure 13, the cylinder head, dp1, is the component of the discharge system with the highest temperature, regardless the compressor operating condition. This is the reason for the higher heat flux found at dp1 in comparison to the locations dm1 and dm2 (Figure 14). It is interesting to observe in Figure 15 that h* is about 1 – 1.5 for the surfaces dm1 and dm2, which are much smaller than the values obtained at the surfaces dm3 and dm4 (in the range 7 - 12). Such a difference can be explained by the proximity of each surface in relation to the crankshaft, as depicted in Figure 3. Specific to this type of compressor, the oil stored in the sump is taken to the upper parts by the centrifugal action of an oil pump that is coupled to the crankshaft. As the shaft spins, the lubricating oil flows inside the pump by centrifugal action until it reaches the other extremity of the crankshaft. Part of the oil that leaves the crankshaft impinges against the surfaces dm3 and dm4, increasing the heat flux there. The highest heat transfer coefficient is found at dm3 because this surface is the closest to the crankshaft. On the other hand, the temperature at dm4 is higher than at dm3, and this explains its greater heat flux.

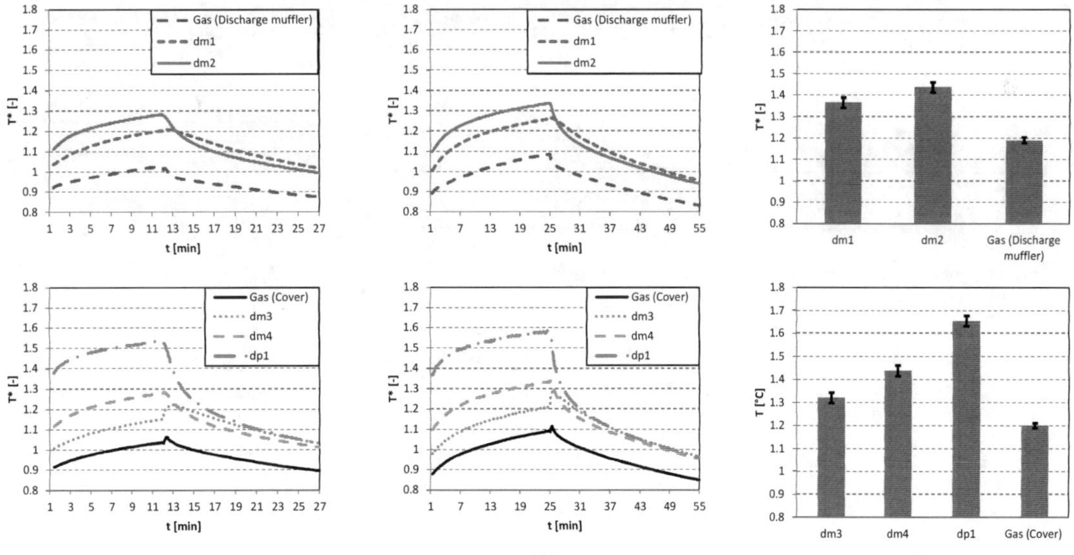

(a) (b) (c)

Figure 13: Temperatures at the discharge system wall: (a) 12 min/15 min; (b) 25 min/30 min; (c) steady state.

The heat transfer enhancement mechanism at the surfaces dm3 and dm4 due the crankshaft spinning suddenly stops when the compressor is switched off (Figures 14 and 15). During the idle period, the heat transfer coefficients for all the compressor surfaces become similar, within a range 0.8 – 1.5. This is a further evidence of the influence of the flow of lubricant oil on the heat transfer in the discharge muffler.

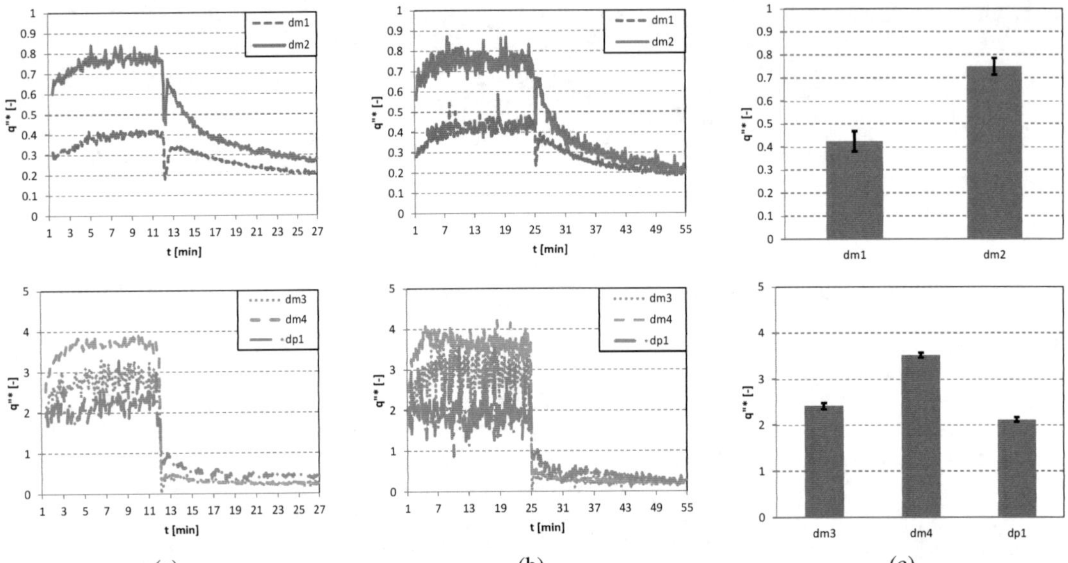

Figure 14: Heat flux at the discharge system wall: (a) 12min/15 min; (b) 25min/30 min; (c) steady state.

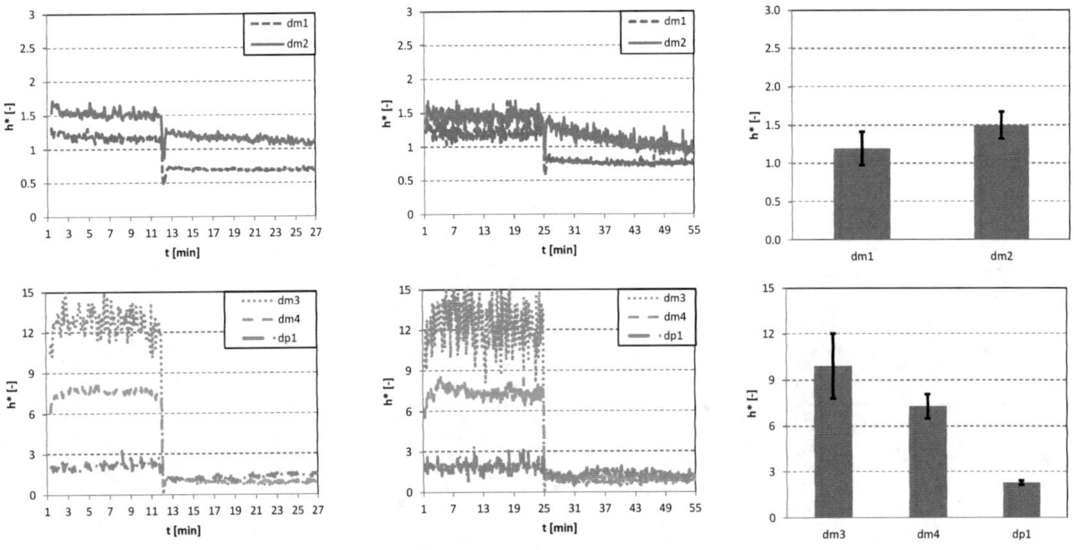

Figure 15: Heat transfer coefficients at the discharge system wall: (a) 12 min/15 min; (b) 25 min/30 min; (c) steady state.

3.5 Comparison of Temperatures and Heat Fluxes between Different Regions

Figure 16 illustrates the temperature difference (T_s - T_∞) at several components of the compressor at the cyclic condition 25 min/30 min. These measurements show the presence of different time scales associated with the heat transfer process in each component. It is interesting to note that the temperature difference between the discharge system components and the refrigerant gas is nearly constant during the ON period, inducing roughly constant heat fluxes (Figure 14), and which do not change regardless the operating condition, as shown in Table 1 for the surfaces dm1 and dp1. On the other hand, the temperature difference between the suction muffler (sm1) surface and the gas increases continuously, bringing about an increase of the heat flux at different operating conditions (Table 1). Moreover, the temperature difference between the electrical motor (em1) and the refrigerant gas first decreases and then increases. Such a slight variation do not affects the heat flux during the cyclic operation, as shown in Table 1. However, if the compressor is kept under operation for a longer period (i.e. until reaching the steady-state condition), the temperature difference continues to increase, bringing about a higher heat flux. It is worthwhile to note from Table 1 that the average heat fluxes at the surfaces sm1 and em1 are 50% higher in the steady-state condition in comparison to the ON period of the 12 min/15 min. This highlights the importance of taking into account thermal transients in analyses of compressor performance.

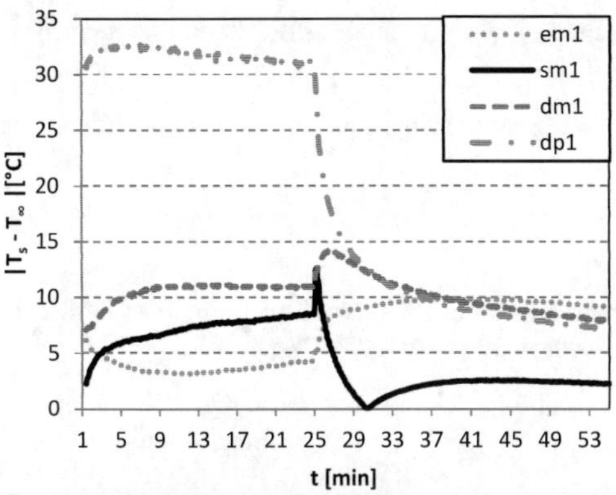

Figure 16: Temperature difference at different components for the 25 min/30 min operating condition.

Table 1: Average heat flux for cyclic and steady state operating conditions.

	Mean heat flux [-]		
	12min/15min	**25min/30min**	**Steady State**
em1	0.23	0.23	0.36
sm1	0.39	0.44	0.60
dm1	0.38	0.41	0.42
dp1	2.17	1.93	2.11

4. CONCLUSIONS

The present paper reported an experimental investigation of heat transfer in a hermetic reciprocating compressor during cyclic and steady-state operating conditions. The compressor was instrumented with heat flux sensors and thermocouples fitted onto the surfaces of the suction muffler, electrical motor and discharge system. The results revealed the presence of two levels of heat transfer coefficients during the compressor operation: (i) 0.8 – 3.5 at most components and (ii) 7 - 12 at the surface of components closest to the crankshaft, brought about by forced convection of lubricant oil and refrigerant gas provided by the crankshaft motion. Furthermore, we found different time scales associated with the heat transfer process of each component, and the heat flux at the suction system and electrical motor during the cyclic condition are quite different of those for the steady state condition. This is an aspect that should be taken into account in the thermal modeling of compressors.

NOMENCLATURE

h	heat transfer coefficient	(W/m².K)	**Subscript**	
h^*	non-dimensional heat transfer coefficient	(–)	s	surface
q"	heat flux	(W/m²)	shell	shell
q''^*	non-dimensional heat flux	(–)	∞	reference
T	temperature	(°C)		
T^*	non-dimensional temperature	(–)		
em	electrical motor	(–)		
dm	discharge muffler	(–)		
dp	discharge plenum	(–)		
sm	suction muffler	(–)		

REFERENCES

Birari, Y. V., Gosavi, S. S., Jorwekar, P. P., 2006, Use of CFD in design and development of R404a reciprocating compressor, *Proc. Int. Compressor Engrg. Conf. at Purdue*, Paper C072.

Dutra, T., Deschamps, C. J., 2013, Experimental characterization of the heat transfer in the components of a small hermetic reciprocating compressor, *App. Thermal Engrg.* 58, pp.499-510.

EN12900:2005, Refrigerant compressors - Rating conditions, tolerances and presentation of manufacturer's performance data presentation.

Hermes, C. J. L., Melo, C., 2008, A first-principles simulation model for the start-up and cycling transients of household refrigerators, *Int. J. of Refrigeration 31*, pp. 1341-1357.

Lohn S. K., Diniz, M. C., Deschamps, C. J., 2015, A thermal model for analysis of hermetic reciprocating compressors under the on-off cycling operating condition, *IOP. Conf. Ser.: Mater. Sci. Eng.*, 90, 012068.

Meyer, W. A., Thompson, H. D., 1990, An analytical model of heat transfer to the suction gas in a low-side hermetic refrigeration compressor, *Proc. Int. Compressor Engrg. Conf. at Purdue*, pp. 898-907.

Ooi, K.T., 2003, Heat transfer study of a hermetic refrigeration compressor, *App. Thermal Engrg.*, 23, pp. 1931–1945.

Raja, B., Sekhar, S. J., Lal, D. M., Kalanidhi, A., 2003, A numerical model for thermal mapping in a hermetically sealed reciprocating refrigerant compressor, *Int. J. of Refrigeration 26 (6)*, pp. 229-236.

Sanvezzo Jr., J., Deschamps, C. J., 2012, A heat transfer model combining differential and integral formulations for thermal analysis of reciprocating compressors, *Proc. Int. Compressor Engrg. Conf. at Purdue*, paper 1343.

Todescat, M. L., Fagotti, F., Prata, A. T., Ferreira, R. T. S., 1992, Thermal energy analysis in reciprocating hermetic compressors, *Proc. Int. Compressor Engrg. Conf. at Purdue*, pp. 1417-1428.

ACKNOWLEDGEMENT

This study is part of a technical-scientific program between the Federal University of Santa Catarina and EMBRACO. Financial support from EMBRAPII Unit POLO/UFSC and CNPq (Brazilian Research Council) is also acknowledged.

Experimental Determination of Correlations for Heat Transfer Coefficients in the Suction Muffler of a Hermetic Reciprocating Compressor

Eduardo ARCENO, Thiago DUTRA, Cesar J DESCHAMPS*

POLO Research Laboratories for Emerging Technologies in Cooling and Thermophysics
Federal University of Santa Catarina
88040-900, Florianopolis, SC, Brazil
earceno@yahoo.com.br, dutra@polo.ufsc.br, deschamps@polo.ufsc.br

* Corresponding Author

ABSTRACT

The efficiency and reliability of hermetic reciprocating compressors adopted for household refrigeration are affected by the temperature distribution of their components. Simulation models developed to assist the thermal design of compressors require heat transfer coefficients in different components, usually estimated from correlations available for simplified geometries. However, the geometries of some components are complex and require more representative heat transfer coefficients. This paper presents experimental results for heat transfer coefficients obtained from measurements of heat flux and temperature in different regions of the suction muffler of a small reciprocating compressor under different operating conditions. A least square method was employed to obtain correlations for local heat transfer coefficients, which were then compared with estimates given by correlations available in the literature for simplified geometries.

1. INTRODUCTION

According to Ribas Jr. (2007), the suction gas superheating is responsible for half of the thermodynamic losses associated with the compression cycle of small reciprocating compressors. Such a superheating is a result of heat sources inside the compressor due to the compression process, electrical motor inneficiency and friction in bearings. Many studies have been carried out to better understand heat transfer between the compressor components. Todescat *et al.* (1992), Rigola *et al.* (2000) and Ooi (2003) proposed simulation models based on lumped formulations to predict the temperature distribution in reciprocating compressors. These models consist in applying energy balances to control volumes which encompass the compressor components, with heat transfer being modeled through heat conductances determined from experimental data or simplified correlations available in the literature.

Chikurde *et al.* (2002), Raja (2003) and Birari *et al.* (2006) carried out three-dimensional simulations of heat conduction and fluid flow in compressors using commercial softwares. Although these models allow a more detailed thermal characterization of the compressor, they are time-consuming and not viable for optimization purpose. Ribas (2007) and Sanvezzo Jr. and Deschamps (2012) circumvented this difficulty of computational cost by proposing the so-called hybrid model in which the benefit of solving heat conduction with a three-dimensional formulation is maintained, but a lumped formulation is adopted for the fluid flow domain. The hybrid model developed by Sanvezzo Jr. and Deschamps (2012) estimates heat transfer at the interface between solid and fluid through convective heat transfer coefficients obtained from the literature.

Heat transfer in reciprocating compressors has also been experimentally investigated (Meyer and Thompson, 1990; Kim *et al.*, 2000), adopting thermocouples to measure the temperature of the gas in different locations. By combining these measurements with energy balances, heat transfer rates for compressor components were determined. Shiva Prasad (1992) carried out instantaneous heat flux measurements in the compression chamber of a 900 rpm air compressor by employing thin-film heat flux sensors (HFSs), with the main objective of analyzing the influence of the regenerative heat transfer on the suction gas superheating. Recently, Dutra and Deschamps (2013) adopted HFSs and thermocouples to measure heat flux at the surfaces of components and estimate local heat transfer coefficients. The authors carried out measurements in the compressor shell, electrical motor, discharge chamber and suction muffler of a single-speed reciprocating compressor and concluded that the lubricant oil flow greatly affects heat transfer inside the compressor and that the local heat transfer coefficients are almost insensitive to evaporating and condensing temperatures.

The present paper reports an experimental characterization of heat transfer coefficients at different locations of the suction muffler of a variable-speed reciprocating compressor. Measurements are carried out for three compressor speeds (1200, 2000 and 4000 rpm) and three pairs of evaporating and condensing temperatures (-15°C/40°C; -25°C/40°C and -25°C/55°C). Based on such measurements, a least square method is employed to obtain correlations for local heat transfer coefficients, which are then compared with estimates of correlations available in the literature for simplified geometries.

2. EXPERIMENTAL PROCEDURE

2.1 Compressor Instrumentation

The suction muffler considered in this work is composed by two independent parts, herein called front and back parts, forming an array of three tubes and two chambers, as depicted in Figure 1. The refrigerant gas enters the muffler through tube 1, being then released into the chamber 1. The gas is then driven to the chamber 2 by flowing in the tube 2. Finally, the gas flows towards the compression chamber through tube 3.

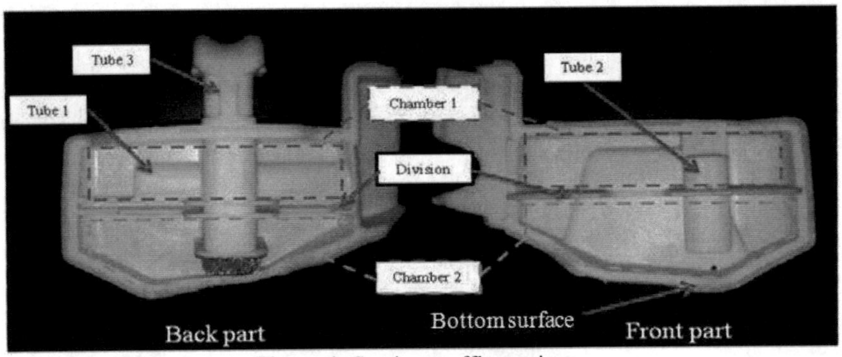

Figure 1: Suction muffler regions.

Six samples of this suction muffler geometry were employed in this study. Three of them, named as group A, were instrumented only with thermocouples and the other three, group B, were instrumented with thermocouples and HFSs. Mufflers of group A were instrumented with thermocouples on their external surfaces (Figure 2) and in convenient positions along the suction path (Figure 3).

Two HFSs were installed in each sample of suction muffler of group B, one positioned on the front wall and the other on the back wall (Figure 4). The working principle of a HFS is based on a self-generated voltage output which is proportional to the heat flux that crosses the sensor. The relationship between output voltage, E, and heat flux, q'', is represented by a linear function ($q'' = E / S$), where S is the sensitivity associated to a specific HFS. The HFS sensitivity is obtained from a calibration procedure and its value is usually specified by the manufacturer. Further details regarding the HFS working principle are available in Dutra and Deschamps (2013).

The HFSs have to be carefully fitted onto the suction muffler walls. Due to the presence of lubricating oil at high temperature inside the compressor, an epoxy-adhesive was used to attach the HFSs to the surfaces. A thin layer of epoxy-adhesive was applied between the HFS and the muffler wall to reduce the contact resistance. This thin layer was thoroughly spread all over the HFS surface for preventing air bubbles from being formed between the sensor and the wall during the fitting process.

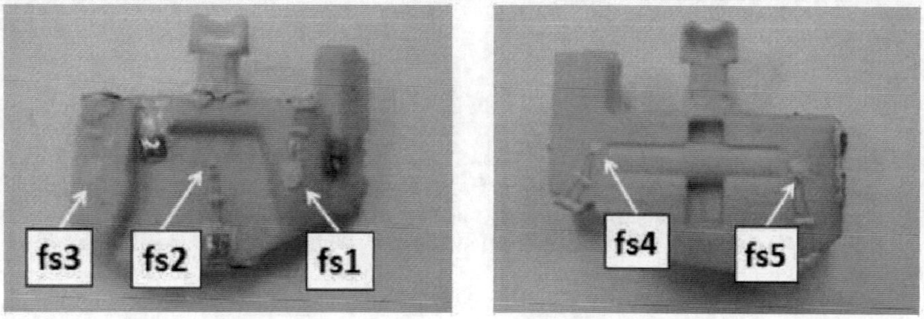

(a) (b)

Figure 2: Thermocouples (fs1 to fs5) located on the (a) front and (b) back surfaces of mufflers of group A.

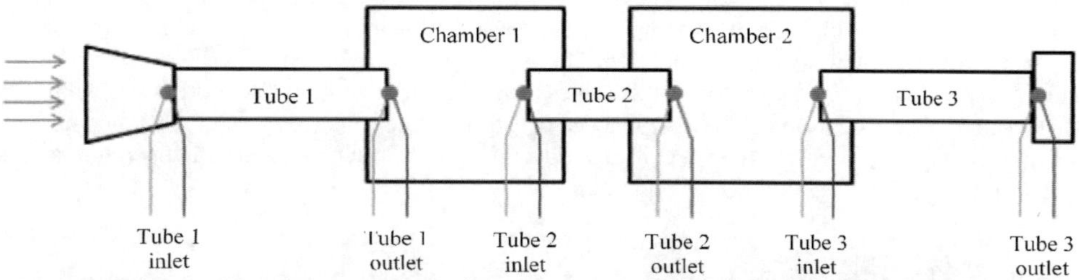

Figure 3: Schematic of the array of tubes and chambers and locations of thermocouples (groups A and B).

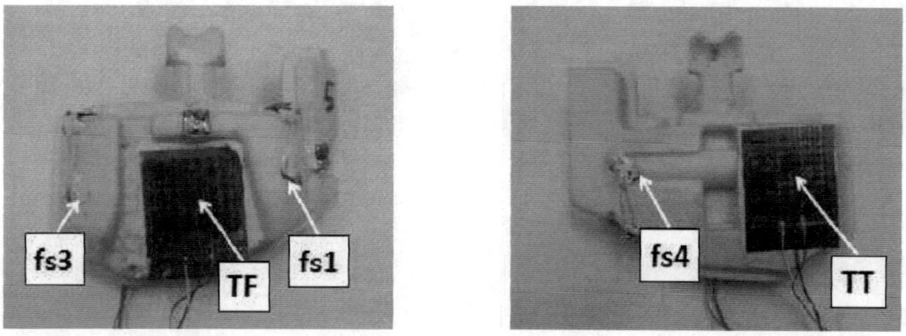

(a) (b)

Figure 4: Thermocouples (fs1, fs3 and fs4) and heat flux sensors (TF and TT) on the (a) front and (b) back surfaces mufflers of group B.

Four thermocouples were installed inside the compressor for measurements of temperature in convenient positions near the muffler external surface. Three of them were positioned near the front surface of the muffler to measure the gas temperature (Ci1, Ci2 and Ci3). The other thermocouple (tc) was installed onto the surface of the electrical motor that is 2 mm apart from the back surface of the muffler. Figure 5 shows a schematic view of the internal components of the compressor, highlighting the locations of the HFSs and thermocouples.

From the temperature measurements in the muffler vicinity, it was possible to establish the reference temperatures, T_∞, to estimate local heat transfer coefficients at the suction muffler surfaces:

$$h = \frac{q''}{T_s - T_\infty} \quad (1)$$

where T_s is the surface temperature measured by a thermocouple in the HFS and q'' is the heat flux measured with the HFS.

2.1 Experimental Facility

The compressor was tested in a calorimeter facility specially designed to establish and control different operating conditions. Figure 6 depicts a schematic of the calorimeter, which is composed of pipelines, control valves (CV), a mass flow meter (FM), heat exchangers (HX), a thermocouple (TC) and pressure transducers (PT). The compressor (C) raises the refrigerant pressure from point 1 up to point 2, which is then cooled by the heat exchanger HX1 at point 3. The refrigerant is then adiabatically throttled to an intermediate pressure level (point 4) by means of a control valve (CV1). A Coriolis mass flow meter (FM) measures the mass flow rate of refrigerant before it is cooled again with another heat exchanger (HX2), reaching point 5. A second expansion device (CV2) takes the refrigerant fluid to the evaporating pressure (point 6). Finally, an electrical heater (EH1) and a thermocouple (TC1) are used to adjust the temperature of the compressor suction line to the required gas superheating condition (point 1), completing the operating cycle.

The first step in experiments is to submit the compressor and the calorimeter pipeline to an adequate vacuum condition, so as to remove air and humidity. Then, a charge of refrigerant is supplied to the system, the operating condition is set and the compressor is switched on. A period of approximately 4 hours is needed until the compressor achieves a thermal stabilized condition. During this process, the control valves in the high and low pressure lines are continuously adjusted to establish the specified suction and discharge pressures within 1% of the set points. The compressor is considered to have reached the steady-state condition when the temperatures at several locations of the compressor vary less than 1°C in one hour. After that, samples of local temperature and heat flux measurements are acquired over a period of 15 minutes.

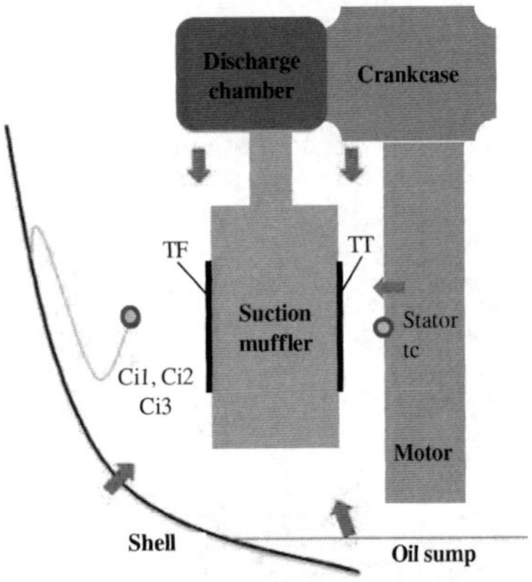

Figure 5: Locations of thermocouples and HFS inside the compressor.

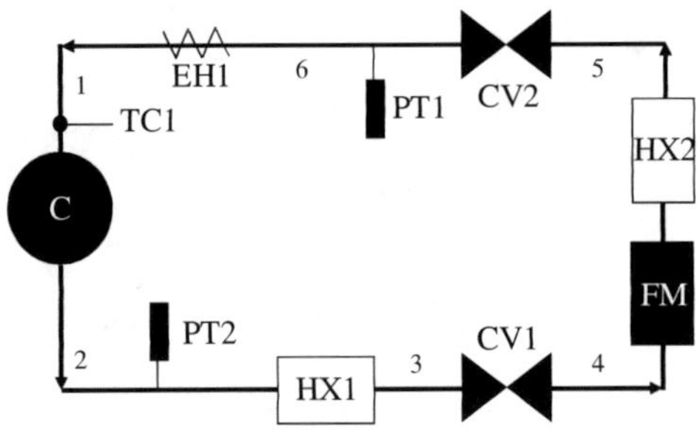

Figure 6: Schematic of the calorimeter facility.

3. RESULTS AND DISCUSSIONS

A variable-speed reciprocating compressor operating with R600a was selected for the analysis. The compressor was tested under three rotational speeds (1200, 2000 and 4000 rpm) and three operating conditions represented by different evaporating and condensing temperatures (T_e, T_c): A (-15°C, 40°C), B (-25°C, 40°C) and C (-25°C, 55°C), corresponding to pressure ratios, $\pi = p_c/p_e = 6.0$, 9.1 and 13.3. Considering that six samples of suction mufflers were employed in the experimental investigation, a total of 54 tests were carried out, providing mean values of local heat transfer coefficients. Such coefficients were made non-dimensional by dividing them by the coefficient h_{shell} associated with convection heat transfer between the shell and the calorimeter environment, i.e. $h^* = h/h_{shell}$, with $15 W.m^{-2}K^{-1} < h_{shell} < 30 W.m^{-2}K^{-1}$.

3.1 Measurements of Heat Transfer Coefficients

Mean values of heat flux and temperature measured at each surface were adopted to estimate the heat transfer coefficients. The reference temperature, T_∞, for the front surface was taken as the average of the values indicated by the thermocouples Ci1, Ci2 and Ci3 (Figure 5). On the other hand, the mean value of temperatures on the motor and muffler back surface was considered as the reference temperature to estimate the heat transfer coefficient on the back surface.

Figures 7 and 8 present results of heat transfer coefficients for the front, h_f^*, and back surfaces, h_b^*, of the suction muffler as a function of pressure ratio. As can be seen, the heat transfer coefficient increases with the compressor speed but decreases with the pressure ratio in most conditions. This is an expected behavior since the increase of speed or the decrease of pressure ratio provide higher mass flow rates for the operating conditions A, B and C, and, therefore, an intensification of the convective heat transfer. It is interesting to note that the heat transfer coefficient on the muffler front surface for the compressor operating at 2000 rpm slightly decreases from $\pi = 6.0$ to $\pi = 9.1$ and then increases as the pressure ratio rises to $\pi = 13.3$. Such a behavior is not trivial but may be related to variation of the lubricant oil flow path inside the compressor.

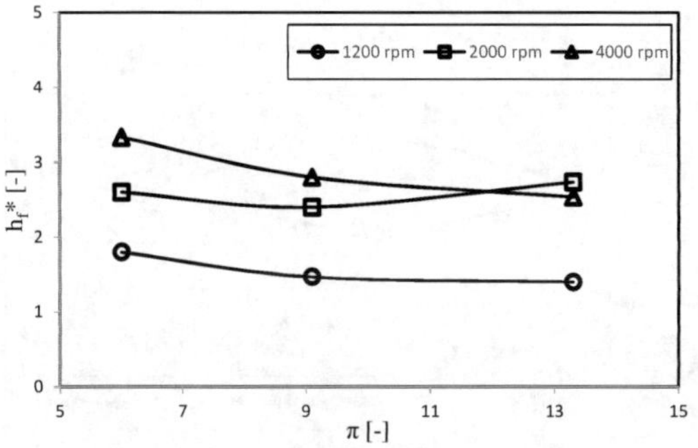

Figure 7: Heat transfer coefficients on the front surface of the suction muffler.

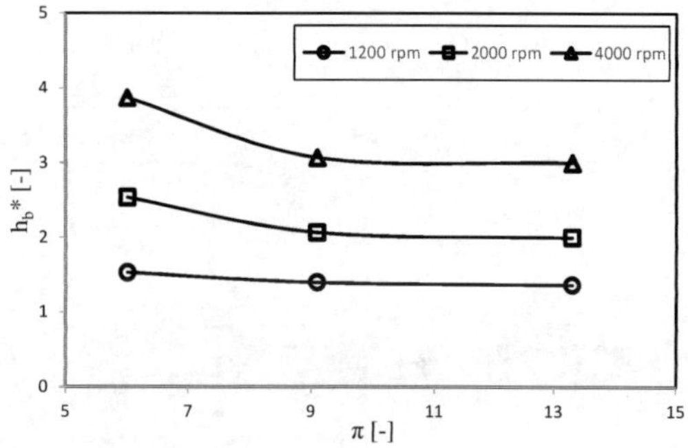

Figure 8: Heat transfer coefficients on the back surface of the suction muffler.

Heat transfer coefficients for the internal fluid flow through tubes and chambers of the muffler are obtained by combining measurements of temperature and heat flux with the hypothesis of one-dimensional heat transfer through the wall:

$$h = \frac{q''k}{k(\overline{T}_{se} - \overline{T}_\infty) - q''L} \tag{2}$$

where k and L are the thermal conductivity and thickness of the suction muffler wall, $\overline{T}_{se}$ is the mean temperature of the external surface and $\overline{T}_\infty$ is the mean temperature of the gas in the component under analysis (tube or chamber).

Tubes 1 and 2 are connected to the back and front surfaces of the muffler, respectively, as shown in Figure 1. Hence, heat flux measurements on these surfaces are used to estimate the local heat transfer coefficients, h, for tubes 1 and 2. This is certainly an approximation that will be verified in future work via energy balance and measurements of temperatures at the inlet and outlet of each tube. The heat transfer coefficient is not evaluated for the tube 3 because a considerable part of this tube is in contact with hot components of the cylinder head. Therefore, the hypothesis of one-dimensional heat transfer would not be applicable to tube 3. Figures 9 and 10 show the results of local heat transfer coefficients, h^*, for the tubes 1 and 2, respectively. Although such heat transfer coefficients are higher than those associated with the muffler external surfaces, their variations with compressor speed and pressure ratio are similar.

The determination of the heat transfer coefficients for the chambers followed a different procedure, since the temperature inside them is affected by heat fluxes through the front and back surfaces. Thus, a mean value of these heat fluxes was used to calculate the local heat transfer coefficients of both chambers, shown in Figures 11 and 12. The values of h^* for chamber 1 present the same levels and trends compared with those measured in the tubes.

However, the values of h^* for chamber 2 are higher than those observed for chamber 1 and show no clear relationship with both the pressure ratio and the compressor speed. This is arguably related to the more complex geometry of chamber 2, which gives rise to large recirculating flow regions inside the chamber that are less sensitive to the mass flow rate. Numerical simulations could assist clarifying this issue.

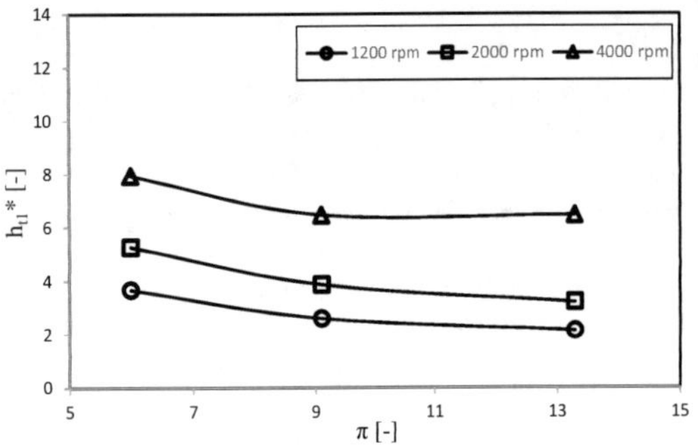

Figure 9: Heat transfer coefficients for tube 1.

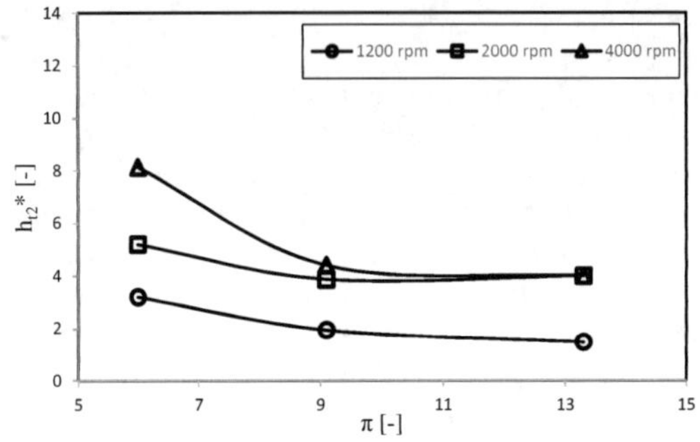

Figure 10: Heat transfer coefficients for tube 2.

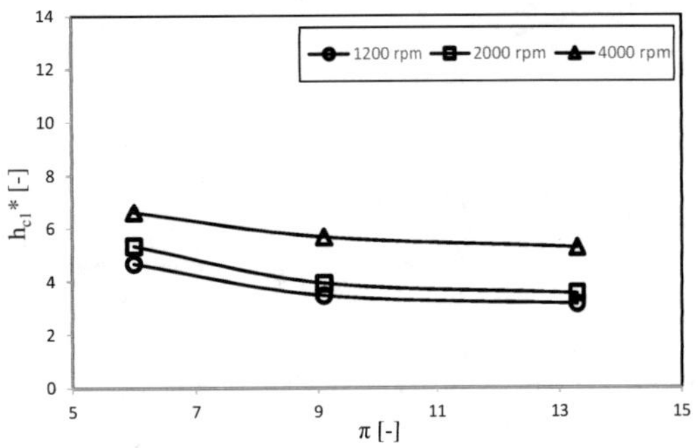

Figure 11: Heat transfer coefficients for chamber 1.

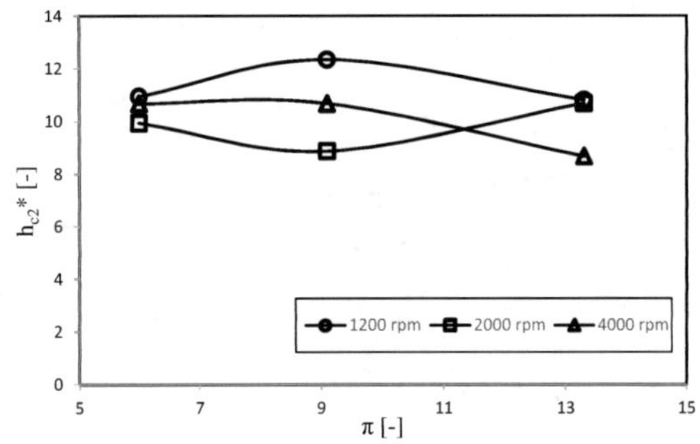

Figure 12: Heat transfer coefficients for chamber 2.

3.2 Heat Transfer Correlations

Heat transfer correlations for the suction muffler are usually expressed through the Nusselt number, Nu, as a function of the Reynolds number, Re, and Prandtl number, Pr, as defined in Equation (3):

$$Nu = \frac{hD}{k_f} \qquad Re = \frac{VD}{\nu} \qquad Pr = \frac{\nu}{\alpha} \tag{3}$$

where ν, α, k_f are the kinematic viscosity, thermal diffusivity and thermal conductivity of the refrigerant gas, respectively, V is the characteristic velocity and D is the characteristic length of the phenomenon. The Prandtl number was found to have a small effect on the Nusselt number for the range of operating conditions tested in our study and, hence, was not included in the heat transfer correlation. Modified expressions for the Nusselt and Reynolds numbers can be written by dividing their original expressions by the characteristic length, D:

$$Re^* = \frac{Re}{D} = \frac{V}{\nu} \qquad Nu^* = \frac{Nu}{D} = \frac{h}{k_f} \tag{4}$$

3.2.1 External surface: The Reynolds number for the external surfaces of the suction muffler was calculated by adopting the crankshaft tangential velocity as the characteristic velocity, V, since its rotational motion promotes the gas flow inside the shell, and as a consequence, around the suction muffler. The Nusselt number was estimated for each operating condition by averaging the heat transfer coefficients measured on the front and back surfaces. Fluid properties were evaluated by using the film temperature.

Reynolds and Nusselt numbers, Re^* and Nu^*, as a function of the compressor speed and pressure ratio are shown in Figures 13 and 14. Both parameters are seen to increase with the compressor speed. The Reynolds number dependence on compressor speed is directly related to the crankshaft tangential velocity. Moreover, the increase of the crankshaft tangential velocity intensifies the flow of gas and oil inside the compressor shell, increasing heat transfer and the Nusselt number. On the other hand, both parameters are found to increase as the pressure ratio decreases. Actually, this dependence on pressure ratio is associated with the evaporating pressure, because both Reynolds and Nusselt numbers vary as π moves from 6.0 (-15°C/40°C) to 9.1 (-25°C/40°C), but remain roughly constant when π changes from 9.1 to 13.3 (-25°C/55°C). In fact, the density of the gas in the suction muffler increases with the evaporating pressure, and hence increasing the Reynolds number and Nusselt number.

Experimental data and two correlations for the Nusselt number as a function of the Reynolds number are shown in Figure 15. Measurements are presented with uncertainty bars corresponding to a 95% confidence interval. A least square method was employed to fit a logarithm function to these measurements:

$$Nu^* = 9.56 \times 10^2 \ln(Re^*) - 1.06 \times 10^4 \tag{5}$$

which is valid for $2x10^5 \leq Re^* \leq 10^6$. The determination coefficient, R^2, is 0.94, indicating that the fitted curve is reasonably correlated with the experimental data. For the purpose of comparison, the curve associated with heat transfer correlation for the flow past a sphere (Whitaker, 1972) is also shown in Figure 15. The good agreement observed between both correlations suggests that a correlation for sphere could be used to estimate heat transfer on the external surfaces of suction mufflers.

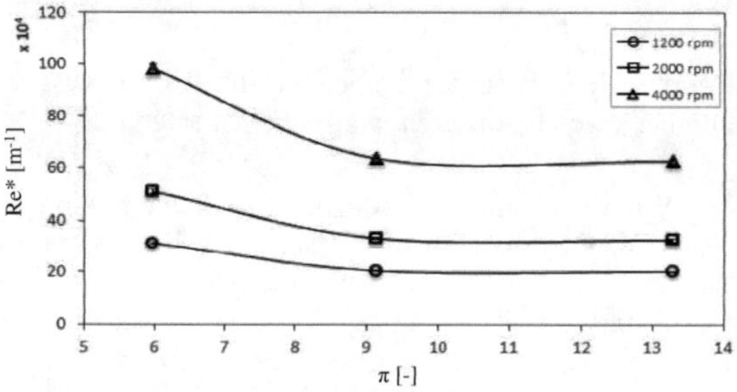

Figure 13: Reynolds number for the external surface of the suction muffler.

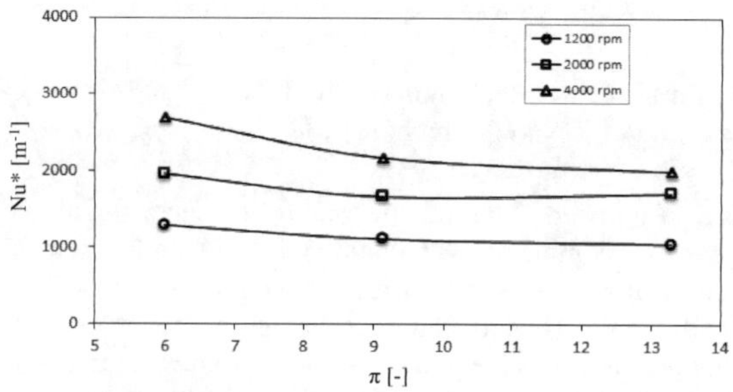

Figure 14: Nusselt number for the external surface of the suction muffler.

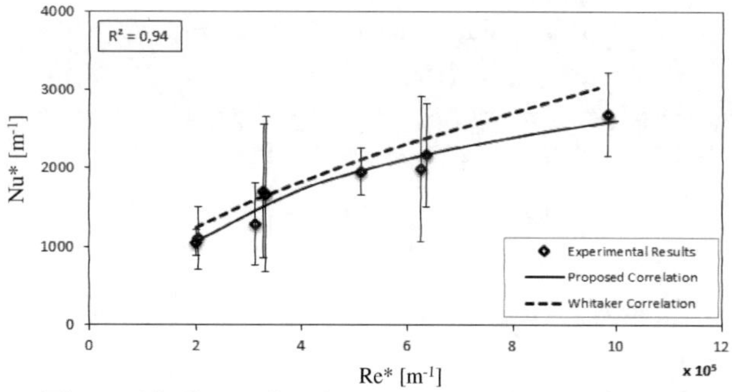

Figure 15: Comparison between correlations at the external surface.

3.2.2 Tubes: The characteristic velocity V necessary to define the Reynolds number is obtained from the mass flow rate, $\dot{m}$, the tube diameter, D, and the gas density in the tube, ρ, ($V = 4\dot{m}/\rho\pi D^2$). Nusselt numbers are evaluated from the heat transfer coefficients presented in Figures 9 and 10.

Figure 16 shows experimental data for Nusselt number as a function of the Reynolds number for tubes1 and 2. As expected, the Nusselt number increases with the Reynolds number for both tubes. Moreover, the results are similar for tube 1 and 2, especially for low Reynolds numbers. A heat transfer correlation for the flow in the suction muffler tubes is proposed by the following logarithm expression:

$$Nu^* = 2.46 \times 10^3 \ln(Re^*) - 3.12 \times 10^4 \quad (6)$$

for a range of $5 \times 10^5 \leq Re^* \leq 3.5 \times 10^6$. Figure 17 shows the fitted curve ($R^2 = 0.85$), with a maximum deviation of 34% between measurements and values given by the correlation. Figure 17 shows a comparison between the experimental data and estimates for Nusselt number in the tubes 1 and 2, given by Equation (6) and the correlations for tubes developed by Dittus and Boelter (1930) and Gnielinski (1976). There is a good agreement between the three correlations for $Re^* < 1.8 \times 10^6$. In the situations of high Reynolds numbers, the estimates from the literature correlations overpredict the heat transfer in the tubes.

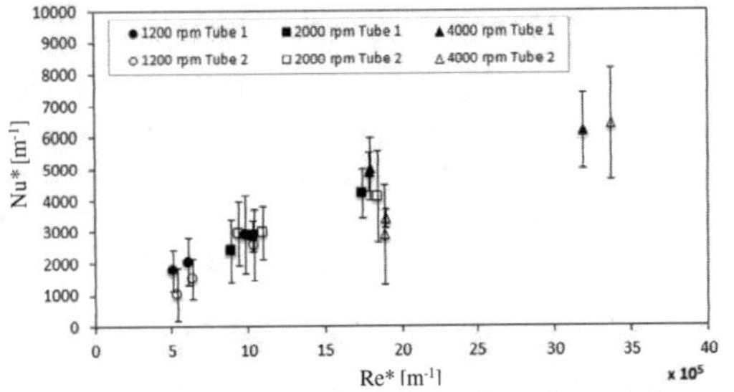

Figure 16: Nusselt number for tubes 1 and 2 as a function of Reynolds number.

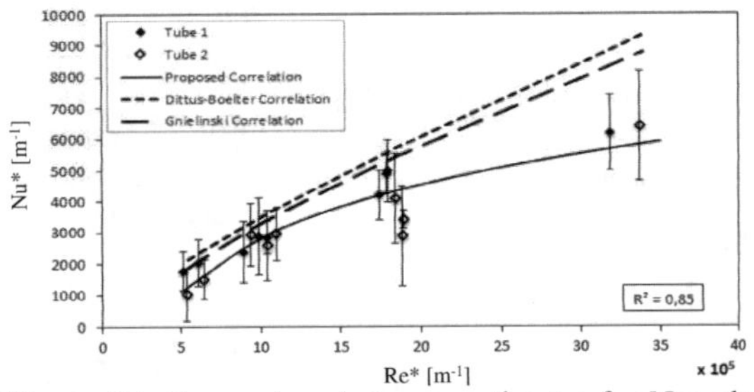

Figure 17: Comparison between estimates for Nusselt number given by different correlations for tubes.

3.2.3 Chambers: The specification of the characteristic velocity V in the chambers is a difficult task, due to the flow geometry complexity. For chamber 1, a hydraulic diameter was defined by using the perimeter of the chamber cross-sectional area immediately after the outlet of tube 1. For chamber 2, the distance between the outlet of tube 2 and the bottom surface of chamber 2 (Figure 1) was adopted as the hydraulic diameter. The characteristic velocity was then evaluated as $V = 4\dot{m}/\rho\pi D_h^2$. The Nusselt numbers are assessed from the heat transfer coefficients presented in Figures 11 and 12.

Figure 18 shows values of Nusselt number for different Reynolds numbers in chambers 1 and 2. The results for chamber 1 are similar to those observed in the tubes. On the other hand, as a consequence of the results shown in Figure 12, there is no clear relationship between Nu^* and Re^* for chamber 2.

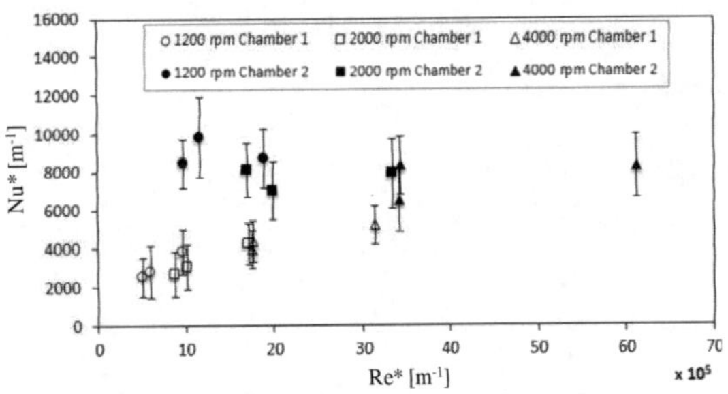

Figure 18: Nusseltnumber for chambers 1 and 2.

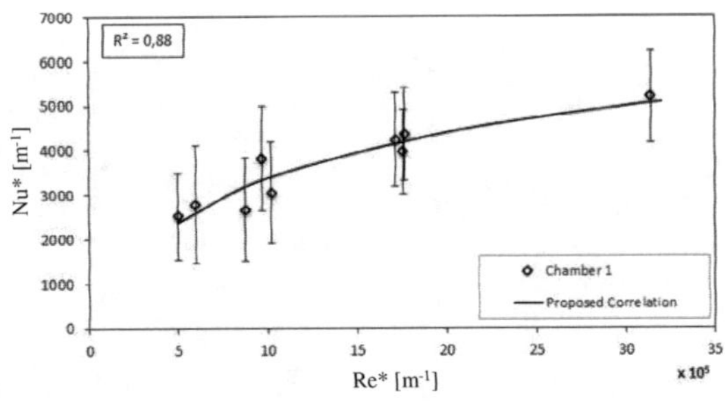

Figure 19: Nusselt correlation for chamber 1.

Since no variation was observed for the Nusselt number in the chamber 2 for different Reynolds numbers, a heat transfer correlation is only proposed for chamber 1,

$$Nu^* = 1.47 \times 10^3 \ln(Re^*) - 1.69 \times 10^4 \quad (7)$$

considering $5x10^5 \leq Re^* \leq 3.2x10^6$. As observed in Figure 19, reasonable fitting was achieved (R^2 = 0.88) and the maximum deviation between the experimental data and correlation is about 27%.

4. CONCLUSIONS

An experimental study of heat transfer in the suction muffler of a variable-speed hermetic reciprocating compressor was carried out. Measurements of temperature and heat flux were carried out in different locations of the suction muffler. Heat transfer coefficients were obtained from such measurements at the external surface of the suction muffler as well as in internal tubes and chambers. Local heat transfer coefficients were found to increase with the compressor speed and evaporating pressure for most of the suction muffler regions. Heat transfer correlations were developed based on the measurements and compared with those available in the literature for simplified geometries. The correlation proposed herein for heat transfer on the external surface of the suction muffler was in good agreement with that obtained by Whitaker (1972) for heat transfer of flow past a sphere. For the suction muffler tubes, the classical correlations of Dittus and Boelter (1930) and Gnielinski (1976) were found to reasonably predict heat transfer for low Reynolds numbers.

NOMENCLATURE

D	characteristic length	(m)	p_e	evaporating pressure	(Pa)
D_h	hydraulic diameter	(m)	q''	heat flux	($W.m^{-2}$)
h	heat transfer coefficient	($W.m^{-2}.K^{-1}$)	Re	Reynolds number	(-)
h^*	non-dimensional heat transfer coefficient	(-)	Re^*	modified Reynolds number	(m^{-1})
h_{shell}	heat transfer coefficient at the compressor shell	($W.m^{-2}.K^{-1}$)	T_s	surface temperature	(K)
k	muffler thermal conductivity	($W.m^{-1}.K^{-1}$)	T_∞	reference temperature	(K)
k_f	gas thermal conductivity	($W.m^{-1}.K^{-1}$)	V	characteristic flow velocity	($m.s^{-1}$)
L	muffler wall thickness	(m)	α	thermal diffusivity	($m^2.s^{-1}$)
$\dot{m}$	mass flow rate	($kg.s^{-1}$)	ν	gas kinematic viscosity	($m^2.s^{-1}$)
Nu	Nusselt number	(-)	π	pressure ratio	(-)
Nu^*	modified Nusseld number	(m^{-1})	ρ	gas density	($kg.m^{-3}$)
p_c	condensing pressure	(Pa)			

REFERENCES

Birari, Y. V., Gosavi, S. S., Jorwekar, P. P., 2006, Use of CFD in design and development of R404a reciprocating compressor, *Proc. Int. Compressor Engrg. Conf. at Purdue*, paper C072.

Cavallini, A., Doretti, L., Longo, G. A., Rosseto, L., Bella, B., Zannerio, A., 1996, Thermal analysis of a hermetic reciprocating compressor, *Proc. Int. Compressor Engrg. Conf. at Purdue*, pp. 535- 540.

Chikurde, R. C., Loganathan, E., Dandekar, D. P., Manivasagam, S., 2002, Thermal mapping of hermetically sealed compressors using computational fluid dynamics technique, *Proc. Int. Compressor Engrg. Conf. at Purdue*, paper C6-4.

Dittus, F. W., Boelter, L. M. K., 1930, Heat transfer in automobile radiators of the tubular type, *Berkeley Publ. Eng.* 2, Univ. California, pp. 443–461.

Dutra, T., Deschamps, C. J., 2013, Experimental characterization of the heat transfer in the components of a small hermetic reciprocating compressor, *App. Thermal Engrg.* 58, pp. 499-510.

Gnielinski, V., 1976, New equations for heat and mass transfer in turbulent pipe and channel flow, *Int. Chemical Engrg*, 16 (2), pp. 359–368.

Kim, S. H., Sim, Y. H., Youn, Y., Min, M. K., 2000, An experimental study on internal temperature cistribution and performance characteristics in a reciprocating compressor for a domestic refrigerator, *Proc. Int. Compressor Engrg. Conf. at Purdue*, pp. 245-252.

Meyer, W. A., Thompson, H. D., 1990, An analytical model of heat transfer to the suction gas in a Low-Side hermetic refrigeration compressor, *Proc. Int. Compressor Engrg. Conf. at Purdue*, pp. 898-907.

Ooi, K. T., 2003, Heat transfer study of a hermetic refrigeration compressor, *App. Thermal Engrg.*, 23, pp. 1931 – 1945.

Raja, B., Sekhar, S. J., Lal, D. M., Kalanidhi, A., 2003, A numerical model for thermal mapping in a hermetically sealed reciprocating refrigerant compressor, *Int. J. of Refrigeration 26 (6)*, pp. 229-236.

Ribas Jr., F. A., 2007, Thermal analysis of reciprocating compressors, *Int. Conf. on Compressors and Their Systems*, London, pp. 277-287.

Rigola, J., Pérez-Segarra, C. D., Oliva, A., 2000, Advanced numerical simulation model of hermetic reciprocating compressors. Parametric study and detailed experimental validation, *Proc. Int. Compressor Engrg. Conf. at Purdue*, pp. 23-30.

Sanvezzo Jr., J., Deschamps, C. J., 2012, A heat transfer model combining differential and integral formulations for thermal analysis of reciprocating compressors, *Proc. Int. Compressor Engrg. Conf. at Purdue*, paper 1343.

Shiva Prasad, B. G., 1992. Regenerative heat transfer in reciprocating compressors, based on the paper presented at *PCRC Int. Reciprocating Machinery Conf. at Denver/USA*.

Todescat, M. L., Fagotti, F., Prata, A.T., Ferreira, R.T.S., 1992, Thermal energy analysis in reciprocating hermetic compressors, *Proc. Int. Compressor Engrg. Conf. at Purdue*, pp. 1417-1428.

Whitaker, S., 1972, Forced convection heat transfer correlations for flow in pipes, past flat plates, single cylinders, single spheres, and for flow in packed beds and tube bundles, *AIChE J.*, 18 (2), pp. 361–371.

ACKNOWLEDGEMENT

This study is part of a technical-scientific program between the Federal University of Santa Catarina and EMBRACO. Financial support from EMBRAPII Unit POLO/UFSC and CNPq (Brazilian Research Council) is also acknowledged.

On the use of Parallel Object-Oriented Code for Solving the Heat Transfer in Hermetic Reciprocating Compressors

López JOAN[1]*, Rigola QUIM[1], Lehmkuhl ORIOL[1], Oliva ASSENSI[1]
[1]CTTC, Polytechnical University of Catalonia,
Terrassa, Spain
joanl@cttc.upc.edu

ABSTRACT

The heat transport phenomenon in a hermetic reciprocating compressor is addressed in this work. This is far from straightforward. It involves several transient physical phenomena interacting with each other. The heat is exchanged between the refrigerant fluid and the solid parts of the compressor (suction muffler, cylinder head, crankcase, etc.). At the same time, the solid parts exchange heat to each other by means of conduction and radiation. Moreover, the phenomenon happens in non-symmetrical complex geometries and the solid parts are made of different materials. This is interesting from both the software engineering and the compressor design viewpoint. A parallel object-oriented software platform for the resolution of multiphysics problems is employed. This platform allows the use of partitioned strategies so that the compressor heat transport problem -a global problem- can be divided into several smaller parts -local problems-. This makes possible the use of multilevel modeling strategies for thermal systems analysis. Furthermore, in order to couple the several sub-problems in an integrated simulation, the platform provides data transfer tools -for matching and non-matching meshes- to exchange sub-domain state information. In particular, the work provides detailed information on the heat distribution and the temperature of the components of a test compressor. By means of comparative studies the thermal properties of some of its components are analyzed. This highlights the importance of choosing proper materials. For example, different suction muffler materials are tested to investigate their influence on the volumetric efficiency. Since the whole compressor is simulated, the consequences of altering specific component properties are also appreciated on the other components. In sum, the work presents illustrative numerical results of the three-dimensional heat transfer in a compressor that show the potential use of computer simulation to support design of components to attain feasibility and energy efficiency.

1. INTRODUCTION

The available computational power has increased significantly in the last decade thanks to the rise of computer clusters. This sets a new paradigm in the development and the usage of software tools for physics simulation. This scenario enables the engineers and scientists not only to incorporate the latest modeling techniques to their codes (LES turbulent flow models, fluid-structure interaction, multiphysics coupling, etc.) but also to exploit classical physical models not feasible before such computing growth (multi-phase modeling, three-dimensional compressible flow, etc.). This work, though an illustrative compressor simulation, aims to show that today's software is a reliable tool to understand better and improve industrial devices and processes. Nowadays, hermetic reciprocating compressors are so optimized that it is becoming increasingly difficult to improve their efficiency. For this reason, one must go to the design details in order to remove the small inefficiencies that prevent improvements in modern compressors. Here, the CFD&HT codes have an essential role. So far, powerful software tools based on one-dimensional mathematical models made it possible optimized compressor designs by means of parametric studies. These tools provide essential global information on the compressor performance such as the cooling capacity and the consumption. Other data such as the transient mass flows in the pipes, the transient temperature maps of the refrigerant, the average temperature of the main solid components, the local pressure perturbation, piston leakage, the sliding-crankshaft mechanism strains and much other local information is also provided indirectly in these one-dimensional simulations (Pérez-Segarra 2004, Rigola 2005).

However, since these models are one-dimensional, their geometry model is too simplified and consequently the effect of the real geometry details on compressor efficiency is not always captured. For instance, the pipes are usually connected to chambers searching an expansion effect of the refrigerant. This expansion depends on the geometries of both the pipe and chamber elements. Although the shape of the pipe is a well-known and simple

canonical geometry, the shape of the chamber can be really complex. Even so, one-dimensional mathematical models for solving this flow transitions only depend on the pipe cross sectional area, its length, and the chamber volume. The mathematical model will provide very realistic results of the flow into the pipe. Nevertheless, the resolution of the flow expansion depends on empirical data, which is not always easy to obtain. Furthermore, some of the common assumptions in expansion-contraction flows are not always reasonable in the context of compressor simulations. Finally, the shape of the chamber is not considered in the model, at least the real one, since it is modeled as a cylinder whose inputs are a diameter and a height.
On the other hand, it is important to note that compressors are particularly complex thermal systems. Not only the time scales of the fluid motion are really small but also all their components are in contact with each other and they are continuously exchanging energy. Moreover, the energy flows in a closed loop due to the presence of the shell which is, excluding the refrigerant itself, the only element that can remove energy to the ambient. Actually, removing heat from a compressor is a difficult matter.
Therefore, understanding the heat transfer details into the compressor device is very important, since most of its inefficiencies are linked to thermal effects (Shiva 1998, Ribas 2008). Both the geometry and the materials of the compressor components (suction muffler, cylinder head, cylinder body, discharge line etc.) are important variables in heat removal. Parameters such as the volume and the thermal diffusivity or the surface size, its roughness and its optical properties are directly related to this process. For this reason, the visualization of the heat fluxes in the compressor components, with real three-dimensional geometries and materials, may help to identify unnecessary effects that produce thermodynamic inefficiencies. In particular, this information could help to reduce the effects of the superheating on the volumetric efficiency of the compressor. Innovative designs of the cylinder head and the cylinder body structures can be analyzed, searching for smart thermal paths aimed to move heat far from the muffler surrounds. For instance, one could ask itself whether the expansion chambers in the cylinder body structure are well positioned or not. By means of multi-dimensional numerical simulation, one could find better positions where to place the chambers.
There are two ways to well understand the thermal behavior of compressor components and its fluid refrigerant in real devices: experimentation and detailed numerical analysis. Experimental techniques, such as the use of thermocouples or infrared thermography may help to obtain compressor temperature maps. However, these techniques are far from easy. They need specific setup for every new prototype, they are time expensive, they are intrusive and some of them lack of detailed information. Moreover, they do not provide information on the heat exchanges between compressor components (Shiva 1998). Alternatively, numerical analysis is much flexible. It can be applied to different designs, with almost no cost, and most times the effort is portable to other devices and technologies since software developers usually focus their codes to general multiphysics purposes, as in the platform is hereby presented.

2. THE NEST SOFTWARE PLATFORM

2.1 Multi-model capability

In the simulation of complex systems it is usually interesting to model each component separately by means of a partitioned approach. This brings some advantages from the point of view of software engineering. A big one is the capability to use legacy codes. Another practical advantage is that replacing or removing some component does not affect the remaining components. Hence, you are able to arrange various compressor configurations in a fast and very flexible way since there is no need to modify the program code.
The platform presented here, called NEST (López 2016), is ready for doing this. Regarding to the simulation of hermetic reciprocating compressors in particular, the heat transport is assessed by splitting the compressor into several parts. The main separation is between the refrigerant fluid and the solid components. However, both the refrigerant as well as the set of solid parts are divided into some more sub-parts at the same time. In order to illustrate this, in the scheme of Figure 1 are depicted the several sub-models used for modeling a compressor. This partitioning is based on the several phases that the fluid is submitted to all along the compression path (suction, muffler, compression and discharge line). On the other hand, the main solid components of the compressor are shown in Figure 2. The several parts have distinct shapes and they are made of different materials. The separate treatment of each part permits a better modeling of the heat transfer, since each sub-model can be customized to the particular mathematical and numerical requirements of each part.

Finally, since the model of a particular element depends on the state of their surrounding elements, NEST provides the necessary tools to exchange information between each other. This is implemented in a transparent way, which allows that the users focus their attention on the physical problem instead of spending time understanding coupling issues and software implementation details.

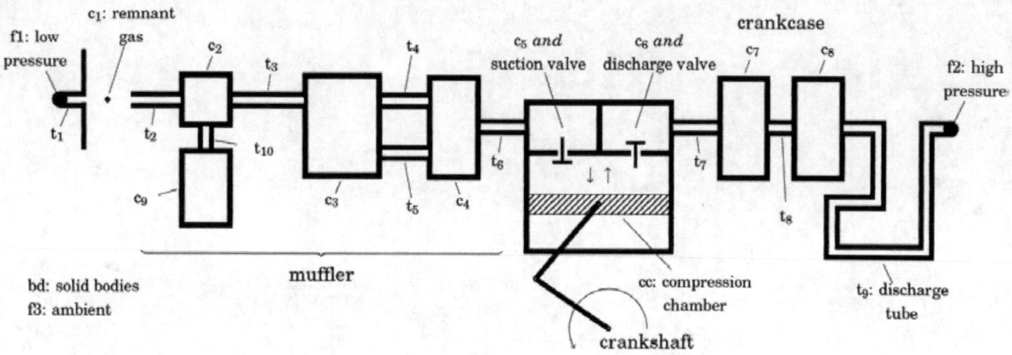

Figure 1: NEST model of a compressor (*c*, *t* and *f* stands for *chamber*, *tube* and *fixed value* respectively).

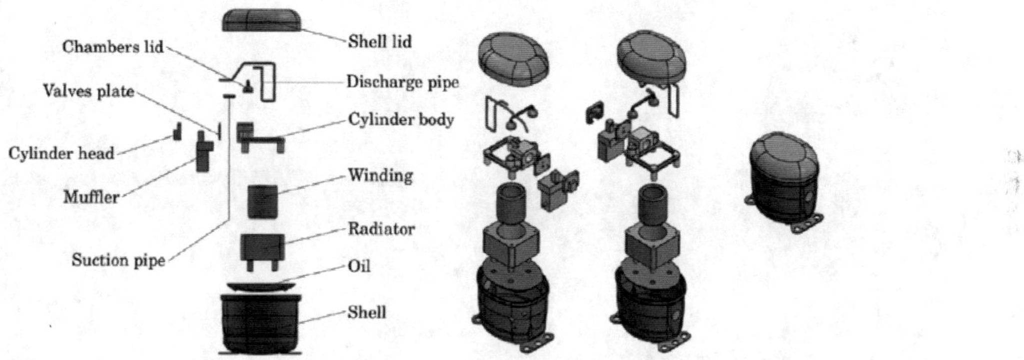

Figure 2: Main solid components of the compressor.

2.1 Parallel capability and non-matching mesh

It is well-known that the computational expenses are very high when detailed simulation by means of CFD&HT analysis is necessary. However, the last advances in parallel computation make it possible to solve complex industrial problems. Parallel computation is an essential software capability to make use of CFD&HT advances in compressors modeling. NEST software platform is also ready for this. When some model is computational expensive it can be executed in parallel using several processors. Of course this does not affect the resolution of the whole compressor system. Such parts that are solved in parallel are still seen as separate software entities system elements).

One important advantage of the partitioned approach is the possibility to use distinct meshes in each multi-dimensional sub-model. This makes the meshing process much easier since each part is treated apart, which significantly simplifies the geometry of the domain to mesh. For example, you can construct orthogonal meshes in the discharge pipeline, whereas the cylinder body structure or the shell can be meshed using an unstructured meshes. In Figure 3 are shown the meshes for some of the compressor parts. An important drawback of this approach is the mismatch of the meshes at the contact boundaries of the components. This makes data exchange very hard since the data must be transferred properly, assuring accuracy and conservation of the transmitted variables. Moreover, this operation should be fast as it must be performed many times during the simulation. Data transfer is not a straightforward problem. Actually, this is considered a challenging problem in multiphysics simulation (Jiao 2004). Nevertheless, the data transfer methods implemented in NEST permit the use of non-matching meshes.

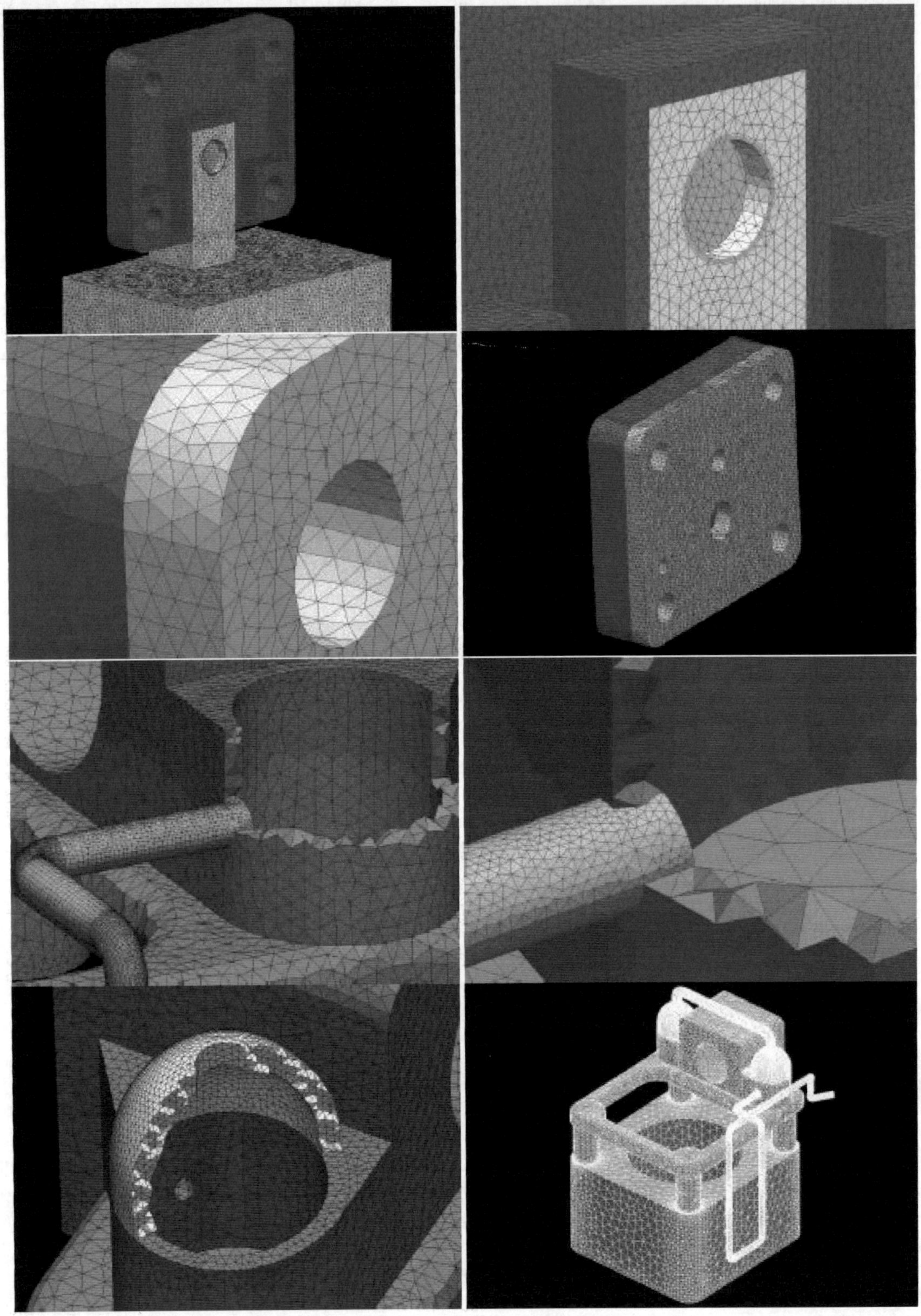

Figure 3: Some views of the meshes used in the numerical simulation. Every component is meshed apart so that the meshes do not match at the interfaces.

3. NUMERICAL EXPERIMENT

In this section the temperature map in a three-dimensional hermetic reciprocating setup is presented. These data are obtained from the numerical resolution of the heat transfer in the compressor components of Figure 2 using NEST software platform. In the background, NEST uses the TermoFluids DiffusionSolver (Lehmkuhl 2009) for resolving the energy equation in the several computational domains. In order to have a complete view of the temperature distribution and the heat flows, the computational domain is composed of all the components where the fluid refrigerant circulates through as well as those are in contact. To get closer to a real-world compressor the parts are made of different materials, for instance the shell is made of steel and the discharge pipeline is made of cooper.
Since the overall compressor model is spitted into several sub-models, every component is discretized apart from the others. Hence, there is one discretization mesh per component. In this way, high quality meshes can be obtained easily. Moreover, this makes it possible to modify, replace or remove compressor components without affecting the meshes of the surrounding components. Information about the mesh of each component is presented in Table 1.

Table 1: Numerical parameters of the distinct meshes.

Mesh	**Cells($x10^{-3}$)**	**Partitions**	**Component**
$mpipe_1$	5	1	Muffler
$mpipe_2$	5	1	Muffler
$mpipe_3$	50	2	Muffler
$mcham_1$	350	12	Muffler
$mcham_2$	250	8	Muffler
cylhead	170	6	Cylinder head
valveplate	120	4	Valve plate
cylbody	300	12	Cylinder body
$ckchamlid_1$	30	2	Crankcase
$ckchamlid_2$	30	2	Crankcase
$ckpipe_2$	10	2	Crankcase
radiator	120	4	Motor
winding	25	2	Motor
oil	85	4	Oil
shell	775	20	Shell
$dtube_{beg}$	30	2	Discharge tub
$dtube_{end}$	20	2	Discharge tub
$dtube_{bends}$	170	6	Discharge tub
$dtube_{extrusions}$	70	4	Discharge tub
SUM	2615	96	Compressor

The boundary conditions for each of these components are fixed using the Newton's approach $q=h(T_{wall}-T_{gas})$. The gas temperature data is based on the refrigerant temperature profile shown in Figure 4. This data is obtained from a simulation of a 9.6cm^3 isobutane compressor using NEST one-dimensional models. The heat transfer coefficients are taken from this simulation as well. They were evaluated using the empirical correlations implemented in the NEST compressors sub-models. The ambient temperature is set at 20C.

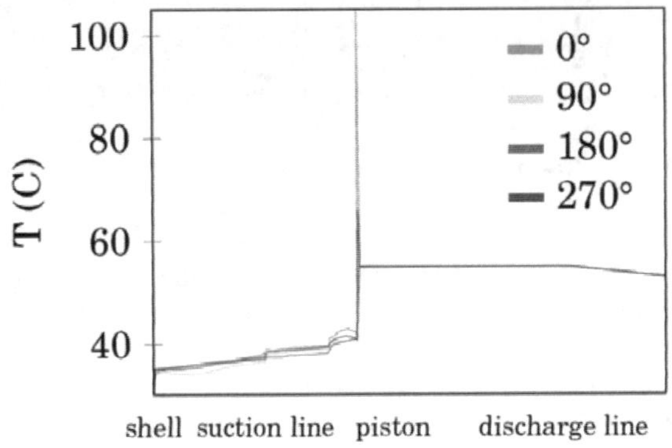

Figure 4: Temperature profile of the refrigerant fluid at four crankshaft positions.

3.2 Numerical results

The following figures show several views of the compressor temperature map. One can clearly identify the cold (dark color) and hot (light color) areas in the general perspective views of Figure 5. As expected, the shell is the coldest part while the cylinder surround is the hottest one. Other interesting effects can be observed here. For instance, the chambers in the cylinder body structure are heated by remnant heat flows coming from the cylinder surface. This is a desired effect since the heat is dissipated from the cylinder surrounds to colder parts of the cylinder body structure, which contributes to create a colder environment around the suction muffler. This is very important since the volumetric efficiency can be reduced significantly if the refrigerant is too heated before entering into the cylinder. Furthermore, the backside of the cylinder body structure is about 15C colder than the cylinder surround (see left-right views in Figure 6). This indicates that there is room for intensifying heat removal from the cylinder. On another hand, the images also show how the heat is transferred to the shell first through the radiator and later through the oil pool. Actually the top part of the shell is colder than its bottom. This means that the heat is mainly dissipated thanks to the oil, being the heat convection though the remnant gas not so relevant. In addition, the simulation also shows how the discharge pipe colds down as it is closer to the shell. This temperature gradient indicates that this pipe is also a heat removal path. However, one cannot forget that the refrigerant leaves the compressor through this pipe. Hence, the state of the discharged gas depends on what happens along this pipe.

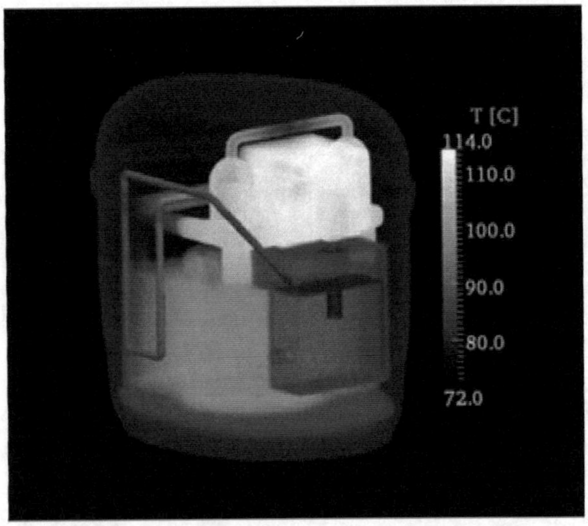

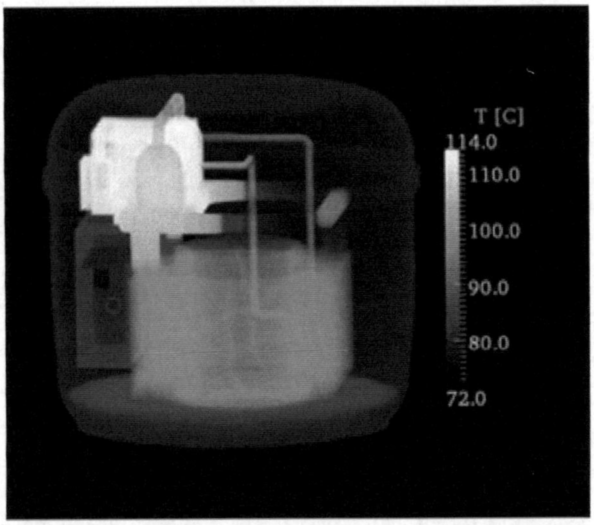

Figure 5: Temperature distribution in the compressor setup. The optical properties of these images have been modified in order to emphasize specific temperature levels. Due to this rendering some temperatures are not appreciated.

Moving towards more lightweight geometries of the cylinder body structure could help to reduce the accumulated heat in the solid parts. Indirectly, this could improve volumetric efficiency since it would reduce the temperature of the refrigerant in the suction area. In a similar way, the choice of the employed materials is also critical. In order to show the effect of changing the suction muffler material, a second simulation has been carried out using exactly the same geometry and boundary conditions. In this case, however, the suction muffler is not made of plastic material as usual. Instead, it is done of a hypothetical material, whose thermal diffusivity is ten times the one of the plastic muffler. Figures 7 and 8 show several comparative views of the obtained numerical results. Remember that in these views, the darker is the color the colder is the material. The main difference relies on the temperature, when the thermal diffusivity is low the muffler is colder. Therefore, this numerical experiment shows why suction mufflers are made of plastic.

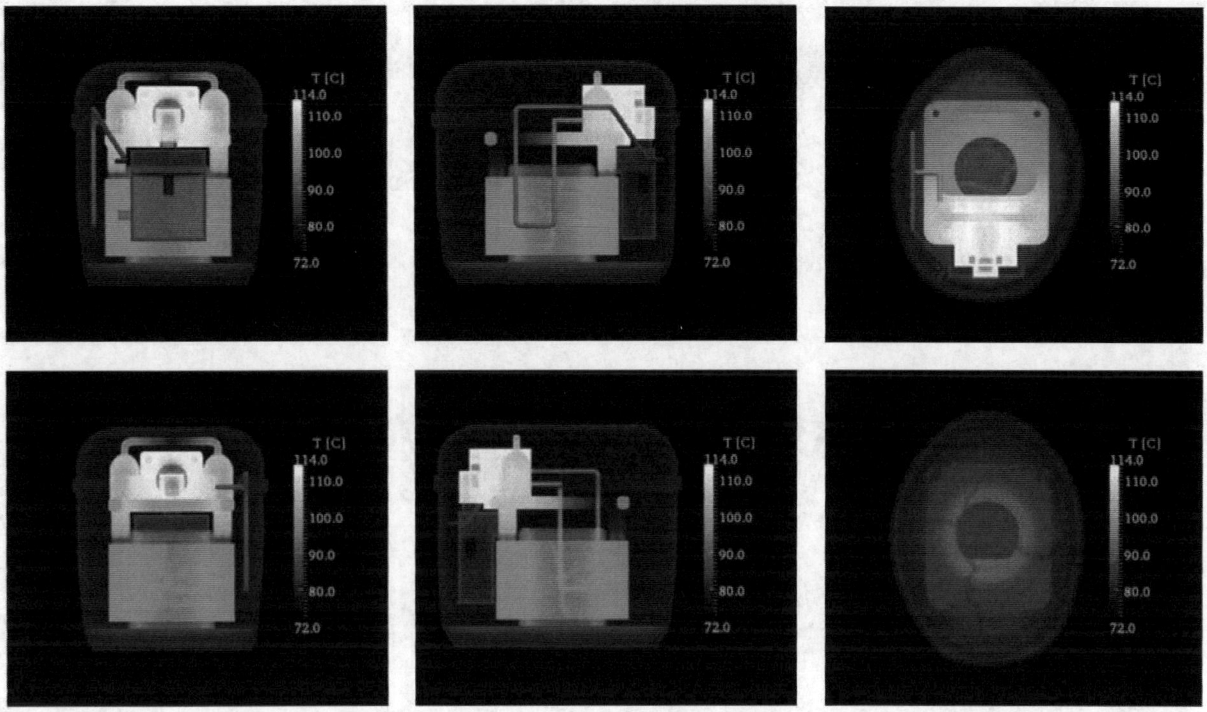

Figure 6: Temperature distribution in the planar views of the compressor setup. The presented images correspond to views from: front and back (left column), left and right (middle column) and top and bottom (right column).

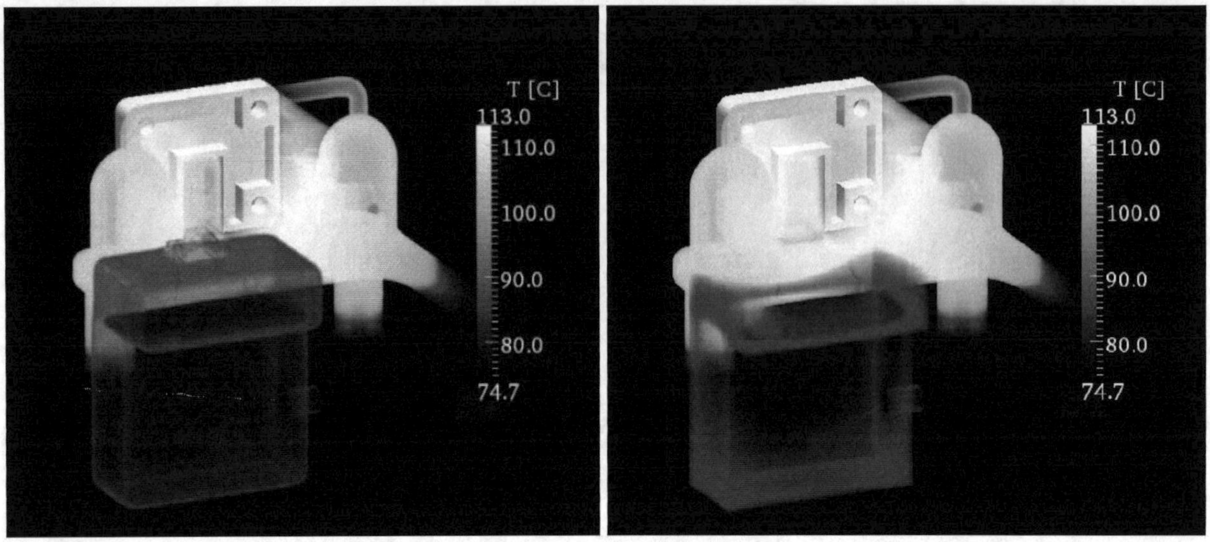

Figure 7: Temperature distribution in general views of suction muffler and its surrounds. Low (left) and high (right) thermal diffusivity material.

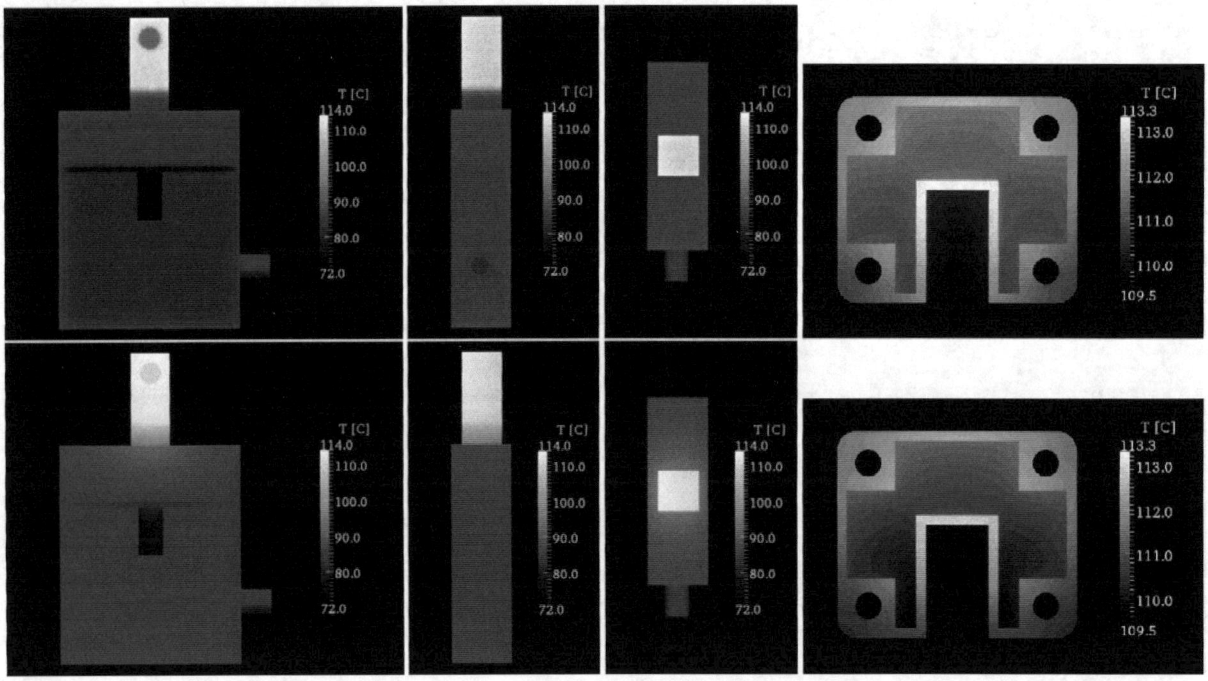

Figure 8: Temperature distribution in the planar views of suction muffler and the cylinder head. Lower thermal diffusivity material (upper row) and higher thermal diffusivity material (lower row).

Figure 9 shows a better perspective of both suction mufflers. In these figures it is easier to see that the outlet pipe of the muffler is at different temperature. In this zone, the heat transfer between the refrigerant and the muffler walls is intensified and it is much important than the one in the previous chambers. This results show that the discharge of the muffler is an important design consideration.

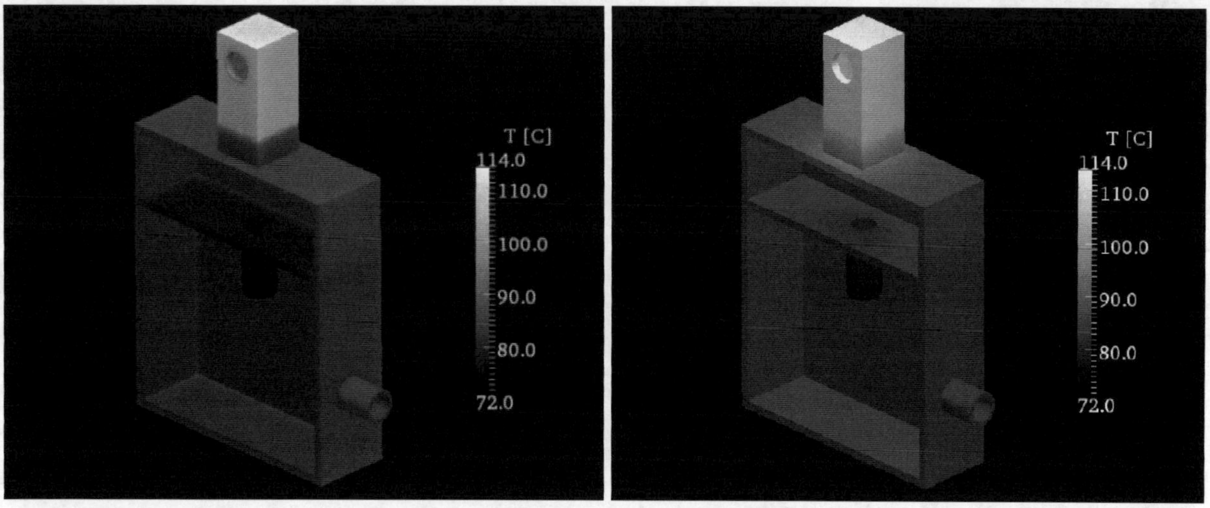

Figure 9: Temperature distribution in perspective views of the suction muffler. Low (left) and high (right) thermal diffusivity material.

4. CONCLUSIONS

The NEST platform, using TermoFluids (Lehmkuhl 2009) software in the background, has been used to solve the heat conduction in a hermetic reciprocating compressor.

Several capabilities of this software have been tested in the simulation. First, the heat transfer has been analyzed in the whole set of compressor components by means of a partitioned approach. This involves the use of several instances of the TermoFluids DiffusionSolver, which exist simultaneously while they are executed on several parallel processes. Using distinct instances has made it possible the use of different meshes and materials. The employed meshes are based on geometric particularities of each component. Consequently, they do not match at the contact boundaries, which permitted to test the NEST data transfer tools also.

Various three-dimensional numerical results of the temperature distribution in the compressor components have been presented and discussed. The obtained numerical results are in good agreement with the expected thermal behavior of a compressor. Although this simulation is far from being a novel contribution to compressors design, it proves that the NEST platform deserves further research. This research can be focus on solving the convection of the remnant fluid refrigerant around the solid components using coupled CFD&HT. In addition, many other compressor details should be considered in this kind of simulations. Some are still far to achieve in numerical simulation, like the oil recirculation or those related with noise generation such as motor vibrations. Some others are already possible, although they make the simulations more computationally expensive. For example, thermal radiation is an important factor in the energy balance and could be introduced by resolving the RTE numerically. Some other details rely on the geometry and the employed materials. Innovative designs pursuing a general cold down of the compressor components could be investigated. This could be focused on many details. For instance, the joining point in the discharge line at the shell side is quite simplified in the simulation model. In real-world compressors this join is usually mechanically fastened using cooper rivets. This is an important consideration because it alters the heat transfer to the shell and the ambient. In sum, the advances in high performance computer and consequently the new possibilities in CFD&HT are not the definitive solution in the improvement of industrial devices. There is still much work to do in this research area and today's simulation tools must be used carefully. Well-thought numerical experiments are necessary to provide substantial information on designs defects and suggest novel enhancements. While the one-dimensional models lack of detail in the provided results, CFD&HT models are still too computationally expensive for using them in industrial problems. The combination of both one-dimensional models -in a low level- and three-dimensional models -in a high level- is the workaround to this problem. When this is possible, the numerical results such as the ones presented in this work can be complemented with estimations of the COP and hence, with an overall compressor performance analysis.

REFERENCES

Jiao, X. and Heath, M.T. (2004). Common-refinement-based data transfer between nonmatching mehses in multiphysics simulations. Int. J. for Numerical Methods in Engineering, 61,2402-2427.

Lehmkuhl, O., Perez-Segarra, C.D., Borrell, R., Soria, M. and A. Oliva (2009). Termofluids: A new parallel unstructured CFD code for the simulation of turbulent industrial problems on low cost PC cluster. Lecture Notes in Computational Science and Engineering, 67:275-282.

López, J. (2016). *Parallel object-oriented algorithms for simulation of multiphysics. Application to thermal systems.* Universitat Politècnica de Catalunya. Thesis.

Perez-Segarra, C.D. et al. (2004). Detailed thermodynamic characterization of hermetic reciprocating compressors. Int. J. Refrig., 28,579-593.

Rigola, J. et al. (2005). Parametric studies on hermetic reciprocating compressors. Int. J. Refrig., 28(2),253-266.

Ribas, F.A. et al. (2008). Thermal analysis of reciprocating compressors - a critical review. In Int. Compr. Eng. Conf., Purdue, IN, USA, Paper 1907.

Shiva, B. G. (1998). Heat transfer in reciprocating compressors - a review. In Int. Compr. Eng. Conf., Purdue, IN, USA, Paper 1349.

Surface Tension of Low-viscous Lubricants in High Pressure Carbon Dioxide Atmospheres

Tobias GÖPFERT[1]*, Christiane THOMAS[1], Ullrich HESSE[1]

[1]Technische Universität Dresden, Bitzer Chair of Refrigeration, Cryogenics and Compressor Technology
Dresden, Saxony, Germany
(Tobias.Goepfert@tu-dresden.de)

* Corresponding Author

ABSTRACT

The optimization of refrigeration systems can be achieved by increasing the isentropic efficiency of the compressor. One possible way is the reduction of the friction and leakage losses which occur at the piston rings. The viscosity and the surface tension of the lubricant must be known for this purpose. From the literature only a small amount of measured values for the surface tension of commonly used lubricants are known. Additionally natural refrigerants like R-744 are more common according to the latest (i.e. European) environmental regulation. The thermophysical properties of lubricant-refrigerant-mixtures can currently only be calculated by generalized calculation methods. For specific mixtures, these equations need to be confirmed by measurements. In this paper the measurement results of the surface tension of different low-viscous lubricants are shown. A test bench was designed to measure the surface tension of different liquids especially under high pressure atmospheres. The investigation was performed by using the pendant drop method and analyzing the results with new algorithms for the solution of the Young-Laplace equation. Effects of the geometry of the used capillaries are discussed which leads to possible optimizations of this measurement method. Dilution effects on the surface tension of the lubricants in carbon dioxide atmospheres are shown and discussed. The effect of higher pressures and different temperatures on the surface tension is shown as well as a comparison of the measurements with generalized methods.

1. INTRODUCTION

With increasing pressure and lower temperatures refrigerants are soluted in the compressor lubricants which leads to a significant change of the thermophysical properties of the lubricant. Especially the transport properties viscosity, thermal conductivity and surface tension are interesting for technical applications. These properties are important for the process of lubrication and sealing in the compressor, i.e. on the piston rings. Studies from (Di Nicola & Moglie, 2011) and (Di Nicola & Pierantozzi, 2013) show that the surface tension of refrigerants can be precalculated by using corresponsing generalized methods. However escpecially the critical parameters of the fluids, the critical temperature, pressure, volume, the normal boiling point and the acentric factor need to be known. (Di Nicola & Pierantozzi, 2013) also show that the surface tension of refrigerant mixtures can be calculated by adjusting one specific parameter of the binary fluid mixture. For this purpose the surface tension and its temperature dependency must be known for both fluids. This isn't usally the case for lubricants. Studies from (Zhelezny P. , Zhelezny, Procenko, & Ancherbak, 2007) and (Zheleznya, Sechenyhb, Ivchenkoa, & Yury, 2014) show that the surface tension of refrigerant-oil-mixtures (ROS) can be calculated when the vapor pressure of the refrigerant lubricant mixture is known for different temperatures and concentrations. These calculations can be used for reduced critical temperatures up to 0.7. This leads to limitations, when fluids like R744, R170, R23, R290, R32 or R1234yf are used, which have critical temperature lower than 373 K. Certain compressor parts and thus the lubricant can have significant higher temperatures in refrigeration or heat pump systems. Also the precalculation according to (Zheleznya, Sechenyhb, Ivchenkoa, & Yury, 2014) leads to problems when the temperature is lower than the critical temperature of the refrigerant but the pressure is higher than the pseudo critical pressure of the lubricant. This is often the case, as the critical pressure of a lubricant is typically rather low, and it can especially be the case for mixtures of R744 with lubricants. On the other side especially for low temperature applications with R744 or R23 low-viscous lubricants with an high viscosity index (VI) must be used to ensure the lubricant transport through the evaporater. To calculate the friction losses and the thickness of the lubricant film inside the evaporator and compressor the surface tension must be known.

2. THEORETICAL APPROACH

2.1 Surface tension of pure fluids calculated with the Young-Laplace equation

When a liquid drop is in equilibrium to its gas phase at a specific pressure and temperature, it can be described by a balance of forces. Referencing the forces to the surface of the drop, the equilibrium can also be described as a pressure equilibrium. The equilibrium of a pendant drop can be described by the Young-Laplace equation (1).

$$\sigma\left(\frac{1}{R_1}+\frac{1}{R_2}\right) = \Delta p \equiv p_1 - p_2 - z \cdot g \cdot (\rho'' - \rho') \tag{1}$$

In equation (1) σ is the surface tension, R_1 and R_2 are the significant radii of the drop, p_1 the internal pressure of the drop, p_2 the pressure in the gas phase, g is the gravitational acceleration and ρ' and ρ'' are the densities of the liquid and gaseous phase. Since the internal pressure difference of the drop Δp is difficult to measure it is more common to measure to geometric parameters of the drop. As shown in publications from (Davida, Sefianea, & Tadristb, 2007), (Saad, Policova, & Neumann, 2011), (Berry, Neeson, Dagastine, Chan, & Tabor, 2015) especially the equivalent spherical radius R_0 can be used. To solve equation (1) it can be assumed that when a specific sphere is found which has the same geometric behavior it should have the same balance of forces which leads to the surface tension. The internal pressure force ($\pi \cdot {R_0}^2 \cdot p_1$), the outer pressure force ($\pi \cdot {R_0}^2 \cdot p_2$) and the interfacial tension ($2\pi \cdot {R_0}^2 \cdot \sigma$) have to cancel each other (2), see Figure 1.

$$\pi \cdot {R_0}^2 \cdot p_1 = \pi \cdot {R_0}^2 \cdot p_2 + 2\pi \cdot {R_0}^2 \cdot \sigma \tag{2}$$

Since equation (2) still needs the difference of the internal and outer pressure of the drop the Bond-Number Bo was defined (3) which can be interpreted as a geometrical index of the drop.

$$\mathrm{Bo} \equiv \frac{(\rho' - \rho'') \cdot g \cdot {R_0}^2}{\sigma} = \mathrm{f}(\phi, s) \tag{3}$$

The solution of equation (3) is shown in detail by (Berry, Neeson, Dagastine, Chan, & Tabor, 2015). Since equation (3) is only a function of the drop shape the detection and calculation of the edge function is essential. Since the density difference $(\rho' - \rho'')$ is not always known by from measurements also the Laplace constant a can be used (4) for comparing the measurements with other tabulated values.

$$\mathrm{a} \equiv \sqrt{\frac{\sigma}{(\rho' - \rho'') \cdot g}} = \frac{R_0}{\sqrt{\mathrm{Bo}}} \tag{4}$$

In the literature like (Zhelezny P. , Zhelezny, Procenko, & Ancherbak, 2007) and (Zhelezny, Semenyuk, Ancherbak, Grebenkov, & Beliayeva, 2007) also the capillary constant α^2 is used (5).

$$\alpha^2 = \frac{2 \cdot \sigma}{(\rho' - \rho'') \cdot g} = 2 \cdot \mathrm{a}^2 = 2 \cdot \frac{{R_0}^2}{\mathrm{Bo}} \tag{5}$$

2.2 Literature survey on surface tension of refrigerants and refrigerant-lubricant mixtures

Only a few measurements and methods for the calculation of the surface tension of lubricants are known from the literature. Generalized and semi-empirical methods are shown by (Reid, Prausnitz, & Poling, 1987) where at least the surface tension of the normal boiling point and the critical parameters of the fluid have to be known. Analysis from (Di Nicola & Moglie, 2011) and (Di Nicola & Pierantozzi, 2013) led to the same results. Since often the database of the lubricants is small these methods have a high uncertainty. Also group contribution methods like

UNIFAC or UNIQUAC cannot be used since the chemical structure of the lubricant and the additives are often unknown.

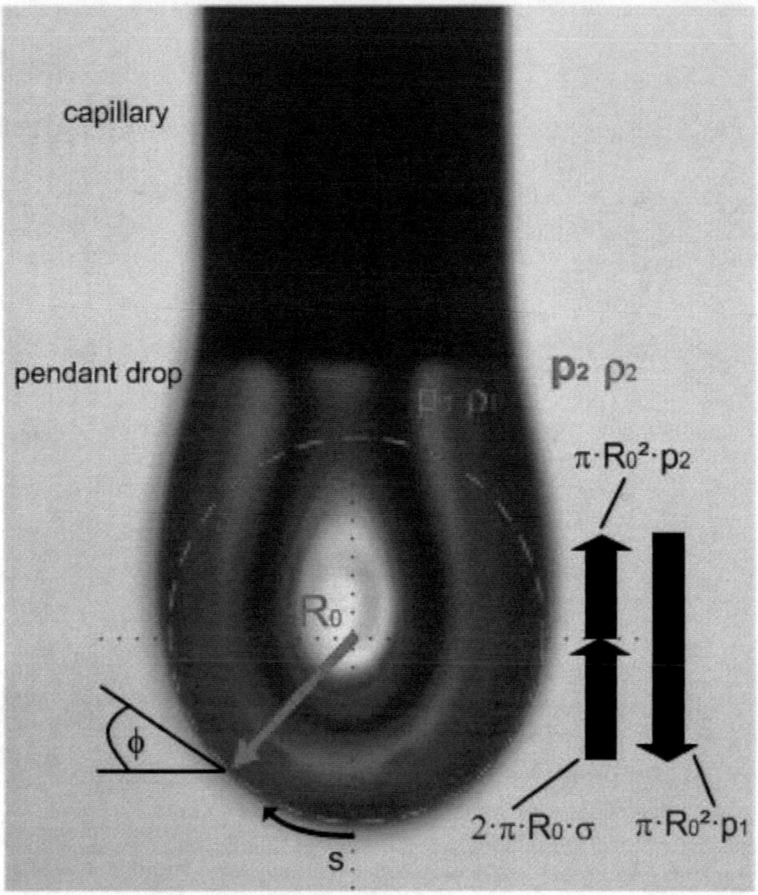

Figure 1 - Forces and geometric values of a pendant drop

A new approach was investigated by (Zheleznya, Sechenyhb, Ivchenkoa, & Yury, 2014) who used a special form of the corresponding state method to calculate the surface tension of lubricants and lubricant refrigerant mixtures. The basic idea was to use the vapor pressure data of the lubricant refrigerant mixtures and approximate the surface tension with universal coefficients and pseudo critical data of the mixtures. Analysis from (Zhelezny P. , Zhelezny, Procenko, & Ancherbak, 2007), (Zhelezny, Semenyuk, Ancherbak, Grebenkov, & Beliayeva, 2007) and (Semenyuk, Zhelezny, Ivchenko, & Geller, 2011) show that these methods can be applied for the surface tension calculation of different refrigerant lubricant combinations. However as discussed by (Vetere, 1992), (Yokozeki, 2001) or (Zhelezny, Zhelezny, & Scripov, 2003) still a lot of data like the liquid density, the compressibility factor, pTx-data of the refrigerant lubricant mixture and others are unknown. The method shown by (Di Nicola & Pierantozzi, 2013) has the advantage that the temperature and liquid phase concentration dependency can be described with simple semi-empirical equations which can be fitted to surface tension measurements and only solubility data are required. However at temperatures or pressures higher than the critical values of the refrigerant, the model reaches its limits considering that the surface tension of supercritical states is not defined.

2.3 High pressure effects on the surface tension of lubricants

Measurements from (Eggers & Jaeger, 1994), (Chun & Wilkinson, 1995) and (Bachu & Bennion, 2009) show that the close to the critical state of carbon dioxide, some fluid mixtures lead to 'special' behavior. This behavior can be explained with the temperature and pressure dependency of the pure fluids. The temperature dependency can be described with equation (6) where the saturation temperature T_S can also be described by a function of the vapor pressure like the Clausius-Clapeyron equation. When the temperature or pressure gets close to the critical state, the surface tension becomes zero, which leads to problems using the equations of (Di Nicola & Pierantozzi, 2013) for binary fluid mixtures. However this problem can be solved when the critical state of the mixture is known as a function of the concentration.

$$\sigma(T_s) = \sigma_{ref} \cdot \left(1 - \frac{T_s(p_s)}{T_c}\right)^n \tag{6}$$

In equation (6) σ_{ref} is the surface tension at a reference state point, T_s is the saturation temperature as a function of the saturation pressure p_s, T_c is the critical temperature and n is a fluid specific exponent. For technical applications it is more common to measure the pressure and temperature and calculate other thermophysical properties. If we consider a system of a lubricant and carbon dioxide without a miscibility gap, the correlation from (Di Nicola & Pierantozzi, 2013) can be used for subcritical temperatures as a function of the concentration x (7).

$$\sigma_{mix}(p,T) = \left(\sigma_1(T) \cdot x_1(p,T)\right)^r + \left(\sigma_2(T) \cdot x_2(p,T)\right)^r \tag{7}$$

The surface tension of the mixture σ_{mix} depends on the temperature (7). With the assumption of an ideal solubility and with the application of Dalton's Law, the pressure dependency can be described. Considering that the lubricant vapor pressure is almost zero the one dimensional equation (8) can be formulated.

$$x(p, T_s) = \frac{p}{p_s(T_s)} \tag{8}$$

Applying equation (8) on (7) for binary systems of pure gas phases and close to an ideal liquid phase equation (9) is resulting.

$$\sigma_{mix}(p,T) = \left(\sigma_1(T) \cdot \frac{p}{p_s(T_s)}\right)^r + \left(\sigma_2(T) \cdot (1 - \frac{p}{p_s(T_s)})\right)^r \tag{9}$$

Using equation (9) and (4) or (5) it can be assumed that also the Laplace constant and the capillary constant should be a function of pressure and temperature. Equation (9) can also be expressed as a polynomial function for the Laplace constant a=f (p, T) because of the following assumptions

- the surface tension is a power function,
- the density of the liquid phase is approximately a linear function of the temperature and
- the vapor density is in first approximation a polynomial function of pressure and temperature.

3. EXPERIMENTAL TEST SETUP AND MEASUREMENT

3.1 Experimental test setup

A test bench was designed to measure the surface tension of lubricants in high pressure atmospheres. It is shown in Figure 3. The key part is a high pressure stainless steel cell. This high pressure cell includes two high pressure sight glasses with an optical distortion of almost zero. From the top of the cell a pipe with a small capillary at the end, two PT-100 resistance thermometers and an additional small diameter pipe are connected to the cell. The drops are generated with by means of the capillary tube. The two resistance thermometers measure the temperature of the lubricant and inside the measuring cell to ensure that they have the same temperature. Through the small pipe gas can flow into the measuring cell to generate a specific pressure. On the one side of the sight glasses a monochrome and collimated LED is used for the lightning of the measuring cell. The whole apparatus is placed in a temperature chamber. On the other side of the sight glasses a camera is installed to photograph the drops. A Nikon D7000 with a macro lens for high quality resolute pictures is used. To reduce effects of scattered light a circular polarization filter was used. Additionally the experimental apparatus includes:

- different valves for controlling the gas and lubricant flow,

- a vacuum pump for degassing the lubricant and removing air and moisture from the measuring cell,
- a pressure tank where the lubricant is stored and
- Two pressure sensors, one for the measuring cell and one for the pressure tank.

The apparatus was designed for pressure from 0 up to 170 bar and a temperature range from 220 up to 400 K. The principle design of the apparatus is shown in Figure 3 and Figure 2.

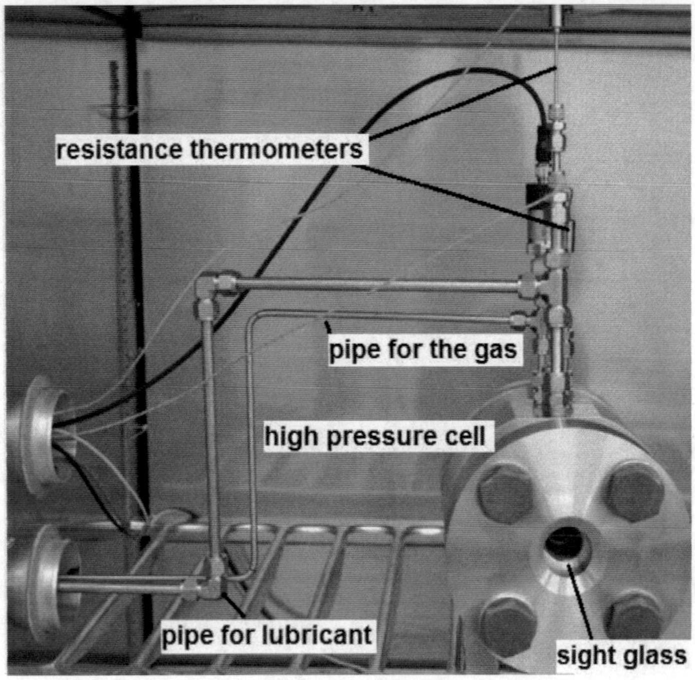

Figure 2 – high pressure cell with connected pipes and wires

3.2 Surface tension measurement procedure

The lubricant is charged to the pressure tank (A) (see Figure 3) by means of a filling spout which is closed afterwards. A vacuum pump evacuates the entire system to pressures lower than 10^{-3} mbar for at least 10 hours. After that, the system is sealed again and the gas from the gas cylinder (C) is charged into the system. For the tests carbon dioxide with a purity of 99.9% was used. The purity was validated by a gas chromatographic analysis. The pressure drop in the system, which is caused by the solution of gas in the lubricant, needs to be considered. After about 90 minutes the lubricant is saturated with the gas. The high pressure cell is located in a temperature chamber. Afterwards the valve (D) in the gas pipe to the high pressure cell is closed. The pressure inside the lubricant tank is slightly increased by letting a small amount of gas into the tank. The pressure difference is about 0.05 to 0.1 bar. After that the flow control valve (E) is opened carefully and the lubricant flows to the high pressure cell due to the small pressure difference. The system is observed by an optical camera until a drop was generated. The drop is held long enough so that the saturation state of the drop is in equilibrium with the gas phase. According to measurements from e.g. (Davida, Sefianea, & Tadristb, 2007) the saturation state is reached in a few seconds. This can be explained by the small volume of the drop where only a small amount of gas molecules can be solved. A picture of the drop is taken and the drop is drained by opening the flow control valve (E) carefully. After that the procedure is repeated until at least 20 drops are generated at a specific pressure and temperature. After that the pressure was changed for the next operation point and the system was given a time of at least 1.5 hours for saturation.

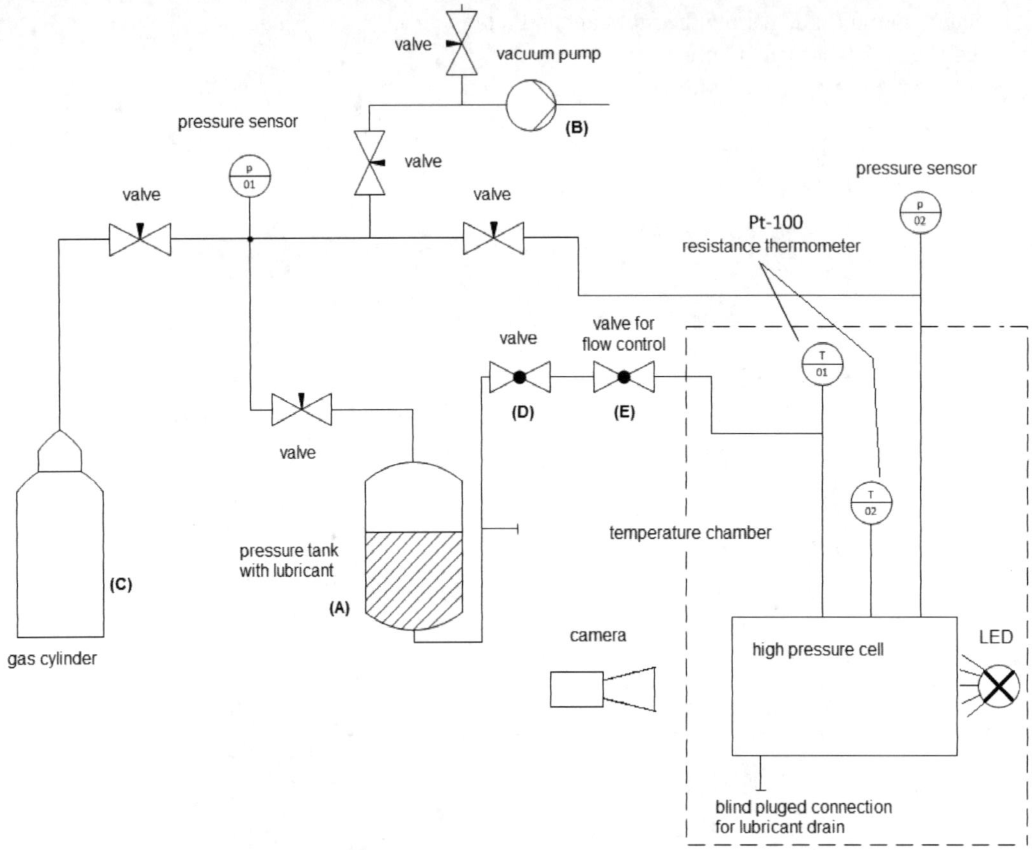

Figure 3 - Principle design of the experimental test bench

3.3 Computational analysis of the surface tension with the pendant drop method

For the analysis of the drops a computational algorithm for the solution of the Young-Laplace equation was used. The solution of the algorithm is well described by (Berry, Neeson, Dagastine, Chan, & Tabor, 2015) who also developed an open source analysis program called "OpenDrop". The basic idea of the program is the detection of the edge of a binarized picture and to calculate the volume and shape of the drop by using dimensionless variables. An example of the analysis is shown in Figure 4. At first the pictures are trimmed, so that the drop and the capillary are shown, see Figure 4, picture 1. The picture is then binarized so that the edges of the drop and the capillary are shown as black contours, picture 2. Since the histograms of the pictures, see Figure 4, picture 1, have two maxima with a few values in between them, the data loss is minimal. Tests have shown that a different threshold value of ±10 % of the original value for the binarisation leads to uncertainties of the analysis lower than 0.5 % of the surface tension.

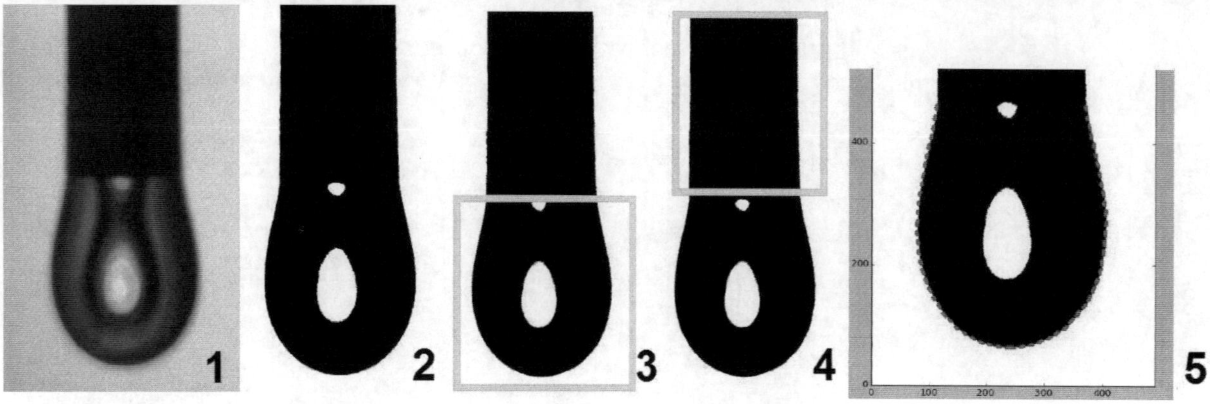

Figure 4 - principle of edge detection and analysis for the drop calculation

After the binarisation the program "OpenDrop" needs the input areas for the drop, see Figure 4, picture 3, and the capillary, picture 4, which are shown as a green rectangles. The outer diameter of the capillary is also an initial value for the calculation of the drop dimensions. Besides this the density values of the liquid and the gas phase have to be known. These parameters however are just used for the calculation of an absolute value of the surface tension of the liquid phase. After all parameters are entered the edge of the drop is detected and the dimensionless parameters are calculated as shown by (Berry, Neeson, Dagastine, Chan, & Tabor, 2015). The calculated shape of the drop is shown to the user as a doted line for control purposes (Figure 4, Picture 5). As shown by the analyses of (Berry, Neeson, Dagastine, Chan, & Tabor, 2015) only drops with a high Bond- and Worthington-Number lead to a low scattering and high certainty of the measured values which means that drops with a high volume have to be produced. Only drops with a Bond-Number higher than 0.5 and a Worthington-Number higher than 0.7 were used for their analysis. For pressures near the critical pressure of carbon dioxide the drops become much smaller so that the Bond-number decreases. That is why also drops with lower Bond-numbers were used in this investigation.

4. EXPERIMENTAL RESULTS

4.1 Surface tension of different lubricants in carbon dioxide atmosphere

As shown in equation (9) the surface tension can be calculated for different boundary conditions. Since especially the effect of higher pressures on the surface tension of mixtures is not well discussed in the literature isothermal measurements of the surface tension for different pressures were initiated. The carbon dioxide had a purity of 99.9%. Three different POE (polyolefin ester) lubricants were used. Lubricant A and C were POE of ISO-VG 32 and lubricant B a POE of ISO-VG 22. The results are shown in Figure 5. The Laplace constant and the capillary constant can be expressed as a second-order polynomial function of the pressure. The extrapolation shows a reasonable behavior so that it can be assumed that the constants can be precalculated with a low number of measured values. As shown in Figure 5 the difference between the three analyzed lubricants is small. The lubricants A and C with a higher viscosity do not tend to a much different Laplace or capillary constant than lubricant B. The variations of the measurements of the Laplace and capillary constant for constant pressures are also small, see Figure 5. The relative differences of the measurements are smaller than 1 % for Lubricant B. For Lubricant A and C especially at pressures higher than 50 bar the differences increase. Lubricant A has a maximum deviation of lower than 1 % for the Laplace constant and smaller than 2 % for the capillary constant for pressure lower than 50 bar. For higher pressures the deviations increase up to 6 % for the Laplace constant and up to 11 % for the capillary constant (see also Figure 5). Lubricant A has also a maximum of the Laplace constant at about 5 bar which is assumed that this caused by the wetting effects of the capillary at this pressure. Lubricant C has a maximum deviation of 0.5 % for the Laplace constant and 0.9 % of the capillary constant at pressures lower than 50 bar. At higher pressures the deviation increases up to 1.3 % for the Laplace constant and 2.6 % for the capillary constant. As shown in equation (9), the behavior of the mixtures can be described with simple polynomial functions. However the temperature effect still needs further investigation, since the solubility of the gas in the lubricant decreases with higher temperatures. The differences of the Laplace constants and capillary constants for the three analyzed lubricants are small. That is why only density data of the lubricant carbon dioxide mixtures is needed to calculate the surface tension of the mixtures, see equation (4) and (5).

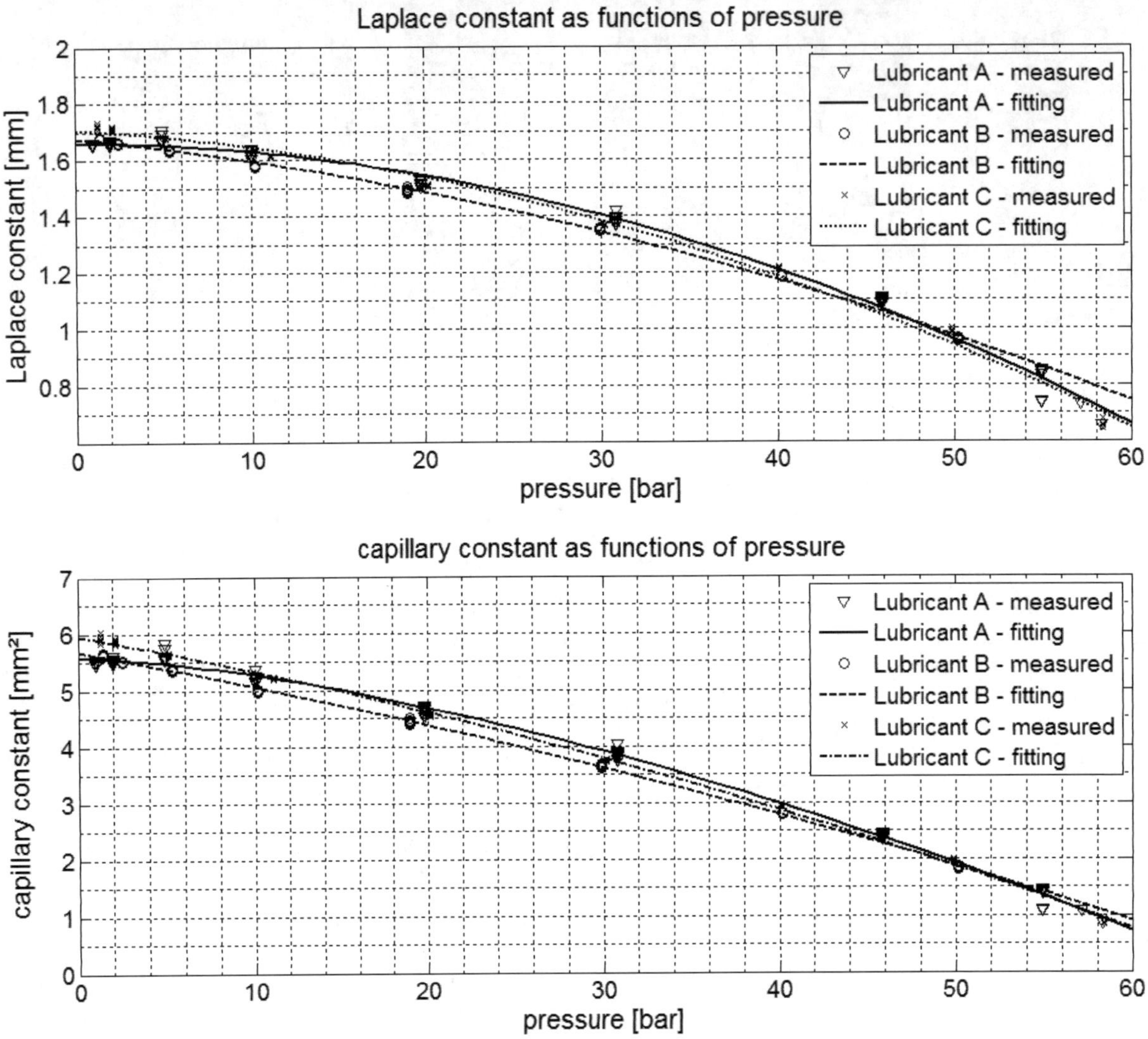

Figure 5 – Laplace and capillary constant as functions of pressure for different pressures at 293.15 K

4.2 Uncertainty of the measurements

Commercial lubricants were used for the measurements. Thus production-related variations of the properties are possible. Additionally the amount of anti-foam additives in the lubricant is not known which means that the surface tension of the base oil can lead to other results. Other effects that were observed while measuring the surface tension of the pure lubricant at different temperatures were the wetting of the capillary, as shown in Figure 6 A to E, and so called "melting" of the drop at low temperatures, as shown in Figure 6 I to V. As shown in A to B the drop is rising on the outside of the capillary upwards. The drop is then growing on the outside of the capillary (C). This leads to problems when analyzing pictures like D or E because the outer diameter of the wet capillary is unknown. When this effect occurred, the capillary was removed and cleaned with a cleaning solvent. At lower temperatures effects like a melting drop, I to V in Figure 6, were observed. The kinematic viscosity is about 50 times higher than at room temperature and the density is about 8 % higher. Since these effects could not be observed at higher temperatures, it can be assumed that at high viscosities the flow resistance increases and internal Reynolds-Number of the drop decreases so that the drop is not stable anymore, see Figure 6. Calibration measurements with water, ethanol and acetone show maximum deviation at different temperatures of about 5% compared to values from (Lemmon, Huber, & McLinden, 2013) for temperatures from 243 up to 373 K. The wetting effects occurred more frequently with small outer diameters of the capillary. Especially at high pressures Bond and Worthington numbers greater than 1 were observed. These values were not used since Worthington numbers are defined as equal or lower than 1.

Another uncertainty is the quality of the edge detection of the drops. To reduce the effect of different Bond numbers drops with different radii R_0 and different Worthington numbers should be analyzed.

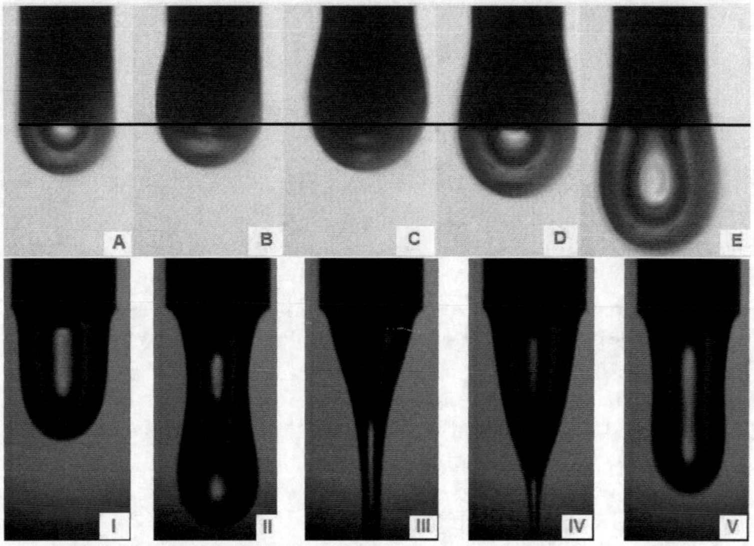

Figure 6 – Wetting and melting effects with the pendant drop method

5. CONCLUSIONS

The measurement and the calculation of the surface tension for different low viscose lubricants in carbon dioxide atmospheres are explained in this paper. The functional dependency of the surface tension and the Laplace constant are discussed and compared to measurements. The measurements show that the Laplace constant and the capillary constant can be easily described for isothermal data sets. With the developed test bench the effects of other gas and lubricant combinations can be analyzed at different temperatures and pressures. Further investigations are required to confirm the dependencies for different boundary conditions. Effects of wetting capillaries and melting drops need further investigation about how to measure the surface tension and of possible effects on the lubrication in compressors.

NOMENCLATURE

a	Laplace constant	(m^2)
α^2	Capillary constant	(mm^2)
Bo	Bond-Number	(-)
ϕ	Drop tangent angel	(-)
s	Arc length	(-)
σ	Surface tension	($N \cdot m^{-1}$)
r	Adjustable parameter for mixtures	(-)
R	Radius	(m)
Δp	Pressure difference (Laplace pressure)	(Pa)
p	Pressure	(Pa)
z	Drop height	(m)
g	Gravitational acceleration	($9.80665\ m \cdot s^{-2}$)
ρ	Density	($kg \cdot m^{-3}$)
T	Temperature	(K)
n	Exponent for surface tension calculation	(-)
x	Mole fraction refrigerant in liquid phase	($mol \cdot mol^{-1}$)
Subscript		
c	critical state point	
mix	mixture	

ref	reference state point
s	saturation state
′	saturated boiling liquid
″	saturated dry steam
0	equivalent spherical values

REFERENCES

Bachu, S., & Bennion, D. (2009). Interfacial Tension between CO2, Freshwater, and Brine in the Range of Pressure from (2 to 27) MPa, Temperature from (20 to 125) °C, and Water Salinity from (0 to 334 000) mg L-1. *J.Chem.Eng.Data, 54*, 765-775.

Berry, J., Neeson, M., Dagastine, R., Chan, D., & Tabor, R. (2015). Measurement of surface and interfacial tension using pendant drop tensiometry. *Journal of Colloid and Interface Science, 454*, 226-237.

Chun, B.-S., & Wilkinson, G. (1995). Interfacial Tension in High-Pressure Carbon Dioxide Mixtures. *Ind.Eng.Chem.Res., 32*, 4371-4377.

Davida, S., Sefianea, K., & Tadristb, L. (2007). Experimental investigation of the effect of thermal properties of the substrate in the wetting and evaporation of sessile drops. *Colloids and Surfaces A: Physicochemical and Engineering Aspects, 298(1-2)*, 108-114.

Di Nicola, G., & Moglie, M. (2011). A generalized equation for the surface tension of refrigerants. *International Journal of Refrigeration, 34*(4), 1098 - 1108.

Di Nicola, G., & Pierantozzi, M. (2013). Surface tension prediction for refrigerant binary systems. *International Journal of Refrigeration*, 562-566.

Eggers, R., & Jaeger, P. (1994). Der Einfluß von Grenzflächenerscheinungen auf den Stoffübergang und die Prozessführung in Gegenstromkolonnen mit überkritischem CO2. *Wärme- und Stoffübertragung, 29*, 373-377.

Lemmon, E., Huber, M., & McLinden, M. (2013). "REFPROP 9.1." NIST Standard Reference Database 23.

Reid, R. C., Prausnitz, J. M., & Poling, B. E. (1987). *The properties of gases and liquids.*

Saad, S., Policova, Z., & Neumann, A. (2011). Design and measurement of pendant drop methods for surface tension measurement. *Colloids and Surfaces A: Physiochem. Eng. Aspects 384*, 442-452.

Semenyuk, Y., Zhelezny, V., Ivchenko, D., & Geller, V. (2011). Refrigerant/Lubricant Mixtures: Problems of Application and Property Research. *In: Proc. 23rd IIR Internal Congress of Refrigeration*, (pp. Paper ID 191, 21-26). Prague, Czech Republic.

Vetere, A. (1992). Again the Rackett equation. *The Chemical Engineering Journal*, 27-33.

Yokozeki, A. (2001). Solubility of Refrigerants in Various Lubricants. *International Journal of Thermophysics, 22(4)*, 1057-1071.

Zhelezny, P., Zhelezny, V., & Scripov, V. (2003). Determination of the pseudocritial parameters of refrigerant/ oil soultions. *Fluid Phase Equilibria, 212*, 285-302.

Zhelezny, P., Zhelezny, V., Procenko, D. A., & Ancherbak, S. N. (2007). An experimental investigation and modelling of the thermodynamic properties of isobutane–compressor oil solutions: Some aspects of experimental methodology. *International Journal of Refrigeration, 30(3)*, 433-445.

Zhelezny, V., Semenyuk, Y., Ancherbak, S., Grebenkov, A., & Beliayeva, O. (2007). An experimental investigation and modelling of the solubility,density and surface tension of 1,1,1,3,3-pentafluoropropane (R-245fa)/synthetic polyolester compressor oil solutions. *Journal of Fluorine Chemistry,128*, 1029–1038.

Zheleznya, V., Sechenyhb, V., Ivchenkoa, D., & Yury, S. (2014). Prediction of the surface tension for refrigerants and refrigerant-oil solutions (ROS). *International Journal of Refrigeration, 40*, 241-245.

The Evolution of Polymer Bearing Processing for Scroll Compressor Performance and Life Enhancement

Michael Kim[1], Derek Marsella[2], Benoit Sidot[3]

1. GGB Bearing Technologies, Thorofare, NJ, USA
2. GGB Bearing Technologies, Annecy, France
3. GGB Bearing Technologies, Clichy, France

ABSTRACT

Polymer bearings with a PTFE-based running surface were first introduced as a breakthrough in scroll compressor performance approximately 30 years ago. These bearings consisted of the polymer surface, porous bronze interlayer and steel supporting layer. That construction has become widespread in compressor applications, as the bearing composition and performance have advanced, with improved environmental impact. Since that introduction, polymer application to the bronze bearing interlayer has typically relied on the deposition of a composite slurry to produce the low-friction, wear resistant surface. While that deposition method has served the compressor application well, bearing life was basically limited by the polymer integrity and layer thickness due to the inherent nature of the slurry-based process.

A new material technology is now available for use in scroll compressor bearing manufacture to provide notably improved performance both in durability and seizure resistance in comparison to the historical slurry-based deposition method. These performance improvements have been documented through wide-ranging tribological testing in general tribological testing conditions as well as methods relevant to scroll compressor conditions.

A comparison of the bearing materials manufactured by multiple process technologies is provided as well as tribological results documenting the performance enhancement. Test data presented include results from the bearing development and manufacture and also qualification tests by a prominent compressor manufacturer, documenting improvements in both durability and seizure resistance.

1. INTRODUCTION

Multilayer PTFE-lined bearing materials have been adopted successfully across a broad application base, primarily because the filled-PTFE running surface provides superior self-lubricating properties. The inherent low friction associated with this polymer has been used to increase efficiency in applications impacted by sliding friction, and to reduce bearing wear in systems that are susceptible to adhesive wear.

The classic three-layer configuration for this bearing type is shown in Figures 1 and 2. This structure consists of a porous bronze inner-layer sintered to a steel backing, and impregnated with a filled-PTFE compound to purposely leave a film of the same composition remaining on top. This upper layer, or "overlay" is fundamental to bearing operation. Figure 1 shows the bearing with a PTFE and lead composition, the original multilayer polymer bearing innovation in scroll compressors. Figure 2 shows a lead-free alternative that has replaced the former in recent years for some applications to comply with increasing environmental requirements.

A longstanding manufacturing approach for this class of materials (1,2) has been to co-coagulate a PTFE-based aqueous dispersion with desired particulate fillers, creating a slurry-based compound that could be impregnated into the porous bronze structure. Subsequent incremental developments around this process led to a dry powder process that together with compositional improvements improved properties such as cavitation resistance and fatigue resistance (3).

Size adjustment after installation for both slurry-based and powder-based liner systems is limited to burnishing for both materials. It was demonstrated that the powder-based liner was able to undergo significant modification through burnishing while maintaining tribological performance. But resizing via machining for both types resulted in a significant life reduction and higher friction through increased exposure of the bronze interlayer.

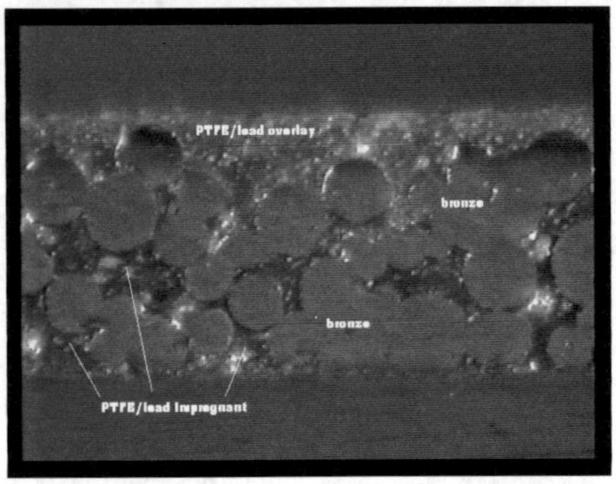

Figure 1: Cross-sectional image showing the multi-layered bearing construction of the. Image shows a bearing of PTFE/lead composition, the original bearing of this type utilized in scroll compressors.

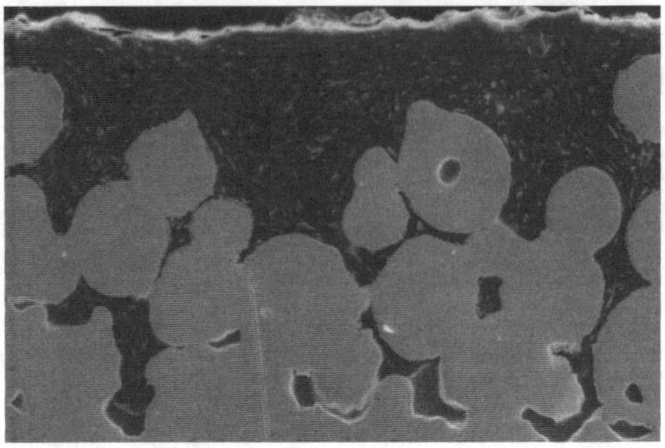

Figure 2: Cross-sectional image of a lead-free composition with MoS_2 filler.

The latest innovation in the manufacture of PTFE-based tribological layers has been through the impregnation of higher-integrity polymer films (5,6). These films provide the ability to build a thicker polymer overlay that delivers increased resistance to key tribological conditions including sliding wear and cavitation erosion. The historical overlay thickness was typically in the range of 5 to 25 µm thickness, with Figure 3 showing such an example with a layer of 17 µm. The enhanced overlay version is nominally over 100 µm thickness, with Figure 4 showing a thickness of 126 µm. This thicker overlay overcame the sizing limitations of its predecessors, permitting machining after bearing installation to a precision equivalent to metallic materials. Additionally, it has also provided favorable tribological improvements in a number of conditions relevant to scroll compressor operation.

Figure 3: Cross-section of historical bearing construction showing limited polymer overlay thickness, in this case 17 µm.

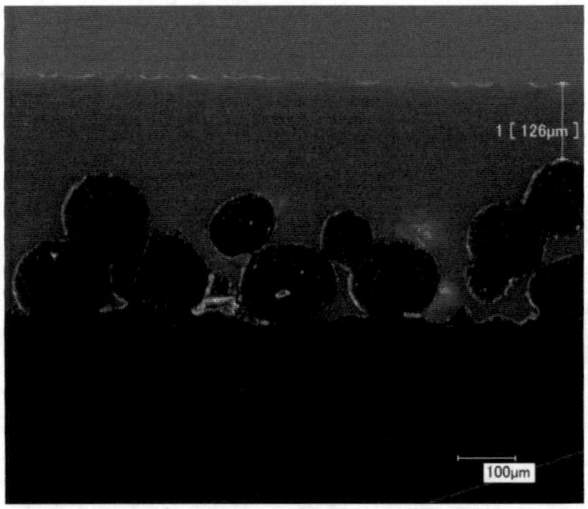

Figure 4: Bearing construction showing machineable polymer overlay, in this case 126 µm.

2. EXPERIMENTAL PROCEDURE

2.1 Materials Tested: Each material consisted of a classic metal / polymer composite structure. That is, a steel backing, porous bronze interlayer and a PTFE-based polymer impregnant, within the porous structure with a polymer overlay as the initial running surface. Three individual materials were tested:

- Slurry-based PTFE layer: MoS_2 particulate filler, balance PTFE; cross-section shown in Figure 2 [product DP10™]
- Powder-based PTFE layer: CaF_2, a second fluoropolymer addition and colloidal Al_2O_3 particulate addition, balance PTFE (3); cross-section shown in Figure 3 [product DP31™]
- Film-based PTFE layer: CaF_2, balance PTFE; cross-section shown in Figure 4 (5,6) [product DTS10™]

Note that the slurry and powder-based polymer layers were produced with a polymer overlay of approximately 25 um, designed to be used in the as-impregnated state. By comparison, the film-based PTFE layer was designed to have the capability to be machined, thus containing a polymer layer of 100 μm or greater. These overlay differences are visible in the cross-sectional images of the various materials. All performance data are presented in the as-formed bearing condition, without secondary sizing operations.

2.2 Test Methods

Two sets of tribological evaluations were conducted. One series was conducted by the bearing manufacturer, GGB Bearing Technologies. A second set was run by compressor manufacturer Daikin Industries, Ltd. As a result, the test descriptions, and results presentation are identified as being by the bearing manufacturer or by the compressor manufacturer.

2.2.1 Bearing Manufacturer Test Conditions:

2.2.1.1 Cavitation Erosion: Testing was conducted on flat samples of composite, polymer-lined discs. The cavitation test rig is shown in Figure 5. Samples were held magnetically in place 1mm below a high frequency sonic horn as shown, immersed in room temperature water. The horn operated at 26 kHz for a pre-determined duration that was determined by the material under test. Material loss was quantified by weighing samples before and after the test. Data are presented as an erosion rate, mg/min.

2.2.1.2 Boundary Lubricated Wear Testing: Wear testing was performed on the rig shown in Figure 6. These tests also used flat samples of the polymer-lined discs as test samples. The test head shown in Figure 2 consists of three hemispherical counterbodies ("shoes") held in a rotating fixture and forced to slide against the flat test specimen. The lubrication regime was altered from boundary lubrication (material / material contact) to a hydrodynamic film (no material contact) based upon speed, load, temperature and oil type. Individual conditions are listed in Table 1.

Table 1: Wear Test Conditions				
Test Regime	Load	Speed	Temperature	Oil
Boundary lubrication	14.5 MPa	0.163 m/s	95°C	Type F

Note that the oil type (Type F transmission fluid) was chosen primarily because of the viscosity and tendency to form a specific type of lubrication regime; not to simulate a specific application.

Figure 5: Cavitation erosion rig

Figure 6: Wear test rig showing the three "shoe" test configuration

2.2.2 Compressor Manufacturer Analysis: The same three materials were also evaluated by using test conditions considered relevant to those in a scroll compressor. The conditions reported here evaluated seizure resistance and general wear resistance. Both seizure and wear tests utilize a bushing configuration of 30 mm ID x 20 mm length.

Table 2: Test parameters	Seizure resistance*	Wear resistance
Load	step load to 31MPa	5.9 MPa
Speed	11.3 m/s	0 m/s (10s) – 2.8 m/s (10s)
Lubricant	24% FVC68D oil w/ R410A	24% FVC68D oil w/ R410A

*Seizure resistance test was stopped at friction coefficient $\geq$ 0.2; temperature $\geq$ 200°C

3. RESULTS

3.1 Tribological Data: Bearing Manufacturer

3.1.1 Cavitation Results: Cavitation testing was conducted on the three materials, and weight loss calculated for each sample grouping. Data are reported in Figure 10 as wear rate (mg / minute) for each of three materials. . The greatest level of erosion loss was measured on the slurry-based liner, followed by the powder-based material. The lowest level of erosion loss was measured on the film based liner. Surface images are also shown in Figures 7 through 9 for examples of each of the three materials at equivalent test duration (6 minutes). Removal of pieces of the overlay may be observed for the slurry-based material in Figure 7. The underlying bronze structure may be seen to have been exposed during the cavitation process. The powder-based liner shown in Figure 8 exhibited reduced exposure of the underlying bronze structure in comparison to Figure 7, and reduced weight loss rate as indicated in Figure 10. The superior cavitation performance of the film-based liner quantified in Figure 10 and supported by the surface image shown in Figure 9. The underlying bronze structure was not exposed, as a result of both the reduced weight loss rate, and the thicker polymer overlay.

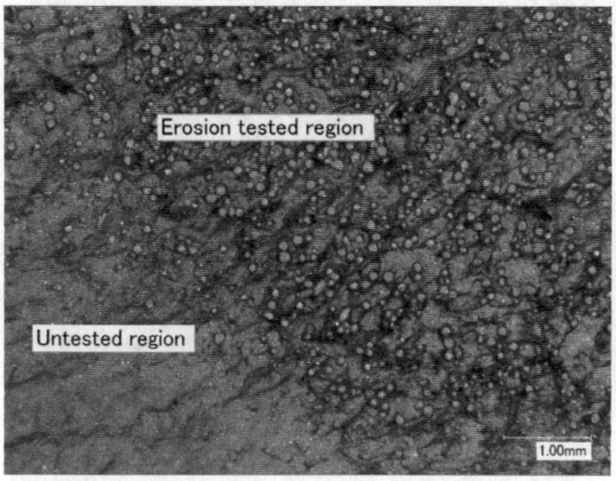

Figure 7: Cavitation erosion test sample, slurry-based liner.

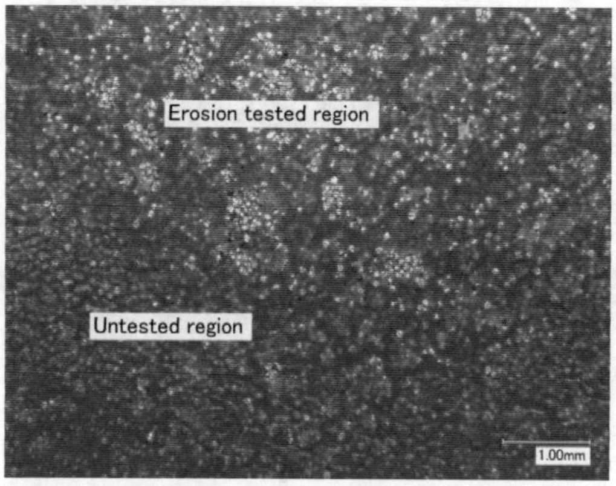

Figure 8: Cavitation erosion test sample, powder-based liner.

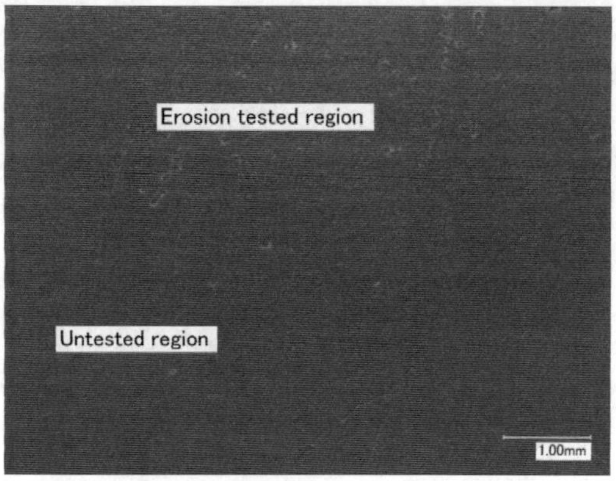

Figure 9: Cavitation erosion test sample, film-based liner.

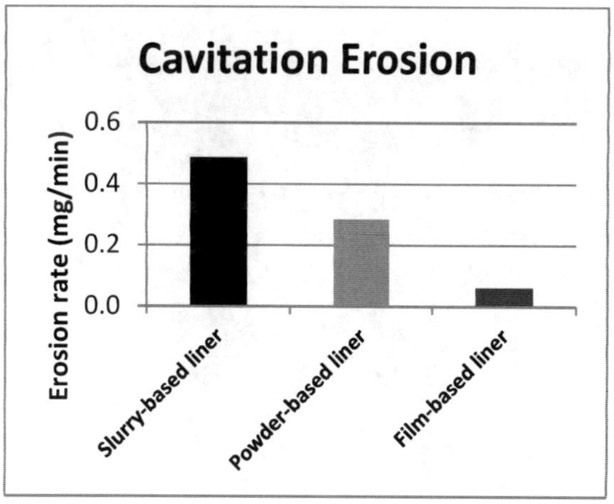

Figure 10: Cavitation erosion data.

3.1.2 **Boundary Lubricated Surface Wear:** Lubricated wear testing was performed on the three sample materials using the three "shoe" test rig shown in Figure 6. Data are shown in Figure 14, presented as wear rate, that is, wear depth (μm) as a function of sliding distance (km). Surface images of the three materials are shown in Figures 11 through 13 indicating the representative surface appearance after a wear test. The overall wear depth for the sample shown is noted in each figure. The slurry-based material experienced the greatest wear rate at 1.9 μm/km. The powder-based product had the next level at 0.66 μm/km with the lowest from the film-based liner at 0.46 μm/km. This lowest wear rate was in addition to having overlay thickness of ≥100 μm, which was significantly greater than either of the other materials (≤ 25 μm). This wear resistance translates directly into extended bearing life in a compressor, with the added advantage of a thicker overlay (even after machining) contributing further to durability.

Surface imagery followed the quantitative wear results in that the material with the greatest wear also showed the greatest disruption of the wear surface. The surface of a tested slurry-based liner in Figure 11 exhibited exposure of the underlying bronze layer. The powder-based liner, shown in Figure 12, exhibited less disruption through wear indicated by only a slight change in bronze exposure between the untested region and the wear track. The film-based product in Figure 13 appeared smooth by comparison with very little disruption following wear testing.

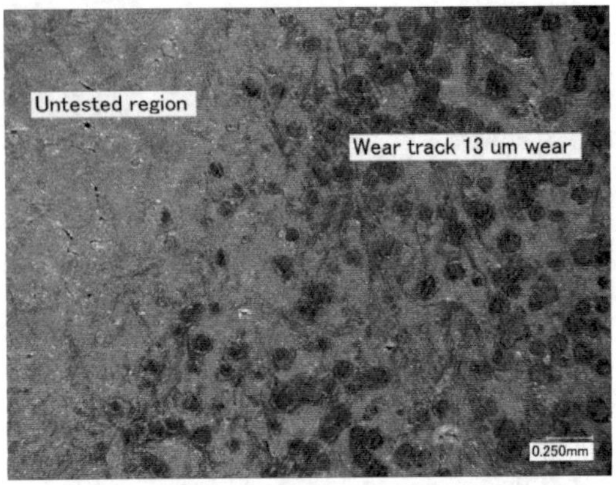

Figure 11: Boundary wear tested sample, slurry-based liner

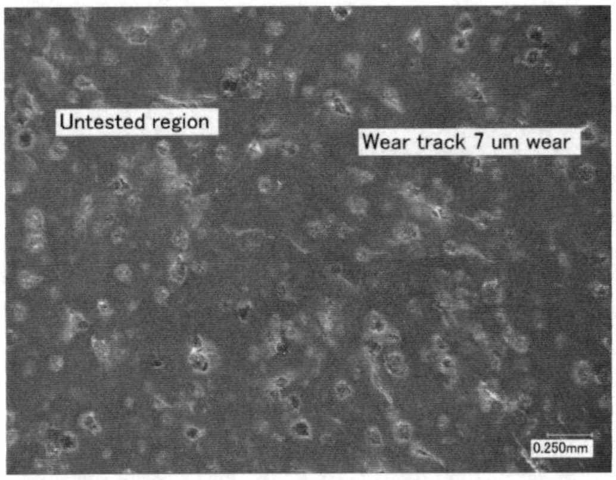

Figure 12: Boundary wear tested sample, powder-based liner

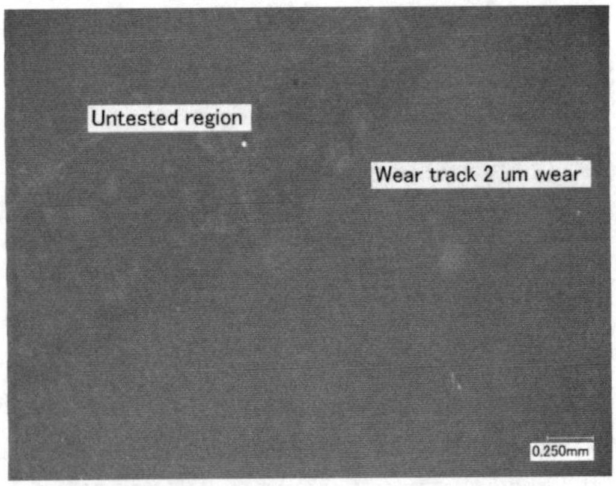

Figure 13: Boundary wear tested sample, film-based liner

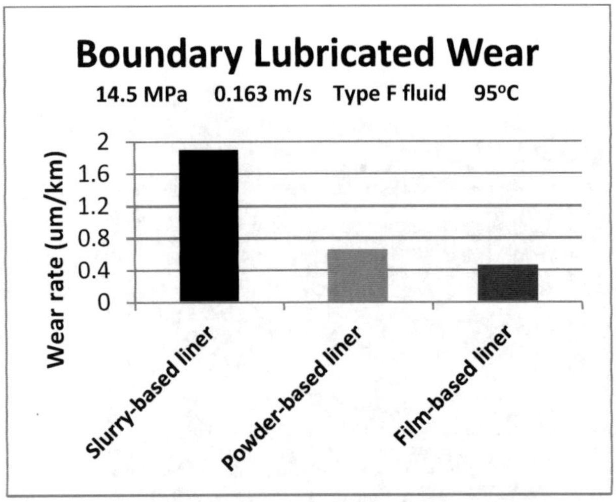

Figure 14: Boundary wear data

3.1.3 **Boundary Lubricated Friction Data:** Representative friction curves are shown in Figure 15 for the three materials. The powder-based and film-based materials show one curve each, as those materials typically exhibit minimal wear at the conditions shown. The slurry-based material, by comparison, shows two friction curves, representing different test responses. The friction curves for the powder-based and film-based materials show that friction was higher initially for those in comparison to the slurry-based liner, but experienced low wear and thus consistent friction. The low wear rate resulted in minimal exposure of the bronze interlayer at the test completion. Figures 12 and 13 show final sample conditions for the powder and film-based samples. In the case of one slurry-based sample higher wear and resulting bronze exposure caused increased friction late in the test. Figures 16 and 17 represent the final condition for two slurry-based samples, with varied levels of bronze exposure. This difference demonstrated the friction increase resulting from bronze exposure compared to the friction of a surface consisting primarily of PTFE.

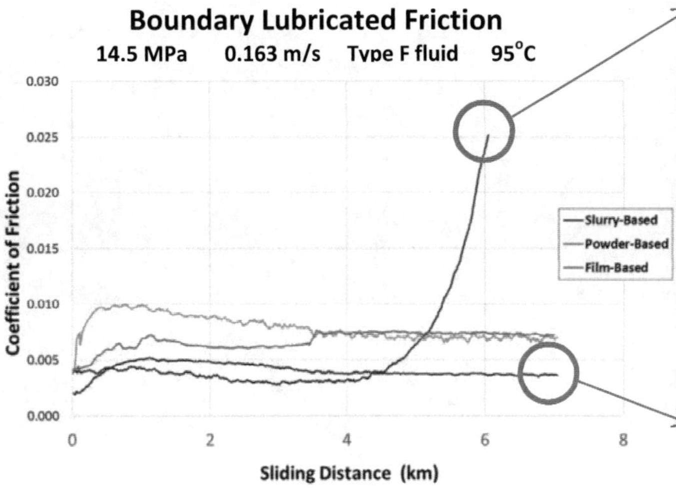

Figure 15: Friction curves from boundary lubricated wear testing

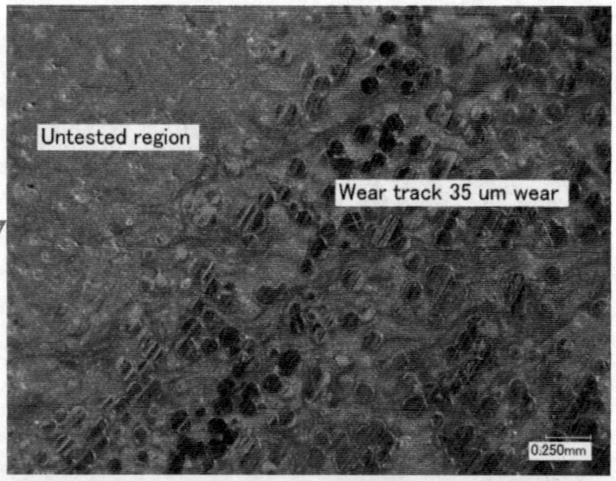

Figures 16 (top) and 17 (bottom) showing surface condition after wear testing for a range of bronze interlayer exposure, and varied friction results.

3.2: Tribological Data: Compressor Manufacturer : Testing was conducted to evaluate both wear resistance and seizure resistance of the three bearing materials listed in Section 2.1. The wear data in Figure 18 generated by compressor manufacturer showed the same order and relative scale of the wear data in Figure 14 by bearing manufcaturer, despite being tested under notably different conditions. Again, the slurry-based material had the highest wear depth and the film-based material the lowest.

Seizure resistance data are shown in Figure 19, presented as the load at which the friction exceeded 0.2 or the temperature exceeded 200°C. The data showed a similar trend as wear and cavitation data with the film-based material exhibiting the most favorable performance.

These data helped to validate test results discussed above, with the film-based liner exhibiting the highest wear resistance. It is also believed that the low and consistent friction over an extended test duration provided the favorable seizure resistance, especially for the film-based bearing material.

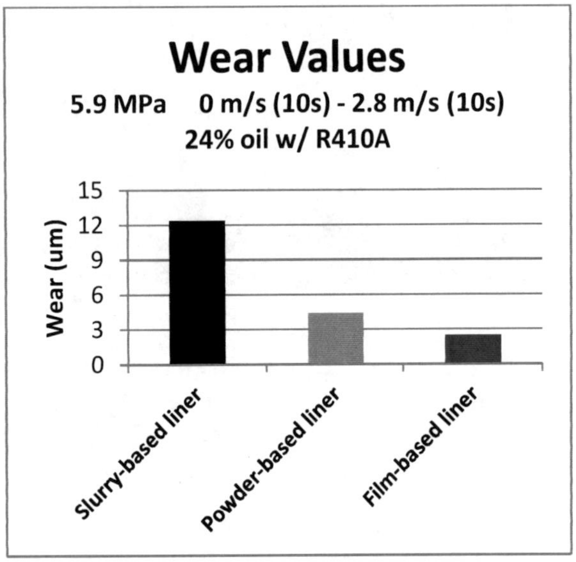

Figure 18: Wear data for compressor manufacturer testing presented as overall wear depth.

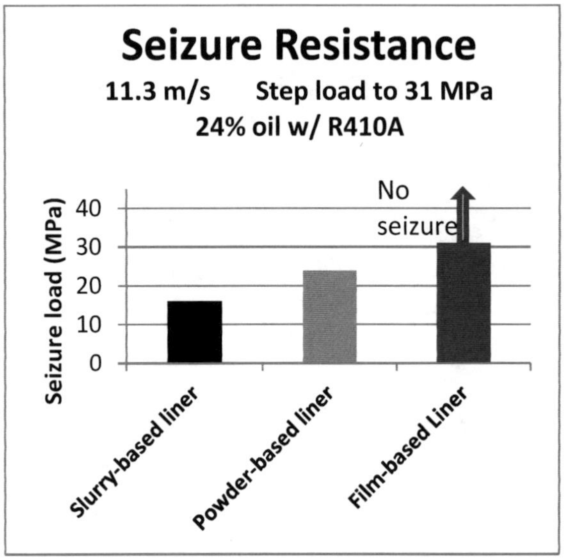

Figure 19: Seizure resistance presented as seizure load. Note that the film-based liner did not seize at the maximum test load.

3.3 Performance Summary: Tribological data were presented comparing bearing materials with differences including manufacturing method, composition and thickness of the polymer wear layer. Wear rate differences were documented, including highly-correlated relative differences between the testing performed in different test locations using varied test methods. Wear rate was lowest for the film-based material, which also had the greatest wear layer thickness. The cavitation erosion rate was also lowest for the film-based material. Seizure resistance was also most favorable for the film-based material. The performance of the powder-based material was favorable to the slurry-based material in each data set and test regime, although wear layer was nominally identical for those two materials.

It is proposed that the manufacturing method by which the polymer layer was produced and applied played a significant role in bearing performance. This enhanced performance would ultimately relate to increased bearing life in scroll compressor applications.

4. CONCLUSIONS

- Three PTFE-lined bearing materials distinguished by their polymer processing methods were tested in wear, cavitation and seizure resistance. The processing method differences concerned the means by which a self-lubricating polymer layer was applied to a supporting porous bronze interlayer. The methods were slurry-based processing, a powder-based process and a film-based method.
- The film-based material had a thicker polymer overlay than the other materials, proving the opportunity for precision machining after installation and/or increased wear life as a result of a greater thickness of the self-lubricating polymer layer.
- Separate tribological evaluations were conducted by both the bearing manufacturer and by a compressor manufacturer. Both sets of results demonstrated the most favorable wear performance for the film-based liner, followed by the powder-based material.
- The film-based polymer liner system exhibited the highest level of wear resistance, cavitation resistance and seizure resistance. The slurry-based material demonstrated the lowest level of resistance to each of the conditions.

5. REFERENCES

1. Davies, G., Moisey, P., Johnston, J., *Plain bearing with polytetrafluoroethylene-based lining*, US Patent 5,911,514 (1999)
2. McMeekin, K., Johnston, J., *Plain Bearing Structure*, US Patent 6,296,392 (2001)
3. McMeekin, K., Johnston, J., *Bearing Materials*, US Patent 6,461,679 (2002)
4. Kim, M., Peng, Y.H., Small, C., *The Tribological Performance of Self-Lubricating Bearings Following Secondary Sizing,* Purdue Compressor Conference (2006)
5. Kim, M., Marsella, D., Johnston, J., *Boreable plain bearing material,* US Patent 7,491,353 (2009)
6. Marsella, D., Belmahdi, S., *Novel Low Friction, Machineable Metal-polymer Bearing for Lubricated Applications (in Fluid Power Systems),* National Conference on Fluid Power (2011)
7. *GGBearing Evaluation Result,* Compressor Engineering and Development Group, Daikin Industries, Ltd. (2015)

ACKNOWLEDGEMENT

The authors would like to thank Daikin Industries, Ltd. for their support in testing and analysis that contributed to the viability of this work.

Evaluating Lubricants for Lower GWP Refrigerant Compressor Operation

Joseph A. Karnaz

CPI Fluid Engineering,
Midland, Michigan, USA
989-496-3780
jska@cpifluideng.com

ABSTRACT

Refrigerant Lubricants are an essential component to refrigerant and air conditioning compressor operation regardless of the refrigerant used. But to maintain reliable and efficient operation, the choice of lubricant can vary depending on the type of compressor, design of the system and especially the kind of refrigerant.
Historically, refrigerants used in applications have changed over the period of time since the invention of refrigeration due to various reasons. Today the global warming potential (GWP) value that exists for each refrigerant are being scrutinized if the GWP value is deemed too high which could result in a negative environmental impact. One of the so called high GWP refrigerants is R-404A which is on the fast track to be minimized through regional and global phase downs and potential phase outs. Various replacement/substitute refrigerants exist today that have been evaluated or are currently being evaluated as lower GWP choices in place of R-404A refrigerant. For R-404A the typical lubricant of choice has been a synthetic lubricant such as a polyol ester (POE) which meets certain criteria for refrigerant and lubricant interaction. Replacement/substitute lower GWP refrigerants for R-404A will vary in the type of lubricant chemistry that is satisfactory for effective operation resulting in numerous lubricant selections.
This paper will evaluate lower GWP refrigerant options and what lubricant chemistry is best suited. Various rules that apply to what testing is needed to make the appropriate conclusions will be outlined and results of refrigerant and lubricant interaction will be presented. This data will then be used to determine how the changes in lower GWP refrigerant properties will dictate these lubricant options. This type of information will allow for maintaining the required compressor performance parameters associated with the industry.

1. INTRODUCTION

Since the beginning of the use of some type of mechanical refrigeration and air conditioning, the working fluid (refrigerant) used has taken center stage in not only the performance of the system but also how the various properties these refrigerant possess will affect the areas it is used in. Early refrigerants used in the late 1800's and early to mid-1900's like Methyl Chloride, Sulphur Dioxide, Carbon Dioxide, Ammonia and Hydrocarbons carried properties like toxicity, flammability and high pressure that brought about safety concerns evolved to ChloroFluoroCarbon (CFC) and HydroChloroFluoroCarbon (HCFC) synthetic refrigerants with less safety concerns. With the industrial revolution came the apprehension of how the environment was being effected by emissions which led to the investigation of CFC and HCFC refrigerants effect on the ozone. These refrigerants deemed as having some level of Ozone Depletion Potential (ODP) were soon mandated by protocols and laws to be phased out or phased down during a certain time period over the course of the late 1980's until today. Once again, despite the benefits of performance properties, these refrigerants displayed the external environmental concerns require other options. Thus HydroFluoroCarbon (HFC) refrigerants which met safety concerns and ODP environmental concerns were developed along with expanded development on so called natural refrigerants (carbon dioxide, ammonia and hydrocarbons) to try to identify refrigerants that would maintain some level of performance while having no ODP. Systems in which these refrigerants are used will have varying levels of refrigerant leaking into the environment which can have a direct effect on the environmental response. Over time studies have claimed that the HFC refrigerants that migrate into the atmosphere have a relationship to the climate change due to the Global Warming Potential (GWP). This now has implemented another change to refrigerants that have a lower GWP which once again include natural refrigerants but also new synthetic refrigerants based on HydroFluoroOlefin (HFO) chemistry. Table 1 shows some of the history of refrigerants based on safety, environmental concerns and other observations.

Table 1: Refrigerant history information

Refrigerant	ODP	GWP	Safety and Other Considerations	Misc.
Methyl Chloride	0.02	9	Toxicity and Flammability concerns.	More recently this refrigerant designated as R-40 has caused some issues due to reactivity.
Carbon Dioxide	0	1	Non-toxic and high pressure gas.	Lower energy efficiency when used at higher ambient temperatures has been a limitation.
HC R290	0	~20	Non-toxic and highly flammable.	The flammability has led to maximum charge limits be set for use in various types of equipment.
CFC R12	0.82	10,900	Non-toxic and non-flammable.	Effect refrigerant based on system operation but ozone depletion potential has led to its removal.
HFC 134a	0	1430	Non-toxic and non-flammable.	Stable refrigerant chemistries but this stability results in too long of atmospheric life.
HFO 1234yf	0	<1	Non-toxic and mild flammability.	Lower GWP values sacrifice some stability when compared to HFC refrigerants.

In addition to the changes in refrigerants over the years, the mechanical mechanism mainly a compressor, has also evolved and expanded to meet certain requirements. Typically these compressors require some type of lubrication to maintain bearing integrity and efficient operation, so various types of lubricants are employed. Designing a lubricant for a particular compressor design and application is basically set by the design and application and may not vary significantly as the same type of compressor changes in size. This cannot be said though for changes to the refrigerant, over the years as we have seen changes to the refrigerants used we have also seen changes to the lubricants required. Lubricant and refrigerant properties like miscibility, solubility, stability are key to the successful and efficient operation of the compressor and system. Some of these lubricant changes over the years caused by refrigerant changes have been small but some have been very significant with an exhaustive amount of evaluation needed. Regardless of the level of change, it is important to understand the requirements needed for each refrigerant to be able to recommend the appropriate lubricants. This concept is no different today with some of the lower GWP refrigerants that are currently being targeted for use.

This paper will investigate some key lower GWP refrigerants either in the market today or being evaluated for future use to identify potential lubricant candidates. Lubricant and refrigerant properties will be discussed and compared to baseline refrigerants and lubricants used today. Other industry work on compressors and systems using some of these lower GWP candidates will also be discussed. The focus will be on alternate lower GWP refrigerants as substitutes/alternatives for HFC refrigerant R-404A.

2. LUBRICANT AND REFRIGERANT INTERACTIONS

Understanding of the lubricant and refrigerant interaction is needed to create an operating system that is both reliable and has maximum energy efficiency. When refrigerants change, a bench study of the interaction is the first start to matching the right lubricant to the refrigerant. The interaction studies can be broken down into various areas of interest each providing both a first look approach and potential final approach to making a lubricant choice. These tests can provide a significant amount of information in a relatively short period of time which will benefit the lubricant supplier along with the compressor and system manufacturer. After bench studies, it is sometimes helpful to obtain other information based on initial compressor or system testing to help decide on direction.

2.1 Lubricant and Refrigerant Interaction - Miscibility

Typically without proper oil separation devices or techniques, lubricants used inside of a compressor sump will migrate into the system by way of the refrigerant or movement (velocity). Once out into the system, it is imperative that the oil move back to the compressor to eliminate any potential bearing lubricant starvation issues and minimizing of heat transfer degradation on the heat exchange surfaces. A quick screening tests can be performed either in sealed glass tubes or windowed cells, for higher pressure situations, to evaluate the interaction of the lubricant with the refrigerant. Lubricants at various concentrations can be viewed over specified temperature ranges to see when the mixture separates into two phases which is considered immiscible or stays one phase which is considered miscible. Figure 1 is a rendering of a typical miscible curve. These evaluations can be made quickly to understand if a current lubricant is going to be acceptable with a new refrigerant or if lubricant modifications are required.

Figure 1: Miscibility of lubricant and refrigerant

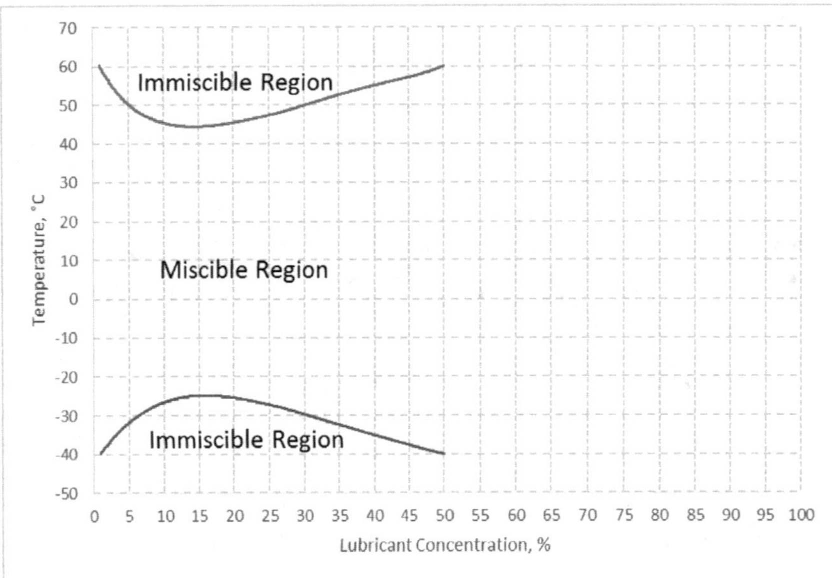

2.2 Lubricant and Refrigerant Interaction - Solubility

Whereas miscibility measurements are usually more of a concern for operations inside the system, solubility is usually a measurement valued inside the compressor or where the oil is being distributed. Concentration levels of refrigerant in the lubricant are measured and then the resulting pressure at varying temperature is measured and plotted. Figure 2 is a representation of a typical pressure vs temperature solubility plot. With this data and knowing operating conditions within a compressor system, the resulting dilution factor the refrigerant will have on the lubricant can be calculated. This dilution factor is important in determining proper lubrication properties that a bearing will experience at conditions. Other dilution information is also important to compressor system design and operation.

Once the solubility is determined, then this will translate into a viscosity of the combination of lubricant and refrigerant sometimes referred to as the "working viscosity". Figure 3 is a typical representation of a plot relating solubility of the refrigerant in the lubricant to working viscosity. The data plotted this way is sometime referred to as a pressure-viscosity-temperature plot or "Daniel Plot". These types of plots are essential to compressor design engineers to get a quick look at what value of film thickness that might be seen at a bearing with a particular lubricant and refrigerant to provide a screen before compressor testing is done. Today sophisticated equipment is used to make measurement of lubricant and refrigerant solubility and working viscosity values from which models can be developed to solve for different operating conditions. Figure 4 shows a picture of the environmental chamber and schematic of the equipment used to measure solubility and working viscosity of refrigerant and lubricant mixtures.

2.3 Lubricant and Refrigerant Interaction – Compressor and System Studies

Screening tests and system tests lasting 2-4 weeks can be used to help make intial studies of not only the candidate refrigerant but also a quick evaluation study of the lubricant. Compressor calorimeter and endurance testing can predict the effective capacity and efficiency that the new refrigerant provides along with expected durability. The lubricant choice can also affect the compressor efficiency and the endurance tests are the best way to identify the optimized lubricant formulation. System testing evaluates the overall benefits or deficiencies that refrigerant testing shows. System testing can also assess how the lubricant is affecting performance based on parameters like oil circulation rates and lubricant-refrigerant interaciton.

Figure 2: Solubility

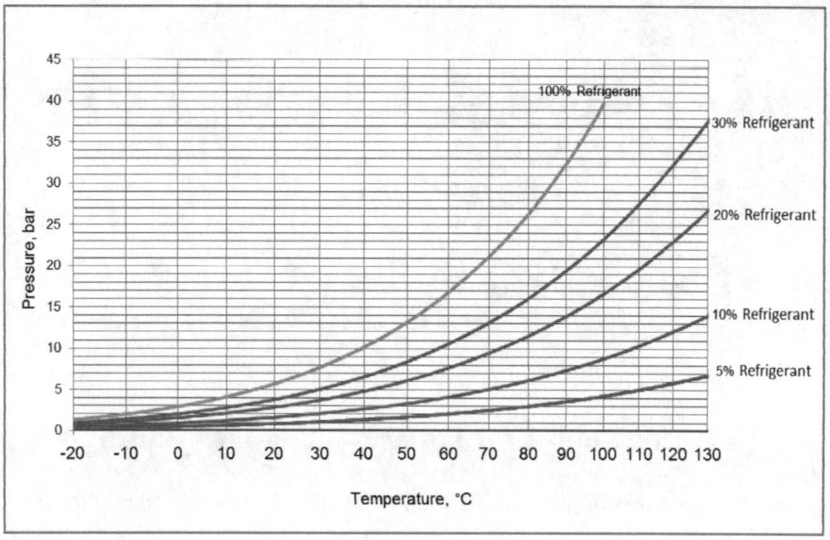

Figure 3: Pressure-Viscosity-Temperature (PVT) curve

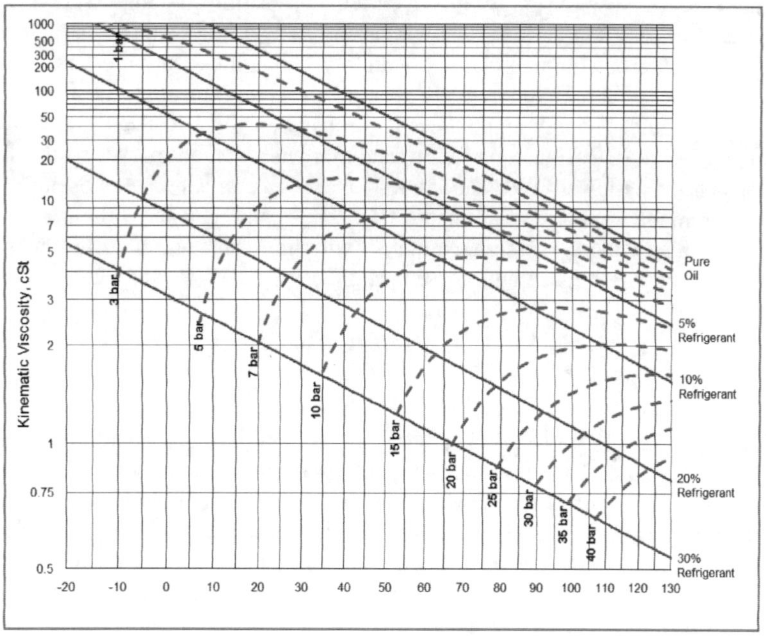

Figure 4: PVT Apparatus

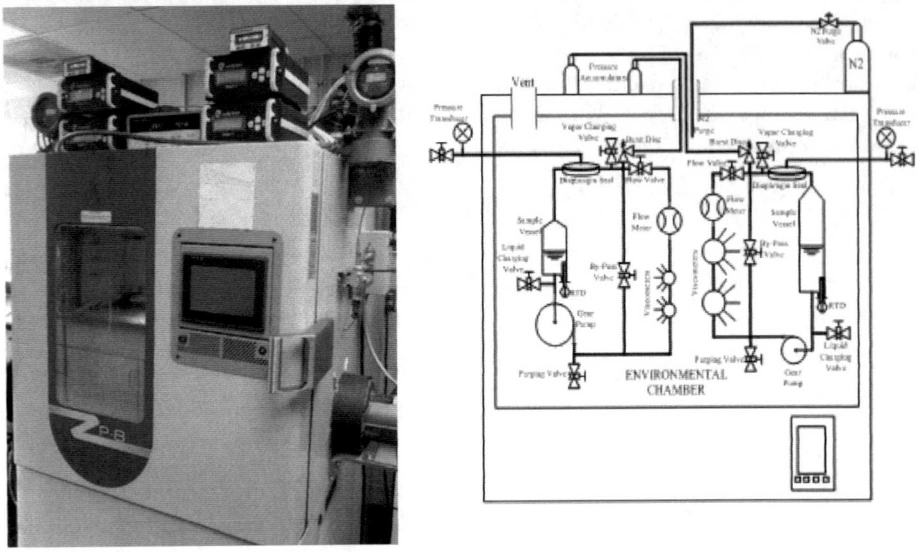

3. Lubricant Studies for Lower GWP R-404A Substitutes

R-404A refrigerant is used in a number of low and medium temperature refrigeration applications ranging from commercial systems using smaller capacity reciprocating compressors to larger semi-hermetic compressors and scrolls used in supermarket rack systems. Since the GWP value of R-404A is one of the largest amongst HFC refrigerants used today, there is pressure to reduce the use of this refrigerant (phase down) to potential phase outs. The European Union has restricted use of refrigerants in certain applications having GWP values above 2500 starting in 2020, the so called F-gas Regulation. The U.S. EPA has also followed similar actions to de-list R-404A as an acceptable refrigerant for certain commercial applications. Given the recent pressure to control the use of R-404A, many compressor and system manufacturers are moving ahead with trying to identify solutions ahead on any phase-down or phase-out schedule. Even though some HFC refrigerants like R-407A and R-407F can meet the anticipated GWP level and R-744 (CO_2) has found more and more use in supermarket applications as a transcritical fluid, this paper will focus on other lower GWP candidates. The refrigerants I have chosen to discuss are three HFO/HFC that are non-flammable, a slightly flammable HFO/HFC candidate and a flammable HC candidate. Investigation will be made on how these refrigerants interact with lubricants and if current marketed lubricants used with R-404A are still good candidates to use with these particular lower GWP candidates.

4. HFO/HFC Blended Refrigerants

4.1 R-448A

This refrigerant is a blend of R32/R125/R134a/R1234yf/R1234zeE at percent concentrations of 26/26/21/20/7 and has an ASHRAE safety designation A1 with a GWP value of 1390. As a first look and comparison, the refrigerant was evaluated for miscibility with a number of lubricant candidates based on polyol ester (POE) chemistry that are used today with R-404A refrigerant. Table 2 is a snapshot of miscibility at 10 and 20 percent lubricant concentration in R-448A. OP in the table designates one phase or miscible and TP designates two phases or when the mixture became immiscible. As a comparative, the POE32 reference in the table has a 10 and 20 percent miscibility profile with R-404A that still maintains one phase below -60°C. Reviewing the data it shows that R-448A will have similar miscibility performance with the same lubricants used with R-404A. Given this similarity it would be predicted that based on the miscibility these lubricants could be used in the same type of applications that R-404A is used in without concern for system operating parameters that may minimize oil return and heat transfer.

Table 2: Miscibility Results of Various POE Lubricant Viscosities with R-448A

	POE 22		POE 32		POE 68A		POE 68B	
	10%	20%	10%	20%	10%	20%	10%	20%
20C	OP	OP	OP	OP	OP	OP	OP	OP
10C	OP	OP	OP	OP	OP	OP	OP	OP
0C	OP	OP	OP	OP	OP	OP	OP	OP
-10C	OP	OP	OP	OP	OP	OP	OP	OP
-20C	OP	OP	OP	OP	OP	OP	OP	TP
-30C	OP	OP	OP	OP	OP	OP	OP	TP
-40C	OP	OP	OP	OP	OP	OP	TP	TP
-50C	OP	OP	OP	OP	OP	OP	TP	TP
-60C	OP	OP	OP	OP	OP	TP	TP	TP

Recently a number of studies have been presented on lower GWP refrigerants through the AHRI Low GWP AREP and other independent manufacturer studies. These compressor and system tests are usually done as drop-in tests or with minimal changes. Sometimes these tests can provide some insight to the potential effect the lubricant might be having with the candidate refrigerant. Recently a couple of studies have been done with R-448A refrigerant in comparison to R-404A. Rajendran (2013) showed in a supermarket mock set-up that at a low temperature application (-22°F/-30°C evap) and at a medium temperature (12°F/-11°C evap) that R-448A when compared to R-404A baseline showed a 3% and 3-8% respectably improvement in energy efficiency. This data was consistent with performance measurements made on this refrigerant in comparison to R-404A. For the system test the scroll compressor POE lubricant used with R-404A was used with the R-448A and based on the system test performance results it would appear that the lubricant had no negative effect. In a separate study by Baba and Yamaguchi (2014), carried out in an inverter controlled unit at constant capacity, R-448A was compared to R-404A at condensing temperatures of 45°C to 55°C. Improvements in volumetric capacity between 6 and 10% and in COP between 4 and 7% were measured. Once again for this test the same POE lubricant used in the standard rotary compressor was used with the R-448A test, and the fact that there was no volumetric capacity loss is a good indication that this oil will not be challenged with solubility changes that could affect performance.

4.2 R-449A

This refrigerant is a blend of R32/R125/R134a/R1234yf at percent concentrations of 24/25/26/25 and has an ASHRAE safety designation A1 with a GWP value of 1400. As a first look and comparison, the refrigerant was evaluated for solubility and working viscosity with a POE lubricant candidate and compared to solubility and working viscosity data of the same POE lubricant with R-404A. Figure 5 shows the Daniel Plot for POE32 lubricant with R-449A.

Figure 5: Pressure-Temperature and Viscosity Data for POE32 and R-448A

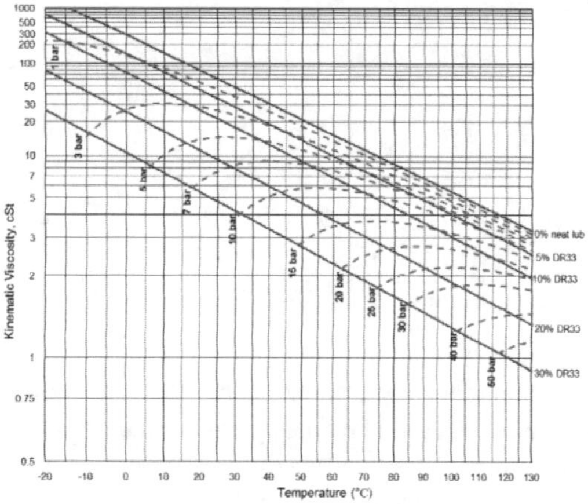

Using this data and Daniel Plot data from POE32 with R-404A a comparison was made at four different pressures and two different temperatures. The comparison data is tabulated in Table 3. Even though there could be some difference in operating conditions between the two refrigerants that might result in slight changes to the numbers, this is a good indication that when using the same POE lubricant for both R-404A and R-449A the working viscosities provided to the bearings will be similar.

Table 3: R-449A and R-404A comparison with same lubricant

40°C	2 Bar	4 Bar	10 Bar	20 Bar	80°C	2 Bar	4 Bar	10 Bar	20 Bar
R-404A					**R-404A**				
Dilution	4%	10%	21%	NA	Dilution	1.6%	5%	8%	20%
Viscosity	24 cSt	15 cSt	6 cSt		Viscosity	8 cSt	7 cSt	5 cSt	2.8 cSt
R-449A					**R-449A**				
Dilution	4%	10%	21%	NA	Dilution	1.5%	4.5%	8%	20%
Viscosity	23 cSt	15 cSt	5.5 cSt		Viscosity	9 cSt	7.5 cSt	4.8 cSt	2.7 cSt

A system study was done with R-449A in comparison to R-404A baseline by Leck and Minor (2014) in a split condensing unit open display food case. A reciprocating compressor using a 32 centistoke POE lubricant was used for both tests changing only to fresh lubricant for the second test after the baseline test. Tests were conducted at a couple different ambient temperatures at both low and medium temperature conditions. The results indicated that R-449A showed 3-4% less energy consumption at low temperature and 8-12% less energy consumption at medium temperature over R-404A tests. These results along with the dilution and viscosity results in Table 3 would suggest that the lubricant has no negative affect on performance when operated with R-449A.

4.3 R-452A

This refrigerant is a blend of R32/R125/R1234yf at percent concentrations of 11/59/30 and has an ASHRAE safety designation A1 with a GWP value of 1945. Initial miscibility and solubility studies were done with the refrigerant and compared to R-404A/Lubricant studies. Current POE chemistry at various viscosities was compared by a manufacturer with R-404A and R-452A to see if the current POE lubricant could be used without complications. Figure 6 is a comparison study with two lubricants and the results indicate that miscibility differences between R-404A and R-452A are small and should not be an issue.

Figure 6: Miscibility study commercial comparison of R-404A and R-452A

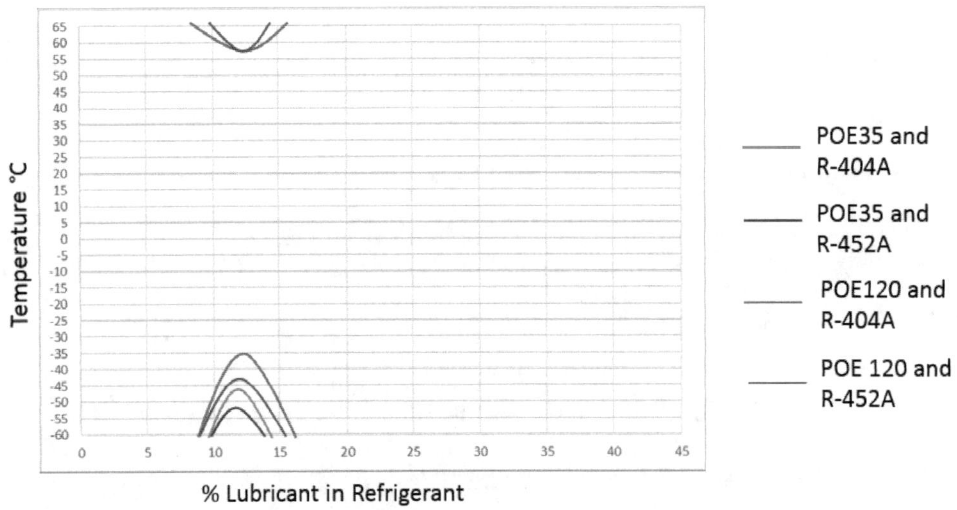

PVT data generated for the two POE lubricants listed in the miscibility tests indicated that these lubricants have similar profiles with the two refrigerants and would be considered similar enough that differences seen in compressor lubrication should also be minimal. Actually, compressor testing was done to confirm what was measured in the PVT data and no issues were reported.

As with the other R-404A substitute refrigerants discussed earlier some studies have been done with R-452A in compressors and systems and reference in literature data. A paper by Kujak *et al.* (2015) showed a system study that was done with a trailer transport refrigeration unit which commercially operates with R-404A. Baseline testing was done with R-404A and after measurements the system was operated with R-452A using the same POE lubricant. Results compared to the baseline study indicate that the volumetric efficiency at higher rpm speed was 2% lower while at lower rpm speed was 0.5% lower. The authors commented that these differences would be considered small enough to say that R-452A has similar performance to R-404A and these results would also indicate that the POE lubricant did not have a negative effect and is an effective product for R-452A.

4.4 R-455A

This refrigerant is a blend of R32/R1234yf/R744 at percent concentrations of 21.5/75.5/3 and has an ASHRAE safety designation A2L with a GWP value of <150. Since this refrigerant has some mild flammability compared to the previous three candidates it is more uncertain at this time about the potential commercialization but most likely will not be an issue. Miscibility tests with R-455A and the same POE32 lubricant mentioned in the previous studies indicates that the combination is fully miscible from 0 to 40% lubricant concentration at temperature measurements from 65°C to -60°C. From experience, lower temperature miscibility usually relates to the refrigerant having an increased level of solubility/dilution, therefore, it was valuable to make comparisons of the solubility/working viscosity of R-455A and the other HFO/HFC blends that were rated A1. Figure 7 are plots showing the working viscosity comparison of R-455A; R-448A; R-449A and R-452A over a range of temperatures at two different pressures with the same POE32 lubricant. The results do indicate that R-455A shows lower working viscosity at some conditions, this does seem to relate to the difference in lower temperature miscibility. There is not enough information to determine if these difference are significant.

Figure 7: Working viscosity differences of various R-404A substitute refrigerants

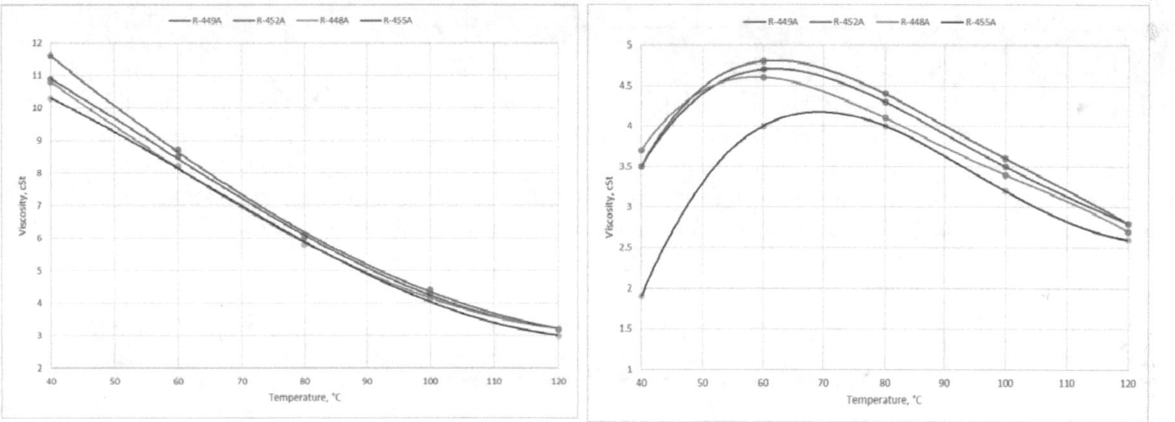

System tests have been conducted with R-455A in a commercial refrigeration unit and in a hermetic compressor. Yana Motta and Spatz (2014) showed test results of a ¾ HP single door reach-in freezer using a reciprocating compressor. No changes were made to the lubricant between tests with only slight changes to the evaporator flow, TXV and charge optimization. Results showed a 6% improvement in compressor efficiency and 3% system efficiency improvement and a 4% capacity loss was measured. These results appear to show no significant effect of the lubricant on performance with R-455A. A slight capacity difference could be from the thermodynamic differences of the refrigerant with possible minor changes related to the solubility factor difference. Zgliczynski and Sedliak (2016) showed results from compressor calorimeter studies on R-455A with a 1900 BTU/hr reciprocating hermetic compressor designed for R-404A. The same ISO 22 POE lubricant was used in both the baseline and alternate refrigerant study at varying evaporator temperatures (-40 to -10°C) and condensing temperatures (30 to 60°C). At the midpoint rating condition, the average capacity loss for R-455A was approximately 2% while the efficiency measured an average gain of approximately 6%. The authors commented that R-455A appears as a reasonable candidate for an R-404A replacement in lower temperature systems. Once again lubricant performance does not seem to be an issue. As an aside for discussion that will take place in the next section, comparison testing was also done in this study with R-290 refrigerant. Once again using the same ISO 22 POE lubricant, the results showed approximately 10% capacity loss and 11% efficiency gain. Solubility of R-290 will be discussed later.

5. HC Refrigerant

R-290 (propane) refrigerant is a single component refrigerant with an ASHRAE safety designation of A3 with a GWP value of <20. R-290 is not really a new refrigerant being investigated since it is already in use in many regions in various applications. Typically in light commercial types of systems, there is a limit to how much R-290 can be used due to the highly flammable nature of the refrigerant which can limit the size of the system in operation. In the previous section, some information was provided which compared the effectiveness of R-290 when compared to R-404A which showed better efficiency but some capacity loss. This loss in capacity could be mainly from the thermodynamics, but since R-290 has a high solubility level in most lubricants, this could also be playing a role. Unlike HFC refrigerants and the HFO/HFC refrigerants discussed earlier, R-290 can be used with mineral type oils but this brings about even higher dilution of refrigerant into the lubricant. Many compressor manufacturers have commercialized R-290 compressors using POE lubricants to help minimize the higher dilution factors that exist with mineral oils that may degrade volumetric capacity. Table 4 is a comparison of R-290 dilution in a mineral oil and a POE lubricant. The results show how dilution increases in the mineral oil.

Table 4: MO and POE dilution comparison of R-290

	4 Bar		8 Bar	
	MO	POE	MO	POE
20C	9.8%	7.5%	47.0%	35.0%
40C	5.4%	4.1%	14.2%	12.1%
60C	3.6%	3.0%	8.2%	6.9%
80C	2.6%	2.3%	5.5%	4.9%

The charge restriction in place for R-290 refrigerant usage minimizes the application and locations that this refrigerant can be used. Since the charge is limited to a certain amount, it is beneficial to understand the regions of a system that the refrigerant resides in to be able to potentially maximize charge usage. Fuentes and Hrnjak (2015) tried to measure and predict the distribution of the R-290 refrigerant in a typical light commercial unit like a beverage cooler. The results indicate that the majority of the charge was found in the condenser concluding the condenser redesign could be beneficial. Also the second highest location of the refrigerant was in the lubricant, attempting to minimize this level would also be helpful to not only charge maximization but also system performance. Lubricant chemistry as seen in Table 4 can play a role in this minimization and further optimization can be done with lubricant change. Figure 8 shows the dilution factor of three different lubricant chemistries.

Figure 8: Dilution factors for various lubricant chemistries at the same condition

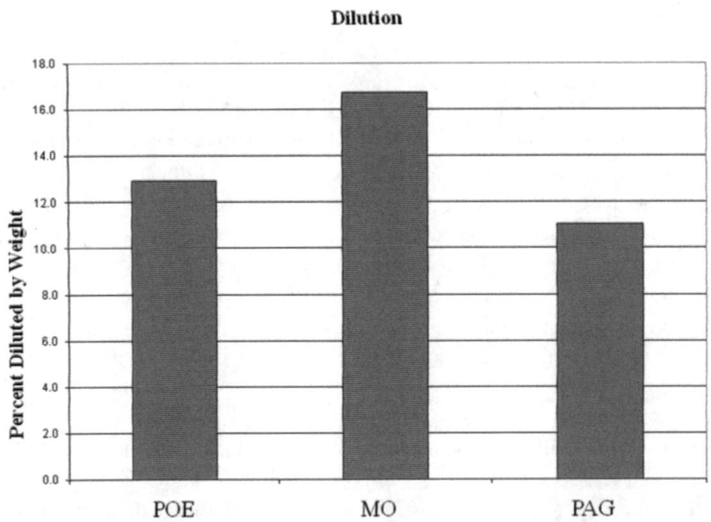

5. CONCLUSIONS

Environmental concerns have moved the HVAC&R industry to evaluate other refrigerant options that have lower GWP values. These refrigerants still need to exhibit adequate to potential better performance because any gains made by the direct environmental effect of a low GWP refrigerant can be lost by the indirect effect of poor performance and reliability. R-404A refrigerant, due to its higher GWP value, is a target that many regions and manufacturers are looking to find alternate replacement refrigerants. Part of this performance and reliability evaluation of the system requires finding the appropriate lubricant which may or may not be the same as what was used with the refrigerant being replaced. This paper evaluated five potential candidates as substitutes for R-404A and the potential impact of lubricant chemistry. For the most part, certain lubricant chemistry currently used with R-404A in compressor operation is found to be satisfactory for use with all the refrigerants investigated by way of evaluating lubricant-refrigerant interaction or system performance studies. Lubricants like mineral oils which typically have been unacceptable for use with R-404A, due to miscibility concerns, are also not miscible with the HFO/HFC refrigerants studied. Since most of the tests outlined in this paper involved either drop-in scenarios or soft optimization techniques, it is always beneficial to determine ways to optimize the performance of the new refrigerant and this can include looking at ways to potentially optimize the lubricant.

REFERENCES

American Society of Heating, Refrigerating and Air Conditioning Engineers, Inc, 2013 Fundamentals Handbook. Section 29 Refrigerants.

Baba, A., Yamaguchi, H., 2014, Performance evaluation of condensing unit using low GWP refrigerants, *The International Symposium on New Refrigerants and Environmental Technology,* Kobe, p. 269-273.

EPA, Final Rule, Federal Register Vol, 80 No. 138, July 20, 2015 Rules and Regulations 42870-42959, Significant New Alternatives Policy (SNAP) Program.

F-gas Regulation (EU) No 517/2014 of the European Parliament Fluorinated Greenhouse Gases.

Fidler, J., 1979, A History of Refrigeration Throughout the World, *International Institute of Refrigeration.*

Fuentes Y., Hrnjak, P., Elbel, S., 2015, Low-Charge Propane Refrigeration Systems Technology for Single and Multi-Door Bottle Coolers, *The 24th IIR International Congress of Refrigeration ICR 2015*, Yokohama, Paper ID 303.

Karnaz, J., 2013, Compressor and System Energy Efficiency Improvements through Lubricant Optimization, *8th International Conference on Compressors and Their Systems,* London, p. 163-171.

Karnaz, J., 2016, Lubricants for Low GWP Refrigerants, Seminar 16 Making the Commercialization of Low-GWP Refrigerants a Reality, *ASHRAE 2016 Winter Conference*, Orlando.

Karnaz, J., Kultgen, D., 2015, Making the Right Lubricant Choices, *The 24th IIR International Congress of Refrigeration ICR 2015*, Yokohama, Paper ID 338.

Karnaz, J., Liu, K., 2013, Lubricant and Refrigerant Properties – The Need for Lubricant Optimization with Various Type of Alternative Refrigerants, *The 5th International Conference on Cryogenics and Refrigerants,* Hangzhou, p. 116.

Kujak, S., Berge, J., Majurin, J., Kolda, M., Crombie, D., 2015, Assessment of Next Generation Refrigerant R452A to Replace R404A for Transport Refrigeration Products, *The 24th IIR International Congress of Refrigeration ICR 2015*, Yokohama, Paper ID 59.

Leck, T., Minor, B., 2014, Experience with Reduced GWP Refrigerants for Commercial Refrigeration and Air Conditioning, *International Symposium on New Refrigerants and Environmental Technology,* Kobe, p. 287-291.

Pottker, G., Yana Mota, S., Spatz M., Becerra, E., Smith, G., 2014, Refrigerants with Low Environmental Impact for Refrigeration Applications, *International Refrigeration and Air Conditioning Conference,* Purdue, Paper 1554.

Rajendran, R., 2013, Promising Lower GWP Refrigerants in Air Conditioning and Refrigeration Systems, *Advancing Ozone & Climate Protection Technologies, Bangkok.*

Yana Mota, S., Spatz M., 2014, Low Environmental Impact Refrigerants for AC and Refrigeration Systems, *The International Symposium on New Refrigerants and Environmental Technology,* Kobe, p. 282-286.

Zgliczynski, M., Sedliak, J., 2016, Compressor Calorimeter Performance Evaluation of Low GWP Alternatives top R404A in Low Temperature Applications, *AHRI Low GWP AREP*, Orlando.

LGWP & HC refrigerant solubility tests performed in running scroll compressor

Pierre GINIES *, Guillaume REBIÈRE; Julie MANDON ; Jean Guillaume CHEURLIN

Danfoss Commercial Compressors,
01600 Trevoux, France
p.ginies@danfoss.com,
rebiere.guillaume@gmail.com; julie.mandon@ecam.fr ;
jean-guillaume.cheurlin@insa-lyon.fr

* Corresponding Author

ABSTRACT

This paper presents the results of investigations on Low Global Warming Potential refrigerants and R290 solubility with different lubricants. Solubility impacts a lot the final lubricant-refrigerant mixture reaching the bearing lubricant inlet and in the same way can impact compressor reliability on wet vapor flowing back to compressor inlet Experimental data, measured in operating compressors, are compared to static steady state values coming from PVT plots. R407C and R410A, two well-known HFC refrigerants, are added as a reference. This work highlights the differences between dynamic solubility (tests performed in a compressor) and static solubility (from PVT plots). A comparison of the impact on the dynamic solubility has been done when a steering feature is added. This work allows us to define the best compromise between different lubricants in these new applications.

1. INTRODUCTION

Recent developments in regulations governing the use of refrigerants with respect to their global environmental impact (2006/842/EC "F-GAS" regulation recently voted in the EU with a new revision coming into force on January 1rst 2015, or 2009/125/EC Eco-design directive about energy performance) imply the introduction of new refrigerants and the adequate lubricant for these new applications. Natural refrigerant R290 and some LGWP refrigerants based on HFO and their blends are part of the solution. For example, R290 has a Global Warming Potential (GWP) equal to 3 which is roughly 500 times lower than R407C and 600 times lower than R410A.
As a consequence, tests with R290 and several HFO based mixtures have been performed in running compressor by sampling a small quantity of lubricant/refrigerant mixture from the oil sump. The mixture was sampled at several saturation suction temperatures and superheat conditions. The compressors used were standard scroll compressors. This article presents the results of these tests and compares them with static steady state solubility values from PVT plots, especially at low superheat.

2. STEADY STATE SOLUBILITY

Steady state values are provided by PVT plots (also named Daniel plots). They indicate the solubility of the refrigerant in the respective lubricant and the viscosity of the mixture in accordance with temperature and pressure. These plots are useful for the first comparisons of refrigerants and lubricants. However, they are more suitable for heat exchangers than compressors. Actually, the accuracy of data for low superheat conditions is unsatisfactory because of the slope of pressure curves (Figure 1). Furthermore, only a few oil ratio values are given (100%, 90%, 80% and 70% in most of plots, each 10% at best).

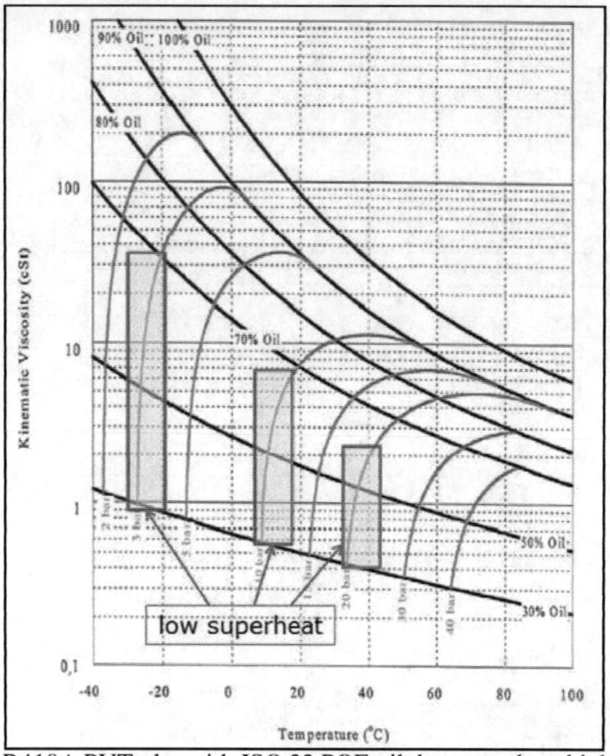

Figure 1 : R410A PVT plot with ISO 32 POE oil, low superheat highlighted

3. EXPERIMENTAL METHOD

3.1 R290 evaluation

Solubility tests must be performed in compressor in order to know the real behavior of lubricant/refrigerant mixtures at low superheat. Furthermore, the PVT plots only provide static steady state values, hence the usefulness of these tests in dynamic conditions corresponding to low super heat conditions.

These tests were run in a refrigeration loop which included a scroll compressor. An oil pump is fixed to the lower end of the crankshaft (Figure 2). Its rotation enables the oil sump lubricant to be stirred. This could homogenize the lubricant/refrigerant mixture and lead to an outgassing process.

The compressor used is a standard model. It is equipped with a valve in order to sample a small quantity of lubricant/refrigerant mixture from the oil sump. These samples are made when the system conditions are stable. Then the solubility is obtained by a similar process to the one of oil circulation rate measurement (ASHRAE standard 41.4-1996).

The tests were performed at several saturated suction temperatures (from -10°C to +15°C) and superheat conditions (5K, 10K and 20K). This allows us to check the lubricant/refrigerant mixture behavior all over the application range of the compressor.

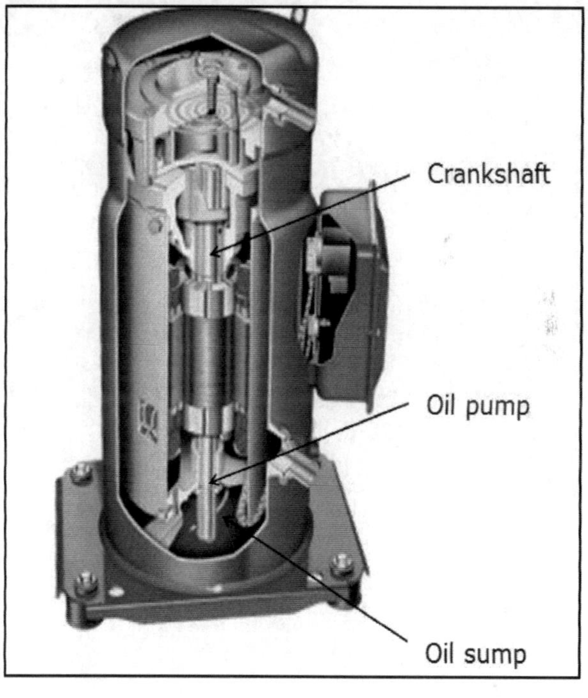

Figure 2 : Type of compressor used for the tests

RESULTS AND DISCUSSION

Five R290-based couples were tested. Two references with R407C and R410A were added. The Table 1 lists the different couples and the viscosity grade of the lubricants used.

Refrigerant	Lubricant	Grade
R407C	POE	32
R410A	POE	32
R290	Lub1	68
R290	Lub2	46
R290	Lub3	68
R290	Lub4	68
R290	Lub5	68

Table 1 : Lubricant/refrigerant couples studied

The Figure 3 presents the static steady state solubility from PVT plots data.

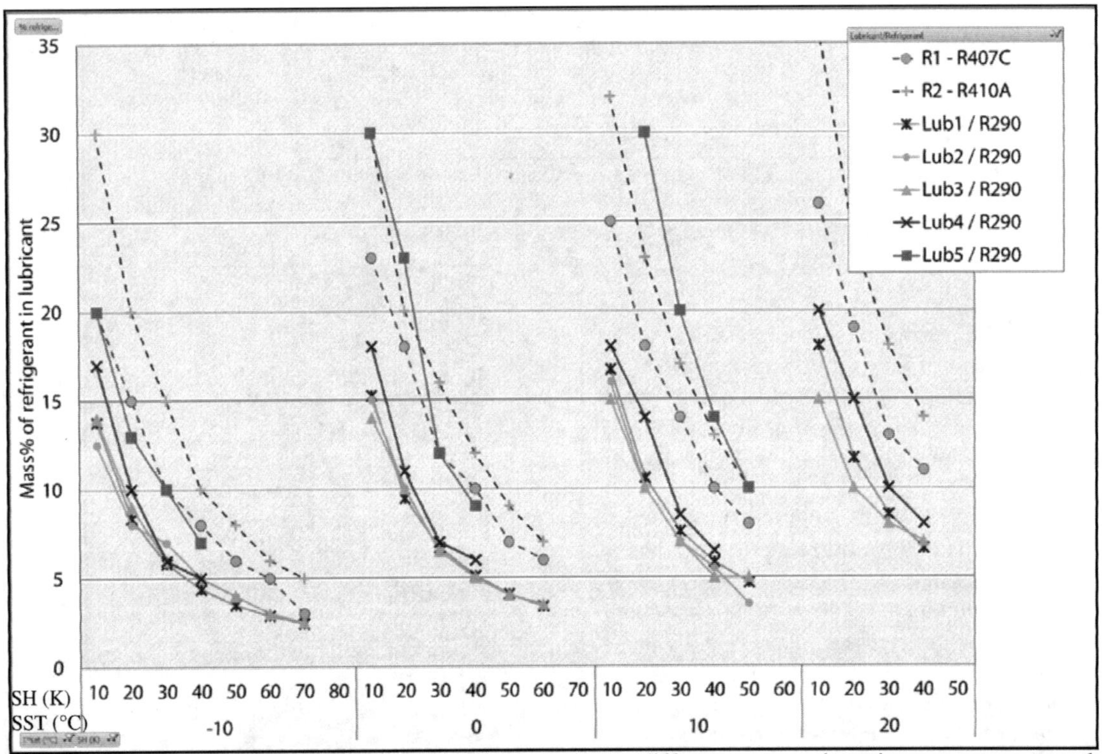

Figure 3 : Steady state solubility provided by PVT plots at different saturated suction temperatures and superheat conditions

We used the oil sump sampling method to compare several lubricants on compressors. The results are shown in Figure 4. We can see 3 different groups of solubility level: 'Lub 5'; 'Lub2' and 'Lub 1-Lub3-Lub4'.

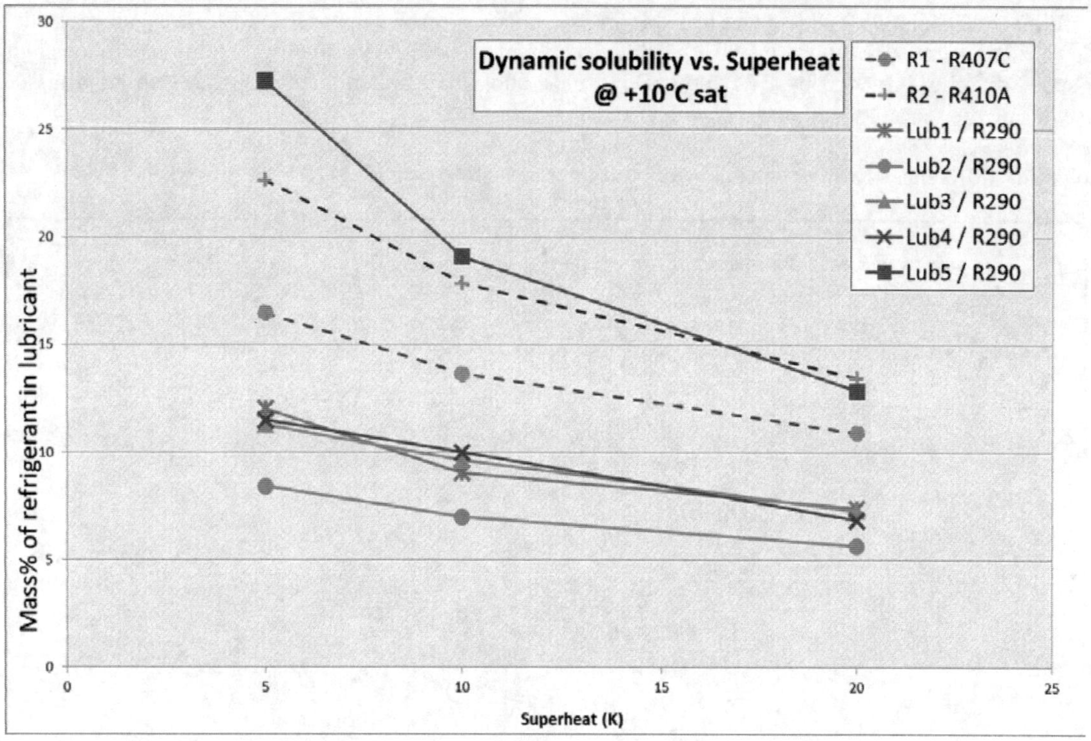

Figure 4 : Dynamic solubility measured in running compressor for T sat = 10°C

Solubility from PVT plots and tests in running compressor can now be compared in order to notice a potential dynamic effect. This comparison at low superheat for T sat = 10°C is presented by Figure 5. R407C and R410A curves are removed in the following graph.

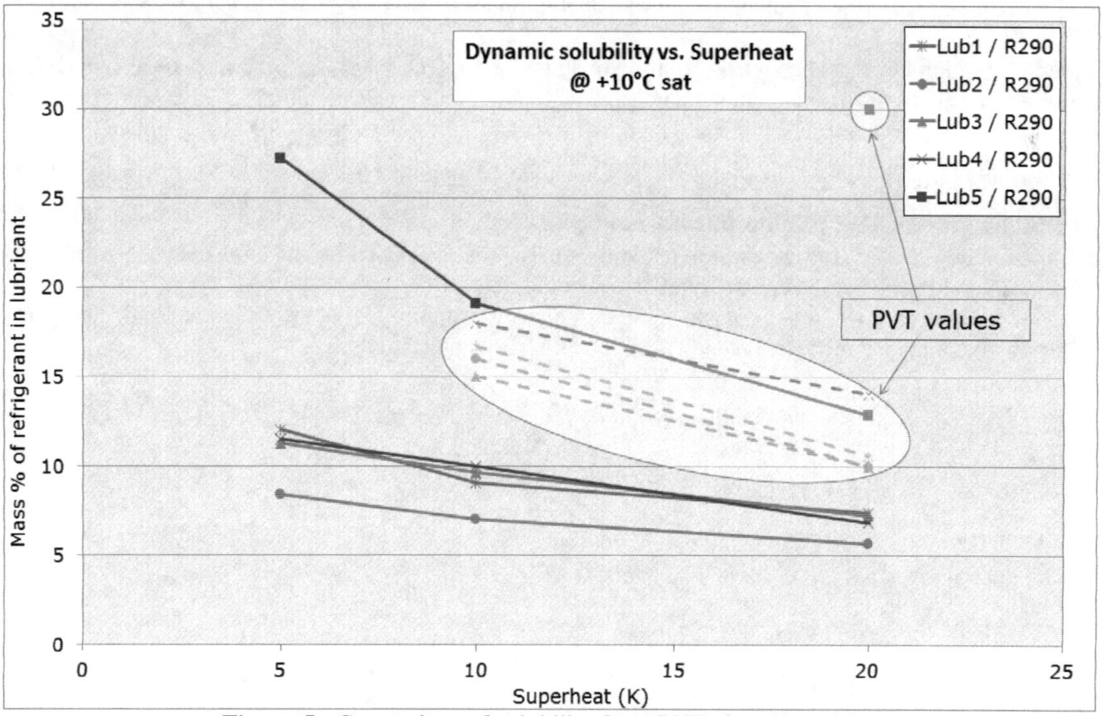

Figure 5 : Comparison of solubility from PVT plots (dotted line)
and dynamic solubility tests performed in compressor

A5% difference between PVT charts and tests performed in compressor values is noticed. In some case this shows that the refrigerant mass is reduce by 70% to 80%. The real solubility (in running conditions) is lower than the steady state solubility provided by PVT plots. P.J. Rubas and C.W. Bullard (1993) already observed a solubility decrease during the startup of compressor.

This solubility difference is also noticeable with both references R407C and R410A (Figure 6).

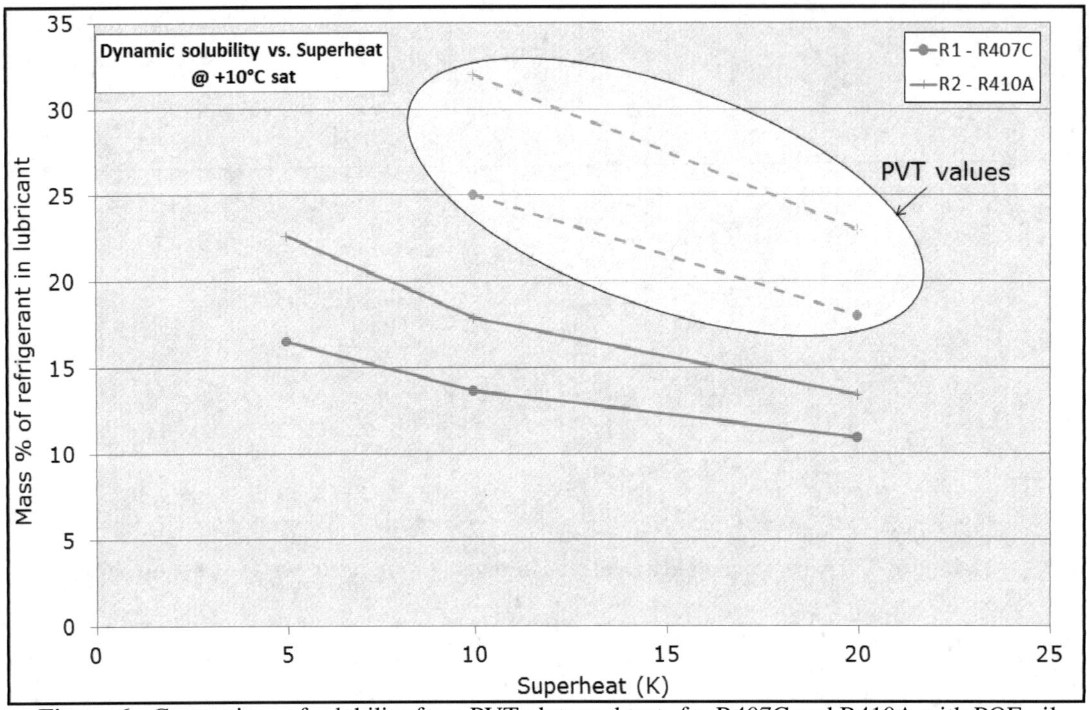

Figure 6 : Comparison of solubility from PVT plots and tests for R407C and R410A with POE oil

This decrease in solubility may be caused by the stirring effect in the oil sump produced by the moving parts, including the crankshaft and the oil pump. Said stirring effect leads to the foaming of the lubricant/refrigerant mixture which brings a decrease in solubility. T. Yanagisawa and T. Shimizu (1985) had early noticed that foaming caused by stirring effect in rolling piston type rotary compressor.

3.2 FGAS: new LGWP refrigerant investigation

On the work for new LGWP refrigerant applications we use similar evaluation method than the one we did for R290. The idea was to see what the existing similarities could be.

Figure 7 shows the results when we compare the Daniel plots solubility data for different POE lubricants with potential replacement LGWP refrigerant candidates versus known applications.

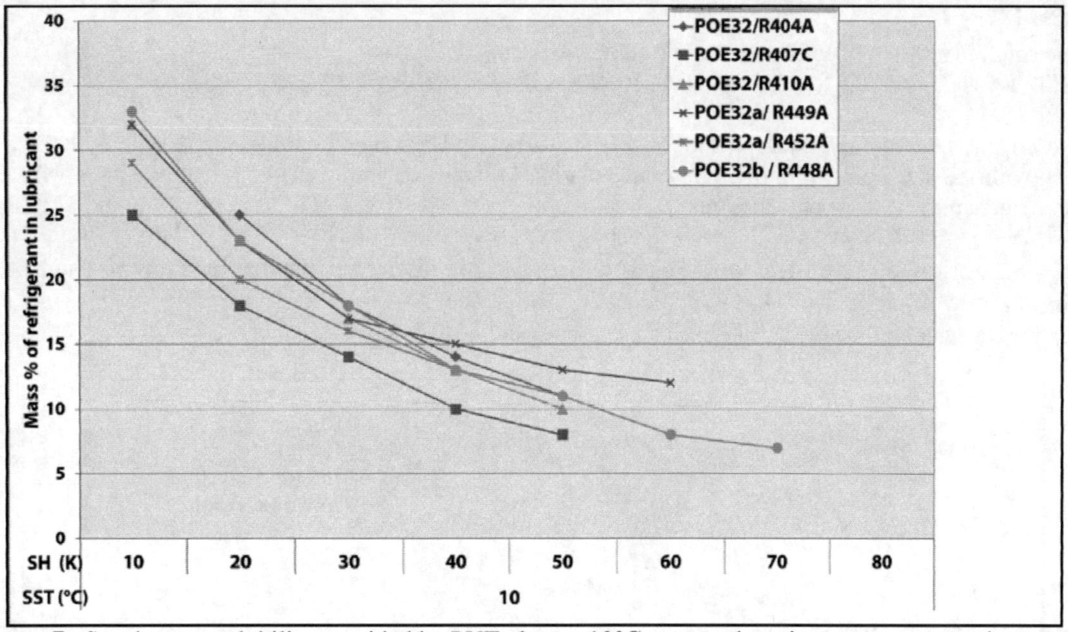

Figure 7 : Steady state solubility provided by PVT plots at 10°C saturated suction temperature and at several superheat conditions

The potential R404A new refrigerant replacement candidates; solubility % on static (Daniel plot) are on the same level than the R404A and R410A.
We can see that the solubility level for R404A and the new refrigerant candidates have a similar percentage value versus SH. We can note that the solubility percentage tends to be higher than the R407C refrigerant solubility in lubricant.

➔ Static solubility: POE32/R407C < POE/R449A ~ POE32/R410A ~ POE32b/R448A

To have a better vision of the behavior on oil sump we need some experimental tests to know the real values on the oil sump.
We perform similar testing process than we carried out for R290 with new LGWP refrigerants comparing the measurements of the refrigerant oil sump solubility % with sampling technique

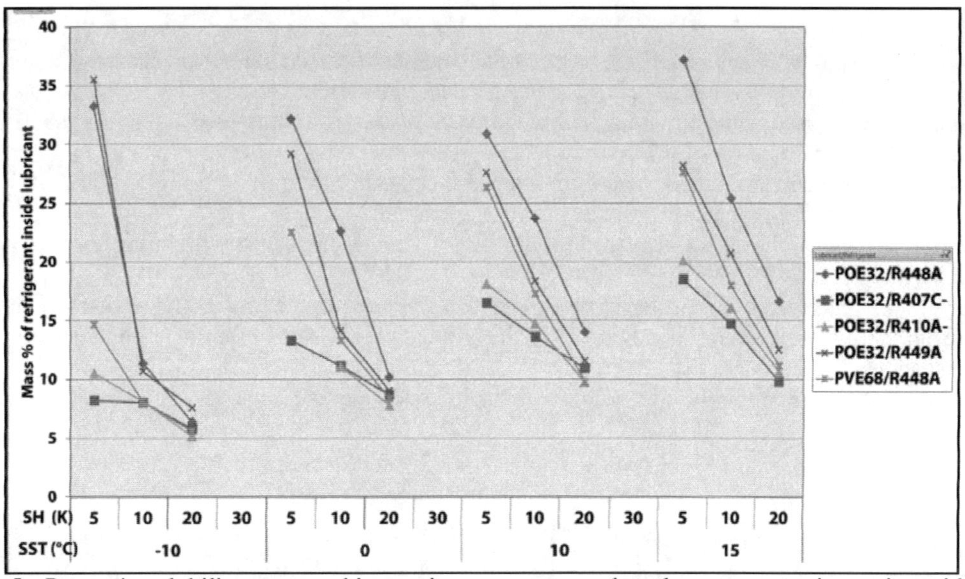

Figure 8 : Dynamic solubility measured in running compressor when the compressor is running with R410A, R407C, R448A and R449A refrigerants

Dynamic solubility with HFO mixtures: can be sorted as followed:
→ POE32/R407C ~ POE32/R410A < POE32/R448A < POE32/R449A < PVE68/R448A

.

We note on this test, that dynamic conditions when using new HFO blends, R448A and R449A, have a higher solubility with the same oil. So, refrigerant concentration on lubricant is higher than what we have with existing R404A refrigerant at the same operating point.

3.2 Evaluation of potential out gassing process on oil sump for reduction of the solubility percentage.

The idea was to evaluate the potential lubricant outgassing on the oil sump.

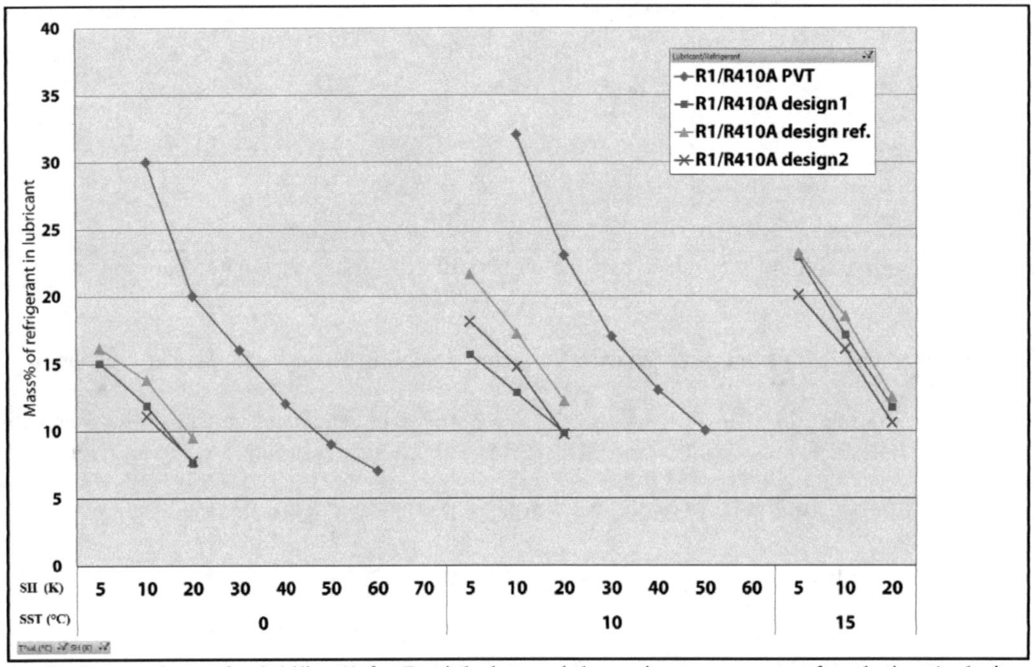

Figure 9 : Comparison of solubility % for Daniel plot, and dynamic measurement for: design 1; design 2 & reference

On Figure 9 we compare the Daniel plot versus sampling measurement on reference compressor and sampling measurement with two oil sump designs to see the changes in steering effect.
The first design has a cage around the oil pump pickup tube. The oil has a slightly lower solubility than in a standard compressor, according to the measures for most of temperatures.
On the second design, we add one stirrer feature on the oil pickup tube. With this design, on the presence of the stirrer, the solubility is higher in each point, whatever the temperature.
Contrary to our expectations, the stirrer increased oil solubility. This result seems to be linked to the stirring action on the oil sump surface level.
However, the oil solubility sampling fitting is placed on the compressor shell at the edge of the housing and is therefore far from the agitation zone. It is possible that the arrangement of the measuring equipment is not adequate with the desired results.
For the next degassing tests it would be wise to place the solubility sampling location as close as possible to the oil pumping area.

3.3 HC/HFO trends

Some of the new LGWP refrigerants are based on HFO. In line with our work done with R290 and some HFO reliability investigations, we want to compare the behavior of HFOs with lubricants in oil sump and compare it to the data measured with R290.

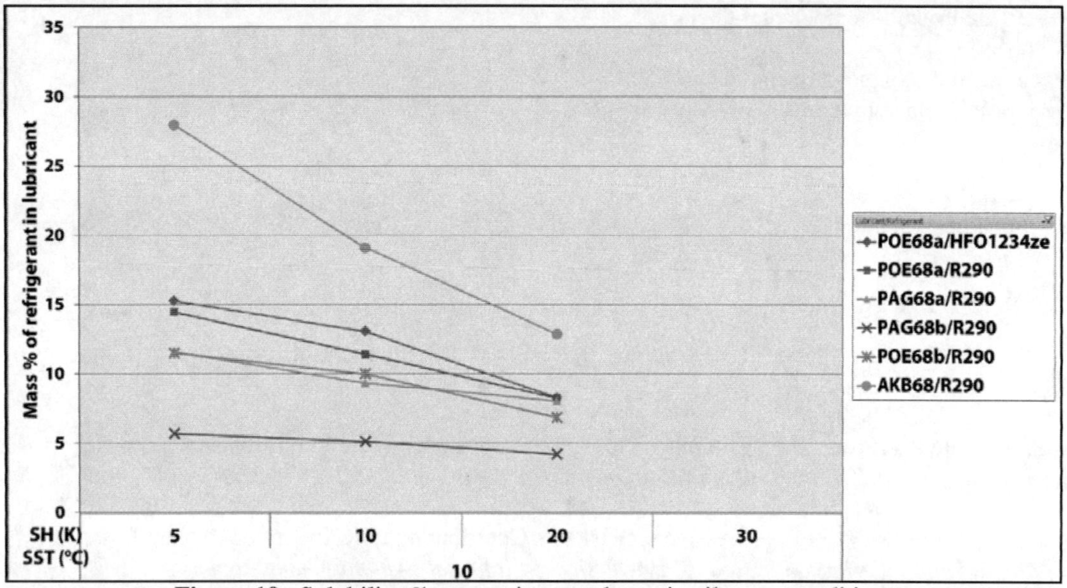

Figure 10 : Solubility % comparison on dynamic oil sump conditions

The results on Figure 10 show that the HFO evaluated behave not that far from R290 on the selected lubricants. This helps us to build up the compressor lubricant selection and qualification program.

4. CONCLUSIONS

R290 solubility tests in running compressor have given lower values than PVT plots. The difference between dynamic and static values reaches 5 to 10 percent of refrigerants in mass of lubricant. This small value brings significant differences. It could be caused by the stirring of the refrigerant/lubricant mixture. Other tests are scheduled in order to study this effect by implementing a stirrer in the oil sump.

The concept of outgassing by steering effect process seems to be promising but the real measurement of lubricant solubility % is not so easy to carry out within the area of the oil pickup tube inlet without disturbing its efficiency.

The solubility test developed for R290 was used on new LGWP refrigerants evaluation and helps to compare properties and this was confirmed during reliability testing.

This measurement principle gives a better knowledge on what happens inside the compressor oil sump. This work will help on the lubricant candidate selection process for compressor on reliability evaluation.

NOMENCLATURE

HFC	hydrofluorocarbon	GWP	global warming potential
POE	polyol ester lubricant	T	temperature
AKB	alkylbenzene lubricant	PAG	polyalkylene glycol lubricant
HFO	Hydro Fluoro Olefin		
SST	saturated suction temperature (°C)		
SH	super heat (K)		
SDT	saturated discharge temperature (°C)		

Daniel plot or PVT pressure temperature viscosity data

Subscripts

Oil	lubricant	Lub	lubricant
Sat	saturation	R	refrigerant

REFERENCES

- ANSI/ASHRAE Standard 41.4-1996 (RA 2006) - Method for Measurement of Proportion of Lubricant in Liquid Refrigerant

- Ginies P., Dewitte P., Terrier M.F., Charni M. (2014) - Review of Hydrocarbon Lubricants Behavior, *2700. Proceedings of the 15th International Refrigeration and Air Conditioning Conference at Purdue*

- Rubas P.J., Bullard C.W. (1993) - Assessment of Factors Contributing to Refrigerator Cycling Losses, *, Air Conditioning and Refrigeration Center, University of Illinois at Urbana-Champaign*
ACRC TR-45 July 1993

- Yanagisawa T., Shimizu T. (1985) - Foaming of refrigerating oil in a rolling piston type rotary compressor, *Shizuoka University*
Rev. Int. Froid 1986 Vol 9 Janvier , 17-20

AKNOWLEDGEMENTS

It would not have been possible to lead this study without the contribution of ATEX lab technicians and manager. In addition I enjoyed a very nice co-authors team work with a trainee engineers.

The work on HC refrigerant is issue from the R&D activities work during the "Next Generation of Heat Pumps working with Natural fluids" EU project. This project has received funding from the European Union's Seventh Program for research, technological development and demonstration under grant agreement No [307169]".

Numerical Modeling of Capillary Compensated Aerostatic Bearing Applied to Linear Reciprocating Compressor

Emílio R. HÜLSE[1]*, Álvaro T. PRATA[2]

[1]Embraco R&D,
Joinville, Santa Catarina, Brazil
(emilio.r.hulse@embraco.com)

[2]Federal University of Santa Catarina,
POLO Research Laboratories for Emerging Technologies in Cooling and Thermophysics,
Florianópolis, Santa Catarina, Brazil
(prata@polo.ufsc.br)

* Corresponding Author

ABSTRACT

Linear compressor uses a linear actuator to drive the piston directly in its reciprocating motion, eliminating several bearings that convert rotational into linear reciprocating motion, which is the common case in conventional reciprocating compressors. Additionally, the presence of side loads is minimized because the main forces in the mechanism are aligned to the axis of motion. Since the only sliding surface is the piston-cylinder interface, lubrication can be achieved using the refrigerant gas.

In the present investigation, the equilibrium condition of capillary compensated gas lubricated piston for a linear compressor is fully explored. The main objective of the present work is to evaluate the performance of a pneumatic bearing and to understand the influence of some design characteristics as radial clearance, diameter of the capillary compensation channels and the position of the feeding ports on the bearing during operation. A mathematical model is developed and then the resulting equations are numerically solved by finite volume methodology.

Results indicated that the linear piston is very stable, and that the gas leakage necessary to lubricate the piston-cylinder interface during the reciprocating motion, is very small. Furthermore, the geometry of the feeding channels to the clearance between piston and cylinder is an important design parameter in achieving a better performance of the aerostatic bearing.

1. INTRODUCTION

Linear compressors are objects of study since the early sixties and one of the first works was presented by Cadman and Cohen (1969) about a double piston electrodynamic oscillating compressor design. Later Polman *et al.* (1978) and Pollak *et al.* (1978) also modeled different linear compressor mechanisms, both showing the potential of such configurations. Recently, the development of linear actuators and magnetic materials stimulated the evolution of this technology and the presence of electronic drives also allowed the precise control of the actuator position as mentioned by Redlich *et al.* (1996).

Due to the constructive aspects of linear compressors, gas bearings have the potential to be used for piston lubrication. The main forces acting on the compressor mechanism are aligned with the main axis of movement and small lateral forces that may arise are only consequences of manufacturing tolerances, either from components or assembling. Van der Walt and Unger (1994) presented a prototype linear compressor with gas bearing technology.

Kazmierski and Trojnarski (1980) investigated the influence of feeding system configuration on externally pressurized radial gas bearings. The governing equations were solved by a finite difference method and the numerical results were compared with experimental results for pocketed and plain orifice feeding systems. Feedholes were treated as source points in the Reynolds' equation. Gommed and Etsion (1993) presented a dynamic analysis of gas lubricated ringless pistons, and included in their model the effects of connecting rod inertia and thermohydrodynamic state of operation; both Reynolds' and the energy equation were solved. Fourka and Bonis (1997) investigated different feeding systems of externally pressurized gas thrust bearings. The influence of quantity and position of pocketed orifices as well as the permeability of porous channels in a porous feeding configuration on the load capacity and stiffness of thrust bearings were explored. Renn and Hsiao (2004) proposed a model to calculate the mass flow through orifice-type restrictors for feeding aerostatic bearings; CFD simulation was used to determine the governing parameters and experimental results were employed to validate the numerical model. Lo *et al.* (2005) analyzed the performance of high speed spindle aerostatic bearings, solving the Reynolds' equation by finite differences and Newton method. Convergence at very thin films was improved by using a modified relaxation method applied to orifice pressure.

2. MODELING OF THE LUBRICATION PROBLEM

2.1 Physical model and local film thickness

A typical configuration with piston and cylinder in a gas bearing system is depicted in Figure 1. The piston-cylinder clearance is fed with refrigerant by a set of orifices regularly arranged in the cylinder surface and connected by feeding channels. The refrigerant plays the role of lubricant fluid, separating the piston and cylinder surfaces through the pressure field generated in the radial clearance.

The coordinate system XYZ used to locate the piston inside the cylinder, as depicted on Figure 2, has its origin in the cylinder top center. The piston can move freely within the clearance and the eccentricities at the piston top $\boldsymbol{e_0}$ and bottom $\boldsymbol{e_1}$, and their components $\boldsymbol{e_{x0}}$, $\boldsymbol{e_{x1}}$, $\boldsymbol{e_{z0}}$ and $\boldsymbol{e_{z1}}$ are used to define its position. Another coordinate system $\xi\eta\zeta$, solidary to the piston and with origin on the piston top center is used to calculate the pressure field. Since the piston-cylinder clearance is much smaller than piston radius, curvature effects can be disregarded. As indicated in Figure 3, for a given axial location, the local film thickness $\boldsymbol{h}$ can be expressed in the $\xi\eta\zeta$ coordinate system as,

$$h = c - e_x \cos\left(\frac{\xi}{R}\right) - e_z \sin\left(\frac{\xi}{R}\right) \tag{1}$$

Also, considering that eccentricity along the η axis varies linearly with the piston length, the general expression for the local film thickness can be written as,

$$h(\xi,\eta) = c - \frac{\eta}{L_{pis}}\left[e_{x1}\cos\left(\frac{\xi}{R}\right) + e_{z1}\sin\left(\frac{\xi}{R}\right)\right] - \left(1 - \frac{\eta}{L_{pis}}\right)\left[e_{x0}\cos\left(\frac{\xi}{R}\right) + e_{z0}\sin\left(\frac{\xi}{R}\right)\right] \tag{2}$$

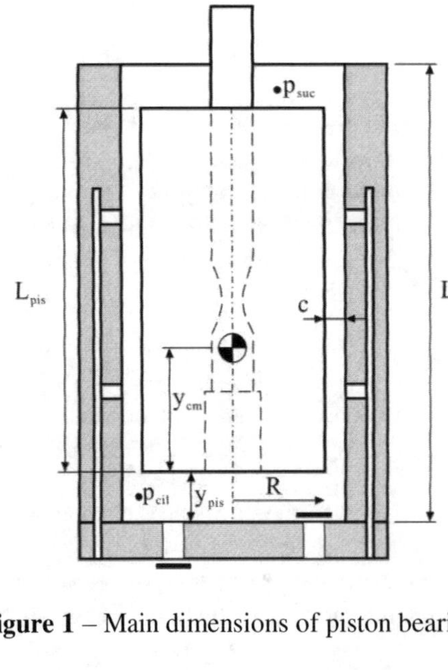

Figure 1 – Main dimensions of piston bearing

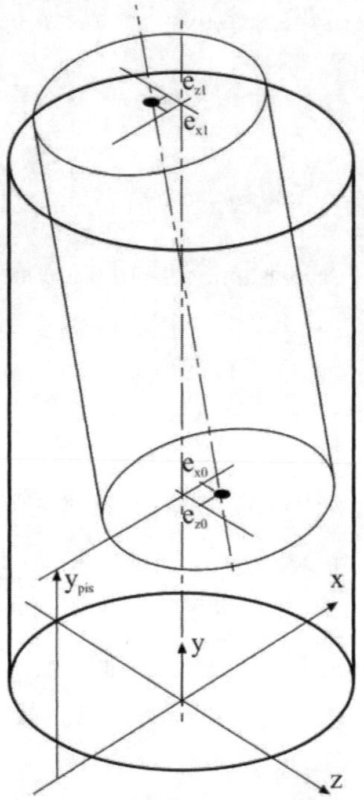

Figure 2 – XYZ coordinate system

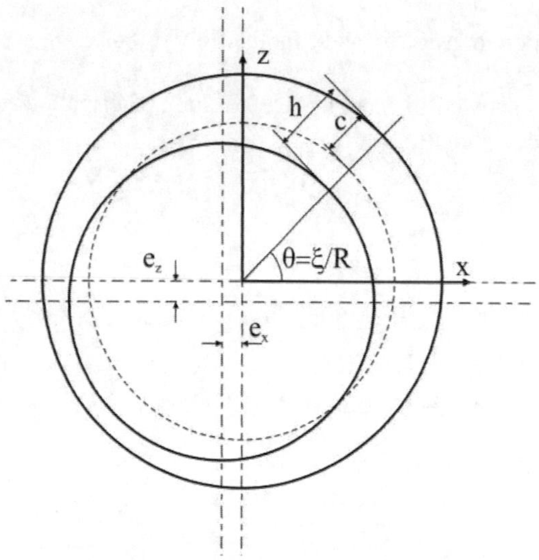

Figure 3 – Local film thickness

2.2 Flow model inside piston-cylinder clearance

The flow between the piston and cylinder surfaces is governed by the Navier-Stokes equations, but according to the lubrication theory these equations can be simplified, once observed the following hypothesis according to Hamrock *et al.* (2004):

a) Field forces are ignored
b) Inertia forces are disregarded
c) Pressure is uniform across the radial clearance
d) Non-slipping flow is observed at the solid interfaces
e) Laminar flow
f) Flow is isothermal and fluid viscosity is constant
g) Lubricant fluid follows ideal gas behavior

Using assumptions (a) to (f), the Reynolds equation for the lubrication flow through the piston-cylinder clearance can be derived; assuming ideal gas behavior density can be expressed in terms of pressure, resulting the following working equation,

$$\frac{\partial}{\partial \xi}\left(ph^3\frac{\partial p}{\partial \xi}\right)+\frac{\partial}{\partial \eta}\left(ph^3\frac{\partial p}{\partial \eta}\right)=-12\mu\left(\frac{V_p}{2}\frac{\partial (ph)}{\partial \eta}+\Re T\frac{\dot{m}}{A_f}-p\frac{\partial h}{\partial t}\right) \tag{3}$$

Boundary conditions for equation (3) are periodic in the ξ direction and prescribed in the η direction,

$$p_{\xi=0}=p_{\xi=2\pi R} \tag{4}$$
$$p_{\eta=0}=p_{cil} \tag{5}$$
$$p_{\eta=L_{man}}=p_{suc} \tag{6}$$

2.3 Flow model through feeding channels

Feeding channels are responsible to conduct the refrigerant to the feeding holes, reducing the supply pressure and creating the compensation effect as described in Hamrock *et al.* (2004). The mass flow rate that is fed into the piston-cylinder clearance by the feeding channels is represented by the term $\dot{m}$ in equation (3) and during the discretization of the Reynolds' equation it becomes a source term. The mass flow rate in each feeding channel can be calculated by as.

$$\dot{m} = \rho_f u_f A_f \tag{7}$$

where density, $\boldsymbol{\rho}$, and average flow velocity, $\boldsymbol{u_f}$, are both evaluated at channel inlet conditions.

As showed in Saad (1992), for an isothermal compressible flow in a pipe, Mach number at inlet conditions can be calculated as,

$$M_f = \sqrt{\frac{1}{\gamma}\frac{1-\left(\frac{p}{p_f}\right)^2}{\ln\left(\frac{p_f}{p}\right)^2 + \left(\frac{4fL}{D_h}\right)_f}} \tag{8}$$

where $\boldsymbol{f}$ is the friction factor calculated from the Reynolds number in the pipe flow. The Mach number is used to calculate the average flow velocity and therefore the mass flow at each feeding channel.

Mass flow leaked from both ends of the piston can be obtained integrating the local mass flux across the piston-cylinder clearance as,

$$\dot{m}_\eta = -\int_0^{2\pi R}\left(\frac{\rho h^3}{12\mu}\frac{\partial p}{\partial \eta} + \frac{\rho V_p h}{2}\right) d\xi \tag{9}$$

2.4 Load capacity of piston bearing

Loads acting on the piston due to driving mechanism lead to piston displacement inside the cylinder clearance. To balance these loads, hydrodynamic forces and moments due wedge, squeeze and mass flow effects arise. Hydrodynamic forces and moments can be determined by the integration of the instantaneous pressure field obtained from the solution of equation (3),

$$F_{rx} = -\int_0^{L_{man}}\int_0^{2\pi R} p\,(\xi,\eta)\cos\left(\frac{\xi}{R}\right) d\xi d\eta \tag{10}$$

$$F_{rz} = -\int_0^{L_{man}}\int_0^{2\pi R} p\,(\xi,\eta)\sin\left(\frac{\xi}{R}\right) d\xi d\eta \tag{11}$$

$$M_{rx} = -\int_0^{L_{man}}\int_0^{2\pi R} p\,(\xi,\eta)(\eta - y_{cm})\sin\left(\frac{\xi}{R}\right) d\xi d\eta \tag{12}$$

$$M_{rz} = \int_0^{L_{man}}\int_0^{2\pi R} p\,(\xi,\eta)(\eta - y_{cm})\cos\left(\frac{\xi}{R}\right) d\xi d\eta \tag{13}$$

3. SOLUTION METHODOLOGY

3.1 Solution of Reynolds equation – pressure field

The finite volume method of Patankar (1980) is used in the present work to solve equation (3) in order to calculate the pressure distribution acting on the piston. Integrating equation (3) over the generic volume in the computational domain represented in Figure 4 will result in,

$$\left[(ph^3)_e\frac{\partial p}{\partial \xi}\bigg|_e - (ph^3)_w\frac{\partial p}{\partial \xi}\bigg|_w\right]\Delta\eta + \left[(ph^3)_n\frac{\partial p}{\partial \eta}\bigg|_n - (ph^3)_s\frac{\partial p}{\partial \eta}\bigg|_s\right]\Delta\xi =$$
$$= -6\mu V_p[(ph)_n - (ph)_s]\Delta\xi - 12\mu\Re T\dot{m}\frac{\Delta\xi\Delta\eta}{A_f} + 12\mu p_P\frac{\partial h}{\partial t}\Delta\xi\Delta\eta \tag{14}$$

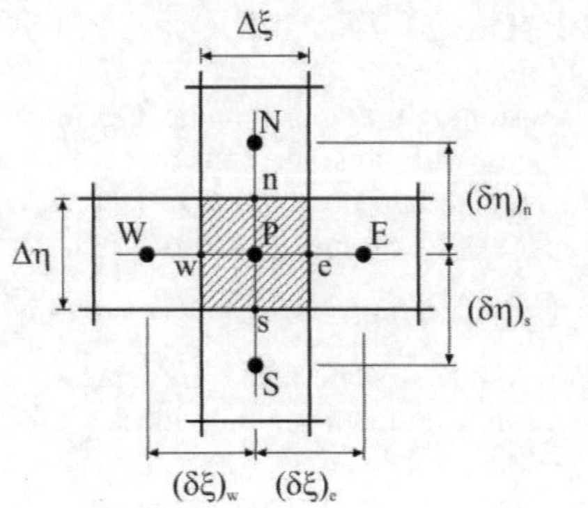

Figure 4 – Typical domain control volume

Derivatives at the volume faces are evaluated linearly using the pressure values at the center of each adjacent volume. Because the derivative coefficients are also function of pressure, an iterative procedure is employed to update these coefficients. The pressure values at the volume faces needed in calculating the derivative coefficients are evaluated linearly from the nodal pressures of previous iteration. The convective terms associated to the pressure at the interfaces are also linearly approximated with values from the current iteration. The mass flux term is calculated only for the volumes coincident with the feeding holes and is linearized because of its dependence on the supply pressure and the volume nodal pressure as indicated in the Mach number given by equation (8). Finally, the local film thickness time derivative is evaluated from (2) substituting the piston eccentricity positions $\boldsymbol{e_{x0}}$, $\boldsymbol{e_{x1}}$, $\boldsymbol{e_{z0}}$ and $\boldsymbol{e_{z1}}$ by their eccentricity velocities $\dot{\boldsymbol{e}}_{x0}$, $\dot{\boldsymbol{e}}_{x1}$, $\dot{\boldsymbol{e}}_{z0}$ and $\dot{\boldsymbol{e}}_{z1}$.

Discretization of equation (14) leads to a linear system of equations solvable by any numerical scheme. Due to the periodical boundary condition, it is advantageous to use cyclic TDMA in the circumferential direction and TDMA in the axial direction. For further details about these methods see Patankar *et al.* (1977).

3.2 Mass flow at piston top and bottom ends

The lubricant leakage at both ends of the piston can be calculated by numerically integrating equation (9), which results in the following equations,

$$\dot{m}_t = -\frac{p_{cil}}{\Re T}\sum_{i=1}^{NI}\left(\frac{h_s^3}{12\mu}\frac{p_{i,1}-p_{cil}}{\Delta\eta}+\frac{V_p h_s}{2}\right)\Delta\xi \tag{15}$$

$$\dot{m}_s = -\frac{p_{suc}}{\Re T}\sum_{i=1}^{NI}\left(\frac{h_n^3}{12\mu}\frac{p_{suc}-p_{i,NJ}}{\Delta\eta}+\frac{V_p h_n}{2}\right)\Delta\xi \tag{16}$$

3.3 Reaction forces and moments

Load reaction of the piston bearing can be calculated by the numerical integration of equations (10) to (13), resulting respectively,

$$F_{rx} = -\sum_{j}^{NJ}\sum_{i}^{NI}\left[\mathrm{p}_{i,j}\cos\left(\frac{\xi_i}{R}\right)\Delta\xi\Delta\eta\right] \tag{17}$$

$$F_{rz} = -\sum_{j}^{NJ}\sum_{i}^{NI}\left[\mathrm{p}_{i,j}\sin\left(\frac{\xi_i}{R}\right)\Delta\xi\Delta\eta\right] \tag{18}$$

$$M_{rx} = -\sum_{j}^{NJ}\sum_{i}^{NI}\left[\mathrm{p}_{i,j}(\eta_j - y_{cm})\sin\left(\frac{\xi_i}{R}\right)\Delta\xi\Delta\eta\right] \tag{19}$$

$$M_{rz} = \sum_{j}^{NJ}\sum_{i}^{NI}\left[\mathrm{p}_{i,j}(\eta_j - y_{cm})\cos\left(\frac{\xi_i}{R}\right)\Delta\xi\Delta\eta\right] \tag{20}$$

4. RESULTS AND DISCUSSIONS

A computational code to solve the piston-cylinder pneumatic bearing model was developed and implemented. Solution were given to static cases (V_p=0), with prescribed piston eccentricity inside piston-cylinder clearance. The influence of design parameters as the number and positioning of feeding holes, feeding channel diameter and radial clearance were investigated in relation to the load capacity and mass flow used in each case.

4.1 Mesh generation

Four feeding holes configurations are evaluated as depicted in Figures 5 and 6. Larger number of feeding holes is not practical due geometrical constrains in manufacturing individual feeding channels.

Mesh in the axial direction is divided into top, middle and bottom regions. Parameters NYT, NYM and NYS are respectively the number of volumes at each region. M parameter is the number of volumes between two feeding holes at each section. Each mesh arrangement is illustrated in Figures 7 and 8. Each feeding port is discretized by one single finite volume. Other geometric characteristics of the linear piston-cylinder are listed on table 1. Working fluid is air with atmospheric pressure at both piston ends and supplying pressure of 6 bar.

Mesh refinement was performed and the results to be presented are based on the configuration 30 x 30 volumes (M=8; NYT=11; NYM=11; NYS=6). This number of control volumes represents a compromise between solving time and accuracy.

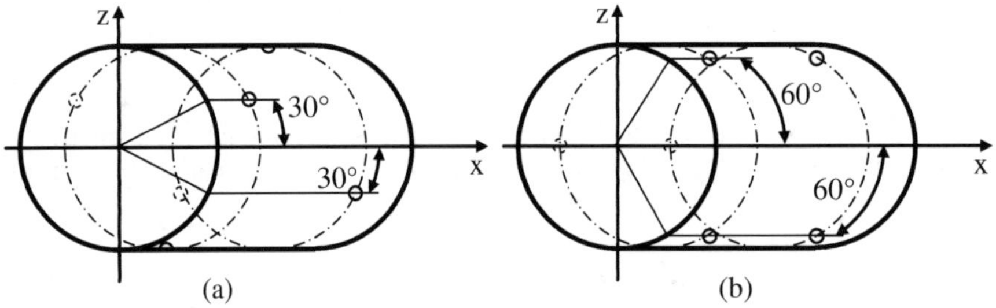

(a) (b)

Figure 5 – Cylinder with 3 feeding holes per section: (a) staggered; (b) aligned

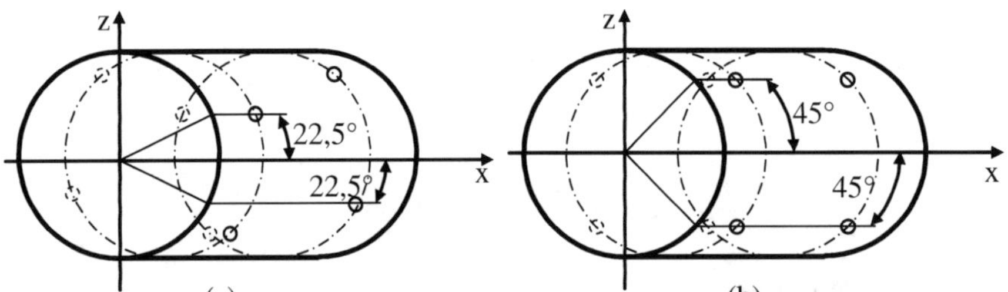

(a) (b)

Figure 6 – Cylinder with 4 feeding holes per section: (a) staggered; (b) aligned

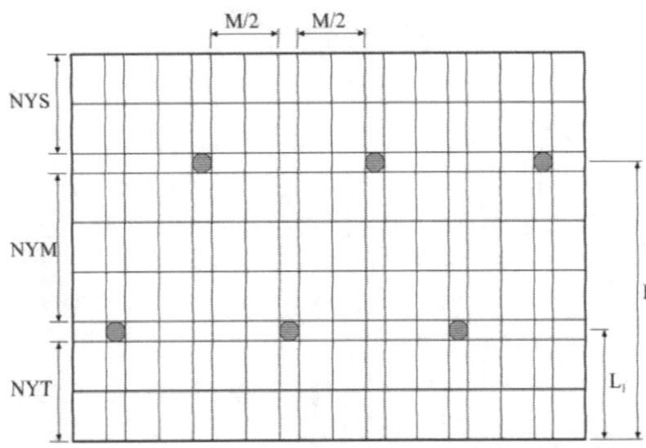

Figure 7 – Mesh configuration for staggered feeding holes

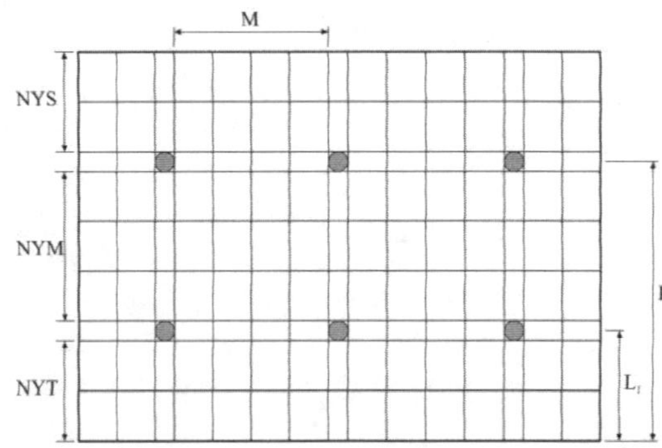

Figure 8 – Mesh configuration for aligned feeding holes

Table 1 – Geometric characteristics

Cylinder diameter / length	0.31
Piston length / cylinder length	0.85
Number of feeding holes per section	3
Feeding hole arrangement	Staggered
Feeding hole diameter / cylinder length	0.016
Position of first section of feeding holes / cylinder length	0.42
Position of second section of feeding holes / cylinder length	0.78
Position of center of gravity / piston length	0.19
Radial clearance / feeding channel diameter	0.0438
Feeding channel length / feeding channel diameter	4.537

4.2 Comparison with experimental results

An experimental set up was developed to measure the load capacity of the pneumatic piston by applying a lateral force while the piston vertically reciprocates inside the cylinder with small speed / frequency. Measuring lateral force (load) and vertical force (friction) that act on the cylinder, maximum bearing reaction force is determined when the friction force deviates from zero.

This specific design uses a radial clearance / feeding channel diameter relation of 0.062 and feeding channel length / feeding channel diameter relation of 4.084. Remaining characteristics are according to table 4.1. Numerical values in general agree with maximum values found in the experiment, as showed in Figure 9. Smaller experimental load capacity values may be caused by manufacturing shape errors in the piston and cylinder, feeding channel with uneven mass flow due to machining variability and piston misalignment during load application.

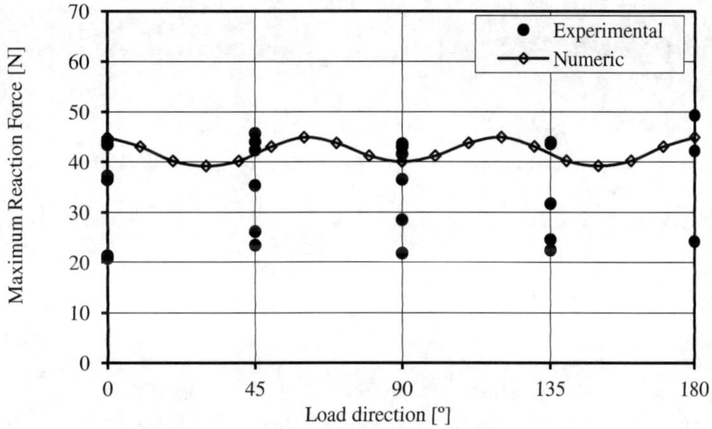

Figure 9 – Maximum bearing reaction force

4.3 Influence of design parameters

Reynolds equation shows a cubic dependence of pressure with the local film thickness; therefore radial clearance is one of the most important parameters in the bearing design. The aerostatic pneumatic piston is not an execution in this regard and it is expected that the load capacity increases with reduction in clearance. In the present configuration, this happens until 3.0μm as showed in Figure 10. For the range of parameters investigated in this work, the optimal radial clearance is between 2.5 and 3.5μm. As observed from Figure 10, total mass flow increases as the radial clearance increases.

The decrease on load capacity for smaller clearances is related to the feeding channel dimensioning, as shown in Figure 11. For a given radial clearance, the maximum bearing reaction force depends on the feeding channel diameter and practically the same level of reaction force can be achieved by an adequate choice of feeding channel diameter. Of course, the smaller the pair radial clearance and feeding channel diameter, the smaller is the mass flow needed to sustain the load, but there is a limit in practical terms related to the components manufacturing and the risk of clogging.

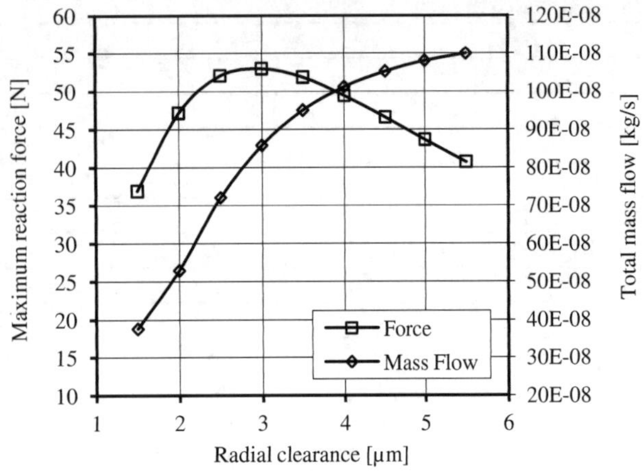

Figure 10 – Effect of radial clearance on maximum bearing reaction force

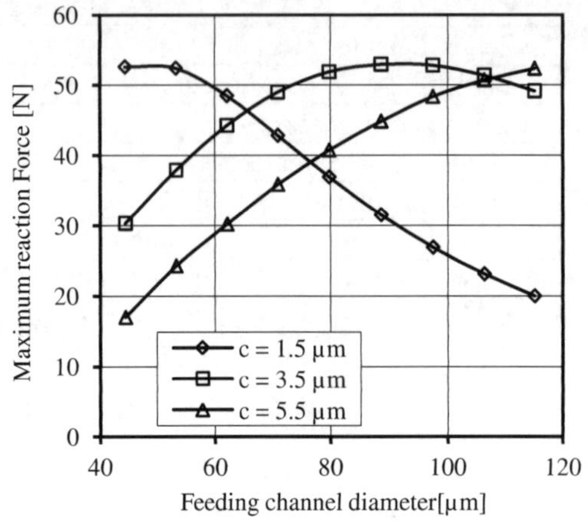

Figure 11 – Effect of feeding channel diameter on maximum bearing reaction force

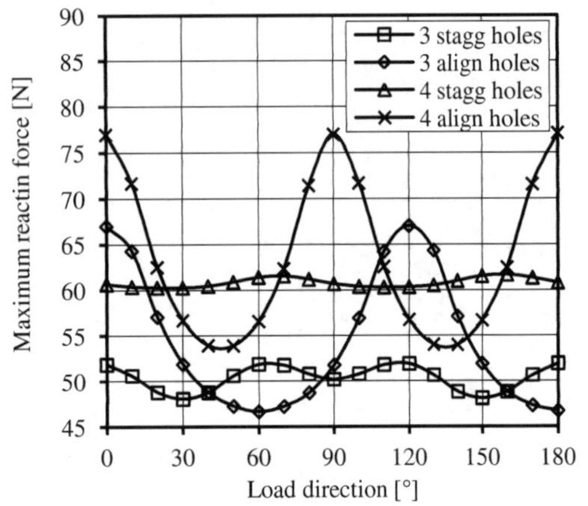

Figure 12 – Effect of feeding hole configuration on maximum bearing reaction force

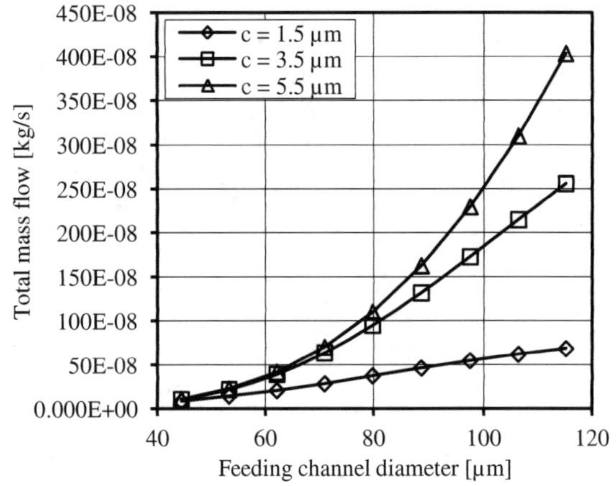

Figure 13 – Effect of feeding channel diameter on total mass flow

Another design parameter investigated was the feeding hole arrangement. Figure 12 shows the effect of four possible configurations. An additional hole in each section increases the maximum reaction force by 20% approximately, but at expenses of 30% more mass flow (not showed). Mass flow was insensitive to feeding hole alignment in the cases investigated.

Staggered feeding hole configuration is less sensitive to the load orientation. Aligned feeding holes provide a greater force when the load direction is at the bisectrix of the angles formed by the axial lines of insufflation, but a much smaller load capacity when the force direction is applied directly over a line of feeding holes.

5. CONCLUSIONS

This work presented a model to explore the influence of design parameters of a gas lubricated bearing for a piston-cylinder interface of a linear compressor. The Reynolds' equation was written for a compressible flow assuming ideal gas behavior and was numerically solved by the volume finite methodology. Feeding channels were modeled as isothermal steady state compressible flow in a constant section duct and inlet and expansion effects were disregarded. Numerical results were in accordance with available experimental data.

Effects of radial clearance, diameter of feeding channels and arrangement of feeding holes on the maximum bearing reaction force and on total mass flow consumption were investigated. Feeding channel diameter is an important design factor and shall be optimized for a given radial clearance range, which is the case for the typical configuration of the linear compressor under analysis.

NOMENCLATURE

A_f	Cross sectional area of feeding holes	(m^2)
c	Radial clearance between piston and cylinder	(m)
e_0	Piston top eccentricity	(m)
e_1	Piston bottom eccentricity	(m)
F_r	Reaction force on the piston due to gas bearing action	(N)
h	Local film fluid thickness	(m)
L_{cil}	Cylinder length	(m)
L_{pis}	Piston length	(m)
L_{man}	Bearing axial length	(m)
L_f	Feeding channel length	(m)
$\dot{m}$	Refrigerant mass flow	(kg/s)
M	Fluid flow Mach number	(–)
M_r	Moment reaction on the piston due to gas bearing action	(N.m)
p	Pressure	(Pa)
p_{cil}	Pressure inside compression chamber	(Pa)
p_{suc}	Pressure inside compressor shell	(Pa)
R	Piston radius	(m)
$\Re$	Ideal gas constant	(J/kg.K)
T	Cylinder temperature	(°C)
y_{pis}	Distance from the piston top to the valve plate	(m)
y_{cm}	Position along y axis of the piston center of gravity	(m)
V_p	Relative speed between piston and cylinder	(m/s)
ρ	Density	(kg/m^3)
ξ	Circumferential coordinate in the bearing domain	(m)
η	Axial coordinate in the bearing domain	(m)
μ	Absolute viscosity of lubricant	(Pa.s)
γ	Ratio between constant pressure and volume specific heat capacities	(–)

Subscript

i,j	Volume index on the computational domain in the circumferential and axial directions respectively
f	Index of feeding hole / channel
P	Central point for the control volume under discretization.
N,S,E,W	Central points for the North, South, East and West volumes of the central volume under discretization
n,s,e,w	Face points to the North, South, East and West volumes of the central volume under discretization

REFERENCES

Cadman, R.V., & Cohen, R., (1969). Electrodynamic oscillating compressors: part 1 – design based on linearized loads. ASME. *Journal of Basic Engineering, 656-670.*

Fourka, M., & Bonis, M. (1997). Comparison between externally pressurized gas thrust bearings with different orifice and porous feeding systems. *Wear, 210, 311-317.*

Gommed, K., & Etsion, I. (1993). Dynamic analysis of gas lubricated reciprocating ringless pistons – basic modeling. ASME. *Journal of Tribology, 115(2), 207-213.*

Hamrock, B. J., Schmid, S. R., & Jacobson, B. O. (2004). *Fundamentals of fluid film lubrication – 2nd edition.* New York, Marcel Dekker Inc.

Kazimierski, Z., & Trojnarski, J. (1980). Investigation of externally pressurized gas bearing with different feeding systems. ASME. *Journal of Lubrication Technology, 102(1), 59-64.*

Lo, C., Wang, C., & Lee, Y. (2005). Performance analysis of high-speed spindle aerostatic bearings. *Tribology International, 38 (1), 5-14.*

Patankar, S. V. (1980). *Numerical heat transfer and fluid flow*. New York, Hemisphere Publishing Corp.

Patankar, S. V., Liu, C. H., & Sparrow, E. M. (1977). Fully developed flow and heat transfer in ducts having streamwise-periodic variations of cross sectional area. *Transactions of ASME, 99, 180-186.*

Pollak, E., Friedlaender, F. J., Soedel, W., & Cohen, R. (1978). Mathematical model of an electrodynamic oscillating refrigeration compressor. *Proceedings of International Compressor Engineering Conference (246-259), University of Purdue. Indiana.*

Polman, J., Jonge, A.K. de, & Castelijns, A. (1978). Free piston electrodynamic gas compressor. *Proceedings of International Compressor Engineering Conference (241-245), University of Purdue. Indiana.*

Renn, J., |& Hsiao, C. (2004). Experimental and CFD study on the mass flow-rate characteristic of gas through orifice-type restrictor in aerostatic bearings. *Tribology International, 37, 309-315.*

Redlich, R., Unger, R., & der Walt, N. R. van (1996). Linear compressors: motor configuration, modulation and systems. *Proceedings of International Compressor Engineering Conference (341-346), University of Purdue. Indiana.*

Saad, M. A. (1992). *Compressible fluid flow – 2nd edition*. New Jersey, Prentice Hall.

der Walt, N. R. van, & Unger, R. (1994). Linear compressor – a maturing technology. *Proceedings of the 45th International Appliance Technical Conference, University of Wisconsin. Madison, Wisconsin.*

ACKNOWLEDGEMENT

This work is part of a technical partnership program between POLO Research Laboratories from Federal University of Santa Catarina and EMBRACO.

Evaluation of Wettability of Solid Surface with Oil/Refrigerant Mixture

Mitsuhiro FUKUTA[1]*, Junki SUMIYAMA[2], Masaaki MOTOZAWA[1]
Akifumi HYODO[2], Tadashi YANAGISAWA[1]

[1]Shizuoka University, Department of Mechanical Engineering,
Hamamatsu, Shizuoka, Japan
+81-53-478-1054, fukuta.mitsuhiro@shizuoka.ac.jp

[2] Shizuoka University, Graduate school of Engineering,
Hamamatsu, Shizuoka, Japan

* Corresponding Author

ABSTRACT

In order to reduce an oil amount discharged with refrigerant from a refrigerant compressor, oil separation from the refrigerant in a compressor shell or an oil separator has been investigated by many researchers. The oil separation in the compressor shell or the separator strongly depends on droplet size of oil mist generated in the compressor and condition of oil film formed on an inner surface of the compressor shell or the separator. The oil mist diameter and generation mechanism of the oil mist are affected by surface tension of the oil, while the condition of the oil film on the inner surface is influenced by wettability of the surface with the oil. In addition, dissolution of the refrigerant into the oil changes the surface tension and the wettability of the surface. In this study, the surface tension of the oil/refrigerant mixture and the wettability of the solid surface with the oil/refrigerant mixture were examined as a first step of the study on the generation mechanism and the droplet behavior of the oil mist in the compressor. The surface tension of PAG oil/CO_2 mixture was measured by a pendant drop method and the wettability of a PTFE solid surface with the mixture was evaluated by measuring a contact angle. It is found that the surface tension of the PAG/CO_2 mixture decreases steeply with increasing the refrigerant concentration in the oil. The droplet of the mixture keeps its shape with equilibrium condition on the PTFE plate. The contact angle on the PTFE surface decreases with the refrigerant concentration in the oil due to reduction of the surface tension.

1. INTRODUCTION

A part of lubrication oil stored in an oil sump of a refrigeration compressor becomes fine oil mist in a compressor shell by motions of sliding parts and a discharge valve, an entrainment induced by gas flow and rotation of a motor rotor. Although some of the oil mist is separated from refrigerant in the compressor shell, a certain amount of the oil mist is discharged with refrigerant into a refrigeration cycle. The oil circulated with refrigerant in the cycle degrades the heat transfer efficiency of heat exchangers, increases pressure drop in lines and the heat exchangers and sometimes reduces reliability of the compressor when an oil return from the cycle to the compressor is insufficient. It is therefore important to reduce the oil discharge from the compressor to the cycle.

There are many researches on the separation of the oil mist in the compressor shell or an oil separator in a discharge line. The oil separation in the compressor shell or the separator strongly depends on droplet size of the oil mist generated in the compressor and condition of oil film formed on an inner surface of the compressor shell or the separator. The oil mist diameter and generation mechanism of the oil mist are affected by surface tension of the oil, while the condition of the oil film on the inner surface is influenced by wettability of the surface with the oil. In addition, dissolution of the refrigerant into the oil changes the surface tension and the wettability of the surface.

Weber number is inversely proportional to the surface tension and is one of the dominant parameters of liquid breakup (Majithia *et al.*, 2008; Sakurai *et al.*, 2012). Khosharay and Varaminian (2014) measured the surface tension of (CH_4+H_2O), (C_2H_6+H_2O), (CO_2+H_2O) and (C_3H_8+H_2O) binary system by a pendant drop method with changing pressure and temperature. The surface tension of the oil/refrigerant mixture was measured by Seeton and Hrnjak (2006) by a maximum bubble pressure method. Jensen and Jackman (1984) measured the surface tension of the mixture by a ring method and proposed a correlation between the surface tension and the refrigerant concentration. Fukuta (2016) also measured the surface tension of the mixture by the maximum bubble pressure method under high temperature pressure condition.

The condition of the oil film formed on the solid surface affects the mist generation due to the entrainment by the flow. It is influenced by the wettability of the surface with the oil and the wettability is evaluated by a contact angle of oil droplet on the surface. It is preferable that the contact angle of the oil on the inner wall of the compressor shell is small to avoid the entrainment of the oil mist by the gas flow. Vadgama and Harris (2007) measured a quasi-static advancing contact angle of refrigerant R134a on copper and aluminum surfaces by a direct optical observation technique where the liquid meniscus at the surface of a vertical plate was captured. It is shown that the contact angle varies between 8.3° and 5.6° for the aluminum and between 5.1° and 6.5° for the copper when the temperature rises from 0 °C to 80 °C. Lin *et al.* (2016) investigated the wetting behavior of refrigerant-oil mixture on copper surface by the capillary rise method. They found that the protuberant liquid film occurs in front of meniscus during evaporation of the refrigerant-oil mixture, and the presence of oil significantly enhances the surface wettability of refrigerant.

In this study, the surface tension of the oil/refrigerant mixture and the wettability of the solid surface with the oil/refrigerant mixture are examined as a first step of the study on the generation mechanism and the droplet behavior of the oil mist in the compressor. The surface tension of the oil/refrigerant mixture is measured by the pendant drop method that can be applicable to a pressurized condition like a refrigerant atmosphere. The contact angle of the mixture on the solid surface is measured by a Sessile Drop Method in parallel with the surface tension measurement. Polyalkylene glycol (PAG) oil and CO_2 are used as refrigeration oil and refrigerant respectively. As the test surface, polytetrafluoroethylene (PTFE) plate is used in the measurement of the contact angle of the mixture on the solid surface.

2. EXPERIMENT

2.1 Experimental Setup

Figure 1 shows a schematic of an experimental setup. A main vessel is designed to withstand pressure up to 10 MPa. Two sight glasses are installed at opposite sides of the main vessel. One is used to light up an inside of the vessel, and the other is used to take pictures of a droplet on a test plate. The oil/refrigerant mixture, i.e. PAG/CO_2 mixture in this study, is prepared in a sub vessel prior to the experiment and fed to the main vessel by a gear pump. The PAG/CO_2 mixture flows through a capillary tube and the droplet drops from the tip of capillary tube on the test plate. Since volume of the droplet in a Sessile Droplet Method is less than 4 μL, the capillary tube of 0.5 mm in outer diameter is used. The tip of the capillary tube is set at 5 mm above the test plate. The test plate is fixed on a rotary stage and the rotary stage is set on a thrust bearing. The rotary stage has two neodymium magnets on a side of the stage. This configuration allows us to rotate the test plate by operating other magnets from outside of the main vessel so that the measurement can be repeated at the same condition with changing the dropping position of the droplet by the rotation of the test plate to confirm reproducibility. Polytetrafluoroethylene (PTFE) plate is used in this study as the test plate. Refrigerant concentration is measured by a sampling method. In the experiment, the measurements of the contact angle, the surface tension and the refrigerant concentration of the mixture are carried out with changing the pressure and the temperature of the PAG/CO_2 mixture. Purity of CO_2 used in this experiment is 99.95%. Viscosity grade of PAG used in this study is 100 mm^2/s at 40 °C.

2.2 Measurement of Contact Angle

The contact angle of the mixture is measured by the Sessile Drop Method for the wettability evaluation. Figure 2 shows a picture of the droplet on the test plate taken by a CCD camera. The contact angle of the droplet is an angle between a horizontal line and a tangential line of the droplet at a contact point of an edge of the droplet. The tangential line is calculated by a curve fitting method. Firstly, detection of droplet boundary is conducted by using an image processing software and *x-y* coordinates (x_i, y_i) of points on the boundary are obtained. Then, an approximation curve of the droplet boundary is calculated by the curve fitting with an assumption that the shape of the droplet boundary is perfect circle when the volume of the droplet is small. Equation (1) is used to obtain the approximation curve of the boundary by a least square method with identifying a coordinate of the center (*a*, *b*), and radius *r* of the circle from the discrete points on the boundary.

$$\begin{pmatrix} -2a \\ -2b \\ a^2+b^2-r^2 \end{pmatrix} = \begin{pmatrix} \sum x_i^2 & \sum x_i y_i & \sum x_i \\ \sum x_i y_i & \sum y_i^2 & \sum y_i \\ \sum x_i & \sum y_i & \sum 1 \end{pmatrix}^{-1} \begin{pmatrix} -\sum\left(x_i^3 + x_i y_i^2\right) \\ -\sum\left(x_i^2 y_i + y_i^3\right) \\ -\sum\left(x_i^2 + y_i^2\right) \end{pmatrix} \tag{1}$$

The broken line in Figure 2 is the approximation curve obtained by Equation (1). The tangential line at the contact point is calculated from the approximation curve and the coordinate at the contact point. The contact angle θ is obtained from the angle between the horizontal line and the tangential line of the approximation curve at the contact point. γ_S, γ_L, and γ_{SL} shown in Figure 2 expresses a balance of surface tension and interfacial tension, and are explained in the following section.

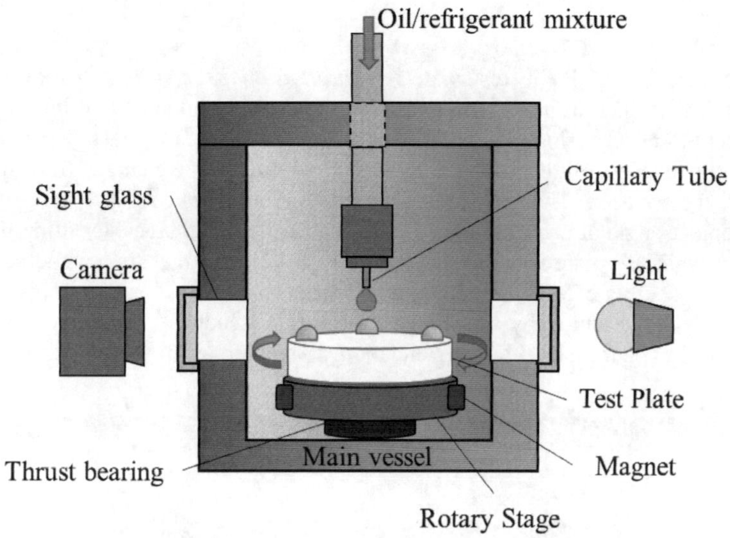

Figure 1 Experimental setup of wettability evaluation

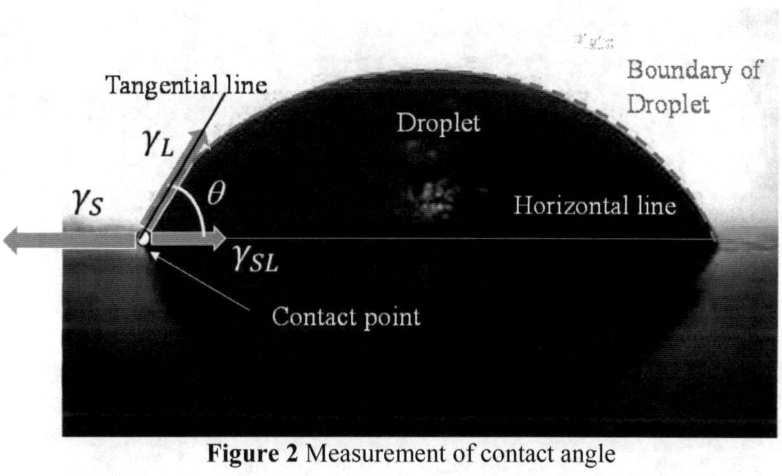

Figure 2 Measurement of contact angle

2.3 Measurement of Surface Tension

Since the surface tension of liquid is closely related to the contact angle of the droplet on the solid surface, the surface tension of the oil/refrigerant mixture is also measured by the pendant drop method in the same experimental setup shown in Figure 1. During dripping the droplet from the capillary tube, the surface tension of the droplet is calculated by the shape of the droplet when the diameter of the droplet becomes the maximum. When the maximum diameter of the droplet is d_e, and the diameter at the point where the height from the bottom of the droplet is equal to d_e is d_s (see upper left in Figure 3), the surface tension of the droplet is calculated from Equation (2).

$$\gamma = J \cdot g \cdot \rho \cdot (d_e)^2 \tag{2}$$

where, γ is surface tension, g is the gravitational acceleration, ρ is density of the oil/refrigerant mixture. J is a correlation factor and calculated by the following equation (Fordham, 1948).

$$J = -5.602x^3 + 17.669x^2 - 19.405x + 7.6496$$
$$(x = d_s / d_e) \tag{3}$$

3. EXPERIMENTAL RESULTS AND DISCUSSIONS

3.1 Contact Angle of Pure PAG on PTFE Plate

The contact angle of pure PAG on the PTFE test plate is measured at first. An observation results of the droplet shape are shown in Figure 3 when the droplet is dripped from the capillary tube and collides against the plate. The temperature condition is kept to 60°C during the experiment. The height from the plate to the capillary tube is set to 5 mm. Time is zero when the droplet collides with the plate. The droplet shape changes from sphere to hemi-sphere on the plate at an instant of the collision. Then, it slightly spreads and becomes equilibrium condition after 5 seconds from the collision. The contact angle at each moment is obtained from the images and time history of the contact angle of the pure PAG on the PTFE plate is shown in Figure 4. It is found that the contact angle decreases at the instant of the collision with the plate and becomes stable after 2 second after the collision. In the following experiment, the contact angle is measured by the image taken at 10 seconds after the collision. In addition, the surface tension of the droplet can be measured by the pendant drop method as previously explained at the section 2.3.

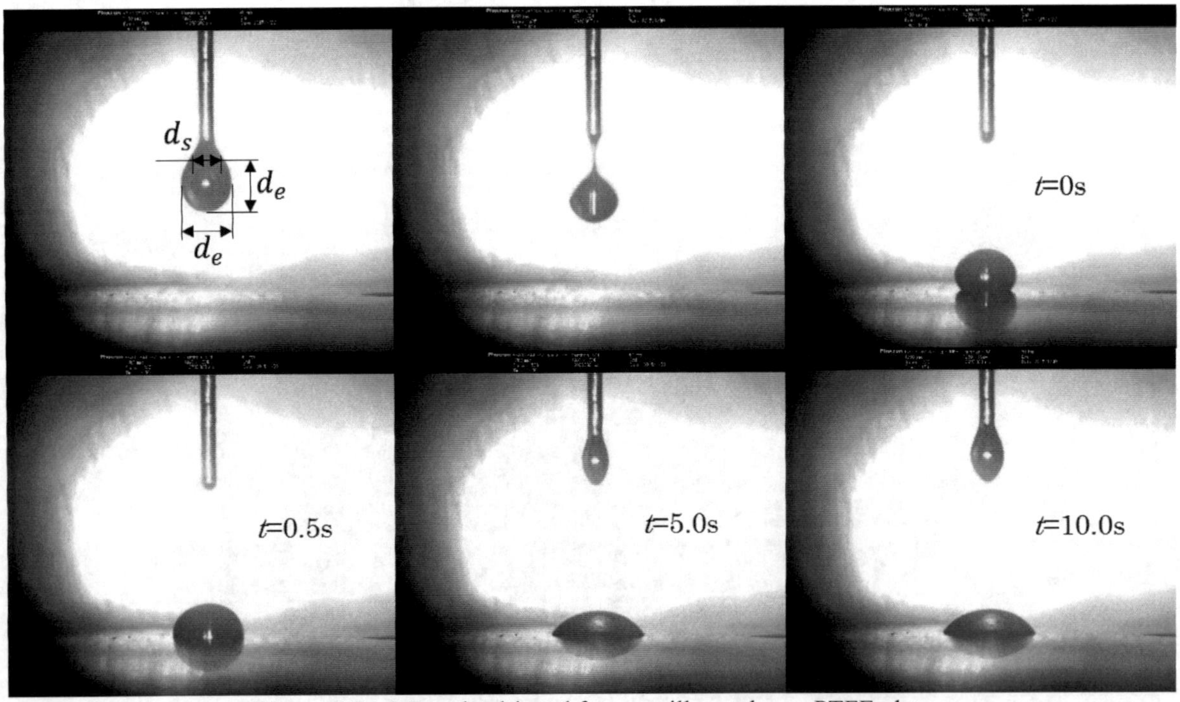

Figure 3 PAG Droplet dripped from capillary tube on PTFE plate

Figure 5 shows the contact angle of PAG on the PTFE plate measured with changing the temperature as a parameter. The contact angle is measured 5 times at each condition, and an average value is plotted against the temperature. The contact angle decreases with increasing the temperature. Under the equilibrium condition, Young's formula expressed by Equation (4) is established as shown in Figure 2.

$$\gamma_S = \gamma_L \cos\theta + \gamma_{SL} \tag{4}$$

where, γ_S is solid surface tension, γ_L is surface tension of liquid phase, γ_{SL} is solid-liquid interfacial tension, and θ is the contact angle. According to Equation (4), the contact angle depends on the liquid surface tension if the solid-liquid interfacial tension are assumed to be constant against the temperature since the solid surface tension hardly changes in this temperature range. The surface tension of PAG measured by the pendant drop method is shown in Figure 6 against the temperature. It is shown that the surface tension of PAG decreases with increasing the temperature, and it causes the reduction of the contact angle shown in Figure 5.

Relationship between the surface tension and the contact angle is shown in Figure 7. It is found that the contact angle increases with the surface tension. Duple's equation shown by the following equation expresses a work of

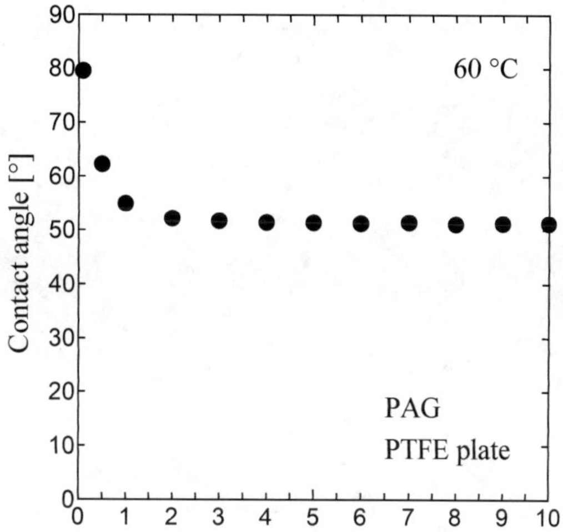

Figure 4 Change of contact angle from collision

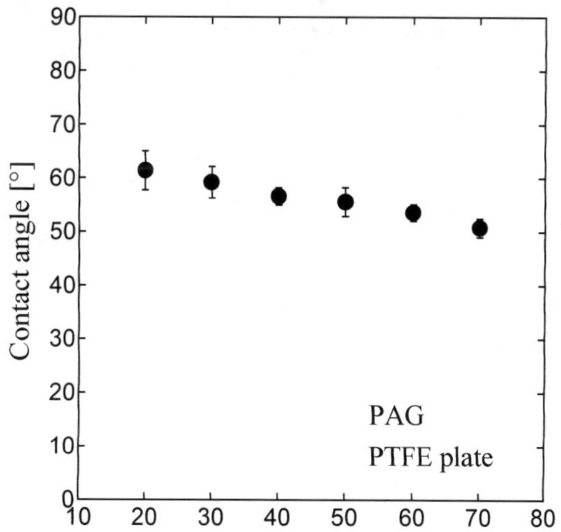

Figure 5 Contact angle of PAG on PTFE versus temperature

adhesion, which is the work per unit area required to separate the drop from the solid surface.

$$W_A = \gamma_s + \gamma_L - \gamma_{SL} \tag{5}$$

The work of adhesion is an index of the wettability and it is indicated that the wettability becomes better with large work of adhesion. With considering that the droplet of the PAG maintains its shape on the PTFE plate with equilibrium condition, Young-Duple equation is derived from Equations (4) and (5) as follows.

$$W_A = \gamma_L (1 + \cos\theta) \tag{6}$$

The work of adhesion under the equilibrium condition is a function of the surface tension of liquid and the contact angle as expressed by Equation (6). The work of adhesion is obtained by the surface tension and the contact angle, and found to be 45.5 mJ/m^2 [mN/m] when the surface tension is 29 mN/m. As a reference value, the work of adhesion of PTFE film with glycerol is shown to be 45.9 mN/m by Gilman (2014). The contact angle calculated by Equation (6) with an assumption of constant work of adhesion of 45.5 mN/m is shown in Figure 7 by solid line. Although the calculated contact angle shows the similar tendency as the experimental one, gradient of the calculated contact angle is larger than experimental one. It may be caused by the fact that the work of adhesion varies with the

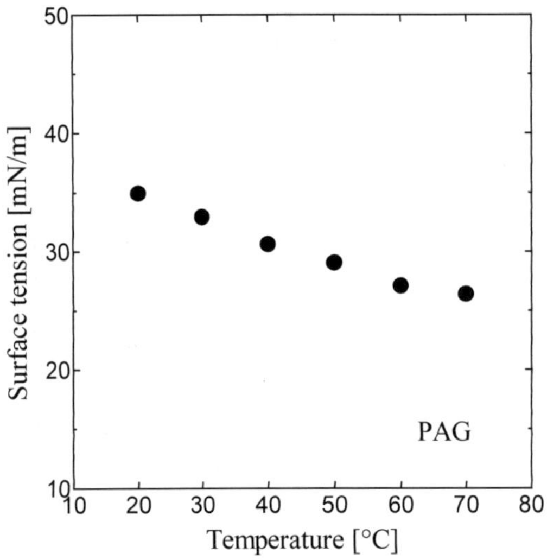

Figure 6 Surface tension of PAG versus temperature

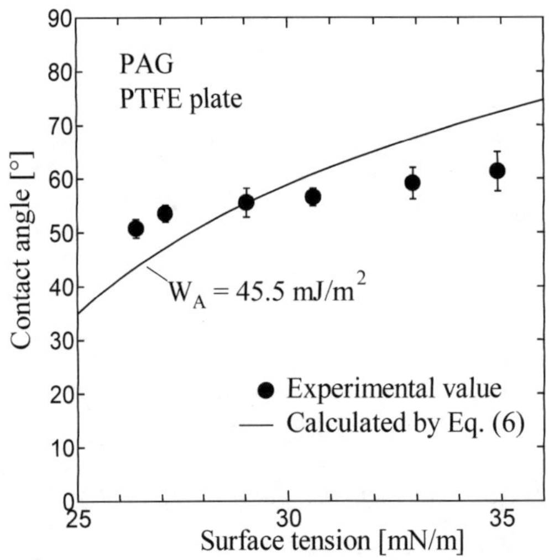

Figure 7 Influence of surface tension on contact angle (PAG on PTFE)

temperature. It is also supposed that contamination on the PTFE surface affects the wetting phenomena.

3.2 Contact Angle of PAG/CO_2 Mixture on PTFE Plate

The influence of refrigerant dissolved into the oil on the contact angle is investigated. Figure 8 shows the relationship between the mass fraction of CO_2 dissolved in the PAG and the surface tension of the mixture measured by the pendant drop method at 30°C. The surface tension of the PAG/CO_2 mixture decreases steeply with increasing the CO_2 concentration in the mixture. The contact angle of the PAG/CO_2 mixture is shown in Figure 9 against the CO_2 concentration in the mixture. It is shown that the contact angle of the PAG/CO_2 mixture on the PTFE plate decreases with increasing the CO_2 concentration in the mixture due to the reduction of the surface tension of the mixture. Therefore, it is supposed that the wettability of the solid surface with the PAG/CO_2 mixture having high CO_2 concentration becomes high. The contact angle is re-arranged against the surface tension as shown in Figure 10. When the surface tension of the mixture becomes smaller than 22.5 mN/m, the contact angle calculated by Equation (6) with the work of adhesion of 45.5 mJ/m^2 becomes zero, which means that the wetting becomes a perfect one in which the droplet spreads and does not keep the droplet shape. In the experiment, however, the droplet of the mixture with the surface tension of 20 mN/m keeps the equilibrium contact angle. This difference may be caused by the fact that the work of adhesion is not constant. Contamination on the PTFE surface is another factor that affects

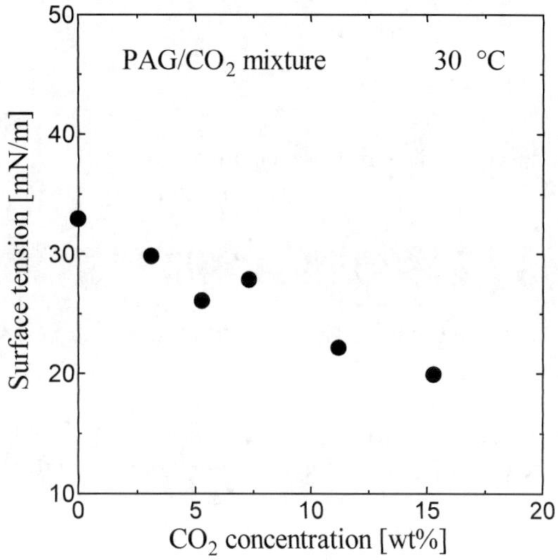

Figure 8 Surface tension of PAG/CO_2 mixture versus refrigerant concentration

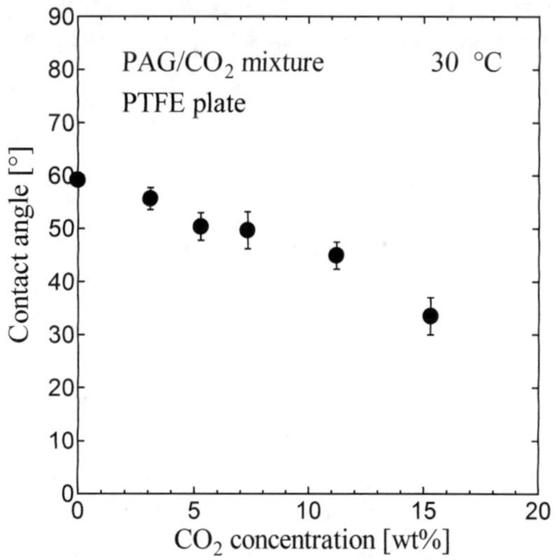

Figure 9 Influence of refrigerant concentration on contact angle of PAG/CO_2 mixture on PTFE

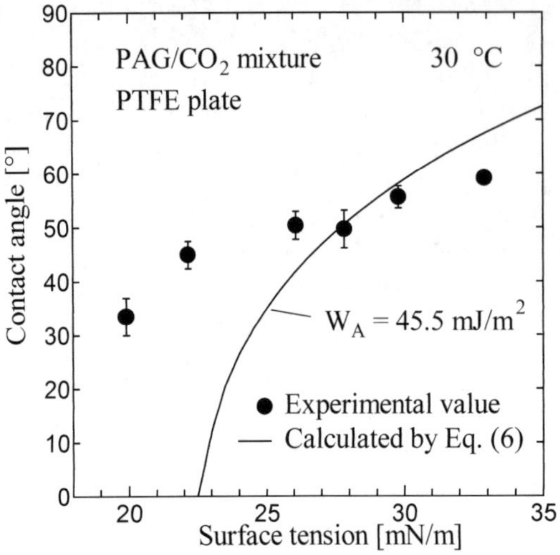

Figure 10 Contact angle of PAG/CO_2 mixture on PTFE versus surface tension

the wetting phenomena.

Since metal materials have much larger solid surface tension or work of adhesion than those of the PTFE, the wetting on the metal surface with the oil/refrigerant mixture will tend to be the perfect one.

4. CONCLUSIONS

In this study, the wettability of solid surface with oil/refrigerant mixture is evaluated experimentaly. The following conclusions are obtained

1. The experimental apparatus developed in this study is applicable to measure the contact angle and the surface tension of oil/refrigerant mixture under high pressure and temperature condition.
2. The droplet of pure PAG on the PTFE plate becomes equilibrium condition after 2 second from the collision with the plate. The contact angle of the pure PAG on the PTFE plate decreases with increasing temperature due to the reduction of the surface tension of the PAG.
3. The droplet of PAG/CO_2 mixture keeps the droplet shape with the equilibrium condition on the PTFE plate, and the contact angle decreases with increasing the CO_2 concentration in the mixture since the surface tension of the mixture steeply decreases with increasing the CO_2 concentration.
4. The wetting of the oil/refrigerant mixture on the metal surface is supposed to be the perfect one due to the much higher solid surface tension of the metal than that of the PTFE.

REFERENCES

Fordham, S. (1948). On the calculation of surface tension from measurements of pendant drops, *Proceedings of Royal society of London*, Volume194, A.

Fukuta, M., Sumiyama, J., Motozawa, M., Yanagisawa, T. (2016). Surface tension measurement of oil/refrigerant mixture by maximum bubble pressure method. *Int. J. Refrig.*, submitted.

Gilman, A., Piskarev, M., Yablokov, M., Kechek'yan, A., Kuznetsov, A. (2014). Adhesive properties of PTFE modified by DC discharge. *J. Physics: Conference Series*, 516, 012012.

Jensen, M.K., Jackman, D.L. (1984). Prediction of nucleate pool boiling heat transfer coefficients of refrigerant-oil mixtures. *J. Heat Transfer*, 106, 184-190.

Khosharay, S., Varaminian, F. (2014). Experimental and modeling investigation on surface tension and surface properties of (CH_4+H_2O), (C_2H_6+H_2O), (CO_2+H_2O) and (C_3H_8+H_2O) from 284.15 K to 312.15 K and pressures up to 60 bar. *Int. J. Refrig.*, 47, 26-35.

Lin, L., Peng, H., Ding, G. (2016). Influence of oil concentration on wetting behavior during evaporation of refrigerant-oil mixture on copper surface, *Int. J. Refrig.*, 61, 23-36.

Majithia, A.K., Hall, S., Bowen P.J. (2008). Droplet breakup quantification and processes in constant and pulsed airflows. *Proceedings of the 22nd European Conference on Liquid Atomization and Spray System*, ILASS08-4-4.

Sakurai, Y., Kobayashi, K., Fujikawa, T., Sanada, T., Watanabe, M. (2012). Observation of deformation of single droplet impact on solid surface (Effects of liquid properties, surface roughness and surrounding gas pressure). *Japanese Journal of Multiphase flow*, Vol.26 No.2, 164-171. (in Japanese)

Seeton, C.J., Hrnjak, P. (2006). Thermophysical properties of CO_2-lubricant mixtures and their affect on 2-phase flow in small channels (less than 1 mm). *Proceedings of the International Refrigeration and Air Conditioning Conference at Purdue*, R170.

Vadgama, B., Harris, K. D. (2007). Measurements of the contact angle between R134a and both aluminum and copper surfaces. *Experimental Thermal and Fluid Science*, 31, 919-984.

ACKNOWLEDGEMENT

The density measurement of the oil/refrigerant mixture was done with help by Mr. Tatsuhiro Suzuki, working at DENSO corp., when he was a student at graduate school of engineering, Shizuoka University, and the authors would like to appreciate for his effort.

Numerical simulation of the flow inside a scroll compressor equipped with Intermediate Discharge Valves

Alain PICAVET, Brian ANGEL

Danfoss Commercial Compressors, 01600 Trevoux ; France

Renuda, Belleville sur Saône ; France

a.picavet@danfoss.com

brian.angel@renuda.com

ABSTRACT

This paper presents the results of CFD simulations of the compression process of a scroll compressor. The compressor geometry accounts for the final scroll gas pocket, two intermediate discharge valves (IDVs), the central discharge zone and the upper shell or high pressure (HP) zone. The numerical model uses a real gas equation of state to determine gas properties during the compression process and accounts for the motion of the orbiting scroll and the IDVs using a mesh smoothing and remeshing algorithm. The IDVs are represented as a spring mass system with their movement controlled via the pressure difference around the valve. Appropriate pressure based boundary conditions are used at entry to the third gas pocket and at the exit of the upper shell. An initial analysis of the results indicates that it is possible to achieve time accurate results of the pressure field throughout the flow domain and also determine the impact of the IDVs on scroll performance. The results will also enable a more thorough analysis of the fluid flow and compression process inside the scroll in order to improve its performance.

1 INTRODUCTION

The use of CFD for modelling scroll compressor fluid flows is now standard practice due to the technical capabilities of commercial CFD codes and the use of high performance machines. Typical modelling processes at Danfoss now account for the scroll compression process using an orbiting scroll, IDVs and flank and tip leakage.

Due to its significant experience in this area, Danfoss uses IDVs to maintain compressor performance when it is operating at low pressure ratios. At these pressure ratios, the pressure in the final gas pockets can be higher than that in the HP zone resulting in lower efficiency and excessive power consumption. The IDVs are used to bleed off this excessive pressure and, as previously mentioned, maintain performance.

Detailed experimental analysis (Giniès et al. and Picavet et al.) and CFD modelling of simplified scroll compressors (Angel et al.) have been previously presented. The work presented here focuses on the CFD modelling of a scroll compressor using two IDVs and the comparison of these results with experimental data.

2 NUMERICAL MODEL

The current numerical model is based on the commercial CFD solver FLUENT. For all simulations, the flow is assumed to be unsteady, three-dimensional, compressible and turbulent with adiabatic walls. The gas is assumed to be a real gas, R410A, modelled using the Soave-Redlich-Kwong equation of state available in FLUENT. The initial properties of R410A to be used in Fluent are computed using REFPROP, the software developed by the NIST.

Turbulence is modelled using the k-ω model. The boundary conditions used are pressure-inlet, pressure-outlet and wall. A second order scheme is used for the discretization of the equations. The SIMPLE algorithm with default relaxation is used to ensure the convergence of the solution in time, using a first order stepping method, with up to 30 iterations per time-step. FLUENT has been set up save flow field data 40 times per revolution (volume or mass averaged data, flow field results and various images on different viewing planes).

The computational domain is shown in Figure 1 and consists of the final gas pocket, the main discharge port, two IDVs and the HP zone of a 2.4 volume ratio machine, the cooling capacity of which being 25 tons. The numerical model uses the mesh motion algorithm and appropriate UDFs to control the movement of the orbiting scroll and the movement of the IDVs. Mesh quality is maintained using the mesh smoothing and remeshing algorithms. The simulations are run using a time step size designed to ensure that the remeshing process does not generate negative volume cells. The initial grids of the pockets of the scrolls are shown in figure 2.

The scroll compressor is running with the refrigerant gas R410A at 2900 rpm with an operating point characterized by an evaporating temperature of 10 °C and a condensing temperature of 40 °C with an superheat of 10 °C. The pressure ratio at this point is 2.23, which ensures that the IDVs will open for each revolution of the orbiting scroll.

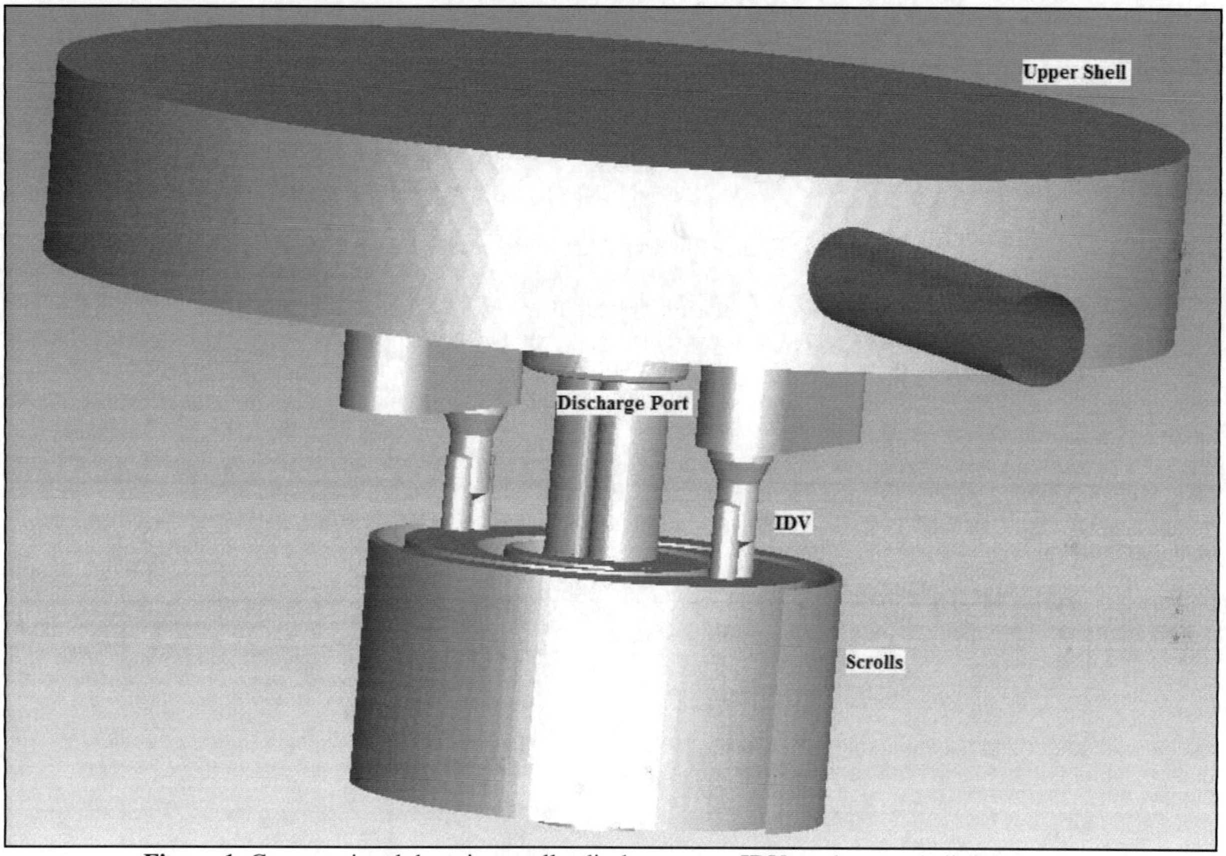

Figure 1: Computational domain: scrolls, discharge port, IDVs and upper shell (HP zone)

Danfoss compressors are equipped with tip seals located on the top of the static and orbiting scroll spirals. The central part of each spiral interacts with the discharge port hence there is no tip seal present due to the risk of the seal being pulled out of its housing. Where present, the seal is assumed to cover the entire width of the spiral so this zone is not meshed. Where the seal is not present in the central zone of each spiral (hence the gas can flow through this tip gap) the tip gap zone is meshed with at least 5 layers of cells for gradient capture. The tip gap mesh does not deform but it rotates with the orbiting scroll.

Flank leakage is controlled via the use of appropriate UDFs which limit the mass flow through flank gap to appropriate flow rates.

Figure 2: Gas pocket initial mesh

Figure 3 presents a sectioned view of a scroll with typical IDV. For the numerical model, the geometry of the Intermediate Discharge Port is respected, however the valve has been simplified to a solid round disc which is subject to the gas pressure in the pocket and the upper shell. The disc movement is vertical and is modeled as a as a spring-mass system. The stiffness and the mass of the real valves are used to model the movement. When the valve is closed, a tiny gap exists between the lower surface of the valve (or disc) and it's housing. This gap is meshed and the cells in this zone are assigned viscosity of the order of 100Pa.s when the valve is in its closed position. This high viscosity ensures a negligible leakage flowrate through this zone. When the valve starts to open, the viscosity is reset to its normal value and the layers of cells are "inflated" via a UDF which is controlled by the dynamic mesh algorithm.

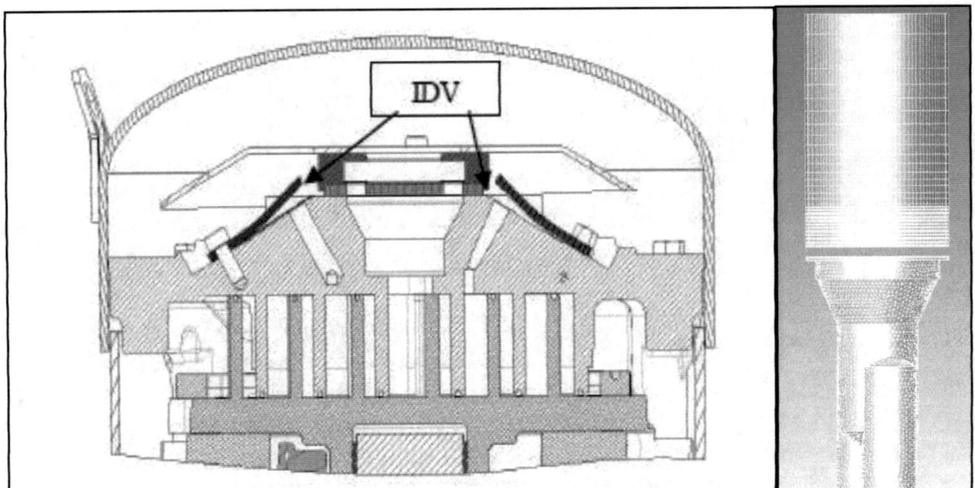

Figure 3: Comparison between real and modelled IDVs

3 RESULTS AND DISCUSSION

3.1 Analysis of pressure curves

Figure 4 presents the pressure in the final gas pockets (red curves) and the discharge port (blue curves) at different monitoring points. Given that only the final gas pockets are modelled, the predicted pressure is only "measured" for a given part of the compression cycle. The initial pressure used at the inlet to each of the final gas pockets is set to the pressure at the end of the compression of the intermediate or second gas pockets. These inlet pressures are calculated using 1D software developed and validated by Danfoss.

Figure 4 shows the pressure increasing as the pocket volume reduces then stabilize as the IDVs open due to the pressure difference between the gas pockets and the HP zone. The discontinuous reduction in pocket pressure is due to the monitoring point moving outside of the flow domain.

In the main discharge port, the pressure can be seen to be oscillatory which is linked to the opening of the scrolls at the beginning of discharge process. This opening means contact between three volumes at different pressures and results in rapid mass transfers of gas and highly turbulent flow between the gas pockets and the discharge port.

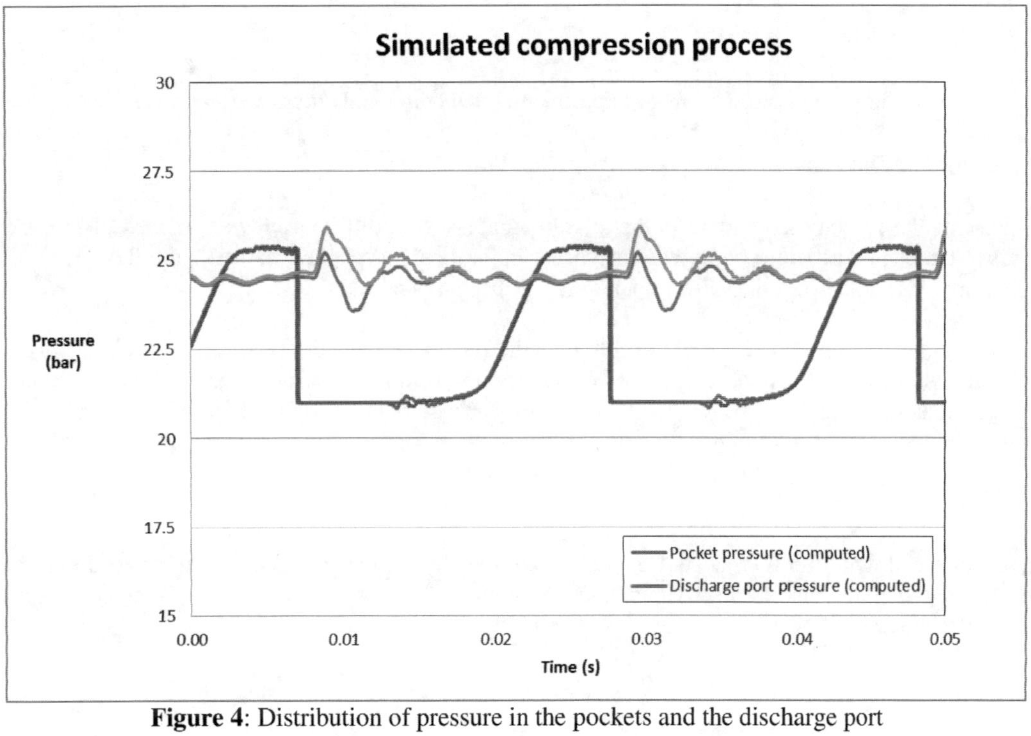

Figure 4: Distribution of pressure in the pockets and the discharge port

Figure 5 compares the pressure distribution and IDV motion measured during a previous test campaign (on a machine with the same volume ratio running at the same operating point) and compares this data with the CFD results. The tests show that the valve starts to move when the pressure difference between the pocket and the discharge port reaches more than 1.5 bar, resulting in a time shift of 3 milliseconds. The simulation shows a pressure difference of less than 0.5 bar before the beginning of IDV motion, with a time shift of 0.5 millisecond and a longer opening duration. The CFD also overestimates the discharge flowrate through the IDV resulting in an underestimated maximum pressure in the pocket. Hence, the CFD model should be tuned to obtain more reliable results.

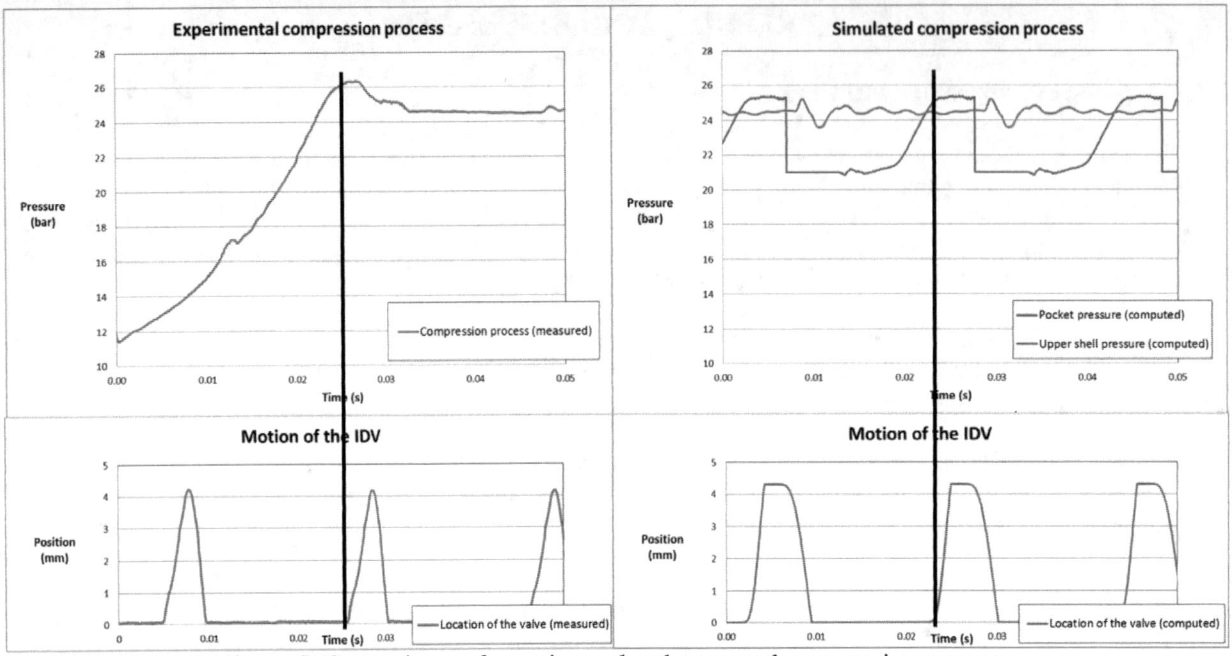

Figure 5: Comparisons of experimental and computed compression process

3.2 Analysis of the flow field

Figure 6 presents the velocity magnitude in the scroll compressor on different viewing planes. Here, the angle 0° corresponds to the beginning of the compression process in the final gas pockets. At 63°, the IDVs start to open and at 144°, they are closed. At 162°, the spirals open onto the discharge port.

The gas velocity can be seen to be in excess of 30m/s in the IDV ducting as the IDVs open and in the discharge port as the gas pockets open onto this zone. The gas velocity in the tip gap can also exceed 30m/s. An analysis of the flow indicates that it is highly turbulent with numerous vortices being generated in the gas pockets and in the discharge port, explaining (at least in part) the oscillations seen in Figure 4. At 144°, high gas velocities are seen in the IDV ducting and this is due to the gas flowing back into the gas pockets due to the pressure in the HP zone being higher than that in the gas pockets.

The analysis of the flow in the IDV ducts has shown that the flow is not identical in each duct due to differences seen in each gas pocket but also locally around each valve in the HP zone. This flow asymmetry is of interest and will be investigated further.

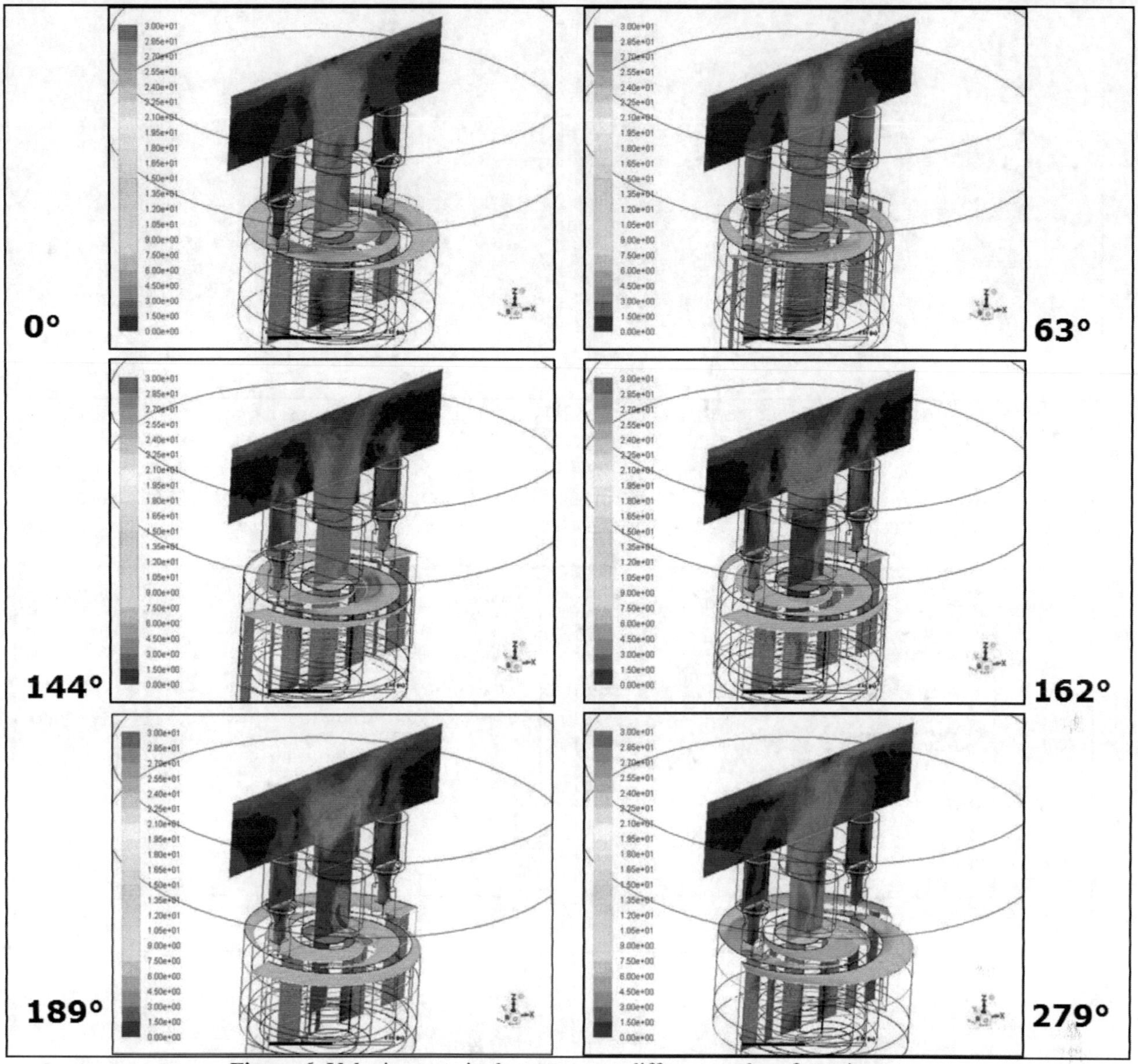

Figure 6: Velocity magnitude contours at different angles of rotation

Figure 7 presents the velocity field around the IDVs and in the discharge port as the spirals open. In the IDV ducting, it can be seen that the gas velocity is of the order of 50m/s and that there is a large zone of recirculating gas downstream of the valve when it is fully open. This will contribute to the overall losses of the scroll compressor hence a possible redesign of the shape of the IDV valve to reduce this loss could be considered (although it should be understood that these losses are only for approximately 1/6th of a complete revolution).

As the spirals open onto the main discharge port, the gas can be seen to be flowing into the port at velocities exceeding 50m/s resulting in highly turbulent flow with large recirculating zones. This will also contribute to the overall losses of the compressor and indicates that the design of the discharge port could be reconsidered in order to improve performance.

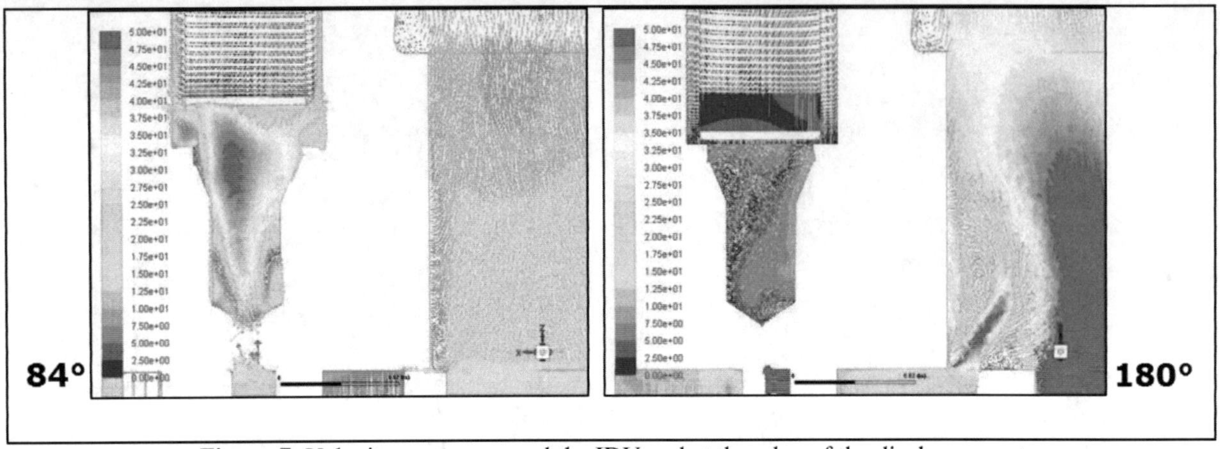

Figure 7: Velocity vectors around the IDV and at the edge of the discharge port

Figure 8 presents the vorticity calculated by FLUENT where there is a significant increase in vorticity as the discharge process starts. This confirms the analysis of the flow field presented in Figures 6 and 7 and indicates that CFD can be used with success to analyse scroll performance and assist in the design process.

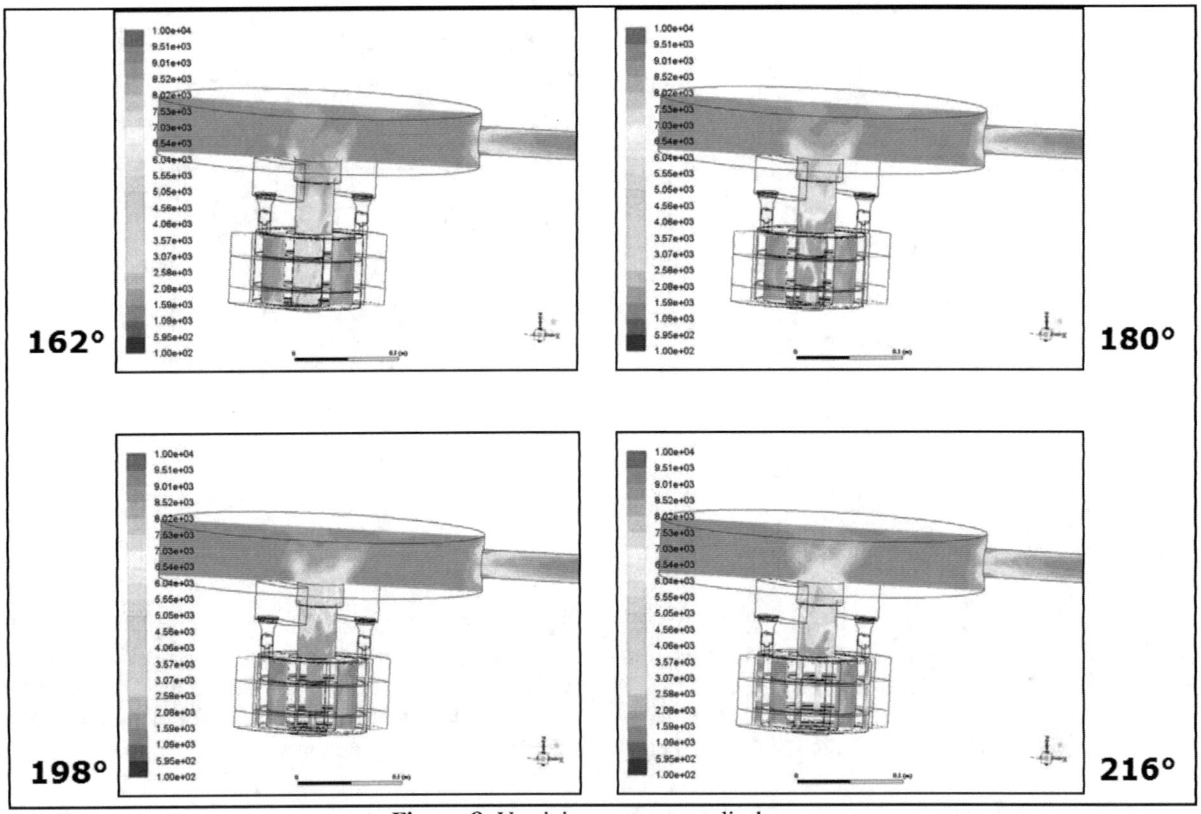

Figure 8: Vorticity contours at discharge

4 CONCLUSIONS

The work presented here has shown that it is possible to simulate, with an acceptable level of accuracy and CPU time, the flow in a scroll compressor whilst accounting for important details such as tip and flank leakage and intermediate discharge valves.. These results of such simulations provide essential details concerning the flow field and how the different components of the compressor interact with one another. These results are also extremely useful to gain an understanding of the unsteady physical phenomena occurring during the compression process, especially in the discharge port, and how it would be possible to improve compressor performance.

REFERENCES

P. Giniès, C. Ancel, D. Gross : Scroll Compressors and Intermediate Valve Ports. *7th International Conference on Compressors and their* Systems, London, *5-6th September 2011.*

B. Angel, P. Giniès, D. Gross, C.Ancel : CFD modelling of scroll compressor intermediate discharge ports. *8th International Conference on Compressors and their Systems, London, 9-10 September 2013.*

Picavet, Alain and Giniès, Pierre : Experimental Pressure-Volume Diagrams of Scroll Compressors. *22nd International Compressor Engineering at Purdue, 14-17th July 2014.*

A Heat Transfer Correlation for the Suction and Compression Chambers of Scroll Compressors

Evandro L.L. PEREIRA[§], Cesar J. DESCHAMPS*

POLO Research Laboratories for Emerging Technologies in Cooling and Thermophysics
Federal University of Santa Catarina
88040-900, Florianopolis, SC, Brazil
evandro.l.lange@embraco.com, deschamps@polo.ufsc.br

* Corresponding Author

ABSTRACT

Heat transfer in the suction and compression chambers of scroll compressors has not been sufficiently studied and typical correlations available in the literature are based on simplified flow conditions. This paper presents the results of a model developed to predict fluid flow and heat transfer inside the suction and compression chambers of scroll compressors. Due to the particular geometry of scroll compressors, an algorithm was developed to automatically adapt the computational mesh for each orbiting angle. Convective heat transfer is strongly affected by the flow in the near-wall region and for this reason a low Reynolds number turbulence model was adopted in the simulations. The study covered a wide range of operating conditions and geometric parameters, allowing the proposal of a new heat transfer correlation for scroll compressors that is compared with other correlations available in the literature.

1. INTRODUCTION

In the suction process of compressors, the refrigerant is heated as it gets into contact with hot components, such as the electric motor, bearings, crankcase, and compressor housing. Some of the heat is rejected to the outside environment through the compressor housing, whereas the remainder ends up heating the refrigerant admitted in the suction line. Figure 1 outlines the interactions between the sources and the heat paths in a typical low-pressure housing scroll compressor. Some gas superheating in the suction process is necessary, either to avoid the presence of liquid in the compression chambers, or primarily to cool the electrical motor. However, suction gas superheating negatively affects the compressor performance. The first effect is the reduction of the volumetric efficiency due to the decrease of gas density, reducing the cooling capacity of the system. The second consequence is the reduction of the isentropic efficiency brought about by the associate increase of specific work of compression (Kremer *et al.*, 2012).

Most simulation models for thermodynamic analyzes of scroll compressors adopt integral formulations (Chen *et al.*, 2002). The accuracy of such models depends on the adequate description of different processes involved, such as suction and discharge, leakage and heat transfer. However, there are few studies (Sunder, 1997; Ooi and Zhu, 2004; Jang and Jeong, 2006) dedicated to convective heat transfer in scroll compressors, greatly due to the level of complexity imposed by the restricted space for measurement and small time scales. As a consequence, there remains much uncertainty about predictions of heat transfer for operating conditions found in actual applications.

Based on experimental results and correlations of heat transfer for turbulent pipe flows, Sunder (1997) pointed out that the physical contact between the stationary and orbiting scrolls would be the primary heat transfer mechanism during the compression process. However, his measurements for the temperature profile of the stationary scroll differs from experimental data of other studies, indicating that this heat transfer mechanism may not always happen.

[§] Present address: Embraco - Research and Development, Joinville, SC, Brazil.

Ooi and Zhu (2004) developed a simulation model for the study of flow and heat transfer in the compression chamber of a scroll compressor. They adopted the standard k-ε turbulence model and wall-functions, in addition to specifying the turbulent Prandtl number equal to 0.6 based on empirical results for Wankel engines. For some operating conditions, their predictions of heat transfer coefficient were much higher than the values indicated by Sunder (1997).

Jang and Jeong (2006) measured the temperature distribution in the stationary scroll and analyzed heat transfer for specific conditions by using an oscillating wall to simulate the orbiting scroll. The authors proposed a correction factor for the Nusselt number based on the Strouhal number to include the effect of the oscillating motion. According to the authors, their correlation was able to predict the discharge temperature better than other correlations, including the correlation for spiral heat exchangers adopted by Chen *et al.* (2002).

The present paper reports a numerical analysis of heat transfer in the suction and compression chambers of scroll compressors. A two-dimensional model was developed to simulate this phenomenon in details, taking into account different geometries, operating conditions and refrigerants. Due to the particular geometry of the scroll wraps, an algorithm was developed to automatically adapt the computational mesh throughout the simulation. A low Reynolds number turbulence model was employed to predict the shear stress and heat transfer at the walls. Based on dimensional analysis and simulations, a new convective heat transfer correlation is proposed and compared to other correlations commonly adopted in simulations of scroll compressors.

2. DIMENSIONAL ANALYSIS

There are invariably many parameters involved in the study of fluid flow and heat transfer. Dimensional analysis allows the reduction of the number of variables in the problem and the effort of numerical simulations. For the specific problem in question, heat transfer between the gas and the walls of the scrolls can be characterized by the Nusselt number as a function other dimensionless parameters:

$$\mathrm{Nu} = f\left(\mathrm{Re}, \mathrm{Pr}, C^*, \frac{h}{a}, \frac{\bar{T}_w}{\bar{T}_g}, \frac{c_p}{c_v}\right) \tag{1}$$

The Nusselt number, Nu, can be written as follows:

$$\mathrm{Nu} = \frac{h_c D_h}{k_f} = \frac{q_w}{\left(\bar{T}_w - \bar{T}_g\right)} \frac{D_h}{k_f} \tag{2}$$

where h_c is the convective heat transfer coefficient, q_{w} is the heat flux at the wall, D_{h} is the hydraulic diameter, k_{f} is the fluid thermal conductivity, $\bar{T}_{\mathrm{w}}$ is the wall mean temperature and $\bar{T}_{\mathrm{g}}$ is the gas mean temperature.

In Equation (1), Re is the Reynolds number of the flow within the compression chamber, and Pr is the fluid Prandtl number, respectively defined by:

$$\mathrm{Re} = \frac{\rho U_s D_h}{\mu} \qquad \mathrm{Pr} = \frac{c_p \mu}{k_f} \tag{3}$$

where ρ is the fluid density, U_s is the characteristic velocity, μ is the fluid dynamic viscosity and c_p is the fluid specific heat at constant pressure.

The Reynolds number requires a characteristic velocity for the fluid flow inside the chamber. Analyzing the compressor mechanism, it is clear that the orbiting scroll drives the refrigerant from the suction port towards the discharge chamber. Therefore, the fluid travels a distance similar to that traveled by the points of conjugacy forming a given chamber. Thus, the characteristic velocity U_s is defined as

$$U_s = \left(\frac{L_{w;int} + L_{w;ext}}{2}\right) f_c \tag{4}$$

where $L_{w;int}$ and $L_{w;ext}$ are, respectively, the lengths of the inner and outer lateral walls forming the chamber in a given instant and f_c is the compressor operating frequency.

The shape of the scroll was modeled herein as an involute of a circle. The scroll is consequently defined by two involutes that develop around a common basic circle of radius a and are offset by a constant distance r_o. From the definition of the involute, the distance L of one point on the involute to its tangent point on the base circle must satisfy the following mathematical relationship:

$$\frac{\partial L(\varphi)}{\partial \varphi} = a \tag{5}$$

where φ is the involute angle. For a scroll with N compression chambers, φ varies from the initial involute angle α to the involute ending angle φ_e $(= 2\pi N + \pi/2\ 2\pi N + \pi/2)$. Therefore, by integrating Equation (5) and assuming a is a constant, the equations describing the pair of involutes in polar coordinates can be obtained (Chen *et al.*, 2002).

Considering only the involute profile for the scroll wraps, it can be shown that

$$U_s = 2\pi a(2\pi N + \pi - \theta^*)f_c \tag{6}$$

where $\theta^* = 2\pi$ if $\theta < 2\pi$, and $\theta^* = \theta$ if $\theta \geq 2\pi$.

The dimensionless curvature of the chamber

$$C^* = \frac{D_h}{R_c} \tag{7}$$

is a function of the hydraulic diameter, D_h, and and average curvature radius, R_c:

$$D_h = \frac{4V}{A_s} \qquad R_c = \frac{1}{2\pi}\left(\frac{L_{w,ext} + L_{w,int}}{2}\right) \tag{8}$$

where V and A_s are the volume and surface area of the control volume.

Based on observations of numerical results, a constant value was assigned to the characteristic velocity of the gas, U_s, during the suction process. However, it should be noted that U_s varies during the compression process ($\theta \geq 2\pi$) even when the compressor speed, i.e., f_c, is kept constant. As shown in Figure 2, the value provided by Equation (6) to U_s agrees satisfactorily with the numerical prediction for the average velocity of the gas in the chamber throughout the processes of suction and compression, being therefore suitable for calculating the Reynolds number.

The geometric similarity of the problem is guaranteed by the relationships shown in Equation (1). Simplifying the problem to a two-dimensional flow implies that $h \gg a$, making the ratio $h/a \to \infty$. For convenience, the solutions at different instants of time will be referenced by the respective orbiting angle to facilitate the analysis of results.

The ratio between the temperature of the wall and the temperature of the gas ($T_w^* = T_w/T_g$) defines a dimensionless boundary condition, comprising gas cooling ($T_w^* < 1$) or gas heating ($T_w^* > 1$). In general, the gas is at a temperature lower than the average surface temperature of the chamber ($T_w^* > 1$) throughout the process of suction and part of the compression process, becoming higher near the discharge process. For typical temperatures of the gas at the suction ($T_{suc} \cong 300$ K) and the wrap ($T_{w,min} - T_{suc} \cong 10$-$20$ K; $T_{w,max} \cong T_{dis}$), T_w^* reaches its maximum value around 1.05 during the suction process and minimum value of 0.95 during the discharge process.

Finally, the specific heat ratio ($\gamma = c_p/c_v$) for most refrigerants is between 1.10 and 1.40, with values usually close to 1.20. In the following sections, we analyze the convective heat transfer during the suction and compression processes by varying the dimensionless parameters indicated in Equation (1), seeking to include the effect of different operating conditions, geometric parameters and refrigerants.

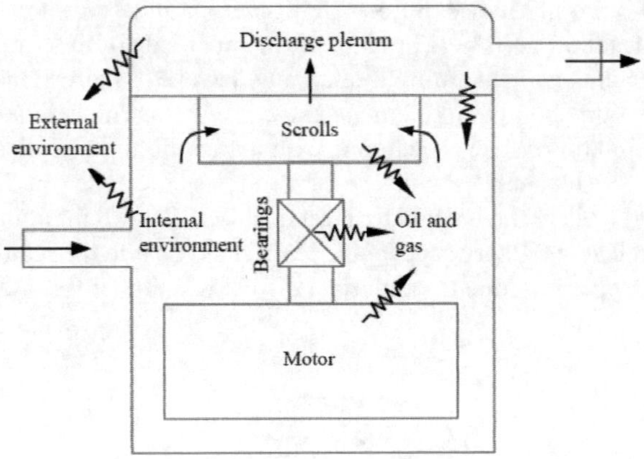

Figure 1. Sources and pathways of heat in a typical low-pressure housing scroll compressor.

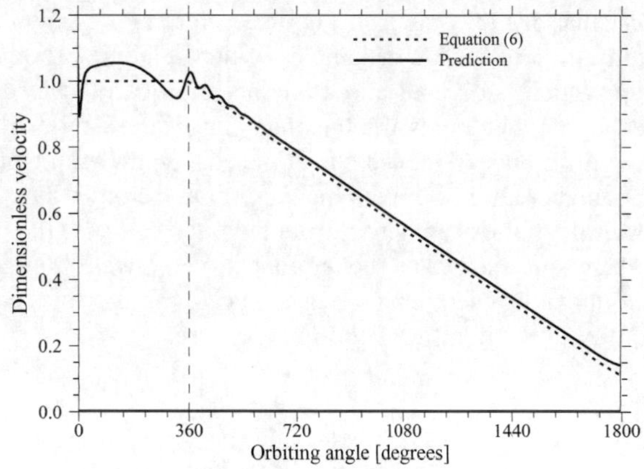

Figure 2. Characteristic velocity of the gas in the chamber during the suction and compression processes.

3. SIMULATION MODEL

The numerical model was developed with a commercial CFD code in which the governing equations of conservation of mass, momentum and energy are solved by the finite volume method, taking into account the unsteady effects due to the motion of the orbiting scroll. The two-dimensional model here developed is similar to those presented by Pietrowicz *et al.* (2002) and Ooi and Zhu (2004), in which only one gas pocket is simulated throughout the compression process. However, the model adopts a different approach to generate and update the solution domain at each time step, and allows the analysis of heat transfer between the gas and scroll wraps also during the suction process.

The compressible turbulent flow that prevails inside the compressor chambers was solved through the concept of Reynolds-averaged quantities, in which the value of a computed variable represents an ensemble average over many cycles at a specified spatial location. The standard k-ε turbulence model is known to be unsuitable for predicting turbulent flow with streamline curvature. Moreover, although the standard wall function has been a common choice for industrial applications, some authors have suggested that a near wall modelling allows a more realistic prediction of convective heat transfer (Launder, 1984). Therefore, the turbulence transport was modeled in the present analysis through the Realizable k-ε model.

The turbulent Prandtl number, $\mathrm{Pr_t}$, is uncertain in the near-wall region and values greater than 0.7 have been reported in the literature. In the present analysis, $\mathrm{Pr_t}$ was estimated via the correlation proposed by Kays and Crawford (1993), as a function of the ratio between the molecular viscosity and the turbulent viscosity, i.e.:

$$\frac{1}{\mathrm{Pr_t}} = \left\{0.5882 + 0.228\left(\frac{\mu}{\mu_t}\right) - 0.0441\left(\frac{\mu}{\mu_t}\right)^2\left[1 - \exp\left(-\frac{5.165}{\mu/\mu_t}\right)\right]\right\} \tag{9}$$

The equation of state of an ideal gas was used to complete the system of equations required to simulate the compression cycle. The fluid viscosity was obtained from the REFPROP software. A second-order upwind scheme was adopted to interpolate the flow quantities needed at the control volume faces. The coupling between the pressure and velocity fields was achieved with the SIMPLEC algorithm and the system of algebraic equations was solved with a segregated implicit procedure.

Due to the unconventional geometry of scroll compressors, an algorithm was developed to adapt the computational mesh throughout the simulation according to the geometric parameters of the compressor. This is needed because the position of each node of the computational mesh must be prescribed for the different orbiting angles in the simulation. This strategy also allows a better control over the quality of the mesh elements. Figure 3 compares the meshes generated by the original mesh generator and by the new code. The meshes shown in Figure 3 were not refined for easier viewing.

The value of the contour in Figure 3 refers to the volume of obliquity index (Squish Cell Index), a measure of grid quality in terms of internal angles of the element. In both codes the grid is generated from the inner side of the scroll wrap, which is divided into a number of equally spaced elements. In generating a quadrangular mesh, the original code automatically divides the outer side of the wrap with the same number of elements, also evenly distributed. Due to the difference in the lengths of the curves, the resultant grid in this case has elements with high obliquity mainly at the contact regions between the wraps. On the other hand, the mesh generated with the new code has the nodes on the outer side of the chamber set from the intersection with normal lines to the inner side in each node, as shown in Figure 3. The result is a mesh of better quality, i.e., lower obliquity index, and lower aspect ratio, especially next to the points of conjugacy. The alignment of the grid with the flow decreases errors due to numerical diffusion and increases the stability of the iterative solution procedure.

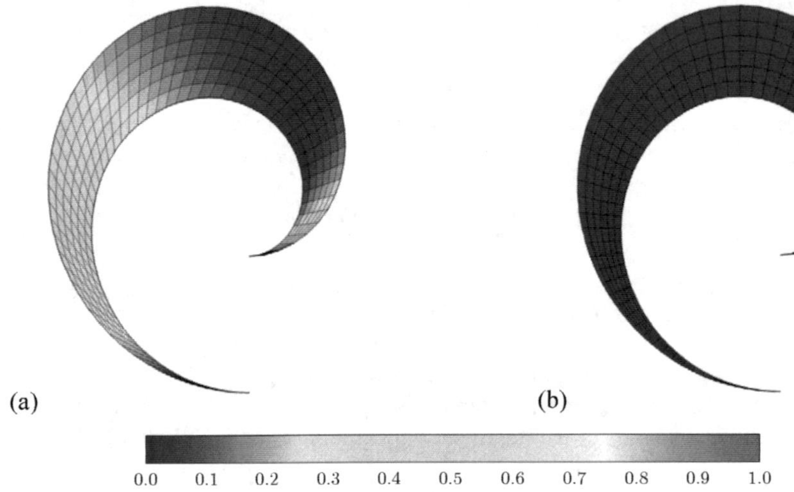

Figure 3. Comparison between the meshes: (a) commercial code (a) developed code. Obliquity index (0 best, 1 worst).

The near wall region plays an important role on the convective heat transfer and hence a low Reynolds number turbulence model was adopted in the simulations. Due to the importance of the mesh refinement near the walls in the solution of the viscous sublayer, the mesh generation followed two criteria: (i) the value of $y+$ in the cells adjacent to the walls was kept on the order of 1, with values of $y+$ greater than 1 being acceptable provided they were within the viscous sublayer ($y+ < 5$); (ii) the viscous layer region ($y+ < 30$) was refined with at least 10 cells to properly estimate the turbulence quantities. The mesh refinement adopted in the circumferential direction was necessary to reduce the aspect ratio of the elements in the region of clearances. The grid nodes are repositioned at each time step to update de chamber geometry, respecting the aforementioned grid requirements. The total number of nodes remained constant throughout the simulation.

The evaporating pressure and suction temperature were employed as boundary conditions at the entrance of the suction chamber. On the other hand, no mass flow rate was considered across the boundaries of the compression chamber, i.e., gas leakage was not taken into account in the compression process. Despite such simplifications, the simulation model was able to provide valuable information on the convective heat transfer during the compression process, as well as during the suction process in which low temperature gas is admitted into the scroll pockets. The simulation procedure employed time steps equivalent to crank angles with one-degree resolution. It is assumed that the initial velocity field is null and the pressure and temperature fields are uniform. It is considered that the solution procedure at each time step is converged when the relative change in the heat transfer rate at the walls is less than 10^{-5} over the last five iterations. The solution procedure is stopped exactly at the discharge angle.

4. RESULTS AND DISCUSSIONS

In the following sections the heat transfer in the suction and compression chamber is analyzed in detail. Initially, we evaluate the relative importance of different parameters on the heat transfer in order to propose a new heat transfer correlation. When not informed, the results are for $Pr = 0.90$ and $\gamma = 1.20$, typical values of refrigerants.

4.1 Numerical Predictions

It is assumed that the temperature of the wraps, $\bar{T}_w$, varies according to the average temperature of the gas, $\bar{T}_g$, such that $T_w^*(=\bar{T}_w/\bar{T}_g)$ is held fixed during the compression process. Thus, for $T_w^* > 1$, the gas temperature is lower than the temperature of the walls and the gas is heated. This occurs in actual operating conditions during the suction process and great part of the compression process. At the end of the compression process, T_w^* is usually smaller than 1, and the gas is cooled. For most applications, it is noted that $0.95 < T_w^* < 1.05$.

As also predicted by the correlation of Dittus-Boelter, heat transfer rate between the gas and the scroll wraps is higher for cooling conditions as shown in Figure 4, for different values of Reynolds number, Re. However, the effect on the Nusselt number, Nu, is smaller than 5% in most of the tested conditions. Figure 5 shows the results Nu as a function of the average curvature of the camera, C^*, for $1000 \leq \text{Re} \leq 240000$, and $\gamma = 1.1$, 1.2 and 1.4. As shown for the temperature ratio T_w^*, Nu is also weakly affected in relation to γ.

Heat transfer along the compression process can be analyzed by varying the curvature of the side walls that form the compression chamber. If the mass and properties of the gas within the chamber and the dimensionless wall temperature are held constant, the variation of the curvature is solely responsible for the variation of the Nusselt number in the compression process. The dimensionless curvature of the camera, C^* is a parameter adopted to ensure the geometric similarity of the problem. However, the geometry of the scrolls in the central region can be formed by different profiles (Bell *et al.*, 2014). As the dimensionless curvature C^* is an average representation of the profile, chambers with different geometries in the central region may have identical values of C^*. To evaluate the impact of C^* on the heat transfer, we analyzed the chamber geometry formed by different involute curves throughout the compression process.

Figures 6 and 7 show the variation of the Nusselt number due to the curvature and orbiting angle for Reynolds numbers equal to 4000 and 120000, respectively. While the scrolls are formed by involute curves, the chambers are identical and there is obviously no difference between the Nusselt numbers calculated (dashed and solid lines). The double-arc meshing profile starts to be formed by circular curves from the points indicated with arrows in Figures 6 and 7. It can be noted that the chambers are formed only by involute curves most of the compression process. Thus, although there is some difference between the values of Nu for the two geometries at the end of compression process, in practical terms, such a difference is small.

The results of the simulation model include instantaneous local properties such as pressure, temperature, density, velocity and turbulent quantities. Overall, the results showed that all properties are not uniform within the chamber, except for the pressure. In the absence of leakages, the pressure field within the chamber is fairly homogenous with variations lower than 1% relative to the average absolute gas pressure, and therefore are not shown herein. Thus, the density inside the chamber varies with temperature only, i.e., the density is low in high temperature regions, and vice-versa.

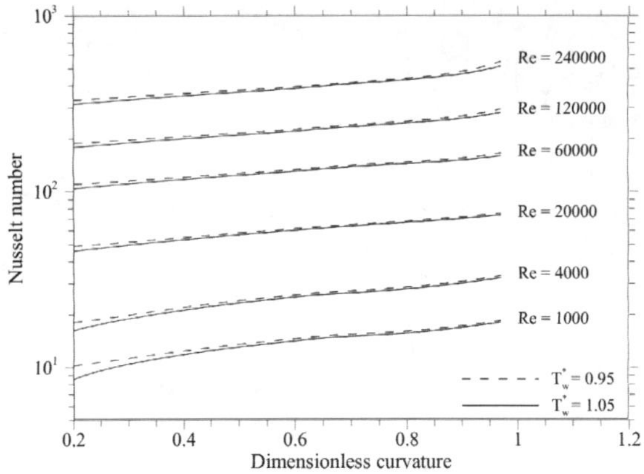

Figure 4. Influence of the heat transfer mode (heating or cooling) on the Nusselt number.

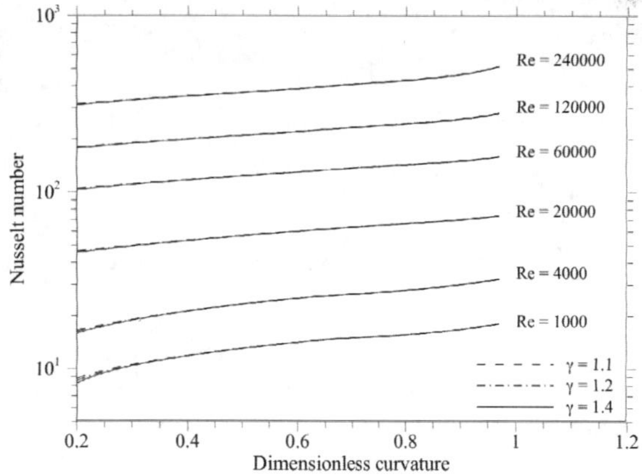

Figure 5. Influence of the ratio of specific heats on the Nusselt number.

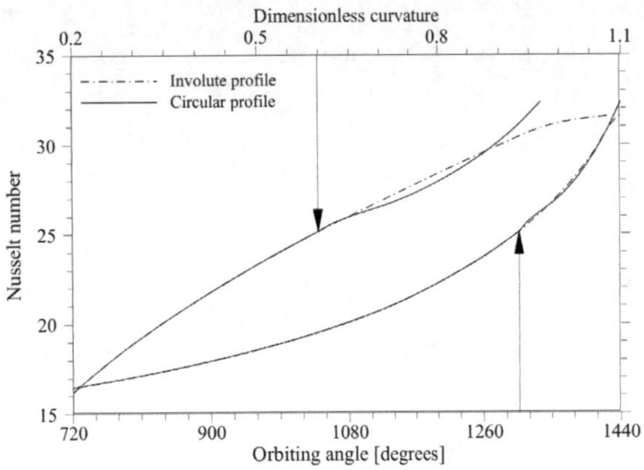

Figure 6. Influence of the wrap curvature in the central region on the Nusselt number for Re = 4000.

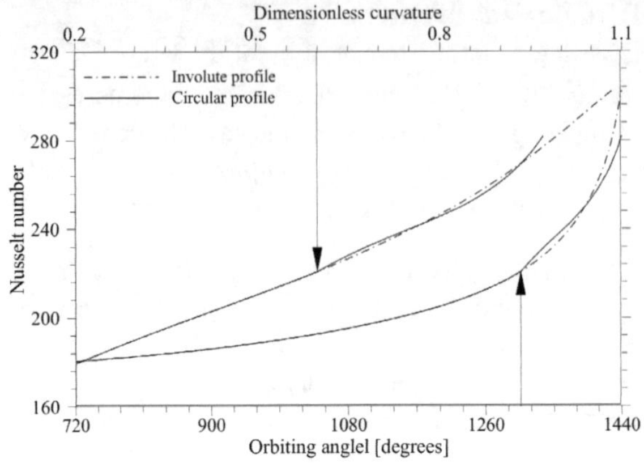

Figure 7. Influence of the wrap curvature in the central region on the Nusselt number for Re = 120,000.

Figure 8 shows the temperature fields associated with the heating condition ($\bar{T}_w/\bar{T}_g$ = 1.05 K) for different Reynolds numbers. The temperature values are made dimensionless by the average temperature of the gas in the chamber at the given time. For completeness, the results for two compression chambers are presented simultaneously in Figure 8. It should be recalled that gas leakage is not considered in the chamber. As can be seen, the variation of the gas temperature is not only brought about by the compression process but also by convective heat transfer between the gas and the scroll wraps, which is responsible for the steep variation of temperature near the walls. In the periphery of the chamber the gas temperature is practically equal to the temperature of the wrap as a result of high heat transfer rates that arise due to high gas velocities near the wall. In the central region, on the other hand, there is a significant temperature difference between the gas and the walls. The influence of the walls on the gas temperature is smaller in the central region due to the greater distance between the walls. The temperature field becomes more homogeneous as the Reynolds number is increased, as a consequence of greater turbulent transport of heat.

4.2 Currently Available Heat Transfer Correlations

The following heat transfer correlations are commonly adopted in lumped models of scroll compressors:

- Correlation of Dittus-Boelter:

$$\mathrm{Nu_{DB}} = 0.023\ \mathrm{Re}^{0.8}\ \mathrm{Pr}^{n} \tag{10}$$

- Correlation for spiral heat-exchanger (Kakac and Shah, 1987):

$$\mathrm{Nu} = \mathrm{Nu_{DB}}\left(1 + 1.77\frac{D_h}{R_c}\right) \tag{11}$$

- Correlation of Jang and Jeong (2006):

$$\mathrm{Nu} = \mathrm{Nu_{DB}}\left(1 + 1.75\frac{D_h}{R_c}\right)\{1 + 8.48[1 - \exp(-5.35\,St)]\} \tag{12}$$

In Equation (10), the exponent n assumes the value of 0.4 when the fluid is being heated, and 0.3 when the fluid is being cooled. The correlation of Jang and Jeong (2006) apply a correction factor to the correlation of spiral heat exchangers as a function of the Strouhal number, St:

$$\mathrm{St} = \frac{f_c D_h/2}{U_s} \tag{13}$$

The results of the two-dimensional model were compared to estimates of heat transfer given by Equations (10)-(12). Besides such correlations, we also considered estimates of Nu from a correlation for turbulent flow between parallel flat plates, in which the Reynolds number is calculated based on the average length of the chamber, L_w. For convenience, the results are presented in terms of heat transfer coefficient, h_c.

The results in Figure 9 demonstrate that the correlation for spiral heat exchangers provides values of h_c close to the predictions of the simulation model for Re = 1000. However, only the correlation for turbulent flow between flat plates provides results in agreement with the predictions for Re = 120000 (Figure 10). The correlation of Jang and Jeong (2006) gives estimates of $h_c$considerably greater than the other correlations in both conditions.

4.3 Proposed Heat Transfer Correlation

Given the differences between the heat transfer coefficients predicted by the simulation model and evaluated with the available correlations, a new heat transfer correlation is proposed for the suction and compression chambers of scroll compressors. Among the different expressions tested to correlate the Nusselt number, Nu, as a function of the Reynolds number, Re, Prandtl number, Pr, and dimensionless curvature, C^*, the following showed the smallest deviations in relation to the numerical results:

$$\mathrm{Nu} = \mathrm{Nu}_{\mathrm{DB}}(c_0 + c_1 C^* + c_2 Pr + c_3\, C^*/\mathrm{Re}) \qquad (14)$$

where $c_0 = 0.4956$, $c_1 = 0.406$, $c_2 = 0.1361$ and $c_3 = 3394$.

Figure 11 shows that the maximum deviation between the values of Nu obtained via simulations and from the correlation proposed herein, Equation (17), is less than 10%. The most significant deviations occur at the end of the compression process, when both the curvature of the wraps and the heat transfer rate between the gas and surfaces are greatly increased.

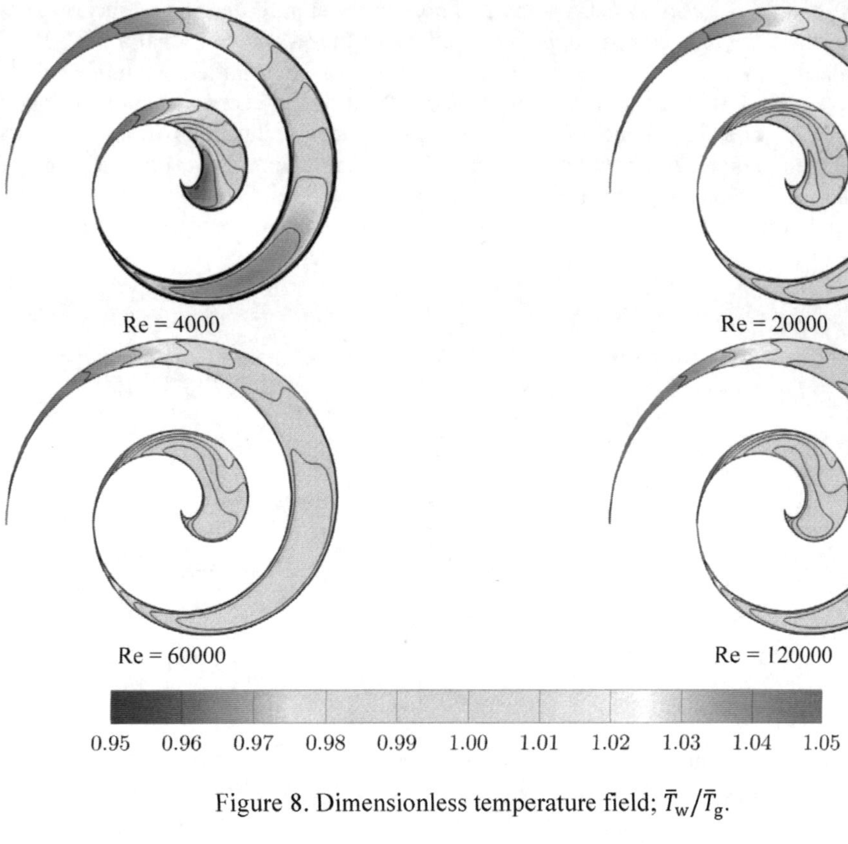

Figure 8. Dimensionless temperature field; $\bar{T}_{\mathrm{w}}/\bar{T}_{\mathrm{g}}$.

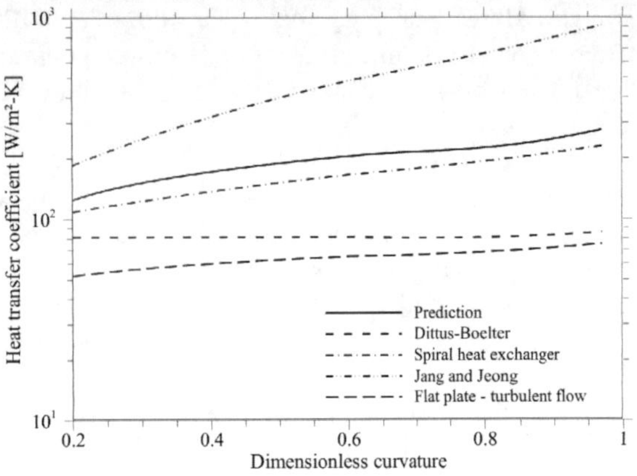

Figure 9. Convective heat transfer coefficients obtained from prediction and different correlations; Re = 1000.

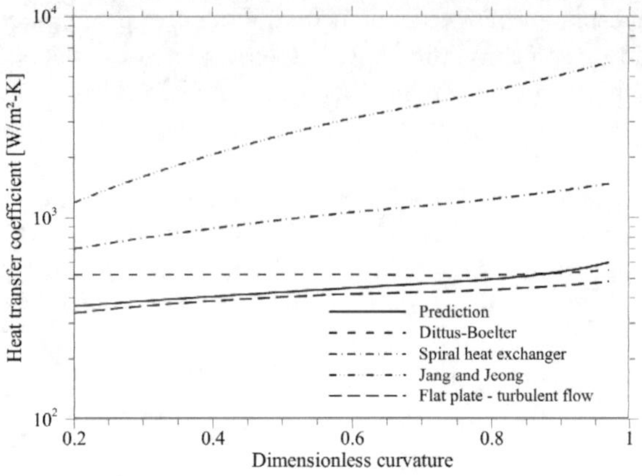

Figure 10. Convective heat transfer coefficients from prediction and different correlations; Re = 120,000.

As previously discussed, the influence of the wall temperature on the Nusselt number is relatively small. Hence, for simplicity, the proposed correlation does not distinguish between gas heating and gas cooling. On average, Equation (14) provides values of Nu around 2% lower than the predictions in the case of gas cooling, and 2% higher in the case of gas heating. It should be noted that the proposed correlation gives the average value of Nu for the side walls of the chamber. Three-dimensional simulations showed that the heat transfer coefficient between the gas and the bases of the scroll wraps is 50% higher than that of the sidewalls. Nevertheless, the correlation proposed herein should provide suitable results for most engineering applications.

Considering the absence of experimental data of heat transfer in the chambers of scroll compressors, the correlation proposed herein was used in a lumped model to predict the gas temperature during the suction and compression processes. Such predictions were then compared with predictions of the two-dimensional model. It is assumed that throughout the suction and compression processes the wraps remain at a constant temperature, such that $\bar{T}_w/\bar{T}_{suc} = 1.05$. Thus, the gas is heated during the suction process and cooled at the end of compression, as occurs in the compressor under common operating conditions. The results of temperature shown in Figure 12 are made dimensionless by the gas temperature at the suction entrance, T_{suc}. For reference, the evolution of the temperature for an adiabatic compression process is also shown in Figure 12. As can be seen, the correlation adopted in the lumped model was able to accurately predict the gas temperature at the end of the suction process and during the compression process.

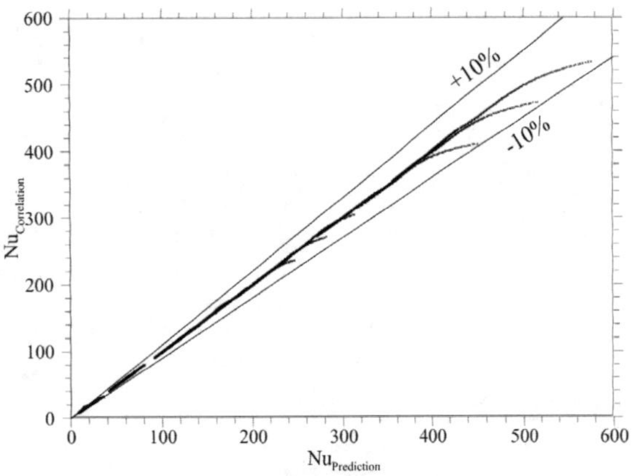

Figure 11. Values of Nusselt number, Nu, obtained from the two-dimensional model and Equation (14).

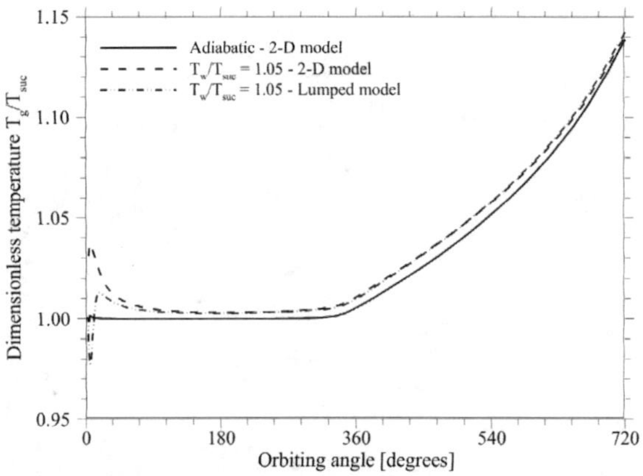

Figure 12. Gas temperature during the suction and compression processes predicted by the two-dimensional and lumped models.

5. CONCLUSIONS

Correlations available in the literature for convective heat transfer in the compression chambers of scroll compressors are based on simplified flow conditions. A numerical model was developed to predict gas flow and heat transfer inside the suction and compression chambers with an algorithm to automatically adapt the computational mesh throughout the simulation. The study covered a wide range of operating conditions, refrigerants and geometries, allowing the proposal of a new correlation for the Nusselt number as a function of the Reynolds number, Prandtl number, and dimensionless curvature of the chamber. In the absence of experimental data, the correlation was adopted in a lumped model to predict temperature of the gas during the suction and compression processes. Such predictions were compared to numerical results of the two-dimensional model developed in the study, and good agreement was observed. None of the correlations available in the literature was able to provide satisfactory estimates of heat transfer in different conditions, although the correlations for spiral heat exchangers and flat plates have shown good agreement with the present numerical results in some specific conditions.

REFERENCES

Bell, I. H., Groll, E. A., Braun, J. E., Horton, W. T., & Lemort, V, (2014). Comprehensive analytic solutions for the geometry of symmetric constant-wall-thickness scroll machines. *Int. J. Refrig.* 45, 223-242.

Chen, Y., Halm, N.P., Groll, E.A., & Braun, J.E, (2002). Mathematical modeling of scroll compressors – part I: overall scroll compressor modeling. *Int. J. Refrig.* 25, 731-750.

Jang, K. & Jeong, S., (2006). Experimental investigation on convective heat transfer mechanism in a scroll compressor. *Int. J. Refrig.,* 29, 744-753.

Kakac S, Shah RK., (1987). *Handbook of single-phase convective heat transfer*. New York, NY: John Wiley & Sons.

Kays, W. M. & Crawford, M. E., (1993). *Convective heat and mass transfer*. New York, NY:McGraw-Hill.

Kremer, R., Barbosa Jr., J. R., Deschamps, C. J., (2012). Cooling of a reciprocating compressor through oil atomization in the cylinder. *HVAC&R Research*, 18, 481-499.

Launder, B. E., (1984). Numerical Computation of Convective Heat Transfer in Complex Turbulent Flows: Time to Abandon Wall Functions?, *Int. J. Heat and Mass Transfer*, 27, 1485-1491.

Ooi, K. T. & Zhu, J., (2004). Convective heat transfer in a scroll compressor chamber: a 2-D simulation. *Int. J. Thermal Science*, 43, 677-688.

Pietrowicz, S., Yanagisawa, T., Fukuta, M., & Gnutek, Z., (2002) Mathematical modeling of physical processes in the scroll compressor chamber. *Proceedings of International Compressor Engineering Conference at Purdue, West Lafayette, USA* (Paper C20-6).

Sunder, S., (1997). *Thermodynamic and heat transfer modeling of a scroll compressor*. Ph.D. Thesis, Massachusetts Institute of Technology, USA.

Zhenquan, L., Guirong, D., Shicai, Y., & Mingzhi, W., (1992) The graphic method of modified wrap of scroll compressor, *Proceedings of International Compressor Engineering Conference at Purdue, West Lafayette, USA* (1099-1106).

ACKNOWLEDGEMENTS

The present study was developed as part of a technical-scientific cooperation program between the Federal University of Santa Catarina and EMBRACO. The authors also acknowledge the support of EMBRAPII Unit POLO/UFSC and and CAPES (Coordination for the Improvement of High Level Personnel).

The Improvement of Motor Cooling Through Stator Profile Optimization Using CFD Analysis in Hermetic Scroll Compressors

Weiping TANG[1], Liu WANG[1], Li YAO[1*], Jiangbo LIN[1]

[1]Danfoss (Tianjin) Ltd., Danfoss Commercial Compressors
Tianjin, P.R. China
Phone: +86 22 82197029, Fax: +86 22 82197799, yaoli@danfoss.com

ABSTRACT

Hermetic compressor motor design face the challenge today that reducing the motor size for cost reason. This causes a higher motor power density and in turns a higher motor temperature. With the help of CFD computation tool, we investigated two ways to improve the motor cooling and bring the motor temperature down. One way is to put rotor ventilation channels to increase the flow in lower pressure side, so that the total percentage of suction gas to cool the motor increased. The other way is to optimize the stator cuts depending on flow rate in each stator cut. By combining these two improvement together, total motor temperature, particularly in stator end wing and rotor lower end ring, was reduced dramatically. Both motor and compressor prototype were tested and results given.

1. INTRODUCTION

With the increased pressure of cost, hermetic compressor sizing become an important part of the design optimization process, while maintain the same performance level in the same time. In order to squeeze the compressor to a minimum size and meanwhile maintain the same performance level, more thorough analysis and sophisticated methods and tools need to be utilized to achieve this goal.

This is also true for hermetic compressor motor design. A smaller motor size will still provide enough torque for the compression process, but with a relatively bigger current feeding to the motor. This leads to extra losses from the motor. From the overall compressor point of view, with this additional losses generated, in a given compressor load condition, when compressor reach thermal stability, motor temperature will become higher and might exceeds the limit required by regulation, or else the design already has some thermal margin to accommodate this increased temperature due to these added motor losses.

It has not been possible to visually check the temperature distribution as well as the gas flow inside the compressor shell, which is sealed against high pressure, so improvement of motor cooling has been a difficult problem for compressor motor designers. Compressor motor cooling improvement using thermal modelling has been investigated before by using thermal network method, Chen *et al.* (2012), with limited accuracy since this do not take all thermal correlation between motor and compressors. In recent years, Computational Fluid Dynamics (CFD) tools have been used to simulate refrigerant gas flow inside the compressor shell, as well as temperature distributions, and this helps designers to optimize the motor cooling.

2. ELECTRIC MOTOR LOSSES AND COOLING

The main losses inside a compressor including motor losses, compression losses, and also heat transfer losses. Here we concentrated on electric motor losses only. Motor losses include electromagnetic losses, and also mechanical losses, which are all dissipated by motor surroundings in the form of heat. The former losses can further be split into eddy current losses, hysteresis losses, copper losses and stray losses. Mechanical losses mainly from frictional losses between bearings and shafts.

Traditionally, core loss P_{Fe} can be divided into two components: hysteresis loss P_h and eddy current loss P_e. According to the Steinmetz equation, measurement and calculation of core losses are normally made with sinusoidal flux density of varying magnitude and frequency. These measurements and calculations are based on the standard coil and frequently modeled by a two term function of the form:

$$P_{Fe} = P_h + P_e = k_h f B^n + k_e f^2 B^2 \qquad (1)$$

where f is the frequency of the external magnetic field, B is flux density, k_h , k_e and n are the coefficients, which depend on the lamination material, thickness, conductivity, as well as other factors. Stator copper losses are:

$$P_{Cu} = m I_1^2 R_1 \qquad (2)$$

Where m is the number of phases of the motor, I_1 is stator current, R_1 is stator resistance per phase. Rotor copper loss is:

$$P_{Al} = I_2^2 R_2 \qquad (3)$$

Where I_2 is rotor current, R_2 is rotor bar resistance include cage bars and end rings.

All these motor losses will turn into the format of heat, and to be dissipated by surrounding area, and reach a certain temperature when the system reaches thermal stability. Hermetic compressor motor is cooled by refrigerant inside the compressor, sometimes with oil also. Stator usually has cuts in the stator outer profile in order to let refrigerant pass through the gap between stator cores and compressor shell to cool the motor, and provide gas return path in the same time. Gas going into the suction port of the compressor is split into two parts, majority of the gas going into scroll set for compressor process, minor part of the gas will go through the stator cuts to low pressure side of the compressor, circulate through the stator lower winding and go back to motor top winding through stator cuts and air gap.

Factors that deciding motor temperatures are: how much motor losses generated and need to be dissipated and how effective this generated heat can be brought away, i.e. the cooling of the motor. Based on this, two major directions to reduce motor temperature can be inferred: (1) increase motor efficiency and reduce losses; (2) keep same motor efficiency level but improve the motor cooling.

Motor cost will be increased if we adding materials to the motor to improve motor efficiency, and this is not the preferred solution. However, it is more interesting to concentrated on optimizing the motor cooling path inside the compressor, such that motor temperature can be reduced but maintain the same cost.

3. COMPUTATIONAL FLUID DYNAMIC MODELE AND HEAT TRANSFER ANALYSIS

3.1 Computational Model

The computational model for inside the compressor of the lower pressure side is shown in Figure 1, where (a) is the compressor model and (b) is the entire mesh of the compressor model. Compressor is modelled from the oil surface to the housing thrust surface, because it is the most important part that related to motor cooling.

The computational model includes fluid domain and solid domain, the solid domain consist of stator, rotor, shaft, balancing counterweights, and suction baffle. The flow path of fluid domain is from the suction inlet of compressor to scroll suction port; while for motor flow path, which includes (1) the stator cut, and (2) the motor air gap which lie in between the circumference of the rotor and the inner circumference of the stator. The motor air gap (2) is very small, roughly 0.5mm per side, so constructing a mesh is difficult. However, since the motor air gap is very important in this motor cooling study, it is important that we include it in the computational model even though huge computation power needed.

The simulation method and calculation conditions are presented in Table 1. We selected FLUENT software to perform the computation, and there are several assumptions we have made also. The refrigerant gas is R410a, and it is considered as an incompressible fluid; oil channel cooling path in the middle of the shaft is neglected, since it is mainly to conduct the oil instead of the refrigeration gas and it is not directly cooling the stator and rotor; oil temperature is assumed to be constant for the computation and only steady state condition is considered, and no radiation heat transfer is considered. The fluid state equation we used in this study is k-ε equation with standard wall function.

Table 1: Simulation method and calculation condition

Computation method	Software	FLUENT
	Turbulent model	k-ε equation with standard wall function
	Mesh number	20,000,000 cells
Operating conditions	Rotation speed	3000 rpm
	Suction inlet pressure	11.5 bar
	Suction inlet temperature	299.7 K
	Discharge mass flow rate	0.1994 kg/s
	Oil temperature	366.5 K

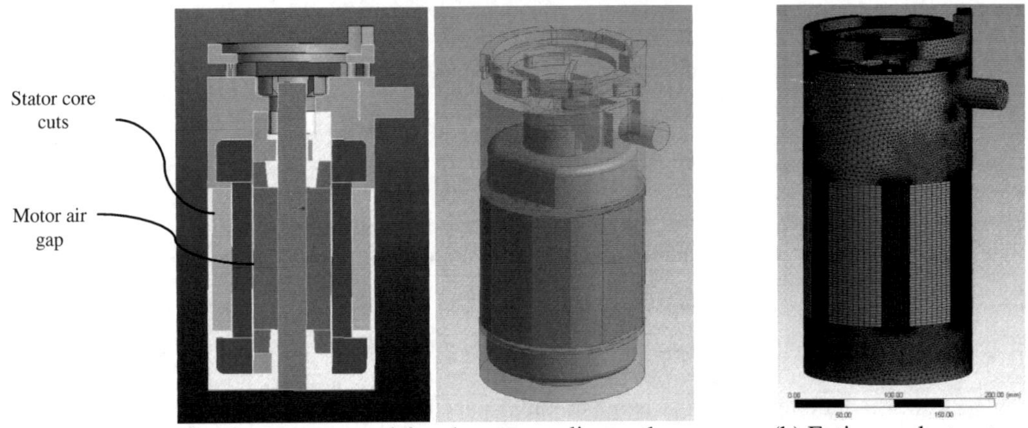

(a) Compressor model and motor cooling path (b) Entire mesh structure

Figure 1: Compressor model

3.2 Gas Flow Path and Heat Transfer Analysis

3.2.1 Refrigeration gas flow path distribution

Figure 2 shows the velocity distribution of the refrigerant gas flow for the standard stator cuts profile model. Figure 2 (a) shows the velocity vector distribution in one section, and (b) shows the gas flow direction. For the amount of suction gas into the compressor inlet, majority of the refrigeration gas goes into the scroll set, while a small portion goes down into stator cuts, particularly true in the stator cuts that adjacent to suction port, as can be seen in (b). These cooling gas then go down to the oil surface and return from the rotor air gap and also through the stator cuts that are opposite to the suction gas. Table 2 shows the percentage of each stator cuts flow in proportion to the total suction gas. Sign in Table 2 represent the direction of the flow; positive sign represent the downward direction, while negative sign represent the upwards flow direction.

Table 2: Percentage of flow in motor cooling path for motor with standard stator cuts

Section	Downward flow rate of cooling path, percentage to total suction gas	Section	Downward flow rate of cooling path, percentage to total suction gas
S1	4.04%	S4	-3.77%
S2	0.25%	S5	-0.16%
S3	-0.20%	S6	0.19%
S0	-0.35%	Total	4.48%

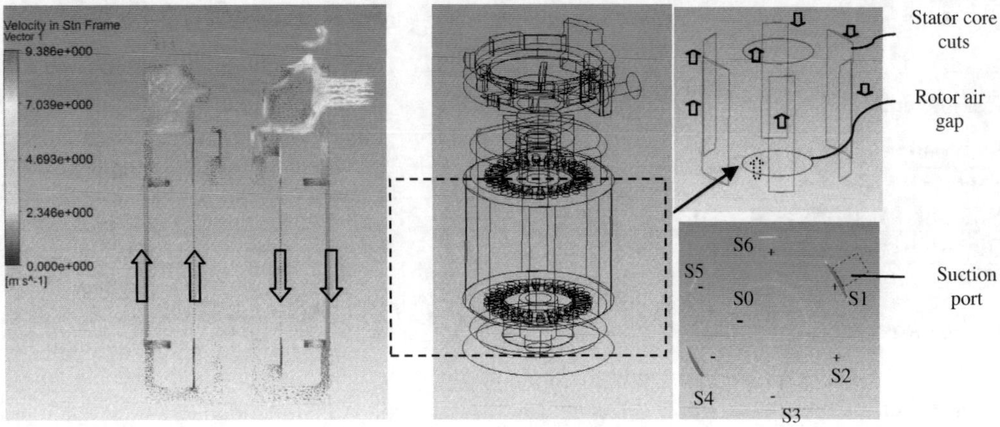

(a) Velocity vector distribution from one section view (b) Flow direction through the motor flow path

Figure 2: Velocity distribution for standard stator cuts

From the simulation result, we can see most of the refrigeration gas goes down to motor through stator cuts that next to suction port S1, and majority of them return through S4, which is the stator cuts that opposite to suction port. The total amount of refrigeration gas return through rotor air gap is small, due to the small air gap. The total amount of cooling gas goes down to cool the motor is 4.48% of the total suction gas, which is really small, and this is one aspects we want to improve.

3.2.2 Motor temperature distribution

Steady state heat transfer computation shows the temperature distribution inside compressor in low pressure side. Figure 3 (a) and (b) shows the temperature distribution cross sectional view, in different direction. Figure 3 (a) shows rotor temperature increases from top to bottom, due to the decreased cooling effects along the flow path. Bottom of rotor has the highest temperature. There is a sharp temperature distribution across the air gap because the existence of air gap and its heat transfer coefficient is much higher than solid. Figure 3 (b) shows the view cut in suction port direction, compare with (a), rotor temperature is with contoured distribution, temperature varies with the distance to suction port, if the distance is more close to the suction port, the temperature will be lower, valid in both axial and radial direction. And it is also the case for stator upper winding temperatures, as shown in Figure 3(c).

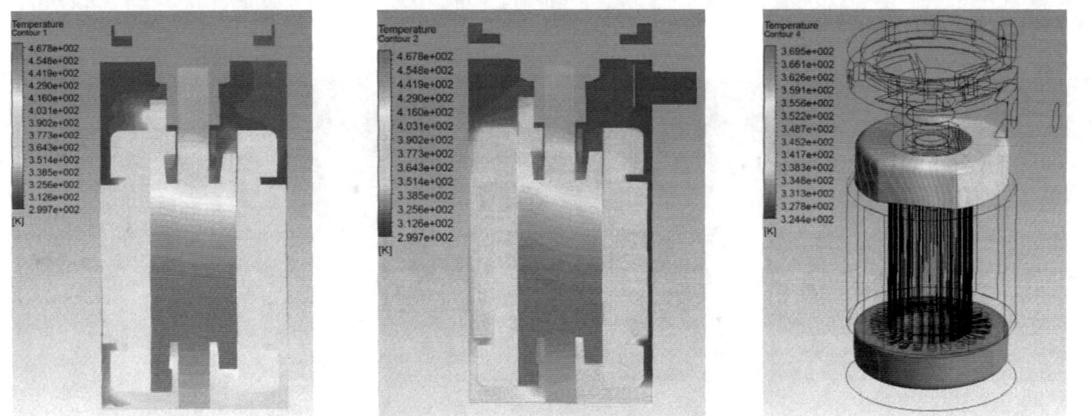

(a) Section view in orthogonal to suction (b) Section view of suction direction (c) stator windings temperature

Figure 3: Temperature distribution for compressor motor

Table 3 shows the temperature of each part of the compressor from CFD computation, in comparison with measurement results for stator upper-winding and stator lower-winding from test. Winding temperatures are measured by inserting thermocouples into the winding. Quite good alignment is found between simulation and test data. This model can be used to optimize motor cooling design. The main issue is that the bottom of motor temperature is too high; there will be insulation system lifetime risk. Further optimization need to be analyzed in order to reach this goal.

Table 3: Temperature for each part of the motor

Components	Temperature Computation (K) Delta to benchmark	Temperature Test (K) Delta to benchmark	Delta between test vs. calculation (K)
Stator upper-winding	benchmark	-15	-15
Stator lower-winding	+47	+39.7	-7.3
Rotor upper end-ring	+44	-	-
Rotor lower end-ring	+112	-	-

4. MOTOR COOLING OPTIMIZATION

The main target is to improve the motor cooling so reduce the motor temperature but maintain the similar motor performance in the same time. Two different cooling methods were investigated for the scroll type compressor motor, i.e. rotor cooling channels and optimized stator cuts profiles. Motor efficiency simulation results as well as CFD results for flow rate and also temperature distribution are given, prototype test also have been done to validate the methods.

4.1 Cooling Optimization on Rotor

Having additional cooling channels on rotor lamination is one of the popular choices for improved motor cooling, and this has been utilized for many other compressor motors, such as by Yokoyama *et al.* (2012). We also want to utilize this improvement for our study here.

4.1.1 Selection and optimization of ventilation channels on rotor
From simulation results in Table 2, only 4.48% of the total suction gas goes down to cool the motor, then flow back to scroll sets through air gap and opposite side of the stator cuts. For some compressor motor with a higher power density, this amount of cooling gas is not enough to bring the motor temperature down to accepted limit. To improve this situation, additional rotor channels are added to improve the cooling. However, the side effects of having channels on rotor lamination will squeeze the flux field (higher flux density) in rotor yoke area thus inducing higher magnetization current, leading to a lower motor efficiency.

Figure 4 shows the flux density impact of having rotor channels in flux distribution map. Flux distribution in rotor yoke increased due to the existence of these cooling channels. Care must be taken when selecting rotor channels not to have significant impact on motor efficiency.

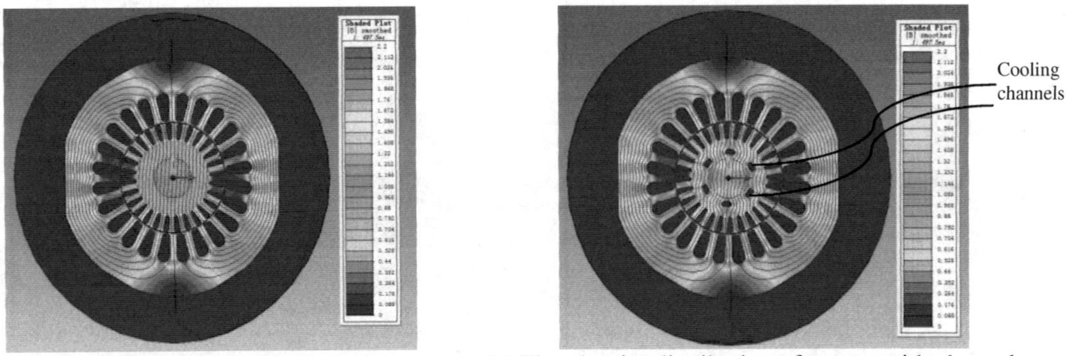

(a) Flux density distribution for a standard rotor (b) Flux density distribution of a rotor with channels

Figure 4: Motor flux distribution with cooling channels

Some important aspects need to be considered when design rotor channels, except the impact to motor efficiency. The motor design has a lower counterweight and end rings, these ventilation holes have to be placed between the inner diameter of the bottom of the end ring and the rotor shaft, to let the refrigeration gas pass through, as shown in Figure 5. From this point of view, there is not enough space to put reasonable sized round cooling holes, so kidney shaped holes are selected in this study. Important factors for selecting cooling holes include minimizing the windage losses and pressure loss but maintaining a maximum ventilation area and also the total circumferential length of the cooling channels, plus mechanical strength of the total rotor lamination with holes. Table 4 listed the key aspects. Figure 5 shows the dimensions of the kidney shape channels. Few optimization candidates are selected, and their area and position are shown in Table 5. We will select one of them to be our solution.

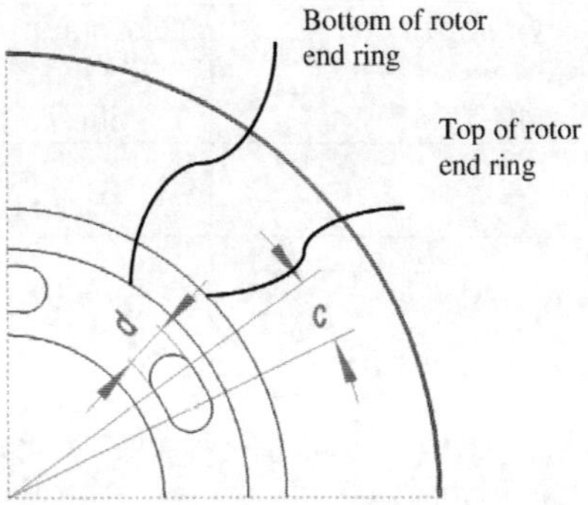

Figure 5: Rotor cooling channel dimensions

Table 4: Important factors to be considered when selecting rotor channels

	Comments
Winding loss	To be minimized
Total ventilation area	To be maximized
Total circumference length	To be maximized
Mechanical strength	Above certain level
Motor Efficiency	Least impact

Where: c: angle between the two center of the kidney shape cooling channel;
d: height of the kidney shape cooling channel in radial direction.

Table 5: Rotor cooling channel total area and circumference length

	Standard	**Design R1**	**Design R2**	**Design R3**
Comments	No holes on rotor	6 Holes on rotor, evenly distributed, 20° separate	6 Holes on rotor, evenly distributed, 10° separate	4 Holes on rotor, evenly distributed, 20° separate
c (degree)	0	20	10	20
d (mm)	0	4	4	4
Total ventilation area (mm^2)	0	229.1	152.3	152.8
Total circumference (mm)	0	152.3	113.8	101.5

4.1.2 Impact to motor efficiency of having ventilation holes
From the motor efficiency simulation results in Figure 6, we can see that with the increased number of cooling holes on rotor, motor efficiency reduced as expected, due to the increased losses. With the same ventilation area, the more the number of holes, the more efficiency reduction. Design R2's motor efficiency is the highest among the three designs, and its total circumference is larger than design R3, while the total ventilation area is the same as design R3. So it is the best option that we will go forward with prototypes and testing.

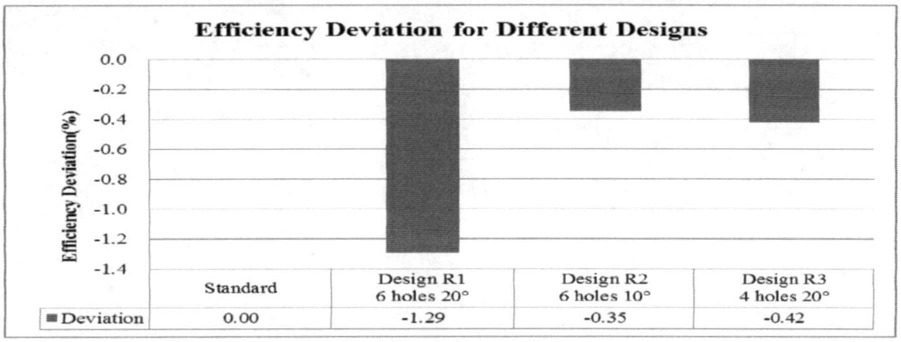

Figure 6: Efficiency comparison for different designs, motor simulation results

A prototype rotor was made with deign R2, 6 holes 10° separate. Motor efficiency is measured with dynamometer. Measurement results in comparison with original motor efficiency benchmark ηs without ventilation holes are shown in Table 6. These tests are performed in rated (ARI) condition torque.

Table 6: Test results comparison between original motor vs. design R2 rotor

	Efficiency [%]				Current [A]			
	Test	Deviation (absolute value)	Calculation	Deviation (absolute value)	Test	Deviation (absolute value)	Calculation	Deviation (absolute value)
Standard	101.72% $*\eta s$	--	ηs	--	97.71% $* Is$	--	Is	--
Design R2 6 holes 10° separate	101.26% $*\eta s$	-0.41%	99.60% $*\eta s$	-0.35%	99.88% $* Is$	0.36A	103.13% $* Is$	0.52A

Motor current increased by 0.36A, and total motor efficiency reduced by 0.41%. From simulation results, motor current increase by 0.52A, and total motor efficiency reduced by 0.35%. They are close enough for us to trust the motor simulations.

4.1.3 CFD simulation results – with ventilation channels on rotor

Table 7 shows the change of flow rate in each part of the cooling path, in terms of percentage to total suction inlet refrigerant gas. The total amount of gas that flow directly down to motor increased from 4.48% to 6.43%, and particularly, the amount of gas flow through S1, which is the stator cuts besides the suction port; S2 and S6, which are the stator cuts close to suction port, have been increased by 1.25%, 0.11%, 0.11% respectively. These flow rates are increased by having these ventilation holes on rotor. Figure 7 shows the position for rotor ventilation holes, i.e. SI1 to SI6. In Table 7, positive sign represent downward flow, while negative sign represent upward flow.

Table 7: CFD results for flow rate between standard designs vs. design R2 in percentage of total suction gas

Section	Downward flow rate of cooling path		Deviation
	Standard	Design R2	
S1	4.04%	5.29%	1.25%
S2	0.25%	0.36%	0.11%
S3	-0.20%	0.09%	0.29%
S4	-3.77%	-2.89%	0.88%
S5	-0.16%	0.07%	0.23%
S6	0.19%	0.30%	0.11%
S0	-0.35%	0.32%	0.67%
SI1	--	-0.57%	-0.57%
SI2	--	-0.58%	-0.58%
SI3	--	-0.58%	-0.58%
SI4	--	-0.59%	-0.59%
SI5	--	-0.60%	-0.60%
SI6	--	-0.61%	-0.61%
Total	4.48%	6.43%	1.95%

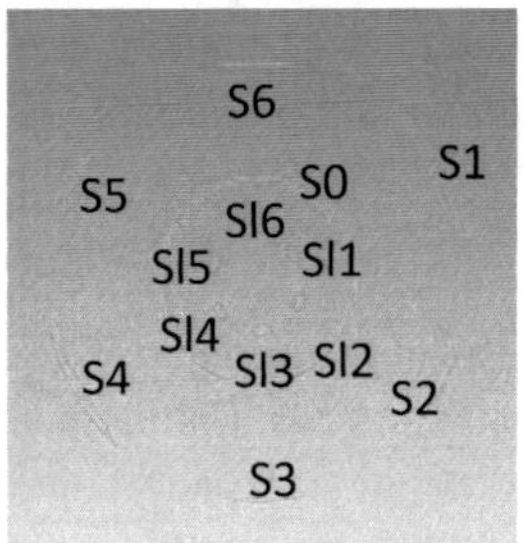

Figure 7: Ventilation holes position

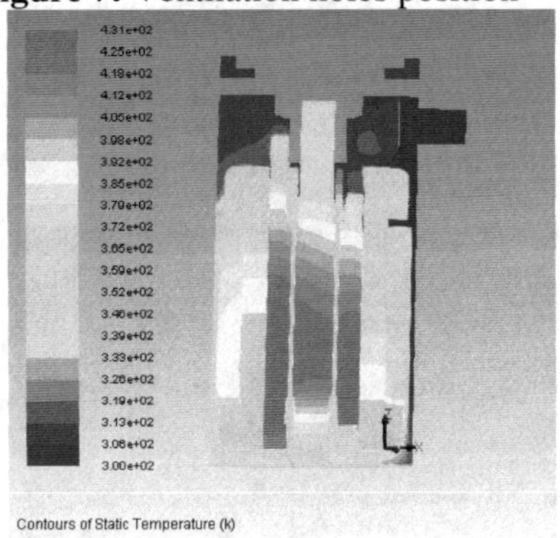

Figure 8: Temperature distribution with ventilation holes

Figure 8 and Table 8 shows the temperature distribution for the motor and also the temperature for each part. We can see that having holes on rotor has a bigger impact on bottom winding of the stator, and even better effects for rotor temperature by itself. This cooling method is preferred if we want to reduce rotor temperature particularly. For our research in this paper, the reduction of stator end winding temperature is still not enough, so further optimization need to be invested as well.

Table 8: CFD simulation result of design R2 in comparison with original design

Component	Temperature reduction from Computation (K) Compare to standard design
Stator upper-winding	-4
Stator lower-winding	-7
Rotor upper end-ring	-14
Rotor lower end-ring	-42

4.1 Cooling Optimization for Complete Motor

With additional ventilation channels on rotor, flow rate on stator cuts also changes, and S1, S2 and S6 flow rate have been increase as the results; but S3, S4 and S5 flow rate reduced, due to the existence of rotor holes. To better accommodate the increased flow down to motor from suction side, we decided to increase the stator cuts in S1, S2, and S6, to allow more gas going through, and keep same stator cuts on S3, S4 and S5, since their flow rate reduced with ventilation holes.

On the other hand, more stator cuts will reduce the yoke thickness of the stator, introducing higher flux density in that area, so to cause higher magnetization current thus more iron losses, especially we already have holes on rotor for ventilation. In order to find the optimized stator cuts, we made several options. Figure 9 shows the dimension of the stator cuts. Cuts parameter *a* vary from 0.5mm to 3mm, 0.5mm per step; parameter *b* vary from 0.5mm to 2mm, 0.5 mm per step. There are 24 combinations in total. Each individual option is evaluated with motor efficiency simulation first.

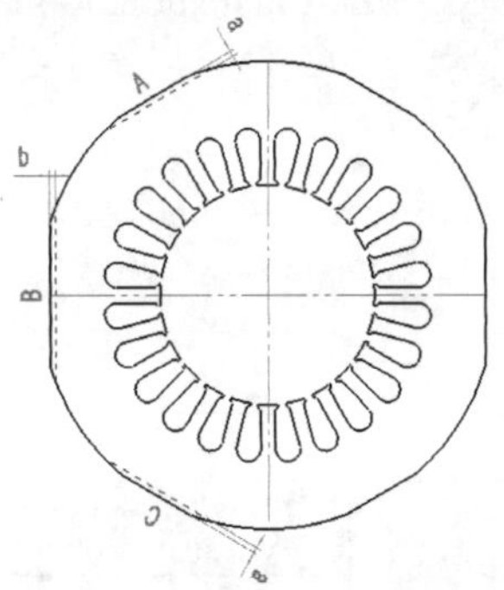

Figure 9: Stator cuts dimensions

In order to select the optimal stator cuts, another important factor we need to consider is the total shrink fit area of the stator with respect to the compressor shell. A certain shrink fit area between stator outer diameter and compressor shell has to be maintained to ensure the stator will not turn inside the shell during operation and transients. Figure 10 show the shrink fit circumference, total ventilation area, and also motor efficiency for different combinations. Motor efficiencies are calculated without ventilations holes on rotor.

For the total ventilation area, we can see that change of parameter *a* has more impact on ventilation area than parameter *b*, and has less adverse impact to motor efficiency as well. Depending on what aspects we want to optimize our design, we can select different combination which suits the design purpose the best. From Figure 10, design *a*=2.5, *b*=1, and *a*=3, *b*=0.5, combinations are selected; they both have the relatively big ventilation area without too much impact to motor efficiency. Next, *a*=2.5 and *b*=1 is selected for final investigation.

In previous text, two different cooling improvement methods have been proposed and validated; on rotor and stator respectively, now we want to combine these two together, to further investigate what is the benefit in terms of motor cooling and the impact to motor efficiency as a whole.

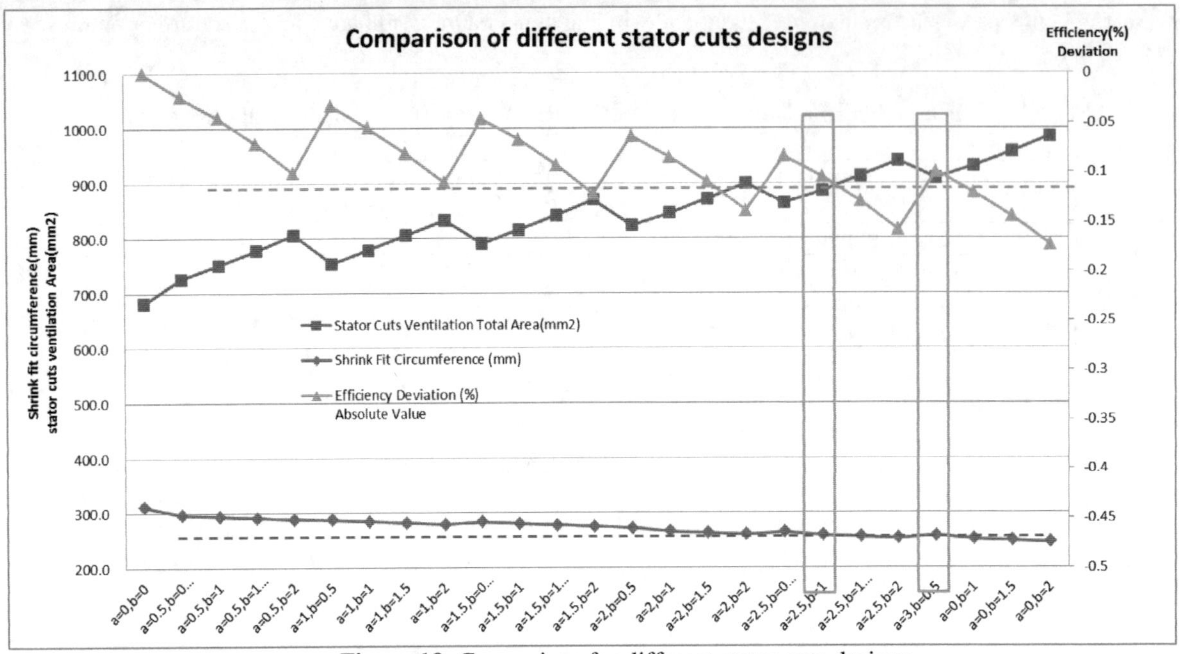

Figure 10: Comparison for different stator cuts designs

4.2.1 Impact to motor efficiency – combination of optimization on stator and rotor

Table 9 shows the simulation and prototype results for design R2, 6 rotor ventilation holes with 10° separate, combined together with stator cuts *a*=2.5, *b*=1 design, taking the calculated value for standard design as the benchmark. Motor efficiency dropped by 0.46% from calculation, this aligns with rotor ventilation holes efficiency impact (-0.35%, Table 6) plus stator cuts efficiency impact (-0.1%, Figure 10). Test results are well accepted within this cooling improvement.

Table 9: Motor test and simulation results with stator cuts and rotor cooling holes @ rated condition compare to standard design

	Efficiency [%]				Current [A]			
	Test	Deviation (absolute value)	Calculation	Deviation (absolute value)	Test	Deviation (absolute value)	Calculation	Deviation (absolute value)
Standard	101.7%* ηs	--	ηs	--	97.7%* *Is*	--	*Is*	--
Design R2 6 holes 10° with stator cuts *a*=2.5, *b*=1	101.3%* ηs	-0.37%	99.5%* ηs	-0.46%	100.7% * *Is*	0.49A	104.1%* *Is*	0.68A

4.2.2 CFD results for optimized combined cooling design

Table 10 shows the CFD computation results for temperatures. We can conclude that:

- The temperature of stator upper-winding decreased about 7 K, this will do benefit for the motor protection selection and matching.
- The temperature of stator lower-winding decreased 22 K, this is really a huge benefit for complete motor temperature, so to insulation life etc.
- For rotor temperature, which is the most difficult part to cool, the optimization seems to be very effective, temperature decreased from 21 K to 60 K, depending on position.

Table 10: Simulated temperature for each part for standard design in comparison to optimized design

Component	Temperature reduction from Computation (K) Compare to standard design
Stator upper-winding	-7
Stator lower- winding	-22
Rotor upper end-ring	-21
Rotor lower end-ring	-60

4.2.3 Compressor test results for the optimized cooling design

Compressor sample was tested with thermocouples inserted into the stator windings. Table 11 shows the comparison results for the optimized design with standard design. Stator upper winding temperature decreased by 17.4 degree; and lower winding temperature decreased by 26.2 degree. This is very promising temperature improvement and can be utilized in future design.

Table 11: Compressor temperature measurement results between final optimized design and standard design

Component	Deviation (K) Test results Optimization vs. standard
Stator upper-wingding	-17.4
Stator lower-winding	-26.2

5. CONCLUSIONS

We investigated how to improve hermetic scroll compressor motor cooling by performing CFD analysis. Flow and temperature analysis were done to help with the investigation. Two methods to improve the motor cooling:

(1) By carefully selecting rotor ventilation channels, considering different factors, total percentage of gas going into low pressure side increased. Stator lower winding temperature reduced by 7 degree from CFD simulation. Rotor lower end ring temperature reduced by 42 degree.
(2) Optimize stator cuts to accommodate flow rate in different cuts position, on top of the rotor ventilation channels optimization, which proved to be very effective. Stator lower winding temperature reduced by 22 degree from simulation, and 26.2 degree from test. Rotor lower end ring temperature reduced by 60 degree from simulation.

Compressor test results confirmed the effectiveness of the combined optimization of motor cooling. Stator upper winding and lower winding temperature were reduced by 17.4 degree and 26.2 degree respectively, which is very promising and exciting way to improve motor cooling for motor and compressor designers.

REFERENCES

Chen, W., Shen J., He J., Wu H., Xing Z., (2012). Thermal Modelling for the Motor in Semi-hermetic Screw Refrigeration Compressor under Part-load Conditions. *Proc. Int. Compressor Engineering Conference at Purdue, C1246.*

Yokoyama, T., Shingu K., Sekiya S., Maeyama H., Nishimura N., (2012). CFD Analysis inside a CO2 Rotary Compressor Shell to Improve Oil Separation and Reduce the Shell Size. *Proc. Int. Compressor Engineering Conference at Purdue, C1230.*

Empirical Calculation Method for Bypass Leakage in Scroll Compressors

Noriaki ISHII[*1], Takuma TSUJI*[2], Eiji NONOGUCHI[1], Keiko ANAMI[1],
Yuya TERADA[1], Takashi MORIMOTO[3], Charles W. KNISELY[4]

[1] Osaka Electro-Communication University, Osaka, Japan.
ishii@isc.osakac.ac.jp

[2] Mayekawa Mfg. Co., Ltd., Ibaraki, Japan
takuma-tsuji@mayekawa.co.jp

[3] Panasonic Corporation Appliances Company, Shiga, Japan
morimoto.takashi@jp.panasonic.com

[4] Mechanical Engineering Department, Bucknell University, USA
knisely@bucknell.edu

ABSTRACT

This study presents a new idea for an empirical method to calculate the bypass leakage mass flow rate along the tip seal in scroll compressors. A bypass leakage test model was precisely developed to be compatible with a practical scroll compressor with large cooling capacity. Detailed tests of pressure decay in a pressurized vessel due to the bypass leakages were conducted with dry refrigerant gas R410A. The measured pressure decay characteristics were subsequently simulated using the Darcy-Weisbach equation with an empirical friction factor determined in our previous study for the leakage flow through axial clearances. In the present simulations of the measured pressure decay, the complicated flow patterns through bypass clearances were decomposed into two thin representative rectangular cross-section leakage passes, one with the effective mean width and another with the effective mean length. Empirical values of the equivalent pass width and length were determined so that the measured pressure decays are well predicted by the calculations. As a result, the bypass leakage flow rate along the tip seal in scroll compressors can be calculated by a very simple scheme introducing the equivalent pass width of one representative leakage pass with a thin rectangular cross-section.

1. INTRODUCTION

In recent years, scroll compressors have become an increasingly popular choice in many fields, because of their low vibration, low noise and high efficiency. Continued performance enhancement is desired, with improvement in volumetric efficiency as one of the key factors. Major factors affecting the volumetric efficiency are the tangential and radial leakages through the passages between the orbiting and fixed scrolls, especially for small cooling capacity scroll compressors. These leakage flows through the small axial and radial clearances between the orbiting and fixed scrolls have been studied extensively by Ishii *et al.* (1996, 2011) and Oku *et al.* (2005). The pressure decay in the pressurized vessel due to each leakage through the axial and radial clearances was measured using a maximum pressure of 3 MPa for CO_2 and 0.6 MPa for R22, and the Darcy-Weisbach equation for incompressible, viscous fluid flow through the thin rectangular cross-section was applied to calculate the leakage mass flow rate, thus simulating the pressure decay characteristics, where the empirical friction factors are determined and plotted on a Moody diagram. As a result, it has been clearly demonstrated that the empirical friction factors for both axial and radial clearance leakage flows take on essentially the same value for both CO_2 and R22, despite the significantly different working pressures.

For a low pressure type scroll compressor with large cooling capacity, in contrast to the small capacity compressors, the axial clearance of the scroll compressor becomes relatively large and hence a large leakage loss due to the radial flow through the axial clearance occurs if designed with no countermeasure. In order to eliminate reduce the impact of this leakage on the volumetric efficiency, tip seals are generally inserted into a tangential slot machined along the end of the scroll wrap. With the installation of tip seals, the radial leakage flow through the relatively large axial clearance can be reduced to a minimum level, but another new leakage flow occurs. It is the so-called "bypass leakage." Bypass leakages appear along the tip seal and has such complicated flow patterns that even the leakage characteristics seems not to be well studied, let alone formulating a method for calculating the bypass leakage mass flow rate (see Youn *et al.,* 2000). The main purpose of this study is to present a new and very simple empirical scheme for calculating the mass flow rate caused by bypass leakages.

In the present study, a new idea is proposed to represent the complicated bypass leakage by a very simple flow model of parallel passages with rectangular cross-section. The complicated flow patterns through bypass clearances are classified into two representative flows through rectangular thin cross-section passes, one with the equivalent width for the radial leakage along the tip seal and the other with the effective mean length for the tangential leakage over the scroll wrap in front of the tip seal. Empirical values of the effective pass mean width and length are so determined such that the measured pressure decays are accurately simulated with calculations. Secondly, a bypass leakage test model is processed precisely, compatible with large-power scroll compressors of about 20 kW class. Two steps of pressure decay tests were conducted in detail, with dry refrigerant gas R410A. The first step was to determine the effective mean length of the rectangular cross-section passage and the second step was to determine the effective mean width. The measured time-dependent curves of pressure decay are simulated by applying the Darcy-Weisbach equation for incompressible and viscous fluid flow, where the empirical friction factor determined in our previous study (Oku et al., 2005) for the leakage flow of dry CO_2 and R22 gas through the axial and radial clearances was applied. Finally, a careful examination of the level of contribution of each representative leakage pass to the resultant leakage was undertaken. Ultimately, the complicated bypass leakage flow can be represented by a single simple leakage flow through a rectangular cross-section passage with an effective mean width empirically determined from the tests presented in this paper

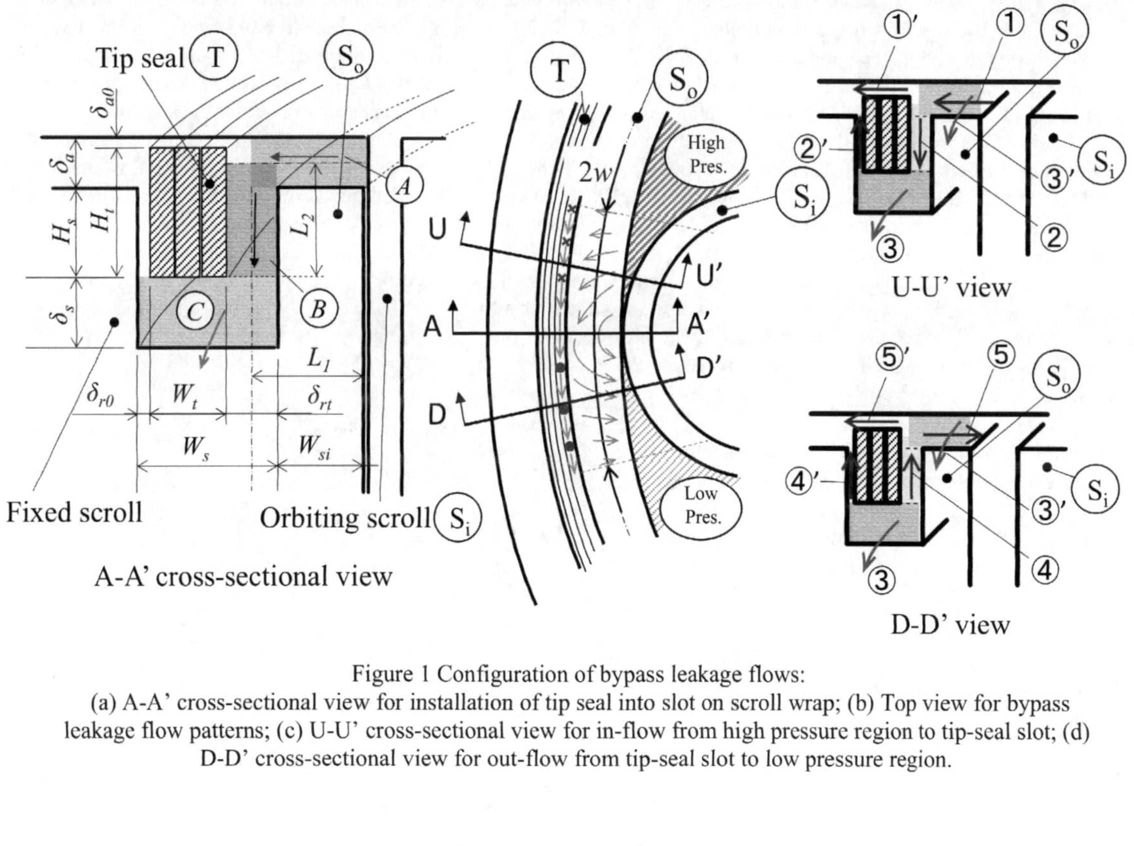

Figure 1 Configuration of bypass leakage flows:
(a) A-A' cross-sectional view for installation of tip seal into slot on scroll wrap; (b) Top view for bypass leakage flow patterns; (c) U-U' cross-sectional view for in-flow from high pressure region to tip-seal slot; (d) D-D' cross-sectional view for out-flow from tip-seal slot to low pressure region.

2. EQUIVALENT LEAKAGE FLOW MODEL WITH PARALLEL PASSAGES WITH RECTANGULAR CROSS-SECTION

2.1 Basic Configuration of Bypass Leakage Flows

The configuration of general installation of the tip seals into a rectangular slot machined on the top of the scroll wrap is shown in Figure 1(a). The right hand side faces a high pressure region and the left hand side faces a low pressure region. The tip seals are pressed against the left hand side wall of the slot and upward onto the orbiting thrust plate, caused by the hydrodynamic effects. However, leakage clearances δ_{r0} and δ_{a0} still remain behind and above the tip seal, respectively. In the present study, however, these leakages through these clearances are assumed to be negligible when compared with the other leakages. The major bypass leakages occur through the clearance δ_a (leakage region "A") over the front scroll wrap, with the length L_1, the clearance δ_{rt} (leakage region "B") in front of the tip seal, with the length L_2 and the clearance δ_s (leakage region "C") with the width W_s. A top view of the bypass leakage flow patterns is illustrated in Figure 1(b). The flow in the region "A" above the scroll wrap consists of radial and tangential flows. The radial flow (labelled ①) from the high pressure region changes its direction downward in front of the tip seal and flows into the region "B" (labelled ②) and then into the slot under the tip seal; furthermore, the flow changes its direction again in the tangential direction along the tip seal slot indicated in region "C" (labelled ③), as shown in Figure 1(c). The out flow (labelled ③ to ⑤) from the tip seal slot into the low pressure region is the opposite of the in-flow (labelled ① to ③), as explained above, and shown in Figure 1(d). The tangential flow in the region "A" (labelled ③') directly flows from the high pressure region into the low pressure region.

2.2 Equivalent Parallel Rectangular-Cross-Section Passages

The in- and out-flows, labelled ① to ⑤, are classified as a leakage flow Model-1, which can be represented as a leakage flow through a simple thin rectangular cross-sectional passage with the effective mean width and an equivalent length. First, the fundamental idea for the effective mean width is presented in Figure 2, in which the radial in-flow velocity distribution along the tangential direction over the scroll wrap is illustrated. The high pressure region between the outer scroll S_o and the inner scroll S_i forms a wedge, and hence the refrigerant gas is concentrated into the tip of the wedge and then flows out in the radial direction over the outer scroll, thus exhibiting an increase in radial flow velocity as it approaches the contact point of the outer and inner scrolls. This kind of flow may be analyzed with FEM computer calculations, but such a time-consuming method is seldom effective in practical calculations of the volumetric efficiency of scroll compressors under a variety of various operating conditions. A very simple scheme to roughly evaluate this leakage flow, which in turn can afford an accurate calculation of the leakage mass flow rate, is absolutely needed. For this purpose, the concept of "effective mean width", represented by w, is introduced, in which the flow velocity is assumed constant. It is suggested here that the effective mean width naturally depends upon the pressure difference between the high and low pressure regions and its empirical value can be easily determined by leakage tests.

With introduction of the effective mean width w, the in- and out-flows, labelled ① to ⑤, can be represented by a very simple step-like passage with the same effective mean width for all the flow steps, as shown in Figure 3, where the clearance and length of each passage is δ_a and L_1 for the flows ① and ⑤, while δ_{rt} and L_2 are for the flows ② and ④. Here, the Darcy-Weisbach equation is applied to each flow labelled ①, ②, ④ and ⑤:

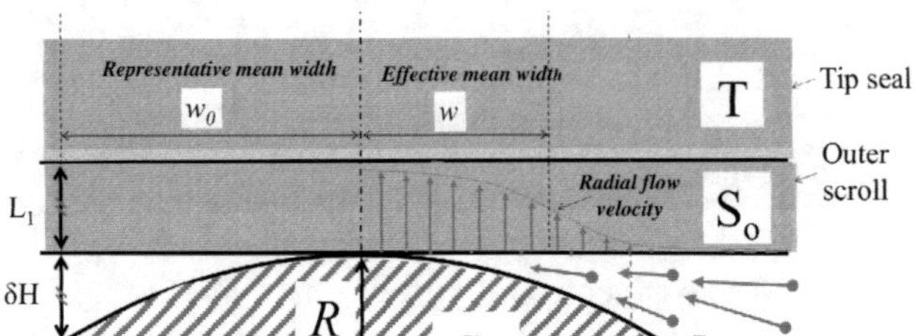

Figure 2 Effective mean width for radial flow over outer scroll (scroll wrap, labelled S_o) and its representative value for non-dimensional form.

$$\frac{P_H - P_{12}}{\rho g} = \lambda \frac{L_1}{2\delta_a} \frac{u_{m1}^2}{2g}, \quad (1)$$

$$\frac{P_{12} - P_s}{\rho g} = \lambda \frac{L_2}{2\delta_{rt}} \cdot \frac{u_{m2}^2}{2g}, \quad (2)$$

$$\frac{P_s - P_{45}}{\rho g} = \lambda \frac{L_2}{2\delta_{rt}} \cdot \frac{u_{m4}^2}{2g}, \quad (3)$$

$$\frac{P_{45} - P_L}{\rho g} = \lambda \frac{L_1}{2\delta_a} \frac{u_{m5}^2}{2g}, \quad (4)$$

where the pressure loss for the flow labelled ③ is disregarded since the clearance height δ_s is generally far larger than δ_a and δ_{rt}. The continuity equations are given by

$$u_{m1}\delta_a w = u_{m2}\delta_{rt} w = u_{m4}\delta_{rt} w = u_{m5}\delta_a w. \quad (5)$$

The summation of each side of the Darcy-Weisbach equations (1) to (4), with consideration of the continuity equation (5), results in the following simple form:

$$\frac{P_H - P_L}{\rho g} = \lambda \frac{l_1}{2\delta_a} \frac{u_{m1}^2}{2g}, \quad (6)$$

where

$$l_1 \equiv 2L_1 + 2L_2\left(\frac{\delta_a}{\delta_{rt}}\right). \quad (7)$$

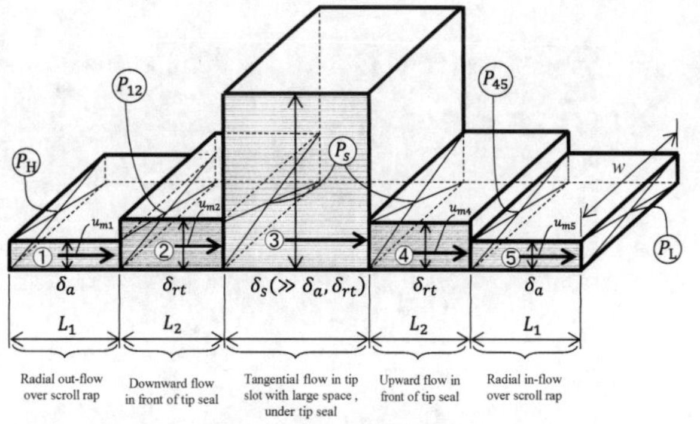

Figure 3 Equivalent steps-like passages with effective width, for flows labelled 1 to 5 of bypass leakage.

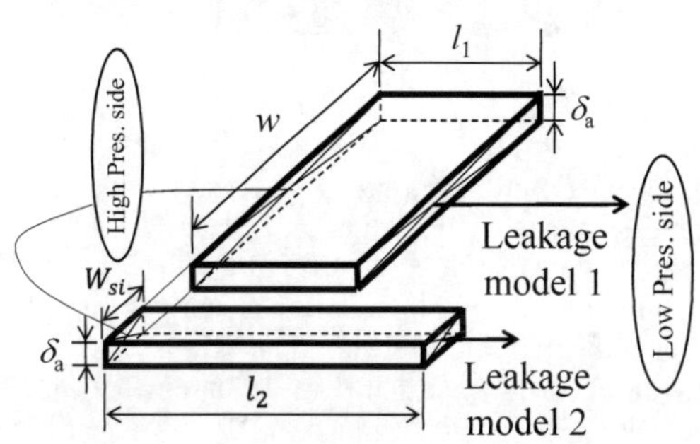

Figure 4 Equivalent flow model of parallel rectangular-cross-section passage, consisted of Model 1 and Model 2, for bypass leakages.

Equation (6) represents the reduction of the step-like passage, shown in Figure 3, to the simplest rectangular cross-section passage, as shown by the leakage Model-1 in Figure 4. The clearance height is δ_a and the equivalent length are l_1, is defined by Equation (7).

One additional leakage exists in addition to this leakage Model-1. This additional leakage is the tangential flow over the scroll wrap, labelled 3', as shown in Figures 1(c) and (d). This tangential flow through the minimal rectangular cross-section with the clearance height δ_a and the width W_{si} can be simply represented by a flow through the rectangular cross-section passage with the effective mean length l_2, as shown by the leakage Model-2 in Figure 4. It is suggested that the empirical value of the effective mean length l_2 can be easily determined by leakage tests. The Darcy-Weisbach equation for this leakage Model-2 is given by

$$\frac{P_H - P_L}{\rho g} = \lambda \frac{l_2}{2\delta_a} \frac{u_{m2}^2}{2g}. \quad (8)$$

3. LEAKAGE TESTS AND DETERMINATION OF EFFECTIVE MEAN LENGTH AND WIDTH OF PARALLEL RECTANGULAR CROSS-SECTION PASSAGES

3.1 Leakage Test Set-Up

Photos of the present leakage test set-up are shown in Figure 5, in which the most important plate S_o is sandwiched between the upper and lower plates. The high pressure region of P_H on the left is connected to the pressurized tank with a volume of 860 cm^3 and the low pressure region of P_S on the right to a refrigerant recovery tank with

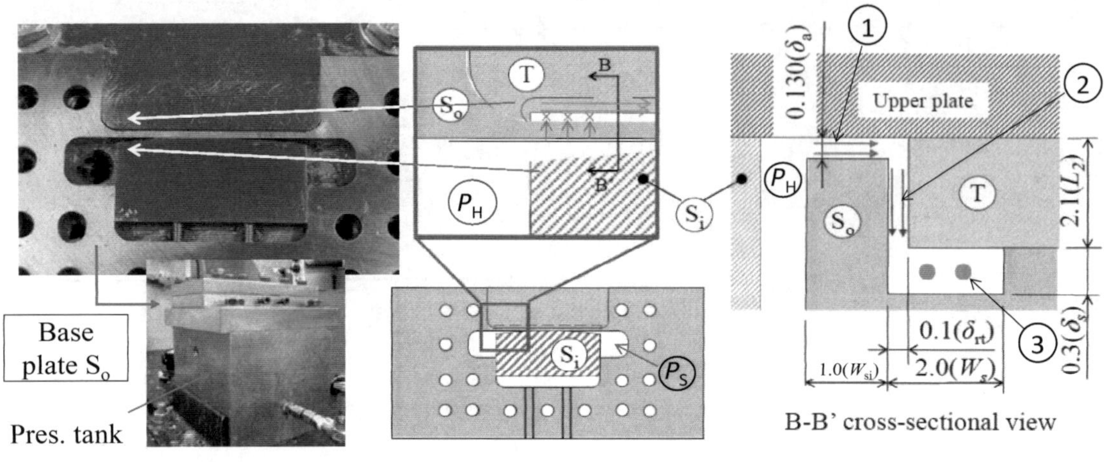

Figure 5 Configuration of lubrication test set-up.

atmospheric pressure, through a release valve. The linear tip seal slot is precisely machined on the base plate S_o, as shown in the B-B' cross-section shown right-most in Figure 5. The width W_s is 2.0 mm and the height is 2.27 mm on the left side and 0.3 mm (δ_s) on the right side, on which the tip seal plate with the height of 2.1 mm (L_2) is mounted so that the axial clearance height δ_a of 0.130 mm is achieved over the left side of the slot. The radial clearance of 0.1 mm is kept between the slot on the left side and the tip seal plate side. The outer scroll S_o has a horizontal length of 70 mm, while the inner scroll S_i with the same length, is shaded in pink, and has a radius R of 30 mm.

Table 1 Major specifications for bypass leakage tests.

Equivalent passage length l_1 [mm]	7.56
Pass length L_1 [mm]	1.05
Pass length L_2 [mm]	2.1
High pressure P_H [MPa]	0.4 ~ 1.2
Low pressure P_S [MPa]	0.1
Inner scroll radius R [mm]	30
Slot width W_s [mm]	2.0
Width W_{si} [mm]	1.0
Axial clearance δ_a [μm]	130
Radial clearance δrt [μm]	100
Slot heghit δs [μm]	300

The refrigerant gas R410A in the high pressure region with P_H flows through the axial clearance over the left side of the slot S_o, labelled ①, and flows downward in front of the tip seal plate left side, labelled ②, and flows tangentially along the tip seal slot, labelled with ③, and then in the opposite direction toward the low pressure region with P_S. The major specifications for the present leakage tests are listed in Table 1. The equivalent passage length l_1 of Model-1 is 7.56 mm, calculated from the defining Equation (7).

3.2 Pressure Decay Test Results and Determination of Effective Mean Width for Leakage Mode-1 and Effective Mean Length for Leakage Model-2

The test itself is very simple and easy to conduct. Initially, the release valve is closed to pull the air from the inside of the whole system with a vacuum pump and then the system is charged with refrigerant gas R410A. The release valve is opened as quickly as possible and the pressure decay in the pressurized tank, due to gas leakage, is measured until the pressure in the high pressure tank decreases to atmospheric pressure. Careful attention was paid to seal perfectly the contact surfaces between the parts, with a fluid gasket, so that the bypass passages are not sealed.

The initial high pressure in the tank, $P_{H,}$ was increased from 0.4 MPa to 1.2 MPa in steps of 0.1 MPa, relative to the atmospheric low pressure P_S of 0.1 MPa. Thus, the pressure difference Δp changed from 0.3 to 1.1 MPa. Test results of the pressure decay due to bypass leakage, detected with a pressure transducer (JTEKT, PMS-5M), are shown by a solid line (black) in Figure 6. Figure 6(a) represents the pressure decay due to the resultant leakage for Models -1&2, while Figure 6(b) represents the pressure decay due to a leakage only for Model-2, where the leakage passages for Model-1 were completely and carefully sealed. The leakage cross-section area for Model-2 is very small compared with that for Model-1 and, hence, the pressure decay shown in Figure 6(b) takes far more time the pressure decay shown in Figure 6(a) for Models-1&2.

The empirical friction factor λ in the Darcy-Weisbach equation was empirically determined by Ishii, et al. (1996, 2011) and Oku et al., (2005). It can be represented by a Nikuradse-style formula

$$\lambda = 0.0032 + 1.5\,\mathrm{Re}^{-0.35}, \text{ where the Reynolds number is } \mathrm{Re} \equiv 2\delta_a \rho u_m / \mu, \quad (9)$$

which is a result for the refrigerant R22 and is shown on the Moody diagram in Figure 7. With the introduction of this empirical friction factor into Equations (6) and (8), the leakage flow velocity u_{m1} and u_{m2} can be calculated and thus the mass flow rate $\dot{m}$ can be calculated by

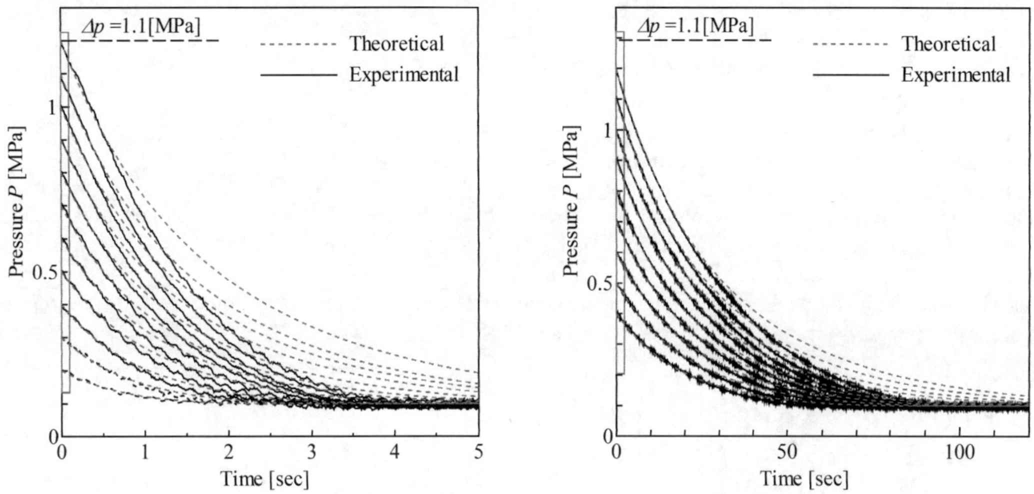

(a) Pressure decay for Model-1&2. (b) Pressure decay for Model-2.

Figure 6 Pressure decay test results and its theoretical simulation for determination of effective mean width w for Model-1 and effective mean length l_2 for Model-2.

$$\dot{m} = \rho\delta_a w u_{m1} \text{ for Model-1} \quad \text{or } \rho\delta_a W_{si} u_{m2} \text{ for Model-2} \quad (10)$$

where the density ρ can be calculated by

$$\rho = \rho_0 (P/P_0)^{1/n}. \quad (11)$$

The initial pressure and density are represented by P_0 and ρ_0, respectively. The residual refrigerant mass in the high pressure chamber, G, can be calculated by subtracting the total leakage mass from the initial refrigerant mass G_0:

$$G = G_0 - \int_0^t \dot{m}dt \quad (12)$$

Finally, assuming a polytropic process with exponent n, the pressure decay ΔP in the high pressure chamber over a small time Δt can be calculated:

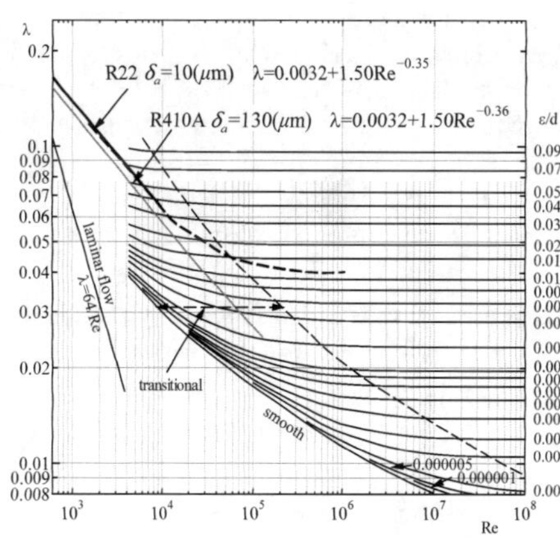

Figure 7 Empirical friction factor on Moody diagram.

$$\Delta P = \frac{P_0}{G_0^{\ n}} nG^{n-1}\dot{m}\Delta t \, . \qquad (13)$$

Calculated results for the pressure decay are shown by a dashed line (red) in Figure 6. During the initial time step, the pressure decay for Model-2, shown in Figure 6(b), was theoretically simulated so that the theoretical curve predicts well the measured pressure decay in the time region immediately after the measurement started, where the effective passage length l_2 was determined first. Similar simulations were conducted for all measured pressure decays and the determined empirical data of l_2 is plotted over the pressure difference ΔP, as shown in Figure 8, exhibiting a linear increase from about 3 mm to about 8 mm, with increasing initial pressure difference ΔP from 0.1 to 1.1 MPa. Adopting this empirical value for l_2 in the next step, the pressure decay for Model-1&2, shown in Figure 6(a) was theoretically simulated so that the theoretical curve again predicted the measured pressure decay over the time increment immediately after the measurement started, where the effective passage width w was determined. The empirical data of w is plotted over the pressure difference ΔP, as shown in Figure 8, exhibiting a quadratic-like decrease from about 22 to about 8 mm, with increasing initial pressure difference ΔP.

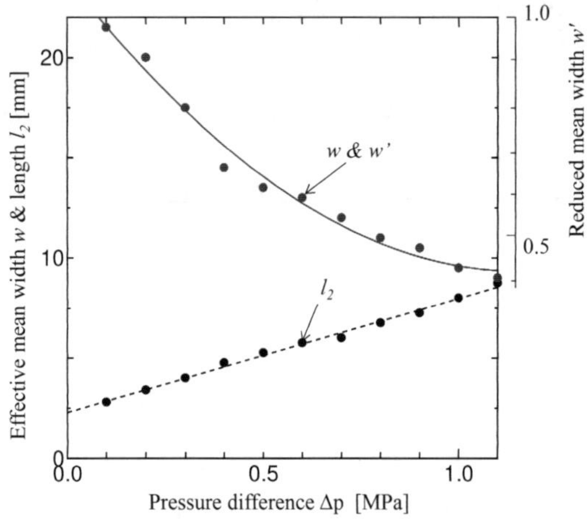

Figure 8 Effective mean width w and length, determined from pressure decay tests.

4. CONFIRMATION OF EMPIRICAL CALCULATION METHOD FOR ITS VALIDITY

In the theoretical simulations shown in Figure 6(a), the effective mean width w and length l_2 were kept at a value determined from the data at the time immediately after the measurement started, for the whole time duration over which the measurements were made. For this reason, it is suggested, the theoretically simulated curves exhibit substantial differences from the measured ones, which increases as time passes.

As a result, the theoretical calculations were conducted again for the whole time duration, adopting the empirical data of w and l_2, shown in Figure 8, depending upon the pressure P. Re-calculation results are all shown by a dashed line (red) again in Figure 9, where the friction factor λ was adjusted slightly to provide a best fit curve, given by

$$\lambda = 0.0032 + 1.5\,\mathrm{Re}^{-0.36} , \qquad (14)$$

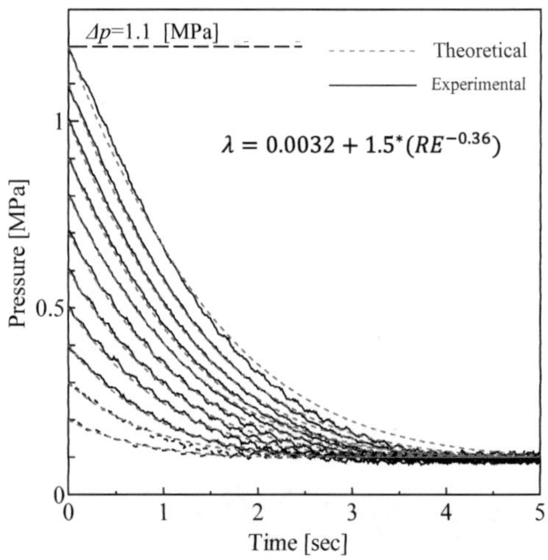

Figure 9 Theoretical simulation of pressure decay for Models-1&2, compared with test results.

which is shown by a red solid line in Figure 7, for comparison with the previous result for the refrigerant R22 and for the radial clearance δ_a of 10 μm. The present empirical result is for the refrigerant R410A and the clearance height of 130 μm which is more than 10 times larger than that for the previous tests. In spite of different refrigerants and different clearances, the empirical friction factor does not change much at all.

The calculated results shown by the red dashed lines are in a close agreement with the test results (solid black lines) for the measured time duration. This drastic improvement from the comparison shown in Figure 6(a) confirms the validity of the present idea of representing the complicated bypass leakage flows by the very simplified flows through the parallel rectangular cross-section passages.

In addition, the effective mean width w is normalized by the representative tangential length w_0, presented in Figure 3, in which special attention is paid to the distance δH between the outer and inner scrolls, increasing with distance from the contact point. The representative mean length w_0 is the horizontal (tangential) length from the contact point where the clearance height δH becomes equal to the outer scroll thickness L_1, which takes on a value of 22.0 mm for the present test set-up. Thus, a normalized quantity, the so-called "reduced mean width w'," is shown on the right hand scale in Figure 8. The reduced mean width $w'(\equiv w / w_0)$ is close to 1.0 when the pressure difference is small, and gradually decreases to about 0.4 at the maximum pressure difference of 1.1 MPa. This result indicate that the present selection of the representative mean width w_0 was physically correct. It is suggested that the empirical result of reduced effective mean width, presented in Figure 8, can be applied to all cases with different radii for the outer and inner scrolls, in calculations of bypass leakages in scroll compressors.

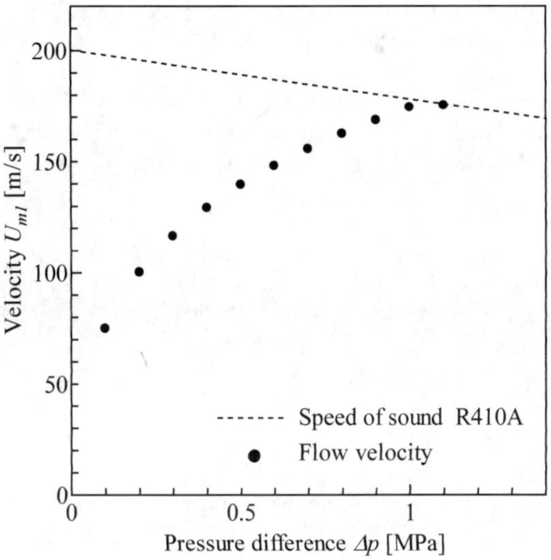

(a) Leakage flow velocity u_{ml}.

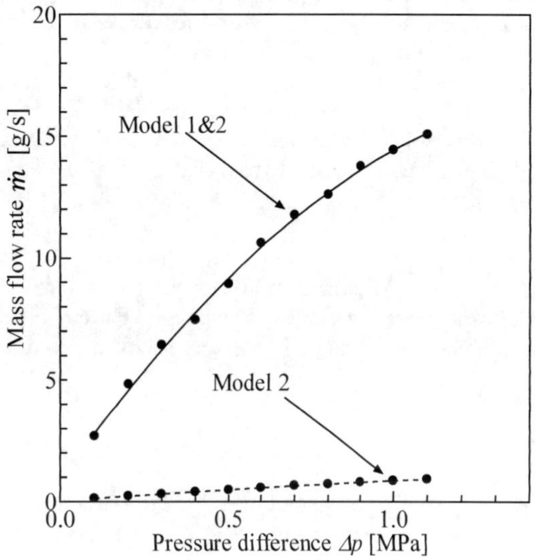

(b) Mass flow rate $\dot{m}$.

Figure 10 Empirical data of leakage flow velocity u_{ml} and mass flow rate $\dot{m}$ vs. pressure difference ΔP.

5. LEAKAGE MASS FLOW RATE AND DOMINANT LEAKAGE FLOW MODEL

In the process of the pressure decay simulations, the leakage flow velocity u_{ml} and mass flow rate $\dot{m}$ have been calculated, as shown in Figure 10, over the pressure difference Δp. All leakage velocity data, except for the single value at Δp = 1.1, are lower than the sound velocity for R410A, shown by the dashed line in Figure 10(a). For this reason the working fluid of refrigerant gas can be treated as incompressible, for the leakage flow through very thin passages. Only one calculated flow velocity data point at Δp =1.1 exceeded the sound velocity. In this case, the choking flow through the passage would force the velocity to be the speed of sound at Δp =1.1.

All data of the mass flow rate $\dot{m}$ align well along the solid and dashed lines, respectively, as shown in Figure 10(b). The upper solid line is for the leakage by Model-1&2, and the lower dashed line is for Mode-2 which is about 5%

relative to that for Model-1&2. The leakage by Model-1 is far dominant. From this result, one may disregard the leakage effect for Model-2, in simulation of bypass leakage for determining the volumetric efficiency of scroll compressors, if 5% error is permitted.

6. CONCLUSIONS

In the present study, the hypothesized treatment of complicated bypass leakage in scroll compressors using a very simple flow model was validated. In the flow model parallel rectangular cross-section passages, classified into Model-1 with the effective mean width for the radial leakage along the tip seal and Model-2 with the effective mean length for the tangential leakage over the scroll wrap in front of the tip seal are used to simulate the bypass leakage flow. The effective mean width and the effective mean length were determined by very simple and easily conducted pressure decay tests, where the measured pressure decays were all successfully simulated by very simple theoretical calculations, based on the Darcy-Weisbach equation for incompressible viscous fluid flow. Furthermore, the level of contribution to the resultant leakage is examined to conclude that the leakage by Model-1 is by far the dominant effect and the complicated bypass leakage can be calculated by just one rectangular cross-section passage represented by Model-1, if an error of about 5% is permissible. Finally, the effective mean width for Model-1 was normalized with the representative tangential length from the contact point of the outer and inner scrolls, where the distance between the outer and inner scrolls becomes equal to the thickness of outer scroll wrap in front of the tip seal. In this way, the reduced effective mean width was formulated, and the present scheme for empirically calculating the complicated bypass leakage was generalized for possible application to all cases of scroll contact area geometry.

The present study was motivated through possible developments of super large scroll compressors with shaft power more than 1000 kW, where the bypass leakage effect on the volumetric efficiency was one unknown factor for the authors. The present scheme for evaluating the complicated bypass leakage will be applied to predict the possible high volumetric efficiency of super large scroll compressors.

NOMENCLATURE

G, G_0	: Refrigerant mass	[kg]	Re	: Reynolds number	[-]
g	: Acc. of Gravity	[m/s^2]	t	: Time	[s]
L_1, L_2	: Length	[m]	u_{m1}, u_{m2}, u_{m3}, u_{m4}	: Mean velocity	[m/s]
l_1, l_2	: Equivalent length	[m]	W_s	: Width	[m]
$\dot{m}$	: Mass flow rate	[kg/s]	δ_a, δ_{rt}, δ_{a0}, δ_{r0}	: Clearance	[m]
n	: Polytropic index	[-]	λ	: Friction factor	[-]
P_H, P_s, P_L, P_{12}, P_{45}	: Pressure	[Pa]	ρ	: Density	[kg/m^3]

REFERENCES

Ishii, N., Bird, K., Sano, K., Oono, M. Iwamura, S. & Otokura, T., 1996, Refrigerant Leakage Flow Evaluation for Scroll Compressors, *International Compressor Engineering Conference at Purdue*, pp.633-638.

Ishii, N., Tsuji, T., Oku, T., Anami, K. Matsui, A. & Knisely, C., 2011, Exploration of Optimal Efficiency Super-Large Cooling Capacity Scroll Compressors for Air Conditioners, *International Conference of ICR at Prague*, ID.: 945.

Oku, T., Anami, K., Ishii, N., Knisely, C., Yasuda, K., Sawai, K. Sano, K. & Morimoto, T., 2005, Gas Leakage in CO2 and R22 Scroll Compressors and Its Use in Simulations of Optimal Performance, *International Compressor Engineering Conference at Purdue*.

Youn, Y., Cho, N., Lee, B. & Man, M., 2000, The Characteristics of Tip Leakage in Scroll Compressors for Air-Conditioners, *International Compressor Engineering Conference at Purdue*, p.797.

ACKNOWLEDGEMENT

This study was initiated through research collaborations with Danfoss Commercial Compressors. The authors express their sincere gratitude to Danfoss people for their kind regards. The authors extend their sincere gratitude to Mr. Isao Tasaka, Director, Air-Conditioning and Cold Chain Development Center, Corporate Engineering Division, and Mr. Akio Kozaki, Director, Refrigeration and Air-Conditioning Devices Business Division, Appliances Company, Panasonic Co., Ltd., for their cooperation in carrying out this work and their permission to publish this study.

CFD SIMULATION OF A DRY SCROLL VACUUM PUMP INCLUDING LEAKAGE FLOWS

Jan HESSE*, Rainer ANDRES

CFX Berlin Software GmbH,
Berlin, Germany
jan.hesse@cfx-berlin.de

* Corresponding Author

ABSTRACT

One challenge for the numerical simulation of a dry scroll vacuum pump is the discretization of the chamber volume, which changes with time. In addition, the flow characteristics are very complex including the leakage flow caused by radial and axial gaps between the rotors and the housing. For engineers and designers it is imperative to understand the influence of such leakage flows on the efficiency of the dry scroll vacuum pump.

This paper describes the workflow for Computational Fluid Dynamics (CFD) of a dry scroll vacuum pump and shows some preliminary results. The computational grids for the time dependent flow volume are generated by the grid generator TwinMesh. The meshing software generates and optimises all necessary grids for each time step prior to the actual simulation and calculates the mesh quality to assure high quality numerical results. Furthermore, the grid generation accurately considers axial gaps.

The transient numerical simulations are performed by the commercial CFD software code ANSYS CFX, which is able to handle complex flow characteristics. The simulations consider compressibility, turbulence, and heat transfer effects using ideal gas properties for the fluid. The pressure ratio between inlet and outlet is varied as well as the gap size. The results are compared with analytical methods. This paper shows time dependent information such as pressure, velocity, and temperature for specific locations in the chamber volume and also integral values such as power, torque, and mass flow. Finally, cross-sectional views are presented for different positions and time steps.

1. INTRODUCTION

Nowadays, dry scroll vacuum pumps (DSVP) are essential devices in many different industrial and academic processes. The electronic and semiconductor industry is the biggest application area of vacuum pumps, for example plasma etching and physical vapour deposition in manufacturing displays or computer chips. Vacuum pumps are also used in food industry for packaging technology or freeze-drying, in metallurgy for degassing of melt or inside a coating line. For research vacuum technology is used for electron microscopes or mass spectrometer. There are several different types of compressor mechanism. The scroll pump is constructively the simplest solution for a vacuum pump. The advantages are high performance down to 10^{-3} mbar, low cost, easy maintenance, low noise and small vibrations compared with other kinds of dry vacuum pumps.

The scroll vacuum pump is quite complex from the fluid dynamics point of view. The driven orbiting scroll wrap changes the position relative to the fixed scroll wrap which results in a time changing working chamber volume. The fluid is compressible and therefore acoustic aspects could be important. Because of the vacuum characteristics the transition from the continuous flow regime to the molecular regime has to be taken into account. Computational Fluid Dynamics (CFD) is a broadly accepted tool, both in academia as well as in industry, to analyze fluid mechanics in complex systems in order to visualize the flow, understand and optimize such systems.

The numerical computation of scroll machines requires accurate grid generation because the fluid flow is transient and depends on the rotor position. Furthermore, the grid needs a high resolution especially in the gaps and their vicinity. Consequently, TwinMesh was developed with the aim to generate high quality grids in a short time. As a result, CFD can now be utilized on a day-to-day basis in industrial applications of PD machines.

This paper is based on the paper of Yue et al. (2015) which includes CFD simulation results of a dry scroll vacuum pump. In addition the axial gaps are included in the simulation domain.

2. GEOMETRY

The working chamber of the scroll pump is a simple cylindrical volume including one fixed and one orbiting scroll wrap. The inlet and outlet volume is simplified as a pipe, which could have an influence on the DSVP performance in comparison to the real geometry. One important geometrical parameter of DSVP is the volume ratio, which is the ratio of the suction chamber volume to the discharge chamber volume. Here a volume ratio of 2.0 is used corresponding to the involute starting angle Φ_S of 3π like in Yue et al. (2015) and Li et al. (2010, Vacuum 84). The pressure inside the working chamber is analyzed on four different locations (orbiting angle of 300°, 570°, 840° and 1110°) next to the fixed scroll wrap, which are the pressure transducer locations of Li et al. (2010, Vacuum 84). The used axial clearance size is 30 micron. The specified axial clearance in the paper of Li et al. (2010, Vacuum 84) of 300 micron must be a mistake.

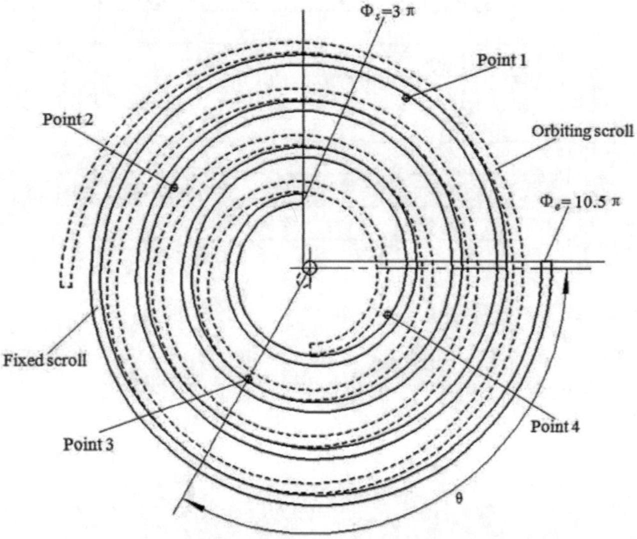

Figure 1: Geometry of the scroll wraps (Yue et al., 2015)

Table 1: Basic parameters of the DSVP

Items	Value
Radius of the basic circle	3 mm
Pitch of involute	18.85 mm
Initial angle of involute	40°
Thickness of scroll wrap	4.19 mm
Height of the scroll wrap	30 mm
Loops of scroll	5.25
Axial clearance	0.03 mm
Minimum radial clearance	0.036 mm

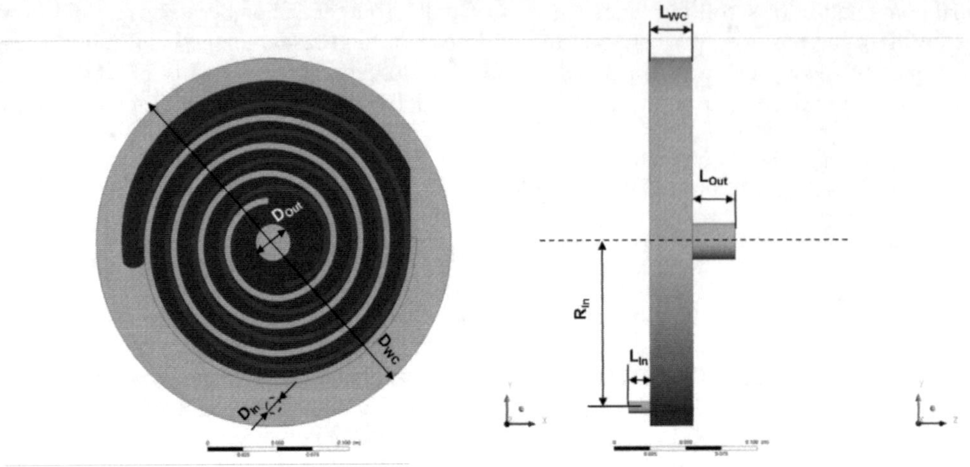

Figure 2: Geometry of the simulation domain

Table 2: Parameters of the simulation domain

Items	Value [mm]
L_{In}	15
L_{WC}	30
L_{Out}	30
R_{In}	112.5
D_{In}	8
D_{WC}	250
D_{Out}	25

3. MESHING

The discretization of the fluid volume is the main challenge to get reliable simulation results. The time changing volume has to be represented by a numerical mesh including all gaps. There are several methods to model the chamber volume with different restrictions.

3.1 Chamber Modeling

The immersed solid method is a comfortable method to setup a simulation including rotor movement. This method needs only a background mesh (without solid of orbiting scroll) and a mesh for scroll wrap solid body. A momentum source forces the fluid to the rotor movement in regions where the solid mesh overlaps with the background mesh. There are a lot of restrictions for the immersed solid method. For example it can only be used for single-phase incompressible fluids. The background mesh needs a fine mesh near the wall to represent the gap region. This method cannot model the compressibility effects of a DSVP.

The remeshing method generates a mesh during the simulation depending on the mesh quality of the deformed mesh for each time step. In comparison to the immersed solid method compressible simulations are allowed also including multi-phase. The main disadvantage is that remeshing is necessary for almost each time step when having small gaps. The time step has to be small to reach an acceptable deformed mesh per time step. This method is not efficient for a CFD simulation of a DSVP.

The method to use pre-generated meshes for each rotor position is the best method for the simulation of a DSVP. This can be done by hand with a meshing tool for structured grids or with special meshing programs like TwinMesh which has a more automatic approach. Structured meshes are generated before doing the simulation and imported later at the beginning of each time step of the simulation. The computational effort is minimized und the mesh quality is high leading to reliable simulation results.

3.2. Simulation Domain

The simulation domain is split into different volumes to use the adequate meshing strategy for each volume (Figure 3). The mesh of the stator domain is generated with ANSYS Meshing including the inlet and outlet pipes and the stationary part of the working chamber and the axial gap of the fixed scroll. The mesh of the rotor domain is created with TwinMesh whereas the axial gap of the rotor is generated with ANSYS Meshing as well.

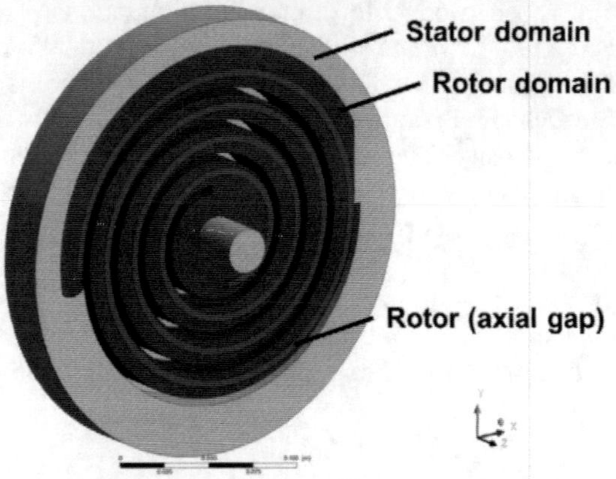

Figure 3: Simulation domains

TwinMesh is a meshing software for PD machines with two axially parallel rotors, with complex rotor geometry, i.e. continuous (e.g. lobe pump), discontinuous (e.g. screw compressor) or with a single rotor (e.g. scroll compressor, eccentric screw pump). CAD data of the rotor and casing curvature can be imported and is used for the structured grid generation. The topology of a scroll is meshed using O-type grid around the rotor curvature. Figure 4 shows the GUI in the case of a scroll pump. The meshing software generates 2D meshes for each rotation angle of the rotor with the same topology and node numbers including a refined boundary layer resolution layers towards rotor and housing walls. The meshes are smoothed with an explicit and iterative method to reach homogeneous node distribution and orthogonality. The resulting mesh of the chamber volume can be analyzed by quality criteria, such as minimum element angle, aspect ratio and volume change of cells. The 2D meshes are used to get the 3D volume mesh. The 3D meshes are exported for each rotor position (i.e. the meshes cover one complete rotation), where an angle increment of 1° is used for the simulation of the dry scroll vacuum pump.

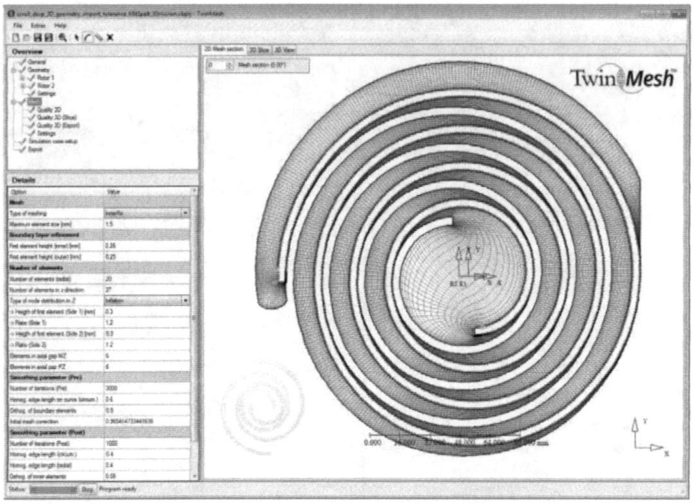

Figure 4: TwinMesh GUI (2D cross section)

4. SIMULATION SETUP

The commercial CFD solver ANSYS CFX is used for the simulation. The meshes for the stator and rotor domain are imported in the setup and connected with a GGI (Generalized Grid Interface) to consider the time step dependent connection, where the flux of mass, momentum, heat and turbulence are conserved. For the definition of the boundary condition the pressure and the temperature at inlet and outlet are specified as well as the rotational speed of the rotating scroll wrap. The wall boundary condition is set to no slip wall, which is correct for viscous flow (Kn<0.01). The wall temperature is set to 65°C for almost all simulations. Three different values for the suction pressure are used (17 kPa, 42 kPa, 95 kPa) and a discharge pressure of 95 kPa to compare the results with the measurement from Li et al. (2010,Vacuum 85). The lowest suction pressure is 17 kPa which results to Kn=0.011. The inlet temperature is 20°C for each simulation. The rotational speed is 1704 rpm. The fluid is defined as ideal gas with constant properties beside the density. The SST turbulence model is used to take turbulence in the free flow into account, whereas the turbulent kinetic energy goes to zero inside the clearances. The advection scheme is high resolution and the transient scheme is second order backward euler.

The mesh deformation due to the orbiting scroll movement is realized by a FORTRAN routine, which imports the point coordinate data (TwinMesh export files) for each node and each time step of the rotor domain. Therefore no interpolation is needed between the time steps leading to a simulation result without interpolation errors.

5. SIMULATION RESULTS

The simulation results are compared with the results of Li et al. (2010, Vacuum 84 and 85) and Yue et al. (2015). First the working process of the DSVP is analyzed. Figure 5 shows the CFD simulation results of Yue et al. (2015) in comparison to the measurements of Li et al. (2010, Vacuum 85) for a rotational speed of 1704 rpm. They use a different discharge pressure for measurement (95 kPa) and CFD simulation (101 kPa) leading a pressure offset for point 4. In the measurement the increase of the pressure with the rotation angle θ is larger than the increase of the CFD pressure curve. In CFD the pressure decrease from the high pressure chamber to the lower pressure chamber (scroll wrap passes the measurement location) is faster than the measured pressure curve. The curves of point 2 and point 3 have a pressure jump at every 360°, which indicates non periodic simulation results of the pressure curve (range of 360° results linked together). It is not clear why the pressure decrease for each measurement point starts before the rotor is passing the pressure transducer position (Rotation angle position of 30° P4, 120° P3, 210° P2, 300° P1). The paper of Li et al. (2010, Vacuum 85) has no information about the triggering of the rotation angle with the measurements. It could be that the pressure transducer information has a wrong link to the rotation angle, which results in an offset along the x-axis or there are other geometrical details (deformation of the scroll) which are important but unknown.

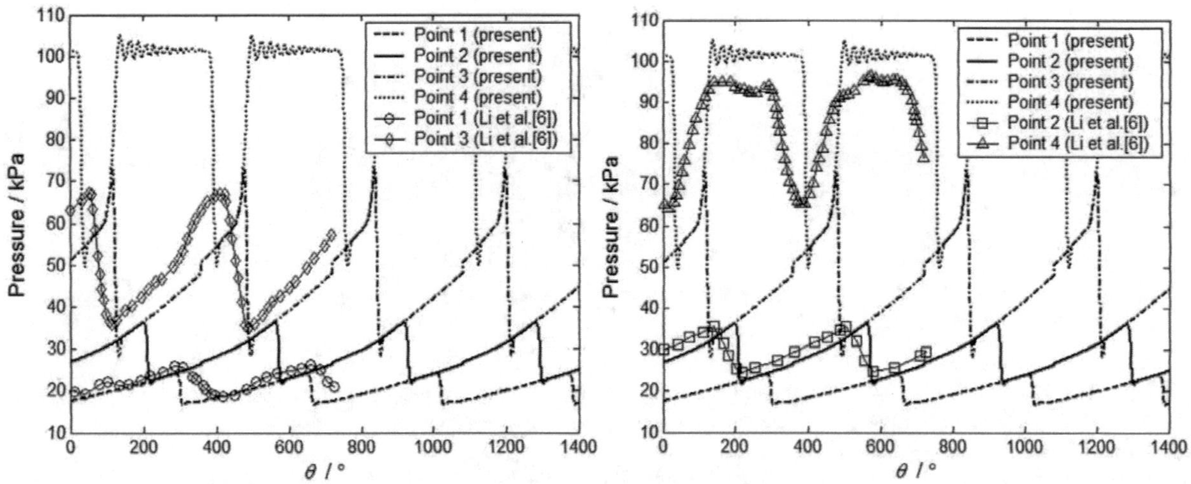

Figure 5: Working process (suction pressure 17kPa, 1704 rpm) (Yue et al., 2015)

5.1 Working Process

The working process of a DSVP describes the pressure increase with the change of the rotation position of the orbiting scroll wrap. The geometry of the DSVP used in this paper is designed for a pressure ratio of 2.64 (Yue et al., 2015) between the inlet and outlet pressure. A higher pressure ratio leads to under-compression (inlet pressure of 17 kPa) whereas a lower pressure ratio leads to over-compression. Figure 6 shows the pressure change at the measurement positions depending on the rotation angle. Our CFD results are compared with the CFD results of Yue at al. (2015). The simulation conditions are nearly the same as good as available in the referenced paper (Yue et al., 2015) except the outlet pressure. The pressure increase of our simulation is a bit slower than described in the paper. The reason could be the non-periodic working cycle of the paper results (more time step needed), differences in the boundary conditions or the different pressure ratio.

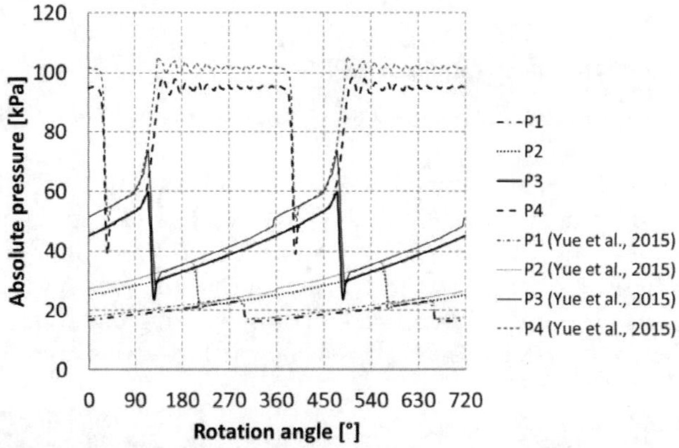

Figure 6: Comparison of our simulation results with the CFD results of Yue et al. (2015)
(no axial gaps, RNG turbulence model, suction pressure is 17 kPa, rotational speed is 1704 rpm)

Based on that setup the axial gaps are included. First we start with the specified unusual axial gap size (300 micron) described in the papers to clarify the possible mistake. Figure 7 shows the working process. The axial gap is too large to handle the pressure ratio. For further investigation we use 30 micron for the axial gap size and switch to the SST turbulence model.

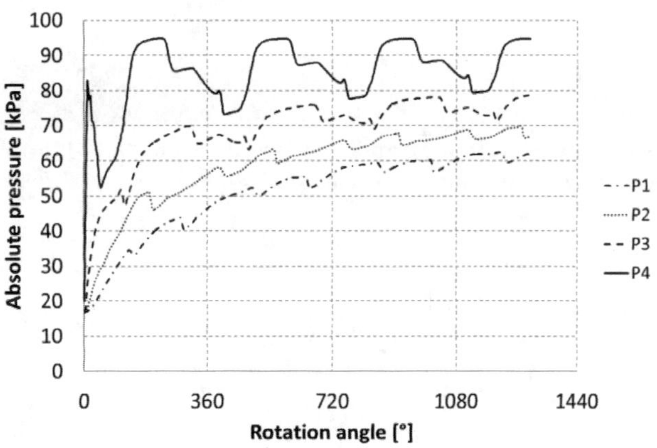

Figure 7: Pressure increase using axial gap size of 300 micron
(suction pressure is 17 kPa, rotational speed is 1704 rpm)

The comparison of our CFD results (considering the axial gaps) with the measurement results including a correction shift along the x-axis shows a good agreement (Figure 8). The correction shift is necessary to have the same pressure drop position where the orbiting scroll wrap passes the pressure transducer positon. The pressure drop in the measurements location of Li et al. (2010, Vacuum 85) and the pitch between the locations were not appropriate to the geometric definition of the transducer position.

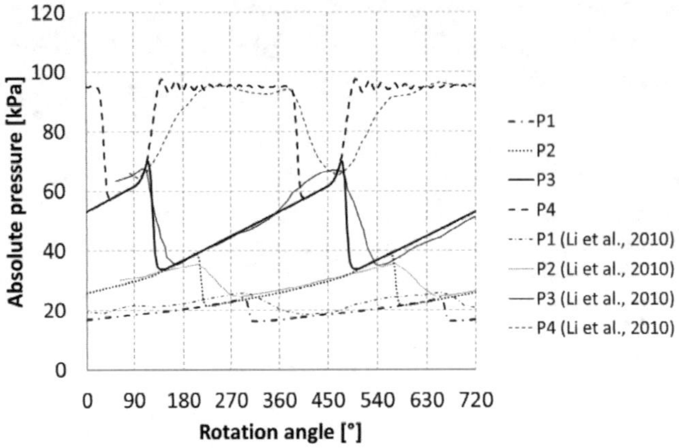

Figure 8: Comparison of our CFD results with measurements of Li et al. (2010, Vacuum 85)

The gradient of the pressure decrease in the measurement data is smaller than in CFD. This is due to the fact that the pressure transducer is connected via a pressure hole in the fixed scroll with a specific diameter and is located in a specific distance to the fixed scroll wrap (Li et al., 2010, Vacuum 85). If the orbiting scroll wrap passes the pressure hole the pressure transducer measures the area averaged pressure in the axial clearance. This leads to a slower pressure decrease. A comparable post processing of the CFD results using the averaged pressure on a circle plane located at a distance half of the scroll wrap thickness gives a comparable working process.

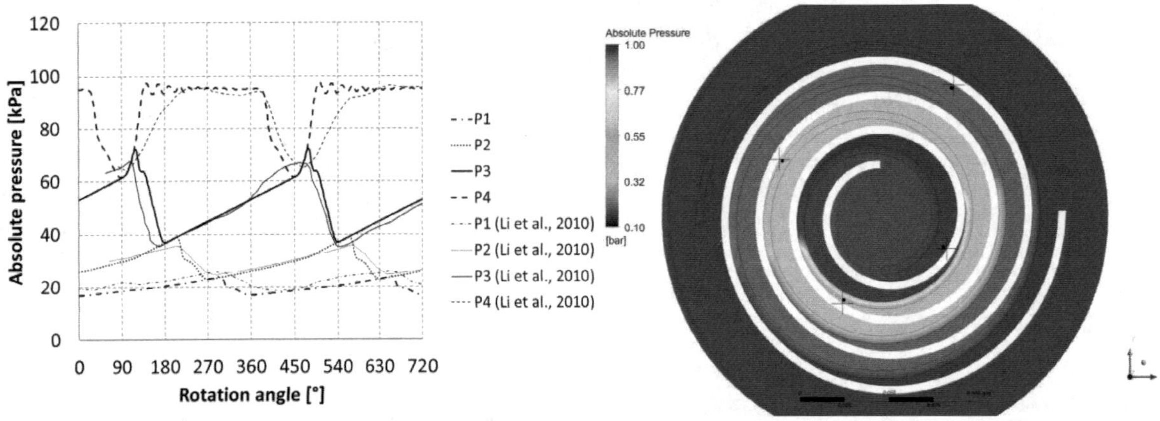

Figure 9: Average pressure calculation (left), slice plane in the axial gap (right) on fixed scroll side with measurement locations (dots) and CFD monitor point (crosses)

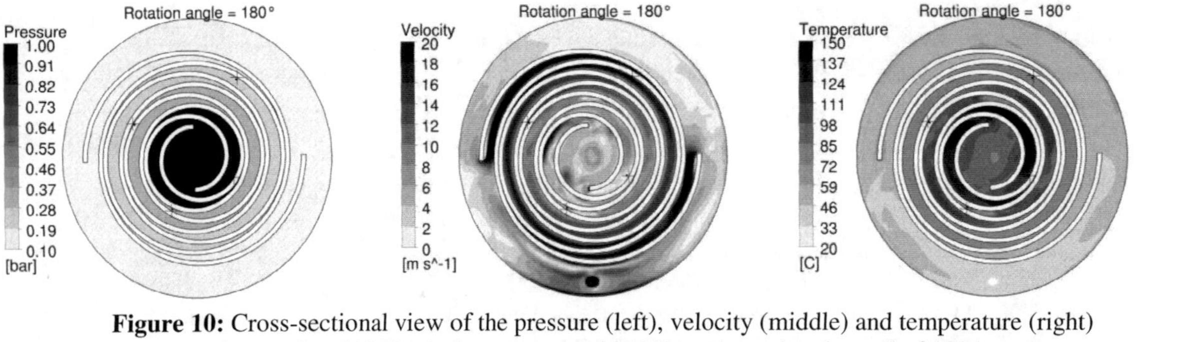

Figure 10: Cross-sectional view of the pressure (left), velocity (middle) and temperature (right) (rotation angle of 180°, suction pressure of 17 kPa and rotational speed of 1704 rpm)

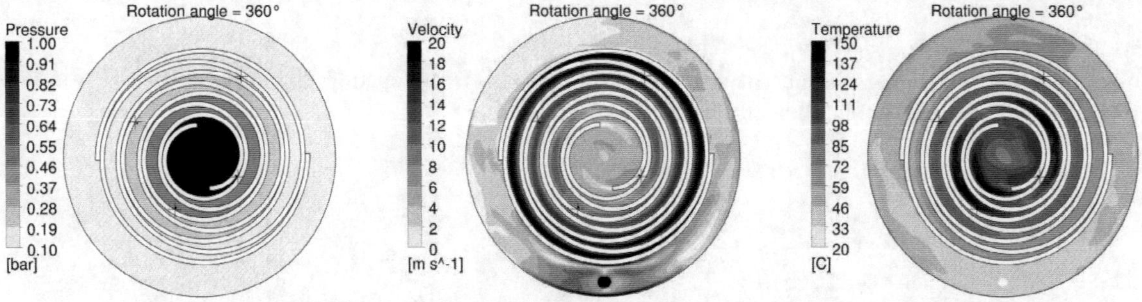

Figure 11: Cross-sectional view of the pressure (left), velocity (middle) and temperature (right) (rotation angle of 360°, suction pressure of 17 kPa and rotational speed of 1704 rpm)

Figure 10 and 11 show the pressure, temperature and velocity for two different rotor positions.

The comparison of our CFD results with the theoretical approach of Li et al. (2010, Vacuum 85) shows a difference in the working process. The pressure increase is faster than in the CFD results. The specification of the temperature behavior has a main influence on the pressure increase.

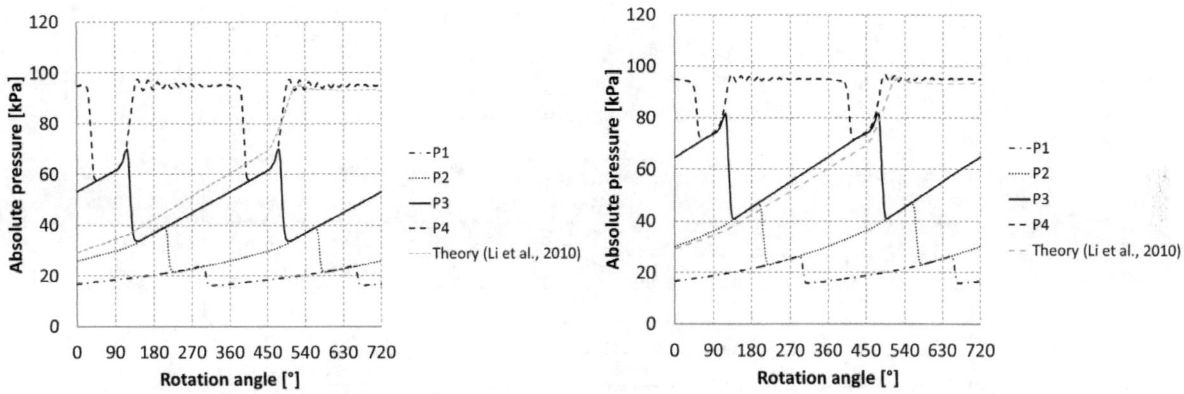

Figure 12: Comparison of theoretical approach of Li et al. (2010, Vacuum 85) with our CFD results using a wall temperature of 65°C (left) and an adiabatic wall (right)

With an adiabatic simulation the pressure increase is faster in comparison to the theoretical process due to the higher temperatures (no cooling over the casing). The correct modelling of the heat transfer from the internal fluid to the ambient is really important to get the correct working process.

5.2 Mass Flow

The mass flow decreases with decreasing suction pressure.

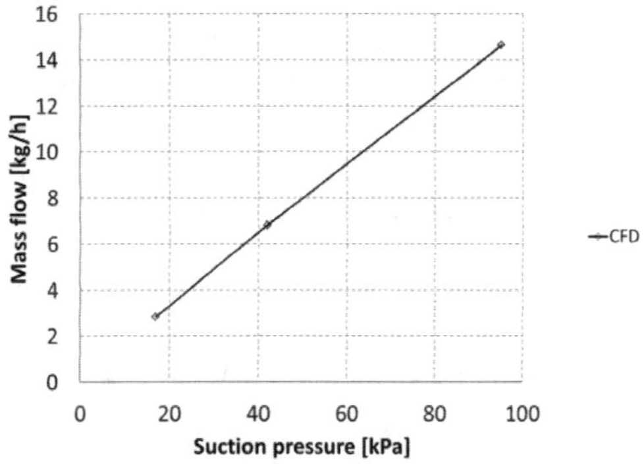

Figure 13: Mass flow from CFD results (rotational speed of 1704 rpm)

5.3 Torque

Figure 14 shows the instantaneous torque of orbiting scroll under different suction pressures calculated from the pressure forces on the orbiting scroll around the rotation axis.

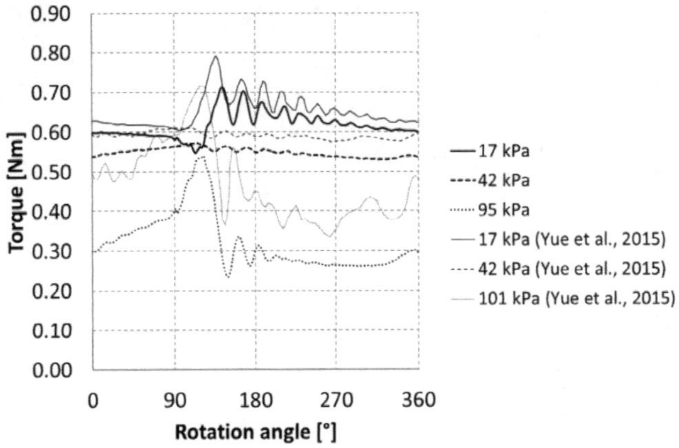

Figure 14: Torque of orbiting scroll from our CFD results in comparison to the CFD results of Yue et al. (2015) (rotational speed of 1704 rpm)

Table 3: Power consumption

Suction pressure kPa	Power consumption W	Reference
17	110.3	CFD results
42	96.4	CFD results
95	54.6	CFD results
31	174	Polytropic process
17	116.6	CFD Yue et al. (2015)
42	105.6	CFD Yue et al. (2015)
101	84.1	CFD Yue et al. (2015)

6. CONCLUSIONS AND OUTLOOK

This paper shows CFD results for a dry scroll vacuum pump including axial gaps. Different suction pressures from 17 kPa to 95 kPa were applied using an inlet temperature of 20°C. The CFD results of the working process are compared with theoretical calculation and experimental data. In addition, average values of the mass flow and power consumption are shown for different suction pressures. For the CFD simulation ANSYS CFX was used and the meshes for the working chamber were created with the meshing software TwinMesh and the stator meshes with ANSYS Meshing.

The comparison of the CFD results with the experimental data indicates that the working mechanism and flow conditions within the dry scroll vacuum pump are well captured by the simulation. The heat transfer from the process gas to the ambient over the casing has a main impact on the increase of the pressure over the rotation angle. The pressure increase of the adiabatic process is faster than specifying a constant wall temperature.

Further investigation will focus on the heat transfer through the casing of the pump, to have the correct solid and fluid temperature depending on the rotational speed. If the boundary conditions including the heat transfer are specified more accurate further investigations are planned for a mesh refinement study also including real gas properties.

NOMENCLATURE

Kn	knudsen number	()
θ	rotation angle	(°)

REFERENCES

Li, Zeyu & Li, Liansheng & Zhao, Yuanyang & Bu, Gaoxuan & Shu, Pengcheng, 2010, Theoretical and experimental study of dry scroll vacuum pump, *Proceedings of the Vacuum 84*, (415-421), Elsevier

Li, Zeyu & Li, Liansheng & Zhao, Yuanyang & Bu, Gaoxuan & Shu, Pengcheng & Lui, Jinping, 2010, Test and analyses on the working process of dry scroll vacuum pump, *Proceedings of the Vacuum 85*, (95-100), Elsevier

Yue, Xiang-Ji & Lu, Yan-jun & Zhang, Ying-Li & Ba, De-chung & Wang, Guang-Yu, 2015, Computational fluid dynamics simulation study of gas flow in dry scroll vacuum pump, *Proceedings of the Vacuum 116*, (144-152), Elsevier

Theoretical and Experimental Analysis of Scroll Expander

Peng Bin[1]*, Zhu Bingguo[2], Vincent Lemort[3]

[1] School of Mechanical and Electronical Engineering, Lanzhou University of Technology.
Lanzhou, Gansu, China
Phone: +86(0)13893459071, E-mail: pengb2000@163.com

[2] School of Mechanical and Electronical Engineering, Lanzhou University of Technology.
Lanzhou, Gansu, China
Phone: +86(0)13919835339, E-mail: 571912509@qq.com

[3] University of Liege, Energy Systems Research Unit,
Liege, 4000, Belgium
Phone: +32(0)43664824, E-mail: Vincent.lemort@ulg.ac.be

* Corresponding Author

ABSTRACT

In this article, both theoretical and experimental analysis on the performance of a new scroll expander is presented and discussed. Various working chamber volumes from suction to discharge are defined. Thereafter, the evolution and derivative of the working chamber volume with respect to the orbiting angle are discussed. In order to investigate the performance of scroll expander, a detail thermodynamic modeling based on energy and mass balances is established. Radial and flank leakage, heat transfer between the working fluid, scroll wraps and plates are considered in the thermodynamic modeling. Volume, pressure, mass flow of working chamber and power are investigated by solving the thermodynamic modeling. A simple experimental rig for scroll expander is set up. From the comparison of the simulated and measured data, it can be seen that the expander model predicts the output power very well (within 5%). So the proposed mathematical modeling can accurately describe all the suction, expansion and discharge processes for scroll expander.

1. INTRODUCTION

In recent years, protecting environment and reducing energy consumption become two major issues for the human being. Many researchers concentrate on the low-grade energy recovery technique in order to economize energy sources. The small scale energy recovery system makes important contribution to energy crisis and environment pollution which is normally used in the vehicle and mobile equipment. As a kind of high efficient positive machine, scroll compressor has many advantages such as simple structure, high-efficiency, low noise, high reliability, low vibration, light weight and small size compared with other types of compressors. It is becoming popular and widely used in refrigeration, air-conditioning, various kinds of gas compression and pressurized pump products, etc (Peng Bin *et al.*, 2016). Scroll expander could be used as the energy recovery machine in the small energy recovery system. In recent years, scroll expander is applied in the low-temperature waste heat recovery system, fuel cell system and carbon dioxide refrigeration system. There are a lot of researches on experimental evaluations and numerical simulations for the performance of scroll expanders (Liu Guangbin *et al.*, 2013. Song P. *et al.*, 2015. Tarique M. A. *et al.*, 2014. Lemort V. *et al.*, 2009. Quoilin S. *et al.*, 2010. Zhu J. *et al.*, 2016. Wu Z. *et al.*, 2015. Quoilin S., 2011). However, for scroll expander, it is worthwhile to get more comprehensive understanding on the various aspects of its working process using the detail mathematical model and quantitative analysis in order to increase the accuracy of simulation and performance prediction. In this paper, effort has been made to present in detail inner working process of scroll expander. This research is expected to provide an insight to understand the qualitative and quantitative characteristics of the working process of scroll expander. So this study develops a new scroll expander and presents a comprehensive simulation model that predicts the performance of a scroll expander. A simple experimental campaign is also conducted to measure expander performance and validate the mathematical model.

2. GEOMETRIC MODEL

The goal of the geometry model is to achieve expressions for the volumes of the different expander chambers as a function of the orbiting angle. The change of the working chamber volume with respect to the orbiting angle is calculated according to one single chamber from suction to discharge. Figure 1 is the structure of scroll expander. The basic parameters of the scroll expander are listed in Table1.

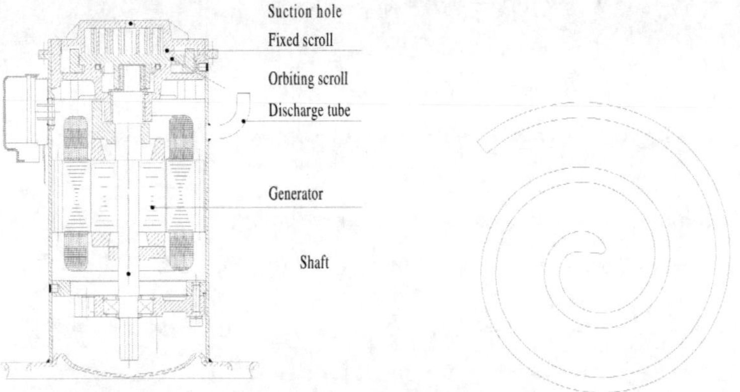

Figure 1: Scroll expander

Table 1: The basic parameters of the scroll expander

Parameters	Value
Base circle radius r_b	2.387mm
Pitch P	15mm
Thickness t	3.6mm
Height h	20mm
Ending angle of the scroll profile φ_e	27.25rad
Radius of big modification arc R_{O1}	4.13mm
Radius of small modification arc R_{O2}	1.02mm
Modification line length L	5.71mm

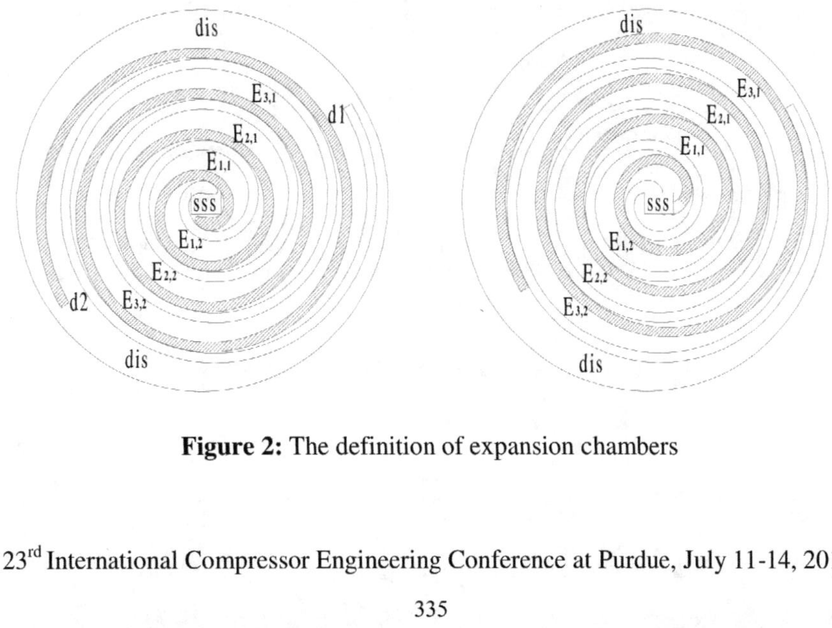

Figure 2: The definition of expansion chambers

Figure 2 is the definition of working chambers. The working chambers $E_{1,1}$, $E_{1,2}$, $E_{2,1}$, $E_{2,2}$, $E_{3,1}$ and $E_{3,2}$ are expansion chambers. The working chambers d_1 and d_2 are discharge chambers. The working chamber sss is suction chamber. The base line method is used to calculate the volume of scroll expander (Peng Bin *et al.*, 2015).
The working chamber volumes $V_{E1,1}(\theta)$ and $V_{E1,2}(\theta)$ can be calculated by:

$$V_{E1,1}(\theta) = h\pi r_b R_{or}(2\theta - (\varphi_{i0} - \varphi_{o0} - \pi)) \quad 0 \le \theta < 2\pi \tag{1}$$

$$V_{E1,2}(\theta) = V_{E1,1}(\theta) \tag{2}$$

The working chamber volumes $V_{E2,1}(\theta)$ and $V_{E2,2}(\theta)$ can be calculated by:

$$V_{E2,1}(\theta) = h\pi r_b R_{or}(2(2\pi - \theta) - (\varphi_{i0} - \varphi_{o0} - \pi)) \quad 0 \le \theta < 2\pi \tag{3}$$

$$V_{E2,2}(\theta) = V_{E2,1}(\theta) \tag{4}$$

The working chamber volumes $V_{E3,1}(\theta)$ and $V_{E3,2}(\theta)$ can be calculated by:

$$V_{E3,1}(\theta) = h\pi r_b R_{or}(2(4\pi - \theta) - (\varphi_{i0} - \varphi_{o0} - \pi)) \quad 0 \le \theta < 2\pi \tag{5}$$

$$V_{E3,2}(\theta) = V_{E3,1}(\theta) \tag{6}$$

The working chamber volumes $V_{d_1}(\theta)$ and $V_{d_2}(\theta)$ can be calculated by:

$$\begin{aligned} V_{d_1}(\theta) &= h r_b R_{or}((2\varphi_e - 2\theta)\varphi_e - (\varphi_e - \theta)^2 - (\varphi_e - \theta)(\varphi_{i0} + \varphi_{o0} + \pi) + \\ & 2(1 - \cos(\varphi_e - \theta)) - 2(\varphi_e - \pi)\sin(\varphi_e - \theta)\frac{\pi}{4}\sin(2(\varphi_e - \theta))) \quad \varphi_e - 2\pi \le \theta < \varphi_e \end{aligned} \tag{7}$$

$$V_{d_2}(\theta) = V_{d_1}(\theta) \tag{8}$$

The working chamber volume $V_{sss}(\theta)$ can be calculated by:

$$V_s(\theta) = h r_b R_{or}\theta(\theta - \varphi_{i0} - \varphi_{o0} + 3\pi) \quad 0 \le \theta < 2\pi \tag{9}$$

The inner volume ratio v_e of scroll expander can be calculated by:

$$v_e = \frac{2\varphi_e - \varphi_{i0} + \varphi_{o0} - 3\pi}{5\pi - \varphi_{i0} + \varphi_{o0}} \tag{10}$$

Figure 3 is the change of the working chamber volume with respect to the orbiting angle. It can be seen that the volume increases in suction process. As expansion volume increases, temperature and pressure decrease. The expansion chambers change into discharge chambers at discharge angle. As the discharge process ends, one entire suction-expansion-discharge process finished.

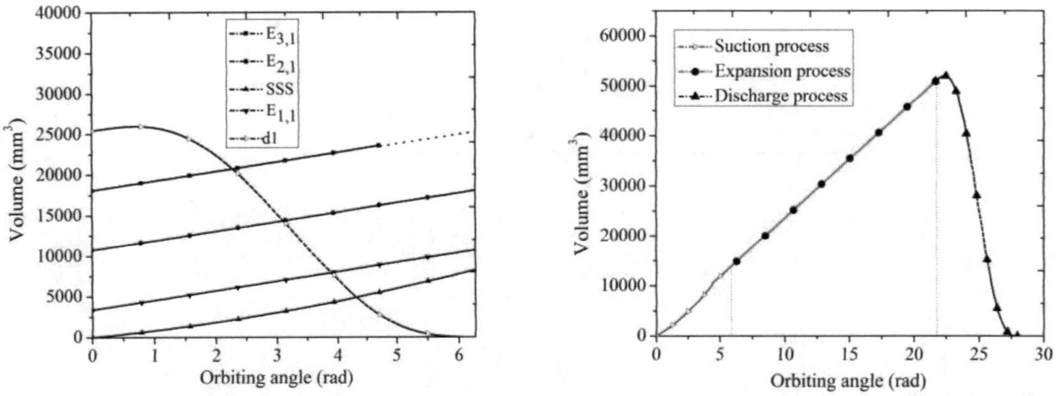

(a) The working chambers volume (b) The expansion process

Figure 3: The change of the working chamber volume with respect to the orbiting angle

3. THERMODYNAMIC MODEL

3.1 Basic equations

Change of temperature, mass and pressure in each expansion chamber with respect to orbiting angle θ can be calculated from the first law of thermodynamics (Equation (11)) for an open control volume in conjunction with an equation of state (13), (14) and the mass balance (Equation (15)). The change of the gas(N_2) temperature with respect to θ can be written as (Halm N.H, 1997. Chen Yu, 2000)

$$\frac{dT}{d\theta}=\frac{1}{mC_v}\left\{-T\left(\frac{\partial p}{\partial T}\right)_V\left[\frac{dV}{d\theta}-\frac{v}{\omega}(m_{in}-m_{out})\right]-\sum\frac{m_{in}}{\omega}(h-h_{in})+\frac{Q}{\omega}\right\} \quad (11)$$

$$\omega=\frac{d\theta}{dt} \quad (12)$$

$$\frac{dp}{dT}=\frac{R}{v},\quad \frac{dp}{dv}=-\frac{RT}{v^2} \quad (13)$$

$$dh=c_p dT\,,\quad c_p=1.11-0.48(\frac{T}{1000})+0.96(\frac{T}{1000})^2-0.42(\frac{T}{1000})^3 \quad (14)$$

$$\frac{dm}{d\theta}=\sum\frac{m_{in}}{\theta}+\frac{m_{out}}{\theta} \quad (15)$$

3.2 Suction gas mass flow

The suction gas mass flow rate is calculated using the flow equation for isentropic flow of a compressible ideal gas, corrected by a flow factor ψ (Fox RW and McDonald AT, 1992)

$$\dot{m}=\psi A_s\sqrt{2p_h\rho_h}\sqrt{\frac{\gamma}{\gamma-1}\left[\left(\frac{p_l}{p_h}\right)^{\frac{2}{\gamma}}-\left(\frac{p_l}{p_h}\right)^{\frac{\gamma+1}{\gamma}}\right]} \quad (16)$$

This flow is restricted by a critical pressure ratio $p_l\,/\,p_h$ for choked flow conditions.

$$\left(\frac{p_l}{p_h}\right)_{crit}=\left(\frac{2}{\gamma+1}\right)^{\frac{\gamma}{\gamma-1}} \quad (17)$$

3.3 Heat transfer model

The expansion process in scroll expander is achieved in pockets between the fixed scroll and the orbiting scroll. The efficiency of scroll expander depends significantly on the convective heat transfer and the leakage between the scrolls and working fluid. Therefore, the accurate understanding of the inside physical condition of gas that is influenced by the convective heat transfer with scrolls, is very important to correctly analyze the expansion process in scroll expander. Heat transfer can occur at several locations in a scroll expander because of the different temperature of ambient, shell, fluid, scroll wraps and plates. Heating during the expansion process causes the expansion to go farther away from the ideal isentropic expansion process.

3.3.1 Heat transfer during suction and discharge

The heat transfer coefficient h_c can be calculated by the following relations (Incropera FP and Dewitt DP, 1996)

$$h_c=0.023\frac{\kappa}{d_p}Re^{0.8}Pr^{0.4} \quad (18)$$

$$Re=\frac{4\dot{m}}{\pi\, d_p\mu} \quad (19)$$

$$Pr=\frac{\mu c_p}{\kappa} \quad (20)$$

The outlet temperature $T_{s,o}$ of the gas can be calculated by

$$T_{s,o}=T_{pipe}-(T_{pipe}-T_i)\exp\left(-\frac{\pi\, d_pL_ph_c}{\dot{m}c_p}\right) \quad (21)$$

The heat flow rate from the pipe to the fluid $\dot{Q}_{pipe}$ is given by

$$\dot{Q}_{pipe} = \dot{m}c_p(T_{pipe} - T_{s,i})\left[1 - \exp\left(-\frac{\pi d_p L_p h_c}{\dot{m}c_p}\right)\right] \quad (22)$$

which is evaluated for both the inlet and outlet paths of the expander.

3.3.2 Scrolls heat transfer

As the fluid is expanded by the scroll expander, it experiences heat transfer from the scrolls, the bottom and top plate. In order to account for these losses, the heat transfer process needs to be modeled and incorporated into the expansion process model.

The correlation h_{cm} for the spiral heat exchanger is as follows (Kakac S and Shah RK, 1987. Rajavel R and Saravanan K, 2008. K Jang and S Jeong, 2006)

$$h_{cm} = 0.023\frac{\kappa}{D_{ef}}Re^{0.8}Pr^{0.4}\left(1+1.77\frac{D_{ef}}{R_{aver}}\right)\left(1+8.48\left[1-\exp(-5.35St)\right]\right) \quad (23)$$

The hydraulic diameter D_{ef} is defined as

$$D_{ef} = \frac{4V}{A} \quad (24)$$

The average radius R_{aver} is defined in this model as

$$R_{aver} = a\left[\frac{(\varphi_k - \pi/2) + (\varphi_{k-1} - \pi/2)}{2}\right] \quad (25)$$

The temperature distribution of the scroll is non-uniform. It is assumed that the temperature along the scroll wraps is linear with the generating angle φ between two generating angles φ_1 at the center of the scroll and generating angle φ_2 at the outer edge of the scrolls. The average temperature of scrolls is at the generating angle of $\frac{\varphi_1+\varphi_2}{2}$ (middle of the scrolls). Then, the temperature distribution in the scrolls can be expressed as (Chen Yu, 2000. Rajavel R and Saravanan K, 2008. K Jang and S Jeong, 2006)

$$T(\varphi) = \frac{\Delta T}{\Delta\varphi}(\varphi - \varphi_1) + T_{s1} \quad (26)$$

The heat exchange rate $\dot{Q}_{scrolls}$ from the scroll walls/plates to the gas in any chamber can now be calculated by an integral method according to the following equation

$$\dot{Q}_{scrolls} = h_c\int_A \left[T(\varphi) - T(k,j)\right]dA \quad (27)$$

3.4 Leakage model

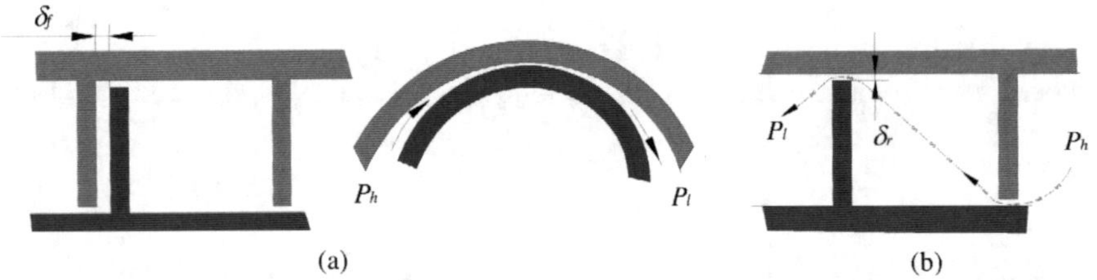

(a) Flank leakage (b) radial leakage

Figure 4: Flank and radial leakage

During expansion process, the pressure difference of adjacent chambers leads to leakage between in the high and low pressure chambers. There are two different paths for leakage in a scroll expander. One is the path that is formed by a gap between the flanks of the two scrolls and is called flank leakage. Another path is formed by a gap between the bottom or the top plates and the scrolls. This kind of leakage is called radial leakage. The two kinds of leakage are illustrated in Figure 4. Flank leakage and radial leakage account for losses during the expansion process as they decrease the volumetric efficiency significantly. Leakages also account for a decrease in specific expansion work rate because gas leaking from high pressure regions back into low pressure regions and needs to be re-expanded. In order to calculate the leakage rate from a higher pressure chamber to one with low pressure, the width of the flow path needs to be evaluated (Halm N.H, 1997).

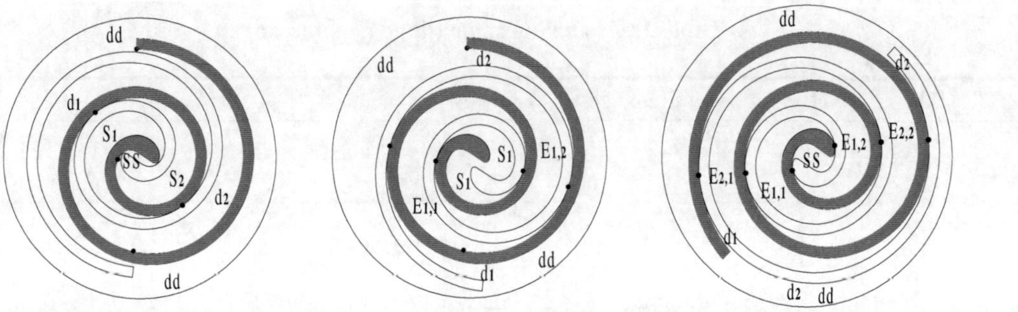

(a) No pair of expansion chamber $N_c = 0$ (b) One pair of expansion chambers $N_c = 1$ (c) Two pairs of expansion chambers $N_c > 1$

Figure 5: Three cases for expansion chambers

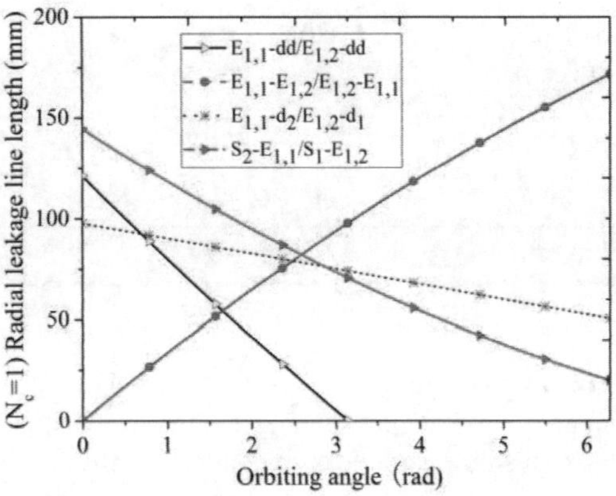

Figure 6: Evolution of the radial leakage line length with respect to the orbiting angle

For the flank leakage, the gas flow area A_f is

$$A_f = h\delta_f \tag{28}$$

According to the scroll expander manufacturer, the flank gap size δ_f can be represented as a linear function of the pressure ratio.

$$\delta_f = -9.615\times 10^{-5}\left(\frac{p_h}{p_l} - 1.67\right) + 20\times 10^{-6} \tag{29}$$

For the radial leakage, the gas flow area A_r is

$$A_r = \delta_r L_r \tag{30}$$

The effect of the expansion ratio on the radial gap size is much less. It is assumed however, that the radial gap increases with increasing expansion ratio. The gap size δ_r is also represented by a linear function.

$$\delta_r = 1.1\times 10^{-6}\left(\frac{p_h}{p_l} - 1.67\right) + 10^{-6} \tag{31}$$

Figure 5 shows three cases for expansion chambers. Table 2 is generating angle of the radial leakage for different cases. The leakage line length L_r is

$$L_r = r_b\left(\frac{1}{2}(\varphi_{max}^2 - \varphi_{min}^2) - \varphi_0(\varphi_{max} - \varphi_{min})\right) \tag{32}$$

φ_{max} and φ_{min} are displayed in the table 2.

Table 2: Generating angle of the radial leakage

N_c	The chambers of leakage	φ_{max}	φ_{min}
$N_c = 0$	$s_2 - d_1 / s_1 - d_2$	φ_{ie}	$\max(\varphi_{ie} - \theta, \varphi_{ie} - \pi)$
	$s_1 - s_2 / s_2 - s_1$	$\varphi_{ie} - \pi$	$\min(\varphi_{ie} - \theta, \varphi_{ie} - \pi)$
$N_c = 1$	$E_{1,1} - dd / E_{1,2} - dd$	$\max(\varphi_{ie} - \theta, \varphi_{ie} - \pi)$	$\varphi_{ie} - \pi$
	$E_{1,1} - E_{1,2} / E_{1,2} - E_{1,1}$	$\max(\varphi_{ie} - \pi - \theta, \varphi_{ie} - \pi)$	$\varphi_{ie} - \theta - \pi$
	$E_{1,1} - d_2 / E_{1,2} - d_1$	$\varphi_{ie} - \theta - \pi$	$\varphi_{ie} - \theta - 2\pi$
	$s_2 - E_{1,1} / s_1 - E_{1,2}$	$\varphi_{ie} - \theta - 2\pi$	φ_{is}
$N_c > 1$	$E_{2,1} - dd / E_{2,2} - dd$	$\max(\varphi_{ie} - \theta, \varphi_{ie} - \pi)$	$\varphi_{ie} - \pi$
	$E_{2,1} - d_2 / E_{2,2} - d_1$	$\max(\varphi_{ie} - \pi - \theta, \varphi_{ie} - \pi)$	$\varphi_{ie} - \theta - \pi$
	$E_{2,1} - E_{2,2} / E_{2,2} - E_{2,1}$	$\varphi_{ie} - \theta - \pi$	$\varphi_{ie} - \theta - 2\pi$
	$E_{\alpha-1,1} - E_{\alpha,2} / E_{\alpha-1,2} - E_{\alpha,1}$	$\varphi_{ie} - \theta - 2\pi(\alpha - 1)$	$\varphi_{ie} - \theta - 2\pi(\alpha - 1) - \pi$
	$E_{\alpha,1} - E_{\alpha,2} / E_{\alpha,2} - E_{\alpha,1}$	$\varphi_{ie} - \theta - 2\pi(\alpha - 1) - \pi$	$\varphi_{ie} - \theta - 2\pi\alpha$
	$s_2 - E_{1,1} / s_1 - E_{1,2}$	$\varphi_{ie} - \theta - 2\pi N_c$	φ_{is}

3.5 Output characteristics of scroll expander

The working process of scroll expander is unsteady state. The parameters change with the orbiting angle. As the shaft rotates, the gas sucked into the expansion chamber is expanded. In the scroll expander structure, the orbiting scroll can move between the fixed scroll and the frame, the gas forces and moments acting on it are finally transferred to the bearings installed in the orbiting scroll and the frame. The tangential and axial gas forces also account for an increase frictional loss. The output torque of scroll expander is mainly caused by the tangential gas force. The main gas forces and moments are showed in Figure 7.

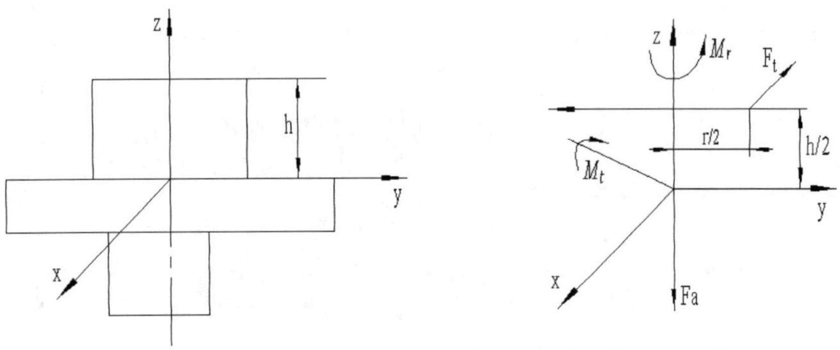

Figure 7: Gas forces and moments acting on the orbiting scroll

The tangential gas force $F_t(\theta)$ can be computed by

$$F_t(\theta) = \sum_{i+1}^{N} P\left(2i - \frac{2\pi - \theta}{\pi}\right) h(p_i - p_{i+1}) \tag{33}$$

The tangential gas moment $M_t(\theta)$ can be computed by

$$M_t(\theta) = F_t(\theta)R_{or} = \sum_{i+1}^{N} P(2i - \frac{2\pi - \theta}{\pi})h(p_i - p_{i+1})R_{or} \tag{34}$$

The output power $W(\theta)$ is given by

$$W(\theta) = M_t(\theta)\omega\eta_m \tag{35}$$

The mechanical efficient η_m is given by (Park Y. C. *et al.*, 2002)

$$\eta_m = 0.868 + 0.0048f - 4.4444\times10^{-5} f^2 \tag{36}$$

4. PERFORMANCE ANALYSIS

4.1 Simulation analysis

The detail thermodynamic model has been implemented into FORTRAN and Matlab computer code. This code calculates the pressure and mass in each of the working chambers as a function of the orbiting angle for an entire revolution. It also calculates the power consumption of the scroll expander for a specified operating condition. Figure 8 is the flowchart for the implementation of the thermodynamic model.

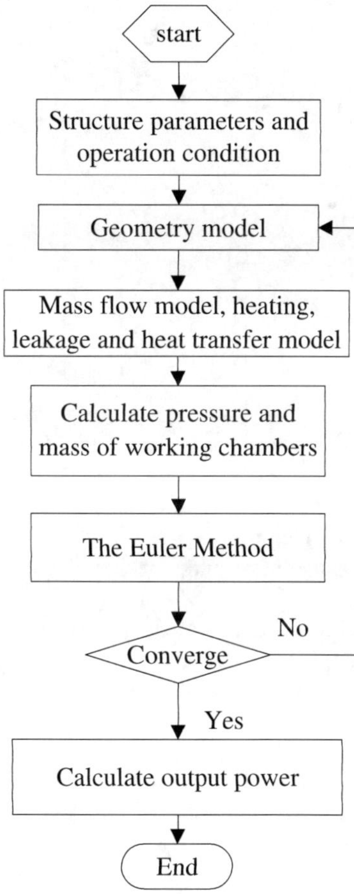

Figure 8: Flowchart of thermodynamic model

Figure 9 is the change of pressure with respect to the orbiting angle and volume. The suction pressure is 683.5kPa and speed is 2500r/min. Due to throttling of suction hole, the pressure has slightly decreased during the suction process. The suction pressure decreases about 77kPa before it enters the expansion chamber. During the expansion stage, the change of pressure is from 606kPa to 143kPa. When the last expansion chamber opens to the discharge chamber the discharge process begins. Because of big discharge hole, during the discharge process the pressure slightly drops, the change of pressure is from 143kPa to 100kPa. During the expansion process, the pressure decreases with the increase of working chamber volume and it changes more and more gently in Figure 9(b).

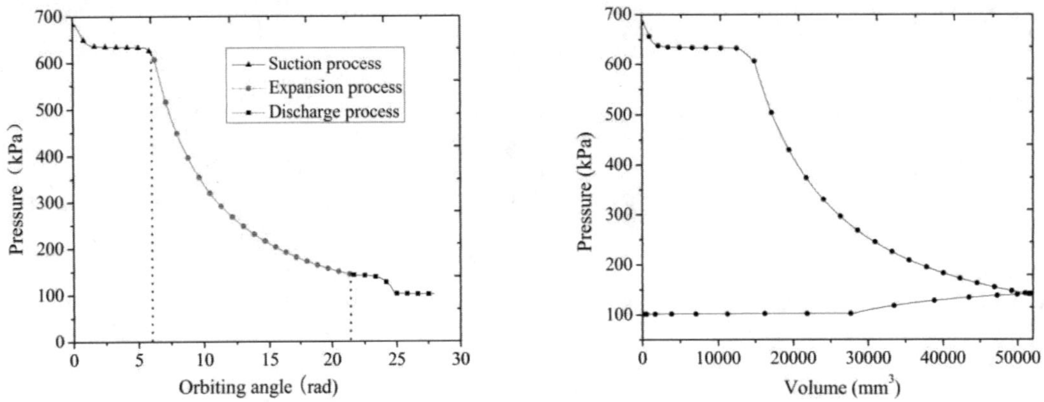

(a) Pressure vs. orbiting angle (b) Pressure vs. volume (p-V)

Figure 9: The change of pressure with respect to the orbiting angle and volume

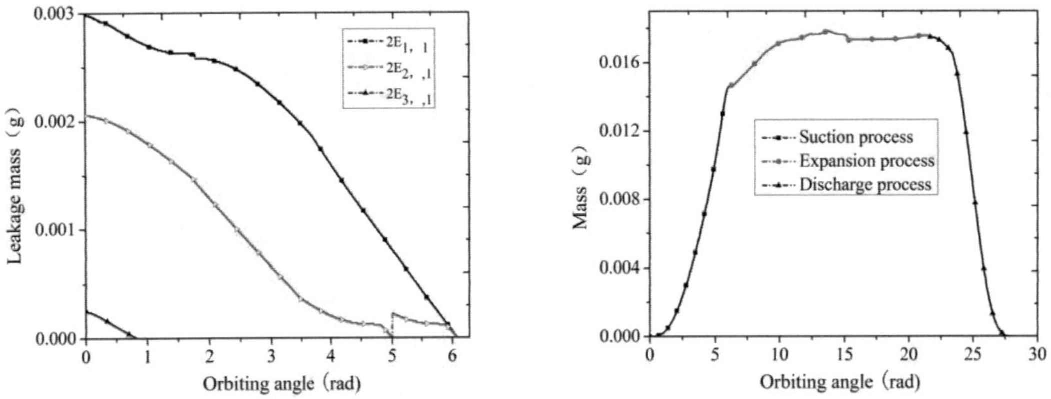

(a) Leakage mass vs. orbiting angle (b) Mass vs. orbiting angle

Figure 10: The change of leakage mass and working chamber mass with respect to the orbiting angle

Figure 10 is the change of leakage mass and working chamber mass with respect to the orbiting angle. The mass first increases until the suction chamber is closed. During the expansion process, the mass first changes sharply and then goes slowly. In the discharge process, the mass starts to decrease until the end of discharge process.

4.2 Model validation

A hermetic scroll expander is used to measure performance and validate model in a simple test rig as seen in Figure 11. The high pressure nitrogen enters the suction chamber through reducing valve, control valve and flow meter. Control valve is used to control suction pressure. The scroll expander is drived by the compressed nitrogen. Electric generator transforms mechanical energy into electricity. The voltage, current and power can be gotten from data acquisition system. Figure 12 is the change of performance with respect to the suction pressure. When the suction pressure reached 1000kPa, the test output power is almost 1242W in Figure 12(b). The absolute error of the output power is from 16W to 45W. By comparing the calculating results with the test results it is proved that the proposed thermodynamic model can simulate the working process of scroll expander accurately.

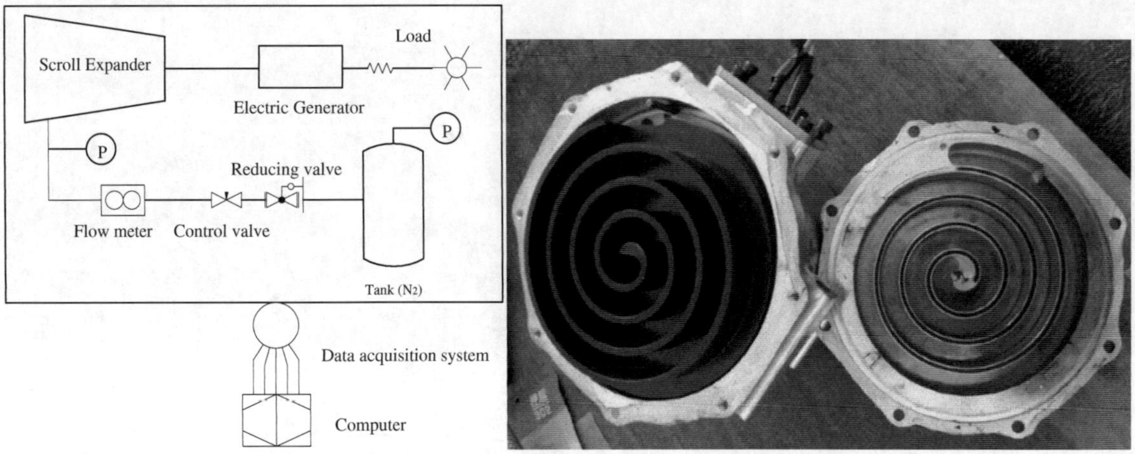

Figure 11: The test rig

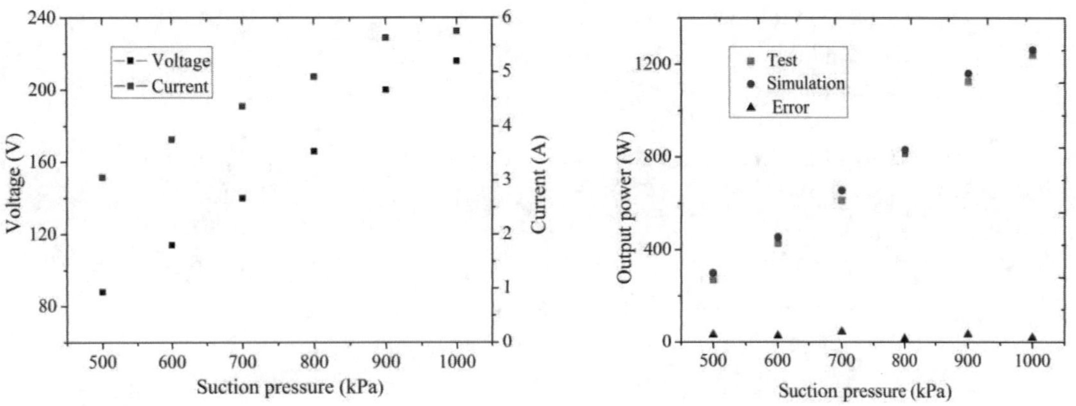

(a) Voltage and current vs. suction pressure (b) Output power vs. suction pressure

Figure 12: The change of performance with respect to the suction pressure

5. CONCLUSIONS

(1) A geometry model of scroll expander is presented. The evolution of the working chamber volume with respect to the orbiting angle has been derived.
(2) Based on energy and mass conservation equations, the thermodynamic modeling of scroll expander is set up. The mass and pressure of gas in working chambers are determined by solving the detail thermodynamic modeling of scroll expander using the Euler explicit method.
(3) The developed scroll expander prototype is used to validate the thermodynamic modeling. The simulation results obtained from the thermodynamic modeling agree very well with experimental results (within 5%).
(4) As future work a more detailed analysis will be done with the mathematical model. And an ORC system will be set up to test different operational conditions of scroll expanders and working fluids.

NOMENCLATURE

φ_{i0}	generating angle of inner scroll	(rad)
φ_{o0}	generating angle of outer scroll	(rad)
R_{or}	turning radius	(mm)
T	temperature	(K)

m	gas mass	(kg)
C_v	specific heat at constant specific volume	(-)
p	gas pressure	(kPa)
v	specific volume	(m^3/kg)
ω	angular speed of expander shaft	(rad/s)
$\dot{m}_{in}$	mass flow rate flowing into control volume	(kg)
$\dot{m}_{out}$	mass flow rate flowing out of control volume	(kg)
h	gas specific enthalpy of the control volume	(kJ/kg)
h_{in}	gas specific enthalpy flowing into the control volume	(kJ /kg)
$\dot{Q}$	heat flow rate flowing into the control volume	(kW)
c_p	gas specific heat ratio	(kJ/kg-K)
p_h	pressure in the high pressure side	(kPa)
A_s	area of the suction chamber opening	(mm^2)
p_l	pressure in the low pressure side	(kPa)
Re	Reynolds number	(-)
T_{pipe}	temperature of the pipe	(K)
d_p	diameter of the pipe	(mm)
Pr	Prandtl number	(-)
St	Strouhal number	(-)
γ	specific heat ratio	(-)
ρ_h	density of the gas in the high pressure side	(kg/m^3)
μ	viscosity	(Pa-s)
κ	conductivity	(kW/m-k)
L_p	length of the pipe	(mm)
T_{s1}	temperature at the generating angle φ_1	(K)
$T(k,j)$	temperature of the gas in the k-th chamber at the angle θ_j	(K)
A	chamber bounding area	(mm^2)
T_{gas}	temperature of the gas in the scroll expander	(K)
dA	heat transfer area of scroll wrap and plate	(mm^2)
f	the frequency of scroll expander	(Hz)

REFERENCES

Chen Yu (2000). *Mathematical modeling of scroll compressors*. Purdue University Doctor article.

Fox RW, McDonald AT (1992). *Introduction to fluid mechanics*. New York: John Wiley & Sons Publishers.

Halm N.H (1997). *Mathematical Modeling of Scroll Compressors*. Purdue University Master article.

Incropera FP, Dewitt DP (1996). *Fundamentals of heat and mass transfer*. New York: John Wiley & Sons Publishers.

K Jang, S Jeong (2006). Experimental investigation on convective heat transfer mechanism in a scroll compressor. *International Journal of Refrigeration*, 29(5): 744-753.

Kakac S, Shah RK (1987). *Handbook of single-phase convective heat transfer*. New York: John Wiley & Sons Publishers.

Lemort V., Quoilin S., Cuevas C., Lebrun J. (2009). Testing and modeling a scroll expander integrated into an organic rankine cycle. *Applied Thermal Engineering, 29*(14), 3094-3102.

Liu Guangbin, Zhao Yuanyang, Yang Qichao, Wang Le, Tang Bin, Li Liansheng(2013). Theoretical and experimental research on scroll expander used in small scale organic rankine cycle system. *Proceedings of the Institution of Mechanical Engineers Part E Journal of Process Mechanical Engineering*, 229(1): 25-35.

Peng Bin, Sun Ying (2015). Investigation of Mathematical Modeling and Experiment for Variable Thickness Scroll Compressor. *Journal of Mechanical Engineering*, 51(14): 185-191.
Peng Bin, Arnaud Legros, Vincent Lemort , Xie Xiaozheng, Gong Haifeng (2016). Recent Advances on the oil-free Scroll Compressor. *Recent Patants on Mechanical Engineering*, 9(1): 37-47.
Park Y. C., Kim Y., Cho, H. (2002). Thermodynamic analysis on the performance of a variable speed scroll compressor with refrigerant injection. *International Journal of Refrigeration, 25*(8), 1072-1082.
Quoilin S. (2011). *Sustainable Energy Conversion Through the Use of Organic Rankine Cycles for Waste Heat Recovery and Solar Applications*. University of Liège (Belgium) Doctor article.
Quoilin S., Lemort V., Lebrun J. (2010). Experimental study and modeling of an organic rankine cycle using scroll expander. *Applied Energy, 87*(4), 1260-1268.
Rajavel R, Saravanan K (2008). Heat transfer studies on spiral plate heat exchanger. Thermal Science, 12(3): 85-90.
Song P., Wei M., Shi L., Danish S. N., Ma C. (2015). A review of scroll expanders for organic rankine cycle systems. *Applied Thermal Engineering, 75*(75), 54-64.
Tarique M. A., Dincer I., Zamfirescu C. (2014). Experimental investigation of a scroll expander for an organic rankine cycle. *International Journal of Energy Research, 38*(14), 1825–1834.
Wu Z., Pan D., Gao N., Zhu T., Xie F. (2015). Experimental testing and numerical simulation of scroll expander in a small scale organic rankine cycle system. *Applied Thermal Engineering, 87*, 529–537.
Zhu J., Chen Z., Huang H., Yan Y. (2016). Effect of resistive load on the performance of an organic rankine cycle with a scroll expander. *Energy, 95*, 21-28.

ACKNOWLEDGEMENT

This work is supported by National Natural Science Foundation of China (Grant No. 51275226), Natural Science Foundation of ZHEJIANG Province (Grant No.LY12E05010) and Natural Science Foundation of GANSU Province (Grant No. 1212RJYA010), China.

Modelling and Preliminary Design of a Variable-BVR Rotary Valve Expander with an Integrated Linear Generator

Sergei GUSEV[1*], Davide ZIVIANI[1], Jasper DE VIAENE[2], Stijn DERAMMELEAERE[2], Martijn VAN DEN BROEK[1]

[1]Ghent University, Department of Flow, Heat and Combustion Mechanics,
Kortrijk, Belgium
[2]Ghent University, Department of Electrical Energy, Systems and Automation,
Kortrijk, Belgium
Contact Information: Servei.Gusev@UGent.be

* Corresponding Author

ABSTRACT

The Organic Rankine Cycle (ORC) is currently one of the most suitable technologies to convert waste heat into mechanical work or electricity. While large and medium scale systems are widely available on the market for various temperature and power ranges, small-scale ORCs below 50 kWe are still in a pre-commercial phase because of the relatively high specific cost per kW and the lack of technologically mature and high efficient expanders. Small-scale ORC installations for automotive applications operate at variable heat source profiles combined with the fluctuating power demand from a vehicle. The prediction of an optimum operating point is challenging.

Exhaust gases are a limited heat source, therefore the more heat is recovered at an optimal cycle efficiency level, the more power is produced. By using advanced cycle architectures (e.g. trilateral ORCs, partial-evaporating ORCs, zeotropic mixture ORCs, etc.) and the right fluids, an optimum can be found. An expander with a variable built-in volume ratio (BVR) can allow to operate at optimal conditions within the whole range of pressures imposed by the variable heat source and heat sink. Adjustable expanders are known but mainly limited to large-scale applications. Neither a positive displacement expander, nor a turbine can provide an optimal expansion of a working fluid in a wide range of operation conditions.

As a response to this challenge, the concept of a variable-BVR piston expander with an integrated linear generator is proposed in this paper. The internal part-load control is based on a rotary valve which controls the suction and discharge processes in the expander. An analytic model has been developed to relate the position of the valve with the motion of the piston.

By means of a deterministic model, the influence of the main design parameters is investigated. A preliminary design based on the expander model results is described and the predicted performance over the operating range of interest is discussed.

1. GENERAL DESCRIPTION

1.1 Mechanical design

The proposed expander is shown in Figure 1 and it consists of the following main parts:

1. Stationary cylindrical housing (shown semi-transparent) with wall openings for intake and discharge of the working fluid. Two working chambers are engaged alternately to expand and to discharge the working fluid. These chambers are formed by two stators located at both sides of the housing;
2. Double acting cylindrical piston, coaxial with the housing, consisting of two skirts connected with each other by a central element. The thickness of this element is defined by the pressure difference, which is dependent on the application, in order to ensure the required mechanical resistance. Permanent magnets are embedded in such a way that the control of the rotor by means of two degrees of freedom (rotation and translation) is possible by the coils. The reciprocation of the piston induces electricity, while the rotation controls the intake and discharge

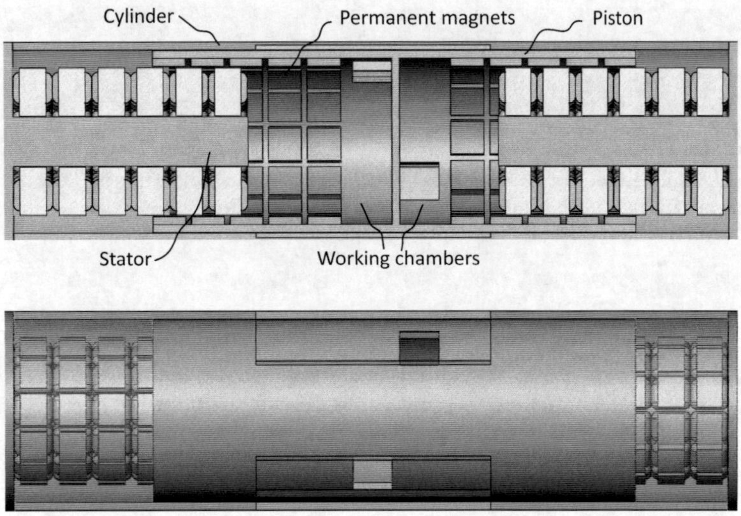

Figure 1: Expander cut-away and with semi-transparent housing

process of the working fluid by opening and closing the intake and the discharge ports. The outer diameter of the piston is matched with the inner diameter of the cylindrical housing with a minimal clearance;

3. Power and control electronics.

1.2 Operation principle

Working chambers are enclosed between the central element, the skirts and the stators. When the piston moves to the right, the volume of the first chamber increases and the volume of the second chamber decreases and vice versa. The skirts with openings act as valves: when the skirt openings converge with the openings in the housing connected to the high pressure side, these openings act as inlet ports. When connected to the low pressure side – as outlet ports. When the skirt opening is not converged with any of the in- and outlet openings of the housing, but facing an inner wall of the cylinder, the working chamber is isolated. Three working phases can identified as function of the piston movement and the angular position: intake, expansion and discharge, which are depicted in Figure 2.

Depending on the piston linear position at the moment of closing of the inlet, the volume V_1 is isolated and further expanded until the volume V_2. The built-in volumetric ratio is defined as:

$$BVR = \frac{V_2}{V_1} = \frac{1}{\chi}, \tag{1}$$

where χ is the piston relative displacement at the moment when the inlet port closes. Figure 3 illustrates the volume ratio of around 10 corresponding to the piston relative displacement of χ=0.1 (position of the right-bottom corner of the skirt opening).

Piston movement is controlled by power electronics for optimal expansion which requires further analysis. In fact, there are no limitations such as volume in function of the crankshaft angle. To adjust the expansion ratio, the intake and the discharge ports are controlled by means of a piston skirt. An additional degree of freedom is added: the piston can rotate closing the intake port quasi at any time independent of the linear position. The volumetric efficiency of such expander is quite high and only defined by machining precision. Ideally, there is no working fluid left in the clearance volume when the piston reaches extreme left or right position. The placement of the in- and outlet ports symmetrically on opposing sides of the cylinder eliminates side forces acting upon the mechanism. The unit can be made fully hermetic with no seals and moving parts except the piston itself.

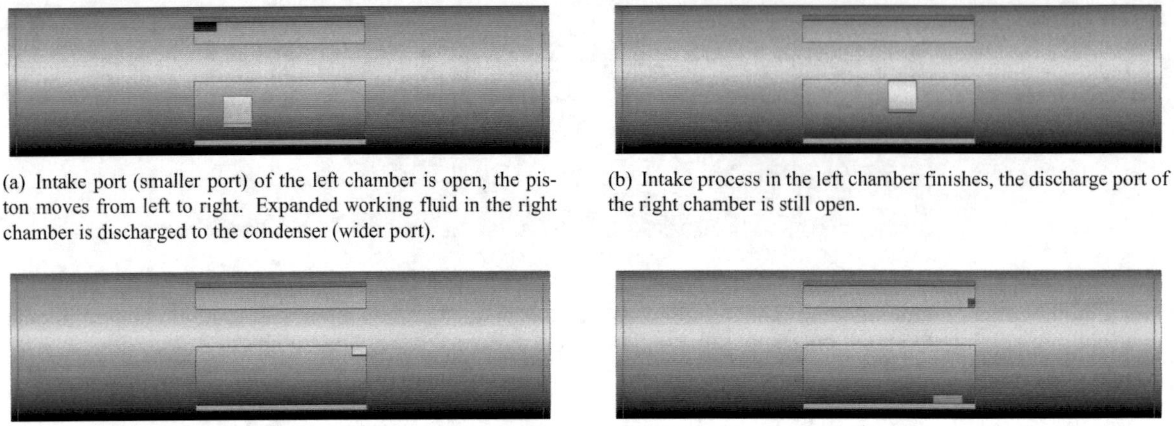

(a) Intake port (smaller port) of the left chamber is open, the piston moves from left to right. Expanded working fluid in the right chamber is discharged to the condenser (wider port).

(b) Intake process in the left chamber finishes, the discharge port of the right chamber is still open.

(c) In the left chamber, the expansion starts: the port is closed, the piston keeps moving to the right.

(d) The left chamber is connected to the condenser and the discharge process starts. The function of the chambers alternates, intake process is now started in the right chamber.

Figure 2: Valve operation principle.

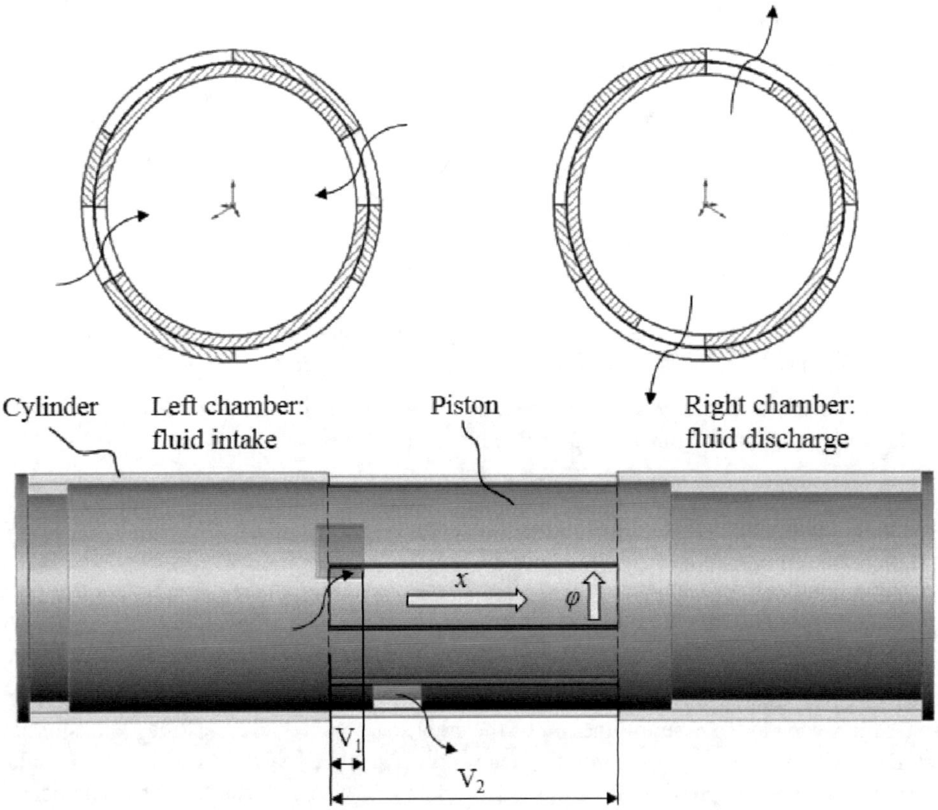

Figure 3: Variable volume ratio

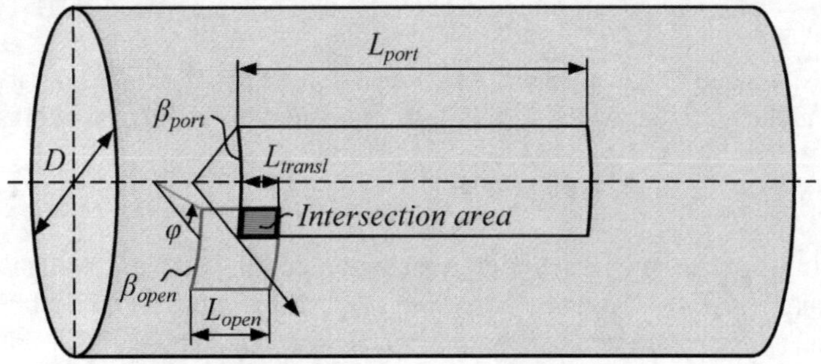

Figure 4: Schematic of the port geometry.

2. THERMODYNAMIC MODEL

An hybrid gray box model adapted from (Glavatskaya et al., 2012) is developed to simulate the expander. The process is split up into infinitesimal steps. In order to calculate the necessary thermodynamic parameters, the CoolProp library connected to Python is used (Bell et al., 2014). Air is chosen as a working fluid in order to avoid complexity at the first stage of the prototype development: leakages in a non-optimized setup are inevitable and the use of expensive working fluids will increase operating costs.

The working process consists of three phases: intake, the inlet pressure assumed constant; expansion, the pressure in the working chamber decreases; discharge, the pressure is equalized with the condenser pressure and remains constant.

2.1 Intake and discharge.

The working medium enters the expander through a rectangular port formed by the openings in the housing and the skirts. This process is modeled as an isentropic expansion. The mass flow rate $\dot{m}$ is dependent on the port area S, the inlet pressure p_{su}, the pressure in the cylinder p_{cyl} and can be calculated using the Fliegner's equation:

$$\dot{m}(t) = C_d \cdot S(t) \cdot \rho_{su} \left(\frac{p_{su}}{p_{cyl}(t)} \right)^{\frac{1}{\kappa}} \cdot \sqrt{\frac{2\kappa}{\kappa - 1} \frac{p_{su}}{\rho_{su}} \left[1 - \left(\frac{p_{cyl}(t)}{p_{su}} \right)^{\frac{\kappa-1}{\kappa}} \right]}. \tag{2}$$

The intersection area is changing according to the rotation and the translation of the piston (Figure 4). Depending on the pressure ratio across the inlet port, the discharge coefficient C_d from 0.5 to 0.9 is interpolated from the Figure 5 of (Novák & Koza, 2013).

The position of the lower left corner of the cylinder wall opening is chosen as the reference point. The start of the intake process corresponds with the position of the piston when the lead-top corner of the skirt opening is at the reference point. A simplified algorithm defining the overlap of two rectangular openings is used to calculate the intersection area:

$$\begin{aligned} S(t) = {} & (max(0, min(L_{transl}(t), L_{port}) - max(0, L_{transl}(t) - L_{open})) \times \\ & \times (max(0, min(\varphi(t), \beta_{port}) - max(0, \varphi(t) - \beta_{open})) \cdot \frac{\pi D}{2} \cdot n, \end{aligned} \tag{3}$$

where n is the number of ports. In current configuration, two axisymmetric ports are used.

A similar model is applied to estimate the mass flow rate during the discharge process. The corresponding pressures p_{cyl} and p_{dis} are used.

2.2 Expansion.

The expansion process is simulated as a sequence of small steps: an isentropic expansion, isochoric heating by friction, the heat transfer from or to the expander walls. The last process is described by a Newtonian model:

$$\dot{Q}_{cyl}(t) = h(t) \cdot A_p(t) \cdot (T_{cyl}(t) - T_p) . \tag{4}$$

In literature, different correlations for the in-cylinder heat transfer coefficient h are available, for example: the Eichelberg correlation (Eichelberg, 1939), the Woschni correlation (Woschni, 1967) and the Hohenberg correlation (Hohenberg, 1979). These correlations applied to a high stroke/diameter ratio reciprocating engine have been compared by Fula et al. (2013). The Woshini correlation have been selected and it is given by:

$$h(t) = 3.26D^{-0.2}\left(\frac{p_{cyl}(t)}{10^5}\right)^{0.8} T_{cyl}(t)^{-0.55}\left(\frac{dx}{dt}\right)^{0.8} . \tag{5}$$

The piston surface temperature is assumed as the average between the in- and the outlet gas temperatures.

$$T_p = \frac{T_{su} + T_{dis}}{2} . \tag{6}$$

This temperature can be adjusted if a water cooling system is used.

2.3 Leakage flow

The leakage flow is dependent on the pressure difference and the gap size between the piston and the cylinder walls. Since the goal of the current study is to optimize the dynamic behavior of the expander, a relatively high leakage flow is permitted in order to minimize the friction forces. The exact flow will be measured during static tests and subtracted from the final results.

3. THE EQUATION OF MOTION

The model of the free-piston expander is developed as suggested by Mikalsen & Roskilly (2008). Since the rotation and translation of the piston have to be synchronized, the model is focused on the accurate definition and control of the piston position in both dimensions. The dynamics of the piston is defined by the Newton's second law.

$$F_{p,cyl} - F_{p,dis} - F_{fr} - F_{el} = m_p \frac{d^2x}{dt^2} , \tag{7}$$

where $F_{p,cyl}$ and F_{dis} are gas forces in the left and right working chambers respectively during the piston motion from left to right. Both gas forces are applied to the same central element and are opposing each other.

The friction force F_{fr} depends on the piston speed, which is the vector sum of the translation and the rotational speeds of the piston. The dynamic friction behaviors of pneumatic cylinders based on the LuGre model is studied in (Tran & Yanada, 2013). The results obtained in this study are used for the simplified friction force correlation for the free-piston model. Due to a continuous rotation of the piston, the resulting speed is non-zero at the extreme piston positions. Therefore, the Stibeck-effect at near-zero speeds can be neglected and a linear dependence of the friction force from the speed can be used.

The electromagnetic force F_{el} applied to the piston is adjusted during the piston motion in order to stop the piston at its extreme left and right positions, so all kinetic energy is absorbed and transformed into electricity. The approach is similar to (Petrichenko et al., 2015). The resulting equation can be written as follows:

$$F_{el}(t) = \frac{\pi D^2}{4}(p_{cyl}(t) - p_{dis}) - F_{fr}\left(\frac{dx}{dt}\right) - m_p \frac{d^2x}{dt^2} . \tag{8}$$

The Equation 8 is used to control the drive of the linear motor, the position sensor is used for the feedback of the piston position.

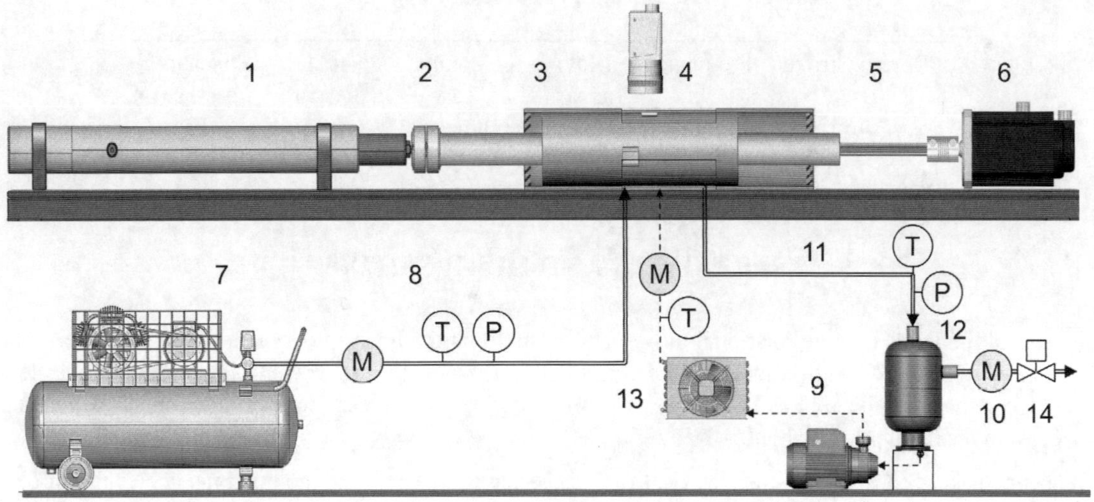

Figure 5: Test setup

4. TEST SETUP

The key feature of this setup is the ability to control both the rotation and translation of the piston. The layout of the setup is shown in Figure 5. While the final goal is a fully integrated system, for convenience and feasibility during the proof of concept phase, both movements are designed to be controlled by a different electromechanical actuators.

4.1 Main Components

The test setup consists of: a linear generator with an integrated position sensor (1) connected to the expander (3) through a coupling (2) transferring only the linear motion and preventing the generator shaft from rotation. The other side of the expander is connected to a servo motor with an encoder by means of a spline shaft (5) which insures the right angle positioning while the expander shaft can freely slide along it.

The expander is driven by compressed air provided by an air compressor unit (7). Temperature (T), pressure (P), and mass flow meters (M) are placed on the air supply line in order to determine the input power.

To ensure the lubrication and the sealing of the expander, an oil injection system is placed. It consists of a circulation pump (9), an oil separator (10) and a cooler (13) to reject the heat produced by friction.

The second flow meter is placed at the air discharge line (12) to measure the leakage flow by calculating the difference between the first flow meter placed at the inlet of the expander. The outlet pressure is controlled by a motorized throttling valve (14) to set up any desired pressure.

A high speed camera (4) is used to observe the inlet port operation.

4.2 Expander sizing

The sizing of the expander for the setup depends on the linear generator available. Since linear generators are still rare on the market, a linear motor operating in a brake mode is found, which fits the requirements. Basing on the characteristics provided by the manufacturer, the maximal static force applied to the generator is 5.18 kN. The nominal inlet pressure of 1 MPa is chosen. These parameters define the diameter of the piston. The speed and the force relationship is dependent on the linear generator used and provided by the manufacturer. This dependence is acquired into the model. The major setup parameters are summarized in Table 1.

Table 1: Main parameters of the designed test setup

Bore (m)	Stroke (m)	Frequency (Hz)	Moving mass (kg)	Volume flow (nm^3/sec)	Inlet pressure (Pa)	Discharge pressure (Pa)
0.08	0.31	5.0	40	0.016	10^6	10^5

5. RESULTS AND DISCUSSIONS

The developed model dictate the rotational movement in line with the linear movement of the expander. The speed and the acceleration of both movements can be controlled independently. In order to lower the pressure loss at the inlet, the linear speed can be kept as low as possible, while during the expansion process the linear movement can be optimized for a maximal power output.

The frequency is kept constant and the electrical force is varied in order to find the maximal achievable stroke length and the maximum power output conditions. The major results of the simulation of the expansion are shown on the Figure 6.

From the simulations it is cleared out that the piston should be forced to move at the beginning of the expansion process as it is shown on Figure 6(a) otherwise it will not be able to reach the maximum speed related to the current electromagnetic force applied (Figure 6(d)). As a result, a much smaller stroke length will be required for the same frequency (Figure 6(c)), which in combination with the diameter of the piston and the working frequency define the swept volume of such an expander and the mass flow rate through the system. The lower mass flow rate will cause a lower power output.

During the market survey, some important limits were figured out. The speed is in close relationship with the moving mass of the translator, which is typically assembled from a number of magnetic elements, and the power applied to the coils. Linear motors, characterized by a high-dynamic behavior, are not able to hold a certain position if a high static force is applied. This is the case at the beginning of the intake process when the piston is exposed to the inlet pressure, which is expected to be relatively high in the target applications. The cycle of acceleration / deceleration is chosen accordingly to the characteristics of the motor available (Figure 6(f)). The resulting average accelerating and the power output are shown on Figure 6(e). The electromagnetic force profile is chosen to be constant for the first simulations but it can be further optimized to allow the generator operate at maximal efficiency.

As it can be seen on Figure 6(b), the pressure drop over the inlet valve is relatively small at the beginning of the intake but at the end it becomes higher. An optimization of the piston rotational speed profile is needed in order to keep the inlet port open as long as it possible.

6. CONCLUSIONS

The dynamics of the piston must be optimized accordingly to the linear generator used. There is a trade-off between the high dynamics of the piston movement and a relatively high static force typical for the piston extreme positions. The crucial requirement is that the piston position must be controlled by the generator electronics at any moment.

The major difference with the most systems studied is that in the current setup the piston is not freely moving but, instead, its movement is fully controlled, so there are no bouncing devices needed. This approach provides potentially a more efficient way to transform the energy of the expanded gas into electricity since irreversibilities in bouncing devices are avoided.

This setup has the advantage of testing various expanders with different sizes, by easily replacing generators and motors to study the dynamics of the whole system. The piston movement can be matched with the maximum efficiency range of the generator used.

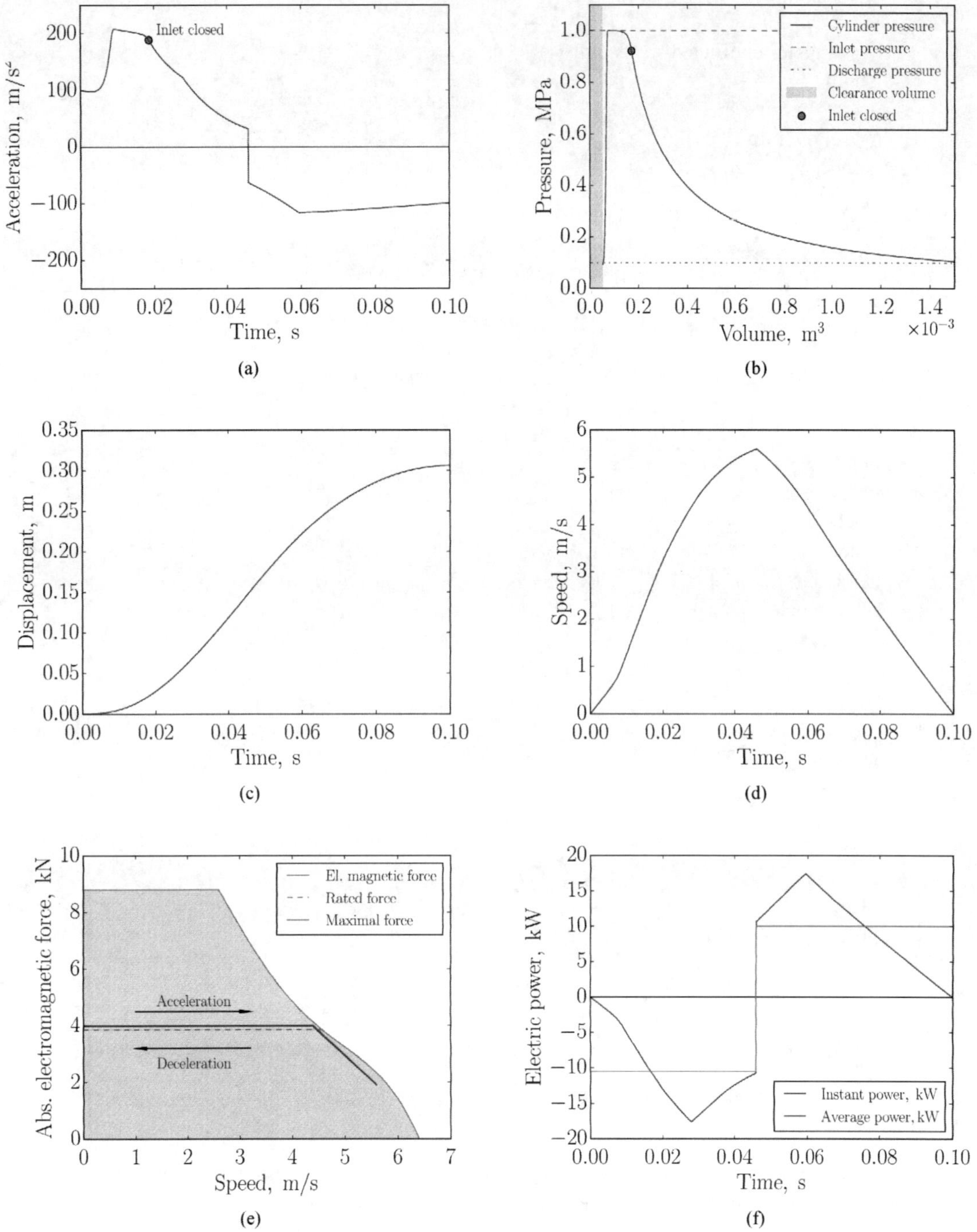

Figure 6: Expansion process.

NOMENCLATURE

A	suface area	(m^2)
C_d	discharge coefficient	(-)
D	diameter	(m)
F	force	(N)
h	heat transfer coefficient	(W/m^2 K)
L	length	(m)
m	mass	(kg)
$\dot{m}$	mass flow rate	(kg/s)
n	number of ports	(-)
p	pressure	(Pa)
$\dot{Q}$	heat flow rate	(W)
S	intersection area	(m^2)
T	temperature	(K)
t	time	(s)
V	volume	(m^3)
x	displacement	(m)

Greek

β	port angle	(rad)
Δ	difference	(-)
κ	isentropic coefficient	(-)
υ	speed	(m/s)
φ	angle of rotation	(rad)
χ	relative displacement	(-)

Subscript

cyl	cylinder
dis	discharge
el	electromagnetic
fr	friction
max	maximal
min	minimal
open	opening
p	piston
port	port
rot	rotation
su	supply
theor	theoretical
transl	translation

REFERENCES

Bell, I. H., Wronski, J., Quoilin, S., & Lemort, V. (2014). Pure and pseudo-pure fluid thermophysical property evaluation and the open-source thermophysical property library coolprop. *Industrial & engineering chemistry research*, *53*(6), 2498–2508.

Eichelberg, G. (1939). Some new investigations on old combustion engine problems. *Engineering*, *148*, 463–446 et 547–560.

Fula, A., Stouffs, P., & Sierra, F. (2013). In-cylinder heat transfer in an ericsson engine prototype. *Renewable Energy and Power Quality Journal, ISSN 2172-038 X*(11).

Glavatskaya, Y., Podevin, P., Lemort, V., Shonda, O., & Descombes, G. (2012). Reciprocating expander for an exhaust heat recovery rankine cycle for a passenger car application. *Energies*, *5*(6), 1751.

Hohenberg, G. (1979). Advanced approaches for heat transfer calculations. *SAE Special Publications*, *SP-449*, 61–79.

Mikalsen, R., & Roskilly, A. (2008). The design and simulation of a two-stroke free-piston compression ignition engine for electrical power generation. *Applied Thermal Engineering*, *28*(5–6), 589–600.

Novák, O., & Koza, V. (2013). Measuring a discharge coefficient of an orifice for an unsteady compressible flow. *Fuels*(1), 21–25.

Petrichenko, D., Tatarnikov, A., & Papkin, I. (2015). Approach to electromagnetic control of the extreme positions of a piston in a free piston generator. *Modern Applied Science*, *9*(1), 119–128.

Tran, X. B., & Yanada, H. (2013). Dynamic friction behaviors of pneumatic cylinders. *Intelligent Control and Automation*, *4*(2).

Woschni, G. (1967). A universally applicable equation for the instantaneous heat transfer coefficient in the internal combustion engine. *SAE*(670931).

CFD Approaches Applied To A Single-Screw Expander

Davide ZIVIANI[1]*, Alessio SUMAN[2], Jacopo GABRIELLONI[2], Michele PINELLI[2], Michel DE PAEPE[1], Martijn VAN DEN BROEK[1]

[1]Ghent University, Department of Flow, Heat and Combustion Mechanics, Ghent, Belgium
davide.ziviani@ugent.be, michel.depaepe@ugent.be, martijn.vandenbroek@ugent.be

[2]Engineering Department in Ferrara, University of Ferrara, Ferrara, Italy
alessio.suman@unife.it, jacopo.gabrielloni@student.unife.it, michele.pinelli@unife.it

* Corresponding Author

ABSTRACT

Organic Rankine Cycle (ORC) systems rely on the expander performance to generate power output in an efficient manner. Especially in the low power range (below 100 kWe), positive displacement (PD) expanders (e.g. scroll, twin-screw, reciprocating, vane, spool, etc.) result to be cost-effective. However, commercially available PD expanders are still limited and, in many cases, the existing PD compressors are operated in reversed mode by introducing design modifications to sealing, bearings, port sizes, lubrication requirements to increase both their performance and reliability. Computational fluid dynamics (CFD) as a design and analysis tool of positive displacement machine has been proven to be viable. Challenges arise when CFD is applied to PD machines due to the dynamics of the expansion (or compression) process, presence of internal leakages and heat transfer mechanisms, as well as deforming working chambers. Different grid generation methods and solution schemes have been successfully implemented to scroll, twin-screw and reciprocating machines (Rane *et al.*, 2012, Rane *et al.*, 2013). The limitation of such methodologies to be applied directly to complex multi-rotor machines has been highlighted by Rane et al. (Rane *et al.*, 2012). In this paper, a single-screw expander is used as benchmark to evaluate different grid generation methodologies (dynamic remeshing and Chimera strategy overlapping grid) and commercial software, in terms of computational resources required, accuracy of the results and limitations. The calculations have been performed with air to reduce the complexity of the problem.

1. INTRODUCTION

Organic Rankine Cycle (ORC) is a proven technology to convert waste heat from several sources and in a wide range of power outputs (Quoilin *et al.*, 2013). Especially in the medium to low power range, volumetric expanders play an important role because of their performance and lower investment costs compared to dynamic expanders (Imran *et al.*, 2016). As a consequence, the efficiency of the expander represents a key parameter to be optimized in order to maximize the power output. Different types of positive displacement machines, such as scroll or twin screw, have been investigated for their use in ORCs. Recently the single screw expander (SSE) is gaining interest as a potential volumetric expander for ORC applications. Similarly to a twin screw expander, the single screw is generated by two meshing profiles: a grooved central rotor engages with the lateral toothed satellites (or gaterotors). The meshing between each groove with the corresponding tooth isolates a working chamber, i.e. a portion of the groove. In comparison with the twin screw, the SSE has longer bearings life due to balanced loading on the main rotor, high volumetric efficiency, low leakage, low noise and vibration and a simplified configuration (Ziviani *et al.*, 2014a). The design of ORC systems requires also the selection of a suitable working fluid. Screening methods have been developed to identify optimal working fluids by taking into account environmental and safety criteria along with thermophysical properties (Saleh *et al.*, 2007). However, using real gas properties in CFD simulations is still a challenging quest due to high computational cost and convergence issues. To the Author best knowledge, CFD analyses applied to single screw machines are rare due to the complexity of the problem especially regarding the mesh generation. The aim of this paper is to evaluate and compare different meshing techniques and solution schemes when applied to a single screw geometry. Pros and cons of commercial software are discussed and the accuracy of the results is analyzed by comparing the results obtained with each of them. Both steady-state and transient analyses have been investigated.

2. SINGLE-SCREW EXPANDER GEOMETRY AND BOUNDARY CONDITIONS

The single-screw expander considered as case study is an 11 kWe air-compressor adapted to be an expander for organic Rankine cycle (ORC) applications. The details of the main dimensions of the expander are listed in (Ziviani *et al.*, 2016). However, the real geometry has been simplified accordingly in order to perform CFD analyses. In particular, the main rotor shaft and its sealing have been removed as well as the bearings. Similar modifications have been applied to the starwheels. Instead, the housing has been kept in its original form to accurately predict the internal flows and the 3D CAD geometry is shown in Figure 1. The same boundary conditions have been applied for each simulation, i.e. an inlet pressure of 1200 kPa, an inlet temperature of 390 K, and an outlet pressure of 200 kPa. The chosen working fluid was air to simplify the analysis, and the equation of state (EOS) used the ideal gas law. The standard turbulence model k-ε combined with standard wall functions was used. This model has proven to be, also in this case, the best compromise between accuracy, robustness and computational effort. The main rotor has a rotation speed of 3000 rpm, while the rotational speed of satellites have been obtained by considering the 6/11 engaging ratio between grooves and teeth.

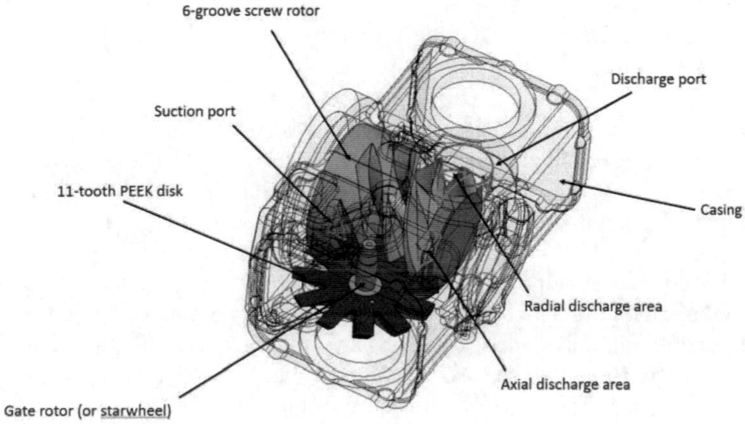

Figure 1: Single-screw expander geometry.

3. MESHING TECHNIQUES

The aim of this section is to evaluate and compare four different meshing techniques applied to a single-screw expander. Since the analyzed methodologies are not available in a single software, each technique is associated to a specific software. The meshing methods and the specific software used (in brackets) are as follows: (i) Dynamic Mesh (DM) (ANSYS Fluent 13.0); (ii) Chimera Strategy (CS) (CD-ADAPCO STAR-CCM+ 10.02); (iii) Mesh Morphing (MM) (PumpLinx 4.0.3); (iv) Key-Frame Re-meshing (KFR) (ANSYS CFX 15.0).

3.1 Dynamic Mesh

By using the DM strategy implemented in ANSYS Fluent 13.0, the mesh inside the screw and satellites profiles is regenerated at each time step to accommodate the shape and size change of the gas pockets. The mesh regeneration could follow element size criteria and/or element quality criteria (such as skweness). The numerical domain of the single-screw expander is discretized through the use of a triangular mesh. In order to to increase the resolution of the mesh close to the walls, a local mesh refining approach has to be adopted. Figure 2 highlights the grid deformations during the rotation of main rotor and corresponding gaterotor. In the flank gap there are at least two triangular mesh elements in agreement with the data reported in literature (Castilla *et al.*, 2010 and Del Campo *et al.*, 2012) for positive displacement machines. One of the CFD simulation was carried out as a steady-state numerical model by using a DM strategy implemented in ANSYS Fluent 13.0. Eventually, a transient analysis is able to reproduce the real operation of the machine through a sequence of relative positions between the central screw and the satellites by imposing an angular increment. For example, the use of 2D CFD analysis by using a DM for studying some particular fluid dynamic phenomena in the gear pumps. In particular, Castilla *et al.* (2010) put the attention to the turbulence structure in the suction chamber, while Del Campo *et al.* (2012) analyzed pressure variations and cavitation phenomena. In these cases, the 2D strategy allows the comprehension of the major fluid dynamic phenomena involved.

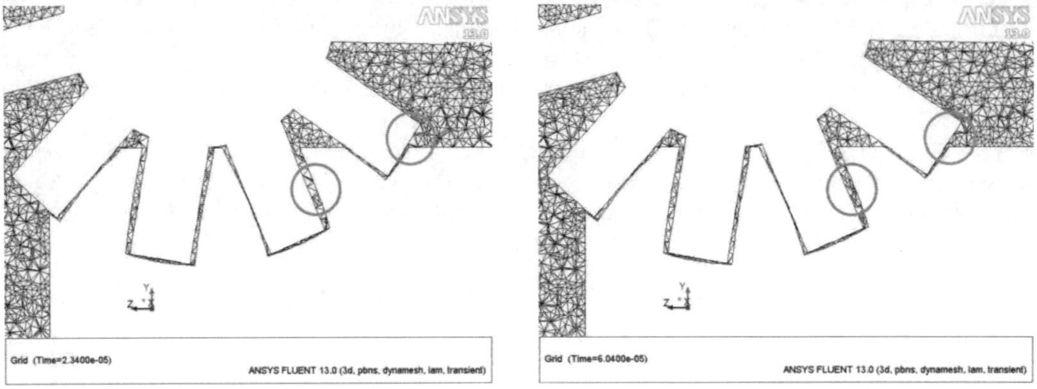

Figure 2: Dynamic tetrahedral mesh evolution through time.

3.2 Chimera Strategy

Overlapping grid, subsequently named the Chimera approach, was first introduced in the early 1980's (Steger and Benek, 1987) when several grid assembly packages have become available (Roget and Sitaraman, 2014). In the present paper, the simulation of the scroll expander with the CS was carried out by means of CD-ADAPCO STAR-CCM+ 10.02. The overset method consist in the contemporary use of active cells and passive cells. In active cells, regular discretized equations are solved, while in passive cells, no equation is solved, they are temporarily or permanently de-activated. Active cells along interface to passive cells refer to donor cells at another grid instead of the passive neighbors on the same grid. The first layer of passive cells next to active cells are called acceptor cells. The solution is computed on all grids simultaneously. Grids are implicitly coupled through the linear equation system matrix. Different interpolation functions can be used to express values at acceptor cells via values at donor cells. The donor cells must be active cells, and the change of cell status is automatically controlled by the solver. Overset grids usually involve one background mesh, which is adapted to environment, and one or more overset grids attached to bodies, overlapping the background mesh and/or each other as reported in Figure 3. Each grid (background and overset) can move according to the motion models implemented in the CFD software. In literature, the comparison between the CS and other CFD strategies are limited. Togasci *et al.*, 2001 and Hoke *et al.*, 2009 report the applications of the overset grid for a rocket booster and a 2D airfoil, respectively. Comparison with experimental data (Togasci *et al.*, 2001) and other CFD solutions (Hoke *et al.*, 2009) demonstrated that the CS is reliable for different fluid dynamics applications. The major difference with respect to the DM approach consists that CS does not deform the mesh during the calculation.

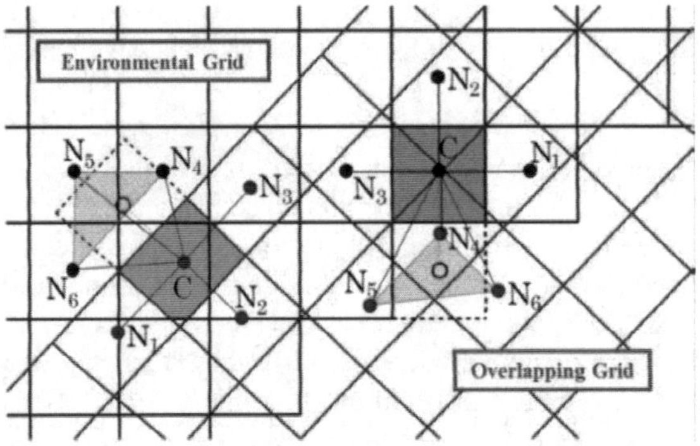

Figure 3: N_1, N_2, N_3 neighbors from the same grid; N_4, N_5, N_6 neighbors from overlapping grid (adapted from Schreck and Peric, 2012).

In the present paper the background mesh and the three overlapping mesh are reported in Figure 4. In this case, each overlapping grid has to interface with the background mesh, moreover there are two additional overset interfaces between each satellite and the central screw. For this particular configuration, the central screw is considered as the background for the two satellites. To obtain a good interface quality, local refinements have been done in the background mesh to lower cell dimensions in the overlapping zone and reasonably increase the element number. Overlapping meshes are made of 5,033,265 elements (839,113 for the screw and 2,097,068 for each satellite), while the background mesh consists of 3,210,621 elements. Overlapping and background mesh are constitute by Cartesian elements and prism layer close to the wall.

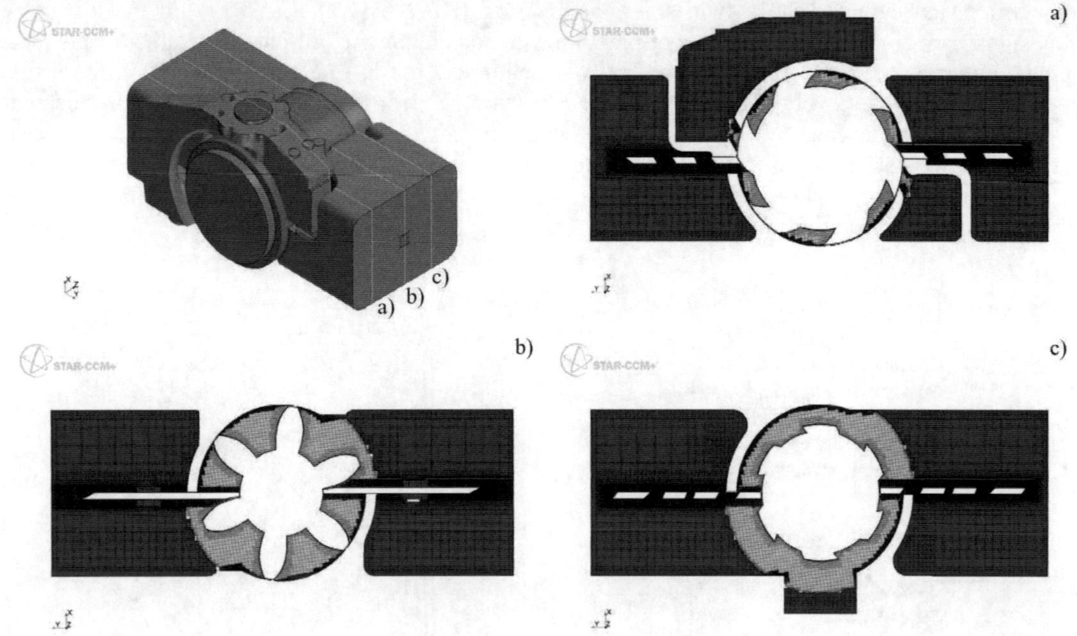

Figure 4: Chimera geometry and planar section at a) inlet ports b) center of the device c) outlet port. It is possible to see the background region (blue) and the overlapping ones (yellow for the screw, grey and orange for the starwheels).

As said before, the presence of three rotating regions implies the presence of three overset mesh boundaries. Moreover, as the gaterotor regions have to communicate between the background and the screw, the number of overset interfaces increases up to five (the three canonical between each rotating region and the background, and the two interfaces between the screw and the gaterotors). This number of interfaces is easily handled by the software, however it is possible that this could lead to higher solution time.

3.3 Mesh Morphing

Metamorphosis (or Morphing) is the process of gradually changing a source object through intermediate objects into a target object (Lee *et al.*, 1999). Morphing a mesh to be used in a numerical simulation is a delicate task, especially for a 3D CFD mesh. In this case, the morphing, also termed smoothing, is not limited to the surface but has to be extended to the entire volume of the mesh and the solver suffers dramatically.. The way the task is accomplished, depends on which smoothing algorithm is selected and on the definition of the control points criteria which can substantially change the result. A key feature that can be used to subdivide them is how they are related to the original mesh. The software in which mesh morphing is implemented and that is used in this paper is PumpLinx. Pumplinx has the capability of meshing pumps, gears and compressors by using this technique under the form of preloaded templates. In particular, the SCORG module has been designed for screw compressors and adapted to screw expanders in this paper. SCORG is a software for the design and analysis of screw compressors, pumps and rotors. The actual limit is given by the number of screw rotors that the SCORG module can deal with. In fact, only compressors with two or more screw rotors can be meshed with this template, which makes it unusable for a SSE application. This method, due to the combination of structured mesh and prescribed motion, sounds to be the best approach to this kind of application. However, as it is not possible to fit the SCORG module to SSE (Kovacevic *et al.*, 2012), at the moment the best compromise is given by the CS.

3.4 Key-frame re-meshing

Key-frame remeshing (KFR) is a meshing method where, after each time step, the volume is checked to see if cell's quality is still good to absorb further deformation without generating negative volume elements. If not, some group of cells are locally remeshed (Rane et al., 2012). This implies that number and connectivity of the cells changes from one step to another. It has been demonstrated that this meshing technique could simulate the complex configuration of screw compressors, however at the moment there are many limitations on pre-processing time, computational resources, results accuracy and difficulties in including features like turbulence or multiphase flows. Two tests were conducted on a piston cylinder and a twin screw compressor (Kovacevic *et al.*, 2012). In the piston cylinder application, significant errors in the prediction of pressure and temperature are present, while all attempts made with a twin screw compressor failed due to the complexity of numerical mesh. The analysis made on the piston gave important information on the level of accuracy of the KFR, with the error that increases with grid refinement. This could be due to mesh replacement at each time step, which means that it is possible that space conservation is violated with KFR. In conclusion, KFR gives inconsistent results, with an error that increases with mesh refinement instead of becoming smaller (Rane *et al.*, 2013). Time step reduction made excessive demands on pre-processing, making this path not suitable at this time.

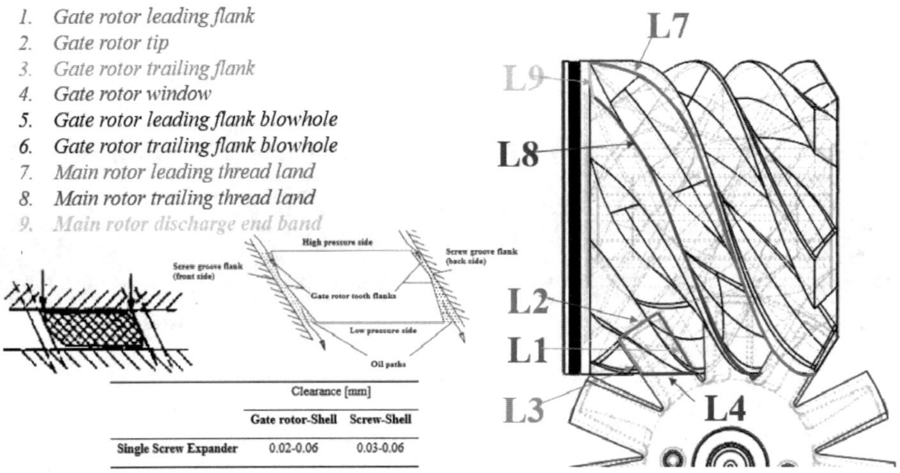

Figure 5: Possible leakage paths in single screw compressors and expanders (Adapted from Ziviani *et al.*, 2014a).

4. SIMULATION ANALYSIS

The prepared simulations were performed under steady-state conditions, with the exception of the CS one which was performed also under transient conditions. Each approach can provide different insights and useful information. The steady-state simulation is useful to evaluate the volumetric efficiency, axial thrusts and therefore a preliminary structural analysis. Moreover, the steady-state approach allows to calculate the flow coefficients (C_d) of the gaps which relates the flow rate through the machine and the pressure drop across the gaps. This is very useful to tune lumped parameter models, where gap flow coefficients are fundamental to improve the model accuracy on the mass flow predictions. At the same time, they are very difficult to be estimated by means of experimental data and/or analytical/empirical relations. For these reasons, CFD models (even steady-state) are increasingly used to determine these crucial coefficients. However, the transient simulation, besides the allows to evaluate the evolution-in-time of parameters of interest such as flow rate, pressure and torque. Since the SSE is a positive displacement machine, its volumetric efficiency is heavily affected by different kind of leakage paths (Ziviani *et al.*, 2014a). The most common leakage paths are shown in Figure 5. In particular, a total of nine sealing lines can be identified: L1 and L3 are related to the gaps which exist between the screw and satellite tooth tip and flank; L2 is satellite tooth tip (its presence is connected to the manufacturing process, i.e. milling); L4 is the gap between housing and screw and it is necessary to permit the screw motion; L7 and L8 are screw leading and trailing edges; L9 is the end curved edge of the groove (even this gap is with the housing); while L5 and L6 are blowholes which existence is due to the imperfect shape of satellite teeth. Both steady and transient simulations consider these leakage paths.

4.1 Steady-state approach

Generally, meshing strategies are useful to simulate moving and deforming fluid domains but at the same time also the computational time required increases. Therefore, in the case of steady-state simulations such approaches are not required.

Each steady-state simulation has been run on a Cartesian grid with slight differences on the grid topology among different solvers. The PumpLinx General Mesher did not allow to set prism layers around the wall boundaries, while the meshes generated in STAR-CCM+, Ansys Fluent and CFX did have the possibility to fully define the prism layers at the walls. In terms of number of elements, the mesh generated in PumpLinx, STAR-CCM+, Fluent and CFX consisted of about 500000, 1500000, 3000000 and 2800000 elements, respectively.

4.2 Transient approach

The transient simulation was performed only by means of the CS implemented in CD-ADAPCO STAR-CCM+ 10.02. Boundary conditions were kept the same, expect for the addition of the time step set to 10^{-4} s. The details of the mesh have been outlined in Section 3.2. For completeness, it is mentioned that all the meshing methods supported by the other software were tested. However, the different tests showed that KFR was not able to converge during early stages of the simulation. MM did not allow for transient simulation and DM required extremely high computational resources.

5. RESULTS

5.1 Steady-state approach results

The results presented in the following are related to the C_d obtained in the four different simulations and are aimed at comparing the different approaches. The C_d is an index which, combined to the volumetric efficiency, can give information on the losses and on the leakage flows which occur in a positive displacement device. It is defined as follows:

$$C_d = \frac{Q}{A\sqrt{\frac{2\Delta p}{\rho}}} \tag{1}$$

where Q is the volumetric flow rare [m^3/s], A is the passage area [m^2], Δp is the pressure drop [Pa] between two sections conveniently chosen upstream and downstream the passage area and ρ [kg/m^3] is the fluid density. The higher this value is, the higher are the losses due to leakage flow. To calculate this factor it is necessary to fix a section of known area A, across which the volumetric flux Q can be calculated by means of CFD simulations, and two arbitrary planes where it is possible to calculate the average pressure and, hence, the pressure drop. Figure 6 shows the section positions, referring to the simulation ran in STAR-CCM+. The calculations performed by means of the other methods rely on the same planes.

Table 1: Comparison between the C_d given by the steady simulations.

Software	C_d – Starwheel 1	C_d – Starwheel 2
Ansys Fluent 13.0	0.402	0.488
Ansys CFX 15.0	0.399	0.463
CD-ADAPCO STAR-CCM+ 10.02	0.402	0.497
PumpLinx 4.0.3	0.364	0.439

The results are reported in Table 1. As can be seen, the C_d values for both Satellite 1 and Satellite 2 are of the same order of magnitude with a maximum difference of about 10%. Moreover, since Satellite 1 has a shorter duct, it was expected that the calculated Cd would be smaller than Satellite 2. The first three calculations surprisingly predicted C_d values in close agreement each other (less than 1% difference). Only the fourth calculation deviates from the other of about 10%. In the case of Satellite 2, almost the same behavior is noticed: the first three calculations predicted C_d values within a 5% difference, while the fourth deviates of about 10%. This behavior could have an explanation by considering that Pumplinx is optimized for transient calculations, and in particular the grid topology is conceived to meet this task. Nevertheless, also this software allowed the obtainment of C_d values which can be considered to achieve an enhanced information with respect to usual practice. As for computational resources, PumpLinx, in particular due to the coarser mesh, reached convergence in less than 1500 iterations and less than an hour. STAR-CCM+ reached convergence in 2000 iterations and some hours of simulation run, while Fluent could not reach convergence before 70000 iterations and 15 days of running simulation. CFX has an intermediate solution time of 15000 iterations and 2 days.

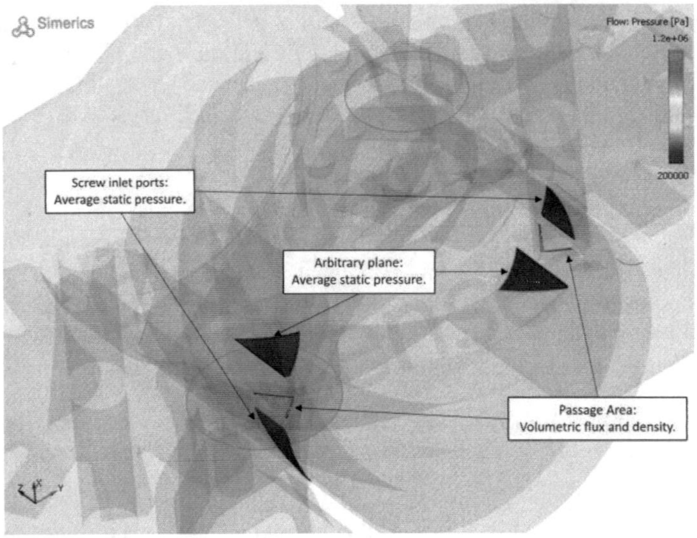

Figure 6: Surfaces employed for the calculation of flow coefficients C_d.

5.2 Transient approach results

A transient analysis provides insights on the dynamic behavior of the expander. As this work presents preliminary results of CFD simulations on a single-screw expander, particular attention is given to verify the reliability of the results obtained. In particular, the inlet and outlet pressure and mass flow rate fluctuations are considered along with the torque generated at the main rotor shaft. Moreover, pressure fluctuations have also been monitored at the screw-expander triangular inlet port closest to the main inlet. The different variables of interest are plotted as a function of the normalized angular position of the screw. The mean values over one full rotation are then compared with the results obtained from the deterministic model (Ziviani *et al*. 2016).

In Figure 7, it is possible to notice the three working chambers that are isolated by the gaterotors on the central screw. As expected, the gaterotors create three sections on the groove, lowering the pressure from the inlet to the discharge pocket. The contour of the pressure is overlaid on the main rotor.

Pressure signals have been monitored circumferentially along the rotor to reconstruct the pressure traces in the grooves. In Figure 8, the main pressure signals are plotted for one revolution. As there are six grooves in the main rotor, six peaks can be identified in the plot.

The 3D CFD model allows to investigate the flow distribution inside the inlet duct. The pressure fluctuations at the triangular inlet ports are shown in Figure 9. The fluctuations present different trends suggesting that the inlet duct induces pressure waves and pressure drops. Such effects cannot be captured by the deterministic model without a coupling with a pressure pulsation model. The pressure fluctuations have direct influence of the mass flow rate as reported in Figure 10a. In fact, each peak of the mass flow rate corresponds to a valley in inlet pressure trend. This behavior can be clearly seen in Figure 10b, where static, dynamic and total pressure traces are overlaid.

The expanding working fluid generates a mechanical torque and therefore produces useful mechanical work. From the CFD model it is possible to extract the actual mechanical torque that can be directly compared to experimental measurements if available. As shown in Figure 8, the value of the shaft torque varies in the range from 12 Nm to 35 Nm, with a mean value of 25 Nm, giving an output power of 7.8 kW (mean value) at 3000 rpm. However, from the deterministic model, the maximum shaft power obtained was in the order of 9 kW. This difference could be given by the presence of enlarged gaps, to help the overset mesh coupling. Similar values were also measured by running the expander with R245fa.

Figure 7: Different working chambers (i.e. portion of the grooves) isolated by the gaterotors.

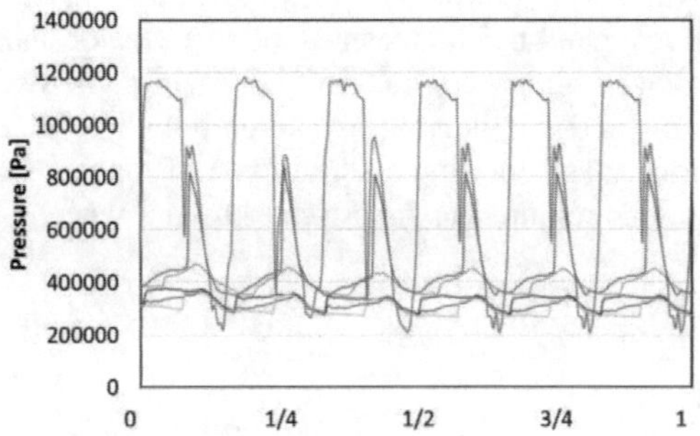

Figure 8: Pressure traces at main screw rotor.

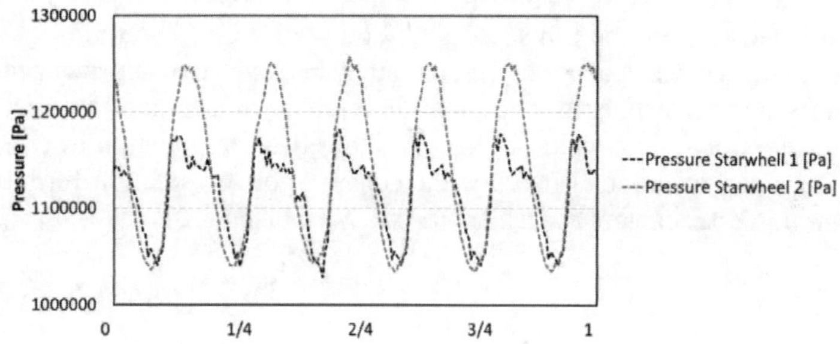

Figure 9: Pressure fluctuations at the main rotor triangular ports.

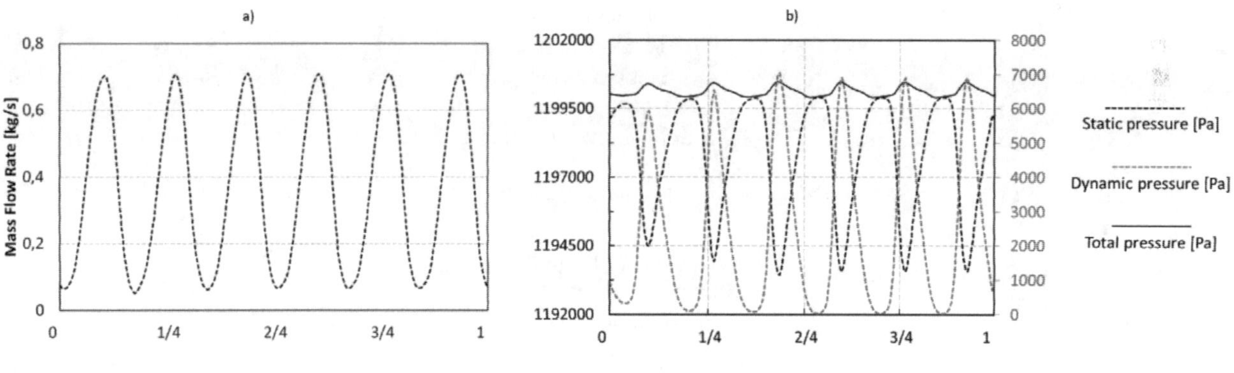

Figure 10: Inlet a) mass flow rate and b) pressure trends.

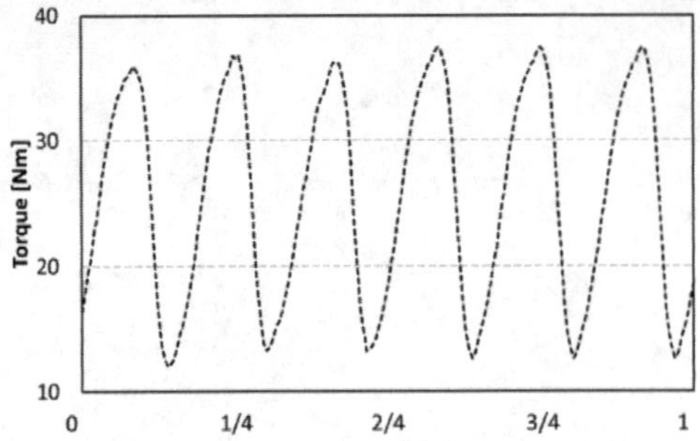

Figure 11: Shaft torque (main screw rotor).

6. CONCLUSIONS

Different CFD approaches applied to a SSE are compared in this paper, both in steady-state and transient conditions. Different commercial software has been considered in order to identify the most suitable to simulate the SSE. All the evaluated software could perform a steady-state simulation to estimate the flow coefficients and a maximum deviation of 10% was obtained, which can be considered acceptable. Among the investigated methods, the transient simulation could be run successfully only with CS. The fluid domain generated by the meshing of screw rotor and two starwheels makes the interpolation between overset and background regions critical and information is lost during the interpolation especially in regions such as gaps between screw and satellites. From the preliminary CFD results it can be concluded that CS showed some potential as meshing strategy for single-screw type of machines. However, in order to improve the converge of the solution further work is required to investigate the influence of meshing parameters as well as the time-step taken.

NOMENCLATURE

Variable	Definition	Units
Q	Volumetric flow rate	[m^3/s]
m	Mass flow rate	[kg/s]
ρ	Density	[kg/m^3]
Δp	Pressure drop	[Pa]
p_{tot}	Total pressure	[Pa]
p_{stat}	Static pressure	[Pa]
A	Passage area	[m^2]
SSE	Single-screw expander	
PD	Positive displacement	
DM	Dynamic Mesh	
CS	Chimera strategy	
MM	Mesh morphing	
SCORG	Screw compressor rotor grid generator	
KFR	Key-frame re-meshing	

REFERENCES

Castilla, R., Gamez-Montero, P. J., Ertürk, N., Vernet, A., Coussirat, M., Codina, E., 2010, Numerical simulation of turbulent flow in the suction chamber of a gear pump using deforming mesh and mesh replacement, Int. J. Mech. Sci., 52, pp. 1334-1342.

Del Campo, D., Castilla, R., Raush, G. A., Gamez-Montero, P. J., Codina, E., 2012, Numerical Analysis of External Gear Pumps Including Cavitation, J. Fluids Eng., 134, p. 081105.

Hoke, C., Decker, R., Cummings, R., McDaniel, D., Morton, S., 2009, Comparison of Overset Grid Deformation Techniques Applied to 2-Dimensional NACA Airfoils, 19th AIAA Computational Fluid Dynamics Conference, 22 - 25 June 2009, San Antonio, Texas, US.

Imran, M., Usman, M., Park B-S, Lee, D.-H., 2016, Volumetric expander for low grade heat and waste heat recovery applications, Renewable and Sustainable Energy Reviews, *57*, 1090-1109

Lee A., Dobkin D., Sweldens W., Schroder P., "Multiresolution Mesh Morphing" (1999), Proceedings of SIGGRAPH 99. Pp. 343-350.

Quoilin, S., van den Broek, M., Declaye, S., Dewallef, P., Lemort, V., 2013, Techno-economic survey of Organic Rankine Cycle (ORC) systems, Renewable and Sustainable Energy Reviews, 2013, *22*, 168-186

Rane S., Kovacevic A., Kethidi M., "CFD Modeling in Screw Compressors with complex multi rotor configurations"(2012), Int. Compressor Engineering Conference at Purdue Univ. Paper 2141.

Rane S., Kovacevic A., Stosic N., Kethidi M., "Grid deformation strategies for CFD analysis of screw compressors", Int. J. Refrigeration, 36(2013), 1883-1893.

Roget, B., Sitaraman, J., 2014, Robust and efficient overset grid assembly for partitioned unstructured meshes, J. Comput. Phys., 260, pp. 1-24.

Saleh, B., Koglbauer, G., Wendland, M., Fischer, J., 2007, Working fluids for low temperature organic Rankine cycles, Energy, 2007, *32*, 1210-1221

Schreck, E., Peric, M., 2012, Overset Grids in STAR-CCM+: Methodology, Applications and Future Developments, STAR Japanese Conference 2012.

Steger, J. L., Benek, J. A., 1987, On the use of composite grid schemes in computational aerodynamics, *Computer Methods Appl. Mech. Eng.*, 64, pp. 301–320.

Togashi, F., Nakahashi, K., Ito, Y., Iwamiya, T., Shimbo Y., 2001, Flow Simulation of NAL Experimental Supersonic Airplane/Booster Separation Using Overset Unstructured Grids, Comput. Fluids, 30(6), pp. 673-688.

Ziviani, D., Bell, I., De Paepe, M., van den Broek, M., 2014a, Comprehensive Model of a Single Screw Expander for ORC-Systems, 22nd International Compressor Engineering Conference at Purdue, Purdue, 14 - 17 July, 2014, Paper n. 1506, pp. 1-10.

Ziviani, D., Suman, A., Lecompt, S., De Paepe, M., van den Broek, M., Spina, P. R., Pinelli, M., Venturini, M., Beyene, A., 2014b, Comparison of a Single-Screw and a Scroll Expander under Part-Load Conditions for Low-Grade Heat Recovery ORC Systems, Proc. of 6th International Conference on Applied Energy (ICAE2014), Taipei, 30 May - 2 June, 2014, Paper n. 78, pp. 1-4.

Ziviani, D., Bell, I., De Paepe, M., van den Broek, M., 2016, Mechanistic model of an oil-flooded single-screw expander, 23rd International Compressor Engineering Conference at Purdue, Purdue, 11 - 14 July, 2016, Paper n. 1486, pp. 1-10.

CFD SIMULATION OF A TWIN SCREW EXPANDER INCLUDING LEAKAGE FLOWS

Rainer ANDRES[1]*, Jan HESSE[1], Haris BABIC[1], Uwe SALECKER[1], Andreas SPILLE-KOHOFF[1]
Alexander NIKOLOV[2], Andreas BRUEMMER[2]

[1]CFX Berlin Software GmbH,
Berlin, Germany
rainer.andres@cfx-berlin.de

[2]TU Dortmund University, Faculty of Mechanical Engineering, Chair of Fluidics
Dortmund, Germany
andreas.bruemmer@tu-dortmund.de

* Corresponding Author

ABSTRACT

The numerical simulation of rotary positive displacement (PD) machines has proven to be a very challenging task characterized by complex geometries with a time dependent change of the chamber volume along with complex flow physics. In addition, the leakage flow caused by housing clearances, front gaps and the intermesh clearance is of particular interest. The resulting losses have a severe impact on the overall machine efficiency.

This paper shows workflow and results for Computational Fluid Dynamics (CFD) simulations of the Screw Expander SE-51.2 of TU Dortmund University. The simulation is performed using ANSYS CFX, where the computational grids for the rotary parts are derived from the grid generator TwinMesh. The SE-51.2 is a gearless, dry running twin screw expander that converts thermal into mechanical energy by decompressing air with pressure-ratios up to 6:1 by a rotational speed range up to 20,000 rev/min. The volume ratio of the expander is 2.5 at a shaft distance of 51 mm with a displaced volume per male rotor revolution of 285 cm³. The numerical model of this machine includes fully resolved clearances along with the meshing of the rotor solid end faces whereas a contact point between the rotors is neglected. For several rotational speeds, CFD results are compared to measurements at 4 bar at the high pressure side, expanding air down to 1 bar.

The validation is based on integral values, e.g. mass flow, torque and power, all averaged over a certain time period for several operating points. Moreover, time-resolved pressure measurements for distinct positions are compared to simulation results. Besides direct evaluation against experimental data, tendencies are analyzed by varying the numerical model with respect to mesh resolution or changing the size of clearances. Apart from local changes in the flow field, it is shown that also global values like torque or mass flow are clearly affected. In addition, flow regimes not captured by measurements, like leakage flows within clearances between rotors and housing, are evaluated and illustrated based on CFD results.

While the simulation allows reliable analysis of dry running screw machines and their complex flow characteristics, oil- or water-injected machines are also commonly used in the industry, allowing higher pressure-ratios. The liquid fluid acts as a sealing for clearances and can also be used as lubricant. From the simulation point of view, multiphase calculations with corresponding multiphase-effects are required in order to cope with injected screw machines. Further investigation also involves heat transfer and resulting deformations of rotary and stationary parts. Clearances can vary due to deformations, eventually resulting in different machine characteristics and efficiency. Thus, fluid-structure interaction is of high demand. For both of these issues, a prospect is given to extend the scope of application, using CFD for screw machine analysis.

1. INTRODUCTION

Different theoretical approaches for the calculation of the operational behavior of rotary PD machines have been reported in the literature. At the Chair of Fluidics, a simulation tool KaSim was developed for the simulation of rotary displacement machines based on a zero dimensional chamber model. The calculation by the time-stepping method is made under consideration of the law of conservation of mass and energy, Janicki (2007). For the purpose of simplicity, it is assumed that spatial gradients in the intensive state variables are negligibly small and the fluid condition within a working chamber can thus be regarded as homogeneous. Hereby, a 3D CFD method is essential, if the spatial distribution of the variables of state is required. By means of an iterative fluid-structural interaction coupling of KaSim und the Finite Element Method (FEM) program ANSYS, the calculation of temperature distribution in the components as well as resulting thermal deformation, was presented by Nikolov et al. (2012). The automation capability is given due to the ANSYS Parametric Design Language (in short: APDL). A dynamic model for the performance prediction of a twin screw expander was presented by Papes et al. (2015). Here, the calculation approach was compared to 3D CFD simulations.

The geometry of the test screw expander SE 51.2 used to validate the simulation model within the framework of this paper is presented by Nikolov and Brümmer (2016). SE 51.2 is a dry-running screw-type expander prototype without timing gear designed at the Chair of Fluidics by Temming (2007). In order to avoid damage in the rotor intermesh due to the lack of lubrication, both rotors are hardened and have a tough wear-protection coating are. Characteristic for SE 51.2 are its relatively small dimensions and high rotational speed up to 20,000 rev/min at operation.

2. CHAMBER MODELING

The modelling of rotation and deformation of the fluid domain between rotors and housing is the most challenging task in order to get a numerical representation of the PD machine. Here, the mesh generator TwinMesh is used to create the numerical grids for the fluid domain. The time dependent change of the chamber volume according to the rotor position is modelled by creating meshes for each rotor position. For the SE-51.2 twin screw expander, the chamber volumes of two axially parallel rotors are meshed with O-type grids for each rotor, using hexahedral elements, as shown in Figure 1. These O-type grids cover the fluid volume which is determined from the 2D rotor and housing contours.

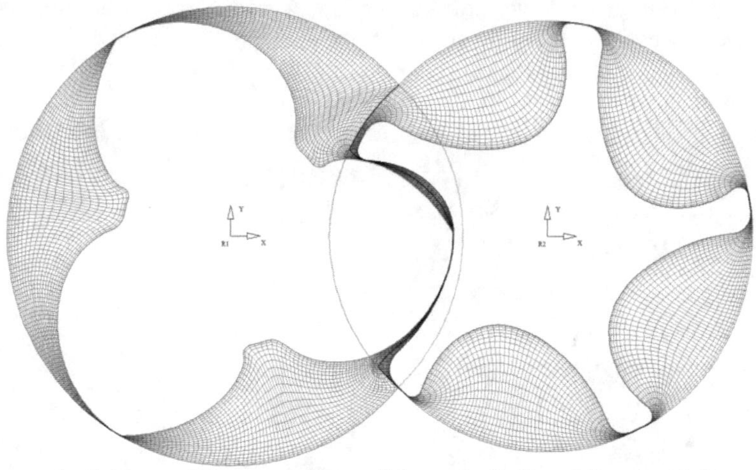

Figure 1: Grid on a 2D cross section of the male (left) and female (right) rotor

The twisted rotor geometry along the axial direction (200° twist angle for male rotor, 120° for female rotor) is modelled by a consecutive staggering of 2D mesh sections. For each axial position, 3D grids are created, which satisfy specific quality criteria. The minimum element angle is above 18°, the aspect ratio below 2000 and volume change below 20. Towards housing and rotor walls, the grid is refined to have an adequate boundary layer resolution. The 3D meshes are exported from TwinMesh for each rotor position (i.e. the meshes cover a periodic section of 120° in this case), where an angle increment of 1° is used for the simulation of the SE-51.2 screw expander. A more detailed description of the mesh generation using TwinMesh is given by Hesse (2016). The stationary fluid domains, i.e. suction and discharge port as well as inlet and outlet pipes are provided as separate meshes. These are generated using ANSYS Meshing and ANSYS ICEM CFD. In this case, hexahedral elements are used for the inlet and outlet pipes, whereas tetrahedral elements are used to mesh the more complex geometry of fluid regions around the suction and discharge port.

3. SIMULATION SETUP

The simulation is carried out in ANSYS CFX, where the simulation setup contains the assembly of all stationary and rotary fluid regions, see Figure 2 (right). The grids for the rotary parts also include the fluid regions for all clearances. All meshes are connected via fluid-fluid interfaces, where the flux of mass, momentum, heat and turbulence is conserved. In addition, appropriate boundary conditions are defined, representing the desired operation point. For the screw expander, pressure and temperature at inlet and outlet are set as well as the rotation speed of the male rotor. Apart from interfaces, inlet and outlet faces, no slip walls are assigned.

For the simulation of rotary PD machines, especially the mesh motion due to the rotor rotation and thus the deformation of the chamber volume is of interest. This is realized by using a FORTRAN routine, which loads the new mesh information (according to the current rotor position) at the beginning of each simulated time step. In the ANSYS CFX solver, the mesh deformation is added to the set of differential equations, so that no interpolation is required between time steps. The constrained pressure at inlet and outlet of the numerical model can lead to an artificial pressure reflection, most distinct at the high pressure side (i.e. inlet side of the expander). These waves affect the pressure and velocity field and represent an undesired discrepancy between numerical and real model. In all simulations performed, the distortion of monitored flow quantities caused by those pressure waves is prevented by using non-reflective boundary conditions, allowing the pressure waves to pass the artificial inlet and outlet boundaries.

In this paper, for a pressure ratio of 4:1 with an air inflow temperature of 90°C, the expander is simulated for several rotational speeds between 1,000 and 16,000 rev/min. For the reference operating point at 4,000 rev/min, a variation of the numerical model is performed. As a geometric variation, one simulation with 25% increase of radial clearances (housing gap, intermesh clearance and blow hole) and one simulation with 25% increase of axial clearances (front gaps) are carried out. Also a simulation with a decreased mesh resolution is performed, decreasing the total amount of elements from about 920,000 down to approximately 52,000 per rotor chamber. Because the sharp and most distinct pressure and velocity gradients are expected to occur within the moving rotor chambers (especially in narrow clearances), the grids for those chambers are assumed to outweigh the stator meshes with regard to their impact on flow characteristics. Thus, rotor meshes are coarsened in the radial, axial and circumferential direction, whereas coarsening of the stator grids is only applied at the interface regions. In this way, discontinuities regarding the element sizes in regions where rotary and stationary parts connect to each other are prevented.

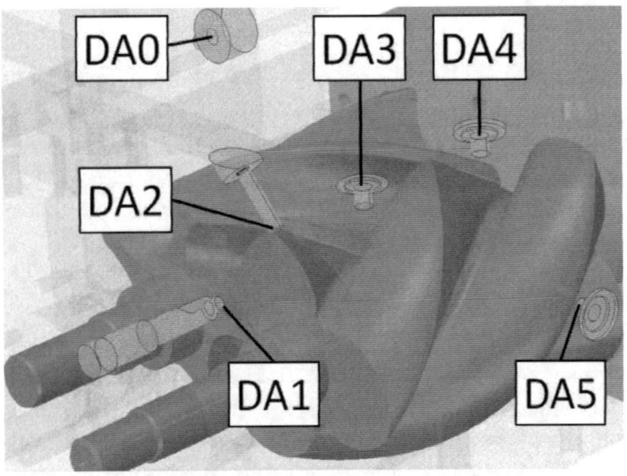

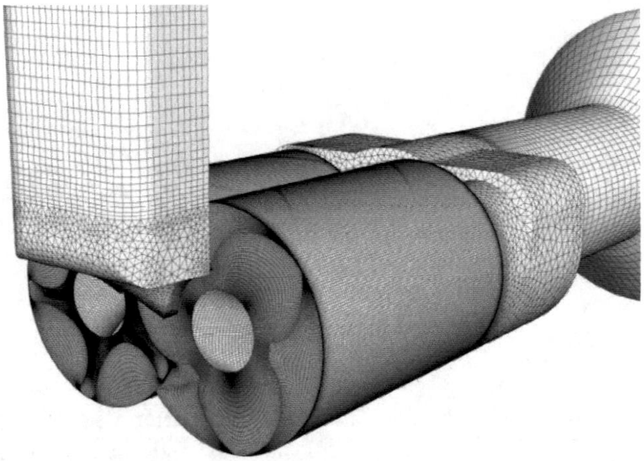

Figure 2: CAD model of the SE-51.2 screw expander including pressure measurement points (left) and assembly of rotary and stationary mesh regions including meshes for the front gap clearances (right)

5. SIMULATION RESULTS

The flow conditions inside the screw expander are visualized by a contour plot of the instantaneous pressure and velocity fields. These spatially resolved results are shown for a specific rotation angle, i.e. for a specific point in time. Time resolved results refer to a periodic solution over time, i.e. for a converged simulation, the monitored flow quantities are periodic over a section of 120° (due to rotational periodicity of the rotors). The gas enters the expander at a pressure of 4 bar, filling the pressure chamber between the flanks of male and female rotor. The screwed rotors form several chambers in axial direction, where the volume of the chamber is increasing with the rotation angle. Because of the pressure inside those chambers acting on the rotor walls, the fluid energy is converted into mechanical energy, represented by the torque of the rotors. This is illustrated in Figure 3 for three different male rotation angles over a periodic section of 120°.

At a rotation angle of 40°, both rotors reach the control edge of the pressure port. At this point in time, the filling process of the preceding chamber (2) has finished, since there is no longer a connection between pressure port and chamber (if clearances are neglected). With further rotation of the rotors, the foremost chamber (1) fills up with air, causing a pressure on the rotor walls of 4 bar, while the chamber volume is increasing simultaneously. At the same time, in chamber (2) a pressure of about 3 to 3.4 bar is calculated.

For a rotation angle of 80° and 120° respectively, the pressure on the rotors in chamber (1) remains nearly constant at 4 bar, because there is an ongoing supply with inflowing air through the pressure port. In chamber (2) on the other hand, pressure is dropping over rotation angle because of the increasing chamber volume. Once the lobes of both rotors reach the control edges again, the air supply of chamber (1) is cut and pressure also starts to decrease, while a new chamber is already formed. Due to the rotational symmetry, the 40° position in Figure 3 can be interpreted as the subsequent position 160°, where the aforementioned pressure drop of chamber (1) can be observed.

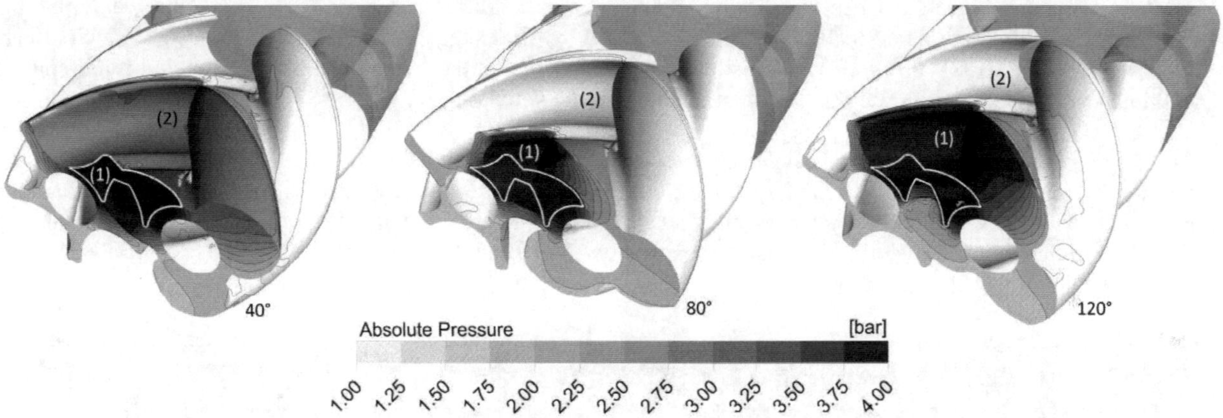

Figure 3: Pressure distribution at 4,000 rev/min on the rotors for male rotation angle of 40°, 80° and 120°

This work mechanism of the screw expander can be also demonstrated in Figure 4. Regarding the averaged chamber pressure over rotation angle for the reference operation point (4,000 rev/min, pressure-ratio of 4:1, air inflow temperature of 90°C), a pressure drop can be noticed at less than 160°, i.e. before the lobes reach the control edge. The cross section area of the pressure port through which chamber (1) is supplied with air is decreasing past 120° of rotation. In dependence of the rotation speed, sufficient replenishment with air to maintain a pressure of 4 bar is not given. This represents a throttle effect with a growing influence on the chamber fill level at increasing rotation speeds. Hence, at a lower rotation speed of 1,000 rev/min, this throttle effect is less pronounced, as the pressure drop is deferred to a higher rotation angle or a higher chamber volume, respectively.

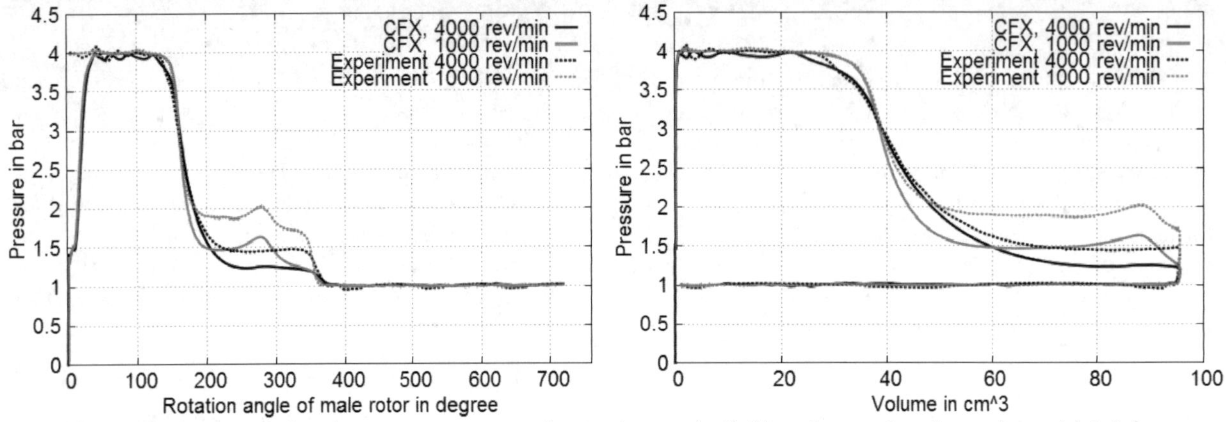

Figure 4: Averaged chamber pressure over male rotation angle (left) and over chamber volume (right) for experimental measurements and simulation results at 1,000 rev/min and 4,000 rev/min

Comparing simulation with experiment, results are qualitatively in good agreement. Nevertheless, chamber pressure measurements only cover male rotation angles higher than 45°. For lower angles, the pressure is estimated, e.g. according to the inlet valve pressure, whereas simulation results cover the entire angle range. Past 45°, filling process and pressure drop pressure due to the expansion of the chamber are in good match as well as the chamber refill seen at 1,000 rev/min for a chamber volume higher than 40 cm³. This effect occurs for lower rotation speeds, where caused by leakage, the chamber is refilled with air from regions with higher pressure, resulting in a pressure rise before discharge. The noticeable difference between simulation and experiment can be seen in the lower pressure level, which is reached through chamber expansion. For a fully expanded chamber at 1,000 rev/min, the difference is approximately 0.4 bar, whereas for 4,000 rev/min the difference is 0.25 bar. It is conceivable that the smaller pressure drop measured in the experiment is caused by the leakage through the rotor bearings, which affects rotation angles from 200° till discharge. This leakage is quantified in the experiment by 7.2 kg/h and represents an additional rotor chamber inflow, which was not modeled in the simulation.

Figure 5 compares the measured and calculated pressure distribution over an angle range of 240° at distinct pressure measuring points. The positions of the individual points can be derived from Figure 2 (left). The calculated results are found to be in very good agreement with the experiment. However, point DA4 shows the abovementioned difference in the pressure level through chamber expansion. The mean deviation at this pressure point is about 0.2 to 0.25 bar, which fits well to the results presented earlier in Figure 4.

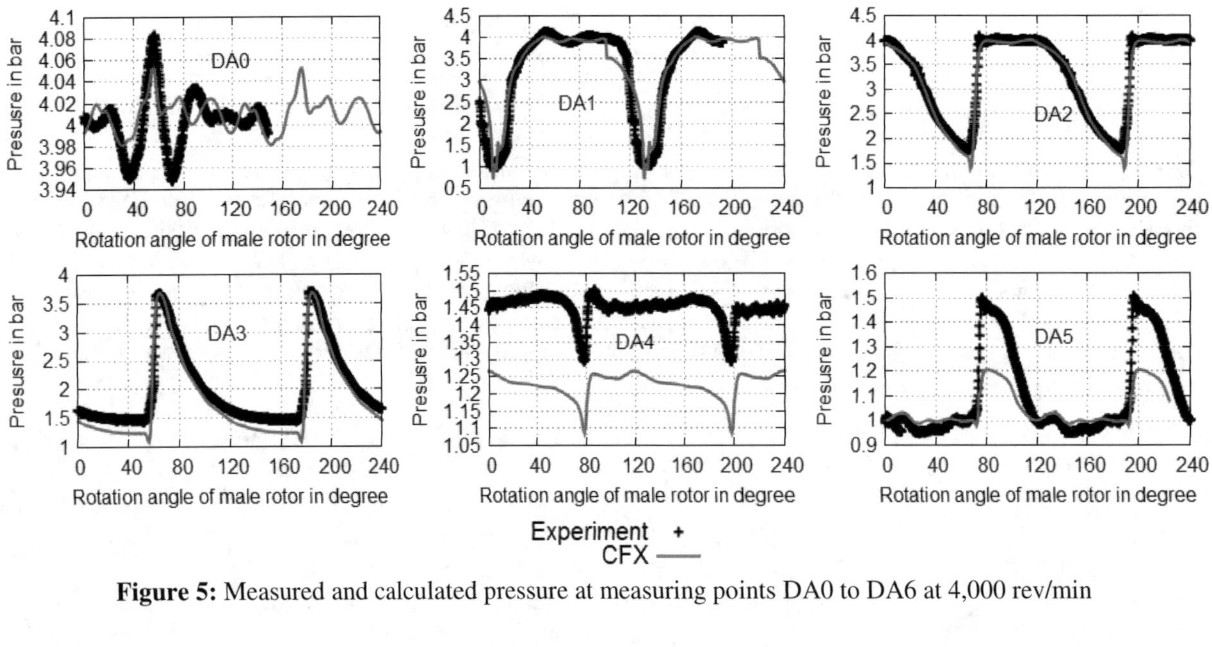

Figure 5: Measured and calculated pressure at measuring points DA0 to DA6 at 4,000 rev/min

The velocity field in Figure 6 shows the flow pattern within rotor chambers and housing. As illustrated, the velocity upstream and downstream of the rotor chamber varies between 10 m/s and 25 m/s. Within the rotor chamber, higher velocities are present while the maxima are located within clearances. Examplified in Figure 7, velocities above 200 m/s up to 500 m/s are illustrated at a wall distance of 0.02 mm from the rotor walls. These high velocities occur primarily within the front and housing gaps at the high pressure side. Also the blow hole causes leackage flow. This can be visualized in Figure 8, where the leackage flow through the blow hole is evident. Caused by the pressure difference between chambers in the the axial direction, a jet of high pressurized air streams into the chamber with a lower pressure level. At the investigated reference operation point, the peak velocity of jet through the foremost blow hole is approximately Ma =1.2.

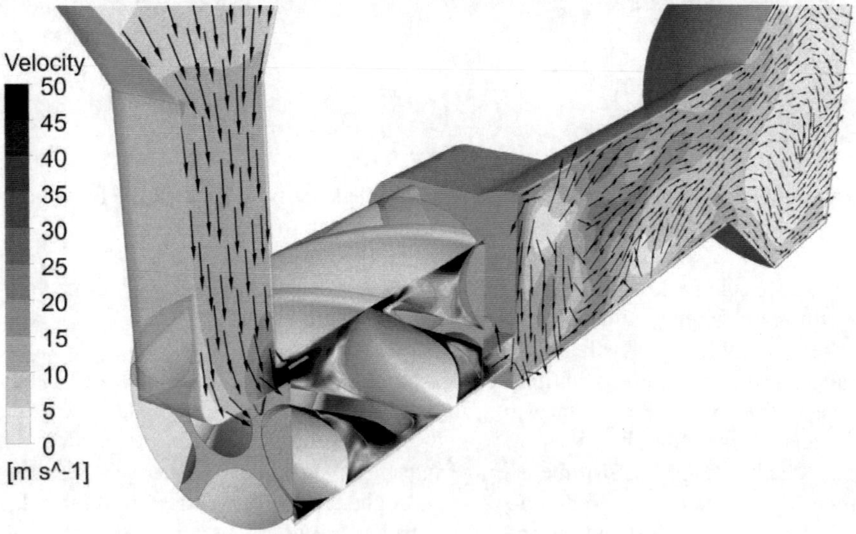

Figure 6: Contour plot of the velocity field at male rotation angle of 0° at 4,000 rev/min

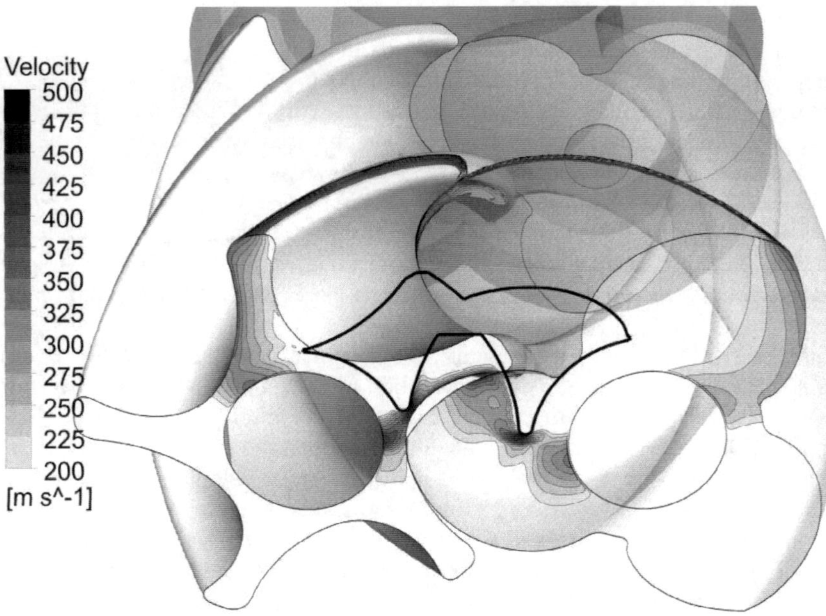

Figure 7: Flow regions with a velocity above 200 m/s at male rotation angle of 0° at 4,000 rev/min

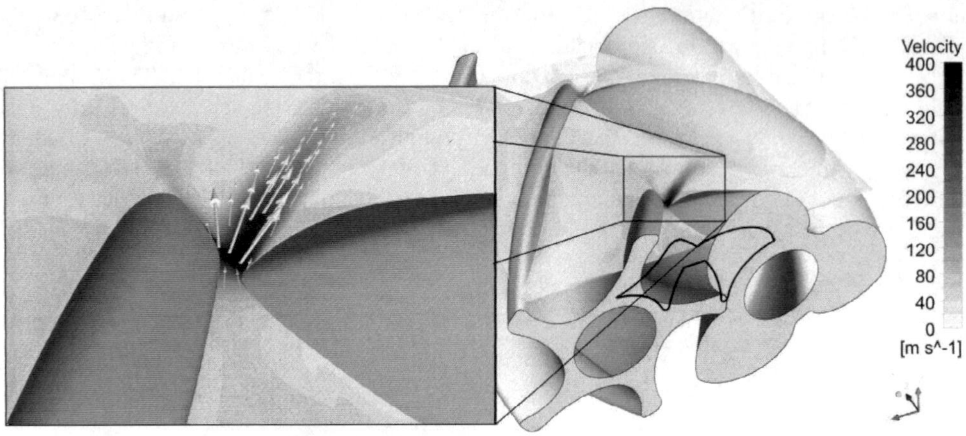

Figure 8: Velocity field on a horizontal plane, illustrating the leakage flow through the blow hole at male rotation angle of 75° at 4000 rev/min

Since leakage flow does partially not contribute to the conversion of thermal energy into mechanical energy, it has an adverse impact on the efficiency and overall performance of the screw expander. In the simulation, it can be demonstrated on the basis of integral values like mass flow (at inlet and outlet) and generated power. Figure 9 depicts the calculated results for these quantities on the basis of varied radial clearances. The case with 80 μm housing gap represents the reference case according to measured gaps on the real machine. As a variation of the model, housing clearance is increased by 25% (i.e. from 80 to 100 μm) for both rotors. Due to a normal rotor scaling of 20 μm to increase the housing gap, also the average intermesh clearance is increased, resulting in a larger blow hole as well. Comparing the results for both gap sizes, the higher mass flow caused by higher leakage is apparent. Averaging the mass flow over one revolution, the case with larger radial clearances exceeds the reference case by approximately 12%. Comparing the same cases in terms of generated power, the difference is smaller but still noticeable. The deviation is about 2 % for total power.

Figure 10 compares averaged results for mass flow and generated power for a rotational speed range between 1,000 and 16,000 rev/min with measurements. Also, the cases with the model variations (i.e. increased clearances, decreased rotor grid resolution) are included. For the reference point at 4,000 rev/min, a mass flow of 180 kg/h was measured, whereas 167 kg/h are calculated by the simulation (without model variation). Concerning the generated power, only the inner power is compared, i.e. bearing friction and friction caused by contact of rotor lobes are neglected. The measured value is 2,973 W, while the calculation yields 2,808 W. Regarding the model variation, the increase of front gaps shows a minor impact on the mass flow, which raised by nearly 3% (compared to 12% caused by increasing radial gaps). This is also the case for the generated power, where the larger front gaps results in only 0.5% (compared to 6.5% for increased radial gaps) change compared to the reference front gap size. The calculation on coarsened rotor grids however shows a significant deviation from both, measurement and reference CFD results. Despite unchanged clearances, the coarsened grids lead to a mass flow of 215 kg/h, which is off by over 28% regarding the fine grid results. As for generated power, the difference is about 12.5% using the coarse rotor grids.

It can be seen that both, mass flow and power are slightly underestimated by the simulation over the investigated range of operation points. However, with regard to the variation of the rotation speed, the trend of resulting mass flow and power shows a very good match with the measurements. The mean deviation between CFD results and measurements can be quantified as approximately -8 % for mass flow and -7% for generated power. Figure 11 presents a quantification of the leakage flow through radial clearances (including housing clearance, intermesh clearance and blow hole) and front gaps for the male rotor over a rotation period of 120° for both, reference and increased radial clearances. For the reference gaps, the average leakage through the front gap (only high pressure side) is approximately 1.2 kg/h, whereas for radial clearances, it is about 10.8 kg/h. As aforementioned, the leakage through the rotor bearing was not included in the simulation, but quantified with 7.2 kg/h, it is near the magnitude of the housing clearance leakage. Thus, it is considered to have noticeable impact on the integral mass flow of the expander, which can contribute to the mass flow underestimation by the simulation compared to the experiment.

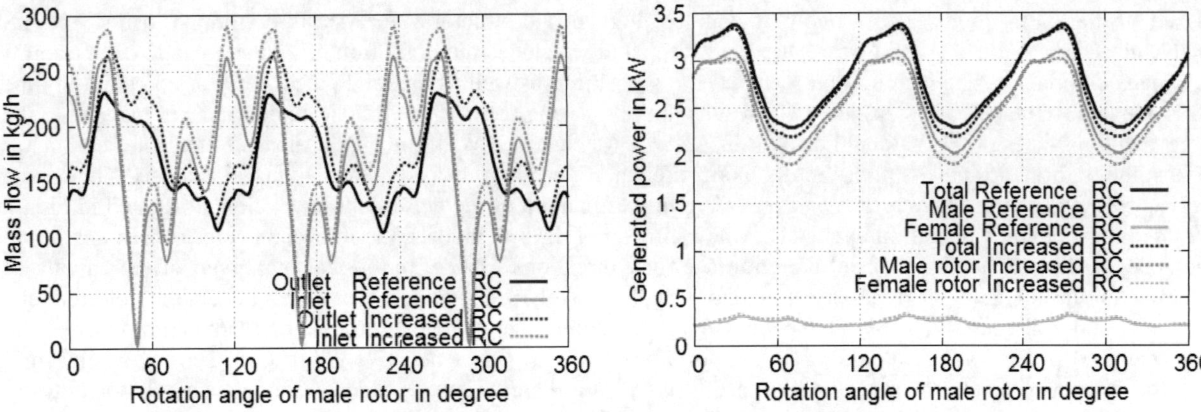

Figure 9: Comparison of Mass flow (left) and generated power (right) at 4,000 rev/min for reference and increased radial clearances (RC) over 360° of male rotor revolution

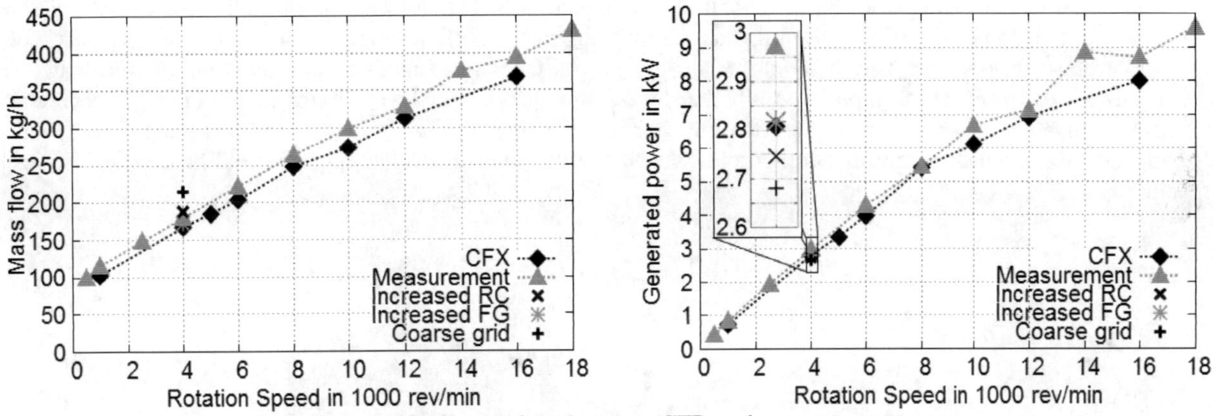

Figure 10: Comparison between CFD and measurements:
Mass flow (left) and generated power (right) for different operation points and for cases with increased radial clearances (RC) and increased front gaps (FG)

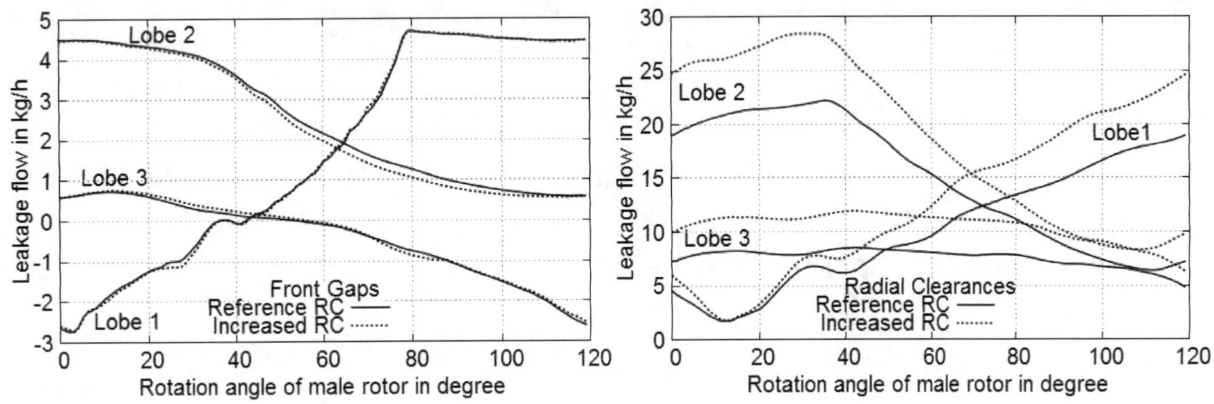

Figure 11: Front gap (left, only front gap at pressure side) and radial clearance (right) leakage flow at 4,000 rev/min for the three male rotor lobes over 120°. Comparison between reference and increased radial clearances (RC)

6. FLUID STRUCTURE INTERACTION AND MULTIPHASE SIMULATIONS

Based on the CFD results, also a one-way coupled thermal and structural analysis is performed, using ANSYS Mechanical. The numerical model for stator and rotor solids includes the definition of clamping and contact points between components, screw pretension forces, bearing positions as well as material properties. For one point in time (here the last simulated time step), the CFD results at the encasing faces of the fluid volume mesh are mapped to the corresponding faces of the stator solid mesh in ANSYS Mechanical. With the applied loads to the rotor and stator solids, the deformations and stresses in the model are calculated from a steady-state structural analysis. The results only correspond to one point in time and provide a first estimation of the deformation, which is illustrated in Figure 12. With the clamping definition applied (i.e. the clamping of the real model), the intended degree of freedom of the expander is directed along the axial direction (i.e. in z-direction). Hence, the global maximum of the calculated deformation occurs along the axial direction and is approximately 0.174 mm. Nevertheless also radial deformations (i.e. in x- and y-direction) are present. The maximum radial deformation of the rotor is about 56.96 µm, whereas the adjacent shell surface of the stator extends for approximately 56.21 µm in negative y-direction. Thus, at this location and for this point in time, the housing clearance is assumed to barely change. The maximum change in the housing gap is calculated approximately at half axial rotor length. Here the deformation of stator and clearance lead to a 13 µm reduction of the housing gap.

The outlined one-way coupling of fluid and structure could be also performed for every time step, so that time resolved deformations and stresses on the structure can be analyzed. This was not part of the current investigation and it is expected to be very time consuming, as the mapping of the CFD results for the steady-state analysis took about four hours for one time step. It should be also noted, that the applied loads do not represent the final state, as the heating of the machine takes place on a higher time frame than covered by the simulation. From experimental results, a temperature increase after more than 30 minutes is still noticeable, whereas the presented CFD results correspond to a simulated time of less than one second (i.e. the start-up of the expander and hence the first few revolutions).

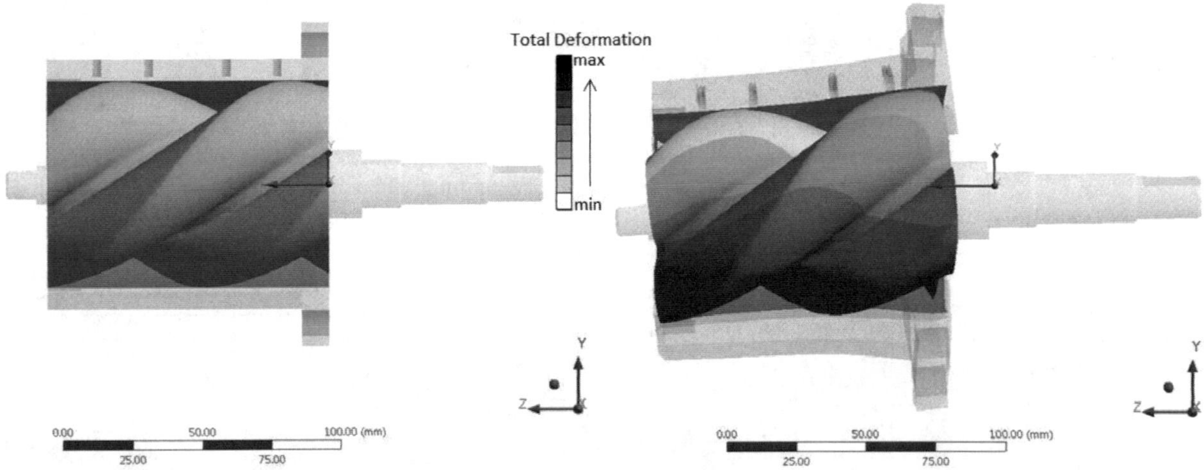

Figure 12: Qualitative deformation at 4,000 rev/min, pressure ratio 4:1 and air inflow temperature of 90°C

Another challenge is the modeling of multiphase flows, e.g. to analyze oil- or water-injected screw machines. Figure 13 shows a qualitative result of a simulation approach of an oil injected screw compressor, where oil is injected through supply valves in both rotor chambers at the pressure side. To model the multiphase flow, the volume of fluid method is used in ANSYS CFX with homogeneous coupling of gaseous and liquid phase. The presented results depict regions with an oil volume fraction higher than 1% after 6 revolutions of the male rotor. The approach demonstrates general feasibility with available multiphase models. Nevertheless, validation of oil distribution within rotor chambers as well as resulting effects like sealing or cooling is desired for well-defined operation points.

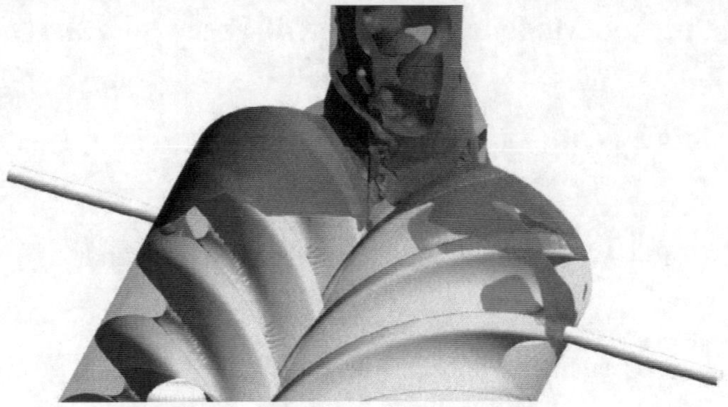

Figure 13: Simulation of an oil injected screw compressor

7. CONCLUSIONS AND OUTLOOK

The presented CFD results of the SE-51.2 twin screw expander in comparison with experimental data indicate that the working mechanism and flow conditions within the expander are well captured by the simulation. It accounts for effects like throttling, chamber refilling or leakage flows and examined flow quantities are overall in good agreement with the measurements. Moreover, the CFD easily allows analyzing tendencies, like e.g. the variation of clearances or changing the operation point. With regard to further investigations, accurate modeling of leakage sources (i.e. all clearances present in the real model) should be treated with particular attention, as it severely affects the machine characteristics. Neglecting the rotor bearing leakage in the presented numerical model is assumed to considerably contribute for fewer losses compared to the experiment, as seen in the pressure volume. Also the separate quantification of the individual leakages through radial clearances (i.e. housing gap, intermesh clearance and blow hole) is of interest. The significance of leakage modeling demands for high quality meshes at high resolutions, as coarse meshes tend to result in a decreasing accuracy regarding geometry approximation, especially the intermesh clearances of rotary PD machines. The significant overestimation of the mass flow derived from the coarsened rotor meshes can be mainly caused by an insufficient spatial resolution of clearances.

Both, fluid structure interactions as well as multiphase flows are a challenge of upcoming numerical investigations, as the numerical effort is clearly increased, especially with the aim of a two-way coupling between fluid and structure. Also procedures with the goal to shorten simulation time are to be elaborated, e.g. applicable initialization of oil initialization or the way of coupling the fluid with the structure. For all issues, there is the claim for reliable experimental results in order to validate the numerical models.

REFERENCES

Hesse, J., Spille-Kohoff, A. (2016). CFD simulation of a screw compressor including leakage flows and rotor heating. *Proceedings of the 3rd International Rotating Equipment Conference (IREC)*, Düsseldorf.

Janicki, M. (2007). Modellierung und Simulation von Rotationsverdrängermaschinen. *Dissertation*, TU Dortmund University, Dortmund.

Nikolov, A., Brümmer A. (2016). Analysis of Indicator Diagrams of a Water Injected Twin-shaft Screw-type Expander. *Proceedings of the 23st International compressor engineering conference at Purdue*, Purdue.

Nikolov, A., Huck, C. & Brümmer A. (2012). Influence of Thermal Deformation on the Characteristic Diagram of a Screw Expander in Automotive Application of Exhaust Heat Recovery. *Proceedings of the 21st International compressor engineering conference at Purdue*, Purdue.

Papes, I., Degroote, J. & Vierendeels, J. (2015). Dynamic model for the performance prediction of a twin screw expander in ORC. *Proceedings of the 3rd International Seminar on ORC Power Systems* (513-521). Brussels.

Temming, J. (2007). Stationärer und instationärer Betrieb eines unsynchronisierten Schraubenladers. *Dissertation,* TU Dortmund University, Dortmund.

Experimental Testing and Modeling of 5 kW Oil-Free Open Drive Scroll Expander Using R245fa

Felipe A. ACCORSI*, Nelson A. JAMES[1], Eckhard A. GROLL[1], James E. BRAUN[1], William T. HORTON[1],

Ray W. Herrick Laboratories, Purdue University
West Lafayette, Indiana, USA

faccorsi@purdue.edu, najames@purdue.edu, jbraun@purdue.edu, groll@purdue.edu, wthorton@ecn.purdue.edu

* Corresponding Author

ABSTRACT

Organic Rankine Cycles (ORC) are thermodynamic power cycles designed to generate work from low temperature heat sources. The low temperature heat input is, typically, between 80 °C to 270 °C. The efficiency of an ORC is highly dependent on the expander characteristics. This article experimentally evaluates the performance of a scroll expander with nominal capacity of 5 kW, built-in volume ratio of 3.5. Tests were conducted in an ORC test-rig using R245fa as the working fluid. The expander was tested over a full performance map consisting of two temperature sources 85°C and 110°C and five expander speeds, from 800 RPM to 3000 RPM.

The scroll expander achieved a maximum isentropic efficiency of 0.58, for a volume ratio of 6.12, at the expander speed of 1600 RPM and temperature source of 110°C. The same temperature source also registered the maximum expander power output, 3.75 kW, at volume ratio of 6.55 and an expander speed of 2500 RPM.

A semi-empirical model was developed to predict the expander output power and isentropic efficiency. The model was in close agreement with the experimental results with a mean deviation of 4.93%, for the expander isentropic efficiency, and 4.23%, for the power output. The semi-empirical model suggests friction and leakage influence efficiency more than any other source.

1. INTRODUCTION

Projections from the Annual Energy Outlook 2015 (AEO 2015) anticipate a 0.3% yearly growth in U.S. energy demand through 2040. However, it predicts that the U.S. total electricity uses grow by an average of 0.8% per year. These projections of rising energy demand are in and of themselves an incentive for research to study ways to increase the efficiency of equipment for power production.

Conventional methods of energy generation are based on the use of limited sources, such as petroleum, natural gas, and coal. Nuclear fission is another traditional source of energy supply, but it has been linked to recent uncertainty and public disapproval regarding the safety of nuclear power plants. In addition, a higher level of awareness among the world population regarding environmental concerns, such as climate change, create considerable pressure for changes in traditional energy conversion, energy efficiency usage and the exploration of renewable sources.

The combination of the projected growth in electricity demand (and energy demand, in general), social concerns regarding climate change and uncertain fluctuations in fossil fuels costs make clear the need for more efficient equipment and an overall improvement in power generation methods. The Organic Rankine Cycle (ORC) is a technology that can use waste heat to produce power and improve the overall efficiency of an existing machine or process.

Rattner and Garimella (2011) assessed the potential of waste heat from U.S. power generation and thermal processes. According to their study, the largest concentration of waste heat is localized in the range of low temperature sources in the transportation sector, followed by power plants.

Internal combustion engines (ICE) are a large source of waste heat having typical 1st Law efficiencies between 30 and 40 % with the remainder of the input energy dissipated as heat. Song et al. (2014) examined waste heat recovery of a marine diesel engine using ORC technology. Two separated ORC systems with R245fa and benzene as the working fluids were studied to recover waste heat from the jacket cooling water system and engine exhaust gas. They estimated that a total net power output 101.1 kW could be reached with an efficiency increase of 10.2% for the marine diesel engine.

Katsanos et al. (2012) conducted a theoretical study to investigate the potential of an ORC coupled with the exhaust of a six-cylinder heavy-duty two-stage turbocharged truck diesel engine. Using the working fluid R245ca, their study showed that the brake specific consumption improvement ranged from 10.2% at 25% engine load to 8.5% at 100% engine load. Battista et al. (2015) investigated the interaction of an ORC mounted on the exhaust line of a turbocharged IVECO F1C engine. Their study reported some difficulties in recovering heat from the exhaust gases in the evaporator but also showed that the gross benefit of the ORC-based unit power was 4% to 5%. The study mentioned that additional improvement is possible with a model-based control and more efficient expander.

ORCs can utilize other heat sources besides waste heat from internal combustion engines. For example, Borsukiewicz-Gozdur et al. (2014), Oudkerk et al. (2013), Peris et al. (2014) studied ORCs for combined heating and power Generation.; Saitoh et al. (2007) coupled an ORC to a solar source; Hettiarachchia et al. (2007) considered a geothermal source; and Uris et al. (2015) studied biomass sources.

Expanders applied to ORCs, in general, are categorized into two types according to Qiu et al. (2011). One is the dynamic (turbo) machine type, such as an axial turbine expander; the other is a volumetric (displacement) type, such as screw expanders, scroll expanders, and reciprocal piston expanders. According to Zanelli and Favrat (1994), volumetric types are preferable for small- and medium-scale ORCs, whereas turbo machines are typically used for large-scale energy conversion due to higher costs and operation speeds.

More recently, Song et al. (2014) conducted a review of the research status of scroll expanders and cited that scroll expanders are still in the early stages of development. Most of the small capacity ORC studies use scroll compressors adapted to run as an expander. The objective of the study presented in this paper was the evaluation of the performance, power output and isentropic efficiency of a 5kW scroll expander operating with temperature sources (85°C and 110°C) and the application of modeling techniques to the expander that are useful in characterizing isentropic efficiency and power output.

2. EXPERIMENTAL SETUP

The operating principle of a simple ORC includes four major processes. The working fluid evaporates at high pressure using the heat transfer from a heat source. Then, the enthalpy of the working fluid is reduced in an expander producing mechanical work that can be used directly or converted into electricity by an electric generator. The low-pressure fluid leaving the expander is passed to a condenser, where its temperature is reduced and changes phase to liquid. From there, the liquid is pumped back to a higher pressure and fed to the evaporator.

A test-rig with a schematic representation in Figure 1 was used to characterize the performance of a scroll expander in this study. The expander tested was an open-drive expander with a nominal power capacity of 5kW. It was equipped with a magnetic coupling to minimize leakage risks. Its built-in volume ratio was 3.5 and it had a displacement volume of 73.6 cm^3 per revolution. The expander was limited to a maximum inlet pressure of 1380kPa (200PSI), a maximum temperature of 175°C, and a speed in the range of 500 RPM to 3600 RPM. The goal of the experimentation was to determine the device's efficiency and power production under oil-free operation. The sensor types and uncertainties for the test rig are summarized in Table 1.

The heat source was simulated by a water boiler and the tests were conducted with source temperatures of 85°C and 110°C. The test matrix is summarized in Table 2. For each expander speed, the tests began with the lowest pump speed, when the pressures and flow were stable, between 80 RPM to 110 RPM, leading to the lowest pressure ratio for that expander speed. The pump speed was increased incrementally over its range leading to increasing evaporating pressure, decreasing expander inlet superheat and increasing the pressure ratio across the expander. The tests for each expander speed were considered ended when one of the three scenarios happened:

- the degree of superheat was lower than 3°C;
- the limit of 23 Nm of the torque sensor was reached; or
- the pressure inlet of the expander was close to 1380kPa (200PSI).

This process was repeated for both temperature sources. Table 2 summarizes all speeds in the test matrix. In addition, tests were performed at the speed of 2400 RPM due to this point being close to a filling factor of 1.

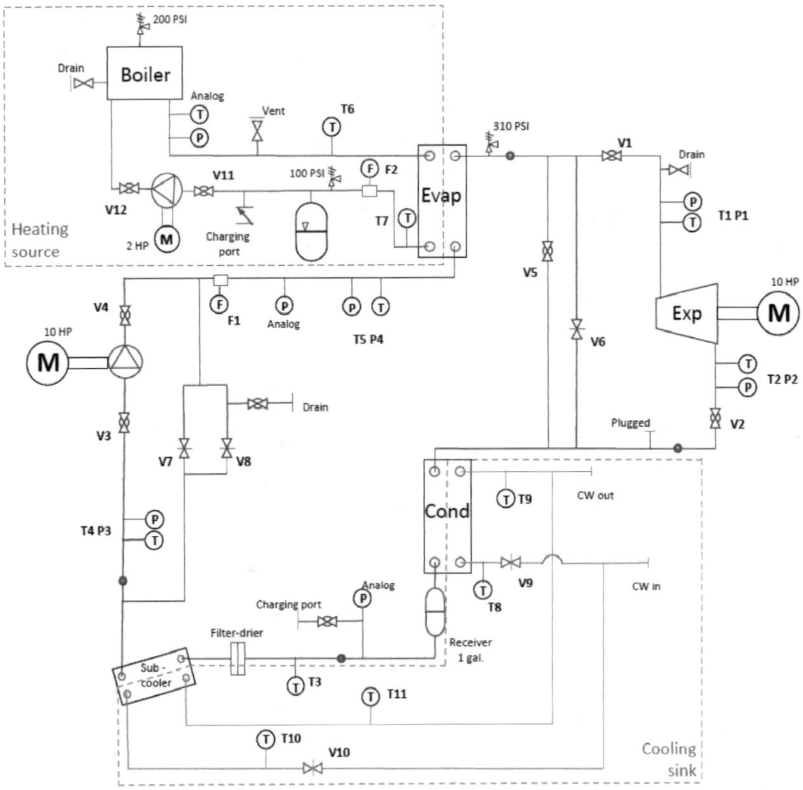

Figure 1: Layout of the ORC test-rig used to test the 5 kW scroll expander.

Table 1: Summary of the sensors and corresponding uncertainties

Measurement	Description	Accuracy
Temperature	Undergrounded T-Type thermocouple	+/- 0.5°C
Pressure (high-pressure side)	Electronic Pressure Transducer 0-500 psig	+/- 5.65 kPa (+/- 0.82 psi)
Pressure (low- pressure side)	Electronic Pressure Transducer 0-250 psig	+/- 4.105 kPa (+/-0.59 psi)
Mass flow (working fluid)	Coriolis mass flow sensor	+/- 0.1 %
Mass flow (hot water fluid)	Coriolis mass flow sensor	+/- 0.5 %
Torque	Rotating torque sensor	+/- 1.5 %
Shaft Speed	VFD Encoder	+/- 6.84 RPM

The determination of a steady state condition was based on the method developed by Woodland et al. (2012). They proposed a method that compares the average values of the measured variables over windows that are separated by 10 minute intervals. The method is illustrated in Figure 2 and the criteria for steady state are given in Table 3.

Table 2: Test matrix of the expander experimental study.

Source Temperature	Expander Speed (RPM)				
85°C	800	1600	2000	2500	3000
110°C	800	1600	2000	2500	3000

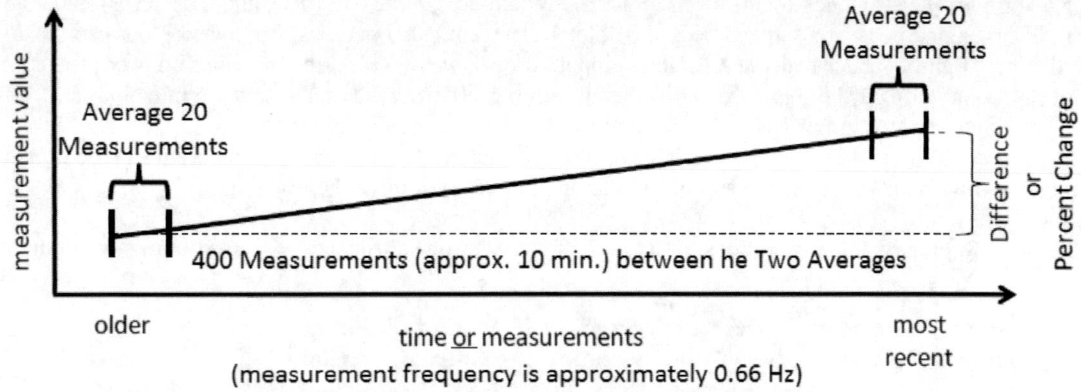

Figure 2: Steady state condition adopted for the experimental evaluation adapted from Woodland et al. (2012).

Table 3: Criteria used in the comparison in each measurement for steady state condition.

Measurement type	Steady State Criteria
Temperature	Difference < 0.5 K
Pressure	Change < 2%
Mass Flow	Change < 2%
Rotating Equipment Speed	Change < 2%

The volumetric performance of the scroll expander is quantified using the filling factor (ϕ) defined as the ratio of the actual flow rate to the theoretical flow rate displaced by the expander.

$$\phi = \frac{\dot{m}_{wf} v_{suc}}{N_{shaft} V_{D,exp}} \tag{1}$$

$V_{D,exp}$ is the theoretical displacement volume of the expander The built-in volume ratio (r_v) is a characteristic of the scroll expander dependent on its geometry.

$$r_v = \frac{V_{D,exp}}{V_{D,comp}} \tag{2}$$

The tested expander has a built-in volume ratio of 3.5. The volume ratio imposed on the expander is the ratio of the working fluid specific volume of the expander outlet by the inlet:

$$V_r = \frac{v_{dis}}{v_{suc}} \tag{3}$$

The imposed pressure ratio is calculated in analogous way:

$$P_r = \frac{P_{dis}}{P_{suc}} \tag{4}$$

The overall performance of the expander is evaluated in terms of the expander isentropic efficiency:

$$\eta_{exp} = \frac{\dot{W}_{shaft}}{\dot{W}_s} = \frac{2\pi N_{shaft} \tau_{shaft}}{\dot{m}_{wf}(h_{suc}(T_{suc}, P_{suc}) - h_{dis}(s_{suc}, P_{dis}))} \tag{5}$$

3. EXPERIMENTAL RESULTS

During the experimental test phase, 75 data points were collected. They consisted of 18 points that were run with the temperature source at 85°C, and 57 points from temperature source at 110°C. Figures 3 shows the isentropic efficiency versus volume ratio and filling factor. The peak efficiency of 0.58 occurred at a volume ratio of about 6 and filling factor of about 1.2.

The highest isentropic efficiency found was 0.58 for a temperature source of 110°C, and 0.51 for a temperate source of 85°C. These conditions are summarized in Table 4. The optimal operating efficiencies correspond to filling factors that are slightly greater than one. For the temperature source of 110°C the maximum power production, 3.75 kW, did not coincide with the maximum expander isentropic efficiency. For the temperature source of 85°C, the point of maximum power and efficiency were found to be at the same operating condition. However, it is possible that a larger pressure ratio was necessary to achieve peak efficiency at this operating temperature. The pressure ratio was limited at this condition due to constraints on the operating condition of the test stand.

Table 4: Summary of the conditions for best expander isentropic efficiency and maximum power output.

Temp. Source	Expander Speed	Degree Superheat	Pressure ratio	Volume Ratio	Filling Factor	Power	Expander Isen. Eff.
[°C]	[RPM]	[°C]	[-]	[-]	[-]	[kW]	[-]
85.20	2000	1.83	4.80	5.02	1.10	**1.64**	**0.51**
110.41	1600	21.74	5.95	6.12	1.20	2.34	**0.58**
110.19	2500	4.20	5.92	6.55	1.00	**3.75**	0.54

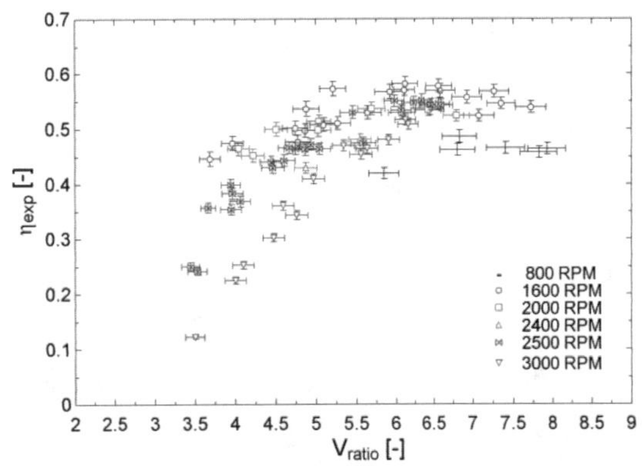

Figure 3: Isentropic Expander Efficiency versus Volume Ratio. Efficiency Peak occurs at volume ratio 6.12. Plot created using all the 75 data points.

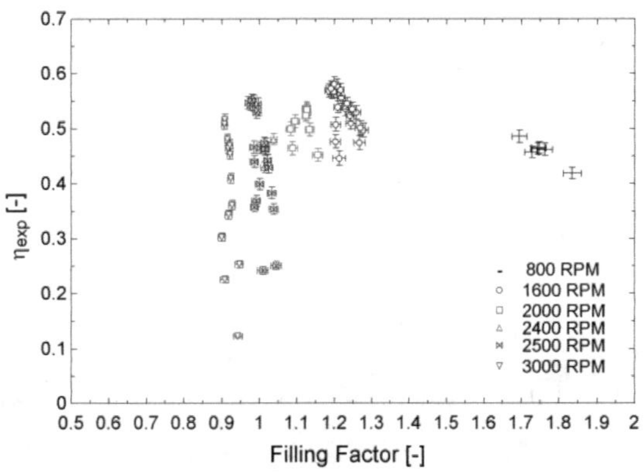

Figure 4: Isentropic Expander Efficiency versus Filling Factor. Efficiency Peak occurs at filling factor of 1.2. Plot created using all the 75 data points.

The expander isentropic efficiency versus pressure ratio is shown for each temperature source in Figure 5 and Figure 6. For the temperature source of 110°C, the expander speed of 1600 RPM consistently shows better efficiency than the other speeds. For the temperature source of 85°C, the expander speed of 2000 RPM yielded higher efficiency than the other speeds. However, close to the expander isentropic efficiency peak of 0.51, the efficiencies at expander speeds of 1600 RPM and 2000 RPM were nearly identical. In both situations, the inlet pressure conditions couldn't be increased due to the degree of superheat at these points. At a speed of 2000 RPM it was 1.83°C and the condensing pressure 155.8 kPa, while for 1600RPM the degree of superheat was 2.4 °C and the condensing pressure 150.2 kPa. However, increasing the pump speed to increase the pressure ratio also elevates the expander inlet pressure and the mass flow rate, which could lead to two phase flow in the expander. An alternative method to test larger pressure ratios for these two speeds would be to lower the condensing pressure. However, the condensing pressure is limited by the sink temperature conditions and since there was no control over the sink water temperature and the flow rate was at a maximum, higher pressure ratios could not be tested.

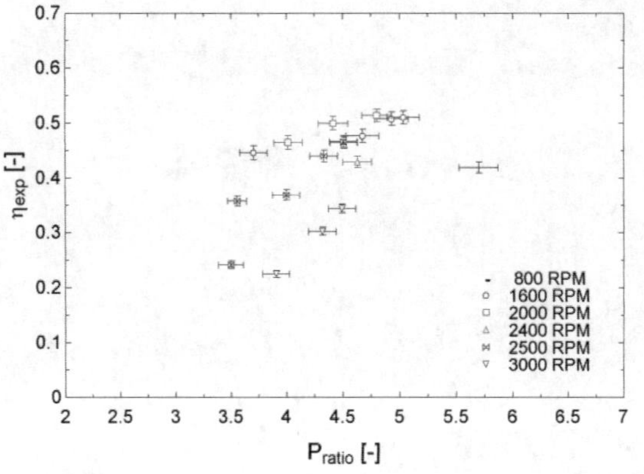

Figure 5: Expander Isentropic Efficiency versus Pressure Ratio for Temperature Source of 85°C.

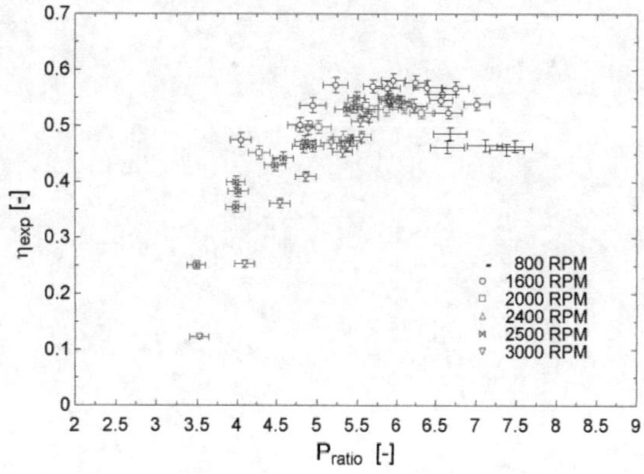

Figure 6: Expander Isentropic Efficiency versus Pressure Ratio for Temperature Source of 110°C.

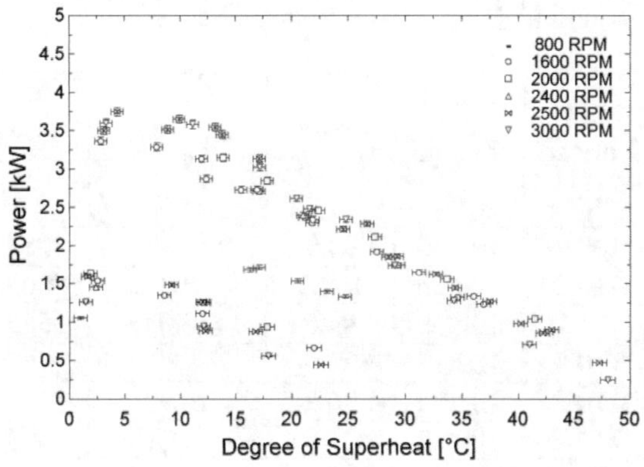

Figure 7: Expander Power Output Versus Degree of Superheat at Expander Inlet. Upper trend represents temperature source of 110°C, while lower trend represents temperature source of 85 °C.

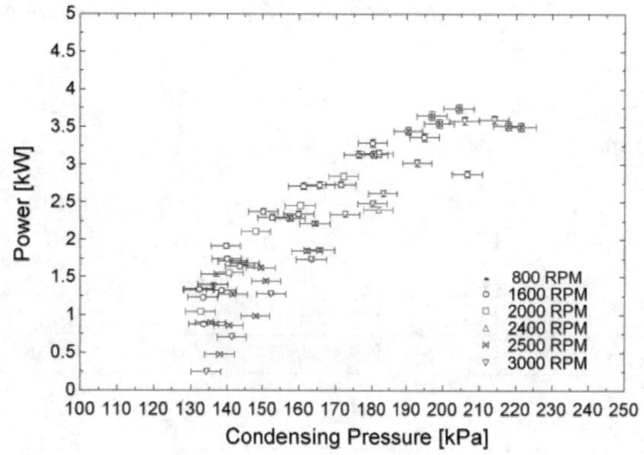

Figure 8: Expander Output Power Versus Condensing Pressure for Temperature Source 110°C.

Figure 7 presents expander power output as a function of inlet superheat. The power output increased with decreasing inlet superheat for all the tested expander speeds but was limited to be greater than zero. Lower inlet superheats are generally associated with higher pump speeds and refrigerant mass flow rates, which lead to higher evaporating pressures. In turn, this leads to higher pressure ratios and density of the working fluid entering this positive displacement expander. Figure 7 shows that the power output is nearly linear with degree of superheat until the superheat approaches zero where power output tends toward a maximum. At this point the pressure ratio across the expander no longer increases with increasing pump speed. This is due to a limitation on the cooling capacity of the test stand. While the evaporation pressure increases due to the increased pump speed and mass flow rate the condensing temperature and hence saturation pressure also increases.

Figure 8 shows power output as a function of condensing pressure for the higher source temperature with a peak output occurring at a condensing pressure of about 200 kPa. This peak power operating point is associated with the maximum pressure ratio over this range of data. There is a maximum pressure ratio for each expander speed for fixed sink conditions (fixed flow rate and condenser inlet temperature). Figure 9 shows that the maximum pressure ratio for the expander speed of 2500RPM occurs at the 200 kPa condensing pressure. The peak occurs because both evaporating and condensing pressures increase with increasing working fluid mass flow rate towards limiting values that depend on the source and sink temperatures but at rates that depend on heat transfer characteristics of the evaporator and condenser.

It should be noted that tests for the expander speed of 800 RPM at a temperature source of 110°C reached the limit of the torque cell. At this point, the superheat was 16.4°C, indicating that higher condensing pressures could have been tested for this speed if higher rated torque cell were employed.

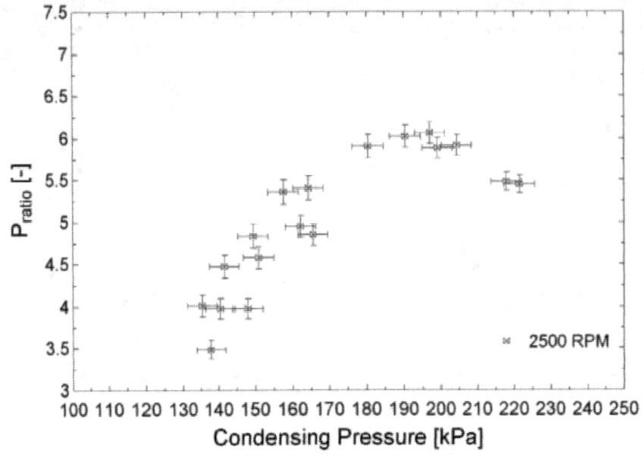

Figure 9: Trend of the Pressure Ratio versus the Condensing Pressure as the Mass Flow Rate Increases.

Table 5: Conditions close to the power limit and minimum degree of superheat.

Expander Expeed	**Suc. Pressure**	**Dis. Pressure**	**P. Ratio**	**Density R245fa**	**Temp. Source**	**Suc. Temp.**	**Suc. Sat. Temp.**	**Superheat**	**Power**
[RPM]	[kPa]	[kPa]	[-]	[kg/m^3]	[°C]	[°C]	[°C]	[°C]	[W]
1600	1262.68	180.15	7.01	68.50	109.8	107.66	99.9	7.74	3285.71
1600	1279.89	194.91	6.57	71.97	110.2	103.23	100.5	2.70	3362.33
2500	1170.88	199.01	5.88	61.20	110.2	109.59	96.6	13.03	3545.15
2500	1208.66	204.32	5.92	66.88	110.3	102.17	98.0	4.20	3745.85
3000	1161.90	205.89	5.64	61.43	110.2	107.22	96.2	10.99	3577.71
3000	1183.83	214.00	5.53	65.85	110.4	100.22	97.0	3.17	3593.62

4. Expander Modeling

To complement the experimental evaluation of the scroll expander, an expander model was developed from the data collected. A semi-empirical model is used to evaluate the expander's performance and estimate the expander losses based on the determined parameters.

4.1 Semi-Empirical Model

This section presents the semi-empirical model introduced by Lemort et al (2009) and its application to the scroll expander tested. The model is able to predict power, isentropic efficiency, mass flow and outlet temperature of the expander for different operating conditions. The model considers that one cycle of the scroll expander includes suction, expansion and discharge processes with the following physical phenomena:

- su → su,1: working fluid suction pressure drop;
- su,1 → su,2: isobaric suction heat transfer;
- su,2 → in: isentropic expansion to the built-in volume ratio;
- in → ex,3: adiabatic expansion at constant machine volume;
- ex,2 → ex,1: isobaric discharge heat transfer;
- ex,1 → ex: discharge pressure drop

The model has ten parameters that need to be determined:

$$AU_{amb},\ AU_{suc.n},\ AU_{ex.n},\ \dot{m}_{r.n},\ V_{s.exp},\ V_{r.in},\ \tau_{loss},\ A_{thr.leak},\ A_{thr.suc},\ A_{thr.dis}$$

The parameters of the model are then identified by minimizing a global error function accounting for errors in the prediction of the main output variables of the model (mass flow rate, shaft power and exhaust temperature):

$$error = \sqrt{\sum_{1}^{N_{data}} \left(\frac{\dot{m}_{calc} - \dot{m}_{meas}}{\dot{m}_{calc}}\right)^2} + \sqrt{\sum_{1}^{N_{data}} \left(\frac{\dot{W}_{calc} - \dot{W}_{meas}}{\dot{W}_{calc}}\right)^2} + \sqrt{\sum_{1}^{N_{data}} \left(\frac{T_{dis,calc} - T_{dis,meas}}{T_{dis,calc}}\right)^2} \quad (6)$$

The generic algorithm implemented by James et al. (2015) was used to perform the error minimization using 10 experimental data points. The resulting model parameters are given in Table 6.

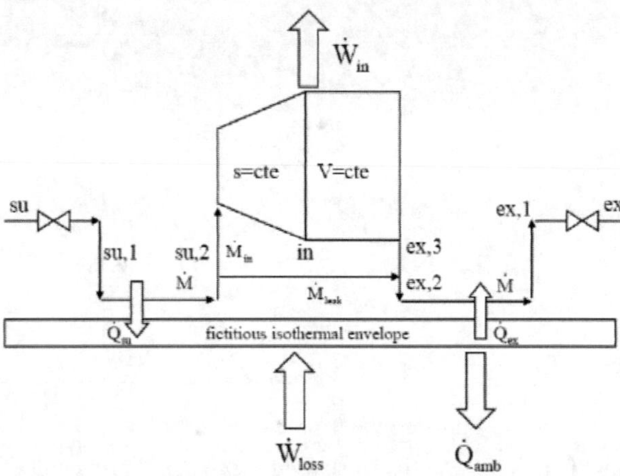

Figure 10: Conceptual scheme of the expander model. Source: Lemort et al (2009)

Table 6: Parameters determined by the minimization of the semi-empirical model.

Semi-Emprical Parameters		
$\dot{m}_{r,n}$	0.1384	[kg/s]
$AU_{suc.n}$	28.3949	[W/K]
$AU_{ex.n}$,	11.7066	[W/K]
AU_{amb}	6.1725	[W/K]
$V_{s,exp}$	8.10E-05	[m^3]
$V_{r,in}$	3.3009	[-]
$A_{thr,suc}$	4.01E-05	[m^2]
$A_{thr,leak}$	7.43E-06	[m^2]
$A_{thr,dis}$	0.000267	[m^2]
τ_{loss}	2.9680	[Nm]

As shown in Figure 11, the semi-empirical model predicts the isentropic efficiency with mean deviation of 4.93 % and a maximum relative error of 17.75% for all the data points. However, the maximum prediction errors for 69 out of the 75 experimental data points are less than 10%. The power output of the expander is predicted with a maximum deviation of 15% and a mean deviation of 4.23%. Figure 12 indicates that the highest relative deviation occurs at low power output of less than 1 kW. For higher output power points, the deviation falls inside 8%.

The mass predictions are in good agreement with the experimental data with a maximum deviation of 8.06% and a mean deviation of 2.2%. However, the model tends to somewhat under predict the mass flow rates at higher values as seen in Figure 13.

Figure 14 shows comparisons of expander outlet temperatures. The model tends to over predict the expander output temperature with a maximum difference of 4.38 K that occurs at the highest temperature. Besides this deviation, all the other 74 data points are predicted with deviation in the interval +3.5 K to -2K. At higher expander speeds the accuracy of the model decreases. High expander outlet temperatures are associated with lower power outlet, and this deviation for high outlet temperatures agrees with the lower accuracy of the power prediction shown in Figure 12.

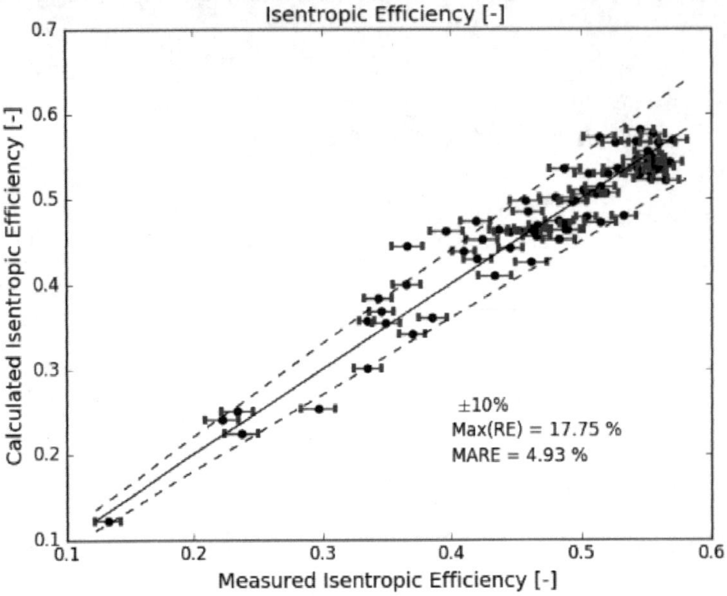

Figure 11: Calculated Isentropic Efficiency versus Measured Isentropic Efficiency of the Semi-Empirical Model with Parameters from Table 6.

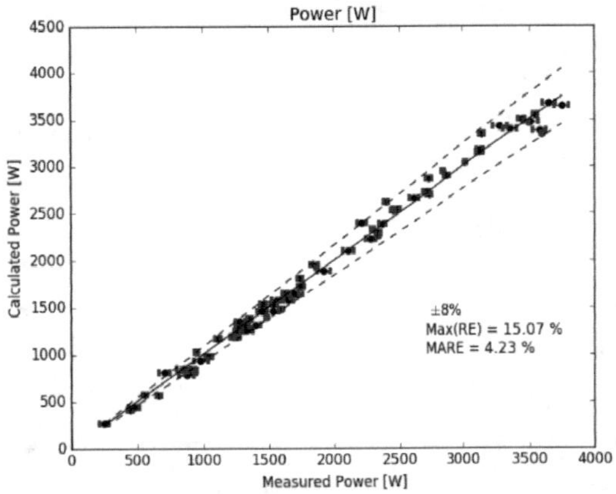

Figure 12: Calculated Output Power versus Measured Output Power of the Semi-Empirical Model with Parameters from Table 6.

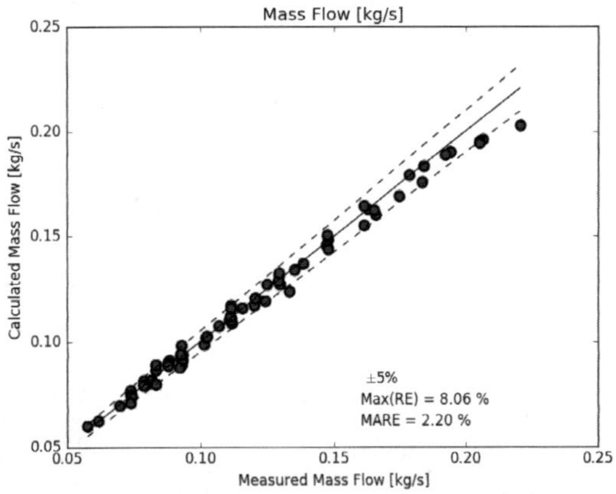

Figure 13: Calculated Mass Flow Rate versus Measured Mass Flow Rate of the Semi-Empirical Model with Parameters from Table 6.

Figure 15 presents model predicted trends of isentropic efficiency for different cases that exclude different types of expander losses. The bottom curve includes all of the losses and then subsequent curves remove sequentially leakage, pressure drop, mechanical, and heat transfer losses. The top curve only includes losses associated with under- or over-compression when the volume ratio is not equal to the built-in volume. For pressure ratios lower than about 5, mechanical losses (friction and bearing losses) have the most significant effect on overall expander performance. These losses cause the location of the expander peak efficiency to shift from its build-in volume ratio to a higher volume ratio that has a corresponding pressure ratio of about 6. The other major source of losses is due to internal leakage, which is leakage between scroll chambers across radial and flank gaps. These losses have the largest relative impacts for pressure ratios above about 5. Losses due to heat transfer are relatively small and could be neglected. Losses due to pressure drop are more important at lower pressure ratios but could be neglected for pressure ratios greater than about 6.

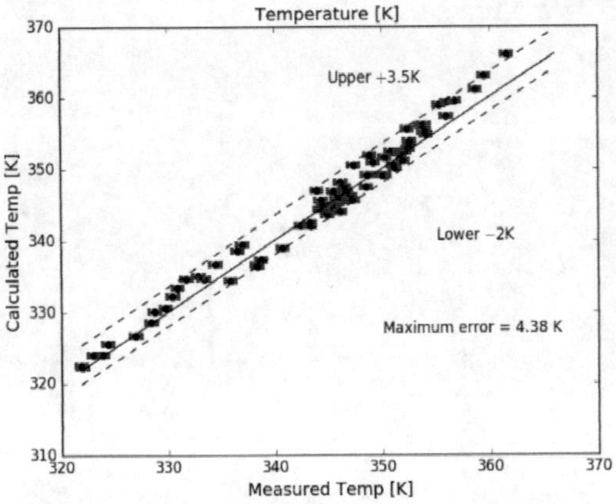

Figure 14: Calculated Outlet Temperature versus Measured Outlet Temperature of the Semi-Empirical Model with Parameters from Table 6.

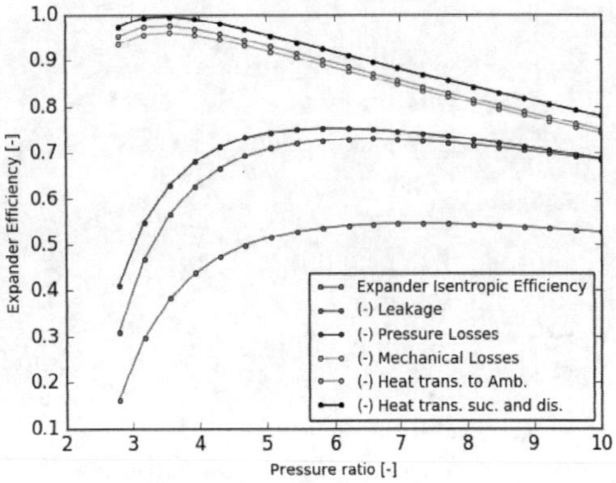

Figure 15: Isentropic Efficiency versus Pressure Ratio. From the Bottom Curve to the Top Curve, The Effects of the Losses Are Shown. Each Curve Accounts for the Loss Effects from its Antecedent Curves.

6. CONCLUSIONS

A set of 75 data points were collected to characterize a 5 kW scroll expander's performance for two temperature sources, 85°C and 110°C. Five expander speeds were considered in the test matrix, Table 2. At each expander speed, pump speeds were varied leading to a range of evaporating pressures, condensing pressures, and expander inlet superheat values with limits set on maximum expander inlet pressure, minimum expander inlet superheat, and maximum shaft torque sensor output.

The conditions of maximum measured isentropic efficiency and power output of the expander are summarized in Table 4. The maximum expander isentropic efficiency of 0.58 was measured for an expander speed of 1600 RPM and temperature source of 110°C. The peak efficiency point can also be seen in Figure 3. For the temperature source of 85°C, the maximum expander isentropic efficiency determined during testing and shown in Figure 5 also corresponded to the maximum power output achieved and occurred at the maximum pressure ratio tested for that operating speed. Therefore, it cannot be affirmed that this is the maximum isentropic efficiency for this expander speed without testing larger pressure ratios. Larger pressures ratios were not possible due to constraints imposed by the minimum inlet superheat and the minimum possible condensing pressure associated with the sink temperature.

Although the expander has a rated nominal maximum capacity of 5 kW, the maximum power output obtained during testing was 3.75 kW. To achieve the nominal power capacity, larger pressure ratios would be necessary. This would require either higher source temperatures or lower sink temperatures. The expander is rated for temperatures up to 175°C and could indeed be tested with higher temperature sources. However, the hot water loop in the current test setup would need to be modified, since the current expansion tank is limited it to 115°C.

An optimum pressure ratio for peak efficiency was found for a source temperature of 110°C, but peak efficiency was not achieved for 85°C with the range of conditions tested. The optimum pressure ratio for the lower source temperature could not be achieved because the maximum pressure ratio for the test stand was limited by the control approach.

The semi-empirical modeling approach of Lemort et al (2009) was applied to the scroll expander and accurately characterizes the performance in terms of mass flow rate, power output, and efficiency. The model was used to analyze the major loss mechanisms. For this expander, leakage and mechanical losses have the most significant impact on overall efficiency and therefore should be addressed first in improving performance.

NOMENCLATURE

h	specific Enthalpy	(J/kg)
$\dot{m}$	Mass flow rate	(kg/s)
N	Expander speed	(rev/sec)
P	Pressure	(kPa)
s	Entropy	(J/ kg °C)
T	Temperature	(°C)
V	Volume	(m^3)
v	Specific Volume	(m^3/kg)
$\dot{W}$	Power	(W)
ϕ	Filling Factor	(-)
η	Expander Isentropic Efficiency	(-)
τ-	Torque	(Nm)
AU	Heat Transfer Coefficient	(W/m)
A	Area	(m^2)

Subscript

amb	Ambient
D, exp	Expander displacement expansion volume
$D, comp$	Expander displacement compression volume
Exp	Expander
r	Ratio
r, in	Built-in
s	Isentropic
suc	Suction
thr	Throat
wf	Working fluid

REFERENCES

Battista, D.; Mauriello, M.; Cipollone, R. (2015): Waste heat recovery of an ORC-based power unit in a turbocharged diesel engine propelling a light duty vehicle. *Applied Energy*, *152*, 109–120.

Borsukiewicz-Gozdur, A.; Wisniewski, S.; Mocarski, S.; M. Bankowski (2014): ORC power plant for electricity production from forest and agriculture biomass. *Energy Conversion and Management, 87*, Pages 1180–1185.

Hettiarachchia, H. D. M., Golubovica, M., Woreka, W. M., Ikegamib, Y. (2007). Optimum design criteria for an Organic Rankine cycle using low-temperature geothermal heat sources. *Energy, 9*,1698-1706.

James, N.; Braun, J.; Groll, E.; Travis, W. (2016): Semi-empirical modeling and analysis of oil flooded R410A scroll compressors with liquid injection for use in vapor compression systems.; *International journal of refrigeration, 66*, 50–63.

Katsanos, C.; Hountalas, D.; Pariotis, E. (2012): Thermodynamic analysis of a Rankine cycle applied on a diesel truck engine using steam and organic medium. *Energy Conversion and Management, 60*, 68–76.

Lemort, V.; Quoilin, S.; Cuevas, C.; Lebrun, J. (2009): Testing and modeling a scroll expander integrated into an Organic Rankine Cycle. *Applied Thermal Engineering, 29*, 3094–3102

Oudkerk, J.; Quoilin, S.; Declaye, S.; Guillaume, L.; Winandy, E.; Lemort V. (2013): Evaluation of the Energy Performance of an Organic Rankine Cycle-Based Micro Combined Heat and Power System Involving a Hermetic Scroll Expander. *Journal of Engineering for Gas Turbines and Power, 135*.

Peris, B.; Navarro-Esbrí, J.; Molés, F.; Martí, J. (2015): Mota-Babiloni, A.: Experimental characterization of an Organic Rankine Cycle (ORC) for micro-scale CHP applications. Applied Thermal Engineering, 79, pages 1-8.

Qiu, G.; Liu, H.; Riffat, S. (2011): Expanders for micro-CHP systems with organic Rankine cycle. *Applied Thermal Engineering, 31*, 3301 to 3307.

Rattner, A.; Garimella, S. (2011): Energy harvesting, reuse and upgrade to reduce primary energy usage in the USA. *Energy, 36*, 6172 to 6183.

Saioth, T.; Yamada, N.; Wakashima, S. (2007): Solar Rankine Cycle System Using Scroll Expander. *Journal of Environment and Engineering, 2*, No. 4.

Song, P.; Wei, M.; Shi, L.; Danish, S.; Ma, C. (2015): A review of scroll expanders for organic Rankine cycle systems. *Applied Thermal Engineering, 75*, 54 to 64.
U.S. Energy Information Administration (20015): Annual Energy Outlook 2015 (AEO2015) – http://www.eia.gov/forecasts/archive/aeo15/
Uris, M., Linares J. I., Arenas, E. (2015): Size optimization of a biomass-fired cogeneration plant CHP/CCHP (Combined heat and power/Combined heat, cooling and power) based on Organic Rankine Cycle for a district network in Spain. *Energy, 9*,935–945.
Woodland, B.; Braun, J.; Groll, E.; Horton, W. (2012): Experimental Testing of an Organic Rankine Cycle with Scroll-type Expander. *International Refrigeration and Air Conditioning Conference at Purdue, July 16-19*, 2012.
Zanelli, R.; Favrat, D. (1994): Experimental Investigation of a Hermetic Scroll Expander-Generator. *International Compressor Engineering Conference*, Paper 1021.

ACKNOWLEDGEMENT

The financial support from the CAPES Foundation, Ministry of Education of Brazil, is greatly appreciated.

Improvement of the Efficiency of Twin-Screw Refrigeration Compressors by means of Dual Lead Rotors

Matthias Utri[1]*, Andreas Brümmer[2]

[1]TU Dortmund University, Chair of Fluidics,
Dortmund, Germany
Phone: +49 231 755 5724, Fax: +49 231 755 5722
E-mail: matthias.utri@tu-dortmund.de

[2]TU Dortmund University, Chair of Fluidics,
Dortmund, Germany
Phone: +49 231 755 5720, Fax: +49 231 755 5722
E-mail: andreas.bruemmer@tu-dortmund.de

* Corresponding Author

ABSTRACT

The paper deals with a theoretical investigation of screw machines applied as refrigerant compressors, focusing on the compression process. The refrigerant used is R227ea with a pressure ratio of four. A wide range of geometric parameters such as rotor length, internal volume ratio and wrap angle are varied in order to show their influence on efficiency, with a view to finding out how energy conversion can be improved by using dual lead rotors. These rotors possess two segments with different rotor lead in order to optimize the progression of chamber volume, clearance size and outlet area. The rotor profile under investigation is an asymmetric SRM-profile whereas the diameter of the rotors as well as the clearance heights remain constant throughout the simulation. The impact of the machine geometry on efficiency is examined by performing a multi-chamber simulation. The improvement in energy conversion that can be achieved by using a dual lead is quantified in terms of isentropic efficiency and delivery rate. Especially for greater length-to-diameter ratios, benefits in efficiency and mass flow rate may be expected.

1. INTRODUCTION

Screw machines are rotary positive displacement machines that are used in various applications for gas compression. The machine basically consists of two counter rotating rotors that are enclosed in a tight-fitting casing. Apart from its physical dimensions, the efficiency of a screw compressor depends to a great extent on its geometrical parameters, e.g. lobe count, length-to-diameter ratio, internal volume ratio, wrap angle and the rotor profile itself due to their influence on the main dissipative effects, throttling and volumetric clearance losses. In particular, the wrap angle needs to be adjusted individually for the intended application because it affects dissipation in two opposing ways: A major twisting of the rotors decreases the time-dependent change in chamber volume and enlarges the port areas of the machine, thus leading to reduced throttling effects, but increasing clearance mass flows due to greater clearance areas and a longer working cycle. The idea of combining the advantages of large and small wrap angles by using non-uniform rotor leads was patent-registered in the 1960's (Gardner, 1969). Rane *et al.* (2014) performed a CFD simulation which presented advantages in energy conversion by varying the rotor lead continuously from the low to the high pressure intersection. Other studies (Utri and Brümmer, 2014; Fost, 2003) focus on the application of screw expanders and also possess potential, especially for high pressure applications, by using two different lead-segments on one rotor ("dual lead rotors").

2. REFRIGERATION AND THE SCREW COMPRESSOR CYCLE

Screw compressors are widely used in refrigeration. The purpose of refrigeration cycles is to refrigerate a cooling chamber by absorbing its heat by means of a refrigerant and thereby changing its state from fluid to vapor (Fig. 1, state 1→2). To be re-liquefied the refrigerant needs to be pressurized to the condenser-dependent temperature (2→3). Next the fluid is condensed by transferring heat to the environment (3→4). Subsequently, the refrigerant is throttled back to the low pressure conditions (4→1).
The change in state which is investigated within the scope of this paper is the compression stroke from state 2 to state 3, which is performed by the screw compressor. The idealized working cycle of this compressor is illustrated in the indicator diagram in Fig. 1. The working cycle can be divided into three characteristic phases: During suction, the working chamber increases in volume and is filled with the refrigerant through the low pressure port area. After reaching maximum chamber volume, the working chamber is disconnected from the low pressure port and the compression phase begins. The refrigerant is now compressed by the decreasing chamber volume. During this phase, the working chamber is almost closed. It only interacts with other chambers and the machine ports through clearances. When it reaches the high pressure control edges, the working chamber connects with the high pressure port and the discharge phase begins. The fluid is forced out through the high pressure port as the chamber volume decreases.

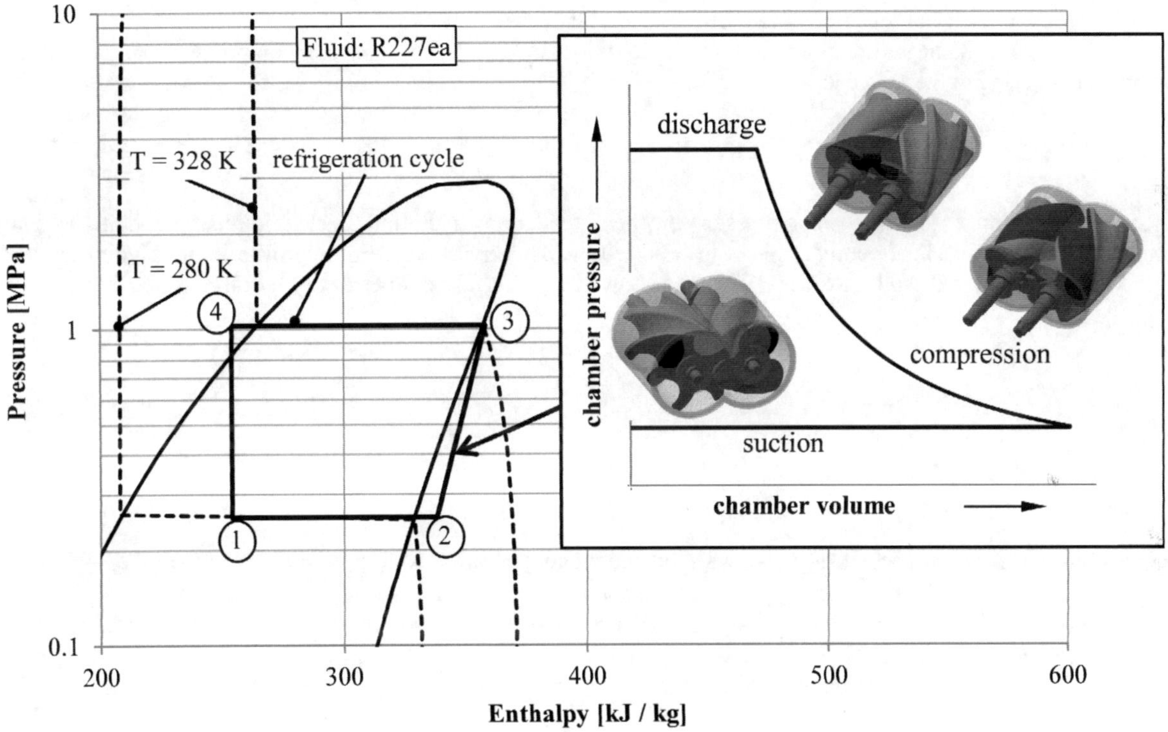

Figure 1: p-h-diagram of refrigeration cycle, idealized working cycle of a screw compressor

3. SIMULATION PRINCIPLE

The thermodynamic simulation of the screw compressor is carried out by performing a multi-chamber simulation mainly based on mass and energy conservation. Multi-chamber simulations are already used to simulate the behavior of displacement machines in different kinds of application (Janicki, 2007; Rohe, 2004) and present an alternative to the more complex CFD simulation that is commonly used for the simulation of turbo machines. In contrast to single-chamber simulation, the fluid state inside all chambers is determined simultaneously. The basis of multi-chamber simulation is a zero-dimensional chamber model which includes all relevant properties of the machine being examined. The relevant parameters are the volume of the working chamber and its connections as a function of the rotation angle. The connections of the working chamber are opening areas, which connect the

working chamber with the high and low pressure ports, and clearance areas, which connect the working chamber with other chambers as well as the ports. The fluid state inside all working chambers is assumed to be homogeneous. The optimization we are aiming for requires a large number of machines to be simulated. A new and rapid approach of chamber model generation has been achieved. For a demonstration machine a complex geometrical analysis has been carried out, supplying all necessary information such as volume curve and clearance areas. Assuming constant rotor profile and diameter, the progression of chamber volume and clearances depends entirely on rotor lead and length. Thus the machine under examination can be used to generate chamber models of machines with different rotor lengths and leads by "recombining" the chamber model of the machine that has already been analyzed. This approach eliminates the major time penalty of chamber model generation and is highly suitable for optimization tasks which necessitate a high number of chamber models.
The thermodynamic simulation was carried out by using the simulation tool "KaSim", which was developed at the Chair of Fluidics at the TU Dortmund University (Kauder *et al.*, 2002; Janicki, 2007; Temming, 2007). The results of the simulation are given as integral values, e.g. for mass flow and internal power, as well as the indicator diagram of the simulated rotary positive displacement machine. The connections maintain an exchange of mass and energy between chambers and ports, thereby influencing pressure and temperature inside all working chambers. In turn, the state inside the working chambers influences the mass flow. Due to this interaction, an iterative calculation of the state inside the working chambers is required. Mass flow is calculated by means of the equation introduced by St. Venant and Wantzel (Sigloch, 2009) which assumes an adiabatic and frictionless flow from a volume with higher pressure through a flow cross-section to a volume with lower pressure. Taking a flow involving friction into account, a flow coefficient α is used to reduce the theoretical mass flow.
Electrical and mechanical losses are not considered in the simulation and the compressor is assumed to be hermetically sealed to the ambience.

4. BOUNDARY CONDITIONS AND VARIATION PARAMETERS

In order to optimize the screw compressor for the intended refrigeration cycle, boundary conditions for the simulation and the machine geometry need to be clarified. The investigated rotor profile is the asymmetric SRM-profile as described in detail by Rinder (1979) and shown in Fig. 2. The diameters of the rotors are equal.

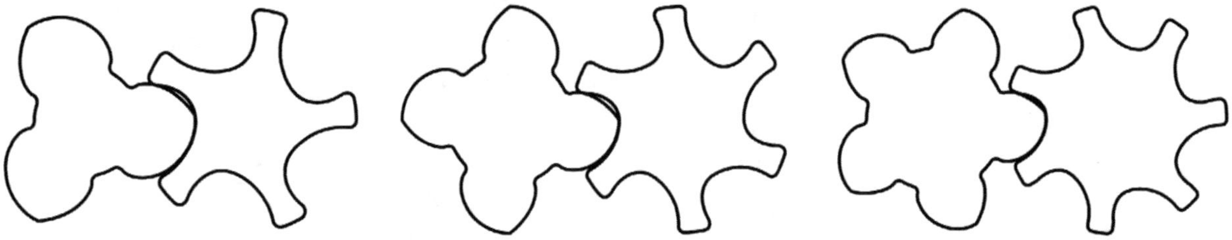

Figure 2: Asymmetric SRM-profile with lobe count combinations of 3+5, 4+6 and 5+7 with $D_{MR} = D_{FR}$

The boundary conditions which remain constant throughout all simulations are summarized in Table 1. The investigated refrigerant is R227ea with a vaporization temperature of 280 K, leading to a low pressure level of 0.25 MPa. The refrigerant temperature in the condenser is set to 328 K, resulting in a high pressure of 1.02 MPa.

Table 1: Constant boundary conditions

Parameter [dimension]	Value
Cycle parameters:	
- Fluid [-]	R227ea
- Compressor inlet temperature [K]	291
- Compressor inlet pressure [MPa]	0.25
- Compressor outlet pressure [MPa]	1.02
Machine parameters:	
- Male rotor crown circle [mm]	144
- Clearance heights [mm]	0.1
Operating parameters:	
- Circumferential speed [m/s]	35

The diameter of the male rotor is set to 144 mm with clearance heights of 0.1 mm for the housing, intermesh and front clearances, whereas the size of the blowhole and the width of the other clearances depend on the lead of the rotors and need to be calculated individually. Liquid-injected compressors are examined within the scope of this paper, so the circumferential male rotor tip speed is set to 35 m/s, leading to a rotational speed of 4642 min^{-1}. The simulation is carried out assuming adiabatic chambers.

5. OPTIMIZATION OF CONSTANT ROTOR LEAD

5.1 Variation parameters for constant lead

To reveal the potential of variable lead rotors the optimal machine for constant rotor lead is also determined using the same simulation tool "KaSim". The varied properties of the machine are shown in Table 2. Three different lobe count combinations are investigated while the length-to-diameter ratio is varied from 0.97 up to 1.8. The internal volume ratio $v_i = V_{max} / V_{compr,end}$ defines the positions of the high pressure control edges, which influence the duration of compression and discharge. The variation up to $v_i = 5$ is large enough to cover the optimal internal volume ratio for all simulated machines. The wrap angle is defined as the degree of twist between the front and the rear surface of the male rotor. The impact of the variations on the geometrical parameters of the machine, e.g. volume curve, opening and clearance areas, are described in detail by Fost (2003) and Utri and Brümmer (2014). These geometrical parameters influence the maximum chamber volume of the machine, affecting mass flow and internal power. With respect to the refrigeration cycle, each machine could realize another cooling capacity. The purpose of the variation is to optimize the screw compressor, not the refrigeration cycle itself. The auxiliary fluid, which partly seals the clearances and opening areas, is taken into account by varying the flow coefficient.

Table 2: Varied parameters for optimization of constant rotor lead

Parameter [dimension]	Minimum / maximum value	Step size
Rotor length L [mm]	140 / 260	60
Internal volume ratio v_i [-]	2 / 5	0.5
Male rotor wrap angle φ_{MR} [°]	200 / 600	50
Lobe count combination [-]	3+5, 4+6, 5+7	-
Flow coefficient α [-]	0.6 / 0.8	0.2

5.2 Simulation results

First results for a constant lobe count combination of 4+6 are shown in Fig. 3. For each point only the optimal internal volume ratio has been evaluated. Internal isentropic efficiency is defined as the ratio of the power which would be necessary in the case of isentropic compression, to the internal power of the compressor determined by multi-chamber simulation:

$$\eta_{is} = \frac{P_s}{P_i} = \frac{\dot{m}_{real} \cdot \Delta h_s}{-\oint p\,dV \cdot n_{MR} \cdot z_{MR}} \tag{1}$$

The results show that an optimal value for the wrap angle exists between 300 and 400 degrees for each rotor length. For small wrap angles, port areas are also small, resulting in throttling effects during discharge, which increase internal power. For high wrap angles throttling effects are minor, but large clearance areas reduce efficiency. This is also indicated by the delivery rate (Fig. 3) which is defined as

$$\lambda_L = \frac{\dot{m}_{real}}{\dot{m}_{theoretical}} = \frac{\dot{m}_{real}}{V_{max} \cdot \rho_{LP} \cdot n_{MR} \cdot z_{MR}}. \tag{2}$$

The delivery rate relates mass flow in the machine to the theoretical mass flow which would be achieved in the case of an ideal machine operating without clearance losses and throttling during suction. Optimal internal isentropic efficiency is achieved for the most efficient compromise between clearance losses and throttling at the outlet port.
Furthermore, the progression in isentropic efficiency indicates that longer rotors are advantageous for energy conversion, which can also be explained with reference to the delivery rate. Assuming a defined constant wrap angle, the length of the machine determines maximum chamber volume whereas the axial outlet area and the duration of the working cycle remain unaffected. Compared with shorter rotors the percentage of mass which is lost during compression due to clearance mass flow, and related to the total mass inside the working chamber, is much smaller and results in higher efficiency in terms of energy.

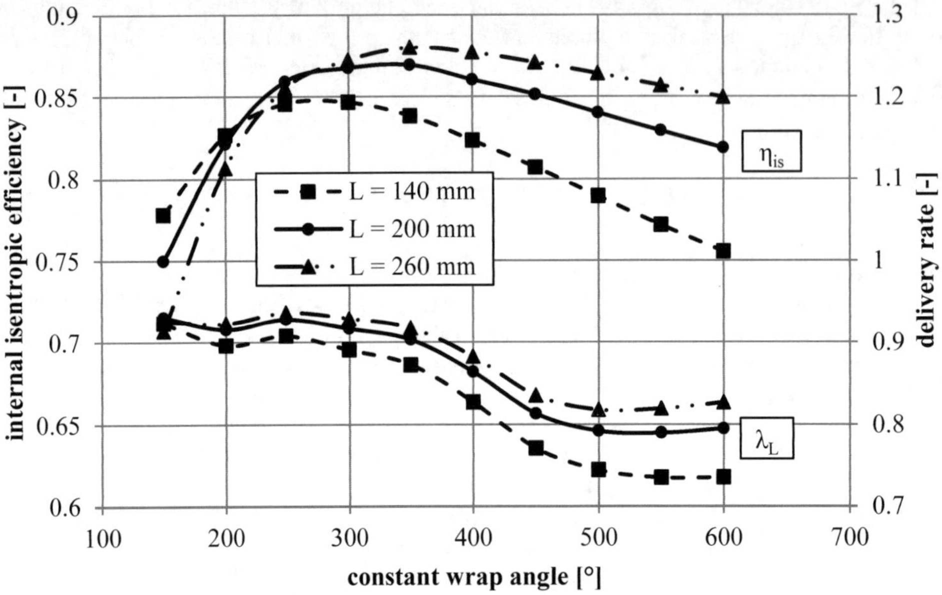

Figure 3: Internal isentropic efficiency and delivery rate as a function of wrap angle for different rotor lengths (lobe count combination 4+6, $\alpha = 0.8$, v_i optimized for each point)

Fig. 4 shows the impact of different lobe count combinations at a constant rotor length. Although the lobe count combination of 3+5 shows maximum values of delivery rate in the optimal range of wrap angle between 300 and 400 degrees, a lobe count combination of 4+6 achieves maximum values of internal isentropic efficiency up to $\eta_{is} = 88$ %. This can be explained in terms of the throttling effects which take place during discharge for the lobe count combination of 3+5, leading to an enlargement of the area in the indicator diagram (not shown), even after optimization of the internal volume ratio. The profile has small sealing lines, but the rapid decrease in chamber volume during discharge combined with a comparatively small outlet area mean that the poor efficiency of this kind of application cannot be improved. Results for other rotor lengths show similar progressions.

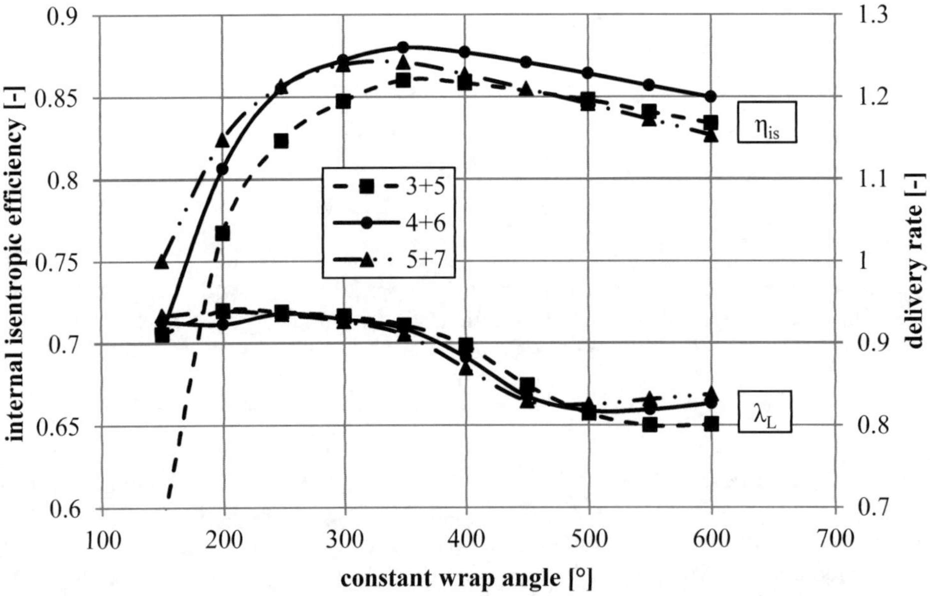

Figure 4: Internal isentropic efficiency and delivery rate as a function of wrap angle for different lobe count combinations ($L_{Rotor} = 260$ mm, $\alpha = 0.8$, v_i optimized for each point)

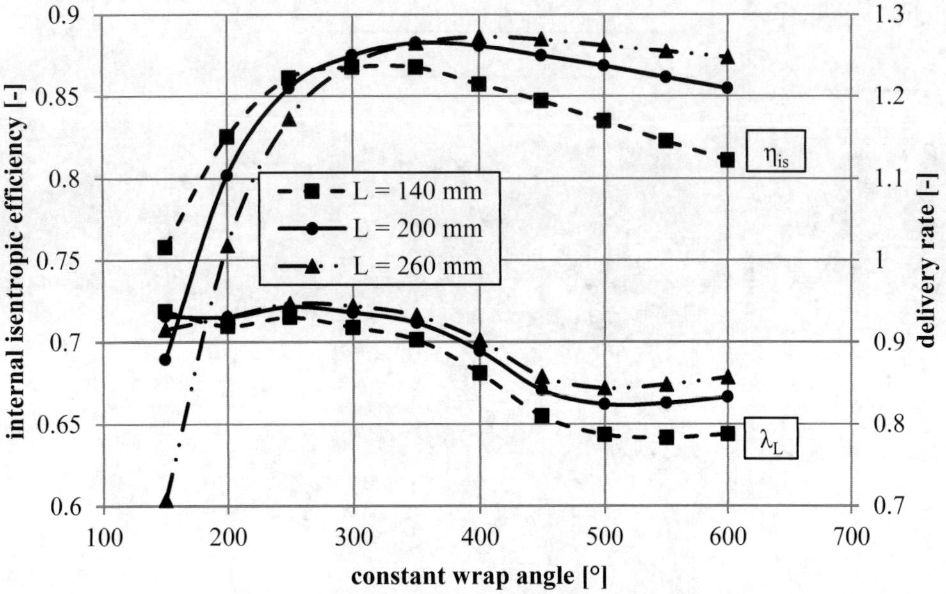

Figure 5: Internal isentropic efficiency and delivery rate as a function of wrap angle for different rotor lengths (lobe count combination 4+6, $\alpha = 0.6$, v_i optimized for each point)

Fig. 5 shows the variation of rotor length for a reduced flow coefficient of $\alpha = 0.6$. Compared with the results for $\alpha = 0.8$ (Fig. 3) the absolute values of delivery rate and internal isentropic efficiency are raised, and maximum efficiency is increased to a value of $\eta_{is} = 88.6\%$ for a rotor length of 260 mm. A lower flow coefficient reduces clearance mass flow but also increases the potential for outlet throttling. This results in higher optimal wrap angles, while, compared with Fig. 3, the influence of the wrap angle is reduced after maximum internal isentropic efficiency is achieved.

6. OPTIMIZATION OF VARIABLE ROTOR LEAD

6.1 Machine parameters for dual lead rotors

The purpose of dual lead rotors is to combine large and small wrap angles on a single rotor as shown in Fig. 6. This makes it possible to reduce the duration of the compression phase - thus decreasing clearance mass flow - and to influence both the duration of the discharge phase and the available port area in order to avoid throttling losses. Compared with constant lead rotors, this kind of geometry provides new parameters which need to be adjusted: the total wrap angle φ_{total}, which is still defined as the degree of twist between the front and rear surface, and the lengths L_{HP} and L_{LP} with their wrap angles φ_{HP} and φ_{LP} on the high and low pressure side (all wrap angles refer to the male rotor). As with the total wrap angle φ_{total} the wrap angle sections φ_{HP} and φ_{LP} are related to the overall length L_{Rotor}, which results in the following equations:

$$L_{Rotor} \cdot \varphi_{total} = L_{HP} \cdot \varphi_{HP} + L_{LP} \cdot \varphi_{LP} \tag{3}$$

$$L_{Rotor} = L_{HP} + L_{LP} \tag{4}$$

Thus, for a defined set of total length L_{Rotor}, wrap angles φ_{total}, φ_{HP} and φ_{LP}, the lengths of the sections L_{HP} and L_{LP} are clearly defined. As for a setup with a constant rotor lead, the total wrap angle together with the total rotor length define maximum chamber volume and the duration of the working cycle. The parameters which are varied to optimize the geometry of dual lead rotors are shown in Table 3. Rotor length, internal volume ratio and flow coefficient are varied according to the optimization of a constant rotor lead. In contrast, only a lobe count combination of 4+6 is examined. With respect to the optimal wrap angle range for a constant rotor lead, the total wrap angle is investigated between 200 and 400 degrees. High and low pressure side wrap angles are adjusted systematically between the lowest wrap angle of 50 and the highest wrap angle of 700 degrees. This guarantees that the optimal balance between clearance and discharge losses for each total wrap angle will be achieved. Considering all parameters, 2160 different machines were simulated for seven internal volume ratios and two flow coefficients, resulting in a total of 30240 multi-chamber model generations and machine simulations.

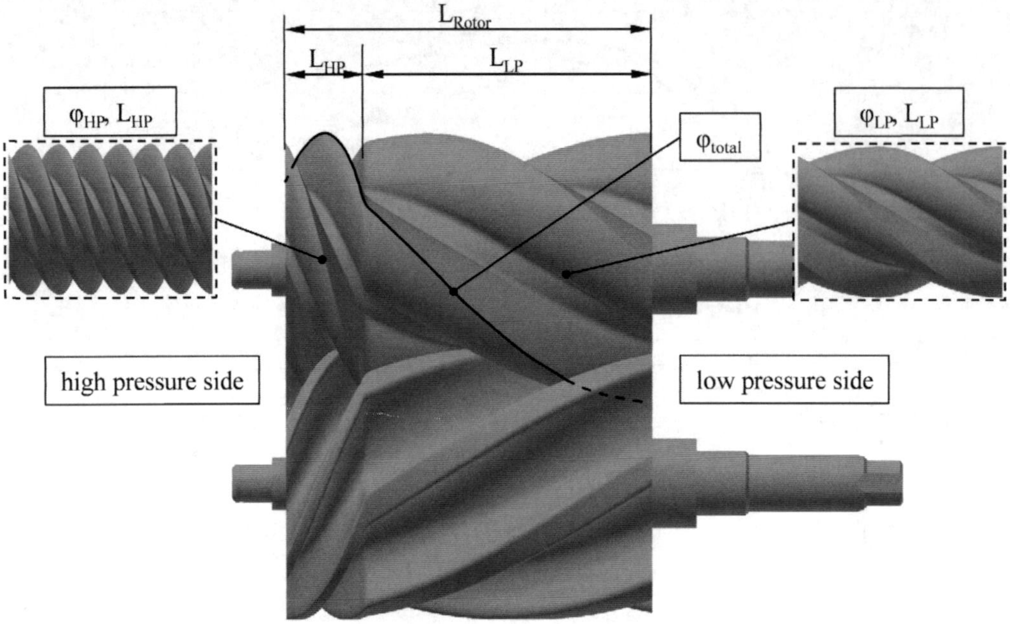

Figure 6: Screw machine with dual lead rotors

Table 3: Varied parameters for the optimization of variable rotor lead

Parameter [dimension]	Minimum / maximum value	Step size
Length L_{Rotor} [mm]	140 / 260	60
Internal volume ratio v_i [-]	2 / 5	0.5
Total male rotor wrap angle φ_{total} [°]	200 / 400	25
High pressure wrap angle φ_{HP} [°]	φ_{total} + 25 / 700	50
Low pressure wrap angle φ_{LP} [°]	50 / φ_{total} - 25	25
Lobe count combination [-]	4+6	-
Flow coefficient α [-]	0.6 / 0.8	0.2

6.1 Simulation results

The internal isentropic efficiency that can be achieved by means of dual lead rotors is shown in Fig. 7. Only maximum values for all investigated machines are shown in the diagram. As for constant lead, the small size of the outlet area for small wrap angles causes outlet throttling. For wrap angles greater than φ_{total} = 275° efficiency cannot be improved because clearance areas increase whereas maximum chamber volume decreases, resulting in a lower mass flow. The curves show significant improvement in efficiency for φ_{total} between 225 and 250 degrees. In comparison with a constant lead (Fig. 3 and Fig. 5) efficiency can be increased, especially for greater length-to-diameter ratios. Maximum efficiency η_{is} can be raised from 88.0 to 89.1 % (α = 0.8) and from 88.6 to 90.6 % (α = 0.6) by using variable lead rotors. Compared to a constant lead setup, maximum values are achieved for lower total wrap angles between 250 and 275 degrees. Dual lead rotors allow the machine to function with larger outlet port areas, which prevents outlet throttling.

The performance map in Fig. 8 shows the influence of the high and low pressure side wrap angle on internal isentropic efficiency, with the total wrap angle remaining constant. Each shell covers a range of internal isentropic efficiency of 0.5 percent. All machines examined in the diagram possess the same maximum chamber volume and length of housing, along with front and intermesh clearance, due to the constant overall wrap angle. There are differences related to the time-dependent emergence of the clearances as well as the sizes of the outlet and blowhole area, which depend to a large extend on the lead progression. The map shows that the optimal machine for the given boundary conditions should possess a wrap angle of 425 degrees on the high pressure side and 160 degrees on the low pressure side, as shown in Fig. 8. For high pressure wrap angles below 400 degrees detrimental outlet throttling occurs, which reduces efficiency.

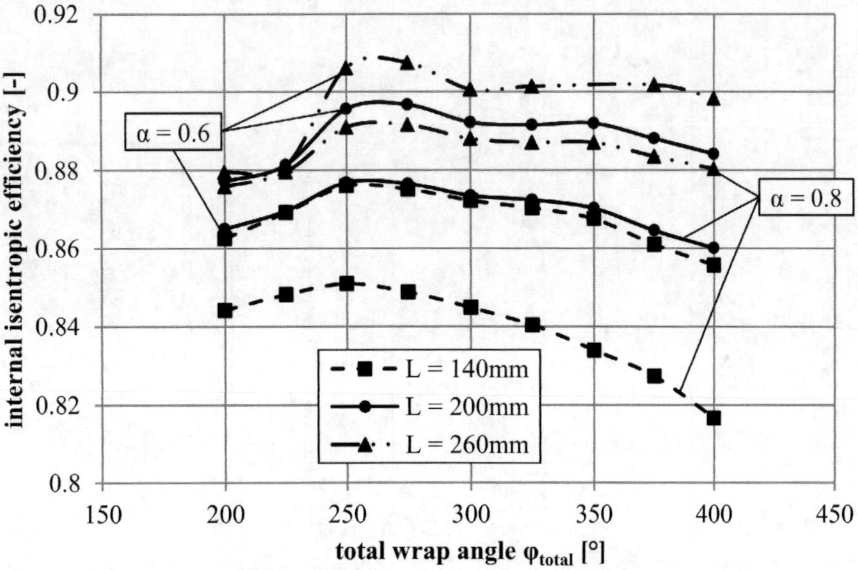

Figure 7: Internal isentropic efficiency as a function of total wrap angle for different rotor lengths and flow coefficients (lobe count combination 4+6, machine geometry optimized for each point)

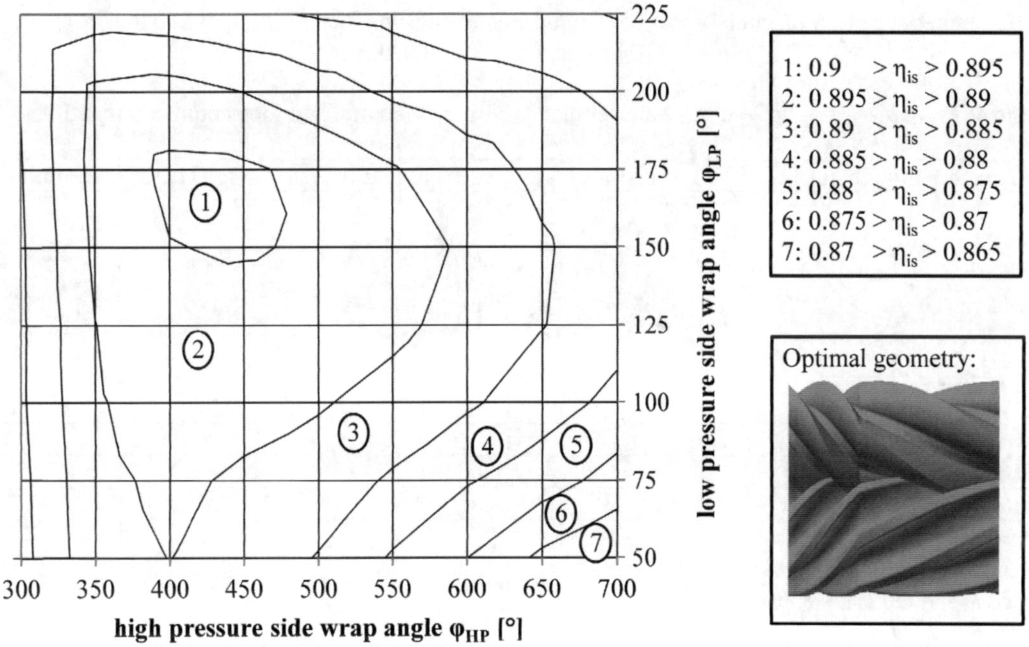

Figure 8: Internal isentropic efficiency as a function of low and high pressure side wrap angle (L_{Rotor} = 260 mm, φ_{total} = 250°, α = 0.8, lobe count combination 4+6)

The decrease in efficiency for high pressure wrap angles above 450 degrees can be explained by referring to the progression in the delivery rate, which is shown in Fig. 9. For higher wrap angles on the high pressure side the length of the low pressure segment L_{LP} increases according to eq. (3) and (4). For small wrap angles on the low pressure side (50 to 125 degrees) the delivery rate decreases, an effect caused mainly by the blowhole, the size of which depends entirely on the rotor lead, which is active for the major part of the working cycle. Due to the high amount of leakage, internal isentropic efficiency diminishes as well. For higher wrap angles on the low pressure side (125 to 225 degrees) the delivery rate is fine, but the high pressure segment is short and causes increased outlet throttling.

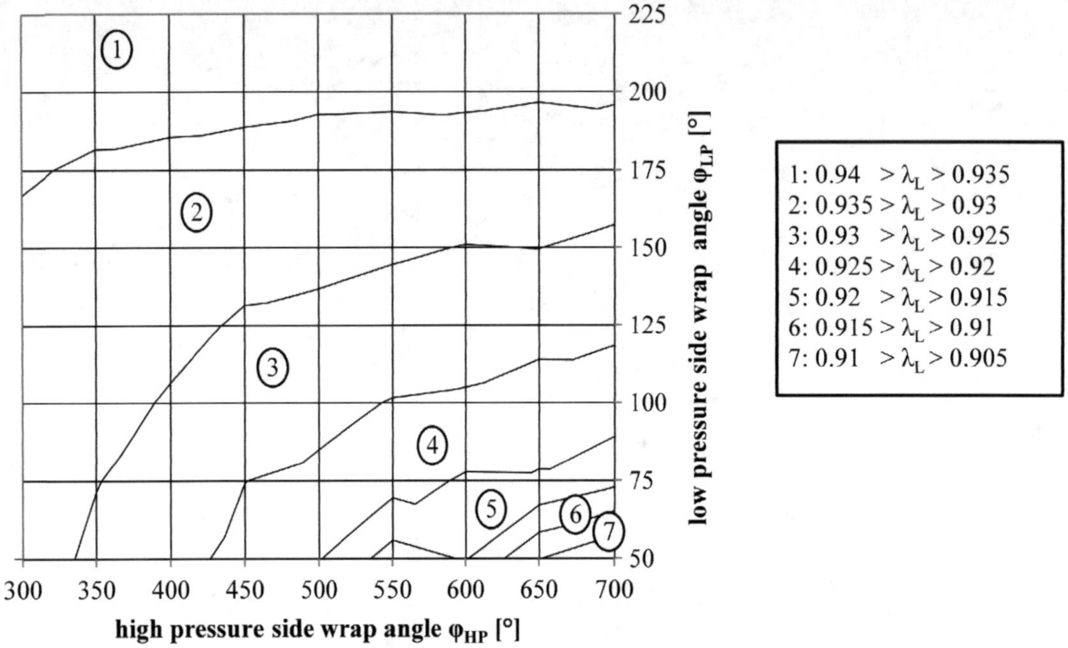

Figure 9: Delivery rate as a function of low and high pressure side wrap angle (L_{Rotor} = 260 mm, φ_{total} = 250°, α = 0.8, lobe count combination 4+6)

Table 4: Properties of optimized machine geometries for L = 260 mm and lobe count combination of 4+6

Parameter [dimension]	Constant lead	Dual lead	Constant lead	Dual lead
Rotor geometry				
Flow coefficient α [-]	0.8		0.6	
Total male rotor wrap angle φ_{total} [°]	350	250	400	250
High pressure wrap angle φ_{HP} [°]	-	400	-	500
Low pressure wrap angle φ_{LP} [°]	-	175	-	150
Internal volume ratio v_i [-]	3.5	3.5	3.5	3.5
Internal Power P_i [kW]	70.2	76.6	64.2	76.5
Internal isentropic efficiency η_{is} [%]	88.0	89.1	88.6	90.6
Relative increase in internal isentropic efficiency [%]	1.25		2.3	
Delivery rate λ_L [%]	91.8	93.4	90.2	94.8
Mass flow $\dot{m}_{real}$ [kg/s]	3.45	3.82	3.18	3.87
Relative increase in mass flow [%]	10.7		21.7	

Optimal machine properties for the assumed flow coefficients are summarized in Table 4. All optimized machines possess an internal volume ratio of 3.5. For a flow coefficient of 0.8 energy-related efficiency can be improved by 1.25 %. Due to the lower total wrap angle maximum chamber volume for dual lead rotors is greater and leads to a higher mass flow rate. This effect is even greater for a flow coefficient of 0.6. As a result of increased outlet throttling, the wrap angle in a constant lead setup, and on the high pressure side of a dual lead setup, is also increased. Efficiency can be enhanced relatively by 2.3 % whereas the mass flow rate is more than 20 % higher.

7. CONCLUSION

For the refrigerant process under examination, energy-related and volumetric efficiency can be enhanced by using dual lead rotors. Wrap angle segments for dual lead rotors should be finely adjusted to improve the efficiency. If the low pressure side wrap angle segment is too long, efficiency decreases due to the large blowhole area, which increases clearance mass flow.

Compared with constant lead rotors, the simulation results show that the optimal total wrap angle of the machine should be reduced, resulting in decreased clearance mass flow, which raises the delivery rate. Furthermore, smaller total wrap angles increase maximum chamber volume. Consequently, dual lead rotors allow much higher mass flow rates on a machine of a given size.

NOMENCLATURE

D	Diameter	(m)
h	Specific enthalpy	(J/kg)
L	Length	(m)
$\dot{m}$	Mass flow	(kg/s)
n	Rotational speed	(1/s)
P	Power	(W)
p	Pressure	(Pa)
s	Specific entropy	(J/kg/K)
V	Volume	(m^3)
v_i	Internal volume ratio	(-)
z	Number of lobes	(-)
α	Flow coefficient	(-)
η_{is}	Internal isentropic efficiency	(-)
Δ	Difference	(-)
φ	Wrap angle	(°)
λ_L	Delivery rate	(-)
ρ	Density	(kg/m^3)

Subscript

compr,end	End of compression
FR	Female rotor
HP	High pressure
MR	Male rotor
max	Maximum
LP	Low pressure
s	Constant entropy
i	Internal

REFERENCES

Fost, C., 2003, *Ein Beitrag zur Verbesserung der Kammerfüllung von Schraubenmotoren*, Dissertation, Universität Dortmund, 168 p.

Gardner, J. W., 1969, *Variable lead compressor*, US patent US3424373 A.

Janicki, M., 2007, *Modellierung und Simulation von Rotationsverdrängermaschinen*, Dissertation, Universität Dortmund, 170 p.

Kauder, K., Janicki, M., Rohe, A., Kliem, B., Temming, J., 2002, Thermodynamic Simulation of Rotary Displacement Machines, *VDI-Berichte 1715*, VDI Verlag, Düsseldorf, p. 1-16.

Rane, S., Kovacevic, A., Stosic, N., Kethidi, 2014, Deforming grid generation and CFD analysis of variable geometry screw compressors, *Computers & Fluids*, vol. 99, p. 124-141

Rinder, L., 1979, *Schraubenverdichter*, Springer-Verlag, Wien New York, 252 p.

Rohe, A., 2005, *Wärmehaushalt von Schraubenspindel-Vakuumpumpen*, Dissertation, Universität Dortmund, 160 p.

Sigloch, H., 2009, *Technische Fluidmechanik*, 7th edition, Springer-Verlag, Heidelberg London New York, 581 p.

Temming, J., 2007, *Stationärer und instationärer Betrieb eines unsynchronisierten Schraubenladers*, Dissertation, Universität Dortmund, 140 p.

Utri, M., Brümmer, A., 2014, A comparative examination of the potential of screw expanders with variable rotor pitch, *International Conference on Screw Machines 2014*, VDI Verlag, p. 249-266.

ACKNOWLEDGEMENT

The work leading to these results has received funding from the European Community's Horizon 2020 Programme (2014-2020) under grant agreement no 678727. The opinions expressed in the document are of the authors only and no way reflect the European Commission's opinions. The European Union is not liable for any use that may be made of the information.

Performance Testing of a Vapor Injection Scroll Compressor with R407C

Thomas W. MOESCH[1], Ammar M. BAHMAN[2*], Eckhard A. GROLL[2]

[1]Technische Universitaet Dresden, Institut fuer Energietechnik,
Bitzer-Stiftungsprofessur fuer Kaelte-, Kryo- und Kompressorentechnik,
Dresden, Germany
thomas.moesch@tu-dresden.de

[2]Purdue University,
School of Mechanical Engineering,
Ray W. Herrick Laboratories,
West Lafayette, Indiana, USA
abahman@purdue.edu, groll@purdue.edu

* Corresponding Author

ABSTRACT

Current studies indicate that the method of economized vapor injection (EVI) increases both cooling capacity and coefficient of performance (COP) of vapor compression systems and enlarges the operating range of compressors by reducing the discharge temperature. The design and analysis of EVI systems require comprehensive and comparable performance data of the compressor. In this work, a thermodynamic model was developed to simulate the potential benefit of EVI systems. Furthermore, the performance of a vapor injection (VI) scroll compressor has been experimentally investigated using a modified compressor calorimeter and the refrigerant mixture R407C. During the experiments, the injection flow was regulated by controlling the injection superheat. The experimental results confirm the predicted tendencies of the EVI model. The investigation also reveals that the injection pressure affects the VI compressor performance and needs to be included in the compressor performance evaluation.

1 INTRODUCTION

The application of lubricated compressors in single-stage vapor compression cycles is limited by the pressure ratio of condensing to evaporating pressures. The main limitation is hereby the maximum compressor discharge temperature at which the lubricant starts to degrade. This limited operating envelope can be extended by refrigerant injection. Dutta *et al.* (2001) showed that using a scroll compressor and injecting liquid refrigerant into the compression pocket reduces the discharge temperature linearly with 1.5 K per percent of injection ratio, which is defined as the ratio of the injection mass flow rate to the mass flow rate through the evaporator. They found that liquid refrigerant injection is not preferable thermodynamically as it reduces the compressor efficiency and thus, increases the power consumption. Winandy and Lebrun (2002) investigated the application of a VI scroll compressor in combination with an internal heat exchanger (IHX) using R22 as a refrigerant. The IHX was used as an economizer, cooling down the refrigerant leaving the condenser while evaporating the injection mass flow at injection pressure. Their experimental results of this EVI system showed that the discharge temperature can be reduced while the cooling capacity is increased, and the systems COP remains unchanged. The design of the VI scroll compressor (Perevozchikov, 2013) allowed an unchanged evaporator mass flow rate as the additional refrigerant was solely injected within the compression process. Wang *et al.* (2009) optimized the EVI process using a detailed VI compressor model for R22. Their investigation revealed that for a fixed injection port location there is an optimal injection pressure for maximizing the cooling capacity based on given condensing and evaporating conditions. Several manufacturers already introduced VI compressors suggesting EVI cycle designs and controls (Emerson, 2015). However, the performance evaluation of VI compressors is not yet standardized. Navarro *et al.* (2013) systematically investigated the performance of a VI scroll compressor using a test setup with IHXs and the refrigerant mixture R407C. The investigation included a variation of the injection pressure for different evaporating and condensing pressures. Their results show a dependence of the heating capacity and heating COP improvement on the intermediate pressure.

This paper presents an experimental investigation of a VI scroll compressor for a HVAC application in high temperature regions using R407C as a refrigerant. Prior to any performance tests, the operational limits of the test setup and the compressor were estimated using a thermodynamic cycle model. The experimental investigation includes performance tests with and without vapor injection and for different injection pressures.

2 EXPERIMENTAL SETUP

A VI scroll compressor with a volumetric flow rate of 14.1 m³/h was tested on an existing compressor calorimeter, which was modified for refrigerant injection as shown in Figure 1. During the test operation, the condensing pressure p_C, the evaporating pressure p_E, the return gas temperature T_{RG}, the compressor ambient air temperature T_{amb}, and the injecting superheat $\Delta T_{SH,inj}$ were adjustable. The condensing pressure was adjusted by PID-1, a PID controller which controlled the process water flow through the shell-and-tube condenser by setting the opening of valve V3 in the process water return line (PWR). The evaporating pressure was adjusted manually by changing the opening of the pneumatically driven expansion valves V1 and V1' at the inlet of the evaporators. Both evaporators sit in a secondary calorimeter filled with R134a as a secondary refrigerant. The return gas temperature was adjusted by PID-3 which controlled a 3 kW on-off heating element in the secondary calorimeter. In addition, three 3 kW and one 6 kW on-off heating elements in the secondary calorimeter were controlled manually. This allowed a wide range of cooling capacities during the compressor testing. The ambient air temperature in the compressor chamber is adjusted by PID-2 which controlled the chamber heater. The control of the injection flow was performed with an electronic expansion valve (EXV). The EXV control was based on the injection superheat which resulted from the temperature and pressure measurement at the injection port. All temperatures were measured with T-type thermocouples. The pressure measurements were conducted with absolute and gauge capacitance pressure transducers. The specifications of all measurement devices are listed in Table 1. The mass flow measurement was confirmed using the secondary calorimeter method according to ANSI/ASHRAE Standard 41.9 (2011).

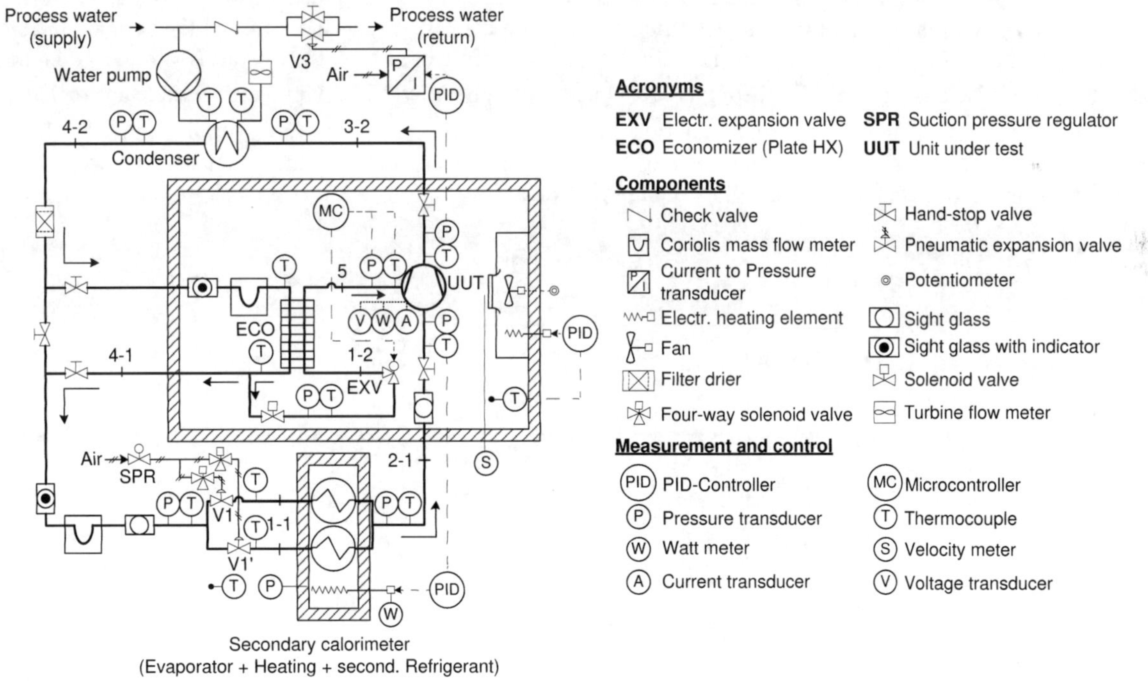

Figure 1: Sketch of modified compressor calorimeter

Table 1: Measurement devices of the test setup

Measurement	Device (Manufacturer, Model)	Accuracy
Current (compr.)	Current transducer (SC[1], 4044-8)	±0.15 % ±0.05 A
Flow rate (Evap.)	Coriolis mass flow meter (MM[2], DS040 & RFT9712)	±0.2 % ±0.24 kg/h
Flow rate (Total)	Coriolis mass flow meter (MM[2], DH025S & IFT9701)	±0.15 % ±0.18 kg/h
Flow rate (H_2O)	Turbine flow meter (EG&G, FT12)	±0.25 %
Power (compr.)	Watt/Watthour transducer (SC[1], DL31K5)	±0.09 % ±2.0 W
Power (Sec. Cal.)	Watt/Watthour transducer (SC[1], DL5C5)	±0.09 % ±1.8 W
Pressure (amb.)	Barom. pressure transducer (Setra, 278)	±0.6 mbar
Pressure (low/Sec.Cal./high)	Absolute pressure transducer (Setra, 204-100/500/1000)	±0.76 / ±3.79 / ±7.58 kPa
Pressure (eco)	Gauge pressure transducer (Setra, 207-500)	±4.48 kPa
Temperature	Thermocouples (Omega, T)	±1 K
Velocity (air amb.)	Air velocity transmitter (Dwyer, 640-1)	±2 %
Voltage (compr.)	Voltage transducer (SC[1], 3588)	±0.25 %

1) Science Columbus, 2) Micro Motion

3 THERMODYNAMIC MODEL

3.1 Model Parameters

The modified calorimeter model is based on an ideal vapor compression cycle and a supplementary refrigerant injection circuit with an IHX economizer as shown in Figure 2. The refrigerant is extracted downstream after the economizer. The calculation of the models state points (SP) requires nine parameters. The main cycle parameters such as the saturated condensing/evaporating temperature $T_{dew,C}/T_{dew,E}$, the high pressure subcooling ΔT_{SC}, and the low pressure superheat ΔT_{SH} resemble the operating conditions of regular compressor performance tests. The supplementary circuit is defined by the injection pressure superheat $\Delta T_{SH,inj}$ and the minimal temperature difference at the economizer ΔT_{min}. The compressor model is reduced to three parameters, the overall isentropic efficiency ε_c, the volumetric efficiency ε_v and the theoretical volume displacement $\dot{V}_{th}$.The volumetric efficiency includes the effect of inner leakages and the internal heat flux to the suction volume flow. The overall isentropic efficiency describes the irreversibility of the compression process. The theoretical volume displacement was the only geometric compressor parameter and given by the manufacturer.

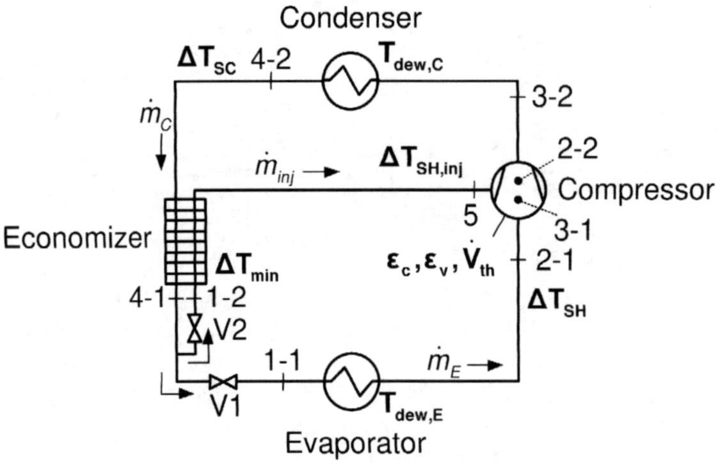

Figure 2: Parameters for EVI cycle model

3.2 Cycle Analysis

The analysis of the thermodynamic model excluded any internal pressure losses that are not associated with the expansion devices, or any external heat losses that are not associated with the heat exchangers. Further, the composition of the refrigerant mixture was assumed to be the same at all SP and the circulation of lubricant was not considered. The EVI system operates at three different pressures levels: the condensing pressure p_C, the evaporating pressure p_E, and the injection pressure p_{inj}. The pressures were derived from their saturated temperatures as stated in Equation (1).

$$p_{C/E/inj} = p_{dew}\left(T_{dew,C/E/inj}\right) \tag{1}$$

The saturated injection temperature $T_{dew,inj}$ was estimated using Equation (2). This empirical correlation was given by Emerson (2015) as part of the economizer sizing process for their VI scroll compressors. The correlation is valid for condensing temperatures $T_{dew,C}$ = 26.6 to 65.6 °C and evaporating temperatures $T_{dew,E}$ = -31.6 to 10.0 °C, and it was based on the assumption of fixed temperature differences $\Delta T_{min} = \Delta T_{SC} = \Delta T_{SH}$ = 5.6 K. The claimed accuracy of this correlation is ± 2.8 K.

$$T_{dew,inj} = 0.8 \cdot T_{dew,E} + 0.5 \cdot T_{dew,C} - \frac{19}{3} K \tag{2}$$

The total mass flow rate through the condenser is divided at the economizer outlet (SP 4-1) which yields to the mass conservation as stated in Equation (3). The mass flow rate through the evaporator was calculated using Equation (4) which is based on the definition of the volumetric efficiency. The energy conservation of the economizer and the assumption of an isenthalpic expansion across V2 yields to the injection ratio μ_{inj} as stated in Equation (5).

$$\dot{m}_C = \dot{m}_E + \dot{m}_{inj} = \dot{m}_E \cdot \left(1 + \mu_{inj}\right) \tag{3}$$

$$\dot{m}_E = \varepsilon_v \cdot \frac{\dot{V}_{th}}{v_{2-1}} = \varepsilon_v \cdot \frac{\dot{V}_{th}}{v\left(p_E, T_{dew,E} + \Delta T_{SH}\right)} \tag{4}$$

$$\mu_{inj} = \frac{\dot{m}_{inj}}{\dot{m}_E} = \frac{\Delta h_{SC,eco}}{h_5 - h_{4-2}} = \frac{\Delta h_{SC,eco}}{h\left(p_{inj}, T_{dew,inj} + \Delta T_{SH,inj}\right) - h\left(p_C, T_{bub,C} - \Delta T_{SC}\right)} \tag{5}$$

Where $T_{bub,C} = T_{bub}(p_C)$ is the bubble point temperature at condensing pressure and $\Delta h_{SC,eco}$ is the specific enthalpy difference that represents the additional subcooling in the liquid line of the economizer as stated in Equation (6). The temperature T_{4-1} is determined iteratively using Equation (7), which assumes an isenthalpic expansion across V2 and includes the constraint ΔT_{min} at the economizer outlet.

$$\Delta h_{SC,eco} = h_{4-2} - h_{4-1} = h\left(p_C, T_{bub,C} - \Delta T_{SC}\right) - h\left(p_C, T_{4-1}\right) \tag{6}$$

$$h\left(p_C, T_{4-1}\right) = \begin{cases} h\left(p_{inj}, T_{4-1} - \Delta T_{min}\right) \text{ when } \mu_{inj} > 0 \\ h_{4-2} \text{ when } \mu_{inj} = 0 \end{cases} \tag{7}$$

The cooling capacity $\dot{Q}_E$ is defined by the enthalpy difference across the evaporator as stated in Equation (8). The assumption of an isenthalpic expansion across V1 and the additional subcooling $\Delta h_{SC,eco}$ yield to Equation (9).

$$\dot{Q}_E = \dot{m}_E \cdot \left(h_{2-1} - h_{1-1}\right) \tag{8}$$

$$\dot{Q}_E = \dot{m}_E \cdot \left(h_{2-1} - h_{4-2} + \Delta h_{SC,eco}\right) = \dot{m}_E \cdot \left[h\left(p_E, T_{dew,E} + \Delta T_{SH}\right) - h\left(p_C, T_{bub,C} - \Delta T_{SC}\right) + \Delta h_{SC,eco}\right] \tag{9}$$

The compression process in this model was divided in a pre-injection compression (SP 2-1 to 3-1) and a post-injection compression (SP 2-2 to 3-2). The pre-injection compression as stated in Equation (10) is assumed to end at the injection pressure whereas the post-injection compression as stated in Equation (11) starts at the injection pressure. The injection process was simplified to an adiabatic and isobaric mixing process at injection pressure and results in the mixing enthalpy $h_{2\text{-}2(s)}$ as stated in Equation (12).

$$h_{3-1} = \frac{h_{3-1s} - h_{2-1}}{\varepsilon_c} + h_{2-1} = \frac{h(p_{inj}, s_{2-1}) - h(p_E, T_{dew,E} + \Delta T_{SH})}{\varepsilon_c} + h(p_E, T_{dew,E} + \Delta T_{SH}) \tag{10}$$

$$h_{3-2} = \frac{h_{3-2s} - h_{2-2s}}{\varepsilon_c} + h_{2-2} = \frac{h(p_C, s_{2-2s}) - h_{2-2s}}{\varepsilon_c} + h_{2-2} \tag{11}$$

$$h_{2-2(s)} = \frac{\dot{m}_E \cdot h_{3-1(s)} + \dot{m}_{inj} \cdot h_5}{\dot{m}_E + \dot{m}_{inj}} = \frac{h_{3-1(s)} + \mu_{inj} \cdot h_5}{1 + \mu_{inj}} \tag{12}$$

The compressor discharge temperature was calculated using the specific enthalpy $h_{3\text{-}2}$ at the end of the post-injection compression. The compressor power consumption was calculated as stated in Equation (14), which results from Equations (10) to (12).

$$T_{dis} = T_{3-2} = T(p_C, h_{3-2}) \tag{13}$$

$$P_{comp} = (\dot{m}_E + \dot{m}_{inj}) \cdot (h_{3-2} - h_{2-2}) + \dot{m}_E (h_{3-1} - h_{2-1}) = \dot{m}_E \cdot \frac{(1 + \mu_{inj}) \cdot h_{3-2s} - \mu_{inj} \cdot h_5 - h_{2-1}}{\varepsilon_c} \tag{14}$$

The software REFPROP (Lemmon *et al.*, 2013) was used to calculate the refrigerant's thermodynamic properties such as the saturated pressures $p_{dew/bub}(T)$, the saturated temperatures $T_{dew/bub}(p)$, the specific volume v = v(p,T), the specific enthalpy h = h(p,T), and the specific entropy s = s(p,T).

3.3 Performance Test Simulation

The model was used to conduct simulations for two different compressor operating conditions, the baseline (BL) operation with $\mu_{inj} = 0$ and the vapor injection (VI) operation with $\mu_{inj} > 0$. For the simulation, $T_{dew,C}$ and $T_{dew,E}$ were varied and the remaining parameters were kept constant as stated in Table 2. The results lead to the definition of a limited test envelope for the BL and VI operation as shown in Figure 3A. The envelope boundaries were based on the limits for both calorimeter and compressor. The restricting parameter for the given calorimeter was the cooling capacity, which was limited to a range of 5 kW to 15 kW. The operation of the scroll compressor was limited to $T_{dis,max}$ = 115 °C and I_{max} = 18 A. The maximum compressor power consumption resulted from Equation (15), which assumes a symmetric power distribution for a three phase power supply. This lead to a maximum power consumption $P_{comp,max}$ = 6.095 kW for a power factor cos(φ) = 0.85 at U_{comp} = 230 V.

$$P_{comp,max} = \cos(\varphi) \cdot \sqrt{3} \cdot I_{max} \cdot U_{comp} \tag{15}$$

The simulation reveals the benefits of the EVI system on the compressor performance. The operating envelope expands towards higher compression ratios through the reduction of the discharge temperature, and the cooling capacity is raised. However, the simulation results also show the impact of the calorimeter limitation, which reduces the amount of possible test points for evaporating temperatures past 4.5 °C.

Table 2: Parameter assumption for the performance test simulation

ΔT_{SH}	ΔT_{SC}	$\Delta T_{SH,inj}$	ΔT_{min}	ε_c	ε_v	$\dot{V}_{th}$
11.11 K	5.56 K	5.56 K	5.56 K	0.7	0.9	14.1 m³/h

4 PERFORMANCE TESTS AND RESULTS

4.1 Test Conditions

Similar to the simulation, the experiment was conducted for two different kinds of compressor operating conditions. During BL operation, the injection port of the scroll compressor was closed and the EVI circuit was bypassed. All steady state test points of the BL operation are plotted in Figure 3A, which also shows the calculated test limits. During VI operation the injection port was opened and the bypass of the EVI circuit was closed. The VI test points were based on two BL conditions at different injection conditions as shown in Figure 3A and Figure 3B. During Condition A ($T_{dew,C}$ = 48.9 °C, $T_{dew,E}$ = 4.5 °C) the impact of two-phase injection was investigated by changing the opening of the EXV between 10 to 30 %. The influence of elevated condensing temperatures on the EVI performance was investigated at Condition B ($T_{dew,C}$ = 60.0 °C, $T_{dew,E}$ = 4.5 °C) while the injection superheat was varied from 0 to 34.6 K. All compressor performance tests were conducted for a suction superheat of 11.1 K and an ambient compressor temperature of 35 °C. During each test condition, the refrigerant charge remained constant, which caused a variation in the subcooling at the condenser outlet (SP 4-2). Depending on the evaporating and condensing conditions, the subcooling varied between 0.5 K and 4.3 K for the BL operation and between 2.3 K and 4.9 K for the VI operation.

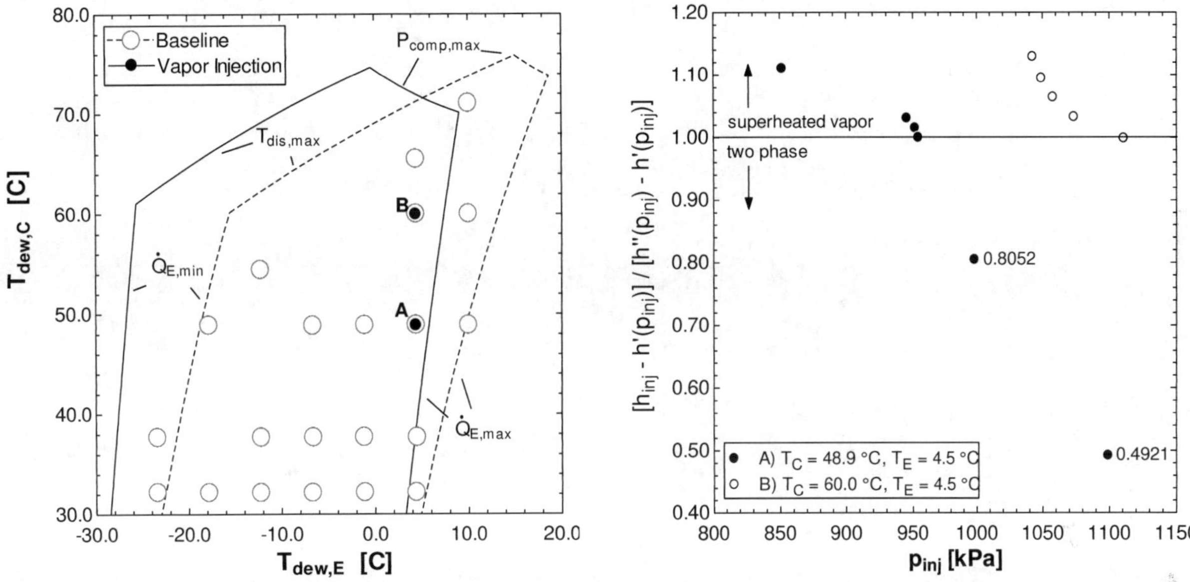

A) Test conditions for BL and VI tests B) Injection conditions during VI tests

Figure 3: Test matrix for performance tests

4.2 Baseline Test Results and Analysis

The results of the baseline performance tests are illustrated in Figure 4A through 4D. The cooling capacity decreased towards increased pressure ratios and ranges from 4 to 15.5 kW as shown in Figure 4A. This was caused by two effects, the reduction of the suction mass flow rate for lower evaporating temperatures and the reduction of the specific enthalpy difference across the evaporator for increased condensing temperatures. The trend of the cooling capacity is in good agreement with the simulation results. The error of ±7% can be explained with the lack of subcooling. The trend of the discharge temperature is shown in Figure 4B. It increased for growing compression ratios reaching values up to 113 °C. The error of the simulated discharge temperature was up to ±10 K for lower evaporating temperatures. This was mainly caused by the decreasing efficiency as illustrated in Figure 4C. The trend of ε_c results from the built-in volume ratio of the scroll compressor. According to Winandy *et al.* (2002), the isentropic efficiency reaches its maximum for an adapted compression ratio and decreases for any external pressure ratio that differs from the inner ratio. By analyzing the trend of the compressor efficiency, the adapted pressure ratio was estimated to be 3.65. The volumetric efficiency of this scroll compressor ranged from 0.86 to 0.96. This rather high ε_v is common for scroll compressors, which generally lack any dead volume and do not encounter re-expansion after the compression process. The reduction of the volumetric efficiency, as shown in Figure 4D, results from increased internal leakages and additional motor losses, which increase the internal suction temperature.

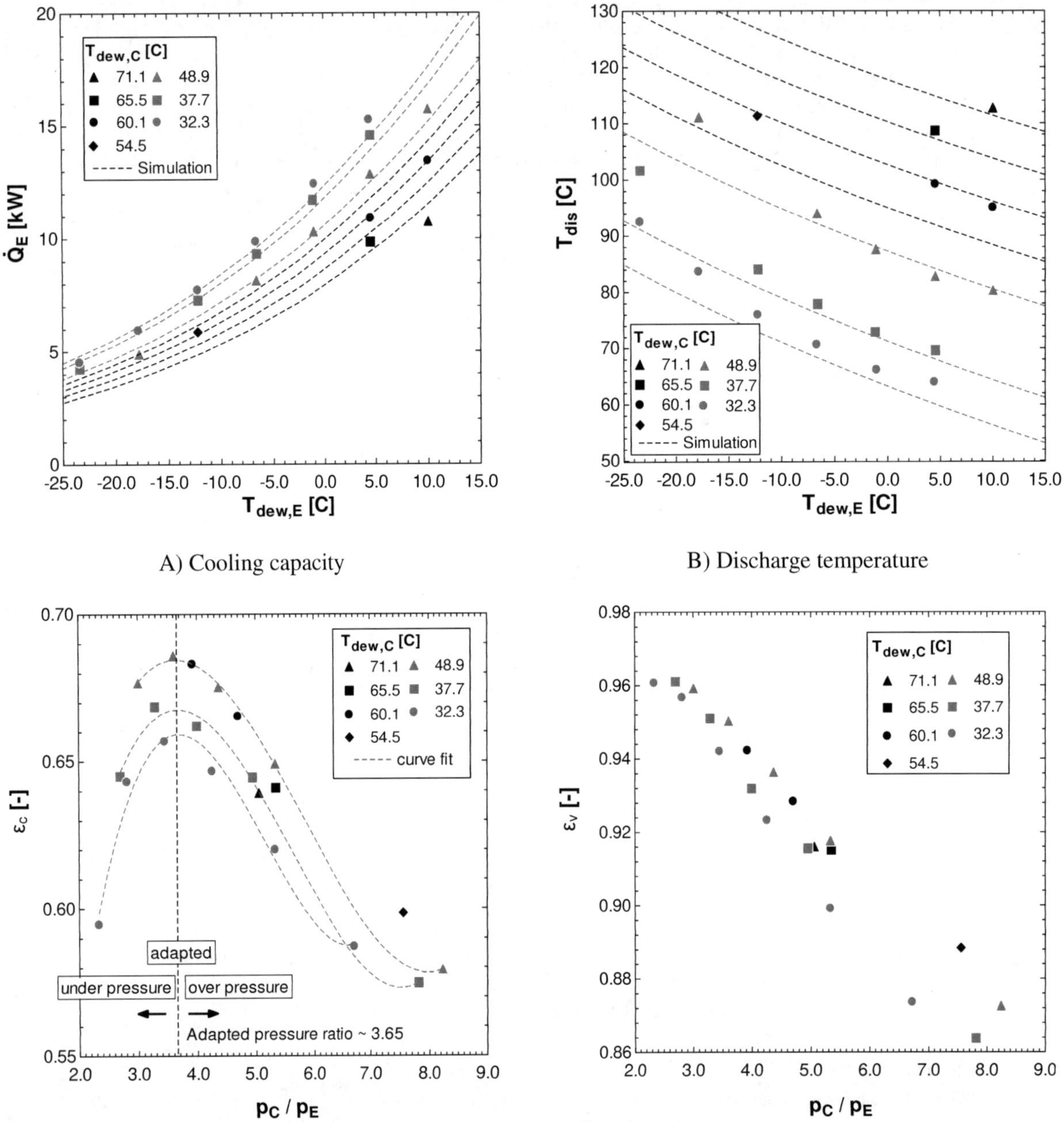

A) Cooling capacity

B) Discharge temperature

C) Overall isentropic efficiency

D) Volumetric efficiency

Figure 4: Performance values during baseline operation

4.3 Vapor Injection Test Results and Analysis

The results of the vapor injection performance tests for the Operating Conditions A and B are illustrated in Figure 5A through 5D. The cooling capacity was raised by 12 to 30% compared to the baseline tests. For vapor injection, the cooling capacity increased with increasing injection pressure until two phase-injection started. The decrease in cooling capacity for wet injection can be explained with an increased saturated temperature on the cool side of the economizer which leads to a reduced subcooling of the liquid refrigerant. The optimum injection pressure for maximum cooling capacity as described by Wang *et al.* (2009) was found to be 94% of the geometric mean of evaporating and condensing pressure at the given conditions. The simulation underestimated the cooling capacity by up to 5%. This underestimation was mainly caused by the assumption of ΔT_{min}, which reached values of 0.4 to 4 K during the VI performance tests. The discharge temperature generally decreased with increased injection pressures. For the two tested conditions, the rate of reduction was found to be dependent on the phase of the injected refrigerant and the condensing temperature as shown in Figure 5B. For Operating Condition A, the reduction rate increased from -3.5 K to -13.1 K per 100 kPa when the injection changed to two-phase injection. During two-phase injection the latent heat of the injected refrigerant causes extra cooling. Compared to Operating Condition A, the reduction during Operating Condition B was more effective with a rate of -9.7 K/100 kPa. This was due to the higher injection pressure, which increased the injection mass flow rate and thus, reduced the mixing temperature after injection. During the VI operation, P_{comp} increased linearly with the injection pressure as shown in Figure 5C. The rate was approximately 204 W/100 kPa for both operating conditions. This leads to an increase in the power consumption of 7.5 to 22.5% compared to the BL operation. The simulation model underestimated the power consumption by 12.5%. The reason for this error is the isentropic compressor efficiency, which was lower than the assumed value of 0.7. Figure 5D shows the reduction of ε_c for increasing injection pressures. For vapor injection the isentropic compressor efficiency decreased by only 0.01 / 100 kPa. However, when the injection turned into wet injection, this rate changed to 0.05 / 100 kPa. During VI operation, ε_c was decreased by -5% to -16.8% compared to the BL operation. The volumetric efficiency remained constant during VI within a range of ±0.005. Compared to BL operation, ε_v was reduced by -3%. The reason for this reduction could be an increase in internal leakages during injection.

5 CONCLUSIONS

A VI scroll compressor has been successfully tested on a modified compressor calorimeter at high temperature range using R407C. The thermodynamic model presented in this paper allowed the creation of a limited test envelope for the performance test of a given compressor. The experimental results indicated that the injection pressure had a significant impact on the performance of the VI compressor. The VI compressor was characterized in regard to its adapted pressure ratio and optimized injection pressure. It was found that the reduction rate of the discharge temperature depends on the phase of the injected refrigerant and the condensing temperature. Further, this paper reveals that any injection pressure might degrade the isentropic and volumetric efficiency of the given compressor compared to an operation without vapor injection. Since the injection pressure influences the degree of the discharge temperature reduction, the cooling capacity improvement, and the isentropic compressor efficiency, their values and variations should be included in future compressor performance tests.

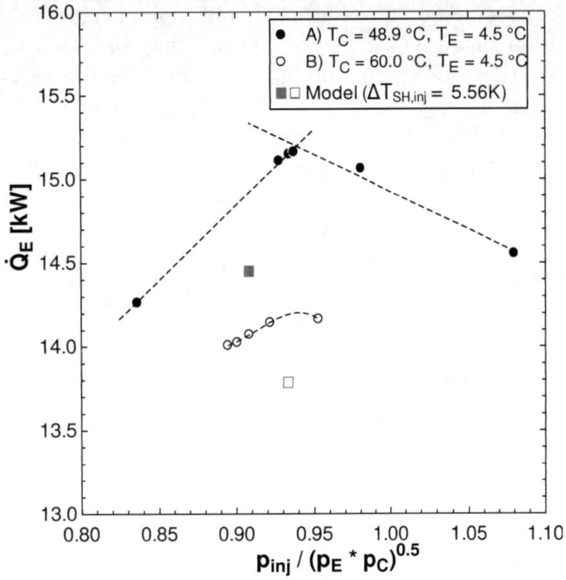

A) Cooling capacity

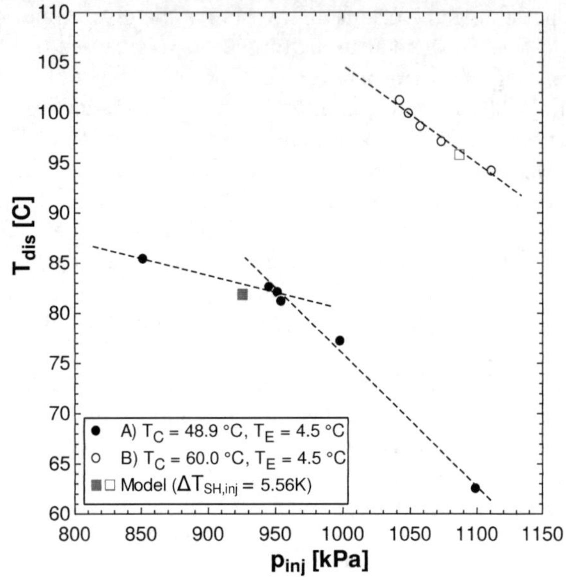

B) Discharge temperature

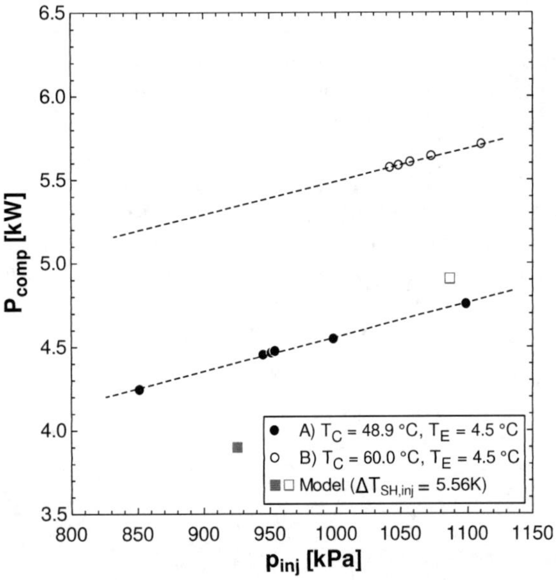

C) Power consumption

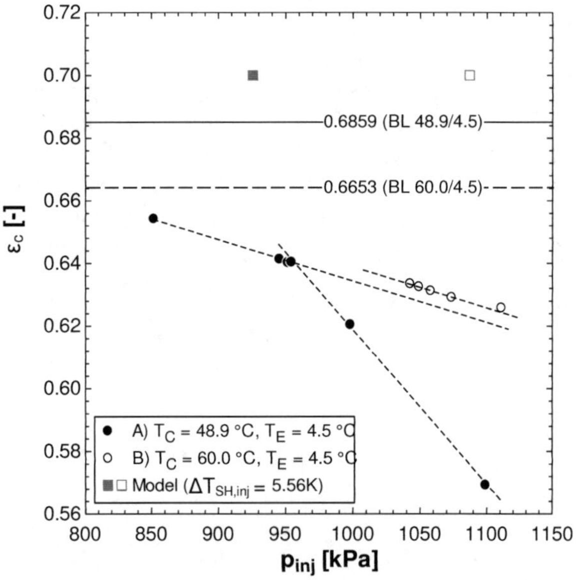

D) Overall isentropic efficiency

Figure 5: Performance values during vapor injection operation

NOMENCLATURE

COP	Coefficient of performance	(-)
h	Specific enthalpy	(kJ/kg)
$\dot{m}$	Mass flow rate	(kg/h)
p	Pressure	(kPa)
$\dot{Q}$	Capacity	(W)
s	Specific entropy	(kJ/kgK)
T	Temperature	(°C)
v	Specific volume	(m^3/kg)
$\dot{V}$	Volume flow rate	(m^3/h)
Greek		
Δ	Difference	(-)
ε_c	Overall isentr. compr. efficiency	(-)
ε_v	Volumetrtic efficiency	(-)
μ	Mass flow ratio	(-)
Subscripts		
1-1...4-2, 5	State points	
bub	Bubble point	
C	Condensing	
comp	Compressor	
dew	Dew point	
dis	Discharge	
E	Evaporating	
eco	Economizer	
inj	Injection	
min	Minimum	
s	Isentropic	
SC	Subcooling	
SH	Superheat	
th	Theoretical	
Superscripts		
'	Bubble point	
"	Dew point	
Acronyms		
BL	Baseline	
EVI	Economized Vapor Injection	
EXV	Electronic Expansion Valve	
IHX	Internal heat exchanger	
PID	Proportional, integration, Differential	
PWR	Process water return	
Sec. Cal.	Secondary calorimeter	
SP	State point	
VI	Vapor Injection	

REFERENCES

ANSI/ASHRAE Standard 41.9-2011, *Standard methods for volatile-refrigerant mass flow measurements using calorimeters*, American Society of Heating, Refrigerating and Air-Conditioning Engineers, Inc., 1791 Tullie Circle NE, Atlanta, GA 30329.

Dutta, A.K., Yanagisawa, T., Fukuta, M., 2001, An investigation of the performance of a scroll compressor under liquid refrigerant injection, *Int. J. Refrig.*, vol. 24, no. 6, p. 577-587.

Emerson Climate Technologies (Ed.), 2015, Economized Vapor Injection (EVI) Compressors, *Bulletin AE4-1327 R12*. Emerson Climate Technologies, Inc., 22p.

Lemmon, E.W., Huber, M.I., McLinden, M.O., 2013, NIST Standard Reference Database 23: Reference Fluid Thermodynamic and Transport Properties (RefProp), Version 9.1, National Institute of Standards and Technology, Standard Reference Data Program, Gaithersburg.

Navarro, E., Redón, A., Gonzálvez-Macia, J., Martinez-Galvan, I.O., Corberán, J.M., 2013, Characterization of a vapor injection scroll compressor as a function of low, intermediate and high pressures and temperature conditions, *Int. J. Refrig.*, vol. 36, no. 7, p. 1821-1829.

Perevozchikov, M.M., 2003, Scroll compressor with vapor injection, U.S. Patent, No. 0133819.

Wang, B., Shi, W., Han, L., Li, X., 2009, Optimization of refrigeration system with gas-injected scroll compressor, *Int. J. Refrig.*, vol. 32, no. 7, p. 1544-1554.

Winandy, E., Saavedra, C., Lebrun, L., 2002, Experimental analysis and simplified modeling of a hermetic scroll refrigeration compressor, *Applied Thermal Engineering*, vol. 22, no. 2, p. 107-120.

Winandy, E.L., Lebrun, J., 2002, Scroll compressors using gas and liquid injection: experimental analysis and modeling, *Int. J. Refrig.*, vol. 25, no. 8, p. 1143-1156.

Analysis of Indicator Diagrams of a Water Injected Twin-shaft Screw-type Expander

Alexander NIKOLOV*, Andreas BRÜMMER

Chair of Fluidics, TU Dortmund University,
Dortmund, Germany
alexander.nikolov@tu-dortmund.de
andreas.bruemmer@tu-dortmund.de

* Corresponding Author

ABSTRACT

Twin-shaft screw-type expanders offer a high potential for energy conversion in the lower and medium power range, for instance as expansion engines in Rankine cycles for exhaust heat recovery. With regard to minimizing internal leakages and lubricating moving machine parts, an auxiliary liquid or liquid working fluid, for example in organic Rankine cycles (ORC)[1], can be fed to the screw expander.

In this paper, indicator diagrams of a twin-shaft screw-type expander prototype SE 51.2 designed at the Chair of Fluidics at TU Dortmund University are presented and analyzed in detail. The experimental investigations are carried out on a hot-air test rig with an expander inlet manifold water injection. The time-dependent working chamber pressure is recorded by means of high-resolution absolute pressure transmitters. Hereby, specific aspects of working chamber pressure measurements are mentioned. Based on the indicator diagrams, relevant influence mechanisms on the expander's operational behavior resulting from water injection are determined. Additionally, the impact of the injected water on the expander's delivery rate and mechanical efficiency is presented.

1. INTRODUCTION

Due to the growing shortage of non-renewable fossil fuel reserves and the resulting increase in primary energy costs, little developed energy potentials increasingly move into the focus of economic interest. Available heat sources in the field of decentralized energy systems of small and medium power ranges, such as industrial exhaust gases or waste heat in vehicles' engines, geothermal or solar thermal energy can be converted into useable mechanical power by means of expanders or turbines within a Rankine cycle.

In this context, twin-shaft screw-type expanders in Rankine cycles possess clear advantages compared with turbo machines or even other expander concepts. In general, twin-shaft screw-type machines are characterized by relatively high energy density and efficiency, good part load behavior, and rather a simple design. Furthermore, screw expanders in Rankine cycles are suitable for wet-vapor operation, so that an overheating system is not essential. On the one hand, the number of ORC components and the overall system costs can be reduced. On the other hand, liquid in screw expanders is even beneficial with regard to minimizing internal leakages and reducing noises or vibrations.

Whereas injecting oil is a common method in screw-type compressors without timing gears, water as an auxiliary fluid offers sufficient potential for investigation. There are commercial water injection concepts for screw-type compressors with timing gears which aim to reduce gas temperature and thermal stress in the working camber. Investigations of water and oil injection into screw-type expanders with timing gears were done by Zellermann (1996). Rinder *et al.* (2004) presented benefits and disadvantages of water compared to oil injection with regard to a screw-type compressor without timing gears. Furthermore, Rinder and Moser (1990) have shown oil distribution in a screw compressor. Kliem (2005) investigated a screw expander application in a trilateral flash

[1] (Organic) Rankine cycles are thermodynamic cycles that operate as a closed loop. In Rankine cycles an operating fluid, water or organic fluid, is continuously pressurized, evaporated, expanded and eventually condensed.

cycle theoretically and experimentally. Here, overheated water is injected into the working chamber where it expands and evaporates. An approach for the theoretical investigation of screw expanders in trilateral flash cycles is presented by Vasuthevan and Brümmer (2016). An application of screw-type expanders in trilateral flash cycles is predicted to be a promising approach in the medium and low power operating range in ORC by Ohman and Lundqvist (2015).

In order to determine hydraulic friction losses in screw expander clearances using an auxiliary liquid, an analytical approach was presented by Gräßer and Brümmer (2014) with regard to water and by Gräßer and Brümmer (2015) to oil. Nikolov and Brümmer (2014) presented a water injected screw-type expander, focusing on an integral investigation of the test expander without differentiation between the impact mechanisms resulting from the auxiliary fluid. The most important expected advantage of water injection into screw-type expanders is sealing of clearances as well as lubrication of the hardened and hard-coated screw rotors during operation. At the same time depending on the liquid injection temperature, the auxiliary fluid lifts the temperature level within the working chamber due to its higher heat capacity. Compared to a dry-running expander, injected water reduces temperature drops during expansion which result in a slight difference between the inlet and outlet temperature of the working fluid, as well as lower thermal stress alongside the expander.

Kovacevic and Rane (2013) presented a 3D analysis of a twin screw expander. Indicator diagrams, mass flow rates and expander power at different operating conditions were estimated and compared with experimental results.

Cao *et al.* (2011) presented the results of an experimental investigation of pressure distribution inside the working chamber of a twin screw compressor for multiphase duties. Additionally, a mathematical model for describing the pressure distribution inside the working chamber is proposed.

Within the framework of the following paper, indicator diagrams of a water injected screw expander are presented. Based on the indicator diagrams, relevant influence mechanisms on the expander's operational behavior resulting from water injection are determined and explained in detail.

2. EXPERIMENTAL SETUP

The experimental rig, measurement devices which were used, and the test screw-type expander are introduced below. In addition, an overview of the test parameters is given.

Diverse measurement devices record information about the fluid state in the rig and the test screw-type expander. Insulated thermocouples and static relative pressure transducers are installed at different points of the test rig. Mean static pressure and temperature of the gaseous working fluid is metered in the test rig before water injection. Inlet and outlet static pressure as well as overall fluid temperature is recorded both in the high and low pressure domains of the test expander. The effective power of the test screw-type expander is calculated by means of rotational speed and driving torque metered using a torque transducer.

Within the framework of this paper, the results of experimental investigations on a water injected screw-type expander SE 51.2 without timing gears (Figure 1) are presented. The expander's geometrical parameters are listed in Table 1. SE 51.2 is a further development of a twin screw charger GL 51 designed at Chair of Fluidics by Temming (2007). The screw rotors are hardened and have a tough wear-protection coating, so that seizure can be avoided at dry running or water injection. Both fixed and loose bearing sets are grease lubricated, so that no oil supply is necessary. To avoid water flow through the fixed bearing sets on the high pressure expander side and protect them from damage, a pressure cushion at constant level near the expander's inlet conditions is set by a pressurized air in the coupling casing.

In the high pressure domain of SE 51.2, a variable plate, which is shaped like the inlet control edge, is used. Thus, the inner volume ratio and the utilizable pressure ratio can be adjusted. Moreover within the framework of experimental investigations, inlet throttling effects can be investigated in detail. However, the inner volume ratio is set constant for the experimental results presented below. Due to its modular structure, the geometry parameters of the screw-type expander SE 51.2 can easily be modified. Apart from front clearance heights, housing and intermesh clearance heights can be varied by changing the corresponding expander module.

In order to have a more detailed look into the physical effects of liquid phase within the working chamber, indicator diagrams for dry- and wet-running operation of SE 51.2 are recorded by means of six high-resolution absolute piezo-resistive pressure transmitters, Figure 1. Hereby, two different sensor types and sizes are used, which are installed flush with the rotor bore surface and at the high-pressure front side. The pressure transmitters of the series M5 manufactured by "KELLER AG für Druckmesstechnik" at positon 0, 1 and 6 record the static pressure at the inlet port (position 0), within a rotational angle in the range between 85° and 130° (position 1) and during the fluid discharge on the low pressure side of the expander (position 6). Each of these three sensors delivers the absolute pressure at the metering point. With regard to the manufacturer's data sheet of these transducers, a resonance frequency of more than 50 kHz can be expected. Nevertheless, a small cavity at sensor tip is constructively available. Therefore especially at water injection, acoustical Helmholtz resonances may occur as a result of the rotor tip gliding at the sensor. In this case, dynamic effects induce an impulse which can excite strong resonance oscillations at frequencies far lower than given natural frequency in the sensor data sheet. This aspect has to be considered within the evaluation of the pressure signal in order to avoid inaccurate interpretation. However, this sensor type is essential for the experiments to record the working chamber pressure for very small chamber volume at sensor position 1 in the high pressure front section, where an extremely narrow mounting space in the expander's casing is available.

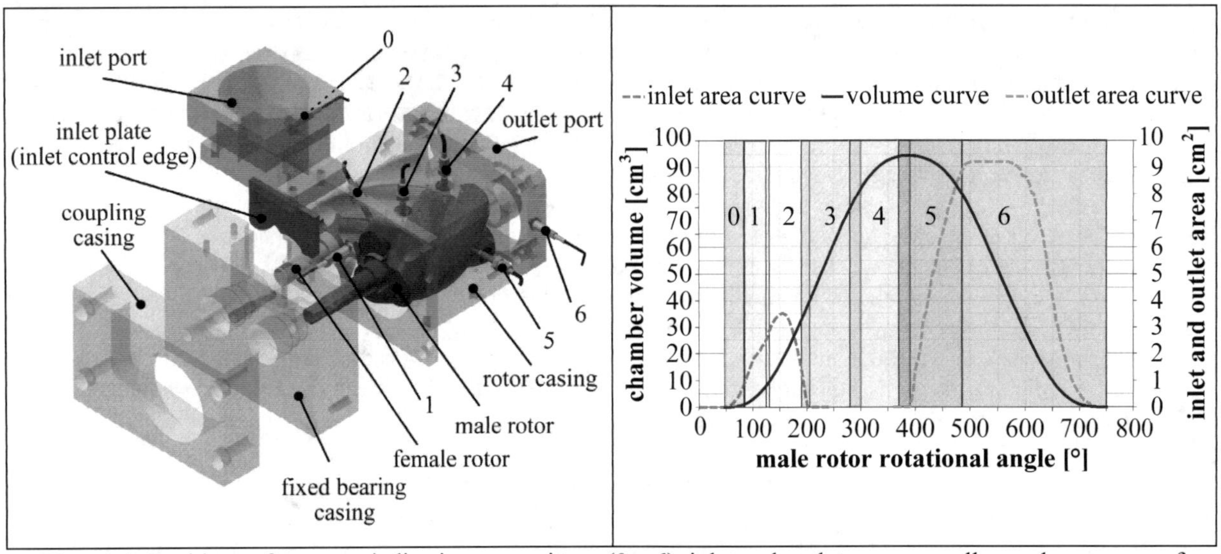

Figure 1: Positions of pressure indication transmitters (0…6), inlet and outlet area, as well as volume curve of the test screw expander SE 51.2

At positon 2 to position 5 XTM-190 pressure transducers manufactured by "Kulite Semiconductor Products, Inc" with natural frequency higher than 425 kHz are mounted flush with the rotor bore surface and have no dead space between their tip and the considered working chamber. Hence, no acoustic issues at the metering point with regard to acoustic Eigen frequencies are expected. Within the calibration of these sensors, the pressure sensitivity proved to be nearly constant. However, a relatively great temperature depending offset could be observed. Hence, insulated thermocouples are installed at each pressure measurement position in order to compensate the impact of temperature changes on the sensor sensitivity.

Table 1: Parameters of the test screw expander SE 51.2

designation	unit		designation	unit	male rotor	female rotor
axis-center distance a	[mm]	51.2	number of lobes z	[-]	3	5
internal volume ratio v_i	[-]	2.5	diameter d	[mm]	71.8	67.5
displaced volume per male rotor rotation V	[cm^3]	286	wrap angle φ	[°]	200	-120
front clearance height $h_{fc,hp}$ (high pressure)	[mm]	0.1	rotor lead s	[mm]	181.8	-303
front clearance height $h_{fc,lp}$ (low pressure)	[mm]	0.17	rotor length l	[mm]	101	
housing clearance height h_{hc}	[mm]	0.08	rotor profile	[-]	modified asymmetric SRM	

Due to relatively high temperature sensitivity and offset of this sensor type, only the relative change in pressure to a reference pressure level is considered. At this point, the reference pressure is determined at the overlapping scope of two adjacent transducers (Figure 1). To begin with, the static pressure in the outlet port of the screw expander is considered as the starting reference pressure.

Within the scope of this indication pressure measurement, a rotational angle range between 130° and 140° cannot be covered due to lack of space in the casing, so the pressure has to be estimated. Preliminary analysis of the recorded pressure delivered a linear pressure distribution in the missing range to be qualitatively and quantitatively sufficiently accurate. For the area which was not surveyed at the beginning of the working cycle from 45° (start of chamber filling) to 85° as shown in Figure 1, the dynamically recorded inlet port pressure is considered identical to the chamber pressure. In this way, the assumed chamber pressure is mapped with regard to an ideal chamber filling without pressure drop over the inlet. In theory, this results in an upper limit for the converted internal working of the screw expander within this working cycle range. Another approach to map the chamber pressure profile in this section is to assume a linear pressure increase starting from the outlet and ending at the pressure level of the transmitter at position 1. This method is more appropriate with respect to increasing inlet throttling losses either at higher expander's rotational speed or increasing amount of injected water. Therefore, the pressure level and the induced inner work within the missing range of the indicator diagram are expected to be lower than the real one during the chamber growth. This approach represents a lower limit for the converted inner work within the first 40° of chamber formation between 45° and 85°.

The relation between the time dependent high-resolution pressure signals and rotational angle and chamber volume is provided by an optical trigger mounted on the male rotor shaft. Thereby, a TTL impulse is generated once per male rotor revolution, which corresponds to a defined position of both rotors in relation to each other and the working chamber volume. Between two trigger signals, an equidistant correlation between time and rotational angle axis is considered. In terms of sufficient high resolution of the time depending signals, a sampling rate of 100 kHz with regard to pressure transducers and trigger signal is provided.

Within the framework of this experimental investigation, both the system and the screw expander's operating parameters were varied. Water was injected at temperature levels of $\vartheta_{i,w} \approx 30$ °C, $\vartheta_{i,w} \approx 60$ °C, and $\vartheta_{i,w} \approx 90$ °C for a constant air inlet pressure of $p_i = 4 \cdot 10^5$ Pa and back pressure of $p_o = 1 \cdot 10^5$ Pa and at hot air temperature of $\vartheta_{i,a} = 90$ °C. The resulting mixture temperature at the expander's inlet is $\vartheta_i < 40$ °C, $\vartheta_i \approx 50$ °C, and $\vartheta_i > 60$ °C respectively depending on water volume flow and expander rotational speed.

With regard to steam or vapour mass fraction, the amount of injected water into the expander's inlet port is represented by the relative air mass fraction related to the overall air and water mass flow:

$$x = \frac{\dot{m}_a}{\dot{m}_a + \dot{m}_w}. \quad (1)$$

Here, measured air mass flow $\dot{m}_a$ (Coriolis mass flow meter) corresponds to almost dry air, since a refrigerant type dryer is used after air compression. Air mass fraction of $x = 1$ corresponds to a dry-running operation and $0 < x < 1$ to water injection operation and a two-phase flow respectively. The air mass fraction x is varied in the range from 1.0 to 0.4 depending on water temperature and expander's rotational speed. For water injection, two nozzle sizes are used at two adjacent water flow ranges with respect to a satisfactory water injection quality at the expander's inlet. The maximum expander rotational speed is $n_{MR} = 18{,}000$ min^{-1}, which equals a male rotor circumferential speed of $u_{MR} = 67.7$ m·s^{-1}.

3. CHARACTERISTIC NUMBERS

The operational behaviour of screw-type expanders is evaluated by means of characteristic numbers. Effects such as the sealing of expander clearances, inlet pressure drops, hydraulic losses, etc., can be evaluated integrally by taking the system and expander operating parameters into account. The characteristic numbers used here are delivery rate and mechanical efficiency.

3.1 Delivery Rate

Delivery rate is a characteristic number which describes different loss mechanisms such as internal or external leakages, inlet pressure drops during the chamber filling, or thermal effects within a displacement machine. For a screw-type expander, delivery rate is defined as:

$$\lambda_L = \frac{\dot{m}_a}{\dot{m}_{th}}. \quad (2)$$

Here, $\dot{m}_{th}$ represents theoretical air mass flow with:

$$\dot{m}_{th} = V_{th,ex} \cdot \rho_{da,i} \cdot n_{MR} \cdot z_{MR}. \quad (3)$$

The parameters z_{MR} and n_{MR} correspond to male rotor lobes and male rotor rotational speed respectively. Displaced chamber volume by the liquid can be ignored due to a relatively high volume fraction of air, at over 98.5 % even at the lowest air mass flow and highest injected water volume flow rate. The parameter $V_{th,ex}$ refers to chamber volume after completing the chamber filling. Moreover, dry air is regarded as an ideal gas and its density $\rho_{da,i}$ is considered with reference to calculated subcooled temperature after water was injected. In terms of delivery rate, measured air mass flow $\dot{m}_a$ does not consider evaporated water steam with respect to saturated humid air after water injection, because the Coriolis mass flow meter is installed upstream of the water injection nozzle. In some cases at high temperature levels and low pressure at the same time, evaporated steam fraction has to be taken into account, since it significantly increases. Within the following investigations, steam mass flow increases up to 4.7 % of metered air mass flow at the highest injected water temperatures. Since steam mass flow is not measured by the Coriolis flow meter, an additional decrease of delivery rate of a maximum of 4.7 % is determined at some operating points.

3.2 Mechanical Efficiency

Mechanical efficiency η_m is a characteristic number, which gives information about the amount of mechanical and hydraulic friction losses in screw expanders relating effective and indicated power to each other:

$$\eta_m = \frac{P_e}{P_{ind}}. \tag{4}$$

Effective power results from the recorded expander's rotational speed and torque:

$$P_e = 2 \cdot \pi \cdot n_{MR} \cdot M_{MR}. \tag{5}$$

Indicated power can be calculated by means of time-dependent working chamber pressure p and volume V, male rotor rotational speed n_{MR} and number of male rotor lobes z_{MR} as follows:

$$P_{ind} = -\, n_{MR} \cdot z_{MR} \cdot \oint p \cdot dV. \tag{6}$$

Within the investigated screw expander, friction losses P_ϕ result from the contact between both screw rotors, in the bearings, in sealing elements, as well as from hydraulic friction of the injected auxiliary fluid within clearances and at rotor surfaces. Friction losses can be calculated by means of Equation (7) as follows:

$$P_\phi = P_{ind} - P_e. \tag{7}$$

4. INDICATOR DIAGRAMS

In the following section, indicator diagrams of the test screw expander are presented for dry-running and water-injection operation. Indicator diagrams deliver an insight into the inner fluid-mechanical and frictional effects of the test screw expander. Statements about inner dissipative mechanisms during each working cycle are thus possible with reference to the chamber pressure distribution per working cycle. Hereby, relevant effects of water injection on the chamber state as well as specific issues concerning the interpretation and evaluation of locally recorded pressure levels are considered. Each indicator diagram, presented in the following, is calculated by averaging the chamber pressure of all resulting working cycles within a given time frame of one second at a rotational speed of 500 min^{-1} and 0.5 s at higher rotational speeds.

4.1 Dry-running Operation

In the following, the chamber pressure dependency on the expander's rotational speed for an inlet pressure of $4 \cdot 10^5$ Pa and temperature of 90 °C is examined, Figure 2. In addition to chamber pressure, the measured signals of the dynamic pressure transducer at position 0 in the inlet port are depicted.

The chamber pressure at male rotor circumferential speeds of $u_{MR} = 9.4\ \text{m} \cdot \text{s}^{-1}$ ($n = 2{,}500\ \text{min}^{-1}$), $u_{MR} = 22.6\ \text{m} \cdot \text{s}^{-1}$ ($n = 6{,}000\ \text{min}^{-1}$), and $u_{MR} = 67.7\ \text{m} \cdot \text{s}^{-1}$ ($n = 18{,}000\ \text{min}^{-1}$) is illustrated. Two significant mechanisms according to the operation of screw expanders can be observed. On the one hand, internal leakages at lower rotational speeds impact the chamber pressure in the range of male rotor rotational angle between nearly 200° and 390°. On the other hand, inlet throttling losses increase at higher rotational speed, whereas the internal leakages decline due to the reduced working cycle period. Internal leakages decrease at operating points with higher rotational speeds because less time is available for the mass exchange between the working chambers.

Particularly remarkable is the decreasing chamber pressure level up to a rotational angle of 200° at high rotational speeds. According to Equation 8, flow velocity has a greater influence on the throttling losses than the inlet density, since the inlet pressure losses do not increase linearly but by the square of flow speed:

$$\Delta p_{\phi} = \zeta \cdot \rho \cdot \frac{c^2}{2}. \quad (8)$$

The relation between increasing inlet throttling and declining internal leakages at high rotor circumferential speed results in overexpansion, which means that the chamber pressure expands lower than the backpressure level.

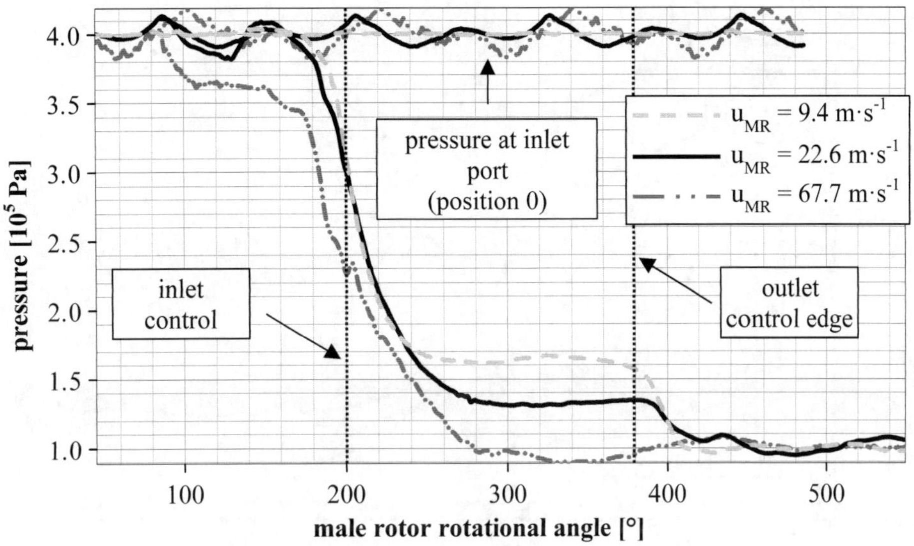

Figure 2: Indicated working chamber and inlet port pressure depending on male rotor rotational angle at an inlet pressure of $4 \cdot 10^5$ Pa and temperature of 90 °C and different male rotor circumferential speeds (no water injection)

Obviously, a periodic pressure pulsation specific for volumetric displacement machines can be detected at the high and low pressure port of the screw expander. The amplitude of the pulsations increases at higher rotational speeds. The recorded pressure pulsations in the inlet port at position 0 are comparable in amplitude and phase with those in the working chamber while it is connected to the inlet port. Thus, an impact of this pressure pulsation can be expected within the working chamber unless it is not disconnected from the inlet port. This effect is known in the literature as resonance charging. Depending on the phase of the oscillations, the chamber filling might be affected differently. Hereby, a local pressure distribution, possibly related to an acoustic standing wave, can be expected. The pressure recorded by the pressure transducers is presumably only a local instantaneous value. Here, a local pressure differing from the mean chamber pressure is possible. Last but not least, the transient inflow into the working chamber and cross-sectional jump over the inlet offer potential for vortices and separated flow regions in the chamber while it is connected to the inlet port. In this way, the mapping of the indicator diagram can lead to deviation from the real inner power, which has to be considered.

4.2 Water Injection

Variation of injected water is carried out in terms of air mass fraction and water temperature. Different water droplet size and thus injection quality, depending on the applied nozzle size, result in changing heat transfer between liquid and gas. At constant injected water volume flow, larger nozzle outlet diameter results in larger droplet size and thus decreasing heat exchange between water and air during the entire working cycle. Hereby at the expander's inlet, the mixture temperature after water injection corresponds to a thermal equilibrium between both fluids.

In Figure 3, indicator diagrams at male rotor circumferential speeds of $9.4\ m \cdot s^{-1}$ and $67.7\ m \cdot s^{-1}$ are presented for different injected water flow with respect to air mass fraction. Hereby, indicator diagrams at air mass fractions of

0.9 and 0.5 are compared with the dry-running operation. In terms of water injection, four significant mechanisms influence the screw expander operation and the pressure situation within the chamber during a working cycle. The first relevant effect on the expander's indicator diagram is the heat transfer from a liquid into an expanding gaseous phase. Due to its higher heat capacity, water increases the pressure level at a constant chamber volume by rising the chamber gas temperature. In Figure 3 considering the expanding chamber after passing over the inlet control edge, the effect of heat transfer can be observed in the indicator diagrams for both low and high male rotor circumferential speeds. Here, the higher the amount of injected water is, the higher the heat transfer and the pressure lift are.

The second significant effect of water injection on the time dependent chamber pressure results from blocking clearances within the screw expander. Theoretically, even low amounts of water partially block the clearances connecting two adjacent volumes in the expander, so they become "sealed" and the internal leakage declines. Hence, a pressure decrease can be expected after finishing the chamber filling, since the expanding chamber cannot be refilled by leakages from the following chamber at a higher pressure level. Nevertheless since the heat transfer effect seems to be more dominant with regard to pressure change, the chamber pressure in general increases for both male rotor circumferential speeds in Figure 3.

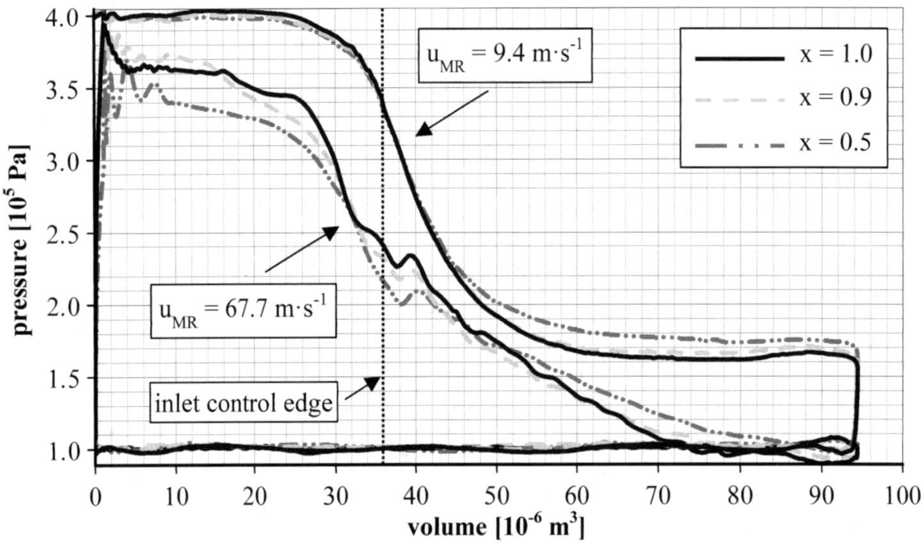

Figure 3: Indicator diagrams at male rotor circumferential speeds of $9.4\ m \cdot s^{-1}$ and $67.7\ m \cdot s^{-1}$, different air mass fractions, inlet pressure of $4 \cdot 10^5$ Pa, air temperature of 90 °C, and water injection temperature of 60 °C

The third relevant influence mechanism is characterized by increasing inlet throttling at higher amounts of injected water due to declining speed of sound in terms of two-phase flow during the chamber filling. At higher male rotor circumferential speeds especially, reduced speed of sound results in a significant drop in pressure in the working chamber while it is connected to the inlet port.

The fourth important influence mechanism of water injection on the indicator diagrams results from hydraulic friction between water and the expander's rotors and casings due to the relatively high viscosity of the liquid compared to air. At this point primarily in the tight expander's clearances, dominant hydraulic frictional losses can be expected. Nevertheless, hydraulic friction losses cannot be directly evaluated by means of the indicator diagram at this point.

With respect to the pressure characteristic at a male rotor circumferential speed of $9.4\ m \cdot s^{-1}$, the pressure drop during the chamber filling at increasing water flow results from a reduced speed of sound at the expander inlet. The adjacent pressure lift during the chamber expansion can be traced back to the increasing heat flow from the liquid into the air which dominates the clearance sealing effect. That means if heat exchange between both fluids can be neglected, the pressure level has, in general, to decrease due to reduced internal leakage.

Considering the indicator diagrams at a male rotor circumferential speed of $67.7\ m \cdot s^{-1}$, a significant pressure drop over the inlet can be observed. Apart from particular inlet blocking by the liquid at the begin and end of chamber filling, a reduced speed of sound of the air-water mixture results in increasing inlet throttling. After the working chamber is disconnected from the inlet port, the chamber state mainly changes due to heat flow from water into air, lifting the chamber pressure level. Since the relevance of internal leakage is utterly low-dominant at high rotational

speeds, no significant pressure drop due to decreased internal leakage in this range can be expected. Due to heat input from the liquid into the air phase, no overexpansion can be observed at a male rotor circumferential speed of $67.7\ m \cdot s^{-1}$.
In Figure 3, a boundary effect of the time-dependent chamber pressure measurement is obvious at a male rotor circumferential speed of $67.7\ m \cdot s^{-1}$. Here at pressure metering position 1 up to a chamber volume of ca. $8 \cdot 10^{-6}\ m^3$, acoustical Helmholtz resonances can be observed due to the sensor geometry as explained above.
The effect of water injection temperature is depicted in Figure 4. The indicator diagrams at both presented male rotor circumferential speeds are influenced by the liquid temperature. At low rotational speed which provides enough time to complete heat exchange between water and air and at which no significant hydraulic friction losses and inlet throttling can be expected, chamber pressure increases due to higher heat transfer from water into air at a constant chamber volume and increasing water temperature. Even if the time for heat flow and complete heat exchange at higher male rotor circumferential speed is not sufficient, higher chamber pressure can be observed at higher water injection temperatures. On the one hand, this can be traced back to declining inlet throttling due to the higher speed of sound at higher temperature levels. On the other hand, heat flow into air increases.

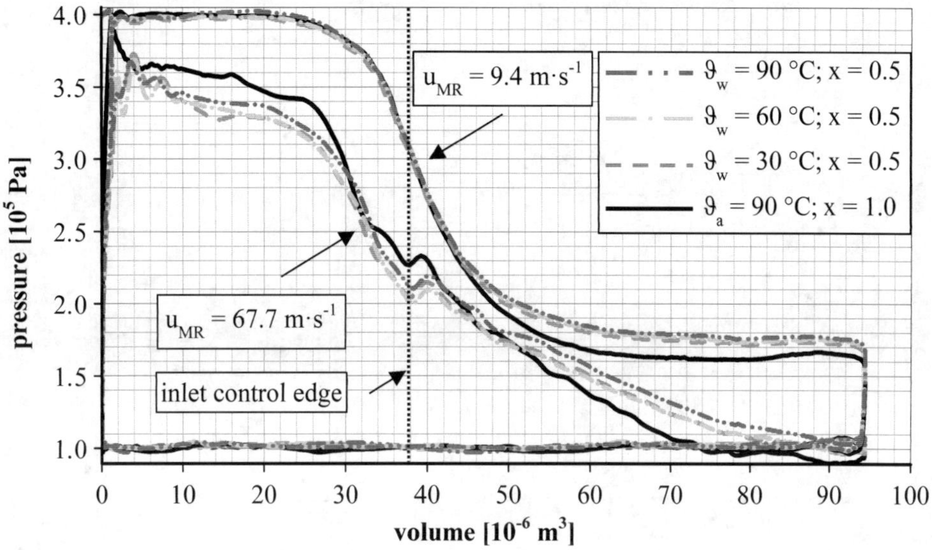

Figure 4: Indicator diagrams at male rotor circumferential speed of $9.4\ m \cdot s^{-1}$ and $67.7\ m \cdot s^{-1}$, inlet pressure of $4 \cdot 10^5$ Pa, air temperature of 90 °C, dry-running operation ($x = 1$), and different water injection temperatures at $x = 0.5$

5. CHARACTERISTIC DIAGRAMS

Characteristic diagrams of SE 51.2 are presented in Figure 5. Here, indicated power, solid and hydraulic friction losses, delivery rate, and mechanical efficiency as a function of male rotor circumferential speed at different air mass fractions and water temperatures are depicted.
With regard to indicated power determined from the indicator diagrams, a nearly linear increase in power up to a male rotor circumferential speed of $30\ m \cdot s^{-1}$ can be observed. Afterward, the power growth flattens out due to progressively increasing inlet throttling. Comparing different amounts of injected water, indicated power first slightly increases in relation to dry-running operation due to a chamber pressure rise resulting from heat input. However, at higher rotational speed and water flow, indicated power decreases since inlet throttling becomes more dominant. Here, an increasing inlet throttling in terms of reduced speed of sound seems to be the more dominant effect. With regard to decreasing water temperatures at an air mass fraction of 0.5, lower indicated power can be observed due to higher inlet throttling and lower heat input.
Friction losses represent the entire losses within the screw expander in terms of the relative movement of the screw expander's components. Here as expected, relatively low frictional losses occur at low rotational speeds independent of whether dry-running operation or water injection is considered and at which temperature water is injected. At increasing injected water flow and high expander's revolution speeds, hydraulic friction losses within the relatively tight clearances increase. In general, solid friction slightly increase at dry running ($x = 1$) and higher

rotational speeds. However, solid body friction between both rotors in the intermesh point seems to slightly decrease in the presence of a small amount of water (x = 0.9) presumably as a result of the more beneficial lubrication by water than dry air. With respect to water temperature, an increase in friction losses at lower water injection temperatures is obvious, which can be traced back to higher liquid's viscosity.

In terms of both dry-running operation and water injection, delivery rate decreases as a function of male rotor circumferential speed due to declining internal leakage and increasing inlet throttling. At the same time, an increase of water flow at low rotational speeds results in a significant drop of delivery rate according to more efficient clearance sealing. In contrast, increasing water flow at high rotational speeds reduces the delivery rate due to increasing inlet throttling. Considering water temperature, delivery rates do not change significantly (not presented here due to clarity reasons).

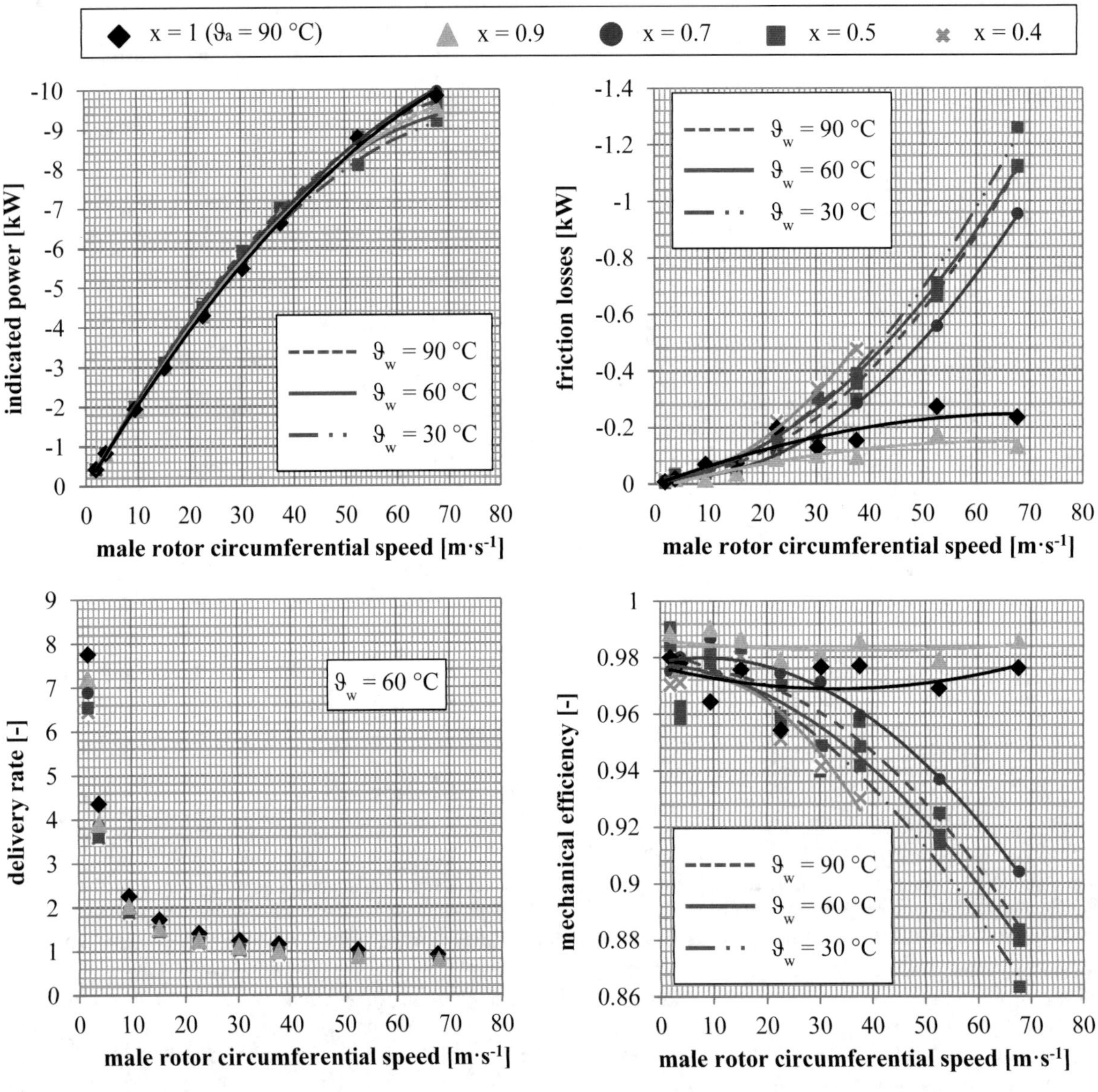

Figure 5: Indicated power, friction losses, delivery rate, and mechanical efficiency as a function of male rotor circumferential speed at different air mass fractions and water injection temperatures ($x = 0.5$ at $\vartheta_w = 30$ °C, $\vartheta_w = 60$ °C, and $\vartheta_w = 90$ °C; $x = 0.9$, $x = 0.7$, and $x = 0.4$ at $\vartheta_w = 60$ °C)

Mechanical efficiency in Figure 5 represents the range of internal losses in the expander, including solid friction between both rotors, in the bearings and contact seals, as well as hydraulic losses due to water injection. With regard to a dry-running expander (x = 1), almost constant mechanical efficiency in the range between 0.97 and 0.99

can be observed. Here, no significant dependency on the rotational speed can be detected. Similar to the friction losses in Figure 5, mechanical efficiency first slightly increases at x = 0.9 and then systematically decreases, particularly at increasing air mass fraction and increasing rotational speeds. At higher water temperatures and a constant air mass fraction of x = 0.5, hydraulic friction declines and, thus, the mechanical efficiency increases moderately. Mechanical efficiency values deviating from the trend line with a second-degree polynomial used to approximate the progress of this characteristic number can be traced back to the mentioned issues evaluating the pressure measurement signals with respect to pressure distribution in the chamber during the chamber filling.
In summary, mechanical efficiency greater than 0.94 and relatively low mechanical losses even at very low air mass fraction and male rotor circumferential speeds up to 30 m/s are detected. In this characteristic operating range for liquid-injected screw expanders, satisfactory sealing of the clearances and heat input from the auxiliary fluid into the gaseous phase during expansion is available. At male rotor circumferential speeds greater than 30 $m \cdot s^{-1}$ hydraulic friction increases, so this range is not appropriate for wet-running expanders at high liquid flows.

5. CONCLUSION

This paper contains the results of an experimental investigation on indicator diagrams of a water injected twin-shaft screw-type expander SE 51.2. By means of the time-dependent chamber pressure distribution, relevant effects on the operational behavior of dry-running and liquid-injected screw expanders are presented.
Experimental investigations on water injected screw-type expanders by means of indicator diagrams show detailed information about the expander's operation with regard to the most relevant mechanisms resulting from water injection. The investigations provide a solid basis for validating and verifying models for hydraulic losses and heat transfer with respect to the operation of screw expanders. The aim of further investigations on liquid injected screw-type expanders is to provide an enhancement of the chamber model simulation tool KaSim developed at the Chair of Fluidics by Janicki (2007) with regard to adequate prediction of the aforementioned mechanisms.

REFERENCES

Cao, F., Gao, T., Li, S. Xing, Z., Shu, P. (2011). Experimental analysis of pressure distribution in a twin screw compressor for multiphase duties. *Experimental Thermal and Fluid Science, Volume 35, Issue 1, January 2011*, 219–225.

Gräßer, M. & Brümmer, A. (2014). An analytical model of the incompressible one-phase clearance flow in liquid injected screw expanders. VDI Bericht 2228, in *9th International Conference on Screw Machines 2014, Dortmund, Germany* (71–89).

Gräßer, M. & Brümmer, A. (2015). Influence of liquid in clearances on the operational behavior of twin screw expanders. In 9th *International Conference on Compressors and their Systems, IOP Conf. Series: Materials Science and Engineering 90 (2015)* 012060.

Janicki, M. (2007). *Modellierung und Simulation von Rotationsverdrängermaschinen*. Dissertation, TU Dortmund University, Germany.

Kliem, B. (2005). *Fundamentals of the Two-Phase Screw-Type Engine*. Dissertation, TU Dortmund University, Germany.

Kovacevic, A. & Rane, S. (2013). 3D CFD analysis of a twin screw expander. *8th International Conference on Compressors and their Systems, City University London.*

Nikolov, A. & Brümmer, A. (2014). Influence of water injection on the operating behaviour of screw machines. VDI Bericht 2228, in *9th International Conference on Screw Machines 2014, Dortmund, Germany* (43–60).

Ohman, H. & Lundqvist, P. (2015), Screw expanders in ORC applications, review and new perspective. In *3rd International Seminar on ORC Power Systems, Brussels, Belgium* (503–512).

Rinder, L., Kuchler, M., & Hackl, H. (2004), Water-Injection or Oil-Injection for Screw-Compressors in Comparison. *Schraubenmaschinen, Nr. 12 - 2004*, 65–74.

Rinder, L. & Moser, I. (1990). Untersuchung der Ölverteilung in den Arbeitsräumen naßlaufender Schraubenkompressoren. *VDI Berichte 859 - 1990*, 1–14.

Temming, J. (2007). *Stationärer und instationärer Betrieb eines unsynchronisierten Schraubenladers*. Dissertation TU Dortmund University, Germany.

Vasuthevan, H. & Brümmer A. (2016). Thermodynamic Modeling of Screw Expander in a Trilateral Flash Cycle. In *23rd International Compressor Engineering Conference at Purdue University*, number 1432.

Zellermann, R. (1996). *Optimierung von Schraubenmotoren mit Flüssigkeitseinspritzung.* Dissertation, TU Dortmund University, Germany.

Performance Investigation on Electrochemical Compressor with Ammonia

Ye Tao[1], Yunho Hwang[1*], Reinhard Radermacher[1], Chunsheng Wang[2]

[1]Center for Environmental Energy Engineering, University of Maryland,
4164 Glenn L. Martin Hall Bldg., College Park, MD 20742, USA
[2]Department of Chemical and Biomolecular Engineering
1223A Chemical & Nuclear Engineering Bldg. College Park, MD 20742, USA
*: Corresponding Author, Tel: (301) 405-5247, E-mail: yhhwang@umd.edu

ABSTRACT

The mechanical compressor on small cooling appliances has considerable energy consumption to the entire vapor compression system. To improve the efficiency of the compressor in the vapor compression system, an electrochemical compressor (EC) was suggested and its performance was investigated. The EC compresses refrigerant by charging an external voltage to the membrane electrolyte assembly (MEA). In this study, ammonia was used as a working fluid due to its high thermodynamic characteristics and electrochemically reactivity with the MEA. This paper mainly focuses on the thermodynamic compression limit at the open loop and the EC performance at the closed loop. The thermodynamic compression limit of the EC at open loop verifies the voltage and compression ratio relationship based on the theoretical Nernst equation. For the closed loop experimental study, the current density of the MEA was investigated with voltage at constant refrigerant charge. The pressure lift across the EC was measured as well. The measured flow rate was in the same order of magnitude of the theoretical values. In conclusion, the ammonia EC can be a new compressor option for the vapor compression system due to its potential energy saving, no moving parts and no lubrication oil.

1. INTRODUCTION

The mechanical compressor used in small cooling appliances such as cooler or freezer has an isentropic efficiency of around 65%. It is also a fairly complicated system with moving parts and lubrication oil. The electrochemical ammonia compressor is proposed to improve the efficiency of vapor compression cycle without using any moving parts and lubrication oil. The electrochemical compressor (EC) pumps gas molecules across a membrane electrolyte assembly (MEA) when charged with an external voltage. The working principle is similar to the reverse of proton exchange membrane fuel cell design. The PEM fuel cell produces electricity by supplying hydrogen and oxygen gases. The EC however, pumps gas by supplying electricity. The electrochemical hydrogen compressor has been built and studied by different groups (Rohland et al., 1998; Strobel et al., 2002; Onda et al., 2007; Gardner and Ternan, 2007; Grigoriev et al., 2011). The focus of this paper is on using the hydrogen transfer feature of proton exchange membrane to let hydrogen drag ammonia across the membrane for compression, which was never been previously reported. The working principle is shown in Figure 1. At the EC suction line, both ammonia and hydrogen are supplied. With the help of Pt catalyst, chemical reactions will take place and the generated NH_4^+ ion will transfer across the membrane with the influence of the electric field. On the other side of the membrane, the NH_4^+ ion will break down to ammonia and hydrogen. The MEA also needs to be humidified to have good NH_4^+ conduction. If the external flow from the EC discharge-side to feed-side is regulated, the gases can be compressed through the EC. Based on the chemical equation shown in Figure 1, the transfer ratio between NH_3 and H_2 is 2:1 due to ionic bonding. The steady transfer rate of NH_3 can be achieved at a constant voltage charge.

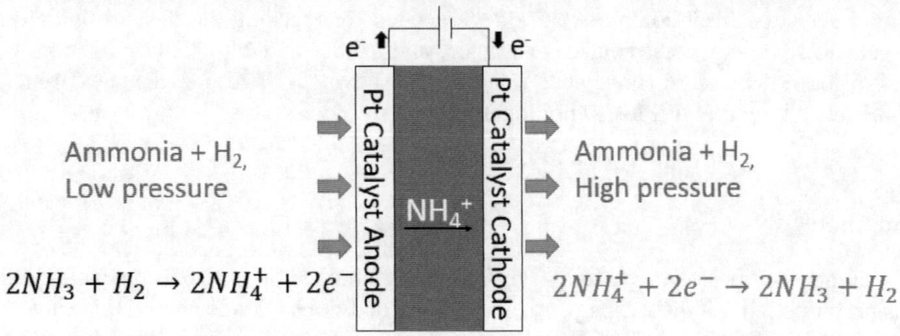

Figure 1: Electrochemical NH_3 compressor using H_2 as carrier gas

2. EXPERIMENTS

2.1 Thermodynamic compression ratio

The EC performance is governed by the Nernst equation, which gives the relationship between pressure ratio and voltage supplied. The discharge pressure is increased by closing the discharge line valve. Therefore, no flow is in the discharge line. The purpose of this experiment is to show that the gas pressure in the suction line and discharge line achieved by the ammonia EC satisfies the Nernst equation. As Figure 2 shows, the EC is charged with constant voltages at 0.05 V, 0.1 V, 0.15 V, and 0.2 V. With the discharge line valve closed, the gas therefore accumulates and pressure builds up. As the pressure increases, the back diffusion across the MEA also occurs at increasing rate. Therefore, eventually the gas transport rate across the MEA will be equal to the back diffusion flow rate and pressure stops building up. The equilibrium pressure and voltage were recorded when pressure in the discharge line was no longer increasing.

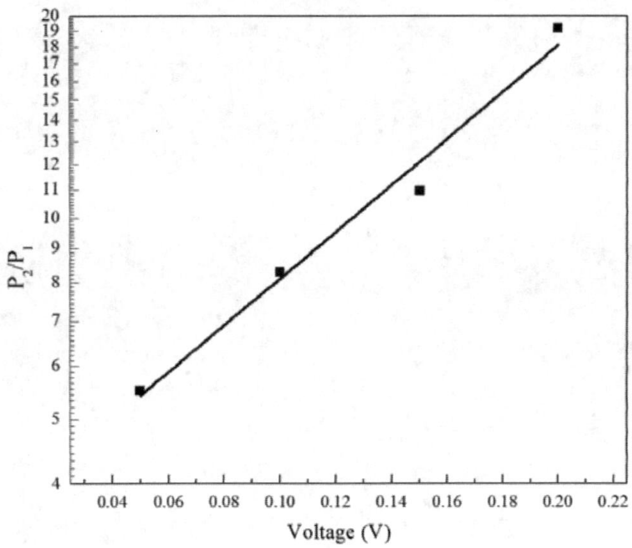

Figure 2: Ammonia EC thermodynamic compression limit

According to the Nernst equation shown in Eq. 1, the log scale of pressure ratio follows linear increase with respect to the cell voltage charge.

$$E = \frac{RT}{nF} \ln\left(\frac{P_d}{P_s}\right) \quad (1)$$

where P_d represents the discharge pressure, and P_s represents the suction pressure, n represents the number of moles of electrons released by each H_2 molecule transferred across the membrane, E represents the EC cell voltage charge.

The Nernst equation gives the ideal compression voltage needed for reaching the desired partial pressure ratio. In reality, the total cell internal resistance, contact resistance, and anode- and cathode-polarization will be taken into account as parasitic loss. Therefore, the linear relationship shown in Figure 2 deviates from the ideal linear relationship, but the overall trend is still following logarithm.

2.2 Closed loop analysis

In order to test the compressor prototype performance for the potential vapor compression cycle application, a closed loop analysis was performed to test the EC performance in real conditions with discharge line valve partially opened. As previously described, the EC builds up pressure by pumping gas molecules across the MEA, therefore regulating the discharge side flow back to the suction line needs to be regulated. The closed loop analysis gives real time performance of the EC. At this moment, the small EC prototype has only very limited flow rate. In order to have flow rate that is measureable, the EC needs to be scaled up by increasing the MEA surface area and multiply stacks. The system was pulled vacuum before charging 2 bar of mixture of H_2 and NH_3 at 1:2 ratio. Figure 3 shows the closed loop design of EC with suction line and discharged line marked. The suction line uses 6.4 mm diameter tubing and 50 cc cylinder for gas reservoir. The discharge line uses 3.2 mm diameter tubing. Figure 4 shows the closed loop schematic of the ammonia EC. Both suction side (anode) pressure and discharge side (cathode) pressure were measured by pressure transducers. As previously discussed in Section 1, the MEA needs to be humidified to be able to transport both H_2 and NH_3, so that the refrigerant loop was heated by a tape heater to keep temperature constant at 60ºC to have reasonable RH of 70% for enough membrane humidification. For the real refrigeration system, the tape heater can be removed and a suction line humidifier would be installed to maintain the humidity of MEA. The humidity is required at a high level in order to keep the ionic conductivity of proton exchange membrane for high ammonia transfer rate. Water is not interacting with the working fluid but only serves as a need for the membrane. Ammonia as the working fluid will be brought over by hydrogen only. The EC performance at different voltage charges was measured in terms of pressure drop and current.

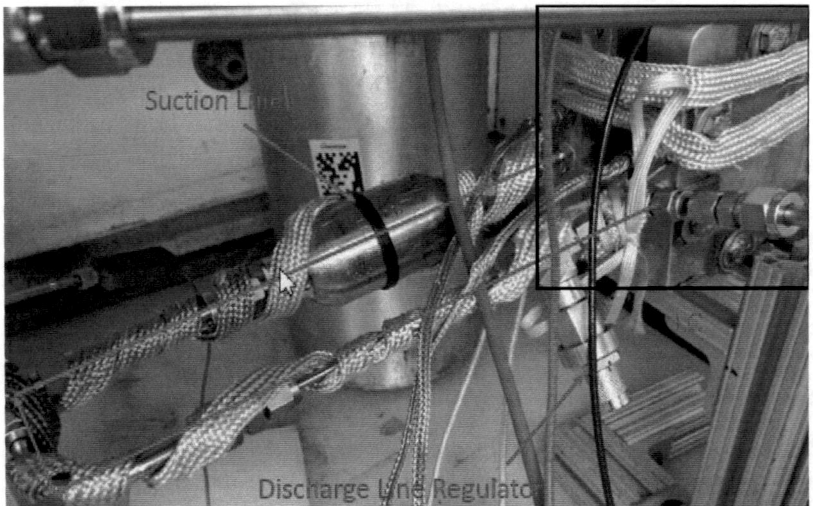

Figure 3: Closed loop analysis with suction and discharge line shown.

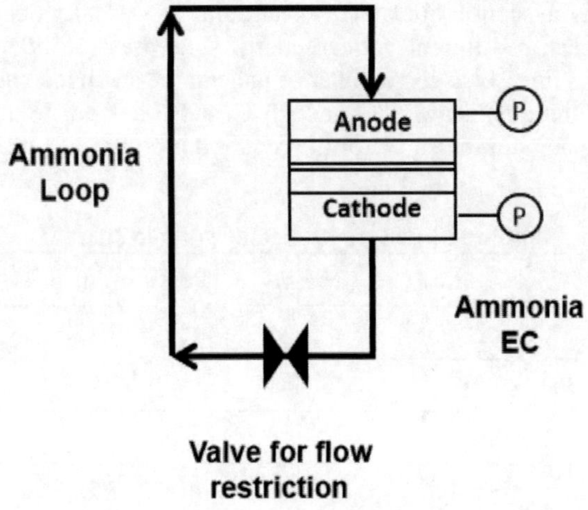

Figure 4: Schematic of EC closed loop

The EC single cell was charged with a voltage for 1,500 seconds and the current was measured as well as the pressure lift from the suction side to the discharge side. Figure 5 shows the current versus voltage graphs. As the voltage charge increases, the current increases as well. The EC was able to maintain constant pressure lift at fixed voltages. The flow rate across the EC can be calculated based on the current by using the Faraday's Law.

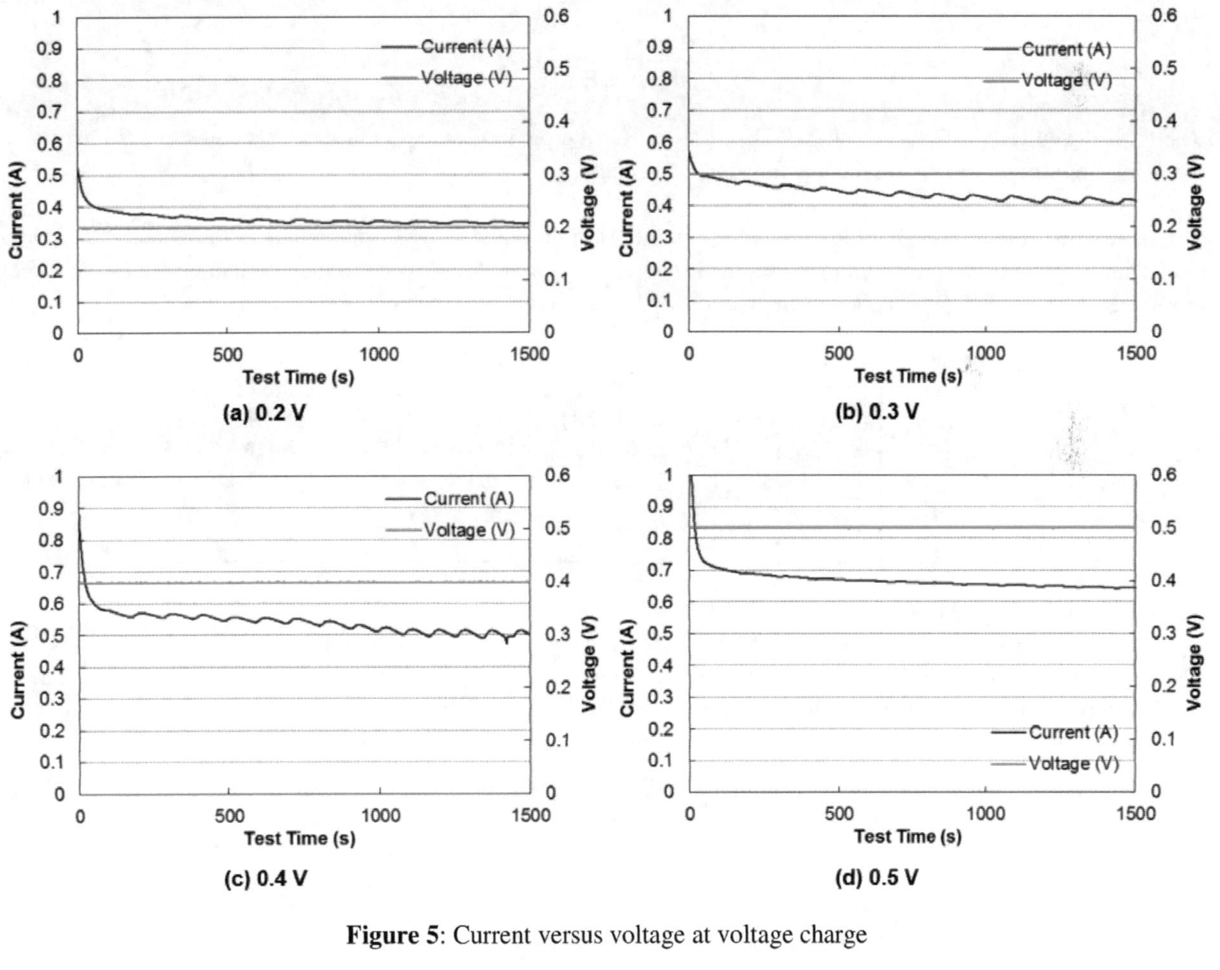

Figure 5: Current versus voltage at voltage charge

The pressure lift was measured at the suction line and discharge line by partially opening the metering valve. Table 1 shows the pressure lift with respect to different voltage charge. The pressure lift of a single EC unit is very low. However, it is expected to increase linearly as the number of units in the stack increases for scale up. The evaporating and condensing temperatures of ammonia are 5°C and 45°C at 5 bar and 18 bar, respectively. For the future development, 1 EC with multiple units in the stack for the required pressure lift will be built and tested for a 200 W cooling system.

Table 1: Pressure lift versus voltage charge

Voltage (V)	Pressure lift (kPa)
0.2	19
0.3	23
0.4	26
0.5	30

2.3 Flow rate measurement

The transfer rate was measured and calculated using titration and results were shown in Table 2. The experimental flow rate was calculated by the amount of hydrogen and ammonia determined by titration over time. The theoretical flow rate was calculated based on the Faraday's Law in Eq. 2.

$$\frac{dn_e}{dt} = \frac{I}{nF} \tag{2}$$

where I is the current, n_e is the number of electrons transferred per molecule, in this case n is equal to 2 for hydrogen and 1 for ammonia, and F represents the Faraday's constant.

The results in Table 2 show that the measured flow rate is in the same order of magnitude with the theoretical flow rate. The discrepancy between the measured and theoretically predicted flow rates may be caused by H_2 and NH_3 cross over, which is a common phenomenon associated with proton exchange membrane.

Table 2: Comparison between measured flow rate and theoretical flow rate

Measured Current (A)	Theoretical flow rate H_2 (g/s)	Theoretical flow rate NH_3 (g/s)	Measured flow rate H_2 (g/s)	Measured flow rate NH_3 (g/s)
0.14	1.45e-06	2.47e-05	1.85e-06	3.22e-05
	1.45e-06	2.47e-05	1.72e-06	3.16e-05
	1.45e-06	2.47e-05	1.78e-06	3.29e-05

2.4 EC scale up calculation

Since the flow rate for a single cell of EC is small, the proper scale up is needed in order to have enough flow rate to provide enough cooling capacity for the size of a cooler. The mass flow rate of EC ($\dot{m}$) can be calculated in Eq. 3.

$$\dot{m} = \frac{dn}{dt} * M_{NH_3} = \frac{I}{nF} * 17 \tag{3}$$

$$I = I_d * np * S \tag{4}$$

where I_d represents the current density of each cell, np represents the number of cells in parallel, S represents the surface area of each cell.

The scale up is going to be based on the stacking up plates design. Based on the theoretical calculation, in order to provide 200 W of cooling capacity, ammonia flow rate needs to be at 0.2 g/s, which requires 70 cells to be connected in parallel, forming an EC stack. The scaled up EC is estimated to have surface area of 100 cm^2 for each cell and the overall size is 10 cm x 10 cm x 70 cm.

3. CONCLUSIONS

A small prototype of the electrochemical compressor working with ammonia was investigated. The open loop thermodynamic compression ratio was measured based on the voltage and compression ratio with discharge line valve closed. The open loop analysis shows the logarithm of compression ratio is in linear relationship with voltage charge. The closed loop analysis gives the actual EC performance when ammonia from the discharge line is fed back to the suction line. Both current and pressure lift increase with voltage charge. The EC can be a new compressor option for the vapor compression cycle.

ACKNOWLEDGEMENT

The authors gratefully acknowledge the support of this effort from the Center for Environmental Energy Engineering (CEEE) at the University of Maryland.

NOMENCLATURE

E	EC cell voltage charge
EC	electrochemical compressor
F	Faraday's constant
I	current
I_d	current density of each cell
$\dot{m}$	mass flow rate
MEA	membrane electrolyte assembly
n	number of moles of electrons released by each H_2 molecule transferred across the membrane
n_e	number of electrons transferred per molecule, in this case n is equal to 2 for hydrogen and 1 for ammonia
np	number of cells in parallel
P_s	suction pressure
P_d	discharge pressure
S	surface area of each cell

REFERENCES

B. Rohland, K. E. (1998). Electrochemical Hydrogen Compressor. *Electrochimica Acta.*

Grigoriev, S., Shtatniy, I., Millet, P., Porembsky, V., and Fateev, V. (2011). Description and characterization of an electrochemical hydrogen compressor/concentrator based on solid polymer electrolyte technology. *International Journal of Hydrogen Energy*, 4148-4155.

Onda, K., Ichihara, K., Nagahama, M., Minamoto, Y., and Araki, T. (2007). Separation and compression characteristics of hydrogen by use of proton exchange membrane. *Journal of Power Sources*, 1-8.

R. Strobel, M. O. (2002). The compression of hydrogen in an electrochemical cell based on a PE fuel cell design . *Journal of Power Sources*, 208-215.

Ternan, M. and Gardner, C. L. (2007). Electrochemical separation of hydrogen from reformate using PEM fuel cell technology. *Journal of Power Sources*, 835-841.

3D-CFD DESIGN STUDY AND OPTIMIZATION OF A CENTRIFUGAL TURBO COMPRESSOR FOR THE OPERATION IN A HYBRID SORPTION/COMPRESSION HEAT PUMP CYCLE

Thomas ECKERT*, Leo DOSTAL, Martin HELM, Christian SCHWEIGLER

Munich University of Applied Sciences,
Department 05: Building Services Engineering
Lothstraße 34, 80335 Munich, Germany
Email: teckert@hm.edu
Phone: +49 89 1265 4298

* Corresponding Author

ABSTRACT

In various applications the use of sorption chillers and heat pumps is limited by the available temperature level of the driving heat source or the heat sink for export of reject heat. These constraints can be overcome by integrating an efficient high-speed transonic turbo-compressor into the internal cycle of a thermally driven water/lithium bromide absorption heat pump. The operation in a hybrid heat pump with the refrigerant water implies specific challenges for the design of the compressor: Saturation pressures in the sub-atmospheric range, low vapor density, high volume flows and a targeted pressure ratio of 3 result in high impeller tip speed up to 660 m/s and transonic flow phenomena in the flow channel of impeller and diffusor. Here the authors present a theoretical design study based on a 3D-simulation of a centrifugal compressor, targeted at the given operating conditions for a hybrid heat pump. Key figures are investigated to figure out the relationship between impeller tip speed, compressor pressure ratio and operating range of the compressor meeting the requirements, wide stable operating range between surge and choke line and appropriate pressure ratio. The optimization of the impeller geometry comprises both fluid dynamic behavior and structural stability.

1. INTRODUCTION

In the field of building energy supply, the reduction of the electric power consumption for heating and cooling is a core issue. The dominating part of the electric demand is to be attributed to the operation of compression chillers serving for cooling and air-conditioning. As an alternative to mechanically driven devices, absorption chillers can be applied, activated by driving heat. Apart from directly fired multi-stage chillers (Ziegler *et al*, 1993), single-stage chillers driven by hot water from a district heating network or supplied by a cogeneration system or a solar thermal installation are available. A major obstacle for the use of these thermally driven chillers is their limited ability for the use of low-grade heat (Schweigler, 1999). For typical operating conditions in building air-conditioning systems the driving hot water can only be cooled down to about 80 °C.

This limitation can be overcome, if a mechanical vapor compressor is integrated into the sorption refrigeration cycle: At the expense of input of mechanical work, additional freedom for the design of the thermal cycle is gained (Alefeld and Ziegler, 1985), (Niebergall, 2011), (Ziegler, 1991) and (Ziegler, 2009). Either the temperature level of the driving heat source can be lowered, or the heat rejection of the chiller can operate at a higher temperature level, allowing for the use of dry cooling towers instead of wet open systems. A third option is to boost the cooling output in order to provide extra capacity during peak demand of the building (Eckert *et al*, 2015) and (Helm *et al*, 2015). Consequently, the application of sorption chillers in conjunction with low-grade heat sources and particularly in small scale systems is facilitated.

There have been some successful applications of vapor compressors using water as refrigerant (Koren and Ophir, 1996), (Madsboell, 2011), (Madsboell *et al*, 2015), (Madsboell and Minds, 1995) and (Paul, 2007). This publication focuses on the design of a small scale radial turbo compressor for the refrigerant water for a hybrid sorption/ compression heat pump system with an evaporator capacity of about 10 kW. The investigation is performed by coupled modeling of the fluid dynamic behavior and the structural-mechanic stability. The focus of this paper is the impeller aerodynamic design and the influence of key design parameters on the shape and the width of the operating envelope of the compressor under varying operating conditions. The characteristic target diagram for the compressor, based on an analysis of the operational envelope of the hybrid cycle heat pump, is being discussed in (Eckert *et al*, 2016).

2. USE OF WATER AS REFRIGERANT IN A HYBRID HEAT PUMP

2.1 Water Vapor Compressor

Comparing water (R718) to other refrigerants, e.g. R134a, it is characterized by a low molecular mass γ, a high sonic velocity c and a high isentropic coefficient κ. The cycle operates at sub-ambient pressures and its p-T curves lie under all common refrigerants (Mueller and Kharazi, 2006) at very low pressures. Thus the specific volumetric cooling capacity of water vapor is very low and therefore the volumetric flowrate for a desired specific cooling capacity and temperature lift can be up to a 100 times higher than the volumetric flowrate of a conventional refrigerant like R134a. Under the same conditions the required pressure ratio is nearly twice as high as for R134a. For a target pressure ratio of 3 a high circumferential Mach Number and isentropic work are required, which lead to a high impeller tip speed of up to 660 m/s, nearly four times higher than for R134a (Eckert *et al*, 2016). The temperature lift for a one-stage water vapor compressor is limited to roughly 20 K. A classical compression refrigeration system will require a two stage water vapor compressor. For a compression chiller using the water as refrigerant, the discharge temperature of the compressor is very high and has a negative influence of the energetic efficiency of the chiller. For water the irreversibility due to de-superheating of the compressed refrigerant vapor is substantially higher than for R134a (Mueller and Kharazi, 2006), but if the turbo compressor is used in a sorption/compression chiller, the superheat can be used for preheating the sorption sub-cycle of the hybrid chiller.

2.1 Hybrid Heat Pump System

To overcome the limitations of the single stage water vapor turbo compressor, a hybrid cycle concept can be applied for the refrigerant water. In this research the combination of a "thermal compressor" with a mechanical compressor is investigated. The so called thermal compressor is a heat driven chiller and the combination of both systems is called hybrid chiller. A jet ejector working in the cycle could also be investigated (Sarevski and Sarevski, 2012), but the available pressure ratio is far below the minimum pressure ratio of 2 which is required for a sound design of the hybrid heat pump.

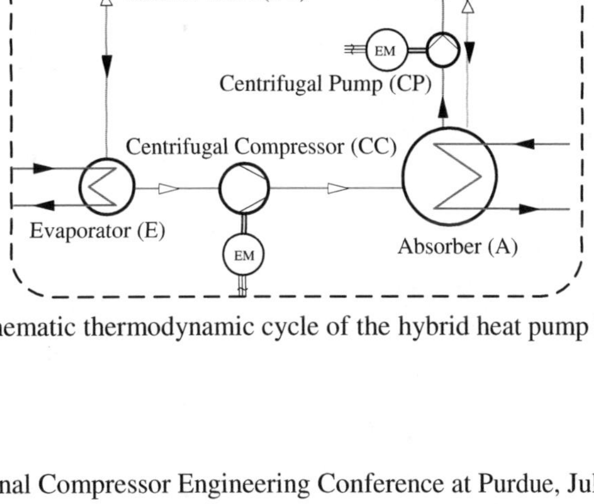

Figure 1: Schematic thermodynamic cycle of the hybrid heat pump in EVA mode

In figure 1 the schematic thermodynamic cycle of the hybrid heat pump in the EVA configuration with evaporator (E), condenser (C), centrifugal compressor (CC), solution heat exchanger (SHX), absorber (A) and desorber (D) is shown. In (Eckert *et al*, 2016) the target compressor map for the turbo compressor for different operating conditions is shown. This map lays out the compressor requirements for various cold water temperatures (6 °C to 15 °C), cold capacity (4 kW to 16 kW) and different settings of the electrical and heat factor EH (0.5 up to 1.5). This diagram is used to compare and evaluate the initial compressor designs.

3. MODELING TECHNIQUE AND BASIC DESIGN

3.1 Discretization and Boundary Conditions of Fluid Analysis

The computational meshes for the rotating and non-rotating fluid domains have been created by use of ANSYS-TurboGrid. An automated topology meshing system with high quality structural hexahedral mesh and local refinements was used. All domains, rotating and non-rotating, are designed as two separate structures and are connected at the interface between rotor and diffusor by usage of a frozen rotor interface. To reduce the computational effort, only one single passage of the flow field is being calculated. The heat transfer was calculated by the total energy heat transfer model. For the turbulence the SST model and for the heat transfer model Total Energy have been chosen. The calculation was carried out with the real gas IAPWS-IF97 water steam model. The use of the high resolution mode for the discretization of the Reynolds-averaged Navier-Stokes (RANS) equations boosted the accuracy in the results for the steady state solution.
The boundary conditions for the rotor are summarized in figure 2 and are set as follows: The flow condition at the impeller inlet was defined as a subsonic flow, by setting total temperature and total pressure input variables. The turbulence intensity in this section has been set to "medium" with an intensity value of 5%. Periodic boundary conditions were defined to allow the reduced calculation of only one passage. The surfaces of hub and blade were defined as adiabatic walls with appropriate rotational velocities. The shroud was defined as a counter-rotating wall in order to reflect the motion of the rotor in relation to the fixed shroud. Because of the low ratio between tip clearance and blade width the return flow between impeller shroud and casing is ignored for the first optimization process. For the domain interfaces between inlet geometry and impeller and between impeller and diffusor "Frozen Rotor" was chosen as domain interface type. This setting allowed for shorter computational time and higher accuracy, as compared to the "Stage" interface mode. The impeller outlet cross section is matched with the diffusor inlet cross section and defined as "Frozen Rotor". The periodic limitations of the diffusor have been set as periodic boundaries. The hub and shroud surfaces were set as adiabatic wall boundaries. At the diffusor exit plane, the subsonic flow velocity was assured by limiting the mass flow rate (setup a). This setup is robust at the surge line, because the pressure ratio exhibits only a moderate increase with falling mass flow. At the choke line, operation of the compressor was characterized by total pressure at inlet and static pressure at outlet (setup b). This setup promises a more robust behavior at the choke line than setup a, where small changes of mass flow induce a strong decrease of the pressure ratio.

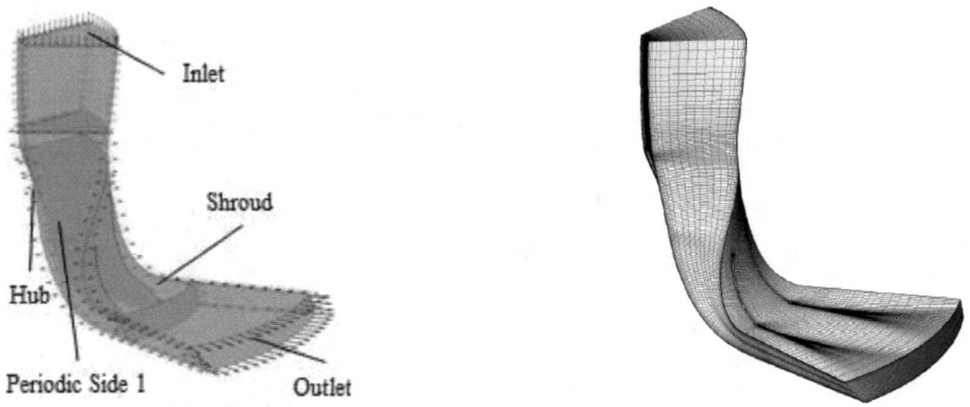

(a) (b)

Figure 2: Fluid flow blade passage with the boundary conditions (a) and structural meshing of the fluid flow passage (b)

3.2 Solver Settings

To achieve an adequate convergence the CFX solver settings had to be adjusted. As the optimal timescale is dependent on several influencing parameters, a fixed given physical timescale seemed unfeasible due to its relation to the rotational speed. Thus the conservative approach of auto timescale was chosen. To shift the simulation focus towards a faster convergence at the expense of stability, the auto timescale value was multiplied by an alterable factor, increasingly varying over the number of iterations. A factor value of 5 for up to 60 iterations followed by a factor of 20 turned out as a good trade-off between a fast and sufficiently robust convergence. To increase the speed of convergence the solver uses the current solver results as starting values, if possible. The calculation of characteristic diagrams therefore yields a so called mass flow ramp, resulting in increasing mass flow over iteration time to grant a higher robustness at the beginning of a simulation. Under consideration of the compressible medium and high stream velocities, the double precision calculation model was enabled to reach a more stable convergence towards the end of the calculation.

3.3 Structural Investigation

In the present modelling the ANSYS structural module is adopted providing a FEM procedure for determination of the von-Mises-stress distribution. The solid geometry for stress analysis is discretized by use of quadratic tetrahedral elements. In general, the centrifugal forces cause about 98% of the overall stress, whereas the forces and stresses caused by temperature and pressure differences have almost no influence. With regard to manufacturing aspects several materials are worth considering. But as the centrifugal force correlates directly with the density of the impeller material, lightweight materials are necessary. Thus for the impeller design only specific materials were considered: high-strength aluminum, titanium and carbon fiber appeared to be adequate choices. In order to reduce the mechanical stress the modification of the following parameters appeared to be the most relevant: rotational speed, blade thickness in means of mass reduction, the shroud radiuses and the blade geometry in terms of wrap angle, high twists and rakes in combination with the grade of tangential displacement in radial direction.

3.4 Basic Compressor Design

In order to fulfill the requirements for structural stability, minimizing the compressor size can be identified as major goal of the rotor design. So the outlet diameter should not exceed a certain given size. Otherwise the centrifugal force F_{fc} and stress σ_{max} induced by the increasing impeller tip speed u_2 would exceed the tolerable limit. One important parameter for a small outlet diameter is a high flow coefficient φ, according to equation 1.

$$D_2 = \sqrt{\frac{V_1}{\pi 4\, u_2 \varphi}} \quad (1)$$

A higher flow coefficient is reached when the specific volume is decreased. But at the same time a higher flow coefficient results in a larger axial compressor length, which induces an increasing bearing span.

$$\frac{p_2}{p_1} = \left[\frac{\Delta h_{tot}}{u_2^2}\,(\kappa - 1)M_{u2}^2 + 1\right]^{\frac{\kappa\eta_p}{\kappa - 1}} \quad (2)$$

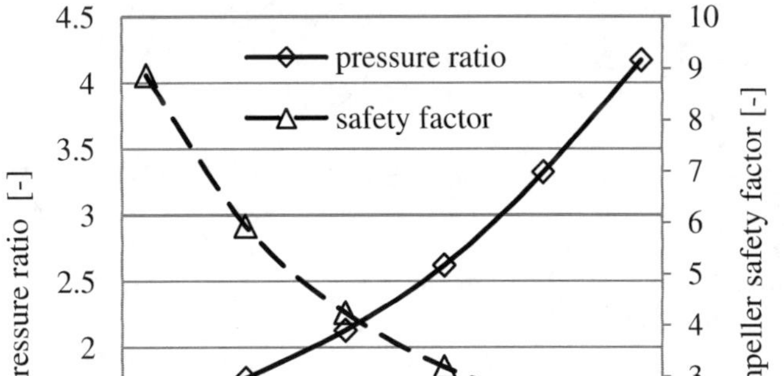

Figure 3: Pressure ratio and impeller safety factor over the tip speed Mach number

In figure 3 the relation between compressor pressure ratio, tip speed Mach number M_{u2} and impeller safety factor for the designed impeller of optimization step 4 is shown. The pressure ratio strongly depends on the tip speed Mach number (equation 2) and the isentropic exponent κ. Also in this diagram the influence of the safety factor is shown for the case of high strength aluminum with a maximum tensile strength (TS) of up to 550 MPa and is calculated according to equation 3. The safety factor of the impeller characterizes the ratio of maximum stress to allowable stress. It is a significant criterion to reach a high durability.

$$s_f = \frac{\sigma_{TS}}{\sigma_{max}} \quad (3)$$

The main goal is to reduce the high Mach number. A high Mach number leads to a narrower operating envelop of the compressor with a surge and choke point near the design point. A compressor with lower M_{u2} has a wider operating range near surge and choke line and has a better efficiency because of lower gas dynamic losses.

4. MEANLINE ANALYSIS AND OPTIMIZATION OF THE ROTOR DESIGN

4.1 Meanline Analysis

The design study for the turbo compressor had been oriented towards the following targets: meet the requirements of the hybrid heat pump, increase the isentropic efficiency, reduce the mechanical stress, increase the volume flow and increase the operating range. The preliminary design of a radial impeller has been calculated by empirical correlations (Pfleiderer, 2005). These empirical correlations had been validated by analysis of performance data of real impellers. As a first step the impeller design from CFturbo has been integrated in ANSYS Vista CCD and then exported into ANSYS DesignModeller. The flow optimization process was done based on the basic impeller dimensioning according to figure 4 (a).

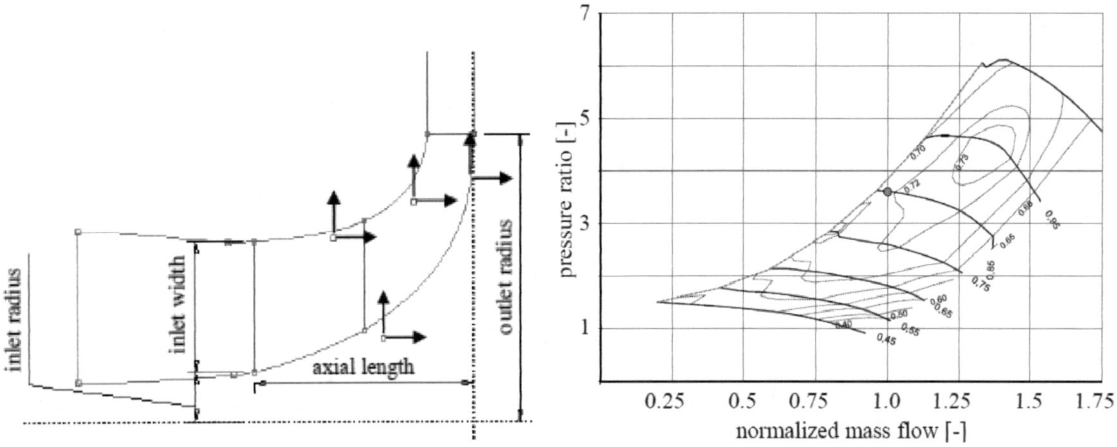

(a) (b)

Figure 4: Impeller Meanline Analysis (a) and radial compressor reference design without paying attention to the occurring stresses (b)

In order to allow a more qualified evaluation of the upcoming optimization results, an optimization in terms of fluid dynamic efficiency without regarding the stresses has been performed. The resulting operational map is shown in figure 4 (b).

The compressor design is limited by stress, vibration and material properties. It is a multi-objective design problem. Vibration induced by fluid dynamic interaction with stator and rotor is not investigated. The focus of this study is the improvement of the fluid dynamic performance, while stress is kept below tensile strength and deformation is minimized. The whole design procedure has been carried out iteratively, with identification of the optimal configuration by systematic variation of the governing design parameters, as described in the following section.

4.2 Coupled Fluid Dynamic and Structural Mechanic Optimization

The optimization is performed by the use of the software optiSLang 4.2.2 and pursued in four steps. The main difference to the meanline analysis is the consideration of stresses, resulting in two-dimensional (2D) optimizations. To get a better basis of assessment a one-dimensional (1D) optimization neglecting the stresses has been carried out in advance, which was expected to bring the best possible results in means of specific compressor work, as described in the previous section (see figure 4 (b)). This design will further be referred to as "reference design". Prerequisite to a parametrical optimization is the parametrization of the model, with the side condition that a lower amount of varied parameters allows a better Coefficient of Prognosis (CoP) for the optimization results. Thus the impeller design created by Vista-CCD had to be parameterized and simplified. This way the impeller could be described by only 25 instead of the original 72 parameters.

1) On these 25 parameters a sensitivity analysis is performed in order to rank their influence on the output values. Based on the results of the sensitivity analysis then a 2D-optimization is performed, letting the objectives pressure ratio and efficiency perform against each other. For verification a compressor map, or compressor characteristic diagram (ChD), is being generated.

2) Additionally the four most influencing parameters are investigated with regard to pressure ratio, isentropic efficiency, choke and surge line. The results are described regarding the pressure ratio and operating range in the next section.

3) The influences of the geometry modifications are applied to the impeller design in order to match the desired operational limits as characterized by the target ChD. Now the design point should lie in the desired position. On the newly chosen design point a 1D-optimization is performed by means of meridional contour and rotational speed to gain a higher efficiency. In this case only a 1D-optimization is necessary as the parameters with the largest influence on the stresses rotational speed and outlet radius are restrained from growing during the optimization. Again a verification by use of an ChD analysis is performed.

4) In the next step the number of blades and their shape has been optimized in a 2D-optimization based on a further sensitivity analysis. Key figures for the optimization are fluid dynamic efficiency and material stress which commonly react contrariwise on the change of the rotor design. Therefore, the best design is deduced by use of a Pareto-diagram. For verification purposes again a ChD is calculated.

5) Following a detailed investigation on undesired behavior has been performed. Based on this knowledge gain an innovative shape for the leading edge has been developed and integrated by use of a final optimization.

6) In the final step, step 2 has been repeated for the final design. The results are shown in the next section as well.

4.2.1 Overall Optimization (step 1): As a result of the multidimensional optimization in step 1, an isentropic efficiency of 51% at the design point has been achieved. It could also be observed that the design point, originally meant to lie at a pressure ratio of 3 at 85,000 RPM, was unintentionally moved up to a pressure ratio of 3.3. For part load operation with lower speed of rotation the characteristic diagram exhibits slightly higher efficiencies up 55 %. The operation with optimum efficiency is found close to the choke line, whereas a steep decrease of efficiency is found for operation with high flow rate close to the surge line.

4.2.2 Dominating Geometric Parameters (step 2 & 6): In step 2 and 6 of the optimization process, the four most influencing parameters, impeller axial length, outlet diameter or outlet radius as well as twist and rake of the blade leading and trailing edges are investigated with regard to the position of choke and surge line. The results of the parameter variations are presented in the following bullet points. For the first parameter variation a design mass flow of 4.0 g/s was chosen at the design speed of 85,000 RPM. The second variation was based on 6.0 g/s and 65,000 RPM

AXIAL LENGTH A shortening of the axial length results in a shift of the characteristic curve towards lower flow rates combined with a slight decrease of the pressure ratio, as shown in figure 5. An enlargement of the axial length results in a shift of the characteristic curve towards higher flow rates with a roughly constant pressure ratio. The behavior appears to be the same for both designs.

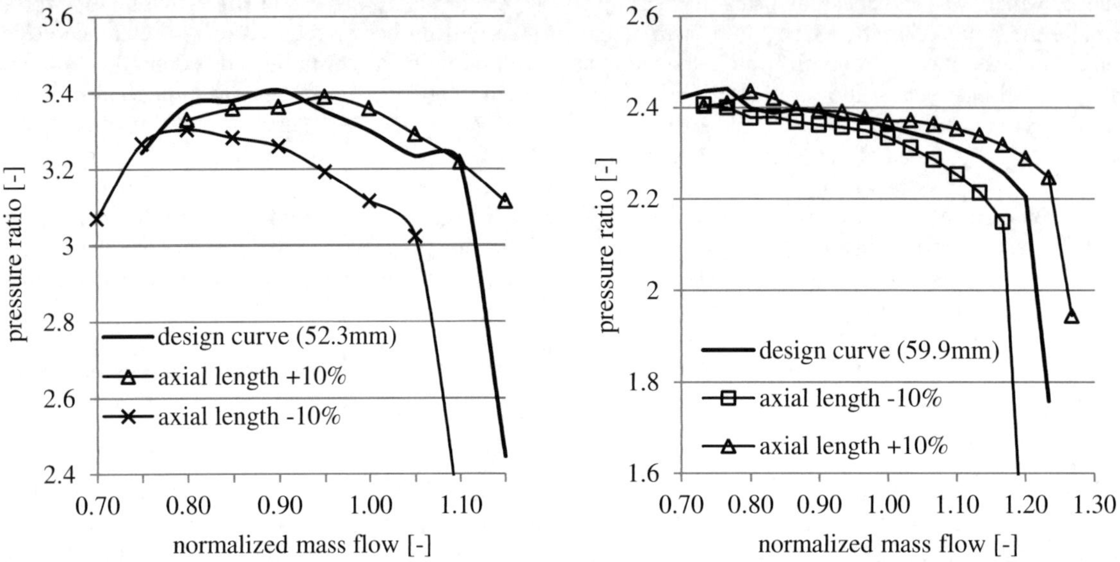

(a) (b)

Figure 5: Influence of the impeller axial length for the first (a) and final (b) design on the compressor design curve

OUTLET RADIUS An increase of the outlet radius – while reducing the outlet height in order to keep the outflow area constant – results in slightly higher flow rates and vice versa, accompanied by a strong influence on the pressure ratio. The effect displayed the same effect for both impeller designs (see figure 6).

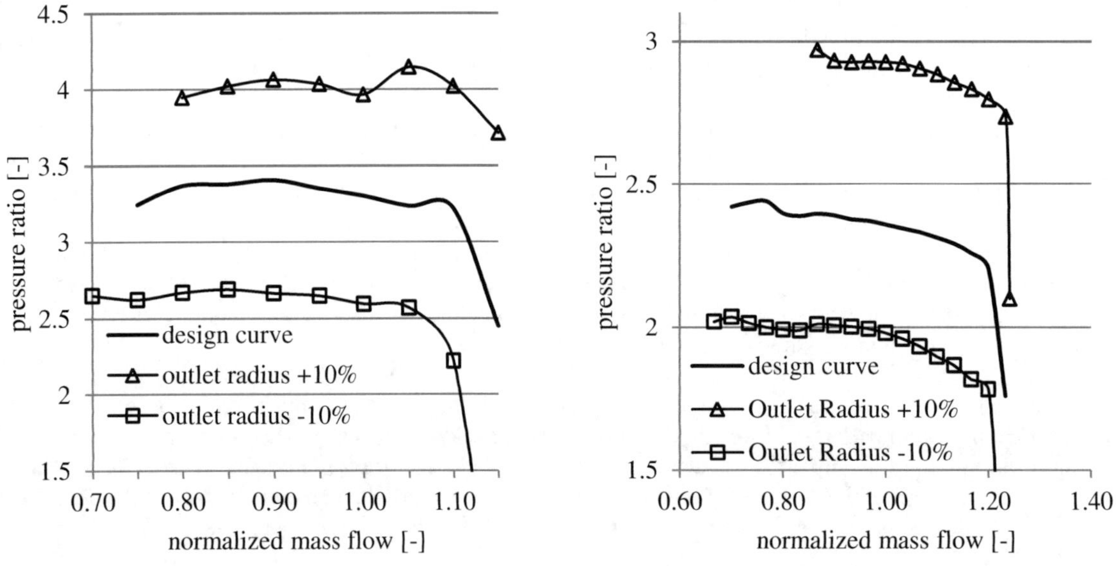

(a) (b)

Figure 6: Influence of the impeller outlet radius on the compressor design curves

TWIST The radial displacement in direction of rotation of the shroud tip at the leading edge compared to the hub tip is called twist. For the first design a positive change of the twist moves the characteristic curve towards lower flow rates and causes a slight increase of the pressure ratio. A negative change of the twist has the opposite effect (figure 7). For the final design the variation of the twist did not show significant results: The cause for this is estimated to lie in the new shape of the leading edge (compare figure 9).

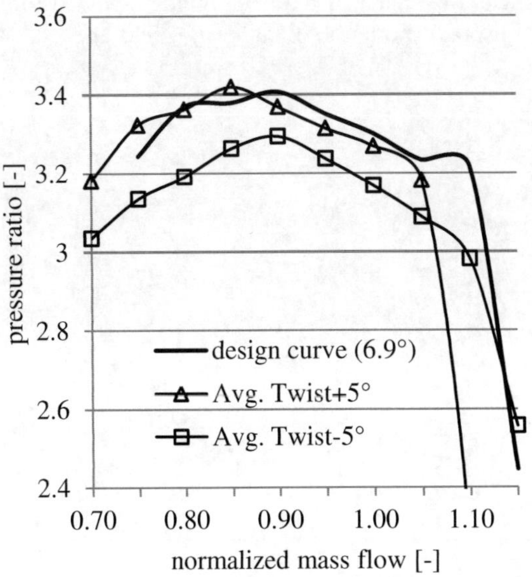

(a)

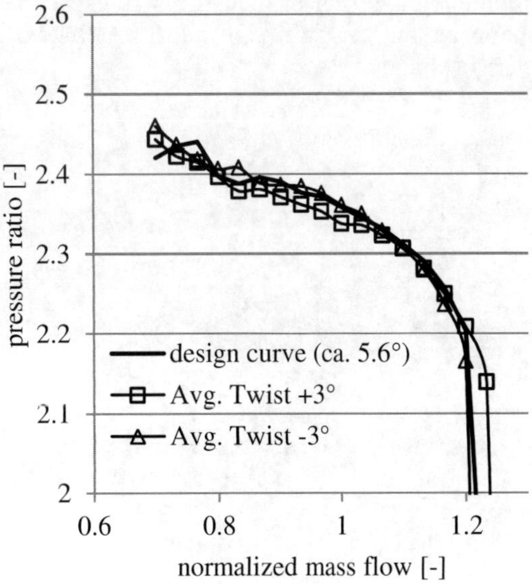

(b)

Figure 7: Influence of the variated twist on the first (a) and the final (b) design.

RAKE The radial displacement in direction of rotation of the shroud tip at the trailing edge of the compressor compared to the hub tip is called rake. For the first design a negative variation of the rake shifts the characteristic curves towards lower flow rates and lowers the pressure ratio. A positive change of the rake seems to have the same effects in a less pronounced manner (figure 8). The reason therefore lies in the previously performed optimization towards a higher pressure ratio. The design curve is already the one causing the highest pressure ratio, so that any modification worsens the result. For the final design, which hadn't been optimized for a maximal pressure raise, it can be seen as well that a lower rake appears to show better behavior in terms of pressure ratio.

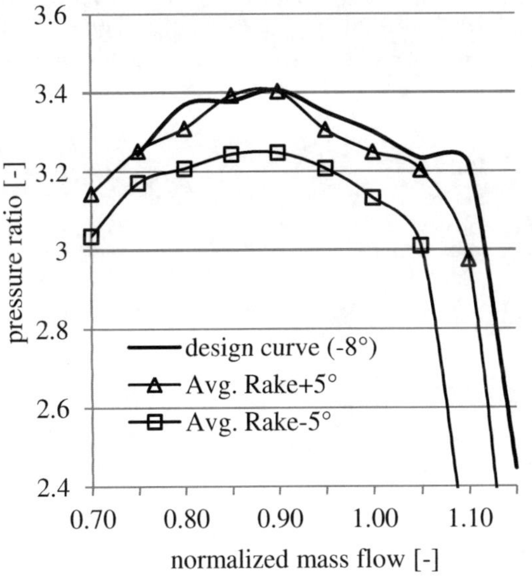

(a)

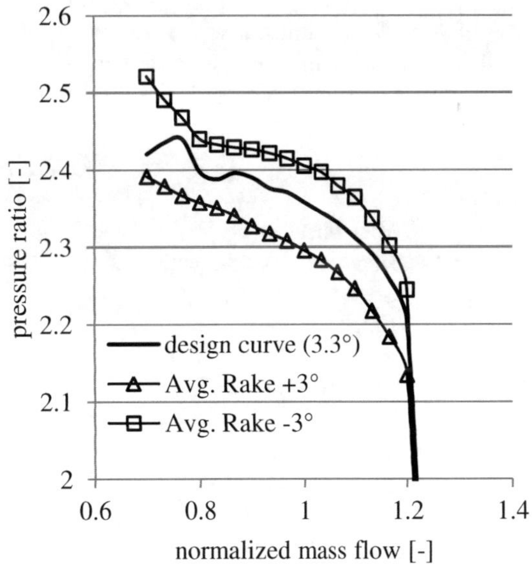

(b)

Figure 8: Influence of the rake variation onto the first (a) and final (b) design on the characteristic curve.

4.2.3 Step 5 and final Results: In step 5 the improved design from step 3 and 4 was analyzed in detail. Renderings of the simulations, both steady state and transient, showed a bad flow distribution within the passage. By interdisciplinary applying methods from aerospace design methods the shape of the leading edge was customized and optimized for the given operating conditions. Figure 9 displays both the final design and the according characteristic diagram.

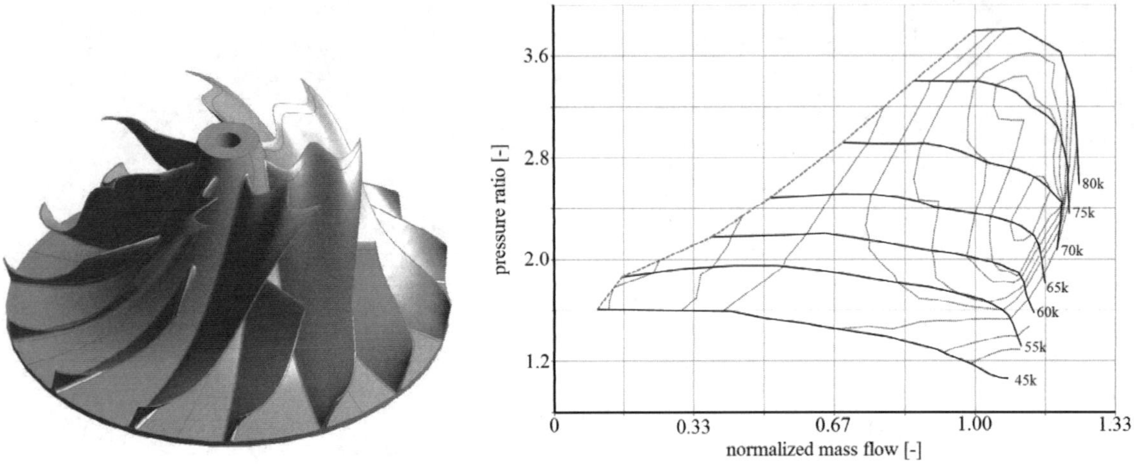

(a) (b)

Figure 9: Shape (a) and characteristic diagram (b) of the final design

5. CONCLUSION AND OUTLOOK

By systematic optimization of the compressor design an effective improvement in operating range and isentropic efficiency has been achieved, developing from the first to the resulting final Characteristic Diagram (ChD). The presented parameter variation showed the expected behavior for the variation of axial length and outlet radius as standard turbomachinery parameters. For the other varied parameters it is interesting to see how the potential for further increase in pressure ratio moved from the leading edge to the trailing edge.

As next step of the development process the thickness distribution along the rotor contour will be optimized for reduction of stress and distortion with regard to mechanical stability. The following steps are the validation of the calculated compressor in the compressor test rig at Munich University of Applied Sciences and the integration of the compressor into the absorption cycle for demonstration of the hybrid concept..

NOMENCLATURE

σ	Von-Mises stress	(MPa)
D_2	Impeller outlet diameter	(m)
c	Speed of Sound	(m/s)
h	Enthalpy	(J/kg)
M_u	Peripheral Mach Number	(-)
RPM	Speed of Rotation	(rev/min)
p	Pressure	(Pa)
T	Temperature	(C, K)
F_{cf}	Centrifugal force	(N)
u_2	Impeller peripheral speed	(m/s)
$\dot{V}_1$	Volumetric flow rate compressor inlet	(m^3/s)
$\dot{m}$	Massflow	(kg/s)
η	Isentropic compressor efficiency	(-)
κ	Conditional isentropic coefficient	(-)
γ	molecular mass	(kg/mol)
Π	Pressure ratio	(-)
s_f	Safety factor	(-)
φ	Flow rate coefficient	(-)

ChD	Characteristic Compressor Diagram
EH	Electrical Heat Coefficient
TS	Tensile Strength
EM	Electric Machine
CoP	Coefficient of Prognosis
CHP	Combined Heat and Power Unit
RANS	Reynolds-averages Navier-Stokes

REFERENCES

Alefeld, G. and Ziegler, F. (1985). Advanced heat pump and airconditioning cycles for the working pair H2O/ LiBr: domestic and commercial applications. *ASHRAE Transaction,* Honolulu

Eckert, T., Helm, M., Grassel, A., and Schweigler, C. (2015) Lithium Bromide/ R718 Hybrid sorption & compression cycle. *International Congress of Refrigeration*, Yokohama

Eckert, T., Dostal, L., Helm, M., Schweigler, C. (2016), Design of a Centrifugal Turbo Compressor with the Working Fluid Water for the Operation in a Hybrid Sorption / Compression Heat Pump Cycle, *ASME Turbo Expo 2016*, Seoul

Helm, M., Eckert, T., and Schweigler, C. (2015). Hybrid Water/ LiBr Absorption Chiller Boosted by High-Speed Turbo Compressor. *6th Int. Conf. Solar-Air-Conditioning,* Rome

Koren, A. and Ophir, A. (1996). Water vapor technology. *Proc. IIR Conf. Applications for Natural Refrigerants*, Aarhus

Madsboell, H. and Minds, G. (1995). A 2 MW Industrial Chiller using Water as Refrigerant. *Proc. IIR Conf. Applications for Natural Refrigerants*, Aarhus

Madsboell, H. (2011) Development of new type of compact, high efficient and cost effective axial compressor for commercial chillers with water as refrigerant. *International Congress of Refrigeration,* Prague

Madsboell, H., Weel, M. and Kolstrup, A. (2015). Development f a water vapor compressor for high temperature heat pump applications. *International Congress of Refrigeration*, Yokohama.

Mueller, N. and Kharazi,A. (2006). Comparing water (R718) to other refrigerants. *ASME Int. Mechanical Engineering Congress and Exposition, IMECE2006-13341*, Chicago

Niebergall, W. (2011). *Sorptions-Kältemaschinen.* Springer, 7th edition.

Paul, J. (2007). State of the art for cooling with water as refrigerant (R718). *Proc. IIR Conf. of Refrigeration,* Bejing

Pfleiderer, C. and Petermann, H. (2005). *Strömungsmaschinen. Klassiker der Technik*. Springer, 7th edition, Berlin.

Sarevski, M.N. and Sarevski, V.N. (2012). Preliminary Study of a Novel Compact R718 Water Chiller with Integration of a Single Stage Centrifugal Compressor and Two-Phase Ejectors. *Int. J. of Refrigeration*

Schweigler C. (1999). *Kälte aus Fernwärme: Konzept, Auslegung und Betrieb der Single-Effect/ Double-Lift-Absorptionskälteanlage: Doktorarbeit*. Technische Universität München, München.

Ziegler, F. (1991). Kompressions-Absorptions-Wärmepumpen, *Volume Nr. 34 of Forschungsberichte des Deutschen Kälte-und Klimatechnischen Vereins DKV*, Stuttgart

Ziegler, F., Kahn, Summerer, F., Alefeld, G. (1993). Multi-effect absorption chillers. *Int. J. of Refrigeration, volume 16.*

Ziegler, F. (2009) Sorption heat pumping technologies: Comparisons and challenges. *Int. J. of Refrigeration, volume 32.*

ACKNOWLEDGEMENT

This project is funded by the German ministry of Education and Research within the program "Forschung an Fachhochschulen" with the grant number 03FH023I3.

Mechanistic model of an oil-flooded single-screw expander

Davide ZIVIANI[1*], Ian H. BELL[2], Michel DE PAEPE[1], Martijn VAN DEN BROEK[1]

[1] Ghent University, Department of Flow, Heat and Combustion Mechanics,
Kortrijk, Belgium
[2] University of Liège, Thermodynamics Laboratory,
Liège, Belgium
Contact Information: Davide.Ziviani@UGent.be

* Corresponding Author

ABSTRACT

The power generated from Organic Rankine Cycles (ORCs) is directly related to the performance of the expander used, which in many cases it is adapted from the original compressor. The efficiency of positive displacement expanders depends on several aspects such as internal volume ratio and pressure ratio (under- and over-expansion), leakages, non-adiabatic expansion and heat losses to the ambient and mechanical losses. Thus, lubrication is often required to reduce friction and to seal leakage gaps. In this paper, the effect of oil-flooding on the performance of a single-screw expander is investigated. By varying the oil fraction, it is possible to analyze the combined effects of reducing the leakages and the possibility to achieve a quasi-isothermal expansion. The flooding of the expander is realized by adding an independent oil loop to the conventional ORC as proposed by Woodland et al. (2013). The mechanistic model developed by Ziviani et al. (2014) for a single-screw expander has been extended to include liquid-flooding. The multi-chamber model with the associated governing equations of conservation of mass and energy for the flooded-expansion process is described along with the overall energy balance model of the expander. Insights on the calculation of the working fluid and lubricant oil mixture properties and the solubility effects are provided. The working fluid considered is R245fa and the lubricant oil is ACD100FY. The mechanistic model is implemented in the Python programming language and the oil-flooded single-screw expander model is integrated in the open-source simulation package PDSim (v3.0).

1. INTRODUCTION

The performance of an organic Rankine cycle depends on the operating conditions of the heat source and cold sink, on the selection of the working fluid to exploit the heat source and on the selection and sizing of the expander. During the last decades, the research on positive displacement machines as expanders for ORC systems is intensified. A comprehensive review of volumetric expanders for low grade heat recovery has been published by Imran et al. (2016). A comparative assessment showed that scroll and screw expanders achieved the highest scores because of the balance between internal losses (mainly leakage and friction losses) and overall performance.

In order to optimize the work extraction for waste heat, different cycle architectures can be investigated Lecompte et al. (2015). As shown by Woodland et al. (2014), ORC with flash expansion from saturated liquid or from a two-phase mixture with a low vapor quality presented a higher Second Law efficiency compared to baseline ORC with a net increase of power conversion percentage of the heat source exergy rate.

Theoretically, the expansion process can also be improved by moving the expansion work path from an adiabatic process toward an isothermal process (Igobo & Davies, 2014). In an organic Rankine cycle, a quasi-isothermal process can be achieved by flooding the expander with a secondary fluid (typically oil) which acts as thermal buffer. Such concept was analyzed by Woodland et al. (2013) and Georges (2012). The scroll expander used had a small internal volume ratio (< 2) which limited the benefits of the flooding process with respect to the effective internal volume ratio. It was suggested that an expander with a larger volume ratio, such as screw-type, could be more suitable.

In this paper a single-screw expander is considered to investigate the benefits of an oil-flooded expansion. A generalized chamber model needs to be developed to improve existing models in the literature to be able to describe any expansion (or compression) processes including flashing and two-phase expansion, oil-flooded, oil-injected, etc.

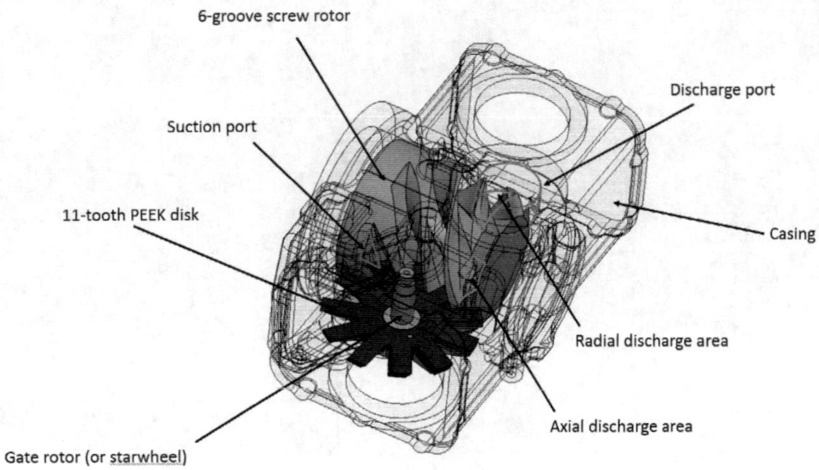

Figure 1: 3D CAD model of the single-screw expander considered in the present work with the different control volumes indicated.

2. SINGLE-SCREW GEOMETRY MODEL

The single-screw expander considered in this work is an oil-injected 11 kW air-compressor adapted to perform as expander. The main geometric characteristics are listed in Table 1. Furthermore, the complete formulation of the geometry model can be found in Ziviani et al. (2014) and Ziviani et al. (2015). In the following subsections, the main improvements on the geometry model are briefly described.

2.1 Extension of the compressor model

The expander model proposed in Ziviani et al. (2014) was derived from the well-known compressor model established by Bein and Hamilton (1982). The expansion process was considered to occur identically in both sides of the rotor which simplified the overall model. The model is extended here to undertake a more comprehensive description of the machine. The single-screw expander and the definition of different control volumes are shown in Figure 1. In particular, both sides of the expander are modeled which allows to a better estimation of the suction losses and to more accurate model of the discharge process as expander. In fact, in the current expander the main housing outlet port is located on one side which means that the effective discharge area is different in the two sides of the expander. As a consequence the volume of the groove needs to be predicted throughout one rotation.

2.2 Volume Curve and Port Sizes

The calculation of the chamber volumes and the porting are done by describing the single-screw geometry by means of polygons, i.e. arrays of node coordinates. Polygon operations are employed to obtain intersections and through Green's theorem areas and volumes are computed, as described in Bein and Hamilton (1982) and Ziviani et al. (2015). While the single-screw compressor models reported in literature start at the end of suction process (beginning of discharge in the case of expander), in the present model, the groove return volume curve is also important to describe the discharge process. The polygon model of the single-screw geometry is shown in Fig. 2. The intersection operation between the starwheels and the main rotor is shaded in grey. At each rotation angle and for each tooth engaged, the area, A_i, and the position of the centroid (c_z, c_r) are calculated and the volume of the groove is obtained as:

$$V(\theta) = \sum_i c_r A_i \Delta\theta \tag{1}$$

The decrease of groove volume after around 180°of rotation is done by subtracting the volume of the expansion chamber increasing on the other side of the rotor. Similar approach is also used to calculate the groove surface.

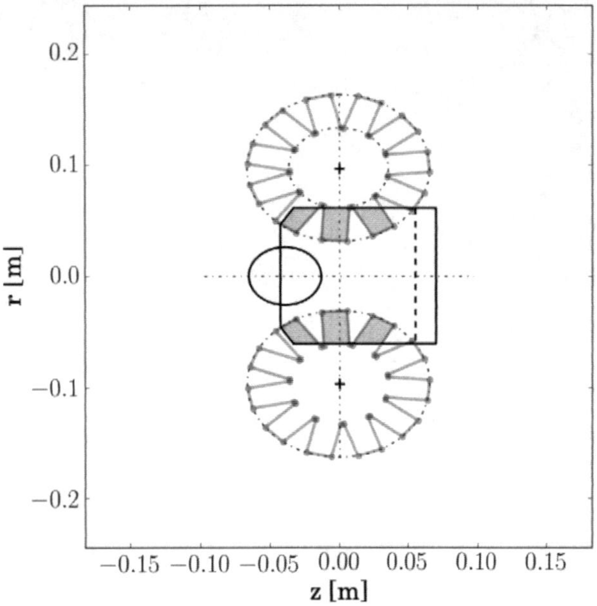

Figure 2: Polygon clipping applied to single-screw expander geometry.

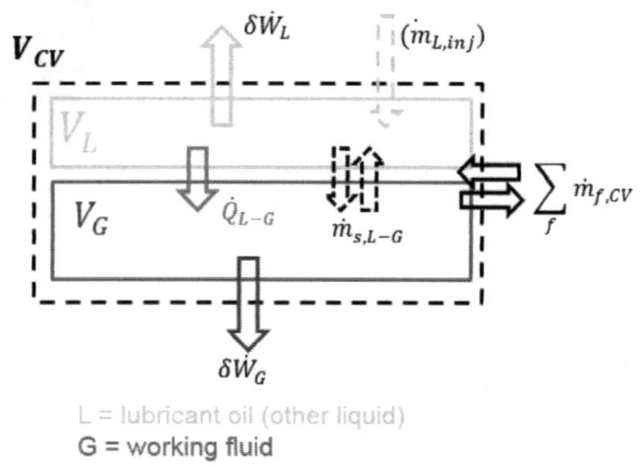

Figure 3: General schematic of a control volume.

The area of suction and discharge ports with respect the crank angle are determined in a similar fashion. The main rotor and the groove are unwrapped and the intersection between the groove with the triangular suction port is obtained through polygon intersection. On both sides of the rotor, the axial discharge port defined by the shape of the end of the groove is the same. As previously mentioned, only on the side with the main outlet of the housing, the groove presents a radial discharge area. By unwrapping the groove, the length of leading and trailing edges can be determined at each crank angle (Ziviani et al., 2015).

3. THERMODYNAMIC MODEL

Different thermodynamic models have been considered in literature to analyze oil-flooded compressors. In particular, Fujiwara et al. (1984) developed an oil-flooded screw compressor model by applying the conservation of mass and energy of an open control volume to both the working fluid and oil. The heat exchange between the working fluid and the oil was also taken into account with a term proportional to the temperature difference between the two fluids. However, the working fluid was treated as an ideal gas and could not change phase. An improvement of such model was done by Stosic et al. (1988) by describing the oil-injection process in details including the influence of the oil droplet size. A significant contribution on handling large amount of liquid in the working chamber was proposed by Bell et al. (2012). In particular, the lubricant oil and the working fluid were considered as a homogeneous mixture and the oil-flooded compressor and expander were modeled by solving the governing equations of conservation of mass of working fluid, oil mass fraction and energy for each working chamber.

Regarding two-phase expansion, Taniguchi et al. (1988) proposed thermodynamic model of a two-phase working fluid in screw expanders by applying conservation of mass and energy to different phases of the expansion process. The specific volume, internal energy and quality were chosen as state variables. The presence of lubricant oil was not taken into consideration. A similar approach was also adopted by Smith et al. (1996) to carry out an extensive study on two-phase expansion in screw machines.

Specifically to single-screw compressor, Wu and Jin (1988) developed a simulation model which accounted for the impact of oil injected into the working chamber and the effects on the leakages. The heat transfer between oil and working fluid was neglected and the working fluid was treated as ideal gas with no phase change. Shear power losses due to the presence of oil were taken into account. A mechanical analysis on the forces and torques was not considered.

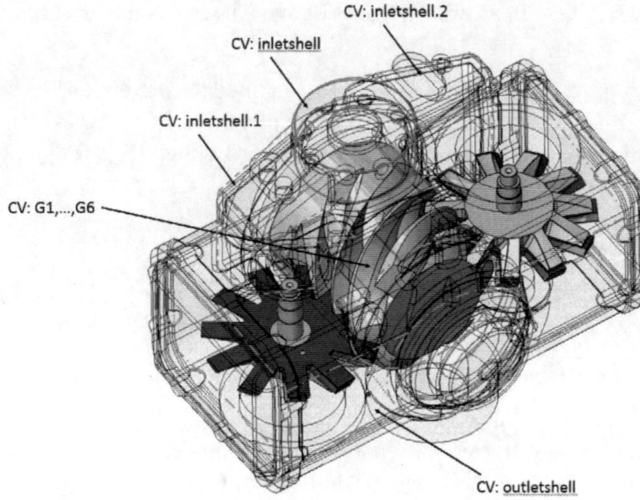

Figure 4: Definition of the different control volumes in a single-screw expander.

A more generalized chamber model is required to model:

- oil-flooded compressors/expanders (two-fluid model);
- phase change of working fluid (flash or two-phase expansion). Compressor slugging can also be modeled;
- allow for liquid/vapor/two-phase injections;
- account for heat transfer between phases when the assumption of thermal equilibrium is not valid (e.g. isothermal expansion);
- working fluid and lubricant oil can leak independently or as mixture.

3.1 Definition of a General Geometric Control Volume

In order to establish a set of governing equations, the structure of a general control volume (CV) needs to be defined. By referring to Figure 3, a general CV, which corresponds to the actual geometry of the working chamber, is divided into two part: working fluid (G) and lubricant oil or other flooding mediums (L). According to Ishii and Hibiki (2011), for a two-fluid model the volume occupied by each fluid can be estimated by introducing the void fraction α:

$$\alpha = \frac{1}{1 + \frac{x_L}{1-x_L} S \frac{v_L}{v_G}} \tag{2}$$

$$V_{CV} = \alpha V_{CV} + (1-\alpha) V_{CV} = V_G + V_L \tag{3}$$

where S is the slip ratio. A slip model between the fluids can be introduced which requires the calculation of the fluid velocities. At first, the slip ratio is assumed unity (mechanical equilibrium). A simplified version of Equation 3 was

Table 1: List of geometric parameters of the single-screw expander.

Parameter	Value
Engaging ratio [-]	11/6
D_{sr} [mm]	122
D_{sw} [mm]	132
$V_{g,max}$ [cm^3]	57.39
$r_{v,built-in}$ [-]	5.3
L_{rotor} [m]	121

adopted by Mathison et al. (2013), where the working fluid volume was estimated by subtracting the volume of oil to the total chamber volume.

The governing equations of conservation of mass and energy are applied to each fluid. In a complete two-fluid model, also the momentum equations should be included. The CVs of identified in a single-screw expander are shown in Figure 4 and they can be distinguished between stationary (inlet and outlet shell CVs) and rotating CVs (grooves G1,...,G6).

3.2 Extension of the Chamber Model

To establish the two-fluid model, a set of differential equations need to be derived. The following assumptions are made:

- the working fluid (G) can change phase and the saturated vapor and liquid phases are denoted as g and l, respectively;
- working fluid and lubricant oil are treated as separate fluids. Oil-refrigerant interaction occurs through heat transfer and mass transfer (solubility effect). The flashing of oil is neglected;
- working fluid and oil thermodynamic states are homogenous throughout the control volume at any moment;
- the kinetic and potential energy term are neglected. Therefore, the total energy of the control volume is $E_{CV} \simeq (mu)_{CV}$.

For each fluid, the conservation of mass and energy are applied. When the working fluid is has a superheated state, density and temperature, i.e. (ρ, T), are chosen as state variables. The rate of change of the working fluid mass in the control volume can be conveniently written as:

$$\rho_G \frac{dV_G}{d\theta} + V_G \frac{d\rho_G}{d\theta} = \frac{1}{\omega}\left[\sum_i \dot{m}_{G,i} \pm (..)\right] \tag{4}$$

Analogously for the secondary liquid phase:

$$\rho_L \frac{dV_L}{d\theta} + V_L \frac{d\rho_L}{d\theta} = \frac{1}{\omega}\left[\sum_i \dot{m}_{L,i} \pm (..)\right] \tag{5}$$

In the RHS of the above equations, additional terms may include working fluid dissolved in the oil or oil vapor in the working fluid. If the secondary liquid phase is expressed as a fraction of the working fluid, Equation 5 can be reworked as proposed by Bell et al. (2012).

The conservation of energy for the two phases is given by:

$$\frac{d(mu)_G}{d\theta} = \frac{1}{\omega}\left[\sum \dot{Q}_{G,b} + \sum_i \dot{m}_{G,i} h_{G,i} + \sum \dot{W}_{G,b} + \dot{Q}_{L-G}\right] \tag{6}$$

$$\frac{d(mu)_L}{d\theta} = \frac{1}{\omega}\left[\sum \dot{Q}_{L,b} + \sum \dot{Q}_{L,mt} + \sum \dot{W}_{L,b} + \sum \dot{W}_{L,mt} - \dot{Q}_{L-G}\right] \tag{7}$$

where:

- $\dot{Q}_b$: heat transfer rate through the boundary of the control volume
- $\dot{Q}_{mt}$: heat transfer rate to the control volume by mass transfer
- $\dot{W}_b$: work rate through the boundary of the control volume
- $\dot{W}_{mt}$: work rate to the control volume by mass transfer
- $\dot{Q}_{L-G}$: heat transfer rate between the two phases.

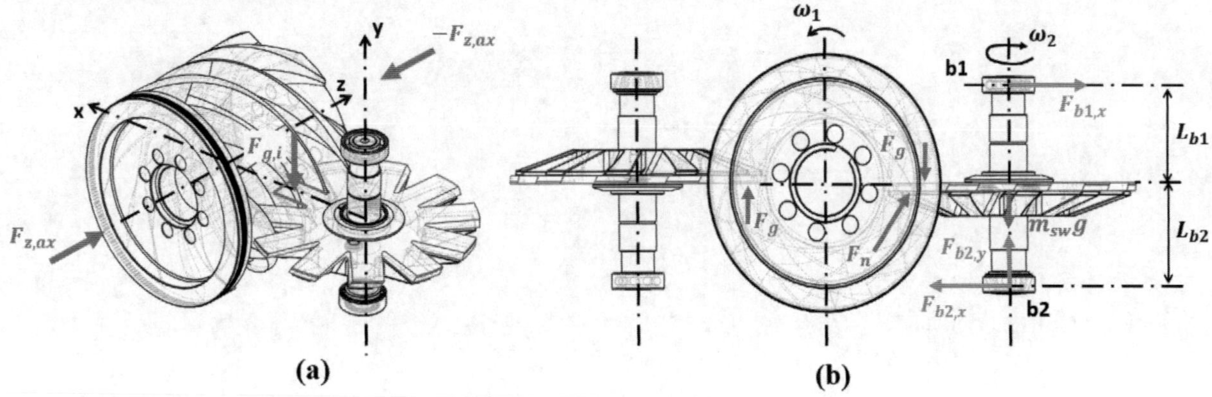

Figure 5: Forces and torques balance on the starwheel.

By reworking Equation 6 and Equation 7, the expressions for $dT_G/d\theta$ and $dT_L/d\theta$ can be obtained.

When the working fluid condition in the control volume is two-phase, temperature and quality need to be considered along with the density in order to accurately define a thermodynamic state. The specific enthalpy of the working fluid is then given by:

$$h_G = x_g h_{G,g} + (1 - x_g) h_{G,l} \tag{8}$$

where the saturated properties are only functions of temperature. By differentiating the specific enthalpy with respect temperature and quality:

$$\frac{dh_G}{d\theta} = \left(\frac{\partial h_G}{\partial T}\right)_{x_g} \frac{dT}{d\theta} + \left(\frac{\partial h_G}{\partial x}\right)_T \frac{dx_g}{d\theta} = \left[x_g \frac{dh_g}{dT} + (1 - x_g)\frac{dh_l}{dT}\right]\frac{dT}{d\theta} + (h_g - h_l)\frac{dx_g}{d\theta} \tag{9}$$

Therefore, the derivative of the specific internal energy becomes:

$$\frac{du_G}{d\theta} = \left[x_g \frac{dh_g}{dT} + (1 - x_g)\frac{dh_l}{dT}\right]\frac{dT}{d\theta} + (h_g - h_l)\frac{dx_g}{d\theta} - \frac{p(\theta)}{m_G}\frac{dV_G}{d\theta} + \frac{pV_G}{m_G^2}\frac{dm_G}{d\theta} \tag{10}$$

By recalling that the specific volume of the two-phase working fluid is equal to the volume to mass ratio of the working fluid, or:

$$x_g v_g + (1 - x_g) v_l = \frac{V_G}{m_G} \tag{11}$$

and by differentiating each side with respect the crank angle θ, the derivative of the quality with respect the crank angle, $dx_g/d\theta$, can be obtained:

$$\frac{dx_g}{d\theta} = \frac{1}{v_{G,g} - v_{G,l}}\left\{\frac{1}{m_G}\frac{dV_G}{d\theta} - \frac{V_G}{m_G^2}\frac{dm_G}{d\theta} - \left[x_g \frac{dv_{G,g}}{dT} + (1 - x_g)\frac{dv_{G,l}}{dT}\right]\frac{dT_G}{d\theta}\right\} \tag{12}$$

Last, the expression of the derivative of the specific energy with respect the crank angle:

$$\frac{du_G}{d\theta} = \left[x_g \frac{dh_{G,g}}{dT} + (1 - x_g)\frac{dh_{G,l}}{dT} - v_{G,g}\left.\frac{dp}{dT}\right|_{sat}\right]\frac{dT_G}{d\theta} + (h_{G,g} - h_{G,l})\frac{dx_g}{d\theta} - \frac{p(\theta)}{m_G}\frac{dV_G}{d\theta} + \frac{pV}{m_G^2}\frac{dm_G}{d\theta} \tag{13}$$

can be solved for $dT_G/d\theta$. The derived set of differential equations are integrated simultaneously by using an adaptive Runge-Kutta 4th/5th order solver. The thermo-physical properties of the working fluids are retrieved from CoolProp (Bell et al., 2014). Furthermore, the first saturation derivatives necessary in the case of two-phase conditions are also obtained from the low-level interface available within CoolProp. A library of lubricant oil has been created taking advantage of the CoolProp *State* class. The properties of ACD100FY are also available in Zhelezny et al. (2007).

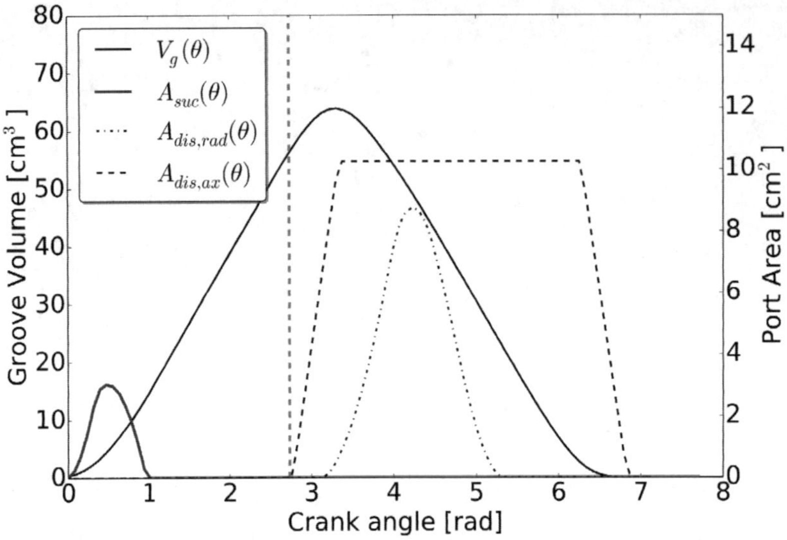

Figure 6: Variation of the groove volume with the crank angle. On the same plot, also the suction port and both radial and axial discharge ports as function of the crank angle are overlaid.

3.3 Overall Energy Balance

The expander model is closed by an overall energy balance applied to the expander shell. A lumped thermal mass model is adopted. The overall energy balance includes the heat fluxes interacting with the shell, the heat transfer loss to the ambient and the mechanical losses. The temperature of the lumped mass, T_{shell}, is obtained by minimizing the energy balance given by:

$$\dot{W}_{mech,loss} + \sum_i \left[\dot{Q}(T_{shell})\right] = 0 \tag{14}$$

In Ziviani et al. (2014), the mechanical losses were expressed by considering a friction torque τ_{loss}. In the present work, a mechanistic model for the mechanical losses is included. In particular, the loads on the starwheel and friction losses between starwheel and main rotor are considered. A schematic of forces on main rotor and starwheels is shown in Figure 5. Once the pressure trace in the working chambers is known at the end of one cycle, the force on each tooth generated by the expansion process is given by:

$$F_{g,i}(\theta) = p_i(\theta)A_i(\theta) \tag{15}$$

and it is applied on the centroid of the intersection tooth area (see Figure 2). As there are three tooth on each side constantly engaged, the total force due pressure loads is given by summing up each contribution, i.e. $\sum_i F_{g,i}, i = 1,2,3$.

The groove applies a force F_n which is perpendicular to the groove surface itself and direct to the tooth flank. As a consequence, F_n changes orientation with the helix angle of the groove during the expansion process. For each tooth, a frictional force component of $F_{n,i}$ can be obtained as:

$$F_{f,i} = \mu_g F_{n,i} \tag{16}$$

where the friction coefficient μ_g depends on the materials of groove and tooth and the presence of oil. The axial and radial force balances on the starwheel are given by:

$$F_{b2,y} - m_{sw}g - F_g + F_{n,y} = 0 \tag{17}$$

$$-F_{b1,x} - F_{n,x} + F_{b2,x} = 0 \tag{18}$$

where m_{sw} is the total mass of the starwheel body plus the toothed PEEK disk. By means of a balance of moments around the upper bearing, the loads on each bearing can be determined and the friction power can also be estimated through the thrust bearing theory.

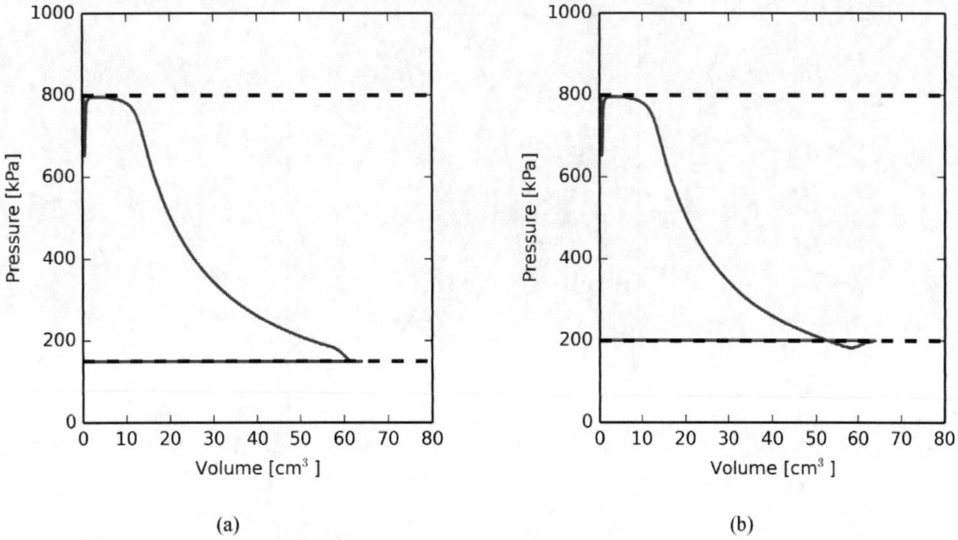

Figure 7: Pressure versus groove volume of single-screw expander (p_{su}= 800 kPa, T_{su}= 100 °C, x_L = 0.01, N = 3000 rpm): (*a*) under-expansion (p_{ex}= 150 kPa); (*b*) over-expansion (p_{ex}= 200 kPa).

4. RESULTS AND DISCUSSION

The complete groove volume curve and the port areas as function of the crank angle are shown in Figure 6. The ability of the geometry model to include also the return part of the groove volume while it starts wrapping on the opposite side of the rotor is important to simulate the under- and over- expansion as well as to quantify the flow losses during the discharge process. In Figure 7, the model has been exercised to obtain the p-V diagrams in both cases of under- and over-expansion. In particular, the inlet pressure and temperature are fixed at 800 kPa and 100 °C, the rotational speed of the main rotor is 3000 rpm and lubricant oil mass fraction is 0.01. Two values of the discharge pressure have been considered: 150 kPa (case with under-expansion) and 200 kPa (case with over-expansion). These values are also representative of the experimental operating conditions (Ziviani et al., 2015). By comparing the two p-V diagrams, it can be noted that the over-expansion, shown in Figure 7(b), is particularly detrimental for the performance of the machine with respect to under-expansion. Under the specified conditions, the slight over-expansion caused a 17% decreased in power output. Due to the fact that the internal volume ratio of the expander is around 5.3, the peak of isentropic efficiency was obtained for an applied pressure ratio between 6 and 8. The gas forces applied to each tooth during one full rotation are shown in Figure 8. The peak of force occurs at the end of the suction process.

The extended chamber model described in Subsection 3.2 has been tested by considering a two-phase expansion of R245fa and lubricant oil ACD100FY. The expander inlet conditions were specified in terms of temperature and quality, 100 °C and 0.6. The lubricant oil mass fraction was set to 0.05. The results of the two-phase expansion are shown in Figure 9. At each step size taken by the solver, the quality of the working fluid is updated by considering the actual volume occupied by the working fluid and its mass:

$$x_g = \frac{\frac{V_G}{m_G} - v_{G,l}}{v_{G,g} - v_{G,l}} \tag{19}$$

If $x_g < 0$ or $x_g > 1$, the working fluid is considered subcooled or superheated, respectively. Once x_g is calculated, its derivative $dx_g/d\theta$ can be evaluated. The evolution of pressure and quality with respect of the crank angle are reported in the top-left chart of Figure 9. During the expansion process, the quality of R245fa increases from 0.60 to 0.87 which corresponds to a specific volume ratio of approximately 17. Thus, the majority of the quality change occurs during the filling process and at the beginning of the closed expansion. As a result, the built-in volume ratio of the machine needs to be lower than the expansion volume ratio to avoid over-expansion and other internal losses. In the case of twin-screw machines, Smith et al. (1996) reported that the built-in volume ratio should be around 20-30% of

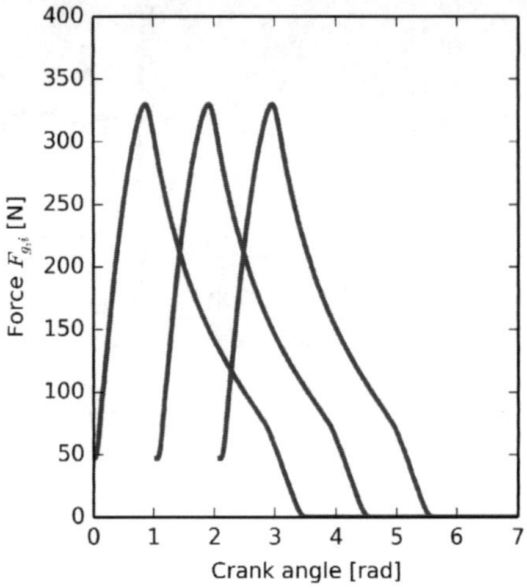

Figure 8: Force induced from the pressure inside the grooves on each tooth engaged during a rotation of the main rotor (p_{su}= 800 kPa, T_{su}= 100 °C, p_{ex}= 200 kPa, x_L = 0.01, N = 3000 rpm).

the overall expansion volume ratio. Under the considered two-phase expansion conditions, shown in the T-s diagram (top-right chart of Figure 9), the built-in volume ratio of the single-screw expander should be between 3.4 and 5.1. This result is in line with the geometry of the existing expander showing potential to explore two-phase expansion and partial-evaporating ORC.

The chamber model has the ability to estimate the temperature evolution of both working fluid and lubricant oil. Different models of heat transfer term $\dot{Q}_{L-G}$ can be applied depending on the data availability, oil-injection parameters etc. (Stosic et al., 1988) . In this case, the oil is pre-mixed with the working fluid prior entering the expander by means of a static mixer. Ideally, the static mixer creates an homogeneous fluid structure and therefore it is reasonable to assume an average oil droplet size to estimate the heat transfer. The heat transfer coefficient at the oil droplet surface is given by:

$$h_{L-G} = \frac{k_L}{D_{d,L}}\left[2 + 0.6 Re_L^{0.6} Pr_L^{0.33}\right] \tag{20}$$

where k_L is the thermal conductivity of the oil, $D_{d,L}$ is mean oil droplet diameter, Re_L and Pr_L represents the Reynolds and Prandtl numbers for the oil. The thermal behaviors of the working fluid and the oil droplet are shown in Figure 9. The thermal inertia of the two fluid is different and the temperature drop experienced by the working fluid is significantly higher under the two-phase conditions. However, the expansion process in the single-screw is quite fast (pressure variations are in the order of 8-9 ms/cycle) and in order to account for phenomena such heat transfer between oil and working fluid as well as oil and the chamber walls, more accurate correlation should be develop based on experimental results. To this end, a dedicated ORC test-rig with independent lubricant oil loop and mixing section have been designed and constructed. The chamber model is general enough to account for primary and secondary thermal effects.

5. CONCLUSIONS

In this paper, a mechanistic model of an oil-flooded single-screw expander has been presented. The main achievements are:

- by describing the main rotor and the starwheels as polygons, boolean operations can be used to obtain the full volume curve of each groove; by unwrapping each groove the suction and discharge porting can be accurately calculated. In a single-screw expander only one side presents a radial groove opening beside the main outlet of the housing;
- a generalized two-fluid chamber model has been implemented which allows to the working fluid to change phase and to account for the presence of oil. The models account for the heat transfer between the fluids;

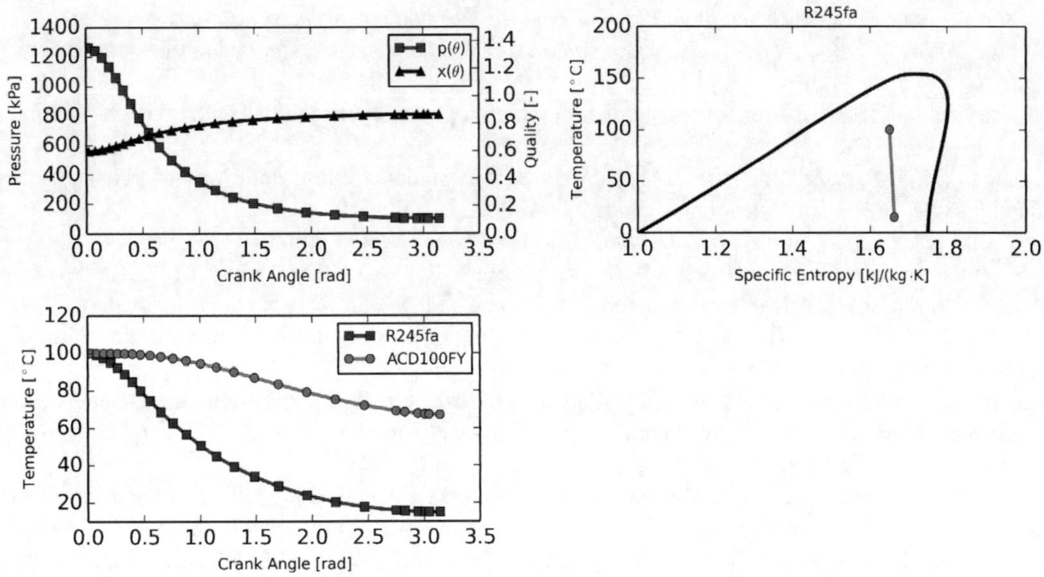

Figure 9: Example of two-phase expansion of R245fa with lubricant oil ACD100FY with the following boundary conditions: T_{su} = 100 °C, $x_{g,su}$=0.6, p_{ex} = 150 kPa, x_L = 0.05. The mean oil droplet size is 50 μm.

- a mechanistic models to compute the mechanical and friction losses on the starwheel has been added. Further work is required to estimate the mechanical torque generated at the main rotor shaft.

NOMENCLATURE

$\dot{m}$	mass flow rate	(kg/s)	**Subscript**	
c	centroid coordinates	(m)	CV	control volume
D	diameter	(m)	ex	exhaust
F	force	(N)	d	droplet
h	specific enthalpy	(kJ/kg)	g	working fluid vapor phase
	heat transfer coeff.	(kW/(m^{2}K))	G	working fluid phase
L	length	(m)	l	working fluid liquid phase
M	force	(Nm)	L	liquid secondary phase
v	specific volume	(m^3/kg)	su	supply
V	volume	(m^3)		
$\dot{W}$	work rate	(kW)		
x	mass fraction	(-)		
α	void fraction	(-)		
ω	angular speed	(rad/s)		

REFERENCES

Bein, T. W., & Hamilton, J. F. (1982). Computer modeling of an oil flooded single screw air compressor. In *International compressor engineering conference.* (Paper 383)

Bell, I. H., Lemort, V., Groll, E. A., Braun, J. E., & Horton, W. T. (2012). Liquid flooded compression and expansion in scroll machines - Part i: Model development. *International Journal of Refrigeration*, *35*, 1878-1889.

Bell, I. H., Wronski, J., Sylvain, Q., & Lemort. (2014). Pseudo-pure fluid thermophysial property evaluation and the open-source thermophysical property library coolprop. *Ind. Eng. Chem. Res.*, *53*, 2498-2508.

Fujiwara, M., Kasuya, K., Matsunaga, T., & Watanabe, M. (1984). Computer modeling for performance analysis of rotary screw compressor. In *International compressor engineering conference at purdue. paper 503.*

Georges, E. (2012). *Investigation of a flooded expansion organic Rankine cycle system* (Unpublished master's thesis). University of Liege.

Igobo, O., & Davies, P. (2014). Review of low-temperature vapor power cycle engines with quasi-isothermal expansion. *Energy*, *70*, 22-34.

Imran, M., Usman, M., B-S, P., & Lee, D.-H. (2016). Volumetric expanders for low grade heat and waste heat recovery applications. *Renewable and Sustainable Energy Reviews*, *57*, 1090-1109.

Ishii, M., & Hibiki, T. (2011). *Thermo-Fluid Dynamics of Two-Phase Fflow.* Springer.

Lecompte, S., Huisseune, H., van den Broek, M., Vanslambrouck, B., & De Paepe, M. (2015). Review of organic Rankine (ORC) architectures for waste heat recovery. *Renewable and Sustainable Energy Reviews*, *47*, 448-461.

Mathison, M. M., Braun, J. E., & Groll, E. A. (2013). Modeling of a novel spool compressor with multiple vapor refrigerant injection ports. *International Journal of Refrigeration*, *36*, 1982-1997.

Smith, I. K., Stosic, N., & Aldis, C. (1996). Development of a trilateral flash cycle system: Part 3: The design of high-efficiency two-phase screw expanders. *Proceedings of the Institution of Mechanical Engineers. Part A: Journal of Power and Energy*, *208(2)*, 135-144.

Stosic, N., Kovacevic, A., Hanjalic, K., & Milutinovic, L. (1988). Mathematical modelling of the oil influence upon the working cycle of screw compressors. In *International compressor engineering conference. paper 645.*

Taniguchi, H., Kudo, K., Giedt, W. H., Park, I., & Kumazawa, S. (1988). Analytical and experimental investigation of two-phase flow screw expanders for power generation. *Journal of Engineering for Gas Turbines and Power*, *110*, 628-635.

Woodland, B. J., Braun, J. E., Groll, E. A., & Horton, W. T. (2014). Methods of increasing net work output of organic Rankine cycles for low-grade waste heat recovery. In *15th international refrigeration and air conditioning conference at purdue. paper 2190.*

Woodland, B. J., Krishna, A., Groll, E. A., Braun, J. E., Horton, W. T., & Garimella, S. V. (2013). Thermodynamic comparison of organic rankine cycles employing liquid-flooded expansion or a solution circuit. *Applied Thermal Engineering*, *61*, 859-65.

Wu, J., & Jin, G. (1988). The computer simulation of oil-flooded single screw compressors. In *International compressor engineering conference.* (Paper 646)

Zhelezny, V. P., Semenyuk, Y. V., Ancherbak, S. N., Grebenkov, A. J., & Beliayeva, O. V. (2007). An experimental investigation and modelling of the solubility, density and surface tension of 1,1,1,3,3-pentafluoropropane (R-245fa)/synthetic polyolester compressor oil solutions. *Journal of Fluorine Chemistry*, *128*, 1029-1038.

Ziviani, D., Bell, I. H., De Paepe, M., & van den Broek, M. (2014). Comprehensive model of a single screw expander for ORC-systems applications. In *22th int.compressor engineering conf. at purdue.*

Ziviani, D., Bell, I. H., De Paepe, M., & van den Broek, M. (2015). Update on single-screw expander geometry model integrated into an open-source simulation tool. In *9th Int. Conf. on Compressors and their Systems, City University of London, London.*

Transient Thermal and Structural Analysis of Cylinder and Bolted Joints on BOG Compressor during Starting Process

Lingxiao Wei[1], Yi Yan[1], Yun Li[1]*, Xueqiang Feng[2]

[1]Xi'an Jiaotong University, School of Chemical Engineering and Technology,
No.28, Xianning West Road, Xi'an, Shaanxi, 710049, China

[2]Zhejiang Qiangsheng Compressor Manufacturing Co., Ltd.
No.161, Economic and Technological Development Zone, Wenzhou, Zhejiang, 325011, China

* Corresponding author, email: yunli@mail.xjtu.edu.cn

ABSTRACT

BOG (Boil off Gas) compressor deals with the gas evaporated from the LNG storage tank. Its working temperature is lower than -120℃. There are two types of starting process for BOG compressor: Direct Starting and No-load Starting. During the starting process, the temperature on cylinder will change rapidly and result in additional thermal stress to some parts on the cylinder, especially to the bolted joints on the cylinder head. In this paper, transient Finite Element thermal analysis is proposed on the cylinder with some improvement of the boundary condition settings, such as the consideration of the ice increase on the cylinder wall. Then, the theoretical and transient FE analysis are proposed subject to the bolted joints of the cylinder head in two starting process. Result shows that the maximum temperature difference on the cylinder is 81.5℃ during direct starting; while it could decrease to 60.9℃in No-Load Starting. Along the bolted joints in the cylinder head, the maximum temperature difference will be up to 52℃ in direct starting and 45℃ in No-Load Starting. The preload increases rapidly over 60% and its mean tensile stress on the bolt is near to the yield strength. Besides, the preload are distributed unevenly during the starting process. The maximum uneven rate is 20% in Direct Starting and 15% in No-Load Starting. It shows that No-Load Starting could offer a more stable starting process. Finally, based on the theoretical analysis, a preload adjustment method is proposed to ensure the safety and validity of bolted joints on BOG compressor.

1. INTRODUCTION

The liquefied natural gas in LNG storage tank will produce evaporated gas when absorbing the incoming heat. Using the compression and reflux after condensation treatment can effectively save these resources. BOG compressor is a kind of equipment that deals with the gas evaporated from the LNG storage tank. Its cylinder components usually work lower than -120℃. Working at such a low temperature, the mechanical behavior of BOG compressor's various parts, especially the bolted joints, will result in great changes. Therefore, it is necessary to analyze its stress and deformation distribution at low temperature to ensure the safely and effectively operation of the compressor at low temperature.

For BOG compressor and many machines, its bolts are tightened in ambient temperature during assembling process, but its operating temperature is far lower. Various components of compressor would face thermal contraction at low temperature. And the thermal contraction of different components varies because of their various materials. For example, BOG compressor cylinder head is made of cast iron (QTiNi35), whose coefficient of thermal expansion is lower than 304 steel, used by cylinder head bolts. The properties of two materials are shown as Table 1.

Table 1: Related properties of two kinds of materials

Materials	Elasticity modulus Pa	Poisson's ratio	Thermal expansion $^{o}C^{-1}$
304 steel	1.95×10^{11}	0.3	16.5×10^{-6}
QTiNi35	2.10×10^{11}	0.3	0.9×10^{-6}

Therefore, the preload of the bolt will change due to the different thermal deformation. In this case, the preload will become larger in low temperature. In other words, even if the bolts were tightened properly at ambient temperature, the preload would be too larger in the low-temperature environment. Accordingly, it is very important to research on the preload characters of the bolts on BOG compressor cylinder head at low temperature and further to guide to select the preload at ambient temperature.

There are mainly two kinds of BOG compressor starting mode. One starts directly at ambient temperature, which is called Direct Starting. The other can be called No-load Starting: during this starting process, cryogenic gas flow into the cylinder and compressor works by keeping the suction valve and discharge valve open until the cylinder temperature is stable. No-load Starting is more conservative than Direct Starting to avoid sharp changes of cylinder temperature when the compressor is in the normal operation and improve the safety. But it needs more time and energy than Direct Starting. At present, both of the starting methods are applied in commercial BOG compressor. One of the tasks of this study is to compare these two starting mode and confirm the advantage of No-load Starting.

The steady and transient thermal characteristics of BOG compressor's starting process has been focused on the steady mechanical behavior of the bolted joints on the cylinder head (Shen *et al.*, 2012) and (Ding *et al.*, 2013). Nevertheless, the transient mechanical behavior of bolted joints during BOG compressor's starting process has not been concerned. Therefore, this paper consists two parts. Based on predecessors' analysis, the boundary conditions of cylinder thermal simulations are improved and the thermal characteristics of Direct Starting in ambient temperature and No-Load Starting are compared in Part I. In Part II, based on the transient thermal condition, theoretical and transient mechanical analysis are proposed in order to research the stress behavior of bolted joints.

2. FE Transient Thermal Analysis

2.1 Model and Analysis Condition

BOG compressor cylinder structure is shown in Figure 1. It is composed by cylinder, cylinder head, piston and bolt on cylinder head. Intake passage is set up on the cylinder top while discharge passage is on the bottom side. The axial dorsal of cylinder is connected with piston rod. In the front, 12 of M36 bolts are arranged uniformly along the cylinder axis direction. Due to the complex structure of the cylinder, all-hexahedral meshing for FE is very difficult. Thus, meshes are components with most of hexahedron and small number of tetrahedral elements. Cylinder is symmetrical along axis. So we only take half of the calculation volume to reduce the computation time. Based on the experimental data (the time when the temperature of cylinder becomes stable), analysis time is set to 10 hour or 36,000 seconds.

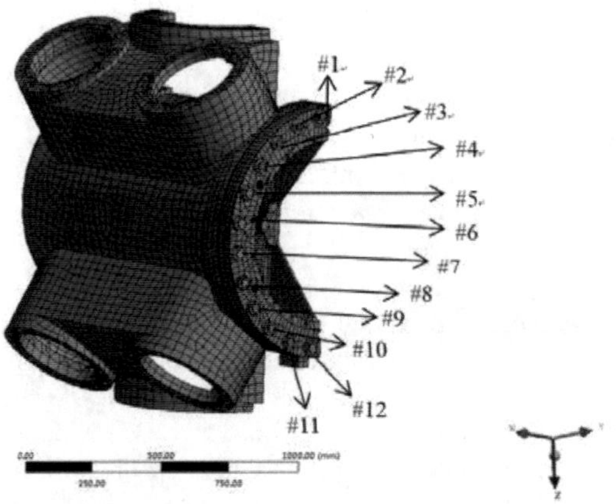

Figure 1: Model and meshing of bog compressor cylinder temperature field simulation

2.2 Boundary Condition of Direct Starting:

For BOG compressor cylinder's thermal analysis, the boundary conditions can be divided into three parts: the outer wall, intake and discharge passage and cylinder compression chamber. For the outer wall, heat can be transferred through the cylinder wall between cylinder and environment. It is considered as a stable natural convection boundary condition (Shen *et al.*, 2012). However, BOG compressor's working temperature is extremely low. Discharge temperature of first stage, which is the maximum in the cylinder, is usually as low as -80℃. So the cylinder will be covered with a thick and increasing ice layer during the starting process. The ice layer will make the heat transfer conditions change, as shown in Figure 2. The ice thickness D increased with time. As a result, the equivalent heat transfer coefficient decreases gradually. Therefore, equivalent heat transfer coefficient of the outer wall of the cylinder h_{eq} is calculated by superposing of ice thermal resistance and natural convection thermal resistance. Its expression is shown as follows:

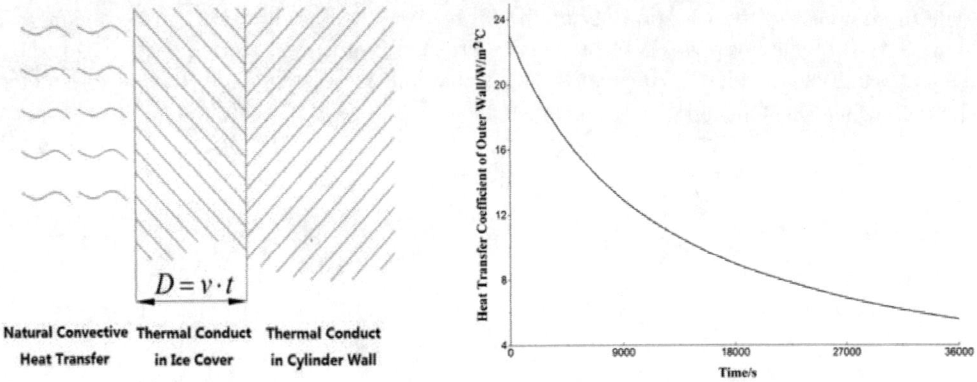

Figure 2: Variation of equivalent heat transfer coefficient

The ice thickness D increased with time. As a result, the equivalent heat transfer coefficient decreases gradually. Therefore, equivalent heat transfer coefficient of the outer wall of the cylinder h_{eq} is calculated by superposing of ice thermal resistance and natural convection thermal resistance. Its expression is shown as follows:

$$h_{eq} = \frac{1}{\frac{1}{h_n} + \frac{vt}{\lambda_i}} \tag{1}$$

Where v represents ice growth rate, which can be obtained by compressor starting test; t is the compressor start time; λ_i is thermal conductivity of ice; h_n is natural convection heat transfer coefficient. In this paper, the growth rate of ice is 8.35×10^{-6} m/s, which is estimated by experimental data. Convective heat transfer coefficient decreases from 23 to 5 with time. It shows that ice growth has significant effect on heat transfer boundary conditions of the outer wall of cylinder.

Heat transfer coefficient of the intake and discharge passage can be calculated by a general correlation for forced convection in tubes. Gnielinski formula (Holman, 2001) is used to apply for the boundary conditions of simulation.

$$Nu = \frac{(f/8)(Re-1000)Pr}{1+12.7\sqrt{f/8}(Pr^{2/3}-1)} \cdot \left[1+(\frac{d}{l})^{2/3}\right] \tag{2}$$
$$f = (1.82\lg Re - 1.64)^{-2}$$

Where Re represents Reynolds number; Pr is Prandtl number; Nu is Nusselt number; d is equivalent diameter; l is passage length.

When compressor is working, the temperature and pressure of the gas in cylinder compression chamber, as well as the convective heat transfer coefficient, is changing with time. However, the convective heat transfer coefficient correlations of the cylinder compression chamber have not been found. In the similar study (Shen *et al.*, 2012), convective heat transfer coefficient correlations of the inner wall of internal combustion engine cylinder are adopted for substitution. This paper adopts the Hohenberg formula (Neshat & Saray, 2014):

$$\alpha_g = 130 \cdot V_g^{-0.06} \cdot P_g^{0.8} \cdot T_g^{-0.4} \cdot (C_m + 1.4)^{0.8} \quad (3)$$

Where P_g and T_g represent the transient pressure and temperature of the gas in cylinder, C_m is the average velocity of piston, and V_g is the volume of compression chamber.

According to the regularities of the compression chamber volume change with the crank motion and the P-V Graph for steady state of compressor, the variation of parameters of formula (3) with time can be obtained. The result is shown in Figure 3. A-B segment represents the expansion process. B-C is the suction phase. C-D is the compression phase, and D-A of the discharge phase. To simplify the calculation, the average convective heat transfer coefficient, $107.67\ \mathrm{W/m^2 \cdot K}$, which is calculated from the results of Figure 3, is used in the following analysis.

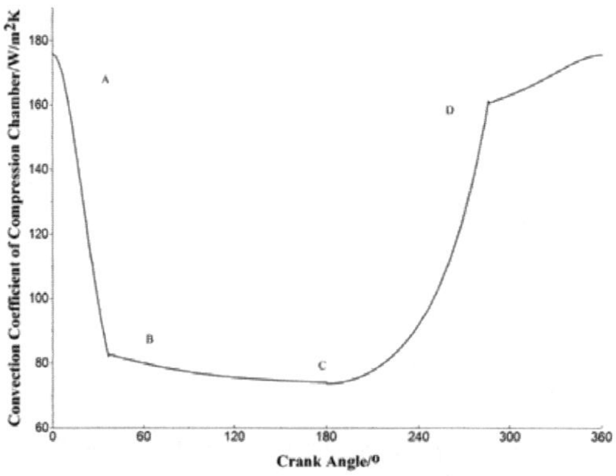

Figure 3: Variation of convection coefficient with crank angle

Besides, during compressor starting process, the gas temperature could be increased because of the heat obtained from cylinder inner wall. That is to say, gas temperature will be higher than its design value at the beginning of the starting process. Therefore, it is unreasonable to set the gas temperature constant. In this study, the temperatures of the gas in intake passage, the compression chamber and the discharge passage decrease with time. Based on the design parameters, an index changes correlation is used to show the compressor chamber average temperature variation ignoring variations over each crankshaft rotation. The expression is shown below.

$$T(t) = \frac{1}{\frac{1}{-T_\infty} + \frac{(\frac{1}{T_\infty} - \frac{1}{T_0})}{(t_{\max})^a} \cdot t^a} + (T_\infty + T_0) \quad (4)$$

Where T_∞ represents steady-state temperature, the design value; T_0 is the gas temperature of initial starting process, which can be obtained by the compressor test data; $t_{\max}$ is the total test time, which is 36000 seconds in this case; a is the exponent refers to the temperature decrease rate, generally taking $a < 1$, the smaller value represents the faster drop of temperature.

2.3 Boundary Conditions of No-load Starting

No-load Starting process is that the valves are forcing-opened and cryogenic gas flow through the cylinder without compression. So the boundary condition setting is similar to Direct Starting with a little difference. Its outer wall boundary condition is almost the same as Direct Starting. Intake and discharge passage boundary conditions use the same model as the Direct Starting: Gnielinski model. The difference is that the heat transfer coefficients of intake passage and discharge passage are equal because there is no compression process. The biggest difference between No-load Starting and Direct Starting is in compression chamber. Absent of the compression process, the boundary conditions of compression chamber can be calculated by Gnielinski model at No-load Starting process. Temperature variations of two starting methods in each boundary are shown in figure 4. Boundary conditions of two starting methods are summarized in Table 3.

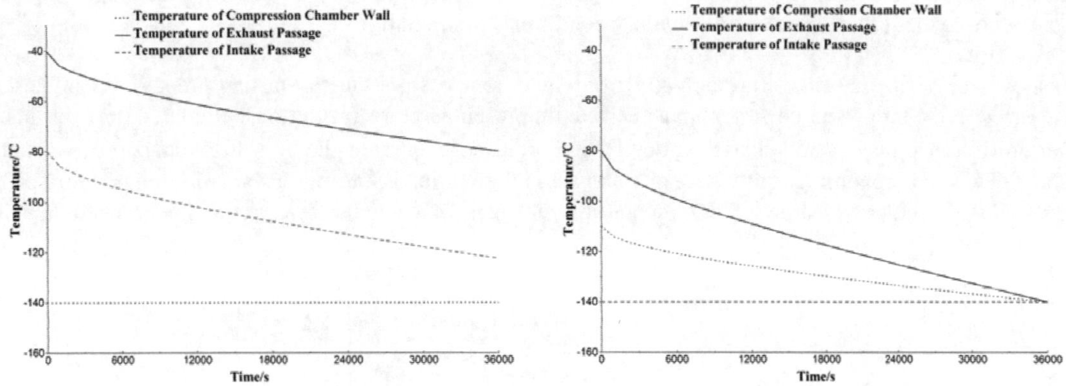

a) Direct Starting b) No-load Starting

Figure 4: Temperature variation in each boundary

Table 3: Summary of temperature field boundary conditions of BOG compressor simulation

Conditions	items	Intake passage	Discharge passage	Compression chamber	Outer wall
Direct Starting	h (W/m$^{2}\cdot$K)	123	103	Fig 6	Fig 4
	T_0 (℃)	-140	-40	-80	33
	T_∞ (℃)	-140	-80	-122.5	33
No-load Starting	h (W/m$^{2741\cdot}$K)	123	123	151	Fig 4
	T_0 (℃)	-140	-80	-110	333
	T_∞ (℃)	-140	-140	-140	33

2.4 Results and analysis

The simulation of cylinder temperature field requires many initial boundary conditions, most of which are hypothetical conditions based on the design parameter. So it is necessary to compare the result with experimental data to verify its reliability. Experimental data used in this section is real-time temperature data in BOG compressor prototype without load test made by Zhejiang Qiangsheng Compressor Manufacturing Co. The compressor prototype's design parameters and structure are consistent with the simulation. Comparative results are shown in the Figure 5.

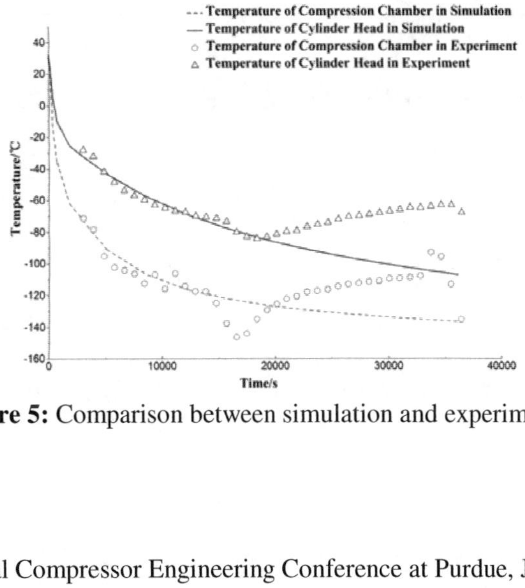

Figure 5: Comparison between simulation and experimental

Firstly, the simulation and experiment value have a good consistency on the trend, especially for the starting half of the data. In the following time, the consistency is lower. It is probably because during the test, the temperature of intake gas is controlled by the proportion of liquid and gas. The flow fluctuation will affect the temperature of the mixed gas. Therefore, the stability of the intake gas temperature is hard to keep stable. In conclusion, the results can preliminarily prove the reasonableness of boundary conditions of simulation.

Figure 6 shows the variation of the extreme value of cylinder temperature during starting process. It can be seen that the minimum temperature dropped faster than the maximum temperature inside the cylinder. Thus, the maximum temperature difference inside the cylinder, around 150℃, did not appear at the end of the start process, but at the earlier stage. Two start conditions show similar trends, and the No-load Starting has smaller temperature difference inside the cylinder. Figure 7 shows the temperatures changes of each bolts at the cylinder head in two start conditions.

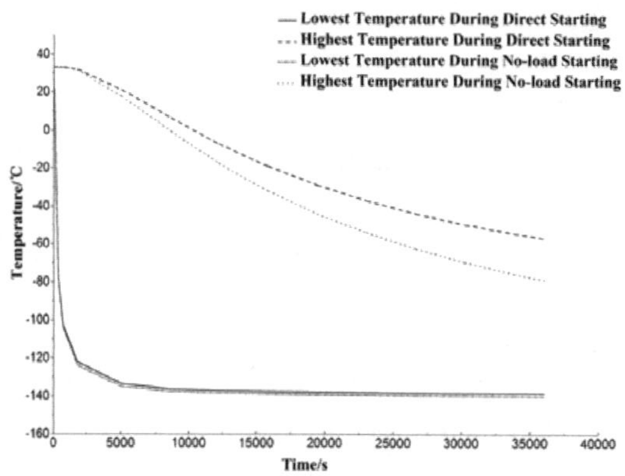

Figure 6: Variation of compressor cylinder extreme temperature during starting process

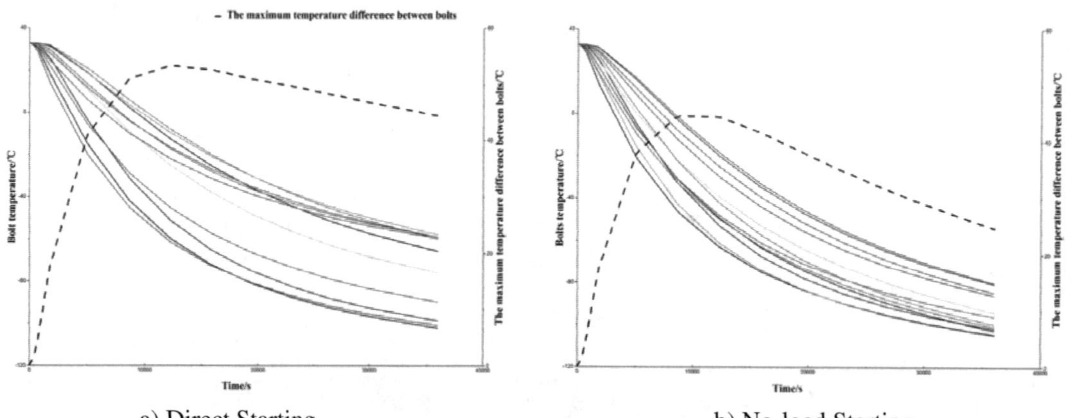

a) Direct Starting b) No-load Starting

Figure 7: Temperature distribution and temperature difference

Due to the obvious temperature gradient in the cylinder, the temperature distributions of each bolt are not uniform. When compressor starts, temperature differences of bolts rises rapidly, and decreases slowly after reaches peak temperature, whose patterns are consistent with the cylinder. There are still some differences of the temperature distribution in two starting conditions. During the Direct Starting condition, the temperatures of the bolts, which are close to the intake side, decrease quickly, but slowly close to the discharge side. Eventually, the steady temperatures difference is about 45℃ between two sides. The maximum temperature difference of bolt is 52℃, but it would decrease to 45℃ during the No-load Starting. The temperatures of bolts are low on the intake side and discharge side, but higher in the middle. Temperature distributions of bolts are more uniform. Different steady temperature indicates that the preload will change in different ranges. When compressor is working at low temperatures, cylinder head will be under an uneven preload state.

3. Structural Analysis of Bolted Joints

During the previous section, BOG compressor starting transient temperature field of cylinder were analyzed. Transient temperature field would be loaded into the transient mechanical analysis in this section. Bolt preload theoretical prediction formula would be applied to determine the cylinder head bolt preload at ambient temperature.

3.1 Introduction of Theoretical Analysis

Preload represents the degree of tightening bolts. Preload value shall comply with the relevant standards. According to the standard (ANSI/ASME B18.2.6, 1996), preload values, which will serve as an initial value in theoretical calculation and finite element calculation, can be represented by the following equation.

$$F_0 \leq (0.5 \sim 0.6)\sigma_s \cdot A_s \quad (5)$$

Where σ_s represents the yield strength of the bolt material; A_s is the minimum cross area of the bolt. The formula shows that the preload should make the average tensile stress within the bolt reaches 0.5 to 0.6 times of the yield stress. Excessive preload value may make the bolt yield even failed, and small preload value cannot guarantee good preloading effect. It is necessary to analysis the bolt preload changing from ambient temperature to working temperature and adjust the assembling preload value to meet the requirements. The preload at ambient temperature is 250kN. The bolt structural sizes are in the Table 4.

Table 4: Original parameters

Structure parameters			
Bolt diameter/mm	32	Thickness of nut/mm	15
Nut outer diameter/mm	52	Thickness of the bolt head /mm	30
Bolt head outer diameter/mm	100	Nut inner diameter/mm	32
Bolt head inner diameter/mm	50	yield strength	640MPa

The theoretical model is taken from the earlier study (Yan *et al.*, 2012). The calculations can be broadly divided into the following steps: The first step is to get the original parameters. Mainly include the structure size parameters, material properties, as well as the working condition.

The second step is to calculate the axial stiffness of each component. For bolt, nut and bolt head, the axial stiffness can be calculated by the following formula:

$$k = \frac{EA}{L} \quad (6)$$

Where E represents the elastic modulus of the material; A is the force area; L is the length along direction of the force.

However, for the clamped objects, such as cylinder and cylinder head, the formula (6) cannot be used due to these complex shapes. Therefore the finite element method is applied to calculate the axial stiffness of connecting pieces in this section, whose result is $2.5 \times 10^9 \mathrm{N/m}$。

Next, calculating the modified stiffness of clamped objects k_c' and the deformation distribution coefficient m:

$$k_c' = \frac{1}{\frac{1}{k_c} + \frac{1}{k_n} + \frac{1}{k_h}} \quad (7)$$

$$m = \frac{k_c'}{k_c' + k_b} \quad (8)$$

Where k_c presents the stiffness of clamped objects, k_h and k_n are the stiffness of bolt head and bolt. Axial stiffness and deformation distribution coefficient of various components are shown in the Table 5:

Table 5: The axial stiffness of several components and the distribution coefficient

Stiffness of each component			
Bolt stiffness k_b / N/m	2.614×10^9	Clamped object stiffness k_c / N/m	2.500×10^9
Nut stiffness k_n / N/m	1.507×10^{10}	Modified clamped objects stiffness k_c'/ N/m	3.822×10^9
Bolt head stiffness k_h / N/m	3.725×10^{10}		
Distribution coefficient m			
0.44			

Finally, the range of bolt preload change could be calculated by following formula:

$$F' = F - m\cdot(\alpha_b - \alpha_c)\cdot \Delta T \cdot L \cdot k_b \quad (9)$$

Where F' represents the preload at low temperature; F is the preload at ambient temperature; α_b is coefficient of thermal expansion of bolt; α_c is coefficient of thermal expansion of connecting piece; ΔT is temperature difference.

3.2 Introduction of the pre-processing of finite element analysis

The analysis in this section is a continuation of the cylinder transient thermal analysis in previous section. So the model and mesh are consistent with Figure 1. In accordance with the spatial position, bolts are numbered along the half circumference of the cylinder head. The preload is simulated by applying an inward compressive force, 250kN, in the middle of cross-section of the bolt. Two steps are taken in analysis. The preload of bolt is loaded in the first step and the temperature field is imported in the second step. The whole analysis time is still 36000s.

3.3 The results and analysis

4 representatives of the bolts along the cylinder head are selected to shown the preload changes in Figure 8.

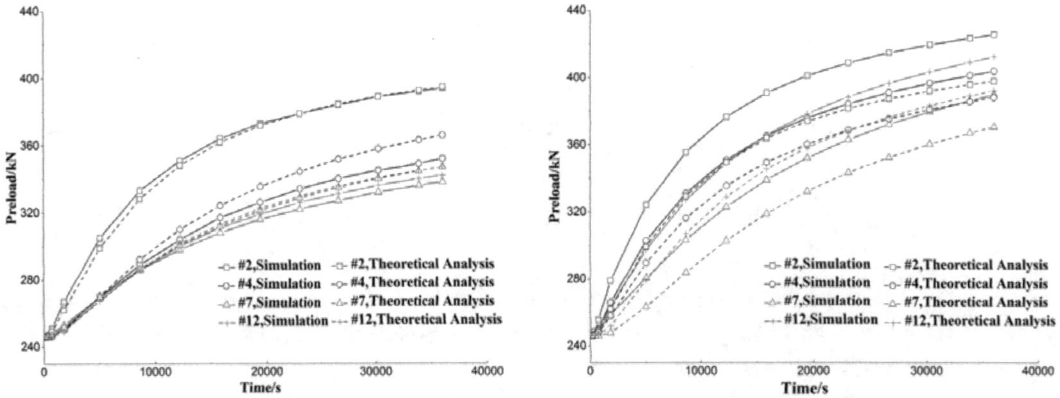

a) Direct Starting b) No-load Starting

Figure 8: Bolt preload change and calculation error distribution

An analysis of the results can be viewed in several aspects: 1) The materials of cylinder and cylinder head (lower expansion coefficient) are different from the bolts (higher expansion coefficient). So the bolt preload will rise at low temperatures. Average tensile stress of bolt section increased from 320MPa in ambient temperature to 468MPa (Direct Starting condition) and 525MPa (No-load Starting condition), equivalent to 73% and 82% of the material yield strength respectively. 2) Errors of bolt preload between theoretical calculation and finite element values are within 10%, which shows good consistency between theoretical calculation and finite element analysis. The maximum error of Direct Starting conditions is 4%, and 8% of the No-load Starting conditions. 3) The bolt preload distribution at low temperatures is no longer uniform because of the significant temperature difference in cylinder head during starting. For Direct Starting condition, the area of cylinder head near the intake passage is at low temperature, but other area near the discharge passage is at higher temperature. Thus, after the start, the maximum preload is 395kN, located at #2 bolts, and minimum preload is 346kN, located at #8 bolts. However, for No-load Starting condition, the area of cylinder near the intake passage and discharge passage head is at low temperature, but other area in the middle is at higher temperature. Thus, after the start, the maximum preload is 398kN, located at #2 bolts, and minimum preload is 371kN, located at #7 bolts. The preload unevenness rate is defined as the following formula:

$$\delta = \frac{\sigma_{max} - \sigma_{min}}{\overline{\sigma}} \times 100\% \tag{10}$$

Where σ_{max} and σ_{min} are the maximum and minimum preload; $\overline{\sigma}$ is average preload. Bolt preload unevenness of two start conditions are shown in Figure 9. Due to the low maximum uniformity of No-load Starting condition, using this method could reduce the stress fluctuation.

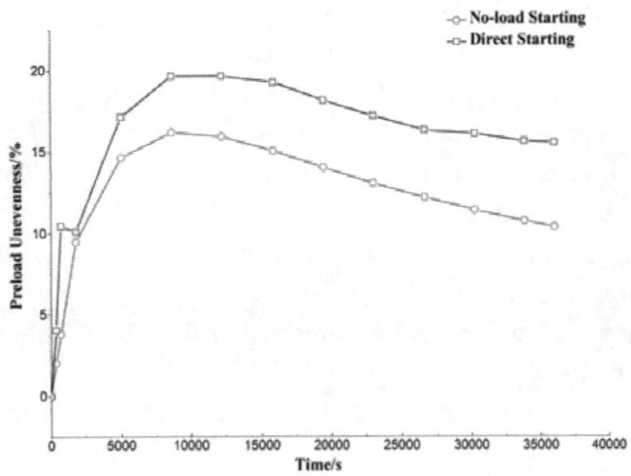

Figure 9: Preload unevenness distributions of two conditions

3.4 Adjustment method for cylinder head bolt assembling preload values

According to the previous section, preloads of BOG compressor cylinder head bolts at the operating temperature are significantly higher than the ambient temperature. Meanwhile, preload distribution is becoming uneven during the starting process. Therefore it is necessary to adjust the value of preload at ambient temperature. Finite element analysis proved the validity of the theoretical formula in previous section. In this section, the theoretical formula is used to deduce the recommended range of bolt preload value in different conditions.

As for the bolts that have great difference between operating temperature and ambient temperature, safety and effectiveness under two conditions should be considered for determine preload values,

$$\begin{cases} b\sigma_s A \le F \le \sigma_s A / a \\ b\sigma_s A \le F - m(\alpha_b - \alpha_c)\Delta T \cdot L \cdot k_b \le \sigma_s A / a \end{cases} \tag{11}$$

Where a represents coefficient of safety, whose value must be bigger than 1; b is coefficient of effect, whose value could be 0.3-0.5 depending on the circumstances, such as requirements for the sealing performance.

The formula (11) can be simplified as follow: If bolt preload rises at the operating temperature, it mustn't be excessive at operating temperature but not too small when assembled at ambient temperature. The expression is as follows:

$$b\sigma_s A \le F \le \sigma_s A / a + m(\alpha_b - \alpha_c)\Delta T \cdot L \cdot k_b \tag{12}$$

Conversely, in the situation that preload decreases at the operating temperature, it must be not too small at operating temperature and not excessive when assembled at ambient temperature. The expression is as follows:

$$b\sigma_s A / a + m(\alpha_b - \alpha_c)\Delta T \cdot L \cdot k_b \leq F \leq \sigma_s A / a \quad (13)$$

In this article, a=1, b=1.5 according to the requirement of the bolt's load condition. The ranges of each bolt preload are shown Table 7. The bolds marks the boundary of the bolt's stress condition.

Table 7: Range of bolt preload

	Preload (kN)		Tensile Stress(MPa)				Yield Ratio (%)			
			Room Temperature		Low Temperature		Room Temperature		Low Temperature	
	Min	Max	Min	Max	Min	Max	Min	Max	Min	Max
#1	148	183	**184.02**	227	289	**324**	**28.75**	35.55	45.16	**50.63**
#4	148	193	**184.02**	240	279	**324**	**28.75**	37.50	43.59	**50.63**
#7	148	226	**184.02**	281	247	**325**	**28.75**	43.91	38.59	**50.78**

3.5 Conclusions

According two starting conditions of BOG compressor, cylinder transient temperature field has been analyzed in this paper. Then theoretical formula and finite element method were used to analyze the preload changes of cylinder head bolts. Finally based on the theoretical calculation formula, adjustment method of bolts preloads at ambient temperature was given. Conclusions are following.

- Temperature distribution of cylinder is more uniform during No-load Starting. compared with the Direct Starting，The maximum temperature difference reduced over 20 ℃.
- Theoretical formula could reflect the variation of bolt preload at low temperature, and agree with the results of finite element simulation well.
- The cylinder head bolt preload will increase substantially at low temperature. The tensile stress reaches 517MPa, close to the yield strength of the material.
- Bolt preload distribution is no longer evenly at low temperatures. The maximum unevenness is 18.8% and ended unevenness is 13.5% during Direct Starting. The maximum unevenness is 16.0% and ended unevenness is 7.1% during No-load Starting.
- Based on the theoretical model, a preload adjustment method is proposed in this study, after the test verification, it can be used to ensure the preload values are even and moderate at operating temperature.

REFERENCES

ANSI/ASME B18.2.6-96 (1996). Fasteners for use in structural applications. American National Standards Institute, Washington, D.C.

Ding, L., Wang, L., & Peng, X. (2013). Bolt Preload Study of BOG Compressor Cylinder Head at Low Temperature. *Compressor Technology*(6), 12-17.

Holman, J. (2001). Heat transfer, Eighth SI Metric Edition: McGraw-Hill Inc.

Neshat, E., & Saray, R. K. (2014). Effect of different heat transfer models on HCCI engine simulation. *Energy Conversion and Management, 88*, 1-14.

Shen, Y., Zhang, B., Xin, D., Yang, D., & Peng, X. (2012). 3-D finite element simulation of the cylinder temperature distribution in boil-off gas (BOG) compressors. *International Journal of Heat and Mass Transfer, 55*(23), 7278-7286.

Yan, Y., Wen, M., & Li, Y. (2015). *FEM Analysis and Preload Evaluation of Bolted Joints Subject to Low Temperature.* Paper presented at the ASME 2015 Pressure Vessels and Piping Conference.

AN INVESTIGATION OF THE HEAT TRANSFER CHARACTERISTICS OF THE INDUCTION MOTOR INSIDE THE HERMETIC RECIPROCATING COMPRESSOR

Haşim Otunç[1]; A. Refik Özdemir[2], Prof.Dr.Hasan Güneş[3]

[1]Arçelik Electrical Motor Plant / R&D Dept., Organize Sanayi Bölgesi, Atatürk Caddesi, 8. Sokak, 59500, Çerkezköy, Turkey
[2] Arçelik Research and Development Center, Tuzla, 34950, Istanbul, Turkey
[3]Istanbul Technical University, Mechanical Eng. Department, Gümüşsuyu, 34437 Istanbul, Turkey

ABSTRACT

In recent years, environmental concerns and energy consumption become more important. Therefore, compressor manufacturers try to design high-efficiency reciprocating compressors. Owing to developing technology, mechanical, electrical and thermodynamic losses of hermetically sealed compressors are reduced. Nevertheless, the effects of electrical motors are significant phenomena. The thermal behavior of an electrical motor is very important for motor performance, but also has a key role for compressor components heat transfer characteristics.

In this study, heat transfer between the components and the electrical motor of the hermetic reciprocating compressor were investigated theoretically and experimentally. The compressor performance, indicator diagram, motor tests and the detailed temperature measurements of electrical motor and compressor components were given. Analytical model simulations of induction motor and compressor components were performed by using commercial softwares. This part describes analytical models to predict the temperatures of motor and the other parts of hermetic reciprocating compressor. The model simulations allow determining component temperatures with different motor temperatures theoretically. In addition to this, conceptual designs, which have affected the heat transfer on the components and performance of the compressor, were also applied during the study.

The results of the theoretical, analytical and experimental studies are used for investigating the network including conduction, convection and radiation forms of heat transfer inside the compressor. Generated heat transfer network helps to characterize the thermal functions of the main components which leads to the new and better compressor designs.

1. INTRODUCTION

Hermetic reciprocating compressors are the main units of the vapor compression refrigeration systems. These compressors are commonly used due to its smallness and compactness. In recent years, environmental concerns and energy consumption become more important. Therefore, compressor manufacturers try to design high-efficiency reciprocating compressors. Therefore performance improvement studies of the compressor play an important role to reduce overall energy consumption of the refrigerators. Electromotor, thermodynamic and mechanical losses are the main three factors which effect the compressor performance.

The performance of a reciprocating compressor is influenced by the electrical losses. The thermal behavior of an electrical motor is very important for general motor performance, but also has a key role for heat transfer characteristics among compressor components. The electrical losses spread as heat energy inside the compressor housing. Therefore thermal behavior of an electric motor has effects on all components of the compressor. Thermodynamic losses of a hermetic compressor occur due to the heat transfer between the components and pressure losses. It is possible to improve performance of a compressor by investigating the heat transfer analysis. Complex heat transfer network was investigated by using different modeling methods and separating compressor component into simple control volumes. The design of the optimum components influences the efficiency of reciprocating compressors. In order to increase the COP of the compressor an efficient electrical motor design which provides optimum motor performance and minimum loss must be used.

In addition to the theoretical analysis, numerical simulations with engineering software coupled with experimental studies have become an essential development tools in the improvement of performance of reciprocating compressors. Positive efforts have also been made for the complexity of the physical phenomenon inside the compressor ambient. A lot of research work dealing with the use of software coupled with experimental studies as a design tool for various compressor components can be found in the literature. There is also some remarkable research in the development of optimum designs.

Dutra et al (Dutra, 2008) developed a thermal-electrical coupling model and used it to predict the volumetric-isentropic efficiencies and temperature distribution. They used equivalent circuit for steady state model of single phase induction motor. Ooi, presented an analytical study (Ooi, 2003) to improve compressor performance and investigate temperature distribution and heat transfer. In this study 46 control volumes were used and complex geometries simplified. Fagotti et al (Fagotti, 1994) developed a model and focused on the heat transfer from gas to component walls. Working characteristics, without experimental data, were determined. Haas et al (Haas, 2013) presented an investigation that thermal network was studied with 63 control volumes. Simulation results were compared with experimental results. Another study was presented by Oner et al. (Oner, 2009) In this study three phased induction motor was modeled and temperature distribution was investigated by using commercial software.

2. EXPERIMENTAL STUDIES

In order to conduct the electrical analysis and separate compression work from input power for numerical calculation, a pV set-up was built and calorimeter measurements were performed. Piezo-resistive miniature pressure transducers flush mounted in the valve plate were used to measure the pressure inside the cylinder for bearing force investigations. Compression work was done on gas by piston during the compression phase of the compressor. The gas force on piston creates reaction forces on crankshaft bearings. These reaction forces must be carried in the bearings by the oil film pressure. For the determination of the cylinder volume an optical encoder was placed on the shaft, from which the position of the shaft can be determined. From the position of the shaft the piston position was calculated and also the momentary cylinder volume. pV measurements were conducted at ASHRAE test conditions (Condensing temperature:54.4 °C, Evaporating temperature: -23.3 °C) to examine the pressure characteristics of the investigated compressor. The results of pV measurements of the reciprocating compressor are shown in Figure 1.

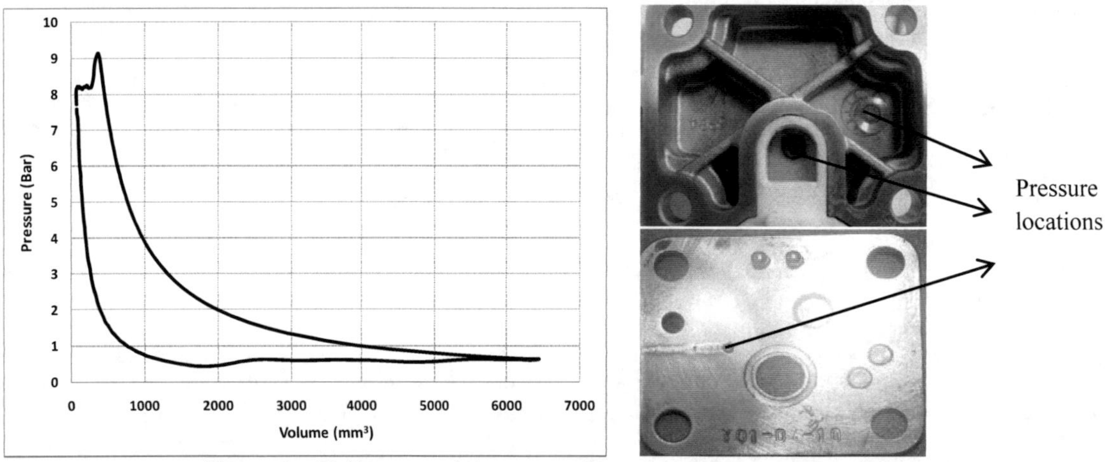

Figure 1: pV-diagram and temperature/pressure measurement locations

Furthermore detailed temperature measurements were also done at various points inside the compressor. Temperature measurement points were given in Table 1 55 thermocouples were mounted on the compressor components walls and flow lines. Besides, 40 thermocouples were used to get single phase induction motor temperatures when calorimeter measurements were performing. Baseline temperatures were collected to compare simulation results. In order to get motor efficiency and working characteristics motor-test system was used. Motor-test system provides torque, speed, power, current, voltage datas for different working conditions.

Table 1. Temperature measurement points

No	Measurement Point	Number	No	Measurement Point	Number
1	Lubrication Oil	4	10	Cylinder Head	2
2	Suction Muffler Backside	4	11	Discharge Muffler	2
3	Suction Muffler Front side	4	12	Discharge Tube Inlet	1
4	Inner Gas	6	13	Discharge Tube Outlet	1
5	Cylinder Surface	4	14	Shell	10
6	Body	8	15	Compressor Outlet.	2
7	Discharge Plenum	2	16	Suction Muffler Inlet	1
8	Compressor Inlet	2	17	Suction Muffler Outlet	1
9	Suction Plenum	1	18	Electrical Motor *	40
		Total:95			

No	Measurement Point	Number	No	Measurement Point	Number
1	Stator Lamination-Left	9	9	Upper Windings- Left	1
2	Stator Lamination-Right	9	10	Upper Windings -Right	1
3	Stator Lamination-Back	7	11	Upper Windings -Back	1
4	Stator Lamination-Front	7	12	Upper Windings -Front	1
5	Lower Windings- Left	1			
6	Lower Windings -Right	1			
7	Lower Windings -Back	1			
8	Lower Windings -Front	1			
* Electrical Motor		Total:40			

3. NUMERICAL ANALYSIS

Analytical models were studied with commercial engineering software. At first, numerical model of the induction motor was used to compare torque results. Numerical model is in good agreement with experimental results. In order to obtain thermal model for induction motor numerical model was imported to commercial engineering software which is dedicated to the electromagnetic performance of motors and the optimization of their thermal calculations. The model is given in Figure 3. The geometric values were edited and external environment of motor was modeled for the conditions like the inside of hermetic compressor housing.
The gas around the motor was modeled as R600a (isobutane) refrigerant and speed was edited at running speed range that the load of induction motor of hermetic compressor which run for ASHRAE condition. The housing was not used and the temperature of the gas was edited as experimental value.

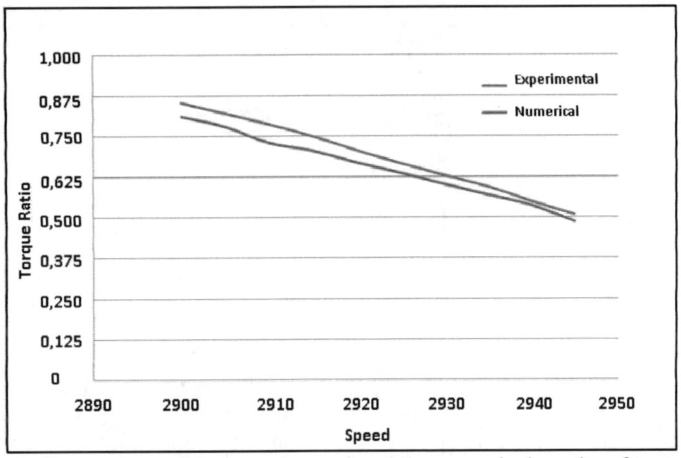

Figure 2: Comparison of the experimental and the numerical results of torque ratio

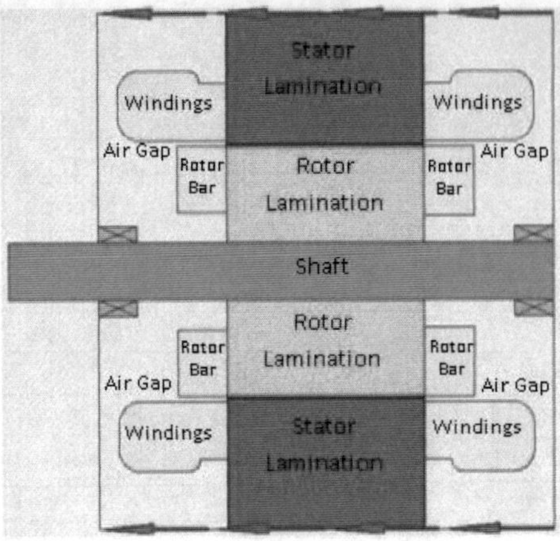

Figure 3 : Simulation model

Schematic thermal network is given in Figure 4. Symbols for Figure 4 explained in Table 2. It shows thermal resistances, interfaces, losses; conduction, convection, radiation heat transfer mechanisms.

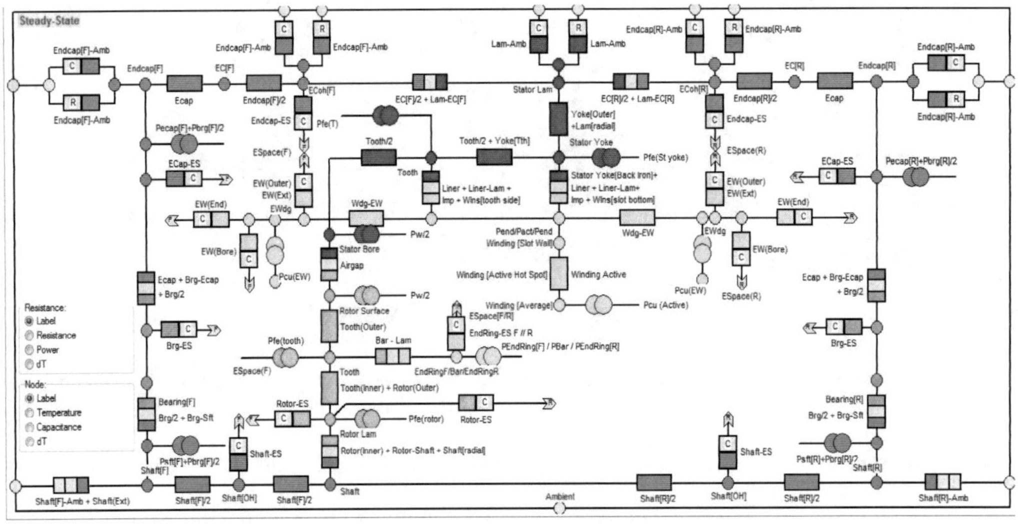

Figure 4: Thermal Network model

Table 2. Symbols for thermal network model

Symbol	Information
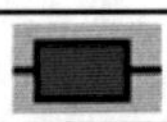	Termal Resistant
	Interface Resistant
	Convection
	Radiation
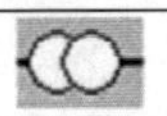	Loss
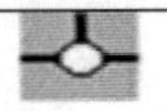	Connection Point

After modeling induction motor for baseline design, the motor cooling methods were used on this model. Spiral water jacket, axial water jacket, spray cooling models were simulated for steady state conditions and according to the parameters temperature changes were investigated.

In addition to analytical models for induction motor, other hermetic compressor components were also studied. Compressor divided into 12 lumped element and energy balance equations were applied to all control volumes. Figure 5 shows these lumped elements: Suction muffler 1st volume, suction muffler channel, suction muffler 2nd volume, suction plenum, cylinder, discharge plenum, discharge pipe, housing, oil, internal environment, body and motor. All suction line and discharge line pressures were taken 0,624 bar and 7,61 bar respectively. Complex geometries were simplified and correlations from the heat transfer network of the compressor (Ozdemir, 2007) were applied. Figure 6 shows solution method for analytical model of lumped parameter method. Simulation starts with geometrical, electrical parameters, operating conditions and initial temperatures. After that step, procedure continuous with calculating fluid properties (dynamic viscosity μ, kinematic viscosity ν, density ρ, specific heat cp, enthalpy h, kinetic energy of turbulance k) for every loop. Energy balances are applied for many times until internal environment temperature converge for all elements. Chart shows suction muffler inlet (T_{in}), 1st volume (T_{sm1}), channel (T_{ch}) and 2nd volume (T_{sm2}); suction plenum (T_{sp}), discharge plenum (T_{dp}), discharge muffler (T_{dm}), discharge pipe (T_{dpip}) compressor outlet (T_{out}) temperatures respectively.

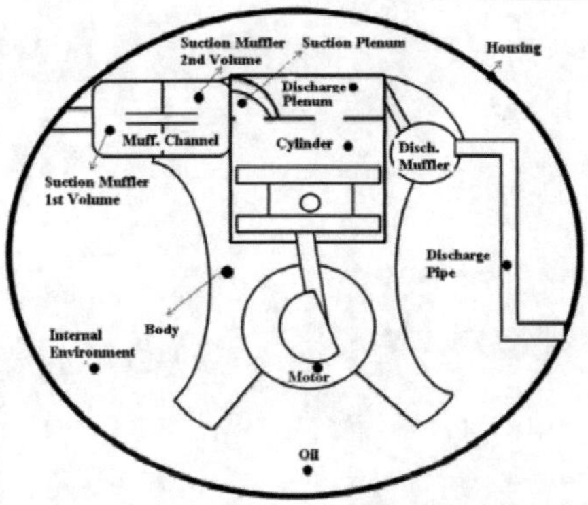

Figure 5: Compressor lumped elements

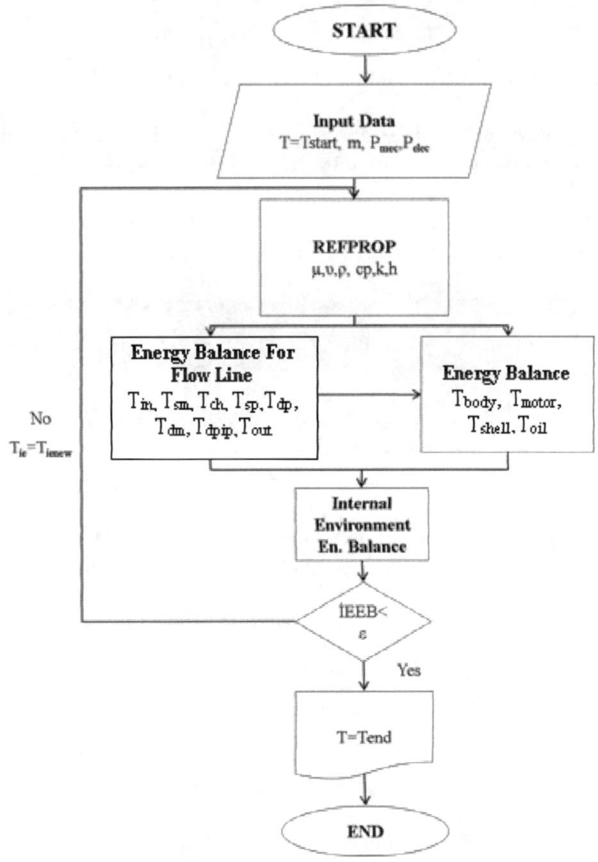

Figure 6: Model algorithm

4. RESULTS AND DISCUSSIONS

The results of experimental studies show that motor losses have significant importance for hermetic compressor performance. Results of compressor performance, indicator diagram, motor tests and the detailed temperature measurements of compressor components proved that the changes in temperature impacts overall performance and losses. Detailed temperature measurements of induction motor are given in Figure 7. It can be easily examined that temperature distribution is nearly homogenous. Average temperature of stator lamination 74,6 °C and there is not remarkable temperature difference between all sides. Average end winding temperature is 74,4 °C. It can be observed from front and back views, due to oil in crankcase, bottom end windings are cooler than upper windings.

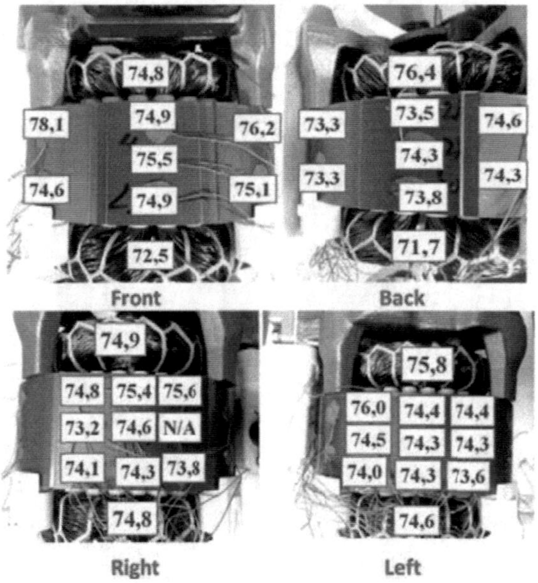

Figure 7: Detailed temperature measurements of induction motor (°C)

Simulation results for baseline model were given in Figure 8. The results are in a good agreement with experimental measurements. There is approximately 3K temperature difference possibly due to unidentified dynamic effects of oil flow over the electrical motor in the model. Simulation results give opportunity to see temperatures of motionless parts of induction motor. Furthermore moving component's temperature distribution which is impossible to measure by thermocouples can be easily found.

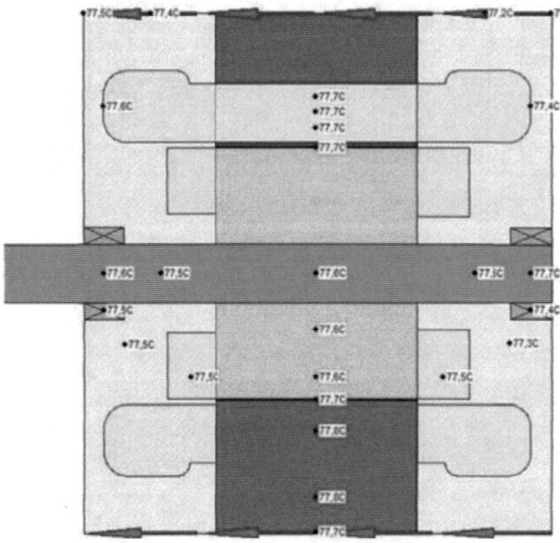

Figure 8: Temperature distribution for simulation model

In simulations temperature measurement points on the electrical motor is divided to different sections. These sections are summarized as front electrical wiring (ew-fr), average electrical wiring (w), stator lamination (slam), stator surface (ssur), rotor surface (rsur),rotor teeth (rteeth), rotor lamination (rlam), rotor bars (rbar), Shaft center (shft), back electrical wiring (ew-bck).

Alternative cooling methods for motor investigated in this study. Spray cooling model simulation results are given in Figure 9. In this cooling method end windings are cooled by oil passing from down to up of crankshaft channel. Inlet temperature was edited 65 °C constant, temperatures all over the motor laminations and windings decreased approximately 6-7 °K.

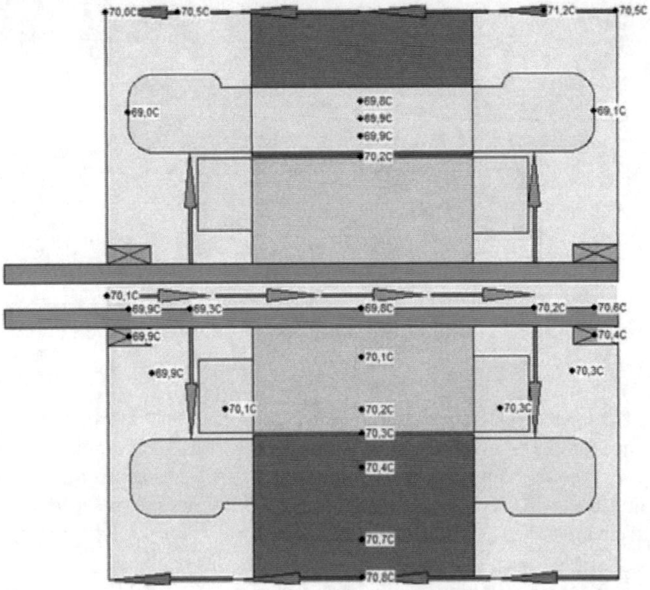

Figure 9: Temperature distribution for simulation model

Spiral water jacked model simulation results are given in Figure 10. In this model, there is a helical channel around stator lamination packet. The fluid inside the channel was modeled as water. Furthermore fluid properties can be set at fixed values and new fluids can be created. For the results which are given below, first, inlet temperature of fluid and channel width were set 32,2°C respectively. For these conditions temperatures were predicted for different flow rates. For constant channel width, simulation results show that, for 1.10^{-7} m^3 flow rate temperature distribution is homogenous. Flow rate from 1.10^{-7} m^3/s to 8.10^{-7} m^3/s decreases the temperatures respectively and homogenous temperature characteristic of induction motor become different.

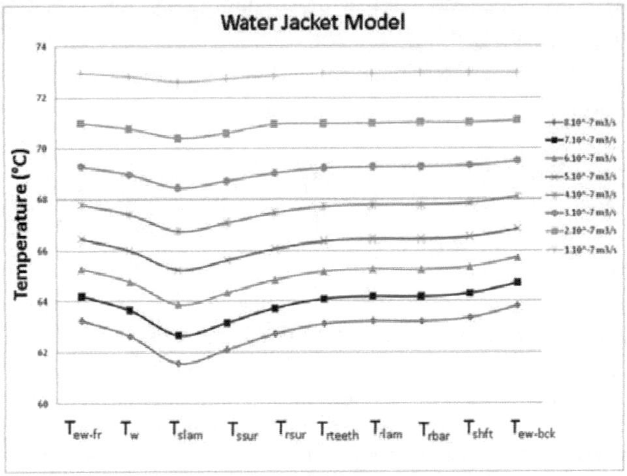

Figure 10: Spiral water jacked model simulation results

In Figure 11 axial water jacket model simulation results are given. For this model, fluid follows axial channels to cool motor lamination. Flow rate is critical parameter. Results show that, motor temperatures decrease for high flow rates.

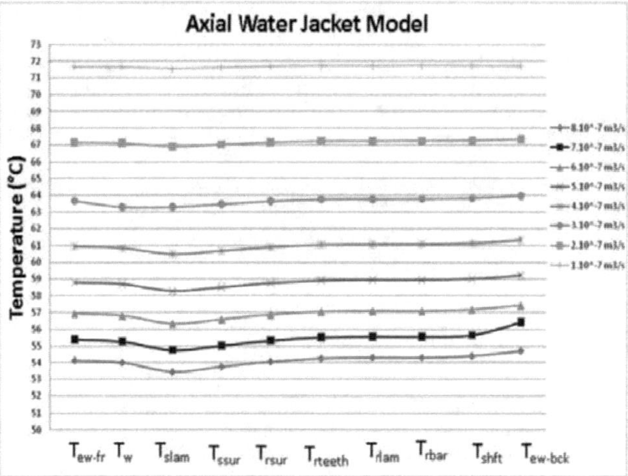

Figure 11: Axial water jacket model simulation results

In addition to the electrical motor's detailed thermal analysis, heat transfer processes between the components of a hermetic reciprocating compressor were also studied. The results of lumped parameter method analytical model for compressor flow line and components are given in Figure 12. According to the temperature results, it can be seen that, flow line and wall temperatures are in a good agreement with experimental measurements. Chart shows suction muffler inlet (T_{in}), 1st volume (T_{sm1}), channel (T_{ch}) and 2nd volume (T_{sm2}); suction plenum (T_{sp}), discharge plenum (T_{dp}), discharge muffler (T_{dp}), discharge pipe (T_{dp}) compressor outlet (T_{out}) temperatures respectively. The other temperatures are summarized as oil (T_{oil}), internal environment (T_{ie}), compressor body (T_{bdy}) and compressor shell (T_h).

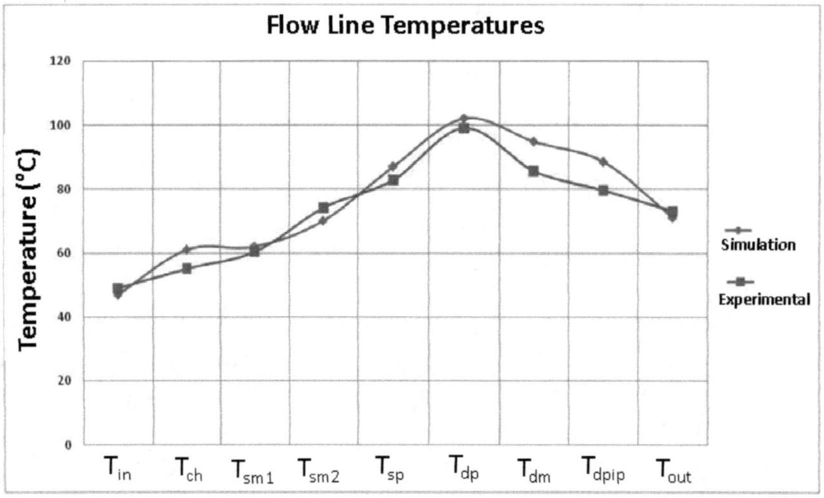

Figure 12: Lumped parameter method analytical model simulation results

General temperatures for different average electrical motor temperatures are given in Figure13 & Figure14. Simulation results show that motor temperatures affect all components directly and they are all dependent motor losses. Decreasing the average temperature of induction motor affects component temperatures. It decreases general temperatures all over the compressor.

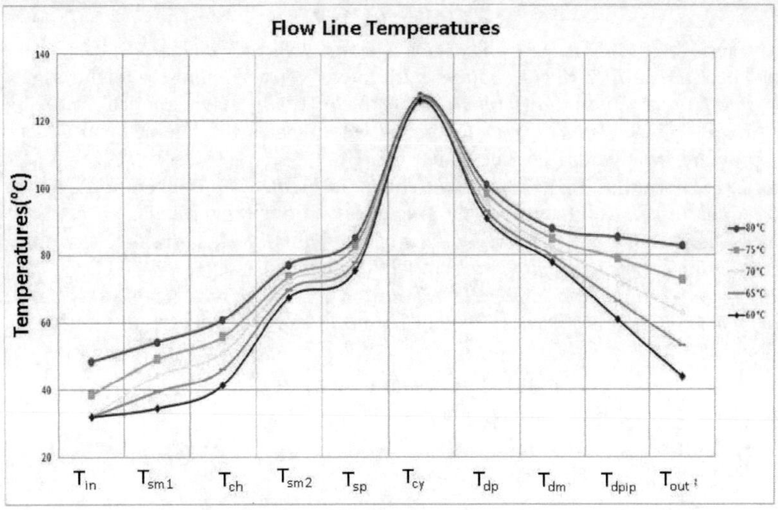

Figure 13: Predicted temperatures for different motor temperature (1)

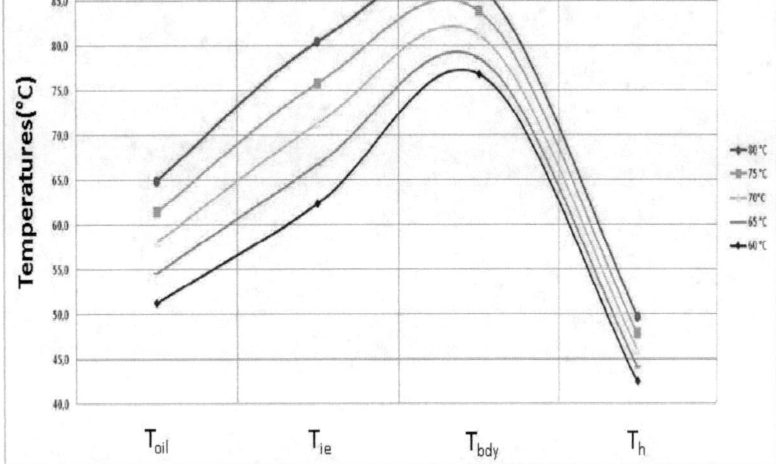

Figure 14: Predicted temperatures for different motor temperature (2)

5. CONCLUSIONS

In this paper, heat transfer between the components and the electrical motor of the hermetic reciprocating compressor were investigated theoretically, numerically and experimentally. The compressor performance, indicator diagram, motor tests and the detailed temperature measurements of electrical motor and compressor components were given. Analytical model simulations of induction motor and compressor components were performed by using commercial software.

- The results of measurements and simulations show that, motor temperatures are nearly homogenous. due to oil in crankcase bottom end windings are cooler than upper windings.
- According to cooling methods motor temperatures can be reduced for different rates. In water jacket applications when flow rate increases, homogenous temperature characteristic of the motor changes.
- According to lumped parameter model that was applied compressor components, motor temperatures directly affect all over the compressor.

The results of the theoretical, analytical and experimental studies are used for investigating the network including conduction, convection and radiation forms of heat transfer inside the compressor. Generated heat transfer network helps to characterize the thermal functions of the main components which leads to the new and better compressor designs.

REFERENCES

Dutra, T. ve Deschamps, C. J. (2014). *Development of a Lumped Paramether Model for Hermetic Reciprocating Compressor with Thermal-Electrical Coupling.* 22th International Compressor Engineering Conference.

Ooi, K. T. (2003). *Heat Transfer Study of a Hermetic Refrigeration Compressor.* Applied Thermal Engineering.

Fagotti, F. (1994). *Heat Transfer Modelling in Reciprocating Compressor.* International Compressor Technology Engineering Conference at Purdue, Purdue University, USA.

Haas, D., Deschamps, C., Diniz, M.(2013) *A Thermal Network to Predict The Temperature Distribution in Household Refrigeration Compressors.* 8th International Conference on Compressors and Coolents, Smolenice-Slovakia.

Öner Y., Gökdemir F., Çetin E. (2009). *Thermal Analysis of The Three-Phase Induction Motor and Calculation of Its Power Loss by Using Lumped-Circuit Model.* IAT'09, Karabük-Turkey.

Özdemir, A.R. (2007). *An Investigation of the Heat Transfer Characteristics Among the Hermetic Reciprocating Compressor Components*, M.Sc . Thesis, Istanbul Technical University. Istanbul. Turkey

ACKNOWLEDGEMENT

The authors would like to express their sincere appreciation to Arcelik A.Ş.

Performance Analysis of a Reciprocating Compressor under Typical Transients of Refrigeration Systems

Marco C. DINIZ[1], Cesar J. DESCHAMPS[1]*

[1]POLO Research Laboratories for Emerging Technologies in Cooling and Thermophysics
Federal University of Santa Catarina
Florianopolis, SC, Brazil
deschamps@polo.ufsc.br

* Corresponding Author

ABSTRACT

The performance of reciprocating compressors is usually evaluated under steady-state operating conditions defined in standards. However, the compressor is submitted to quite different conditions in actual refrigeration systems, such as transients associated with the on/off cycling operation. This paper presents a simulation model developed to evaluate the performance of reciprocating compressors during typical transients of refrigeration systems. The model is formed by the coupling of two sub models, one for the compression chamber and the other to calculate the temperatures and mass flow rates in components other than the compression chamber. As the model only simulates the compressor, experimental data was required to provide input from the remaining components of the system. The model was validated by comparing predictions and experimental data of discharged mass flow rate and power consumption of a compressor operating under two transient system conditions. Then, the simulation model was employed to predict the variation of the compressor efficiencies and temperatures during the typical transients analyzed.

1. INTRODUCTION

About 1.4 billion household refrigerators and freezers are in use worldwide (Barther and Götz, 2012) and virtually all of these refrigerators operate based on the vapor compression refrigeration cycle. The performance of these appliances depends on the efficiency of each of its components, but also on the manner they interact with each other.

Despite the advent of variable speed compressors, the vast majority of the compressors employed on refrigerators are still single speed. In this case, the refrigerator operation is characterized by alternate periods at which the compressor is either operating (on) or idle (off). The analysis of the transient operating conditions of refrigerators has been the subject of several researches during the last 30 years. However, most of such studies have focused their attention either on the heat exchangers or on the throttling device.

Regarding the compressor operation during the refrigerator transients, two main aspects must the addressed. First, and most important, is the variation of the pressure ratio to which the compressor is submitted. Unlike the conditions during the refrigerator pulldown, when the compressor is started in a typical cycling condition of the refrigerator, the time to reach a stabilized pressure ratio is usually short. However, this may differ from refrigerator to refrigerator due to aspects such as the volume of the refrigerated compartment, volume of the heat exchangers, compressor displacement volume, throttling device and refrigerant charge. The second aspect is related to the transient behavior of the compressor temperatures during the on/off cycling conditions, which may vary even when the pressure ratio is already stabilized.

Few studies have been developed regarding the compressor performance under transient refrigerator conditions. Some of these studies focused on the crankshaft mechanism and/or bearing system during the first seconds after startups and shutdowns (Dufour *et al.*, 1995; Link and Deschamps, 2011). In other studies, the objective was to predict mass flow rate and power consumption (Porkhial *et al.*, 2002; Hermes and Melo, 2008; Ndiaye and Bernier, 2010; Negrao *et al.*, 2011) or heat transfer phenomena (Lohn *et al.*, 2015). None of such works has focused on understanding the phenomena that influence the efficiency of the compressor during typical transients of refrigeration systems.

This paper presents a simulation model developed to characterize the performance of reciprocating compressors under transient system conditions. It consists of two sub models: one to predict the compression chamber and the other to calculate the variation of the temperatures and mass flow rates in components other than the compression chamber. Some input data required for the model were obtained from measurements, which were also used to validate the model predictions.

2. EXPERIMENTAL PROCEDURE

The model developed in this work only simulates the compressor. Therefore, the input parameters that would be calculated by models for the other components of the system had to be experimentally obtained. The required parameters are the evaporating and condensing pressures and the gas temperature at the compressor inlet. Additionally, several parameters were also measured in the experiments to validate model predictions, including the mass flow rate, indicated power, electrical power consumption and compressor temperatures.

The tested system was a one door typical frost-free 300-L refrigerator, in which the compressor is on-off controlled by the freezer temperature, while a mechanical damper is responsible to establish the temperature in the fresh food compartment. The original compressor employed in the refrigerator was substituted by an instrumented compressor with 30% less volumetric displacement than the original and the refrigerant charge (R600a) was set to 27g. The tests were performed with the refrigerator in a controlled environment with temperature of 25°C and 50% of relative humidity. The refrigerator temperature setting was set to its medium value, at which the freezer temperature reached -16°C by the time the compressor was switched off.

Two different "on" periods were analyzed in this study. The first was typical of the periodic cyclic operation of the refrigerator. In this case, the equalizing pressure was around 0.8 bar and the "on" cycle had a duration of around 15 minutes. The second period, usually termed recovery on period, takes place immediately after refrigerator defrost. In this case, the equalizing pressure before the compressor is switched on was around 2.6 bar and the "on" period presented a duration of approximately 60 minutes. The transient pressures and suction line temperature that were used as input for the simulation model will be presented and discussed further.

3. MATHEMATICAL MODEL

The model consists of two sub models: one to simulate the compression chamber and the other to calculate the variation of compressor temperatures and mass flow rates. The time scale of the compression cycle (milliseconds) is considerably smaller than that of the compressor thermal transients (minutes), allowing the compression chamber to be simulated using a quasi-steady state approach. The coupled solution procedure will be detailed in section 3.3.

3.1 Compression chamber

The model for the compression chamber follows a transient lumped formulation based on four groups of equations (Todescat *et al.*, 1992). The first group allows the calculation of the instantaneous volume occupied by the gas inside the compression chamber. The second group is the conservation equations applied to the compression chamber. The mass conservation equation is given by:

$$\frac{dm_g}{dt} = \dot{m}_{sv} - \dot{m}_{sv,b} - \dot{m}_{dv} + \dot{m}_{dv,b} - \dot{m}_{pc} \tag{1}$$

In Equation (1), $\dot{m}_{sv}$ and $\dot{m}_{dv}$ are the instantaneous mass flow rates through the suction and discharge valves, while $\dot{m}_{pc}$ represents the instantaneous leakage through the piston cylinder gap and the subscript b indicates backflow in both valves. With the instantaneous mass and volume of refrigerant inside the compression chamber it is possible to calculate its density.

The instantaneous temperature of the refrigerant inside the compression chamber is obtained by applying the energy conservation equation, given by:

$$\frac{dT_g}{dt} = A - BT_g \tag{2}$$

where

$$A = \frac{1}{m_g c_{v}}\left[H_g A_w T_w - h_g \frac{dm_g}{dt} + (\dot{m}_{sv} h_{sc}) - (\dot{m}_{sv,b} h_g) - (\dot{m}_{dv} h_g) + (\dot{m}_{dv,b} h_{dc}) - (\dot{m}_{pc} h_g)\right] \tag{3}$$

$$B = \frac{1}{m_g c_{v}}\left[H_g A_w + \left.\frac{\partial p_g}{\partial T_g}\right|_{v} \frac{d\forall_{cc}}{dt} - \left.\frac{\partial p_g}{\partial T_g}\right|_{v} v_g \frac{dm_g}{dt}\right] \tag{4}$$

In the above equations H_g is a convective heat transfer coefficient calculated using the correlation proposed by Annand (1963). The method to estimate the mass flow rates in the above equations will be explained shortly. The temperatures of the suction chamber (T_{sc}), discharge chamber (T_{dc}) and cylinder wall (T_w) are calculated by the model that will be described in section 3.2. With the temperature calculated using Equation (2) and the density of the refrigerant it is possible to estimate the pressure inside the compression chamber by means of a state equation.

The third group of equations is used to estimate the dynamics of the suction and discharge valves. A single degree-of-freedom mass-spring model is employed, with effective force areas being used to characterize the force acting on the valve. Finally, the fourth group of equations predicts the mass flow rates that cross the control volume surface. Mass flow rate through the valves is estimated using the concept of effective flow area to correct the theoretical model for the isentropic flow through a convergent nozzle. Leakage through the piston cylinder gap is estimated by considering a fully developed laminar Couette-Poiseuille flow (Ferreira and Lilie, 1984).

The parameters associated with the compression chamber that are used as input in the other sub model are calculated as shown in Table 1. The mechanical losses $\dot{Q}_b$ and the electrical efficiency η_{ele} were specified based on experimental data previously obtained for the compressor under analysis.

Table 1 Output from the compression chamber model.

Parameter	Calculation
$\dot{W}_{ind}$	$f\oint p_g d\forall_{cc}$
$\dot{m}_s$, $\dot{m}_d$, $\dot{m}_l$, $\dot{m}_{s,b}$ and $\dot{m}_{d,b}$	$f\oint \dot{m}_i dt$, where $i = sv$, $i = dv$, $i = pc$, $i = sv,b$ or $i = dv,b$
T_d	$\frac{1}{\dot{m}_d}\oint \dot{m}_{dv} T_g dt$
T_l	$\frac{1}{\dot{m}_l}\oint \dot{m}_{pc} T_g dt$
$T_{s,b}$	$\frac{1}{\dot{m}_{s,b}}\oint \dot{m}_{sv,b} T_g dt$
$\dot{Q}_{mot}$	$(\dot{W}_{ind} + \dot{Q}_b)/\eta_{ele} - (\dot{W}_{ind} + \dot{Q}_b)$

3.2 Temperatures and mass flow rates through control volumes

To calculate the temperature and mass flow rates inside components other than the compression chamber, transient energy and mass balances were applied to the control volumes presented in Figure 1. The energy balances applied to each control volume are presented in Table 2 in the Appendix. The mass balances can also be deduced through the equations presented in Table 2.

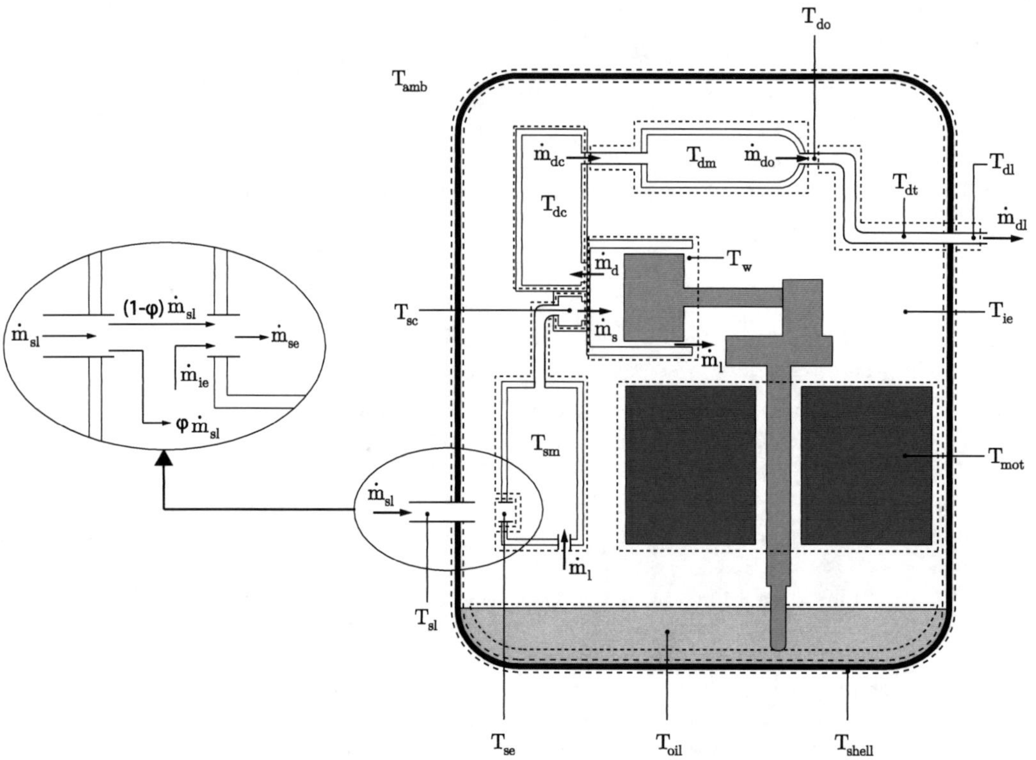

Figure 1: Control volumes for the thermal analysis.

In the compressor employed in this work, part of the gas in the suction line flows directly to the suction muffler, while the other share fills the internal environment (Figure 1). In the model, this flow bifurcation is represented by a factor φ. The φ factor and the heat conductances between the components were obtained by adjusting the model to a measured steady state temperature distribution in a calorimeter, a procedure that has been previously adopted by several other authors, such as Todescat *et al.* (1992). Naturally, the transient terms in the system of equations were not taken into account in this adjustment procedure. The condition chosen for the adjustment consisted on an evaporating temperature $T_e = -28\,°C$ and a condensing temperature $T_c = 40\,°C$ with ambient temperature $T_{amb} =$ 25 °C. The adjusted heat conductances and the φ factor were maintained constant throughout the simulation of the compressor under transient conditions.

The nonlinear system of equations represented in Table 2 is solved by means of a Newton-Raphson method. Of especial interest are the suction chamber, cylinder and discharge chamber temperatures, which are used as input data in the compression chamber model.

3.3 Solution procedure

The solution procedure is based on exchange of information between the two previously discussed models. Figure 2 presents the flowchart that characterizes the coupled solution for the overall model. The instantaneous operating conditions are given by the system condensing and evaporating pressures and compressor inlet temperature, obtained from the experimental tests of the refrigerator.

Since the compression chamber and the thermal model require input data from each other, inside each time step an iterative procedure is necessary to verify convergence, which is considered to be reached when the suction chamber temperature and the inlet mass flow rate present a variation between successive iterations of less than 0.1%.

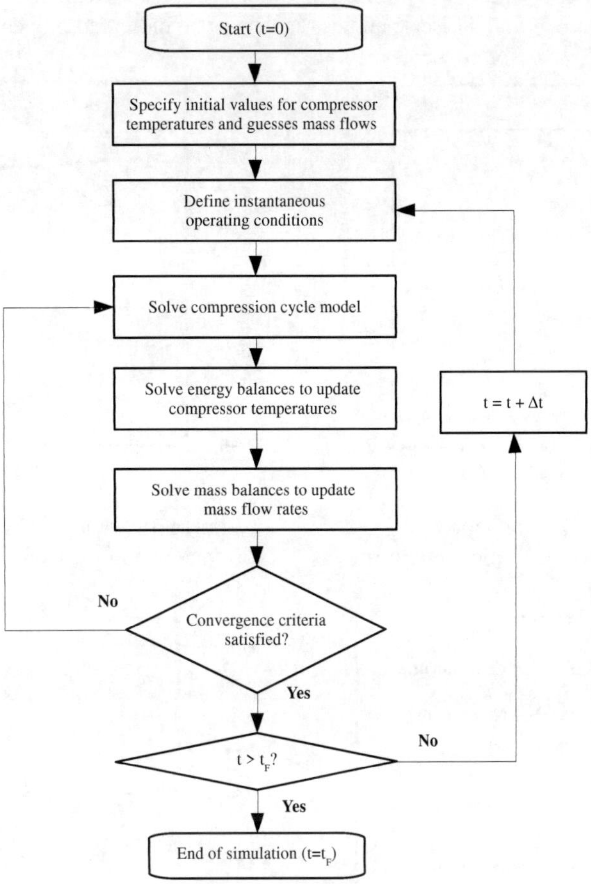

Figure 2: Flowchart of the solution procedure.

4. RESULTS

Figures 3 and 4 present the measured suction and discharge pressures during the two "on" periods considered in this study. The data is normalized by the total time of each period (15 min. and 60 min.). It can be observed that for the typical cycling condition the suction pressure rapidly approaches its stabilized value. On the other hand, the transient associated with the defrost recovery is quite different, and resembles more a refrigerator pulldown.

The transient evaporating and condensing pressures shown in Figures 3 and 4 and the compressor inlet temperature were used as input data for the compressor model. Figures 5 and 6 show the comparison between experimental and numerical results of mass flow rate in the two transients (cycling and defrost recovery), with good agreement being observed for both cases. The discrepancy observed during the first 20 to 30 seconds, only visible in Figure 5, cannot be fully explained at this stage of our study but is more likely to be a result of measurement errors due to the presence of some liquid droplets inside the measurement section of the flow meter. Considering the cycling condition, the results show that the variation of mass flow rate is not significant after a normalized time around 0.2. On the other hand, this variation of mass flow rate is slightly more intense for the defrost recovery condition due to the variation of the suction pressure, as shown in Figure 4.

Figures 7 and 8 present the evolution of electrical power consumption and indicated power during the typical cycling and defrost recovery conditions. As observed for the mass flow rate, the trends of variation are different, with the defrost recovery showing a considerable variation during the whole period. Some discrepancies between predictions and measurements can be observed for the indicated power, which can partially be attributed to absence of models to predict the pressure pulsations in the suction and discharge systems, as well as to errors in the leakage and heat transfer models. Furthermore, the more significant discrepancies observed for the electrical power consumption can be attributed to lack of appropriate models for the mechanical losses and electrical motor efficiency, but also to the aforementioned discrepancies regarding the indicated power.

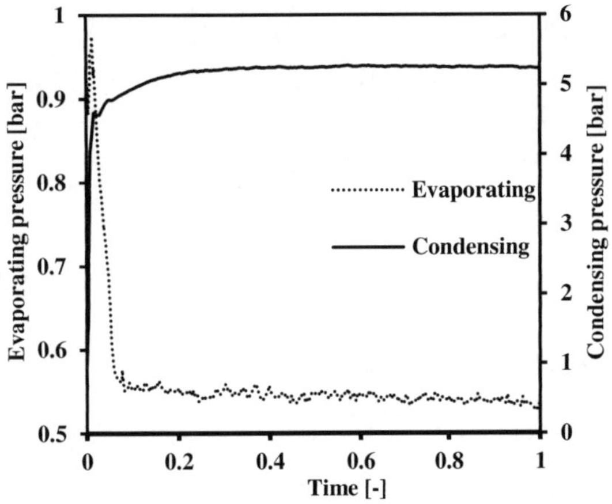

Figure 3: Transient variation of evaporating and condensing pressures for the typical cycling

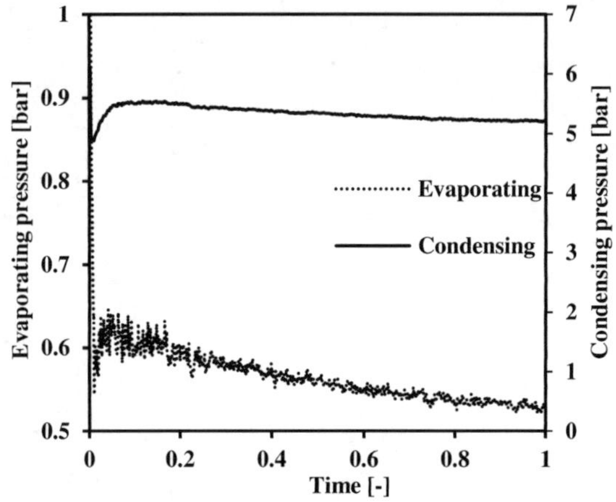

Figure 4: Transient variation of evaporating and condensing pressures for the defrost recovery

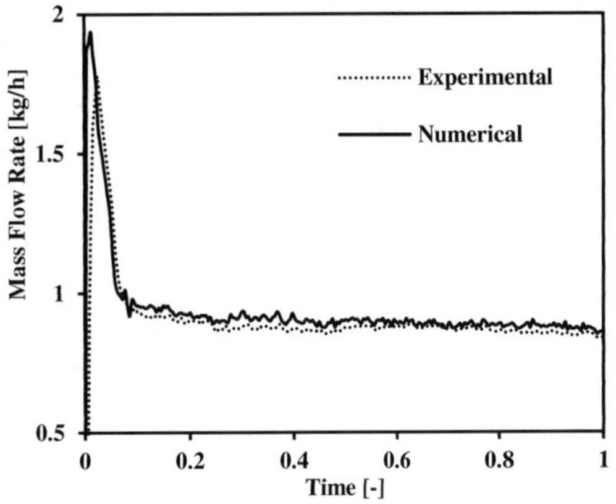

Figure 5: Numerical and experimental results of mass flow rate for the typical cycling

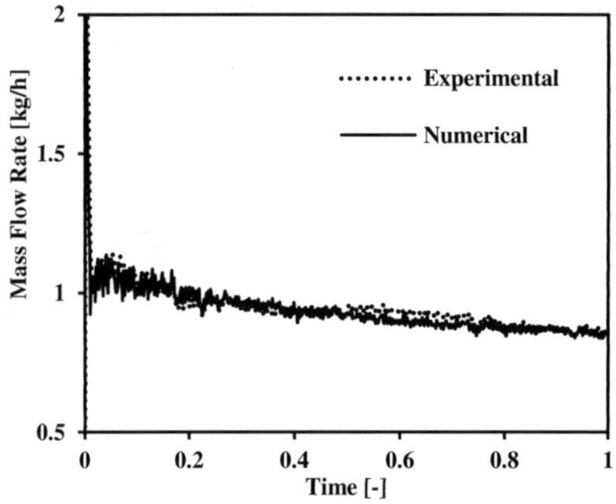

Figure 6: Numerical and experimental results of mass flow rate for the defrost recovery

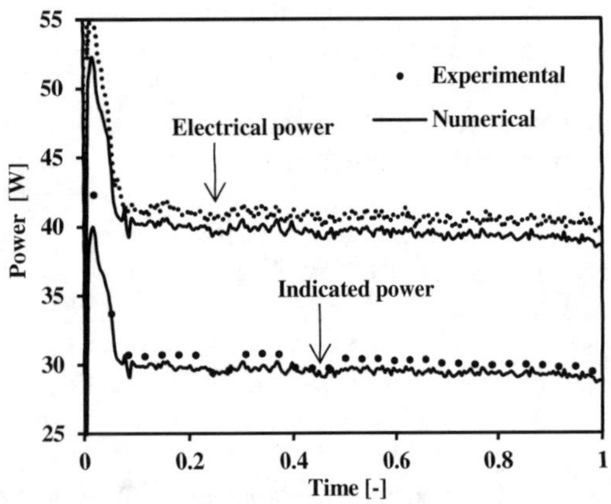

Figure 7: Numerical and experimental results of electrical and indicated power for the typical cycling

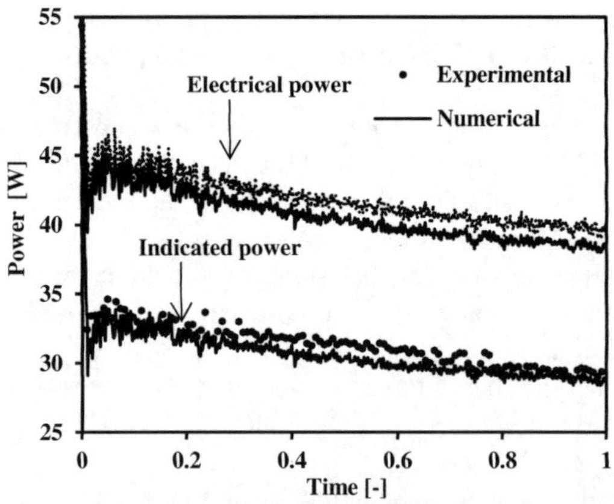

Figure 8: Numerical and experimental results of electrical and indicated power for the defrost recovery

Figure 9 and 10 show the transient variation of the temperatures in suction chamber and at the cylinder wall. Despite the deviations, the trends are well captured by the simulation model, which predicts the temperatures in both conditions with an average error of around 2ºC. In part, the deviations can be attributed to the fact that the heat conductances and the φ factor were maintained constant throughout the transient simulation. These coefficients are expected to vary, especially during the first minutes of compressor operation, due to the considerable variation in the mass flow rate. Therefore, improvements in this methodology will be incorporated in future work.

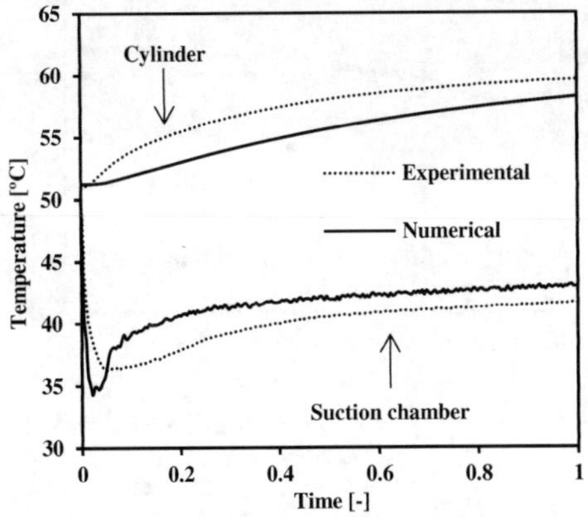

Figure 9: Numerical and experimental results of temperatures in the suction chamber and at the cylinder wall for the typical cycling

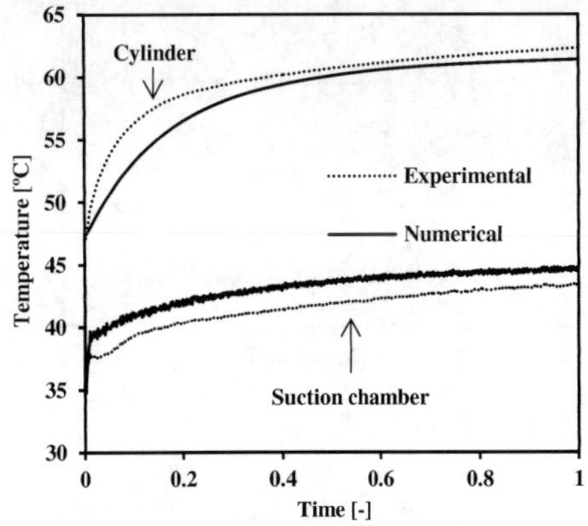

Figure 10: Numerical and experimental results of temperatures in the suction chamber and at the cylinder wall for the defrost recovery

Figures 11 and 12 presents the pressure ratio to which the compressor is submitted and the numerical results of thermodynamic, volumetric and global the efficiencies for the two conditions analyzed. For the cycling condition (Figure 11), the thermodynamic and global efficiencies approach a minimum value in the beginning of the period due to aspects such as suction superheating and losses in the suction and discharge processes. A local minima for these efficiencies is also observed in the defrost recovery moments after the compressor is turned on. However, the global minima for this condition are observed immediately before compressor shut down, mainly as a result of the increase in compressor temperatures. On the other hand, for both conditions, the volumetric efficiency presents a monotonic behavior that is mainly governed by the pressure ratio.

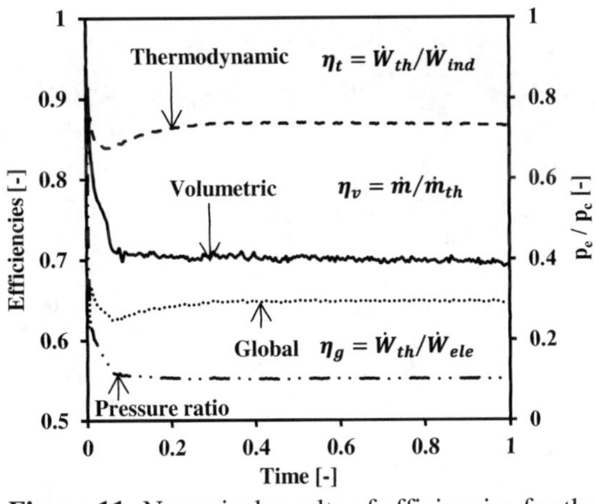

Figure 11: Numerical results of efficiencies for the typical cycling

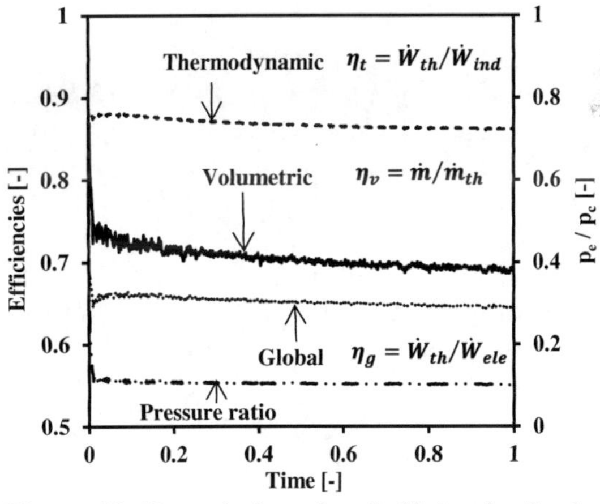

Figure 12: Numerical results of efficiencies for the defrost recovery

Figures 13 and 14 present the effect of suction superheating (ΔT_{sup}) on the thermodynamic power (ϕ) and mass flow rates (ψ) during both transients. The peak of suction superheating reached soon after the compressor is switched on is more intense for the cycling condition. This increases the power necessary the compress the gas by almost 10%, which explains in part the minimum value for the thermodynamic efficiency (Figure 11). The mass flow rate is also reduced from its theoretical value, but this impact is smaller than that due to the pressure ratio. The results for defrost recovery condition show that the negative effect of the suction superheating increases with time.

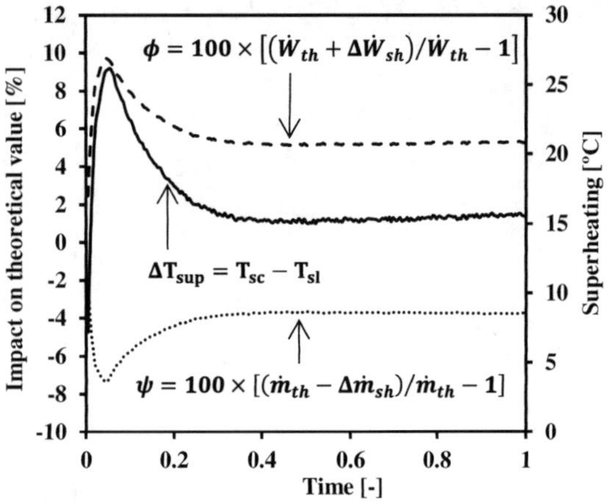

Figure 13: Impact of suction superheating during the cycling on

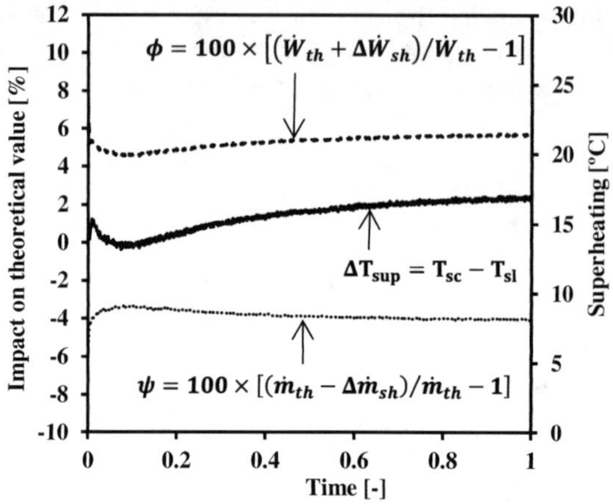

Figure 14: Impact of suction superheating during the recovery on

6. CONCLUSIONS

This paper presented a simulation model to calculate the performance of reciprocating compressors under two transient conditions found in refrigerators, represented by standard cycling and defrost recovery conditions. The model consists of two sub models, one for the compression chamber and the other to predict the compressor temperature distribution and mass flow rates through components other than the compression chamber. A compressor was tested in a refrigerator and several parameters were measured (mass flow rate, indicated power, electrical energy consumption, temperatures). Some of these measurements were used as input data and others to validate the simulation model. Overall, the model was shown to be capable of predicting the transient performance of a compressor in a refrigerator. The results showed that the volumetric efficiency, which is closely related to the pressure ratio, presents a monotonic variation during both transient transients. The thermodynamic and global efficiencies presented a minimum value in the typical cycling condition immediately after the compressor start up. This minimum value was shown to be strongly related to the suction gas superheating. For the recovery on the minimum value of thermodynamic and global efficiencies were found to occur immediately before compressor shutdown.

NOMENCLATURE

General symbols

A	Area	[m^2]	$\dot{Q}_{mot}$	Motor losses	[W]
f	Compressor speed	[Hz]	T	Temperature	[ºC]
H	Heat transfer coefficient	[W/m^{2}K]	T_d	Average discharge temperature	[ºC]
h	Enthalpy	[J/kgK]	UA	Heat conductance	[W/K]
$\dot{m}$	Mass	[kg]	$\forall$	Volume	[m^3]
$\dot{m}$	Mass flow rate	[kg/s]	$\dot{W}_{ind}$	Indicated power	[W]
$\dot{m}_d$	Mass flow rate at discharge orifice	[kg/s]	η_g	Global efficiency	[-]
$\dot{m}_s$	Mass flow rate at suction orifice	[kg/s]	η_t	Thermodynamic efficiency	[-]
p	Pressure	[Pa]	η_v	Volumetric efficiency	[-]
$\dot{Q}$	Heat transfer rate	[W]	φ	Semi-direct suction factor	[-]
$\dot{Q}_b$	Mechanical losses	[W]			

Subscripts

amb	External environment	l	Leakage
b	Related to backflow	mot	Electrical motor
c	Condensing	oil	Lubricant oil

cc	Compression chamber	pc	Piston cylinder gap
d	Averaged value at discharge orifice	s	Averaged value at suction orifice
dc	Discharge chamber	sc	Suction chamber
dl	Discharge line	se	Suction muffler inlet
dm	Discharge muffler	$shell$	Compressor shell
do	Discharge muffler outlet	sl	Suction line
dt	Discharge tube	sm	Suction muffler
dv	Discharge valve	sv	Suction valve
e	Evaporating	th	Theoretical value
ie	Compressor internal environment	w	Compression chamber wall
g	Gas inside compression chamber		

REFERENCES

Annand, W. J. D. (1963). Heat transfer in the cylinders of reciprocating internal combustion engines. *Proc. I. Mech. Eng.* ***117*** *973-96.*

Barther, C. and Götz, T. (2012). *The overall worldwide saving potential from domestic refrigerators and freezers.* Wuppertal, Germany. Wuppertal Institute for Climate, Environment and Energy.

Dufour, R., Hagopian, J. D., & Lalanne, M. (1995). Transient and steady state dynamic behavior of single cylinder compressors: prediction and experiments. *J. Sound Vib.*, 181(1), 23-41.

Hermes, C. J. L. & Melo, C. (2008). A first-principles simulation model for the start-up and cycling transients of household refrigerators. *Int. J. Refrig.*, 31, 1341-1357.

Ferreira, R. T. S. & Lilie, D. E. B. (1984). Evaluation of the leakage through the clearance between piston and cylinder in hermetic compressor. *In: Proc. Int. Compressor Engineering Conference at Purdue University*, (paper 424). West Lafayette, USA.

Lohn, S. K., Diniz, M. C. & Deschamps, C. J. (2015). A thermal model for analysis of hermetic reciprocating compressors under the on-off cycling operating condition. *IOP. Conf. Ser.: Mater. Sci. Eng.*, 90, 012068.

Link, R., & Deschamps, C. J. (2011). Numerical modeling of startup and shutdown transients in reciprocating compressors. *Int. J. Refrig.*, 34, 1398-1414.

Ndiaye, D. & Bernier, M. (2010). Dynamic model of a hermetic reciprocating compressor in on–off cycling operation (Abbreviation: Compressor dynamic model). *Appl. Therm. Eng.*, 30, 792-799.

Negrão, C. O. R., Erthal, R. H., Andrade, D. E. V. & Silva, L. W. (2011). A semi-empirical model for the unsteady-state simulation of reciprocating compressors for household refrigeration applications. *Appl. Therm. Eng.*, 31, 1114-1124.

Phorkial, S., Khastoo & Razavi, M. R. M. (2002). Transient characteristic of reciprocating compressors in household refrigerators. *Appl. Therm. Eng.*, 22, 1391-1402.

Todescat, M. L., Fagotti, F., Prata, A. T. & Ferreira, R. T. S. (1992). Thermal energy analysis in reciprocating hermetic compressors. *In: Proc. Int. Compressor Engineering Conference at Purdue University*, (paper 936). West Lafayette, USA.

ACKNOWLEDGEMENT

The present study was developed as part of a technical-scientific cooperation program between the Federal University of Santa Catarina and EMBRACO. The authors also acknowledge the support of EMBRAPII Unit POLO/UFSC and CAPES (Coordination for the Improvement of High Level Personnel).

APPENDIX

Table 2 Energy balances of thermal hydraulic sub model

CV	$\frac{d}{dt}(mu)$	$\sum mh\|_{in}$	$\sum mh\|_{out}$	$\sum Q$	$\sum W$
1	–	$m_{ie}h_{ie} + (1-\varphi)m_{sl}h_{sl}$	$m_{se}h_{se}$	–	–
2	$m_{sm}\frac{u_{sm}^{t+\Delta t} - u_{sm}^{t}}{\Delta t} + u_{sm}\frac{m_{sm}^{t+\Delta t} - m_{sm}^{t}}{\Delta t}$	$m_{se}h_{se} + m_{s,b}h_{s,b} + m_l h_{ie}$	$m_s h_{sc}$	$-UA_{ie-sm}(T_{ie} - T_{sm})$	–
3	$m_w cp_w \frac{T_w^{t+\Delta t} - T_w^t}{\Delta t}$	$m_s h_{sc} + m_{d,b}h_{dc}$	$m_{s,b}h_{s,b} + m_d h_d + m_l h_l$	$UA_{w-oil}(T_w - T_{oil})$	$W \quad + Q_b$
4	$m_{dc}\frac{u_{dc}^{t+\Delta t} - u_{dc}^{t}}{\Delta t} + u_{dc}\frac{m_{dc}^{t+\Delta t} - m_{dc}^{t}}{\Delta t}$	$m_d \bar{h}_d$	$m_{dc}h_{dc} + m_{d,b}h_{dc}$	$UA_{dc-ie}(T_{dc} - T_{ie})$	–
5	$m_{dm}\frac{u_{dm}^{t+\Delta t} - u_{dm}^{t}}{\Delta t} + u_{dm}\frac{m_{dm}^{t+\Delta t} - m_{dm}^{t}}{\Delta t}$	$m_{dc}h_{dc}$	$m_{dm}h_{do}$	$UA_{dm-ie}(T_{dm} - T_{ie})$	–
6	$m_{dt}\frac{u_{dt}^{t+\Delta t} - u_{dm}^{t}}{\Delta t} + u_{dt}\frac{m_{dt}^{t+\Delta t} - m_{dt}^{t}}{\Delta t}$	$m_{dm}h_{do}$	$m_{dl}h_{dl}$	$UA_{dt-ie}(T_{dt} - T_{ie})$	–
7	$m_{mot}cp_{mot}\frac{T_{mot}^{t+\Delta t} - T_{mot}^{t}}{\Delta t}$	–	–	$UA_{mot-ie}(T_{mot} - T_{ie}) - Q_{mot}$	–
8	$m \quad cp \quad \frac{T_{oil}^{t+\Delta t} - T_{oil}^{t}}{\Delta t}$	–	–	$UA_{oil-sh} \quad (T_{oil} - T_{sh} \quad)$ $- UA_{w-} \quad (T_w - T \quad)$	–
9	$m_{sh} \quad cp_{sh} \quad \frac{T_{sh}^{t+\Delta t} - T_{shell}^{t}}{\Delta t}$	–	–	$UA_{shell-amb}(T_{shel} - T \quad)$ $- UA_{oil-sh} \quad (T_{oil} - T_{shell})$ $- UA_{ie-shell}(T_{ie} - T_{shell})$	–
10	$m_{ie}\frac{u_{ie}^{t+\Delta t} - u_{ie}^{t}}{\Delta t} + u_{ie}\frac{m_{ie}^{t+\Delta t} - m_{ie}^{t}}{\Delta t}$	$\varphi m_{sl}h_{sl} + m_l h_l$	$m_l h_l + m_{ie}h_{ie}$	$UA_{ie-shell}(T_{ie} - T_{sh} \quad)$ $+ UA_{ie-sm}(T_{ie} - T_{sm})$ $- UA_{dc-ie}(T_{dc} - T_{ie})$ $- UA_{dm-ie}(T_{dm} - T_{ie})$ $- UA_{dt-ie}(T_{dt} - T_{ie})$ $- UA \quad _{-ie}(T \quad - T_{ie})$	–

Integral and Differential Model of Hermetical Compressor Heat Losses Including Experimental Validation

Maria GOOSSENS[1&2*], Philippe RIVIERE[1], Odile CAURET[2], Cedric TEUILLIERES[2], Dominique MARCHIO[1]

[1] Mines ParisTech, PSL Research University, Centre for Energy efficiency of Systems (CES), Paris, 75006, France
Phone: +33 1 40 51 90 80, Fax: +33 1 40 51 94 49, E-mail: philippe.riviere@mines-paristech.fr

[2] EDF R&D, Energy in Buildings and Territories Department, Ecuelles, 77250, France
Phone: +33 1 60 73 67 10, E-mail: maria.goossens@edf.fr

* Corresponding Author

ABSTRACT

The current evaluation of heating and cooling system performances has become a major issue in European buildings sector. A promising method for measuring heat pump performances on-field has been previously developed and published. However, the accuracy of the method is strongly dependent on the evaluation of heat transfer from compressor towards the ambient air. In this paper a developed compressor heat transfer model that couples integral and differential formulations is presented. The model consists of three fundamental steps: thermodynamic analysis of the compression process, detailed thermophysical flow analysis using a CFD software, and finally compressor thermal network analysis consisting of integral equations that describe heat transfer of solid-fluid interfaces. Temperature distribution of the compressor shell is the main output of the model. Internal and external thermal profiles obtained from the numerical model and experimental measurements are compared in one operating condition at two compressor speeds. The RMS errors of external profiles are 3.20 °C and 1.96 °C, at 30 rps and 60 rps, respectively. The model can be used to determine the minimum number of surface temperature sensors and their locations on the compressor shell necessary to measure heat losses on-field. More accurate evaluation of compressor heat losses towards the ambient air will improve the accuracy of the performance measurement method.

1. INTRODUCTION

In an increasingly demanding regulatory setting, current methods to evaluate performances of heating and cooling systems have become problematic in European buildings sector (Ertesvag, 2011). Accurately measuring performances of air-to-air heat pumps (HPs) on-field is particularly challenging (McWilliams, 2002). A method based on mass/energy balances of the refrigeration cycle components has been adapted to measure the heat pump performances on-field (Tran *et al.*, 2013). This method integrates a simplified model of compressor heat transfer towards the ambient air, herein after referred to as the compressor heat losses. It has been established that compressor heat losses influence significantly the overall accuracy of the method: Tran *et al.* (2013) showed that the sensitivity index of the uncertainty of heating power predicted by the internal refrigerant method is 40%. Thus, a more precise evaluation of compressor heat losses is required in order to improve the HP performance analysis method.

Modeling compressor heat transfer is a challenging task. The complexity of compressor geometry hinders simplifications and contributes to various complex flow and heat transfer phenomena. Experimental and numerical approaches are typically used to model compressor thermal behavior. A short literature review in Section 2 highlights the state-of-the-art of compressor heat transfer modeling. The purpose of literature review was to aid in determining the most adequate modeling technique.

In this paper numerical model that evaluates compressor thermal behavior and heat transfer from the shell towards the ambient air is presented. Hybrid modeling approach, where differential and integral formulations are coupled, was adapted to be able to provide results with better accuracy than simply integral models, yet remaining relatively simplified, with low computational costs, and more flexible in terms of compressor design applicability, in comparison with differential models. This modeling approach is applicable to scroll and rotary compressors. Thermal behavior of scroll compressors is modeled. Thermal profiles obtained from the numerical model were compared to the ones obtained from a scroll compressor test bench. The model can be used to determine the minimum non-intrusive instrumentation, *i.e.* the minimal number of surface temperature sensors and their locations on the compressor shell, necessary to measure heat losses on-field.

2. STATE-OF-THE-ART OF COMPRESSOR HEAT TRANSFER MODELING

2.1 Correlations based on experimental measurements

Experimental techniques, where thermocouples are placed inside and/or outside the compressor, are one of the most straightforward and conventional ways to determine the thermal behavior of compressors. This approach is intrusive and the location of sensors has to be considered very carefully for the flow dynamics and the heat transfer inside the compressor to remain unaffected. Another technique is to use heat flux sensors and infrared cameras (Ribas *et al.*, 2008). Experimental data gives an insight into the main heat transfer mechanisms in each internal domain of the compressor (Jang & Jeong, 2005). Furthermore, the most significant heat transfer zones between component surfaces and surrounding fluid can be quickly identified (Dutra & Deschamps, 2013).

Correlations for compressor heat losses and component thermal conductances can be derived with the acquired experimental data. Heat transfer correlations are typically functions of variables that can be measured on-field, for instance condensation temperature. The correlations can also be used in some numerical models (Diniz *et al.*, 2012). However, these correlations are based on the data obtained from a limited number of operating conditions, therefore, their accuracy and overall validity is limited/unknown outside the operating range or when compressor parameters, such as the compressor layout or refrigerant type, are modified (Kim & Bullard, 2002). On the other hand, more accurate experimental models based on a great deal of operating conditions typically involve too many variables that are hard to measure on-field, such as the volumetric and isentropic efficiencies (Duprez *et al.*, 2009).

2.2 Numerical models

Numerical models can be divided into three different approaches: integral, differential, and hybrid. The choice of the modeling approach depends on the desired level of complexity and applicability of the model.

Integral models consist of building a system of steady or unsteady state energy balance equations (integral formulations) for each control surface. They are often referred to as lumped conductance or thermal network (TNW) models. Control surfaces are usually limited by the geometrical boundaries of the compressor solid components. With appropriate energy balances and heat transfer coefficients the compressor components are thermally interconnected, thus forming a network. Heat transfer coefficients can be adapted from the correlations already available in literature, like in the works of Sim *et al.* (2000) and Ooi (2003). Another approach is to use experimental data to calibrate thermal conductances, such as in the model developed by Todesca *et al.* (1992). The main advantage of the integral method is its relative simplicity and flexibility to adapt to different compressor types and internal component layouts. However, the main disadvantage of this modeling approach is its poor accuracy.

Differential models solve flow equations, such as Navier-Stokes equations, for each control volume with the aid of a computational fluid dynamics (CFD) program. Such models are able to predict flow and thermal profiles in a specific compressor design (Almbauer *et al.*, 2006). In the works of Chikurde *et al.* (2002), Birari *et al.* (2006), Pereira & Deschamps (2012), and Raja *et al.* (2003) steady-state thermal behavior of a hermetical reciprocating compressor is simulated, where different compressor component zones are assigned as heat sources/sinks resulting from electrical motor losses, frictional losses and heat losses due to a non-isentropic compression process. Such models provide a very accurate and detailed thermal analysis for a variety of operating conditions, and there is no need to prescribe thermal conductances. However, the developed model is usually compressor design specific. Furthermore, the computational processing time is quite high and requires powerful computers. Also, some complex physical phenomena inside the compressor, such as the flow of lubricating oil, complicate significantly the modeling strategies.

Hybrid models resolve integral and numerical formulations in a coupled manner: thermal network models are used to optimize simplified CFD models (Diniz *et al.*, 2012). Thus, hybrid models are a compromise between differential and integral models, as seen in the works of Almbauer *et al.* (2006), Sanvezzo & Deschamps (2012), and Ribas (2007). Similarly to integral models, hybrid methods assign heat transfer coefficients and/or other necessary information as boundary conditions to represent the effects of the flow dynamics on the heat transfer between solid and fluid interfaces (integral formulations). This eliminates the need to create complex, computationally expensive and detailed CFD models. For instance, modeling in detail rotations and other movements of interior components and their effects on the flow and heat transfer can be avoided in the CFD geometry, mesh and setup. In this case the differential formulations resolve the evolution of flow inside the compressor and its effects on the component wall temperature with relatively low computational costs. Such models are more accurate than the integral models. Yet, they remain relatively simplified and more flexible in terms of compressor design applicability in comparison with differential models (Ribas et al., 2008). Heat transfer from compressor shell towards ambient air was not a primary interest of the hybrid models found in literature. Also, most hybrid models consider reciprocating compressors. Diniz *et al.* (2012) developed a hybrid thermal model for scroll compressors. However, the model integrates temperature distribution along the scroll wraps; thus, it is adapted only for scroll compressors.

3. COMPRESSOR HEAT LOSS MODEL

Figure 1 is a schematic representation of the computational domain for scroll compressors. The model is simplified by removing unnecessary fillets and parts in order to facilitate the meshing in the preprocessing stage and the convergence in CFD solver. Compressor is surrounded by ambient air with enclosure walls maintained at a constant temperature. Depending on the configuration compressor shell can either be exposed to forced or natural convection. The geometrical domain consists of compression chamber, crankshaft, compressor shell, oil sump at the bottom, rotor, and stator. The listed components were considered as the most essential from the heat transfer point of view.

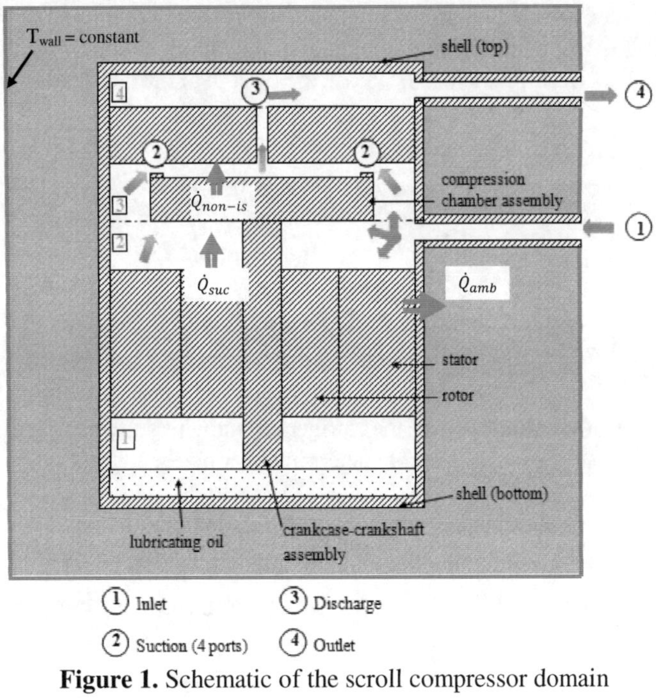

Figure 1. Schematic of the scroll compressor domain

The calculation procedure can be divided into three fundamental steps: thermodynamic cycle analysis, detailed thermophysical flow analysis, and compressor thermal network analysis, as depicted in Figure 2.

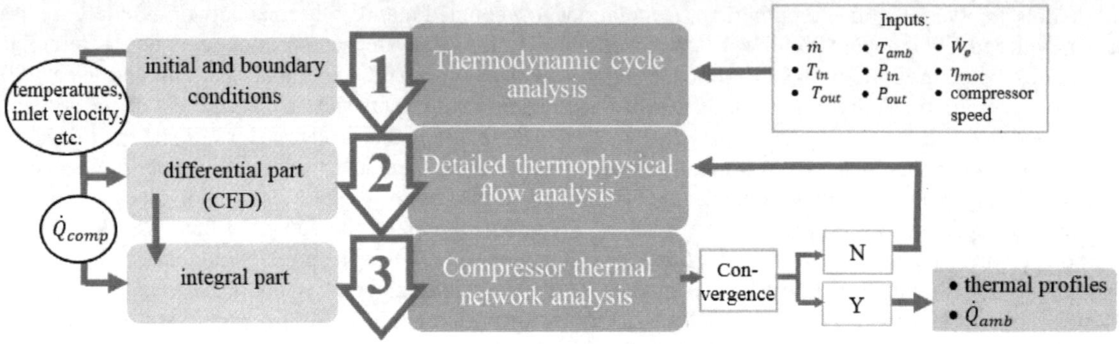

Figure 2. Schematic of the model calculation steps

3.1 Thermodynamic cycle analysis

The first part of the model analyzes the thermodynamic compression cycle in order to evaluate the initial and boundary conditions, such as temperatures, inlet velocity, and heat sources, used in integral and differential parts of the model, steps two and three, respectively. In this part heat released by the compression process is calculated. This variable is used in the third step of the model to determine whether the model has converged.

It is assumed that the heat released by the components, $\dot{Q}_{comp}$, is partially absorbed by the fluid and partially transmitted towards the ambient air, and can be calculated from Equation (1):

$$\dot{Q}_{comp} = \dot{Q}_{suc} + \dot{Q}_{non-is} \tag{1}$$

where $\dot{Q}_{suc}$ is the superheat of the refrigerant gas passing from the compressor inlet to the compression chamber, and $\dot{Q}_{non-is}$ is the heat released due to a non-isentropic compression process. Gas superheat is assumed to be equal to electrical losses dissipated as heat from the motor, located at the bottom of the hermetical compressor (Figure 1), given in Equation (2):

$$\dot{Q}_{suc} = (1-\eta_{mot})\dot{W}_e \tag{2}$$

where η_{mot} is the motor efficiency and $\dot{W}_e$ electrical power consumption. In order to calculate $\dot{Q}_{non-is}$, the isentropic compression power, given in Equation (3), must be known:

$$\dot{W}_{is} = \dot{m}\left(h_{dis,is} - h_{suc}\right) \tag{3}$$

where $\dot{m}$ is the mass flow rate of the fluid, h_{suc} is the refrigerant enthalpy at the suction chamber cavity, and $h_{dis,is}$ is the discharge enthalpy of an isentropic compression determined from Equation (4):

$$h_{dis} = h\left(s\left(T_{suc}, P_{in}\right), P_{dis}\right) \tag{4}$$

where P_{dis} is the discharge pressure and $s(T_{suc}, P_{in})$ is the fluid entropy at the suction chamber cavity. Consequently, $\dot{Q}_{non-is}$ is obtained from Equation (5):

$$\dot{Q}_{non-is} = \eta_{mot}\dot{W}_e - \dot{W}_{is}\,. \tag{5}$$

Figure 3 illustrates the thermodynamic cycle of heat pumps and associated power distribution. This step is coded in a computational code, MATLAB. The required refrigerant gas properties, such as enthalpies, are obtained from the libraries of REFPROP (NIST, 2008).

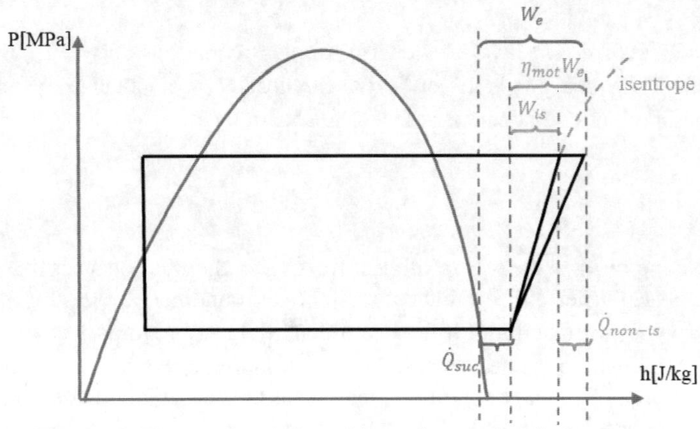

Figure 3. Thermodynamic refrigeration cycle of a heat pump

3.2 Detailed thermophysical flow analysis

The second step of the model is executed in ANSYS FLUENT, a CFD code that numerically solves a set of governing differential equations (conservation of mass, momentum and energy). The outcomes of this part of the model are refrigerant thermal and flow profiles, as well as solid zone temperatures. This information is subsequently transmitted to the final step of the model in order to perform a TNW analysis that verifies the CFD simulation results.

The domain geometry of the CFD model is decomposed into solid and fluid zones. The sub domains are oil continuum (on the bottom of the compressor cylinder), discrete refrigerant fluid zones, surrounding air continuum, and solid zones bounded by simplified geometrical boundaries of the components. The domain mesh consists of $3.40 \cdot 10^6$ tetrahedral elements. Denser mesh is imposed in narrow passages and regions with more significant curvature. Steady state, conduction between solid-solid and convection and radiation between fluid-solid interfaces is enabled in the domain. The solver type is segregated and pressure-based. Gravitational acceleration is enabled in the domain. Standard k-epsilon model, was chosen as the turbulence model for its robustness, economy, and reasonable accuracy. Velocity inlet and outflow were set at the inlet and outlet boundary conditions, respectively. The walls of the air enclosure surrounding the compressor are isothermal. Boussinesq approximation was applied for the air density. Volumetric and surface heat fluxes were added to the computational domain.

The CFD model performs exclusively a thermal analysis and, therefore, no compression is assumed to take place in this part of the model. To represent an increase in temperature from T_{suc} to T_{dis}, temperatures of the refrigerant at suction and discharge, respectively, due to compression and non-isentropic heat, a heat flux is introduced in the compression chamber, $\dot{Q}_{comp-cham}$ (Figure 4). Oil sump and top of the crankshaft temperatures are set as constant and calculated from a multivariate correlation for oil temperature. The correlation was derived from experimental data over a range of operating conditions, where the difference between condensation and evaporation temperatures varied from 25 to 60 °C. Also, multiple compressor speeds and ambient temperatures were considered.

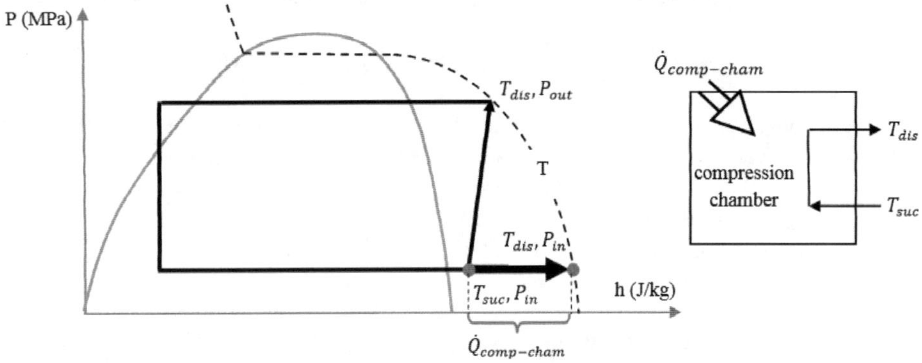

Figure 4. Required heat to reach T_{dis} (temperature of the compressed fluid at discharge)

3.3 Compressor thermal network analysis

The final step of the model involves a network of energy transfer equations between the solid-fluid interfaces by convection. The goal of the TNW analysis is to verify that the heat, $\dot{Q}_{comp}$, calculated from Equation (1), equals to the heat dissipated from compressor components, as listed in Equation (6):

$$\dot{Q}_{comp} = \dot{Q}_{rotor} + \dot{Q}_{stator} + \dot{Q}_{crank} + \dot{Q}_{comp-cham} \tag{6}$$

The total heat flux of each component is the sum of all heat fluxes of individual surfaces that constitute the component in question. For instance, heat transferred from the rotor to the surroundings is the sum of heat fluxes between the fluid and rotor surface in contact with the fluid. If the heat released due to an imperfect compression process, $\dot{Q}_{comp}$, obtained from Equation (1), is equal to the heat released by the compressor components, obtained from Equation (6), the model converged. If not, more boundary conditions may be assigned in CFD model to guide the differential step more effectively, until convergence is achieved.

In order to obtain the net energy released by the components, the temperatures of the solid components and surrounding fluid, as well as the refrigerant velocities in proximity to the solid components must be known. These values are obtained from the CFD model. Consequently, convective heat transfer coefficients are evaluated. Nusselt number correlations for different flow regimes and geometries were taken from literature. Correlations for internal and external flows for horizontal and vertical, rotating and static plates and cylinders were considered. Equation (7) describes the relation between the Nusselt number and the convective heat transfer coefficient:

$$Nu_x = h_x x / k \tag{7}$$

where x is the characteristic length, which depending on the geometry of the object and flow characteristics can be a diameter, radius or length, k is the thermal conductivity of the fluid, and h_x is the convective heat transfer coefficient (Incropera, 2002).

As the convergence in step three is achieved, $\dot{Q}_{amb}$ values are extracted directly from the CFD model. Numerical results of $\dot{Q}_{amb}$ presented in Table 1 were obtained at T_{cond}=60 °C, T_{evap}=0 °C, and T_{amb}=10 °C at two compressor speeds, 30 rps and 60 rps, with R407C as the refrigerant fluid.

Table 1. Numerical results of $\dot{Q}_{amb}$ expressed as a percentage of compressor power input at different speeds

Speed (rps)	Numerical (%)
30	10.84
60	7.36

This part established the hybrid model. The results in Table 1 are expressed in percentages for confidentiality reasons. The second step (CFD model) took approximately 4 hours to converge, and for the third step to converge, the value obtained in Equation (6) had to fall within the ±10% range of the value obtained from Equation (1).

4. EXPERIMENTAL COMPARISON

Numerical results were compared to experimental values obtained from a scroll compressor test bench. Compressor component layout was identical to the one of the numerical model (Figure 1). Air temperature in the climate chamber was maintained at 10 °C, and the evaporation and condensation temperatures were 0 °C and 60 °C, respectively, at 30 rps and 60 rps. The tested compressor had no insulation, and R407C was used as the refrigerant fluid. In the experimental setup 40 thermocouples were placed along the compressor shell and on the internal components, such as the motor and discharge baffle. Refrigerant mass flow rate, oil concentration in the refrigerant circuit, compressor inlet and outlet temperatures, condensation and evaporation pressures, and compressor power input were also measured. All values were registered in steady operating state, about 10 minutes after start up. Experimental values of compressor heat losses were calculated from the compressor energy balance equation given in Equation (8):

$$\dot{Q}_{amb} = \dot{W}_{comp} - \dot{m}\left[\left(1-C_g\right)\left(h_{r,comp,out} - h_{r,comp,in}\right) + C_g \Delta h_o^{T_{comp,in} \to T_{comp,out}}\right] \tag{8}$$

where $h_{r,comp,out}$ and $h_{r,comp,in}$ are the refrigerant enthalpies at compressor inlet and outlet, respectively, $\dot{W}_{comp}$ is the compressor power input, $\Delta h_o^{T_{comp,in} \to T_{comp,out}}$ is the enthalpy change of oil, and C_g is the oil mass fraction. Error propagation formula, listed in Equation (9), was used to evaluate the absolute error of experimental heat losses:

$$\sigma_{\dot{Q}_{amb}} = \sqrt{\sum_i \left(\frac{\delta \dot{Q}_{amb}}{\delta n_i}\right)^2 \sigma_{n_i}^2} \tag{9}$$

where n_i is the measured variable, and σ_{n_i} is the absolute uncertainty of the measured quantity. Wattmeter, flow meter, pressure sensor, and surface temperature sensor uncertainties were taken into consideration.

In Figure 5, numerical and experimental temperatures, expressed in dimensionless values for confidentiality reasons, distributed along the exterior compressor shell are compared at two compressor speeds. Measuring points 1-7 represent the locations of thermocouples along the shell, from the bottom to the top. The RMS errors of external profiles are 3.20 °C and 1.96 °C, at 30 rps and 60 rps, respectively. Numerical and experimental temperature values of the internal components are compared in Figure 6. Compressor heat losses obtained from the numerical model and experimental tests are presented in Table 2.

In hermetical compressors lubricating oil prevents potential damage of internal components due to friction. In addition to oil being stored in the sump at the bottom, a thin oil film is present over many components inside the compressor, such as the crankcase, stator and the internal surface of the compressor shell. The oil pump is coupled with the crankshaft; as the shaft spins, the lubricating oil flows inside the pump through the crankshaft by centrifugal forces. As it reaches the other extremity of the shaft assembly, part of it is projected as a jet hitting the upper surface of the compressor shell along which it later flows as a thin oil film. The other part of the projected oil returns to the sump by flowing over the compressor crankcase. Oil forms an interface between the refrigerant fluid and the surface of the solid components. Thus, it can affect the heat transfer inside the compressor (Dutra & Deschamps, 2013). In the numerical model, oil is still and present only at the bottom. This could be the reason for the discrepancies between the experimental and numerical results.

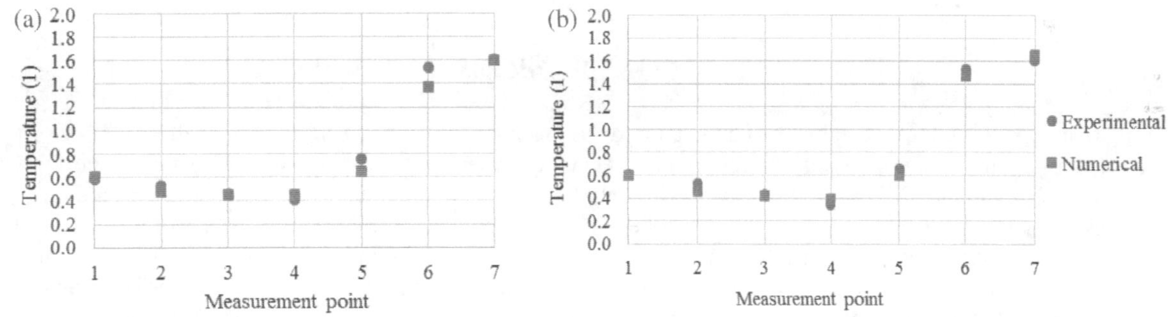

Figure 5. Shell temperature distribution (dimensionless values) at 30 rps (a) and 60 rps (b)

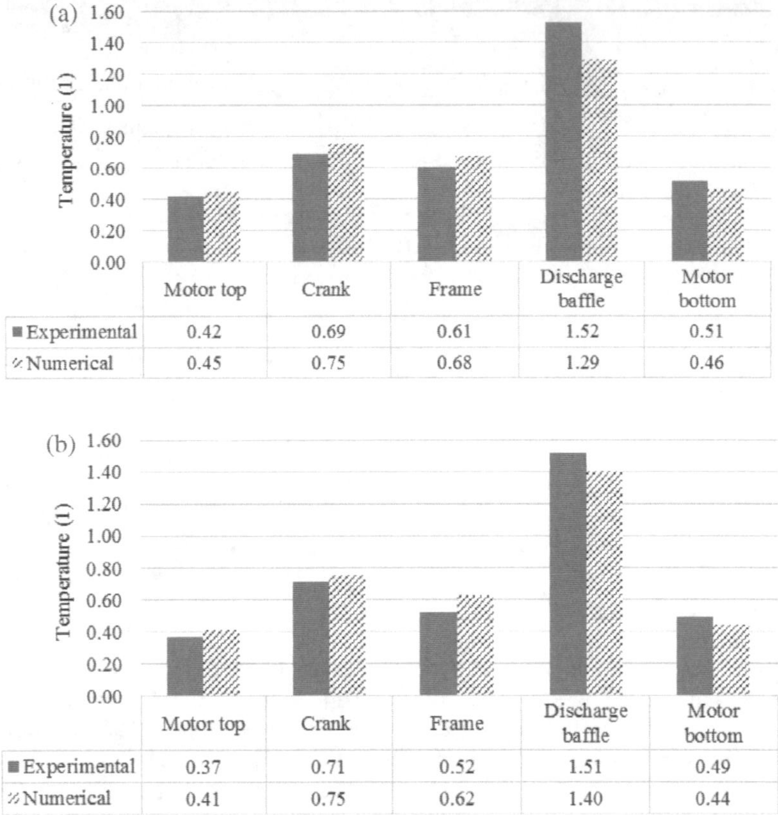

Figure 6. Dimensionless temperature values of internal compressor components at 30 rps (a) and 60 rps (b)

Table 2. Experimental and numerical $\dot{Q}_{amb}$ values expressed as a percentage of compressor power input

Speed (rps)	Experimental (%)	Numerical (%)
30	11.48±1.71	10.84
60	8.31±1.78	7.36

5. CONCLUSIONS

A numerical model used to determine the temperature distribution of the compressor exterior shell was developed. A code was written in a computing language representing the thermal network of the compressor, which is coupled with a commercially available CFD code that solves differential equations in order to perform a more elaborate flow and heat transfer analysis in the compressor domain. Model outputs are compared with experimental results obtained from a hermetical scroll compressor that has been accurately instrumented and tested in a compressor test bench. Internal and external thermal profiles are compared in one operating condition at two different speeds. The RMS errors of the shell temperatures at different locations are 3.20 °C and 1.96 °C at compressor speeds of 30 rps and 60 rps, respectively. Compressor heat losses calculated from the compressor energy balance using the measured data were compared with the heat losses obtained from the numerical model. Numerical model can be used to determine the locations of surface temperature sensors used to evaluate heat losses of scroll compressors on-field.

NOMENCLATURE

C_g	oil mass concentration	(-)
h	specific enthalpy	(J kg^{-1})
h	heat transfer coefficient	(W m^{-2} °C^{-1})
k	thermal conductivity	(W m^{-1} °C^{-1})
m	number of thermal zones	(-)

$\dot{m}$	mass flow rate	(kg s^{-1})
n	measured variable	(-)
Nu	Nusselt number	(-)
P	pressure	(Mpa)
$\dot{Q}$	heating power	(W)
s	specific entropy	(J kg^{-1} °C^{-1})
T	temperature	(°C)
UA	thermal conductance	(W °C^{-1})
$\dot{W}$	power consumption	(W)
x	characteristic length	(m)
η_{mot}	motor efficiency	(-)
σ	standard deviation	(-)

Subscript

in	inlet
out	outlet
comp	component
comp-cham	compression chamber
amb	ambient
suc	suction
non-is	non-isentropic
e	electric
dis	discharge
is	isentropic
o	oil
r	refrigerant

Abbreviations

CFD	computational fluid dynamics
HP	heat pump
RMS	root-mean-square
R407C	refrigerant type
TNW	thermal network

REFERENCES

Almbauer, R. A., Burgstaller, A., Abidin, Z., & Nagy, D. (2006). 3-Dimensional simulation for obtaining the heat transfer correlations of a thermal network calculation for a hermetic reciprocating compressor. *Proc. of the Int. Compressor Engineering Conference at Purdue*, Paper C079.

Birari, Y. V., Gosavi, S. S., & Jorwekar, P. P. (2006). Use of CFD in design and developement of R404a reciprocating compressor. *Proc. of the Int. Compressor Engineering Conference at Purdue*, Paper C072.

Chikurde, R. C., Loganathan, E., Dandekar, D. P., & Manivasagam, S. (2002). Thermal mapping of hermetically sealed compressors using computational fluid dynamics. *Proc. of the Int. Compressor Engineering Conference at Purdue*, Paper C6-4.

Diniz, M. C., Pereira, E. L., & Deschamps, C. J. (2012). A lumped-parameter thermal model for scroll compressors including the solution for the temperature distribution along the scroll wraps. *Int. J. Refrig.*, *53*(2015), 184-194.

Duprez, M.-E., Dumont, E., & Frère, M. (2009). Modeling of scroll compressors - Improvements. *Int. J. Refrig.*, *33*(4), 721-728.

Dutra, T., & Deschamps, C. J. (2013). Experimental characterization of heat transfer in the components of a small hermetic reciprocating compressor. *Appl. Therm. Eng.*, *58*(1-2), 499-510.

Ertesvag, I. S. (2011). Uncertainties in heat-pump coefficient of performance (COP) and exergy efficiency based on standardized testing, *Energy and Buildings*, *43*, 1937-1946.

Incropera, F. P. (2002). *Fundamentals of heat and mass transfer*. New York, USA: John Wiley & Sons.

Jang, K., & Jeong, S. (2005). Experimental investigation on convective heat transfer mechanism in a scroll compressor. *Int. J. Refrig.*, *29*(5), 744-753.

Kim, M.-H., & Bullard, C. W. (2002). Thermal performance analysis of small hermetic refrigeration and air-conditioning compressors. *JSME International Journal*, *45*(4), 857-864.
McWilliams, J. (2002). *Review of Airflow Measurement Techniques*. Contract for U.S. Department of Energy, Contract No. DEAC03e76SF00098
NIST, National Institute of Standards and Technology (2008). *Refprop – Reference Fluid Thermodynamic and Transport Properties*, vol. 8. USA.
Ooi, K. T. (2003). Heat transfer study of a hermetic refrigeration compressor. *Appl. Therm. Eng.*, *23*(15), 1931-1945.
Pereira, E. L., & Deschamps, C. J. (2012). A numerical study of convective heat transfer in the compression chambers of scroll compressors. *Proc. of the Int. Compressor Engineering Conference at Purdue*, Paper 1274.
Pizarro-Recabarren, R. (2009). Modelling the influence of the lubricant oil on the heat transfer in hermetic reciprocating compressor. *Proc. of the Int. Conference on Compressors and Their Systems* (365-374). London.
Raja, B., J., Sekhar. S. J., Lal, D. M., & Kalanidhi, A. (2003). A numerical model for thermal mapping in a hermetically sealed reciprocating refrigerant compressor. *Int. J. Refrig.*, *26*(6), 652-658.
Ribas, F. A. (2007). Thermal analysis of reciprocating compressors. *Proc. of Int. Conference on Compressor and Their Systems* (227-287). London.
Ribas, F. A., Deschamps, C. J., Fagotti, F., Morriesen, A., & Dutra, T. (2008). Thermal analysis of reciprocating compressors - critical review. *Proc. of the Int. Compressor Engineering Conference at Purdue*, Paper 1306.
Sanvezzo, J., & Deschamps, C. J. (2012). A heat transfer model combining differential and integral formulations for thermal analysis of reciprocating compressors. *Proc. of the Int. Compressor Engineering Conference at Purdue*, Paper 1343.
Sim, Y. H., Youn, Y., & Min, M. K. (2000). A study on heat transfer and performance analysis of hermetic reciprocating compressors for refrigerators. *Proc. of the Int. Compressor Engineering Conference at Purdue*, (229-236).
Todesca, M. L., Fagotti, F., Prata, A. T., & Ferreira, R. T. (1992). Thermal energy analysis in reciprocating hermetic compressors. *Proc. of the Int. Compressor Engineering Conference at Purdue* (1417-1428).
Tran, C. T., Rivière, P., Marchio, D., & Arzano-Daurelle, C. (2013), In situ measurement methods of air to air heat pump performance. *Int. J. Refrig.* *35*(5), 1442-1455.
White, F. (2005). *Viscous Fluid Flow*. New York, USA: McGraw-Hill Science.

ACKNOWLEDGEMENT

The authors would like to thank Mitsubishi Heavy Industries (MHI) for providing experimental data from conducted compressor tests. This study was also made possible with the means provided by EDF R&D and Centre for Energy efficiency of Systems (CES) of Mines ParisTech.

Heat Transfer Analysis in the Cylinder of Reciprocating Compressor

Jan Tuhovcak[1]*, Jiri HEJCIK[1], Miroslav JICHA[1]

[1]Brno University of Technology, FME, Energy Institute
Brno, Czech Republic
jan.tuhovcak@vutbr.cz

* Corresponding Author

ABSTRACT

Heat transfer inside the cylinder of reciprocating compressor causes decrease of volumetric efficiency due to superheating of the working fluid. Understanding heat transfer phenomenon inside the cylinder is necessary to achieve higher efficiency. There are several empirical correlations available in the literature to date, which are employed in simulation tools due to their easy-to-use properties and they provide fast results with solid accuracy compared to more complex numerical models for heat transfer. Nevertheless, there is a problem with validation these empirical correlations. Solution of this problem could be found in CFD simulation tools. This paper presents the CFD analysis of heat transfer in order to compare heat fluxes calculated by numerical tools with those obtained by empirical correlations. Distribution of transferred heat among the main typical surfaces inside the cylinder is described in this paper. The analysis was done on simplified 3D geometry including the movement of valves and remeshing through the simulation to guarantee mesh quality. Based on the results of numerical analysis a proposal to increase the accuracy of heat transfer prediction is suggested.

1. INTRODUCTION

Superheating of a gas in reciprocating compressors causes a considerable decrease in the overall efficiency. Increase in the gas temperature is caused by heat transfer from the walls along the suction line and inside the cylinder from surrounding walls – piston, liner and cylinder head. As it was proved by Almbauer *et al.* (2010) and Morriesen and Pereira (2009), increase of the gas temperature before the beginning of the compression has critical impact on the volumetric efficiency of the compressor. In fact, increase in gas temperature by 1 °C at the beginning of compression causes decrease of efficiency around 0.3%. Moreover, there is a potential risk of ignition of a lubricating oil or rapid material degradation of compressor components. Therefore a good knowledge of heat transfer inside the compressor as well as a development of new materials could prevent the gas superheating and increase the overall compressor efficiency. This paper presents analysis of heat transfer inside the cylinder of reciprocating compressor using two different approaches and the results are used to propose an improvement of heat transfer correlation.

2. HEAT TRANSFER MODELS

There are two mathematical approaches used to determine the heat flux. The first and simpler possibility is to use semi-empirical correlation usually based on non-dimensional numbers. The advantage of using this approach is low computational demand. Second option is more advanced numerical simulation. The results are more accurate, however the complexity of the CFD simulation together with high computational demands make it less convenient and useful. In the following chapters there is an overview of both approaches. Many authors used CFD to introduce new correlations, e.g. Disconzi *et al.*(2012) and Aigner and Steinrück (2007) as they have an important role in the compressor analysis. Many researches use integral correlation to describe heat flux from cylinder in thermal analysis of the whole compressor body. Balduzzi *et al.* (2014) used Annand (1963) correlation to describe heat flux from a compression chamber in his analysis of cylinder cooling. He employed 0d numerical model to obtain heat transfer coefficient and bulk temperature for compression chamber. Based on these results he carried out a CFD analysis of cooling system. Another thermal analysis was performed by Dutra and Deschamps (2013), who applied the model of Annand (1963) inside the cylinder of hermetic reciprocating compressor.

As it was pointed out by Disconzi *et al.* (2012) and Tuhovcak *et al.* (2015), there are significant differences between the results of heat transfer correlations, which can lead to a drop in compressor efficiency. According to Adair (1972), due to the heat transfer correlations accuracy the compressor efficiency could be decreased by as much as three percent. Moreover, the volumetric efficiency could decrease by 10 – 20 percent due to heat transfer regardless on correlation. Some aspects affecting the heat transfer are neglected in this paper, especially oil film on the cylinder walls and heat generation caused by friction between piston and cylinder liner. These aspects are difficult to implement in current heat transfer modelling and will be examined in future works.

2.1 Integral correlations

The literature survey shows that most of the heat transfer models for reciprocating compressors were developed for internal combustion engines, which have different operation conditions although the geometry could be similar. First attempts were done by Annand (1963) and Woschni (1967), later by Adair *et al.* (1972) and most recently by Disconzi *et al.* (2012) and Aigner and Steinrück (2007). The last two authors focused directly on reciprocating compressors. They used correlations based on non-dimensional Nusselt number and Reynolds number. The main differences between all the correlations are in coefficients *a*, *b* and *c* and also in the expression of characteristic velocity for Reynolds number, see the Table 1. Aigner in his correlation equation uses the Stanton number that can be directly related to the friction coefficient.

$$Nu = a\,\mathrm{Re}^b\,\mathrm{Pr}^c \tag{1}$$

$$\mathrm{Re} = \frac{u\,D}{\nu} \tag{2}$$

$$St = f\left(C_f\right) \tag{3}$$

where D is the cylinder diameter, u is characteristic velocity and ν is kinematic viscosity of the gas.

Table 1: Empirical correlation for heat transfer prediction

Woschni	$Nu = 0.035\,\mathrm{Re}^{0.7}$	$\mathrm{Re} = \frac{u\,D}{\nu}$	$u = 6.618 \cdot u_p$ - opened valves $u = 2.28 \cdot u_p$ - closed valves
Annand	$Nu = 0.7\,\mathrm{Re}^{0.7}\,\mathrm{Pr}^{0.7}$	$\mathrm{Re} = \frac{u\,D}{\nu}$	u_p
Adair	$Nu = 0.053\,\mathrm{Re}^{0.8}\,\mathrm{Pr}^{0.6}$	$\mathrm{Re} = \frac{u\,D_e}{\nu}$, $D_e = \frac{6 \cdot Cylinder\ volume}{Cylinder\ area}$	$u = 2\omega\left[1.04 + \cos\left(2\alpha\right)\right]$ $\frac{3}{2}\pi < \alpha < \frac{1}{2}\pi$ $u = \omega\left[1.04 + \cos\left(2\alpha\right)\right]$ $\frac{1}{2}\pi < \alpha < \frac{3}{2}\pi$
Disconzi *Compression*	$Nu = 0.08\,\mathrm{Re}^{0.8}\,\mathrm{Pr}^{0.6}$	$\mathrm{Re} = \frac{u\,D}{\nu}$	$u = u_p$
Disconzi *Discharge*	$Nu = 0.08\,\mathrm{Re}^{0.8}\,\mathrm{Pr}^{0.6}$	$\mathrm{Re} = \frac{u\,D}{\nu}$	$u = u_p + u_p^{0.8} u_{c(t)}^{0.2}$
Disconzi *Expansion*	$Nu = 0.12\,\mathrm{Re}^{0.8}\,\mathrm{Pr}^{0.6}$	$\mathrm{Re} = \frac{u\,D}{\nu}$	$u = u_p$
Disconzi *Suction*	$Nu = 0.08\,\mathrm{Re}^{0.9}\,\mathrm{Pr}^{0.6}$	$\mathrm{Re} = \frac{u\,D}{\nu}$	$u = u_p + u_p^{-0.4} u_{c(t)}^{1.4}$
Aigner	$St = \frac{h}{\rho c_p u} = \frac{Nu}{\mathrm{Re}\,\mathrm{Pr}}$	$St = \frac{C_f}{2\,\mathrm{Pr}^{2/3}}$	$C_f = 0.078\,\mathrm{Re}^{-0.25}$ $C_f = 0.046\,\mathrm{Re}^{-0.2}$

where u_p is the mean piston speed, α is the crank angle and ω is the angular velocity of crankshaft.

Adair was one of the first authors, who focused directly on reciprocating compressors. His correlation for heat transfer differs from other correlations mainly in characteristic diameter for Reynolds number and in calculation of characteristic gas velocity. He uses the variable equivalent diameter D_e, instead of cylinder diameter and the approximated swirl velocity with angular speed of a crankshaft, ω, which has a larger magnitude than linear speed of the piston.

Disconzi *et al.* (2012) divided the whole process inside the cylinder into the four basic processes: suction, compression, discharge and expansion. For each of them she proposed a new correlation with different constants a, and exponents b and c in the general equation (1) for Nusselt number, see Table 1. Characteristic velocity is formed from the mean piston speed and the velocity of exchanging gas, equation (4).

$$u_c = \frac{\left|\dot{m}_{(t)}\right|}{\rho_{(t)} \cdot A_c} \tag{4}$$

where $\rho_{(t)}$ is the density of gas in the cylinder and A_c is the cylinder cross-sectional area. The correlation was developed using a CFD simulation of simplified compressor geometry and verified for several working conditions. The approach of Aigner was later excluded from this work as it requires a deeper analysis of friction coefficient. One possibility to calculate Stanton number using skin friction coefficient C_f could be found in Bejan (2013), however even Aigner adjusted Stanton number according to CFD simulations. There are two formulas to calculate friction coefficient in Table 1. First one is for lower Reynolds number, second one for higher Reynolds number, which could be useful when compression/expansion process and suction discharge process are distinguished .

2.2 Numerical modelling of heat transfer

Rakopoulos *et al.* (2010) examined and compared four commonly used heat transfer models in commercial CFD tools concluding that none of these models is capable of predicting heat transfer accurately for combustion engines. He also developed a new model which provided more acceptable results than the four previous ones. Nevertheless, this effort was done on combustion engines, which have different operating conditions compared to reciprocating compressors. Recent investigations in compressor field were done by Pereira *et al.* (2010), who studied heat transfer inside the cylinder using two dimensional model. The purpose was to compare results from CFD with integral correlation obtained from literature. Author uses RNG k-ε turbulence model and wall function based on Launder-Spalding within commercial CFD code Ansys FLUENT. His work shows that integral correlation predicts lower heat flux than CFD simulation. Later, Disconzi *et al.* (2011) found out that the use of Launder-Spalding wall function provides lower heat flux at cylinder walls compared to near-wall modelling approaches. In following work of Disconzi *et al.* (2012), enhanced wall treatment based on Kader (1981) was employed instead of Launder-Spalding. RNG k-ε turbulence model was used on simplified two dimensional geometry again. Based on the results, the author proposed new integral correlations for heat transfer, which have the main improvement in the characteristic velocity of Reynolds number. Most of the authors used only mean piston velocity, but Disconzi *et al.* (2012) takes into account also the flow through the valves, which considerably influences the flow field inside the cylinder.

Most of the aforementioned works were focused either on combustion engines or refrigeration compressors. Literature survey shows only a few publications focused directly on air compressors. Regarding the CFD simulations with particular focus on heat transfer there is a work of Müllner and Bielmeier (2008), however he did not state exactly which type of compressor he used in his analysis. He also used a commercial CFD tool to investigate heat fluxes in the cylinder of compressor for different valve configurations. Another study of heat transfer was done by Aigner and Steinrück (2007), who used commercial CFD tool for a full three dimensional model in order to catch the heat flux inside the cylinder and he reconstructed it using non-dimensional Stanton numbers.

3. COMPRESSOR MODELS AND SIMULATION PROCEDURE

3.1 Matlab model

The basis of the Matlab - 0D model is the energy balance over a control volume calculated using the 1st Law of Thermodynamics, equation (5). Both pressure and temperature are considered as uniform in the control volume, which is the whole volume of cylinder. The flow inside the control volume is neglected.

$$dW + dQ + \sum_i dm_i \cdot h_i = dU \tag{5}$$

where dQ stands for heat transfer from/to the cylinder, $dm_i h_i$ is inflow or outflow of gas through the valves, dU is the change of internal energy and dW is the piston work.

Solution procedure of this equation is similar to the one presented by Farzaneh-Gord *et al.* (2014). Discrete values of temperature, pressure and the rest of the fluid properties are obtained by solving the differential equation (5) with the finite difference method for a given time-step as follows

$$u_{t+1} = \frac{1}{m}\left\{\dot{Q} + (\dot{m}_S h_S)_t - (\dot{m}_D h_D)_t - p_t \frac{\Delta V}{\Delta t} - \frac{\Delta m}{\Delta t}\right\} + u_t \tag{6}$$

Continuity equation is solved similarly

$$\frac{dm_{cyl}}{dt} = \dot{m}_{in} - \dot{m}_{out} \tag{7}$$

where m_{in} is the mass flow coming into the cylinder, m_{out} is the flow escaping through the discharge valve and dm_{cyl} is the change of mass inside the cylinder over a time step. The rest of the thermodynamic properties are obtained by employing the CoolProp library Bell *et al.* (2014).

Valve behavior is the main factor affecting the mass flow coming into or out of the cylinder. A mathematical description of valve movement was introduced by Costagliola (1950) as a mass-spring system. This description uses the single degree of freedom equation (8) solved using the 4th order Runge-Kutta method. This approach assumes the valve to be a rigid object.

$$m \cdot \ddot{x} + d \cdot \dot{x} + k \cdot x = \sum_i F_i \tag{8}$$

where m is the mass of the valve, k is the spring stiffness and d is the damping constant. The exact values used in the presented work can be found in Table 2. Valve position, velocity and acceleration are expressed by x and its first and second derivative. A flow area, S_{flow}, is calculated from the valve geometry and substituted in Fliegner's equation (9) to evaluate flow through a valve:

$$\dot{m}_i = S_{flow} \cdot \zeta \cdot \rho_1 \cdot \left(\frac{p_2}{p_1}\right)^{\frac{1}{\kappa}} \sqrt{\frac{2\kappa}{\kappa - 1}\frac{p_1}{\rho_1}\left(1 - \left(\frac{p_2}{p_1}\right)^{\frac{\kappa-1}{\kappa}}\right)} \tag{9}$$

The loss coefficient ζ is the function of valve position and must be determined experimentally or from the CFD simulations. This is one of the most important steps in the procedure, as this value cannot be subtracted from primary parameters of a compressor. The loss coefficient depends on valve geometry, type of fluid used in analysis and manifold geometry in front of and behind the valve. In this case we used CFD.

3.2 CFD model

The majority of the previous studies using commercial CFD tools have been done in FLUENT using k-ε turbulence model. The results of present work were computed in Ansys CFX using SST k-ω turbulence model. The model includes automatic wall function for Near Wall Modelling and it is also recommended by Ansys for resolving details of boundary layer. This function switches automatically from wall-function to a low Reynolds near wall formulation as the mesh is refined. Advantage of this automatic wall function is lower sensitivity to mesh quality and therefore it should be useable for a wide interval of y^+ (equation (10)). However, the grid should have y^+ value around 1 for highly accurate simulations, such as heat transfer predictions,

$$y^+ = \frac{u_{fric} \cdot \Delta n}{\nu} \tag{10}$$

where Δn is the distance between the first and second grid points off the wall, u_{fric} is the friction velocity and ν is kinematic viscosity.

Thermal boundary layer is solved by Kader (1981) function, where temperature distribution is modeled by blending the viscous sublayer and the logarithmic law of the wall, eq. (11).

$$T^{+} = \Pr y^{*} e^{-\Gamma} + \left[2.12 \ln\left(y^{*}\right) + \beta\right] e^{\frac{-1}{\Gamma}} \tag{11}$$

$$\beta = \left(3.85 \Pr^{1/3} - 1.3\right)^{2} + 2.12 \ln\left(\Pr\right) \tag{12}$$

$$\Gamma = \frac{0.01\left(\Pr y^{*}\right)^{4}}{1 + 5\Pr^{3} y^{*}} \tag{13}$$

where y* is dimensionless coordinate.

3.3 Compressor specification

The simplified geometry of compressor was used for simulation purposes. The three dimensional model of compressor was created in ANSYS Workbench 16.1, see the Figure 1.

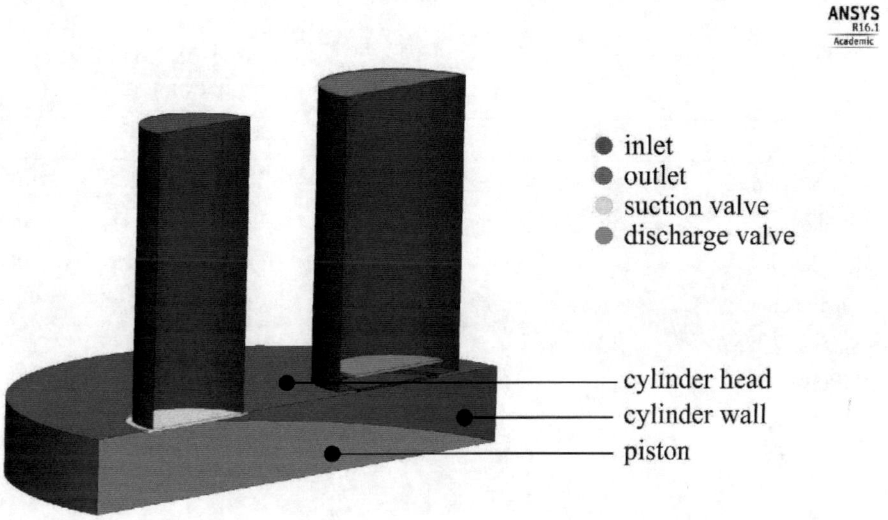

Figure 1: Overview of compressor geometry

Simplifications were also made in the modelling of valves. They are simulated as rigid bodies without any deformation during the solution process. The valves are allowed to move only in Z axis. CFX has a feature which is used for this kind of a simulation – Rigid Body solver. Another challenging task for reciprocating compressor simulation is the conservation of acceptable mesh during the solution. The movement of piston stretches the cells during the solution, which can be in some cases acceptable deformation with no need to remesh the geometry. However, the valve movement causes strong deformation around the valve geometry, which could result in inaccurate results. Therefore, a remeshing method using ICEM was implemented in this case. After the signal from user expression, the solver stops the solution, sends the geometry with current valve position to ICEM, where the remeshing script is activated. The quality of cells can be seen in Figure 2, where the valve is in a different positions.

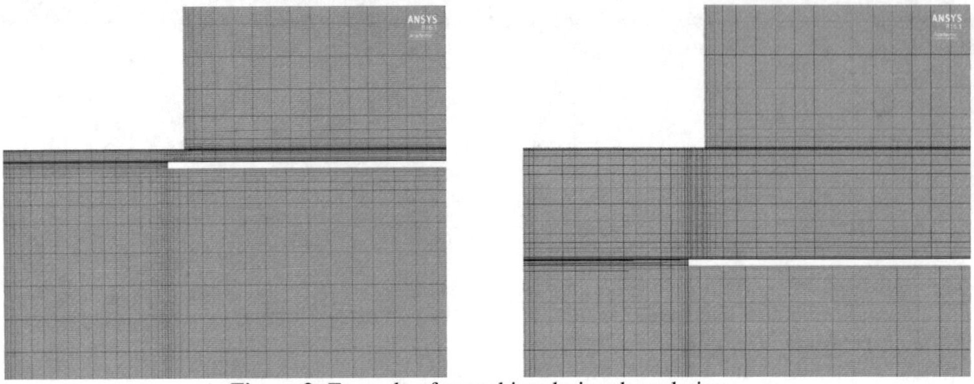

Figure 2: Example of remeshing during the solution

Table 2: Compressor specification.

Parameter	Value	Unit
Cylinder bore	20	mm
Stroke	20	mm
Rod length	55	mm
Clearance	0.2	mm
Suction valve mass	1	g
Spring stiff. - suction	250	N
Discharge valve mass	0.1	g
Spring stiff. - discharge	300	N

Table 3: Boundary conditions used in the simulation tool.

Boundary condition	Value	Unit
Suction pressure	1.00	bar
Discharge pressure	10.0	bar
Inlet temperature	300	K
Wall temperature	363	K

3.4 Boundary conditions

Temperature and pressure were prescribed for inlet and outlet boundary condition according to Table 3. The boundaries were set to an opening condition, which allows flow in both directions instead of simple inlet (or outlet) condition, which would allow no flow in opposite direction. The movement of the piston was prescribed by a user function based on crankshaft mechanism equation. The valves are solved as rigid bodies, as it was mentioned above. In addition, a symmetry condition was used to save the number of cells and the duration of calculation. Non-zero wall velocity component was prescribed for moving boundaries – piston and valves. During the suction and discharge processes the velocity component at wall refers to the movement of valve boundaries. The same applies for piston boundary for the entire simulation. Zero velocity component is defined for the rest of the boundaries.

The simulation consisted of three cycles of the piston in order to provide enough time for flow development inside the cylinder. We used air in ideal gas model as a working fluid.

4. RESULTS

Heat transfer changes its direction during the cycle of crankshaft depending on the gas temperature inside the cylinder. When the gas temperature is lower than the wall temperature, heat flux is positive, which means that the heat is transferred from the wall into the gas. When the compressed gas reaches the same temperature as the wall, heat flux is zero and after that it changes direction, meaning the heat is transferred from the gas into the wall. This change of direction occurs not just during the compression, but also during expansion process (Figure 3). The negative effect of heat transfer inside the cylinder is the superheating of the sucked gas, resulting in decrease of compressor efficiency. Magnitude of this decrease is mostly determined by the wall temperature, therefore it is desirable to calculate it

correctly, especially when the transient simulation of compressor is the point of interest. Most of the previous models have used one correlation to predict the heat transfer inside the cylinder for all surfaces. A distribution of heat flux depends usually on the actual heat transfer area of the cylinder. The heat transfer area is constant for the piston and the cylinder head, but for cylinder wall it is based on the actual piston position.
Another important phenomenon inside the cylinder, which influences the heat flux, is the fluid flow. High velocities close to the wall increase the heat transfer coefficient and consequently the heat flux, however this is included in simplified correlations for heat transfer. Therefore, a more detailed numerical tool was used to study the distribution of heat flux inside the cylinder.

The simulation procedure started with the expansion process moving the piston from top dead center towards bottom dead center. When the pressure force acting on suction valve overcomes the spring force, suction process begins. Usually the suction process is ended when the piston is in its bottom dead center followed by the compression and discharge. Movement of the discharge valve is again controlled by pressure and spring force. A special care must be taken when the valves are opening and closing in numerical simulation. Keeping the mesh in the gap between the valve and the seat is fundamental for CFD simulation and therefore it is not possible to close/open valves continuously, but when the gap is less than 0.01 mm, suction or discharge process is terminated.

Numerical simulation of heat transfer requires high quality mesh, especially close to the walls, where the heat transfer coefficient is calculated. The value of y^+ should be kept around 1, however, when the fluid flow is not static, it is difficult to keep the high mesh quality and not to enormously increase the computational requirements. In this simulation the value of y^+ was checked on each surface during the whole simulation and the average value was mostly below ten. Also the convergence of the momentum equations, continuity equation and energy equation did not overcome the 10^{-4} criteria.

The analysis of Müllner and Bielmeier (2008) shows that most of the heat flows between the gas and piston (around 50 %), e.g. it is 35 % for cylinder head and less than 5 % goes into the cylinder wall. The rest of the heat is transferred in out-walls, which are walls around the valve. Similar results were obtained in this paper despite the fact that boundary conditions and settings were not described in Müllner and Bielmeier (2008). The results in Table 4 show the area integrated heat fluxes during one cycle of crankshaft.

Table 4: Percentage distribution of total heat flux from one cycle of crankshaft

	piston	wall	head
full cycle	67	-4	37

If the compressor works in steady conditions, after a certain point the temperature of all walls should reach certain value, which depends on the heat flow inside the cylinder and outside as well. Heat flux distribution shows that these temperatures may not be equal.

Nevertheless, the results in Table 4 take into account both positive and negative values of heat flux, which could bring some inaccuracy. The point of direction change is different for different boundary conditions. To overcome this problem, it is essential to divide the heat flux by the points of direction change. This approach makes it possible to understand the heat flux distribution during each phase of crankshaft cycle.

Heat flux history through the one cycle of crankshaft for a simplified geometry is displayed in Figure 3. Suction process and discharge process are positive and negative, respectively. There is no change of heat flux direction. Another situation occurs for compression and expansion process, when the direction changes, hence the process was divided in two parts – positive and negative.

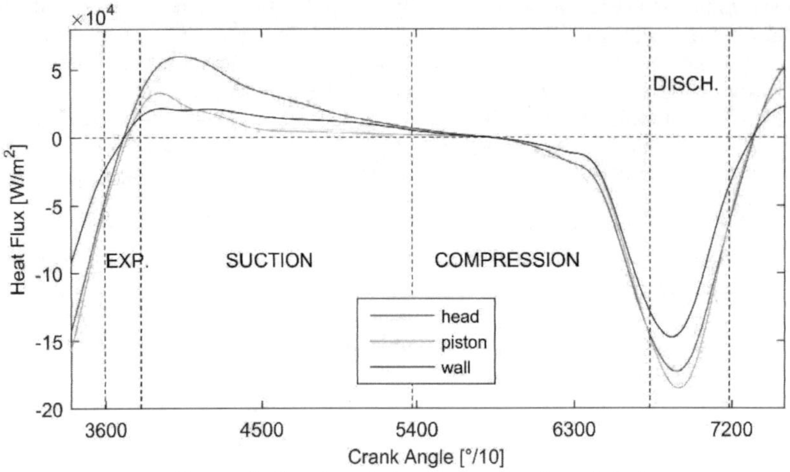

Figure 3: Heat flux during one cycle of crankshaft

The results in Table 4 emphasize the importance of the surface of piston and cylinder head at the expense of cylinder wall surface. Yet the surface of cylinder wall is particularly important not just during the suction phase, but also during the compression. The reason of that is the largest heat transfer area. As the flow velocity increases around the valves, it brings more importance to cylinder head as well.

When the heat flux direction is changed, the importance of cylinder wall decreases rapidly. The least heat is transferred through the cylinder wall during the expansion, which is expected as a consequence of piston motion. Cylinder head plays a stable part in heat flux distribution as the amount of transferred heat is always around 40 % except for positive compression. Heat flow through the piston wall contributes mostly during the discharge and expansion phase, when the velocities are considerably high in a region close to the piston surface. When the piston is moving towards bottom dead center, it has the biggest distance to the port of incoming gas. Moreover, it moves in the same direction as the incoming gas reducing the velocity of gas in a region close to the surface.

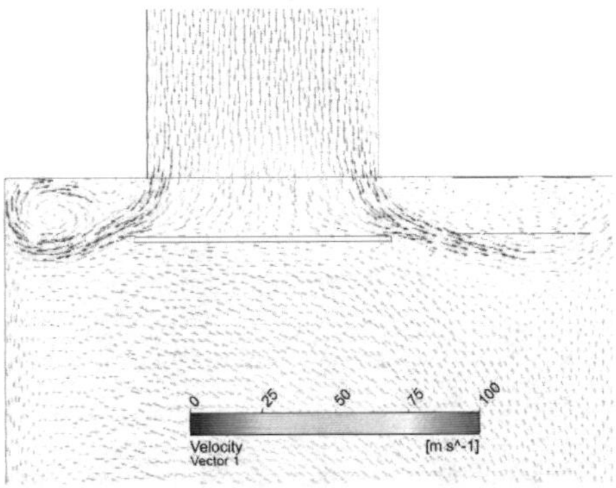

Figure 4: Velocity inside the cylinder (Crank A. 803°)

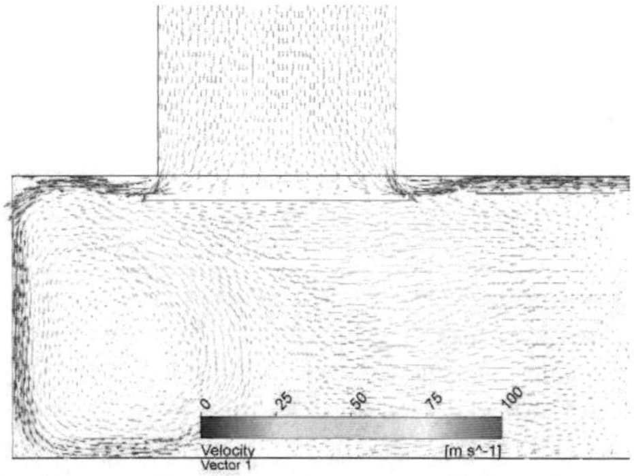

Figure 5: Velocity inside the cylinder (Crank A. 775°)

CFD analysis provides with more detailed flow field inside the cylinder including the velocity and temperature distribution, which helps to divide the heat flow more accurately among the surfaces. On the other hand, the computational cost is enormous, which is not convenient during the development process. Due to the complicated use of CFD, the objects of interest are simplified tools, such as the one presented in section 3.1. The results of complex analysis can be used in 0d model, which uses integral correlations to predict heat flux. However, it is distributed just according to the current heat transfer area. As the heat transfer area is the same for all the models, distribution of heat flux is consequently almost the same as well. The values used in Table 5 are the results of Disconzi correlation.

Table 5: Percentage distribution of heat flow among the surfaces from numerical simulation

	CFD simulation				0d simulation		
	piston	wall	head		piston	wall	head
expansion							
positive	56	6	38		46	13	41
negative	55	1	45		51	3	46
suction	17	44	40		17	69	15
compression							
positive	8	73	19		10	80	9
negative	29	32	39		25	53	22
discharge	53	10	37		40	24	36
TOTAL +	19	43	39		16	69	15
TOTAL -	45	17	38		36	31	33

Good agreement was found for the piston surface, where the same trend can be observed – minor share on total heat flux during suction and compression and higher during discharge and expansion process. The heat flux through the cylinder wall is overestimated. The surface of cylinder wall does not play such an important role as it is predicted by 0d model, especially during the suction process. CFD predicts lower heat flux through the cylinder wall than 0d simulation and vice versa for cylinder head. During discharge process the heat flux from cylinder wall is higher again at the expense of the piston. All of this could be the consequence of high velocities of gas around the cylinder head and piston, which increase the heat flux on these surfaces.

The comparison of results from the 0d model and the CFD analysis shows that one integral correlation for all surfaces inside the cylinder is not sufficient to properly describe distribution of heat flux. The same idea of using different correlation for each process of compressor was already applied by Disconzi *et al.* (2012), therefore it seems reasonable to use different correlation for each surface inside the cylinder as well. Usually there are three main surfaces, as it is displayed in the Figure 1. The surfaces of valves can be neglected as they are much smaller than other surfaces. Models mentioned in section 2.1 show important differences, as it was mentioned in Tuhovcak *et al.* (2015) and finding correct one is a difficult task. However, some correlations show better agreement with numerical simulations than others for particular surface. Therefore, the combination of well-known correlations could help to achieve higher accuracy in heat flux predictions for simplified models used for compressors.

In the pictures below (Table 6), a comparison of particular correlations with CFD results can be found. The comparison is divided according to heat transfer areas – head, piston and wall and shows area integrated heat fluxes. Integral correlations offer only one calculation of heat transfer coefficient for all surfaces, therefore the curve produced by each correlation is the same for head, piston and wall. On the other hand, the results from numerical analysis show area integrated heat fluxes at each surface calculated from local heat transfer coefficient. Discontinuities occurring at curves from CFD are the consequence of remeshing and interpolation results from previous time step on new mesh. Even with more iteration within one time step they did not disappear.

Most of the models underestimate the heat fluxes, especially during the suction phase and at cylinder head. The same problem occurs for the piston surface, only the model of Annand predicted results in reasonable agreement with CFD simulation. For the surface of cylinder wall, only the model of Disconzi shows a sufficient accuracy. The model of Adair predicts much lower heat flux for all surfaces, especially during the suction and discharge phase.

Table 6: Comparison of heat flow for particular empirical correlations and numerical simulation

	Head	Piston	Wall
Adair	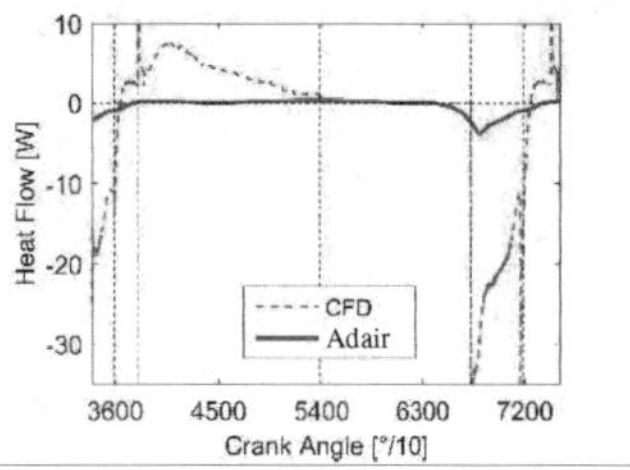	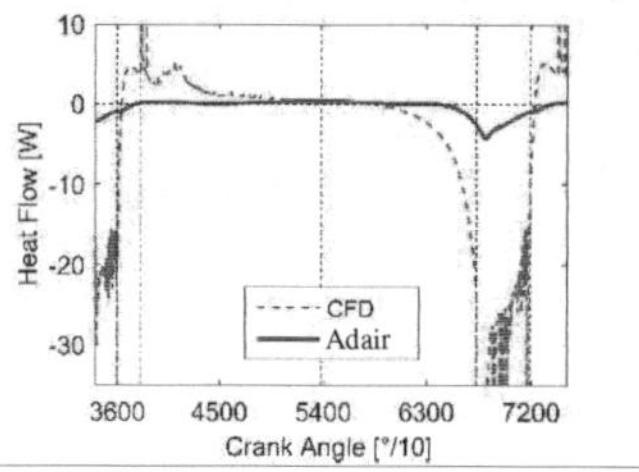	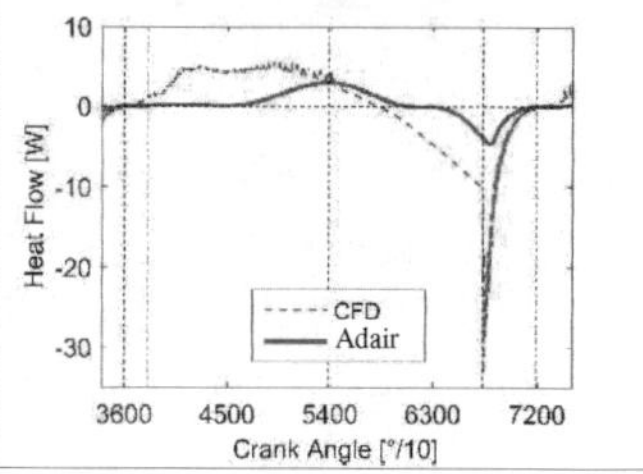
Annand	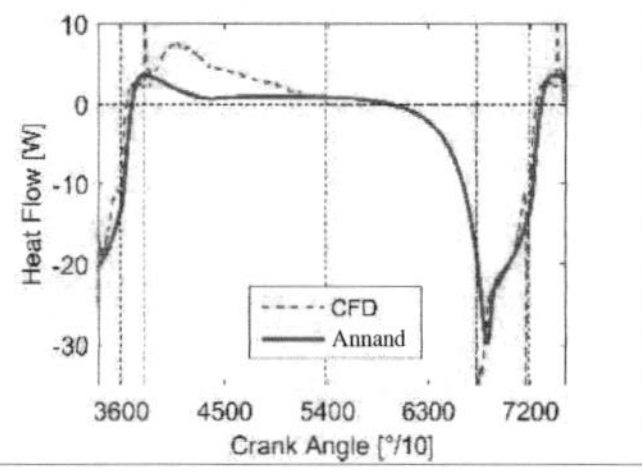	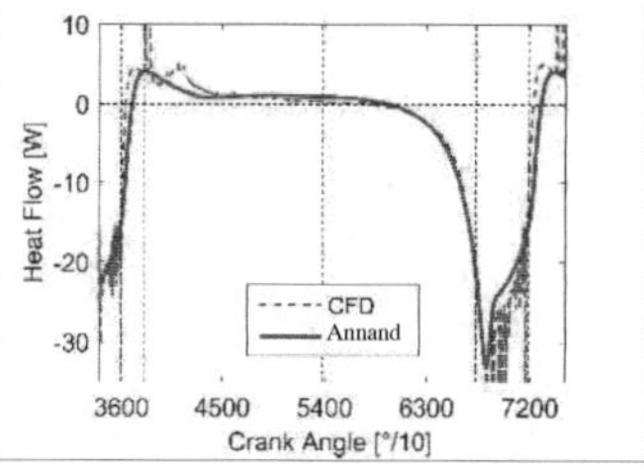	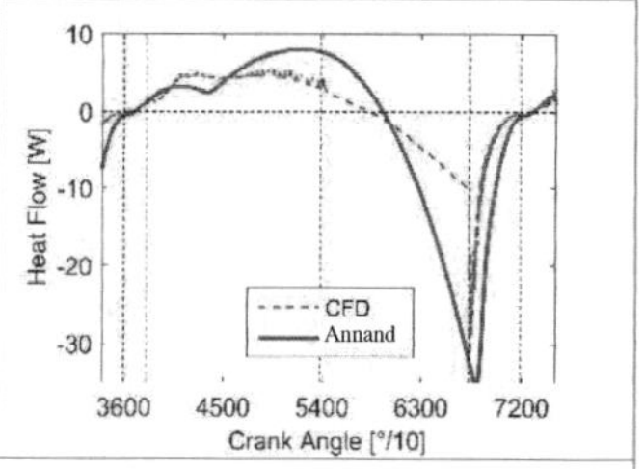
Disconzi	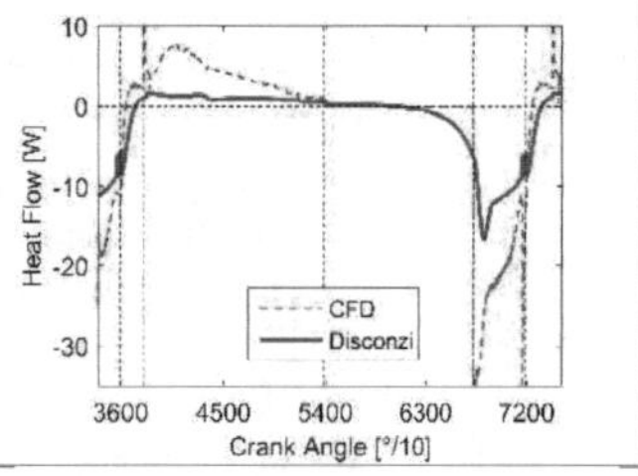	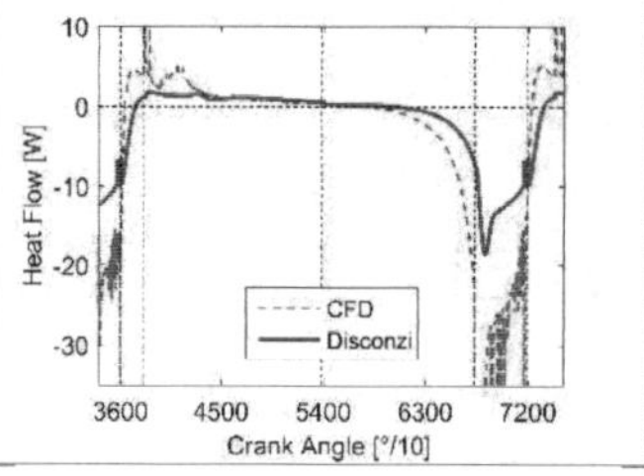	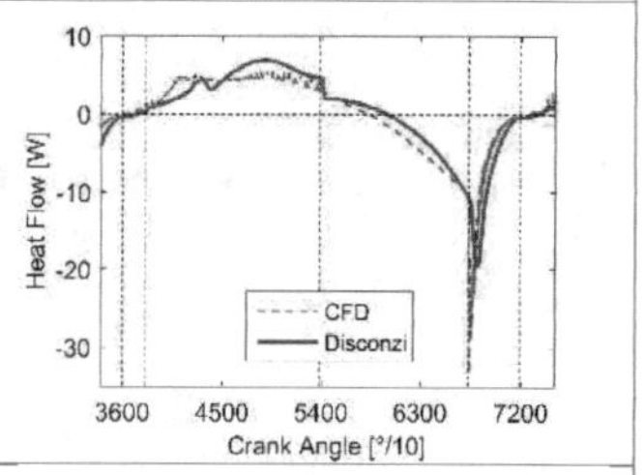
Woschni	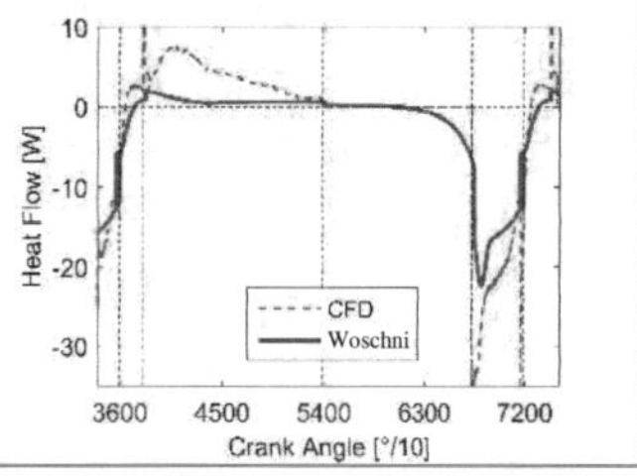	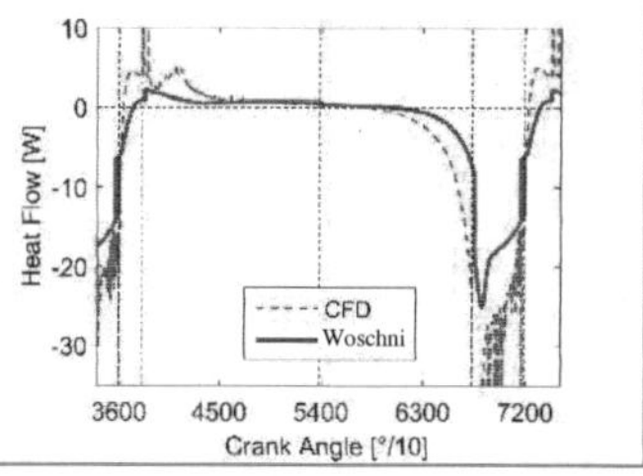	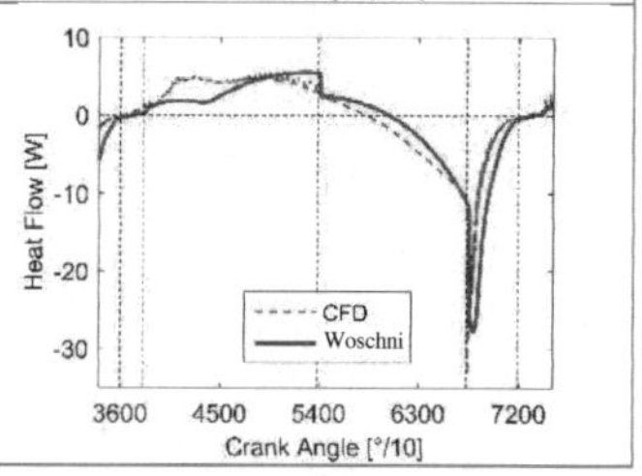

As it was mentioned before a combination of known models could help to increase the accuracy of heat flux prediction in simplified tools. Each process and surface would have its own correlation, based on the comparison with CFD numerical simulation. The combination of models based on presented results can be found in Table 7. On the left side there is a ratio of heat flux predicted by correlations and CFD. On the right side there are models which showed best agreement with CFD simulation. In most of the cases the models of Annand and Disconzi appear, however there is also contribution of Adair and Woschni. The results show a big difference between integral correlations and numerical model for heat transfer prediction in CFD. More investigation is necessary, especially in CFD, to perform simulation with higher accuracy.

It must be pointed out, that all of the presented models were developed either for combustion engines (Woschni, Annand) or for refrigeration compressor (Disconzi, Adair). The compressor used in this paper was working with air, which could have influenced the results. Another important thing is the CFD simulation, particularly the simulation of heat transfer, which is very demanding on mesh quality close to heat transfer surface.

Table 7: Heat flux ratio of empirical correlations and numerical simulation

	piston	wall	head		piston	wall	head
expansion				**expansion**			
positive	68	166	90	*positive*	Annand	Annand	Annand
negative	99	134	110	*negative*	Woschni	Adair	Woschni
suction	88	110	33	**suction**	Annand	Disconzi	Annand
compression				**compression**			
positive	148	121	69	*positive*	Disconzi	Disconzi	Woschni
negative	145	90	96	*negative*	Annand	Disconzi	Annand
discharge	77	134	100	**discharge**	Annand	Disconzi	Annand

5. CONCLUSION

Heat transfer inside the cylinder effects the efficiency of compressor. Most of the simplified simulation tools use integral correlation to predict heat flux. This paper presented a comprehensive comparison of most common models with more complex numerical simulation, which includes also the solution of velocity field inside the cylinder. Velocity close to the heat transfer areas directly affects the heat transfer coefficient and consequently the heat flux. The compressor cylinder was divided into three main surfaces in order to see the distribution of transferred heat among them. The results show, that this distribution based just on the area of particular surface could decrease the accuracy. Solution of this problem can be found in different correlations for each surface in the cylinder and each process of the compressor (expansion, suction, compression and discharge). This would increase the number of equations for heat transfer prediction, however these equations are solved analytically and therefore, their computational costs are very low. More simulations are needed for verifying suggested combination of correlations, especially for different fluids and working conditions.

NOMENCLATURE

A_c	cross section of the cylinder	[m^2]
C_f	friction coefficient	[-]
D	characteristic dimension	[m]
De	equivalent diameter	[m]
d	damping constant	[-]
F	force acting on valve	[N]
h	specific enthalpy	[J/kg]
k	spring stiffness	[N/m]
m	mass	[kg]
p	pressure	[Pa]
Q	heat transfer rate	[J]
S	flow area	[m^2]
u	velocity	[m/s]
u	specific internal energy in eq.6	[J/kg]
u_c	velocity through the valve	[m/s]
u_{fric}	friction velocity	[m/s]
u_p	mean piston velocity	[m/s]
U	internal energy	[J]
V	volume	[m^3]

x	valve position	[m]
y^+	distance from the wall	[-]
y^*	dimensionless coordinate	[-]
w	specific work	[J/kg]
W	work	[J]
α	crank angle	[°]
κ	adiabatic exponent	[-]
ρ	density	[kg/m^3]
ζ	loss coefficient	[-]
ν	kinematic viscosity	[m^2/s]

Subscript

cyl	cylinder

REFERENCES

Aigner, R., & Steinrück, H. (2007). Modelling Fluid Dynamics , Heat Transfer and Valve Dynamics in a Reciprocating Compressor. In *5th Conference of the EFRC* (pp. 171–180).

Almbauer, R., Lang, W., Nagy, D., & Berger, E. (2010). Application of Computational Fluid Dynamics for the Thermodynamic Development of a New Generation of Hermetic Reciprocating Compressor. In *International Compressor Engineering Conference* (pp. 1–7).

Bejan, A. (2013). *Convection heat transfer* (4th ed.). New Jersey: John Wiley & Sons.

Bell, I. H., Wronski, J., Quoilin, S., & Lemort, V. (2014). Pure and pseudo-pure fluid thermophysical property evaluation and the open-source thermophysical property library coolprop. *Industrial and Engineering Chemistry Research, 53*(6), 2498–2508. doi:10.1021/ie4033999

Costagliola, M. (1950). The theory of spring-Loaded Valve for Reciprocating Compressor. *Journal of Applied Mechanics, 17*, 415–420.

Disconzi, F. P., Pereira, E. L. L., & Deschamps, C. J. (2012). Development of an In-Cylinder Heat Transfer Correlation for Reciprocating Compressors. In *International Compressor Engineering Conference* (pp. 1–10).

Farzaneh-Gord, M., Niazmand, A., Deymi-Dashtebayaz, M., & Rahbari, H. R. (2014). Thermodynamic analysis of natural gas reciprocating compressors based on real and ideal gas models. *International Journal of Refrigeration*. doi:10.1016/j.ijrefrig.2014.11.008

Kader, B. A. (1981). Temperature and concentration profiles in fully turbulen boundary layers. *Int.J.Heat Mass Transfer, 24*(9), 1541–1544.

Morriesen, A., & Pereira, E. L. L. (2009). Numerical Prediction of Superheating in the Suction. In *20th International Congress of Mechanical Engineering*.

Müllner, T., & Bielmeier, O. (2008). Heat transfer in small piston compressors. In *International Compressor Engineering Conference* (pp. 1–8).

Tuhovcak, J., Hejcik, J., & Jicha, M. (2015). Modelling of fluid flow and heat transfer in a reciprocating compressor. In *IOP Conference Series: Material Science and Engineering* (Vol. 012018). doi:10.1088/1757-899X/90/1/012018

ACKNOWLEDGEMENT

The presented research was supported by the Brno University of Technology project FSI-S-14-2355.

A 3-D Transient CFD Model of a Reciprocating Piston Compressor with Dynamic Port Flip Valves

Sujan DHAR[1]*, Hui DING[1], Jônatas LACERDA[2]

[1]Simerics, Inc,
Bellevue, WA, USA
Phone – 425-502-9978, Fax – 425-502-9978, E mail – sd@simerics.com

[2]Tecumseh Products Company,
São Carlos, SP, Brazil

* Corresponding Author

ABSTRACT

In this paper a full transient 3-D CFD model of a complete single piston compressor system along with dynamic inlet and outlet port flip valves is presented. The model includes the motion of the cam driven piston, and also predicts the motion of the dynamic flip valves. All activities related to model creation, simulation and post-processing were performed within the simulation software Simerics–PD (also known as PumpLinx).

The mesh creation and mesh motion for the piston chamber and the flip valve volumes were performed using meshing algorithms implemented within the simulation software. In addition, a novel mesh motion algorithm was implemented to enable a fully coupled simulation of the inlet port flip valve and the piston surface: which have intersecting swept volumes and may interfere during the operation of the compressor. The entire simulation was performed in a fully coupled manner in which the motion of the valves is calculated within Simerics–PD during the transient CFD simulation. This is achieved by a Fluid Structure Interaction (FSI) methodology where the pressure distribution acting on the valve is used to calculate the valve motion using a robust ordinary differential equation (ODE) solver. Important input parameters of the rotational dynamics (such as torsion constant) of the flip valve were calculated from the beam bending theory. In addition a Full Closure model for flip valves implemented within Simerics–PD ensures that there is no leakage through the valves when they are in the fully closed position, which are essential for accurate compressor simulation.

In the present work, detailed descriptions are provided of the model creation and simulation methodology for an industrial single piston compressor system. The model creation was performed on the real world geometry without simplifications and real-gas fluid properties were used for the simulation. The fully coupled 3-D transient CFD simulation of the piston compressor and the flip valves is shown to have excellent simulation speed and several relevant results are presented. The model set-up and performance of the simulation are demonstrated to be robust and user friendly and the algorithms developed can be readily applied to single piston compressor systems with dynamic port valves.

1. INTRODUCTION

Single piston compressors are popular devices used in refrigerators and air conditioners. While known to be high efficiency machines, international regulations have created a need for the efficiency and performance of such machines to be further improved. Traditional methods of improving the design and efficiency of such machines depend on a trial and error approach in which several iterations of design improvement and experimental testing are performed. However, such a method can be expensive as well as time-consuming and recent effort to minimize the environmental impact of refrigerators and air conditioners as well as an increase in available computing power has led to a focus on computational techniques to improve the design of such machines. Most of the effort in the modeling of reciprocating compressor systems has been using one or zero dimensional models. Soedel (1972) and Ussyk (1984) studied the reciprocating compressor using an integral formulation to represent the cylinder fluid properties. The valves dynamics were calculated using free modes superposition and the mass flow through the valves using discharge coefficients obtained experimentally. Further improvements in representing the valve movement was conducted by Lopes (1996) and Matos et al. (2000, 2006) modeling a radial diffuser as a one- degree of freedom mass-spring model and the fluid domain with a two-dimensional compressible finite volume methodology also including turbulence effects. The suction reed valve, in particular, has been studied in individual experimental work such as Arantes et al. (2013) as well as computational work like by Bhakta et al. (2012), Suh et al. (2006) and Nagy et al. (2008). The valves have also been studied using an immersed boundary method approach by Lacerda and Gasche (2010), Rodrigues and Gasche (2011) and Barbi et al. (2012) where a virtual physical model was created using a two-dimensional fixed walls radial diffuser, a two-dimensional moving wall radial diffuser and three-dimensional moving reed valve.

However, since the compressor and port valves interact with each other during its operation and the valve dynamics have a large role to the play in the efficiency and performance of the compressor system, to have a useful simulation model the piston compressor and the valves should be simulated in a fully coupled manner. Very few such CFD models of a combined piston compressor valve system exist in literature. In 2006, Birari et al. published a 3D CFD model of a reciprocating compressor, without considering the inlet or the exit valves or their dynamics. Pereira et al. (2010) presented a 2D axisymmetric model of the compressor along with suction and discharge valves, and in 2012 the same group published a 3D CFD model of a simplified compressor with only the suction valve modeled in a simplified form. In 2010, Kinjo et al published a CFD model of a reciprocating compressor where both the suction and discharge valves were modeled dynamically. However, the geometry for this problem was simplified to a 2D approximation of the real geometry.

While the above studies approach a complete high fidelity model of the reciprocating compressor system, all of them have significant approximations which limit their usability to study real-world industrial problems. In addition, the CFD setup for reciprocating compressor simulations can be quite involved using complex mesh motion algorithms as well as variable time step models in order to preserve the stability of the simulation which makes them quite cumbersome for CFD users. Over the past several years the commercial CFD software Simerics-PD (also known as PumpLinx) has been developed keeping in mind positive displacement (PD) fluid machines such as pumps and compressors, and several 3D CFD transient models of PD compressors have been published over the last few years such as scroll compressors (Gao and Jiang, 2014), rolling piston compressors by (Ding and Gao, 2014) and twin screw compressors by (Kovacevic et al., 2014). The present work addresses the two issues discussed above by implementing a template based 3D CFD meshing and solution approach for reciprocating compressors. In order to be useful for industrial compressor design, the present work considers the entire compressor system with full geometric fidelity, and both port valves are modeled dynamically using a fluid structure interaction (FSI) algorithm. The model formulation, meshing technique and simulation results will be discussed in the following sections.

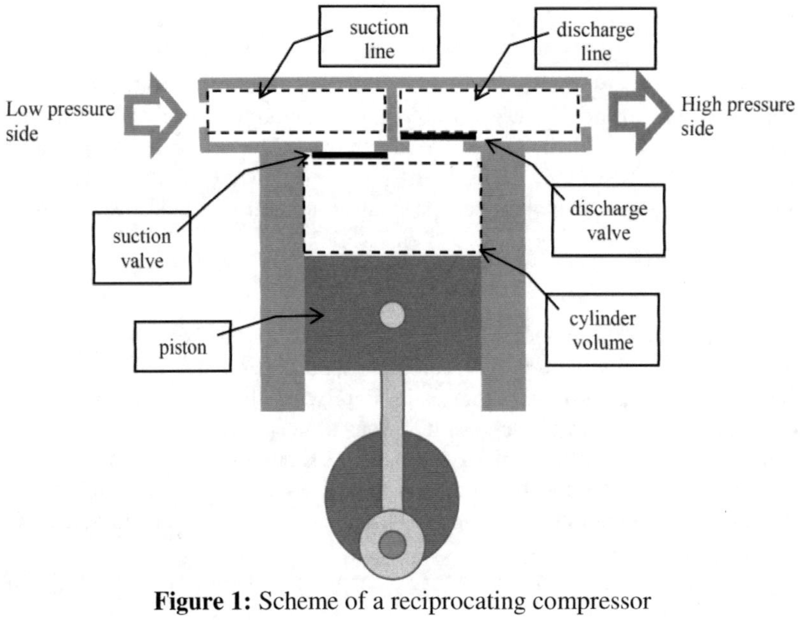

Figure 1: Scheme of a reciprocating compressor

2. RECIPROCATING COMPRESSOR MODEL

In the present section the working principle of the full reciprocating compressor will be described and details of various steps involved in setting up the model such as mesh creation and calculating dynamic parameters for the valves will be discussed.

2.1 Working Principle of the Reciprocating Compressor

The reciprocating compressor typically has three individual moving parts: the piston which has a well-defined prescribed kinematic, and the inlet and the outlet valves which open dynamically based on the pressure forces acting on them. Reciprocating compressor volumetric and energetic efficiency is very dependent on the valve system design. Usually, it uses a reed valve whose dynamics are controlled by the balance between the opening force caused by the pressure difference upstream and downstream the valve and the closing force, like a spring restitution, due to its deflection. A good valve design should have quick response, high mass flow, low pressure drop, avoid backflow and be reliable during extreme operation conditions. Since the flow field affects the structure of the valve and vice-versa, it is considered a fluid-structure interaction problem. Besides the valve system with the suction and discharge valves, the reciprocating compressor is composed by a cylinder, a suction line and a discharge line, as shown in Fig. 1. The cylinder volume variation is due the linear displacement of a piston connected in a crank slider mechanism moved by an electrical motor. The compressor efficiency is the ratio between the cooling capacity and the consumption energy necessary to perform the work.

2.2 Model Creation

The CFD model of the compressor is created starting from the complete solid CAD of the compressor, including all inlet and outlet porting as well as real valve geometry as shown in Fig. 2. The fluid volume was extracted, and the resultant volumes were used for mesh creation.

Figure 2: The CFD model is created based on the solid CAD model of the compressor (top). The extracted fluid volume includes the suction and the discharge valves (shown in green and red).

A view of the entire domain mesh is shown in Fig. 3. For the ports, an unstructured 3D hex-dominant Cartesian mesh was created using Simerics Binary Tree mesh generator. Further details follow about the mesh of the outlet valve, as well as the meshes of the suction valve and the piston chamber.

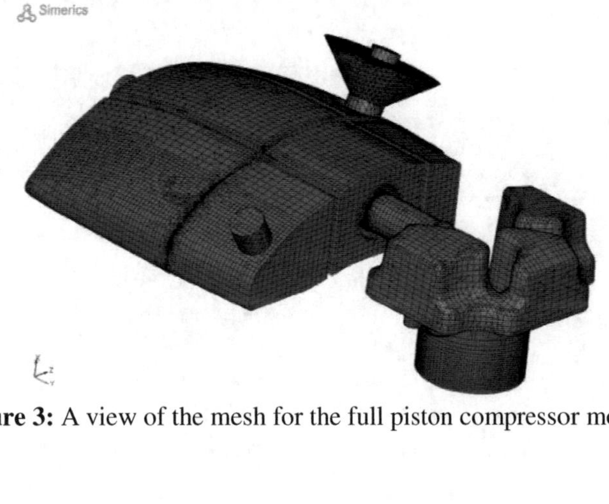

Figure 3: A view of the mesh for the full piston compressor model.

2.3 Piston Compressor, Suction and Discharge Valve Mesh

The discharge valve mesh was created using the specialized Simerics valve template mesher previously described in Ding and Gao (2014). The template mesher creates a structured mesh considering the possible path of motion of the valve, and during simulation the mesh is deformed appropriately to account for the valve motion as shown in Fig. 4.

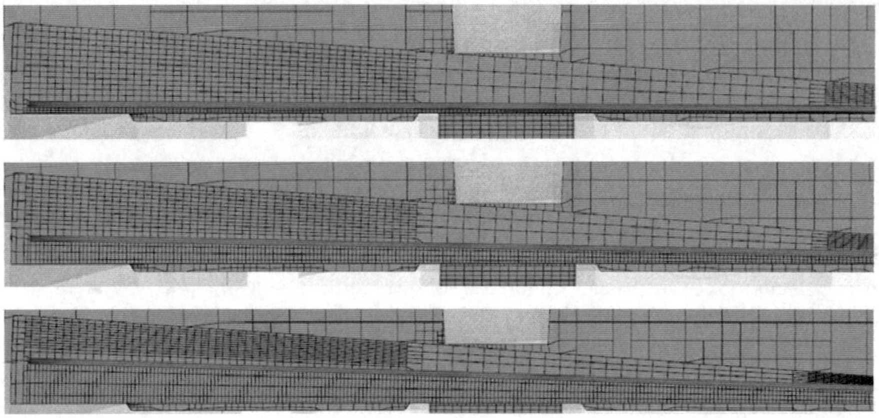

Figure 4: A view of a section through the discharge valve with the valve (in red) at different positions of opening

In reciprocating compressors the swept volumes of the suction valve and the piston head intersect since the suction valve opens into the compression chamber. In order to model this correctly a mesh deformation algorithm was implemented within Simerics-PD. The initial mesh for the valve and the remainder of the chamber volumes were created separately and connected using an implicit mesh interface. During the operation of the compressor the piston motion is well defined, but the valve motion is determined by the dynamics of the machine. In order to ensure that the both motions are captured accurately two mesh deformation zones are defined. These are shown in Fig. 5: Zone1 deformation captures the valve motion, while Zone 2 deformation captures the motion of the piston. The location of separation between the two zones is represented by the yellow line and this separation location is updated automatically during the simulation based on the separation between the suction valve and the piston.

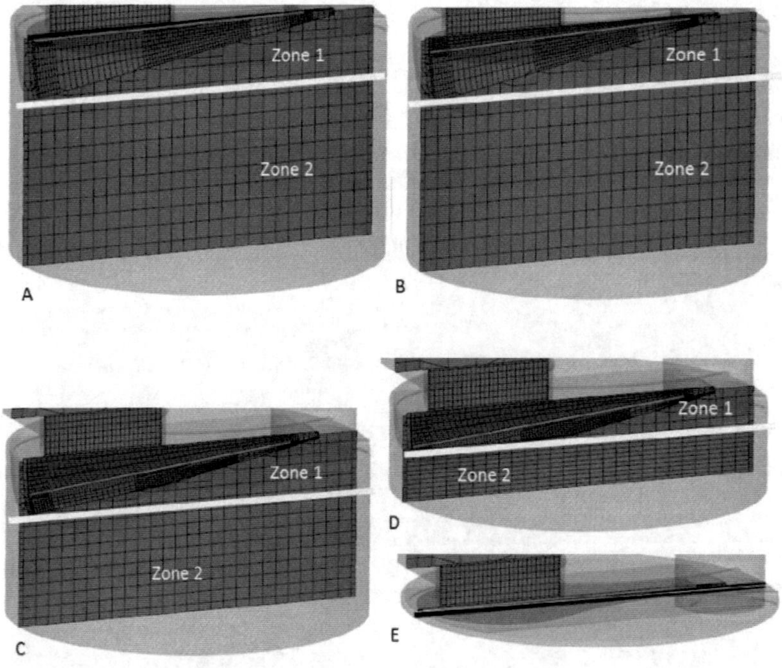

Figure 5: A section view through the combined mesh of the suction valve and piston compressor chamber. The mesh deformation algorithm is demonstrated.

In Fig. 5a the valve is in the closed position while the piston is close to the bottom dead center. Through Fig. 5b to d the gap between the piston and the valve starts to reduce but the mesh separation into zones ensures that both motions are captured fully. Finally in Fig. 5e the piston is essentially at the top dead center, with the valve fully closed and the gap between the piston and the valve is of the order of microns. Thus, it can be seen that due to the mesh deformation algorithm implemented for reciprocating compressors within Simerics-PD, the intersection of the swept volumes of the suction valve and the piston can be modeled. It must be noted at this point that the initial mesh created is only deformed to simulate the machine and that there is no additional re-meshing step which would add to the simulation time and reduce simulation efficiency.

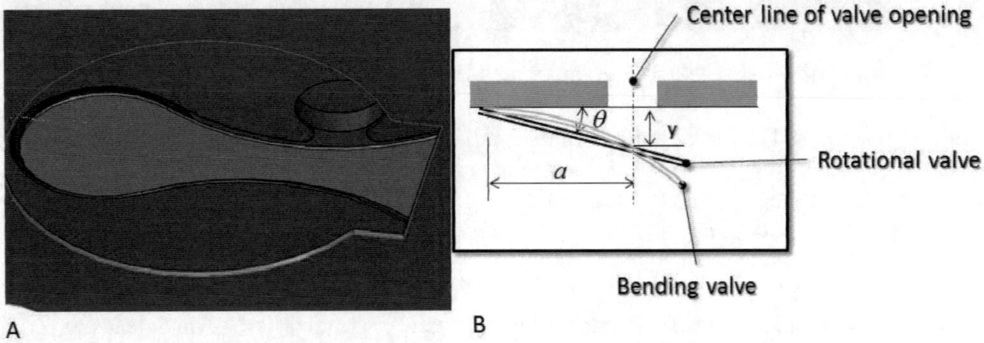

Figure 6: A) Suction valve geometry and B) rotation vs bending of the reed valve

2.4 Discharge and Suction Valve Dynamics

The suction reed valve is shown in Fig. 6: it is made of a thin sheet of metal with one end fixed and the other end free. The valve will bend under fluid force, and create an opening for the fluid to flow. Due to strong interaction force and small inertia of the reed valve the coupling between the piston compressor and reed valve becomes a very stiff FSI system to solve. A semi-implicit proprietary procedure is applied to eliminate the requirement of extremely small time steps of explicit coupling methods. In the present work both the suction and the discharge valve are modeled as reed valves, where the opening of the valve is captured using an equivalent rigid body rotation for the bending motion.

For the valves the dynamics are solved using a torsional mass and spring system using the following ODE:

$$I\frac{d^2\theta}{dt^2} + C\frac{d\theta}{dt} + k\theta = \tau(t) \quad (1)$$

Where θ is the valve opening angle starting from the valve closed position, I is the moment of inertia, C is the rotational friction, k is the coefficient of torsion spring, τ is the torque from fluid forces, and t is time. For the suction valve the values of I and k need to be chosen based to obtaining an equivalent solution to the analytical solution for cantilever beam bending as in Ding and Gao (2014). This is a reasonable simplification, because for a reed valve that covers a small port opening, we can assume it to be a cantilever beam with a concentrated load at the center of the opening. In this case, the deflection at the opening is given by

$$y = \frac{Pa^3}{3EI} \quad (2)$$

where P is the load, a is the length from fixed end to the center of the opening, E is the modulus of elasticity, and I is the area moment of inertia of the reed valve. From Fig 6b, we can see that $\theta \approx \frac{y}{a}$ and the torque from the load at the opening can be calculated by $T = Pa$, finally leading to

$$k = \frac{3EI}{a} \quad (3)$$

This value, in addition to the moment of inertia of the valve (can be calculated from any CAD package) can be then used in Eq.1. By using the above method, with carefully chosen dynamic parameters of the flip valve, the port opening center can have the same deflection as a bending valve under the same load.

For the discharge valve a similar procedure is followed, but with some differences since the discharge valve is fixed at two ends due the presence of a retainer, as shown in Fig.7. Therefore, for this valve the values for Eq.1are chosen carefully to correspond with the analytical solution of a beam supported at both ends, rather than the cantilever beam bending solution.

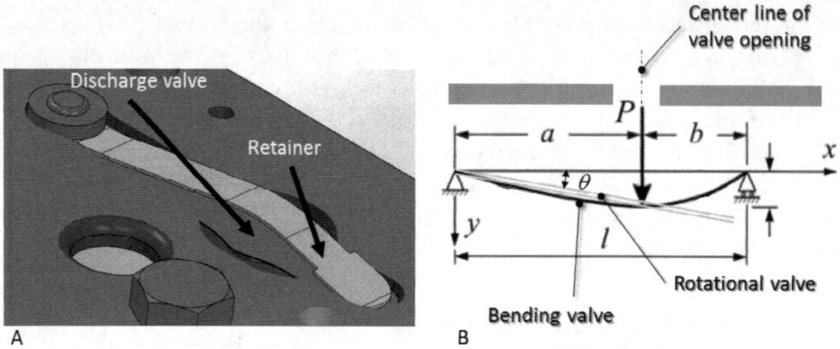

Figure 7: A) Discharge valve geometry and B) rotation vs bending of the reed valve

As for the suction valve, the valve opening area is quite small, and the force on the valve can be approximated as a point load P acting through the center of the valve opening area at a distance a from one end as shown in Fig.7b. The resultant deformation at the point of load application is

$$y = \frac{Pba}{6lEI}(l^2 - a^2 - b^2) \tag{4}$$

where l is the total length of the valve, and $b=l$ - a. Since the opening is quite small, $\theta \approx \frac{y}{a}$ and the torque from the load at the opening can be calculated by $T= Pa$, and k for Eq.1 can be written as $k=\frac{T}{\theta}$ which leads to,

$$k = \frac{6EIl^3}{ab^2(l-a)} \tag{5}$$

As before, this calculated value along with the moment of inertia of the valve can be then used in Eq.1 to provide a realistic dynamic simulation of the valve where the opening of the reed valve along the line of load application is same as that of the bending valve.

3. CASE STUDY

An industrial reciprocating piston compressor from Tecumseh Products Company with both port valves was simulated to demonstrate the capability of the approach described in the previous sections. The simulation was performed using the real gas properties of the refrigerant R134a. Conservation equations of mass, momentum, and energy of a compressible fluid were solved using a finite volume approach. The standard $k - \varepsilon$ two-equation model (Launder and Spalding, 1974) was used to account for turbulence. In addition all the properties of the R134a: density, viscosity, heat capacity and conductivity are considered as a function of pressure and temperature.

The equation for the kinematics of the piston was set as follows:

$$x = -\sqrt{(r1 + r2)^2 - e^2} - r1\cos(\omega t) - \sqrt{(r2^2 - (r1\sin(\omega t) - e)^2} \tag{6}$$

Where r1 is the crank eccentric length, r2 is the con-rod length and e is the eccentric diameter. The rotational speed of the piston ω, is set as 3600 RPM. The boundary conditions of the model were set as shown in Fig.8.

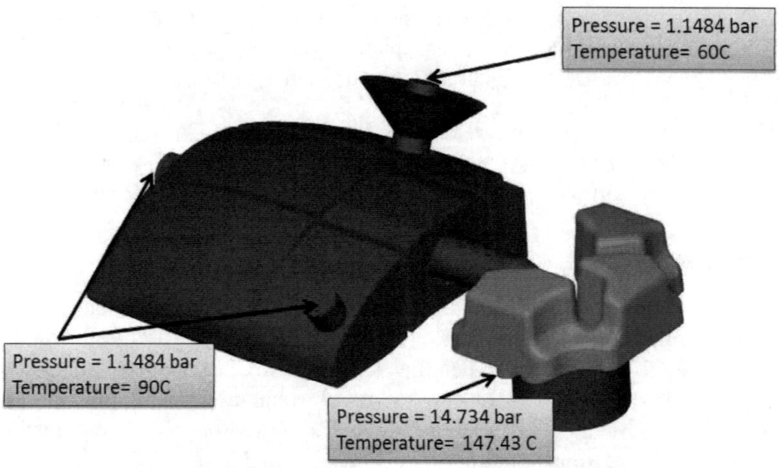

Figure 8: Model boundary conditions

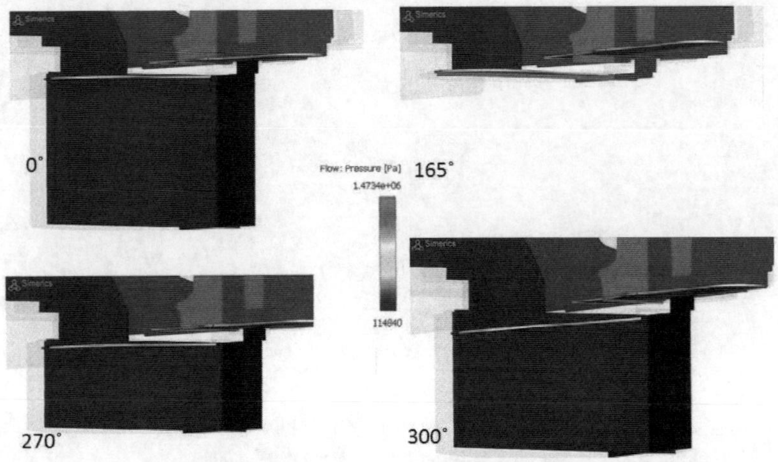

Figure 9: Pressure inside the compressor at different point during one cycle of the piston movement. The discharge and the suction valve solids are both shown in yellow.

A transient simulation was performed and the pressure inside the compressor at different crank angle is shown in Fig.9 with the help of two section planes which pass through the suction and the discharge valves. It can be seen that at 0°, the piston is at bottom dead center and both the suction valve is still slightly open (but just about to close) and the discharge valve is shut. At close to top dead center 165°the chamber is at high pressure and the discharge valve is open. As the piston goes through its suction stroke at 270° and 300° the pressure in the chamber falls and the suction valve opens. A similar view of the temperature inside the compressor is shown in Fig.10. It can be generally seen that as the piston gets closer to the top dead center, the temperature in the chamber rises, while the temperature remains low during the suction stroke.

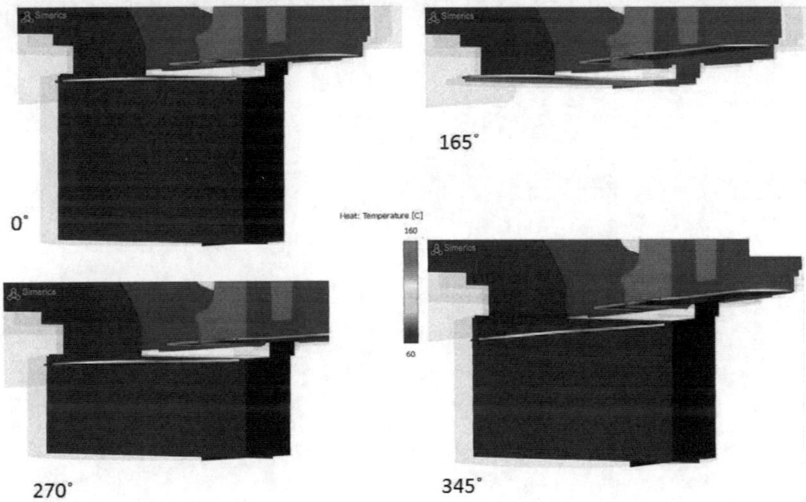

Figure 10: Temperature inside the compressor at different point during one cycle of the piston movement. The discharge and the suction valve solids are both shown in yellow.

The pressure-volume (P-V) diagram for the compressor is shown in Fig.11. The suction and the discharge pressures along with the overpressure and the underpressure are also represented. The underpressure and overpressure areas represent the spent energy in the suction and discharge process respectively, while the P-V diagram can be used to find the efficiency and the power consumption of the compressor. At piston TDC the gap between the piston and the cylinder head is 100 microns, and this small dead volume is reflected on the P-V diagram.

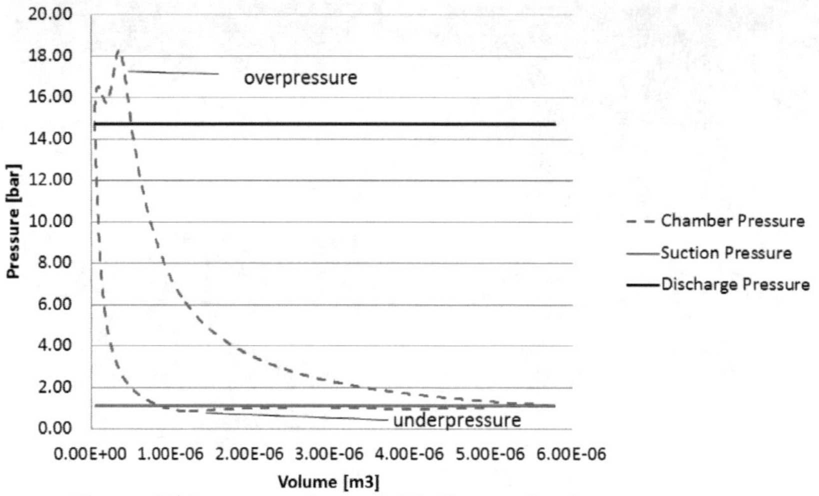

Figure 11: Pressure-volume (P-V) diagram for the compressor

The motion of the suction and the discharge valves as predicted by the model (according to the dynamic force balance in Eq.1) along with the chamber volume are plotted in Fig.12. As can be seen, the results are periodic between the cycles. The trends which were observed in Fig. 9 and 10 can be examined now in more detail and it can be seen that the outlet valve opens for around last 30 degrees of the compression half of the cycle, and is closed by the time the suction stroke starts. During the suction stroke, the suction valve opens about 40 degrees of crank rotation after the top dead center. The suction valve exhibits a valve flutter during the suction stroke, and closes about 30 degrees into the compression stroke. In general it was seen that due to the presence of the retainer the discharge valve shows a motion with much higher stiffness, which makes it open lesser than the suction valve and also closes faster. The suction valve has a much lower stiffness, and therefore can open a much higher degree, and also exhibits a slow valve oscillation.

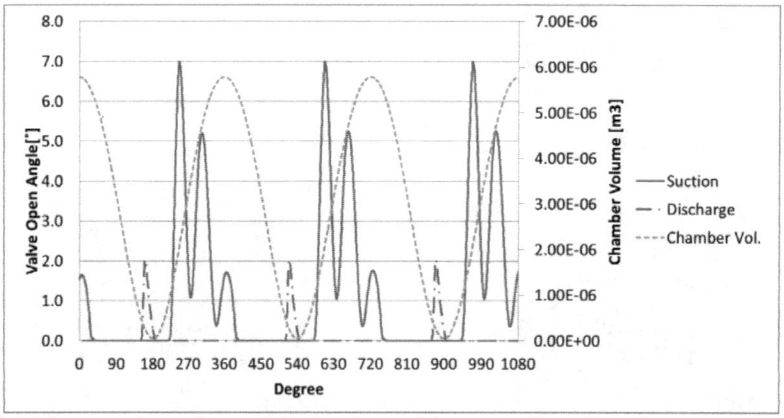

Figure 12: Valve openings along with chamber volume over 3 cycles of compressor simulation

Finally, the discharge and the suction flow rates are plotted in Fig. 13. As expected the flow rates correspond very closely with the valve opening angles with the discharge flow rates showing up in the form of a sharp spike, whereas the suction flow rates are more evenly distributed along the suction stroke. Another important point to note is that the flow rates go to zero when the valves are closed, since the simulation is able to handle valve full closure with essentially zero gaps between the valves and their seats when they are predicted to be fully closed by the dynamics.

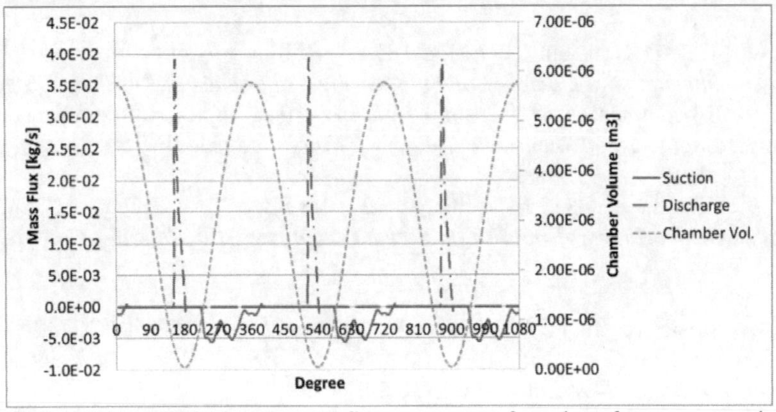

Figure 13: Suction and Discharge mass flow rates over 3 cycles of compressor simulation

Starting from a properly prepared CAD geometry, the meshing and the setup of the simulation takes less than an hour with the help of the Valve templates within Simerics-PD. Simulation time is about 3 hours for one compressor cycle on a standard 8GB Windows laptop with quad-core 2.7GHz Intel i7 processor. Simulation results were seen to be periodic within 3 cycles of the piston.

4. SUMMARY AND CONCLUSIONS

A novel 3D CFD, transient model of a reciprocating single piston compressor was presented in this work. The compressor simulation included fully coupled FSI simulations of the suction and the discharge flip valves. A methodology for creating the single piston compressor model has been outlined. In particular, the meshing of the discharge valve using the Valve template mesh, and the mesh deformation algorithm used to resolve the suction valve piston swept volume interference has been detailed. The procedure to calculate the parameters for the dynamic FSI simulation of both the valves is also outlined. In the simulation all the geometrical details of the compressor are captured and the physical system is modeled accurately without any simplifications unlike in 0D/1D models. In addition, in the current approach the dynamics of the flip valves take into account the accurate pressure distribution on the valve surfaces which are expected to provide much greater fidelity for the valve motion prediction as well as compressor performance and efficiency prediction. The developed methodology was demonstrated on an industrial Tecumseh reciprocating compressor, and the simulation results and the predicted valve motions were found to be physically realistic. Due to the template based meshing techniques and the robustness of flow solver, the setup and simulation is easy and fast. The ease-of-use, quick simulation turn-around time, and valuable information that can be extracted from such a transient 3D simulation makes this novel model a useful tool for the design and analysis of reciprocating single piston compressor systems.

5. REFERENCES

Arantes, D.M., Andreotti, T., and Gasche, J.L. (2013). Experimental Analysis of the Fluid-Structure Interaction in a Model of Refrigeration Compressor Valve. *Proceedings of the International Compressor Engineering Conference at Purdue*. West Lafayette, IN, USA.

Barbi F., Gasche J. L. & Villar, M. M. (2012). An efficient immersed boundary method for solving the unsteady flow through actual geometries of reed valves. *Proceedings of the International Compressor Engineering Conference at Purdue.* West Lafayette, IN, USA.

Bhakta, A., Dhar, S., Bahadur, V., Angadi, S., and Dey, S., (2012). A Valve Design Methodology For Improved Reciprocating Compressor Performance. *Proceedings of the International Compressor Engineering Conference at Purdue*. West Lafayette, IN, USA.

Birari, Y. V., Gosavi, S. S., & Jorwekar, P. P. (2006). Use of CFD in design and development of R404a reciprocating compressor. *Proceedings of the International Compressor Engineering Conference at Purdue.* West Lafayette, IN, USA.

Ding, H., & Gao, H. (2014). 3-D Transient CFD Model For A Rolling Piston Compressor With A Dynamic Reed Valve. *Proceedings of the International Compressor Engineering Conference at Purdue.* West Lafayette, IN, USA.
Gao, H. and Jiang, Y. (2014). Numerical simulation of unsteady flow in a scroll compressor. *Proceedings of the International Compressor Engineering Conference at Purdue.* West Lafayette, IN, USA.

Kinjo, K., Nakano, A., Hikichi, T., & Morinishi, K. (2010). Study of CFD considering valve behavior in reciprocating compressor. *Proceedings of the International Compressor Engineering Conference at Purdue.* West Lafayette, IN, USA.

Kovacevic, A., Rane, S., Stosic, N., Jiang, Y., Lowry, S., & Furmanczyk, M. (2014). Influence of approaches in CFD Solvers on Performance Prediction in Screw Compressors. *Proceedings of the International Compressor Engineering Conference at Purdue.* West Lafayette, IN, USA.

Lacerda, J. F. & Gasche J. L. (2010). Analysis of the flow in hermetic compressor valves using the immersed boundary method. *Proceedings of the International Compressor Engineering Conference at Purdue.* West Lafayette, IN, USA.

Launder, B. E., & Spalding, D. B. (1974). The numerical computation of turbulent flows. *Computer methods in applied mechanics and engineering*, *3*(2), 269-289.

Lopes, M. N. (1996). Uma metodologia numérica para a análise do comportamento dinâmico de válvulas tipo palheta em escoamentos periódicos. Masters dissertation, Federal University of Santa Catarina, Florianopolis/SC, Brazil.

Matos, F. F. S., Prata A. T. & Deschamps, C. J (2000). A numerical methodology for the analysis of valve dynamics. *Proceedings of the International Compressor Engineering Conference at Purdue*. West Lafayette, IN, USA.

Matos, F. F. S., Deschamps, C. J. & Prata A. T (2006). A two-dimensional simulation model for reciprocating compressors with automatic valves. *Proceedings of the International Compressor Engineering Conference at Purdue*. West Lafayette, IN, USA.

Nagy , D., Almbauer, R. A., Lang, W., and Burgstaller, A. (2008). Valve Lift Measurement for the Validation of a Compressor Simulation Model. *Proceedings of the International Compressor Engineering Conference at Purdue.* West Lafayette, IN, USA.

Pereira, E. L., Deschamps, C. J., & Ribas, F. A. (2010). Numerical analysis of heat transfer inside the cylinder of reciprocating compressors in the presence of suction and discharge processes. *Proceedings of the International Compressor Engineering Conference at Purdue.* West Lafayette, IN, USA.

Pereira, E. L., Santos, C. J., Deschamps, C. J., & Kremer, R. (2012). A Simplified CFD Model for Simulation of the Suction Process of Reciprocating Compressors. *Proceedings of the International Compressor Engineering Conference at Purdue.* West Lafayette, IN, USA.

Rodrigues, T. T. & Gasche J. L. (2011). Simulation of the flow through automatic valves of hermetic compressor by the immersed boundary method approach. *Proceedings of the International Conference on Compressors and Their Systems at London* (181-189). London, UK.

Suh, K., Heo, D., & Kim, H. (2006). CAE/CFD application for linear compressor. *Proceedings of the International Compressor Engineering Conference at Purdue*. West Lafayette, IN, USA.

Modeling Reciprocating Compressors Using A Cartesian Cut-Cell Method With Automatic Mesh Generation

David ROWINSKI*, Kenneth DAVIS

Convergent Science, Inc.
Madison, WI, USA
david.rowinski@convergecfd.com

* Corresponding Author

ABSTRACT

Computational Fluid Dynamics (CFD) models can offer great insight into flow phenomena and complex fluid-structure interactions present in reciprocating compressors. This is often achieved, however, at large computational cost and considerable user setup time. In this study, a Cartesian cut-cell finite-volume method is applied to model a small displacement refrigeration compressor. The cut-cell method has the key feature of representing discrete cell volumes exactly without requiring the computational grid to coincide with the bounding geometry. Additionally, the grid is dynamically generated at each time step based on the instantaneous boundary positions and is automatically refined based on gradients of local flow variables. These two features make this method ideal for modeling the deformation of the valves, the motion of the piston, and the complex geometries of the suction and discharge mufflers. The model is validated against experimental data. The sources of numerical error in the model are assessed, including the spatial and temporal discretization error and the model treatment for valve opening, closure, and contact. Lastly, several automated grid generation strategies are presented to establish guidelines for balancing cost and accuracy. The model formulation highlights the ease of incorporating complex and moving geometries characteristic of reciprocating compressors into a CFD model at a reasonable cost.

1. INTRODUCTION

Understanding the nature of the flows in reciprocating compressors is vital in engineering work involving their design and analysis. While models of various levels of complexity have been developed, three-dimensional time-accurate computational fluid dynamics (CFD) provides some key advantages. These advantages include the identification of flow structures important to viscous losses, local heat transfer, mixing, and the elucidation of the interaction of the flow with the geometry. Furthermore, these models are more complete in the sense that they are mostly free from empirical corrections which may be valid over small ranges of operating conditions. As Prasad (2004) points out in his review, though reciprocating compressors are typically viewed as thermodynamic devices, most of the important losses are fluid dynamic in nature, making understand the flow of critical importance.

Over the last twenty years, considerable advancement has been made in the application of CFD models to reciprocating compressors. Cyklis (1994) applied a commercial CFD package to model two-dimensional flow through a suction valve. Fagotti and Possamai (2000) applied commercial CFD tools to model flows in compressor components, and noted that one important aspect of the problem setup was the user selection of grid size. Ottitsch (2000) applied CFD to model a variety of different valve types in steady state incompressible flow. Chikurde (2002) presented an analysis of a full compressor geometry and noted that it was critical to apply sufficient grid resolution in regions of important flow features. In addition, complex flow features such as rivets and fillets were removed to simplified the meshing procedure. In more recent studies, Kim (2006) applied two-way coupled fluid-structure interaction (FSI) to model the behavior of valves, and Pereira (2007) applied CFD to model the full three-dimensional domain of a refrigeration compressor. Pereira (2008) compared models of various levels of complexity and found advantage of the CFD's predictive nature, despite finding the three-dimensional model costs too high for use in optimization. Mistry (2012) and Rigola (2012) explored three-dimensional modeling of the flow over valves, among other topics such as turbulence model effects. Rodrigues (2014) applied a variety of turbulence models to full three-dimensional compressor models.

One of the major limitations to the adoption of CFD in the routine design and analysis process for reciprocating compressors has been the bottleneck of grid generation, as pointed out by Prasad (2004) and others. Reciprocating compressors typically feature complex geometries in the suction and discharge mufflers, as well as moving parts of the piston and valves. Furthermore, the types of valves and other moving parts used may vary widely between different reciprocating compressor types, especially when prototyping new design concepts.

In this study, an alternate approach to modeling reciprocating compressors is presented. The majority of previous works on CFD studies of reciprocating compressors have featured models requiring the construction of a numerical grid. Here the authors present the use of a Cartesian cut-cell method in which the grid is dynamically generated at every time step based on exact and instantaneous geometry definition. The purpose of this work is to investigate the numerical accuracy of this cut-cell based method, specifically in its application to modeling reciprocating compressors.

2. METHODOLOGY

The Cartesian cut-cell method described by Senecal et al. (2007) is employed in this work. The model begins from a triangulated surface which describes the volume of a computational domain. The volume is filled by an orthogonal base grid of a user-specified size. Wherever the triangulated surface intersects the base grid, the base grid cells are cut into arbitrary sided polyhedra. Thus the volume represented in the cut-cells is exactly the same as the volume of the original triangulated surface, independent of the base grid size used. Because the process of forming the mesh is entirely automatic, it can easily accommodate complex boundary shapes and complex boundary motion without any user intervention. Because the method is based on standard orthogonal grids, it is computationally simple and demonstrates good numerical stability.

The conservation equations are solved using the finite-volume method. All values in the model are collated at the center of the computational cells. The algorithm of Rhie and Chow (1983) is used to prevent checker-boarding in the velocity and pressure fields. In the present study, second-order spatial discretization is used for solving all the conservation equations. To maintain stability, the time accuracy is set to a fully implicit first-order scheme for this work. The transport equations are solved using a modified Pressure Implicit with Splitting of Operators (PISO) method from Issa (1985). A variable time-step approach is used to ensure stability and to increase the time step when possible. The time step is controlled through the Courant-Friedrichs-Lewy (CFL) numbers based on local cell sizes, velocity, speed of sound, and diffusivity. Turbulence is modeled using the RNG k-epsilon model of Yahkot (1986) with standard model constants.

All simulations are run in parallel using Message Passing Interface (MPI). The automatic domain decomposition technique of Karypis (2013) is used to control the load balance, which is not trivial for this method. Because of change to the computational volume over time (for example, by moving boundaries), or addition of cells through Automatic Mesh Refinement (AMR), there may be a significant non-uniformity to the distribution of cells in space.

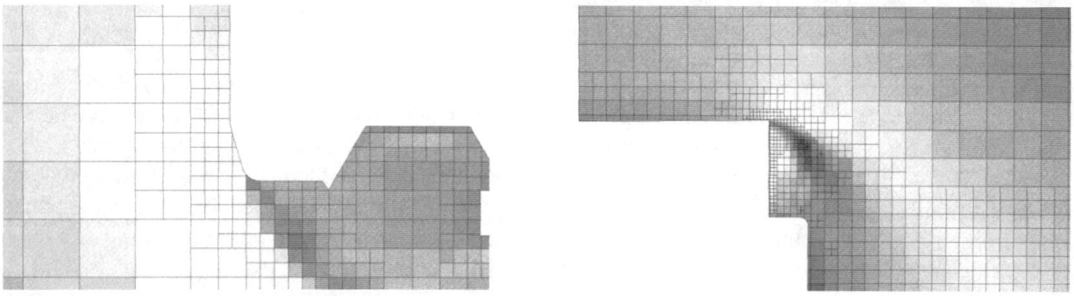

Figure 1: (Left) A cut-plane showing the base grid cells, far to the left, and the smaller embedded cells which are cut by the boundary into arbitrary sided polyhedra. (Right) A cut-plane showing smaller cells which have been created automatically through a refinement based on the curvature of instantaneous field variable values.

It is essential to ensure that the grid resolution at any location is sufficiently small to have minimal impact on the solution of the flow field. In most cases, it is very difficult to determine beforehand the locations and times over which a finer grid resolution is required. Automatic Mesh Refinement (AMR) is employed in these models to provide a way to handle this problem, as demonstrated in Figure 1. Ideally, a good AMR method should add the cells where the simulation is most under-resolved, or where the sub-grid scale variation of a field variable is the largest. As shown by Pomraning (2000) and others, any sub-grid scalar field can be expressed as a series sum. So here, the refinement criteria are based on the instantaneous curvature of the solution fields, or the first term in the series for the sub-grid scale term.

The equation of state used in this study is that of Peng and Robinson (1976), in which the critical temperature, critical pressure, and acentric factor of the working fluids must be provided. The standard departure functions described by Poling (2001) are used to compute real-gas properties for enthalpy, entropy, and specific heat capacity. The transport properties of viscosity and conductivity are described as tabulated functions of temperature and pressure based on the data provided by Perkins (2000) and Tillner-Roth (1994).

In this particular study, moving and deforming solids are incorporated into the model using an explicit solid-fluid coupling scheme. The classical Euler-Bernoulli beam equation as described by Timoshenko (1953), is discretized over the length of the solid reeds. A finite-difference scheme is used to determine the local beam velocity and displacement. The scheme is spatially second-order accurate, and first-order accurate in time through a fully implicit solution in time. The local pressure and viscous forces over the solid are evaluated and fed into the beam equation as a spatially varying load. The model can account for variation in the cross-section area, density, thickness, elastic modulus, and moment of inertia across the length of the beam. The model assumes the beam is subject to lateral loads only and that the displacement is small. To handle contact along the beam, an implicit penalty force method is used. If contact is determined, the time step is re-evaluated with a force applied constraining the node displacement.

3. TEST CASE DESCRIPTION

The method described has been used often in the modeling of internal combustion engines, for example by Senecal (2007) and Pomraning (2014). This application area shares several important similarities and differences from the application presented here. In this section, the physical test case studied in this work is described, the computational model inputs are presented, and the numerical test cases are outlined.

3.1 Description of Physical Compressor Studied

The compressor studied in this work is a 13.0 cubic centimeter displacement hermetic refrigeration compressor. All information, including the geometry, operating conditions, and experimental data has been provided to the authors. The working fluid is R-134a. The geometry consists of the components shown in Figure 2, and extends from the input of the suction muffler to the end of the discharge line. The operating conditions for the case considered here is a low back pressure (LBP) with a condensing temperature of 54°C. The nominal speed of the compressor is 3000 revolutions per minute (RPM). The cylinder pressure is measured, as are the lifts of the suction and discharge reeds.

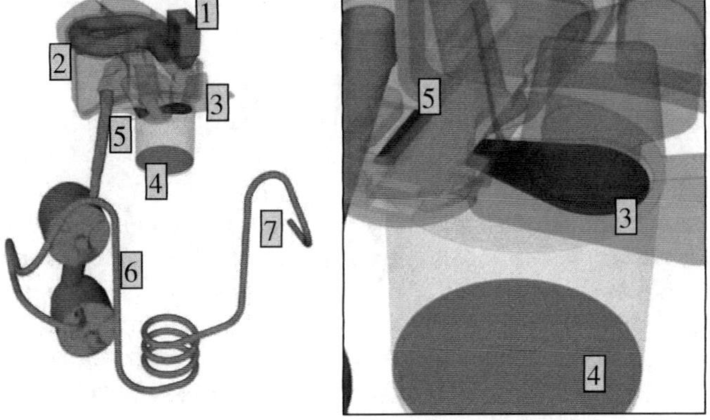

Components of Compressor Assembly in CFD Model
1. Inflow
2. Suction Muffler and Port
3. Valve Plate and Suction Reed Gasket
4. Piston
5. Discharge Reed, Booster, and Limiter
6. Discharge Muffler and Lines
7. Outflow

Figure 2: The geometry of the test case compressor.

3.2 Computational Model Setup

The computational model is constructed first by extracting a wetted surface geometry of the compressor components in contact with the working fluid. This surface contains all the necessary information concerning the geometry. The suction muffler inlet is set as a total pressure inflow boundary condition, and the measured temperature at that inlet as well. At the discharge line outlet, the discharge pressure is set as a static pressure boundary condition. At the walls, standard two-layer law-of-the-wall models are used for velocity and temperature, and a measured wall temperature is used around the cylinder to model the heat transfer.

The motion of the piston is set according to a standard crank-slider mechanism using the geometric parameters of the compressor's crankshaft, connecting rod, and wrist pin offset. The motion of the suction valve, discharge valve, and discharge booster are set according to the beam model mentioned in Section 2. The authors emphasize that all these boundary motions are easily incorporated into the model without the need for any user intervention concerning how the motion will affect the computational grid.

4. Results and Discussion

As the purpose of this work is to ascertain the numerical errors in the Cartesian cut-cell model, a number of test cases are performed to target specific parts of the model which may be prone to numerical error. First, a base case is run using an intermediate grid resolution. Those calculations begin at bottom dead center with the cylinder region initialized at the conditions of the suction muffler. The model is run until cyclic convergence is reached. Figure 3 shows the cooling capacity (normalized by the measured cooling capacity), and the power (normalized by the measured power) for the last three cycles. As can be seen from the data, there is less than 1% variation in these measures by the third cycle. The pressure-volume (PV) diagram is shown in Figure 3 for the last three cycles. In this figure, the pressure is normalized to P^* by the difference between suction and discharge pressures, so that the normalized pressure, P^*, is zero at suction conditions and unity at discharge. The volume is also normalized to V^* by the displacement.

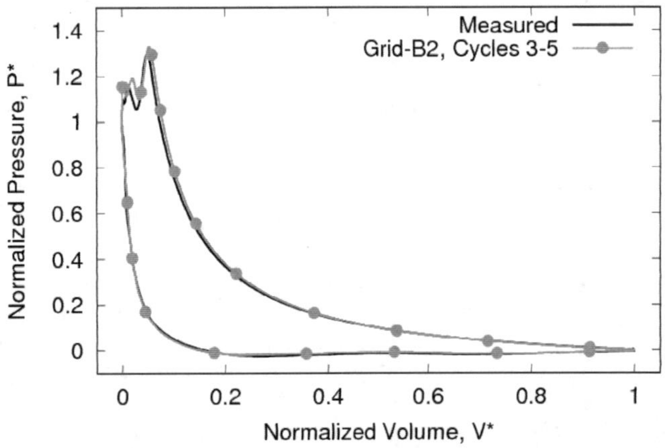

Case	Norm. Cooling Capacity	Norm. Power (Point)	Norm. Power (Mean)
Measured	1	1	N/A
Model, Cycle #3	0.994	1.019	1.037
Model, Cycle #4	0.992	1.019	1.037
Model, Cycle #5	0.995	1.020	1.038

Figure 3: PV Diagram (left) for base case for cycles three to five along with experimental measurements, and table (right) of normalized cooling capacity and power statistics.

4.1 Effects of the spatial accuracy of the model

Once cyclic convergence is reached, a series of grid resolution studies are performed for both suction and discharge events. These simulations begin approximately 20 degrees before the valves open, and are initialized with conditions from the previous model. The authors note that each grid resolution study is performed by changing just a single set of parameters in the model inputs. For two of the grids, Grid-A0 and Grid-B0, there is no automatic mesh refinement (AMR) activated. For the remaining grids, AMR is activated with various levels of refinement allowed. An AMR scale value means that cells are allowed to be divided recursively by that number of times. (For example, a scale of 2 means that a base grid cell can be subdivided into eight grid cells, which can again be divided into a total of sixty-four grid cells.) The details of all seven grids studied are shown in Table 1.

Table 1: Details about the seven grids used in the various grid refinement studies.

Grid Name	Number of Base Grid Cells Across Bore	AMR Scale	Base Grid Size (mm)	Finest Grid Size (mm)	Max Cell Count (MCell)
Grid-A0	10	0	2.6	0.65	0.047
Grid-B0	20	0	1.3	0.33	0.18
Grid-B2	20	2	1.3	0.33	0.24
Grid-B3	20	3	1.3	0.16	0.48
Grid-C2	40	2	0.65	0.16	1.3
Grid-C3	40	3	0.65	0.08	2.1
Grid-C4	40	4	0.65	0.08	6.3

To compare the cases, two additional metrics are introduced: Y^*, the delivered mass fraction, and M^*, the normalized mass flow rate. For the suction event, Y^* is the mass fraction of fresh suction gas in the cylinder, and M^* is the mass flow rate into or out of the cylinder normalized by the characteristic mass flow rate, defined by the mass at suction conditions occupying the displacement volume per time length of a cycle. For the discharge event, Y^* is defined as the mass in the discharge port which originated from the cylinder, normalized by the mass of the cylinder prior to discharge. These quantities are all tracked in simulation by solving transport equations for two additional passive scalars.

Figure 3: Cut-planes along the center of the suction reed, colored by normalized velocity for three of the grids studied. From left to right, the grids are Grid-A0, Grid-B2, and Grid-C3.

Three representative grids from the suction event grid refinement studies are shown in Figure 3. In the leftmost figure is Grid-A0, the coarsest grid with no AMR. In the center figure is the intermediate Grid-B2. One of the finest grids is shown to the right. As can be seen from the figure, AMR adds the cells in the regions where the critical flow structures are present.

The grid resolution study for the suction process showed relatively little sensitivity of either M^* or Y^* on all but the coarsest grid. The results were encouraging in that they indicated numerical grid convergence of the solution on relatively coarse meshes. The error between the measured values and the modeled values is far greater than the grid discretization error on the finest six grids. Figure 4 shows the PV diagram, the suction reed lift (normalized by the valve length), and the mass flow rates for four of the grids. The similarity in the solution for the six finest grids can be seen in all four of these plots.

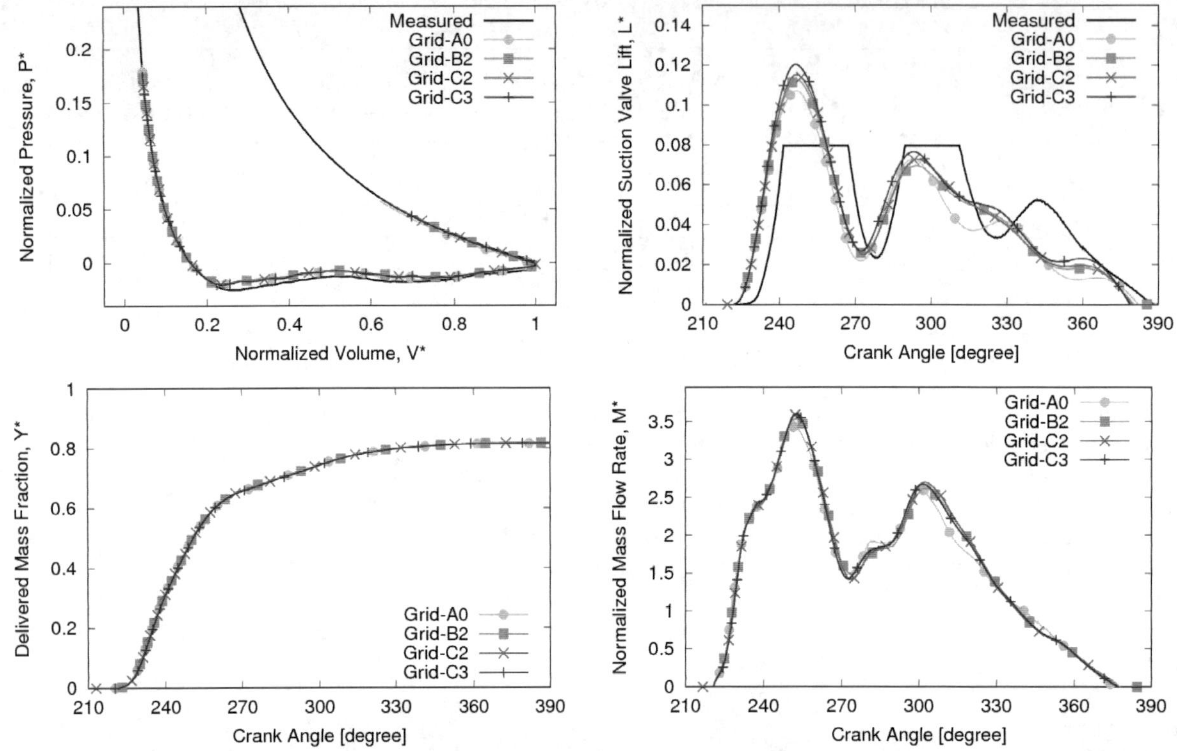

Figure 4: Calculated quantities during the suction process for four of the grids studied and experimental measurements. Top left shows the pressure vs. volume. Top right shows the suction reed lift, bottom left shows the delivered mass fraction, and bottom right shows the mass flow rate into the cylinder all as functions of crank angle.

The discharge process showed slightly more grid sensitivity than did the suction process. Figure 5 shows two representative cut-planes from during the discharge event. The bending of the reed valve and booster can be seen in these figures, as can the large amount of cells that are added adaptively near the valve opening. Figure 6 shows the PV diagram, discharge lift (normalized by the valve length), and Y* and M* evolution. The only real outlying point is the coarsest grid, Grid-A0, and the other grids indicate reasonable convergence of these quantities. The data from all grid convergence studies is summarized in Table 2.

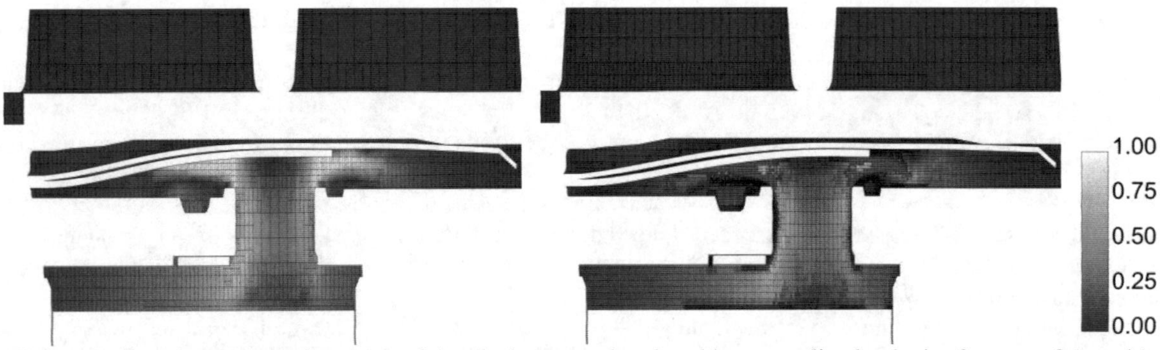

Figure 5: Cut-planes along the center of the discharge reed, colored by normalized velocity for two of the grids studied. The grids are Grid-B3 and Grid-C4 from left to right. The view is skewed to exaggerate the bending.

The results of the grid resolution studies for both suction and discharge events are summarized by Table 2. Global quantities of Y^* at the end of the cycle showed very little variation on any of the grids, save the coarsest Grid-A0. The maximum flow rate, M^*, showed slightly more variation, however, which does indicate important grid effects in resolving the temporally accurate and spatially accurate mass flow through the valves.

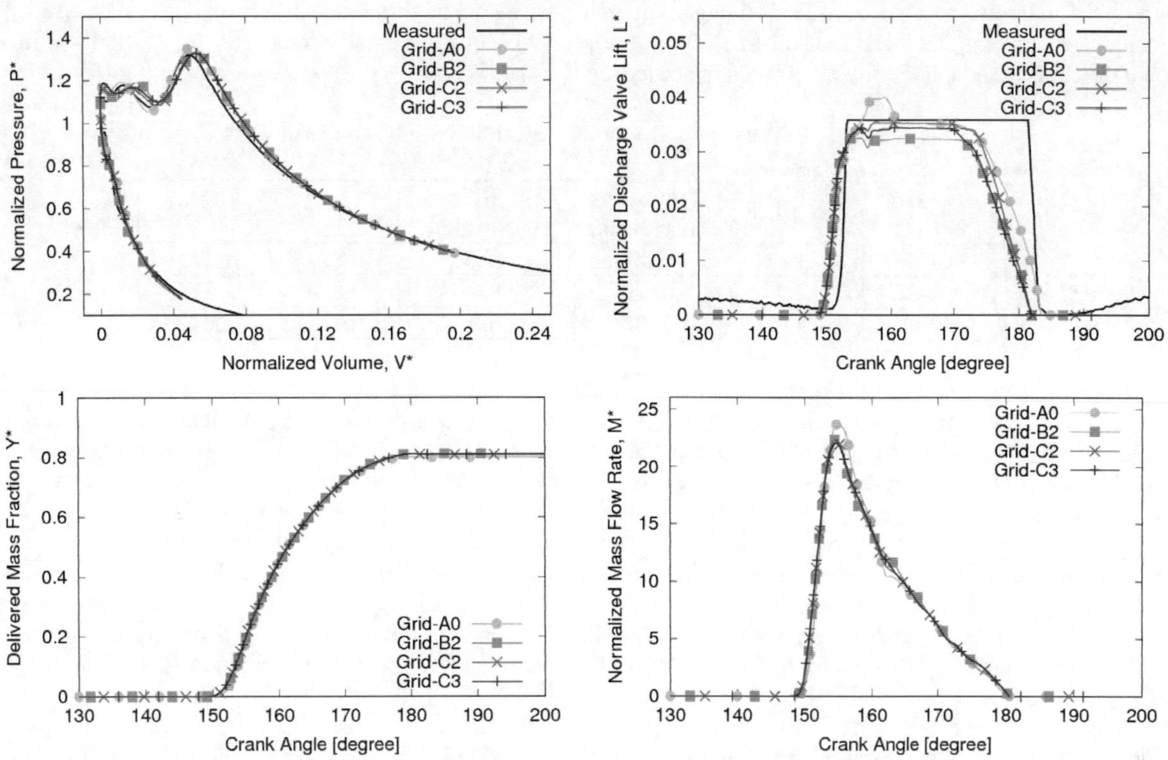

Figure 6: Calculated quantities during the discharge process for four of the grids studied and experimental measurements. Top left shows the pressure vs. volume. Top right shows the suction reed lift, bottom left shows the delivered mass fraction, and bottom right shows the mass flow rate into the cylinder all as functions of crank angle.

Table 2: Results from Grid Resolution Studies

	Suction				Discharge			
Grid Name	*Y**	**Max(*M**)**	**Wall Clock Time (hr)**	**CPU Time (core hr)**	*Y**	**Max(*M**)**	**Wall Clock Time (hr)**	**CPU Time (core hr)**
Grid-A0	0.816	3.43	0.9	3.6	0.802	23.9	0.5	2.1
Grid-B0	0.817	3.53	3.3	13	0.812	22.4	1.5	5.8
Grid-B2	0.817	3.54	5.0	20	0.812	22.4	4.0	16
Grid-B3	0.817	3.56	8.1	65	0.810	22.8	3.4	27
Grid-C2	0.817	3.59	10.2	163	0.810	22.4	3.7	58
Grid-C3	0.817	3.60	26.4	423	0.811	22.1	8.1	130
Grid-C4	0.817	3.57	73.8	1180	0.811	22.9	27.0	432

4.2 Effects of temporal discretization

As the grid is refined, the time-step also adjusts accordingly based on the imposed CFL restrictions. It is possible, however, to run the model at larger time steps by increasing the CFL number. If the CFL number becomes too large, there may be significant temporal error in the solution fields. Grid-C2 has been assessed at larger CFL numbers of 2.0 and 4.0. The results in Table 3 show measures for a complete cycle.

The suction process is virtually unaffected by running at larger time steps, however the discharge process shows slightly greater sensitivity, affecting the maximum pressure by around 2% when a CFL number as large as 4.0 is used. It is recommended, therefore, to allow a higher CFL number during suction. The tests show that a large improvement in run-time, near 50% savings, can be achieved by using the higher CFL numbers. The run-time improvement is very good from CFL of 1.0 to 2.0, but the improvement is marginal when CFL is increased to 4.0

due to the extra work required when solving each transport equation to the specified tolerance. The use of the biconjugate gradient stabilized method for solving the pressure equation greatly improves the method's ability to solve the pressure equation efficiently when the time step is large.

Table 3: Results from Temporal Resolution Studies on Grid-C2

Grid	CFL	Timesteps (per cycle)	Wall Time (hr/cycle)	Y*	Max(P*)
Grid-C2	1.0	13000	15.5	0.824	1.311
Grid-C2	2.0	6500	8.4	0.824	1.315
Grid-C2	4.0	3800	6.0	0.824	1.322

4.3 Effects of valve sealing model

The numerical modeling of the sealing of the valves is another important part of the model that must be scrutinized for its effect on the accuracy. When the valve lift at the center of the seat is below a user-specified threshold, the flow between the valve and port is treated as perfectly sealed. When this lift exceeds the threshold, flow passes freely through the valve opening. By default, this tolerance is set to be very small, at 10^{-8} m. If it is too small, the simulation may be expensive to due high velocities and small cells. If it is too large, the simulation will suffer from the numerical artifact of flow being restricted when the valve is open. Values of 10^{-8}, 10^{-7}, 10^{-6}, and 10^{-5} m are tested here and found to have negligible effect on the maximum flow rate and the total mass flow. In Figure 7 (a), the artifact caused by late opening is seen. Despite the small effect on the global statistics, it is recommended to use a small value around 10^{-8} to 10^{-7} m for this parameter, as the additional run time to the simulation is also small.

A second numerical effect in the valve modeling is the modeling the seat contact. The current approach carries the limitation that the surfaces describing the computational volume are not allowed to intersect. Therefore, where valves contact seats there exists a thin gap in the model. Gaps of 10^{-4}, 10^{-5}, and 10^{-6} m were investigated in this study. It was found that a gap of 10^{-5} m performed best. If the gap were too small, the time step restriction is quite severe; if the gap were too large, the flow areas were slightly over-estimated.

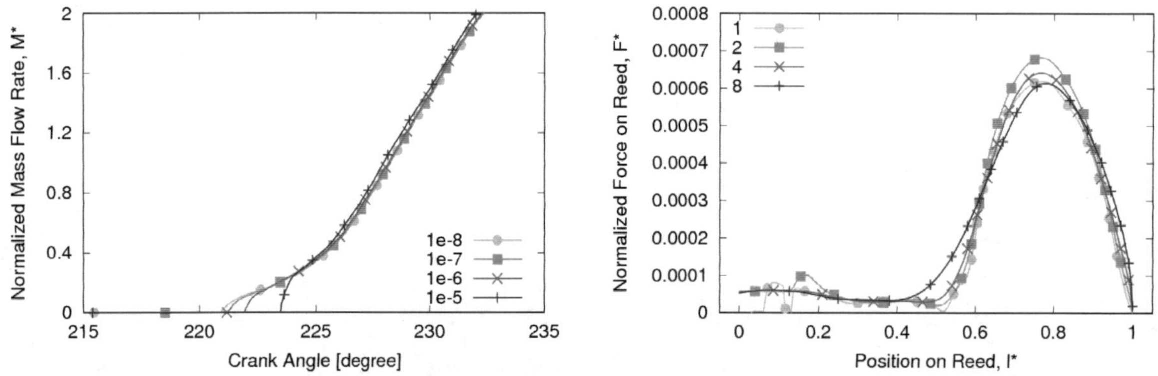

Figure 7: Mass flow right vs. crank angle (left) for varying lift openings, and pressure force along reed (right) for varying pressure window sizes.

4.4 Effects of pressure coupling between fluid and solid

The final numerical aspect of the model investigated here is the window size over which the pressure is evaluated on the solid beam. By default, a value equal to the base grid size is used. A tent function is employed to average the pressure from the fluid cells onto the beam. When the window size is increased, there may be a smearing of pressure data over the beam which can affect the motion. If the window size become two small, there may be nodes on the beam around which no fluid cell is close enough to evaluate. Figure 7 (b) shows how the normalized pressure force on the beam varies with window sizes of 1, 2, 4, and 8 factors of the base grid size. Because of the tent-function averaging, the smearing effect is somewhat reduced. There are no obvious changes to the model results over this range of variation of the pressure window.

6. CONCLUSIONS

In this study, we have investigated the numerical accuracy of applying a Cartesian cut-cell finite-volume method to reciprocating compressors. The method was found to be ideally suited to handling the moving complex geometries associated with the compressor tested. The numerical error showed that grid convergence can be achieved on relatively coarse grids, in part due to the small length scales and velocity scales in this problem. As the Reynolds number of the problem increases, the grid may show more sensitivity. Running at a higher CFL number, especially during the suction process, can significantly reduce the run time, with very minor effects on the accuracy. Recommended values for the valve height at which to open the flow and the gap to leave in the seat were reported. It is important to consider that the metrics used here for numerical convergence were primarily global quantities; there almost certainly more variation among the different grids when comparing local quantities. The numerical settings used in any calculation should be determined so that the main statistics of interest, whether they be global or local quantities, are sufficiently converged.

Overall, the amount of numerical uncertainties in the model were quite small compared to the difference between the experimental and model metrics of pressure and valve lift. This is demonstrative of two things: Firstly, the numerical technique applied in this work demonstrates numerical accuracy over the range of values tested; secondly, there is more uncertainty in the physical models themselves than in the numerics. Specifically, implementing realistic models for phenomena like leakage and stiction represent important physics which are lacking in the current model. There is considerable research effort devoted to these two topics already. The numerical method described here has been demonstrated as a platform in which these models could be tested further.

REFERENCES

Chikurde, R. C., Loganathan, E., Dandekar, D. P., Manivasagam, S. (2002). Thermal Mapping of Hermetically Sealed Compressors Using Computational Fluid Dynamics Technique. *Proceedings of the 16th International Compressor Engineering Conference at Purdue* (Paper 1520). West Lafayette, USA: Purdue University.

Cyklis, P. (1994). CFD Simulation of the Flow through Reciprocating Compressor Self-acting Valves. *Proceedings of the 12th International Compressor Engineering Conference at Purdue* (Paper 1016). West Lafayette, USA: Purdue University.

Fagotti, F., Possamai, F. C. (2000). Using Computational Fluid Dynamics as a Compressor Design Tool. *Proceedings of the 15th International Compressor Engineering Conference at Purdue* (Paper 1377). West Lafayette, USA: Purdue University.

Issa, R. I, (1985). Solution of the Implicitly Discretised Fluid Flow Equations by Operator-Splitting. *Journal of Computational Physics*, 62.

Karypis, G. (2013). A Software Package for Partitioning Unstructured Graphs, Partitioning Meshes and Computing Fill-Reducing Orderings of Sparse Matrices.

Kim, J., Wang, S., Park, S., Ryu, K., La, J. (2006). Valve Dynamic Analysis of a Hermetic Reciprocating Compressor. *Proceedings of the 18th International Compressor Engineering Conference at Purdue* (Paper 1805). West Lafayette, USA: Purdue University.

Mistry, H., Bhakta, A., Dhar, S., Bahadur, V., Dey, S. (2012). Capturing Valve Dynamics in Reciprocating Compressors through Computational Fluid Dynamics. *Proceedings of the 21st International Compressor Engineering Conference at Purdue* (Paper 1210). West Lafayette, USA: Purdue University.

Ottitsch, F. (2000). CFD: A Viable Engineering Tool for Compressor Valve Design or Just a Toy? *Proceedings of the 15th International Compressor Engineering Conference at Purdue* (Paper 1417). West Lafayette, USA: Purdue University.

Peng, D. Y. and Robinson, D. B. (1976). A New Two-Constant Equation of State. *Ind. Eng. Chem. Fund.*, 15, 59-64.

Pereira, E.L.P, Deschamps, C.J., Ribas Jr., F.A. (2007). Performance analysis of reciprocating compressor through CFD simulation. *Proceedings of the International Conference on Compressors and their Systems* (pp. 310-318). London, England: City University of London.

Pereira, E.L.P, Deschamps, C.J., Ribas Jr., F.A. (2008). A comparative analysis of numerical simulation approaches for reciprocating compressor. *Proceedings of the 19th International Compressor Engineering Conference at Purdue* (Paper 1303). West Lafayette, USA: Purdue University.

Perkins, R.A., Laesecke; A., Ramires, M.L.V., Gurova, A.N., Cusco, L. (2000). Experimental thermal conductivity values for the IUPAC round-robin sample of 1,1,1,2-tetrafluoroethane (R134a), *NIST*, Gaithersburg, MD.

Poling, Prausnitz, O'Connell. (2001). The Properties of Gases and Liquids, 5th Ed., McGraw-Hill, p. 6.5.

Pomraning, E. (2000). Development of Large Eddy Simulation Models. *Ph.D. Dissertation, The University of Wisconsin-Madison.*

Pomraning, E., Richards, K., and Senecal, P. (2014). Modeling Turbulent Combustion Using a RANS Model, Detailed Chemistry, and Adaptive Mesh Refinement. *SAE Technical Paper* 2014-01-1116.

Prasad, B. G. (2004). CFD for Positive Displacement Compressors. *Proceedings of the 17th International Compressor Engineering Conference at Purdue* (Paper 1689). West Lafayette, USA: Purdue University.

Rhie, C. M. and Chow, W. L. (1983). Numerical Study of the Turbulent Flow Past an Airfoil with Trailing Edge Separation. *AIAA J.*, 21, 1525-1532.

Rigola, J., Lehmkuhl, O., Ventosa, J., Pérez-Segarra, C.D., Oliva, A. (2012). Numerical simulation of the turbulent fluid flow through valves based on Low Mach models. *Proceedings of the 21st International Compressor Engineering Conference at Purdue* (Paper 1375). West Lafayette, USA: Purdue University.

Rodrigues, T. T., Da Silva, D. L. (2014). Turbulence Modelling Evaluation for Reciprocating Compressor Simulation. *Proceedings of the 22nd International Compressor Engineering Conference at Purdue* (Paper 2297). West Lafayette, USA: Purdue University.

Senecal, P. K., Pomraning, E., Richards, K. J., Briggs, T. E., Choi, C. Y., McDavid, R. M., Patterson, M. A., Hou, S., & Shethaji, T. (2007). A New Parallel Cut-Cell Cartesian CFD Code for Rapid Grid Generation Applied to In-Cylinder Diesel Engine Simulations. *SAE Technical Paper*, #2007-01-0159.

Tillner-Roth, R.; Baehr, H.D. (1994). An international standard formulation of the thermodynamic properties of 1,1,1,2-tetrafluoroethane (HFC-134a) covering temperatures from 170 K to 455 K at pressures up to 70 MPa. *J. Phys. Chem. Ref. Data*, 23, 5, 657-729.

Wang, Y., Krishan, K. and Dennin, M. (2007). Limits of the equivalence of time and ensemble averages in shear flows. *Phys. Rev. Lett.*, Vol. 98, Issue 22.

Yakhot, V. and Orszag, S. A. (1986). Renormalization Group Analysis of Turbulence. I. Basic Theory. *J. Sci. Comput.*, 1, 3.

ACKNOWLEDGEMENT

The authors acknowledge the support of Embraco for providing the compressor geometry, boundary conditions, and experimental data, as well as for their guidance in setting up and assessing the models.

Study of Valve Motion in Reciprocating Refrigerator Compressors Based on the 3-D Fluid-Structure Interaction Model

Weifeng Wu[1,2*], Zhifang Jian[1], Mingyi Song[1], Zhao Zhang[1]

[1]School of Energy and Power Engineering, Xi'an Jiaotong University
Xi'an, Shaanxi, 710049, China
86-29-82665526, weifengwu@xjtu.edu.cn

[2]Faculty of Engineering and Architecture, Department of Industrial System and Product Design, Ghent University, Kortrijk 8500, Belgium
weifeng.wu@ugent.be

* Corresponding Author

ABSTRACT

In this paper, valve motion in the refrigerator reciprocating compressor is studied based on a 3-D fluid-structure interaction model. Experiment of the P-V graph was carried out to validate the numerical model. Simulations showed that increased thickness of the valve would finally prolong delayed closure. It seems a little different from the traditional prediction that increased thickness would result in increased strength and advanced closure.

1. INTRODUCTION

The compressor is core of a refrigerating facility. Its reliability is very important to the whole system. It is well known that valves are main wearing parts in a reciprocating compressor (Danqing W., Jingtong C., 1979). In published references, many different methods have been adopted to simulate the working process and the valve motion of the reciprocating refrigerating compressors. Most of the estimation of inner state parameters of the compressor is based on energy balance and spring model of the valve motion (Junlong Z. *et al.*, 2012; Satyam B., 2008).One-dimensional method has been applied to simulate and design the compressor including the valve (Mithraratne P. *et al.*, 2000; Lee L. and Randall J., 2003). Experience parameters are necessary for the simulation program. And the one dimensional model fails to predict the detail of the motion of the valve. Multi-dimensional method is also used to simulate the compression and valve motion (Derek S. and Sorin M., 2002; Kim J. *et al.*, 2006). It is complex and costs big computer resources.
In this paper, a refrigerator compressor with refrigerant R600a is adopted to carry out the simulation of the valve motion based on a 3-D fluid-structure interaction model.

2. THE SIMULATION MODEL

The geometric model is established using Unigraphics NX8 software, and the simulation is carried out based on the ADINA. Thermal physical parameter of the Refrigerant comes from NIST database.

2.1 The geometric Model

Geometric model of a refrigerator compressor is established. Displacement of the compressor is 15.3 cm^3. Parameters of the compressor are shown in Table 1. Geometric model of the compressor is shown in Figure 1.

Table 1: Parameters of the compressor

Parameters	Value
Diameter of the cylinder/mm	31
Stroke/mm	24
Diameter of the suction valve hole/mm	8.2
Diameter of the discharge valve hole/mm	5.6

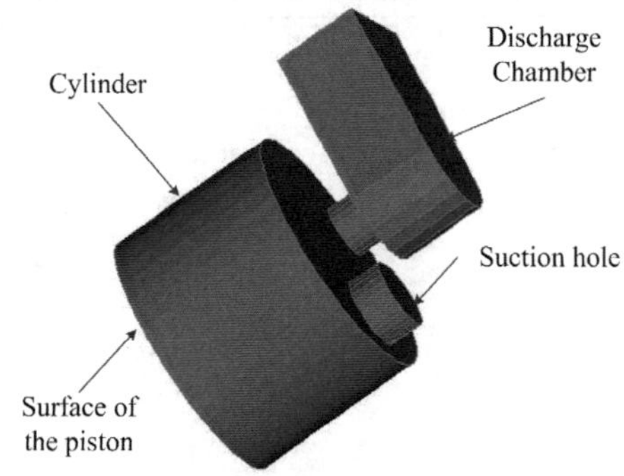

Figure 1: Geometric model of the simulation model

The solid models of the suction valve and the discharge valve (Mei L. and Yongzang Y., 2000) are shown in Figure 2.

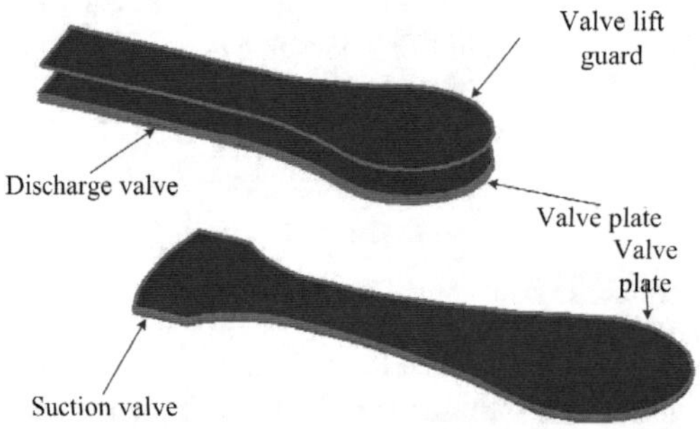

Figure 2: Solid model of the suction valve and the discharge valve

Grid of the model is shown in Figure 3.

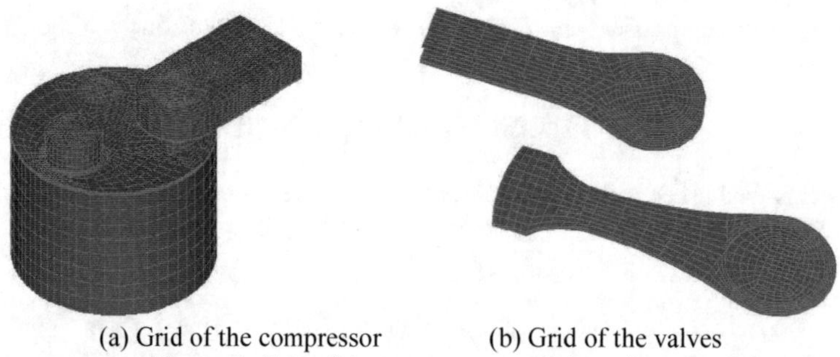
(a) Grid of the compressor (b) Grid of the valves

Figure 3: Grid of the compressor and the valves

2.2 Boundary Conditions

Temperature and pressure conditions of the suction gas are set as 305K and 0.068MPa. For the discharge port, pressure pulsation recorded by experiments is input into the simulation model as the boundary condition, to ensure the accuracy of simulation results of the valve motion.

Gap condition is set for the suction valve and the discharge valve. The piston surface is set as the moving wall. Eccentric installation of the cylinder is considered in the simulation model. The motion equation of the piston is

$$x = R[(1-\cos\alpha)+\frac{\lambda}{4}(1-\cos 2\alpha)-\lambda\eta\sin\alpha+\frac{\lambda^2\eta^2}{2(1+\lambda)}]. \quad (1)$$

Here, λ is the ratio of rotating radius, R, of the crankshaft to the length, L, of the rod. η is the eccentric ratio of the eccentric distance to the rotating radius, R.

To simulate the deformation and motion of the valve, the leader-follower condition is set between the valve and its contact surface as shown in Figure 4.

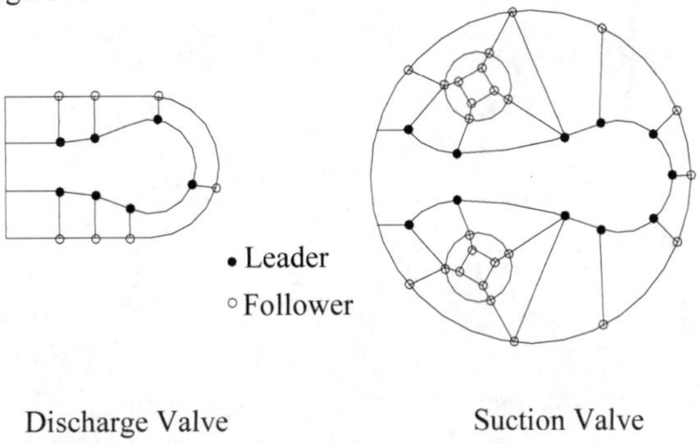

Figure 4: Leader-follower condition of the valve

3. VALIDATION OF THE 3-D FLUID-STRUCTURE INTERACTION MODEL WITH EXPERIMENTS

To validate the 3-D fluid-structure interaction model, experiments of the inner working process was carried out.

3.1 The experiment system

An experiments system of the inner working process of the compressor was established as shown in Figure 5.

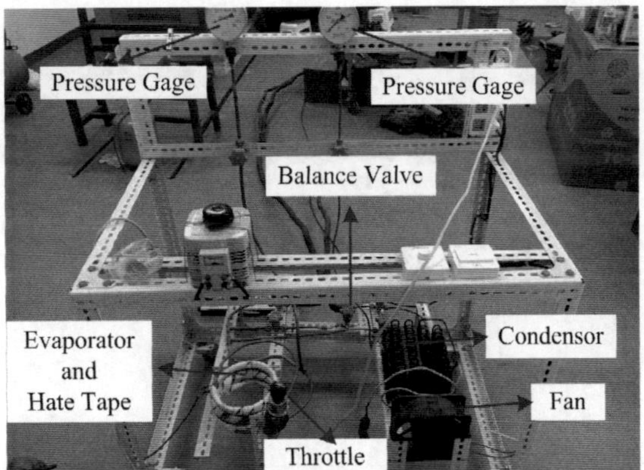

Figure 5: Experiment system of the refrigerator compressor

The inner pressure is tested using a kulite pressure sensor with the type of XTL-AC-190M. Rotating angle of the crankshaft is tested using a Proximity Switch produced by Omron. All the data is recorded by a NI system.

3.2 Comparisons of the simulation results to the experiment results

A comparison of the working process obtained by simulation and experiment is shown in Figure 6. The rotation speed of the compressor is 2970 rpm, the suction pressure is 0.068MPa, and the discharge pressure is 0.78MPa. Well consistency between the simulation results and the experiment results could be found.

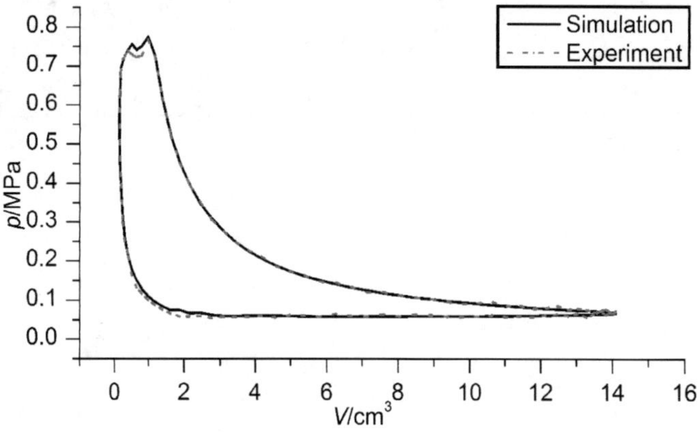

Figure 6: P-V diagram obtained by the experiment and simulaton

The difference of the power between the simulation and the experiment is 1.9%. It means 3-D fluid-structure interaction model proposed in this paper is valid.

4. SIMULATION RESULTS OF THE VALVE MOTION

The 3-D fluid-structure interaction model could predict the three dimensional motion of the valve. For the reed valve in the refrigerator compressor, a decline of the valve would result in difference lift at different place of the valve. Thus motion of the center point, which is located at the center of the area covering the suction hole or discharge hole on the valve, is chose as motion of the valve. Motion of the valves is then shown by the lift-angle line in the Figure 7.

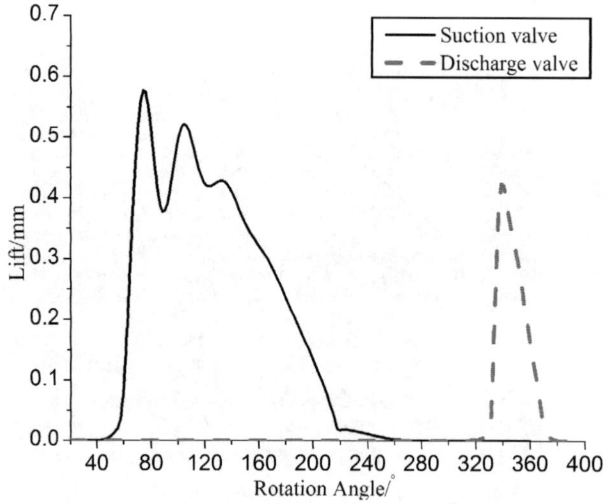

Figure 7: The lift-angle line of the valve motion

It could be found that, there are time delayed closure, both of the suction valve and the discharge valve. It is normal for a small reciprocating compressor. To obtain the highest efficiency, the suction should be as full as possible, and the discharge should be as empty as possible. A little delayed closure could enlarge the lift-angle section, which means low suction resistance. And such a small delay results in almost no leakage. It is similar for the discharge valve.

Simulation of the valve with different thickness shows that an increasing thickness of the valve would result in prolong the delayed time of the closure.

4.1 Influence of the thickness on the closure of the suction valve

Motions of the suction valve with three different thicknesses are shown in Figure 8. It could be found that the lift-angle section of the valve motion is decreased with the increasing of the valve thickness. That means increased resistance of the suction. However, the delayed time of the closure point is prolonged with the increasing of the valve thickness. For the thickness of 0.152mm, 0.203mm and 0.254mm, the closure point of the valve located at the angle of 203.4°, 214.2° and 219.6° respectively. This conclusion seems conflict to traditional predictions that the increased strength of the valve would result in advanced closure.

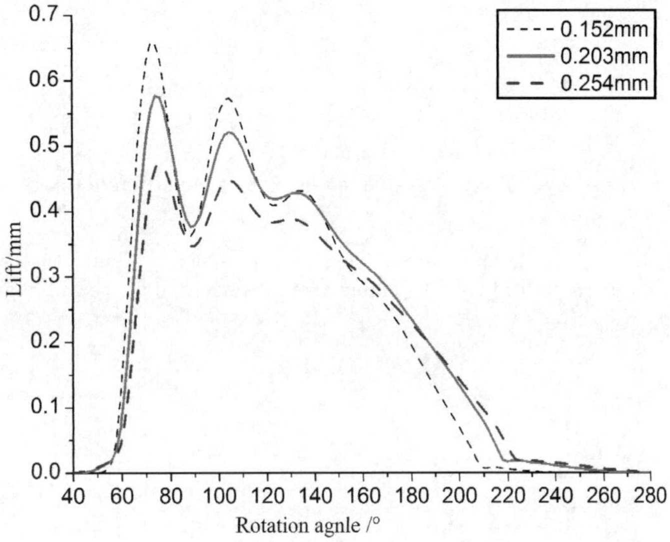

Figure 8: Motion of the suction valve with different thickness

This motion could be explained by investigating the forces acting on the valve. The simulation results show that higher gas load acting on the thicker valve. For this suction valve, the closure point is after the downside dead point, which means the closure is after the 180°, where the piston is moving toward the suction valve. This could result in quick close. However, the most important reason is the lift of the valve. For a thinner valve, the valve lift is bigger. That means a bigger deformation. During the close process, this bigger deformation means quicker recovery. Furthermore, the lift of the valve is longer, which could result in a higher final velocity of the valve. Because of these reasons, the thinner valve results in an advanced closure than that of the thicker valve. Using traditional one dimensional method fails to get this conclusion. This is the advantage of the 3D method because of its details in calculating the gas force on the valve plate, the bending of the valve plate and its related rebound force which drives the valve plate fall back on the valve hole.

However, the COP of the compressor is decreased with the increasing of the valve thickness.

4.2 Influence of the thickness on the closure of the discharge valve

For the discharge valve, the situation is similar. The valve motions with three valve thickness are shown in Figure 9.

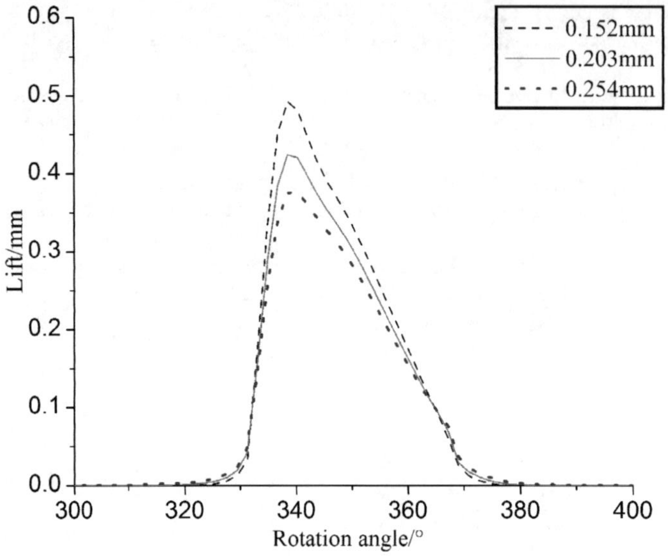

Figure 9: Discharge valve motion with three different thickness

The enlarged figure of the lift-angle line at the closure point is shown in Figure 10. A clearly delayed closure is shown. For different thickness of 0.154mm, 0.203mm and 0.254mm, the crankshaft angle of the closure point are 379.8°, 387° and 408.6° respectively.

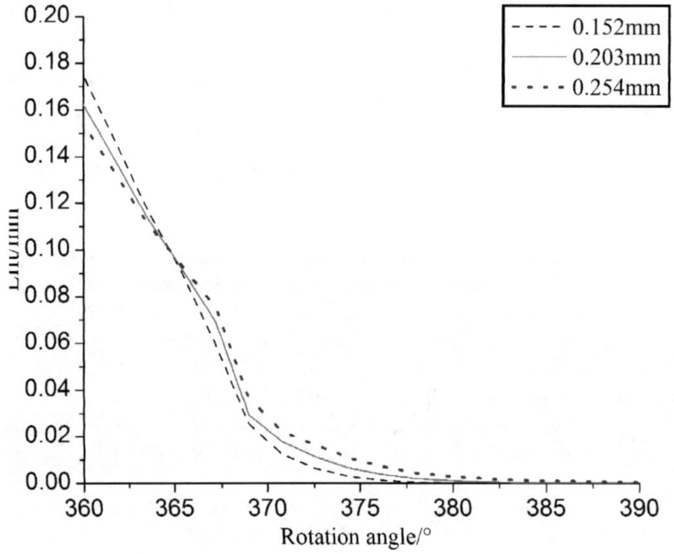

Figure 10: Enlarged figure of the lift curve at the closure point

It is same with the suction valve that the large lift of the valve with the thinner thickness finally results in its quick closure.

5. CONCLUSIONS

A three dimensional fluid-structure interaction model of small refrigerator reciprocating compressor was established in this paper. The simulation model was validated by experiments of investigating its inner working process. Then simulation of the motion of both the suction valve and the discharge valve was carried out.

Obtained results show that increased thickness of the valve could result in delay of its closure. The situation is same for the suction valve and the discharge valve. Reason for this situation is a little complicated. It is a comprehensive result of the motion of the valve plate, due to the changing rebound force, the gas force on the valve plate and the lift. Overall, the higher lift of the thinner valve is the main reason for the quicker recover of the valve plate. This is advantage of the 3-D Fluid-Structure Interaction simulation method.

NOMENCLATURE

P	Pressure	(MPa)
V	Stroke Displacement Volume	(cm^3)
R	Crankshaft Rotating radius	(m)
α	Rotation angle	($^\circ$)
L	length of the rod	(m)
λ	R/L	
η	ratio of the eccentric distance to R	
x	Displacement of the piston	(m)

REFERENCE

Danqing W., Jingtong C. (1979). *Numerical simulation and design for reed valve in compressors*, Beijing, China: China Machine Press. (In Chinese)

Junlong Z., Fangyuan Z., Weifeng W. and Kun Z. (2012). Improvement design of the oil pump system in household inverter refrigerating compressors. *Compressor Technology*, (01), 18-22.

Satyam B. (2008). A comparison of moving-boundary and finite-volume formulations for transients in centrifugal chillers. *International Journal of Refrigeration*, 31(8), 1437-1452.

Mithraratne P., Wijeysunderaa E. and Bongb Y. (2000). Dynamic simulation if a thermostatically controlled counter-flow evaporator. *International Journal of Refrigeration*, 23(4), 174-189.

Lee L. and Randall J. (2003). An immersed interface method for incompressible navier stokes equations. *Society for industrial and applied mathmatics*, 25(3), 832-856.

Derek S. B. and Sorin M. (2002). Wave propagation method for conservation laws and balance laws with spatially varying flux functions. *Society for industrial and applied mathmatics*, 24(3), 955-978.

Kim J., Wang S., Park S., et al. (2006). Valve dynamic analysis of a hermeic reciprocating compressor. *18th International Compressor Engineering Conference, Purdue University*(107.1-107.6). USA.

Mei L., Yongzang Y. (2000). *Manual for displacement compressors* (328-368), Beijing, China: China Machine Press.(In Chinese).

Linear Compressor Discharge Manifold Design for High Thermal Efficiency

Dietmar E. B. LILIE[1]; Rinaldo PUFF[2*]; Marcos G. D. de BORTOLI[3]

Embraco – R&D
Joinville, Santa Catarina, Brazil

[1]++55 47 3441 2348, dietmar.lilie@embraco.com
[2]++55 47 3441 2044, rinaldo.puff@embraco.com
[3]++55 47 3441 2783, marcos.g.bortoli@embraco.com

* Corresponding Author

ABSTRACT

In the design of hermetic compressors for household refrigeration, one important part of the overall efficiency of the machine comes from an optimized thermal efficiency of the discharge manifold. It is aimed that the heat generated during the compression process and heats up gas in the discharge chamber does not return to the compression chamber walls. Heating up the pump implies reducing the volumetric efficiency. Current technologies, like on/off and variable speed compressors, normally make use of metallic parts in the discharge chamber, such parts having high thermal conductivity coefficients, jeopardizing the compressor efficiency. Linear compressor technology, together with the absence of oil, allows the improvement of the discharge manifold system, increasing the thermal insulation of the heat from discharge chamber to the compression chamber, improving overall efficiency. This article aims to show the development of this improved discharge system, tests performed, and results obtained.

1. INTRODUCTION

The development of efficient machines requires from researchers a lot of work, creativity and use of computational and experimental techniques, aiming the achievement of engineering solutions proper to each application. Puff (2008) reported a new solution for a regular hermetic reciprocating compressor that improved thermal efficiency. In the industry of hermetic compressors for household and commercial refrigeration, recent advances lead to the development of new compressors, with lower size and that does not require the use of lubricating oil. The following three points are certainly the ones with major effort spent in this industry field, every time aiming to increase energy efficiency of the compressors used.

- Increase in the electrical motor's efficiency;
- Friction and wear reduction;
- Increase in the thermal efficiency;
- Reduce pressure loss in the valves.

Going a little bit more deeply in the third one, it is well known that one of the greatest heat generation points takes place inside the cylinder, during the compression. One part of the generated heat is transmitted to the discharge chamber wall (*Wb*), and the other part is carried out by the discharge gas (*Wa*). A fraction of the heat that the gas transmits to the discharge chamber wall is transmitted to the internal compressor environment by a convection phenomenon, and the other part to the cylinder wall (*Wc*), by conduction, increasing the cylinder wall temperature and reducing the compression efficiency.

These heat fluxes can be seen in the Figure 1, for linear compressor technology with suction through the piston.

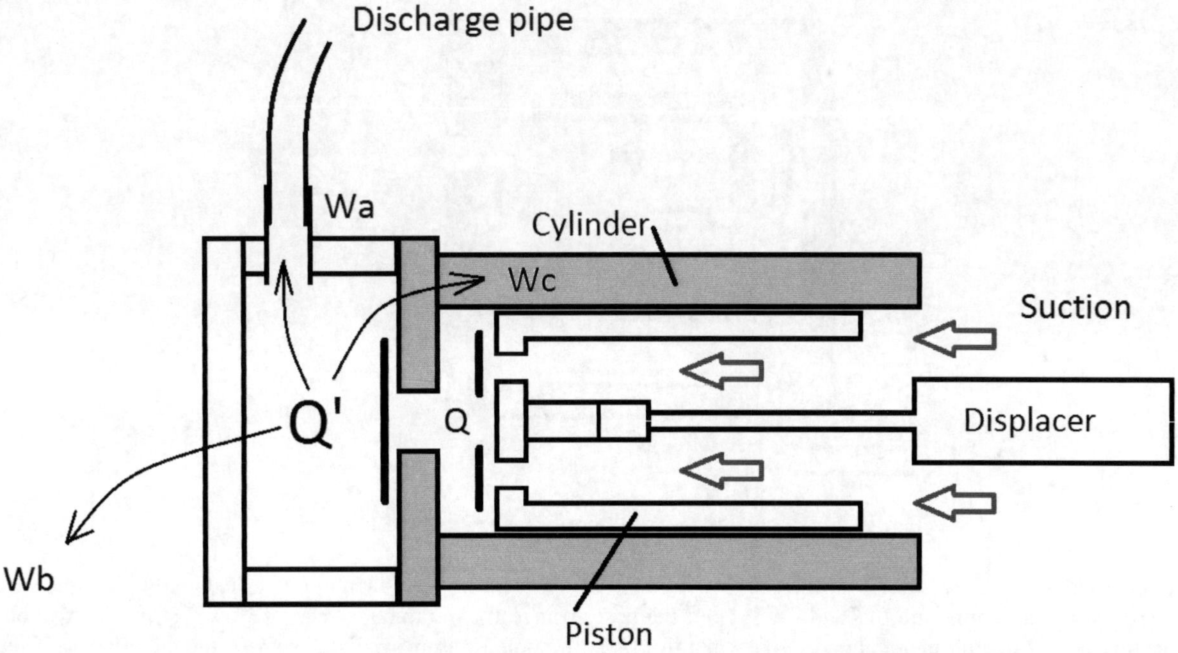

Figure 1: Heat generation and fluxes in the cylinder-manifold system.

Considering the compression efficiency reduction due to this fraction of the heat generated during the compression process which returns to the block and cylinder wall, a new design has been developed in the discharge manifold in order to reduce *Wb* and *Wc*, increasing the amount of heat that is carried by the gas (*Wa*), in that way increasing compressor efficiency.

With relation to the discharge valve pressure drop, the main solution to reduce it, is working on the geometry, flow area and number of valves. This article presents the gains coming from the discharge valve and manifold optimization as well.

2. IMPROVEMENTS IN THE DISCHARGE GAS FLOW

The reference manifold for this part of the analysis is the initial version of the linear compressor technology and which is composed by 4 discharge valves. The vapor is discharged directly into a first volume (*V1*), passing through a tube (*T1*) to the second volume (*V2*). In this reference manifold, the first volume is composed by a metallic material. From the second volume, gas is discharged to the discharge tube (*Td*), and from it to the system discharge line, out of compressor shell. This reference discharge manifold is depicted in Figure 2.

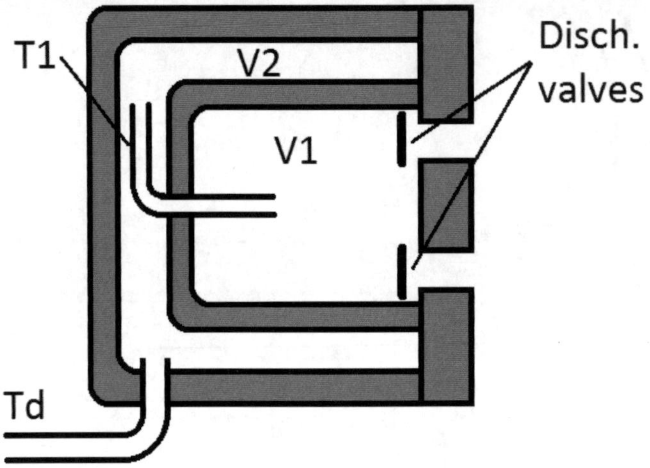

Figure 2: Schematic view of the reference discharge manifold.

The optimization work was done for the gas flow initially by performing analytic and numerical analyses in the way of testing different configurations and finding out the best arrange for minimizing the gas flow restrictions without impacting sound quality characteristics. The manifold was modeled parametrizing the main characteristics of the volumes and tube lengths. This model was implemented into an optimization procedure, using ModeFrontier commercial package and many alternatives could be simulated, aiming the minimization of the pressure pulsation and restriction for the gas flow. In addition, the experimental technique was used to measure the pressure profile in the first volume right after the valves chamber. The experiment consists in installing a pressure transducer and measuring the pressure during compressor operation. Some of the alternatives were experimented and tested using this method.

The analysis and optimization resulted in a configuration composed by 6 discharge valves. Vapor is discharged and passes through a funnel that concentrates the flux to a first volume (*V1*) of the discharge manifold. From *V1*, the vapor passes through a tube (*T1*), directed to a second volume (*V2*). From *V2*, vapor is discharged to the system's discharge line through a discharge tube (*Td*). The third volume (V3) is semi-hermetic and has thermal function as is going to be described in the next section. Figure 3 shows schematically the optimized discharge manifold configuration for this linear compressor generation.

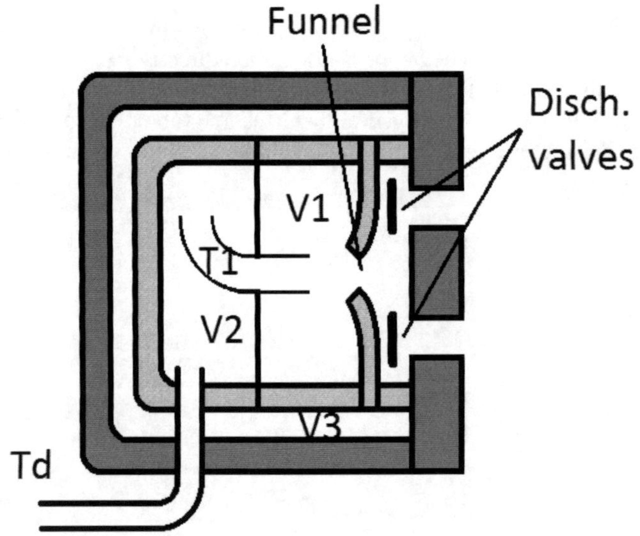

Figure 3: Schematic view of the optimized discharge manifold.

The gain obtained with the use of this optimized gas flow in discharge manifold can be observed in the graphic of Figure 4, that shows the pressure profile right ahead the discharge valves. It can be observed that the integration of the pressure for this manifold represents a reduction in discharge power of 1.2%, with relation to the reference.

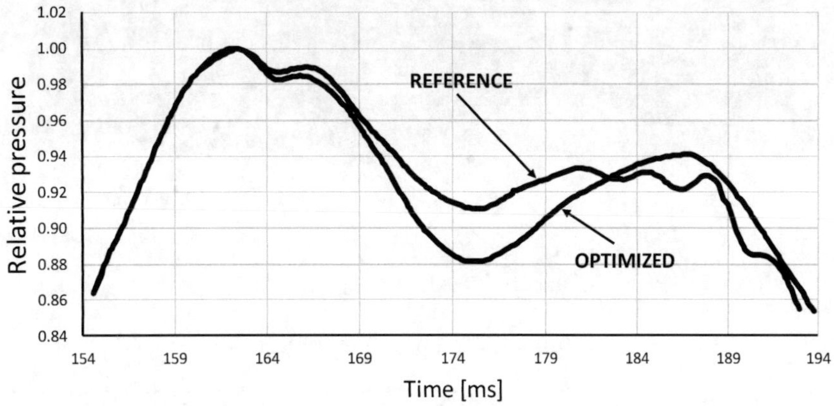

Figure 4: Pressure profile comparison between optimized discharge manifold and reference.

3. IMPROVEMENTS IN THE THERMAL EFFICIENCY

Shiva Prasad (1998) presented an excellent review of research on the effect of superheating in the reduction of thermal efficiency of compressors. The author included in his work an account about major developments of theoretical, numerical and experimental methodologies, part of them used in this work.

Ribas et al. (2008) performed a critical review of different approaches for reciprocating compressors thermal analysis they showed that gas superheating may account for up to half of the thermodynamic losses in a small reciprocating compressor. A good understanding of the compressor thermal dynamics is a requirement for the development of high efficiency compressors.

In this work, some of techniques pointed in the previous articles were used. The strategy adopted for the thermal optimization was initially mapping the compressor temperatures through thermocouples. In an equivalent way, Thermal simulation was used in order to evaluate the temperature distribution inside the compressor, in order to map the hot spots of the machine. Reference configuration is the same shown in Figure 2. It this case, the first volume is composed by a metallic material, as the second volume, which is the cylinder head wall. Both have poor thermal insulation characteristics and the amount of heat that flows to the internal compressor shell volume is high.

The absence of oil in this compressor allows, beside other features, new geometric solutions, because there is no risk that the lubricating oil becomes trapped in specific geometric points. Considering this benefit, many alternatives could be modeled and simulated using a commercial package, making it possible to find a configuration that could fit the initial purposes.

In that way, the proposed solution as well illustrated in Figure 3 for reducing pressure drop. In terms of superheating reduction, the improvement is achieved by using *V1* and *V2* manufactured in polymeric material, with low thermal conductivity. Beside that fact, there is a third volume (*V3*) semi-hermetic, with an additional thermal insulation between volumes *V1* and *V2* vs. cylinder head cover. Hot gas from the compression process is conducted through the discharge tube to the system, out of the compressor.

The optimization result was evaluated, as mentioned, by the measurement of the temperature profiles of the compressor and compared to the reference. Figure 5 illustrates in a compressor scheme the measured points, and Table 1 illustrates the experimental results for the optimized configuration.

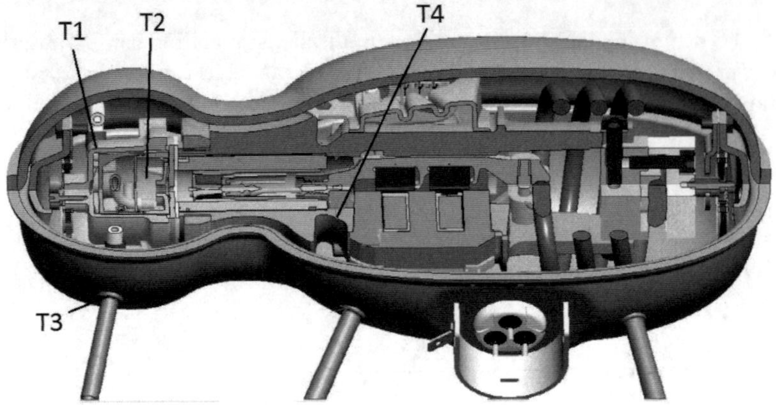

Figure 5: Thermocouples distribution inside compressor.

Table 1: Temperature results.

Thermocouple	Position	Reference	Optimized	Improvement
T_1	Cylinder head wall	109°C	85°C	- 24°C
T_2	First volume (*V1*)	119°C	119°C	0°C
T_3	Discharge pipe	74°C	96°C	+ 22°C
T_4	Block wall	51°C	54°C	+ 3°C

A final evaluation was done in terms of energy efficiency. This evaluation showed an increase of around 1% due to the new temperatures distribution, which is result of the new heat balance in the optimized discharge manifold. It can be observed that the cylinder head walls temperature (T_1) decreased 24°C, even internal volume V_1 temperature (T_2) remaining the same. This indicates that the amount of heat transmitted to the internal shell volume *Wb* is lower. In the other hand, the temperature on the discharge pipe (T_3) increased 22°C, indicating that the gas is carrying more heat *Wa* to outside the discharge line. Finally, the amount of heat transmitted to the block and cylinder did not increase significantly, indicated by the block temperature (T_4) that increased only 3°C.

4. RESULTS AND DISCUSSION

The design of the new discharge manifold for the linear compressor was performed taking into account the absence of lubricating oil. This characteristic allowed the use of new concepts in terms of materials, gas flow with lower pressure drop, and thermal barriers.

The optimization of the gas flow, with 6 valves instead of 4, and the use of the discharge funnel ahead of the first volume *V1* resulted in a decrease of 1.2% in the discharge power.

New materials like the use of polymers could be applied, with lower thermal conductivity, as well as the application of semi hermetic volumes, aiming to avoid heat conduction to the internal volume of the shell. This result is a consequence of the increase of heat carried by the gas to outside the compressor through the discharge line, and the reduction of the heat transmitted to the internal shell volume.

These thermal concepts resulted in an increase of around 1% in the thermal efficiency of the compressor.

REFERENCES

Puff, R., 2008, Hermetic Compressor Manifold analysis with the Use of the Finite Element Method, 19th *International Compressor Engineering Conference at Purdue*, Proceedings of 2008 International Compressor Engineering Conference at Purdue.

Shiva Prasad, B.G., 1998, Heat Transfer in Reciprocating Compressors - A Review, 14th *International Compressor Engineering Conference at Purdue*, Proceedings of 1998 International Compressor Engineering Conference at Purdue.

Ribas Jr, F. A., Deschamps, C. J., Fagotti, F., Morriesen, A., Dutra, T., Thermal Analysis of Reciprocating Compressors – A Critical Review, 18th *International Compressor Engineering Conference at Purdue*, Proceedings of 2006 International Compressor Engineering Conference at Purdue.

ACKNOWLEDGEMENT

A special acknowledgement to the researcher Talita Wajczyk for the extensive analysis performed during this research.

Numerical Simulation for the Internal Flow Analysis of the Linear Compressor with Improved Muffler

Hyunjoo OH[1]*, Sungchul KONG[1], Wonsik OH[1], Kyeongbae PARK[1]

[1]LG Electronics, L & A Research Center
Seoul, Republic of South Korea
hyunjoo.oh@lge.com
wonsik.oh@lge.com
sungchul.gong@lge.com
kb.park@lge.com

ABSTRACT

The numerical simulation is frequently used to estimate experimental results in fluid flow dynamics. In this paper, we focus on increasing the efficiency of linear compressor by applying modified muffler. Furthermore, we analyze a fluid flow in a suction line of linear compressor by using a computational fluid dynamics. We assume an unsteady flow, compressible, and two dimension axis-symmetric condition. In numerical simulation models, all processes of linear compressor, i.e., compression, discharging, expansion and suction, are continuously performed. Numerical simulation results are compared with experimental data to validate an accuracy of simulation. Furthermore, we show that numerical simulation results are well converged according to time. The results present that the mass compressor efficiency increases when the pressure field in the muffler exit is highly maintained during the suction process.

1. INTRODUCTION

In recent world, many countries have been strengthening the regulation of energy savings. Global home appliance markets interested on energy consumption have been also growing. Especially, because household refrigerator consumes about 5% (RECS, 2009) of using home energy, numerous manufacturers related household refrigerator have made a great effort to increase the energy efficiency. As example, the compressor accounting for up to 90% (LG homepage, 2016) of household refrigerator efficiency plays an important role to save energy in home appliance.

The efficiency of compressor can be improved by optimizing muffler design and by decreasing superheating loss and pressure loss. The muffler design especially affects efficiency of compressor by changing the flow field in suction line of compressor. Therefore, the suction muffler in compressor has been actively studied to improve energy efficiency of compressor. Furthermore, the computation fluid dynamics (CFD) has been used to expect results of compressor performance with respect to suction muffler, because the CFD commercial tool has been developed and we can obtain more exact results in short time. Hence, many works for muffler design have been studied by using CFD. Kemal *et al.* (2012) studied the refrigerant flow inside the suction muffler of compressor by using CFD. The results of numerical analysis shown the mass flow rate calculated by CFD is good agreement with the experiment results. C. J. *et al.* (2002) presented muffler affection in the compressor performance. This numerical model was applied one dimensional, compressible and thermal flow. Kwangha *et al.* (2006) developed the research of compressor by computational simulation. The compressor shell's modal simulation was conducted by CAE, and model with muffler and suction valve in suction line was analyzed by CFD with FSI. Umut and Hasan (2013) investigated change of the mass flow rate and pressure with respect to muffler shape. They solved simulation model of three dimension, unsteady, and validated experimental data by comparing CFD data.

In this paper, we analyze the internal flow of the linear compressor with improved muffler by using CFD. The simulation of compressor is carried out two dimensional axis-symmetry, unsteady, compressible and turbulence flow.

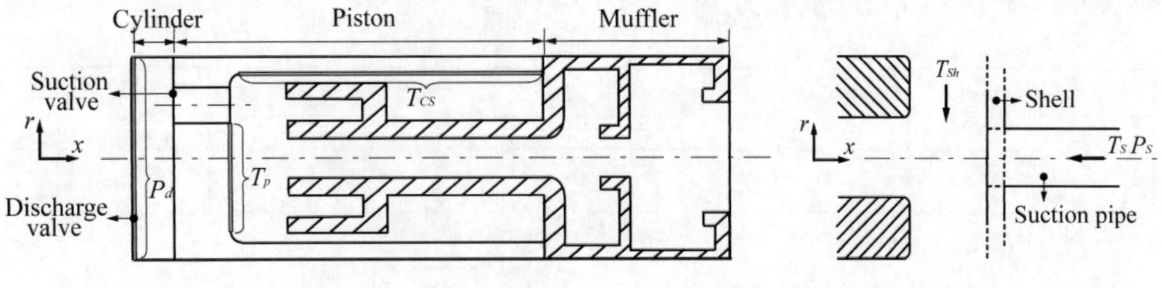

(a) Piston, cylinder and muffler (b) Suction part around shell

Figure 1: Schematic in suction system of the linear compressor and temperature condition on boundary

We considered four cases to evaluate the effect of suction muffler. Furthermore, we optimize suction muffler by analyzing change of flow field in suction muffler exit and suction muffler entrance of linear compressor.

2. NUMERICAL METHODOLOGY

In this section, we define initial condition, boundary condition and simulation method for computational model of linear compressor. Figure 1 shows schematic in suction system of the linear compressor and boundary conditions for temperature and pressure. As shown in Figure 1(a), we denote temperature of the inner piston's cylindrical surface and the inner piston's surface on suction port entrance as T_{cs} and T_p, respectively, which is about 333 Kelvin and 336.5 Kelvin. Moreover, we denote compression pressure applied on discharge port as P_d. We denote boundary conditions in suction pipe and around muffler entrance connected shell space, respectively, as T_s, P_s and T_{sh} in Figure 1(b). T_s, which is about 298 Kelvin and P_s respectively represent the temperature and the pressure of refrigerant flowing from evaporator. We involve shell space in simulation domain that connected muffler entrance. The boundary temperature on shell space is about 310.7 Kelvin, is denoted as the T_{sh}. The T_{cs}, T_p and T_{sh}, which are already mentioned, are obtained by experiment, and they are constant values in according with time.

The number of mesh of simulation model to finite volume method is about 65000. We generate rectangular meshes by using ANSIS ICEM program, and simulation domain is solved unsteady, compressible, viscous flow with ANSYS FLUENT V.15 commercial program. We set the simulation fluid as a real-gas-isobutan based on physical property in REFPROP. To implement motion of linear compressor, we use dynamic mesh including layering and re-meshing method. SIMPLE scheme is used by pressure-velocity coupling algorithm. We adopt second order upwind to interpolation between the nodes and nodes. The time step size to run calculation and time step to converge are set as $5xe^{-05}$ and 50, respectively.

In this CFD simulation, whole process of the linear compressor, i.e., compression, discharging, expansion, suction is continuously generated by using user defined function and scheme-language until the cycle converges. Figure 2 shows the pressure-volume diagram of whole processes. The compression process is defined that pressure of area-weighted average in cylinder increases from P_s to P_d as a polytropic process. In the discharging process, the pressure in cylinder is affected by refrigerant flowing out cylinder through discharse valve until the piston reaches to top dead center. After that, the pressure in cylinder decreases from P_d to P_s as a polytropic process. We call this process as the

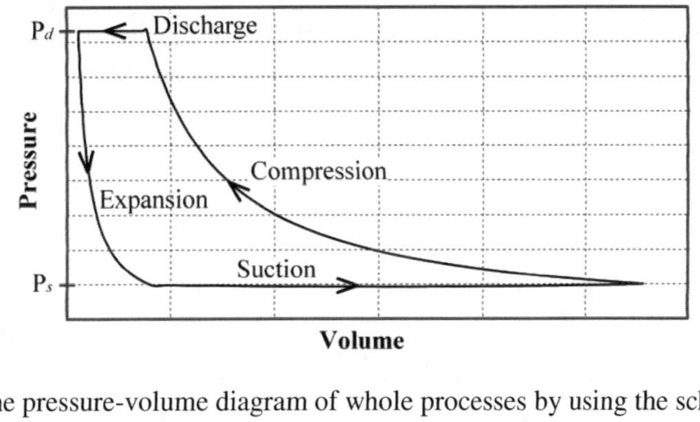

Figure 2: The pressure-volume diagram of whole processes by using the scheme language

Table 1: The four cases of modified suction muffler

Applied item	case1	case2	case3	case4
Partition wall	none	none	applied	applied
Cylindrical projection	none	applied	none	applied

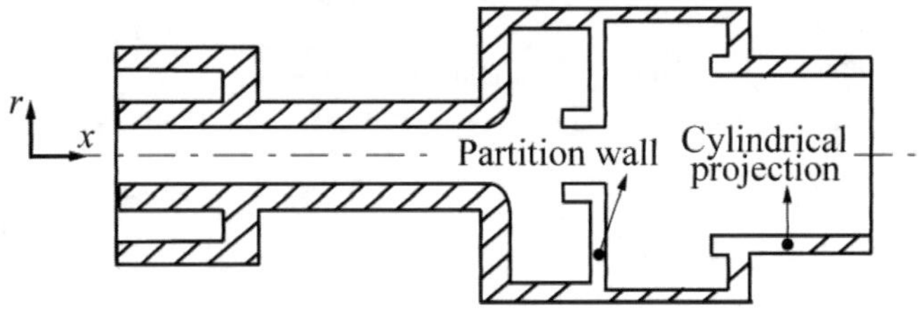

Figure 3: Schematic of the modified suction muffler

expansion process. Finally, the suction process is affected by refrigerant flowing in cylinder through suction valve. Last process is continued until pressure in cylinder reaches the P_s. Hence, valves of linear compressor only controlled by pressure difference between cylinder and muffler exit area and by simulation time. The suction valve and discharge valve are assumed as ideal valve. Therefore, valve motion is carried out to fully-open, fully-close. Because the simulation model is solved in two dimensional axis-symmetry method, it is assumed to be three dimensional structures based on the rotation axis. Therefore, suction valve is made by donut shape.

3. RESULTS AND DISCUSSION

3.1 Validation of the Computational Simulation

The internal flow of the linear compressor is affected by suction muffler structure existed to suction line from suction pipe to suction valve. Therefore, we analyze the internal flow of the linear compressor with improved muffler by using computational simulation.

We consider four cases of modified suction muffler in Table 1. Figure 3 shows schematic of the modified suction muffler. As shown in Figure 3, the partition wall is located between muffler entrance and exit, and the cylindrical projection is connected to the entrance of the muffler. The internal and an external diameter of partition wall is 0.008m, 0.025m, respectively. The height and diameter of cylindrical projection are 0.015m. In Table 1, case1 indicates base model without partition wall and cylindrical projection. case2 is applied cylindrical projection only.

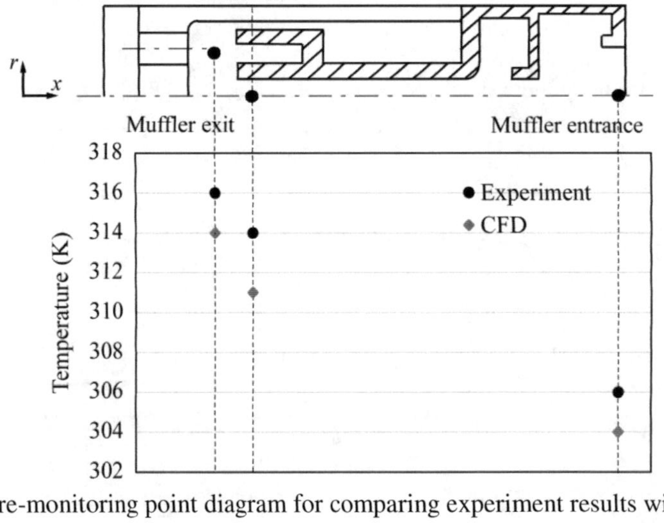

Figure 4: Temperature-monitoring point diagram for comparing experiment results with CFD data in case1

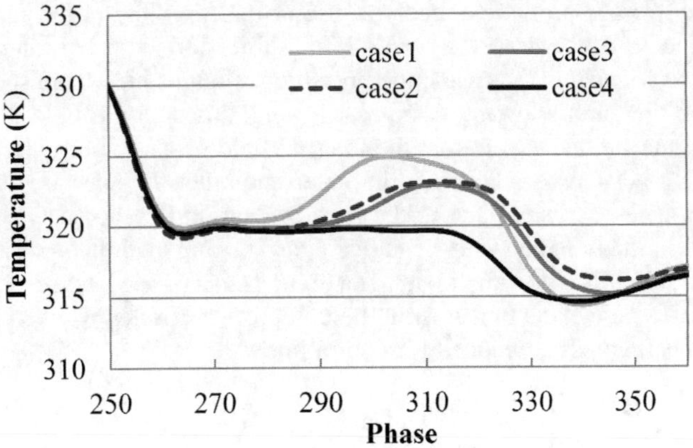

Figure 5: The temperature-phase diagram in muffler exit in case1, case2, case3 case4 during the suction process

On the other hand, case3 is applied only partition wall and the last one is applied all of them. We compare experimental data with CFD results to validate an accuracy of simulation. Figure 4 shows the temperature of each monitoring point in model with partition wall without cylindrical projection. The experiment and CFD data is monitored at the muffler exit, the muffler entrance and the distance of 0.01m from the muffler entrance. The results show a good correlation. The temperature difference of each monitoring point is less than 1%. Therefore, there results verify that CFD results are sufficiently reliable.

3.2 Comparative Analysis of Modified Muffler

In this paper, we analyze the internal flow in the muffler and modify the muffler to increase the compressor's efficiency. In Figure 5, we show the temperature-phase diagram at muffler exit during the suction process. The suction valve opens and refrigerant goes in the cylinder due to the low pressure field in cylinder, and this refrigerant temperature in the muffler exit of linear compressor represents the valve between 315 and 330 Kevin in all cases. In case 1 in which the partition wall and cylindrical projection are not adopted, the temperature increases about 325 Kelvin in about 300 degree of phase and decreases 314.5 Kelvin in about 340 degree of phase. In case2 and case3, as in case1, similar results are shown in which the temperature increases about 323 Kelvin, and decreases 316 and 314 Kelvin, respectively. The temperature in case3 is lower than that in case2 by difference between 0.5 and 1. However, the trends of temperature are greatly similar. In case4 in which the partition wall and cylindrical projection are adopted, the temperature is about 320 Kelvin in about 313 degree of phase and decreases about 314.5 Kelvin in about 338 degree of phase.

We investigate the temperature contour to explain the result aforementioned. In Figure 6, we show the temperature in case3 and case4 at the muffler entrance. We provide the temperature contour in case 1 and case4 in Figure 7. The cylindrical projection interrupts the diffusion of cool refrigerant going out in the shell space. Hence, in case4, the temperature can be lower than that in case3. Furthermore, in case1, 2, and 3, the temperature is higher than that in case4 due to the effect of the hotter refrigerant coming from the upper side of the muffler.

In Figure 8, we show the pressure-phase diagram in muffler exit of all cases in the whole process of compressor. In case2 and case3, the pressures are similar in the compression, discharge, and suction processes. Furthermore, the pressure of case1 is lower than that of case2, 3, and 4 in about 300 degree of phase.

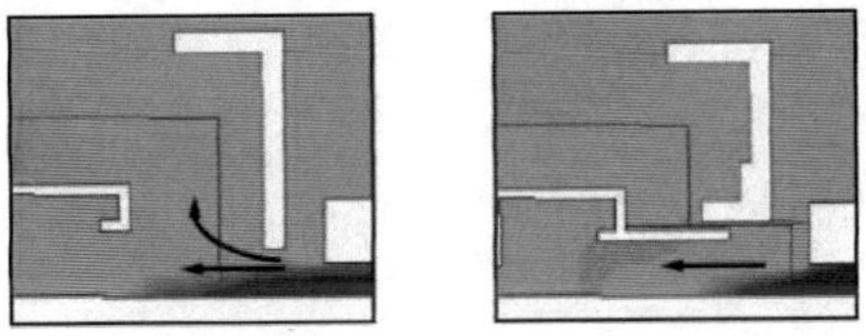

Figure 6: The temperature contour of case3(left) and case4(right)

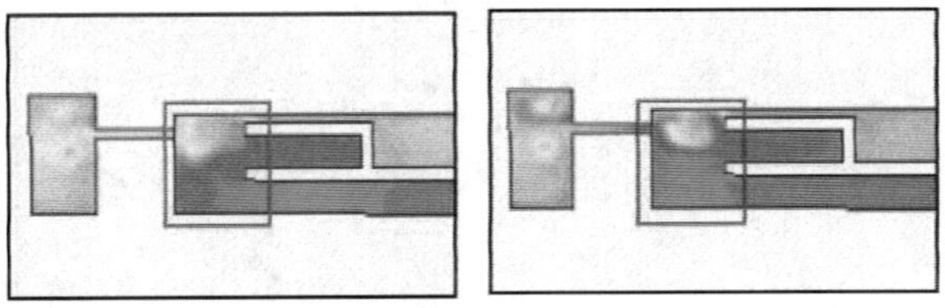

Figure 7: The temperature contour of case1(left) and case4(right) in the suction process

In Figure 9, we show the time average pressure at each position of suction line in the suction process. In case4, the narrow suction line made in the muffler entrance by cylindrical projection and partition wall, and that increases the pressure field in the suction line. Hence, in the case4, the pressure maintains highest in suction line, the pressures are similar in case2 and 3, and the pressure is lowest in the case1. And, this pressure field develops pressure difference of between inner cylinder and muffler exit. Note that the internal field of the cylinder has relatively low pressure in the suction process. The pressure differences between the inner and outer cylinder are about 146Pa, 157Pa, 176Pa, and 188Pa in case1, 2, 3, and 4, respectively. Due to the high pressure difference between the muffler exit and the cylinder, the cool refrigerant of suction line can be moved fast in the suction process, and it can interrupt hot refrigerator of muffler upper side with flowing in muffler exit. Hence, case4 represents the highest compressor efficiency as a result in Figure 10. We expect increasing the compressor efficiency when the pressure can be high at the muffler exit and the superheating loss is reduced in suction line.

4. CONCLUSIONS

We can estimate compressor efficiency in according with muffler design by using CFD tool. Through this study, we can obtain conclusions as in the following.

- We suggested improving suction muffler of the linear compressor by using CFD.
- The refrigerant of low temperature flows into the cylinder when the pressure in muffler exit is highly maintained during the suction process.
- The efficiency of compressor improved when refrigerant flows in the suction line without superheating loss.

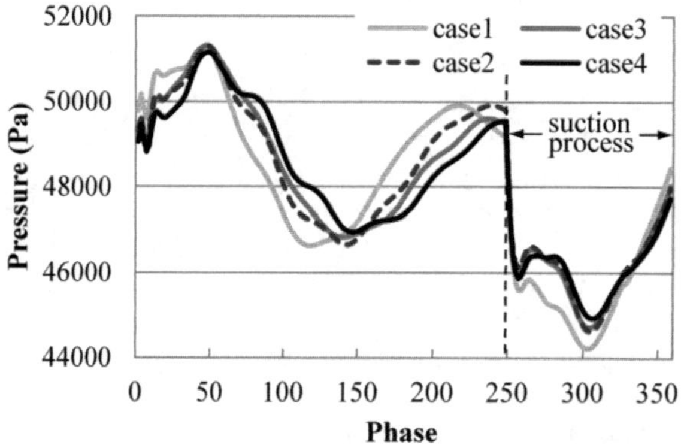

Figure 8: The pressure-phase diagram in muffler exit in case1, case2, case3 and case4 in the whole process

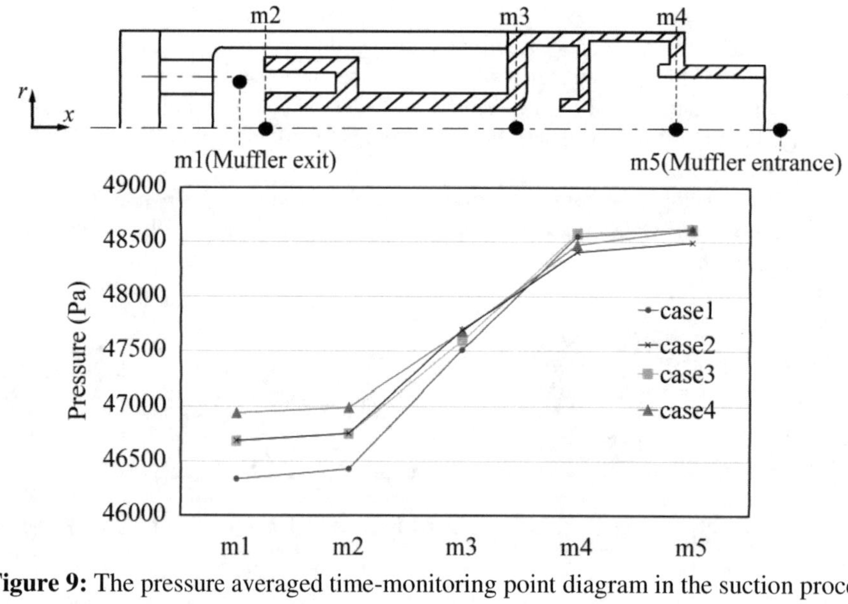

Figure 9: The pressure averaged time-monitoring point diagram in the suction process

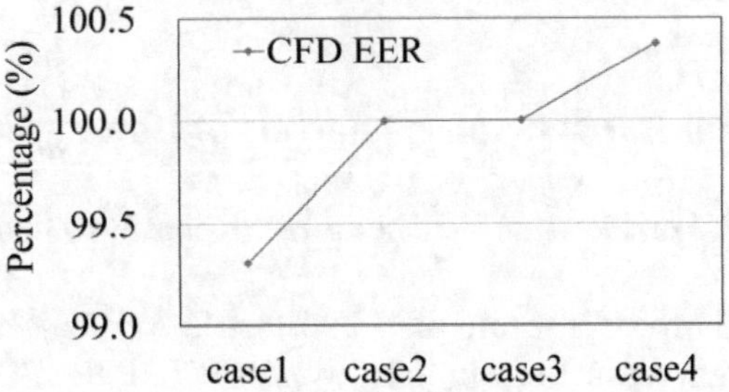

Figure 10: The percentage of the compressor EER for comparing experiment with CFD data in case3 and case4

NOMENCLATURE

T	temperature	(K)
P	pressure	(Pa)
r	radial direction	(m)
x	x direction	(m)

Subscript

cs	cylindrical surface
p	piston surface
s	suction
sh	shell
d	discharge

REFERENCES

U.S. Energy Information Administration, Residential Energy Consumption Survey (RECS) 2009, How Energy Is Used in Homes (2009).

www.lg.com, Discover LG Refrigerator Technology, Guaranteed Reliability.

Kemal, S., Ahmet, O., Emre, O., & Atilla, K. (2012), An Experimental and Numerical Analysis of Refrigerant Flow Inside the Suction Muffler of Hermetic Reciprocating Compressor, *International Compressor Engineering Conference at Purdue.*

Kwangha, S., Daenyoung, H., & Hyeongseok, K. (2006), CAE/CFD Application for Linear Compressor, *International Compressor Engineering Conference at Purdue.*

C, J. D., F, C. P., E, & L. L. P. (2002), Numerical Simulation of Pulsating Flow in Suction Mufflers, *International Compressor Engineering Conference at Purdue.*

Umut, C. C., Hasan, G., & Kemal, S. (2013), Numerical Investigation of Suction Muffler in Household Refrigerator Compressor, *International Conference on Mechanics, Fluids, Heat, Elasticity and Electromagnetic Fields.*

ACKNOWLEDGEMENT

This research was supported by L & A Research Center of LG Electronics.

Mechanical Impact Noise Analysis of Rotary Compressor

Jinquan Zhang *, Peizhen Que, Yusheng Hu, Rongting Zhang, Yuanjie Shang

Compressor and Motor Institute of Gree Electric Appliances, Inc. of Zhuhai,
Jinji West Rd., Zhuhai City, 519070, P. R. China
Phone: +86-756-8669017, Fax: +86-756-8668386,
E-mail: popway@163.com

* Corresponding Author

ABSTRACT

In the noise and vibration test of 8HP rotary compressor, the peak value of noise and vibration is especially high at frequency range between 600~1000Hz. The sound pressure level (SPL) exceeds the enterprise criterions and the sound quality is so tough that persons cannot bear such annoying noise. Two experiments, pressure pulsation above/ below compressor motor and mechanical impacts of compressor pump-rotor unit, are particularly carried out in order to identify the noise sources. The test results confirm that the axial up-down movement of pump-rotor unit is the primary reason, causing the abnormal noise and vibration. Several measures, including partially cutting the circumferential edge of the stator and increase the height differential between rotor and stator, have been taken in order to reduce pressure pulsation above/below motor and restrain the axial movement of shaft-rotor unit. With these measures, the noise problem of 8HP rotary compressor has been solved successfully. Finally the sound pressure level decreases about 4dB and the sound quality is comfortable and good to hear.

1. INTRODUCTION

Traditionally, usage of multiple scroll compressors is widely applied for general variable refrigerant flow (VRF) system at more than 5HP capacity because of its performance and large capacity advantages. But usage of multiple scroll compressors have problems of controlling cost, which greatly limits the further expanding of commercial VRF system market. Fortunately, with the development of technology and progress of manufacture equipment, inverter rotary compressors are used not only in house-hold air conditioner market but also in commercial VRF market because of its efficiency, comfort, simple structure, low cost, etc. Recently, 5HP~12HP inverter rotary compressors, which can operate in a wide running range from low speed to high speed and from low load to high load, are developed successfully to replace similar capacity scroll compressors due to the need of cutting cost.

What is more, with the capacity getting larger and larger, more attentions should be paid to noise and vibration controlling, so how to diagnose the abnormal noise rapidly and correctly is an urgent issue for large capacity rotary compressor. In addition, the strong requirements for low noise product of customers have also enforced many compressor makers worldwide to focus on development of low noise compressor. As compressor is the main noise source of air conditioner, therefore the noise and vibration is a big concern over the course of development [1].

In this study, noise, vibration, various forces, pressure pulsation, mechanical impacts were investigated of 8HP rotary compressor. In the noise and vibration test of the compressor, the peak value was especially high at frequency range between 600~1000Hz. The sound pressure level (SPL) achieved 80.3dB with its operating speed of 60Hz, which exceeded the enterprise criterions. The sound quality was so tough that persons can't bear such annoying noise. Two experiments, pressure pulsation above/below compressor motor and mechanical impacts of compressor pump-rotor unit, were particularly carried out for the purpose of identifying the noise source. The results confirmed that the axial up-down movement of pump-rotor unit was the primary reason, causing the abnormal noise.

According to forces analysis of pump-rotor unit and two experiments, several measures had been taken, especially the height differential between rotor and stator was increased to improve the axial magnetic force dramatically, which could overcome the discharge gas force. With these measures, the noise problem of 8HP rotary compressor has been solved successfully. As a result, the peak value disappeared at frequency range between 600~1000Hz and the sound pressure level reduced approximately by 4.1dB.

2. NOISE PROBLEM DESCRIPTION

In order to diagnose the abnormal sound, noise and vibration were measured of the 8HP, large capacity compressor in hemi-anechoic room. Figure 1, Figure 2 shows the noise spectrum and vibration spectrum respectively, from which several key characteristics of the noise problem can be observed.

Firstly the high peak values mainly concentrate at frequency range between 600~1000Hz both in noise and vibration spectrum. Secondly the most serious noise problem occurs at around the operating speed of 60Hz, in which the peak value reaches to about 64.4 dB. At last, the vibration spectrum is particularly remarkable. Unlike the spectrum of X, Y direction of compressor pump, the spectrum of Z direction is extraordinary high at 600~1000Hz. Thus a conclusion can be made, that is the abnormal vibration of Z direction is the prime reason for the noise problem. This study is designed to investigate the source of the vibration issue and focus on reducing it.

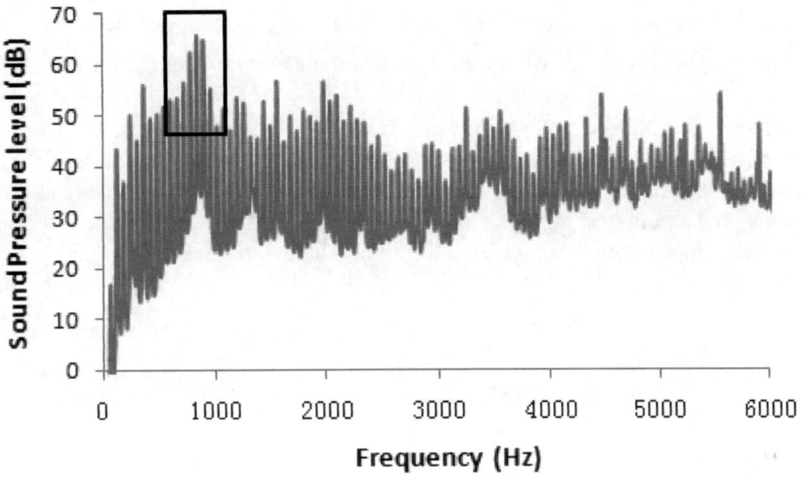

Figure 1: Noise spectrum of 8HP compressor at 60Hz running

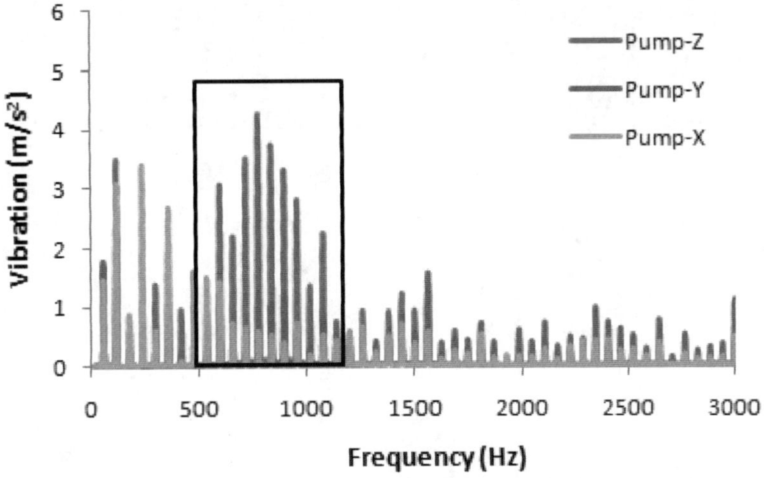

Figure 2: Vibration spectrum of 8HP compressor at 60Hz running

3. FORCE ANALYSIS AND PRESSURE PULSATION EXPERIMENT

According to the results of above noise and vibration measurement, the pump-rotor unit may move up and down in cylinder under some rather unique condition. Of all components for pump-rotor unit, the unstable movement of shaft-rotor most probably causes axial displacement up and down, which could result in serious axial impact vibration [2]. In order to investigate the mechanism and relationship between axial movement of shaft-rotor unit and the abnormal noise problem, a detailed analysis is taken on correlative forces of shaft-rotor unit, including pressure pulsation, axial magnetic force, gravity, damping effect.

3.1 Axial Force Analysis of Shaft-Rotor Unit

In discharge period of rotary compressor, the refrigerant flows into the chamber under the motor, and then reaches the chamber above the motor through motor passages. The pressure in chambers above and under the motor is different from each other, which may cause axial displacement of the shaft-rotor unit when the pressure differential force is larger than self-weight of the shaft-rotor unit and axial magnetic force of the motor. The axial motion of the shaft-rotor unit shown in Figure 3 is expressed by the following equation:

$$m\frac{d^2X}{dt^2} = \Delta PS_r - mg - C\frac{dX}{dt} - F_m \qquad (1)$$

Where, m :mass of the shaft-rotor unit, X :axial displacement, ΔP :differential pressure ($\Delta P = P_2 - P_1$), S_r :sectional area of the rotor, C :damping coefficient of vibration, F_m :axial magnetic force of the motor.

The axial magnetic force is determined by the height differential Δh between rotor and stator shown in Figure 3. In addition, simulations of axial magnetic force have been completed to estimate the effect of height differential and motor torque. Fig.4 shows that the axial magnetic force increases greatly with the increasing of height differential between rotor and stator, but the increasing trend is not linear relationship. Fig.5 indicates that the axial magnetic force increases linearly with motor torque for a certain 4mm height differential.

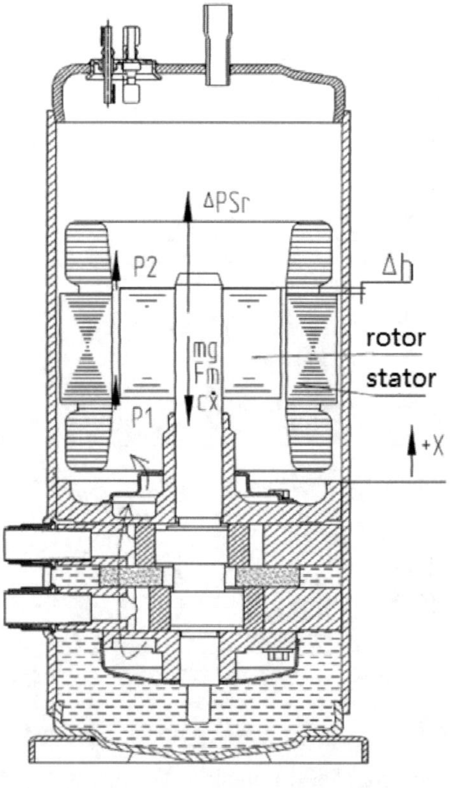

Figure 3: Axial force analysis of shaft-rotor

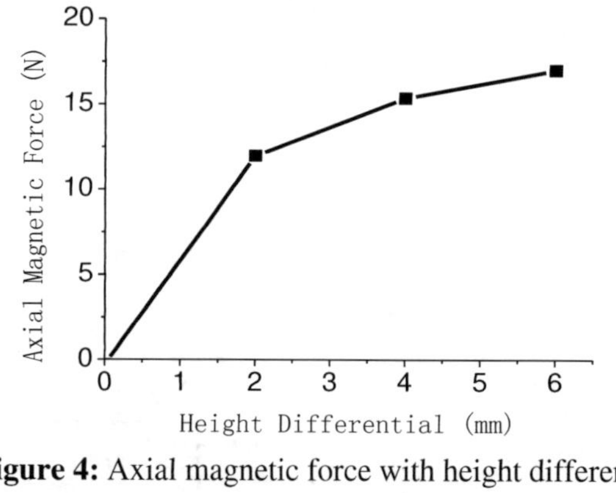

Figure 4: Axial magnetic force with height differential

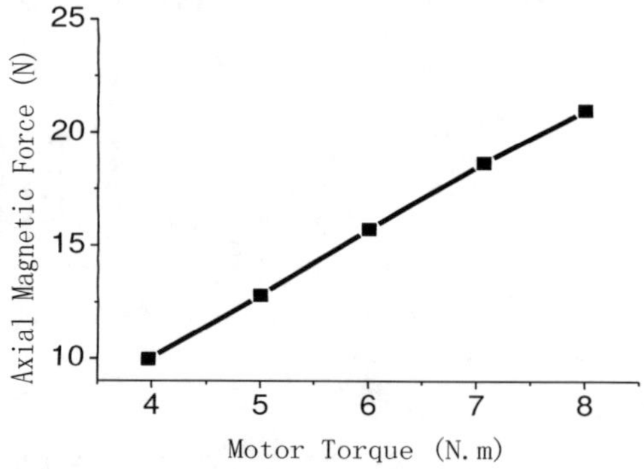

Figure 5: Axial magnetic force with motor torque

3.2 Pressure Pulsation Experiment

In recent years, with the development of high speed, large capacity compressor, the discharge pressure pulsation becomes more and more serious, which may induce various problems. What's more, due to the intermittent discharge, pressure pulsation has a great influence on the axial movement of shaft-rotor unit, especially for two-cylinder, large capacity rotary compressor. Experiment of pressure pulsation was carried out with two pressure sensors in chambers above and below the compressor motor.

Fig.6 shows the experimental results of the differential pressure ΔP between the lower chamber pressure P_1 and the upper chamber pressure P_2. As shown in Fig.6, the maximum differential pressure approaches 0.02MPa, which may introduce great impact force on pump-rotor. The axial discharge force depends on pressure pulsation in chambers above and below the motor. When the lower chamber pressure P_1 overcomes the upper chamber pressure P_2, the pressure differential force will push the shaft-rotor unit move upward, otherwise the pressure differential force will push the shaft-rotor unit move downward. However, the axial up-down displacement most probably takes place when the load condition is very tough and the pressure differential force exceeds self-weight of the shaft-rotor unit. In such situation, abnormal vibration of Z direction would be quite obvious and the noise problem emerges at the same time.

By the way, the damping force will not be discussed in detail of this paper, because the damping coefficient depends on the thickness of oil film, which is hard to provide. Although there have been some investigations on the value of the flow damping coefficient, but their results cannot been applied directly to this study. Concerning damping force of the vibration, squeeze force of oil film on the shaft thrust is superior to shearing force of oil film on the shaft journal.

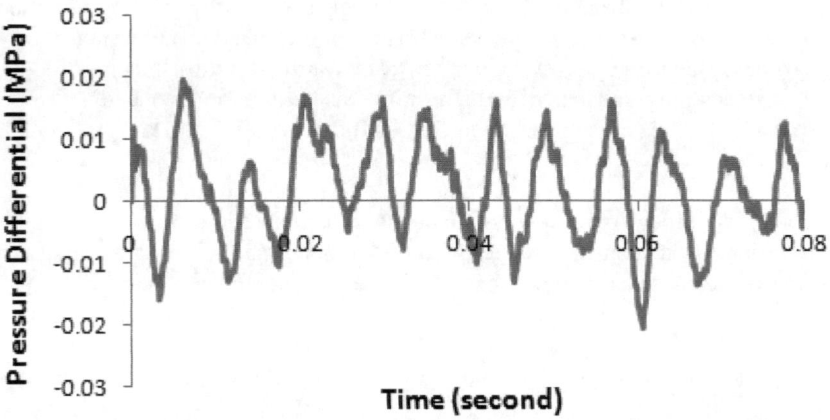

Figure 6: Pressure differential of chambers above/under motor at 60Hz running

4. MECHANICAL IMPACT EXPERIMENT

For the purpose of confirming and identifying the noise source of impact noise, experiment of mechanical impacts of compressor pump-rotor unit was particularly designed and carried out, which would help engineers comprehend the noise mechanism more clearly. As shown in Fig.7, the measurement equipments mainly consist of a pump-rotor unit, a microphone, and corresponding signal sampling & analyzer system.

In the experiment, hold the rotor with a hand and push the shaft-motor unit rotate intermittently and move up and down. This kind of movement not only brings axial up and down displacement but also operates as a normal working compressor. In this way, the unstable movement, causing abnormal vibration of Z direction and noise problem, is simulated outside the hermetic compressor. In addition, the signals of components in compressor pump impact each other are received with a microphone. Thus, the sound response of the unstable movement can be obtained easily.

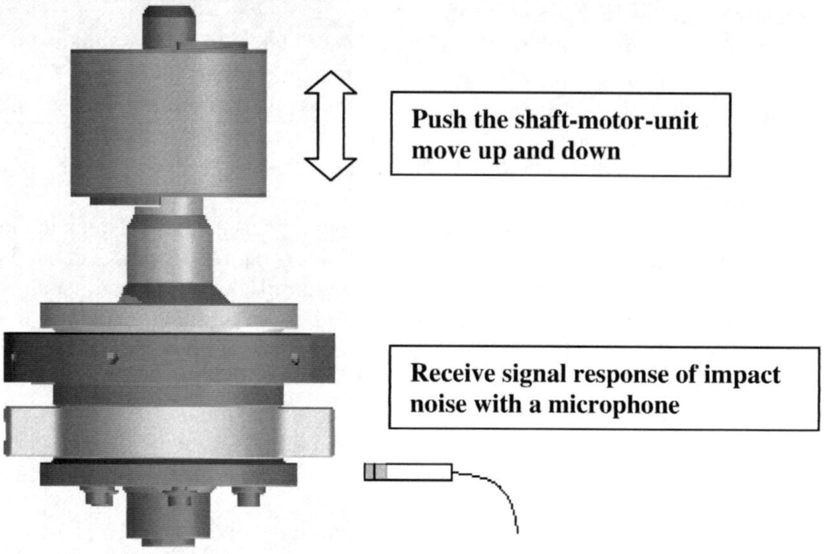

Figure 7: Experiment of mechanical impacts of pump-rotor unit

With FFT analysis, the received sound response can be transformed to frequency domain. One important thing should be mentioned, as the frequency distribution is the core interesting point of the impact response, therefore the magnitude of the response curve has been modified for easy comparison with noise and vibration spectrum. The red curve in Fig.8 and Fig.9 describes the frequency characteristics of the sound signal response for unstable movement of shaft-rotor unit. In the response curve, the maximum response mainly distributes at frequency range between 600~1100Hz, which has a strong coherence with the noise and vibration frequency distribution. And there is relatively larger response at frequency range between 1800~2100Hz, 2300~2500Hz, 3200~3500Hz, which also need more attentions.

As shown in Fig.8 and Fig.9, through comparison with the noise and vibration spectrum, it can be seen that there is a good agreement of the interested frequency range between the abnormal noise/vibration spectrum and the impact response curve. Moreover, the impact response has great effect on compressor noise at frequency range between 600~1000Hz, 1800~2100Hz, 3200~3500Hz. In addition, the shape of the noise spectrum has changed by the influence of the corresponding frequency range in Figure 8. The results confirm that mechanical impacts of shaft-rotor unit with the other components in pump are the primary reason, causing the abnormal vibration of Z direction and noise problem.

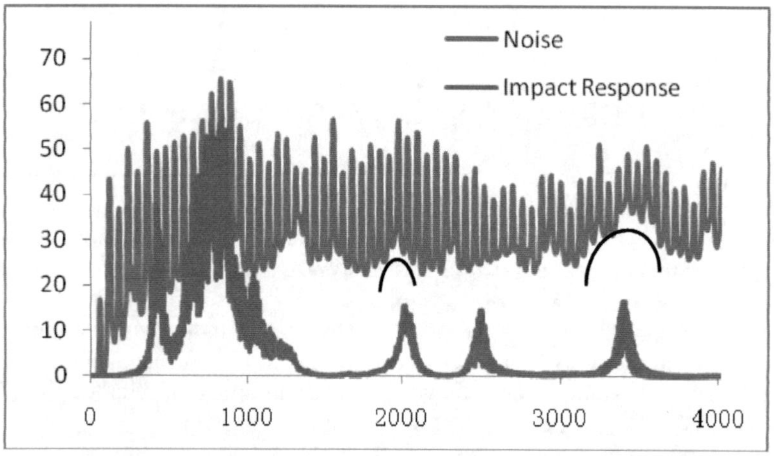

Figure 8: Comparison of noise and impact spectrum

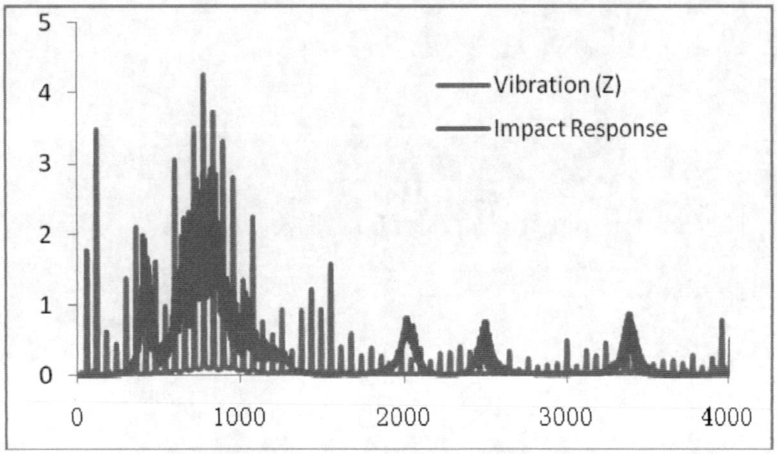

Figure 9: Comparison of vibration and impact spectrum

5. NOISE REDUCTION

From the above consideration, it is known that the shaft-rotor unit moves up and down as rigid motion, then corresponding noise and vibration are brought by being excited up-down movement of the pump. In order to solve the problem, several measures have been taken as follows:

(1) Reduction of pressure pulsation above and below motor:

A. Setting several holes passing through the rotor.

B. Partially cutting the circumferential edge of the stator.

(2) Restraint of axial movement of the shaft-rotor unit

Increase the height differential Δh between rotor and stator to improve the axial magnetic force. The direction of the magnetic force must point to the ground. Together with self-weight of the pump-rotor unit and damping force, the axial magnetic force could overcome the discharge gas force so that the shaft could be pushed down stably.

With these measures, sound pressure level of the modified compressor has been decreased from 80.3dB to 76.2dB at the operating speed of 60Hz. Compared with the original compressor, the sound pressure level is reduced approximately by 4.1dB and the peak value has been decreased approximately by 20.2dB at frequency 840Hz as shown in Table 1. The final results indicate that the measures applied take great significance for the noise problem brought by up-down movement of shaft-rotor unit.

Table 1: Comparison of sound pressure level and peak values

	Original Compressor	**Modified Compressor**	**Reduction (dB)**
Sound Pressure Level	80.3	76.2	4.1
Peak value	64.4 / 840Hz	44.2 / 840Hz	20.2

In detail, noise and vibration spectrum are compared for checking the frequency difference. As shown in Fig.10, the noise spectrum of the modified compressor has been decreased dramatically at frequency range between 600~1020Hz, 1800~2520Hz, which has a good coherence with the frequency distribution of impact response curve. In addition, the vibration spectrum of pump-Z direction has also been decreased greatly at frequency range between 300~1080Hz for the modified compressor as shown in Fig.11, which has a good agreement with the impact response curve. Fig.12 shows vibration spectrum of the modified compressor at location of pump. Obviously the vibration value of pump is low enough for an 8HP capacity compressor. As a result, the noise problem of 8HP rotary compressor has been solved successfully and the peaks disappear at frequency range between 600~1000Hz.

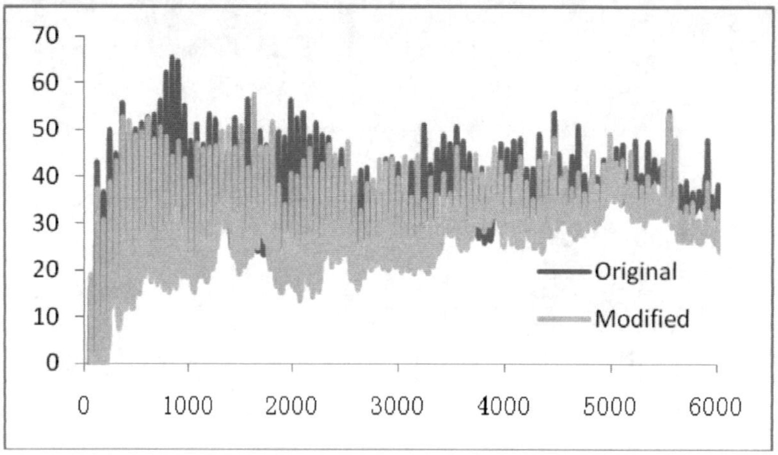

Figure 10: Noise spectrum of original and modified compressor

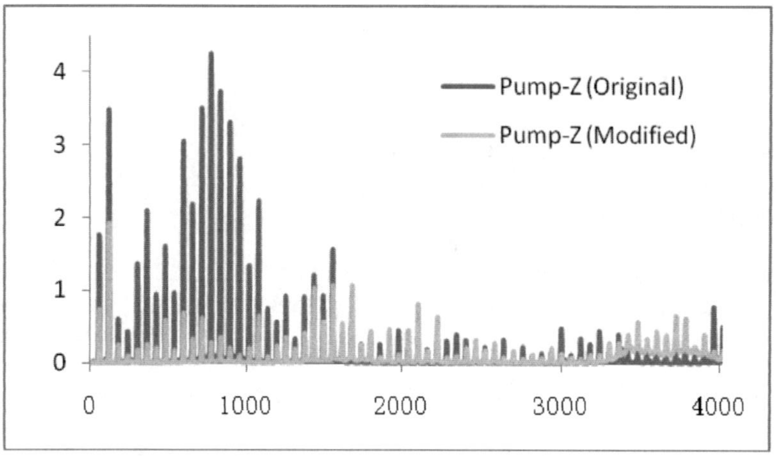

Figure 11: Vibration spectrum of original and modified compressor (Pump-Z)

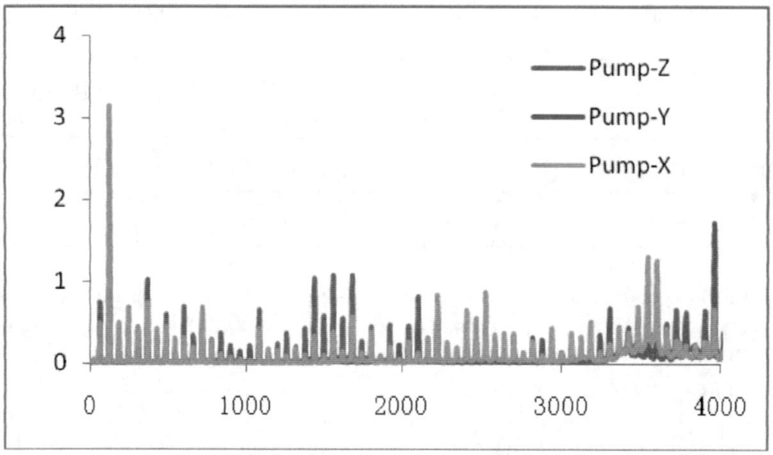

Figure 12: Vibration spectrum of modified compressor

6. CONCLUSIONS

In this paper, the noise problem of 8HP rotary compressor was investigated in order to reduce sound pressure level and improve the sound quality with large capacity compressor. According to the abnormal frequency spectrum of noise and vibration, correlative forces of shaft-rotor unit were made a careful analysis, including pressure pulsation, axial magnetic force, gravity, damping effect. Two experiments, pressure pulsation above/below compressor motor and mechanical impacts of shaft-rotor unit, were carried out for the purpose of identifying the noise source. The test results confirmed that this noise problem was originated from the axial up-down movement of the pump-rotor unit. What's more, the frequency characteristics of impact response were obtained, which had a good coherence with the frequency distribution in abnormal noise and vibration spectrum. Several measures had been taken as shown in chapter 5 to reduce pressure pulsation above/below motor and restrain the axial movement of shaft-rotor unit. As a result, the noise problem had been solved successfully and the sound pressure level decreased about 4dB and the sound quality was comfortable and good to hear.

REFERENCES

[1] Jinquan Zhang, Rongting Zhang, Huanhuan Gu, etc. (2012). Abnormal compressor noise diagnosis using sound quality evaluation and acoustic array method. 2012 Purdue Compressor Technology Conference.

[2] Tadashi YANAGISAWA, Takashi SHIMIZU, Mitsuhiro FUKUTA, Motohiko UEDA. (1992). Pressure pulsation in hermetic casing of refrigerating rotary compressor. 1992 Purdue Compressor Technology Conference, 743-750.

Research on low frequency vibration of rotary compressor

Zhiming Wen, Rongting Zhang, Yaosi Zhang, Huanhuan Gu, Yusheng Hu

Compressor and Motor Institute of Gree Electric Appliances, Inc. of Zhuhai,
Jinji West Rd., Zhuhai City, 519070, P. R. China
Phone: +86-756-8669017, Fax: +86-756-8668386,
E-mail: madicus@163.com

ABSTRACT

The abnormal noise of an outdoor domestic air-conditioner operating at low speed is experimentally analyzed. The structure-borne noise which passes through the mounting system is confirmed to be the main source of the abnormal noise due to the large low frequency vibration on compressor foot. Then the characteristic of low frequency vibration of rotary compressor including the dynamic model, exciting forces and dynamic response is researched. Based on this, mounting system including compressor foot and rubber grommet is optimized to solve this problem, more than 8dB reduced.

1. INTRODUCTION

With the popularization of the inverter-driven air-conditioner and the requirement of comfort and energy saving, the operating frequency of inverter-driven air-conditioner is lower and lower, even to 1Hz. At the same time, the concern about the noise of air-conditioner is higher and higher. So noise level is the important factor of the air-conditioner performance [1].

Generally, rotary compressor is a main noise source of air-conditioner outdoor unit, which is composed of two components, airborne noise and structure-borne noise. The airborne noise is directly radiated from the compressor shell. The structure-borne noise is caused by the vibration transmitted through mounting system, suction pipes and discharge pipes that are linked to the compressor.

To a certain domestic air-conditioner, when the compressor operates at 1440-1980 RPMs, the noise level of outdoor unit increases obviously comparing to other RPMs. The structure-borne noise caused by the mounting system of compressor is confirmed by experimental results. First, it was found from the experimental data that the low frequency vibration of compressor plays an important role in the abnormal noise. Next, the low frequency vibration of rotary compressor is analyzed. Finally, the mounting system of the compressor is optimized and a decrement of 8.9 dB in the sound pressure is achieved.

2. ABNORMAL NOISE AND IDENTIFICATION

2.1 Experiment Environment

Noise and vibration experiments are taken to the outdoor unit in semi-anechoic room. Two microphone sensors are set at two points as shown in Figure 1.

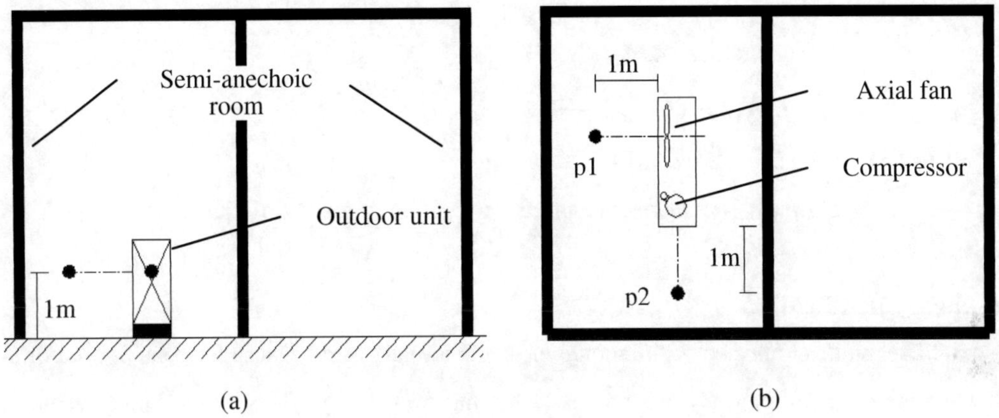

Figure 1: (a) Experiment environment. (b) Top view of (a).

2.2 Characteristic of the abnormal noise

The sound pressure level is measured from 8 to 90Hz, which is the rotation frequency of the rotary compressor. As show in Figure2, the sound pressure levels both at p1 and p2 increase sharply when the compressor operating from 24 to 33Hz (1440-1980 RPMs).

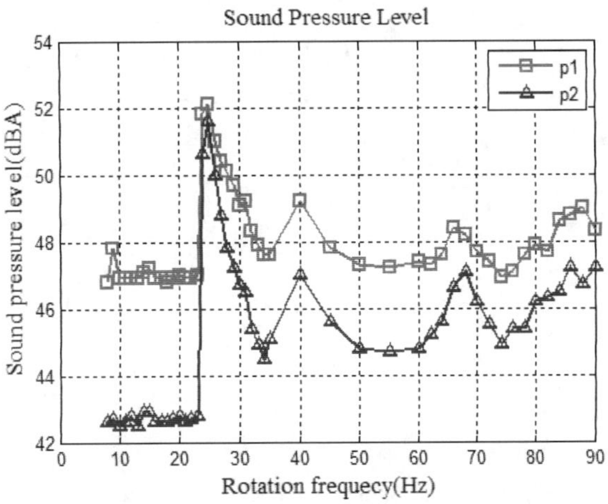

Figure2: Measured sound pressure level at p1 and p2

It can be confirmed that the abnormal noise is a problem of broad band spectrum upwards 200Hz when comparing the sound pressure spectrum at rotation frequency 23Hz with 24Hz, as shown in Figure3.

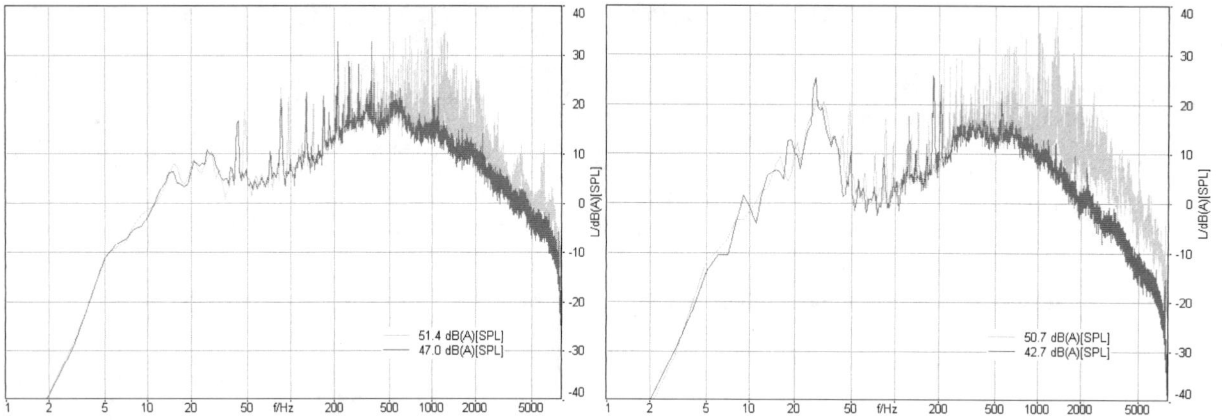

(a) p1: — 24Hz — 23Hz. (b) p2: — 24Hz — 23Hz.

Figure3: Sound pressure spectrum at different rotation frequency.

2.3 Noise source identification

Firstly, the axial fan can be eliminated as the source of the abnormal noise because the rotary speed of axial fan do not change when the rotation frequency of compressor turning from 23 to 24Hz. By the same reason, the airborne noise of compressor will not change so sharply just when the rotation frequency of compressor increases 1Hz. Thus, the structure-borne noise is the only possible reason. Finally, the structure-borne noise caused by mounting system of compressor is confirmed as the real reason by checking the vibration of pipes and unit base.

As shown in Figure 4, the acceleration at unit base increases obviously just like the characteristic of sound pressure spectrum when the rotation frequency of compressor turning from 23 to 24Hz.

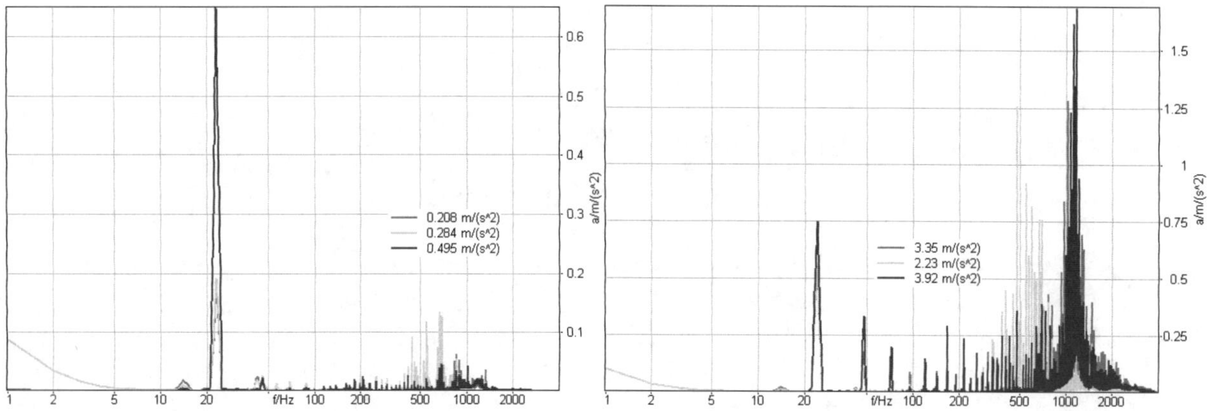

(a) 23Hz (b) 24Hz

Figure 4: Measured acceleration at unit base

(— Radial direction — Axial direction — Tangential direction)

Furthermore, the 1st order tangential vibrating displacement at compressor foot increases obviously when the rotation frequency of compressor turning from 23 to 24Hz as shown in Figure 6. A rigid link between compressor foot and bolt will be formed if the low frequency tangential vibrating displacement is bigger than the actual radial gap between rubber grommet and bolt as shown in Figure 5. High frequency vibration of compressor will be directly transmitted to the unit base through the bolt while rigid link formed between compressor foot and bolt.
The low frequency vibration of compressor plays a very important role in the sound pressure level of outdoor unit.

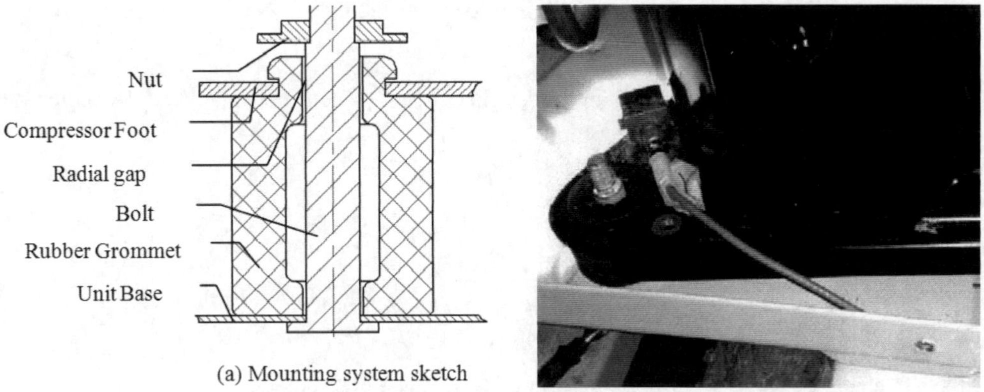

(a) Mounting system sketch (b) Measured unit

Figure 5: Compressor mounting system

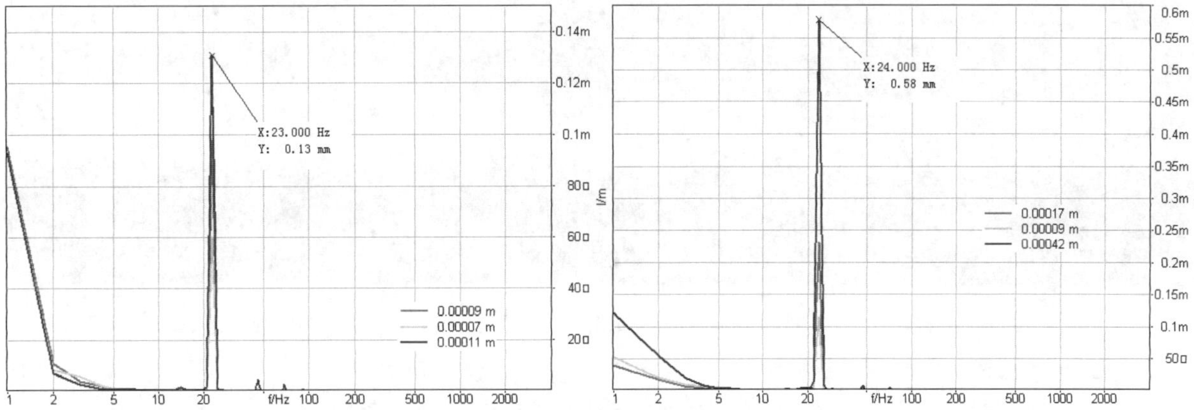

(a) 23Hz (b) 24Hz

Figure 6: Measured displacement at compressor foot

(— Radial direction — Axial direction — Tangential direction)

3. LOW FREQUENCY VIBRATION ANALYSIS

3.1 Dynamic model

To a dynamic system, the dynamic characteristic can be described by equation (1).

$$[M]\{\ddot{q}\}+[C]\{\dot{q}\}+[K]\{q\}=\{f\} \tag{1}$$

Generally, the first flexible mode of a rotary compressor is about 200Hz. So the low frequency vibration is nearly decided by the rigid modes of a compressor. As we all know, there are 6 rigid modes which can be calculated by equation (2) and (3) when ignoring the damping of a dynamic system.

$$[M]\{\ddot{q}\}+[K]\{q\}=0 \tag{2}$$

Eigen function can be expressed as follow:

$$M^{-1}K\Phi=\lambda\Phi \tag{3}$$

where M and K are the global stiffness and mass matrices, respectively; λ is the eigenvalues, and $\mathbf{\Phi}$ is the eigenvectors.

Then, low frequency vibration response under a certain force $\{f\}$ can be calculated by equation (4) when considering modal damping coefficient c_i.

$$\{X\}=\sum_{i=1}^{6}\frac{\{\phi_i\}\{\phi_i\}^T\{f\}}{\left(k_i+j\omega c_i-\omega^2 m_i\right)} \tag{4}$$

where $\{\phi_i\}$ is the i-th eigenvector; k_i and m_i is the i-th modal stiffness and mass.

Six rigid modes are identified by impact testing of the rotary compressor as shown in Figure 7. The theoretical calculated result is accurate enough comparing to the experiment result as shown in Table 1.

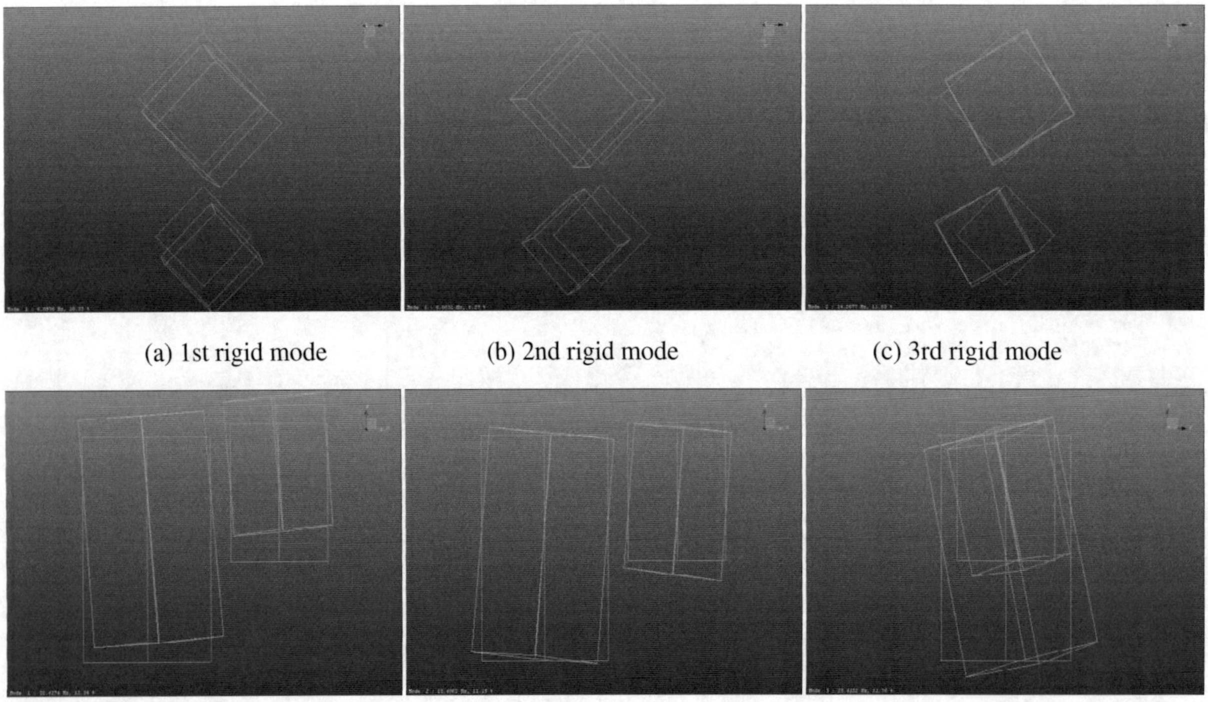

(a) 1st rigid mode (b) 2nd rigid mode (c) 3rd rigid mode

(d) 4th rigid mode (e) 5th rigid mode (f) 6th rigid mode

Figure7: Measured 6 rigid modes of the rotary compressor.

Table 1: Rigid modes of a certain rotary compressor (Hz)

Rigid mode	Calculated	FEM	Measured
1	6.7	6.7	6
2	6.8	6.7	6.1
3	14.2	15.0	14.3
4	20.1	20.0	20.4
5	23.1	23.0	23.4
6	24.7	24.1	25.4

3.2 Low Frequency Forces

The low frequency forces should be discussed firstly in order to calculate the low frequency vibration. The main forces such as gas force and moment, centrifugal inertia forces, magnetic force and torque are considered based on the structure of a rotary compressor. Finally, it is found that the unbalanced axial moment is the main force deciding the low frequency vibration.

(1) Gas forces and moment

Considering the model shown in Figure 8 (a), where F_h is the gas force on blade; F_q is the gas force on cylinder; F_g is the gas force on crankshaft. As a rigid body, moving these forces to the centre of cylinder and considering the forces concentrated on cylinder as shown in figure 8 (b). The F_g do not generate an additional moment on cylinder because only radial force can be transmitted through the top and bottom bearings when taking no account of friction. After calculation, it is found that the radial gas forces concentrated on cylinder are balanced forces and only the unbalanced moment M_h is left. Furthermore, M_h is equal to the gas moment on crankshaft with opposite direction.

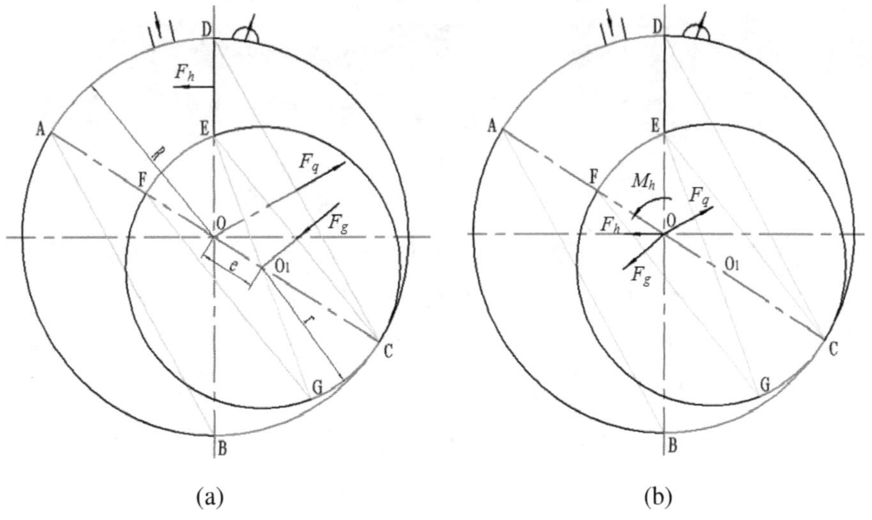

Figure 8: Simplified sketch of gas force

(2) Centrifugal inertia forces [2]:

Centrifugal inertia forces caused by rotational eccentric parts can be calculated by equation (5). Generally, these forces can be neglected when the rotation speed is not too high because they are designed to be a nearly balanced force.

$$F_i = m_i \omega^2 r_i \tag{5}$$

where m_i is mass of the eccentric parts (a pair of balancers, a rolling and a crankshaft eccentric part), ω is speed of rotation, r_i is the distances from the mass centre of the eccentric parts to the axis of revolution.

(3) Magnetic force and torque:

Theoretically, no dynamic concentrated radial magnetic force exists if no dynamic eccentricity existing between the axes of rotor and stator. Also the concentrated radial magnetic forces on stator and rotor are a pair of balanced forces to a rigid compressor when considering the low frequency vibration.

The magnetic counter torque on stator is a constant value and will not influence the low frequency vibration if no torque compensation algorithm is taken to the invert controller as shown in Figure 9.

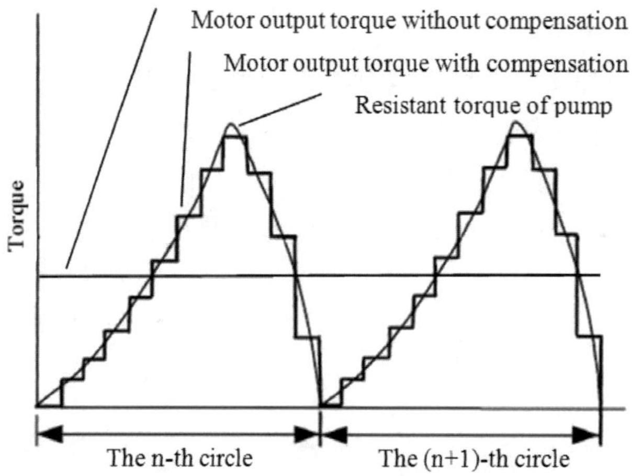

Figure 9: A sketch of torque compensation control [3]

As discussed above, the main force which influencing the low frequency vibration is the unbalanced moment. It can be calculated more accurate when using a commercial program to take account of other factors such as friction etc.

A theoretical comparison of unbalanced moment between single and dual cylinder rotary compressor is presented in Figure 10. The unbalanced moment of a signal cylinder compressor is even bigger than a dual cylinder compressor which owning double displacements and its frequency is half lower at the same rotation frequency. Single cylinder compressor is used widely in the domestic air-conditioner because of its lower cost although it may cause worse vibration problem when the unbalanced moment value is bigger and the frequency is lower.

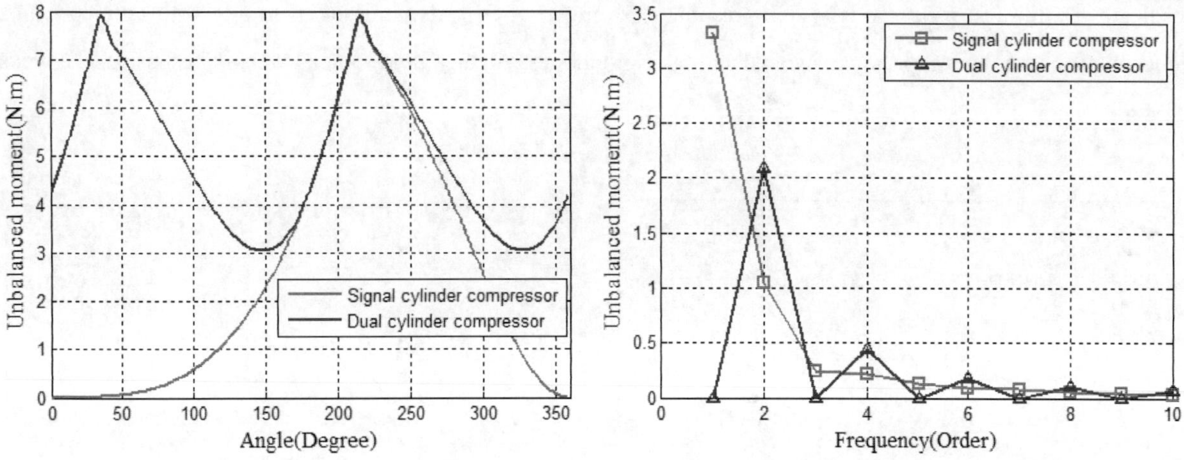

(a) Unbalanced moment curve (b) FFT analyzed spectrum

Figure 10: Theoretical calculated unbalanced moment

3.3 Characteristic of low frequency vibration

Firstly, the coordinate is defined as shown in Figure 11 where the *w* and *Z* axes are decided by the right hand rule.

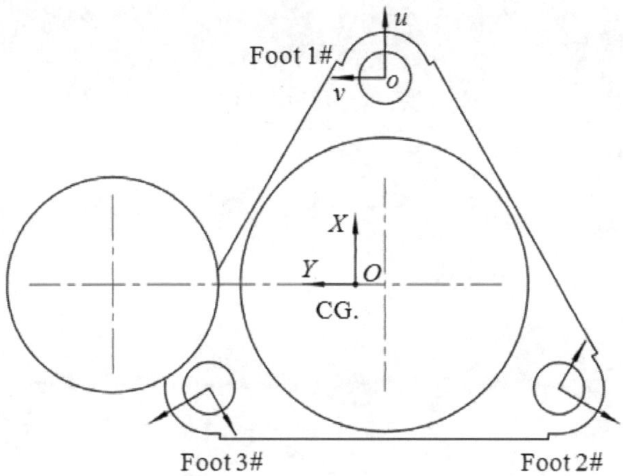

Figure 11: Coordinate definition

The vibrating displacement responses at three feet under the load of unit axial torque are presented in Figure 12(a), (b) and (c). Three response peaks appear at about 7, 14 and 25Hz which are near the natural frequency of the 2nd, 3rd and 6th rigid modes. The tangential vibrating displacement at each foot and at each frequency is nearly the biggest response against the other two directions. The maximum value of tangential vibrating displacement appears at 14Hz which is the natural frequency of the 3rd rigid mode.

Furthermore, the characteristic of vibration under unit axial torque can be explained by discussing the data given in Table 2. The 3rd rigid mode has the largest energy distribution on the θ_Z direction which is the direction of the loaded torque. Also, the 3rd rigid mode is coupled with the 2nd and 6th rigid modes weakly under the defined coordinate. So the 3rd rigid mode will be resonated strongly when a dynamic axial torque with corresponding frequency acting on the compressor. The 2nd and 6th rigid modes will also be resonated weakly because of the weak coupling.

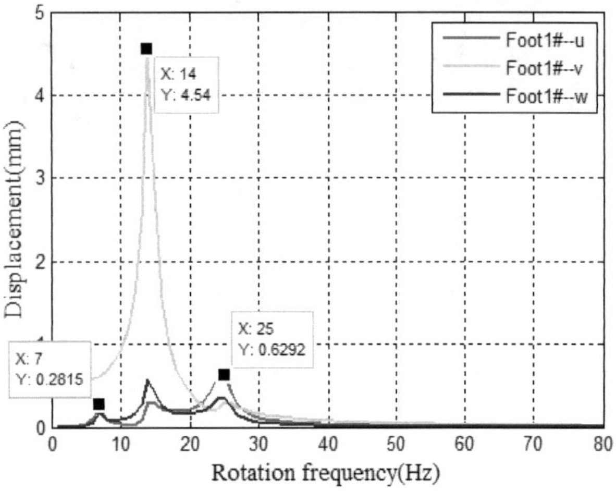

(a) Response at Foot 1#

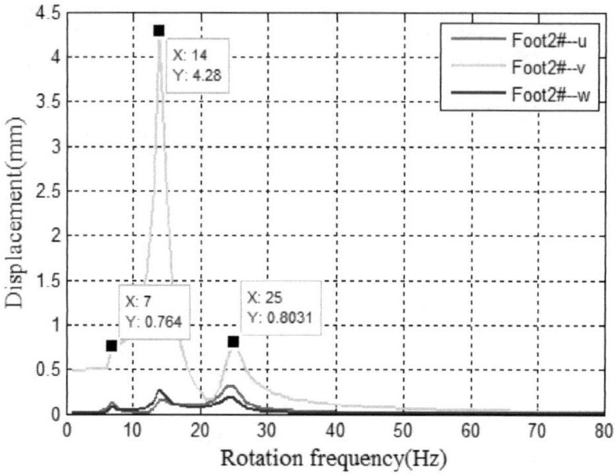

(b) Response at Foot 2#

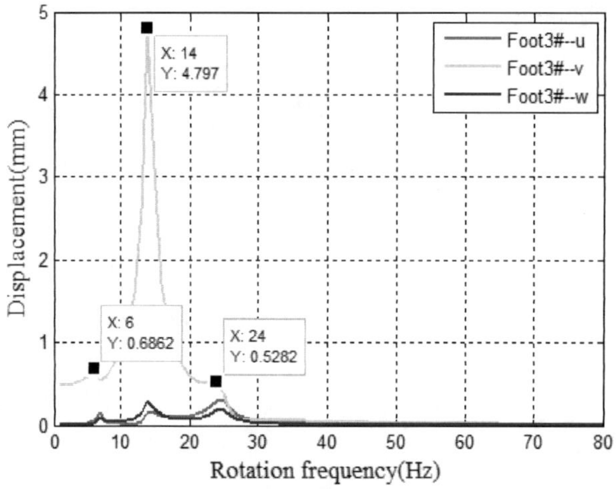

(c) Response at Foot 3#

Figure 12: Vibration response under unit axial torque

Table 2: Natural frequency and energy distribution of rigid modes

Natural Frequency (Hz)		**6.7**	**6.8**	**14.2**	**20.1**	**23.1**	**24.7**
Energy Distribution (%)	***X***	0.04	92.11	0.23	0.00	0.00	7.61
	Y	89.52	0.04	0.00	5.41	5.03	0.00
	Z	0.27	0.00	0.00	63.15	36.56	0.01
	θ_X	10.16	0.00	0.00	31.43	58.38	0.02
	θ_Y	0.00	7.84	5.31	0.00	0.03	86.82
	θ_Z	0.00	0.00	94.46	0.00	0.00	5.54

4. IMPROVEMENT AND VALIDATION

Finally, the mounting system is optimized which including the structure of compressor foot and rubber grommet considering the abnormal noise as well as the operational stress level of suction and discharge pipes. The sound pressure levels of optimized unit is improved obviously as shown in Figure 13, 4.9 and 8.9 dB reduced at measured points p1 and p2 when the compressor operating at 25Hz.

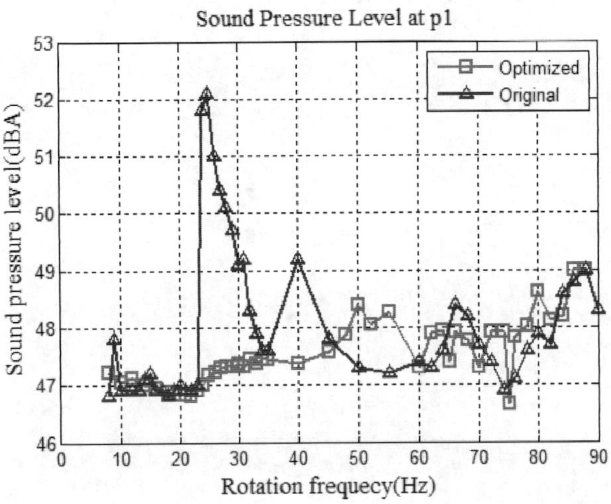

(a) p1 (b) p2

Figure 13: Sound pressure level comparison

5. CONCLUSIONS

In this paper, the abnormal noise of a domestic air-conditioner outdoor unit when the compressor operating at certain low RPMs is experimentally analyzed. The structure-borne noise which passes through the mounting system is confirmed to be the main source of the abnormal noise due to the large low frequency vibration on compressor foot. Characteristic of low frequency vibration is researched including the dynamic model, low frequency forces and dynamic response. Based on this， mounting system including compressor foot and rubber grommet is optimized to solve this problem, 4.9 and 8.9 dB reduced at measured points p1 and p2 when the compressor operating at 25Hz.

REFERENCES

[1] Huanhuan Gu, Rongting Zhang, Yusheng Hu. (2014). The Mechanism Discuss of Periodic Sound in Rolling Piston Compressor under Low Operating Frequency in Air-conditioner System. Proceedings of the 22nd International Compressor Engineering Conference at Purdue: 1214:p1.

[2] Lingchao Kong, Liping Ren, Jia Xu, Yusheng Hu. (2014). The Finite Element Analysis of the Deflection of the

Crankshaft of Rotary Compressor. Proceedings of the 22nd International Compressor Engineering Conference at Purdue: 1211:p2-3.
[3] Kazuhiro Kageyama, Toshiyuki Shikanai, Toshiyuki Gotoh, etc. (2002). Development of High-Efficiency Low-Vibration Rotary Compressor for HFC-410A Air-Conditioner. Proceedings of the 16th International Compressor Engineering Conference at Purdue: C5-1:p4-7.

AUTHOR INDEX

AUTHOR INDEX

AUTHOR INDEX

AUTHOR INDEX

23rd International Compressor Engineering Conference at Purdue 2016

West Lafayette, Indiana, USA
11 - 14 July 2016

Volume 2 of 2

ISBN: 978-1-5108-2872-8

Printed from e-media with permission by:

Curran Associates, Inc.
57 Morehouse Lane
Red Hook, NY 12571

Some format issues inherent in the e-media version may also appear in this print version.

Copyright© (2016) by Perdue University
All rights reserved.

Printed by Curran Associates, Inc. (2016)

For permission requests, please contact Ray W. Herrick Laboratories
at the address below.

Ray W. Herrick Laboratories
School of Mechanical Engineering
Purdue University
177 S. Russell Street
West Lafayette, IN 47907-2099, USA.

Phone: +1 (765) 494-2132
Fax: +1 (765) 494-0787

rhlab@ecn.purdue.edu

Additional copies of this publication are available from:

Curran Associates, Inc.
57 Morehouse Lane
Red Hook, NY 12571 USA
Phone: 845-758-0400
Fax: 845-758-2633
Email: curran@proceedings.com
Web: www.proceedings.com

Ray W. Herrick Laboratories
Center for High Performance Buildings

Table of Content

C-01: Screw Compressors I

Time: Monday July 11, 2016: 1:00 PM - 3:00 PM — *Location*: 202

Session Chair: Matt Cambio

C-02: Valves I

Time: Monday July 11, 2016: 1:00 PM - 3:00 PM — *Location*: 206

Session Chair: Cesar J. Deschamps

C-03: Novel Compressors

Time: Monday July 11, 2016: 3:30 PM - 5:30 PM — *Location*: 202

Session Chair: Margaret Mathison

C-04: Heat Transfer in Compressors I

Time: Monday July 11, 2016: 3:30 PM - 5:30 PM — *Location*: 206

Session Chair: Vincent Lemort

C-05: Tribology and Lubrication I

Time: Tuesday July 12, 2016: 9:45 AM - 11:45 AM — *Location*: 202

Session Chair: Joseph Karnaz

C-06: Scroll Compressors I

Time: Tuesday July 12, 2016: 9:45 AM - 11:45 AM — *Location*: 206

Session Chair: Gene Fields

C-07: Expanders I

Time: Tuesday July 12, 2016: 1:30 PM - 3:30 PM — *Location*: 206
Session Chair: Martijn van den Broek

C-SPC: Student Paper Competition

Time: Tuesday July 12, 2016: 1:30 PM - 3:30 PM — *Location*: 278

Session Chair: Gene Fields

C-08: Heat Transfer in Compressors II

Time: Tuesday July 12, 2016: 4:00 PM - 6:00 PM — *Location*: 202

Session Chair: Joe Orosz

C-09: Linear and Reciprocating Compressors

Time: Tuesday July 12, 2016: 4:00 PM - 6:00 PM — *Location*: 206

Session Chair: Craig R Bradshaw

C-10: Noise, Vibration, & Harshness I

Time: Wednesday July 13, 2016: 9:45 AM - 12:00 PM — *Location*: 202

Session Chair: Stuart Bolton

C-11: Compressor Modeling

Time: Wednesday July 13, 2016: 9:45 AM - 12:00 PM — *Location*: 206

Session Chair: Craig R Bradshaw

C-12: Expanders II

Time: Wednesday July 13, 2016: 1:00 PM - 3:00 PM — *Location*: 202

Session Chair: Jason Hugenroth

C-13: Scroll Compressors II

Time: Wednesday July 13, 2016: 1:00 PM - 3:00 PM — *Location*: 206

Session Chair: Bryce Shaffer

C-14: Tribology and Lubrication II

Time: Wednesday July 13, 2016: 3:30 PM - 5:30 PM — *Location*: 202

Session Chair: Michael Perevozchikov

C-15: Rotary Compressors I

Time: Wednesday July 13, 2016: 3:30 PM - 5:30 PM — *Location*: 206
Session Chair: Doug Collings

C-16: Testing and Measurement

Time: Thursday July 14, 2016: 9:45 AM - 12:00 PM — *Location*: 202

Session Chair: Stefan Bertsch

C-17: Centrifugal Compressors

Time: Thursday July 14, 2016: 9:45 AM - 12:00 PM — *Location*: 206

Session Chair: Christian Bach

C-18: Noise, Vibration, & Harshness II

Time: Thursday July 14, 2016: 1:00 PM - 3:00 PM — *Location*: 202

Session Chair: Patricia Davies

C-19: Rotary Compressors II

Time: Thursday July 14, 2016: 1:00 PM - 3:00 PM — *Location*: 206

Session Chair: Margaret Mathison

C-20: Valves II

Time: Thursday July 14, 2016: 3:30 PM - 5:30 PM — *Location*: 202

Session Chair: Akash Bhatia

C-21: Scroll Compressors III

Time: Thursday July 14, 2016: 3:30 PM - 5:30 PM — *Location*: 206
Session Chair: Kirill M. Ignatiev

ID: 1079

Experimental Investigations on the Performance Improvement of Oil-Gas Separator in Electric Driven Scroll Compressor for Eco-Friendly Vehicles.....1066

Donglim Nam[1,4], Poyoung Lee[1], Seungbin Jung[1], Geonho Lee[2], Yunki Kwon[3], Jinho Lee[4]

[1]R&D Center of Doowon Heavy Industrial Co., Ltd., Korea, Republic of (South Korea); [2]Department of Building Engineering Services, Doowon Technical University College; [3]Department of Computer Aided Design, Doowon Technical University College; [4]School of Mechanical Engineering, Yonsei University

Keywords: Electric Driven Compressor, Oil Separator, Vortex Finder, Pressure Drop

ID: 1344

Influencing Factors Study of the Variable Speed Scroll Compressor With EVI Technology.....1076

Xiaoli Kang, Yusheng Hu, Caixia Shan, Yun Liu, Xiaojun Gao, Gang Lv

GREE ELECTRIC APPLIANCES, INC. OF ZHUHAI, China, People's Republic of

Keywords: Enhanced Vapor Injection; Scroll Compressor;the Area of Supplementary Channel;the Position of Supplementary Inlet;gas Force

ID: 1497

Lubrication Tests to Support Optimal Performance Design Guidelines for Thrust Slide-Bearings in Scroll Compressors.....1084

Noriaki Ishii[1], Eiji Nonoguchi[1], Keiko Anami[5], Atsushi Sakuda[2], Yuya Terada[1], Yusuke Imai[2], Kiyoshi Sawai[3], Charles William Knisely[4]

[1]Osaka Electro-Communication University, Japan; [2]Panasonic Corporation Appliances Company; [3]Hiroshima Institute of Technology; [4]Bucknell University, USA; [5]Ashikaga Institute of Technology, Japan

Keywords: Scroll Compressor, Thrust Slide-Bearing, Optimization, Dsign Guideline, Friction Power Loss

ID: 1613

Visualization and Simulation of Oil Flow in a Scroll Compressor Plenum.....1093

Jiu Xu[1], Pega Hrnjak[1,2]

[1]Air Conditioning and Refrigeration Center, University of Illinois at Urbana-Champaign; [2]Creative Thermal Solutions Inc., Urbana, Illinois

Keywords: Compressor, Oil, CFD, Flow Visulization, Droplet

ID: 1634

Numerical Simulation and Experimental Examination of an Oldham Coupling.....1101

Hang Ye[1], Zhigang Huang[1], Jinduo Ye[2]

[1]Danfoss (Tianjin) R&D, Tianjin, PR. China; [2]School of Mechanical, Tianjin University of Technology, Tianjin, PR.China

Keywords: Scroll Compressor, Oldham Coupling, Numercial, Experimental, Prediction

Rotary Compressor Noise Analysis Using Mechanisms and Electromagnetics Coupled Approach

Jinah Chung [1*], Uiyoon lee[1], Jeongbae Lee[1], Unseop Lee[1],Eunsil Han[2], Jinhwan Yoon[2]

[1]Samsung Electronics Co., Ltd, Compressor & Motor Business Team
129, Samsung-ro, Yeongtong-gu, Suwon-si, Gyeonggi-do, 16677, Korea
(Office: +82 -31-8062-7939, Fax: +82-31-8062-9316, E-mail: jinah.chung@samsung.com)

[2]Taesung S&E, Inc
27, Seongsui-ro 7-gil, Seongdong-gu, Seoul, Korea
(Office : +82-2-3888-3156, Fax: +82-2-2117-0017, E-mail : ehan@tsne.co.kr)

*Corresponding Author

ABSTRACT

This research is conducted to investigate noise source and design low noise compressors. For improving energy efficiency, there is expansion usage of high performance variable speed brushless DC motor for rotary compressor. However brushless DC motor makes more vibration of compressor than constant speed motor compressor in high speed operating condition. Therefore it is necessary to reduce noise and vibration for improving air conditioner quality. In this study, compressor's noise and vibration are simulated using structural and electromagnetics coupled methods. From the rotor motion simulation result, precession motion of rotor is applied for simulating actual motor movement.

1. INTRODUCTION

Recently home appliance companies make an effort to improve energy efficiency of their product, air conditioning system and refrigerator. Compressor is most energy consuming part of appliances. Rotary compressors in air-conditioner have been considered for energy efficiency enhancement and noise reduction. For reducing energy consumption of compressor, Kageyama, K.et al(2002) studied about brushless DC(BLDC) motor type compressors. BLDC type compressor can drive with optimized speed depending on cycle conditions like setting temperature, environment temperature, etc. However increased power of motor makes compressor vibration more than constant speed type compressor.

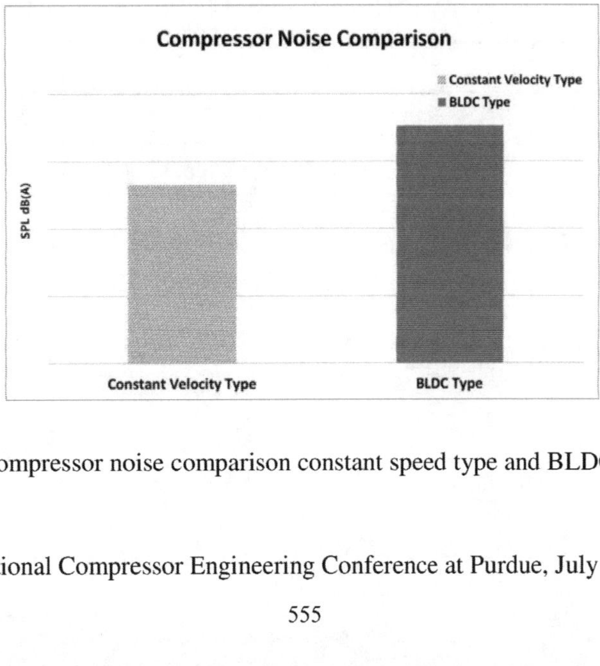

Figure 1: Compressor noise comparison constant speed type and BLDC type motor

Sungtae woo. et al(2008) studied about improving compressor's noise and vibration. Noise sources of compressor are electromagnetic noise, structure-born noise and air-born noise. J Lee and U Yoon Lee (2012) showed optimized flow path designs of accumulator to reduce air-born noise. For reducing structure-born and electromagnetic noise, structural improvement is needed. The goal of study is conducting structural simulation, and designing low noise and vibration compressor. In this study, structural and electromagnetics coupled simulation is carried out for increasing simulation accuracy. Motor force is set as noise and vibration source because motor directly excites compressor parts on operating condition. Electromagnetics simulation results are mapped to stator teeth, and then structural analysis is performed. From result, acoustic field analysis is carried out. Finally simulation result is analyzed with noise measurement test.

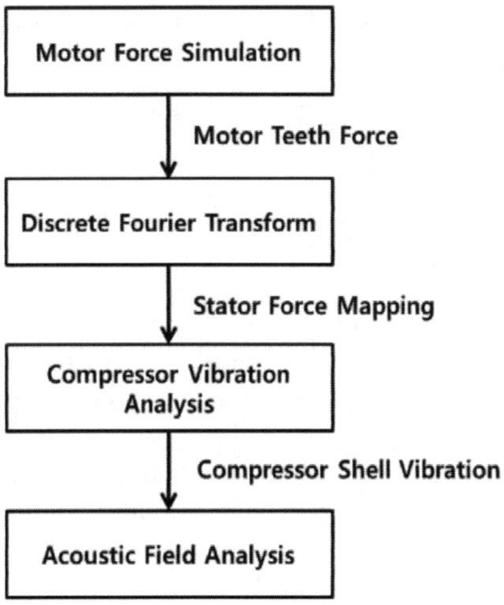

Figure 2: Compressor noise analysis flow

2. COMPRESSOR SIMULATION

2.1 Simulation Model

For simulation, compressor model is selected. Twin rotary compressor is applied for simulation. Motor is 6poles, 9slots and compressor cooling capacity is 20000BTU/hr.

Table 1: Compressor specification

Specifications	Value	Feature
Refrigerant	R410a	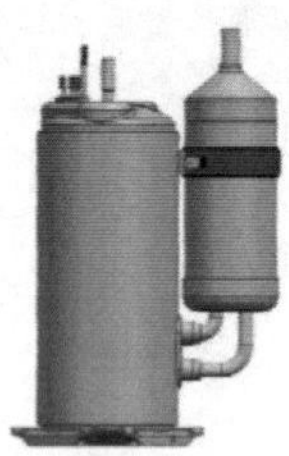
Motor	Brushless DC Motor, 6 Pole 9Slot	
Magnet	Neodymium	
Type	Twin Rotary	
Cooling Capacity	20000 Btu/hr	

2.2 Motor simulation

In this paper motor force is main noise and vibration source of rotary compressor. In rotary compressor, stator is usually assembled to case by press fit, and motor vibration directly transferred to other parts of compressor.

Motor behavior is analyzed prior to compressor structural simulation. In rotary compressor, rotor is supported overhanging. Due to structure, rotor's whirling motion can be shown when compressor is operating. Then forces on stator tooth are not same because of air gap imbalance in real whirling motion. Force imbalance between teeth makes vibration on compressor. However, it is difficult to realize whirling motion with air gap imbalance. For realizing similar motor behavior, centrifugal, eccentric, and precession motor motion simulations are conducted. Motor force simulation is carried out using ANSYS Maxwell Software. Maxwell software is commercial program and used to simulate low frequency electromagnetic field.

2.2.1 Centrifugal Motion

Centrifugal motion means rotor is rotating ideally without revolving and every stator tooth has same size of air gap. All forces on stator teeth are same by air gap size. Simulation result shows motor pole's effect only. Figure 4 shows 348Hz is highest force. 348Hz is multiplying of 58 rps driving speed and motor poles number. In the other frequencies, except multiple of operating speed and motor poles number, there are no significant forces.

Figure 3: Centrifugal motion

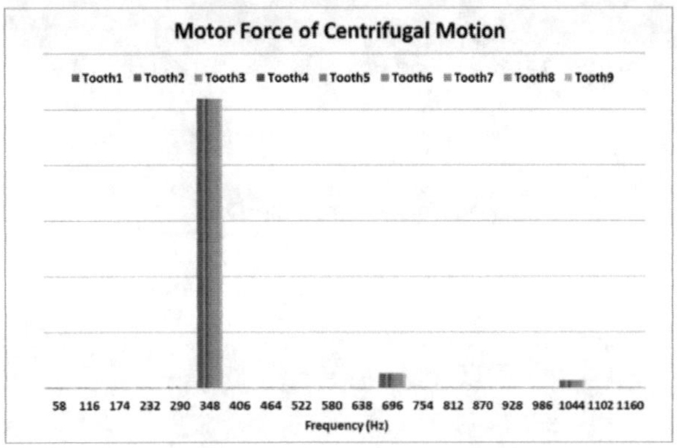

Figure 4: 58 rps centrifugal motion force result

2.2.2 Eccentric Motion

Eccentric motion center of rotating is moving to one side. In eccentric motion, one of air gap is very smaller than the others. Then rotor does not revolve along stator. For simulation, the minimum air gap value refers to assembly specification. In eccentric motion simulation, result represents air gap imbalance, but cannot shows whirling motion of rotor.

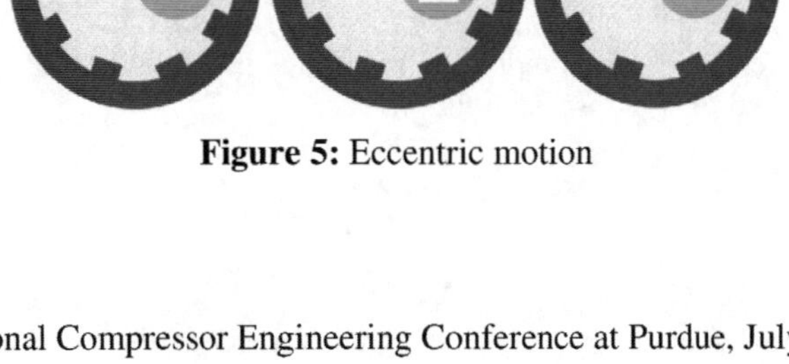

Figure 5: Eccentric motion

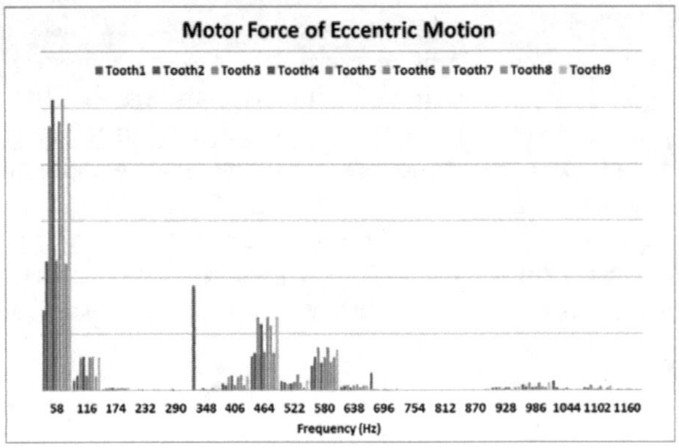

Figure 6 : 58 rps eccentric motion force result

2.2.3 Precession Motion

In rotary compressor, precession motion always appears in case parts and assembly tolerance. When rotor is rotating and revolving around stator. Rotating speed and rotating speed are same. Revolving radius is subtracting the minimum air gap value from the stator inner radius. In the precession motion simulation, all air gaps are same and rotor whirling motion can be shown.

Figure 7: Precession motion

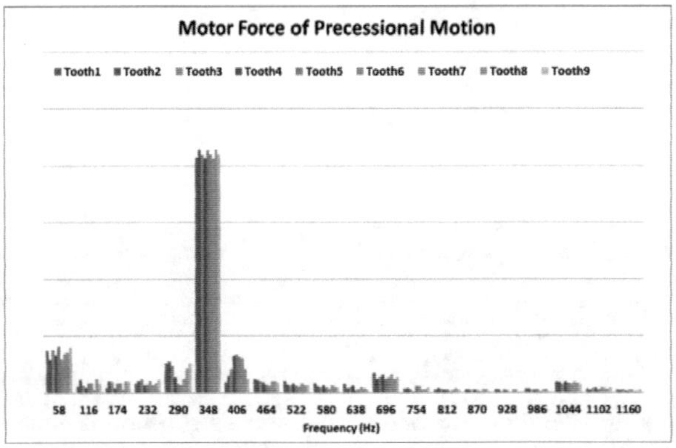

Figure 8: 58 rps precession motion force simulation result

2.2.4 Result Analysis

Centrifugal motion simulation result shows only motor poles multiple components. In this case, frequencies except motor poles multiple components cannot be used for structural analysis. In eccentric motion simulation, result shows that tooth forces have a big difference by air gap difference. But result cannot represent motor poles number at all. Precession motion simulation result shows compressor vibration characteristics well. Stator forces are uniformly distributed at each tooth, and result shows compressor operating speed and motor poles number also. Therefore, precession motion result is applied to structural simulation.

2.3 Structure Simulation

2.3.1 Force setting

From motor simulation results, stator forces are divided into Fx, Fy and transferred as tangential and radial force on stator teeth. For vibration analysis, discrete fourier transform is applied. Transformed tangential force and radial force are mapped teeth1 to teeth9, and these forces are used for exciting force of compressor vibration.

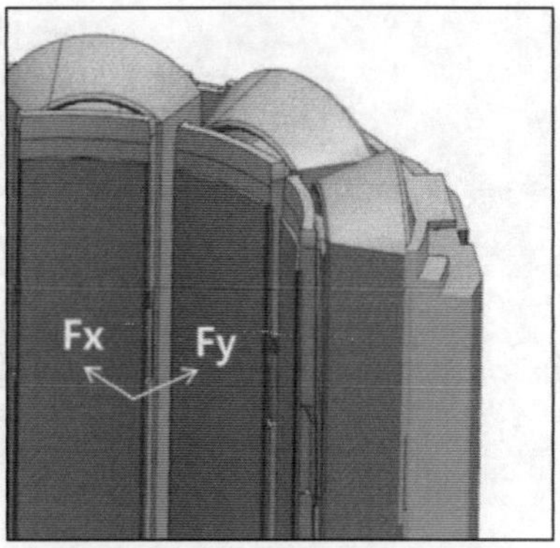

Figure 9: Teeth force setting

2.3.2 Modal Analysis

Prior to structural simulations, modal parameters are examined with modal test. From modal test results, natural frequency, damping ratio, and modal shape can be detected. Experiment modal analysis (EMA) is impacts object using hammer and detects modal parameters. In this research, EMA is applied for verifying modal value of compressor parts. There are difference between modal test and simulation result. For correlating modal test and simulation result, elastic modulus and density of simplified parts are slightly adjusted.

Table 2: Mode shape of FRF test and simulation

	Shall + Bottom Cap	**Shall + Stator**	**Shall + Stator +Core**
Model			
Test			
Simulation	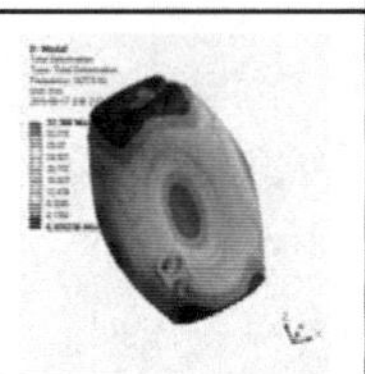		

2.4 Acoustic Simulation

In this paper, semi spherical acoustic field is applied reflecting experiment environment. ANSYS Workbench is used for acoustic simulation. It is assumed that vibration on compressor surface excites surrounding air. Noise signals are measured using 4 microphone settings which are set every 90dregrees at 900mm from compressor surface.

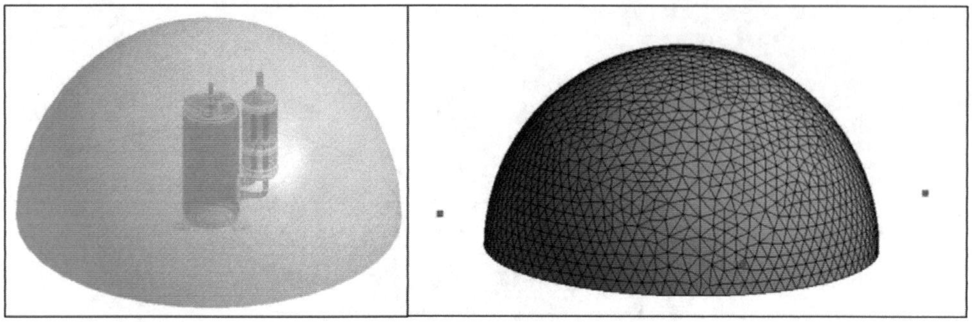

Figure 10: Acoustic field and microphones setup. Red points indicate microphone.

3.RESULTS

3.1 Results analysis

For analysis simulation, the results of the compressor noise are expressed through the sound pressure level. Between 4 microphones, the maximum value of noise is used for comparing frequency behavior. First, 58 rps operating speed of compressor condition is simulated. From simulation, overall value of simulation is 49.5dB(A) and, measurement is 56.1dB(A). Its overall frequency range is 0~2KHz. Results show that overall values are roughly similar, and peak level have similar trend, too. Additionally other operating speeds are applied to simulation for verifying compressor simulation accuracy. Operating frequency is 72 rps, 86 rps and 94 rps for comparing with measurement results. Simulation results are shown in table 3, and maximum error is 11.9%. Results show higher simulation accuracy than without coupling simulation of structural and electromagnetic.

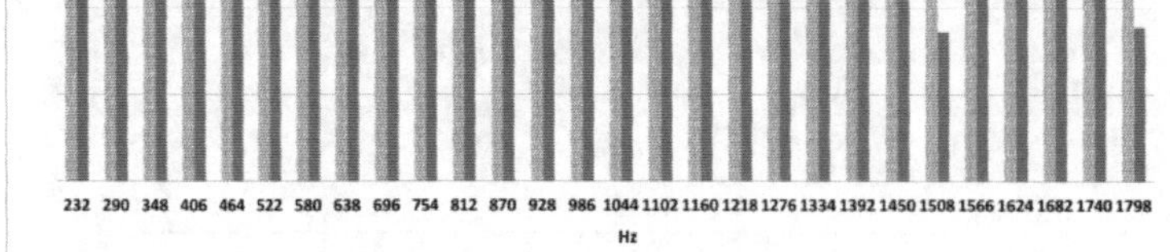

Figure 11: 58 rps analysis result

Table 3: Comparison of overall noise

Operating Frequency	Overall dB(A) (0~2KHz)		
	Measurement	Simulation	Error (%)
58 rps	56.2	49.5	11.9
72 rps	56.5	52.7	6.7
86 rps	58.9	54.5	7.5
94 rps	59.7	54.9	8.0

4. CONCLUSIONS

The purpose of study is noise analysis of compressor using coupled structural and electromagnetics simulation. From research, simulation results are very similar with measurements, and simulation can be assumed that it has high accuracy by comparison with independent acoustic simulations without motor electromagnetic coupling. In terms of motor simulation, the real motion of rotor is assumed as combination of eccentric motion and precession motion. However, due to limitation of simulation, only precession motion motor simulation has been conducted for getting more accuracy result.

The compressor air-born noise can be observed at accumulator, muffler and other cavities can make sharp changing in flow energy. Moreover there is refrigerant oil effect, also. Such a refrigerant flow is air-born noise source of compressor. The reason why this study does not conduct flow simulation have a lot of time cost in 3D flow analysis. In future research, the study of coupled simulation with flow noise will be conducted for better analysis accuracy.

REFERENCES

Kageyama, K. et al(2002). Development of high-efficient low-vibration rotary compressor for HFC-410A air conditioner: Proceedings of International Compressor Engineering Conference at Purdue, Paper 1511

Sungtae woo. et al (2008). Noise reduction in two-cylinder rotary compressor : Proceedings of International Compressor Engineering Conference at Purdue, Paper1987

J Lee, U Yoon Lee (2012).Design optimization of an accumulator for reducing rotary compressor noise. Int. J. of Process Mechanical Engineering, vol. 226, no.4 285-296

Stick-Slip Squeal in a Dry Scroll Vacuum Pump

John CALHOUN[1]*, James MOORE[2], Ronald FORNI[3]

[1, 3]Agilent Technologies, Vacuum Products Division
Lexington, Mass., USA
Phone: 781-861-7200, Email: john.calhoun@agilent.com, ronald.forni@agilent.com

[2]Acentech Incorporated
Cambridge, Mass., USA
Phone: 617-499-8011, Email: jmoore@acentech.com

* Corresponding Author

ABSTRACT

Vacuum pumps have been described as compressors for rarefied gases. For more than twenty-five years the scroll compressor principle has been used in designing vacuum pumps that do not rely on oil to provide their pumping action. The dry scroll vacuum pump has been particularly successful as a source of rough vacuum for analytical instruments such as liquid-chromatography or gas-chromatography mass spectrometers. Such instruments are typically located in quiet laboratory environments where the scroll vacuum pump may be the only significant source of noise. Consequently, quiet operation is a key requirement of dry scroll vacuum pumps. This paper discusses an occurrence of stick-slip squeal that developed during the design of a new model, low noise, dry scroll pump. Stick-slip phenomena and the associated noise are the result of "self-excited" oscillations related to the difference between static and sliding friction. Rapidly oscillating friction forces occurring during a portion of the orbital cycle can result in induced vibration of the participating surfaces and structure, which radiate high-pitched noise.

In the case described, considerable effort was expended in developing a quiet pump design. However, a loud "chirping" noise found in a few late-stage prototypes caused a good deal of concern. It was necessary quickly to fully characterize the vibration behavior within the pump related to the noise and to develop design modifications to ensure that the noise was eliminated. The characteristics of the noise are described for the scroll pump along with the methods utilized to eliminate it from the design.

1. INTRODUCTION

1.1 Pump Observations

A dry scroll vacuum pump is typically fitted with tip seals in both the fixed and orbiting scrolls as shown in Fig. 1. The tip seals are designed to float in their grooves with little or no contact pressure between the mating surfaces. Due to manufacturing tolerances and thermal expansion in operation, one or both seals can be compressed between the bottom of the groove and the mating surface. The seals are a firm (thermoplastic) material, so a small variation in axial dimensions can cause a large variation in contact (normal) force.

The noise that caused this investigation was usually characterized as a "chirp". The noise was heard as a high pitched noise delivered in rapid pulses (one or two per revolution, at 1500-1800 rpm). It was observed early on that the noise occurred more often with new seals before they were worn in. Once worn in the noise occurred less often. Furthermore, changing the cooling fan speed would often eliminate the chirp noise. Based on the investigation, it was determined that the most likely cause of the noise was stick-slip motion between the tip seal and its mating surface.

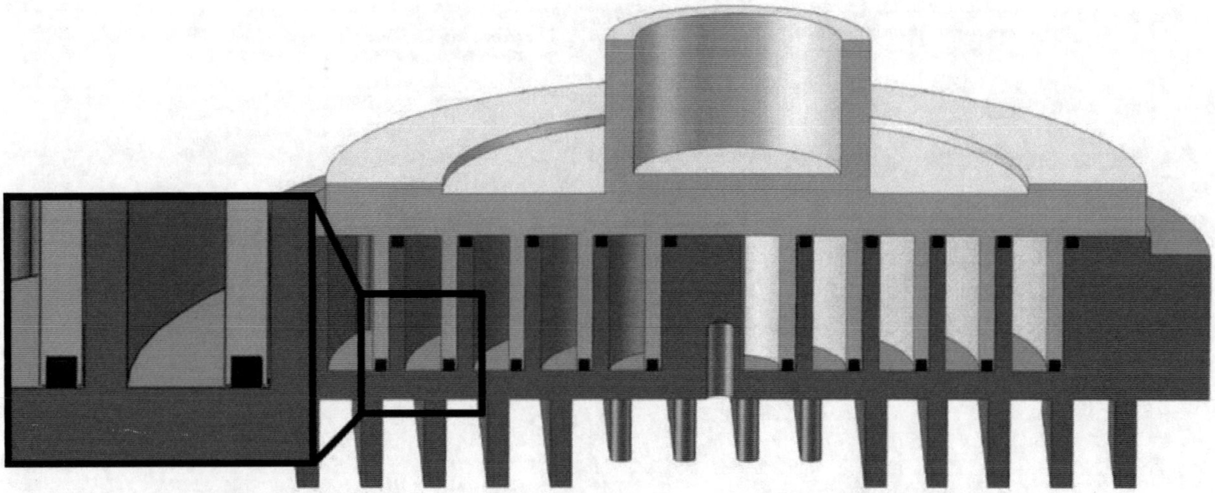

Figure 1: Longitudinal section view of scrolls and tip seals in working position

1.2 Theory of the Noise

Stick-slip motion can occur, if the conditions of normal force and relative velocity are right, when the tip of the vane-seal assembly becomes momentarily fixed - due to static friction - relative to the base plate. As the motion of the orbiting vane continues, the deformation of the vane increases to the point that the internal forces of deformation in the vane are sufficient to overcome static friction and the vane releases, resulting in a drop in friction force which is now governed by the lesser coefficient of sliding friction. Because more energy is built up in static deflection than is dissipated in sliding motion, the resonant response of the vane builds until there is significant vibration transmission to the external surfaces of the pump.

It is believed that the stick-slip motion occurs during the circular orbit when the relative motion is substantially perpendicular to the vane. The stick-slip motion excites vibration resonances within the vane, of a frequency much higher than the rotational frequency of the mechanism. These resonances primarily involve flexural motion perpendicular to the vane surface. This behavior should happen to a greater extent at the outer turns of the spiral vane because its curvature is less there and the vane is less stiff in flexure than in the inner turns. The flexural motion should be greatest at the vane tip. During the part of the orbital cycle where the relative motion is along the vane, only sliding motion without stick-slip oscillation occurs and the friction forces act to dissipate vibration energy in the vane. As a result, the high frequency vibration and noise occur as a series of pulses rather than a continuous tone, resulting in the description of the noise as "chirping".

We were able to start and stop the phenomenon at will, and to study the onset of the noise, by varying the cooling airflow and thus the component temperatures. It is assumed that differential thermal expansion of the various pump components results in changes in contact pressure between the seals and base plates.

2. MEASUREMENTS AND NOISE CHARACTERISTICS

Figure 2a is a time plot of vibration of the orbiting scroll housing as measured by an accelerometer. This figure shows the development of the chirping noise, as seen in the vibration time signal, versus time, over an approximately 2.5 minute period. Before the start of the plot, the pump is running without noticeable noise and with the cooling fan turned off. When the cooling fan is turned on and the pump components begin to cool unevenly, the signal begins to grow due to the chirping noise; then a sudden increase or jump occurs, followed by a steeper growth in the amplitude. Eventually the vibration and noise level out to a steady state, though this is not shown on the plot. Figure 2b shows a 0.07 sec duration time slice of the same vibration signal, after the chirping noise is well established. Note that pulses, approx. 0.01 sec. long, occur every 0.036 sec. which is the rotational time period of the pump/scroll at 1680 rpm or 28 Hz. The content of each pulse is a tone whose fundamental component is centered around 2000 Hz.

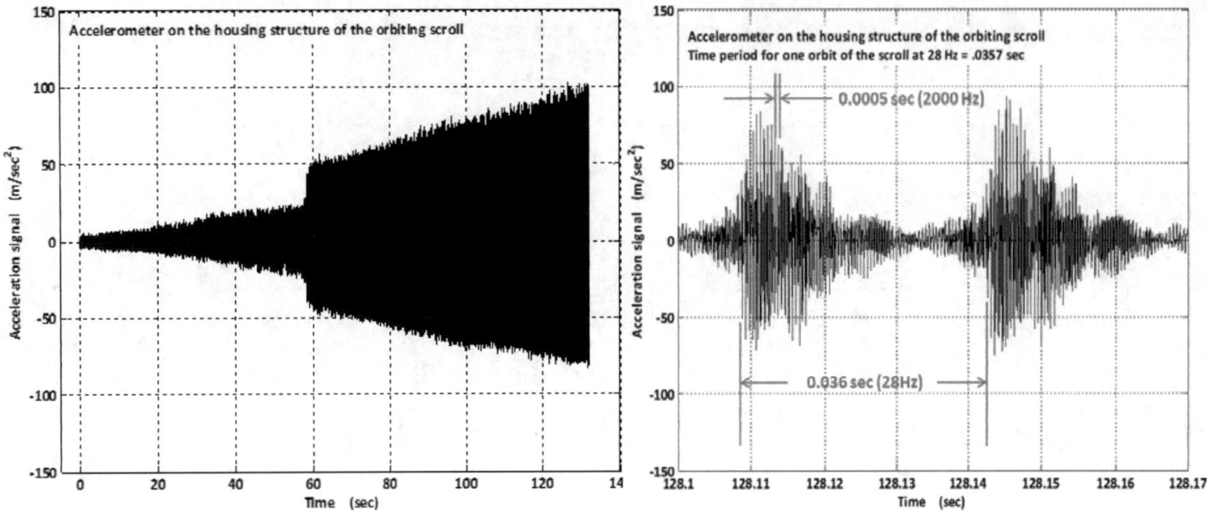

(a) **(b)**

Figure 2: Housing acceleration signal: (a) showing evolution of "chirping" noise; (b) 0.07 sec. time slice showing the noise/vibration when well established

Fig. 2c is a noise spectrum of the pump with the noise well established, showing broad peaks at 2000 Hz and multiples of 2000 Hz.

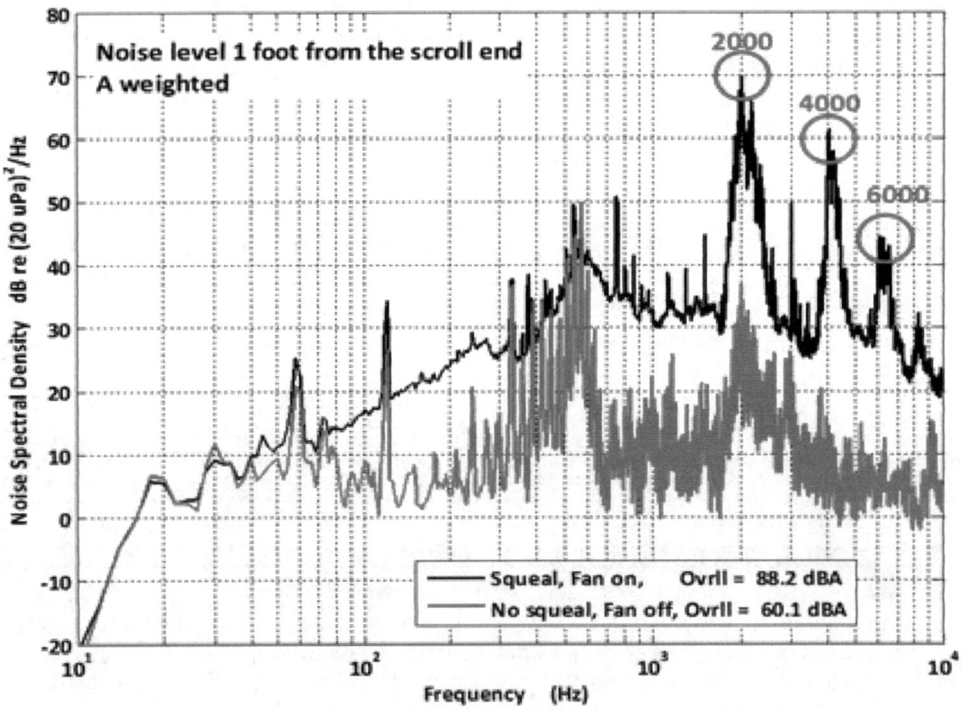

Figure 2c: Audible spectrum with and without "chirping" noise

Figures 2d-g show shorter slices during the initial increase in amplitude as the chirping noise develops, but before the sudden increase in amplitude. (The y axis range of 2d, 2e is different than that of 2f, 2g).

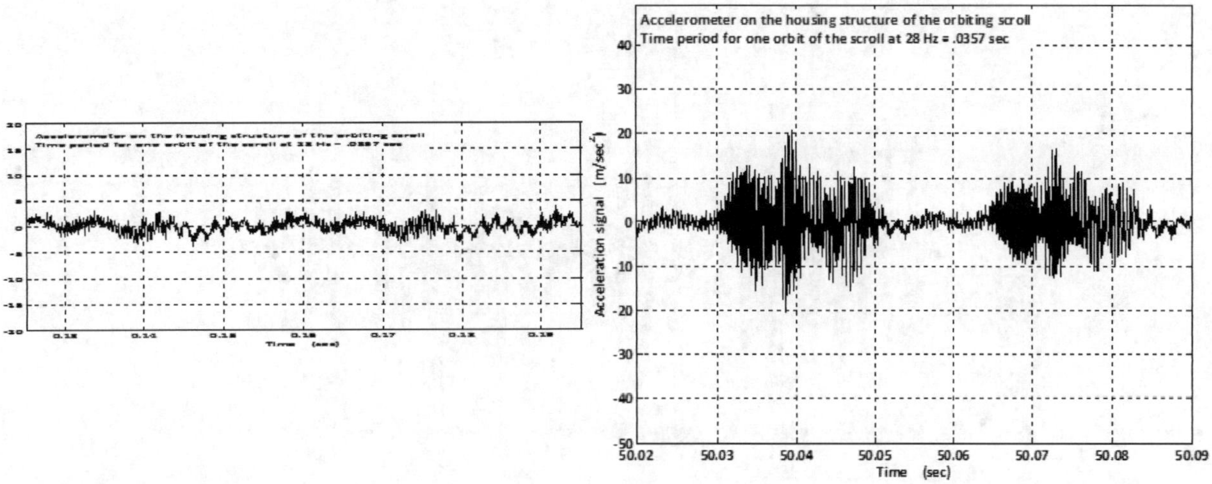

(d) (e)

Figure 2 (cont'd): Housing accceleration signal, short time slices: d) just before "chirping" noise became noticeable; e) just before the sudden jump in amplitude

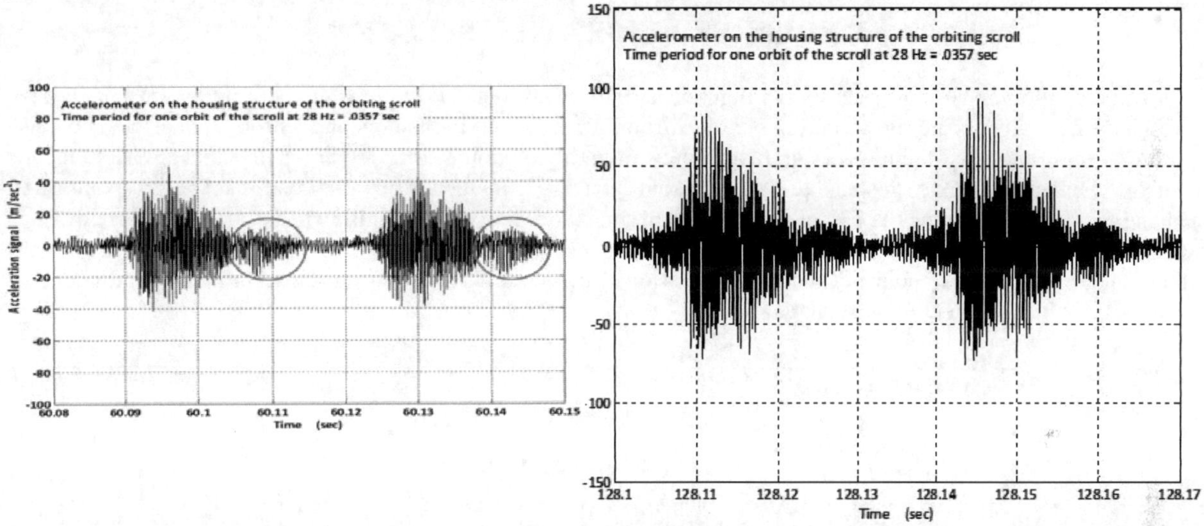

(f) (g)

Figure 2 (cont'd): Housing acceleration signal, short time slices: (f) after the sudden jump in amplitude - red circles indicate a possible second pulse; (g) near the end of the plot in Fig. 2a

We did not have a position marker to relate the timing of the pulses to the actual position of the orbiting scroll, but the results do imply that the pulse of noise occurs primarily during one direction of relative motion and less intensely (if at all) during the other direction of relative motion. (In Figure 2f the red ovals indicate a possible, smaller, second disturbance at 180 degrees orbital position from the main pulse.) Due to manufacturing tolerances in the pump drivetrain, the movement of the orbiting scroll will not be perfectly parallel to the surfaces of the fixed scroll, thus the contact force is likely to generate stick-slip behavior at only one point in the orbiting path.

Figure 2h shows a different vibration measurement on the same pump over many rotations of the orbiting scroll. An interesting pattern was observed in that the pulses appear to come in groups of approximately four pulses; the first in the group has the largest amplitude and the subsequent pulses in the group have decreasing amplitude, until a pulse with large amplitude occurs again. The number of pulses in a group was not consistent: generally four pulses to a group, but groups were observed with three to seven pulses. This pattern wasn't always seen during retesting of the same pump and the behavior remains unexplained.

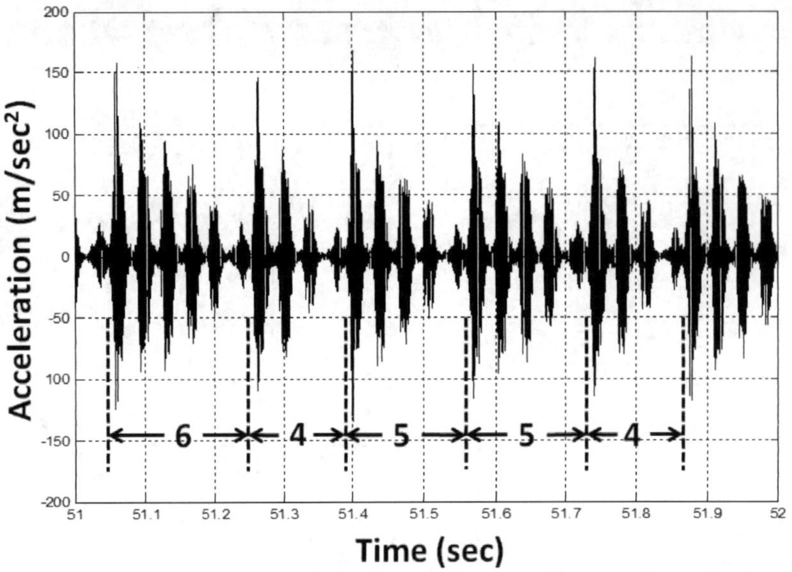

Figure 2h: Longer time slice of acceleration for same unit; note varying numbers of pulses between large pulses

3. NOISE AND VIBRATION SPECTRA

Narrowband vibration spectra are shown in Figures 3a, b for measurement locations on the orbiting scroll and fixed scroll housings. Curves are shown during steady chirping with the cooling fan on, and with the fan off and no noise. During the chirping noise there were significant bands of high vibration near 2, 4, 6 and 8 kHz that are harmonically related. The bands do not correspond to fixed sinusoids at a single frequency but are broad, likely corresponding to a changing frequency during the part of the orbit when squeal occurs. Other vibration peaks at lower frequencies were relatively consistent with the fan on or off, i.e. with squeal or no squeal, respectively. These likely correspond to the fundamental and harmonics related to the rotation of the motor. The overall linear weighted vibration levels were nearly 25dB greater during squeal.

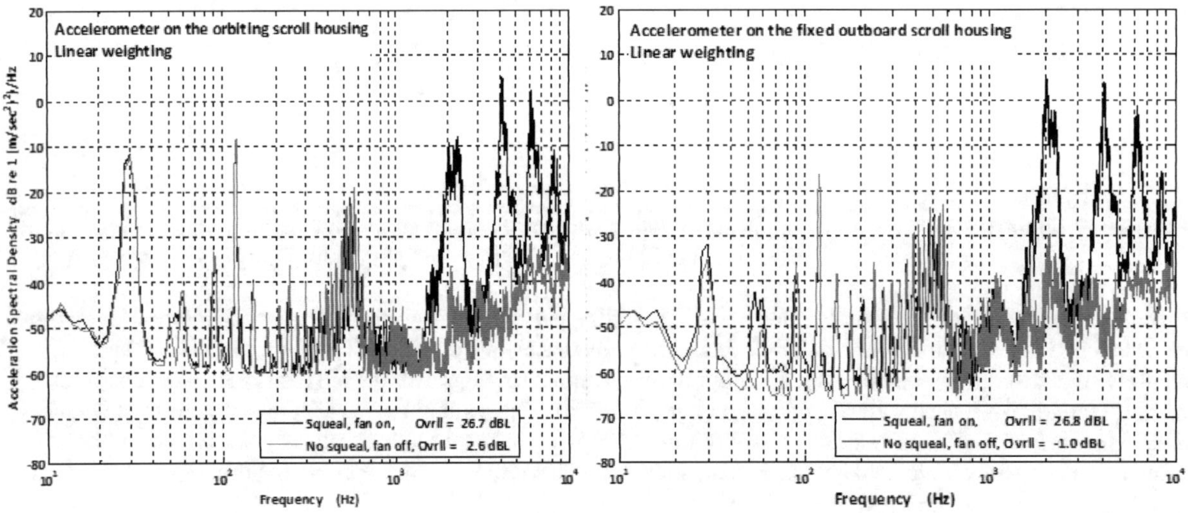

(a) (b)

Figure 3: Acceleration spectra of (a) orbiting scroll housing; (b) fixed scroll housing: "chirping" noise vs. no noise.

Noise spectra are shown in Figures 4a, b for microphone locations approximately 1 foot from the scroll and motor ends of the pump. The harmonically related bands are again seen in the noise at higher frequencies where the overall A weighted noise levels were also greater by nearly 25 dBA with chirping noise (fan on) than without chirping noise (fan off). At the motor end of the unit, Figure 4a, the larger broad spectrum levels at low frequencies were due to wind noise as the microphone was positioned within the cooling airflow.

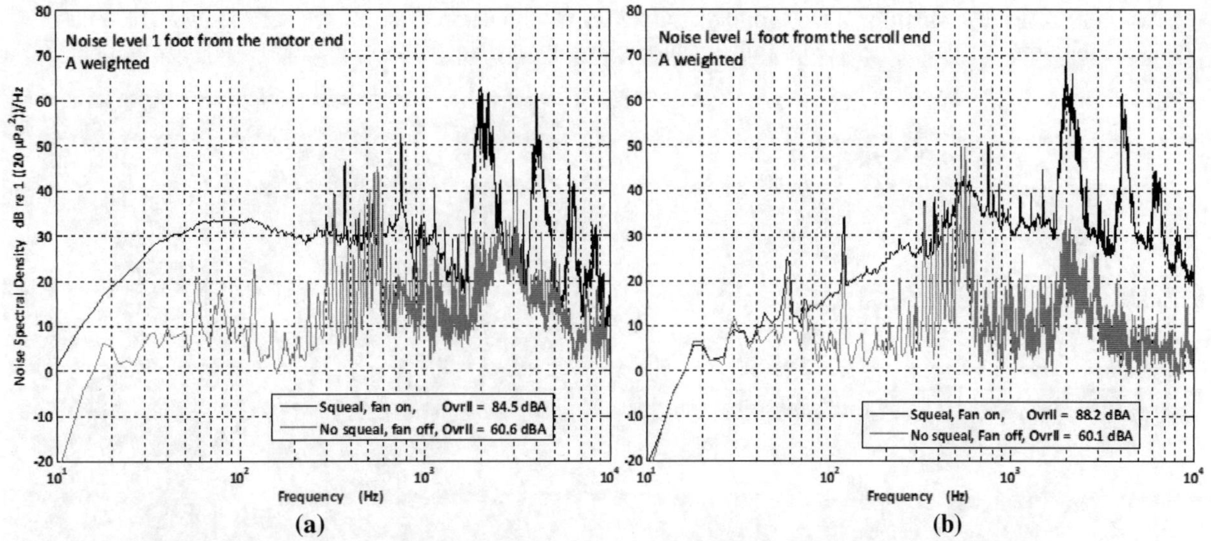

Figure 4: Noise spectra from (a) scroll end; (b) motor end: "chirping" noise vs. no noise.

4. RESONANT MODES OF THE SCROLL ASSEMBLIES

The conjecture is that the chirping noise is related to an oscillation of the vane that occurs primarily at the outer end of the vane with greatest motion at the seal edge furthest from the base plate. This motion is driven by fluctuating friction forces during the stick-slip process and excites resonances in the scroll assemblies that are dominated by motion of the vanes.

To investigate the resonant behavior of the vanes, an impulsive force was applied with an impact hammer to the end of the orbiting scroll vane at the outer turn, this being the location where the vane is least stiff in flexure. The impact was applied at the top of the vane near the tip seal. A small, lightweight, 0.5 gm accelerometer was attached to the vane at the same location and a microphone was placed approximately one inch from the vane near the impact point. Transfer function results from force to vane acceleration ($m/sec^2/N$) and noise (20 μPa/N) are shown in Figure 5. For these tests the fixed outboard scroll housing was removed in order to access the orbiting scroll. The intent was to identify whether there was significant resonant activity in the vane in the primary frequency where chirping noise occurs, *i.e.*, near 2 kHz.

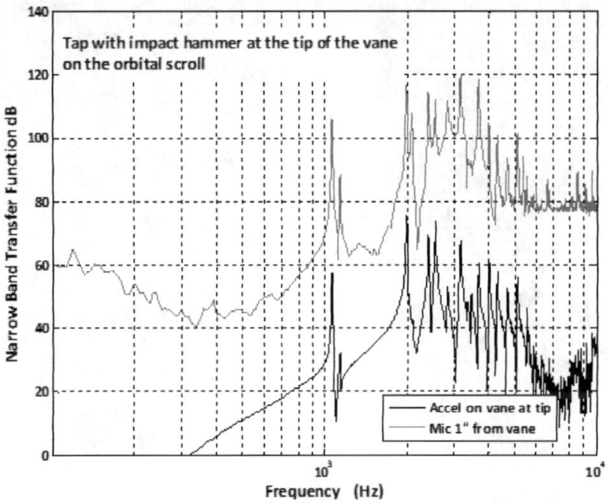

Figure 5: Transfer function from impulsive force to scroll vane

The first resonance occurs near 1 kHz and is characterized, based on finite element models, by the motion of the orbital scroll assembly on its support relative to the pump shaft. Resonances that involve significant motion of the vane relative to the base plate structure occur starting near 2 kHz, consistent with the primary noise frequency. Fig. 6 shows an FEA result, a vibrational mode of the orbiting scroll with natural frequency near 2000 Hz. FEA also indicated vibrational modes in the fixed scroll structure with natural frequencies near 2000 Hz as shown in Fig. 7.

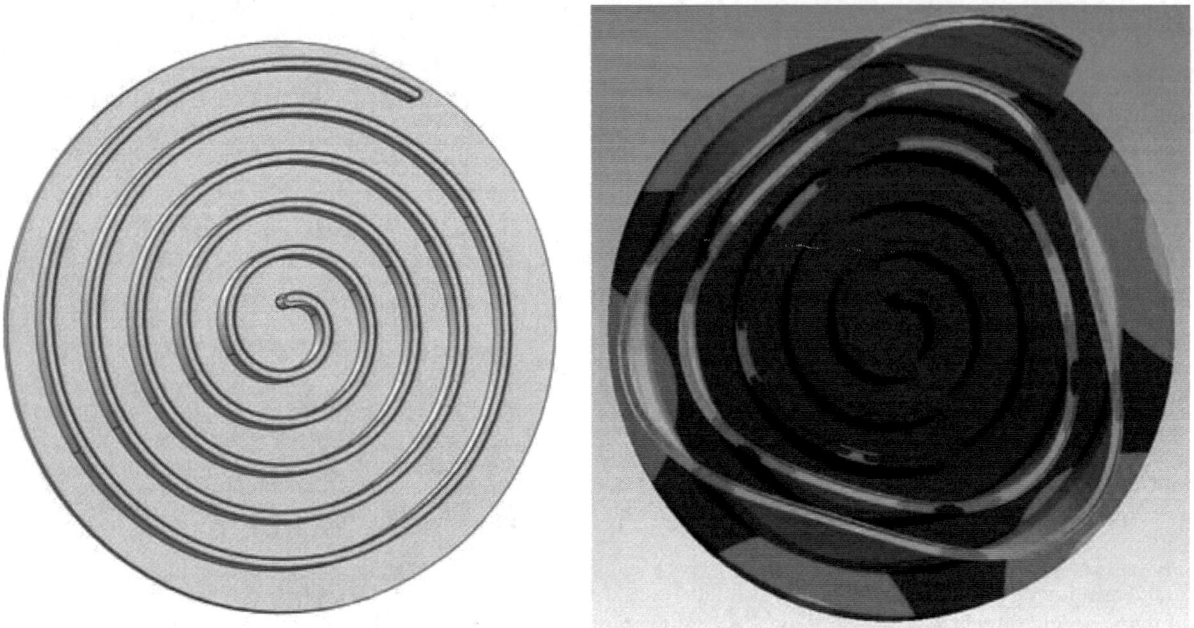

Figure 6: Typical deflection mode near 2000 Hz for orbiting scroll

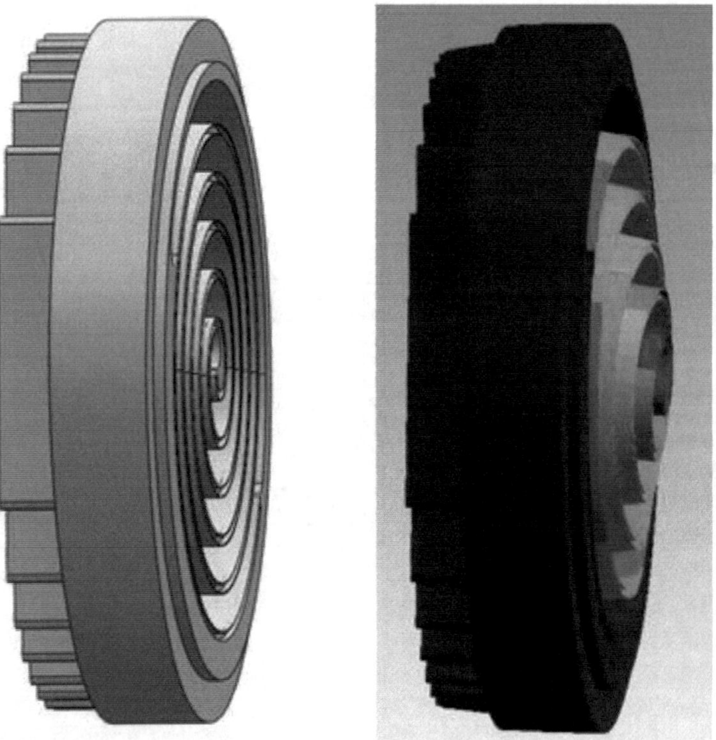

Figure 7: Typical deflection mode near 2000 Hz for fixed scroll

The stick-slip behavior ***seeks*** out resonant behavior within the mechanical system that involves an interaction between contacting parts with sliding friction to establish the stick-slip oscillation. It is not simply an excitation at a particular frequency based on external factors. The frequency of the stick-slip oscillation doesn't simply match the resonance frequency of the individual component alone, as the contact condition between the mating surfaces is at times during the oscillation fixed and at other times sliding. This establishes a limit cycle behavior with a more complicated relationship between the primary stick-slip frequency and the frequency of the mechanical resonance and also the generation of significant harmonic content in the oscillation. It is believed that the stepwise increase in amplitude of the vibration signal, seen in Fig. 2a, may be caused by a sudden large increase in the resonant response of the fixed scroll and its structure to the motion of the orbiting scroll..

5. MITIGATION OF THE NOISE

As described above, the stick-slip phenomenon requires a certain normal force applied between the surfaces of concern. If the dimensions of the tip seal and its groove are chosen so that there is always axial clearance between the top surface of the tip seal and the mating surface, the normal force cannot reach the level needed to induce the stick-slip mechanism. Based on the finite element analysis of orbiting scroll vibrational modes, described above, and the expectation that the scroll vane is most flexible near the periphery, it was determined to alter the detail design of the scrolls to increase axial clearance of the tip seal in the outermost portion of the vane. It was hoped that, even if rigid axial contact between the tip seal and its mating surface were to occur, it would occur near the center of the scroll, and that the higher vane stiffness near the center would prevent the stick-slip noise. The validation testing described below would thus also serve as confirmation of the location of the problem.

Validation of the changes consisted of building a number of pumps, with the new design and a variety of tolerance stacks, and monitoring using accelerometers and vibration monitoring software to detect whether the vibration characteristic of the "chirp" noise was detected. One pump did experience the vibration. Upon investigation it turned out that this unit had been machined out of tolerance, *i.e.*, to the old dimensions. No units with the corrective action properly applied experienced the vibration or noise of concern.

6. CONCLUSIONS

Machinery designers should take note that wherever sliding contact occurs between components that are relatively rigid in the normal direction and relatively elastic in a direction parallel to the motion, there is a risk of stick-slip noise. In addition, such noises may not be present on all units, as small variations in normal force or sliding speed can cause the noise to appear or disappear. This should be taken into account especially in design of low-noise products.

REFERENCES

Allgaier, R., *et al.* 2002, A Study on Brake Squeal Using a Beam-on-Disc Model. *Proceedings of the 2002 IMAC-XX,* Society for Experimental Mechanics, pp. 528-534.

Ghazaly, N., El-Sharkawy, M., Ahmed, I. 2013, A Review of Automotive Brake Squeal Mechanisms. *J. Mechanical Design and Vibration*, vol. 1, no. 1, pp. 5-9.

Ghazaly, N., Mohammed, S., Abd-El-Tawwab, A.M. 2012, Understanding Mode-Coupling Mechanism of Brake Squeal Using Finite Element Analysis. *Int. J. Engineering Research and Applications*, vol. 2, issue 1, pp. 241-250.

Kim, U., Krousgill, C., Mongeau, L. 2005, Friction-Induced Vibrations of Window Sealing Systems. *SAE 2005 Noise and Vibration Conference and Exhibition,* Society of Automotive Engineers, no. 2005-01-2540.

Oberst, S., Lai, J.C.S., Marburg, S. 2013, Guidelines for Numerical Vibration and Acoustic Analysis of Disc Brake Squeal Using Simple Models of Brake Systems. *J. Sound and Vibration*, no. 332, pp. 2284-2299.

Soedel, W. 2006, *Sound and Vibrations of Positive Displacement Compressors.* CRC Press, Boca Raton, Fla., 343 pp.

Stoll, A., Cavalcante, M. 2012, Stick-Slip Characteristics of Leather/Artificial Leather, *In:* Trapp, M., Chen, F. *Automotive Buzz, Squeak, and Rattle: Mechanisms, Analysis, Evaluation and Prevention*. Butterworth-Heinemann, Waltham, Mass., pp. 63-98.

Investigation of the Effects of Coil Spring on Discharge Tube

*Ergin TASDELEN[1], Atilla KAYA[2]

[1]Arcelik AS, Compressor Plant, R&D Department
26110 Eskisehir, Turkey
Phone: +90 222 213 43 43, Fax: +90 222 213 43 44,
Email: ergin_tasdelen@arcelik.com

[2]Arcelik AS, Compressor Plant, R&D Department
26110 Eskisehir, Turkey
Phone: +90 222 213 43 43, Fax: +90 222 213 43 44,
E-mail: atilla.kaya@arcelik.com

*Corresponding Author

ABSTRACT

In this paper, effects of coil spring and coil spring parameters on vibration and acoustic characteristics of compressors are studied. Different tube and spring combinations are prepared and analyzed in order to compare the frequency response function of tube as well as no spring situation. Vibration responses of coil springs are investigated by experimental studies. Results are compared with numerical model. The numerical methods can be used to study the vibration characteristics of a compressor and decreases the need of preparing several prototypes.

1. INTRODUCTION

In hermetic compressors, one of the vibration transmission paths from main body to the shell is discharge tube. The main parts of discharge tube are tube part, coil spring on the tube and resonator cap. The end which is mounted to main body is excitation point. The point that tube is fixed to the shell is evaluation point to investigate the transmitted vibration energy towards compressor shell.

The coil spring on the tube has two main effects on tube. One of them is the mass effect. The usage of coil spring creates additional mass to the discharge tube and it causes change of natural frequency of discharge tube. The other one is the damping effect. Coil spring on discharge tube creates impact damping and reduces vibration kinetic energy of discharge tube.

The coil spring parameters effect vibration and acoustics characteristics of compressor therefore these parameters should be determined carefully.

2. EFFECTS OF COIL SPRING

Discharge tubes should provide two main functionalities in compressor. One of them is discharging of the pressurized refrigerant gases and the other one is reduction of vibration which is transmitted from main body through the shell. For vibration reduction function, coil spring on discharge tube has importance. Discharge tube is a vibration transmission path and because of its complexity of shape, design is performed most of times by trial and error method with some good engineering judgment.

Coil spring on discharge tube adds nonlinear characteristics to discharge tube. This causes modeling of this system analytically very challenging job. Some previous studies tried to model this complex system, and Wang and Park (2006) are modeled this system by using 2-degree of freedom impact damper system. Wang and Park (2006) stated that coil spring can have two different effects on vibration characteristics of discharge tube according to phase synchronization of coil spring and discharge tube. When out-of-phase, because of impact mechanism, momentum transfer occurs. However, in-phase case coil spring shows mass loaded effect.

Coil spring have some design parameters. These parameters are length of spring, spring constant, clearance between spring and discharge tube, position of spring, spring constant and etc. All these parameters should be selected carefully during design processes.

3. EXPERIMENTAL STUDY

To understand effects of coil spring and also coil spring parameters, an experimental study is done. Hammer test is used to obtain Frequency Response Functions of specimens. In real working conditions, discharge tube is excited from resonator cap by main body and it transmits the vibration energy to the shell from the other end of tube. Discharge tube is welded to shell by using an interconnecting tube. To simulate this configuration in experimental study, resonator cap is at free condition and other end of discharge tube is fixed to ground. Excitation forces are generated in compressors by piston mechanism and unbalanced inertia. In experiment, a hammer is used to excite specimen. Evaluation point is selected a near point to shell end, from this point by using accelerometer response of system is measured. Experiments are made multiple times for each specimen to check stability of experiment system.

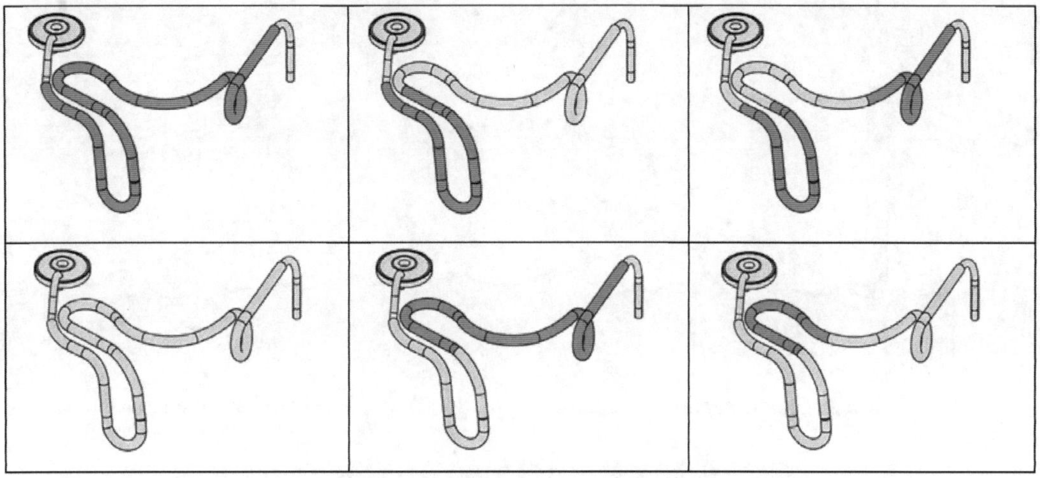

Figure 1 : Experimental Testing Specimens

In this experimental study, spring length and spring position parameters are variable parameters. By chancing them, six different type specimen sets are obtained. One of type does not have any coil spring on it. This type is needed to create a baseline to understand effects of coil spring in detail.

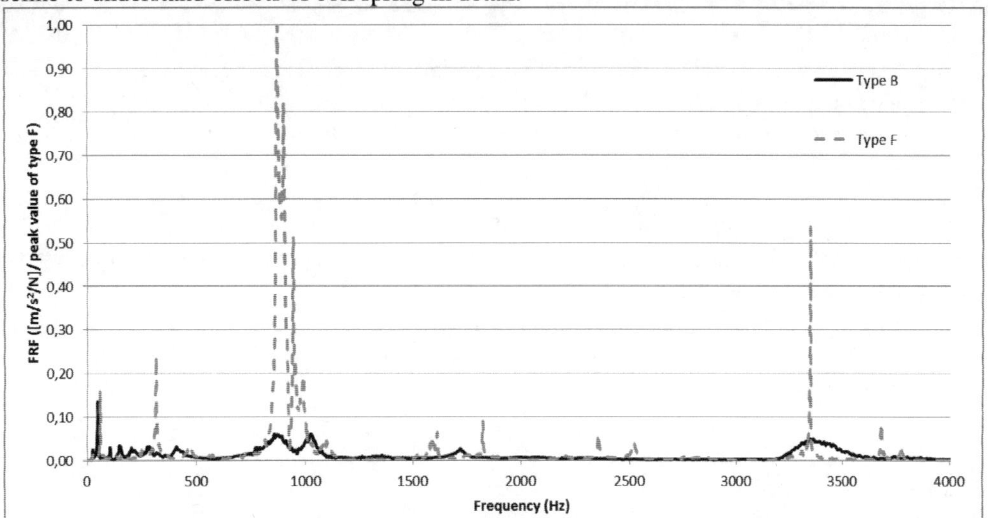

Figure 2 : Experimental FRF results of Type B and Type F

Figure 2 shows that experimental test results of type B and F. Type F does not have any spring on discharge tube. In higher frequencies, especially after 500 Hz, damping effect of spring comes up. Effects of coil spring on discharge tube observed similarly in previous studies. (Wang *et al.*, 2004)

The Figure 3 shows that lower frequency responses of type B and F. In this figure, mass loading effect of coil spring causes a shifting to lower frequencies of peaks of type B with respect to peaks of type F.

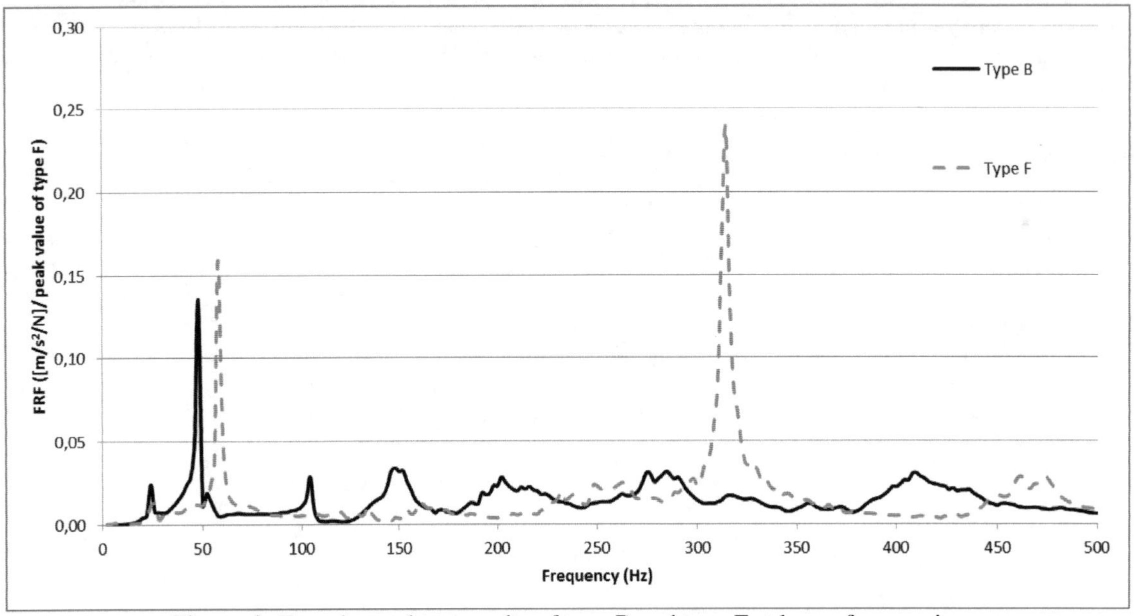

Figure 3: Experimental test results of type B and type F at lower frequencies

4. NUMERICAL ANALYSIS

Discharge tube has complex shape and also nonlinear characteristic because of coil spring on it. Therefore, analytical analysis and also a complete numerical analysis are very competitive duties. Samples or prototypes should be prepared for experiments. Tests should be done multiple times with multiple samples, so all of these take times. However, numerical analysis can be done easily and after verification of model is done, it needs less time than experiments.

A numerical model is created to investigate effects of coil spring on discharge tube. In this numeric analysis, main area of concern is mass effect of coil spring. Six different spring and tube combinations were created in experimental tests. Same six combinations are modeled numerically. Additional masses are created on discharge tube at specific locations to simulate experiment specimens.

To obtain same results with experiments, boundary conditions of numerical analysis are selected as same and also material properties are defined with respective to real material properties. Investigated frequency range is also defined same as experimental test results.

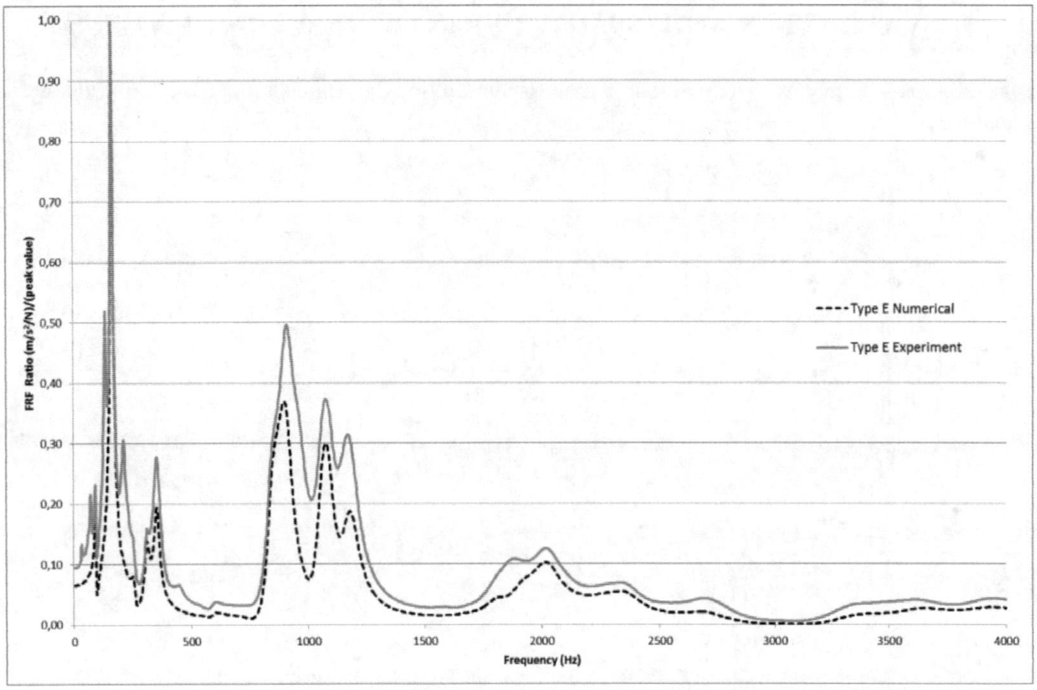

Figure 4 : Experimental tests and Numerical analysis results of Type E

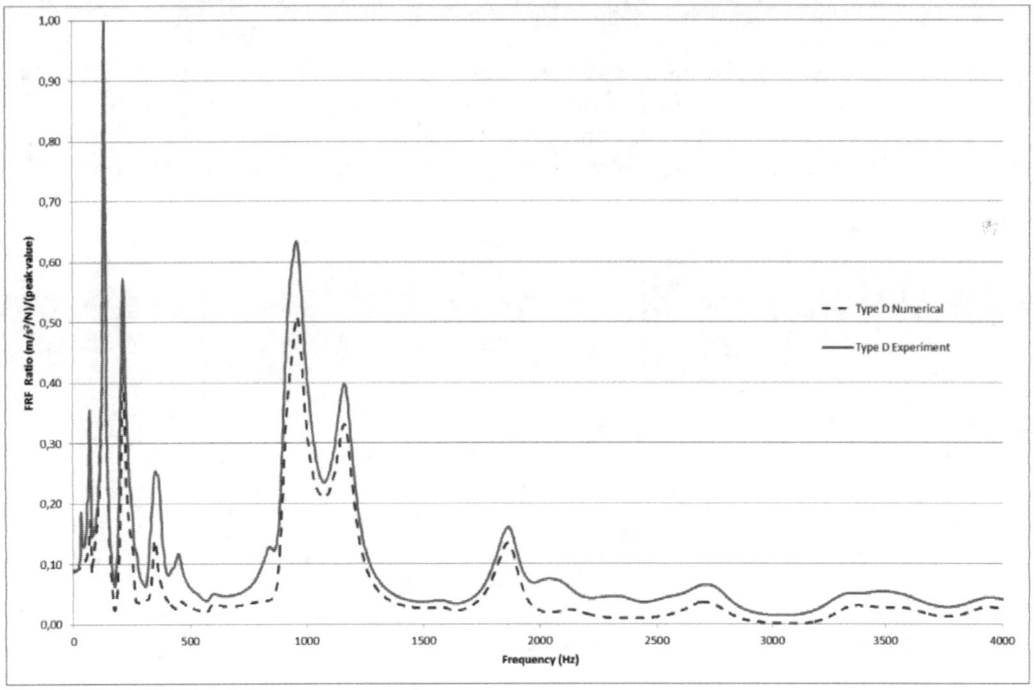

Figure 5 : Experimental tests and Numerical analysis results of Type D

Figure 4 and Figure 5 show that numerical analysis results are suitable with experimental results at both lower and higher frequencies. Peaks occur at same frequencies in results of numerical and experimental studies. According to these results, numerical analysis can be useful for design processes.

5. EVALUATION AND COMPARISION OF RESULTS OF TYPES

Numerical results are suitable with experimental results therefore types of samples can be evaluated by using numerical analysis.

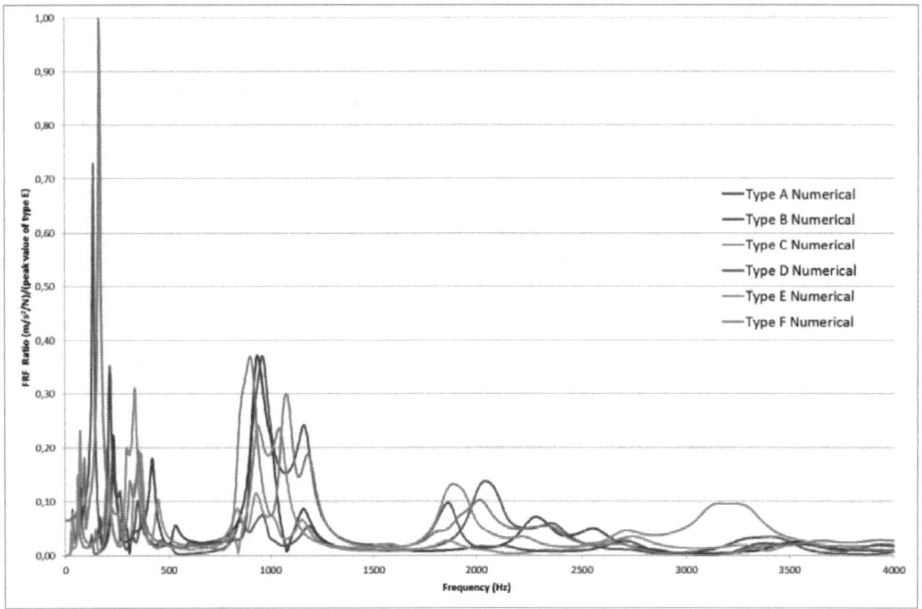

Figure 6 : Numerical analysis results of six types

Figure 6 shows that frequency response function graph of different coil spring and discharge tube combinations. Type F does not have any spring on tube; as a result of this it has high vibration amplitudes at frequencies between 3000 Hz and 3500 Hz. Figure 7 shows this trend obviously. The other types have spring on different locations with different length. These types do not have high vibrations at high frequencies. Spring provides positive effect at that frequency range.

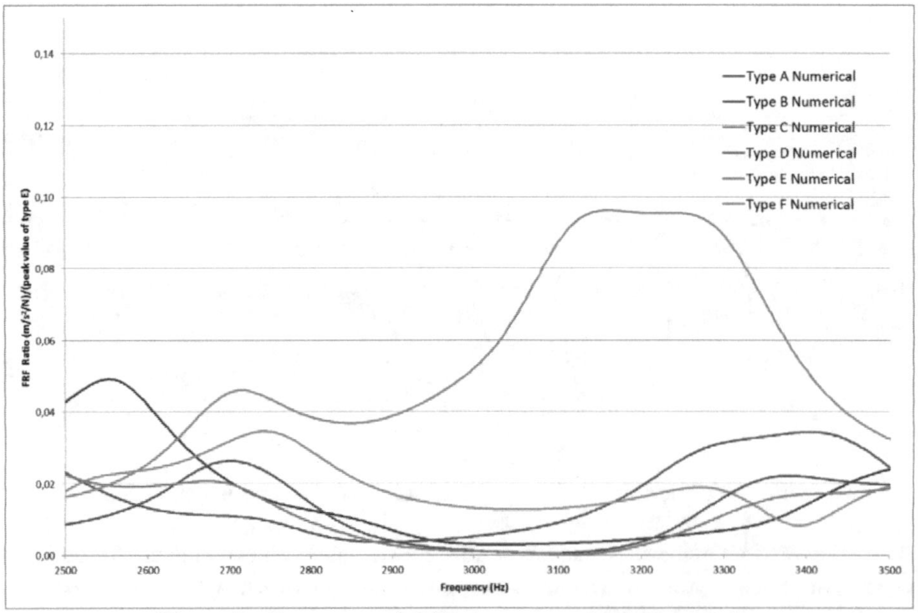

Figure 7 : Responses of different types at high frequencies

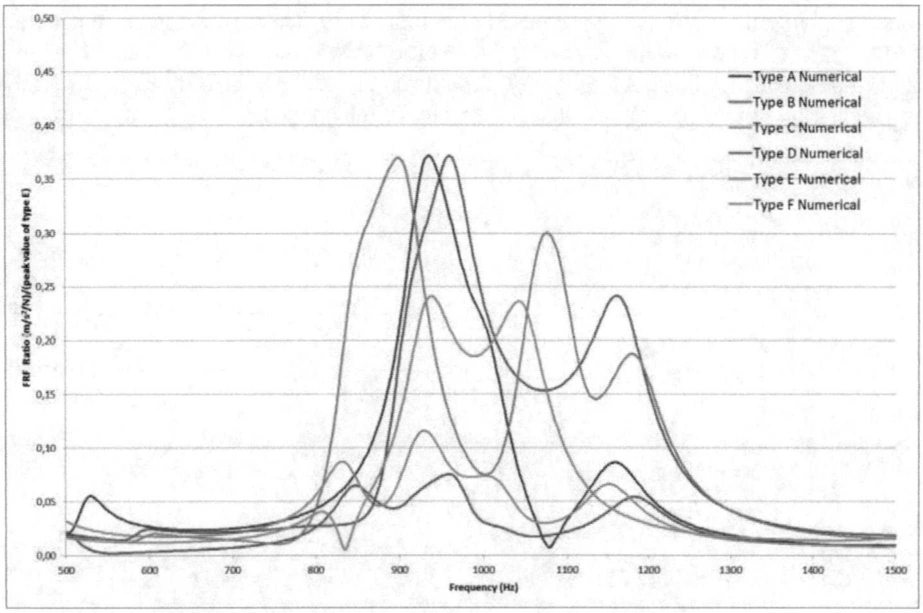

Figure 8 : Responses of different types at mid frequencies

Figure 8 shows that type B and type C have lower amplitudes than others at frequency range between 500 Hz and 1500 Hz. At lower frequencies between 100 Hz and 200 Hz, type D and type E vibrate high amplitudes. In contrast, vibration of type F has very low vibrations at same range which is shown in Figure 9. As a result of this, coil spring causes negative effect for these low frequencies.

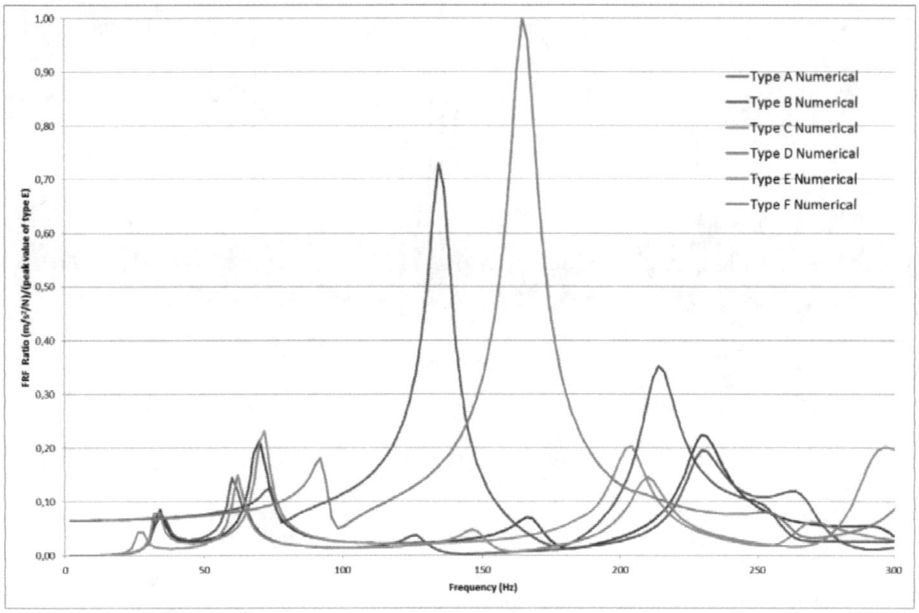

Figure 9 : Responses of different types at low frequencies

Prototype compressors are prepared to see effects of these different vibration characteristics on compressors. Vibration and sound measurements of all of prototypes are performed. In below Figure 10, first graph shows that numerical analysis results of type B, C and F discharge tubes. The second graph indicates that vibration measurement of compressor which is prototyped by using these three types and last graph shows sound measurement of compressors. Type F has higher vibration amplitudes than type B and C at frequencies about 3000 Hz. This reflects negatively on sound and vibration results of type F used in compressors. There is a resonance peak between 800 Hz and 1200 Hz in FRF graph of type F discharge tube. As a result of this, sound pressure level of this type compressor is higher than other types about 1250 Hz. All of these do not have any peak about 1600 Hz and sound level of all are low; therefore, spring has no effect on compressor vibration and sound characteristics about this frequency.

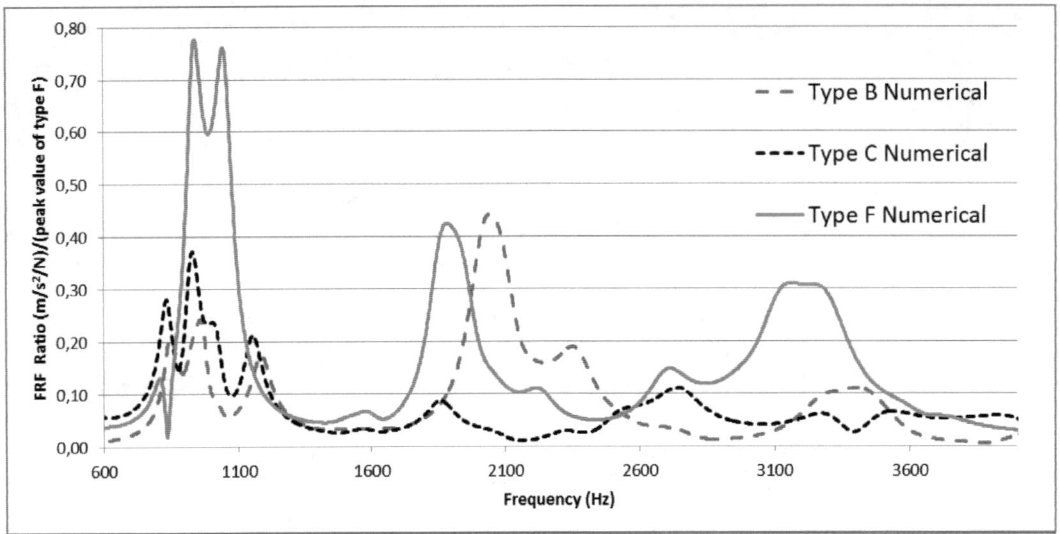

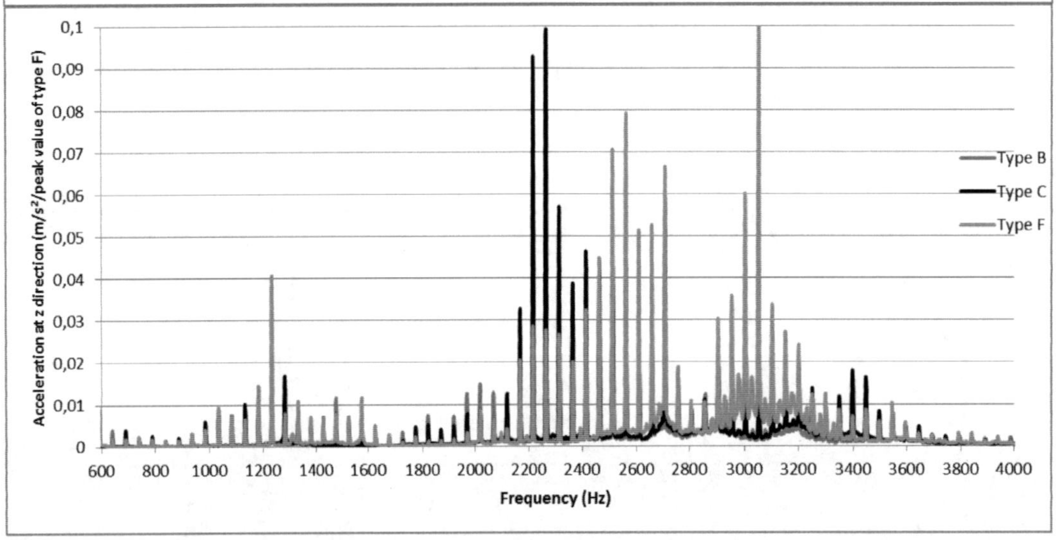

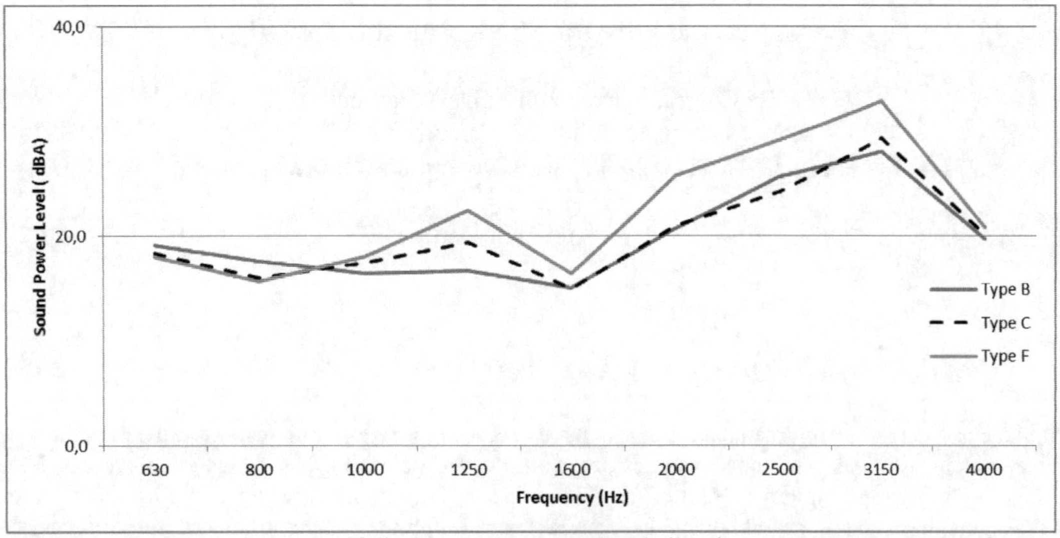

Figure 10 : Numerical analysis results of discharge tubes, sound and vibration measurement results of compressors with type B, C and F samples

6. CONCLUSIONS

Discharge tube on compressors has geometrical complexity and coil spring on it causes nonlinearity therefore analytical analysis of discharge tube with coil spring became very hard and it needs detailed study. Experiments can be used to overcome this problem and obtain desired designs by changing parameters. However, most of time experiments are needed so much time and repetitive tests. Numerical analysis is another design tool for engineers but numerical solutions should be verified initially.

In this paper, by changing length and position of coil spring on discharge tube different groups are created. By making hammer test, frequency response functions of specimens are obtained. After that, numerical model is prepared by using mass effect of coil spring. Numerical analysis is performed for six different types. Results of numerical and experimental studies are compared and numerical results show similarity with experiments. After numerical model is verified, by using numerical results effects of coil spring are investigated. These effects are observed on compressor by using sound pressure level and vibration measurements. The peaks which are observed on numerical and experimental analysis reflect negatively on sound measurements of compressor. As a result of this, numerical model is useful for pre-design processes of discharge tube with coil spring of compressors.

REFERENCES

S. Wang & J. Park, 2006, "*Noise Reduction Mechanisms of a Spring Wound LDT for a Reciprocating Compressors*", ICSV13, Vienna, Austria.

S. Wang, J.Kang, J. Park, C. Kim, 2004, "*Design Optimization of a Compressor Loop Pipe using Response Surface Method*", Purdue.

Singiresu S. Rao, 2007, *Vibration of Continuous Systems*, John Wiley & Sons.

ACKNOWLEDGEMENT

The authors gratefully acknowledge to Associate Professor Ender Cigeroglu, Middle East Technical University for valuable discussions about nonlinear vibrations.

Analysis of Vibration Isolators for Hermetic Compressors

Carlos Eduardo VENDRAMI*, Claudio PELLEGRINI

Embraco, Research & Development
Joinville, SC, Brazil
Phone: +55 47 3441-2650, carlos_vendrami@embraco.com.br

* Corresponding Author

ABSTRACT

Recently, the customers of refrigerators and freezers have become more exigent in respect to the acoustic comfort of these home appliances. Among the noise sources, there are the refrigerant flow, the fans and the compressors. The compressors might be responsible for a part of the cabinet low frequency noise, which could be generated either by the gas pulsation that excites the tubing or the compressor low frequency vibration that transfers energy to the cabinet base plate and tubing.
The compressor low frequency vibration is a function of the mechanism loads and the gas pressure that acts on the piston and it is modulated by its running frequency and harmonics. Besides, at low frequencies, the compressor behaves as a rigid body assembled on isolators or grommets, which link the compressor to the cabinet base plate. If one knows these rigid body natural frequencies, it is possible to optimize the refrigerator project, providing a better acoustic performance for manufacturers and final customers.
Based on this context, this paper aims to present a study of the compressor grommets, usually made of rubber. First of all, an investigation about the materials is carried out using a technique with a single degree-of-freedom system. Secondly, an analysis of the static behavior of an isolator using the rubber properties is performed. Finally, a numerical model to predict the modal behavior of a compressor assembled on a rigid base is created and experimentally validated.

1. INTRODUCTION

The low frequency noise radiated by a cabinet can have some sources, such the gas flow, the fans and the compressor. Among the sources related to the compressor itself, there are the gas pulsation and the compressor low frequency vibration, which are created by mechanism loads and the gas pressure that acts on the piston.
At low frequencies the compressor behaves as a rigid body, because it is assembled on isolators or grommets which are components, normally made of rubber, that link the compressor base plate to the appliance base plate. Figure 1 presents an example of grommet assembled in a compressor and an example of grommet.

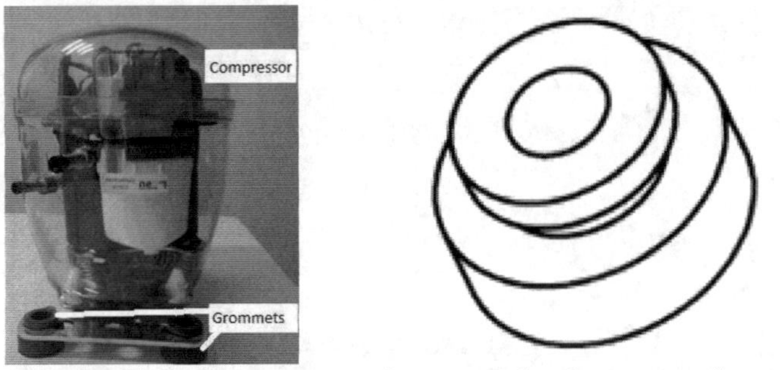

Figure 1: Compressor assembled on grommets and an example of grommet. Source: Vendrami, 2013.

Therefore, grommets are components that transfer some energy from the compressor to the appliance and because of this they can generate noise problems in low frequency. These problems are more frequent when a rigid body natural frequency of the compressor assembled on the base plate coincides with the compressor running frequency or any other critical frequency. Figure 2 shows how it is possible to optimize the low frequency noise with a good choice of grommets for a specific appliance.

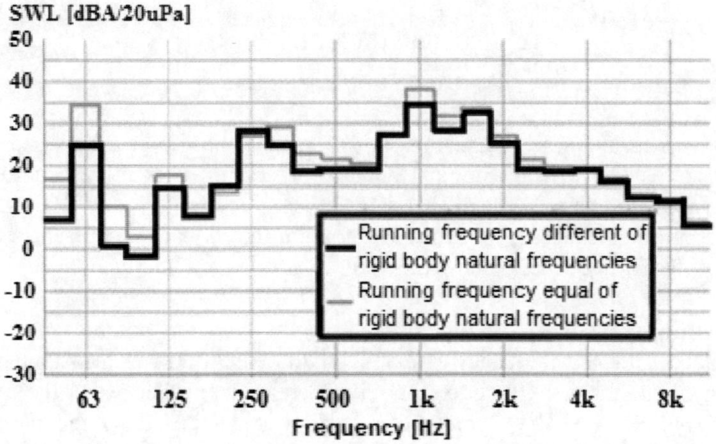

Figure 2: Low frequency noise optimization with a good choice of grommets. Source: Porto, 2010.

Knowing that nowadays customers are more and more exigent about the noise radiated by their home appliances, this work aims to present a study about the grommets, presenting some basic properties of the rubber and some ways to find out other properties. Besides, it presents some numerical static and dynamic analysis of these components as well as some validation of these simulations.

2. THE RUBBER AND ITS PROPERTIES

Rubbers are materials which have low plastic deformation, low Young Modulus (E) and are able to support deformations up to 1,000%. Besides, they have high resilience, high internal damping and do not follow the Hooke's law under deformation.

According to Istvan (2006), rubber behavior varies with the temperature. At low temperatures, the molecules are almost static and there is small relative movement between them, which means that rubbers have high Young Modulus (E), close to 0.5 (0.499 according to Jones, 2001) and small damping factor (η), respectively.

At high temperatures, the relative velocity between the molecules is higher, which means the Young Modulus tends to be smaller. As the interaction between molecules is lower, thus the loss factor is also lower.

Finally, at mid temperatures, there is small Young modulus but the loss factor is high.

Figure 3 shows the dependency of the Young Modulus and the Loss factor with the temperature for viscoelastic materials, like rubber.

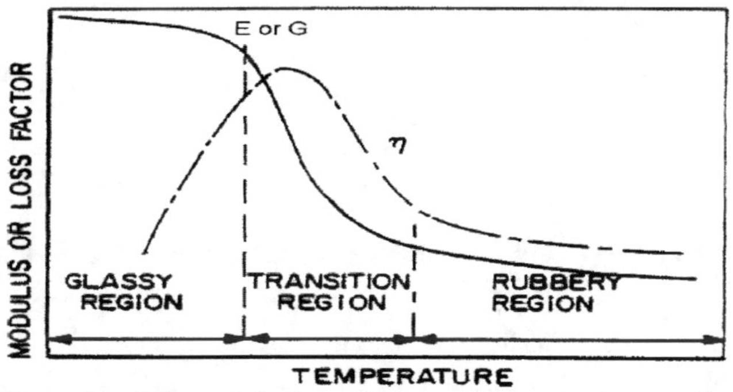

Figure 3: Variation of the Young Modulus and the Loss Factor with the temperature. Source: Istvan, Beranek, 2006.

The effect of the frequency is the opposite of the effect of the temperature: bigger the frequency of a harmonic excitation that acts on the rubber, the higher the Young Modulus (Jones, 2001).

Another property largely used to characterize a sample of rubber is the hardness, that is the resistance to the plastic deformation when an external load is applied (Almeida, 2009). Some authors created calculations and graphs which allows the association of the hardness of the rubber to its Young Modulus. One of them was proposed by Gent (1958) and correlates the Young Modulus *E*, in MPa, to Shore A hardness (*S*):

$$E = \frac{0.0981(56+7.62336S)}{0.137505(254-2.54S)} \quad (1).$$

Next sections will show two techniques used to measure rubbers Young modulus and loss factor: Oberst Beam, presented in ASTM Standards (2005), and Single Degree of Freedom. It is important to mention that for both techniques, the system temperature must be controlled.
The last section will present the Reduced Frequency Nomogram that is used to present Young modulus and loss factor data, obtained for different frequencies and temperatures, in only two curves.

2.1 Oberst Beam Technique

According to this technique, first the frequency response of the base beam is measured using the measurement system exemplified in Figure 4A, at specific temperatures which are controlled when the apparatus is placed in an environmental chamber (Figure 4B). The Oberst beam, which is composed by the base beam and a layer of viscoelastic material, is presented in Figure 4C.

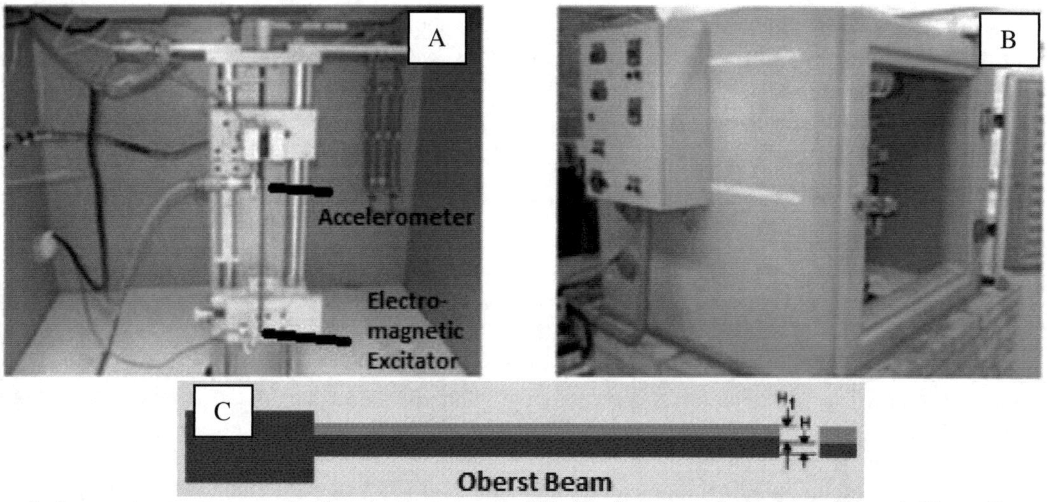

Figure 4: Apparatus to measure the frequency response (A), Environmental chamber (B) and Oberst Beam (C).
Source: Stumpf, 2013.

According to ASTM Standards (2005), Young Modulus E_b and loss factor ηb of base beam material are calculated:

$$E_b = \frac{12\rho_b . L^4 {f_n}^2}{H^2 {C_n}^2} \quad (2),$$

$$\eta_b = \frac{\Delta f_n}{f_n} \quad (3).$$

where C_1 =0,55959, C_2 =3,5069, C_3 =9,8194 and $C_n = (\pi/2)(n - 0,5)$ for n>3, f_n are the resonance frequencies of n modes, Δf_n is the half power band ($\Delta f_n = \frac{f_{2n} - f_{1n}}{2f_n}$, where f_{2n} and f_{1n} are the two frequencies closer to f_n which the energy is the half part of the energy of the peak given for f_n, considering $f_{1n} < f_n < f_{2n}$), ρ_b is the material density, H is the thickness and L is the length.
After measuring the frequency response of the base beam, the damped beam response is also measured. This damped beam is basically composed by the base beam and a layer of rubber which is glued on the metal and its natural frequencies, the rubber Young Modulus E_v and the rubber loss factors η_v must be obtained at the same temperatures the base beam was tested, according to equations (4) and (5).

$$E_v = \frac{E_b}{(2R^3)}\left[(\alpha - \beta) + \sqrt{\{(\alpha - \beta)^2 - 4R^2(1-\alpha)\}}\right] \quad (4),$$

$$\eta_v = \eta_c\left[\frac{(1+MR)(1+4MR+6MR^2+4MR^3+M^2R^4)}{(MR)(3+6R+4R^2+2MR^3+M^2R^4)}\right], \quad (5),$$

where c is an index, f_n is the resonance frequency for mode n of base beam, f_c is the resonance frequency of the mode c of the composed beam, Δf_c is the half power band of mode c of the composed beam (calculated as Δf_n), H_1 is rubber layer thickness, $M = \frac{E_v}{E_b}$, $R = \frac{H_1}{H}$, $\alpha = (f_c/f_n)^2\left(1 + \frac{\rho_v}{\rho_b} R\right)$, $\beta = 4 + 6R + 4R^2$, $\eta_c = \Delta f_c/f_c$ is the composed beam loss factor and ρ_v is rubber density. This technique is indicated specially for determining properties of materials with Young modulus over than 100MPa (ASTM Standards, 2005), value normally higher than rubbers commonly used in grommets.

2.2 Single Degree of Freedom Technique

Another technique that might be used for the analysis of the rubber properties is the single degree of freedom technique. In this case, the dynamic properties are evaluated in tests which a cylindrical rubber sample is glued on a flat surface that needs to be sufficiently stiff. On the rubber sample upper surface, it is necessary to glue a steel cylindrical block (rubber sample and steel block must have the same central axis) with a mass much bigger than the rubber sample mass.

Then, this system is then placed inside an environmental chamber like the one showed in Figure 4B at a specific temperature and must be excited with a sweep sine signal, using a shaker. In this case, the rubber sample behaves as a spring and the steel block has a point mass behavior.

The frequency response of the point mass related to the base is obtained using two accelerometers (both *B&K* type 4517) as showed in Figure 5.

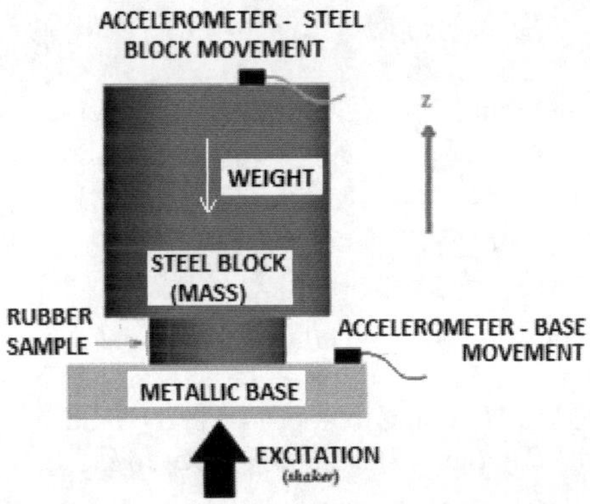

Figure 5: Measurement system for single degree of Freedom Technique. Source: Vendrami, 2013.

Considering the same rubber sample and three different steel blocks with masses A, B and C and using the measurement system presented in Figure 5, curves like the ones presented in Figure 6 are obtained.

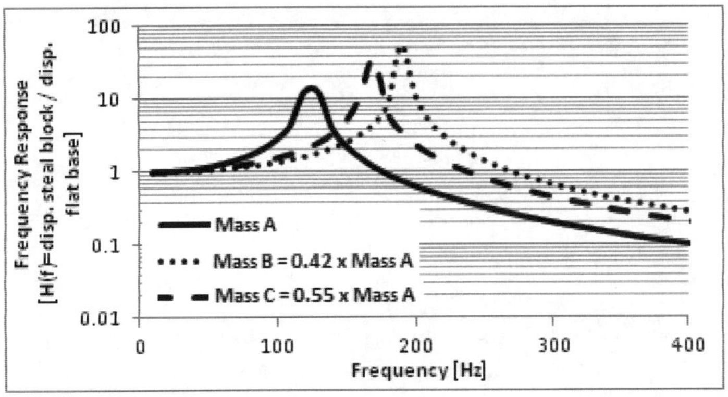

Figure 6: Example of experimental curves obtained with Single Degree of Freedom Technique.

The rubber loss factor normally is obtained directly from the curves at the peaks. For example, *B&K Pulse 12.0* uses half Power band to calculate the damping factor ζ, which is the half part of the loss factor.
On the other hand, the rubber Young modulus needs to be obtained adjusting a numerical model to the experiment. For this purpose, it is necessary to set in the numerical model the density and the Poisson's ratio of the rubber sample material as well as the density, the Poisson's ratio and the Young modulus of the steel.
Using all these properties, some numerical modal analysis can be done, varying in the software the value of rubber Young modulus until the system first natural frequency results similar (difference between the experiment and the simulation lower than 0.1%) to the frequency of the peak obtained during the experiments. To do this numerical modal analysis, one must consider fixed all the nodes of rubber lower surface and common nodes at the interface between the rubber sample and the steel block.
This technique is interesting to calculate the dynamic properties of the rubbers commonly used in grommets, which the Young Modulus for frequencies up to 400Hz is lower than 50 MPa.

2.3 Reduced Frequency Nomogram

As already commented, the Reduced Frequency Nomogram is a way to present Young Modulus and Loss Factor magnitudes, obtained at different temperatures and frequencies, using only two curves. This fact allows extrapolation of these parameters to frequency and temperature ranges that were not evaluated (ASTM Standards, 2005).
This Nomogram is obtained from the Superposition Principle that affirms the dynamic properties as E and η of the rubber are function of the product $f \times \alpha_T(T)$ where f is the frequency and α_T is the shift factor, that is a function of the temperature T. The product $f \times \alpha_T(T) = f_R$, where f_R is also known as reduced frequency.
Based on these concepts, the Young Modulus and the Loss Factor can be defined as follows:

$$E = E(f,T) = E\big(f.\alpha_T(T)\big) = E(f_R) \tag{6}$$

$$\eta = \eta(f,T) = \eta\big(f.\alpha_T(T)\big) = \eta(f_R) \tag{7}$$

Finally, for Reduced Frequency Nomograms consolidation, it is necessary to calculate the shift factor $\alpha_T(T)$. Williams, Landel and Ferry (1955) proposed an WLF equation to do this calculation:

$$log[\alpha_T(T)] = -C_1 \frac{T-T_0}{B_1+T-T_0} \tag{8}$$

where B_1 and C_1 are constants which must be determined. Osswald (2010) suggests some values for these constants, and *T0* is the temperature where there is the transition from glassy region to transition region of the material.
Based on the loss factors measured and Young Modulus and Reduced frequencies calculated at different temperatures, it is possible to create the nomograms as the one showed in Figure 7.

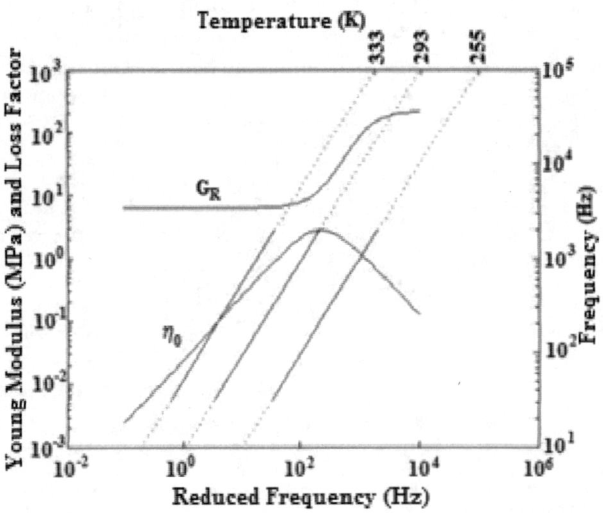

Figure 7: Reduced Frequency Nomogram Example. Source: Medeiros Júnior, 2010

Further recommendation to create and to use Reduced Frequency Nomograms, which are necessary to correctly set rubber dynamic properties for grommet analysis and projects, are found in ASTM Standards (2005).

3. GROMMET STATIC ANALYSIS

The first analysis done to understand the characteristics of a grommet was a static analysis. This analysis is important to understand how much the grommet structure deforms because of compressor weight.

3.1 Numerical Model

A FEM numerical model can be defined to understand grommet behavior under the compressor's weight load.
To set the properties to do this type of analysis, the construction of Reduced Frequency Nomograms is not necessary, but it is mandatory to consider the Poisson's ratio, the density and the rubber static Young Modulus E_0. This value can be obtained by several ways, especially from the rubber material hardness, using calculations like the one proposed by Gent (1958).
The geometric model must consider only grommet utile part for the point of view of the stiffness which is the part that is under compressor base plate as it is showed in Figure 8.

Figure 8: Schematic representation of the analyzed grommet (L) and its functional part (R). Source: Vendrami, 2013.

For the model, a hexahedrical mesh (element SOLID185) was created and using *ANSYS 11.0*, all the nodes of the grommet lower surface are considered fixed. On the upper surface, a displacement from 0.02 up to 0.5mm in intervals of 0.02mm was imposed, knowing the fact that 0.5mm is bigger than the deformation imposed by most of compressors (Porto, 2010).
The goal of the simulation is to calculate the reaction force, in newtons, at the lower surface for each increment of displacement, as the outline presented in Figure 9 proposes.

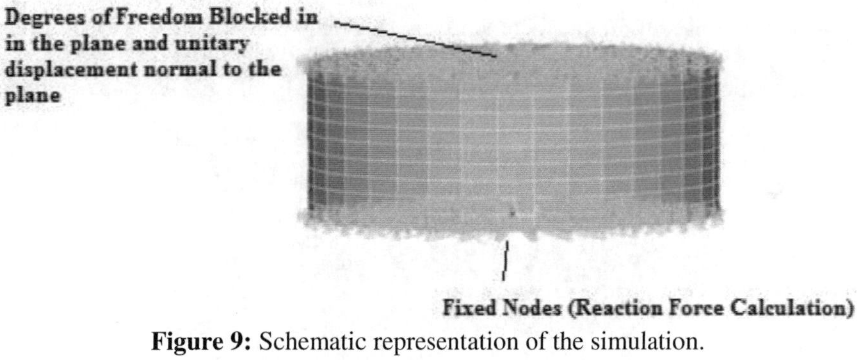

Figure 9: Schematic representation of the simulation.

The result for the proposed grommet showed a straight line in a chart Force vs. Displacement, where its inclination represents the grommet stiffness given in N/mm.

3.2 Model Validation

For the validation of the static numerical model, a Universal Testing Machine (model EMIC DL2000) was used in order to obtain a curve of the reaction force, in N, at the lower surface as a function of deformation imposed at the upper surface, in mm. The speed of the grommet deformation was set as 1mm/min.

Figure 10 presents the equipment and the position of the grommets for the described test and how the grommet sample was prepared.

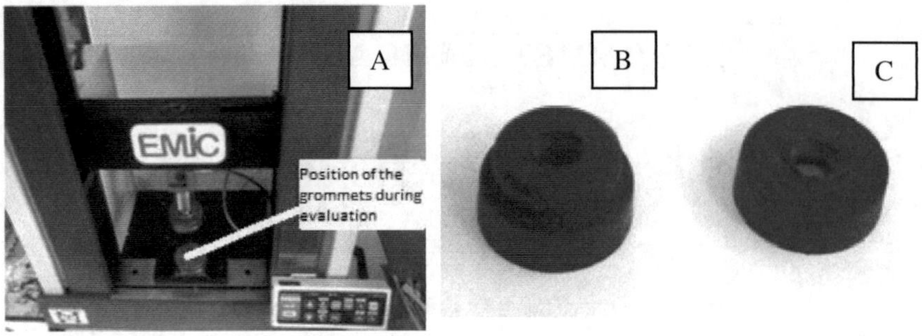

Figure 10: Equipment used for the experiments (A), entire grommet used in the analysis (B) and its utile part (C).

A comparison of the numerical and experimental curves (Figure 11) suggests a good correlation between the results due to a difference lower than 3% in the reaction force, with higher forces for the numerical curve.

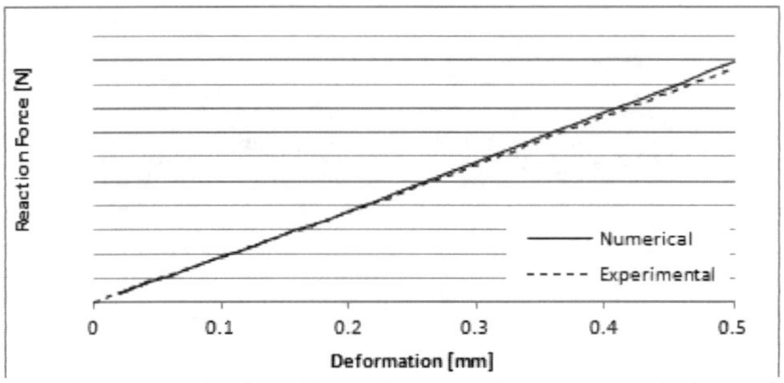

Figure 11: Numerical vs. Experimental Grommet static behavior.

4. MODAL ANALYSIS: COMPRESSOR ASSEMBLED ON GROMMETS

4.1 Simplified Model

In order to simplify the model, focusing on the grommet model (because the emphasis of this work is in frequencies up to 400Hz), the compressor geometry was simplified, representing it with steel block with total mass and *W* and *LB* dimensions similar to the ones of a typical compressor. Dimension *H* is around 20% of a typical reciprocating compressor.

Figure 12 shows the simplified model where two base plates are welded, with grommets assembled to them.

Figure 12: Steel block used to simplify the compressor geometry.

4.2 Numerical Model

For the modal analysis to determine the natural frequencies and modes of the system presented in Figure 12, a commercial software can be used. For the proposed range of frequencies, it is expected to find out six modes: in

three of them the dominant movement is a translation (in three orthogonal directions) and in three of them the dominant movement is a rotation (the three rotation axis should be orthogonal).
The first parameters that must be set are the material properties: for the block and base plates, standard steel properties are used; for the grommets, the properties defined for a static condition (commented on session 3.1) and dynamic condition (from nomograms like the one showed in Figure 7) are used.
The second step is to create the system mesh: the steel block can be modeled as a rigid body, while the base plates are modeled with quadratic tetrahedral elements and the grommets with quadratic hexahedral elements.
After finishing the mesh, the boundary conditions and joints are defined: the nodes of the lower grommet surfaces are fixed and fixed-fixed joints are created between the grommets and the base plates as well as between the base plates and the steel block.
The first analysis is a static analysis (static properties for the grommet), used to calculate the grommet deformation due to the gravity acceleration which is considered 9.8m/s^2.
Considering the deformed mesh obtained from the static analysis, a modal analysis with *Block Lanczos* algorithm is launch. The following steps were followed:

a) The first 6 natural frequencies are calculated for Young modulus E_{Mat} at 10Hz and 25°C in order to have a first approach for all of the rigid body frequencies.
b) For each of the six natural frequencies, Young Modulus E_{Mat} is recalculated, using the equation that defines it at 25°C. This equation is based on the rubber Reduced Frequency Nomogram and is defined as it follows:

$$E_{Mat}(f) = C1\ln(f) + C2 \quad (9),$$

where $C1$ and $C2$ are constants obtained from the nomogram and f is each one of the natural frequencies obtained from the software. The modal analysis must be redone for all the six natural frequencies, varying the dynamic Young Modulus, up to the value of E_{Mat} at equation (9) is equal to the value of dynamic Young Modulus set.

The results for the dimensionless natural frequencies are in Table 1 and two mode shapes are presented in Figure 13. The dimensionless natural frequency f^*_n of mode n is defined by equation (10):

$$f^*_n = {f_n}/{f_1} \quad (10),$$

where f_n is the value of the natural frequency of mode n and f_1 is the natural frequency of the first mode.

Table 1: dimensionless natural frequencies of Figure 14 system.

Number of mode	#1	#2	#3	#4	#5	#6
f^*_n	1.00	1.06	1.78	2.34	2.46	4.78

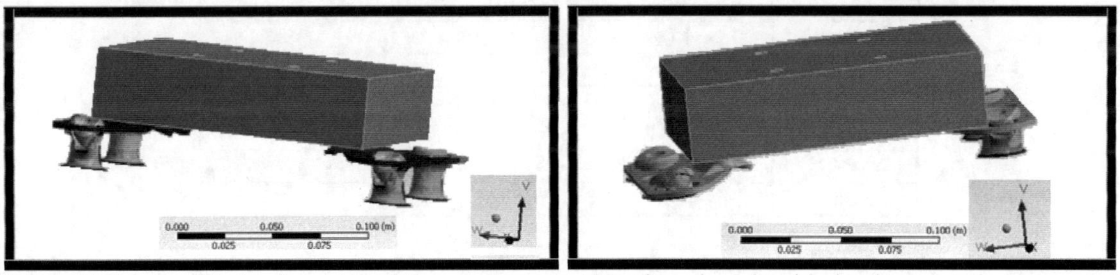

Figure 13: Mode with dominant translation movement at V axis and mode with dominant rotation movement around X direction.

4.3 Model Validation

The steel block presented in Figure 12 was used to do the validation as well as an impact hammer (*B&K* type *8202*) with rubber tip and a uniaxial accelerometer (*B&K* type 4397). Glue connected the grommets to the base plate and the grommets to the rigid base where the experiments toke place.
For the hammer force transducer, a transient window was set (the FFT must contain only the force peak generated by the hummer excitation), while for the accelerometer an exponential window was defined (the sinusoidal response

decreases with the time because of the rubber internal damping). With these settings using *B&K Pulse 12.0*, it was possible to obtain all the six rigid body natural frequencies, considering the range of interest until 400Hz.
The accelerometer and the impact position with the hammer are important to excite the rigid body modes. Figure 14 shows how to excite and measure the response of two different modes and a typical response curve.

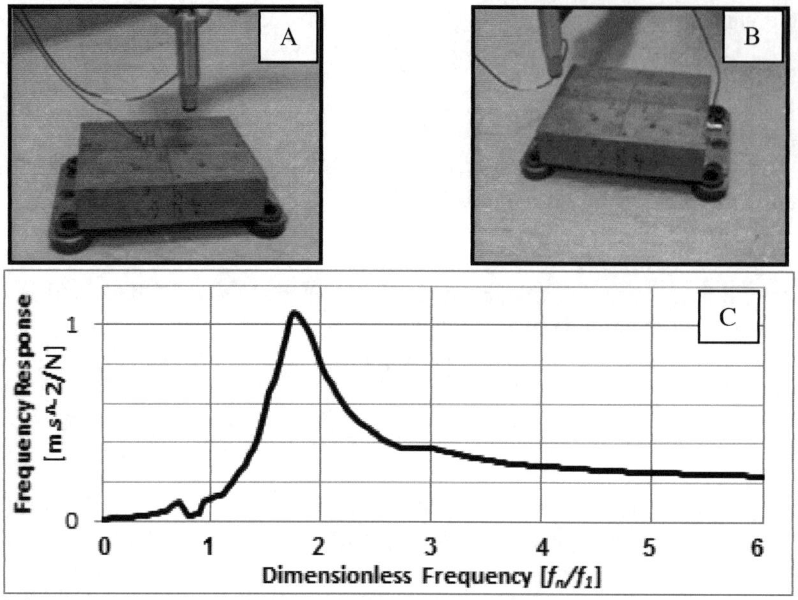

Figure 14: Excitation and measurement of a mode with dominant translation movement at V axis (A) and rotation movement around X direction (B) and typical response curve (C).

Table 2 shows a comparison between numerical and experimental results in terms of natural frequencies. The *Difference*, in %, is calculated according to equation (11), where *Num* represents the natural frequency calculated by *ANSYS* and *Exp* is the one obtained from the experiments.

$$Difference[\%] = \left(\frac{Exp-Num}{Num}\right).100\% \qquad (11).$$

Table 2: Comparison between numerical and experimental dimensionless natural frequencies.

Mode	**Numerical**	**Experimental**	**Difference**
	f^*_n	f^*_n	%
#1	1.00	1.04	3.8
#2	1.06	1.15	8.9
#3	1.78	1.84	3.6
#4	2.34	2.23	-4.5
#5	2.46	2.54	3.2
#6	4.78	4.53	-5.2

It is observed from Table 2 that numerical and experimental results are very similar, therefore validating the numerical model.
Based on this model, it is possible to analyze, design and optimize grommets.

5. CONCLUSIONS

- This paper presented the importance of vibration isolators, also known as grommets, and how a good grommet choice or project may optimize appliance noise in the range from 0 to 400Hz third octave bands.

- The main properties of the rubber, the material which usually is used to made grommets, were presented as well as Oberst Beam and Single Degree of Freedom Techniques. It was shown Single Degree of Freedom Technique is a good approach to determine rubber dynamic properties. With this technique it is possible to create a Reduced Frequency Nomogram, a graph that shows the Young Modulus and the Loss Factor as a function of the temperature and the frequency

- It was presented a grommet static analysis and its validation; the results showed a difference of around 3% between numerical and experimental results.

- Based on the static analysis and using a modal analysis with grommet pre-deformation, a simplified model of a compressor assembled on grommets was analyzed, showing its six rigid body natural frequencies and some of its modes. This model was validated experimentally, showing differences lower than 10% between numcrical and experimental results.

- Obtaining the rubber properties from the Single Degree of Freedom technique and using the static and the dynamic models presented in this paper, it is possible to analyze, design and optimize vibration isolator for refrigeration compressors.

REFERENCES

Almeida, C. M. (2009). *Caracterização de defeitos mecânicos produzidos por nanoindentação no InP*. Rio de Janeiro, Brazil: Rio de Janeiro Catholic University.

American Society for Testing and Materials International (2005). *ASTM E0756 – 05: Standard Test Method for Measuring Vibration-Damping Properties of Materials*. West Conshohocken, PA.

Gent, A. N. (1958). On the relation between indentation hardness and Young's modulus. *Rubber Chemistry and Technology, 31(4),* 896-906.

Istvan, L., Beranek, L. (2006). *Noise and Vibration Control Engineering - Principles and applications*: John Wiley & Sons.

Jones, D. I. G. (2001). *Handbook of Viscoelastic Vibration Damping*. Chichester: John Wiley & Sons.

Medeiros Júnior, W. B. (2010). *Caracterização Dinâmica Integrada de Elastômeros via Derivadas Fracionárias e Método GHM*. 2010. Curitiba, Brazil: Paraná Federal University.

Osswald, T. (2010). *Understanding Polymer Processing – Processes and Governing Equations*: Munich, Germany: Hanser Publishers.

Porto, L. L. R. (2010). *Simulação Numérica dos Esforços Dinâmicos de Baixas Frequências Transmitidos por Compressores Recíprocos*. Florianópolis, Brazil: Federal University of Santa Catarina.

Stumpf, C. (2013). *Determinação de Propriedades Mecânicas de Materiais Viscoelásticos através de Ajuste de Modelo*. Florianópolis, Brazil: Federal University of Santa Catarina.

Vendrami, C. E. (2013). *Análise de Isoladores de Vibração usados em Refrigeradores*. Florianópolis, Brazil: Federal University of Santa Catarina.

Williams, M. L., Landel, R. F., Ferry, J. D. (1955). The Temperature Dependence of Relaxation Mechanisms in Amorphous Polymers and Other Glass-forming Liquids. *J.Amer. Chem. Soc., 77 (14)*, 3701-3707.

Study on Vibrations of Inverter Refrigerator Compressor

Shoufei WU*, Jun LUO, Yuqian LI

Jiaxipera Compressor Co., Ltd., Technology Research Department,
Jiaxing, Zhejiang, People's Republic of China
shoufei.wu@qq.com; +86 573 8258 6106

* Corresponding Author

ABSTRACT

Vibration is a very important characteristic of inverter compressor, and it's critical for refrigerator running speeds selection. First, the reasons why the compressor vibrates are discussed, unbalanced forces of comprssor moving parts are the source of the vibration. The flexible parts in the vibration transmission route are also very critical, they are inner discharge tube, suspension springs, rubber grommets, and so on. When the natural frequencies of the system are similar to the source excitation frequencies, resonance will happen, vibrations will be very large. Second, one inverter compressor vibrations were tested, and the results show that the vibrations varied a lot in different speeds and directions. Then the reasons why the inverter compressor vibrations are very large in some speeds are analyzed in the third part. The natural frequencies of shell and grommets system and inner discharge tube were got by experiment or simulation. The results show that those natural frequencies are too close to the soure excitation frequencies in some speeds, that is the reason why the compressor vibrates so large. By the grommets and inner discharge tube redesigns, the natural frequecies are moved far away from excitation frequencies, the compressor vibrations are reduced. The methods suggested in this research are helpful in reducing inverter compressor vibrations.

1. INTRODUCTION

Vibration is one of the most important characteristics of compressor, it also can make big influence on refrigerator noise and reliability. Compressor is connected to the cooling system by pipes, when compressor vibrates too large, it will cause large stress and strain of the connection pipes, and may cause the pipes to break. When compressor vibrations are transferred to other parts of cooling system, they will produce noise and even resonances. Recently, great attention is being paid to compressor vibrations. Dugast (2008) studied a method of compressor noise level calculation from vibrations measurement results. Bratti *et al.* (2012) identified the vibrations energy path in the interaction between compressor and refrigerator. Gu *et al.* (2012) researched the compressor vibrations identification based on angle domain analysis method. Kuan and Kim (2014) did a theretical analysis of revolving compressor vibrations.

Compared to fixed speed compressor, inverter compressor has advantages in many aspects, such as precise temperature control, food preservation, energy conservation, etc. More and more inverter compressors are to be used in refrigerators. For vibrations control, inverter compressor is much more difficult than the fixed speed compressor. Inverter compressor has many speeds, we have to research the vibrations at each speed that may be used in the refrigerator.

In the second part, factors that affect inverter compressor vibrations are discussed. In the thrid part, inverter compressor vibrations in the whole speed range were tested. In the fourth part, the cause of large vibrations in inverter compressors is explored. The methods suggested are helpful in reducing inverter compressor vibrations.

2. FACTORS AFFECTING INVERTER COMPRESSOR VIBRATIONS

Most refrigerator compressors are reciprocating compressors that use piston driven by a crankshaft to deliver gases at high pressure. The reciprocating movement of piston and rotation of crankshat produce unbalanced forces, unbalanced forces cause compressor to vibrate, and then vibrations transmit from inner of compressor to the outside. Unbalanced forces are the source of compressor vibrations.

The formula for calculating the reciprocating unbalanced force I_s is,

$$I_s = m_s r\omega^2 \cos\omega t + \lambda m_s r\omega^2 \cos 2\omega t \tag{1}$$

m_s is reciprocating unbalanced mass, r is radius of rotation, ω is crankshaft rotation speed, λ is the ration of r and rod length.

Formula for calculating rotational unbalanced force I_r is:

$$I_r = m_r r\omega^2 \tag{2}$$

m_r is rotational unbalanced mass. Formula 1 shows that, reciprocating unbalanced force can be separated into two parts, one is the same frequency as the shaft rotation, the other is two times frequency of the shaft rotation. The unbalanced force is proportional to the unbalanced mass and square of rotating speed.

From vibrations tranmission view, vibrations transmit from crank to shell mainly by two paths. One is inner discharge tube, the other is suspension spring. So they are critical parts for compressor vibrations isolation.

When the compressor is installed in refrigerator, shell vibrations are transferred to the refrigerator mainly through three paths, one is suction connection pipe, the second is discharge connection pipe, and the third is rubber grommets. The vibrations transmission of the compressor can be shown as Figure 1.

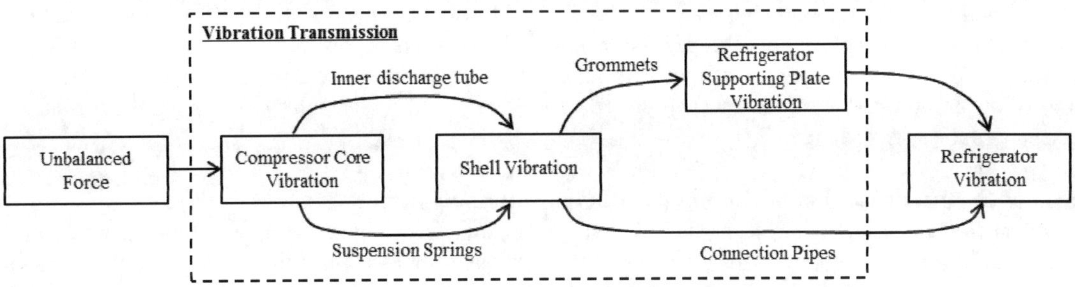

Figure 1: Compressor vibrations analysis

The unbanlaced forces are the fundamental reason for the vibrations. The inner discharge tube, suspension springs, connection pipes, grommets are the main paths for vibrations transmission.

3. INVERTER COMPRESSOR VIBRATIONS TEST

The invert compressor vibrations were tested in speeds of 1200 rpm to 4500 rpm. The compressor was put on the rubber grommets, three accelerometers were attached to the shell in three directions. The running speed was adapted by the inverter board. The vibration data and directions are shown in Figure 2. X direction is the direction of piston moving, Y direction is normal to X in the horizontal plane, Z direction is the vertical direction.

Figure 2 shows that there are four peak values in the whole speed range in three directions. There is one peak at about 38 Hz in X direction. There are two peaks in Y direction, one is at 22 Hz, the other is at 66 Hz. There is one peak at about 72 Hz in Z direction.

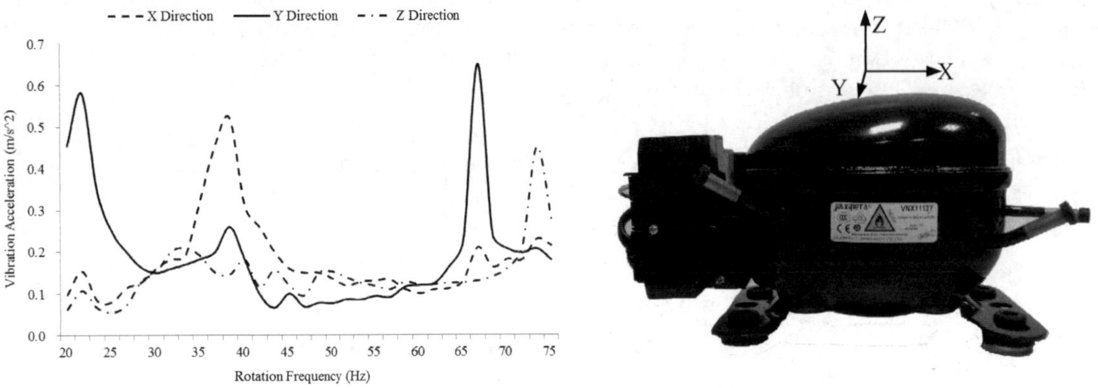

Figure 2: Inverter compressor vibrations in three directions

4. INVERTER COMPRESSOR VIBRATIONS ANALYSIS AND OPTIMIZATION

To reduce the vibrations of the compressor, to find the reasons why there are large vibrations is the most important. Usually, the compressor was put on four rubber grommets, the shell and grommets composed a spring-mass system. In each direction there is a natural frequency, that is to say, it may cause one peak value in each direction. But in Y direction, there are two peaks. We think that the second peak in Y direction may come from the inner discharge tube. Because, the stiffness of inner discharge tube is not very high, the natural frequencies may be close to the excitation frequency, and the vibrations that inner discharge tube transfers to shell are mainly in y direction.

To verify the conclusion, the natural frequencies of the shell and grommets system and inner discharge tube were tested. When doing the test, a impact hammer was used to knock one end of the tube, and an triaxial accelerometer was used to pick the response at the other end. Then we got the transmission functions from the test system, and resonance frequecies were identified, mostly they are the natural frequencies of the tube. The test system we used is PULSE from Bruel & Kjær, the impact hammer is 8206-002, the accelerometer is 4525-B-001. The test results showed that there was a resonance frequency at 132 Hz, it is just two times of the shaft rotation frequency at the speed of 66 Hz, so we conclude that the peak value in Y direction at 66 Hz comes from the resonance of inner discharge tube. To avoid the resonance of inner discharge tube, the shape of the discharge tube was redesigned. The vibrations of the compressor with new inner discharge tube were tested. The results showed that the resonance of the inner discharge tube was disapperaed.

4.1 Natural Frequencies Test of the Shell and Grommets System

To test natural frequencies, Bruel & Kjær PULSE test system was used. Impact hammer 8206-002 was used to give a excitation, and accelerometer 4514-001 was used to pick the response. By doing classical modal analysis, the modal parameters can ben identified. When doing this test, compressor shell was put on the grommets, MTC simple hammer method was used to do the modal analysis. Hammer was used to knock shell in x direction, the accelerometer was put in x direction. The transfer function can be got from the test. The same tests had been done in y direction and z direction. The tested transfer functions in three directions were shown in Figure 3.

Figure 3 shows that, in x direction, there is a natural frequency at about 22 Hz, the x direction vibrations also has a peak value at about 22Hz; in y direction, there is a natural frequency at about 38 Hz, the y direction vibrations also has a peak value at about 38Hz; in z direction, there is a natural frequency at about 73 Hz, the z direction vibrations also has a peak value at about 72 Hz. The three natural frequencies are consistent with the peak value frequencies in the vibrations test. It is concluded that the natural frequency is the reason for the peak value in the vibrations test. Except the three peak values, there is another peak value at about 66 Hz, the reason might be the natural frequency of inner discharge tube.

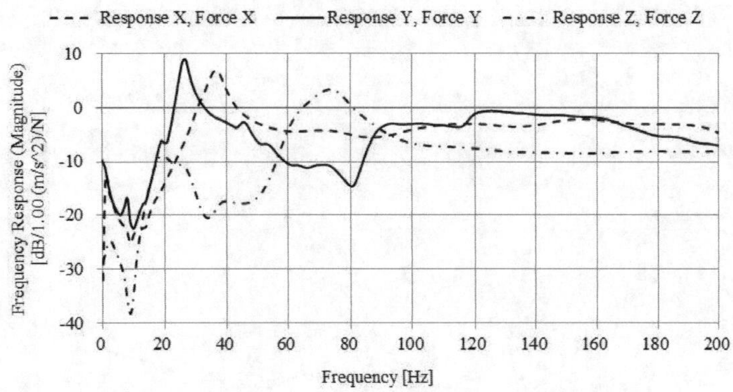

Figure 3: Transfer functions of shell and grommets system

4.2 Inner Discharge Tube Natural Frequency Simulation and Optimization

The modal analysis simulation of inner discharge tube was done by using finite element method, shown in figure 4.

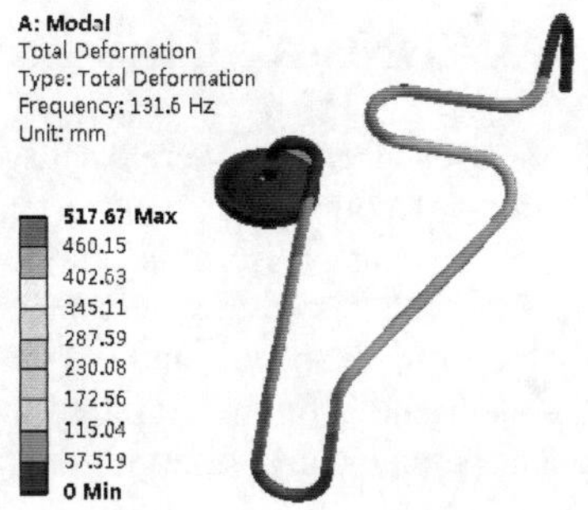

Figure 4: Modal analysis of inner discharge tube

Its first six order natural frequencies are shown in Table 1. The first order natural frequency is 132 Hz. In the vibrations test, there was a peak at about 66 Hz, 132 Hz is just two times of 66 Hz. According to section 2, when the compressor is running at 66 Hz, the unbalanced force also has a component at 132Hz, so there is a excitation force at 132Hz. When the excitiation frequency is similar to the natural frequency, resonance happens, and then the compressor vibrations are large.

Table 1: Six order natural frequencies (Hz)

First Order	Second Order	Third Order	Fourth Order	Fifth Order	Sixth Order
132	167	211	350	414	626

To avoid the inner discharge tube resonance, we redesigned the tube and the 1 order natural frequency was improved to 160Hz.

4.3 Vibrations Test and Verifications

To verify the influence of the inner discharge tube, the compressor vibrations with the redesigned inner discharge tube was tested. The vibrations of the compressor are shown in Figure 5.

Compared with the previous vibrations test results in Figure 2, the peak value at 66 Hz is disappeared, the above conclusion is verified. The peak values of the other three points are also reduced after using new grommets. That is to say, vibration transmission parts can make great influence on compressor vibration, and to aviod resonance is a basic and effective way for solving vibration problems. Especially for inverter compressor, it has many speeds, we should be more careful for the resonance problems.

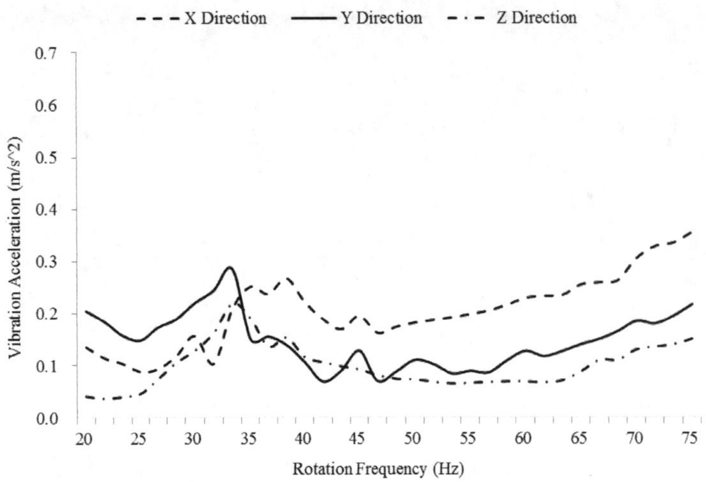

Figure 5: Compressor vibrations after modification

5. CONCLUSIONS

The factors which influence the inverter compressor vibrations were discussed, and the vibrations transmission paths were analyzed. It shows that unbalanced forces of comprssor moving parts are the source of the vibration. The flexible parts such as inner discharge tube, suspension springs, rubber grommets are critical for vibrations transmission. The inverter compressor vibrations were tested in whole speed range from 1200 rpm to 4500 rpm, the results show that compressor vibrates very large in some speeds, and the reasons were analyzed. The most important reason is that the natural frequencies of the shell and grommets system and inner discharge tube are too close to excitation frequencies, and resonances happen. After inner discharge tube and grommets redesigns, the compressor vibrations are reduced.

REFERENCES

Bratti GioVanni, Santini Junior Otavio, Baars Edmar, Barbieri Renato, Pellegrini Claudio, Silva Olavo, Lenzi Arcanjo, 2012, Identification of Energy Path in the Interaction between Compressor and Refrigerator, *International Compressor Engineering Conference*, paper 1302.

Dugast, 2008, New Portable Hybrid method of Compressor Noise Level Calculation from Vibrations Measurements, *International Compressor Engineering Conference*, paper 1261.

Huanhuan Gu, Rongting Zhang, Jia Xu, 2012, Identifying Noise and Vibrations of the Discharge Stage in the Rotary Compressor Based on Angle Domain Analysis Method, *International Compressor Engineering Conference*, paper 1199.

Kuan Thai Aw, Kim Tiow Ooi, 2014, Theoretical Analysis of Revolving Vane Compressor Vibrations, *International Compressor Engineering Conference*, paper 1216.

A Calibration Procedure for Compressor Simulation Models using Evolutionary Algorithm

Ernane SILVA, Willian M. FERREIRA, Cesar J. DESCHAMPS*

POLO Research Laboratories for Emerging Technologies in Cooling and Thermophysics
Federal University of Santa Catarina
88040-900, Florianopolis, SC, Brazil
deschamps@polo.ufsc.br

* Corresponding Author

ABSTRACT

Comprehensive models are widely adopted to predict the performance of compressors due to their low computational cost and acceptable prediction capability. In general, the accuracy of such models depends strongly on the correct adjustment of some parameters that are of difficult to determine both analytically and experimentally. However, due to the nonlinearities of the compressor model, the tuning of such parameters affects many output variables and hence can be very challenging and time consuming. In this paper, we consider this procedure of adjustment of parameters as a multiple objective optimization problem that can be solved by using an elitist non-dominated sorting genetic algorithm (NSGA-II). The parameters of two simulation models are adjusted following this new procedure. In the first model the suction muffler was neglected and a mass-spring-damper system was adopted to predict the suction valve dynamics. The second model solves the suction valve dynamics by the finite element method and the flow in the suction muffler with the finite volume method. In both models the clearance volume and parameters associated with the suction valve were chosen to be adjusted while the deviations between predictions and measurements for the mass flow rate, indicated power, and suction valve dynamics were defined as the objective functions to be minimized.

1. INTRODUCTION

Numerical simulation has been widely adopted in the development of refrigeration compressors, providing detailed results that complement experimental data. Simulation models are less time consuming and less expensive than measurements and, with the current development of computers and numerical methods, can be applied to predict complex physical phenomena in compressors. For instance, valve dynamics and fluid flow through valves can be solved in a coupled manner to analyze the performance of compressor valves (Kim *et al.*, 2006).

From the thermodynamic point of view, low-cost simulation models (Link and Deschamps, 2011) are based on the unsteady formulation of the conservation equations for mass and energy in the compression chamber with sub-models for different physical phenomena. The accuracy of such models depends strongly on the adjustment of some parameters that are difficult to determine both analytically and experimentally. A clear example of these parameters is the clearance volume in the compressor chamber, which varies significantly around its nominal value according to the assembling process and the manufacturing tolerances of the compressor components. However, due to the nonlinearities of the simulation model, the tuning of these parameters is very challenging and time consuming since it affects many output variables, such as mass flow rate, indicated power, and valve dynamics.

In this paper, we propose a procedure to adjust parameters of compressor simulation models following a multiple objective optimization problem that is solved via an elitist non-dominated sorting genetic algorithm, NSGA-II (Deb *et al.*, 2002). To illustrate this, the procedure is applied to two versions of a simulation model. In the first model the suction muffler is neglected and a mass-spring-damper system is adopted to model the suction valve dynamics. The second model solves the suction valve dynamics via the finite element method and the flow in the suction muffler with the finite volume method. In both models the clearance volume and parameters associated with the suction valve were chosen to be tuned while the deviations between predictions and measurements for the mass flow rate, indicated power, and suction valve dynamics were defined as the objective functions to be minimized.

2. SIMULATION MODEL

The unsteady-state equations for the conservation of mass and energy are applied in conjunction with an equation of state to predict the thermodynamic state of the gas inside the compression chamber. The mass flow rates of gas through the suction and discharge valves are calculated with reference to the isentropic flow in a convergent nozzle and the introduction of the effective flow area coefficient, C_{ep}:

$$\dot{m} = C_{ep} A_o p_{up} \sqrt{\frac{2k}{RT_{up}(k-1)}} \sqrt{\Pi^{2/k} - \Pi^{(k+1)/k}} \tag{1}$$

where A_o is the area of the orifice, p_{up} is the upstream pressure, k is the ratio of specific heats, R is the specific gas constant, T_{up} is the absolute upstream temperature, and Π is the pressure ratio. The leakage of gas through the piston-cylinder gap is obtained by assuming a fully developed Couette-Poiseuille laminar flow.

Two models are available to solve the valve dynamics. The first one considers the valve as a mass-spring-damper system with a single degree of freedom moving perpendicularly to the valve seat. In the second model the valve is discretized in beam elements and solved via the finite element method (Silva *et al.*, 2012). The governing equation of both models can be written as:

$$[M]\{\ddot{x}\} + [C]\{\dot{x}\} + [K]\{x\} = \{F\} \tag{2}$$

where $[M]$, $[C]$, and $[K]$ stand for the mass, damping, and stiffness matrices associated to the valve, $\{x\}$, $\{\dot{x}\}$, and $\{\ddot{x}\}$ represent the vectors of displacement, velocity, and acceleration of the valve nodes, and $\{F\}$ is the vector of forces applied to the valve nodes. In the finite element method, the Rayleigh damping model (Cook *et al.*, 1989) is used:

$$[C] = \alpha[K] + \beta[M] \tag{3}$$

where α and β are the coefficients of the model. If the valve is characterized as a mass-spring-damper system, Equation (2) is reduced to a single scalar equation. The resulting force acting on the valve may result from oil stiction, F_S, preload imposed on the valve assembly, F_P, and gas pressure load, F_v. The magnitude of the latter is evaluated through the concept of effective force area coefficient, C_{ef}:

$$F_v = C_{ef} A_o \Delta p \tag{4}$$

where Δp is the pressure difference across the valve. In the present model, the magnitude of F_S and F_p were assumed constant and prescribed as input parameters of the simulation.

When applicable, the pulsating compressible flow through the suction and discharge mufflers is modeled via the conservation equations for mass, energy, and momentum following a one-dimensional formulation. Such equations are solved using the finite volume method in a coupled manner with the valve dynamics, and taking into account heat transfer and viscous effects. Details of this modeling approach can be obtained in Deschamps *et al.* (2002).

3. EXPERIMENTAL SETUP AND PROCEDURE

Measurements of different parameters were carried out in a compressor placed in a calorimeter, so as to allow tests for operating conditions found in refrigeration systems. As can be seen in the schematic representation of the calorimeter (Figure 1a), the experimental setup is composed of pipelines, control valves (CV1 and CV2), a mass flow meter (FM), heat exchangers (HX1 and HX2), a thermocouple (TC), and pressure transducers (PT1 and PT2). The calorimeter is designed in a way that refrigerant fluid can flow through the high and low pressure lines in the superheated state, as indicated in the pressure–enthalpy diagram of Figure 1b. In this experimental setup, refrigerant from the suction line (1) is admitted in the compressor C and compressed to the condensing pressure, p_C, of the system (2). Pressure transducers PT1 and PT2 monitor pressures in the suction and discharge lines, respectively. The temperature of the fluid is then reduced in the heat exchanger HX1, reaching condition (3), and is expanded adiabatically through control valve CV1 to an intermediary pressure (4). Following the measurement of the mass flow rate at the flow meter FM, the temperature of the fluid is reduced again in the heat exchanger HX2, reaching condition (5), and expanded adiabatically to the evaporating pressure (6), p_E, through control valve CV2. Finally, with the use of an electrical heater EH and a thermocouple TC, the temperature in the suction line is controlled to reach condition (1) again.

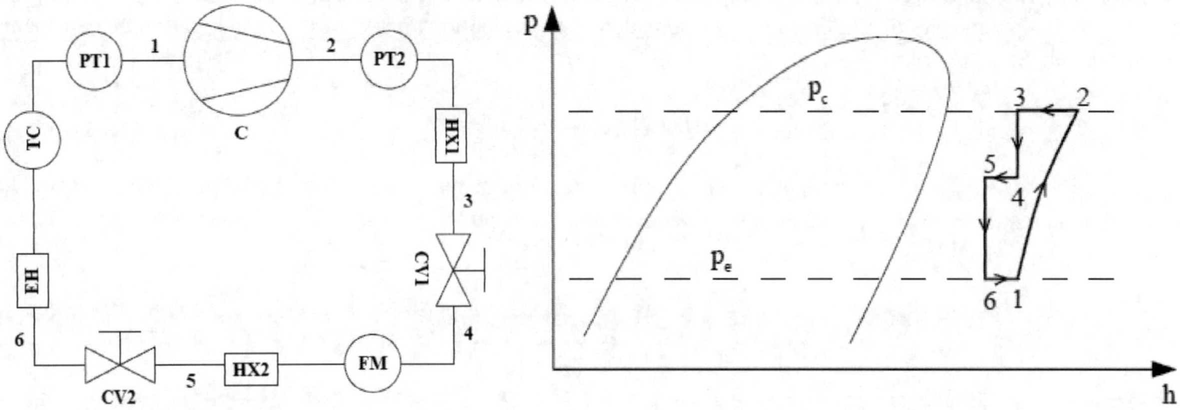

Figure 1: Hot gas test bench mechanical scheme (a) and pressure-enthalpy diagram of the thermodynamic cycle (b).

The adjustment of the amount of refrigerant in the system, the opening of the control valves, and the amount of thermal energy exchanged by the fluid in the heat exchangers and with the electrical heater, allow the control of the thermodynamic conditions in the suction (1) and discharge (2) lines and, consequently, the operating conditions to which the compressor is subjected. In addition to mass flow rate, the following parameters were also measured during compressor operation: crankshaft angle, pressures in the suction and compression chambers, temperatures in the suction chamber and cylinder wall, and the suction valve opening. The experimental procedure consisted in mounting the compressor in the system and submitting it to vacuum, allowing the desired amount of refrigerant to enter the system, running the compressor, and adjusting the control valves to reach the desired operating condition.

The measurements were conducted after the compressor had been running for at least 2 hours in order to achieve a fully cyclic operating condition. Then, measurements were carried out during a 30 minute interval. The compressor was evaluated under two operating conditions: A (T_e = - 28 ºC; T_c = 40 ºC) and B (T_e = 10 ºC; T_c = 70 ºC), where T_e and T_c stand for evaporating and condensing temperatures, respectively. The test of each condition was repeated 3 times and global parameters of the compression cycle, such as mass flow rate, indicated power, and local temperatures, were established in terms of a mean value. The parameters described as a function of the crankshaft angle, such as pressures in the suction and compression chamber and valve opening, were obtained from the cycle in which the global parameters were closer to the mean value.

4. CALIBRATION PROCEDURE

The accuracy of models developed to simulate refrigeration compressors depends strongly on the correct adjustment of some parameters that are difficult to determine both analytically and experimentally. Some of these parameters are adjusted in an indirect manner in order to reduce deviations between numerical and experimental results. In view of the complex physical interactions between different phenomena that take place inside the compressor and the consequent nonlinearity present in the models, this is a time-consuming task based on trial-and-error. Considering the importance and the difficulty of this activity, it is more convenient to consider the calibration procedure as an optimization problem. In this sense, the parameters to be adjusted become the design variables and the deviations between numerical and experimental results become the objective functions to be minimized. Therefore, the calibration procedure can be stated as:

$$\text{minimize: } f_i = f_i(x_1, x_2, \dots, x_j, \dots, x_{m-1}, x_m),\ i = 1 \dots n \tag{5}$$

$$\text{subjected to: } x_j^l \leq x_j \leq x_j^u \tag{6}$$

where f_i represents one of the n objective functions of the problem and x_j represents one of the m design variables, with x_j^l and x_j^u as its lower and upper bounds, respectively.

Two different types of variables can be used as objective functions: global parameters of the compression cycle, which are scalar variables, such as the mass flow rate, and parameters that vary with the crankshaft angle, such as the valve opening. When parameters of the first type are considered, the deviations between the numerical and experimental results are calculated as:

$$f_i = \left|\varphi_{num} - \varphi_{exp}\right| \tag{7}$$

where φ_{num} and φ_{exp} represent the numerical result predicted by the compressor model and the experimental measurement, respectively. On the other hand, the deviations for parameters described as a function of the crankshaft angle, θ, are calculated as

$$f_i = \frac{\int_{0^\circ}^{360^\circ} \left|\varphi_{num} - \varphi_{exp}\right| d\theta}{\int_{0^\circ}^{360^\circ} \left|\varphi_{exp}\right| d\theta} \tag{8}$$

The genetic algorithm NSGA-II (Elitist Non-Dominated Sorting Genetic Algorithm) developed by Deb *et al.* (2012) was used to solve this optimization problem, since it is commonly used in problems with multiple objectives and real-valued variables. In this algorithm the ranking of the feasible solutions is based on the fitness and density of solutions in the objective space and the maintenance of the optimum solutions as the analysis evolves is based on the evaluation of the combined population of solutions formed by solutions belonging to the parental population, that is, the group of optimum solutions from the last iteration, and the offspring population, which represents the population of optimum solutions of the current iteration. In the current analysis, the adopted stopping criterion was based on the number of generations of the algorithm. The algorithm was implemented in the programming language C++ and parallelized using the application programing interface OpenMP to reduce the processing time of the calibration procedure.

5. RESULTS

In the present analysis, we used two different simulation models. In the first one, the presence of the suction muffler was neglected and a mass-spring-damper system was adopted to predict the suction valve dynamics. On the other hand, the second model solves the suction valve dynamics by the finite element method and the flow in the suction muffler with the finite volume method. In both models the discharge valve dynamics was solved considering a mass-spring-damper system and the presence of a discharge muffler. The preload force acting on the suction valve, F_p, was considered negligible and the value of the stiction force, F_s, was adjusted together with the clearance volume, V_c, and parameters of the suction valve.

The parameters of the suction valve adjusted in the first model were its mass and stiffness, without considering any damping effect. In the second model, only the parameters α and β associated with the modeling of the valve damping were considered, since the mass and stiffness matrices were obtained directly from the valve geometry. In both cases the absolute error of the mass flow rate, $\Delta\dot{m}$, indicated power, $\Delta\dot{W}$, and suction valve opening, Δd, in relation to experimental data, were defined as the objective functions to be minimized.

Both models were calibrated in the condition A ($T_e = -28$ °C; $T_c = 40$ °C) and validated in the condition B ($T_e = 10$ °C; $T_c = 70$ °C), which is quite different from the one used in the calibration. The optimization algorithm adopted the number of generations equals to 20, population size 80, mutation probability 0.375, and recombination probability 0.9. These parameters were defined based on a sensitivity analysis for the specific case considered.

The results of the calibration procedure for the first model at condition A are presented in terms of the objective functions in Figure 2. In this figure, the solutions that provide the minimum deviation in relation to the measurements for each of the objective functions considered are highlighted. Since each generation forms a Pareto frontier in the NSGA-II algorithm, the results obtained from the last iteration form a group of non-dominated solutions, that is, all solutions are equally optimal. Therefore, an additional criterion has be adopted in order to select a single solution. In our study, we selected the solution that produced the smaller deviation for the suction valve dynamics, since its correct prediction affects the compressor efficiency and reliability. The deviations of the predictions in comparison to experimental measurements are presented in Table 1 for both operating conditions. It is possible to observe that the simulation model provided satisfactory results for indicated power and mass flow rate even in the operating condition far from the one used for calibration.

Figure 3 presents the comparison between numerical and experimental results for the suction valve displacement, s, as a function of the crankshaft angle, θ. It is worth mentioning that the adopted measurement technique did not allowed measurements of valve displacements greater than 2 mm. Based on the curves, it is possible to verify that the numerical model reproduced the valve displacement with good accuracy in the calibration condition A, with the opening and closing points matching the measurements. However, predictions of the valve opening and closing points do not match so closely the measurements for condition B, though the opening amplitudes seem to be well predicted. Such discrepancies may be associated with the consideration of a constant stiction force, independent of the operating condition selected. Therefore, the use a model for the stiction force based on physical reasoning could improve the predictions of valve dynamics in operating conditions far from that used in the model calibration.

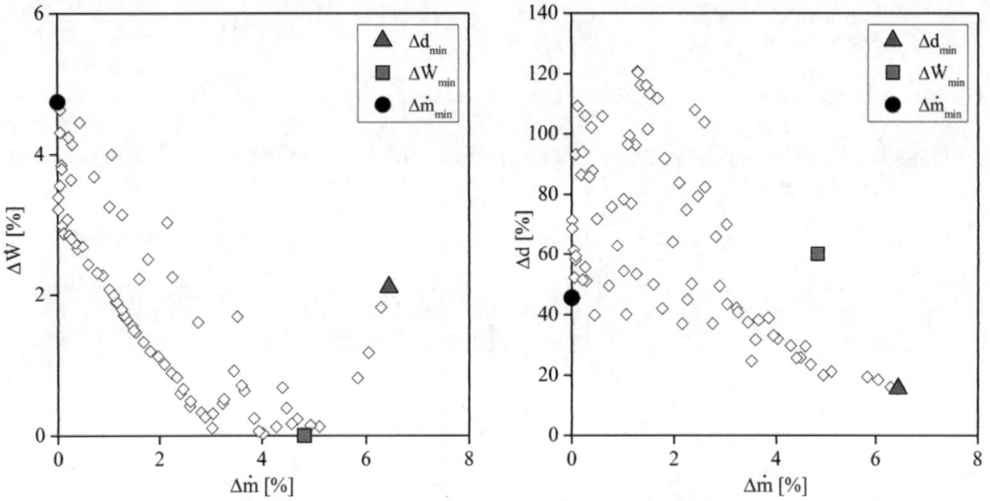

Figure 2: Optimal solutions in the solution domain.

Table 1: Deviations between predictions and measurements of mass flow rate and indicated power.

Parameter		Operating conditions	
		A (T_e = - 28 °C; T_c = 40 °C)	B (T_e = 10 °C; T_c = 70 °C)
1st model	$\Delta\dot{m}$ [%]	6.4	3.3
	$\Delta\dot{W}$ [%]	2.1	2.3
2nd model	$\Delta\dot{m}$ [%]	4.0	2.6
	$\Delta\dot{W}$ [%]	1.0	1.8

The results obtained in the calibration of the second model are presented in the solution domain in Figure 4, while the deviations for mass flow rate and indicated power are presented in Table 1. As can be seen, this model presents smaller deviations when compared to experimental measurements as one would be expected, since it provides more complete descriptions of the valve dynamics and fluid flow in the suction muffler. Again, larger deviations are observed in the predictions for the opening and closing of the valve (Figure 5) at the operating condition B (T_e = 10 °C; T_c = 70 °C). Lack of modeling for the stiction force and variation of the damping matrix $[C]$ with the operating conditions are possible sources of such deviations. In spite of that, the model predicts the valve displacement in reasonable agreement with the measurement.

In terms of computational time, 1600 simulations were run with both models to complete the calibration procedure. The calibration of the first model took about 4 hours and 15 minutes on a computer with 2.93 GHz Intel Core i7 processor, 6 GB DDR3 RAM and 64-bit Windows 7 operating system. The parallel processing allowed us to reduce the computational time by a factor of almost 5 in comparison to serial processing.

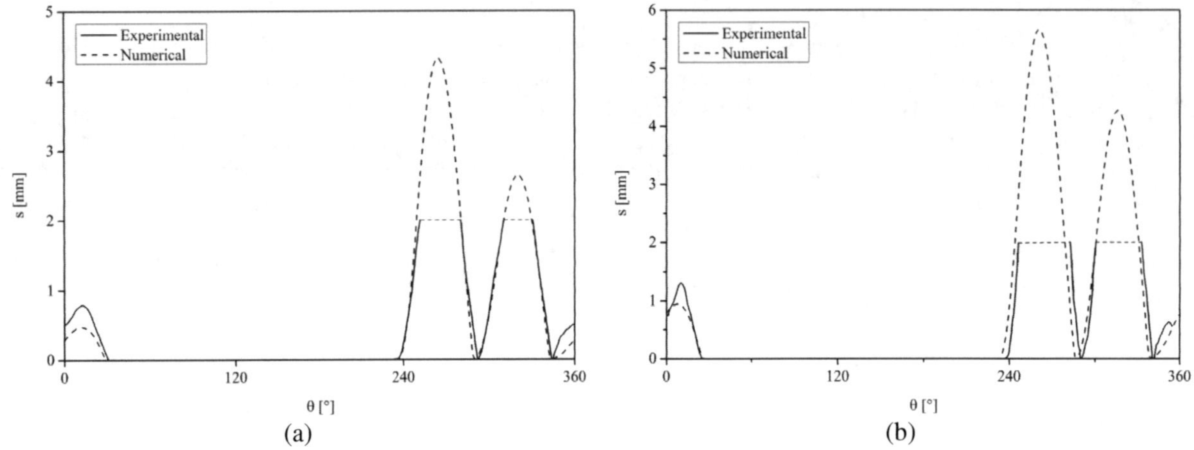

Figure 3: Comparison of numerical predictions and experimental measurements of valve displacement in the conditions: (a) (T_e = - 28 °C; T_c = 40 °C); and (b) (T_e = 10 °C; T_c = 70 °C).

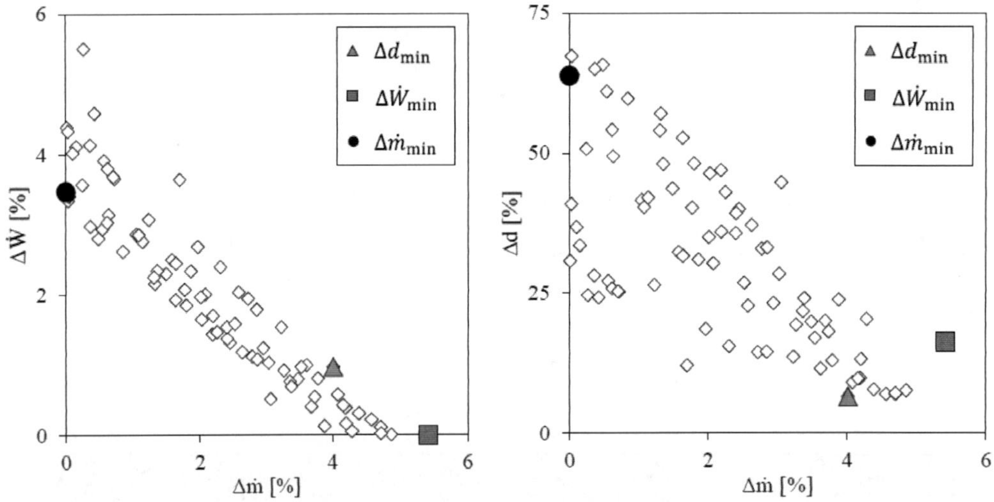

Figure 4: Optimal solutions in the solution domain.

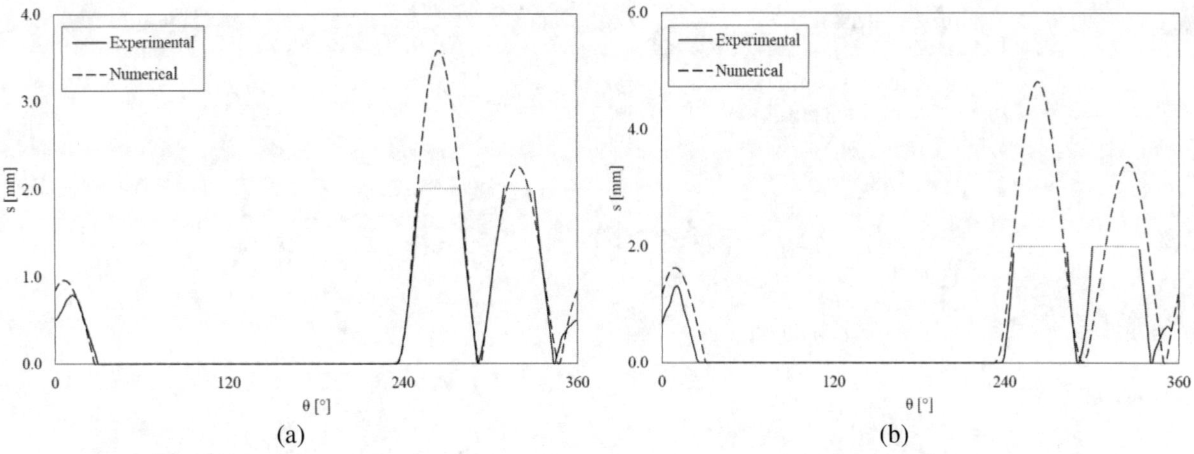

Figure 5: Comparison of numerical predictions and experimental measurements of valve displacement in the conditions: (a) (T_e = - 28 °C; T_c = 40 °C); and (b) (T_e = 10 °C; T_c = 70 °C).

6. CONCLUSIONS

A calibration procedure was developed to adjust input parameters of compressor simulation models that are difficult to obtain from measurements, such as clearance volume and stiction force. The procedure consists of a multi-objective optimization problem that is solved with an evolutionary algorithm. The parameters to be adjusted are considered the optimization variables while the deviations between predictions and measurements of output quantities, such as mass flow rate, indicated power, and valve displacement, were considered the objective functions. The procedure was applied to the calibration of two different simulation models. The two calibrated models were seen to predict mass flow rate, input power and valve displacement in close agreement with experimental data, even in operating conditions far from the one used for calibration. Nevertheless, some discrepancies were observed between predictions and measurements for the opening and closing crankshaft angles of the suction valve and attributed in part to lack of modeling of the stiction force. Overall, the calibration procedure showed to be effective in reducing the time required to calibrate simulation models, which give predictions in closer agreement with experimental data than models adjusted by the trial-and-error method.

NOMENCLATURE

A	area	(m^2)
C_{ef}	effective force area coefficient	(–)
C_{ep}	effective flow area coefficient	(–)
f	generic objective function	
F	force	(N)
k	ratio of specific heats	(–)
$\dot{m}$	mass flow rate	(kg/s)
p	pressure	(Pa)
R	specific gas constant	(J/kg K)
s	valve displacement	(mm)
T	temperature	(K or °C)
x	generic variable	
$\{F\}$	force vector	(N)
$\{x\}$	displacement vector	(m)
$\{\dot{x}\}$	velocity vector	(m/s)
$\{\ddot{x}\}$	acceleration vector	(m/s^2)
$[C]$	damping matrix	(N s/m)
$[K]$	stiffness matrix	(N/m)

$[M]$	mass matrix	(kg)
α	1st damping coefficient	(s)
β	2nd damping coefficient	(s^{-1})
Δd	deviation of valve dynamics	(%)
$\Delta \dot{m}$	deviation of mass flow rate	(%)
Δp	pressure difference	(Pa)
$\Delta \dot{W}$	deviation of indicated power	(%)
φ	generic value	
Π	pressure ratio	(–)
θ	cranckshaft angle	(°)

Subscript

c	condensing
e	evaporating
exp	experiment measurement
l	lower bound
num	numerical prediction
o	orifice
p	preload
s	stiction
u	upper bound
up	upstream
v	gas pressure load

REFERENCES

Cook, R. D., Malkus, D. S., Plesha, M. E. (1989). *Concepts and applications of finite element analysis*. New York: John Wiley & Sons.

Deb, K., Pratap, A., Agarwal, S., Meyarivan, T. (2012). A fast eletist multi-objective genetic algorithm: NSGA-II. *IEEE Transactions on Evolutionary Computation 6* (2), 182–197.

Deschamps, C. J., Possamai, F. C., Pereira, E. L. L. (2002). Numerical simulation of pulsating flow in suction mufflers. *Proceedings of the International Compressor Engineering Conference*. West Lafayette: Purdue University.

Kim, J., Wang, S.,Park, S.,Ryu, K., La, J. (2006). Valve dynamic analysis of a hermetic reciprocating compressor. *Proceedings of the International Compressor Engineering Conference*. West Lafayette: Purdue University.

Link, R., Deschamps, C.J. (2011). Numerical modeling of startup and shutdown transients in reciprocating compressors, *International Journal of Refrigeration* 34, 1398-1414.

Silva, E., Deschamps, C. J., Fancello, E. A. (2012). A procedure to optimize reed type valves considering efficiency and bending fatigue. *Proceedings of the International Compressor Engineering Conference*. West Lafayette: Purdue University.

ACKNOWLEDGEMENT

The present work was developed with the support of CNPq, National Council for Scientific and Technological Development – Brazil, and is part of a technical-scientific cooperation program between the Federal University of Santa Catarina and EMBRACO. The authors also thank the financial support of EMBRAPII Unit Polo/UFSC and CAPES (Coordination for the Improvement of High Level Personnel).

Structural Analysis of Reciprocating Compressor Manifold

Marcos Giovani Dropa de BORTOLI

EMBRACO, R&D
Joinville, Santa Catarina, Brazil
Phone: 0055 47 34412783
E-mail: marcos.g.bortoli@embraco.com

ABSTRACT

The reciprocating compressor manifold design is very important for the compressor to achieve the goals of energy efficiency, acoustic and reliability. The main components involved in the manifold design are the cylinder, piston, valve plate, valves, cylinder head, bolts and gaskets. The design of a reciprocating compressor manifold is a multifunctional activity, which involves knowledge from thermal, acoustic, structural and tribology areas. The objective of this paper is to present the main aspects related with the structural analysis of the manifold, with special attention to gasket and valve plate deformation. The analysis is performed by the finite element method, and it involves highly non-linear aspects because of the gasket behavior. The gasket properties are determined by a special experimental procedure. Experimental results are presented to verify the precision of the methodology used.

1. INTRODUCTION

Normally, the performance of a refrigeration compressor is evaluated by the coefficient of performance (COP). COP is defined by the rate between the refrigerating capacity and power consumption. Just one part of the energy provided for the compressor is transformed in work to compress the gas, because there are losses during the compressing process. The losses are related to thermodynamic process, mechanic aspects and electrical motor efficiency.

A very important point to be considered in the COP evaluation is the compressor volumetric efficiency. Volumetric efficiency is the ration between the actual volume flow rate in the compressor inlet and the displacement rate of the compressor (maximum theoretical flow rate). Ideally, volumetric efficiency should be equal to one. However, several aspects reduce the volumetric efficiency, such as: re-expansion of gas in clearance volume; pressure drops in valves and flow passages; leakage from compression chamber; internal superheat and back-flow through valves. Equation 1 presents the volumetric efficiency for a reciprocating compressor, where is possible to observe the influence of the clearance volume.

$$\eta_v = 1 - \frac{V_c}{V_{sw}} \left(\left(\frac{P_d}{P_s} \right)^{\frac{1}{\kappa}} - 1 \right) \tag{1}$$

The main components of the reciprocating compressor cylinder head system are showed in Figure 1. Clearance volume is defined by the empty space at the top cylinder region, when the piston reaches its top-most position. It is defined by the piston, crankcase, cylinder gasket, valve plate and suction valve. The objective of this work is to evaluate the main factors that affect the clearance volume. It is important to mention that a large clearance volume causes a reduction of refrigeration capacity and COP, whereas a small volume could cause the hit of the piston against the valve plate, reducing the compressor reliability and increasing the total noise.

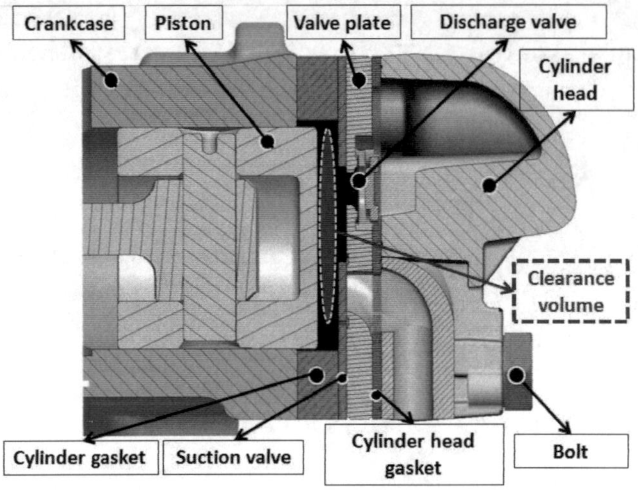

Figure 1: Schematic view of main components of the cylinder head system and the clearance volume definition.

2. CLEARANCE VOLUME

The main factors that affect clearance volume size are: the distance between the top piston face and the top cylinder face (when the piston is at top dead center); the cylinder gasket thickness; the cylinder head system components deformation, mainly the valve plate, after the bolts torque application; empty volumes due to the suction valve and discharge hole; and of course, the piston diameter. This work will evaluated the piston position related to the cylinder top, cylinder gasket and the cylinder head system components deformation after the bolts torque application (the assembly process).

2.1 Cylinder Gasket

For reciprocating compressor, the cylinder head system is where the suction and the compression of the refrigerant fluid happen. This area is subject to high pressure gradients. As the cylinder head system is formed by a series of components fixed by bolts, and forming a type of sandwich, the high pressure inside the cylinder head system tends to force the refrigerant fluid to leak among the components to a lower pressure area. Therefore, the main reason for using gaskets, alternating them with the other components, is to keep the insulation between the discharge and suction volumes. The gasket is also responsible for absorbing the imperfections at the contact surfaces among the components, such as flatness errors and roughness at the surface.

A second aspect for using gaskets is to increase a control over the clearance volume, because of the variations that happen in the maximum positioning of the piston top in relation to the cylinder top. This positioning is the function of a large chain of tolerances with the participation of all the compression mechanism components. This distance between the piston top and the cylinder top is known as piston boss. Figure 2 shows two cylinder head system cases: the first with a big piston boss and the second, with a small one. In the first case, there is a small clearance volume and in the second, a large volume. Clearance volume correction can be made by using a thinner gasket for the second case. Thus, a second function of the cylinder is, through the known piston boss, use gaskets of different thicknesses to absorb larger tolerances from the compression mechanism components. This process guarantees a clearance volume optimization, increasing the volumetric efficiency of the compressor, without the necessity to improve the dimensional quality of the mechanism components.

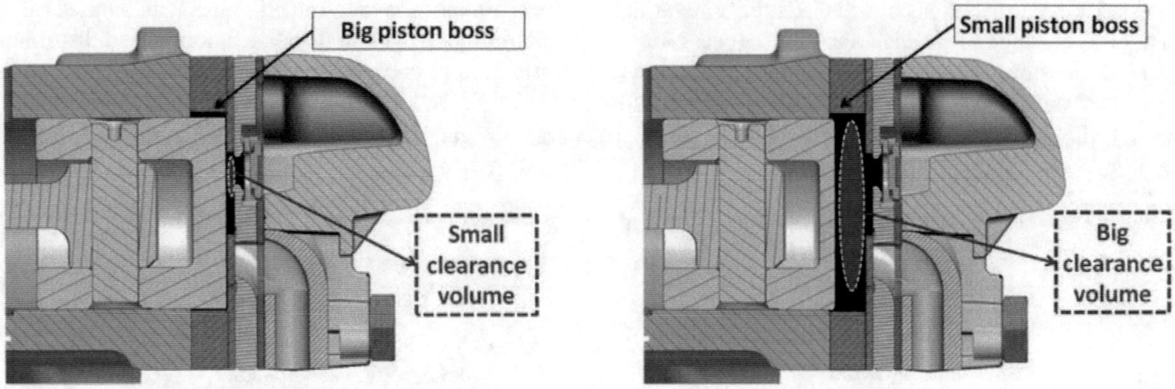

Figure 2: Influence of the piston boss size on the clearance volume.

2.2 Piston Boss

Piston boss is function of a chain of dimensions of the compression mechanism, which are related to manufacturing tolerances and specifications of hydrodynamic bearing clearances. Figure 3 shows the chain of dimensions involved in determining the piston boss, represented in the figure as R_1. There are two possibilities of value distribution for R_1, which are referenced as R_{1e} and R_{1f}. The first possibility, R_{1e}, known as 'empty', happens when there are no pressure charges on the piston top, and only the inertial forces act on the components. In this situation, all the clearances add to the salience. The second possibility, R_{1f}, and known as 'full', happens when there are pressure charges on the piston top, causing all the clearances to be subtracted from the salience.

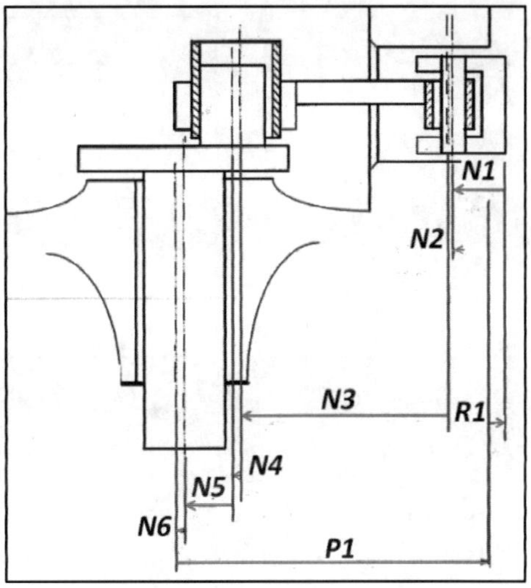

Figure 3: The chain of tolerances to evaluate the piston boss.

The best way to evaluate R_1 is through statistical analysis, where the manufacturing tolerances of the components and the clearances among the components are considered. The statistical method presents more realistic results than using other methods, as for example, the worst case methodology. R_{1e} and R_{1f} are calculated through Equation 2 and 3. This methodology is called *vector loop.*

$$R_{1e} = (N_1 + N_2 + N_3 + N_4 + N_5 + N_6) - P_1 \quad (2)$$

$$R_{1f} = (N_1 + N_3 + N_5) - (N_2 + N_4 + N_6 + P_1) \quad (3)$$

The statistical distributions of each independent variable of the chain must be determined or estimated, based on real data, or knowledge of the manufacturing process, or from similar projects. Figure 4 shows a normalized distribution for N_2 dimension. Through statistical methods, like Monte Carlo, it is possible to generate data in a pseudo random form, and statistically create the distribution, according to Figure 5. The distributions of R_{1e} and R_{1f} are obtained through the statistical analysis of the universe of data generated by the results of the mathematical operations, as Figure 6 and 7 show.

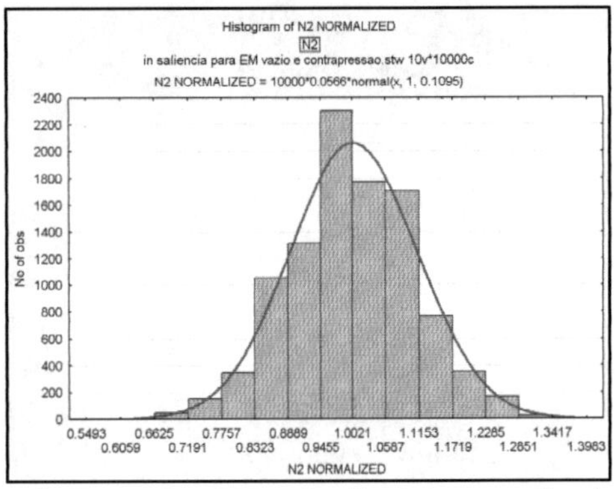

Figure 4: Example of N_2 - distribution normalized.

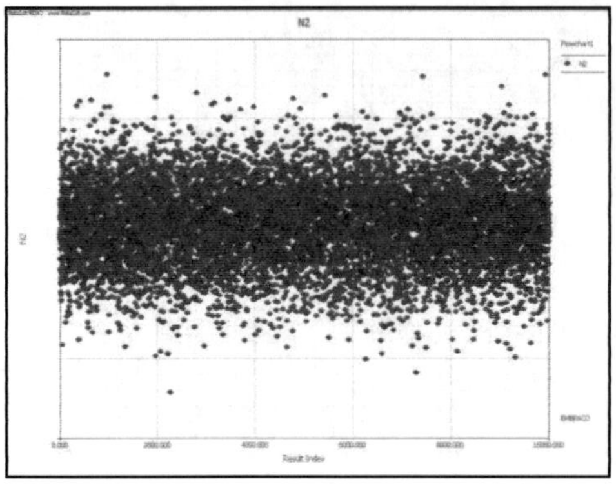

Figure 5: Example of N_2 - distribution created by Monte Carlo method.

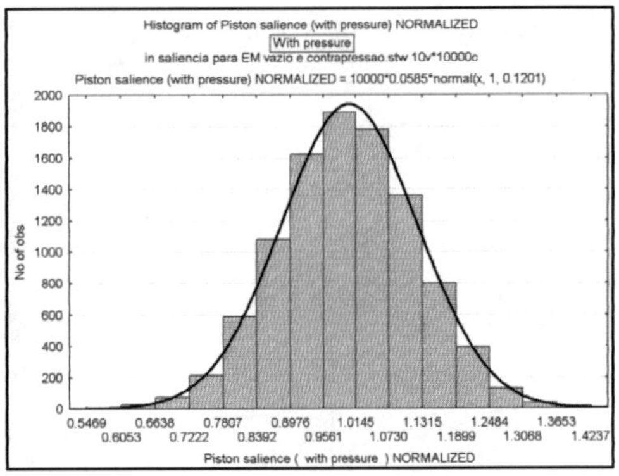

Figure 6: $\boldsymbol{R_{1f}}$ distribution with pressure (normalized).

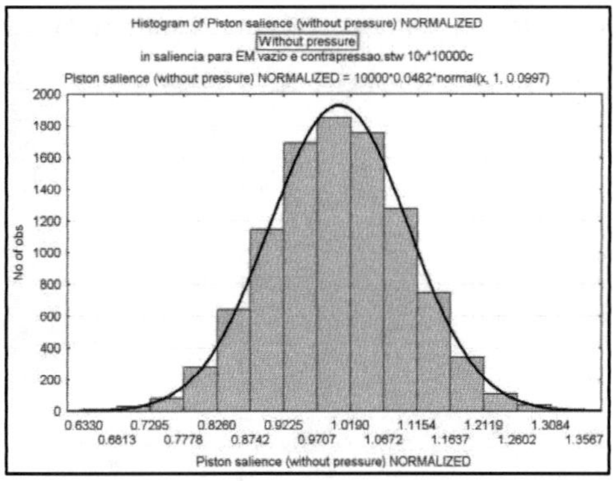

Figure 7: $\boldsymbol{R_{1e}}$ distribution without pressure (normalized).

2.3 Cylinder Head System Deformation

All the cylinder head system components are fixed by four bolts and torque application. An undesirable consequence of this procedure is the deformation of all the components, which influences the clearance volume. Figure 8 shows a hypothetical deformation of a compressor cylinder head system, where mainly it is possible to see the valve plate behavior. In the situation presented, the excessive deformation of the valve plate is harmful to the performance of the compressor, as the deformations happen by invading the area that was defined by the salience to be the clearance volume. If this situation is not corrected through a thicker gasket, the piston may hit the valve plate. A thicker gasket causes an increase in the compressor clearance volume, with all the negative consequences already presented at the beginning of this article.

An efficient way to predict cylinder head system deformations is by a numerical simulation through the finite element method, where the main factors involved in the deformation can be considered in the analysis. In the following topic, the main aspects to be considered for a proper numerical modeling of the cylinder head system will be presented.

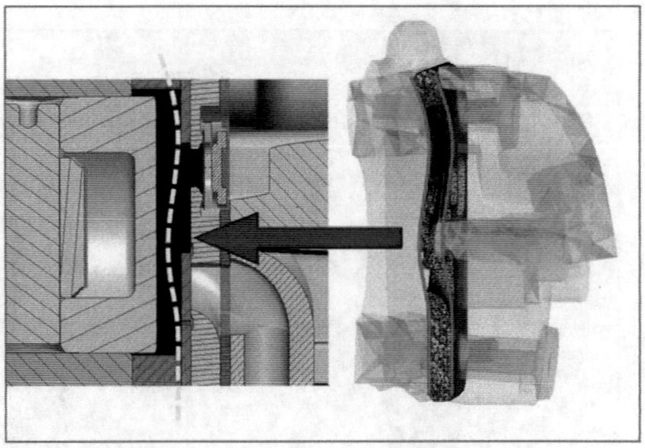

Figure 8: Manifold deformation after bolts assembly.

3. NUMERICAL SIMULATION FOR CYLINDER MANIFOLD SYSTEM

For the finite element model of cylinder-manifold system, the main aspects to be considered are: the geometry of the components, material property and boundary conditions, mainly forces applied over the structure. The main difficulty to create a precise numerical to predict the deformations is primarily related to the variety of materials that form the cylinder head system, where the gaskets and the interaction among the other components have a highly nonlinear behavior.

3.1 Geometry and Mesh

The geometrical modeling of the components, with the creation of the respective finite element meshes, is topics of the numerical simulation that had great advances in the last decades, and made the numerical simulation of very complex structures feasible. With the total integration between CAD and CAE systems, along with the availability of robust algorithm for mesh creation, the task of creating the finite element models presented a significant increase in productivity. Puff (2008) presented a finite element model which will be basically considered in this analysis.

3.2 Gasket Mechanical Properties

Gaskets are mechanical elements used to promote insulation among the several components that form the cylinder head system. In the case of alternating compressors, two gaskets are used: one between the valve plate and the cylinder top and another between the valve plate and the cylinder cover. The two main materials used in this type of application are hydraulic cardboard and metallic sheets coated with polymeric material. These materials show a nonlinear behavior when under a charging, and are also strongly influenced by thickness. The mechanical properties of the gaskets are rarely provided by their suppliers, which creates the necessity to develop specific tests to evaluate them. The gaskets compressibility property (deformation versus load) may be determined by a special compression experiment, as shown in Figure 9.

Figure 10 shows the results of load versus deformation for two gaskets made of the same material, but with different thicknesses. The gasket on the right is 30% thicker than the one on the left. Another phenomenon observed from the results, through the application and removal of the charge, is a material hysteretic behavior, and with different results for every operation cycle. In this case presented here, five cycles of load and relief were applied, where a convergence of results is observed for the last charge cycles. This is an important topic because, due to the running of the compressor, the cylinder head system suffers large variations of temperature, on account of the release of heat during the refrigerant fluid compression process. The thermal expansion of the components, and consequently a significant increase in load over the gaskets, causes different deformations for the cylinder head system. Understanding the curves load versus deformation for the gaskets is the vital importance for the cylinder head system numerical simulation model.

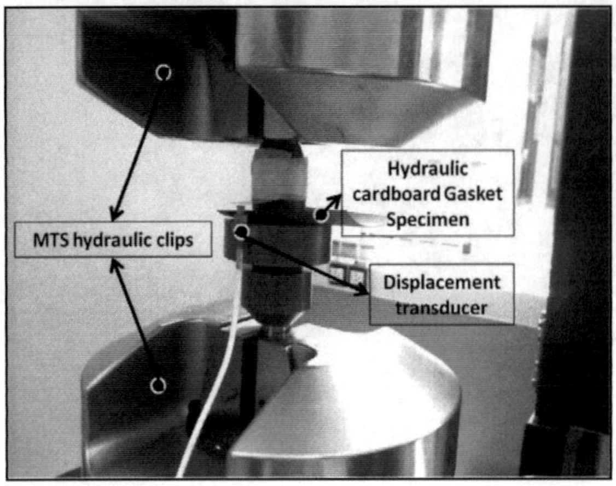

Figure 9: Compressibility gasket property test.

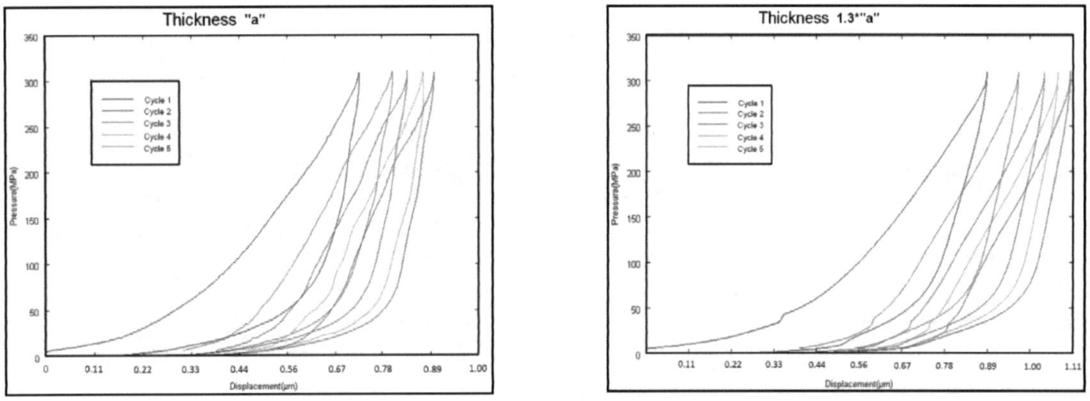

Figure 10: Gasket compressibility properties for two different thicknesses.

3.3 Bolts Force

All the cylinder head system components are fixed by four bolts, with the application of their respective torques. For the numerical model, it is necessary to know precisely the relation between the applied torque and the effectively resulting force over the components. According to Figure 11, these compression forces are the boundary conditions for the evaluation of the components deformations, either by experiments or by numerical simulation. Force is the function of the torque and the friction between the surfaces on the head and the thread of the bolt. According to Mischke and Shigley (1989), it is possible to evaluate the portion of torque dispersed in friction in a simplified way by Equation 4.

$$T = KF_b\, d \tag{4}$$

As there is a portion referring to the friction between the bolt and the other components, an experiment is necessary to verify or calibrate the relation between the torque and the applied force. Figure 12 shows an experiment where through the charge cells, it is possible to evaluate experimentally the behavior of the force as function of the applied torque. Figure 13 shows the results found for a given cylinder head system configuration.

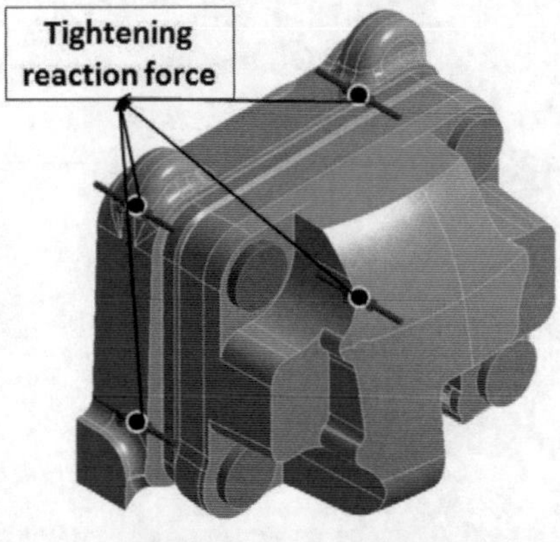

Figure 11: Bolt forces.

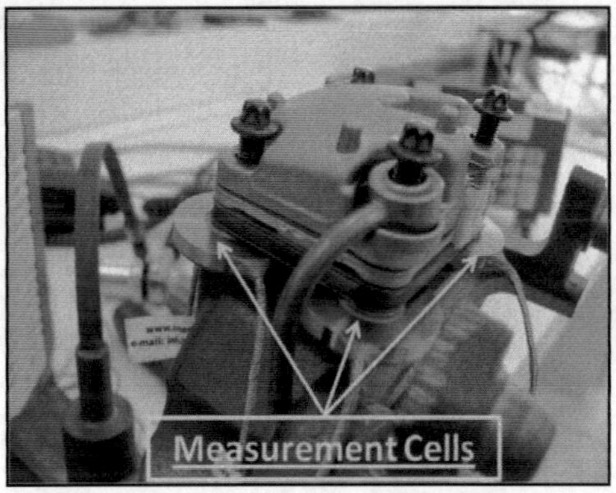

Figure 12: Experimental test for measurement the effective bolt force.

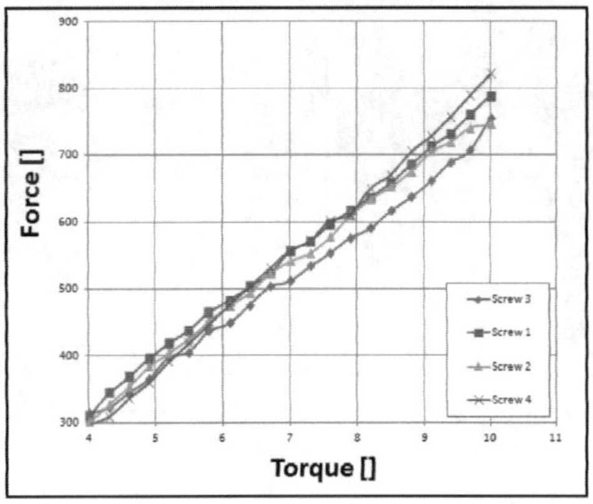

Figure 13: Relationship between force versus torque.

3.4 Numerical Analysis

Figure 14 shows the deformation gradient of the valve plate after applying the force charge resulting from the torque, as shown in Figure 13. What is relevant for this analysis is how much the valve plate invades the hot chamber, and for that, the best analysis is through the deformation profile towards the cylinder, as shown with an enlarged scale in Figure 15.

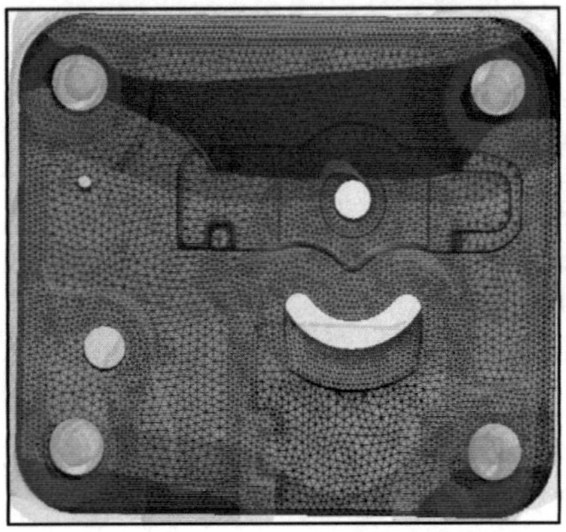

Figure 14: Valve plate deformation.

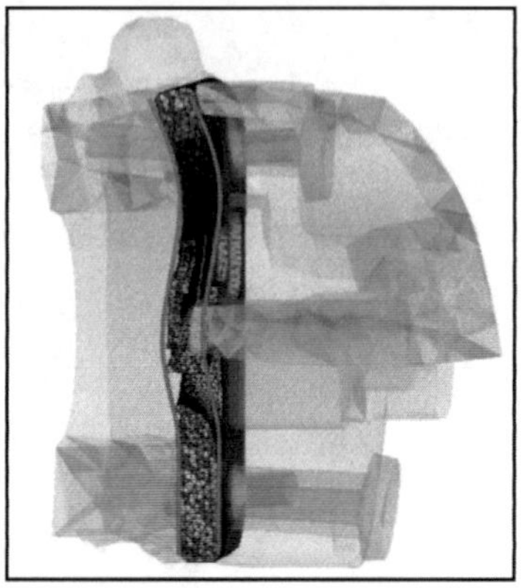

Figure 15: Clearance volume and valve plate deformation.

4. NUMERICAL VERSUS EXPERIMENTAL RESULTS

Accuracy of the obtained results with the numerical models was done through experimentation by laser measurements, where after the torque application; the valve plate deformation was measured by the inner side of the cylinder, as Figure 16 shows. In order to compare the results, the same region scanning by the laser was created in the numerical model (Figure 17).

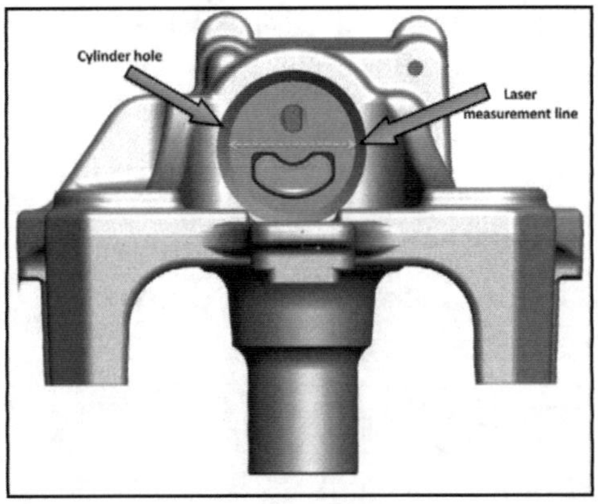

Figure 16: Measurement laser line.

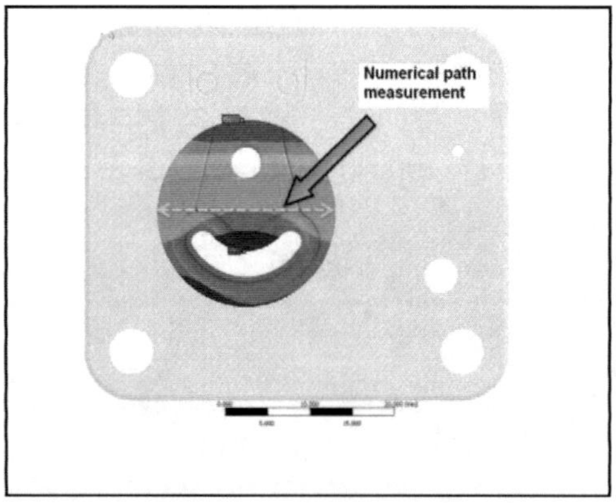

Figure 17: Numerical path displacement.

Figure 18 shows the comparison of the displacement results measured in the valve plate through the experimental and the numerical processes. The maximum difference between the numerical model and the experiment is around 5%.

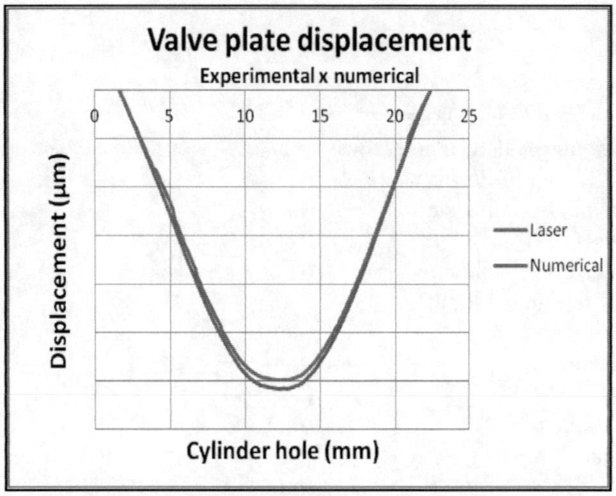

Figure 18: Comparison experimental and numerical results for valve plate deformation.

5. CONCLUSIONS

The reciprocating compressor manifold design is very important for the compressor to achieve the goals of energy efficiency, acoustic and reliability. The main aspects related with the structural analysis of the manifold are:

- The use of gasket is the fundamental importance for absorbing the imperfections at the contact surfaces among the components, such as flatness errors and roughness at the surface. The second aspect for using gaskets is to increase a control over the clearance volume, because of the variations that happen in the maximum positioning of the piston top in relation to the cylinder top.
- It is the fundamental importance to use statistical tools to define the gasket thickness according the variation of piston boss, because this dimension is resultant of several variables.
- The use of special experiment to evaluate the mechanical properties for the gaskets is crucial for the success of numerical model (non-linear properties).
- Experimental procedure for calibration the relation torque versus force for the bolts is important to find the bolt constant.
- The finite element method is a powerful tool to have a good design for the manifold, considering the deformation of the components after the bolts assembly. This tool reduces the development time, with an optimized design.
- The success of the numerical model to predict the manifold deformation, beyond of numerical capabilities, is the quality of the input data, as material properties and forces. Just in this way it is possible to have a maximum difference between the numerical model and the experiment around 5%. It is a very good result due the complexity of the model.

NOMENCLATURE

CAD	Computer Aided Design	
CAE	Computer Aided Engineering	
COP	Coefficient of Performance	(-)
d	nominal bolt diameter	(mm)
F	force	(N)
K	torque coefficient	(-)
k	polytrophic coefficient	(-)
N	dimension	(mm)
P	dimension	(mm)
P	pressure	(N/mm^2)
R	piston boss	(mm)
T	torque	(N.mm)
V	volume	(mm^3)
η	efficiency	(-)

Subscript

b	bolt
c	clearance
d	discharge
s	suction
sw	swept
v	volumetric
1 ... 5	indexes
1e	piston boss empty
1f	piston boss full

REFERENCES

Mischke, C. R., Shigley, J. E., 1989, *Mechanical Engineering Design,* McGraw-Hill, Inc., USA, p.346

Puff, Rinaldo, 2008, Hermetic Compressor Manifold Analysis With the Use of the Finite Element Method, 2008 *International Compressor Engineering Conference. Paper 1843.*

ACKNOWLEDGEMENT

Thanks to the Eng. Julio Cesar da Silva for his contribution to this paper.

Refrigeration Cycle and Compressor Performance for Various Low GWP Refrigerants

Jit Guan Edwin ONG [1a*], Kim Tiow OOI[1b],

[1]School of Mechanical and Aerospace Engineering, Nanyang Technological University,
50 Nanyang Avenue, Singapore 639798.

[a]e-mail: A140003@e.ntu.edu.sg
[b]Phone: +65 6790 5511, fax: +65 6791 1859, e-mail: mktooi@ntu.edu.sg

* Corresponding Author

ABSTRACT

The recent meetings of the global leaders in COP21 have gathered global consensus to fight the global warming. One of the important approaches in aiding toward fighting global warming is to use environmental friendly refrigerants. Recently a few new low GWP refrigerants have been introduced. In this paper, we will discuss the energy efficiencies of some new low GWP refrigerants as compared to the existing ones.

Cycle performance calculations assuming an ideal vapor compression cycle from a pressure-enthalpy diagram and those from a comprehensive compressor mathematical model will be shown and compared. Refrigerants which are considered are R32, R134A, R404A, R407C, R1234ZE and R1234YF, and the results quoted will be normalized using data from R410A, which is currently a widely used refrigerant.

The results shows that, when compared to the predictions from the compressor mathematical model, the results from the P-h diagram using the ideal cycle may overestimate the COP by up to 2 times, and this is caused by the up-to 50% of underestimation on the compressor work input.

1. INTRODUCTION

The COP21 United Nations climate summit that was held in Paris November 2015, after 13 days of negotiations by negotiators representing 195 countries, reached an agreement on the effort on climate change which covers both the developed and the developing countries. One of the agreements made is to limit the average rise in global temperature to be 2 degrees Celsius above pre-industrial times. One of the causes of the global warming is due to the use of refrigerants, hence many new green and environmentally more friendly refrigerants had been introduced. In this paper, we will compare the energy efficiencies and refrigerant cycle performance when using these new low GWP refrigerants with the existing ones. The basis of the comparison will be the ANSI/AHRI 540 testing standards, which is suitable for positive displacement refrigerant compressors. The testing cycle for the refrigerant will be based on the idealised vapor-compression refrigeration cycle. With these conditions imposed on all the refrigerants tested, the coefficient of performance (COP), cooling capacity, work input required, as well as the working pressure of the refrigerants will be compared and discussed. The effects of these new low GWP refrigerants and their properties on the compressor performance will be evaluated, discussed and compared. Compressor performance parameters such as work input, discharge pressure and temperature, volumetric efficiency, mechanical efficiency will also be shown and discussed.

2. MATHEMATICAL MODEL

There are two approaches used in this paper to compare the performance of these refrigerants. One approach is to use an idealized vapor compression cycle on a P-h diagram and the other is to use a more comprehensive mathematical simulation model for a rolling piston compressor.

This brief description for the compressor mathematical model will be presented below for completeness, detailed model can be found in ref [3]. The compressor used is rolling piston type, the schematic of which is shown in fig. 1. The volume $V(\theta)$ of the chamber of the rolling piston compressor can be expressed in terms of the length of the compressor l, radii of cylinder R_c and rotor R_r, the rotational angle θ and vane thickness t_v as below.

$$V(\theta) = f(l, R_c, R_r, \theta, t_v) \tag{1}$$

The variation of the fluid properties in the chamber can be obtained by applying First Law of Thermodynamics,

$$\dot{E}_{in} - \dot{E}_{out} = \frac{d(mu)_c}{dt} \tag{2}$$

where $\dot{E}_{in}$, $\dot{E}_{out}$ and $(mu)_c$ the energy in and out of the chamber as well as the change in internal energy of the working chamber, respectively.

Real gas properties of the refrigerant that relates enthalpy h_c of the working fluid with the pressure P and specific volume v,

$$h_c = f(P, v) \tag{3}$$

The conservation of mass in the working chamber,

$$\Sigma\frac{dm_i}{dt} - \Sigma\frac{dm_o}{dt} = \frac{dm_c}{dt} \tag{4}$$

where m is the mass of the working fluid in chamber and subscript i, o, c represents in, out and chamber respectively.

Assuming that the flow through the valves is steady one-dimensional and adiabatic, the mass flow can be expressed as,

$$\frac{dm_2}{dt} = \frac{C_d A_2\sqrt{2(h_1-h_2}}{v_{s2}} \tag{5}$$

where C_d indicates non-isentropic and flow losses, A is the flow area, v_s the specific volume of refrigerant and h enthalpy of the refrigerant. The subscript 1, 2 and s represents upstream, downstream and isentropic conditions respectively. The model also includes kinematic, roller dynamics, thermodynamics, valve dynamics, in-chamber heat transfer, mechanical frictional and lubrication [2,3,4].

The model is written in Fortran programming language, solving simultaneous equations using 4th order Runge-Kunge numerical integration method. The model has been verified using R22 as working fluid operating at -23.3°C and 54.4°C at 2875 rev/min, see fig. 1.

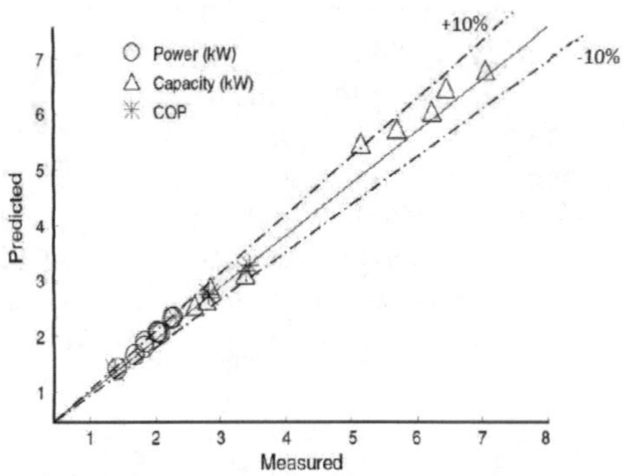

Figure 1: Comparison between measured and predicted results

Fig. 1 shows that the discrepancies between measured values and prediction are within 10%. The model also includes effects of in-chamber heat transfer, internal leakages and frictional losses [3].

3. REFRIGERANTS AND THEIR PROPERTIES

Properties of the refrigerants such as the GWP, critical temperature and pressure are listed in table 1. As we can observed, the GWP of newer refrigerants are lower, with R1234YF has only 0.2% of that of the R410A. The table also shows that the critical temperatures are all well above the condensing temperature in the applications and hence the refrigeration cycle will operate at the subcritical regions.

In this paper, the performance of various refrigerants shown in Table 1 will be compared using ANSI/AHRI 540 2015 standard operating condition, henceforth refers to as standard condition. Two approaches are used: (i) employing the ideal vapor compression cycle as shown in fig. 2 and, (ii) using a more comprehensive compressor model as described in section 2.

Table 1 Properties of refrigerants

Refrigerant	Refrigerant type	GWP normalised *	Critical Temperature (°C)	Critical Pressure (kPa)
R32	HFC	0.32328	78.11	5782
R134A	HFC	0.68487	101.06	4059.3
R404A	HFC	1.87835	72.12	3734.9
R407C	HFC	0.84962	86.14	4639.4
R410A	HFC	1.00000	71.34	4901.2
R1234ZE	HFO	0.00287	109.36	3634.9
R1234YF	HFO	0.00192	94.70	3382.2

*GWP values shown are normalized with respect with R410A

Figure 2 shows the idealized refrigeration cycle on a P-h diagram operating under the standard conditions. Under this condition, the refrigerant enters the compressor at 11°C of super-heat and leaving the condenser at 46°C as saturated liquid with no subcooling. The compression is assumed isentropic and there is no pressure loss at the evaporator, condenser and along the pipelines. The refrigerant will be throttled through the expansion valve and enter the evaporator at 10°C.

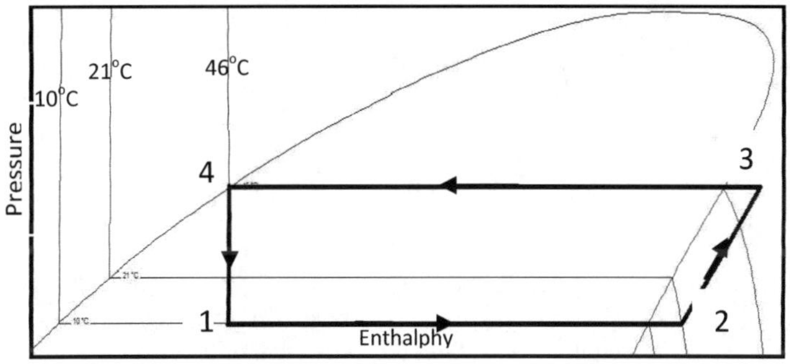

Figure 2: Ideal Refrigeration Cycle on a P-h diagram

4. RESULT AND DISCUSSIONS

Calculations from the idealized cycle on P-h diagram (henceforth refers to as "**calculation**") and predictions from the compressor model (henceforth refers to as "**prediction**") have been made using operational conditions spelt out in ANSI/AHRI 540 2015. The results are normalized, with respect to R410A, a commonly used refrigerant in the refrigeration system.

Figure 3 shows that all the **predicted** Coefficient of Performance (COP) are lower than that of the **calculated**. This is because the prediction includes mechanical losses and the internal leakages. The difference between the two varies significantly from 35% to more than 60%.

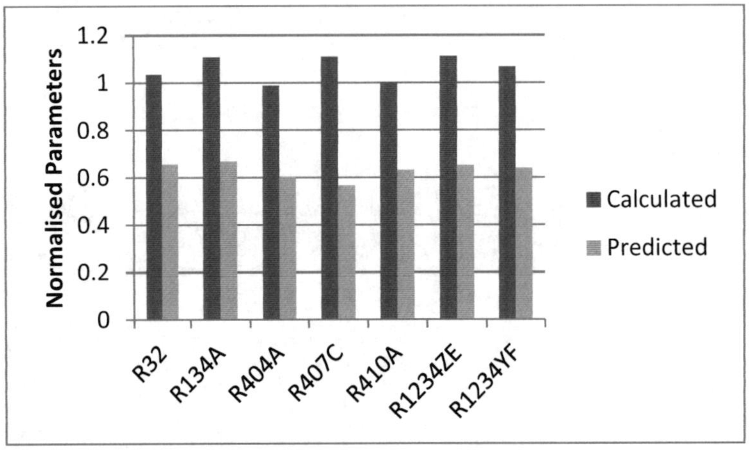

Figure 3: COPs of various refrigerants from calculation and predicted

The cooling capacity of the refrigerant depends on the refrigerant mass flow rate. For given displacement volume and operating conditions, the mass flow rate depends on the specific volume of the refrigerant. Fig. 4 shows the specific volume alongside with the mass flow rate. As expected, the higher the specific volume, the lower the mass flow rate. R404A and R410A has similar mass flow rate as they have similar specific volume.

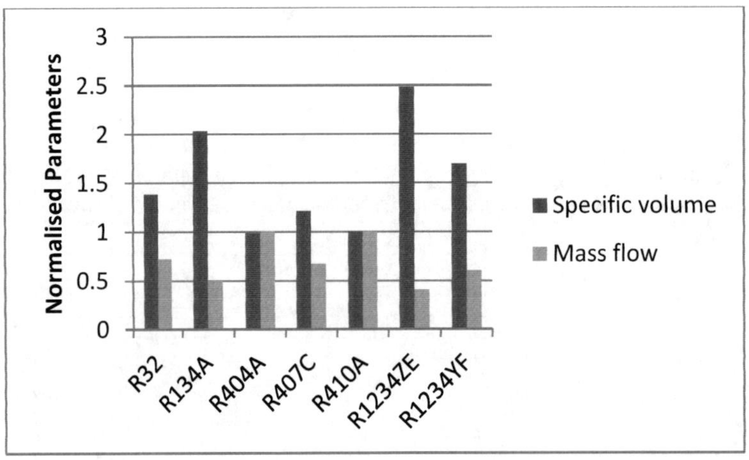

Figure 4: Specific Volume and the predicted mass flow rates

Another factor that will affect the cooling capacity is the latent heat of the refrigerant at the evaporating condition, in this case 10°C. Figure 5 presents the refrigerating capacity, the mass flow rate and the latent heat of the refrigerant at the evaporator condition. It shows that R32 gives the highest refrigerating capacity, which is 10% more than R410A while R1234ZE is the lowest, which is more than 60% lower than R410A.

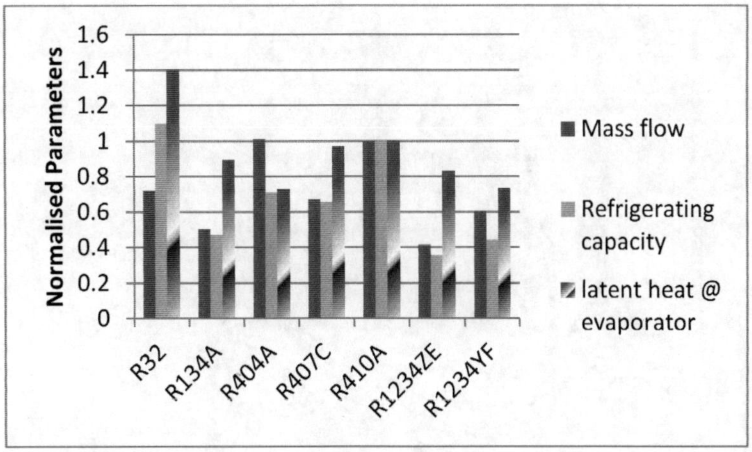

Figure 5: Variations of mass flow, refrigerating capacity and latent heat at evaporator

The second parameter that affects the COP of the refrigeration cycle is the work input. Fig. 6 shows the comparison between the calculated and the predicted work. As seen from the figure, the predicted work input is higher than that of the calculated and the difference can be easily more than 50%. This is due to flow losses, volumetric and mechanical efficiencies considered in the predicted values.

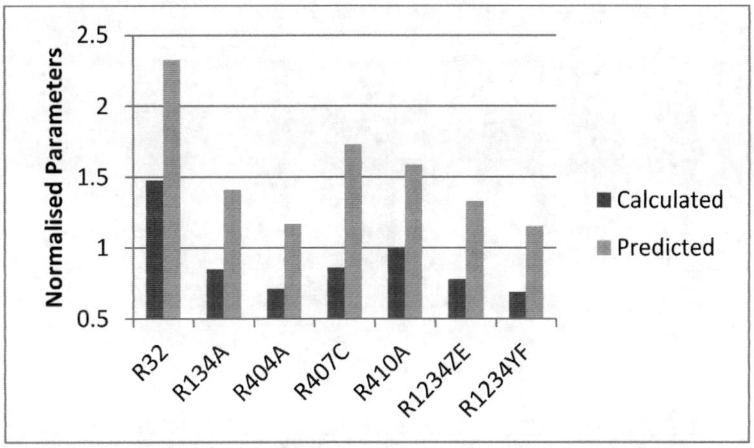

Figure 6: Variations of work input from calculated and predicted

In the rotary vane compressor, the friction loss occurs mainly at 6 rubbing areas, namely the eccentric and inner surface of the roller, surface between roller and cylinder, eccentric face and cylinder head face, vane tip and roller as well as vane and slot. As seen from fig. 7, for a given refrigerant, when the pressure difference between suction and discharge (P_d-P_s) is high, the contact force between the surfaces will be large and hence results in a higher friction loss.

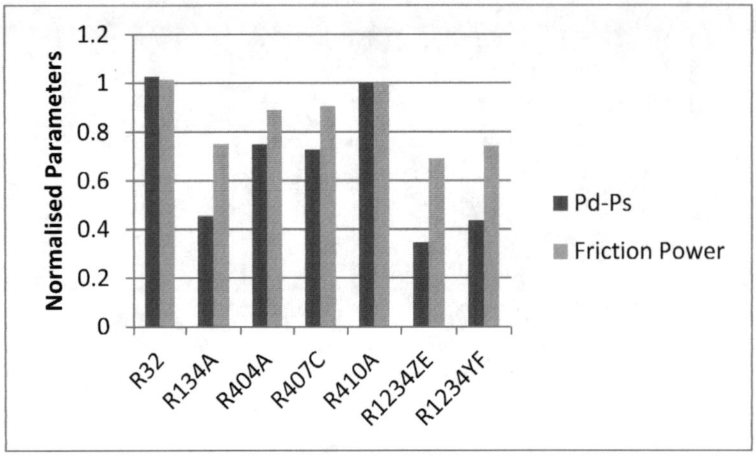

Figure 7: Variations of P_d-P_s and friction power (predicted)

Figure 8 shows the variation of the predicted mechanical efficiencies for various refrigerants. It shows that R410A and R32 have the highest values. It would expect that for a higher P_d-P_s the mechanical efficiency will be lower, however the influence of the indicated work will also affect the mechanical efficiency.

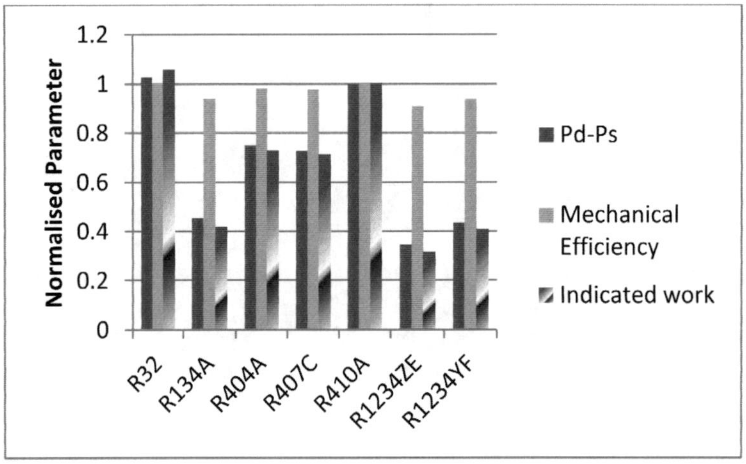

Figure 8: Variations of P_d-P_s, mechanical efficiency (predicted) and indicated work for various refrigerants.

Figure 9 shows that the volumetric efficiency for various refrigerants are very similar to each other, because the refrigerant's experiencing similar suction valve losses with suction heating effect was not considered. The marginal differences in volumetric efficiencies were caused by the differences in operating pressures.

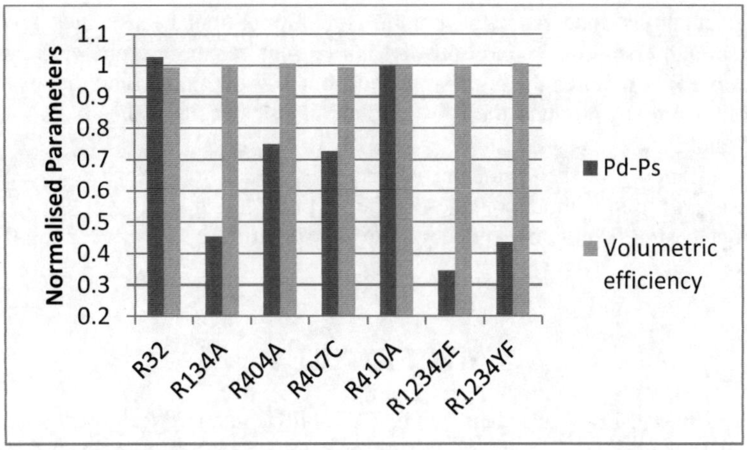

Figure 9: Variations of Pd-Ps and volumetric efficiency (predicted)

Figure 10 shows clearly that the torques of the motor shaft for various refrigerants are dependent on the differences between P_d and P_s.

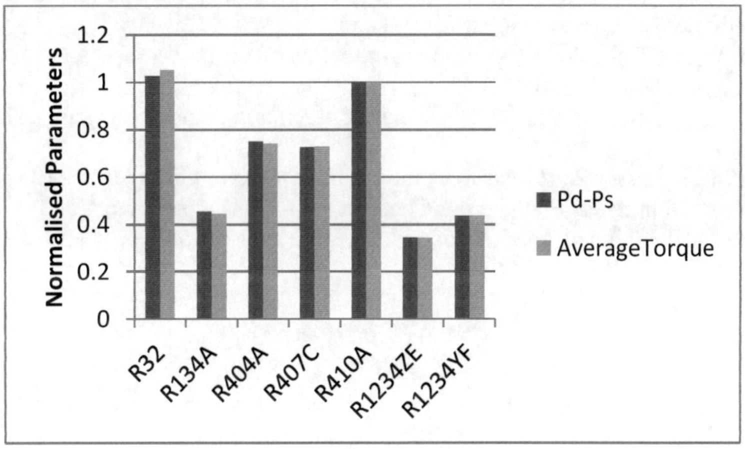

Figure 10: Variations of Pd-Ps and average torque (predicted)

5. CONCLUSION

This paper presents comparison of performance for various refrigerants when applied to a vapour compression cycle. Calculations of the performance data assuming a simplified cycle using a textbook based P-h diagram (hence forth refers to as calculations) have been presented alongside with the predictions from a more comprehensive compressor mathematical model (henceforth refers to as predictions) together with operational data for condenser and evaporator sides using ANSI/AHRI 540 2015 standard operating condition. The refrigerants tested are R32, R134A, R404A, R407A, R1234ZE and R1234YF. The results show that:

1. Performance data obtained from the textbook based P-h diagram using an idealized refrigeration cycle deviated significantly from those predicted using a more comprehensive mathematical model. It is therefore concluded that P-h diagram based calculations are not suitable for accurate performance calculation estimation.
2. These “calculations” may over estimate COP values from 30% to 60%, when compared with the more comprehensive COP predications.
3. For a given compressor displacement volume, the cooling capacity for R32 is 10% higher than that of R410A and the lowest is the R1234ZE, which is 60% lower than R410A.

4. P-h diagram calculations underestimated compressor power input by as much as 40% to 80%. For the same compressor and operational conditions, R1234YF requires the least power input, which is 40% lower than R410A, while R32 requires the highest power input of 70% higher than that of R410A.
5. Refrigerant R32 seems to produce the most frictional loss which is due to its large pressure difference between the P_d and P_s.
6. The volumetric efficiency of the various refrigerants tested are not differ significantly, this is because the same compressor was used without considering the suction heating effect.
7. The motor torque is significantly dependent on the pressure difference between the P_d and P_s.

REFERENCES

1) Lemmon, E.W., Huber, M.L., McLinden, M.O., 2007. NIST Standard Reference Database 23: Reference Fluid Thermodynamic and Transport Properties REFPROP, Version 8.0. National Institute of Standards and Technology, Standard Reference Data Program, Gaithersburg.
2) Linde Group (2007) Refrigerants Environmental Data retrieved from http://www.linde-gas.com/internet.global.lindegas.global/en/images/Refrigerants%20environmental%20GWPs17_111483.pdf
3) Ooi, K.T., 2008, Assessment of a rotary compressor performance operating at transcritical carbon dioxide cycles, Applied Thermal Engineering 28 1160–1167
4) Ooi, K.T. and, Wong, T.N., 1997, A computer simulation of a rotary compressor for household refrigerators, Appl. Therm. Eng. 17 (1) 65–78.
5) Ooi, K.T., 2003. Heat transfer study of a hermetic refrigeration compressor. Appl. Therm. Eng. 23 (15), 1931-1945.
6) Ooi, K.T, 2012 Compressor Performance Comparison When Using R134 and R1234YF as Working Fluids, 21st International Compressor Engineering Conference at Purdue, paper no. 2040
7) T. Yanagisawa and T. Shimizu, 1985, Friction losses in rolling piston type rotary compressors. III (164-165)

ACKNOWLEDGEMENT

We wish to acknowledge the funding for this project from Nanyang Technological University under the Undergraduate Research Experience on CAmpus (URECA) programme.

Comparative Analysis of Two Types of Positive Displacement Compressors for Air Conditioning Applications

Marco C. DINIZ[1], Cesar J. DESCHAMPS[1]*

[1]POLO Research Laboratories for Emerging Technologies in Cooling and Thermophysics
Federal University of Santa Catarina
Florianopolis, SC, Brazil
deschamps@polo.ufsc.br

* Corresponding Author

ABSTRACT

Different types of positive displacement compressors are employed in low-capacity air conditioning applications. Understanding the advantages and drawbacks of each of these technologies is important in order to identify the ranges of cooling capacities at which a specific compression technology is more efficient. This paper presents a comparative analysis between two positive displacement compressors (rolling piston and scroll) designed for air conditioning applications. The compressors operate with R-22 and present a nominal cooling capacity of 18000 BTU/h at the AHRI-A checkpoint condition. First, the compressors were tested in a hot-cycle test bench to measure the mass flow rate and power consumption for MBP and HBP operating conditions. The results showed that the most efficient compressor depends on the operating condition. After this, simulation models for each compressor were validated against the experimental data, showing good agreement. Therefore, through the simulation models, it was possible to evaluate the effect of different aspects, such as leakage, heat transfer and discharge process, on the efficiency of each compressor.

1. INTRODUCTION

A significant percentage of the energy used in buildings is related to air conditioning systems, especially in tropical climates (Chua *et al.*, 2013). Most of these systems employ the vapor compression refrigeration cycle, in which the compressor is the main component when it comes to electrical power consumption. Therefore, to improve the performance of the whole system it is necessary to employ efficient compressors.

The first generations of air conditioning systems used in small and medium size applications employed reciprocating compressors. Gradually, with the advent of rotary compressors, reciprocating compressors were increasingly replaced by rolling piston and scroll compressors, which presented higher efficiency and reliability. Nowadays, rolling piston and scroll compressors are widely spread and dominate the market for small and medium size air conditioning systems. Nevertheless, it should also be mentioned that some researchers (Ooi, 2014; Orosz *et al.*, 2014) are continuously working to develop new types of compressors that overcome the main drawbacks associated with rotary compressors.

The global efficiency of hermetic compressors is related to its electrical, mechanical and thermodynamic performances. The electrical motor design is one of the most important aspects when defining the compressor operating envelope and the research on this subject is usually focused on motor reliability. The studies related to the mechanical behavior of rolling piston and scroll compressors dominated the research efforts in the beginning of their technical development, when the objective was to develop durable and reliable compressors. As the electrical and mechanical improvements of the compressors are usually limited by cost, the detailed analysis of the thermodynamic phenomena is fundamental in order to identify opportunities to increase compressor global efficiency.

Several works have been developed regarding the thermodynamic performance of both rolling piston (Ooi, 2012) and scroll compressors (Maertens and Richardson, 1992), usually with focus on the impact of using different refrigerants. On the other hand, few works (Gomes and Deschamps, 2007) have been conducted with the objective of comparing the thermodynamic efficiency of these compressors. However, the comparative thermodynamic analysis of these compressors is important to identify the advantages and drawbacks of each technology.

This work presents a comparative analysis of the thermodynamic efficiency of a rolling piston and a scroll compressor operating under MBP and HBP applications. Two compressors with similar volumetric displacements were tested in a hot-cycle test bench in nine operating conditions. Simulation models were adjusted for each compressor, validated against experimental data, and used to evaluate and compare their energy and mass losses under different operating conditions.

2. EXPERIMENTAL PROCEDURE

The rolling piston and scroll compressors evaluated in this work were tested with their original shells, i.e., no internal measurements were possible. Externally, the compressors were instrumented with a thermocouple to measure the shell temperature, which was used to establish a criterion for thermal stabilization during the measurements. Both compressors used refrigerant R22 and operated at 60 Hz. The displacement volume of the rolling piston compressor was 23.9 cm^3, while for the scroll the displacement volume was 25.2 cm^3.

The compressors were tested in a hot-cycle test bench, in which all the thermodynamic processes occur with the gas in the superheated vapor state. In this facility, a set of micrometric valves is adopted to submit the compressor to the required pressure ratio, while the temperature at the compressor suction line is controlled through heat exchangers. A coriolis mass flow meter and a power transducer available in the test bench allowed the measurements of mass flow rate and power consumption, respectively. Details about the operation of such a test bench are available in Dutra and Deschamps (2013).

The tests were performed under nine operating conditions defined by the combination of three evaporating temperatures (-1.1ºC, 7.2ºC and 12.8ºC) and three condensing temperatures (37.8ºC, 48.9ºC and 60.0ºC). For all conditions, the temperature in the compressor inlet was set to be 8.3ºC higher than the evaporating temperature. Each test was stopped when the compressor was considered thermally stabilized. This condition was achieved when during 30 minutes the temperature of the compressor shell showed a maximum variation of ± 0.5°C. Then, the next 30 minutes of the test were used to calculate the average mass flow rate and power consumption. Three repetitions were conducted for each operating condition, showing good repeatability. With the average mass flow rate, power consumption and temperature at the compressor inlet, the main performance parameters for the nine operating conditions can be calculated.

Tables 1 and 2 present two dimensionless important performance indicators of rolling piston and scroll compressors. Table 1 presents the volumetric efficiency $\eta_v = \dot{m}/\dot{m}_{th}$, where $\dot{m}$ is the measured mass flow rate and $\dot{m}_{th}$ is the theoretical mass flow rate. Table 2 shows the global efficiency $\eta_g = \dot{W}_{th}/\dot{W}_{ele}$, where $\dot{W}_{th}$ is the theoretical power of isentropic compression, estimated taking into account the measured mass flow rate, and $\dot{W}_{ele}$ is the measured electrical power consumption.

Table 1 – Experimental results of volumetric efficiency for rolling piston and scroll compressors

Volumetric efficiency [%]		T_e [°C]								
		-1.1			7.2			12.8		
		Rolling piston	Scroll	Difference [%]	Rolling piston	Scroll	Difference [%]	Rolling piston	Scroll	Difference [%]
T_c [°C]	37.8	90.6	90.8	+0.2	91.3	92.8	+1.6	93.0	94.0	+1.1
	48.9	87.4	88.4	+1.1	88.6	90.7	+2.4	89.6	92.0	+2.7
	60.0	83.1	81.2	-2.3	84.9	87.4	+2.9	85.8	89.3	+4.1

Table 2 – Experimental results of global efficiency for rolling piston and scroll compressors

Global efficiency [%]		T_e [°C]								
		-1.1			7.2			12.8		
		Rolling piston	Scroll	Difference [%]	Rolling piston	Scroll	Difference [%]	Rolling piston	Scroll	Difference [%]
T_c [°C]	**37.8**	65.2	64.5	-1.1	63.2	64.7	+2.4	61.1	61.5	+0.7
	48.9	64.5	61.8	-4.2	64.8	66.2	+2.2	65.0	67.1	+3.2
	60.0	59.7	52.4	-12.2	61.4	61.4	+0.0	62.9	64.9	+3.2

As expected, it can be observed that the operating condition is important to define the most efficient compressor. The scroll compressor is more efficient than the rolling piston compressor both in terms of volumetric and global efficiency in the two higher evaporating temperatures analyzed, especially closer to its internal pressure ratio. However, for the lower evaporating temperature, the rolling piston becomes more competitive even overcoming the performance of the scroll compressor. This aspect is more pronounced at the condition with the higher pressure ratio, where the global efficiency of the rolling piston is more than 10% higher than that of the scroll compressor. These experimental results are useful to establish which compressor presents the best performance in each condition. However, as internal measurements were not performed, it was not possible to clearly identify the reasons for the differences of thermodynamic performance between the two compressors. In order to allow such an analysis, two simulation models were employed and will be described in the next section.

3. MATHEMATICAL MODELS

The models used to simulate the compression cycles of both rolling piston and scroll compressors are based on a transient lumped formulation, which allows the calculation of the thermodynamic properties of the gas throughout the suction, compression and discharge processes. This procedure is based on a method widely employed in the literature, where four groups of equations are used to characterize the whole compression cycle. The solution procedure for each model will be briefly described herein. Further information regarding the models can be found in Brancher (2013) for the rolling piston compressor and Pereira and Deschamps (2010) for the scroll compressor.

The first group of equations is used to calculate the instantaneous volume of the suction, compression and discharge chambers, which depend solely on the geometrical parameters of each compression mechanism. For the rolling piston compressor these parameters include the cylinder, piston and blade dimensions, whereas for the scroll compressor the number of turns of the scroll wrap and its dimensions are required. To estimate these parameters both compressors were opened and the important data were measured with appropriate instruments.

The second group of equations is the conservation equations for the volumes formed by the compressor chambers. As the instantaneous volume of the chambers is known, to estimate the instantaneous density a mass balance is necessary:

$$\frac{dm_g}{dt} = \sum \dot{m}_{in} - \sum \dot{m}_{out} \tag{1}$$

where the summations of instantaneous mass flow rates include those in the suction and discharge orifices and also the ones due to leakage through clearances. An energy balance, Equation (2), is used to calculate the instantaneous temperature during the compression cycle. With the density and the temperature being evaluated via Equations (1) and (2), it is possible to obtain the refrigerant instantaneous pressure by employing an equation of state.

$$\frac{dT_g}{dt} = \frac{1}{m_g c_v}\left\{-T_g\left(\frac{\partial p_g}{\partial T_g}\right)_{\forall} - \frac{1}{\rho_g}\frac{dm_g}{dt} - \sum \dot{m}_{in}\left(h_g - h_{in}\right) + H_g\left(T_w - T_g\right)\right\} \tag{2}$$

In Equation (2), H_g and T_w are, respectively, the convective heat transfer coefficient and the temperature at the cylinder wall (for the rolling piston) or at scroll wrap (for the scroll), while h_{in} is the enthalpy of the gas entering the chamber. During the formation of the suction chamber of both compressors, this temperature is very important to characterize the so-called suction gas superheating, which is known to affect compressor efficiency.

For the rolling piston compressor, H_g was calculated using the correlation employed by Liu and Soedel (1992). The cylinder temperature was considered equal to the shell temperature, for which experimental measurements in the nine different conditions were available. This decision was considered acceptable, as the cylinder and shell temperature are quite similar for a high pressure shell rolling piston compressor (Padhy, 1994). The suction temperature was estimated by calculating the heat transfer in the suction passage with the $\epsilon - NTU$ method, in which the cylinder was considered as the heat source.

For the scroll compressor model, H_g is estimated using a correlation developed from the numerical simulations presented by Pereira and Deschamps (2012), whereas the suction temperature was calculated following the procedure presented by Diniz and Deschamps (2014). The temperature of the scroll wraps was estimated by considering a linear profile from its most outer point up to the most central point. Based on the results of Jang and Jeong (2004), the lowest temperature of the scroll was considered to be 15ºC higher than the suction temperature and the highest temperature of the scroll was set to be 10ºC lower than the average discharge temperature of the gas.

The third group of equations is required to predict valve dynamics. A single degree-of-freedom mass spring model was employed to calculate the discharge valve dynamics of the rolling piston. Effective force areas were previously obtained through CFD simulations to predict the force acting on the valve during the compression cycle.

Finally, the fourth group of equations predicts the mass flow rate through the suction and discharge orifices and the gas leakage through clearances. For the rolling piston, the mass flow rate through the suction and discharge orifices was estimated with reference to an isentropic flow through a convergent nozzle, which was corrected via the concept of effective flow areas. Different models were employed to calculate leakage through the clearances of the rolling piston compressor, and details can be obtained in Bezerra Jr *et al.* (2013). For the scroll compressor, the mass flow into the suction pocket was estimated using the correlation developed by Pereira (2012), while the mass flow through the discharge orifice was calculated with the concept of effective flow areas. The leakages through the flank and top clearances were predicted by using the correlations presented by Pereira and Deschamps (2015).

Both models were simulated for a certain number of cycles until the compressor operating conditions were cyclically repeated, allowing the predictions of the main performance parameters and the inventories of energy and mass losses.

4. RESULTS

Before using the models to predict the mass and energy losses of the compressors, some numerical predictions were validated through comparisons with the available experimental data. Table 3 presents the differences between predictions and measurements of mass flow rate, showing that the error is smaller than 2% in most of the conditions analyzed. To achieve such an agreement for the mass flow rate it was necessary to adjust the clearances of the scroll for each of the conditions.

Table 3 – Errors associated with predictions of mass flow rate.

Error in mass flow rate [%]		T_e [ºC]					
		-1.1		7.2		12.8	
		Rolling piston	Scroll	Rolling piston	Scroll	Rolling piston	Scroll
T_c [ºC]	37.8	-1.66	-0.11	-0.04	+0.40	-0.47	+0.14
	48.9	-0.85	+0.12	+0.51	+0.64	+1.03	+0.79
	60.0	+0.93	+4.25	+1.80	+0.38	+2.64	+0.48

Table 4 shows the errors associated with the power consumption prediction. This parameter was numerically estimated by using the numerical indicated power and values of electromechanical efficiency adjusted for each condition for both compressors. The agreement between experimental and numerical results can be considered satisfactory.

Table 4 – Errors associated with predictions of power consumption

Error in power consumption [%]		T_e [°C] -1.1 Rolling piston	-1.1 Scroll	7.2 Rolling piston	7.2 Scroll	12.8 Rolling piston	12.8 Scroll
T_c [°C]	**37.8**	0.64	-2.21	-0.64	-0.23	+1.25	-0.20
	48.9	0.46	-0.65	0.17	+0.18	+0.21	+0.89
	60.0	0.87	-1.04	0.00	+1.02	-0.73	+2.33

To extrapolate the analysis to other operating conditions, it was necessary to obtain curves that estimate the variation of the aforementioned adjusted parameters within the envelope. For each operating condition of the rolling piston, the electromechanical efficiency and the cylinder temperature (necessary to calculate the suction temperature and to estimate heat transfer inside the chamber) were expressed as a function of the evaporating and condensing temperatures. A similar approach was employed for the scroll compressor in order to predict the clearances and the electromechanical efficiency throughout the application envelope. The results shown hereafter were obtained by using such curves.

Figures 1 and 2 present the suction and discharge temperatures of the rolling piston and scroll compressors at different condensing temperatures, for the three evaluated evaporating temperatures. As in the scroll compressor the gas that enters the shell fills the internal environment before flowing to the suction chamber, its suction temperature is higher than that of the rolling piston. However, the discharge temperatures of the rolling piston are higher than those of the scroll compressor for most of operating conditions.

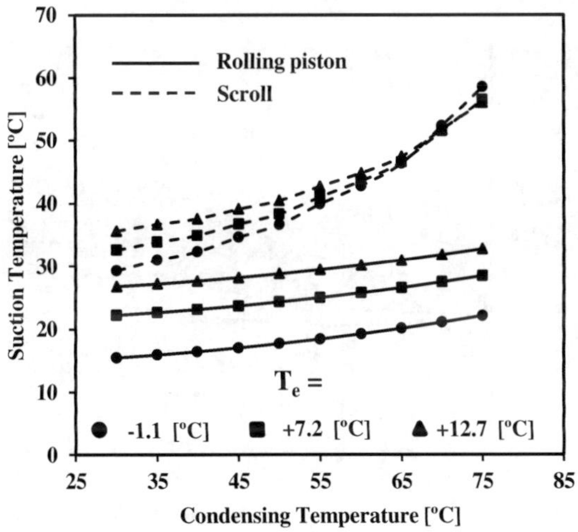

Figure 1: Suction temperature of rolling piston and scroll compressor

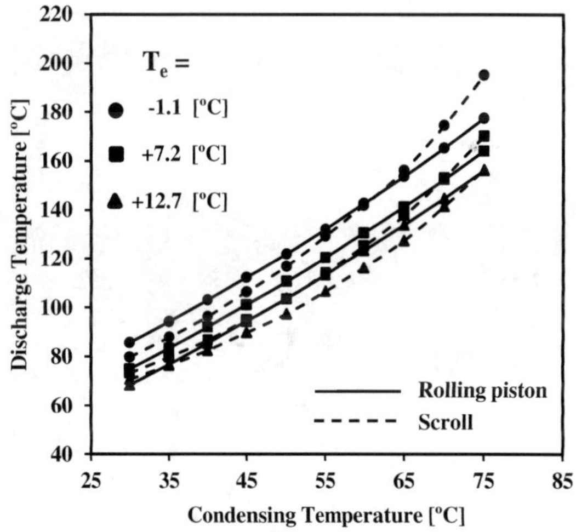

Figure 2: Discharge temperature of rolling piston and scroll compressor

Figures 3 and 4 present the indicated power and the thermodynamic efficiency of the rolling piston and scroll compressors at different condensing temperatures for the three evaluated evaporating temperatures. The indicated powers of both compressors have the same order of magnitude, with a monotonic increase with the pressure ratio π (= p_c/p_e). However, Figure 4 show that the thermodynamic efficiency of the scroll compressor is much more sensitive to changes in the operating condition, with a considerable decrease of efficiency when applied far from its design operating condition, which corresponds the condition in which the system pressure ratio matches the internal pressure ratio of the compressor. In fact, the efficiency of the scroll compressor is higher than that of the rolling piston in the design condition. On the other hand, the rolling piston compressor presents a more flat behavior for the thermodynamic efficiency throughout the operating conditions analyzed.

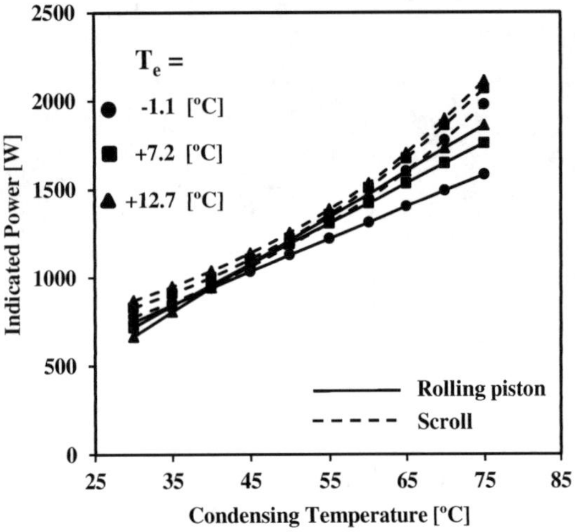

Figure 3: Indicated power of rolling piston and scroll compressor for different evaporating temperatures.

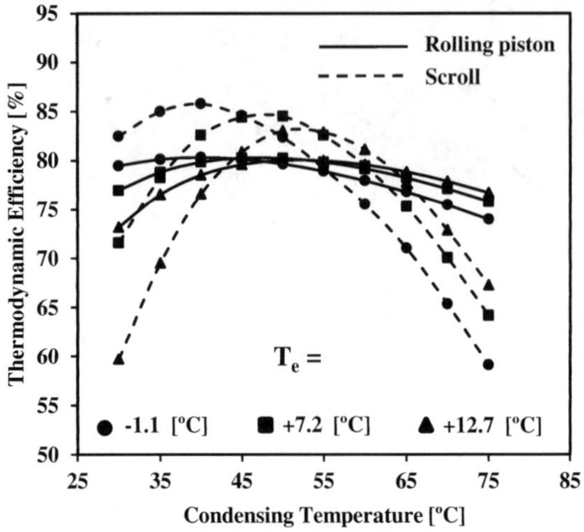

Figure 4: Thermodynamic efficiency of rolling piston and scroll compressor for different evaporating temperatures.

The inventory of energy losses is a convenient way of identifying the reasons for the trends shown in Figure 4. Figures 5 and 6 present the energy losses associated with the compression cycle of the rolling piston for evaporating temperatures of T_e = -1.1°C and T_e = +12.7°C at different condensing temperatures. The energy loss due to viscous friction in the suction orifice was not significant and therefore was neglected.

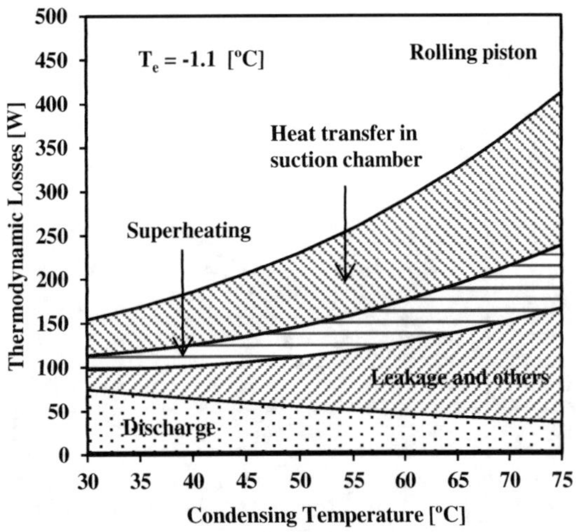

Figure 5: Thermodynamic losses for the rolling piston compressor at T_e = -1.1 °C

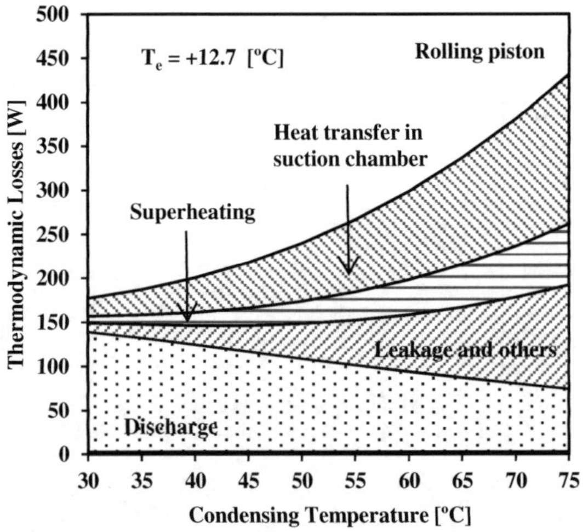

Figure 6: Thermodynamic losses for the rolling piston compressor at T_e = 12.7 °C

It can be observed that total thermodynamic loss is more dependent on the condensing temperature than on the evaporating temperature. The impact of heat transfer was divided in two contributions: the loss due to the increase in refrigerant temperature from the suction line until the suction port (superheating) and the loss associated with the increase in refrigerant temperature from the suction port until the beginning of the compression process (heat transfer in suction chamber). In the rolling piston, due to the small path of the suction tube, the increased amount of work due to suction superheating is small when compared to the increase caused by heat transfer during the formation of the suction chamber, which is mainly associated with the adopted high-pressure shell design. As expected, the energy losses due to leakage are more pronounced at higher pressure ratios. The losses in the discharge process are more significant at higher evaporating and low condensing temperatures because under these conditions the mass flow rate is higher, increasing the flow velocity and viscous losses in the discharge valve.

The diagrams depicting the scroll compressor thermodynamic losses are shown in Figures 7 and 8. For this compressor, the thermodynamic losses are considerably influenced by under and over compression. Such losses occur because the initial pressure in the discharge pocket is governed by the suction pressure and the compressor internal pressure ratio, regardless of the condensing pressure. The under compression loss occurs if the discharge pocket initial pressure is lower than that of the condenser, resulting in backflow from the compressor discharge plenum to the discharge pocket. When this high pressure refrigerant re-enters the discharge pocket, the volume of the chamber is still decreasing and therefore this gas is compressed again. This process will continue until the gas in the discharge pocket reaches a slightly higher pressure than that of the discharge plenum, when the gas is finally discharged. This phenomenon was incorporated into the leakage contribution of the thermodynamic losses and is more pronounced at higher condensing temperatures, especially for the lower evaporating temperature (Figure 7). A solution commonly employed to minimize the under compression loss is to use a discharge valve, although it increases the viscous losses of the discharge process.

If the pressure in the discharge pocket is above the condensing pressure, then the compressor is operating with over compression as the compression process required more energy than the one that would be necessary to match the system pressure ratio. This loss, identified as a discharge loss in Figure 8, is considerable for the lower condensing temperatures. By comparing Figures 7 and 8 it is possible to identify that the point of lowest thermodynamic losses occurs at higher condensing temperatures when the evaporating temperature is increased, following the compressor internal pressure ratio. These aspects show that the design of scroll compressors is strongly dependent on the operating conditions determined by the system.

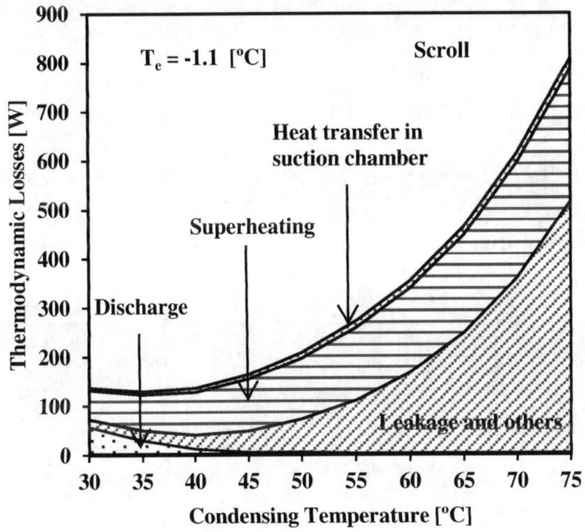

Figure 7: Thermodynamic losses for scroll compressor at T_e = -1.1 ºC

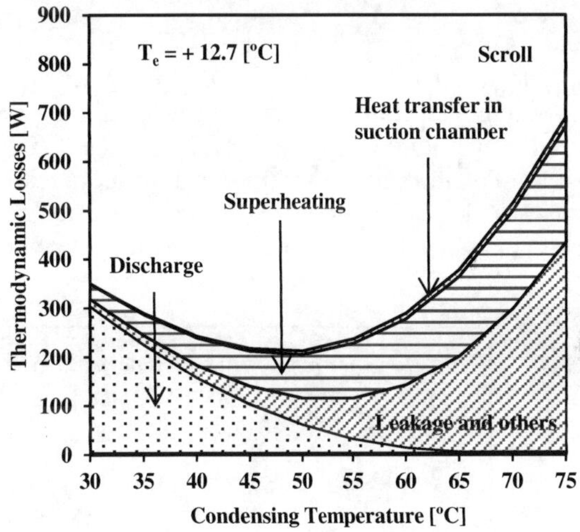

Figure 8: Thermodynamic losses for scroll compressor at T_e = 12.7 ºC

The diagrams of energy losses for the rolling piston and scroll compressors (Figures 5 to 8) clarify the reasons for the trends observed for the thermodynamic efficiencies shown in Figure 4. The rolling piston compressor presents a monotonic behavior for the total thermodynamic loss as the condensing temperature increases. On the other hand, the scroll compressor presents an inverted U-shaped curve for the thermodynamic efficiency mainly due to under and over compression losses. Figures 5 to 8 also show that, regarding heat transfer, the main loss for the rolling piston compressor is related to the heating of the gas that occurs during the formation of the suction chamber, whereas for the scroll compressor the main loss related to heat transfer is associated with suction superheating. This is a result of the differences between the layouts of the rolling piston and scroll compressor shells.

Figures 9 and 10 present the mass flow rate and volumetric efficiency for the rolling piston and scroll compressors at different operating conditions. It can be noticed that the mass flow rate of both compressors is reduced as the condensing temperature increases, regardless of the evaporating temperature. This reduction of mass flow rate is more significant for the scroll compressor, especially for higher condensing temperatures. For this reason, the scroll compressor presents a higher volumetric efficiency than the rolling piston at lower condensing temperatures, but it eventually becomes lower as the condensing temperature is increased.

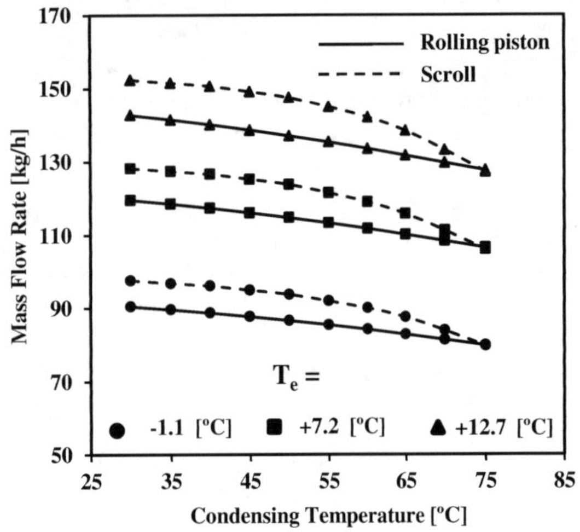

Figure 9: Mass flow rate for scroll and rolling piston compressors for different evaporating temperatures.

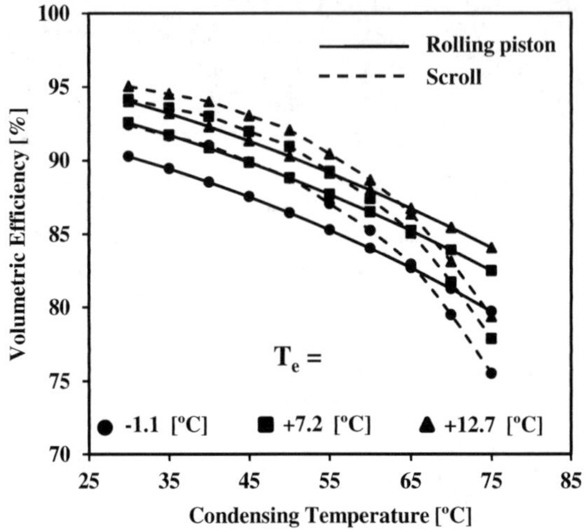

Figure 10: Volumetric efficiency of rolling piston and scroll compressor for different evaporating temperatures.

Opposite to what occurs for the energy losses, the mass losses inventories do not vary significantly with the evaporating temperature for the scroll and rolling piston compressors. For this reason, only the mass loss inventory for the evaporating temperature of T_e = +7.2 °C is shown for both compressors in Figures 11 and 12. It can be observed that the mass losses present a linear increase for the rolling piston, while a non-linear pattern is observable for the scroll compressor, which is directly related to the increase of leakage and superheating at higher condensing temperatures. This explains the decrease of volumetric efficiency observed in Figure 10 for the scroll compressor..

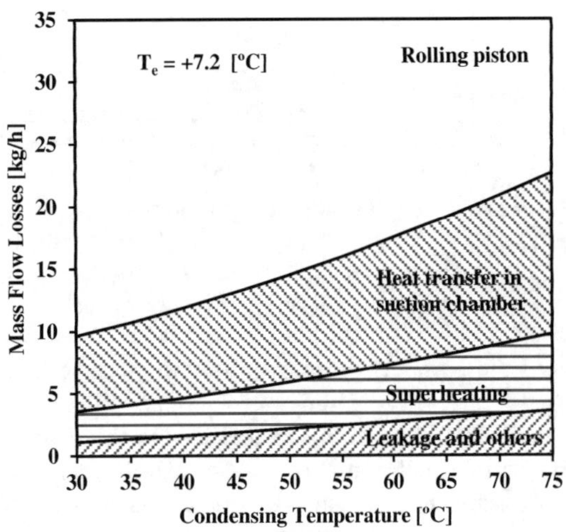

Figure 11: Mass losses for rolling piston at T_e = +7.2

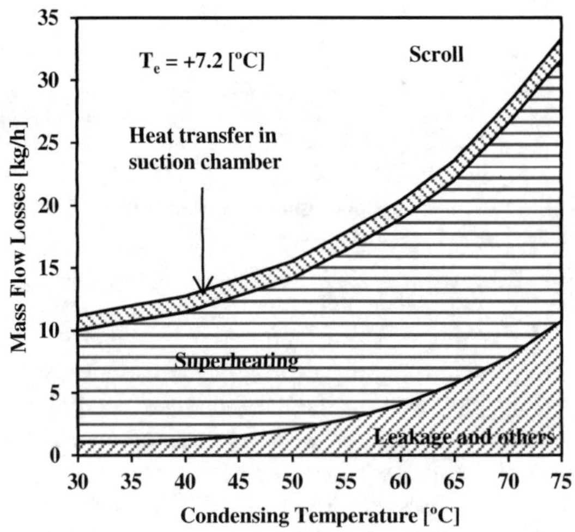

Figure 12: Mass losses for scroll at T_e = +7.2

6. CONCLUSIONS

This paper presented a comparative analysis of the thermodynamic efficiency of rolling piston and scroll compressors for air conditioning applications. To this extent, two compressors designed to operate at AHRI-A condition (T_e = 7.2 ºC and T_c = 54.4 ºC) were compared in different operating conditions that included MBP and HBP applications. Both compressors were initially experimentally analyzed and calculations of volumetric and global efficiency showed that the most efficiency compressor depends on the operating condition. Two validated simulation models were then adjusted and used to evaluate the thermodynamic and mass losses of each compressor. The analysis showed that the scroll compressor presents a better performance than the rolling piston compressor close to the design pressure ratio, which is related to the scroll geometry. Away from these operating points the scroll compressor shows a considerable decrease of efficiency, mainly due to under and over compression effects. It was also possible to identify that at higher evaporating temperatures the rolling piston compressors presents significant losses in the discharge process due to the higher mass flow rate. Finally, heat transfer was seen to significantly decrease the efficiency of both compressors.

NOMENCLATURE

General symbols

A	Area	[m^2]	T	Temperature	[ºC]
H	Heat transfer coefficient	[W/m^{2}K]	T_e	Evaporating temperature	[ºC]
$\dot{m}$	Mass	[kg]	T_c	Condensing temperature	[ºC]
$\dot{m}$	Mass flow rate	[kg/s]	$\forall$	Volume	[m^3]
h	Enthalpy	[J/kgK]	η_g	Electromechanical efficiency	[-]
p	Pressure	[Pa]	η_g	Global efficiency	[-]
$\dot{Q}$	Heat transfer rate	[W]	η_v	Volumetric efficiency	[-]

REFERENCES

Bezerra Jr, J. G., Deschamps, C. J., Silva, E. & Brancher, R. D. (2013). Thermodynamic performance of reciprocating and rolling piston compressors for heat pump application. *In: Proc. Int. Congress on Compressors*. Papiernicka, Slovak Republic.

Brancher, R. D. (2013). Development and validation of a model to predict the performance of rolling piston compressors (in Portuguese). *M.Sc. Thesis*, Federal University of Santa Catarina, Brazil.

Chua, K. J., Chou, S. K., Yang, W. M. & Yan, J. (2013). Achieving better energy-efficient air conditioning – A review of technologies and strategies. *Appl. Energy.*, 104, 87-104.

Diniz, M. C. & Deschamps, C. J. (2014). A NTU-based model to estimate suction superheating in scroll Compressors. *In: Proc. Int. Compressor Engineering Conference at Purdue University*, (paper 2325). West Lafayette, USA.

Dutra, T. & Deschamps, C. J. (2013). Experimental characterization of heat transfer in the components of a small hermetic reciprocating compressor. *Appl. Therm. Eng..*, 58, 499-510.

Gomes, A. R. & Deschamps, C. J. (2007). A thermodynamic analysis of scroll, rolling piston and reciprocating compressors for commercial refrigeration. *In: Proc. Int. Congress of Refrigeration*. Beijing, China.

Jang, K. & Jeong, S. (2006). Experimental investigation on convective heat transfer mechanism in a scroll compressor. *Int. J. Refrig.*, 29, 744-753.

Liu, Z. & Soedel, W. (1992). Modeling temperatures in high speed compressors for the purpose of gas pulsation and valve loss modeling. *In: Proc. Int. Compressor Engineering Conference at Purdue University*, (paper 932). West Lafayette, USA.

Maertens, M. J. & Richardson, H. (1992). Scroll compressor operating envelope considerations. *In: Proc. Int. Compressor Engineering Conference at Purdue University*, (paper 850). West Lafayette, USA.

Ooi, K. T. (2012). Compressor performance comparison when using R134 and R1234yf as working fluids. *In: Proc. Int. Compressor Engineering Conference at Purdue University*, (paper 2040). West Lafayette, USA.

Ooi, K. T. (2014). Twenty Years of Compressor Innovation at NTU, Singapore. *In: Proc. Int. Compressor Engineering Conference at Purdue University*, (paper 2384). West Lafayette, USA.

Orosz, J., Bradshaw, C. R., Kemp, G. & Groll, E. A. (2042). An update on the performance and operating characteristcs of a novel rotating spool compressor. *In: Proc. Int. Compressor Engineering Conference at Purdue University*, (paper 2327). West Lafayette, USA.

Padhy, S. K. & Dwivedi, S. N. (1994). Heat transfer analysis of a rolling-piston rotary compressor. *Int. J. Refrig.*, 17, 400-410

Pereira, E. L. L. (2012). Modeling and analysis of the thermodynamic performance of scroll compressors (in Portuguese). *Ph.D. Thesis*, Federal University of Santa Catarina, Brazil.

Pereira, E. L. L. & Deschamps, C. J. (2010). A lumped thermodynamic model for scroll compressors with special attention to the geometric characterization during the discharge process. *In: Proc. Int. Compressor Engineering Conference at Purdue University*, (paper 2007). West Lafayette, USA.

Pereira, E. L. L. & Deschamps, C. J. (2012). A numerical study of convective heat transfer in the compression chambers of scroll compressors. *In: Proc. Int. Compressor Engineering Conference at Purdue University*, (paper 2191). West Lafayette, USA.

Pereira, E. L. L. & Deschamps, C. J. (2015). Prediction of gas leakage through clearances in scroll compressors. *In: Proc. Int. Congress of Refrigeration*, (paper 2191). Yokohama, Japan.

ACKNOWLEDGEMENT

The present study was developed as part of a technical-scientific cooperation program between the Federal University of Santa Catarina and EMBRACO. The authors also acknowledge the financial support of EMBRAPII Unit POLO/UFSC and CAPES (Coordination for the Improvement of High Level Personnel).

Design Methodology Improvements of a Rotating Spool Compressor using a Comprehensive Model

Craig R. BRADSHAW[1*], Greg KEMP[1], Joe OROSZ[1], Eckhard A. GROLL[2]

[1] TORAD Engineering,
Cumming, GA, USA
ph. 770-889-8400, craig.r.bradshaw@gmail.com

[2] Ray W. Herrick Laboratories, School of Mechanical Engineering at Purdue University,
West Lafayette, IN, USA
groll@purdue.edu

* Corresponding Author

ABSTRACT

An improvement to the design process of the rotating spool compressor is presented. This improvement utilizes a comprehensive model to explore two working fluids (R410A and R134a), various displaced volumes, at a variety of geometric parameters. The geometric parameters explored consists of eccentricity ratio and length-to-diameter ratio. The eccentricity ratio is varied between 0.81 and 0.92 and the length-to-diameter ratio is varied between 0.4 and 3. The key tradeoffs are evaluated and the results show that there is an optimum eccentricity and length-to-diameter ratio, which will maximize the model predicted performance, that is unique to a particular fluid and displaced volume. For R410A the modeling tool predicts that the overall isentropic efficiency will optimize at a length-to-diameter ratio that is lower than for R134a. Additionally, the tool predicts that as the displaced volume increases the overall isentropic efficiency will increase and the ideal length-to-diameter ratio will shift. The result from this study are utilized to develop a basic design for a 141 kW (40 tonsR) capacity prototype spool compressor for light-commercial air-conditioning applications.

1. INTRODUCTION

The rotating spool compressor is a novel rotary compressor mechanism most similar to the sliding vane compressor. Primary differences are described by Kemp et al. (2008) and include three key differences from a sliding vane compressor, as shown in Figure 1.

- The vane is constrained by means of an eccentric cam allowing its distal end to be held in very close proximity to the housing bore while never contacting the bore.
- The rotor has affixed endplates that rotate with the central hub and vane forming a rotating spool.
- The use of dynamic sealing elements to minimize leakage between the suction and compression pockets as well as between the process pockets and the compressor containment.

The movement of the rotor is purely rotary with only the vane and tip seals performing any oscillating movement. The eccentric cam will force the movement of the vane to oscillate by twice the eccentricity during a single rotation. The tip seals will oscillate relative to the vane two times per rotation by an amount proportional to the ratio of diameters of the rotor to the housing bore. The tip seal movement amount is roughly an order of magnitude smaller than the eccentricity and follows a sinusoidal path. Analytical details regarding the geometry, including the mathematical expressions describing the chamber volumes, is presented in Bradshaw and Groll (2013).

1.1 Previous Work on the Spool Compressor

Since its introduction in 2008 the spool compressor has been improved and better understood. A comprehensive model was developed to deepen the understanding of the device holistically, as a system (Bradshaw and Groll, 2013). This

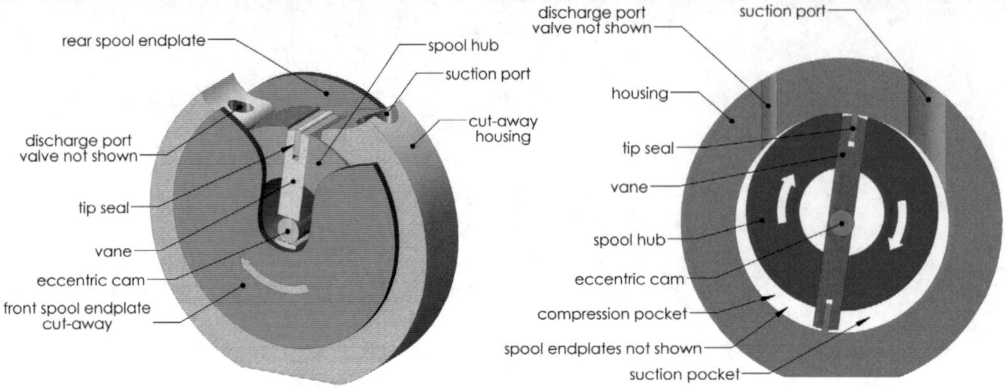

Figure 1: Cutaway view of rotating spool compressor mechanism with key components highlighted.

holistic approach has allowed for better prediction of non-intuitive behavior of the spool compressor especially applied to applications and working fluids without experimental experience. Figure 2 shows a schematic of mass and heat flows represented by the comprehensive model. The model includes 10 leak paths, 8 heat flow paths, port geometries, as well as oil solubility and frictional component sub-models.

Additionally, sub-components of the spool compressor have been studied in detail. Bradshaw (2013) presented a study on the analytical model developed to represent the tip seal in the spool compressor. The model was developed based on the tip seal dynamic model coupled with hydrodynamic lubrication theory. Kemp et al. (2012) presented iterations on the spool seal design and described the critical nature of the design of this component.

Bradshaw et al. (2016) recently presented the development of a loss analysis to describe the 5th generation spool compressor. This analysis utilized high-speed pressure measurements to determine the indicated, or flow, losses within the device. Coupled with analysis of the frictional losses an complete loss Pareto was generated. This analysis has resulted in a deep understanding of the location of potential improvements as well as an improved comprehensive modeling tool which accounts, more accurately, for the various frictional losses within the device. The resulting improvements using this analysis were also presented which showed a significant improvement in performance between the 5th and 6th generation compressor. These results echo the improved performance presented by Orosz et al. (2014).

This work will utilize the comprehensive model from Bradshaw and Groll (2013), and its improvements, to study the spool compressor operating on R134a with a variety of displacements to determine an appropriate size and shape for a R134a air-conditioning compressor. Additionally, a comparison of designs for R410A and R134a is presented which highlights the utility of the comprehensive design tool for various applications. This is accomplished by examining how the efficiencies of the compressor change with changes in the fundamental geometric parameters which define the shape of the spool compressor; the eccentricity ratio and the length-to-diameter ratio.

2. PARAMETRIC ANALYSIS

The basis of a spool compressor design starts with the selection of two quantities, the eccentricity ratio (ε), and the length to diameter ratio($^L/_D$). The eccentricity ratio is defined as,

$$\varepsilon = \frac{R_r}{R_s} \tag{1}$$

while the length to diameter ratio is defined as,

$$\frac{L}{D} = \frac{h_{stator}}{2R_s}. \tag{2}$$

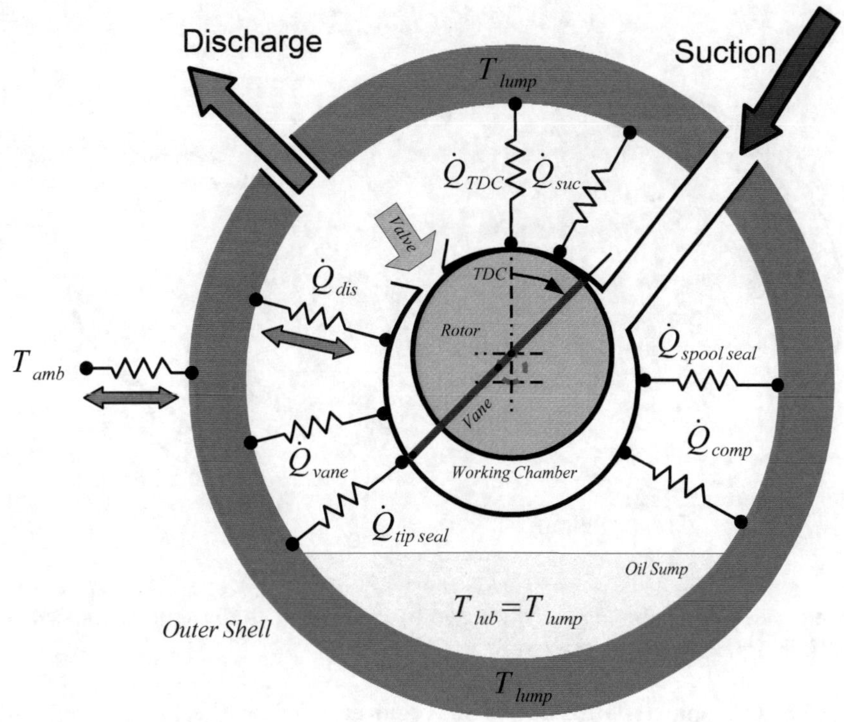

Figure 2: Schematic showing the mass and heat flows in the comprehensive spool compressor model as explained in Bradshaw and Groll (2013).

Figure 3 shows the basic dimensions of the spool compressor and two examples of the basic spool compressor geometry which have the same displaced volume. For a fixed displaced volume, the basic shape of the spool compressor can change depending on what values of eccentricity ratio and length to diameter ratio are selected, as shown in Figure 3.

2.1 Compressor Displacement

The eccentricity and length-to-diameter ratio are not sufficient to define a basic compressor shape, a displaced volume is also required. The displacement is a function of the working fluid and cooling capacity required at a particular operating condition. This study will compare the model predicted efficiencies of a fixed displaced volume on two working fluids, R410A and R134a, as well as 5 displaced volumes on R134a to study the influence of the displaced volume on the compressor efficiency.

For all studies the operating conditions will be fixed at an evaporating temperature of 40F (4.4C) and a condensing temperature of 100F (38C). Table 1 provides the calculated pressure ratios and calculated displaced volumes for the proposed compressors assuming a 90% volumetric efficiency. This figure represents the volume required to process the required volumetric flow rate of working fluid to generate the desired cooling capacity. This assumes an ideal refrigeration cycle with 10R (5.6K) of subcooling and 15R (8.3K) of compressor superheat. The volumetric efficiency and speed are fixed and used as an estimate though it is acknowledged that these values will vary based on operating condition and compressor geometry.

2.2 Manufacturing Considerations

One of the key leakage paths in the spool compressor is the Top Dead Center (TDC) gap. This is the gap formed by the small gap between the rotor and cylinder block. To improve leakage, and reduce manufacturing variability, the rotor is inset into the cylinder. This inset requires that a portion of the cylinder is manufactured with a discontinuous profile. The portion of the profile adjacent to the rotor is cut with a radius which is equal to the rotor. This is highlighted in Figure 4 which shows how the rotor is inset into the cylinder to increase the leakage path length of the TDC region.

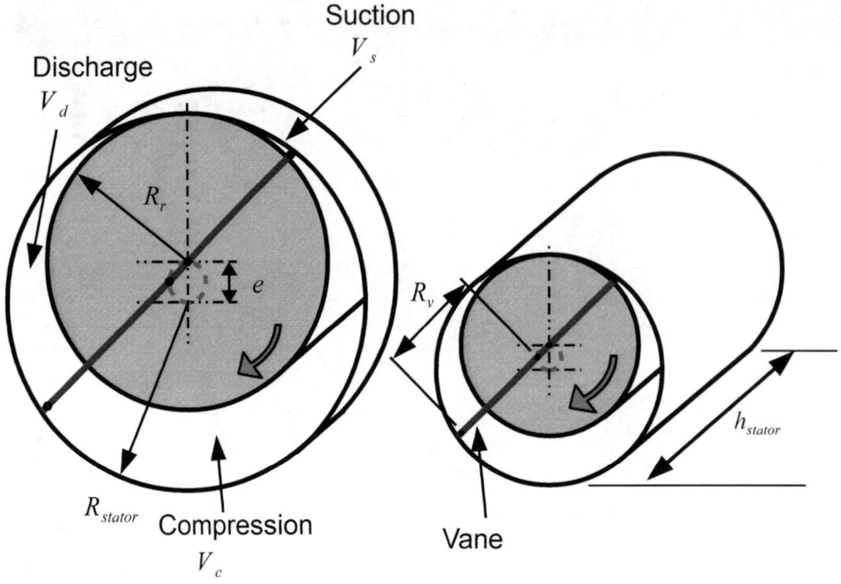

Figure 3: Spool compressor basic dimensions shown on two designs with the same displaced volume and eccentricity ratio with different length to diameter ratio.

Table 1: Table of calculated compressor displacements required to achieve various cooling loads used for study.

Fluid	**PR**	$\dot{Q}_{cool}$	η_{vol}	**Speed**	$V_{disp,tot}$	V_{disp}
-	-	**tonsR (kW)**	-	**rpm**	cm^3/rev	$cm^3/pocket$
R410A	2.49	5 (17.6)	0.9	1750	112	56
R134a	2.79	5 (17.6)	0.9	1750	254	127
R134a	2.79	20 (70.3)	0.9	1750	1014	507
R134a	2.79	40 (141)	0.9	1750	2029	1014
R134a	2.79	80 (281)	0.9	1750	4057	2029

The remaining portion of the cylinder is cut at a different radius. The section of the cylinder with the same radius as the rotor creates a region where the two components come in close contact. However, it elongates the leak path so it can be assumed the leak path is closer to two long plates in close contact opposed to a tangential line contact without this feature.

While this inset improves the leakage performance for a given compressor it is subject to the ability to manufacture the parts. This manufacturing ability changes with the eccentricity and length-to-diameter ratio selected. The length-to-diameter ratio has the most significant influence on this gap. More specifically the stator height, h_{stator}, as the stator radius has minimal influence. As the stator height increases the ability to maintain the squareness of the rotor, relative to its centerline, and the total indicated runout become more challenging. Similarly, when boring the cylinder block the ability to maintain the squareness of the TDC scallop relative to its centerline becomes more challenging as the stator height increases. The result is, for a given displacement, the TDC leakage gap will increase as the length-to-diameter increases. The amount of the deviation varies with manufacturing process used and thus is also a function of the volume of parts produced. Assuming the rotor is turned and ground and the cylinder is bored it is estimated that the TDC leakage gap will vary based on the following expression:

$$g_{TDC} = a + bh_{stator} \tag{3}$$

where a is $0.0005in$ ($12.7\mu m$) and b is $0.0001in/in$ ($1\mu m/cm$) and h_{stator} is in inches (cm).

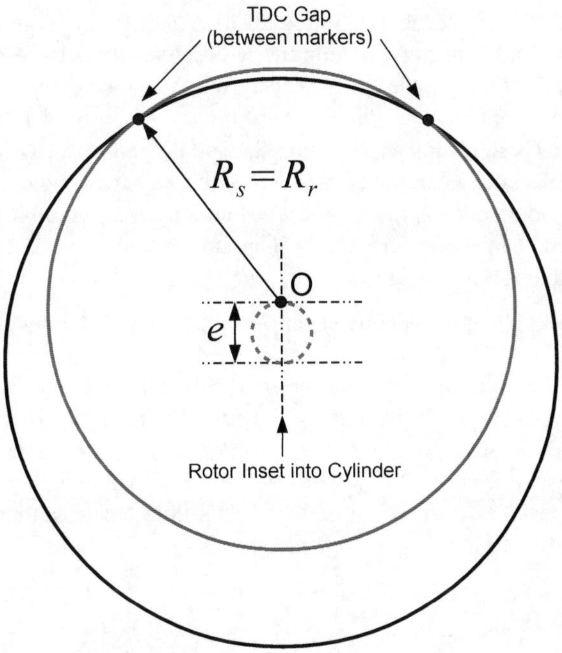

Figure 4: A rotor and cylinder schematic of the spool compressor highlighting the TDC region and the inset feature of the rotor into the cylinder.

3. COMPARING R410A TO R134a

Using the displaced volumes presented above in Table 1 for the 5 ton (17.3 kW) compressors a comparison between an optimum design on R410A and R134a is presented. For each displacement the model is exercised for ε values from 0.81 to 0.92 and $^L/_D$ values from 0.4 to 3 with fixed operating conditions described in Section 2.1. Figure 5 shows the overall isentropic efficiency predicted by the model for the 5 ton (17.3 kW) displacements on both R410A and R134a for all eccentricity ratios and length-to-diameter ratios.

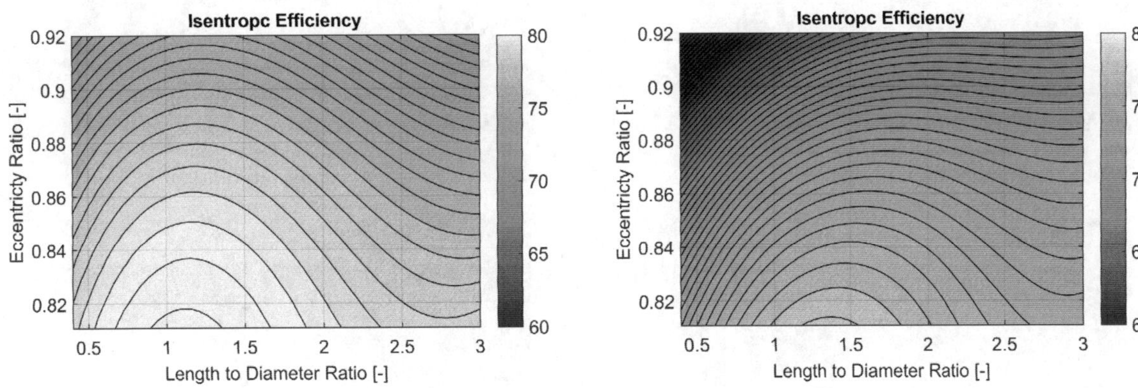

Figure 5: Overall Isentropic Efficiency of 5 ton (17.3 kW) compressor on R410A (left) and R134a (right) for various eccentricity and length-to-diameter ratios.

The overall isentropic efficiency is significantly higher for the R410A simulation at its peak (~5% points). However, it should be noted that the peak efficiency locations are at different length-to-diameter locations. For R410A the peak efficiency is predicted at length-to-diameter ratio of about 1.05 whereas the R134a peak is predicted to be at roughly 1.4 length-to-diameter ratio. Both fluids are predicted to have the highest efficiency at the lowest eccentricity possible (0.81).

Looking at the figures from the lowest eccentricity ratios to the highest there is generally a decrease in efficiency. The eccentricity ratio has minimal influence on the volumetric efficiency (i.e. leakage) so the decrease in efficiency is a result in changes of frictional losses. At the lowest eccentricity ratio the cylinder diameter (and thus spool seal) diameter is small but the movement of the vane is high. At the highest eccentricity ratio the cylinder diameter is large but the vane movement is low. These results suggest that the spool compressor is more sensitive to the spool seal friction compared with the vane friction. Therefore, based on the results here it suggests for this displacement it would be advantageous to design a compressor with the smallest eccentricity ratio possible. The R134a has a much larger range of efficiencies across a change in eccentricity ratios. This suggest the R134a design is more sensitive to changes in frictional loads compared with a R410A design.

Examining each figure from lowest length-to-diameter ratio to the highest the tradeoff between friction and leakage can be seen. At the lowest length-to-diameter ratios the cylinder diameter (and thus spool seal diameters) are forced to increase non-linearly. This creates a lot of friction per unit displacement and thus to minimize seal friction the suggestion would be to increase the length-to-diameter ratio as high as possible. However, as the length-to-diameter ratio is increased the influence of the TDC manufacturing considerations become apparent and the increase in leakage begins to overwhelm the decrease in friction. This causes a peak value within the ranges explored which is different for each fluid. For R410A the peak being lower suggests that this fluid creates an increase in sensitivity to leakage. Conversely, R134a is less sensitive to the length-to-diameter leakage which pushes the peak to a higher length-to-diameter peak value.

4. SCALING ANALYSIS

Extending the previous analysis for various displacements on R134a is presented. The displacements were previously provided in Table 1 and operating conditions defined in Section 2.1. The objective is to further understanding of the tradeoffs in design when scaling to various sizes of compressor. Figure 6 shows the results of the study of the four displaced volumes.

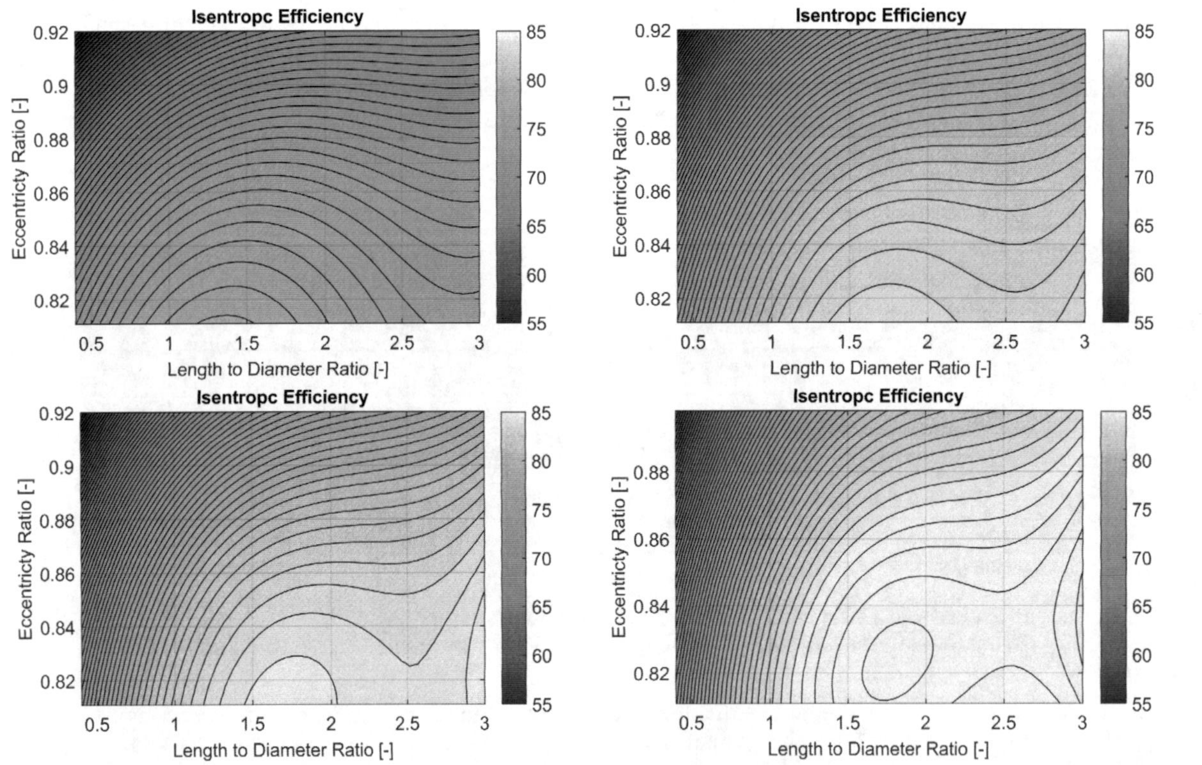

Figure 6: Overall Isentropic Efficiency of R134a compressors at 5 ton (17.3 kW), 20 ton (70.3 kW), 40 ton (141 kW), and 80 ton (281 kW) at various eccentricity and length-to-diameter ratios.

The trends are similar to the trends presented in the previous section for any given displacement. However, as the displaced volume increases the overall isentropic efficiency increases and the optimum location changes. The overall isentropic efficiency increases from a peak of roughly 77% at 5 tons (17.3 kW) to a peak of 86% at 80 tons (281 kW). This is a result of processing dis-proportionally more volume per unit frictional load as the compressor increases in displacement. While this is typical of most compressor technologies, the amount of improvement varies. The optimum location shifts from a length-to-diameter ratio of roughly 1.4 to 1.85 as the displaced volume increases. This is a result of a reduction in the sensitivity to frictional components at this displacement compared with the smaller displacements. However, as the displacement increases the peak eccentricity ratio increases, suggesting the vane friction becomes more important as the displacement increases. The peak overall isentropic efficiency and accompanying length-to-diameter ratio and eccentricity ratio are tabulated in Table 2.

Table 2: Final estimated peak efficiencies and key geometric parameters.

$\dot{Q}_{cool}$	$^{L}/_{D}$	ε	$\eta_{o,is}$
tonsR (kW)	**-**	**-**	**%**
5 (17.6)	1.4	0.81	77%
20 (70.3)	1.75	0.81	83%
40 (141)	1.82	0.815	85%
80 (281)	1.85	0.825	86%

5. DESIGN OF R134A PROTOTYPE COMPRESSOR FOR LIGHT-COMMERCIAL AIR-CONDITIONING

Using the analysis from the previous section a design for a light-commercial air-conditioning compressor is presented. The 40 ton (141 kW) size was selected as its overall isentropic efficiency is predicted to be reasonably high and the displaced volume represents a displaced volume that is difficult to find efficient products in the market. The typical alternatives, screw compressors are generally not as efficient at this displaced volume and scroll compressors are not either (Kreith, 2000).

From the data above the most efficient design for a 40 ton (141 kW) compressor is provided in Table 3.

Table 3: Final preliminary design parameters for light-commercial compressor prototype.

Parameter	Units	Value
ε	-	0.811
$^{L}/_{D}$	-	1.82
R_r	in (cm)	2.21 (5.60)
R_s	in (cm)	2.72 (6.9)
h_{stator}	in (cm)	9.890 (25.1)
$\eta_{o,is}$	%	84.6
η_{vol}	%	92.6
Discharge Temp	F (C)	128 (53.3)
Massflow	lbm/min (kg/min)	118 (53.5)
Shaft Power	kW	23.5
% Carryover Vol.	%	1.76

6. CONCLUSIONS

A design methodology utilizing a comprehensive compressor model is presented which culminates in a preliminary design of a 40 ton (141 kW) air-conditioning compressor. The study examined the utility of the approach when applied to different refrigerants as well as examining various displaced volumes. It was found that the tool provided great insight into the appropriate basic dimensions of a compressor that would have otherwise been difficult to obtain through trial and error.

NOMENCLATURE

g_{TDC}	Gap between rotor and cylinder at TDC region	(m)
h_{stator}	Cylinder height	(m)
$^{L}/_{D}$	Length-to-diameter ratio	(-)
PR	Pressure Ratio	(-)
$\dot{Q}_{cool}$	Cooling capacity	(kW)
$\dot{Q}_{dis}$	Discharge process heat transfer	(W)
$\dot{Q}_{vane}$	Frictional loss from vane	(W)
$\dot{Q}_{tipseal}$	Frictional loss from tip seal	(W)
$\dot{Q}_{comp}$	Compression process heat transfer	(W)
$\dot{Q}_{spoolseal}$	Frictional loss from spool seal	(W)
$\dot{Q}_{suc}$	Suction process heat transfer	(W)
$\dot{Q}_{TDC}$	Frictional loss from TDC	(W)
R_r	Rotor radius	(m)
R_s	Cylinder radius	(m)
T_{amb}	Ambient temperature	(K)
T_{lub}	Lubricant temperature	(K)
T_{lump}	Compressor temperature	(K)
$V_{disp,tot}$	Total compressor displaced volume for single shaft revolution	(cm^3/rev)
V_{disp}	Compressor displaced volume for single pocket	$(cm^3/pocket)$

Greek

ε	Eccentricity ratio	(-)
η_{vol}	Volumetric efficiency	(-)
$\eta_{o,is}$	Overall isentropic efficiency	(-)

REFERENCES

Bradshaw, C. (2013). Spool compressor tip seal design considerations. In *8th International Conference on Compressors and their Systems*, page 341.

Bradshaw, C. R. and Groll, E. A. (2013). A comprehensive model of a novel rotating spool compressor. *International Journal of Refrigeration*, 36(7):1974 – 1981. New Developments in Compressor Technology.

Bradshaw, C. R., Kemp, G., Orosz, J., and Groll, E. A. (2016). Development of a loss pareto for a rotating spool compressor using high-speed pressure measurements and friction analysis. *Applied Thermal Engineering*, 99:392–401.

Kemp, G., Orosz, J., Bradshaw, C., and Groll, E. A. (2012). Spool seal design and testing for the spool compressor. In *International Compressor Engineering Conference at Purdue University*, number 1259.

Kemp, G. T., Garrett, N., and Groll, E. A. (2008). Novel rotary spool compressor design and preliminary prototype performance. In *International Compressor Engineering Conference at Purdue University*, number 1328.

Kreith, F. (2000). *The CRC handbook of thermal engineering.* Springer Science & Business Media.

Orosz, J., Kemp, G., Bradshaw, C. R., and Groll, E. A. (2014). An update on the performance and operating characteristics of a novel rotating spool compressor. In *International Compressor Engineering Conference at Purdue University*, number 1378.

Systematic calculation of the piston rod unit

Vasillaq Kacani

Leobersdorfer Maschinenfabrik GmbH
Leobersdorf, Low Austria, Austria
Vasillaq.kacani@LMF.at

ABSTRACT

In this article a modern and economic method for the strength calculation of the piston rod unit and its components under different operating conditions will be presented. Therefor the commercial FEA - Software will be linked with the company-owned calculation tools. The parametric user input will be followed by an automatic Pre- and Postprocessing. Afterwards the strength calculation is processed on all critical points of the piston rod connection, assisted by an extra module, based on general standards and special codes for reciprocating compressors. In this process most arrangements of the piston rod unit as well as the special geometries of the single-components (piston, piston rod and piston nut) can be considered easily. In this article the modeling of the notches, especially on the piston rod, piston as well as the piston nut will be covered in detail.

1. INTRODUCTION

The paper investigates the parametric design and calculation of all components of piston rod unit considering the contact properties of all clamped parts and different piston design. The piston rod unit generally includes piston, piston nut, ring and piston rod. On crankcase side the piston rod is connected to the crosshead.
These components are under extremely high loads during the operation of the reciprocating compressor. The forces acting on the components are the pre-load force, the gas force, the thermal related forces as well as the mass force. During one revolution of the crankshaft the gas and mass forces are changing their value and the direction. Additionally for each crankshaft angle the forces are different on different sections of the piston rod unit. That means the strength calculation must be performed for each angular step with the actual pressure and acceleration at the local section and for each notch on all components of the piston rod unit.
Furthermore the boundary conditions such as geometry, embedding of the contact surfaces, materials, strength grade, tight techniques as well as different operating cases of the reciprocating compressor are into consideration.
In this paper a systematic design and calculation of the piston rod unit will be presented. This consists of three main steps: structure modelling, (preprocessing), definition of boundary condition, meshing, loads apply, calculation for each crank angle and, post-processing (results). All steps are realized automatically using manufacturer calculation/design tools, commercial programming software VB6, C++, FEA-Software[3], Scripting Language APDL[3] (ANSYS Parametric Design Language) as well as guideline and standards.
The strength verification will be performed according the different guidelines[1,2,4,5].
A calculation software was developed to perform an automatic strength analysis of the components of the piston rod unit. The special features of the software are:

- Reduction of the simulation time
- Increasing the quality of calculation

- Calculation of stress curves at each notch along the notch angle α
- Systematic calculation/design and fault prevention through certified software.

.

2. PRE-PROCESSING

The software consists of several calculation modules and their interfaces between the components. The compressor data (like type, frame size, piston rod design, crosshead, connection rod etc) material-data of the main components (modulus of elasticity, ultimate tensile strength, yield strength, Poisson ratio, fatigue stress values, thermal expansion coefficient etc) can be generated automatically and internally from machine- and material databank.
The basic module is the manufacturer program for the design/calculation of the reciprocating compressors. The interface enables every user after the selection and design of a specific compressor, to use the software for the piston unit calculation.
The main data/parameters are a result of the previous compressor selection:

- Stroke, type of piston (one piece, two piece, welded piston etc), compressor speed..
- Actual pressure variation over the crankshaft angle, in each cylinder compartments (HE, CE, idle or active compartment, SV-Unloader .).
- Piston rod parameters: size, neck diameter, thread diameter, surface roughness, heat treatment..
- Piston nut data..

Data can be input/edited interactively by the user. Fig. 1 shows a screen shot of the user input form with the sketch of the two pieces piston and various input parameters.

- Notch geometry, radius, element type, element size or number of elements. Distance Δs normal to the notch surface for the calculation of the stress gradient required for the fatigue analysis.

- Geometry of contact area between piston, piston rod and piston nut. Type and size of contact elements. Friction coefficient.
- Piston geometry (2D, 3D-modell)
- Welded piston with ribs
- Temperature distribution.
- The assembly preload force of the unit

Hence the geometry and the loads of the piston rod unit are determined. The data are written to a temporary file. The next steps are the generating the FE-model, the definition of boundary condition,
applying of the loads, the calculation, as well as the stress evaluation. All this steps are effected automatically. A special interface for the FEA-simulation in the script language (APDL) was generated.

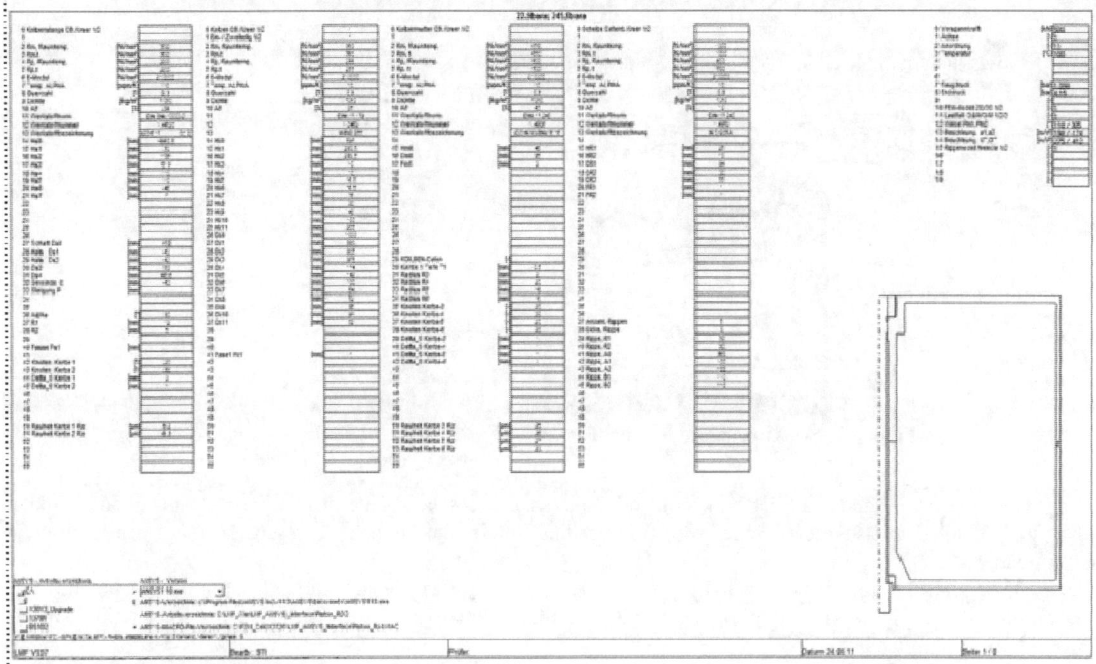

Figure 1. User Input Form

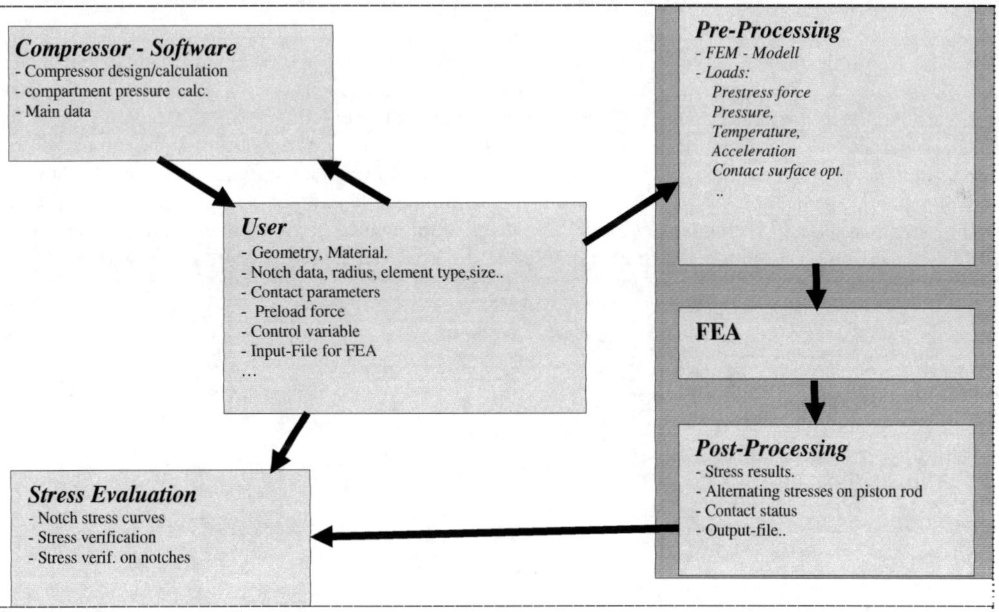

Figure 2. Work process flowchart

3. CRITICAL STRESS ZONES AND CONTACT AREAS

Figure 3 shows the critical high stress zones 1 to 9 and the contact areas I, II, III,…VII.

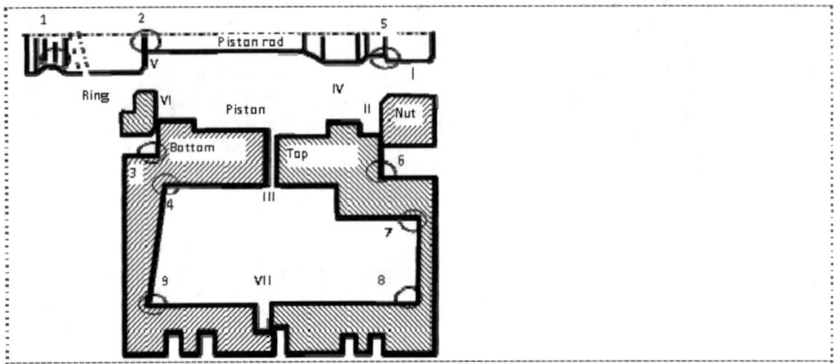

Figure 3. Stress zones and contact areas.

The critical notches 1, 2 and 5 are in the piston rod. Notch 1 on crosshead side of piston rod is a conical shoulder with radius. Notch 2 is a square shoulder with fillet (radius) in circular shaft. 5 represents a V-notch in a circular shaft (thread).

Notches 3, 4, 9 are located in bottom half of the piston 6, 7, 8 are in the top half of piston. These notches are representing changes on cross-section of piston with fillet of radius r. For welded piston the fillet radii shall be assumed to be 1mm (Radaj[5]).

4. LOADS AAPPLIED TO THE STRUCTURE

The first load step is the preload force on the piston rod due to tightening of the piston nut. This load can be applied in ANSYS[3] with pretension elements. This is the basic load. All other loads are superimposed on the basic load. When specifying the minimum axial assembly forces the loss of the preload force[1] due to the plastic deformation of the contact surfaces must be taken into consideration. In API 618 the recommendation for the minimum preload is 1.5 times the maximum allowable continuous piston rod load. The preload force must be checked for each application under the actual operating loads: pressure, acceleration and temperature. Often the required preload force is much higher than the API recommendation. In The status of all contact areas, pressure and gap must be checked carefully for all operation condition. The prestress force has to prevent lifting or slide of all contact surfaces.

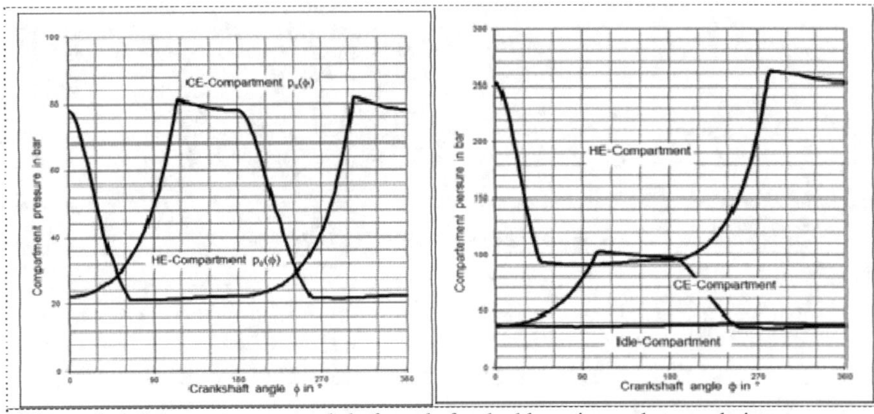

Figure 3, 4. Pressure versus crankshaft angle for double acting and stepped piston

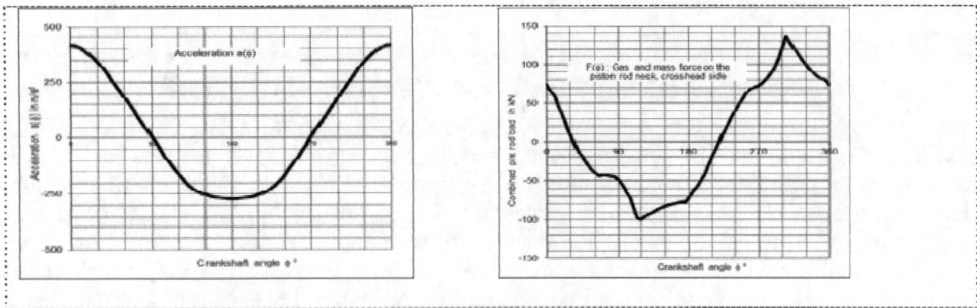

Figure 5, 6. Acceleration and combined rod load at the piston rod neck

Figure 3 and 4 shows the typical pressure curves $p_o(\phi)$, $p_u(\phi)$ on CE-, HE- and idle-compartment for double acting and for stepped piston. These loads are applied on the corresponding areas of the structures (Fig. 7). Figure 5 represents the acceleration and Figure 6 shows the resulting combined rod load at the piston rod neck crosshead side.

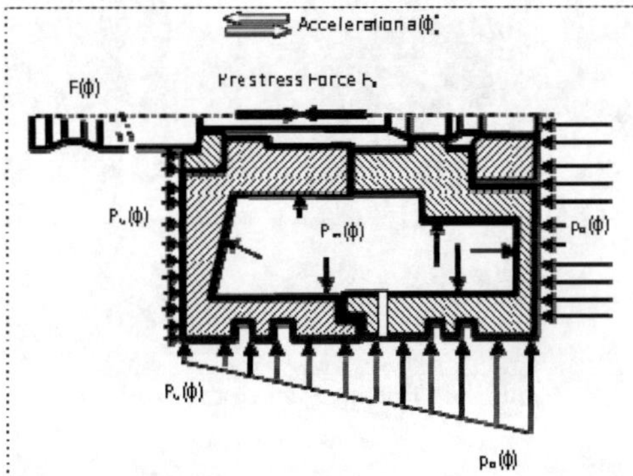

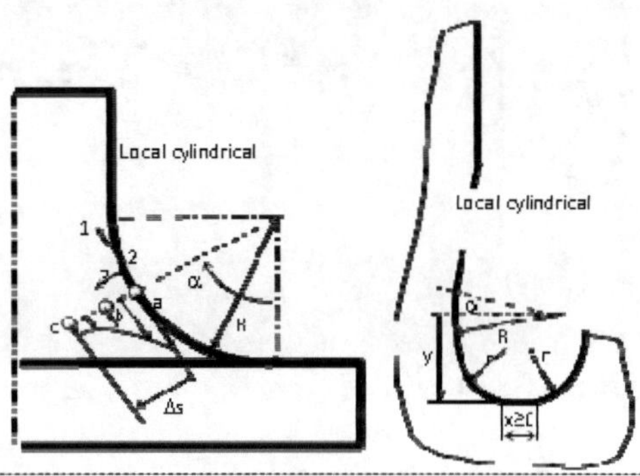

Figure 7. Loads on the piston rod unit: Preload Fv, pressure $p_u(\phi)$, $p_o(\phi)$ and acceleration $a(\phi)$

Figure 8. Typical notch shape and required parameters

5. NOTCH GEOMETRY, PARAMETERS AND STRESS EVALUATION

The main parameters of notches are (Fig.8):

- Shape of notch, local coordinate systems
- Geometry parameters: R, r, α, x, y
- Depth in vertical direction Δs
- Number of elements in circumferential direction 1
- Number of elements in radial direction 3

The principal stresses in the local coordinate system are calculated in three directions 1, 2, and 3 along the path (angle α) and for each step of crankshaft ($\Delta\phi$ acc. [1] is 5°). The direction 1 and 2 are always parallel to the free surface. The third direction is perpendicular to free surface of the component.

The stresses are calculated on three points: position a on the free surface $\sigma_a(\alpha,\phi)=(\sigma_{1a},\sigma_{2a},\sigma_{3a})$, position b in a depth of $\Delta s/2$ $\sigma_b(\alpha,\phi)=(\sigma_{1b},\sigma_{2b},\sigma_{3b})$, as well as position c in a depth of Δs $\sigma_c(\alpha,\phi)=(\sigma_{1c},\sigma_{2c},\sigma_{3c})$, (Fig.8). The stress value for σ_1 in perpendicular direction (Δs) to the free surface can be approximated as a polynom (second order): $\sigma_1=A_0 + A_1.s + A_2.\ S^2$. The coefficients A_0, A_1 and A_2 can be calculated easy from the three stress values (position a,b,c) established by the FEA-Analysis. Finally the stress gradient can be determined, as required in [2,5], for the strength verification. The formulas are listed in appendix. The equation for the gradient is (1):

$$\frac{d\sigma_1}{ds} = A_1 + 2 \cdot A_2 \cdot s \qquad (1)$$

6. MESHING STRATEGY OF THE STRUCTURE

The meshing of the structure must be practice-oriented and the model should have a reasonable size. On the one hand the meshing of the components should be as coarse as possible to reduce the simulation time and on the other hand as fine as necessary with reference to the engineering tasks. For notches (Fig. 10) and the contact areas a fine controlled meshing is used to get the correct information for the required stresses, contact status as well as the contact pressure. For other areas of the structure a free coarse mesh can be used. The element type and size can be adjusted by user. The following figure shows the different piston design (Fig.9): stepped piston 2D-model, two pieces double acting piston 2D and 3D- model, double acting welded piston with ribs 3D-model. All these models can be calculated automatically with the computer program.

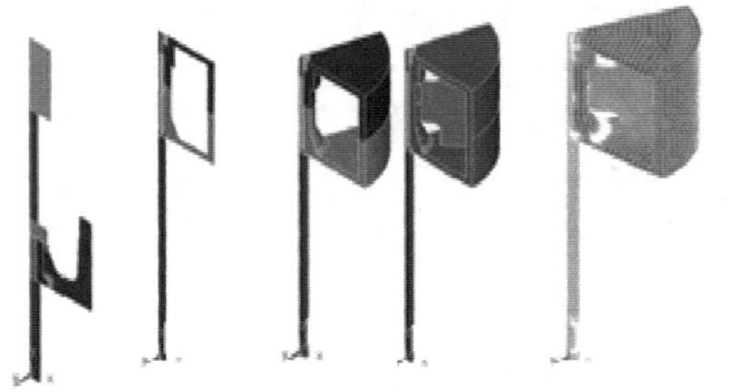

Figure 9. Piston designs

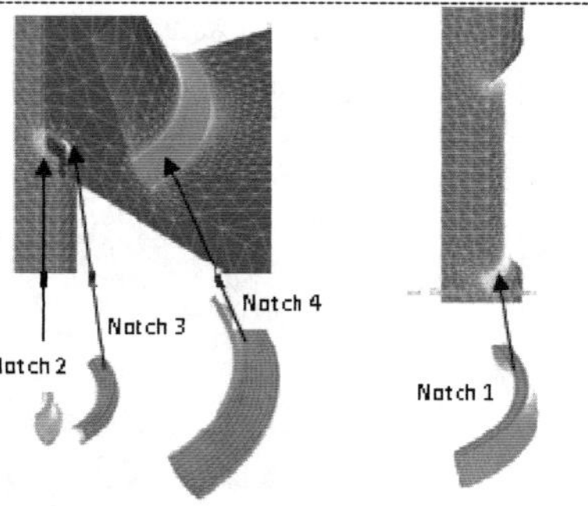

Figure 10 . Controlled fine and coarse mesh.

7. FINITE ELEMENT ANALYSIS AND RESULTS

The first simulation is the calculation of the pre-stressed structure under the preload force F_v. The next calculations are successively under operating condition: pressure, temperature as well as acceleration for each crankshaft angle. The main results of the Finite Element Analysis are the forces / stresses at the piston rod neck, thread (piston side and crosshead side) and the stresses on all notches.

The strength verification of pre-stressed connection piston rod, ring, piston and piston nut is performed according [1,4] guideline. The reduction of preload force of bolted joint due to the deformation of the contact surfaces is taken into consideration. From the FEA the actual preload force, the amplitude of the alternating force/stresses, maximum forces/stresses in the piston rod neck /thread as well as the resilience of the superimposed clamped parts and piston rod can be determined. From these values the safety factors are calculated. Furtherore the status and the pressure on all contact areas is checked.

The next step is the stress evaluation on the notches. From the FEA results the distribution (stress curves) of stresses on the free surface, on the depth $\Delta s/2$ as well as on the depth Δs along the angle α for each notch and crankshaft angle ϕ are calculated. Further the required alternating stress amplitude σ_a, average stress σ_m as well the stress gradient $d\sigma/ds$ for the fatigue strength verification according [2] are calculated.

Figure 11 to 14 shows the stress tangential to the free surface (direction 1) along the angle α in Notch 1, 2, 3 and 4 for two piston positions: Bottom-Dead-Center BDC and Top-Dead-Center TDC.

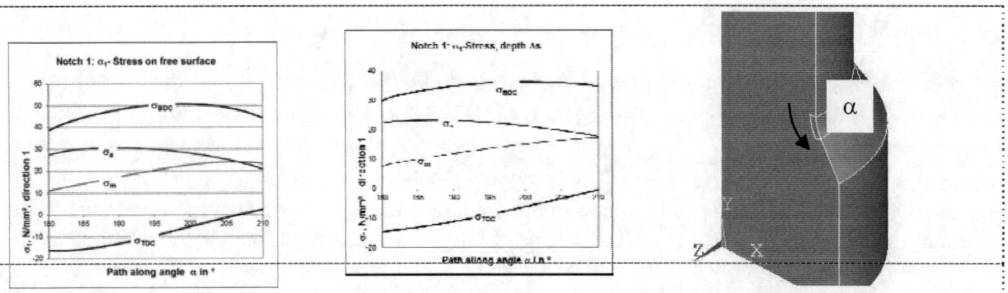

Figure 11. Notch 1, Tangential stress curves at BDC, TDC, amplitude and average stress on free surface and in depth Δs.

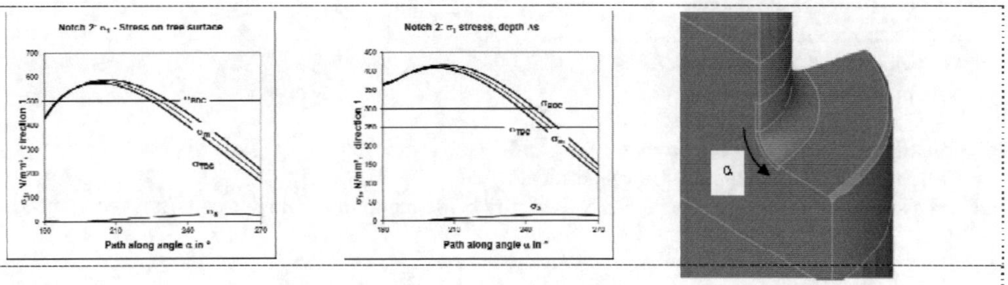

Figure 12. Notch 2, Tangential stress curves at BDC, TDC, amplitude and average stress on free surface and in depth Δs.

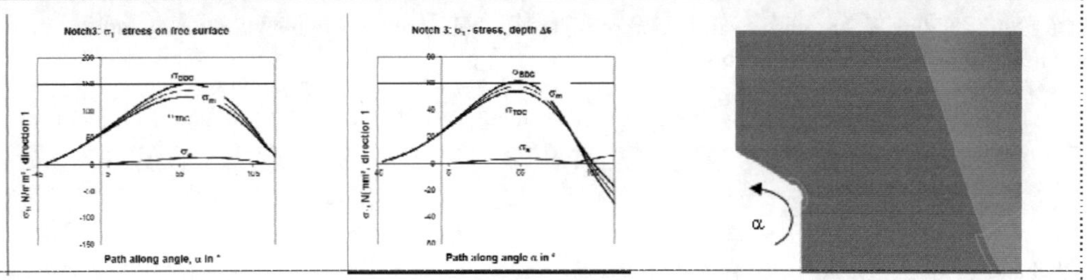

Figure 13. Notch 3, Tangential stress curves at BDC, TDC, amplitude and average stress on free surface and in depth Δs.

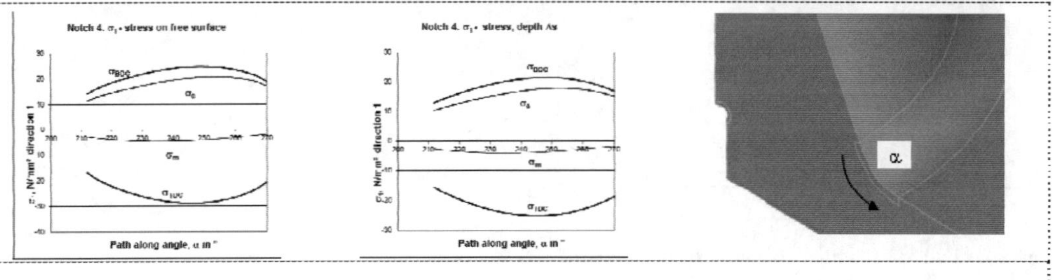

Figure 14. Notch 4, Tangential stress curves at BDC, TDC, amplitude and average stress on free surface and in depth Δs.

8. CONCLUSION

The components of the piston rod unit are under very high dynamic loads due to: pressure, temperature and acceleration. Depending on the operation mode of the reciprocating compressor, normal operation, SV-Unloaders, VSD, variable operation conditions the load profile of piston rod unit changes. The individual single calculation for each load step requires an enormous effort. The traceability and documentation of analysis would be also a big effort. The method presents a so-called expert system connects different tested and proved modules like company tools, standards, company owned software as well as commercial software together. The calculation is carried out automatically. The main steps of the calculation are:

- Compressor calculation
- User input data
- Modelling of the structure
- Finite Element Analysis
- Verification of the preload force
- Status of the contact surfaces
- Stress verification on the relevant critical notches.

Instead of the suggested factor 1.5 in API 618 5th ed., para.6.10.1, a minimum factor of 1.75 is recommended.

REFERENCES

[1] VDI 2230 Richtlinien. Systematic calculation of high duty bolted joints. Joints with one cylindrical bolt. Verein Deutscher Ingenieure, Düsseldorf, Februar 2003

[2] Rechnerischer Festigkeitsnachweis für Maschinenbauteile aus Stahl, Eisenguss- und Aluminiumwerkstoffe. FKM-Richtlinie, VDMA Verlag 5. Ausgabe 2003

[3] ANSYS , Simulation Software.

[4] API Standard 618 5th Edition, December 2007. Reciprocating Compressors for Petroleum, Chemical, and Gas Industry Services.

[5] Hobbacher. Empfehlungen zur Schwingfestigkeit geschweißter Verbindungen und Bauteile. DVS Verlag, IIW-Dokument XIII-1536-96 /XV-845-96, The International Institute of Welding .1997

APPENDIX, STRESS VERIFICATION ACCORDING FKM [2]

<u>Static stress verification with local stresses</u>
-Material Properties

Ultimate and yield strength of material from standard: $R_{m,N}$, $R_{p,N}$.

Technology Factor: $K_{d,m}, K_{d,p}$

Constants: $a_{d,m}$, $a_{d,p}$

Anisotropy factor: K_A

Effective diameter: $d_{eff}, d_{eff,N,m}, d_{eff,N,p}$

Ultimate and yield strength of prefab. piece from the draw: $R_M = K_{d,m} \cdot K_A \cdot R_{m,N}, R_p = K_{d,p} \cdot K_A \cdot R_{m,p}$

Tension and shear stress factor : f_σ, f_τ

Compressive strength, shear strength : $R_{c,m} = f_\sigma \cdot R_m, R_{c,p} = f_\sigma \cdot R_p \quad R_{s,m} = f_\tau \cdot R_m, R_{s,p} = f_\tau \cdot R_p$

-Characteristic construction values

Plastic stress concentration factor: $K_{p,\sigma_1}, K_{p,\sigma_2}, K_{p,\tau}$

Plastic notch sensitivity factor: $n_{pl,\sigma_1}, n_{pl,\sigma_2}, n_{pl,\tau}$

Stress concentration factor squer shoulder: $K_{t,ZD}, K_{t,b}, K_{tt}$

Construction factors for block-shaped pieces:

$$K_{SK,\sigma_1} = \frac{1}{n_{pl,\sigma_1} \cdot K_{NL}} \quad K_{SK,\sigma_2} = \frac{1}{n_{pl,\sigma_2} \cdot K_{NL}}, K_{SK,\sigma_3} = 1$$

-Component strength

Static local strength:

$$\sigma_{SK,1} = \frac{f_\sigma \cdot R_m}{K_{SK,\sigma_1}}, \sigma_{SK,2} = \frac{f_\sigma \cdot R_m}{K_{SK,\sigma_2}}, \sigma_{SK,3} = \frac{f_\sigma \cdot R_m}{K_{SK,\sigma_3}}$$

-Safety factors

Overall safety factor: j_{Ges}

-Stress characteristics

The stress characteristics are result of FEA. The maximum extreme stresses will be evaluated from the stress curves in each notch Fig.11, 12, 13, 14 $\sigma_{Max,ex,1}, \sigma_{Max,ex,2}, \sigma_{Max,ex,3}$

-Stress verification

Static degree of utilization in three directions:

$$a_{SK,\sigma_1} = \left| \frac{\sigma_{\max,ex,1}}{\sigma_{SK,1}/j_{ges}} \right| \le 1 \quad a_{SK,\sigma_2} = \left| \frac{\sigma_{Max,ex,2}}{\sigma_{SK,2}/j_{ges}} \right| \le 1 \quad a_{SK,\sigma_3} = \left| \frac{\sigma_{Max,ex,3}}{\sigma_{SK,3}/j_{ges}} \right| \le 1$$

<u>Fatigue strength calculation with local stresses</u>

-Stress characteristics

The stress characteristics are result of FEA. The stress values: principal, maximum, minimum, mean as well as amplitude will be evaluated from the stress curves in each notch Fig.11, 12, 13, 14

Principal stresses in three directions: $\sigma_{1,ZD} = \sigma_1, \sigma_{2,ZD} = \sigma_2, \sigma_{3,ZD} = \sigma_3$

Maximum and minimum stresses: $\sigma_{\min,1}, \sigma_{\min,2}, \sigma_{\min,3} \quad \sigma_{\max,1}, \sigma_{\max,2}, \sigma_{\max,3}$

Stress amplitude, mean stresses in three directions, i=1,2,3 :

$$\hat{\sigma}_{a,i} = \frac{\sigma_{\max,i} - \sigma_{\min,i}}{2} \quad \sigma_{m,i} = \frac{\sigma_{\max,i} + \sigma_{\min,i}}{2}$$

Stress ratio: $$R_{\sigma,i} = \frac{\sigma_{\max,i}}{\sigma_{\min,i}}$$

-Material characteristics

Tension, compressive and shear fatigue strength factor: $$f_{W,\sigma} \quad f_{W,\tau} \quad \tau_{W,Zd} = f_{W,\tau} \cdot \sigma_{W,Zd}$$
$$\sigma_{W,Zd} = f_{W,\sigma} \cdot R_m$$

Polynomial function for the stress in direction i=1,2,3: $$\sigma_i(s) = A_{0,i} + A_{1,i} \cdot s + A_{2,i} \cdot s^2$$

-Construction characteristics

Stress gradient on free surface of the notch, direction i=1,2,3: $$\overline{G}_{\sigma,i} = \frac{d\sigma_i}{ds} = A_{1,i}$$

Constants, mean arithmetic roughness $$a_G, b_G \qquad A_{R,\sigma},\ R_{m,N,Min} \qquad R_Z \qquad \tilde{K}_f,\ K_{NL,E}$$

Notch sensitivity factor
$$n_{\sigma,i} = \begin{cases} 1 + \overline{G}_{\sigma,i} \cdot 10^{-\left(a_G - 0.5 + \frac{R_m}{b_G}\right)} & \text{für} \quad \overline{G}_{\sigma,i} \leq 0.1 \\ 1 + \sqrt{\overline{G}_{\sigma,i}} \cdot 10^{-\left(a_G + \frac{R_m}{b_G}\right)} & \text{für} \quad 0.1 \leq \overline{G}_{\sigma,i} \leq 1 \\ 1 + \sqrt[4]{\overline{G}_{\sigma,i}} \cdot 10^{-\left(a_G + \frac{R_m}{b_G}\right)} & \text{für} \quad 1 \leq \overline{G}_{\sigma,i} \leq 100 \end{cases}$$

Roughness factor for normal and shear stresses
$$K_{R,\sigma} = 1 - A_{R,\sigma} \cdot \log(R_Z) \cdot \log\left(\frac{2 \cdot R_m}{R_{m,N,\min}}\right)$$
$$K_{R,\tau} = 1 - f_{W,\tau} \cdot A_{R,\sigma} \cdot \log(R_Z) \cdot \log\left(\frac{2 \cdot R_m}{R_{m,N,\min}}\right)$$

Surface layer factor, coat factor construction factor
$$K_{WK,\sigma_i} = \frac{1}{n_{\sigma_i}} \cdot \left[1 + \frac{1}{\tilde{K}_f} \cdot \left(\frac{1}{K_{R,\sigma}} - 1\right)\right] \cdot \frac{1}{K_V \cdot K_S \cdot K_{NL,E}} \qquad K_V, K_S$$

-Component strength

Alternate fatigue strength direction i =1, 2, 3 $$\sigma_{WK_i} = \frac{\sigma_{W,zd}}{K_{WK,\sigma_i}}$$

Residual stress factor, constants, mean stress sensitive factor $K_{E,\sigma}$ a_M, b_M

$$M_\sigma = a_M \cdot 10^{-3} \cdot R_m + b_M, M_\tau = f_{W,\tau} \cdot M_\sigma$$

Mean stress factor

$$K_{AK,\sigma_i} = \begin{cases} \frac{1}{1 - M_\sigma} & Zone\ \ I\ R_\sigma \succ 1 \\ \frac{1}{\left(1 + M_\sigma \cdot \frac{\sigma_{m_i}}{\sigma_{a_i}}\right)} & Zone\ \ II\ -\infty \le R_\sigma \le 0 \\ \frac{\left(\frac{1 + M_\sigma /3}{1 + M_\sigma}\right)}{\left(1 + \frac{M_\sigma}{3} \cdot \frac{\sigma_{m_i}}{\sigma_{a_i}}\right)} & Zone\ \ III\ , 0 \prec R_\sigma \prec 0.5 \\ \frac{3 + M_\sigma}{3 \cdot (1 + M_\sigma)^2} & Zone\ \ IV\ R_\sigma \succ 0.5 \end{cases}$$

Fatigue strength amplitude $$\sigma_{AK_i} = K_{AK,\sigma_i} \cdot K_{E,\sigma} \cdot \sigma_{WK_i}$$

Endurance strength factor $$K_{BK,\sigma_i}\ \ (K_{BK,\sigma} = 1\ f\ddot{u}r\ N = \infty)$$

Endurance strength amplitude $$\hat{\sigma}_{BK_i} = K_{BK,\sigma_i} \cdot \sigma_{AK_i}$$

-Safety factors

Safety factor j_D

- Fatigue stress verification

Degree of utilization in three directions i=1,2,3 $$a_{BK,\sigma_i} = \frac{\hat{\sigma}_{a_i}}{\left(\frac{\sigma_{BK_i}}{j_D}\right)} \le 1$$

Investigation of High Pressure Fluid Circuit Gas Flow Dynamics in a Hermetic Compressor Using CFD to Improve Qualitative Flow Understanding

David N. Halbrooks[1*]

[1] Bristol Compressors International, LLC,
Bristol, Virginia, USA
Email: davidhalbrooks@gmail.com, Phone: 615.519.6430

ABSTRACT

While many approaches have been investigated to identify areas for hermetic compressor efficiency improvements over the past few decades, there remains no conclusive method to gain a holistic view of the system domain to help identify potential target areas to guide product research direction. Hence, modern tools, methods, and problem perspectives are needed to develop new insight into mature problem-sets on a system level to satisfy increasingly stringent efficiency mandates. Reasonably mature simulation tools currently exist that enable scientists and engineers to gain a new perspective into system domains. One such tool is Computational Fluid Dynamics (CFD). The results of the following investigation will show how CFD, coupled with the knowledge of its inherent strengths and weaknesses, can be leveraged to allow scientists and engineers to view mature problem-sets from a qualitative perspective to yield different results by using the high-pressure fluid circuit of a hermetic compression system as a simulation test-bed.

1. INTRODUCTION

Traditional Hermetic Compressor research has largely centered around classic trial-and-error approaches based on years of industry knowledge. A byproduct of this approach is the production of a results-centered mindset in respect to product development, which in turn brings about new opportunities and challenges. The prime challenge is being able to gain insight into the dynamic operating conditions of the Hermetic Compressor. This challenge is exacerbated by the fact that accurate internal system measurement data is very difficult to acquire due to the hermetic conditions in which the refrigerant must be maintained. That said, taking random measurements can often prove to be an exercise in futility itself since repeatability can also be a function of measurement location. Sometimes trial-and-error research yields very good results, while at other times it can yield less-than-optimal results. Regardless of the view, new efficiency requirements being mandated require old problems to be viewed through a different lens to find answers to the question of where to concentrate future development efforts. The following study addresses this need through use of a macro-level CFD analysis approach.

2. LITERATURE REVIEW

Numerical analysis has began to take a more substantial role with respect to research in the HVAC&R industry during the last decade. Current fluid modeling techniques comprise of 1-D Lumped Conservation Models, 2-D Computational Fluid Dynamics, 3-D Computational Fluid Dynamics, and hybrid approaches containing elements of each mentioned method. Due to the large computational requirements of Computational Fluid Dynamics (CFD), compressor systems have been segregated to enable detailed study into smaller sub-systems that can be categorized into four general sub-systems: suction (low-side) fluid circuit, suction valving, discharge valving, and discharge (high-side) fluid circuit.

The suction (low-side) fluid circuit has been a recent hot topic of study due to its reduced gas complexity with respect to thermodynamics coupled with the inherent need for improvements. Rigola et al. (2014) investigated the effects of suction mufflers with respect to system pulsations, using modular sub-systems to simulate the suction side of the system. Mayer et al. (2014) compared strengths and weaknesses of 1-D lumped modeling to CFD simulation for reed valve motion using the suction valve as the center of study. Pereira et al. (2012) recently investigated the use of a hybrid modeling approach that consisted of a 1-D lumped model in conjunction with a 1-D valve motion model and a 3-D CFD model to predict suction pulsation with some success. Novak (2014) investigated the effects of geometry simplification for CFD analysis and the potential impact for quantitative result accuracy with respect to low-side compressor shell

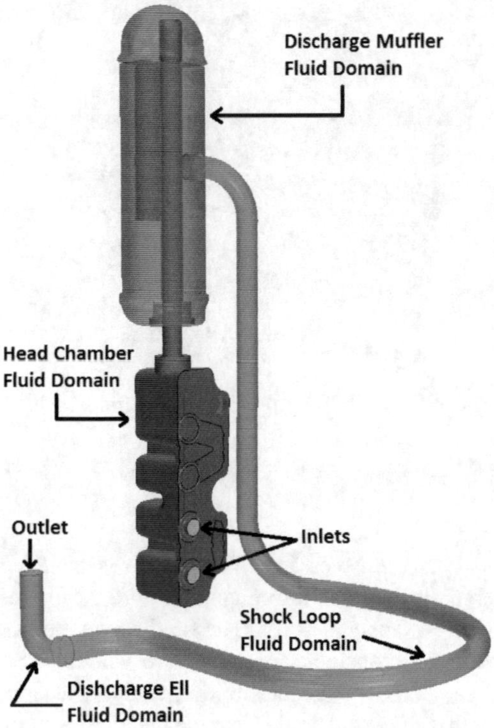

Figure 1: High Pressure Fluid Domain

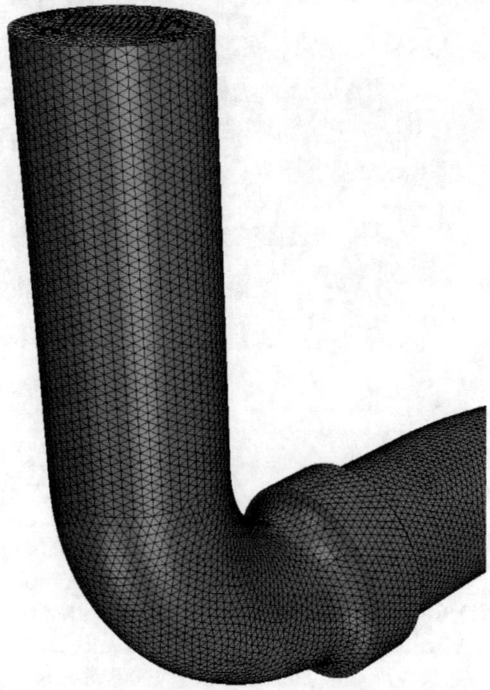

Figure 2: 2-D Surface Mesh

volume. Kerpicci and Oguz (2006) also investigated system influences on numerical simulations with respect to results for a suction valve analysis. Done et al. (2014) have recently leveraged CFD to successfully present some interesting conclusions with respect to suction muffler design approaches.

The discharge (high-side) fluid circuit has not been as actively researched as the suction side of the compressor system. However, Oliveira and Real (2010) recently applied a 1-D analytical model to investigate discharge pressure pulsations to further investigate electronic valving. Park et al. (2008) present the use of a 1-D lumped model to investigate low frequency compressor noise due to discharge pulsations.

Current CFD quantitative state-of-the-art accuracy for simulation systems with unknown test data can be implicitly gathered through the work of Liu et al. (2012). The work of Liu et al. (2012) investigated pressure loss coefficients for converging and diverging duct sections in a competitive study solicited by ASHRAE to conduct a blind comparison between CFD and experimental results. Liu et al. (2012) presented the most accurate results to ASHRAE in response to the project titled, "CFD Shootout Contest - Prediction of Duct Fitting Losses".

3. SYSTEM OF INTEREST

The system of interest for this numerical investigation is the high-pressure side of the hermetic compressor. This includes all compressed gas after the discharge valve up to (and including) the discharge ell. The complete simulation system consists of three general flow domains: the head chamber, discharge muffler, and discharge shock loop that includes the discharge ell as seen in Figure 1. All fluid volume after the inlet ports up to the discharge muffler entrance is included in the head chamber fluid domain. The discharge muffler fluid domain entrance begins inside the head chamber and encompasses the fluid volume up to the discharge shock loop entrance. Lastly, gas flow from the discharge muffler is carried through the shock loop and discharge ell fluid domains to the outside the compressor shell. It should be noted that all three flow domains are simulated as one domain for the purposes of this investigation. Valving has not been included in this analysis due to the simplification intent of the study.

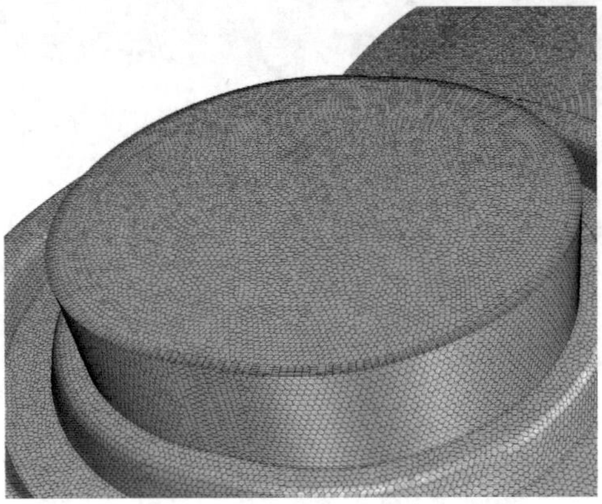

Figure 3: Polyhedral Inlet Volume Mesh

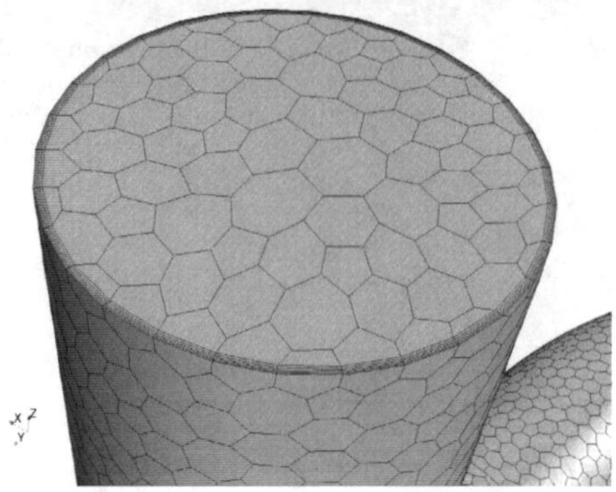

Figure 4: Polyhedral Outlet Volume Mesh

4. SIMPLIFYING ASSUMPTIONS

To adequately study the system of interest, simplifying assumptions were necessary. The system is modeled in a steady-state condition. Dynamic valve operation is not considered since it should be considered as a micro-level transient event. Structural interaction is neglected. Oil suspension is not accounted for. The fluid is modeled as pure R410A with no impurities. All fluid walls are treated as adiabatic and considered to have zero velocity. Surface roughness is not considered. The valve is fixed at maximum lift (against the valve stop) for the steady-state simulated condition. Finally, the inlet massflow rate is considered uniform for this study. It should be noted that these simplifying assumptions are not valid for a micro-level analysis. However, they are conducive to positive results on a macro-level as is the case for this analysis.

5. NUMERICAL MODELING APPROACH

While many numerical methods for CFD simulation exists, the Finite Volume (FV) method was employed for this simulation. Continuity, momentum, and energy equations were solved using a pseudo-transient, density-based implicit coupled flow model through use of a second-order upwind discretization scheme. Buoyancy and density changes are accounted for due to compression and expansion. The equations-of-state for density is modeled using the Ideal Gas Law due to its simplicity coupled with its ability to predict reasonable qualitative results (Branch, 2014). For brevity, the governing equations for the numerical models are not presented. That said, general scalar transport equations, equations-of-state, and discretization schemes are covered in detail in the text published by Ferziger and Perič (2001).

6. DOMAIN DISCRETIZATION

For this study, an airtight 2-D triangular surface mesh (see Figure 2) is used to generate the base surface mesh to enable clean volume mesh generation. From the generated 2-D surface mesh, a tetrahedral volume mesh is created to enable generation of a polyhedral mesh based on tetrahedral cell centroids. The polyhedral volume mesh used for numerical discretization can be seen in Figures 3 and 4. Mesh cross-sections for the head cavity region can be seen in Figures 5 and 6.

7. FLUID PROPERTIES

The fluid of interest for this study is R410A. As stated in the simplifying assumptions, oil suspension is neglected and the fluid is modeled as pure. Hence, fluid properties were extracted from RefPROP v9.1 for this simulation (Lemmon et al., 2010). A truncated table of example fluid properties can be seen in Table 1 on page 4.

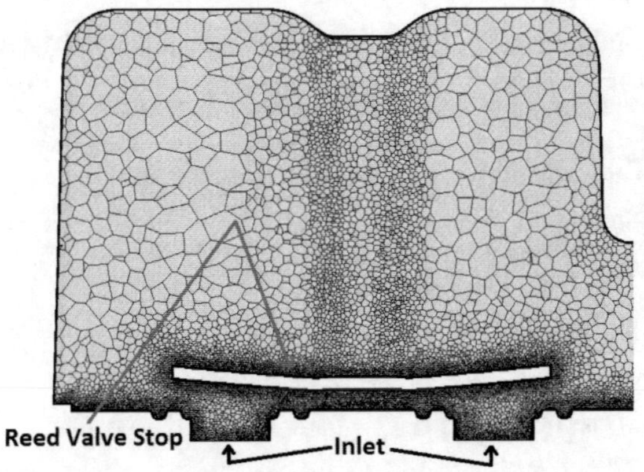

Figure 5: Mesh Density Inlet Cross Section One

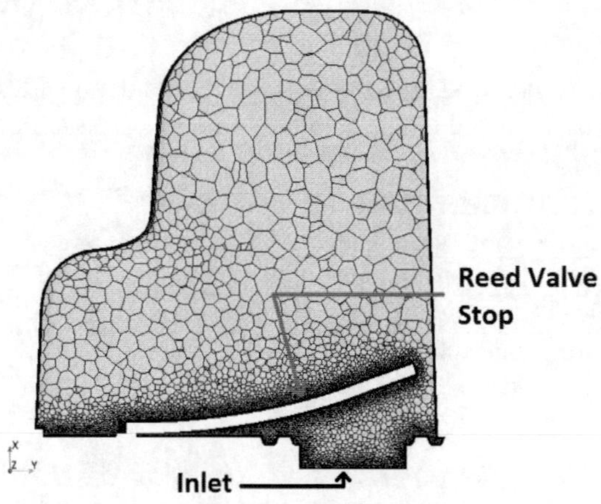

Figure 6: Mesh Density Inlet Cross Section Two

Table 1: R410A Example Fluid Properties

T (K)	P (MPa)	$\rho\left(\frac{kg}{m^3}\right)$	$Cp\left(\frac{J}{kg-K}\right)$	$\left(\frac{Cp}{Cv}\right)$	c $\left(\frac{m}{s}\right)$	Molar Mass	k $\left(\frac{W}{m-K}\right)$	μ (Pa-s)	Pr
310.8	2.3	989.76	1889.9	1.9881	354.4	72.585	0.082181	9.9244E-5	2.2823
320.8	2.3	86.824	1434.7	1.575	165.64	72.585	0.01797	1.5039E-5	1.2007
330.8	2.3	79.691	1286.5	1.4579	174.14	72.585	0.018039	1.542E-5	1.0997
340.8	2.3	74.292	1205.9	1.3872	181.29	72.585	0.018394	1.5819E-5	1.0371
350.8	2.3	69.948	1157.3	1.3391	187.55	72.585	0.018914	1.6227E-5	0.99291

8. TURBULENCE MODEL SELECTION

Many models that enable closure of the Reynolds Averaged Navier-Stokes (RANS) equations exist. These models are used to statistically provide closure the the Reynolds stresses (Pope, 2000). While it is not mandatory that RANS modeling be employed, it is much more computationally efficient than employing Direct Numerical Simulation (DNS) to model the flow field on a macro-level using current computing platforms. Three approaches for selecting a suitable turbulence model as well as general thought process strategies are discussed below.

A first approach in simulating an unfamiliar system should be to perform a literature review to identify similar simulations that have had extensive parametric studies performed in respect to turbulence models. This often gives a good indication on where to begin with respect to model selection. More importantly, this approach aids in identifying potential shortcomings based on past turbulence model results.

The most reliable approach is to explicitly perform a parametric study using the simulated model as a test-bed for model selection. This often reveals strengths and weaknesses in different turbulence modeling approaches. However, it should be understood that many turbulence models are also grid dependent which makes a parametric study that accounts for grid independence a necessity. Re-circulation zones, surface attachment, and surface detachment based on the fluid domain can have a significant impact on the turbulence model selection.

The last method involves model selection based on modeling experience. It should be noted that this method can be somewhat misleading since problem extrapolation is not always as straight-forward as it may appear. Parametric turbulence model studies can often reveal weaknesses within a turbulence model application that previous experience has not yet discovered.

Turbulence model selection is generally based on trade-offs between performance, robustness, and accuracy. With robustness and computational performance, and macro-level simulation targets in-mind, the realizable, two-layer wall treatment k-ε turbulence model is selected to model turbulent kinetic energy (k) and the dissipation rate (ε) based on a combination of literature review and past experience (Rodrigues, 2014; Pereira et al., 2012; Liu et al., 2012).

9. BOUNDARY CONDITIONS

Simplified boundary conditions for the simulation are set according Table 3. It should be noted that all boundary conditions are based on the problem's simplifying assumptions listed in section 5 on page 3.

Table 2: Discretization Domain Details

Discretization Region	# of Elements
Head Cavity	7.2 Million
Muffler, Shock Loop, & Discharge Ell	653 Thousand

Table 3: Simulation Boundary Conditions

Inlets	Mass Flow Inlet	$0.0726 \left(\frac{kg}{s}\right)$
Outlet	Pressure Outlet	311 K
All Walls	No Slip, Smooth, & Adiabatic	N/A

10. SOLUTION CONVERGENCE

Solution convergence for this investigation is monitored based on three separate methods. The first is through solution residuals. Residuals allow real-time solution monitoring to allow an insight to be established in respect to convergence as well as divergence. However, residuals do not always tell a complete story in respect to convergence. Changing CFD solution variables such as the under-relaxation factor can affect the solution residual presentation. In practice, residuals are better served as a solution performance metric to answer the question: Is the solution diverging?

Hence, for this study, the pressure drop and mass conservation from the inlet to the discharge outlet is also monitored to ensure solution convergence is adequately determined. Even using these two additional solution monitoring methods, careful consideration must be paid to convergence. Figure 7 shows massflow convergence somewhere around eight to nine hundred iterations. However, 8 shows that the pressure drop (inlet to outlet) does not show convergence until around fourteen hundred iterations.

11. GRID INDEPENDENCE

It should be noted that grid independence is not rigorously tested in this study due to the nature of the investigation intent. To maintain computational efficiency, the mesh density is normally decreased in areas that have little effect on the results once results are discovered. In contrast, mesh refinement (mesh density increase) can be executed in areas of high pressure, temperature, and velocity gradient. Figures 5 and 6 show a cross-section showing mesh density in the head cavity fluid region. Careful consideration based on prior simulation experience can be leveraged in areas of large gradient expectations, such as the reed valve stop walls, the head cavity floor, and inlet wall regions. A general element size matrix can be seen that describes the mesh densities in Table 2 on page 5.

12. QUALITATIVE RESULTS

The purpose of this investigation is to gain qualitative insight into the energy flow-field on a macro-level. That said, the amount of interesting qualitative information must be broken into digestible chunks. In that regard, the results presented have been parsed down to the head cavity section to facilitate a quality discussion. Furthermore, the information will be segregated into three qualitative sections: Velocity, Pressure, and Temperature.

Figure 9 shows a two-dimensional cross section of the velocity vectors for the simulation. Of particular interest is the apparent velocity "bubble" that forms on each face of the discharge valve in addition to a "jet-effect" in the valve curtain area. It can be seen in Figure 10 that multiple re-circulation zones exist throughout the head chamber domain. The head chamber fluid domain also shows an interesting "bow" in the flow-field as the fluid exits the head chamber and enters the muffler inlet tube. Figures 11 and 12 show velocity vectors acting on the face and the back side of the valve stop as well as a substantial re-circulation zone located on the back side of the valve stop. High velocity gradients are often an indicator of potential energy losses.

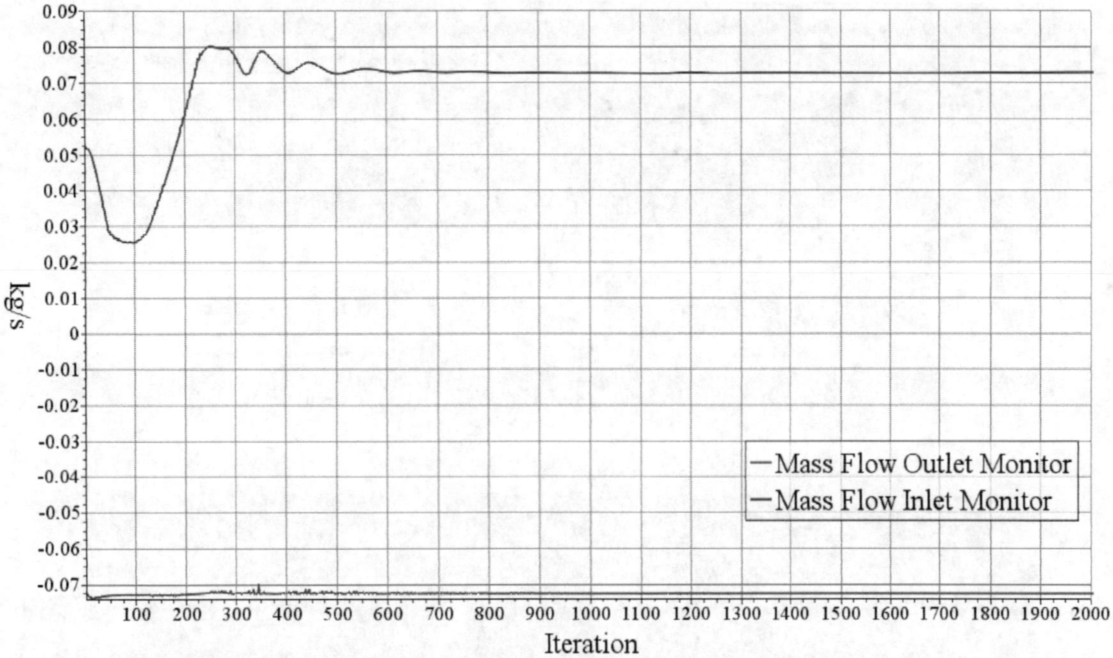

Figure 7: Mass Flow Balance

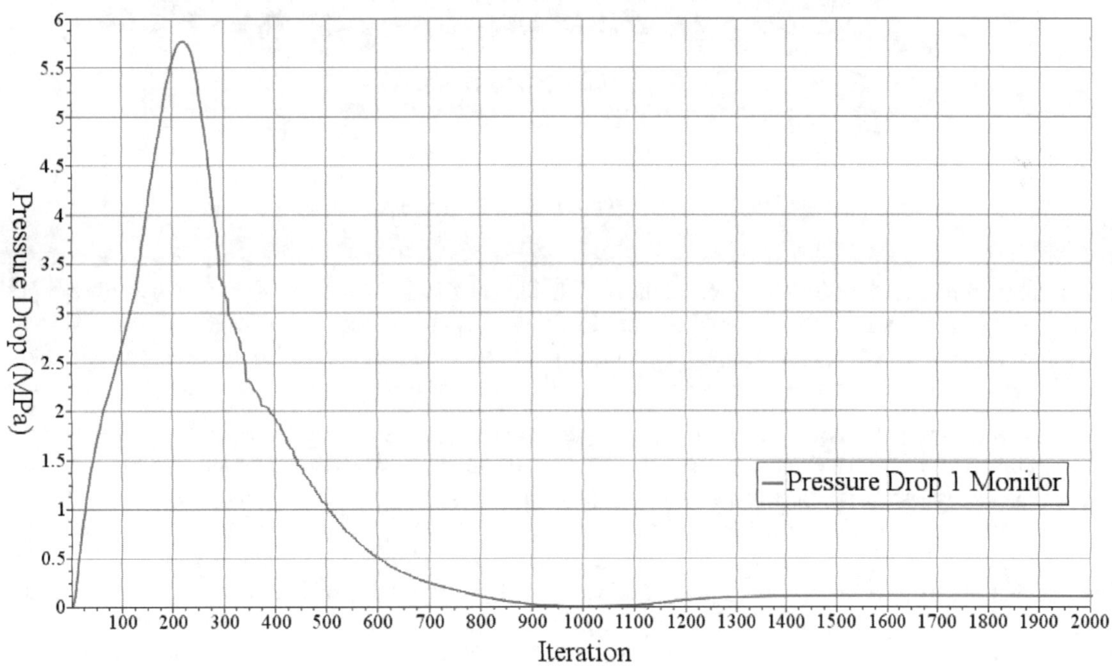

Figure 8: Pressure Drop Convergence

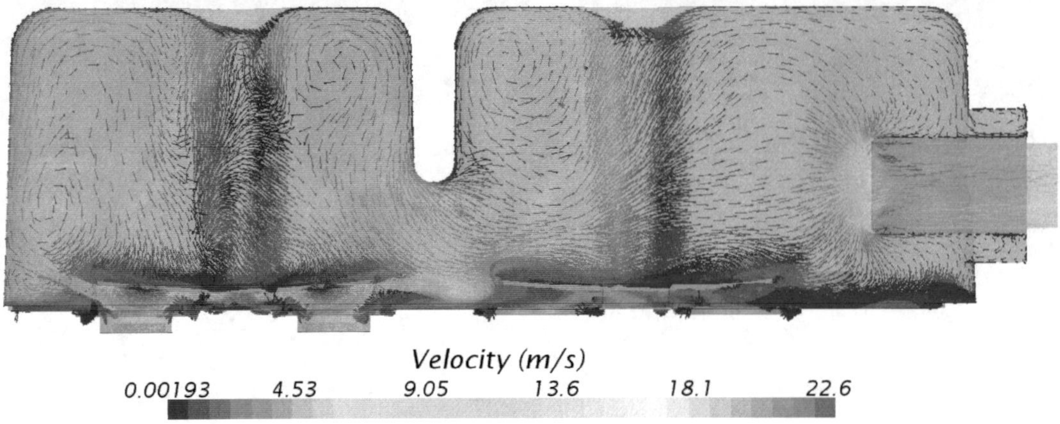

Figure 9: Velocity Vectors of Head Cavity Cross-Section

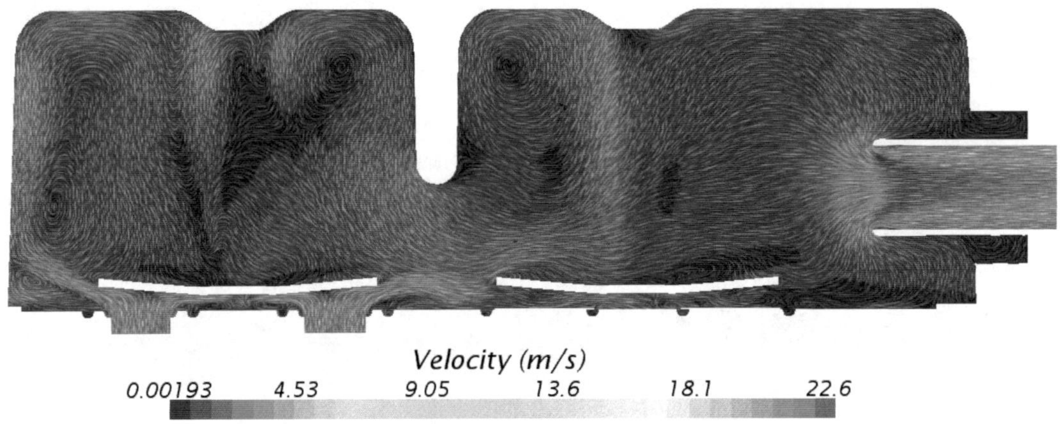

Figure 10: Velocity Stream of Head Cavity Cross-Section

Figure 13 shows three very interesting aspects of the flow-field. The first is the high pressure that the valves experience. Of interest, is the pressure differential from the front of the valve stop to the back side of the valve stop. Next, areas of high pressure present themselves on the upper left corner of the fluid domain. This also indicates an area of low fluid velocity. Insight such as this can be leveraged to investigate flow improvement design strategies.

Lastly, the temperature field is queried for insight into general energy transfer characteristics for the flow domain. An HVAC&R compressor's primary purpose is to increase the refrigerant's delivered temperature to the heat exchanger (via pressure increase) to enable heat transfer to take place in the condenser and evaporator, thus satisfying the thermodynamic refrigeration cycle. That said, temperature is a form of energy and understanding it's flow path can be beneficial. Figures 14 and 15 show some interesting trends as the gas expands into the head chamber volume. It is well understood that energy flows from hot-to-cold under ideal conditions. With that said, it can be seen that the temperature contours are very discreet, particularly in the cross-section depicted in Figure 15. An apparent area of interest is seen in the valve and valve-stop region. The temperature gradient is very pronounced on each side of the valve stop due to the gas expansion path governed by the geometric flow-path of the design.

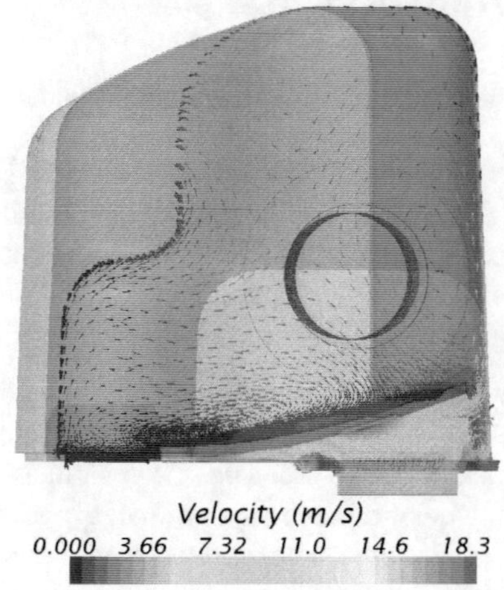

Figure 11: Velocity Vectors Section View Two

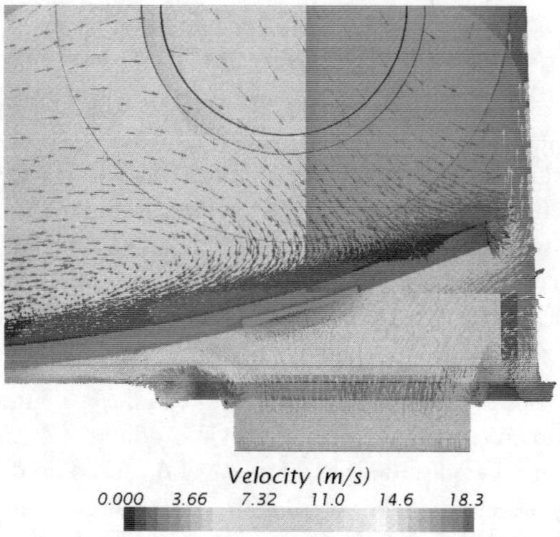

Figure 12: Velocity Vectors Section View Two Zoomed

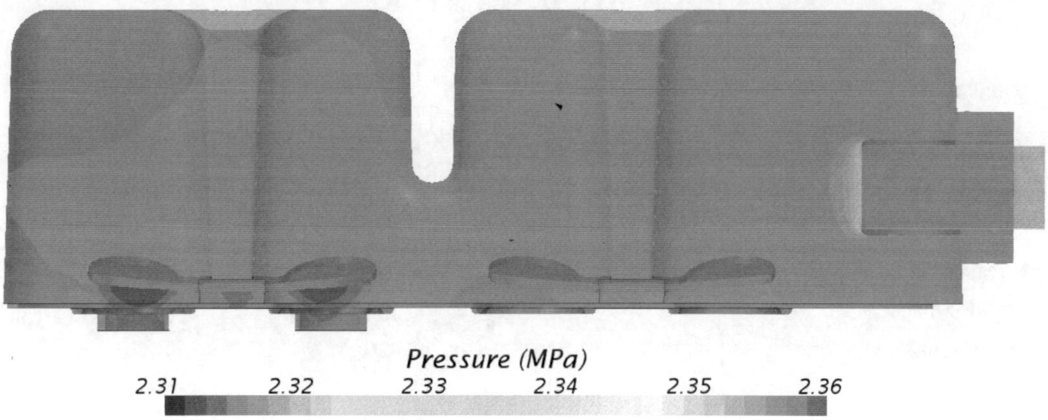

Figure 13: Head Cavity Cross-Section Pressure Contours

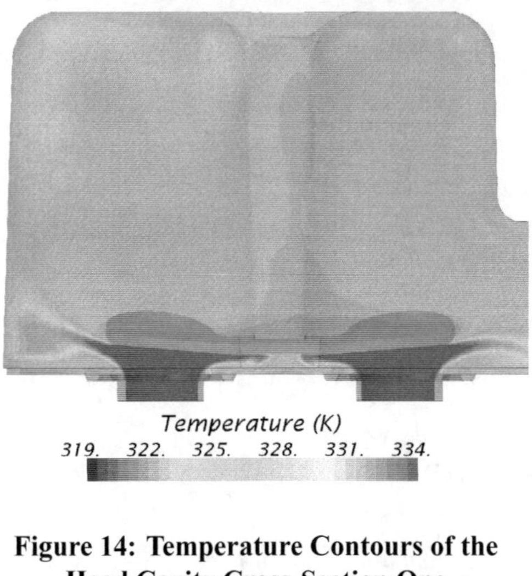

Figure 14: Temperature Contours of the Head Cavity Cross-Section One

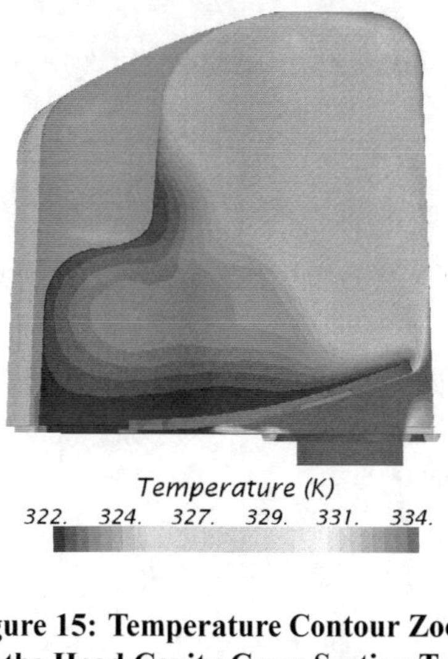

Figure 15: Temperature Contour Zoom of the Head Cavity Cross-Section Two

13. CONCLUSIONS AND RECOMMENDATIONS

Upon conclusion it can be seen that well thought-out steady-state simulation simplification strategies can be applied to enable the study of a transient, dynamic system through the use of CFD numerical simulations. Results presented in this study are able to provide a virtual window into the head chamber's tortuous flow domain on a macro-scale, thus enabling the ability to gain a holistic insight into system optimization direction. That said, computational limitations still exist that must be considered. Hence, the need for practical simplifying assumptions. Once computational and numerical limitations are identified for a simulation, they can be exploited to aid research direction. Many times, the solution isn't known... nor is it necessary to be known as shown through this investigation.

Future work could include parametric studies that vary the mass flow rate to study system trends as the compressor experiences different system operating conditions. Furthermore, transient micro-scale sub simulations could be employed to enhance accuracy and provide additional energy transfer understanding. That said, the general energy-flow trends discovered through this investigation could hold true across many conditions the system will experience in respect to the energy flow-field. Lastly, results discovered during this investigation can be used to enhance experimental measurements based on optimum measurement locations or perhaps even correction factor applications that can be understood and accounted for. Hence, computational effort on the front-end can be used to augment the measurement process... creating an iterative loop for computational analysis and experimental measurements to improve research and design efficiency.

NOMENCLATURE

ρ	Density	$\left(\frac{kg}{m^3}\right)$
ε	Turbulence Dissipation	$\left(\frac{m^2}{s^3}\right)$
μ	Dynamic Viscosity	$\left(\frac{kg}{m-s}\right)$
V	Velocity	$\left(\frac{m}{s}\right)$
P	Pressure	(MPa)
T	Temperature	(K)
k	Turbulent Kinetic Energy	$\left(\frac{m^2}{s^2}\right)$
kg	Kilogram	(N/A)
m	Meter	(m)
ASHRAE	American Society of Heating, Refrigerating, and Air-Conditioning Engineers	(N/A)
CFD	Computational Fluid Dynamics	(N/A)
RANS	Reynolds Averaged Navier Stokes	(N/A)
DNS	Direct Numerical Simulation	(N/A)
Pr	Prandtl Number	(N/A)
k	Thermal Conductivity	$\left(\frac{W}{m-K}\right)$
c	Speed of Sound	$\left(\frac{m}{s}\right)$
s	Second	(s)
W	Watts	(W)
Cp	Specific Heat	$\left(\frac{J}{kg-K}\right)$

REFERENCES

Branch, S. (2014). Methods of fluid properties for compressible refridgerant cfd analysis. In *International Compressor Engineering Conference*. Paper 2289.

Done, V., Balide, V., Tamma, B., Soni, K., and Dey, S. (2014). Muffler design for a refridgerator compressor. In *International Compressor Engineering Conference*. Paper 2262.

Ferziger, J. and Periĉ, M. (2001). *Computational Methods for Fluid Dynamics*. Springer, 3 edition.

Kerpicci, H. and Oguz, E. (2006). Transient modeling of flows through suction port and valve leaves of hermetic reciprocating compressors. In *International Compressor Engineering Conference*. Paper 1806.

Lemmon, E. W., Huber, M. L., and McLinden, M. O. (2010). Nist standard referencedatabase 23: Reference fluid thermodynamic and transport properties /- refprop. v.9.1.

Liu, W., Long, Z., and Chen, Q. (2012). A procedure for predicting pressure loss coefficients of duct fittings using cfd (rp-1493). *HVAC&R Research*, 18(6):1168--1181.

Mayer, J., Bjerre, P., and Brune, F. (2014). A comparative study of different numerical models for flapper valve motion. In *International Compressor Engineering Conference*. Paper 2309.

Novak, K. (2014). Influence of shell volume on pressure pulsations in a hermetic reciprocating compressor. In *International Compressor Engineering Conference*. Paper 2272.

Oliveira, S. and Real, M. (2010). Analytical and experimental study of discharge flow behavior provided by electronically controlled valves in hermetic compressors. In *International Compressor Engineering Conference*. Paper 1941.

Park, J., Kang, J., Choi, J., Jang, P., Yoon, M., Kwon, D., and Oh, I. (2008). Reduction of the low frequency noise due to the discharge pulsation of a reciprocating compressor. In *International Compressor Engineering Conference*. Paper 1898.

Pereira, E., Santos, C., Deschamps, C., and Kremmer, R. (2012). A simplified cfd model for simulation of the suction process of reciprocating compressors. In *International Compressor Engineering Conference*. Paper 2086.

Pope, S. (2000). *Turbulent Flows*. Cambrigdge University Press, 1 edition.

Rigola, J., López, J., Papakokkinos, G., Lehmkuhl, O., and Oliva, A. (2014). Numerical analysis of suction mufflers. In *International Compressor Engineering Conference*. Paper 2381.

Rodrigues, T. (2014). Turbulence modelling evaluation for reciprocating compressor simulation. In *International Compressor Engineering Conference*. Paper 2297.

ACKNOWLEDGMENT

This research was made possible through funding provided by the National Institute of Science and Technology, The Virginia Tobacco Region Revitalization Commission, and Bristol Compressors International, LLC. A special thanks is also due to GENEDGE alliance for working with Bristol to secure funding and manage implementation of the required research resources to enable Bristol to "look forward". Lastly, the author would like to extend significant gratitude to Professor Steve Idem of Tennessee Technological University for providing thought-provoking feedback in regards to presentation context.

Integral Compressor / Generator / Fan Unitary Structure.

Dr. Nelik DREIMAN

Consultant, P.O.Box 144, Tipton, MI 49287 USA
E-mail: ndreiman@frontier.com

ABSTRACT

A compact, space saving single structure of short axial length have been developed by radial integration of a revolving piston rotary compressor, a motor-generator and a fan, to replace a plurality of separate units (compressors, generators, fans, etc.) employed, as usual, in contemporary air conditioning and refrigeration systems. A dual drive mechanism of the structure permits to operate it in generating mode when a motive power supplied by an external mechanical force or to utilize an electric motor mode to run all integrating entity when an electric power is available. A suction gas delivery scheme excludes a direct distribution of a refrigerant to the suction chamber, includes double-stage liquid-gas separation to prevent slugging, provides cooling of the motor–generator, following by supercharge of the refrigerant into the suction chamber. An additional liquid-gas separation is performed on a discharge side of the structure before expelling of the gas. The structure can be employed in a single standby condenser/generator units in residential, commercial, health facilities air conditioning systems to replace separately mounted condensers and standby generators or can be used as a space and cost saving alternative in RV, transport refrigeration units and, with some modification, in vehicle air conditioning systems.

1. INTRODUCTION

Since the dawn of air conditioning and refrigeration engineering, compressor and generator have generally been two separately spaced individual electric machines which are especially adapted to their particular functions in residences, retail facilities, commercial and public buildings, vehicles, transport refrigeration, etc..

In remote air conditioning systems, the refrigeration equipment (outdoor air-cooled condensing units equipped with a compressor driven by an electric motor, a cooling fan, etc.) is located away from the space to be conditioning and mounted at ground level, on the roof, on the outside wall. A portable generator or an independently installed standby, driven mechanically by an engine generator is used to provide backup power during line power outages.

In vehicles systems, a compressor and a generator (alternator) are belt-driven from the prime engine c'shaft. Generators and air conditioning compressors used in hybrid electric vehicles are also segregate entities and can be driven as by an auxiliary engine so by an electric motor. When the prime electric motor is stopped, including traffic light stops, an auxiliary electric motor powered by the battery drives the air conditioning compressor.

Transport refrigeration systems have employed belt driven, mechanically linked shaft driven or electric motor driven compressor and airflow fans. Alternatively, various types of on-board generators driven by auxiliary engine have provided the power required by individual entities of the refrigeration system. Typically, the size of the equipment used in truck trailer refrigeration units is restricted to the limited space bounded by the tractor swing radius and the trailer front wall. Use of a tandem mounted auxiliary engine and generator or (and) utilization of compact components such as, for example, a synchronous generator may help in meeting required space limitation. These arrangements, however, will not change the number of separate entities each of which requires individual mounting space and have to be assembled with the predetermined distances there between.

The present paper contemplates one compact, single unit (*the structure*) formed by integration of a modified rolling piston rotary compressor, Dreiman (2014), a generator and an airflow fan components mounted on a stationary crankshaft of the compressor.

2. KINEMATICS OF THE STRUCTURE

2.1 Motion analysis

The construction of the compressor used as a base element of the structure comprises, in combination: a revolving piston cylinder rotatably mounted on opposed ends of the stationary crankshaft, a rotor block cylinder spaced eccentrically inside of the revolving piston cylindrical cavity and a vane rigidly fixed axially in the wall of the revolving piston. The rotor block and the revolving piston of the structure can be considered, consequently, as internal cylinder (driver) and external cylinder (follower) having rolling contact at point P (see Figure 1) with transmission of the motion from the driver to the follower through the rigid link L (coupler) and a force developed at a line of contact. As shown in Figure 1, the line connected piston center O_P and point P_R is considered as the part of rotational link with $O_P P_R = d$, rotor cylinder center O_R and the piston center O_P are individually taken as the two immovable points related to the stationary crankshaft of the mechanism with $O_R O_P = e$ (fixed link). For plane triangle $\nabla O_P P_R O_R$ (see Figure 1), as per the Law of Sines and the Law of Cosines

$$e \sin\theta = f \sin\phi; \; d = e \cos\theta + f \cos\phi \tag{1}$$

Due to the fact that the rotor-cylinder-vane is integral part and rotates as one solid body around *z*-axis, there is no reciprocating movement of the vane. The flats of the bushing implanted in the roller slide along the suction and discharge sides of the vane with change of the rotor-cylinder turning angle θ. The part of the vane **L** (θ) exposed to the suction and discharge pressure inside of the working chamber will be

$$L(\theta) = P_R P_P = R_P - d, \tag{2}$$

where radius of the cylinder $R_C = R_R + e$ and radius of rotor $R_R = f$. After substitution and transformation

$$L(\theta) = e(1 - \cos\theta) + R_R \left[1 - \sqrt{1 - (\lambda \sin\theta)^2}\right], \tag{3}$$

where $\lambda = e / R_R$. We can use binominal series shown below for the last term of the Eq.3

$$(a + b)^n = a^n + n a^{n-1} b + n(n-1)/2! \, [a^{n-2} b^2] + n(n-1)(n-2)/3! \, [a^{n-3} b^3] \; .... \tag{4}$$

with a = 1, $b = -(\lambda \sin\theta)^2$ and n = ½. After expanding the root into a binominal expansion with inclusion of the first two terms as the series rapidly converges, we obtain

$$L(\theta) = e\,(1 - \cos\theta + \lambda/2 \sin^2\theta - \lambda^3 \sin^4\theta / 8 + ...) \tag{5}$$

Thus the length of the exposed part of the vane very approximately (error 0.15%) will be:

$$L(\theta) = e\,(1 - \cos\theta + \lambda/2 \sin^2\theta) \tag{6}$$

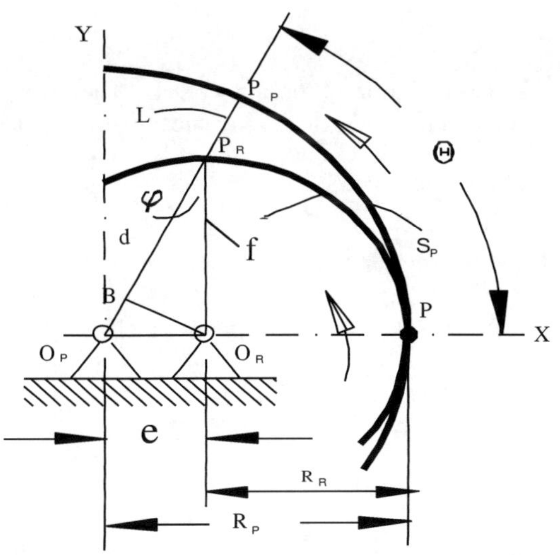

Figure 1. Diagram of rolling piston rotary compressor mechanism.

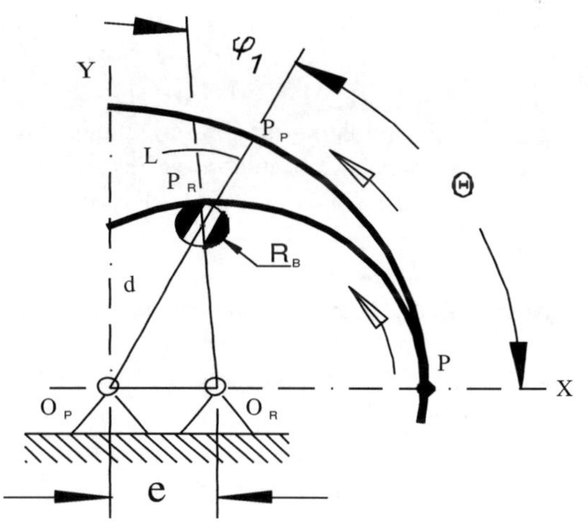

Figure 2. Displacement volume of a rolling piston rotary compressor.

The bushing mounted in the rotor block slides radially against the vane sides and turns to accommodate the vane with changing of the rotor block angular position. Differentiating Eq. (6) and denoting $d\theta / dt = \omega$ we obtain the sliding speed V_{BV} between the bushing and the vane with change of the turning angle.

$$V_{BV} = e\,\omega\,(\sin\theta + \lambda^2/2 \sin 2\theta) \qquad (7)$$

where ω is angular velocity of the rotor block. The bushing turning angle φ_1 (see Figure 2) with change of the vane rotational angle θ will be

$$\phi_1 = \arcsin\,[e \sin\theta /(R_R - R_B)\,] \qquad (8)$$

where R_B is the radius of the bushing. The values of the bushing turning angle ϕ_1 and exposed to suction and discharge pressures vane length $L(\theta)$ with change of the rotational angle θ in the limits 0-2π (single revolution) are illustrated in Figure.3.

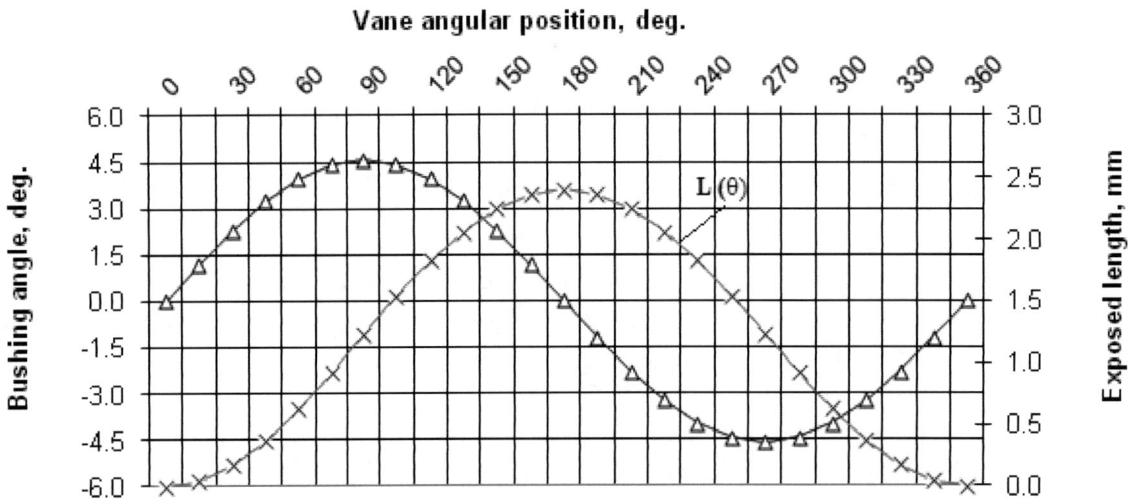

Figure 3. Turning angle of vane bushing and exposed length of the vane with change of the rotational angle.

The bushing has ±4.5° oscillating motion in the axial cavity per single revolution of the rotor block. The extreme values of the turning angle have been observed at 90° and 270° of the working cycle. The vane's maximum exposed length has been recorded at 180° of the rotor block circular position.

2.2 Piston- swept volume

The swept volume **W** (θ) of the working chamber at an arbitrary angle of the vane is

$$W(\theta) = A(\theta)\,h\,, \qquad (9)$$

where h is the height of the rotor block and A (θ) is the area of compression chamber at an arbitrary angle θ of the vane position as shown in Figure 4.

$$A(\theta) = \pi \int_0^\theta (R_P^2 - d^2)\,d\theta = \pi \int_0^\theta [R_R^2 (1+\lambda)^2 - d^2]\,d\theta\,, \qquad (10)$$

where

$$d = R_P - L(\theta) = R_R (1+\lambda\cos\theta - \lambda^2/2 \sin^2\theta), \qquad (11)$$

or

$$d^2 = R_R\,(1 + 2\lambda\cos\theta + \lambda^2 \cos 2\theta), \qquad (12)$$

if we neglect the terms containing λ^3 and λ^4 due to the small value and put $\cos 2\theta =. \cos^2\theta - \sin^2\theta$. Refer back to Eq. 9 and Figure 4 the compression chamber volume trapped within piston, vane and rotor block at an arbitrary angle θ

$$W(\theta) = \pi \lambda h R_R^2 / 2 [(1+\lambda)\theta - 2\sin\theta - \lambda/2 \sin 2\theta] , \tag{13}$$

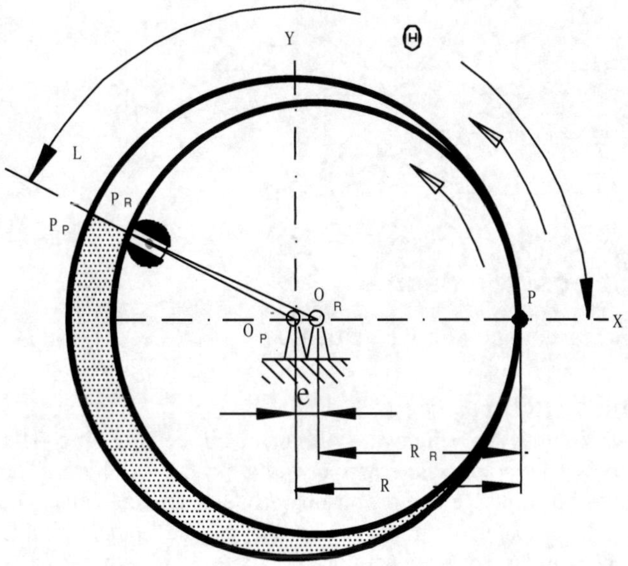

Figure 4. Displacement volume of the revolving piston compressor

In this calculation we have neglected the volume occupied by the vane. The compression volume given by Eq. (13) has to be reduced by the volume of the exposed in the working chamber part of the vane:

$$W_1(\theta) = W(\theta) - h\,t\,L(\theta) = W(\theta) - e\,h\,t\,(1 - \cos\theta + \lambda^2/2 \sin^2\theta), \tag{14}$$

where t is the thickness of the vane.

2.3. Piston-rotor block relative velocity

Motion of internal cylinder (rotor block) is transmitted to the external cylinder (piston) through the driving link **L.** If driver link **L** rotates counterclockwise with constant angular velocity, the follower will revolve in the same direction at a varying (accelerating and decelerating) speed. The relative velocity between the cylinders is a function of the driver angle θ, radii of the contact cylinders, centers distance. The linear velocities of the piston v_P and rotor $\boldsymbol{v}_R$ are shown below:

$$v_P = dS_P / dt = \omega\, dS_P / d\theta = \omega R_P \tag{15}$$

$$v_R = dS_R / dt = \omega\, d\left[\int_0^\theta d\, d\theta\right] / d\theta = \omega\, d \tag{16}$$

where S_P and S_R are revolving arcs of the cylinder and roller. The *relative* velocity of the roller to the cylinder

$$\Delta v = \omega (R_P - d) \tag{17}$$

Taking derivative we find that

$$dv_R/dt = \omega [(dd/d\theta)(d\theta/dt)] = \omega^2 (dd/dt) = -\omega^2 \lambda R_R \sin\theta\, (1+\lambda/2 \cos\theta) \tag{18}$$

The extreme values of the roller relative velocity are shown below:

$$\Delta v_{max} = 2\omega (R_P - R_R) \quad \text{and} \quad \Delta v_{min} = 0 \tag{19}$$

Effect of the R_P / R_R ratio on the values of the relative velocity between the roller and the cylinder at different vane angular positions is illustrated in Figure 5. It is shown that rather smaller ratio give us lower relative velocity between rotating in one direction roller and cylinder

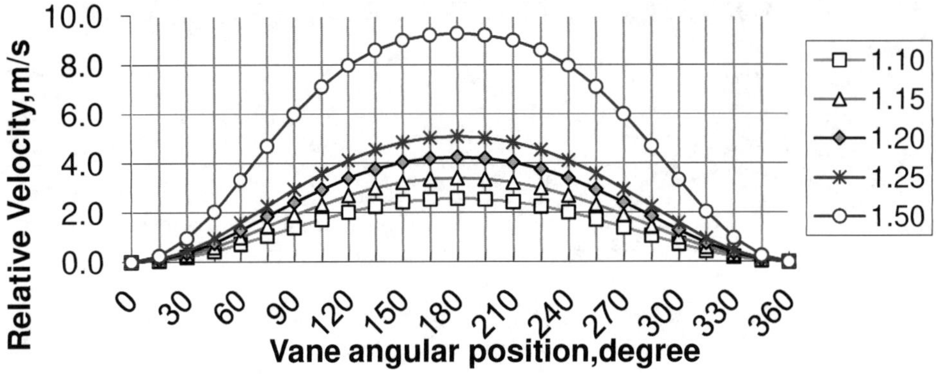

Figure 5. Effect of the radii ratio on the relative velocity.

2.4. Drive and power generating system of the structure.

Typically, the compressors, generators, airflow fans which may be employed within an air conditioning or a refrigerant unit, have to be selectively driven either by an electric motor or alternatively by a mechanical driver. The electrical motors of each component may be powered by utility line, generator, on-board battery or fuel cells. The mechanical driver can be a self-contained gasoline, diesel, propane or natural gas engine, but any other suitable source of mechanical power may be employed. The expression "motor"," generator" brings to mind two electrical machines with two separate rotational shafts, two rotors, stators, double numbers of bearings, etc.. Figure 6 is diagrammatical representation of some possible motor and generator arrangements in the structure.

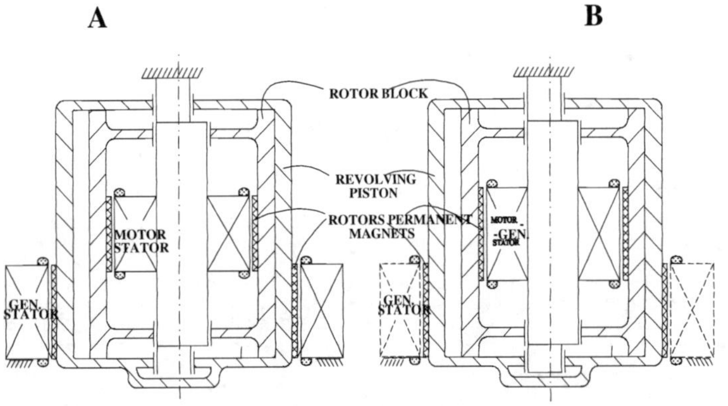

Figure 6. Schematic diagram of a motor, a generator (A) and a motor-generator unit (B) integrated with the compressor components.

A disadvantage of the structure construction shown in Figure 6A is that an electric motor can be used for running of the compressor (lane power is available) or for starting of the engine but has no useful function during power generation process when compressor is driven by an engine, thus motor represents “dead weight”. The structure assembly shown in Figure 6B (chosen for the structure) employs a single electrical motor-generator (MG) integrated with components of a revolving piston rotary compressor. The structure comprises a rigidly fixed on the stationary crankshaft MG stator, and a rotor rigidly fixed inside of the rotor block circumferentially to the stator. MG has ability to generate electric power from motive power supplied by an external mechanical force or can operate in a motor mode wherein motive power developed from electrical power supplied to the compressor. Either, the rotor block or the revolving piston, can be the driven by the motor, but the mechanical driving can be applied only to the piston due to the stationary crankshaft of the compressor. A power distribution capacity of the structure can be increased by positioned additional generator stators circumferentially around the piston external wall of which holding a plurality of permanent magnets. The multiple generator sets can be wiring in series or parallel to sum the power output or selectively distribute the power to individual electrical systems.

2.5 Modes of operation

The structure with combine motor –generator (Figure 6, B) can operate at following modes: motor mode to transfer motive power to the compressor (line power is available); auxiliary engine starting mode at which the motor motive power transfers to the engine (motor powered by a battery); power generating mode (compressor will be driven by an engine, the structure operates as a generator). Referring now to Figure 7, there is illustrated a flow diagram of the structure's dual drive mechanism operation for selectively driving a refrigerant compressor by either an electric motor or alternately by an internal combustion engine.

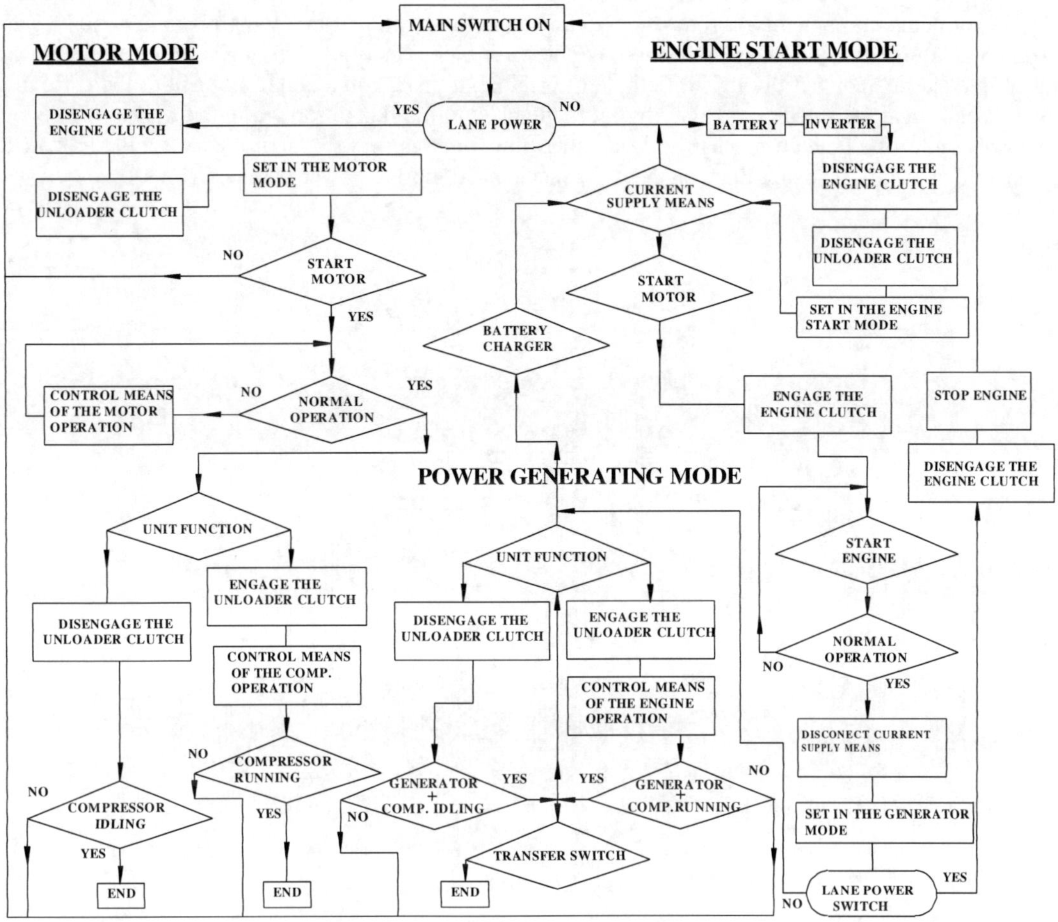

Figure 7. A flow diagram of the structure operations

The flow diagram does not include details of electronic units controlling start and running conditions of the compressor or the engine due to the specific requirements for every entity. An electromagnetically operated unload mechanism located on a suction side of the structure serves substantially to start up the compressor under no load, or to switch the compressor to an idling operation during power generation mode where the structure can function only as a generator or can generate electric power and simultaneously perform compression. The piston and the rotor block couplet by the vane forms high rotational inertia unit supported by bearings located symmetrically on opposite ends of the stationary crankshaft. Such rotational unit performs as a flywheel and can be used to start the engine when there is disconnection from the grid. In this case the MG does not start the engine itself, but rather accelerates the rotational unit (in the beginning decoupled from the engine). Upon reaching a sufficiently high speed, the rotational unit is coupled to the crankshaft of the engine by means of a frictional coupling. The rotational energy stored in the rotational unit then turns over the engine. In generator mode, the electric machine is permanently coupled to the engine by the frictional coupling

3. COMPONENTS OF THE STRUCTURE

An integral compressor / generator / fan unitary structure 1 (see Figure 8) comprises, in combination, the following: a revolving piston rotary compressor pump 2; an internal suction gas delivery system which excludes a direct distribution of a refrigerant to the suction chamber; the MG 3 rigidly fixed on a stationary crankshaft 4, Dreiman (2016). The compressor pump components--a rotor block 5 and an eccentrically fit revolving piston assembly 6 are unidirectional spinning around the stationary crankshaft 4. The rotor block and the revolving piston assembly have no radial clearance internal line contact through which the rotor block transfers an angular moment to the revolving piston assembly. A centrifugal fan impeller wheel 7 has been attached to the flange formed in the wall of the piston. An electromagnetically controlled clutch mechanism 8 coupled the compressor with an auxiliary engine to start it, activate energy generation process and simultaneous transfer mechanical motive power to the compressor.

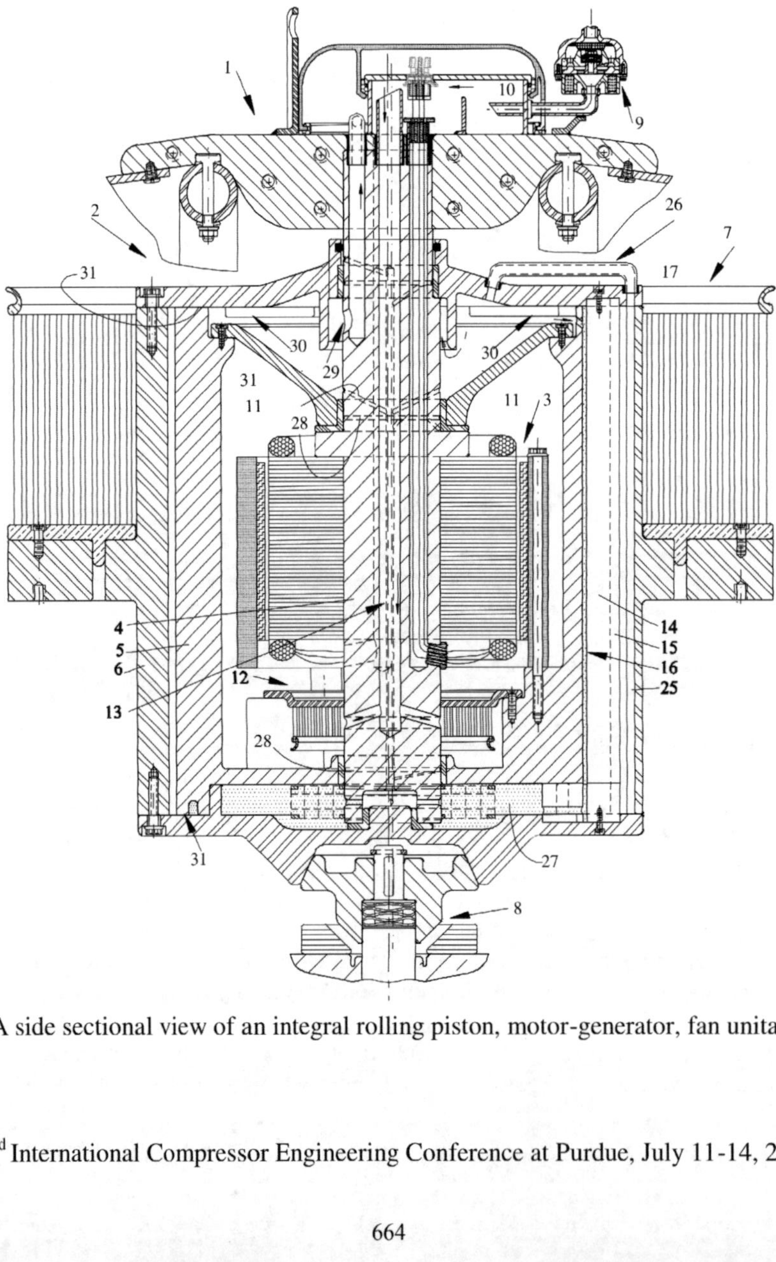

Figure 8. A side sectional view of an integral rolling piston, motor-generator, fan unitary structure

3.1 Suction System

The suction system of the novel rotary compressor consists of an intake adapter 9 equipped with an electromagnetically activated a stepless unloader. The unloader serves substantially to start up the compressor under no load, or to switch the compressor to an idling operation. The suction system further including an input suction cavity 10, the MG compartment 11, an impeller 12 rigidly fixed to the revolving rotor block, formed in the crankshaft a suction channel 13 through which refrigerant will be delivered to the MG compartment. The incoming refrigerant separates from liquid in the input suction cavity 10 (due to gravity) and in the MG compartment 11 (due to a centrifugal force), compressed (due to action of the impeller), cooling motor-generator 3 and delivers under higher pressure into the suction chamber.

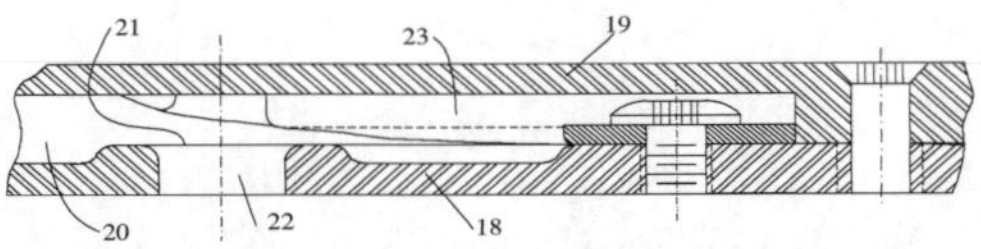

Figure 9. A partial sectional view of the vane-valve assembly.

3.2 Vane assembly and discharge system

The vane 14 of the compressor separates suction chamber and compression chamber (see Figure 8) and has an axial edge 15 rigidly fixed without operating clearance in a an axial slot formed in the piston 6 with opposite axial edge 16 fitted in between sliding segmental bushings mounted in the rotor block. The vane edges 17 are radially rigidly fitted in between the revolving piston heads without operating clearances. It eliminates frictional losses and also blocks high-low side leaks. Integration of the vane with the revolving piston excludes frictional and leakage losses observed in contemporary rotary compressors where a vane and roller slide against facing surfaces of the stationary cylinder block and stationary heads. There is also no" grinding" interference between vane and roller (roller and spring are eliminated) and, in addition, nose of the vane (axial edge 16 does not required any special treatment, such as precision machining, TiN coating, special vane tip seal, etc., recommended and typically used in rotary compressors. The vane consists of a held together discharge side plate 18 and suction side plate 19 (see Figure 9); a discharge valve cavity 20; at least one reed type valve 21 clamped to the back of the discharge side plate against discharge port opening 22. The suction side plate 19 is formed integrally with a valve stop 23 which prevents excessive movement of the reed valve. A cone-shaped discharge gas expansion cavity 24 is in fluid communication with the valve cavity 22 through an axial discharge channel 25 formed in the piston wall and a bypass connecting discharge passage 26.

3.3 Lubrication

The lubrication system (see Figure 8) comprises an oil rotary type pump spaced below the rotor block, plurality of axial channels and radial passages 28 formed in the stationary c'shaft through which oil flows to the bearings and delivered to the cone- shaped discharge cavity 24. During operation of the compressor the accumulated in the cavity 24 oil and liquid/gas mixture from the incoming discharge are thrown against the wall of the cone-shaped reservoir by action of centrifugal force. This causes a boundary layer to be formed between the oil and relatively lighter gas in axial region of turbulence. The volume $\boldsymbol{v}$ of the vapor separated from oil depends upon the angular velocity $\boldsymbol{\omega}$ and will be

$$v = \int_0^H \frac{R_0^2}{H} x\,dx - \pi H R_{cs}^2 = \pi H/2\,(R_B^2 - R_{cs}^2) \qquad (20)$$

where

$$H = \omega^2 R_B^2 / 2g \qquad (21)$$

Here H is the axial depth, R_B is the radius of the rotor block, Rcs is the radius of the crankshaft**,** and g is the acceleration of gravity. The reservoir has an axial gas outlet 29 formed at the top portion of the stationary crankshaft for discharge of the vapor and plurality of oil outlets 30 connected to plurality of vertical channels in the wall of the rotor block through which oil delivered by centrifugal force flows downwardly under gravity into an intake cavity of the oil pump. The oil thrown outward by centrifugal force forms also an annular seal of liquid at the periphery 31of the chamber and along the axial edge16 of the vane. Any leakage that may take place past the sides of the vane or past the periphery 31 of the rotor block facing piston heads will be leakage of the lubricant inward into the crescent-shaped space, as this oil is under the delivery pressure (discharge + pumping) in addition to the pressure due to centrifugal force.

4. A STANDBY OUTDOOR CONDENSER-GENERATOR UNIT

A standby outdoor air-cooled condenser - generator unit has been formed by combining the structure with residential (commercial) outdoor condenser. A natural gas, gasoline, or propane auxiliary engine may be used to drive a motor-generator of the compressor and to provide backup power during line power outages. The Figure 10 illustrates the system with the engine placed below the unit which utilizing an impeller wheel of a centrifugal fan.

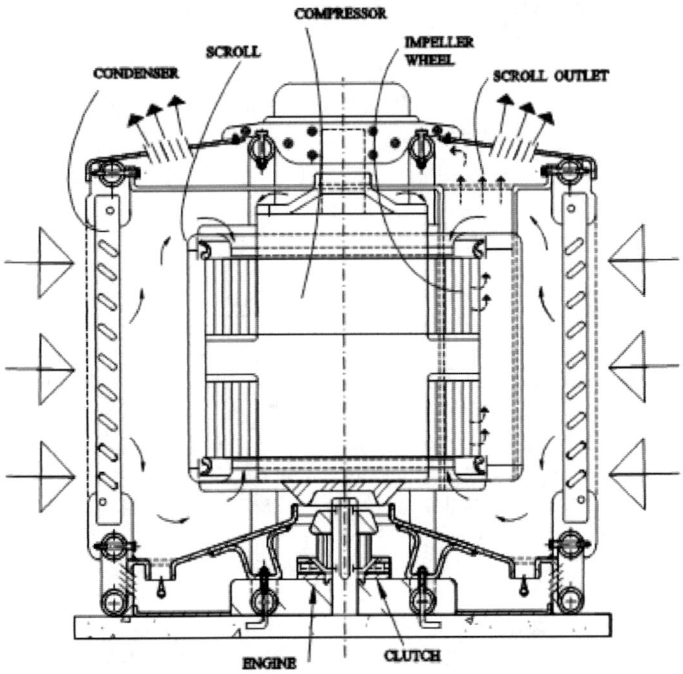

Figure 10. A schematic vertical cross section of standby outdoor condenser-generator unit utilizing the compressor/generator/fan structure

In this arrangement suction input, discharge output, the motor/generator electric power input/output terminals are spaced above the structure. Disposed around and extending axially along the compressor is a scroll type design housing for efficient conversion of velocity pressure to static pressure. It is fixed to the frame of the structure and serves also as a safety screen surrounding a spinning assembly. The engine, compressor, optional generators, cooling fan, and condenser coils combined with associated hardware and electronic control means are located in a common weatherproof and noise reduction enclosure. It is possible to reverse the location of the inputs/outputs by position the engine at the top of the structure. An axial fan propeller or the impeller wheel can be attached to the upper portion of the piston head and equipped with the electromagnetically controlled clutch to cycling the fan in response to changes of outdoor ambient temperature.

5. CONCLUSIONS

- A single structure has been formed by radial integration of a revolving piston rotary compressor and electrical motor-generator, fan components.
- A dual drive system permits to run the structure in generating mode when a motive power supplied by an external mechanical force or to utilize an available electric power to operate it. The design of the structure and the drive system excludes multiple conversions of transferred energy observed for the plurality of the individual units employed before.
- The compressor components with attached impeller wheels are spinning around their center of gravity –stationary c'shaft with rigidly fixed on it stator. Short length of the c'shaft is due to the radial integration of the motor –generator and compressor components. Large diameter of the c'shaft makes

it possible to accommodate inside electrical power conduits, oil and refrigerant channels. The bearings supported all rotational parts of the structure are positioned symmetrically on the stationary c'shaft below and above the fixed stator.

- The single structure replaces the plurality of separate entities each of which has been an individual contributor to the overall noise level and vibrations. Additional reduction of noise and vibration has been achieved by the symmetrical arrangement of the close-spaced bearings. Such arrangement combined with the dimensionally parameters mentioned above practically eliminates deflection of the c'shaft, and improve reliability of the structure.
- A suction gas delivery system excludes a direct distribution of a refrigerant to the suction chamber and includes double-stage liquid-gas separation to prevent slugging, provides cooling of the motor-generator following by supercharge of the refrigerant into the suction chamber due to action of an internal impeller. An additional liquid-gas separation is performed on a discharge side of the structure before expelling of the gas
- A power distribution capacity of the structure can be increased by positioned additional generator stators circumferentially around the piston external wall of which holding a plurality of permanent magnets. The multiple generator sets can be wiring in series or parallel to sum the power output or selectively distribute the power to individual electrical systems.
- A compact, therefore space saving unitary structure employs single motor-generator to operate compressor and fan or compressor, fan and generator during electric power outages. Utilization of the structure in air conditioning and refrigeration systems will reduce cost due to elimination duplicity of such parts as crankshafts, stators, rotors, bearings, belts, etc..It will simplify wiring procedure and reduces assembly and service time.

REFERENCES

Dreiman N.I., (2014). Rotary Compressor with the Stationary Crankshaft. *Int. Comp.Eng.Conf. at Purdue*

Dreiman N.I., (2016). Revolving Piston Rotary Compressor with Stationary Crankshaft. *US Patent Office, Publication No.: 2016/0032925A1. Filed Apr.2, 2014.*

Study on gas-liquid two-phase flow patterns and pressure drop in a helical channel with complex section

Bo CAI[1]*, Guo-Dong XIA[2], Jing-Fu WANG[3]

[1]Key Laboratory of Enhanced Heat Transfer and Energy Conservation, Beijing University of Technology, Beijing, China
E-mail: caibo2014@emails.bjut.edu.cn

[2,3]Key Laboratory of Enhanced Heat Transfer and Energy Conservation, Beijing University of Technology, Beijing, China

ABSTRACT

Single screw expansion is widely used in petroleum and chemical department, energy recovery engineering, power engineering and many practical applications. Yet the problems of gas leakage in the single screw expansion seriously affect its efficiency. An experimental investigation of gas-liquid two-phase flow patterns and pressure drop inside the single screw expansion has been carried out. The flow patterns are distinguished and recorded by high speed camera using the backlight imaging tomography. The test section made of plexiglass is manufactured according the screw of single screw expander prototype whose diameter and length of are 155mm and 112mm. There are five flow patterns in the screw channel: bubbly flow, dispersed bubbly flow, slug flow, churn flow and annular flow. The flow pattern map of the test channel is presented and compared with the flow pattern map of typical available models. The effect of Re_L and Re_G on pressure drop is presented. Pressure frequency of four flow patterns and their PDF function are presented. Based on the experimental data a new correlation to predict the liquid multiplier factor of test channel is purposed.

Key words: screw channels; two-phase flow; pressure drop; flow patterns

1. INTRODUCTION

Organic Rankine Cycle (ORC) has been found in wide applications in metallurgical, electric, the natural gas transmission and so on, which is an effective way of recovering the waste heat and pressure resources. In these industries, a large amount of waste heat at low temperature (below 230 ℃) has been underused because traditional steam turbines are not available to convert heat energy into electrical energy. The single screw expander is a kind of volumetric expansion machine with the power from 1kW to 1000 kW. This type of expander applied in ORC system extends the range of heat source temperature. As the working fluid (dry/wet vapor) expands in the single screw expander the lubricating flow together with the gas. It is noted that the amount of gas were decreased in leak by 25 to 30 percent even in the oil feeding of 100g/h (Schels C, 1984). Then the development of prime movers and used in low temperature source is more significant.

Two-phase flow through the channels of single screw expander is greatly more complex substantially than that of general helical coiled tubes. The complexity performance exists in two aspects. The first is two-phase flow interaction effect between two adjacent screw channels. The other one is continuous changes in cross section area and shape alone the axial direction of the screw channel. It is important to assess the irreversible loss for precisely designing single screw expander. Liu X F and Xia G D (2014) simulated the two-phase flow frictional pressure drop and liquid holdup under different torsions and inlet sectional liquid holdup in screw channels. The results showed that decreasing the pitch diameter or increasing the inlet liquid holdup would bring about the increase of pressure drop. They fitted the pressure drop correlations in the consideration of pitch diameter.

The experimental researches on pressure drop for helically coiled channels are large, but they are questionable if the channels are tested in vertical orientation. Some of works discussing the pressure drop characteristics and the correlations have been proposed by Ripple et al. (1966), Owhadi et al. (1968), Akagawa et al. (1972), Xin et al (1996). Others carried out air-water flow experiments in helically tubes and analyzed differential pressure fluctuation characteristics through diverse analytical tools. Among the former works, the Lockhart-Martinelli (1949) method has been generally used to predict the two phase flow pressure drop for helical channels in vertical orientation. However, the screw channels of single screw expander are worked in horizontally orientation. Chen and Guo (1999) found that the relationship between two-phase multiplier and the reciprocal of the Martinelli parameter X is not monotonous in helical tubes at horizontally orientation. So they modified Chisholm method which showed that the maximum deviation between experimental data and predicted values is within 30%. Awwad (1995) also indicated that L-M correlation was not valid in the prediction except in the high flow rates and modified the L-M correlation by firstly induced the Froude number and dimensionless parameter d/D in the expression. In spite of Awwad (1995) indicated that helix angle (0-20°) had no effect on pressure drop, R. C. XIN (1996) induced the inclined angle β into proposed correlation to predict the pressure drop experiment data. Sailendra M and Sudip D (2002) carried out the experimental for gas-Newtonian liquid through 24 helical coils. They suggested the correlations to predict the pressure drop with acceptable accuracy based on the Awwad (1995) model. Asit B (2007) evaluated the two-phase frictional pressure drop for air non-Newtonian liquid through 36 different helical coils and correlated the experimental data by dimensional analysis.

For assessing the irreversible loss and implementing real-time monitor of the screw channel, a lot of researches on differential pressure fluctuations were reported. Pinto and Campos investigated the behavior of Taylor bubble by measurement the pressure fluctuation in vertical tubes. Bai (2002) conducted the two-phase flow differential pressure fluctuations experiment and analyzed the change of the average, root mean square, and fractal dimension with different gas and liquid superficial velocity. Zhou (2003) extracted the probability density function of differential pressure fluctuations in different two-phase flow patterns and concluded that it could identify the two-phase flow patterns. Shaband (2014) studied the power spectral density function of two-phase flow differential pressure fluctuations characteristics. Literature review showed that the previous works of research subject mainly questioned on upward two-phase flow in vertical straight tubes. The works on the helical tube are rather scarcity in spite of the various applications the research.

This work focuses on the pressure drop characteristics of air-water two-phase flow in screw channels. Five two-phase flow patterns were identified in screw channels. Differential pressure fluctuations characteristics in different superficial gas and liquid velocity for five flow patterns are discussed and analyzed by extracting their probability density function. The frictional pressure drop data of screw channel are compared with the correlations of Lockhart-Martinelli (1949). The correlation of the pressure drop is proposed for screw channel.

2. EXPERIMENT SETUP

The air-water two-phase flow experimental apparatus including the gas and water supplying system, measuring apparatus, test section and data acquisition system is shown schematically in Fig.1. In this flow loop, water stored in a storage tank with an internal volume of 1 m^3 and driven by a stainless steel centrifugal pump in the loop. Air was supplied by an air-compressor and then flows through the air measuring station. The gas flow rate was measured by a thermal mass flow meter with an accuracy of ±1.0%. The water flow rate was measured by an electromagnetic flow meter with an accuracy of±0.5%. Then these two fluids entered the test section after a mixing chamber. The flow patterns are distinguished and recorded by high speed camera using the backlight imaging tomography. The pressure measurement probes were fixed near the inlet and outlet on the text channel and they were statistical to calculate the pressure drop in the screw channel.

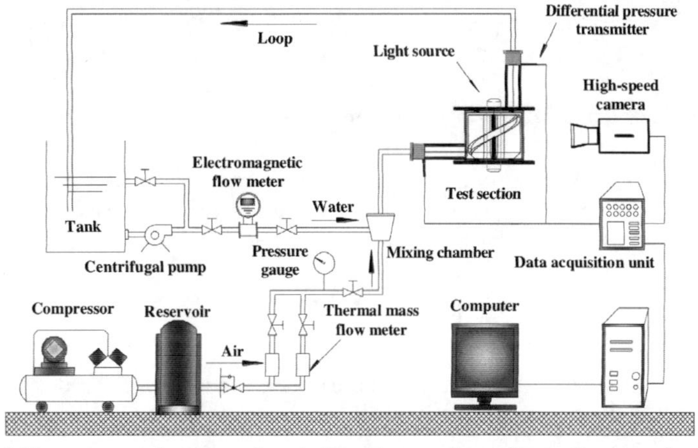

Fig.1 Experimental system for air-liquid two-phase flow of screw channel

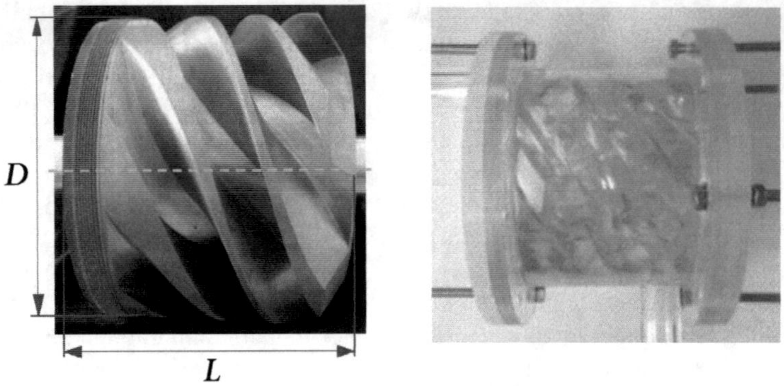

Fig.2 Single screw expander prototype and the test section

Diagram of single screw expander prototype and the test section is presented in Fig. 2. The test section made of plexiglass is manufactured according the screw of single screw expander whose diameter (*D*) and length (*L*) of are 155mm and 112mm. Diagraph of the a test channel model in the screw is presented in Fig.3. The screw channel is a helical channel with variable quadrilaterals cross-sections. And the screw channel is characterized that the largest flowing area is in the 90° turn relative to the inlet and slightly decrease to the outlet and inlet.

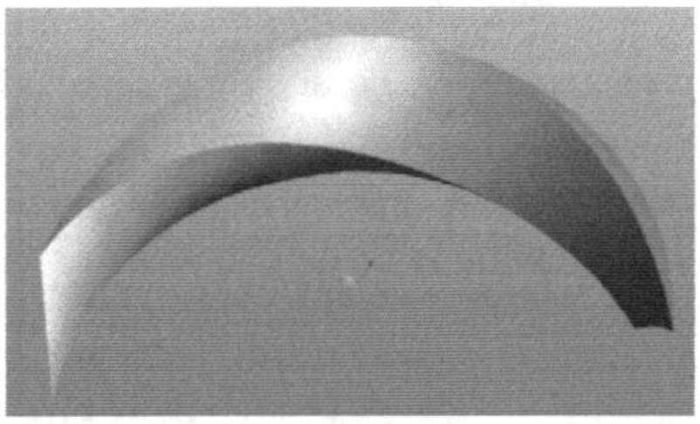

Fig.3 The model of test channel in single screw expander prototype

3. RESULTS AND DISCUSSIONS

3.1 The flow pattern map of screw channel

Two typical models of flow pattern maps for different tube structures were used for exploring the screw channel flow pattern transitions mechanism. Flow pattern map of Taitel et al. (1983) for upward two-phase flow has been selected, since they have been successfully applied to many flow configurations. The flow pattern map of Guo (1987) was developed for air-water two-phase flow in the helically coiled tubes. Fig.3 (a)-(b) compares the flow pattern map of upward flow section in screw channel with the selected models.

A good agreement is found between the present data and the Taitel et al. (1983) model for flow pattern transition from bubbly flow to slug flow. Taitel et al. model is based on a critical void fraction, with predicted values of 0.25. The gas superficial velocity U_{sG} value at slug to churn transition lines in screw channel is lower than that in vertical tube. It is can be seen the consistency of flow pattern transition between the text channel and the Taitel et al. model is not very well because the different geometry structure between these two channels.

In the screw channel, dispersed flow emerges at the liquid superficial velocity U_{sL} value increase to more than 1.2m/s. The consistency between Guo et al. model and present date is very well for bubble to dispersed transition. In addition, a good agreement between the experimental data and Guo et al. model for churn to annular flow pattern transition can be found in Fig.3 (b).

3.2 Two-phase pressure drop frequency

The probability density function applied to character the pressure frequency is defined as:

$$P_{\mathrm{i}} = \frac{n_{\mathrm{i}}}{N} \quad (1)$$

Where n_i is the number of amplitude value in the *i* region, *N* is the total number of the measurement data.

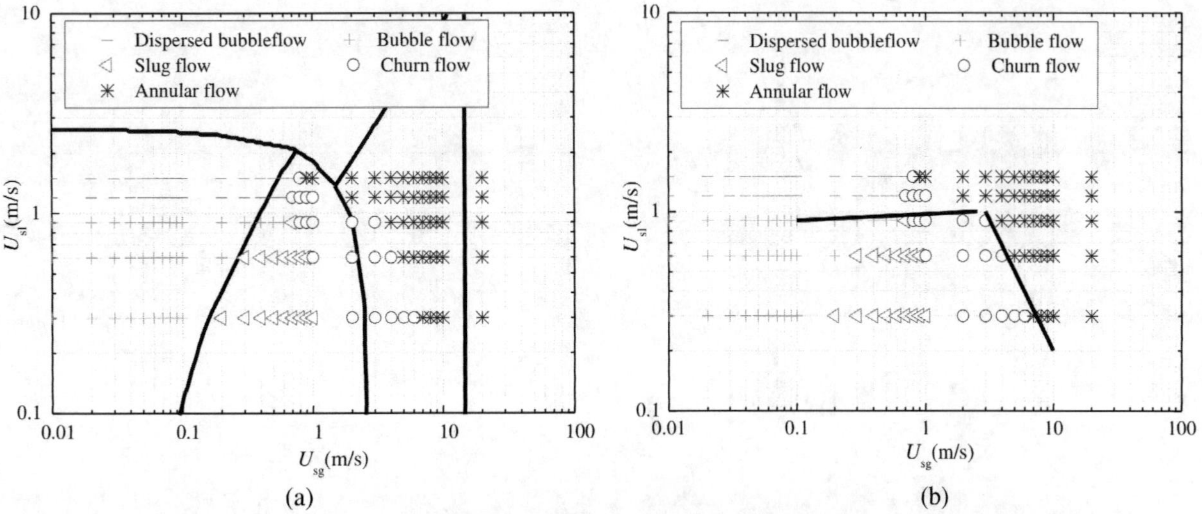

Fig.3 Comparison between experimental data and existing flow pattern transition models for: (a) Taitel et al. (b) Guo et al.

The fluctuations signal of differential pressure with different U_{sL} and U_{sG} of bubbly flow, dispersed bubble flow, slug flow and annual flow in screw channel are shown in Fig. 4, 6, 8 and 10, respectively. The range of fluctuations for bubbly flow is smallest and for annular flow is largest among these flow patterns. The fluctuations of differential pressure were in the range of 0.05-0.4 kPa and 0.6-6.5 kPa for bubbly flow and annular flow. The fact is bubbly flow occurred in the low U_{sL} and U_{sG} with low turbulence intensity. And the annular flow always happened in high U_{sG} and accompanied by strong swirling flow of gas. There are obvious periodic oscillations in the fluctuations signals of slug flow with the range of -0.3-0.3kPa. The fluctuations signal of differential pressure of dispersed bubble flow with the range of 0.7-2.2kPa. Dispersed bubble flow happened in high U_{sL} with strong liquid turbulence intensity. The increasing of U_{sL} enlarged the liquid turbulence intensity which made the bubbly flow transferred to the dispersed bubble flow.

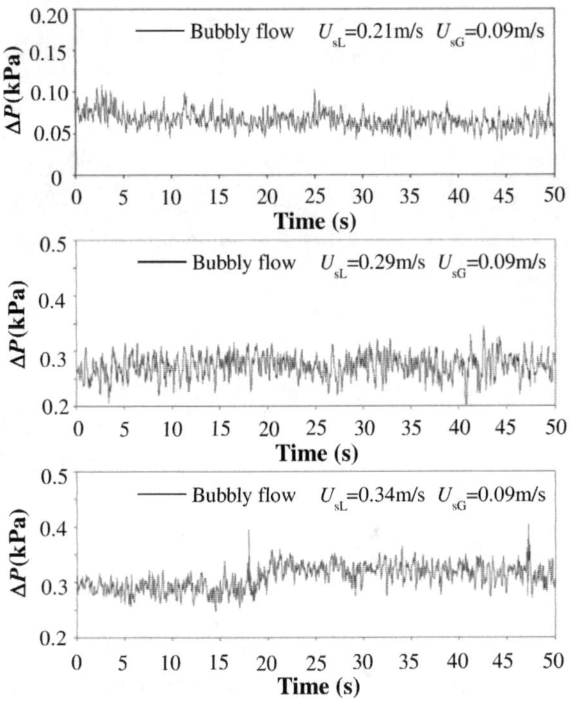

Fig. 4 Pressure fluctuations in bubbly flow

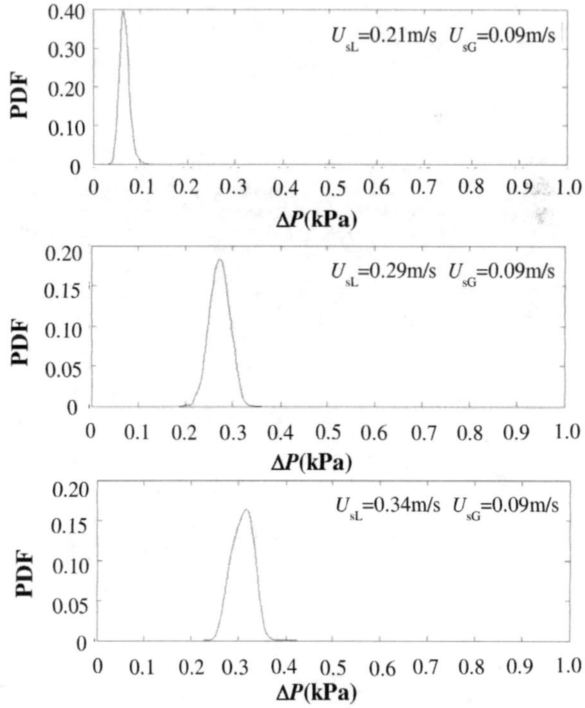

Fig. 5 Probability density function in bubbly flow

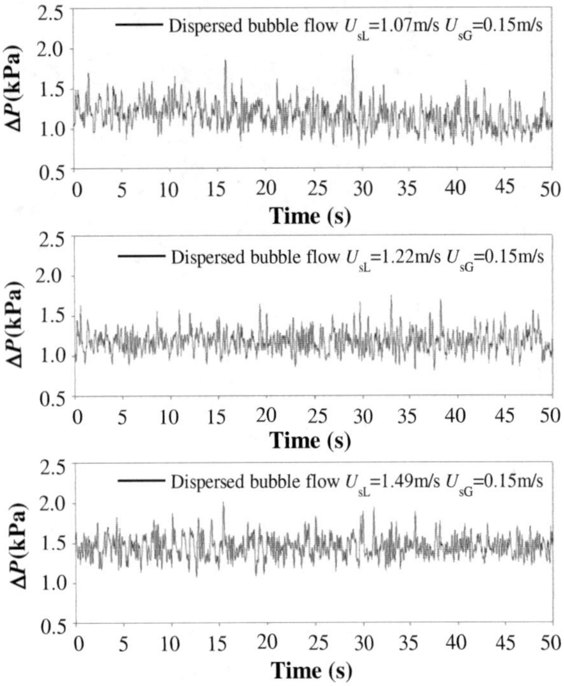

Fig. 6 Pressure fluctuations in dispersed bubble flow

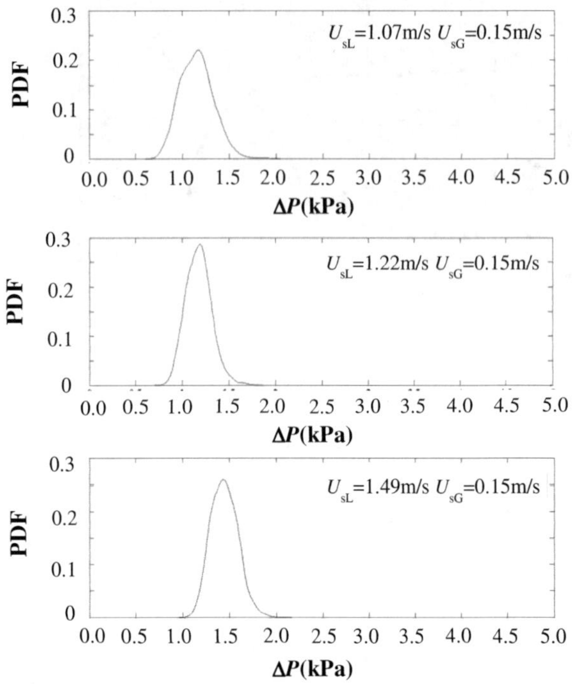

Fig. 7 Probability density function in dispersed bubbly flow

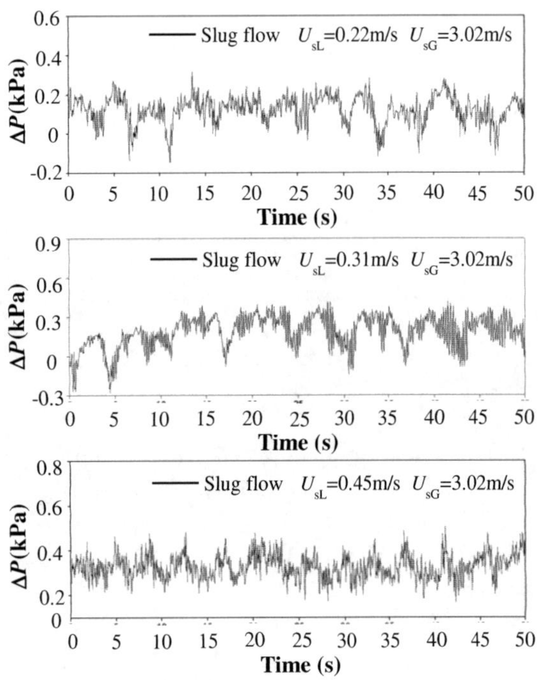

Fig. 8 Pressure fluctuations in slug flow

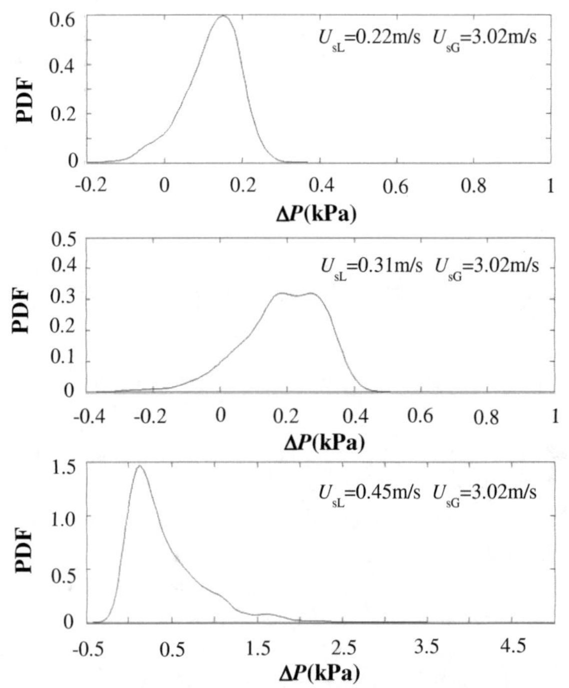

Fig. 9 Probability density function in slug flow

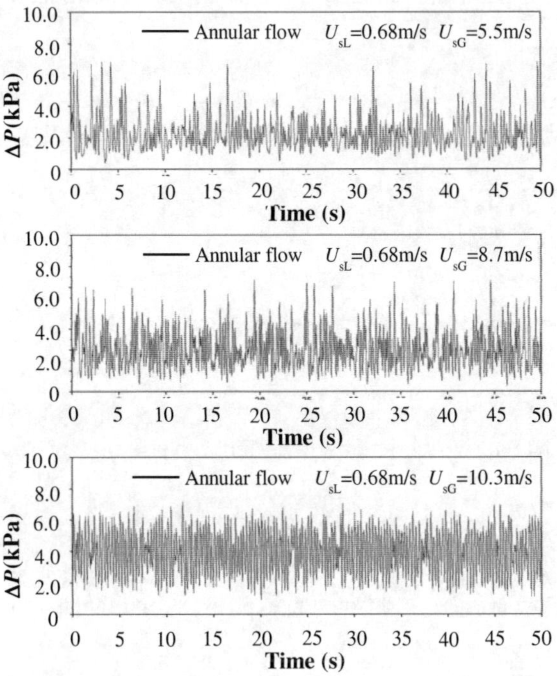

Fig. 10 Pressure fluctuations in annular flow

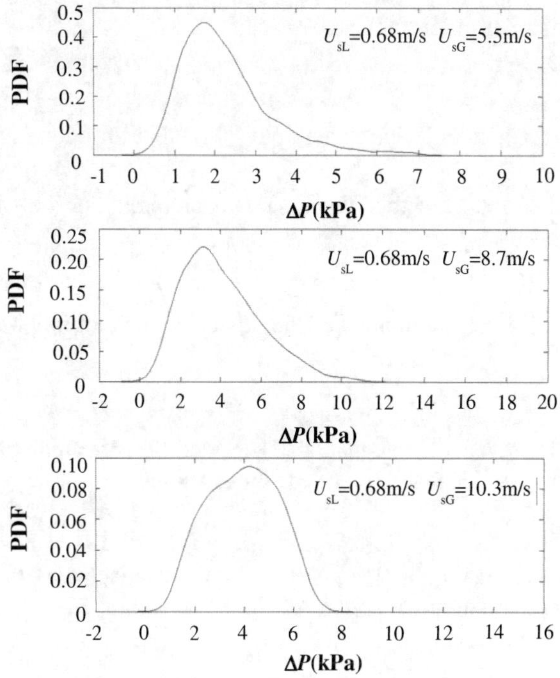

Fig. 11 Probability density function in annular flow

For the purpose of more information extraction from differential pressure fluctuations signal probability density function of these four flow regimes are shown in Fig. 5, 7, 9 and 11. Transverse amplitude of the probability density function represented the amplitude of fluctuation. Probability density curve in the bubbly flow was forward-peaked distribution with narrow transverse amplitude. The mean value (U_{sL}= 0.21m/s, 0.29m/s, 0.34m/s) are 0.066 kPa, 0.273 kPa, 0.309 kPa, and standard deviation are 0.003 kPa, 0.009 kPa, 0.013 kPa. With the increase of liquid flow rate the mean value and standard deviation of the differential pressure fluctuations signal are increasing. This may due to the increasing U_{sL} would enlarge the liquid disturbance, which made the energy fluctuation increase. For the dispersed bubble the probability density curve resembled normal distribution flow and its transverse amplitude was wider than that of bubbly flow. The mean value (U_{sL}= 1.07m/s, 1.22m/s, 1.49m/s) are 0.122 kPa, 0.136 kPa, 0.156 kPa, and standard deviation are 0.041 kPa, 0.225 kPa, 0.015 kPa. The increasing mean value and standard deviation for dispersed bubble flow due to the strong turbulence intensity. Probability density curve of this flow regime was abnormal distribution and exist the double peak distribution. The mean value (U_{sL}= 0.22 m/s, 0.31 m/s, 0.45 m/s) are 0.182 kPa, 0.256 kPa, 0.352 kPa, and standard deviation of slug flow are 0.041 kPa, 0.225 kPa, 0.015 kPa. Standard deviation of slug flow is smaller than that of dispersed flow because of periodic oscillations in the slug flow. The mean value of annular flow (U_{sL}= 5.5 m/s, 10.3 m/s, 8.7 m/s) are 1.854 kPa, 3.759 kPa, 4.066 kPa, and standard deviation are 0.671 kPa, 0.873 kPa, 2.874 kPa. Transverse amplitude of probability density curve for annular flow was the widest.

3.3 Friction pressure drop prediction of the screw channel

The Lockhart-Martinelli method is the effective approach to harmonize data of two-phase flow. Yet Lockhart-Martinelli method applied for non-straight tubes have lower accuracy (Awwad (1995), Chen (1999)). Measured pressure drop of two phase flow is generally expressed as follows:

$$\Delta P_{\mathrm{TP}} = \Delta P_{\mathrm{TPf}} + \Delta P_{\mathrm{TPa}} + \Delta P_{\mathrm{TPg}} \quad (2)$$

Where ΔP_{TPf} is the frictional pressure drop, ΔP_{TPa} is the acceleration pressure drop and ΔP_{TPg} is the gravitational pressure drop. The acceleration effect can be neglect because of no phase change in two-phase flow. The gravitational pressure drop has been considered as zero because the inlet and outlet of screw channel were in the same height from ground level. Then the frictional pressure drop can be expressed by

$$\Delta P_{\mathrm{TPf}} = \Delta P_{\mathrm{TP}} \tag{3}$$

The experiment data are reduced with the help of liquid friction multiplier Φ_L and gas friction multiplier Φ_{G}, their definitions are as follows:

$$\Phi_{\mathrm{L}} = \sqrt{\Delta P_{\mathrm{TPf}}/\Delta P_{\mathrm{L}}} \tag{5}$$

$$\Phi_{\mathrm{G}} = \sqrt{\Delta P_{\mathrm{TPf}}/\Delta P_{\mathrm{G}}} \tag{6}$$

The liquid friction multiplier Φ_L and gas friction multiplier Φ_{G} related the Lockhart-Martinelli parameter, X

$$X = \sqrt{\Delta P_{\mathrm{L}}/\Delta P_{\mathrm{G}}} \tag{4}$$

For L-M method the liquid friction multiplier $\Phi_{\mathrm{L,C}}$ is calculated by:

$$\Phi_{\mathrm{L,L-M}} = \left(X^{4/(\mathrm{n}-5)} + 1\right)^{(5-\mathrm{n})/4} \tag{7}$$

$$\Phi_{\mathrm{G,L-M}} = \left(X^{4/(5-\mathrm{n})} + 1\right)^{(5-\mathrm{n})/4} \tag{8}$$

For Chisholm method the liquid friction multiplier $\Phi_{\mathrm{L,C}}$ is calculated by:

$$\Phi_{\mathrm{L,C}} = \sqrt{1 + \frac{c}{X} + \frac{1}{X^2}} \tag{9}$$

Where n and c in equation (7), (8) and (9) is related with the state of two-phase flow. The single-phase flow pressure drop is calculated by the equations below:

$$\Delta P_{\mathrm{L}} = 2 f_{\mathrm{L}} \rho_{\mathrm{L}} U_{\mathrm{L}}^2 L_{\mathrm{S}}/d \tag{10}$$

$$\Delta P_{\mathrm{G}} = 2 f_{\mathrm{G}} \rho_{\mathrm{G}} U_{\mathrm{G}}^2 L_{\mathrm{S}}/d \tag{11}$$

Where L_{S} is the length of the screw channel, d is the hydraulic diameter of the screw channel. The single-phase pressure drop friction factors in the helical tubes for laminar flow are acquired from the expression by Manlapaz and Churchill (1980):

$$\frac{f}{f_s} = \left[1 + \left(1 + \frac{d/D}{3}\right)^2 \left(\frac{De}{88.33}\right)\right]^{0.5} \tag{12}$$

$$f_s = 16/Re \tag{13}$$

Where D is the coil diameter, De is the Dean number. For turbulence flow, the friction factor is calculated from the expression by Ito (1959):

$$f\left(\frac{D}{d}\right)^{0.5} = 0.00725 + 0.076\left[Re\left(\frac{D}{d}\right)^{-2}\right]^{-0.25} \tag{14}$$

Fig.12 shows the influence of Re_{G} on the two-phase frictional pressure drop of the screw channel various Re_{L}. It is can be found that the pressure drop of the screw channel is increases with increasing the Re_{G} and Re_{L}. Fig. 13-14 shows the measured gas and liquid pressure drop multiplier factors for the screw channel in terms of L-M parameter.

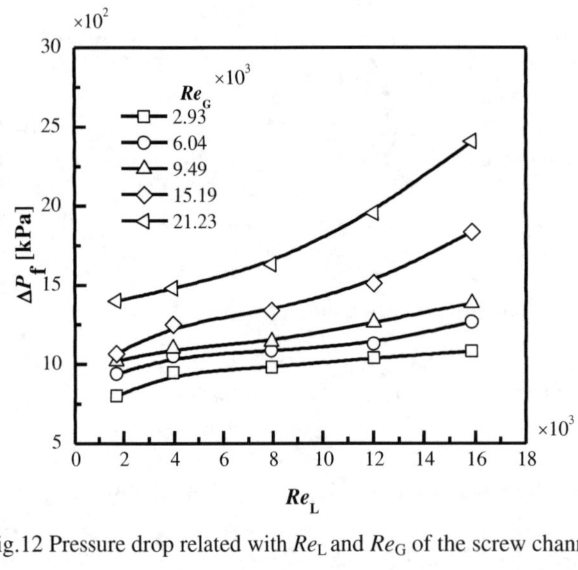

Fig.12 Pressure drop related with Re_{L} and Re_{G} of the screw channel

In addition, the L-M correlations for two phase flow in screw channel are presented with imaginary line. The friction multiplier factors are closed to the L-M correlations. However, the pressure drop of screw channel is also related to the liquid flow rate.

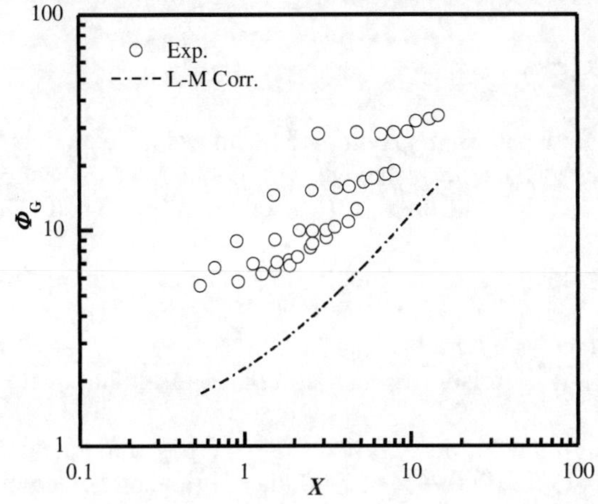

Fig.13 Gas multiplier factor of the screw channel compared with L-M correlation

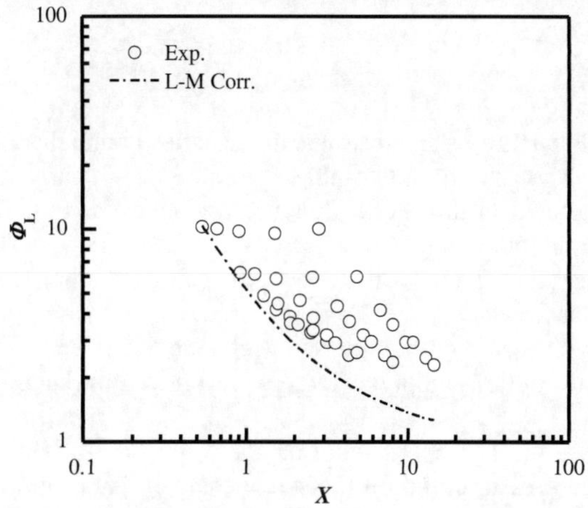

Fig.14 Liquid multiplier factor of the screw channel compared with L-M correlation

A new correlation predicting the pressure drop in screw channel based on the expressions by Xin (1996) is presented to fit with experimental data. The new correlation is terms of *Fr*, *D/d* and *X*, which considering the inertia force, liquid gravity force and centrifugal force effect. The model applied to predict the pressure drop is as following:

$$\Phi_L = \Phi_{L,C}\left(1 + \frac{X}{CF_S^{0.6}}\right) \tag{15}$$

Where for F_S<0.1, C=7.4; 0.1< F_S <0.5, C=9.6, F_S >0.5, C=15.6. F_S is defined as

$$F_S = \frac{U_{sL}^2}{Gd}\left(\frac{d}{D}\right)^{0.5} \tag{16}$$

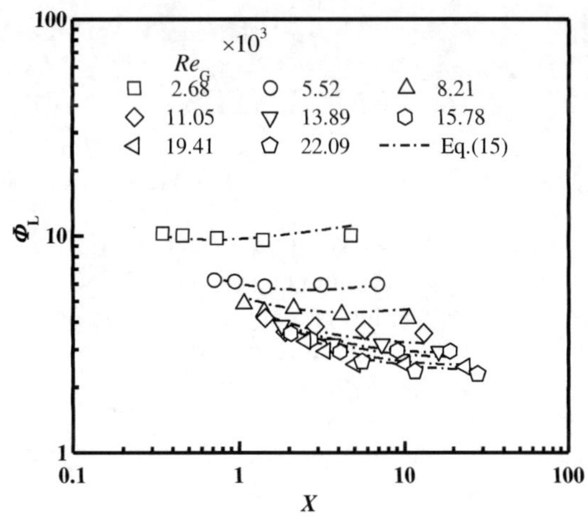

Fig.15 Friction pressure drop prediction of the screw channel compared with experimental data

The maximum error of the liquid multiplier factor Φ_L between the equation (15) and the experimental data is 15%. The pressure drop experimental data of screw channel compared with the new correlation is showed in fig.15. It is shown that the new correlation have a good agreement with the experimental data.

CONCLUSION

Five flow patterns are observed in the screw channel and the flow pattern map compared with available models is analyzed. The gas-liquid two-phase pressure drop characteristics in the screw channel with complex cross-section are investigated in this work. Pressure frequency of four flow patterns and their PDF function are presented. The results are as following:

(1) There are five flow patterns in the screw channel: bubbly flow, dispersed bubbly flow, slug flow, churn flow and annular flow.

(2) The pressure drop of the screw channel is increases with increasing the Re_G and Re_L.

(3) The friction multiplier factors are not only related the L-M correlations, but also related to the liquid flow rate.

(4) Base on the experimental data and the existing predicting model a new correlation is purposed to calculate the friction pressure drop for the screw channel. And for $0<Re_G<2100$, $1000<Re_L<18000$, there is a good agreement between the experimental data and the correlation.

NOMENCLATURE

D	diameter of the screw	(mm)
d	hydraulic diameter of the screw channel	(mm)
L	length of the screw	(mm)
L_S	length of the screw channel	(mm)
U_{sL}	liquid superficial velocity	(m/s)
U_{sG}	gas superficial velocity	(m/s)
Re_L.	liquid Reynolds number	
Re_G	gas Reynolds number	
n_i	number of amplitude value in the *i* region	
N	total number of the pressure data	
Φ_L	liquid friction multiplier	
Φ_G	gas friction multiplier	
De	Dean number	

REFERENCES

Akagawa K T. Study on gas-liquid two-phase flow in helically coiled tubes. Bull JSME. 1971,14.64-71.

Awwad A, XIN R C, Dong Z F, et. al. Measurement and correlation of the pressure drop in ,air-water two-phase flow in horizontal helicoidal pipes. Int. J. Multiphase Flow. 1995. 21(4): 607-619.

Bi Q C, Chen T K, Luo Y S, Zheng J X, Jing J G. Frictional pressure drop of steam-water two-phase flow in helical coils with small helix diameter. Process of the 3rd International Symposium on Multiphase Flow and Heat Transfer. Xi'an Jiaotong University Press and Begell House, 1994, 1,498-505.

Chisholm D. A theoretical basis for the Lockhart–Martinelli correlation for two-phase flow. Int. J. Heat Mass Transfer. 1967, 10, 1767–1778.

Chisholm D. Pressure gradients due to friction during the flow of evaporating two-phase mixtures in smooth tubes and channels. Int. J. Heat Mass Transfer .1973, 16, 347–348.

Chen X J, Guo L J. Flow patterns and pressure drop in oil-air-water three-phase flow through helically coiled tubes. International Journal of Multiphase Flow. 1999,25: 1053-1072.

Fang X H, et. al. Evaluation of frictional pressure drop correlations for two-phase flow in pipes. Nuclear Engineering and Design. 2012, 253: 86-97.

Guo L J, Feng Z P, Chen X J. An experimental investigation of the frictional pressure drop of steam-water two-phase flow in helical coils. International Journal of Heat and Mass Transfer,2001, 14(44),2601-610.

Guo L J, Feng Z P, Chen X J. An experimental investigation of the frictional pressure drop of steam-water two-phase flow in helical coils. International journal of heat and transfer.2001,44:2601-2610.

Hong W P, Teng F Y. The Flow Pattern and Differential Pressure Fluctuations of transition Flow across Tube Bundles. Energy Procedia, 2012, 1(17): 1507-1512.

Ito H. Frictional factors for turbulent flow in curved pipes. Journal of Engineering, 1959, 2(8),123-134.-488.

Lockhart R W, Martinelli R G. Proposed correlations for isothermal two-phase two-component flow in pipes. Chemical Engineering Progress, 1949, 45, 39-48

Santini L, Cioncolini A. Two-phase pressure drops in a helically coiled steam generator. International Journal of Heat and Mass Transfer, 2008,51:4926–4939.

Manlapaz R L, Churchill S W. Fully developed laminar flow in a helically coiled tube of finite pitch. Chem. Engng Commun. 1980 ,7, 57-78.

Mandal S N, Das S K. Gas-Liquid flow through helical coils in horizontal orientation. R&D Note.2002, 80: 979-983.

Owhadi A, Bell K J. Force convection bolling inside helically- coiled tubes. Heat and mass transfer.1968, 11, 1779-1793.

Ripple G R, Eidt C M. Two-phase flow in a coiled tubes. Ind. Eng. Chem. Proc. Des. Dev. 1996,32-39.

XIN R C, AWWAD A. An investigation and comparative study of the pressure drop in air-water two-phase flow in vertical helicoidal pipes. Int. J. Heat Mass Transfer. 1996, 39(4):735-743.

Xin R C, Awwad A, Z. F. Dong and M. A. An experimental study of single-phase and two-phase flow pressure drop in annular helicoidal pipesInt. J. Heat and Fluid Flow. 1997, 18: 482

Schels A H. et. al. Leakage calculation on Spray type rotary compressor. Foreign shipstechnology, 1984, 33-35.

Mandal S N, Das S K. Gas-Liquid Flow through Helical Coils in Vertical Orientation. Ind. Eng. Chem. Res. 2003, 42, 3487-3494.

Shaban H, Tavoularis S. Identification of flow regime in vertical upward air–water pipe flow using differential pressure signals and elastic maps. International Journal of Multiphase Flow, 2014, 5(61): 62-72.

Srinivasan P S, Nandapurka S S, Holland F A. Pressure drop and heat transfer in coils. ChemicalEngineering, 1968, 2(18),113-119.

Liu X F, Xia G D. 2014. Numerical analysis the two phase flow distribution and pressure characteristic in the screw channel. CIESC Journal. 59, 2784-2791.

ACKNOWLEDGEMENTS

This work is supported by the Beijing Natural Science Foundation (3142004) and the National Natural Science Foundation of China (No. 51576005).

Development of a general numerical methodology for CFD analyses in sliding vane machines and application on a mid-size oil injected air compressor

Giuseppe BIANCHI[1,2]*, Sham RANE[2], Ahmed KOVACEVIC[2],
Roberto CIPOLLONE[1], Stefano MURGIA[3], Giulio CONTALDI[3]

[1]University of L'Aquila,
Department of Industrial and Information Engineering and Economics,
L'Aquila, Italy
(giuseppe.bianchi@univaq.it, roberto.cipollone@univaq.it)

[2]City University London,
Centre for Compressor Technology,
London, United Kingdom
(sham.rane@city.ac.uk, a.kovacevic@city.ac.uk)

[3]Ing. Enea Mattei S.p.A.,
Vimodrone, Italy
(stefano_murgia@mattei.it, giulio_contaldi@mattei.it)

* Corresponding Author

ABSTRACT

The current work presents the development of a numerical methodology to investigate sliding vane rotary machines by means of advanced design tools such as the Computational Fluid Dynamics ones. Although highly limited in this topic, literature shows that the major constraint for the employment of such approaches is the deformation and motion of the rotor mesh, i.e. the computational grid related to the fluid volume between stator, rotor and blades of the positive displacement vane device. To address these issues, a novel grid generation approach is herein proposed and accomplished through a series of steps: geometrical 2D modeling of the machine cross section profile, boundary generation of the rotor mesh and, eventually, distribution of computational nodes using algebraic algorithms with transfinite interpolation, post orthogonalization and smoothing. This methodology was subsequently tested on an industrial vane compressor comparing the results of oil free and oil injected simulations set up in the ANSYS CFX solver. Results show angular pressure evolution inside the compressor vanes, a recirculation region induced by the clearance between the vane tip and the stator as well as the cooling effects of the oil entrained in the cells during the compression phase.

1. INTRODUCTION

Compressed air is an indispensable utility and, at the same time, one of the factors that highly affects the energy costs of industrial facilities. On a global scale, compressed air production accounts for 4.2% of electricity consumptions, 710 Mtonne/yr in absolute terms (IEA, 2015). In order to lower this share and the corresponding CO_2 emissions, the scientific and industrial communities have been developing energy saving and recovery strategies in compressed air systems. In particular, it has been estimated that more efficient compressors might contribute for up to 10% of the energy saving potential (Saidur et al., 2010). Although current air compressor market for industrial applications is led by screw machines, other positive displacement technologies are also available. Among them, sliding vane rotary compressors benefit of intrinsic geometrical and operating features that might forecast a breakthrough in compressed air systems.
The lower industrial interest in the development of advanced sliding vane machines, in turn, motivates the limited scientific literature about this research topic. Among the available works, Badr et al. (1985a, 1985b) firstly

performed a series of studies concerned to the development of a comprehensive modeling approach for sliding vane expanders including geometry, vane kinematics, blade dynamics and friction phenomena as well as energy balances. Modeling and experimental studies in steady state conditions were also carried out by Tramschek and Mkumbwa (1996) on circular vane compressors using radial and non-radial blade arrangements inside the rotor slots. Additionally, Al-Hawaj (2009) modeled a dual stage elliptical sliding vane compressor for refrigeration applications. Bianchi and Cipollone (2015a, 2015b) eventually presented comprehensive numerical and experimental methodologies to address mechanical and thermodynamic improvements in sliding vane machines, with particular focus on air compressors.

The modeling approaches proposed in all the above mentioned research works mostly belong to lumped parameters or one-dimensional (1D) categories. These simplified methods assume uniform or 1D spatial distributions of the physical quantities involved during mass and energy exchanges that highly reduce the computational cost of the simulation. However, they are not able to represent real machine geometry and effects like fluid leakages which are of foremost importance in devices like positive displacement machines. Furthermore, although the experimental validation of these mathematical models allowed to address and estimate further performance improvements of industrial machines, zero and one-dimensional approaches are not suitable to detect failures on existing devices or to properly support the development of future sliding vane prototypes at an industrial level. In order to accomplish studies of greater complexity and accuracy on real industrial products, very recently scientists from academia and industry started to employ state of the art Computational Fluid Dynamics (CFD) tools on sliding vane machines, namely expanders for energy recovery systems based on Organic Rankine Cycles (ORC) and automotive oil pumps.

Montenegro et al. (2014), investigated a sliding vane ORC expander with elliptic stator and working with R245fa using the OpenFOAM CFD solver. Due to the lack of an automatic grid generator, the vane mesh was created relying upon the built-in Cartesian mesh tool in OpenFOAM. To speed up the computations, the momentum equation was modified by the introduction of a Darcy-Forchheimer type source term, which takes into account both the viscous and inertial effects occurring in a tiny gap; this resistance source term was locally applied only on the cells located at the flat shaped vane tip. Discrepancies on the output power were up to 12%.

Kolasiński and Błasiak (2015) performed experimental and numerical studies on a micro ORC sliding vane expander with maximum power of 600W using R123. In their modeling activity, to avoid negative volumes in the mesh, the expander geometry was modified reducing vane thickness and increasing the vane tip clearance, whose profile was still flat. This latter adjustment yielded to a modified eccentricity. Movement and deformation of the vane grid were accomplished using an embedded tool of ANSYS suite. In particular, internal vane nodes were relocated according to a displacement diffusion model.

Zhang and Xu (2014) investigated the role of vane tip radius on local cavitation phenomena in an oil pump. In particular, the Authors developed a parameterized vane grid generator coupled with the software ICEM-CFD. Because of grid deformation, to avoid generation of artificial mass sources during the numerical solution, an additional equation, namely the space conservation law proposed by Demirdzic and Peric, had to be simultaneously satisfied with the other conservation equations. As concerns the turbulence model set in the Star-CD solver, k-ε was considered in the free stream regions while k-ω SST in regions of adverse pressure gradients and separating flow. To account for cavitation phenomena, Rayleigh model and Volume Of Fluid method were eventually considered.

The bibliographic survey clearly shows that generation and handling of rotor mesh, i.e. the computational grid that discretizes the fluid volume between stator, rotor and blades, are undoubtedly the most challenging tasks for CFD analyses on sliding vane machines and, actually, on any positive displacement device. The limitations of commercial CFD solvers in analyzing positive displacement machinery is additionally mentioned in Prasad (2004). Rane et al. (2013) compared most of the available approaches for grid deformation, namely diffusion equation mesh smoothing, user defined nodal displacement and key-frame re-meshing. These methods were firstly tested on a simple piston test case and further assessed using a real twin screw compressor geometry. Among the three approaches, only the user defined nodal displacement could be successfully applied to the complex twin screw geometry. This fact allowed to state the need of customized grid generators for numerical analyses in positive displacement machines.

In order to overcome all the assumptions and the limitations considered in previous research, the current work proposes the development of a general methodology for grid generation in sliding vane machines based on analytical models to reproduce the boundaries of the rotor mesh and on the meshing algorithms which have been developed at City University London for screw machines over the last two decades. As reference test case, an industrial oil injected sliding vane air compressor was considered. In particular, suction, discharge and oil injection routes were meshed using the embedded tool of the ANSYS suite and subsequently coupled with the rotor mesh generated with the novel grid generation procedure by means of non-conformal sliding interfaces. Transient three-dimensional (3D) multiphase simulations were carried out using the pressure coupled solver of ANSYS CFX. This procedure allowed to retrieve remarkable information on the flow field developing inside the compressor vane throughout the whole operation of the machine.

2. GEOMETRICAL MODELING

In geometrical terms, the core of a sliding vane machine is essentially composed of two cylinders which are eccentrically mounted one inside the other and tangent along a generatrix. The inner cylinder has a circular cross section and presents a series of slots that host an equal number of parallelepipedic blades. The outer cylinder may even have an elliptical cross section and it usually is a motionless component called stator, unlike the inner one that rotates around its axis and it is consequently called rotor. As the inner cylinder rotates, the blades slide along the rotor slots and build up the cells, whose volume changes over a full revolution because of the mounting eccentricity. To minimize friction losses, blade tip usually has a circular profile with circle center on the blade axis such that the contact between blades and stator theoretically occurs along a single generatrix.
Since the core of a sliding vane machine has a constant cross section profile with respect to the rotation axis, the boundaries of the fluid volume enclosed in the core were parameterized using an analytical bi-dimensional (2D) modeling approach that tracks the trajectories of all the relevant points displayed in Figure 1. This method, proposed in Bianchi and Cipollone (2015a), was herein improved to account for finite blade thickness and circular blade tip profile.

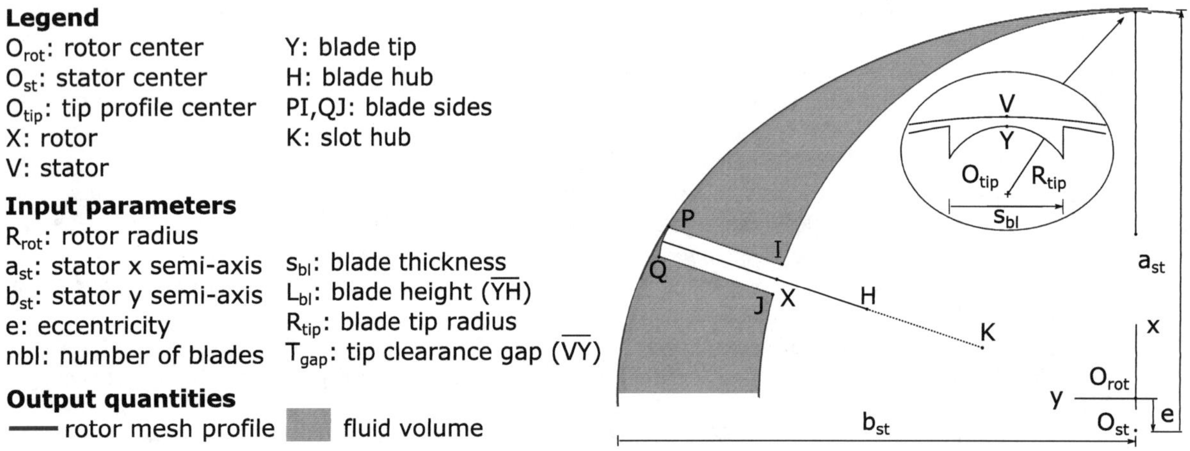

Figure 1: Schematics of the parametric geometrical modeling for sliding vane machines with inputs and outputs

A major assumption in the 2D modeling of the sliding vane machine core and, in turn, in the resulting rotor grid was to consider a constant minimum clearance between blade tip and stator along a full vane revolution. The minimum tip clearance gap is shown in Figure 1 as T_{gap}. This choice is a trade-off between the ideal tangency described above and real operating conditions, whereas blades' tips slide along a fluid layer (usually oil) that prevents a dry contact with the stator. The first limiting case would have indeed prevented the usage of a single rotor O-grid while the calculation of the instantaneous minimum liquid film thickness between blade tip and stator would have required either the usage of simplified hydrodynamic approaches, as the one in Bianchi and Cipollone (2015a), or complex fluid-structure computations. Additionally, any leakage phenomenon between consecutive vanes at the clearance between rotor slots and blade walls as well as the ones occurring across the end wall plates were neglected due to the tight gaps and slow flow velocities at those locations.
The resulting 2D boundary of the fluid volume enclosed in the core of a sliding vane machine is eventually composed of the stator profile, and the rotor profile with all the blades outside the rotor slots such that the minim gap between blades' tips and the stator occurs only once per tip.

3. GRID GENERATION AND MOTION

The gas domain of the compressor is the volume entrained between the housing, the rotor and the end plates. During operation, as the rotor turns, each vane divides the full volume into successive chambers that undergo suction, compression and discharge phases. The CFD model of the compressor represents these chambers as fluid volumes that change or deform with flow time. Additionally, at the tip of each vane there is a thin gap that needs to be treated as the leakage volume. Even though the change in volume of this gap is very small compared to the core region, because of the eccentricity between rotor and stator, the relative shape of these gaps change with rotor positions. The topology of each compression chamber and every connecting leakage gap is a rectangle in the computational domain. Figure 2 shows the domain transformations for the core and the leakage areas when they are treated separately. Although several algorithms for the structured meshing of such rectangular domains are available in practice, Kovacevic et al. (2007) showed that algebraic algorithms with transfinite interpolation, post orthogonalization and smoothing are the most efficient ones for applications like twin screw compressors where a number of mesh files are required to be generated and supplied to the CFD solvers in order to treat mesh deformation. Similar algebraic grid generation principle was used in this work for the vane compressor rotor domain mesh.

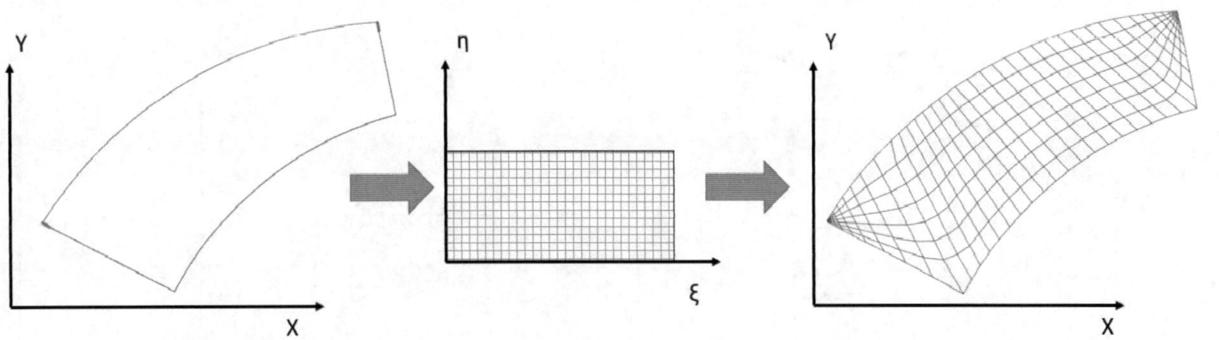

Figure 2: Domain transformation from the core region to computational region and then to the grid

Flow chart in Figure 3 lists the main stages of the grid generation procedure. The geometry program described in Section 2 is used to produce the rotor profile along with the vane at the initial rotor position. This forms an input to the meshing algorithm. Successive rotor positions appear to be rigid body rotations for the rotor while for the fluid volume that is deforming, it is a complex boundary motion where rotor rotates, stator is stationary and each of the vane is undergoing a general body motion. For simplification, blade tip clearance gap changes were not accounted in the presented analysis, but the meshing algorithm allows the possibility to further introduce a gap variation function.

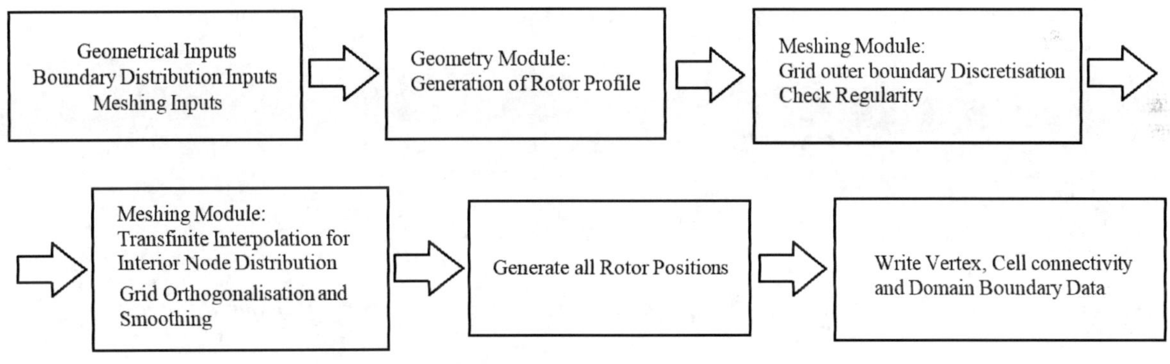

Figure 3: Vane compressor mesh generation procedure

The first stage in the meshing procedure is to discretize the boundaries. At this stage, inputs regarding the number of nodes and parameters for various control functions are taken from the user. The boundary discretization produces the distribution into the topology of an 'O' grid, as shown in Figure 4. An 'O' structure for the rotor mesh was purposely selected because it avoids any inaccuracies that can be introduced due to a non-matching mesh connection between core and leakage regions. The sudden transition from the leakage gaps to the core was handled by the introduction of stretching functions that gradually flare the radial mesh lines from leakage gap into the core as seen in views A and C in Figure 4.

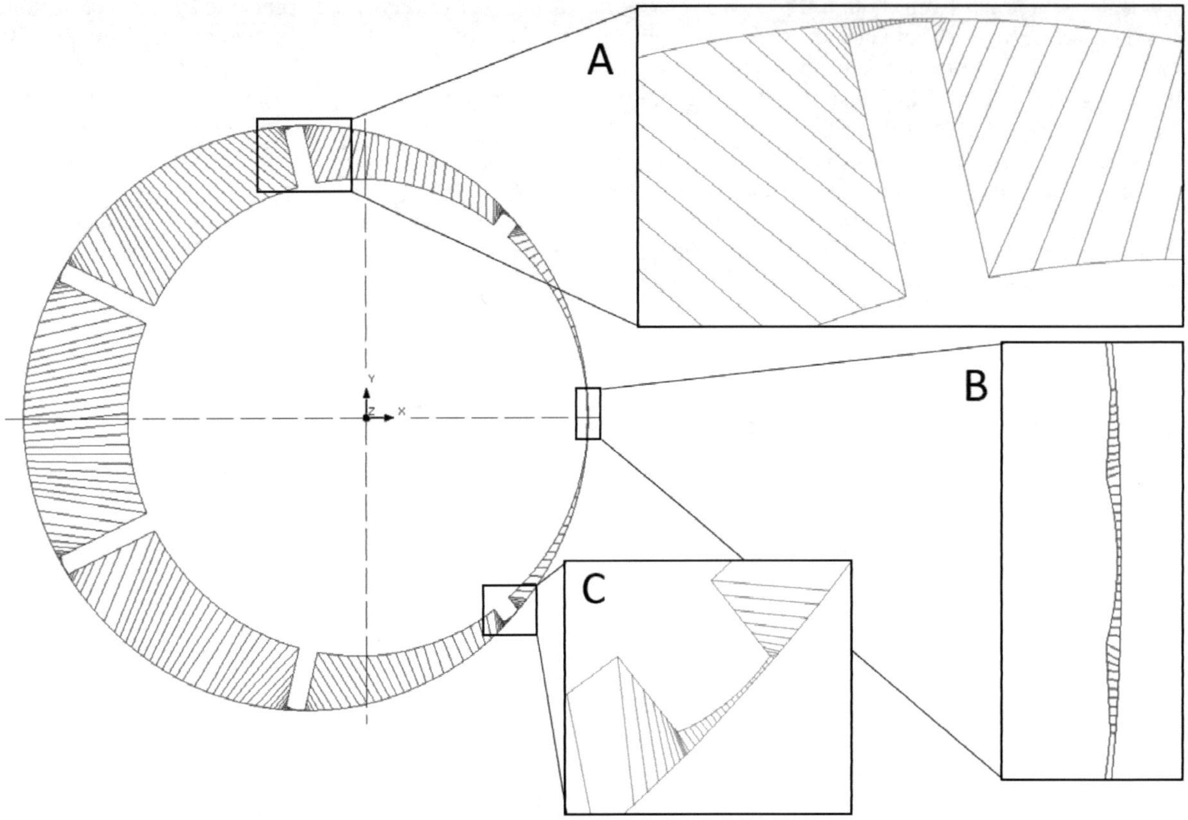

Figure 4: Boundary discretization as an 'O' Grid and transition function

At the position where the core region topology does not remain rectangular, as it happens when the vane slides inside the rotor, a simplification was introduced to convert the vane side walls into short segments on the rotor surface. This is seen in the view B of Figure 4 and it is anticipated that the leakage flow will not be influenced by this simplification as the gap clearance does not change. Nevertheless, this will have an influence on the solver stability due to sudden change in topology and could demand higher solver relaxation parameters.

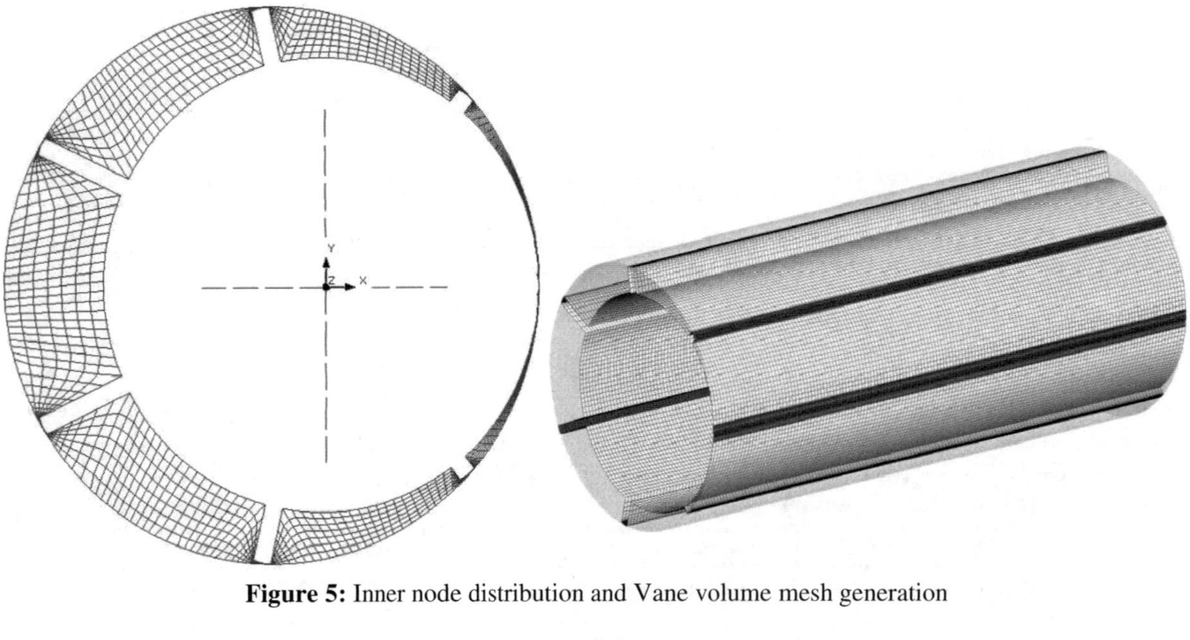

Figure 5: Inner node distribution and Vane volume mesh generation

The second stage in mesh generation is distribution of interior nodes. Transfinite interpolation, orthogonalization and smoothing as described in Kovacevic et al. (2005) were used in these procedures to get a good quality hexahedral cell structure. Once the 2D mesh is generated for all rotor positions, the mesh is assembled into 3D volume in the required format of the solver (Figure 5). In the current implementation there is possibility to export the mesh in solvers such as ANSYS CFX, ANSYS FLUENT, STAR CCM+ and Simerics PumpLinx. The entire meshing procedure was implemented in the software package SCORG™ developed at City University London.
The rotor domain is eventually connected to suction and discharge ports via non-conformal interfaces, as usually occurs in CFD analysis on positive displacement machines (Kovacevic et al., 2005). Ports meshing is usually carried out using standard tools and with reference to a tetrahedral cell structure to better capture the real geometry.
In the presented analysis, ANSYS CFX solver was used to calculate the oil free and oil injected flow fields in the compressor. This solver was selected mainly because of the Authors' experience gained with oil injected twin screw machines. In particular, ANSYS CFX provides a model called Junction Box Routine that is a user defined library to specify mesh deformation from custom applications such as SCORG. The solver updates the nodes coordinates from set of pre-generated files after every crank angle step (or its submultiples). Time step finally results from crank angle step and revolution speed.

4. TEST CASE

As reference application to test the numerical methodology just presented, an industrial mid-size oil injected air compressor was considered. As shown in Figure 6, in this machine the suction process occurs both through axial and radial ports: the first ones are located on the end wall plates of the compressor while the radial suction port is composed of three slanted slots along the stator. On the other hand, the discharge process takes place only through a series of radial slots located on the stator. At a specified position of the rotor and compression process, oil is injected in the compressor cells through a series of calibrated orifices displaced along the axial direction of the machine and fed by a common rail whose cross section is represented by the yellow dot in Figure 6. The operating conditions that were considered as benchmark for the numerical results were 980 RPM as revolution speed and 8.7 bar_a as outlet pressure (Table 1).

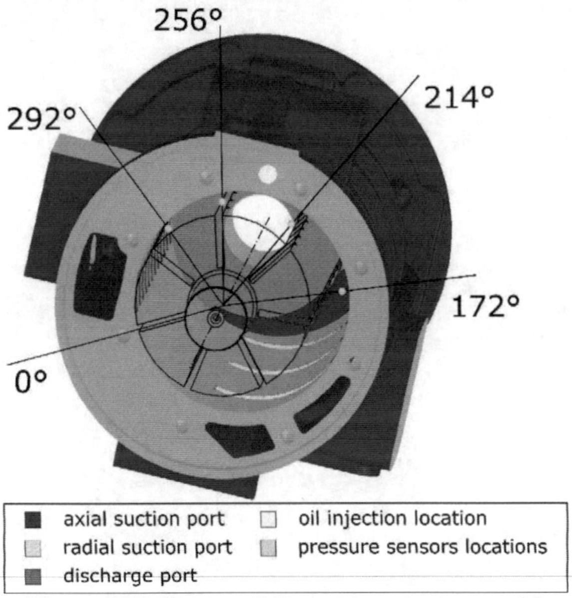

Figure 6: Reference sliding vane compressor with indication of ports and sensors locations

Inlet pressure	1.01	bar_a
Inlet temperature	25.4	°C
Outlet pressure	8.7	bar_a
Outlet temperature	88.7	°C
Revolution speed	980	RPM
Oil mass flow rate	2.9	kg/s
Oil injection pressure	6.6	bar_a
Oil injection temperature	65.0	°C
Air mass flow rate	0.203	kg/s
Indicated power	68.2	kW
Shaft power	82.8	kW

Table 1: Experimental data for the CFD model

4.1 Experimental setup

The compressor was instrumented with pressure, temperature and flow rate sensors across the machine and along the oil injection circuit. Furthermore, a torque meter allowed to retrieve the mechanical power while a set of four piezoelectric pressure transducers were installed along the compression and the discharge phases. On the other hand, intake process was reasonably considered as isobaric. The in-vane pressure data were processed according to the methodology proposed in Bianchi and Cipollone (2015b) and allowed to reconstruct the indicator diagram of the compressor and, in turn, to calculate the indicator power which is listed in Table 1 together with all the quantities measured.

4.2 Numerical setup

In order to evaluate the sealing effect of the oil, two compressor configurations were considered, namely oil free and oil injected. In its complete layout, the computational domain was composed of four main zones: suction line, rotor mesh, oil rail and discharge line. Grids' sizes, expressed in terms of millions of cells, are 1.4, 0.3, 0.2 and 0.4 respectively. As shown in Figure 7, the first and latter domains not only include axial or radial ports but also the intake valve and the discharge pipe. Apart from fluid domain enclosed among stator, rotor and blades, all the fluid zones were retrieved using CAD models of the industrial machine and were further discretized using tetrahedrons. On the other hand, the grid generation of the rotor mesh was accomplished using the methodology presented in the previous sections and with reference to actual machine geometry but using a conservative value for the minimum tip clearance gap (T_{gap}) of 200 μm.
ANSYS CFX is a pressure based coupled CFD solver. Air was modeled as ideal gas while oil as a fluid whose thermophysical properties were customized with reference to proprietary data. Pressure boundary conditions were considered both at air and oil inlets as well as at the compressor outlet. The specification of pressure at the discharge boundary and at the oil injection rail is a major factor influencing the numerical stability of the solver. Hence time dependent functions were used to gradually ramp the boundary pressure as shown in Figure 8. Mass and energy exchanges between fixed and rotating grids occurred through non-conformal sliding interfaces circumferentially located along the stator surface (e.g. suction and discharge ports as well as oil orifices) or on the end wall plates for the axial suction ports.

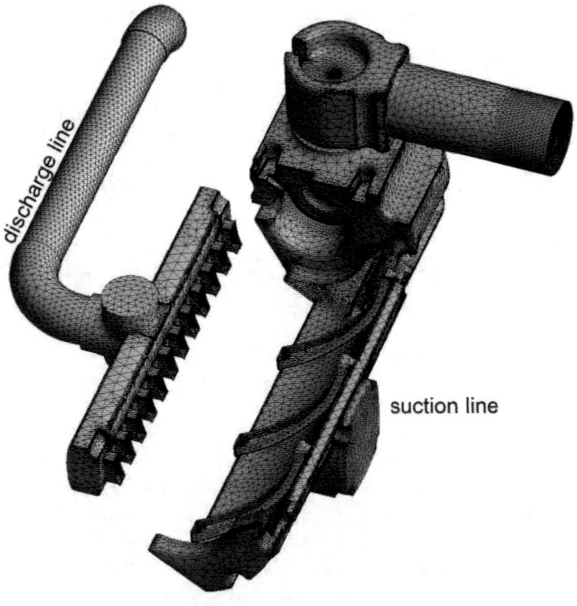

Figure 7: Tetrahedral mesh in the Suction and Discharge lines

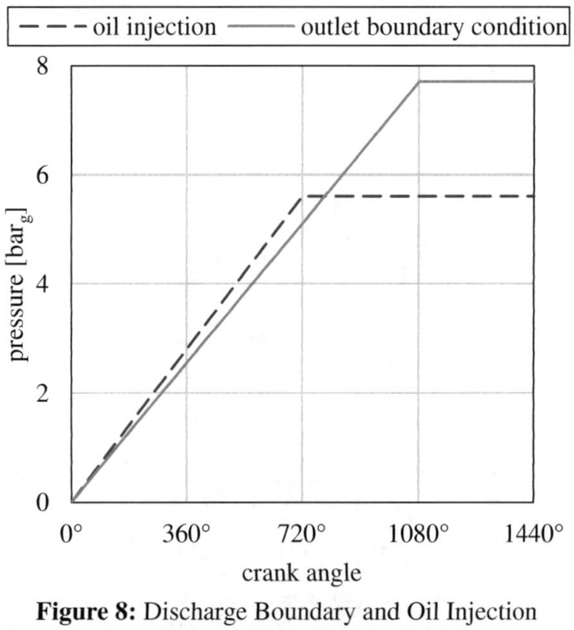

Figure 8: Discharge Boundary and Oil Injection Pressure ramp functions

5. RESULTS

A first attempt in the numerical investigation of the sliding vane air compressor was performed considering a set of single phase simulations, whereas air was solely taken into account. Results of these calculations after 3 full revolutions are presented in Figures 9 and 10. In particular, Figure 9 reports the pressure contours in a cross section of the machine close to mid length, namely at 285 mm with respect to a full axial length of 540 mm. In this Figure, velocity field is additionally superimposed and normalized. A virtual pressure probe rotating with the vane further allowed to retrieve the angular pressure evolution over a full cycle that is reported in Figure 10. The x-axis of this Figure has been intentionally divided in 7 ranges in order to be in perfect agreement with Figure 9. Angular displacements of intake and discharge ports are reported in both Figures with cyan and magenta lines while dashed blue line in Figure 10 indicates the experimental outlet pressure and, in turn, the outlet boundary condition after the ramp initialization. Due to the relatively large gap at the tip clearance, a strong backflow could be noticed between the discharge phase and the suction process. Indeed, the clearance at the tangency between stator and rotor (0° in Figure 10) acts as an orifice which drops the pressure inside the first vane from the discharge value to more than 3 bar. The vane volume increase from the first to the second vane and its interaction with the suction port further contribute to expand the fluid up to the inlet pressure. Nevertheless, the temperature level of the backflow driving the expanded air inside the suction ports is greater than the fresh charge coming from the inlet and mixes with it as occurs in Figure 13, but with a less relevant extent, for the oil-injected case. In turn, the starting temperature of the compression process, that takes place from 186° to 305°, leads the fluid to an over-compression up to 11 bar before being suddenly damped as the discharge begins. This trend is in disagreement with the experimental one which did not show any over-compression. Furthermore, the backflow towards the suction ports led to a mass unbalance in the simulations even though the pressure coupled solver handled the continuity equation in direct way. These remarks contribute to state the need of taking into account the oil presence in the simulations but, more importantly, show that oil injection in sliding vane air compressors is indispensable for sealing across any clearance. Moreover, additional features of the lubricant injection are friction power reduction and air cooling.

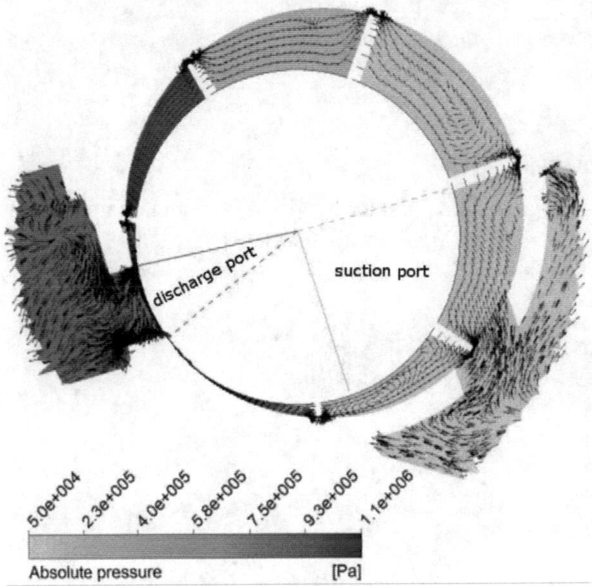

Figure 9: Absolute pressure contours and velocity field in a cross section close to mid length

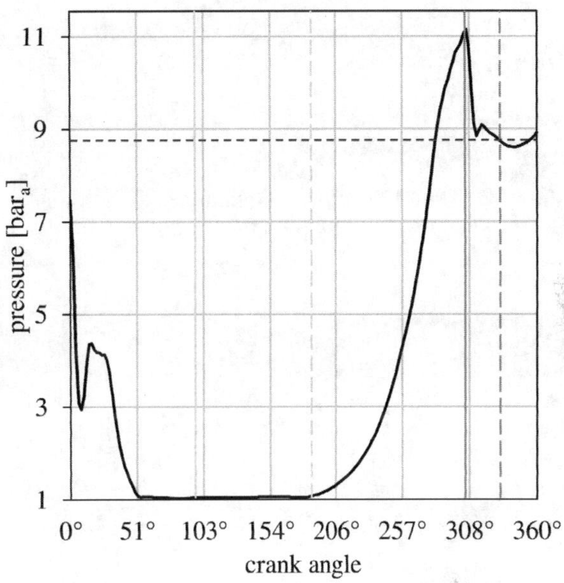

Figure 10: Vane pressure evolution with angular references of ports and outlet pressure

Figure 11 presents a detailed view of the air velocity field across the tip region in oil injected simulations with reference to the vane downstream the oil injection process (206°-257°). Despite the sealing action of the lubricant, the clearance at the vane tip induces a backflow driven by adverse pressure gradient in opposite direction to the counter clockwise revolution speed sense. In turn, the overall flow field in the compressor vane is affected by a recirculation region whose magnitude is gradually lowered as the leakage flow interacts with the main one (which is aligned with the rotation sense). This phenomenon contributes to enhance the heat transfer between the compressing air and the oil that, after the radial injection shown in Figure 12, spreads along the leading side of the vane and mixes with air. The oil accumulation on the walls of the rotor grid, i.e. on the metallic surfaces which define the vane in reality, can be seen in the bottom left of Figure 12. These oil layers allow to lower friction losses between surfaces in relative motion, namely stator with blade tip and blade side walls with rotor slots.

The cooling effect of the oil is additionally noticeable in Figure 13, which reports the total temperature contours in the same mid cross section at 285 mm. The temperature increase during the first stages of the compression process is indeed mitigated in the vane where the oil supply takes place. Nevertheless, as the compression proceeds, air temperature keeps increasing up to 100 °C-120 °C, as it can be seen in cell whose angular range is 257°-308°. Finally, at the discharge manifold, an intense heat transfer takes place and leads the temperature of the air-oil mixture back to 90 °C.

The thermal power gained by the lubricant during the convective heat transfer with air increases oil temperature and, in turn, lowers its viscosity. This phenomenon is presented in Figure 14 using an isosurface of points having an oil volume fraction equal to 2.5%. The variation of oil dynamic viscosity along the axial direction is essentially due to discrepancies in oil flow rate supplied by each injector because of the single central point inlet. Hence, air cooling is more effective in the central region of the compressor core. On a global scale, oil dynamic viscosity at the discharge phase has lowered 4.4 times with respect to the value at the injection. This effect has a detrimental drawback in the hydrodynamic capabilities that the oil layer must have to sustain the blade tip and to prevent any dry contact with the stator surface.

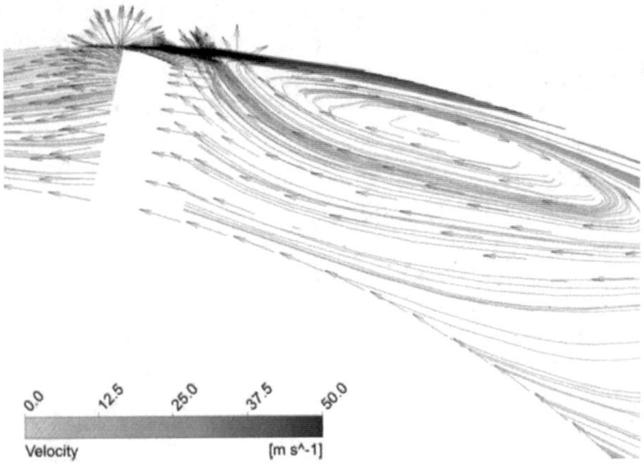

Figure 11: Detail of velocity streamlines and vectors in the tip region

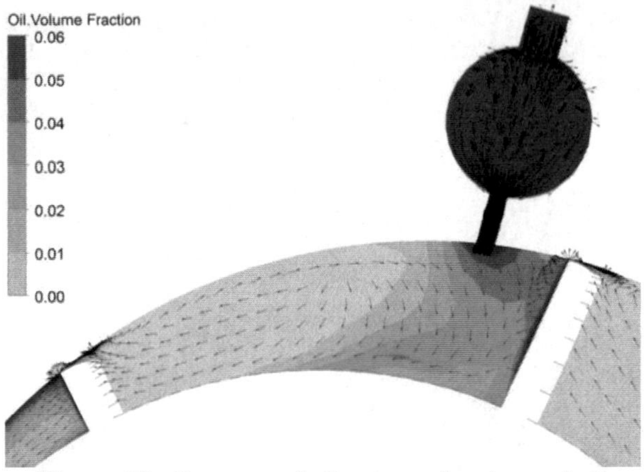

Figure 12: Contours of oil volume fraction and oil flow field during the injection process

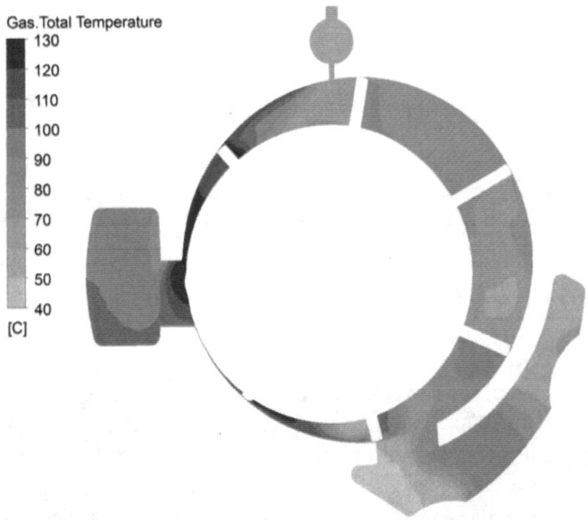

Figure 13: Air total temperature contours in a cross section close to mid length

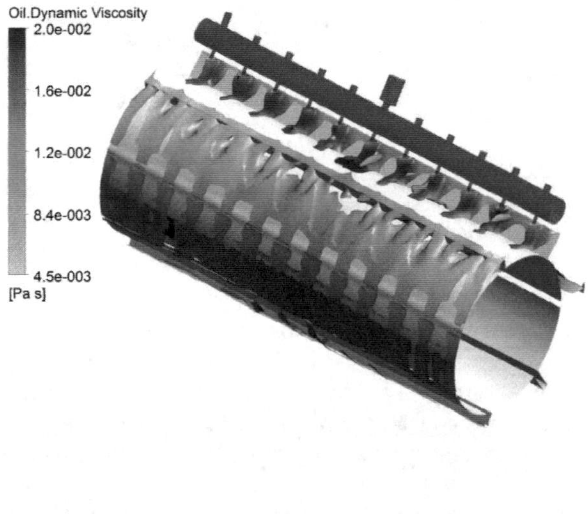

Figure 14: Variation of oil dynamic viscosity along an isosurface having a volume fraction of 2.5%

6. CONCLUSIONS

The research contents presented in this paper attempted to address the need of advanced designed methodologies in positive displacement vane machines using commercial CFD tools at the state of art. The challenges involved to proper handle computational grids that move and deform during the numerical simulations, such as the ones that enclose the fluid volume inside a vane device, is still an open issue in literature and industry. For this reason, an innovative grid generation methodology for general vane machines was herein developed using a parametric geometrical modeling of the rotor fluid volume and the discretization algorithms successfully implemented for screw machines. This numerical methodology was eventually tested with reference to an industrial sliding vane air compressor using the ANSYS CFX solver. Preliminary single phase simulations showed a high leakage flow at the clearance between stator and blades' tips. These phenomena were particularly severe when the vane was approaching the tangency region between rotor and stator. Large backflows at high pressure and temperature indeed destabilized the mass balance and led to an over-compression not revealed experimentally. Three-dimensional transient multiphase simulations were consequently performed to take into account the injection of oil during the compression process. This upgrade allowed to track a recirculation region induced in the vane flow field by the leakage flow through the clearance at the tip. Furthermore, the convective heat transfer with air showed the cooling capabilities of the lubricant even when supplied into the vanes as liquid jets. On the other hand, the oil temperature rise decreased its dynamic viscosity and, in turn, its hydrodynamic potential to prevent dry contact between blades and stator. The effects of a variable tip clearance due to oil temperature and blade dynamics will be taken into account in future upgrades of the numerical methodology.

REFERENCES

O. Al-Hawaj (2009), Theoretical modeling of sliding vane compressor with leakage, International Journal of Refrigeration, Volume 32, Issue 7, Pages 1555-1562, ISSN 0140-7007, DOI:10.1016/j.ijrefrig.2009.07.005.

ANSYS 14.0 (2011), ANSYS CFX Manual

O. Badr, P.W. O'Callaghan, S.D. Probert, (1985a), Multi-vane expanders: Geometry and vane kinematics, Applied Energy, Volume 19, Issue 3, Pages 159-182, ISSN 0306-2619, DOI:10.1016/0306-2619(85)90006-6.

O. Badr, S.D. Probert, P. O'Callaghan, (1985b), Multi-vane expanders: Vane dynamics and friction losses, Applied Energy, Volume 20, Issue 4, Pages 253-285, ISSN 0306-2619, DOI:10.1016/0306-2619(85)90018-2.

G. Bianchi and R. Cipollone, (2015a), Theoretical modeling and experimental investigations for the improvement of the mechanical efficiency in sliding vane rotary compressors, Applied Energy, Volume 142, Pages 95-107, ISSN 0306-2619, DOI: 10.1016/j.apenergy.2014.12.055

G. Bianchi and R. Cipollone, (2015b), Friction power modeling and measurements in sliding vane rotary compressors, Applied Thermal Engineering, Volume 84, Pages 276-285, ISSN 1359-4311, DOI: 10.1016/j.applthermaleng.2015.01.080

International Energy Agency (IEA), (2015), Key world energy statistics, 2015.

A. Kovacevic, (2005), Boundary adaptation in grid generation for CFD analysis of screw compressors, International Journal for Numerical Methods in Engineering Volume 64, Issue 3, Pages 401–426, DOI: 10.1002/nme.1376

A. Kovacevic and N. Stosic and I. Smith, (2007), Screw Compressors - Three Dimensional Computational Fluid Dynamics and Solid Fluid Interaction, Springer-Verlag Berlin Heidelberg, New York, ISBN 3-540-36302-5.

P. Kolasiński and P. Błasiak, (2015), Numerical and experimental investigation on the rotary vane expander operation in micro ORC system, 3rd International Seminar on ORC Power Systems, Brussels, Belgium

G. Montenegro and A. Torre et al., (2014), Evaluating the Performance of a Rotary Vane Expander for Small Scale Organic Rankine Cycles Using CFD tools, Energy Procedia, Volume 45, Pages 1136-1145, ISSN 1876-6102, DOI:10.1016/j.egypro.2014.01.119.

Prasad, B. G. Shiva, (2004), CFD for Positive Displacement Compressors, International Compressor Engineering Conference at Purdue University, West-Lafayette (IN), United States, URL: docs.lib.purdue.edu/icec/1689/

S. Rane, A. Kovacevic, N. Stosic, & M. Kethidi, (2013), Grid deformation strategies for CFD analysis of screw compressors, International Journal of Refrigeration, Volume 36, Issue 7, Pages 1883-1893, ISSN 0140-7007, DOI: 1016/j.ijrefrig.2013.04.008.

R. Saidur, N.A. Rahim, M. Hasanuzzaman, (2010), A review on compressed-air energy use and energy savings, Renewable and Sustainable Energy Reviews, Volume 14, Issue 4, Pages 1135-1153, ISSN 1364-0321, DOI: 10.1016/j.rser.2009.11.013.

Tramschek, A. B., and M. H. Mkumbwa, (1996), Mathematical modelling of radial and non-radial rotary sliding vane compressors, International Compressor Engineering Conference at Purdue University, West-Lafayette (IN), United States, URL: docs.lib.purdue.edu/icec/1151/

Q.Zhang and X. Y. Xu, (2014), Numerical Simulation on Cavitation in a Vane Pump with Moving Mesh, 5th International Conference on Computational Methods, Cambridge, England

ACKNOWLEDGEMENT

The work documented in this report has been funded by the 2015 Scholarship of the Knowledge Center on Organic Rankine Cycle technology (www.kcorc.org), the organization formed by the members of the ORC Power Systems committee of the ASME International Gas Turbine Institute (IGTI).

Thermodynamic Modeling of Screw Expander in a Trilateral Flash Cycle

Hanushan VASUTHEVAN[1]*, Andreas BRÜMMER[2]

[1]TU Dortmund University, Chair of Fluidics
Dortmund, Germany
Phone: +49 231 755 5723, Fax: +49 231 755 5722, E-mail: hanushan.vasuthevan@tu-dortmund.de

[2]TU Dortmund University, Chair of Fluidics,
Dortmund, Germany
Phone: +49 231 755 5720, Fax: +49 231 755 5722, E-mail: andreas.bruemmer@tu-dortmund.de

* Corresponding Author

ABSTRACT

The present paper focuses on the thermodynamic modeling of screw expanders in a trilateral flash cycle. In this process a liquid is pumped from low to high pressure and then heated up close to saturation point. Instead of a vaporous working fluid as usual, a hot liquid is filled in the working chamber of the screw expander. During the filling process the pressure of the liquid drops below the saturation pressure while the temperature remains almost constant due to a fast process. Hence the liquid is in a metastable state. The liquid aspires a stable state and therefore is vaporized during the filling and expansion process due to its increasing chamber volume. Thus a two-phase mixture, vapor and liquid, exists in the working chamber. The simulation presented includes the calculation of fluid states within each chamber to be in thermodynamic equilibrium, which assumes that a sufficient heat transfer exists between the phases to reach a stable state within a time step. Thermodynamic simulations are carried out using water as the working fluid for an exemplary screw expander geometry. The conclusive assessment of the thermodynamic model is demonstrated by the comparison of the simulation results with available experimental measurement results for a corresponding two-phase screw expander.

1. INTRODUCTION

Increasing requirements in compliance with environmental regulations and laws, particularly reduction in CO_2 emissions, lead to concepts to improve the efficiency in energy conversion systems. In modern power plants, in the high power range, combined gas and steam processes are applied to increase efficiency. In this process the steam is heated by the exhaust gases to a temperature of approximately 850 K. Contrarily, efforts are made to increase the efficiency in power recovery from low-temperature heat in the 373 K – 573 K temperature range. Typical examples for low-temperature heat are geothermal sources or industrial waste heat.

The trilateral flash cycle which basically contains liquid heating only and vaporizing during expansion is a means of power recovery for aforementioned low-temperature sources (Smith, 1993). A screw expander is suited for the trilateral flash cycle process as this type of displacement machine is able to expand working fluids with a high liquid content (Kliem, 2002). At the beginning two-phase screw expanders with flash vaporization were essentially examined experimentally. The application of a two-phase screw expander with flash vaporization from a geothermal well was mentioned in a patent for the first time by Sprankle (1973), which was realized with geothermal water for power generation subsequently (Sprankle and McKay, 1974). First experiments of a screw expander in a trilateral flash cycle test rig were carried out by Steidel et al. (1982). In these experiments the flash vaporization with water was realized with a throttle valve on the high pressure side in front of the machine such that a two-phase mixture with a high liquid content enters the screw expander. The maximum measured effective isentropic efficiency was 53%.

First approaches for a theoretical analysis of a two-phase screw expander in comparison to experimental results were made by Taniguchi et al. (1988). For the theoretical analysis a one chamber model was used, which was based on the conservation of mass and energy. In the experiments the screw machine was used as a replacement for a throttle in a heat pump with the aim to increase efficiency and to decrease power consumption of the heat pump. As a working fluid the refrigerant R12 was used. The maximum measured effective isentropic efficiency was 50%. An overview for the development of a trilateral flash cycle system is given by Smith et al. (1993), (1994), and (1996). With thermodynamic cycle simulations the working fluids R12, R113 and n-pentane were examined. For the experimental investigation the working fluid R113 was chosen. The maximum measured effective isentropic efficiency amounts to 70%. Kliem (2005) examined a two-phase screw expander experimentally with a focus on a novel filling system with rotating short nozzles in which hot water was accelerated and injected as a liquid jet in the working chamber. In these tests a maximum effective isentropic efficiency of 55% was measured.
With the evaluation of the available literature it can be said, that there are no detailed methods or models for the design of a two-phase screw expander for a trilateral flash cycle although this technology seems to have a high efficiency potential. A first step in this direction is made with this paper, which presents a method to model a two-phase screw expander. In contrast to the approach of Taniguchi et al. (1988), the conservation of mass and energy in the present work will be derived separately for liquid phase and gas phase. This approach allows a detailed examination of the gas and liquid phase in thermodynamic non-equilibrium. Furthermore detailed heat flow models between phases within working chambers and during leakage flows can be developed and applied on the presented model, which will be realized in future works.
Nevertheless in this paper, as a first step, calculations are done in thermodynamic equilibrium. Taniguchi et al. (1988) and Smith et al. (1996) demonstrated that the assumption of the thermodynamic equilibrium provide acceptable results. The simulations are carried out for an exemplary screw expander geometry, in which a hot liquid is injected through a nozzle, and compared with available experimental results (Kliem, 2005).

2. FUNDAMENTALS

2.1 Trilateral Flash Cycle

The ideal trilateral flash cycle is shown in Figure 1 in a Ts-diagram. The liquid working fluid is pumped isentropic to a higher pressure p_1 and is then heated isobaric to the saturation temperature T_2 related to the pressure p_1. From the saturation point the hot liquid is expanded isentropic within the expander into the wet vapour region to the pressure p_3. Expansion and phase change take place at the same time. Depending on boundary conditions the phase change can occur explosively which is referred to as flash vaporization. The cycle is closed with condensation of the two-phase mixture.

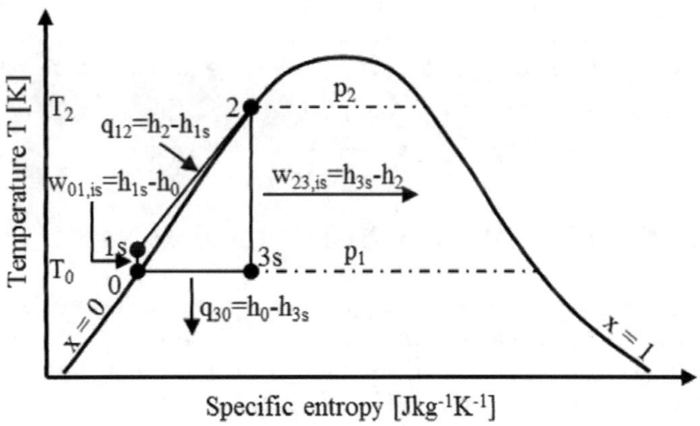

Figure 1: Ts-diagram of the ideal trilateral flash cycle

2.2 Screw Expanders

Screw expanders are rotary displacement machines with spiral shaped intermeshing rotors enclosed in a housing. The gaps of the rotors with the enclosed housing form the working chambers. The periodical change of the working chamber volume in a working cycle takes place due to the rotation of the rotors. Figure 2 qualitatively presents the progression of the chamber volume as well as the inlet and the outlet area as functions of the phase angle for one working chamber of a screw expander. A phase angle of $\varphi = 1$ is equal to the tooth pitch angle of the male rotor which depends on the number of lobes. The working cycle is divided into three steps. The first step is the filling process, in which the working chamber is filled with the working fluid from the high pressure side while the volume increases. With further rotation the rotor edges passes the control edges which are within the housing. At this moment the connection between the high pressure side and the working chamber over the inlet area is separated. The filling process ends and the working chamber -except of clearances- is enclosed from the environment. At this moment the expansion process begins at the theoretical expansion volume $V_{ex,th}$. The expansion process proceeds with further rotating of the rotors while pressure and temperature drops. This process ends when the working chamber reaches the control edge on the low pressure side at its maximum chamber volume V_{max}. The working chamber is then connected to the low pressure side through the outlet area. At this point the discharge of the working fluid begins while the chamber volume decreases, which is the third step of the working cycle. The working cycle ends with the disappearing of the chamber volume.

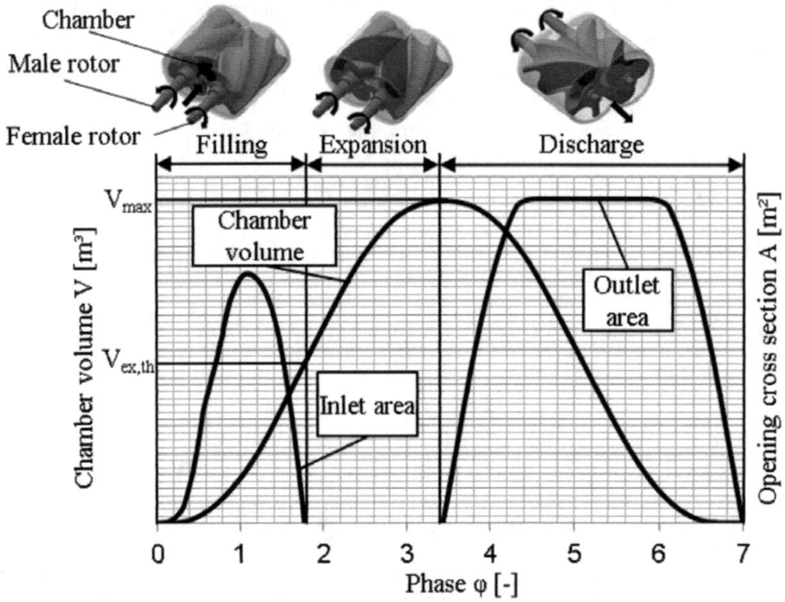

Figure 2: Volume curve and chamber openings for a workings cycle of a screw expander

For a given screw expander geometry with a maximum chamber volume the initiation of the theoretical expansion depends on the internal volume ratio v_i which is defined in equation (1). Thus the internal volume ratio affects the inlet area of a screw expander. The inlet area decreases with increasing internal volume ratio.

$$v_i = \frac{V_{max}}{V_{ex,th}} \tag{1}$$

3. SIMULATION PROCEDURE AND THERMODYNAMIC MODELLING

In this chapter the procedure of the simulation and its thermodynamic modelling are described. The simulation tool used for a multi-chamber model is mainly based on Janicki (2007). The essential requirement for the simulation is a multi-chamber model that presents the geometrical description of the angle dependent chamber volumes and gap areas between chambers. Every working chamber and its connections over gap areas are calculated simultaneously within one tooth pitch angle of the male rotor. The calculations are mainly based on the conservation of mass and energy which are considered separately for each phases. This approach offers the option to consider thermodynamic non-equilibrium states within working chambers.

3.1 Conservation of mass

In equation (2) and equation (3) the conservation of mass for each phase is shown. The change of the mass over time in a gas or liquid control volume depends on the mass flow over the surface of the control volumes including the phase changes. The function $\dot{m}_{phase}$ describes the mass flow which vaporizes or condenses from the corresponding phase. If $\dot{m}_{phase}$ is a positive value the mass flow is vaporized from the liquid phase which is subtracted from the liquid mass flow and added to the gas mass flow. Vice Versa if $\dot{m}_{phase}$ is a negative value the mass flow condenses from the gas phase.

$$\frac{dm_G}{dt} = \sum_f (\dot{m}_G) + \dot{m}_{phase} \tag{2}$$

$$\frac{dm_L}{dt} = \sum_f (\dot{m}_L) - \dot{m}_{phase} \tag{3}$$

3.2 Conservation of energy

The conservation of energy for the gas and liquid phase is shown in equation (4) and equation (5). The left-hand-side of equation (4) and equation (5) is the change of the energy in the control volume over the time of the gas phase and liquid phase. It is assumed that kinetic and potential energy due to the gravitation can be neglected in the control volumes. Hence, only the internal energy will be considered in the control volumes.

On the right-hand-side of equation (4) and equation (5) the energy flows over the control volume surfaces are described. The terms $\dot{Q}_{SG}$ and $\dot{Q}_{SL}$ are the heat flow between a solid part and the gas or liquid phase. The heat flow between the phases is described with $\dot{Q}_{LG}$. It is assumed that the pressure-volume work of the whole chamber W_{Ch} is only performed by the gas phase and not from the liquid phase. The calculation of the pressure-volume work W_{Ch} by means of equation (6) includes the change of the whole chamber volume over time, which depends on the volume curve of the geometry. Independent of the pressure-volume work of the whole chamber the control volume of the gas phase can be deformed due to a volume change of the liquid. The volume of the liquid can be considered as incompressible and occupy the whole chamber volume in parts. If the control volume of the gas phase is deformed by the liquid phase to a smaller volume, pressure-volume work W_L is performed on the gas phase. This energy is taken from the liquid phase because the control volume of the gas phase is reduced by the liquid phase. All inlet and outlet total-enthalpy flows over the control volume surfaces of each phase are described by the summation of $\dot{m}_G h_{t,G}$ and $\dot{m}_L h_{t,L}$. The total enthalpy is that of the upstream state point of the flow. That means if the flow is into the control volume, the total-enthalpy is from the chamber the mass flow came from. If the flow is out of the control volume, the total-enthalpy is that of the considered control volume.

$$\frac{dE_G}{dt} = \dot{Q}_{SG} + \dot{Q}_{LG} + \frac{dW_{Ch}}{dt} - \frac{dW_L}{dt} + \sum_f \dot{m}_G h_{t,G} \tag{4}$$

$$\frac{dE_L}{dt} = \dot{Q}_{SL} - \dot{Q}_{LG} + \frac{dW_L}{dt} + \sum_f \dot{m}_L h_{t,L} \tag{5}$$

$$\frac{dW_{Ch}}{dt} = -p\frac{dV_{Ch}}{dt} \tag{6}$$

3.3 Chamber filling

The filling process is the flow from the high pressure side into a chamber volume. It is assumed that the working chamber is filled with a hot liquid over an injector nozzle instead of a conventional inlet area to improve flash vaporization of the fluid in the working chamber. The calculation of the mass flow rate is carried out with the Bernoulli equation for unsteady and incompressible flows from point 1 (reservoir upstream of the nozzle) to point 2 (outlet of the nozzle) along a streamline s, which is given in equation (7) neglecting the geodetic height differences between 1 and 2 and friction losses.

$$\frac{p_1}{\rho} + \frac{c_1^2(t)}{2} = \frac{p_2(t)}{\rho} + \frac{c_{2s}^2(t)}{2} + \int_1^2 \frac{\partial c(s,t)}{\partial t} ds \tag{7}$$

Point 1 can be a reservoir where the kinetic energy can be neglected and point 2 is the position at the nozzle outlet. At point 2 the pressure p_2 is assumed to be equal to the pressure of the working chamber which is to be filled. The pressure of the working chamber changes dependently on rotor angle or time.

The conversation of mass for a streamline from point 1 to point 2 is given in equation (8).

$$c(s,t) \cdot A(s) = c_{2s}(t) \cdot A_2(t) \tag{8}$$

With the assumption that the outlet area A_2 is constant and time-independent, equation (8) can be differentiated with respect to the time and integrated along the streamline, which yields to equation (9)

$$\int_1^2 \frac{\partial c(s,t)}{\partial t} ds = \frac{dc_{2s}(t)}{dt} \int_1^2 \frac{A_2}{A(s)} ds \tag{9}$$

Combining equation (7) and (9) yields to the following differential equation (10) with the stationary velocity $c_{stat,2}$, which is depending on the time-dependent pressure difference between point 1 and the filled working chamber:

$$\frac{dc_{2s}(t)}{dt} = \left(\frac{c_{stat,2s}^2(t)}{2} - \frac{c_{2s}^2(t)}{2}\right) \cdot \frac{1}{\int_1^2 \frac{A_2}{A(s)} ds} \tag{10}$$

The solution of the differential equation (10) for a constant and time-independent steady state velocity $c_{stat,2s}$ is given in equation (11). The time-dependent velocity $c_{2s}(t)$ is a hyperbolic tangent function which will lead to the stationary velocity $c_{stat,2s}$ for infinite time. The parameter D is determined by the initial condition and is zero for $c_{2,s}(0) = 0$. Furthermore the velocity at the nozzle outlet is depending on the geometry along the streamline.

$$c_{2s}(t) = c_{stat,2s} \cdot \tanh\left(\frac{c_{stat,2s} \cdot \left(D \int_1^2 \frac{A_2}{A(s)} ds + t\right)}{2 \int_1^2 \frac{A_2}{A(s)} ds}\right) \tag{11}$$

During the filling process the steady state velocity $c_{stat,2s}$ changes at the nozzle outlet which is not considered in the solution of equation (11). In the simulation the changeable steady state velocity at the nozzle outlet will be applied in equation (11), which is an approximation.
With equation (11) the theoretical maximum mass flow rate of the liquid is determined with the density ρ and the nozzle exit area A_2. Due to losses along the streamline the actual mass flow rate will be smaller. Therefore the losses are considered with a time dependent flow coefficient $\alpha(t)$. The actual mass flow rate $\dot{m}_L(t)$ for the filling of the working chamber is given in equation (12).

$$\dot{m}_L(t) = \alpha(t)\dot{m}_{th}(t) = \alpha(t)\rho A_2(t) c_{2,s}(t) \tag{12}$$

3.4 Leakage Calculation
For the calculation of the leakages a homogenous distribution of the two-phase mixture in the working chamber is assumed. This is a simplification as the leakage flow is essentially depending on the amount of liquid in the working chamber due to its sealing properties concerning the gap flows. Furthermore the gap flow is calculated in thermodynamic equilibrium which is also a simplification.
In Figure 3 an exemplary situation of a housing gap flow is shown. The calculation of the velocity from a chamber volume in the gap plane is done with the conservation of energy. At position 1 (chamber) the state is known. For the position in the gap plane it is assumed that the pressure p_{gap} equals to the pressure p_2 of the underlying working chamber. Furthermore assuming isentropic flow from chamber 1 to the gap plane the fluid state in the gap plane is known too.

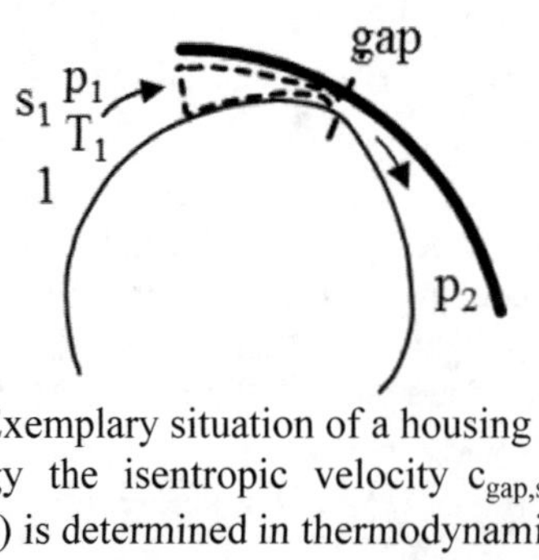

Figure 3: Exemplary situation of a housing gap flow

By means of the conservation of energy the isentropic velocity $c_{gap,s}$ in the gap plane is calculated with equation (13). The enthalpy in equation (13) is determined in thermodynamic equilibrium for the two-phase mixture.

Furthermore it is assumed that the velocity from the gas and liquid is equal in the gap plane. In the gap plane the vapor content x_{gap} will be evaluated in thermodynamic equilibrium to determine the liquid and gas mass flow as shown in equation (14) and equation (15). The calculated gas mass flow in the gap plane, which is inserted in chamber 2, is not equal to the change of gas mass within chamber 1, because during the flow between chamber 1 and the gap plane, phase change occurs. This also applies for the liquid mass flow. The gas and liquid mass taken out of the chamber 1 must be calculated with the vapor content x_1 in chamber 1. This must be taken into account by use of the conservation of mass and energy.

$$c_{gap,s} = \sqrt{2 \cdot \left(h_1(p_1, s_1) - h_{gap,s}(p_2, s_1)\right)} \tag{13}$$

$$\dot{m}_{G,s} = \dot{m}_{gap,s} x_{gap} = c_{gap,s} \rho_{gap} A_{gap} x_{gap} \tag{14}$$

$$\dot{m}_{L,s} = \dot{m}_{gap,s}(1 - x_{gap}) = c_{gap,s} \rho_{gap} A_{gap} (1 - x_{gap}) \tag{15}$$

Due to the assumption that the pressure in the gap plane equals the underlying working chamber, a choked flow may occur. This has to be checked by calculation of the critical pressure $p_{gap,crit}$. If the pressure p_2 is smaller than the pressure $p_{gap,crit}$ the assumed mass flux in the gap plane equals the maximum mass flux is $(\rho c)_{crit}$, as shown qualitatively in Figure 4.

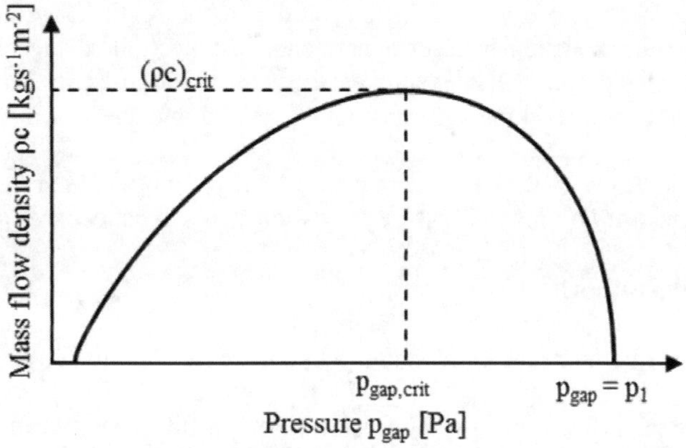

Figure 4: Mass flux depending on pressure $p_2 = p_{gap}$ for constant pressure p_1

The calculation of the critical mass flux is done iteratively by varying the pressure p_{gap} for constant pressure p_1. The criterion for the critical mass flux in the gap plane is given in equation (16). To limit the required calculation time for the iteration the first step is to calculate the derivative of the mass flux at the known pressure p_2. If the gradient is negative the flow is subcritical and the determination of the critical mass flux is not necessary.

$$\left(\frac{d(\rho c)}{dp}\right)_{gap} = 0 \tag{16}$$

Furthermore losses have to be considered and thereby the calculation of the actual mass flow rate can be treated in the same way as for the filling process with a flow coefficient as shown in equation (12).

3.5 Solution Procedure

For a given time step or rotor angle which is dependent on the angular frequency the geometry will be considered as a frozen model. It therefore follows, that chamber volumes and gap areas between chambers are defined. The basic simulation steps for this frozen model are shown in Figure 5. The first step is the calculation of the pressure-volume work due to the volume change compared to the time step before. The second step is the calculation and the transfer of heat flows between phases in a chamber volume. Thermodynamic equilibrium can be assumed if a sufficient heat transfer exists to reach the stationary stable state. Then the state can be calculated directly by means of an equation of state. Another approach is to calculate the heat flows with suitable heat transfer and vaporization models with consideration of the distribution of liquid and gas in the chamber which is not considered and not used here but scheduled for future work. In the present calculations the fluid states within the chambers are calculated in thermodynamic equilibrium. The last step is the calculation and the exchange of mass and energy due to fluid flows (gap, filling, discharge flows and phase changes) and heat flows (between solid parts and fluid). Subsequently the next time step can be calculated. The steps are performed successively within a small time increment to minimize the error. The calculation of all time steps in a working cycle is referred to as iteration. The stationary solution is achieved when the deviation of mass and energy in all time steps and all chamber volumes between two iterations is lower than a predefined deviation.

Current time step
↓
pressure-volume work
↓
heat flows between phases
↓
flows
↓
Next time step

Figure 5: Calculation procedure within a time step

4. COMPARISON OF SIMULATION AND EXPERIMENTAL RESULTS

Kliem (2005) examined a two-phase screw expander with a novel filling system in a trilateral flash cycle with the working fluid water. The filling system consists of a disk with short nozzles, which is fixed on the high pressure side of the female rotor. The inlet nozzles are located in the middle of the tooth space area. In the housing an inlet segment (elongated hole) is built in which has got a permanent connection to the high pressure side. During a specific rotating angle the nozzles passes the inlet segment and a connection exists between the high pressure side and a nozzle. In this configuration a working chamber will be filled and the hot liquid will be vaporized in the chamber.

In the experimental investigation flow coefficients were measured depending on the rotor angle for steady state flow conditions without the machine. Furthermore a pressure indication of the expander was performed.

4.1 Data of Geometry and Simulation

In Table (1) data of geometry and simulation are presented. The gap heights of the screw expander are estimated because these are unknown. Furthermore the fluid states per chamber are calculated in thermodynamic equilibrium and no heat transfer between solid parts and fluid is considered (adiabatic). The flow coefficient for all gaps are set constant to $\alpha = 0.8$ neglecting the influence of the rotor speed. For the filling system measured flow coefficients are used which are depending on the rotor angle. The geometric parameter for the unsteady flow of the filling system is estimated rudely from the available data due to lack of detailed information of the geometrical dimension of the flow path.

Table 1: Geometry and Simulation Data

	Geometry and simulation data
Rotor profile	Asymmetric SRM profile
Number of lobes – male/female	5 / 7
Crown circle – male/female	0,166 m / 0,154 m
Maximum chamber volume	$6{,}64 \cdot 10^{-4}$ m^3
Internal volume ratio	4,1
Profile gap height	$5 \cdot 10^{-5}$ m
Front gap height/ housing gap height	$8 \cdot 10^{-5}$ m
Heat flow model between phases	Thermodynamic equilibrium
Flow coefficient gap	0,8
Flow coefficient filling	Flow coefficients measured for steady state flow conditions by Kliem (2005)
Geometry parameter – filling: $\int_1^2 \frac{1}{A(s)} ds$	500 m^{-1}

4.2 Mass Flow Rate

In the experiment the mass flow rate was measured on the high pressure side in front of the machine. Due to the relative rotation between the rotating disk and the inlet segment a gap exists which results to a leakage mass flow.

This leakage flow is not considered in the simulation. Therefore it is to be expected that the simulated mass flow rate is lower than the measured mass flow rate. In Figure (6) the mass flow rate of simulation and experiment are presented depending on the male rotor circumferential speed for different inlet temperatures with corresponding inlet pressures. The symbols are measured points from the experiment and the curves are the results from the simulation.

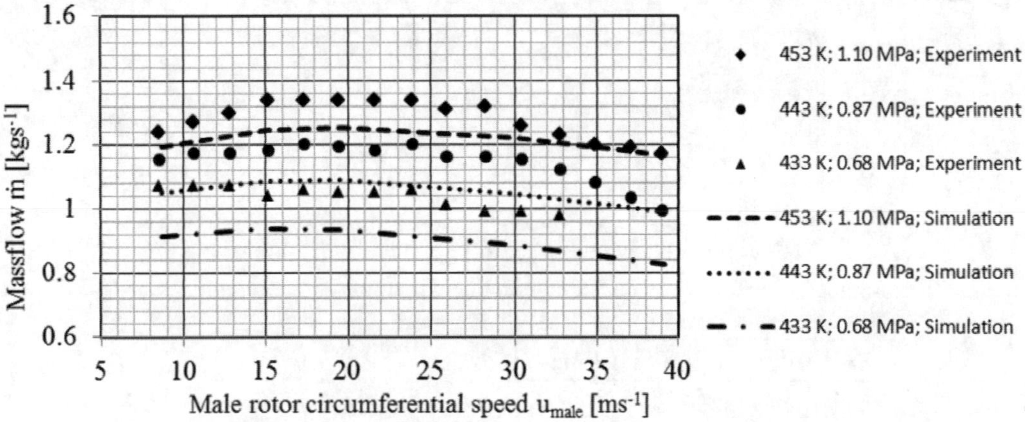

Figure 6: Mass flow rate depending on the male rotor circumferential speed

The mass flow rate has got a minor dependency of the circumferential speed. For high inlet temperatures a maximum of the mass flow rate can be identified which is decreasing with further circumferential speed. The reason for the decreasing mass flow rate is the unsteady filling. The liquid has got to be accelerated once a connection to the nozzle exists and does not reach the steady state velocity within the filling phase at high circumferential speed. The mass flow rates of the simulation are smaller than measured values due to the leakage mass flow between the inlet segment and the rotating disk as expected. The deviation can be interpreted as an additional not simulated leakage. With increasing circumferential speed the deviation between simulation and experiment decreases. A possible reason for this is a higher unsteady loss mechanism with increasing circumferential speed which decreases the mass flow rate in the experiment. Another reason is that the additional not simulated leakage mass flow rate in the experiment is decreasing with increasing circumferential speed.

4.3 Internal Work and Indicator Diagrams

In Figure (7) the internal work (pressure-volume work of working cycle) depending on the male rotor circumferential speed is shown. The symbols are measured points from the experiment and the curves represent the simulation results. The internal work is calculated from the indicator diagrams. The simulation provides an acceptable estimation of the internal work. The maximum deviation for the examined range is 17,5% at a temperature of 453K and a circumferential speed $u_{male} = 13{,}9\ ms^{-1}$.

A closer examination can be done with an indicator diagram. In Figure (8) the indicator diagrams for a temperature of 453K and three circumferential speeds are shown. For a circumferential speed of $u_{male} = 8{,}71\ ms^{-1}$ the pressure of the simulation decreases during the filling while the pressure of the experiment is nearly constant. It is assumed that the pressure decreases due to the simplified simulation of the leakage flow. In the real case there is the possibility that leakages are sealed with the liquid phase.

For a higher circumferential speed of $u_{male} = 19{,}6\ ms^{-1}$ the time for the leakage flow is reduced such that the pressure of the simulation during the filling stays nearly constant. During the expansion process the pressure of the simulation decreases faster than the pressure of the experiment. The reason for this is again the modeling of the leakage flow whereby sealing effects of gaps and distribution of the liquid phase or the gas phase in the chamber are not considered. With increasing circumferential speed the stationary velocity during the filling cannot be reached. The pressure of the simulation at the circumferential speed $u_{male} = 39{,}2\ ms^{-1}$ does not reach the pressure of the experiment during the beginning of the filling. A possible reason is that the approximation for the calculation of the chamber filling at higher circumferential speed is inaccurate. In the expansion phase the curve progression of simulation and experimental are nearly identical. At this circumferential speed the impact of the leakage flow seems to be minor.

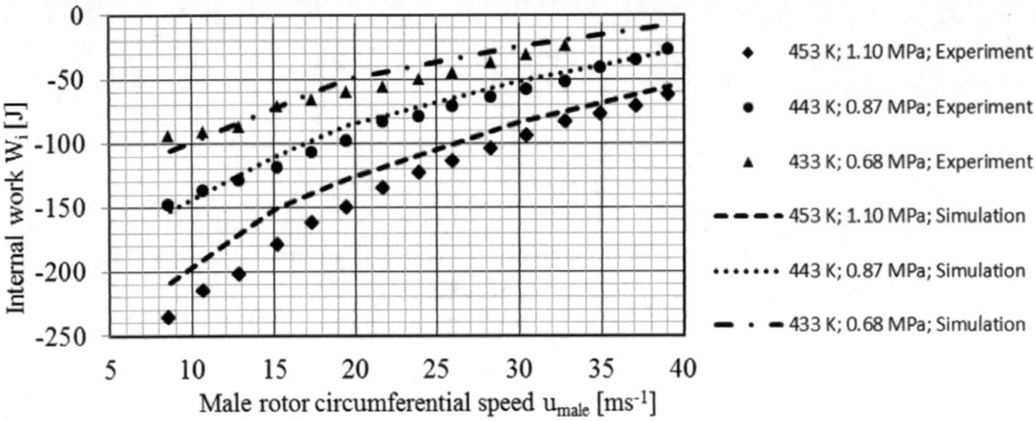

Figure 7: Internal work depending on the male rotor circumferential speed

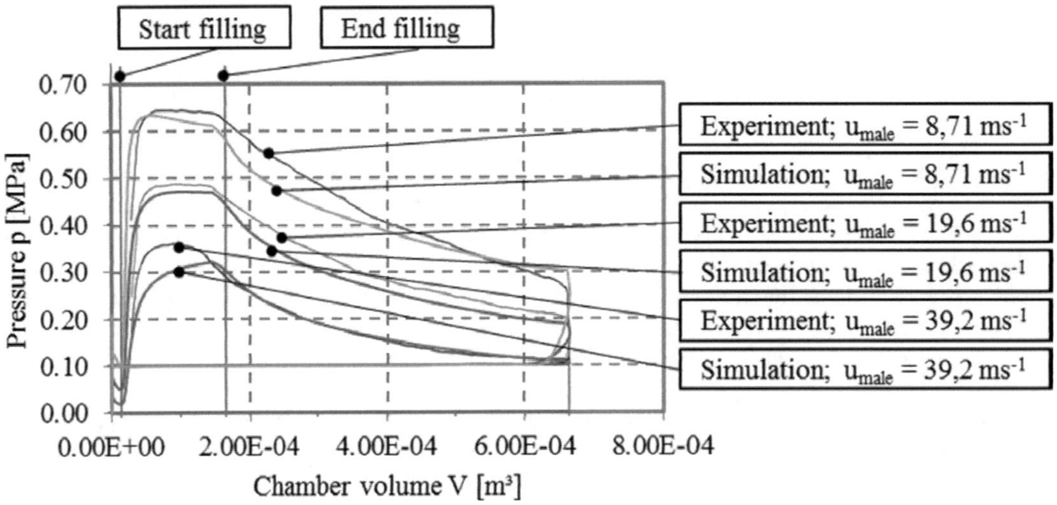

Figure 8: Indicator diagrams for different circumferential speeds at T_{inlet} = 453K and p_{inlet} = 1,1MPa

6. CONCLUSIONS

The presented simulation model to calculate a screw expander with hot liquid injection and flash vaporization is an acceptable approach for a first estimation of mass flow and internal power. This is validated by comparison of simulation and experimental results. A potential reason for existing deviations between simulation and experiment is the assumption of the thermodynamic equilibrium for the calculation of chamber states and gap flows.
Consideration of a detailed atomization and vaporization model for the filling process may improve the simulation model. Additionally a model of the distribution of liquid and gas within the chambers is required in order to model more realistic gap flows.

NOMENCLATURE

A	cross section	(m^2)
c	velocity	(m/s)
h	specific enthalpy	(J/kg)
$\dot{m}$	mass flow	(kg/s)
p	pressure	(Pa)
$\dot{Q}$	heat flow	(J)

t	time	(s)
T	temperature	(K)
u	circumferential speed	(m/s)
v_i	internal volume ratio	(-)
V	volume	(m^3)
W	work	(J)
x	vapor content	(-)
α	flow coefficient	(-)
ρ	density	(kg/m^3)
φ	phase	(-)

Subscript

ch	chamber
crit	critical
ex	expansion
G	gas
L	liquid
max	maximum
S	solid
s	isentropic
stat	stationary

REFERENCES

Janicki, M., (2007), *Modellierung und Simulation von Rotationsverdrängermaschinen*, Dissertation, Universität Dortmund

Kliem, B., (2002), The Two-Phase Screw-Type Engine with Flash Evaporation, *Schraubenmaschinen 2002 VDI-Berichte 1715*, Düsseldorf, VDI Verlag, 167-177

Kliem, B., (2005), *Fundamentals of the Two-Phase Screw-Type Engine*, Dissertation, Universität Dortmund

Smith, I. K., (1993), Development of the trilateral flash cycle system: Part 1: fundamental considerations, *Journal of Power and Energy*, 207(3),179 – 194.

Smith, I. K., Marques da Silva, R.P., (1994), Development of the trilateral flash cycle system: Part 2: increasing power output with working fluid, *Journal of Power and Energy*, 208(2),135-144.

Smith, I. K., Stosic, N., Aldis, C. A., (1996), Development of the trilateral flash cycle system: Part 3: the design of high-efficiency two-phase screw expanders, *Journal of Power and Energy*, 210(1), 75-93.

Sprankle, R. S.,(1973), *Electrical Power Generating System*, patent specification, United States, publication number: 3,751,673

Sprankle, R. S., McKay, R.A., (1974), *Experience, Plans, and a Mixed Flow Expander*, Hydrothermal Power, Ltd, California, Pasadena

Steidel, R. F., Weiss, H., Flower, J.E., (1982), Performance characteristics of the Lysholm engine as tested for geothermal power applications in the Imperial Valley, *Journal of Engineering for Power,*104(1), 231-240

Taniguchi, H., Kudo, K., Giedt, W. H., Park, I., Kumazawa S., (1988), Analytical and experimental investigation of two-phase flow screw expanders for power generation, *Journal of Engineering for Gas Turbines and Power,*110(4), 628-635

Theoretical Analysis of the Impact of an Energy Recovery Expansion Device in a CO_2 Refrigeration System

Nicholas CZAPLA[1], Harshad INAMDAR[2], Riley BARTA[3]*, Eckhard GROLL[4]

Purdue University, School of Mechanical Engineering,
West Lafayette, Indiana, United States
(nczapla, hinamdar, bartar, groll)@purdue.edu

ABSTRACT

Carbon dioxide (CO_2) is being widely used as a refrigerant in HVAC&R applications due to its low Global Warming Potential (GWP). There are many aspects of CO_2 systems that make it unique to other traditional refrigerants in that it has higher pressure levels and typically operates at transcritical levels. Up to 40% of the losses in a transcritical CO_2 cycle can be due to throttling across the expansion valve. These significant losses make CO_2 systems ideal for installing an energy recovery expansion device that consists of a nozzle, micro-turbine and a generator. The expander has been developed by Regal Beloit Corporation and will be referred to as the Viper Expander. The device functions by using a nozzle to convert the high pressure of the refrigerant into a high velocity jet that is directed into the impeller of the micro-turbine. The turbine impeller then spins a shaft that is coupled with a generator to generator electrical energy. The Viper Expander is to replace the standard passive expansion device. Experimental testing of this device with R410A indicates that the device performs better for refrigerant with low viscosity and high pressure. For these reasons, the implementation of this energy recovery device in a CO_2 refrigeration system has been investigated. The results of this paper quantify the potential impact that this device could have in the system by predicting theoretical recoverable power and improvements to the coefficient of performance (COP). The analysis predicts that the Viper Expander can improve the COP from 1 - 11% and indicates that further experimental investigation should be conducted to verify the results.

1. INTRODUCTION

In 1987 the Montreal Protocol ruled to begin phasing out ozone depleting substances like chlorofluorocarbons (CFCs). The results of this international agreement led to the discovery of non-ozone depleting hydrofluorocarbons (HFCs) such as R134a and R410A. Even though HFCs are not contributors to ozone depletion, they do have significant global warming potentials. As the need to reduce the impact on climate change grows, efficient systems using alternative refrigerants must be developed. Since the natural refrigerant carbon dioxide (R744) is currently being advocated to replace CFCs and HFCs, there has been a renewed interest to develop more efficient CO_2-based system. The reason for this is that CO_2 has many advantages when compared to standard refrigerants and system components are now able to be manufactured to properly operate with CO_2.

Carbon dioxide has a global warming potential (GWP) of 1 and acts as the basis of comparison for all other substances. It is a non-toxic, colorless and odorless substance that is the by-product of many industrial applications. Therefore as significant amounts of CO_2 are being produced, it is a more cost-effective solution than traditionally manufactured refrigerants. Additionally, CO_2 also has many unique thermophysical properties that make it favorable as a natural refrigerant. These include a high heat transfer coefficient, very low viscosity as well as having low pressure drop in piping and heat exchangers. It has also been shown that CO_2 has a lower compression ratio of 20-50% less than HFCs and ammonia, which leads to increased volumetric efficiencies. The operating pressures of CO_2 are also much higher than traditional refrigerants and its critical point is very close to normal operating conditions. Therefore such systems are typically run transcritically and have different designs than pure subcritical systems. The system components must be designed and constructed out of materials to handle the high pressures.

Current commercial designs of CO_2 systems include transcritical and subcritical or cascade systems. One of the main applications for transcritical CO_2 systems is in commercial refrigeration for supermarkets. In the United States, roughly 40% of the electricity consumption is due to refrigeration (Energy Star, 2008). Therefore, there is a huge potential for savings if energy efficient CO_2 systems are designed. A year-long study funded by the U.S. Department of Energy in 2015 showed that a transcritical CO_2 system consumed 6-21% less electricity than an HFC unit during the winter and spring months. A milestone was reached in January of 2014 when The Coca-Cola Company announced the installation of its 1 millionth natural refrigerant CO_2 cooler. The result of this venture for more sustainable refrigeration has resulted in preventing the emission of 5.25 million metric tons of CO_2 into the Earth's atmosphere. By switching to CO_2 as the working fluid for coolers has assisted in allowing The Coca-Company to improve the energy efficiency of its systems by 40% (The Coca-Cola Company, 2014).

Aside from these applications, numerous other uses for CO_2 as a natural refrigerant have been found. As stricter energy regulations and emissions standards are being imposed on the automotive industry, many automotive manufactures are focusing on designing CO_2 based air-conditioning systems for their vehicles. The U.S. Army has invested in research of military environmental control units (ECUs) with CO_2. Li and Groll (2004), discovered that CO_2 based ECUs demonstrate both a higher cooling COP and a higher cooling capacity as compared to traditional R22-based ECUs.

Due to the potential for an energy recovery, a considerable amount of research has been conducted. For example, Yang et al., (2009) performed an experimental investigation of installing a double acting rotary vane expander into a transcritical CO_2 cycle. An electrical generator was connected to the shaft in order to consume the recovered power. Tests were conducted at expansion inlet pressures ranging from 7.85 – 8.35 MPa and the power output from the expander ranged from 348 – 379 W. This corresponded to an increase in COP of 11.1 – 14.2%. For this expander the isentropic efficiency was recorded to be between 19.1 – 22.63%.

Another work output expansion device for transcritical CO_2 systems was developed by Baek and Groll (2005). The device was chosen to be a piston cylinder expander and is known as the Expansion Device-With Output Work (ED-WOW). Three different conditions were tested with indoor and outdoor room temperatures of 20°C and 35°C, respectively The increase in COP ranged from 7.07% to 10.46% which includes the increased cooling capacity resulting from the non-isenthalpic expansion process.

2. THE TRANSCRITICAL CO_2 SYSTEM

The fundamentals of the transcritical system CO_2 refrigeration system investigated will now be presented. The system is comprised of an evaporator, multi-stage compressor, gas cooler with intercooling, two TXVs as well as a flash tank. The schematic of this cycle can be seen below in Figure 1 (left) and the respective log p-h diagram in Figure 1 (right). Detailed specifications of the system's components have been held confidential at the request of the manufacturer.

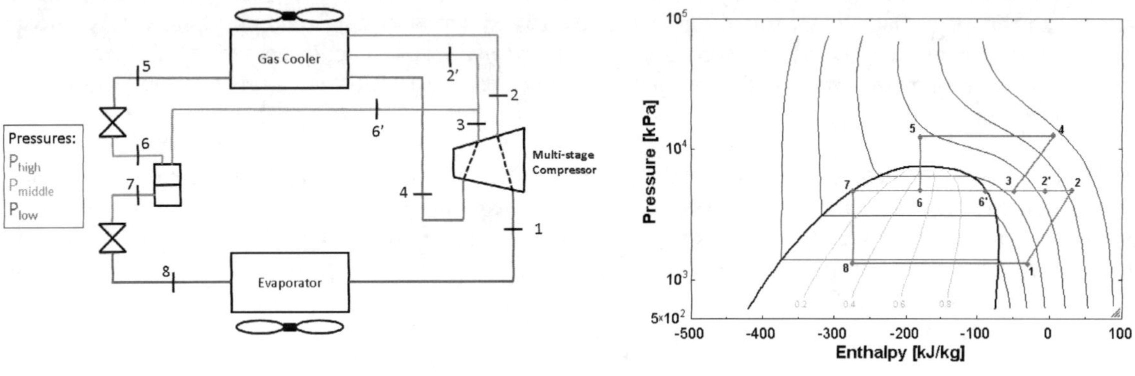

Figure 1: Schematic of transcritical CO2 cycle (left) and log p – h diagram (right). State points: 1- 1st stage suction, 2- 1st stage discharge, 2'- Intercooler exit, 3- 2nd stage suction, 4- 2nd stage discharge, 5- Gas cooler exit, 6- 2nd expansion exit, 6'- Separator exit (sat. vapor), 7- Separator exit (sat. liquid), 8- 1st stage expansion outlet/evaporator inlet

The CO_2 starts by entering the multi-stage compressor at State Point 1 and is compressed to the middle system pressure at State Point 2. Then the refrigerant enters the intercooling portion of the gas cooler which reduces the temperature to State Point 2'. The superheated vapor at State Point 2' is then mixed with saturated vapor at the same middle pressure. This mixing further reduces the temperature of the refrigerant that is to enter the 2nd stage of the compressor. This process is important because it reduces the discharge temperature from the 2nd stage of the compressor. The temperature at state point 4 cannot exceed certain values because the system may not be rated to handle such high temperatures. From State Point 4 to State Point 5 the CO_2 exchanges heat with the surroundings in the gas cooler at a transcritical state. This is ideally a constant pressure heat exchange process. State Point 5 is the inlet to the high side expansion process. The standard unit uses a TXV and isenthalpically expands from State Point 5 to 6. The liquid and gaseous CO_2 is then separated at State Point 6. The saturated vapor is considered to be 6' and the saturated liquid is 7. The saturated vapor is mixed with State Point 2' to determine State Point 3 and the saturated liquid at 7 goes through the low side expansion process to State Point 8. The cycle is then completed as the refrigerant exchanges heat with the ambient air in the evaporator and is heated to the suction temperature at State Point 1.

2.1 Analysis of Transcritical CO_2 System

The analysis will be performed over a range of operating conditions. Five different operating conditions are chosen for the investigation. These five operating points offer a wide range of system performance which will allow for the Viper's potential impact to be thoroughly investigated. The high and low side temperatures of each chosen operating conditions are shown in Table 1.

Table 1: Operating points for analysis

Operating Point (OP)	Low-Side Temperature		High-Side Temperature	
	[°F]	[°C]	[°F]	[°C]
1	-22	-30.0	77	25.0
2	30	-1.11	77	25.0
3	32	0.00	100	37.8
4	-20	-28.9	100	37.8
5	-22	-30.0	122	50.0

The baseline system data for all operating conditions is shown in Table 2. The data in this table was used to find the main system state points that are necessary to construct the cycle in a log p-h diagram. Both temperature and pressure have been experimentally measured for each of the five system operating conditions. Only three pressures are reported: high, middle and low. The pressure difference between all the state points of a particular pressure level is considered to be negligible. Additionally the total system power consumption is reported. These values will be used when determining the COP.

In order to fully construct the cycle, the inlet to the 2nd stage of the compressor (State Point 3) must be calculated. Applying an energy balance to the mixing process at the 2nd state compressor inlet will enable State Point 3 to be determined. Refrigerant from the gas cooler exit (State Point 2') is mixed with saturated vapor from the flash tank (State Point 6'). This mixture then enters the compressors as the 2nd stage suction (State Point 3). The process is shown in the log p–h diagram in Figure 2.

Table 2: Set of given data for all operating conditions

	Operating Point	Units	OP1	OP2	OP3	OP4	OP5
	Test Condition Temperatures	C	-30.0/25.0	-1.11/25.0	0.0/37.8	-28.9/37.8	-30.0/50.0
Press	High-Side Discharge	kPa	7496.3	7717.9	12071.6	9989.1	12398.8
	Middle	kPa	3568.9	4405.5	6590.0	4194.4	4845.1
	Low-Side Suction	kPa	1170.2	2978.0	2718.4	1115.7	1068.5
	Total Power	W	3256.6	2730.7	9254.8	8431.2	9326.2

The first law of thermodynamics states that energy must be conserved for a thermodynamic system. By assuming steady state, adiabatic process, no work and neglecting potential and kinetic energy change, the first law of thermodynamics for the mixing process at State Point 3 reduces to:

$$\dot{m}_{6'}h_{6'} + \dot{m}_{2'}h_{2'} = \dot{m}_3 h_3 \tag{1}$$

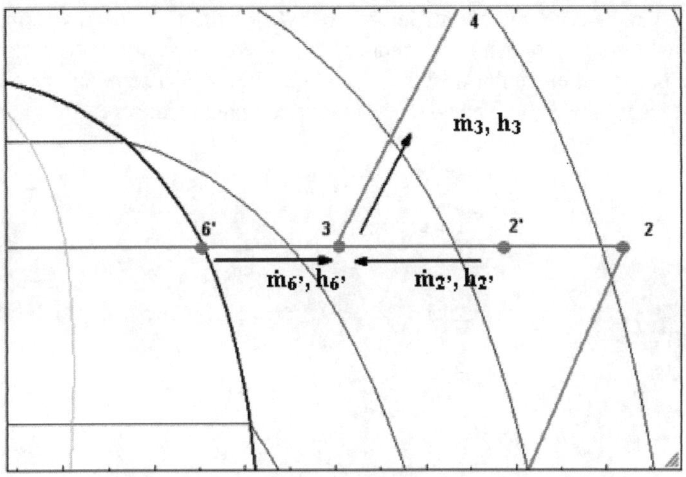

Figure 2: Energy balance for determining State Point 3

Therefore the enthalpy at the inlet to the 2nd stage of the compressor can be solved and the complete cycle can now be determined. In order to understand the potential impact of an energy recovery expansion device, a 2nd law analysis needs to be performed on both the high and low side expansion processes. The maximum available power can be predicted by assuming the expansion processes occur isentropically. This means the inlet entropy will be equal to the outlet entropy.

$$s_{out,is} = s_{in} \tag{2}$$

The outlet isentropic enthalpy is then calculated with the outlet pressure and the outlet entropy:

$$h_{out,s} = f\left(p_{out}, s_{out,is}\right) \tag{3}$$

The total available power from the throttling process can now be calculated as the product of the mass flow rate and difference in enthalpy values for the isentropic process:

$$\dot{W}_{is} = \dot{m}\left(h_{out,is} - h_{in}\right) \tag{4}$$

3. THE VIPER EXPANDER

The Viper Expander is a power generating device that consists of a nozzle, turbine and a generator. The purpose of the expander is to harvest the normally wasted work potential during the expansion process of heat pumps or refrigeration systems. The device is to replace the system's thermostatic expansion valve (TXV) or electronic expansion valve (EXV). The nozzle is used to convert the high pressure energy of the refrigerant at the outlet of the condenser into kinetic energy in the form of low pressure, high velocity jet. The refrigerant will impinge on the impeller causing it to spin and the generator to produce a 3-phase alternating current (AC). The current is then to be rectified to direct current (DC) and fed to an electronically commutated motor (ECM). ECMs operate with direct current and have many advantages over standard AC motors like higher efficiency and lower cost. The power from the Viper Expander will be used to partially power the fan and thus reduce the total system power consumption. Additionally, the work extracted from the Viper Expander will result in a lower outlet enthalpy than the inlet to the nozzle. This reduction in enthalpy is due to the expansion process now being pushed towards partially isentropic which causes an increase in the cooling capacity. Both the reduced power consumption and added cooling capacity should increase the system COP.

The particular Viper Expander design to be tested with the transcritical CO_2 system is based off of an impulse turbine design. For pure impulse turbines, all of the pressure energy is converted into kinetic energy across the nozzle, which correlates to the refrigerant exit velocity. Since the power corresponds to the kinetic energy of the refrigerant impinging on the impeller, the goal is to ensure that as much pressure drop as possible will occur across the Viper nozzle. Typical designs for such impulse impellers are Pelton wheels which have two symmetric buckets with a splitter blade in the center. The splitter blades act to balance the fluid forces acting on the impeller and the buckets cause the fluid flow direction to turn. This magnitude of the change in flow direction relates to the rotational speed of the impeller. The proposed expander design is shown in Figure 3. The nozzle is on the left, the housing is in the middle and the impeller is on the right. The nozzle is a separate piece and screws into the housing.

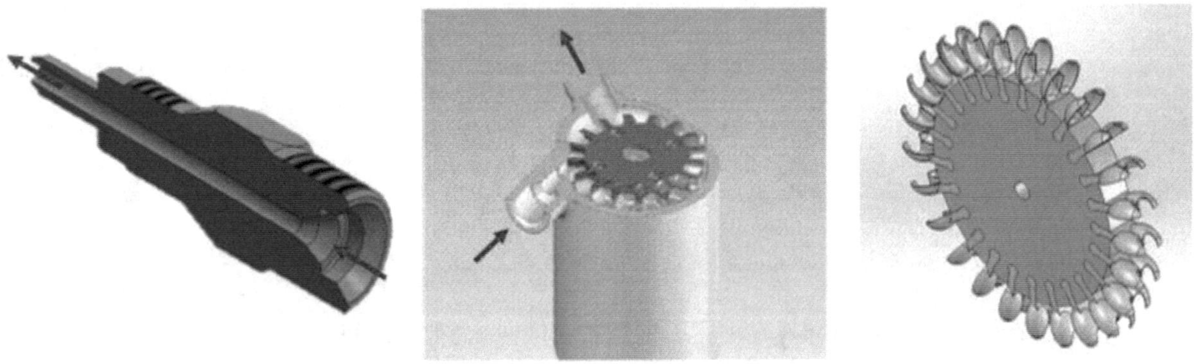

Figure 3: The nozzle (left), housing (middle) and impeller (right) for the expander

The nozzle shown in Figure 3 (left) has a very short converging portion followed by a long constant diameter cross-sectional area. Since this is a two-phase expansion process, the flow will continue to accelerate through the constant diameter section due to the decreasing density as the fluid expands into the two-phase dome. The velocity of the refrigerant can be calculated as follows:

$$V = \frac{\dot{m}}{\rho A} \tag{5}$$

The power generated from the Viper Expander can then be determined by the following equation:

$$\dot{W}_{Viper} = \eta_{Viper}\left(\frac{1}{2}\dot{m}V^2\right) \tag{6}$$

Additionally, the power generated from the Viper Expander can also be determined from the enthalpy change across the expansion process:

$$\dot{W}_{Viper} = \dot{m}\left(h_{out} - h_{in}\right) \tag{7}$$

Since this is a transcritical system with two expansion valves, there are three different possibilities to install the expander in the system. The first possibility is to install the device on the high-side, the second possibility is to purely install it on the low-side and the third possibility is to install it on both the high and low-side. Figure 4 shows the expander installation options in more detail.

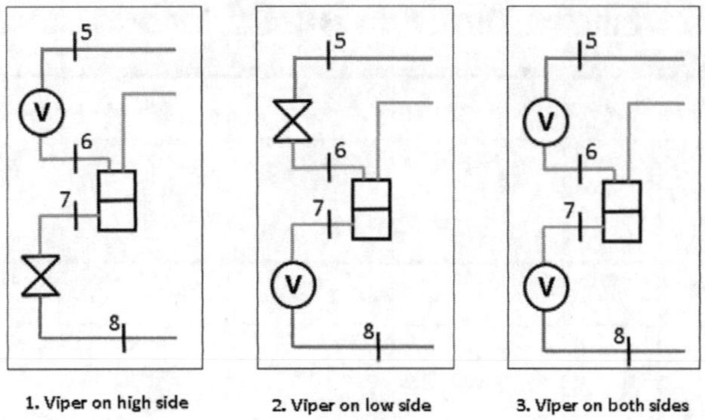

Figure 4: The three possibilities for installing the expander

4. RESULTS

The results of implementing each of the three Viper Expander installation options with all five of the operating conditions were analyzed. Before presenting the impact of the expander on the cycle, it is important to understand how much power is actually available to be harvested. Figure 5 shows isenthalpic and isentropic expansion processes for both the high and low side expansion.

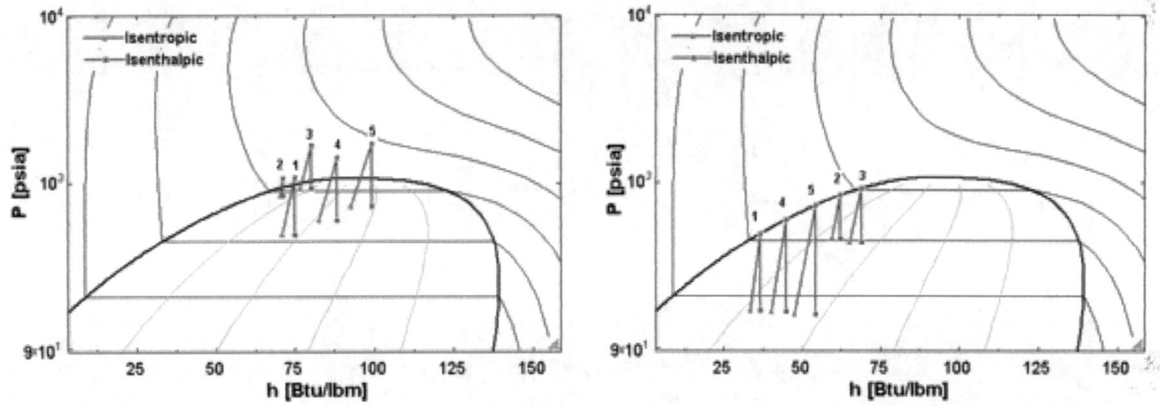

Figure 5: High expansion process (left) and low expansion process (right)

The estimated power to be recovered from the Viper Expander can be found by multiplying the available power during the isentropic expansion process by the target efficiency of the device. The power output can be determined as follows:

$$\dot{W}_{Viper} = \eta_{Viper}\dot{W}_{is} \tag{8}$$

Based on prior experimental data, the maximum achievable value of the expander efficiency has been predicted to range from 30 – 50%. For the sake understanding the theoretical power to be harvested, both the high and low side expander efficiencies were assumed to be 50%. These are just estimates to gather a general idea of how the Viper Expander may perform in the actual CO_2 system. Table 4 shows the available power for each operating point and how much power can potentially be recovered if the aforementioned efficiencies are achieved.

Table 1: Comparison between available power and potential recovered power by Viper Expander

Side	Power	Unit	OP1	OP2	OP3	OP4	OP5
High Side	Available	W	182.5	43.2	606.6	563.2	859.2
	Viper $\eta = 50\%$	W	91.2	21.6	303.3	281.6	429.6
Low Side	Available	W	85.2	86.1	530.5	241.7	356.3
	Viper $\eta = 50\%$	W	42.6	43.0	265.2	120.8	178.2
	Viper $\eta = 30\%$		25.6	25.8	159.2	72.5	106.9

Figure 6 shows the high and low-side expansion processes for standard isenthalpic process, the isentropic process as well as the Viper process. This processes depicts the values of OP1.

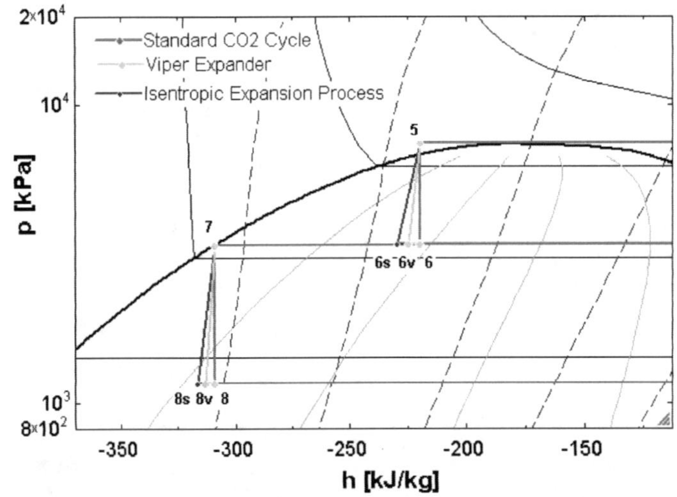

Figure 6: Isenthalpic, isentropic and Viper expansion processes

Observing the power output from the different operating conditions, it is seen that the operating condition with the largest potential power to be recovered is OP5 (50.0°C/-28.9°C) followed by OP3 (37.8°C/0.0°C) and then OP4 (37.8°F/-28.9°F). The available power for these operating points ranges from 563 to 859 W, which correlates to 6 to 9% of the total system power consumption. Also in comparing the high and the low side expansion processes, it is seen that the most power can be generated from the high side expansion in 4 of the 5 cases with the exception of OP2 (25°C/-1.1°C). The reason for this is because of the very low pressure ratio on the high side. Aside from observing the potential power that can be recovered, the impact on the COP will also be compared for the standard system and the system with the Viper Expander. The refrigerant-side COP can be calculated with the following equation:

$$COP_{ref} = \frac{\dot{Q}_C}{\dot{W}_{net}} \quad (9)$$

Table 5 displays the impact on COP for all operating points, Viper efficiencies and Viper installation options. The percentage improvement to the COP, ξ, is also displayed and has been calculated as:

$$\xi=\left(\frac{COP_{Viper}}{COP_{ref}}-1\right)\times 100\% \tag{10}$$

Table 2: Viper COP Improvement in CO_2 with Viper Efficiencies of 50% and 50%.

OP	Baseline	Viper Installed on:					
	COP	High Side		Low Side		Both Sides	
		COP	ξ	COP	ξ	COP	ξ
1	0.955	0.977	2.30	0.983	2.96	1.006	5.36
2	1.198	1.210	1.00	1.235	3.07	1.247	4.11
3	1.249	1.293	3.50	1.336	6.92	1.384	10.81
4	0.608	0.643	3.23	0.633	4.07	0.653	7.48
5	0.526	0.552	5.02	0.555	5.46	0.583	10.86

Upon investigating the predicated improvement in reported in Tables 5, it is shown that there is a significant potential for the Viper Expander in this transcritical CO_2 system. The impact of the Viper being installed on just the high side could be as high as 5% improvement and the Viper being installed on just the low side could range from 3 to 7% improvement. Additionally, the impact of installing two Vipers into the system would offer the largest improvement. These predicted improvement values range from 4 to11% improvement in COP.

5. CONCLUSIONS AND FUTURE WORK

This paper presented a theoretical analysis of installing an energy recovery expansion device into a transcritical CO_2 refrigeration system. The Viper Expander is a device that consists of a nozzle, impeller and a generator which connects to an ECM of a fan to reduce the power consumption. The device is to replace either the TXV or the EXV of the system. Experimental baseline data from 5 different operating conditions were used to investigate the potential power available and to estimate the impact that the Viper Expander could have in the system. The results show that the system COP can be improved from 1-11% by installing one or two Vipers into the system. The predicted results indicate that further experimental research with the device is worth pursuing.

The future work of this project is to manufacture the described Viper Expander and test it at Purdue University's Ray W. Herrick Laboratories. The Viper will be installed in a CO_2 test stand and the five operating conditions will be simulated. The power output from the Viper Expander will be compared to the theoretical and the device efficiency will be determined.

NOMENCLATURE

A	area	(m^3)
H	specific enthalpy	(kJ/kg)
$\dot{m}$	mass flow rate	(kg/s)
P	pressure	(kPa)
s	specific entropy	(kJ/kg-K)
$\dot{W}$	power	(W)
V	velocity	(m/s)

Greek symbols

H	efficiency
P	density
Ξ	improvement to COP

Subscript

1,2,3…	state points
C	cooling
In	inlet
Is	isentropic
Net	net total value
Out	outlet
Ref	refrigerant
V	Viper Expander
Viper	Viper Expander

Acronyms

AC	Alternating Current
DC	Direct Current
CFC	Chlorofluorocarbon
COP	Coefficient of Performance
ECM	Electronically Commutated Motor
ED-WOW	Expansion Device - With Output Work
EXV	Electronic Expansion Valve
GWP	Global Warming Potential
HFC	Hydrofluorocarbon
TXV	Thermostatic Expansion Valve
OP	Operating Point

REFERENCES

Baek, J.S., Groll, E.A., P.B. Lawless (2005): Piston-Cylinder Cork Croducing Expansion Device in a Transcritical Carbon Dioxide Cycle. International Journal of Refrigeration.

Energy Star. (2008). Energy Star Building Manual. 11. Facility Type: Supermarkets and Grocery Stores.

Li, D. and E. Groll. (2004). Theoretical Performance Evaluation of a Carbon Dioxide based Environmental Control Unit (ECU) with Microchannel Heat Exchangers. 6th IIR-Gustav Lorentzen Conference on Natural Working Fluids. Glasgow, Scotland.

The Coca-Cola Company. (2014). Coca-Cola Installs 1 Millionth HFC-Free Cooler Globally, Preveting 5.25MM Metric Tons of CO2. http://www.coca-colacompany.com/press-center/press-releases/coca-cola-installs-1-millionth-hfc-free-cooler-globally-preventing-525mm-metrics-tons-of-co2/.

United States Department of Energy. (2015). Case Study: Transcritical Carbon Dioxide Supermarket Refrigeration Systems. Energy Efficiency & Renewable Energy.

Yang, B., Peng, X., Sun, S., Guo, B. and Z. Xing. (2009). Study of a Rotary Vane Expander for the Transcritical CO_2 Cycle – Part I: Experimental Investigation. HVAC&R Research, 15:4, 673-688.

ACKNOWLEDGEMENT

The authors would like to thank Regal Beloit Corporation for the support of this project and the faculty of Ray W. Herrick Laboratories.

High Efficiency Inverter Scroll Compressors

Yasuhiro MURAKAMI[1*], Hiroshi KITAURA[2], Kazuhiko MATSUKAWA[3],
Katsumi KATOU[4], Yoshitomo TSUKA[5], Yasuo MIZUSHIMA[6]

[1]Compressor Development and Engineering Group, Daikin Industries, LTD
3-12 Chikko-shinmachi, Nishi-ku, Sakai, Osaka, Japan
Tel: +81-72-241-1261, Fax: +81-72-245-3740, E-mail:Yasuhiro.murakami@daikin.co.jp

[2] Compressor Development and Engineering Group, Daikin Industries, LTD
3-12 Chikko-shinmachi, Nishi-ku, Sakai, Osaka, Japan
Tel: +81-72-241-1261, Fax: +81-72-245-3740, E-mail:Hiroshi.kitaura@daikin.co.jp

[3] Compressor Development and Engineering Group, Daikin Industries, LTD
3-12 Chikko-shinmachi, Nishi-ku, Sakai, Osaka, Japan
Tel: +81-72-241-1261, Fax: +81-72-245-3740, E-mail: kazuhiko.matsukawa@daikin.co.jp

[4] Compressor Development and Engineering Group, Daikin Industries, LTD
3-12 Chikko-shinmachi, Nishi-ku, Sakai, Osaka, Japan
Tel: +81-72-241-1261, Fax: +81-72-245-3740, E-mail: katsumi.katoh@daikin.co.jp

[5] Compressor Development and Engineering Group, Daikin Industries, LTD
3-12 Chikko-shinmachi, Nishi-ku, Sakai, Osaka, Japan
Tel: +81-72-241-1261, Fax: +81-72-245-3740, E-mail: yoshitomo.tsuka@daikin.co.jp

[6] Compressor Development and Engineering Group, Daikin Industries, LTD
3-12 Chikko-shinmachi, Nishi-ku, Sakai, Osaka, Japan
Tel: +81-72-241-1261, Fax: +81-72-245-3740, E-mail: yasuo.mizushima@daikin.co.jp

* Corresponding Author

ABSTRACT

We have to consider the global warming and the energy saving to reduce the environment load when the new product is going to be developed.

Improving the compressor efficiency is the essential factor for air conditioning for buildings, which are widely spreading now, to reduce the environment load.

Needless to say, it is very important to improve not just the efficiency of compressor at the rated condition but also at the small load condition.

In this paper we explain the technic of inverter scroll compressor, aiming the dramatic improvement of efficiency compared with the one of the traditional compressors, by employing the back pressure control structure and injection mechanism with check valve.

1. INTRODUCTION

The protection of environment is becoming a big issue from the destruction of the ozone layer, the global warming and the energy saving point of view.
On the other hand, there is expectation of improving the comfort of the room environment as well, so that we have to contribute the environment protection by developing the air conditioning technic.
The improvement of energy saving technic is now also expected for the widely spreading air conditioning for buildings. In Figure 1 it shows the load ratio and the operation time of outdoor unit for an office.
About 90% of the yearly operation time is at 50% or less load rate, which means that for the further energy saving, it needs to improve the efficiency of the operation at the low load rate.
When the load rate becomes low, the operation compression ratio also becomes low, which leads the low rotation speed of the inverter compressor. The traditional inverter scroll compressor had huge drop down of its efficiency due to the leak loss when the load rate becomes less than about 30% with its low compression ratio and low rotation speed.
We contributed to the energy saving of air conditioning for buildings by improving the efficiency at the low compression ratio, low rotation speed and even under the comparatively high load rate condition and contributed.
We have developed high efficiency and high reliability, high-pressure dome type inverter scroll compressors, which are for the building air conditioning use.
With the new technologies: back pressure control structure and injection mechanism with check valve, as compared with the conventional scroll compressor, in a wide operation range from a low compression ratio operation to high compression ratio operation and from low-speed operation to high speed operation, efficiency and reliability were significantly improved.

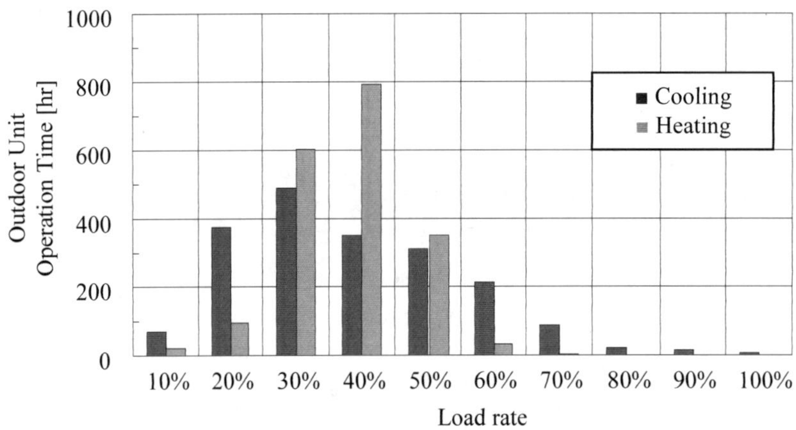

Figure 1: Load Rate / Operation Time
(Office in building : "airnet" data of Daikin)

2. STRUCTURE OF COMPRESSOR

The traditional technology, which we employed to achieve high efficiency and high reliability, such as,
- The high-low pressure dome structure to suppress the suction overheat
- The high thrust mechanism to suppress the increase in the surface pressure by introducing the high pressure oil between thrust sliding surface of the fixed scroll and the orbiting scroll.

In addition to these technologies we employed the following technology as well.
- The back pressure control in order to prevent the turnover of the orbiting scroll even at a low compression ratio operation,
- The injection mechanism with check valve to achieve high efficiency both at rated condition and a low compression ratio condition by minimizing the dead volume of the compression chamber.

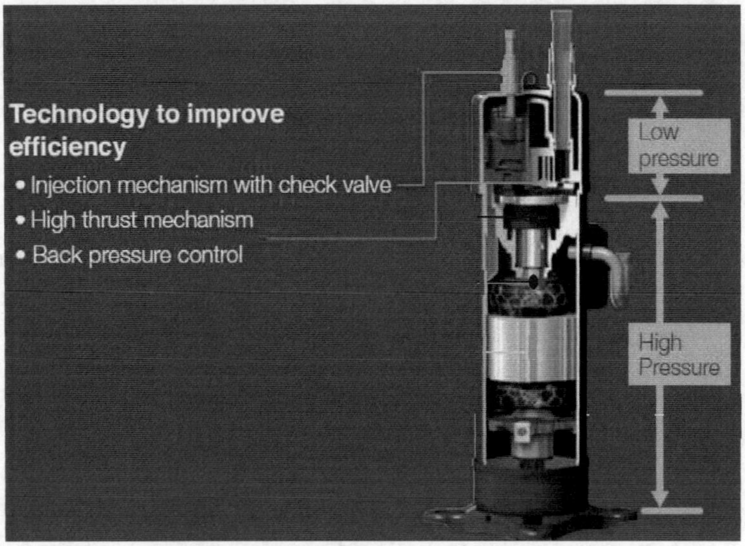

Figure 2: The structure of new model compressor

3. HIGH EFFICIENCY TECHNOLOGY

3.1 Back Pressure Control

There is a problem that efficiency goes down by increasing leak loss, which is caused by the orbiting scroll's overturn at the low compressing ratio operation. It happens because the compression chamber is formed with the orbiting scroll pressed against the fixed scroll.

The gas during compression is employed to the back surface of the orbiting scroll to solve this problem and the orbiting scroll is pushed by the pressure of the discharge gas and the arbitrary gas pressure during compression (Middle pressure).

Figure 3 shows the structure of back pressure control.

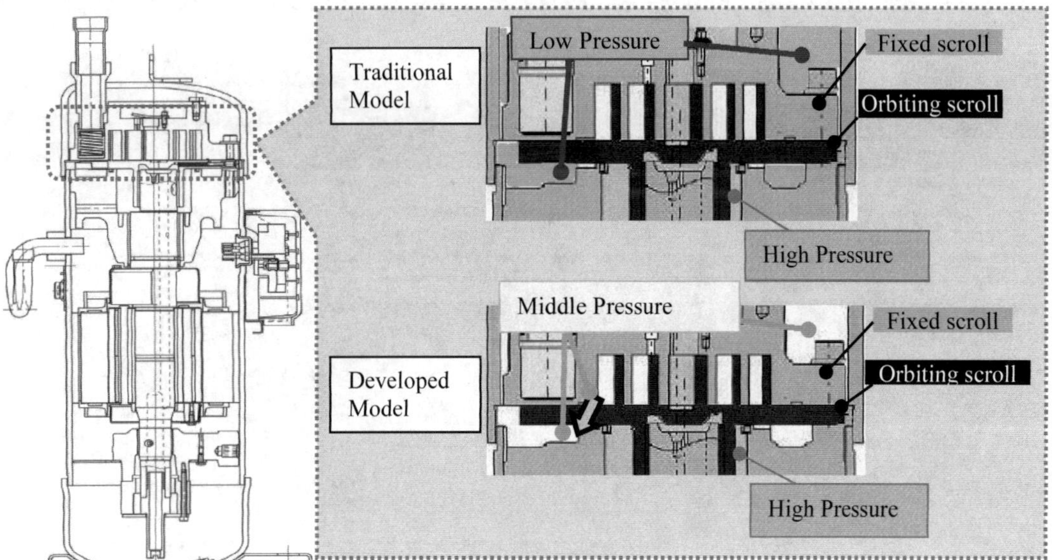

Figure 3: The structure of back pressure control

To adjust the value of the middle pressure, we have examined the pressure which makes the compressor input become smaller.
Figure 4 is the result of the test which we evaluated the compressor input by changing the value of middle pressure. It shows that the compressor input at each condition become smaller when the ratio of middle pressure to suction pressure is between 1.4-1.6. We have improved efficiency of the developing model by setting the middle pressure within this area.

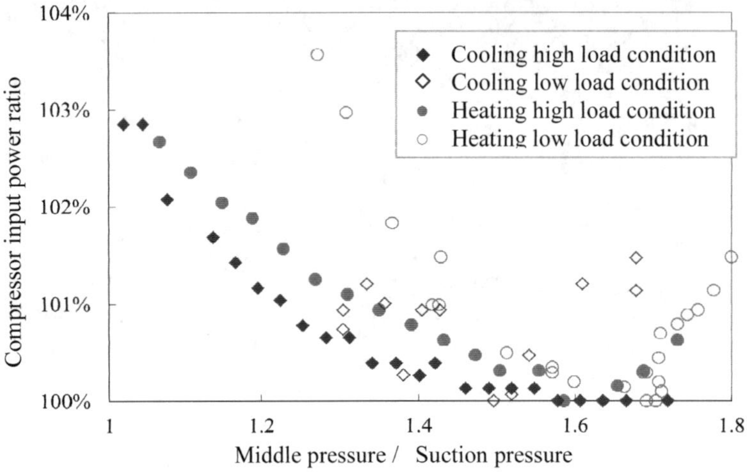

Figure 4: Relation between middle pressure and compressor input ratio

3.2 Injection mechanism with check valve

Traditionally the middle pressure injection structure has been employed for scroll compressors to improve the heating capacity and also system efficiency under the rated condition or high compressing ratio. But when the air conditioning load is small such as under the low compressing ratio and the injection is not employed, the injection structure between the system and the compressing chamber becomes the dead volume, which ends up the dramatic decrease of the compressor efficiency. To employ the injection structure with check valve, which minimizes the dead volume, we can suppress the efficiency drop at low compressing ratio condition without gas injection.

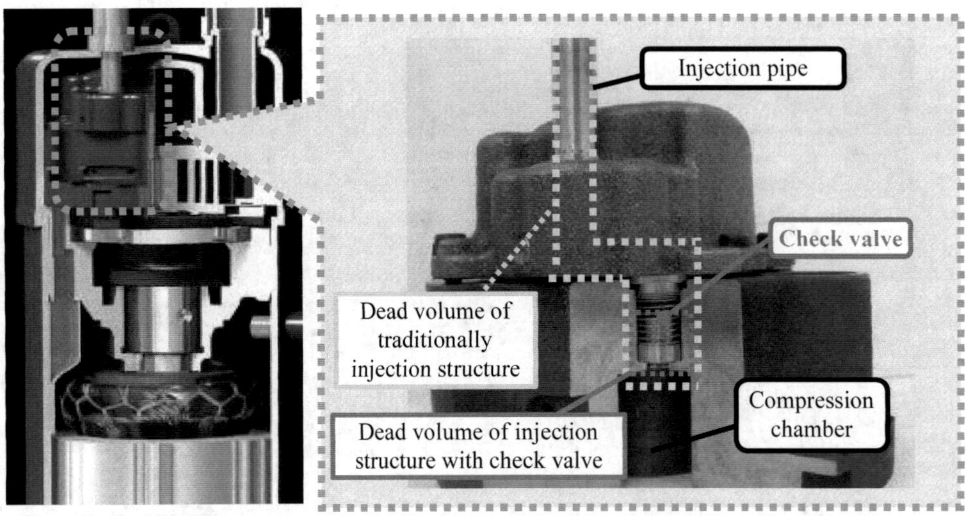

Figure 5: The structure of injection mechanism with check valve

Figure 6 shows the calculation of dropping amount of indicated efficiency ratio to the suction volume ratio of dead volume under the condition where the compression ratio is 1.8, which is comparatively low compressing ratio. Compared with the dead volume of the traditional injection structure, which is considered as the cause of efficiency drop, the dead volume becomes now less than 1/20 by employing the new structure with the check valve set nearby the compressing chamber.

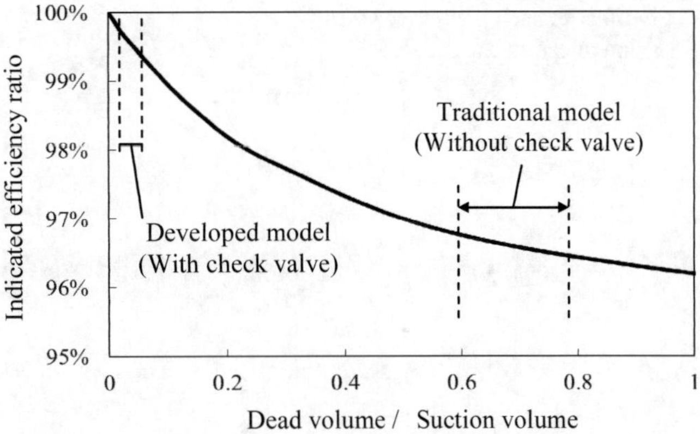

Figure 6: The Relation between dead volume and indicated efficiency

Figure 7 shows the test results of the compressor efficiency change which has injection structure. Under the condition where the compression ratio is 1.8 without gas injection, the compressor efficiency was improved 2.2pt by employing the injection structure with check valve.

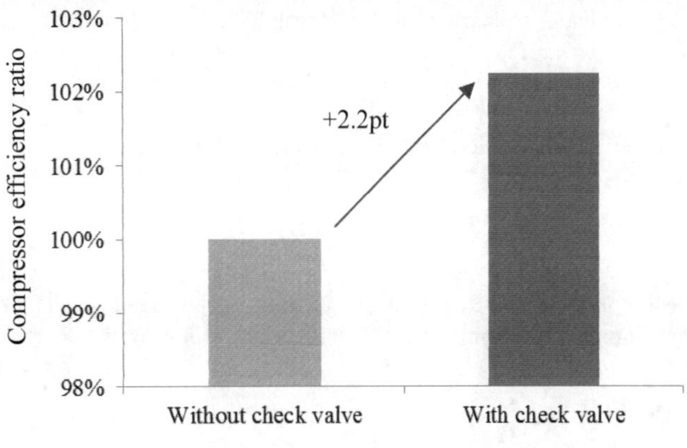

Figure 7: Efficiency ratio between employing injection with/without valve

3.3 Efficiency Improvement Result

Figure 8 shows the efficiency of inverter scroll compressor employing back pressure control and middle pressure injection with check valve.

Compared with the traditional model, the efficiency of the developed model was improved at whole operation area by improving the efficiency at low load ratio employing the back pressure control and the injection structure with check valve, and also by improving the efficiency at high load ratio employing the middle pressure injection.

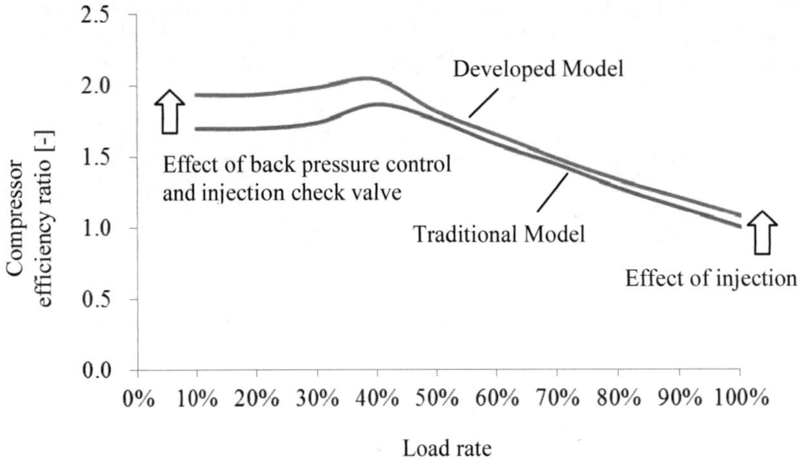

Figure 8: Efficiency of back pressure control and injection with check valve

4. CONCLUSIONS

Here is the summary of improvement efficiency for inverter scroll compressor employing the back pressure structure and injection mechanism with check valve.

- Compared with the traditional model under small load operation condition, the efficiency was dramatically improved by controlling back pressure of middle pressure as the input becomes optimum at each condition.
- By employing the injection structure with check valve, the efficiency was improved at high load operation condition by injection and the efficiency drop down was suppressed at small load operation condition.
- Compared with the traditional model, the efficiency of whole operation area was improved by back pressure control structure and injection mechanism with check valve.

REFERENCES

K. Matsukawa, et, al. , "Development of the New Capacity Control Technique for the High Efficiency Scroll Compressor.", International Compressor Engineering Conference at Purdue, 2008

Development of Compact and High-efficient Scroll Compressor with Novel Bearing Structure

Sungyong AHN[1], Junchul OH[1*], Junghoon PARK[1], Seheon CHOI[1],
Byeongchul LEE[1], Hyunwoong CHO[2], Jeonghun KIM[2]

[1]Air Conditioning Compressor Team, LG Electronics Inc.
Seoul, Korea
E-mail: sungyong.ahn@lge.com

[2]Air Conditioning Compressor Development Team, LG Electronics Inc.
Changwon, Gyeongnam, Korea

* Corresponding Author

ABSTRACT

High-Side Shell(HSS) scroll compressors have been widely used for Variable Refrigerant Flow(VRF) system which is a powerful solution for the cooling and heating of commercial buildings. In order to improve the characteristics of the VRF system, a new HSS scroll compressor has been developed with the outer bearing structure. The core elements of the outer bearing structure are an outer-type bush bearing mounted on an orbiting scroll and a female-type eccentric journal inside a shaft. The outer-type bush bearing which is made of engineering plastic without a back steel layer has been newly developed. The new HSS scroll compressor employing the outer bearing structure has a compact size, high efficiency, and low noise level compared to a conventional HSS scroll compressor. In order to confirm the advantages of the new HSS scroll compressor, basic tests and theoretical analysis have been performed in this study.

1. INTRODUCTION

The compression part of a scroll compressor is composed of two scrolls. One is a Fixed Scroll(F/S) which is usually clamped onto a Main Frame(M/F). The other is an Orbiting Scroll(O/S) that has orbit motion by an eccentric journal and an Oldham's couple ring. The scroll compressor has relatively a low noise and vibration level, and large capacity among positive displacement compressors because multi-chambers for compression are formed between two scrolls at the same time. Characteristics described above are the main reasons to use a scroll compressor in various heating, ventilating, and air conditioning systems. In general, the scroll compressor are classified into two types, Low-Side Shell(LSS) type and High-Side Shell(HSS) type, according to a pressure level in a compressor shell below the M/F. Variable Refrigerant Flow(VRF) systems used for cooling and heating of commercial buildings usually employ the HSS scroll compressor because that has good volumetric efficiency and the ease of oil management in addition to basic good characteristics of the scroll compressor.

Recently, the main issues of compressors for the VRF system are compact size and high efficiency. The high efficiency of VRF system is essential because regulations for building energy consumption have been reinforced with the environment issue such as global warming. Compact compressors can enhance the space utilization and cost competitiveness of the VRF system. Additionally, a heating capacity in cold weather and a noise level are basic characteristics which must meet the needs of customers using the VRF system. We have developed a new HSS scroll compressor with the outer bearing structure to raise the competitiveness of the VRF system. The core of the outer bearing structure is an outer-type bush bearing driving an O/S. The newly developed HSS Scroll compressor is represented in Fig. 1 with a conventional one, and the specifications of the two scroll compressors are listed in Table 1. The two scroll compressors have the same orbiting radius and suction volume. Furthermore, a compression volume ratio and O/S mass are very similar comparing the two scroll compressors. That is, the two compressors are identical in terms of a capacity and theoretical motor torque required for a compression process. The main difference between the two compressors is the design of an O/S journal bearing. The O/S journal bearing of the conventional scroll compressor is designed with a male-type eccentric journal and a traditional Polytetrafluoroethylene(PTFE) bush bearing. Meanwhile, the O/S journal bearing of the new one is modeled with a female-type eccentric journal inside a shaft and an outer-type bush bearing made of engineering plastic without a steel layer. These changes in the new HSS scroll compressor bring many advantages such as compact size, high efficiency, and low noise level.

At the beginning of scroll compressor history, Inaba *et al.*[3] and Yamamora Muramatsu *et al.*[4] studied for similar type scroll compressors to the new scroll compressor. Moreover, some manufacturers produced scroll compressors which adopt the female-type eccentric journal. However, they didn't use the outer-type bush bearing. The new scroll compressor is the world's first one of which O/S journal bearing is designed with the outer-type bush bearings. In this study, basic test items of a compressor such as efficiency and a noise level are evaluated to compare the two scroll compressors each other. The lubrication characteristics of two bush bearings are also compared by bench tests. Furthermore, a theoretical analysis for shaft dynamics and the lubrication characteristics of journal bearings is performed to confirm shaft motion and journal bearing loads.

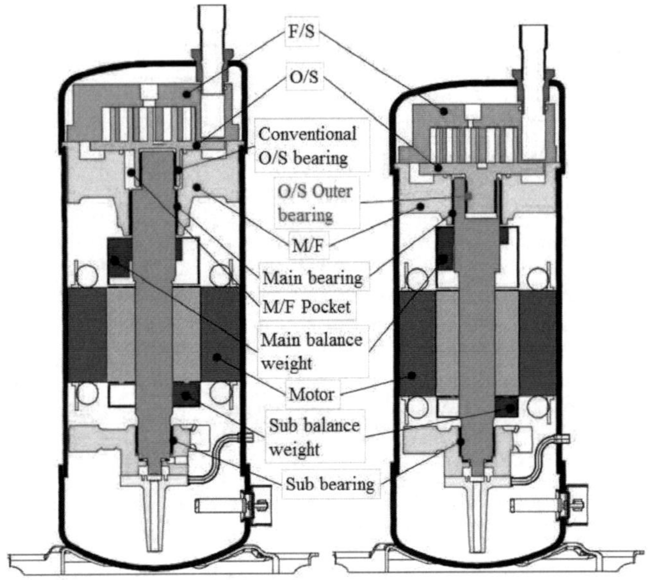

(a) Conventional (b) Newly developed

Figure 1: Comparison of scroll compressors

Table 1: Spec. of scroll compressors

Item	Conventional	Newly developed
O/S bearing type (Lubricating surface)	Inner	Outer
Weight	38kg	32kg (84%)
Height	460mm	440mm (96%)
Inner diameter	160mm	147mm (92%)
Operating speed	15~150Hz	12~165Hz

2. NEW OUTER-TYPE BUSH BEARING

While the shaft of a scroll compressor rotates synchronized with a rotor, the O/S of a scroll compressor has orbit motion. That means the loading point of a gas force is not changed in the circumferential direction of an eccentric journal as Fig. 2. The certain area of the eccentric journal sweeps all surfaces of an O/S bearing circumferentially, although the minimum clearance point on the eccentric journal is varied by shaft motion and an attitude angle. The operating mechanism of the new scroll compressor is equal to that of the conventional one. Therefore, the outer-type bush bearing should be applied to the O/S.

It is also possible to insert a bush bearing into the female-type eccentric journal in terms of manufacture. However, there are two problems in that case. One is the one-point wear of the bush bearing due to the operating mechanism of a scroll compressor mentioned above. The other is the excessive diameter of a main journal. The diameter of the main journal must be grown by twice the thickness of the bush bearing in the case. In a design using the outer-type bush bearing, the main journal already has slightly larger diameter than that of the conventional scroll compressor to secure shaft stiffness of the minimum thickness point by the female-type eccentric journal. The diameter of the main journal has to be minimized in an available design range to reduce bearing loss. Therefore, the outer-type bush bearing made of engineering plastic is mounted on the O/S like as Fig. 3(b). The traditional PTFE bush can't be used as the outer-type bearing because that has a slit.

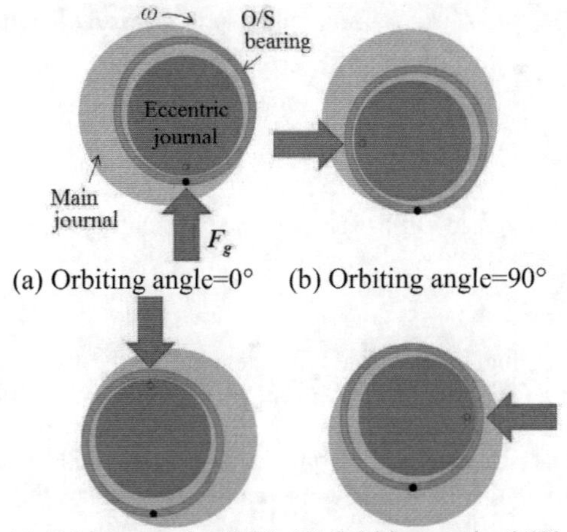

(a) Orbiting angle=0° (b) Orbiting angle=90°

(c) Orbiting angle=270° (d) Orbiting angle=360°

Figure 2: Acting direction of gas force

(a) Inner-type bush bearing (Conventional)

(b) Outer-type bush bearing (New)

Figure 3: Bush bearings for O/S journal bearing

All development processes of the outer-type bush bearing are focused on its reliability for high temperature oil and refrigerant in a compressor. Additives and thermal treatment processes have a key role in the reliability. Figure 4 shows the test results of bush bearings using a bench tester for the journal bearing of a scroll compressor. The new bush bearing made of an engineering plastic is compared with two kinds of bush bearings to confirm its feasibility. One is the PTFE bush bearing used for the conventional scroll compressor. The other is an Aluminum(Al) bush bearing which is well known for its low friction coefficient in a hydrodynamic lubrication region. Figure 4(a) represents the Stribeck curves of the three bush bearings from a step load test (60Hz, μ=0.005Pa·s). The curve of the new bush bearing is located between two curves of the PTFE bush and Al bush bearing. That means the new bush bearing has a medium characteristics between the PTFE and Al bush bearing in terms of friction loss. The oil block test results are presented in Fig. 4(b). Oil supply is suddenly blocked during a normal operation (60Hz, 3000N, μ=0.005Pa·s). While the Al bush bearing has short duration time after oil block, the new bush bearing and PTFE bush bearing has quite larger duration time than the Al bush. From the results, it is confirmed that the new bush bearing has typical good characteristics of two existing bearings.

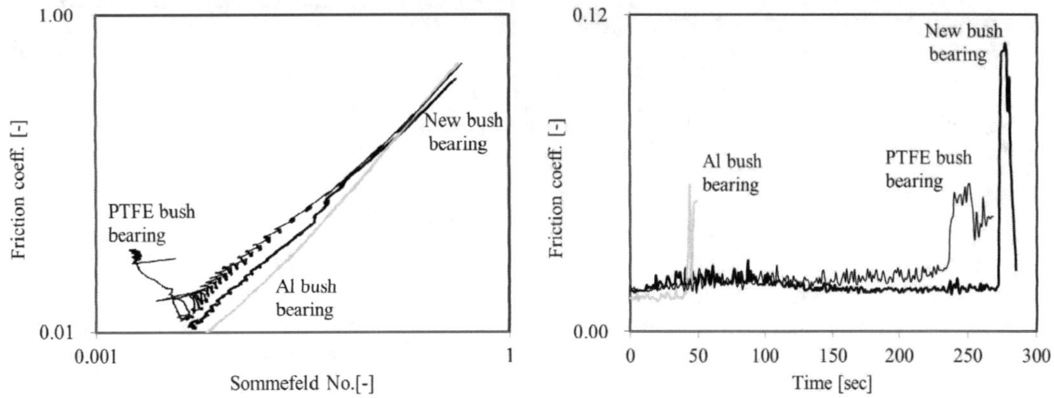

(a) Stribeck curve (step load test) (b) Oil block test

Figure 4: Results of bearing tests

The Sommerfeld No. in Fig. 4 (a) is a non-dimensional bearing load. That is defined as follows:

$$\text{Sommerfeld No.} = \frac{2\mu NRL}{F}\left(\frac{R}{c}\right)^2 \qquad (1)$$

Where, μ is oil viscostiy, N is the revolutions of a journal per second., R is a bearing radius, L is a bearing length, F is a bearing load, and c is a raidal clearance.

3. ADVANTAGES AND TECHNOLOGIES

The main advantages of the new HSS scroll compressor are introduced in this section. The two scoll compressors illustrated in Fig. 1 are compared to each other.

3.1 Reduced journal bearing loads

As mentioned in a chapter 1, the two sources of journal bearing loads are almost identical in the two scroll compressors. One is a gas force which is the driving force of an O/S. The other is a centrifugal force by O/S orbit motion. However, main and sub journal bearing loads by the two forces are clearly different in the two scroll compressors. The simple theories of statics are used to evaluate the journal bearing loads. In this analysis, we assume that all resultant forces act on the mid-point of bearing length, and a F/S wrap completely supports the O/S centrifugal force equally over wrap height.

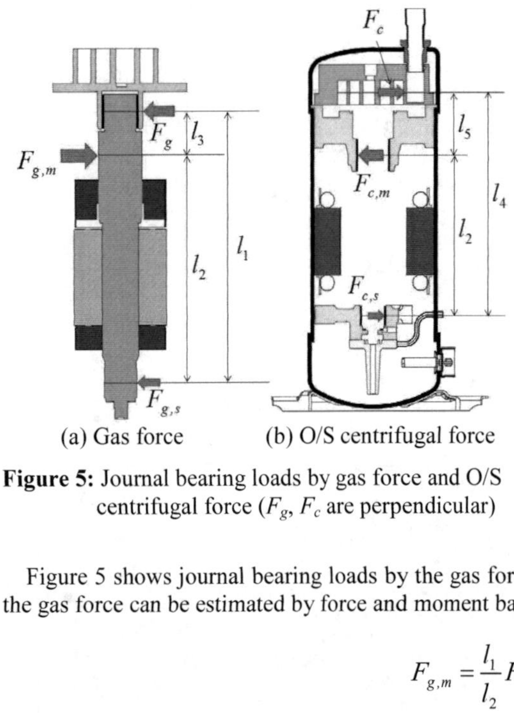

Figure 5: Journal bearing loads by gas force and O/S centrifugal force (F_g, F_c are perpendicular)

Table 2: Journal bearing loads by gas force

Item	Load (normalized by O/S bearing load)	
	Conventional	New
O/S (Gas force)	1.00	
Main	1.18	1.02 (86%)
Sub	0.18	0.02 (11%)

Table 3: Journal bearing loads by O/S centrifugal force

Item	Load (normalized by O/S centrifugal force)	
	Conventional	New
O/S centrifugal force	1.00	
Main	1.39	1.23 (88%)
Sub	0.39	0.23 (59%)

Figure 5 shows journal bearing loads by the gas force and the O/S centrifugal force. Journal bearing loads due to the gas force can be estimated by force and moment balances of a shaft as follows:

$$F_{g,m} = \frac{l_1}{l_2}F_g,\ F_{g,s} = F_{g,m} - F_g \qquad (2)$$

Where, $F_{g,m}$ is a main journal bearing load and $F_{g,s}$ is a sub journal bearing load by the gas force. The definition of each length which represents distance between the mid points of bearing length is illustrated in Fig. 5 (a). The good designed balancing system for a compressor also should be met the force and moment balances of the stationary part of a compressor. Therefore, journal bearing loads by the O/S centrifugal force can be expressed as follows:

$$F_{c,m} = \frac{l_4}{l_2}F_c,\ F_{c,s} = F_{c,m} - F_c \qquad (3)$$

Where, $F_{c,m}$ is a main journal bearing load and $F_{c,s}$ is a sub journal bearing load by the O/S centrifugal force.

From Eqns. (2) and (3), the length l_3 and l_5 in Fig. 5 have a key role in journal bearing loads due to the gas force and the O/S centrifugal force. The two lengths are reduced by the outer bearing structure in the new scroll compressor. Tables (2) and (3) list calculation results of the journal bearing loads using Eqns. (2) and (3). All journal bearing loads of the new scroll compressor are reduced compare to the conventional scroll compressor. That has very important meaning because the reduced journal bearing loads should be helpful for all characteristics of the new scroll compressor.

3.2 Compact design

In a chapter 3.1, the O/S centrifugal force is reviewed with the force and moment balances of the stationary part without balance weights. Considering the equilibrium of the moving part, the balance weights of the new scroll compressor have a smaller unbalanced mass than the balance weights of the conventional one. There are two main reasons. One is a female-type eccentric journal. It can offset some unbalanced mass of an O/S, while the mass of a male-type eccentric journal is added to the unbalanced mass of the O/S. The other is reduced distance between the mass center of the O/S and that of a main balance weight, which is similar to the principle of reduced journal bearing loads in Eqn. (2). The reduced unbalanced mass of balance weights has a key role in the compactness of a scroll compressor when the outer radius of a balance weight has to be decreased.

Some technologies including the outer bearing structure are adopted to the new scroll compressor to implement its compactness. Those are listed in Table 4 compared to the conventional scroll compressor.

Table 4: Technologies for compact scroll compressor

Part	Item	Specification (%) (Normalized by conventional value)		Technology
		Conventional	New	
O/S	Plate diameter	100%	94%	Hybrid wrap design
F/S	Thrust area	100%	80%	Optimization of back pressure mechanism (Outer bearing structure)
Shaft	Length	100%	92%	Outer bearing structure
Main balance weight	Unbalanced mass	100%	71%	Outer bearing structure
Sub balance weight		100%	63%	Outer bearing structure
Motor	Volume	100%	94%	High power motor design

In Table 4, The outer bearing structure has a key role in the optimization of back pressure mechanism because a back pressure seal size can be reduced. While the back pressure seal size of the conventional scroll compressor is constrained by a M/F pocket size, that of the new scroll compressor is constrained by a main bearing size. The available mininum seal diameter of the new scroll compressor is 89% of that of the conventional one. The M/F pocket is denoted in Fig. 2.

3.3Low noise

The noise level of the new scroll compressor is lower than the conventional scroll compressor in all operating conditions. In particular, the noise level is reduced by 4.0dB(sound power level) at 150Hz heating condition. Although both the reduced journal bearing loads and the compact size of the new scroll compressor have good effects on the noise, we think those are not main effects of the improvement. we compare the noise level of the new scroll compressor with another conventional type scroll compressor. The capacity of that is two-third of the new scroll compressor, and the size of that is equal to the new scroll compressor. The noise level of the new scroll compressor is slightly lower than the another conventional type scroll compressor. That means the new scroll compressor has a certain effect which compensate for the increase of compressor capacity in terms of noise.

The main reason for noise reduction would be lower exciting moments produced by the gas force and the O/S centrifugal force. Figure 6 represents the exciting moments produced by the gas force in the two scroll compressors. The gas force, F_g, acts on the F/S and O/S journal bearing simultaneously, and the main journal bearing should support a load by the gas force, $F_{g,m}$. These two forces cause a huge moment because the F/S and M/F is combined as one body. Furthermore, the moment excites compressors. If a M/F and compressor shell are combined by welding points, the moment would be a reason for fatigue fracture of the welding points. In Fig. 6 (a), l_5 can be considered by a moment arm which is coupled with F_g and $F_{g,m}$. The ratio l'_5 in Fig. 5 (b) to l_5 is 0.62, and $F'_{g,m}$ to $F_{g,m}$ is 0.86.

Consequently, The ratio of the compressor exciting moment due to the gas force, $M_g^{'}$ to M_g, is estimated at 0.57. A compressor exciting moment by the O/S centrifugal force also reduced by the same principle as described earlier. This effect has been a main reason for the development of the new scroll compressor because it is essential to reduce compressor noise in high speed operations.

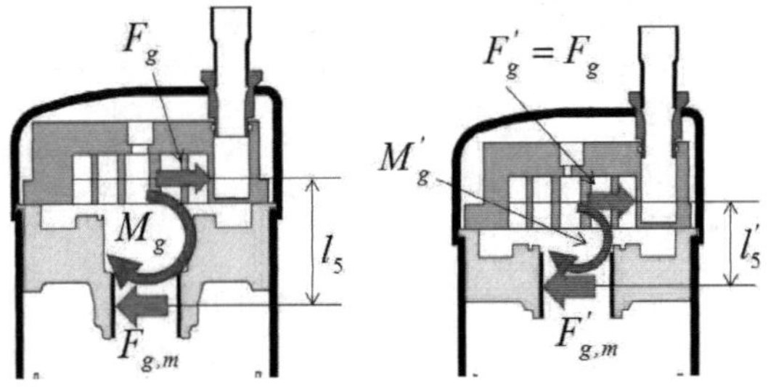

(a) Conventional (b) New

Figure 6: Exciting moment due to O/S centrifugal force

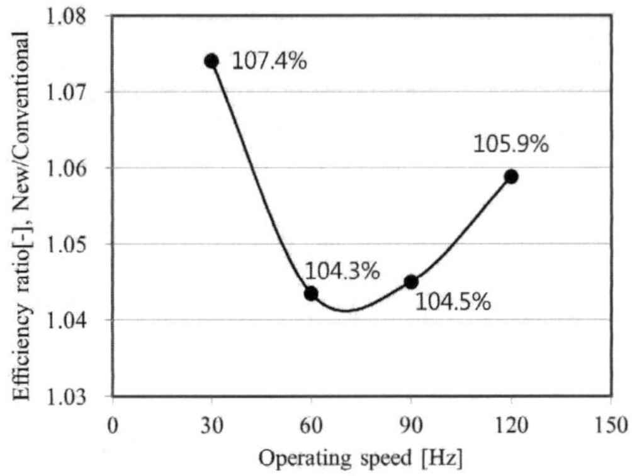

Figure 7: Compressor efficiency at rated condition (Evaporating temp.=7.2°C, Condensing temp.=54.4°C)

3.4 Efficiency improvement

Figure 7 shows the efficiency improvement of the new scroll compressor at a rated condition. The efficiency of the new scroll compressor is improved more than 5% compared to the conventional one. Especially, the efficiency improvement is remarkable at low speed and high speed conditions.

Even if we can list all items of the efficiency improvement, it is very difficult to quantify each item because of interactions between the items. However, some roles of the outer bearing structure for the efficiency improvement can be explained qualitatively. The reduced bearing loads which is explained in a chapter 3.1 clearly helpful in efficiency improvement in all operating conditions, and the outer-type bush bearing has a lower friction coefficient than the traditional PTFE bush bearing. The stirring loss in the M/F pocket which is usually filled with an oil is also eliminated by the outer bearing structure. Finally, the outer bearing structure helps the optimization of a thrust bearing as reducing back pressure seal size.

4. ANALYSIS OF SHAFT DYNAMICS AND BEARINGS

Shaft dynamics and journal bearings are analyzed to confirm the eccentricities and loads of journal bearings. In this analysis, it is assumed that F/S wrap completely support the O/S centrifugal force, and O/S tilting motion is synchronized with a shaft. Therefore, the O/S journal bearing is analyzed with the most favorable condition, while main and sub journal bearings should support all centrifugal forces. Figure 8 shows forces and coordinate systems applied to this analysis. Centrifugal forces by the unbalanced mass of shaft, main balance weight, and sub balanced weight are considered in this analysis. The weight of shaft assembly is also considered to confirm its z direction motion. The gas force which is equally applied to the two compressors in this analysis is represented in Fig. 9. The gas force is calculated by theoretical analysis with the rated operating condition denoted in Fig. 7. The definition of orbiting angle is represented in Fig. 2.

In order to determine the pressure distribution of journal bearings, the Reynolds equation is solved in θ-z coordinate systems:

$$\frac{1}{R^2}\frac{\partial}{\partial\theta}\left(h^3\frac{\partial p}{\partial\theta}\right)+\frac{\partial}{\partial z}\left(h^3\frac{\partial p}{\partial z}\right)=6\mu\omega\frac{\partial h}{\partial\theta}+12\mu\frac{\partial h}{\partial t} \tag{4}$$

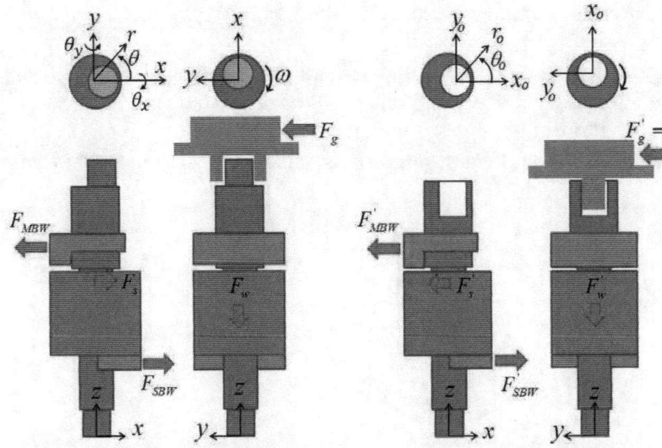

(a) Conventional (b) New

Figure 8: Forces and coordinate systems

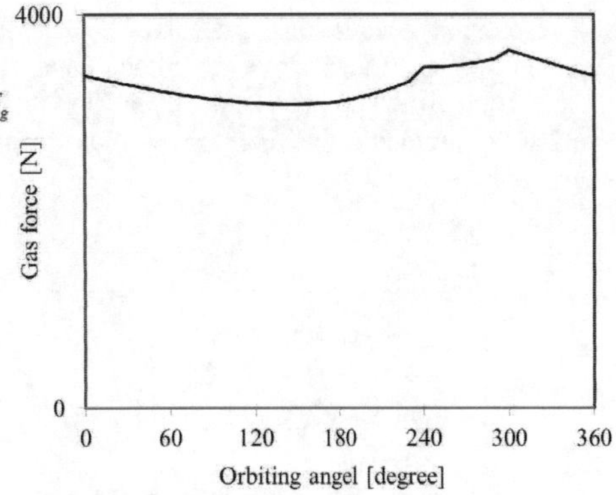

Figure 9: Gas force varations at rated condition

Where, h is clearance, p is pressure, t is time, and ω is the angular velocity of a shaft. The clearacne distribution of main and sub journal bearings can be expressed as follows:

$$h = c - u_x \cos\theta - u_y \sin\theta + z(\alpha \sin\theta - \beta \cos\theta) \tag{5}$$

Where, u_x and u_y denote shaft translation motion in x and y directions , and α and β represent shaft tilting motion in θ_x and θ_y directions respectively. The clearance distribution of an O/S journal bearing is given by:

$$\begin{aligned} \text{Conventional}: h &= c - a_x \cos\theta_o - a_y \sin\theta_o \\ \text{New}: h &= c + a_x \cos\theta_o + a_y \sin\theta_o \end{aligned} \tag{6}$$

A new coordinate system of which origin is the O/S journal center is needed for the O/S journal bearing analysis. The new coordinate system have orbit motion synchronized with a shaft rotation. θ_o in Eqn. (6) denotes the circumferential direction of the new coordinate system. a_x and a_y represent O/S translation motion in x_o and y_o directions in the new coordinate system respectively. The subscript o means the new coordinate system. The Reynolds equation for the O/S journal bearing is not represented because its form is the same as Eqn. (4).

Thrust bearing to support the weight of shaft assembly is also analyzed to consider z direction motion of a shaft. The Reynolds equation and clearance distribution of the thrust bearing can be expressed as follows:

$$\begin{aligned} &\frac{1}{r^2}\frac{\partial}{\partial\theta}\left(h^3\frac{\partial p}{\partial\theta}\right) + \frac{1}{r}\frac{\partial}{\partial r}\left(rh^3\frac{\partial p}{\partial r}\right) = 6\mu\omega\frac{\partial h}{\partial\theta} + 12\mu\frac{\partial h}{\partial t} \\ &h = u_z + r(\alpha \sin\theta - \beta\cos\theta) \end{aligned} \tag{7}$$

Where, u_z is the translation motion of the shaft in z direction.

Dynamic equations for the five-degrees-of-freedom motion of a shaft are given by:

$$\begin{aligned} m\ddot{x} &= \sum F_x = \sum F_{x,action} + \sum F_{x,reaction} & I_{xx}\ddot{\theta}_x &= \sum M_{\theta_x} = \sum M_{\theta_x,action} + \sum M_{\theta_x,reaction} \\ m\ddot{y} &= \sum F_y = \sum F_{y,action} + \sum F_{y,reaction} & I_{yy}\ddot{\theta}_y &= \sum M_{\theta_y} = \sum M_{\theta_y,action} + \sum M_{\theta_y,reaction} \\ m\ddot{z} &= \sum F_z = \sum F_{z,action} + \sum F_{z,reaction} \end{aligned} \tag{8}$$

Where, m is the mass of shaft assembly. I_{xx} and I_{yy} are the moment of inertia of shaft assembly. Subcript *reaction* represents the oil film pressure of bearings.

The Reynolds equations are solved by the finite difference method, and dynamic equations in Eqn. (8) are calculated by the Euler explicit method. Total calculation is over when the relatively error of shaft motion between the previous cycle and present cycle is below 0.5% for all orbiting angles. One orbiting cycle(orbiting angle from 0° to 360°) is divided into 20,000 steps. Oil viscosity is set to 0.005Pa·s, and c/R is fixed as 0.0015 for all journal bearings in this analysis.

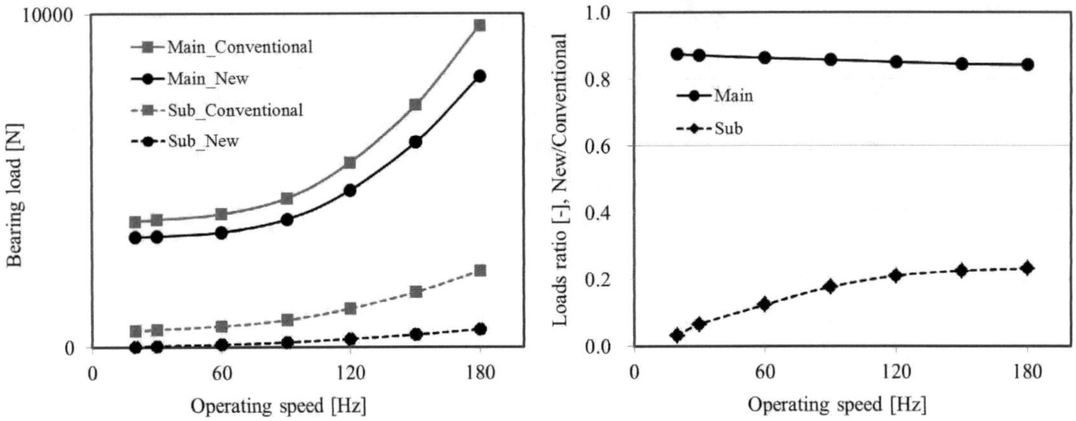

(a) Bearing loads (b) Loads ratio

Figure 10: Comparison journal bearing loads

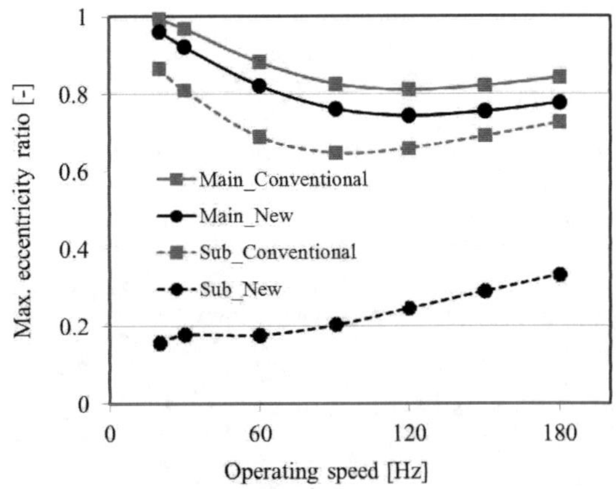

Figure 11: Max. eccentricity ratio

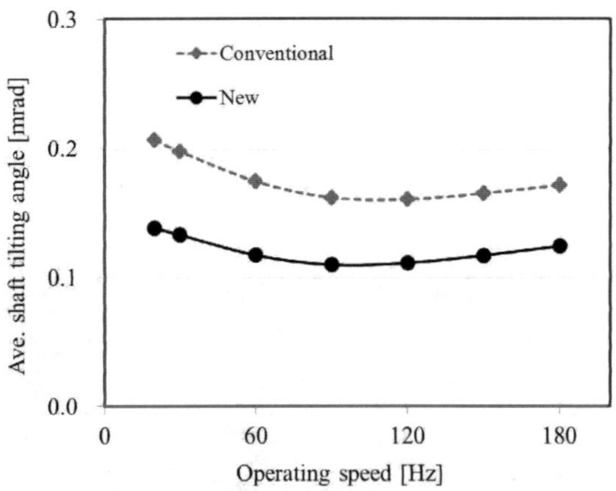

Figure 12: Ave. shaft tilting angle

Analysis results for the bearing loads are denoted in Fig. 10. Both main bearing load and sub bearing load of new socrll compressor are lower than the conventional one especially sub bearing load of the new one is dramatically decreased. In Fig 10 (b), the loads ratio of main bearings is about 0.87 at 20Hz. That is correspond to the result in Table 2. The gas force is dominant in the low speed conditions. The loads ratio of main bearings at 180Hz is 0.84 in Fig 10 (b). That is quite different with the result in Table 3. This difference may be caused by the assumption for a supporting point of the centrifugal force in a chapter 3.1.

Figure 11 shows the maximun eccentricity ratio of journal bearings during one rotation of a shaft. The definition of the maximun eccentricity ratio is given by:

$$\varepsilon_{max} = 1 - (h_{min} / c) \quad (9)$$

The sub journal bearing of the new scroll compressor does not significantly deviate from a concentric point in Fig. 11, and the main journal bearing of the new one also has relatively larger minimum clearance compared to the conventional one. Consequently, shaft tilting angle of the new scroll compressor is quite reduced in comparision with the conventional one as Fig.12. The reduced shaft tilting angle is good effect on the side contacts of journal bearings. The shaft tilting angle is defined as follows:

$$\text{Shaft tilting angle} = \sqrt{\alpha^2 + \beta^2} \quad (10)$$

Although the results of this analysis don't explain all advantages of the outer bearing sturcture which are came up in a chapter 3, the validity of the outer bearing structure for efficiency and wide operating speed is confirmed.

5. CONCLUSION

The new HSS scroll compressor with the outer bearing structure has developed for the VRF system. Compactness, high efficiency, and a low noise level are implemented by various technologies in the new scroll compressor. The outer bearing structure has a key role in the improvements.

1) The size of the new scroll compressor is about 80% of the conventional one. The outer bearing structure reduces shaft length and balance weight size.
2) The efficiency of the new scroll compressor improves more than 5% compared to the conventional one. Reduced journal bearing loads due to the outer bearing structure improve compressor efficiency in all operating conditions.
3) The outer bearing structure decreases the noise level by reducing the compressor exiting moment caused by the O/S centrifugal force and the gas force.
4) The maximum operating speed of the new scroll compressor is expanded by 165Hz. The reduced journal bearing load and low noise level have a key role in the expansion of operating speed.

NOMENCLATURE

Roman

a	Linear displacement of orbiting scroll [m]
c	Radial clearance of journal bearing [m]
F	Force [N]
F'	Force in the scroll compressor [N]
h	Clearance of bearing [m]
L	Bearing length [m]
l	Distance [m]
l'	Distance in new scroll compressor [N]
M	Moment [N m]
M'	Moment in new scroll compressor [N]
m	Mass [kg]
N	Revolution speed [rev s^{-1}]
p	Pressure [Pa]
R	Bearing radius [m]
u	Linear displacement of shaft [m]
I_{xx}	Moment of inertia in θ_x direction [kg m^2]
I_{yy}	Moment of inertia in θ_y direction [kg m^2]

Greek

α	Angular displacement in θ_x direction [rad]
β	Angular displacement in θ_y direction [rad]
ε	Eccentricity ratio of journal bearing [-]
μ	Viscosity [Pa s]
ω	Angular velocity [rad s^{-1}]

Sub script

1	Between sub bearing to O/S bearing
2	Between sub bearing to main bearing
3	Between main bearing to O/S bearing
4	Between sub bearing to F/S wrap
5	Between main bearing to F/S wrap
c	Centrifugal
g	Gas
m	Main bearing
s	Sub bearing

REFERENCES

[1] C. Kim, S. Choi, Y. Cho et al., 2010, Development of High-Side Shell Scroll Compressor with Novel Oil Return Mechanism, *International Compressor Engineering Conference at Purdue*, C-15, No. 1412

[2] B. Yoo, S. Choi, I. Won et al., 2012, Development of New High-Side Shell Scroll Compressor with Real-Time Oil Level Sensor, *International Compressor Engineering Conference at Purdue*, No. 1418

[3] T. Inaba, M. Sugihara, T. Nakamura, T. Kimura, E. Morishita, 1986, A Scroll Compressor with Sealing Means and Low Pressure Side Shell, *International Compressor Engineering Conference at Purdue*, No. 577

[4] M. Yamamora Muramatsu, Y. Kojima, S. Yamamoto, S. Kawahara, N. Ishii, 1990, Compact Type Scroll Compressor for Air Conditioners, *International Compressor Engineering Conference at Purdue*, No. 699

CFD Simulation of An Oil Flooded Scroll Compressor Using VOF Approach

Hui Ding*, Yu Jiang

Simerics Inc.,
1750 112th Ave NE Ste A203,
Bellevue, WA 98004, USA
Phone – 425-502-9978, Fax – 425-502-9978, E mail – hd@simerics.com

* Corresponding Author

ABSTRACT

Liquids have been commonly used in different types of compressors to cool the compressed gas and to seal the leakage gaps in order to increase the efficiency. CFD simulation provides valuable insights to help design engineers to verify, to analyze, and to improve the performance of a compressor. However two phase flow with moving parts and small gaps is a very challenging CFD problem. For compressor simulation, thermal effects and heat transfer are also essential. Therefore simulation of liquid flooded compressors is extremely difficult.

In this paper a full 3D transient CFD model for a generic oil flooded scroll compressor will be described in detail. Volume Of Fluid (VOF) multiphase approach will be used to model gas and liquid phases. Effects of flooded oil will be evaluated by comparing the simulation results for the cases with and without oil. Simulation will also demonstrate that the approaches used in the paper are robust, fast, and user friendly, and can be readily applied to industrial compressor systems.

Keywords: oil flooded, scroll compressor, CFD, two phase flow, VOF.

1. INTRODUCTION

Compressors are designed to compress gases. However, liquids, such as oil and water, are often introduced to compression process for various purposes. For example, oil in compressor not only lubricates the moving components, but also cools the compressed gas, and seals the leakage gaps to help improve the efficiency of the process (Bell, 2011).

In recent years, CFD has been widely used to analyze performance of varies Positive Displacement (PD) compressors. Those models cover different types of PD compressors including rolling piston compressor (Lenz and Cooksey, 1994, Geng *et al.,* 2004, Liang, *et al.,* 2010, Ding and Gao, 2014,), scroll compressor (Feng et al., 2004, Cui, 2006, Gao and Jiang 2014, Gao, *et al.*, 2015), twin screw compressor (Voorde *et al.*, 2005, Kovacevic *et al.,* 2007, Pascu *et al.*, 2012, Kovacevic *et al.,* 2014), and reciprocating compressor (Birari *et al.*, 2006, Pereira *et al.,* 2010, Dhar *et al.*, 2016). Intake and/or discharge valves are also included in some of the models. However almost all of the models are running as a "dry" process, meaning no liquid modeled in the compression chamber. The lack of CFD model of liquid flooded PD compressors is mainly due to the difficulties in CFD multiphase flow simulation.

Even for a relatively simple two phase problem, high density ratio between liquid and gas, sophisticated interaction among the phases, and the interface tracking with complex shape make the flow difficult to solve. It becomes even more difficult to solve a multiphase flow in a PD compressor. In such a case, gas phase has to be treated as compressible, heat transfer is also essential, and interface tracking has to be done in moving, deforming volumes. The major issues users experience with many CFD solvers in multiphase simulations are poor convergence, very long simulation time, and unsatisfactory mass/energy conservation.

Simerics-PD (also known as PumpLinx) has been designed to model PD machines including PD compressors. Its VOF based multiphase model has also been validated with many different industrial applications including, liquid ring vacuum pump (Ding et al., 2015), and oil pump priming (Kucinschi and Shieh, 2016). Recently, heat transfer capability was added to Simerics-PD's VOF multiphase model. Oil flooded PD compressor is one of the targeted application for this new capability.

Scroll compressors are widely used in many industries, such as refrigeration, air-conditioning and automotive. It is believed that scroll compressors have the advantages of high efficiency, lower noise and vibration levels. Scroll compressor can run dry or oil flooded, and it was chosen as the candidate for this study.

This paper presents a full 3D transient CFD model for a generic oil flooded scroll compressor. Due to the time limitation, the focus of this paper will be on the demonstration of the new capability with emphasis on the qualitative trends revealed from simulation results, and the conservation of mass and energy in the results.

2. CFD SOLVER AND GOVERNING EQUATIONS

2.1 Conservation Equations For Gas Liquid Mixture

The CFD package used in this study solves conservation equations of mass, momentum, and energy of a compressible fluid using a finite volume approach. Those conservation laws can be written in integral representation as

$$\frac{\partial}{\partial t}\int_{\Omega(t)} \rho d\Omega + \int_{\sigma} \rho(\boldsymbol{v} - \boldsymbol{v}_{\sigma}) \cdot \boldsymbol{n} d\sigma = 0 \tag{1}$$

$$\frac{\partial}{\partial t}\int_{\Omega(t)} \rho \boldsymbol{v} d\Omega + \int_{\sigma} \rho\big((\boldsymbol{v} - \boldsymbol{v}_{\sigma}) \cdot \boldsymbol{n}\big)\boldsymbol{v} d\sigma = \int_{\sigma} \tilde{\boldsymbol{\tau}} \cdot \boldsymbol{n} d\sigma - \int_{\sigma} p\boldsymbol{n} d\sigma + \int_{\Omega} f d\Omega \tag{2}$$

$$\frac{\partial}{\partial t}\int_{\Omega(t)} \rho E d\Omega + \int_{\sigma} \rho\big((\boldsymbol{v} - \boldsymbol{v}_{\sigma}) \cdot \boldsymbol{n}\big)E d\sigma = \int_{\sigma} k\nabla T \cdot \boldsymbol{n} d\sigma - \int_{\sigma} p\boldsymbol{v} \cdot \boldsymbol{n} d\sigma + \int_{\sigma} (\boldsymbol{v} \cdot \tilde{\boldsymbol{\tau}}) \cdot \boldsymbol{n} d\sigma + \int_{\Omega} \boldsymbol{f} \cdot \boldsymbol{v} d\Omega \tag{3}$$

The standard $k - \varepsilon$ two-equation model (Launder and Spalding, 1974) is used to account for turbulence,

$$\frac{\partial}{\partial t}\int_{\Omega(t)} \rho k d\Omega + \int_{\sigma} \rho\big((\boldsymbol{v} - \boldsymbol{v}_{\sigma}) \cdot \boldsymbol{n}\big)k d\sigma = \int_{\sigma} (\mu + \frac{\mu_t}{\sigma_k})(\nabla k \cdot \boldsymbol{n}) d\sigma + \int_{\Omega} (G_t - \rho\epsilon) d\Omega \tag{4}$$

$$\frac{\partial}{\partial t}\int_{\Omega(t)} \rho \varepsilon d\Omega + \int_{\sigma} \rho\big((\boldsymbol{v} - \boldsymbol{v}_{\sigma}) \cdot \boldsymbol{n}\big)\varepsilon\, d\sigma = \int_{\sigma} (\mu + \frac{\mu_t}{\sigma_\varepsilon})(\nabla\epsilon \cdot \boldsymbol{n}) d\sigma + \int_{\Omega} (c_1 G_t \frac{\varepsilon}{k} - c_2 \rho \frac{\varepsilon^2}{k}) d\Omega \tag{5}$$

Together with equation of state, where properties are functions of temperature and pressure, to form a closed system:

$$\rho = f(p, T) \tag{6}$$

In the solver, each of the fluid properties can be a function of local pressure and temperature, and can be input as a formula or in a table format.

2.2 VOF Model for Multiphase

VOF models are widely used in simulation of two phase flow (Ubbink 1977, Hirt and Nichols, 1981) . VOF solves a set of scalar transport equations representing the fraction of the volume each fluid component occupies in every computational cell. In the region close to a sharp interface, reconstruction will be used to determine the shape of the interface. The transport equation of the volume fraction for each fluid component can be written as:

$$\frac{\partial}{\partial t}\int_{\Omega(t)} \rho_i F_i d\Omega + \int_{\sigma} \rho_i(\boldsymbol{v} - \boldsymbol{v}_{\sigma}) \cdot \boldsymbol{n} F_i d\sigma = 0 \tag{7}$$

Where F_i is the volume fraction of the *ith* fluid component, and ρ_i is the local density of *ith* fluid component. The weighted mixture density of the fluid in equation (1) to (5) are then calculated as:

$$\rho = \sum \rho_i F_i \tag{8}$$

Both implicit and explicit methods are implemented to solve this equation. Close to the sharp phase interface, high resolution scheme was implemented for reconstruction of the interface.

This software package has been validated against many different types of compressors including: centrifugal compressor, lobe compressor, twin screw compressor (Kovacevic1 *et al.*, 2014), scroll compressor (Gao and Jiang, 2014, Gao et al., 2015), rolling piston (Ding and Gao, 2014), and reciprocating compressor (Dhar et al., 2016) for single phase compression of air, refrigerants, and other type of gases. The VOF model has also been validated against many industrial applications (Ding et al., 2015, Kucinschi and Shieh, 2016) for multiphase flow without solving heat transfer directly.

3. SCROLL COMPRESSOR TEST CASE

A generic scroll model was used to demonstrate the functionality and capability of proposed approach. The complete system includes an inlet port, a scroll, and an outlet port. Mesh of the scroll was created using Simerics-PD Scroll Template (Gao and Jiang, 2014). The rest of fluid volumes are meshed using Simerics binary tree unstructured mesher. All the fluid volumes are connected together using Miss Matched Grid Interface (MGI). The total number of cells is around 0.3 million. Figure 1 shows the complete fluid domain. Figure 2 shows the mesh in a cutting plane.

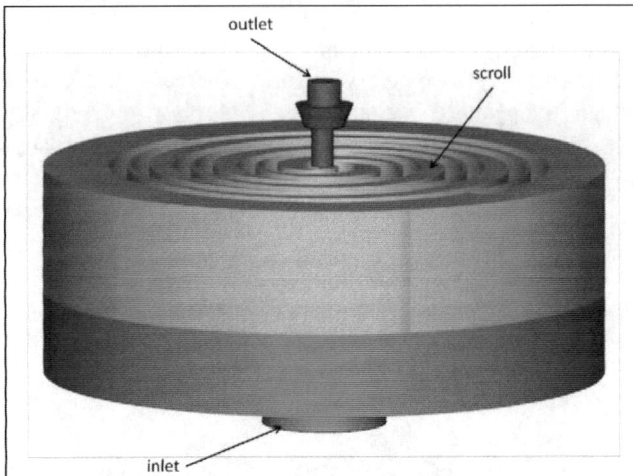

Figure 1: All fluid volumes

Figure 2: Mesh in a cutting plane

The inlet is set to a fixed pressure, fixed temperature boundary condition. The outlet is set to a fixed pressure boundary condition. The refrigerant is R410a, modeled using ideal gas law. The molecular weight of R410a is 72.63 g/mole, and the heat capacity is 1035 J/kgK. The compressor rotation speed is 3500 RPM. The oil was assumed to be incompressible with a density of 800 Kg/m3, and a heat capacity of 1670 J/kgK. In order to demonstrate the effects of the flooded oil, a similar case with the same parameters but without oil was also simulated for comparison.

Start from properly prepared CAD geometry, the meshing and the setup of the simulation take less than half an hour with the help of the Scroll template. Simulation time is about 1.5 hours per revolution for oil flooded simulation and about 20 minutes per revolution without oil on a PC with quad-core Intel Xeon CPU at 2.67GHz.

4. RESULTS AND DISCUSSION

In the simulation, the inlet and the outlet pressure are set to 1MPa and 3.4MPa respectively. The inlet temperature is set to 300K. The oil is assumed uniformly mixed with refrigerant at the entrance. Two oil concentrations were simulated. The oil mass fraction were set to 2.4% and 18.1% respectively. The corresponding inlet oil volume fraction were about 0.086% and 0.65%. Simulation results start to stabilize after around 6 revolutions. Table 1 shows the mass and energy imbalance for the three simulations.

Table 1 Conservation of simulation results

	Oil flooded (2.4%)	Oil flooded (18.1%)	Dry
Gas mass imbalance	1.2%	2.0%	0.2%
Oil mass imbalance	0.7%	1.2%	N/A
Energy imbalance	1.3%	1.8%	0.5%

Figure 3 shows typical pressure contour at 4 different crankshaft angles. The pressure in each isolated fluid "pocket" keeps increasing with crankshaft angle due to the continuous volume reduction of the pocket till it reaches outlet.

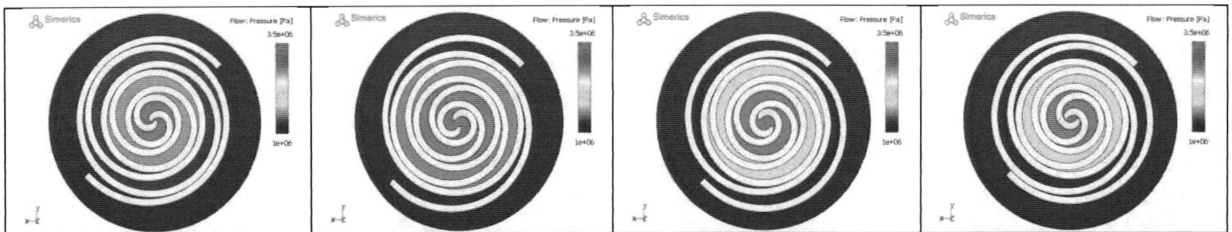

90 degree	180 degree	270 degree	360 degree

Figure 3: Pressure contour at different crankshaft angles

Figure 4 shows typical temperature contour at 4 crankshaft angles. Temperature in the pocket follows a similar trend. However unlike the pressure, the temperature inside each pocket is not very uniform. The non-uniformity is caused by the leakage flow from the high pressure/high temperature region towards the low pressure/low temperature region. Pressure propagates with pressure wave in the speed of sound, while temperature propagates with much slower convection and diffusion process. It is expected that the pressure in each chamber equalizes much fast than the temperature.

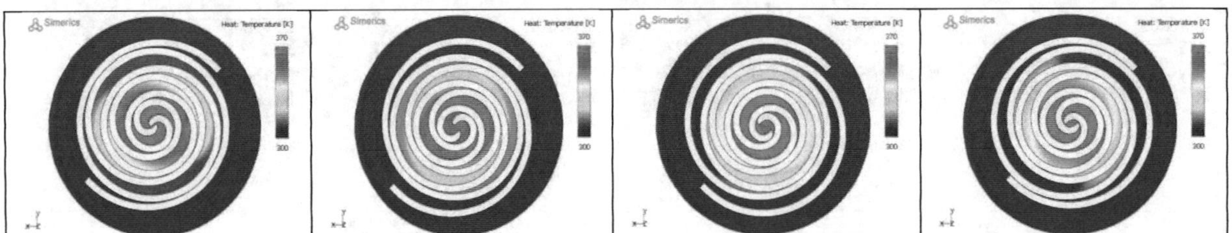

90 degree	180 degree	270 degree	360 degree

Figure 4: Temperature contour at different crankshaft angles

Figure 5 shows oil volume fraction contour for 2.4% oil case at 4 crankshaft angles. When pocket pressure increases, volume of the gas phase will decrease, and the volume fraction of oil will increase. Therefore oil volume fraction also increase when pocket moves towards the center.

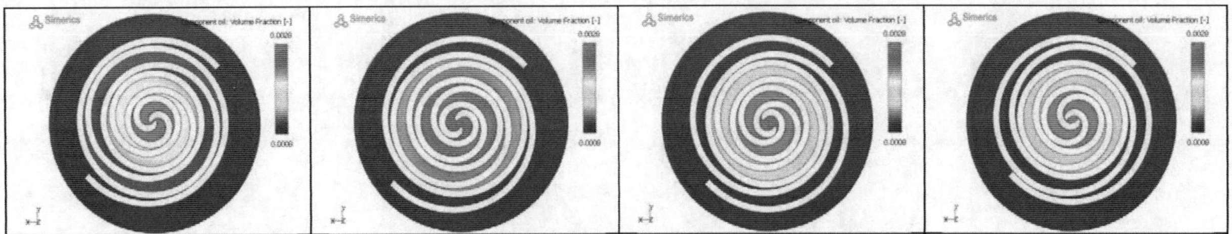

90 degree	180 degree	270 degree	360 degree

Figure 5: Oil volume fraction at different crankshaft angles

Figure 6 and Figure 7 compare the pressure and the temperature for the cases with and without oil at the same crankshaft angle. Although the pressure distribution looks very similar, the temperature for the dry compression is apparently higher.

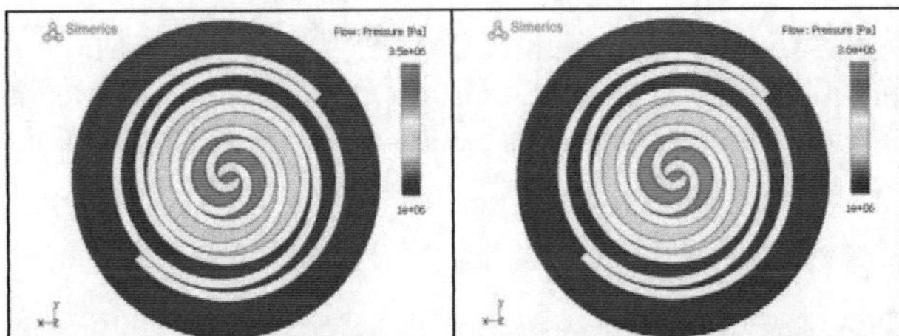

Oil flooded	Dry

Figure 6: Pressure with/without oil

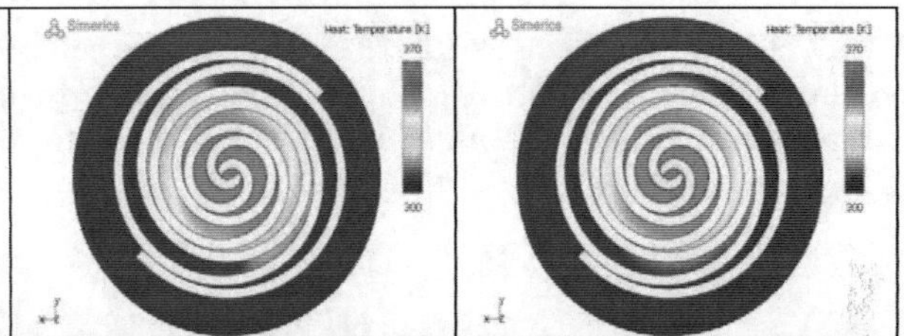

Oil flooded	Dry

Figure 7: Temperature with/without oil

Figure 8 and Figure 9 plot the pressure and the temperature history of a single fluid pocket for three simulated cases. Those data are read from monitor points moving together with the pocket. From the plots, the pressure history are similar for all the cases, and the 2.4% oil flooded case is almost identical to the dry case. But there are much more differences in temperature history. The temperature for the case with 18.1% oil dropped about 20 degree at the outlet.

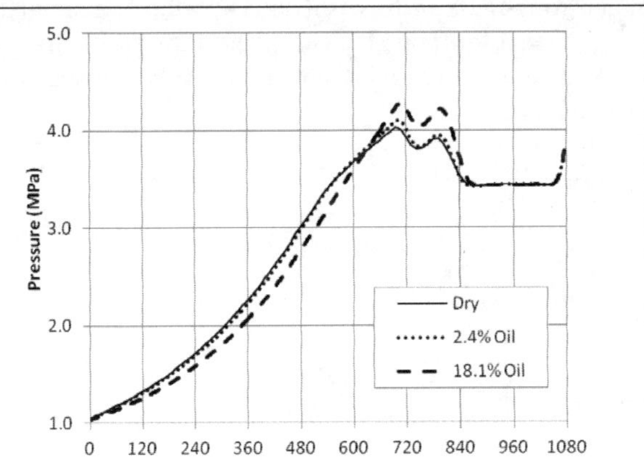

Figure 8: Pressure history of a pocket

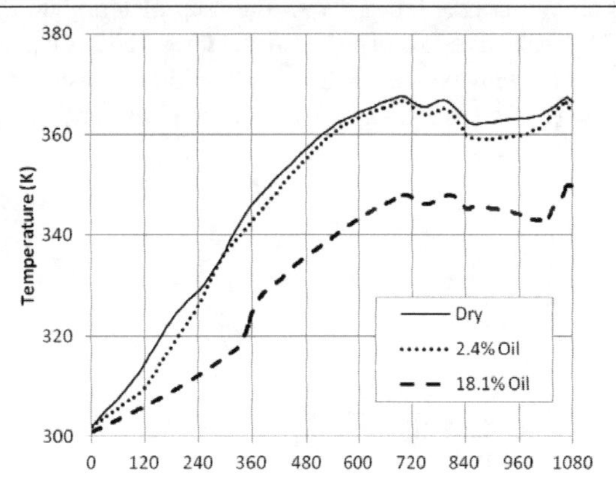

Figure 9: Temperature history of a pocket

Figure 10 compares the leakage velocity for the cases with 2.4% oil and without oil at the same crankshaft angle. The leakage velocity for the dry compression is higher than the oil flooded one.

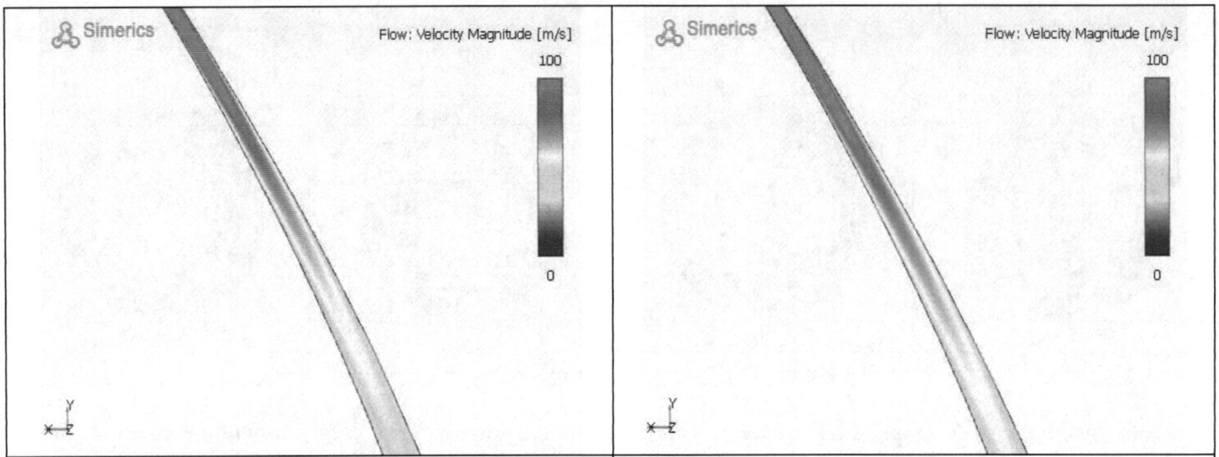

Oil flooded	Dry

Figure 10. Leakage flow velocity with/without oil

Table 2 shows the major performance difference when compressor run dry or with oil flooded. The results show a gradual increase in refrigerant mass flow rate, a gradual decrease in power consumption, and a significant temperature reduction when the oil contents in compressor increases. Those trends match the expected oil cooling and sealing effects.

Table 2 Differences in compressor performance

Oil flooded/Dry	2.4% Oil	18.1% Oil
Gas mass flow rate	101.3%	107.1%
Outlet temperature rise	90.4%	63.9%
Rotor power	99.5%	97.6%

5. CONCLUSIONS AND FUTURE WORK

A newly improved VOF multiphase model has been successfully applied to an oil flooded scroll compressor. Simulation results predict correctly the oil cooling and sealing effects on compression process, and show good mass and energy conservation. With the help of template design and robustness of flow solver, the setup and simulation are also easy and fast. The new model has demonstrated great potential for modeling two phase flow problems in PD compressors. Next step will be validation against available test data. Further improvement of simulation speed, and inclusion of phase change model are in the future development plan.

NOMENCLATURE

C_1	Turbulence model constant		T	Temperature	K
C_2	Turbulence model constant		t	Time	s
C_μ	Turbulence model constant		$\boldsymbol{v}$	Velocity vector	
$\boldsymbol{E}$	Total energy	J	VOF	Volume of fluid	
$\boldsymbol{F_i}$	Volume fraction of ith component		ε	Turbulence dissipation	m^2/s^3
$\boldsymbol{f}$	Body force	N	μ	Fluid viscosity	Pa-s
G_t	Turbulent generation term		μ_t	Turbulent viscosity	Pa-s
k	Heat conductivity	W/m-K	ρ	Fluid density	kg/m^3
k	Turbulence kinetic energy	m^2/s^2	σ	Surface of control volume	
$\boldsymbol{MGI}$	Mis-matched grid interface		σ_k	Turbulence model constant	
$\boldsymbol{n}$	Surface normal		σ_ε	Turbulence model constant	
PD	Positive displacement		$\boldsymbol{\tau}$	Stress tensor	

p	Pressure	Pa	Ω	Control volume
RPM	Revolution per minute			

REFERENCES

Bell, I., 2011. Theoretical and Experimental Analysis of Liquid Flooded Compression in Scroll Compressors. Ph.D. thesis, Purdue University. Full-text: http://docs.lib.purdue.edu/herrick/2/

Birari, Y. V., Gosavi, S. S., & Jorwekar, P. P. (2006). Use of CFD in design and development of R404a reciprocating compressor. *Proceedings of the International Compressor Engineering Conference at Purdue.* West Lafayette, IN, USA.

Cui, M.M., 2006, "Numerical Study of Unsteady Flows in a Scroll Compressor", 2006, *J. Fluids Eng.* Vol.128 no.5, p. 947-955.

Dhar, S., Ding, H., and Lacerda, J., 2016, "A 3-D Transient CFD Model of a Piston Reciprocating Compressor with Dynamic Port Flip Valves," *23rd Int. Compressor Engrg. Conf. at Purdue*. Paper 1392.

Ding, H. and Gao, H., 2014,"3-D Transient CFD Model For A Rolling Piston Compressor With A Dynamic Reed Valve", 2014, *22nd International Compressor Engineering Conference*, Paper 1548.

Ding, H., Jiang, Y., Wu, H., and Wang, J., 2015, "Two Phase Flow Simulation of Water Ring Vacuum Pump Using VOF Model," ASME/JSME/KSME 2015 Joint Fluids Engineering Conference, Seoul, South Korea, July 26–31, 2015

Feng, J., Qu, Z., and Lin, X., "Three-Dimension Numerical Simulation of Discharge Flow in a Scroll Air Compressor", 2004. *International Compressor Engineering Conference*. Paper 1647.

Gao, H., and Jiang, Y., 2014, Numerical Simulation of Unsteady Flow in a Scroll Compressor, *Int. Compressor Engrg. Conf. at Purdue*.

Gao H., Ding, H., and Jiang, Y., 2015 "3D Transient CFD Simulation of Scroll Compressors with the Tip Seal,"IOP Conference Series: Materials Science and Engineering. Vol. 90. No. 1. IOP Publishing, 2015.

Geng, W., Liu, C., and Wang, Y., 2004, The Performance Optimization of Rolling Piston Compressors Based on CFD Simulation, *Int. Compressor Engrg. Conf. at Purdue*, pp. 1488-1492.

Hirt, C.W. and Nichols, B.D., 1981, Volume of fluid (VOF) method for the dynamics of free boundaries. J. Comput. Phys., Vol. 39, p. 201-225.

Kinjo, K., Nakano, A., Hikichi, T., and Morinishi, K., 2010, Study of CFD Considering Valve Behavior in Reciprocating Compressor, *Int. Compressor Engrg. Conf. at Purdue*. Paper 1975

Kovacevic A., Stosic N. and Smith I. K., 2007, Screw compressors - Three dimensional computational fluid dynamics and solid fluid interaction, ISBN 3-540-36302-5, Springer-Verlag Berlin Heidelberg New York.

Kovacevic, A., Rane, S., Stosic, N., Jiang, Y., and Lowry, S., 2014, Influence of approaches in CFD Solvers on Performance Prediction in Screw Compressors, *Int. Compressor Engrg. Conf. at Purdue*.

Kucinschi, B. and Shieh, T., "Estimation of Oil Supply Time during Engine Start-Up at Very Low Temperatures," SAE Int. J. Fuels Lubr. 9(2):2016, doi:10.4271/2016-01-0893.

Launder, B.E., and Spalding, D.B., 1974, The numerical computation of turbulent flows, Comput. Methods Appl. Mech. Eng., 3, pp. 269-289.

Lenz, J. R., Cooksey, E. A., 1994, Application of Computational Fluid Dynamics to Compressor Efficiency Improvement, *Int. Compressor Engrg. Conf. at Purdue*, pp. 441-446.

Liang, S., Xia, S., Kang, X., Zhou, P., Liu, Q., and Hu, Y., 2010, Investigation of Refrigerant Flow Simulation and Experiment of Rolling Piston, *Int. Compressor Engrg. Conf. at Purdue*, pp.

Machu, G.,; Albrecht, M., Bielmeier, O., Daxner, T., and Steinruck, P., 2004, A Universal Simulation Tool for Reed Valve Dynamics, *Int. Compressor Engrg. Conf. at Purdue*. Paper 1716

Pascu M., Kovacevic A., and Udo N., 2012, Performance optimization of Screw Compressors based on numerical investigation of the flow behaviour in the discharge chamber. Int Compressor Conf at Purdue, Purdue, p. 1145.

Pereira, E. L., Deschamps, C. J., & Ribas, F. A. (2010). Numerical analysis of heat transfer inside the cylinder of reciprocating compressors in the presence of suction and discharge processes. *Proceedings of the International Compressor Engineering Conference at Purdue.* West Lafayette, IN, USA.

Prasad, B. G. S., 2004, CFD for Positive Displacement Compressors, *Int. Compressor Engrg. Conf. at Purdue*. Paper 1689

Silva, E., Deschamps, C. J., and Fancello , E. A., 2012, A Procedure to Optimize Reed T ype V alves Considering Efficiency and Bending Fatigue, *Int. Compressor Engrg. Conf. at Purdue*. Paper 2106.

Ubbink, O., 1977. "Numerical prediction of two fluid systems with sharp interfaces". Ph.D. Thesis, Department of Mechanical Engineering, Imperial College, University of London.

Voorde John Vande and Vierendeels Jan, 2005, A grid manipulation algorithm for ALE calculations in screw compressors. 17th AIAA Computational Fluid Dynamics Conference, Canada, AIAA 2005-4701.

A Numerical Investigation of the Oil Pump Suction Behaviour in a Hermetic Reciprocating Compressor

Stefan POSCH[1]*, Johann HOPFGARTNER[1], Martin HEIMEL[1], Erwin BERGER[1], Raimund ALMBAUER[1], Peter SCHÖLLAUF[2]

[1] Institute of Internal Combustion Engines and Thermodynamics, Graz University of Technology, Inffeldgasse 19, 8010 Graz, Austria

posch@ivt.tugraz.at, +43 316 873 30233
hopfgartner@ivt.tugraz.at, +43 316 873 30240
heimel@ivt.tugraz.at, +43 316 873 30235
berger@ivt.tugraz.at, +43 316 873 30234
almbauer@ivt.tugraz.at, +43 316 873 30230

[2] Secop Austria GmbH, Jahnstraße 30, 8280 Fürstenfeld, Austria
p.schoellauf@secop.com, +43 3382 5010 984

* Corresponding Author

ABSTRACT

In addition to the adequate lubrication of the moving parts, the oil flow in a hermetic reciprocating compressor has a significant influence on the thermal characteristics of a compressor. The present work is concerned with the investigation of the oil pump system of a reciprocating hermetic compressor used in household refrigeration appliances. The considered oil pump system consists of a centrifugal pump immersed in the oil sump of the hermetic compressor and a helical groove machined on the crankshaft. The focus of this work lies on the immersed part of the centrifugal oil pump and its interaction with the oil in the oil sump. To analyse the flow in the immersed area of the oil pump, the commercial computational fluid dynamics (CFD) software ANSYS Fluent is used. The free surface of the oil flow is modelled with the Volume of Fluid (VOF) method. A numerical investigation is used to study the influence of the immersion depth and the oil pump design on oil mass flow rate and flow field at the oil pump intake. To evaluate the oil pump regarding the applicability in variable speed compressors, the influence of the rotational speed on the oil mass flow is also explored.

1. INTRODUCTION

The most common compressor type used in domestic refrigeration appliance is the hermetic reciprocating compressor. This kind of compressor uses a reciprocating piston driven by a slider-crank mechanism for suction, compression and discharge of the refrigerant. Small-scale hermetic reciprocating compressors have a crankshaft which is directly driven by the electric motor. In addition to the energy transfer from the motor to the piston, the crankshaft acts also as oil pump device. Depending on the compressor type and the required amount of oil, different oil pump designs are used which differ in the arrangement of centrifugal and helical pumps. Several studies are available which deal with the investigation of single pumping parts or the entire oil pump, respectively. A numerical study of a compressor oil pump using commercial CFD software can be found in Lückmann *et al.* (2009) and Kerpicci *et al.* (2013). The authors used the Volume of Fluid (VOF) technique to model the two-phase flow in three dimensions. Besides the calculation of the oil mass flow rate, specific focus was laid on the determination of the oil climbing time. To keep the number of cells and therefore the computational time under a tolerable limit, several simplifications like the neglect of the gap between shaft and crankcase in the helical grove have to be made. Kim *et al.* (2002) used the analogy between the oil supply system of a reciprocating compressor and an electric circuit. The authors described the pumping parts by equivalent electric elements to get a mathematical model of the system. A validation of the simulation results by experimental data showed good agreement in terms of the oil flow rate. The fluid flow in helical pumps using analytical models is analysed in the works of Alves *et al.* (2009 and 2010). Thereby, the Navier-Stokes equations were adapted to a finite channel with specific boundary conditions. The gap between shaft and crankcase was neglected in these studies. Different techniques like generalized integral transform technique have been presented to solve the resulting equation. A more detailed investigation of helical groove oil pumps was

presented by Posch *et al.* (2015). In this study, the oil flow in channel direction is assumed to be fully developed and the 2d flow in the channel cross section has been calculated considering the gap between shaft and crankcase. The pressure gradient in channel direction can be considered to get the ability to couple the helical groove with other pumping parts. In Tada *et al.* (2014) an uncoupled simulation of the helical pump of a reciprocating compressor using commercial CFD software is shown. The method has been utilized to analyse the applicability of an oil pump for the use in variable speed compressors. The authors calculated the volumetric oil flow depending on the rotational speed and immersion depth of the shaft in the compressor oil sump. Another example of the use of commercial CFD software to analyse the performance of a compressor lubrication system can be found in Ozsipah *et al.* (2014). In addition to the investigation of the influence of geometric parameters on the pumped oil mass flow, the suction behaviour of the oil pump was analysed. Therefore, the oil sump of the compressor was also modelled, assuming that the oil expelled from the oil pump does not flow back into the oil sump. A calculation of the oil mass flow rate, transient oil distribution and pressure fluctuations at the pump inlet, depending on geometrical parameters, was carried out.
The present paper focuses on the investigation of the suction behaviour of a hermetic reciprocating compressor oil pump. Therefore, an uncoupled simulation of the immersed centrifugal pump interacting with the oil sump using commercial CFD software ANSYS Fluent is carried out. To increase the accuracy of the simulation, the expelled oil is assumed to be transported back into the oil sump. The approach is utilized to get an insight in the suction behaviour of the oil pump in dependence on the immersion depth, pump design and rotational speed. Additionally, an assessment according to the oil mass flow rate of the different cases is presented.

2. MODELLING

2.1 Theoretical background

Computational fluid dynamics (CFD) simulation is based on the solution of the flow governing equations using numerical methods. The governing equations for mass and momentum are as follows (Moukalled *et al.*, 2015):

$$\frac{\partial \rho}{\partial t} + \nabla \cdot [\rho \boldsymbol{v}] = 0 \tag{1}$$

$$\frac{\partial}{\partial t}[\rho \boldsymbol{v}] + \nabla \cdot \{\rho \boldsymbol{v} \boldsymbol{v}\} = -\nabla p + \mu \nabla^2 \boldsymbol{v} + \boldsymbol{f}_b \tag{2}$$

The simulation domain is divided into a finite number of cells, the so called control volumes. The governing equations are spatial discretized on the control volumes using the finite volume method to get a linear system of equations which is solved by mathematical methods. Commercial CFD software like ANSYS Fluent, which is used in the present study, executes these steps automatically.
Two phase flows with sharp interface are calculated with the volume of fluid approach (VOF) by Scardovelli and Zaleski (1999) which is implemented in ANSYS Fluent. The VOF approach models two or more immiscible fluids by tracking the volume fraction of each fluid in the domain, solving the volume fraction equation (3). Fluid properties like density or viscosity can be determined by summing up the volume fraction weighted values (4 and 5).

$$\frac{\partial \chi}{\partial t} + \frac{\partial (u_i \chi)}{\partial x_i} = 0 \tag{3}$$

$$\rho = \rho_g \chi + \rho_l (1 - \chi) \tag{4}$$

$$\mu = \mu_g \chi + \mu_l (1 - \chi) \tag{5}$$

Further information about the VOF method can be found in the work of Scardovelli and Zaleski (1999) and the Fluent User Guide (2011).

2.2 Simulation domain

The investigated compressor oil pump consists of three single pump parts: a helical pump and two centrifugal pumps at the inlet and outlet of the crankshaft, respectively. To increase the oil mass flow rate, the considered oil pump is designed with a spiral sheet (catcher) at the pump inlet. The aim of the present study is to investigate the suction behaviour of the oil pump and the flow conditions at the pump inlet. Therefore, an uncoupled simulation of the centrifugal pump at the crankshaft inlet in combination with a simplified oil sump is carried out. The computational mesh is created with about 250.000 cells for the pump and about 400.000 cells for the sump, respectively. The mesh is refined at interfaces, inlet and outlet to increase simulation accuracy. Figure 1 shows the position of the centrifugal pump in the crankshaft (a) and the computational mesh (b).

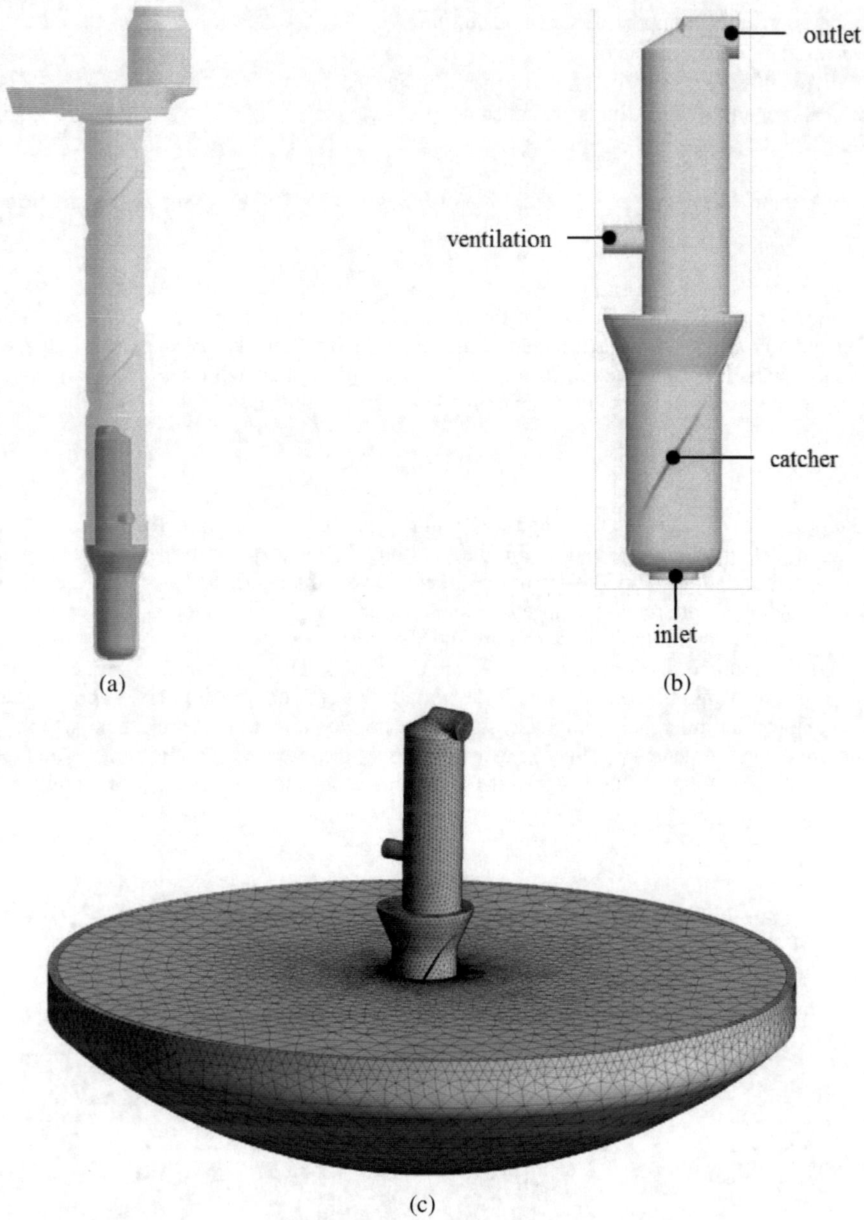

Figure 1: Position of the centrifugal pump (a), geometrical domain (b) and computational mesh (c)

To fulfil the overall mass balance, the oil at the outlet of the pump is filled back into the oil sump at the outer area of the oil sump surface to reproduce the wall oil flow. A mass flow boundary condition which is linked to the pump outlet via user defined function (UDF) is applied on the outer area of the oil sump surface. On the rest of the oil sump surface, a pressure inlet boundary condition with zero relative pressure is applied. At pump outlet and ventilation hole, a pressure outlet boundary condition is used with also zero relative pressure. To decrease computational time, the pump is filled with oil according to the current immersion depth at the initialization process. The simulation is carried out until pressure at the pump/sump interface and oil mass flow at the pump outlet show a steady-state behaviour.

2.3 Assumptions and Settings

The following assumptions are made for the simulation of the centrifugal pump and the compressor oil sump:

- Isothermal flow.
- Acceleration at the compressor start-up is infinite.
- Air is used for the gaseous phase.
- Dissolving of gaseous phase in liquid phase is neglected.
- Physical properties of the fluids are constant.
- Laminar flow.

The oil density is set to 832 kg m^{-3} and the oil viscosity is set to 6.7 x 10^{-3} Pa s. Surface tension is considered and is set to 0.028 N m^{-1}.

The simulation is carried out using implicit VOF method in combination with bounded second order implicit time formulation. The spatial discretization of the momentum equation is utilized with a second order upwind scheme (Moukalled *et al.*, 2015) and, for the volume fraction equation, the compressive scheme is used. The pressure-velocity coupling is executed with the PISO (pressure-implicit with splitting of operators) algorithm (Isaa, 1986). Time step is set to 10^{-4} s and double precision mode is utilized.

3. RESULTS

The important properties to value the functionality of oil pumps for the usage in hermetic compressors are the oil mass flow rate, the oil climbing time and the oil distribution. To quantify the influence of the immersion depth, the rotational speed and the pump design on the mentioned parameters, a basic configuration with 7 mm immersion depth and 3000 rpm rotational speed in combination with the original geometry is used.
Figure 2 shows the oil volume fraction depending on the immersion depth at different time steps during the compressor start-up until and after reaching stead-state conditions. The figures show the development of the oil parabola due to the centrifugal forces. Although the variant with the highest immersion depth gives the highest static pressure at the pump inlet due to the hydrostatic pressure, the oil parabola sinks nearly to the pump inlet at the beginning of compressor start-up. This can be explained by the increasing friction forces on the outer surface which induce centrifugal forces. The centrifugal forces reduce the static pressure at the pump inlet. If the oil parabola would reach the pump inlet, the oil flow would break down immediately.

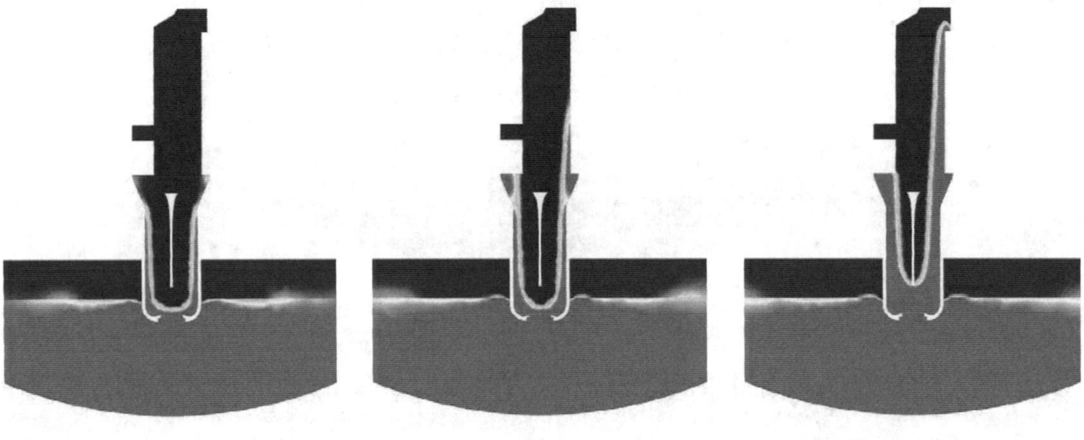

0.12s (a) | 0.24s (b) | 1s (c)

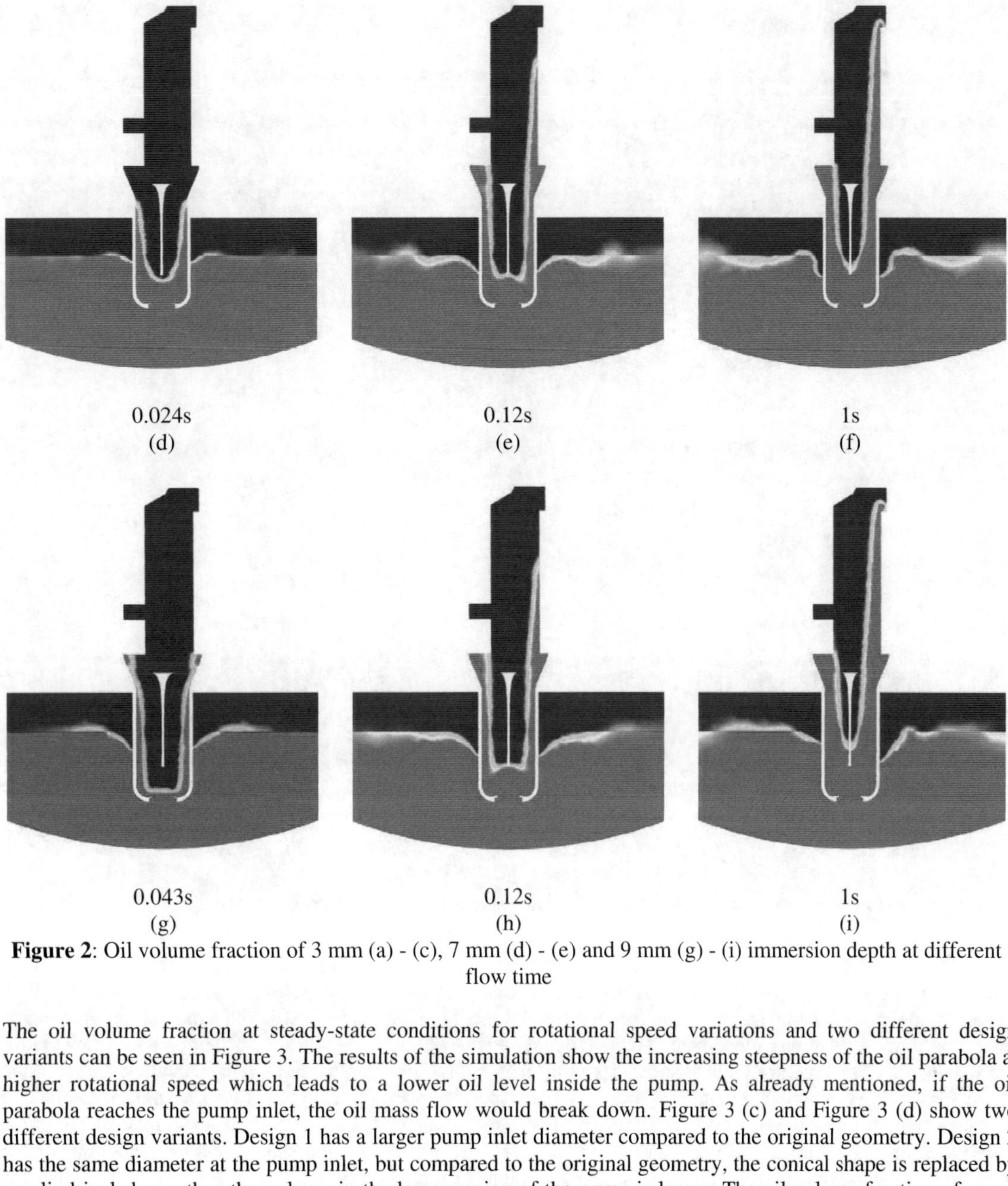

Figure 2: Oil volume fraction of 3 mm (a) - (c), 7 mm (d) - (e) and 9 mm (g) - (i) immersion depth at different flow time

The oil volume fraction at steady-state conditions for rotational speed variations and two different design variants can be seen in Figure 3. The results of the simulation show the increasing steepness of the oil parabola at higher rotational speed which leads to a lower oil level inside the pump. As already mentioned, if the oil parabola reaches the pump inlet, the oil mass flow would break down. Figure 3 (c) and Figure 3 (d) show two different design variants. Design 1 has a larger pump inlet diameter compared to the original geometry. Design 2 has the same diameter at the pump inlet, but compared to the original geometry, the conical shape is replaced by a cylindrical shape, thus the volume in the lower region of the pump is larger. The oil volume fraction of pump design 1 shows an uneven shape of the oil parabola due to the influence of the spiral sheet which interacts with the oil in the sump. A comparison between design 2 and the original geometry shows similar shape of the oil parabola and an increased amount of oil in the catcher due to the higher volume of design 2.

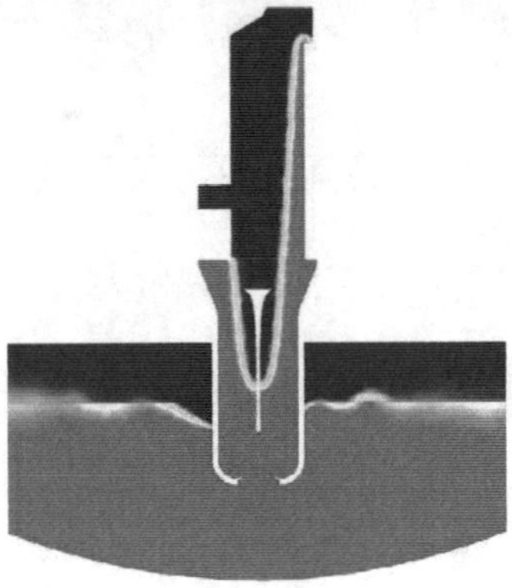

2500 rpm
(a)

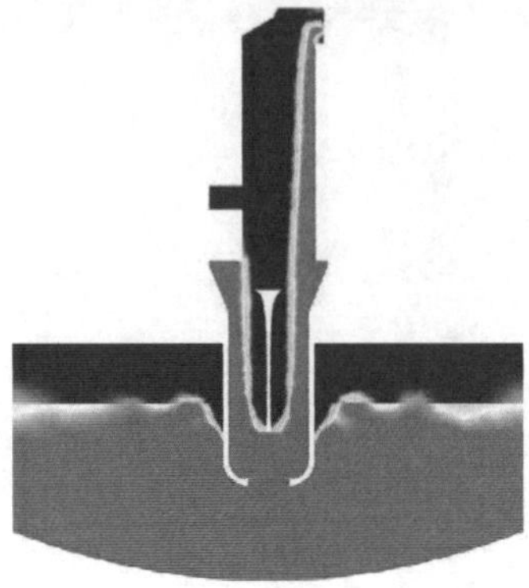

4000 rpm
(b)

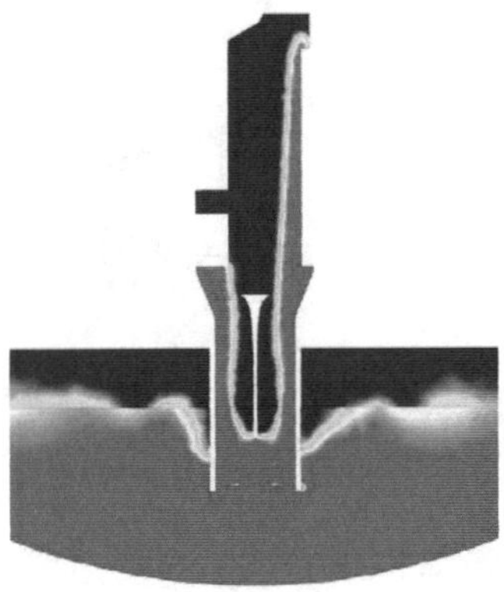

design 1
(c)

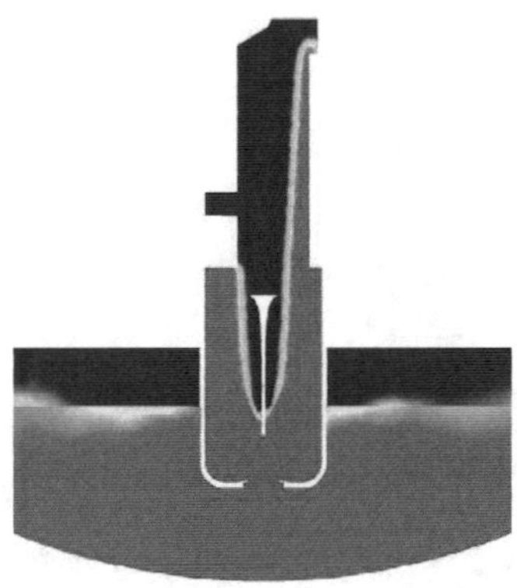

design 2
(d)

Figure 3: Steady-state oil volume fraction

Figure 4 shows the pressure distribution and the flow field of the basic configuration in the vicinity of the pump inlet. It can be seen that the main flow of the oil is situated around the centreline of the pump until the flow develops the oil parabola. Apart from the centreline, hardly oil is transported, thus the oil in that region undergoes only a rotational movement. The tangential velocity at the outer surface of the pump is a function of the distance from the centreline and increases with growing radius. Due to the higher velocities, the pressure decreases in the vicinity of the surface and oil is transported to low pressure regions causing vortexes. The vortexes spread out in the oil sump and can cause acoustic problems. High velocities at the inlet of the pump result in a decrease of the static pressure. Table 1 gives an overview of the static pressure reduction for the observed variants compared to the analytically determined geodetic pressure. It can be pointed out, that the major part of the static pressure reduction occurs due to the flow in axial direction. The influence of tangential and radial flow on the reduction of the static pressure can be considered as efficiency loss. Especially higher rotational speed induces higher tangential and radial velocities at the pump inlet due to higher frictional forces.

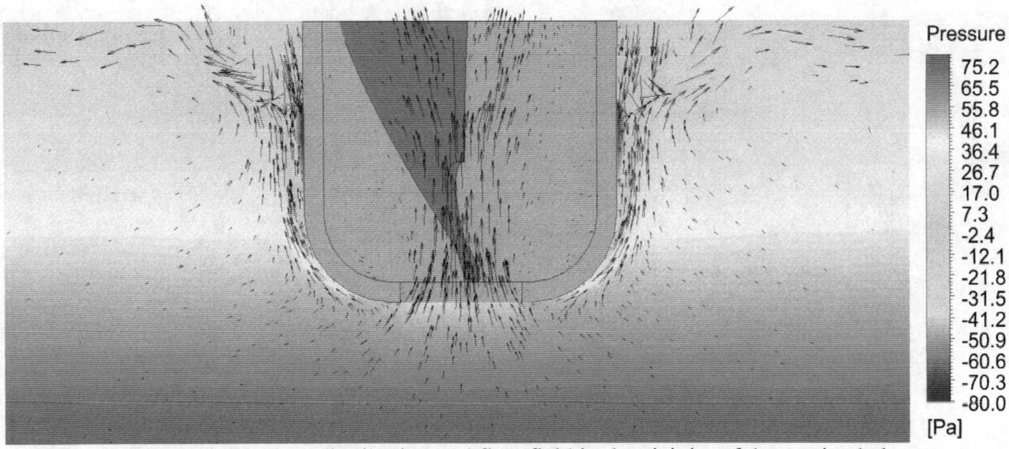

Figure 4: Pressure distribution and flow field in the vicinity of the suction hole

The Figures 5, 6 and 7 illustrate the mass flow rate at pump outlet as a function of time of the investigated variants. Simulated mass flow rates are normalized to the basic configuration. Variations of the immersed depth of the oil pump show the dependence of the oil mass flow rate on the immersion depth and therefore on the static pressure. Reduction of the immersion depth of ~50 % yields a decrease of the mass flow rate of 20 % and an increase of the oil climbing time of 0.4 s. The results also show that higher immersion depths do not affect the oil climbing time significantly but yield higher oil mass flow rates as expected. The comparison between the original geometry, design 1 and design 2 shows deviations of the oil mass flow rate of about 10 %. A shorter oil climbing time could be achieved with design 1 compared to the original geometry. Design 2 requires significant longer time for the oil reaching the pump outlet. This can be explained due to the higher volume in design 2 which has to be filled with oil until it is furtherly transported.

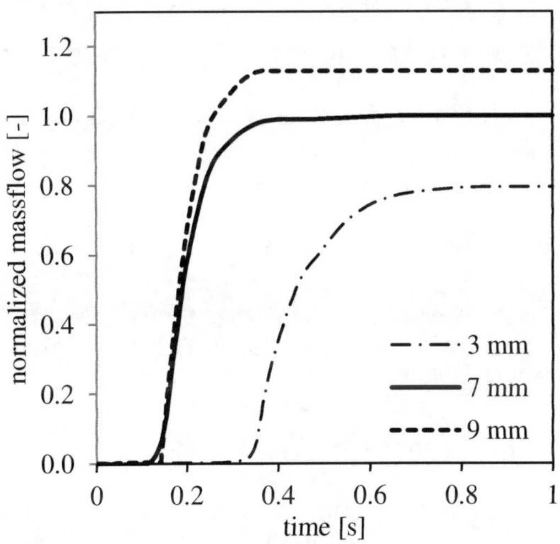

Figure 5: Normalized oil mass flow as a function of immersion depth

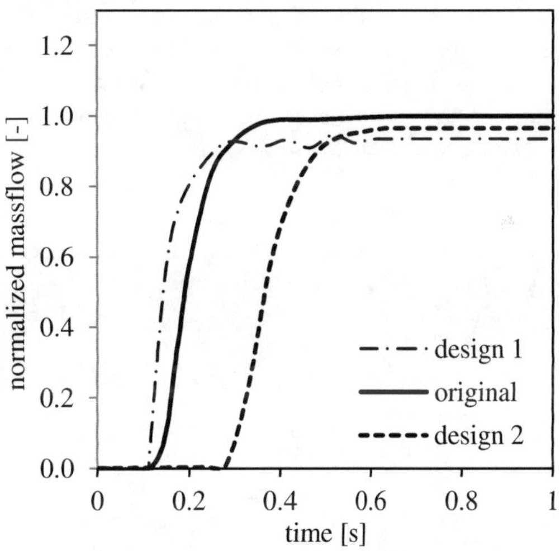

Figure 6: Normalized oil mass flow as a function of pump design

Rotational speed variations influence the oil mass flow rate and the oil climbing time in the same intent. Increasing the rotational speed by a third yields an increase of the oil mass flow rate by a factor of 2.2. The results also show the behaviour of the oil pump at low rotational speeds. A reduction of 500 rpm compared to the basic configuration results in an oil mass flow decrease of 60%. Simulations with reduced rotational speed show that no oil is transported under 2000 rpm, thus the use of the investigated pump design in variable speed compressors is unsuitable.

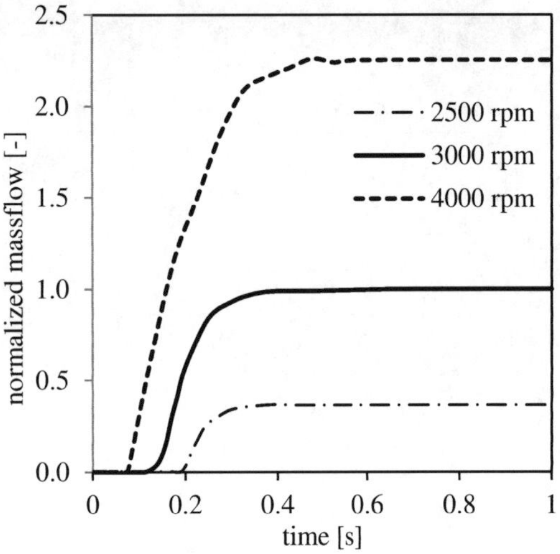

Figure 7: Normalized oil mass flow as a function of rotational speed

Table 1: Reduction of the static pressure between analytic (geodetic pressure) and numerical (CFD) calculation

Variant	Value [%]
basic model	21.0
3 mm immersion depth	45.5
9 mm immersion depth	24.1
2500 rpm	9.2
4000 rpm	71.2
design 1	55.3
design 2	52.5

4. CONCLUSION

A numerical investigation of the suction behaviour of the oil pump of a hermetic compressor is shown. The simulations are carried out using the commercial software ANSYS Fluent and they concentrate on the immersed centrifugal pump and its interaction with the oil sump. Special focus lies on the oil volume fraction in the pump, the oil mass flow rate, the oil climbing time and the static pressure at pump inlet. After analysing the simulation results the following conclusions can be drawn:

- Especially in the start-up phase of the compressor, the oil parabola can sink significantly which has to be taken into account in the pump design.
- Frictional forces reduce the static pressure at the pump inlet due to high radial and tangential velocities which occurs at high rotational speed.
- Vortexes are developed on the immersed pump surface which are spread out in the oil sump and can lead to acoustic problems.
- Volumes in the lower area of the pump should be kept small to decrease the oil climbing time considering the development of the oil parabola.
- The investigated oil pump is not suitable for the use in variable speed compressors because the oil transport fails below 2000 rpm.

NOMENCLATURE

$\boldsymbol{f}_b$	body force vector	(N/m³)
p	pressure	(N/m²)
t	time	(s)
u_i	velocity component	(m/s)
$\boldsymbol{v}_i$	velocity vector	(m/s)
x_i	coordinate component	(m)

Subscripts

g	gaseous phase	
l	liquid phase	

Greek symbols

μ	dynamic viscosity	(N s/m²)
ρ	density	(kg/m³)
χ	gas volume fraction	(-)

REFERENCES

Alves, M.V.C., Barbosa Jr, J.R., Prata, A.T. (2009). Analytical solution of single screw extrusion applicable to intermediate values of screw channel aspect ratio. *Journal of Food Engineering*, Volume 92, pages 152-156.

Alves, M.V.C., Barbosa Jr, J.R., Prata, A.T., Ribas Jr, A.T. (2010). Fluid Flow in a Screw Pump Oil Supply System for Reciprocating Compressors. *Proceedings of the International Compressor Engineering Conference at Purdue*, Purdue University, IN, USA, paper 1326.

Fluent 14 (2011). *User Guide*. Fluent Inc., USA, 2498.

Isaa, R.I. (1986). Solution of the Implicitly Discretised Fluid Flow Equations by Operator-Splitting. *Journal of Computational Physics*, Volume 62, pages 40-65.

Kerpicci, H., Alper, Y., Onbasioglu, S.U. (2013). Investigation of Oil Flow in a Hermetic Reciprocating Compressor. *International Journal of Refrigeration*, Volume 36, pages 215-221.

Kim, H.J., Lee, T.J., Kim, K.H., Bae, Y.J. (2002). Numerical Simulation of Oil Supply System of Reciprocating Compressor for Household Refrigerators. *Proceedings of the International Compressor Engineering Conference at Purdue*, Purdue University, IN, USA, paper 1533.

Lückamnn, A.J., Alves, M.V.C., Barbosa Jr, J.R. (2009). Analysis of Oil Pumping in a Reciprocating Compressor. *Applied Thermal Engineering*, Volume 29, pages 3188-3123.

Moukalled, F., Mangani, L., Darwish, M., (2015). *The Finite Volume Method in Computational Fluid Dynamics*. Springer, Switzerland, 791.

Ozsipah, M., Cadirci, S., Gunes, H., Sarioglu, K., Kerpicci, H. (2014). A Numerical Study on the Lubrication System for a Hermetic Reciprocating Compressor used in Household Refrigerators. *International Journal of Refrigeration*, Volume 48, pages 210-220.

Posch, S., Berger, E., Heimel, M., Hopfgartner, J., Almbauer, R., Schöllauf, P. (2015). Fluid Flow in the Oil Pumping System of a Hermetic Compressor. *Proceedings of the International Conference on Compressors and their Systems*, City University, London, GB, paper 012028.

Scardovelli, R., Zaleski, S. (1999). Direct numerical simulation of free-surface and interfacial flow. *Annual Review Fluid Dynamics*, Volume 31, pages 567-603.

Tada, M.P., Hoffmann, T., Couto, P.R., Manke, A.L., de Bortoli, G.D. (2014). Numerical and Experimental Examination for Oil Pump System Using a Simplified Uncoupled Simulation. *Proceedings of the International Compressor Engineering Conference at Purdue*, Purdue University, IN, USA, paper 1248.

ACKNOWLEDGEMENT

This work has been carried out within the framework of ECO-COOL, a research project initiated and funded by the FFG (Austrian Research Promotion Agency). Furthermore the authors particularly acknowledge the technical support by Secop Austria GmbH, formerly ACC Austria GmbH and Liebherr-Hausgeräte Lienz GmbH.

A Numerical Friction Loss Analysis of the Journal Bearings in a Hermetic Reciprocating Compressor

Stefan POSCH[1]*, Johann HOPFGARTNER[1], Martin HEIMEL[1], Erwin BERGER[1], Raimund ALMBAUER[1], Peter SCHÖLLAUF[2]

[1] Institute of Internal Combustion Engines and Thermodynamics, Graz University of Technology, Inffeldgasse 19, 8010 Graz, Austria

posch@ivt.tugraz.at, +43 316 873 30233
hopfgartner@ivt.tugraz.at, +43 316 873 30240
heimel@ivt.tugraz.at, +43 316 873 30235
berger@ivt.tugraz.at, +43 316 873 30234
almbauer@ivt.tugraz.at, +43 316 873 30230

[2] Secop Austria GmbH, Jahnstraße 30, 8280 Fürstenfeld, Austria
p.schoellauf@secop.com, +43 3382 5010 984

* Corresponding Author

ABSTRACT

In addition to the electrical and the thermodynamic losses in hermetic compressors, mechanical losses have a significant influence on the performance of the compressor. In the present paper the friction losses in the journal bearings of a hermetic reciprocating compressor are investigated using numerical methods. A dynamic model is set up to solve the Reynolds equation using a finite volume approach to calculate the pressure field in each journal bearing. The calculation of the fluid film thickness is accomplished with the formulas of the parallel gap. The resulting hydrodynamic forces are equated with forces obtained by a dynamic multibody model of the compressor crank drive to calculate the transient orbit movement of the bearing. Based on the movement of the crankshaft at steady-state conditions, the shear stresses in the gap between crankshaft and housing can be calculated. Thus the cycle averaged friction power loss can be determined. To consider effects such as surface roughness of the bearings or possible contacts between the solids, correlations found in literature are implemented. The present method is used to assess the friction power loss of the journal bearings during the operation with different oil viscosities. The simulated data is verified by simple analytical friction loss calculations based on shear stresses in the Couette flow between bearing and housing.

1. INTRODUCTION

Hermetic compressors either running in ON-OFF or variable speed mode are widely used in domestic refrigeration appliances. The hermetic design of the compressors requests a durability of all the compressor parts of at least 15 years, which is specified by the producer of the cooling appliance. Another difficulty caused by the hermetic design is the complex interaction between electrical, mechanical and thermodynamic losses. The identification of the individual losses is one of the main tasks in the development process of modern hermetic reciprocating compressors. In addition to challenging experimental investigations, simulation models are increasingly used. An example of simulation application in compressor development is the quantification of friction losses in moving parts, especially in journal bearings. Due to the low absolute level of friction power losses (relative level with regard to compressor power is about 15 %) and the hermetic design, classical strip-down methods which are used in engine development are not suitable for the investigation of friction losses in reciprocating compressors for domestic refrigeration appliance. Several studies dealing with the modelling of the journal bearing dynamic behaviour are available. The studies resemble each other concerning the procedure to solve the friction loss problem of journal bearings. The procedure can be summarized as follows: (i) a multi body simulation of the compressor crank drive is carried out to get the dynamic loads on the journal bearings, (ii) the Reynolds equation is solved using numerical schemes to calculate the hydrodynamic forces in the bearings, (iii) a Newton-Raphson algorithm is used to get the movement of the shaft orbit and (iv) based on the shaft movement the friction losses can be determined. A distinction between these studies can be made concerning the modelling of the Reynolds equation or the model accuracy. The use of the short bearing approximation can be

found in e.g. Estupinan and Santos (2009) or Kim *et al.* (2012). A comparison between results of the short bearing approximation and the finite bearing model was carried out by Chieh *et al.* (2007). Duyar and Dursunkaya (2002 and 2006) analysed the dynamic behaviour of compressor journal bearings considering the elastic deformation of the shaft using finite element discretization. An investigation of compressor journal bearings considering mixed lubrication was presented in the work of Matsui *et al.* (2010).
The present paper deals with the investigation of the friction losses in compressor journal bearings for being used as input data for a holistic thermal model of a hermetic compressor. The method is similar to the previous explained procedure using finite bearing model for the solution of the Reynolds equation. Furthermore, mixed lubrication models found in the literature are considered. The gap in the bearing is approximated as parallel. The simulation is used to calculate the influence of the oil viscosity and gap width on the friction losses. The simulated data is compared to simple analytical friction loss calculations based on shear stresses in the Couette flow to give an assessment of the required accuracy of the model to fulfil the requirements of a thermal compressor model.

2. KINEMATICS AND DYNAMICS

The friction analysis of the compressor journal bearings requires the determination of the bearing forces. For this purpose a multibody dynamics model of the piston-conrod-crankshaft system is developed. A detailed description of the friction behaviour of the junctions between piston and conrod respectively conrod and crankshaft is not carried out in the present study. Moreover, the misalignment of the piston is not considered. Figure 1 shows the coordinate systems used for the kinematic description of the crank mechanism parts. The inertial coordinate system is located in the static position of bearing A. The misalignment of the crankshaft due to the bearing eccentricity is considered in the distance and rotating vector of the crankshaft regarding to the initial coordinate system.

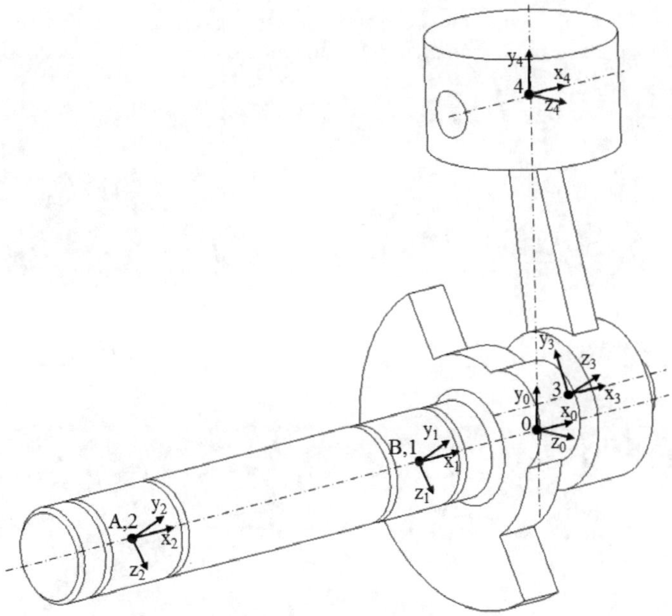

Figure 1: Coordinate systems

Using the Newton-Euler method the equations of motion can be expressed for each part.

Piston:

$$m_{Pi} \cdot {}_I\boldsymbol{a}_{Pi} = {}_I\boldsymbol{F}_p - {}_I\boldsymbol{F}_4 + {}_I\boldsymbol{F}_{Cyl} \tag{1}$$

Conrod:

$$m_{Cr} \cdot {}_I\boldsymbol{a}_{Cr} = {}_I\boldsymbol{F}_4 - {}_I\boldsymbol{F}_3 \tag{2}$$

$${}_I\Theta_{Cr} \cdot \dot{\omega}_{Cr} + m_{Cr} \cdot {}_I\boldsymbol{r}_{3-Cr} \times {}_I\boldsymbol{a}_{Cr} = {}_I\boldsymbol{r}_{3-4} \times {}_I\boldsymbol{F}_4 \tag{3}$$

Crankshaft:

$$m_{Cs} \cdot {}_I\boldsymbol{a}_{Cs} = {}_I\boldsymbol{F}_A + {}_I\boldsymbol{F}_B + {}_I\boldsymbol{F}_3 \tag{4}$$

$${}_I\Theta_{Cs} \cdot \dot{\omega}_{Cs} + \omega_{Cs} \times \left({}_I\Theta_{Cs} \cdot \omega_{Cs} \right) + m_{Cs} \cdot {}_I\boldsymbol{r}_{1-Cs} \times {}_I\boldsymbol{a}_{Cs} = {}_I\boldsymbol{r}_{1-2} \times {}_I\boldsymbol{F}_A + {}_I\boldsymbol{r}_{1-3} \times {}_I\boldsymbol{F}_3 + \boldsymbol{T} \tag{5}$$

The set of equations gives the time dependent functions of the reaction forces in the main journal bearings according to the cylinder pressure. The cylinder pressure can be determined by experiments or computational fluid dynamics (CFD), respectively. In the present study the cylinder pressure is given by CFD simulation of the compressor gas line.

$${}_I\boldsymbol{F}_p = \begin{pmatrix} 0 \\ p_{CFD} \cdot A \\ 0 \end{pmatrix} \tag{6}$$

3. HYDRODYNAMIC FORCES

The modelling of the hydrodynamic forces in the journal bearings is carried out by solving the Reynolds equation which links the hydrodynamic pressure and the fluid film thickness.

$$\frac{\partial}{\partial x}\left(\frac{\rho\, h^3}{12\eta}\frac{\partial p_h}{\partial x}\right) + \frac{\partial}{\partial y}\left(\frac{\rho\, h^3}{12\eta}\frac{\partial p_h}{\partial y}\right) = \frac{(u_I + u_{II})}{2}\frac{\partial(\rho h)}{\partial x} + \frac{\partial(\rho h)}{\partial t} \tag{7}$$

The terms on the left-hand side of the Reynolds equation represent the Poiseuille flow, the terms on the right-hand side represent the Couette flow and the displacement flow, respectively. To solve this kind of partial, inhomogeneous elliptic differential equation, numerical methods have to be applied. In the present study the Reynolds equation is solved using the finite volume approach. The fluid film area in each bearing is divided in a certain number of cells and the terms of the Reynolds equation can be discretised. To avoid negative fluid film pressures, the Gümbel boundary condition is used which sets negative pressure values to zero. Figure 2 shows the hydrodynamic pressure distribution in bearing A without (a) and with (b) Gümbel boundary condition.

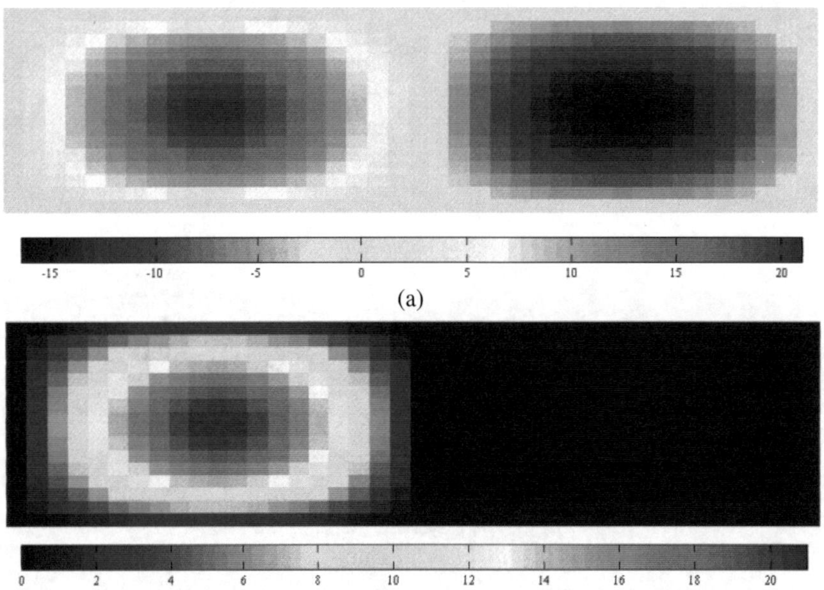

Figure 2: Hydrodynamic pressure in bearing A without (a) and with (b) Gümbel boundary condition [MPa]

To create a link between the eccentricity of the bearing and the fluid film thickness, the gap in the bearing is assumed to be parallel. According to the geometric relations shown in Figure 3 the fluid film thickness can be expressed as follows (Woschke, 2013):

$$h(\varphi) = (r_I - r_{II}) - ex \cdot \cos(\varphi - \xi) \tag{8}$$

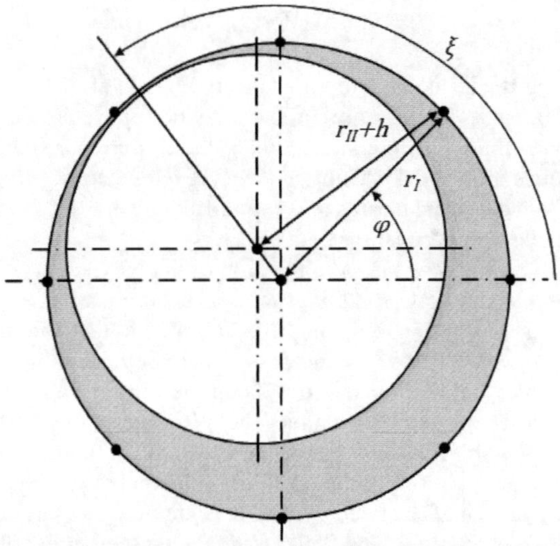

Figure 3: Determination of the fluid film thickness

The basic form of the Reynolds equation (7) considers only the macro geometry of the bearing parts. Patir and Cheng (1978 and 1979) introduced an extended form of the Reynolds equation to consider also micro effects due to surface roughness. The authors used flow factors which depend on the roughness of the bearing surface and called the new form Average Reynolds Equation.

$$\frac{\partial}{\partial x}\left(\phi_x \frac{\rho\, h^3}{12\eta}\frac{\partial \overline{p_h}}{\partial x}\right)+\frac{\partial}{\partial y}\left(\phi_y \frac{\rho\, h^3}{12\eta}\frac{\partial \overline{p_h}}{\partial y}\right)=\frac{(u_I+u_{II})}{2}\left(\frac{\partial\left(\rho\overline{h_T}\right)}{\partial x}+\sigma_\delta\frac{\partial \phi_S}{\partial x}\right)+\frac{\partial\left(\rho\overline{h_T}\right)}{\partial t} \quad (9)$$

For a more detailed description of the flow factor determination the interested reader is referred to the primary works by Patir and Cheng (1978 and 1979). To calculate the possible solid contact between the surfaces the model of Greenwood and Tripp (1970) is used in the present study.

4. NUMERICAL ALGORITHM

The solving procedure starts with the input of the geometric data of the compressor crank mechanism parts. Initial values of eccentricity *ex* and minimum fluid film thickness angle ξ are set to zero as well as their time derivatives. Time resolution is fixed for the whole procedure and corresponds to a crank angle of two degrees. Starting at a certain position of the crankshaft (here crankshaft angle is zero) the cylinder pressure of the CFD calculation is read into the bearing simulation program. Solving the multibody dynamic system, the bearing forces can be calculated for the current crank angle. The calculation of the hydrodynamic bearing forces starts with the determination of the fluid film thickness according to (8) with the defined start values of the new time step. The start values of the new time step are the calculated eccentricity *ex* and minimum fluid film thickness angle ξ of the previous time step or the initial values if it is the first time step, respectively. To get the orbit of the crankshaft for the current time step, a two-dimensional Newton algorithm depending on *ex* and ξ is used (10). In each iteration step the Averaged Reynold Equation (9) is solved numerically for both bearings and the hydrodynamic bearing forces are compared with the calculated bearing forces of the multibody system regarding absolute value and force direction. If the deviation between hydrodynamic and multibody forces in both bearings is higher than the convergence criteria β_F the iteration procedure is repeated, otherwise the algorithm proceeds to the next time step. The numerical algorithm is carried out until the deviation of the time-dependent shaft orbit between the current and the previous cycle is lower than the convergence criteria β_C and steady-state conditions are reached.

$$\begin{gathered}\boldsymbol{x}=\binom{ex}{\xi}\\ \boldsymbol{f}(\boldsymbol{x})=\boldsymbol{F}_{MBS}-\boldsymbol{F}_{Rey}\\ \boldsymbol{x}_{n+1}=\boldsymbol{x}_n-\left(\boldsymbol{J}_f(\boldsymbol{x}_n)\right)^{-1}\boldsymbol{f}(\boldsymbol{x}_n)\end{gathered} \quad (10)$$

5. RESULTS

The present numerical bearing model is used to simulate hydrodynamic bearings of a hermetic reciprocating compressor running at 3000 rpm. The pressure in the cylinder is determined by CFD simulation of the compressor gas line considering fluid-structure interaction of the compressor valves. Operating conditions are set to -23 °C for evaporating temperature and 45 °C for condensing temperature (R600a), respectively. The baseline value for oil viscosity is 8 cSt and bearing clearance is 5 µm. Bearing width of bearing A is one quarter less than bearing B, whereas the shaft diameter is equal for both bearings.

Figure 4 shows the shaft orbits of both bearings as a function of the oil viscosity. The illustrated curves represent one crankshaft revolution after steady-state conditions are reached which is the case after 7-8 cycles depending on the considered configuration. According to the higher forces due to the compression process at bearing B the eccentricity ratio of bearing B is higher than bearing A. An increase of the oil viscosity results in a decrease of the shaft deflection. Furthermore, higher oil viscosities damp the relative motion between the bearing parts and thus the crankshaft orbit is focused on a smaller region. The influence of the shaft orbit on the bearing clearance is shown in Figure 5. The curves show an increase of the eccentricity ratio of both bearings with an increased bearing clearance. This behaviour is a result of the nonlinear relation between the bearing pressure and the local oil film thickness. Small bearing clearance already yields in higher oil pressure at small shaft movement, so the deflection of the crankshaft is kept at low values. This effect can be seen in the shape of the crank orbit which is smoother at smaller bearing clearance values.

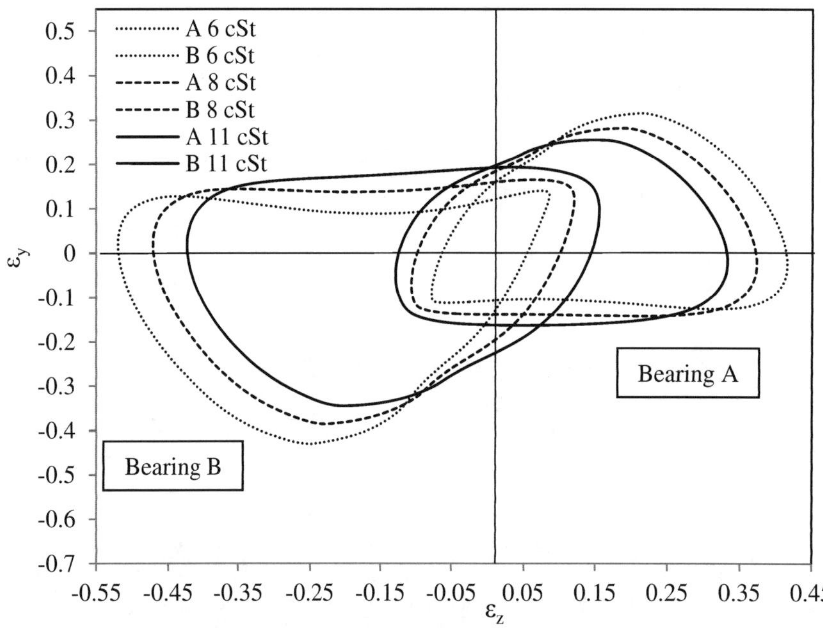

Figure 4: Shaft orbits with different oil viscosity

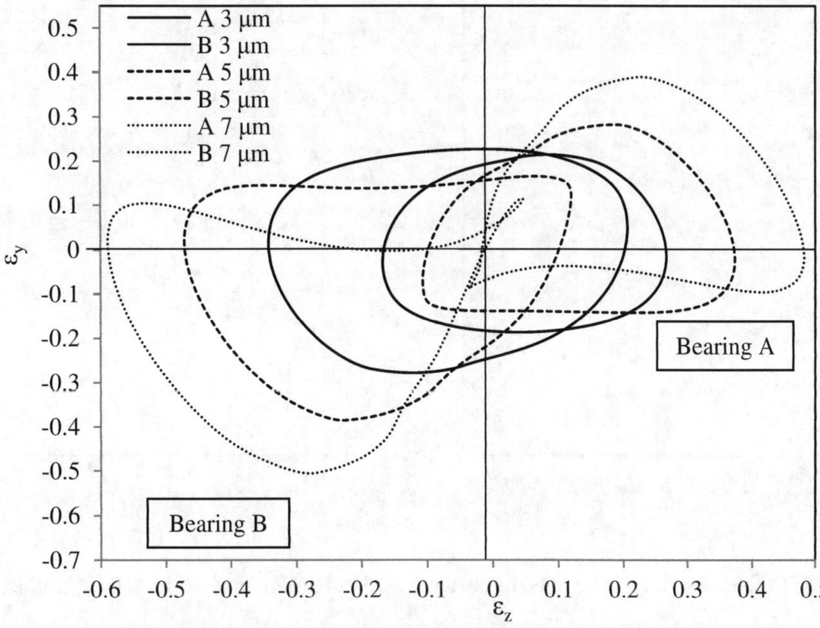

Figure 5: Shaft orbits with different bearing clearances

An addition illustration of the simulated shaft movement in terms of the absolute eccentricity ratio ε over the crank angle φ can be seen in Figure 6 and 7. According to the shaft orbit curves the eccentricity ratio of bearing B is higher than of bearing A. The figures show the peak in the eccentricity curve in the area of the top-dead centre of the compressor piston and the resulting high reaction forces in the piston-conrod-crankshaft system. The nonlinear dependence between bearing pressure and local oil film thickness (and its time derivatives) can also be seen in the eccentricity ratio curves. Curves of different viscosity or gap width can intersect especially in regions of low reaction forces and small eccentricity ratios.

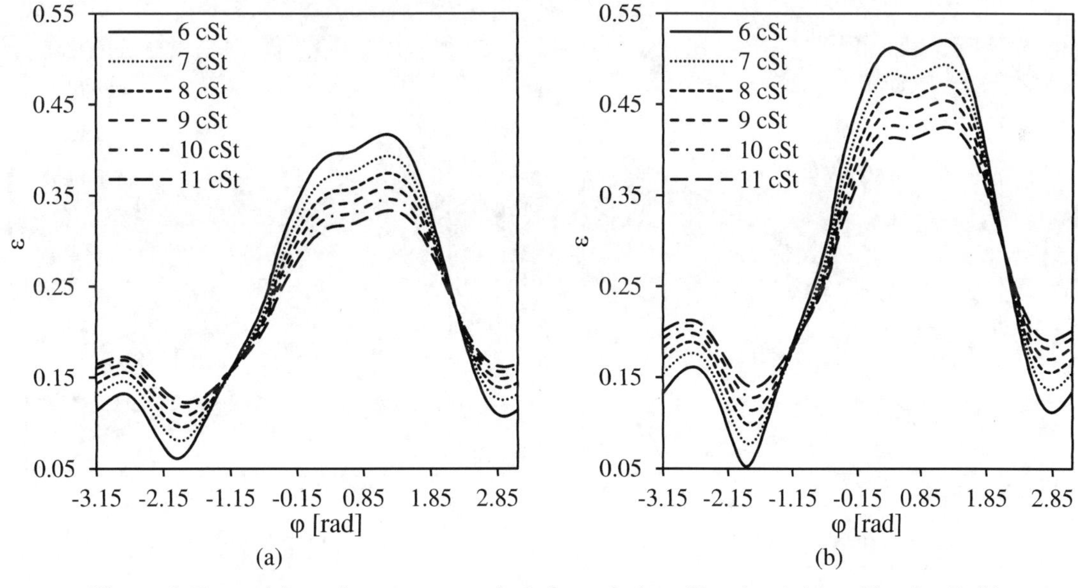

Figure 6: Eccentricity ratio over one crank shaft revolution of bearing A (a) and bearing B (b)

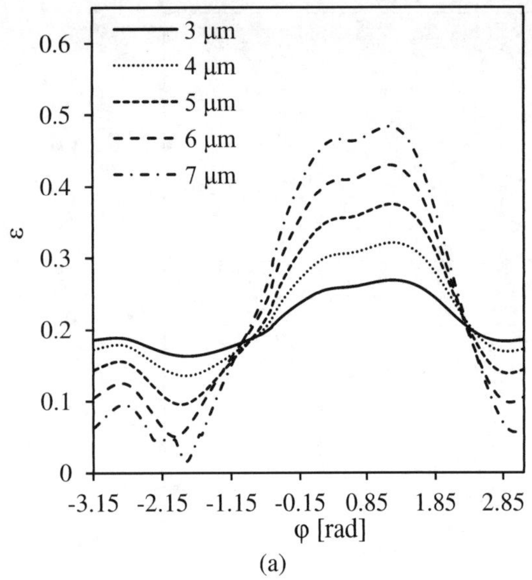

Figure 7: Eccentricity ratio over one crank shaft revolution of bearing A (a) and bearing B (b)

Results of the determination of the friction power loss in the bearings are shown in Figure 8. Friction power loss values of bearing B are higher than of bearing A. This is because of the higher reaction forces in addition with the larger bearing width of bearing B which results in higher friction surface. Viscosity variations show a linear dependency of the friction power loss increasing with higher viscosities. The numerical simulation of the friction power loss with different gap widths shows a stronger power loss increase at smaller gap widths. A comparison between numerical results and simple analytical Couette flow calculation show constant curve offset for a viscosity variation. For gap width investigation, the offset between numerical and analytical results increases for higher gap widths. Generally, the deviation between numerical and analytical results is significantly higher for higher loaded bearings like bearing B in the present study.

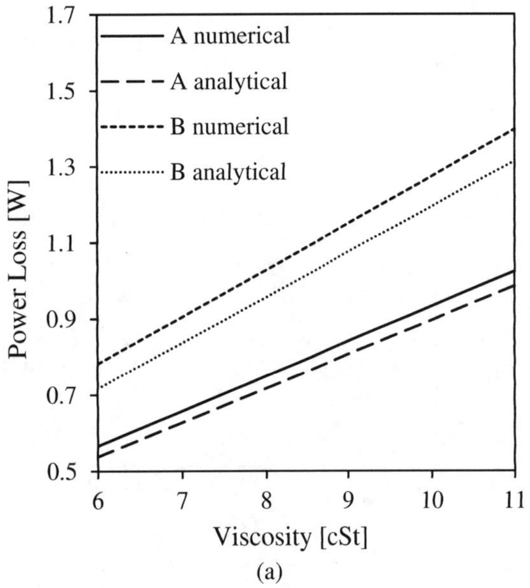

Figure 8: Power loss as function of viscosity (a) and gap width (b)

A comparison between the simulations with and without the usage of Patir and Cheng's (1978 and 1979) Average Reynolds Equation yields in friction power loss deviations below 1 %. It should be mentioned, that the maximum determined eccentricity ratios of both bearings in the present study are significant smaller than 1 so the influence of the surface roughness on the friction power loss is also small. In simulations of compressor operating points with higher cylinder pressure which results in higher reaction forces in the bearings the eccentricity ratio will increase and thus the surface roughness influence on the friction power loss will also increase. Similar considerations can be made regarding the influence of the shaft movement in the multibody dynamics. Stronger shaft movement influences the dynamics of the piston-conrod-crankshaft system and consequently the hydrodynamics of the bearings. In the present study the influence of the shaft movement consideration on the friction power loss is below 1 %.

6. CONCLUSION

A numerical investigation of the hydrodynamic bearing system of a hermetic reciprocating compressor is shown in this study. The model is based on the solution of the Average Reynolds Equation of Patir and Cheng (1978 and 1979) to compare hydrodynamic bearing forces with reaction forces gained by multibody dynamics calculation which results in a two-dimensional Newton algorithm problem for the fluid film properties ex and ξ. The presented algorithm is used to calculate the dynamic behaviour of the journal bearings of the compressor operating at -23 °C evaporating temperature and 45 °C condensing temperature (R600a). A parameter study for several oil viscosities and gap widths are carried out to show the abilities of the method for the compressor development. Additionally, the present algorithm is used to calculate the friction power losses of the journal bearings. In terms of the friction power loss modelling, the following conclusions can be emphasized:

- The comparison between numerical and analytical (simple Couette flow) simulations show a similar dependence of the parameters, but analytical simulation underestimates friction power losses especially at higher loaded bearings.
- The use of the complex Averaged Reynolds Equation of Patir and Cheng (1978 and 1979) can be avoided at lower loaded bearings with small shaft movements. In this case, the classical Reynolds equation can be used.
- A similar conclusion can be made for the consideration of the shaft movement in the multibody dynamics. Small shaft movements can be neglected here.

NOMENCLATURE

$\boldsymbol{a}$	acceleration vector	(m/s²)	Cyl	cylinder	
A	area	(m²)	I	inertial system	
ex	eccentricity	(m)	I, II	shaft, bearing	
$\boldsymbol{F}$	force vector	(N)	MBS	multi body system	
h	fluid film thickness	(m)	Pi	piston	
m	mass	(kg)	Rey	Reynolds forces	
p	pressure	(N/m²)	**Greek symbols**		
$\boldsymbol{r}, \boldsymbol{x}$	distance vector	(m)	β	convergence criteria	(-)
t	time	(s)	ε	relative eccentricity	(-)
$\boldsymbol{T}$	torque moment vector	(N m)	η	dynamic viscosity	(N s/m²)
u	velocity	(m/s)	Θ	inertia tensor	(kg m²)
x, y, z	distance components	(m)	ξ	minimum fluid film thickness angle	(°)
Subscripts			ϕ_x, ϕ_y, ϕ_s	flow factors	(-)
0, 1, 2, 3, 4	coordinate system index		φ	attitude angle	(°)
A, B	bearing index		ω	rotational speed	(s^{-1})
Cr	conrod		ρ	density	(kg/m³)
Cs	crankshaft		σ_δ	standard deviation of the combined roughness	(m)

REFERENCES

Chieh, H., Couto, P.R.C., Prata, A.T. (2007). An Integral Analysis and Simulation of the Complete Bearing System in Reciprocating Hermetic Compressors. *Proceedings of the International Conference on Compressors and their Systems*, City University, London, GB, paper 658054.

Duyar, M., Dursunkaya, Z. (2002). Design Improvement of a Compressor Bearing Using an Elastohydrodynamic Lubrication Model. *Proceedings of the International Compressor Engineering Conference at Purdue*, Purdue University, IN, USA, paper 1537.

Duyar, M., Dursunkaya, Z. (2006). Design Improvement Based on Wear of a Journal Bearing Using an Elastohydrodynamic Lubrication Model. *Proceedings of the International Compressor Engineering Conference at Purdue*, Purdue University, IN, USA, paper 1821.

Estupinan, E.A., Santos, I.F. (2009). Modelling Hermetic Compressors Using Different Constraint Equations to Accommodate Multibody Dynamics and Hydrodynamic Lubrication. *Journal of the Brazilian Society of Mechanical Science and Engineering*, Volume 31, pages 35-46.

Greenwood, J.A., Tripp, J.H. (1970). The contact of two nominally flat rough surfaces. *Proceedings of the Institution of Mechanical Engineers*, Volume 185, pages 635-633.

Kim, J., Park, S., Cho, I., Baek, I., Jo, J., Jung, J. (2012). Journal Bearing Lubrication Characteristics in a Reciprocating Compressor. *Proceedings of the International Compressor Engineering Conference at Purdue*, Purdue University, IN, USA, paper 1263.

Matsui, M., Kitsunai, Y., Inagaki, K. (2010). High Efficiency Development of a Reciprocating Compressor by Clarification of Loss Generation in Bearings. *Proceedings of the International Compressor Engineering Conference at Purdue*, Purdue University, IN, USA, paper 2025.

Patir, N., Cheng, H.S. (1978). An average model for determining effects of three-dimensional roughness on partial hydrodynamic lubrication. *Journal of Lubrication Technology*, Volume 100, pages 12-17.

Patir, N., Cheng, H.S. (1979). Application of average flow model to lubrication between rough sliding surfaces. *Journal of Lubrication Technology*, Volume 101, pages 220-230.

Woschke, E. (2013). Simulation gleitgelagerter Systeme in Mehrkörperprogrammen unter Berücksichtigung mechanischer und thermischer Deformation. *PhD Thesis*, Otto von Guericke University Magdeburg, GER.

ACKNOWLEDGEMENT

This work has been carried out within the framework of ECO-COOL, a research project initiated and funded by the FFG (Austrian Research Promotion Agency). Furthermore the authors particularly acknowledge the technical support by Secop Austria GmbH, formerly ACC Austria GmbH and Liebherr-Hausgeräte Lienz GmbH.

Effect of Refrigerant Gases (HFC134a and HC600a) on the Tribological behaviour of a Multifunctional DLC Coating.

Marcio SILVERIO[1*], Roberto BINDER[1], Emilio Rodrigues Hulse[1], José Daniel BIASOLI DE MELLO[2]

[1*] Embraco, R&D
Joinville, Santa Catarina, Brazil
(+55 47 3441-2828, marcio.silverio@embraco.com)
* Corresponding Author

[1] Embraco, R&D
Joinville, Santa Catarina, Brazil
(+55 47 3441-2066, roberto.binder@embraco.com)

[1] Embraco, R&D
Joinville, Santa Catarina, Brazil
(+55 47 3441-2414, emilio.r.hulse@embraco.com)

[2] Universidade Federal de Santa Catarina, Laboratório de Materiais
Florianópolis, Santa Catarina, Brazil
(+55 48 3721-9268, d.mello@labmat.ufsc.br)

ABSTRACT

Multifunctional DLC, Diamond-like Carbon coatings have been extensively studied and recognized as a promising solution to avoid wear and friction problems. However, the effect of the environment on the tribological behaviour, in particular in the refrigeration industry, is little studied yet. The present work aims to study the influence of refrigerant gases (HFC134a and R600a) on the tribological behaviour of multifunctional CrN-Si rich Diamond-like Carbon coatings, in particular on the tribo-chemical reactions. Tribological behaviour was evaluated using constant and incremental load reciprocating sliding tests. The constant load tests were conducted to access the friction coefficient and wear rates of specimen and counter-bodies whereas the incremental load tests (increments of 2 N at 15 min. intervals) allowed the assessment of the scuffing resistance of the coatings. The scuffing resistance was defined as the work (N.m) at which the value of the friction coefficient first rose above 0.20 (lubricity effect). The characterization of wear scars (tribolayers) was performed by optical interferometry, energy dispersive X-ray spectrometry associated with scanning electron microscopy, and micro Raman spectroscopy. It was verified a strong influence of the atmosphere via the formation of a tribolayer at the interface between body and counter body. This influence resulted in some disturbance of friction coefficient and contact resistance values. Tests performed in R600a atmosphere presented lower (36%) friction coefficient and lower (40%) scuffing resistance (surface durability) than those performed in HFC134a refrigerant gas. The observed differences were justified in terms of the tribo-chemical reaction between the multifunctional coating, the counter body and the refrigerant gases. Both results presented traces of Oxygen, but in the tests with HFC134a gas it was observed the presence of Fluorinated compounds in the tribo-layer. In the tests performed with R600a gas atmosphere it was found only carbon and Silicon, indicating different interactions of the environment. It is worth noting that these results can serve as a guide to the application: the low friction induced by the presence of R600a recommend it for applications where sustainability and energy saving are required while the use of HFC134a is recommended for applications requiring increased reliability (longer life).

1. INTRODUCTION

In the last years, academic and governmental entities have developed several projects in search for energy saving technologies. Many of these studies focused on surface engineering and related topics. In this context, surface

treatments and surface coating modifiers have faced the important challenge of reducing friction coefficient, wear losses and, consequently, energy consumption. For decades, the refrigeration industry has investigated and proposed various new technologies aiming energy consumption reduction. Only in the USA, Accordingly to United States Environmental Protection Agency, the refrigeration industry is responsible for approximately of 17.2% of the domestic energy consumption, and therefore plays a very important role in saving energy and natural resources.
In a refrigerator, the compressor is the main responsible for energy consumption and numerous emergent technologies are being studied and implemented each year. The great majority of compressors nowadays use fluid lubrication to reduce friction and wear. However, environmental factors suggest the reduction or even elimination of lubricating oils. In this sense, oil-less tribo systems represent a new challenge for surface science and engineering, stimulating new projects related to coatings and self-lubricant materials. According to Erdermir (2001), the development of advanced technology for coatings with low friction coefficient and high wear endurance became a high source of activities and researches around the world in the tribological field. Amorphous carbon coatings, known as diamond-like carbon (DLC), have been an alternative in many technological applications, such as hard coating for tools, automotive parts, computer hard drives and micro electro mechanical systems (MEMS). The applications are mainly associated with its high hardness, high chemical inertness and solid lubrication capacity Viana et al (2010). The low friction coefficients associated with Diamond-like Carbon coating is often associated with tribolayer formation, which is a result mainly of DLC transfer from to the counter body and vice versa Ronkainen et al (1996), Gril (1993), Lifang and Guang (2008), Erdemir et al (1995). The genesis of these tribo layers is strongly dependent on the environment. For atmospheres such as air, oxygen or nitrogen, the increase of humidity usually reduces the friction coefficient of Diamond-like Carbon-steel contacts Lifang and Guang (2008), Erdemir et al (1995). In a study performed by De Mello et al. (2009), it was shown that the presence of protective atmosphere (R600a) modifies the chemical structure of the tribolayer generated on the counter body, reducing the friction coefficient and the wear rate of the system (DLC / steel AISI 52100). In another work performed by Demas and Polycarpou (2006) pin-on-disc tests were performed with gray cast iron (pin and disc) without lubrication and different atmospheres (O_2, CO_2, N_2, Ar and R134a). Under O_2 atmosphere, the wear was ruled by the oxidation that occurs on the wear track, whereas in air and R134a it also contained an adhesive component. Under N_2 atmosphere, the wear was dominated by adhesion, while in CO_2 a slight polishing on the surface occurred, removing only the most prominent asperities, thus leading to the best tribological performance among the tested atmospheres. The settling of tribo layers on the real contact area rule the tribological behaviour of tribo systems. Their dimensions are reduced (nanometres), and this factor hinders to understand their formation and stabilization. Tribo layers are formed from physicochemical interactions between the surfaces in contact and relative motion, which ranges from mutual transfer of materials and reactions between atmosphere, lubricants and contaminants present in the contact Luo et al (2009). This work presents a surface treatment that has been used with great success in oil free hermetic compressors, consisting of a multi-functional coating composed of an outermost Diamond-like Carbon film applied onto a low Carbon steel substrate with an intermediate Chromium base layer used as a load bearing layer. The Diamond-like Carbon family consists of amorphous carbon hydrogenated alloys (a-C:H), as shown in the ternary diagram on Figure 1, where different families of DLC's are characterized by the percentage of hybridizations sp^2, sp^3 and amount of hydrogen Robertson (2002). In addition, doping elements can be incorporated into Diamond-like Carbon coatings to modify their properties, such as silicon (a-C:H:Si), which decreases the free energy and the residual stress of the coating, modifying its tribological behaviour Donnet (1998).
According to Bowden and Young (1951) the low friction coefficient of diamond films is due its inertness nature and/or passivation of contact interfaces. In this way, some researchers such as Miyake et al (1995), Smentkowski et al (1996, 1997) e Molian et al (1993) developed different routes for the surface passivation of Carbon films, adding different chemical elements to reach ultra low friction coefficients. Other studies investigated the presence of fluorine in the Diamond-like Carbon coatings to reduce friction coefficients and wear rates (Miyake et al, 1995, Smentkowski et al, 1996; Molian et al, 1993). Dugger et al (1992) showed that adsorbed gases at the contact interface have great influence on wear rates due to the bonding changes caused by those gases. De Mello et al (2009) showed the tribological behaviour of multi component Diamond-like Carbon coating in different refrigerant gases used by the refrigeration industry. They presented different friction coefficients and wear rates for each atmosphere including the influence of Oxygen contamination on tribological properties. The present study aims to identify the influence of two of the most used refrigerant gases in the refrigeration industry, HFC134a and HC600a refrigerant gases, on the frictional behaviour and wear of CrN-SiDLC multifunctional coatings.

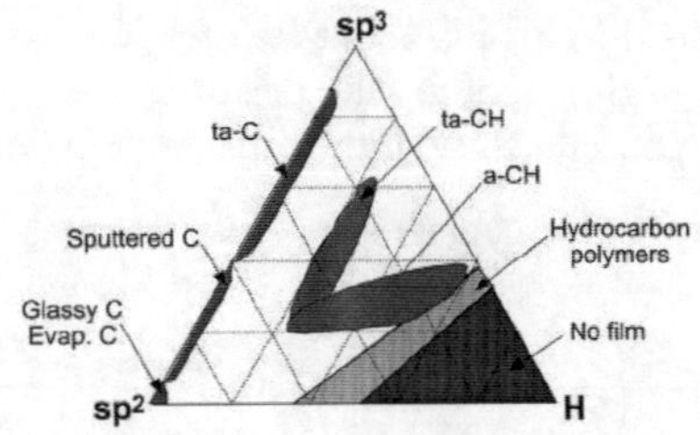

Figure 1: Ternary phase diagram for hydrogenated amorphous carbon Robertson (2002).

2. MATERIALS AND METHODS

The multifunctional Diamond-like Carbon coating used in this work was a proprietary multifunctional Chromium base plus a DLC coating, where an intermediate layer composed of chromium nitride is responsible for the load bearing and an uppermost a-C:H layer deposited by PACVD (Plasma Assisted Vapour Deposition) is used as a self lubrication film. The substrate was an AISI 1020 low carbon steel. The aim of this configuration was to avoid the so called "egg shell effect", which may cause coating spalling due differences in mechanical and physical properties of the Diamond-like Carbon layer (high hardness) and the soft substrate. The samples were polished prior to coating deposition in order to standardize the surfaces.

2.1 Tribological Evaluation

The tribological tests were conducted using a ball-on-flat contact geometry in a micro tribometer UMT-01 CETR under reciprocating sliding configuration, as indicated in Figure 2-a using two types of experiments:

i- Wear tests: Reciprocating sliding tests conducted at constant normal load to access the friction coefficients and wear rates of specimen and counter-bodies.

ii- Durability tests: Reciprocating sliding tests carried out using an incremental loading mode. The scuffing resistance was determined by increasing the normal load in increments of 2 N at every 15 min. (up to 91N, which was the limit of the load cell used in this test configuration). In this study, the scuffing resistance was defined as the work (N.m) at which the value of the friction coefficient first rose above 0.20 (lubricity effect) as proposed by De Melo and Binder, 2006 and illustrated in Figure 2-b.

In both types of experiments, a hard Si_3N_4 ball (diameter 2.381 mm) was fixed onto a pivoted arm and rested against the specimen surface, which moved at constant stroke (5 mm) and frequency (2 Hz). The ball surface was used in the as-received condition and a new surface region was used for each test. Contact potential and friction force were continually logged (acquisition rate of 1 Hz). The tests were conducted under controlled relative humidity (50 %) and temperature (22±2 °C). The results are the averages from, at least, five experiments for each analysed condition. Wear scars were analyzed by using SEM-EDX as well as FEG-SEM; micro Raman spectroscopy and white light interferometry.

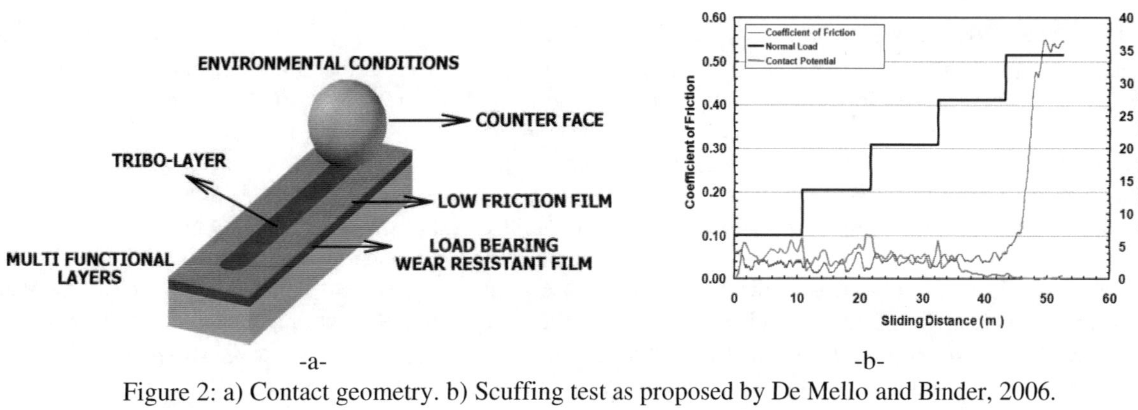

-a- -b-

Figure 2: a) Contact geometry. b) Scuffing test as proposed by De Mello and Binder, 2006.

A semi-hermetic chamber, Figure 3, was developed to vary the test environment (refrigerant gas atmospheres) and avoid humidity and air contamination. Light vacuum was performed before each test and slight positive pressure was assured during the whole tests by a constant flow of the refrigerant gases.

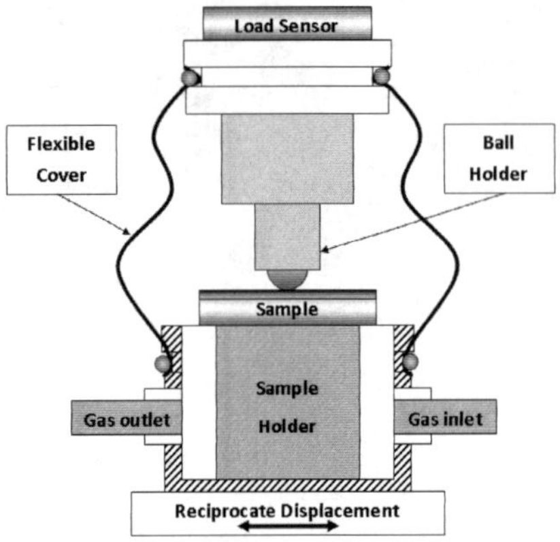

Figure 3: Semi-hermetic chamber.

The surface topography of the specimens before and after the tests was assessed using white light interferometer WYKO NT9000 and Vision software provided by Veeco®, which was used to obtain the topographic parameters, as well as measurements of wear volume of the Diamond-like Carbon coating samples, as exemplified in Figure 4.

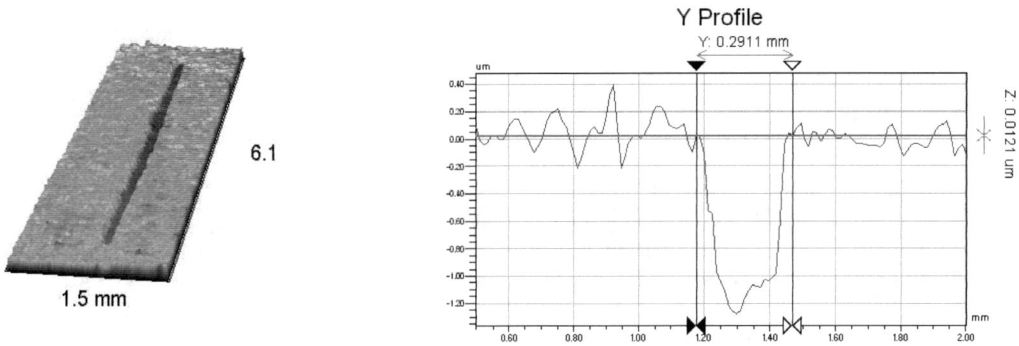

-a- -b-

Figure 4: a) 3D Topographic analysis; b) Profile of the wear scar.

The wear rates of counter-bodies were calculated measuring the wear cap diameter on the spheres using the volume of the removed material and the wear rate were calculated according to G99-95 ASTM (2010) assuming flatness of the cap.

2.2 Tribochemical Analysis

The characterization of the tribochemical layers in the wear tracks produced in the Diamond-like Carbon coatings and counter-bodies was performed using EDS mapping elementary analysis by energy dispersive spectrometer coupled to a SEM – Scanning Electron Microscope, and Raman scattering spectroscopy using a Renishaw - InVia Raman Microscope system with an Ar+ ion laser (λ=514 nm).

Recent studies showed that Raman scattering spectroscopy is a powerful and safe methodology to characterize Diamond-like Carbon films, which presents two bands in the 800 - 2000 cm^{-1} range. The D band centred at 1360 cm^{-1} and the G band at 1560 cm^{-1}. In amorphous carbon the I_D/I_G index is measured to identify the size of the sp^2 phase organized in rings. If I_D/I_G is negligible, the sp^2 phase is mainly organized in chains, or, even if rings are present, the π bonds are not fully delocalized on the rings Staturi et al (2009), Cassiraghi et al (2005a). Hydrogen content indicates the density of sp^3 fraction in the bonding parameters of Diamond-like Carbon films. The ratio between the slope, m, of the fitted linear background and the intensity of the G peak, m/I(G), can be empirically used as a measure of the bonded H content Cassiraghi et al (2005b). The calculation of H content follows the equation (Casiragui et al, 2005):

$$H(at\%) = 21{,}7 + \log\left\{\frac{m}{I(G)}[\mu m]\right\} \quad (1)$$

3. RESULTS

Figure 5-a presents the typical aspect of the multifunctional coatings. The cross section clearly shows the presence of the columnar Chromium base layer and the uppermost Si rich a-C:H.

The deposition process had little influence on surface topography, since the grinding marks before deposition were still visible, as indicated by the arrows in Figure 5-b. In fact, the grinding marks were still present, and even more evident, within the wear scars.

Figure 5: Typical aspects of the multifunctional coating. a) Cross section. b) Wear scar (top view)

Figure 6-a presents the effect of the different atmospheres on the evolution of friction coefficient with sliding distance and applied normal load.

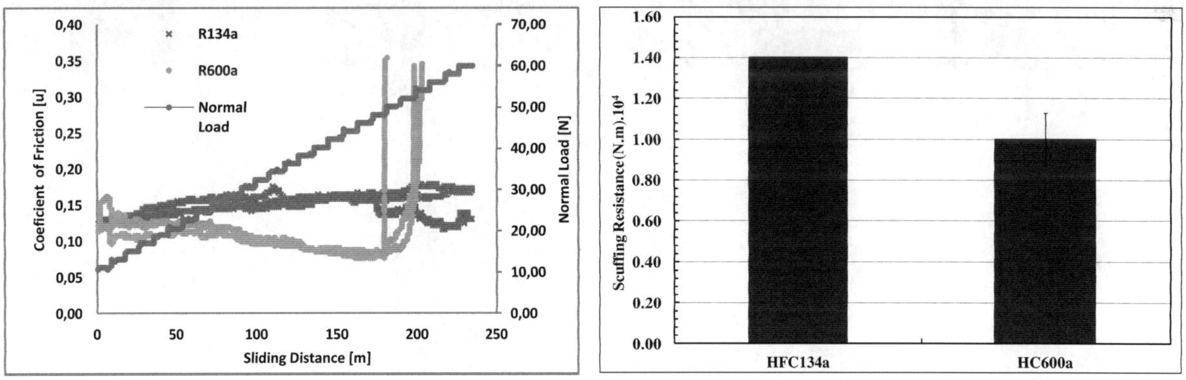

a) b)

Figure 6: Durability tests. a) Evolution of friction coefficient with sliding distance and applied normal load.; b) Surface durability.

There is a transient associated with the onset of contact between specimen and counter-body at the beginning of tests before a steady state is reached. It is reasonable to suppose that the stabilisation of the friction coefficient after the

transient is related to the generation of a protective tribolayer, where the transient corresponds to the kinetics of formation of the layer. As long as the formation rate of the tribolayer is different from the degradation rate, friction coefficient will vary. Once this formation rate is equal to or greater than the degradation rate, a steady state is reached Balu (2005), Balu (2009). The tribological tests performed with refrigerant gas R600a presented lower average steady state friction coefficient when compared to those obtained in the tests using R134a, Figure 6-a. However, in R134a atmosphere the multifunctional coatings presented higher durability (approximately 40%), Figure 6-b. It is worth noting that the R600a tests were interrupted when the coatings collapsed ($\mu \geq 0.2$), whereas the tests conducted in R134a were finalized only when the maximum load supported by the tribometer was applied.
A synthesis of the tribological behaviour (constant loads wear tests) is presented in Figure 7. It is possible to identify different results for each refrigerant gas for all parameters analyzed: friction coefficient, specimen wear rate and counter-body wear rate. Although the tests performed with R134a gas showed a higher friction coefficient (36%), they presented lower wear rates for both specimens (50%) and counter-bodies (one order of magnitude) than those conducted in R600a. In order to understand the differences in tribological behaviour the wear scars present on the specimens and counter bodies were analysed by micro-Raman spectroscopy (Figure. 8). The differences between the peak positions of the Raman spectra for the specimen wear scars are not significant, Figure 8-a and Figure 8-c. They present two typical DLC broad peaks (G-band at approximately 1572±12 cm^{-1} and D-band at 1360±5cm^{-1}). However, the intensity ratio of the D-band to G-band, I_D/I_G (I_D/I_{G134a}= 0.92±0.04; I_D/I_{GR600a}= 0.39±0.02), as well as the hydrogen content ($H\%_{134a}$= 33±4 $H\%_{R600a}$= 22±4) varies according to the atmosphere indicating tribochemical reaction of the specimens with the surrounding atmosphere. It is also noticeable that, for the case of the tests conducted under R600a environment, the spectra of the tribo layers presented in the wear scars of the specimens (Figure 8-c) or in the counter-bodies (Figure 8-d) are almost identical and indicate the beneficial presence of Diamond-like Carbon coating on both sides of the tribo pair. In this sense, the transfer of Diamond-like Carbon film to the counter-body may contribute for its consumption and its higher wear rate when compared to those presented by the tests conducted under R134a environment.

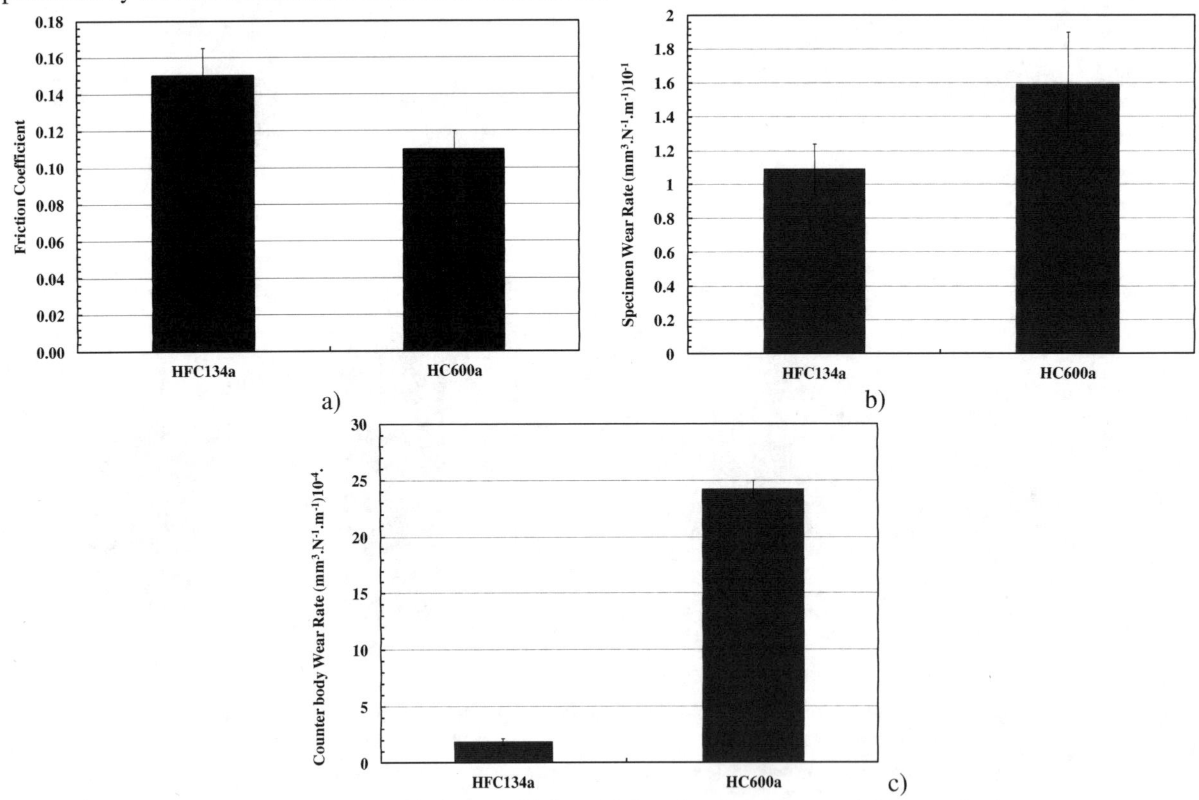

Figure 7: Constant load wear tests. a) Friction coefficient. b) Specimen wear rate. c) Counter body wear rate.

On the contrary, the spectra found in the wear track of the tests in R134a atmosphere changed according to their position in the tribo pair. In the wear scar, the spectra were similar to those found in the previous case with R600a, Figure 8-a, whereas those found in the wear marks of the counter bodies were radically different, Figure 8-b. The spectra presented smaller bands at lower frequencies. The positions of the bands match fairly well those observed for iron oxide phases De mello et al (2009) and their origin can be attributed to the formation of iron oxide by tribochemical reaction of the counter body with the surrounding and is, probably, a consequence of the high energy availability associated with high friction coefficients. Sei J. Oh et al (1998), Ouyang et al (1995a,1997b), Scharf et al (2003), De Mello et al (2009). It is reasonable to suppose that the presence of iron oxides is likely to induce higher friction coefficient. Figure 9 shows the typical SEM and elemental mapping images of the specimen wear scar for samples tested under R134a and R600a atmospheres. It is clearly visible that there was material transfer between specimen and counter-body as well as tribochemical reactions with the surrounding atmosphere. Oxygen was found on both wear scars, as determined by the distribution map of this element indicating a contamination of the chamber. In a recent paper, Barbosa et al (2015) clearly evidenced that this contamination originates from the refrigerant gases, in particular R134a. For the tests conducted under R134a atmosphere it is noticeable the presence of fluorine as indicated in the correspondent elemental map. Fluorine is adsorbed onto the surface of Diamond-like Carbon coating, forming a passivation layer with low friction coefficient and wear explaining the superior tribological performance for the tests under R134a environment. Indeed, some researchers such as Smentkowski et al (1996) and Molian (1993) developed effective ways to passivate these films, incorporating fluorine into the Diamond-like Carbon coating in the deposition step.

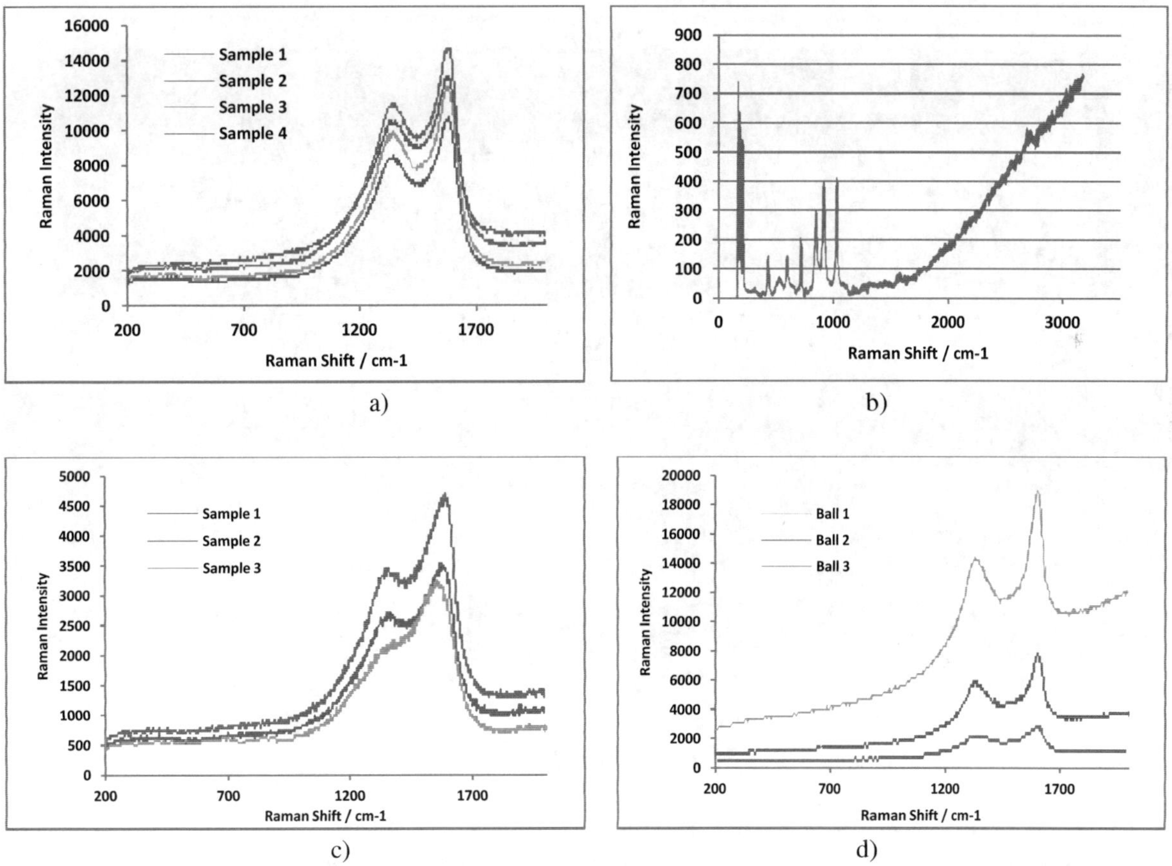

Figure 8 - Raman spectra. a) Specimen, R134a. b) Counter-body, R134a. c) Specimen, R600a. d) Counter-body, R600a.

On the other hand, the EDS elemental mapping analysis of the wear marks produced with R600a, Figure 10-b, presented carbon, silicon and oxygen, indicating only contamination of the test chamber with air.

CONCLUDING REMARKS

1. The surrounding atmosphere had a strong influence on the tribological behaviour of the multifunctional coating.
2. In R134a atmosphere the multifunctional coatings presented higher scuffing resistance for this experiment (approximately 40%). The tests performed with R134a gas although presenting a higher friction coefficient (36%) presented higher wear rate for both specimens (50%) bodies and counter-bodies (one order of magnitude) than those conducted in R600a.
3. For the case of the tests conducted under R600a environment the spectra of tribo layers in the wear scars of the specimens and in the counter-bodies are almost identical and indicate the beneficial presence of Diamond-like Carbon films on both sides of the tribo pair.
4. On the contrary, the spectra found in the wear track of the tests in R134a atmosphere changed according to their position in the tribo pair. Spectra found in wear marks of the counter bodies were radically different: They presented smaller bands at lower frequencies which were attributed to the formation of iron oxide by tribochemical reaction of the counter body with the surrounding.
5. In tests performed in an R134-a atmosphere, fluorine is adsorbed on the surface of Diamond-like Carbon coating, forming a passivation layer with low coefficient of friction and wear, explaining the superior tribological performance of the R134a tests

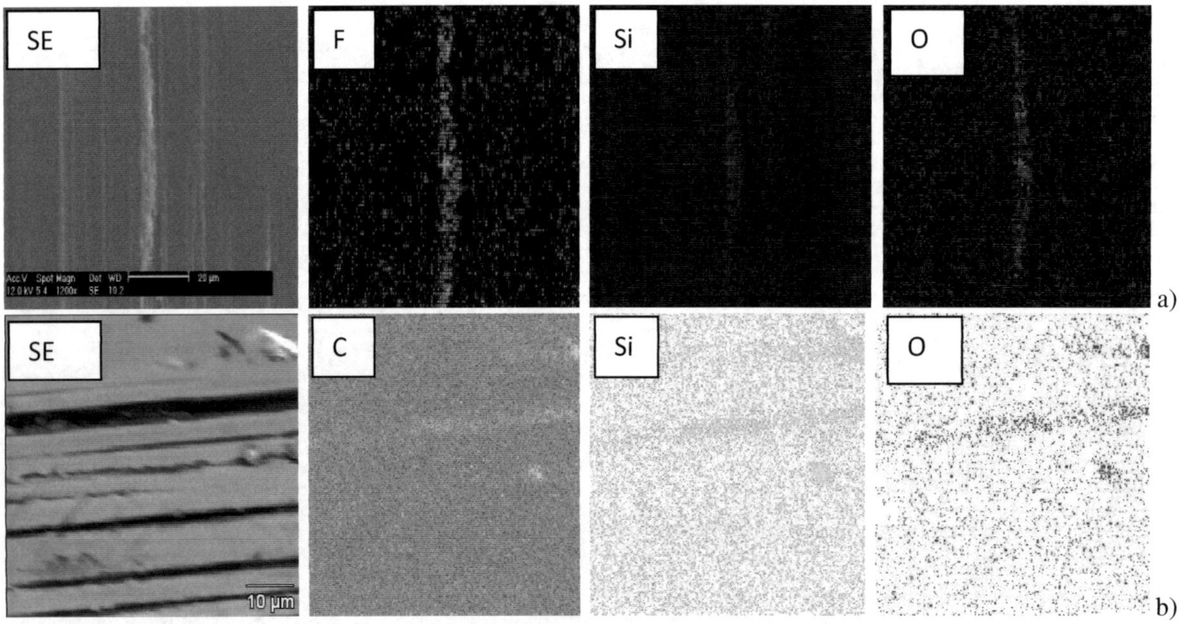

Figure 9: Elemental mapping of wear scars. a) R134a. b) R600a.

REFERENCES

ASTM G99 - 05(2010) Standard Test Method for Wear Testing with a Pin-on-Disk Apparatus.

Barbosa, M.V, Hammes, G., Binder, C., Klein, A.N., de Mello, J.D.B. Physicochemical characterisation of tribolayers by micro-Raman and GDOES analyses. Tribology International, (2015), v.81, pp.223 - 230.

Blau, P.J., "Friction science and technology: From Concepts to Applications", (2009), 2nd ed., CRC Press, 400p.

Blau, P.J., , "On the nature of running-in", Tribology International, (2005)38, pp. 1007-1012.

Casiraghi, C.; Ferrari, A. C.; Robertson, J. Raman spectroscopy of hydrogenated amorphous carbons. PHYSICAL REVIEW B 72, 085401. 2005a.

Casiraghi, C.; Piazza, F.; Ferrari, A.C.; Grambole, D.; Robertson, J. Bonding in hydrogenated diamond-like carbon by Raman spectroscopy. Diamond & Related Materials v. 14, p. 1098– 1102. 2005b.

Chandraseka, S. and Bhushan, B. The role of environment in the friction of diamond for magnetic recording head applications, Wear, 153, 79-89. 1992.

De Mello, J. D. B.; Binder, R. A Methodology to Determine Surface Durability in Multifunctional Coatings Applied to Soft Substrates. Tribology international, v. 39, p. 769-773. 2006.

De Mello, J. D. B.; Binder, R.; Demas, N.G.; Polycarpou, A. A. Effect of the actual environment present in hermetic compressors on the tribological behaviour of a Si-rich multifunctional DLC coating. Wear, v 267, p. 907–915. 2009

Demas, N.G., Polycarpou, A.A., Tribological investigation of cast iron air-conditioning compressor surfaces in CO_2 rfrigerant, Tribol Lett (2006), pp. 271–278

Donnet, C., Recent progress on the tribology of doped diamond-like and carbon alloy coatings: a review, Surf Coat Technol (1998), pp. 180–186

Dugger, D.; Peebles, E.; Pope, L.E. Counterface material and ambient atmosphere: role in the tribological performance of diamond films, in Surface Science Investigations in Tribology, Experimental Approaches, Chung, Y.-W., Homolo, A.M., and Street, G.B. (Eds.), ACS Symposium Series 485, American Chemical Society, Washington, D.C., 72-102. 1992.

Erdemir, A. Bindal, C. Pagan, J. Wilbur, P., Characterization of transfer layers on steel surfaces sliding against diamond-like hydrocarbon films in dry nitrogen, Surf Coat Technol (1995), pp. 559–563

Erdemir, A. Design criteria for super lubricity in carbon films and related microstructures. Tribology International, v. 37, p. 577-583, 2004.

Grill, A., Review of the tribology of diamond-like carbon, Wear (1993), pp. 143–153

Hayward, I.P.; Field, J.E. Friction and wear of diamond, Int. Conf. On Tribol., 50 Years on, IMechE, C159/87, 205-209. 1987.

Lifang, X., Guang, L., The frictional behavior of DLC films against bearing steel balls and Si3N4 balls in different humid air and vacuum environments, Wear (2008), pp. 1077–1084

Luo, Q., Zhou, Z., Rainforth, W.M., Bolton, M., Effect of tribofilm formation on the dry sliding friction and wear properties of magnetron sputtered TiAlCrYN coatings ,Tribol Lett (2009), pp. 113–124

Miyake, S.; Miyamoto, T.; Kaneko, R.; Miyazaki, T. Macro- and micro-tribological properties of polished CVD diamond films and trial processing diamond, IEICE Trans. Electron., E78, 180-185. 1995.

Miyoshi, K.; Wu, R.L.C.; Garscadden, A.; Barnes, P.N.; Jackson, H.E. Friction and wear of plasma-deposited diamond films, J. Appl. Phys., 74, 4446-4450. 1993.

Molian, P.A.; Janvrin, B.,; Molian, A.M. Laser chemical vapor deposition of fluorinated diamond thin films for solid lubrication, Wear, 165, 133-140. 1993.

Ouyang, M.; Hiraoka, H. Preparation and characterization of iron oxide films by the excimer laser ablation of poly(ferric methacrylate). Materials Science and Engineering B34. p. 188-191.1995.

Ouyang, M.; Hiraoka, H. Structure and magnetic properties of iron oxide films deposited by excimer laser ablation of a metal-containing polymer. Materials Research Bulletin.Vol.32.No.8. 1099 - 1107.1997.

Robertson, J., Diamond-like amorphous carbon, Mater Sci Eng (2002), pp. 129–281

Ronkainen, H, Likonen, J, Koskinen,J., Varjus S.,Effect of tribofilm formation on the tribological performance of hydrogenated carbon coatings, Surf Coat Technol (1996), pp. 87–94

Scharf, T.W.; Singer, I.L., Thickness of diamond-like carbon coatings quantified with Raman spectroscopy. Thin Solid Films 440. p. 138–144. 2003.

Sei J. Oh, D.C. Cook and H.E. Townsend. Characterization of Iron Oxides Commonly Formed as Corrosion Products on Steel. Hyperfine Interactions Journal, V. 112, p. 59-66, 1998.

Smentkowski, V.S; Yates, J.T., Jr. Fluoroalkyl iodide photodecomposition on diamond (100) — An efficient route to the fluorination of diamond surfaces, MRS Symp. 416, 293-298. 1996.

Statuti, R.P.C.C.; Radi, P.A.; Santos, L.V.; Trava-Airoldi; V.J.. A tribological study of the hybrid lubrication of DLC films with oil and water. Wear 267 p. 1208–1213. 2009.

Viana, G.A., Motta, E.F. da Costa, M.E.H.M. Freire Jr., F.L. Marques, F.C., Diamond-like carbon deposited by plasma technique as a function of methane flow rate, Diam Relat Mater (2010), pp. 756–759

Wear Volume Equations, Appendix 1, Wear Control Handbook, American Society of Mechanical Engineers, p. 449-457. 1980

ACKNOWLEDGEMENT

The authors are grateful to the Federal University of Santa Catarina for the collaboration programme and the BNDES – Brazilian National Bank for Economic and Social Development for sponsoring this study.

Lubrication Analysis of Journal Bearings in R410A Rotary Compressor

Yongjun Fu[1], Xingbiao Zhou[1]*, Hong Guo[1], Peipei Mei[1]

[1]* Research and Development Center, Guangdong Meizhi Compressor Limited, Shunfengshan Industrial Development Zone, Foshan city, GuangDong Province, 528333, P.R. China
zhouxb3@chinagmcc.com

ABSTRACT

Understanding the state of lubrication in rotary compressors is of paramount importance for their reliability. This paper refers to an improvement of the lubrication in R410A rotary compressor. The lubricating conditions on the journal bearings can be evaluated by metallic contact that consists of measuring the electrical resistance between lubricated surfaces. The contact resistance can be used to indicate the degree of metal-to-metal contact between the sliding surfaces in the compressor. And some parameters of them are numerically analyzed. It is elucidated from the investigation that when unfavorable condition occurs, the metal contact between the journal and bearings become severe. And lubricating condition on the bearings in the compressor using high viscosity oil is better than that for the compressor using low viscosity.

1. INTRODUCTION

As alternative refrigerant of R22, R410A rotary compressors are widely used in room air conditioning units. Fig.1 shows the cross section of the rotary compressor. The rotary compressor is a high side compressor with the lubricant subjected to discharge conditions. And the crankshaft is under a large dynamic load including gas force and unbalanced mass forces of comprising mechanical parts (crank, roller, and balancers on motor rotor).As a result, lubrication conditions at bearings would become severe owing to the deformation of crankshaft. Therefore, to ensure rotary compressor operating well, the dynamic behavior of crankshaft at various conditions should be analyzed (Fei Xie *et al.*, 2006) (Zengli Wang *et al.*, 2013) (Haifeng Zhang *et al.*, 2014). The discussions of the lubrication problem, which were reported previously (Masao Ozu *et al.*, 1980) (Satoru Fujimoto *et al.*, 1984) (Takashi Fukuda *et al.*, 1996), have become increasingly important in designing of the rotary compressor.

In this paper, the dynamic behavior of the crankshaft for R410A rotary compressor were analyzed considering both unbalanced mass force and gas force. The influence of oil viscosity on the reliability is investigated. And we have tried to solve the problem with a method of electrical resistance between the lubricated surfaces.

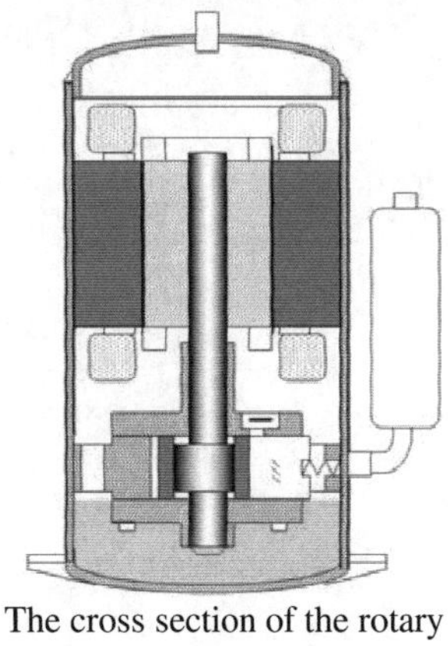

Figure 1: The cross section of the rotary compressor

2. NUMERICAL ANALYSIS

The loads acting on the crankshaft are load acting on the crank due to refrigerant pressure difference, centrifugal forces produced by the shaft itself and other attached parts such as upper and lower balance mass and a magnetic attractive force generated between the rotor and stator. And the shaft is supported by support forces of the bearings. At a rotation angle, the shaft is supposed to deform due to these forces.

2.1 Load Acting On The Crank

The crank is supported by the roller as shown in Fig.2a. The loads acting on the roller are defined as following (Katsuya, M.*et al.*, 1998)

$$F_x = F_{gx} + F_{vx} + F_{ux} \tag{1}$$

$$F_y = F_{gy} + F_{vy} + F_{uy} \tag{2}$$

Where F_{gx} and F_{gy} are the gas forces, F_{ux} and F_{uy} are the unbalanced forces, F_{vx} and F_{vy} are the forces at the contact point with the vane. F_{vx} and F_{vy} are defined as following(Katsuya, M.*et al.*, 1998).

$$F_{vx} = F_{sx} + F_{px} + F_{fx} \tag{3}$$

$$F_{vy} = F_{sy} + F_{py} + F_{fy} \tag{4}$$

Where F_{sx} and F_{sy} are the vane spring forces, F_{px} and F_{py} are the forces produced by the difference of gas pressures around the vane, F_{fx} and F_{fy} are the frictional forces between the vane and the vane slot as shown in Fig.2b.

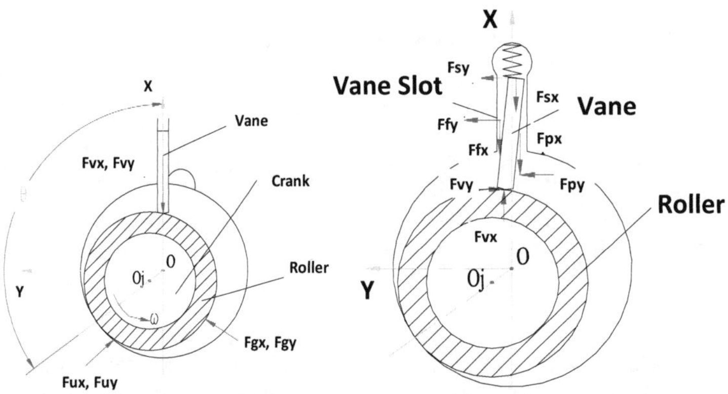

(a) The loads acting on the roller. (b) The loads acting on the vane.

Figure 2: Cross section of the cylinder

2.2 Support Force Of The Bearing

The dynamic bearing loads of the main bearing and sub bearing requires numerical solution to the Reynolds equations based on the finite length journal bearing theory as given by the following equations (Fei Xie *et al.*, 2006) (Haifeng Zhang *et al.*, 2014).

$$\frac{1}{R^2}\frac{\partial}{\partial\theta}\left(\frac{h^3}{\mu}\frac{\partial P_b}{\partial\theta}\right)+\frac{\partial}{\partial z}\left(\frac{h^3}{\mu}\frac{\partial P_b}{\partial z}\right)=6\frac{U}{R}\frac{\partial h}{\partial\theta}+12\frac{\partial h}{\partial t} \tag{5}$$

$$\begin{Bmatrix} F_{bx} \\ F_{by} \end{Bmatrix} + R\int_0^L \int_{\theta_1}^{\theta_2} P_b \begin{Bmatrix} \cos\theta \\ \sin\theta \end{Bmatrix} d\theta dz = 0 \tag{6}$$

Where P_b is the oil pressure between the bearing (the main bearing or sub bearing) and the shaft. It also means the bearing pressure. And the bearing forces F_{bx} and F_{by} are calculated by equation (6).

3. EXPERIMENT

As shown in Fig.1, there are two bearings in rotary compressor, such as main bearing and sub bearing. The two electrodes are fitted in the bearings to detect the contact. The position of the electrodes is decided on the basis of the system dynamic analysis and several reliability test results. In this evaluation, the electrical resistance change between sliding parts is monitored as the potential drop. The test system is shown in Figure 3. And Figure 4 explains the method. The electrodes are metallic pin, fitting in the bearings while insulated from the bearing around it. Electrical resistance of oil film between lubricated surfaces can be measured when the two surfaces are electrified with direct current. Differences in resistance caused by metallic contact between sliding surfaces, can be analyzed from the output voltage. When the lubricating condition is good enough, there is hydrodynamic lubricating oil film that can insulate between the bearing and crankshaft. However, as the condition becomes worse and the oil film is broken, the electrode and the crankshaft make direct contact, and the potential of electrode becomes the same as that for other parts. So, by monitoring the electrode potential as the voltage output, the lubricating condition changes during the operation can be monitored instantly (Masao Ozu *et al.*, 1980) (Satoru Fujimoto *et al.*, 1984) (Takashi Fukuda *et al.*, 1996).

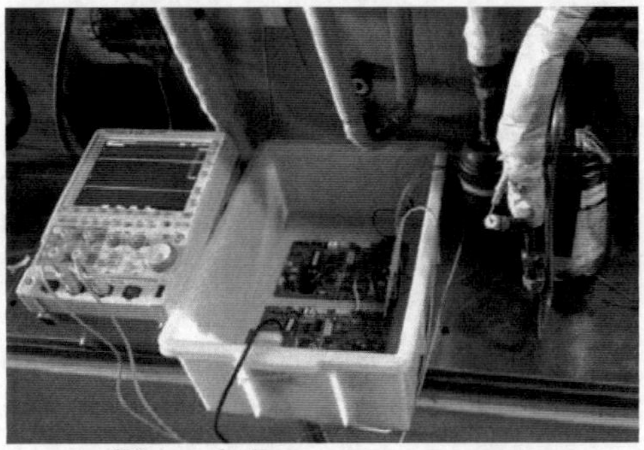

Figure 3: Test system

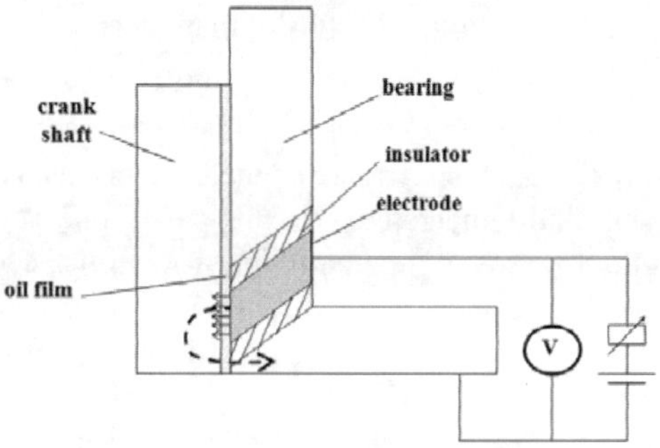

Figure 4: Electrical contact detection system

4. RESULTS AND DISCUCCION

4.1 Numerical results

The dynamic behaviors of the crankshaft in R410A rotary compressor are analyzed in this paper. The suction pressure is 1.2 MPa and the discharge pressure is 4.4 MPa. As a result of the elasticity of crankshaft, wear-out between journal and bearing usually occurs on four vertical planes, namely MBT (main bearing top), MBB (main bearing bottom), SBB (sub bearing bottom) and SBT (sub bearing top), as shown in Fig. 5. Besides, the impact between motor rotor and motor stator may occur at the section of MTop (the top rotor motor).

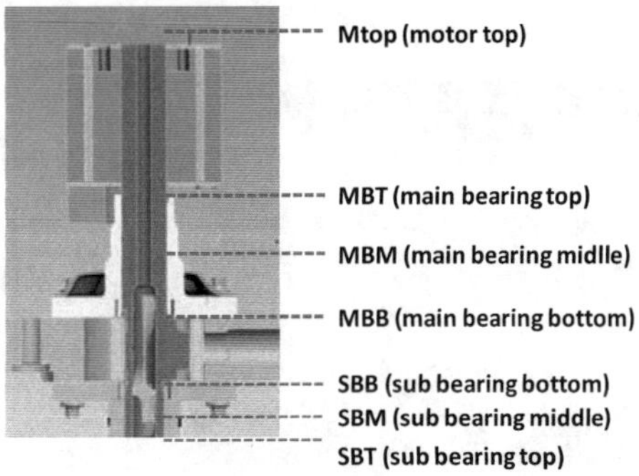

Figure 5: Vertical planes of the shaft

Journal orbit plays a key role in the dynamics of rotary compressor. The Journal orbits of the bearings are shown in Fig. 6. The journal orbits at SBB and MBB are relatively larger. And the wear-out may occur in these sections where oil film is much thinner.

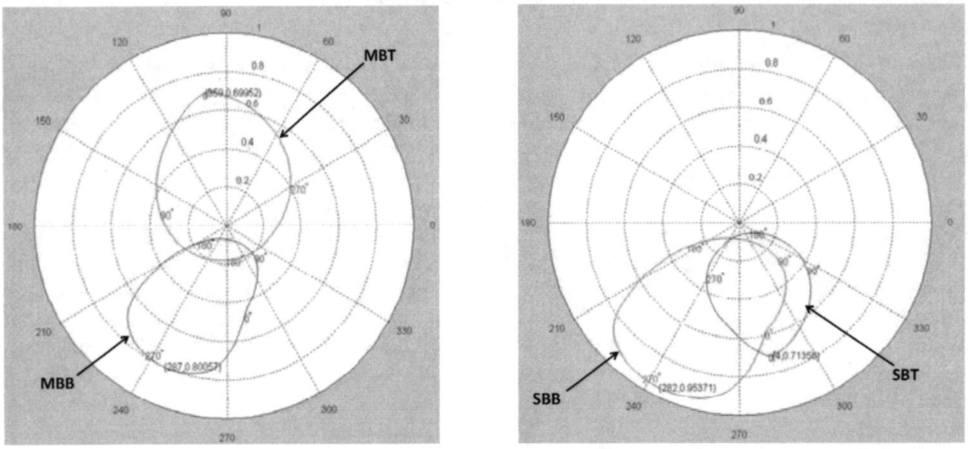

(a) Journal orbits of main bearing (b) Journal orbits of sub bearing

Figure 6: Journal orbits of the bearings

And the analysis was examined under two kind of oil. The viscosity of oil is given by discharge pressure and temperature. In this paper, one is 3 cP. Another is 10 cP. Table 1 shows the minimum oil film thicknesses of the bearings. At low oil viscosity of 3 cP, the wear-out may occur in SBB where oil film is much thinner.

Table 1: Minimum oil film thicknesses

Viscosity/cP		3	10
Main bearing	Top	44%	100%
	Bottom	29%	74%
Sub bearing	Bottom	6%	48%
	Top	37%	81%

4.2 Experiment results

A test model ran under the condition (the suction pressure is 1.2 MPa and the discharge pressure is 4.4 MPa). The metallic contact signals change during the test of MBB and SBB are shown in Fig. 7. It shows that the condition becomes worse and the oil film is broken, the electrode and the crankshaft make direct contact. And it may indicate that the lubricating oil film condition on MBB is a little better than that on SBB. Because the voltage drop of SBB is larger than MBB. It agrees with the numerical results shown in Fig. 6 and Table 1.

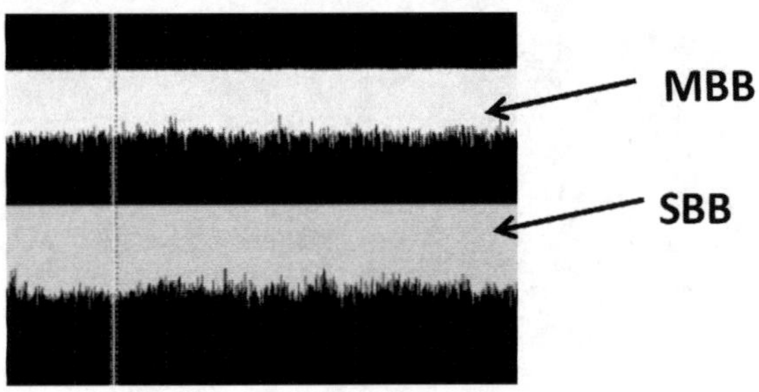

Figure 7: Metallic contact in MBB and SBB

5. CONCLUSIONS

In this paper, the dynamic behaviour of the crankshaft in R410A compressor has been studied. The vulnerable sections of crankshaft vary with the viscosity of oil. At low oil viscosity, the wear-out may occur in SBB. And the lubricating conditions on the journal bearings were evaluated by metallic contact that consists of measuring the electrical resistance between lubricated surfaces. The experiment results show that the condition of MBB and SBB becomes worse and the oil film is broken. It agrees with the numerical results.

REFERENCES

Xie, F., et al., 2006. Dynamic analysis of a rotor-journal bearing system of rotary compressor. In: *Proceedings of International Compressor Engineering Conference at Purdue.*

Wang, Zengli, et al., 2013. Dynamic analysis for the rotor-journal bearing system of a variable speed rotary compressor. *Int. J. Refrigeration*, vol. 36, 1938 - 1950.

Haifeng Zhang et al., 2014. Dynamic behaviors of the crankshafts in single cylinder and twin-cylinder rotary compressors. *Int. J. Refrigeration*, vol. 47, 36 - 45.

Katsuya, M.et al., 1998. Lubrication Analysis of Rotary Compressors for HFC Refrigerants. In: *Proceedings of International Compressor Engineering Conference at Purdue.*

M Kalin, I Velkavrh, 2009, The Stribeck curve and lubrication design for non-fully wetted surfaces, *Wear*, vol. 267, p. 1232–1240.

Dufour, et al., 1998. Dynamics prediction of refrigerant rotary compressor crankshaft. In: *Proceedings of International Compressor Engineering Conference at Purdue.*

T Hirayama, K Miura, 2006. Numerical analysis for mixed lubrication in journal bearings of rotary compressors. In: *Proceedings of International Compressor Engineering Conference at Purdue.*

Yoko Kitsunai, Masaru Matsui, et al. 2010. High efficiency development of a rotary compressor by clarification of its shaft dynamic motion. In: *Proceedings of International Compressor Engineering Conference at Purdue.*

H Hattori, N Kawashima, 1990. Dynamic analysis of a rotor-journal bearing system for twin rotary compressors. In: *Proceedings of International Compressor Engineering Conference at Purdue.*

Masao Ozu, Tsugio Itami, 1980. Some electrical observations of metallic contact between lubricated surfaces under dynamic conditions of rotary compressor, *Proceedings of International Compressor Engineering Conference at Purdue*, pp. 105-111.

Satoru Fujimoto, Katsumi Sakitani, Masahiro Watada, 1984. Tribology analysis in rolling piston type compressor. *Proceedings of International Compressor Engineering Conference at Purdue*, pp. 377-382.

Takashi Fukuda, Makoto Hayano, 1996, HFC/POE Lubricity Evaluation on the Rotary Compressor in System Operation, *Proceedings of International Compressor Engineering Conference at Purdue*, pp. 121-126.

Oil Flow Measurement at the Compressor Discharge

Jiu XU[1], Pega HRNJAK[1,2,*]

[1]University of Illinois at Urbana-Champaign, Mechanical Science and Engineering, Urbana, IL, USA

[2]CTS, Creative Thermal Solutions, Urbana, IL, USA

jiuxu2@illinois.edu, pega@illinois.edu
[*]Corresponding author

ABSTRACT

Oil circulating in the vapor compression system has a significant impact on the system performance. This paper presents a method to calculate the oil retention and oil circulation ratio (OCR) based on oil flow parameters, including oil film thickness and velocity, oil droplet size and velocity, and system mass flow rate. Oil flow parameters are determined by visualization using high-speed camera and video processing program. The estimated oil retention and oil circulation ratio results are compared with the experimental sampling results under different compressor speed. The agreement between video results and sampling results verifies the accuracy of the optical method, which can be applied in other annular-mist flow analysis. It also shows that most of the oil exists in film by mass while oil droplets contributes more to the oil mass flow rate because oil droplets travel in a much higher speed.

1. INTRODUCTION

Oil presence in the vapor compression system ensures the lubrication of the compressor moving parts, evacuation of the heat generated by frictions and sealing between the compression. Oil also affects the performance of the system besides the compressor. Despite few researchers indicating small amount of lubricant has positive effect in some cases on heat transfer (Nidegger et al., 1997)(Zürcher et al., 1998)(Kim and Hrnjak, 2012)(Hrnjak and Kim 2013), refrigerant distribution or capacity (Li and Hrnjak, 2014) (Li and Hrnjak, 2015), adverse effects of oil on system performance has been identified by many other researchers (McMullan et al.,1992)(DeAngelis and Hrnjak, 2005a)(DeAngelis and Hrnjak, 2005b).

The origin of oil circulating in the system is the compressor so research on oil flow starts with the compressor discharge. Toyama et al. (2006) concluded that oil flow inside scroll compressor is mainly mist flow. The mean diameter of the oil droplets decreases as the flow speed of the refrigerant gas increases. Zimmerman and Hrnjak (2014) showed with visualization that the form of the flow in the compressor plenum is misty. Oil enters the discharge chamber during the opening and closing process of the discharge valve. Therefore, discharge valve is the gateway for the oil to leave the compressor. Wujek and Hrnjak (2011) showed the form of the oil-refrigerant flow is developing mist-annular flow in the compressor discharge tube. The oil mist generated by the periodic movement of the discharge valve flows to the discharge tube. Oil droplets may collide, coalesce or splash during the process and then part of oil is entrained by the high gas flow rates. Therefore, annular mist flow in a horizontal tube consists of two forms of oil: oil film spread around the inner wall and oil droplets in the vapor core.

With the importance of oil flow in the refrigeration systems, it raises the interest to develop different techniques to quantify the oil flow parameters. Wada et al. (1992) used a novel approach to measure the in-line OCR based on the ultra-violet light absorption behavior of the oil. Fukuta et al. (2004) used reflective index of oil-refrigerant mixture to measure the oil concentration of the mixture. Peuker and Hrnjak (2010) presented three different techniques to determine the oil distribution in the major components of a R134a automotive system.

Xu and Hrnjak (2016) measured the OCR in a non-invasive way based on the real-time images and videos of the refrigerant-oil mixture in the compressor discharge tube of an automotive AC system. The main purpose of this research is quantifying the oil flow with developed visualization techniques and verifies the method by experimental results for both swashplate compressor and scroll compressor. Furthermore, in order to ensure the reliability of compressor and reduce the negative effect of oil on the system performance, it is important to separate the oil from the refrigerant vapor and drain it back to the compressor. A better understanding of flow parameter of oil at the compressor discharge can provide guidelines for oil separator design.

2. FACILITIES

Both scroll compressor (for residential systems) and swashplate compressor (for automotive systems) driven by variable frequency drive motor are used as research objectives. PVE 32 oil is used for the scroll compressor while PAG 46 is used for the swashplate compressor. The refrigerant used for both systems is R134a. As shown in Figure 1, two environmental chambers maintain the desired pressure and temperature conditions at condenser and evaporator. In order to achieve different system mass flow rate as well as vapor velocity at compressor discharge, the motor speed is varied from 1200~4000 rpm. The compressor pressure ratio is maintained the same (around 3) through the experimental by varying the working condition at condenser and evaporator. The mass flow rate of the system is measured by a Coriolis-effect flow meter.

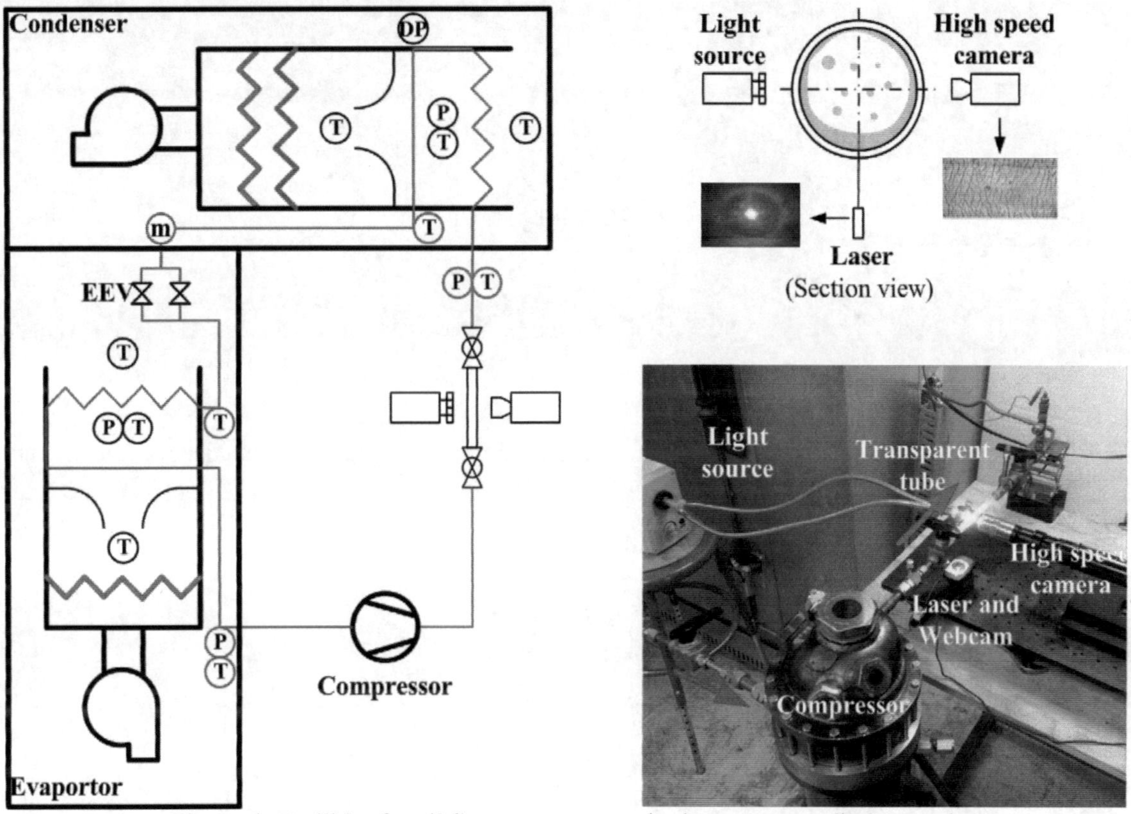

Figure 1: Facilities for oil flow measurement in the compresor discharge tube

A section of perfluoroalkoxy alkane (PFA) tube is used as the transparent observation window at the discharge of the compressor. The inner diameter of the PFA tube is 6 mm (1/4 inch) while the wall thickness is 1.5 mm (1/16 inch), which is the same as the copper tube size used for the compressors. For video capture, a Phantom high speed camera is used to catch the flow image at the compressor discharge. An optical platform offers the precise adjustment of horizontal and vertical position of the camera. The resolution of the position adjustment is 0.02 mm. Halogen lamps are used for back-light of the video, which can offer enough brightness while prevent adding extra heat to the flow.

To measure the oil retention by sampling, two valves are used to cut off the flow at a steady working condition. The compressor was shut down at the same time as the two valves are closed. The section of transparent tube is then taken out of the system and weighed by an analytical balance with a readability of 0.1 mg. Another refrigerant and oil mixture sample is taken in the liquid line (between condenser outlet and throttling valve) to measure the OCR. Both oil retention and OCR sampling results are compared to the video results under the same working condition.

3. METHODS

3.1 Analysis of oil retention and OCR

Refrigerant-oil mixture flow in the compressor discharge tube is mist-annular flow (Wujek and Hrnjak, 2011). In order to quantify the oil retention and OCR, oil droplets and oil film need to be considered separately. Oil retention is a measurement of mass staying in a certain section while OCR is a measurement of mass flow rate circulating in the system. The relation between these two parameters and different physical quantities are shown in Figure 2.

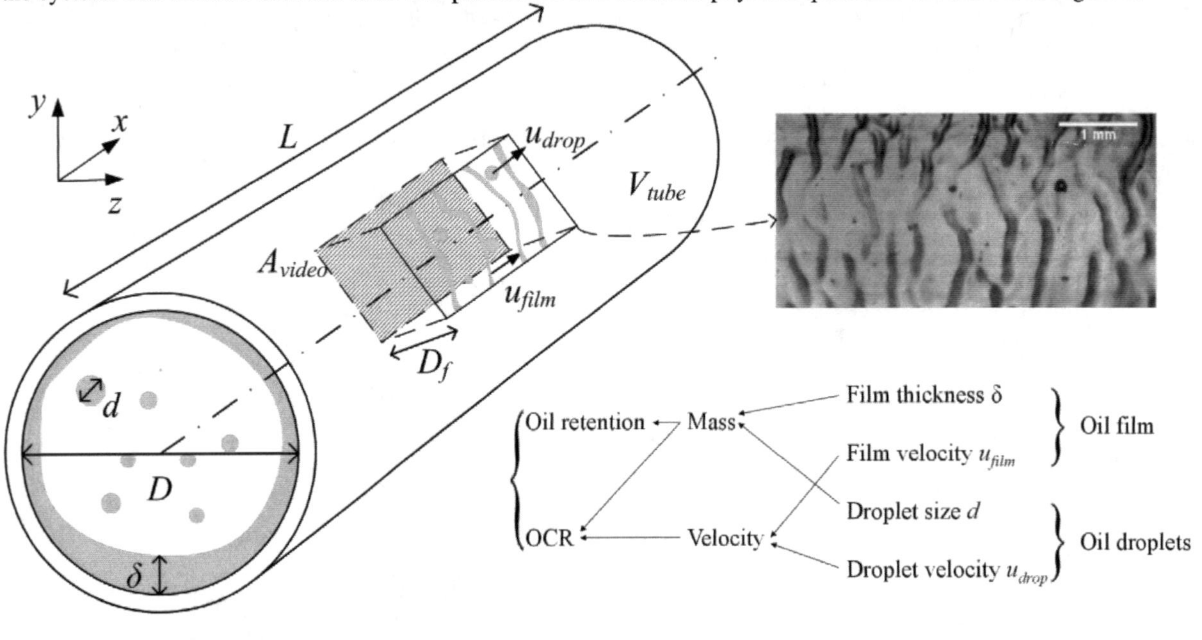

Figure 2: Annular-mist flow in compressor discharge tube and break-down of the contrubutions of film in annulus and droplets in mist on oil retention and OCR

Oil retention in the form of the film can be calculated as

$$M_{film} = \rho_{oil} V_{film} = \rho_{oil}\pi(D-\delta)\delta L \tag{1}$$

Similarly, OCR contributed by the oil film can be calculated as

$$OCR_{film} = \frac{\dot{m}_{oil}}{\dot{m}_{oil} + \dot{m}_{ref}} = \frac{\rho_{oil}\pi(D-\delta)\delta u_{film}}{\dot{m}_{oil} + \dot{m}_{ref}} \tag{2}$$

where ρ_{oil} is the density of the oil, D is the inner diameter of the tube, δ is the average thickness of the oil film, L is the length of the tube, u_{film} is the average velocity of the oil film, $\dot{m}_{oil}$ and $\dot{m}_{ref}$ are the mass flow rate of oil and mass flow rate of refrigerant, respectively.

In the same way we can derivate the equations for oil retention and OCR in the form of droplets. Since the video only represents a small part of the tube, the ratio of tube volume over video volume needs to be multiplied to the droplet totally volume acquired by the video. For the video control volume, the projection area is easy to estimate but the depth of focus D_f, is determined by the method which will be discussed in 3.4. In short, the total volume of droplets in the tube is the sum of the volume of each droplet; the total mass flow rate contributed by oil droplets is the sum of the mass flow rate of each droplet.

$$M_{drop} = \rho_{oil} V_{drop,video} \frac{V_{tube}}{V_{video}} = \frac{\pi^2 \rho_{oil} D^2 L \sum d^3}{24 A_{video} D_f} \tag{3}$$

$$OCR_{drop} = \frac{\dot{m}_{drop}}{\dot{m}_{oil} + \dot{m}_{ref}} = \frac{\rho_{oil} V_{drop,video}}{\left(\dot{m}_{oil} + \dot{m}_{ref}\right) \Delta t_{drop}} \frac{V_{tube}}{V_{video}} = \frac{\pi^2 \rho_{oil} D^2 \sum (d^3 u_{drop})}{24(\dot{m}_{oil} + \dot{m}_{ref}) A_{video} D_f} \quad (4)$$

where d is the equivalent diameter of the droplets recognized by the video processing, u_{drop} is the velocity of oil droplets, D_f is the depth of focus (will be discussed in 3.4), A_{video} is the video projection area.

From equation (1)-(4), it can be concluded that we need determine the oil film thickness and velocity, oil droplet diameter and velocity in order to calculate the oil retention and OCR from the video. The details of these measurements will be discussed in the following sections.

3.2 Oil film thickness

Shedd and Newell (1998) used an optical method to determine the liquid film thickness of annular flow in a horizontal tube. Xiao and Hrnjak (2015) used the same method to estimate the liquid film thickness for refrigerant condensation in a horizontal tube. The method utilized the light reflected from the surface of a liquid film flowing over a transparent wall to generate an image. The image can be processed to determine the position of reflected light rays and calculate the film thickness without intruding into the tube.

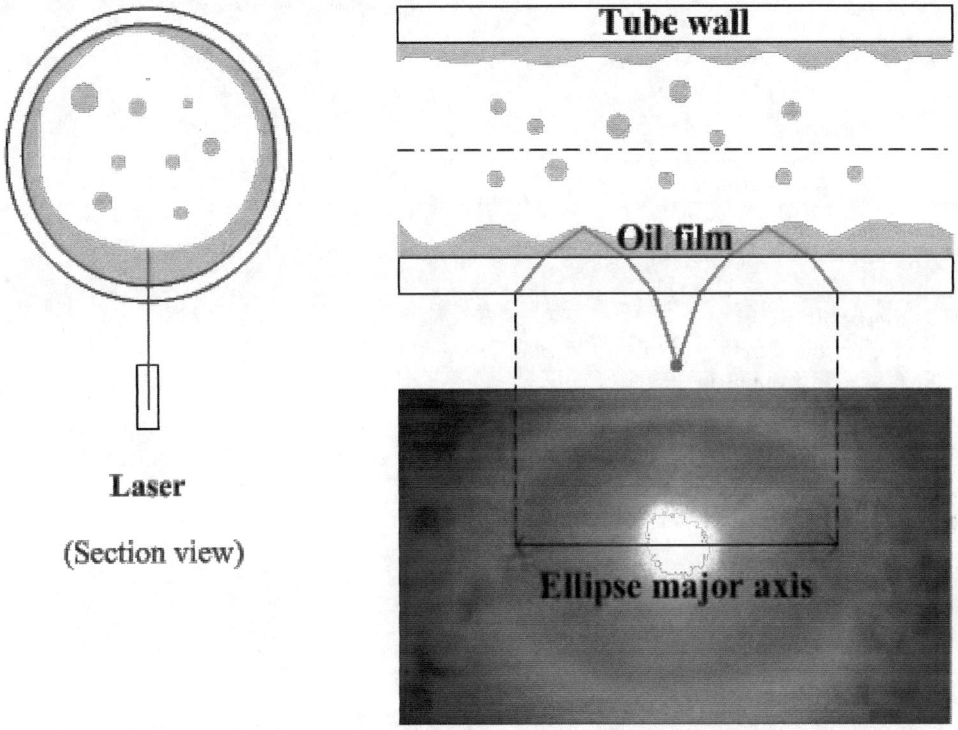

Figure 3: Total reflection method for oil film thickness measurement in a round tube

As shown in Figure 3, a laser beam shoots to the transparent tube wall to generate an ellipse pattern. The ellipse pattern is formed by the reflection and deflection of the laser on curved surfaces and oil-tube, oil-refrigerant interfaces. A webcam provides the image of ellipse pattern. Numerical model shows that with oil film thickness inside the tube getting larger, the size of the ellipse will grow almost linearly. The angle between the laser beam and the camera can never be zero but the error caused by a small camera view angle can be neglected.

Calibration is necessary in order to get the correlation between the oil film thickness and the ellipse major axis. A half tube is used to hold the oil whose film thickness is measured by a point tip micrometer. The difference between a half tube and a full tube is negligible because the reflection and deflection from the upper part has a minor impact on the image pattern. Therefore, the calibration result of a half tube is considered as the same with that of a full tube. Multiple data points at different liquid level can fit into a linear curve. It is important to mention that the relative error of oil film thickness measurement may be considerable when the thickness is very close to zero.

Oil film circumferential distribution inside horizontal tube is considered to be uneven because of the effect of gravity and surface tension. Hurlburt and Newell (2000) developed a model to predict the circumferential film thickness distribution in horizontal annular gas-liquid flow. The model fits well with experimental results in our case. The ratio between oil film thickness at different circumferential location and oil film thickness at bottom of the tube is shown in Figure 4. Higher superficial gas velocity will lead to a more uniform film distribution. With a relatively high superficial gas velocity in our case, it is reasonable to use the average of oil film thickness measured at θ=0°, 60°, 120°, 180° to estimate the volume of oil film in the tube.

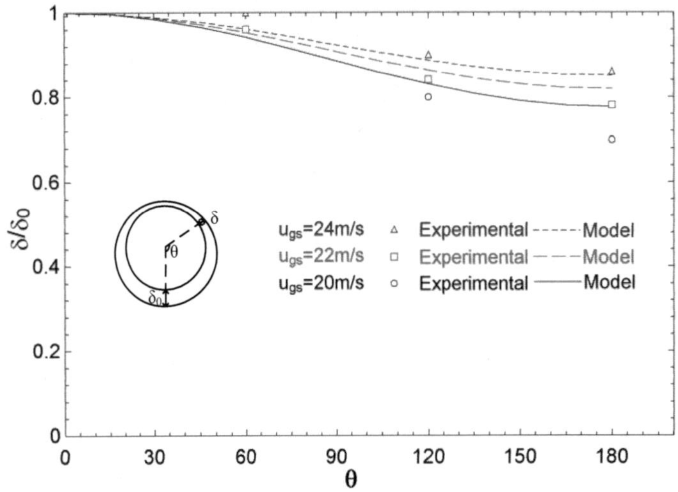

Figure 4: The circumferential distribution of oil film in the horizontal tube

3.3 Oil film velocity

From the video taken at the transparent tube, it is clearly shown that oil film with wavy pattern spread on the tube inner wall and the discrete droplets in the vapor core.

In order to calculate the mass flow rate contributed by the oil film, the oil film average velocity needs to be determined. Since the oil is viscous, the axial film velocity is not the same along the radical direction of the tube. Hurlburt and Newell (2000) developed a model for horizontal annular flow to estimate the average velocity of the liquid film by mass and momentum conservation. In this research, as shown in Figure 5, average axial velocity in the film is a function of gas superficial velocity and oil film thickness. Therefore, the oil mass flow rate contributed by the oil film can be calculated using equation (2).

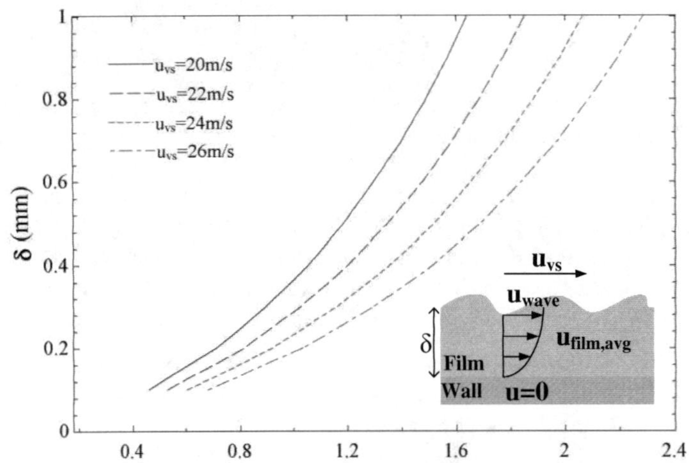

Figure 5: Average oil film velocity given by annular mist flow model in the horizontal tube

3.4 Oil droplets size measurement

From the same video taken at the discharge tube, the size and velocity of oil droplets in the vapor core can be calculated from video processing. Wujek and Hrnjak (2011) developed a video processing program which can eliminate the droplet image based based on the steepest grayscale gradient method.

The basic idea of the steepest grayscale gradient method is to determine the boundary of droplets by the sharp grayscale transition. 8-bit grayscale allows 256 different color intensities which defines pure black as 0 while pure white as 255. Therefore, one frame of the video can be converted to a matrix where each element represents the pixel grayscale.

The grayscale gradient in a 2D picture is defined as

$$\nabla G = \sqrt{\left(\frac{\partial G}{\partial x}\right)^2 + \left(\frac{\partial G}{\partial y}\right)^2} \tag{5}$$

For an image matrix shown in Figure 6, the value of grayscale is shown in each grid element. The boundary detected by the video processing is where steepest grayscale gradient takes place (red line in Figure 7).

One of the limitations of grayscale gradient is that the video processing cannot tell if the droplet is in-focus or not. As shown in Figure 6, if we adjust the distance between the camera and a known size particle, the image captured by the camera will be blurry when it is too close to or too far away from the particle.

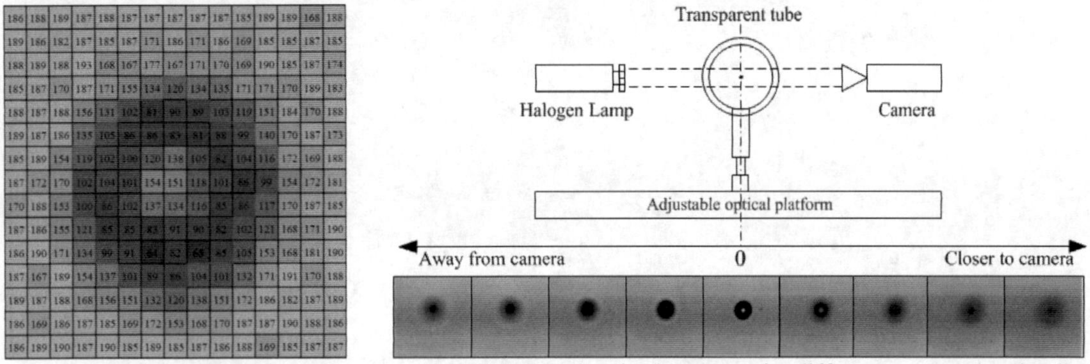

Figure 6: Droplet image matrix for boundary detection and droplet images under different camera-droplet distances

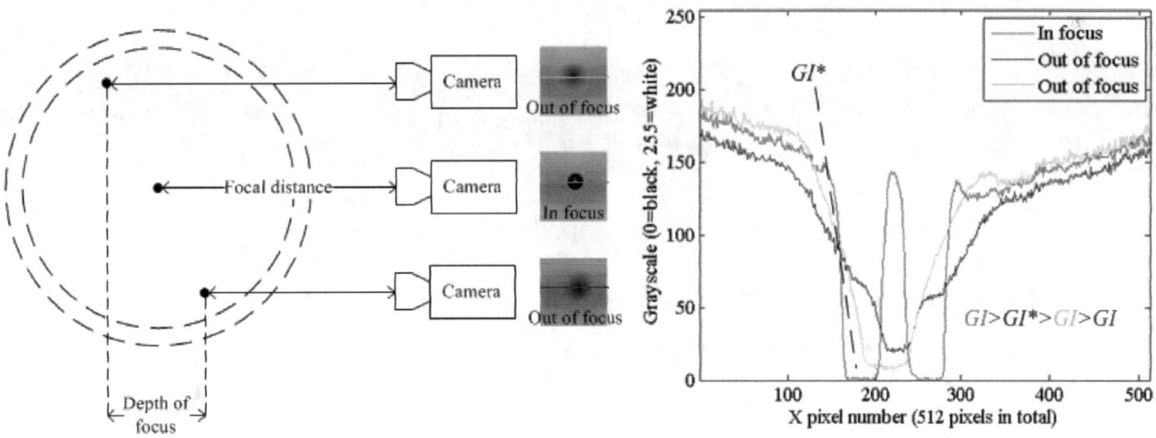

Figure 7: In-focus criteria defined by critical grayscale indicator *GI**

Koh et al. (2001) used the grayscale gradient indicator (*GI*) to determine the in-focus criteria for a processed image of a specific particle. Grayscale gradient indicator is defined as the local grayscale gradient divided by the difference between maximum grayscale and minimum grayscale in the image.

$$GI = \frac{\nabla G}{(G_{max} - G_{min})/2\Delta x} \tag{6}$$

Therefore, the steepest grayscale gradient of droplets in focus on the boundary will be higher than that of droplets out of focus. In order to tell in-focus image from out-of-focus, *GI** is defined as in-focus criteria. All the droplets with boundary *GI* less than *GI** are excluded out of the visualization control volume. In Figure 7, the blue and green lines show the *GI* less than *GI** which is considered as out-of-focus while the red line has a steeper gradient represents in-focus. A pratical *GI** value is given by Koh et al. (2001) as

$$GI^* = GI_{max} - 0.22 \tag{7}$$

The value of *GI** could be adjusted according to the image quality. A higher *GI** will include less in-focus droplets with sharper boundaries.

With the in-focus criteria, we can estimate the depth of focus for different particles and build a correlation between the particle size and depth of focus. The correlation is calibrated by four known size particles and compared with the results in other literatures, as shown in Figure 8.

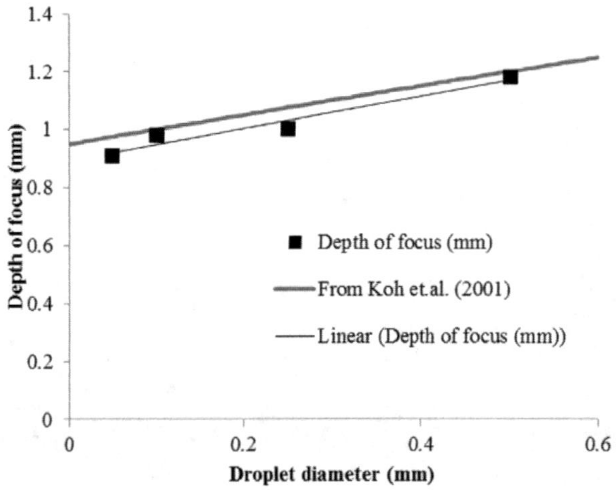

Figure 8: Correlation between depth of focus and droplet diameter

Since depth of focus is different for different particle diameters, Equation (3) and (4) can be rewritten as

$$M_{drop} = \rho_{oil}\frac{V_{drop}}{V_{video}}V_{tube} = \frac{\pi^2 D^2 L}{24A_{video}}\sum\frac{d^3}{D_f} \tag{8}$$

$$OCR_{drop} = \frac{\dot{m}_{drop}}{\dot{m}_{oil} + \dot{m}_{ref}} = \frac{\rho_{oil}V_{drop,video}}{\left(\dot{m}_{oil} + \dot{m}_{ref}\right)\Delta t_{drop}}\frac{V_{tube}}{V_{video}} = \frac{\pi^2 \rho_{oil} D^2}{24(\dot{m}_{oil} + \dot{m}_{ref})A_{video}}\sum\frac{d^3 u_{drop}}{D_f} \tag{9}$$

In this way, a more precise estimation of total mass of oil droplets in the discharge tube can be achieved.

3.5 Oil droplets velocity

Wujek and Hrnjak(2011) developed a video processing method similar to the particle image velocimetry(PIV) in order to determine the droplets moving velocity in axial direction. Oil droplets, instead of seeding particles, are used directly to determine the velocity field of the gas core misty flow. Two images of the flow are taken in rapid succession with a high speed camera to make sure the same droplets must appear in consecutive frames. The video processing principle is shown in Figure 9 in the following steps:

1. Wave elimination by subtracting the average pixel value in period of time;
2. Droplet size determination with steepest grayscale gradient;

3. Comparison of two consecutive images in small sections to interrogate potential displacement ;
4. Velocity determination with PIV algorithm.

Numerous erroneous velocity vectors will be generated throughout this process and the two-dimensional cross-correlation can be used to screen out true velocity vectors.

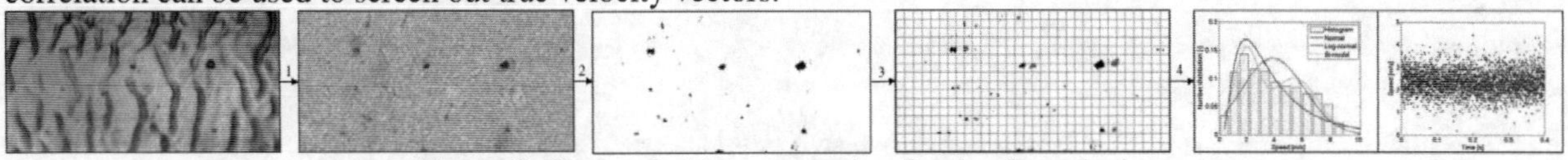

Figure 9: Steps of video processing for droplet velocity

4. RESULTS

4.1 Oil retention

Figure 10(a) shows that oil film thickness gets thinner when gas velocity goes up because of the higher shear stress. The mass of oil droplets gets higher because of the higher compressor speed.

If the oil retention in the form of droplets and that in the form of film are added together, we can compare the oil retention measured by the video to sampling results, as shown in Figure 10(b). Although the uncertainty of oil retention by video is relatively large because of thin film thickness and rough estimation of video control volume, the results can still be compared to oil retention by sampling. It means the optical method is acceptable as a non-invasive way to measure the oil retention.

If we compare the results of the compressor without oil sump and the compressor with oil sump, it can be concluded that from oil retention point of view, oil sump has a minor impact on the amount of oil in the compressor discharge tube.

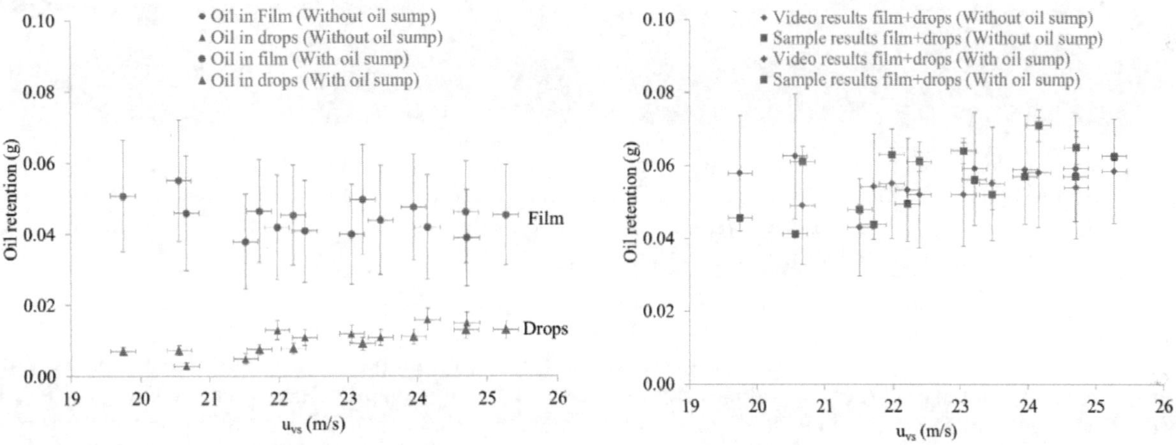

(a) (b)

Figure 10: (a) Oil retention contributed by oil film or oil droplets in the discharge tube (b) Oil retention in the discharge tube under different vapor superficial velocity (u_{vs}) measured by both video and sample

4.2 Oil in circulation

Both OCR contributed by oil film or oil droplets gets larger, when the gas velocity goes up. This is because higher compressor speed leads to more oil injected by the discharge valves. It is interesting to see that the distribution of OCR in droplets and in film is very different from the distribution of oil retention. OCR is mainly contributed by oil droplets because oil droplets travel in a much higher speed than oil film does. For OCR, both mass and velocity need to be taken into consideration, which means fast-moving oil droplets can contribute more than slow moving oil film even though they are minority by mass. Similarly, Figure 11(b) shows that the OCR measured by video can be comparable with OCR measured by sampling in an acceptable accuracy.

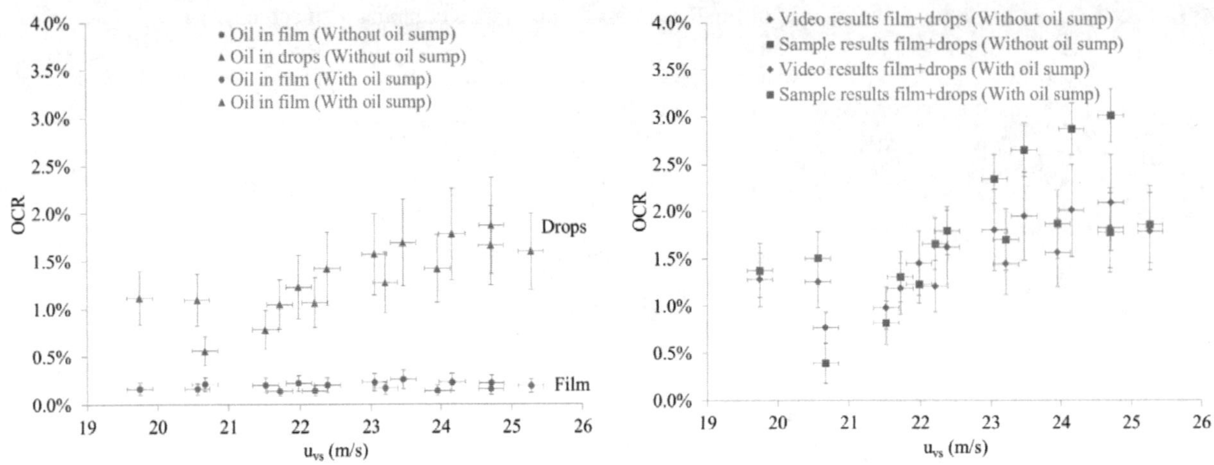

(a) (b)

Figure 11: (a) Oil retention contributed by oil film or oil droplets in the discharge tube (b) Oil retention in the discharge tube under different vapor superficial velocity (u_{vs}) measured by both video and sample

The effect of the oil sump becomes essential when it comes to OCR. Compressor without oil sump cannot hold a lot of oil in the compressor so that higher compressor speed tends to increase the OCR in the system dramatically. The different trends in Figure 10(b) and Figure 11(b) clearly show that compressor without oil sump generates a much higher OCR than compressor with oil sump.

SUMMARY

This paper provides a validated non-invasive method to measure the oil retention and OCR at the compressor discharge tube. The video processing method can give more information, like the oil droplet velocity and size, than the traditional sampling method. Furthermore, the visualization method can be applied to any similar annular-mist multiphase flow in order to quantify the flow parameters.

The distribution of the oil in the annular-mist flow at the compressor discharge is researched at different gas superficial velocity and different compressor types. In the compressor discharge tube, most of the mass is in the form of oil film while most of the mass flow rate is in the form of oil droplets. Velocity difference between the refrigerant vapor core, the oil droplets and the oil film causes the distribution difference of oil retention or OCR. In order to eliminate the oil circulating in the system, it is important to separate oil droplets travelling in the refrigerant-oil mixture annular-mist flow at the compressor discharge.

REFERENCES

DeAngelis, J., & Hrnjak, P. S. (2005). Development of transcritical R744 systems for small commercial applications. In *Pre-prints of the IIR International Conference on Commercial Refrigeration, Vicenza, Italy* (p. 167e178).

DeAngelis, J. M., & Hrnjak, P. S. (2005). Experimental Study of System Performance Improvements in Transcritical R744 Systems with Applications to Bottle Coolers, *ACRC Contract Report CR-57, University of Illinois at Urbana-Champaign.*

Fukuta, M., Yanagisawa, T., Miyamura, S., & Ogi, Y. (2004). Concentration measurement of refrigerant/refrigeration oil mixture by refractive index. *International journal of refrigeration*, *27*(4), 346-352.

Hrnjak, P., & Kim, S. (2013, July). Oil Effects on In-Tube Evaporation of CO2 by Altering Flow Regime and Properties. *In ASME 2013 Heat Transfer Summer Conference collocated with the ASME 2013 7th International Conference on Energy Sustainability and the ASME 2013 11th International Conference on Fuel Cell Science, Engineering and Technology* (pp. V002T04A012-V002T04A012). American Society of Mechanical Engineers.

Hurlburt, E. T., & Newell, T. A. (2000). Prediction of the circumferential film thickness distribution in horizontal annular gas-liquid flow. *Journal of fluids engineering*, *122*(2), 396-402.
Kim, S., & Hrnjak, P. S. (2012). Effect of oil on flow boiling heat transfer and flow patterns of CO2 in 11.2 mm horizontal smooth and enhanced tube.
Koh, K. U., Kim, J. Y., & Lee, S. Y. (2001). Determination of in-focus criteria and depth of field in image processing of spray particles.*Atomization and Sprays*, *11*(4).
Li, H., & Hrnjak, P. (2014). Lubricant Effect on Performance of R134a MAC Microchannel Evaporators (No. 2014-01-0692). *SAE Technical Paper*.
Li, H., & Hrnjak, P. (2015). An experimentally validated model for microchannel heat exchanger incorporating lubricant effect. *International Journal of Refrigeration*, *59*, 259-268.
McMullan, J. T., Hewitt, N. J., Masson, A. J., & Murphy, N. E. (1992). The influence of oil viscosity and refrigerant quality on evaporator performance. *International journal of energy research*, *16*(7), 567-581.
Nidegger, E., Thome, J. R., & Favrat, D. (1997). Flow boiling and pressure drop measurements for R-134a/oil mixtures Part 1: evaporation in a microfin tube. *HVAC&R Research*, *3*(1), 38-53.
Peuker, S., & Hrnjak, P. S. (2010). Experimental Techniques to Determine Oil Distribution in Automotive A/C Systems. *International Refrigeration and Air Conditioning Conference.* Paper 1011
Shedd, T. A., & Newell, T. A. (1998). Automated optical liquid film thickness measurement method.*Review of Scientific Instruments*, *69*(12), 4205-4213.
Toyama, T., Matsuura, H., & Yoshida, Y. (2006). Visual techniques to quantify behavior of oil droplets in a scroll compressor. *International Refrigeration and Air Conditioning Conference.* Paper 2787
Wada, A., Nomura, M., Tsuboi, K., Kutsuna, K., & Nabeta, T. (1992). A novel approach to instrumentation and application for OCR measurement in refrigeration system. *International Refrigeration and Air Conditioning Conference.* Paper 840
Wujek, S., & Hrnjak, P. (2011). Mist to Annular Flow Development Quantified by Novel Video Analysis Methods. *ACRC Technical Report TR-285*, University of Illinois at Urbana-Champaign.
Xiao, J., & Hrnjak, P. (2015). Condensation in Horizontal Smooth Round Tube. *ACRC Technical Report TR-321*, University of Illinois at Urbana-Champaign.
Xu, J., & Hrnjak, P. (2016). Refrigerant-Oil Flow at the Compressor Discharge (No. 2016-01-0247). SAE Technical Paper.
Xu, J. (2016). Oil Flow from Compressor Valve to Discharge Tube. *Master Thesis*, University of Illinois at Urbana-Champaign.
Zimmermann, A. J. P., & Hrnjak, P. S. (2014). Visualization of the Opening Process of a Discharge Reed Valve in the Presence of Oil. *International Refrigeration and Air Conditioning Conference.* Paper 1569.
Zimmermann, A. J., & Hrnjak, P. S. (2015). Oil flow at discharge valve in a scroll compressor. *Proceedings of the Institution of Mechanical Engineers, Part E: Journal of Process Mechanical Engineering*, 229(2), 104-113.
Zürcher, O., Thome, J. R., & Favrat, D. (1998). In-tube flow boiling of R-407C and R-407C/Oil Mixtures Part II: Plain tube results and predictions. *HVAC&R Research*, *4*(4), 373-399.

ACKNOWLEDGEMENT

The authors would like to acknowledge the support of the member companies of the Industrial Advisory Board for the Air Conditioning and Refrigeration Center at the University of Illinois through ACRC project 341.

Study on Flow Characteristics of Oil Viscosity Pump for Refrigerant Compressors

Kiyoshi SAWAI[1]*, Manabu DOI[1], Norimaki ISHII[2], Noboru IIDA[3], Kenji KINJO[3]

[1]Hiroshima Institute of Technology, Dept. of Mechanical Engineering
Hiroshima, Japan
Phone: +81-82-921-5475, E-Mail: sawai@me.it-hiroshima.ac.jp

[2]Osaka Electro Communication University, Osaka, Japan
Tel/fax: +81-72-820-4561

[3]Panasonic Corporation, Appliances Company, Corporate Engineering Division
Kusatsu, Shiga, Japan
Tel: +81-77-562-5658, fax: +81-77-563-1967

* Corresponding Author

ABSTRACT

Reciprocating compressors are commonly used in household refrigerators, and lubricants is supplied to the compression mechanism by the oil viscosity pump. While it is expected that this lubricants has a great influence on the compressor's performance and reliability, its flow characteristics is does not mean elucidated. In this study, we examined the flow characteristics of the oil viscosity pump that had a spiral groove on the shaft surface opposite to the bearing surface, using both theoretical and experimental study. The results revealed followings:

(1) When oil flow in the oil groove is assumed to be a "uniform, two-dimensional flow which is a steady and sufficiently developed laminar flow," the theoretical flow rate in the oil viscosity pump and the pump head can be expressed with a linear function formula. It was also confirmed in experiments that both were in a linear function relationship.

(2) Whether the theoretical flow rate formula can be applied does not depend on the oil groove angle, but on the oil groove length. It is possible to estimate the flow characteristics within an average relative error of 30% if the oil groove length is 60 mm or larger.

(3) Whether the theoretical flow rate formula can be applied depends on the oil groove depth if the oil groove has a rectangular cross section. It is possible to estimate the flow characteristics within an average relative error of 30% if the aspect ratio of the oil groove is 0.5 or smaller.

1. INTRODUCTION

Recently, refrigerators account for 14% of the total electricity consumed by households in Japan. In addition, the compressor accounts for 80% of the electricity consumed by a refrigerator. As shown in Figure1, suction and compression of refrigerants are conducted as the piston makes a reciprocating motion within the cylinder, and reciprocating compressors which have relatively high efficiency are commonly used in refrigerators. In the reciprocating compressors, lubricants is supplied to the compression mechanism from the bottom of the shell by the oil pump system. This oil pump system is comprised of [1] a centrifugal pump comprising the cylindrical part and a cap attached at the bottom end of the shaft, [2] a reed pump with a blade placed in the cylinder, and [3] the oil viscosity pump which supplies oil using the shearing stress generated on the bearing's surface as a spiral groove is formed on the surface of the shaft opposite to the bearing surface. The lubricants supplied by this oil pump system is expected to have a great influence on the performance and reliability of the compressor, then its flow characteristics have been studied by several researchers. Asanuma (1951) made a theoretical analysis on the flow characteristics of the oil viscosity pump, but the influence of gravity is not taken into consideration. Weifeng *et al.* (2010) made a numerical simulation of the centrifugal pump of the reciprocating compressor. Kim *et al.* (2002) and Luckmann *et al.* (2009) made a numerical analysis of centrifugal and oil viscosity pump in reciprocating compressor. Alves *et al.* (2010) made a numerical analysis of screw pump system attached to shaft-end for reciprocating compressor. However, in these studies, flow characteristics (relationship between oil flow rate and pump head) of the oil pump system has not been clear. In this study, we focused on the oil viscosity pump which is expected to have the largest influence on the flow characteristics out of all the oil pump system's components, and examined its flow characteristics using both theoretical and experimental study.

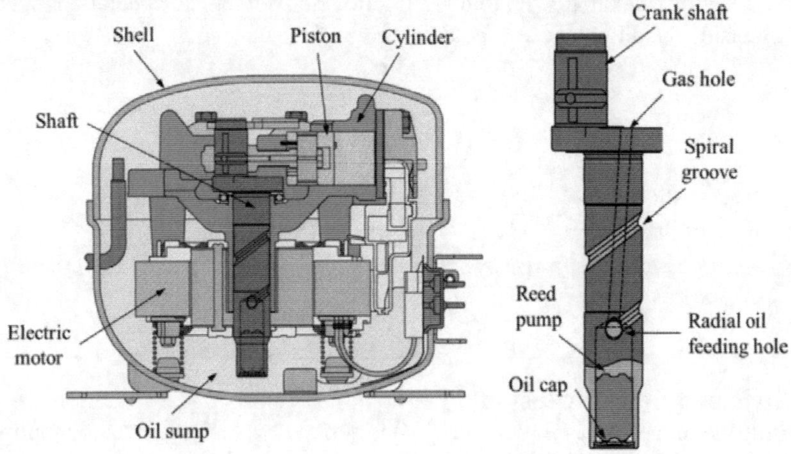

Figure 1: Compressor for household refrigerator

2. THEORETICAL STUDY OF OIL VISCOSITY PUMP

2.1 Overview of and coordinate system for oil viscosity pump

Figure 2 shows a schematic diagram of the oil viscosity pump. It shows that the shaft has a spiral groove with a rectangular cross section. A schematic view of the oil groove is also shown in Figure 2. The bearing surface moves relatively to the right direction when the shaft rotates to the left. In this case, a viscous shearing force acts on the oil inside the oil groove on the surface in contact with the inner wall of the bearing, which makes the oil climb up along the oil groove as a consequence.

The length of this oil groove which functions as a viscosity pump was designated as L, the length of the oil groove in the direction of the shaft axis L_s, and the angle of the oil groove attachment β. In addition, the oil groove was formed with a rectangular cross section with a shallow depth h and width b. And as the coordinate system for theoretical study, the direction of the oil flow in the oil groove was defined as the x axis, the direction of depth in the oil groove as the y axis, and the direction of width in the oil groove as the z axis.

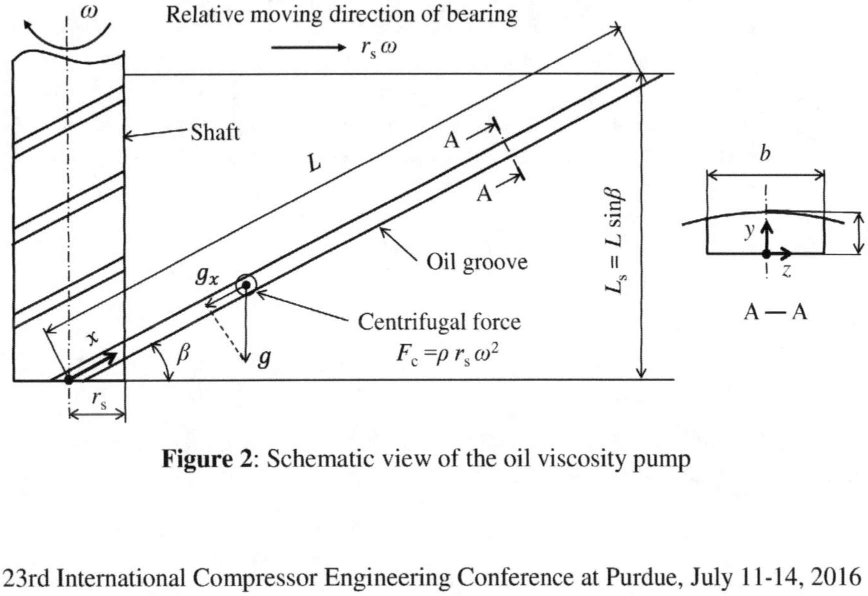

Figure 2: Schematic view of the oil viscosity pump

2.2. Fundamental equations for oil flow in spiral groove

In theoretical study, we introduced two assumptions to the oil flow inside the groove. The first assumption was that the oil flow was a steady and sufficiently developed laminar flow in the direction of the x axis. The second assumption was that the oil flow had a velocity component in the z axis direction $w = 0$ on each cross section of the oil groove, and was a uniform $x - y$ flow in z direction. In this theoretical study, it was also assumed that the leakage through the clearance could be neglected as the clearance between the shaft surface and bearing was sufficiently small compared to the oil groove depth h. The equation of continuity in Eq.(1) and Navier-Stokes formula in Eqs.(2) and (3) were applied to this $x - y$ flow in oil groove. In this case, it was taken into consideration that the centrifugal force F_c was applied to the oil inside the oil groove as a body force.

$$\frac{\partial u}{\partial x} + \frac{\partial v}{\partial y} = 0 \tag{1}$$

$$\frac{\partial u}{\partial t} + u\frac{\partial u}{\partial x} + v\frac{\partial u}{\partial y} = g_x - \frac{1}{\rho}\frac{\partial p}{\partial x} + \nu\left(\frac{\partial^2 u}{\partial x^2} + \frac{\partial^2 u}{\partial y^2}\right) \tag{2}$$

$$\frac{\partial v}{\partial t} + u\frac{\partial v}{\partial x} + v\frac{\partial v}{\partial y} = g_y + \frac{F_c}{\rho} - \frac{1}{\rho}\frac{\partial p}{\partial y} + \nu\left(\frac{\partial^2 v}{\partial x^2} + \frac{\partial^2 v}{\partial y^2}\right) \tag{3}$$

2.3. Oil flow Velocity *u* and flow rate formula

Oil flow velocity u inside the oil groove as shown in Eq.(4) is derived by applying two assumptions in the previous section to Eqs.(1), (2) and (3) and integrating Eq.(2) twice:

$$u = \frac{y}{h}\left\{U - \frac{h^2}{2\mu}\left(\frac{\partial p}{\partial x} + \rho g \sin\beta\right)\left(1 - \frac{y}{h}\right)\right\} \tag{4}$$

Figure 3 illustrates the oil flow velocity u expressed by Eq.(4). The bottom end ($y = 0$) in Figure 3 indicates the bottom of the oil groove, and the top end ($y = h$) the bearing inner wall. The broken line corresponds to the Couette-flow, and the solid line shows velocity distribution u that is suppressed by the pressure gradient $\partial p/\partial x$ and gravity g acting on the flow compared to the Couette-flow.

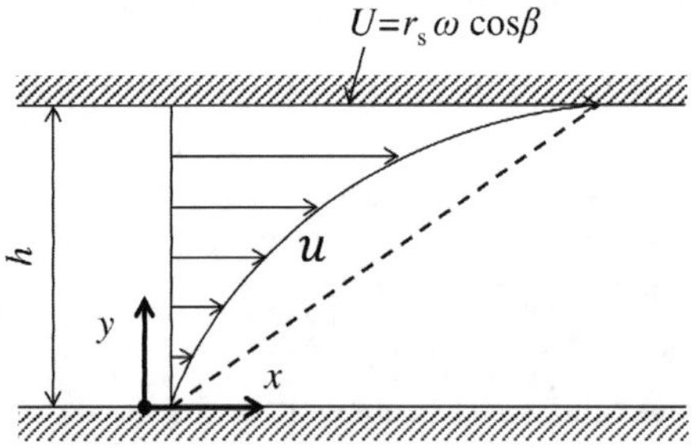

Figure 3: Velocity distribution in spiral groove

Here, pressure gradient $\partial p/\partial x$ in Eq.(4) is expressed by Eq.(5) using pump head H. Therefore, by substituting it in Eq.(4), integrating Eq.(4) until y=0 → h, and multiply the oil groove width b, Eq.(6) can be used to express the theoretical flow rate of oil Q_t flowing inside the oil groove.

$$\frac{\partial p}{\partial x} = \frac{\rho g H}{L} \tag{5}$$

$$Q_t = \frac{bh}{6}\left\{-\frac{h^2 g}{2\nu L}H + 3U - \frac{h^2}{2\nu}g\sin\beta\right\} \tag{6}$$

Based on Eq.(6), it was revealed that there was a linear functional relationship between the theoretical flow rate Q_t and pump head H.

3. OUT LINE OF THE EXPERIMENT

3.1. Experimental apparatus

We prepared some evaluation apparatus to examine the oil viscosity pump's flow characteristics as shown in Figure 4. In this experiment, lubricants with a viscosity close to the actual operation viscosity on refrigeration cycle at room temperature was selected and stored in the oil bath. Shafts equipped with a spiral groove on the surface are prepared to suit the purpose of each experiment. When this shaft is rotated, oil is pumped up along the oil groove formed on the shaft surface. By pulling out the oil into a tube from the top of the bearing and giving resistance to the flow with a valve installed in the middle of the pipe line, pump head H was applied to the oil viscosity pump in this system. During experiments on oil flow characteristics, pump head H was read on a scale set up at the head-measuring section, and the experimental oil flow rate Q_e was measured using an electronic balance.

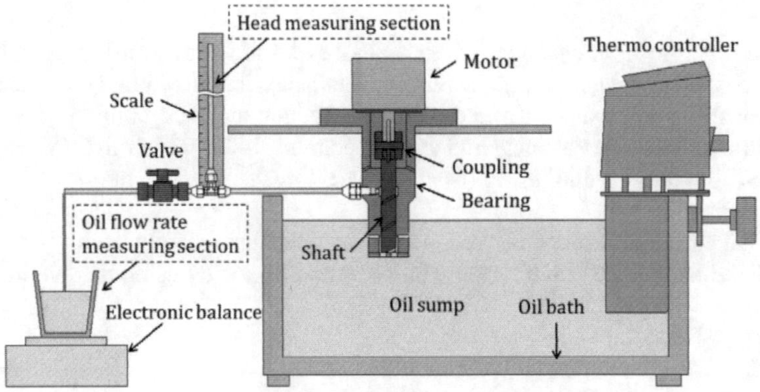

Figure 4: Experimental apparatus

3.2. Description of experiments

To find out the applicable range for the theoretical flow rate formula for the oil viscosity pump in Eq.(6), we conducted three types of experiments as shown in Tables 1 to 3. The purpose of experiments in Tables 1 and 2 was to study whether Eq. (6) was applicable or not when what value of oil groove length L was ensured, and flow characteristics is studied by varying each of the shaft length L_s and oil groove angle β. In the experiment in Table 3, we studied the relationship between the oil groove aspect ratio h/b and the possibility of application for Eq.(6) by varying the oil groove width b and depth h while keeping the oil groove's cross section constant. Furthermore, a shaft with a $\phi 18mm$ diameter was used in all experiments.

Table 1: Shaft specifications 1 (various oil groove length)

No.	b, h [mm]	β [°]	L_s [mm]	L [mm]
1	3.0, 1.0	72	100	105.2
2			75	78.9
3			50	52.6
4			30	31.5

Table 2: Shaft specifications 2 (various oil groove angle)

No.	b, h [mm]	β [°]	L_s [mm]	L [mm]
3	3.0, 1.0	72	50	52.6
5		61		57.2
6		54		61.8
7		43		73.3
8		30		100.0

Table 3: Shaft specifications 3 (various rectangular cross-section of oil groove)

No.	b [mm]	h [mm]	β [°]	L_S [mm]	L [mm]
9	3.5	0.85			
6	3.0	1.0			
10	2.5	1.2	54	50	61.8
11	2.0	1.5			
12	1.5	2.0			

4. EXPERIMENTAL RESULTS

As typical experimental results, the flow characteristics of Shaft No.8 are shown in Figure 5. The blue line in the figure indicates the experimental flow rate Q_e, and the red line the theoretical flow rate Q_t calculated using Eq.(6) for the same head. Based on this result, it was confirmed that the tendency matched both the theoretical flow rate and experimental flow rate and that there was a linear functional relationship between oil flow rate and pump head. It showed that the theoretical flow rate formula in Eq.(6) expressed flow characteristics of the oil viscosity pump that were close to the actual situation.
In this study, we also evaluated the difference between the flow characteristics obtained in the experiment and the flow characteristics calculated from the theoretical flow rate formula Eq.(6) by using the average relative error E_{ave} defined in Eq.(7).

$$E_{ave} = \left(\sum_{i=1}^{n} \frac{|\delta_{e,i} - \delta_{t,i}|}{\delta_{e,i}} \right) \Big/ n \times 100 \ [\%] \tag{7}$$

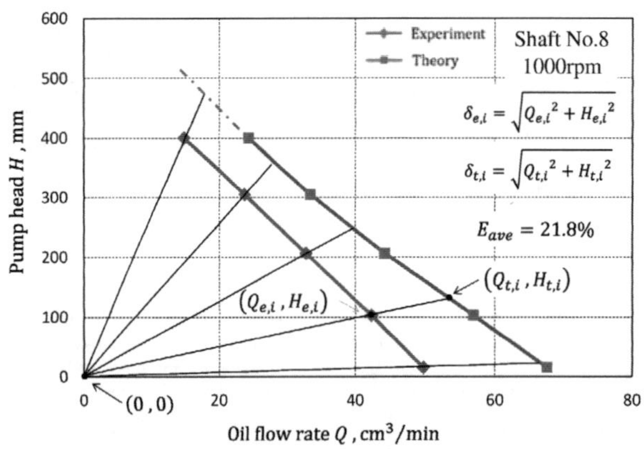

Figure 5: Experimental results (representative example)

$\delta_{e,i}$ in Eq.(7) indicates the distance between the origin $(0, 0)$ and experimental value $(Q_{e,i},\ H_{e,i})$ in Figure 5. In addition, $\delta_{t,i}$ indicates the distance between the origin $(0, 0)$ and the point $(Q_{t,i},\ H_{t,i})$ that is the intersection between the line connecting the origin $(0, 0)$ and experimental value $(Q_{e,i},\ H_{e,i})$, and the theoretical flow rate characteristic line. In addition, n is the number of samples used for calculating the average relative error δ_{ave}, and it was set to the same value as the number of plots in experiment. It was also supposed that theoretical flow rate formula Eq.(6) was applicable when $E_{ave} \leqq 30\%$.

4.1. Experiment varying the oil groove length L

Figure 6 shows the relationship between the oil groove length L and average relative error δ_{ave} derived from the results of the experiment which evaluated the flow characteristics by varying the oil groove length L and calculated each average relative error δ_{ave}. Based on Figure 6, it was found that the difference between the experimental values and theoretical values became larger at low rotational speed when the oil groove length L was shorter, although they showed a relatively good match at high rotational speeds. These phenomena indicate that Assumption (1), set up when the theory was discussed, holds for a shorter oil groove length L at higher rotational speeds than lower ones. It was also found based on Fig. 6 that application of the theoretical formula Eq.(6) was possible with $E_{ave} \leqq 30\%$ when the oil groove length $L \geqq 60\text{mm}$.

4.2. Experiment varying oil groove angle β

Figure 7 shows the relationship between oil groove angle β and average relative error δ_{ave} derived from the results of the experiment which evaluated the flow characteristics by varying the oil groove angle β. Based on Figure 7, it was found that the experimental value and theoretical value was approximately matched in the high rotational speed from low rotational speed. While Figure 7 also describes oil groove length L, it was found that $E_{ave} \leqq 30\%$ was obtained for oil groove length $L \geqq 60\text{mm}$, making it possible for a theoretical flow rate formula Eq.(6) to be applied. Based on these, it was found that whether the theoretical flow rate formula Eq.(6) is applicable or not is dependent on the oil groove length L, not on the oil groove angle β.

4.3. Experiment varying the oil groove cross section shape

Figure 8 shows the relationship between oil groove aspect ratio h/b and average relative error δ_{ave} derived from the results of the experiment which evaluated the flow characteristics by varying oil groove width b and depth h with the oil groove cross section kept constant. Based on Figure 8, it was found that the difference between the experimental value and theoretical value became larger when the oil groove aspect ratio h/b was larger at any rotational speed N. It is assumed that this was because the oil flow came under a larger influence of the side wall when the oil groove width b was smaller if aspect ratio h/b was larger, making it impossible to apply Assumption (2), in which the oil flow is uniform *x-y* flow in z direction. It was also found that application of the theoretical formula was possible at $E_{ave} \leqq 30\%$ when aspect ratio was $h/b \leqq 0.5$.

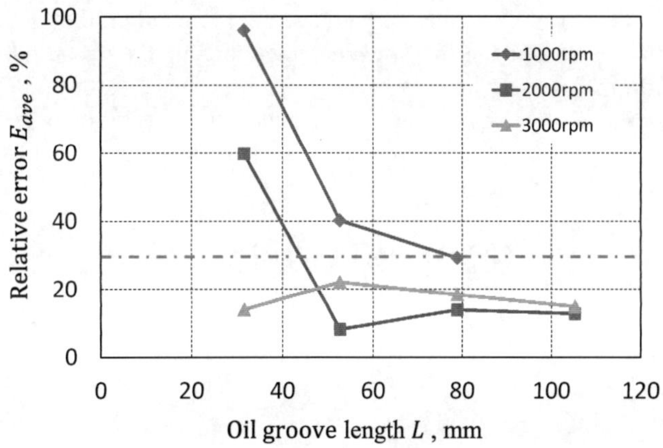

Figure 6: Relationship between oil groove length and average relative error

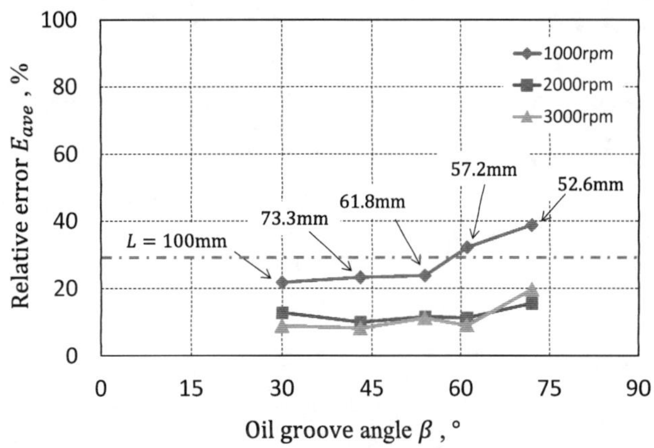

Figure 7: Relationship between oil groove angle and average relative error

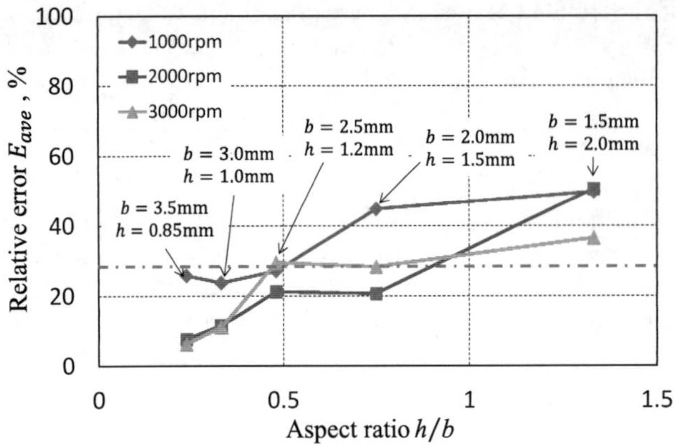

Figure 8: Relationship between oil groove aspect ratio and average relative error

5. CONCLUSION

We conducted a theoretical study of fluid dynamics on the flow characteristics of the oil viscosity pump with a spiral groove on the surface of the shaft opposite to the bearing and verification experiments using a model test machine, and found the followings:

(1) When the oil flow in a spiral groove is assumed to be a "uniform, two-dimensional flow which is a steady and sufficiently developed laminar flow," the theoretical flow rate Q_t in the oil viscosity pump and the pump head H can be expressed with a linear function formula. This relationship was also verified in experiments.

(2) Whether the theoretical flow rate formula can be applied does not depend on the oil groove angle β, but on the oil groove length L. It is possible to estimate the flow characteristics with average relative error $E_{ave} \leqq 30\%$ if the oil groove length $L \geqq 60\text{mm}$.

(3) Whether the theoretical flow rate formula can be applied depends on the oil groove depth h if the oil groove has a rectangular cross section. It is possible to estimate the flow characteristics with average relative error $E_{ave} \leqq 30\%$ if the oil groove aspect ratio $h/b \leqq 0.5$.

NOMENCLUTURE

b	Oil groove width	(m)
E_{ave}	Average relative error	(%)
F_c	Centrifugal force	(N)
g	Gravity	(m/s^2)
h	Oil groove depth	(m)
H	Oil pump head	(m)
L	Oil groove length	(m)
L_s	Shaft length	(m)
n	Sample number	(−)
N	Shaft rotating speed	(rpm)
Q	Oil flow rate	(m^3/s)
r_s	Shaft radius	(m)
u	Oil velocity of x direction	(m/s)
U	Relative velocity of bearing wall	(m/s)
v	Oil velocity of y direction	(m/s)
β	Oil groove angle	(°)
δ	Distance in Figure 5	(mm)
ν	Oil kinematic viscosity	(m^2/s)
ρ	Oil density	(kg/m^3)
Subscript		
e	Experimental	
t	Theoretical	

REFERENCES

Tsutomu Asanuma. (1951). A Study on sealing action by viscous fluid, *Transactions of the JSME*, Vol.17, (No.60), pp.119-125.

Weifeng Wu, Jian Li, Longquan Lu, Quanke Feng. (2010). Analysis of Oil pumping in the Hermetic Reciprocating for Household Refrigerators, *International Compressor Engineering Conference at Purdue*, Paper 1168

H.J. Kim, T.J. Lee, K.H. Kim, Y.J. Bae. (2002). Numerical Simulation of Oil Supply System of Reciprocating Compressor for Household Refrigerators, *International Compressor Engineering Conference at Purdue*, Paper 1553.

Antonio J. Luckmann, Marcus Vinicius C. Alves, Jader R. Barbosa Jr. (2009). Analysis of pumping in a reciprocating compressor, *Applied Thermal Engineering,* Vol.29, pp.3118-3123.

Marcus V.C. Alves, Jader R. Barbosa Jr., Alvaro T. Prata, Fernando A. Ribas Jr. (2010). Fluid Flow in a Screw Oil Supply System for Reciprocating Compressor, *International Compressor Conference at Purdue,* Paper 1326.

Performance Comparison between a Single-stage and a Two-stage Hermetic Rotary CO_2 Compressor

Li ZHANG[*] , Min YANG, Xiaolong HUANG

R&D Centre, Shanghai Hitachi Electrical Appliances Co., Ltd.,
Shanghai, 201206, P.R. China
Email: zhangli@shec.com.cn Phone:86-21-50554560

*Corresponding Author

ABSTRACT

As a natural refrigerant, CO_2 gets more and more attention in recent years. The CO_2 system runs at a trans-critical cycle. This leads to a high running pressure and high pressure difference for the CO_2 compressor. In this paper the single stage and two stage rolling piston compressors with same displacement have been developed for CO_2 heat pump water heater. And a comprehensive performance comparison has been made with these two kinds of compressor at different pressure ratio and superheat. With these results, a better understanding for the single stage and two stage CO_2 compressors could be got.

1. INTRODUCTION

Carbon dioxide (CO_2) as a nature refrigerant has attracted many attentions again because of global warming issue. As a natural refrigerant, CO_2 has been used in the past and proved to be environmentally friendly. The physical properties of CO_2 make it particularly suitable for heat pump water heaters. Compared with the traditional Freon refrigerants heat pump water heater, CO_2 heat pump water heater has higher energy efficiency, higher outlet water temperature (up to 90℃ in winter), and lower performance degradation in low temperature environment.

The research and development of CO_2 refrigeration system have been paid more attention in recent years. The compressor is a key component of CO_2 refrigeration system. Due to the high pressure difference for CO_2 compressor, the initial products mainly were two-stage compressor. With the in-depth of development, single-stage compression products have gradually been adopted. The performance comparison between single-stage and two-stage CO_2 compressor, especially at a wide running condition, were mentioned less in previous literature.

In this paper, a single-stage and a two-stage hermetic rolling piston CO_2 compressor with same displacement were developed. The performance of the two types of CO_2 compressor at different pressure ratio and superheat were tested and compared. With these results, a better understanding for the single-stage and two-stage CO_2 compressor could be got.

2. DEVELOPMENT OF CO_2 COMPRESSOR

In this paper, a single-stage and a two-stage hermetic rolling piston CO_2 compressor were developed.

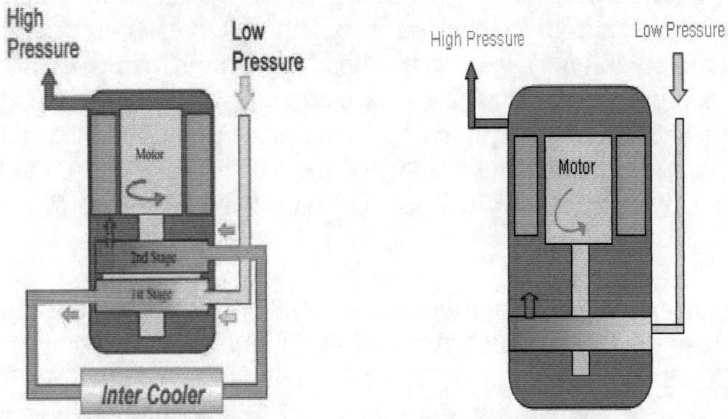

(a) two-stage compressor (b) single-stage compressor

Figure 1: The schematic view of the compressors

Figure 1(a) shows the schematic view of the two-stage compressor in conjunction with the refrigerant flow path. The low pressure refrigerant flows into the first-stage compression unit located in the lower part of the whole compression unit. And then it is compressed to an intermediate pressure determined by the suction pressure and the structure of the compressor. After that, the refrigerant is directly discharged out of the compressor from the first-stage compression unit. After a cooling process in the intercooler, the refrigerant enters into the second-stage compression unit located in the upper part of the compression unit. In this compression unit, the refrigerant is compressed to high pressure, and then discharged into the shell of the compressor. Flowing through the gaps of motor on the upper of the compressor, the refrigerant is discharged out of the compressor to gas cooler of CO_2 trans-critical refrigeration cycle.

Obviously, the discharge pressure in the shell of the two-stage compressor is much higher than that of the intermediate pressure compressor. According to this special compression arrangement, the oil on the bottom of the compressor can be easily pumped into the clearance between the movement parts. The lubrication becomes better and compressor reliability is improved. In addition, before being discharged out of the compressor, the refrigerant flowing through the gaps of motor can reduce the oil flow rate ratio (defined as the ratio of the oil flow rate to the refrigerant mass flow rate) and improve the efficiency of the refrigeration system.

Figure 1(b) shows the schematic view of the single-stage compressor in conjunction with the refrigerant flow path. It's simpler than the two-stage compressor. The low pressure refrigerant flows into the compression unit located in the lower part of the compressor. In the compression unit, the refrigerant is compressed to high pressure, and then discharged into the shell of the compressor. The high pressure refrigerant flows through the gaps of motor on the upper of the compressor, and then is discharged out of the compressor.

Table 1 shows the basic parameters of these two kinds of compressors.

Table 1: Basic Parameters of Compressors

Structure of compressor	Rolling piston compressor	
Stage	Single-stage	Two-stage
Displacement	$3.26cm^3/rev$	$3.26cm^3/rev$
Power	220～240V/50Hz	220～240V/50Hz
Motor	Single phase asynchronous	Single phase asynchronous

3. PERFORMANCE TEST

3.1 Performance Test Rig

Figure 2 shows the schematic view of the test rig. The test rig consists of the main components such as tube-in-tube gas cooler, evaporator, intercooler, electronic expansion valve and super heater and auxiliary components such as oil separator, filter and vapor-liquid separator. A water tank whose temperature can be controlled is used as heat source and heat sink for evaporator and gas cooler. The mass flow meter was installed at the outlet of the gas cooler where the density of the refrigerant is the highest. During the tests, the suction pressure of compressor is controlled by adjusting the water flow rate flowing through the evaporator. The super heater is used to fine-tune the suction temperature and the discharge pressure of the compressor is controlled by the opening degree of the electronic expansion valve.

The measurement range and uncertainties for pressure, temperature, mass flow rate and power consumption are summarized in Table 2. The resulting uncertainties for volumetric efficiency and isentropic efficiency were about ±0.5% and ±0.6% respectively.

Table 2: Uncertainties for measurements

Parameter	Measurement Range	Uncertainty
Pressure	0～16MPa	±0.1%
Temperature	-20～150℃	±0.15℃
Mass flow rate	0～200kg/h	±0.5%
Power consumption	0～2000W	±0.25%

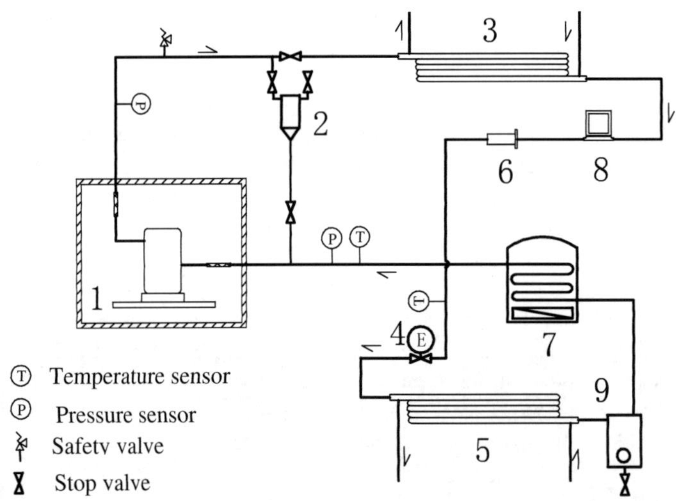

1—compressor, 2—oil separator, 3—gas cooler, 4—electronic expansion valve
5—evaporator, 6—filter, 7—suction superheater, 8—mass flow meter, 9—gas-liquid separator

Figure 2: The performance test rig for hermetic CO_2 compressor

3.2 Compressor Efficiency Calculation

In this paper, the volumetric efficiency η_v and isentropic efficiency η_i were used to evaluate the performance of the compressor. The formulas are as following.

$$\eta_v = q_{v,r} / q_{v,t} \tag{1}$$

$$\eta_i = \frac{q_m(h_2 - h_1)}{3.6P} \quad (2)$$

Where, $q_{v,r}$ is measured volumetric flow rate, and $q_{v,t}$ is the theoretical volumetric flow rate, and the unit is m^3/h. q_m is measured mass flow rate, and the unit is kg/h. h_1 is the specific enthalpy of the refrigerant at inlet of the 1st stage of the compressor, and the unit is kJ/kg. h_2 is the theoretical specific enthalpy of the refrigerant at the discharge pressure after a isentropic compression from the suction state of the compressor, and the unit is kJ/kg. P is the measured input power of the compressor, and the unit is W.

3.3 Compressor Operating Conditions for Performance Test

For the purpose of getting a comprehensive performance comparison between the single-stage compressor and two-stage compressor, the two kinds of compressors were tested at different operating conditions. The suction pressure range is 3.5~4.5MPa, the discharge pressure range is 8.0~11.0MPa, and suction superheat range is 10~20K.

For the tested two-stage compressor, the discharge pipe of the first stage was connected directly to the suction pipe of second stage, so there is no intercooler in the test.

4. RESLUTS AND ANALYSIS

4.1 Effect of Pressure Ratio on Volumetric Efficiency

Figure 3 shows the relationship between volumetric efficiency η_v and pressure ratio (defined as the ratio of the discharge pressure to the suction pressure of the compressor) under different discharge pressure for tested compressors. It can be seen that the volumetric efficiency of single-stage CO_2 compressor is higher than that of two-stage CO_2 compressor at lower pressure ratio. With the increase of pressure ratio, the volumetric efficiency will decrease, and the dropping speed of η_v of single-stage CO_2 compressor is faster than that of two-stage CO_2 compressor. The volumetric efficiency gaps between the two kinds of compressors become small. When the pressure ratio is closed to 2.8, the η_v of two-stage CO_2 compressor will be higher than that of single-stage CO_2 compressor. Furthermore, there is an optimal pressure ratio for two-stage CO_2 compressor, at that pressure ratio the η_v reaches a peak.

For a two-stage compressor, the pressure ratio of the first stage (defined as the ratio of the discharge pressure of the first stage to the suction pressure of the first stage) mainly is determined by the displacement ratio of the first stage to second stage which is a structure parameter. For the developed two-stage compressor, this pressure ratio is about 1.8. When the pressure ratio of the compressor is small, the load coming from the pressure difference and pressure ratio mainly acts on the first stage. At this situation, the second stage almost doesn't play the role of compression unit. However the frictional loss and heat loss coming from the second stage remains, so the performance of the compressor will be affected. Only when the first stage and the second stage can undertake the load evenly, the compressor can get a higher performance. That's why there is an optimal pressure ratio for the two-stage compressor.

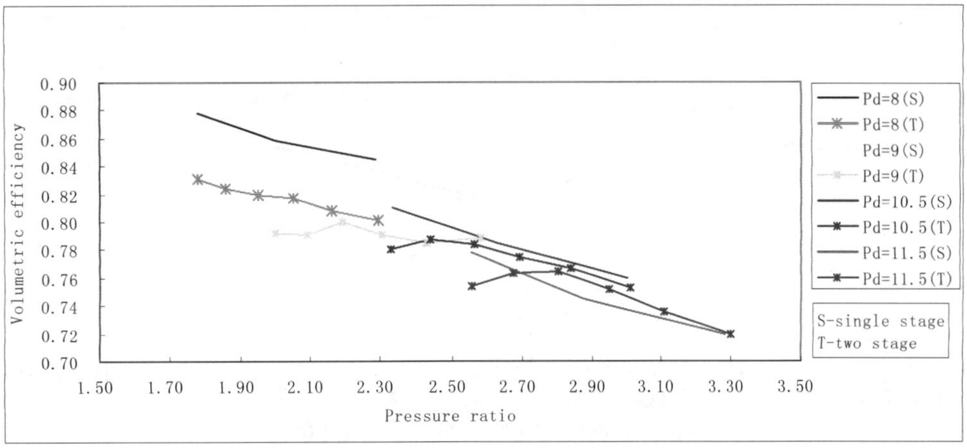

Figure 3: The η_v at different pressure ratio

4.2 Effect of Pressure Ratio on Isentropic Efficiency

Figure 4 shows the relationship between isentropic efficiency η_i and pressure ratio under different discharge pressure for tested compressor. The trend is similar to volumetric efficiency. At lower pressure ratio, the isentropic efficiency of single-stage CO_2 compressor is higher than that of the two-stage CO_2 compressor. With the increase of the pressure ratio, the isentropic efficiency will decrease, and the dropping speed of η_i of single-stage CO_2 compressor is faster than that of two-stage CO_2 compressor. The isentropic efficiency gaps between the two kinds of compressors become small. Similarly there is an optimal pressure ratio for two-stage CO_2 compressor; at that pressure ratio the η_i reaches a peak. The reason is same as the volumetric efficiency as mentioned above.

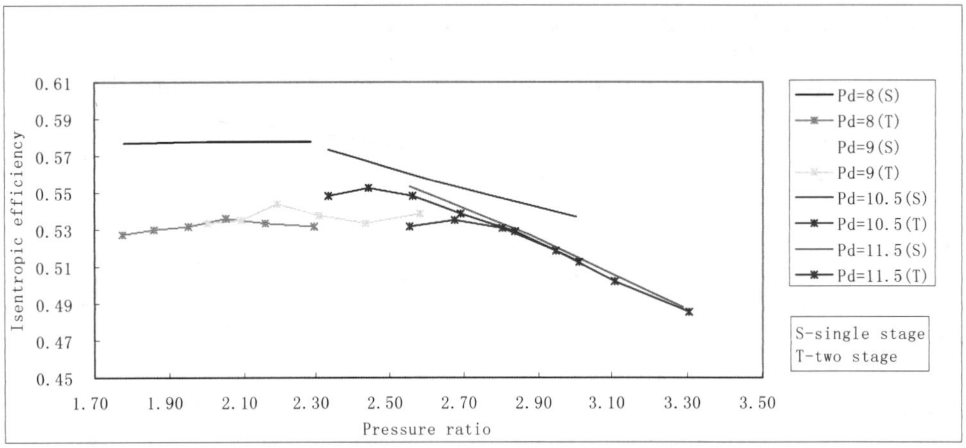

Figure 4: The η_i at different pressure ratio

4.3 Effect of Superheat on Volumetric Efficiency and Isentropic Efficiency

Figure 5 shows the trend of η_v and η_i with suction superheat for the tested compressor. It can be seen that the superheat has very small effect on the volumetric efficiency and isentropic efficiency under the testing conditions whether for the single-stage compressor or two-stage compressor.

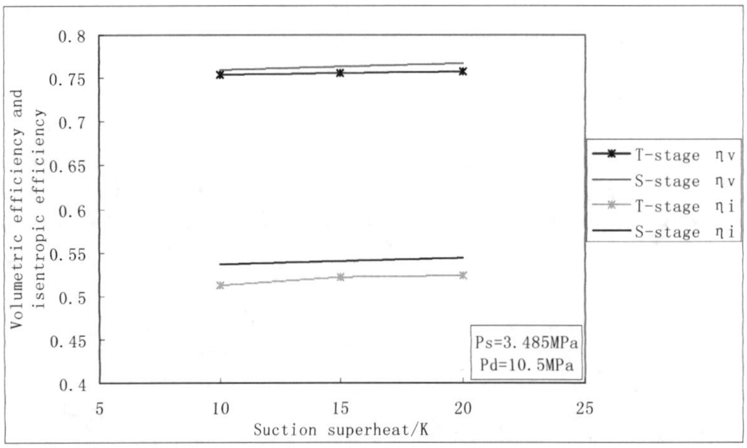

Figure 5: The η_v and η_i at different superheat

5. CONCLUSIONS

A single-stage and a two-stage CO_2 compressor used for heat pump water heater with same displacement have been developed. A test rig for performance test of the CO_2 compressor has been built and the two kinds of compressors were experimentally compared. Based on the tested results, the effects of pressure ratio and superheat on performance have been analyzed. It is shown that the volumetric efficiency and isentropic efficiency of single-stage

CO_2 compressor are higher than that of the two-stage CO_2 compressor at lower pressure ratio. With the increase of pressure ratio, the volumetric efficiency and isentropic efficiency will decrease, and the gaps of efficiencies between the two kinds of compressors become small. Furthermore, there is an optimal pressure ratio for two-stage CO_2 compressor, at that pressure ratio the volumetric efficiency or isentropic efficiency reaches maximum. And the superheat has very small effect on the volumetric efficiency and isentropic efficiency under the testing conditions whether for the single-stage compressor or two-stage compressor.

REFERENCES

Dreiman, N. (2004). Two-stage Rolling Piston Carbon Dioxide Compressor [C] // *Proc. International Compressor Engineering Conference*. C010, West Lafayette, Purdue University.

Hideaki M, Tetsuhide Y, Hideto N. (2006). Development of the Compressor for CO_2 Heat Pump with the Single Rotary Mechanism [C] // *Proc. International Compressor Engineering Conference*. C056, West Lafayette, Purdue University.

Takeshi O, Hiroshi H, Atsuo O, et al.(2006). Reduction of Oil Discharge for Rolling Piston Compressor Using CO_2 Refrigerant. *Proc. International Compressor Engineering Conference*. C095, West Lafayette, Purdue University.

YANG Jun，LU Ping，ZHANG Li, et al. (2008). Development and Performance Measurements of a New Hermetic Rotary CO_2 Compressor. *Journal of Shanghai Jiaotong University*, 42(3). China.

Takayuki Kobayashi, et al. (2012). Commercial Use CO_2 Heat Pump Water Heater with Scrotary Compressor. *The International Symposium on New Refrigerants and Environmental Technology*. Japan.

ZHANG Li, HUANG Xiaolong. (2015). Development and Performance Measurements of a single stage hermetic rotary CO_2 Compressor. *Academic Annual Conference of Chinese Association of Refrigeration*. China.

YANG Jun, QI Zhaogang, et al. (2012). Experimental Study on a Two-stage Rolling PistonCO2 Compressor Based on P-V indicator diagrams. Journal of Mechanical engineering science. vol 226 no. 4 995-1003.e. vol 226 no. 4 995-1003.

CFD Analysis and Experiment Study of the Rotary Two-Stage Inverter Compressor with Vapor Injection

Liying Deng, Shebing Liang, Jia Xu, Yusheng Hu

Compressor and Motor Institute of Gree Electric Appliances, Inc. of Zhuhai,
Jinji West Rd., Zhuhai City, 519070, P. R. China
Phone: +86-756-8668720, Fax: +86-756-8668386,
E-mail: haitunsai@163.com

ABSTRACT

The offset angle of the upper and lower part of the crankshaft will affect the resistance of discharge of low pressure stage cylinder and suction of high pressure stage cylinder in the rotary two-stage inverter compressor with vapor injection, and then affect the performance of compressor. This paper presents the performance of the rotary two-stage inverter compressor in different bias angle of the crankshaft, which calculated by CFD soft and compared with the experimental results. Under the operation of close vapor injection and open vapor injection, the performance of compressor can be improved by optimize the bias angle of crankshaft。

1. INTRODUCTION

Under the extremely atrocious weather conditions for a traditional rolling piston rotary compressor in home air-conditioner, the efficiency of refrigeration or thermal energy usually become worsen due to lack of motivation. The rotary two-stage inverter compressor can solve this problem by open vapor injection.

Huifang Luo et al.(2014) studied the optimal displacement ratio of rotary two-stage inverter compressor with vapor injection based on theory of thermodynamics and experiments. Sicao Zhang et al.(2014) discussed the effect of intermediate pressure gas compensating parameters on the thermal performance of the system. The intermediate pressure gas compensating parameters are changed primarily for exhaust of low pressure stage cylinder and suction of high pressure stage cylinder.

This paper study the influence of offset angle of the upper and lower part of the crankshaft on performance of rotary two-stage inverter compressor by CFD and experimental analysis.

2. CFD Model of Rotary Two-Stage Inverter Compressor

The discharge gas of low pressure stage cylinder mixed with intermediate pressure vapor which injected from flash tank passes through interior channels arrive the gas suction of the high pressure stage cylinder, which form two stage compression. The vapor from flash tank can improve the efficiency of the compressor by cooling the discharge gas which from the low pressure stage cylinder and increasing the suction gas volume of high pressure stage cylinder. On the other hand it causes more exhaust resistance loss by increasing the discharge pressure of low pressure stage cylinder.

Figure 1 shows the mesh model of rotary two-stage inverter compressor which contains suction pipe, low pressure stage cylinder, intermediate pressure chamber, flash tank, high pressure stage cylinder and exhaust chamber. The calculation model use hexahedral grids and total number of mesh is 511,119. Sliding mesh model is adopted, in which the cylinder parts are moving grids, and the rest parts are stationary. Event file is used to connect the adjacent attach boundary and invoke .cgrd file controls the moving mesh.

The turbulent model use K-Epsilon/RNG model, standard wall function is adopted in the near wall treatment. The physical properties of R410A are used as ideal gas. Operation frequency of the rotary two-stage inverter compressor is 60HZ. The inlet of suction pipe, flash tank and the outlet of exhaust chamber all use pressure boundary.

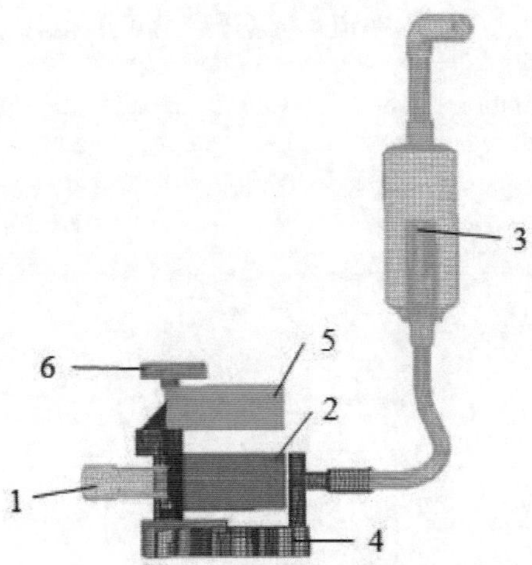

1. Suction pipe 2. Low pressure stage cylinder 3. Flash tank 4.Intermediate pressure chamber 5. High pressure stage cylinder 6. Exhaust chamber

Figure 1: Mesh model of rotary two-stage inverter compressor

This paper has three calculation schemes; crankshaft and main mesh part of each scheme are showed in figure 2:
Scheme 1: Use the crankshaft of traditional rotary double cylinder compressor, the offset angle of upper and lower part of the crankshaft is 180 degrees. (Fig 2(a))
Scheme 2: On the basis of Scheme1, the offset angle of upper and lower part of the crankshaft is counter-clockwise 120 degrees. (Fig 2(b))
Scheme 3: On the basis of Scheme1, the offset angle of upper and lower part of the crankshaft is clockwise 120 degrees. (Fig 2(c))

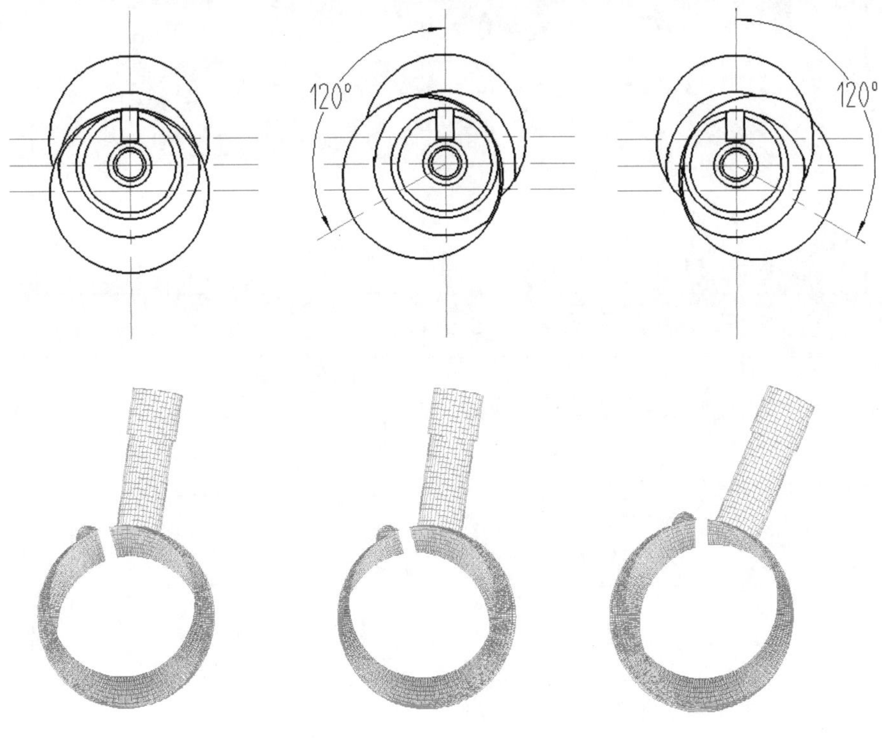

(a) Scheme 1 (b) Scheme 2 (c) Scheme 3

Figure 2: Crankshaft and main mesh part of each scheme

3. Results of CFD Analysis

The first three period results of calculation are listed in figures as calculation become stability from the third period. The low pressure stage cylinder is tagged as lower and the high pressure stage cylinder is tagged as upper in result figures.

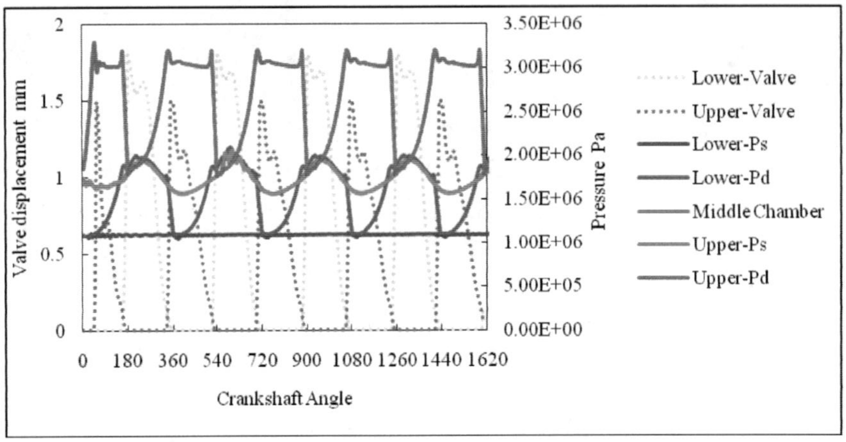

(a) Scheme 1

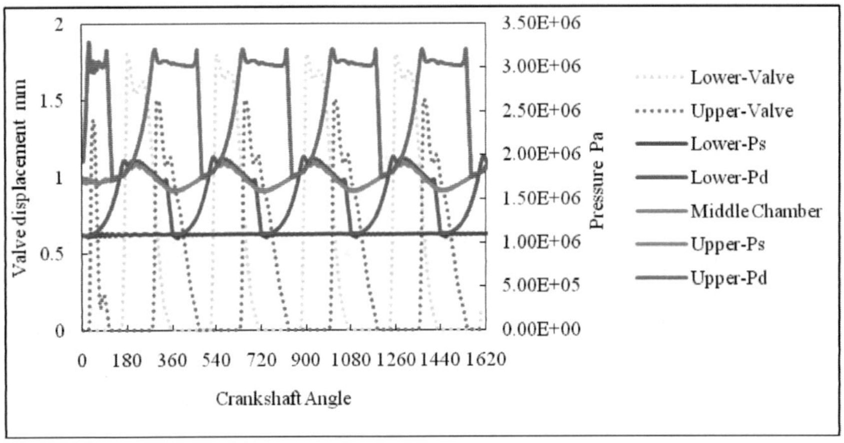

(b) Scheme 2

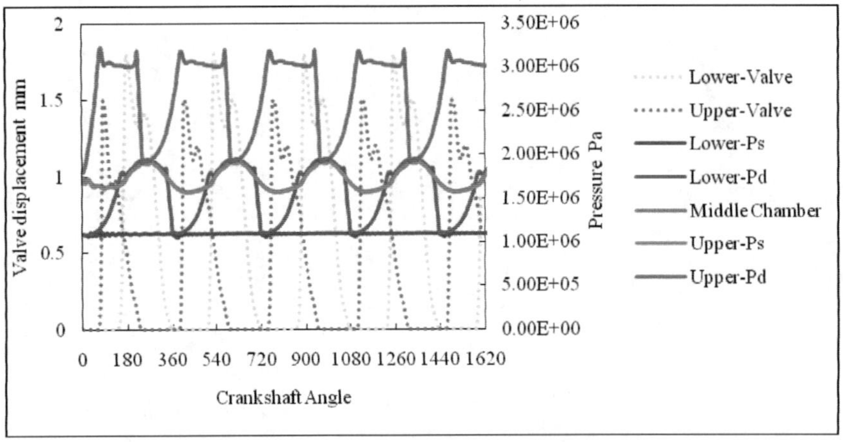

(c) Scheme 3

Figure 3: Valve displacement and pressure translation curves of each scheme

Figure 3 shows the displacement curves of valves and pressure translation curves of each chamber. Horizontal axis is Crankshaft angle, Vertical axis on the left is valve displacement and on the right is pressure.
As figure 3(a) shows, the crankshaft angle of upper valve open is near to the close angle of lower valve. The upper cylinder starts a new round compression and suction after the valve of lower cylinder has opened about 30 degrees.
In figure 3(b), the crankshaft angle of upper valve open is before the close angle of lower valve. And a new round compression of the upper cylinder is before the gas discharge of lower cylinder. The interval angle is about 30 degrees.
In figure 3(c), the starting point of a new round compression of upper cylinder is at the middle point of discharge gas line of the lower cylinder.
Figure 4 shows the flux curves of the lower cylinder, the flash tank and the upper cylinder of each scheme, the peak exhaust gas flow rate of lower cylinder is according decreasing order of scheme 2, scheme 1and scheme 3; The inflow rate peak of flash tank is almost equivalent for scheme 1 and scheme 3 which both greater than scheme 2; But the reflux of scheme 1 and scheme 3 also greater than scheme 2 significantly, and the greatest reflux is scheme 1.

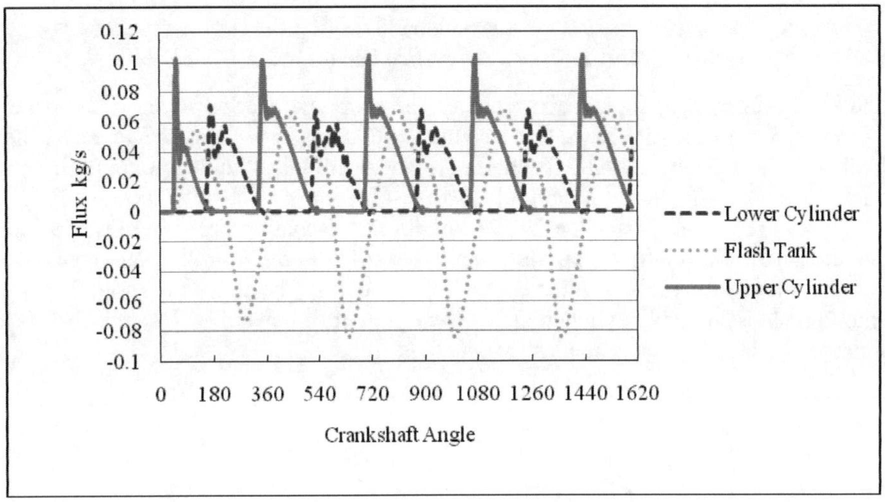

(a) Scheme 1

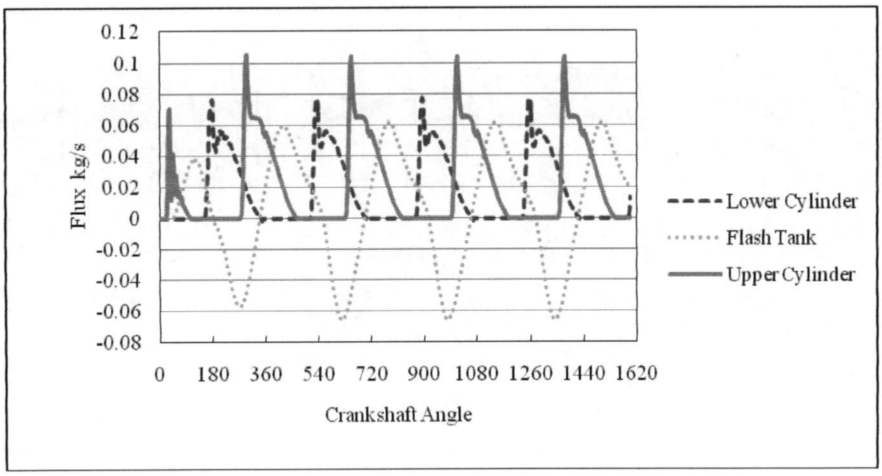

(b) Scheme 2

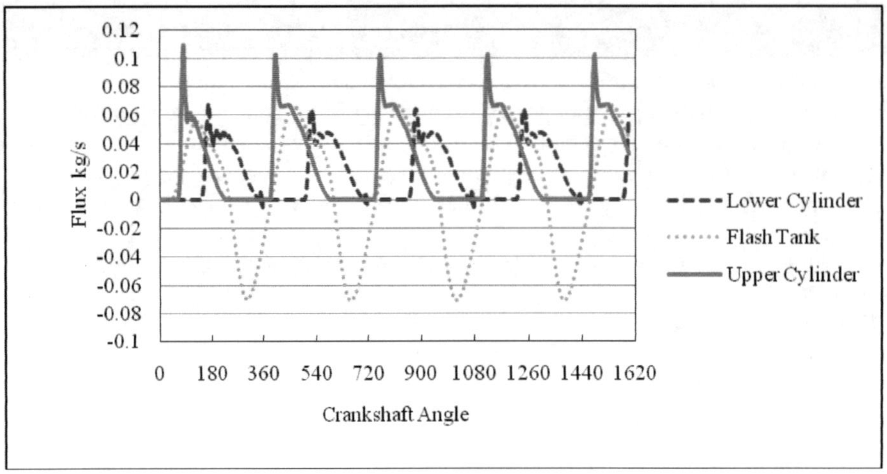

(c) Scheme 3

Figure 4: Flux curves of each scheme

Figure 5 shows the PV curves comparison of each scheme. The lower cylinder's PV curves almost overlap during the compression process. Significant difference between the lower cylinder's PV curves of each scheme appeared in the stage of gas exhaust. The exhaust pressure of scheme 2 is the maximal at the beginning of discharge, when compression volume change smaller than 4000mm^3, the exhaust pressure of scheme 1 become bigger than scheme 2 gradually. After compression volume is smaller than 2300mm^3, the discharges pressure of scheme 2 is becoming the smallest while the scheme 3 is becoming the biggest. The upper cylinder's PV curves both have obviously difference during either the suction process or the compressing process. The area which contains discharge loss of lower cylinder and suction loss of upper cylinder enclosed by pressure curves of each scheme are listed on the right of figure 5 and the area of scheme 2 is the smallest among these schemes.

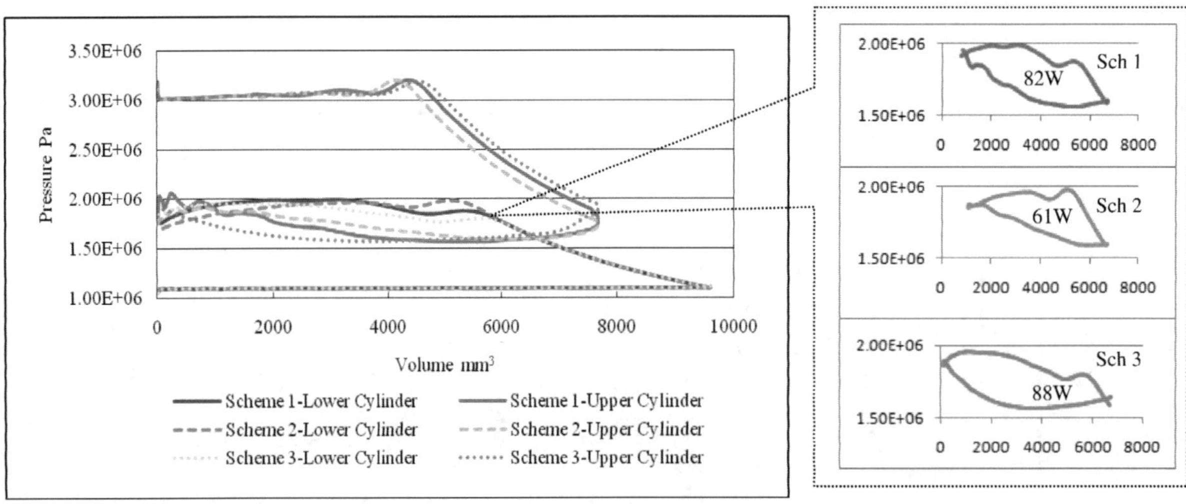

Figure 5: PV curves comparison of each scheme

Table 1: Exhaust flux and indicated power of each scheme

	Flux(Kg/h)				PV(W)			
	Lower Cylinder	Flash Tank	Upper Cylinder	Difference	Lower Cylinder	Upper Cylinder	Total	Difference
Scheme1	65.39	16.31	81.70	/	355.99	508.93	864.92	/
Scheme2	65.29	15.68	80.97	-0.90%	349.15	472.60	821.75	-4.99%
Scheme3	65.09	17.21	82.29	0.72%	341.15	544.76	885.92	2.43%

According to figure 4 and figure 5 statistical exhaust flux and indicated power of each scheme are shown in table 1 which use scheme 1 as the original contrast data. Discharge flux from upper cylinder of scheme 2 has decreased 0.9% due to the flux of lower cylinder and flash tank both have decreased slightly, while in scheme 3 the flux of the upper cylinder has increased 1.78% . The total indicated power of scheme 2 has decreased 4.99% while in scheme 3 have increased 2.43%.
By comprehensive comparison, in the case of little change in the amount of cooling, the scheme 2 can be used as optimal plan due to the power consumption has significantly reduced.

4. Results of Experimental Research

Experiments were made to compare the performance of each scheme under the conditions of close vapor injection and open vapor injection which results have been shown in figure 6.The operating frequency is 60HZ and 90HZ. Horizontal axis is operating frequency and name of each scheme, vertical axis on the left is energy and power while on the right is COP. The differences of energy and power between three schemes are not very distinct. But the COP of the scheme 2 is the best both under the operating condition of close vapor injection and open vapor injection.

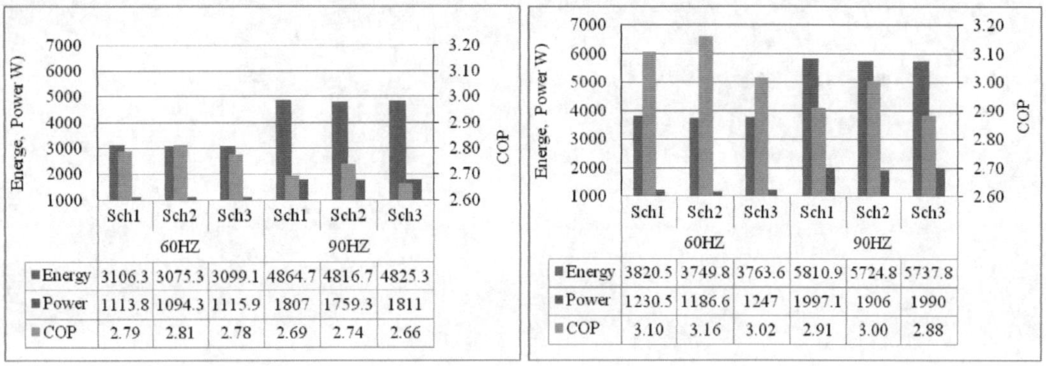

Figure 6: Performance compared with scheme 1

Table 2 list the percent of performance difference compared with scheme 1. The energy both have decreased for scheme 2 and scheme 3 whenever vapor inject is close or open under operating frequency of 60HZ and 90HZ. The power consumption have decreased obviously for scheme 2 while increased slightly for scheme 3 except under open vapor inject operating condition at 90HZ.
Over all, the scheme 2 is the best in performance due to the reduction of power consumption. Scheme 3 has no obvious advantages in both the energy and the power consumption. The calculated results are in agreement with the experimental results in prediction the optimal scheme.

Table 2: Percent of performance difference compared with scheme 1

Scheme	vapor inject	Frequency	Energy	Power	COP
Sch2	Close	60 HZ	-1.00%	-1.75%	0.77%
		90 HZ	-0.99%	-2.64%	1.70%
	Open	60 HZ	-1.85%	-3.57%	1.78%
		90 HZ	-1.48%	-4.56%	3.23%
Sch3	Close	60 HZ	-0.23%	0.19%	-0.42%
		90 HZ	-0.81%	0.22%	-1.03%
	Open	60 HZ	-1.49%	1.34%	-2.79%
		90 HZ	-1.26%	-0.36%	-0.91%

5. CONCLUSIONS

The CFD model of the optimal offset angle between upper part and lower part of the crankshaft of rotary two-stage inverter compressor with vapor injection has been developed and verified with the experimental data in this paper. From above analysis and comparison, changing offset angle between upper part and lower part of the crankshaft have slightly influence on energy of rotary two-stage inverter compressor, but the optimal offset angle between upper part and lower part of the crankshaft can improve the performance by decrease power consumption of rotary two-stage inverter compressor obviously especially with vapor injection.

REFERENCES

Huifang Luo et al. 2014, Theoretical and Experimental Research on The Optimal Displacement Ratio of Rotary Two-Stage Inverter Compressor With Vapor Injection. International Compressor Engineering Conference at Purdue, July 14-17, 2014, Paper 1262.

Sicao Zhang et al.2013, Experimental investigation on working performance of two-stage compression heat pump system with vapor injection. Cryo. & Supercond. 2013, 41(5) : 40-43.

Comparative Research on Air Conditioner with Gas-injected Rotary Compressor through Injection Port on Blade

Xingru LIU[1], Baolong WANG[2]*, Wenxing SHI[3]

Department of Building and Science, School of Architecture,
Tsinghua University, Beijing, 100084, China
[1] Tel.: Tel.: +86 18711346668; Fax: +86 10 62786571; E-mail: liuxr11@yeah.net
[2] +86 10 62786571; Fax: +86 10 62786571; E-mail: wangbl@tsinghua.edu.cn
[3] Tel.: +86 10 62796114; Fax: +86 10 62786571; E-mail: wxshi@mail.tsinghua.edu.cn
* Corresponding Author

ABSTRACT

Economizer technology has become a critical method to improve the performance of air conditioners with rotary compressor in low ambient temperature. A novel vapor injection structure on the blade of the rotary compressor has been proposed in previous research to overcome the drawback of the traditional cylinder injection structure. Based on a verified numerical system model, the performance of an air conditioner adopting different economized rotary compressors has been investigated. The results indicate that: compared to the single cylinder rotary compressor with traditional injection structure, the rotary compressor with blade injection structure can enhance the heating capacity and COP by 12.9%~15.7% and 1.5%~4.1% in low ambient temperature, respectively. And compared to the twin-cylinder rotary compressor, the rotary compressor with blade injection structure has a competitive performance.

1. INTRODUCTION

Economizer technology has previously been used in centrifugal compressors, screw compressors and scroll compressors. Many investigations indicated the economizer technology could effectively improve the compressor performance in low ambient temperature [1-3].However, the application of economizer in single-cylinder rotary compressors, which is also called vapor injection into the rotary compressor, is not as wide as twin-cylinder rotary compressors and scroll compressors [4-5]. The main barrier constraining the application of economizer in single-cylinder rotary compressors is its inherent compressor structure. For the single-cylinder rotary compressor with conventional injection structure, whose injection port is opened in the cylinder and close to the discharge port, considerable injected refrigerant gas directly flows back into the suction tube. The back-flowing will reduce the practical suction refrigerant from the suction tube and decrease the volumetric efficiency of the compressor. On the other side, constrained by the limited space for injection port and considering to avoid a larger back-flowing, the area of the injection port could not be enlarged. Finally, back-flowing of the injected refrigerant and limited injection port area leads to the constricted injection mass flow-rate and constricted enhancement on the performance. That is the main reason why the economizer technology is not well developed in single-cylinder rotary compressor, not like in twin-cylinder rotary compressors and scroll compressors.

However, compared to twin-cylinder rotary compressors, the single-cylinder rotary compressor has some advantages, such as simpler structure, lower cost and better reliability. So if the gas injection performance of the single-cylinder rotary compressor can be greatly improved, it will produce great value. According to above, a novel vapor injection structure on the blade for the single-cylinder rotary compressor has been proposed [6]. Preliminary research indicated that the proposed vapor injection structure can eliminate the back-flowing of the injected refrigerant completely. Meanwhile, the novel blade injection structure can enhance the heating capacity and COP of the compressor under the constant evaporating temperature and the constant condensing temperature by 23.1~48.9% and 3.2~8.0%, respectively, compared to the traditional injection structure.

To evaluate the system performance of single-cylinder rotary compressor with the new blade injection structure, a numerical study on an air conditioner equipped with rotary compressor is conducted in this research. Three kinds of rotary compressors with same suction volume, including a twin-cylinder rotary compressor, a single rotary compressor with traditional injection structure, and a single rotary compressor with the blade injection structure, are respectively adopted in the air conditioner for comparison.

2. AIR CONDITIONERS WITH DIFFERENT ECONOMIZED ROTARY COMPRESSORS

Figure 1 shows system schematics of air conditioners with different rotary compressors. Figure 1(a) present an air conditioner with twin-cylinder rotary compressor. Figure 1(b) present an air conditioner with single-cylinder rotary compressor using traditional injection structure. Figure 1(c) shows the air conditioner with a rotary compressor adopting the new blade injection structure. The high pressure refrigerant gas coming from discharge port of compressor is condensed in the condenser into high pressure liquid, then throttled by the first expansive valve and flows into the flash tank. In the flash tank, the saturated refrigerant gas at the top of tank goes through the injection pipe, the injection passage and the injection port on the blade and finally is injected into the compression pocket. The other refrigerant in the flash tank, saturated liquid refrigerant, is throttled by the second expansive valve and flows into the evaporator. The refrigerant is sucked by the compressor into the suction pocket after heated in the evaporator into a superheated gas.

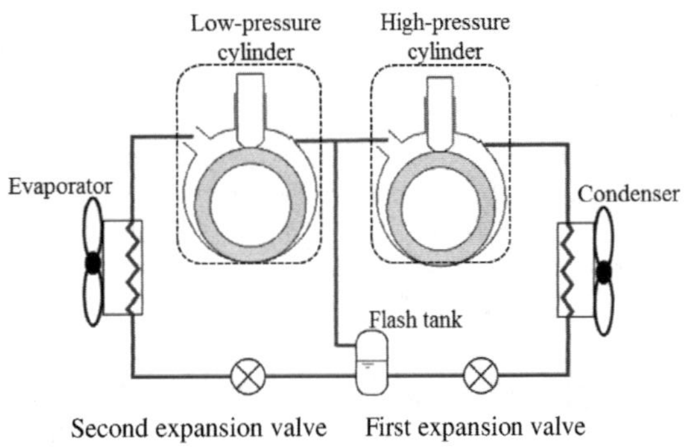

(a) With twin-cylinder rotary compressor

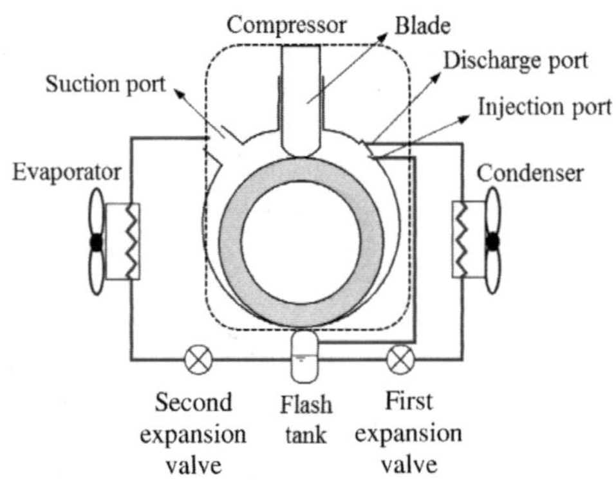

(b) With rotary compressor using cylinder injection

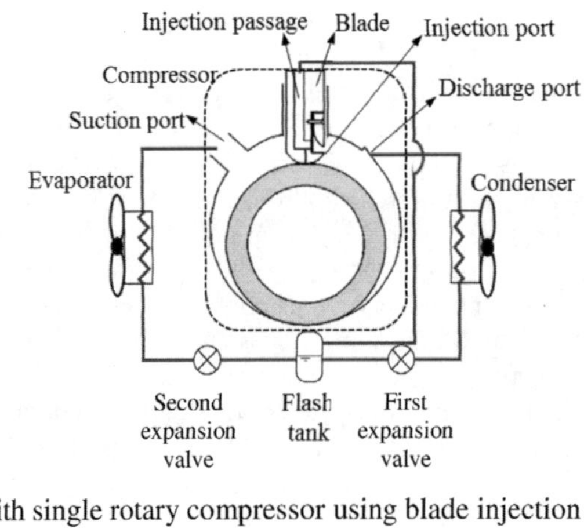

(c) With single rotary compressor using blade injection

Fig. 1 Air conditioners with different rotary compressors

The difference between Figure 1(b) and Figure 1(c) is the position of injection port. The injection port of the rotary compressor with traditional injection structure is opened in the cylinder as Figure 1(b), in which the back-flowing of the injected refrigerant gas into the suction tube is inevitable.

3. MODELLING

In order to investigate the performance of the air conditioner with different economized rotary compressors, the detailed distributed-parameter models of three kinds of rotary compressors and systems are built as follows.

3.1 Compressor

Except for the injection model, the models of the gas-injected rotary compressors, including the rotary compressor with traditional cylinder injection structure and the rotary compressor with blade injection structure, are completely same. The models of these two kinds of gas-injected rotary compressor have been established in previous research [6]. For the twin-cylinder rotary compressor, the working principle actually is two-stage compression process. It should be noted that, because the flash tank is used as economizer, the injected gas refrigerant mass flow-rate must equal to the gas generated in the flash tank, which is one of the important convergence index.

3.2 Heat exchanger, flash tank and expansive valve

A distributed-parameter model is developed to describe the characteristics of the condenser and the evaporator. Gnielinsk [7] correlations is adopted to calculate the in-tube convective heat transfer coefficient of refrigerant at the single-phase region. Gungor-Winterton [8] correlations are used to calculate the in-tube convective heat transfer coefficient of refrigerant at two-phase zone of the evaporator. Dobson-Chato [9] model is used to calculate the in-tube convective heat transfer coefficient of refrigerant at two-phase zone of the condenser. Friede [10] model is applied to calculate the pressure drop of refrigerant in tube.

For the flash tank, the inlet refrigerant enthalpy and the total outlet enthalpy both are equal to the refrigerant enthalpy at the outlet of the condenser. And the outlet gas and the outlet liquid both are saturated.

3.3 System model

C++ was used to solve equations because the solution procedure includes many submodules. And the method of variable step size was applied during calculating to reduce the calculation time. The method of 4th Order Runge-Kutta was used to improve calculation accuracy and accelerate the convergence process. Besides, the REFPROP, which is an international common software, was called to calculate refrigerant properties.

3.4 System model verification

To verify the accuracy of the model, the experimental data from a public papers [11] are collected to compare with simulation results. Figure 2 illuminates the simulation results of the air source heat pump with cylinder gas-injected rotary compressor compared to experimental results, which indicates that the heating capacity error is within 6% and the power consumption error is within 7%. Figure 3 presents the simulation results of air source heat pump with a two-stage rotary compressor compared to experimental results, which indicates the heating capacity error is within 7% and the power consumption error is within 6%. Generally speaking, the model have relative high accuracy and can be used in the further research.

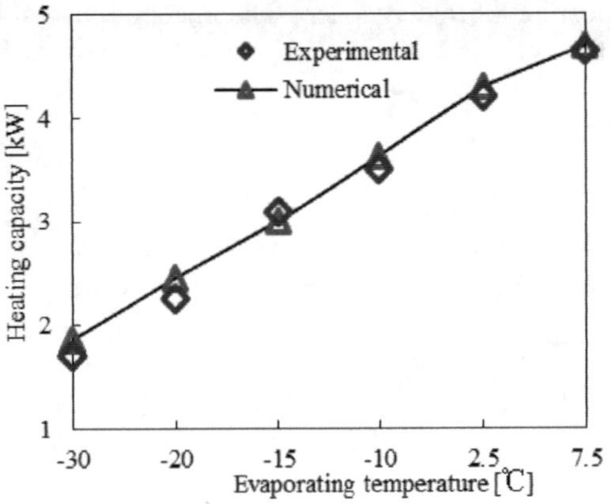

Fig. 2. Rotary compressor with traditional injection structure

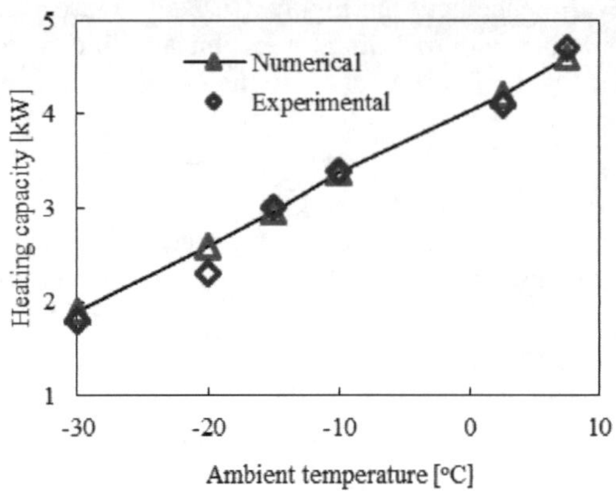

Fig. 3. Twin-cylinder rotary compressor

4. RESEARCH SCENARIO

Just as aforementioned, the gas injection will enhance the system heating performance under low ambient temperature conditions. So the heating performance of the air conditioners under low ambient temperature conditions are investigated in this paper. The suction volumes of single-cylinder rotary compressors and the low-pressure cylinder of the twin-cylinder rotary compressor keep same. The suction volume ratio of the high-pressure cylinder to the low-pressure one of the twin-cylinder is selected as 0.7 according to previous optimal research [4].

5. RESULTS AND DISCUSSION

Figure 4 shows the refrigerant flowrate in the evaporator and condenser. According to Figure 4(a), the refrigerant mass flowrate of twin-cylinder system in evaporator is greater than the air conditioner with single rotary compressor as same suction volume and close evaporating temperature for three kinds of compressors and the refrigerant injection has a clear negative effect on the suction of the rotary compressor, which can be attributed to the increase of the inner leakage of blade injection compressor[6]. The mass flowrate for the air conditioner with the blade injection is greater than the cylinder injection system as the heavy back-flowing of injected refrigerant in cylinder injection compressor [6].

As an addition of refrigerant flowrate in evaporator and injected refrigerant flowrate, the condenser mass flowrate of the two-stage rotary compressor is greater than the single-cylinder rotary compressor with the blade injection structure and the formers both are greater than the rotary compressor with traditional injection structure as Figure 4(b).

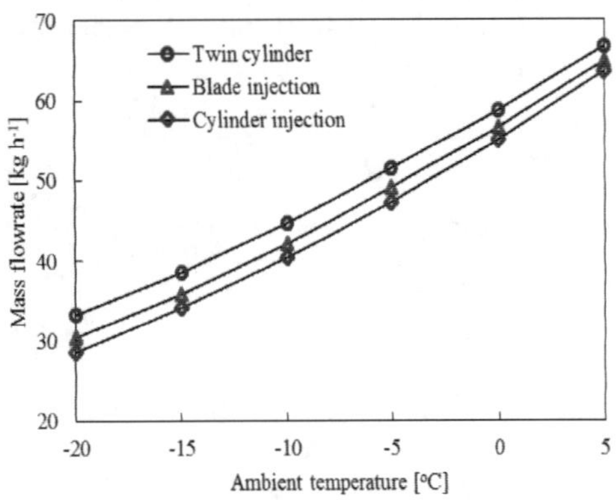

(a) Evaporator side

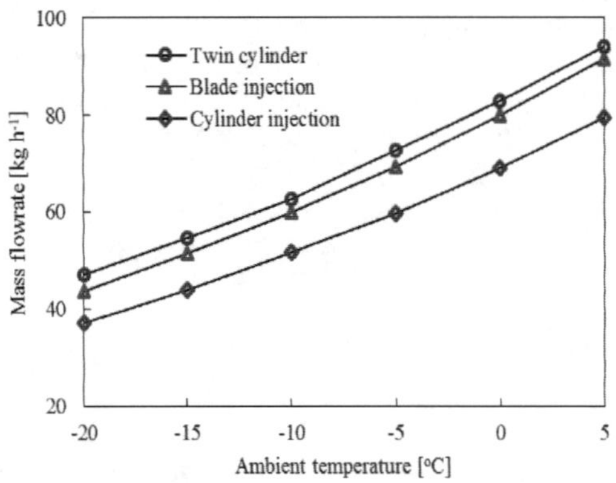

(b) Condenser side

Fig. 4 comparison of system refrigerant mass flowrates

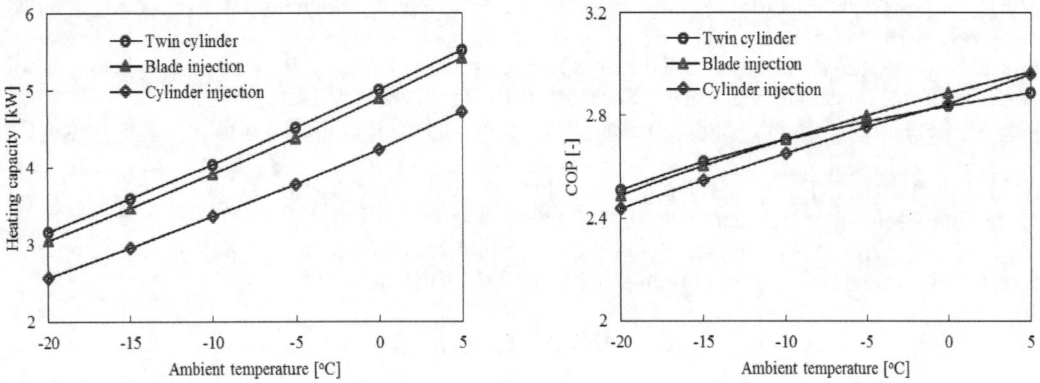

(a)Heating capacity (b)COP

Fig. 5 System performances of different air conditioners with different economizer

Figure 5(a) and Figure 5(b) show the heating capacity and COP of air conditioner with different economizer. The heating capacity of the rotary compressor with the novel injection structure is slightly smaller than the one with the two-stage rotary compressor under all the working conditions. The blade injection system shows a slight improvement in COP relative to the twin-cylinder rotary compressor when the ambient temperature is higher than -10°C. But generally, the air conditioner with blade injected rotary compressor has a competitive performance with the air conditioner with twin-cylinder rotary compressor. On the other side, compared to the rotary compressor with cylinder injection structure, the heating capacity and COP with the new blade injection structure is enhanced by 12.9%~15.7% and 1.5%~4.1%, respectively.

6. CONCLUSIONS

Based on a verified numerical model, air conditioners with different rotary compressors, including a two-stage rotary compressor, a single rotary compressor with traditional injection structure and a single rotary compressor with the novel injection structure, are investigated and the results indicate:

(1)The air conditioner with blade injected rotary compressor has a competitive performance with the air conditioner with twin-cylinder rotary compressor, and the rotary compressor with the blade injection structure could be considered to replace the two-stage rotary compressor.

(2)Compared to the air conditioner with cylinder injection, the blade injection structure can enhance heating capacity and COP by 12.9%~15.7% and 1.5%~4.1%, respectively.

(3)The heating capacity of the rotary compressor with the blade injection structure is slightly smaller than the one with the two-stage rotary compressor under all the working conditions. The blade injection system shows a slight improvement in COP relative to the twin-cylinder rotary compressor when the ambient temperature is higher than -10°C.

REFERENCES

[1] B. L. Wang, W. X. Shi, X. T. Li, Numerical analysis on the effects of refrigerant injection on the scroll compressor. Applied Thermal Engineering 29 (2009) 37–46.

[2] G. Y. Ma, Q.Chai, Characteristics of an improved heat-pump cycle for cold regions. International Journal of Refrigeration 26 (2003) 12–18.

[3] J.Heo, M. W. Jeong,C. Baek, Comparison of the heating performance of air-source heat pumps using various types of refrigerant injection. International Journal of Refrigeration 34 (2011) 444–453.

[4] S. X. Xu, G. Y. Ma, Experiment study on two-stage compression refrigerant/heat pump system with dual-cylinder rolling piston compressor. Applied Thermal Engineering, 62 (2014) 803–808.

[5] Y. Ko, S. Park, S. Jin, The selection of volume ratio of two-stage rotary compressor and its effects on air-to-water heat pump with flash tank cycle. Applied Energy, 104 (2013) 187–196.

[6] X. R. Liu, B. L. Wang, W. X. Shi, P. L. Zhang, A novel vapor injection structure on the blade of a rotary compressor, Applied Thermal Engineering (2016), http://dx.doi.org/doi: 10.1016/j.applthermaleng.2016.02.098.

[7] Gnielinski V. New equations for heat mass transfer in turbulent pipe and channel flows. International Chemical Engineering, 1976, 16: 359-368.
[8] Gungor K E, Winterton R H S. Simplified general correlation for saturated flow boiling and comparison of correlations with data. Chemical Engineering Research & Design, 1987, 65: 148-156
[9] Dobson M K, Chato J C. Condensation in Smooth Horizontal Tubes. ASME Journal of Heat Transfer 1998, 120: 193–213
[10] Didi M B O, Kattan N, Thome J R. Prediction of two-phase pressure gradients of refrigerants in horizontal tubes. International Journal of Refrigeration, 2002, 25: 935–947
[11] Q. L. Jia, L. W. Feng, G. Yan, Experimental Research on Heating Performance of Rotary Compression System with Vapor Injection. Journal of Refrigeration 36 (2015) 65–70. (In Chinese)

ACKNOWLEDGEMENT

This research is funded by Innovative Research Groups of the National Natural Science Foundation of China (grant number 51521005).

Development of a New Dual-Cylinder Rotary Compressor for VI System

Guoyong Yang[1*], Cheng Zhang[2], Yanping Wu[3], Siqing Liao[4]

Guandong Meizhi Compressor Co. Ltd. R&D Section,
Foshan City, Guangdong Province, China, 528333

[1]Phone: +86-757-22325058, E-mail: yangguoy@chinagmcc.com
[2]Phone: +86-757-22329089, E-mail: zc@chinagmcc.com
[3]Phone: +86-757-22325073, E-mail: wuyp2@chinagmcc.com
[4]Phone: +86-757-22329158, E-mail: liaosq@chinagmcc.com

ABSTRACT

On the vapor compression refrigeration system, Vapor injection (VI, the phase separator type injection or the internal heat exchanger type injection) compression cycle's superiority over non-injection cycle has been well known. VI system produces the high heating/cooling capacity, and its power consumption is less than the non-injection system. But if a VI compression cycle uses a single rotary compressor, there is the problem that refrigerant injection increases the indicated power by mixture loss. If we use a two-stage rotary compressor, indicated power also increases because of its two times exhaust process. To solve these problems, we developed a new dual-cylinder rotary compressor for VI systems, one of the cylinders is used to compress the gas from the evaporator, and the other is used to compress the gas from the phase separator (flash-tank). Its design method is discussed and its performance under different conditions is analyzed.

1. INTRODUCTION

The demands of people about the efficient energy consumption are increasing continuously by reason of the exhaustion of fossil fuel and global warming. To meet these demands, most governments are strengthening the regulation of the CO2 emission and standard about certification of products and each company is spurred to the development of new technology for the high efficiency. Also, the demand for high efficiency in heating and cooling system is increasing by a growing customer's recognition about the energy conservation. On the vapor compression refrigeration system, Vapor injection compression cycle's superiority over non-injection cycle has been well known. VI system produces the high heating/cooling capacity, and its power consumption is less than the non-injection system. Unfortunately, other problems arise, such as mixture loss on refrigerant injection into the compression chamber of a single-stage compressor. If a two-stage compression cycle uses a single rotary compressor, there is the problem that refrigerant injection increases the indicated power by mixture loss (Sekiya *et al.*, 2005), compared to that of ideal isentropic compression.

Another solution for the two-stage compression cycle is the use of a two-stage compressor. A two-stage flash tank cycle (FTC) contains two additional components compared to the standard vapor compression cycle, namely an additional expansion valve and a flash tank separator. The two-stage flash tank cycle plotted on a P-h diagram is shown in Figure 1. In a two-stage flash tank cycle the condensed liquid leaving the condenser is first throttled through the first stage expansion device and the two-phase refrigerant (state point 4) enters in the flash tank separator. In the flash tank the liquid and vapor components of the two-phase mixture are separated. The saturated vapor refrigerant (state point 5) is injected at an intermediate compressor suction port and the saturated liquid refrigerant (state point 6) is throttled through the second stage expansion device prior to entering the evaporator. After entering the compressor, the saturated vapor refrigerant mixing with the exhaust gas form the first stage, and then was compressed by the second stage together. In a two-stage flash tank cycle, usually we use a two-stage rotary compressor, because most of the gas is compressed two times, the indicated power increases because of its two times exhaust process.

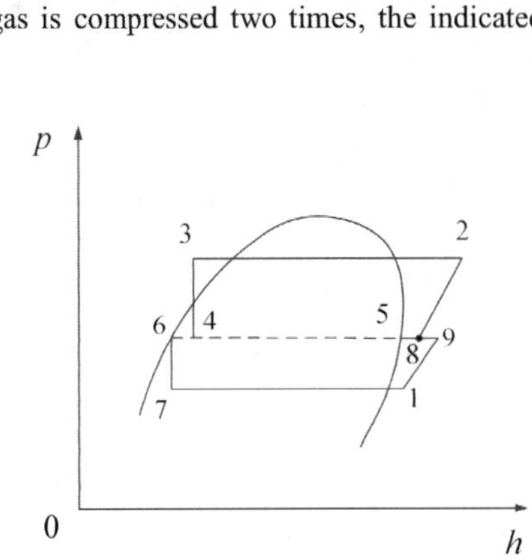

Figure 1: Two-stage compression refrigeration/heat pump system

2. MODEL OF NEW DUAL-CYLINDER ROTARY COMPRESSOR VI SYSTEM

The new dual-cylinder rotary compressor in this paper is a rotary compressor provided with two cylinders, each of which is provided with an independent suction port. The two cylinders woke synchronously by the same electric motor.

The schematic diagram of the new dual-cylinder rotary compressor VI system plant and its corresponding thermodynamic diagram are shown in Figure 2. In this new dual-cylinder rotary compressor VI system, the condensed liquid (state point 3) leaving the condenser is first throttled through the first stage expansion device and the two-phase refrigerant (state point 4) enters in the flash tank separator. In the flash tank the liquid and vapor components of the two-phase mixture are separated. The saturated vapor refrigerant (state point 5) is sucked in by one of the cylinders (the subordinate cylinder) and compressed directly to the exhaust pressure (state point 8) and the saturated liquid refrigerant (state point 6) is throttled through the second stage expansion device prior to entering the evaporator (state point 7), and then, after evaporation (state point 1), is sucked into the other cylinder (the main cylinder) and compressed to the exhaust pressure (state point 9). The gas discharged from the two cylinders is discharged after mixing in the compressor (state point 2), and then is condensed (state point 3). Because the gas is compressed independently, mixed losses and two exhaust losses are avoided. The state point 3' is assumed to be obtained by direct throttling of the state point 3, in fact it does not exist in this system. By assuming the state point 3', we get the basic cycle (state point 1-9-3-3') under the same conditions.

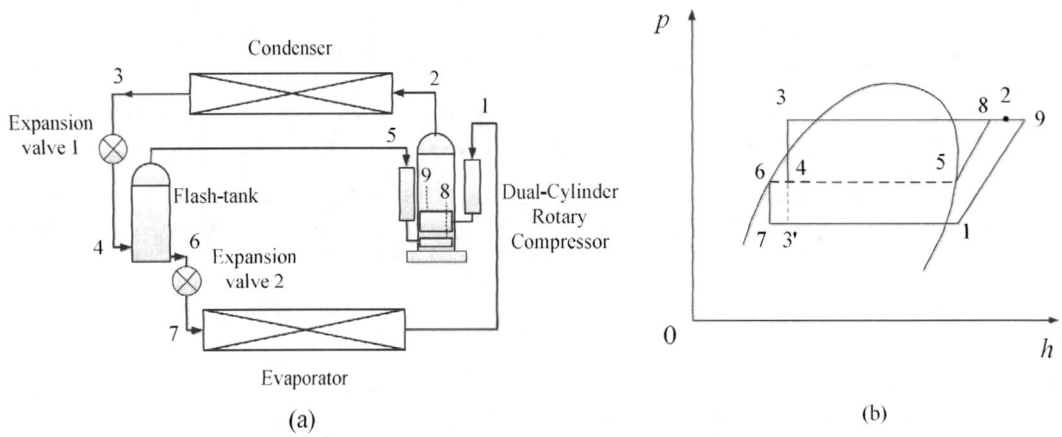

Figure 2: New dual-cylinder rotary compressor VI system

3. THERMODYNAMIC ANALYSIS

For system shown in Figure 2, mathematical is presented as below:
i. The refrigerating capacity per unit weight in evaporator is

$$q_o = h_1 - h_7 \tag{1}$$

ii. The power consumption for the main cylinder is

$$w_{1-9} = \frac{h_{9s} - h_1}{\eta_{s,1\text{-}9}\eta_{\mathrm{m}}\eta_{\mathrm{mo}}} \tag{2}$$

$\eta_{s,1\text{-}9}$ is the isentropic efficiency of compression process 1-9, h_{9s} is the enthalpy assuming that compression process 1-9 is isentropic, η_{m} is the mechanical efficiency of the compressor, η_{mo} is the electrical efficiency of the compressor.
iii. The power consumption for the subordinate cylinder is

$$w_{5-8} = \frac{(m_5 / m_1)(h_{8s} - h_5)}{\eta_{s,5\text{-}8}\eta_{\mathrm{m}}\eta_{\mathrm{mo}}} \tag{3}$$

$\eta_{s,5\text{-}8}$ is the isentropic efficiency of compression process 5-8, h_{8s} is the enthalpy assuming that compression process 5-8 is isentropic.

iv. The cooling COP of the new dual-cylinder rotary compressor VI system is

$$\mathrm{COP_L} = \frac{q_o}{w_{1-9} + w_{5-8}} \tag{4}$$

v. As contrast, the cooling COP of the basic refrigeration cycle (base system) is

$$\mathrm{COP_{BASE}} = \frac{h_1 - h_{3'}}{w_{1-9}} = \frac{h_1 - h_3}{w_{1-9}} \tag{5}$$

In equations (2) and (3), $\eta_{s,1\text{-}9}$ and $\eta_{s,5\text{-}8}$ were calculated by the follow formula (Xu Shuxue and Ma Guoyuan, 2014)

$$\eta_s = 1 - 0.6[1 - (p_d / p_s)^{-0.3}] \tag{6}$$

p_d is the discharge pressure, and p_s is the suction pressure. The efficiency ($\eta_{\mathrm{m}}\eta_{\mathrm{mo}}$)for the compressor take a value of 0.70.
The following assumptions have been made to simplify the analyses:
(1) System operates in steady state.
(2) Heat transfer with the surroundings is negligible.
(3) No pressure drop occurs in evaporator, condenser, flash tank economizer, and their connecting pipelines.

Based on Equations (1) -(6) and the properties of the refrigerant, R410A, the variations of cooling performance with the intermediate temperature T_m (temperature of flash tank/state point 6) is calculated. The calculation conditions were: condensing temperature, T_k, of 45℃, evaporating temperature, T_o, of 3℃, 5℃, 7℃ under cooling condition, the superheat is 5℃ and the sub cooling is 5℃; The calculation results are plotted in Figure 3.

As shown in Figure 3, the cooling COP increase firstly and then decrease with the increasing of the intermediate temperature T_m. The intermediate temperature T_m which makes the VI system has the highest COP_L called the optimal T_m. The variations of the optimal T_m with the evaporating temperature T_o is plotted in Figure 4 (the condensing temperature T_k= 45℃, the superheat = 5℃ and the sub cooling = 5℃). As shown in Figure 4, the optimal T_m almost increased linearly with the increasing of the intermediate temperature T_m. Under the optimum intermediate temperature T_m, the variations of the COP_L/COP_{BASE} (the COP ratio of the new VI system to the base system) with the evaporating temperature T_o is plotted in Figure 5(the condition is same as Figure 4), and we can get that the lower the evaporation temperature, the greater the COP improvement.

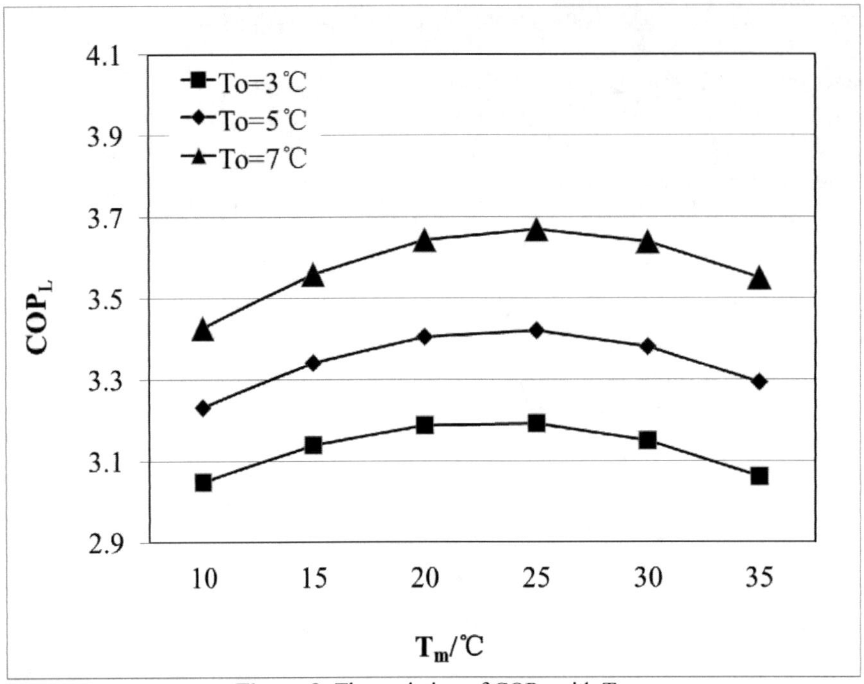

Figure 3: The variation of COP_L with T_m

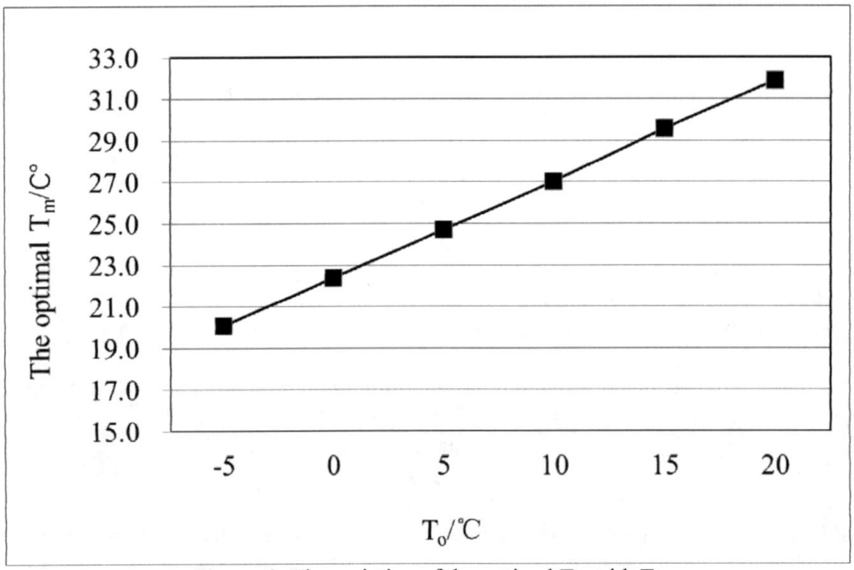

Figure 4: The variation of the optimal T_m with T_o

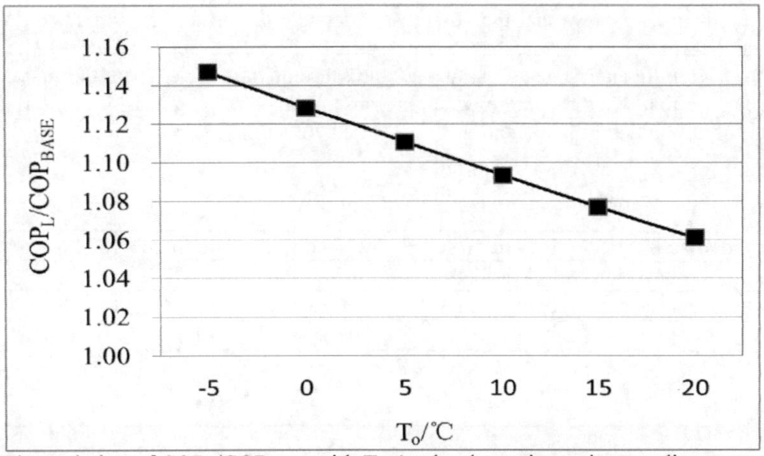

Figure 5: The variation of COP_L/COP_{BASE} with T_m (under the optimum intermediate temperature T_m)

4. DESIGN OF THE NEW DUAL-CYLINDER COMPRESSOR

Based on the above analysis and the refrigerant R410A, a new dual-cylinder compressor used in this VI system is designed, for a residential air conditioner with 3500W target ability at rated condition. Consider the Chinese Annual Performance Factor (APF) requirements, the compressor is designed based on 5 conditions, showed in Table 1.

By the previous analysis, the optimum intermediate temperature of each working condition is different, and the design of the compressor is based on the comprehensive energy efficiency of this 5 conditions. The comprehensive energy efficiency is the weighted average COP_L of the 5 conditions, the contribution ratio is also showed in Table 1 respectively. The design result of displacement of the compressor is shown in Table 2.

Table 1: The refrigeration operating conditions for evaluating performance

Operating condition	The first	The second	The third	The fourth	The fifth
Condensing temperature (℃)	45.6	39.6	43.6	31.5	46.5
Evaporating temperature (℃)	10.7	17.8	0.8	2.9	-4.7
Superheat (℃)	9.3	7.2	5.3	5.1	6.7
Sub cooling (℃)	6.1	1.0	7.0	4.0	9.9
Ambient temperature (℃)	35	35	35	35	35
Contribution ratio	25%	33%	13%	10%	19%

Table 2: The design result of displacement

Parameters	The main cylinder	The subordinate cylinder
Displacement (cc)	10	0.9

5. EXPERIMENTAL RESULT OF THE NEW DUAL COMPRESSOR

Based on the design result, a new dual-cylinder rolling piston compressor was manufactured. At the same time, a comparative traditional compressor for the basic refrigeration cycle is also processed as a base compressor with a similar structure, and then we the tested this these two types of compressors. The test results of cooling COP of the 5 conditions are shown in Table 3, and we can get that the cooling COP increase 4.6%-9.1% under this 5 conditions. The comparison between theory and experiment of COP_L/COP_{BASE} is shown in Figure 6, the experimental results are in agreement with the theory, it shows that the compressor achieve the desired performance goals. In the system operation, the actual intermediate pressure has a great influence on the performance of the compressor. The variations of the COP_L with the intermediate pressure p_m (pressure of flash tank) about the first condition is plotted in Figure 7, it can be seen that when the intermediate pressure exceeds the optimum value, the COP_L is decreased dramatically because of the suction of liquid and this should be avoided in the application.

Table 3: The test result of cooling COP

Operating condition	**The first**	**The second**	**The third**	**The fourth**	**The fifth**
COP_{BASE} of the base compressor	3.824	6.597	2.897	4.951	1.847
COP_L of the new dual compressor	4.033	6.916	3.134	5.181	2.016
COP_L/COP_{BASE}	105.5%	104.8%	108.2%	104.6%	109.1%

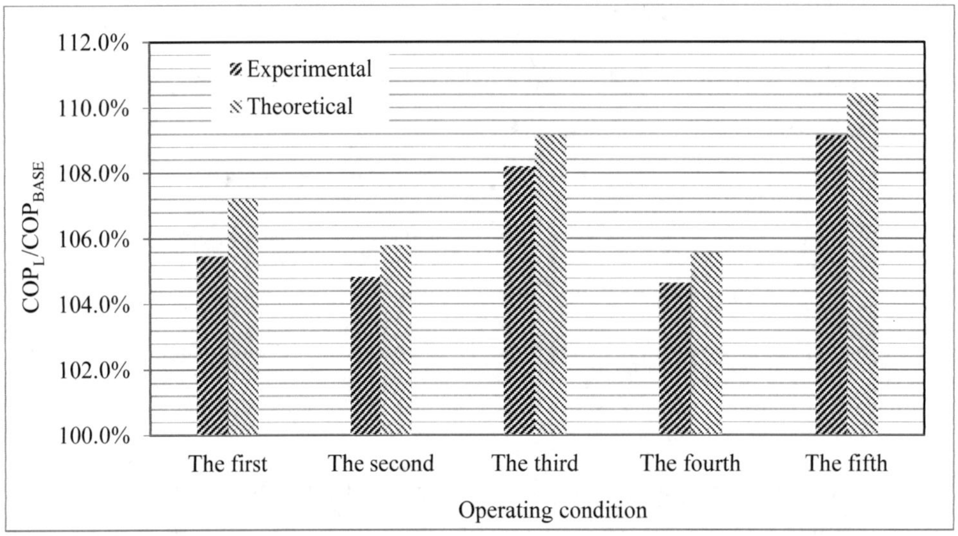

Figure 6: The Comparison between theory and experiment of COP_L/COP_{BASE}

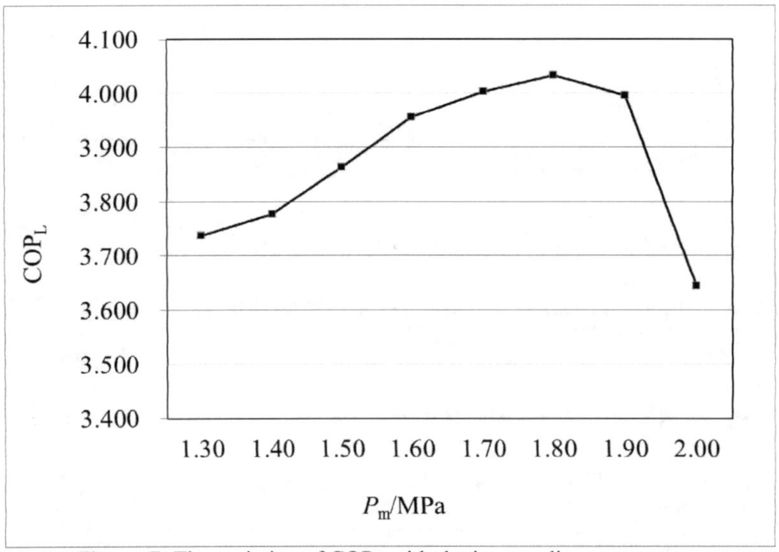

Figure 7: The variation of COP_L with the intermediate pressure p_m

6. CONCLUSIONS

A new dual-cylinder rotary compressor for VI systems is developed, one of the cylinders is used to compress the gas from the evaporator, and the other is used to compress the gas from the phase separator (flash-tank). Its design method is discussed and its performance under different conditions is analyzed:
(1) the cooling COP increase 4.6%-9.1% under the test conditions. The comparison between theory and experiment of COP_L/COP_{BASE} indicates the experimental results are in agreement with the theory.
(2) The variations of the COP_L with the intermediate pressure p_m (pressure of flash tank) indicates that when the intermediate pressure exceeds the optimum value, the COP_L is decreased dramatically because of the suction of liquid and this should be avoided in the application.

NOMENCLATURE

h	enthalpy	(kJ/kg)
m	mass flow rate	(kg/s)
p	pressure	(MPa)
p_d	discharge pressure	(MPa)
p_m	intermediate pressure	(MPa)
ps	suction pressure	(MPa)
T_k	condensing temperature	(℃)
T_o	evaporating temperature	(℃)
T_m	intermediate temperature	(℃)
w	compressor work	(kJ/kg)
COP_L	coefficient of cooling performance	(–)
COP_{BASE}	coefficient of heating performance	(–)
η_i	isentropic efficiency	(–)

Subscript

1, 2, 3, ...	state points shown in the figures

REFERENCES

Sekiya, S., Kakuda, M., Koda, T., Yoshiyasu, H., (2005), *Proceedings of the 39th Japanese Joint Conference on Air conditioning and Refrigeration: 42*, p.155-158.

Xu Shuxue, Ma Guoyuan, (2014). Experimental study on two-stage compression refrigeration/heat pump system with dual-cylinder rolling piston compressor. *Applied Thermal Engineering, 62(2014)*, 803–808

Prediction of Leakage Flow of Radial Clearance in a Rolling Piston Rotary Compressor

Geon-woo Kim[1a], Ki-youl Noh[1b], Byung-chae Min[1c], Sang-jin Song[1d], Sang-kyung Na[1e], Tae-seung Yoon[1f], Kenichiro teshima[2], Jang-sik Yang[3], Gyung-min Choi[4a*], Duck-jool Kim[4b]

[1]Graduate School of Mechanical Engineering, Pusan National University, Busan, Korea
(Tel: +82-51-510-3072, E-mail : [a]kwkim@pusan.ac.kr, [b]nky0317@naver.com, [c]bcmin@pusan.ac.kr, [d]songsj@pusan.ac.kr, [e]naskng@pusan.ac.kr, [f]tsluv89@pusan.ac.kr)

[2]Interdisciplinary Graduate School of Engineering Sciences, Kyushu University, Kyushu, Japan
(E-mail : teshima@phase.cm.kyushu-u.ac.jp)

[3]Rolls-Royce University Technology Centre, Pusan National University, Busan, Korea
(E-mail : yangjs@pusan.ac.kr)

[4]Department of Mechanical Engineering, Pusan National University, Busan, Korea
(E-mail : [a*]choigm@pusan.ac.kr, [b]djkim@pusan.ac.kr)

* Corresponding Author

ABSTRACT

The functional formula of the flow coefficients for the radial clearance in a rolling piston rotary compressor was developed by the experiments and Computational Fluid Dynamics (CFD) to improve the accuracy of a compressor simulation program. The experiments and CFD simulation were conducted to obtain the flow coefficients at the radial clearance under the assumptions of incompressible flow and steady state. The experiments were carried out with the variation of pressure ratio and low upstream pressure using the nitrogen as a working fluid. The CFD simulation was adopted to calculate the flow coefficients at the radial clearance with change of the radial clearance and high upstream pressure. The results of CFD calculation were compared with the experimental results for verifying the reliability of CFD simulation. The functional formula obtained from the results of the CFD calculation was applied to compressor performance simulation for calculating the leakage flow through the radial clearance. The results of compressor simulation with the functional formula showed tolerable errors with the experimental data of the compressor calorimeter in terms of EER (Energy Efficiency Ratio). Thus, the function of the flow coefficients may be acceptable for prediction of leakage through the radial clearance under the various conditions.

1. INTRODUCTION

A rolling piston type rotary compressor has been widely used as a core component in the air-conditioning and refrigeration system. This is mainly due to the advantages like design simplicity, compactness, light weight, low cost and high performance. However, the steep rise in concern for energy conservation and environmental problem demanded the development of a higher efficiency compressor. Among the influenced factors for performance of the rolling piston type rotary compressor, the refrigerant leakage is very important from the point of view of volumetric efficiency. Especially, the leakage through the radial clearance constitutes the major contributions (Pandeya, P and Soedel, W. 1978).

Previous researches show that the leakage through the radial clearance has been calculated using simplified model. This model was applied to calculate the leakage through the radial clearance and predict the performance of a rolling piston rotary compressor.

The base model for predicting the leakage in the radial clearance is the incompressible flow of a gas through a convergent divergent nozzle with isentropic flow. This isentropic model has been applied to calculate the mass flow rate through the radial clearance.

In order to consider the fluid friction loss, several researches modeled this case of leakage assuming compressible flow of pure refrigerant gas. Yanagisawa and Shimisu (1985) designed the radial clearance between the rolling piston and the cylinder wall as a convergent nozzle and a straight friction channel. This paper suggests the general flow coefficient (0.3) to calculate the mass flow rate in the radial clearance. Ian H. Bell et al (2013) studied leakage through the radial clearance by applying the frictional correction term to the isentropic nozzle model with the Reynolds number.

Furthermore, previous researches established model of mass flow rate through the radial clearance with oil-refrigerant mixture (Ferreira, R.T.S et al., 1984), (Dias, J.P. et al, 2011), (Gasche, J.L. et al, 2012).
The flow models of these researches are good to predict the mass flow rate through the radial clearance, but these of model were relatively complicated to apply to compressor simulation and insufficient consideration for operating condition of compressor. B.C. Min et al (2014) developed the simple functional formula of flow coefficient for a discharge valve system with consideration about diameter of discharge port and lift height of the discharge valve to simplify the prediction for the mass flow rate in the discharge valve system for simulating the performance of compressor.

In this work, the functional formula of the flow coefficients at the radial clearance was derived from experiments and CFD simulation under a wide range of operating conditions and developed for simplifying the application to compressor simulation. The comparison between functional formula and 0.3 (Yanagisawa and Shimisu (1985) for predicting mass flow rate through the radial clearance was done with the simulation of compressor performance. Consequently, functional formula of the flow coefficient could be helpful to predict the performance of a compressor for various operating conditions.

2. NUMERICAL ANALYSIS

In this paper, the model of mass flow rate through the radial clearance can be described by the simple orifice flow. The flow through the radial clearance is assumed as the isentropic and one-dimensional flow. Eq. (2) and Eq. (3) were used to calculate the mass flow rate through the radial clearance. The mass flow rate through the nozzle can be classified into two divisions with criteria of the critical pressure ratio as Eq. (1). At first, when the pressure ratio is bigger than the critical pressure ratio, mass flow rate is calculated by using the Eq. (2) under the no-choke condition. However, when the flow reaches choke condition, Eq. (3) is used to calculate the mass flow rate through the radial clearance.

$$P_{cr} = \left(\frac{2}{k+1}\right)^{\frac{k}{k-1}} \tag{1}$$

For $P_{cr} < P_d/P_u$

$$\dot{m} = C_v P_u A \sqrt{\frac{2k}{(k-1)RT_u}\left[\left(\frac{P_d}{P_u}\right)^{\frac{2}{k}} - \left(\frac{P_d}{P_u}\right)^{\frac{k+1}{k}}\right]} \tag{2}$$

For $P_{cr} \geq P_d/P_u$

$$\dot{m} = C_v P_u A \sqrt{\frac{k}{RT_u}} \left(\frac{2}{k+1}\right)^{\frac{k+1}{2(k-1)}} \tag{3}$$

C_v is flow coefficient in each equation. The major influence factors for the actual mass flow rate are presented as the flow area, pressure and temperature condition in this equation. To develop the functional formula of flow coefficient, these factors were considered.

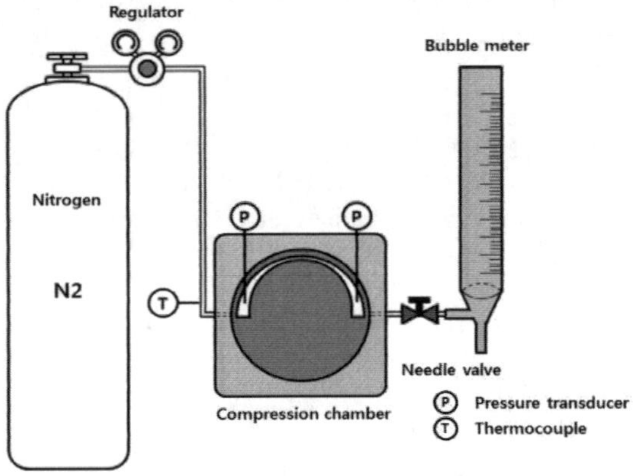

Figure 1: Schematic diagram of experimental set-up

Figure 2: Compression chamber simulator

3. EXPERIMENTAL SET-UP

The experiment variables were determined based on the equation of the isentropic nozzle flow. Figure 1 and Figure 2 show the experimental apparatus used to measure the mass flow rate, pressure and temperature through the radial clearance. The roller is fixed in the cylinder to maintain the gap of clearance, δ. The radial clearance is confirmed by clearance gauge and SEM (Scanning Electron Microscope). The side of top and bottom in cylinder is packed by using the hard sheet packing. A. Levy (1964) investigated accuracy of bubble meter according to the range of mass flow rate. Since the mass flow rate under the 1 L/s was measured within 1% error, the mass flow rate was measured by bubble meter in this experiment. The upstream pressure was controlled by using the regulator with the pressure transducer. And, the downstream pressure was controlled by using the needle valve with the pressure transducer. The ambient temperature was not changed during the experiments. Nitrogen is used as a working fluid.

4. CFD SIMULATION

In this study, CFD simulation was done as a tool to investigate the mass flow rate through the radial clearance. Since the width of the radial clearance is quite small (typically, 10-60μm), it is difficult to measure the actual area and mass flow rate in this experiment. Therefore, CFD simulation was conducted using Fluent 14.5 to obtain the mass flow rate with various operating conditions. Table 1 shows the conditions of CFD simulation. Figure 3 shows geometry of radial clearance. In order to make the pipe flow fully developed, the length of pipe was sufficiently long.

5. RESULTS AND DISCUSSION

5.1 Validation of CFD simulation

The CFD simulation results were compared with the experimental results to validate the reliability. In the condition of high upstream pressure (30 bar), the measurement of mass flow rate with the bubble meter may have large uncertainty. Thus, the validated CFD simulation under the condition of low upstream pressure was used for predicting the flow coefficient through the radial clearance. Figure 4 shows the comparison results of mass flow rate at the radial clearance with variable upstream pressure condition. The maximum error of comparison result is within 10%, when the upstream pressure is upper than 8 bar. Although the maximum error of mass flow rate is 10%, the difference of mass flow rate is near the 0.01 g/s in the each case.

The mass flow rate through the radial clearance obtained by CFD simulation has acceptable reliability for applying to prediction of the mass flow rate through the radial clearance.

Thus, the mass flow rate through the radial clearance under the various operating conditions was obtained by the CFD simulation. The functional formula is derived from the results of CFD simulation.

Table1: Conditions for the CFD simulation

	Incompressible steady-state
Viscous model	Realizable k-ε model
Inlet boundary condition	Pressure inlet
Outlet boundary condition	Pressure outlet

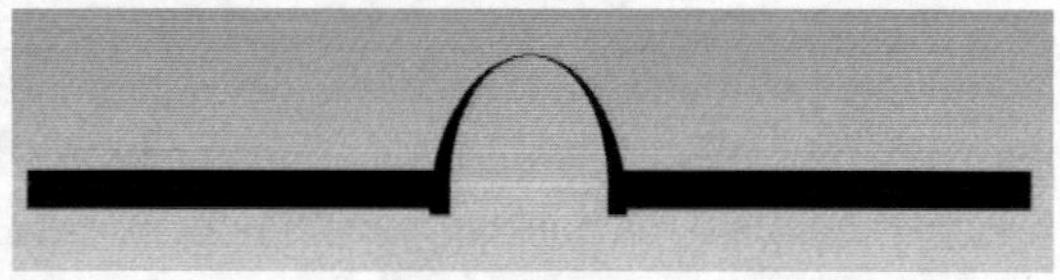

Figure 3 Geometry for the CFD simulation

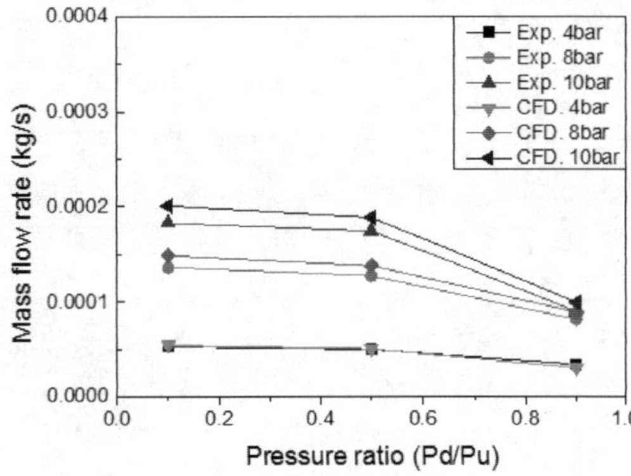

Figure 4 Comparison of mass flow rates from experimental and CFD simulation

5.2 Functional formula of flow coefficients

The influence factors for determining the flow coefficient are area of radial clearance, pressure ratio and upstream pressure. Figure 5 shows the flow coefficient with a change of each influence factor. The trend of flow coefficient is directly proportional to the width of radial clearance and upstream pressure. The rate of change of flow coefficient is different with choke condition. However, the flow coefficient is not affected by the height of radial clearance, even though the area of the radial clearance was changed. It seems that the effective area with the change of height in radial clearance has a linear correlation with the mass flow rate through the radial clearance.

Thus, the functional formula was developed by considering the width of radial clearance, upstream pressure and pressure ratio. Eq. (4) was derived from the results of CFD simulation to obtain the flow coefficients under the various conditions.

This equation of the flow coefficients was applied to compressor simulation in order to calculate the mass flow rate through the radial clearance. The mass flow rate was calculated by Eq. (2) and Eq. (3) with consideration of choke condition as Eq. (1). Where, P_r is the pressure ratio, w is the width of a radial clearance and P_u is the upstream pressure.

$$C_v = f_1(w) + f_2(w)P_{ub} + f_3(w)P_r + f_4(w)P_{ub}^2 + f_5(w)P_r^2 + f_6(w)P_{ub}P_r \quad (4)$$

$$f_n(w) = aw^4 + bw^3 + cw^2 + dw \ \ (n = 1, 2, 3, 4, 5, 6\,) \quad (5)$$

Table 2: Constants for Eq. (5)

	1	2	3	4	5	6
a	1.7E-07	-5.3E-08	3.1E-07	1.2E-09	-1.5E-07	1.6E-09
b	-2.0E-05	6.9E-06	-3.2E-05	-1.5E-07	1.2E-05	-5.5E-07
c	6.1E-04	-2.9E-04	8.3E-04	5.9E-06	1.4E-04	3.9E-05
d	7.E-03	4.8E-03	4.2E-04	-8.9E-05	-2.4E-02	-6.0E-04

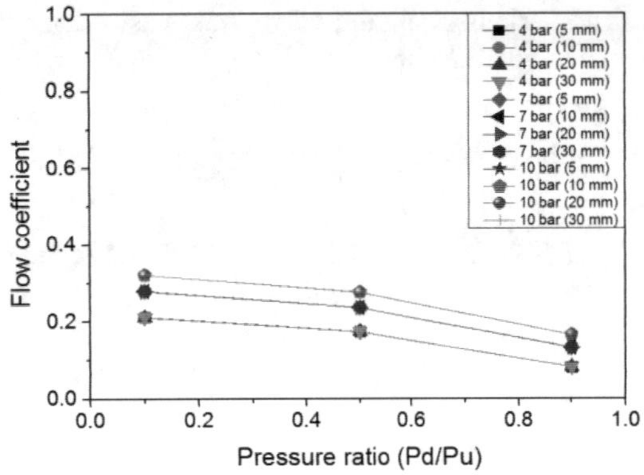

(a) : 10 μm

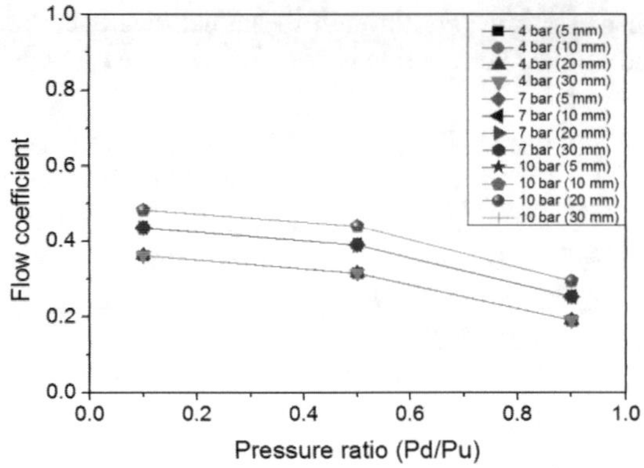

(b) : 20 μm

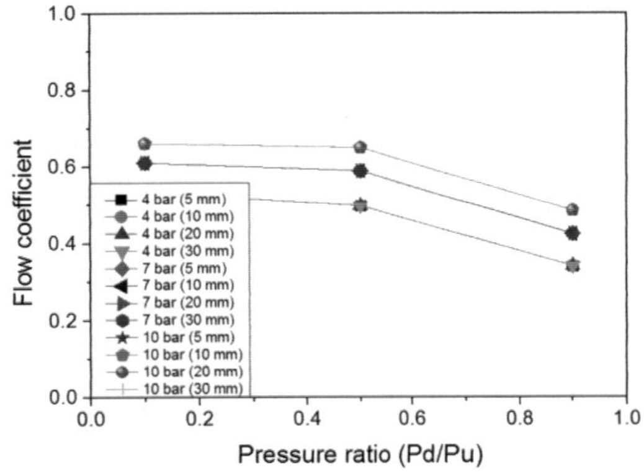

(c) : 40 μm

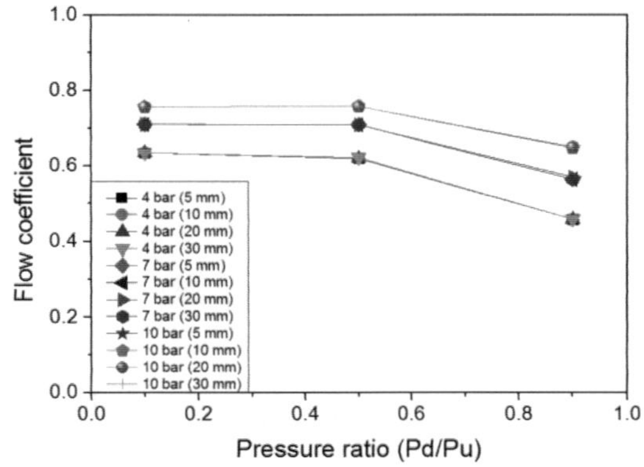

(d) : 60 μm

Figure 5 Flow coefficient in the radial clearance for each width of radial clearance

5.3 Application of functional formula to performance analysis of compressor

The compressor performance simulation with the functional formula was conducted the several operating conditions. This simulation data was compared with the results of simulation with $C_v = 0.3$ (Yanagisawa, T. and Shimisu, T., 1985) for calculating the mass flow rate and the results of calorimeter experiments. Table 3 shows the operating conditions and specifications of compressor for using in compressor performance simulation and experiments. Figure 7 shows the Energy Efficiency Ratio with varying operating conditions.

The comparison results of experiments and compressor performance simulation with the functional formula and 0.3 showed tolerable errors of 3 % in the case 1 and 3 except for 80 Hz condition. Since over compression in the compression chamber occurred the insufficient the leakage flow rate through the radial clearance, the results of high frequency of 80 Hz in the case 1, 3 have a higher error than 3 %. The simulation results applying the functional formula and 0.3 showed 3.7 % and 6.7 % errors respectively with comparison to results of experiment under the 80 Hz of operating condition.

In the case 2, the results of simulation with functional formula and 0.3 showed acceptable error of 3 % with the results of calorimeter experiments. The pressure ratio of case 4 is different from other case. The error between functional formula and 0.3 was upper than 5% under the various Hz condition. The simulation results applying the functional formula and 0.3 showed 7 % and 9 % errors respectively with comparison to results of experiment in the case 4. Although the compressor input power was increased with increase of the leakage through the radial clearance, it seems that the cooling capacity was more increased with the increase of the volume efficiency according to the compressor performance simulation.

The functional formula may be acceptable for predict the performance of compressor because that of equation considered for high upstream pressure and various width of radial clearance.

Table 3: Operating conditions and specifications of compressor for numerical simulation

	Case 1	Case 2		Case 3	Case 4
Suction pressure (bar, absolute)	9.95	9.95		9.95	10.15
Discharge pressure (bar, absolute)	33.8	33.8		33.8	23.37
Capacity of Comp. (cc)	10.2	24	42	14.1	14.1
Comp. type	Single rotary	2stage rotary		Twin rotary	Twin rotary

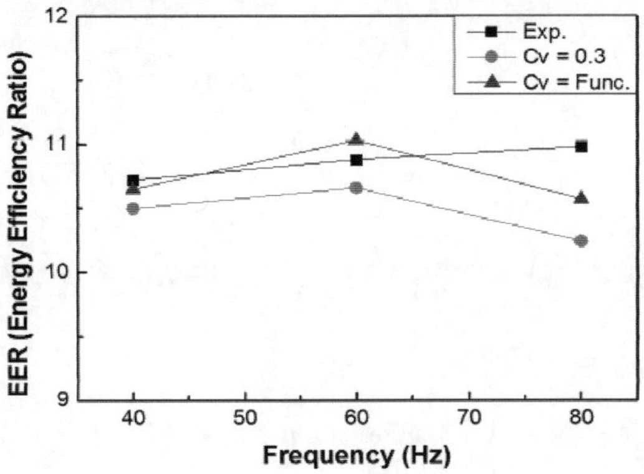

(a) case 1

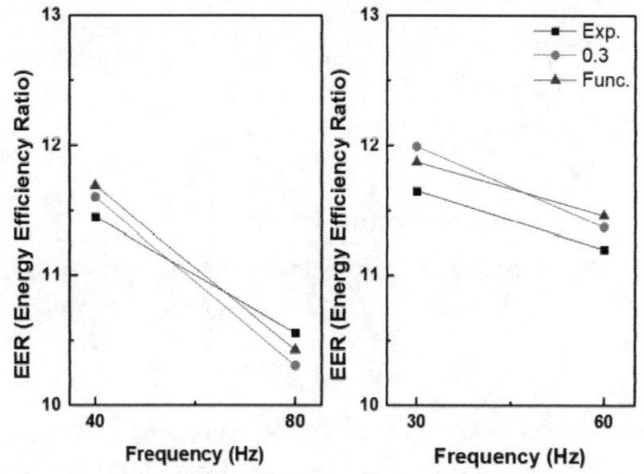

(b) case 2

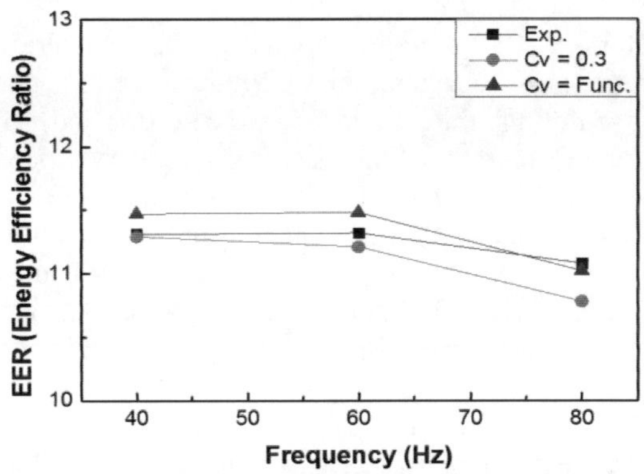

(c) case 3

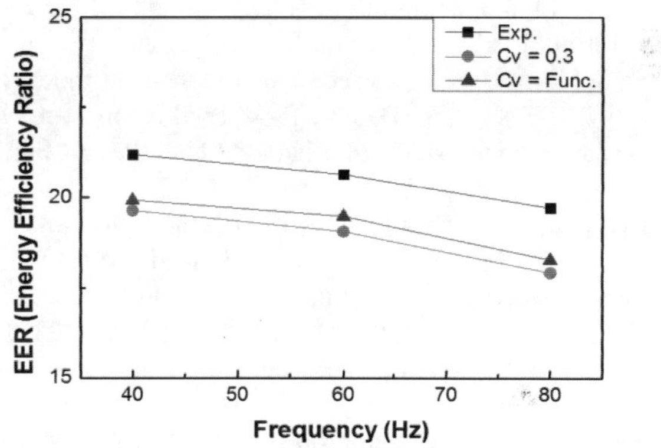

(d) case 4

Figure 6 The results of numerical compressor simulation

6. CONCLUSIONS

In this work, the flow coefficient through the radial clearance in the rolling piston type rotary compressor was functionalized with various operating condition. The results are summarized as follows.

The functional formula of flow coefficient (= C_v) in the radial clearance consist of the width of radial clearance, pressure ratio and upstream pressure. The range of flow coefficient is from 0.08 to 0.85 under the various width of radial clearance (10 μm - 60μm) and upstream pressure (4 bar – 30 bar). When the width of radial clearance was the 20 μm with the various upstream pressure and height, the average value of flow coefficient was 0.3. But, flow coefficient was different from 0.3 under the various operating condition. The results of compressor performance simulation with the functional formula showed good agreement with the results of experiments.

The developed functional formula is helpful for predict the performance of compressor in the design process of novel compressor under the various operating condition.

NOMENCLATURE

A	Area of the radial clearance	(m^2)
C_v	Flow coefficient of the radial clearance	(−)
k	Specific heat ratio	(-)
$\dot{m}$	Mass flow rate	(kg/s)
P_{cr}	Critical pressure ratio	(-)
P_d	Downstream pressure	(Pa)
P_r	Pressure ratio	(-)
P_u	Upstream pressure	(Pa)
P_{ub}	Upstream pressure	(bar)
R	Gas constant	((N*m)/(kg*K))
T_u	Upstream temperature	(K)
w	Width of the radial clearance	(μm)

REFERENCES

Pandeya, P., Soedel, W., 1978, Rolling piston type compressors with special attention to friction and leakage, International Compressor Engineering Conference at Purdue, pp 209-218.

Yanagisawa, T., Shimisu, T., 1985. Leakage losses with a rolling piston type rotary compressor. I. Radial clearance on the rolling piston. Int. J. Refrigeration 8 (2), 75-84.

Ian H. Bell, Eckhard A. Groll, James E. Braun, W. Travis Horton, 2013, A computationally efficient hybrid leakage model for positive displacement compressors and expanders, Int. J. Refrigeration 36, 1965-1973.

Ferreira, R.T.S., Lilie, D.E.B., 1984, Evaluation of the leakage through the clearance between piston and cylinder in hermetic compressors, International Compressor Engineering Compressor Engineering Conference at purdue, 424.

Dias, J.P., Gasche, J.L., Seixlack, A.L., 2011, Mathematical modeling of the Ester oil-refrigerant R134a mixture two-phase flow with foam formation through a small diameter tube. J. Braz. Soc. Mech. Sci. Eng. 33, 314e323

Gasche, J.L., Andreotti, T., Roberto Macedo Maia, C., 2012, A model to predict R134a refrigerant leakage through the radial clearance of rolling piston compressors, Int. J. Refrigeration 35, 2223-2232.

Ren, J.C., Hsiao, C.H., 2004, Experimental and CFD study on the mass flow-rate characteristic of gas through orifice-type restrictor in aerostatic bearings, Tribology International 37, 309-315

Min, B.C., Noh, K.Y., Yang, J.S., Choi, G.M., Kim, D.J., 2014, Prediction of refrigerant leakage for discharge valve system in a rolling piston rotary compressor, International Compressor Engineering Conference, paper 2346

A.LEVY, 1964, The accuracy of the bubble meter method for gas flow measurements, Journal of Scientific Instruments, Volume 41, Number 7.

ACKNOWLEDGEMENT

This work was supported by the Human Resources Development program (No. 20144010200780) of the Korea Institute of Energy Technology Evaluation and Planning (KETEP) grant funded by the Korea government Ministry of Trade, Industry and Energy.

The Influence of Main Bearing Parameters on the Bearing Wear in Rotary Compressor

Lingchao Kong, Qingfu Zhao, Liping Ren, Jia Xu, Xiaotong Cheng, Yusheng Hu

Compressor and Motor Institute of Gree Electric Appliances, Inc. of Zhuhai,
Jinji West Rd., Zhuhai City, 519070, P. R. China
Phone: +86-756-8974645, Fax: +86-756-8668386,
E-mail: klc03@163.com

ABSTRACT

The main bearing and sub bearing which support the crankshaft rotation often have wear in the rotary compressor, and the exceptional wear will cause a series of problems which contain the vibration, noises, frictional power rising and reliability reduced. The process of improving the actual exceptional wear problems of bearings is analyzed. And based on the finite element method (FEM), the results of the original and improved bearings are compared with each other; contact stress is chosen to be used to evaluate the wear condition of bearings. Then the influence of height, diameter and clearance of main bearing on the wear of the bearing is analyzed by the accelerated life test and the FEM simulation, and the feasibility of the bearing contact stress to evaluate the wear condition of bearings is further verified, at the same time it provides a theoretical basis for the design of compressor bearing.

1. INTRODUCTION

Internal parts cannot be removed and replaced in hermetic rotary compressor. As shown in Figure 1, the main bearing supports the upper part of the crankshaft, while the sub bearing supports the lower part of the crankshaft. Due to the long high-speed operation, improper design may cause the bearing wear, and cause a series of problems of compressor.

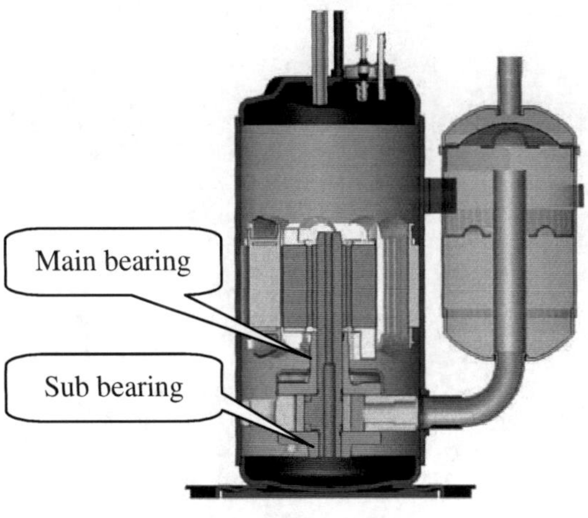

Figure 1: Rotary compressor

On the one hand, wear results in the change of bearing clearance which influence the stability of compressor pump, and then the noise and vibration problems will appear.

On the other hand, wear will make the friction power of bearings increase. As an important part of mechanical power in compressor, friction power increase will reduce the efficiency of the compressor.

Therefore, the effective evaluation method of wear is very important for the design of compressor.

Hydrodynamic lubrication theory is generally used for the analysis of the compressor journal bearing. The oil film pressure and thickness of the bearing which are used to evaluate the bearing are obtained by solving the Reynolds equation. Because the compressor bearing environment is complex, it is difficult to obtain accurate results using this method without overall consideration. In addition, the Reynolds equation needs to be solved by computer programming. In this paper, the contact stress between the crankshaft and bearing is obtained to evaluate the bearing wear by structural simulation analysis based on commercial finite element software. This method is simple and feasible, and it is very suitable for the application of engineering personnel.

2. EXCEPTIONAL WEAR ANALYSIS

In the process of R & D (Research & Development) of the compressor, there are two different types of compressor appearing large power, downtime and other anomalies. Abnormal wear of the crankshaft was found by dissecting, as shown in Figure 2.

Figure 2: Photo of the wear of the crankshaft

The crankshaft has had a severe adhesive wear. It is well known that adhesive wear is mainly affected by pressure and temperature. When the surface pressure reaches a certain critical value, the adhesion occurs (Figure 3). When the temperature reaches the critical value, the lubrication film is invalid and the adhesion is produced (Figure 4). The main influences on the temperature characteristics are surface pressure and sliding velocity. Therefore, controlling the pressure at a certain level is conducive to avoiding adhesive wear.

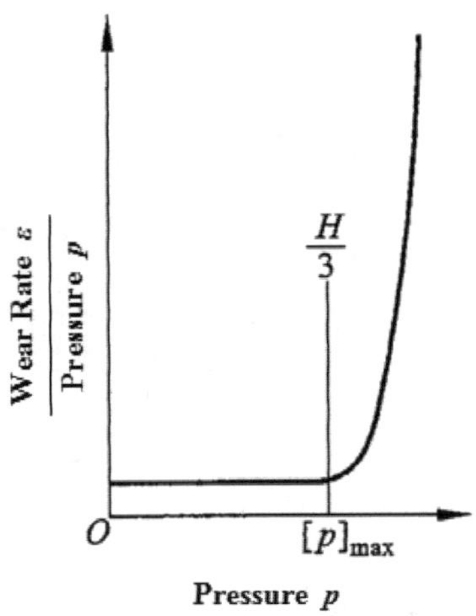

Figure 3: $\frac{\varepsilon}{p} - p$ Curve

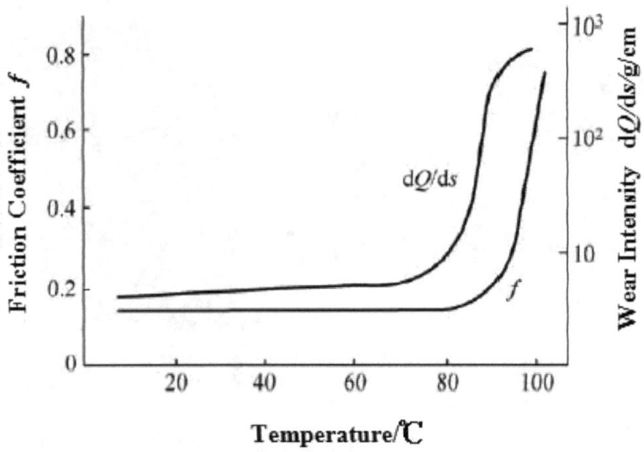

Figure 4: Influence of temperature on adhesive wear

After many studies, finally it is found that the design of the flexible structure of bearing can effectively solve the problem of abnormal wear of the crankshaft, as shown in Figure 5.

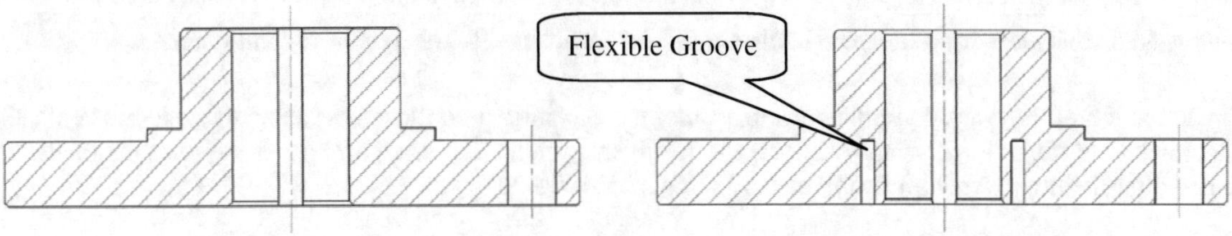

(a) Before improvement (b) After improvement

Figure 5: The structure of the bearing

The flexible groove is easy to deform due to its low stiffness. It can reduce the pressure between the crankshaft and the bearing, and then eliminate the adhesive wear.

3. FEM ANALYSIS

Based on the FEM, the static calculation is carried out before and after the improvement of the bearing. The FEM analysis model is built up, as shown in Figure 6. It contains the crankshaft, the main bearing and sub bearing. Because of not taking into account the oil film, the clearances between the crankshaft and bearings are cancelled.

Figure 6: FEM analysis model

The loads that are applied on the crankshaft contain centrifugal inertia forces and gas force. These forces can be calculated by the equations as follows.

1) Gas force F_g :

$$F_g = RL\left(1-\frac{e}{R}\right)(P_d - P_{s0})\sqrt{2(1-\cos\varphi)+\frac{e}{R-e}(1-\cos 2\varphi)} \tag{1}$$

where e is eccentricity of the crankshaft, R is radius of the cylinder, L is height of the rolling, P_{s0} is suction pressure, P_d is discharge pressure, φ is angle of starting to discharge.

2) Centrifugal inertia forces F_i :

$$F_i = m_i \omega^2 r_i \tag{2}$$

where m_i is mass of the eccentric parts (a pair of balancers, a rolling and a crankshaft eccentric part), ω is speed of rotation, r_i is the distances from the mass centre of the eccentric parts to the axis of revolution.

The top deflection of the crankshaft, the deformation of the eccentric part, the equivalent stress and the contact stress are calculated. Before and after improvement, the result maximums are compared, and the changes of the results before and after the improvement are obtained, as shown in Table 1.

Table 1: Computed results before and after improvement

	Top deflection	**Deformation of eccentric**	**Equivalent stress**	**Contact stress**
Original	0.0596mm	0.029mm	86.5MPa	99.8MPa
Improved	0.061mm	0.0317mm	86.4MPa	34.9MPa
Change	2.35%	9.31%	-0.12%	-65.03%

From Table 1, compared with the original scheme, the top deflection of the scheme with flexible groove has increased a little, and the deformation of eccentric increases by 9.31%, while the equivalent stress is basically unchanged, and the contact stress is greatly reduced by 65.03%.
The contact stress reflects the bearing surface pressure. The flexible groove effectively reduces the surface pressure of bearing. It is consistent with the front analysis.
The distributing state of the contact stress of the crankshaft is shown in Figure 7, and it is compared with the actual exceptional wear (Figure 2). It is found that the locations of the maximum contact stress and the actual wear are consistent. And the contact area becomes larger after improvement. It is one of the reasons for the contact stress reduced.

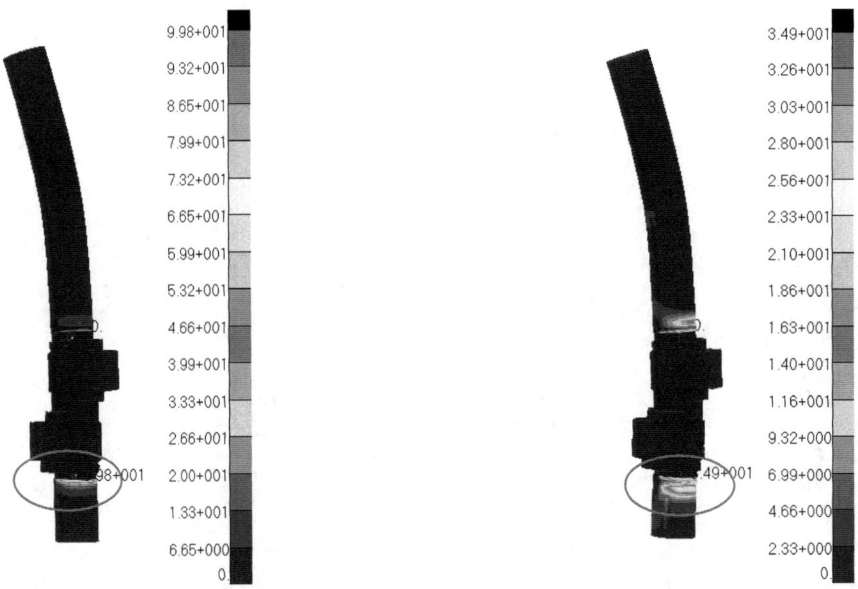

(a) Before improvement (b) After improvement

Figure 7: Contact stress fringe result

On the basis of the previous analysis the contact stress can be used to evaluate the bearing wear.

4. INFLUENCE FACTORS AND EXPERIMENTS

For much better design of bearing, the influence of main bearing parameters on bearing wear is analyzed by the forenamed method. And experiments are performed to validate the influence.

4.1 Influence Factors

The main parameters of the bearing design include the height, diameter and clearance. Because the clearance is not considered in the calculation method, the effect of height and diameter on the wear of the bearing is analyzed. Changing the height of the main bearing and the diameter of the crankshaft, the changing trend of the bearing contact stress is shown in Figure 8, 9.

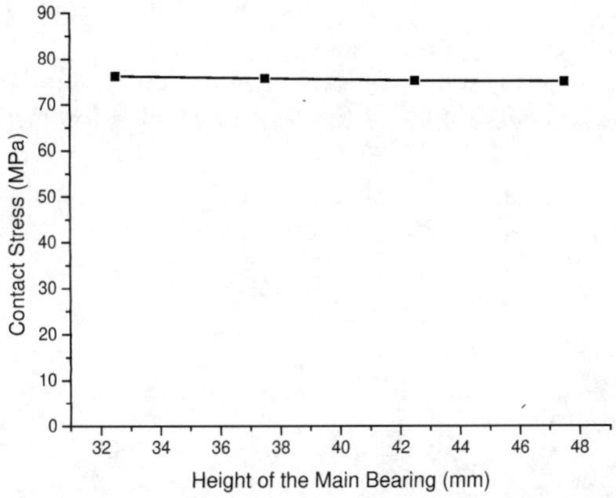

Figure 8: Contact stress vs. height of main bearing

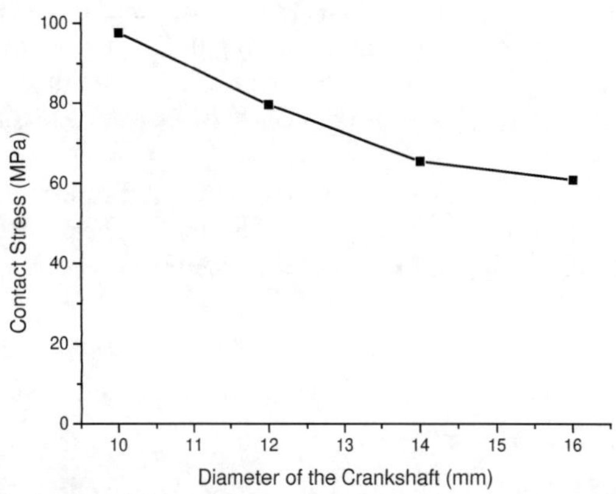

Figure 9: Contact stress vs. diameter of the crankshaft

These results show that:

1) With the increase of the height of the main bearing, the contact stress is basically unchanged. It indicates that the main bearing height has little influence on the contact stress, which has little effect on the bearing wear.
Because the bearing height is limited, it cannot represent the results of other limit values.

2) With the increase of the diameter of the crankshaft, the contact stress gradually decreases, and the decreasing tendency becomes slow. It indicates that the larger the bearing diameter, the better the improvement of bearing wear. But the bearing diameter increases to a certain extent, the effect of improvement become weak.

4.2 Experiments

According to the previous analysis, two schemes have been designed, one: reducing the height of the main bearing, and the other: reducing the diameter of the crankshaft. Through the accelerated life test, the wear condition of the crankshaft is obtain, and compared with the normal compressor.
The anatomical images of crankshaft wear of different schemes are shown in Figure 10, 11, 12.

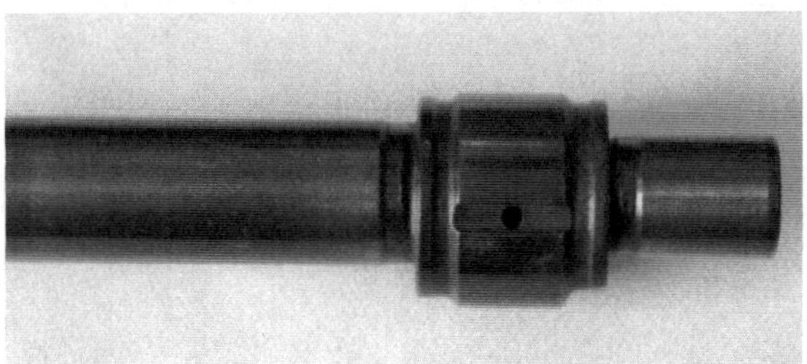

Figure 10: Wear of the crankshaft of the normal

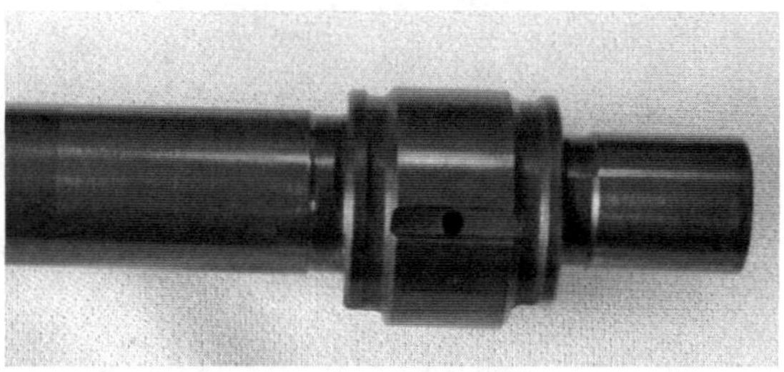

Figure 11: Wear of the crankshaft with short main bearing

Figure 12: Wear of the crankshaft with small diameter of crankshaft

Although the height of the main bearing decreases, the crankshaft wear state is consistent with the normal compressor. The height of main bearing has little effect on the bearing wear.
As the diameter of the crankshaft decreases, the crankshaft wear increases. The small diameter of the crankshaft will increase the bearing wear. Reducing the diameter of the crankshaft will aggravate the wear of bearings.
The test results of two schemes are all consistent with the calculation results. The influence of height and diameter of main bearing on bearing wear is further demonstrated, at the same time the feasibility of the evaluation method is confirmed.
In addition, because the calculation method cannot deal with the clearance of bearing, the influence of the clearance on the bearing wear is analyzed by the accelerated life test. The scheme with small clearance bearing is designed. And its anatomical image of crankshaft wear is shown in Figure 13.

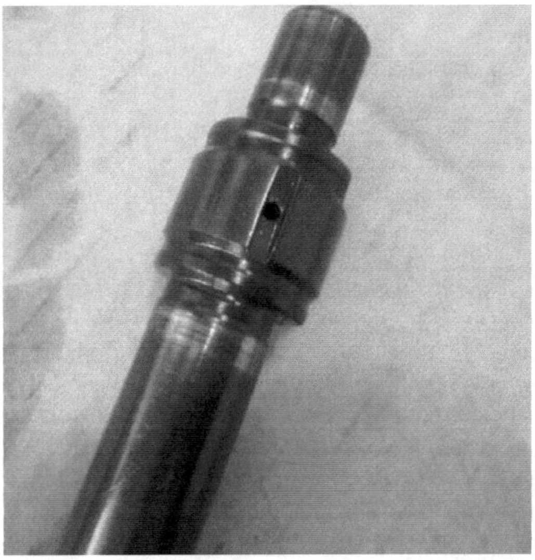

Figure 13: Wear of the crankshaft with small clearance bearing

A severe adhesive wear is found on the crankshaft when the clearance of bearing decreases. This shows that the clearance of bearing has considerable influence on bearing wear. The decrease of the clearance possibly changes the lubricant state, thus causing exceptional wear.

5. CONCLUSIONS

A method of using contact stress to evaluate the bearing wear has been put forward on the basis of FEM. By the experiments, it is validated that the method has certain feasibility. But there are also limitations.

The influence of several parameters on bearing wear is analyzed, it shows that: the height of main bearing has little effect on the bearing wear; the decrease of the diameter of crankshaft will aggravate the wear of bearings; small clearance of bearing will cause exceptional wear.

REFERENCES

Yoko KITSUNAI, Masaru MATSUI, Singo OYAGI (2010). High efficiency development of a rotary compressor by clarification of its shaft dynamic motion. *International Engineering Compressor Conference at Purdue.*

Junghyoun Kim, Seongwoo Park, Ihnsung Cho, Ilhyun Baek, Jaecheon Jo, Jaeyoun Jung (2012). Journal bearing lubrication characteristics in a reciprocating compressor. *International Engineering Compressor Conference at Purdue.*

A. Refik Özdemir, Erhan Kasapoğlu, Bilgin Hacıoğlu, Mustafa Duyar (2014). An investigation on the bearing design and friction characteristics of a hermetic reciprocating compressor. *International Engineering Compressor Conference at Purdue.*

Shizhu Wen, Ping Huang (2002). *Principles of Tribology*. beijing: Tsinghua University Press.

Lingchao Kong, Liping Ren, Jia Xu, Yusheng Hu (2014). The finite element analysis of the deflection of the crankshaft of rotary compressor. *International Engineering Compressor Conference at Purdue.*

Design and Realization of an Automated Test-Stand for Variable Capacity Household Compressors

Bernhard VETSCH[1]*, Gabriel FEICHTER[1]
Adrian H. BACHMANN[2], Stefan S. BERTSCH[1]

[1]University of Applied Sciences of Technology, NTB
Institute for Energy Systems
Werdenbergstrasse 4
9471 Buchs, Switzerland
bernhard.vetsch@ntb.ch, gabriel.feichter@ntb.ch, stefan.bertsch@ntb.ch

[2]V-ZUG Cooling Technology LTD
Zelgstrasse 3
9320 Arbon, Switzerland
adrian.bachmann@vzug-kt.ch

* Corresponding Author

ABSTRACT

In Switzerland the efficiency of household refrigerators has to be proven by the manufacturer to get the European energy label for the devices. The efficiency is tested in the beginning of the lifetime of the fridge. However, the compressor might reach its best performance only after some hours of operation. Knowing precisely the compressor characteristics is necessary for the manufacturer to optimize the products. Therefore, an automated compressor test-stand was designed and built to test variable speed compressors on their evolution of isentropic and volumetric efficiencies over time.

The test-stand features three expansion valve sections to control the high- and low-pressure states, and the suction gas temperature of the compressor. The test setup was programmed to be automatically controlled. Additional heaters can balance the non-isenthalpic pressure drop through the different components (mass flow meter and expansion valves) and the large internal heat exchange of the compressor enabling high enough suction gas temperatures. Furthermore, the compressor is installed in a temperature-controlled chamber. The power consumption of the compressor is measured with and without the inverter to get additional information. The testing range is 20 to 200 W cooling capacity with pressures of the refrigerant R600a down to its corresponding evaporation temperature (-35°C) and up to its condensing temperature (55°C). The test-stand is equipped with high precision sensors to ensure uncertainties of volumetric and isentropic efficiencies smaller than 5%. This paper reports the design of the test-stand, the control of the system, as well as the first results.

1. INTRODUCTION AND MOTIVATION

The EU directive 1060/2010 (European Commission, 2010) demands that all energy-related products are labeled with an indication of, among others, their energy consumption. Domestic household refrigerators and freezers were the first products that have been subjected to this directive.
The energy label initially classified the energy efficiency from class G (low efficient) to class A (highly efficient). Various improvements of components and materials gradually led to additional efficiency classes currently standing at A+++ for the highest efficiency. Nowadays, Swiss regulation requires that every new fridge sold in Switzerland have at least an energy consumption that corresponds to the class A++ (EnV, 1998). Consequently, the sales figures of high efficient domestic household refrigerators were steadily increased while low efficient products phased out in the past years, as shown in Figure 1 (FEA, 2015).

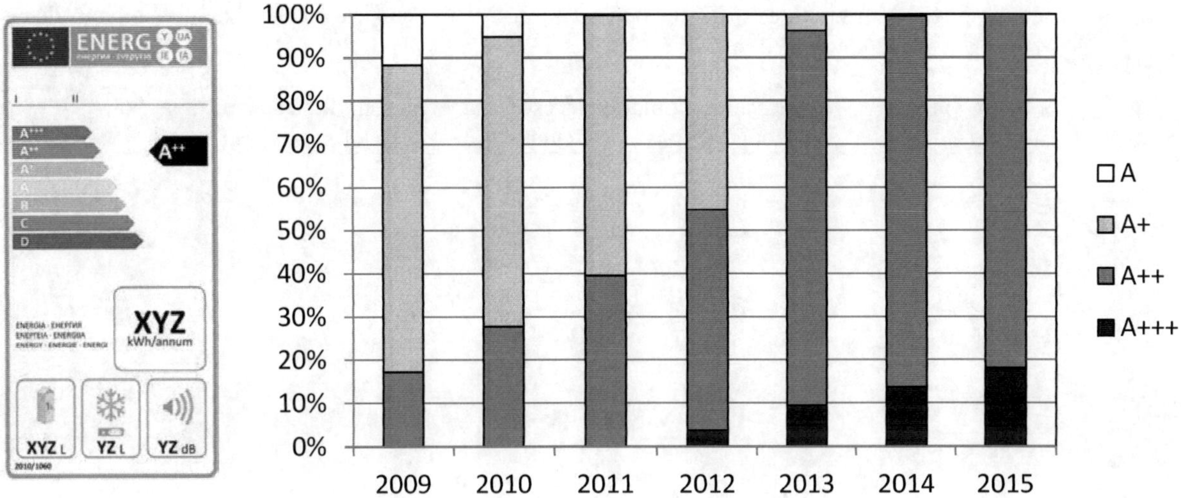

Figure 1: Energy label and development of sales figures of household refrigerators with different energy labels.

For example, an A+++, double-door, two-compartment household refrigerator with storage volumes of 203 liters (fridge: fresh food) and 73 liters (freezer: frozen food) must not consume more than 155 kWh/year (climate class tropical) at 5°C/-18°C (fridge/freezer) storage temperatures. This results in an average power consumption of 17.7 W. In order to fulfill this requirement, this specific appliance combines many system improvements stated by Bansal *et al.* (2011) as:

- advanced insulation by implementing vacuum insulation panels (VIP),
- isobutane (R600a) as refrigerant, and a
- variable speed compressor

The transition of existing, integrated household refrigerators to a better efficiency class is challenging since the outer dimensions are given. Consequently, increasing the insulation thickness directly reduces the storage volume which ultimately lowers the allowed energy demand to obtain the label. Therefore, this is not a preferred approach.

One key component of the appliance is the variable speed compressor. In order to increase the quality and efficiency of existing household refrigerators and to get additional design information, a compressor test-stand was designed and realized. Information about the transient evolution of volumetric and isentropic efficiencies during the first few hundred hours of operation, optimal conditions for high efficiencies, as well as inverter performances are key results obtained by the test-stand.

Compressor test-stands are commonly built to map and test compressors during development. This will also be possible in the presented test-stand. However, the main focus is to monitor the transient behavior of the efficiency over long periods of time and to analyze the effect of various parameters such as the suction gas temperature.

2. EXPERIMENTAL SET-UP

2.1 Test-stand design

The main specifications for the compressor test-stand are the following:

- Testing of compressors with cooling capacity ranging from 20 to 200 W
- Monitoring isentropic and volumetric efficiencies with uncertainty below 5%
- Testing conditions according to EN 12900 Household/CECOMAF
    - Ambient temperature 32°C
    - Suction gas temperature 32°C
- Condensing temperatures from 35 to 55°C
- Evaporation temperatures from -35 to -10°C

In order to fulfill these requirements three different options to design the compressor test-stand were considered. These concepts are schematically presented in Table 1.

Table 1: Different approaches of compressor test-stands with simplified schematics

Calorimetric test-stand	Hot gas bypass test-stand	Hot gas bypass
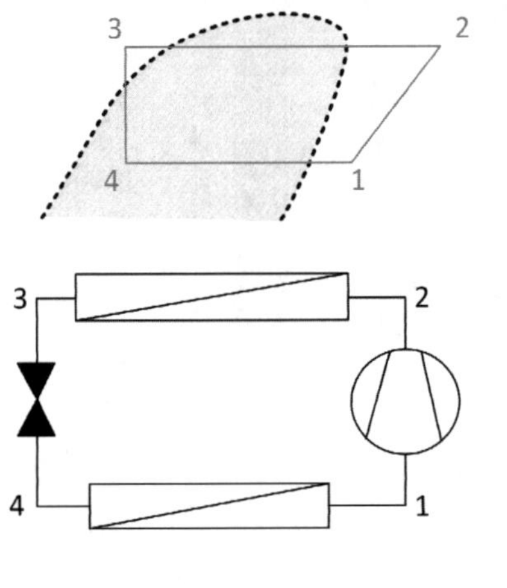	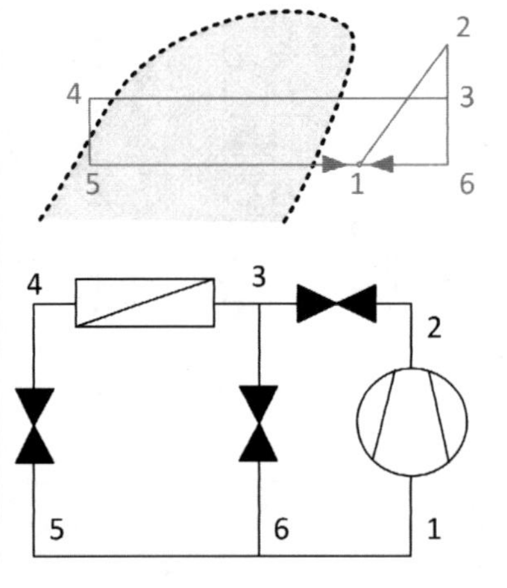	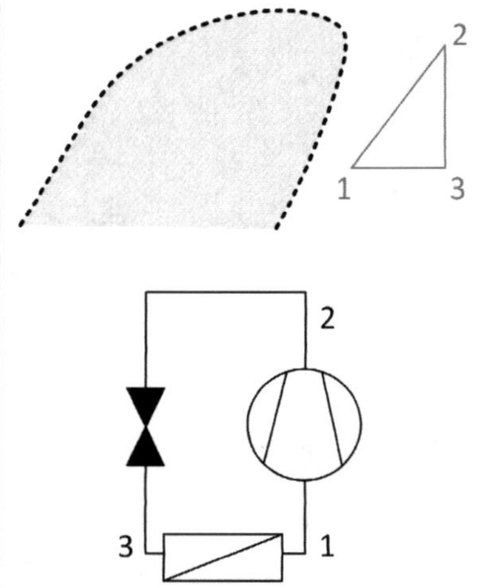
Duggan *et al.* (1988)	Christen *et al.* (2006)	Zingerli and Ehrbar (2000)

The hot gas bypass test-stand (Christen *et al.*, 2006) was chosen since this concept features advantages considering versatility, controllable range of the suction gas temperature, refrigerant inventory, and low thermal inertia.
The first household compressor under test showed a significant internal desuperheating of the discharged gas. At some operating conditions, the refrigerant reached a two-phase state at the compressor outlet, as indicated in Figure 2. In addition, an external source of heating energy became necessary due to heat losses in the expansion processes through the valve sections 2-3, 3-6, and the Coriolis type mass flow meter (Figure 3).

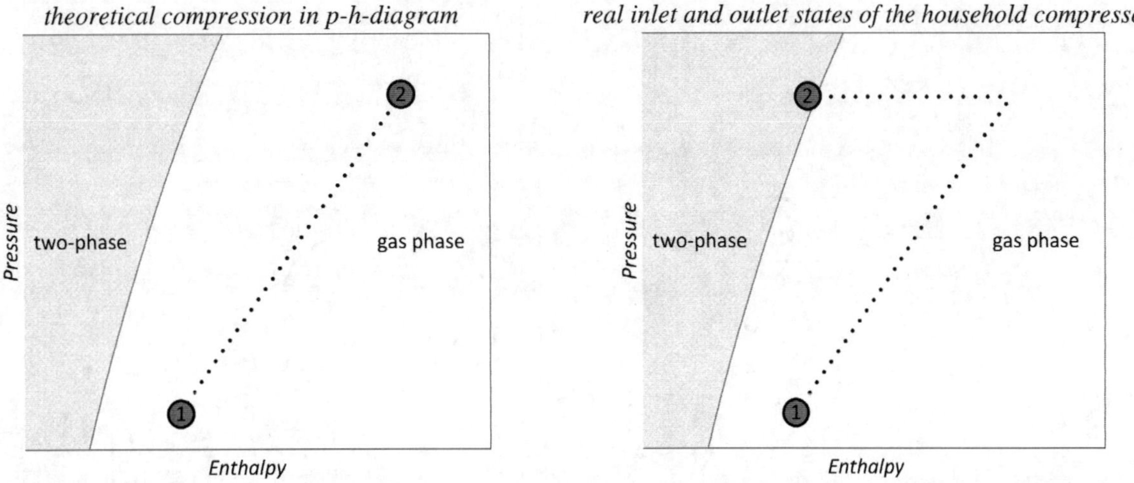

Figure 2: Compression only compared to real inlet (1) and outlet (2) states of a household compressor.

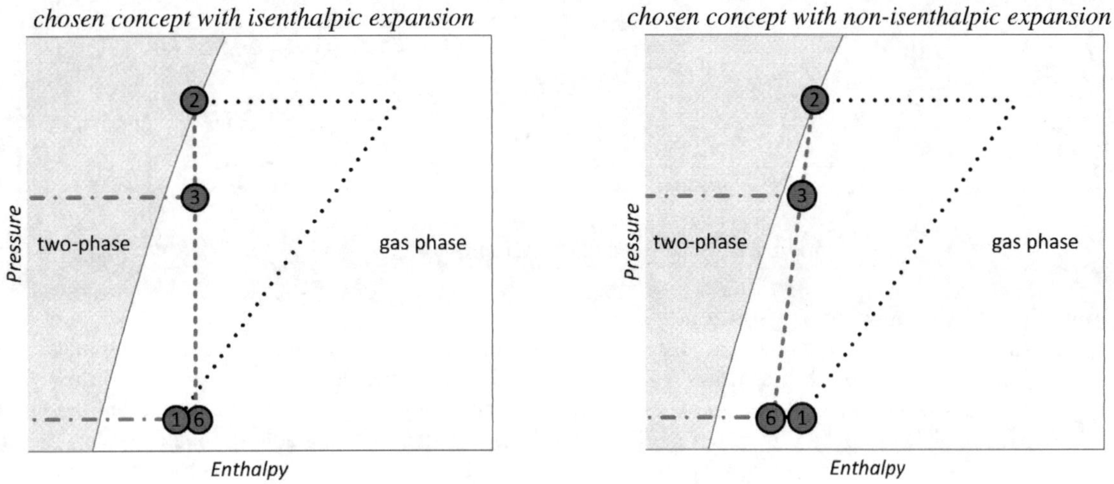

Figure 3: Influence of non-isenthalpic expansion through valves on the concept

Figure 4 shows the schematic of the final test-stand. The mass flow sensor measuring the refrigerant flow rate of the compressor was positioned before the split in the refrigerant line in point 3. The mass flow meter was originally giving inconsistent results due to an inhomogeneous flow at the sensor inlet which was caused by the two-phase state of the refrigerant. The refrigerant was in this state because the compressor has a large internal desuperheating capacity. Furthermore, due to the combination of the desuperheating and the non-isenthalpic expansion, the temperature at point 6 was lower than the target suction temperature. This made the implementation of heating stages necessary. As can be seen in the schematic, heating stages were integrated at the compressor outlet and upstream of the expansion valve (ExpV2-section). The heaters have a maximum heating capacity of 100 W each. Additionally, the high temperature section has been insulated and the mass flow meter itself was equipped with a 15 W electric heater. These modifications were necessary to ensure proper measurement of the mass flow rate. The heating stages 1 and 2 were realized using cartridge heaters that were clamped onto the refrigerant tube within the available space. In order to prevent the decomposition of the oil the maximum clamp-on-element temperature was limited to 105°C.

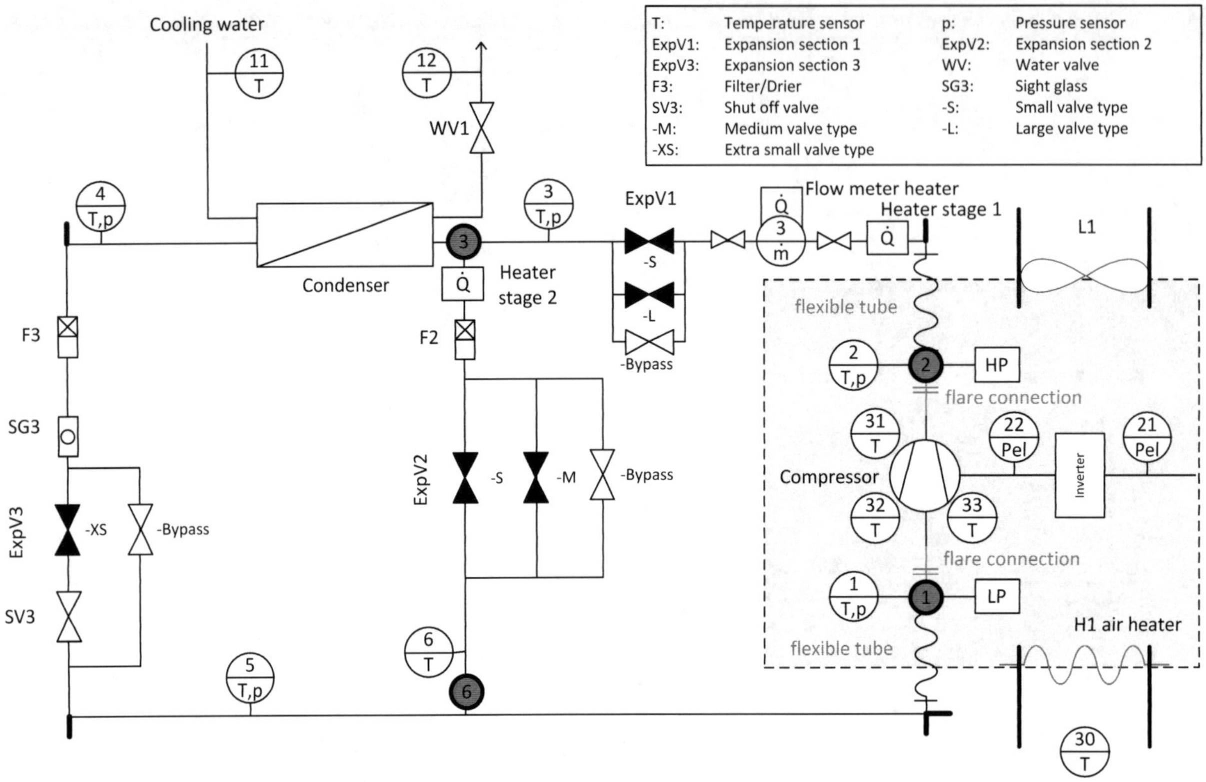

Figure 4: Schematic of the compressor test-stand

2.2 Operation modes and key components

The compressor test-stand can operate in two different modes: the gas cycle mode or the split stream mode. For low cooling capacities, with refrigerant mass flows below 0.3 g/s, and suction gas temperatures at 32°C the flow through the ExpV3-section is about 1/10 of the total mass flow, when operating in split stream mode. Even the smallest electronic expansion valve found in the actual market is too large and therefore not able to reliably control the mass flow. This led to severe fluctuation of the suction gas temperature. Hence, the condensing section is shut off and the suction gas temperature is controlled using the electric heater in stage 2 for small refrigerant flow rates.

The compressor is installed in an insulated compartment. Ambient air is drawn into the compartment by a fan, where its temperature is being controlled using two electric air heaters with 200 W capacity, each. The temperature of the air surrounding the compressor shell is calculated using the average of three temperature probes. To enable different compressor types to be easily placed in the compartment, the compressor is connected to flexible tubes using flare connections. The discharged gas passes the first heater stage and the mass flow is measured by the mass flow meter. Then, the refrigerant passes the ExpV1-section that controls the high pressure at the compressor outlet, since the pressure after the ExpV1-section is anchored by the condensing temperature. The secondary fluid in the condenser is tap water. The refrigerant flow may split up (depending on the operation mode) and one fraction is passing the condenser and expanded from liquid state to two-phase state by the ExpV3-section that controls the suction gas temperature in split stream mode. The rest of refrigerant is heated again by heater stage 2 and throttled by passing the ExpV2-section that controls the suction pressure of the compressor.

2.3 Controls

In order to meet the capacity of the compressor each expansion section features an array of expansion valves with different sizes. Each section may use the valves individually or any combination thereof to control the system. The KV-values of all expansion valves are summarized in Table 2 including controller settings calculated according to Chien, Hrones, Reswick (Haager, 2008).

Table 2: Control devices and settings

section	control device	size	Controller settings		
			P-gain	**I-gain**	**D-gain**
ExpV1	S L Needle valve (bypass)	KV: 0.009 m^3/h, at OD 100% KV: 0.024 m^3/h, at OD 100% KV: 0.33 m^3/h, at OD 100%	-0.0001	0.05	-
ExpV2	S M Needle valve (bypass)	KV: 0.009 m^3/h, at OD 100% KV: 0.014 m^3/h, at OD 100% KV: 0.08 m^3/h, at OD 100%	0.09	0.5	-
ExpV3	XS	KV: 0.003 m^3/h, at OD 100%	-0.0005	0.009	-
Heater stage 2	Cartridge heater	100 W	3.3	80	-
Air heater	Finned heating coil	2x200 W	7	5	-

The compressor is set to a constant rotational speed according to the test point. As input signal to the inverter a frequency-modulated signal corresponding to the desired rotational speed is generated using an Arduino Uno Board. A conversion uncertainty of 1% was taken into account for uncertainty analysis.

2.4 Instrumentation and measurement

In order to measure the compressor's isentropic and volumetric efficiency under 5%, the precisions of the probes are crucial.

Table 3: Measurement device uncertainty

Measurement point	Measurement device	Absolute uncertainty	Relative uncertainty
T_1	PT100, 1/3 DIN	0.1 K	0.0017 K/°C·|T|
T_2	PT100, 1/3 DIN	0.1 K	0.0017 K/°C·|T|
p_1	Piezo resistive pressure transducer	0.0015·300 kPa	
p_2	Piezo resistive pressure transducer	0.0015·1000 kPa	
$\dot{m}_3$	Coriolis mass flow meter		0.5%
$P_{el,21}$	Power Analyzer	0.364 W	(0.2+0.05/PF·f)%
n	Conversion of rotational speed (DAQ-System – Arduino Uno – Inverter)		1%

As can be seen from Figure 5 the temperature probes for ambient temperature measurement (T_{31}, T_{32}, T_{33}) are inserted in flexible coolant pipes to enable the individual placement. This is necessary because the outer dimensions of the different compressor types do vary significantly. At the compressor's suction and discharge line the temperature measurement is taken using an in-tube PT100-temperature sensor and the pressure is measured close to that location. Figure 5 shows the flare connections as well as the temperature and pressure measurements at the discharge line with insulation removed.

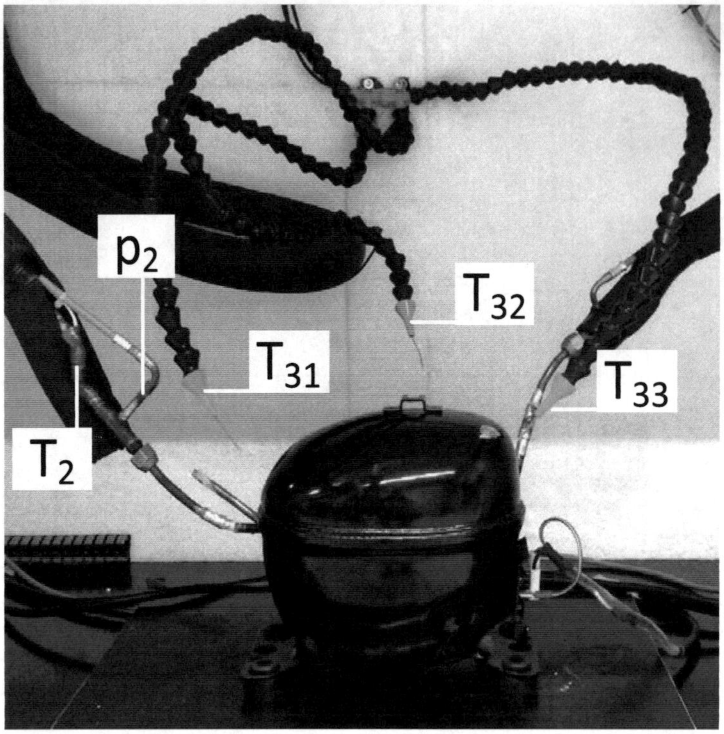

Figure 5: Compressor and sensors in insulated test compartment

In addition to the overall power consumption $P_{el,21}$, the power input to the compressor motor only ($P_{el,22}$) was acquired by a three channel power analyzer to rate the efficiency of the inverter.

3. METHODOLOGY AND EXPERIMENTAL RESULTS

3.1 Uncertainty Analysis

Uncertainty analysis was based on the method outlined in Taylor and Kuyatt (1994) applying the law of propagation of uncertainty. The following contributions to the uncertainty were considered:

- Distribution of fluctuations of the measurement values in terms of the one sided 95% confidence interval of the one sided t-distribution.
- Sensor accuracy according to Table 3 and propagation of sensor accuracy through calculated properties.
- Uncertainty of rotational speed setting of 1%.

The following contributions to the uncertainty were neglected:

- Uncertainty of the property calculation routines (Coolprop, Bell, 2011).
- Manufacturing tolerances of the displacement of the compressor.

3.2 Checking the uncertainty of the compressor efficiencies

A compressor using R600a was tested at several operating conditions (Table 4) from low to high cooling capacity.

Table 4: Uncertainty check: Test runs

Test run	1	2	3	4	5	6	7	8	9	10
n [s^{-1}]	20.0	20.0	26.7	26.7	26.7	26.7	33.3	33.3	33.3	50.0
p_1 [kPa]	88.8	36.6	36.6	88.8	36.6	88.8	88.8	36.6	88.8	36.6
p_2 [kPa]	464.5	464.5	464.5	464.5	772.4	772.4	464.5	772.4	772.4	772.4

Test run	11	12	13	14	15	16	17
n [s^{-1}]	50.0	50.0	50.0	75.0	75.0	75.0	75.0
p_1 [kPa]	88.8	36.6	88.6	36.6	36.6	88.8	88.8
p_2 [kPa]	464.5	464.5	772.4	464.5	772.4	772.4	604

Using this data the isentropic and volumetric efficiency was calculated according to equations (1) and (6) and the law of propagation of uncertainty was applied.

$$h_1 = f(R600a, p_1, T_1) \tag{1}$$
$$s_1 = f(R600a, p_1, T_1) \tag{2}$$
$$\rho_1 = f(R600a, p_1, T_1) \tag{3}$$
$$h_{2s} = f(R600a, p_2, s_1) \tag{4}$$
$$\eta_{isentropic} = \frac{h_{2s} - h_1}{P_{el,21}} \tag{5}$$
$$\lambda_{volumetric} = \frac{\dot{m}}{V_{disp} \cdot \rho_1 \cdot n} \tag{6}$$

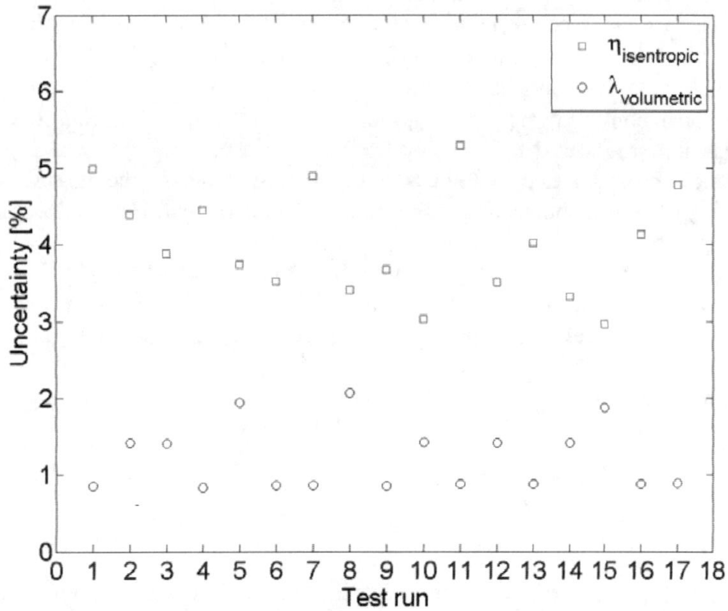

Figure 6: Resulting relative uncertainty considering the propagation of all uncertainties according to Table 3

The resulting relative uncertainties are shown in Figure 6. The uncertainties are under 5% for 94% of the values for the isentropic efficiency and under 3% for every value of the volumetric efficiency measured.

3.2 Comparison of measurements to the check point data of the data sheet

A new compressor has been installed and run for 250 hours for the quality assessment of the test-stand. Afterwards the check point conditions according to the compressor manufacturer data sheet (refer to Table 5) were tested. The resulting power consumption and mass flow rate were compared to the data sheet values (Table 6).

Table 5: Check point data

Check point	**1**	**2**	**3**	**4**	**5**
n [s^{-1}]	20.0	26.7	33.3	50.0	75.0
T_{evap} [°C]	-25	-25	-25	-25	-25
T_{cond} [°C]	55	55	55	55	55

Table 6: Check point conditions: Test results and comparison

-25°C/55°C		**Data sheet**		**Measurement**		**Relative deviation**	
		$P_{el,total}$ [W]	**$\dot{m}$ [kg/h]**	**$P_{el,total}$ [W]**	**$\dot{m}$ [kg/h]**	**$P_{el,total}$ [W]**	**$\dot{m}$ [kg/h]**
1	20.0 s^{-1}	25	0.45	24.4	0.458	-2.2%	1.9%
2	26.7 s^{-1}	32	0.61	32.0	0.626	0.0%	2.6%
3	33.3 s^{-1}	40	0.77	39.9	0.797	-0.2%	3.5%
4	50.0 s^{-1}	61	1.16	60.7	1.206	-0.5%	4.0%
5	75.0 s^{-1}	89	1.65	94.3	1.742	5.9%	5.5%

The measurement matches well with the check point data of the compressor. At the highest rotational speed a larger deviation can be noticed. However, since both the mass flow rate and the power consumption deviate in the same direction the coefficient of performance will not be negatively affected.

5. CONCLUSIONS

The design of a compressor test-stand for low capacity, variable speed household compressors is described. It has been noticed that various issues had to be taken into account when testing highly efficient household compressors.

- The compressor features large internal desuperheating of the discharged gas, which can lead to a refrigerant in two-phase condition at the compressor outlet.
- The initial superheat controlling approach is not applicable to some testing conditions due to the non-isenthalpic expansion processes.

Therefore, two electric heating stages have to be added to the test-stand and the mass flow sensor itself is heated to a temperature higher than the condensing temperature of the refrigerant. Hence, the compressor test-stand can operate in two modes:

- Gas cycle mode : High superheat conditions are controlled by the power of the heating stages.
- Split stream mode: Moderate and low superheat conditions are controlled by the fraction of the refrigerant passing by the condenser and expanding before being mixed back to the rest of the hot gaseous refrigerant.

The first test results show relative uncertainties of the isentropic and volumetric efficiencies of around 5% and 2% respectively. The check point data were similar to the data sheet values.

6. OUTLOOK

The next step is to rate the compressor test-stand using calorimetrically mapped compressors from the manufacturer. Then, various tests will be carried out in order to get data of the transient evolution of the compressor efficiencies and to further optimize the application in household refrigerators.

NOMENCLATURE

f	Frequency	kHz	**Subscript**	
h	Enthalpy	kJ/kg	1...33	Index/location in schematic
KV	Flow coefficient	m^3/h	2s	State point after isentropic compression
$\dot{m}$	Mass flow	kg/h	cond	Condensation
n	Rotational speed	s^{-1}	disp	Displacement
OD	Opening degree	%	evap	Evaporation
p	Pressure	kPa		
P_{el}	Power (electric)	W		
PF	Power factor	-		
T	Temperature	°C		
V	Volume	cm^3		
η	Isentropic efficiency	-		
λ	Volumetric efficiency	-		
ρ	Density	kg/m^3		
LP	Low pressure switch			
HP	High pressure switch			

REFERENCES

Bansal P., Vineyard E., Abdelaziz O., 2011, Advances in household appliances – A review, Applied Thermal Engineering 31, p. 3748-3760.

Bell I.H., 2011, Coolprop, https://sourceforge.net/projects/coolprop/

Christen T., Hubacher B., Bertsch S.S., Groll E.A., Experimental Performance of Prototype Carbon Dioxide Compressors (2006). International Compressor Engineering Conference. Paper 1730.

Duggan M.G., Hundy C.F., Lawson S., 1988, Refrigeration Compressor Performance Using Calorimeter and Flowrater Techniques, International Compressor Engineering Conference, Paper 622.

EnV, 1998, Energieverordnung 730.01, https://www.admin.ch/opc/de/classified-compilation/19983391/index.html#app10, 22.04.2016.

European Commission, 2010, Commission Delegating Regulation (EU) No 1060/2010, Official Journal of the European Union.

FEA, 2015, Fachverband Elektroapparate für Haushalt und Gewerbe Schweiz, www.fea.ch, April 2015.

Haager W., 2008, Regelungstechnik, 2. Auflage, Nachdruck, Verlag Hölder-Pichler-Tempsky, Wien.

Taylor, B. N., Kuyatt, C. E., 1994, NIST technical note 1297, 1994 edition: Guidelines for evaluating and expressing the uncertainty of NIST measurement results. Technical report, National Institute of Standards and Technology.

Zingerli A.P., Ehrbar M., 2000, Charakteristiken von Vielzellen- und Scrollkompressoren, Etappe 1: Vielzellen-kompressor EUROVANE 140, Swiss Federal Office of Energy.

ACKNOWLEDGEMENT

We gratefully acknowledge the support of the sponsors of this project: V-ZUG Cooling Technology LTD and the Commission for Technology and Innovation CTI.

Experimental Study on the Thermal Behavior of a Domestic Refrigeration Compressor during Transient Operation in a Small Capacity Cooling System

Johann HOPFGARTNER[1]*, Martin HEIMEL[1], Erwin BERGER[1], Stefan POSCH[1], Raimund ALMBAUER[1], Stefan STANGL[2]

[1]Institute of Internal Combustion Engines and Thermodynamics, Graz University of Technology, Inffeldgasse 19, 8010 Graz, Austria

hopfgartner@ivt.tugraz.at, +43 316 873 30240
heimel@ivt.tugraz.at, +43 316 873 30235
berger@ivt.tugraz.at, +43 316 873 30234
posch@ivt.tugraz.at, +43 316 873 30233
almbauer@ivt.tugraz.at, +43 316 873 30230

[2]Secop Austria GmbH, Jahnstraße 30, 8280 Fürstenfeld, Austria
s.stangl@secop.com, +43 3382 5010 824

* Corresponding Author

ABSTRACT

Generally, domestic refrigerators and freezers are running in non-continuous operation mode most of the time, which is a necessity to match cooling capacity to thermal loads. In currently available domestic appliances it can be observed, that this matching is mainly realized in two different ways: On the one hand, a simple on/off control of the hermetic compressor is installed in lower priced appliances with limited energy efficiency for the mass market. On the other hand, modern top efficiency class appliances have a variable frequency controlled compressor installed. Both control strategies have a repetitive and transient change of thermodynamic states of the refrigerant in common.

For better understanding of these cyclic patterns in terms of internal temperature distribution, a state of the art domestic refrigeration compressor with a displacement of approximately 6 cubic centimeters is integrated in a commercial freezer. The compressor which has an on/off control is equipped with extensive measurement instrumentation. Several temperature probes are inserted and temperatures on surfaces inside and outside the compressor as well as refrigerant temperatures are logged for both cyclic and steady-state behavior. Finally, a comparison between transient experimental data and steady-state data from a standardized calorimeter test bench is done.

1. INTRODUCTION

Since a respectable part of the worldwide energy consumption can be attributed to household refrigerators and freezers, it is a necessity to increase the overall efficiency of cooling systems. These systems consist of several parts whereat the compressor plays the key role concerning the energy consumption. The energy consumption of hermetic compressors is mainly influenced by three aspects: the electrical, the mechanical and the thermodynamic efficiency. The electrical efficiency of the motor is usually about 90%, the mechanical efficiency, which is essentially affected by the bearing system, is also mostly around 90% and the thermodynamic efficiency is usually about 80 to 83% (Ribas *et al.,* 2008). Therefore, the thermodynamic analysis of reciprocating compressors for domestic refrigerators was the main focus of several different studies which were published in the last few years. Most of them were based on a quasi-static behavior of the hermetic compressor under certain standardized conditions.

Kara and Oguz (2010) investigated the thermal behavior of a small hermetic reciprocating compressor. They measured the gas and surface temperature on several positions while the compressor was tested in a calorimeter that enabled stabilized operating conditions (ASHRAE conditions) and compared the experiments with results by numerical models. Another work, wherein a thermal analysis of a hermetic reciprocating compressor was made, was published by Cavallini *et al.* (1996). The authors investigated several computational models and compared them with experimental data. These experimental measurements had been carried out on a hermetic compressor which also was operating under steady conditions (ASHRAE conditions). In the field of thermal analysis of reciprocating compressors, transient conditions represent very often only the unsteady phenomena during one revolution as in the study of Longo and Caracciolo (2002). Similar investigations were done by Negrao *et al.* (2010).

To investigate the complex thermal phenomena inside reciprocating compressors, experiments and numerical calculations are almost always carried out at stabilized operating conditions with standardized boundary conditions. These conditions can differ from the real world operating conditions especially for non-variable speed compressors. If a compressor should be improved in order to increase the overall efficiency and therefore, to reduce the annual energy consumption of the entire cooling system, measuring data from a calorimeter test bench could lead to misleading potentials for improvements. Therefore, this work compares steady-state and transient temperature measurements of a fixed speed compressor (50 Hz) operating at different conditions.

2. EXPERIMENTAL SETUP

Experimental measurements are carried out with a commercial hermetic reciprocating compressor for domestic refrigerators with a displacement of approximately 6 cubic centimeters. The refrigerant flows through the suction pipe into the shell of the compressor. There it enters the suction muffler and flows through a reed valve into the cylinder. The compressed refrigerant is then discharged and leaves the shell, passing the cylinder head, discharge mufflers and the serpentine.

The investigated compressor, running with R600a, is equipped with extensive measurement instrumentation. In order to measure the temperature profile during a duty cycle, which is in a range of several minutes, T-type thermocouples are used. Thirteen thermocouples are distributed on several positions inside the compressor to get a meaningful statement of the thermal behavior. The description of the locations of these thermocouples is given in Table 1. Additionally, the temperatures of the suction pipe (T14) and discharge pipe (T15) outside the compressor are measured. To get the outer surface temperature distribution of the compressor, another six thermocouples are placed around the shell. These measuring positions are described in Table 2. Furthermore, two pressure transducers are placed at the cylinder head (P1) and the suction muffler neck (P2). Figure 1 shows drawings of the compressor, wherein the temperature and pressure measuring positions are depicted.

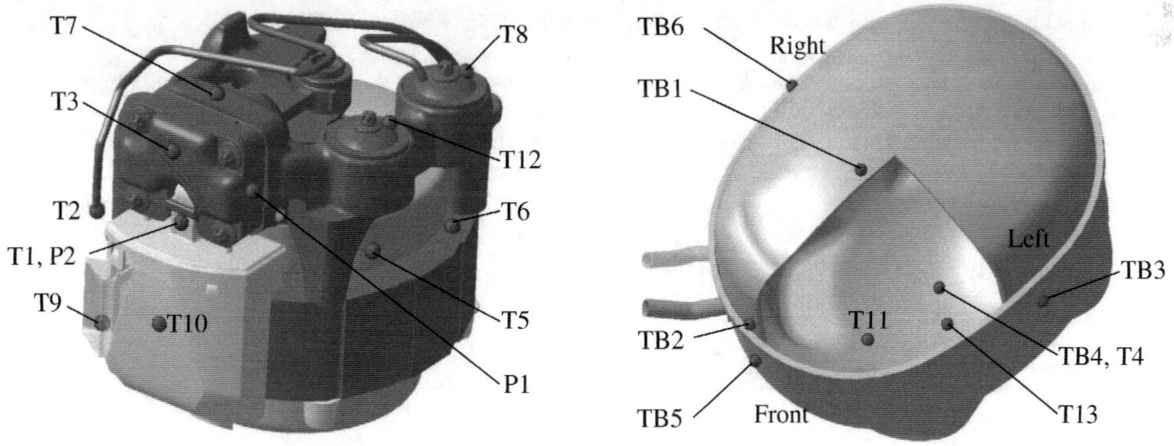

Figure 1: Temperature and pressure measuring positions

Due to the requirement of hermetical tightness of the compressor, every cable-feedthrough represents a potential vulnerability. Therefore, the Teflon®-insulated sensor wires are passed through the shell with a specially sealed feedthrough. The left side of Figure 2 shows the compressor without the upper part of the shell. One of the challenges is to arrange all the measurement instrumentations in a way that the influence on the behavior of the compressor is as small as possible. Thermocouples which measure a surface temperature inside the shell are fixed with epoxy resin which can be seen in Figure 2 (right).

Table 1: Thermocouple measuring positions at the compressor

No.	Temperature measuring position description	Location
T1	Suction muffler neck (gas)	-
T2	Serpentine end (surface)	-
T3	Cylinder head outside (surface)	Front
T4	Shell inside (surface)	Bottom
T5	Stator windings (surface)	Top
T6	Stator lamination (surface)	Top
T7	Cylinder (surface)	Top
T8	Discharge muffler-2 (surface)	Top
T9	Suction muffler intake tube (gas)	-
T10	Suction muffler wall (surface)	Centre
T11	Oil	Centre
T12	Discharge muffler-1 (surface)	Top
T13	Shell (gas)	-

Table 2: Thermocouple measuring positions at the shell (outside)

No.	Temperature measuring position description	Location
TB1	Shell cover (surface)	Middle
TB2	Shell cover (surface)	Front
TB3	Shell (surface)	Left
TB4	Shell (surface)	Bottom
TB5	Shell (surface)	Front
TB6	Shell cover (surface)	Right

To get steady state-data, the compressor is running on a fully automated calorimeter test bench under four different stabilized operating conditions. Thereby, ambient temperature is set to 32 °C and the adjusted evaporation and condensation temperature can be seen in Table 3. Each test takes several hours, whereat only the data which is recorded during the last 50 minutes is analyzed to ensure steady-state conditions.

The transient measurements are done with the same compressor installed in a commercial freezer with a net capacity of approximately 100 liters. Several different measurements are carried out such as varying the ambient temperature, varying the cycle duration or varying the duty-cycle ratio. Table 4 describes the operating conditions of the respective transient measurements whereat evaporation and condensation temperatures refer to the ON-period of the compressor.

Table 3: Evaporation and condensation temperature of calorimeter measurements

Measurement	Evaporation temperature T_{eva} [°C]	Condensation temperature T_{con} [°C]
Calo -23/45	-23.3	45.0
Calo -23/55	-23.3	55.0
Calo -37/49	-37.0	49.0
Calo -34/50	-34.0	50.0

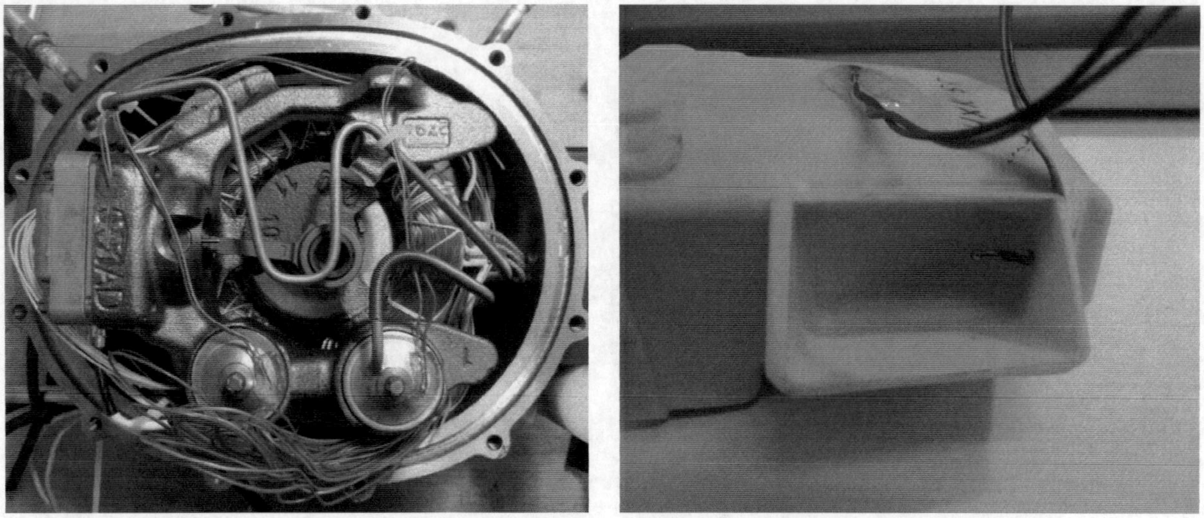

Figure 2: Compressor without upper shell (left), suction muffler details (right)

Table 4: Operating conditions of transient measurements

Measurement ($T_{amb}/t_{cyc}/r$)	Ambient temperature T_{amb} [°C]	Cycle duration t_{cyc} [min]	Duty-cycle ratio r [%]	Evaporation temperature T_{eva} [°C]	Condensation temperature T_{con} [°C]
Real 32/30/67	32	30	67	-33.8	50.0
Real 32/45/67	32	45	67	-34.9	50.0
Real 32/105/80	32	105	80	-37.2	49.1
Real 25/30/65	25	30	65	-31.2	41.2

3. RESULTS AND DISCUSSION

In the following chapter, the experimental data is presented and discussed. All measurements are carried out with the same hermetic compressor which is either running on a calorimeter test bench or in a commercial household freezer. Several operating conditions of the compressor are tested and effects on temperature distribution are investigated. Finally, a comparison between steady-state and transient operating behavior is done.

3.1. Steady-state data of the calorimeter test bench

Figure 3 shows a chart where the arithmetic mean values of the measured temperatures can be seen. Considering the outer surface temperatures of the shell (TB1 – TB6), it can be observed that the left (TB3) and the right side of the shell (TB6) show the highest temperatures. The lowest temperatures of the shell occur at the middle of the shell cover (TB1). One would expect that the temperatures of the three mentioned positions are more similar and higher than the other ones (TB2, TB4 and TB5) because they are close to the discharge line which has the highest temperature level inside the compressor. One explanation for this phenomenon could be the influence of the air flow inside the calorimeter test bench on the heat transfer between compressor shell and ambient air. When comparing the calorimeter-datasets among each other, it should be kept in mind that different operating conditions lead to different mass flow rates.

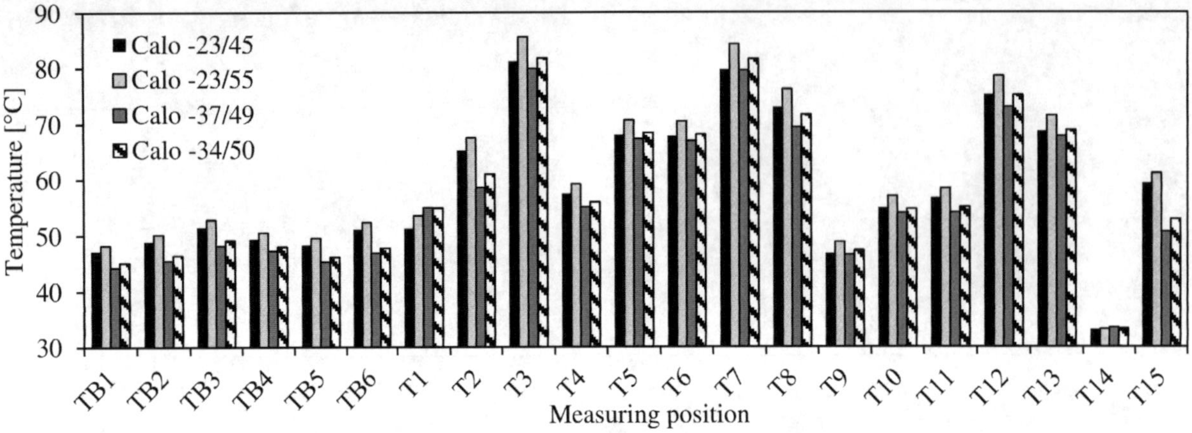

Figure 3: Arithmetic mean values of temperatures at different measuring positions - calorimeter data

The highest temperatures occur along the discharge line. From the cylinder head (T3) to the second discharge muffler (T8) a temperature drop of approximately 10 K can be observed for all four operating conditions. Along the serpentine, the temperature drops for another 8-10 K. The temperature of the gas inside the shell (T13) and temperatures of the electric machine (T5 and T6) are at similar levels, leading to the assumption that measuring position T13 is influenced by the motor. Due to the limited space inside the compressor, it is hardly possible to find a position, where influences from other parts could be fully excluded. The heating of the suction gas (T9, T1) is about 4.5 K for the measurements "Calo -23/45" and "Calo -23/55", and approximately 8 K for the other two operating conditions. This discrepancy results from the different mass flow rates.

3.2 Transient experimental data

To gain transient experimental data, the hermetic compressor is installed in a commercial freezer with a net capacity of approximately 100 liters. Thereby, cycle duration and duty-cycle ratio are varied. In this work, the duty-cycle ratio is defined as the duration of the ON-cycle divided by the total cycle duration. Additional to the three measurements at ambient temperature of 32 °C, another measurement at 25 °C is done to investigate the influence of different ambient temperatures.

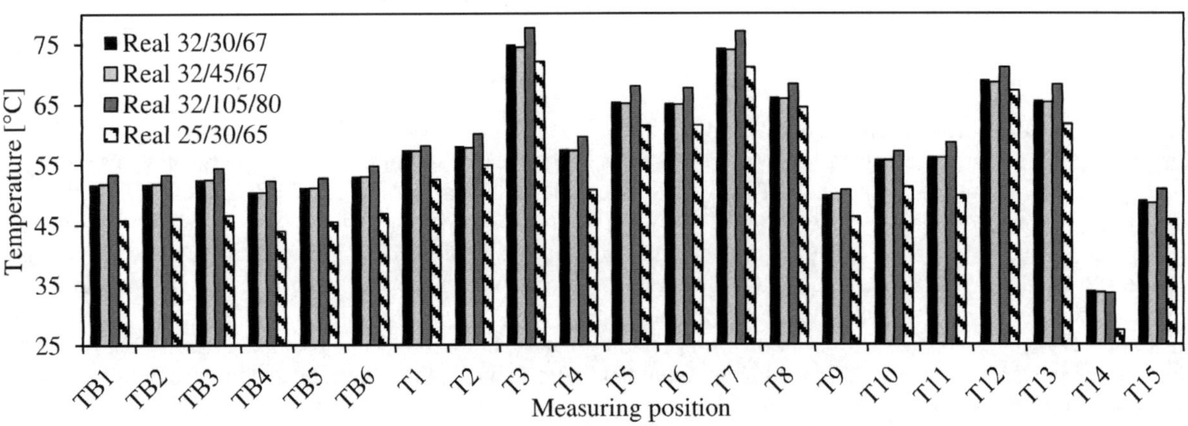

Figure 4: Arithmetic mean values of temperatures at different measuring positions (transient data, total cycle duration)

Figure 4 shows measured temperatures which are averaged over the total cycle duration. A closer look at the outer surface temperatures of the shell (TB1-TB6) shows that the lowest temperature occurs at the bottom (TB4) and the highest at the right and the left side of the shell (TB3 and TB6). This temperature distribution indicates the influence of the hot and cold compressor components on the surface shell temperature more pronounced than those determined by the steady-state measurements on the calorimeter test bench.

Similar to the calorimeter measurements, it can be seen, that the temperature drops approximately 8.5 K between the cylinder head (T3) and the second discharge muffler (T8) and another 8 K along the serpentine. The suction gas temperature increases by 7.5 K from the intake tube to the neck of the suction muffler for measurements at an ambient temperature of 32 °C and increases by 6.5 K for the measurement at 25 °C ambient temperature. Comparing the real-world measurements among each other, rather small differences can be seen when varying the cycle duration (Real 32/30/67 and Real 32/45/67). A similar thermal behavior but at higher temperature level (1.5-2.5 K higher) can be found when increasing the duty-cycle ratio up to 80 percent (Real 32/105/80). Due to the control logic of the commercial freezer, the total cycle duration has also increased. Decreasing the ambient temperature from 32 °C to 25 °C leads to an approximately 6 K lower shell surface temperature. However, the temperature level reduction of the discharge line, the suction gas and the electrical machine is only about 1.7-3 K, 4.5 K and 4 K, respectively.

Figure 5 illustrates differences between temperatures averaged over the ON-period, OFF-period and total cycle, on the basis of the transient measurement "Real 32/30/67". Concerning the outer surface temperature of the shell, it can be noted, that the deviation between ON- and OFF-period is smaller than 1 K except TB4 (shell bottom) which seems to correlate stronger with the oil temperature (T11). Inside the compressor, the same small temperature deviation can be observed for components with a high thermal mass such as the electric machine and the two discharge mufflers. Temperatures at measuring positions T3 and T7 (cylinder head and cylinder) show a significant deviation between ON- and OFF-period because they are directly affected by the heat generation resulting from the compression of the refrigerant. The temperature drop of the serpentine (T2) is similar to that of T3 and T7. Further considerable differences appear at measuring positions T1, T9 and T10 (suction muffler neck, suction muffler intake tube and suction muffler wall) which show the heating of the suction line during the OFF-period.

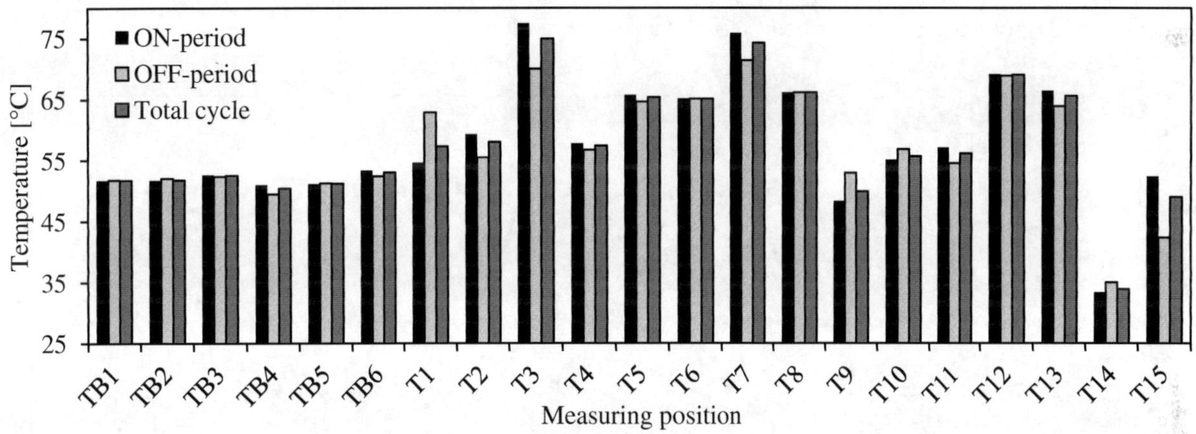

Figure 5: Averaged temperatures of transient measurement "Real 32/30/67" – comparison between ON- and OFF-period

3.3. Comparison between steady-state and transient data

In this section, a comparison between steady-state and transient measurement data is done for two different operating conditions. The adjusted evaporating and condensing pressure is 0.386 bar and 7.02 bar for the measurements "Real 32/30/67" and "Calo -34/50", and 0.386 bar and 6.68 bar for the measurements "Real 32/105/80" and "Calo -37/49", respectively. Figure 6 shows the time averaged temperatures of the measurements mentioned before. Thereby, the temperatures for the transient measurements are averaged over the entire cycle duration. Comparing the outer surface temperatures of the shell (TB1-TB6), a difference of approximately 5-7 K can be observed for both operating conditions. Based on the assumption of a similar input power to the compressor for transient and steady-state measurements, nearly the same or even a lower temperature level of the transient measurements could have been expected. The main reason for this discrepancy is the air flow over the compressor in the calorimeter test bench, which is needed to keep the ambient temperature constant at the desired level. During the transient measurements, the compressor is located in a small recess at the backside of the freezer. There, the air temperature close to the compressor is higher than the ambient temperature and heat transfer only occurs by natural convection and heat radiation.

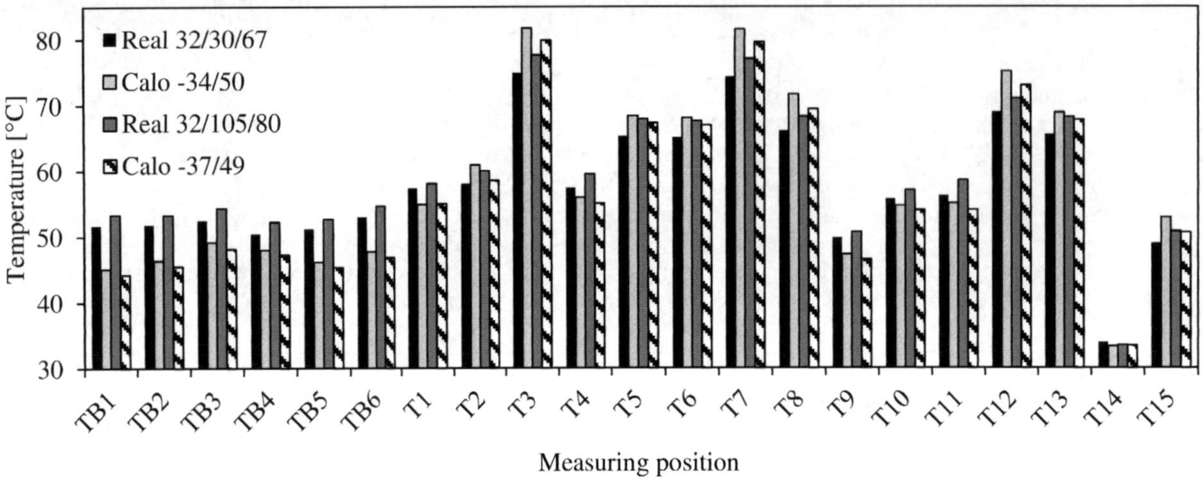

Figure 6: Comparison between transient and steady-state temperature measurements

Inside the compressor, higher temperatures along the discharge line (T3, T7, T8, T12) can be observed for steady-state measurements compared to transient ones, which is expected due to the OFF-period of the compressor. Similar differences can be seen for temperatures influenced by the cold suction gas. T1, T9, and T10 show values which are significant lower for steady-state measurements. Worth mentioning is that the average temperature level at operating condition 0.386 bar/7.02 bar (Real 32/30/67) is lower when comparing both transient measurements, whereas it is higher (Calo -34/50) when comparing both steady-state measurements.

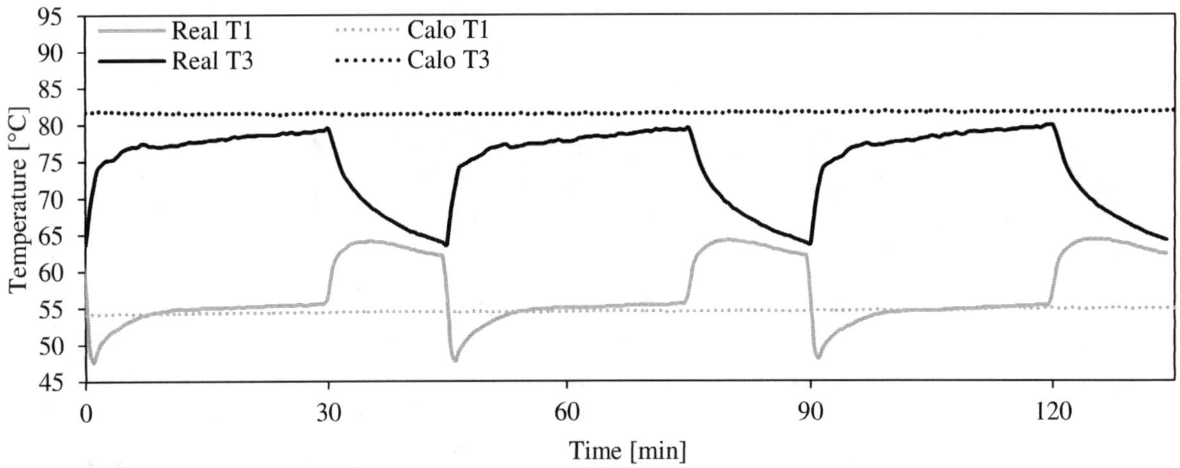

Figure 7: Temperature profile of measuring positions T1 and T3 – transient versus steady-state operating conditions

Figure 7 shows the profile of the gas temperature at the suction muffler neck (T1) and the surface temperature at the cylinder head (T3) of the transient (Real 32/30/67) and the steady-state (Calo -34/50) measurement. Therein it can be seen, that the deviation between steady-state and transient temperature decreases during the ON-period of the compressor and reaches a value of approximately 1-2 K shortly before the compressor stops. The deviations of the time averaged temperatures are of course much more significant which can be seen in Figure 6. When the compressor is switched on, temperature T1 decreases very fast due to the cold refrigerant which is sucked in. Afterwards, the valve plate is heated and influences the gas at the muffler neck which can be seen in the rather fast increase of T1 until a state of equilibrium is reached. When the compressor stops and, therefore, the refrigerant mass flow is zero, the remaining gas in the neck is heated very fast and then cools down together with the surrounding solid parts.

4. CONCLUSIONS

This work deals with the investigation of the thermal behavior of a state-of-the-art compressor used in domestic refrigerators during steady-state and transient operating conditions. Due to the difficulty of sensor-signal feedthrough during operation of hermetic compressors, this investigation is not a standardized measurement and provides valuable temperature data. To gain measuring data, the non-variable speed compressor was operated in a fully automated calorimeter test bench and in a commercial freezer. The temperature was measured at nineteen measuring positions distributed around the shell and inside the compressor. Influences of the duty-cycle ratio, the cycle duration or the ambient temperature on the temperature level of the compressor were investigated at transient operating conditions. Furthermore, calorimeter measurements at four different stabilized operating conditions were compared. Finally the data of transient and steady-state measurements at two different operating conditions were compared. Based on these results the following major conclusions can be drawn:

- The temperature drop along the discharge line (without serpentine) is in the same range for all eight measurements and is almost independent of the refrigerant mass flow rate. On the contrary, the influence of different operating conditions on the heating of the suction gas and the cooling along the serpentine is more significant.
- The influences of the total cycle duration on the temperature levels are rather small. An increase of the duty-cycle ratio results in a similar thermal behavior but generally at a higher temperature level, whereas a variation of the ambient temperature results also in a slightly different thermal behavior.
- The comparison between transient and steady-state measurements shows, that the temperature level differs significantly due to different environmental conditions. Inside the calorimeter test bench, a forced air circulation keeps the exact ambient temperature, whereas inside the recess at the backside of the freezer, where the compressor is located, only buoyancy driven flow occurs. This leads to lower heat transfer coefficients and a higher air temperature close to the compressor.

REFERENCES

Cavallini, A., Doretti, L., Longo, G. A., Rossetto, L., Bella, B., Zannerio, A. (1996). Thermal analysis of a hermetic reciprocating compressor, Proceedings of the International Compressor Engineering Conference at Purdue, West Lafayette, USA, 535-540.

Kara, S., Oguz, E. (2010). Thermal analysis of a small hermetic reciprocating compressor, *Proceedings of the International Compressor Engineering Conference at Purdue,* West Lafayette, USA, paper 1307.

Longo, G. A., Caracciolo, R. (2002). Unsteady state analysis of a hermetic reciprocating compressor: heat transfer inside the cylinder and valve dynamics, Proceedings of the International Compressor Engineering Conference at Purdue, West Lafayette, USA, paper C4-4.

Negrao, C. O. R., Erthal, R. H., Andrade, D. E. V., Silva, L. W. (2010). An algebraic model for transient simulation of reciprocating compressors, Proceedings of the International Compressor Engineering Conference at Purdue, West Lafayette, USA, paper 1486.

Ribas, F. A., Deschamps, C. J., Fagotti, F., Morriesen, A., Dutra, T. (2008). Thermal analysis of reciprocating compressors - a critical review, *Proceedings of the International Compressor Engineering Conference at Purdue,* West Lafayette, USA, paper 1306.

ACKNOWLEDGEMENT

This work has been carried out within the framework of ECO-COOL, a research project initiated and funded by the FFG (Austrian Research Promotion Agency). Furthermore the authors particularly acknowledge the technical support by Secop Austria GmbH, formerly ACC Austria GmbH and Liebherr-Hausgeräte Lienz GmbH.

A Methodology for Characterization of Vapor-injection Compressors

Fernando M. TELLO OQUENDO[1]*, Emilio NAVARRO-PERIS[1], José GONZÁLVEZ-MACIÁ[1]

[1]Institute for Energy Engineering, Universitat Politècnica de València, Camino de Vera s/n, Valencia, 46022, Spain.
fertelo1@upvnet.upv.es, emilio.navarro@iie.upv.es, jgonzalv@ter.upv.es
Phone: +34 963 879 120
Fax: +34 963 879 126

* Corresponding Author

ABSTRACT

Vapor-injection compressor technology is being used in refrigeration systems and heat pumps working at large differences between evaporating and condensing temperatures. However, there are no published standards for the characterization of this kind of compressors. Calorimetric benches are widely used to characterize compressors, nevertheless, the calorimetric benches, as known, cannot be used directly to characterize vapor-injection compressors, since the refrigerant conditions at the injection port are defined by the injecting temperature and the intermediate pressure. In addition, these intermediate conditions depend critically on the way in which the injection is performed (economizer, flash tank, liquid injection, etc.). Therefore, the characterization process of this type of compressors is more complex. In this paper, a methodology for characterizing vapor-injection compressors is proposed. The method allows measurement of vapor-injection compressors in a typical calorimetric test bench after implementing some minor modifications in order to incorporate the injection line. With these modifications, the system is capable of independently controlling the intermediate pressure and the injection temperature. The proposed methodology has been evaluated with a vapor-injection scroll compressor (of 17.1 m^3/h) using R407C as a refrigerant. The measurement results and the performance parameters of the vapor-injection scroll compressor are presented. A simple correlation to estimate the injection mass flow rate as a function of the intermediate pressure has been found, that can be used to correlate the experimental results with high accuracy for several operating conditions. The advantages of the proposed methodology are the better control of the intermediate compression conditions and the possibility of testing vapor-injection compressors in a wide range of operating conditions.

1. INTRODUCTION

Cooling systems and heat pumps present several limitations working at high-pressure ratios: high compressor discharge temperature, cooling/heating capacity loss, rapid decrease of the COP and the like (Bertsch and Groll, 2008). A common solution is a two-stage compression with vapor-injection. The important advantages of this type of compression are the capacity improvement in severe climates; the system capacity can be varied by controlling the injected refrigerant mass flow rate, which permits energy savings by avoiding the intermittent operation of the compressor and the compressor discharge temperature is lower than that of a conventional single-stage cycle (Xu *et al.*, 2011).

To our best knowledge, there are no published standards for characterization of vapor-injection compressors. However, several recent research have been developed about vapor-injection compressors; most of them mainly focused on the experimental study of the heat pump system with economizer using vapor-injection scroll compressors (Bertsch and Groll, 2008), (Ma *et al.*, 2003), (Ding *et al.*, 2004), (Ma and Chai, 2004), (Wang *et al.*, 2009a), and (Wang *et al.*, 2009b). Regarding compressor characterization, Navarro *et al.* (2013) presented a test campaign of a vapor-injection scroll compressor considering a wide range of nominal operating conditions. The system used was air to water refrigerant injection circuit installed in a climatic chamber. The economizer was composed of three heat exchangers in order to adjust the economizer size. The intermediate superheat was defined by the intermediate pressure and by the economizer size (UA).

In Europe, manufacturers characterize single-stage compressors based on the Standard UNE-EN 13771-1 (2003). The standard proposes several procedures for testing compressors. These procedures require the definition of three external conditions: evaporating pressure, condensing pressure and superheat at the compressor inlet. In these conditions, the mass flow rate and the power consumption have to be measured. The characterization of vapor-injection compressors is more complex because there are two additional degrees of freedom, the intermediate pressure, and the injection temperature. These parameters depend on how the system is designed internally, that is the way in which the injection is performed (economizer, flash tank, liquid injection, etc.). For a full characterization of these compressors, it is necessary to measure the injection mass flow rate.

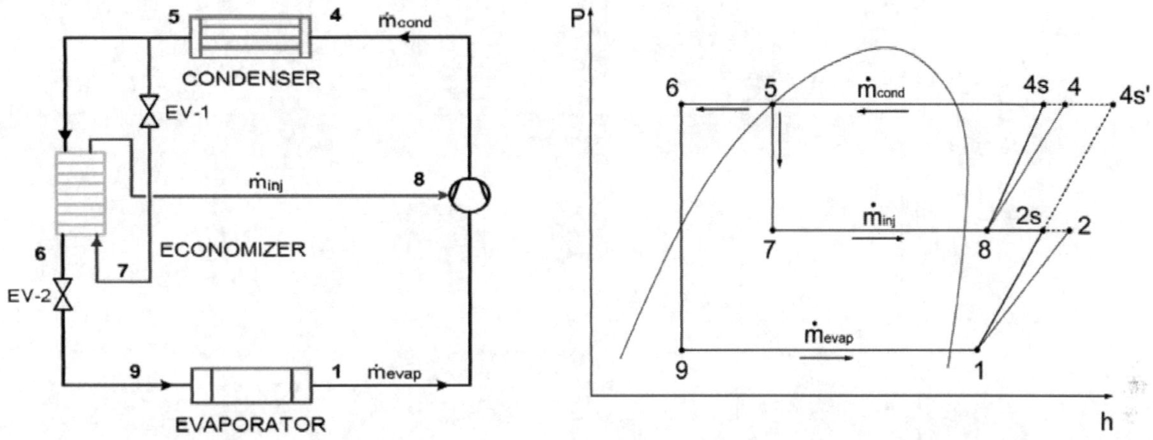

Figure 1: Schematic of a vapor-injection cycle with economizer and p-h diagram

For a given test matrix, when including the two additional parameters in the system, the number of experimental points increase considerably because the intermediate pressure can take several values for each operating point. However, not all points of the resultant test matrix describe the behavior of the compressor working in real operating conditions, since the compressor only works with a single intermediate pressure when the injection mechanism is set (injection cycle and control algorithm).

Figure 1 shows one of the most typical vapor-injection cycles. The system uses a heat exchanger (economizer) to vaporize the injection mass flow rate. The intermediate conditions are set from the economizer size and the chosen mechanism of control, which is usually a thermostatic expansion valve. Therefore, in this configuration, for each compressor size, a determined heat exchanger size has to be selected to define the different operating points of the compressor, which means having a set of heat exchangers (economizers) to characterize the compressor, hence the costs of the test bench increases dramatically.

In several compressor catalogs, the vapor-injection compressors are characterized considering a superheat at the compressor inlet of 5K; an injection superheat of 5K and a temperature approach in the economizer (T_6 -T_7 in Figure 1) of 5K. Therefore, for each operating point, many heat exchangers are needed in testing in order to maintain the temperature approach in the economizer.

This paper presents a methodology for characterization of vapor-injection compressors using a calorimetric bench. For the characterization, a system with economizer was simulated (see Figure 1). The economizer effect in the system is defined by the temperature approach in the economizer (T_6 -T_7). Fixing of this temperature approach is equivalent to set the economizer size. When the temperature approach in the economizer is set, it is possible to find a single intermediate pressure for each operating point (evaporating temperature, condensing temperature and superheat); therefore, the number of test points can be reduced dramatically. The advantages of the methodology are the better control of the intermediate compression conditions and the possibility of simulate several sizes of economizer, depending on the operating conditions.

2. EXPERIMENTAL SETUP

The compressor rating procedure was performed according to the European Standard UNE-EN 13771-1. Based on this standard, the refrigerant mass flow rate is the determining parameter to be measured, and primary and confirming measurements have to be made. The primary test procedure chosen is the secondary refrigerant calorimeter method. A Coriolis-type mass flow meter was used as the confirming test method. In all cases, confirming tests were carried out simultaneously with the primary mass flow rate determination.
Figure 2 shows the scheme of the calorimetric bench used for the compressor characterization.

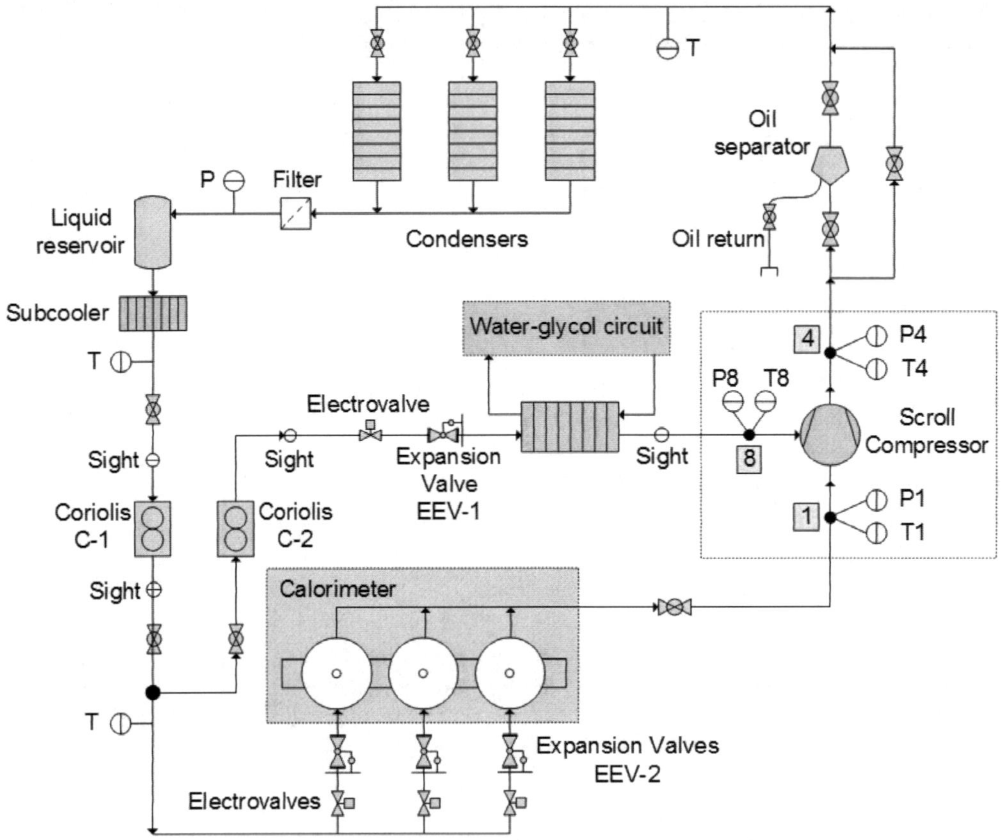

Figure 2: Calorimetric bench schematic

The condenser mass flow rate is directly measured using a Coriolis-type (Fisher–Rosemount Micro-Motion CMF025M), C-1 in Figure 2. Several PID control loops were incorporated to allow a precise adjustment of the refrigerant conditions at compressor inlet (evaporating temperature and superheat) and outlet (condensing temperature) with a precision of 1 kPa. The calorimetric bench is fully automated, and designed to reach any allowable test conditions without manual adjustments. The instrument accuracies of pressure transmitter (Fisher–Rosemount 3051) and temperature transmitter (RTD-PT 100) are 0.02% and 0.05°C, respectively.

In order to test the vapor-injection compressors, the calorimetric bench was modified to simulate the injection conditions and the interaction with the economizer. Part of the liquid (injection mass flow rate) is derived from the condenser outlet and is expanded to the intermediate pressure in an expansion valve (EEV-1 in Figure 2). After the expansion valve, the injection mass flow rate is vaporized in a heat exchanger using a secondary circuit of a water-glycol mixture. Electric resistors control the temperature of the water-glycol mixture in order to fix the injection superheat. With this arrangement, the system is capable of independently controlling the intermediate pressure and the injection temperature. The injection line is also equipped with a Coriolis-type mass flow meter with uncertainty of ±0.025 g/s (C-2 in Figure 2), a pressure transducer with a precision of 0.2%, a RTD with a precision of 0.1K, visors and an electrovalve located before the expansion valve (EEV-1 in Figure 2).

The evaporator mass flow rate is calculated with equation (1) and is compared with the secondary refrigerant calorimeter based result.

$$\dot{m}_{evap} = \dot{m}_{cond} - \dot{m}_{inj} \quad (1)$$

The calorimetric bench was designed to control the operating conditions of the vapor-injection compressor at the points: (1), (4) and (8) (see calorimetric bench schematic in Figure 2). The compressor used for the characterization was a vapor-injection scroll compressor (SCVI), model ZH18KVE-TFD of 17.1 m^3/h (swept volume). The SCVI was tested with R407C as a refrigerant.

3. CHARACTERIZATION PROCEDURE

In vapor-injection cycles with economizer (see Figure 1), for a given compressor size, the intermediate pressure is defined by the heat transfer in the economizer once the injection superheat is supplied. Therefore, for a given pressure ratio and an economizer size (UA), the intermediate pressure and the injection mass flow rate are fixed.

In the proposed calorimetric bench configuration (see Figure 2), there is no economizer because the evaporator mass flow rate does not interact with the injection mass flow rate into a heat exchanger. Therefore, to characterize the compressor, the working conditions of a real economizer cycle will be reproduced with the calorimetric bench using the following methodology.

The main objective of the methodology is to determine the corresponding intermediate pressure in order to have an injection mass flow rate. This injection mass flow rate must be able to produce a heat transfer in the virtual economizer which allows maintaining a temperature approach (T_6 -T_7) of 5K (see Figure 1). In that sense, this configuration would be equivalent to set the real economizer size, which would have in these working conditions.

The procedure of characterization begins with the setting of the condensing pressure, evaporating pressure and the superheat at the evaporator outlet acting on the flow rate of the water condenser, valves EEV-2 and resistors of the calorimeter, respectively. The intermediate pressure is regulated by the electronic expansion valve (EEV-1). The injection superheat is fixed with the water-glycol temperature through a heat exchanger. An injection superheat of 5K was chosen for testing, since in the majority of systems, the intermediate pressure control is performed with a thermostatic expansion valve. This valve needs a minimum superheat to regulate properly and to ensure that no liquid is injected, therefore, an intermediate superheat of 5K is appropriate. Furthermore, the injection superheat should be as low as possible to reduce the discharge temperature of the compressor. For all evaporating and condensing temperatures, the values of the considered parameters in the characterization are the superheat at the compressor inlet of 5K and the injection superheat of 5K. The cooling capacity and the COP are reported with a subcooling of 0K.

At this point, the injection mass flow rate and the intermediate pressure with which the compressor will work are not known. Thus, the methodology proposes an iterative process in order to find the intermediate conditions. The iterative process involves the progressive change of the intermediate pressure. The iterations end when the resultant injection mass flow rate produces a temperature approach of 5K in the virtual economizer of the cycle.

The iterative process begins by setting a starting intermediate pressure equal to the geometric mean of the evaporating and condensing pressures. Once the system is in equilibrium, the total mass flow rate ($\dot{m}_{cond}$), the injection mass flow rate ($\dot{m}_{inj}$), the injection temperature (T_8) and the condenser outlet temperature (T_5) are measured. The thermophysical properties of the refrigerant at points (5), (7) and (8) of Figure 1, are calculated with the NIST database (Lemmon *et al.*, 2010), and the evaporator mass flow rate is obtained from equation (1).

In the simulated cycle with an economizer (Figure 1), there is a heat exchange between the injection mass flow and the mass flow through the evaporator. Therefore, the following energy balance must be verified:

$$Q_{econo} = \dot{m}_{inj}.(h_8 - h_7) = \dot{m}_{evap}.(h_5 - h_6) \quad (2)$$

where Q_{econo} represents the economizer capacity; h represents the enthalpy corresponding to the different points described in the schematic of the vapor injection cycle of Figure 1. Solving the equation (2) the enthalpy and the temperature at the point (6) are calculated. At this point, the temperature approach in the economizer (T_6 -T_7) must be verified. If the temperature approach is 5K, then we have found the intermediate pressure for those working conditions. Otherwise, it must assume another value of intermediate pressure, and the iterative process is repeated. The new intermediate pressure (P_{int}*) is calculated assuming a constant injection ratio (X_{inj} = $\dot{m}_{inj}/\dot{m}_{evap}$) and a temperature approach in the economizer of 5K. All the iterative process is represented in detail in the flowchart of Figure 3.

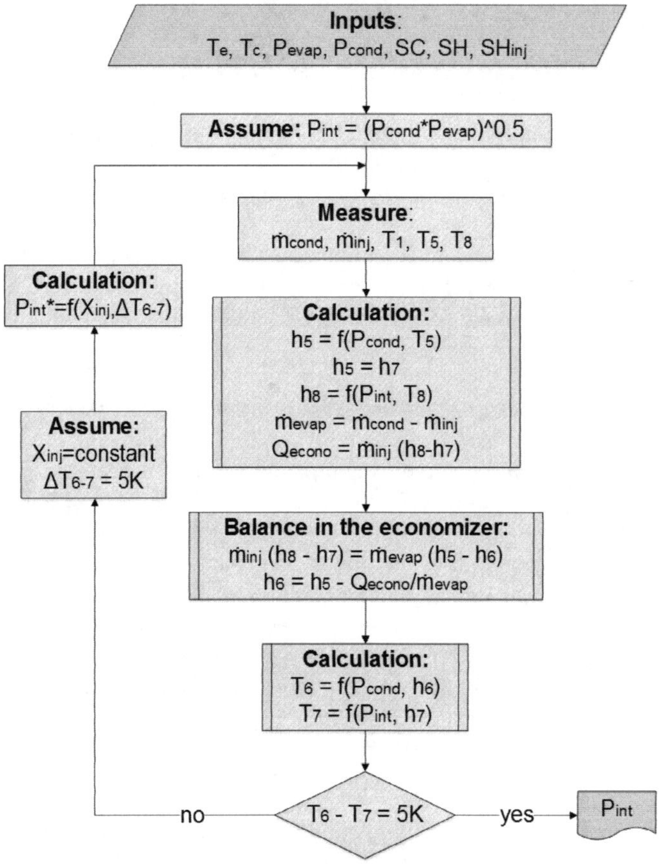

Figure 3: Flowchart for determining the intermediate conditions in a vapor-injection cycle with an economizer

4. RESULTS AND DISCUSSION

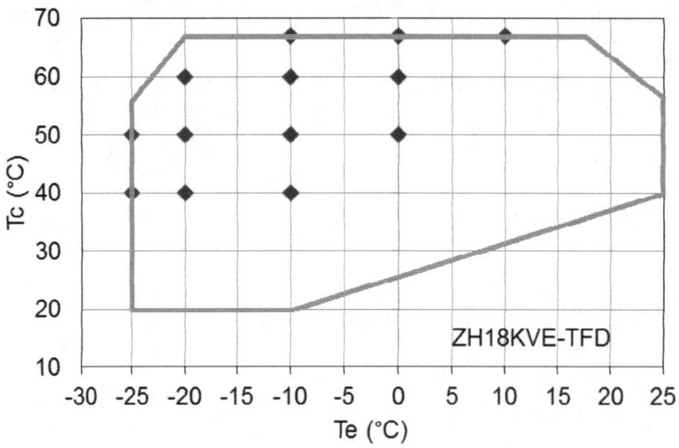

Figure 4: Working envelope and test matrix of the vapor-injection scroll compressor

Using the proposed methodology, vapor-injection compressors can be measured under the conditions of superheat at compressor inlet of 5K, injection superheat of 5K and temperature approach in the economizer of 5K. The methodology described above was applied to characterize a vapor-injection scroll compressor ZH18KVE-TFD at several working conditions. Figure 4 shows the working envelope of the compressor and the test matrix.

Table 1 shows the results of the compressor characterization working at evaporating temperature of -10ºC and a condensing temperature of 60ºC with a superheat at the compressor inlet of 5K. This table illustrates the iterative process shown in Figure 3 including all the performed iterations.

Table 1: Results of the compressor characterization working at the point (-10ºC, 60ºC)

P_{int} (kPa)	$\dot{m}_{evap}$ (kg/s)	$\dot{m}_{inj}$ (kg/s)	T_5 (ºC)	T_8 (ºC)	$h_5 = h_7$ (kJ/kg)	h_8 (kJ/kg)	$\dot{Q}_{econo}$ (kW)	h_6 (kJ/kg)	T_6 (ºC)	T_7 (ºC)	$T_6 - T_7$ (ºC)
843.18	0.05412	0.02581	55.85	22.88	287.53	422.31	3.48	223.24	16.03	14.71	1.33
780.33	0.05381	0.02269	55.85	21.10	287.52	422.00	3.05	230.81	21.17	12.21	8.96
792.42	0.05395	0.02331	55.85	21.47	287.52	422.08	3.14	229.39	20.21	12.70	7.51
803.90	0.05404	0.02388	55.85	22.04	287.52	422.40	3.22	227.93	19.23	13.16	6.06
812.67	0.05369	0.02428	55.85	22.38	287.52	422.54	3.28	226.46	18.23	13.51	5.01

Normally, it is necessary to make 5 iterations for each operating point. The intermediate pressure as a function of the temperature approach in the economizer is plotted in Figure 5; it can be seen that the temperature approach shows a linear dependence with the intermediate pressure. This fact can be an advantage at the time of reducing the number of iterations of intermediate pressure required to reach a temperature approach of 5K in the economizer. Therefore, for other working points, it will be necessary to iterate twice to get a linear equation relating the intermediate pressure with the temperature approach. With this equation, it is possible to estimate the desired intermediate pressure in the third iteration.

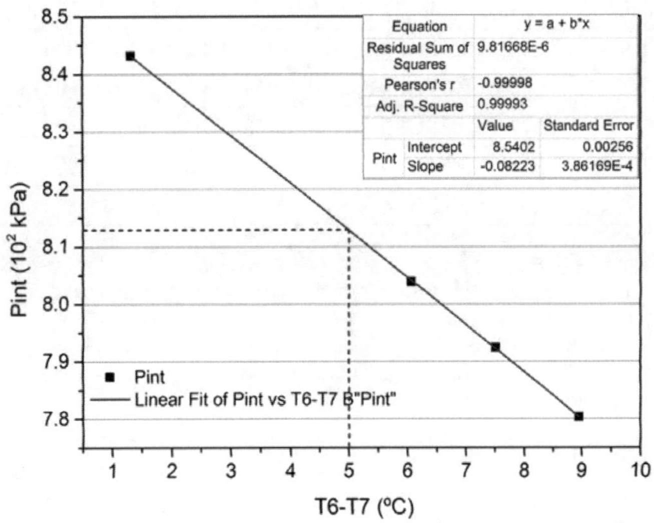

Figure 5: Intermediate pressure as a function of the temperature approach in the economizer, point (-10ºC, 60ºC)

Once the intermediate compression conditions are known, the cooling and heating capacity, the COP and the efficiencies are calculated. The volumetric and compressor efficiencies are defined by the equations (3) and (4) respectively. The results are shown in Table 2.

$$\eta_v = \frac{\dot{m}_{evap}}{\rho_1 \times \dot{V}_1} \qquad (3)$$

$$\eta_c = \frac{\dot{m}_{evap}(h_{4s'} - h_1) + \dot{m}_{inj}(h_{4s} - h_8)}{P_a} \quad (4)$$

Table 2: Compressor performance parameters for the point (-10ºC, 60ºC)

Q_c (kW)	Q_h (kW)	P_a (kW)	COP_c	COP_h	η_c	η_v
9.802	15.975	6.441	1.522	2.480	0.544	0.840

In the iterative process, the linear dependence between the intermediate pressure and temperature approach in the economizer was verified for all measured points; however, the adjustment coefficients of the resulting linear equation are different for each operating point. Table 3 shows the results of the compressor characterization for each working condition of the test matrix (Figure 4).

Table 3: Compressor performance parameters at several working conditions

T_e (ºC)	T_c (ºC)	P_a (kW)	Q_h (kW)	Q_c (kW)	COP_h	COP_c	P_{econo} (kW)	$\dot{m}_{evap}$ (kg/s)	$\dot{m}_{inj}$ (kg/s)	P_{int} (kPa)	P_{evap} (kPa)	P_{cond} (kPa)	η_c	η_v
-25	40	3.817	9.762	6.206	2.557	1.626	1.731	0.03016	0.01082	390.15	172.84	1540.66	0.539	0.849
-20	40	3.978	11.461	7.695	2.881	1.935	1.948	0.03799	0.01202	456.86	215.03	1540.28	0.570	0.869
-10	40	4.280	15.362	11.252	3.589	2.629	2.291	0.05720	0.01380	607.42	319.87	1544.96	0.606	0.895
-25	50	4.600	10.174	5.861	2.212	1.274	2.010	0.02931	0.01386	455.38	173.34	1987.80	0.512	0.823
-20	50	4.805	11.781	7.213	2.452	1.501	2.276	0.03679	0.01548	525.27	215.08	1987.15	0.543	0.841
-10	50	5.234	15.660	10.644	2.992	2.033	2.839	0.05581	0.01879	697.36	320.35	1988.08	0.587	0.872
0	50	5.502	20.325	15.012	3.694	2.728	3.144	0.08185	0.02033	885.11	461.50	1990.36	0.619	0.899
-20	60	5.846	12.252	6.710	2.096	1.148	2.577	0.03568	0.01964	615.67	215.19	2527.89	0.505	0.815
-10	60	6.441	15.975	9.802	2.480	1.522	3.278	0.05369	0.02428	812.67	320.02	2528.32	0.544	0.840
0	60	6.827	20.520	13.920	3.006	2.039	3.868	0.07938	0.02795	1027.30	461.25	2528.52	0.587	0.871
-10	67	7.480	16.228	9.082	2.170	1.214	3.523	0.05168	0.02869	911.51	320.15	2972.13	0.504	0.808
0	67	8.030	20.569	12.839	2.562	1.599	4.246	0.07604	0.03370	1149.98	461.83	2972.30	0.541	0.835
10	67	8.318	25.723	17.693	3.092	2.127	4.883	0.10872	0.03798	1409.89	645.18	2971.81	0.577	0.859

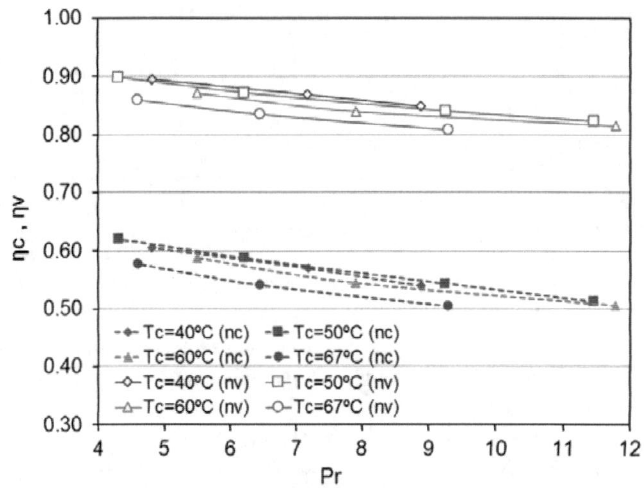

Figure 6: Compressor efficiency and volumetric efficiency as a function of pressure ratio at several condensing temperatures

Figure 6 depicts the compressor and volumetric efficiency of the SCVI as a function of the pressure ratio for several condensing temperatures. At lower condensing temperatures, the compressor efficiency, and volumetric efficiency are higher. The SCVI presents high volumetric efficiency values (above 0.8) for any operating point, because the compressor doesn´t have undesirable dead space and no inlet and outlet valves, the contact between the flanks of scrolls and their bases and upper edges is almost perfect and constant; thus, it has very good axial and radial compliance. The compressor efficiency varies from 0.5 to 0.62. At lower condensing temperatures (40ºC), the compressor efficiency is greater and for higher condensing temperatures and pressure ratios (around 12), the compressor efficiency decreases to 0.5.

Figure 7 illustrates the heating capacity (a) and the heating COP (b) as a function of evaporating temperature for several condensing temperatures. At low condensing temperatures, the COP_h is higher since the temperature focuses are closer. For a given evaporating temperature, the COP_h decreases as the condensing temperature increases, since the pressure ratio is greater and the compressor efficiency is reduced as seen in Figure 6. For a given condensing temperature, the heating capacity and the COP_h decrease when the compressor works with lower evaporating temperatures.

a) b)

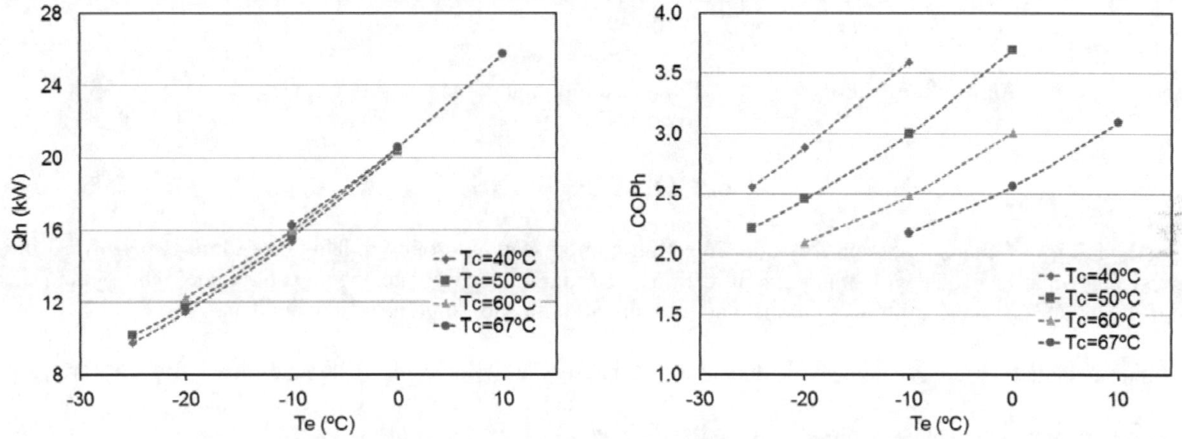

Figure 7: a) Heating capacity as a function of evaporating temperature. b) Heating COP as a function of evaporating temperature at several condensing temperatures

4.1 Injection mass flow rate correlation

In the compressor characterization process, the number of iterations is a very important factor, which should be minimized. One way of minimizing the number of iterations is using a cycle model to estimate the intermediate conditions. Because this type of cycle has two additional degrees of freedom (the intermediate pressure and the injection mass flow rate), it is necessary to have an expression that relates these two parameters to close the system.

Figure 8 represents the evolution of the injection ratio ($\dot{m}_{inj}/\dot{m}_{evap}$) to the pressure ratio (P_{int}/P_{evap}). This figure shows a linear tendency between the both plotted ratios. Based on Figure 8, the injection mass flow rate can be correlated with the intermediate pressure. The correlation is presented in equation (5). By linear regression, the coefficients A and B are obtained with a correlation factor higher than 0.99.

$$\frac{\dot{m}_{inj}}{\dot{m}_{evap}} = A + B\frac{P_{inj}}{P_{evap}} \quad (5)$$

A= -0.366; B= -0.322

To define the correlation at least 12 points of the compressor working envelope have been measured. Once the correlation is defined, the intermediate pressure can be calculated directly for different working conditions. Consequently, the time required to characterize a compressor is dramatically reduced.

The compressor manufacturer should provide this kind of correlations in order to define the intermediate conditions depending on the working conditions.

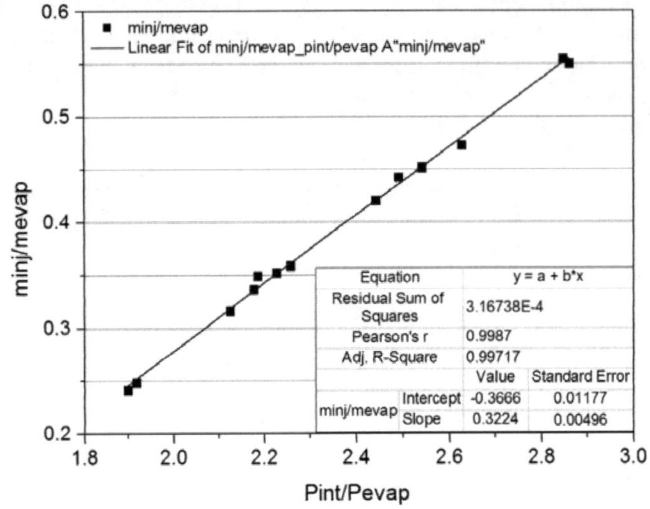

Figure 8: Injection ratio ($\dot{m}_{inj}/\dot{m}_{evap}$) as a function of the pressure ratio (P_{int}/P_{evap})

6. CONCLUSIONS

A methodology of characterization for vapor-injection compressors is proposed. This procedure is performed in a typical calorimetric bench with minor modifications in order to incorporate the injection line. The system can control the intermediate conditions (intermediate pressure and injection superheat) independently.

The intermediate pressure is calculated with an iterative process until the resulting injection mass flow rate produces a temperature approach in the economizer of 5K. A linear dependence between the intermediate pressure and the temperature approach in the economizer is found, which reduces the number of iterations.

With the proposed methodology, a vapor-injection scroll compressor was characterized in a wide range of operating conditions. A simple correlation to estimate the injection mass flow rate as a function of the intermediate pressure has been found, and it is able to correlate the experimental results with high accuracy for several operating conditions. This correlation can be used for modeling the cycle and to define easily the intermediate conditions.

The advantages of the proposed methodology are the better control of the intermediate compression conditions and the possibility of testing vapor-injection compressors in a wide range of operating conditions with no need of having heat exchangers (economizers) of different sizes for each working point; therefore, the costs of the test bench are reduced dramatically.

NOMENCLATURE

COP	coefficient of performance	(–)
h	enthalpy	(kJ/kg)
$\dot{m}$	mass flow rate	(kg/s)
P	pressure	(kPa)
P_a	compressor consumption	(kW)
P_r	pressure ratio	(–)
Q_c	cooling capacity	(kW)
Q_{econo}	economizer capacity	(kW)
Q_h	heating capacity	(kW)
SC	subcooling	(ºC)

SCVI	scroll compressor with vapor-injection	
SH	superheat	(ºC)
T	temperature	(ºC)
Greek symbols		
ρ	density	(kg/m^3)
η_c	compressor efficiency	(–)
η_v	volumetric efficiency	(–)

Subscript

c	condensing
cond	condenser
e	evaporating
econo	economizer
evap	evaporator
inj	injection
int	intermediate
s	isentropic
1	compressor inlet

REFERENCES

Bertsch, S. & Groll, E. (2008). Two-stage air-source heat pump for residential heating and cooling applications in northern U.S. climates. *Int. J. Refrig., 31*, 1282-1292.

Ding, Y., Chai, Q., Ma, G., & Jiang, Y. (2004). Experimental study of an improved air source heat pump, *Energ. Convers. Manage., 45*, 2393-2403.

Lemmon, E., Huber, M., & McLinden, M., (2010). NIST Standard Reference Database 23: Reference Fluid Thermodynamic and Transport Properties-refprop. Version 9.0., *National Institute of Standards and Technology*, Standard Reference Data Program, Gaithersburg.

Ma, G., Chai, Q., & Jiang, Y. (2003). Experimental investigation of air-source heat pump for cold regions, *Int. J. Refrig., 26*, 12-18.

Ma, G., & Chai, Q. (2004). Characteristics of an improved heat-pump cycle for cold regions, *Appl. Therm. Eng., 77*, 235-247.

Navarro, E., Redón, A., Gonzálvez, J., & Martinez, I. (2013). Characterization of a vapor injection scroll compressor as a function of low, intermediate and high pressures and temperature conditions, *Int. J. Refrig., 36,* 1821-1829.

UNE-EN 13771-1. (2003). Compressors and condensing units for refrigeration – Performance testing and test methods.

Wang, B., Shi, W., Han, L., & Li, X. (2009a). Optimization of refrigeration system with gas-injected scroll compressor, *Int. J. Refrig., 32,* 1544-1554.

Wang, X., Hwuang, Y., & Radermacher, R. (2009b). Two-stage heat pump system with vapor-injected scroll compressor using R410A as a refrigerant, *Int. J. Refrig., 32,* 1442-1451.

Xu, X., Hwang, Y., & Radermacher, R. (2011). Refrigerant injection for heat pumping/air conditioning systems: literature review and challenges discussions, *Int. J. Refrig., 34,* 402-415.

ACKNOWLEDGEMENT

Fernando M. Tello Oquendo acknowledges the financial support provided by the "CONVOCATORIA ABIERTA 2013-SEGUNDA FASE" program, which was funded by the SENESCYT ("Secretaría Nacional de Educación Superior, Ciencia, Tecnología e Innovación") (Grand No 2015-AR37665) of Ecuador.

Characterization and Performance Testing of Two-Stage Reciprocating Compressors using a Hot-Gas Load Stand with Carbon Dioxide

Xinye Zhang[1*], Bin Yang[1], Orkan Kurtulus[1], Eckhard A. Groll[1], Andres Osorio[2], Dylan Bethel[2]

[1] Purdue University, Ray W. Herrick Laboratories, West Lafayette, IN, USA
xinye@purdue.edu, yang62@purdue.edu, orkan@purdue.edu, groll@purdue.edu

[2] BlackPak Inc. San Leandro, CA, USA
aosorio@blackpaktech.com, dbethel@blackpaktech.com

* Corresponding Author

ABSTRACT

This paper presents performance measurements of a carbon dioxide two-stage reciprocating compressor system using a hot-gas bypass load stand. The experimental data collected was used to characterize the compressors' performance and evaluate oil management between the two compressors. A closed loop compressor load stand has been redesigned to conduct the compressor performance tests. Two hermetic reciprocating compressors, including two oil separators, two intercooling fans, and safety control valves, were connected to the load stand for testing. The first series of compressor tests were conducted for varying suction pressures and discharge pressures. In a second series of tests, the running time was varied and the suction pressure kept constant. Based on the acquired test measurements, compressor performance metrics, such as the volumetric and overall isentropic efficiencies, have been determined. Furthermore, compressor initial clearance factors were calculated, based on available compressor maps, which could be useful for future carbon dioxide system modeling. Also, a simulation model to predict two-stage compressor performance is presented. The entire process can be simulated to provide compressor performance data for different working conditions. Finally, the predicted performance was validated using the acquired test data.

1. INTRODUCTION

Many numerical and experimental studies of carbon dioxide trans-critical and conventional sub-critical refrigerating systems have recently received increased attention as a possible replacement for vapor compression cycle technologies that use fluorocarbon-based refrigerants. Aprea and Maiorino (2008) evaluated energy performance using an internal heat exchanger. The experimental plant employed a semi-hermetic compressor, plate-finned tube type heat exchangers, an electronically controlled back pressure valve, and an expansion valve. Rigola et al. (2005) designed a small cooling capacity unit with an evaporation temperature of 0 °C. A detailed numerical simulation model for hermetic reciprocating compressor performance, validated using conventional refrigerants, was extended to numerically obtain the CO_2 compressor prototypes behavior.

Hubacher et al. (2002) presented several performance measurements of a prototype carbon dioxide compressor using a hot-gas bypass compressor load stand. The compressor was a semi-hermetic, two-piston, single-stage, reciprocating compressor with an estimated cooling capacity of 3 tons. However, Hubacher mainly focused on how to find the volumetric efficiency. Christen et al. (2006) described performance measurements of a prototype carbon dioxide compressor using a compressor load stand based on a hot-gas bypass design were five types of compressors were used for testing. However, there were no simulation results presented in Christen's study to validate the experiment results.

With that said, the purpose of this paper is to fill the gap in open literature with experimental validation for the sub-critical cycle using hermetic compressors. To accomplish this, more detailed experimental results of a closed loop system will be analyzed. Additionally, a suitable intermediate pressure will be found using the approach of pressure recalculation once test conditions are defined. Furthermore, the simulation model presented is expected to predict a two-stage compressor systems' performance.

2. EXPERIMENTAL RESULTS

2.1 Description of system and experiments

The schematic of the closed loop test stand setup is shown in Figure 1. CO_2 was used as the working fluid during this phase of testing. The closed loop was divided into three fluid circuits: the oil return line, the bypass line, and a reduced temperature hot gas line. In the experiments presented in this paper, the reduced temperature hot gas line was used for testing only. However, due to the nature of the system design, the oil return and hot-gas bypass lines are introduced as follows.

In the oil return line, an oil separator having one inlet and two outlets, was used to separate oil that was discharged with the compressed CO_2 by the compressor system. The purified compressed CO_2 was then allowed to continue into the remaining fluid circuits. The separated oil resulting from the oil separation process, was routed to the oil return line for re-entrance to the compressor system. To ensure oil starvation did not occur, an oil measuring gauge was added to monitor oil levels in the test system to indicate if additional oil was needed.

In the bypass line, two metering valves of different sizes were used to control the CO_2 flow rate, which will affect the compressor test conditions accordingly. Coarse adjustment was conducted by a metering valve (NV106) installed in the larger size tube. Flow rate fine tuning was accomplished using a smaller needle valve (NV105).

The heat exchangers installed in the system were used to reject energy received during the compression process. Two heat exchangers were installed in parallel to enable variable cooling capacity. Both heat exchangers used tap water to carry the excess energy away from the system. A membrane tank filled with the system working fluid was also used in the reduced temperature hot gas line. The pressure inside the membrane tank was held at a constant intermediate pressure by enforcing an appropriate back pressure on the membrane. This enforced intermediate pressure provided a reliable anchor point to the cycle. If any system working gas happened to be lost through leakage of the pressure membrane, the gas inside the tank was able to continue holding the test cycle at stable operating conditions. Pressure variation in the membrane tank was not large enough to significantly change the intermediate system pressure. Thus, results from the controlled membrane will not be discussed in this paper.

Pressure relief valves were installed in each section to protect the test stand from uncontrolled pressure increases. A mass flow meter was installed to enable measurement the systems' mass flow rate. A power meter was also installed to measure the electric power input. All measured data was obtained and displayed using an appropriate DAQ system.

The compressor box of interest, consisting of two hermetic reciprocating compressors used in series to achieve the target pressure in the discharge line, was connected to the test stand. An electronic control box is installed inside the compressor box. The installed electronic control box contains an ON/OFF switch and status indicator lights.

Two adjustable pressure switches are located in the compressor box: one is installed immediately after the first stage compressor and a second is installed in the final discharge line of the compressor box. The first pressure switch is used to control activation of the second-stage compressor. Once the pressure measured by the pressure switch immediately after the first stage compressor reaches a predefined set pressure, the second compressor begins contributing to the system pressure rise. That said, during steady state testing, the initial pressure will be higher than the pressure set value for two-stage compressor activation; leading to both compressors starting and operating simultaneously. The second pressure switch, located in the final compressor box discharge line and used for safety purposes, is charged with the task of shutting the test system down after reaching a predetermined discharge pressure. There are two additional pressure switches, which are used to prevent over pressurization of the first and second compressor stages. These two additional switches are normally closed and connected the electronic box interlock circuit. Whenever pressure exceeds 4800 kPa in discharge pressure or 1100 kPa in intermediate pressure, the interlock circuit opens and the compressors shut down. Two Oil separators and filters have been used after each compressor to provide access for system oil management.

2.2 Test stand instrumentation

All measuring instrumentation is indicated in the schematic shown in Figure 1 . Temperatures were measured with T-type thermocouples with ±0.25 ºC accuracy. The inlet pressure of the open loop test stand was measured with an absolute pressure sensor that has an advertised accuracy of ±0.25%. Other pressures were measured with a pressure sensor gauge that has an accuracy of ±0.13% of full scale. A Coriolis-effect mass flow meter with an advertised accuracy of ±0.1% was also installed in the open loop test stand to measure mass flow rates of the gas. Electric power consumption of the compressors and the cooling fans were measured with power meters with an accuracy of ±0.04%. A data acquisition system was used to convert the incoming voltages from the measuring instrumentation to digital signals for transfer to a personal computer. The computer uses a proper data reduction program for further data analysis.

2.3 Test matrix

A test matrix outlining the operating conditions for the closed loop system is shown in Table 1. The maximum attainable high-side pressure of the closed loop test stand without compressor failure was obtained using CO_2 as the working fluid. Differences in temperature, mass flow rate, and oil slippage as a function of run time were evaluated during these tests.

Table 1: Test matrix of steady state testing with closed loop test stand

Compression Ratio	Low Side Pressure (kPa)	High Side Pressure (kPa)	Runtime (min)	Recording Data
20	105	2100	120	Temperature
30	105	3150	120,240	Pressure, Power
40	105	4200	240,360	Mass Flow Rate
43	105	4550	240,360	Oil Slippage

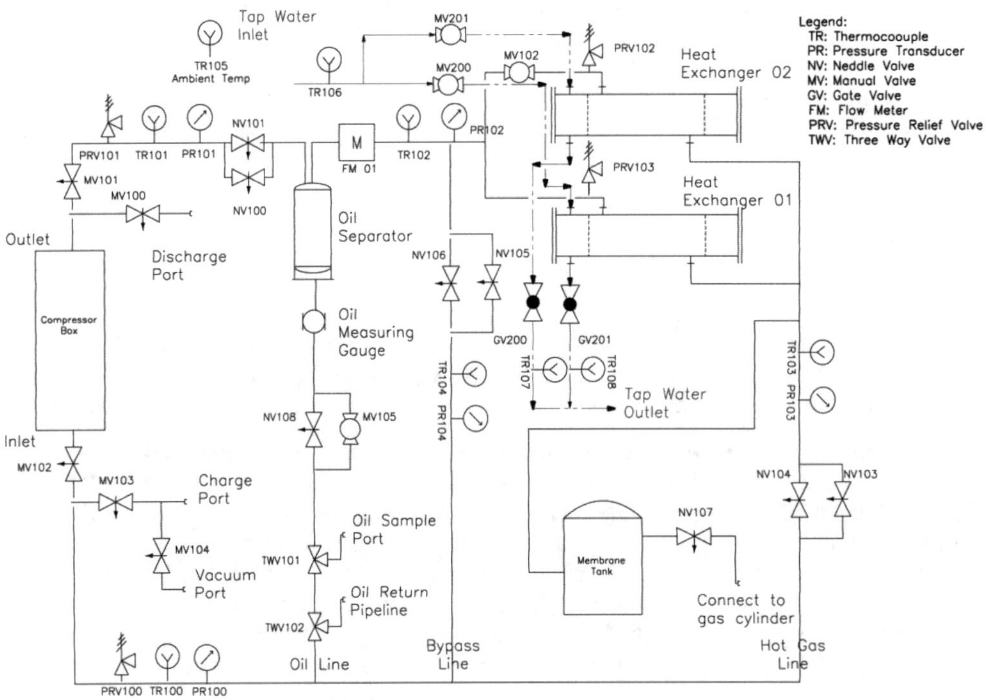

Figure 1: Schematic of the closed loop system

2.4 Experimental results

During the presented tests, the following compressor measurements were recorded: compressor mass flow rate, suction and discharge temperatures, suction and discharge pressures, intermediate pressure and temperature, and compressor power consumption. Using these measurements, the two-stage compressor system performance was evaluated. While multiple tests were conducted under various pressure levels and time durations, this paper only presents the results of the 2100 kPa testing over the 120 min testing period. The other test results lead to similar conclusions.

The steady state test with 2100 kPa high side pressure was conducted and the total run time for complete test execution was 140 minutes. The total testing time included 2 minutes for releasing pressure, 10 minutes for charging process, 120 minutes for steady state testing, and 8 minutes for shut-down testing. During the pressure release stage, only the cooling fans were set to operate. Both compressors were shut down simultaneously during the shutdown stage. Figure 3 shows the variation of the CO_2 pressures as a function of time in the closed loop test stand. It can be noticed that the initial pressure in system was approximately 1050 kPa, releasing to 630 kPa or lower to satisfy the prerequisite condition for compressor start-up.

After both compressors began to operate, it took several minutes to reach steady state. Discharge valve manipulation (MV101) was used to set the target high side pressure. Once steady state with high side pressure 2100 kPa was acquired, the intermediate pressure was set to approximately 840 kPa.

Figure 2: Closed loop system test stand

After two-hours of testing, the compressor box was shut down manually and all pressures in system returned to an equilibrium state. The new equilibrium state was higher than the initial equilibrium system pressure due to the refrigerant charge. That said, it was necessary to release the increased system pressure first before moving on to the next test to enable compressor start-up.

Figure 4 shows the variation of measured CO_2 temperature as a function of time during closed loop, steady state testing. After the compressor system was started, it was found that the 1st stage discharge temperature increased significantly to a value of 125 °C by the end of the test. The compressor thermal protector after 1st stage compressor is engineered to shut down if the temperature exceeds 130 °C. The 2nd stage discharge temperature, increased significantly, but to a much lower overall value than the 1st stage discharge temperature. As is shown in Figure 4, differences between these two discharge temperatures became larger during testing. It can be noticed that the temperature variations after the heat exchanger in both stages were relatively small due to the intercooler. Note that the intercooler was at a temperature close to ambient temperature during testing. Also, a small jump of the 2nd stage discharge temperature can be noticed in the right corner of Figure 4. This small temperature jump was due to the cooling fans being shut off with heat remaining in the compressor system after shut-down.

Figure 5 shows variation of the pressure ratio, i.e. the ratio of discharge pressure over suction pressure for each stage as a function of time for the closed loop system. It can be noticed that the pressure ratio of the 1st stage compressor was reached immediately after both compressors started working. This pressure ratio dropped during the test and was relatively stable at approximately 6. However, the pressure ratio of the 2nd stage compressor continued a gradual increase until the system was turned off. The 2nd stage pressure ratio was smaller than the 1st stage pressure ratio, which resulted in a higher discharge temperature in 1st stage compressor, as shown in Figure 4. The total pressure ratio was stable at approximately 18 in steady state. As discussed before, the compressors hold residual pressures at the end of the test. These residual pressures are why the 1st stage pressure ratio is larger than 1.

Measured data in the test stand side is also analyzed to allow a better understanding of the different measurements in the closed system loop. Figure 6 shows the variation of CO_2 pressures as a function of time in the test stand. Two pressures: high side pressure and low side pressure existing in this plot. During the presented two-hour test, both pressures appeared to be very stable. That said, expected pressure drops existed in Coriolis-effect mass flow meter and pipelines.

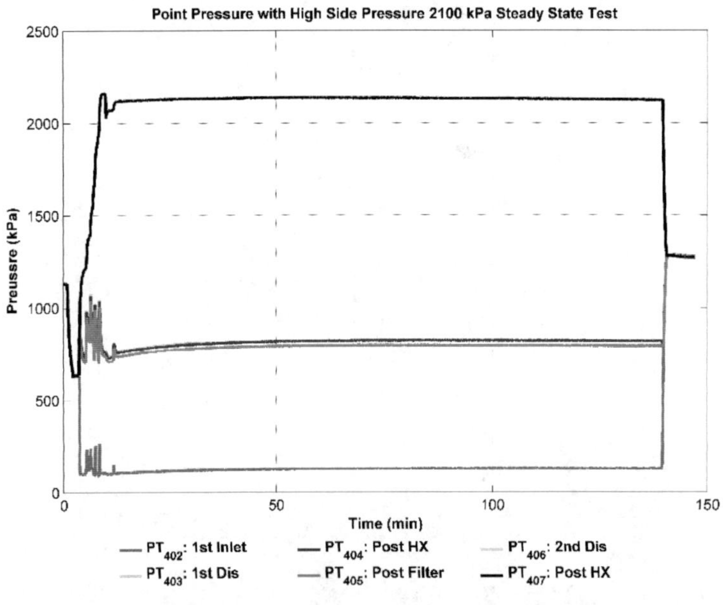

Figure 3: Point pressure with high side pressure 2100 kPa steady state test

Figure 7 shows variations of the measured temperatures as a function of time in the test stand for multiple measurement locations. All temperatures, except for TR_{100} and TR_{103}, appear to have stabilized at approximately 23 °C. Note that 23 °C was close to ambient temperature during testing. The two heat exchangers, used to reject heat from the hot discharge gas using 6 gal/min tap water, caused the post-membrane (TR_{103}) temperature measurement to decrease.

2.5 Oil management

An oil management analysis was conducted during each test. The differences of the oil weight in the discharge filters during each test were recorded to develop a general understanding of how oil migrated within the system during testing, as well as study general factors that affect oil slippage. This analysis helped to indicate how much oil was discharged with the compressed CO_2 and if additional oil was needed. To facilitate the slippage study, two filters were installed after each compressor stage. Oil weight data of the various tests performed are shown in Table 2.

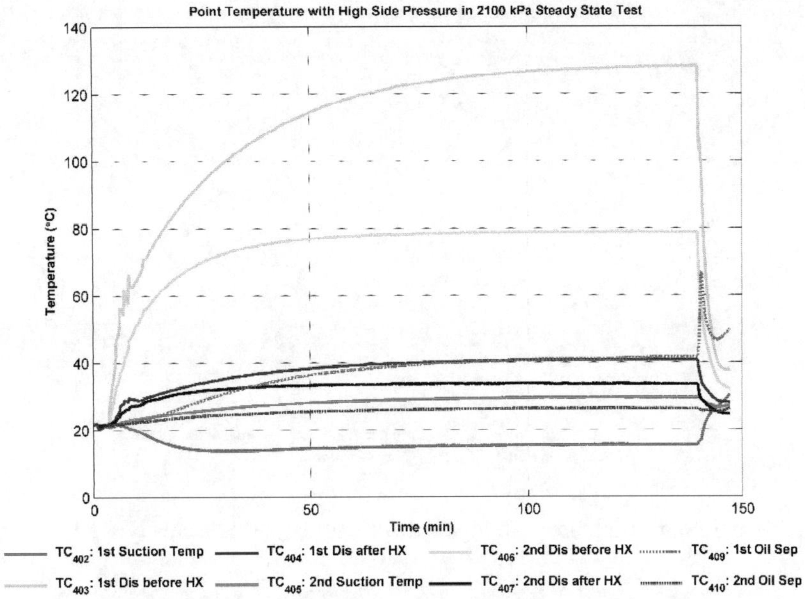

Figure 4: Point temperature with high side pressure 2100 kPa steady state test

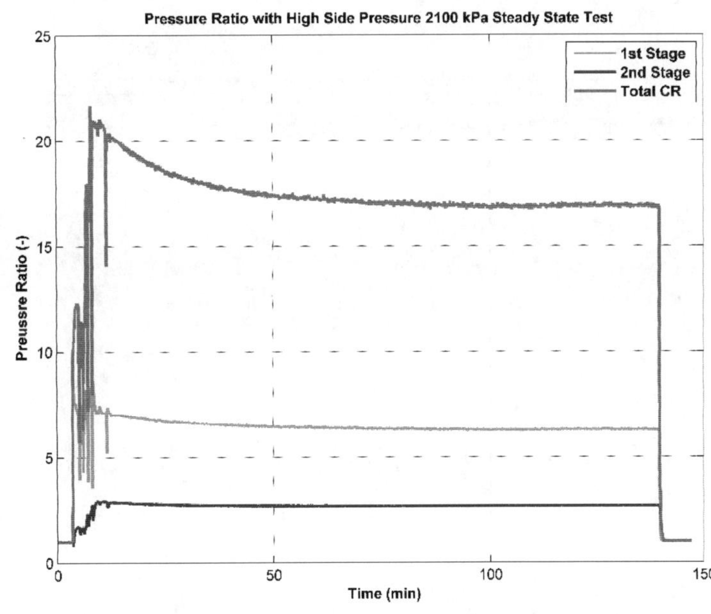

Figure 5: Pressure ratio with high side pressure 2100 kPa Steady State Test

A majority of the oil was discharged by the 1st stage compressor. Also, as expected, more oil was collected by the filters during the tests that required higher target pressures. That said, the importance of analyzing the measured data to estimate which factors affect oil slippage the most becomes apparent. It should be noted that most of the oil was collected by the first filter, which is not shown in Table 2.

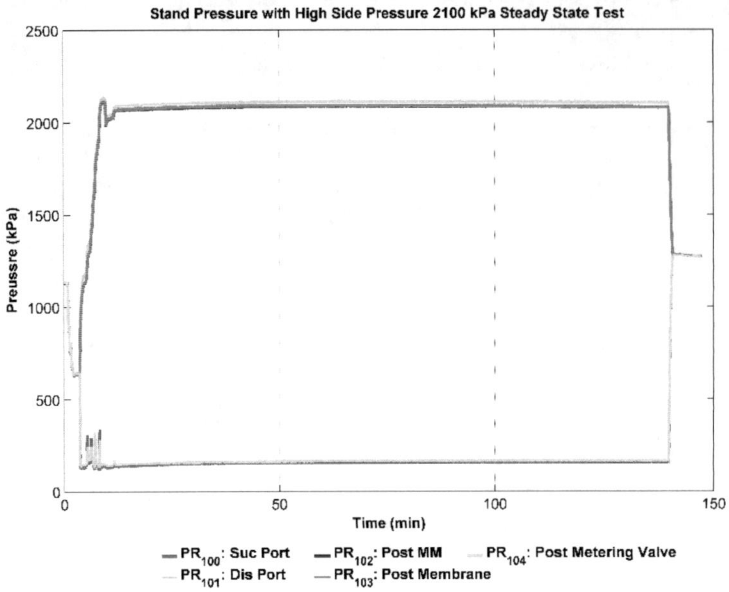

Figure 6: Stand pressure with high side pressure 2100 kPa steady state test

Based on the testing performed, it can be concluded that more oil is collected during testing in a two-stage compressor system with higher target high-side pressures. However, if the same target pressures are compared, more oil will be discharged in the lower pressure range by the 1st stage compressor than discharged in the 2*nd* stage compressor.

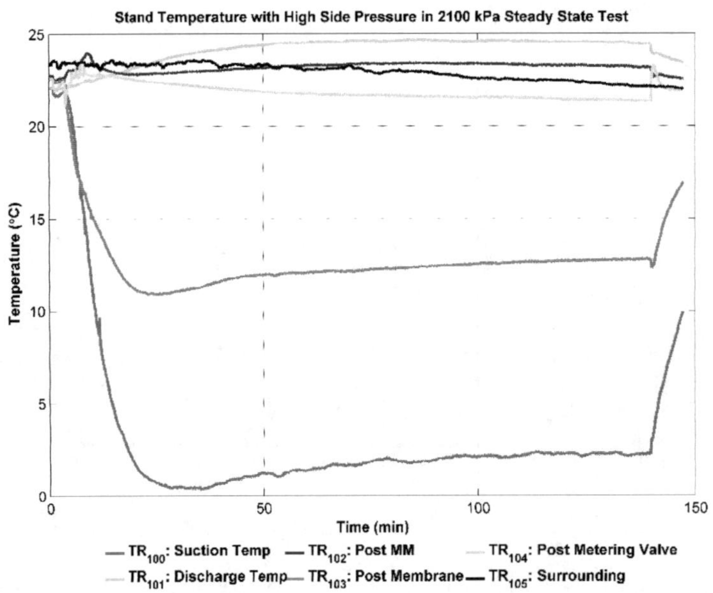

Figure 7: Stand temperature with high side pressure 2100 kPa steady state test

Table 2: Oil slippage analysis

1st Stage (g)	2nd Stage (g)	Total (g)	High Side Pressure (kPa)
9.1	0.4	9.5	2100
11.9	1.6	13.5	3150

3. SIMULATION MODEL

A simulation model to predict the two-stage compressors' performance has been developed. In this model, the initial clearance factor for the compressor was calculated based on available compressor maps that were used as an input to the steady state model. The entire compression process has been simulated to provide the compressor performance data as a function of time. Predicted performance will be validated using test data.

3.1 Clearance volume factor

The clearance volume factor is a significant value, which affects the performance of the compressors. It must be noted that the clearance volume factor is an attribute of the compressor itself and can be found through experimentation. The approach used here is to find an estimated value based on published compressor performance maps for use in the simulation model.

Compressor map data given by the compressor manufacturer provides the relation between mass flow rate, suction, and discharge conditions. Equation (1) is used to calculate the clearance volume factor. The clearance volume factor C is the slope of the function curve between the volume efficiency and the pressure ratio.

$$\eta_v = 1 - C\left[\left(\frac{P_d}{P_s}\right)^{1/m} - 1\right] = \frac{\dot{m}_e}{\dot{m}_{th}} \tag{1}$$

3.2 Intermediate pressure recalculation

In a two-stage compressor system, intermediate pressure observes the principle that the discharging gas volume from the 1st compressor will completely transfer to the 2nd compressor. However, the corresponding intermediate pressure strongly depends on the relationship between two compressor displacements. It should be noted that when the discharge condition of 1st stage compressor changes, the gas volume of 2nd compressor and the intermediate pressure will response accordingly. Therefore, the intermediate pressure value must be iteratively adjusted in the simulation to satisfy both compressor displacements. The intermediate pressure between any two stages in a two-stage compressor system can be calculated using Equation (2):

$$V_{hj}\frac{\eta_{vj}\lambda_{Tj}\lambda_{Pj}\lambda_{lj}}{\lambda_{\varphi j}}\frac{P_j}{P_1}\frac{T_1}{T_j} = V_{hj+1}\frac{\eta_{v,j+1}\lambda_{T,j+1}\lambda_{P,j+1}\lambda_{l,j+1}}{\lambda_{\varphi,j+1}}\frac{P_{j+1}}{P_1}\frac{T_1}{T_{j+1}} \tag{2}$$

where j is referred to the 1st stage and $j+1$ is referred to 2nd stage in a two-stage system.

Given parametric data for one compressor, the recalculation method was used to find the corresponding intermediate pressure for the two-stage compressor system. Based on the calculated intermediate pressure value; power consumption, and gas flow rate can be updated accordingly in each loop. The recalculation process is as follows:

1) Related Coefficient:
   In the recalculation method, all related coefficients are assumed to be a constant value for both compressors respectively, as shown in Table 3.

Table 3: Compressor coefficient

Stage	λ_p	λ_T	λ_l	$\lambda_0 = \lambda_p \lambda_T \lambda_l$
1st	0.96	0.96	0.92	0.848
2nd	0.98	0.97	0.92	0.874

2) Displacement guess value:
Define a displacement guess value to be used for estimating the suction pressure and pressure ratio in each stage. Assume 3 °C increase in the 2nd stage due to incomplete intercooling.

$$V_{hj,guess} = V_{hj} \frac{T_1}{T_j} \lambda_{0j} \tag{3}$$

where $\lambda_{0j} = \lambda_{Pj} \lambda_{Tj} \lambda_{lj}$ and T_j is the suction temperature in each stage.

3) Equation (4) is used to update pressure and the volumetric coefficient with a guess value. Results are shown in Table 5:

$$P_{sj} = P_{s1} \frac{V_{h1,guess}}{V_{hj,guess}} \frac{\eta_{v1}}{\eta_{vj}} \tag{4}$$

Table 4: Displacement guess value

Stage	$V_{hj}(10^{-5}m^3)$	$T_j(K)$	$\lambda_{0j}(-)$	$V_{hj,guess}(10^{-5}m^3)$
1st	3.436	291	0.848	2.91336
2nd	0.424	294	0.874	0.36703

Table 5: Update pressure and volumetric coefficient

Stage	$P_{sj}(kPa)$	$P_{dj}(kPa)$	$\varepsilon_j(-)$	$\eta_{vj}(-)$
1st	112	886	7.9375	0.6766
2nd	886	2101.5	2.3708	0.8707

4) Convergence Criteria:
Based on Equation (2), the suction gas volume in each stage was converted to the volume at the suction condition in the 1st stage, which is shown in Equation (5):

$$V_j = \frac{P_{sj}}{P_{s1}} V_{hj,guess} \eta_{v1} \tag{5}$$

In all results of V_j for each stage, if $V_{min}/V_{max} > 0.97$, the recalculation process stops and results converged.

Table 6: Pressure recalculation in each stage

Iteration	Stage	$P_{sj}(kPa)$	$P_{dj}(kPa)$	$\varepsilon_{j,update}(-)$	$V_j(10^{-5}m^3)$	$V_{min}/V_{max}(-)$	Converged (N/Y)
1	1st	112	688.7	6.168	2.257	0.8151	N
	2nd	688.7	2101.5	3.051	1.8397		
2	1st	112	845.4	7.597	2.079	0.8699	N
	2nd	845.4	2101.5	2.487	2.391		
…..							
5	1st	112	757.7	6.786	2.177	0.9562	N
	2nd	757.7	2101.5	2.773	2.083		
6	1st	112	768.6	6.884	2.165	0.9793	Y
	2nd	768.6	2101.5	2.734	2.121		

Final results from the last iteration were used for cycle calculations, as shown in Table 7 :

Table 7: Final converged results

Stage	$P_{sj}(kPa)$	$P_{dj}(kPa)$	$\varepsilon_j(-)$
1st	112	768.6	6.884
2nd	768.6	2101.5	2.734

Polytropic coefficients for two compressors operating in series can be obtained using experimental data. These are a function of the pressure ratio and experiment time, as shown in Equation (6) and (7). Equation (8) was used to predict the discharge temperature by using the pressure ratio and polytropic coefficients. P-h and T-s diagram, shown in Figure 8 and 9, can then be derived to show the entire gas cycle. In the cycle diagram, all temperature and pressure in each point, at 100 min running time, can be seen clearly and validated by the measurement data discussed above.

$$n_{T,1} = 1.0013 + 2.6944e^{-4} \times t + 7.4677e^{-4} \times PR - 1.2135e^{-7} \times t^2 - 1.9413e^{-5} \times t \times PR \quad (6)$$

$$n_{T,2} = 0.0023416 - 0.016275 \times t + 3.2629 \times PR - 3.9045e^{-5} \times t^2 + 0.019146e^{-5} \times t \times PR \quad (7)$$

where t is duration of time and PR is pressure ratio in each stage.

$$T_{out} = T_{in} \left(\frac{P_{out}}{P_{in}} \right)^{\frac{n_T - 1}{n_T}} \quad (8)$$

where n_T is polytropic coefficient

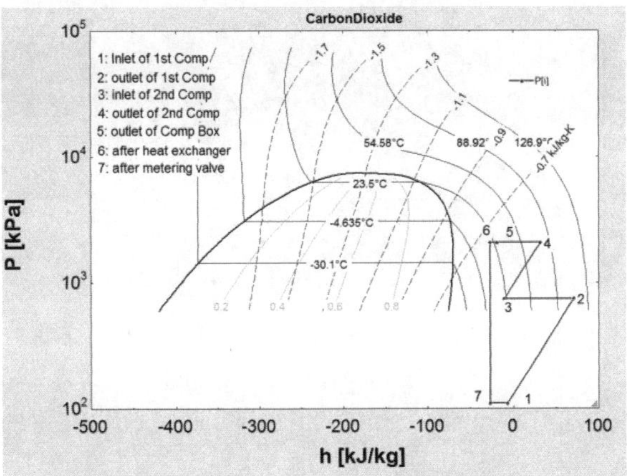

Figure 8: P-h diagram in closed loop system

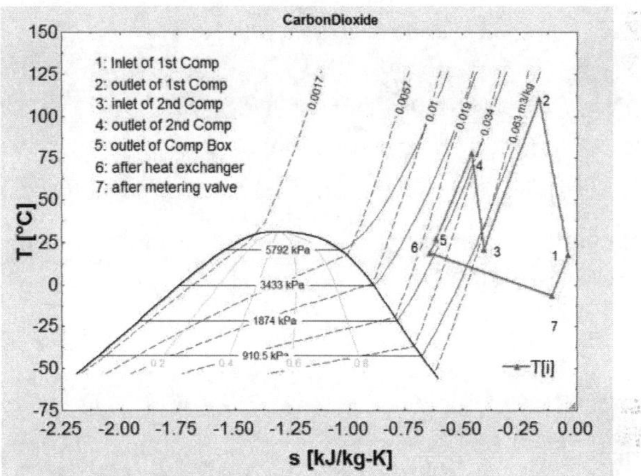

Figure 9: T-s diagram in closed loop system

4. CONCLUSION AND FUTURE WORK

Several experimental measurements with different target pressures and run times have been conducted using CO_2 as the working fluid. The following parameters have been recorded: pressures, temperatures, surface temperatures, and pressure ratios.

At the 2100 kPa steady state test condition, the intermediate pressure is 750 kPa and the maximum discharge temperature is approximately 125 °C.

Oil management studies have been conducted to track oil slippage between the two compressor stages at different target operating pressures and run times. The dominant factor affecting oil slippage in a two-stage compressor system is the pressure ratio. In lower pressure ranges, high pressure ratios will lead to more significant oil slippage.

A simulation model has been created to predict compressor performance and model predictions were validated through the use of the acquired test data. The simulation studies presented in the paper employed a pressure recalculation method and correlation equations which were derived using experimental data. The resulting model is used to analytically predict temperature which potentially leads to some inaccuracy. This inaccuracy is likely a consequence of the temperature calculation being somewhat inaccurate due to a lack of accounting for the 1st to 2nd stage intercooler, as well as heat transfer effects between the compressor shells and gas. Therefore, future work should include usage of a more accurate differential model. Additionally, heat transfer and back pressure calculations should be investigated to form a more complete understanding of the actual two-stage compressor steady-state process.

NOMENCLATURE

η_v	volumetric efficiency	(–)
n	speed of compressor	(r)
m	expansion efficiency	(–)
V_h	displacement	(m^3)
λ_p	pressure coefficient	(–)
λ_l	leakage coefficient	(–)
λ_T	temperature coefficient	(–)
λ_ϕ	condensation coefficient	(–)
C	clearance volume factor	(–)
ρ	density	(kg/m^3)
u	specific internal energy	(kJ/kg)
t_{step}	time in each loop	(s)
h	convection coefficient	($kJ/m^2 - K$)
A	heat transfer area	(m^2)

Subscripts

d	discharge
dis	discharge
s	suction
e	experimental
th	theoretical
L	entering tank

REFERENCES

Aprea, C. and Maiorino, A. (2008). An experimental evaluation of the transcritical co 2 refrigerator performances using an internal heat exchanger. *International journal of refrigeration*, 31(6):1006–1011.

Christen, T., Hubacher, B., Bertsch, S. S., and Groll, E. A. (2006). Experimental performance of prototype carbon dioxide compressors.

Hubacher, B., Groll, E., and Hoffinger, C. (2002). Performance measurements of a semi-hermetic carbon dioxide compressor. *2006 Purdue Conference*.

Rigola, J., Raush, G., Perez-Segarra, C., and Oliva, A. (2005). Numerical simulation and experimental validation of vapour compression refrigeration systems. special emphasis on co 2 trans-critical cycles. *International Journal of Refrigeration*, 28(8):1225–1237.

ACKNOWLEDGEMENT

The authors would like to thank BlackPak Inc. for sponsoring and providing the compressor box for testing.

Characterization and Performance Testing of Two-Stage Reciprocating Compressors during the Dynamic Charging of a Tank with Air

Xinye Zhang[1*], Bin Yang[1], Orkan Kurtulus[1], Eckhard A. Groll[1], Andres Osorio[2], Dylan Bethel[2]

[1] Purdue University, Ray W. Herrick Laboratories, West Lafayette, IN, USA
xinye@purdue.edu, yang62@purdue.edu, orkan@purdue.edu, groll@purdue.edu

[2] BlackPak Inc. San Leandro, CA, USA
aosorio@blackpaktech.com, dbethel@blackpaktech.com

* Corresponding Author

ABSTRACT

Relatively little information is available in the literature with respect to performance of compressors used during the dynamic charging process of a tank. Therefore, work presented in this paper focuses on experimental performance testing of a two-stage reciprocating compressor system using air as the working fluid during the dynamic charging of a tank. A new open loop system test stand, has been designed and built to conduct the dynamic compressor performance tests. Testing was conducted from an initial pressure of 101.3 kPa to a final tank pressure of 3600 kPa. To achieve the final pressure of 3600 kPa, two hermetic reciprocating compressors were used in series. Due to the nature of the charging process, initially only the first stage compressor was activated until the tank pressure reached a level that required activation of the second stage compressor.

Along with the experimental effort, a simulation model to predict compressor dynamic performance has been developed. In this model, the initial clearance factors of the two compressors were calculated based on available compressor maps. The calculated clearance factors were then used as inputs to the dynamic model. The entire charging process was then simulated to provide the compressor performance data as a function of time. Finally, the predicted performance was validated using the acquired test data also presented in this paper.

1. INTRODUCTION

Various methods have been proposed to model the dynamic process of the reciprocating compressor using air as the working fluid. Elhaj et al. (2008) developed a two-stage reciprocating compressor model performing under various conditions for the development of diagnostic features for predictive condition monitoring. Ndiaye and Bernier (2010) presented a dynamic model of a hermetic reciprocating compressor in on–off cycling operation that accounts for the important phenomena influencing the suction and discharge mass flow rates and the electrical power drawn by the compressor using readily available data from manufacturers. MacLaren and Kerr (1969) created a model that could be applied to a two cylinder high speed refrigeration compressor that accounted for unsteady flow pressure prediction in the valve chamber (for capacity and energy consumption considerations), and at the manifold end (for muffling effectiveness consideration). The model developed by MacLaren and Kerr (1969) had experimental data validation but focused on the discharging process only.

However, little work has been found in open literature relative to experimental validation of the dynamic charging process and charging times. Also, there is little information about how to find the intermediate pressure when the 2nd stage compressor begins contribution to the dynamic process. The purpose of this paper is to fill this gap in the open literature using experimental data to find the maximum attainable pressure and temperature of the compressor system and acquire charging times for different target pressures. Furthermore, a simulation model has been formulated to predict the charging process. Experimental data are needed to validate the model predictions.

2. EXPERIMENTAL RESULTS

2.1 Description of system and experiments

The schematic of an open loop system test stand is shown in Figure 1. The test stand is designed to withstand pressures of up to 14 MPa (140 bars) by using stainless steel piping. A silencer and strainer, filled with desiccant, is used at the inlet of the open loop and installed before the compressor box to avoid moisture and dust contamination of the air. One manual ball valve is located after the dryer to shut off the air flowing into the compressor box, if necessary. After the compressor box, a sight glass is used to check for oil in the compressed air stream. A check valve is also used to prevent any air backflow from the pressurized tank. Furthermore, a mass flow meter is installed to measure the mass flow rate of air after the compressor box. A pressure gauge with a 0 – 7 MPa pressure range indicates the high side pressure directly to the operator for control and safety purposes. Also, a pressure relief valve with a 6 MPa cracking pressure is used to protect the tank from uncontrolled pressure increases. At the end of the open loop, a compressed natural gas (CNG) tank with a service pressure ceiling of 25 MPa is used for storage of the compressed air. A power meter was installed to measure the electric power input and all measured data is obtained and displayed using a DAQ-PC system.

One compressor box is connected to the test stand that leverages two hermetic reciprocating compressors in series to achieve the system target pressure in the storage tank. An electronic box is installed inside the compressor box, which contains an ON/OFF switch, status indicator lights, and a remote control button. Two adjustable pressure switches are located downstream of the first-stage compressor and in the discharge line of the compressor box. The first pressure switch is used to control the activation of the second-stage compressor. Once the system reaches a predefined set pressure, the second compressor begins contributing to the system pressure rise. The second pressure switch is used for safety purposes to shut down the whole system after reaching a predetermined discharge pressure. Two intercoolers, located after each compressor, are used to cool down the discharge gas temperatures. Two oil separators and filters were also used after each compressor to provide access to conduct the oil measurements.

2.2 Test stand instrumentation

All measuring instrumentation is indicated in the schematics shown in Figure 1 . All temperatures were measured with T-type thermocouples with ±0.25 ºC accuracy. The inlet pressure of the open loop test stand was measured with an absolute pressure sensor that has an advertised accuracy of ±0.25%. Other pressures are measured with a gauge pressure sensor that has an accuracy of ±0.13% of full scale. A Coriolis-effect mass flow meter with an advertised accuracy of ±0.1% was installed in the open loop test stand to measure mass flow rates of the gas. Electrical power consumptions of the compressors and the cooling fans were measured with power meters with an accuracy of ±0.04%. A data acquisition system was used to convert the incoming voltages from the measuring instrumentation to digital signals for transfer to a personal computer. The computer uses a proper data reduction program for further data analysis..

2.3 Test matrix

A test matrix outlining the operating conditions for the open loop system, is shown in Table 1. The maximum attainable pressure inside the discharge tank was obtained using air as the working fluid. Each test was conducted twice; to ensure the results were repeatable. Tests were conducted using a bisection method with a target pressure resolution of less than 170 kPa. Once the maximum attainable pressure was found, two tests at P_{max}+170 kPa and five additional tank charging tests at P_{max} were conducted.

2.4 Experimental results

During the tests, the following compressor measurements were recorded: compressor mass flow rate, suction and discharge temperatures, suction and discharge pressures, intermediate pressure and temperature, and compressor power consumption. Using these measurements, the compressor performance was evaluated.

A dynamic test with a 2800 kPa target pressure was conducted first as a shakedown test. Figure 2 shows the variation of air pressures as a function of time. It can be noticed that the intermediate pressure, measured as P_{42} and P_{43}, increases significantly when only the 1st stage compressor is in operation. This increment rate became smaller once the 2nd compressor began to contribute to the system pressure rise. The discharge pressure of the compressor box, increases significantly until reaching the target pressure. Additionally, pressure drop between the discharge pressure and the tank pressures due to the filters, the Coriolis-effect mass flow meter, and the pipe connections can be seen in Figure 2. Once the desired pressure of 2800 kPa is reached, the slope of the decrease in discharge pressure and tank pressure is slightly different. This difference is due to backflow, resulting from unavoidable leakage in the compressors, which leads to a small reduction of the 2nd stage discharge pressure. However, once the discharge valve is open, these two pressures decreased at the same slope. Also, it is found that the final 2nd stage discharge pressure taken immediately after shut-down is larger than the tank pressure at ambient temperature. The higher discharge pressure is due to the compressors' in-head temperature environment being higher than the holding tank temperature.

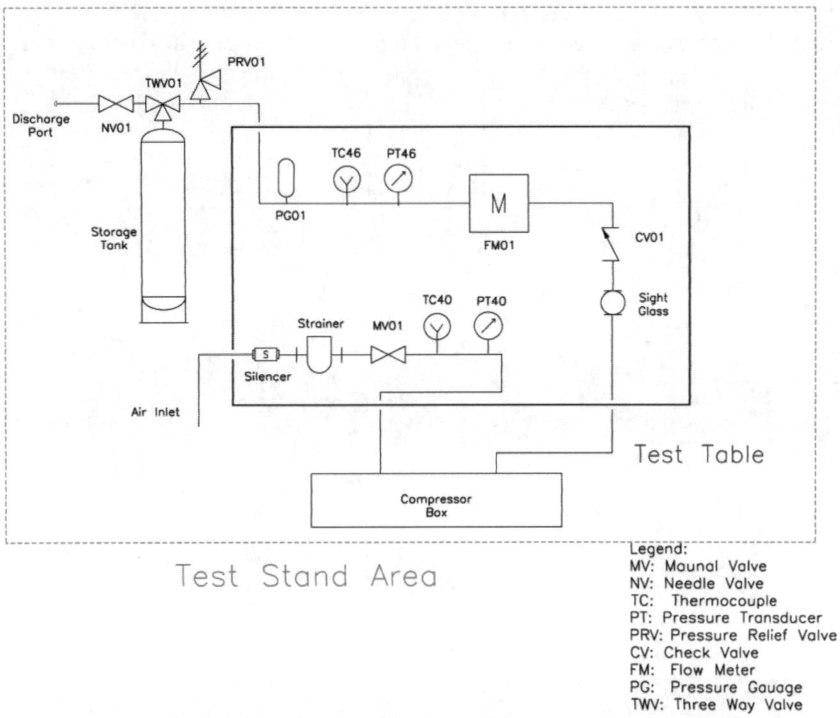

Figure 1: Schematic of the open loop system

Table 1: Test matrix of dynamic testing with open loop test stand

Test Number	Target Pressure (kPa)	Failure	Comments
1	2100	N	Initial low boundary
2	4900	Y	Initial high boundary
3	3500	N	Average of test 1 and 2
4	4200	N	Average of test 2 and 3
5	4550	Y	Average of test 2 and 4
6	4375	N	Average of test 4 and 5. Temporary P_{max} found.
7	4550	Y	1st repetition at P_{max}+25
8	4550	Y	2nd repetition at P_{max}+25
9	4375	N	1st repetition at P_{max}
10	4375	N	2nd repetition at P_{max}
11	4375	N	3rd repetition at P_{max}

Figure 3 shows the variation of the measured compressed air temperature as a function of time during the open loop testing. Total run time was 30 minutes, which included 3 minutes for warm-up testing, 25 minutes for dynamic testing, and 15 minutes for shut-down testing. During the warm-up stage, only the cooling fans were set to operate. During the shut-down stage, both compressors were shut down at the same time and the discharge valve (NV01) was opened to gradually release the compressed air from tank. In addition, the cooling fans were set to operate until the maximum temperature (2nd stage discharge temperature) dropped to approximately 35 °C.

After the compressor system started, it was found that the 1st stage discharge temperature, measured as T_{41}, increased significantly once the 1st stage compressor started working. Once the 2nd stage compressor started working, the discharge temperature of the second compressor increased significantly until it reaches to a final temperature of 75 °C. Another interesting conclusion that can be drawn from Figure 3 is the 1st stage discharge temperature was larger than the 2nd stage discharge temperature until the run time reached approximately 15 minutes. At that point, the 2nd stage discharge temperature climbed dramatically in Figure 3. It can be noticed that the temperature variations after the heat exchanger in both stages were relatively small, which were close to the ambient temperature during the testing. In addition, a small jump of the 2nd stage discharge temperature can be seen in the right corner of Figure 3. This small temperature jump is due to the cooling fans being shut off with heat remaining in the compressor system after shut-down.

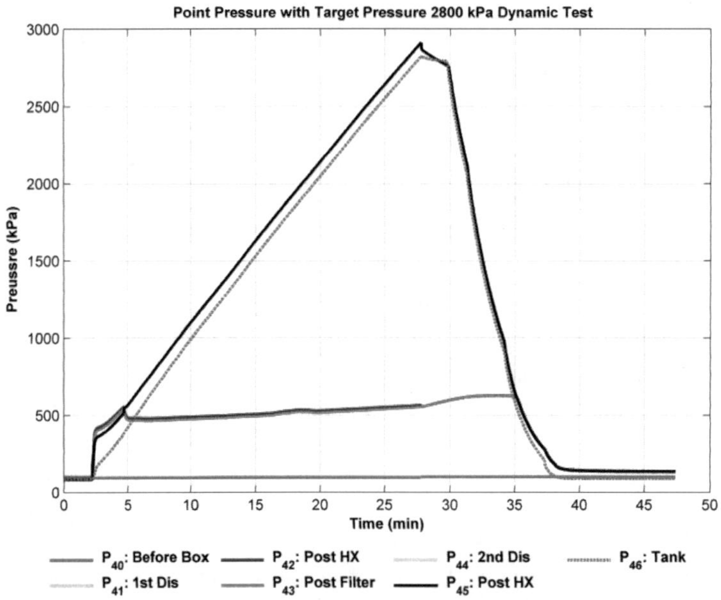

Figure 2: Point pressure with target pressure 2800 kPa dynamic test

Figure 4 shows the variation of the pressure ratio, i.e. the ratio of discharge pressure over suction pressure for each stage as a function of time for the tested system. It can be noticed that the pressure ratio of the 1st stage compressor is reached immediately once the 1st stage compressor starts operating. This pressure ratio is relatively stable at approximately 5. However, the pressure ratio of the 2nd stage compressor increased gradually until the system was turned off. The 2nd stage pressure ratio was smaller than the 1st stage pressure ratio. Also, it can be seen that the backflow previously discussed has great effects on both pressure ratios by enhancing the intermediate pressures. With increasing intermediate pressure, the 1st stage pressure ratio is shown to increase accordingly until the 2nd stage pressure ratio drops to approximately 1, which means there is no obvious pressure difference between the inlet and the outlet of the 2nd stage compressor. As discussed before, the compressors will hold some pressures at the end of the test, which is the reason why the 1st stage pressure ratio is shown to be larger than 1.

To find the maximum attainable pressure inside the discharge tank, several tests were conducted as shown in the test matrix on Table 1. Two selected informative tests will be discussed in this section.

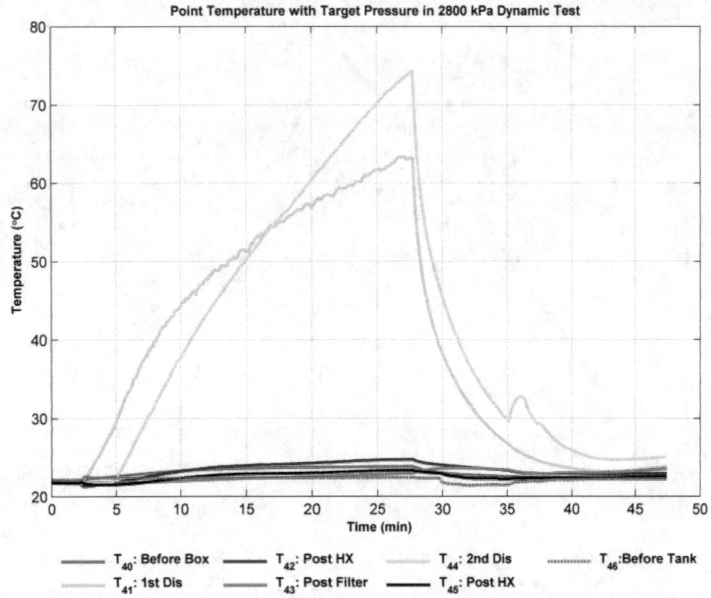

Figure 3: Point temperature with target pressure 2800 kPa dynamic test

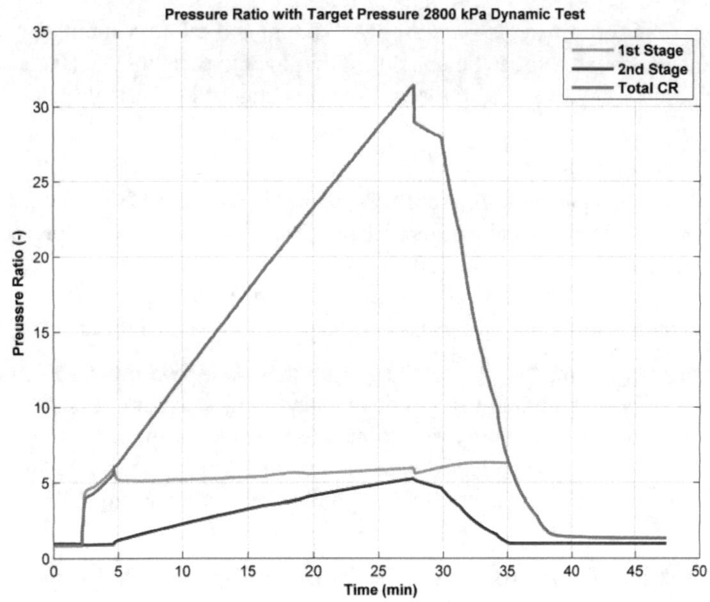

Figure 4: Pressure ratio with target pressure 2800 kPa dynamic test

Figure 5 shows the variation of the air pressures as a function of time with a target pressure of 4550 kPa which was set by the pressure switch in the compressor box. It can be noticed that the tank pressure, measured as P_{46}, only reached a pressure of 4375 kPa, which is lower than the target pressure. Compared to results in 4200 kPa case, the maximum attainable temperature obtained in the 2nd stage reached a value of 95 °C, which was slightly higher than the maximum temperature during the 4200 kPa test. The reason why the target pressure of 4550 kPa, was not reached is due to the thermal protection installed in the compressor box, which protects the compressor from burning out. Once the low-side operating temperature is higher than the pre-set critical value, the system will automatically shut down due to the thermal protector's engineered design.

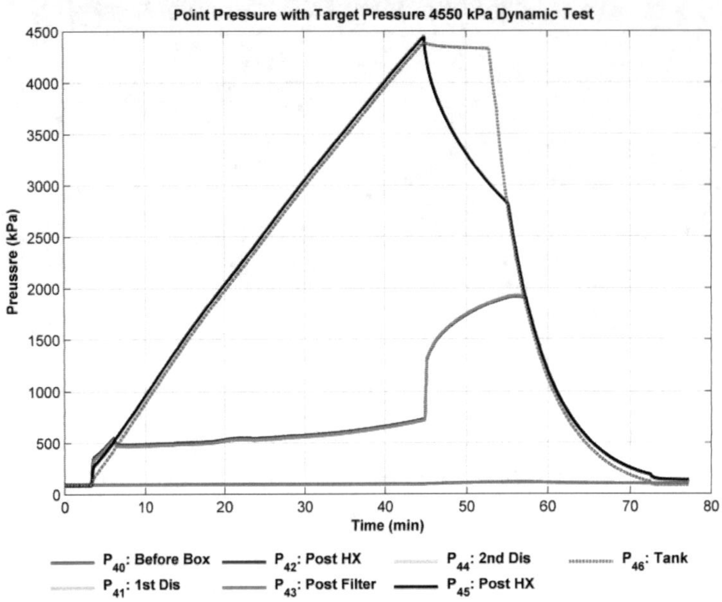

Figure 5: Point pressure with target pressure 4550 kPa dynamic test

Based on the discussion above, 4375 kPa is subsequently referred to as the maximum attainable pressure for the system tested. The charging time for this target pressure was 42 min and the maximum discharge temperature was found to be 93 °C at the end of the tests. It can be concluded that the temperature needs to be monitored during the test to ensure the system doesn't shut down due to thermal overloading of the thermal protector.

2.5 Oil management

An oil management analysis was conducted during each test. And the differences of the oil weight in the discharge filters during each test were recorded, to gain a general understanding of how oil migrated within the system during testing, as well as general factors that affect oil slippage. This analysis also helped to indicate how much oil was discharged with the compressed air and if additional oil was needed. Three filters were installed after each compression stage. Oil weight data of the various tests performed are shown in Table 2.

It can be noticed that the majority of the oil was discharged by the 1st stage compressor. Also, as expected, more oil was collected during the tests with higher target pressures and longer running time. With that information in-hand, the conclusion can be drawn that the compressors must be recharged with oil after a certain run time, or an oil separator with an automatic oil return must be installed in the system to avoid compressor failure. Furthermore, it should be noted that most of the oil was collected by the first filter which is not shown in Table 2.

Table 1: Test matrix of dynamic testing with open loop test stand

1st Stage (g)	2nd Stage (g)	Total (g)	Target Pressure (kPa)	Running Time (min)
3.2	0.1	3.3	2100	19
5.1	0.3	5.4	2800	25
9.9	0.2	10.1	3500	30
12.1	0.6	12.7	4340	39

Figure 6 shows the variation of the oil slippage as a function of pressure ratio for the 1st and 2nd stage compressors. It can be seen that the oil slippage of the 1st stage compressor increases significantly with higher pressure ratio. This relation can be estimated as a linear curve with an R value of 0.98, as shown in Equation (1). This equation is used to predict oil slippage in the 1st stage compressor. Oil slippage of the 2nd stage compressor shows only a minor increase as a function of pressure ratio. The 2nd stage compressor appears to be relatively stable when compared with oil slippage of the 1st stage compressor.

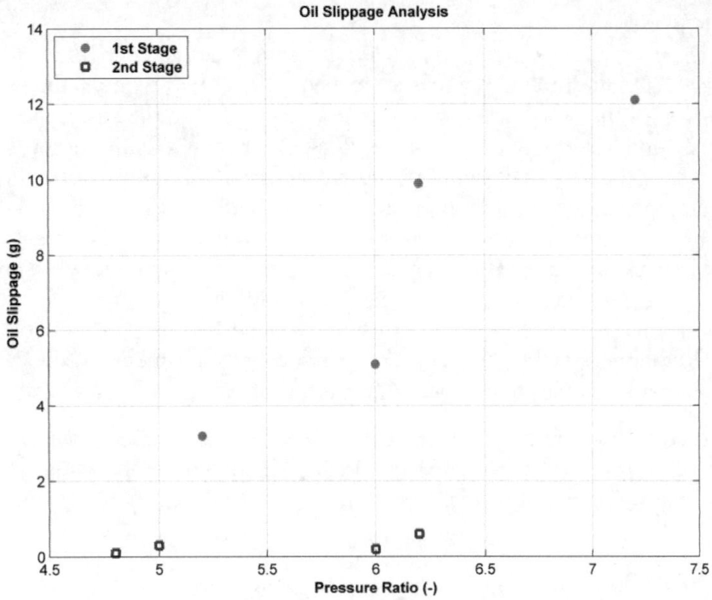

Figure 6: Oil slippage with different pressure ratio

$$oil\_slippage = 3.75 \times PR - 0.97 \quad (1)$$

where PR is referred to pressure ratio which should be larger than 1.

Based on the testing performed, it can be concluded that more oil is collected during tests in a two-stage compressor system with higher target pressure. However, if the same target pressures are compared, more oil will be discharged in the lower pressure range by the 1st stage compressor and a smaller amount of oil will be discharged in the 2nd stage compressor. In the lower pressure range, high pressure ratios will lead to more significant oil slippage.

3. SIMULATION MODEL

A simulation model to predict the compressor dynamic performance has been developed. In this model, the initial clearance factor for the compressor was calculated based on available compressor maps used as an input to the dynamic model. The entire compression process has been simulated to provide compressor performance data as a function of time. Predicted performance will be validated using test data.

3.1 Clearance volume factor

The clearance volume factor is a significant value, which affects the performance of compressors. It must be noted that the clearance volume factor is an attribute of the compressor itself and can be found based on experimental results. The approach used here was to find an estimated value based on published compressor performance maps for use in the simulation model.

Compressor map data given by the compressor manufacturer provides the relation between mass flow rate, suction, and discharge conditions. Equation (2) is used to calculate the clearance volume factor. The clearance volume factor C is the slope of the function curve between the volumetric efficiency and the pressure ratio.

$$\eta_v = 1 - C\left[\left(\frac{P_d}{P_s}\right)^{1/m} - 1\right] = \frac{\dot{m}_e}{\dot{m}_{th}} \tag{2}$$

3.2 Critical intermediate pressure

Due to the nature of the charging process, the intermediate pressure is not achieved immediately. Initially, only the 1st stage compressor is activated until the tank pressure reaches a level that requires the activation of the 2nd stage compressor. At the beginning of the process, the system pressure is equal to the atmospheric pressure. In two stage compressor systems, the suction volume of the 1st stage compressor is inherently larger than that of the 2nd stage compressor due to the nature of the system design. For the system presented in this work, the discharge gas of the 1st stage compressor flows through the 2nd stage compressor without additional compression occurring and arrives at the storage tank when only the 1st stage compressor runs. The storage tank is used as a back pressure of system. As a response to the gradually increasing back pressure produced by the storage tank, the pressure inside the compressor system as well as the intermediate pressure will increase. In addition, the increasing intermediate pressure will decrease the discharge gas volume. Once the discharge gas volume of the 1st stage compressor is equal to the displacement of 2nd stage compressor, the 2nd stage compressor starts to operate. After this critical condition is achieved, both compressors work in series to compress the gas until the target pressure in storage tank is reached.

Equations (3) to (5) are used to find the critical intermediate pressure value. Since there is no condensation in this system, $\lambda_{\varphi 1} = \lambda_{\varphi 2} = 1$. Due to initial system conditions being set for no compression to occur in the *2nd* stage compressor, $\varepsilon_2 = 1$, $\eta_{v2} = 1$, $\lambda_{T2} = 1$ in Equation (3).

$$\frac{P_2}{\eta_{v1}} = \frac{V_{h1}}{V_{h2}} \frac{\lambda_{P1}}{\lambda_{P2}} \frac{\lambda_{l1}}{\lambda_{l2}} \frac{\lambda_{T1}}{\lambda_{T2}} \frac{\lambda_{\varphi 1}}{\lambda_{\varphi 2}} \frac{T_2}{T_1} \frac{P_1}{\eta_{v2}} \tag{3}$$

$$\eta_v = 1 - C[(\frac{P_2}{P_1})^{1/m} - 1] \tag{4}$$

by Combining equation (3) and (4), the simplified Equation (5) can be obtained as follows:

$$\frac{P_2}{1 - C\left[\left(\frac{P_2}{P_1}\right)^{1/m} - 1\right]} = \frac{V_{h1}}{V_{h2}} \frac{\lambda_{P1}}{\lambda_{P2}} \frac{\lambda_{l1}}{\lambda_{l2}} \frac{T_2}{T_1} P_1 \lambda_{T1} \tag{5}$$

Parametric data for these equations can be found in the compressor data sheet, and the clearance volume factor can be obtained from Section 3.1. In addition, it can be assumed that $\lambda_{P1} = 0.97$, $\lambda_{P2} = 0.98$, $\lambda_{l1} = \lambda_{l2}$, *and* $\lambda_{T1} =$ 0.97. Armed with those assumptions, the critical intermediate pressure has been calculated to be 490 kPa for the given system.

3.3 Dynamic charging process

In two-stage compressor systems, the intermediate pressure observes the principle that the discharging gas volume from the 1st compressor will be completely transferred to the 2nd stage compressor. However, the corresponding intermediate pressure depends on the relationship between the two compressor displacements. It should be noted that when the discharge condition of the 1st stage compressor changes, the gas volume of 2nd compressor and the intermediate pressure will respond accordingly. Therefore, the intermediate pressure needs to be adjusted in the simulation to satisfy both compressor displacements. The intermediate pressure between any two stages in a multi-stage compressor system can be calculated using Equation (6) :

$$V_{hj} \frac{\eta_{vj}\lambda_{Tj}\lambda_{Pj}\lambda_{lj}}{\lambda_{\varphi j}} \frac{P_j}{P_1} \frac{T_1}{T_j} = V_{hj+1} \frac{\eta_{v,j+1}\lambda_{T,j+1}\lambda_{P,j+1}\lambda_{l,j+1}}{\lambda_{\varphi,j+1}} \frac{P_{j+1}}{P_1} \frac{T_1}{T_{j+1}} \tag{6}$$

where *j* is referred to the 1st stage and *j* + 1 is referred to 2nd stage in a two-stage system

Once the intermediate pressure is found, Equations (7) to (9) can be used to simulate the dynamic compression process and predict all parameters for every time step until the target pressure is achieved. The volumetric efficiency (η_v) can be obtained from Equation (4).

$$\dot{m}_{dis} = \eta_v V_h \rho n / 60 \tag{7}$$

$$m'_{\text{tank}} = m_{\text{tank}} + \dot{m}_{dis} t_{step} \tag{8}$$

where t_{step} is the time step in each loop.

$$P_{\text{tank}} = \frac{m_{\text{tank}} R_g T_{\text{tank}}}{V_{\text{tank}}} \tag{9}$$

It must be noted that the tank temperature, T_{tank}, which is a key parameter used for calculation of the tank pressure is still unknown in Equation (9). Therefore, the storage tank is modeled as a lumped system within the simulation model to calculate the temperature during the charging process as shown in Equations (10) and (11).

$$Q_{loss} = (T_{\text{tank}} - T_{sur}) \times \rho_{air} \times C_{p,air} \times V_{\text{tank}} \left[1 - \exp\left(\frac{-hA}{C_{p,air} V_{\text{tank}}} t \right) \right] \tag{10}$$

where T_{sur} is the surrounding temperature of the storage tank.

$$m'_{\text{tank}} u'_{\text{tank}} - m_{\text{tank}} u_{\text{tank}} = m_{\text{tank}} Q_{loss} + \dot{m}_{dis} t_{step} h_L \tag{11}$$

where h_L is the air entering enthalpy to tank and $u_{\text{tank}} = f(T_{\text{tank}})$

Figure 7 shows the variations of pressure with time as predicted by the simulation model. When the target pressure is 2800 kPa, the total charging time is predicted to be approximately 20 minutes, which is validated by the experimental data as discussed above. It is noticed that the 2nd stage compressor does not work until the intermediate pressure reaches 525 kPa, which satisfies mass flow rate matching between the two compressors. In addition, the maximum pressure ratios were 7.2 and 3.91 of the 1st stage and 2nd stage compressors, respectively.

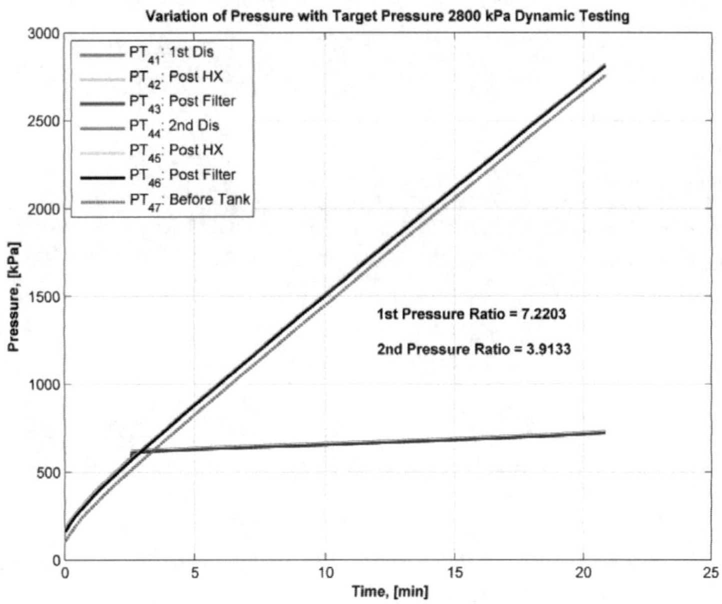

Figure 7: Variation of pressure with target pressure 2800 kPa charging process

4. CONCLUSION AND FUTURE WORK

This paper presents the experimental performance testing of a two-stage reciprocating compressor systems using air as the working fluid during the dynamic charging of a tank. In addition, a simulation model to predict the compressor performance is discussed as well. The following conclusions can be drawn.

- Several results with different target pressures and run times have been conducted. The following parameters have been recorded: pressures, temperature, surface temperatures, and pressure ratios. In addition, the maximum attainable pressure and temperature in the storage tank were also found as: 4375 kPa and 93 °C. Oil measurements have also been conducted to track oil slippage between the two stages with different target pressures and run times. Experimental testing has shown that the dominant factor affecting oil slippage in a two stage compressor system is pressure ratio. A simulation model has been developed to predict the compressor performance. The model predictions have been validated by using the test data.
- The simulation studies presented in this paper employed a simple element approach and lump system method to model the compressor dynamic process and pressurized tank. This leads to some inaccuracy, since it is difficult to predict the mass flow rate accurately using this approach. In addition, the temperature calculations are not as accurate as desired due to the existence of the intercooler and heat transfer between the compressor shells and the surrounding air. Future work should therefore include the usage of a more accurate differential model. Additionally, the heat transfer and back pressure calculations should be investigated, which will lead to a better understanding of the actual two stage compressor dynamic process.

NOMENCLATURE

η_v	volumetric efficiency	(–)
n	speed of compressor	(r)
m	expansion efficiency	(–)
V_h	displacement	(m^3)
λ_p	pressure coefficient	(–)
λ_l	leakage coefficient	(–)
λ_T	temperature coefficient	(–)
λ_ϕ	condensation coefficient	(–)
C	clearance volume factor	(–)
ρ	density	(kg/m^3)
u	specific internal energy	(kJ/kg)
t_{step}	time in each loop	(s)
h	convection coefficient	($kJ/m^2 - K$)
A	heat transfer area	(m^2)

Subscripts

d	discharge
dis	discharge
s	suction
e	experimental
th	theoretical
L	entering tank

REFERENCES

Elhaj, M., Gu, F., Ball, A., Albarbar, A., Al-Qattan, M., and Naid, A. (2008). Numerical simulation and experimental study of a two-stage reciprocating compressor for condition monitoring. *Mechanical Systems and Signal Processing*, 22(2):374–389.

MacLaren, J. and Kerr, S. (1969). Paper 3: An analytical and experimental study of self-acting valves in a reciprocating air compressor. In *Proceedings of the Institution of Mechanical Engineers, Conference Proceedings*, volume 184, pages 24–33. SAGE Publications.

Ndiaye, D. and Bernier, M. (2010). Dynamic model of a hermetic reciprocating compressor in on–off cycling operation (abbreviation: Compressor dynamic model). *Applied Thermal Engineering*, 30(8):792–799.

ACKNOWLEDGEMENT

The authors would like to thank BlackPak Inc. for sponsoring and providing the compressor box for testing.

A Comprehensive Evaluation of Regression Uncertainty and the Effect of Sample Size on the AHRI-540 Method of Compressor Performance Representation

Vikrant AUTE[1]*, Cara MARTIN[2]

[1]University of Maryland College Park, Center for Environmental Energy Engineering,
College Park, MD, USA
(301-405-8726, vikrant@umd.edu)

[2]Optimized Thermal Systems, Inc.
Beltsville, MD, USA
(866-485-8233 x27, cmartin@optimizedthermalsystems.com)

* Corresponding Author

ABSTRACT

AHRI Standard 540 (AHRI, 2015) is the current standard defining the methods for representing compressor performance data. The standard is widely used across the industry and uses a 10-coefficient third order polynomial equation to represent compressor published ratings. The accuracy of the data representation can be affected by multiple factors including measurement uncertainty, regression uncertainty, compressor to compressor variation, and operation outside of the normal operating envelope (extrapolation). In addition, the number and location of points in the operating envelop also affects the accuracy of the resulting 10-coefficient polynomial. The measurement uncertainty is well known and can be factored into the data reduction. However, the measurement uncertainty is generally not propagated into the regression uncertainty and hence the overall uncertainty in prediction using the polynomial is not known. This uncertainty also changes according to the number of samples used for developing the polynomial.

As a first step of the evaluation, a regression uncertainty analysis was conducted using a Monte Carlo simulation method. Results showed that the average uncertainty in mass flow rate prediction can be as high as 4% and that in power prediction can be as high as 5%. Error in predicted power and mass flow rate is higher for larger capacity compressors. For most compressors, the high errors occur in the region of the envelope with low suction and low discharge dew point temperatures.

A study of sampling considering different sample sizes and multiple sampling methods was conducted. Two additional methods of compressor performance representation were also analyzed. This analysis was presented with several challenges, particularly since the compressor operating envelope is a non-rectangular domain. A sampling method using Latin Hypercube Sampling (LHS) design and a proposed alternative sampling method based on polygonal design of experiments (PDOE) were evaluated. The resulting models were validated against a measured data set of more than 600 points encompassing the operating envelope for each compressor. In general, both the LHS and PDOE methods yielded similar errors in mass flow rate for samples sizes of 12, 14 and 16. Thus, for mass flow rate, it is possible to build a model with 12 systematically selected test points. For power prediction, the average error for the LHS and PDOE methods using AHRI Standard 540 and two other methods was lower than 2% for all sample sizes.

1. INTRODUCTION

The current performance rating standard for positive displacement compressors requires manufacturers to report tabular performance data over a specified operating envelope. The data is usually generated using a 10-coefficient third order polynomial equation. The coefficients for the equation are derived from measured and/or extrapolated values using the method of "Least Squares". This equation is also frequently used by manufacturers to represent the tabular data in a form suitable for use in simulation programs.

The original intent of this reporting method was to provide a consistent and uniform way of presenting performance data to enable manufacturers to evaluate different compressors for a particular application. Manufacturers also desire to use compressor performance with simulation tools to calculate system performance values for rating purposes. The current equation that is used to represent the data does not, however, always provide sufficient accuracy to meet the need for product rating. Furthermore, due to the costs involved with collecting the measured data, there is a desire to use the fewest number of data points. No specific information is provided to understand the impact of the number and distribution of measured test points to achieve a particular level of accuracy. Other sources of uncertainty are associated with the variation of measurement reproducibility across the application envelope as well as the compressor to compressor variation. An improvement in the method for representing data will likely enable an improvement in the accuracy of the product ratings without increasing the cost of generating the data.

2. DATA USED FOR ANALYSIS

A total of 29 sets of raw compressor data were used to evaluate the current method of developing a representative compressor performance map from experimental data. As can be seen from Table 1, data accounted for three types of compressors (reciprocating, rotary and scroll) over a complete capacity range for both pure fluids (R134a and R22) and mixed refrigerants (R404A and R410A). Data for multiple compressors of the same model were also acquired to evaluate the impact of unit to unit variation. Data generally included compressor geometry, operating conditions (pressure, temperature, mass flow rate, superheat, etc.), performance, instrument uncertainties, and Standard 540 coefficients.

Table 1: Data Set Used for Project Analysis

Compressor Type	Application	Refrigerant	Capacity Range	Number of Data Sets	Number of Units Per Model
Reciprocating	A/C & HP	R410A	670 – 19,079 W (2,285 – 65,099 Btu/hr)	9	2-3
		R22	1,487 – 10,574 W (5,075 – 36,079 Btu/hr)	2	3
	Low Temp.	R404A	532 – 3,028 W (1,816 – 10,331 Btu/hr)	1	1
	Medium Temp.		751 – 3,951 W (2,564 – 13,483 Btu/hr)	1	1
	High Temp.	R407C	1,644 – 11,988 W (5,611 – 40,906 Btu/hr)	1	1
		R134a	858 – 3,733 W (2,929 – 12,739 Btu/hr)	1	1
Scroll	A/C & HP	R410A	1,762 – 227,460 W (6,013 – 776,125 Btu/hr)	8	1
	Low Temp.	R404A	1,420 – 5,853 W (4,844 – 19,970 Btu/hr)	1	1
Rotary	A/C & HP	R410A	1,125 – 13,020 W (3,838 – 44,427 Btu/hr)	2	1
		R134a	446 – 5,479 W (1,523 – 18,695 Btu/hr)	2	1
		R407C	1,411 – 6,111 W (4,813 – 20,851 Btu/hr)	1	1
Total Complete Data Sets Available				29	

3. REGRESSION UNCERTAINTY ANALYSIS

Experimental measurements involve uncertainty due to various sources and it is important to characterize the effect of measurement uncertainties on the regression coefficients (Coleman and Steele, 2009). This helps us to better understand the effect of measurement uncertainty on the values predicted by the compressor map and the corresponding uncertainty when the map is used for system rating. The uncertainty information can then be used in understanding the effect of tolerances used in the various rating and enforcement tests. The typical data used in the map are measured temperatures, mass flow rate and power consumption. These are all measured quantities (and not derived) and hence no uncertainty propagation is required. A Monte Carlo simulation method was used to conduct a regression uncertainty analysis for the provided compressor map data.

3.1 Approach

In the case of compressor testing, the measured quantities include the suction and discharge dew point temperatures, the power consumption and the mass flow rate. The ASHRAE Standard 23 provides the required measurement accuracies for these quantities, which are depicted in Table 2.

Table 2: Measurement accuracies

Quantity	Accuracy
Temperature	0.5°F
Power	1% of measured value
Mass flow rate	1% of the measured value

Random errors also occur during the course of actual experiments. These errors are included in the measurements, but are not quantitatively identified. For the purposes of the Monte Carlo analysis, the various measurement errors were assumed to be normally distributed with a mean value as the measured value reported in the data sets and the standard deviation as the accuracy value reported in Table 2. This practice is not normally advised since the actual standard deviation will be lower, but it was used here to account for the overall measurement uncertainties including both systematic and random errors.

The following steps describe the Monte Carlo simulation carried out for each of the 29 data sets shown in Table 1. Assume that there are 50 data points in a single data set.

1. Generate 25,000 data sets, based on normal distribution, each comprising of 50 data points. 25,000 sets was found to be a reasonably good number to minimize computational time while maintaining a good level of accuracy.
2. Develop a regression model (curve fit) for each data set, based on the 10-coefficient polynomial. One model each is developed for power consumption and mass flow rate, as a function of Te and Tc.
3. Predict each point in the original dataset using the regression models. Thus, for each of the 50 points, there are 25,000 predicted values for power consumption.
4. Compute the mean and the standard deviation of the 25,000 predicted values. The difference between this mean value and the measured value is termed as the Error in Mean. The ratio of the standard deviation and the mean is termed as the coefficient of variance (COV) and is expressed as a percentage of the mean.
5. For each of the 25,000 predicted values, compute the individual errors, i.e., the difference between predicted power consumption at that point (i.e., [Te,Tc]) and the measured value. The maximum absolute percent error (MAPE), which indicates the worst case prediction error in mass flow rate (or power) over the entire operating envelop, can then be computed for each of these 50 points.
6. The error in mean and the maximum absolute error can then be plotted on [Te,Tc] plot representing the compressor operating envelope. Thus for each compressor dataset, we can generate four plots (i.e., two each for mass flow rate and power consumption).

3.2 Results

A sample of the plots generated for each data set is shown in Figure 1. Results are summarized in Table 3. Based on the error metrics and the individual analysis of each of the plots generated for all data sets, the following general conclusions can be drawn:

1. The minimum absolute error in the predicted mean power is 0.2% and the maximum error in the predicted mean power is 5%. The maximum COV is 1.7%.

2. The minimum absolute error in predicted mean mass flow rate is 0.4% and the maximum error in the predicted mean mass flow rate is 4.3%. The maximum COV is 3.5%.
3. The worst case maximum absolute error in predicted power is 9%.
4. The worst case maximum absolute error in predicted mass flow rate is 17%.
5. In general, the maximum (worst case) errors in mass flow rate prediction are much higher than those in power prediction.
6. For most compressors, the highest errors occur in Quadrant-3, i.e., low suction dew point and low discharge dew point region. These are the compressors that have "sufficient" number of data points distributed across the entire operating envelope.
7. Refrigerants R134a, R404A and R407C exhibit higher errors compared to R410A, but the corresponding datasets do not have 'sufficient' points to begin with. As such these errors could be due to insufficient data available for regression.
8. Larger compressors (i.e., high power and mass flow rate values) exhibit higher errors.

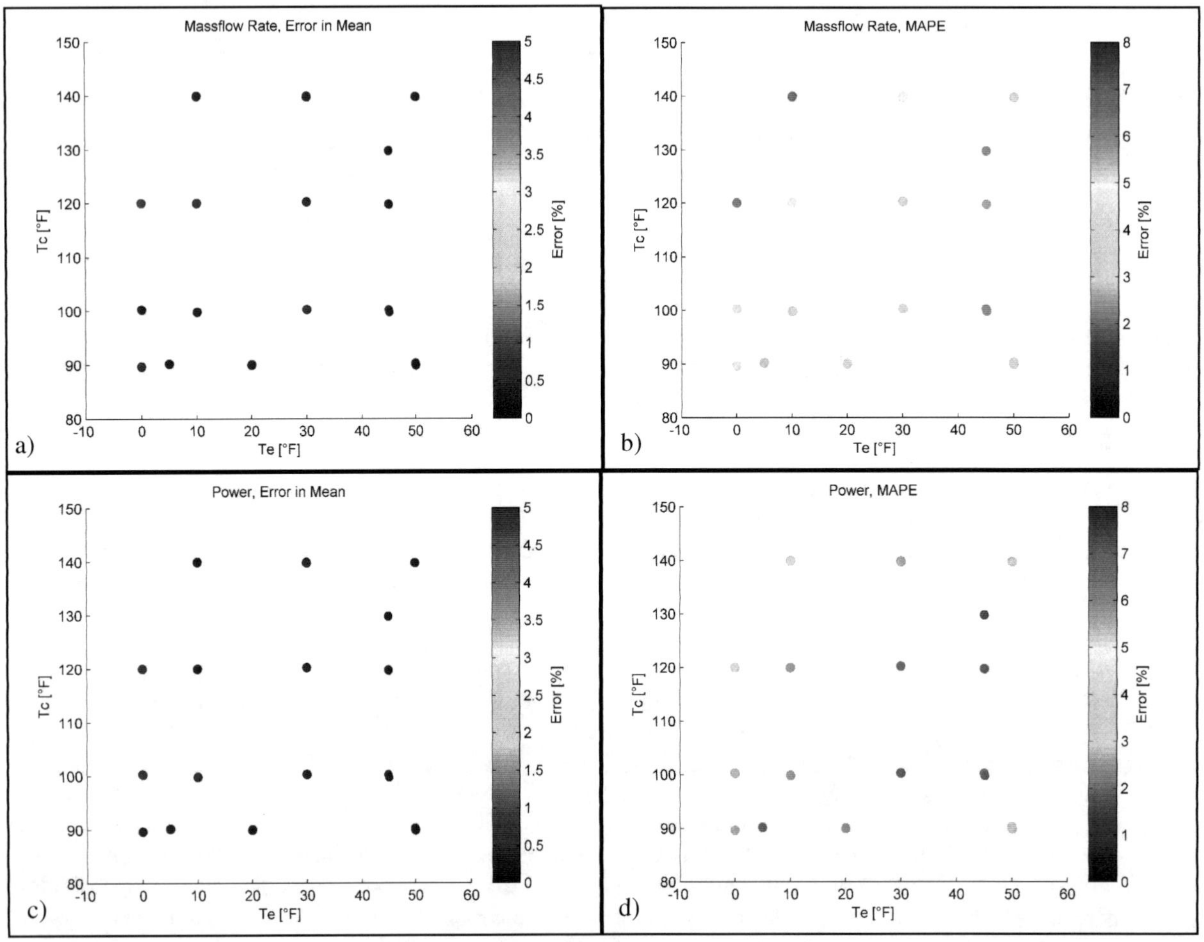

Figure 1: Monte Carlo simulation results for R22 compressor showing (a) error in mean mass flow rate, (b) maximum absolute percent error in mass flow rate, (c) error in mean power, and (d) maximum absolute percent error in predicted power

Table 3: Monte Carlo simulation results

Fluid	Test Points	Power Max. Error in Mean [%]	Power COV [%]	Power MAPE [%]	Mass Flow Max. Error in Mean [%]	Mass flow COV [%]	Mass flow MAPE [%]
R134a	17	1.09	1.18	5.82	1.12	1.82	9.03
R134a	17	0.56	1.25	6.00	0.80	2.10	12.49
R134a	17	1.87	1.29	6.16	0.68	2.00	9.21
R22	48	0.63	0.77	3.31	0.96	1.39	6.87
R22	48	0.98	0.79	3.78	2.81	1.50	7.54
R22	49	0.64	0.76	3.73	1.40	1.44	7.15
R22	48	0.38	0.76	3.06	0.67	1.39	6.22
R22	49	0.56	0.75	3.57	0.75	1.28	6.22
R22	49	0.45	0.74	3.74	4.33	1.19	9.03
R404A	17	2.04	1.53	8.60	3.16	2.55	16.08
R404A	22	3.38	1.21	7.66	2.23	1.94	8.98
R404A	21	2.27	1.31	7.62	1.70	2.50	13.02
R407C	19	0.76	1.22	5.86	4.36	2.28	15.31
R407C	18	2.81	1.46	7.31	4.19	2.36	14.02
R410A	21	5.09	1.27	7.07	1.13	1.90	7.77
R410A	16	0.21	1.09	5.03	1.89	1.92	10.27
R410A	16	0.88	1.73	9.10	1.88	3.51	17.40
R410A	17	0.75	1.31	5.86	0.98	2.62	12.65
R410A	17	2.52	1.47	7.25	1.67	2.98	13.01
R410A	21	1.44	1.32	5.63	2.79	2.05	13.51
R410A	19	0.41	1.13	4.94	0.41	1.54	6.72
R410A	20	12.28	1.12	15.37	3.06	1.48	7.01
R410A	22	0.34	1.15	6.06	0.99	1.71	8.23
R410A	21	1.19	1.16	6.60	0.59	1.57	7.58
R410A	23	0.90	1.21	5.99	0.47	1.64	8.68
R410A	21	1.34	1.15	5.50	0.73	1.58	7.70
R410A	49	1.17	1.28	6.16	1.47	2.90	12.96
R410A	50	0.79	1.03	5.10	1.61	2.22	11.16
R410A	49	1.40	1.05	5.12	2.20	2.46	11.33
R410A	49	0.71	0.80	3.54	0.79	1.17	5.55
R410A	49	0.52	0.78	3.44	0.72	1.22	5.45
R410A	50	0.62	0.76	3.95	1.78	1.26	6.97
R410A	49	0.86	1.04	5.02	1.96	2.57	11.97
R410A	47	1.09	1.14	4.55	2.28	2.51	12.01
R410A	48	0.36	0.87	3.83	1.70	1.98	9.13
R410A	52	0.63	0.77	3.58	2.15	1.66	7.89
R410A	52	0.48	0.84	3.89	3.53	1.78	9.38
R410A	52	0.54	0.88	4.05	3.09	1.84	8.46
R410A	49	0.37	0.79	3.38	0.65	1.25	5.88
R410A	49	0.36	0.80	3.78	1.00	1.25	6.23
R410A	49	1.14	0.79	3.36	1.07	1.22	6.66
R410A	48	0.70	0.80	3.59	0.47	1.22	5.76
R410A	48	0.76	0.79	3.83	0.59	1.19	5.33
R410A	48	0.91	0.81	3.96	0.96	1.26	5.87

4. SAMPLING STUDY

A study was conducted to evaluate and understand the influence of different sampling methods. Based on feedback received from manufacturing partners, there is no well-established methodology used in selecting the sample points for developing the 10-coefficient map. Two approaches were considered, a conventional approach using a Latin Hybercube Sampling (LHS) design, and a Polygon Design of Experiments (PDOE) method taking into account non-rectangular domains.

4.1 Conventional Sampling Using the Latin Hybercube Sampling (LHS) Design

A sampling study using the LHS method was conducted. The map data set contains a total of 16 points such that 15 points are selected for use in the analysis. The sampling study consisted of the following steps:

1. Use the provided map data to develop a 10-coefficient polynomial model for mass flow rate and power consumption. Verify these models against the comprehensive data set.
2. Using the comprehensive verification dataset, develop a regression surface using the Kriging metamodel. The Kriging metamodel does not require a functional form and is capable of exactly reproducing the source data.
3. Select 15 points from the operating envelope.
4. Compute the mass flow rate and power consumption using the Kriging metamodel for the sampled points.
5. Develop a 10-coefficient polynomial model based on the data from Step-4.
6. Verify the model from Step-5 against the comprehensive data set.

Figure 2 shows sample plots generated for the errors in predicted mass flow rate for a given compressor data set. Similar plots resulted for predicted power. Based on these and other results, the 10-coefficient polynomial obtained using the LHS samples is more accurate (lower errors) in terms of both mass flow rate and the power consumption over the entire operating envelope, including some extrapolation. Based on this preliminary analysis, the LHS sampling may be better suited for use with the 10-coefficient map. The challenge with LHS sampling, however, is that it is based on a random number generator. As such, the samples can change from run to run and the accuracy may not be reproducible. Furthermore, conventional sampling techniques are designed to work with square or rectangular domains. Due to availability of over 600 data points encompassing the entire operating envelope, it was possible to use the LHS method for sampling for use with the AHRI-540 model in this particular analysis. Ultimately, for the analysis conducted, the LHS method yielded average errors lower than 2% for all sample sizes.

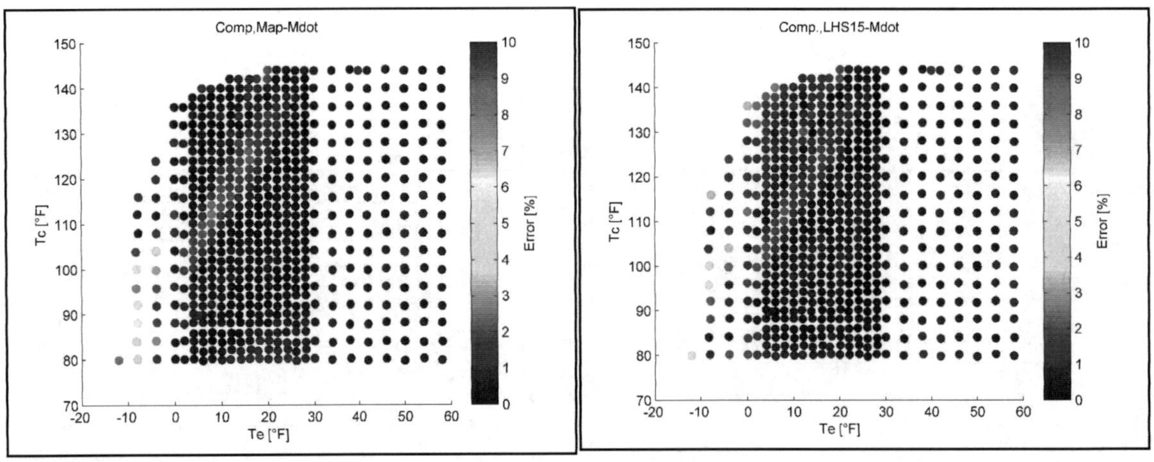

Figure 2: Comparison of errors in predicted mass flow rate for map data vs. LHS samples

4.2 Non-Rectangular Domains

The compressor operating envelope is non-rectangular, and hence, conventional sampling techniques are not directly applicable for selecting test points. A mathematically sound approach is required to handle non-rectangular domains. The two main methods published on space-filling designs for non-rectangular domains are Draguljic et al. (2012) and Lekivetz and Jones (2014). There is no open-source implementation available for either of the two methods, and hence, the method by Lekivetz and Jones (2014) was implemented with some simplifications for use in the conducted analysis. Consider a non-rectangular 2-D domain such as a compressor operating envelope represented on a [Te,Tc] plot. Assume it is desired to sample *n* points inside this domain. The steps in the approach are as follows:

1. Start with the non-rectangular domain.
2. Generate a large number of candidate sample points, either randomly or based on a uniform grid. These candidate sample points are superimposed on the design domain. Using the conventional methods, such points are generated on a rectangular domain, hence these points will span beyond the operating envelope (or domain).
3. Filter the candidate sample points that are outside the operating domain.
4. Use a clustering algorithm to generate *n* clusters based on a Euclidean distance metric.
5. Compute the convex hull for each of these clusters. Note that some of the polygons corresponding to the convex hulls overlap. This is due to the way the clustering technique works.
6. Use the convex hull information to find the centroids of these clusters. The *n* centroids represented as the (x,y) coordinates are the desired sample points.

While the above method is an approximate implementation, it has several advantages as follows:

1. It retains the space-filling properties;
2. The computational cost is very low compared to a true implementation that would require one to solve successive optimization problems for each set of samples;
3. The use of the clustering followed by choosing centroid points guarantees that the samples are 'inside' the domain and sufficiently away from each of the vertices; and,
4. Reproducibility.

4.4 Polygon Design of Experiments (PDOE) Method

An alternate method to the LHS design was developed to account for a non-rectangular domain, given the reasons noted above. This method is referred to as the Polygon Design of Experiments (PDOE) method for brevity. The following sampling strategy was implemented, assuming that the compressor operating envelope is given and *n* total samples are desired:

1. Initialize empty sample set *D*.
2. Start with the compressor operating envelope represented as a polygon with *k* vertices.
3. Add these *k* vertices to *D*.
4. Sample the remaining (*n-k*) points using the non-rectangular sampling method described in the previous section.

The data representation methods are meant to be used for interpolation purposes. As such, adding the *k* initial vertices is the simplest and the logical starting point. The AHRI Standard 540 and four other alternate data representation methods found in the literature were analyzed with regards to the number of samples used in generating the performance map. The number of samples that were tried included [8, 10, 12, 14, 16]. Appropriate models were chosen for the different number of samples. For example, the AHRI Standard 540 model requires a minimum of 11 points and so it cannot be used with 8 and 10 samples. Furthermore, comprehensive data sets were available for three separate compressors. These datasets were analyzed separately from the normal (limited) data sets.

Results indicated that the LHS method provides a much better average prediction performance for a given number of samples, but it is quite possible that this performance may change for a different run of the LHS design. While the overall prediction using the LHS design is better, there are areas within the map where in the local errors are unacceptably high. In general, the AHRI Standard 540 model shows average errors below 1%. The maximum absolute errors, on the other hand, are relatively higher. Irrespective of the number of samples and the sampling method, the MAPE in predicted power is less than 5% for the AHRI Standard 540 and two other evaluated models.

5. CONCLUSIONS

There are several sources of uncertainty in the prediction of compressor performance using maps or models. The most important amongst them are the uncertainty due to measurement and the uncertainty due to regression during model development. The measurement uncertainty has been quantified in the literature. The regression uncertainty was studied in the present work for the AHRI Standard 540 method. The worst case maximum absolute error in predicted mass flow rate across all data sets was 17% and that for power was 9%. For most compressors, the high errors occur in the region of the envelope with low suction and low discharge dew point temperatures. The average uncertainty in power prediction can be as high as 5% and that in mass flow rate prediction can be as high as 4%.

A study was conducted to evaluate the effect of sample size on the prediction capabilities for the various models. One of the challenges was the selection of samples from the operating envelope. The conventional LHS method was used as well as a new method (PDOE) from the literature for sampling on non-rectangular domains. For the case of mass flow rate, it was found that the LHS method yielded average errors lower than 2% for all the sample sizes. However, the maximum absolute errors in mass flow rate were higher when there were a lower (<= 10) number of samples. In general, both the LHS and PDOE methods yielded similar errors for all models for samples sizes of 12, 14 and 16. Thus, for mass flow rate, it is possible to build a model with 12 systematically selected test points. For power prediction, the average error for LHS and PDOE methods using AHRI Standard 540 and two other evaluated methods was lower than 2% for all sample sizes. In general, the errors were of similar magnitude for 12, 14 and 16 samples. For some cases with the LHS method, the errors were unusually high, due the random nature of the algorithm.

Based on the results of the analysis conducted, the following observations and recommendations are made to improve compressor data representation:

1. Reducing the measurement uncertainty is important. Particular attention must be paid during measurements involving low suction and low discharge dew point temperatures.
2. The regression uncertainty has an additive effect on the overall model prediction when the measurement uncertainty is factored into the overall model uncertainty. As such, it is possible to define a lower bound on the expected uncertainty in model prediction. Higher bounds may be possible depending on the availability of data.
3. In order to reduce the regression uncertainty, numerically stable and linearly regressed models should be selected, though this is not always possible for a physics-based power prediction model. Mass flow models can be adapted to linear regression.
4. There is potential for reducing the number of tests used to develop the performance map for a compressor. In general, for the models analyzed as a part of this work, a sample size of 10 or 12 is recommended. This requires additional verification.
5. The use of a systematic DOE method is recommended for selection of samples once an operating envelope is determined. It should be noted, however, that most DOE methods are designed to work with rectangular-domains.

NOMENCLATURE

T	Temperature	(°C/°F)

Subscripts

c	condenser
e	evaporator

Abbreviations

A/C	Air Conditioning
AHRI	Air Conditioning, Heating, and Refrigeration Institute
ASHRAE	American Society of Heating, Refrigerating, and Air-Conditioning Engineers
COV	Coefficient of Variance
DOE	Design of Experiments
HP	Heat Pump
LHS	Latin Hypercube Sampling
MAPE	Maximum Absolute Percent Error
PDOE	Polygon Design of Experiments

REFERENCES

ANSI/AHRI., 2015. Standard for Performance Rating of Positive Displacement Refrigerant Compressors and Compressor Units. Standard 540. Arlington, VA: AHRI.

Coleman, H. W., and Steele, W. G., 2009, Experimentation, Validation and Uncertainty Analysis for Engineers, Third Edition, John Wiley and Sons.

Draguljic D, Santner TJ, Dean AM. 2012, Noncollapsing Space-Filling Designs for Bounded Nonrectangular Regions, Technometrics 2012; 54(2):169–178.

Lekivetz, R. and Jones, B., 2014, Fast Flexible Space-Filling Designs for Nonrectangular Regions, Research Article, Quality and Reliability Engineering International, DOI:10.1002/qre.1640.

ACKNOWLEDGEMENT

The authors would like to acknowledge AHRI and the members of the committee for Project 8013 for support of this analysis effort.

Experimental Study on Noise Characteristics of Centrifugal Compressor Surge

Qichao YANG, Yuanyang ZHAO*, Yue SHU, Xiaosa LI, Liansheng LI

State Key Laboratory of Compressor Technology,
Hefei General Machinery Research Institute
230031, Hefei, P. R. China
(+86-551-65335667, yuanyangzhao@163.com)

* Corresponding Author

ABSTRACT

The centrifugal air compressor test rig is designed and established. The experimental study was carried out on the surge characteristics of centrifugal compressor including the pressure in the pipe and the noise under different rotation speed. The tested results showed that both the suction pressure and discharge pressure fluctuation increase with the rising of speed under surge condition. The amplitude of discharge pressure fluctuation is significantly higher than that of suction pressure. In addition, the sound pressure level near the inlet pipe also fluctuates under surge condition. The fluctuation amplitude of sound pressure level is increases with the rising of rotation speed. It is relative to the variation of discharge pressure of centrifugal compressor. The total sound pressure level under surge conditions is larger than that of steady working conditions. The change of sound pressure level reflects the changing of field flow in centrifugal compressor. The signals of sound could be used as a method to predict the occurring of surge. The experimental results will lay the foundation for future research on the monitoring surge of centrifugal compressor and the development of the surge control method.

1. INTRODUCTION

Centrifugal compressor is widely used for various kinds of industrial fields, such as chemical processes, gas transmission, refrigeration, engines. When centrifugal compressor operates at a lower flow rate, the instability phenomena such as rotating stall and surge will occur, which limits the safe operating region and the reliable performance of the centrifugal compressor. Especially, the surge may cause critical operating conditions and strong fluctuations and limit cycle oscillations can generate large vibrations to fatigue and damage of the compressor systems and its components. Therefore, the occurrences of surge or rotating stall have to be prevented in industrial application. The research on the surge mechanism and experimental investigation on surge characteristics are the base to develop the anti-surge control methods and to enhance the stable working range of centrifugal compressor.

In the past, theoretical and experimental study on surge has been done by many researchers (Galindo et al., 2006, Gravdahl et al., 2004, Hansen et al., 1980, Liu and Zheng, 2008, Zheng and Liu, 2015). Arnulfi G L, et al. (2006) reports the results of an extensive study concerning surge instability in an industrial compression system and its dynamic suppression. Both numerical simulations and experiments have been applied to analyze the system behavior under unstable operating conditions and to compare different control devices for surge suppression. Galindo J., et al. (2008) carried out the experimental work on the measurement of centrifugal compressor behavior within the surge zone by means of a specifically designed facility. A model based on a non quasi-steady hypothesis has also been proposed. Kin Tien Lim et al. (2011) study the surge phenomenon in a centrifugal compression system and to investigate a novel method of surge control by active magnetic bearing servo actuation of the impeller axial tip clearance. Hideaki Tamaki (2008) study the effect of piping system on surge in centrifugal compressor and confirmed that the lumped parameter model was a very useful tool to predict the effect of piping systems on surge characteristics in centrifugal compressor. Guillou E. et al. (2011) investigate experimentally on the steady and unsteady flow phenomena at the inlet of the radial compressor of a turbocharger typically used on diesel engines. Particle Image Velocimetry (PIV) was used to identify the aerodynamic changes occurring in the flow field while surge is developing. Digital Particle Imaging Velocimetry (DPIV) is used in conjunction with dynamic pressure transducers to capture transient velocity and pressure measurements simultaneously in the unsteady flow field during compressor surge. (Mark P. Wernet, 2001).

However, most of the research work mainly focused on the numerical analysis on the flow instability especially the surge phenomena. The noise characteristics of centrifugal compressor during the surge condition were less reported. The purpose of this study is to investigate the performance characteristics of centrifugal compressor surge through experiments. The change properties of suction pressure and discharge pressure are obtained. In addition, the noise characteristics under different working conditions and rotational speed are described.

2. EXPERIMENTAL FACILITIES

Figure 1 shows the test apparatus of the experimental system including the centrifugal compressor employed in this research and the photos of the test rig are shown in Figure 2. It consists of the driving system, the compressor, lubricating oil system, cooling system, data acquisition and control system. The motor driven by a frequency converter is a variable frequency motor that is rated for 250 kW. The maximum speed of the compressor rotor can be reached 30000 r/min. The compressor rotor is supported by a radial bearing and a thrust bearing and consist of a 13-bladed impeller and vaneless diffuser. The impeller was originally designed and the design rotational speed of the compressor is 14250 r/min. The rotational speed and the torque were measured through the torque meter, which was installed between the motor and gearbox. Temperature sensor and pressure sensor were installed at the inlet pipe and discharge pipe to measure the temperature and pressure of the compressor, respectively. The electrically throttle valve located at the discharge pipe was employed to adjust the discharge pressure and the operating point of the centrifugal compressor. In order to investigate the performance and monitor the working conditions of centrifugal compressor, a data collection board providing enough channels of simultaneous sampling analog to digital conversion are employed. The data acquisition and control system based on virtual instrument technology was programed to record the experimental data and adjust the rotational speed and throttle valve precisely. The steady characteristics and unsteady flow phenomena of the centrifugal compressor under different rotational speed with 10725 rpm, 12830 rpm and 15282 rpm. The microphone was located near the inlet pipe of the compressor test rig to evaluate the inlet noise at different working conditions.

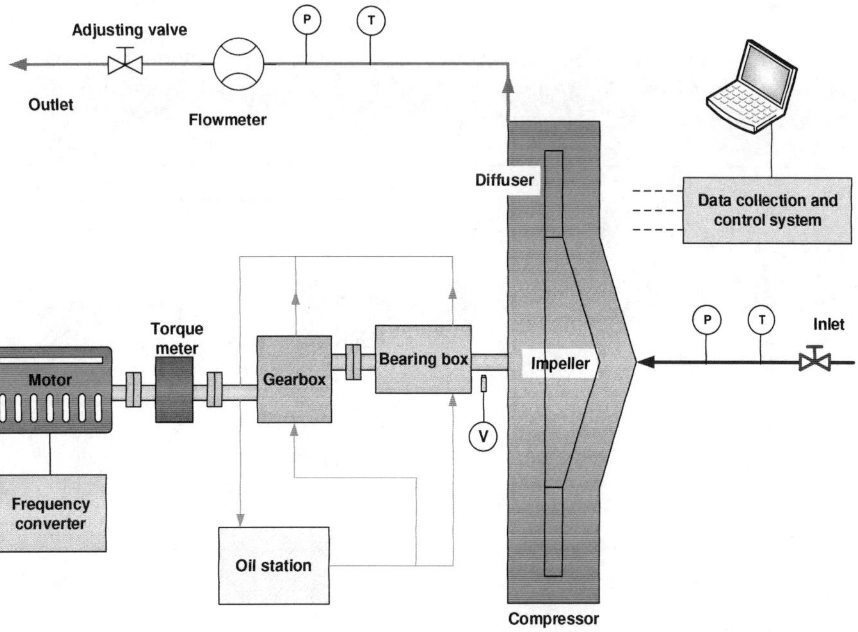

Figure 1: Schematic diagram of centrifugal compressor test system

Figure 2: Photos of centrifugal compressor test rig

3. EXPERIMETNAL RESULTS AND ANALYSIS

3.1 Pressure characteristics

Figure 3 shows the comparison of the inlet and discharge pressure of the centrifugal compressor under different working conditions and different rotational speeds. Figure 3(a) indicates the changing of suction and discharge pressure of the compressor. Suction pressure at three rotational speeds are very similar, therefore, there is only one line at 15282 rpm in Fig. 3(a). With the rising of the rotational speed, the discharge pressure is also increasing and the amplitude of pressure fluctuation is relatively small for different rotational speeds. Figure 3 (b), Figure 3(c) and Figure 3 (d) shows the fluctuation of suction and discharge pressure under surge condition at the rotational speed of 10725 rpm, 12830 rpm and 15282 rpm, respectively. The large fluctuation of pressure occurred under surge conditions. It is obvious that the amplitude of discharge pressure fluctuation is larger than that of suction pressure. Both the fluctuation amplitude of discharge pressure and suction pressure are increasing with the rising of rotational speed. For example, the peak-to-peak value of discharge pressure is about 16.85 kPa and the peak-to-peak value of suction pressure is about 5.15 kPa at 10725 rpm. When the rotational speed is increases to 12830 rpm and 15282 rpm, the peak-to-peak value of discharge pressure is increases to 22.54 kPa and 29.63 kPa, the peak-to-peak value of suction pressure is increases to 6.97kPa and 8.85 kPa. The frequency of pressure fluctuation under surge conditions is low compared to the fundamental frequency of the centrifugal compressor and is decreased with the rising of rotation speed of the compressor. The surge frequency is about 6.3Hz at the rotational speed of 10725 rpm. When the rotational speed is increases to 12830 rpm and 15282 rpm, the pressure fluctuation frequency under surge condition is reduces to approximately 5.95Hz and 5.5 Hz.

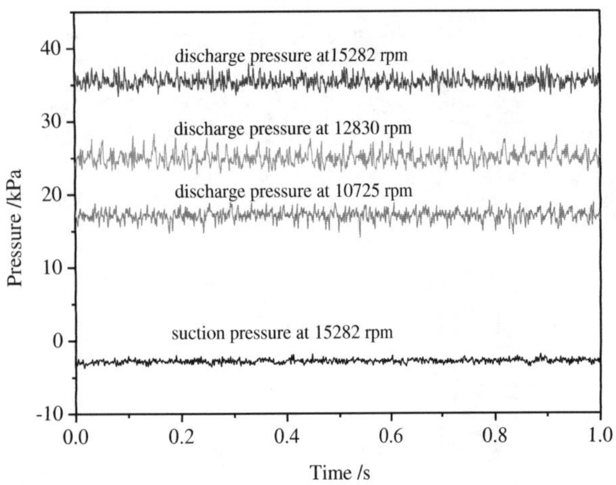

(a) Steady working condition

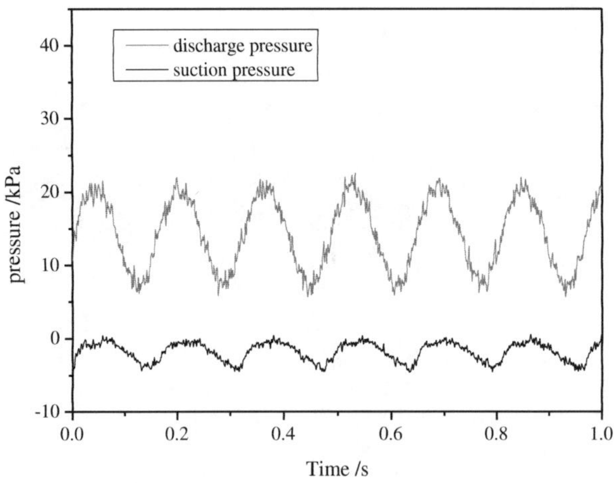

(b) Surge working condition (10725 r/min)

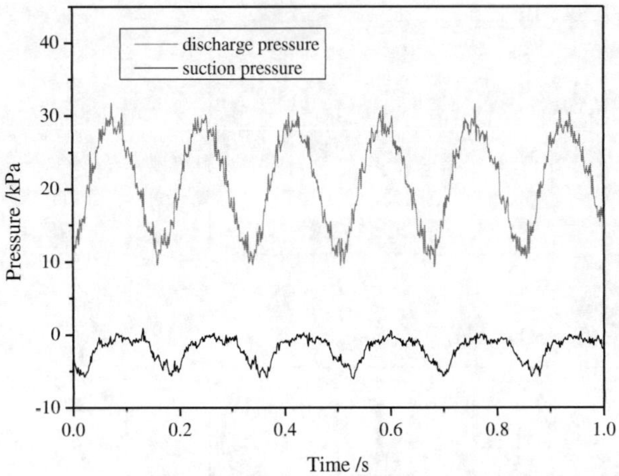

(c) Surge working condition (12830 r/min)

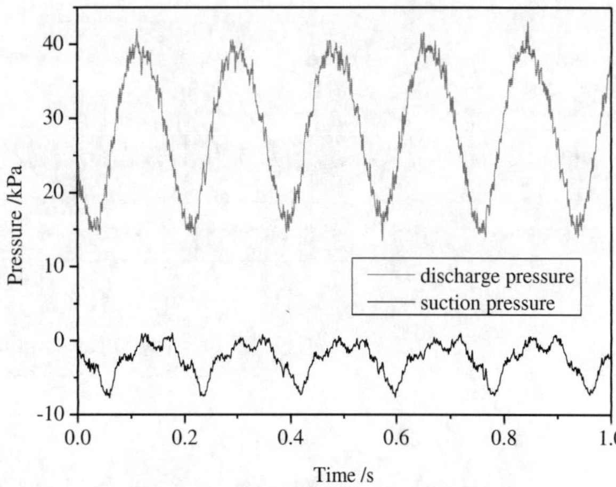

(d) Surge working condition (15282 r/min)

Figure 3: The change of pressure under different speed and working conditions

3.2 Noise characteristics

The fluctuations of sound pressure level of centrifugal compressor under the steady working condition and surge condition are shown in Figure 4. It can be seen from Figure 4 (a) that the noise level of the compressor in the time domain under the steady working condition is stable for any rotational speed of the compressor. While under the surge conditions, the sound pressure level will fluctuate and the fluctuation amplitude of sound pressure level is increasing with the rising of rotation speed as showed in Figure 4 (b), Figure 4 (c) and Figure 4 (d).

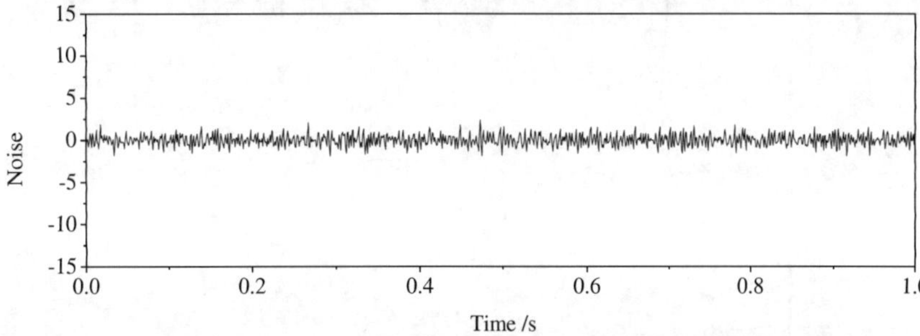

(a) Steady condition (15282 rpm)

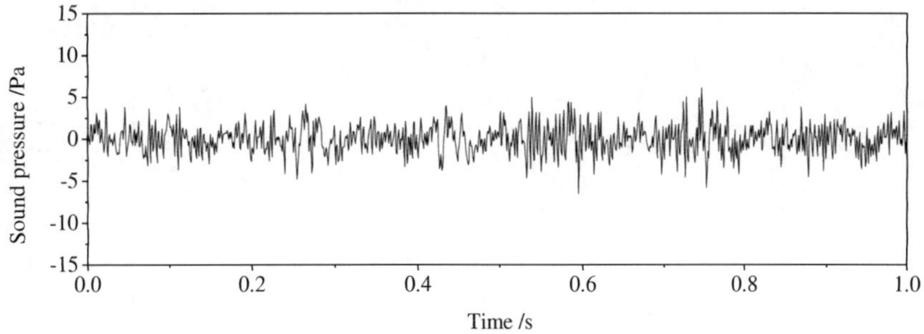

(b) Surge condition (10725 r/min)

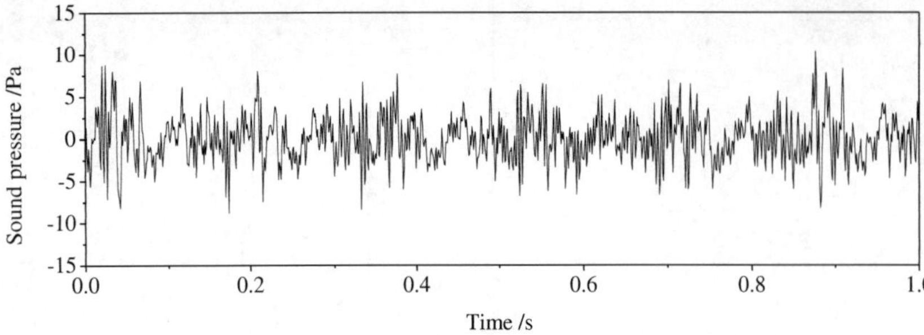

(c) Surge condition (12830 r/min)

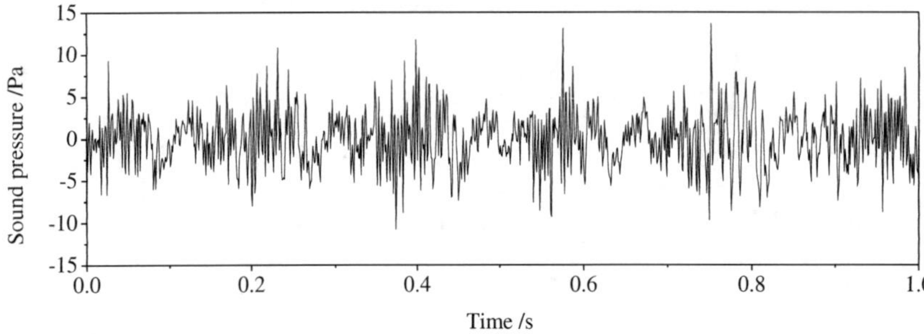

(C) Surge condition (15282 r/min)

Figure 4: The change of noise under different speed and working conditions

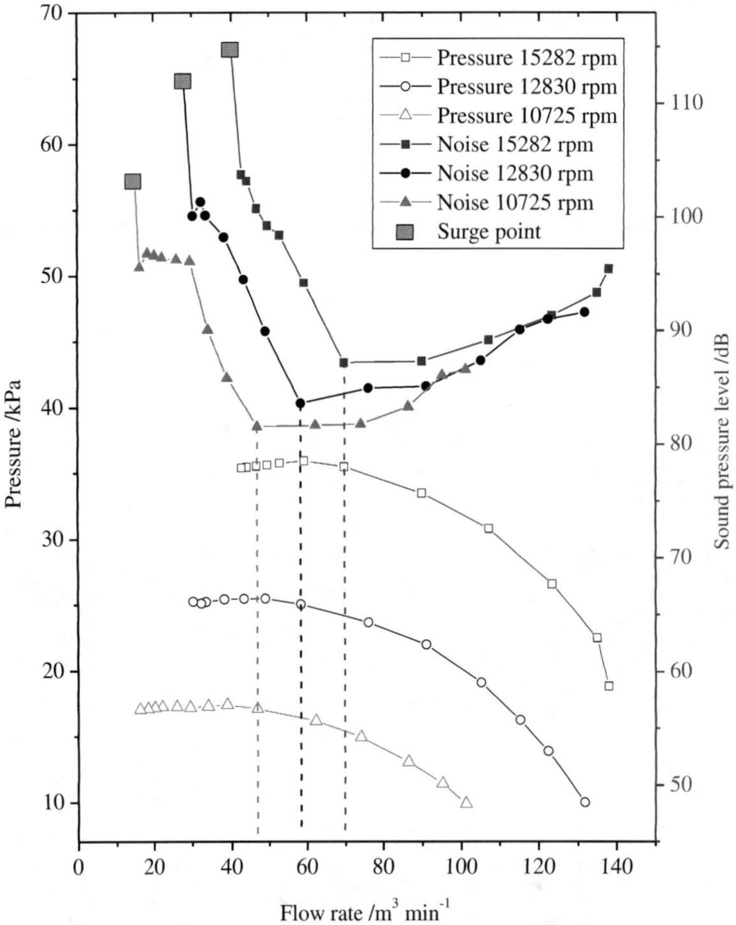

Figure 5: Performance characteristics of centrifugal compressor under different speed

Figure 5 shows the performance characteristics including the noise properties of the centrifugal compressor under different rotational speed of 10725 rpm, 12830 rpm and 15282 rpm. It can be found that the changing of sound pressure level of the compressor is relative to the changes of the discharge pressure of the compressor. In general, the discharge pressure of centrifugal compressor is increasing with the decreasing of the flow rate under the constant rotational speed and then decreasing. Maximum discharge pressure exists for the compressor. The tested results on sound pressure level of this open cycle compressor system indicate that there exists a minimum sound pressure level that is corresponding to the maximum discharge pressure. At the beginning, the sound pressure level is decreasing with the decreases of the flow rate of the compressor, and then increasing rapidly. At the surge conditions, the sound pressure level reached the maximum value. For example, at the rotational speed of 15282 rpm, when the flow rate is 59.29 m^3/min, the discharge pressure of the centrifugal compressor reached the maximum value of 35.94 kPa and the sound pressure level can achieve the lowest value of 87.19 dB. Although the error exists between the changing point of the discharge pressure and sound pressure level, the tested results still proven that the rotation stall occurred after the maximum discharge pressure. The total sound pressure level under surge conditions is larger than that of steady working conditions. At the surge conditions, the sound pressure level can reach to a very high value. For example, the sound pressure level under the surge condition at 15282 rpm can got to 114.76 dB, which is higher than the sound pressure level value of 103.73 dB at the near surge point and is much higher than the sound pressure level value of 87.19 dB at the design working conditions. The same change tendency was found at other rotational speed of 10725 rpm and 12830 rpm. Therefore, the change of sound pressure level during the operation of centrifugal compressor could be used as a characteristic signal to predict the occurring of rotation stall and surge phenomenon.

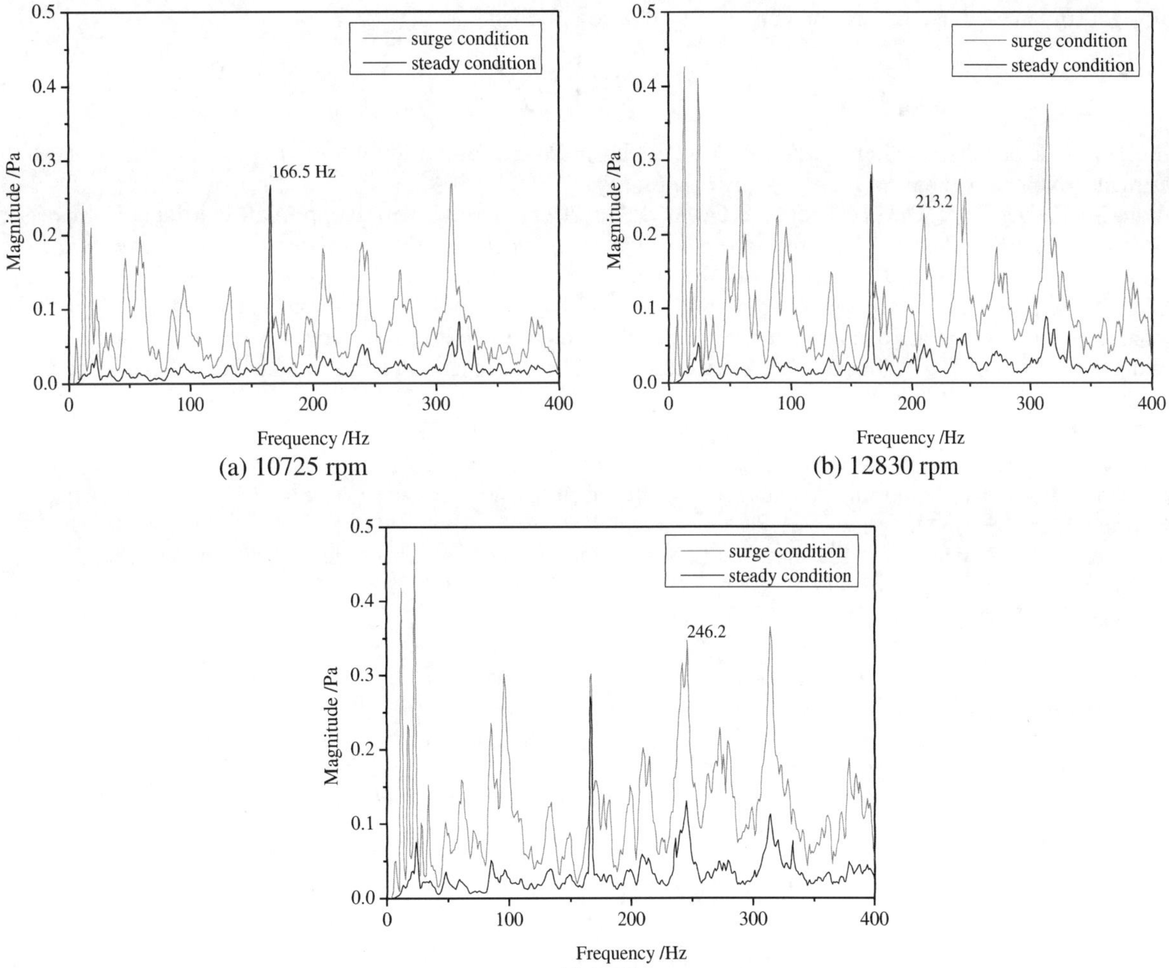

(a) 10725 rpm

(b) 12830 rpm

(c) 15282 rpm

Figure 6: Frequency spectrum at different rotational speed

The results of the frequency analyses on the sound pressure level signals are shown in Figure 6. Since the frequency of surge is often low, analysis on experimental tests focus on the frequency range less than 400 Hz. Experimental tests were carried out at three speeds to evaluate the effect of speed change on the observed dominant frequencies. Figure 6(a) shows the results at 10725 rpm. Figure 6(b) shows the results at 12830 rpm, Figure 6(c) shows the results at 15282 rpm. In Figure 6 (a), the 166.5 Hz are the fundamental frequency of the tested centrifugal compressor. At steady working conditions, the dominant frequency is the fundamental frequency related to the rotation speed of centrifugal compressor. The other harmonic frequency is rather low. However, the low frequency appears and becomes strong under the surge condition and the harmonic frequency are also becoming larger.

4. CONCLUSIONS

The centrifugal air compressor test rig is designed and established. The experimental study was carried out on the surge characteristics of centrifugal compressor including the pressure in the pipe and the noise under different rotation speed. The tested results showed that both the suction pressure and discharge pressure fluctuation increase with the rising of speed under surge condition. The amplitude of discharge pressure fluctuation is significantly higher than that of suction pressure. In addition, the sound pressure level near the inlet pipe also fluctuates under surge condition. The fluctuation amplitude of sound pressure level is increased with the rising of rotation speed. It is relative to the variation of discharge pressure of centrifugal compressor. The total sound pressure level under surge conditions is larger than that of steady working conditions. The change of sound pressure level reflects the changing of field flow in centrifugal compressor. The signals of sound could be used as a method to predict the occurring of surge phenomenon. The experimental results will lay the foundation for future research on the monitoring surge of centrifugal compressor and the development of the surge control method.

REFERENCES

Arnulfi, G. L., Blanchini, F., Giannattasio, P., Micheli, D., & Pinamonti P. (2006). Extensive study on the control of centrifugal compressor surge.*Journal of Power and Energy*, (220):289-304

Guillou, E., Gancedo, M., DiMicco, R.G., & Gutmark, E. (2011). Surge Investigation in a Centrifugal Compressor by Stereoscopic PIV. *49th AIAA Aerospace sciences meeting including the new horizons forum and aerospace exposition*. Orlando, Florida. (AIAA 2011-742) Reston, USA: AIAA.

Galindo, J., Serrano, J.R., Guardiola, C., & Cervello´, C. (2006). Surge limit definition in a specific test bench for the characterization of automotive turbochargers. *Experimental Thermal and Fluid Science*, 30(5): 449-462.

Galindo, J., Serrano, J.R., Climent, H., & Tiserira, A. (2008). Experiments and modelling of surge in small centrifugal compressor for automotive engines. *Experimental Thermal and Fluid Science*. (32):818-826.

Gravdahl, J.T., Willems, F., Jager, B. & Egeland, O. (2004). Modeling of surge in free-spool centrifugal compressors: Experimental validation. *Journal of propulsion and power*, 20(5): 849-857.

Hideaki, T. (2008). Effect of piping systems on surge in centrifugal compressors. *Journal of Mechanical science and technology*. (22):1857-1863.

Hansen, K.E., Jorgensen, P., & Larsen, P.S. (1980). Experimental and Theoretical Study of Surge in a Small Centrifugal Compressor. *Journals of Fluids Engineering*, No.81-FE-1. (103) 391-395.

Kin, T. L., Se, Y. Y., Christopher, P. G., Lin, Z. & Allaire, P. E. w(2011). Design and characterization of a centrifugal compressor surge test rig. *International Journal of Rotating Machinery*, 738275. 1-8.

Liu, A.X. & Zheng, X.Q. (2008). Methods of surge point judgment for compressor experiments. *Experimental Thermal and Fluid Science,* (51):204-213.

Wernet, M.P., Bright, M.M., & Skoch, G.J. (2001). An investigation of surge in a high-speed centrifugal compressor using digital PIV. *Journal of Turbomachinery*, 123(4) 418-428.

Zheng, X. Q., & Liu, A.X. (2015). Experimental Investigation of Surge and Stall in a High-Speed Centrifugal Compressor. *Journal of Propulsion and Power*, 31(3): 8215-525.

ACKNOWLEDGEMENT

The authors would like to thank to the support by the National Key Basic Research Program of China (2012CB026000), the Science and Technological Fund of Anhui Province for Outstanding Youth (1508085J05), and the Science and Technology Plan of Anhui Province (1501041128).

The Effects of Blade Fillets on Aerodynamic Performance of a High Pressure Ratio Centrifugal Compressor

Justin (Jongsik) Oh

Danfoss Turbocor Compressors, Inc.
Tallahassee, FL
jongsik.oh@danfoss.com

ABSTRACT

In most cases of aerodynamic design of centrifugal compressors, the effects of impeller blade fillets on performance are not considered in the process, but could be estimated later from some limited studies. As one of numerical investigations for the effects in centrifugal compressors, the Krain backswept impeller was modeled with and without blade fillets on the hub. A vaneless diffuser and impeller tip clearances were included in the steady state analysis using a commercial CFD code. Over the range of flows at design speed, the case with blade fillets showed a slight drop in the pressure ratio, the efficiency, the choke flow and the range of operation, relative to the case of clean blades. A more detailed look into three-dimensional flow structure inside the impeller shows that a small scraping vortex, developed in the case of clean blades at the corner of the hub pressure surface, disappears in the case with blade fillets due to a local flow acceleration produced by the fillet. As a result of balancing forces acting in the impeller passage, it was observed that the shroud passage vortex in the case with blade fillets grows toward the impeller exit with a higher vortex core than that in the case of clean blades, which means that there are more wake flows with blade fillets, leading to aerodynamic performance drops.

1. INTRODUCTION

Fillets around the interface of blades with endwalls are generally required in the turbomachinery manufacturing process such as the 5-axis machining or a simple welding. Even some cases require them for structural integrity in both rotor and stator rows in order to ensure mechanical strength at higher tip speeds. For centrifugal compressors, however, the blade fillets are rarely considered in the conventional aerodynamic design process. Their effects on aerodynamic performance could be later assumed from some limited studies, but most of the findings were with axial-flow machines. Curlett(1991) found from a low speed axial cascade test that the addition of a fillet in the CDA(Controlled Diffusion Airfoil) blades increased secondary flow and profile losses, but in the DCA(Double Circular Arc) blades the loss increases were not significant, and rather an extended range of operation was expected with blade fillets. Calvert et al.(1999) claimed that the fillet radius needs to be larger in a transonic axial-flow compressor for aerodyanmic reasons. Hoeger et al.(2006) found that fillets removed corner stall for high incidences. Kuegler et al.(2008) showed in a numerical investigation of an axial-flow 15-stage compressor that a simulation with blade fillets had a reduced choke flow and a lower efficiency but an extended operability at design speed. The wider working range was explained with the impact of the blade fillets on the secondary flow which contributed to an accumulative effect on the multistage blade rows by reducing corner stall at the rotor hub and stator tip. As for a radial machine, Syka et al.(2015) numerically investigated the performance of a low pressure ratio shrouded centrifugal compressor with and without blade fillets on the impeller hub and shroud. Reduced choke flows, a lower pressure ratio and a lower efficiency were predicted with fillets, but the effect on the range of operation could not be identified.

Based on such limited information, the effects of blade fillets on aerodynamic performance look beneficial in the range of operation for multistage axial-flow compressors, despite reductions in choke flows and efficiencies. The effect on the pressure ratio is still not clear because it could be dependent of the level of corner stall in axial-flow compressors without fillets. For centrifugal compressors it would be too soon to mention about the effects due to lack of relevant studies. The present paper deals with the influence of blade fillets on the impeller hub in a high pressure ratio centrifugal compressor on aerodynamic performance. A numerical CFD investigation was performed on a well-known public impeller with test results available.

2. CENTRIFUGAL COMPRESSOR IMPELLER

Detailed flow field measurement inside a centrifugal compressor impeller was obtained by Krain(1988). The public impeller, shown in Fig.1, has been frequently cited in many other studies probably because it is (30 deg) backswept at the exit with a working range of higher pressure ratios (i.e., an impeller design total pressure ratio of 4.7). Table 1 shows a brief summary of the impeller specifications.

Table 1: Krain Impeller Specifications

Parameter	Unit	Value
Impeller Exit Diameter	mm	400
Impeller Exit Height	mm	14.7
Design Speed	rpm	22,360
Design Mass Flow (of Air)	kg/s	4.0
Blade Count	-	24

Fig.1 Photo of Krain Impeller

3. NUMERICAL MODELING

A commercial CFD code, ANSYS CFX R15.0, was used in the present study. For a fair comparison between clean blades (without fillets) and blades with fillets, the same type and approximately the same node number of grids were used. A constant fillet radius of 3.7 mm was applied all the way around the blade hub, which size was found the maximum available inside the blade pitch. Fig.2 shows two numerical grids near the blade leading-edge hub. A variable tip clearance, from 0.8 mm at the leading-edge to 0.6 mm at the trailing-edge, was included in the impeller grids, and a vaneless diffuser with a radius ratio of 1.95 (relative to the impeller exit radius) was added downstream of the impeller in the analysis. About 1.2 million nodes of grids were generated for both cases including the impeller upstream region and the vaneless diffuser.

The standard air conditions were used to specify total pressure and total temperature at the inlet boundary, and an averaged static pressure was given at the exit boundary with the k-epsilon turbulence model for a steady state analysis.

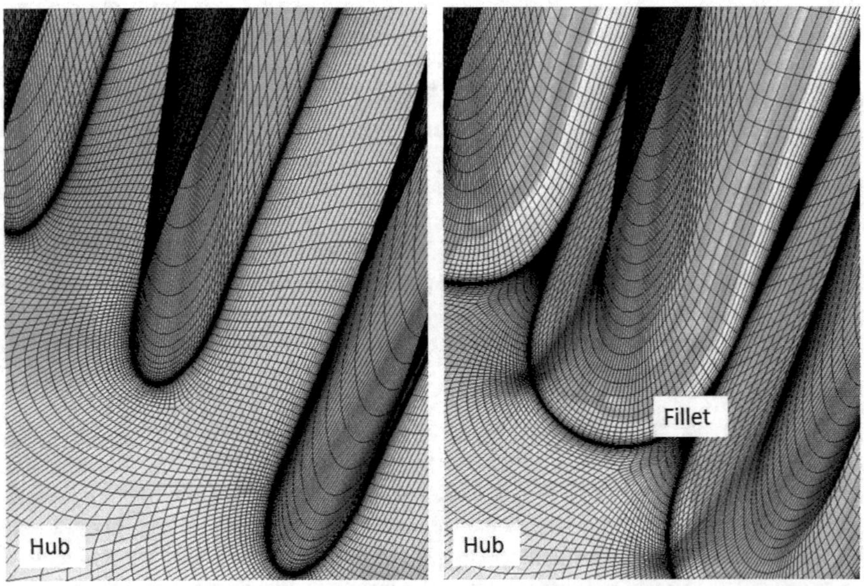

(a) Clean blade (b) Blade with fillets

Fig.2 Structured grids for CFD near blade leading edges (with only wall surface grids shown)

4. RESULTS AND DISCUSSION

4.1 Code Validation

An interesting comparison was made between the present predictions with clean blades and some of the published test results by Krain(1988) in order to make sure that the numerical approach is reliable. Fig.3 presents a compressor map including the vaneless diffuser with changes of rotational speed and flow for the measurements and the present analysis. Mass-averaged total pressure and total temperature were used in the predicted compressor map. Near the stall/surge flows the computation was carried out through increasing the exit static pressure by small amounts until the steady-state solution failed to converge. A good agreement on choke and stall flows was found at the three selected speeds, but the magnitude gap of pressure ratio and efficiency became larger at higher speeds, showing over-predicted pressure ratios and under-predicted efficiencies.

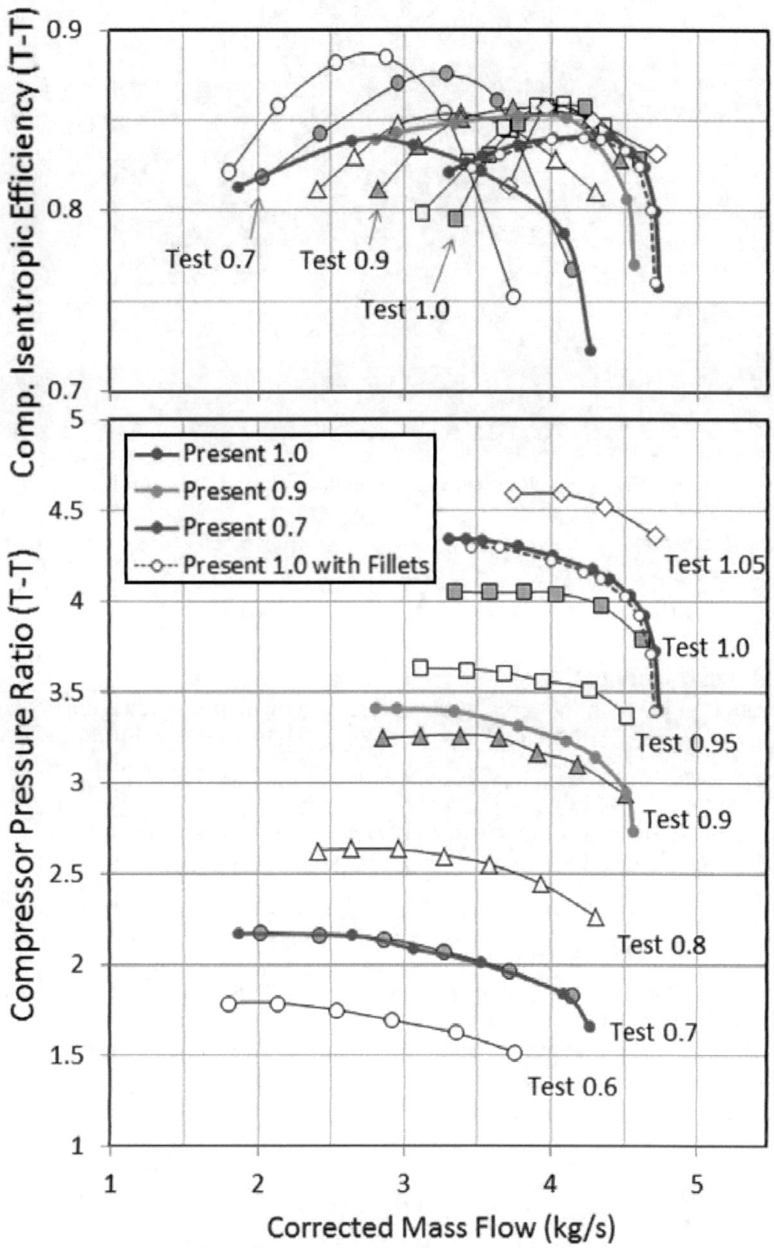

Fig.3 Compressor map with vaneless diffuser

Fig.4 shows meridional velocity contours, normalized by the impeller tip speed, on two orthogonal sections at design flow, which generally provide the development of secondary flow in the impeller passage. An impressive agreement was observed on both planes, and especially at Section V, where very strong movements of secondary flow normally happen, a reliable prediction was found with the pattern of the tip leakage vortex and the passage vortex near the suction surface.

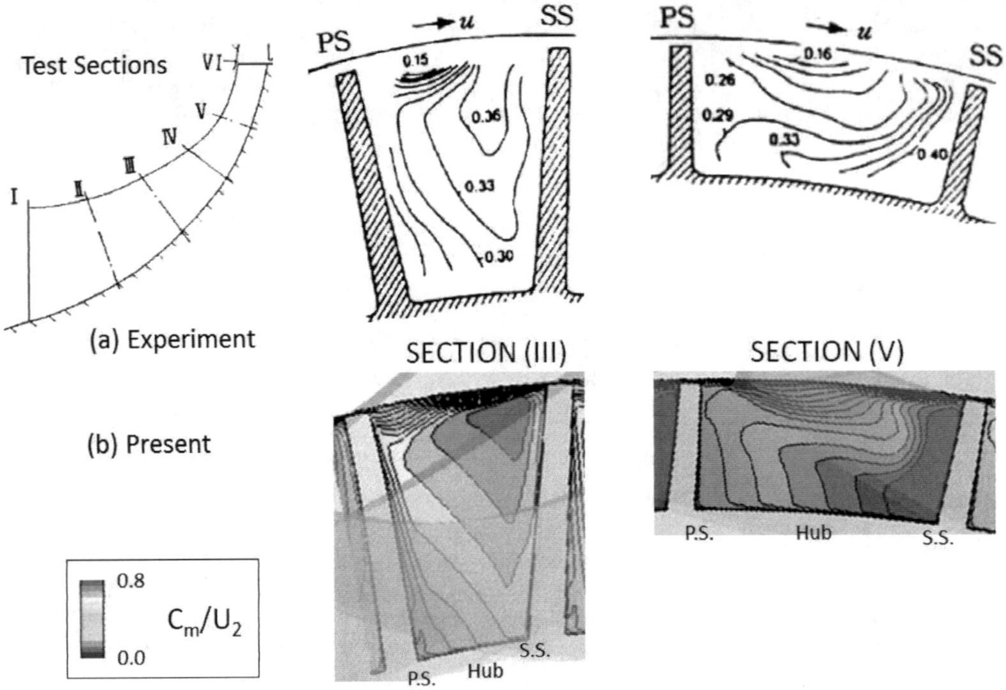

Fig.4 Normalized meridional velocity contours at design flow

4.2 Effects of Blade Hub Fillet

4.2.1 Overall Performance

At design speed, the case with the 3.7 mm blade fillets on the hub was run from choke to stalled minimum flow in the same manner, and the performance was plotted in Fig.3 together with the case with clean blades. A zoomed-in version is given in Fig.5 where every aerodynamic performance, such as the choke margin, the range of stable operation, the pressure ratio and the efficiency, clearly dropped relative to the case of clean blades without fillets. The reduction of choke flow is of course predictable because of a smaller geometric area from fillets, but it is of concern that both pressure ratio and efficiency were down over the range and even the compressor operability was decreased. One finding, not too bad, is that the efficiency drop was minimized at design flow.

4.2.2 Secondary Flow Development

By such a small change of the impeller geometry through blade fillets it was observed that overall compressor performance was affected to some degrees that cannot be ignored, while blade fillets are still in most cases neglected in the aerodynamic design process. Looking for backgrounds needs the investigation of the secondary flow development inside the impeller passage which is generally related with the causes of losses. The normalized relative helicity, H_n, is quite useful in understanding the vortex structure in the blade passage, which is defined as,

$$H_n = \frac{\vec{\zeta} \cdot \vec{W}}{\left|\vec{\zeta}\right| \cdot \left|\vec{W}\right|} \tag{1}$$

where W is the relative velocity in the impeller, and ζ is the relative vorticity. The normalized relative helicity means the cosine of the angle between the relative vorticity and the relative velocity vectors, the magnitude of which approaches unity in the vortex's core, with its sign indicating the vortex swirl direction relative to the streamwise velocity component.

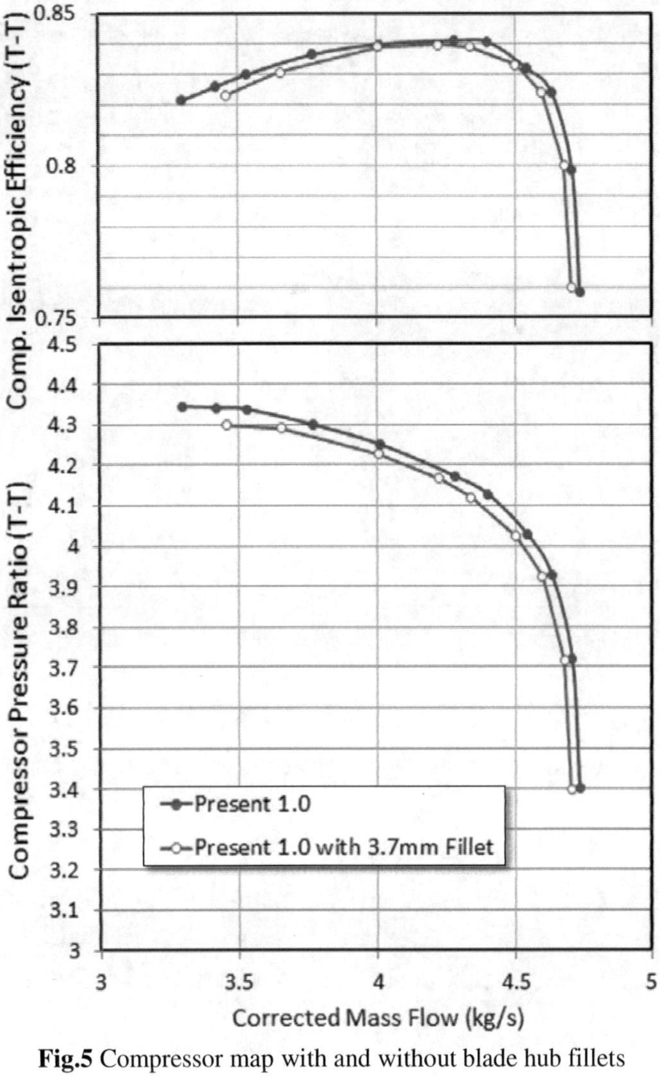

Fig.5 Compressor map with and without blade hub fillets

Fig.6 shows CFD predicted contours of normalized meridional velocity and normalized relative helicity at Section (IV), whose position is depicted in Fig.4, at design flow for both cases. According to Zangeneh et al.(1997), the following incompressible equation can describe the development of main secondary flows in rotating systems.

$$\vec{W}\cdot\nabla(\vec{W}\cdot\vec{\zeta}) = 2\vec{\zeta}\cdot(\vec{W}\cdot\nabla)\vec{W} + \vec{\zeta}\cdot(2\vec{\omega}\times\vec{W}) \tag{2}$$

where $\vec{\omega}$ is the rotational velocity. The left-hand side of equation (2) is the streamwise component of the relative vorticity, that is, the secondary flows. The first term of the right-hand side denotes a component of flow acceleration or diffusion due to streamline curvatures, and the second term means the Coriolis force in the direction of the relative vorticity. The equation tells that the secondary flows are created where flow turns in either meridional or blade-to-blade planes due to vorticities in the wall boundary layers, and where Coriolis acceleration becomes effective in the wall boundary layers. In the relative helicity contours of Fig.6, it is clearly shown that two blade

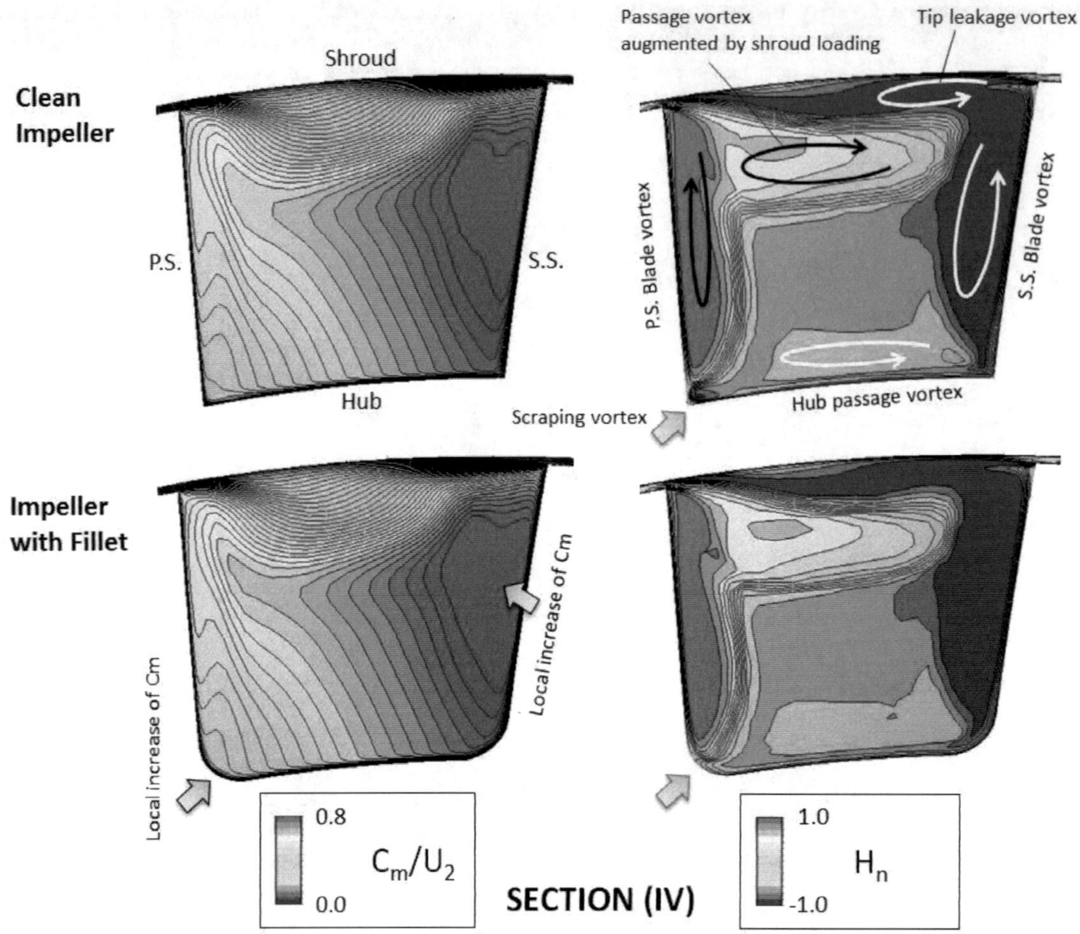

Fig.6 Meridional velocity and relative helicity contours at Section (IV) at design flow

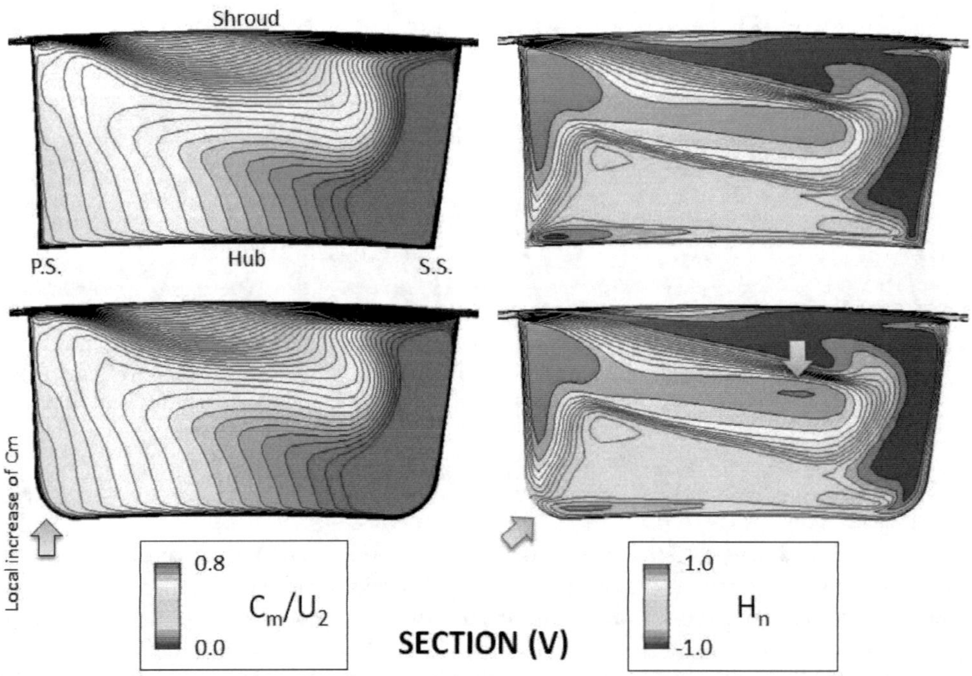

Fig.7 Meridional velocity and relative helicity contours at Section (V) at design flow

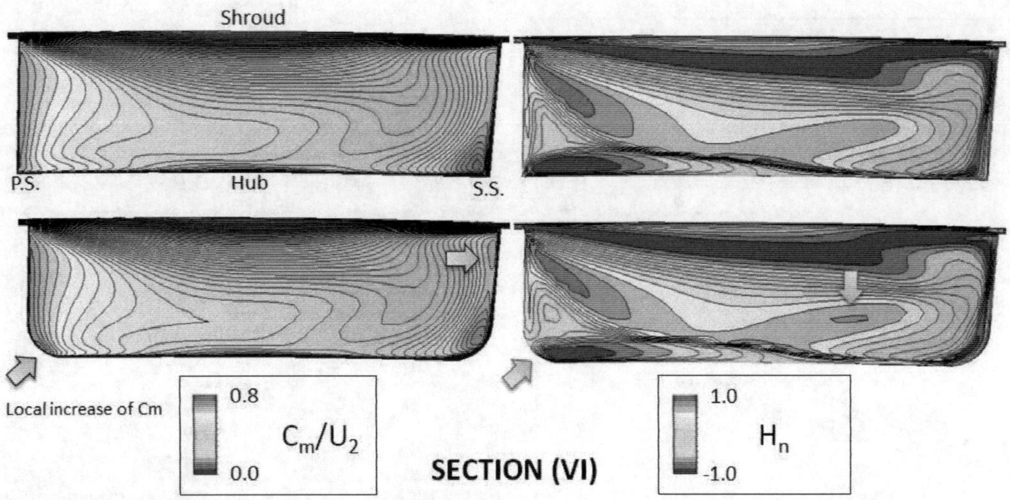

Fig.8 Meridional velocity and relative helicity contours at Section (VI) at design flow

vortices are developed along the suction and the pressure surface of the blade heading to the shroud, and another two passage vortices are created near the hub and the shroud directing from the pressure surface toward the suction surface. The counter-clockwise blade vortex near the suction surface is dominating, reinforced by the hub passage vortex and the tip leakage vortex. The shroud passage vortex, augmented with blade loading and the Coriolis passage vortex, is pushed down by the tip leakage vortex. In the meridional velocity contour, a large low-momentum area is formed near the shroud which is caused by the spreading-out tip leakage vortex interacting with other secondary flow vortices. In the case with blade fillets, compared with the case of clean blades, a local flow acceleration is found at the corner of the hub pressure surface, caused by the fillet, and another increase of the meridional velocity is seen near the suction surface. Such a local acceleration can be also confirmed from a small scraping vortex, formed at the corner with a counter-clockwise rotation in the clean blades, which is however removed with blade fillets.

At Section (V), Fig.7, the shroud passage vortex, growing supported by blade loadings, the Coriolis vortex and the blade vortex near the pressure surface, starts to occupy most area of the section with a clockwise rotation. Its core is filled with the accumulated low-momentum fluid which becomes a main source of losses. The blade vortex near the suction surface, carrying the high-momentum fluid, is pushed toward the hub, but the tip leakage vortex still stays near the shroud. The scraping vortex at the corner of the hub pressure surface in the clean blades is growing, however, it is obviously suppressed with blade fillets by the local flow acceleration. In the case with blade fillets, as a result of different balancing forces acting in the passage, the extended shroud passage vortex is observed with a higher core than the case of clean blades, which means more wake flows.

At the impeller exit, Section (VI) of Fig.8, the shroud passage vortex covers most part of the section, while the high-momentum fluid region becomes confined near the hub toward the pressure surface corner. Due to an interaction of passage vortices with the strong tip leakage vortex, the wake is positioned apart from, the so-called, near shroud suction surface. In the case with blade fillets, the core strength of the extended shroud passage vortex is still slightly higher than the case of clean blades. Another local flow acceleration is seen near the shroud suction surface in addition to that at the corner of the hub pressure surface from the meridional velocity contours.

The local flow acceleration around the hub blade fillet can be visualized through blade loadings and streamline plots, as shown in Fig.9 and Fig.10. Such a reduction of blade loadings at 5% span near the hub contributes to the drop of pressure ratios, while the increased core strength of the shroud passage vortex, already observed in Fig.7 and Fig.8, would lead to the drop of compressor efficiency and operability, found in Fig.5.

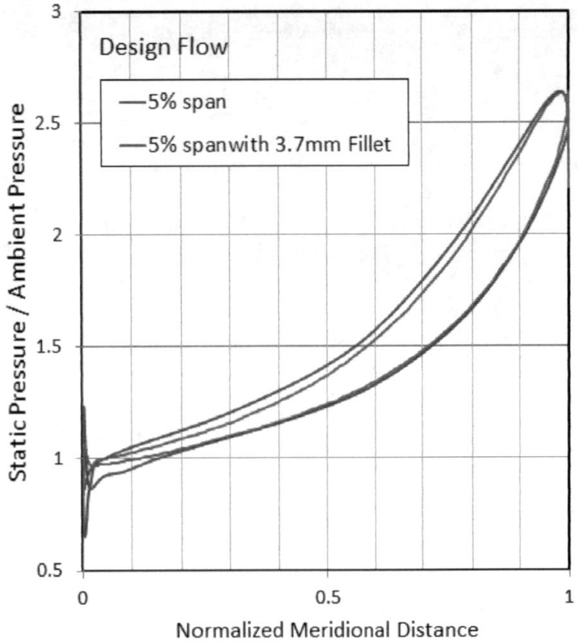

Fig.9 Blade loadings at 5% span near hub

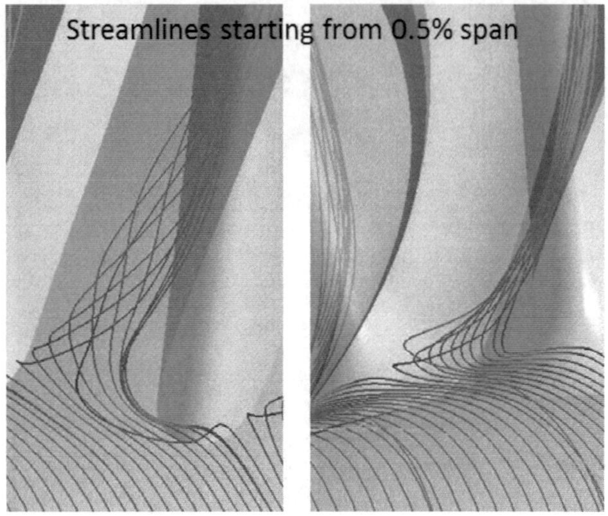

Fig.10 Streamlines near hub leading-edge

6. CONCLUSIONS

As one of numerical investigations for the effects of blade fillets on aerodynamic performance in centrifugal compressors, the Krain backswept impeller was modeled with and without blade fillets on the hub. A uniform fillet radius of 3.7 mm was applied around the impeller blade hub in the case with blade fillets. A vaneless diffuser and impeller tip clearances were included in the steady state analysis using a commercial CFD code.

- Over the range of flows at design speed, the case with blade fillets showed a slight drop in the pressure ratio, the efficiency, the choke flow and the range of operation, relative to the case of clean blades.
- Blade loading near the hub was reduced in the case with blade fillets, due to a local flow acceleration around the fillet, leading to the drop of pressure ratios.
- A small scraping vortex, developed in the case of clean blades with a counter-clockwise rotation at the corner of the hub pressure surface in the impeller passage bend, disappears in the case with blade fillets due to the local flow acceleration.
- As a result of different balancing forces acting in the impeller passage, the shroud passage vortex in the case with blade fillets grows toward the exit with a higher vortex core than that in the case of clean blades, which means more wake flows.
- At the impeller exit in the case with blade fillets, another local flow acceleration is seen near the shroud suction surface in addition to that at the corner of the hub pressure surface, which supports the raised level of flow non-uniformity and therefore the drop of efficiencies and operability.

NOMENCLATURE

W	Relative velocity	m/s
ω	Rotational speed	rad/s
ζ	Relative vorticity	1/s
C	Absolute velocity	m/s
U	Blade speed	m/s
H	Relative helicity	m/s^2

Subscript

2	Impeller exit
m	Meridional component
n	Normalized

REFERENCES

Calvert, W.J. et al., 1999, Transonic Fan and Compressor Design, *Proceedings of the Institution of Mechanical Engineers, Part C: Journal of Mechanical Engineering Science*, Vol.213, pp.419-436

Curlett, B.P., 1991, The Aerodyanmic Effects of Fillet Radius in a Low Speed Compressor Cascade, NASA TM-105347

Hoeger, M. et al., 2006, Impact of a Fillet on Diffusing Endwall Flow Structure, ISROMAC 2006-057

Krain, H., 1988, Swirling Impeller Flow, *ASME Journal of Turbomachinery*, Vol.110, pp.122-128

Kuegeler, E. et al., 2008, Influence of Blade Fillets on the Performance of a 15 Stage Gas Turbine Compressor, ASME GT2008-50748

Syka, T. et al., 2015, Hub and Shroud Fillets Influence on the Radial Compressor Stage Efficiency, EPJ Web of Conference 92, 02090

Zangeneh, M. et al., 1997, On the Design Criteria for Suppression of Secondary Flows in Centrifugal and Mixed Flow Impellers, ASME Paper 97-GT-393

ACKNOWLEDGEMENT

The author gratefully acknowledges the support of Danfoss Turbocor for the present study and its publication.

Experimental Research on Surge and Stability Enhancement of Centrifugal Compressor

Zhao Yuanyang *, Yang Qichao, Li Liansheng, Xiao Jun, Liu Guangbin, Wang Le

State key laboratory of compressor technology, Hefei General Machinery Research Institute, Hefei 230031, P. R. China

+86-055165335667， yuanyangzhao@163.com, yyzhao2@gmail.com

* Corresponding Author

ABSTRACT

Centrifugal compressors are wildly used in many process industries. The stability is one of the most important characters of centrifugal compressor. When the compressor operates in the condition of small volume flow rate, the working conditions of rotating stall and surge will occur, which leads to the unstable condition for centrifugal compressor.

The signals of compressor are tested and analyzed when surge condition occurs in this paper. In addition, a new method to improve the compressor stability is proposed. It is called the active control casing treatment (ACCT) system. The flow in the compressor impeller is changed by the ACCT system and the stability of compressor is improved. The experimental researches have been done in this paper. The test results of ACCT system are also discussed in this paper.

1. INTRODUCTION

Centrifugal compressors are widely used in many industry fields, such as petrochemical, coal chemical, and refrigeration industries. Surge is one of the most important characters of centrifugal compressors. When the centrifugal compressors operate at the margin of the surge line, the compressors will be unstable. In the real industry systems, the anti-surge control systems are always used to make sure that compressors operate at the stable working conditions. There are some kinds of methods to enlarge the stable working condition range of the centrifugal compressors, such as inlet guide vane (M Coppinger& E Swain, 2000). In addition, some casing treatment methods have been researched. Fisher (1988) first researched a recirculation casing treatment in a turbocharger compressor. The surge margin of compressor was extended without much decrease of efficiency. L. Ding (2013) investigated the holed casing treatment (self-recirculation) in an industrial centrifugal compressor. The surge margin of the compressor is increased by about 10%Q_{design} with negligible decrease in efficiency.

This paper present a stability enhancement method based on casing treatment, which is called Active Control Casing Treatment (ACCT). All parts of this system are introduced in details in this paper. A test system is set up to verify the function and performance of the ACCT system. Moreover, the compressor performance is tested when using the ACCT system. At the same time, the character signals of surge are tested in this paper.

2. TEST SYSTEM

Figure 1 shows the schematic diagram of test system. The driving system includes motor, frequency converter, torque meter, gearbox, and bearing box and oil station. At the inlet and outlet pipes, there are temperature, pressure and vibration sensors. During the test process, the rotating speed of compressor is controlled by the frequency converter. The ratio of compressor speed and motor speed is 10. The real-time speed and power is tested by torque meter. The discharge pressure is controlled by the adjusting valve in the outlet pipes. The flow rate of compressor is tested by the flowmeter. The transient pressure and vibration of compressor are recorded by the data acquisition and control system.

Figure 2 shows the main schematics of ACCT system. In principle, the unsteady flow in impeller leads to the rotating stall and surge of centrifugal compressor. The unsteady flow is usually located at the throat of the impeller. For the surge working condition, the flow separation occurs at the throat because the decrease of the mass flow rate. In the moment of surge, some high-speed gas is injected into the impeller by ACCT system to change and depress the unsteady flow. The ACCT subsystem includes the control valve, small compressor and test sensors. The transient pressure, temperature and volume flow rate at the ACCT subsystem are also recorded by the data acquisition and control system.

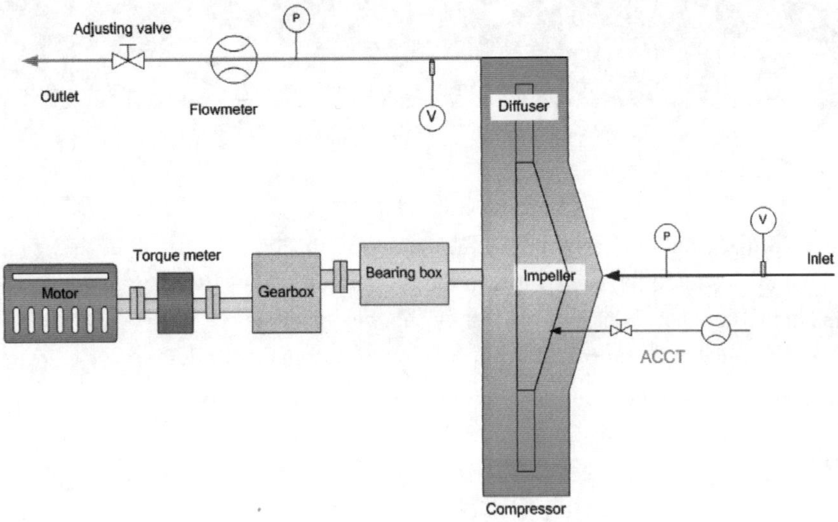

Figure 1: Schematic diagram of test system

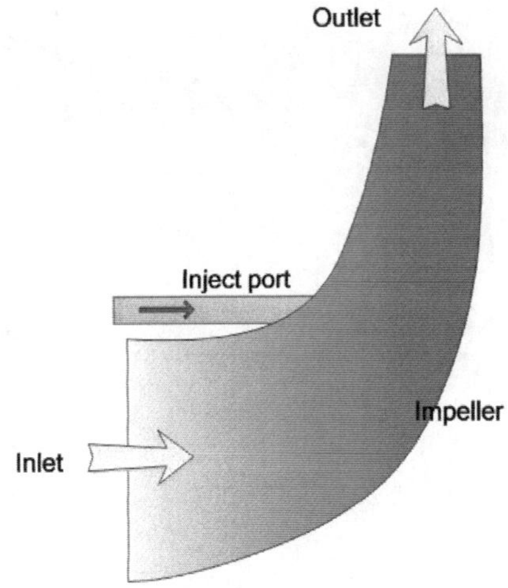

Figure 2: Schematics of ACCT

Figure 3 shows the photograph of the test system. In this picture, all the details of the test system can be seen, such as gas inject pipe (red pipes), compressor and driven system, and many sensors installed in the test rig to monitor the working conditions of the whole system.

Figure 3: Photograph of the test system

3. SURGE TEST

To capture the surge working condition accurately, the surge of the compressor was test firstly. When the compressor is operating at a speed, the adjusted valve in the discharge pipe is closed by degrees. The pressures of inlet and outlet pipe and the vibration of inlet pipe are test during this process.

Figure 4-6 shows the fast Fourier transform (FFT) results of vibration signals of inlet pipe. The signals were gotten in the normal and surge working conditions. The results also compared in the figures.

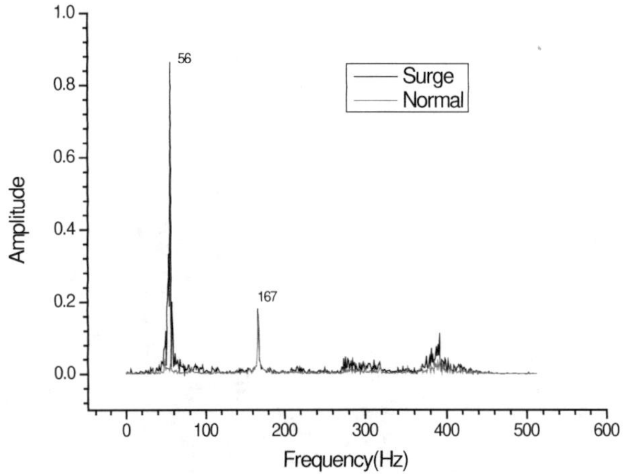

Figure 4: Vibration test of inlet pipe at 10000rpm (166.7Hz)

From the figures it can be seen that the maximum amplitude is at the working frequency of the centrifugal compressor when the compressor operates in the normal working conditions, such as 166.7Hz (10000rpm), 200Hz (12000rpm), and 233.3Hz (14000 rpm). When the compressor operates in the surge working conditions, the amplitudes at working frequency are almost same as that in the normal working conditions. However, the maximum amplitude in the surge working conditions is at the lower frequency and it is about four times of that at working frequency. For different working speed, the maximum amplitude is at 56 Hz. This may relates to the character of the pipe system.

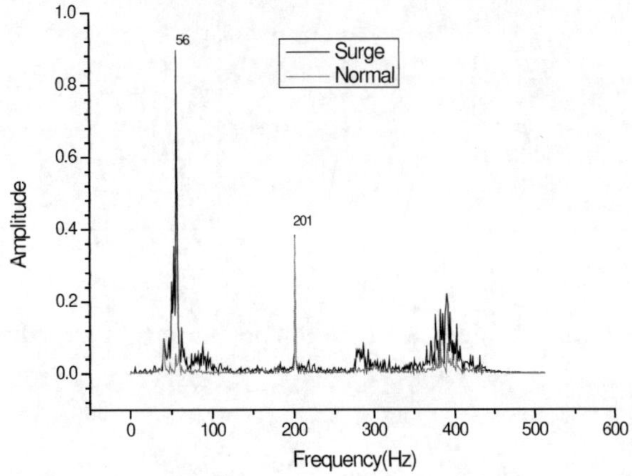

Figure 5: Vibration test of inlet pipe at 12000rpm (200Hz)

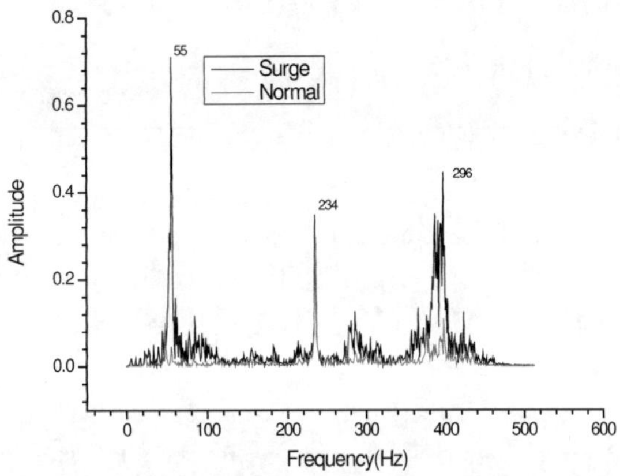

Figure 6: Vibration test of inlet pipe at 14000rpm (233.3Hz)

Figure 7-8 show the suction and discharge pressure and the FFT analysis when the compressor operates in the surge working conditions. It can be seen that when the compressor is operating at the surge condition, the pulse of pressure is increasing. Moreover, the pulse of the discharge pressure is bigger than that of suction pressure. The suction pressure is even smaller than the atmospheric pressure at some moments. From the FFT analysis, it can be seen that the frequency of the pressure pulse is 6Hz, which is about 0.03 of the rotating speed frequency.

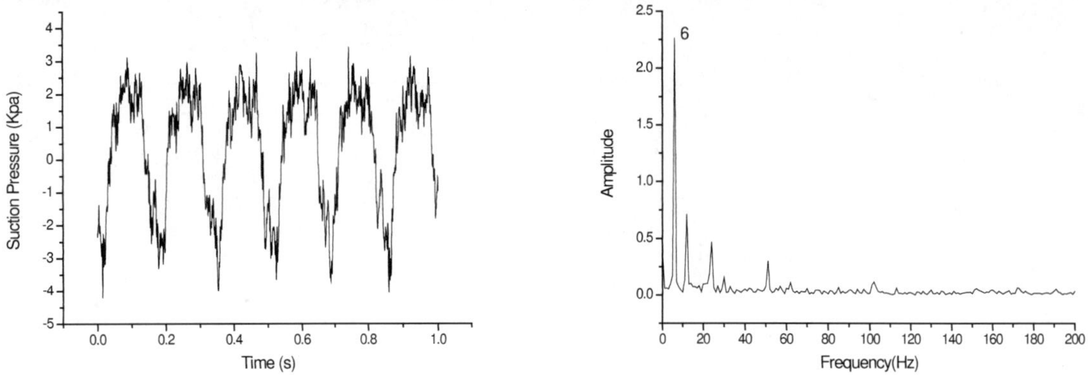

Figure 7: Suction pressure of the compressor in surge condition

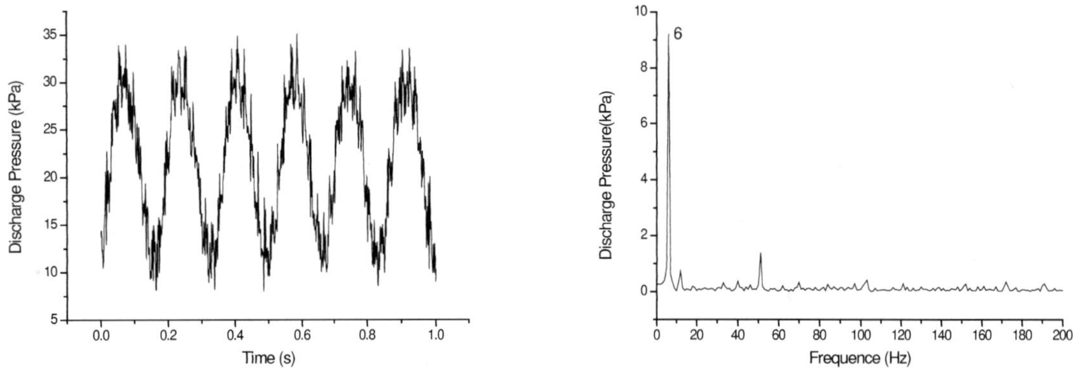

Figure 8: Discharge pressure of the compressor in surge condition

4. STABILITY ENHANCEMENT

Figure 9 shows the performance test results of the centrifugal compressor. The centrifugal compressor was tested under three rotating speeds. The flow rate of the injecting gas by ACCT system is 2.5 m^3/min. During the test process, the rotating speed is changed at one fixed value (about 10000, 12000, 13000 rpm) using the frequency converter. Then the control valve is closed gradually by the data acquisition and control system. The relationship between the discharge pressure and volume flow rate can be gotten. When the ACCT system operates, the test process is repeated. The difference of these two curves is the effect of ACCT on the performance of test centrifugal compressor.

The compressor performance is almost not being changed by using the ACCT system in the normal working conditions. That means when the flow rate of compressor is bigger than the smallest flow rate, the differences of performance of compressor are very little for ACCT condition and traditional condition. The effect of ACCT on the compressor in the normal working condition can be ignoring.

However, the smallest flow rate at the surge condition obviously decreases when the ACCT system is used. The black solid line is the test surge line when the compressor is the original (traditional) structure. The red dot line is the surge line when ACCT system is used. That means the surge line of the test centrifugal compressor is moved from the right of the compressor performance map to the left. The stability of the compressor under the small flow rate is enhanced.

The test results show the improvement of surge by ACCT is increasing when the compressor speed decrease. When the flow rate of injecting gas is fixed, the decrease of compressor speed means the proportion of injecting gas on the

compressed gas by the centrifugal compressor is increasing. That is the main reason for the difference of affection of ACCT on the different compressor speeds.
When the compressor is operating around the condition of surge, the flow in the impeller is unstable. There are many low speed eddies at the throat of compressor impeller. The phenomena causes the unstable operate of compressor. When the ACCT used, the high-speed gas is injected into the throat of impeller, the energy is added to the low speed eddies. The location and scale of eddies in impeller are changed by the ACCT system. The flow in compressor is improved. Hence, the operational stability of compressor is improved too. The surge of centrifugal compressor is moved from right to left when using the ACCT system.

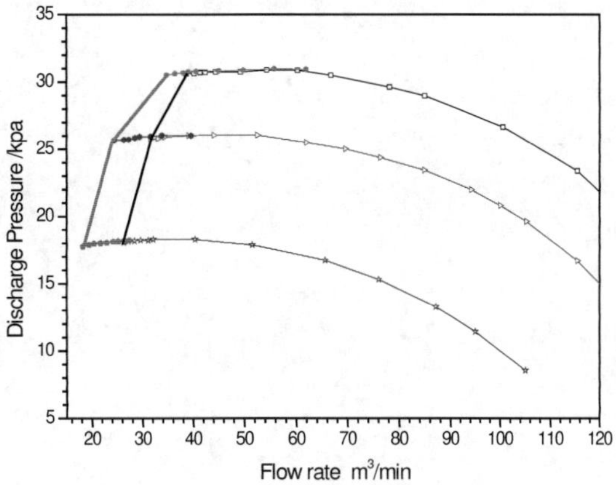

Figure 9: Test results of compressor

5. CONCLUSIONS

The surge of the centrifugal compressor is researched by experiment. The suction and discharge pressures are tested when the compressor operates in the surge working conditions. The signals of pressure are analyzed by FFT method. The pulse frequency of the pressures is very small (6Hz), and it is about 0.03 of the rotating speed frequency. At the same time, the vibration of the inlet pipes is also researched as the character of the surge of compressor. The vibration test results show that it can be as a signal to estimate the surge working condition. When the surge occurs, the vibration amplitude at the low frequency is bigger, and it is about four times of that at the working frequency. However, the maximum amplitude is not change by the rotating speeds. The effect of the ACCT system is tested on the experimental rig. The results show that the surge line can be moved from right side to the left side in the compressor performance map. That means the ACCT system can enhance the range of stable working conditions. In addition, the ACCT is a potential method to use as the anti-surge technology.

REFERENCES

Fisher, F. B. (1988). Application of Map Width Enhancement Devices to Turbocharger Compressor Stages. *SAE paper,* No. 880794.

Liang, D., Tong, W., Bo, Y., et al.(2013). Experimental investigation of the casing treatment effects on steady and transient characteristics in an industrial centrifugal compressor. *Exp. Therm Fluid Sci., 45*, 136-145.

M Coppinger, E Swain. (2000). Performance prediction of an industrial centrifugal compressor inlet guide vane system. *P I MECH ENG A-J POW, 214*, 153-164.

ACKNOWLEDGEMENT

The work was supported by the National Key Basic Research Program of China (2012CB026000), the Science and Technological Fund of Anhui Province for Outstanding Youth (1508085J05), and the science and technology plan of Anhui province (1501041128).

Performance Gain for Multiple Stage Centrifugal Compressor by Using Non-Equal Impeller Configuration

Yuanjie WU[1]*, Chris THILGES[2], Philippe GUILLEROT[3]

Ingersoll Rand - Trane
Compressor Development and Technology
3600 Pammel Creek Road; La Crosse, WI 54601 USA

[1]Email: jim.wu@irco.com, Phone: 608-787-3765
[2] Email: cthilges@trane.com, Phone: 608-787-2768
[3] Email: pguillerot@trane.com, Phone: 608-787-4050

* Corresponding Author

ABSTRACT

Multi-stage centrifugal compressors used for air-conditioning applications are traditionally designed to have near-identical impeller tip diameters. This kind of design can ensure even work delivered into refrigerant gas by each stage. This design simplification may have profound drawbacks on compressor efficiency, especially for chillers using an economizer cycle, where each stage of compression deals with an increasing mass flow rate. For those multiple stage centrifugal compressors connected with economized cycle chillers, the flow of the later stage of the compressor is merged with the flow from the economizer. Therefore, the flow rate for each stage will be much different at any given time. Non-equal impeller diameters will thus show performance advantages. A possible reason for benefit of non-equal diameters is to put more work into higher efficiency stages.

This paper will show that, by removing this design constraint and judiciously size each stage tip diameter, compressor efficiency can be improved by up to 5% or on average level around 2.4%. Analytical studies, based on Euler power, and laboratory experiments have both validated these gains. By extensively using geometric similarity, the proposed methodologies are also applied to other compressors in the same family and similar performance improvement can be achieved. Since the non-equal impeller designs extensively use the existing impeller map data its design lead time is considerably short.

1. INTRODUCTION

The head of the centrifugal compressors in comfortable air conditioning application is decided by difference of saturation pressures for both evaporator and condenser shells. The mass flow rate, or equally volumetric flow rate, is basically decided by the chiller cooling capacity. The compressor intakes refrigerant vapor that is usually without super heating from evaporator shell (Ingersoll Rand, 2008). The refrigerant isentropic enthalpy rise from evaporator to condenser is the required specific work for air conditioning process. Considering the aero performance efficiency the centrifugal compressor must deliver more work than that required one. A single stage compressor usually cannot provide enough energy and thus multiple-stage compressor is required.

The compressor configuration design is a process that uses existed compressor performance maps to determine impeller tip sizes based on customer's requirements which are weather conditions and cooling capacity. The weather conditions directly affect saturation pressures of evaporator and condenser shells, through which the minimum enthalpy rise (head) of the compressor can be calculated. The cooling capacity not only directly determines compressor flow rate, but also partially influences on the compressor enthalpy rise. Proper size for each impeller will provide required head. In order to determine impeller size one common method is to set all impellers to same tip diameters. This algorithm is easy to be implemented and widely being used in commercial air conditioner manufactures. However it is well know that, for the multiple stage centrifugal compressors, each stage possesses different head/flow and efficiency/flow maps. Impellers with same diameter are therefore not necessary to be the configuration through which the best overall chiller performance can achieve.

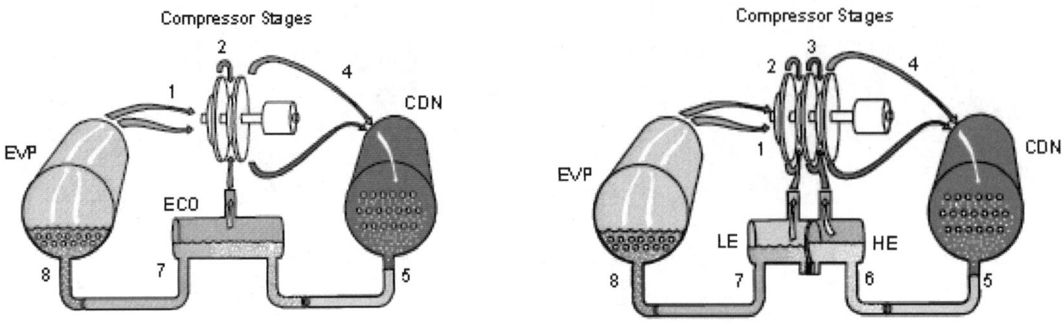

1a: Two-stage unit –one economizer   1b: Three-stage unit–two economizers

Figure 1: Design schematics for centrifugal chiller with economizers

This paper proposes a methodology of how to optimally configure impeller sizes for multiple stage centrifugal compressors to achieve better overall performance of chillers comparing with traditional equal impeller configurations. The gain can be directly reflected on the chiller manufacturing cost saving for equivalent performance. Since the non-equal impeller designs extensively use the existed impeller map data, its design lead time is also considerably shorter.

2. THEORY AND METHODOLOGY

2.1 Impeller power capability

Both head (isentropic enthalpy rise) and flow rate prescribe the minimum energy that refrigerant vapor must absorb from compressor stage. The specific work delivered from each stage to operating gas can be given by Euler turbine equation (Dixon, 1998):

$$w = U_2V_{\theta 2} - U_1V_{\theta 1} \tag{1}$$

For the chillers working at their full capacities, their inlet guide vanes (IGV) are always fully opened. Therefore the second term of the Euler equation can be ignored, which causes:

$$\mathrm{w} = \mathrm{U_2V_{\theta 2}} \propto \mathrm{U_2^2} \propto \mathrm{N^2D_2^2} \tag{2}$$

Where D2 is the impeller tip diameter. For multiple stage compressors the specific work is proportional to the summation of impeller tip diameter square:

$$w \propto N^2 \sum D_{2i}^2 \tag{3}$$

The above relationships ignore a few impacts, such as the slip influence, but are good enough to describe the application. If the compressor is the fixed speed directly driving, the shaft speed can be even omitted and get

$$w \propto \sum D_{2i}^2 \tag{3'}$$

2.2 Impeller size determination

The performance of chillers with multiple stage centrifugal compressor can be improved by 5-10% if proper designed economizer tank is hooked between compressor stages. Figure 1 below shows centrifugal chiller design schematics (Wu and Thilges, 2014). The compressor aero performance maps include head coefficient-Q/N map and adiabatic efficiency-Q/N map for several selected impeller tip sizes of each stage. Those maps were tested and saved in database. Maps for any size impeller can be interpolate or extrapolate from those tested maps.

The compressor configuration includes deciding impeller sizes for each stage based on customer's weather condition and required cooling capacity. After impeller sizes is decided, the compressor performance, chiller performance and chiller cost can thus be calculated from the tested maps.

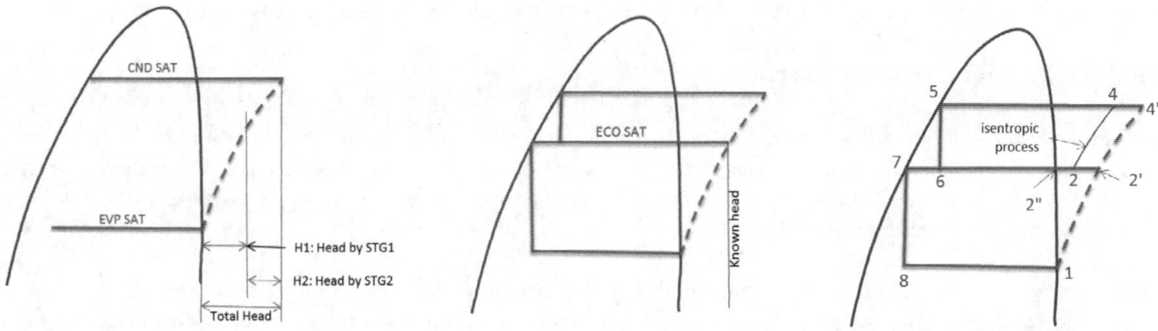

2a: Head distribute 2b: Economizer pressure calculation 2c: Mass flow calculation

Figure 2: P-h Diagram for impeller configuration procedure

Figure 2a-2c are shown the configuration principle for 2-stage centrifugal chiller, whose principle is same as that of other compressors whose stage is more than 2. The principle is illustrated on refrigerant p-h chart. The customer's weather condition can be translated into saturation pressures of evaporator and condenser shells which are compressor suction and discharge pressures, respectively. Thus the isentropic enthalpy rise (head) of entire compressor can be calculated, as shown in Figure 2a. The next step is to distribute the head to each stage. The economizer saturation pressure can be calculated based on stage enthalpy rise and isentropic process, as shown in Figure 2b. The last step of the configuration is to calculate the mass flow rate and real head for each stage. As shown in Figure 2c, known h5=h6, h7=h8, and h1 and h2", the mass flow rate for the first stage of the compressor is

$$\dot{m}_1 = cooling\ capacity/(h_1 - h_8) \tag{4}$$

The economizer mass flow

$$\dot{m}_{ec} = \dot{m}_1 \frac{h_5 - h_7}{h_{2"} - h_5} \tag{5}$$

The next stage mass flow:

$$\dot{m}_2 = \dot{m}_{ec} + \dot{m}_1 \tag{6}$$

Using the mass flow relations, i.e. equation (5) and (6), the enthalpy at the inlet of the next stage is

$$h_2 = \frac{(h_5 - h_7)h_{2"} + (h_{2"} - h_5)h_{2'}}{h_{2"} - h_7} \tag{7}$$

Where the h2' can be found from entropy s1 and economizer pressure, see Figure 2b. Entropy s2 can be found from h2 and economizer pressure. Enthalpy h4 can be found from s2 and condenser pressure, see Figure 2a. Therefore the real head for each stage is

$$Head1 = h_{2'} - h_1 \quad (8a)$$

$$Head2 = h_4 - h_2 \quad (8b)$$

Both head and flow together can be used to access the pre-tested stage performance maps to retrieve the stage efficiency to calculate chiller overall performance.

From the above discussion one can find out that different head distribution strategy results in different stage efficiency and thus different chiller performance which will result in different manufacturing cost for equivalent performance. The traditional way is to evenly distribute the total head to each stage which can get same impeller tip diameter for each stage. However this plan cannot always ensure the compressor stage to be at the optimal efficiency. Or there may exist other plans through which the better compressor efficiency can be retrieved comparing to the equal distribution plan that will be taken as the baseline in following discussions.

2.3 Data mining and Cartesian sweep

The amount of energy that impeller should deliver depends upon mass flow rate, specific work and stage efficiency. As discussed above the specific work is related to the summation of square of impeller size. Mass flow rate that mainly decided by the cooling capacity is also affect the impeller size. Stage efficiency with respect to mass flow rate is the character of each compressor stage.

Cartesian sweep is used to enumerate and simulate all possible combinations of variables { head, flow rate, impeller sizes} over pre-existed compressor performance maps. The simulation returns chiller's overall performance for each combination, either based on full load or part load. Further, for a given {head, flow rate}, there always exists a group of impellers through which the overall performance is the best. Record down this combination: {given {head, flow rate}, best impeller size}. Apparently from this record the diameter ratios, R1=D1/D2 and R2=D2/D3 (for 3-stage compressor only) can be simply calculated. For same given {head, flow rate} the equal impeller configuration, {given {head, flow rate}, equal impeller size} which is the baseline, is recorded as well. Since the summation of the square of impeller size can be related to the {head, flow rate}, one can create a relation from the baseline to optimized one by using data fitting method. Similarly relations from baseline summation of impeller size square to diameter ratios can be created too, as shown in Figure 3 which shows typical 2-stage (Figure3a) and 3-stage (Figure 3b) compressor configuration optimization models based on Cartesian sweeps.

Under the premise that the equal impeller configuration, or baseline, is known, the summation of square of impeller size and diameter ratios can be calculated through their data fitting formula indicated in the Figure 3. Therefore all impeller sizes can be calculated and the performance thus is able to be simulated.

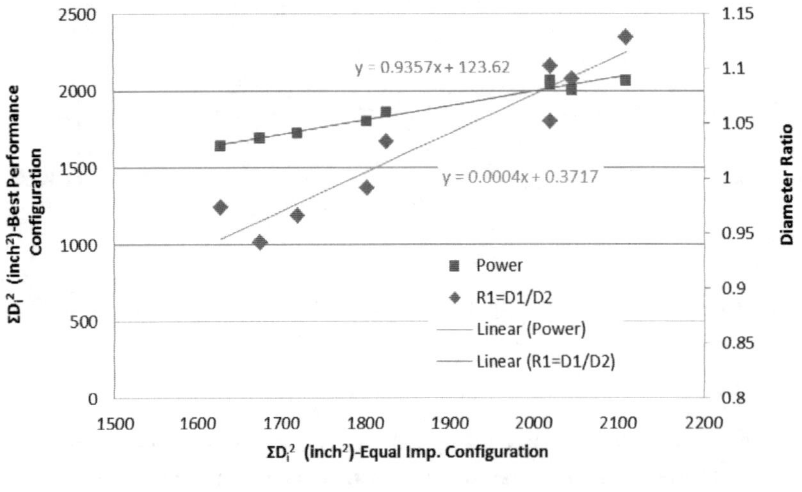

3a: Models for 2-stage compressor

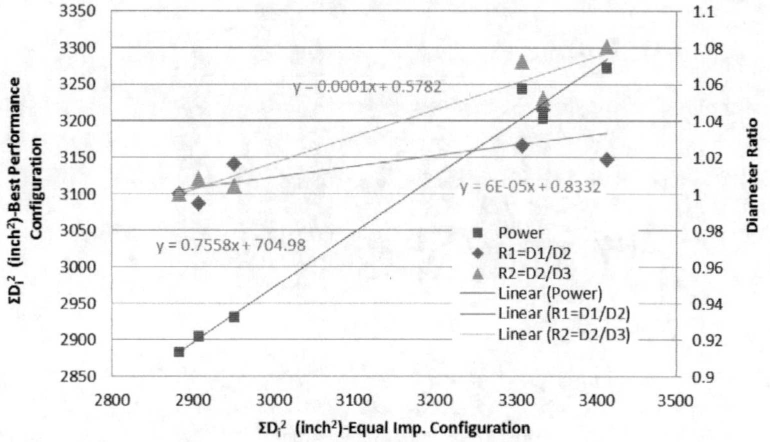

3b: Models for 3-stage compressor

Figure 3: Models for configuration optimization based on Cartesian sweeps

3. RESULTS AND DISCUSSIONS

The non-equal impeller configuration method described in previous sections are successfully applied to a few centrifugal chillers that work with refrigerant R1233zd(E) and CTV centrifugal chillers that work with refrigerant R123. Its algorithm has been implemented into the Ingersoll Rand-Trane's commercial ECTV configuration system. Comparing to the traditional equal impeller configuration, a typical 4% manufacturing cost can be saved due to the improved performance.

Figure 4, 5 and 6 show the performance gain of non-equal impeller configuration for selected ECTV centrifugal chillers. They are CVHH 1700, chiller with 2-stage centrifugal compressor, shown in Figure 4; CVHH 950 chiller with 3-stage centrifugal compressor, shown in Figure 5; CVHH 900 chiller with 2-stage centrifugal compressor that is driven in both fixed speed and variable speed modes, shown in Figure 6. The target of the configuration optimization can be either full load performance, chiller overall kW/ton, or part load performance that is based on weighted average at selected evaluating load percentages prescribed in ARI standard 550/590. In most cases significant part load performance improvement can also be gained even the optimization target is set to be the full load performance, and vice versa.

Generally when the fixed speed driving chiller works at high capacity and high head conditions, it requires high impellers, under which the non-equal impeller configuration shows more performance improvements optimized on both full load and part load, for both 2-stage compressors chiller, Figure 4a, 4b , 6a and 6b, and 3-stage ones, Figure 5a, 5b.

The performance improvements for the fixed speed 2-stage chillers are much higher than the 3-stage ones. The Figure 4a shows the average full load performance improvement is 4.7% for high impeller, 2.1% for medium impeller 0.5% foe low impeller. However, for 3-stage chiller shown in Figure 5a those numbers are 1.8%, 1.3%, and negligible, respectively. Comparing to Figure 4b and 5b, similar conclusions can be obtained. There may existed a few reasons. One of them is that the impeller size difference obtained through optimization calculation for 3-stage compressor is bigger than that of the 2-stage compressors. Therefore 3-stage optimization is easily suffer to the restrictions of mechanical structure designs, such as rotor dynamics limits, and the configuration system had to discard optimized configuration but picked up other configurations. This seems to provide some hints of cooperation between designers in different area.

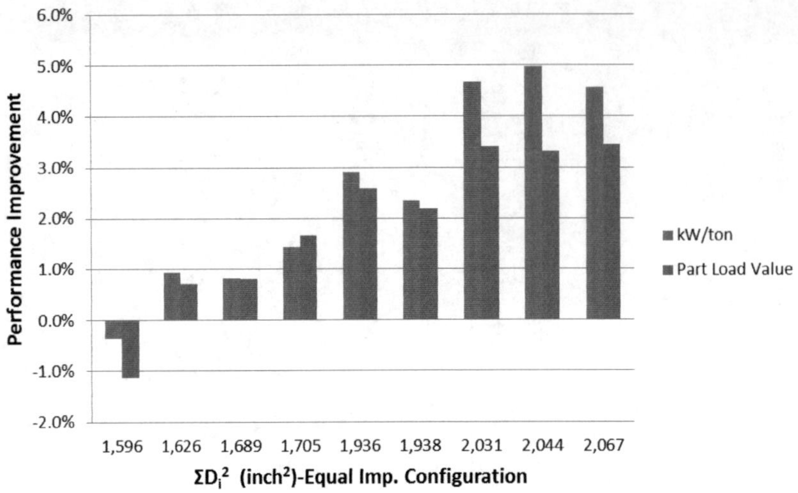

4a: Optimized to full load performance

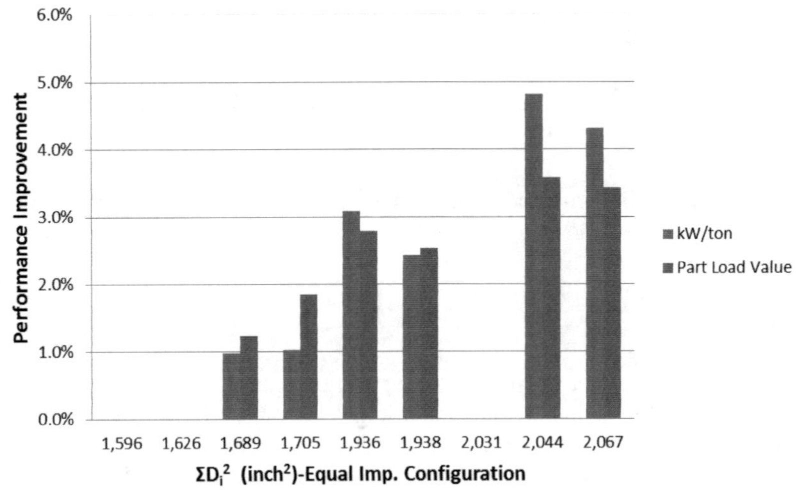

4b: Optimized to part load performance

Figure 4: Optimization results for 2-stage compressor, CVHH 1700.

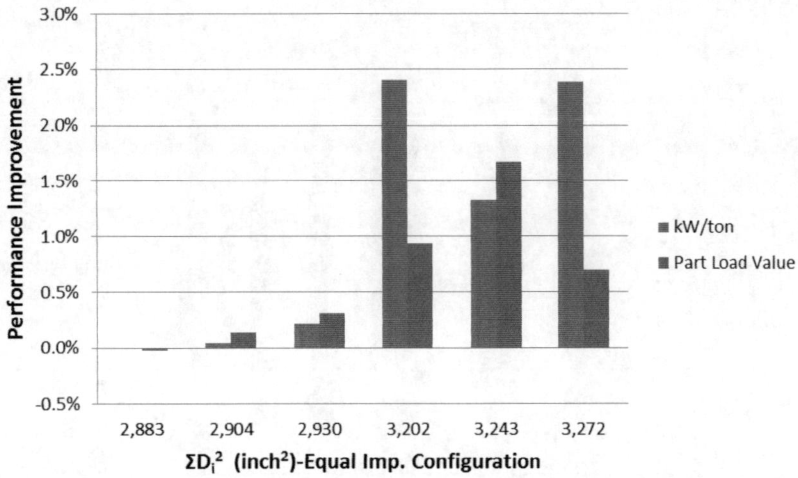

5a: Optimized to full load performance

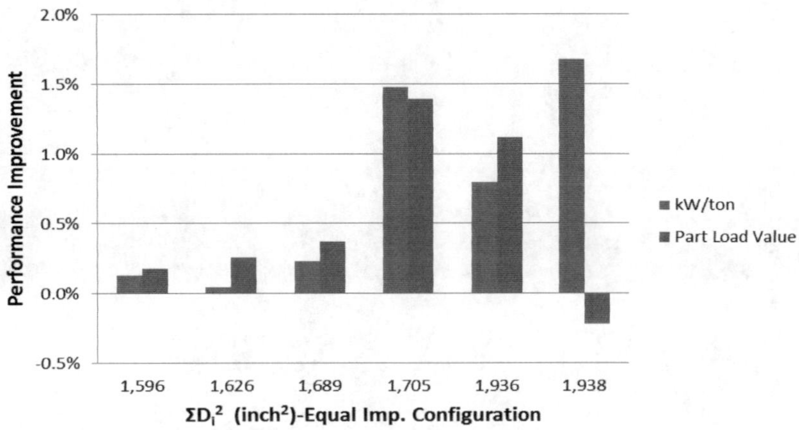

5b: Optimized to part load performance

Figure 5: Optimization results for 3-stage compressor, CVHH 950.

Figure 6 compares the improvements for fixed speed and variable speed driving of 2-stage compressor chiller. For the fixed driving, the pattern of the performance improvement is similar to those for CVHH 1700 and CVHH 950 described above. However the substantial benefit from the variable speed driving compressor is on the part load performance improvement. The benefit exists on two aspects. First it is spread to entire range, i.e. from low impeller to high impeller, rather than only on medium and high impeller, which is typical in fixed speed driving. The second the magnitude of the part load performance improvement is munificently greater than that of in fixed speed driving. The reason is perceptible that is variable speed driving provides more optimization manipulation space through which better optimization may be reached.

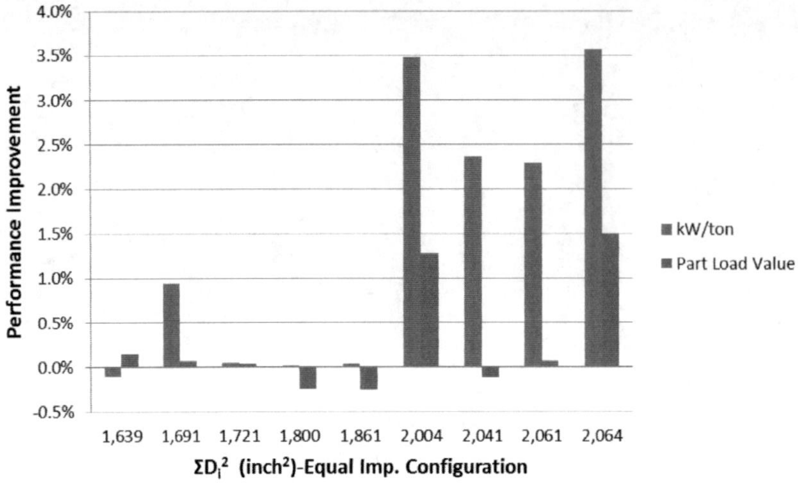

6a: Fixed speed and optimized to full load performance

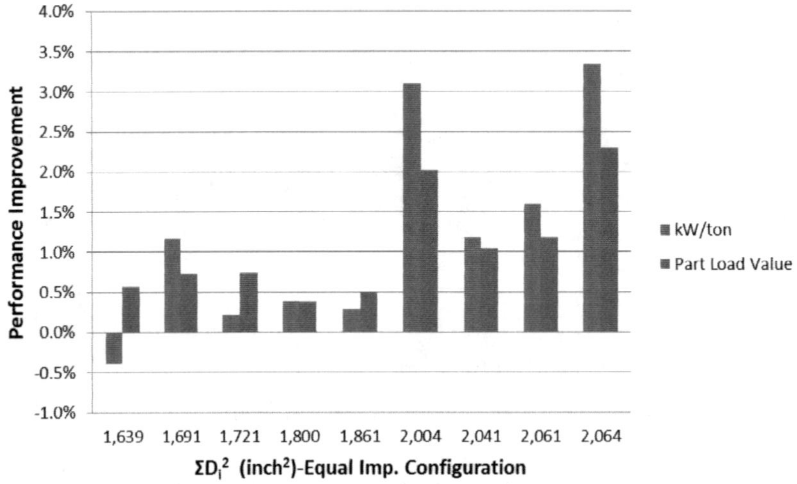

6b: Fixed speed and optimized to part load performance

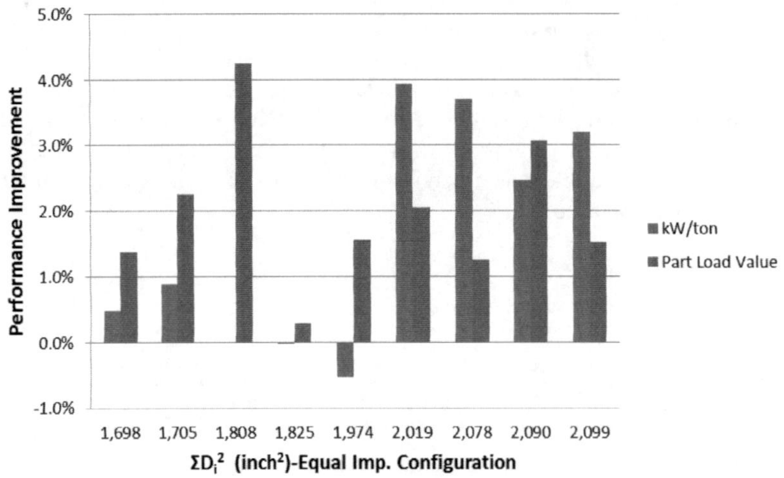

6c: Variable speed and optimized to full load performance

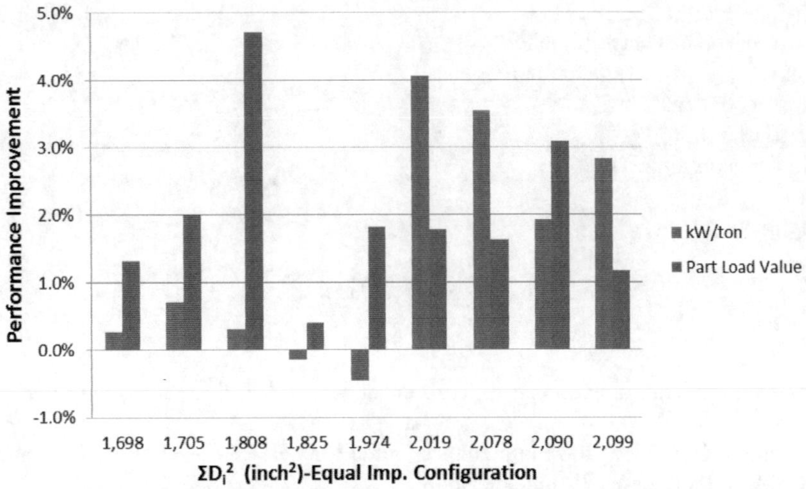

6d: Variable speed and optimized to part load performance.

Figure 6: Optimization results for both fixed and variable speed compressors

4. CONCLUSIONS

Comparing to the equal impeller configuration, the non-equal impeller configuration can optimally allocate the head duty for each stage of a centrifugal compressor, which reveals a feasible way to save manufacturing cost without deteriorating performance.

Impeller configuration optimization can be either on full load performance or part load performance. In most cases they are consistent, i.e. the part load performance improvement can also be reached even the optimization target is full load performance, and vice versa.

For the fixed speed driving compressors, the improvements for both full load performance and part load performance are mainly on high and medium impellers. However, the variable speed driving compressors, the improvements of part load performance can be achieved through entire range, i.e. from low impeller to high impeller. Moreover the magnitude of the improvement is bigger than that of in fixed speed driving compressor.

NOMENCLATURE

D	Impeller blade tip diameter	(inch)
$\dot{m}$	Mass flow rate	(lbm/sec)
N	Impeller shaft speed	(RPM)
U	Impeller tangential velocity	(foot/sec)
V	Impeller absolute velocity	(foot/sec)
w	Specific work	($foot^2/sec^2$)
h	Enthalpy	(BTU/lbm)

Subscript

1	Location of centrifugal compressor stage inlet Impeller inlet
2	Location of centrifugal compressor stage exit; Impeller exit; The 2nd stage impeller inlet
2'	Isentropic exit of the 1st stage impeller

2"	Gas status of the economizer
4	Isentropic exit of the last stage impeller
4'	Isentropic exit of the centrifugal compressor
5	Liquid status of condenser
6	Economizer flash point
7	Liquid status of economizer
8	Evaporator flash point
θ	Tangential direction

REFERENCES

Dixon, S.L. Fluid Mechanics , Thermodynamics of Turbomachinery, 4th Edition Pergamon Press Ltd 1998.

Ingersoll Rand, Installation, Operation, and Maintenance Guide for CVHE, CVHF, CVHG Water-Cooled CenTraVac™ Chillers, La Crosse Wi, January 2008.

Wu, Yuanjie, and Thilges, Chris, Centrifugal Compressor Performance Deviations with Various Refrigerants, Impeller Sizes and Shaft Speeds, *Proceedings of International Compressor Engineering Conference*, Purdue University, West Lafayette, IN, USA – July 2014.

Design of Oil-Free Turbocompressors for a Two-Stage Industrial Heat Pump under Variable Operating Conditions

Adeel JAVED[1*], Cordin ARPAGAUS[2], Stefan BERTSCH[2], Jürg SCHIFFMANN[1],

[1] Laboratory for Applied Mechanical Design (LAMD)
École Polytechnique Fédérale de Lausanne (EPFL)
Rue de la Maladière 71b
Neuchâtel 2002, Switzerland

[2] NTB University of Applied Sciences and Technology Buchs
Institute of Energy Systems
Werdenbergstrasse 4
9471 Buchs, Switzerland

* Corresponding Author, E-mail: adeel.javed@epfl.ch

ABSTRACT

A pair of mechanically driven small-scale turbocompressors running on gas lubricated bearings have been designed for a two-stage heat pump application functioning under variable operating conditions. Novelty in the present two-stage heat pump system lies in the application of oil-free turbocompressor technology and the introduction of unused secondary heat from various sources. Managing the operational deviations and the secondary heat during off-design heat pump operation is challenging for the turbocompressors. The turbocompressors can potentially exceed their operating range defined by the surge and choke margins, and the maximum rotational speed limit set by the structural and rotordynamic considerations. A wide operating range is, therefore, a prerequisite design condition for the turbocompressors. The present paper will guide the readers through different stages of the design process of such turbocompressors subjected to various operational and design constraints. Moreover, a stochastic evaluation on the influence of variable operating conditions on the heat pump off-design performance will be detailed.

1. INTRODUCTION

1.1 Motivation

Heat pumps have been identified as a key technology for reducing the exergy losses as compared to conventional boiler systems (Favrat et al., 2008). One of the significant developments has been the realization of multi-stage heat pumps (Favrat et al., 1997; Zehnder, 2004). Studies have shown that domestic-scale multi-stage heat pumps are feasible and achieve higher coefficient of performance (COP) than single-stage types, but suffer from oil migration issues. Subsequently, a comprehensive design and experimental investigation of an oil-free gas bearing supported turbocompressor unit for a domestic two-stage heat pump has been made (Schiffmann & Favrat, 2009; Schiffmann, 2013, 2015). The turbocompressor has seen application as a compressor-turbine unit in the experimental investigation of a thermally driven heat pump cycle with promising results (Demierre et al., 2014).

In this paper, a 6.5kW two-stage heat pump concept driven by oil-free turbocompressors with two heat sources at different temperature levels is proposed for decentralized domestic application (Granwehr & Bertsch, 2012; Arpagaus et al., 2016). A secondary heat from different unused sources such as waste heat, liquid from cooling circuits or process steam has been added to the thermodynamic cycle to increase the COP. Figure 1 illustrates the thermodynamic cycle of the heat pump concept along with various operating conditions and their expected deviations. As the heat pump shifts to an off-design operating condition, the turbocompressors can potentially surge or choke depending on the magnitude of deviations and the secondary heat. Hence, miniature turbocompressors designed for wide operating range satisfying the manufacturing and assembly constraints are necessary for this application.

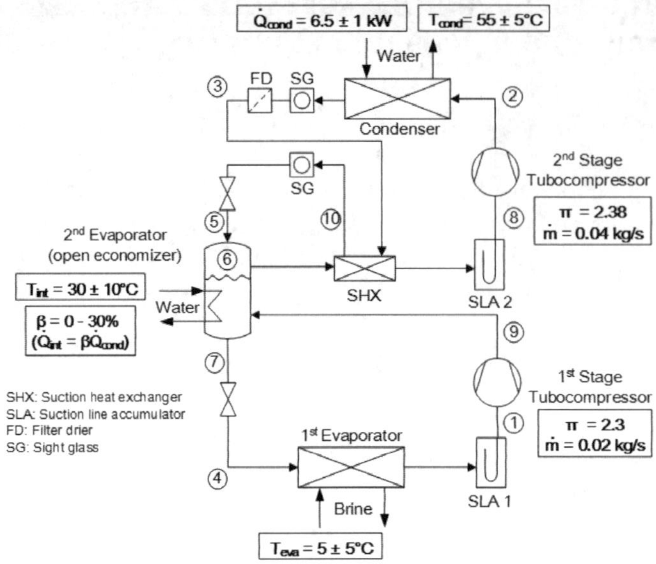

Figure 1: Schematic diagram of the two-stage heat pump cycle

1.2 Goals and Objectives

The goal of this study is to achieve viable turbocompressor designs by using a mix of 1D and 3D design methods. The main objectives are: (1) define the compressor specifications, (2) assess the effects of operational uncertainties on the heat pump performance, and (3) devise unified design guidelines for optimum impeller blading design.

1.3 Scope of the Paper

A comprehensive methodology has been applied to design the turbocompressors. An in-house Matlab based two-stage heat pump model has been used to calculate the steady-state system performance for the given operating conditions with 20% β secondary heat. The model calls the real gas refrigerant database Refprop (Lemmon et al., 2002) to estimate the R134a refrigerant fluid properties. A 1D centrifugal compressor design has been made based on the heat pump requirements. The 1D design specifies the preliminary geometry and performance maps of the two turbocompressor stages. The 1D maps are then integrated in the heat pump model to simulate off-design heat pump operation. A stochastic evaluation using the Monte Carlo method has been made to estimate the heat pump performance statistics under operational deviations. Finally, a 3D impeller blading optimization is performed using comprehensive CFD modeling. The influence of blading on the compressor fluid physics is studied and a design guideline is presented.

2. 1D TURBOCOMPRESSOR DESIGN

The turbocompressor design process begins with 1D meanline modeling of different components as illustrated in Fig 2. A commercial 1D model COMPAL ® from Concepts NREC has been used. The model is based on the two-zone methodology (Dean, 1974; Japikse, 1985, 1996). The design process begins at the inducer (station 1), proceeds to the exducer with primary zone, secondary zone and mixed-out state calculations, and then to the stationary components comprising a vaneless diffuser (station 5) and the discharge volute (station 6 and 7).

2.1 Design Selections

A set of inputs is required during the design process. Selection of these inputs is based on designer's experience, performance requirements and any design constraints.

Rotational speed selection: The $n_s d_s$ diagram for single stage compressors (Balje, 1981) has been used to obtain a conservative estimate of the turbocompressor rotational speed at design point. The maximum rotational speed and transmission torque are limited to 280,000 rpm and 45 mNm, respectively. For the first stage turbocompressor, n_s = 0.62

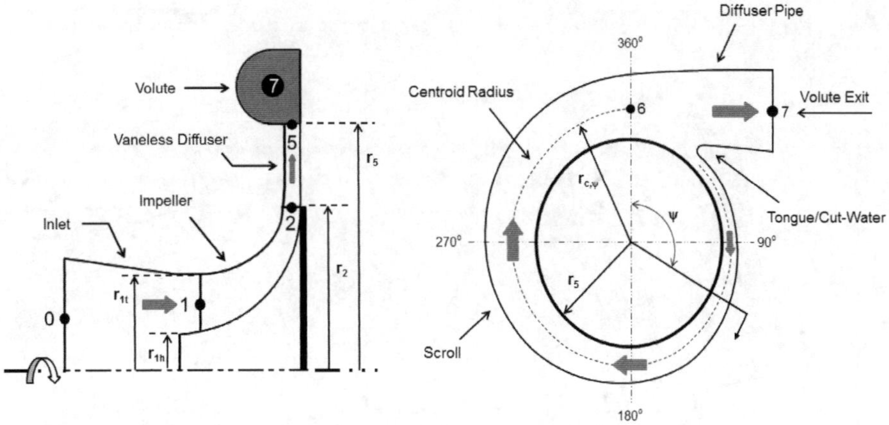

Figure 2: The centrifugal compressor stage and volute with station numbers and main dimensions

has been selected, which complies to the 80% compressor efficiency island for a radial-type compressor. This results in an impeller with 15.2 mm tip diameter with design rotational speed of approximately 227,000 rpm at a torque of 21 mNm. For the second stage turbocompressor, a small trade-off has been made by selecting a comparatively lower specific speed n_s = 0.57 to gain more maximum speed margin. Consequently, an impeller with 14.8 mm tip diameter rotating at approximately 235,000 rpm at a torque of 39 mNm is obtained from 1D calculation at design point.

Inducer design: The impeller will be mounted on a separate titanium shaft fused to the main rotor. Since the shaft can be designed independently of the main rotor and the torque transmission is well within the torsional rigidity limit, r_{1h}= 1.5 mm is selected for both turbocompressors following a parametric study.

Blading and tip-clearance: A total of 16 blades have been specified for both turbocompressor impellers with half of the blades allocated to splitter blades. A blade thickness t_b= 0.25 mm is selected considering manufacturability with aluminum. The tip-clearance t_{clr}= 80 μm is defined initially based on the manufacturing and assembly limitations, and bearing clearances with an intention to reduce it in subsequent evaluations.

Exducer design: The two-zone modeling is case-specific in terms of computation of the mass flow distribution between the primary and secondary flow zones represented by the secondary zone mass flux fraction χ (where $\chi = \dot{m}_{sec}/\dot{m}$). Since data is not available for small-scale impellers in the range of 15 mm tip diameter, a value χ = 0.2 has been specified for the 1D designs. The use of blade backsweep β_{2b} extends the compressor operating range, improves impeller exit flow uniformity, and increases the overall pressure recovery in the compressor (Japikse, 1996; Cumpsty, 2004; Hildebrandt & Genrup, 2007). For the turbocompressors, β_{2b} = 45° has been selected. The exducer depth or height b_2 has been defined by the exit swirl parameter $\lambda_{2m} = c_{\theta 2m}/c_{m2m}$ (also a measure of the absolute flow angle $\alpha_{2m} = tan^{-1}\lambda_{2m}$). Following the guideline provided in Japikse (1996), λ_{2m} = 4.0 (α_{2m} = 76°) and λ_{2m} = 4.2 (α_{2m} = 76.6°) have been specified for the first and second stage turbocompressors, respectively. Finally, the diffusion properties in the primary flow zone are estimated from the two-elements-in-series model (TEIS) by defining the effectiveness parameters η_a= 0.4 and η_b= 0.2 suggested for small-scale impellers (Japikse, 1984, 1996).

Vaneless diffuser design: Vaneless diffusers provide a wide operating range. For the impellers having design point exit swirl parameter λ_{2m} in the range of 4.0 ($\alpha_{2m} \approx 76°$), the diffuser is likely to experience a rotating stall triggered by local flow reversal. For wide-range and peak efficiency at design point, λ_{2m} should ideally be approximately 2.0–2.5 (α_{2m} between 63° and 68°) (Japikse, 1996; Jansen, 1964; Johnston & Dean, 1966; Rodgers & Sapiro, 1972). Hence, a frontal pinch is applied to reduce the diffuser swirl to a more desirable value. For the first stage turbocompressor, the front pinch reduces the exit height of the vaneless diffuser b_5 to 0.65 times the impeller exit height b_2 (b_5/b_2 = 0.65) resulting in λ_5 = 2.25 (α_5 = 66°) at diffuser exit. For the second stage turbocompressor, the front pinch reduces the exit height of the vaneless diffuser b_5 to 0.8 times the impeller exit height b_2 (b_5/b_2 = 0.8) resulting in λ_5 = 2.9 (α_5 = 71°) at diffuser exit. Both diffusers have the exit radius r_5 located at 1.65 times the impeller exit radius r_2 (r_5/r_2 = 1.65), while the pinch is located at about 1.1 times the impeller exit radius r_2 (r_{pinch}/r_2 = 1.1).

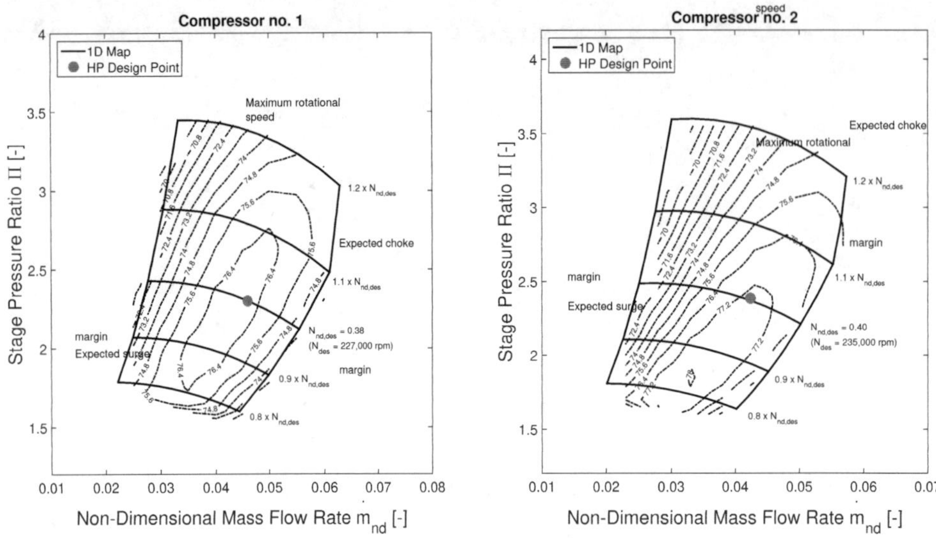

Figure 3: 1D compressor maps for first and second stage turbocompressors with isentropic efficiency contours

Volute design: An external overhung volute design with machined tunnel-type cross-sections has been opted for the two turbocompressors. The cross-sectional areas A_ψ at different azimuth angles ψ have been sized following a simple correlation derived from the principles of angular momentum and mass conservation as $A_\psi = \psi \times SP \times \frac{r_{c,\psi} b_5}{\lambda_5}$. Here SP represents the sizing parameter used to scale the volute through an iterative procedure, $SP = 0.85$ has been found to be an appropriate scale for both turbocompressors to maintain almost constant $c_{\theta 5} \approx c_6$ at design point.

2.2 1D Turbocompressor Performance

Following the extensive design exercise, 1D designs are obtained for the two turbocompressor stages. Table 1 compiles the main compressor geometric and the design point performance data. Considering the geometry, the two turbocompressors will be one of the smallest of their kind with tip diamaters d_2 in order of 15 mm once manufactured and tested. Despite their small size and relatively large tip-clearance (relative tip-clearance $\varepsilon = t_{clr}/b_2$ is approximately 13–15%), the compressors are able to achieve a healthy isentropic efficiency in the range of 77% at design point. Off-design performance is equally important for the turbocompressors. Therefore, non-dimensionalized 1D performance maps have been obtained, as shown in Fig. 3. By converting the compressor performance to non-dimensional rotational speed $N_{nd} = Nd_2/a_{00}$ and non-dimensional mass flow rate $\dot{m}_{nd} = \dot{m}/(\rho_{00} a_{00} d_2^2)$, the influence of varying inlet conditions with a real gas is removed (Dixon, 2005). Empirical stall checks provided in the 1D design tool for different stage components have been monitored at low mass flow rate operations. Considering the stable operating range $\mathrm{R} = \frac{\dot{m}_{choke} - \dot{m}_{surge}}{\dot{m}_{choke}}$ at design rotational speed, the two turbocompressors are in a relatively high range category with R reaching 50% for moderately low flow coefficient compressors having a design pressure ratio around 2.3–2.5 (Japikse, 1996).

2.3 Off-Design Heat Pump Performance Evaluation

The 1D compressor performance maps for the two turbocompressor stages are primarily used to interpolate of isentropic efficiencies at heat pump off-design operating conditions by using the triangulation-based scattered interpolation func-

Table 1: Geometric specifications and design point performance calculated by the 1D meanline model

Parameter	Compressor no. 1	Compressor no. 2
Inducer shroud radius, r_{1t} (mm)	3.62	3.43
Tip radius (diameter), r_2 (d_2) (mm)	7.62 (15.2)	7.4 (14.8)
Tip height, b_2 (mm)	0.62	0.53
Tip-clearance (relative), t_{clr} (ε) (mm)	0.08 (13%)	0.08 (15%)
Blade angle at inducer tip, β_{1b} (deg)	57°	57°
Backsweep angle, β_{2b} (deg)	45°	45°
Diffuser radius, r_5 (mm)	12.57	12.22
Diffuser height, b_5 (mm)	0.405	0.42

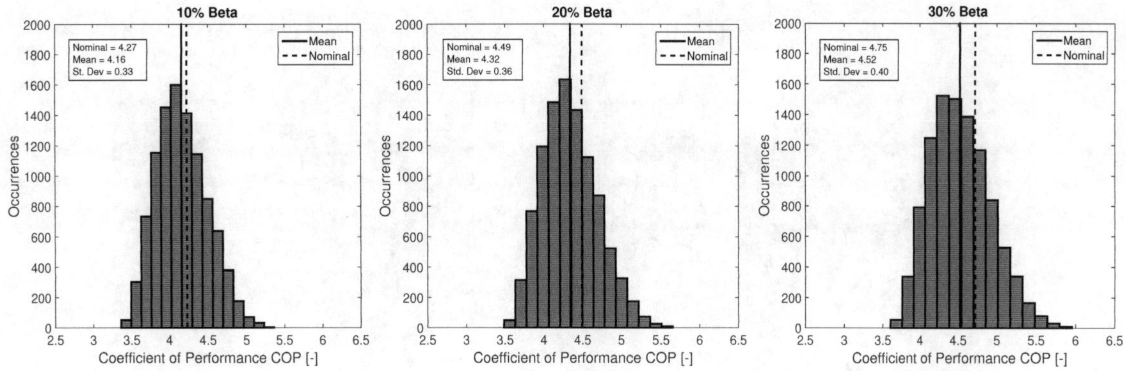

Figure 4: Heat pump performance under deviations in operating conditions

tion provided in Matlab. The method is efficient and particularly advantageous due to its extrapolation functionality (Amidror, 2002). A total of 10,000 Monte Carlo samples of uniformly distributed random input parameter variations have been simulated with the heat pump model. The heat pump input parameters and their respective deviations are summarized in Table. 2. Two assumptions have been made during the stochastic evaluation: (1) the temperature drops across source and sink fluid streams are kept constant by adjusting their respective mass flow rates for more stable regulation of pressures in the heat exchangers, and (2) the turbocompressors are able to operate beyond their expected operating ranges (data is thus extrapolated) since the actual operating ranges are not known at this point. These aspects will be studied in more detail during the experimental phase.

The off-design performance of the two-stage heat pump is statistically summarized as COP probability distributions in Fig. 4 for different β levels. The nominal COP increases by 10% from 4.27 to 4.75, while the mean COP increases by 8% from 4.16 to 4.52 with 10-30% β second source heat addition. Furthermore, the COP standard deviation increases from 0.33 to 0.40. Note that the mean COP levels are actually lower than the nominal COP levels, thereby indicating a systematic heat pump performance shift under stochastic input parameter deviations. This systematic shift also increases in magnitude with increase in β levels.

3. 3D COMPRESSOR IMPELLER DESIGN

Many important analytical and experimental studies have been made to quantify the flow field in centrifugal compressors (Johnson, 1978; Eckardt, 1975, 1976; Krain, 1981; Hirsch et al., 1996; Casey et al., 1992). According to these studies, the flow field in a centrifugal compressor stage is highly complex and subjected to local accelerations due to centrifugal forces (curvature effect) and Coriolis forces (rotational effect). The resulting pressure gradients generate secondary flow in shape of streamline vorticity along the end-walls (hub and shroud surfaces) from pressure side (PS) to the suction side (SS) due to blade-to-blade curvature, secondary flow along the blade surfaces (span-wise direction) due to meridional curvature, and the secondary flow generated by the Coriolis forces mainly in the radial parts of the impeller. The tip-leakage flow due to migration of high pressure flow from PS to SS of the blades through the tip-clearance gap supplements the secondary zone by forming a large vortex that stretches stream-wise in the blade passages causing a considerable performance loss. An appropriate 3D impeller blade design suppressing the secondary and tip-leakage flow is, therefore, very important. An observation based approach has been applied to design the turbocompressor impellers using steady-state CFD modeling with Ansys CFX. Note that the first stage turbocompressor impeller has been optimized. Similar blading is to be applied for the second stage turbocompressor impeller.

Table 2: Heat pump operating parameters and their deviations

Parameter	Nominal	Deviation
1st source temperature, T_{eva} [°C]	5	±5
2nd source temperature, T_{int} [°C]	30	±10
Sink temperature, T_{cond} [°C]	55	±5
Condenser heating capacity, $\dot{Q}_{cond}$ [kW]	6.5	±1
2nd source heat capacity, $\dot{Q}_{int}$ [kW]	10-30% β	±0.5

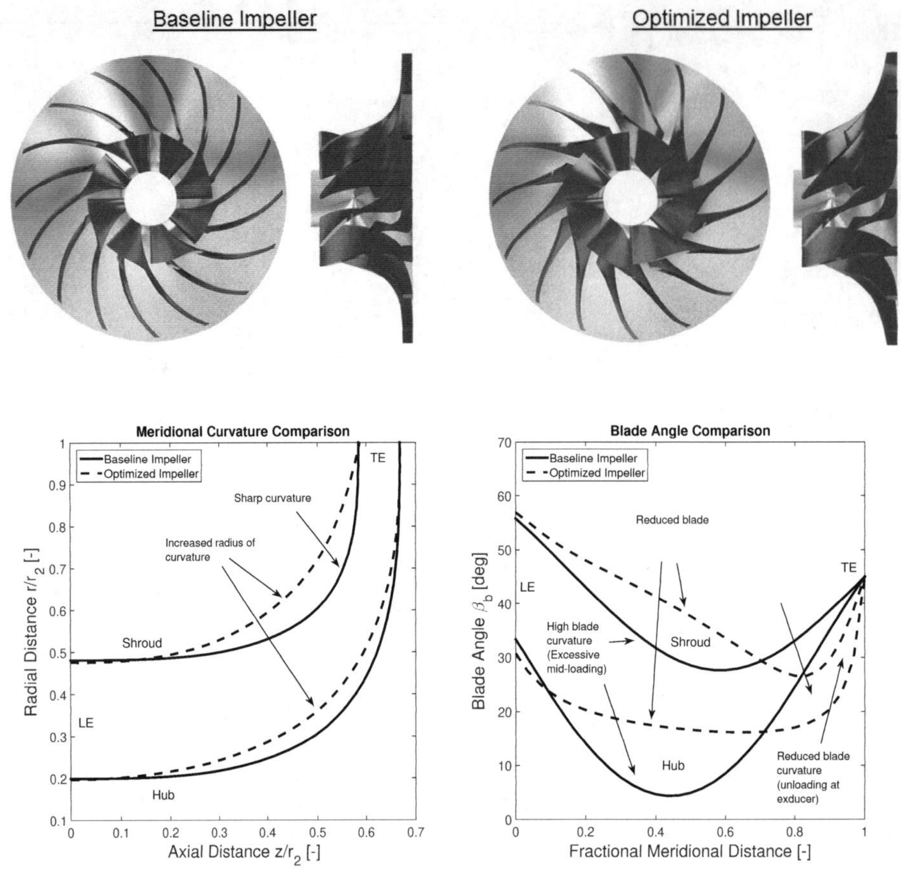

Figure 5: Comparison between the baseline and the optimized impeller blade designs

3.1 CFD Setup

The following steps have been taken to set up an appropriate CFD model of the turbocompressor stage:

3D geometric model: The 1D geometric data has been used to construct a 3D geometric model using ANSYS Blade-Gen. The model consists of a straight inlet duct, impeller and the vaneless diffuser in a single passage fluid domain. The impeller meridional hub and shroud profiles, and blade shape have been defined by Bézier curves of 5th order (six control points). The splitter blades begin from a meridional distance of approximately 40% from the leading edge (LE) and follow the blade angle and thickness distributions of the full blades till the trailing edge (TE). Note that the volute is not included in the present CFD study to reduce the computational cost.

Grid processing: A 3D structured grid has been created using the H-Grid and O-Grid topologies in ANSYS TurboGrid. The tip-clearance gap is also introduced in the grid processing tool by trimming down the impeller shroud profile. A grid independence study has been made by simulating three different grid sizes containing 0.28 million, 1.2 million and 4.0 million hexahedral elements at design point. The grid with 1.2 million elements is found to be a suitable choice by comparing compressor pressure ratio and isentropic efficiency convergence with increase in grid size.

Solver setup: Since a single passage domain simulation is considered, the symmetric surfaces are defined as periodic boundaries. The interfaces between the inlet duct, impeller and the vaneless diffuser are defined as stage/mixing plane. All the wetted surfaces have been considered as hydraulically smooth; hence surface roughness is not defined and skin friction loss is only caused by fluid viscosity. The total inlet conditions have been defined at the inlet boundary, while design mass flow rate is defined at the outlet boundary of the domain. Refrigerant fluid R134a has been set as the fluid medium. The k-ω Shear Stress Transport (SST) turbulence model (Menter, 1993) has been specified for all the CFD analyses.

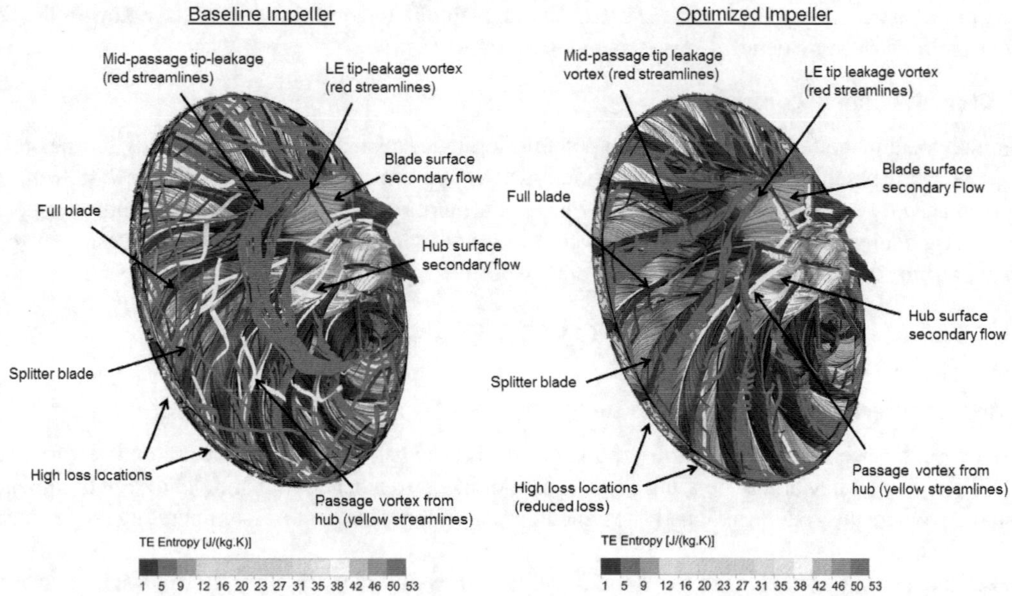

Figure 6: Comparison of secondary and tip-leakage flow between the baseline and optimized impeller designs

3.2 Methodology for optimized blade design

Although each impeller design is a new problem, a good blade passage can be designed by controlling the meridional and blade-to-blade curvatures in order to have a uniform velocity or pressure distribution in the blade passages, thus reducing the secondary and tip-leakage flows. Figure 5 compares the baseline and optimized impeller geometries in detail. Following steps have been taken to iteratively optimize the impeller design:

- The excessive meridional curvature has been reduced at the hub and shroud to have a more circular axial-to-radial bend. The small inducer hub diameter facilitates the design of a smooth circular channel.
- The mean blade loading, generally defined as blade-to-blade loading coefficient $\Lambda = \frac{w_{ss}-w_{ps}}{\bar{w}}$ (where $\bar{w} = \frac{w_{ss}+w_{ps}}{2}$), has been reduced in the middle section of the impeller passages by reducing the blade curvature (approximately 10° higher mean blade angle compared to the baseline impeller). Furthermore, the blade curvature is maximum between 60% and 80% of the passage length, which ensures an aft-loaded blading.
- The influence of Coriolis force (dominant in the radial part of the impeller) has been reduced by unloading the exducer through lowering the blade curvature from 80% of the passage length till the TE of the impeller.
- The splitter blades further helps in reducing the average loading per blade and reduce the tip-leakage flow.

The end result is a more organized impeller design with smoother blade passages compared to the baseline impeller.

3.3 Internal flow comparison

Figure 6 presents a detailed picture of the tip-leakage flow for the baseline and optimized impeller designs. Looking at the baseline impeller flow field, the blade passages are engulfed with large amount of vorticity originating mainly from the tip-leakage flow. A large tip-leakage vortex initiates from the LE shroud, which moves towards the splitter blade SS and stretches to the TE forming a large loss region indicated by the TE entropy contours. Due to high mid-loading, an excessive tip-leakage also appears and swirls in the mid-passage region. Fig. 6 also shows the end-wall and blade surface secondary flows. As expected, the secondary flow moves from the PS to SS across the hub surface (blade-to-blade) and span-wise from hub to shroud. In comparison, the optimized impeller shows an improved vorticity structure in the blade passages. Similar to the baseline impeller, the tip-leakage initiates at the LE shroud and propagates to the splitter blade SS mixing with the tip-leakage flow originating from the splitter blade. Unlike the baseline impeller, this tip-leakage vortex is more aligned with the SS of the splitter blades. Some more leakage flow is seen in the aft-sections of the blade passages. Overall, the flow field contains lesser vorticity, which results in a more uniform flow and lesser entropy production at the TE compared to the baseline impeller. The end-wall and blade surface secondary flows appear to be improved slightly. However, it is extremely challenging to control the secondary flow in the presence of large tip-leakage flow. A more detailed survey is required in this case.

3.4 Overall performance comparison

With an improvement in the flow field, the design point performance also increased significantly for the optimized impeller compared to the baseline. The pressure ratio increased by approximately 15% from 2.13 to 2.48, while isentropic efficiency increased by 3.5% points from 76% to 79.5%. Furthermore, approximately 16% higher choke mass flow rate then the 1D model prediction is expected from the optimized compressor design, thus increasing the compressor operating range R to almost 57% at the design rotational speed.

4. CONCLUSIONS

The following conclusions are drawn from this study:

- The stochastic performance evaluation exposes the difference between nominal heat pump performance and the performance under deviating operating conditions. An integrated heat pump and 1D design optimization could be studied, where the system is robustly designed for maximizing the compressor operating range and heat pump performance in general.
- Smaller inducer hub diameter improves the impeller and overall compressor performance by allowing a lower relative velocity at the tip. The blade design can also benefit from a more circular channel and thus improved passage flow.
- For a wide operating range, a careful balance between the work transfer in the impeller and vaneless diffuser performance is necessary. A pinched vaneless diffuser can improve the work transfer in the impeller, while allowing the diffuser to operate stably at off-design conditions.
- For small-scale compressors, the large tip-clearance ($\varepsilon > 5\%$) and the associated tip-leakage flow dominate the overall flow field in the impeller passages. A moderate blade loading and more aft-loaded distribution is found to be most favorable in the present case.
- Miniature inlet guide vanes and low-solidity vaned diffusers could be studied in a future project to enhance the operating range and peak efficiency of the first stage turbocompressor.

NOMENCLATURE

a	speed of sound	(m/s)
A	area	(mm^2)
b	height or width	(mm)
c	absolute velocity	(m/s)
d	diameter	(mm)
d_s	specific diameter	(-)
$\dot{m}$	mass flow rate	(kg/s)
n_s	specific speed	(-)
N	rotational speed	(rpm)
$\dot{Q}$	heat capacity	(kW)
r	radius	(mm)
R	range	(-)
SP	sizing parameter	(-)
t	thickness/tip-clearance	(mm/mm)
T	temperature	(K)
w	relative velocity	(m/s)
z	axial distance	(mm)
α	absolute angle from the meridional axis	(deg)
β	secondary heat fraction/relative angle from the meridional axis	(-/deg)
χ	secondary mass flux fraction	(-)

ε	relative tip-clearance	(-)
η	effciency	(%)
λ	swirl parameter	(-)
Λ	blade loading coefficient	(-)
ρ	density	(kg/m^3)
Π	pressure ratio	(-)
ψ	azimuthal angle	(deg)

Subscripts

b	blade
c	centroid
clr	clearance
$cond$	condenser
des	design
eva	evaporator
h	hub
int	intermediate
m	meridional/mixed-out
nd	non-dimensional
t	tip
θ	tangential

Subscript

$\bar{}$	average or mean value

REFERENCES

Amidror, I. (2002). Scattered data interpolation methods for electronic imaging systems: A survey. *Journal of Electronic Imaging*, *11(2)*, 157-176.

Arpagaus, C., Bertsch, S., Javed, A., & Schiffmann, J. (2016). Two-stage heat pump using oil-free turbocompressors–system design and simulation. In *Proceedings of 16th international refrigeration and air conditioning conference* (pp. 1–10). Purdue, USA: Paper ID 2101.

Balje, O. (1981). *Turbomachines, a guide to design, selection and theory*. John Wiley & Sons.

Casey, M., Dalbert, P., & Roth, P. (1992). The use of 3d viscous flow calculations in the design and analysis of industrial centrifugal compressors. *ASME Journal of Turbomachinery*, *114*, 27–37.

Cumpsty, N. (2004). *Compressor aerodynamics*. Krieger Drive, Malabar, Florida 32950: Krieger Publishing Company.

Dean, R. (1974). *The fluid dynamic design of advanced centrifugal compressors* (Lecture Series). Von Karman Institute for Fluid Dynamics (VKI).

Demierre, J., Favrat, D., Schiffmann, J., & Wegele, J. (2014). Experimental investigation of a thermally driven heat pump based on a double organic cycle and an oil-free compressor-turbine unit. *International Journal of Refrigeration*, *44*, 91–100. doi: 10.1016/j.ijrefrig.2014.04.024

Dixon, S. (2005). *Fluid mechanics, thermodynamics of turbomachinery*. Oxford, UK: Elsevier Inc.

Eckardt, D. (1975). Measurements in the jet-wake discharge flow of a centrifugal compressor impeller. *ASME Journal of Engineering for Power*, *97(3)*, 337–346.

Eckardt, D. (1976). Detailed flow investigations within a high speed centrifugal compressor impeller. *Journal of Fluids Engineering*, 390–402.

Favrat, D., Marechal, F., & Epelly, O. (2008). The challange of introducing an exergy indicator in a local law on energy. *Journal of Energy*, *33*, 130–136. doi: 10.1016/j.energy.2007.10.012

Favrat, D., Nidegger, D., Reymond, D., & Courtin, G. (1997). Comparison between a single-stage and a two-stage air to water domestic heat pump with on variable speed compressor. In *Proceedings of the iir conference on heat pump systems, energy efficiency and global warming.* Linz, Austria.

Granwehr, E., & Bertsch, S. (2012). *Heat pump* (Patent No. WO 2012/040864 A1).

Hildebrandt, A., & Genrup, A. (2007). Numerical investigation of the effect of different back sweep angle and exducer width on the impeller outlet flow pattern of a centrifugal compressor with vaneless diffuser. *ASME Journal of Turbomachinery*, *129*, 421–433. doi: 10.1115/1.2447873

Hirsch, C., Kang, S., & Pointel, G. (1996). A numerically supported investigation of the 3d flow in centrifugal impellers - part ii: Secondary flow structure. In *Proceedings of asme international gas turbine and aeroengine congress and exhibition.* Birmingham, UK: 96-GT-152.

Jansen, W. (1964). Rotating stall in a radial vaneless diffuser. *ASME Journal of Basic Engineering*, *119*, 750–758.

Japikse, D. (1984). A critical evaluation of stall concepts for centrifugal compressors and pumps - studies in component performance part 7. In *Proceedings of asme winter annual meeting.* New Orleans, USA: American Society of Mechanical Engineers (ASME).

Japikse, D. (1985). Assessment of single- and two-zone modeling of centrifugal compressors. studies in component performance: Part 3. In *Proceedings of gas turbine conference and exhibit.* Houston, Texas, USA.

Japikse, D. (1996). *Centrifugal compressor design and performance*. Concepts ETI.

Johnson, M. (1978). Secondary flow in rotating bends. *ASME Journal of Engineering for Power*, *100*, 553–560.

Johnston, J., & Dean, J., R.C. (1966). Losses in vaneless diffusers of centrifugal compressors and pumps. *ASME Journal of Engineering for Power*, 49–60.

Krain, H. (1981). A study on centrifugal impeller and diffuser flow. *ASME Journal of Engineering for Power*, *103*, 688–697.

Lemmon, E., McLinden, M., & Huber, M. (2002). *Nist reference fluid thermodynamic and transport properties* (NIST Standard Reference Database No. 23). National Institute of Standards and Technology.

Menter, F. R. (1993). Zonal two-equations k-ω turbulence models for aerodynamic flows. In *Proceedings of the aiaa 24th fluid dynamics conference.* Orlando, Florida, USA: AIAA 93-2906.

Rodgers, C., & Sapiro, L. (1972). Design considerations for high-pressure-ratio centrifugal compressors. In *Proceedings of asme gas turbine and fluids engineering conference.* San Francisco, USA: 72-GT-91.

Schiffmann, J. (2013). Enhanced groove geometry for herringbone grooved journal bearings. *Journal of Engineering for Gas Turbines and Power*, *135*(10), 102501-1–102501-8. doi: 10.1115/1.4025035

Schiffmann, J. (2015). Integrated design and multi-objective optimization of a single stage heat-pump turbocompressor. *Journal of Turbomachinery*, *137*, 071002-1–071002-9. doi: 10.1115/1.4029123

Schiffmann, J., & Favrat, D. (2009). Experimental investigation of a direct driven radial compressor for domestic heat pumps. *International Journal of Refrigeration*, *32*, 1918–1928. doi: 10.1016/j.ijrefig.2009.07.006

Zehnder, M. (2004). *Efficient air–water heat pumps for high temperature lift residential heating, including oil migration aspects*. École Polytechnique Fédérale de Lausanne (EPFL).

ACKNOWLEDGMENT

The authors would like to thank the Swiss Competence Center for Energy Research on Efficiency of Industrial Processes (SCCER EIP) and the Swiss Federal Office for Energy (SFOE) for their financial support, and Prof. Michael Casey and Dr. Daniel Rusch for their useful discussions and insights.

Vertical-Vibration Suppressing Design of Accumulator with New Vibration-Measuring Method

Hikaru Wada[1]*, Hideki Matsuura[2], Satoru Takanezawa[3], Ayumi Ogawa[4]

[1]*Technology and Innovation Center, Daikin Industries, Ltd.
1-1, Nishi-Hitotsuya, Settsu, Osaka, 566-8585, Japan
(Phone: +81-6-6195-7074, hikaru.wada@daikin.co.jp)

[2]Technology and Innovation Center, Daikin Industries, Ltd.
1-1, Nishi-Hitotsuya, Settsu, Osaka, 566-8585, Japan
(Phone: +81-6-6195-7074, hideki.matsuura@daikin.co.jp)

[3]Technology and Innovation Center, Daikin Industries, Ltd.
1-1, Nishi-Hitotsuya, Settsu, Osaka, 566-8585, Japan
(Phone: +81-6-6195-7074, satoru.takanezawa@daikin.co.jp)

[4]Daikin Device Suzhou Co., LTD.
256, Chang Yang Street Suzhou Industrial Park,
Suzhou, 215126, China
(Phone: +86-512-6283-8668, ayumi.ogawa@daikin.co.jp)

*Corresponding Author

ABSTRACT

A compressor was developed using R32 which was a low GWP R32 refrigerant for coping with global warming. However, there is a tendency for vibrations of a compressor to also increase because the circulation of refrigerant increases according to specification for using R32 refrigerant. Since large vibrations of a compressor causes outdoor units to generate solid propagation sound, there is a need for a technology that can reduce vibration of a compressor. However, a vibration-measuring method to analyze such vibration had not been fully developed and it was difficult to specify the cause. Accordingly, a new vibration-measuring method was developed specifically for compressors. The use of this measuring method allowed to discover that one of the unresolved problems with sound of outdoor units was the vertical vibration of the accumulator housed in the compressor. Moreover, it was found that the accumulator vibrates vertically due to the acoustic resonance inside the casing, which led to develop a new (accumulator) design with a focus on resonance. There is a type of spatial resonance inside an accumulator that has a phase difference with the antinode of sound pressure appearing at the upper and lower ends of the space. The vertical force caused by the difference in the sound pressure becomes excitation force, which then causes vertical vibration of the accumulator. Therefore, preventing this resonance from occurring can help suppress the vertical vibration. In order to prevent the generation of resonance, a plate-shaped reflection member was placed inside the accumulator, and it proved to be effective in suppressing the vertical vibration.

1. INTRODUCTION

A compressor was developed using R32 which was a low GWP R32 refrigerant for coping with global warming. There is a tendency for vibrations to also increase because the circulation of refrigerant is increase according to specification for using R32 refrigerant, so there is a need for a technology that can reduce such vibration. However, it was difficult to identify the cause because a method to support compressors during measurement of vibrations had not been fully developed.

Accordingly, a measuring method that can analyze vibrations was developed that does not inhibit the compressor movement via the use of a flexible support. In this report, we describe the method of measuring compressor vibrations that uses a flexible support and introduce a case that used this new method to develop a vibration-suppression seed.

2. VIBRATION-MEASURING METHOD FOR COMPRESSORS AND THE ISSUE

One of the issues with the outdoor unit was the sound caused by vibrations of the compressor housing. For example, it is shown that the same compressor was mounted in an outdoor unit to concurrently measure the vibration of the compressor and the sound from the outdoor unit. Figure 1 shows a contour map of 1/3 octave band sound from the outdoor unit and Figure 2 shows the overlaid acceleration spectrums of acceleration vertical vibration of the accumulator. The accumulator is hereafter referred to as "Accm.". Figure 2 shows a graph with a focus on the vertical vibration of the Accm, on which the vibration-acceleration spectrum per compressor rotation was overlaid on a single sheet. In this case, the outdoor unit generated a loud sound in the 500 Hz band, which indicated that the Accm vertical vibration was the cause of the sound from the outdoor unit.

Meanwhile, before we conducted the sound and vibration test on the outdoor unit, also conducted the vibration test on the compressor alone. Figure 3 is a picture showing the state of vibration of a compressor alone being evaluated with the conventional method. Suction piping was connected to the copper facility piping and discharge piping to the iron refrigeration piping, and the legs were placed on rubber support legs. A handy vibrometer and acceleration sensor were placed so that they come in contact with the compressor and piping surfaces to measure vibrations. Figure 4 shows a graph with a focus on the vertical vibration of the Accm, on which the vibration-acceleration spectrum per compressor rotation was overlaid on a single sheet. Unlike the vibration test on the outdoor unit, the vertical vibration significantly decreased in the 500 Hz band when the compressor was evaluated alone.

This suggests that there were cases in which the compressor vibration, the cause of the sound from the outdoor unit, could not be properly evaluated with the conventional vibration-measuring method for a compressor alone.

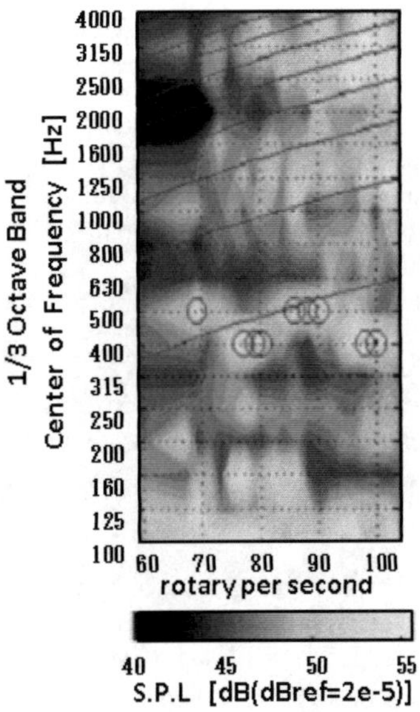

Figure 1: Contour map of 1/3 octave band sound from an outdoor unit

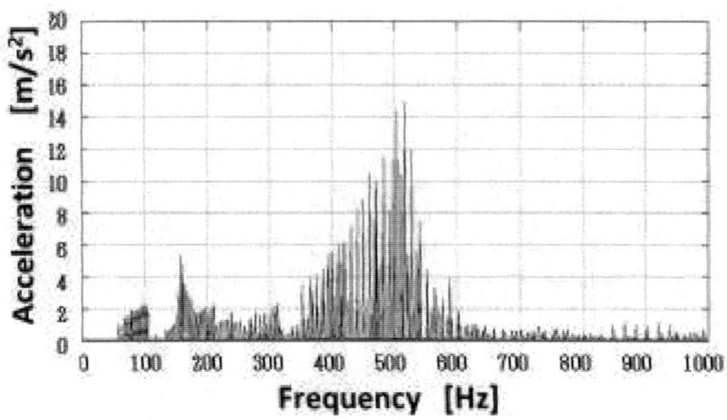

Figure 2: Overlaid acceleration spectrums of Accm vertical vibration when mounted in an outdoor unit

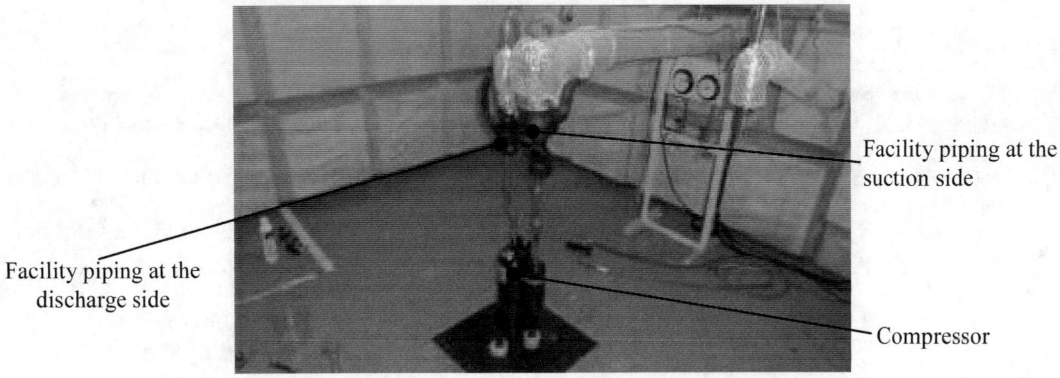

Figure 3: Evaluation of vibrations of a compressor alone: Conventional method

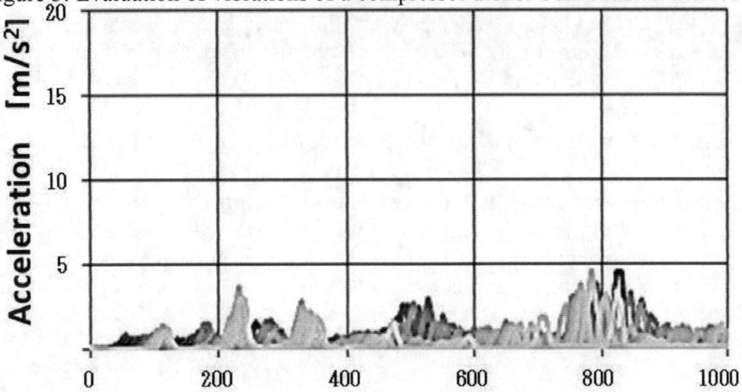

Figure 4: Overlaid acceleration spectrums of Accm vertical vibration of a compressor alone

3. Vibration-measuring method for compressors

3.1 Flexible support structure

As the reason for the difference described in Section 2, we considered smaller compressor vibrations due to the large piping mass and high support rigidity associated with the equipment to measure the vibrations of the compressor alone. This led to develop a new vibration-measuring method that incorporates a flexible support. Figure 5 shows vibration of a compressor being evaluated by the new vibration-measuring method.

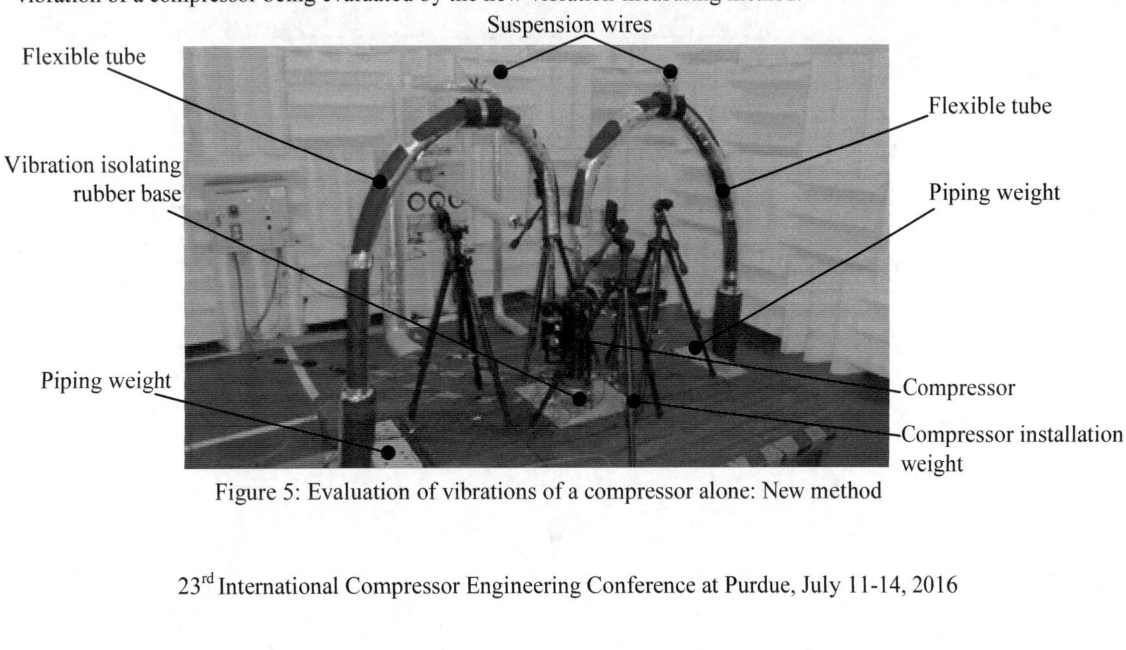

Figure 5: Evaluation of vibrations of a compressor alone: New method

The following describes the effect of each part.

- Flexible tube: Supports the compressor flexibly.
- Piping weight: Places approximately 150 kg of mass onto the connection between the flexible tube and system piping in order to isolate the vibrations from the system piping.
- Suspension wire: Prevents the gravity of the flexible tube from being applied onto the compressor. Has a tension adjustment mechanism to allow for the evaluation of compressors with different heights.
- Compressor installation weight: A foundation with approximately 200 kg of mass to prevent the ground vibrations from being transmitted to the compressor.
- Vibration-isolating rubber base: An increase in the compressor vibrations due to the use of a flexible support may cause the compressor to fall down. Therefore, the base is fixed to the installation weight to allow for testing to be conducted safely.

3.2 Confirmation of the effects

As described in Section 2, the vertical vibration was measured while flexibly supporting the compressor. Figure 6 shows the overlaid acceleration spectrums of the Accm vertical vibration. The flexible support enabled the measurement of previously difficult-to-measure vertical vibrations of an Accm.

4. Development of a compressor that suppresses the vertical vibration of an Accm

4.1 Assumption of a mechanism that generates vertical vibration

In order to specify the cause of 500 Hz vertical vibration of an Accm examined in Section 3, an acoustic eigenvalue analysis was conducted on the Accm internal space. Figure 7 shows the results of the eigenvalue analysis on the Accm internal space. It was found that there were acoustic modes in the 500 Hz band with the antinode appearing at the top and bottom and the node appearing in the center of the Accm.

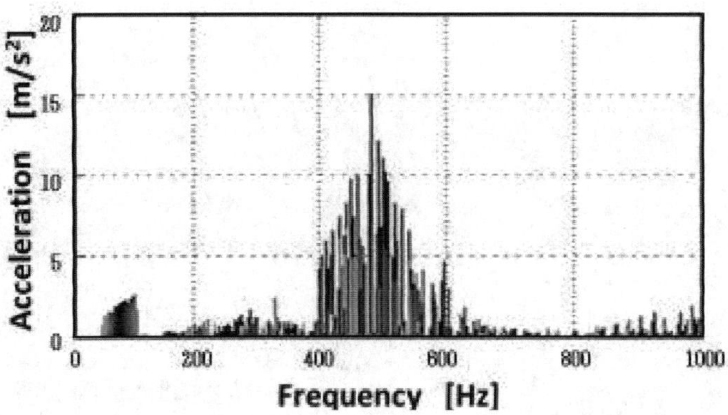

Figure 6: Overlaid acceleration spectrums of the Accm vertical vibration of a compressor alone: New evaluation method

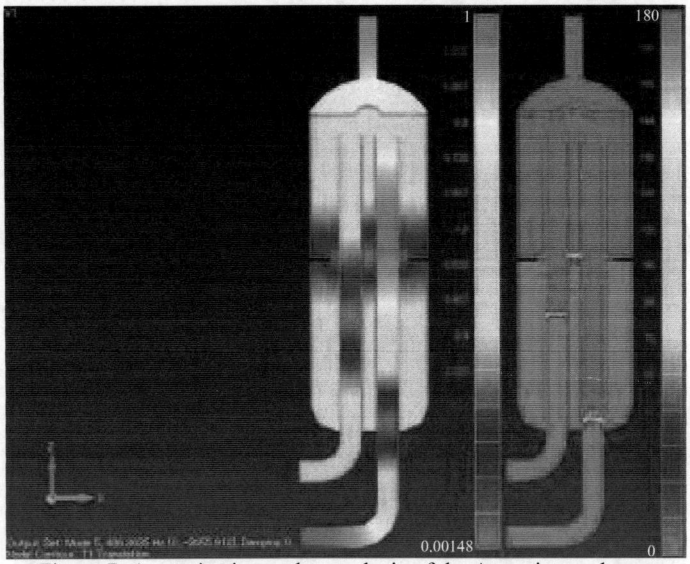

Figure 7: Acoustic eigenvalue analysis of the Accm internal space

In addition, the vibration and internal pressure of the Accm were concurrently measured. For the measurement of internal pressure, pressure sensors were inserted on the top and bottom of the Accm, and the pressure difference was calculated to determine the excitation component that vibrates the entire casing. Figure 8 shows the overlaid acceleration spectrums of the Accm vertical vibration and Figure 9 shows the overlaid pressure difference from top of Accm to bottom of Accm spectrums and the phase difference. These figures show that the peak of the vertical vibration roughly coincides with the peak of the pressure difference. They also show that the phase lag occurs in the band frequencies where the peaks are observed, which demonstrates that the spatial mode exists as indicated by the results of the eigenvalue analysis.
Therefore, it was shown that the pressure difference due to the acoustic mode that exists in the 500 Hz band acts as excitation, causing the vertical vibration.

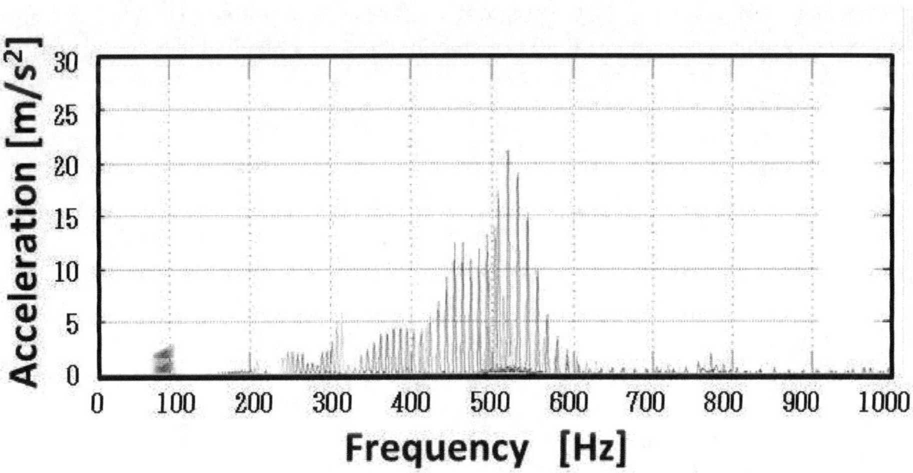

Figure 8: Overlaid acceleration spectrums of the Accm vertical vibration

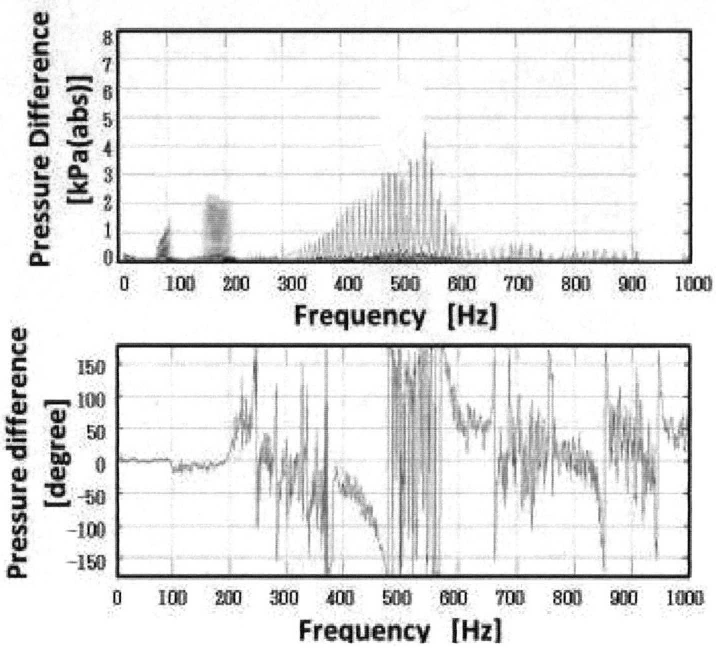

Figure 9: Pressure differences between the top and bottom of the Accm
(Upper: Overlaid amplitude differences; Lower: Phase differences)

4.2 Accm with a countermeasure against vertical vibration

Figure 10 shows the cross-sectional views of a conventional Accm and an Accm equipped with a countermeasure against vertical vibration of the Accm. In consideration of reduction of the mode excitation, a part that prevents pressure fluctuations ("countermeasure part" in the figure) was attached to the antinode of the acoustic mode inside the Accm.

To verify the effect (of the countermeasure part), the acoustic response analysis was conducted on the conventional Accm and countermeasure Accm. Figure 11 shows the excitation of the acoustic pressure point and response point at the acoustic response analysis and Figure 12 shows the results of the analysis.

It was shown that the pressure fluctuations caused by the acoustic mode in the 500 Hz band at the suction side was reduced in the countermeasure Accm to approximately one-third that of the conventional Accm. The vertical vibration was also measured while the compressor was being supported flexibly. Figure 13 shows the vertical-vibration acceleration spectrums of the Accm. It was shown that the peak from 400Hz to 600Hz of the conventional Accm was approximately 19m/s2 and the peak of the countermeasure Accm was approximately 14m/s2. It confirmed that the vertical vibration has been reduced by approximately 27 percent.

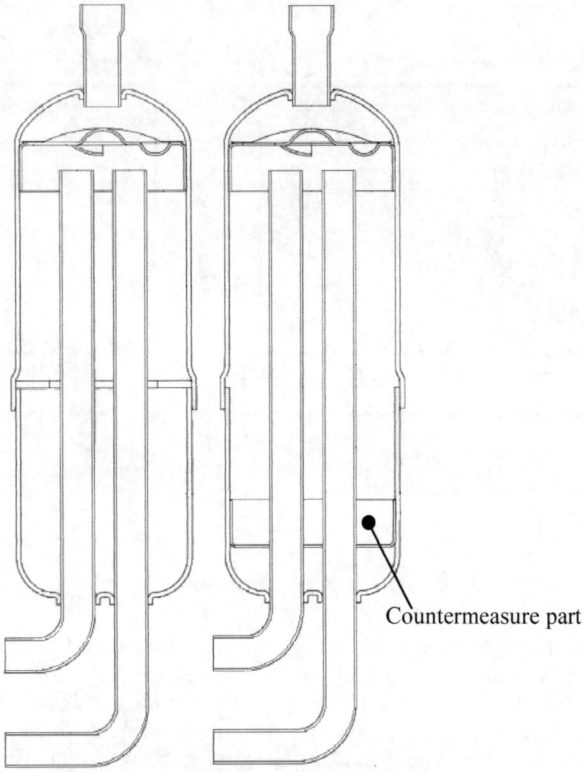

Figure 10: Cross-sectional views of Accms
(Left: Conventional Accm; Right: Countermeasure Accm)

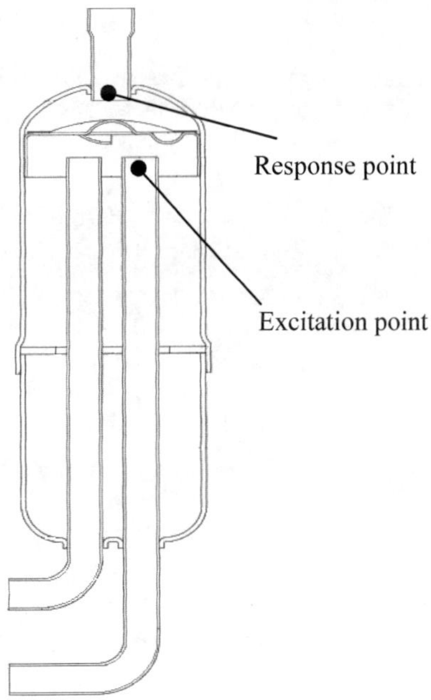

Figure 11: Excitation and response points
in the acoustic response analysis of the Accm internal space

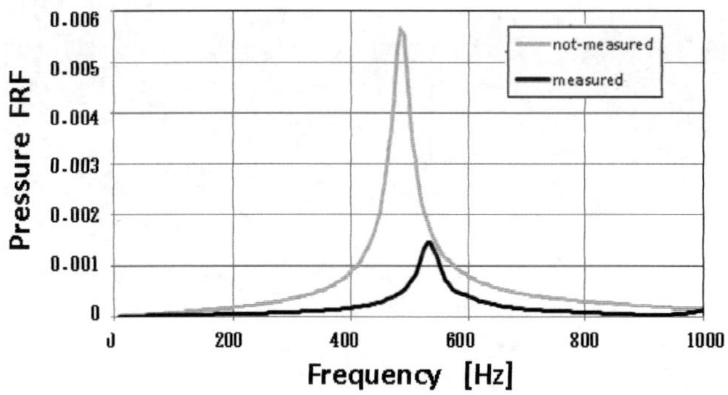

Figure 12: Acoustic response analysis of
the Accm internal space: FRF

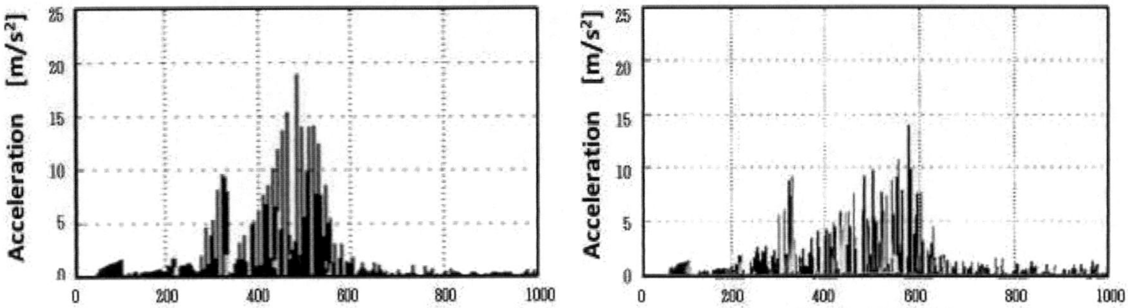

Figure 13: Overlaid acceleration spectrums of Accm vertical vibration of a compressor alone: New evaluation method
(Left: Conventional Accm; Right: Countermeasure Accm)

5. Conclusion

In this report, we introduced a method of measuring compressor vibrations that uses a flexible support in order to examine the compressor vibrations that cause the outdoor unit sound. The following is the summary of the report.

- The conventional vibration-measuring method for a compressor alone was unable to evaluate the vertical vibration of the Accm that can cause the outdoor unit sound. Accordingly, a vibration-measuring method that utilizes a flexible support was developed, which enabled the evaluation of vertical vibration of the Accm.
- As the new method allows for the evaluation of vertical vibration, the mechanism generating such vibrations was examined. Based on the concurrent measurement of vibration and pressure as well as the acoustic eigenvalue analysis, it was shown that the acoustic mode in the 500 Hz band with the antinode appearing at the top and bottom and the node appearing in the center of the Accm inside the Accm was the cause of the vertical vibration.
- In consideration of reduction in the acoustic mode excitation, a part that prevents pressure fluctuation was attached to the antinode of the acoustic mode. Based on the acoustic response analysis as well as the vibration measurement that utilized a flexible support, it was confirmed that the vertical vibrations had been reduced by approximately 27 percent.

Study on the Contribution of Compressor noise to Refrigerator Overall Noise

Shoufei WU*, Jiayou SONG, Gaowei SHEN

Jiaxipera Compressor Co., Ltd., Technology Research Department,
Jiaxing, Zhejiang, People's Republic of China
shoufei.wu@qq.com; +86 573 8258 6106

* Corresponding Author

ABSTRACT

Recently, greater attention is being paid to refrigerator noise. As we know, compressor is an important part of refrigerator, and it's also a very important noise soure. Based on comparison between compressor noise spectrum and refrigerator noise spectrum, the contributions of compressor noise to the refrigerator overall noise are discussed. Besides compressor direct radiation noise, compressor pressure pulsation and vibration also make great influence on refrigerator noise. A suction and discharge pressure pulsations test bench was built, and then the compressor pressure pulsation data were tested. By compressor suction and discharge pipelines optimization, the pressure pulsations and refrigerator noise were reduced. Vibration transmission path from compressor to refrigerator was investigated, and resonance frequencies on the vibration transmission path were identified. By refrigerator connection pipe optimization, refrigerator noise was reduced.

1. INTRODUCTION

Noise is one of the most easily perceived refrigerator characteristics. Recently, more and more attentions are being paid to the noise. Refrigerator noise is a complicated problem, and it can be affected by many aspects. Refrigerator noise is not only related to its own design, but also affected by the application environment and user preferences. Refrigerator produces different sounds in different seasons and regions. Different user preferences can also make the noise different. The complication also performs as that the noise is not stable and static. When compressor is starting the noise may be a little bit higer, and when it is stopping, maybe a sudden and short abnormal noise appears. Also the noise is not objective, different persons have different impressions on the same noise, and sometimes we have to make a subjective evaluation.

Compressor as one of the core components of refrigerator is closely related to the refrigerator noise.It is one of the most important noise and vibration sources. Baratti *et al.* (2012) identifed the vibration energy and acoustic energy paths in interaction between compressor and refrigerator. Olavo *et al.* (2014) researched the compressor supporting plate optimization for reducing the vibration energy transmission. Jeong *et al.* (2008) proposed the methods to reduce the low frequency noise of a reciprocating compressor in a refrigerator. Elvedin *et al.* (2008) researched the influence of the discharge port shape on screw compressor acoustic performance.

In this paper, the influence of the compressor on the operating noise of the refrigerator was discussed. Refrigerator noise spectrum and compressor noise spectrum were compared, and the reasons for the differnce were discused. Besides the compressor direct radiation noise, compressor suction and discharge pressure pulsations and vibration can also make great influence on the refrigerator noise. The compressor pressure pulsations were tested, and the refrigeator noises with different compressors were compared. The vibration transfer paths from compressor to refrigerator were analyzed, and the inherent frequencies of vibration transmission parts were indentified. After vibration transmission path redesign, the refrigerator noise was reduced.

2. REFRIGERATOR NOISE AND COMPRESSOR NOISE COMPARISON

Typical noise spectrums of compressor and refrigerator are shown in Figure 1. The compressor noise was tested using the load system. To make sure that the spectrums are compariable, the suction and discharge pressures of compreesor were tested when it was running in refrigerator. When testing the compressor noise in the lab, the same suction and discharge pressures were applied.

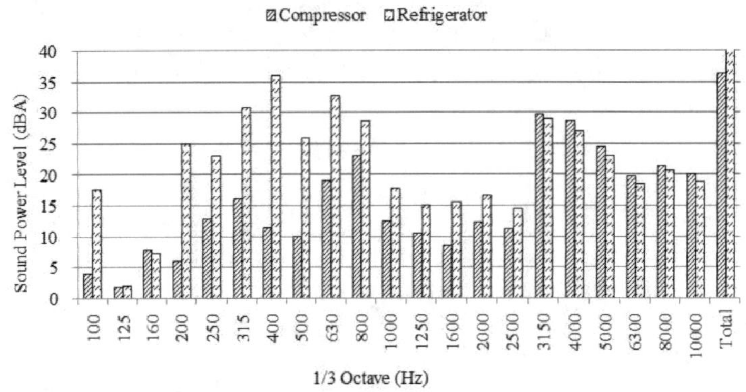

Figure 1: Compressor and refrigerator noise spectrums comparison

Figure 1 shows that there are great differences between the compressor noise and refrigerator noise: a) The refrigerator noise is mainly at the low frequencies, less than 1000 Hz in 1/3 octave; the compressor noise is mainly at the high frequencies, higher than 2500 Hz in 1/3 octave. b) For the high frequency noise (higher than 2500Hz), refrigerator noise spectrum is similar to the compressor noise spectrum, just the sound power level value is a litter smaller. c) For the low frequency noise (lower than 1000Hz), refrigerator noise is obviously higher than the compressor noise nearly in each 1/3 octave band.

Refrigerator compressor is hermetic; most parts are closed in the shell. Most noise is gererated form the inner of the compressor, the noise has to go out through the shell, in other words, most of the compressor noise propagates by the shell radiation. The shell has sound insulation effect, especially for the noise which frequency is lower than the first order natural frequency of the shell. Usually the size of refrigerator compressor shell is small, the first order natural frequency is higher than 3000 Hz. Therefore, the shell can make more sound insulations for the lower frequency noise.

Compared to compressor, refrigerator has much bigger size, it has foam layers, pipes, plates, and so on. Usually the stiffness of those parts are not high, the inherent frequencies of them are lower than 1000 Hz, so refrigerator is easy to produce low frequency noise. In addition, many refrigerator parts have damping, they are not good for high frequency noise propagation, so most refrigerator system noise is low frequency, the refrigerator's high frequency noise (higher than 3000Hz) comes from the direct radiation of the compressor noise.

Besides compressor direct radiation noise, suction and discharge pressure pulsations and vibration contribute to refrigerator noise. During suction and discharge phases, compressor suction and discharge pressure pulses respectively, the pulsations make involved pipes vibrating and generate extra noise. Section 3 addresses the contribution of compressor suction and discharge pressure pulsations to refrigerator overall noise. And section 4 focuses on the influence of compressor vibration on it.

3. INFLUENCE OF COMPRESSOR PRESSURE PULSATIONS ON REFRIGERATOR NOISE

Refrigerator compressor is the reciprocating compressor. Due to the periodical suction and exhaust, it caused suction and discharge pressure fluctuations. The compressor suction and exhaust pressure pulsations test bench was built, the pulsations were tested. Different suction and exhaust designs were discussed. When installing the different compressors in regrigerator, the refrigerator noise was tested. The influence of compressor pressure pulsation on refrigerator noise was analyzed.

3.1 Compressor Suction and Discharge Pressure Pulsations Test

Compressor suction and exhaust pressure pulsation test bench is shown in Figure 2. It also can be used to test the pressure pulsations in the refrigerator. The dynamic pressure sensors were fixed in suction pipe and discharge pipe, the pressure sensor sensitivity is 450mV/bar, the sensor can catch the pressure pulsation upto 3000Hz. Data was collected by a NI PCI-4472 8 channels acquisition card, a data analysis program was coded by Labview. After the compressor operation was stable, the pressure data was collected. By Fast Fourier Transform (FFT), the spectrum of pressure pulsations were got.

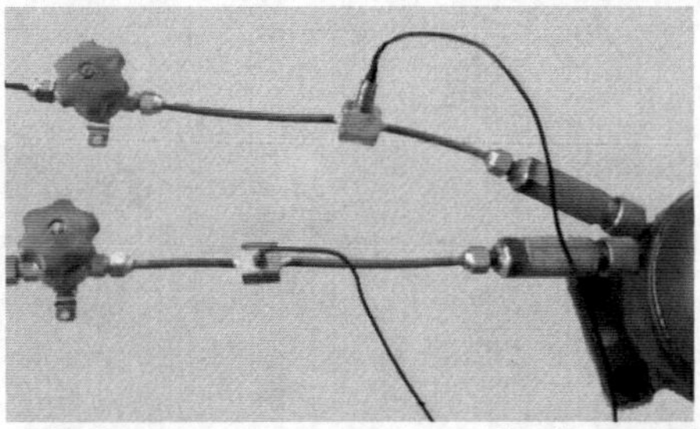

Figure 2: Compressor pressure pulsation test bench

The suction pressure pulsation data is shown in Figure 3. Figure 3.a) is the FFT data of pressure pulsation, Figure 3.b) shows the pressure pulsation in 1/3 octave, the reference pressure is 2E-5 Pa.

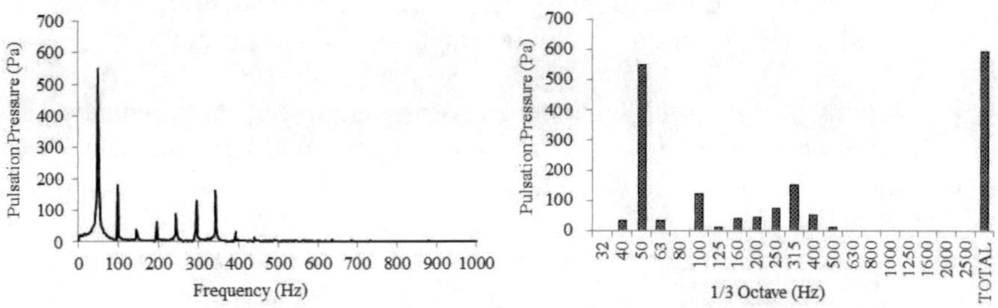

a) FFT data b) 1/3 Octave data

Figure 3: Compressor suction pressure pulsation

Figure 3 shows that most of the suction pressure pulsation frequecies are lower than 1000Hz, and the frequencies are harmonic, they are multiples of the fundamental frequency of the compressor. The maximum value appears in the compressor working frequency 50Hz, that is easy to understand. The total suction pulsation pressure is 500Pa.

The discharge pressure pulsation data was listed in Figure 4. Figure 4.a) is the FFT data, Figure 4.b) is the 1/3 octave data, the reference pressure is also 2E-5 Pa.

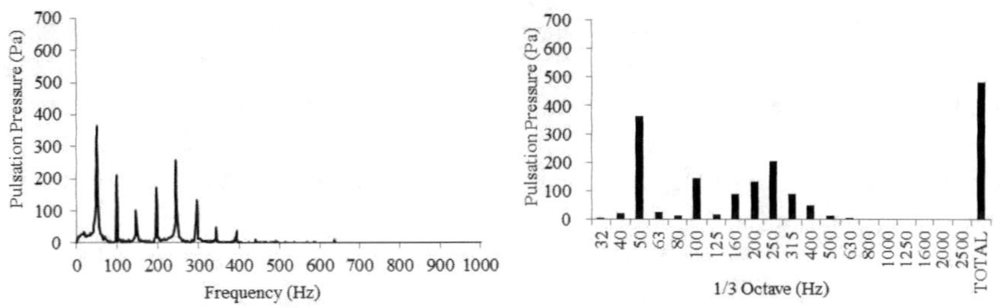

a) FFT data b) 1/3 Octave data

Figure 4: Compressor discharge pressure pulsation

Figure 4 shows that the peak valve of the compressor exhaust pressure pulsation is also at the working frequency

50Hz, and the pressure pulsation is harmonic, the pulsation frequencies are less than 1000Hz. The total discharge pulsation pressure is 800Pa.

3.2 Suction Pressure Pulsation Optimization and Refrigerator Noise Test

For the reciprocating compressor, the suction pressure pulsation is unavoidabel, but we can optimize the pressure pulsation by structure redesign. Figure 5 shows two typical designs for compressor suction system, one is direct suction compressor, the other is semi-direct suction compressor. The main difference is that there is a rubber connection between the shell and suction muffler in the direction suction compressor, the semi-direct suction doesn't have this part. For the direct suction compressor, the suction muffler is directly connected with the shell, and the refrigerant gas in the suction pipe is directly sucked into the muffler. For the semi-direct suction compressor, the refrigerant gas goes into the shell cavity, then be sucked from the cavity. The shell cavity acts as a big cushion chamber.

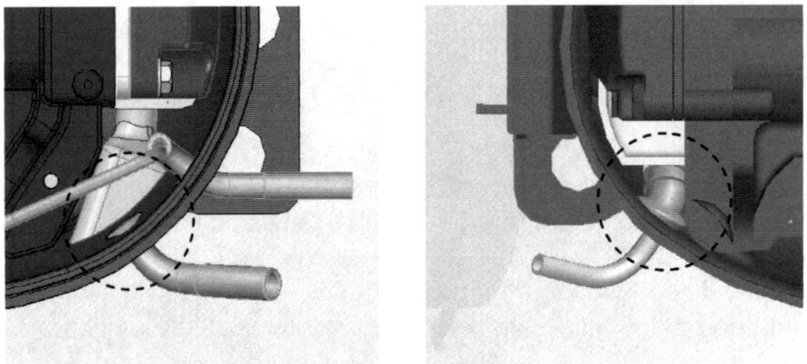

a). Semi-direct suction b). Direct suction

Figure 5: Semi-direct suction and direc suction structures

The suction pressure pulsations of the two different compressors were tested; the results are listed in Figure 6. It shows that the prssure pulsation of semi-direct suction compressor is much smaller than that of direct suction compressor. Press pulsation is reduced from 500Pa to 60 Pa. And in 315 Hz frequency band, 400Hz frequency band, pressure pulsations are reduced obviously.

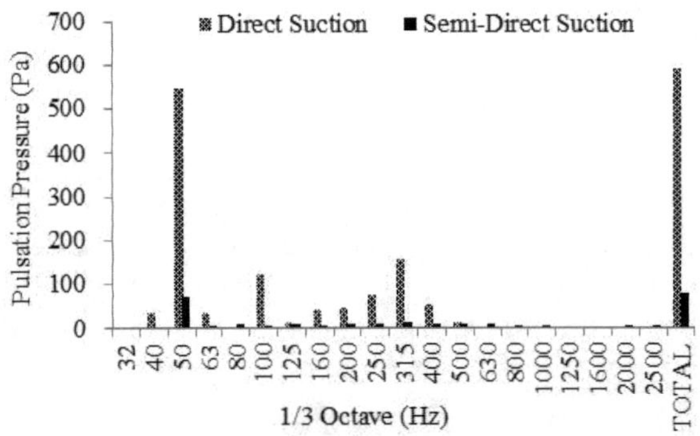

Figure 6: Suction pressure pulsation of direct suction and semi-direct suction compressors

Installing the above two compressor in the same refrigerator separately, the refrigerator noise was tested, and result is shown in Figure 7. It shows that, for the refrigerator with semi-direct suction compressor, the noise in 50Hz, 315Hz and 400Hz frequency bands reduced a lot. Compared to the pressure pulsation test results in Figure 6, we can find that the frequency bands where the pressure pulsation reduced are corresponding to frequency bands where the refrigerator noise reduced. So the pulsation pressure results in refrigerator additional noise in the corresponding frequency bands.

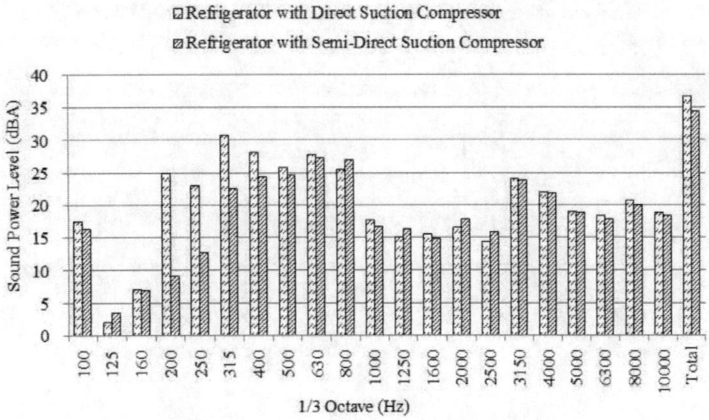

Figure 7: Refrigerator noise with direct suction and semi-direct suction compressors

3.3 Discharge Pressure Pulsation Optimization and Refrigerator Noise Test

Compressor also can produce pressure pulsation in the discharge side as in the suction side. The pressure pulsation can be optimized by discharge line redesign. Figure 8 shows a discharge line of a compressor.

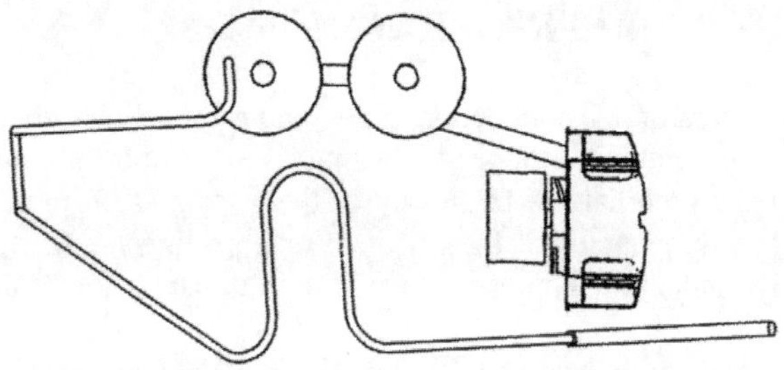

Figure 8: Discharge line of compressor

In order to reduce the discharge pressure pulsation, two buffer cavities had been arranged in the exhaust line, and the throttle orifice was arranged in the pipe between the two cavities. The inner discharge pipe was designed to be long and thin. To reduce the pulsation furthermore, each buffer cavity volume was extended to two times of the original. And the discharge pressure pulsation was tested. The data is shown in Figure 9.

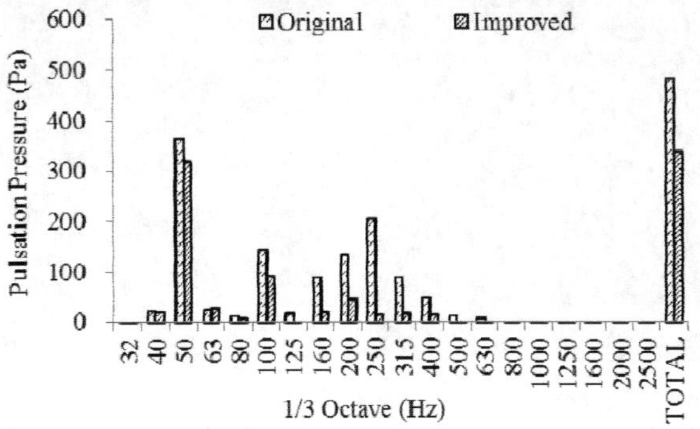

Figure 9: Discharge pressure pulsation of two designs

Figure 9 shows that the compressor discharge pressure pulsation is reduced significantly by doubling the volume of the buffer cavities. The compressors were installed on the refrigerator, the refrigerator noise was tested, shown in Figure 10.

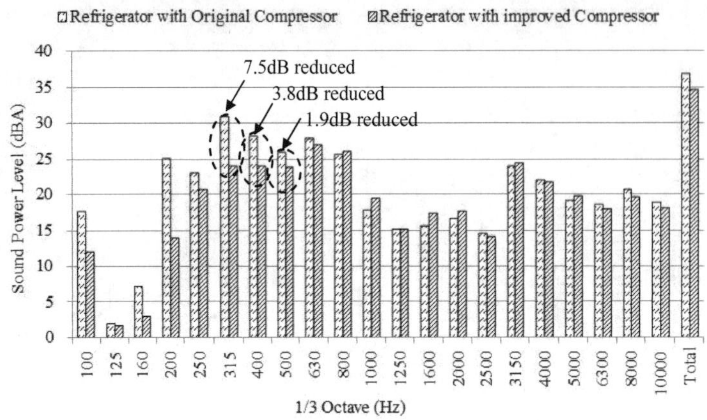

Figure 10: Refrigerator noise comparison

Figure 9 and Figure 10 show that, when the discharge pressure pulsation is reduced in the 315Hz, 400Hz, 500Hz frequency bands, the noise of the refrigerator in those frequency bands is also reduced.

4. INFLUENCE OF COMPRESSOR VIBRATION TRANSMISSION ON REFRIGERATOR NOISE

Compressor is connected to refrigerator by connection pipes and supporting rubber grommets. The compressor vibration can be transmitted to refrigerator by them. According to the vibration theory, when the frequency of the vibration is the same as or similar to the natural frequency of the transmission path, it will cause resonance, noise will be very large. The resonance frequencies of the refrigerator pipe lines were studied. A case showed that the refrigerator noise was reduced by reducing the compressor vibration transmission of connection pipes.

4.1 Vibration Transmission Path Analysis and Transmission Functions Test

Many parts of refrigerator are made by pipes, for example evaporator and condenser, most of them are not completely fixed, fixation stiffness is not high, and their inherent frequencies are low. They are easy to transmit vibration, cause resonance, and produce noise. Indentifing the inherent frequencies, and then avoiding them similar to source's vibration frequency, is a good way to avoid refrigerator additional noise. Suction pipeline inherent frequencies were tested, as shown in Figure 11.

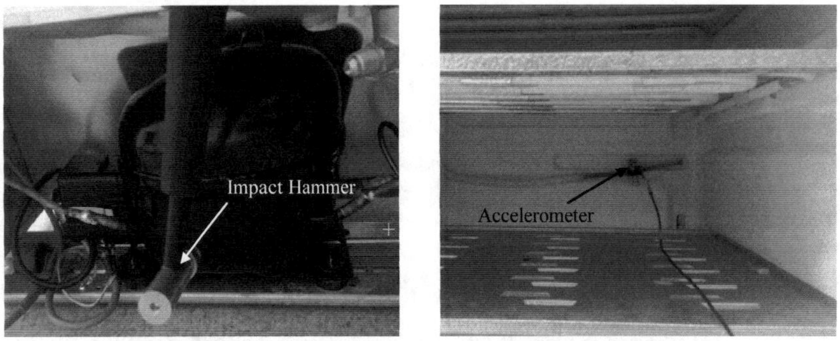

Figure 11: Vibration transmisson functions test

The accelerometer was installed on the pipeline, the hammer was used to knock the compressor to give a excitation. The transmission functions from hammer to the accelerometer can be got from the test; the inherent frequencies can be identified. One transmission function is shown in Figure 12. Figure 12 shows that, several inherent frequencies are lower than 1000Hz for the suction pipe line. When the excitation frequency is close to the inherent frequency, it will produce resonance and refrigerator noise.

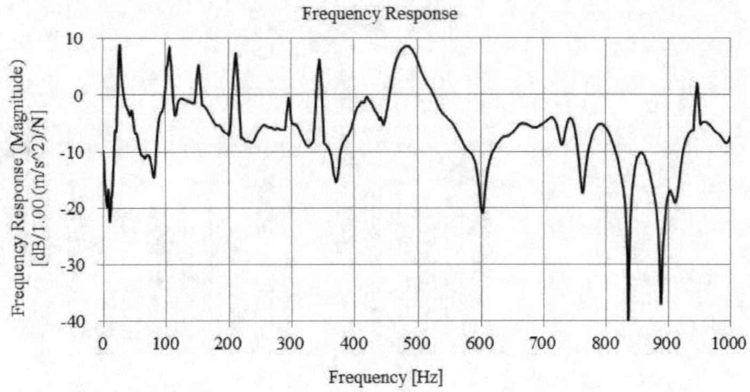

Figure 12: Vibration transmission function

4.2 Connection Pipe Optimization and Refrigerator Noise Comparation

Figure 13 shows one refrigerator connecting pipe. When using this pipe, the pipe and compressor vibrations are great. The pipe was modified and the inherent frequencies were optimized by using Finite Element Method (FEM), shown in Figure 14. The modified pipe was installed in the refrigerator, and the refrigerator noise was tested, the result was shown in Figure 15. The experiment shows that, optimizing the connection pipes frequencies to reduce the vibration transmission can be helpful for reducing refrigerator noise.

Figure 13: Refrigerator connecting pipes

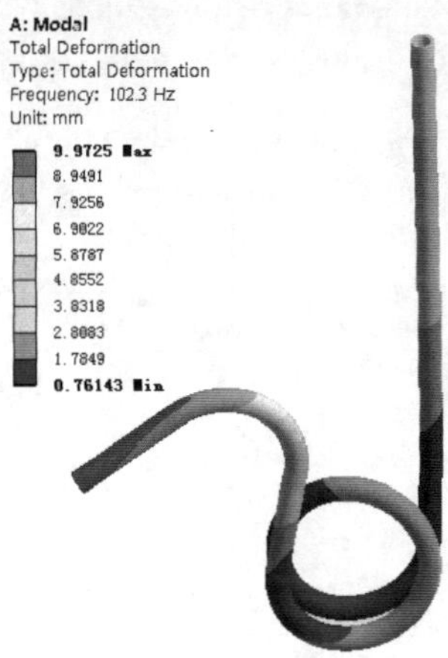

Figure 14: Connection pipe modal analysis using FEM

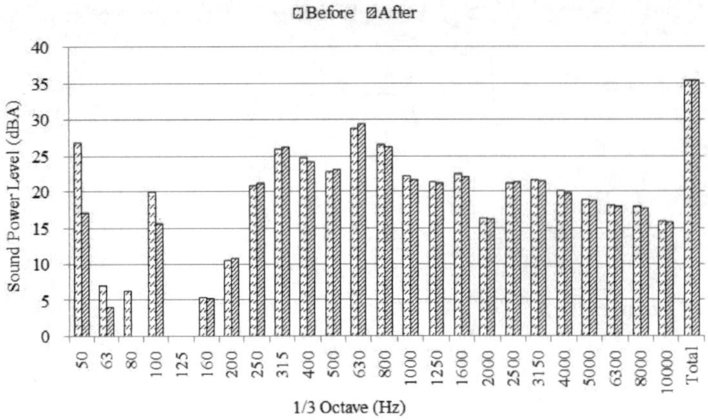

Figure 15: Refrigerator noise comparison

5. CONCLUSIONS

The compressor and refrigerator noise spectrums were compared, the differences between of them were discussed. The compressor noise is mainly at the high frequencies for the reason of the shell sound insulation; the refrigerator noise is mainly at the low frequencies for that, the inherent frequencies of most refrigerator parts are not high, these parts are easy to produce low frequency noise.

Besides the compressor radiation noise, compressor suction and discharge pressure pulsations and compressor vibration also make great influence on the refrigerator noise. A suction and discharge pressure pulsation test bench was built, and then the compressor pressure pulsation data were tested. By compressor suction and discharge pipelines optimization, the pressure pulsations and refrigerator noise were reduced. Vibration transmission path from compressor to refrigerator was investigated, and resonance frequencies were identified. By refrigerator connection pipe shape optimization, refrigerator noise was reduced.

REFERENCES

Bratti GioVanni, Santini Junior Otavio, Baars Edmar, Barbieri Renato, Pellegrini Claudio, Silva Olavo, Lenzi Arcanjo, 2012, Identification of Energy Path in the Interaction between Compressor and Refrigerator, *International Compressor Engineering Conference*, paper 1302.

Elvedin Mujie, Ahmed Kovacevie, Nikola Stosic, Ian K. Smith, 2008, The Influence of the Discharge Port Shape on Screw Compressor Performance and Gas Pulsations, *International Compressor Engineering Conference*, paper 1170.

Jeong Il Park, Jeong Hoon Kang, Ji Hoon Choi, Phil Soo Jang, Min Chul Yoon, Dae Il Kwon, Il Kyo Oh, 2008, Reduction of the Low Frequency Noise due to the Discharge Pressure Pulsation of a Reciprocating Compressor, *International Compressor Engineering Conference*, paper 1335.

Olavo M. Silva, Thiago M. Guesser, Igor M. Guesser, Glaudio de Pellegrini, Carlos E. Vendrami, Arcanjo Lenzi, 2014, Shape Optimization of a Compressor Supporting Plate Based on Vibration Modes, *International Compressor Engineering Conference*, paper 2312.

Analysis and Control of Severe Vibration of a Screw Compressor Outlet Piping System

Ying ZHAO[1], Bin ZHAO[1], Qiang ZHOU[1], Xiaohan JIA[1,2], Jianmei FENG[1,2]*, Xueyuan PENG[1,2]

[1] Xi'an Jiaotong University, Xi'an, Shannxi, China
[2] National Engineering Research Center of Fluid Machinery and Compressors, Xi'an, Shannxi, China
E-mail: jmfeng@mail.xjtu.edu.cn

* Corresponding Author

ABSTRACT

The severe vibration of a screw compressor outlet piping system caused the fatigue failure of some thermowells and the unscheduled shut down of the system. The main reasons of the abnormal vibration in the outlet piping system were investigated by developing an acoustic model to simulate the gas pulsation and establishing two finite element models to conduct the mechanical vibration analyses. The acoustic analysis results showed that the pulsation amplitudes of most nodes in the outlet piping system exceeded the allowable values. The results of mechanical vibration analyses indicated that the insufficient stiffness of the outlet piping system and the first-order structure resonance occurred on thermowells were also the key factors inducing vibration. Several methods were put forward to attenuate vibration amplitude of the outlet piping system as well as the thermowells. A new pulsation attenuator was installed and the piping layout was rearranged to reduce pulsation amplitudes and shaking forces of all nodes in the outlet piping system. Several reasonable supports were added to improve the stiffness of the outlet piping system. After reinforcing the thermowells, the first-order natural frequency of the thermowells increased from 207.4Hz to 280.7Hz, away from the excitation frequency of 196.67Hz. The field measurement results showed that vibration amplitude and the vibration velocity decreased significantly after modifications.

1. INTRODUCTION

Compressors are widely utilized in the petrochemical industries. Piping vibration is a potential threat to the stability and reliability of the compressor system. Gas pulsation is one of the most important sources of piping vibration. Therefore, it is of great significance to investigate on gas pulsation and piping vibration of the compressor piping system.

The gas pulsation and piping vibration have drawn much attention since 1950s. The studies concerning gas pulsation and piping vibration mainly include theoretical research and engineering application. Kinsler and Frey (1962) deduced the equation of plane wave theory. Sakai *et al.* (1973a and 1973b) calculated the gas column natural frequencies of complex pipe system by the transfer matrix method. Benson (1972) proposed to simulate the gas pulsation in the piping system by using the one-dimensional non-isentropic flow theory. Soedel *et al.* (1978) investigated the acoustic characteristics of pulsation attenuator. Jia *et al.* (2015a) studied the influence of the orifice plate on gas pulsation in a reciprocating compressor piping system. Liang *et al.* Liu *et al.* (2013) investigated the acoustic attenuation and flow resistance characteristics of perforated tube silencers though numerical simulation. (2015b) estimated the main causes of the abnormal vibration in a reciprocating compressor inlet piping system through acoustic analysis, vibration analysis and field measurements, and took relevant treatments to eliminate the vibration. These research lays a good foundation for the analysis and control of vibration problems of the compressor piping systems.

This paper is devoted to investigate the main causes and elimination of the vibration problem of a screw compressor outlet piping system. An acoustic model and two mechanical models were established to simulate the gas pulsation and structure vibration of the screw compressor outlet piping system. The main causes of abnormal vibration in the outlet piping were investigated and relevant elimination treatments were put forward. The good performance of the improvement measures were verified by field vibration measurements.

2. FIELD VIBRATION PROBLEM

The serious vibration of a screw butadiene compressor outlet piping system caused the unscheduled shut down of the system. The compressor operates at rated condition most of the time. The rated speed is 2950 rpm, rated motor power is 1500 kW, rated flow rate is 15277 $m^3.h^{-1}$, the inlet pressure and temperature are 159 kPa and 318 K, and the outlet pressure and temperature are 530 kPa and 374 K, respectively. The vibration velocity at the first elbow after the outlet silencer was 38.8mm/s, much exceeding the allowable value. Worse still, the small-bore appendages attached to the mainline vibrated violently, and the fatigue fracture happened on the junction of some thermowells and mainline, as shown in Figure 1. Although the temporary measure of changing the elastic support near the outlet silencer to guide support was adopted, there were still severe vibrations. As a result, vibration caused analysis and eliminating measures are urgently needed to ensure the safety and reliability of the system.

Figure 1: Fatigue fracture of the thermowells

3. VIBRATION CAUSES ANALYSIS

3.1 Acoustic Analysis

The gas pulsation is an important excitation source of vibration in the compressor piping system. Eliminating the gas pulsation in the piping system within the allowable range is an effective measure to control the vibration in the compressor piping system. The plane wave theory and transfer matrix method were both applied to analyze the gas pulsation of the outlet piping system. Accordingly, the piping system can be divided into several basic elements such as pipe, valve, tee and volume. Each element has a transfer matrix *M*. For example, the transfer matrix of the uniform section straight pipe is given by Equation (1). The transfer matrix of the overall piping system is formulated by multiplying the transfer matrixes of all the piping elements as Equation (2).

$$M = \begin{bmatrix} \cos(\omega l / a) & -\rho_0 a \sin(\omega l / a) \\ -\frac{1}{\rho_0 a}\sin(\omega l / a) & \cos(\omega l / a) \end{bmatrix} \tag{1}$$

$$\begin{bmatrix} p_{end} \\ u_{end} \end{bmatrix} = M_n \cdot M_{n-1} \cdots M_2 \cdot M_1 \cdot \begin{bmatrix} p_{start} \\ u_{start} \end{bmatrix} \tag{2}$$

The acoustic natural frequency can be obtained by applying relevant boundary conditions of the piping system. Then, the pressure unevenness defined by Equation (3) and corresponding shaking forces can be calculated.

$$\delta = \frac{(p_t)_{\max} - (p_t)_{\min}}{\frac{1}{2}\left[(p_t)_{\max} + (p_t)_{\min}\right]} \tag{3}$$

The acoustic model of the outlet piping system was shown in Figure 2. The outlet piping system was divided into 84 elements and 103 nodes. The compressor boundary conditions were applied to node 1, 4, 7, and 9. The non-reflecting boundary condition was applied to node 45, and the boundary conditions of thermowells and pressure gauge branches were applied to node 46-58.

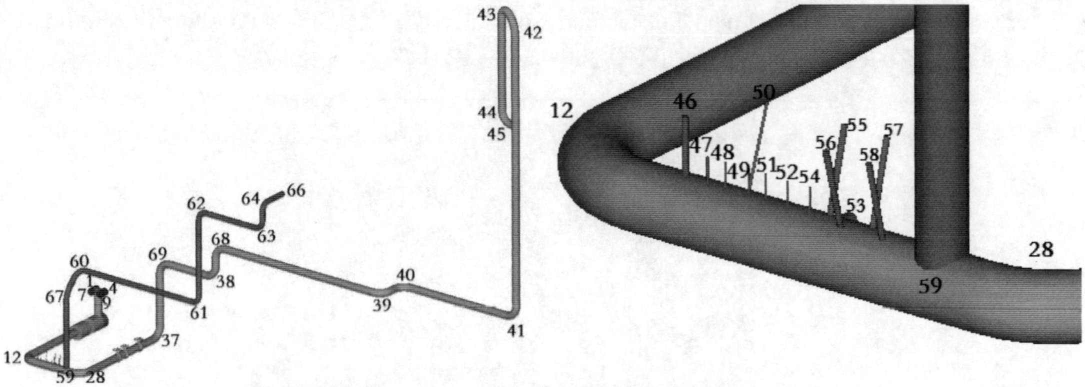

Figure 2: Acoustic model of the outlet piping system

The acoustic analysis results showed that the pressure unevenness of most nodes exceeded the allowable value of 2% under the operating condition. The maximum pressure unevenness was 15.5%, located on a pressure gauge branch (node 53). Therefore, the first step in controlling the severe vibration is to eliminate the gas pulsation of the piping system.

3.2 Mechanical Vibration Analyses

Piping system is a continuous elastic structure system, the structure resonance will be induced when the excitation frequency is within ±10% of the mechanical natural frequency of the piping system. Therefore, the mechanical vibration analysis plays an important role in the design of piping systems. Modal analysis is an effective method to investigate natural frequencies and mode shapes of a structure. Dynamic response analysis can be used to evaluate the piping system design due to the operating load cases. Since the mainline as well as small-bore appendages had severe vibration, the vibration analysis of overall outlet piping system and the thermowells were performed, respectively.

3.2.1 Vibration analysis of overall outlet piping system

The mechanical model of the outlet piping system was established, and the modal analysis and harmonic response analysis were performed. The pipes were modeled with beam elements, and the valves were simplified as rigid elements. Small-bore appendages were modeled and analyzed separately in the following and not included in the overall pipeline analysis.

Modal analysis results showed that the first order natural frequency of the outlet piping system was only 0.9Hz; the modal shape of the natural frequency 198.8Hz, close to the excitation frequency of 196.67Hz, occurred mainly on the horizontal pipe behind the first elbow that the small-bore appendages attached, as shown in Figure 3. The results indicated that the outlet pipeline was prone to vibrate due to insufficient stiffness.

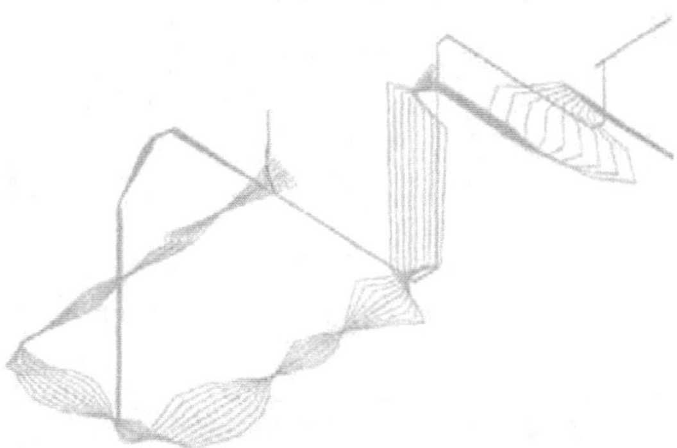

Figure 3: The mode shape at the MNF of 198.8Hz

The harmonic response analysis was performed under the action of calculated shaking forces by the acoustic analysis. The vibration velocities under the excitation frequency of 196.67Hz were calculated and compared with the field measured data, as shown in Figure 4. It can be seen that the calculated values agreed well with the measured values. Accordingly, the mechanical model of the outlet piping system and the shaking forces were reliable, which indirectly verified the reliability of acoustic model of the piping system.

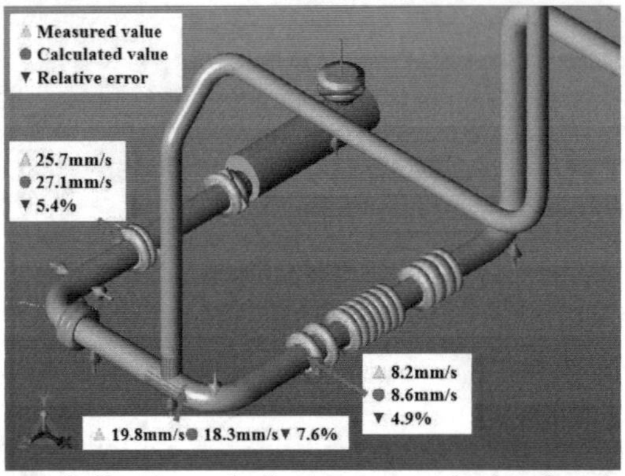

Figure 4: Comparison of vibration velocity between calculated and measured values

3.2.2 Vibration analysis of thermowells

Failures of small-bore appendages due to alternating stresses caused by vibration are typically occurred in compressor piping systems. Shifting the MNFs of the appendages away from the excitation frequency will go a long way toward avoiding such failures. A mechanical model including a thermowell, a section of the mainline and the connection flanges between them was established and the modal and harmonic response analysis were performed.

The modal analysis results showed that the first order natural frequency of the thermowell was 207.4Hz, near the excitation frequency of 196.67Hz, which means the first order structure resonance will occur on the thermowell. The harmonic response analysis results showed that the maximum equivalent stress of the model under the excitation frequency of 196.67Hz was 44.4MPa, appeared at the root of the thermowell, and prone to induce fatigue fracture of the thermowell, as shown in Figure 5.

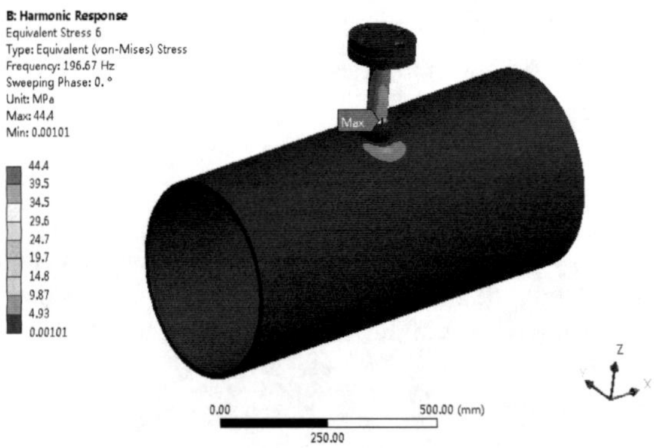

Figure 5: Stress distribution at the excitation frequency of 196.67Hz

4. VIBRATION ELIMINATING TREATMENTS

4.1 Pulsation Reduction Measures

At first, the pulsation reduction measure of adding a new pulsation attenuator was put forward. The 1650-milimeter-long straight pipe after the original discharge silencer was replaced by a pulsation attenuator of the same length, and the outer diameter of the new pulsation attenuator was 800mm, as shown in Figure 6. Acoustic analysis results of the modified model showed that the pulsation of the pipe after the new pulsation attenuator was significantly decreased, but the pressure unevenness of some small-bore appendages cannot meet the requirements of API 619.

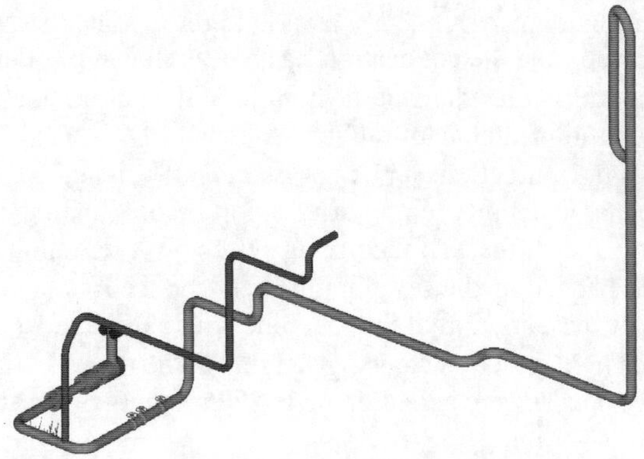

Figure 6: The first pulsation eliminating scheme

In order to eliminate the pressure unevenness and shaking forces at the small-bore appendages, another treatment including adding a new pulsation attenuator, adjusting the mainline, reflux pipeline and instrument branches was put forward. On the basis of the first scheme, the horizontal pipe after the first elbow was extended by 4000mm, the reflux pipeline was moved 1000 mm after the second elbow, and the layout of the instrument branches was adjusted, as shown in Figure 7.

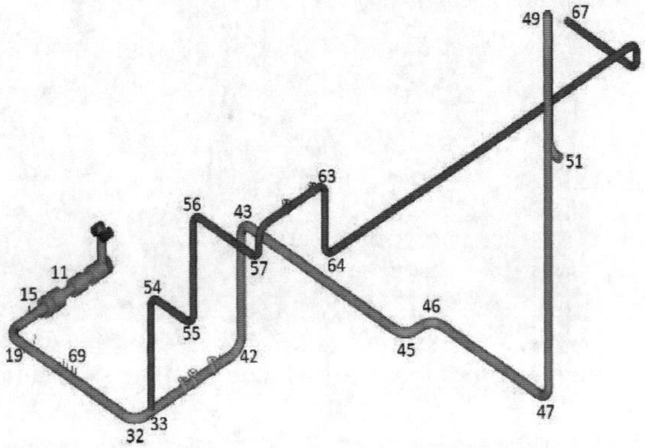

Figure 7: The second pulsation eliminating scheme

After modifications, both the pressure unevenness and shaking forces all over the piping system under designed operating conditions can meet the requirements of API 619; the maximum pressure unevenness was 1.53%, occurred on a thermowell (node 70), as shown in Figure 8.

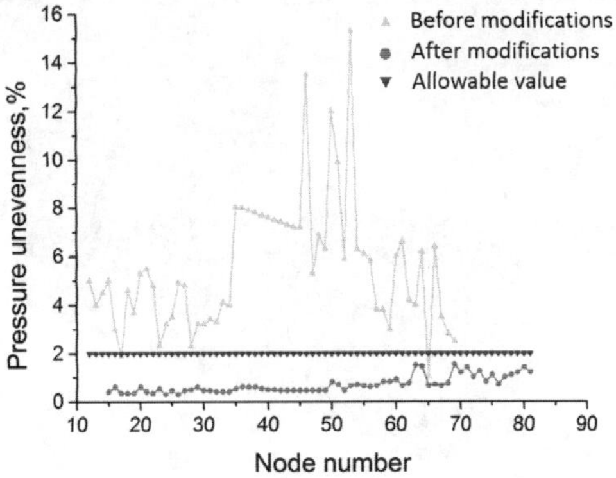

Figure 8: Comparison of the pressure unevenness before and after modifications

4.2 The piping vibration control measures

The vibration analysis results showed that the outlet pipeline with poor stiffness was prone to vibrate, and the fatigue fracture caused by the first-order structure resonance occurred on thermowells. Some supports were added to improve the stiffness of the overall piping system, as shown in Figure 9. The thermowells were reinforced by adding four gussets between the thermowells and the connection flanges, as shown in Figure 10.

- Two variable spring supports were added at the bottom of the pulsation attenuator (node 2022) to support the weight of the new pulsation attenuator (about 2 tons);
- Since the shaking forces in X direction can be generated at the first and the second elbow after the new pulsation attenuator, the poor stability temporary support before the first elbow (node 50) was changed to a guide support to improve the stiffness of the piping system in X direction.
- Two guide supports were added on the straight pipe between the first and the second elbows (node 200 and node 20260) to reduce the base motion of the mainline where the small-bore appendages attached.
- Since the shaking forces in Y direction can be generated at the tee which connected the outlet pipe and reflux pipe, the bearing support were added on node 250 to improve the stiffness of the piping system in Y direction.
- A guide support was added to the vertical pipe of reflux pipeline (node 565) to improve the stiffness of the reflux pipeline.
- A guide support was added to node 442 to improve the stiffness of the mainline.

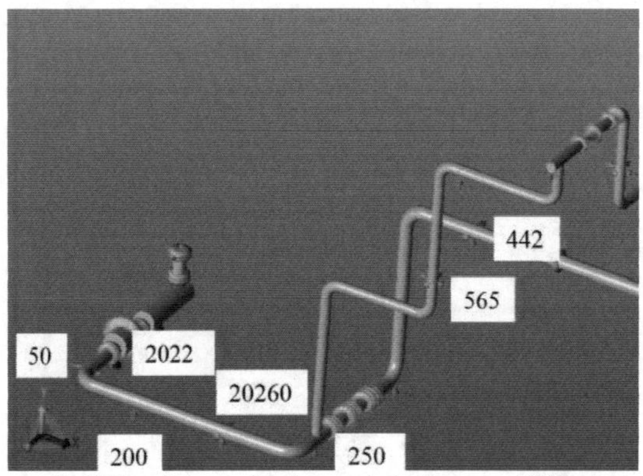

Figure 9: Vibration analysis model after modifications

After above modifications, the first order natural frequency of the outlet piping system improved from 0.9Hz to 1.67Hz, and the calculated maximum vibration velocity was 11.4 mm/s, far less than the allowable value, which indicated that the stiffness of the outlet piping system increased. The first order natural frequency of the thermowell improved from 207.4Hz to 280.7Hz, away enough from the excitation frequency of 196.67Hz, and the first order structure resonance could be avoided. The maximum equivalent stress of the modified model at the excitation frequency of 198Hz was 2.65 MPa, occurred on the welding location between a gusset and the thermowell, as shown in Figure 10, which indicated that the fatigue life of the thermowell was improved.

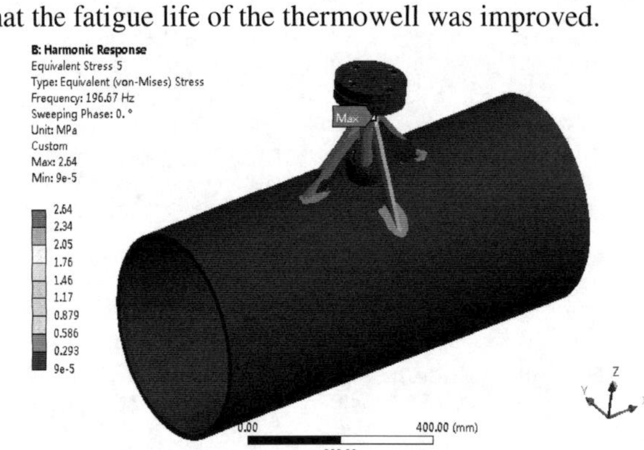

Figure 10: Stress distribution at the excitation frequency of 196.67Hz after modifications

4.3 Field Measured Results

The comparison of the field measured data before and after modifications is shown in Table 1. It can be seen that the maximum vibration amplitude dropped from 0.07mm to 0.017mm and decreased by 76%, and the maximum vibration velocity dropped from the 38.8mm/s to the 10.6mm/s and decreased by 73%. The vibration of the outlet piping system meet requirements of API 619, which indicated that the gas pulsation eliminating treatments and vibration control measures were effective.

Table 1: Comparison of the field vibration measurements data before and after modifications

	Before modifications	**After modifications**	**Decreasing amplitude**
The maximum horizontal vibration amplitude	0.046 mm	0.012 mm	72%
The maximum vertical vibration amplitude	0.07 mm	0.017 mm	76%
The maximum horizontal vibration velocity	21.4 mm/s	6.8 mm/s	68%
The maximum vertical vibration velocity	38.8 mm/s	10.6 mm/s	73%

5. CONCLUSIONS

The most important conclusions of the present paper were as follows:

- The main vibration causes of the severe vibration problem of the outlet piping system were investigated through the acoustic analysis, mechanical vibration analyses and field measurements. It was found that the excessive pulsation, low stiffness of the overall outlet piping system and the first-order structure resonance occurred on the thermowelles mainly contributed to the severe vibration.
- To control the abnormal vibration, some practical elimination treatments including adding a pulsation attenuator and some supports, changing the arrangement of the outlet piping system and reinforcing the thermowelles were carried out. The results of numerical analyses and field vibration measurements showed that the pressure pulsation, vibration amplitude and velocity after modifications satisfied well with standard requirements, which indicated that the treatments adopted indeed eliminated the vibration sources.

NOMENCLATURE

M	transfer matrix	(-)
a	sound speed	(m/s)
l	length	(m)
t	time	(s)
p	pressure	(Pa)
ω	angular frequency	(rad/s)
ρ	density	(kg/m^3)

Subscripts

start	the start node
end	the end node
max	the maximum value
min	the minimum value

REFERENCES

API619. Rotary type positive displacement compressors for petroleum chemical gas industry services. 2010.

Benson, R.S. (1972). Numerical solution of one-dimensional non-steady flow with supersonic and subsonic flows and heat transfer. *International Journal of Mechanical Sciences*, *14*(10):635-642.

Jia, X., Liu, B., Feng, J., et al. (2015a). Influence of an orifice plate on gas pulsation in a reciprocating compressor piping system. *Proceedings of the Institution of Mechanical Engineers, Part E: Journal of Process Mechanical Engineering, 221*(1):64-77.

Kinsler, L. E., Frey, A. R. (1962). Fundamentals of Acoustics. *Applied Science*, 56-78.

Liu, C., Ji, Z. L. (2013). Computational fluid dynamics-based numerical analysis of acoustic attenuation and flow resistance characteristics of perforated tube silencers. *Journal of Vibration and Acoustics, 136*(2):021006.

Liang, Z., Li, S., Tian, J., et al. (2015b). Vibration cause analysis and elimination of reciprocating compressor inlet pipelines. *Engineering Failure Analysis, 48*:272-282.
Sakai, T., Saeki, S. (1973a). Study on pulsations of reciprocating compressor piping systems: 2nd report, model experiment of natural frequency. *Bulletin of the Japan Society of Mechanical Engineers*, *16*(91):63-68.
Sakai, T., Saeki, S. (1973b). Study on pulsations of reciprocating compressor piping systems: 1st report, calculation of natural frequency of complicated piping systems. *Bulletin of the Japan Society of Mechanical Engineers*, *16*(91):54-61.
Soedel, W. (1978). Designing simple low-pass flitter mufflers for small two-cycle engines. *Noise Control Engineering, 10*(2):60-66.
Wiggert, D. C., Tijsseling, A. S. (2001). Fluid transients and fluid-structure interaction in flexible liquid-filled piping. *Applied Mechanics Reviews*, *54*(5): 455-481.
Yu, D., Wen, J., Zhao, H., et al. (2008). Vibration reduction by using the idea of phononic crystals in a pipe-conveying fluid. *Journal of Sound and vibration*, *318*(1): 193-205.

Prediction of compressor muffler frequency response function using CFD

Tadeu Tonheiro RODRIGUES[1]*, Cristiano STUMPF[2], Ricardo Mikio DOI[3]

Embraco - R&D
Joinville, Santa Catarina, Brazil

[1]tadeu.t.rodrigues@embraco.com
[2]cristiano.stumpf@embraco.com
[3]ricardo.m.doi@embraco.com

* Corresponding Author

ABSTRACT

The acoustic filters of hermetic reciprocating compressors, also called mufflers, are usually developed through acoustic simulation solving the discretized wave equation to obtain the Frequency Response Function, which translates the acoustic response of the muffler. Nonlinear effects are neglected in this approach, which are attributed to flow patterns, as turbulence phenomena, which occur in the contractions, expansions and changing directions within the geometry. The main aim of this work is to investigate the influence of non-linear effects in the acoustic response of mufflers, solving the flow field by computational fluid dynamics (CFD). A discharge acoustic filter design was simplified for the study purpose and simulated using both CFD and Linear Acoustic techniques; the difference in the two approaches is made by comparing the Frequency Response Function (FRF). The flow effects are analyzed varying the compressor piston displacement and operating conditions. FRF predicted by CFD presents reasonable agreement with acoustics approach for lower frequencies identifying resonances and anti-resonances. It was observed increased disagreement for higher mass flow rates due to the predominance of flow effects over acoustics vibrations modes.

1. INTRODUCTION

Reactive mufflers are widely used in reciprocating compressor design for control and reduction of noise (Soedel, 2006) originated from the intermittent flow through valves, which generates gas pulsation in suction and discharge lines. Engineering approaches to design reactive mufflers are diverse and embrace experimental, analytical and numerical methods.

Lumped formulations were the first known method to evaluate mufflers until the introduction of advanced numerical methods based on the solution of the three-dimensional wave equation to obtain the harmonic response of reactive mufflers. Examples of application of lumped formulation are found in Soedel (2006) and more recently in Park *et al.* (2008), which demonstrated the usefulness of reducing low frequency noise due to discharge pulsation in a reciprocating compressor.

Application of numerical methods is presented in Vendrami *et al.* (2014), which optimizes the suction muffler design by acoustic harmonic analysis, reducing the irradiated noise. Designing reactive mufflers for reciprocating compressors requires multidisplinary approach, since them imply head losses which jeopardizes the compressor efficiency. The optimization of mufflers aiming improvements in noise and efficiency commonly follows a decoupled approach between thermal and acoustic methodologies as presented in Done *et al.* (2014).

Kim *et al.* (2015) applied CFD simulations to improve both acoustics and flow performance of a suction muffler, for these purposes the flow performance was evaluated in terms of mass flow rate increase while the acoustics performance was verified observing the Transmission Loss curves. The authors validated the muffler improvement directly in the compressor assembly by standard performance and acoustic tests, with reasonable agreement with numerical predictions.

The previous review show a lack of comparison between the decoupled and coupled approaches, despite the attractiveness of CFD simulations for obtaining reliable data, one can observe that transient simulations are far more costly than acoustic simulation of the wave equation, which is performed in the frequency domain.

The main aim of the present work is to present the results of a coupled evaluation through CFD technique, investigating the influence of non-linear effects related to pump size and operating conditions and comparing the results with linear acoustics simulations. For this purpose, the FRF was chosen as muffler system identity parameters. The discharge muffler was chosen due its importance in the total noise generation of reciprocating compressors, as reported by Bratty *et al.* (2012).

2. NUMERICAL PROCEDURE

2.1 Geometry simplification

The internal geometry of a real discharge system compressor presented in Figure 1.a was simplified to geometry of equivalent dimensions, as illustrated in Figure 1.b; the objective is to obtain a simpler geometry for mesh generation still representative for the purpose of this work.

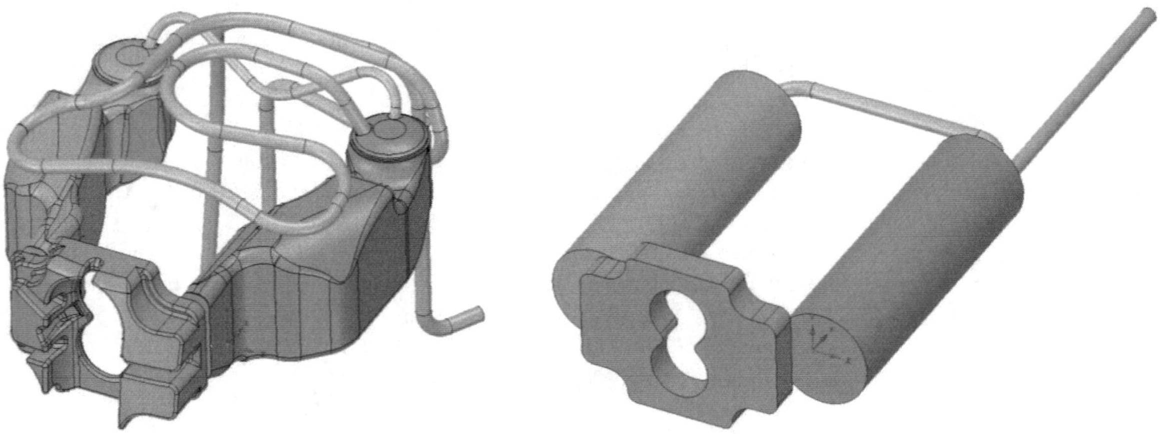

(a) Real geometry. (b) Simplified geometry.

Figure 1: Discharge muffler internal geometry.

2.2 CFD approach

The computational domain was discretized in tetrahedrons, observing skewness quality parameter; at the wall inflation layers were applied for better representation of boundary layer, resulting in computational mesh of 260,000 nodes, presented in Figure 2.

The unsteady, compressible, turbulent fluid flow through the discharge system was solved using the commercial code ANSYS CFX® release 15, based on the element-based finite volume method to discretize the partial differential equations of the mass conservation, momentum, energy and turbulence.

Advection terms were discretized using a High Resolution Scheme and the temporal discretization obtained by a second order backward Euler scheme. The Shear Stress Transport model was set to bridge the turbulent effects to the mean flow, which has been proven as a suitable model for compressor simulation (Rodrigues and Da Silva, 2014). The resultant system of equations is solved through Incomplete LU (ILU) decomposition, algebraic multigrid method and coupled strategy. The pressure evaluation correlation is obtained from the Redlich-Kwong library, available in the numerical tool.

The boundary conditions are illustrated next in Figure 2, with inlet at the discharge ports and outlet at the end of the discharge tube. The boundary conditions are based on performance rating points ARI LBP (Low Back Pressure) and

MBP (Medium Back Pressure) for propane refrigerant. Mass flow rate profile is prescribed at the inlet according to the crank angle (Figure 3), obtained from results of in-house compressor simulation software (Ussyk, 1984) for 60 Hz application, while at the outlet the condensing pressure is applied. At the inlet are set constant temperature, obtained from experimental measurements and turbulence intensity of 5%, which is the recommended value for tubular geometries.

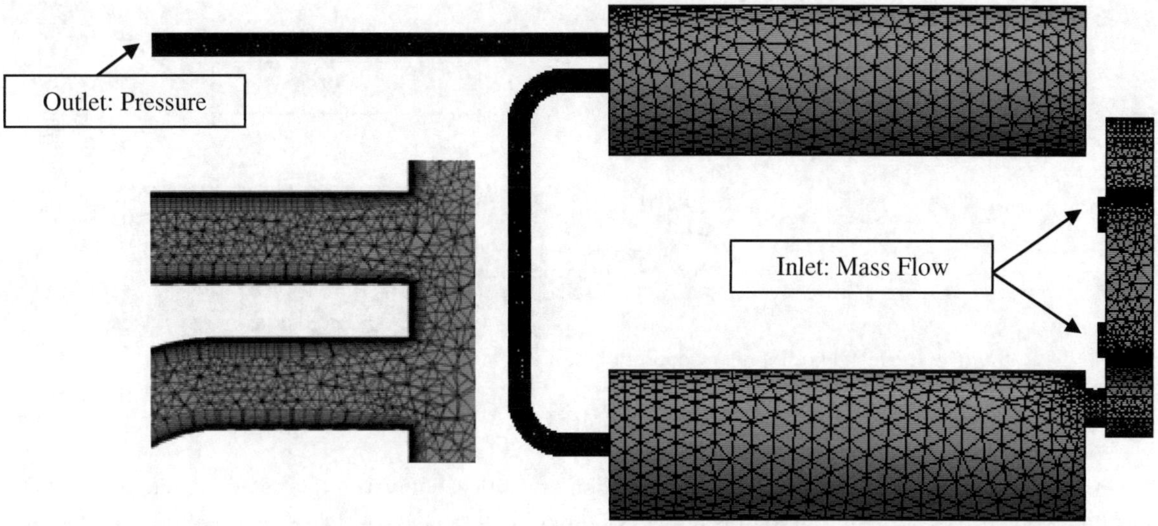

Figure 2: Mesh and boundary conditions for CFD simulation.

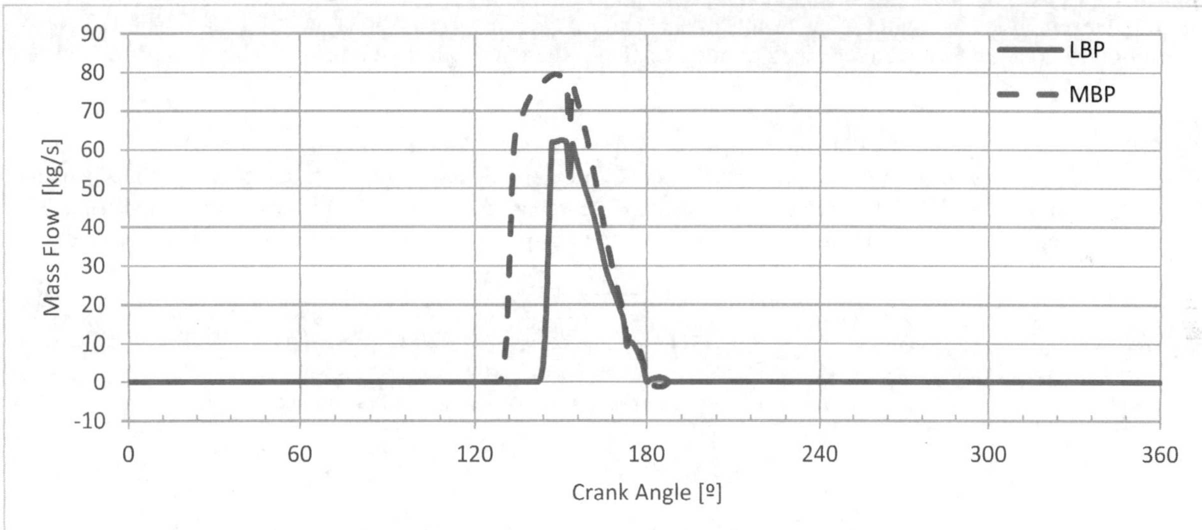

Figure 3: Mass flow rate profiles for inlet boundary condition.

2.3 Acoustics approach

A harmonic response simulation of the acoustic domain was performed in ANSYS Mechanical® release 15 for obtaining the FRF, based on the Finite Element Formulation for solving the governing equation of the physics chosen, in this case, the acoustic wave equation. The physical domain is discretized with fully hexahedral mesh, presented in Figure 4, which is compounded with 9070 second order elements and 41462 nodes. The fluid density and sound speed are set based on experimental measurement of temperature and pressure in the discharge system. At the fluid flow inlet, the system is excited with mass flow, while at the outlet is prescribed an acoustic absorption surface, as illustrated in Figure 4.

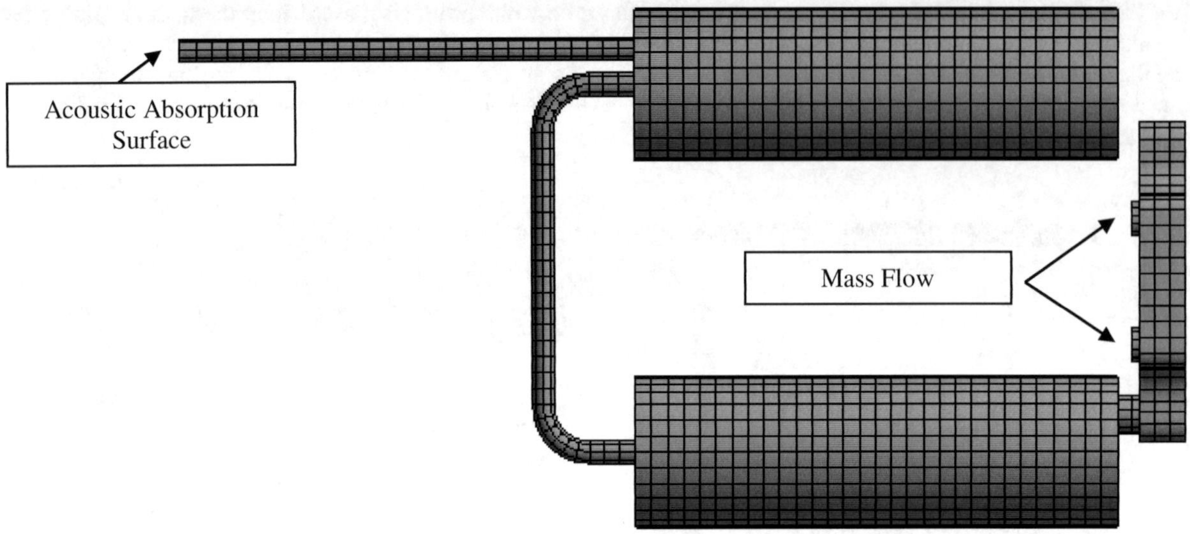

Figure 4: Mesh and boundary conditions for harmonic acoustics simulation.

3. RESULTS

FRF results are obtained by CFD with temporal discretization of $2x10^{-5}$s, equivalent to a sampling rate of 50000Hz, the simulations were performed with static flow initial conditions until obtained a statistical stable flow, in this case the sixth cycle. In order to observe nonlinear effects due to the influence of mass flow, the discharge system was simulated for ARI LBP and MBP rating points, each one for three pump sizes requiring correspondent mass flow rate profile exemplified in Figure 3. The results of each pump according to size order are referred as CFD1, 2 and 3. Acoustics results of the FRF were obtained through single simulation for the mentioned rating points, differentiated by the fluid properties set in the acoustics solver.

The FRF calculation through CFD requires temporal sampling of the excitation (inlet) and the resulting (outlet) pressures from the flow field, an example of the sampled pressure signals is illustrated in Figure 5. The sampled pressure signals are treated by Fast Fourier Transform (FFT) method to obtain the pressure magnitude spectrum. The FRF is calculated using the FFT results, based on the estimator provided in Equation 1 for deterministic methods:

$$FRF(f) = \frac{P_{out}(f)}{P_{in}(f)} \quad (1)$$

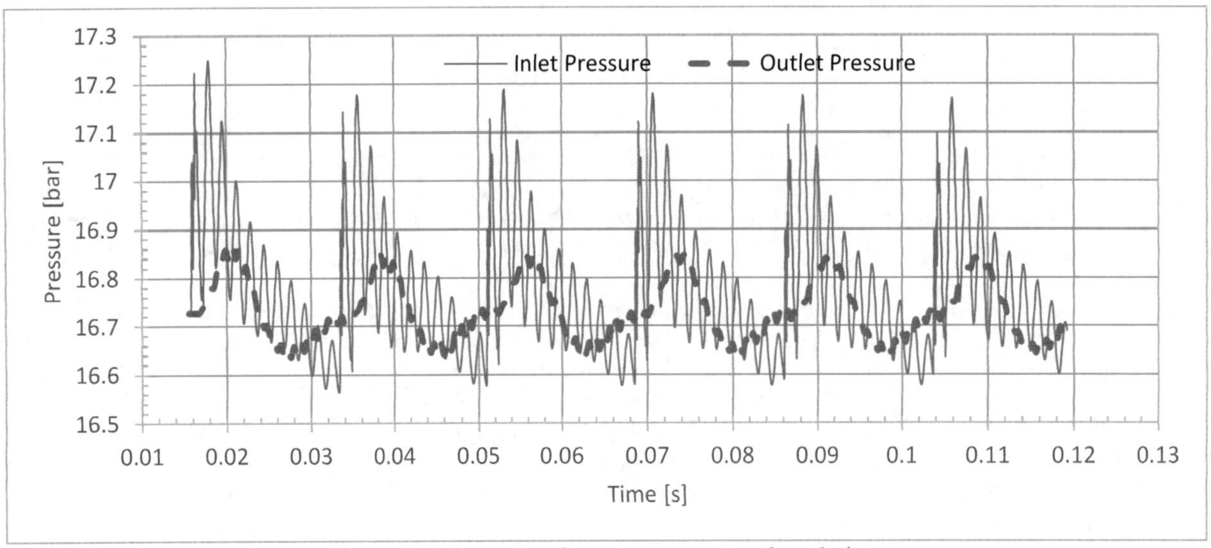

Figure 5: Inlet and outlet pressure temporal evolution.

Pressure spectrum magnitude results are presented in Figures 6 and 7, for inlet and outlet, respectively. One can note good agreement among all conditions simulated up to 1500Hz for inlet and 1400Hz for outlet, especially concerning identification of the main frequencies. Significant difference is found for the outlet pressure spectrum magnitude results for frequencies higher than 1400Hz, mainly related to the operating point. Compared to LBP results, the MBP results presents the main peaks anticipated and an overall increase of the spectrum due to the significant increase of the mass flow.

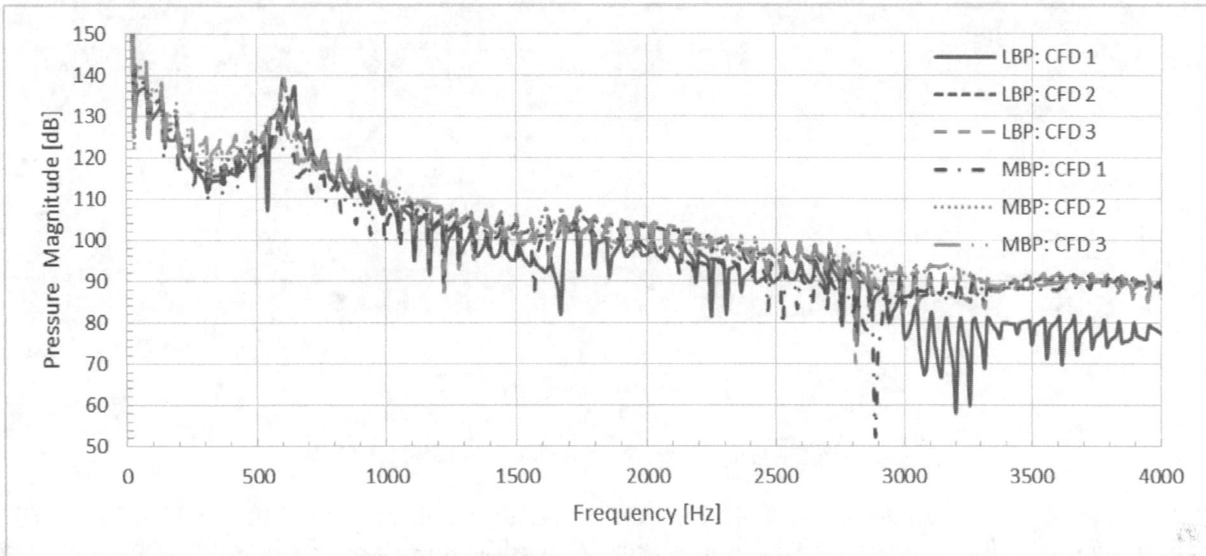

Figure 6: Spectrum magnitude of inlet pressure.

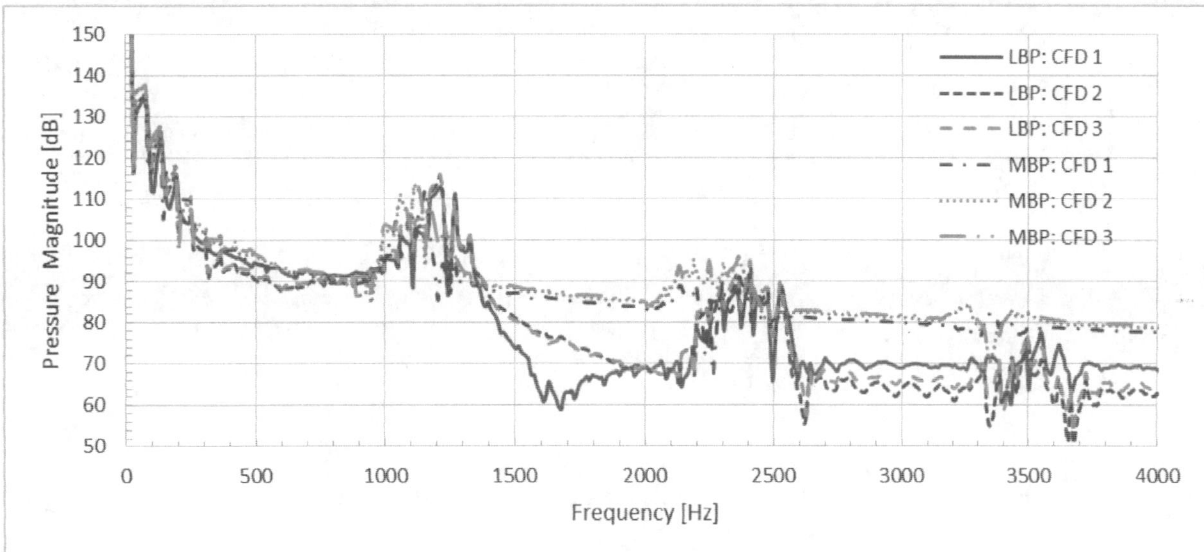

Figure 7: .Spectrum magnitude of outlet pressure.

Results of the FRF's obtained through CFD are compared to acoustics results in Figures 8 and 9, for LBP and MBP rating points, respectively. For LBP, the main profile shape agrees reasonably with the acoustics results until 1500Hz, then the agreement decreases until 2500Hz, from this frequency on the CFD results does not emulate any characteristics observed in the acoustics results.

In the range until 1500Hz, there is no identification of the first two resonances, one possible reason is due their proximity to the average compressor frequency. The first anti-resonance agrees well while the group of resonances around 1250Hz, the CFD approach identifies only the intermediate. Despite the effects of mass flow rate at LBP conditions, it was not found relevant differences among pump sizes.

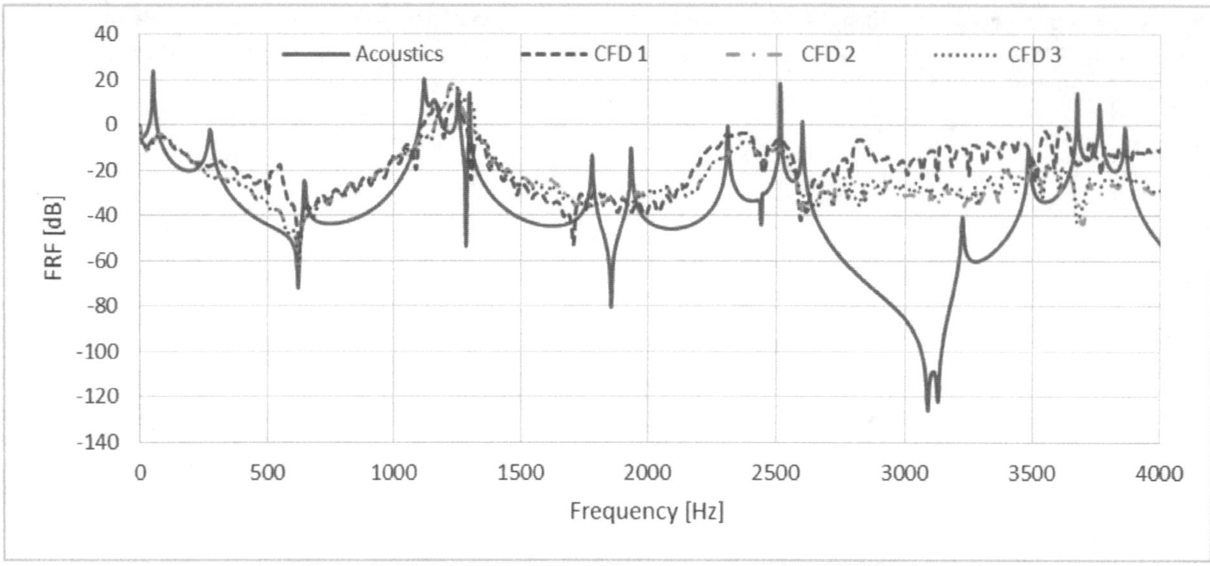

Figure 8: Frequency response function results for ARI LBP.

It is observed higher level of disagreement between CFD and acoustics approaches for MBP condition. In a similar way, it is possible identify the first anti-resonance and the intermediate resonance around resonance group till 1250 Hz, however the values are significantly damped compare to the acoustics results and the ones found for LBP condition.

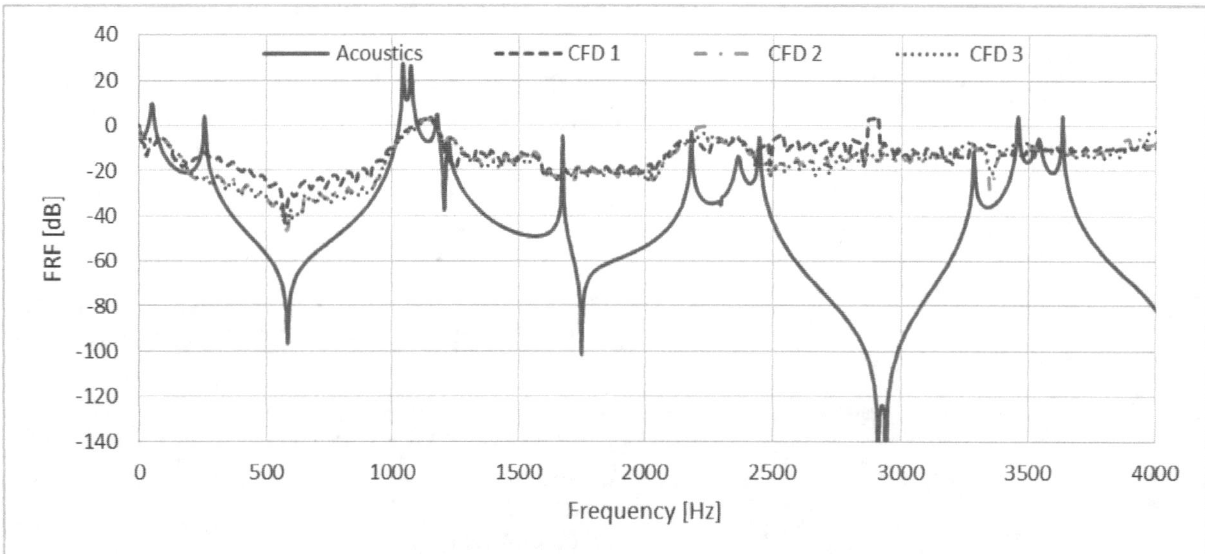

Figure 9: Frequency response function results for ARI MBP.

The interpretation of the results obtained for the rating point conditions indicates relevant influence of mass flow at expressive increase levels, for example, ideally the mass flow rate at MBP condition is 85% higher than LBP. For the pump sizes approached at each rating condition the mass flow rates vary within 36%.

A complementary comparison through visualization of solution fields is performed in order to check the consistency of the FRF results and presented only for LBP. The identification of the first six acoustic modes is presented in Figure 10 as acoustics pressure contours.

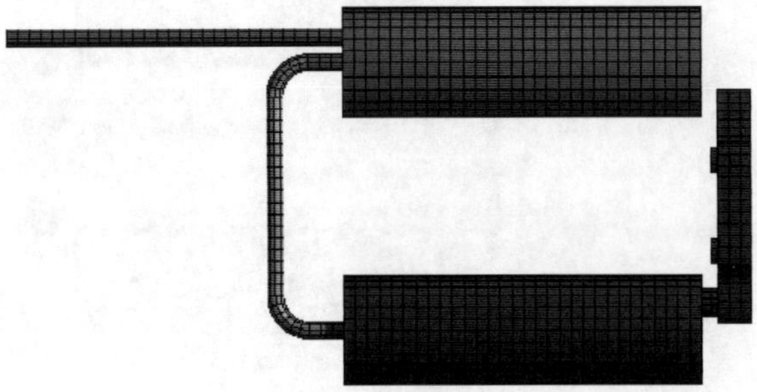

(a) First Mode.

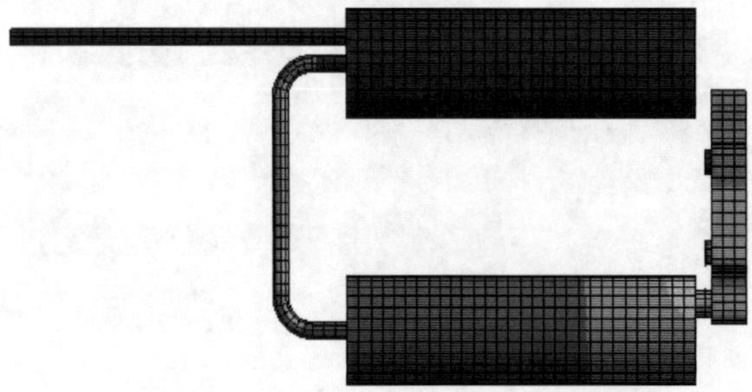

(b) Second Mode.

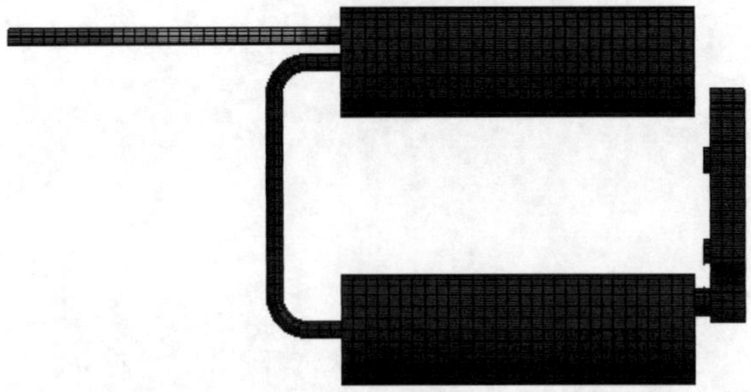

(c) Third Mode.

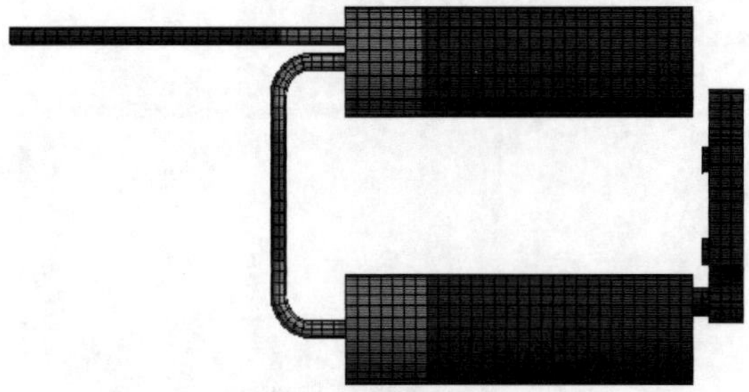

(d) Fourth Mode.

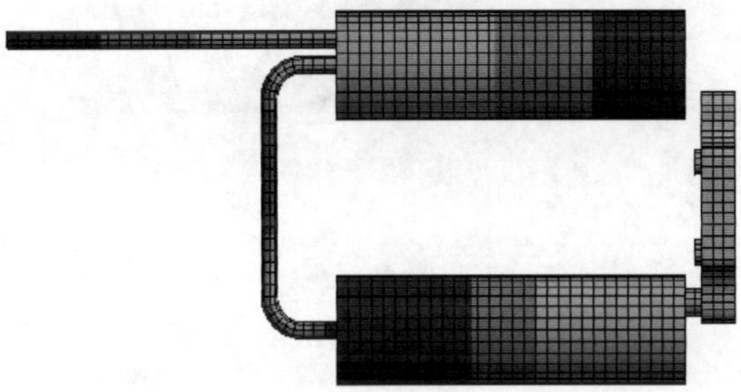

(e) Fifth Mode.

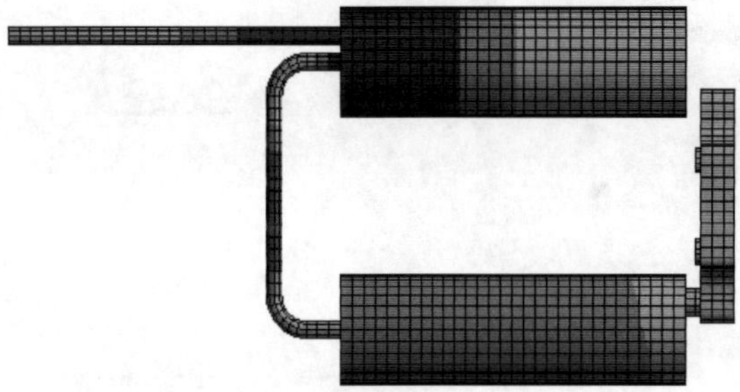

(f) Sixth Mode.

Figure 10: Acoustic mode shapes in pressure contours.

Based on the frequencies and acoustics pressure contours of the acoustic modes the pressure field obtained via CFD is analyzed for identification of similar patterns, depicted in Figures 11 and 12.

Equivalent periodic patterns to the second and third modes are clearly identified in Figure 11, however a deeper insight reveleals the equivalent second mode patterns its fully correlated with the mass flow pulse of 60 Hz frequency, which corroborates the absence of the first and second modes identications in FRF results. The remaining acoustic modes are very similar in patterns, a possible reason why the CFD approach does not dintingush them, due to the mass flow effects, and capture one intermediate mode in FRF, which is presented as time snapshots in Figure 12.

Observing practical applicability the CFD computational cost is significant higher compared to acoustics simulation, it was observed a total calculation time up to 120 times longer. The differences found does not indicate the acoustic method is unsuitable to predicted tendencies, based on this one can conclude is still more effective for large optimization computations.

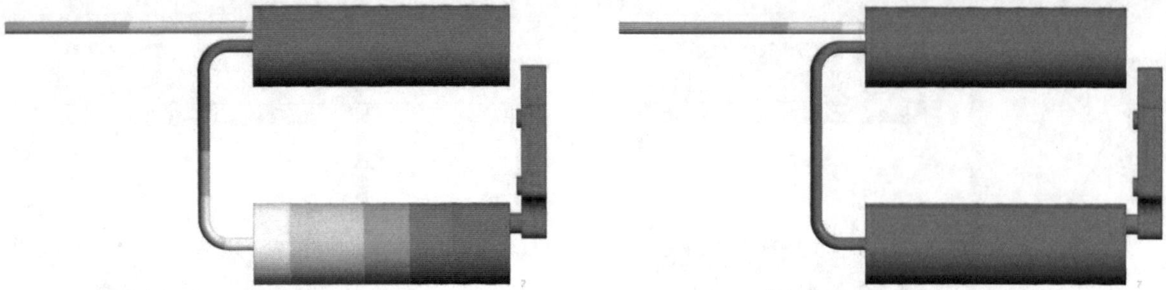

(a) Second mode pattern. (b) Third mode pattern.

Figure 11: Pressure flow patterns equivalence to acoustic modes.

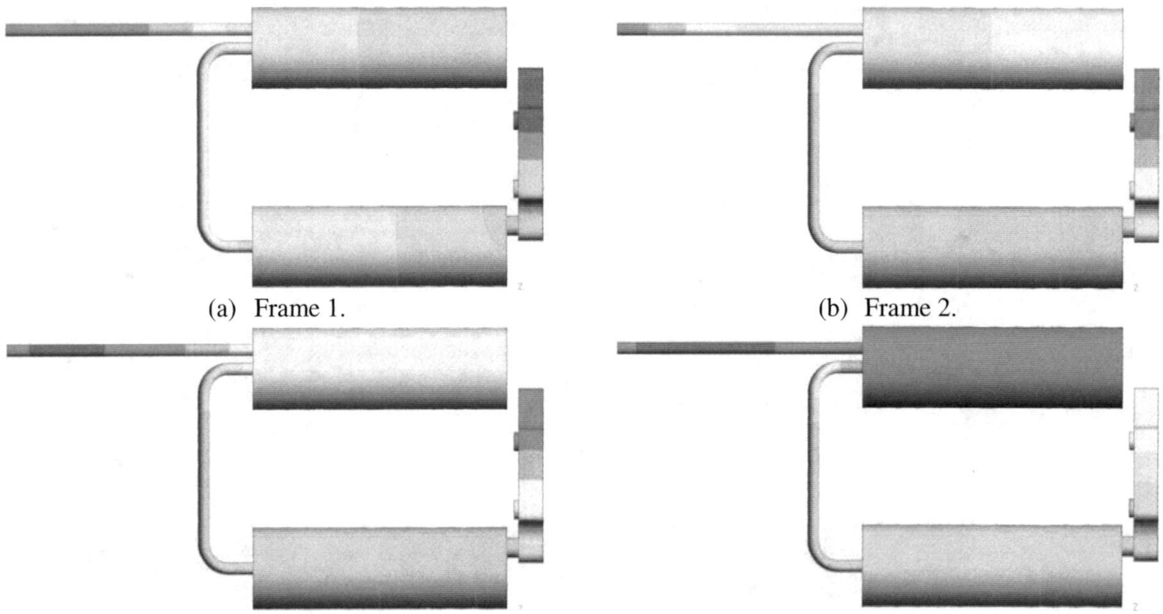

(a) Frame 1. (b) Frame 2.

(c) Frame 2. (d) Frame 4.

Figure 12: Pressure flow patterns equivalence to fourth-sixth acoustic modes.

4. CONCLUSIONS

In the presented work, an evaluation of CFD approach as coupled technique for analysis of acoustic performance is presented; for this purpose the FRF of a simplified discharge system was obtained through acoustic and CFD numerical methods. It was observed the CFD approach is able to predict FRF with similar profiles to the acoustics method and includes mass flow effects, which was evidenced that for higher mass flow rates the FRF obtained from CFD solution tends to distance from acoustic FRF, as the mass flow rate increases. The FRF results were confirmed observing the pressure patterns in both acoustics and flow solution field. Although the CFD approach is useful to perform acoustics evaluation and it is able to include mass flow effects in the analysis, the computational cost is more expensive compared to acoustics approach, becoming unsuitable for large computation in optimization procedures. On the other hand. due to the wider inclusion of physical phenomena, its applicability is appropriated as complementary evaluation of the acoustics approach in order to understand the effects of the mass flow on the muffler´s FRF previously to performing experiments.

NOMENCLATURE

CFD	Computational Fluid Dynamics	(–)
FFT	Fast Fourier Transform	(–)
FRF	Frequency Response Function	(–)
ILU	Incomplete LU factorization	(–)

LBP	Low Back Pressure	(–)
MBP	Medium Back Pressure	(–)
SST	Shear Stress Transport	(–)
P	Pressure spectrum	(Pa)

Subscript

in	inlet
out	outlet

REFERENCES

Bratti, G., Santini Junior, O., Baars, E., Barbieri, R. & Pellegrini, C. (2012). Identification of Energy Path in the Interaction between Compressor and Refrigerator. *Proceedings of the 21nd International Compressor Engineering Conference at Purdue* (Paper 2091). West Lafayette, USA: Purdue University.

Done, V., Balide, V., Tamma, B., Soni, K., Dey, S., Angadi, S. & GP, V. (2014). Muffler Design for a Refrigerator Compressor. *Proceedings of the 22nd International Compressor Engineering Conference at Purdue* (Paper 2262). West Lafayette, USA: Purdue University.

Kim, S., Cheong, C., Park, J., Kim, H. & Lee, H. (2015). Numerical and Experimental Investigation into Flow and Acoustic Performances of Muffler for a Hermetic Reciprocating Compressor. *Proceedings of the 22nd International Congress on Sound and Vibration.* Florence, Italy: IIAV.

Park, J., Kang, J., Choi, J., Jang, P., Yoon, M., Kwon, D., & Oh, I. (2008). Reduction of the Low Frequency Noise Due to the Discharge Pressure Pulsation of a Reciprocating Compressor. *Proceedings of the 19nd International Compressor Engineering Conference at Purdue* (Paper 1898). West Lafayette, USA: Purdue University.

Rodrigues, T. T., Da Silva, D. L. (2014). Turbulence Modelling Evaluation for Reciprocating Compressor Simulation. *Proceedings of the 22nd International Compressor Engineering Conference at Purdue* (Paper 2297). West Lafayette, USA: Purdue University.

Soedel, W. (2006). *Sound and Vibrations of Positive Displacement Compressors*. Boca Raton, FL: CRC Press.

Ussyk, M.S., "Numerical Simulation of Hermetic Reciprocating Compressor Performance", (in Portuguese), M.Sc. Dissertation, Federal University of Santa Catarina, 1984.

Vendrami, C. E.; Pellegrini, C., Hattori, M. A. & Climaco, D. (2014). Acoustic Improvements for a New Generation of Variable Speed Compressor. *Proceedings of the 22nd International Compressor Engineering Conference at Purdue* (Paper 2308). West Lafayette, USA: Purdue University.

ACKNOWLEDGEMENT

The authors acknowledge the EMBRACO technical and financial support

3D Compressible Simulation Of a Muffler With Pseudosound Prediction Levels

Jesús RUANO[1*], Joan LÓPEZ[1], Oriol LEHMKUHL[1], Joaquim RIGOLA[1], Carles David PÉREZ SEGARRA[1]

[1]Heat and Mass Transfer Technological Center (CTTC)
Universitat Politècnica de Catalunya – Barcelona TECH (UPC),
Colom, 11, E-08222, Terrassa (Barcelona), Spain
Tel. +34 93 739 81 92
Email: cttc@cttc.upc.edu, http://www.cttc.upc.edu

* Corresponding Author

ABSTRACT

The main objective of this paper is to present a numerical resolution of a suction muffler configuration by using an in-house object oriented CFD & HT code TermoFluids (Lehmkuhl et al. 2007), able to handle tridimensional geometries, unstructured meshes with intrinsic parallelization. This code has been adapted to be able to resolute 3D Navier-Stokes equations in their compressible form and coupled with the numerical resolution of the whole compressor domain by means of a parallel and object-oriented called NEST (Lopez, 2016).

The numerical results aim to study the influence of the suction muffler inner geometry in the fluidynamic behavior inside the muffler while considering how this internal geometry affects the global performance of the compressor. Hence, the inlet and outlet boundary conditions at the muffler are obtained from the numerical simulation of the whole compressor using NEST, while the fluid behavior inside the muffler is numerically simulated by means of detailed analysis. The paper presents a methodology that handles with Large Eddy Simulation (LES) models for the turbulent motion of fluid inside the muffler, the formulation of Navier-Stokes in their compressible form, dealing with numerical problems derived from the compressible part, and the coupling of the whole compressor simulation to set boundary condition.

1. INTRODUCTION

The understanding of the physical phenomena happening inside a reciprocating compressor is a must to improve its design and efficiency. However, nowadays is still a challenging problem due the fluid inside the compressor is submitted to periodic changes mainly in its pressure and velocity and consequently to all the thermo-dynamical values associated with them. Hence, a numerical code used to simulate compressors must be able to handle with the compressible nature of the refrigerant and the periodicity of the variations in the main values of the fluid.

One-dimensional models are one of the most used tools to numerically analyze compressors. These models offer a faster simulation rather than conventional CFD & HT simulations, where the three dimensionality of the problem is taken into account. The great improvement in speed makes possible the comparison of different geometries and configurations in a relative short period (Rigola, 2002).

While one-dimensional models have been used to study mufflers since more than 40 years, three-dimensional simulations had recently begun to be adopted by the community. While several authors have relied on constant boundary conditions (Choi, 2002), Sarioglu (2012) simulated a full cycle inside the muffler using time dependent boundary conditions extracted from an empirical study. Done (2014) simulated different muffler geometries imposing acoustic pressure at inlet and measuring the outlet signal. His study also shown although fluid-dynamics can be correctly computed with one-dimensional models; acoustics at medium-high frequencies are wrongly computed if three-dimensional models are not taken into account.

The present paper is focused on the numerical resolution of a muffler using 3D unstructured, parallel and object oriented CFD & HT code TermoFluids (Lehmkuhl et al. 2007). This in-house code uses Large Eddy Simulation (LES) turbulence models instead of classical Reynolds Averaged Navier-Stokes equations (RANS). As a post process, the spectrum of the pressure at both inlet and outlet will be analyzed in order to understand muffler's behavior towards pseudonoise generation and propagation trough the component.

2. NUMERICAL METHOD

The numerical simulation of the suction muffler is based on a 3D explicit finite volume method adapted to compressible flows and integrated in time by using a second order Adams-Bashford time scheme. Turbulence modeling is an extension of Wall-Adapting Local Eddy-viscosity (WALE) (Nicoud, F. and Ducros., 1999) to unstructured meshes. These methods have been implemented in an in-house code TermoFluids (Lehmkuhl *et al.*, 2007).

2.1 Navier-Stokes equations

The main features of the discretized Navier-Stokes equations in its compressible form will be presented here. The compressible form of Navier-Stokes equations takes into account the effect of density variations, which are of high importance inside a compressor. If these variations want to be taken into account then momentum equations alone are not enough to simulate the problem; density and energy equations as well as a state equation are necessary to be added to the set of equations to correctly simulate the problem.

$$\Omega_c \frac{\partial \rho_c}{\partial t} + M u_c = 0 \tag{1}$$

$$\Omega_c \frac{\partial (\rho_c u_c)}{\partial t} + C(\rho_c u_c) u_c + D u_c + \Omega_c G p_c = 0 \tag{2}$$

$$\Omega_c \frac{\partial (\rho_c h_c)}{\partial t} + C(\rho_c u_c) h_c + D T_c + \tau : G u_c = 0 \tag{3}$$

The modeled substance is assumed to follow real gas law, being the selected equation the ideal gas law with compressibility factor calculated via acentric factor:

$$p = z \rho R T \tag{5}$$

$$z = z_0 + \omega \cdot z_1 \tag{6}$$

$$z_0 = 1 + (0.083 - 0.422/T_r^{1.6}) \cdot P_r/T_R \tag{7}$$

$$z_1 = (0.139 - 0.172/T_r^{4.2}) \cdot P_r/T_r \tag{8}$$

The acentric factor ω determines which substance is being simulated. Due a large amount of domestic compressors use isobutene as refrigerant, the selected acentric factor has been the corresponding to this fluid (0.18).

2.2 Turbulence modeling

The intrinsic turbulence associated to the chaotic fluid movement inside the muffler is modeled using TermoFluids (Lehmkuhl *et al.*, 2007), which has an extended library of turbulence models developed. The chosen one for this article is the Wall-Adapting Local Eddy-viscosity model (WALE) (Nicoud, F. and Ducros., 1999), which has been demonstrated to have a proper behavior near walls even with coarser grids compared with other models.

$$\mu_{SGS} = \rho \left(C_\omega V^{1/3} \right)^2 \frac{tr(S \cdot S^T)^{3/2}}{tr(\bar{S} \cdot \bar{S}^T)^{5/2} + tr(S \cdot S^T)^{5/4}} \tag{9}$$

$$S = \frac{1}{2}(G(\bar{u}')^2 + G^T(\bar{u}')^2) - \frac{1}{3} I (G(\bar{u}')^2) \tag{10}$$

$$\bar{S} = \frac{1}{2}(G(\bar{u}') + G^T(\bar{u}')) \tag{11}$$

Where the operator $G(\varphi)$ is the gradient matrix operator presented previously and I is the identity matrix.

2.3 Solver properties

Summarizing, the properties which the used solver has are:

- The convective and diffusive terms have been discretized using second order schemes. These schemes have been chosen to ensure the preservation of the kinetic energy (Jameson 2008, Baez et al. 2014, Pedro et al. 2015).
- At inlet and outlet, two buffer zones are used in order to assure stability during the simulation.
- Temporal evolution of the equations is computed via Runge-Kutta integration method, RK3. Time step is computed at step with a CFL 0.5. We remark the requirement that the used time step has to be able to compute not only the hydrodynamic behavior of the simulation but also the acoustic; hence, the maximum velocity of the simulation is the speed of sound (no supersonic Mach numbers are expected).
- LES turbulence model with WALE model which has produced overall good results without the additional effort to include new scalar equations to be resolved such as k-ε or k-ω models.
- Gas properties have been calculated at each time step using ideal gas equation with compressibility factor correction.

3. COMPUTATIONAL DOMAIN AND BOUNDARY CONDITIONS

3.1 Boundary conditions

The detailed simulation of the whole compressor with a three dimensional modeling is still not possible even with High Performance (HPC) facilities. Hence, the muffler has to be simulated as a part of the whole system but, at the same time, respecting the effect it produces to the whole compressor.

For this reason, first a 1D simulation of the whole compressor is performed in order to extract the values at both inlet and outlet of the muffler, in order to obtain the boundary conditions which couple the element with the whole system.

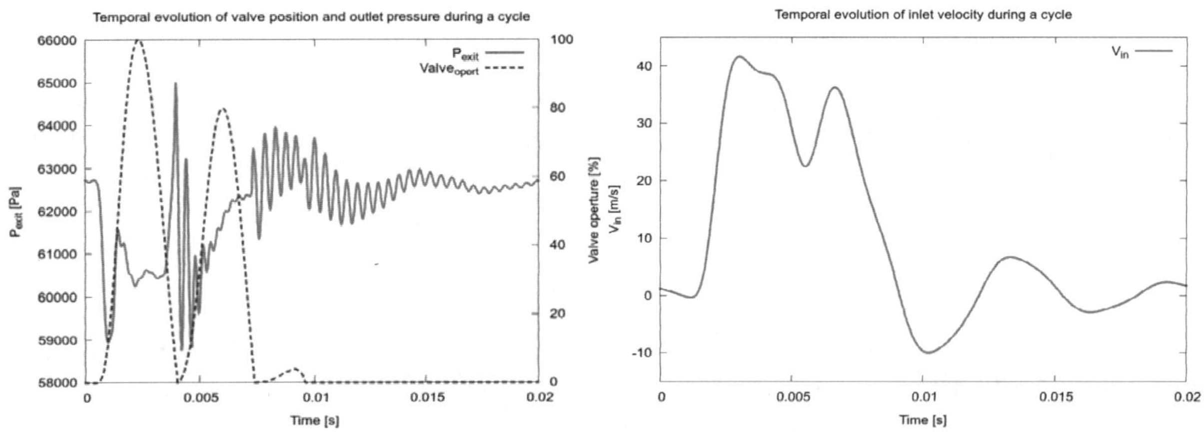

Figure 1: Boundary conditions at a) Outlet: Pressure and valve position and b) Inlet: velocity profile

Using the previous values as boundary conditions at both inlet and outlet, the simulation of the muffler becomes a reality. Hence, the velocity is imposed at the inlet of the element while a mixed boundary condition imposing valve position and pressure is set at the outlet. This mixed condition behaves as a non-slip wall boundary condition or as an outflow boundary condition by imposing the pressure depending of the valve's position. Summarizing, these both boundary conditions, at inlet and outlet, extracted from a one-dimensional simulation of the whole compressor, make feasible the simulation of a single element but taking into account, to some degree, its effect to the whole system. Finally, the condition applied at the inner walls of the muffler is an adiabatic non-slip boundary condition.

3.2 Computational domain

The domain selected to perform the simulation enlarges the muffler geometry as well as two buffer zones in order to attenuate spurious oscillations which can occur during simulation and ensure numerical stability during the computation.

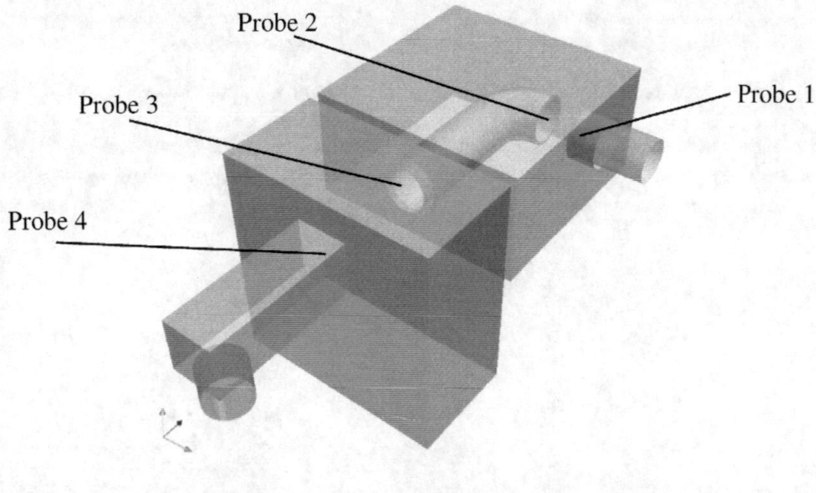

Figure 2: Computational domain and probes location

Inside the domain, where maximum precision is mandatory, schemes which preserve the kinetic energy, such as Symmetry Preserving schemes, are used in the discretization whereas up-winding techniques like upwind discretization schemes are used in buffer layers as has been previously commented. The use of a hybrid discretization is able to ensure numerical stability due to domain truncation at boundaries while in the inner domain the energy, and consequently noise and pressure perturbations, are not damped by the discretization.

Figure 3: Isometric and front view of the used mesh

A tetrahedral unstructured mesh of approximately 2 Million Control Volumes is used. In order to correctly capture fluid expansion as well as pressure gradient at walls, the mesh is refined at the beginning and the end of all the tubes inside the muffler as well as at the walls of the computational domain. Furthermore, the quality of the inner cells has been set by iteration as a minimum aspect ratio value of 0.4 in order to improve the overall results of the simulation.

4. COMPUTATIONAL RESULTS

4.1 Hydrodynamic results

In this section, the results obtained from the transient three-dimensional numerical simulation of the studied muffler are commented. First, the velocity magnitude inside the muffler will be analyzed at several time localizations in order to check muffler's behavior qualitatively.

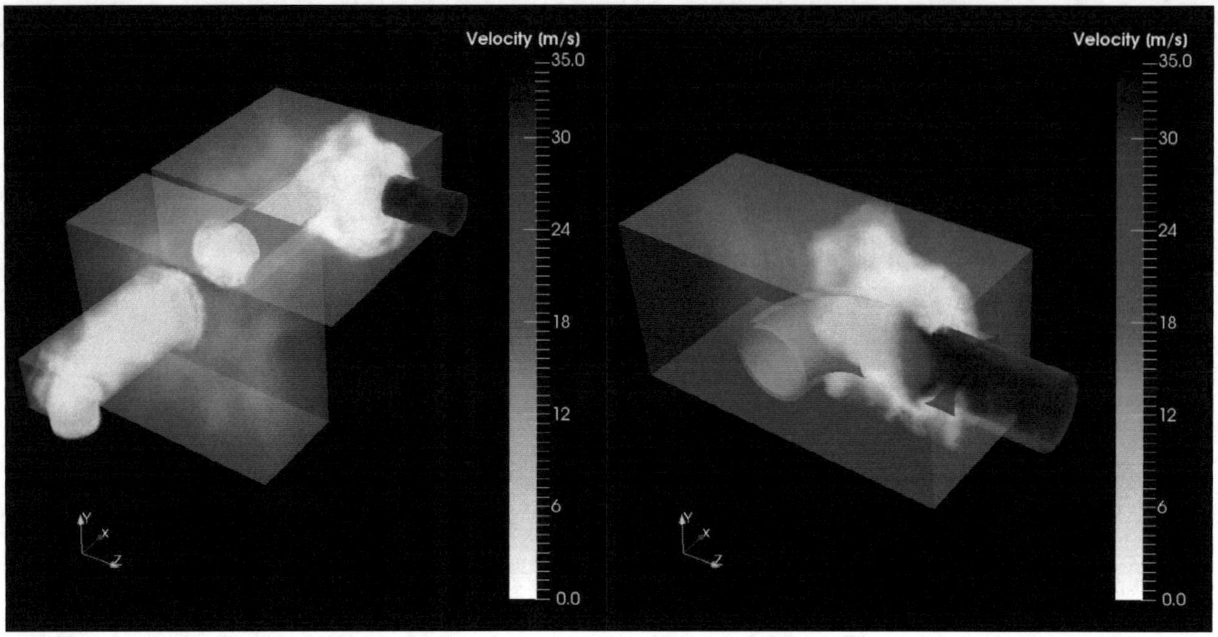

Figure 4: Velocity magnitude at valve maximum opening in a) the entire muffler and b) end of the inlet pipe.

The previous pictures show the velocity magnitude inside the muffler when the valve is fully open (0.0023 sec). At this point of the cycle the inlet flow has not reached the outlet, hence the outflow consists in the remnant fluid of the previous cycle. This could be observed in the expansion of the fluid at the end of the inlet pipe; if the fluid is still expanding this means it clearly hasn't reached the outlet. The maximum velocity during all the cycle is achieved during this period of time, when the outflow reaches approximately 40 m/s.

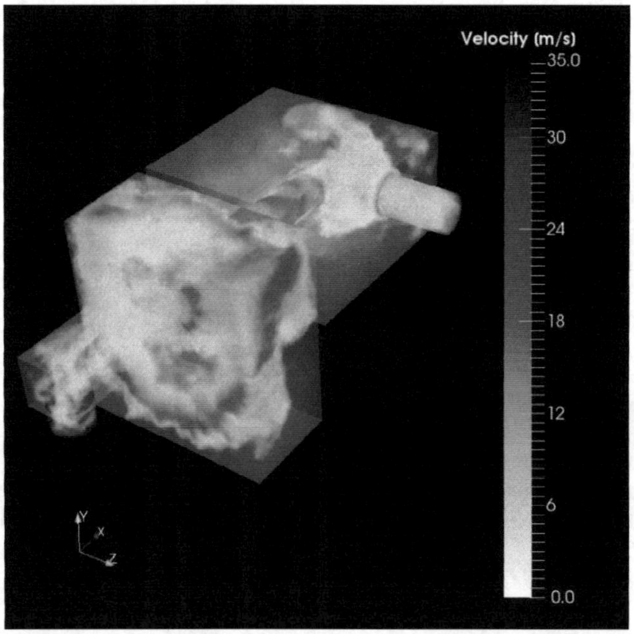

Figure 5: Velocity magnitude at valve closure in the entire muffler.

Figure 5 shows the velocity field when the valve is closed, at 0.0095 seconds. As can be seen, the velocity magnitude at the exit is practically 0, consisting in just small fluctuations. However, the rest of the domain has a clearly non null value. At the first expansion chamber could be observed how the fluid is still moving towards the muffler due to the form the expansion jet has. Thus, the chamber is accumulating mass when the valve has already

closed. Finally, is important to notice the velocity at the first chamber has only noticeable values at the near zone of the expansion tube, while the fluid at the rest of the chamber behaves as a fluid at rest state.
On the other hand, the second expansion chamber shows how the fluid is still moving and expanding. This happens as a conjunction of the previous facts: the valve has closed and consequently the fluid cannot go to the compression chamber but at the same time there is fluid coming from the inlet and first chamber.

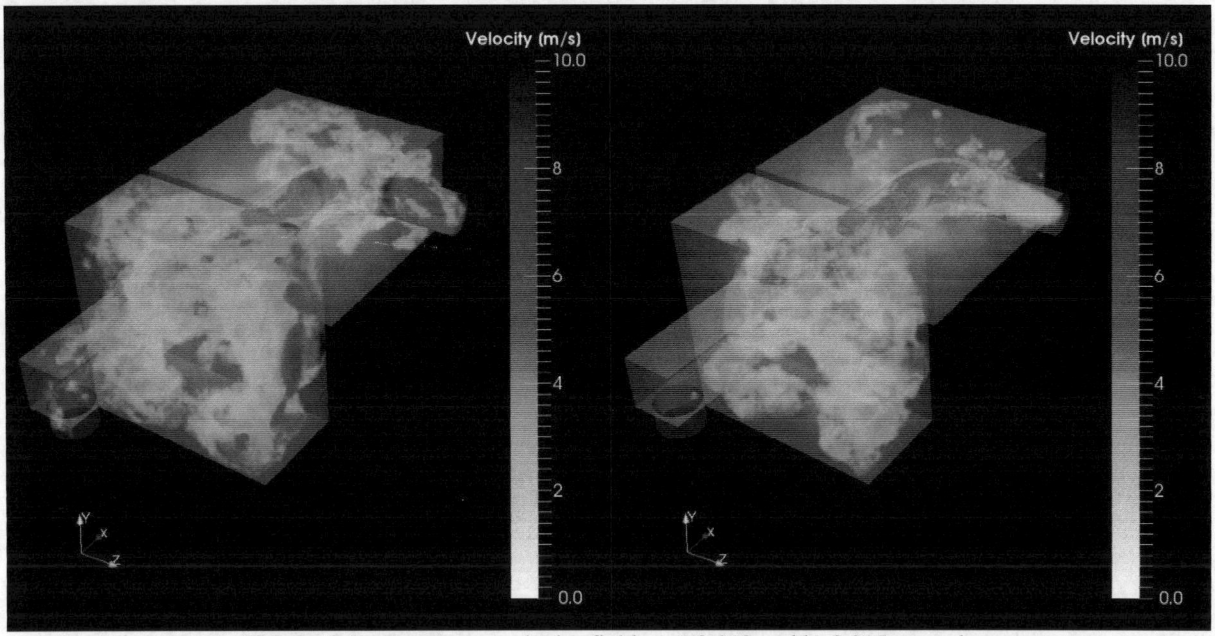

Figure 6: Instantaneous velocity field at a) 0.012 and b) 0.015 seconds

The previous pictures show the velocity field at 0.0125 and 0.015 seconds. These two instants have been selected because two facts could be observed: first, how the velocity magnitude starts to decrease (note the selected scale); second, how the maximum value is located at the inner pipe which communicates both chambers. This last fact comes from recirculation of fluid that lives between the first chamber and the second one during the part of the period outlet is closed; this back and forth movement is slowly decreased until the fluid stays at rest.

Apart from the instantaneous velocity fields, the pressure at several inner points of the muffler has been saved during the simulation, in order to study its temporal evolution which will be useful in a posterior acoustic analysis of the results. These points can be observed in figure 2.

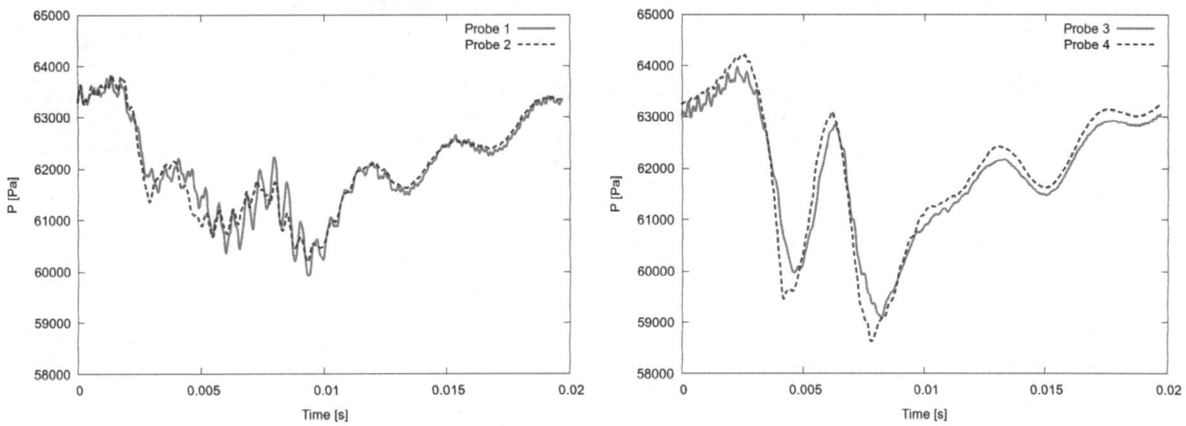

Figure 7: Temporal evolution of pressure at several locations

Both figures show a similar pattern: an initial pressure drop, with more or less superimposed oscillations until half cycle approximately, and a rise in pressure during the second half of the cycle until achieving initial values. The initial pressure drop is explained by the mass outgoing of the element; while the rise in pressure during the second half is explained by positive mass flow entering the muffler.

In figure 7, it can be observed how, apparently, probe 1 has larger oscillations than probe 2 even when the main purpose of the muffler is to attenuate the oscillations at the inlet. This happens due fluid behavior at the end of the inlet pipe is similar to a jet and consequently oscillations which hadn't been taken into account appear. This is checked during the second half of the cycle, when this jet-like behavior stops and these oscillations disappear.

The second set of probes doesn't show the commented oscillations, mainly because velocity magnitude at the second expansion is lower than in the first one. This set of probes show how the peaks of the oscillations are attenuated in the second chamber.

Furthermore, the pressure drop across the muffler during the cycle as well as the ratio between pressures at inlet and outlet can been computed. A pressure ratio similar to 1, or equivalently a pressure drop closer to 0, will indicate a good performance of the muffler from an hydrodynamic point of view due the presence of this element has as objective reduce pressure fluctuations without introducing additional losses, such as additional pressure drop, to the compressor.

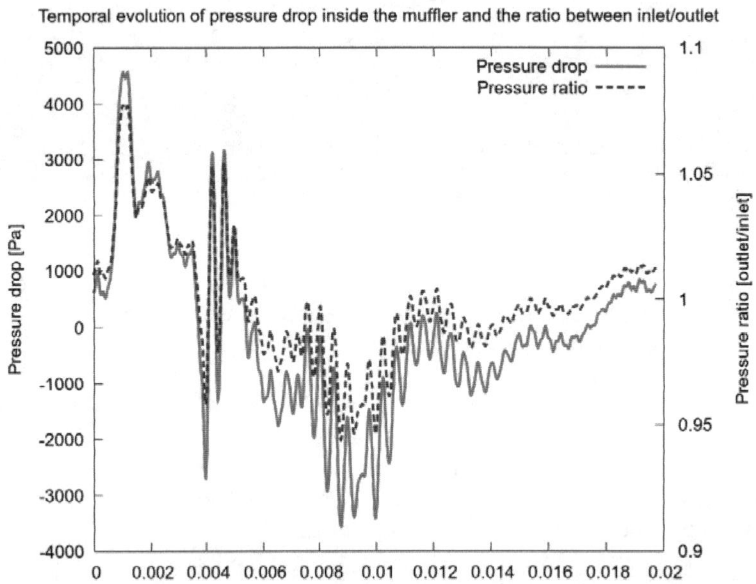

Figure 8: Pressure drop and ratio inside the muffler during a cycle

The previous figure shows how the pressure drop oscillates between 4500 and -3500 Pascals. Negative values are no surprise due as has been commented previously, fluid started to accumulate near the exit and hence the pressure is expected to grow more than in inlet when this happens. On the other hand, pressure ratios values oscillate near 1, which means the studied muffler doesn't affect the overall performance of the compressor.

4.2 Acoustic results

The acoustic behavior of the simulated muffler is studied via transforming the obtained pressure signals in the temporal domain to the spectral domain. This methodology permits studying which frequencies are being attenuated as well as if several resonances exist inside the muffler. By using a CFD simulation instead than simulating just the propagation of acoustic waves, two different acoustic waves exist:

- Waves which exist at the outlet of the muffler and are propagated through the inlet.
- Waves which are generated inside the muffler as a flow-induced noise.

Hence, the key point of this methodology is that is possible to take into account the noise generated by the muffler itself rather than just considering the muffler as an element which propagates noise but doesn't generate.

Figure 9 shows the spectrum of the pressure signals at both inlet and outlet of the muffler obtained after a Fourier Transform of the time dependent data. The ratio between the two signals is known as transmission losses (TL) of the muffler that shows how the element is capable to reduce pressure oscillations from the outlet (acoustic inlet) to the inlet (acoustic outlet); according to Lee (2002) the obtained parameter in this simulation would be more approximate to the insertion losses (IL) of the muffler instead of TL due the flow induced noise of the muffler is being taken into account. Regarding this measure, the obtained results in this simulation show good agreement with the ones of Lee (2002).

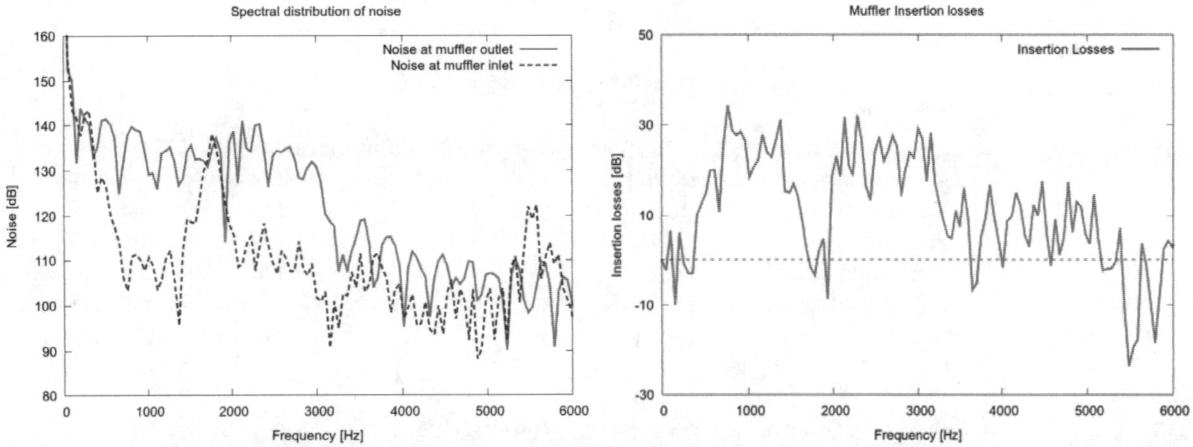

Figure 9: Noise at inlet and outlet and Insertion Losses of the muffler

Additionally, the spectrum at the different probes has also been computed to study how the different elements inside the muffler affect noise reduction. Moreover, the equivalent noise at each probe has been calculated to understand how it is reduced. The obtained results show good performance of the simulated muffler in the different studied positions, being the larger levels of noise those near the outlet of the muffler.

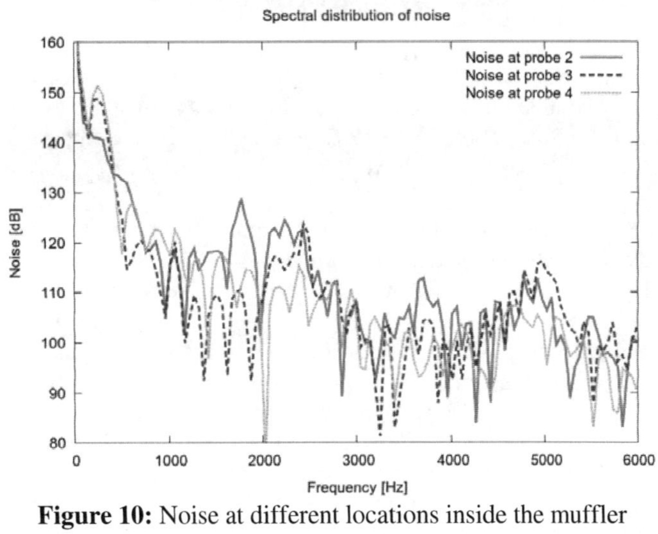

Figure 10: Noise at different locations inside the muffler

Table 1: Equivalent noise at different locations inside the muffler

Probe	1	2	3	4	Outlet
Noise [dBA]	153.086	153.24	155.588	156.845	157.367

5. CONCLUSIONS

The conclusions which can be extracted from the three-dimensional simulation of an academic muffler are:

- The present article uses a code able to predict numerically hydrodynamic and acoustic behavior in mufflers.
- The overall pressure inside the compressor drops after the valve opening. During the second half of the cycle the pressure rises achieving the values at the initial cycle time.
- During the second part of the cycle the total net mass flow entering the muffler is positive. However, several backflows occur.
- Jet-flow like behavior at the end of the inlet tube generates additional oscillation frequencies related to self-induced noise inside the muffler.
- The acoustic results show muffler is able to attenuate oscillations and good agreement with other authors.

NOMENCLATURE

Variables

ρ	Density	(kg/m^3)
u	Velocity vector	(m/s)
p	Pressure	(Pa)
T	Temperature	(K)
R	Ideal gas constant	(J/mol·K)
ω	Acentric factor	(-)
z	Compressibility factor	(-)
μ_{SGS}	Subgrid-scale viscosity	(Pa·s)
C_{ω}	WALE coefficient	(-)
$V^{1/3}$	Filter length	(m)

Subscript

r	Reduced value

Operators and matrices

Ω_c	Mass discrete operator
M	Divergence discrete operator
C	Convective discrete operator
D	Difusive discrete operator
G	Gradient discrete operator
τ	Shear stress tensor
I	Identity matrix

Superscript

$-$	Filtered value
$'$	Small scale component
T	Transposed matrix

REFERENCES

Baez, A., Pedro, J., Lehmkuhl, O., Trias, X., Perez Segarra, C.D. 2014. A filtered kinetic energy preserving methodology for large eddy simulations of compressible flows on unstructured meshes. . Kaliningrad.

Baez, A., Pedro, J., Lehmkuhl, O., Rodriguez, I., Perez Segarra, C.D. 2014. Comparing Kinetic energy preserving and Godunov schemes on the flow around a NACA0012. *11th European Conference on Computational Fluid Dynamics*. Barcelona.

Choi, J.M., Joo, J.M., Oh, S.K., Park, S.W, 2000, Smart suction muffler design for a reciprocating compressor. *International Compressor Engineering Conferece*. Purdue.

Done, V., Ventakesham, B., Tamma, B., Soni, K., Dey, S., Angadi, S., GP, V. 2014 Muffler Design for a Refrigerator Compressor. *International Compressor Engineering Conferece*. Purdue.

Jameson, A. 2008. Formulation of Kinetic Energy Preserving Conservative Schemes for Gas Dynamics and Direct Numerical Simulation of One Dimensional Viscous Compressible Flow in a Shock Tube Using Entropy and Kinetic Energy Preserving Schemes, *J. Sci. Comput.*, v.34, pp.188-208.

Lee, J.H., An, K. H., Lee, I. S. 2002, Design Of The Suction Muffler Of A Reciprocating Compressor. *International Compressor Engineering Conferece*. Purdue.

Lehmkuhl, O., Perez-Segarra, C.D., Soria, M., Oliva, A. 2007, TermoFluids: A new Parallel unstructured CFD code for the simulation of turbulent industrial problems on low cost PC cluster. *Proceedings of the Parallel CFD 2007 Conference*, Springer, pp. 275–282.

Lopez, J. 2016, *Parallel object-oriented algorithms for simulation of multiphysics. Application to thermal systems.* Universitat Politècnica de Catalunya.

Lopez, J., Ruano, J., Lehmkuhl, O., Rigola, J., Oliva, A. 2015, Compressible 1D-3D simulation of a muffler with pseudosound prediction levels. *24th International Congress of Refrigeration*. Japan.

Pedro, J., Baez, A., Lehmkuhl, O., Perez, C., Oliva, A. 2015. On the extension of RANS/LES methods from incompressible to compressible transonic turbulent flows with SBLI. *8th International Symposium on Turbulence, Heat and Mass Transfer*. Sarajevo.

Poinsot, T.J. 1992, Boundary conditions for direct simulations of compressible viscous flows, *Journal of Computational Physics*. 101 (1): 104 – 129.

Rigola, J. 2002, *Numerical Simulation and Experimental Validation of Hermetic Reciprocating Compressors. Integration in Vapour Compression Refrigerating Systems*. CTTC, Terrassa, 112 p.

Roe, P.L. 1981, Approximate riemann solvers, parameter vectors and difference schemes, *Journal of Computational Physics*. 43 (2): 357 – 372

Sarioglu, K., Ozdemir, A.R., Orguz, M., Kaya, A., 2012, An experimental and numerical analysis of refrigerant flow inside suction muffler of hermetic reciprocating compressor. *International Compressor Engineering Conference*. Purdue.

Soedel, W. 1992, *Mechanics, simulation and design of compressor valves, gas passages and pulsation mufflers*. Purdue University, 262 p.

Trias, F.X., Lehmkuhl, O., 2011. A self-adaptive strategy for the time integration of Navier–Stokes equations. *Numerical Heat Transfer*, Part B 60, 116–134.

ACKNOWLEDGEMENTS

The authors acknowledge the collaboration between CTTC-UPC and Termo Fluids, S.L. company (C06550).
This work was supported by the Agència de Gestió d'Ajuts Universitaris i de Recerca (AGAUR), Catalonia, grant FI-DGR 2015.

Gas Pulsation Control by a Shunt Pulsation Trap with Perforated Tubes and an Optional Absorptive Silencer

Paul Xiubao HUANG[1]*, Sean YONKERS[2], David HOKEY[3]

[1, 2] Hi-Bar MC Technologies, LLC.
Fayetteville, GA, USA
(Tel: 678-817-1497, E-mail: paulxbh@yahoo.com)

[3] Howden-Roots
Connersville, IN, USA

* Corresponding Author

ABSTRACT

This paper is a continuing work from the same authors on the same topic of gas pulsation control using a shunt pulsation trap method by Huang (2012a, 2014). Traditionally, a serial pulsation dampener/muffler, often a reactive type, is connected AFTER the discharge of a positive displacement compressor or engine. It is capable of reducing pressure pulsation level up to 10-20 folds or 20-40 dB but is bulky and suffers sizable back pressure losses. It has been demonstrated in the previous theoretical and experimental investigations by Huang (2012a, 2014) that gas pulsations can be as effectively controlled by an alternative method, shunt pulsation trap (SPT) of a parallel configuration, which is more compact and tackles the gas pulsations BEFORE the compressor or engine discharge, hence not suffering any back pressure losses. However, it is observed that the dampening element inside SPT by a perforated plate (p-plate) induces a severe "drum effect" with enhanced vibration and noise.

This paper investigates experimentally the effect of a new dampening element inside the SPT by using circular perforated tubes (p-tubes) with minimum trap volume. It is found that p-tubes can effectively resolve the adverse "drum effect" of p-plate with improved pulsation/vibration/noise reduction. Furthermore, the use of an absorption type silencer is beneficial to further reduce noise level by 10-15 dB in addition to the p-tubes under Exhauster and Blower modes. However, this type of silencer will not have the sizable back pressure drop associated with reaction type silencers and therefore will not affect system efficiency. The integration of absorptive silencer into a p-tube based SPT would provide an optimal design choice for size/weight and pulsation/noise reduction, energy saving and for improving equipment fatigue life. The potentials for applying the SPT to screw and scroll compressors are enormous in the future since a significant portion of their operation is under off-design conditions of either an under-compression (UC) or over-compression (OC).

1. INTRODUCTION

1.1 Gas Pulsations vs. Noises (Acoustic Waves)

Gas pulsations commonly exist in HVACR, energy and automotive industry. They are believed to be a major source for system inefficiency, vibrations, noises and fatigue failures. It has been widely accepted that gas pulsations mainly take place at the discharge side of a positive displacement (PD) type compressor such as a screw, scroll or Roots as discussed in detail by Soedel's book (2007), and Mujiü *et al.* (2007), Koai and Soedel (1990), Sangfors (1999), Wu *et al.* (2004) and Gavric and Badie-Cassagnet (2000), Price and Smith (1999), Tweten *et al.* (2008) and Peters (2003). Recently by using the shock tube analogy, Huang (2012a) revealed that the nature of gas pulsations is a composition of compression waves (CW), expansion waves (EW) and an induced fluid flow (IFF). Moreover, the dynamics of gas pulsations is such that the pressure pulsations consisting of CW (coalescing into a quasi-shockwave) and EW travelling in opposite directions are the forces driving the unidirectional flow pulsation as strong pressure waves sweep across the gas at the speeds of waves.

Though pressure pulsations travel so rapidly, they can be quantitatively measured by modern high response pressure transducers and their magnitude is typically ranging from 0.002 – 2 bar (0.03 - 30 psi). By comparison, pressure fluctuations of acoustic waves are typically less than 0.0002 bar (0.003 psi) or equivalent to the pressure level of 120 dB according to Beranek (1988). Table 1 lists the corresponding magnitudes of pressure pulsations and acoustic waves in bar, psi and dB. It can be seen that the pressure level of gas pulsations is much higher and well beyond the linear pressure range of classical acoustics. That is, the "real life" pressure pulsations are non-linear waves with finite amplitude and changing wave form.

Table 1: Magnitude of acoustic waves in comparison with industrial gas pulsations (Huang, 2012a)

	Acoustic Waves ⟺ Gas Pulsations						
Pressure Pulsation, bar	0.000002	0.00002	0.0002	0.002	0.02	0.2	2
Pressure Pulsation, psi	0.00003	0.0003	0.003	0.03	0.3	3	30
Sound Pres. level, L_p, dB	80	100	120	140	160	180	200

On the other hand, the frequency of gas pulsations are relatively low, ranging from 10-500 Hz while acoustic noise's frequency typically ranges above 500 up to 10,000 Hz.

1.2 Traditional Gas Pulsation and Noise Control Methods: Reactive vs. Absorptive Silencers

Traditionally, a serial pulsation dampener, often a reactive type silencer, is connected AFTER the PD compressors for gas pulsation control as shown for a screw compressor in Figure 1.

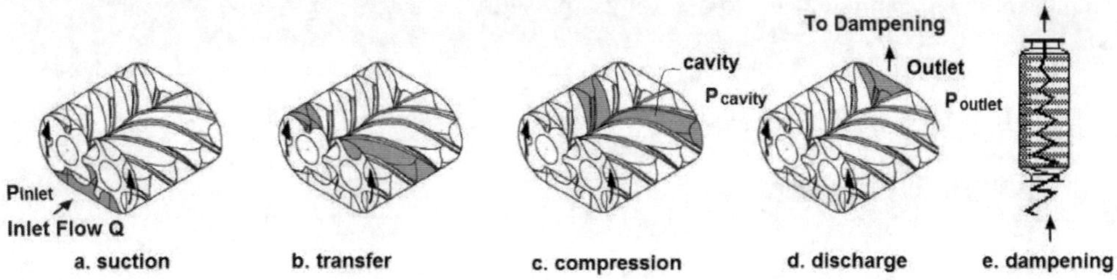

Figure 1(a-e): Traditional screw compression cycle with a serial dampener at discharge

A reactive type silencer, shown in Figure 2a, works by employing a series of chambers connected with a number of perforated tubes. Its effectiveness in pulsation control depends on the size and number of the chambers or stages of dampening. By increasing the size and number of stages attenuation is increased. An ideal design can achieve a reduction of 20 plus dB, but results in a large size which creates other problems such as inducing more noises and vibration due to the increase of surface area and supported weight. Sometimes internal dampener structure fatigue failure results in possible catastrophic damage to downstream components and equipment. Another major defect associated with a reactive type silencer is the static pressure loss that reduces the compressors overall efficiency. The pressure loss is directly proportional to the attenuation for a fixed size silencer and typically ranges from 0.4 psi up to 5 psi or even higher.

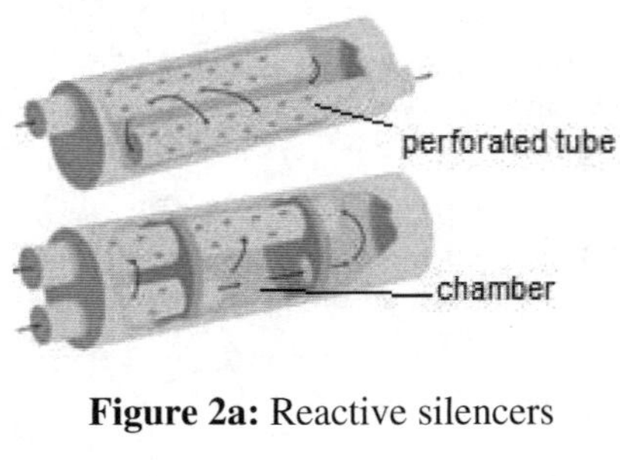

Figure 2a: Reactive silencers

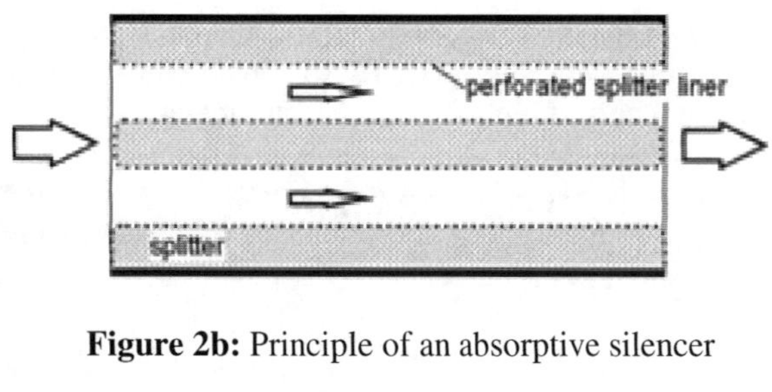

Figure 2b: Principle of an absorptive silencer

On the other hand, absorptive or dissipative silencers, shown in Figure 2b, are configured in a parallel baffle arrangement and lined with sound absorbing materials. They are commonly used in HVAC duct systems after dynamic type blowers or compressors that generate gas pulsations (acoustic waves) with smaller amplitudes and higher frequencies than PD types. They can effectively attenuate pressure fluctuations up to 20 plus dB while suffer much less pressure loss, at least an order of magnitude lower than reactive types.

1.3 Gas Pulsation Control Methods: Serial vs. Parallel (SPT) Dampening

Traditionally, serial dampening configuration of either reactive or absorptive types has been and perhaps still is the dominant method tackling gas pulsations. However, an alternative method called a shunt pulsation trap (SPT) uses a parallel configuration which tackles the gas pulsations BEFORE the compressor discharge. The SPT method is based on the shock tube theory (Huang, 2012a & 2012b) that predicts that the most dominant sources of gas pulsations are the direct results from either an Over Compression (OC) or an Under Compression (UC) suddenly discharging at the compressor outlet when running at off-design conditions.

As a brief description of the principle of a shunt pulsation trap, Figure 3 shows a Roots type blower, PD compressor at 100% UC, with a shunt pulsation trap near the compression cavity. In principle, a SPT is used to both TRAP and ATTENUATE of all three components of gas pulsation CW-IFF-EW at the same time. For comparison, a traditional discharge dampener is connected in series with the cavity, compression chamber, AFTER the discharge port, as shown in Figures 1e and 4a, and through which both the cavity flow Q and the gas pulsation flow IFF pass. While a shunt pulsation trap is connected in parallel with the cavity BEFORE the discharge port through which only the gas pulsation flow IFF passes, shown in Figure 4b. During the discharging phase of the operation, the flow cavity is pre-opened to the "trap inlet" which connects the cavity to the pulsation trap, which in turn communicates with the compressor outlet through the "trap outlet". Between the trap inlet and trap outlet and within the shunt pulsation trap, there is a pulsation dampening means to control gas pulsations energy.

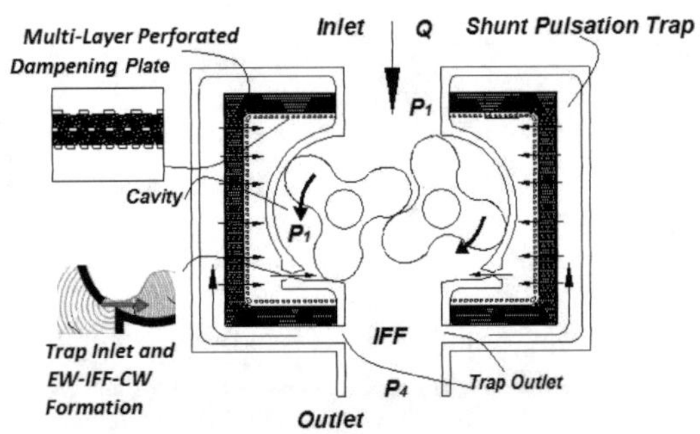

Figure 3: Roots blower with a Shunt Pulsation Trap for UC mode

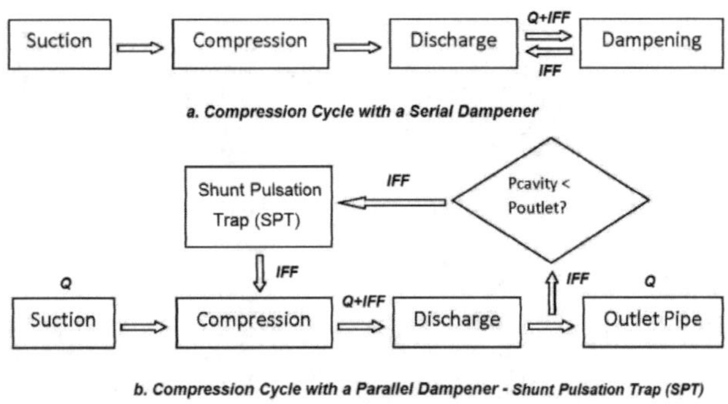

Figure 4(a-b): Comparison of a Shunt Pulsation Trap with traditional serial configuration shown for UC mode

The difference between serial and parallel configurations is fundamental and significant here as there are several distinct advantages associated with the shunt pulsation trap when compared with the traditional serially connected pulsation dampener. First of all, the attenuation of the flow pulsation IFF is separated from the main cavity gas flow Q so that an effective IFF dampening will not affect the main cavity gas flow Q, resulting in both higher system efficiency (no discharge pressure loss) and attenuation efficiency. In a traditional serially connected dampener, both the pulsating flow IFF and cavity gas flow Q travel together through the dampener where a better attenuation always comes at a cost of higher static pressure losses. So a compromise is often made in order to reduce the back pressure loss by sacrificing the degree of pulsation dampening or have to employ a very large volume and costly dampener in a serial setup. Secondly, the parallel pulsation trap attenuates the gas pulsation closer to the source of pulsation generation and in a shape more compact and conformal to the cavity shape than a serial one, resulting in smaller size and less weight.

2. EXPERIMENTAL INVESTIGATION OF SPT WITH P-PLATE AND P-TUBES

2.1 Pulsation Control Testing Using a SPT with Perforated Plate (p-plate)

To validate the shunt pulsation trap (SPT) concept experimentally, two prototypes of 75 HP and 350 HP Roots type blowers were built and tested with SPT equipped with a perforated plate (p-plate) as shown in Figure 5a (refer to Huang, 2014 for more details). The previous experimental investigations have demonstrated that a 10-fold or 20 dB pressure pulsation reduction can be as effectively achieved by the shunt pulsation trap (SPT) as a traditional reactive type silencer as shown in Figure 5b. However, SPT is much more compact in size and is not suffering any back pressure losses. The energy saving of SPT without any back pressure that compressor has to overcome with a serial silencer is significant, as shown in principle by P-V diagram in Figure 6 that was validated by tests on a 350 HP prototype with efficiency gain across the whole flow range by 5- 16% (Huang, 2014).

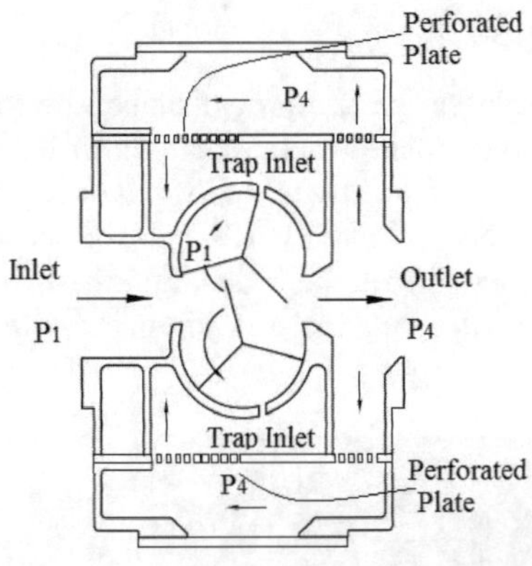

Figure 5a: SPT design with perforated plate (p-plate)

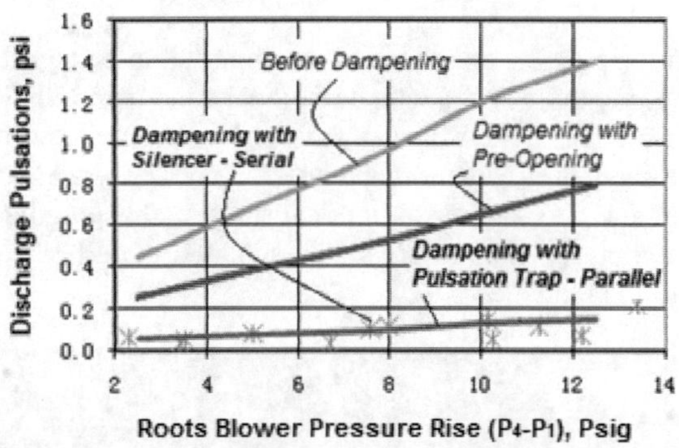

Figure 5b: SPT test results vs. serial silencer

However, there is a downside for p-plate SPT. The most serious one is the generation of a wide spectrum of secondary vibration and noise by the dampening plate itself as they are excited by the strong EW waves and induced fluid flow (IFF) inside the shunt pulsation trap by the perforated plate (p-plate), the so called "drum effect".

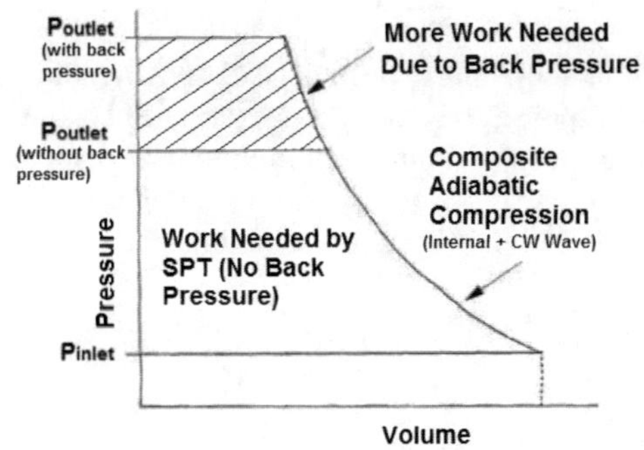

Figure 6: SPT energy saving principle on P-V

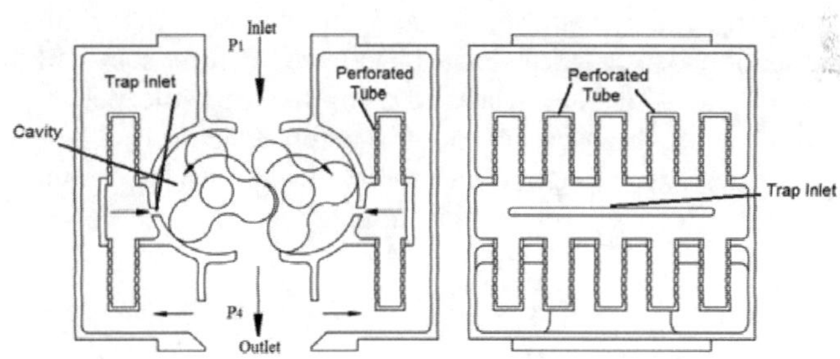

Figure 7: SPT testing with perforated tubes (p-tube)

2.2 Pulsation Control Testing Using a SPT with Perforated Tubes (p-tubes)

To overcome the "drum effect", a new round of testing is conducted on the same 75 HP prototype as the above by using circular shaped perforated tubes (p-tubes) inside the SPT instead of perforated plate (p-plate), as shown in Figure 7. The cylindrical perforated tube offers the advantage of a balanced out aerodynamic forces induced by IFF and EW or CW waves inside a shunt pulsation trap to undergo self-cancellation thus exciting no or little secondary vibration and noise.

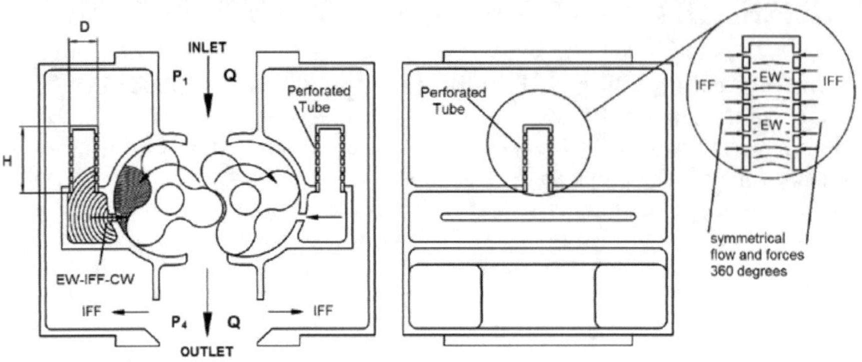

Figure 8: Working principle of perforated tube (p-tube) shown for UC mode

Pay attention to the exploded view in Figure 8 showing a SPT with at least one cylindrical shaped perforated tube (p-tube) per compressor cavity. In comparison with a SPT with perforated plate (p-plate) shown in Figure 5a, the induced flow IFF by EW waves through the holes of the p-tube are now symmetrical and the induced forces that excite the secondary vibration and noise are hence cancelled out throughout the entire perforated surface. Moreover, the perforated tube is of the cylindrical shape with a length (H) to diameter (D) ratio at least larger than 2 to 1 so that the total cylindrical surface area is much larger than the flat top end area that acts like a drum, hence eliminating so called "drum effect".

Figure 9: Comparison of prototype setup of perforated tubes (p-tubes) with a perforated plate (p-plate)

To experimentally compare the performance of a p-tube with p-plate, new p-tube tests are conducted with the same test setup and instruments as p-plate tested before, shown in Figure 9, following ASME PTC-9 specification and pressure pulsations are measured using the same piezoelectric high response dynamic pressure transducers at the same inlet and discharge locations as before. The measured pulsation results are summarized in Figure 10 comparing various designs of p-tubes with p-plate while measured total noise level is shown in Figure 11. It is found that p-tubes can effectively resolve the "drum effect" of p-plate and improve pulsation, vibration and noise reduction.

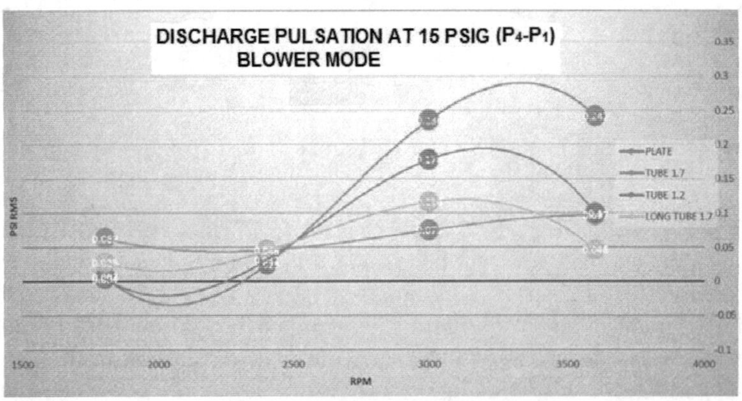

Figure 10: Comparison of measured pulsations of perforated tubes (p-tubes) with a perforated plate (p-plate)

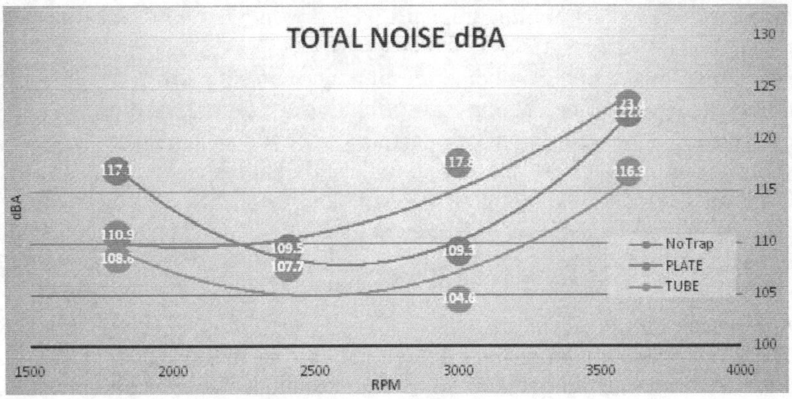

Figure 11: Comparison of measured noises of perforated tubes (p-tubes) with a perforated plate (p-plate)

2.3 Further Noise Control Testing Using an Absorptive Silencer

During previous discussed testing in section 2.2, it is observed that even with very low discharge pulse levels using SPT with p-tubes, around 0.05 - 0.2 psi as shown in Figure 10, there still appears to be a lot of noise coming from the discharge. There could be explained by Table 1 that indicates 0.03 - 0.3 psi pressure pulsations is corresponding to 140-160 dB in acoustic noise. In another words, SPT is effective with a 10-fold large pulsation reduction, say from psi to sub-psi level, the problem remains as how to further reduce the sub-psi level smaller pulsations, hence the induced noise and vibrations?

The upper limit of the maximum pulsation reduction by a SPT for Roots type is discussed by Huang (2015) in the previous theoretical and experimental investigations that gas pulsations can be generated either by a sudden velocity change (ΔU) say from a piston or lobe movement, or by a sudden opening of a pressure difference ($\Delta p41$) such as during the discharge phase of a rotary PD compressor (screw, scroll or Roots) under UC (Under-Compression) or OC (Over-Compression) conditions. The test results confirm that the dominant source of the gas pulsations for the 100% UC case is mainly from the sudden release of a pressure difference (Δp_{41}) while the non-uniform rotor movement (ΔU) induced pulsation is about one order of magnitude lower. In another word, a different scheme is needed to tackle the sub-psi level smaller pulsations at discharge in order to further reduce noises.

A traditional absorption type silencer is used at discharge together with p-tube equipped SPT prototype. Preliminary test results show the beneficial trend to further reduce total noise level by 10-15 dB in addition to the p-tubes under Exhauster and Blower mode. However, this type of silencer will not have the back pressure drop associated with reaction type silencers and therefore will not affect system efficiency (the cost of power wasted). The integration of an absorptive silencer into a p-tube based SPT would be an optimal design choice in both pulsation and noise reduction.

3. CONCLUSIONS AND FUTURE RECOMMENDATIONS

3.1 Test Conclusions

The control of large amplitude gas pulsations and the induced NVH have been a continuing challenge for over a hundred years due to its importance to the reliability and efficiency for widely used internal combustion engines and PD type compressors. Instead of following the path of traditional serial configuration, an alternative parallel approach called SPT (shunt pulsation trap) is devised and validated by experiments in this paper and the previous one (Huang, 2014) based on a new pulsation generation theory - the shock tube mechanism. It is concluded:

1. SPT (shunt pulsation trap) concept has been experimentally validated by perforated plate (p-plate) as well as perforated tubes (p-tubes) on a 75 HP Roots blower, demonstrating that it is capable of a 10-fold plus or 20 dB plus pressure pulsation reduction, as effective as a traditional reactive type silencer. However, SPT is more compact in size and is not suffering any back pressure losses as traditional reactive silencer. The

energy saving potential of SPT by eliminating any back pressure is significant with potential efficiency gain across the whole flow range by 5- 16%;

2. SPT with p-tubes does not suffer the adverse "drum effect" and is more compact and effective than p-plate with overall improved performance of pulsation, vibration and noise reduction;
3. Furthermore, the use of an absorption type silencer is beneficial to further reduce noise level (sub-psi pressure pulsations) up to 10-15 dB in addition to the SPT with p-tubes without suffering the sizable back pressure drop associated with reaction type silencers and therefore will not affect system efficiency. The integration of absorptive silencer into a p-tube based SPT would provide an optimal design for both size/weight and pulsation/noise reduction, energy saving and for improving equipment fatigue life.

3.2 Future Recommendations

Though tests so far have only been conducted for the special case of a 100% under compression for Roots type blower, it is believed that the principle of SPT control strategy could also be applied to other types of PD compressors without automatic discharge valves such as screw and scroll types that suffer even more pulsation problems under over compression (OC) or under compression (UC) than Roots type (Soedel, 2007). The shunt pulsation trap concept (SPT) with p-plate or/and p-tubes configurations could be applied in schemes as shown in Figure 12 for example. More examples and embodiments could be found in SPT patents listed in the references of this paper.

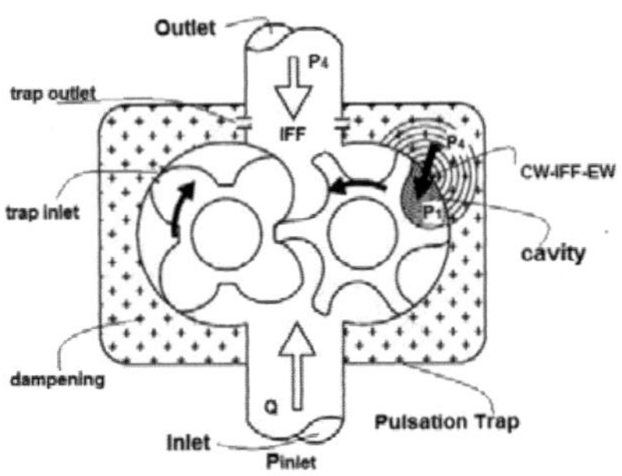

Figure 12a: Screw compressor with a Shunt Pulsation Trap for UC mode

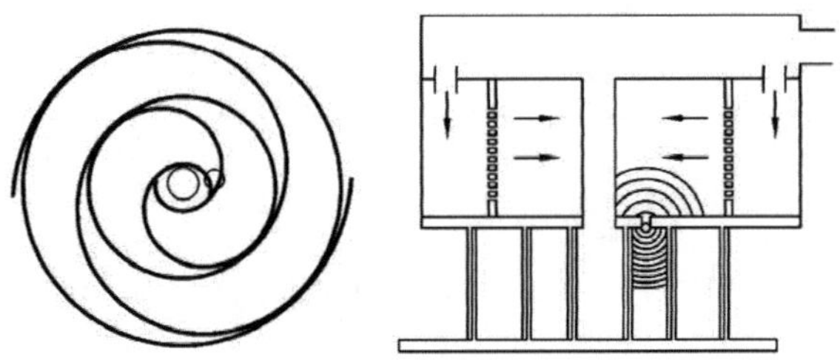

Figure 12b: Scroll compressor with a Shunt Pulsation Trap for UC mode

Finally, we hope, in our continued search to understand and control gas pulsations, more experiments based on other types of PD compressors incorporating SPT principle could be conducted under both UC and OC conditions to explore the enormous potential for this new approach. With more research and development resources devoted by both academia and industry, it is anticipated that future generation of PD compressors and Internal Combustion Engines can be designed to be even simpler in structure, smaller in size and smoother in running than those used today.

NOMENCLATURE

CW	compression wave pulsation component
EW	expansion wave pulsation component
IFF	induced fluid flow pulsation component
OC	over compression
p	gas pressure pulsation
P	absolute gas pressure
PD	positive displacement
Q	inlet volume flow rate
SPT	shunt pulsation trap
UC	under compression
ΔU	IFF velocity
V	PD compressor volume

Subscripts

1	initial low pressure in shock tube, or inlet pressure in Roots blower
2	pressure after shockwave in shock tube, pressure after CW in PD compressor
4	initial high pressure in shock tube or outlet pressure in Roots blower
Cavity	compressor cavity
Inlet	compressor inlet
Outlet	compressor outlet

REFERENCES

Beranek, L., 1988, *Noise & Vibration Control*, Inst. of Noise Control Engineer, New York, p. 396.

Gavric, L., Badie-Cassagnet, A., *Measurement of gas pulsations in discharge and suction lines of refrigerant compressors*, The 2000 International Compressor Engineering Conference at Perdue, p. 627-634, 2000.

Huang, P., *Gas Pulsations: A Shock Tube Mechanism.* The 2012 International Compressor Engineering Conference at Perdue, 2012a.

Huang, P., *Under Compression: An Isochoric or Adiabatic Process?* The 2012 International Compressor Engineering Conference at Perdue, 2012b.

Huang, P., Yonkers, S., Hokey, D., *Gas Pulsation Control Using a Shunt Pulsation Trap.* The 2014 International Compressor Engineering Conference at Perdue, 2014.

Huang, P., Yonkers, S., *Rotary Lobe Pump or Vacuum Pump with Shunt Pulsation Trap*, US Pat# 9,140,260, 2015.

Huang, P., Yonkers, S., *Shunt Pulsation Trap for Cyclic PD Compressors*, US Pat# 9,140,261, 2015.

Huang, P., Yonkers, S., *Screw Compressor with a Shunt Pulsation Trap*, US Pat# 9,151,292, 2015.

Huang, P., Yonkers, S., Hokey, D., *The Origin of Gas Pulsations in Rotary PD Compressors: ΔP or ΔU?* The 2015 International Compressor Engineering Conference at City University, 2015

Huang, P., Yonkers, S., *Shunt Pulsation Trap For PD Gas Machinery*, US20150361982 A1, 2015.

Koai, K., Soedel, W., *Determination of port flow and interpretation of periodic volume source*, The 1990 International Compressor Engineering Conference at Perdue, p. 369-377, 1990.

Mujiü, E., Kovaþeviü, A., Stošiü, N., Smith, I., *The influence of port shape on gas pulsations in a screw compressor discharge chamber*, International Conference on Compressors and their Systems 2007, London, UK, p. 213-223, 2007.

Peters, R., *Damping of low frequency noise in piping systems by means of perforated plates*, International Conference on Compressor and Their Systems, London, p.545-558, 2003.

Price, S., Smith, D., *Sources and Remedies of Hi-Frequency Piping Vibration and Noise*, Proceedings of The 28^{th}Turbomachinery Symposium, p. 189-212, 1999.

Sangfors, B., *Computer simulation of gas-flow noise from twin-screw compressors*, International Conference on Compressor and Their Systems 1999, London, p.707-716, 1999.

Soedel, W., *Sound and Vibrations of Positive Displacement Compressors, CRC Press*, p. 16-18, 2007.

Tweten, D., Nored, M., Brun, K., *The Physics of Pulsations*, Gas Machinery Conference, 2008.

Wu, H., Xin, Z., Peng, X., Shu, P., *Simulation of discharge pressure pulsation within twin screw Compressors*, International Conference on Compressors and their Systems 2007, London, UK, 2004.

ACKNOWLEDGEMENT

The authors would like to pay special tribute to Dr. Joost Brasz, who passed away suddenly late last year, for his enduring encouragement at the early stage of the project, for introducing Purdue and City University Conferences to the author and whose long time mentoring has always proved to be valuable. Moreover, we would like to thank Howden-Roots (formerly GE Oil & Gas), whose special grant made the prototype testing possible.

Development of a Novel Structure Rotary Compressor for Separate Sensible and Latent Cooling Air-Conditioning System

Chunhui LIU*, Haifeng ZHANG, Lei ZHANG, Jin PAN
R&D Center, Shanghai Hitachi Electrical Appliances Co., Ltd
888 Ningqiao Rd., Pudong Jinqiao, Shanghai, China, 201206
Office: +86 -21-50554560, Fax: +86-21-58991084, E-mail: liuchh@shec.com.cn

*Corresponding Author

ABSTRACT

The energy efficiency of separate sensible and latent cooling (SSLC) air-conditioning system is higher than traditional one, since SSLC system uses separate cycles which can raise the evaporating temperature of sensible cycle. Separate cycles often include two separate compressors, which makes the initial investment of SSLC cycle too expensive to use in room air conditioning. A novel structure rotary compressor for SSLC system is proposed in this paper. Using just one novel compressor, SSLC system can realize separate cycles. This paper introduces the schematic of new compressor, shows how the compressor works and the good performance of system. Furthermore, the advantage of this novel compressor is used in two evaporating temperature system.

1. INTRODUCTION

The evaporator of conventional air conditioning systems undertakes the load of the cooling,(i.e. meet the sensible load) and removing moisture (i.e. meet the latent load). In order to remove the moisture, the return air temperature has to be below the dew point temperature of the indoor air. And the evaporating temperature has to be much lower than the dew point temperature. The lower evaporator temperature leads to higher pressure ratio of the compressor. Thus, this doesn't make the running air conditioning systems economical and the supply air of low temperature makes people uncomfortable.

The Constant Temperature and Humidity air-conditioner in present market can meet the demand of constant humidity. The supply air is heated by the condensation part of the indoor after it is dehumidified, and the supply air of higher temperature makes people comfortable, but it isn't economical.

Some SSLC (separate sensible and latent cooling) system uses two independent Separate cycles with two separate compressors, which makes the initial investment of SSLC cycle too expensive to use in room air conditioning.

In some Multi-split Air-Conditioning systems, for example, one outdoor drives two indoors， it can realize two different room temperature. But there is only one evaporate temperature in this system. The evaporate temperature is low in order to meet the lower room temperature. The higher pressure ratio of the compressor leads to the low COP of the Multi-split Air-Conditioning system.

Ling et al.(2009) proposed an idea of using the separate sensible and latent cooling (SSLC) system. In SSLC system, one of the vapor compressions dealt with sensible load only, and it is operating at a higher evaporating pressure than the conventional air conditioning system, which reduces pressure ratio. The reduced pressure ratio brings the benefit of better compressor efficiency and less power input than the conventional system. The other vapor compression dealt with both latent load and a small amount of sensible load. Under the ASHRAE standard ambient condition（35℃，44% RH）, the energy consumption of such SSLC system was reduced by 30% compared with that of a conventional system, and the saving was reported to be up to 50% under hot and dry condition（37℃，15% RH） (Ling et al, 2010).

A novel structure rotary compressor has developed using for the SSLC system, which includes two cylinders and two accumulators, which can meet the sensible load and the latent load separately. It can also apply the two different evaporate temperature at the same time. Its compact structure and high efficiency is suitable for the SSLC system.

2. DESIGN OF SSLC COMPRESSOR

2.1 Compressor structure features

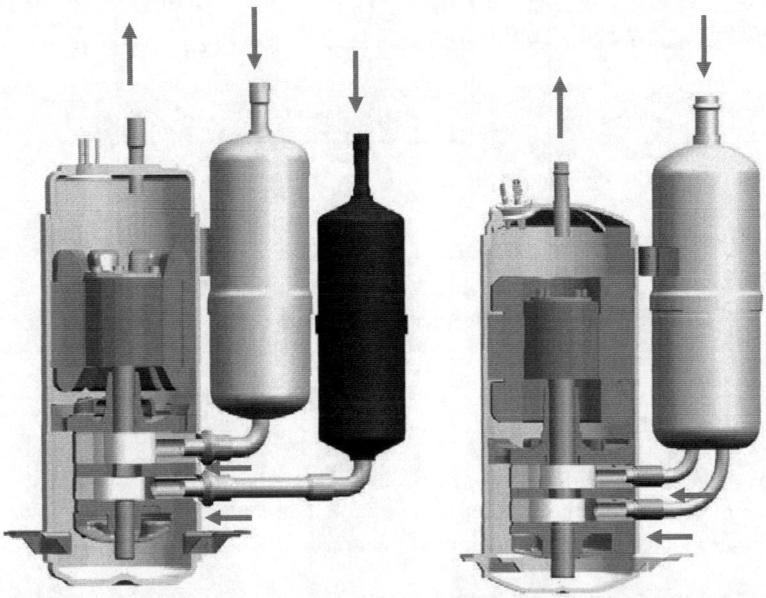

(a) SSLC compressor (b) conventional twin rotary compressor

Fig1: Comparison of compressor structure

Fig 1 (b) shows the conventional twin rotary compressor with double suction structure and single accumulator. The flow of refrigerant in the suction side is as follows: First, the refrigerant from one or two evaporators is mixed in the accumulator of the suction side of compressor, and then the refrigerant is evenly distributed into the two cylinders to be compressed. After being compressed, the refrigerant is mixed in the exhaust chamber of the compressor, then it is exhausted from the discharge pipe into the condenser of the air-conditioning system.

The SSLC compressor is different from the conventional twin rotary compressor. Fig 1 (a) shows its structure features with two cylinders, two suction structures of two different sizes of accumulators. The flow of refrigerant distribution in the suction side is as follows: First, the refrigerant from one or two evaporators enters the desired accumulator of the suction side independently, then the refrigerant enters the desired cylinder to be compressed. After being compressed independently, the refrigerant is mixed in the exhaust chamber of the compressor, then it is exhausted from the discharge pipe into the condenser of the air-conditioning system.

Compared with the traditional twin rotary compressor, the two cylinders of SSLC compressor use independent suction structures which can control two different evaporating temperatures independently because of the same exhaust chamber, the two cylinders may have different suction pressure but the same discharge pressure. Each cylinder's displacement is determined by its cooling capacity requirement and operation condition, so they are maybe different. The two cylinders use the same crankshaft as shown in Fig 1 (a), so they work in the same time and have the same rotational speed.

In SSLC (separate sensible and latent cooling) system, the SSLC compressor uses the two cylinders to deal with the sensible load and the latent load independently. We can raise the evaporate temperature of the sensible load after meeting the latent load, which can improve the overall energy efficiency of the air conditioning system.

2.2. Theoretical performance analysis of SSLC compressor

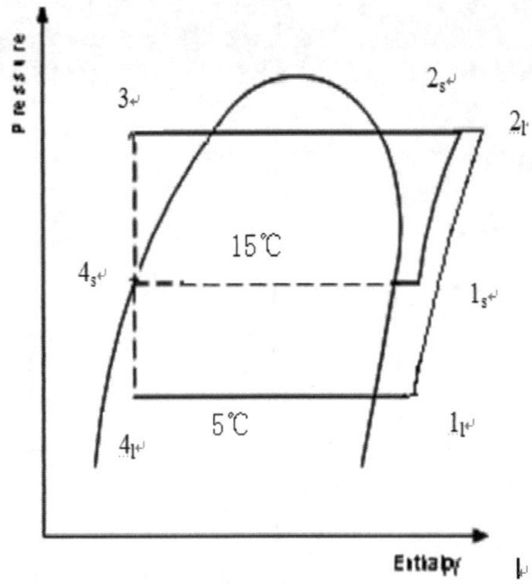

Fig2: the enthalpy-pressure diagram of SSLC compressor

Fig2 shows the enthalpy-pressure diagram of the vapor compression cycle of SSLC compressor and conventional compressor.

The cycle of the conventional air conditioning system as follow: the suction 1l--the discharge 2l--the outlet of condenser 3-- the outlet of throttle 4l-- the suction 1l. The evaporate temperature of the VCC（Vapor Compression Cycle） is set 5℃ in order to meet the need of dehumidification.

In SSLC system, the sensible evaporate temperature is set 15℃ in order to meet the thermal comfort of human body, which mainly deals with the sensible load of the room air conditioner. The latent evaporate temperature is set 5℃ in order to meet the need of dehumidification, which mainly deals with both the latent load and a small amount of sensible load.

The cycle of the sensible VCC of higher evaporate temperature as follow: the mixture of 2s and 2l of discharge -- the outlet of condenser 3-- one of the outlets of throttle 4s—one of the suctions 1s—discharge 2s. The cycle of the latent VCC of lower evaporate temperature as follow: the mixture of 2s and 2l of discharge -- the outlet of condenser 3-- one of the outlets of throttle 4l—one of the suctions 1l—discharge 2l.

The Operating conditions are list in the table 1.

According to the ratio of sensible load and latent load under most conditions for room air conditioners, the volume ratio of the two cylinders of Compressor is set to 1.5:1. That is to say, the displacement of the sensible cylinder is 15.2cc and the displacement of the latent cylinder is 10.0cc.

The calculating formulas for cooling capacity and power are as follows：

$$Q_0 = q_m \times (h_1 - h_3) \qquad (1)$$

$$W_0 = q_m \times (h_2 - h_1) \qquad (2)$$

$$q_m = \rho_1 \times v \times n / 60 / 1000 \qquad (3)$$

Q_0 --- cooling capacity W

W_0 --- Power W

q_m --- mass flow rate kg/s

h_1 --- enthalpy at suction status kJ/kg

h_2 --- enthalpy at discharge status kJ/kg

h_3--- enthalpy before expansion throttling valve kJ/kg

ρ_1--- density at suction status kg/m^3

v--- displacement of compressor cylinder dm^3

n--- speed of compressor rpm

The Comprehensive cooling capacity is the total of the capacity of the two cylinders.

The Comprehensive power is the total of the compression work of the two cylinders.

The performance calculation result of the theoretical cycle is list in the table1.

Table1: comparison of performance calculation result of the theoretical VCC of SSLC compressor and conventional compressor

	Mode of AC system	**Conventional AC system**	**SSLC AC system with two evaporate temperatures**	
Theoretical calculation conditions	Evaporate temperature ℃	5	15	5
	Suction temperature ℃	10	20	10
	Condensation temperature ℃	45	45	45
	Inlet temperature of expansion valve ℃	36.7	36.7	36.7
theoretical calculation performance of vapor compression cycle	Unit mass cooling capacity kJ/kg	**168.61**	**171.33**	**168.61**
	Unit mass compression work kJ/kg	**30.06**	**21.21**	**30.06**
	Unit volume cooling capacity kJ/m^3	**5853.40**	**8067.30**	**5853.40**
	Unit volume compression work kJ/m^3	**1043.50**	**998.90**	**1043.50**
	Speed rpm	3500	**3500**	3500
	Displacement cc	**25.2**	**15.2**	**10**
	Theoretical cooling capacity W	**8604.427**	**7153.001**	**3414.46**
	Comprehensive cooling capacity W		10567.46（↑22.8%）	
	Power W	**1534.014**	**885.7144**	**608.736**
	Comprehensive power W		1494.4504	
	Comprehensive COP	**5.61**	**7.07**	
	Comparison of COP		**↑26.02%**	

According to calculated result in Table 1, it can be seen that the cooling capacity of the SSLC system is increased by 22.8% than that of the conventional compressor. The theoretical COP of the SSLC system is increased by 26.02% than that of the conventional compressor. The energy saving effect is remarkable.

2.3. Analysis of test result of SSLC compressor

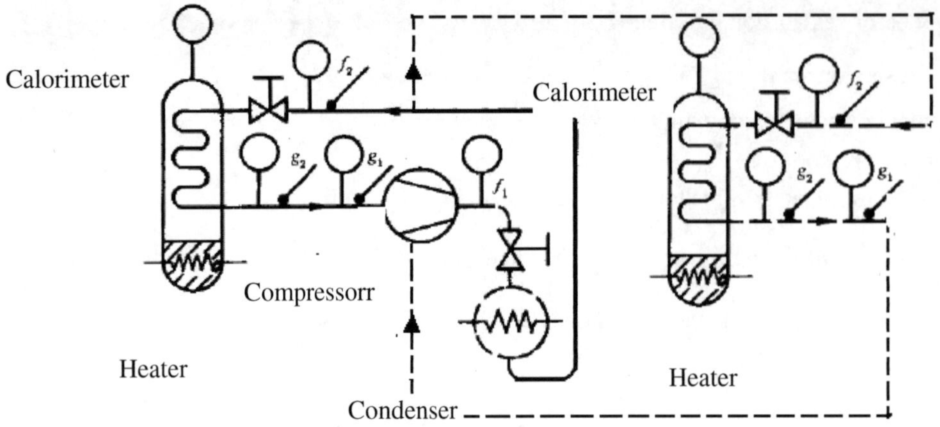

Fig 3: Illustration of performance test for SSLC compressor

SSLC compressor can achieve two different evaporation temperature at the same time. But the conventional performance test equipment of compressor provides only one evaporation temperature. Therefore, we have developed the multifunctional performance test equipment for SSLC compressor.

Fig 3 shows the principle of the performance test equipment for SSLC compressor:

We adopt two sets of separate calorimeter system at the part of evaporation side, which can test the two parts of cooling capacity under two different evaporation temperatures respectively.

Using the multifunctional performance test equipment for SSLC compressor, we get the performance of the SSLC compressor listed in Fig 4.

Fig 4 shows the cooling capacity of the SSLC compressor is increased by 26.4% than that of the conventional compressor with the same displacement under the rated cooling conditions listed in table 1, and the COP of the SSLC compressor is increased by 29.5%. The efficiency gains of the SSLC compressor is consistent with the theoretical performance of SSLC compressor calculated in table 1 under the rated cooling conditions.

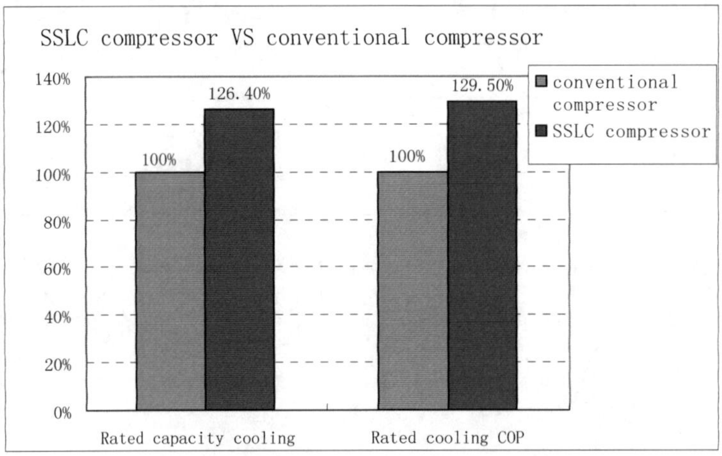

Fig 4: comparison of the performance of compressor

3. TEST AND ANALYSIS OF PERFORMANCE OF SSLC COMPRESSOR USED IN AIR CONDITIONING SYSTEM

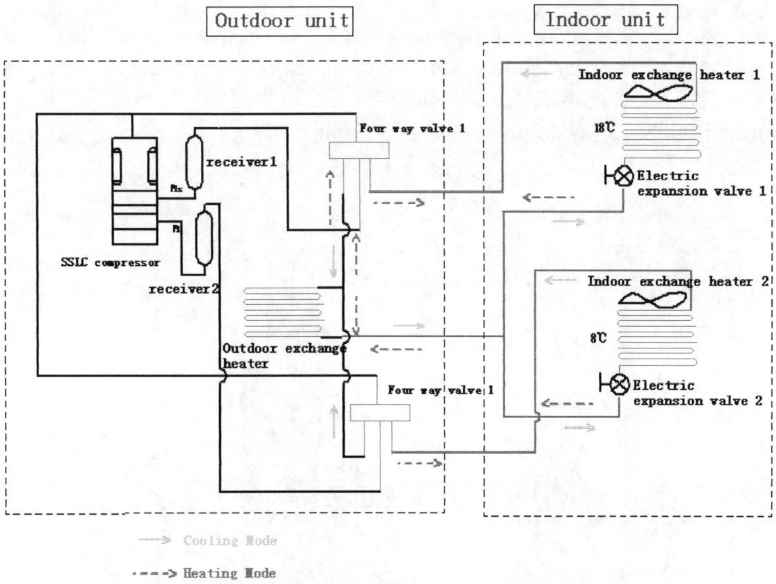

Fig 5: Illustration of the air conditioning system using SSLC compressor

According to the structural characteristics of SSLC compressor, two different evaporation temperatures can be realized in the air conditioning system. We used a one-driving-two air conditioning system to evaluate the performance of the SSLC compressor. The capacity of the one-driving-two air conditioning unit is 7500W. The capacity of the indoor unit 1 is 5000W（cooling capacity）, and the capacity of the indoor unit 2 is 2500W（cooling capacity）.

First, we made the test of baseline using the conventional compressor in order to compare with the SSLC system. There is only one evaporation temperature in the one-driving-two air conditioning system using the conventional compressor. Then we replaced the SSLC compressor to test the performance of the one-driving-two air conditioning system with two different evaporation temperatures. The illustration of the air conditioning system using SSLC compressor is shown in Fig 5.

The cooling mode of the AC system using SSLC compressor is as follows:
In order to meet both cooling and dehumidification, two different evaporation temperatures should be adopted. The refrigerant divided into two parts enters into the respective four-way reversing valve after being exhausted from the SSLC compressor. Then the refrigerant converges before the outdoor condenser. The refrigerant is divided into two parts again from the outdoor condenser in order to enter into the respective electronic expansion throttling valve. The two parts of the refrigerant enter into the respective indoor unit with different evaporation temperature after being throttled. The refrigerant enters into the respective four-way reversing valve after being evaporated from the different indoor unit. Finally, the refrigerant return to the respective accumulator to enter the respective cylinder to be compressed.

The heating mode of the AC system using SSLC compressor is as follows:
The dehumidification needn't be considered in heating mode. So there is only one evaporation temperature in the one-driving-two air conditioning system. The refrigerant divided into two parts enters into the respective four-way reversing valve after being exhausted from the SSLC compressor. Then the refrigerant enters into the respective indoor unit to be condensed. The refrigerant converges before the outdoor evaporator after being throttled in the respective electronic expansion throttling valve. Then the refrigerant enters the outdoor evaporator to be evaporated. The refrigerant is divided into two parts again from the outdoor unit to enter into the respective four-way reversing valve. Finally, the refrigerant return to the respective accumulator to enter the respective cylinder to be compressed.
Then the refrigerant converges before the outdoor condenser. The refrigerant is divided into two parts from the outdoor condenser in order to enter into the respective electronic expansion throttling valve. The two parts of the refrigerant enter into the respective indoor unit with different evaporation temperature after being throttled. The refrigerant enters into the respective four-way reversing valve after being evaporated from the different indoor unit. Finally, the refrigerant returns to the respective accumulator to enter the respective cylinder to be compressed.

The heating mode is the same as that of the conventional AC system. We only evaluate the performance of the cooling mode of the one-driving-two air conditioning system using the SSLC compressor. The rated cooling condition is as follows: outdoor dry bulb 35℃/ wet bulb 24℃, indoor dry bulb 27℃/ wet bulb 19℃. The test results of the performance of the SSLC system are shown in Figure 6-7.

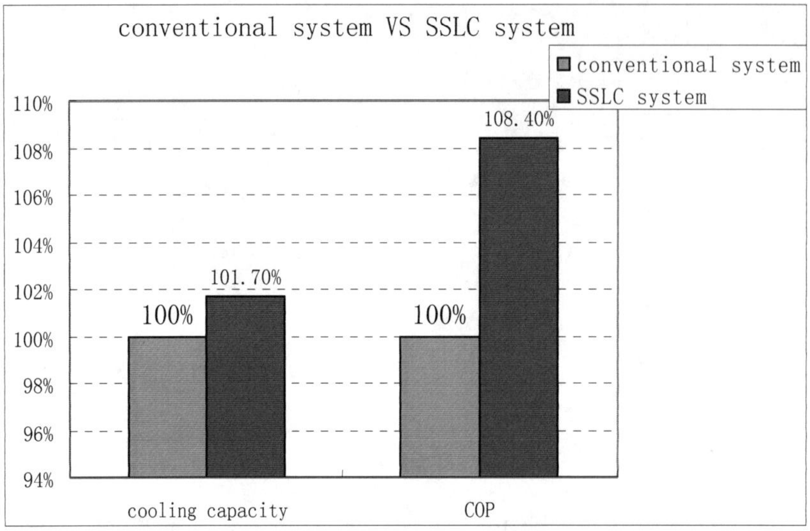

Fig 6: comparison of performance of cooling mode

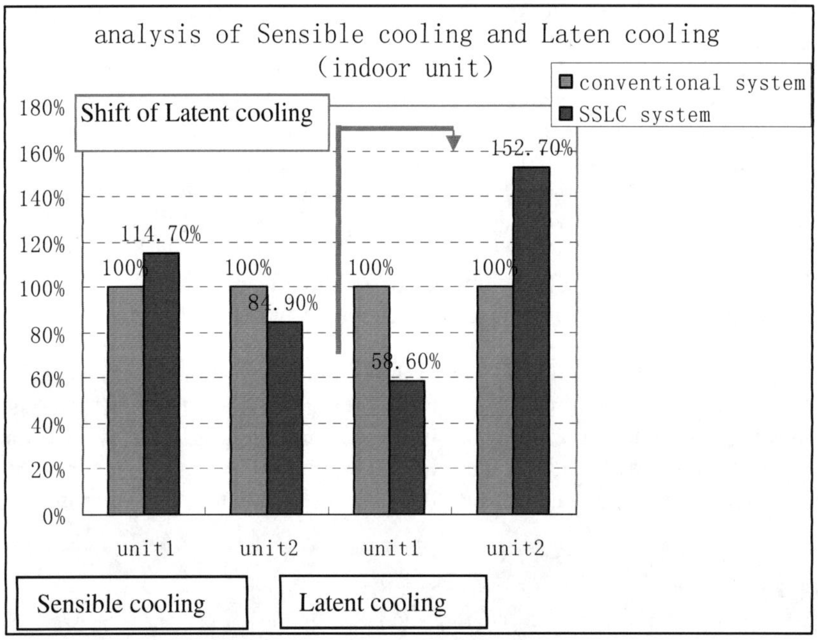

Fig 7: comparison of the shift of sensible cooling and latent cooling

3.1 Higher efficiency in the SSLC (separate sensible and latent cooling) mode

Fig 6 shows that the two indoor units undertake the sensible load and latent load respectively by adopting the SSLC mode with two different evaporation temperatures. The indoor unit 1 undertakes the main sensible load due to the more corporation of the total load. The comprehensive efficiency of the SSLC system is much higher in SSLC mode because of the higher evaporation temperature of the indoor unit 1. The cooling capacity is increased by 1.70% in SSLC mode. And the energy efficiency is increase by 8.40% in SSLC mode.

These results are only the drop-in tests. Though the energy efficiency of the SSLC system is higher than that of the conventional AC system, there is a wide gap between the theoretical calculation results and actual test results. We will optimize the SSLC system in the next work in order to achieve the full energy efficiency.

3.2 Achieving the effect of SSLC (separate sensible and latent cooling) by the one-driving-two air conditioning unit

Fig 7 shows the latent load of the indoor unit1 is decreased by 58.6% in SSLC mode than in conventional mode under rated cooling conditions. And the latent load of the indoor unit2 is increased by 52.7% in SSLC mode than in conventional mode. That is to say, the shift of the part of the latent load from indoor unit1 to indoor unit2 can be realized in SSLC mode. The temperature difference between sensible load and latent load can be realized by adjusting the air volume of the indoor units and the refrigerant flow. Thus the control of the separate sensible and latent cooling can be achieved.

4. CONCLUSIONS

（1）The SSLC compressor adopts the novel structure with two cylinders, two suction structures of two different sizes of accumulators to realize the two different evaporation temperatures independently. The energy efficiency of the air conditioning system using the SSLC compressor is much higher than that of the conventional air conditioning system.

（2）According to the structural characteristics of SSLC compressor，we have developed a multifunctional performance test equipment for SSLC compressor. The test results show the cooling capacity of the SSLC compressor is increased by 26.4% than that of the conventional compressor under the rated cooling conditions and the COP of the SSLC compressor is increased by 29.5%. The energy efficiency is much more remarkable.

（3）Test and analysis of performance of SSLC compressor used in air conditioning system：

- The energy efficiency is increase by 8.40% in SSLC mode than that of the conventional mode under rated cooling conditions. We will optimize the SSLC system in the next work in order to achieve the full energy efficiency.
- The indoor unit 1 undertakes the main sensible load . The indoor unit 2 undertakes the main latent load. The shift of the part of the latent load from indoor unit1 to indoor unit2 can be realized in SSLC mode.

5. REFERENCES

Jiazhen Ling, Osamu KUWABARA, Yunho HWANG, Reinhard RADERMACHER(2010), Enhancement of the Separate Sensible and Latent Cooing Air-Conditioning System, *Proceedings of the International Refrigeration and Air Conditioning Conference at Purdue 2251（1-11）*. West Lafayette, American: Purdue-Pubs

Ling, J, Hwang, Y, Rademacher, R,(2009), Theoretical Study on Separate Sensible and Latent Cooling Air-conditioning System. Int. J. Refrigeration 33(2010)., 510-520

A Study on High Efficiency Wing-vane Compressor
- Part 1: A Simulation Analysis of Dynamic Model -

Raito KAWAMURA[1]*, Shin SEKIYA[2], Tatsuya SASAKI[3], Hideaki MAEYAMA[4],
Shinichi TAKAHASHI[5], Kanichiro SUGIURA[6]

[1, 2, 3] Mitsubishi Electric Corporation, Advanced Technology R&D Center,
8-1-1 Tsukaguchi-Honmachi, Amagasaki, Japan
Phone: +81-6-6497-7173, Fax: +81-6-6497-7285,
[1] Email: Kawamura.Raito@cj.MitsubishiElectric.co.jp
[2] Email: Sekiya.Shin@eb.MitsubishiElectric.co.jp
[3] Email: Sasaki.Tatsuya@ea.MitsubishiElectric.co.jp

[4] Mitsubishi Electric Corporation, Living Environment Systems Laboratory,
3-18-1 Oshika, Suruga-ku, Shizuoka, Japan
Phone: +81-54-287-3053, Fax: +81-54-287-3056,
Email: Maeyama.Hideaki@dx.MitsubishiElectric.co.jp

[5, 6] Mitsubishi Electric Corporation, Shizuoka Works,
3-18-1 Oshika, Suruga-ku, Shizuoka, Japan
Phone: +81-54-287-3112, Fax: +81-54-287-3127,
[5] Email: Takahashi.Shinichi@bk.MitsubishiElectric.co.jp
[6] Email: Sugiura.Kanichiro@ay.MitsubishiElectric.co.jp

ABSTRACT

Low global warming potential refrigerants such as HFO (hydro fluoro olefin) -1234yf are attractive as alternative working fluids in air-conditioning and cooling systems to address global warming. However, larger compressor sizes were required due to the lower fluid density compared with conventional refrigerants such as R410A. Thus, a new wing-vane compressor which has no contact between the vane and cylinder has been developed to prevent on increase in size without performance degradation. The wing-vane type has higher compression efficiency compared to the conventional sliding-vane type as it has no friction loss on the vane tip and it can significantly reduce the mechanical loss for the conventional sliding-vane type. Since the structure of the wing-vane type is different from the conventional vane type, the load characteristics on the rotor and vanes are completely different from the conventional vane type. Therefore, the compression characteristics and the load characteristics were investigated by creating a dynamic model of the wing-vane type. The results are as follows: (1) Characteristics of torque and load are equivalent to those of the conventional vane type. (2) Lubrication of the newly added partial arc guide bearings is important to ensure reliability.

1. INTRODUCTION

Low global warming potential refrigerants such as HFO-1234yf are attractive as alternative working fluids in air-conditioning and cooling systems to address global warming. However, larger compressor sizes were required due to the lower fluid density compared with conventional refrigerants such as R410A. Thus, a new wing-vane compressor which has no contact between the vane and cylinder has been developed to prevent an increase in size without performance degradation. In this paper, we introduce the basic structure and the operation principles of the wing-vane type, and describe the load and torque characteristics revealed by the simulation analysis of a dynamic model. In addition, the features and problems of the wing-vane type are described at the end.

2. BASIC STRUCTURE

Figure 2.1 shows the basic structure of wing-vane type. In order to compare it with the conventional sliding-vane type, the number of vanes in the wing-vane type was selected as two. Figure 2.1a is a longitudinal-section view and Figure 2.1b shows the disassembly view of the compression mechanism. The compression mechanism is accommodated under the shell, and it is composed of a frame, cylinder, cylinder-head, rotor, shaft, four bushes, and two wing-vanes. A wing-vane is composed of one vane and two vane-guides and the vane-guides are fitted into the groove in the frame and the cylinder-head. The outer periphery of the groove acts as a bearing to support the vane-guide, and its center coincides with the center of the cylinder. Therefore, the vane rotates around the center of the cylinder. Incidentally, the vane is inserted into the rotor to rotate with the rotor and the vane sides are supported with bushes which can change their position to face the center of cylinder.

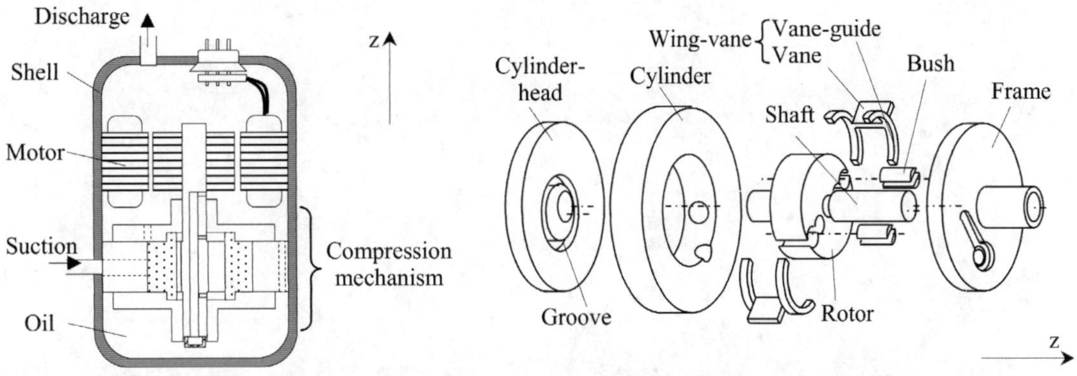

(a) Longitudinal section **(b)** Disassembly view of compression mechanism

Figure 2.1: Basic structure of wing-vane compressor

Figure 2.2 shows a horizontal cross-sectional view of the compression mechanism of the wing-vane type. It is compared the compression mechanisms of the rotary type, and the conventional sliding-vane type. Since in the rotary type, houses the vane in the cylinder, the cylinder outer diameter is larger than the other types of compressors. Since the sliding-vane type houses the vanes in the rotor, the cylinder outer diameter can be made smaller than the rotary type. However, since the tip of the vane rotates at high speed in contact with the inside of the cylinder, the performance is lower than the rotary type due to the increase of frictional losses. Further, since the vane doesn't rotate around the center of the cylinder, it is necessary to decrease the curvature of the vane tip rather than the cylinder inside. In the case of the wing-vane type, the vane is housed in the rotor and the tip of vane does not contact with the cylinder inside with the same curvature of vane tip. Therefore, the wing-vane type can achieve a high efficiency and small size using this structure.

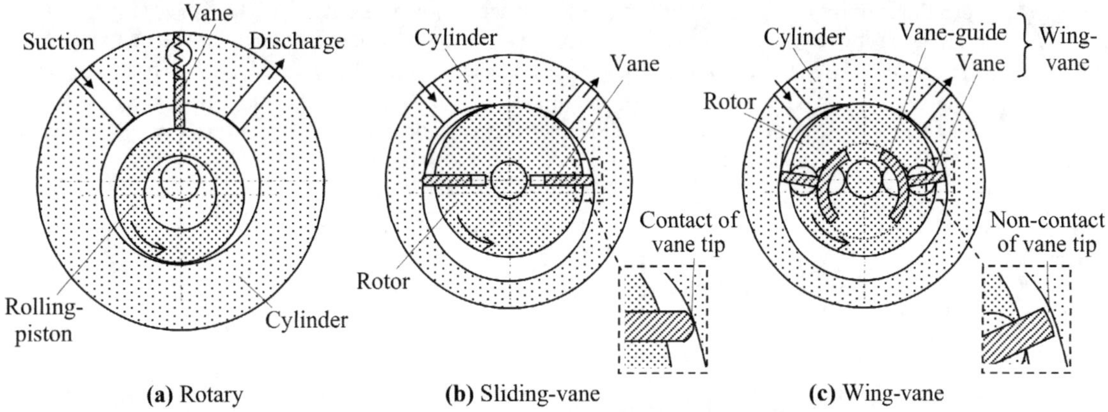

(a) Rotary **(b)** Sliding-vane **(c)** Wing-vane

Figure 2.2: Horizontal cross-sectional view of compression mechanisms

3. OPERATING PRINCIPLE

Figure 3.1 shows the compression process of the wing-vane type. Figure 3.1a shows the overall operation and Figure 3.1b shows the vane and the vane-guide operation. Because there are two vane, the compression process is completed at 180 degrees. The center chamber at 90 degrees is the timing when the suction process is completed, and then the compression process and the discharge process are started. As shown, the vane rotates around the center of the cylinder, so it is possible to keep the gap between the vane tip and the cylinder constant.

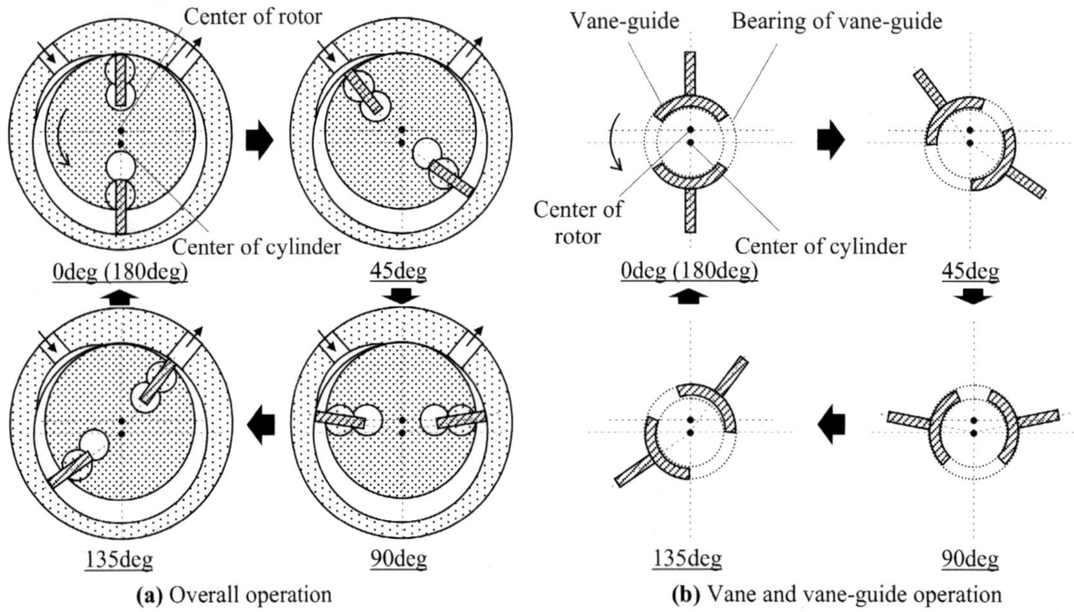

Figure 3.1: Compression process of the wing-vane compressor

4. CHARACTERIZATION OF TORQUE AND ROTOR LOAD

4.1 Calculation of compression chamber volume change

Figure 4.1 shows the calculation model of the volume change in the compression chamber (with two vanes). This figure shows the symbols below: a: cylinder radius, b: rotor outside radius, e: cylinder eccentricity (= a-b), r_s: distance between center of rotor and the bush(ON), L (θ): distance between M and Q, L_1 (θ): distance between O_c and N, L_2 (θ): distance between O_c and M, O: center of rotor, O_c : center of cylinder, P: proximity point, Q: tip of vane, M: intersection of line O_cQ and outside of rotor, N: intersection of line O_cQ and center line of bush, θ: ∠PON rotor rotation angle (P is set to 0 [deg]), φ: ∠PO_cM, and ψ: ∠POM. For simplicity, the thickness of the vane is not taken into account in the drawing. The area A (θ) of the space enclosed between the 1st vane and point P can be obtained by the following equation (4.1),

$$A(\theta) \; = A_1 - A_2 - A_3 \tag{4.1}$$

Here, A_1: area of sector PO_cQ, A_2: area of sector POM, and A_3: area of ΔOO_cM. Each of the areas is given by equation (4.2),

$$A_1 = \frac{a^2\phi}{2},\; A_2 = \frac{b^2\varphi}{2},\; A_3 = \frac{be\sin(\varphi)}{2} \tag{4.2}$$

Therefore, A (θ) is given by equation (4.1) and equation (4.2)

$$A(\theta) = \frac{a^2\phi}{2} - \frac{b^2\varphi}{2} - \frac{be\sin(\varphi)}{2} \tag{4.3}$$

Then, we find the angle φ and ψ. First, from the cosine theorem of ΔOO_cN,

$$L_1(\theta)=\sqrt{r_s^2+e^2+2r_se\cos(\theta)} \qquad (4.4)$$

$$r_s^2=L_1(\theta)^2+e^2-2L_1(\theta)e\cos(\phi) \qquad (4.5)$$

From equations (4.4) and equation (4.5), φ is as follows,

$$\phi=\cos^{-1}\left(\frac{L_1(\theta)^2+e^2-r_s^2}{2L_1(\theta)e}\right)\ (0<\theta<\pi) \qquad (4.6)$$

$$\phi=2\pi-\cos^{-1}\left(\frac{L_1(\theta)^2+e^2-r_s^2}{2L_1(\theta)e}\right)\ (\pi<\theta<2\pi) \qquad (4.7)$$

On the other hand, from the cosine theorem of ΔOO_cM,

$$L_2(\theta)=e\cos(\phi)+\sqrt{b^2-e^2\cos(\phi)^2} \qquad (4.8)$$

$$L_2(\theta)=\sqrt{b^2+e^2+2be\cos(\varphi)} \qquad (4.9)$$

From equations (4.8) and equation (4.9), ψ is as follows;

$$\varphi=\cos^{-1}\left(\frac{L_2(\theta)^2-b^2-e^2}{2be}\right)\ (0<\theta<\pi) \qquad (4.10)$$

$$\varphi=2\pi-\cos^{-1}\left(\frac{L_2(\theta)^2-b^2-e^2}{2be}\right)\ (\pi<\theta<2\pi) \qquad (4.11)$$

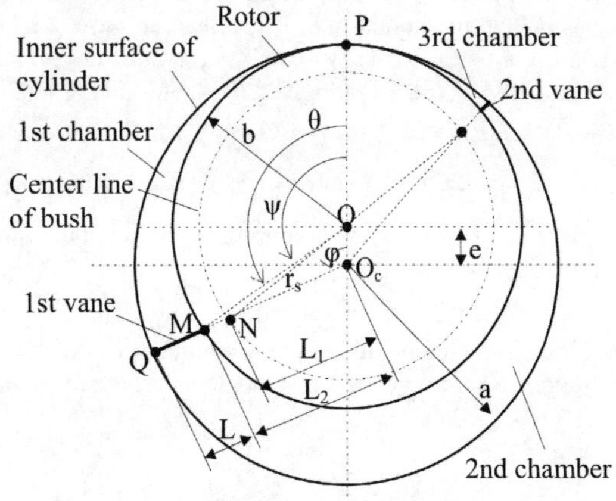

Figure 4.1: Calculation model of volume change in compression chamber

The areas of the 1st, 2nd, and 3rd chambers are S_1, S_2, and S_3, respectively and S_1, S_2, and S_3 are given by the following equation (4.12),

$$S_1=A(\theta),\ S_2=A(\theta+\pi)-A(\theta),\ S_3=\pi(a^2-b^2)-A(\theta+\pi) \qquad (4.12)$$

The maximum compression chamber area S_{max} is given by $\theta=\pi/2$ in S_2 as follows,

$$S_{max}=A(3\pi/2)-A(\pi/2) \qquad (4.13)$$

Here, if the cylinder height is h, the suction volume V_{st} is given by the following equation (4.14),

$$V_{st}=S_{max}h \qquad (4.14)$$

When considering the vane thickness t_v, the suction volume V_{st}^* is given by the following equation (4.15),

$$V_{st}^*=[S_{max}-\frac{t_v}{2}\{L(\frac{3\pi}{2})+L(\frac{\pi}{2})\}]h \qquad (4.15)$$

From Figure 4.1, L(θ) is given by the following equation (4.16),

$$L(\theta)=a-L_2(\theta) \qquad (4.16)$$

It should be noted that in internal compression interval $\theta=\pi/2\sim\pi$, the internal volume ratio ε is given by the following equation (4.17),

$$\varepsilon=\frac{S_{max}}{S_2(\pi)-S_{max}} \qquad (4.17)$$

4.2 Calculation of torque and load

Next, we will calculate the applied torque of vane T_r and the applied load of rotor F_r. In Figure 4.2, $P_i(\theta)$ is the pressure applied on i-th chamber and F_{pi} is the load applied on the i-th vane by the pressure difference, and F_{si} is the applied load of the bush. F_{p1} and F_{p2} are as follows,

$$F_{p1}=L(\theta)h\{P_2(\theta)-P_1(\theta)\} \qquad (4.18)$$

$$F_{p2}=L(\theta+\pi)h\{P_3(\theta)-P_2(\theta)\} \qquad (4.19)$$

The applied torque of vane T_r is as follows,

$$T_r=F_{p1}\{a-\frac{L(\theta)}{2}-e\cos(\phi)\}+ F_{p2}\{a-\frac{L(\theta+\pi)}{2}-e\cos(\phi+\pi)\} \qquad (4.20)$$

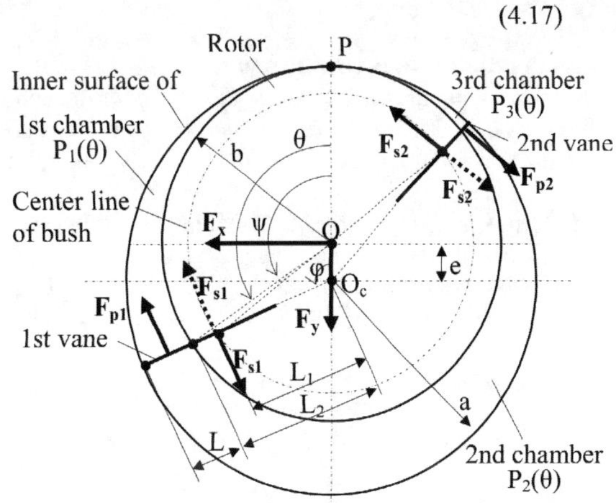

Figure 4.2: Calculation model of torque characteristics and the rotor load characteristics

$P_1(\theta)$, $P_2(\theta)$ and $P_3(\theta)$ have the following relationship,

$$P_2(\theta) = P_1\{\theta + \pi\}, \quad P_3(\theta) = P_1\{\theta + 2\pi\} \tag{4.21}$$

Next, we calculate the gas load applied on rotor F_r. Here F_r is divided into the gas load applied on rotor F_e and the gas load applied on bush F_s, where F_s is generated by the vane load.

In Figure 4.2, F_{exi}, F_{eyi} are the x-direction and y-direction components of the load applied on the rotor in each i-th chamber, F_{exi} and F_{eyi} are given by the following equations (4.22),

$$F_{ex1} = -bhP_1(\theta)\{1-\cos(\varphi)\}, \quad F_{ey1} = bhP_1(\theta)\sin(\varphi), \quad F_{ex2} = -2bhP_2(\theta)\cos(\varphi), \tag{4.22}$$

$$F_{ey2} = -2bhP_2(\theta)\sin(\varphi), \quad F_{ex3} = bhP_3(\theta)\{1+\cos(\varphi)\}, \quad F_{ey3} = bhP_3(\theta)\sin(\varphi)$$

Here F_{ex}, F_{ey} are the x-direction and y-direction components of F_e and $F_{ex,}$ and F_{ey} is given by the following equations (4.23),

$$F_{ex} = F_{ex1} + F_{ex2} + F_{ex3}, \quad F_{ey} = F_{ey1} + F_{ey2} + F_{ey3} \tag{4.23}$$

Next, we calculate the gas load applied on the rotor (bush) through the vanes F_{s1} and F_{s2}. From the balance of the moment around the center of the cylinder, the following equations (4.24) are obtained,

$$F_{s1} = F_{p1}\frac{\{a - \frac{L(\theta)}{2}\}}{L_1(\theta)}, \quad F_{s2} = F_{p2}\frac{\{a - \frac{L(\theta+\pi)}{2}\}}{L_1(\theta+\pi)} \tag{4.24}$$

Here, F_{s1} and F_{s2} are applied on the rotor and bush. The x-direction and y -direction components of this load are F_{sxi} and F_{syi} respectively, and F_{sxi} and F_{syi} are given by following equations (4.25) - (4.28),

$$F_{sx1} = -\frac{L(\theta)}{L_1(\theta)}\{a - \frac{L(\theta)}{2}\}h\{P_2(\theta) - P_1(\theta)\}\cos(\phi) \tag{4.25}$$

$$F_{sy1} = -\frac{L(\theta)}{L_1(\theta)}\{a - \frac{L(\theta)}{2}\}h\{P_2(\theta) - P_1(\theta)\}\sin(\phi) \tag{4.26}$$

$$F_{sx2} = -\frac{L(\theta+\pi)}{L_1(\theta+\pi)}\{a - \frac{L(\theta+\pi)}{2}\}h\{P_3(\theta) - P_2(\theta)\}\cos(\phi+\pi) \tag{4.27}$$

$$F_{sy2} = -\frac{L(\theta+\pi)}{L_1(\theta+\pi)}\{a - \frac{L(\theta+\pi)}{2}\}h\{P_2(\theta) - P_1(\theta)\}\sin(\phi+\pi) \tag{4.28}$$

The x-direction and y -direction components of F_s are F_{sx} and F_{sy}, respectively, and F_{sx} and F_{sy} are given by the following equations (4.29),

$$F_{sx} = F_{sx1} + F_{sx2}, \quad F_{sy} = F_{sy1} + F_{sy2} \tag{4.29}$$

Here, F_r is the applied load of the rotor, and F_{rx} and F_{ry} are the x-direction and y-direction components of F_r. So F_{rx}, F_{ry} and F_r are given by the following equations (4.30),

$$F_{rx} = F_{ex} + F_{sx}, \quad F_{ry} = F_{ey} + F_{sy}, \quad F_r = \sqrt{F_{ey}^2 + F_{sy}^2} \tag{4.30}$$

4.3 Characteristics comparison

Next, we compare the characteristics of the torque and the rotor load with the conventional sliding-vane type. For the comparison, it is assumed that the suction volume and cylinder inside diameter and the rotor outside diameter are the same. The basic dimensions are below: suction volume: V_{st} = 28.5 [cc / rev], cylinder radius: a = 30 [mm], rotor outside radius: b = 25.5 [mm] and the vane thickness is ignored. The maximum chamber area of the wing-vane type is S_{max}, this is slightly larger than that of the sliding-vane type S_{max}*, S_{max} / S_{max}* = 1.017. The compression chamber volume has almost identical characteristics, and is not shown. Accordingly, the characteristics of the torque and the rotor load have almost identical characteristics as shown in Figure 4.3 and Figure 4.4. These figures are the characteristics under specific operating conditions (evaporation temperature 52℃, condensation temperature 5℃).

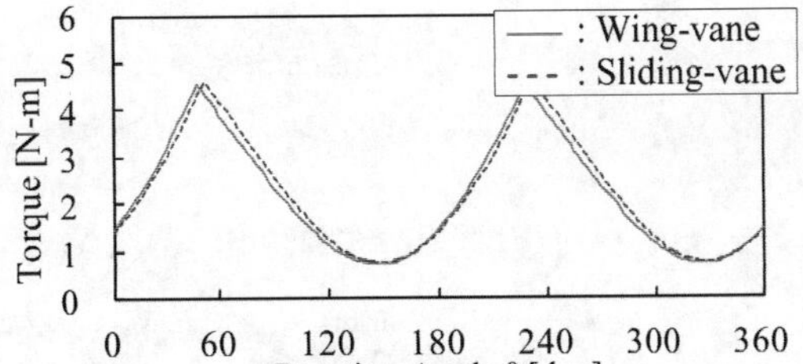

Figure 4.3: Characteristics of torque with respect to rotational phase

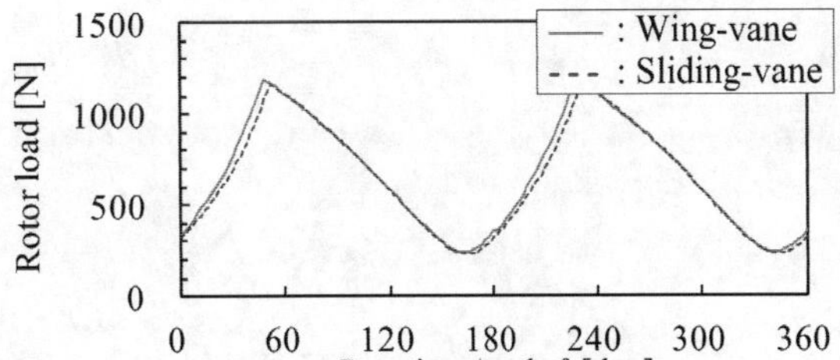

Figure 4.4: Characteristics of rotor load with respect to rotational phase

5. CHARACTERIZATION OF VANE LOAD

5.1 Constraints of arc angle of vane-guide

We now calculate the applied load of the vane-guide. The applied load of the vane is shown in Figure 5.1a, and the applied load of the vane-guide is shown in Figure 5.1b. In order to determine the load applied on only the 1st vane, the subscript is omitted. In Figure 5.1b, the arc of the vane-guide is asymmetric with respect to the vane, α_1 is the arc angle from the center line of the vane measured in the rotational direction, and α_2 is the arc angle from the center line of the vane measured in the counter-rotational direction.

Here, as described in Section 2, the circular arc angle α_1 and α_2 is limited, so this should be invastigated first. As shown in Figure 3.1b, the two vane-guides are closest to each other when θ is 90 degrees. In this phase, the angle between the 1st vane and the 2nd vane β is given by following equation (5.1),

$$\beta = 2\pi - \{\phi(\frac{3\pi}{2}) - \phi(\frac{\pi}{2})\} \tag{5.1}$$

On the other hand, since the portion of arc angle α_1 and the portion of arc angle α_2 are in the angular range of β, it is necessary that following relation (5.2) is established for non-contact,

$$\alpha_1 + \alpha_2 < \beta \tag{5.2}$$

From equations (5.1) and (5.2), the arc angle of a vane-guide has the following constraints (5.3),

$$\alpha_1 + \alpha_2 < 2\tan^{-1}(\frac{r_s}{e}) \tag{5.3}$$

5.2 Calculation of bearing load of vane-guide

In Figure 5.1, F_b is the radial-direction load applied on the vane from the pressure difference between the back and the tip, F_{cv} is the centrifugal force applied on the vane, F_{ca} is the centrifugal force applied on the vane-guide, M is the longitudinal-direction of the vane, N is the orthogonal-direction of the vane, m is the central-direction of the vane-guide, n is the direction perpendicular to m, and F_m and F_n are the n-direction and m-direction loads applied on the bearing of the vane-guide. When ζ is the angle of m-direction and M-direction, F_m and F_n are as follows,

$$F_m = F_{ca} + \frac{F_b + F_{cv}}{2}\cos(\zeta) + \frac{F_s - F_p}{2}\sin(\zeta)\ , \ F_n = -\frac{F_b + F_{cv}}{2}\sin(\zeta) + \frac{F_s - F_p}{2}\cos(\zeta) \tag{5.4}$$

Here, ζ is given by the following equation (5.5),

$$\zeta = \frac{\alpha_1 - \alpha_2}{2} \tag{5.5}$$

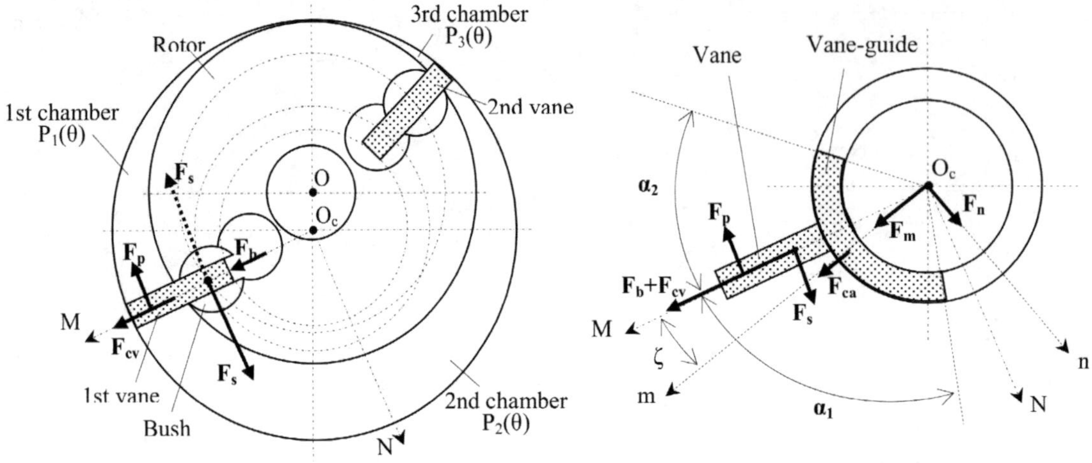

(a) Load applied on vane and bush **(b)** Load applied on vane-guide

Figure 5.1: Calculation model of vane-guide characteristics

The bearing of the vane-guide is a partial bearing which is different from an ordinary bearing. Therefore, the calculation method corresponding to a partial bearing is needed to calculate the characteristics of the vane-guide. The results of the calculation of partial bearing characteristics using a new calculation method are reported in Part 2

In the case of a partial bearing, when the ratio of F_n for F_m becomes larger, it can't be supported by the bearing surfaces of the vane-guide. In this situation, there is a risk that the wing-vane will be unstable. In this paper, a simple means of calculating the limit angle where the wing-vane is unstable is described. In Figure 5.2, F_1 and F_2 are the loads applied on the circumferential ends of the vane-guide. The vane-guide is assumed to be supported at two points. From the balance of forces, F_1 and F_2 are represented by the following equation (5.6),

$$F_1 = \frac{F_m \sin(\frac{\alpha_1+\alpha_2}{2}) + F_n \cos(\frac{\alpha_1+\alpha_2}{2})}{\sin(\alpha_1+\alpha_2)}, \quad F_2 = \frac{F_m \sin(\frac{\alpha_1+\alpha_2}{2}) - F_n \cos(\frac{\alpha_1+\alpha_2}{2})}{\sin(\alpha_1+\alpha_2)} \tag{5.6}$$

It is necessary for $F_2 > 0$ to prevent stable operation. Therefore, if following relation (5.7) is satisfied, the vane will be stable.

$$\frac{F_n}{F_m} < \tan(\frac{\alpha_1+\alpha_2}{2}) \tag{5.7}$$

Figure 5.3 shows a calculation example of F_1 and F_2. If both loads are greater than 0, the vane-guide can be stable and maintain reliability.

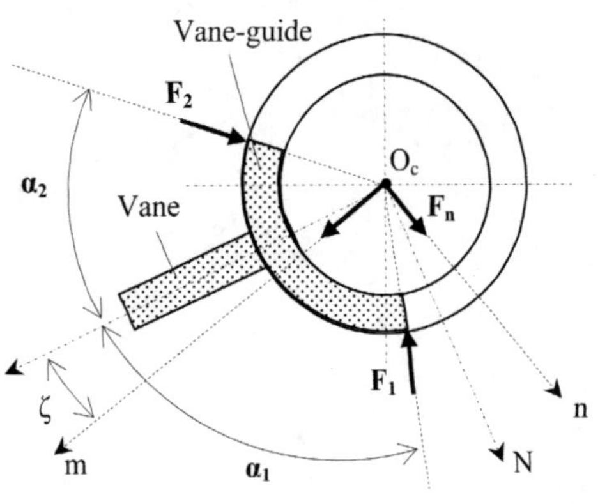

Figure 5.2: Calculation model of the stability of the vane-guide

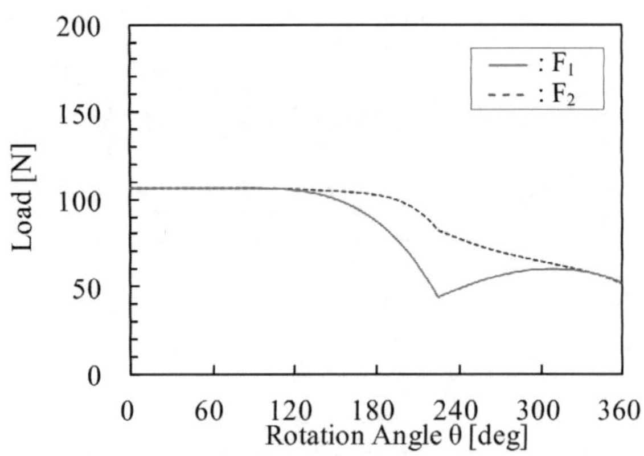

Figure 5.3: Characteristics of the stability of the vane-guide with respect to rotational phase

6. FEATURES AND PROBLEMS ON STRUCTURE

Next, we compare several features of the conventional sliding-vane type and the wing-vane type. We describe the comparison of the noteworthy matters as follows.

6.1 Compression Process, Torque, Rotor Load Characteristics

As indicated in Section 4.2, the characteristics of the chamber volume, torque, and rotor load are almost the same as the sliding-vane type. However, as will be described later, since the eccentricity must be smaller than the sliding-vane type, the rotor load tends to slightly increase over that of the sliding-vane type.

6.2 The size of the compression mechanism

Basically, since the center-direction of the vane tip curvature and the longitudinal-direction of the vane are constant, the operation of the wing-vane type is similar to that of the sliding-vane type. Therefore, it is necessary to increase the eccentricity which is smaller than that of the sliding-vane type which has longer vanes by deflecting the vane to the center of rotor (vane of general sliding-vane type is deflected). For this reason, the wing-vane type is slightly larger in the radial-direction than the sliding-vane type.

6.3 Performance

In the wing-vane type, the friction loss of the vane tip does not occur. However, a bearing loss in the vane-guide and a leakage loss in the vane tip clearance are now generated. If the sum of these losses is equal to the sum of the bearing loss inside the rolling-piston and the friction loss of the vane tip in the rotary type, the efficiency is expected to be the same as the rotary type.

6.4 Reliability

Since the vane-guide bearing and the bush are new frictional parts, it is necessary to ensure their reliability. In particular, as described above, since the vane-guide has a partial bearing, the stability of wing-vane operation is important. Therefore, it is necessary to understand the characteristics of the vane-guide bearing. The bush is used in the conventional swing compressor, but it was never used incorporated in the rotating body. Therefore, it is necessary to consider the frictional properties of the bush.

7. CONCLUSIONS

In this paper, a wing-vane compressor, which can achieve both high efficiency and downsizing, has been proposed and the characteristics of torque, rotor load and vane load were calculated. The characteristics of the wing-vane type is as follows.

- The striking feature of the wing-vane type is that there is no friction loss at the vane tip . Therefore, it eliminates the largest disadvantage of the sliding-vane type.
- However, a leakage loss in the vane tip clearance and a bearing loss in the vane-guide are now generated. If the sum of these losses is equal to the sum of the bearing losses of the rolling-piston and the friction loss in the vane tip of the rotary type, the efficiency is expected to be the same as the rotary type.
- The characteristics of the compression process, torque, and rotor load, are almost the same as the sliding-vane type. However, since it is difficult to handle a large eccentricity, the applied load of the rotor is larger than the sliding-vane type.
- Lubrication of the newly added vane-guide and bush is important to ensure reliability.

A Study on High Efficiency Wing-Vane Compressor -Part 2: Lubrication Characteristic of The Partial Arc Guide Bearing-

Tatsuya SASAKI[1]*, Shin SEKIYA[2], Raito KAWAMURA[3], Hideaki MAEYAMA[4], Shinichi TAKAHASHI[5], Kanichiro SUGIURA[6]

[1,2,3]Mitsubishi Electric Corporation, Advanced Technology R&D Center,
8-1-1, Tsukaguchi-Honmachi, Amagasaki-City, Hyogo, Japan
Phone: +81-6-6497-7582, Fax: +81-6-6497-7285,
[1]E-mail: Sasaki.Tatsuya@ea.MitsubishiElectric.co.jp
[2]E-mail: Sekiya.Shin@eb.MitsubishiElectric.co.jp
[3]E-mail: Kawamura.Raito@cj.MitsubishiElectric.co.jp

[4]Mitsubishi Electric Corporation, Living Environment Systems Laboratory,
3-18-1, Oshika, Suruga-ku, Shizuoka-City, Shizuoka, Japan
Phone: +81-54-287-3053, Fax: +81-54-287-3056,
E-mail: Maeyama.Hideaki@dx.MitsubishiElectric.co.jp

[5,6]Mitsubishi Electric Corporation, Shizuoka Works,
3-18-1, Oshika, Suruga-ku, Shizuoka-City, Shizuoka, Japan
Phone: +81-54-287-3112, Fax: +81-54-287-3127,
[5]E-mail: Takahashi.Shinichi@bk.MitsubishiElectric.co.jp
[6]E-mail: Sugiura.Kanichiro@ay.MitsubishiElectric.co.jp

ABSTRACT

Low global warming potential refrigerants such as HFO-1234yf are attractive as alternative refrigerants in air-conditioning and cooling systems. However, larger compressor sizes were required due to the lower fluid density compared with R410A. Thus, a new wing-vane compressor which has no contact between the vane and the cylinder has been developed to prevent an increase in the size without performance degradation. The rotary motion of the vane is supported by the partial arc vane guide on the upper and lower ends of the vane. Discovering the lubrication properties of the vane guide was one of the issues required in development. So we analyzed the friction coefficient between the vane guide and the guide bearing with numerical analysis, and measured the friction coefficient by experiment. We confirmed that the friction coefficient of the analysis and experiment result are in good agreement, and can be organized in a Stribeck curve as well as full journal bearing in the fluid lubrication region.

1. INTRODUCTION

Low global warming potential refrigerants such as HFO (hydro fluoro olefin) -1234yf are attractive as alternative working fluids in air-conditioning and cooling systems to address global warming (Minor, 2008). However, larger compressor sizes were required due to the lower fluid density compared with conventional refrigerants such as R410A (Akasaka, 2010). Thus, a new wing-vane compressor which has no contact between the vane and the cylinder has been developed to prevent an increase in the size without performance degradation. The rotary motion of the vane is supported by the partial arc vane guide on the upper and lower ends of the vane. Discovering the lubricating properties of the vane guide was one of the issues required in development. So we clarified the friction coefficient between the vane guide and the guide bearing by lubrication analysis and experiment, and evaluated the stability of the vane guide attitude by motion analysis.

2. COMPRESSOR STRUCTURE AND LUBRICATION PROBLEM

Figure 1 shows the basic structure of the wing-vane compressor. The wing-vane compressor consists chiefly of a shell, a motor, and a compression mechanism. Low pressure refrigerant taken in from the suction port is compressed to high pressure in the compression mechanism, and discharged to the refrigerant pipe after passing through the shell and the discharge port. The power of the motor is transmitted to the compression mechanism via the shaft. The vane which divides the compression chamber has a partial arc vane guide on its upper and lower ends. The vane guide slides over the guide bearing which has the same curvature radius. The vane guide is a new sliding part not found in a conventional rotary compressor, and we had to clarify the lubricating properties of the vane guide to develop the wing-vane compressor.

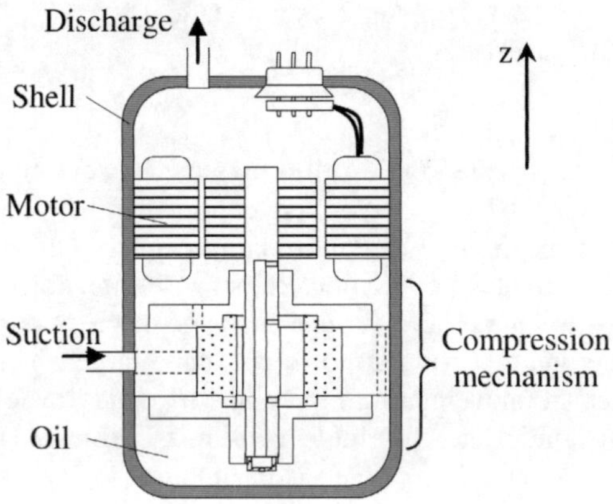

(a) Longitudinal section

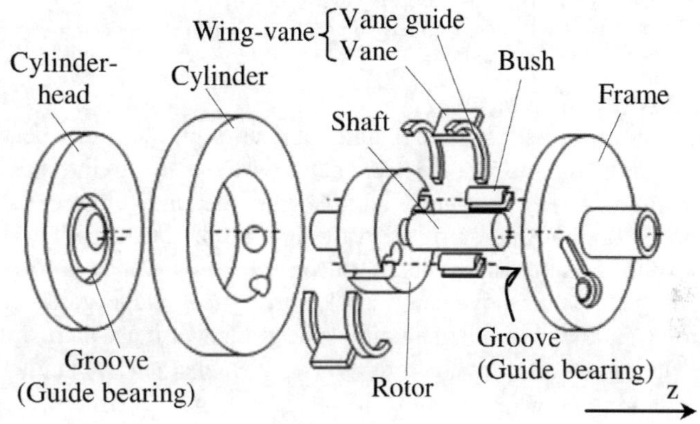

(b) Disassembly view of compression mechanism

Figure 1: Basic structure of wing-vane compressor

3. LUBRICATION PROPERTY OF GUIDE BEARING

We predicted the friction coefficient between the vane guide and the guide bearing by lubrication analysis with a static load. The validity of the analytical model was verified by comparing the experimentally measured friction coefficient by simulating the sliding between the vane guide and the guide bearing.

3.1 Guide Bearing Load

Figure 2 shows the calculation model of the vane guide characteristics, where F_b is a radial direction force acting on the vane due to the pressure difference between the vane back and the vane tip, and F_{cv} and F_{ca} are the centrifugal forces of vane and vane guide respectively. Here, F_p is the gas load caused by the pressure difference between the adjacent compression chambers, and F_s is bush reaction force. The result of these forces acting on the vane is applied to the vane guide, Figure 2b, where M is the vane longitudinal direction coordinate, N is the vane orthogonal direction coordinate, m is the center of the vane guide direction coordinate, and n is the coordinate at right angle to m coordinate. Here, F_m and F_n are the vane guide force of the m direction and n direction respectively. The result of F_m and F_n is represented as F. The angle of the F direction and m coordinates is defined as ϕ.

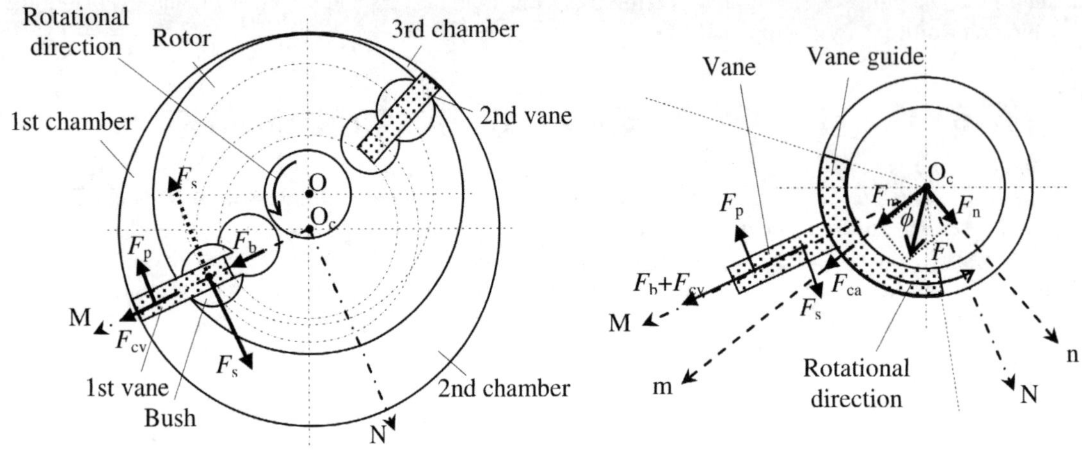

(a) Load applied on vane and bush (b) Load applied on vane guide

Figure 2: Calculation model of vane guide characteristics

3.2 Analysis Model

In order to satisfy hydrodynamic lubrication in the guide bearing, we analyzed the oil film pressure by solving the Reynolds equation of oil film between the vane guide and the guide bearing. We first considered the sliding condition of the vane guide and the guide bearing. Figure 3 shows a magnified view of around the bush. The vane moves by the rotation of the rotor through the bush. When the rotor rotates in the counterclockwise direction, the bush (a) and the vane side are in contact. A clearance is formed between the bush (b) and the vane of the opposite side. With this movement the clearance of the vane guide and the guide bearing at point B expand larger than at point A. Figure 4 is a diagram of the guide bearing which simulates the movement in Figure 3, where the Reynolds equation of oil flowing through the clearance between the vane guide and the guide bearing is expressed by Equation (1), where h is oil film thickness, p is oil film pressure, η is oil viscosity, and v_j is sliding velocity. The Reynolds equation was solved numerically by the finite difference method.

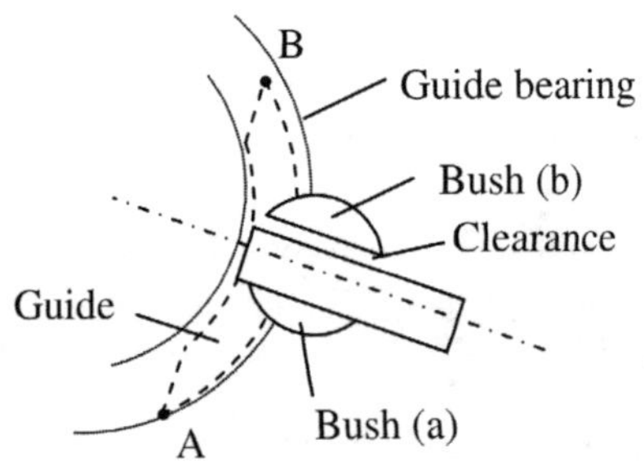

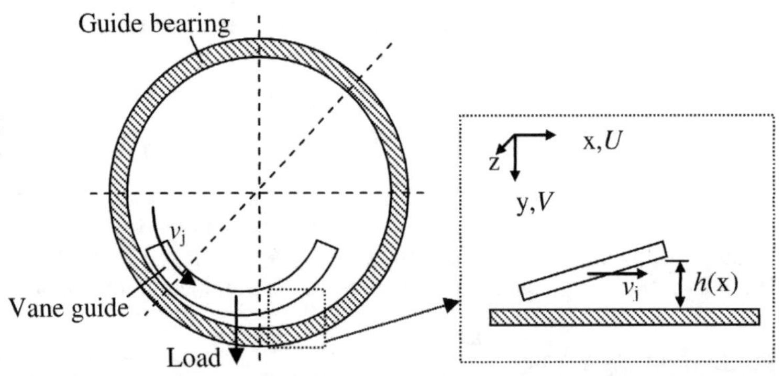

(a) Overall view (b) Magnified view of clearance

Figure 3: Magnified view of around bush

Figure 4: Diagram of guide bearing

$$\frac{\partial}{\partial x}\left(h^3\frac{\partial p}{\partial x}\right)+\frac{\partial}{\partial z}\left(h^3\frac{\partial p}{\partial z}\right)=-6\eta v_j\frac{\partial h}{\partial x} \qquad (1)$$

3.3 Experiment

The friction coefficient was measured by the test apparatus simulating the sliding of the vane guide and the guide bearing in order to verify the validity of the analysis results, and the presence or absence of contact was confirmed.

3.3.1 Experimental apparatus: Figures 5 and 6 show a schematic view of the experimental apparatus and a perspective view of the test vane guide. The test apparatus has a configuration which loads and slides the test vane guide to the rotating test guide bearing in an atmosphere. The driving force of the motor is transmitted to a spindle which drives the test guide bearing via a belt. The test vane guide is attached to the shaft through a bush. The shaft is supported by the base. When the loading shaft is strained, a load is applied to the test vane guide by the spring. The frictional force acting on the test vane guide is measured by the load cell facing the torque arm attached to the upper end of the shaft. Lubricating oil is dropped from the oil supply path of the upper end of the shaft, when the test guide bearing rotates, oil is supplied to the sliding surface by centrifugal force. To measure the temperature of the lubricating oil, thermocouples inserted in the oil supply path are fixed to the lower end of the shaft. The sliding surface roughness of the test guide bearing and the test vane guide is 0.8 μm in a ten-point average roughness. Using a contact electric resistance method, contact between the sliding surfaces was examined (Yano, 2015). The potential difference between the test vane guide and the test guide bearing was measured

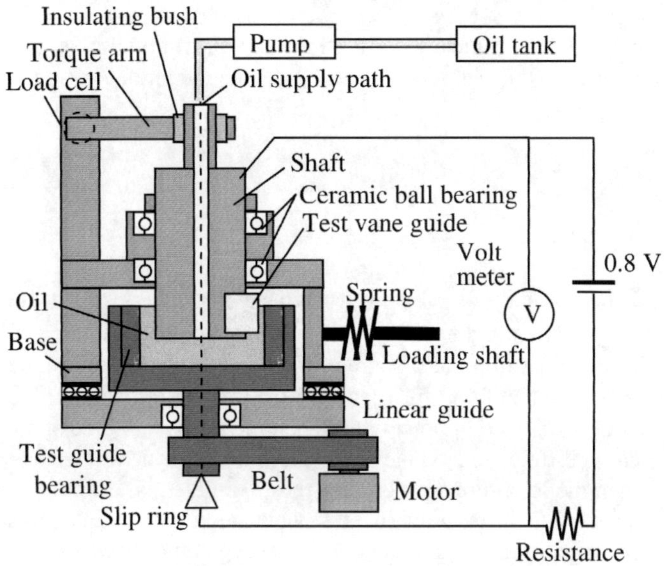

Figure 5: Schematic view of experimental apparatus

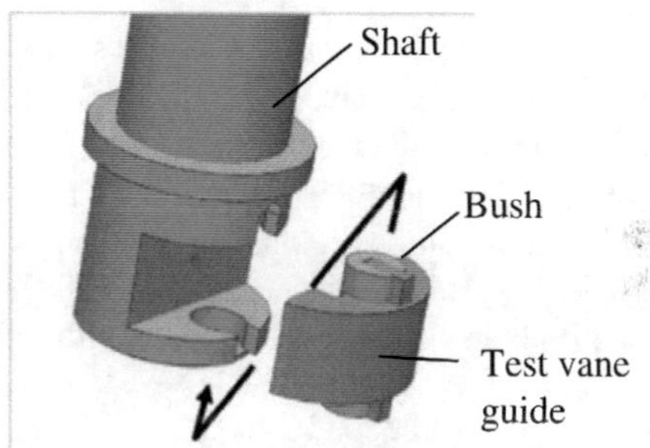

Figure 6: Perspective view of the test vane guide

3.3.2 Experimental conditions: The parameters that affect the lubricating properties of the guide bearing are the load direction, the arc angle, the load value, the viscosity of the lubricating oil, the rotational speed, the bearing radius, and the radial clearance. These parameters other than the load direction and the arc angle can be organized in a Sommerfeld number represented by Equation (2), where N_b is the rotational speed, F is the load acting on the vane guide, L is the guide bearing width, R is the guide bearing radius, and C is the radial clearance. Table 1 shows the test conditions. Figure 7 shows the load direction on the test vane guide. The load is applied in the range of 0-40 degrees toward the counter-rotation direction of the test guide bearing. The rotational speed was constant, the load was increased by 10 N up to 10-100 N, and the contact resistance and the friction coefficient were measured for 60 seconds.

$$S = 2\eta \frac{N_b}{F} LR\left(\frac{R}{C}\right)^2 \qquad (2)$$

Table 1: Test conditions

Load	10-100	N
Load Direction	0-40	deg.
Rotating Speed	20	rps
Arc Angle	30-95	deg.
Sommerfeld Number	0.1-5	-
Oil Viscosity @ 313 K	3.4×10^{-3}	Pa·s

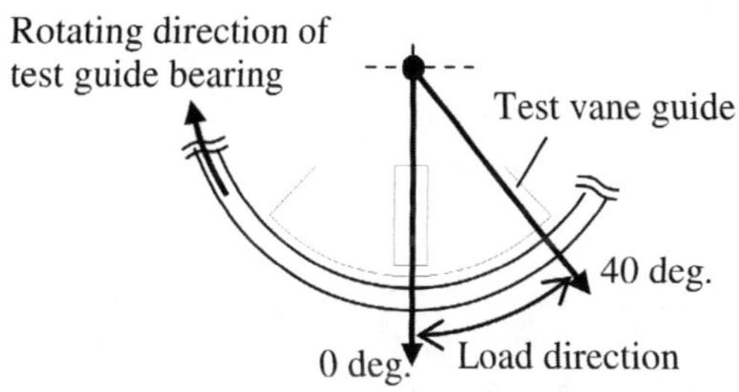

Figure 7: Load direction on the test vane guide

3.4 Analysis and Experimental Result

3.4.1 The effect of the load direction on the friction coefficient: Figure 8 shows the relationship of the friction coefficient and load direction at the arc angle of 95 degrees. In the load direction of 0 degrees, the experimental and the analytical results show a good agreement in the test range. The friction coefficient becomes lower as the Sommerfeld number becomes smaller, it can be seen that the fluid lubrication region is maintained. It was confirmed that the friction coefficient of the guide bearing is organized in a Stribeck curve in the same way as the full journal bearings. When the load direction is not less than 25 degrees, a minimum point exists on the friction coefficient in the region where the Sommerfeld number is smaller than the minimum point of the friction coefficient. With a decreasing Sommerfeld number the friction coefficient increases. Because of increasing the load the oil film thickness is reduced and the test vane guide has contact with the test guide bearing. When the load direction is 40 degrees, it is considered the oil film was broken throughout the test range. The index showing the state of surface contact is generally expressed by Equation (3), where Λ is the film thickness ratio, h is the oil film thickness, and σ is the root-mean-square roughness of each surface (Kimura, 2001). Subscripts 1 and 2 denote the guide bearing and guide.

$$\Lambda = \frac{h}{\sqrt{\sigma_1^{\ 2} + \sigma_2^{\ 2}}} \tag{3}$$

When the projections of the roughness are assumed to be normally distributed, it is considered that $\Lambda > 3$ in fluid lubrication region, and $\Lambda \leqq 3$ in mixed lubrication region. The roughness of the test vane guide and the test guide bearing are 0.8 μm in ten-point mean roughness. In the load direction of 30 degrees, the minimum oil film thickness is 3.5 μm, and the thickness ratio is 3.1, suggesting that it is a condition for transition from fluid lubrication region to mixed lubrication region. Figure 9 shows the results of difference in potential. When the test vane guide and the test guide bearing are separated by an oil film, the potential difference indicates equivalent to the input voltage 0.8 V. When both are in contact the voltage drops lower. When the load direction is 0 degrees, the potential difference in the tested range is constant at 0.8 V, indicating that the test guide bearing and the test vane guide are separated by an oil film. When the load direction is 25 degrees, the reduction ratio of the potential difference is increased when the Sommerfeld number is below 0.3, and it was found that this corresponds to the friction coefficient of Figure 8. When the load direction is 40 degrees, the potential difference has a low value over the entire test range, consistent with the trend of the friction coefficient.

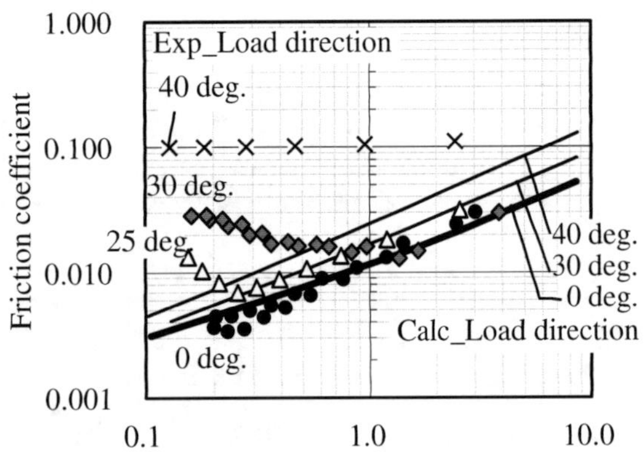

Figure 8: The relationship of friction coefficient and load direction

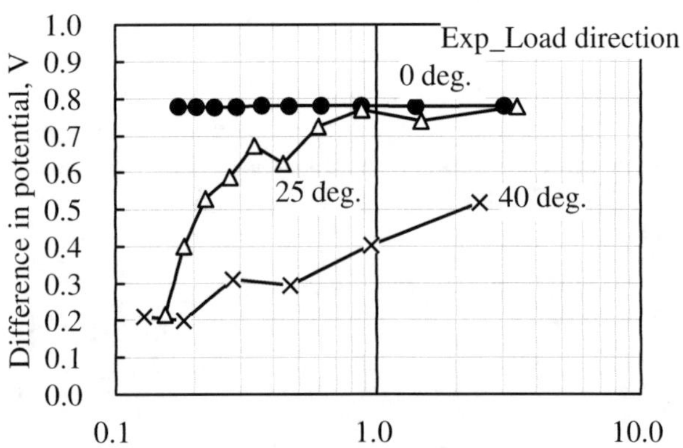

Figure 9: Result of difference in potential

3.4.2 The effect of the arc angle on the friction coefficient: Figure 10 shows the relationship of the friction coefficient and the arc angle of the guide. The load direction is constant at 0 degrees. When the arc angle is 95 degrees and 50 degrees, the friction coefficient is consistent with the analysis result. It can be seen that the sliding condition is a fluid lubrication region. When the arc angle is 40 degrees, the sliding state transited from the fluid lubrication region to the mixed lubrication region with the Sommerfeld number being smaller. When the arc angle was 30 degrees, it was impossible to maintain the fluid lubrication state in the test range. The Sommerfeld number at the time of transition to the mixed lubrication region increases as the arc angle becomes smaller.

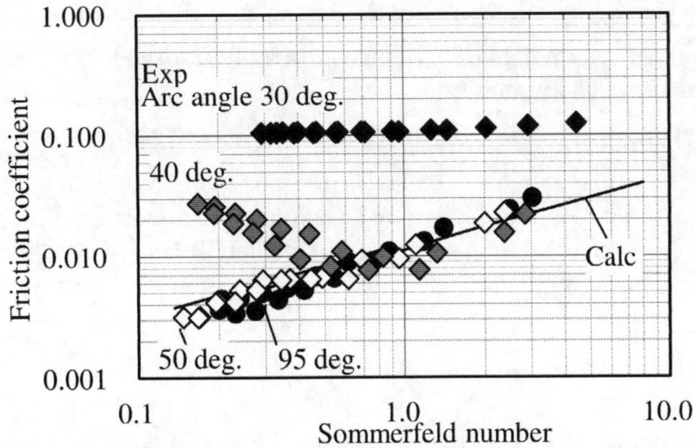

Figure 10: The relationship of arc angle of guide and friction coefficient

4. STABILITY ANALYSIS (DYNAMIC CHARACTERISTICS)

4.1 Motion Equation

During the actual operation of the compressor the load of vane guide varies in magnitude and direction. To predict the operating conditions in which the vane guide can be maintained stably, a vane guide motion analysis was conducted. Figure 11 shows the vane guide load under the low load conditions (evaporation and t condensation temperature is 308 K and 298 K respectively). Figure 12 is a diagram of the vane guide load when the load direction is on the outside of the arc angle. In the figure, the clearance between the vane guide and the guide bearing becomes a reverse wedge shape and oil film pressure does not occur. At this time, the clockwise moment N around point A acts on the vane guide. As the clearance between the vane guide and the guide bearing at point B expand, the frictional force acts on the bush plane and the bush cylindrical surface inhibits the separation movement of the vane guide.

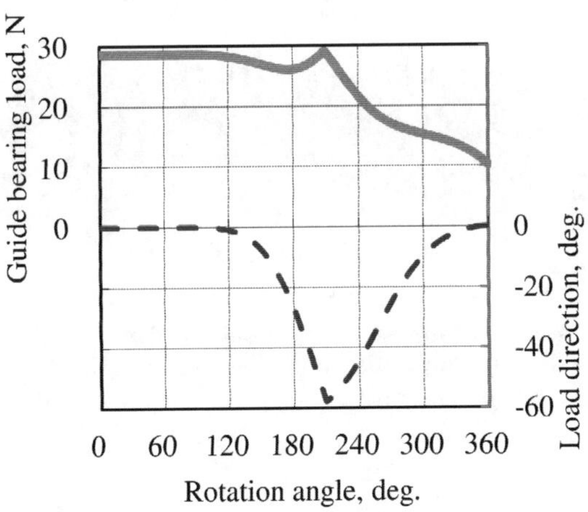

Figure 11: Vane guide load

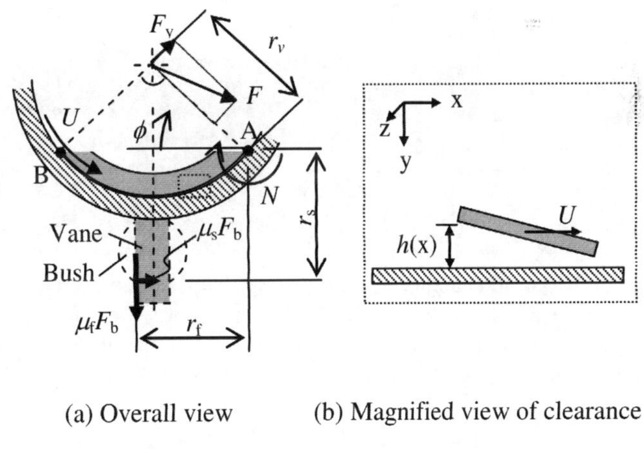

Figure 12: Diagram of vane guide load

The motion equation is expressed by Equation (4), where I is the inertia moment about point A of the vane guide, ϕ is the displacement angle around point A of the vane guide, t is time, and N is the sum of the moment. Here, ϕ is defined as 0 degrees when point B of the vane guide is in contact with the guide bearing. The moment M^+ where the vane guide is separated off from the guide bearing is expressed by Equation (5). When the vane guide starts to separate from the guide bearing the reaction force acts from the bush to the vane, friction force $\mu_f F_b$ acts between the vane and the bush plane, the same as $\mu_s F_b$ which acts between the holder and the bush cylindrical surface, where μ_s is friction coefficient of the vane and bush plane, and μ_f is friction coefficient of the bush holder and bush cylindrical surface. Moment M^- with these forces is expressed by Equation (6). The length of the moment arm is represented by r_v, r_f, r_s. In addition, the derivative term of Equation (4) is represented as shown in Equation (7), where ω is the angular velocity of the vane guide movement direction. Moment N is the sum of Equations (5) and (6), and Equation (8) is obtained by substituting Equations (5)-(7) in Equation (4). Since F_y and F_b are a function of time they can be represented as $F_y = F_y(t)$, $F_b = F_b(t)$. Further, dimensionless $\bar{t}$ and $\bar{\omega}$ are represented by Equations (9) and (10), where ω_v is the angular velocity of the vane guide along the guide bearing. Equation (11) is obtained by substituting Equations (9) and (10) into Equation (8). Equations (12) and (13) are obtained by integrating Equation (11) twice. Since solving the integral calculation analytically is difficult, the dimensionless angular velocity and the displacement angle are calculated with the Riemann integral shown in Equations (12) and (13).

$$I\frac{d^2\phi}{dt^2} = N \tag{4}$$

$$M^+ = F_y r_v \tag{5}$$

$$M^- = -\mu_f F_b r_f - \mu_s F_b r_s \tag{6}$$

$$\frac{d^2\phi}{dt^2} = \frac{d\omega}{dt} \tag{7}$$

$$\frac{d}{dt}\omega = \frac{1}{I}\left(F_y r_v - \mu_f F_b r_f - \mu_s F_b r_s\right) \tag{8}$$

$$\bar{t} = \omega_v t \tag{9}$$

$$\bar{\omega} = \frac{\omega}{\omega_v} \tag{10}$$

$$\frac{d\bar{\omega}}{d\bar{t}} = \frac{1}{I\omega_v^{\,2}}\left\{r_v F_y\left(\bar{t}\right) - \left(\mu_f r_f + \mu_s r_s\right)F_b\left(\bar{t}\right)\right\} \tag{11}$$

$$\bar{\omega} = \frac{1}{I\omega_v^{\,2}}\int\left\{r_v F_y\left(\bar{t}\right) - \left(\mu_f r_f + \mu_s r_s\right)F_b\left(\bar{t}\right)\right\}d\bar{t} \tag{12}$$

$$\phi = \frac{1}{I\omega_v^{\,2}}\int\int\left\{r_v F_y\left(\bar{t}\right) - \left(\mu_f r_f + \mu_s r_s\right)F_b\left(\bar{t}\right)\right\}d\bar{t}\,d\bar{t} \tag{13}$$

4.2 Analysis Conditions

Table 2 shows the analysis conditions. So that the load direction was outside of the vane guide arc, we set the guide arc angle to 50 degrees. As shown in Figure 11 the vane guide become unstable statically under the low load conditions(evaporation and condensation temperature is 308 K and 298 K respectively) when the load direction is outside of the vane guide arc in the rotational angle range of 178-269 degrees.

Table 2: Analysis conditions

Rotating speed	20	rps
Arc angle of guide	50	deg.
Friction coefficient of bush	Variable	-

4.3 Analysis Result

Figure 13 shows the analysis results. It shows a separation range analyzed by the static analysis in the figure too. The separation range becomes narrow as the friction coefficient of bush becomes large. With a bush friction coefficient of 0.1, it was found that separation of the vane guide from the guide bearing was suppressed. In conditions where the load direction becomes outside of the vane guide arc, as the friction coefficient of the bush increase, the stability of the vane improves. The actual compressor verification results are reported in Part 3.

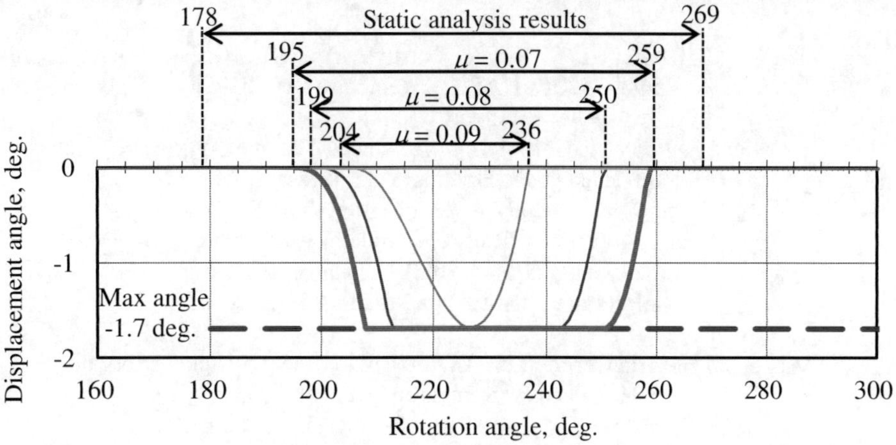

Figure 13: Analysis results

5. CONCLUSIONS

To establish a design technology for the guide bearing which is specific to the wing-vane compressor, the static lubricating analysis of the guide bearing was considered and compared to the friction coefficient by experiment. It was subjected to motion analysis in consideration of the fluctuating load to clarify the behavior and lubrication performance of the guide bearing in the region where the load direction is out of the vane guide arc and the oil film pressure does not occur. The results are shown below.

- When the load acts on the center of the arc angle of the vane guide, the analysis results and the test results of the friction coefficient match and it is revealed that lubrication analysis of guide bearing according static load is appropriate.
- If the guide bearing is satisfied by fluid lubrication region, the friction coefficient can be organized in a Stribeck curve as well as full journal bearings.
- When the load direction is equal to or larger than 25 degrees, lubrication condition changes from fluid lubrication region to mixed lubrication region as the Sommerfeld number is reduced, and as the load direction is large the Sommerfeld number at the boundary of mixed lubrication region and fluid lubrication region is large.
- Where the arc angle of the vane guide is small, the lubrication state is likely to transition from fluid lubrication region to mixed lubrication region.
- As a result of motion analysis it was revealed that the displacement of the vane guide can be reduced by considering the frictional force of the bush in the region where the load direction is out of the vane guide arc.

REFERENCES

Barbara, M., & Mark, S. (2008). HFO-1234yf low GWP refrigerant update. *International Refrigeration and Air Conditioning Conference at Purdue* (2349). Indianapolis, IN: Purdue University.

Ryo, A., Yohei, K., Katsuyuki, T., & Yukihiro, H. (2010). *JSRAE Thermodynamic Table Vol.3 HFO-1234yf.* Japan: Japan Society of Refrigerating and Air Conditioning Engineers.

Akihiko, Y., Eiichi, I., Masaki, M., & Sadao, Y. (2015). Study on the load carring of sliding bearing lubricated by synthetic ester oils, *Tribology Online*, *10*(5), 377-389.

Yoshitsugu, K. (2001). *Tribology Handbook.* Japan: Yokendo.

A Study on High Efficiency Wing-vane Compressor
- Part 3: Experimental Evaluation of The Prototype -

Raito KAWAMURA[1]*, Shin SEKIYA[2], Tatsuya SASAKI[3], Hideaki MAEYAMA[4],
Shinichi TAKAHASHI[5], Kanichiro SUGIURA[6]

[1, 2, 3] Mitsubishi Electric Corporation, Advanced Technology R&D Center,
8-1-1 Tsukaguchi-Honmachi, Amagasaki, Japan
Phone: +81-6-6497-7173, Fax: +81-6-6497-7285,
[1] Email: Kawamura.Raito@cj.MitsubishiElectric.co.jp
[2] Email: Sekiya.Shin@eb.MitsubishiElectric.co.jp
[3] Email: Sasaki.Tatsuya@ea.MitsubishiElectric.co.jp

[4] Mitsubishi Electric Corporation, Living Environment Systems Laboratory,
3-18-1 Oshika, Suruga-ku, Shizuoka, Japan
Phone: +81-54-287-3053, Fax: +81-54-287-3056,
Email: Maeyama.Hideaki@dx.MitsubishiElectric.co.jp

[5, 6] Mitsubishi Electric Corporation, Shizuoka Works,
3-18-1 Oshika, Suruga-ku, Shizuoka, Japan
Phone: +81-54-287-3112, Fax: +81-54-287-3127,
[5] Email: Takahashi.Shinichi@bk.MitsubishiElectric.co.jp
[6] Email: Sugiura.Kanichiro@ay.MitsubishiElectric.co.jp

ABSTRACT

Low global warming potential refrigerants such as HFO (hydro fluoro olefin) -1234yf are attractive as alternative working fluids in air-conditioning and cooling systems to address global warming. However, larger compressor sizes were required due to the lower fluid density compared with conventional refrigerants such as R410A. Thus, a new wing-vane compressor which has no contact between the vane and the cylinder has been developed to prevent an increase in size without performance degradation. The structure of the wing-vane compressor is different from a conventional sliding-vane compressor. Therefore, a prototype of the wing-vane compressor was operated using R134a refrigerant to simulate HFO-1234yf refrigerant. The results were as follows: (1) the APF (annual performance factor) ratio of the prototype of the wing-vane compressor was 100.2% with respect to the twin-rotary compressor. (2) Stable operation was confirmed without problems in the process from suction to discharge. (3) There were no serious abrasions and scratches in the friction parts of the wing-vane compressor after 100 hours of operation under high load operating conditions. (4) The vane-guide, whose arc angle is over 60 degrees, operates stably in low load conditions which are the most worrying conditions for instability of a vane-guide.

1. INTRODUCTION

Low global warming refrigerants such as HFO-1234yf are attractive as alternative working fluids in air-conditioning and cooling systems to address global warming. However, larger compressor sizes were required due to the lower fluid density compared with conventional refrigerants such as R410A. Thus, a new wing-vane compressor which has no contact between the vane and the cylinder was developed to prevent an increase in the size without performance degradation. The structure of the wing-vane compressor is different from a conventional vane compressor. The basic structure and dynamics model were described in Part 1, and the lubrication characteristic of the partial arc guide bearing of the vane-guide in Part 2. In this paper, a prototype of the wing-vane compressor was operated using R134a refrigerant to simulate HFO-1234yf refrigerant. The results of the experimental evaluation of the prototype are compared with the conventional twin-rotary compressor.

2. SPECIFICATION OF THE PROTOTYPE

Figure 2.1 shows the overall configuration of the 1st prototype of the wing-vane compressor. Figure 2.1a shows the longitudinal section of entire structure and Figure 2.1b shows the horizontal cross-section of the compression mechanism. Figure 2.2 shows the appearance of 1st prototype. Table 2.1 shows the basic specifications of the 1st prototype. The refrigerant is R134a to simulate HFO-1234yf, and the stroke volume of the compressor is 28.5 cc. The number of vanes is two, and the shell is a high-pressure type. The flow of the refrigerant is described as follows. First, low-pressure refrigerant is sucked into the compression mechanism. The refrigerant is pressurized from low pressure to high pressure in the compression mechanism and it is discharged into the interior of the shell. Next, the refrigerant is discharged to the outside of the compressor through the gap of the motor.

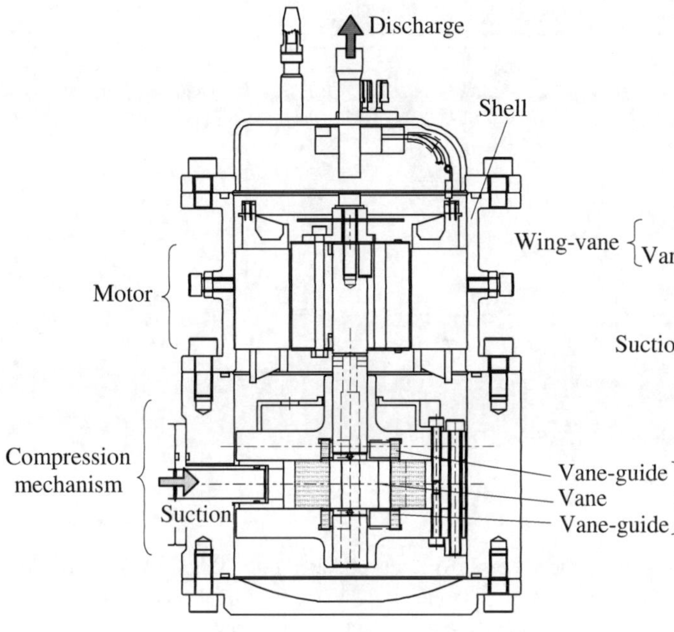

(a) Longitudinal section of entire structure

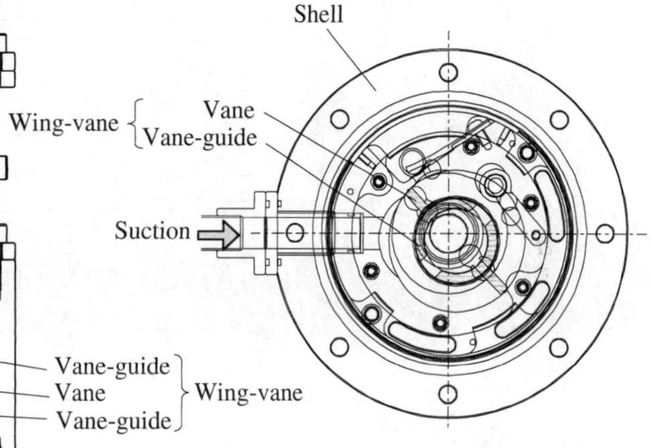

(b) Horizontal section of compression mechanism

Figure 2.1: Overall configuration of 1st prototype

Table 2.1: Basic specifications of 1st prototype

Refrigerant	R134a
Stroke Volume	28.5cc
Number of vanes	2

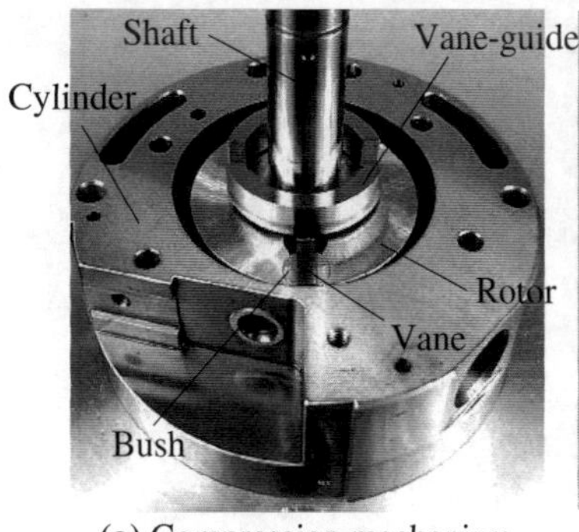

(a) Compression mechanism

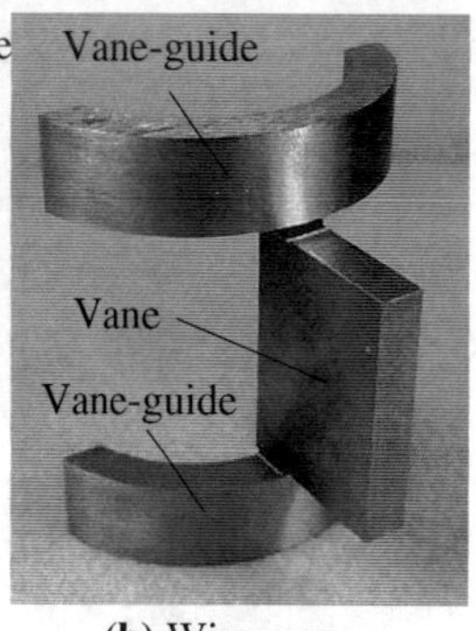

(b) Wing-vane

Figure 2.2: Appearance of 1st prototype

3. EVALUATION OF BASIC RELIABILITY

3.1 Operation check test

In order to confirm the stable operation of compressor, the pressure waveform of the prototype was measured under the specific operating conditions (evaporation temperature 52℃, condensation temperature 5℃, rotational speed 30rps). Figure 3.1 shows the pressure waveform measured in the above condition. As a result of the measurement, stable operation was confirmed without problems in the process from suction to discharge. It should be noted that the range without a waveform (over 320 degrees) is out of measurement range.

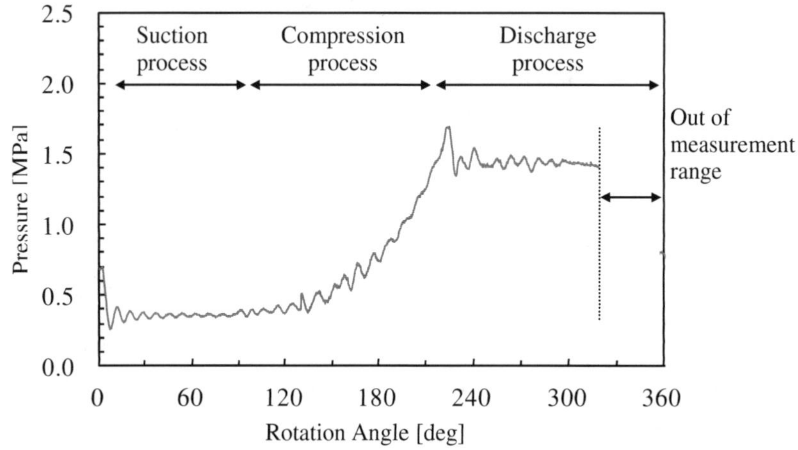

Figure 3.1: Pressure waveform of compression chamber

3.2 Durability test of the friction parts

In order to confirm the durability of the friction parts, the 1st prototype was operated for 100 hours under high load operating conditions (evaporation temperature 68℃, condensation temperature 12℃, rotational speed 60rps). Figure 3.2 shows the typical friction parts of the wing-vane compressor after operation under the above conditions. As shown in Figure 3.2, there are no serious abrasions and scratches. Therefore, it can be said that the structure of the wing-vane has high reliability under normal operating conditions.

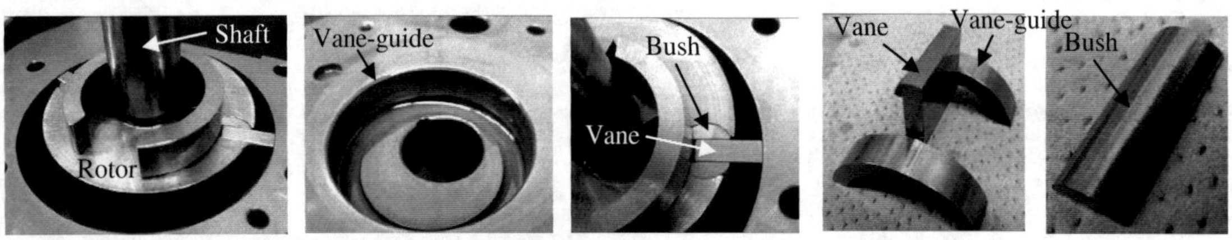

Figure 3.2: Appearance of friction parts after operation

3.3 Stable Operation of Vane

The stability of the vanes described in Part 2 were evaluated in the actual 1st prototype compressor. To confirm the stability limits of the arc angle of the vane-guide, the pressure waveform was compared in the arc angle range from 50 degrees to 135 degrees. From the evaluation, when the arc angle was over 60 degrees, the vane-guide operated stably under low load conditions (evaporation temperature 35℃, condensation temperature 20℃, rotational speed 20rps) which is the most worrying condition of the unstableness of the vane-guide.

Figure 3.3 shows the pressure waveform in the unstable specification of the vane-guide (arc angle was 50 degrees). If the arc angle of the vane-guide is small, the vane-guide operates unstably. In this case, the vane-guide separated from the bearing and the vane separated from the cylinder inside. For this reason, the compression chamber is no longer sealed and the chamber pressure is reduced to the range between 250 deg to 280 deg shown in Figure 3.3.

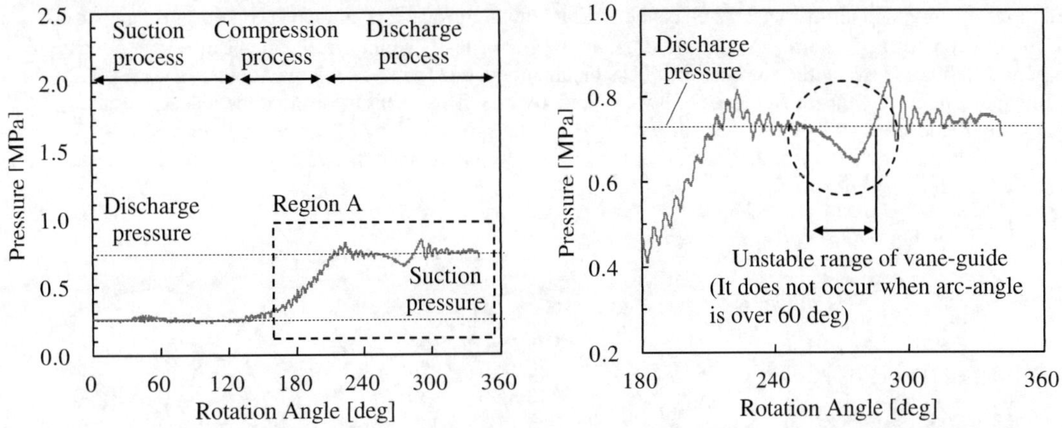

(a) Whole trend **(b)** Magnification of Region A

Figure 3.3: Pressure waveform in the unstable condition of the vane (arc angle is 50 deg)

4. EXPERIMENTAL EVALUATION OF PERFORMANCE

4.1 Evaluation of 1st Prototype

Figure 4.1 shows the performance evaluation results of the 1st prototype in comparison with the twin-rotary type. The stroke volume of the twin-rotary type is 28.5cc which is the same as the wing-vane 1st prototype. According to the results of the evaluation, the APF (annual performance factor) ratio of the 1st prototype is 88.4% with respect to the twin-rotary type. The APF is a closer condition to the actual usage of air-conditioners in Japan.

Figure 4.2 shows the loss analysis of the 1st prototype in comparison with the twin-rotary type under specific operating conditions (evaporation temperature 52℃, condensation temperature 5℃, rotational speed 60rps). In comparison with the twin-rotary type, 1st prototype has large indicative losses. A pressure loss and a leakage loss accounts for the vast majority of them. It is necessary to reduce these losses to improve the performance of the wing-vane compressor. The improved items for reducing these losses are described in the next section.

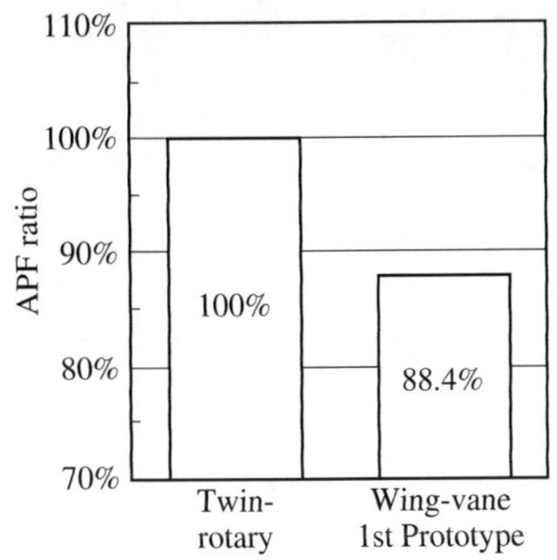

Figure 4.1: Performance evaluation results of 1st prototype

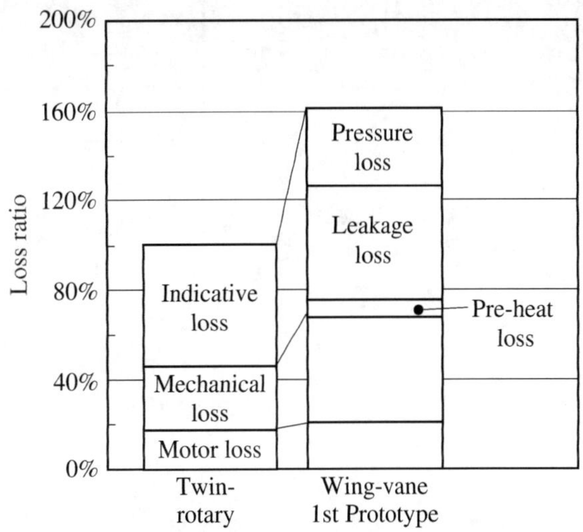

Figure 4.2: Loss analysis of 1st prototype

4.2 Improvement for Loss Reduction

(1) Selection of the number of vanes

In order to reduce the leakage loss, the influence of the number of vane against leakage loss was examined in actual evaluation. As shown in Figure 4.3, the performance (compression efficiency) was compared by changing the number of vanes between 2 and 4. Figure 4.4 shows the influence of the number of vane under specific operating conditions (evaporation temperature 52℃, condensation temperature 5℃, rotational speed 60rps). The performance was improved by using 3 vanes and 4 vanes compared with 2 vanes. This improvement is caused by reducing the pressure difference between the adjacent compression chambers due to increasing the number of vanes. There is no significant difference in the performance of 3 vanes and 4 vanes. This result means that the leakage loss is sufficiently reduced using 3 vanes. Therefore, we selected the specification of 3 vanes using less parts than 4 vanes. In addition, the arc angle of the vane-guide was 135 degrees in the 2-vane type, 95 degrees in the 3-vane type, and 65 degrees in the 4-vane type.

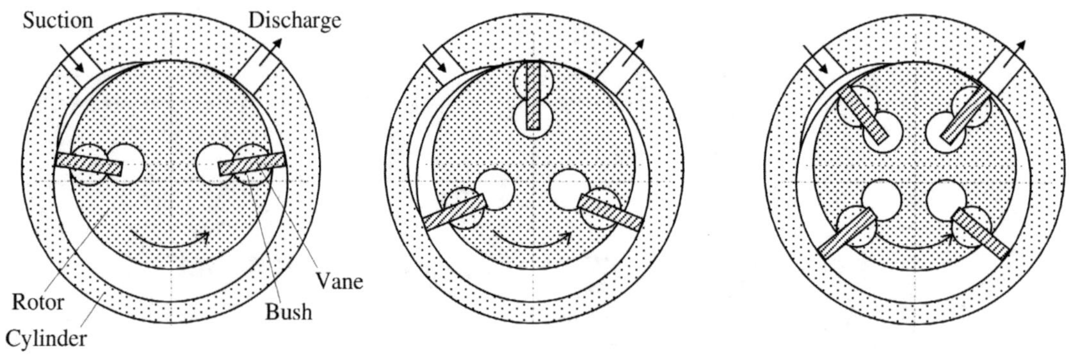

(a) 2-vane type 1st Prototype **(b)** 3-vane type **(c)** 4-vane type

Figure 4.3: Compressor arrangement when changing the number of vanes

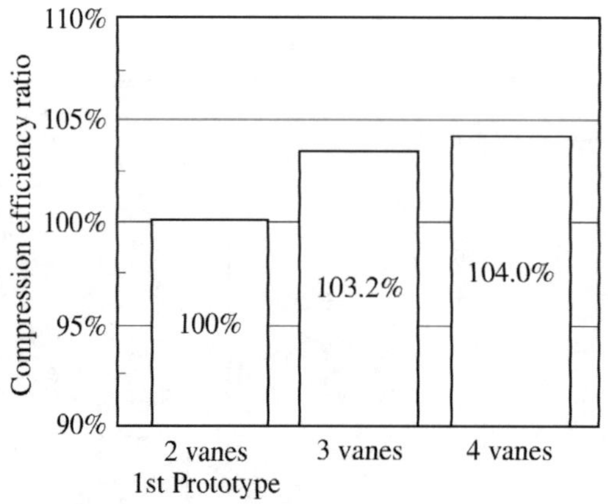

Figure 4.4: Influence of the number of vanes

(2) Adoption of a two-step discharge method

In order to reduce the discharge pressure loss, a new discharge method for the compression chamber was devised for dividing the discharge process into two steps. In the sliding-vane type and the wing-vane type, there is a problem that the flow passage area in the compression chamber is decreased at the end of the compression process and the discharge pressure loss is increased. Therefore, the 1st discharge port was placed at the point with a large flow passage area and the 2nd discharge port was placed behind the 1st discharge port to discharge the rest of the refrigerant with a small flow passage area. Figure 4.5 shows the structure of two-step discharge method applied to the sliding-vane type and the wing-vane type. In the sliding-vane type, since the vane tip curvature is small, a leakage loss occurs at the timing when the vane passes the 1st discharge port as show in Figure 4.5a. In the wing-vane type, since the vane tip curvature is large and it is the same as the curvature of cylinder inside, the leakage loss is smaller than the sliding-vane type as show in Figure 4.5b.

Figure 4.6 shows the pressure loss improvement by the two-step discharge method under specific operating conditions (evaporation temperature 52℃, condensation temperature 5℃, rotational speed 60rps). Figure 4.6a shows the pressure waveform of the conventional discharge method (1 port), and Figure 4.6a shows that of the two-step discharge method. In Figure 4.6, the hatching range over the discharge pressure is the discharge pressure loss. By applying the two-step discharge method, it was possible to suppress the discharge pressure loss to about 1/3.

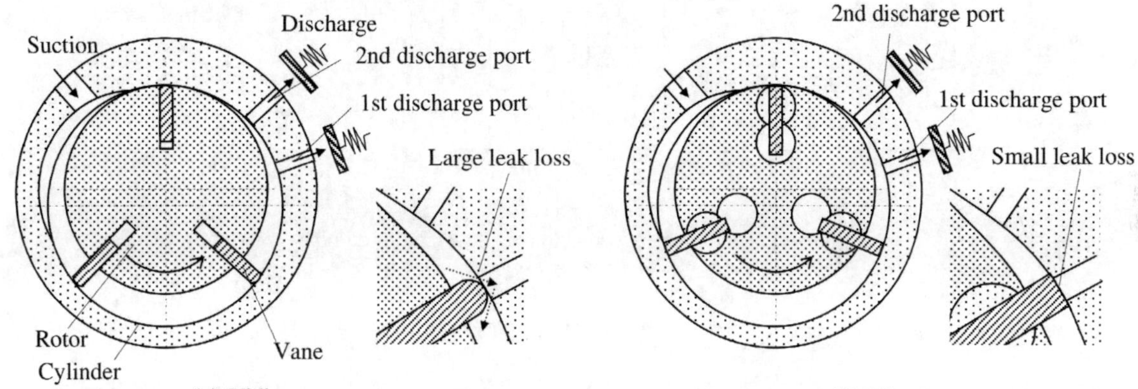

(a) Sliding-vane **(b)** Wing-vane

Figure 4.5: Structure of two-step discharge method

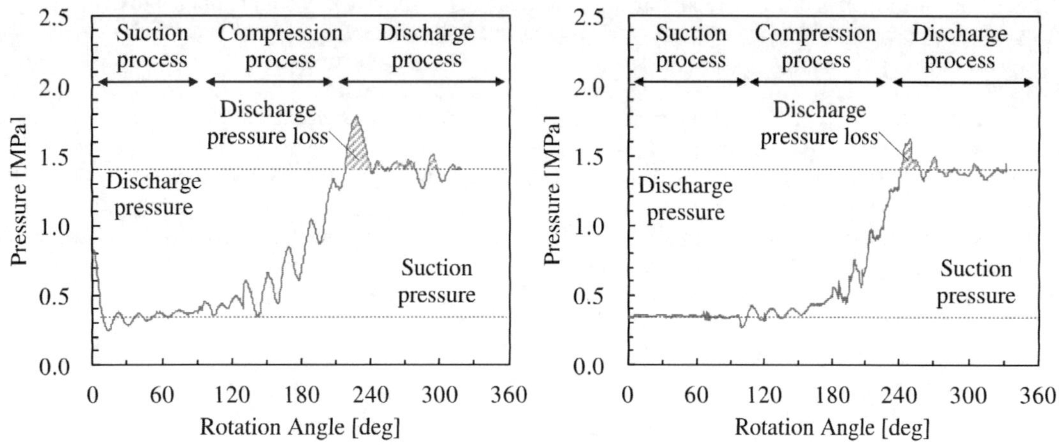

(a) Conventional discharge method (1 port) **(b)** Two-step discharge method

Figure 4.6: Pressure loss improvement by two-step discharge method

4.3 Evaluation of 2nd Prototype

Figure 4.7 shows the performance evaluation results of the 2nd prototype mainly applying the improvement items in the previous section. The Stroke volume of the 2nd prototype is 28.5 cc which is the same as the 1st prototype. The APF ratio of the 2nd prototype is improved by 11.8% from the 1st prototype. In the breakdown of this improvement, 5.1% is due to the two-step discharge method, 4.1% is due to the optimization of the number of vanes, and 2.6% is due to other improvements. As the result of these improvements, the APF ratio of the 2nd prototype is 100.2% with respect to the twin-rotary type. In other words, the wing-vane type can achieve both high performance and miniaturization.

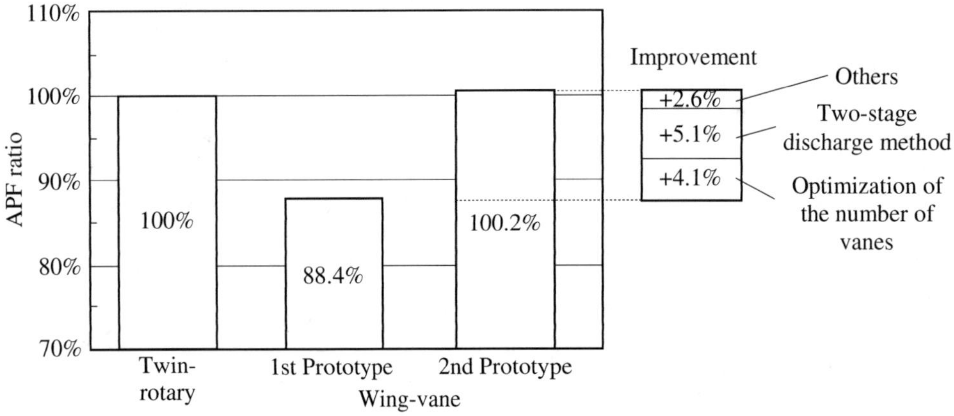

Figure 4.7: Performance evaluation results of 2nd prototype

5. CONCLUSIONS

A wing-vane compressor which can achieve both high efficiency and downsizing has been proposed. In this paper, prototypes of the wing-vane compressor were operated using R134a refrigerant to simulate HFO-1234yf refrigerant. The evaluation results in the prototype are as follows.

(1) Performance

By optimizing the number of vanes and improving the discharge method, the APF ratio of the prototype of the wing-vane compressor is 100.2% with respect to the twin-rotary compressor.

(2) High Reliability

・From the pressure waveform, stable operation was confirmed without problems in the processes from suction to discharge

・There were no serious abrasions and scratches in the friction parts of the wing-vane compressor after 100 hours of operation under high load operating conditions.

・The vane-guide, whose arc angle is over 60 degrees operates stably under low load conditions which is the most worrying condition for instability of the vane-guide.

Optimal Structural Design of Swing Double-Vane Compressor

Junjie MA [1*], Xiang CHEN [1], Xu YANG [1], Zongchang QU [1]

[1]School of Energy and power Engineering
Xi'an Jiaotong University,Xi'an, Shaanxi, China
Phone:+8618292876783, Fax:+86(29)82668691
E-mail: amazing_ma@163.com

ABSTRACT

In this paper, a novel rotary compressor, namely Swing Double-vane Compressor, was introduced. Based on the working principle of the compressor, thermodynamic model and kinetic model of upper and down vanes, rotor and eccentric shaft were established. Combined with each model, with the working efficiency as the objective function, the primary structural sizes were optimally designed for the compressor by optimization algorithm, after which the working efficiency can be to 0.8757 and its theoretical displacement has improved by 66.94% compared with the single-vane compressor. According to this optimal design, it will provide theoretical foundation to improve the performance of the Swing Double-vane Compressor.

1. INTRODUCTION

In order to solve the problems of working face frictional wear and leakage for the high speed of small rotary compressor, professor Qu, Z. C. (2011) put forward a novel swing vane compressor. Due to lower speed of translation and rotation, the friction loss was reduced and the mechanical efficiency was effectively improved for the hinged structure of the compressor with the better sealing. Li, Q. C. (2009) has carried on the simple analysis of the thermodynamics, leakage and kinetic for the swing vane compressor, compared with the rolling rotor compressor and synchronal rotary compressor, he found that the swing vane compressor has lower speed and smaller acceleration which change slowly, and the friction power is relatively small with a more stable operational condition. Based on the compressor geometry model, thermodynamics model and kinetic model, Li, X. L. (2009) has conducted viable optimizing analysis in structural parameters and verified the feasibility of improved compressor in actual operation, after a relative test of the performance parameters. Teh, Y. (2002) presented a designed improvement to reduce the wear and friction at the vane side effectively by achieving a 30% reduction over its predecessor. Zong, W. B. (2015) established the dynamic model of translational rotary compressor to discuss the natural frequency and vibration type by comparing the critical and actual speed.

In this paper, on the basis of previous studies of swing vane compressor, increase a vane is added into the lower end of the rotor to form double vane structure, its movement mechanism basically consistent with compressor. Based on the analysis of the machine geometry model, thermodynamic model and dynamics model, the mathematical optimization design is established with the EER of compressor defined as the objective function. The state parameters of thermodynamics of compressed gas are obtained using the optimization algorithm, and the optimal structure parameters can be carried out by analyzing the calculation results.

2. THE ESTABLISHMENT OF THE MATHEMATICAL MODEL

2.1 The Geometric Model

With working principle of new double swing vane compressor shown in Fig.1, the tip of the main board is embedded into the cylinder by the way of hinged point and the end inserted in the groove of the rotor. Due to eccentric motion of rotor, the vane is swinging around its tip, while the end is dong linear motion in the groove. The down vane which can move randomly in the slot is embedded in the symmetry side of the rotor, and the tip of the down vane is tightly pressed on the inner wall of the cylinder for the centrifugal force and the high pressure of the lubricating oil. Because of the two vanes between rotor and cylinder and simultaneously operation bilaterally of two working chambers, the compressor has completed two cycles when the eccentric shaft rotates a full revolution.

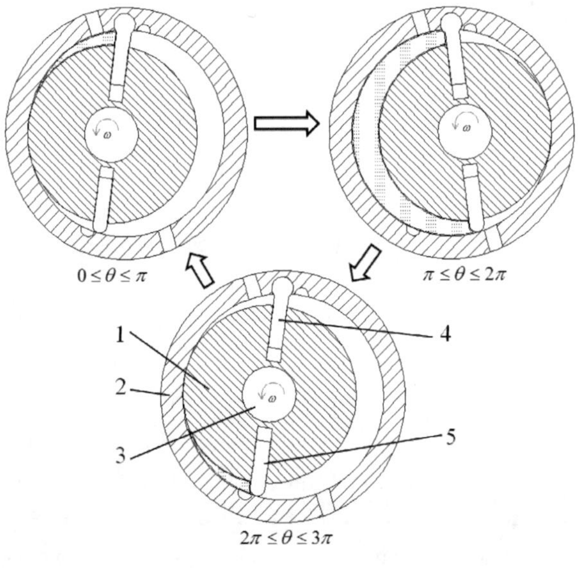

1-rotor 2-cylinder 3-eccentric shaft 4-upper vane 5-down vane

Figure 1: The working process of double swing vane compressor

Taking the left-side working chamber as an example, The chamber volume of the compressor calculated by the method of Ma, J. J. (2015).can be expressed as shown below:

$$V_1(\theta) = H \cdot [\frac{R^2}{2}\theta - \frac{r^2}{2}(\theta + \arcsin\frac{e\sin\theta}{\sqrt{R^2+e^2-2e\mathrm{R}\cos\theta}}) - \frac{eR\sin\theta}{2} - (\sqrt{R^2+e^2-2e\mathrm{R}\cos\theta} - \mathrm{r}) \cdot \frac{b_1}{2}] \quad (1)$$

When $\pi \le \theta \le 2\pi$, the relative geometric parameters are given by,

$$L_0 = \sqrt{e^2 + R^2 - 2\cdot e\cdot R\cdot \cos\theta} \tag{2}$$

$$\alpha = \arcsin[\frac{e}{L_0}\cdot \sin(2\pi - \theta)] \tag{3}$$

$$L_1 = 2R\cdot \cos\alpha - L_0 - r \tag{4}$$

Taking the center of the rotor as the origin of the polar coordinates system, its polar coordinates is expressed as $(e, \theta - \pi)$ and the value of radius is R. Then Equation (5) is given by cosine theorem,

$$R^2 = \rho^2 + e^2 - 2\cdot \rho\cdot e\cdot \cos[\gamma - (\theta - \pi)] \tag{5}$$

Equation (6) is simplified as follows,

$$\rho = e\cdot \cos[\gamma - (\theta - \pi)] + \sqrt{(R^2 - e^2) + e^2\cdot \sin^2[\gamma - (\theta - \pi)]} \tag{6}$$

According to the equations above, the volume of working chamber is described in the formula,

$$V_2(\theta) = H\cdot [\frac{1}{2}\int_{\alpha}^{\pi+\alpha}(\rho^2 - r^2)d\gamma - (L_0 - r)\cdot \frac{b_1}{2} - L_1\cdot \frac{b_2}{2}] \tag{7}$$

When $2\pi \le \theta \le 3\pi$, as shown in Fig.2c, the calculation formula of stroke volume is basically similar to that of $\pi \le \theta \le 2\pi$ (Equation (3)~(6)),

$$\alpha = \arcsin[\frac{e}{L_0}\cdot \sin(\theta - 2\pi)] \tag{8}$$

$$\rho = e\cdot \cos[\gamma - (\theta - \pi)] - \sqrt{(R^2 - e^2) + e^2\cdot \sin^2[\gamma - (\theta - \pi)]} \tag{9}$$

The volume of working chamber is represented as follows,

$$V_3(\theta) = H\cdot [\frac{1}{2}\int_{\theta-2\pi}^{\pi-\alpha}(\rho^2 - r^2)d\gamma - L_1\cdot \frac{b_2}{2}] \tag{10}$$

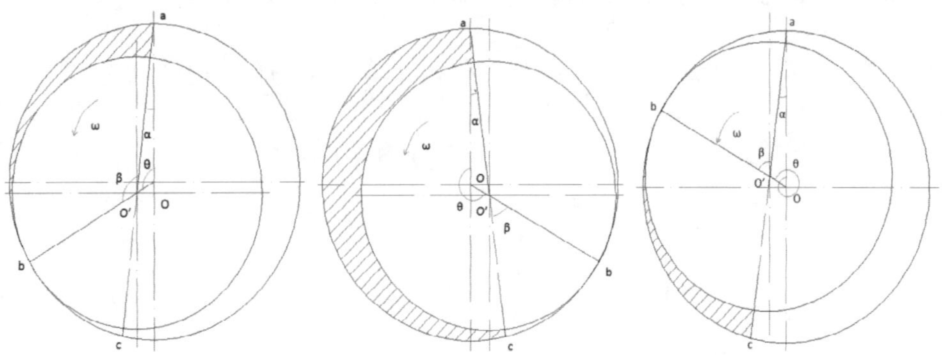

(a) $0 \le \theta \le \pi$ (b) $\pi \le \theta \le 2\pi$ (c) $2\pi \le \theta \le 3\pi$

Figure 2: The diagram of working volume of compressor

Due to the symmetry of the structure, the right working volume of compressor can be defined as:

$$V(\theta)=\begin{cases}V_3(4\pi-\theta) & \pi\le\theta\le 2\pi\\ V_2(4\pi-\theta) & 2\pi\le\theta\le 3\pi\\ V_1(4\pi-\theta) & 3\pi\le\theta\le 4\pi\end{cases} \tag{11}$$

2.2 The Thermodynamic Model

The change of chamber volume will cause the pressure, temperature and mass of fluid to change during the working process, if regarding the chamber volume of compressor as control volume and the medium in the homogeneous state, the equation of energy conservation is established according to the law of thermodynamics as:

$$\frac{d(\mathrm{mu})}{dt}=\frac{\mathrm{dQ}}{dt}+\frac{\mathrm{dE}_i}{dt}-\frac{dE_0}{dt}+\frac{dW}{dt} \tag{12}$$

Assume that the speed of the compressor is constant, and the angle of the shaft is $\theta=\omega t$, namely $dt=d\theta/\omega$, then Equation(13) is given according to the law of thermodynamics $h=u+pv$, $v=V_c/m$ as follows:

$$\frac{d(\mathrm{mu})}{d\theta}=\frac{mdh+hdm-(\mathrm{pdV}_c+\mathrm{V}_c\,\mathrm{dp})}{d\theta}=m\frac{dh}{d\theta}+h\frac{dm}{d\theta}-p\frac{dV_c}{d\theta}-V_c\frac{dp}{d\theta} \tag{13}$$

The total derivative of the pressure is obtained:

$$\frac{\mathrm{d}p}{\mathrm{d}\theta}=(\frac{\partial p}{\partial v})_T\frac{\mathrm{d}v}{\mathrm{d}\theta}+(\frac{\partial p}{\partial T})_v\frac{\mathrm{d}T}{\mathrm{d}\theta} \tag{14}$$

The Equation (15) is the deduction of derivative of specific volume:

$$\frac{\mathrm{d}v}{\mathrm{d}\theta}=\frac{\mathrm{d}(V_c/\mathrm{m})}{\mathrm{d}\theta}=\frac{1}{m}\frac{\mathrm{d}V_c}{\mathrm{d}\theta}-\frac{V_c}{m^2}\frac{\mathrm{dm}}{\mathrm{d}\theta} \tag{15}$$

And the expressions of pressure and temperature in the control volume can be written as:

$$\frac{\mathrm{d}p}{\mathrm{d}\theta}=\frac{\frac{1}{v}[(\frac{\partial h}{\partial v})_T-(\frac{\partial h}{\partial T})_v(\frac{\partial p}{\partial v})_T/(\frac{\partial p}{\partial T})_v]\frac{\mathrm{d}v}{\mathrm{d}\theta}-\frac{1}{V_c}\{\sum[\frac{\mathrm{d}m_i}{\mathrm{d}\theta}(h_i-h)]+\frac{\mathrm{d}Q}{\mathrm{d}\theta}\}}{1-(\frac{\partial h}{\partial T})_v/[v(\frac{\partial p}{\partial T})_v]} \tag{16}$$

$$\frac{dT}{d\theta}=\frac{[\frac{1}{v}(\frac{\partial h}{\partial v})_T-(\frac{\partial p}{\partial v})_T]\cdot\frac{dv}{d\theta}-\frac{1}{V_c}\{\sum[\frac{dm_i}{d\theta}(\mathrm{h}_i-\mathrm{h})]+\frac{dQ}{d\theta}\}}{(\frac{\partial p}{\partial T})_v-\frac{1}{v}(\frac{\partial h}{\partial T})_v} \tag{17}$$

2.3 The Kinetic Model

The input energy of the swing double-vane compressor mainly consists of two parts, namely the indicated work and consumption of friction. And on the basis of the analysis of thermodynamics, the indicated work of a whole working cycle is formulated as:

$$W_i = \frac{n}{60}\int_0^{2\pi} V(\theta)dp \tag{18}$$

And the indicated power of compressor can be obtained by the equation:

$$P_i = \frac{n}{60} W_i \tag{19}$$

One of the principal factors impacting on the performance of the compressor is the friction losses of motion parts, in the calculation mainly taking the five kinds of losses into consideration, and they are the friction of sliding vanes; the friction between vane-end and cylinder; the friction of the end faces of rotor and cylinder; the friction of the bearing of rotor and shaft; the friction between the rotor and eccentric shaft.

Generally, the calculation method of friction consumption can be summed up as:

$$L_f = \begin{cases} \mu Fs \\ \omega M \end{cases} \tag{20}$$

According to the Ma, J. J. (2015), the sum of five friction losses is respectively the total loss by the definition:

$$P_f = \frac{n}{60}\sum L_{f,i} \tag{21}$$

And the total power consumption can be exactly given as:

$$P_{tot} = P_i + P_f \tag{22}$$

3. RESEARCH OF OPTIMIZATION

In the stage of design and development of the compressor, the method of genetic optimization is adopted to improve the efficiency and reliability of the compressor, combining with the analysis of thermodynamics and kinetics. That is the EER of the compressor is defined as objective function to find out the optimal structural parameters of compressor Manole, D. (2002).

The structural parameters of the compressor contain the outer radius of rotor, the inner radius of cylinder, the length of vane, the revolving speed, the axial length of cylinder and so on. If considering too many design parameters, the burden of computation can be increasing so that in this paper, three of the structural parameters, namely the radius of rotor r, the axial length of cylinder H_c and the length of vane L_s, are chosen to be design variables. And the other structural parameters are assigned appropriate constants, for example, the assignment of the inner radius of cylinder R is 50mm, the inner diameter of rotor d is 40mm, the width of vane is 6mm and the material density of rotor and vane is about 7800kg·m^3.

Thus the design variable of the compressor is expressed as:

$$X = [r, H_c, L_s]^T \tag{23}$$

In this paper, in order to obtain maximum EER of the compressor, the EER of any design variables is defined as the objective function as the following equation:

$$F_{obj} = EER = \frac{Q_0}{P_{tot}} \tag{24}$$

In the simulation, the environmental parameters of the compressor are set: suction pressure-0.1MPa, suction temperature-293K, discharge pressure-0.6MPa, revolving speed-2000r· min^{-1} and the physical parameters of the working medium is regarded as ideal gas.

In the iterative computations, when the calculation error is less than 10^{-4}, the optimization design has been considered to be converged. According to the simulation, the optimal structural parameters can be obtained when the number of genetic optimization iteration is 44, they are the rotor radius-42.4mm, axial length of rotor-46.2mm and the length of vane -20.3mm, with the process of iteration shown in Fig.3. Fig.4 shows that the average fitness of the generation has a obvious change in the iteration and finally, the corresponding EER reaches to 0.8757.

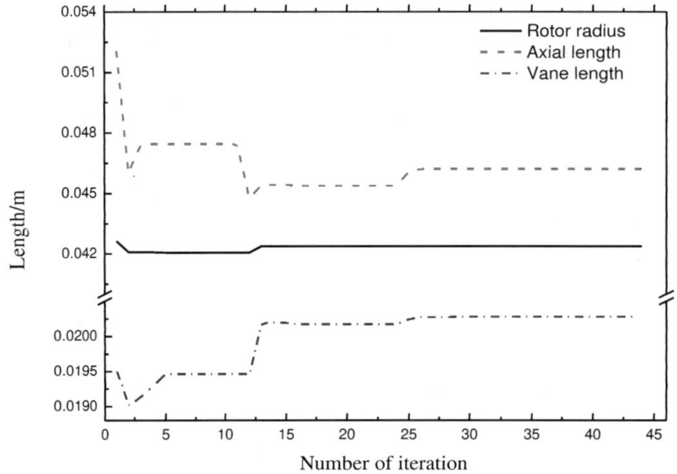

Figure 3: The process of parameters optimization

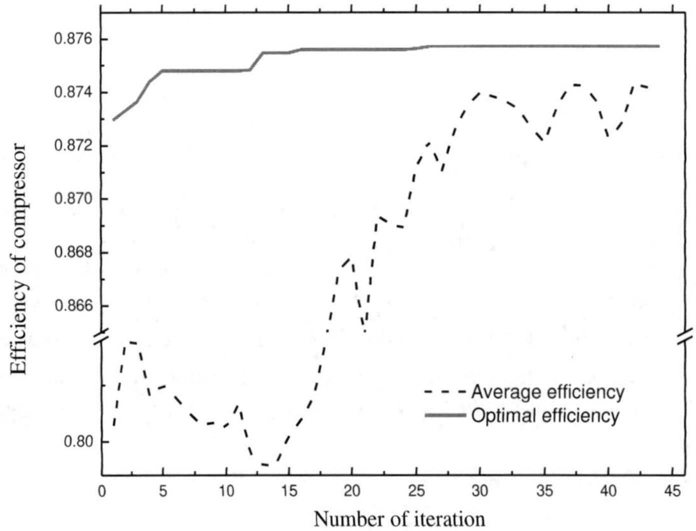

Figure 4: The process of efficiency optimization

4. SIMULATION ANALYSIS

According to the established thermodynamic model, numerical simulation of the double-vane compressor has completed and subsequently has drawn the analysis and research in detail Ooi, K. T., & Wong, T. N. (1997). In the model, the structural parameters are chosen to be design variable, Table 1 shows the initial structural parameter values. The feasible initial parameter values can be random set between the upper and lower limits of the feasible region.

Table 1: The structural parameters of the compressor

Main parameters	Values
Cylinder diameter/mm	100
Outside diameter of rotor /mm	90
inside diameter of rotor /mm	52
Eccentric distance/mm	5
Thickness of vane/mm	6
Length of vane/mm	20
Axial length of cylinder/mm	60
Diameter of eccentric shaft/mm	35

The working chamber of the compressor is surrounded by inner surface of cylinder, external surface of rotor and sides of vane, thus Fig.5 shows how the working chambers of the double-vane compressor and the single-vane compressor change with the angle of eccentric shaft.

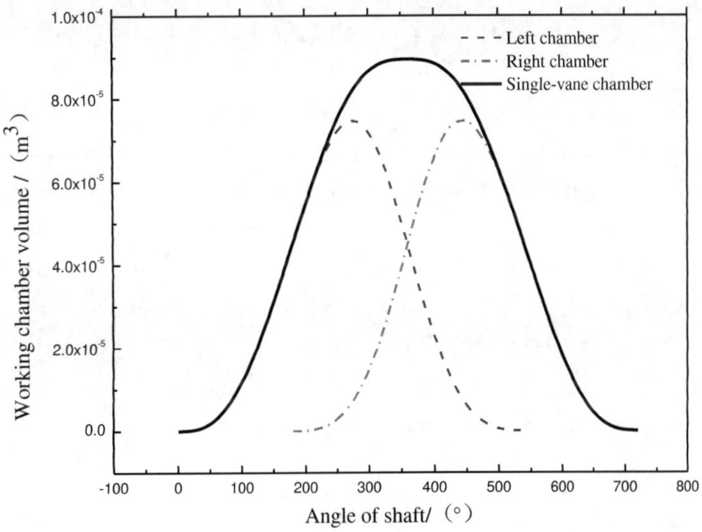

Figure 5: Change of working chamber volume

As indicated in Fig.5, by the time the eccentric shaft completes two revolutions, each chamber in the compressor will have gone through one stroke of suction-exhaust respectively with theoretical displacement being 7.47×10^{-5} m^3, while the value of single-vane compressor is 8.95×10^{-5}m^3. Thus the double-vane compressor possesses an obvious advantage: its theoretical displacement has improved by 66.94% compared with the single-vane compressor.

Taking the left working chamber as research object, after shaft revolution about 540°, an operation circulation has completed with the temperature and pressure of the chamber changing as shown in Fig.6. Pressure varies with the rotation Angle and pressure fluctuation in the process of suction and exhaust is the result of the movement characteristic of valve plate. And the diagram demonstrates that the changing characteristic of temperature is tolerably consistent with the pressure, which corresponds to the law of thermodynamics, it is verified that the thermodynamic model established can be feasible.

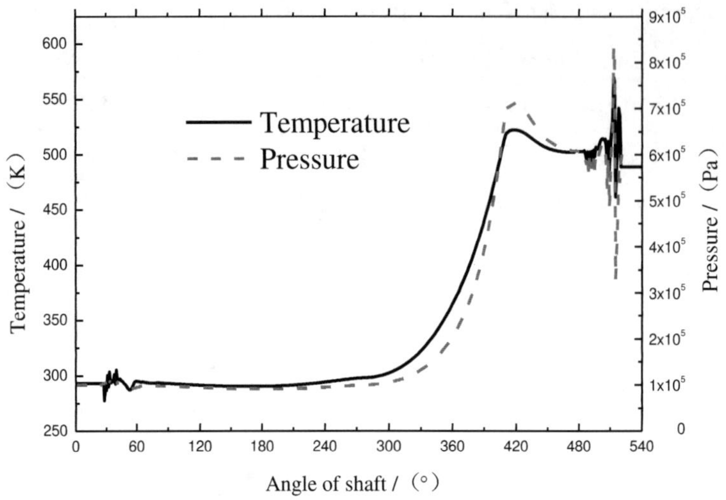

Figure 6: Change of chamber temperature and pressure

6. CONCLUSION

According to the basic structure of the swing double-vane compressor and the relevant theory of thermodynamics and kinetics, the basic optimization parameters of the compressor can be determined. After choosing the appropriate objective function and the method of genetic optimization under the restricted reasonable conditions, mathematical model of optimum design is established. Friction loss of optimized compressor will be reduced dramatically, on the basis of simulation results, we can draw that the optimal structural parameters are the rotor radius-42.4mm, axial length of rotor-46.2mm and the length of vane -20.3mm, and energy efficiency ratio reaches to 0.8757, its theoretical displacement has improved by 66.94% compared with the single-vane compressor, which will provide theoretical foundation to improve the performance of the novel compressor.

NOMENCLATURE

A	valve flow area	(m^2)
b	thickness of vane	(m)
C	discharge coefficient	
e	eccentric distance	(m)
F	force	(N)
H	height of rotor	(m)
L	length	(m)
L_f	frictional loss	(W)
m	mass of fluid	(Kg)

n	rotational speed of compressor	(rev/min)
p	pressure	(pa)
Q	heat transfer rate	(W)
Q_0	cooling capacity	(W)
R	radius	(m)
T	temperature	(K)
U	velocity	($m \cdot s^{-1}$)
V	volume	(m^3)
v	specific volume	(m^3/kg)
W_i	indicate work	(J)
δ	clearance	(m)
θ	drive angle	(rad)
ε	bearing eccentric ratio	
η	kinetic friction coefficient	
μ	dynamic viscosity of lubricant	(pa·s)
ω	angular velocity	($rad \cdot s^{-1}$)

REFERENCE

Hu, X., & Qu, Z. C. (2011). Geometric theory for a swing vane compressor. *Chinese Journal of Mechanical Engineering*, 22, 1896-1900.

Li, Q. C. (2009). *Research on the Mechanism and Test of the Planar Rotary Compressors*. Xi'an, Shaanxi: Xi'an Jiaotong University.

Li, X. L. (2009). *Simulation and Research of Optimization of the Structural Parameters on Plane Rotary Compressor for Air Conditioning*. Xi'an, Shaanxi: Xi'an Jiaotong University.

Ma, J. J. (2015). Structural optimization research of swing vane compressor. *Compressor technology, 4*, 1-5.

Manole, D. (2002). Refrigerant dependent optimum clearance for rotary compressors. *International Compressor Engineering Conference*(1512-1520).

Ooi, K. T., & Wong, T. N. (1997). A computer simulation of a rotary compressor for household refrigerators. *Applied Thermal Engineering, 17*(1), 65-78.

Teh, Y. (2002). Design and friction analysis of the improved revolving vane (rv-i) compressor. *Behavioural Brain Research, 128*(2), 161–168.

Zong, W. B. (2015). The force analysis and study of the critical speed for sliding vane of translational rotary compressor. *Compressor technology, 3*, 11-16.

ACKNOWLEDGMENTS

Many thanks to my supervisor: Professor Qu for his advice and to my teammates: Cang, Rong;Chen, Xiang:Yang, Xu for their work.

Modeling And Testing The Thermal Effect Of Lubricating Oil Sprayed In Sliding-Vane Air Compressors Using Pressure-Swirl Nozzles

Gianluca VALENTI[1,3]*, Stefano MURGIA[2], Ida COSTANZO[2],
Giulio CONTALDI[2], Alessandro VALENTI[3]

[1]Politecnico di Milano, Dip. Energia,
Milano, Italy
Contact Information (+39-02.2399.3845, gianluca.valenti@polimi.it)

[2]Ing. Enea Mattei S.p.a.,
Vimodrone (Milano), Italy
Contact Information (+39-02.2530.5283, stefano_murgia@mattei.it)

[3]Valenti Energie S.r.l.,
Segrate (Milano), Italy

* Corresponding Author

ABSTRACT

A number of studies report the benefit of proper oil spraying in positive-displacement compressors. This work presents a thermodynamic model for simulating sliding-vane air compressors that employ an injection system for the lubricating oil based on both plain orifices and pressure-swirl nozzles to exploit the thermal effect of sprayed oil inside the compression chambers. An experimental campaign on a large-size compressor is conducted measuring, among all, the pressure within a chamber. This measurement is used to validate the model. Oil droplet diameters depend strongly on the injection technology, nozzles generating a finer atomization than orifices, as well as the differential pressure. In any case, the proper positioning of the nozzles allows for an effective heat transfer from the air to the oil.

1. INTRODUCTION

Positive-displacement compressors and, among them, sliding-vane machines are widely used in the compressed air industry. As in other fields, the efficient utilization of energy is a major goal. Hence, the aim of this work is the numerical modeling and the experimental testing of the thermal effect due to spraying the lubricating oil inside sliding-vane air compressors using pressure-swirl nozzles. This effect can be exploited to reduce the compression work.

The benefit of proper oil spraying in positive-displacement compressors has been reported by a number of studies, starting from Singh and Bowman (1986) and from Stosic *et al.* (1988 and 1992) who investigated numerically and experimentally the potential advantage in screw compressors. Shortly later, Fujiwara and Osada (1995) confirmed those positive result still on screw machines. Recently, Valenti *et al.* (2013 and 2014) and Bianchi *et al.* (2014) have focused on sliding-vane compressors, highlighting the higher potential in this kind of machines thanks to their cylindrical shape that allows for a very effective oil spraying parallel to the axis of the compressor itself.

The novelty of this work resides in the extension of the previous model to describe more accurately the quantity and the diameter distribution of the oil droplets and, consequently, to predict more precisely the heat transfer between the oil and the air within a compression chamber. Importantly, *this extension allows not using any tuned parameter*. The model is implemented in a MATLAB code and employed to simulate an existing large-size sliding-vane air compressor that is equipped with both plain orifices and pressure-swirl nozzles. Simultaneously, an experimental campaign is conducted on that compressor measuring the pressure within the chambers by piezoelectric transducers and to reconstruct the pressure-angle diagram. The pressure measurement is used to validate the numerical results.

2. THERMODYNAMIC MODEL

A *pressure-swirl nozzle* is an atomizer that generates liquid droplets imparting a swirl motion to the fluid prior to entering an orifice. The centrifugal forces break the liquid as soon as it leaves the orifice, producing droplets of different dimensions (Lefebvre, 1989). To model this phenomenon, during compression process, a thermodynamic model is developed here as an extension of a previous work (Valenti *et al.*, 2013). The extended model takes into account the *number* and the *position* of the injectors as well as their *jet characteristics*. The number influences the quantity of injected oil and the position influences the jet properties, while both influence the compression process.

When an *oil jet* enters a compression chamber, the related injector is said *activated* for that chamber. Moreover, the jet splits into a number of droplets of different dimensions. The fraction of the jet that breaks up into droplets of the similar size is named *class*. At the end of the compression process, all droplet classes of all injectors are present inside the chamber, assuming that the droplets do not collide against each other or against the chamber walls.

2.1 Process mathematical model

The analysis of the heat transfer between gas and liquid during the compression process is based on the following assumptions, which are the same as those employed in the previous work (Valenti *et al.*, 2013):

- the gas behaves as an ideal gas and the liquid as an incompressible fluid, both having a constant specific heat;
- the gas is non-soluble into the liquid and the liquid non-volatile into the gas;
- the compression is adiabatic toward the surroundings;
- the droplets are non-deformable spheres and the coalescence of droplets is neglected;
- the temperature variation inside the droplets is neglected.

Regarding the last assumption on the temperature within a droplet, its variation can be practically neglected, as explained by Incropera and DeWitt (1996), if the following condition on the *Biot number* Bi [-] is satisfied:

$$Bi = \frac{h\,d}{k_l} \leq 0.1 \qquad (1)$$

where h is the convective heat transfer coefficient at the droplet surface [W m^{-2} K^{-1}], d is the droplet diameter [m], k_l is the liquid thermal conductivity [W m^{-1} K^{-1}]. In its turns, h can be evaluated from the *Nusselt number* Nu_d [-]:

$$Nu_d = \frac{h\,d}{k_g} \qquad (2)$$

where k_g is the gas thermal conductivity [W m^{-1} K^{-1}]. In this study, Nu_d is always taken equal to 2 because this value corresponds, from the point of view of heat transfer, to the case of *pure conduction*, which is a conservative approach.

The *instantaneous energy balance* applied to a *single compression chamber* is:

$$\frac{dU_g}{dt} + \sum_{i=1}^{K} \frac{dU_{l,i}}{dt} = \dot{L} - \dot{Q}_{amb} \qquad (3)$$

where U_g is the gas while U_l the liquid internal energy [J] and $\dot{L}$ the chamber compression power [W], taken positive if transferred to the chamber itself. $\dot{Q}_{amb}$ is the thermal power exchanged with the surroundings [W], neglected as a first approach so that the compression is considered adiabatic. K is the number of all liquid droplet classes present inside the chamber at any time step, reaching the maximum value at the end of the process when all injectors are activated. Furthermore, the instantaneous energy balance applied to the *liquid phase* is:

$$\frac{dU_{l,i}}{dt} = m_{l,i} c_l \frac{dT_{l,i}}{dt} = -h_i A_i \left(T_{l,i} - T_g\right) \qquad \forall i \in [1, K] \qquad (4)$$

where i identifies each droplet class, while the instantaneous energy balance applied to the *gas phase* is:

$$\frac{dU_g}{dt} = m_g c_{v,g} \frac{dT_g}{dt} \qquad (5)$$

In equations (4) and (5) m is the mass [kg], T the temperature [K], A the class heat transfer surface [m^2], given by the product of the number N of droplets and the surface of a single droplet for each class, c_l the liquid while $c_{v,g}$ the gas isochoric specific heat [J K^{-1} kg^{-1}].

The *chamber compression power* is equal to the gas compression power as the liquid is assumed incompressible. Thus:

$$\dot{L} = -p_g \frac{dV_g}{dt} \tag{6}$$

where p_g is the gas pressure [Pa] and V_g the gas volume [m^3]. In its turn, V_g is computed as:

$$V_g(t) = V(t) - V_l(t) \tag{7}$$

where V is the chamber volume [m^3] and V_l the liquid volume [m^3]. The chamber volume V is known, as it is calculated by geometrical means as a function of the angular position. Here, the angular position of a chamber is defined by the position of its trailing vane and starting from the tangent point between rotor and stator.

Finally, the *chamber pressure* is related to the volume and the temperature of the gas by the ideal gas equation of state:

$$p_g V_g = m_g R_g T_g \tag{8}$$

2.2 Process numerical model

The system of equations presented above is implemented using a *backward Euler method*. Rearranging the equations:

$$\begin{cases} m_g c_{v,g}\, T_g^{\,n+1} + \sum_{i=1}^{K} m_{l,i} c_l T_{l,i}^{\,n+1} = m_g c_{v,g}\, T_g^{\,n} + \sum_{i=1}^{K} m_{l,i} c_l T_{l,i}^{\,n} - p_g\left(V_g^{\,n+1} - V_g^{\,n}\right) & (9) \\ -h_1 A_1 dt T_g^{\,n+1} + (m_{l,1} c_l + h_1 A_1 dt) T_{l,1}^{\,n+1} = m_{l,1} c_l T_{l,1}^{\,n} & (10) \\ \vdots & \\ -h_K A_K dt T_g^{\,n+1} + (m_{l,K} c_l + h_K A_K dt) T_{l,K}^{\,n+1} = m_{l,K} c_l T_{l,K}^{\,n} & (11) \\ p_g^{\,n+1} = \dfrac{m_g R_g T_g^{\,n+1}}{V_g^{\,n+1}} & (12) \end{cases}$$

Equations (9) to (11) can be seen as a *linear system of equations*, with unknowns $T_g^{\,n+1}$ and $T_{l,i}^{\,n+1}$. In matrix form:

$$\underline{\underline{A}}\, \underline{T} = \underline{b} \tag{13}$$

where $\underline{\underline{A}}$, $\underline{T}$ and $\underline{b}$ change size every time an injector is activated. As seen, when all injector are activated, K is equal to the total number of droplet classes (sum of all classes of all injectors) and the linear system reaches the maximum dimension. In particular, $\underline{\underline{A}}$ dimension is [1+ K; 1+ K] while $\underline{T}$ and $\underline{b}$ dimension is [K;1]. Mathematically:

$$\underline{\underline{A}} = \begin{bmatrix} m_g c_{v,g} & m_{l,1} c_l & m_{l,2} c_l & \cdots & m_{l,K} c_l \\ -h_1 A_1 dt & m_{l,1} c_l + h_1 A_1 dt & 0 & 0 & 0 \\ -h_2 A_2 dt & 0 & m_{l,2} c_l + h_2 A_2 dt & 0 & 0 \\ \vdots & 0 & 0 & \ddots & 0 \\ -h_K A_K dt & 0 & 0 & 0 & m_{l,K} c_l + h_K A_K dt \end{bmatrix} \tag{14}$$

as well as:

$$\underline{T} = \begin{bmatrix} T_g^{\,n+1} \\ T_{l,1}^{\,n+1} \\ \vdots \\ T_{l,K}^{\,n+1} \end{bmatrix} \tag{15}$$

and:

$$\underline{b} = \begin{bmatrix} m_g c_{v,g} T_g^{\,n} + \sum_{i=1}^{K} m_{l,i} c_l T_{l,i}^{\,n} - p_g\left(V_g^{\,n+1} - V_g^{\,n}\right) \\ m_{l,1} c_l T_{l,1}^{\,n} \\ \vdots \\ m_{l,K} c_l T_{l,K}^{\,n} \end{bmatrix} \tag{16}$$

Solving numerically the system, gas and liquid temperatures at any time step n can be calculated. Ultimately, the gas pressure at each step is computed via Equation (12).

2.3 Pressure-swirl nozzle and plain orifice models

For *pressure-swirl nozzles*, the jet is described by the *Rosin-Rammler relation* as detailed by Mugele and Evans (1951):

$$f(d) = q\frac{d^{q-1}}{SMD^q}e^{-\left(\frac{d}{SMD}\right)^q} \tag{17}$$

where SMD is the *Sauter Mean Diameter*, defined as the diameter of a sphere that has the same volume-to-surface area ratio of the entire droplet sample, and q is the shape parameter providing a measure of the spread of droplet sizes. In its turn, q is taken equal to 3.5 (Lefebvre, 1989) and SMD computed via the relation by Wang and Lefebvre (1987):

$$SMD = 4.52\left(\frac{\sigma_l\, \mu_l{}^2}{\rho_g\, \Delta p^2}\right)^{0.25}\left[2.7\left(\frac{D_{or}\, \dot{m}_l\, \mu_l}{\rho_l\, \Delta p}\right)^{0.25} cos\frac{\gamma}{2}\right]^{0.25} + 0.39\left(\frac{\sigma_l\, \rho_l}{\rho_g\, \Delta p}\right)^{0.25}\left[2.7\left(\frac{D_{or}\dot{m}_l\, \mu_l}{\rho_l\, \Delta p}\right)^{0.25} cos\frac{\gamma}{2}\right]^{0.75} \tag{18}$$

where σ_l is the liquid surface tension [mN m^{-1}], Δp the differential pressure [Pa], ρ the density [kg m^{-3}], D_{or} the nozzle/orifice diameter [m], γ the cone spray angle [rad], $\dot{m}_l$ and μ_l are the liquid mas flow rate [kg s^{-1}] and viscosity [Pa s]. The droplet diameter depends strictly on the nozzle differential pressure, which is calculated as the difference between the injection pressure and the pressure inside the compression chamber at the moment of injection. The *injection pressure* is the same for all nozzles and equal to the compressor delivery pressure minus the pressure drop along the oil circuit. Figure 1 shows the effect of the differential pressure on SMD. The pressure depends strictly on the injector position: the closer the nozzle to the discharge port (i.e. Nozzle 4, as described in Section 3), the smaller the differential pressure and, thus, the bigger the droplet diameter. Lastly, the *mass flow rate through the nozzle* is derived, considering the liquid density, from a correlation provided by the manufacturer for the volume flow rate:

$$\dot{V}_l = F\, \dot{V}_2 = F\, \dot{V}_1\left(\frac{\Delta p_2}{\Delta p_1}\right)^{0.4} \tag{19}$$

where $\dot{V}_l$ is the volume flow rate [l min^{-1}] and Δp is the pressure difference across the nozzle [Pa]. Subscripts 1 and 2 identify two different test conditions: the first is associated to the manufacturer test condition with water while the second to the actual one with lubricant oil. F is a density conversion factor, in this case equal to 1.06.

For *simple plain orifices*, only one droplets class is considered because this kind of atomizers is not able to produce a fine atomized jet. The SMD of this class is evaluated via the correlation by Varde *et al.* (1984):

$$SMD = 8.7\, D_{or}\, (Re\, We)^{-0.28} \tag{20}$$

where Re [-] is the Reynolds and We [-] the Weber number. Lastly, the *mass flow rate through the orifice* is derived from Bernoulli equation taking a discharge coefficient C_d of 0.772, calculated according to Lichtarowicz *et al* (1965):

$$C_d = \frac{1}{\frac{1}{C_{du}} + 20\frac{(1 + 2.25\frac{L}{D_{or}})}{Re}} \tag{21}$$

where L [m] is the orifice length and C_{du} the ultimate discharge coefficient, defined as:

$$C_{du} = 0.827 - 0.0085\frac{L}{D_{or}} \tag{22}$$

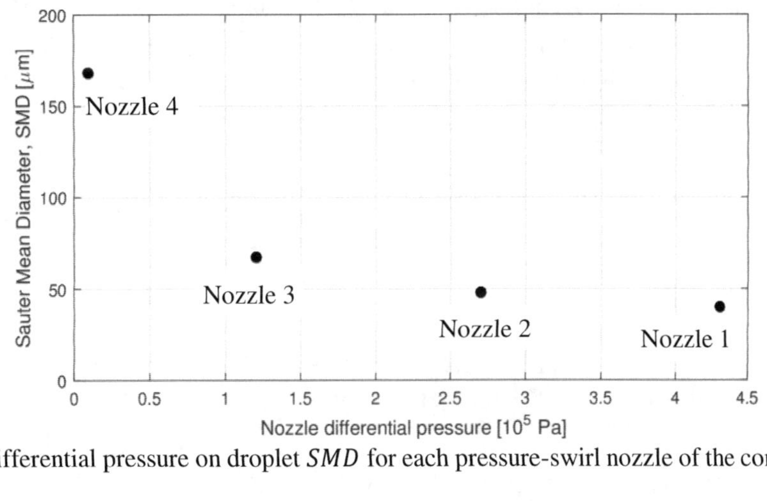

Figure 1: Effect of differential pressure on droplet SMD for each pressure-swirl nozzle of the considered compressor.

2.4 Lubricant oil property models

The thermo-physical properties of the liquid phase, which is the lubricating oil, are evaluated using semi-empirical correlations tuned with experimental data from the oil supplier. However, one parameter rarely available is the surface tension σ_l [mN m^{-1}], which is here calculated as (Brock and Bird, 1955):

$$\sigma = p_c^{2/3}\, T_c^{1/3} Q \left(1 - \frac{T}{T_c}\right)^{11/9} \tag{23}$$

in which:

$$Q = 0.1196 \left[1 + \frac{\frac{T_b}{T_c} \ln\left(\frac{p_c}{1.01325}\right)}{1 - \frac{T_b}{T_c}}\right] - 0.2793 \tag{24}$$

where p_c [bar] and T_c [K] are respectively pressure and temperature at the critical point of the oil, and T_b [K] temperature at the normal boiling point (NBP). These can be evaluated according to the methodology proposed by Conde (1996), which makes use of two equations simultaneously. The first equation is that by Nokay (1959):

$$T_c = 15.2762\, S^{0.2985} T_b^{\;0.62164} \tag{25}$$

and the second that by Kesler and Lee (1976):

$$T_c = 189.8 + 450.6\, S + (0.4244 + 0.1174\, S)\, T_b + (0.1441 - 1.0069\, S)\, \frac{10^5}{T_b} \tag{26}$$

where S is the specific gravity, equal to the ratio between the liquid density and that of pure liquid water at 60°F (15.56°C). Solving these two equations simultaneously, T_b and T_c can be evaluated. Kesler and Lee (1976) proposed also the following equation for the critical pressure:

$$\begin{aligned} \ln p_C = {} & 17.2019 - 0.0566\, S^{-1} - (0.4364 + 4.1216\, S^{-1} + 0.21343\, S^{-2})\, 10^{-3}\, T_b \\ & + (4.7579 + 11.819\, S^{-1} + 1.5302\, S^2)\, 10^{-7} T_b^{\;2} - (2.4505 + 9.9009\, S^{-2}) 10^{-10} T_b^{\;3} \end{aligned} \tag{27}$$

2.5 Diameter and probability of droplet classes

Each droplet class is characterized by a diameter and a probability. From the Rosin-Rammler droplet size distribution seen in Equation (17), the mean α and the variance β^2 of the function are calculated as (Montgomery *et al.*, 2011):

$$\alpha = SMD\, \Gamma\left(1 + \frac{1}{q}\right) \tag{28}$$

and:

$$\beta^2 = SMD^2 \left\{\Gamma\left(1 + \frac{2}{q}\right) - \left[\Gamma\left(1 + \frac{1}{q}\right)\right]^2\right\} \tag{29}$$

where Γ is the Gamma function. For each nozzle, its droplet classes are identified by a vector $\underline{\varphi}$, determined from α and β: each element of the vector $\varphi(i)$ is a representative diameter for all droplets whose diameters fall in the interval $\varphi(i) \pm \beta/2$. In the case here adopted of five classes, the diameters of these classes are:

$$\underline{\varphi} = \{\alpha - 2\beta;\ \alpha - \beta;\ \alpha;\ \alpha + \beta;\ \alpha + 2\beta\} \tag{30}$$

The probability of a class of a nozzle $\varphi(i)$, that is the fraction of the mass flow rate though that nozzle breaking into droplets of diameters in that class interval, is determined exploiting the property of a probability density function:

$$\int_{-\infty}^{+\infty} f(x)\, dx = 1 \tag{31}$$

Geometrically, the probability of the class is the area of a rectangle centered in $\varphi(i)$ with base β. The heights of the second to the last intervals are calculated according to the *mean value theorem for integrals*, while the heights of first and last intervals, which are identical, are taken so that the sum of all rectangle areas is unity.

3. EXPERIMENTAL TEST

Simultaneously to the development of the thermodynamic model, an experimental campaign is conducted on a *large-size sliding-vane air compressor* that adopts *two oil injection systems* in order to validate the model:

- plain orifices on the lateral surface of the stator along the axis and positioned close to the discharge port;
- four couples of pressure-swirl nozzles on both end plates distributed from the suction to the discharge port.

The *test rig* for the experimental campaign comprises an absolute pressure transducer to measure pressure of the ambient air and a group of T-type thermocouples placed on the compressor stator near the inlet valve to measure the initial compression temperature (Figure 2). An additional T-type thermocouple indicates the injected oil temperature. Four piezoelectric transducers, placed circumferentially along one end plate, measure the pressure along the process (Figure 3). The measurement uncertainties of the adopted sensors are listed in Table 1. Ultimately, the pressure data acquired from the piezoelectric transducers are post-processed to reconstruct *the pressure-angle diagram* following the methodology defined by Cipollone et al. (2009).

The *test case* is characterized by ambient pressure of 100325 Pa and injected oil temperature of 60°C. The compressor delivery pressure is 7.5 bar(g) and the angular speed is 990 rpm.

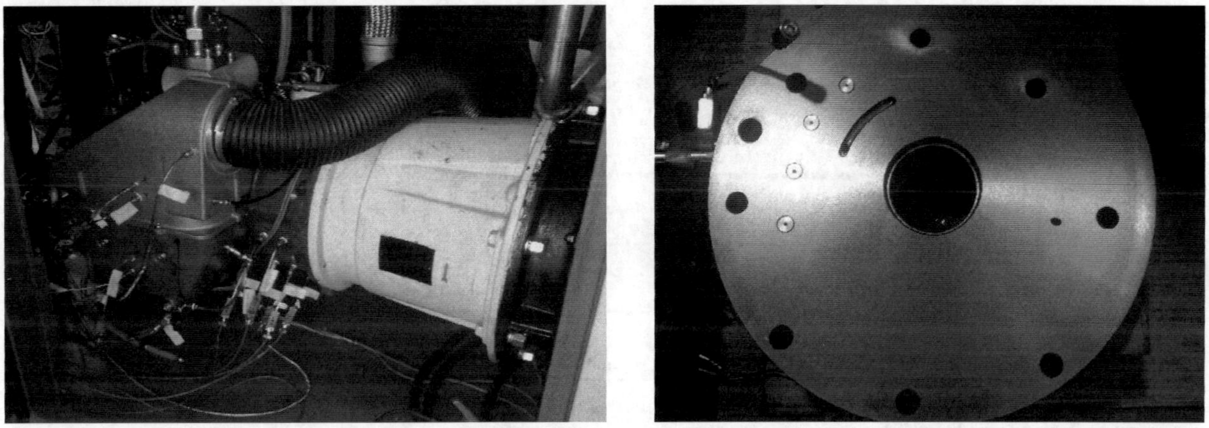

Figure 2: The considered large-size sliding-vane air compressor. Left: group of T-type thermocouples (see blue connectors) placed on the stator near the inlet valve. Right: four pressure-swirl nozzles placed on one end plate.

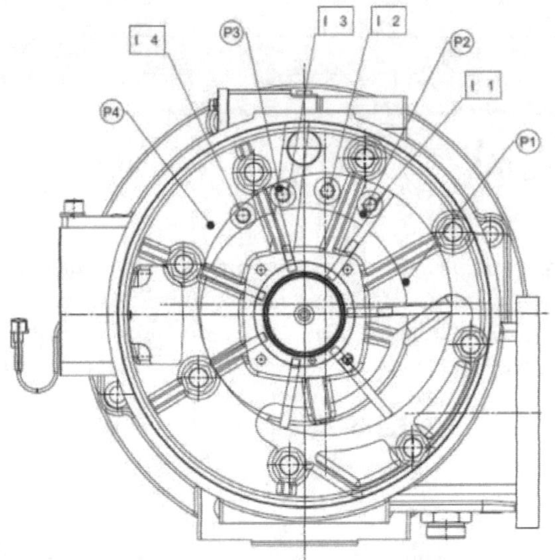

Figure 3: Position of the pressure-swirl nozzles and of the piezoelectric pressure transducers on one end plate. "I" indicates nozzles while "P" transducers.

Table 1: Measurement sensors and uncertainties.

Instrument type	Quantity, unit	Absolute uncertainty
T-type thermocouple	Temperature, °C	0.5
Absolute pressure transducer	Pressure, bar	0.08
Piezoelectric pressure transducer	Pressure, bar	0.01

4. RESULTS AND VALIDATION

The geometrical features of the considered compressor are utilized to determine the chamber volume V as a function of the angular position. Next, the process in the compressor is simulated with the thermodynamic model described in Section 2 and the thermo-physical properties for air and lubricant oil listed in Table 2. In particular, each of the four pressure-swirl nozzle jets is described dividing the droplets into five classes, as presented in Section 2.4, to take into account their fine nebulization, whereas plain orifice jets only in a single class, yielding 21 total classes.

The resulting class diameters for each pressure-swirl nozzle jet are shown in Figure 4. Nozzle 1, which is the closest to the suction port exploiting the highest differential pressure Δp, is characterized by very fine dimensions, from 13 μm to 59 μm. Nozzle 5, instead, from 55 μm to 247 μm. In contrast, the plain orifices are characterized by a class of 800 μm. Figure 5 shows the temperature of all classes during the compression process. Temperature of bigger droplets (Class 5) increases by only a fraction of degree, but that of smaller ones (Class 1) increases even by over 50°C thanks to a very effective heat transfer from the air being compressed to the droplets, resulting in a non-adiabatic air compression. The calculated average oil temperature increase, which is the mass-averaged of all classes, is 12.1°C.

Additionally, Figure 6 shows the comparison between the measured and the calculated pressure within a compression chamber. The absolute value of the relative error is always below 5% and it is averagely equal to 2.3%. The relative error reaches higher values in the last part of the process, where the numerical model underestimates the real process. This is likely due to two effects. First, neglecting the coalescence of droplets and the collision on the chamber walls, the model assumes that all droplets are active until the end of the compression process, which is improbable. Second, the last part may be affected by leakages into the chamber from the discharge port, here not yet considered.

Table 2: Assumed parameters for the thermo-physical properties of the gas (air) and the liquid (lubricating oil).

Parameter	**Symbol**	**Unit**	**Value**
	Gas (Air)		
Name			Air
Gas specific constant	R_g	$\mathrm{J\,kg^{-1}\,K^{-1}}$	287.1
Isochoric specific heat	$c_{v,g}$	$\mathrm{J\,kg^{-1}\,K^{-1}}$	720
	Liquid (Lubricating oil)		
Name			Oil
Density	ρ_l	$\mathrm{kg\,m^{-3}}$	930
Viscosity at 40°C	μ_l	Pa s	0.05
Isobaric specific heat	c_l	$\mathrm{J\,kg^{-1}\,K^{-1}}$	1842
	Compression		
Inlet gas temperature	$T_{g,0}$	K	338.15
Inlet liquid temperature	$T_{l,0}$	K	333.15

5. CONCLUSIONS

This work presents a thermodynamic model developed for positive-displacement compressors, in particular air sliding-vane machines, that employ a lubricant oil injection system based on plain orifices and/or pressure-swirl nozzles to exploit the thermal effect of spraying the oil inside the compression chambers. An experimental campaign is conducted measuring, among all, the pressure within a chamber, used for the validation. Conclusions are as follows.

- Diameters of droplet classes depend strongly on the injection technology and on the differential pressure, which depends on the position of the orifice or nozzle. Class diameters can be as small as 13 μm for the nozzle at the highest differential pressure, as large as 247 μm for that at the lowest, or 800 μm for the orifices.
- Heat transfer from the air to the lubricating oil droplets depends strongly on the droplet diameters. Temperature of the bigger droplets increase by a fraction of degree, but that of the smaller ones increases even by over 50°C. Average temperature increase is 12.1°C, resulting into a non-adiabatic air compression.
- The model is validated comparing the measured and the calculated pressure within a compression chamber. The relative error is within 5% and averagely equal to 2.3% despite the model does not use tuned parameters.

In short, *air is being cooled by the oil*. The *future work* will investigate further the phenomena that have a negative impact on the heat transfer, as the coalescence of droplets and their collision on the chamber walls, as well as leakages.

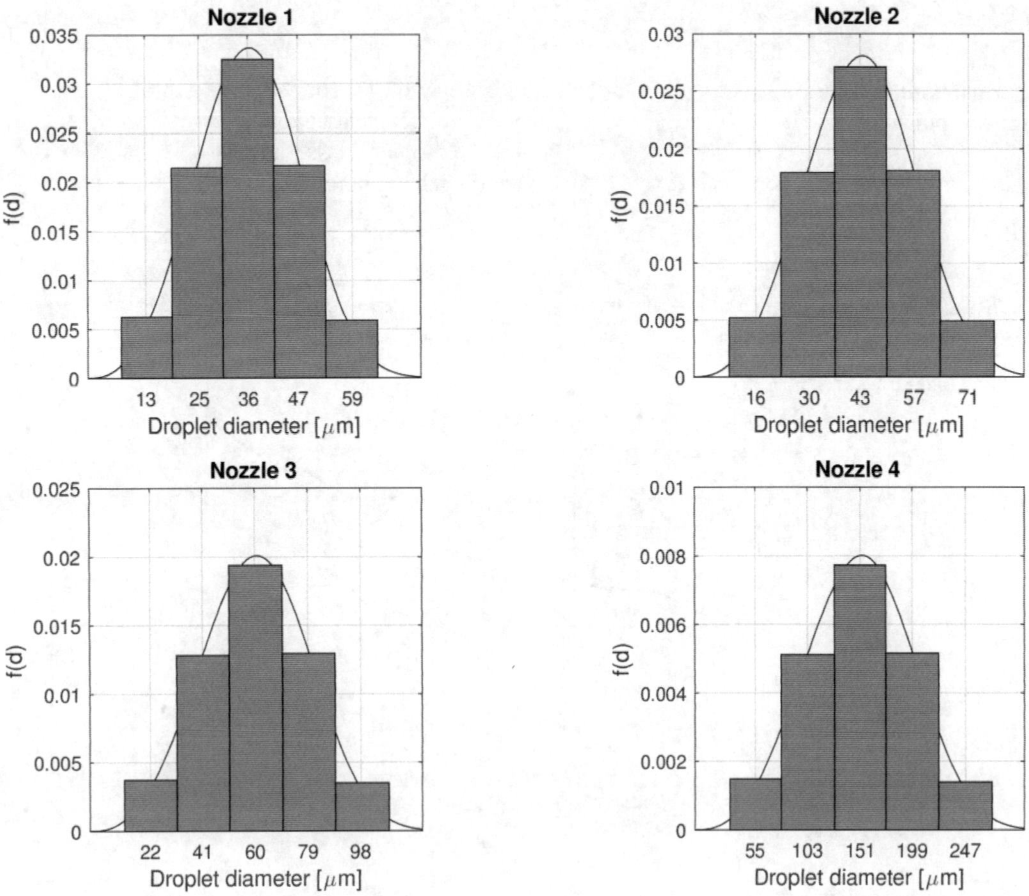

Figure 4: Rosin-Rammler probability density functions and calculated classes of the four pressure-swirl nozzles.

NOMENCLATURE

Symbols

Γ	Gamma function	(-)	Re	Reynolds number	(-)
Δ	difference	(-)	S	specific gravity	(-)
α	mean	(m)	SMD	Sauter Mean Diameter	(m)
β^2	variance	(m^2)	T	temperature	(K)
γ	cone spray angle	(rad)	$\underline{T}$	vector of temperatures	(K)
μ	viscosity	(Pa s)	U	internal energy	(J)
ρ	density	(kg m^{-3})	$\dot{V}$	volumetric flow rate	(l min^{-1})
σ	surface tension	(mN m^{-1})	V	volume	(m^3)
$\underline{\varphi}$	vector of droplet diameters	(m)	We	Weber number	(-)
A	heat transfer area	(m^2)	$\underline{b}$	vector	(J)
$\underline{\underline{A}}$	matrix of coefficient	(J K^{-1})	c	mass specific heat	(J kg^{-1} K^{-1})
Bi	Biot number	(-)	d	droplet diameter	(m)
C_d	discharge coefficient	(-)	h	heat transfer coefficient	(W m^{-2} K^{-1})
F	density correction factor	(-)	k	thermal conductivity	(W m^{-1} K^{-1})
K	droplets classes number	(-)	m	mass	(kg)
$\dot{L}$	compression power	(W)	n	time step	(-)
L	orifice length	(m)	p	pressure	(Pa)
N	number of droplets	(-)	q	size distribution parameter	(-)
Nu_d	Nusselt number	(-)	t	time	(s)
$\dot{Q}$	thermal power	(W)	x	generic variable	(-)
R	gas constant	(J kg^{-1} K^{-1})			

Subscript

amb	ambient	*or*	orifice
b	at normal boiling point	*p*	at constant pressure
c	at critical point	*v*	at constant volume
g	gas	*i*	droplets class identifier
l	liquid	0	inlet condition

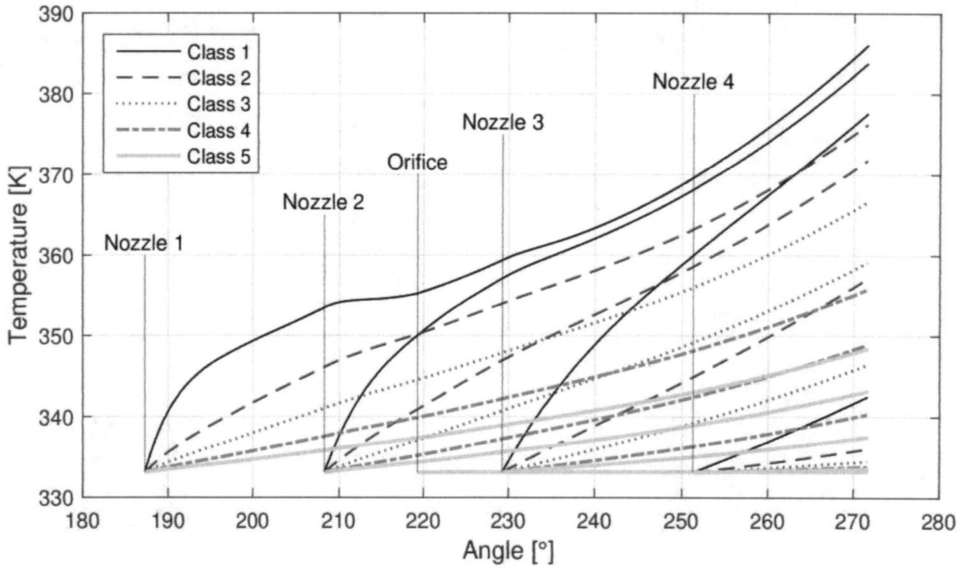

Figure 5: Calculated oil temperature as function of the angular position. Each injector is activated at a different angular position. In particular, it is possible to distinguish pressure-swirl nozzle jets, plain orifice jet, and the evolution of temperature of each class of droplets (Class 1 is the smallest for each nozzle, while Class 5 the biggest).

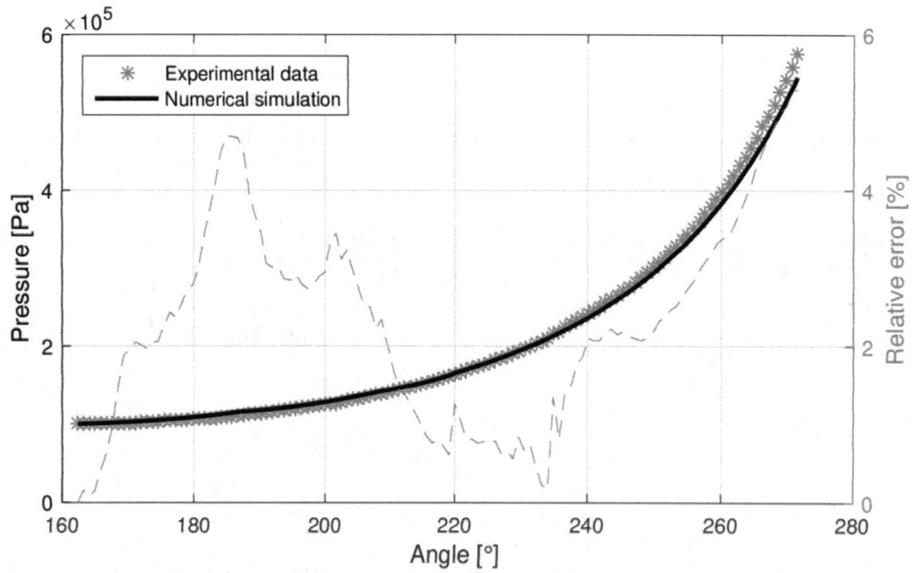

Figure 6: Numerical model validation. Comparison between the measured and the calculated pressure (left axis) and the absolute value of the relative percent error (right axis).

REFERENCES

Bianchi, G., Cipollone, R., Murgia, S., & Contaldi, G. (2014). Performance enhancement in sliding vane rotary compressors through a sprayed oil injection technology. In *Proceeding of the International Compressor Engineering Conference*. Paper Purdue, Lafayette, IN, USA.

Brock, J.R. & Bird, R. B. (1955). Surface tension and the principle of corresponding states. *AIChE Journal, 1*(2), 174-177. doi:10.1002/aic.690010208

Cipollone, R., Contaldi, G., Capoferri, A., & Valente, R. (2009). Theoretical and experimental study of the p-V diagram for a sliding vane rotary compressor, In *Proc. IMechE International Conference on Compressors and Their Systems (279-289)*, London, UK.

Conde, M. R. (1996). Estimation of thermo physical properties of lubricating oils and their solutions with refrigerants: An appraisal of existing methods. *Applied Thermal Engineering, 16*(1), 51–61. doi:10.1016/1359-4311(95)00011-2

Fujiwara, M., & Osada, Y. (1995). Performance analysis of an oil-injected screw compressor and its application. *International Journal of Refrigeration, 18*(4), 220-227. doi:10.1016/0140-7007(95)00008-Y

Incropera, F.P., & DeWitt, D.P. (1996). *Introduction to heat transfer.* New York, NY, USA: John Wiley & Sons.

Kesler, M.G., & Lee, B.I. (1976). Improved prediction of enthalpy of fractions. *Hydrocarbon Proc. 55*(3), 153-158.

Lefebvre, A.H. (1989). *Atomization and Sprays*. New York, NY, USA. Hemisphere Publishing Corporation.

Lichtarowicz, A.K., Duggins, R.H., & Markland, E. (1965). Discharge Coefficients for Incompressible Non-Cavitating Flow Through Long Orifices. *Journal of Mechanical Engineering Science. 7*(2), 210-219. doi:10.1243/JMES_JOUR_1965_007_029_02

Montgomery D.C., Runger G.C., & Hubele N.F. (2011). *Engineering statistics.* New York, USA: John Wiley & Sons.

Mugele, R.A., & Evans, H.D. (1951). Droplet Size Distribution in Sprays. *Ind. Eng. Chem., 43*(6), 1317–1324. doi:10.1021/ie50498a023

Nokay, R. (1959). Estimate petrolchemical properties. *Chem. Eng., 66*(2), 147-148.

Singh, P.J., & Bowman, J.L. (1986). Heat transfer in oil-flooded screw compressors. In *Proceedings of the International Compressor Engineering Conference (135-152).* Purdue, Lafayette, IN, USA.

Stosic, N., Kovacevic, A., Hanjalic, K., & Milutinovic, L. (1988). Mathematical modelling of the oil influence upon the working cycle of screw compressors. In *Proceeding of the International Compressor Engineering Conference (354-361)*. Purdue, Lafayette, IN, USA.

Stosic N., Milutinovic, L., Hanjalic, K., & Kovacevic, A. (1992). Investigation of the influence of oil injection upon the screw compressor working process. *Int. J. of Refrigeration, 15*(4), 206-220. doi:10.1016/0140-7007(92)90051-U

Valenti, G., Colombo, L., Murgia, S., Lucchini, A., Sampietro, A., Capoferri, A., & Araneo, L. (2013). Thermal effect of lubricating oil in positive-displacement air compressors. *Applied Thermal Engineering, 51*(1-2), 1055-1066. doi:10.1016/j.applthermaleng.2012.10.040

Valenti, G., Murgia, S., Contaldi, G., & Valenti, A. (2014). Experimental evidence of the thermal effect of lubricating oil sprayed in sliding-vane air compressors. *Case Studies in Thermal Engineering, 4*, 113-117. doi:10.1016/j.csite.2014.08.001

Varde, K., Popa, D., & Varde, L. (1984). Spray Angle and Atomization in Diesel Sprays, *SAE Technical Paper, 841055*. doi:10.4271/841055

Wang, X.F., & Lefebvre, A.H. (1987). Mean Drop Sizes from pressure-swirl nozzles. Journal of Propulsion and Power, 3(1), 11-18. doi:10.2514/3.22946

Flap-X: Investigation of residual stresses and tumbling.

Alexander LOF[1], Chris MILLWARD[1], Azhar NAWAZ[1, 2*]

[1, 2*]voestalpine Precision Strip AB, Research & Development,
Munkfors, Sweden
[1] (Phone: +46 56316134, E-mail: alexander.lof@voestalpine.com)
[1] (Phone: +46 732717642, E-mail: chris.millward@voestalpine.com)
[1, 2*](Phone: +46 734618240, E-mail: azhar.nawaz@voestalpine.com)

* Corresponding author

ABSTRACT

Compressive residual stresses are an important influence on the fatigue life of flapper valves in compressor applications. The new material, Flap-X, has optimised residual stresses for a longer fatigue life reed valve. Residual stresses were measured using XRD for various materials before and after tumbling. Hardened strip samples of various thicknesses and steel grades were stamped to produce valve samples. Tumbling trials were performed with different durations. The edge radius and the surface roughness were measured for each thickness to understand if there was any correlation between them. Compressive residual stresses increase with an increase of tumbling time. Residual stresses before and after tumbling of various steel grades vary.

1. INTRODUCTION

It has previously been shown that tumbling, shot peening and other similar edge and surface preparation methods improves the fatigue life of strip material (Dusil, 1978). These improvements in fatigue properties are achieved by reducing the effect of edge and surface defects and by introducing compressive residual stresses, both at the edges and on the surfaces (Dusil 1978). For the new alloy Flap-X, this work aims to reproduce some of the investigations reported on Blanking, Tumbling and Shot-Peening of Compressor Valves (Dusil 1978). The work also aims to expand on the previous knowledge by adding investigations of a new material and varying sample thicknesses.

2. MATERIALS AND SAMPLES

The reference tests were undertaken for two common flapper valve steel grades. One was a modified version of the AISI 420 standard martensitic stainless steel referred to as"420" in this report. The second was a high carbon steel grade which is similar to the AISI 1095, referred to as "1095 in the current study. Both of these steel grades were investigated along with three other materials. One was a new improved version of the AISI 420 standard called "Flap-X". The second addition was a high carbon martensitic stainless steel grade here referred to as "13Cr0.7C" in this report and the third addition was a hot-work Tool steel grade referred to as "Tool steel". The samples used the tests were produced by blanking, with all samples having the same shape. The three thicknesses investigated were "thin" (0.203 mm), medium (0.305 mm) and the "thickest" (0.508 mm)

.

3. TUMBLING

The tumbling of the samples was done in a rotating lab tumbling machine using triangular ceramic chips (6 mm) called WXC-6x6 D. A grinding compound 9A (SiC) was added along with water and a liquid inhibitor. The speed of rotation was 40 RPM. The samples were tumbled for various durations between zero and 40 hours in different intervals (see table 1)

In the first stage, the materials were tumbled on two different occasions, called Batch A and Batch B, see table 1. Batch A consisted of the following materials.

1. Flap-X
2. 1095 (20C)

Batch B consisted of the following materials.

1. 420 (SS716)
2. 13Cr0.7C (AEB-L)
3. Tool steel (S-Coat L+)

In the second stage, all of the samples were not loaded in the tumbling barrels at the same time. They were, instead divided into three time groups; depending on the length of time they were to be tumbled as shown in table 1.
In the third stage, the tumbling media from one group was reused without any changes for next tumbling.

Table 1: Tumbling setup

<table>
<tr><th>Tumbling times</th><th colspan="2">Condition of the tumbling media at start of each tumbling time.</th></tr>
<tr><td>hours</td><td>Batch A</td><td>Batch B</td></tr>
<tr><td>1</td><td rowspan="4">Used 15 hours</td><td rowspan="4">Fresh</td></tr>
<tr><td>3</td></tr>
<tr><td>5</td></tr>
<tr><td>10</td></tr>
<tr><td>15</td><td>Fresh</td><td>Used 10 hours</td></tr>
<tr><td>20</td><td rowspan="2">Fresh</td><td rowspan="2">Fresh</td></tr>
<tr><td>40</td></tr>
</table>

4. EDGE PROFILE

The edge profiles of the samples were investigated in cross-sections in the same way that they were in the original report (R. Dusil, 1978) as shown in Fig. 1 and 2. During comparison, it is found that the samples from Batch A showed a much more rapid edge rounding than the samples from Batch B.
The Batch A materials showed notable rounding of the edge corners after only one hour tumbling as can be seen in the example in Fig. 1. The Batch B materials needed almost 3 hours to show the same amount of edge rounding. As illustrated in Fig. 2.
After 10 hours of tumbling the edges of all the 0.203 mm thin samples were well rounded with a single radius equal to about half the thickness (see Fig. 1 and 2). Increasing tumbling time only resulted in reducing the radius of the edge and increasing the stock removal from the samples as can be seen as a narrowing of the edge profiles after 40 hours tumbling in fig. 1 and 2.

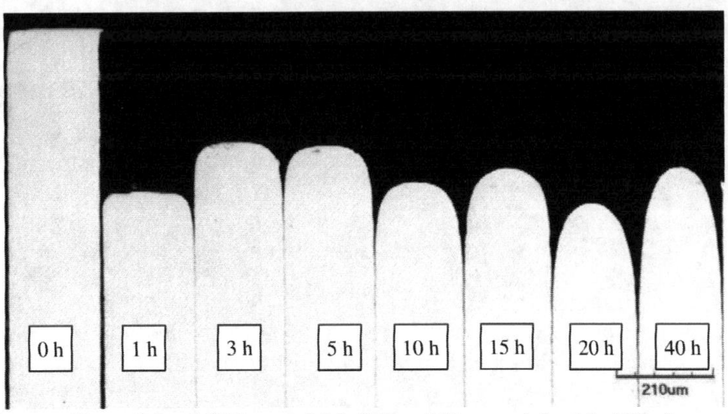

Figure 1: The edge profiles of the 0.203 mm thick "Flap-X" material with differing tumbling durations.

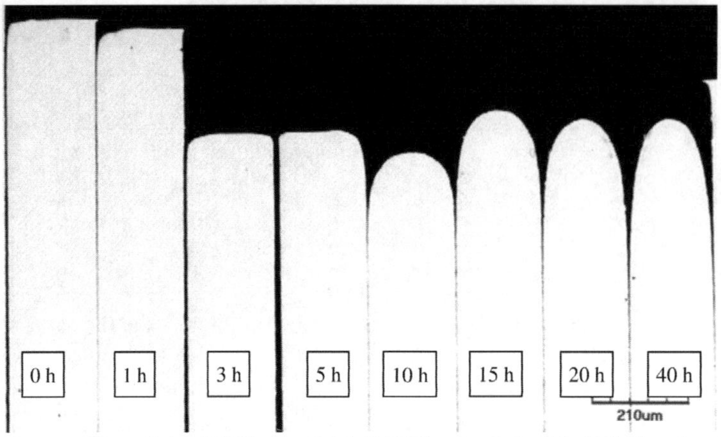

Figure 2: The edge profiles of the 0.203 mm thick "420" material with differing tumbling durations.

The tumbling duration necessary to achieve a single rounded edge was progressively longer for the 0.305 and 0.508 mm thick samples. With a single rounded edge achieved within 15 hours tumbling on the 0.305 mm thick samples and the 0.508 mm thick samples needing more than 20 hours of tumbling for all tested materials.

5. SURFACE APPERANCE

Both surfaces of all samples were photographed using a stereo microscope as shown in figure 3. This examination clearly showed that the 1 to 10 hour tumbling performed on the Batch A affected the sample surfaces more rapidly than the Batch B as illustrated in Fig. 3. It is clear that one hour of tumbling had a greater effect on the surface of the Flap-X than it did on the 420 samples. For "420" samples, after three hours in tumbling, it is still possible to see the traces of the original brush lines while these traces are completely eliminated after one hour of tumbling on the Flap-X samples.

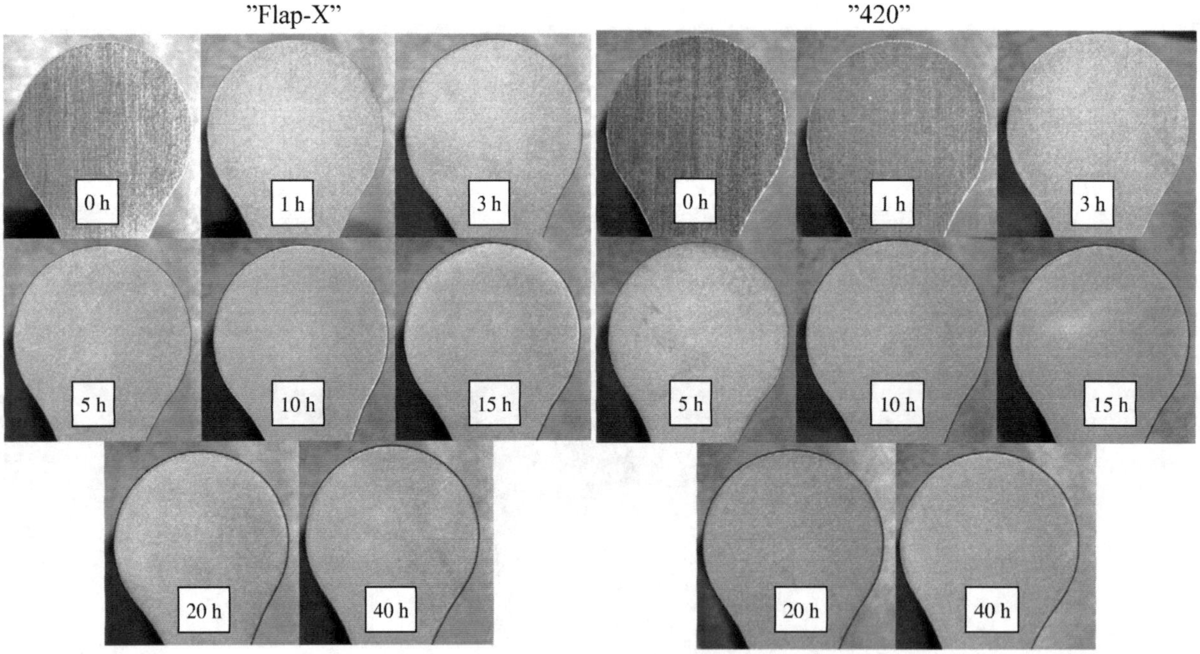

Figure 3: Showing a comparison between the surfaces of 0.203 mm thick Flap-X (left set of images) and 420 samples (right set of images) after different tumbling times.

6. SURFACE ROUGHNESS

The 2D surface profile of the samples was measured using Precision Measurement system called Form Talysurf 60. This was done by randomly measuring a line on one surface along the rolling direction on all of the samples.
The results of the measurements were used to calculate a range of different roughness parameters for each material, sample thickness and tumbling duration as shown in figure 4. Note that it is the average Rz values that are presented for each tumbled Batch.
The graphs show that the samples in Batch A had a rougher surface, at all tumbling times, for one to five hours of tumbling compared to the samples in Batch B. They also show that the Batch A samples had a smoother surface after 10 hours of tumbling when compared to the other materials. After 15 hours of tumbling it appears that the surface roughness of the all materials is roughly the same. Between 15 and 40 hours of tumbling there appears to be no consistent difference between the tested materials.

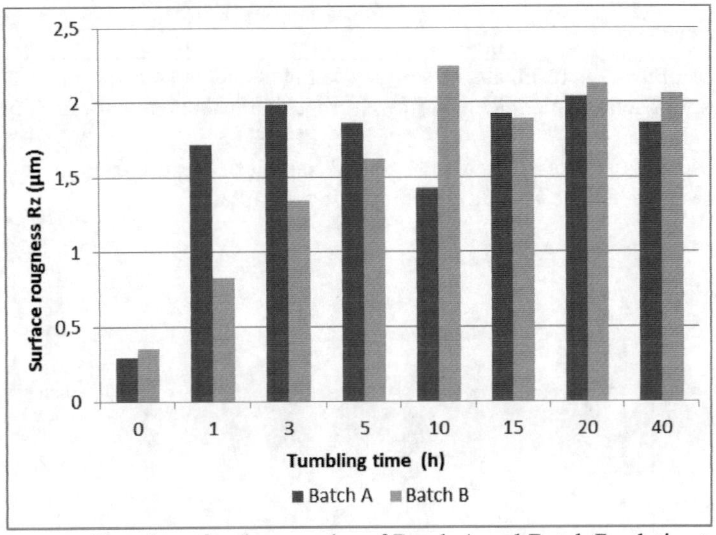

Figure 4: Showing the average Rz values for the samples of Batch A and Batch B relative to the tumbling duration.

7. RESIDUAL STRESSES

A sample from each material and tumbling duration was measured for residual stresses using XRD with one measurement on each side of the sample. The XRD measurements were made using Cr-Ka1 radiation, Bragg angle 156 degrees and lattice planes (2 1 1). The full measurement was made in oscillation and with nine scans each. Note that, when discussing surface residual stress measurements, focus should be on the overall difference between results rather than the absolute values. This is due to the fact that actual numbers in residual stress measurements are closely linked to the material constants. These constants are used during raw data analysis. The setup of the instrument also affects the measurements. In this way, the difference between the materials and tumbling times basically stays the same independent of other variables.
In the diagram in figure 5 a) its shown that Flap-X achieved approximately 33 % greater compressive residual stresses as compared to the "420" material. The higher residual stresses contribute to higher fatigue life in the final application. The"13Cr0.7C" material had slightly less residual stresses than"420" but higher than the Tool steel samples.
The results in the 5b) figure show that thicker samples had a higher residual stresses after tumbling compared the thinner ones. The 0.508 mm thick samples had on average 20 % high residual stresses than the 0.203 mm samples.
When residual stresses are compared with the tumbling time, the results are in agreement with previous work (Dusil 1978) which shows a rapid initial increase for the first 10 hours of tumbling and then levelling off after 15 to 20 hours, as seen in Fig. 5c).

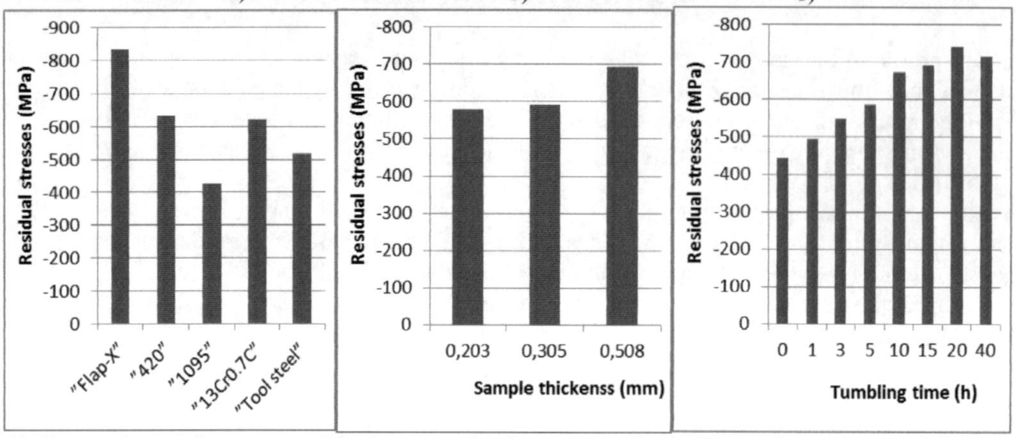

Figure 5: Comparisons of average residual stresses (a) Average residual stresses for each material. (b) Average residual stresses for each sample thickness. (c) Average residual stresses for each time in tumbling.

The rate of change of residuals stresses for batch A and B samples for different tumbling times is shown in the following figure 6. The results are mixed for different tumbling durations.

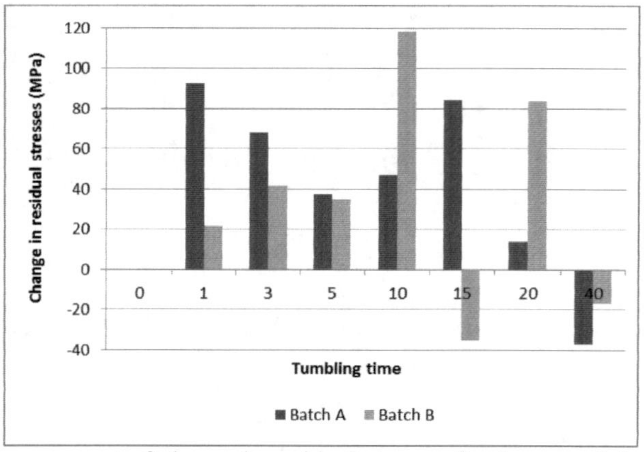

Figure 6: Showing the average amount of change in residual stresses for the materials in each Batch relative to the tumbling duration.

8. DISCUSSION

The results of the edge profile, surface appearances and roughness examinations points towards an initial difference in the effect of the tumbling between the different tumbling batches. The samples for which the 10 hour batch tumbling began with a used tumbling media have rounder edges, more rapidly affected surface appearances and rougher surface profiles for up to 5 hours of tumbling.

For tumbling times longer than 10 hours it is clear that the condition of the tumbling media at the start of the tumbling has little effect on the final outcome.

It appears that the differences in tumbling between the two batches of material have had an effect on the residual stresses as well as the surfaces and edges.

The residual stress measurements indicate that tumbling for more than 20 hours will result in increased compressive residual stresses. They also show that between 10 and 20 hours the increase in residual stresses is comparatively lower while it appears to be possible to modify how rapidly the residual stresses increase for the first few hours of tumbling, with relatively small modifications to the tumbling media.

(Dusil 1978) showed that the "420" and "1095" materials have significantly different amounts of residual stresses after tumbling and that this difference had a clear effect on the fatigue properties of the materials.

Based on this the "Flap-X" material is likely to have notably better fatigue life after tumbling as compared to the current standard flapper valve steel grades included in this investigation.

The study also reemphasises how critical it is for valve manufacturers to measure different metallurgical aspects of reed material before and after tumbling.

9. CONCLUSIONS

- The composition of the tumbling media can have a significant effect on the rate of edge rounding.
- The composition of the tumbling media can have a significant effect on the duration required for the material surface to be modified and also how smooth it becomes.
- The composition of the tumbling media appears to have an effect on the increase in residual stresses for short tumbling times.
- The tumbling time necessary to achieve maximum compressive residual stresses is between 15 and 40 hours.
- The ”Flap-X” material has notably higher residual stresses compared to the other materials investigated.
- The”1095” carbon strip material has comparatively less residual stresses compared to the other materials investigated.
- The thickness of the material appears to have a notable effect on the amount of residual stresses in the surface after tumbling.
- ”Flap-X” is likely to have improved fatigue properties after tumbling as compared to the other flapper valve steel grades in the investigation.
- The findings of this investigation provide a valuable reference to allow valve manufacturers to understand how to optimise their tumbling process.

REFERENCES

Dusil, R. (1978). On Blanking, Tumbling and Shot-Peening of Compressor Valves. *International Compressor Engineering Conference at Purdue,* 256.

Analysis of Viscous Slip at the Wall in Gas Flows of R134a and R600a through Metallic Microtubes

Ernane SILVA[1], Murilo F. NICOLUZZI[1], Marcos ROJAS-CARDENAS[2], Cesar J. DESCHAMPS[1]*

[1]POLO Research Laboratories for Emerging Technologies in Cooling and Thermophysics
Federal University of Santa Catarina
88040-900, Florianopolis, SC, Brazil
deschamps@polo.ufsc.br

[2]Institut Clément Ader UMR CNRS 5312
Université de Toulouse, INSA de Toulouse
Toulouse, France
marcos.rojas@insa-toulouse.fr

* Corresponding Author

ABSTRACT

The leakage of gas between the reed valve and the valve seat may significantly affect the efficiency of compressors adopted for household refrigeration. Such leakages are induced by differences of pressure between the compression chamber and the suction/discharge chamber and occur through small gaps in the order of micrometer, formed as a result of surface imperfections. Given the small dimensions of these clearances, a rarefied flow of refrigerant can occur under some operating conditions of the compressor. In this case, non-equilibrium phenomena, such as viscous slip between the fluid and solid boundaries, appear and they cannot be predicted by the classical fluid-mechanics continuum assumptions. The tangential momentum accommodation coefficient (TMAC) is a flow parameter that must be considered in order to correctly describe these rarefied gas flows, characterizing the exchange of momentum between the gas molecules and the surface. Many studies provide the value of TMAC, but usually the results are limited to nitrogen and noble gases for glass and silicon microchannels. The present paper reports measurements of mass flow rates and TMAC values for R134a and R600a through metallic microtubes (stainless steel and copper). The results show that viscous slip can occur even in flows of heavy polyatomic molecules typical of gases used in the refrigeration industry.

1. INTRODUCTION

Gas leakage is a well-known source of inefficiency in positive displacement compressors and can significantly reduce the volumetric and isentropic efficiencies. Such leakages can occur between the valves and valve seats as a result of surface irregularities and misalignment, which are responsible for the incomplete sealing. This can be especially important in oil-free compressors, since the presence of lubricating oil tends to improve sealing. Despite its importance, few studies in the literature have reported this problem. Machu (1990) conducted a theoretical analysis of a double acting cylinder compressor while Silva and Deschamps (2015) developed a numerical model to estimate the leakage in reed valves. Machu (1990) indicated that the tightness of the valve has a great effect on the efficiency in some operating conditions, while Silva and Deschamps (2015) showed that valve leakage may significantly deteriorate the compressor performance even for very small clearances.

Because of its micrometric scale, the gap formed between the valve and valve seat causes the flow to be rarefied under some operating conditions of the compressor (Silva and Deschamps, 2015). In this flow situation, non-continuum effects, such as viscous slip, thermal transpiration, and temperature jump at the wall, are significant and must be taken into account. Different theories are used to describe the flow according to the level of rarefaction, which is measured by means of the Knudsen number (Kn), defined as the ratio of the mean free path of the gas (λ) and the characteristic dimension of the system (L):

$$Kn = \frac{\lambda}{L} \tag{1}$$

The characteristic dimension of the system (L) can be calculated locally based on the gradients of macroscopic parameters of the flow. However, it is usually assumed to be a constant value, especially in the case of channels with constant cross-section. For microtubes, the characteristic dimension of the system is normally assumed to be the internal diameter (D), as considered in this work. An expression for the mean free path depends on the models considered to describe the molecular interactions between the gas molecules. For simplicity purposes, we adopt the formula proposed by Maxwell (1879):

$$\lambda = \sqrt{\frac{\pi}{2}}\frac{\mu\sqrt{RT}}{P} \tag{2}$$

For values of Kn around 10^{-3}-10^{-2}, or greater, the flow is considered to be rarefied, invalidating the flow description via the classical continuum hypothesis with the use of no-slip boundary conditions at the walls. Under such circumstances, the interaction of the gas molecules with the channel surfaces plays an important role in order to model properly the physical behavior of the flow. The tangential momentum accommodation coefficient (TMAC) is normally adopted to describe statistically such interactions, and can be understood as the fraction of molecules (α) with diffuse reflection at the walls. Conversely, $1 - \alpha$ represents the fraction of molecules with specular reflection, that is, the fraction of molecules that have the same tangential momentum before and after the collision against the wall. The value of TMAC may vary with the chemical composition of the gas, surface material, roughness and level of contamination, as well as gas temperature (Agrawal and Prabhu, 2008).

The TMAC has already been experimentally determined for a variety of gases and materials. Nevertheless, the majority of the experiments have been performed with nitrogen and noble gases flowing through silicon and/or silica microchannels or microtubes (Maurer *et al.*, 2003; Colin *et al.*, 2004; Graur *et al.*, 2009). Some exceptions are the works of Arkilic *et al.* (2001), Hadj Nacer *et al.* (2014), Yamaguchi *et al.* (2011, 2012), Szalmas *et al.* (2010), and Pitakarnnop *et al.* (2010), which have also considered flows of carbon dioxide, oxygen and gas mixtures. To our knowledge, the only efforts to test metallic surfaces were undertaken by Hadj Nacer *et al.* (2011), who covered the internal surfaces of the microchannel with a thin gold coat, and by Yamaguchi et al. (2012) and Hadj Nacer *et al.* (2014), who studied rarefied gas flows in commercially available stainless microtubes. More recently, Silva *et al.* (2016) presented experimental results of TMAC for the refrigerant gases R134a and R600a flowing through a commercially available stainless steel microtube. In this paper, the analysis has been extended to different metallic materials, such as stainless steel and copper. The mass flow rate of R134a and R600a were measured by means of the dynamic constant-volume technique and compared to analytical expressions in the range of *Kn* up to 0.3 in order to extract the corresponding values of TMAC. Although the geometry of the microtubes does not correspond to the geometry of the gap between the valve and valve seat, these experiments allow the determination of the TMAC, which is independent of the flow geometry and a necessary parameter in models adopted to predict valve leakage.

2. EXPERIMENTAL APPARATUS AND METHODOLOGY

2.1 Experimental Apparatus

The measurement of mass flow rates of gases through single microtubes is usually a difficult task given the small magnitudes involved. Different methods have been employed, with the direct measurement using flow meters being the simplest one (Jang and Wereley, 2004). Since this method is usually limited to mass flow rates higher than 10^{-8} kg/s (Ewart *et al.*, 2006), the constant-volume (Arkilic *et al.*, 1997; Ewart *et al.*, 2006; Pitakarnnop *et al.*, 2010) and the drop tracking methods (Colin *et al.*, 2004; Ewart *et al.*, 2006), which are indirect measurement techniques, have become the most commonly used techniques for measuring small mass flow rates of gases. In the present study, we adopted a dynamic measurement technique based on the constant-volume method, which was an alternative version of the well-known constant-volume proposed by Rojas-Cárdenas *et al.* (2011) for the measurement of mass flow rates induced by temperature gradients. Silva *et al.* (2016) applied recently the same technique for the case of pressure-driven isothermal flows. The experimental setup (Figure 1) consists of a test section with two rigid reservoirs (R1 and

R2) equipped with pressure (PT1 and PT2) and temperature (TT1 and TT2) transducers. These reservoirs are connected by the micro-device to be analyzed. The inlet and outlet valves (A and B) are used to isolate the test section from the other components of the system when necessary.

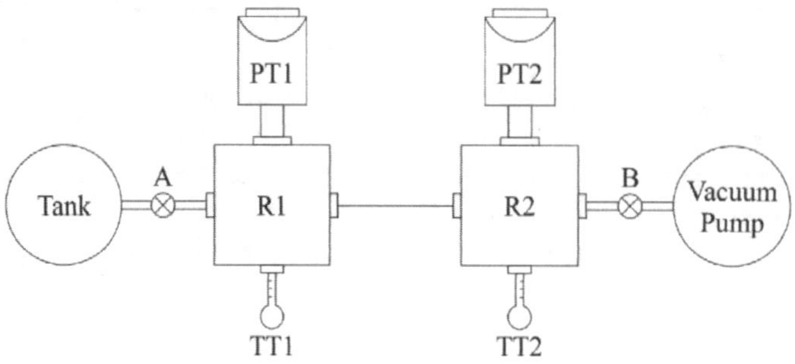

Figure 1: Experimental setup.

Capacitance diaphragm gauges with full scales of 133Pa or 1333Pa and accuracy 0.2% of reading were used to measure pressure in the upstream and downstream reservoirs, while T-type thermocouples were used for temperature measurements. The experiments were conducted through a stainless steel microtube (D = 438.6±4.5μm, L = 92.22±0.01mm) and a copper microtube (D = 444.17±4.5μm, L = 92.44±0.01mm).

The volumes of the upstream (R1) and downstream (R2) reservoirs were measured to be 181.1±0.9 ml and 28.5±0.8 ml, respectively. A tank with the gas to be analyzed was connected to the upstream reservoir (R2). Commercially available R134a and R600a gases were used in the experiments and their viscosities were obtained from the correlations developed by Huber *et al.* (2003) and Younglove and Ely (1987), respectively. The properties of the gases are provided in Table 1.

Table 1: Properties of the gases R134a and R600a.

Gas	$\mu \times 10^{-5}$ [Pa s]*	R [J kg-1 K-1]
R134a	1.181	81.50
R600a	0.7415	143.06

* at 297.15K

Even though the microtubes used in the experiments have small diameters, the dimensions of the gaps between the valve and valve seat are much smaller, on the order of micrometer (Rezende *et al.*, 2016). Therefore, with the purpose of reproducing the rarefied flow conditions that occur in the flow between the valve and the valve seat of compressors, it was necessary to lower the pressure inside the test section by means of a vacuum pump connected to the reservoir downstream of the micro-device (R2), obtaining a similarity between the flows in terms of the Knudsen number (Kn).

Tests were carried out by setting a uniform pressure in the test section and monitoring it for long periods, much longer than the duration of a single experiment, in order to assure that the test-section was leakage-free. In all cases, the difference between the highest and lowest pressures measured during a typical test of two hours did not exceed the uncertainty of the pressure sensors, meaning that air from the outside ambient was not able to leak towards the experimental test-section, at least in a measurable proportion.

2.2 Mass Flow Rate Measurements

The dynamic constant-volume technique consists in tracking the pressure variation with time inside the reservoirs R1 and R2 when a pressure difference is established between both ends of the micro-device. In this case, both valves A and B must be closed and the volumes of the reservoirs must be known. Thus, the variations of pressure can be associated with the mass flow rate of gas flowing through the micro-device by assuming a quasi-steady process.

By using the equation of state for an ideal gas, the mass flow rate through the micro-device, $\dot{M}$, can be related to the pressure variation with time in any of the two reservoirs (R1 or R2) as

$$\dot{M} = \frac{dM_i}{dt} = \frac{V_i}{RT_i}\frac{dP_i}{dt}(1-\varepsilon) \quad ,i = 1,2 \tag{3}$$

where M_i is the mass of gas in the reservoir, V_i is the volume of the reservoir, R is the specific gas constant, t is the time, and P_i and T_i are, respectively, the pressure and temperature of the gas in the reservoir. The term

$$\varepsilon = \left(\frac{dT_i}{dt}/T_i\right)\Big/\left(\frac{dP_i}{dt}/P_i\right) \tag{4}$$

takes under account the variation of temperature in relation to the variation of pressure and gives an indication of how closely the experiment follows the isothermal condition. For all experimental cases considered it was assured that $\varepsilon \leq 0.01$ and hence the mass flow rate was calculated based solely on the pressure variation with time (Equation (3), with $\varepsilon = 0$), considering a maximum relative standard uncertainty of 1% related to the temperature variation.

Each experimental campaign was performed via the following steps: a) the entire system was pumped down for long periods in order to assure that no residual gases were left, after which valves A and B were closed (Figure 1); b) valve A was opened and gas from the high pressure tank was allowed to enter the system until a desired upstream pressure was achieved, when then valve A was closed again; c) valve B was opened in order to release some gas from reservoir R2 into the vacuum pump, creating thus a pressure difference between the two tanks; d) valve B was closed and the experiment initiated with the pressure relaxation process taking place in both reservoirs until a final equilibrium pressure was reached in the entire test section (Figure 2a). Steps from "b" to "d" were repeated for different initial pressures until all the rarefaction range of interest was analyzed.

The dynamic constant-volume technique relies on the assumption that the pressure variation with time inside the reservoirs can be described by a fitting function $P_i(t)$. Then, the mass flow rate is calculated from the time derivative of this function (Equation (3)). The main advantage of the dynamic technique in respect to the classical constant-volume method is that a large data set is obtained from a single experiment. In fact, during a single experiment, a wide range of rarefaction conditions can be investigated, since both the pressures in the upstream and downstream reservoirs vary with time (Figure 2b). The rarefaction condition is measured by means of the average Knudsen number (Kn_m), which is calculated from Equation (1) based on the average pressure along the tube given by $P_m = (P_1 + P_2)/2$.

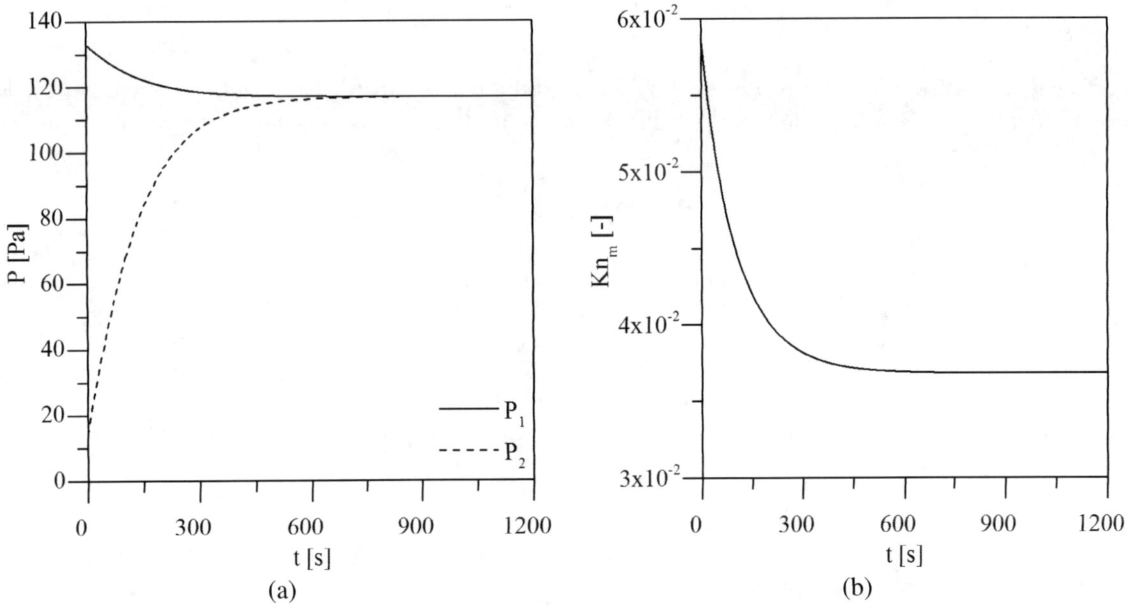

Figure 2: Data from a single experiment with R600a in a copper microtube: (a) variation of pressure in the reservoirs and (b) variation of Kn_m.

2.3 Determination of TMAC

For slightly rarefied gas flows ($Kn \leq 0.1$), the mass flow rate of gas through a tube of diameter D can be obtained directly from the solution of the Navier-Stokes equation provided that the boundary conditions are modified to account for velocity slip at the wall. Several authors (Maurer *et al.*, 2003; Colin *et al.*, 2004; Ewart *et al.*, 2006; Yamaguchi *et al.*, 2011) have proposed the use of second-order slip boundary conditions to extend the validity of the continuum equations for moderately rarefied gas flows ($Kn \leq 0.3$). In this case, the slip velocity reads as follows:

$$u_s = \pm A_1 \lambda \left(\frac{\partial u}{\partial r}\right)_w - A_2 \lambda^2 \left[\frac{1}{r}\left(\frac{\partial}{\partial r} r \frac{\partial u}{\partial r}\right)\right]_w \tag{5}$$

where A_1 and A_2, according to the molecular collision model considered herein, are defined as

$$A_1 = \frac{2}{\sqrt{\pi}} \sigma_P \quad , \quad A_2 = \frac{4}{\pi} \sigma_{2P} \tag{6}$$

In this equation σ_P and σ_{2P} stand for the first and second viscous slip coefficients, respectively. The TMAC (α) can be related to σ_P following the Maxwellian model that describes the gas/surface interaction, i.e.:

$$\sigma_P = \frac{\sqrt{\pi}}{2} \frac{2 - \alpha}{\alpha} \tag{7}$$

Taking into account the velocity slip at the wall (Equation (5)), an analytical solution for the mass flow rate can be easily obtained for the case of a long tube

$$\dot{M} = \frac{\pi D^4 \Delta P P_m}{128 \mu R T L}\left(1 + 8 A_1 K n_m + 32 A_2 \frac{P_m}{\Delta P} \ln \Pi \; K n_m^2\right) \tag{8}$$

where $\Delta P = P_1 - P_2$ is the pressure difference between inlet and outlet reservoirs, $\Pi = P_1/P_2$ is the pressure ratio, $P_m = (P_1 + P_2)/2$ is the mean pressure, and L is the length of the tube. An expression for the non-dimensional mass flow rate, S, is obtained by dividing Equation (8) by the Poiseuille analytical solution for the hydrodynamic regime (i.e., $Kn_m \to 0$):

$$S = \frac{128 \mu R T L}{\pi D^4 \Delta P P_m} \dot{M} = 1 + 8 A_1 K n_m + 32 A_2 \frac{P_m}{\Delta P} \ln \Pi \, K n_m^2 \tag{9}$$

The coefficients of Equation (9) can be obtained by curve fitting of the experimental data of mass flow rate, as described by Maurer *et al.* (2003), with the slip coefficients and the TMAC being obtained from Equations (6) and (7), respectively.

3. RESULTS

Four series of experiments were conducted to cover a wide range of flow rarefaction for the gases and materials considered. In all experiments the pressure variation with time in each reservoir was fitted and the mass flow rate was obtained from Equation (3). The mass flow rates extracted from the first 400s of pressure variation presented in Figure 2 can be seen in Figure 3a. In this case, both the results obtained from the variations of pressure in the upstream and downstream reservoirs are presented and show an excellent agreement, within the uncertainty range, proving that the technique is solid and accurate. The relative uncertainty in the mass flow rate measurements obtained from the downstream reservoir is estimated to be 3% and this margin of uncertainty is represented in the graph by the thinner solid lines. Since the downstream reservoir (R2) has a smaller volume, the variations of pressure are higher in respect to the variations of temperature, and the isothermal condition is more properly satisfied (Figure 3b). For this reason, the mass flow rate obtained from the pressure measurement in the downstream reservoir could be considered to be more accurate and were used to extracted the TMAC.

The non-dimensional mass flow rate, S, given by Equation (9) and presented as a function of Kn_m in Figure 4, can be regarded as a measure of the rarefaction effect. The higher is the value of S the higher is the influence of rarefaction on the mass flow rate. When S equals to unity no rarefaction effect is observed. Figure 4 shows a slightly higher velocity slip for the flows in the microtube of copper, even though the differences are inside the uncertainty range. It is also possible to note in Figures 4b and 4d that rarefaction increases the mass flow rate by approximately 10% ($S =$ 1.1) in both microtubes when $Kn_m = 10^{-2}$. This mass flow rate is further increased with Kn_m, reaching almost 50% ($S = 1.5$) when $Kn_m = 0.05$. Values of Kn_m in this range were predicted by Silva and Deschamps (2015) for a gap of 0.25μm between the valve and valve seat. Recent results from Rezende *et al.* (2016) have shown that gaps on this order of dimension are likely to be found in compressor valves and justify the consideration of rarefaction effects in the modeling of leakage.

Results of TMAC extracted from the curve fittings of the non-dimensional mass flow rate shown in Figure 4 are presented in Table 2 together with the respective coefficients of determination. In all cases, coefficients of determination were higher than 0.999, justifying the adoption of a second-order slip boundary condition. An incomplete accommodation ($\alpha < 1$) was observed for all cases and smaller TMAC (α) values were found for the copper in relation to the stainless steel microtube, even though the differences in the mass flow rate measurements are within the uncertainty range.

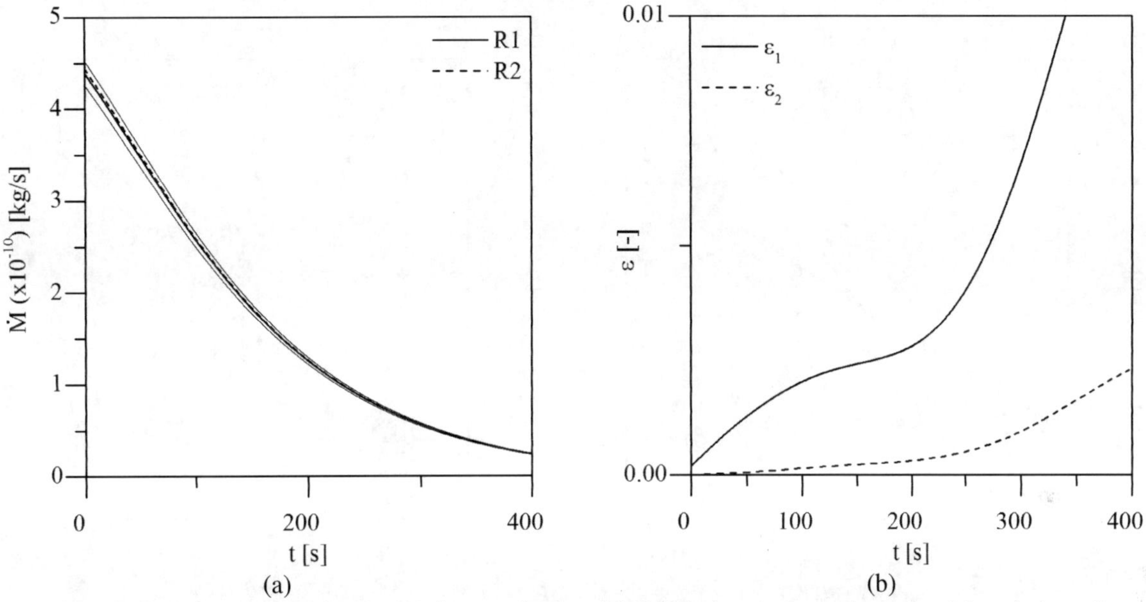

Figure 3: Data extracted from a single experiment with R600a in a copper microtube: (a) variation of the mass flow rates through the microtube and (b) variation of ε.

Table 2: TMAC and coefficient of determination of the curve fitting of *S*.

Microtube	Gas	α	R^2
Stainless Steel	R134a	0.956	0.999
	R600a	0.974	0.999
Copper	R134a	0.917	1.000
	R600a	0.938	1.000

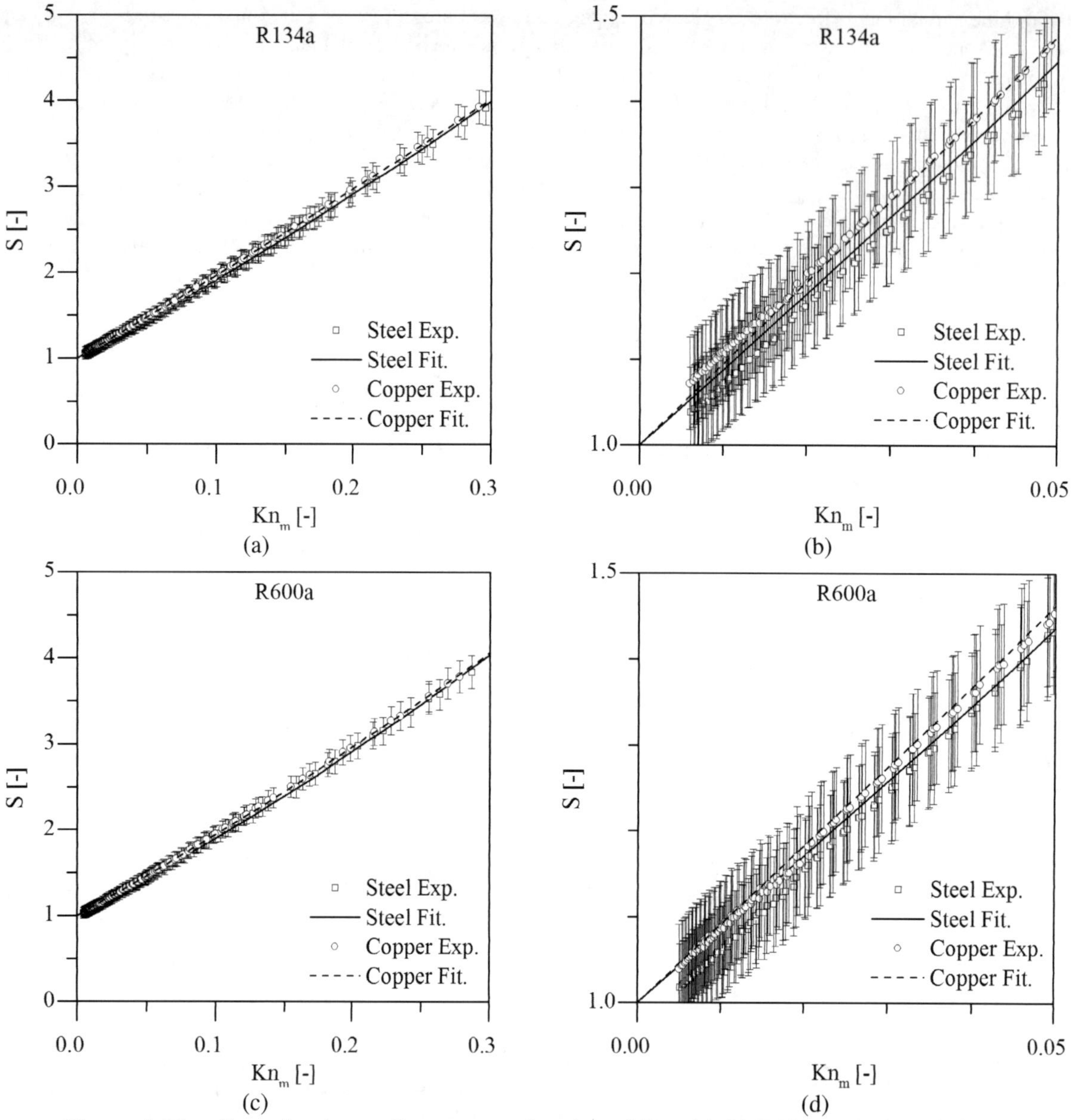

Figure 4: Non-dimensional mass flow rate as a function of Kn_m: (a)-(b) R134a and (c)-(d) R600a.

4. CONCLUSIONS

This paper reported measurements of mass flow rate of rarefied gases through metallic microtubes made of stainless steel and copper, carried out with the dynamic constant-volume method. Flows of gases commonly employed in the refrigeration industry, namely R134a and R600a, were considered and it was shown that rarefaction effects can occur even in such gases of complex polyatomic molecules. Tangential momentum accommodation coefficients (TMAC) were found in the range of 0.917 to 0.974 for the two gases and microtubes investigated, confirming the existence of an incomplete accommodation of the molecules at the walls ($\alpha < 1$). Our results also indicated that the metallic material of the microtube does not have great influence on TMAC when there is no special surface treatment, usually referred to as engineering surfaces, even though the results suggest a higher velocity slip in the copper microtube. Future investigations will be focused on the effect of the surface roughness on the TMAC. It is expected that the roughness of the surface plays a more significant role than the surface material when rough surfaces are considered. Probably the surface material will be more relevant for atomically smooth surfaces, but such a surface finishing is not found in compressors.

NOMENCLATURE

A_1	1st coefficient of the velocity slip boundary condition	(-)
A_2	2nd coefficient of the velocity slip boundary condition	(-)
D	diameter	(m)
Kn	Knudsen number	(-)
L	length	(m)
$M, \dot{M}$	mass, mass flow rate	(kg, kg s^{-1})
$P, \Delta P$	pressure, difference of pressure	(Pa, Pa)
r	radial coordinate	(m)
R	specific gas constant	(J kg^{-1} K^{-1})
S	non-dimensional mass flow rate	(-)
t	time	(s)
T	temperature	(K)
u	velocity	(m s^{-1})
u_s	slip velocity	(m s^{-1})
V	volume	(m^3)
ε	relative variation of temperature	(-)
λ	mean free path	(m)
μ	dynamic viscosity	(Pa s)
Π	pressure ratio	(-)
σ_p	1st order slip coefficient	(-)
σ_{2p}	2nd order slip coefficient	(-)

Subscript

i	generic reservoir
m	mean
w	wall

REFERENCES

Agrawal, A., Prabhu, S. V. (2008). Survey on the measurement of tangential momentum accommodation coefficient. *J. Vac. Sci. Technol. A 26(4)*, 634-645.

Arkilic, E. B., Schmidt, M. A., Breuer, K. S. (1997). Gaseous slip flow in long microchannels. *J. Microelectromech. Syst. 6*, 167-178.

Arkilic, E. B., Breuer, K. S., Schmidt, M. A. (2001). Mass flow and tangential momentum accommodation in silicon micromachined channels. *J. Fluid Mech. 437*, 29-43.

Colin, S., Lalonde, P., Caen, R. (2004). Validation of a second-order slip flow model in rectangular microchannels. *Heat Transfer Eng. 25*, 23-30.

Ewart, T., Perrier, P., Graur, I., Méolans, J. G. (2006). Mass flow rate measurements in gas micro flows. *Exp. Fluids 41*, 487-498.

Graur, I. A., Perrier, P., Ghozlani, W., Méolans, J. G. (2009). Measurements of tangential momentum accommodation coefficient for various gases in plane microchannel. *Phys. Fluids 21*, 102004.

Hadj Nacer, M., Graur, I., Perrier, P. (2011). Mass flow measurement through rectangular microchannel from hydrodynamic to near free molecular regimes. *Houille Blanche 4*, 49-54.

Hadj Nacer, M., Graur, I., Perrier, P., Méolans, J. G., Wuest, M. (2014). Gas flow through microtubes with different internal surface coatings. *J. Vac. Sci. Technol., A 32*, 021601.

Huber, M. L., Laesecke, A., Perkins, R. A. (2003). Model for the viscosity and thermal conductivity of refrigerants, including a new correlation for the viscosity of R134a. *Ind. Eng. Chem. Res. 42*, 3163-3178.

Jang, J., Wereley, S. T. (2004). Pressure distributions of gaseous slip flow in straight and uniform rectangular microchannels. *Microfluid Nanofluidics 1*, 41-51.

Machu, E. (1990). How leakages in valves can influence the volumetric and isentropic efficiencies of reciprocating compressors, *Proceedings of the International Compressor Engineering Conference* (482-494). West Lafayette, USA.

Maurer, J., Tabeling, P., Joseph, P., Willaime, H. (2003). Second-order slip laws in microchannels for helium and nitrogen. *Phys. Fluids 15*, 2613-2621.

Maxwell, J. C. (1879). On stresses in rarefied gases arising from inequalities of temperature. *Philosophical Transactions of the Royal Society of London 170*, 231-256.

Pitakarnnop, J., Varoutis, S., Valougeorgis, D., Geoffroy, S., Baldas, L., Colin, S. (2010). A novel experimental setup for gas microflows. *Microfluid Nanofluidics 8*, 57-72.

Rezende G. C., Silva, E., Deschamps, C. J. (2016). A hybrid experimental-numerical procedure to estimate the leakage gap in compressor valves, *Proceedings of the International Compressor Engineering Conference* (1-10). West Lafayette, USA.

Rojas-Cárdenas, M., Graur, I., Perrier, P., Meolans, J. G. (2011). Thermal transpiration flow: a circular cross-section microtube submitted to a temperature gradient. *Phys. Fluids 23*, 031702.

Silva, L. R., Deschamps, C. J. (2015). Modeling of gas leakage through compressor valves. *Int. J. Refrigeration 53,* 195-205.

Silva, E., Rojas-Cárdenas, M., Deschamps, C. J. (2016). Experimental analysis of velocity slip at the wall for gas flows of nitrogen, R134a, and R600a through a metallic microtube. *Int. J. Refrigeration,* In Press.

Szalmas, L., Pitakarnnop, J., Geoffroy, S., Colin, S., Valougeorgis, D. (2010). Comparative study between computational and experimental results for binary rarefied gas flows through long microchannels. *Microfluid Nanofluidics 9*, 1103-1114.

Yamaguchi, H., Hanawa, T., Yamamoto, O., Matsuda, Y., Egami, Y., Niimi, T. (2011). Experimental measurement on tangential momentum accommodation coefficient in a single microtube. *Microfluid Nanofluidics 11*, 57-64.

Yamaguchi, H., Matsuda, Y., Niimi, T. (2012). Tangential momentum accommodation coefficient measurements for various materials and gas species. *J Phys.: Conf. Ser. 362*, 012035.

Younglove, B. A., Ely, J. F. (1987). Thermophysical properties of fluids. II. Methane, ethane, propane, isobutane and normal butane. *J. Phys. Chem. Ref. Data 16*, 577-798.

ACKNOWLEDGEMENT

The present work was developed with the support of CNPq, National Council for Scientific and Technological Development – Brazil, and is part of a technical-scientific cooperation program between the Federal University of Santa Catarina and EMBRACO. The authors would like to express their gratitude to the invaluable support of the Center of Reference in Innovative Technologies (CERTI) in providing the measurements of the microtube diameters. The authors also thank EMBRAPII Unit Polo/UFSC and CAPES (Coordination for the Improvement of High Level Personnel) for the financial support.

Experimental Investigation of Damping Coefficient for Compressor Reed Valves

Sérgio K. LOHN[1]*, Evandro L. L. PEREIRA[1], Humberto F. da CÂMARA[1],
Cesar J. DESCHAMPS[2]

[1]Embraco, R&D
Rua Rui Barbosa, 1020
89219-901, Joinville, SC, Brazil
Sergio.K.Lohn@embraco.com

[2]POLO Research Labs for Emerging Technologies in Cooling and Thermophysics
Federal University of Santa Catarina
88040-535, Florianopolis, SC, Brazil
deschamps@polo.ufsc.br

* Corresponding Author

ABSTRACT

Reed valve dynamics has a major role on the performance and reliability of compressors and has been the subject of many studies over the years. In terms of mathematical modeling, most of these studies describe valve dynamics through a mass-spring-damper system in which a damping coefficient has to be empirically adjusted. This is a consequence of little knowledge about the effect of different parameters on valve damping. For instance, in spite of great effort to understand viscous effects on the pressure distribution acting on the valve surface, very few works have addressed such effects on valve damping, especially for valve geometries found in compressors. The present study aims to experimentally quantify valve damping under controlled conditions considering the effects of clamping geometry, clamping force, gasket thickness and the presence of fluid around the valve. Finally, experimental data of damping coefficients are used to predict valve dynamics and compressor efficiency. We found that the damping coefficients of typical valves adopted in small reciprocating compressors have negligible effect on compressor efficiency and valve bending stress, but can change by up to 17% the valve impact velocity against the seat.

1. INTRODUCTION

The dynamics of reed valves can be described in a simplified way with a single degree-of-freedom model:

$$m_{eq}\ddot{x} + c\dot{x} + kx = F_{valv} - F_{ot} \tag{1}$$

where m_{eq}, c and k are the reed equivalent mass, damping coefficient and stiffness, respectively. On the other hand, F_{valv} is the force induced by the pressure load on the reed surface and F_{ot} can represent any other force, such as reed pre-tension or stiction force due to the presence of a lubricating oil film between the reed and the valve seat (Khalifa and Liu, 1998). Finally, x, $\dot{x}$ and $\ddot{x}$ are the instantaneous reed lift, velocity and acceleration, respectively.

Valve dynamics is affected by dissipative (damping) forces, leading to smaller valve lifts and valve velocities. Smaller valve lifts increase viscous loss in the suction and discharge processes, reducing the compressor efficiency. On the other hand, small lifts reduce both the valve bending stress and the valve impact velocity, increasing valve reliability. Valve damping can be attributed to three different dissipative sources: i) material damping due to complex interactions in the crystalline structure and internal imperfections; ii) clamping damping due to the friction force of micro-slips between the surfaces that are used to fasten the valve; iii) viscous damping due to the presence of fluid surrounding the valve during its displacement. Prater and Hnat (2003) carried out experiments to characterize reed valve damping in rotary compressors. Material, clamping and viscous damping were obtained by measuring the valve free vibration and its oscillation decay with an optical sensor to avoid any interference in the valve dynamics.

The present paper reports the results of an experiment investigation similar to that of Prater and Hnat (2003) with the aim of determining the damping coefficient of a reed type valve of a small reciprocating compressor. The influence of clamping geometry, fastening force and density of the fluid surrounding the valve are investigated. Additionally, simulations are carried out to estimate the effect of valve damping on the compressor efficiency and valve impact velocity.

2. EXPERIMENTAL SETUP

The method applied to measure the valve damping follows the procedure described by Granick and Stern (1965), but we consider actual valve geometries instead of a simplified beam geometry. The experiment takes place inside a small vessel (Figure 1) with absolute pressure below 2 mbar and temperature equal to 25 °C. At such condition the air density was approximately 0.002 kg/m³, hence minimizing viscous damping.

In order to measure the valve damping the reed valve was allowed to oscillate freely at its natural frequency after being excited by an automatic device mounted inside the vessel. Valve oscillation was measured using two coils of a fast linear displacement transducer (FLDT), as shown in Figure 2. The use of more sophisticated devices, such as a laser displacement meter, was not possible due to space limitation inside the vessel. The FLDT coils were design to accurately measure small displacements and to compensate errors caused by temperature change in the coils. The test bench also had a signal processor and a data acquisition system with sampling rate of 100 kHz.

The inductive coils to measure valve displacement were calibrated for each valve assembly. The calibration procedure consisted in moving the valve from its rest position to its maximum displacement, with the displacement being simultaneously measured by a high-accuracy laser sensor and the coils connected to the signal sensor. A calibration curve was then made available via a transfer function between the voltage signal from the two coils and the valve displacement measured with the laser system. Figure 3 presents a diagram of the calibration apparatus.

The damping coefficient was measured for a typical suction valve of a small reciprocating compressor used in household refrigeration systems. Measurements were repeated four times for each configuration of interest, including valve reassembly, following a procedure of six steps:

i. Assembling the reed valve in the in the measuring device formed by the two coils (Figure2);
ii. Calibrating the valve displacement transducer (FLDT coils);
iii. Mounting the measuring device inside the vessel (Figure1);
iv. Establishing vacuum in the vessel with a vacuum pump;
v. Moving the reed valve to its maximum displacement (x_0) with a mechanical device (Figure2);
vi. Releasing the reed valve and measuring its oscillating motion.

Figure 1: Vessel with rarefied atmosphere.

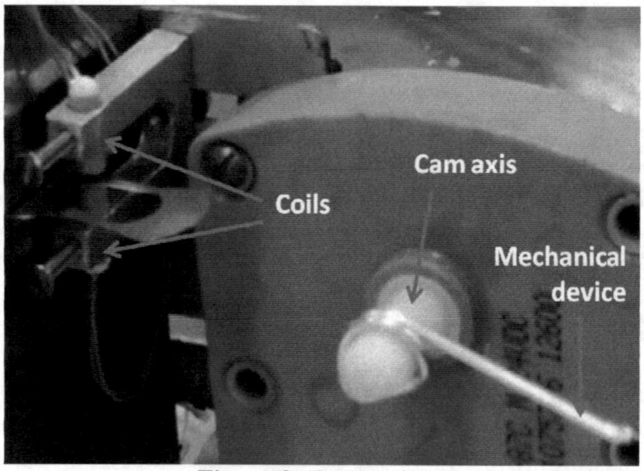

Figure 2: Test section.

The valve damping coefficient can be expressed as:

$$c = 2\gamma m_{eq} \quad (2)$$

The decay parameter γ in Equation 2 is estimated by adjusting its value in Equation (3) so as to obtain the actual decay of the valve oscillating motion observed experimentally (green line), as shown in Figure 4.

$$x_{valv} = x_0 e^{(-\gamma t)} \quad (3)$$

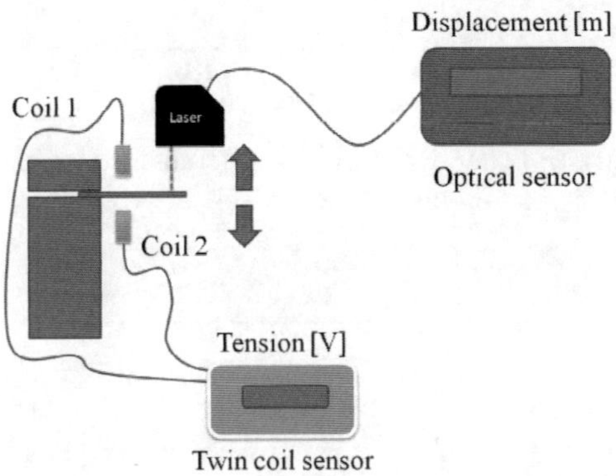

Figure 3: Calibration apparatus.

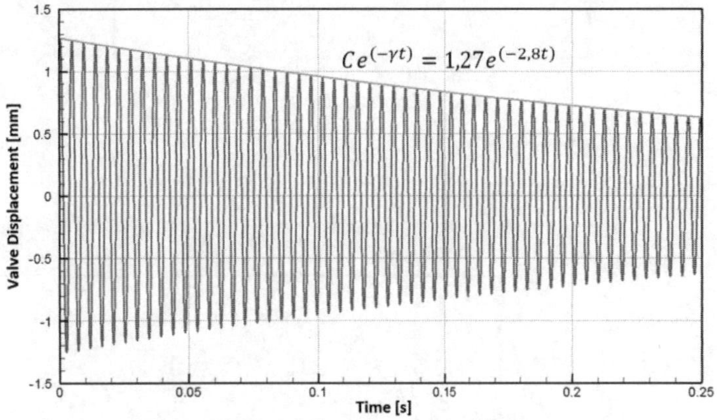

Figure 4: Determination of γ coefficient.

3. EXPERIMENTAL RESULTS

Valve damping was initially analyzed in relation to valve clamping with and without gasket. When the valve fastening system adopts a gasket the maximum torque is applied to the bolts, which represents the situation in the early life of the compressor. In the absence of a gasket, the valve is fastened in place against the other metallic surface by using the same torque. This configuration is expected to result in smaller dissipation force at the clamp region. Figure 5 shows the average values and dispersions of measurements for the damping ratio with and without gasket. As can be noted, the dispersion of both results are similar and small enough to allow the conclusion that the damping ratio for the clamping with gasket is approximately 60% greater than that without gasket. In fact, the damping ratio for the clamping without gasket is quite small and can be attributed solely to the material damping.

When the compressor is in the field, i.e., operating in a refrigeration system, the clamping torque is much lower than the original torque applied in the compressor assembly due to torque relaxation in extreme temperatures. Hence, we decided to analyze also the damping coefficient as a function of two different torques (T1 = 9 Nm; T2 = 4.5 Nm) and two gasket thicknesses (t1 = 0.25 mm; t2 = 0.48 mm), resulting in four test configurations (T1t1, T2t1, T1t2, T2t2). Figure 6 shows a clear increase in the valve damping when the torque is decreased, probably because the reduction of the fastening force makes the gasket less rigid. Additionally, the valve damping seems to increase with the gasket thickness for the same reason.

Nashif *et al.* (1985) analyzed the effect of the beam natural frequency f_n on the valve damping. We explored this aspect by comparing the dynamics of a second valve with f_n = 280 Hz with the dynamics of the original valve whose natural frequency was f_n = 200 Hz. Both natural frequencies are typical of valves used in compressors designed for domestic refrigeration. Figure 7 presents prediction and measurement for the decay of the valve oscillating motion with f_n = 280 Hz. It should be mentioned that the material damping specified in the simulation model was equal to the value measured for the valve with f_n = 200 Hz. Both results are quite similar and show that the natural frequency is of secondary importance in the valve material damping.

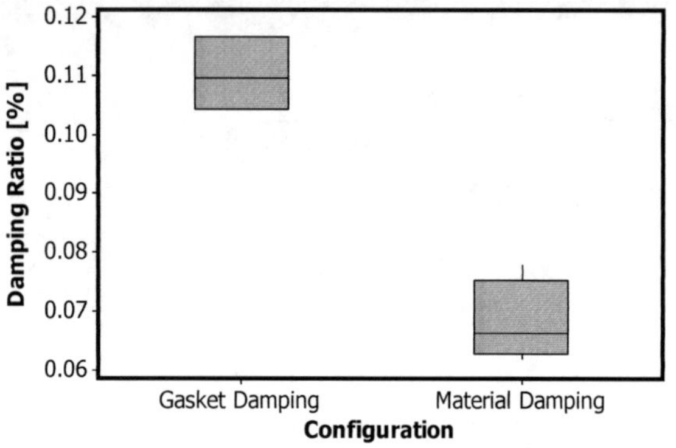

Figure 5: Damping ratios for valves with and without gasket.

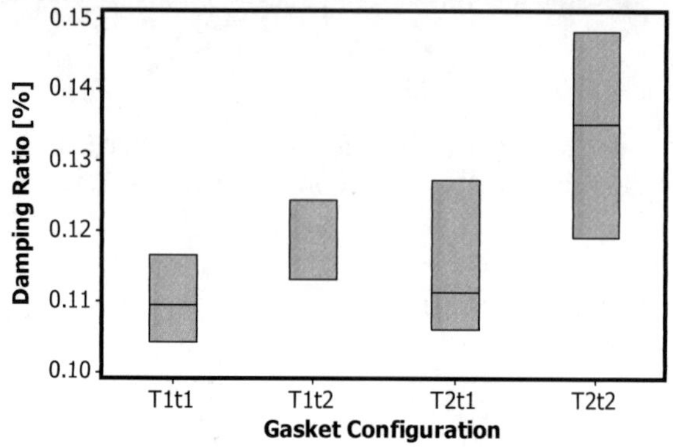

Figure 6: Damping ratios for different torques and gasket thicknesses.

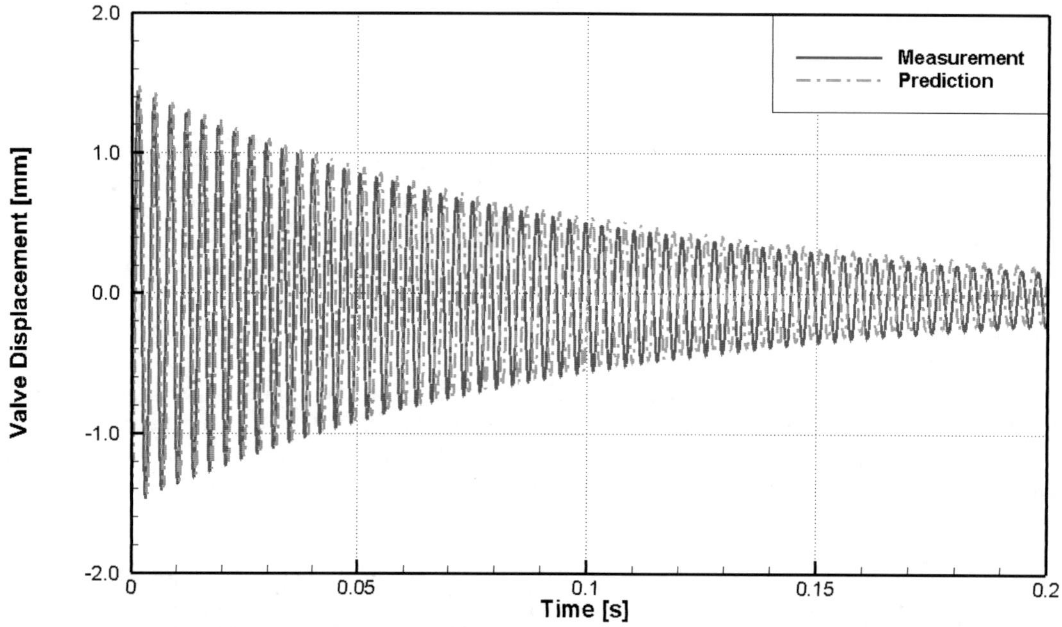

Figure 7: Comparison between numerical and experimental results for valve displacement (without gasket).

4. NUMERICAL ANALYSIS

The experimental setup only allows tests with air at pressures lower than or equal to the atmospheric pressure, hence not being able to cover the entire range of pressure levels experienced by the compressor during its compression cycle. In order to investigate the viscous damping for different gases and pressures we decided to adopt two auxiliary simulation models, one based on the finite volume method (FVM) for the fluid flow through the valve and the other following the finite element method (FEM) to solve the valve displacement.

As shown in Figure 8, the computational domain for the fluid flow is formed by a large volume of fluid around the valve to minimize the effect of the adopted boundary condition (total non-reflective pressure) on the solution. Figure 9 shows the computational grid of the solution domain for the valve dynamics via FEM. In this fluid-structure interaction analysis, the valve dynamics and the time dependent flow field are coupled and have to be solved simultaneously. For instance, the fluid flow pressure field causes the reed to deform, which in turn affects the fluid flow solution. Further details of this fluid-structure model can be obtained in Shiomi *et al* (2009).

The simulation procedure consists in applying an initial impact on the valve and then registering the decay of its oscillating motion similarly as carried out in the experimental setup. We tested different fluid densities in the simulations, since Prater and Hnat (2003) reported this parameter strongly affects the viscous damping. The results allowed us to obtain an expression for the damping ratio, ξ $(= c/c_c)$, as a function of the fluid density (Equation 4). In accordance to the operating conditions of refrigeration compressors, the density was varied from 1kg/m³ to 70kg/m³, leading to changes up to 18% in the viscous damping ratio.

$$\xi = 4.785 \times 10^{-6}\rho^2 + 2.256 \times 10^{-3}\rho - 9.377 \times 10^{-4} \quad (4)$$

For validation purpose, the damping ratio, ξ, given by Equation 4, was adopted in Equation 1 to predict the valve oscillation decay when left to oscillate freely on air at atmospheric pressure. The material damping previously measured was adopted in the model. As shown in Figure 10, close agreement was found between measurement and prediction, validating Equation 4.

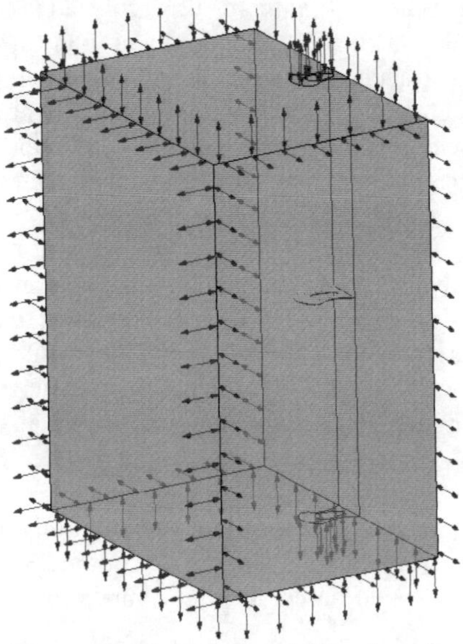

Figure 8: Fluid flow solution domain.

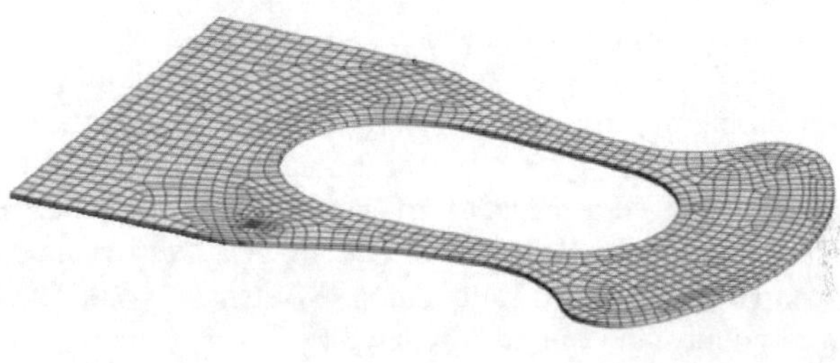

Figure 9: Computational grid of the structural domain.

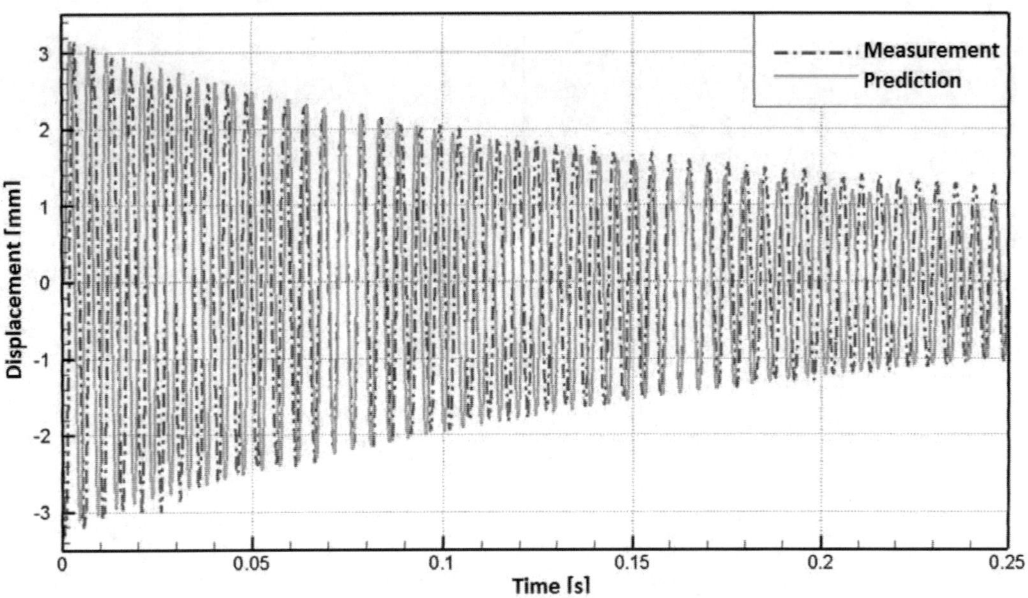

Figure 10: Comparison between numerical and experimental results for valve displacement considering air as the working fluid.

5. EFFECT OF DAMPING COEFFICIENT ON COMPRESSOR PERFORMANCE

Figure 11 shows the contributions of material, gasket, fluid and the total damping ratio, ξ, for the suction valve of a compressor operating with R-600a. As can be noted, the total damping coefficient is not high, being just 0.4% of the critical damping, c_c. The contribution of the fluid surrounding the valve on the damping is similar to the other two contributions (material and gasket). The total damping associated with the discharge valve of the same compressor is 2.6% of the critical damping (Figure 12) and it is essentially due the fluid. Naturally, the damping coefficient is also a function of the valve geometry and Equation 4 should be used with caution in the simulation of considerably distinct valve designs.

5.1. Compressor Efficiency

The effect of the damping coefficient on the efficiency of a small reciprocating compressor was analyzed taking into account the dynamics of suction and discharge valves. A common value attributed to the damping ratio of suction valves is $\xi \cong 0.1$. Figure 13 shows the increase of energy loss in the suction process due to valve damping in comparison to a baseline situation without damping. As can be seen, a damping ratio of 0.1 would increase the suction loss by 5.8%, whereas the actual damping ratio of 0.004 (0.4%) is responsible by an increase of just 0.2% in the suction loss. Therefore, the assumption of no damping for the suction valve would not bring any significant error in predictions of the compressor efficiency.

As seen before, the damping ratio of the discharge valve is greater than that of the suction valve. Moreover, the discharge process takes place in a much shorter period than the suction process. Figure 14 shows that the valve damping leads to a reduction of just 2.5% in the discharge loss. Therefore, the damping coefficient could also be neglected in the modeling of the discharge valve dynamics. It is a common practice to model the dynamics of the discharge valve as a critically damped system ($\xi = 1$). The results in Figure 14 show that this assumption leads to an error of almost 26% in the discharge loss.

Overall, high damping coefficients for both valves ($\xi = 0.1$ and 1) can lead to reductions up to 0.6% in the COP of typical household refrigeration systems. On the other hand, the damping coefficients obtained from our measurements would decrease COP by less than 0.1%, which is again evidence that valve damping is not important when the compressor efficiency is the main interest of the analysis.

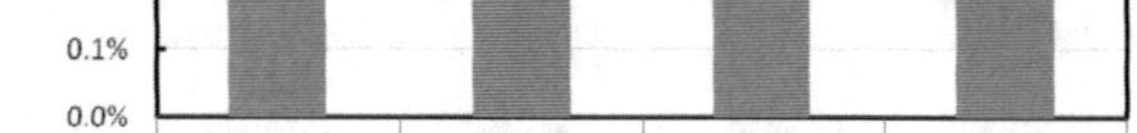

	Material	Gasket	Fluid	Total
Suction	0.09%	0.13%	0.19%	0.41%

Figure 11: Damping ratio for the suction valve of a compressor operating with R-600a.

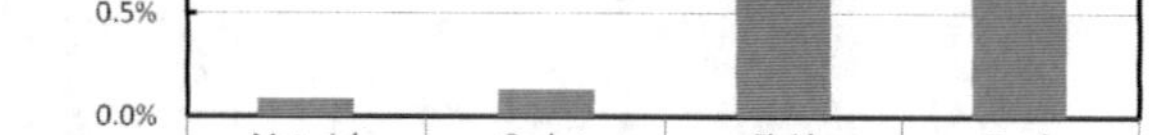

	Material	Gasket	Fluid	Total
Discharge	0.09%	0.13%	2.38%	2.60%

Figure 12: Damping ratio for the discharge valve of a compressor operating with R-600a.

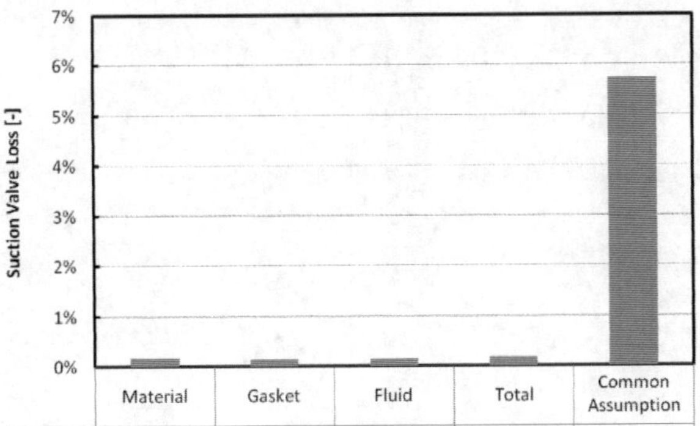

	Material	Gasket	Fluid	Total	Common Assumption
Suction	0.16%	0.15%	0.15%	0.20%	5.75%

Figure 13: Suction loss due to suction valve damping.

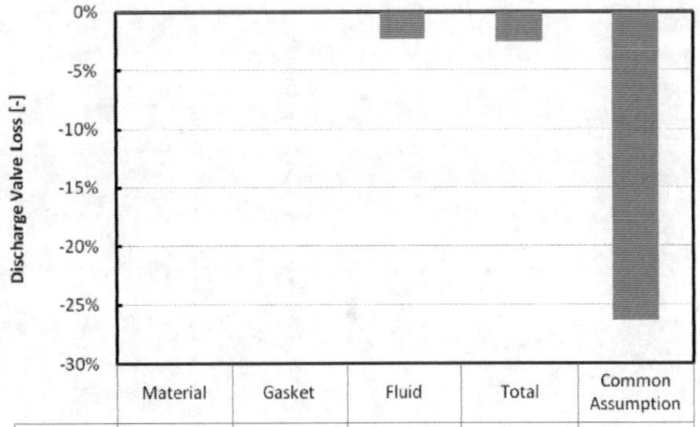

	Material	Gasket	Fluid	Total	Common Assumption
Discharge	-0.04%	-0.07%	-2.29%	-2.54%	-26.34%

Figure 14: Discharge loss change due to discharge valve damping.

5.2. Valve Impact Velocity

In this section we explore the effect of the damping coefficient on the valve maximum opening, which defines the bending stress and, to some extent, the valve impact velocity. The analysis considers the compressor under an extreme operating condition that may occur during its duty cycles. Figure 15 shows that the effect of the damping coefficient on the suction valve opening is small, but it can also be observed that the common assumption of $\xi \cong 0.1$ leads to considerable error in the result. Figure 16 considers the influence of the damping on the suction valve impact velocity. As can be seen, despite its negligible effect on the valve opening, the actual damping affects the valve impact velocity by 17%, which is quite significant in terms of compressor reliability.

The discharge valve usually adopts a stopper device to limit its maximum opening. In the present analysis, the damping force increases the discharge valve impact velocity by 1.1%. On the other hand, the assumption of a critically damped system ($\xi = 1$) for the valve dynamics would overestimate the valve impact velocity by almost 28%. This occurred in our simulation because the high damping coefficient delayed the valve closing. As a consequence, the piston had already started its motion towards the bottom dead center position just before the valve was closed, suddenly increasing the pressure load on the valve and its final velocity.

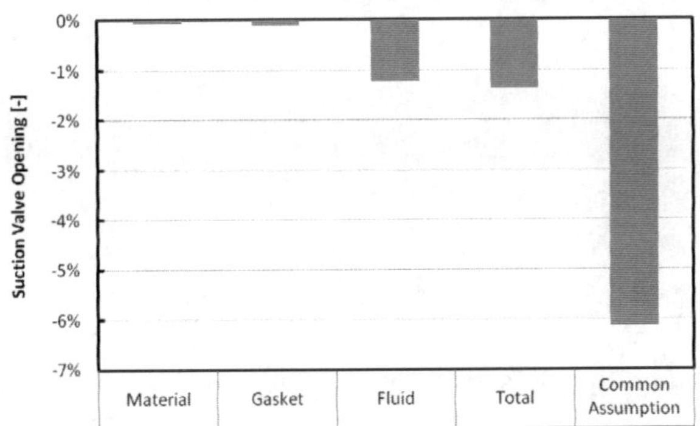

	Material	Gasket	Fluid	Total	Common Assumption
Suction	-0.06%	-0.10%	-1.22%	-1.37%	-6.14%

Figure 15: Effect of the damping coefficient on the suction valve opening.

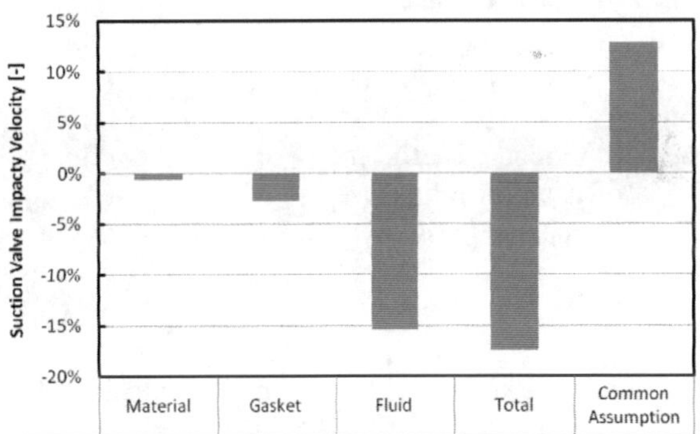

	Material	Gasket	Fluid	Total	Common Assumption
Série1	-0.63%	-2.73%	-15.45%	-17.40%	12.77%

Figure 16: Effect of the damping coefficient on the suction valve impact velocity.

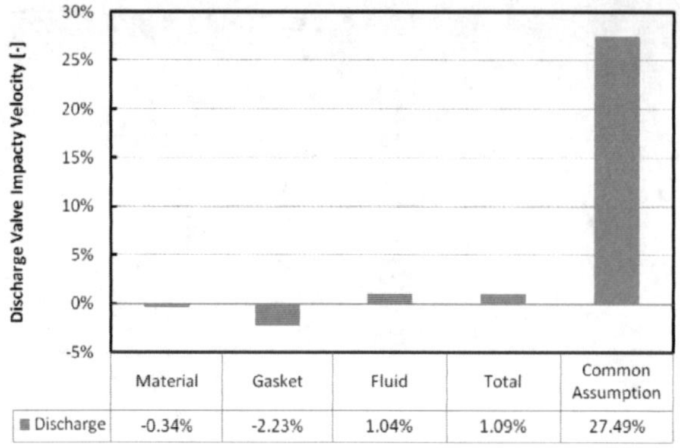

Figure 17: Effect of the damping coefficient on the discharge valve impact velocity.

6. CONCLUSIONS

The present investigation experimentally quantified valve damping under controlled conditions considering the effects of clamping geometry, clamping force, gasket thickness and the presence of fluid around the valve. The results showed that the damping due to the material and clamping is much less important than that associated with the fluid. It was observed that valve damping can be neglected without incurring significant errors in predictions of compressor efficiency. We found that the damping coefficients of valves of small reciprocating compressors have negligible effect on the valve bending stress. However, the damping force can change the valve impact velocity by 17%, which is important in the assessment of compressor reliability.

NOMENCLATURE

x	valve displacement	(m)	**Subscript**	
$\dot{x}$	valve velocity	(m/s)	*Valv*	valve
$\ddot{x}$	valve acceleration	(m/s²)	0	initial condition
F	force over the valve	(N)	ot	other
m	valve mass	(kg)	eq	equivalent
k	valve stiffness	(N/m)	n	natural
f	valve frequency	(Hz)		
γ	half damping and mass ratio	(1/s)		
t	time	(s)		
c	damping coefficient	(kg/s)		
ξ	damping ratio	(-)		
ρ	specific mass	(kg/m³)		

REFERENCES

Granick, N., Stern, J. (1965). Material Damping of Aluminum by a Resonant-Dwell Technique, NASA Technical Note, D-2893, Washington, D.C., USA.

Khalifa, H. E., Liu, X. (1998). Analysis of Stiction Effect on the Dynamics of Compressor Suction Valve, In: *Proc. Int. Compressor Eng. Conference at Purdue, West Lafayette, USA* (87-92).

Nashif, A. D., Jones, I. G., Henderson, J. P. (1985). Vibration Damping, Wiley, New York.

Prater Jr., G., Hnat, W.P. (2003), Optical Measurement of Discharge Valve Modal Parameters for a Rolling Piston Refrigeration Compressor, Measurement, Vol. 33, p. 75-84.

Shiomi, F. H., Takemori, C. K., & Deschamps, C. J. (2009). Comparative Analysis of Different Modeling Approaches for Valve Dynamics of Small Reciprocating Compressors. In: *Proc. 2009 IIR/IIF Int. Congress on Compressors, Papiernicka, Slovakia* (8 p.).

A Combined Experimental-Numerical Procedure to Estimate Leakage Gap of Compressor Valves

Gustavo C. REZENDE, Ernane SILVA, Cesar J. DESCHAMPS*

POLO Research Laboratories for Emerging Technologies in Cooling and Thermophysics
Federal University of Santa Catarina
88040-900, Florianopolis, SC, Brazil
deschamps@polo.ufsc.br

* Corresponding Author

ABSTRACT

Leakage through valves can significantly reduce the volumetric and isentropic efficiencies of compressors. Despite its importance in compressor design, the dimensional characterization of leakage gaps is not a trivial task. In this paper, we present a combined experimental-numerical method developed to estimate the leakage gap of compressor valves. Measurements of leakage were carried out via the constant volume method, which is widely employed in the analysis of gas flow through microchannels. Additionally, predictions were obtained with a one-dimensional flow model, taking into account viscous friction, slip at the walls, and gas compressibility. The leakage gap was adjusted in the simulation model so that predictions matched the measurements of leakage for different pressure differences. The procedure was applied for the analysis of three valve designs of refrigeration compressors.

1. INTRODUCTION

The efficiency of reciprocating compressors adopted in household refrigerators has dramatically increased in the last decades (Possamai and Todescat, 2004). However, further development is still necessary to allow the production of refrigeration systems in compliance with the ever-increasing energy-saving requirements. The main energy losses in compressors are usually classified in three groups: electrical losses, associated with inefficiencies of the electrical motor; mechanical losses, due to friction in bearings; thermodynamic losses, brought about by irreversibilities in the compression cycle. Leakage is a significant source of thermodynamic inefficiency and occurs in components of the compressor where high-pressure gas is not completely sealed.

Leakage in reciprocating compressors occurs through two main paths: (i) the clearance between the piston and the cylinder; (ii) the clearance between the valve and the seat. Most studies in the literature focus on leakage through the piston-cylinder clearance (Pandeya and Soedel, 1978; Ferreira and Lilie, 1984; McGovern, 1990). However, valve leakage is becoming increasingly important as compressors are manufactured with smaller dimensions.

This paper investigates leakage in valves that takes place through microchannels formed by geometric irregularities, which may result from macrogeometric characteristics, such as alignment and flatness; or from microgeometric aspects, such as surface roughness. The driving force for leakage in valves is the pressure difference between the compression and suction/discharge chambers. In addition to the valve-seat gap geometry, other aspects may contribute to leakage, namely the working fluid and valve/seat materials.

Some studies have demonstrated the relevance of valve leakage in compressor performance. For instance, Machu (1990) developed a model to account for the influence of leakage in a double acting cylinder compressor. By varying the leakage area in the suction and discharge valves, he obtained two main conclusions: (i) the volumetric efficiency can significantly drop with valve leakage; and (ii) superheating increases with valve leakage, decreasing the isentropic efficiency. Fujiwara (1998) evaluated the performance of an oil-free microcompressor and showed that a 1 μm gap in the valve would reduce the discharge pressure to half of the value that could be obtained with a completely sealed valve. Silva and Deschamps (2014) developed a simulation model to predict valve leakage in reciprocating compressors and its impact on the compressor volumetric and isentropic efficiencies. For a particular household compressor considered in their study, the authors showed that a valve-seat gap of 1 μm would reduce the volumetric and isentropic efficiencies by 2.7% and 4.4%, respectively.

Leakage in the valve-seat gap is associated with mass flow rates on the order of 10^{-5} kg/s. Under such small mass flow rates, the direct measurement of leakage is challenging and a more convenient approach is to employ accurate indirect measurement techniques used to analyze gas flow through microchannels. The three most used indirect measurement techniques are the drop method (Pitakarnnop *et al.*, 2010; Ewart *et al.*, 2006), the constant pressure method (Jousten *et al.*, 2002; McCulloh *et al.*, 1987) and the constant volume method (Yamaguchi *et al.*, 2011; Pitakarnopp *et al.*, 2010).

In this paper, we report an experimental setup based on the constant volume method to measure leakage in the valve-seat gap, $\delta(r)$, of reciprocating compressors, as depicted in Figure 1. As shown by Silva and Deschamps (2014), the varying gap between the valve and the seat, $\delta(r)$, can be expressed as a function of the valve bending due to pressure load and the reference gap at the edge of the valve port, δ_e, which will be denominated hereafter as 'leakage gap'. An experimentally validated model is used to estimate δ_e and hence characterize valve sealing. This combined procedure is used to analyze leakage in three valve designs under different pressure differences.

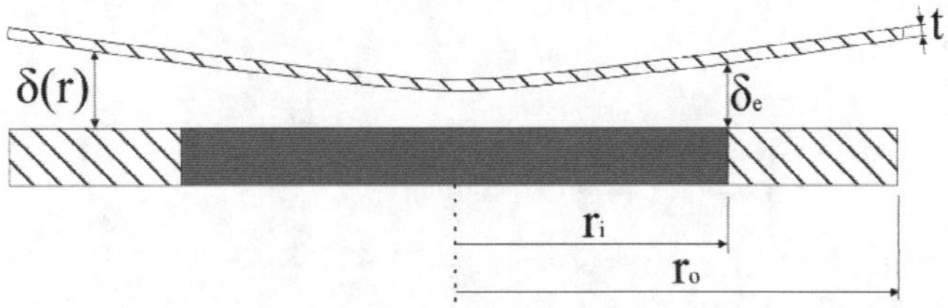

Figure 1: Cross-section of the valve system.

2. EXPERIMENTAL SETUP AND PROCEDURE

2.1 Constant Volume Method

We chose the constant volume method to measure valve leakage, since it is easier to implement than both the constant pressure method and the drop method. This method consists in monitoring the pressure p_r and temperature T_r inside a reservoir of known fixed volume V_r from which an ideal gas leaks. Hence, the mass of gas inside the reservoir, m_r, can be calculated from the equation of state for an ideal gas as a function of pressure and temperature:

$$m_r = \frac{p_r V_r}{R T_r}, \tag{1}$$

where R is the gas constant. By differentiating Equation (1) with respect to time, the mass flow rate of gas leaving the reservoir can be determined by:

$$\frac{dm_r}{dt} = -\frac{V_r}{RT_r}\frac{dp_r}{dt} + \frac{p_r V_r}{RT_r^2}\frac{dT_r}{dt}. \tag{2}$$

The equation above can be rewritten as

$$\frac{dm_r}{dt} = -\frac{V_r}{RT_r}\frac{dp_r}{dt}(1-\varepsilon), \tag{3}$$

where

$$\varepsilon = (dT_r / T_r)/(dp_r / p_r). \tag{4}$$

The parameter ε is a measure of how much the actual leakage process deviates from an isothermal process, i.e., $\varepsilon = 0$. Since very small temperature variations were observed in our experiments ($\varepsilon < 0.01$), Equation (3) was simplified to

$$\frac{dm_r}{dt} = -\frac{V_r}{RT_r}\frac{dp_r}{dt}. \tag{5}$$

It should be noted that dm_r/dt in the previous equations represent the gas leakage, which will be denoted hereafter by $\dot{m}$.

2.2 Experimental Setup

As shown in Figure 2, the test rig is composed by the tank TK, pressure gauge P, thermocouple T and five control valves V1-V5. The compressor mechanical kit with the valve plate contains the compression chamber CC, suction chamber SC and discharge chamber DC, as well as the compressor suction valve SV and discharge valve DV. The tubes connecting the components of the test rig were made of copper alloy. The volume formed by the tank, tubes, discharge and compression chambers of the mechanical kit will be referred herein as the reservoir R1, being subjected to pressure decrease due to the leakage through the valve-seat gap.

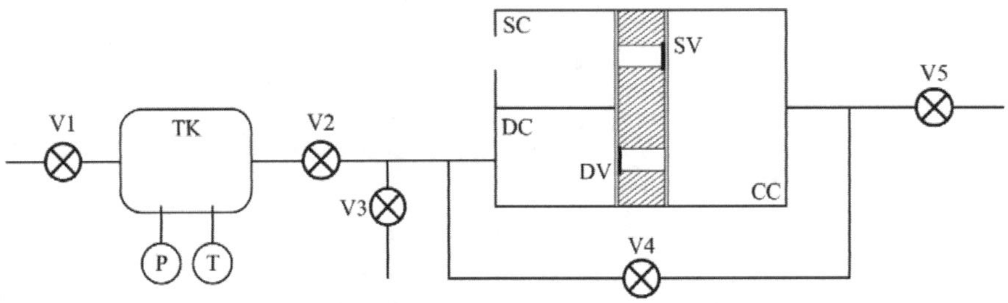

Figure 2: Experimental setup.

Leakage can be measured in one of the compressor valves (SV or DV) at a time, by modifying the setup of the test rig. The pressure in the discharge chamber (DC) does not depend on the test setup and is always equal to the pressure in the reservoir R1, varying from p_{max} to the atmospheric pressure. On the other hand, the pressure in the compression chamber (CC) can be adjusted to the pressure in the reservoir or to the atmospheric pressure, depending on the required setup. The suction chamber (SC) pressure is independent from the setup, since it is always open to the atmosphere.

The effective volume V_r of the reservoir R1 varies according to the valve being analyzed (SV or DV) and must be determined in each case. Accordingly, an auxiliary indirect measurement method was used to calculate V_r, which follows five steps: (i) Connect a second reservoir (R2) of known volume to the control valve V3; (ii) Set the pressure in the reservoirs R1 and R2 to pressure p_1; (iii) Close the control valve V3 to maintain the reservoir R2 at p_1 isolated from the reservoir R1; (iv) Increase the pressure in the reservoir R1 to p_2; (v) Open the control valve V3 to connect the reservoirs R1 and R2 again and let the system to establish the equilibrium pressure p_3 ($p_2 > p_3 > p_1$). By using Equation (1) and the aforementioned procedure, the volume of the reservoir R1 can be determined for the suction and discharge setups. Naturally, this procedure has to be applied to all suction/discharge compressor systems analyzed, since their dimensions are different.

2.3 Experimental Procedure

The control valves (V1-V5) are used to adjust the required pressure difference ($\Delta p = p_r - p_{atm}$) between the compressor chambers (SC, DC and CC). Each test consists in pressurizing the reservoir to p_{max} and then letting the gas to leak until the pressure in the reservoir $p_r = 2\ p_{atm}$. Table 1 shows the open/closed settings of the control valves for the suction and discharge setups. As can be seen, adjustments are required only in valves V4 and V5.

We took special care to manipulate the compressor components, especially valves and valve plates, since small scratches or oxidation on their surfaces may affect microscopic characteristics and, hence, the gas leakage. Therefore, these components were stored in recipients with silica gel and handled with rubber gloves.

Table 1: Control valve settings for the analysis of leakage in the suction and discharge valves.

Control valve	Suction setup	Discharge setup	Control valve	Suction setup	Discharge setup
V1	Closed	Closed	V4	Open	Closed
V2	Open	Open	V5	Closed	Open
V3	Closed	Closed			

Two types of tests were carried out: (a) total leakage; and (b) secondary leakage. The first procedure measures the total leakage, i.e., leakage in the compressor valve ($\dot{m}_l$) and unwanted leakages in the experimental workbench. In fact, the compressor mechanical kit is composed of many assembled parts and leakages through different paths are prone to happen. Such leakages will be referred as secondary leakage ($\dot{m}_s$). To mitigate secondary leakage, we tightened the screws always with an adequate torque and used actual compressor gaskets with silicon treatment. To measure the secondary leakages, we removed the compressor valves from the valve plate and sealed completely the suction and discharge orifices.

New samples of valve assemblies were used to measure the total leakage ($\dot{m}$). We noticed that leakage decreased as the tests were repeated and attributed this effect to deformation of the valve assembly, which resulted in better sealing. To mitigate this effect, we repeated the tests until the measurements yielded similar results. When this condition was met, five measurements of total leakage tests were carried out and valve leakage was subsequently obtained by subtracting the secondary leakage from the average value of total leakage, i.e.,

$$\dot{m}_l = \dot{m} - \dot{m}_s. \tag{6}$$

2.4 Measurement Uncertainties

The standard uncertainty of the valve leakage is a function of uncertainties associated with the volume V_r, temperature T_r and pressure drop rate dp_r/dt in the reservoir:

$$u^2(\dot{m}) = \left(\frac{\partial \dot{m}}{\partial V_r}u(V_r)\right)^2 + \left(\frac{\partial \dot{m}}{\partial T_r}u(T_r)\right)^2 + \left(\frac{\partial \dot{m}}{\partial (dp_r/dt)}u(dp_r/dt)\right)^2 \tag{7}$$

The reservoir volume V_r was evaluated from a series of indirect measurements, and its uncertainty is associated with test repeatability. The three valve designs used in the present investigation resulted different volume standard uncertainties: $u(V_{r1}) = 8.69\times10^{-6}$ m³; $u(V_{r2}) = 2.86\times10^{-5}$m³; $u(V_{r3}) = 1.34\times10^{-6}$m³. The temperature T_r was taken as the average value along the test, in line with the hypothesis of an isothermal process. The main uncertainty source of T_r arises from oscillations in the ambient temperature and the calibration of the thermocouples, resulting a standard deviation $u(T_r) = 6.2\times10^{-2}$ °C.

From experimental data of pressure in the reservoir as a function of time, local linear curve fitting was used to determine dp_r/dt for different pressure ranges. Therefore, the uncertainty of dp_r/dt is associated with this linear curve fitting with a coefficient of determination r^2> 0.99. The highest value found for $u(dp_r/dt)$ was 107 Pa/s.

As suggested by Equation (6), the valve leakage uncertainty is also affected by the secondary leakage uncertainty. Therefore,

$$u^2(\dot{m}_l) = u^2(\dot{m}) + u^2(\dot{m}_s). \tag{8}$$

Once the standard uncertainty is obtained, the expanded uncertainty $U(\dot{m}_l)$ is evaluated through a t-Student distribution:

$$U(\dot{m}_l) = t\,u(\dot{m}_l). \tag{9}$$

where t is the Student factor for a 95% confidence level with a number of degrees of freedom (ν) obtained from the Welch-Satterthwaite equation.

3. NUMERICAL MODEL

We adopted the simulation model developed by Silva and Deschamps (2014) to estimate the leakage gap at the edge of the valve port, δ_e. This section briefly explains this model and its validation based on comparisons between predictions and experimental data for flow of different gases through a metallic microtube (Silva *et al.*, 2016).

The model calculates leakage in compressor valves given the pressure difference across the valve and gas properties at the inlet. It solves the conservation equations for an ideal gas, assuming one-dimensional, steady, adiabatic, laminar flow through a channel of variable cross-sectional area. Reed valves are thin structures and the model calculates valve bending due to pressure load, which is required to define the flow geometry, as represented in Figure 1. The valve-seat gap can be regarded as a radial nozzle in which $\delta(r)$ is a function of both the local radius and the pressure load. On the other hand, the gap at the edge of the valve port, δ_e, can be regarded as the valve leakage gap. The valve-seat gap $\delta(r)$ is usually on the order of micrometer and rarefaction effects may affect the flow when $\delta_e < 0.25$ μm, as shown by Silva and Deschamps (2014).

Since the model was originally developed to predict leakage in compressor valves, we had to modify some equations in the model to allow its validation by comparing predictions with measurements of mass flow rate in a microtube. Moreover, the fanning friction factor required in the model was corrected for slip flow condition via the following expression:

$$C_f = \left(\frac{16}{\mathrm{Re}}\right)\left(\frac{1}{1+8\left[(2-\sigma_v)/\sigma_v\right]Kn}\right), \tag{10}$$

where the tangential accommodation coefficient σ_v was obtained from Silva *et al.* (2016), Re is the Reynolds number and Kn is the Knudsen number used to characterize flow rarefaction.

The experimental setup developed by Silva *et al.* (2016) allows measurements of mass flow rate in microchannels, including tests in the continuum and slip flow regime. The experimental rig consists of two reservoirs connected by a stainless steel microtube (diameter 438.6 ± 4.5 μm and length 9.22 ± 0.01 mm).

Pressure and temperature inside the reservoirs are monitored with pressure transducers and thermocouples. In order to reproduce rarefaction flow conditions that occurs in valve leakage, the pressure inside the test section has to be decreased. This is achieved by means of a vacuum pump connected to the reservoir located downstream of the microtube. In the reservoir located upstream a tank with gas at high pressure was connected.

Numerical and experimental results of mass flow rate were compared for three different gases (Nitrogen, R134a and R600a) and pressure ratio $\Pi = 3$. The relative difference between both results is given by

$$\Delta = \left|\frac{m_{\exp} - m_{mum}}{m_{\exp}}\right|, \tag{11}$$

where $\dot{m}_{exp}$ and $\dot{m}_{num}$ are the mass flow rates from measurements and predictions, respectively.

Figure 3 shows the relative difference between the results as a function of the Knudsen number. It should be mentioned that Equation (10) is valid for the continuum flow regime ($Kn < 10^{-2}$) and slip flow regime ($10^{-2} < Kn < 10^{-1}$). The agreement between numerical and experimental is acceptable for these range of Knudsen numbers, with a maximum difference of approximately 8%, and the model was assumed to be validated.

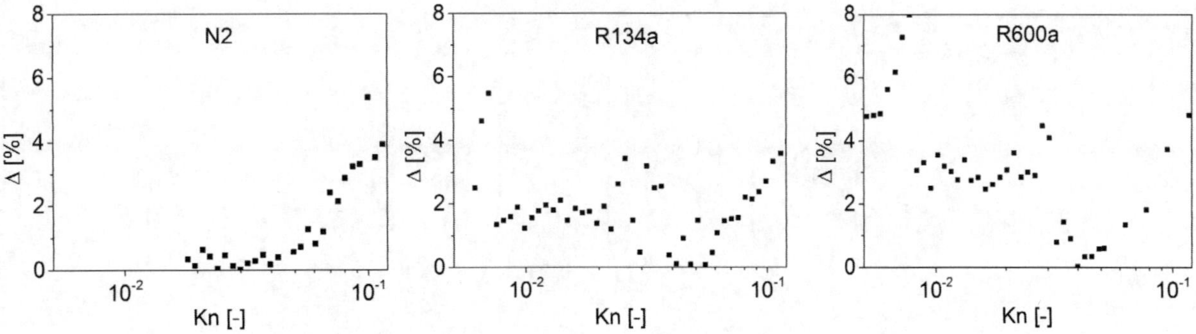

Figure 3: Difference between experimental and numerical results for mass flow rate in microtube.

4. RESULTS

We tested three valve system designs (D1, D2 and D3). The valve system D1 is adopted in a high-capacity compressor, and the valve systems D2 and D3 are employed in low-capacity compressors. We used nitrogen as the working fluid at the ambient temperature (T = 23 °C). The pressure differences set across the valves correspond to conditions typically found in refrigeration compressors.

4.1 Mass Flow Rate Measurements

Figures 4 to 7 present experimental (black squares) and numerical (gray circles) results for valve systems D1, D2 and D3. Measurements of leakages were made dimensionless with respect to the maximum leakage, $\dot{m}_1^*$, measured in the valve system D1. Figure 5 shows that the secondary leakages of valve system D1 is small compared to the valve leakage itself, not exceeding 7% of the valve leakage.

Figure 6 indicates that the leakage in the valve system D2 is much higher than the leakages in the other two valve systems (D1 and D3). Naturally, leakage has a greater impact on the efficiency of low-capacity compressors. Since the valve system D2 has the worst sealing and is adopted in a low-capacity compressor, a significant reduction of efficiency is expected.

An interesting behavior is observed in the measurements of leakage for the valve system D3. As can be seen in Figure 7, after reaching a local maximum at $\Delta p/\Delta p_{max} \cong 0.3$, the leakage is decreased to a local minimum at $\Delta p/\Delta p_{max} \cong 0.5$, before increasing again with the pressure difference. This behavior may be the result of combined effects in the valve system D3, such as valve bending and microgeometry interactions between the valve and seat surfaces.

4.2 Estimate of Leakage Gap

The gray circles in Figures 4, 6 and 7 show estimates of the leakage gap (δ_e) in the valve systems for different pressure differences. Such estimates were obtained by adjusting δ_e in the simulation model so as to make the predicted leakage equal to the measurement. The leakage gap can be regarded as a convenient parameter to assess the sealing of valve systems. However, caution should be taken when comparing valve systems of different dimensions, since the leakage area is as function of the valve orifice diameter.

The smallest leakage gaps were found for the valve system D1 while the valve system D2 had the greatest leakage gaps. Hence, it is natural the higher levels of leakage in the valve system D2, $\dot{m}_{L_2}$, in comparison to that of the valve system D1, $\dot{m}_{L_1}$, as shown in Figure 6. For instance, considering the same pressure difference ($\Delta p/\Delta p_{max} \cong 0.4$), the ratio between the leakage gaps of these two valve systems $(\delta_e)_2/(\delta_e)_1 \cong 8$ and $\dot{m}_{L_2} \cong 20\ \dot{m}_{L_1}$. It is worthwhile to note that the leakage gap decreases in all three valve systems as the pressure difference increases, suggesting that valve sealing is improved with the pressure load on the valve.

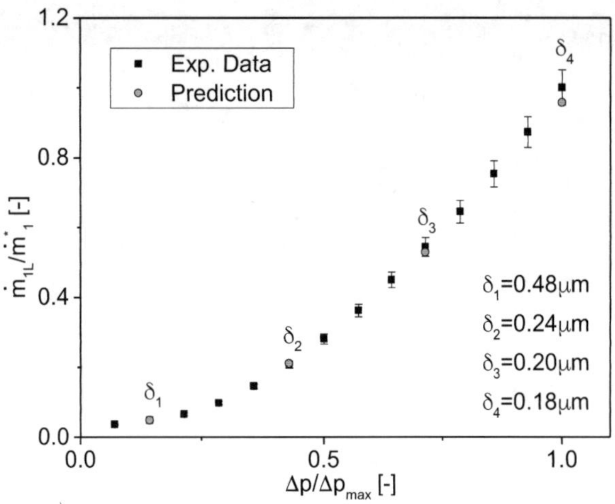

Figure 4: Leakage in the valve system D1.

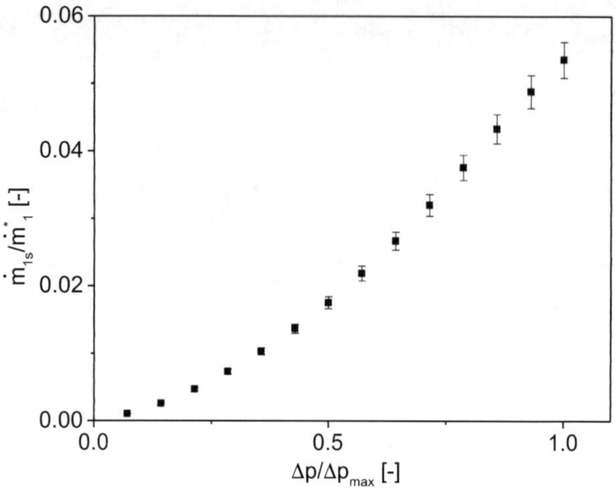

Figure 5: Secondary leakage in the valve system D1.

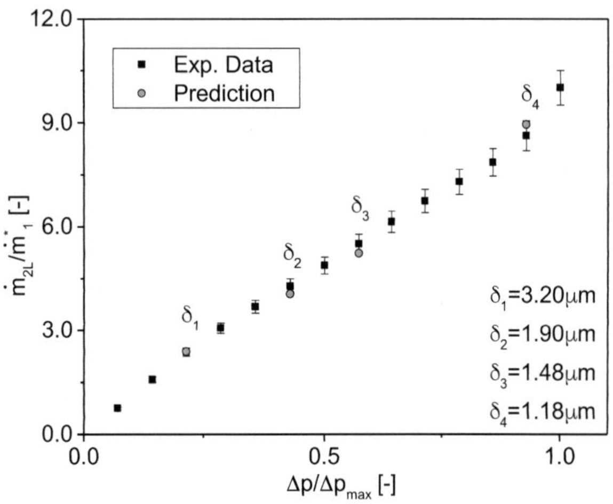

Figure 6: Leakage in the valve system D2.

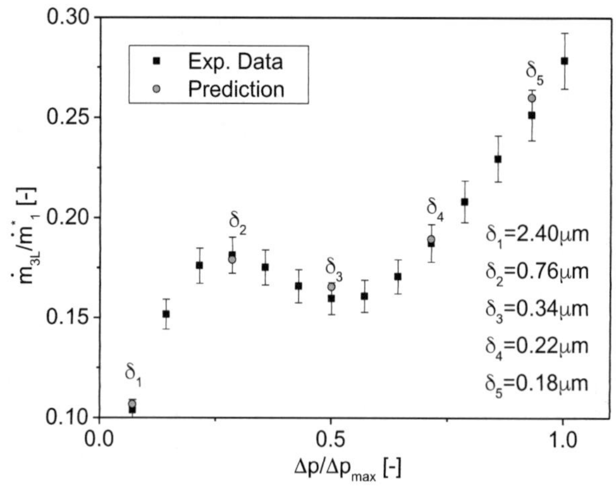

Figure 7: Leakage in the valve system D3.

5. CONCLUSIONS

Valve leakage can significantly reduce the efficiency of small reciprocating compressors, but the dimensional characterization of leakage gaps is not a trivial task. We developed and applied a combined experimental-numerical procedure to estimate the leakage gap of compressor valves. Measurements of leakage were carried out via the constant volume method and predictions of leakage gap were obtained with an experimentally validated simulation model, considering viscous friction, slip at the walls, and gas compressibility. The procedure was applied for the analysis of three valve designs, with leakage gaps being found in the range from 0.18 to 3.2 μm. We found that the leakage gap can be strongly affected by valve design and is reduced with the pressure load on the valve, probably due to valve bending and interactions between the valve and seat surfaces.

REFERENCES

Ewart, T., Perrier, P., Graur, I., & Méolans, J.G. (2006). Mass flow rate measurements in gas micro flows. *Experiments in fluids*, Vol. 41, p. 487-498.

Ferreira, R.T.S., & Lilie, D.E.B. (1984). Evaluation of the Leakage Through the Clearance Between Piston and Cylinder in Hermetic Compressors. *Proceedings of the International Compressor Engineering Conference at Purdue* (424).

Fujiwara, M., & Kazama, T. (1998). Performance Improvement of a Reciprocating Air Microcompressor. *Proceedings of the International Compressor Engineering Conference at Purdue* (1328).

Jousten, K., Menzer, H., & Niepraschk (2002). A new fully automated gas flowmeter at the PTB for flow rates between 10e-13 mol/s and 10e-6 mol/s. *Metrologia*, Vol. 39, p. 519-529.

Machu, E. (1990). How Leakages in Valves Can Influence the Volumetric and Isentropic Efficiencies of Reciprocating Compressors. *Proceedings of the International Compressor Engineering Conference at Purdue* (739)

McCulloh, K.E., Tilford, C.L., Ehrlich, C.D., & Long, F.G. (1987). Lowrange flowmeters for use with vacuum and leak standards. *Journal of Vacuum Science & Technology A*, Vol. 5, p. 376.

McGovern, J.A. (1990). Utilization of Volumetric Displacement in Reciprocating Compressors. *Proceedings of the International Compressor Engineering Conference at Purdue* (713).

Pandeya, P.N., & Soedel, W. (1978). A Generalized Approach Towards Compressor Performance Analysis. *Proceedings of the International Compressor Engineering Conference at Purdue* (257).

Pitakarnnop, J., Varoutis, S., Valougeorgis, D., Geoffroy, S., Baldas, L., & Cólin, S. (2010). A novel experimental setup for gas microflows. *Microfluid Nanofluid*, Vol. 8, p. 57-72.

Possamai, F.C., & Todescat, M.L. (2004). A Review of Household Compressor Energy Performance. *Proceedings of the International Compressor Engineering Conference at Purdue* (C067).

Silva, R.L., & Deschamps, C.J. (2014). Modeling of gas leakage through compressor valves. *International Journal of Refrigeration*, Vol. 53, p. 195-205.

Silva, E., Rojas-Cardenas, M., & Deschamps, C.J. (2016). Experimental analysis of velocity slip at the wall for gas flows of Nitrogen, R134a, and R600a through a metallic microtube. *International Journal of Refrigeration*, Vol. 66, p. 121-132.

Yamaguchi, H., Hanawa, T., Yamamoto, O., Matsuda, Y., Egami, Y., & Niimi, T. (2011). Experimental measurement on tangential momentum accommodation coefficient in a single microtube. *Microfluid Nanofluid*, Vol. 11, p. 57-64.

ACKNOWLEDGEMENT

The present study was developed as part of a technical-scientific cooperation program between the Federal University of Santa Catarina and EMBRACO. The authors also thank the support of CAPES and EMBRAPII Unit Polo/UFSC.

Integrating Numerical Models for Efficient Simulation of Compressor Valves

Ajay PARIHAR[1], David MYSZKA[2]*, Benjamin ROBINET[1], Thomas HODAPP[1]

[1]Emerson Climate Technologies, Applied Mechanics
Ajay.Parihar@Emerson.com, Benjamin.Robinet@Emerson.com, Thomas.Hodapp@Emerson.com

[2]University of Dayton, Department of Mechanical and Aerospace Engineering
Dayton, OH, USA
dmyszka@udayton.edu

* Corresponding Author

ABSTRACT

The motion of automatic, self-acting valves is a primary aspect in achieving superior compressor reliability and performance. Simulation of compressor valve mechanics involves a complex set of interactions that include characteristics of the compression chamber, thermodynamics, gas flow, valve motion, and pressure pulsations through the valve passages. One-dimensional (1D) lumped simulation models that encompass these interactions have been formulated and refined over the past several decades. During that same time period, finite element theory for fluid-structural-interaction (FSI) has been developed. FSI provides three-dimensional (3D) results across the entire fluid and solid domain and is well-suited for compressor simulation. However, the usage of FSI on real problems has not been adopted until much more recently, as sufficient computational resources became available. Still, the high computational expense and significant run times create a practical barrier to using FSI as a routine design tool.

This paper presents techniques for integrating 1D-lumped models with 3D-FSI models. Methods to properly formulate the 1D-models are discussed. Once formulated, these 1D-models provide quick and accurate results that are used to narrow various design alternatives. Additionally, the 1D-models provide initial conditions for the higher resolution 3D-FSI models in critical regimes of operation. Lastly, experimental data is shown to confirm the techniques.

1. INTRODUCTION

The design of valve systems has a strong influence on the cost, performance and reliability of compressors. The primary function of a valve is to permit gas flow and prevent reverse flow. For compressor efficiency, self-acting valves should have a quick response and not impede the forward flow. The durability of valves is threatened as they are subjected to reversed stress and impacts, repeated with each rotation of the crank shaft.

Numerical valve simulation has long been the focus of research projects. Over thirty years ago, Soedel (1984) compiled a short course that presented simulation strategies for the compression chamber, valve dynamics and gas pulsations through the suction and discharge passages. These models predict aggregate measures of compressor performance including mass flow, capacity, power consumption and efficiency. Additionally, the models predict pressure pulsation levels, valve stresses and impact velocities. These simulations are termed one-dimensional (1-D) lumped models as they assume one-dimensional flow or refrigerant, consider the average properties within the compression chamber and use minimal degrees of freedom to track the motion of the valves.

Several subsequent developments have been made to the original 1-D-lumped models outlined by Soedel. Fagotti et al. (1996) increased the degrees of freedom of reed valve motion model by adopting a finite element approach. Bukac (2002) accounted for the flexibility of the valve lift-limiter (stop) and seat. Khalifa and Liu (1998) derived an expression for adhesion force caused by the interfacial surface tension of the lubricating oil film between the valve and seat, which is often termed a stiction force. Pizarro-Recabarren et al. (2012) expanded on that expression to consider the existence of a finite amount of oil between the valve and the seat.

A second, more recent, option for compressor valve simulation is a finite element analysis scheme that is adapted for systems that involve fluid-structural-interaction (FSI). FSI is encountered when fluid flow causes deformation of a structure, which alters the boundary conditions of the fluid flow. With the development of high performance computing, it has become possible to analyze complex systems with a coupling of fluid and structural dynamics. For a compressor, FSI offers a refined examination of the three-dimensional (3D) variation of mechanical properties across the entire fluid and solid domains.

Several studies using FSI models to simulate compressor behavior are found in the literature. Bassiouny and O'Neal (2004) first demonstrated the use of FSI analysis on refrigerant flow through a short flexible tube. Kim et al. (2008) expanded an FSI to a reciprocating compression chamber with a focus on discharge valve impacts. Wu et al. (2012) showed the importance of including the suction and exhaust systems within the FSI analysis.

While the fidelity of the FSI results are superior, the computational expense prevents it being used as a routine design tool. A pragmatic approach is to use a combination of the FSI and 1-D lumped models to explore compressor design options. Such an approach advocates that the 1-D lumped model be used for broad design studies, while the FSI model is used to make final adjustments. Mayer et al. (2014) conducted a study to compare the results from an FSI model and a basic 1-D lumped model. This paper presents strategies for both the FSI and 1-D lumped models to facilitate convergence between the methods and agreement with experimental results. The remainder of the paper is organized as follows. Section 2 presents the 1-D lumped models used by the authors. Section 3 shows an overview of FSI. Section 4 presents the modeling techniques used in three examples that illustrate correlation between 1-D lumped models, FSI, and experimental results.

2. ONE-DIMENSIONAL LUMPED SIMULATION MODEL

A schematic of a 1D-lumped model for a single cylinder reciprocating compressor is shown in Fig. 1. Since many versions of a 1-D lumped models have been implemented and discussed in the literature, the methods used by the authors is detailed in the following sections. While a reciprocating compressor is illustrated, the modeling approach applies to any compression technologies.

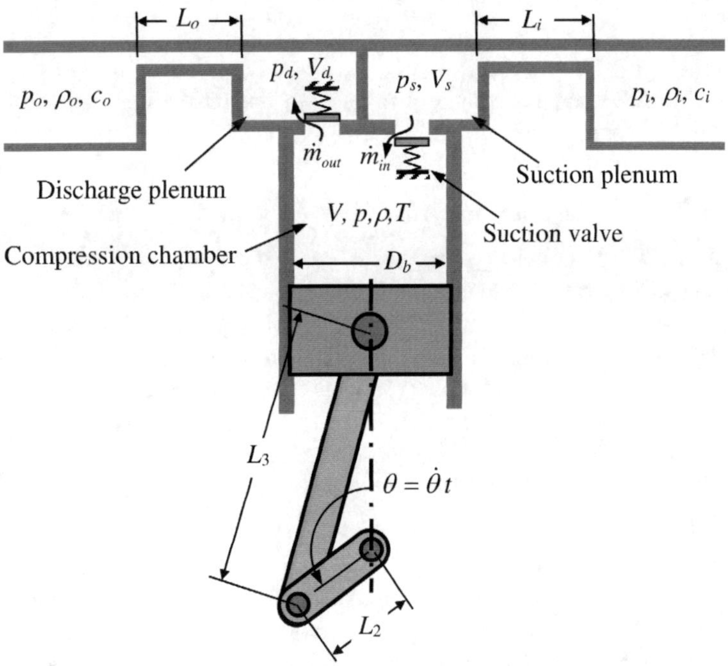

Figure 1: Schematic of a 1D-lumped model for a single cylinder reciprocating compressor.

2.1 Compression Chamber Volume

Closed form equations for the volume of a compression chamber should be generated. For the case of a reciprocating compressor, the volume at a crank angle θ with a bore diameter D_b, crank length L_2, connecting rod length L_3, and clearance volume V_c is

$$V = V_c + \frac{\pi D_b^2}{4}\left[L_2(1-\cos\theta) + L_3\left(1-\sqrt{1-\left(\frac{L_2}{L_3}\sin\theta\right)^2}\right)\right] \quad (1)$$

The time derivative of Eq. (1) is a necessary value and is

$$\frac{dV}{dt} = \frac{dV}{d\theta}\frac{d\theta}{dt} = \frac{\pi\,\dot{\theta}\,D_b^2\,L_2}{4}\left(\sin\theta + \frac{L_2\sin\theta\cos\theta}{\sqrt{L_3^2 - L_2^2\sin^2\theta}}\right) \quad (2)$$

where $\dot{\theta}$ is the speed of the crank shaft.

2.2 Compression Chamber Density

Conservation of mass within the compression chamber is used to derive an expression for the density ρ. Considering the change of mass $dm/dt = d(\rho V)/dt$, and expanding gives

$$\frac{d\rho}{dt} = \frac{1}{V}\left[\dot{m}_{in} - \dot{m}_{out} - \rho\frac{dV}{dt}\right] \quad (3)$$

2.3 Compression Chamber Pressure and Temperature

Energy balance models are commonly used to generate expressions for the temperature T within the compression chamber (Chen et al., 2002). However, obtaining reliable heat transfer coefficients through the cylinder walls can prove to be problematic. Alternatively, the polytropic compression model, $p/\rho^{\gamma} = \text{constant}$, can be used to develop an expression for the pressure p within the compression chamber,

$$\frac{dp}{dt} = \frac{p\gamma}{\rho}\frac{d\rho}{dt} \quad (4)$$

where γ is a polytropic exponent that can be judiciously predicted (Lenz, 2002). The corresponding temperature within the compression chamber can be determined from the density and the pressure by implementing equations of state, with queries to a refrigerant property database, such as REFPROP (Lemmon et al., 2010), or lookup tables (Laughman et al., 2012).

2.4 Flow through Suction and Discharge Ports

A one-dimensional, isentropic flow model is commonly applied to generate expressions for the flow through the ports. The mass flow through the suction port is

$$\dot{m} = \rho_u\,(CA)\,c\sqrt{\frac{2}{\kappa-1}\left[\left(\frac{p_u}{p_d}\right)^{1-\kappa/\kappa} - 1\right]} \quad (5)$$

where p_u is the upstream pressure, p_d is the downstream pressure, C is the flow coefficient, A is the effective flow area, c is the speed of sound in the refrigerant, and κ is the ratio of specific heats (c_v/c_p). Schwerzler and Hamilton (1972) developed a widely-adopted method for determining appropriate flow area CA values for the ports as a function of valve displacement. Equation (5) is applied to both the suction and discharge valves.

2.5 Valve Motion

Some compressors use translating poppet valves that can be accurately modeled as a single degree of freedom. The displacement w is governed by

$$M\,\ddot{w} + C_1\,\dot{w} + C_2\left|\dot{w}\right|\dot{w} + K\,w = F(t) \quad (6)$$

where K is the stiffness of a valve spring, M is the equivalent valve mass (which includes a portion of the valve spring), $\dot{w}$ is the valve velocity, and $\ddot{w}$ is the valve acceleration. Damping is partitioned into a linear term to simulate structural or viscous damping and a second-order term to simulate high-velocity flow damping. Both terms are represented by coefficients C_1 is C_2, respectively. As with CA, Schwerzler and Hamilton (1972) developed an expression for the force on a valve as

$$F(t) = FA\left(p_u - p_d\right) \quad (7)$$

and outlined a process for determining effective force area FA values for the ports as a function of w. Equation (6) is applied to both the suction valve (with displacement w_s) and discharge valve (with w_d).

Translating disc style valves have a single degree-of-freedom (dof) and naturally behave according to Eq. (6). Flexible valves can be modeled as an equivalent single dof system, or with a finite element approach as discussed below. Expressing the valve velocity as $\dot{w}$, constraints must be applied to enforce that the valve does not penetrate the valve seat or stop, if one is used. Thus, $0 < w < w_{max}$, where w_{max} is the maximum valve displacement permitted by a valve stop. Additionally, it is common to assume that the valve bounces off the seat or backer with a coefficient of restitution C_R. Thus, if $w = w_{max}$ or $w = 0$, then $\dot{w} = -C_R\,\dot{w}$. The conditions are enforced for the suction and discharge valves.

A finite element approach can be used to more accurately model the distributed mass and elasticity of the flexible valves. This approach is particularly useful when the boundary conditions of the attachments and backer are difficult to approximate with a single degree of freedom. In this case, the coefficients of Eq. (6) become matrices. A two-node plane Bernoulli-Euler beam element that has length ℓ_v^e, thickness t_v, average element width b_v^e and uniform mass density ρ_v has element mass and stiffness matrices as

$$M^e = \frac{\rho_v t_v b_v^e \ell_v^e}{420}\begin{bmatrix} 156 & 22\ell & 54 & -13\ell \\ 22\ell & 4\ell^2 & 13\ell & -3\ell^2 \\ 54 & 13\ell & 156 & -22\ell \\ -13\ell & -3\ell^2 & -22\ell & 4\ell^2 \end{bmatrix} \qquad K^e = \frac{E_v b_v^e (t_v)^3}{12(\ell_v^e)^3}\begin{bmatrix} 12 & 6\ell & -12 & 6\ell \\ 6\ell & 4\ell^2 & -6\ell & 2\ell^2 \\ -12 & 13\ell & 12 & -6\ell \\ 6\ell & 2\ell^2 & -6\ell & 4\ell^2 \end{bmatrix} \tag{8}$$

The element matrices in Eq. (8) are assembled into a global set of equations and reduced according to boundary conditions.

2.6 Plenum Pulsations

With a finite volume of the suction and discharge plenum and intermittent flow through the ports, a time-varying pressure pulse is created behind the valves. A common model for the passages consists of a cavity connected by a relatively short tube into a larger collecting tank with steady conditions. The Helmhotz principle can be used to model the motion of a slug of gas within the tube (Soedel, 1973). Considering the discharge passage with discharge volume V_d, a tube length L_o, area A_o terminating with steady pressure, density and speed of sound conditions of p_o, c_o, ρ_o, the resonator equation is

$$L_o A_o \rho_o \ddot{\xi} + D_o \dot{\xi} + \frac{c_o^2 \rho_o A_o^2}{V_d}\xi = \frac{c_o^2 A_o}{V_d}\int_0^t \dot{m}_{out}\,dt \tag{9}$$

where ξ is the movement of gas through the tube. Equation (9) is also written for the suction passage and can be solved having the mass flow rate through the valves in Eq. (5). The time-varying pressure behind the valve is obtained once the movement of gas through the passage ξ is known.

$$p_d = \frac{c_o^2}{V_d}\int_0^t \dot{m}_{out} - \frac{c_o^2 \rho_o A_o}{V_d}\xi + p_o \tag{10}$$

2.7 Valve Stiction

Adhesion (or stiction) is a source of thermodynamic and volumetric losses in the compressor as a higher pressure difference is required to open the valve, which subsequently delays in the valve opening. Stiction is attributed to the presence of a lubricant oil film between the valve and the seat. The stiction force results from a combination of interfacial tension, capillarity effects and deformation (viscous flow) of the oil film. An expression (Khalifa and Liu, 1998) depends on the valve lift, valve velocity, lubricant viscosity μ, effective inner and outer contact radii of the valve and seat, r_i and r_o.

$$F_{st} = \mu (r_o - r_i)^3 (r_o + r_i)\frac{\dot{w}}{w^3} \tag{11}$$

The stiction force of Eq. (11) will appear as an additional term on the right side of Eq. (7).

2.8 Solution

Equations (3), (4), (7) and (9) are time-based, ordinary differential equations (ODEs) of the form $dx(t)/dt = f(x(t))$, where $x(t) = [\rho, p, w_s, \dot{w}_s, w_d, \dot{w}_d, \xi_s, \dot{\xi}_s, \xi_d, \dot{\xi}_d,]$. Explicit numerical methods are commonly used to solve ODEs of this form. To start the solution process, a set of initial conditions $x(0)$ must be specified. For reciprocating compressor simulation, TDC is commonly used. The Runge-Kutta technique is a commonly used explicit method to progressively step through a time interval by determining a new set of simulation values $x(t_{n+1})$ using the set of values already known $x(t_n)$. The routine iterates through multiple crank rotations, checking whether the conditions at 0° are equal to those at 360°, $x(t_{0°}) = x(t_{360°})$. Additionally, the total mass flow leaving the compressor must equal to the total mass flow entering the compressor.

3. THREE DIMENSIONAL, FLUID-STRUCTURAL INTERACTION MODELS

In FSI analysis, fluid forces are applied onto the solid and the solid deformation changes the fluid domain. The computational domain is typically divided into separate discretized fluid and solid domains. The fluid and solid models are defined respectively, through their material data, boundary conditions, etc. Fig. 2 shows a flow chart of the solution process.

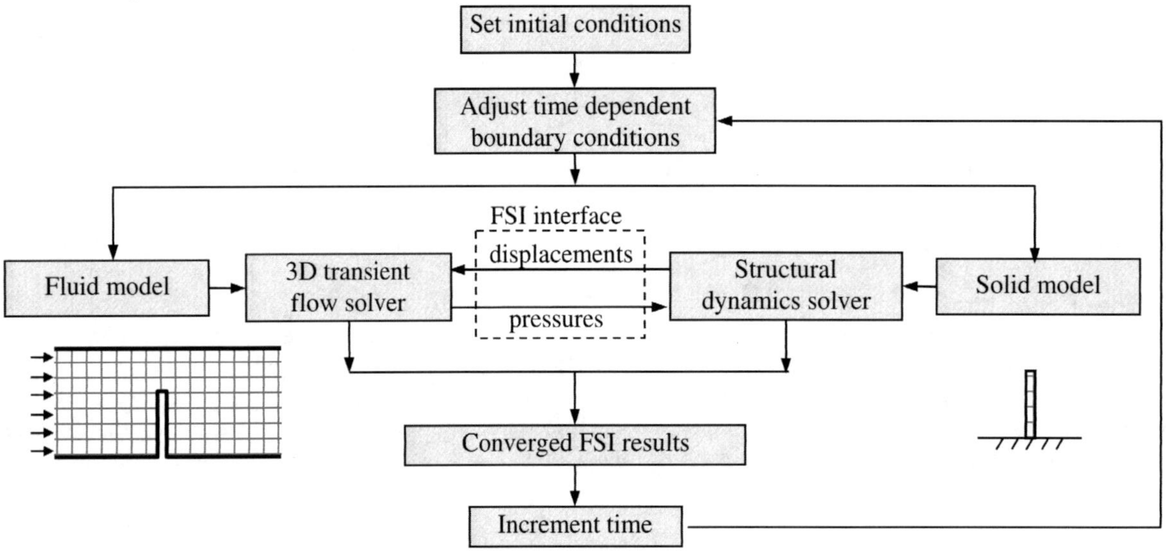

Figure 2: FSI simulation flow chart.

3.1 Transient Fluid Dynamics

The computational fluid dynamics (CFD) model is defined with a spatial fluid domain, wall boundary conditions, and the fluid-structure interface condition. In addition to the interface condition, the fluid domain may have prescribed changes with time (ie., a moving mesh).

Fluid solver involves the Navier-Stokes equations, which are expressed as a continuity equation, three directional momentum equations and an energy equation. In order to obtain a closed system for solution variables, additional equations, such as equations of state and turbulence models for the fluid under consideration must be provided.

3.2 Structural Dynamics

The structural finite element analysis (FEA) model involves the valve components, including the reed, attachment hardware, and backer. The structural dynamics solver involves a traditional, transient dynamic finite element method similar to that shown in Eq. (8).

3.3 Solution Parameters

Most commercial packages use an iterative coupling approach as shown in Fig. 2, where the equations governing the flow and the displacement of the structure are solved separately with two distinct solvers. Apart from standard boundary conditions applied, an FSI boundary is declared at the interface of fluid and solid models to transfer the loads in two-way coupled fashion. FSI simulation requires smaller time step size to resolve the flow gradients and contact convergence on solid side for valve applications. Sufficient number of iterations is given to converge the solution per time step as well as to converge load transfer across two solvers. Higher order discretization schemes in space and time is used for better accuracy of results.

4. TECHNIQUES DEVELOPED FOR USING THE MODELS

Strategies for using the 1-D lumped model and FSI model to facilitate convergence between the methods and agreement with experimental results is presented.

4.1 Techniques for using the 1D-Lumped Model

As an application example, a three-cylinder compressor shown in Fig. 3 contains a flexible suction valve and translating discharge valve. When the objective of the simulation is to determine the performance implications of valve size and arrangements, an equivalent single-dof valve model is deemed sufficient.

Figure 3: Reciprocating compressor with a flexible discharge valve and translating (Discus) discharge valve.

For flexible reed valves, the centroid of the port is used as the reference point for displacement coordinate. Thus, all distributed properties are lumped to that point on the valve.

The equivalent valve stiffness can be determined through classical beam theory, finite element analysis, or experimentally if physical parts are available. Experimental determination of stiffness is preferred when stacked valve components are used, as there is uncertainty in the friction values between the stacks. The equivalent valve mass can be determined through energy-equivalency methods or FEA modal analysis.

Contact with valve stops alter the boundary conditions for the valve dynamics. The equivalent single-dof lumped valve stiffness and lumped mass is determined within the bounds of the various boundary conditions. A lookup table is generated to provide appropriate values at increments of valve lift, to distinguish changes in the constraints.

The effective flow area *CA* and effective force area *FA* approach is incorporated. Instead of using an experimental setup described by Schwerzler and Hamilton (1972), CFD models are used to determine the flow area *CA* and the force area *FA* at various valve displacements. A steady-state analysis is performed on a fluid domain constructed upstream, downstream and through the port, with the valve set at a prescribed displacement. An arbitrary mass flow and downstream pressure is specified. Using the CFD results, the upstream pressure and force exerted onto the valve model is tabulated. Values for *CA* and *FA* at that prescribed displacement are determined from Eqs. (5) and (7).

Using the approach above, aggregate measures of compressor performance, namely mass flow and power, have typically been within 3%. Figure 4 illustrates comparison of the 1-D model with a compressor that was instrumented with a pressure transducer. It is noted that for compressors with low volumetric clearance, the volume under the valve as it displaces is included in the compression chamber, which improve correlation.

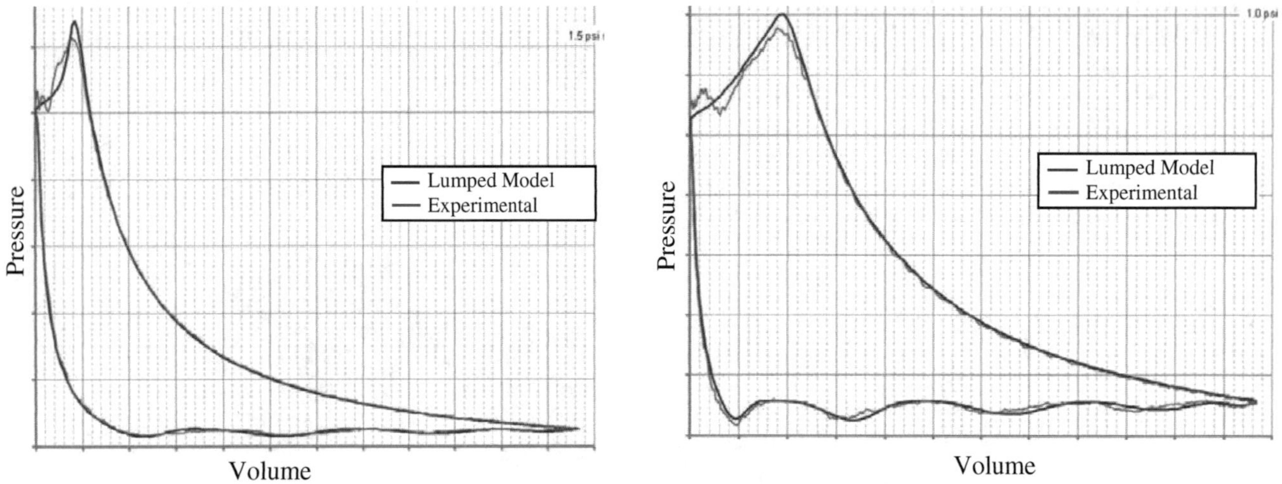

(a) High pressure ratio condition (b) Medium pressure ratio condition

Figure 4: 1-D lumped model comparisons with experimental pressure-volume data.

To more accurately determine stress and impact velocities of flexible valves, a finite element approach of Eq. (8) replaced the single dof valve model. For Eq. (7), the forcing function was distributed across the area projected by the port. Comparison of a multiple dof valve within the lumped model is provided in the following section.

4.2 Techniques for using the FSI Model

A second application example is shown in Fig 5, containing a four-cylinder compressor with two flexible suction and two flexible discharge valves in each cylinder. The focus of the study is to predict dynamic stresses and impact of the suction valve by performing three-dimensional transient FSI simulation

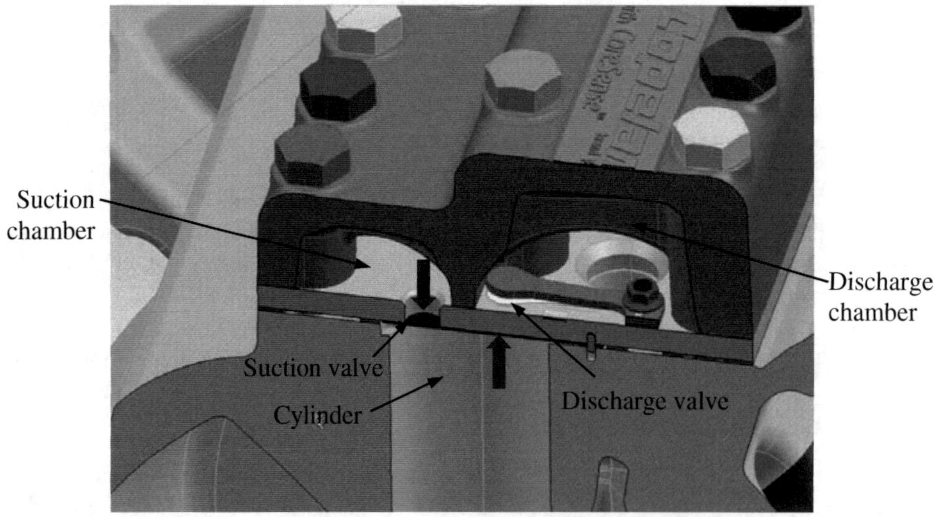

Figure 5: Refrigeration compressor arrangement of cylinder, suction and discharge valves and passages.

Due to symmetry, half a cylinder was modeled. A fluid domain included the inlet plenum chamber, valve passge, and compression chamber as shown in Fig. 6a. The compression chamber had a time-varying moving mesh that was synchronized with movement of the top of the cylinder. With a symmetry model, a single suction reed valve was model, as shown in Fig. 6b. Boundary conditions included a fixed end and valve stop. As is typical for a compressor FSI model, the fluid domain included nearly 50 times more elements as the solid mesh. The surfaces exposed to the fluid were defined as the FSI interface. Zero shear was specified at the symmetry face. All other faces were defined as no-slip boundary condition.

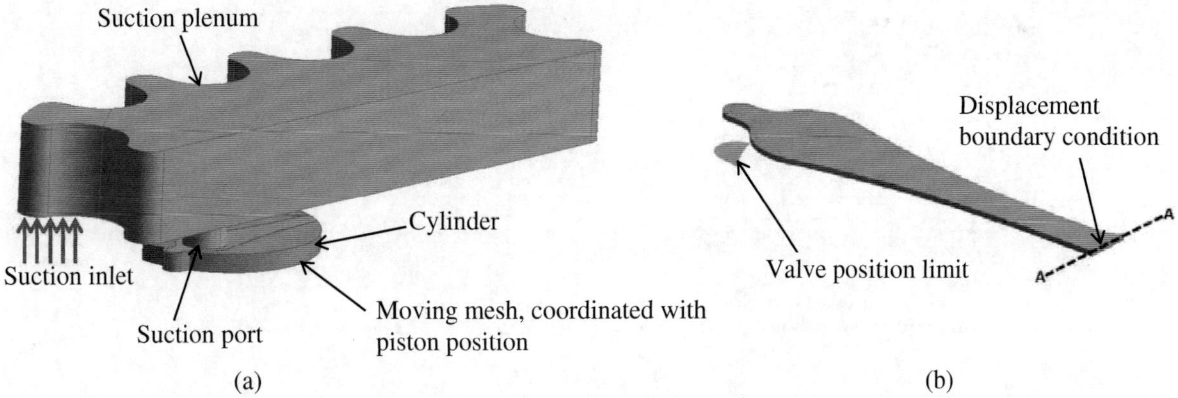

Figure 6: Fluid (a) and structural (b) models for compressor FSI analysis

Suction pressure and temperature were specifed as inlet boundary condition as shown in Fig 6a. The top dead center position of the piston served as the initial condition. The suction pressure and temperature were used with REFPROP to provide initial refrigerant properties. Flow was modeled as turbulent using standard k-epsilon model along with Redlich-Kwong, real gas density formulation (Redlich, Kwong, 1949). The SIMPLE scheme (Patankar et. al., 1972) is used for pressure-velocity coupling, whereas mesh displacement is solved using a direct sparse solver (Angeleri et al., 1989) on the structural side. To speed up the overall simulation time, a variable time step size is used by keeping smaller time step size at the time of valve opening and closing to resolve the steep flow gradients.

A lumped model of the compressor was also constructed using the techniques described in the previous section. A finite element approach (ie., Eq. (8)) was used to model the valve, using 10 nodes. The displacement of each node was limited to motion between valve plate and stop. Figure 7 illustrates comparison of valve displacement and velocity from the FSI model and the lumped model.

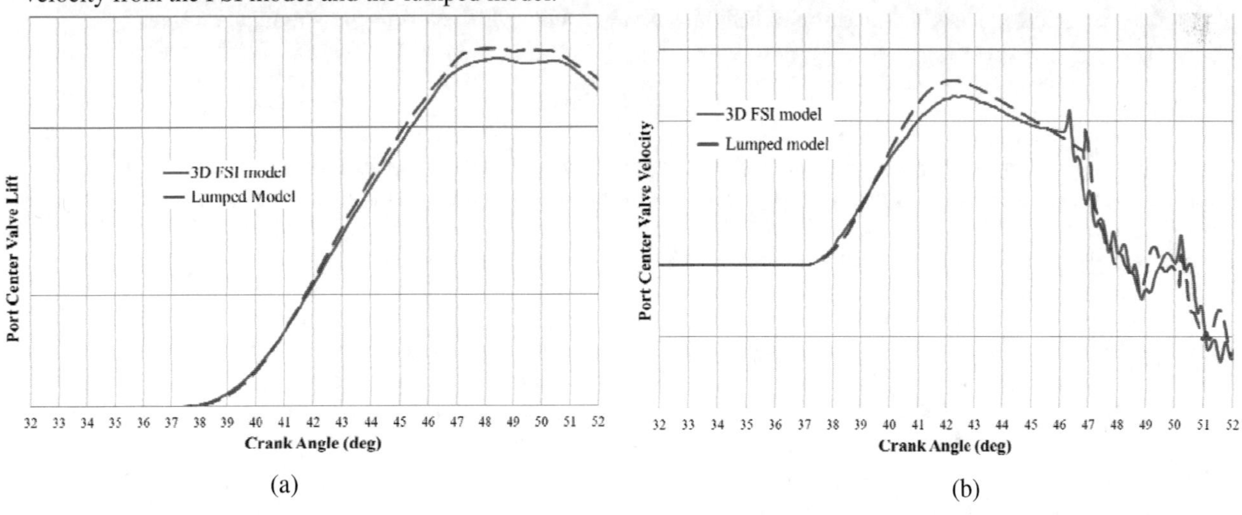

Figure 7: Comparison of (a) displacement and (b) velocity results for lumped and FSI models.

The FSI methodology described above was used to study the flexible discharge reed dynamics. FSI result comparisons were made with a compressor equipped with strain gages on the discharge reed. Three gages were placed (1) at the port center, (2) at mid-length on the edge of the reed, and (3) at the clamped end on the edge of the reed. The comparison of the FSI and strain gage results are shown in Fig 8. The FSI model results were within 3%, 5%, and 7%, respectively.

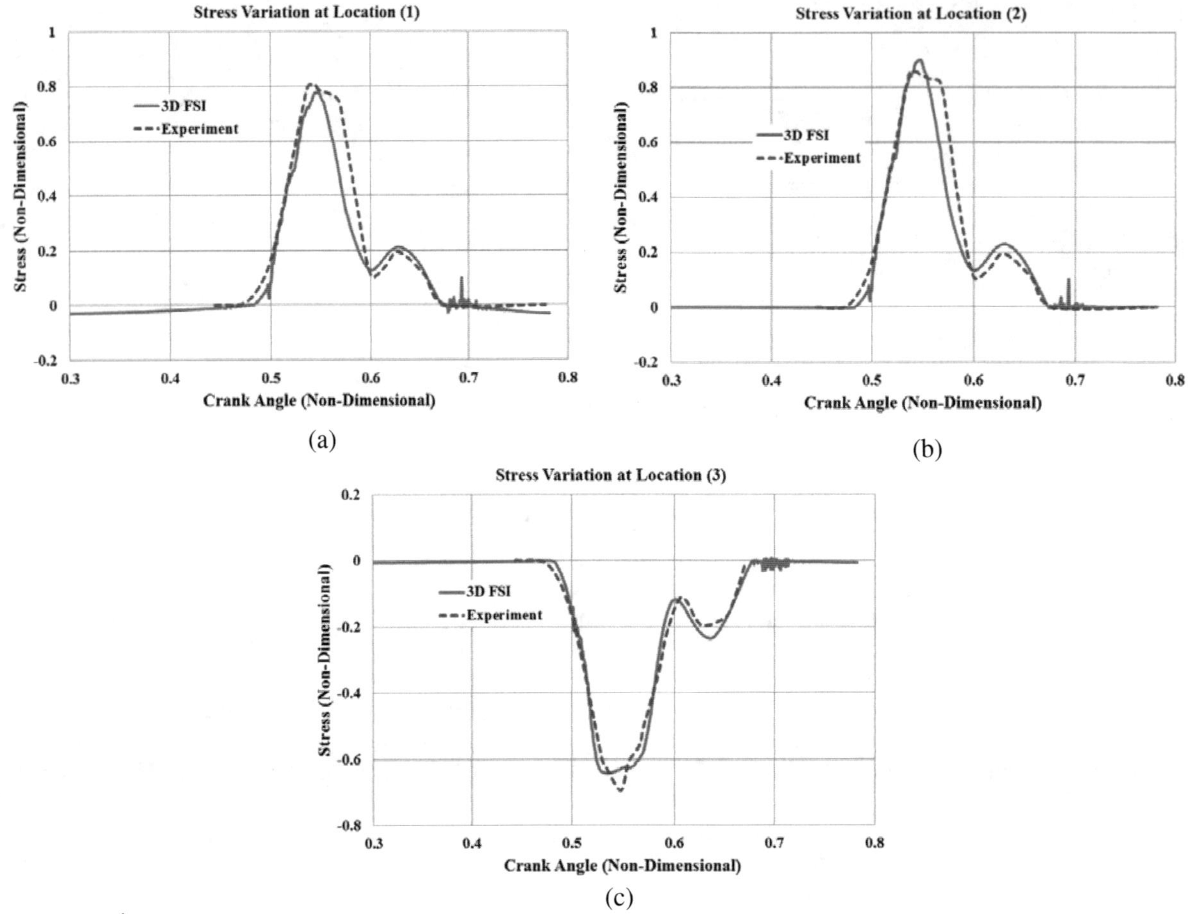

Figure 8: Comparison of FSI valve strain results with experimental strain gage values

5. CONCLUSIONS

This paper presented techniques for compression, flow and valve simulations using both 1D-lumped and 3D-FSI models. Detailed equations and methods to effectively formulate the lumped models are provided. Approaches for using commercial FSI software are discussed. Correlations between the lumped model, 3D-FSI model, and experimental data are shown to confirm the techniques.

This paper illustrates that both models can be used in complimentary fashion within a commercial product development environment. The FSI models provide a superior fidelity, but come at a computational cost. At this time, it is unreasonable to complete a full cycle FSI analysis on numerous design alternatives. Properly formulated, lumped models provide quick and accurate results that can be used to narrow various design alternatives. Additionally, the lumped models can be used to provide initial conditions for the higher resolution 3D-FSI models to be solved in critical regimes of operation.

REFERENCES

Angeleri, F., Sonnad, V., Bathe, K-J. (1989) Studies of finite element proceduresan evaluation of preconditioned iterative solvers, *Computers and Structures*, Vol 32, No. 3-4, pp. 671-677.

Bukac, H. (2002) Understanding Valve Dynamics, *Proc. of the Int. Compressor Eng. Conf. at Purdue*, Paper No. 1564.

Baassiouny R., O'Neal, D. (2004) Analysis Of Refrigerant Flow and Deformation for a Flexible Short-Tube using a Finite Element Model, *International Journal of Refrigeration*, Vol. 27, No. 1, pp. 176-183.

Chen, Y., Halm, N.P., Groll, E.A., Braun, J.E. (2002) Mathematical modeling of scroll compressors - part i: compression process modeling, , *International Journal of Refrigeration,* Vol. 25, No. 6, pp. 731-750.

Fagotti, F., de Bortoli, M., Barbieri, R. (1996) A Finite Element Approach to Compressor Valves Motion Simulation, *Proc. of the Int. Compressor Eng. Conf. at Purdue*, Paper No. 1132.

Khalifa, H., Liu, X. (1998) Analysis of Stiction Effect on the Dynamics of Compressor Suction Valve, *Proc. of the Int. Compressor Eng. Conf. at Purdue*, Paper No. 1221.

Laughman, C., Zhao, Y., Nikovski, D. (2012) Fast Refrigerant Property Calculations Using Interpolation-Based Methods, *Proc. of the Int. Compressor Eng. Conf. at Purdue*, Paper No. 1344.

Lenz, J. (2002) Polytropic Exponents for Common Refrigerants, *Proc. of the Int. Compressor Eng. Conf. at Purdue*, Paper No. C9-3.

Lemmon, E.W., Huber, M.L., McLinden, M.O. (2010) *NIST Standard Reference Database 23: Reference Fluid Thermodynamic and Transport Properties-REFPROP*, Version 9.0, National Institute of Standards and Technology.

Mayer, J., Bjerre, P., Brune, F. (2014) A Comparative Study Of Different Numerical Models For Flapper Valve Motion, *Proc. of the Int. Compressor Eng. Conf. at Purdue*, Paper No. 2309.

Patankar, S., Spalding, D. (1972), "A calculation procedure for heat, mass and momentum transfer in three-dimensional parabolic flows", *Int. J. of Heat and Mass Transfer*, Vol. 15, No. 10, pp. 1787-1806.

Pizarro-Recabarren, R., Barbosa, J., Deschamps, C. (2012) Modeling the Stiction Effect in Automatic Compressor Valves, *Proc. of the Int. Compressor Eng. Conf. at Purdue*, Paper No. 2044.

Redlich, O., Kwong, J. (1949). On Thermodynamics of Solutions, *Chemical Reviews,* Vol. 44, No. 1, pp. 233–244.

Soedel, W. (1984) *Mechanics, Simulation and Design of Compressor Valves, Gas Passages and Pulsation Mufflers,* Ray W. Herrick Laboratories, School of Mechanical Engineering, Purdue University.

Soedel, W. (1984) On Helmholtz resonator effects in the discharge system of a two-cylinder compressor, *MJ. of Sound and Vibration,* Vol. 30, No. 3, pp. 263-277.

Schwerzler, D., Hamilton, J. (1972) An Analytical Method for Determining Effective Flow and Force Areas for Refrigeration Compressor Valving Systems (1972). *Proc. of the Int. Compressor Eng. Conf. at Purdue*, Paper No. 5.

Kim, H., Ahn, J., Kim, D. (2008) Fluid Structure Interaction and Impact Analyses of Reciprocating Compressor Discharge Valve, *Proc. of the Int. Compressor Eng. Conf. at Purdue*, Paper No. 1936.

Wu, S., Wang, Z. (2014) A Numerical Simulation of Fluid-Structure Interaction for Refrigerator Compressors Suction and Exhaust System Performance Analysis, *Proc. of the Int. Compressor Eng. Conf. at Purdue*, Paper No. 1131.

ACKNOWLEDGEMENT

The authors wish to acknowledge Ramesh Behera and Sanket Junghare at Emerson Climate Technologies for their contribution the FSI analyses. Also, Keting Pan and Xinyu Cui from the University of Dayton efforts in developing the finite element portion of the lumped models is recognized.

Experimental investigations on the performance improvement of oil-gas separator in electric driven scroll compressor for eco-friendly vehicles

Donglim NAM[1,4], Poyoung LEE[1] , Seungbin JUNG[1], Geonho LEE[2], Yunki KWON[3], Jinho LEE[4]

[1] R&D Center of Doowon Heavy Industrial Co., Ltd.,
Ansung city, Republic of KOREA (+823180567324, +823180567352,dlnam@doowonhi.com)
[2] Department of Building Engineering Services, Doowon Technical University College,
Ansung city, Republic of KOREA
[3] Department of Computer Aided Design, Doowon Technical University College,
Ansung city, Republic of KOREA
[4] Department of Mechanical Engineering, Yonsei University,
Seoul city, Republic of KOREA

ABSTRACT

Experimental research on the oil-gas separator was conducted to improve the performance of electric driven scroll compressor used in eco-friendly HEV/EVs. The compressor used in the test was "back pressure" method type by oil and an oil separator was made in the discharge chamber to manage the adequate back pressure. The pressure drop could be occurred when the refrigerant passing to oil separator. It makes input power increase and COP decrease. Various parameters of the oil separator— the length of vortex finders; installation angle; and the inlet, outlet diameters— related to the pressure drop were considered. The installation angle and the outlet diameters had no significant effect on the pressure drop; however, the pressure drop decreases as to the length variation of the vortex finder and the inlet diameter. As to the vortex finder length decreased and the inlet diameter increased, the input power of the compressor is decreased about 4.12% and the COP is increased about 2.66%.

1. INTRODUCTION

Since the scroll compressor was introduced in 1983, the demand and application range of scroll compressor has been increased due to the high efficiency and low noise level compared to the other types of compressors. Currently, in response to global environmental problems, the research and development has been focused on hybrid and electric vehicle along with advanced air conditioning systems. Unlike the conventional gasoline engine, the electric driven scroll compressor should be used for the air conditioning system in the HEV or EV. (Nam et al. 2012)

There are some factors to be determined the performance of the compressor. The oil management is one of them. It is necessary for the compressor to circulate the oil of the appropriate amount to lubricate the friction parts. The oil in the compressor prevents the wear of each part and the leakage of the refrigerant. Particularly, in the case of the back pressure type scroll compressor, the oil management is important factor depend on performance and reliability as to control the hermetic sealing about the axial leakage gap with the back pressure force of the orbiting scroll bottom.

However, when the gas of the compressor is discharged to the system after the refrigerant gas is compacted, oil is flowed out to the system along with the discharge gas. The oil from the compressor flows into the evaporator of the system, then the heat transfer coefficient of the heat exchanger is decreased and the overall performance of the refrigeration system is decreased. We installed the oil separator in the compressor in order to reduce the effect of such oil. The oil separator is installed in the compressor discharge port and the separated oil is circulated into the compressor through the back pressure chamber. The oil to circulate the compressor is the lubrication in the compressor and seal the leakage gaps of the scroll. However, the oil separator for this purpose causes a pressure drop and performance degradation because of the flow path resistance in the discharge chamber. Accordingly, this study measured the pressure drop about the each case in which a change of inlet diameter, the mounting angle, the vortex finder length, and the outlet diameter of the oil separator. And it was confirmed the improved performance.

2. EXPERIMENTAL FACILITY AND METHODS

2.1 Electric driven scroll compressor

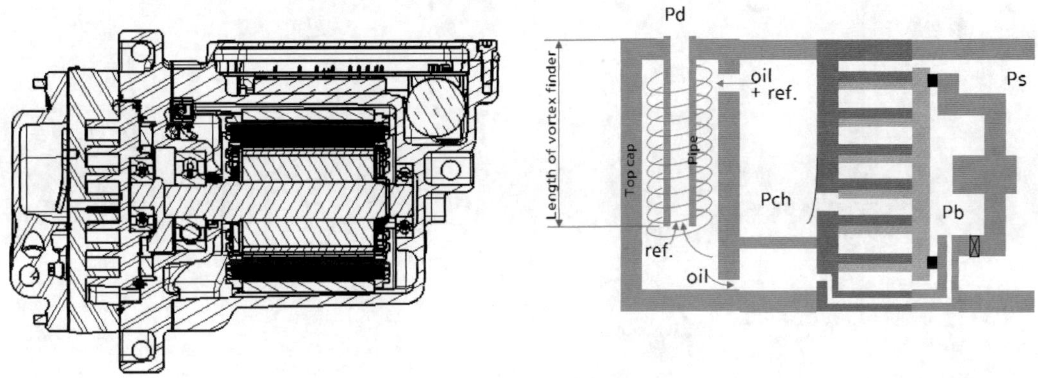

Figure 1. Section view of electrically driven scroll compressor and schematic diagram of oil separator

Figure 1 shows the electrically driven scroll compressor and oil separator in discharge chamber that is used in this research. The compressor is composed of the compression part, the motor and the inverter. The scroll compressor has an axial and radial leakage gap. It is one of hot issues and many approaches are being to try to minimize this gap. The compressor in this research applied the oil back pressure to minimize the axial gap. Using the oil separator, the gas and oil exerted from the compression part is separated. Then the high-pressure oil is sent to the orbiting scroll lower part, and the orbiting scroll adheres closely to the fixed scroll. This method could be minimized the axial gap. In addition, it lubricates the bearings in the back pressure chamber and improves the compressor durability. The R134a and POE were used for the refrigerant and oil respectively.

2.2 Experimental System

Figure 2 displays the schematic diagram of the bench system of the vehicle air conditioning system for the teat. As described above, the electrically driven scroll compressor is used, which includes an AC/DC converter to operate the compressor. The suction, discharge and back pressures were measured with pressure transducers and the accuracy of the pressure measurement was ± 0.08 kgf/cm^2 full scale. The temperatures were measured with k, t type thermocouples and the accuracy was ±0.75°C. The condenser and evaporator of the actual vehicle were also used and an expansion valve (TXV) was used. In addition, a sensor to measure the refrigerant temperature was installed on the front side of the expansion valve. The temperature measurement module was mounted at the evaporator's front and back to measure the refrigerant temperature. And it was configured to can install the OCR meter between the mass flow meter and expansion valve. To measure the OCR, a sound velocity transducer(maker : SensoTech) was used. It measures the temperature and the velocity of sound in the liquid. It has the full range of 150 to 3000m/s and the accuracy of ±0.1m/s. Measurements can be performed in the temperature range of -20°C to 120°C and at pressures up to 16000kPa. For the measurement of sound velocity in refrigerant and refrigerating oil mixtures.

The digital signals from thermocouples, pressure transducers, and mass flow meters were collected and processed by a data acquisition.

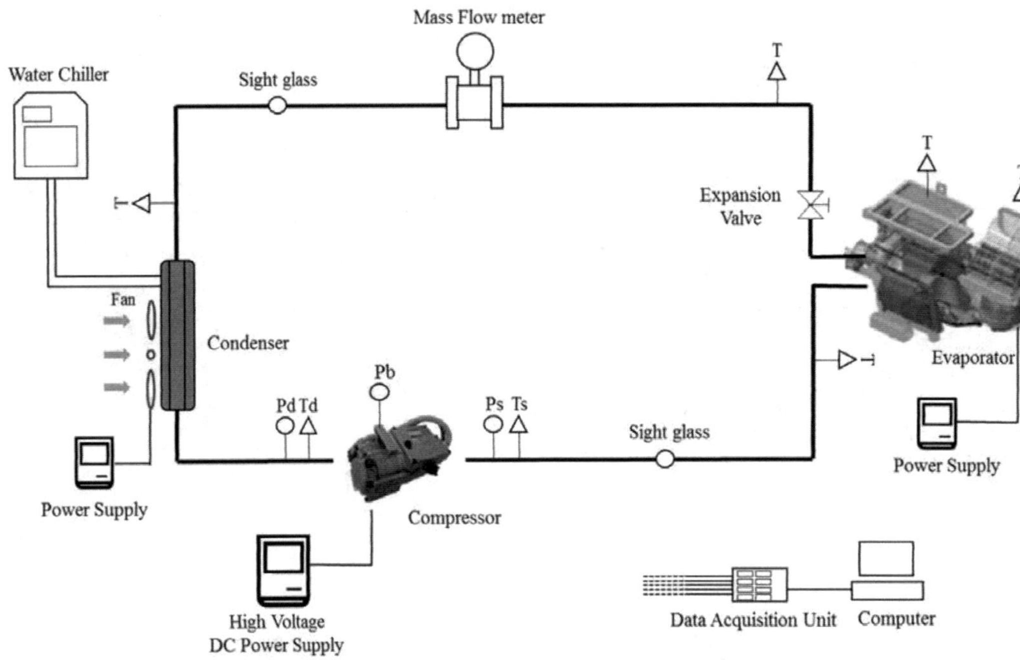

Figure 2. Schematic diagram of the bench system

2.3 Experimental procedure

2.3.1 System refrigerant charge test

At first, a test was carried out to determine the optimal amount of the refrigerant charge on the test equipment. For this, the temperature-cross-over method was used proposed by Wandell al.(1997) in ACRC TR-128. This method adds 10% to the refrigerant amount and then measures the refrigerant temperature of the evaporator inlet and outlet where they intersect. For the test condition, the fan speed of the condenser and evaporator was set to keep the performance test within a temperature range of 40~43°C and humidity within 17~21%. In addition, the EEV opening degree was fixed at that time. The refrigerant temperatures on the front and back of the evaporator were measured at 25g intervals, starting with 600g of refrigerant with fixed boundary conditions.

As shown in figure 3, the refrigerant temperatures of the evaporator front and back intersect at the refrigerant amount of 775g. In addition, the refrigerant was confirmed to be at liquid status in the sight glass in the condenser backend.

Therefore, the refrigerant charge resulted in 852.5g, a 10% increase over the amount of 775g. Finally, the amount of refrigerant was determined to be 860g for the optimum experiment.

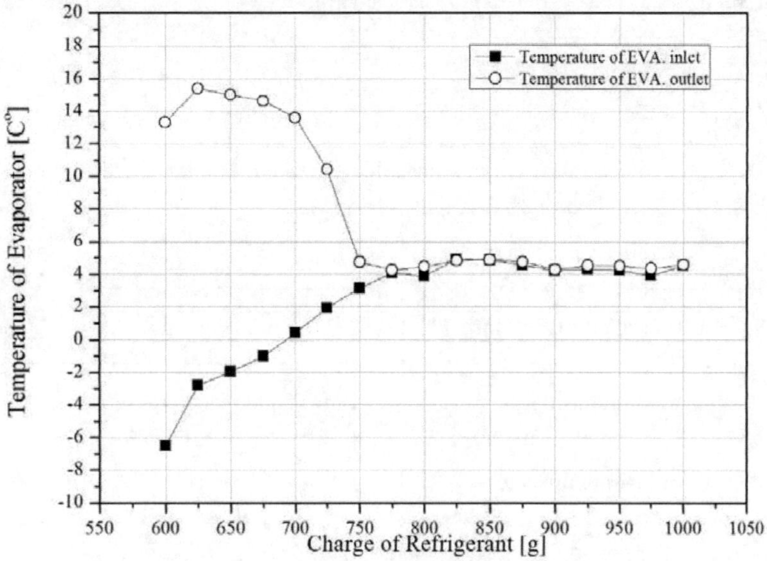

Figure 3. Temperature of evaporator

2.3.2 Experimental conditions

A test was conducted about the pressure drop by vortex finder length in the oil separator. The length of vortex finder is a factor which depends on the pressure drop. The length at the initial design is 50mm. To determine the change in length, tests was conducted to the variation of lengths, 35mm, 40mm, 45mm, 50mm, 55mm, 60mm, and 65mm.

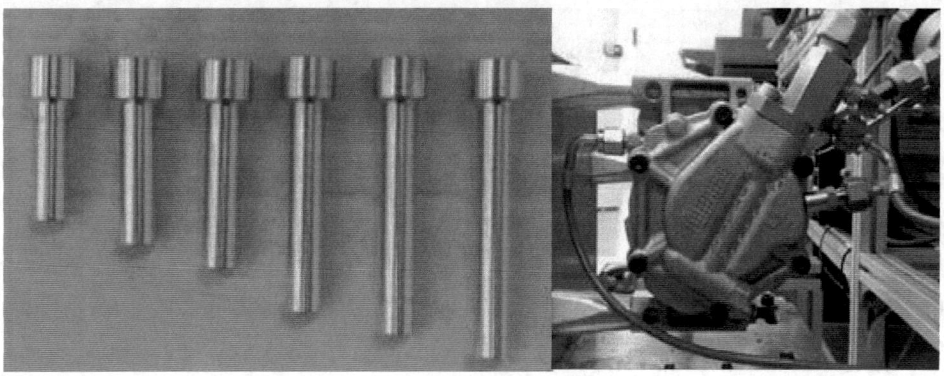

Figure 4. Vortex Finder

The compressed refrigerant and oil discharged through the oil separator. Therefore, the pressure drop was measured by the pressure before and after oil separator through the length of vortex finder. The pressure transducers was added in the discharge chamber. The test condition is shown in below table 1. The suction and discharge pressure fixed 2.04kgf/cm^2 and 14.28kgf/cm^2. It is performance test condition at the refrigerant compressor for real vehicles. And compressor rotation speed was increased from3000 rpm to 8600rpm.

Table. 1 vortex finder length test conditions.

Suction Port Pressure	2.04kgf/cm^2
Discharge Port Pressure	14.28kgf/cm^2
Rotation speed	3000 ~ 8600 rpm
Ambient temperature	20 ~ 24℃
Ambient humidity	14 ~ 26 %

Second test is parameter study on the vortex finder angle and length, inlet and outlet diameter. The inlet diameter is 6mm and 8mm. The outlet diameter is 3mm and 5.5mm. The vortex finder angle is 60 and 90 degree. The vortex finder length is 25mm and 50mm. Table 2 shown the each case through the various parameters. Test condition was shown table 3. The suction and discharge pressure is same to the condition of performance test. And compressor rotation speed is 8000rpm.

Table.2 Test case through each parameter

Test case	**Base**	**#1**	**#2**	**#3**	**#4**	**#5**	**#6**	**#7**	**#8**	**#9**	**#10**	**#11**
VF length [mm]	50	50	50	50	50	50	25	25	25	25	25	25
VF angle [mm]	60	90	60	90	60	90	60	90	60	90	60	90
I.D [mm]	6	6	8	8	8	8	6	6	8	8	8	8
O.D [mm]	3	3	3	3	5.5	5.5	3	3	3	3	5.5	5.5

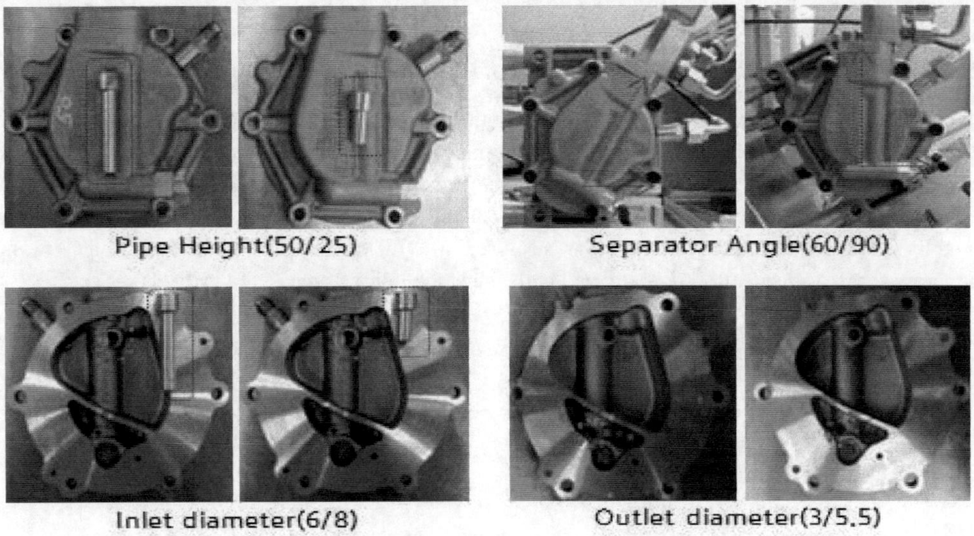

Figure 5. Vortex finder angle, inlet and outlet diameter test parameters

Table. 3 Vortex finder angle, inlet and outlet diameter test conditions.

Suction Pressure	2.04 kgf/cm^2 (const.)
Discharge Pressure	14.28 kgf/cm^2 (const.)
Rotation speed	8600 rpm
Ambient temperature	19 ~ 24 ℃
Ambient humidity	15 ~ 25 %

3. EXPERIMENTAL RESULTS AND DISCUSSION

3.1 Vortex finder length test

Figure 6 shows a pressure drop in the vortex finder length. The pressure drop increases to the faster rotational speed of the compressor. And the longer length of the vortex finder also makes pressure drop increased caused the flow resistance. When the length of the Vortex finder 65mm, the pressure drop was measured by 0.275kgf/cm² at 3000rpm, the pressure drop was measured by 1.035kgf / cm² at 8600rpm. And when its length is 35mm, the pressure drop is measured by 0.218kgf/cm² at 3000rpm, and was measured by 0.788kgf/cm² at the 8600rpm. In the low-speed, the pressure drop was measured by 0.057kgf/cm² and in the high-speed, it was measured by 0.247kgf/cm².

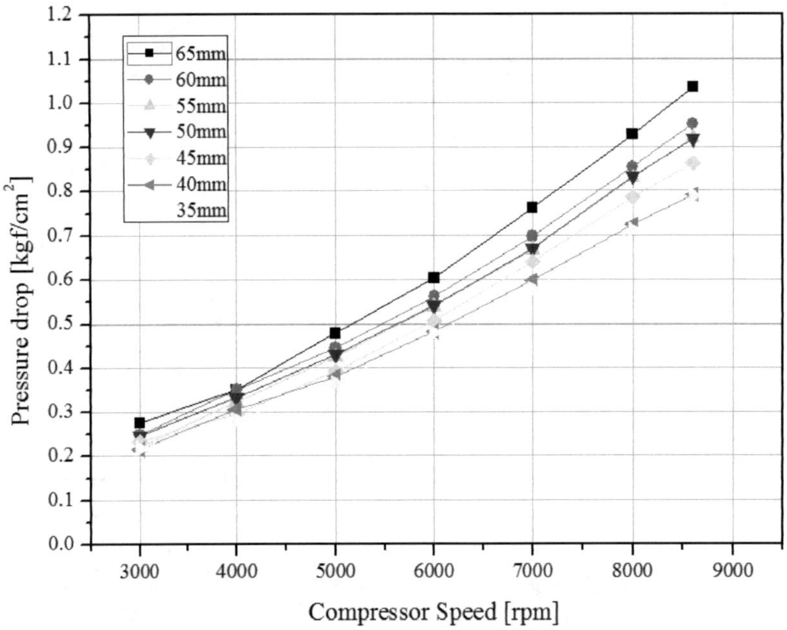

Figure 6. Pressure drop according to the length of vortex finder

3.2 Vortex finder angle, inlet, outlet diameter test

Depending on the length of the Vortex finder angle, inlet / outlet diameter, and the vortex finder, the compared evaluation was carried out. Table.4 shows the result of the pressure drop to each case. First, according to the result of the vortex finder length, it is equal to the previous test results. The pressure drop due to the angle of installation of the Vortex finder was below which means not meaningful. However, the size of the inlet diameter that is the inlet of the refrigerant and oil mixture in the oil separator is measured to affect the pressure drop. Depending on the size of the inlet diameter, increases of pressure drop was reduced up to 26%. The size of the separated oil outlet diameter in the oil separator was determined to have no effect on the pressure drop.

Table. 4 Test Results

Test case	Base	#1	#2	#3	#4	#5	#6	#7	#8	#9	#10	#11
VF length [mm]	50	50	50	50	50	50	25	25	25	25	25	25
VF angle [°]	60	90	60	90	60	90	60	90	60	90	60	90
I.D [mm]	6	6	8	8	8	8	6	6	8	8	8	8
O.D [mm]	3	3	3	3	5.5	5.5	3	3	3	3	5.5	5.5
Pressure drop [kgf/cm^2]	0.9	0.9	0.67	0.66	0.67	0.67	0.75	0.75	0.59	0.59	0.58	0.59
Power [kW]	2.67	2.67	2.59	2.6	2.59	2.58	2.63	2.61	2.58	2.58	2.57	2.56
COP	1.88	1.87	1.91	1.9	1.91	1.92	1.88	1.89	1.91	1.92	1.94	1.93

The power of the test case was decreased with decreases in the pressure drop. The power was measured by 2.67kW in the base case, but it was measured by 2.56kW in case 11. It was reduced about 4.12%. And the coefficient of performance were up 2.66% at 1.88 to 1.93.

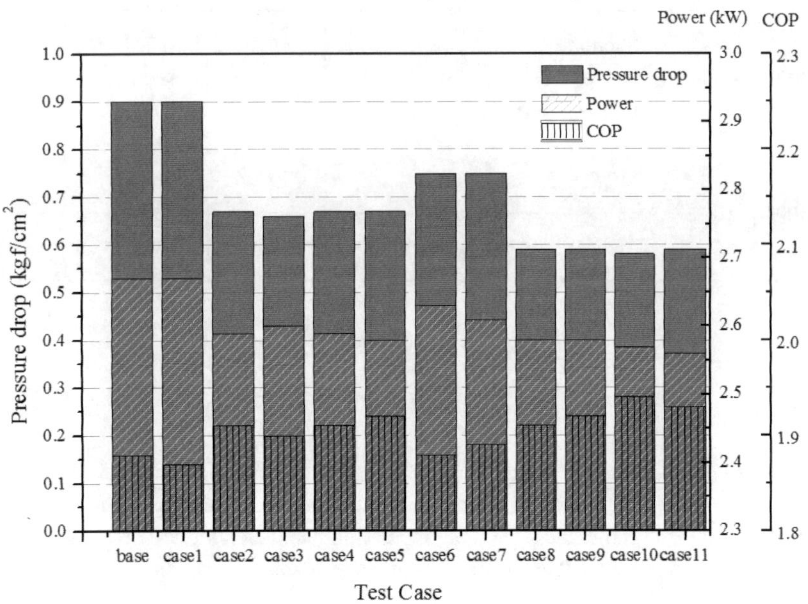

Figure 7. Test results

3.3 OCR and NVH test about base case and improved case

Parameter study results shown that the improvement of 10, 11 case appear larger than the base case. So it measured to the change of OCR in the base case against improved case. Measurement results, the OCR of improved case is measured with the equivalent as compared with the OCR of base case. The OCR of base case was measured in 3.5%, the OCR of improved case was measured in 3.4%. Also, the back pressure was measured in 5.195kgf/cm2 at the base case and 5.18kgf/cm2 at the improved case. This indicates that the separated oil flows equally to go the back pressure chamber. Therefore, it shows that the oil separation efficiency is not difference between the two cases.

And the sound level and pulsation was measured between the two cases through the NVH test. There are measured according to specifications of the car maker. Table. 5 shown the results of NVH test. Measurement results, the sound level of improved case is measured with the equivalent as compared with the base case. Suction pulsation was measured with the equivalent about two cases, but discharge pulsation was measured in improvement case larger than base case. But the discharge pulsation of improved case is applicable because of the satisfaction with the standards by the car maker.

Table. 5 NVH test results

Division			Base case	Improved case	Deviation	Note
Noise	3000rpm	RMS dB(A)	54.8	54.8		
		Peak dB(A)	43.5	45.1		
Pulsation	3000rpm	Suction Pulsation, mbar	16	18		
		Discharge Pulsation, mbar	152	178	8.7% ↑	
	4000rpm	Suction Pulsation, mbar	19	18		
		Discharge Pulsation, mbar	206	237	10.3% ↑	

4. CONCLUSIONS

The study was tested by changing the design parameters of the compressor discharge oil separator room for the performance upgrading of the electric driven compressor. First, to measure the pressure drop along the length of the Vortex finder. And the test has been carried out about the design parameters of various oil separator depend on the number of case. Finally, through the OCR and NVH tests, it was possible to verify that it was applied. The test results are as follows.

- The test results according to the length of the vortex finder, the longer length of the vortex finder is increased pressure drop. And the faster rotational speed of the compressor increases the pressure drop.

- The test results of the various case, the pressure drop by the installed angle of vortex finder and the size of the separated oil outlet diameter in the oil separator was determined to have no effect.

- The test results of the various case, the length of vortex finder and the size of the inlet diameter that is the inlet of the refrigerant and oil mixture in the oil separator was measured to affect the pressure drop. The pressure drop was decreased by max. 23.9% and the power was decreased by 4.12%. So COP was increased by 2.66%.

- Test results about the OCR of base case and improved case was measured with the equivalent. It shows that the oil separation efficiency is not difference between the two cases.

- The sound level and pulsation was measured between the two cases through the NVH test. Base case. , the sound level of improved case is measured with the equivalent as compared with the base case. Suction pulsation was measured with the equivalent about two cases, but discharge pulsation was measured in improvement case larger than base case. But the discharge pulsation of improved case is applicable because of the satisfaction with the standards by the car maker.

REFERENCES

1. Murakami, Hiroki, Shinichi Wakamoto, and Osamu Morimoto. (2006), "Performance prediction of a cyclone oil separator".

2. D.L.Nam, P.Y.Lee, G.H.lee, Y.K.Kwon and J.H.Lee, (2015), "Optimization of an oil charge amount on electric driven scroll compressor for eco-friendly vehicle.", *Int. J. Refrig.* 57, 54-61.

3. Xiong, Zhiyi, Zhongli Ji, and Xiaolin Wu, (2014), "Development of a cyclone separator with high efficiency and low pressure drop in axial inlet cyclones." *Powder Technology* 253, 644-649.

4. Ficici, Ferit, Vedat Ari, and Murat Kapsiz, (2010), "The effects of vortex finder on the pressure drop in cyclone separators." *Int J Phys Sci* 5.6, 804-813.

Influencing Factors Study of the Variable Speed Scroll Compressor with EVI Technology

Xiaoli Kang, Yusheng Hu, Caixia Shan, Yun Liu, Xiaojun Gao, Gang LV,
Compressor and Motor Institute of Gree Electric Appliances, Inc. of Zhuhai,
Jinji West Rd., Zhuhai City, 519070, P. R. China
Phone: +86-756-8974653, Fax: +86-756-8668386
E-mail: 84070191@qq.com

ABSTRACT

Heating capacity of general air-condition system at low ambient temperature has been heavily attenuated. To solve the problem stated, the variable speed Scroll compressor with enhanced vapor injection (EVI) becomes the focus of the study in recent years. Influencing factors of the capacity and energy efficiency of the scroll compressor with EVI technology have been introduced in paper, including the area of Supplementary channel, the parameters of orbiting and fixed scroll profile, the position and number of Supplementary inlet. The different scheme of prototype is tested and the effect of the scheme is compared and analyzed. With the vapor injected into compress chambers, the Gas forces of working chambers change, and the characteristics of gas force about the scroll compressor with EVI Technology have also been analyzed.

1. INTRODUCTION

Heating mode is bounded by the Qinling Mountains In winter of china, centralized heating has been adopted by using coal, heavy oil and natural gas as the heat source in north of the Qinling Mountains. People hope to use new energy as a heat source because of increasingly severe environmental problems. South of the Qinling Mountains, including Jiangsu, Zhejiang, Hunan and Hubei areas currently adopt heaters and other equipment group, which cannot meet the needs of most people. Ordinary heating capacity of heat pump system at low ambient temperature has been heavily attenuated by the characteristics of the refrigerant, and reliability is also affected.

A variable speed Scroll compressor with EVI can solve the problems, which greatly enhance cooling capacity in winter by increasing mass flux of refrigerant. There is enough oil and gas to lubricate the pump when gas is injected into the scroll, so the compressed gas inside Compression chamber can be cooled to reduce the temperature of the exhaust gas and improve the reliability of the compressor. Therefore, Scroll compressor with EVI becomes the focus of the study in recent years. HuiXia Zhao（2005）has established theoretical model of supplement gas process and compared the heating capacity and energy efficiency of the different position of supplementary inlet by theoretical calculation.

In this paper, we analyze and summarize the factors of the scroll compressor with EVI, and then Gas forces of working chambers are compared by switching vapor injection channel at the end of the article.

2. ENHANCED VAPOR INJECTION PROGRESS

There is symmetric and asymmetric scroll profile for scroll compressor. Compared with symmetrical profile, asymmetric profile can reduce suction superheat and exhaust pressure loss, and easily use a single supplementary channel. At a word, the compressor with using Asymmetric profile has excellent efficiency, simple Structure and high reliability. As shown in figure 1, with the movement of orbiting scroll which is driven by the crank shaft, the scroll wrap of orbiting scroll covers the supplementary inlet when orbiting scroll rotate to the position in figure 2(a), and the orbiting scroll continues to rotate to the position in figure 2(b), gas is Injected to the inner chamber, orbiting scroll wrap covering the inlet again at the position in figure(c), the outer chamber Inject gas at the end, a cycle ends.

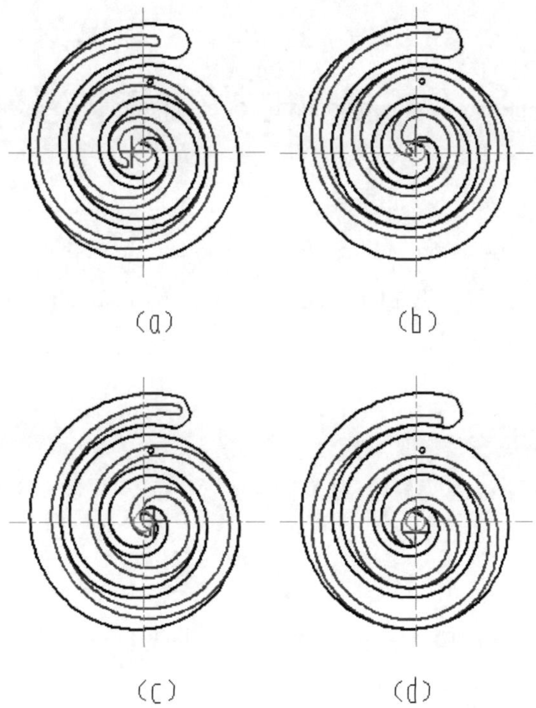

Figure 1: The vapor inject process

Economy of heat pump system which Uses Scroll compressor with EVI has two options which including flash-tank and sub cooler. The principle of heat pump system with flash-tank is introduced as an example, and the pressure enthalpy diagram is shown in Figure 2.liquid (point 9) becomes a gas-liquid mixture (point 10) into the flash-tank through the expansion valve .In the flash-tank, vapor in the upper part is sucked into supplementary inlet, Sub cooled liquid is throttled to the evaporation pressure Pe again and enters the evaporator. Low pressure gas (point 1) enters the compressor through the suction port, and is compressed to a certain pressure (point 2), which is mixed with gas through supplementary inlet in the compression chamber, and is discharged through further compression (point 7).

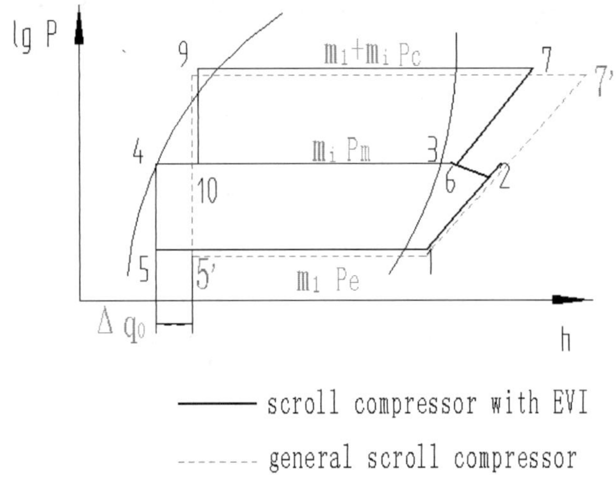

Figure 2: the pressure-enthalpy diagram of heat pump system with flash-tank

Cooling capacity is determined by enthalpy difference between points 1 and point 5 and mass flux of refrigerant which enters the evaporator. When the intermediate pressure is lower, the enthalpy difference is the greater between two points. If the pressure is too low, it will affect mass flux of refrigerant which enters the evaporator. Therefore, the optimal intermediate pressure value exists. To increase refrigerant mass flow of gas which enters condenser can enhance the heat capacity.

3. ANALYSIS OF INFLUENCING FACTORS

3.1 Supplementary channel

Reducing pressure loss of supplementary channel will help enhance the mass flux of supplementary gas, and increase heat capacity. Geometry model of supplementary channel is shown in Figure 3, the author uses CFD software to study the pressure loss of supplementary channel, and the pressure distribution is shown in Figure 4. The resistance loss mainly locates at both ends of the supplementary inlet, because the cross-section area suddenly changes, which resulting in local resistance losses. The effect of the diameter of supplementary channel is relatively small. It should focus on reducing losses here. We choose two different areas of radial channels to test, and the diameter is 5 mm and 7 mm. test results of the two programs are very similar, experimental and simulation results are consistent.

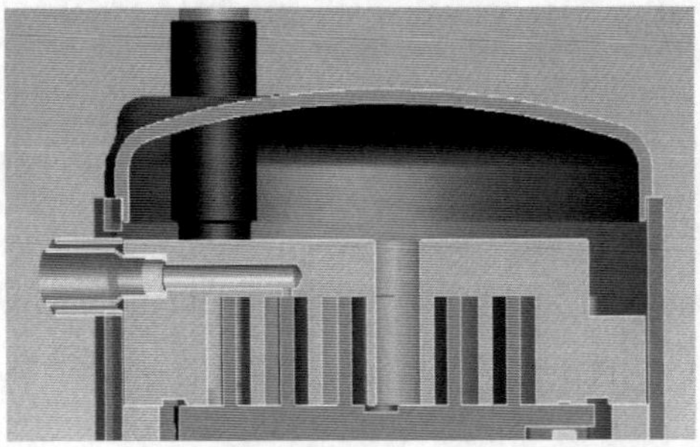

Figure 3: geometry model

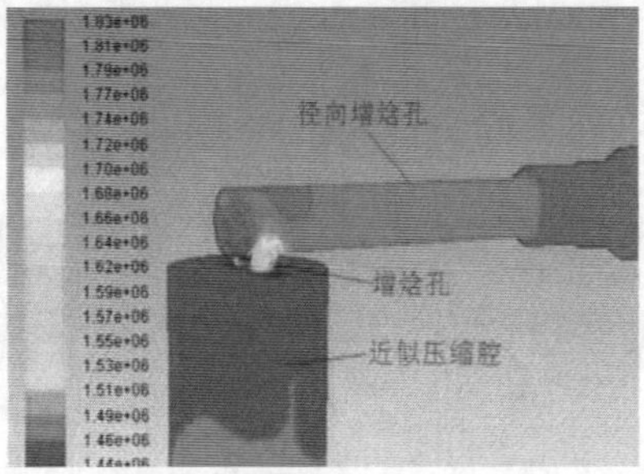

Figure 4: pressure cloud

3.2 Parametric of scroll profile

As the compressor with EVI, we need to consider how to choose the various parameters or equations, so that the volume of the intermediate compression chamber is large, and pressure is small. The rate of volume is determined by different parametric of scroll profile. For the same displacement, the compression chamber pressure is different with different parametric which including the pitch, wall thickness and wrap height. Therefore, in view of the above analysis, we designed two different parameters. By comparison, compression chamber pressure in option 1 is lower than option 2, so profile parameter chooses option 1, the result is shown in figure 5.

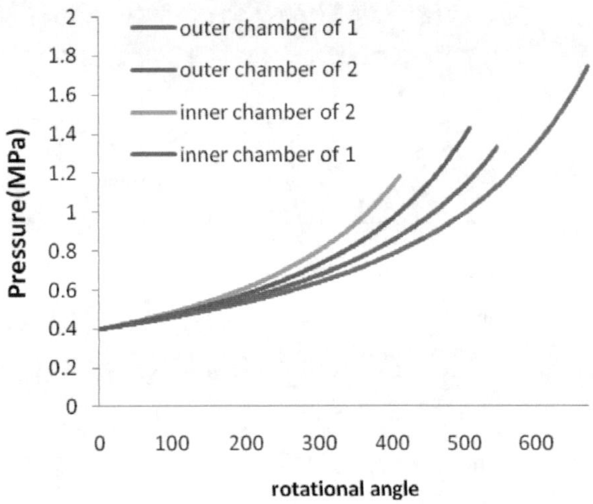

Figure 5: the curves of pressure according to the crank shaft

3.3 The position of supplementary inlet

Different position haves the greatest impact on pressure in the compression chamber. When position of supplementary inlet is near the suction finished chamber, pressure of the compression chamber is the lowest. We design the some different position of supplementary inlet, the paper chooses only two options as an example to analyze data. The starting step 0° is defined as the moment when the outermost or inner chambers are sealed off. The first scheme supplementary inlet is away from the closed position, and the second scheme is near the closed position, the angle of the crank shaft of the first program is 102-277 when vapor inject, the second one is 1-174.

Compared without EVI, the increases of cooling capacity and efficiency with EVI are distinctly shown in fig.7. Cooling capacity of the first scheme is improved between 3% -14% except IPLV50%, the effect in Cryogenic conditions is best, and the improved mount achieves 23%; Cooling capacity of the second scheme is improved 5%-26%, and cooling efficiency of the first one is improved between 5%-15.85%.

Compared the first scheme , the increase curve of cooling capacity and efficiency of the second scheme with EVI is shown in figure 8, Cooling capacity and efficiency of the second one is higher than 2%-12% and 0.6%-11.08% than first respectively. So the effect the second program is better than the first one.

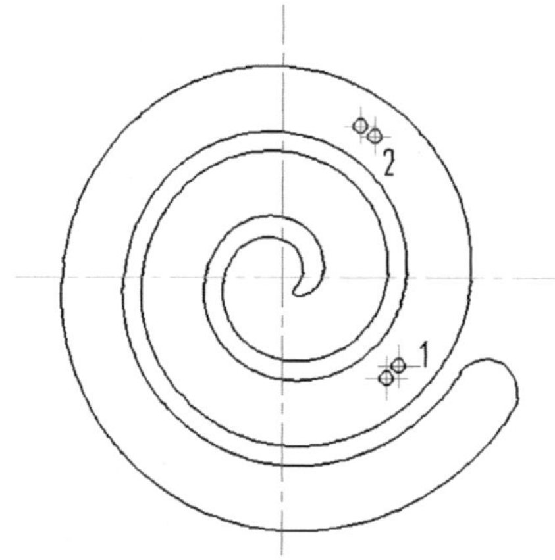

Figure 6: the position of supplementary inlet

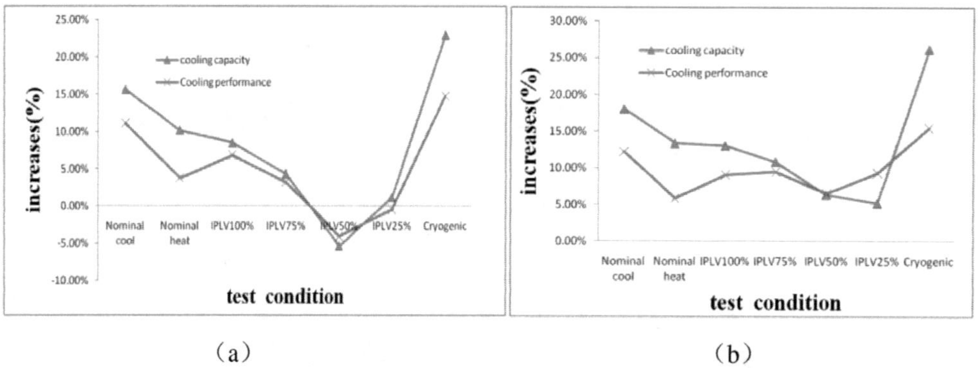

Figure 7: increases of cooling capacity and cooling efficiency with EVI:

(a) the first scheme, (b) the second scheme,

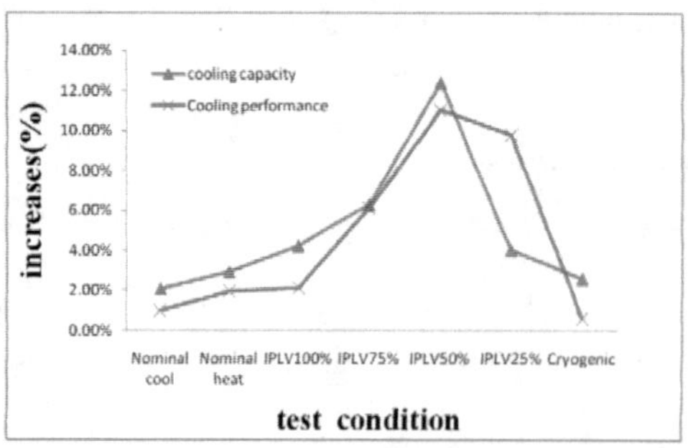

Figure 8: increases of cooling capacity and cooling efficiency between the second and first scheme

As can be seen from the above test data, improvement effect of cooling capacity and energy efficiency is remarkable including a light load conditions. Effect of the second program near the closed position is better.

3.4 Number of Supplementary inlet

In order to seal the compression chamber, enthalpy diameter of the hole is limited by the thickness of the wrap, and cannot be too large. We can only increase the number of Supplementary inlet to reduce fluid resistance loss, the author designed two programs, including single and double holes. On the basis of the same position, the three low temperature conditions, the evaporation temperature is respectively -7, -15 and -20 degrees centigrade. Heating capacity and efficiency is tested in the compressor efficiency test rig with flash evaporator. The results of the comparison are shown in figure 9. At the three low-temperature conditions, heating capacity and energy efficiency of the compressor which adopt double holes have been enhanced in contrast with single hole.

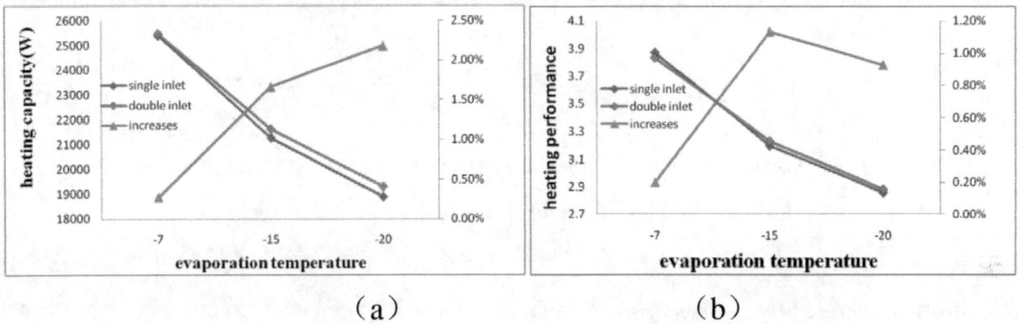

Figure 9: comparison of single inlet and double inlet: (a) heating capacity, (b) heating efficiency

4. ANALYSIS OF GAS FORCE

Orbiting scroll of scroll compressor is floating, seal of the pump is controlled by back pressure force of orbiting scroll and pressure of the compression chamber. Change of pressure in compression chamber will affect the stability and movement of orbiting scroll. If back pressure force is too small, the leakage of the scroll pump have important influence on capacity and energy efficiency, and the back plate of orbiting scroll can wear. If back pressure force is too big it will increase the clamping force between orbiting scroll and fixed scroll, increasing power consumption. So measurement of pressure of compression chamber is essential for efficiency and reliability of the compressor.

In order to measure pressure in the compression chamber, four piezoelectric pressure sensors were arranged in different position of the fixed scroll. The pressure can be monitored throughout the compression process, while the

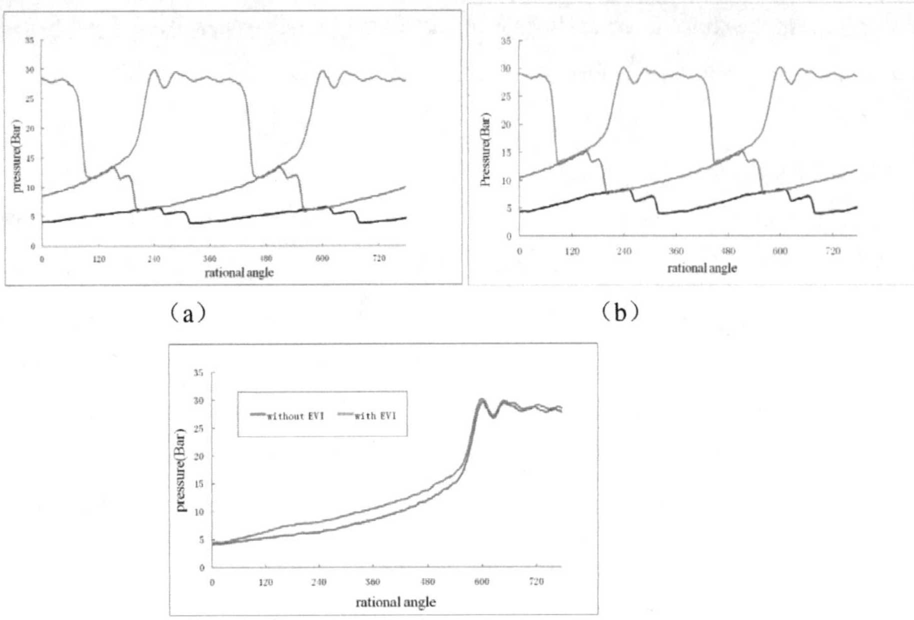

(a) (b)

(c)

Figure 10: Variation Curve of Pressure in compression chamber:
(a) without EVI,(b) with EVI, (c)comparison result

Crank angle also is monitored by displacement sensor. The curve of pressure in compression chamber with crank angle is exported. For all operating conditions the pressure shows similar pattern. As an example, pressure at low temperature condition has been analyzed. Variation curve of pressure is shown in figure 10.

When vapor is injected, pressure in the compression chamber becomes higher. Pressure compression chamber at the end of vapor inject is slightly less than the intermediate pressure, and the pressure of the compression chamber depending on the intermediate pressure. Therefore, with conventional compressor, sealing problems should be different for scroll compressor with EVI.

5. CONCLUSIONS

Based on theoretical analysis and experimental validate, the influencing factors of capacity and energy efficiency of the scroll compressor with EVI technology have been analyzed. We can summarize the following:

- Different profile parameters will affect the structure and effect of scroll compressor with EVI. Asymmetric profile is adapted by scroll compressor with EVI.
- Profile parameter is designed to reduce pressure of the compression chamber. The optimal position of supplementary inlet is near position of suction finished. Supplementary channel is designed rationally according to the volume of fixed scroll. Effect of double supplementary inlets is superior to the single one.
- Pressure in compression chamber at the end of vapor inject is slightly less than the intermediate pressure, and the pressure of the compression chamber depending on the intermediate pressure.

REFERENCES

Huixia Zhao.(2005).Theoretical Analysis and Experimental Study On Heat Pump System With Flash-Tank Coupled with Scroll Compressor. *Beijing University of Technology.*

Lubrication Tests to Support Optimal Performance Design Guidelines for Thrust Slide-Bearings in Scroll Compressors

Noriaki ISHII[1], Eiji NONOGUCHI[1], Keiko ANAMI[1], Atsushi SAKUDA[2]
Yuya TERADA[1], Ysuke IMAI, Kiyoshi SAWAI[3] and Charles W. KNISELY[4]

[1]Division of Mechanical Engineering, Osaka Electro-Communication University,
Neyagawa-City, Osaka, Japan
Tel: +81-72-820-4561; E-mail: ishii@isc.osakac.ac.jp

[2] Air-Conditioning And Cold Chain Development Center, Corporate Engineering Division,
Appliances Company, Panasonic Corporation
Kusatsu-City, Shiga, Japan, E-Mail: sakuda.atsushi@jp.panasonic.com

[3] Department of Mechanical Engineering, Hiroshima Institute of Technology,
Hiroshima, Japan, Phone: +81-82-921-5475, E-Mail: sawai@me.it-hiroshima.ac.jp

[4] Mechanical Engineering Department, Bucknell University,
Lewisburg PA 17837, USA, E-Mail: knisely@bucknell.edu

ABSTRACT

This study focuses on experimental verification of the optimal performance design guideline in lubrication for the thrust slide-bearings in scroll compressors. The auhtors' previously derived theoretical analysis using the average Reynolds equation and the solid contact theory, when applied to a cylindrical rigid thrust plate model with a small fixed wedge angle at its pheriphery, predicted a dramatic decrease in the friction power loss with increasing outer-to-inner radius ratio and determined a minimum value at a certain value of the radius ratio. For experimental confirmation of this predicted optimal lubrication performance, a cylindrical-model of the thrust slide-bearing, submerged in a refrigerant oil POE (polyol ester), was operated under a pressurized condition of 0.6 MPa using R410A as the pressurizing gas, over a wide range of orbiting speeds from 1200 rpm to 6000 rpm for a fixed orbiting radius of 3.0 mm. The resulting measured data agreed well with the predicted tendency that the lubrication of the thrust slide-bearing is substantially improved by increasing the outer radius. As was theoretically predicted, the minimum loss with a radius ratio of about 2.1 at an operating speed of 3600 rpm, a significant 80% reduction relative to the usual conventional design ratio.

1. INTRODUCTION

Scroll compressors are widely used in room air-conditioners. In these compressors a thrust bearing is used to maintain the stable orbiting motion of the orbiting thrust plate as it is firmly pressed against the fixed thrust plate. There are several types of thrust bearings, including the slide bearing type, the ball bearing type, and so on. Among these, the thrust slide-bearing is used most commonly becasue of its low mechanical friction losses and low noise generation. In a thrust slide-bearing, the flat orbiting thrust plate is firmly pressed against the fixed thrust plate. The thrust slide-bearing supports a large thrust force and does not require high pressure lubrication from a dedicated high pressure oil pump. Despite the lack of a high pressure oil pump, the thrust slide-bearing generally does not exhibit any significant lubrication problems that would result in seizure of the sliding surfaces. The thrust slide-bearing, in fact, exhibits better performance than might be expected. The excellent lubrication performance was traced to the formation of a fluid wedge between the friction surfaces, as experimentally found by Ishii *et al.* (2008a). The thrust plate elastically deforms due to axial loadings caused by the pressure difference between the outer and inner regions of the thrust slide-bearing, thus forming a fluid wedge between the sliding surfaces, as is illustrated in Figure 1.

A theoretical analysis for the excellent lubrication of the thrust slide-bearing was conducted by Oku et al. (2008), where the thrust plate with elastic deformation is represented by a rigid, fixed thrust plate with an equivalent

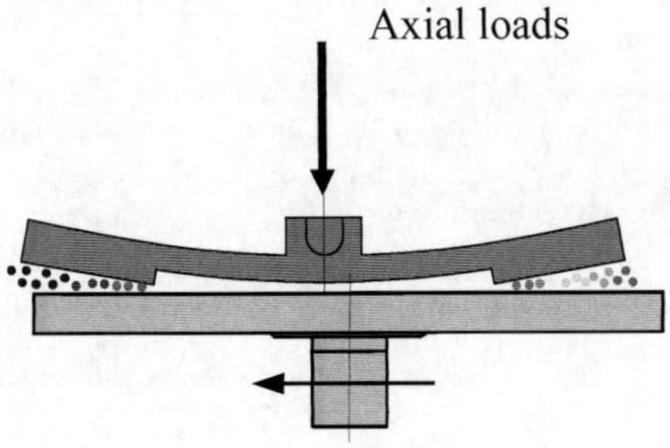

Figure 1 Elastic deformation of thrust plate, forming a fluid wedge between friction surfaces, due to pressure difference between the outer and inner regions.

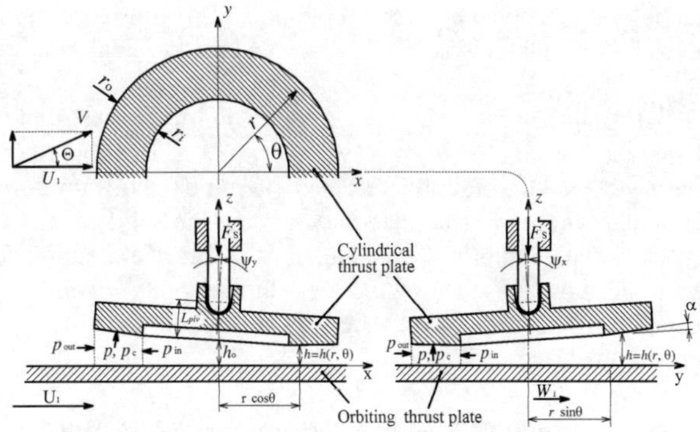

Figure 2 Mathematical model of a rigid thrust slide-bearing for theoretical analysis of fluid lubrication, after Oku *et al.* (2008).

geometrical wedge at its periphery, as shown in Figure 2. In this model, the second flat and rigid plate is forced to perform the orbiting motion. The model replicated the hypothsized outstanding improvement in the oil film lubrication performance. The validity of the fluid wedge mechnism was substantiated with detailed experiments.

Furthermore, the same theoretical development was extended to evaluate the role of lubrication in the optimal performance design guidelines for the thrust slide-bearings in scroll compressors. The theoretical analysis by Ishii *et al.* [2008] showed that the minimum friction power loss in a thrust slide-bearing occurs at a certain outer radius of the friction surface. The outer radius of the friction surface was varied for a fixed inner radius, where the fluid wedge angle between the sliding surfaces was assumed to be constant at a small value. The average Reynolds equation by Patier & Cheng (1978, 1979) and the solid contact theory by Greenwood & Williamson (1966) were applied to calculate the resultant lubrication performance and, finally, the friction power loss at the sliding surface. The model showed that the friction power loss drastically decreases and then gradually increases, with an increasing outer radius of the sliding surface. The minimum value of friction power loss was about 80% lower than that of the conventionally designed thrust slide-bering.

The primary purpose of the present study is to provide further experimental verification of the theoretical prediction for the outstanding optimal performance design guideline of thrust slide-bearing in scroll compressors. This study presents an overview of the theoretical analysis and resulting friction power loss characteristics. Subsequently, experimental confirmation of the predicted optimal lubrication performance is presented, where a thrust slide-bearing cylindrical-model submerged in a refrigerant oil POE was operated under pressurized conditions using R410A as the pressurizing gas. The pressure difference across the friction surface of the thrust bearing was fixed at 0.6 MPa, corresponding to the rated operation condition of a small cooling capacity scroll compressor. In the experiments, a special device was fabricated to maintain a constant fluid wedge angle between the friction surfaces due to the net pressure-induced elastic deformation of the thrust plate. The friction power loss at the friction surface was measured over a wide range of orbiting speeds from 1200 rpm up to 6000 rpm for a fixed orbiting radius of 3.0 mm.

2. THEORETICAL RESULTS FOR OPTIMAL PERFORMANCE IN LUBRICATION

The oil-film pressure $p(r,\theta)$ generated in the oil film for the model of the rigid thrust slide-bearing, shown in Figure 2, can be numerically calculated from the average Reynolds equations by Patir and Cheng (1978, 1979) developed for a rough surface slide-bearing with isothermal and incompressible fluid. Integrating $p(r,\theta)$ over the whole bearing surface, the resultant oil film force can be calculated. In addition, the oil film viscous force on the bearing surface with random roughness can be calculated. On the other hand, using the solid contact theory by Greenwood and Williamson (1966), the local real contact area between the sliding surfaces can be calculated from the clearance height distribution. As a result, the solid contact force and the solid shearing force can be calculated. The resultant frictional force is given by the sum of oil film shearing force and solid shearing force, while the resultant thrust force is given by the sum of axial spring force and the nominal gas thrust force.

First, assuming an initial attitude of the cylindrical thrust plate for a given values of the mean clearance height h_0 and the rotation angles Ψ_x and Ψ_y, the average Reynolds equations were solved numerically by the method of Successive Over-Relaxation (SOR) to determine the oil film forces for given boundary conditions. Subsequently, the solid contact theory was applied to determine the solid contact and friction forces. Then, the calculated results were fed back to the equilibrium equations for the forces and moments on the cylindrical thrust plate to determine a more correct attitude of the cylindrical thrust plate. This calculation procedure was repeated several times until the solution converged.

Table 1 Major specifications for theoretical calculations

Surface roughness R_a	Orbiting thrust plate [μm]	3.0
	Cylindrical thrust plate [μm]	0.056(in) ~ 0.27(out)
Standard deviation of surface roughness	Orbiting thrust plate σ_1 [μm]	1.458
	Cylindrical thrust plate σ_2 [μm]	0.188(in) ~ 1.15(out)
Bearing dimension	Radius ratio γ	1.2 ~ 3.0
	Outer radius r_o [mm]	36 ~ 90
	Inner radius r_i [mm]	30
Wedge angle tanα (×10^{-6})		80
Pivot height L_{piv} [mm]		15.0
Plastic flow pressure p_c [MPa]		1600
Shearing strength τ [MPa]		240
Surface density of asperities η [mm^{-2}]		400
Asperity summits radius β [μm]		8.0
Inner pressure p_{in} [MPa]		0.5
Outer pressure p_{out} [MPa]		1.1
Pressure difference Δp [MPa]		0.6
Axial spring force F_s [N]		600
Orbiting speed N [rpm]		300～6000
Orbiting radius r_{obt} [mm]		3.0
Oil viscosity μ^* [Pa s]		0.013
Sliding velocity V [m/s]		0.0942～2.26
Number of grids	Radial	30
	Circumferential	180

Calculations were conducted for a thrust slide-bearing model with the major specifications presented in Table 1. The wedge angle α on the friction surface of the rigid thrust plate model is fixed at 4.8 mili-degrees, based on FEM analysis of the elastic deformation of orbiting thrust plate with a radius of 66 mm and a thickness of 10.4 mm, axially loaded by the rated pressure difference Δp of 0.6 MPa. The surface roughnesses on the fixed and orbiting plates are assumed to have a Gaussian distribution and are characterized by a factor σ which was introduced by Oku *et al.* (2008), to which the reader is referred for further details. The oil viscosity μ^* is assumed to take on a value of 0.013 Pa·s for the friction surface temperature T_f of 70°C, measured in experiments. The orbiting speed N is varied from 300 rpm to 6000 rpm with an orbiting radius of 3.0 mm, resulting in a bearing surface sliding speed V from 0.0942 to 2.26 m/s. Numerical calculations by the method of SOR were conducted with 180 lattice divisions in the tangential direction and 30 in the radial direction.

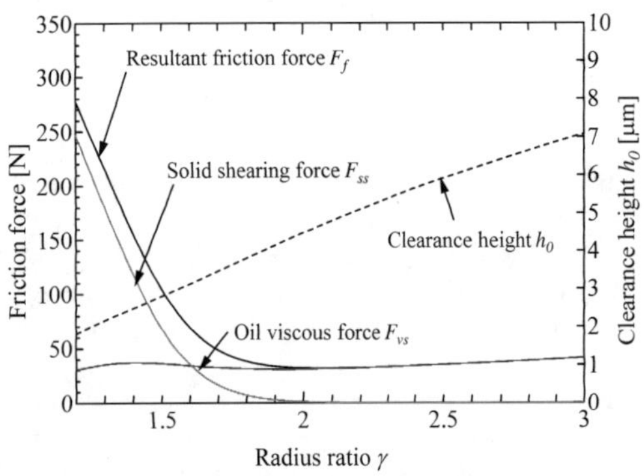

Figure 3 Calculated results of friction force F_{ss}, F_{vs}, F_f and average oil film thickness h_0 vs. radius ratio γ (for pressure difference Δp=0.6 MPa and orbiting speed N = 3600 rpm).

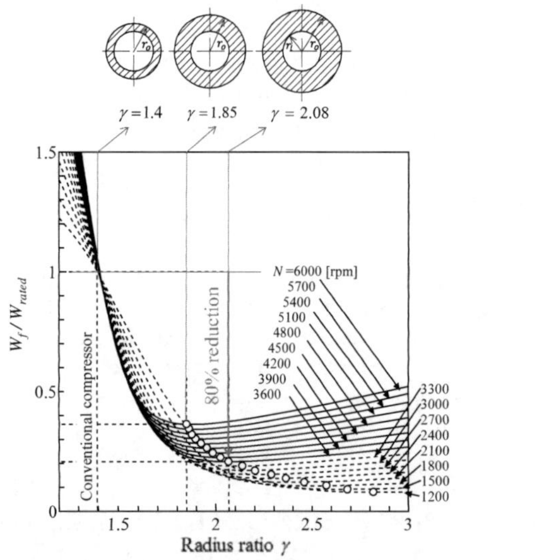

Figure 4 Friction power loss ratio W_f/W_{rated}, relative to rated condition (Δp = 0.6 MPa), with orbiting speed as parameter

Representative calculated results for the friction forces and the average clearance height (oil film thickness) are shown in Figure 3 for an orbiting speed N of 3600 rpm. The solid shearing force F_{ss}, oil viscous shearing force F_{vs} and resultant friction force F_f are shown by solid lines, while the mean clearance height h_0 (mean oil film thickness) is indicated by the dashed line. With increasing radius ratio γ, the oil film thickness h_0 increases and consequently the solid shearing force F_s rapidly decreases, while the oil viscous force F_{vs} increases. Consequently, the resultant friction force F_f rapidly decreases initially and then increases with increasing radius ratio γ, thus exhibiting a minimum at a certain value of γ. The dominant friction power loss was calculated from the resultant friction force F_f, and is shown in Figure 4, in which the ordinate is friction power loss normalized by the rated power loss W_{rated} for the actual scroll compressor with $\gamma = 1.4$. The filled circle is the condition for which the friction loss is equal to the rated loss. The minimum friction power loss is plotted as an open circle for each orbiting speed. At the rated orbiting speed of $N = 3600$ rpm, the minimum friction power loss of slightly more than 0.20 occurs at a radius ratio of $\gamma = 2.08$, corresponding to a drastic reduction of approximately 80% relative to that in the actual current scroll compressor design.

3. EXPERIMENTAL LUBRICATION TESTS TO VERIFY THEORETICAL RESULTS

3.1 Set-up of Thrust Slide-Bearing Model and Test Conditions

In the experimental setup, as shown in Figure 5, the orbiting thrust plate slides over the fixed thrust plate, forming the blue-shaded thrust slide-bearing area in Figure 5(a). The simplified model of the thrust slide-bearing, as portrayed in Figure 5(b), was developed. The orbiting scroll thrust plate is replaced by a cylindrical thrust plate, which is positioned on the upper side and fixed through a pivot bearing, while the fixed scroll thrust plate is replaced by a rigid flat thrust plate, which is positioned on the lower side and driven by the motor in orbiting motion. The material of the thrust plate is aluminum alloy for the fixed cylindrical thrust plate (upper side) and cast iron for the orbiting thrust plate (lower side), as listed in Table 2. The surface roughness R_a is carefully adjusted to 0.5 μm and 1.7 μm, respectively, at the initial stage of testing. Photos of the friction surface side of the thrust plates are shown in Figure 5(c).

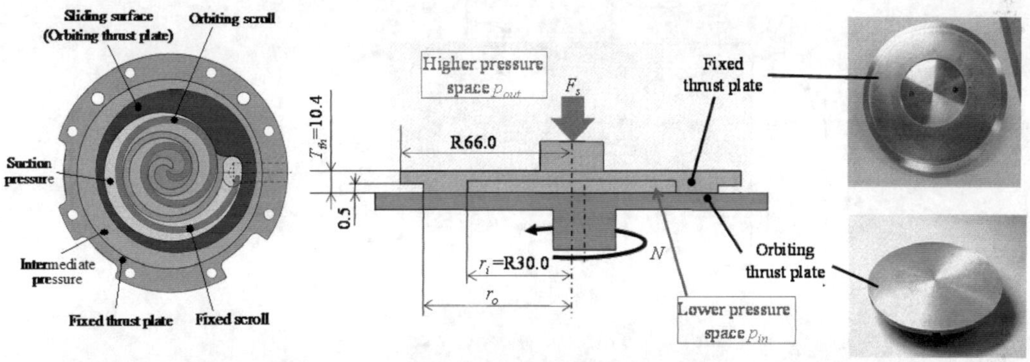

(a) Thrust slide-bearing area (b) Fixed and orbiting thrust plate (c) Test pieces

Figure 5 Thrust slide-bearing.

Table 2 Specifications of thrust slide-bearing model.

Material	Fixed cylindrical plate	Aluminum alloy (Al 90% ,Si 10%)
	Orbiting plate	Cast iron (C 0.25%)
Surface roughness R_a [μm]	Fixed cylindrical plate	0.5
	Orbiting plate	1.7

The lubrication test conditions are listed in Table 3. The cylindrical thrust plate was fixed at the outer radius of 66 mm and the inner radius of 30 mm. The thickness was also fixed at 10.4 mm. In order to decrease the outer radius of the friction surface, the lower and outer surfaces of the thrust plate were machined with the depth of 0.5 mm maintaining a constn degree of elastic deformation of the thrust plate. As a result, the wedge angle between the friction surfaces, α, was kept at a constant value of 4.8 mili-degrees for all outer radii of the friction surface, r_o. The resultant axial loading was 3144 N, consisting of the gas force F_p of 2544 N caused by the pressure difference Δp of 0.6 MPa and the spring force F_s of 600 N. The outer radius r_o was changed in 5 steps from 66 mm to 60, 54, 48 and 42 mm, corresponding to outer-to-inner radius ratios, γ, of 2.2, 2.0, 1.8, 1.6 and 1.4, respectively.

Table3 Major parameters for the lubrication tests.

Outer radius r_o [mm]	42~66	Gas thrust force F_p [N]	2544
Inner radius r_i [mm]	30	Spring thrust force F_s [N]	600
Radius ratio γ	1.4~2.2	Resultant thrust force F_t (=F_s +F_p) [N]	3144
Outer pressure p_{out} [MPa]	1.1	Orbiting speed N [rpm]	1200~3600
Inner pressure p_{in} [MPa]	0.5	Orbiting radius r_{obt} [mm]	3.0
Pressure difference Δp [MPa]	0.6	Refrigerant oil	RB68A
		Refrigerant	R410A

This thrust slide-bearing model is built into a specially-developed tribo-tester pressurized at 1.1 MPa with R410A, as shown in Figure 6. The upper side view the fixed thrust plate is shown in Figure 7(a). Two capillary tubes are attached through the fixed thrust plate into the inner region of the thrust slide-baering: one is for the pressure release and other is for pressure measurement of the inner region. The positioning thrust shaft is fitted to the pivot bearing at the center of the fixed thrust plate, which is supported with a ball slide bush bearing and instrumented with strain gauges to measure the friction forces on the fixed thrust plate, as shown in Figure 7(b). The thrust load spring presses down on the top of the positioning thrust shaft, which is instrumented with strain gauges to measure the axial spring force, as shown in Figure 7(c). The fixed thrust plate is entirely submerged in the refrigerant oil RB68A. The orbiting thrust plate was operated over a speed range from 1200 rpm to 6000 rpm.

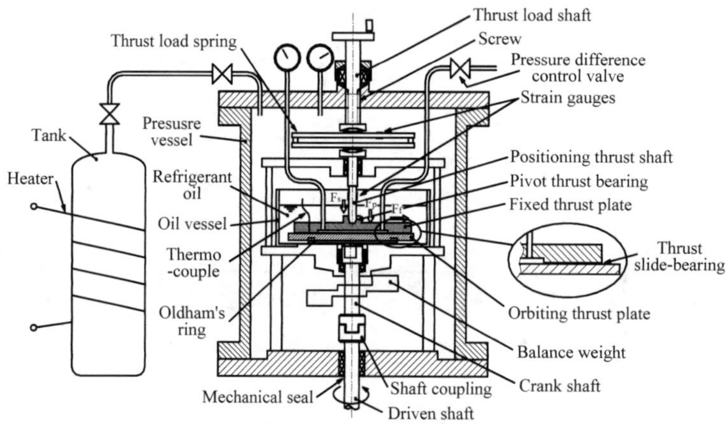

Figure 6. Cross-sectional view of tribo-tester.

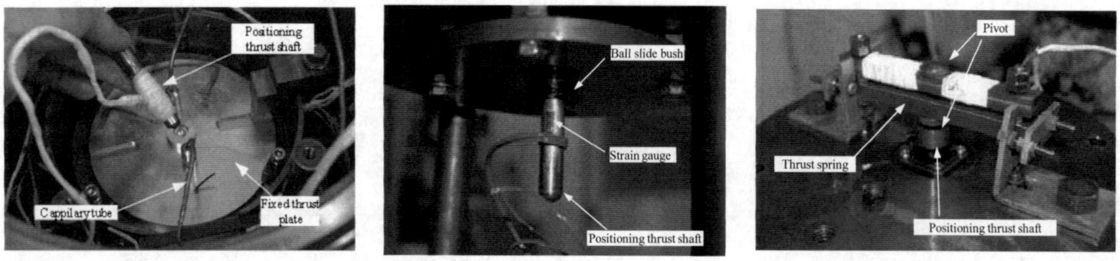

(a) Fixed thrust plate. (b) Positioning thrust shaft. (c) Thrust load spring.

Figure 7 Sequential build-up photos of test apparatus.

The pressure in the inner region of the thrust slide-bearing was decreased to 0.5 MPa, through the pressure difference control valve. A representative example of the measured friction forces F_f on the fixed thrust plate with the outer-to-inner radius ratio γ of 2.2 is shown in Figure 8, for orbiting speeds of N = 1200, 2100, 4200 and 5400 rpm. These measured data were analyzed with a frequency analyzer and the normalized values of amplitude of the friction forces were measured from each spectral peak value. With increasing operating speed N, the amplitude initially decreases from 47 N to 26, and then gradually increases to 35 N and further to 45 N. These friction force F_f, data, labeled with the circled numbers 1 to 4, are plotted in Figure 9 with the orbiting speed N on the abscissa and the radius ratio γ as the parameter. Similar detailed measurements and data analyses were undertaken for operating speeds N from 1200 rpm to 6000 rpm. All friction force F_f data are also plotted in Figure 9. The solid lines represent the theoretical predictions. The measured friction force F_f data agree well with the theoretical results, except for the data for radius ratio γ=1.4. The friction area is the smallest for the radius ratio γ=1.4, and thus the experiments were very unstable, resulting in relatively large error in friction force measurement.

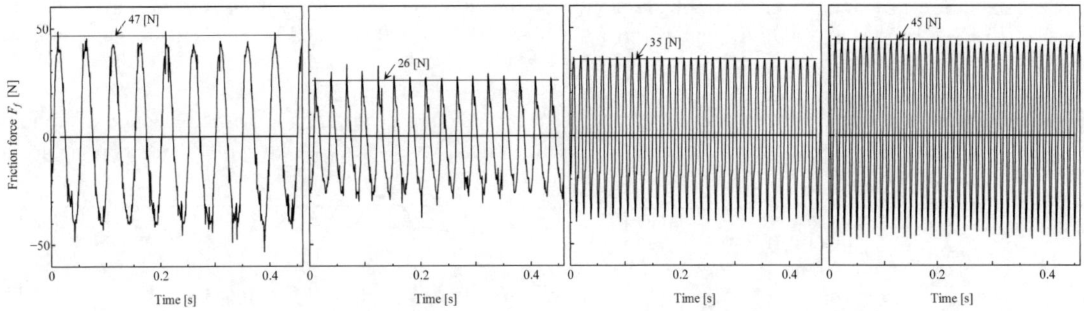

(a) 1200 [rpm] ; ①　(b) 2100 [rpm] ; ②　(c) 4200 [rpm] ; ③　(d) 5400 [rpm] ; ④

Figure 8 Representative example of measured friction forces (F_f) on the fixed thrust plate with a radius ratio γ=2.2.

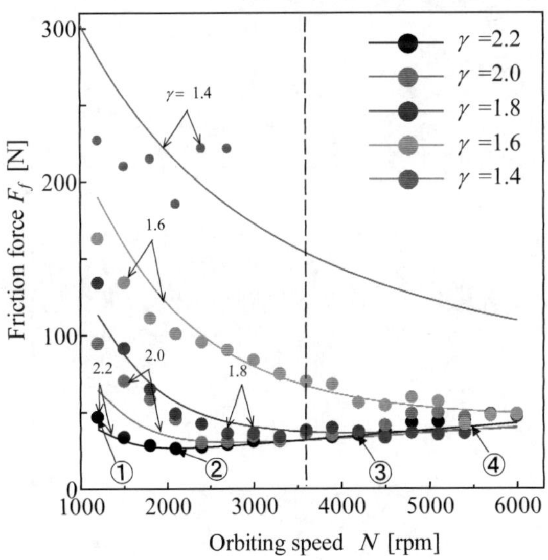

Figure 9 Measured friction forces F_f on fixed thrust plate vs. orbiting speed N, compared with theoretical predictions, for radius ratios γ =1.4, 1.6, 1.8, 2.0 and 2.2.

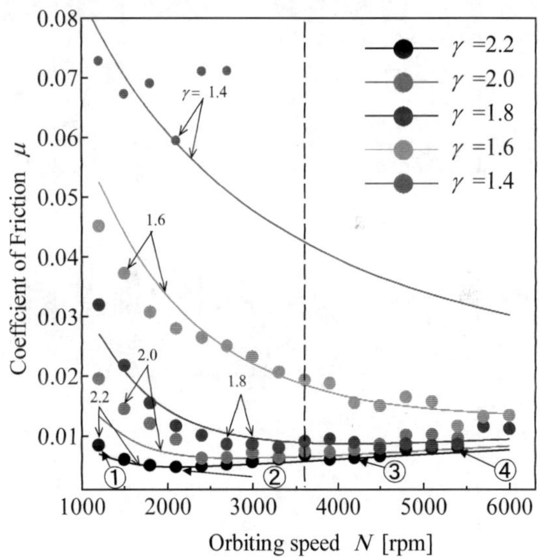

Figure 10 Friction coefficient μ of cylindrical thrust slide-bearing model vs. orbiting speed N, compared with theoretical predictions, for radius ratios γ =1.4, 1.6, 1.8, 2.0 and 2.2.

The friction coefficient μ can be calculated by the ratio of the friction force F_f to the resultant axial load F_t of 3144 N, and is plotted in Figure 10. The theoretical predictions, shown by the solid lines, portray a drastic decrease in friction coefficient with increasing radius ratio, for example, a reduction by more than a factor of 10 (from about 0.07 to about 0.007) for the lowest orbiting speed. This theoretical prediction was confirmed by the measured data which fall very close to the theoretical curves.

From data obtained in this study, optimal performance design guidelines can be formulated. The dominant friction power loss was calculated from the measured resultant friction force F_f, and is plotted over the outer-to-inner radius ratio γ, with the orbiting speed N as a parameter, in Figure 11. The theoretical predictions are shown by solid lines with the calculate minimum plotted as open (while) circles for each orbiting speed. In this figure, the ordinate is friction power loss normalized by the rated power loss W_{rated} for the actual scroll compressor with $\gamma = 1.4$ which, for instance, takes on a value of 131 watts in the theoretical calculations for an orbiting speed of $N = 3600$ rpm. The measured data for the friction power loss ratio are plotted as filled circles. Figure 11(a) is for $N = 1200$ to 2100 rpm, Figure 11(b) for $N = 2400$ to 3300 rpm, Figure 11(c) for $N = 3600$ to 4500 rpm and Figure 11(d) for $N = 4800$ to 6000 rpm.

The plotted experimental data fall very close to the theoretical predictions, indicating a drastic reduction of friction power loss with increasing outer-to-inner radius ratio γ, as theoretically predicted by Oku *et al.* (2008), is possible.

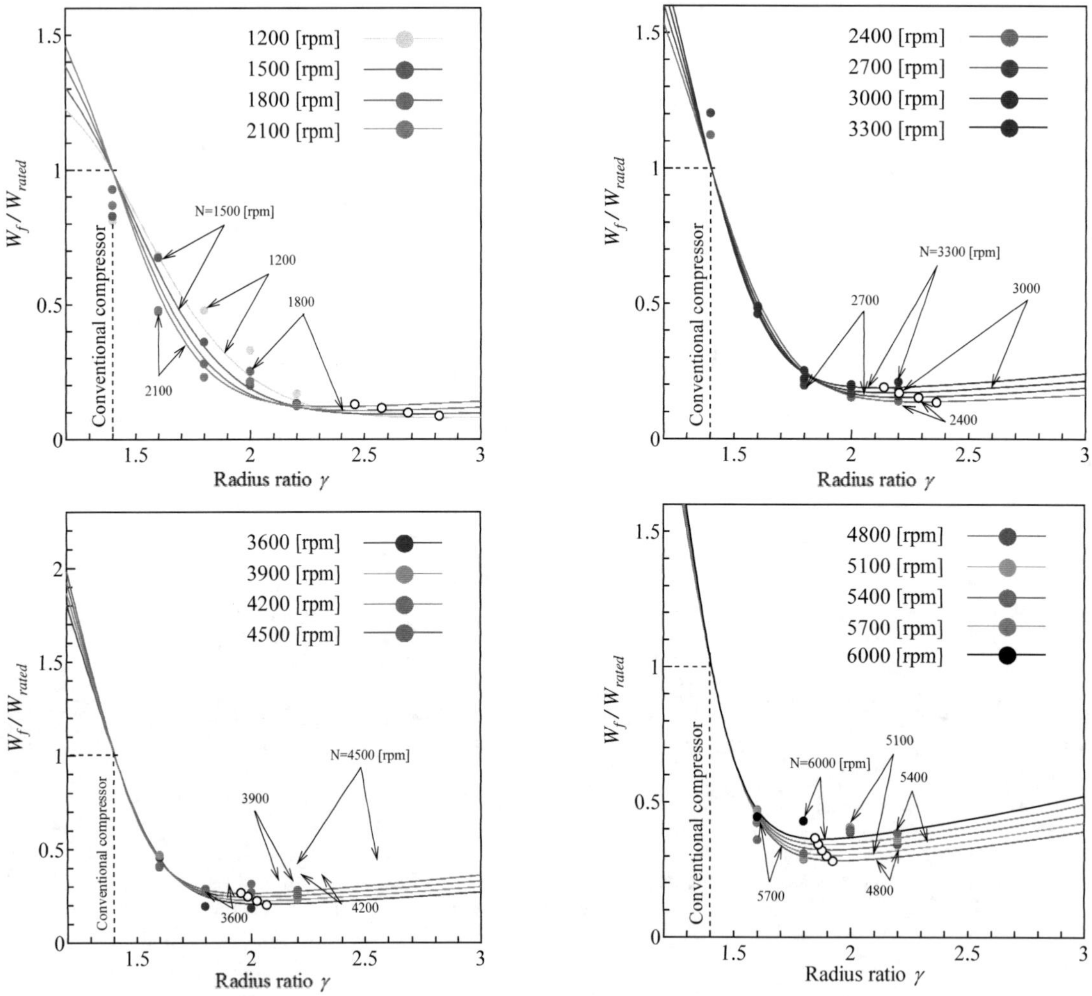

Figure 11 Measured friction power loss ratio W_f / W_{rated}, compared with theoretical predictions: (a) $N = 1200$ to 2100 rpm; (b) $N = 2400$ to 3300 rpm; (c) $N = 3600$ to 4500 rpm; (d) $N = 4800$ to 6000 rpm.

4. CONCLUSIONS

Fundamentally, the inner radius of the thrust slide-bearing is determined by the refrigeration power of the compressor, for example, by executing optimal design analyses presented by Ishii *et al.* (2002). However, a significant concern remaining for the engineer is the determination of the outer radius. There was no design criterion to determine the outer radius of the thrust slide-bearing, until 2008. At that time our previous study (Oku *et al.*, 2008) was presented to provide a theoretical prediction of the optimal performance design method for the thrust slide-bearing in scroll compressors to maximize efficiency. Our theoretical predictions suggested a drastic improvement in lubrication performance of the thrust slide-bearing, by increasing outer radius of the thrust plate. It seems that few designers have accepted these theoretical predictions. However, the present experimental study has entirely confirmed the theoretical predictions.

Designers of scroll compressors are reluctant to adopt a larger outer radius for the thrust plate design. It is, however, recommended without reservation for designers to specify an outer radius of the thrust plate so that the outer-to-inner radius ratio takes on a value of at least 1.8. With this recommendation, the lubrication performance of the thrust slide-bearing would be substantially improved. The friction power loss at the thrust slide-bearing can be decreased by about 80%, compared with that for the conventional design with the outer-to-inner radius ratio of 1.4. Scroll compressor designers should immediately check their design value of the outer-to-inner radius ratio of the thrust slide bearing.

NOMENCLATURE

			r_{obt}	orbiting radius	(mm)
F_f	resultant frictional force	(N)	T_f	friction surface temperature	(°C)
F_{OIL}	oil film force	(N)	V	boundary velocity	(m/s)
F_p	nominal gas thrust force	(N)	W_f, W_{rated}	friction loss energy	(W)
F_s	axial spring force	(N)	α	wedge angle	(rad)
F_{sc}	solid contact force	(N)	β	asperity summits radius	(μm)
F_{ss}	solid shearing force	(N)	Δp	pressure difference	(Pa)
F_T	resultant thrust force	(N)	γ	radius ratio of thrust bearing	(-)
F_{vs}	oil viscous force	(N)	η	surface density of asperities	(m^{-2})
L_{piv}	pivot height	(mm)	μ	frictional coefficient	(-)
N	orbiting speed	(rpm)	$\mu*$	oil viscosity	(Pa s)
p	oil film pressure	(Pa)	σ_1, σ_2	standard deviations	
p_{out}, p_{in}	boundary pressure	(Pa)		of surface roughness	(μm)
r_o, r_i	plate radius	(mm)	τ	shearing strength	(Pa)

REFERENCES

Ishii, N., Yamamoto, S., Sano, K., Sawai, K. Hiwata, A. & Nakamoto, T., 2002, Efficiency Simulations of A Compact CO2 Scroll Compressor and Its Comparison with Same Cooling Capacity R410A Scroll Compressor, *International Compressor Engineering Conference at Purdue*, C22-2.

Ishii, N., Oku, T., Anami, K., Knisely, C., Sawai, K., Morimoto, T. and Iida, N., 2008a, Experimental Study on Lubrication Mechanism at Thrust Slide-Bearing of Scroll Compressors, *ASHRAE Journal of HVAC&R Research*, Vol 14, No.3, pp.453-465.

Ishii, N., Oku, T., Anami, K., Knisely, C., Sawai, K., Morimoto, T. and Iida, N., 2008b, Optimal Performance Design Guidelines for Thrust Slide-Bearings in Scroll Compressors for Maximum Efficiency, *International Compressor Engineering Conference at Purdue*, C1428.

Oku, T., Ishii, N., Anami, K., Knisely, C., Sawai, K., Morimoto, T. and Hiwata, A., 2008, Theoretical Model of Lubrication Mechanism in the Thrust Slide-Bearing of Scroll Compressors, *ASHRAE Journal of HVAC&R Research*, Vol. 14, No.2, pp.239-258.

Patir, N. & Cheng, H. S., 1978, An average flow model for determining effects of three dimensional roughness on partial hydrodynamic lubrication, *Transactions of the ASME* , Vol. 100, pp.12-17.

Patir, N. & Cheng, H. S., 1979, Application of average flow model to lubrication between rough sliding surfaces, *Transactions of the ASME* , Vol. 101, pp.220-230.

Greenwood, J.A. & Williamson, J.B.P., 1966, Contact of nominally flat surfaces, Burndy Corporation Research Division, Norwalk, Connecticut, USA, pp.330-319.

ACKNOWLEDGEMENT

The authors extend their sincere gratitude to Mr. Isao Tasaka, Director, Air-Conditioning and Cold Chain Development Center, Corporate Engineering Division, and Mr. Akio Kozaki, Director, Refrigeration and Air-Conditioning Devices Business Division, Appliances Company, Panasonic Co., Ltd., for their cooperation in carrying out this work and their permission to publish this study.

Visualization and Simulation of Oil Flow in a Scroll Compressor Plenum

Jiu XU[1], Pega HRNJAK[1,2,*]

[1]University of Illinois at Urbana-Champaign, Mechanical Science and Engineering, Urbana, IL, USA

[2]CTS, Creative Thermal Solutions, Urbana, IL, USA

jiuxu2@illinois.edu, pega@illinois.edu

[*]Corresponding author

ABSTRACT

Discharge valves located in the compressor plenum are the gateway for the lubrication oil to leave the compressor. The compressor plenum space has the potential to separate the oil droplets and decrease the oil circulation ratio. In this paper, a video processing method is used to capture the oil droplet flow near the reed valve and in the discharge tube in a modified scroll compressor. Also, discrete phase model in computational fluid dynamics is used as the numerical approach to study the oil droplet trajectory inside the compressor plenum. In order validate the CFD simulation, size distribution from video processing at the valve is used as the input of the calculation and the calculation results are compared to size distribution in the discharge tube. Calculation shows how oil droplet size distribution affects the oil flow in the plenum.

1. INTRODUCTION

Oil existence in the compressor is necessary while oil circulating in vapor compression system is known to have adverse effects on system heat transfer and pressure drop. Therefore, it is preferable to reduce the oil circulation ratio (OCR) in the system and keep the oil in the compressor.

Separating the oil from the refrigerant vapor and draining it back to the compressor is a feasible way to achieve low OCR. For a scroll compressor, it has the space to integrate oil separation structure into the discharge plenum. Oil flow parameters, like oil droplet size and velocity, are very important for oil separator design. This justifies a closer look at the oil flow inside the compressor plenum as well as the discharge tube.

Many researchers have studied the oil flow and distribution in the compressor by visualization or numerical simulation. Toyama et al. (2006) described methods to quantify oil droplet behavior using high-speed photography in the scroll compressor shell via sight glasses. Besides the size distribution of oil droplets, it is concluded that the mean diameter of oil droplets decreases as the flow speed of the refrigerant gas increases.

Yokoyama et al. (2012) analyzed the oil flow in a CO_2 rotary compressor shell by CFD. Numerical simulations shows that flattening the lower balance weight or installing a rotating disk over upper balance weight can reduce the OCR and the results are verified by system experimental results.

Wujek and Hrnjak (2011) developed visualization techniques to quantify the annular mist flow at compressor discharge tube. The size and velocity of each oil droplet captured by the high speed camera can be determined by video processing. It shows that the droplet population periodicity at the discharge of a swash plate compressor with R134a and PAG46 oil.

Zimmermann and Hrnjak (2014a, 2015) showed that the most significant source of oil droplets in a scroll compressor is the break-up process of oil film between valve and valve seat during the opening and closing. The visualization of the reed valve in the scroll compressor shows the valve periodical movement and oil atomization process.

Xu and Hrnjak (2016) measure the OCR in a non-invasive way based on the oil annular mist flow visualization in the compressor discharge tube. Camera focus effect is considered into the video processing which gives a better estimation of oil droplet amount in a specific control volume.

Therefore, previous research has shown 1) the origin source of oil droplets in the scroll compressor; 2) the flow pattern of refrigerant-oil mixture at the compressor discharge 3) the visualization used to quantify the oil flow. 4) Numerical model used to simulate the oil flow.

The motivation of this research is to link the visualization with simulation in order to study the oil droplet flow in a scroll compressor plenum. Discrete phase model in CFD is used to predict the oil droplet trajectory in the plenum. Also the simulation results will be validated with experimental results from flow visualization. Furthermore, based on the validated simulation, it is meaningful to explore the effect of oil droplet size and potential oil separation structures that can be integrated into scroll compressor plenum.

2. OIL FLOW VISUALIZATION

2.1 Compressor modification

The oil flow visualization is realized inside the compressor plenum based on the video captured by a high speed camera. In order to get clear images of the oil droplets inside the plenum, side glasses and a transparent ring are added to the upper part of the compressor, as it is shown in Figure 1.

Figure 1: Modification of a scroll compressor plenum for visualization

The transparent ring is made of polycarbonate which can hold the high side pressure of the compressor. The transmittance of the transparent ring is highly dependent on the surface finishing and polishing process. A flange structure, together with O-ring and bolts is used to seal the chamber and make sure it's air tight under high pressure.

A plastic screen is placed in front of the compressor valve in order to eliminate the misty flow in front of the valves so that a clear vision can be achieved when capturing the video of oil near the reed valve. The compressor has three discharge valves but in this work, two auxiliary valves are forced to close intentionally for easier visualization.

2.2 Videos of oil droplets

Videos are taken by a Phantom camera with different lens at three locations: the reed valve, the valve vicinity and the discharge tube, as shown in Figure 2.

The video of the reed valve shows the valve opening sequence and how oil film between valve and valve seat breaks up. A cloud of small oil droplets flows from the valve edge as the valve opens and the oil film breaks. Zimmermann and Hrnjak (2014b) gave a method to quantify the initial velocity of oil droplets injected by the film break-up process.

The video near the valve is processed to give the information of oil droplet size distribution. In the plenum chamber, the oil droplets are moving in all directions at a relatively low velocity. The droplet size can be better estimated than droplet velocity because it is difficult to give the three dimensional velocity by a series of planar images. Therefore, oil droplet size distribution is the key parameter that is used to compare visualization results and simulation results.

The video in the discharge tube is taken at a section of transparent Polytetrafluoroethylene observation tube. Compared to the video near the valve, it is easier to process to get the droplet velocity since the axial velocity along the tube is much larger than the radical one. The effect of camera depth of focus is also considered to estimate the oil droplet size.

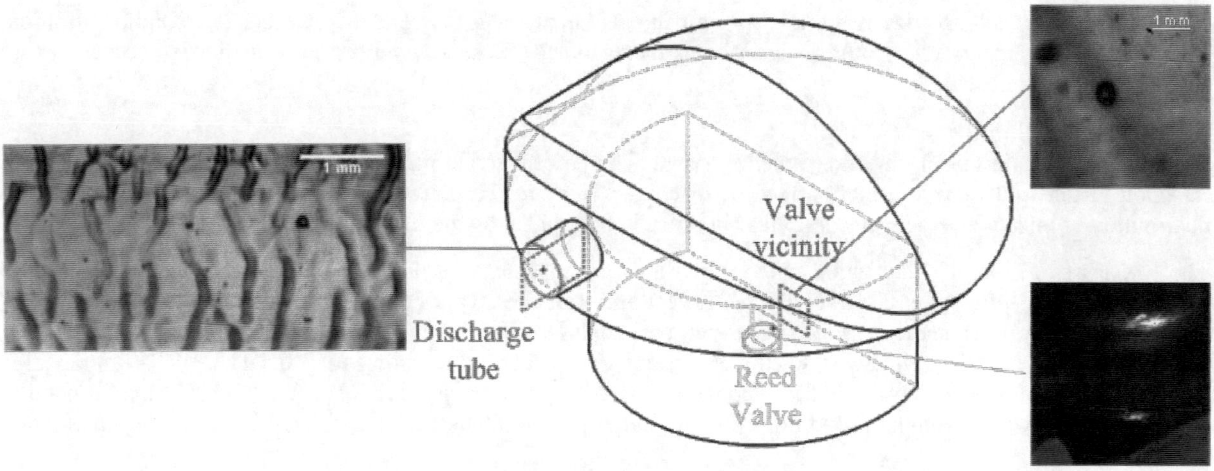

Figure 2: Locations of video taken in the modified compressor

3. OIL FLOW CFD SIMULATION

3.1 Discrete Phase Model

Discrete phase model (DPM) in computational fluid dynamics is usually used to simulate one fluid phase dispersed in another continuous fluid phase in a Lagrangian frame of reference. Discrete phase model is commonly used to simulate the liquid/solid particle motion in a continuous fluid. An important preliminary assumption made in the discrete phase model is that the dispersed phase occupies a low volume fraction. In this research, oil flow inside compressor discharge chamber is considered to be misty and oil volume ratio is always less than 5%, which is consistent with the DPM assumptions.

In our case, we use the Euler-Lagrange approach. The fluid phase is treated as a continuum by solving the time-averaged Navier-Stokes equations, while the dispersed phase is solved by tracking a large number of droplets through the calculated flow field. The dispersed phase can exchange momentum with the fluid phase.

3.2 Geometry and mesh

The research object of CFD simulation in this research is limited to the modified compressor plenum and discharge tube. The surface swept by the edge of the reed valve is defined as the inlet of refrigerant vapor flow and oil droplet injections while the discharge tube is defined as the outlet of the continuous fluids. Figure 3 shows the geometry and mesh of the modified compressor plenum used for calculation. Besides the original plenum space, the additional space surrounded by transparent polycarbonate ring and transparent screen is also included to make sure the simulation results can be comparable with the experimental results. Half of the geometry is actually used for CFD simulation in order to save the calculation time and the boundary of the center plane is set as "symmetry".

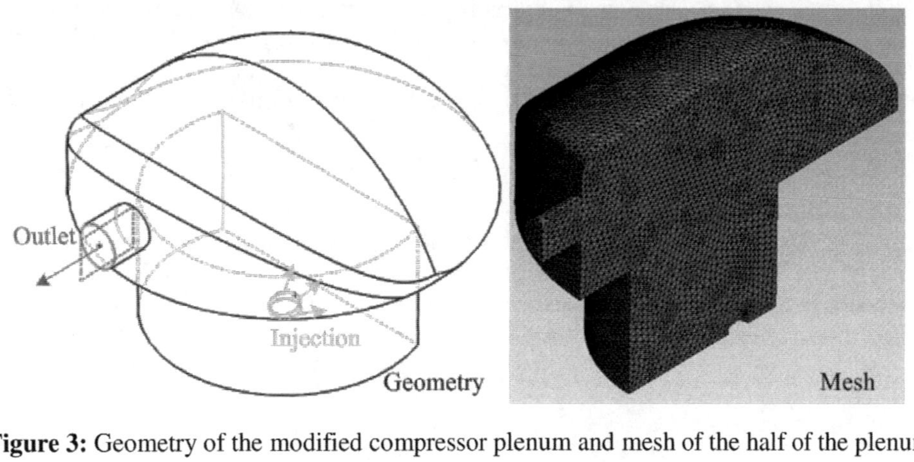

Figure 3: Geometry of the modified compressor plenum and mesh of the half of the plenum

For meshing, the max face size is set as 1 mm for the 3-D model and the elements number is around one million. The mesh around the valve is refined because of finer geometry details. Overall mesh quality is checked before running the CFD calculation.

3.3 Boundary Condition

In the discrete phase model, injection function set the source of discrete phase (usually droplets or particles). The injection is released from the inlet surface of discharge valve in the direction normal to the inlet surface. The distribution of injection velocity and size distribution can be decided by the results from video processing.

The physics of oil droplet interaction with solid wall is complicated. In this research, the boundary condition is simplified to four different cases: trap, escape, reflect and wall film. Trap condition means that the droplet will be absorbed by the wall while escape condition allows the droplet to pass through when they hit the boundary. A reflect wall will give an opposite velocity vector to the incoming droplet based on momentum balance. Wall-film model is most realistic case which allows a single component liquid drop to impinge upon a boundary surface and form a thin film. Table 1 shows the boundary conditions for continuous phase and discrete phase that are used in the simulation.

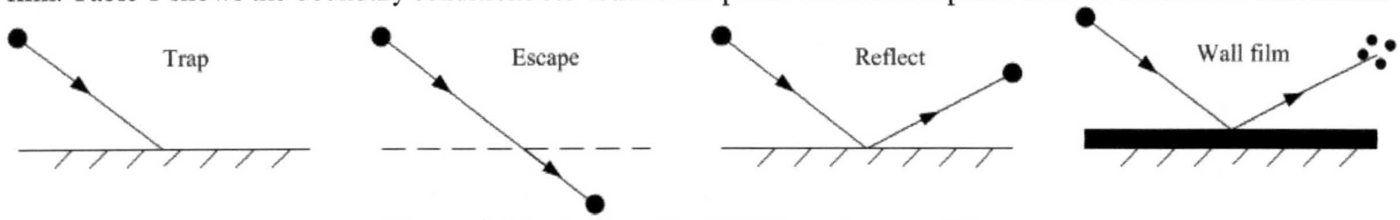

Figure 4: Mechanisms for DPM boundary conditions

Surface	Continuous phase boundary condition	Discrete phase boundary condition
Inlet	Velocity inlet	Injection
Valve	No-slip wall	Reflect
Wall	No-slip wall	Wall-film
Bottom	No-slip wall	Wall-film
Outlet	Outflow	Escape

Table 1: Boundary conditions defined for each

3.4 Droplet Injection

As mentioned in 3.3, injection is the discrete phase boundary condition for the inlet. The information of the droplet injection from video can be specified in the CFD calculation. The distribution of droplet diameter is fitted by Weibull distribution to be used as the injection parameter at the inlet. A typical droplet size distribution curve can be seen in Figure 8.

For the real case, the geometry should be a moving mesh because valve opens and closes from time to time. However, since the volume of valve moving space is relatively small compared to the whole plenum space, the valve is set to be open all the time. Instead of the moving mesh, pulsating flow at the inlet is used to simulate the dynamic mass flow discharged by the reed valve. The valve opening and closing period is decided by video taken at the reed valve and the mass flow rate is given by the mass flow meter in the system. As shown in Figure 5, the pulsating flow of the vapor injected by the discharge valve is simplified to a square wave and the vapor velocity and valve opening displacement are linked with mass flow rate conservation.

3.5 Numerical Settings

The effect of turbulence is considered and k-ε model is used in the flow calculation. The simulation takes SIMPLE as the scheme and uses second order for pressure discretization and second order upwind for momentum discretization. The time step for transient calculation should be as small as possible but here it is set as 0.001s for reasonable computational time.

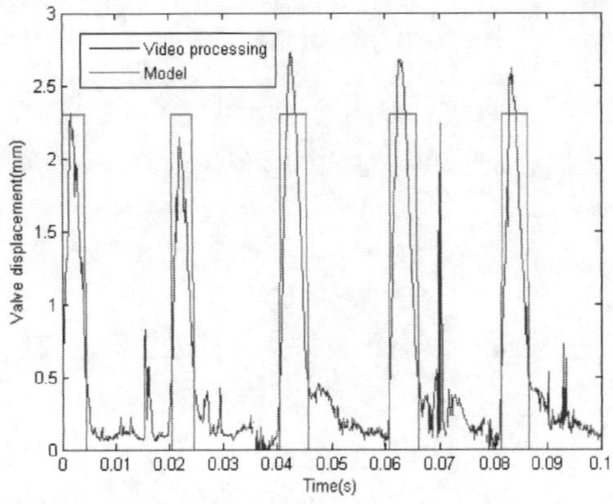

Figure 5: Pulsating flow to simulate reed valve opening and closing at 50 Hz

4. VALIDATE SIMULATION WITH VISUALIZATION RESULTS

4.1 The effect of oil droplet size

The trajectories of oil droplets can be calculated by DPM model in Fluent. When the oil droplets hit the boundary, they will either bounce back or splash into more droplets or be absorbed by the wall. Some of the oil droplets can leave the plenum and recorded by the monitor at the compressor discharge tube. Figure 6 shows the droplets in the free stream and bottom wall-film, colored by droplet diameter. It can be concluded that larger droplets tend to settle down in the wall film at the bottom of the compressor while smaller droplets tend to flow along with the vapor free stream.

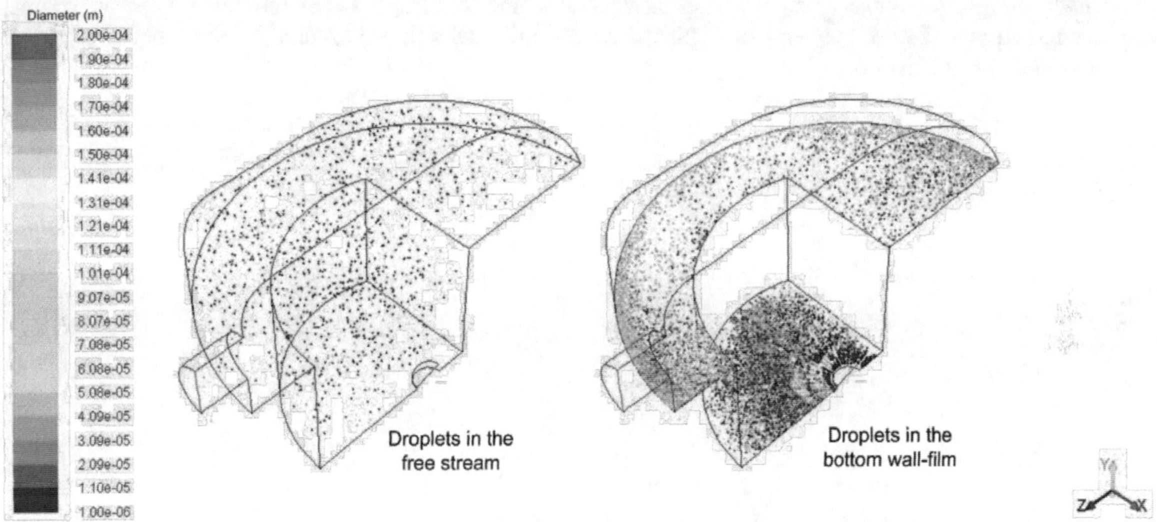

Figure 6: Tracked droplets position at t =1s, colored by droplet diameter

4.2 Oil droplet size distribution under different compressor speed

In order to validate the CFD simulation, video taken near the valve is used to give the as the input parameters for injection, like oil droplet size and velocity. The vapor velocity is given by the pulsating flow model mentioned in 3.4. Based on the input parameters, transient CFD calculation can monitor the oil droplets that escaping out of the compressor plenum and get a droplet size distribution. On the other side, the video taken at the discharge tube is processed to give a droplet size distribution. These two distributions are compared to see if the CFD results can agree with the video results. The process of validation is shown graphically in Figure 7.

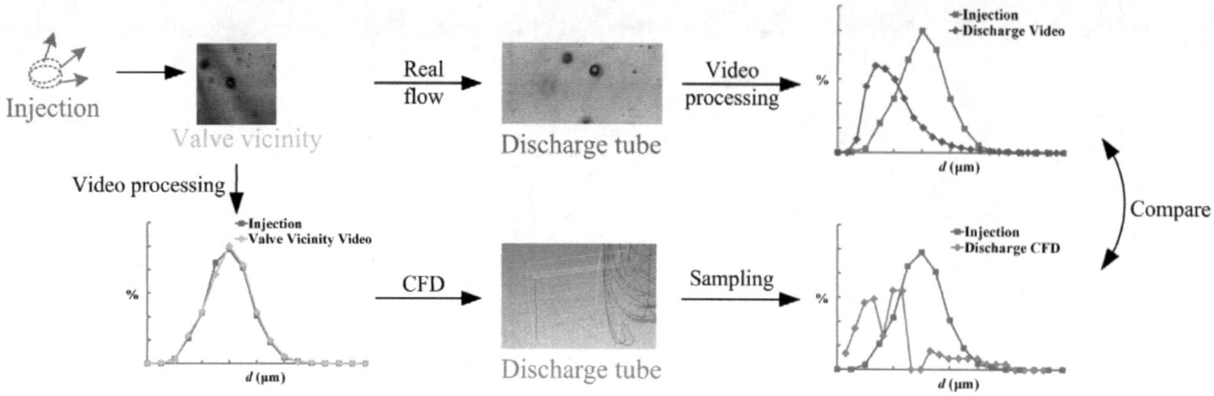

Figure 7: Comparison of visualization results and simulation results under different compressor speed

Both experiments and simulation are carried out at different compressor speed in steady state. Figure 8 to Figure 10 show the comparison of droplet size distribution between CFD results and video results. The red curve represents the injected oil droplet size distribution, which is the input information to the model. Experimental flow visualization gives the green dash line while the simulation gives the blue dot line. Both of them represent the escaped oil droplet size distribution.

The results show that CFD simulation can give a qualitatively prediction about the droplet size distribution at the compressor outlet. From the comparison between droplet size distribution of the injection and that of the discharge, it is clear that large droplets are more likely to hit the wall and splash on the wall film to generate more small droplets. Smaller droplets are less affected by gravity and inertia so that they can catch up with the vapor flow to escape the compressor and form the annular-mist flow in the discharge tube.

It is also interesting to see higher compressor speed generates finer oil droplets near the valve when we compare between the red curve in Figure 8, 9, and 10. This trend also agrees with the conclusion given by other researchers (Zimmerman and Hrnjak, 2014a).

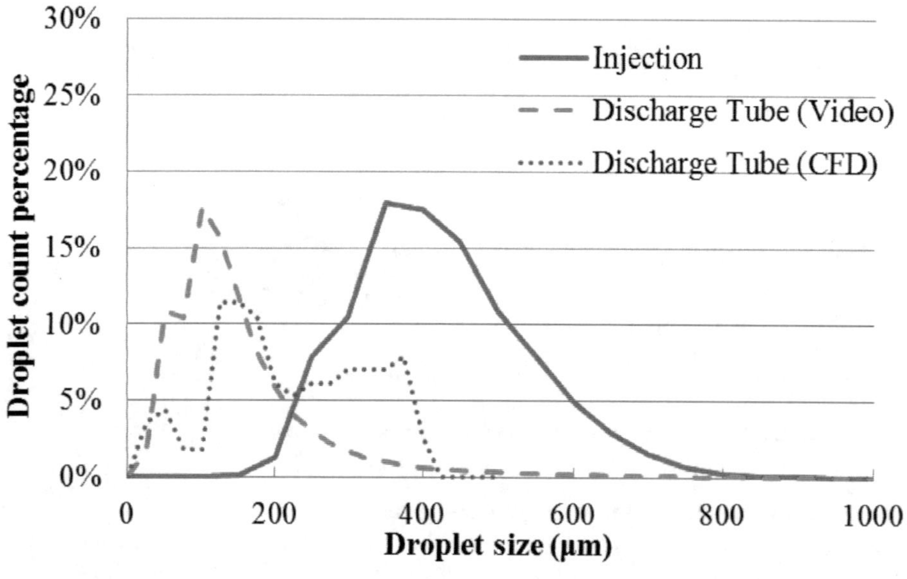

Figure 8: Comparison of oil droplet size distribution by video or CFD for compressor running at 40 Hz

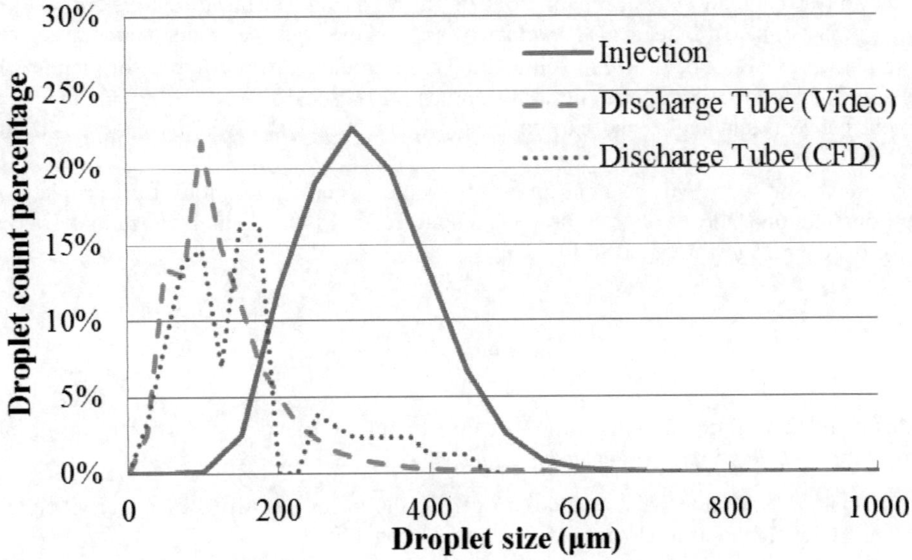

Figure 9: Comparison of oil droplet size distribution by video or CFD for compressor running at 50 Hz

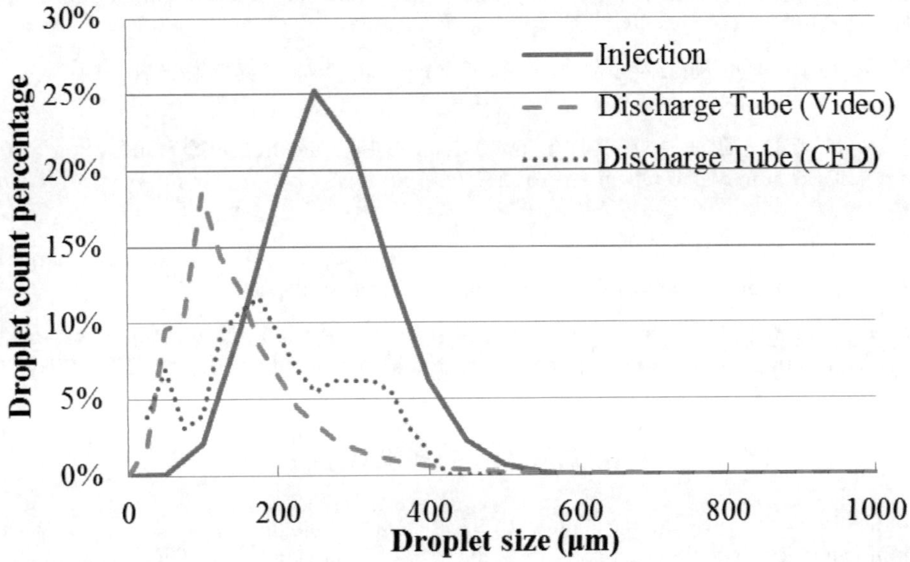

Figure 10: Comparison of oil droplet size distribution by video or CFD for compressor running at 60 Hz

This research is limited to the top plenum of the high side of a scroll compressor. For the start-up process of a compressor, oil droplets are generated by the opening and closing process of the discharge valve and some of the oil will retain in the plenum while the other will escape from the discharge tube. With more and more oil retained in the plenum, the re-entrainment effect becomes more and more significant and finally the oil mass flow rate entering the plenum will equal to the oil mass flow rate leaving the plenum. Therefore, at steady state, the oil flow rate at the valve is the same as the oil flow rate in the tube. The difference is oil droplet size distribution. The results shown in this research also verifies the idea that oil separating structure can be designed in the compressor plenum for easier separation where most of the oil exists in larger size.

5. CONCLUSIONS

Both visualization and simulation method are applied to research the oil flow inside compressor discharge plenum. From the comparison between visualization results and simulation results, we can conclude that:

- Visualization of the compressor plenum shows that the source of oil in the system is oil droplets injected by the reed valve opening and closing process.

- Videos of oil droplet flow are taken in front of the valve and in the discharge tube. Video processing techniques can estimate droplet size and velocity based on the high speed video captured.
- Discrete phase model can be utilized to simulate the oil droplets inside compressor plenum. The simulation results are compared to the video results at the compressor discharge under different compressor speed. The agreement between simulation and experiments validate the CFD model used for oil droplets trajectory prediction.
- Higher compressor speed will generate more oil droplets. The distribution of oil droplets size fulfills the Weibull distribution statically. Larger droplets tend to retain in the plenum while smaller droplets are more likely follow the refrigerant vapor flow.

REFERENCES

Toyama, T., Matsuura, H., & Yoshida, Y. (2006). Visual techniques to quantify behavior of oil droplets in a scroll compressor. *International Compressor Engineering Conference*. Paper 1788.

Wujek, S. S., & Hrnjak, P. S. (2011). Mist to annular flow development quantified by novel video analysis. ACRC Report TR285, University of Illinois at Urbana-Champaign, Urbana, IL.

Yokoyama, T., Shingu, K., Sekiya, S., Maeyama, H., & Nishimura, N. (2012). CFD Analysis inside a CO2 Rotary Compressor Shell to Improve Oil Separation and Reduce the Shell Size. *International Compressor Engineering Conference*. Paper 2074.

Xu, J., & Hrnjak, P. (2016). Refrigerant-Oil Flow at the Compressor Discharge (No. 2016-01-0247). SAE Technical Paper.

Zimmermann, A. J. P., & Hrnjak, P. S. (2014). Source Identification and In Situ Quantification of Oil-Refrigerant Mist Generation by Discharge Valve Opening Process. *International Compressor Engineering Conference*. Paper 2367.

Zimmermann, A. J. P., & Hrnjak, P. S. (2014). Visualization of the Opening Process of a Discharge Reed Valve in the Presence of Oil. *International Compressor Engineering Conference*. Paper 2369.

Zimmermann, A. J., & Hrnjak, P. S. (2015). Oil flow at discharge valve in a scroll compressor. Proceedings of the Institution of Mechanical Engineers, Part E: Journal of Process Mechanical Engineering, 229(2), 104-113.

ACKNOWLEDGEMENT

The authors would like to acknowledge the support of the member companies of the Industrial Advisory Board for the Air Conditioning and Refrigeration Center at the University of Illinois through ACRC project 341.

Numerical Simulation and Experimental Examination of an Oldham Coupling

Hang YE[1]*, Zhigang HUANG[1], Jinduo YE[2]
[1]Danfoss, R&D
Tianjin, P. R. China
Phone: +86-22-82196789, Fax:+86-22-72197299
yehang@danfoss.com
huangzhigang@danfoss.com

[2]Tianjin University of Technology, School of Mechanical & Engineering
Tianjin, P. R. China
jinduoye@126.com
Phone: +86-22-23363367, Fax:+86-22-23363367

*Corresponding Author

ABSTRACT

In scroll compressors, Oldham couplings (OCs) have typical been formed of aluminum alloy and have often broken issues due to the high loads which are transmitted from the orbiting scroll. This problem was especially pronounced during flooded start or slug conditions. In this paper, five pcs mass production OCs (standard OC), which are broken during a special slug test are investigated by classifying the ring broken and pin broken. During the investigation, both the quantities and the probabilities of each failure position are counted. Then, Finite Element Method (FEM) is employed in strength calculation for the standard OC and an optimized OC. The Finite Element Analysis (FEA) results explain that the cause of the ring's breakage is due to high tensile stress and the pin's breakage is due to high shear stress. After simulation, the authors design a series of strength experiments and DOE tests, with a great number of samples, which includes standard full treatment OC, CNC machined OC, half treatment OC and optimized OC, to verify the FEA results and repeat the failures in slug tests. Further, the author investigates the experiments results and compares the errors group by group. Finally, under the help of numerical simulation and experiment results, two new numerical models for OC strength prediction are proposed and validated by both strength tests and slug tests. The implementation of the prediction models are not only benefit to obtain a stronger OC or ensure it to reach a high reliability level, but also helps the develop engineers to shorten the new OC's development duration and reduce the fix expense in research activities.

1. INTRODUCTION

In scroll compressors, it is typical to employ a crankshaft to drive an Orbiting Scroll in orbiting motion with respect to a Fix Scroll (FS) or housing. In order to prevent relative rotation between OS and stationary body, OC has normally been implemented. Generally, the OC incorporates two pairs of pins, each pair of pins projecting in the opposite direction from an annular ring with one pair of pins engaging slots in the OS and the other pair of pins engaging either slots in FS or housing. Figure 1 shows a typical assembly method of an OC. An OC is mounted on the upper side of housing. Above it, located an OS which is driven by a crankshaft.

In the prior art, the Oldham couplings have typical been formed of aluminum, and have often breakage issue. This problem was especially pronounced in some larger scroll compressors and in particular during flooded start operation or slug condition. Under such worse conditions, the force transmitted to the coupling by the orbiting scroll can be substantially higher than during normal operation, causing the Oldham coupling to break.

As a result, to overcome this problem and deliver a robust design becomes an important work to develop engineers. Several approaches are tried to produce a strong OC. For example, it has been proposed in the past that some engineers use cast iron to form the OC. Although it becomes more stronger than before, the OC would be unduly heavy and cause excessive vibration issues. Another example is Titanium and Titanium alloys have been utilized in OC design and manufacture, which provides additional strength and resistance to breakage as compared to other more conventional materials such as Al alloys or cast iron.(Alexander Lifson. 2006) However, Ti or Ti Alloy OC may result in another issue – cost increase in mass production.

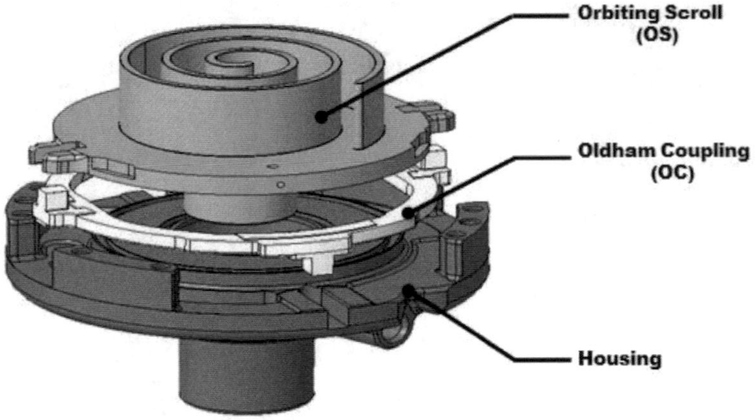

Figure 1: Assembly relationship of an OC

Moreover, the validation of a new OC often includes three main steps: First, 3D design (including FEA); Second, samples manufacture; Third, qualification tests. For each step, it needs about 4-10 weeks to complete corresponding tasks. For example, suppliers needs 6-8 weeks to prepare a new die and engineers need 8-10 weeks to finish all qualification tests. If any failure occurs in the qualification tests phase, engineers must go to the start point, or supplier prepares a new die for another 6-8 weeks. In this case, total development duration and fix expense will increase greatly.

Therefore, providing a robust design and improving OS' reliability level in critical conditions with minimum expense and shorten lead-time is a significant and meaningful task for all design engineers.

2. STASTISTIC AND NUMERCIAL SIMULATION OF OLDHAM COUPLINGS

2.1 Compressor slug test and its results

In order to check and to know X compressor's reliability level during slug conditions, five pcs X series compressors are tested by injecting more liquid step by step during their suction phase. All compressors can run well until the liquid is injected 130~150% times than safety level. Since the purpose of slug test is to check the strength of OC and wrap, there must be some damages in OC or warps. Tear down compressors proves the deduction that five pcs OCs are broken. It also means that current OC becomes the bottleneck of increasing X compressor's liquid handling ability. For this reason, an optimized OC with robust design is need for X compressors.

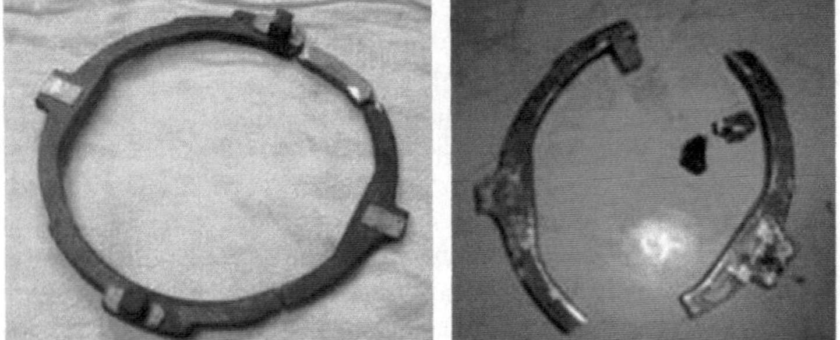

Figure 2: Two broken OC

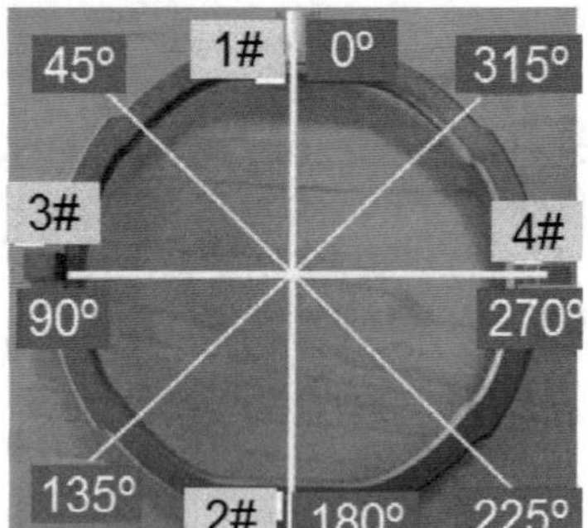

Figure 3: Statistic of failure zone

Two broken OCs are shown in Figure 2. From the pictures, it can be seen that breakage occurs in both pins and the ring. If take OC assembly relationship into consideration, the breakage of the ring is mainly due to high tensile stress, which is caused by OS when the compressor is compressing liquid refrigerant. The breakage of the pin is mainly due to the high shear stress applied on the contact surface through the slots. There is no doubt that breakage always starts from the most weak point or position. Thus, a further statistic is completed on those broken OCs.

The pins that contact with OS are specified as the first pair of pins and numbered 1# and 2#, while the other two pins are regarded as the second pair of pins and are numbered 3# and 4#. Second, four cross lines divides the ring into eight equal zones, see Figure 3. Pin 1# is defined as 0° position and the remaining positions are named from 0° to 315° in 45° increment along a counter-clockwise direction. This operation will be very helpful in finding out the greatest weakness zone(s) in standard OC.

Table 1 and Table 2 list the statistic results of rings and pins. "NOK" means corresponding rings or pins are broken while "OK" means they are still intact after slug tests. It can be seen from Table 1 that rings are more likely to break at 315° and 135° positions, because there are five pcs and three pcs of standard OC, respectively, broken here. For the remaining positions, the failure rate seems average. If looking at Table 2, the first pair of pins, compare to the second pair of pins, breaks more.

Table 1: Statistic of Ring broken

Compressor	0°	45°	90°	135°	180°	225°	270°	315°
x-1	OK	NOK	OK	OK	OK	OK	NOK	NOK
x-2	NOK	OK	NOK	NOK	OK	OK	OK	NOK
x-3	OK	NOK	NOK	OK	OK	NOK	NOK	NOK
x-4	OK	OK	OK	NOK	NOK	OK	OK	NOK
x-5	OK	OK	OK	NOK	OK	NOK	OK	NOK
Total of NOK (pcs)	1	2	2	3	1	2	2	5

Table 2: Statistic of Pin broken

Compressor No.	1#	2#	3#	4#
X-1	NOK	NOK	OK	OK
X-2	NOK	NOK	OK	NOK
X-3	NOK	NOK	OK	NOK
X-4	NOK	NOK	NOK	OK
X-5	OK	NOK	OK	OK
Total of NOK (pcs)	4	5	1	3

2.2 Numerical simulation on OC

Two kinds of 3D models, standard OC and optimized OC are investigated via ANSYS 14.0 software. Both the ring's tensile stress and pin's shear stress are simulated. OC's material property, which is used for simulation, is obtained by using Zwick/Roell Z100 test bench (see Figure 4) and three pcs Al alloy specimen (see Figure 5).

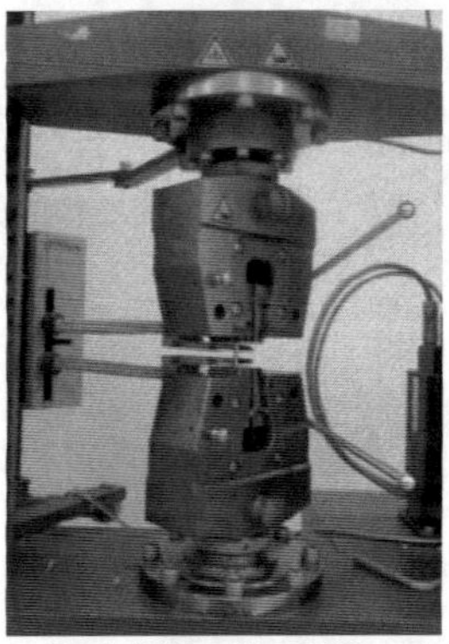

Figure 4: Zwick/Roell Z100 test bench

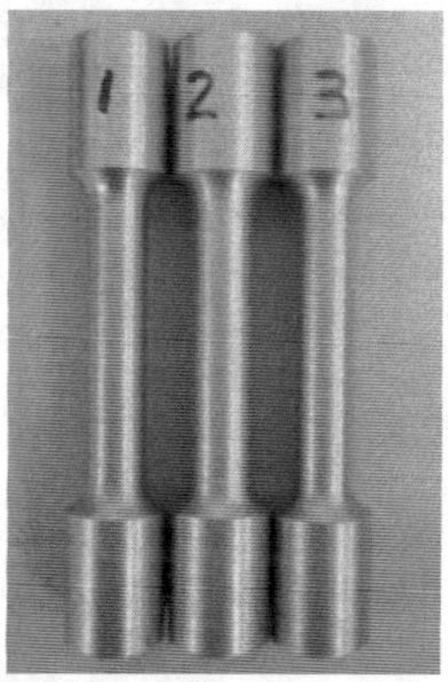

Figure 5: Specimen for material property test

When conducting the ring analysis, a full 3D OC model is implemented (see Figure 6). According to OC assembly relationship and its constraints when the compressor is running, a unit torque and an inertial force is applied on the first pair of pins. Then, corresponding directional zero displacements are applied on the second pair of pins. Instead of using a full 3D mode, a simplified 3D model is implemented in the pin's stress analysis. Figure 7 shows that a part of the pin is taken from the full OC model; meanwhile, a block is created to simulate contact surface in the slot. Then, a unit load is applied according to the red arrow to calculate the pin's stress. After simulation, the high stress fields and high deformation zones are compared between standard OC and optimized OC.

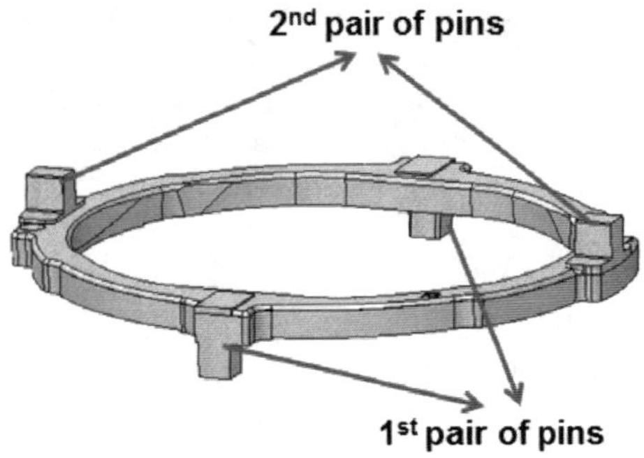

Figure 6: Ring stress analysis

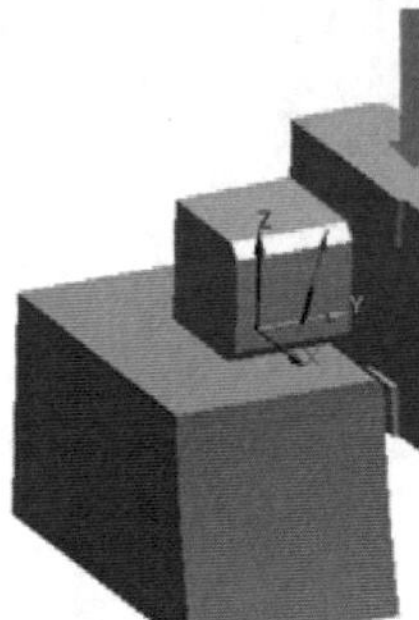

Figure 7: Pin shear stress analysis

Figure 8: Von Mises stress of ring

Figure 9: Von Mises strsss of pin

Figure 8 and Figure 9 show the FEA results of an OC. In Figure 8, although high stress zones are located at 45°, 135°, 335° and 315° positions, which are the same in both standard OC and optimized OC, the maximum stress values have been decreased in optimized model. In Figure 9, high shear stress zones are around the root of pins, which is not changed in optimized model, however, the max stress decreased in optimized OC. Meanwhile, the distribution of high stress zones also explains why there are so many breakages on OC.

Table 3: Summary of numerical simulation

Sample Type	**std. OC (MPa/MPa)**	**Optimized OC (MPa/MPa)**	**Delta Δ (%)**
Ring	1	0.816	18.4%
Pin	1	0.871	12.9%

Table 3 shows the results of standard OC and optimized OC. Here, results are normalized using denominator values, which are acquired from standard OC. After optimization, under same load, there will be 18.4% and 22.9% strength improvement in ring and pin.

3. CORRELATION BETWEEN EXPERIMENTAL EXAMINATION AND NUMERCIAL SIMULATION

In modern research and development activities, Finite Element Method (FEM) tools have wide applications and enjoy extensive utilization in the structural, thermal and fluid analysis fields. However, it is also important to recognize the limitations of FEM - that this method can reduce product testing, but it cannot totally replace it. In other words, numerical results still need tests to validate. For this reason, a series of tests have been conducted on several samples.
The objective of these tests is not only to verify FEA results, but also to understand the correlation between numerical simulations and experimental results, especially the relationship between numerical model and mass production parts. Furthermore, the outputs from correlation analysis can be used as another inputs for the qualitative and quantitative analysis which can help to clarify the factors that affect the correlations or cause the deviations between FEA and experiments.

3.1 Test preparation

In this section, six groups of samples are implemented in strength experiments (see Table 4), which includes standard OC with mass production process (including rings and pins); standard OC with CNC machining and optimized OC with CNC machining. It is easy to understand the strength improvements can be obtained by testing standard OC and optimized OC, but the reason of involving two kinds of standard OC samples (mass production parts and CNC machined parts) is because the strength of an OC with CNC machining can represent the results that gained from FEA software much better than the mass production parts. This is because some significant surface treatments have been involved in mass production OC, which bring some benefits in OC strength, but unfortunately, the benefits is hard to be simulated in FEA tools. Thus, by using these tests samples, the relative strength improvement results in FEA can be verified via standard CNC machined OC and optimized OC. Meanwhile, absolute deviation between numerical simulations and experiments can be obtained via standard OC with mass production process and standard OC with CNC machining.
Strength tests are conducted by using Zwick/Roell Z100 test bench. Corresponding results are recorded until the samples are broken.

Table 4: List of test specimen

No.	Name of specimen	Qty (pcs)	Note
1	std. OC	4	Mass production part
2	CNC std. OC	3	std. OC w CNC machining
3	CNC Optimized OC	3	Optimized OC w CNC machining
4	std. pin	4	mass production part
5	CNC std. pin	3	std. pin w CNC machining
6	CNC optimized pin	3	Optimized pin w CNC machining

3.2 Results of experimental examination

In order to obtain a better comparison, test results are normalized using denominator value of an avg. force, which is obtained through standard mass production OC.
Test results are shown in Table 5 and Table 6. In Table 6, breakage force of CNC machined OC is just about three-quarters of the mass production OC. Even if it has been optimized, the strength is still 12.3% less than mass production parts. Since mass production OC and standard CNC machined OC are same in dimension, excluding the system errors, these gaps may due to the different manufacture methods. If looking at pin tests results, deviations still exist; but seems smaller. In Table 5, the avg. shear stress of standard CNC machined OC is about 89% than that

of mass production OC. Even after optimization, it only increases 3% if comparing with standard mass production OC.

Table 5: Results of Ring strength tests

No. of specimen	Sample Type	Avg. Force (kN/kN)	Delta Δ (%)
P01 ~ P04	std. ring	1	-
S01 ~ S03	CNC std. ring	0.732	-26.8%
O04 ~ O06	CNC optimized ring	0.877	-12.3%

Table 6: Results of Pin's shear tests

No. of specimen	Sample Type	Avg. Force (kN/kN)	Delta Δ (%)
M01 ~ M04	std. pin	1	-
C01 ~ C03	CNC std. pin	0.887	-11.3%
C04 ~ C06	CNC optimized pin	1.030	+3%

Table 7: Comparison of std. OC and optimized OC

Pin	Avg. Force (kN/kN)	Improvement (%)
CNC std. Ring	0.732	14.5%
CNC optimized Ring	0.877	
CNC std. Pin	0.887	14.3%
CNC optimized Pin	1.03	

Nevertheless, if only comparing the results of CNC machined samples, it is obvious that the optimization is effective. In Table 7, there are 14.5% and 14.3% strength increments on ring and pin. Moreover, these results are well aligned with 18.4% and 12.9% Von Mises stress decrease in simulation.

3.3 Correlation between numerical simulation and experimental examination

If the experiments help research engineers to validate the FEA results of an optimized model and let them know that there will be some deviations when the samples are manufactured by different surface treatment methods, then the correlation analysis will help to complete the quantitative analysis for the factors that generated the deviations and corresponding weights. Thus, in this paper, correlation analysis contains two aspects. The first one is to compare the relative design improvements, which are obtained from numerical simulation and experiment examination by using standard OC and optimized OC. In this way, engineers can know whether current approaches, such as a simplified OC 3D model, material properties, load application method and their values are correct or not. The second target is to learn the difference between two different machining methods, which will help engineers to understand how to estimate the influence weights of the factors that cause the deviations in experimental tests. In addition, there might be more benefit in further investigations.

As discussed in section 3.2, FEA results align with test results within about <10% error. Therefore, it can be said that the simplified 3D model, material properties, load application method, and, FEA solution settings can be implemented in other OC design or optimization tasks.

Although the optimization has been validated through tests, there is still an obvious deviation between mass production OC and CNC machined OC. Thus, understanding how the gap is generated and closing it by giving corresponding weights accurately become an interesting work. However, before giving the weights, knowing why mass production parts have better strength performance is more important. Except the minor difference in dimension for example, the variance of tolerance and errors in the tests system, the key difference between two samples is the process. Unlike CNC machined parts, before delivery, mass production OC are treated by two significant surface treatment methods: aging and shooting. Aging is to heat the casting parts to certain temperature and then to keep them for a certain period in order to achieve a high tensile strength, or good plastic/ductility performance. Meanwhile, shooting, as another important treatment method, can introduce very good finish to final components. More, they also increase OC tensile and fatigue stress capacity by conducting high compressive residual stress on the surfaces (Björn Aurén, Guocai Chai. 2002).

3.4 Inspection of two surface treatments

According to preliminary study, since the minor differences and system errors exist in all test samples, the influences of these factors become normal, or can be considered as a constant value. However, the influence of two surface treatments, aging and shooting—or maybe three factors if considering their interaction effects, are variable. For this reason, a two-level factorial DOE test has been conducted in Minitab for the investigation. Thanks to this test, several interesting questions, such as which factor (or factors) play(s) a significant role in strength improvement, and whether the two factors are independent or interactives can be confirmed.

Aging and shooting are assigned as two discrete factors while the samples' breakage forces are set as the response in a four replicates DOE tests. According to the tests plan, four groups of samples are implemented, including standard samples with full treatments, samples without any treatments, samples only with aging and only with shooting. For each group, four pcs samples (pin and ring) are prepared. All samples are tested until broken, then, corresponding forces are recorded for factorial design analysis.

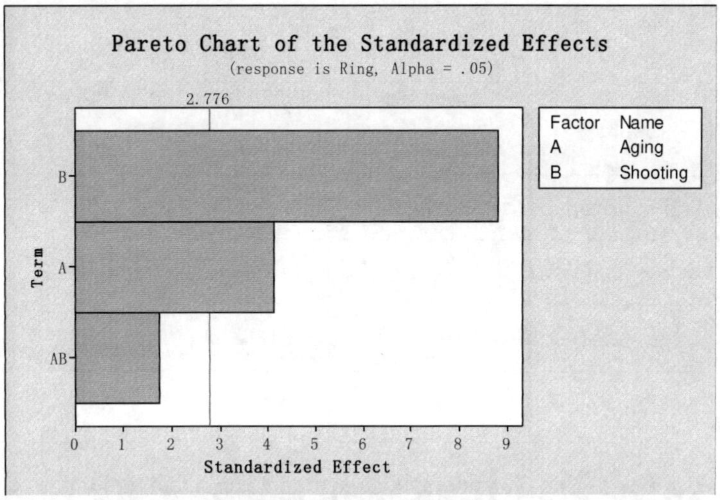

Figure 10: Pareto chart of ring

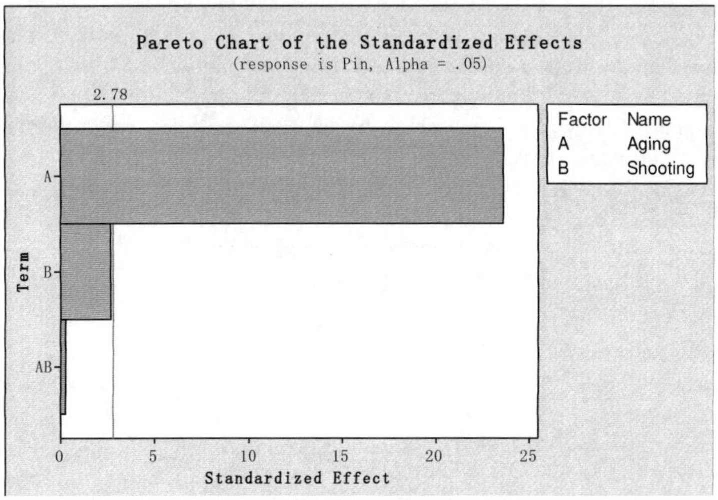

Figure 11: Pareto chart on pin

Figure 10 and Figure 11 show the Pareto charts of two tests. It is obvious that in Figure 10 that both aging and shooting are significant factors to ring's strength, while the interaction effect is not. However, in Figure 11, only aging is the key factor. Meanwhile, shooting and the interaction impacts become less important factors. Since the interactive influence is not a significant factor in two charts, aging and shooting, can be considered as independent factors.

The data, which are used in influence weights analysis, are taken from DOE tests. Here, test results are still normalized using as denominator value of average results obtained from mass production parts. Now, it is more easier to understand how the gaps are generated in section 3. Take the ring's results as an example. As discussed in section 3.2, a big gap (26.8% ring strength deviation) is found between mass production parts and CNC machined parts. Then, in Table 8, Row 2, after removing all surface treatments, the ring's strength will drop about 23%. Or, it can be said that ring's strength can increase at least 20~25% if proper surface treatments can be implemented in OC production. Then, a further analysis shows that if only one surface treatment method is introduced into mass production, the increments of corresponding strength is +5% for aging and +19% for shooting, which aligns with the outputs in two-level factorial DOE tests that only aging and shooting are significant factors for ring's strength. If looking the pin's strength, there are similar results. The difference between treatment parts and non-treatment parts is 10%, which aligns with Table 6 that pin's strength -11.3% on CNC machined parts. If introducing the aging process in production, there will be about 8% improvement. Although shooting also donates 2%, its influence is very limited. Now, the influence weight becomes clear that current aging treatment can bring +5% in ring strength and +8% in pin strength, while shooting can contribute +19% in ring strength and +2% in pin strength. If taking system error (about 5~10%) into account, there will be a slight variance for the weight factors.

Table 8: Influence weight analysis

Code of specimen	**Note**	**Rings avg. Broken Force (kN/kN)**	**Delta Δ (%)**	**Pins avg. Broken Force (kN/kN)**	**Delta Δ (%)**
P01 ~ P04	Full treatments	1	-	1	-
1# ~ 4#	w/o treatment	0.77	-23%	0.90	-10%
5# ~ 8#	Aging only	0.82	-18%	0.98	-2%
9# ~ 12#	Shooting only	0.95	-5%	0.92	-8%

4. STRENGTH PREDICTION MODEL AND VALIDATION TESTS

According to traditional products development procedure, research engineers cannot validate a new design until it is manufactured and passes enough qualification tests. For example, the development of an OC. First, qualification tests are required to be conducted on mass production OC or at least die made OC, which means engineers have to wait at least for 8-10 weeks to have final parts after design frozen. Second, the qualification tests, such as performance, vibration and critical life tests mean high fix expense and long test durations. Third, if there are any failures during the qualification tests phase, unfortunately, engineers must go back to step 1 and step 2. This procedure is very common in R&D; however, under the background of less expense, shorten new product launch time and obtain more achievements; current product development procedure becomes a kind of waste. Thus, to develop a model, which can be used for new component performance prediction before its manufacturing or help to shrink the lead-time for new product validation becomes a challenge and meaningful task.

4.1 Strength prediction

Based on the results in section 3 and section 4, the prediction model for mass production OC stress σ_m can be given by:

$$\sigma_m = \sigma_c * (1 + \delta\%) * (1 \pm C\%) \quad (1)$$

Where σ_c is max tensile stress which can be tested via CNC machined samples, $\delta\%$ represents the overall impact weight of two significant surface treatment methods, which includes two parts:

$$\delta\% = \alpha_a\% + \beta_s\% \tag{2}$$

where $\alpha_a\%$ and $\beta_s\%$ are the corresponding influence weights of aging and shooting. Once the process parameters are changed—for example, temperature or time, there will be some variance in two parameters. In this paper, two parameters are valued as α_a=5% and β_s=18% according to test results, so $\delta\%$ equals 23%. $C\%$ can be regarded as a sum of all system errors, a constant value, which is in a result of the minor differences in samples dimension, tests system, etc. In the view of previous analysis, C is assigned as $\pm10\%$.

The prediction model for mass production OC pin's stress τ_m can be given by:

$$\tau_m = \tau_c * (1 + \mu\%) * (1 \pm C\%) \tag{3}$$

Where τ_c is max shear stress that is tested via CNC machined samples, $\mu\%$ is the influence weight of surface treatment. Be different with tensile stress, only aging is the significant factor for shear stress, while shooting and interaction effect are not. Thus, according to section 3.4 results, μ equals 8%. The last term $C\%$ represent system errors which have the same value as equation (1).
According to the prediction model, new OC tensile strength and pin's shear strength will increase to 1.15MPa/MPa and 1.16MPa/MPa with 5% tolerance.

4.2 Prediction of optimized OC and test validation

Some optimized OCs, which are same as the one shown in Figure 12, are made according to standard mass production process. Then, two kinds of tests are conducted for design validation. In the strength tests, five pcs optimized OCs are tested. Prediction results and test results are compared in Table 9. According to the prediction model, the optimized OC's max tensile and max shear stress will increase to 1.1MPa/MPa to 1.2MPa/MPa. Physical tests show that that ring's maximum tensile strength increase to 1.2MPa/MPa while pin's max shear strength is up 1.17MPa/MPa, which are in the range of prediction.

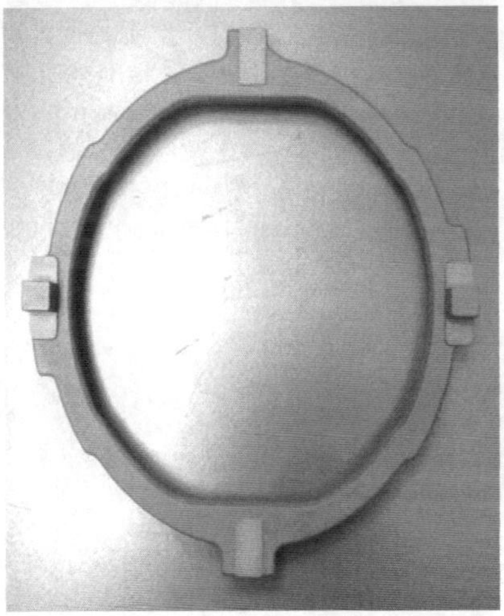

Figure 12: Optimized OC with full treatments

Figure 13: Slug tests

Table 9: Results of strength validation test

Group of samples	Note	Tensile Strength (MPa/MPa)	Shear Strength (MPa/MPa)
I	Current design (Full treatments, 3pcs)	1	1
II	Optimized design (Prediction)	1.08~1.21	1.11~1.20
III	Optimized design (Full treatments, 5pcs)	1.20	1.17

Beside strength tests, five pcs compressors are constructed with optimized OC and subjected to slug tests (see Figure 13). Final results indicate that after optimization, the new OC's slug performance will increase at least 50% than current OC.

5. CONCLUSIONS

- The approach which is implemented in the investigation and optimization of X compressor failure OC is proved to be feasible. Results of FEA, prediction results and physical tests results are aligned well. Although there are some deviations, the gaps are under control. Furthermore, both the factors and gaps can be identified and closed by qualitative and quantitative analysis.
- Several groups of tests prove that aging and shooting are two significant surface treatment methods, and plays a significant role in OC production. In future design activities, beyond adding more materials to obtain stronger components, engineers should consider to introduce more useful surface treatment methods to achieve a robust design.
- A prediction model has been given and validated by strength and slug tests. By using this method, both the new OC product development lead-time and fix expense can be controlled under a very small level. Furthermore, a similar model can be developed and implemented in other components design or optimization works.

NOMENCLATURE

The nomenclature should be located at the end of the text using the following format:

σ	Tensile stress	(MPa)
τ	Shear Force	(F)
δ	Influence weight	(%)
μ	Weight of process (shear)	(%)
C	System error	

Subscript

m	Manufacture parts
c	CNC machined parts
f	FEA result
a	Aging
s	Shooting

REFERENCES

Alexander Lifson, 2006, Scroll compressor With Titanium Oldham Coupling, *US patent:* 2008/008524A1, 2006
Björn Aurén, Guocai Chai., 2002, Effect Of Material Properties And Surface Treatment On The Performance Of Stainless Flapper Valve Steel For Compressors, *International Compressor Engineering Conference at Purdue,* C13-1
ANSYS WORKBENCH User guide, 2014
Minitab User guide, 2014

ACKNOWLEDGMENT

The authors would like to thank Kui Su, Xiaokun Ji, Dandan Liu and Pengfei Zhao for the special help in this study.

AUTHOR INDEX

AUTHOR INDEX

AUTHOR INDEX

AUTHOR INDEX